Genes IX

GENES IX

Benjamin Lewin

JONES AND BARTLETT PUBLISHERS
Sudbury, Massachusetts
BOSTON TORONTO LONDON SINGAPORE

World Headquarters
Jones and Bartlett Publishers
40 Tall Pine Drive
Sudbury, MA 01776
978-443-5000
info@jbpub.com
www.jbpub.com

Jones and Bartlett Publishers Canada
6339 Ormindale Way
Mississauga, Ontario L5V 1J2
CANADA

Jones and Bartlett Publishers International
Barb House, Barb Mews
London W6 7PA
UK

Jones and Bartlett's books and products are available through most bookstores and online booksellers. To contact Jones and Bartlett Publishers directly, call 800-832-0034, fax 978-443-8000, or visit our website, www.jbpub.com.
Substantial discounts on bulk quantities of Jones and Bartlett's publications are available to corporations, professional associations, and other qualified organizations. For details and specific discount information, contact the special sales department at Jones and Bartlett via the above contact information or send an email to specialsales@jbpub.com.

Production Credits

Chief Executive Officer: Clayton Jones
Chief Operating Officer: Don W. Jones, Jr.
President, Higher Education and Professional Publishing: Robert W. Holland, Jr.
V.P., Design and Production: Anne Spencer
V.P., Manufacturing and Inventory Control: Therese Connell
V.P., Sales and Marketing: William J. Kane
Acquisitions Editor, Science: Cathleen Sether
Managing Editor, Science: Dean W. DeChambeau
Editorial Assistant, Science: Molly Steinbach
Senior Production Editor: Louis C. Bruno, Jr.
Production Assistant: Jennifer M. Ryan
Marketing Manager: Andrea DeFronzo
Interactive Technology Manager: Dawn Mahon Priest
Book Designer: Anne Spencer
Cover Designer: Kristin E. Ohlin
Art Director: Jan VanAarsen
Illustrations: Imagineering Media Services, Inc.
Photo Development and Research Manager: Kimberly Potvin
Composition: Shepherd Inc.
Printing and Binding: Courier Kendallville
Cover Printing: Courier Kendallville
Cover Image: © Professor Oscar Miller/Photo Researchers, Inc.

About the cover: DNA transcription by mRNA. Colored transmission electron micrograph of DNA and messenger RNA (mRNA) molecules forming a feather-like, transcriptionally active structure. This DNA is from the nucleus of an amphibian egg. The backbone of the feather, running down the image, is a long strand of DNA coated with protein. Numerous mRNA molecules extend in clusters from the DNA strand. Transcription of genetic information begins at one end of the gene, with the mRNA molecules growing longer as they approach completion. Transcription is the first step in protein synthesis. Magnification: approximately ×30,000.

Library of Congress Cataloging-in-Publication Data

Lewin, Benjamin.
Genes IX / Benjamin Lewin.
p. ; cm.
Includes bibliographical references and index.
ISBN-13: 978-0-7637-4063-4 (alk. paper)
ISBN-10: 0-7637-4063-2
1. Genetics. 2. Genes. I. Title. II. Title: Genes 9. III. Title: Genes nine.
[DNLM: 1. Genes—physiology. 2. DNA—genetics. 3. Genetic Processes. 4. Genome. 5. Proteins—genetics. 6. RNA—genetics. QU 470 L672g 2006]
QH430.L4 2006
576.5—dc22 2006010787
6048

Printed in the United States of America
12 11 10 09 08 10 9 8 7 6 5 4 3 2

Brief Contents

Contents

31 Epigenetic Effects Are Inherited 818

Preface

Science is a wonderfully resilient venture. There are new and interesting developments to report in each revision of this book, and this revision includes much updated material to account for new findings in molecular research. The general organization of material in this edition has been revised along the same lines as *Essential Genes,* making it easier to use the two books in conjunction. With increasing size becoming a problem, the content has been more sharply focused on genes and their expression by eliminating the chapters dealing with the consequences of gene expression for cell biology. Striking changes occur in the first part of the book, dealing with genomes, resulting from the success of many genome sequencing projects. The importance of RNA as a regulator has become increasingly evident and now can be seen to extend across all levels of gene expression in both prokaryotes and eukaryotes. Somewhat of a "missing link," it casts further light on how the current apparatus for gene expression must have evolved from the early RNA world.

My policy in this book has been to cite research and review articles that I believe readers will reasonably be able to access. My preference is for articles that are free after six months; where that is not possible, the publication should be widely available.

I thank the following individuals who served as proofreaders and consultants for this revision:

Elliott Goldstein	University of Arizona, Tempe
Jocelyn Krebs	University of Alaska, Anchorage
Kathleen Matthews	Rice University, Houston

Benjamin Lewin
January 2007

Organization

The new organization of *GENES IX* allows instructors and students to focus more sharply on genes and their expression with expanded coverage of key topics. The number of chapters and the order of topic coverage remains the same; however, several chapters were expanded into two or more chapters. These changes are as follows:

Chapter 1 in *GENES VIII,* Genes are DNA, is expanded to two chapters in *GENES IX.* Basic information on DNA structure, replication, and mutation remains in Chapter 1, whereas the discussion of the gene's function as the unit of heredity appears in the new Chapter 2, Genes Code for Proteins.

Chapter 3 in *GENES VIII,* The Content of the Genome, becomes two chapters in *GENES IX.*

Chapter 4, The Content of the Genome, includes information on DNA sequences, genome mapping, and DNA in organelles.

Chapter 5, Genome Sequences and Gene Numbers, now contains genome size and expression information for a number of organisms, as well as new material on genes in the Y chromosome.

The new Chapter 12, The Operon, comprises *GENES VIII* Chapter 10, as well information on regulation of transcription and translation from *GENES VIII* Chapter 11, Regulatory Circuits. Material on regulatory RNA is now found in Chapter 13.

The material in *GENES VIII* Chapter 13, The Replicon, is expanded in three chapters in *GENES IX.* Chapter 15, The Replicon, covers the structure and function of the replicon, as well as replication origins. Chapter 16, Extrachromosomal Replicons, contains material on terminal proteins, rolling circle replication, plasmids, and T-DNA. Information on how bacterial replication is connected to the cell cycle is found in Chapter 17.

Recombination and Repair, Chapter 15 in *GENES VIII,* is now covered in two chapters in *GENES IX.* Chapter 19 covers homologous and site-specific recombination, and Chapter 20 covers the repair systems, including new information on excision-repair pathways in mammalian cells.

Chapter 23, Controlling Chromatin Structure, in *GENES VIII* is now Chapter 30, discussing the relation between chromatin structure and gene expression.

Chapter 31, Epigenetic Effects are Inherited, details the causes and mechanisms of epigenetic inheritance.

Art Program and Design

GENES IX has a new, contemporary look. Both the design and art program for this edition were updated and revised to facilitate student learning. In addition, the style and design for *GENES IX* intentionally matches that of Lewin's new cell biology text, *CELLS*, which allows students and instructors to easily utilize both texts.

Supplements to the Text

For the Student

The web site developed exclusively for the ninth edition of this text, http://biology.jbpub.com/book/genes/, offers a variety of resources to enhance understanding of molecular biology.

Laboratory Investigations in Molecular Biology, by Williams, Slatko, and McCarrey, presents well-tested protocols in molecular biology that are commonly used in active research labs. The experiments are designed to guide students through realistic research projects conducted in modern research laboratories.

For the Instructor

Compatible with Windows and Macintosh platforms, the Instructor's ToolKit—CD-ROM provides instructors with the following traditional ancillaries:

- The *Test Bank* is available as straight text files.
- The *PowerPoint® Lecture Outline Slides* presentation package provides lecture notes, and images for each chapter of *GENES IX*. Instructors with the Microsoft PowerPoint® software can customize the outlines, art, and order of presentation.
- The *Image Bank* provides the critical art and tables in the text to which Jones and Bartlett Publishers holds the copyright or has digital reprint rights. The image library enables instructors to project images from the text in the classroom, insert images into PowerPoint® presentations, or print overhead acetates.

Genes IX

1

Genes Are DNA

CHAPTER OUTLINE

Continued on next page

1.1 Introduction

The hereditary nature of every living organism is defined by its **genome,** which consists of a long sequence of **nucleic acid** that provides the *information* needed to construct the organism. We use the term "information" because the genome does not itself perform any active role in building the organism; rather it is the sequence of the individual subunits (bases) of the nucleic acid that determines hereditary features. By a complex series of interactions, this sequence is used to produce all the proteins of the organism in the appropriate time and place. The proteins either form part of the structure of the organism, or have the capacity to build the structures or to perform the metabolic reactions necessary for life.

The genome contains the complete set of hereditary information for any organism. Physically the genome may be divided into a number of different nucleic acid molecules. Functionally it may be divided into **genes.** Each gene is a sequence within the nucleic acid that represents a single protein. Each of the discrete nucleic acid molecules comprising the genome may contain a large number of genes. Genomes for living organisms may contain as few as <500 genes (for a mycoplasma, a type of bacterium) to as many as >25,000 for a the human being.

In this chapter, we analyze the properties of the gene in terms of its basic molecular construction. FIGURE 1.1 summarizes the stages in the transition from the historical concept of the gene to the modern definition of the genome.

A genome consists of the entire set of chromosomes for any particular organism. It, therefore, comprises a series of DNA molecules (one for each chromosome), each of which contains many genes. The ultimate definition of a genome is to determine the sequence of the DNA of each chromosome.

The first definition of the gene as a functional unit followed from the discovery that individual genes are responsible for the production of specific proteins. The difference in chemical nature between the DNA of the gene and its protein product led to the concept that a gene *codes* for a protein. This in turn led to the discovery of the complex apparatus that allows the DNA sequence of a gene to generate the amino acid sequence of a protein.

Understanding the process by which a gene is expressed allows us to make a more rigorous

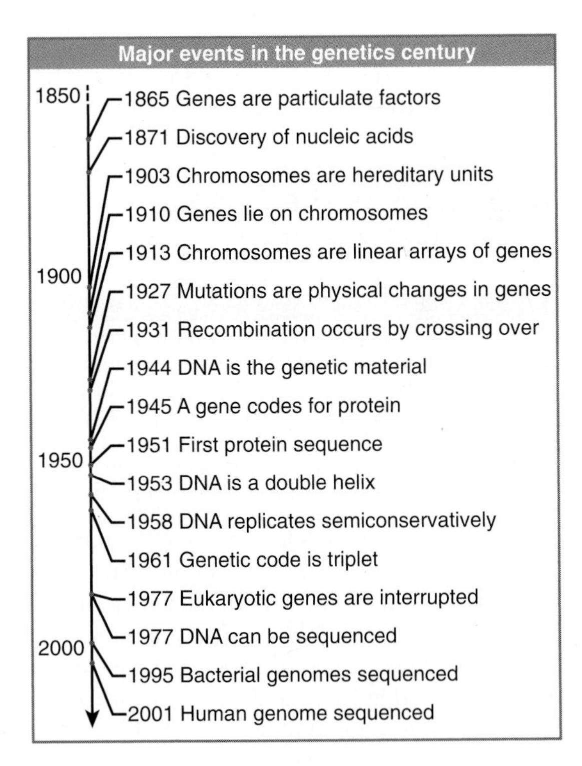

FIGURE 1.1 A brief history of genetics.

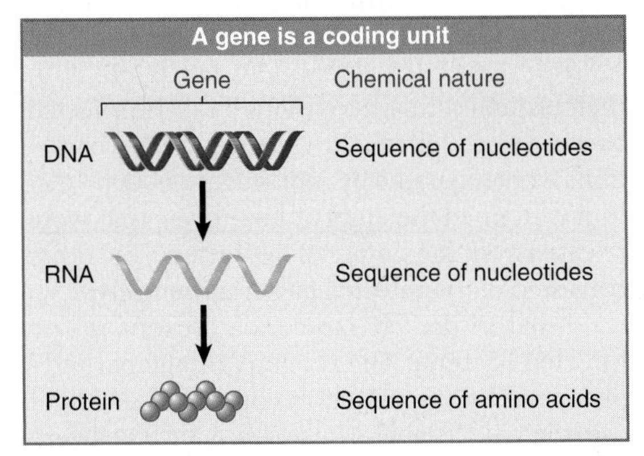

FIGURE 1.2 A gene codes for an RNA, which may code for protein.

definition of its nature. FIGURE 1.2 shows the basic theme of this book. A gene is a sequence of DNA that produces another nucleic acid, RNA. The DNA has two strands of nucleic acid, whereas the RNA has only one strand. The sequence of the RNA is determined by the sequence of the DNA. (In fact, it is identical to one of the DNA strands.) In many, but not all, cases, the RNA is in turn used to direct production of a protein. *Thus a gene is a sequence of DNA that codes for an RNA; in protein-coding genes, the RNA in turn codes for a protein.*

From the demonstration that a gene consists of DNA, and that a chromosome consists of a long stretch of DNA representing many genes, we move to the overall organization of the genome in terms of its DNA sequence. In Chapter 3, The Interrupted Gene, we take up in more detail the organization of the gene and its representation in proteins. In Chapter 4, The Content of the Genome, we consider the total number of genes, and in Chapter 6, Clusters and Repeats, we discuss other components of the genome and the maintenance of its organization.

1.2 DNA Is the Genetic Material of Bacteria

Key concept

- Bacterial transformation provided the first proof that DNA is the genetic material of bacteria. Genetic properties can be transferred from one bacterial strain to another by extracting DNA from the first strain and adding it to the second strain.

The idea that genetic material is nucleic acid had its roots in the discovery of **transformation** in 1928. The bacterium *Pneumococcus* kills mice by causing pneumonia. The virulence of the bacterium is determined by its *capsular polysaccharide.* This is a component of the surface that allows the bacterium to escape destruction by the host. Several types (I, II, and III) of *Pneumococcus* have different capsular polysaccharides. They have a smooth (S) appearance.

Each of the smooth *Pneumococcal* types can give rise to variants that fail to produce the capsular polysaccharide. These bacteria have a *rough* (R) surface (consisting of the material that was beneath the capsular polysaccharide). They are **avirulent.** They do not kill the mice, because the absence of the polysaccharide allows the animal to destroy the bacteria.

When smooth bacteria are killed by heat treatment, they lose their ability to harm the animal. But inactive heat-killed S bacteria and the ineffectual variant R bacteria together have a quite different effect from either bacterium by itself. FIGURE 1.3 shows that when they are jointly injected into an animal, the mouse dies as the result of a *Pneumococcal* infection. Virulent S bacteria can be recovered from the mouse postmortem.

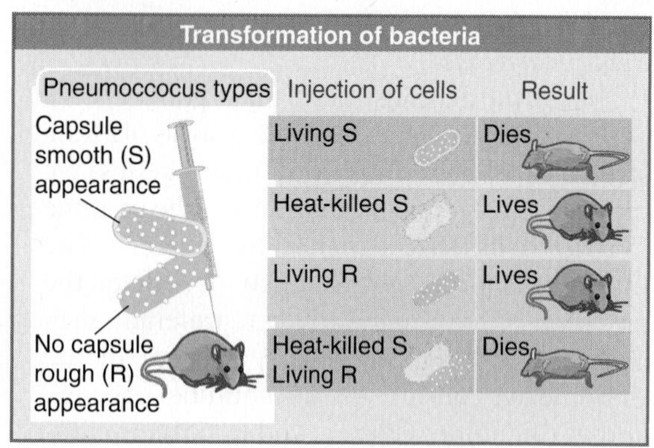

FIGURE 1.3 Neither heat-killed S-type nor live R-type bacteria can kill mice, but simultaneous injection of both can kill mice just as effectively as the live S-type.

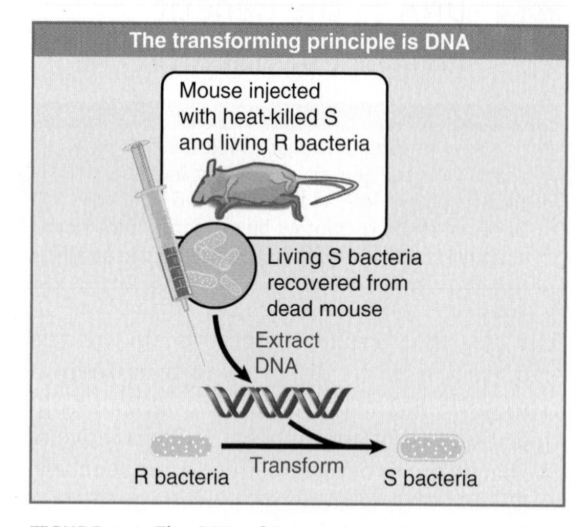

FIGURE 1.4 The DNA of S-type bacteria can transform R-type bacteria into the same S-type.

In this experiment, the dead S bacteria were of type III. The live R bacteria had been derived from type II. The virulent bacteria recovered from the mixed infection had the smooth coat of type III. So some property of the dead type III S bacteria can *transform* the live R bacteria so that they make the type III capsular polysaccharide and as a result become virulent.

FIGURE 1.4 shows the identification of the component of the dead bacteria responsible for transformation. This was called the **transforming principle.** It was purified by developing a cell-free system, in which extracts of the dead S bacteria could be added to the live R bacteria before injection into the animal. Purification of the transforming principle in 1944 showed that it is **deoxyribonucleic acid (DNA).**

1.3 DNA Is the Genetic Material of Viruses

Key concept

- Phage infection proved that DNA is the genetic material of viruses. When the DNA and protein components of bacteriophages are labeled with different radioactive isotopes, only the DNA is transmitted to the progeny phages produced by infecting bacteria.

Having shown that DNA is the genetic material of bacteria, the next step was to demonstrate that DNA provides the genetic material in a quite different system. Phage T2 is a virus that infects the bacterium *Escherichia coli.* When phage (virus) particles are added to bacteria, they adsorb to the outside surface, some material enters the bacterium, and then ~20 minutes later each bacterium bursts open (lyses) to release a large number of progeny phage.

FIGURE 1.5 illustrates the results of an experiment in 1952 in which bacteria were infected with T2 phages that had been radioactively labeled *either* in their DNA component (with ^{32}P) *or* in their protein component (with ^{35}S). The infected bacteria were agitated in a blender, and two fractions were separated by centrifugation. One fraction contained the empty phage coats that were released from the surface of the bacteria; the other consisted of the infected bacteria themselves.

Most of the ^{32}P label was present in the infected bacteria. The progeny phage particles produced by the infection contained ~30% of the original ^{32}P label. The progeny received very little—less than 1%—of the protein contained in the original phage population. The phage coats consist of protein and therefore carried the ^{35}S radioactive label. This experiment therefore showed directly that only the DNA of the parent phages enters the bacteria and then becomes part of the progeny phages, which is exactly the pattern of inheritance expected of genetic material.

A phage reproduces by commandeering the machinery of an infected host cell to manufacture more copies of itself. The phage possesses genetic material whose behavior is analogous

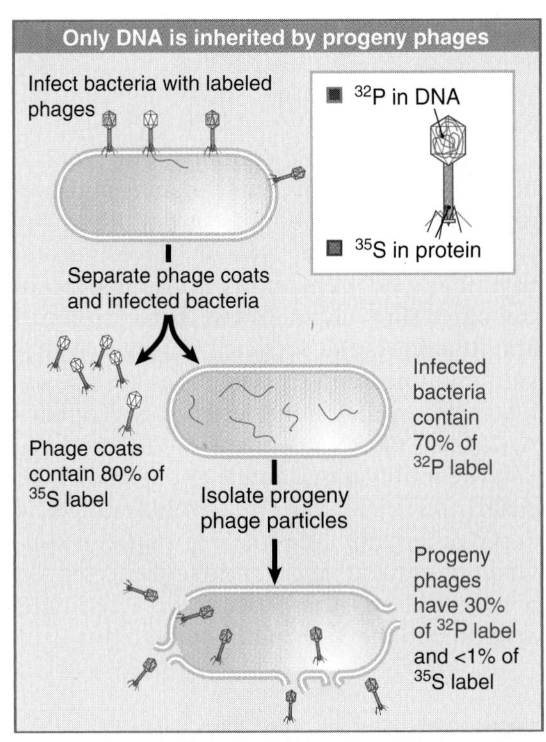

FIGURE 1.5 The genetic material of phage T2 is DNA.

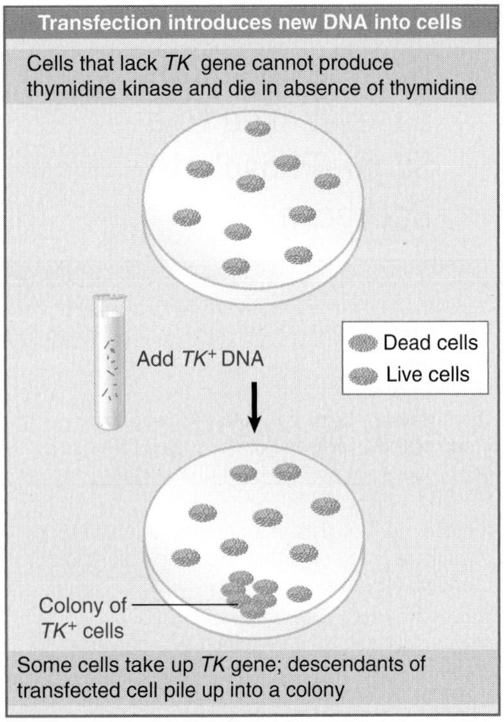

Transfection introduces new DNA into cells

Cells that lack *TK* gene cannot produce thymidine kinase and die in absence of thymidine

Add *TK*$^+$ DNA

Dead cells

Live cells

Colony of *TK*$^+$ cells

Some cells take up *TK* gene; descendants of transfected cell pile up into a colony

FIGURE 1.6 Eukaryotic cells can acquire a new phenotype as the result of transfection by added DNA.

to that of cellular genomes: its traits are faithfully reproduced and are subject to the same rules that govern inheritance. The case of T2 reinforces the general conclusion that the genetic material is DNA, whether part of the genome of a cell or virus.

1.4 DNA Is the Genetic Material of Animal Cells

Key concepts

- DNA can be used to introduce new genetic features into animal cells or whole animals.
- In some viruses, the genetic material is RNA.

When DNA is added to populations of single eukaryotic cells growing in culture, the nucleic acid enters the cells, and in some of them this results in the production of new proteins. When a purified DNA is used, its incorporation leads to the production of a particular protein. FIGURE 1.6 depicts one of the standard systems.

Although for historical reasons these experiments are described as **transfection** when performed with eukaryotic cells, they are a direct counterpart to bacterial transformation. The DNA that is introduced into the recipient cell becomes part of its genetic material and is inherited in the same way as any other part. Its expression confers a new trait upon the cells (synthesis of thymidine kinase in the example of Figure 1.6). At first, these experiments were successful only with individual cells adapted to grow in a culture medium. Since then, however, DNA has been introduced into mouse eggs by microinjection, and it may become a stable part of the genetic material of the mouse.

Such experiments show directly not only that DNA is the genetic material in eukaryotes, but also that *it can be transferred between different species and yet remain functional.*

The genetic material of all known organisms and many viruses is DNA. However, some viruses use an alternative type of nucleic acid, *ribonucleic acid (RNA),* as the genetic material. The general principle of the nature of the genetic material, then, is that it is always nucleic acid; in fact, it is DNA, except in the RNA viruses.

1.5 Polynucleotide Chains Have Nitrogenous Bases Linked to a Sugar–Phosphate Backbone

Key concepts

- A nucleoside consists of a purine or pyrimidine base linked to position 1 of a pentose sugar.
- Positions on the ribose ring are described with a prime (′) to distinguish them.
- The difference between DNA and RNA is in the group at the 2′ position of the sugar. DNA has a deoxyribose sugar (2′–H); RNA has a ribose sugar (2′–OH).
- A nucleotide consists of a nucleoside linked to a phosphate group on either the 5′ or 3′ position of the (deoxy)ribose.
- Successive (deoxy)ribose residues of a polynucleotide chain are joined by a phosphate group between the 3′ position of one sugar and the 5′ position of the next sugar.
- One end of the chain (conventionally the left) has a free 5′ end and the other end has a free 3′ end.
- DNA contains the four bases adenine, guanine, cytosine, and thymine; RNA has uracil instead of thymine.

The basic building block of nucleic acids is the nucleotide, which has three components:

- a nitrogenous base,
- a sugar, and
- a phosphate.

The nitrogenous base is a purine or pyrimidine ring. The base is linked to position 1 on a pentose sugar by a glycosidic bond from N_1 of pyrimidines or N_9 of purines. To avoid ambiguity between the numbering systems of the heterocyclic rings and the sugar, positions on the pentose are given a prime (′).

Nucleic acids are named for the type of sugar; DNA has 2′–deoxyribose, whereas RNA has ribose. The difference is that the sugar in RNA has an OH group at the 2′ position of the pentose ring. The sugar can be linked by its 5′ or 3′ position to a phosphate group.

A nucleic acid consists of a long chain of nucleotides. FIGURE 1.7 shows that the backbone of the polynucleotide chain consists of an alternating series of pentose (sugar) and phosphate residues. This is constructed by linking the 5′ position of one pentose ring to the 3′ position of the next pentose ring via a phosphate group. So the sugar–phosphate backbone is said to consist of 5′–3′ phosphodiester linkages. The nitrogenous bases "stick out" from the backbone.

Each nucleic acid contains four types of base. The same two purines, adenine and guanine, are present in both DNA and RNA. The two pyrimidines in DNA are cytosine and thymine; in RNA uracil is found instead of thymine. The only difference between uracil and thymine is the presence of a methyl substituent at position C_5. The bases are usually referred to by their initial letters. DNA contains A, G, C, and T; RNA contains A, G, C, and U.

The terminal nucleotide at one end of the chain has a free 5′ group; the terminal nucleotide at the other end has a free 3′ group. It is conventional to write nucleic acid sequences in the 5′to3′ direction—that is, from the 5′ terminus at the left to the 3′ terminus at the right.

1.6 DNA Is a Double Helix

Key concepts

- The B-form of DNA is a double helix consisting of two polynucleotide chains that run antiparallel.
- The nitrogenous bases of each chain are flat purine or pyrimidine rings that face inward and pair with one another by hydrogen bonding to form A-T or G-C pairs only.
- The diameter of the double helix is 20 Å, and there is a complete turn every 34 Å, with ten base pairs per turn.
- The double helix forms a major (wide) groove and a minor (narrow) groove.

The observation that the bases are present in different amounts in the DNAs of different species led to the concept that the *sequence of bases is the form in which genetic information is carried.* By the 1950s, the concept of genetic information was common: the twin problems it posed were working out the structure of the nucleic acid and explaining how a sequence of bases in DNA could represent the sequence of amino acids in a protein.

Three notions converged in the construction of the double helix model for DNA by Watson and Crick in 1953:

- X-ray diffraction data showed that DNA has the form of a regular helix, making a complete turn every 34 Å (3.4 nm), with a diameter of ~20 Å (2 nm). Since the distance between adjacent nucleotides is 3.4 Å, there must be 10 nucleotides per turn.

- The density of DNA suggests that the helix must contain two polynucleotide chains. The constant diameter of the helix can be explained if the bases in each chain face inward and are restricted so that a purine is always opposite a pyrimidine, avoiding partnerships of purine–purine (too wide) or pyrimidine–pyrimidine (too narrow).
- Irrespective of the absolute amounts of each base, the proportion of G is always the same as the proportion of C in DNA, and the proportion of A is always the same as that of T. So the composition of any DNA can be described by the proportion of its bases that is G + C. This ranges from 26% to 74% for different species.

Watson and Crick proposed that the two polynucleotide chains in the double helix associate by *hydrogen bonding between the nitrogenous bases.* G can hydrogen bond specifically only with C, whereas A can bond specifically only with T. These reactions are described as **base pairing,** and the paired bases (G with C, or A with T) are said to be **complementary.**

The model proposed that the two polynucleotide chains run in opposite directions **(antiparallel),** as illustrated in FIGURE 1.8. Looking along the helix, one strand runs in the 5′ to 3′ direction, whereas its partner runs 3′ to 5′.

The sugar–phosphate backbone is on the outside and carries negative charges on the phosphate groups. When DNA is in solution *in vitro,* the charges are neutralized by the binding of metal ions, typically by Na^+. In the cell, positively charged proteins provide some of the neutralizing force. These proteins play an important role in determining the organization of DNA in the cell.

The bases lie on the inside. They are flat structures, lying in pairs perpendicular to the axis of the helix. Consider the double helix in terms of a spiral staircase: the base pairs form the treads, as illustrated schematically in FIGURE 1.9. Proceeding along the helix, bases are stacked above one another like a pile of plates.

Each base pair is rotated ~36° around the axis of the helix relative to the next base pair. So ~10 base pairs make a complete turn of 360°. The twisting of the two strands around one another forms a double helix with a **minor**

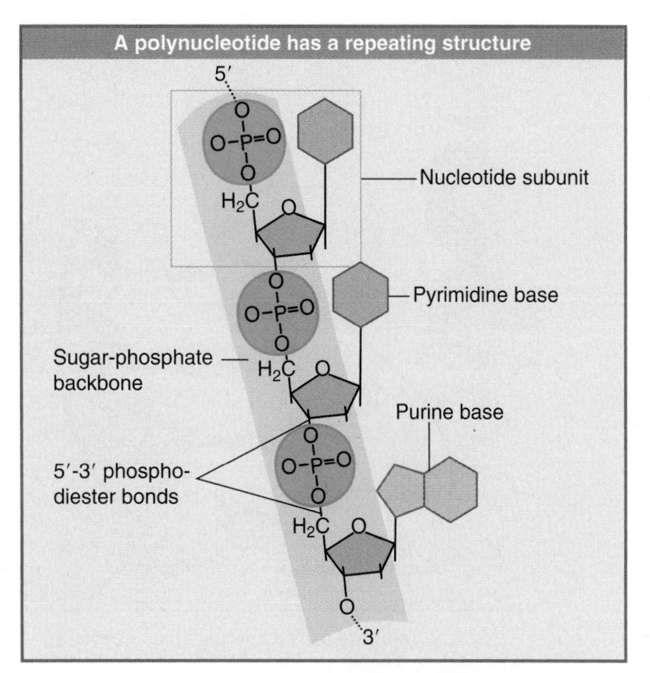

FIGURE 1.7 A polynucleotide chain consists of a series of 5′–3′ sugar-phosphate links that form a backbone from which the bases protrude.

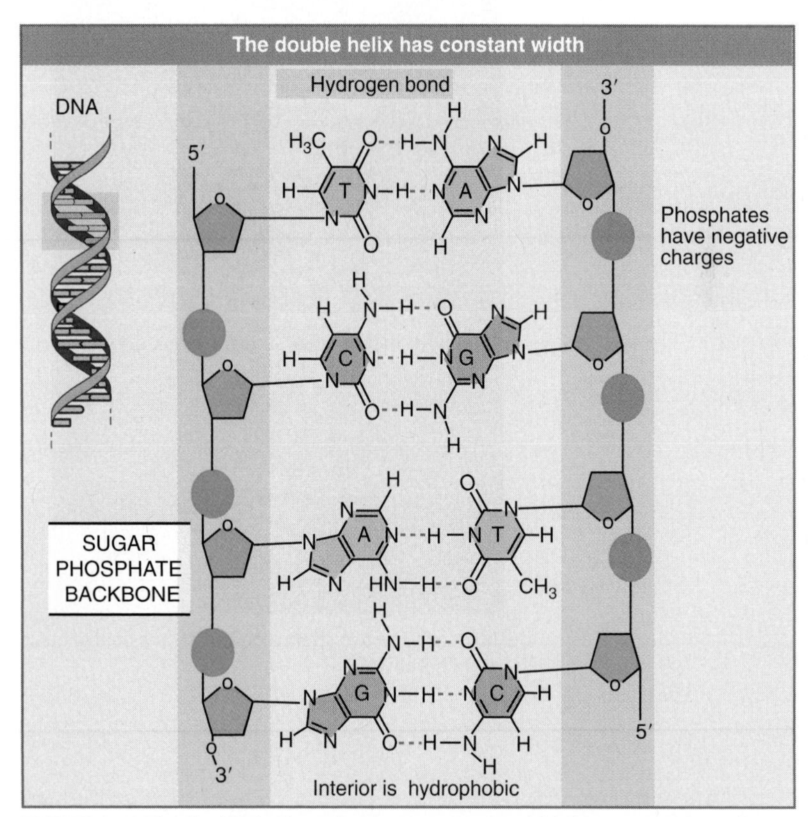

FIGURE 1.8 The double helix maintains a constant width because purines always face pyrimidines in the complementary A-T and G-C base pairs. The sequence in the figure is T-A, C-G, A-T, G-C.

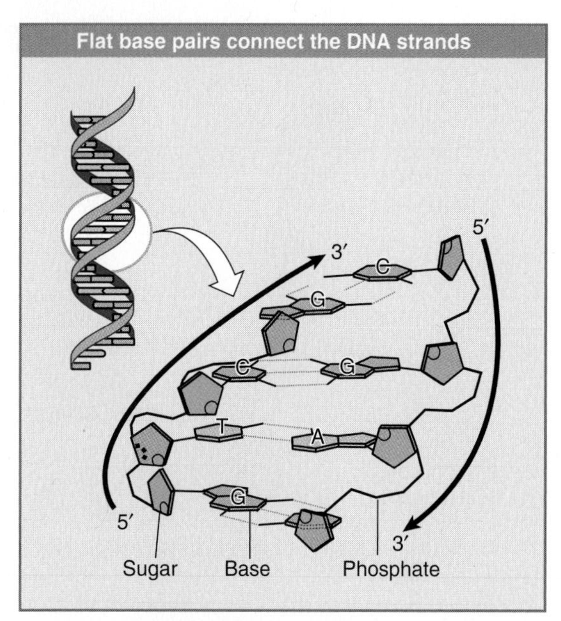

FIGURE 1.9 Flat base pairs lie perpendicular to the sugar-phosphate backbone.

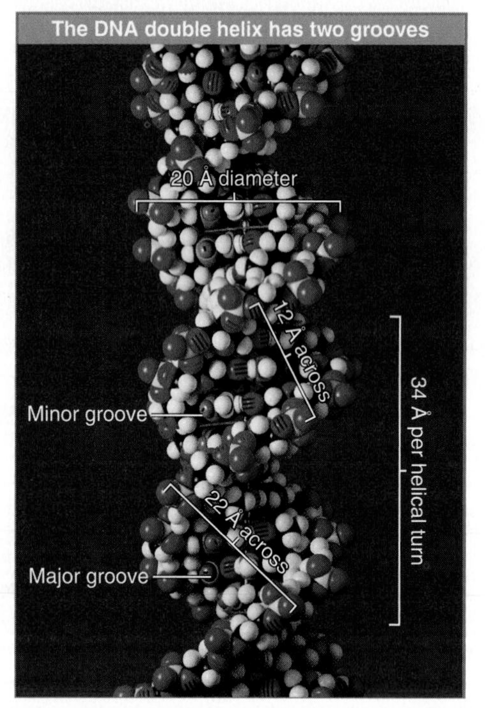

FIGURE 1.10 The two strands of DNA form a double helix. Photo © Photodisc.

groove (~12 Å across) and a **major groove** (~22 Å across), as can be seen from the scale model of FIGURE 1.10. The double helix is **right-handed;** the turns run clockwise looking along the helical axis. These features represent the accepted model for what is known as the **B-form** of DNA.

It is important to realize that the B-form represents an *average,* not a precisely specified structure. DNA structure can change locally. If it has more base pairs per turn it is said to be **overwound;** if it has fewer base pairs per turn it is **underwound.** Local winding can be affected by the overall conformation of the DNA double helix in space or by the binding of proteins to specific sites.

1.7 DNA Replication Is Semiconservative

Key concepts

- The Meselson–Stahl experiment used density labeling to prove that the single polynucleotide strand is the unit of DNA that is conserved during replication.
- Each strand of a DNA duplex acts as a template to synthesize a daughter strand.
- The sequences of the daughter strands are determined by complementary base pairing with the separated parental strands.

It is crucial that the genetic material is reproduced accurately. The two polynucleotide strands are joined only by hydrogen bonds; thus they are able to separate without requiring breakage of covalent bonds. The specificity of base pairing suggests that each of the separated **parental** strands could act as a **template strand** for the synthesis of a complementary **daughter** strand. FIGURE 1.11 shows the principle that a new daughter strand is assembled on each parental strand. The sequence of the daughter strand is dictated by the parental strand; an *A* in the parental strand causes a *T* to be placed in the daughter strand, a parental *G* directs incorporation of a daughter *C,* and so on.

The top part of Figure 1.11 shows a parental (unreplicated) duplex that consists of the original two parental strands. The lower part shows the two daughter duplexes that are being produced by complementary base pairing. Each of the daughter duplexes is identical in sequence

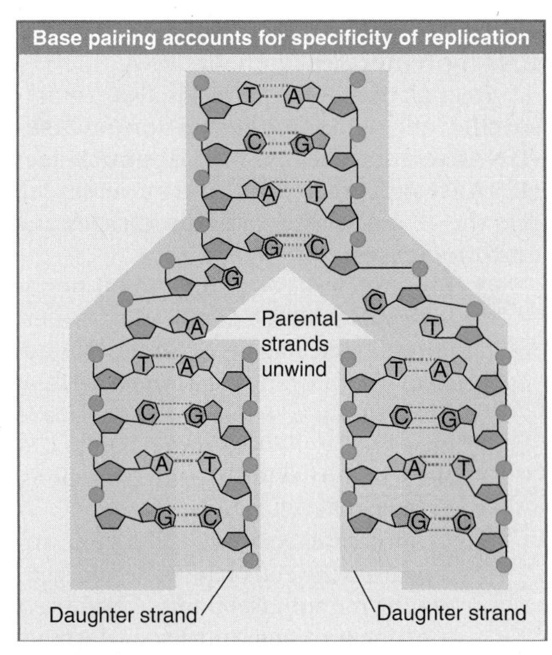

FIGURE 1.11 Base pairing provides the mechanism for replicating DNA.

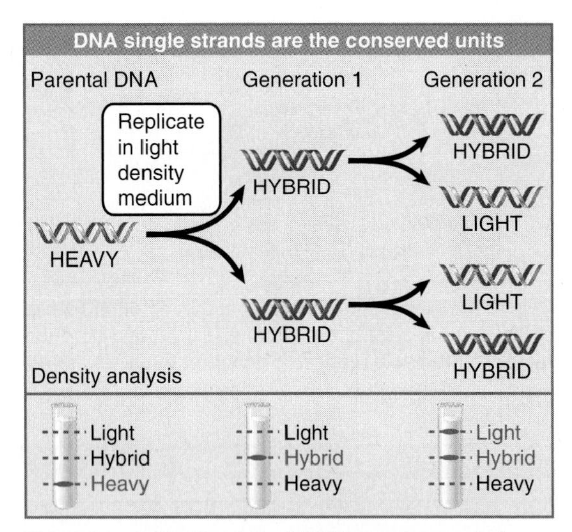

FIGURE 1.12 Replication of DNA is semiconservative.

with the original parent and contains one parental strand and one newly synthesized strand. *The structure of DNA carries the information needed to perpetuate its sequence.*

The consequences of this mode of replication are illustrated in FIGURE 1.12. The parental duplex is replicated to form two daughter duplexes, each of which consists of one parental strand and one (newly synthesized) daughter strand. *The unit conserved from one generation to the next is one of the two individual strands comprising the parental duplex.* This behavior is called **semiconservative replication.**

Figure 1.12 illustrates a prediction of this model. If the parental DNA carries a "heavy" density label because the organism has been grown in medium containing a suitable isotope (such as ^{15}N), its strands can be distinguished from those that are synthesized when the organism is transferred to a medium containing normal "light" isotopes.

The parental DNA consists of a duplex of two heavy strands (red). After one generation of growth in light medium, the duplex DNA is "hybrid" in density—it consists of one heavy parental strand (red) and one light daughter strand (blue). After a second generation, the two strands of each hybrid duplex have separated. Each gains a light partner, so that now one half of the duplex DNA remains hybrid and the other half is entirely light (both strands are blue).

The individual strands of these duplexes are entirely heavy or entirely light. This pattern was confirmed experimentally in the Meselson–Stahl experiment of 1958, which followed the semiconservative replication of DNA through three generations of growth of *E. coli*. When DNA was extracted from bacteria and its density measured by centrifugation, the DNA formed bands corresponding to its density—heavy for parental, hybrid for the first generation, and half hybrid and half light in the second generation.

1.8 DNA Strands Separate at the Replication Fork

Key concepts

- Replication of DNA is undertaken by a complex of enzymes that separate the parental strands and synthesize the daughter strands.
- The replication fork is the point at which the parental strands are separated.
- The enzymes that synthesize DNA are called DNA polymerases; the enzymes that synthesize RNA are called RNA polymerases.
- Nucleases are enzymes that degrade nucleic acids; they include DNAases and RNAases and can be divided into endonucleases and exonucleases.

Replication requires the two strands of the parental duplex to separate. However, the disruption of structure is only transient and is reversed as the daughter duplex is formed. Only

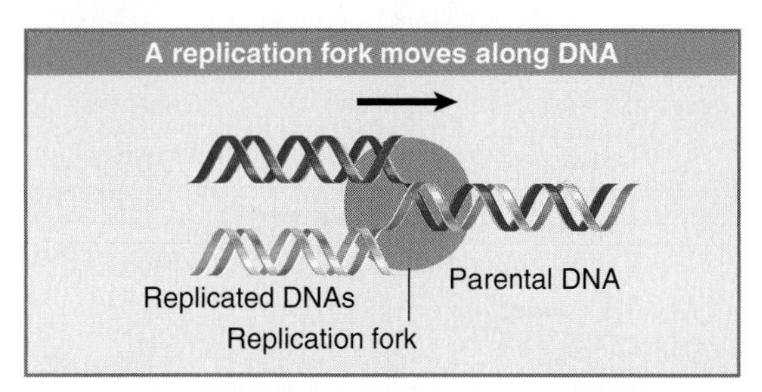

FIGURE 1.13 The replication fork is the region of DNA in which there is a transition from the unwound parental duplex to the newly replicated daughter duplexes.

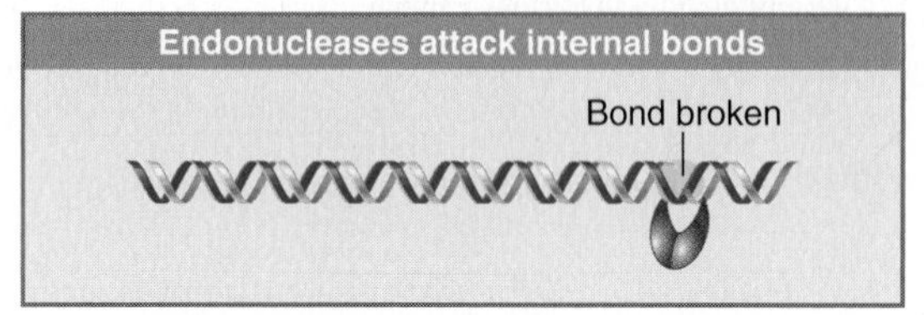

FIGURE 1.14 An endonuclease cleaves a bond within a nucleic acid. This example shows an enzyme that attacks one strand of a DNA duplex.

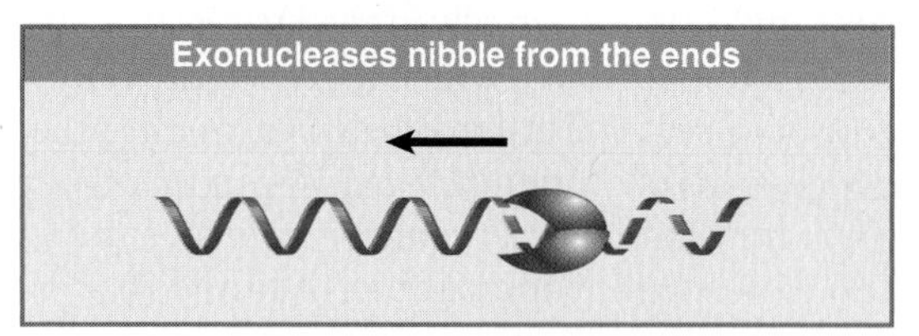

FIGURE 1.15 An exonuclease removes bases one at a time by cleaving the last bond in a polynucleotide chain.

a small stretch of the duplex DNA is separated into single strands at any moment.

The helical structure of a molecule of DNA engaged in replication is illustrated in FIGURE 1.13. The nonreplicated region consists of the parental duplex, opening into the replicated region where the two daughter duplexes have formed. The double helical structure is disrupted at the junction between the two regions, which is called the **replication fork.** Replication involves movement of the replication fork along the parental DNA, so there is a continuous unwinding of the parental strands and rewinding into daughter duplexes.

The synthesis of nucleic acids is catalyzed by specific enzymes, which recognize the template and undertake the task of catalyzing the addition of subunits to the polynucleotide chain that is being synthesized. The enzymes are named according to the type of chain that is synthesized: **DNA polymerases** synthesize DNA, and **RNA polymerases** synthesize RNA.

Degradation of nucleic acids also requires specific enzymes: **deoxyribonucleases (DNAases)** degrade DNA, and **ribonucleases (RNAases)** degrade RNA. The nucleases fall into the general classes of **exonucleases** and **endonucleases:**

- Endonucleases cut individual bonds *within* RNA or DNA molecules, generating discrete fragments. Some DNAases cleave both strands of a duplex DNA at the target site, whereas others cleave only one of the two strands. Endonucleases are involved in cutting reactions, as shown in FIGURE 1.14.
- Exonucleases remove residues one at a time from the end of the molecule, generating mononucleotides. They always function on a single nucleic acid strand, and each exonuclease proceeds in a specific direction, that is, starting either at a 5′ or at a 3′ end and proceeding toward the other end. They are involved in trimming reactions, as shown in FIGURE 1.15.

1.9 Genetic Information Can Be Provided by DNA or RNA

Key concepts

- Cellular genes are DNA, but viruses and viroids may have genomes of RNA.
- DNA is converted into RNA by transcription, and RNA may be converted into DNA by reverse transcription.
- The translation of RNA into protein is unidirectional.

The **central dogma** defines the paradigm of molecular biology. Genes are perpetuated as sequences of nucleic acid, but function by being expressed in the form of proteins. Replication is responsible for the inheritance of genetic information. Transcription and translation are responsible for its conversion from one form to another.

FIGURE 1.16 illustrates the roles of replication, transcription, and translation, viewed from the perspective of the central dogma:

- *The perpetuation of nucleic acid may involve either DNA or RNA as the genetic material.*

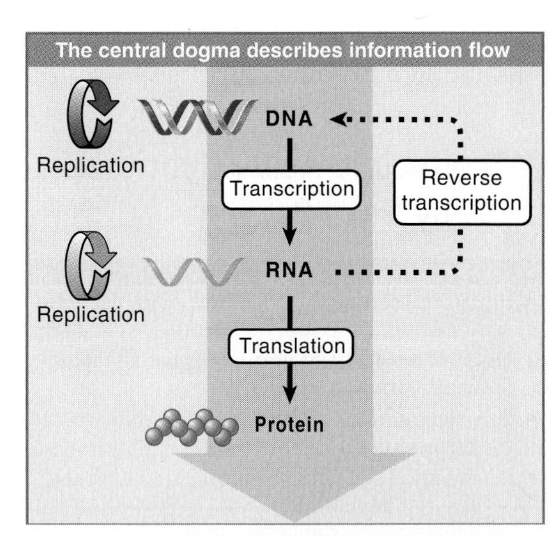

FIGURE 1.16 The central dogma states that information in nucleic acid can be perpetuated or transferred, but the transfer of information into protein is irreversible.

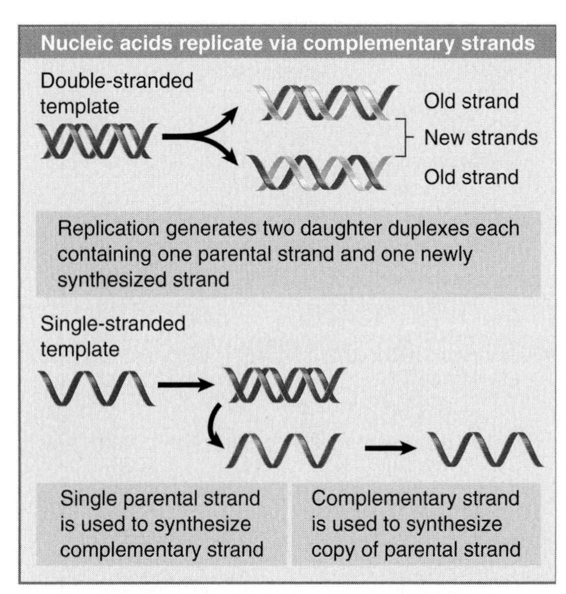

FIGURE 1.17 Double-stranded and single-stranded nucleic acids both replicate by synthesis of complementary strands governed by the rules of base pairing.

Cells use only DNA. Some viruses use RNA, and replication of viral RNA occurs in the infected cell.

- *The expression of cellular genetic information usually is unidirectional.* Transcription of DNA generates RNA molecules that can be used further *only* to generate protein sequences; generally they cannot be retrieved for use as genetic information. Translation of RNA into protein is always irreversible.

These mechanisms are equally effective for the cellular genetic information of prokaryotes or eukaryotes and for the information carried by viruses. The genomes of all living organisms consist of duplex DNA. Viruses have genomes that consist of DNA or RNA, and there are examples of each type that are double-stranded (ds) or single-stranded (ss). Details of the mechanism used to replicate the nucleic acid vary among the viral systems, but the principle of replication via synthesis of complementary strands remains the same, as illustrated in FIGURE 1.17.

Cellular genomes reproduce DNA by the mechanism of semiconservative replication. Double-stranded virus genomes, whether DNA or RNA, also replicate by using the individual strands of the duplex as templates to synthesize partner strands.

Viruses with single-stranded genomes use the single strand as a template to synthesize a complementary strand; this complementary strand in turn is used to synthesize its complement, which is, of course, identical with the original starting strand. Replication may involve the formation of stable double-stranded intermediates or use double-stranded nucleic acid only as a transient stage.

The restriction to unidirectional transfer from DNA to RNA is not absolute. It is overcome by the **retroviruses,** whose genomes consist of single-stranded RNA molecules. During the infective cycle, the RNA is converted by the process of **reverse transcription** into a single-stranded DNA, which in turn is converted into a double-stranded DNA. This duplex DNA becomes part of the genome of the cell and is inherited like any other gene. *So reverse transcription allows a sequence of RNA to be retrieved and used as genetic information.*

The existence of RNA replication and reverse transcription establishes the general principle that *information in the form of either type of nucleic acid sequence can be converted into the other type.* In the usual course of events, however, the cell relies on the processes of DNA replication, transcription, and translation. But on rare occasions (possibly mediated by an RNA virus), information from a cellular RNA is converted into DNA and inserted into the genome.

Genomes vary greatly in size		
Genome	Gene Number	Base Pairs
Organisms		
Plants	<50,000	$<10^{11}$
Mammals	30,000	$\sim 3 \times 10^9$
Worms	14,000	$\sim 10^8$
Flies	12,000	1.6×10^8
Fungi	6,000	1.3×10^7
Bacteria	2–4,000	$<10^7$
Mycoplasma	500	$<10^6$
dsDNA Viruses		
Vaccinia	<300	187,000
Papova (SV40)	~6	5,226
Phage T4	~200	165,000
ssDNA Viruses		
Parvovirus	5	5,000
Phage fX174	11	5,387
dsRNA Viruses		
Reovirus	22	23,000
ssRNA Viruses		
Coronavirus	7	20,000
Influenza	12	13,500
TMV	4	6,400
Phage MS2	4	3,569
STNV	1	1,300
Viroids		
PSTV RNA	0	359

FIGURE 1.18 The amount of nucleic acid in the genome varies over an enormous range.

Although reverse transcription plays no role in the regular operations of the cell, it becomes a mechanism of potential importance when we consider the evolution of the genome.

The same principles are followed to perpetuate genetic information from the massive genomes of plants or amphibians to the tiny genomes of mycoplasma and the yet smaller genetic information of DNA or RNA viruses. FIGURE 1.18 summarizes some examples that illustrate the range of genome types and sizes.

Throughout the range of organisms, with genomes varying in total content over a 100,000-fold range, a common principle prevails: *The DNA codes for all the proteins that the cell(s) of the organism must synthesize, and the proteins in turn (directly or indirectly) provide the functions needed for survival.* A similar principle describes the function of the genetic information of viruses, whether DNA or RNA: *The nucleic acid codes for the protein(s) needed to package the genome and also for any functions additional to those provided by the host cell that are needed to reproduce the virus during its infective cycle.* (The smallest virus—the satellite tobacco necrosis virus [STNV]—cannot replicate independently. It requires the simultaneous presence of a "helper" virus—the tobacco necrosis virus [TNV], which is itself a normally infectious virus.)

1.10 Nucleic Acids Hybridize by Base Pairing

Key concepts

- Heating causes the two strands of a DNA duplex to separate.
- The T_m is the midpoint of the temperature range for denaturation.
- Complementary single strands can renature when the temperature is reduced.
- Denaturation and renaturation/hybridization can occur with DNA–DNA, DNA–RNA, or RNA–RNA combinations and can be intermolecular or intramolecular.
- The ability of two single-stranded nucleic acid preparations to hybridize is a measure of their complementarity.

A crucial property of the double helix is the ability to separate the two strands without disrupting covalent bonds. This makes it possible for the strands to separate and reform under physiological conditions at the (very rapid) rates needed to sustain genetic functions. The specificity of the process is determined by complementary base pairing.

The concept of base pairing is central to all processes involving nucleic acids. Disruption of the base pairs is a crucial aspect of the function of a double-stranded molecule, whereas the ability to form base pairs is essential for the activity of a single-stranded nucleic acid. FIGURE 1.19 shows that base pairing enables complementary single-stranded nucleic acids to form a duplex structure.

- An intramolecular duplex region can form by base pairing between two complementary sequences that are part of a single-stranded molecule.
- A single-stranded molecule may base pair with an independent, complementary single-stranded molecule to form an intermolecular duplex.

Formation of duplex regions from single-stranded nucleic acids is most important for RNA, but single-stranded DNA also exists (in the form of viral genomes). Base pairing between independent complementary single strands is not restricted to DNA–DNA or RNA–RNA, but can also occur between a DNA molecule and an RNA molecule.

The lack of covalent links between complementary strands makes it possible to manipu-

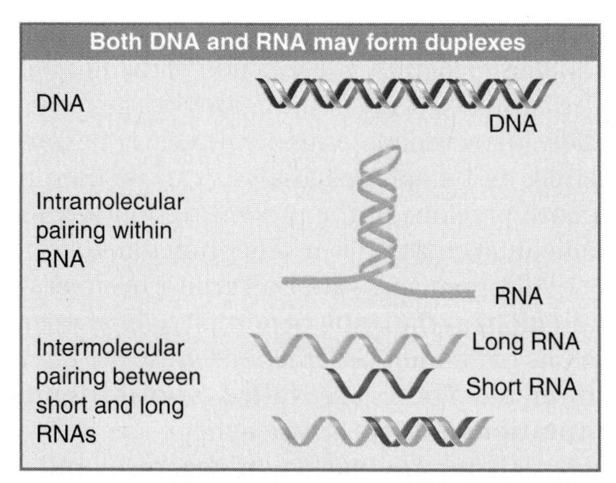

FIGURE 1.19 Base pairing occurs in duplex DNA and also in intra- and inter-molecular interactions in single-stranded RNA (or DNA).

late DNA *in vitro*. The noncovalent forces that stabilize the double helix are disrupted by heating or by exposure to low salt concentration. The two strands of a double helix separate entirely when all the hydrogen bonds between them are broken.

The process of strand separation is called **denaturation** or (more colloquially) *melting*. ("Denaturation" is also used to describe loss of authentic protein structure; it is a general term implying that the natural conformation of a macromolecule has been converted to some other form.)

Denaturation of DNA occurs over a narrow temperature range and results in striking changes in many of its physical properties. The midpoint of the temperature range over which the strands of DNA separate is called the *melting temperature* (T_m). It depends on the proportion of G-C base pairs. Because each G-C base pair has three hydrogen bonds, it is more stable than an A-T base pair, which has only two hydrogen bonds. The more G-C base pairs are contained in a DNA, the greater the energy that is needed to separate the two strands. In solution under physiological conditions, a DNA that is 40% G-C—a value typical of mammalian genomes—denatures with a T_m of about 87° C. So duplex DNA is stable at the temperature prevailing in the cell.

The denaturation of DNA is reversible under appropriate conditions. The ability of the two separated complementary strands to reform into a double helix is called **renaturation.** Renaturation depends on specific base pairing between the complementary strands. FIGURE 1.20 shows that the reaction takes place in two stages. First, single strands of DNA in the solution encounter

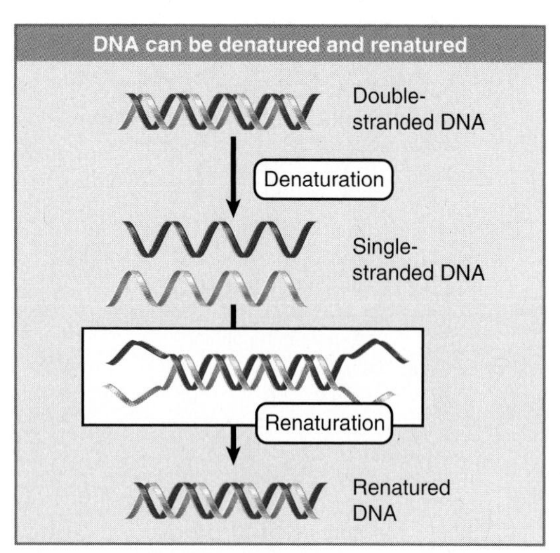

FIGURE 1.20 Denatured single strands of DNA can renature to give the duplex form.

one another by chance; if their sequences are complementary, the two strands base pair to generate a short double-helical region. Then the region of base pairing extends along the molecule by a zipper-like effect to form a lengthy duplex molecule. Renaturation of the double helix restores the original properties that were lost when the DNA was denatured.

Renaturation describes the reaction between two complementary sequences that were separated by denaturation. However, the technique can be extended to allow any two complementary nucleic acid sequences to react with each other to form a duplex structure. This is sometimes called **annealing,** but the reaction is more generally described as **hybridization** whenever nucleic acids of different sources are involved, as in the case when one preparation consists of DNA and the other consists of RNA. *The ability of two nucleic acid preparations to hybridize constitutes a precise test for their complementarity because* only *complementary sequences can form a duplex structure.*

The principle of the hybridization reaction is to expose two single-stranded nucleic acid preparations to each other and then to measure the amount of double-stranded material that forms. FIGURE 1.21 illustrates a procedure in which a DNA preparation is denatured and the single strands are adsorbed to a filter. Then a second denatured DNA (or RNA) preparation is added. The filter is treated so that the second preparation can adsorb to it only if it is able to base pair with the DNA that was originally adsorbed. Usually the second preparation is

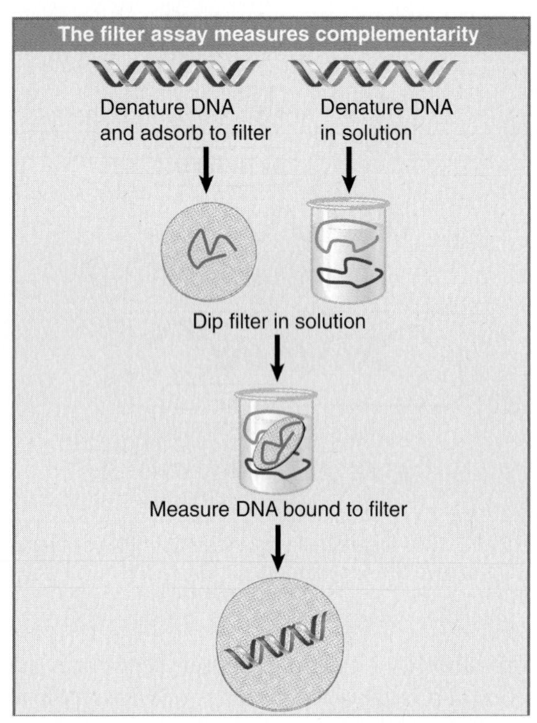

FIGURE 1.21 Filter hybridization establishes whether a solution of denatured DNA (or RNA) contains sequences complementary to the strands immobilized on the filter.

radioactively labeled, so that the reaction can be measured as the amount of radioactive label retained by the filter.

The extent of hybridization between two single-stranded nucleic acids is determined by their complementarity. Two sequences need not be *perfectly* complementary to hybridize. If they are closely related but not identical, an imperfect duplex is formed in which base pairing is interrupted at positions where the two single strands do not correspond.

1.11 Mutations Change the Sequence of DNA

Key concepts

- All mutations consist of changes in the sequence of DNA.
- Mutations may occur spontaneously or may be induced by mutagens.

Mutations provide decisive evidence that DNA is the genetic material. When a change in the sequence of DNA causes an alteration in the sequence of a protein, we may conclude that the DNA codes for that protein. Furthermore, a change in the phenotype of the organism may allow us to identify the function of the protein. The existence of many mutations in a gene may allow many variant forms of a protein to be compared, and a detailed analysis can be used to identify regions of the protein responsible for individual enzymatic or other functions.

All organisms suffer a certain number of mutations as the result of normal cellular operations or random interactions with the environment. These are called **spontaneous mutations;** the rate at which they occur is characteristic for any particular organism and is sometimes called the **background level.** Mutations are rare events, and of course those that damage a gene are selected against during evolution. It is therefore difficult to obtain large numbers of spontaneous mutants to study from natural populations.

The occurrence of mutations can be increased by treatment with certain compounds. These are called **mutagens,** and the changes they cause are referred to as **induced mutations.** Most mutagens act directly by virtue of an ability either to modify a particular base of DNA or to become incorporated into the nucleic acid. The effectiveness of a mutagen is judged by how much it increases the rate of mutation above background. By using mutagens, it becomes possible to induce many changes in any gene.

Spontaneous mutations that inactivate gene function occur in bacteriophages and bacteria at a relatively constant rate of $3\text{–}4 \times 10^{-3}$ per genome per generation. Given the large variation in genome sizes between bacteriophages and bacteria, this corresponds to wide differences in the mutation rate per base pair. This suggests that the overall rate of mutation has been subject to selective forces that have balanced the deleterious effects of most mutations against the advantageous effects of some mutations. This conclusion is strengthened by the observation that an archaeal microbe that lives under harsh conditions of high temperature and acidity (which are expected to damage DNA) does not show an elevated mutation rate, but in fact has an overall mutation rate just below the average range.

FIGURE 1.22 shows that in bacteria, the mutation rate corresponds to $\sim 10^{-6}$ events per locus per generation or to an average rate of change per base pair of $10^{-9}\text{–}10^{-10}$ per generation. The rate at individual base pairs varies very widely, over a 10,000-fold range. We have no accurate measurement of the rate of mutation in eukary-

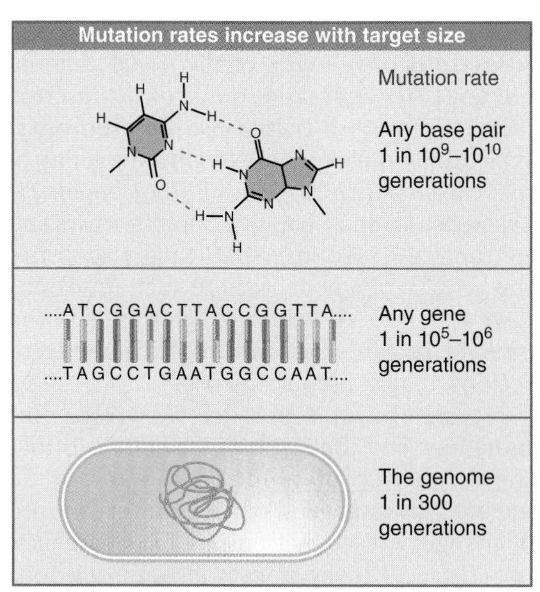

FIGURE 1.22 A base pair is mutated at a rate of 10^{-9}–10^{-10} per generation, a gene of 1000 bp is mutated at $\sim 10^{-6}$ per generation, and a bacterial genome is mutated at 3×10^{-3} per generation.

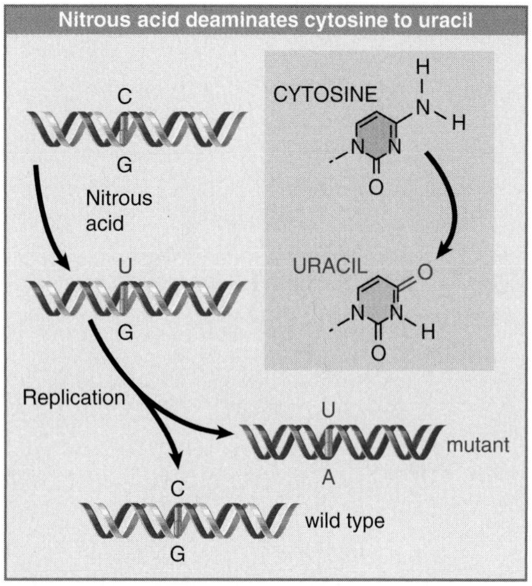

FIGURE 1.23 Mutations can be induced by chemical modification of a base.

otes, although usually it is thought to be somewhat similar to that of bacteria on a per-locus per-generation basis.

1.12 Mutations May Affect Single Base Pairs or Longer Sequences

Key concepts

- A point mutation changes a single base pair.
- Point mutations can be caused by the chemical conversion of one base into another or by mistakes that occur during replication.
- A transition replaces a G-C base pair with an A-T base pair or vice versa.
- A transversion replaces a purine with a pyrimidine, such as changing A-T to T-A.
- Insertions are the most common type of mutation and result from the movement of transposable elements.

Any base pair of DNA can be mutated. A **point mutation** changes only a single base pair and can be caused by either of two types of event:

- Chemical modification of DNA directly changes one base into a different base.
- A malfunction during the replication of DNA causes the wrong base to be inserted into a polynucleotide chain during DNA synthesis.

Point mutations can be divided into two types, depending on the nature of the change when one base is substituted for another:

- The most common class is the **transition,** comprising the substitution of one pyrimidine by the other, or of one purine by the other. This replaces a G-C pair with an A-T pair or vice versa.
- The less common class is the **transversion,** in which a purine is replaced by a pyrimidine or vice versa, so that an A-T pair becomes a T-A or C-G pair.

The effects of nitrous acid provide a classic example of a transition caused by the chemical conversion of one base into another. FIGURE 1.23 shows that nitrous acid performs an oxidative deamination that converts cytosine into uracil. In the replication cycle following the transition, the U pairs with an A, instead of with the G with which the original C would have paired. So the C-G pair is replaced by a T-A pair when the A pairs with the T in the next replication cycle. (Nitrous acid also deaminates adenine, causing the reverse transition from A-T to G-C.)

Transitions are also caused by **base mispairing,** when unusual partners pair in defiance of the usual restriction to Watson–Crick pairs. Base mispairing usually occurs as an aberration resulting from the incorporation into DNA of an abnormal base that has ambiguous pairing properties. FIGURE 1.24 shows the

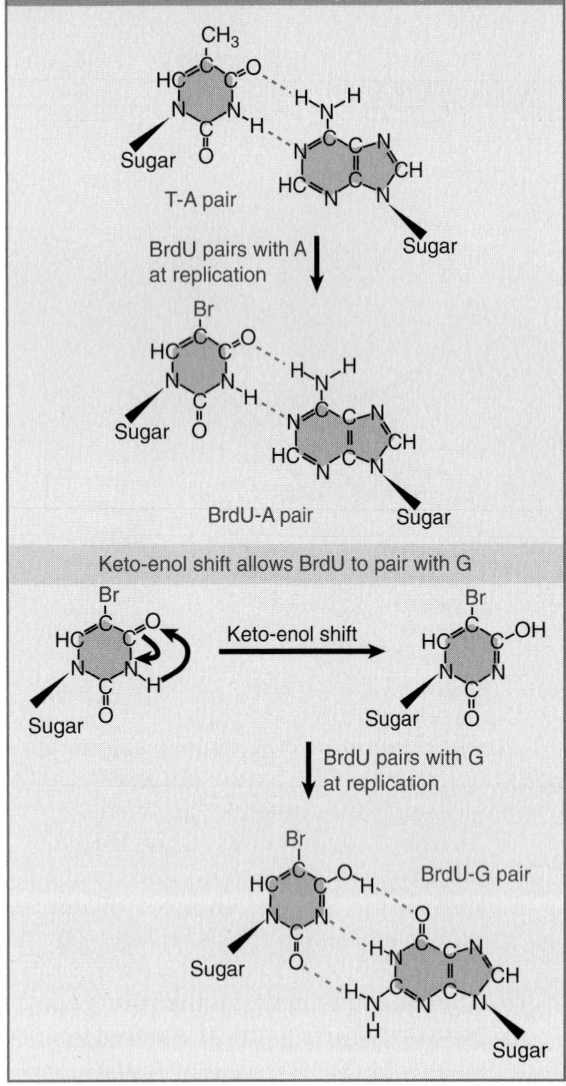

FIGURE 1.24 Mutations can be induced by the incorporation of base analogs into DNA.

example of bromouracil (BrdU), an analog of thymine that contains a bromine atom in place of the methyl group of thymine. BrdU is incorporated into DNA in place of thymine. However, it has ambiguous pairing properties, because the presence of the bromine atom allows a shift to occur in which the base changes structure from a keto (=O) form to an enol (–OH) form. The enol form can base pair with guanine, which leads to substitution of the original A-T pair by a G-C pair.

The mistaken pairing can occur either during the original incorporation of the base or in a subsequent replication cycle. The transition is induced with a certain probability in each replication cycle, so the incorporation of BrdU has continuing effects on the sequence of DNA.

Point mutations were thought for a long time to be the principal means of change in individual genes. However, we now know that **insertions** of stretches of additional material are quite frequent. The source of the inserted material lies with **transposable elements,** which are sequences of DNA with the ability to move from one site to another (see Chapter 21, Transposons, and Chapter 22, Retroviruses and Retroposons.) An insertion usually abolishes the activity of a gene. Where such insertions have occurred, **deletions** of part or all of the inserted material, and sometimes of the adjacent regions, may subsequently occur.

A significant difference between point mutations and the insertions/deletions is that the frequency of point mutation can be increased by mutagens, whereas the occurrence of changes caused by transposable elements is not affected. However, insertions and deletions can also occur by other mechanisms—for example, those involving mistakes made during replication or recombination—although probably these are less common. In addition, a class of mutagens called the acridines introduce (very small) insertions and deletions.

1.13 The Effects of Mutations Can Be Reversed

Key concepts

- Forward mutations inactivate a gene, and back mutations (or revertants) reverse their effects.
- Insertions can revert by deletion of the inserted material, but deletions cannot revert.
- Suppression occurs when a mutation in a second gene bypasses the effect of mutation in the first gene.

FIGURE 1.25 shows that the isolation of **revertants** is an important characteristic that distinguishes point mutations and insertions from deletions:

- A point mutation can revert by restoring the original sequence or by gaining a compensatory mutation elsewhere in the gene.
- An insertion of additional material can revert by deletion of the inserted material.
- A deletion of part of a gene cannot revert.

Mutations that inactivate a gene are called **forward mutations.** Their effects are reversed by **back mutations,** which are of two types: true reversion and second-site reversion.

An exact reversal of the original mutation is called **true reversion.** So if an A-T pair has

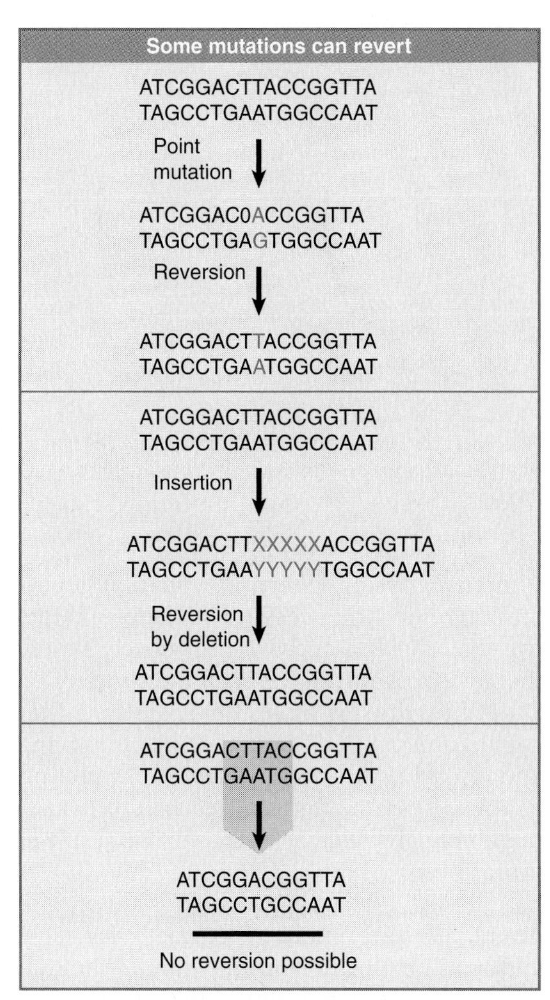

FIGURE 1.25 Point mutations and insertions can revert, but deletions cannot revert.

been replaced by a G-C pair, another mutation to restore the A-T pair will exactly regenerate the wild-type sequence.

The second type of back mutation, **second-site reversion,** may occur elsewhere in the gene, and its effects compensate for the first mutation. For example, one amino acid change in a protein may abolish gene function, but a second alteration may compensate for the first and restore protein activity.

A forward mutation results from any change that inactivates a gene, whereas a back mutation must restore function to a protein damaged by a particular forward mutation. So the demands for back mutation are much more specific than those for forward mutation. The rate of back mutation is correspondingly lower than that of forward mutation, typically by a factor of ~10.

Mutations can also occur in other genes to circumvent the effects of mutation in the original gene. This effect is called **suppression.** A locus in which a mutation suppresses the effect of a mutation in another locus is called a **suppressor.**

1.14 Mutations Are Concentrated at Hotspots

Key concept

- The frequency of mutation at any particular base pair is determined by statistical fluctuation, except for hotspots, where the frequency is increased by at least an order of magnitude.

So far we have dealt with mutations in terms of individual changes in the sequence of DNA that influence the activity of the genetic unit in which they occur. When we consider mutations in terms of the inactivation of the gene, most genes within a species show more or less similar rates of mutation relative to their size. This suggests that the gene can be regarded as a target for mutation, and that damage to any part of it can abolish its function. As a result, susceptibility to mutation is roughly proportional to the size of the gene. But consider the sites of mutation within the sequence of DNA: are all base pairs in a gene equally susceptible, or are some more likely to be mutated than others?

What happens when we isolate a large number of independent mutations in the same gene? Many mutants are obtained. Each is the result of an individual mutational event. Then the site of each mutation is determined. Most mutations will lie at different sites, but some will lie at the same position. Two independently isolated mutations at the same site may constitute exactly the same change in DNA (in which case the same mutational event has happened on more than one occasion), or they may constitute different changes (three different point mutations are possible at each base pair).

The histogram of FIGURE 1.26 shows the frequency with which mutations are found at each base pair in the *lacI* gene of *E. coli.* The statistical probability that more than one mutation occurs at a particular site is given by random-hit kinetics (as seen in the Poisson distribution). So some sites will gain one, two, or three mutations, whereas others will not gain any. Some sites gain far more than the number of mutations expected from a random distribution; they may have 10× or even 100× more mutations

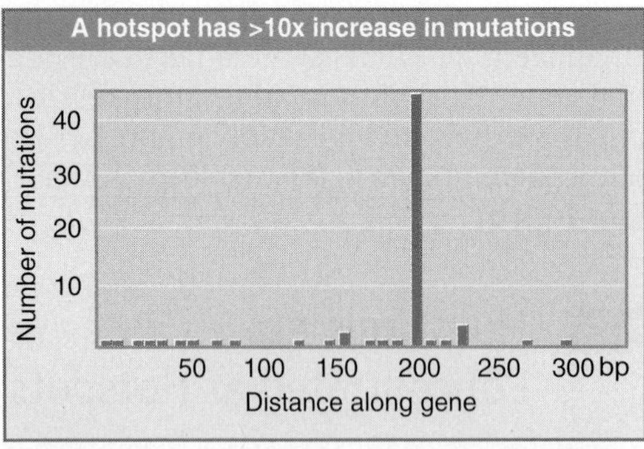

FIGURE 1.26 Spontaneous mutations occur throughout the *lacI* gene of *E. coli* but are concentrated at a hotspot.

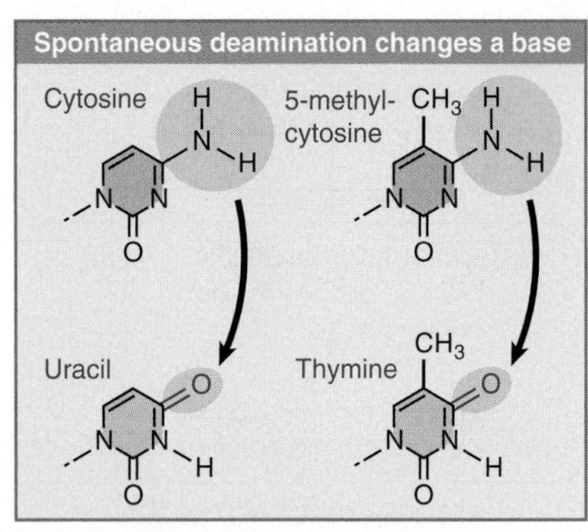

FIGURE 1.27 Deamination of cytosine produces uracil, whereas deamination of 5-methylcytosine produces thymine.

than predicted by random hits. These sites are called **hotspots.** Spontaneous mutations may occur at hotspots, and different mutagens may have different hotspots.

1.15 Many Hotspots Result from Modified Bases

Key concept

- A common cause of hotspots is the modified base 5-methylcytosine, which is spontaneously deaminated to thymine.

A major cause of spontaneous mutation results from the presence of an unusual base in the DNA. In addition to the four bases that are inserted into DNA when it is synthesized, **modified bases** are sometimes found. The name reflects their origin; they are produced by chemically modifying one of the four bases already present in DNA. The most common modified base is 5-methylcytosine, which is generated by a methylase enzyme that adds a methyl group to certain cytosine residues at specific sites in the DNA.

Sites containing 5-methylcytosine provide hotspots for spontaneous point mutation in *E. coli*. In each case, the mutation takes the form of a G-C to A-T transition. The hotspots are not found in strains of *E. coli* that cannot methylate cytosine.

The reason for the existence of the hotspots is that cytosine bases suffer spontaneous deamination at an appreciable frequency. In this reaction, the amino group is replaced by a keto group. Recall that deamination of cytosine generates uracil (see Figure 1.23). FIGURE 1.27 compares this reaction with the deamination of 5-methylcytosine where deamination generates thymine. The effect in DNA is to generate the base pairs G-U and G-T, respectively, where there is a **mismatch** between the partners.

All organisms have repair systems that correct mismatched base pairs by removing and replacing one of the bases. The operation of these systems determines whether mismatched pairs such as G-U and G-T result in mutations.

FIGURE 1.28 shows that the consequences of deamination are different for 5-methylcytosine and cytosine. Deaminating the (rare) 5-methylcytosine causes a mutation, whereas deamination of the more common cytosine does not have this effect. This happens because the repair systems are much more effective in recognizing G-U than G-T.

E. coli contains an enzyme, uracil-DNA-glycosidase, that removes uracil residues from DNA (see Section 20.5, Base Flipping Is Used by Methylases and Glycosylases). This action leaves an unpaired G residue, and a "repair system" then inserts a C base to partner it. The net result of these reactions is to restore the original sequence of the DNA. This system protects DNA against the consequences of spontaneous deamination of cytosine. (This system is not, however, active enough to prevent the effects of the increased level of deamination caused by nitrous acid; see Figure 1.23.)

Note that the deamination of 5-methylcytosine leaves thymine. This creates a mismatched base pair, G-T. If the mismatch is not corrected before the next replication cycle, a mutation results. At the next replication, the bases in the mispaired G-T partnership separate, and then they pair with new partners to

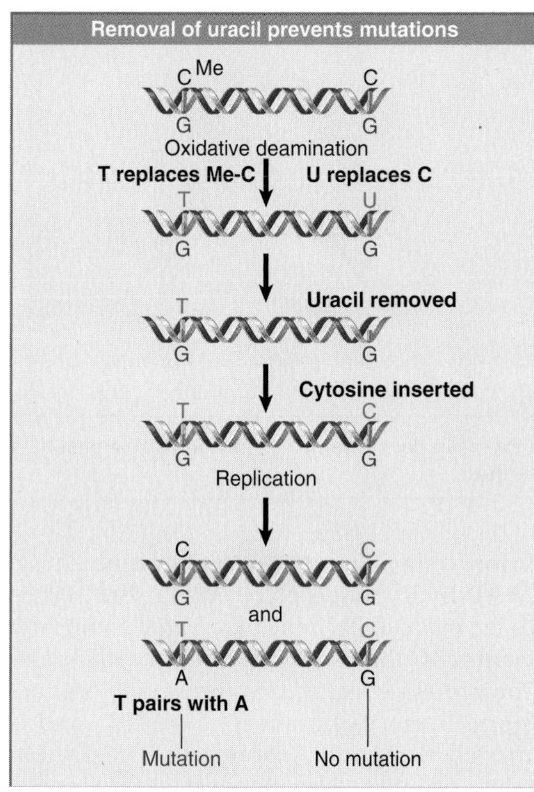

FIGURE 1.28 The deamination of 5-methylcytosine produces thymine (by C-G to T-A transitions), while the deamination of cytosine produces uracil (which usually is removed and then replaced by cytosine).

produce one wild-type G-C pair and one mutant A-T pair.

Deamination of 5-methylcytosine is the most common cause of production of G-T mismatched pairs in DNA. Repair systems that act on G-T mismatches have a bias toward replacing the T with a C (rather than the alternative of replacing the G with an A), which helps to reduce the rate of mutation (see Section 20.7, Controlling the Direction of Mismatch Repair). However, these systems are not as effective as the removal of U from G-U mismatches. As a result, deamination of 5-methylcytosine leads to mutation much more often than does deamination of cytosine.

5-methylcytosine also creates hotspots in eukaryotic DNA. It is common at CpG dinucleotides that are concentrated in regions called CpG islands (see Section 24.19, CpG Islands Are Regulatory Targets). Although 5-methylcytosine accounts for ~1% of the bases in human DNA, sites containing the modified base account for ~30% of all point mutations. This makes the state of 5-methylcytosine a particularly important determinant of mutation in animal cells.

The importance of repair systems in reducing the rate of mutation is emphasized by the effects of eliminating the mouse enzyme MBD4, a glycosylase that can remove T (or U) from mismatches with G. The result is to increase the mutation rate at CpG sites by a factor of 3×. (The reason the effect is not greater is that MBD4 is only one of several systems that act on G-T mismatches; we can imagine that elimination of all the systems would increase the mutation rate much more.)

The operation of these systems casts an interesting light on the use of T in DNA compared with U in RNA. Perhaps it relates to the need of DNA for stability of sequence; the use of T means that any deaminations of C are immediately recognized because they generate a base (U) not usually present in the DNA. This greatly increases the efficiency with which repair systems can function (compared with the situation when they have to recognize G-T mismatches, which can be produced also by situations where removing the T would not be the appropriate response). Also, the phosphodiester bond of the backbone is more labile when the base is U.

1.16 Some Hereditary Agents Are Extremely Small

Key concept

- Some very small hereditary agents do not code for protein but consist of RNA or of protein that has hereditary properties.

Viroids are infectious agents that cause diseases in higher plants. They are very small circular molecules of RNA. Unlike viruses—for which the infectious agent consists of a **virion,** a genome encapsulated in a protein coat—*the viroid RNA is itself the infectious agent.* The viroid consists solely of the RNA, which is extensively but imperfectly base paired, forming a characteristic rod like the example shown in FIGURE 1.29. Mutations that interfere with the structure of the rod reduce infectivity.

A viroid RNA consists of a single molecular species that is replicated autonomously in infected cells. Its sequence is faithfully perpetuated in its descendants. Viroids fall into several groups. A given viroid is identified with a group by its similarity of sequence with other members of the group. For example, four viroids related to PSTV (potato spindle tuber viroid)

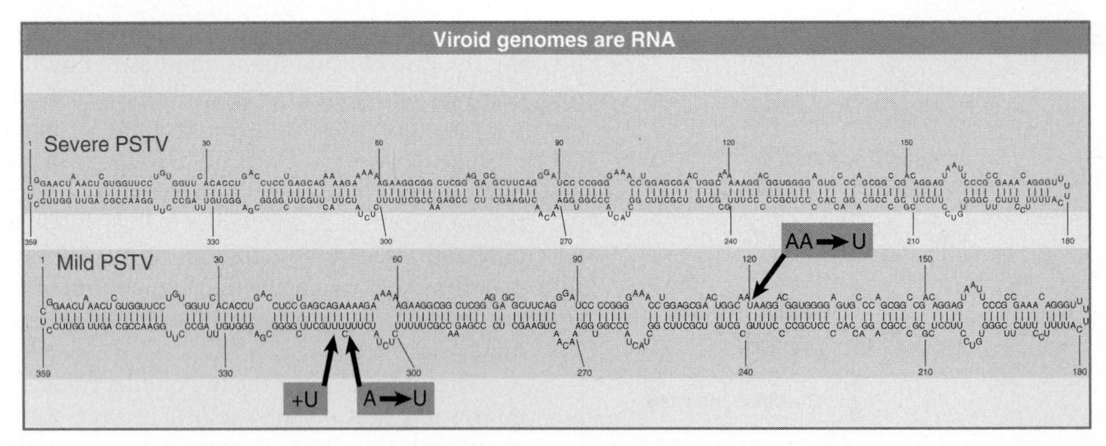

FIGURE 1.29 PSTV RNA is a circular molecule that forms an extensive double-stranded structure, interrupted by many interior loops. The severe and mild forms differ at three sites.

have 70%–83% similarity of sequence with it. Different isolates of a particular viroid strain vary from one another, and the change may affect the phenotype of infected cells. For example, the *mild* and *severe* strains of PSTV differ by three nucleotide substitutions.

Viroids resemble viruses in having heritable nucleic acid genomes. They fulfill the criteria for genetic information. Yet viroids, which are sometimes called **subviral pathogens,** differ from viruses in both structure and function. Viroid RNA does not appear to be translated into protein, so it cannot itself code for the functions needed for its survival. This situation poses two questions: How does viroid RNA replicate, and how does it affect the phenotype of the infected plant cell?

Replication must be carried out by enzymes of the host cell, subverted from their normal function. The heritability of the viroid sequence indicates that viroid RNA provides the template.

Viroids are presumably pathogenic because they interfere with normal cellular processes. They might do this in a relatively random way, for example, by sequestering an essential enzyme for their own replication or by interfering with the production of necessary cellular RNAs. Alternatively, they might behave as abnormal regulatory molecules, with particular effects upon the expression of individual genes.

An even more unusual agent is **scrapie,** the cause of a degenerative neurological disease of sheep and goats. The disease is related to the human diseases of kuru and Creutzfeldt–Jakob syndrome, which affect brain function.

The infectious agent of scrapie does not contain nucleic acid. This extraordinary agent is called a **prion** (proteinaceous infectious agent). It is a 28 kD hydrophobic glycoprotein, **PrP.** PrP is coded by a cellular gene (conserved among the mammals) that is expressed in normal brain. The protein exists in two forms: The product found in normal brain is called PrP^{c} and is entirely degraded by proteases. The protein found in infected brains is called PrP^{sc} and is extremely resistant to degradation by proteases. PrP^{c} is converted to PrP^{sc} by a modification or conformational change that confers protease-resistance, and which has yet to be fully defined.

As the infectious agent of scrapie, PrP^{sc} must in some way modify the synthesis of its normal cellular counterpart so that it becomes infectious instead of harmless (see Section 31.12, Prions Cause Diseases in Mammals). Mice that lack a PrP gene cannot be infected to develop scrapie, which demonstrates that PrP is essential for development of the disease.

1.17 Summary

Two classic experiments proved that DNA is the genetic material. DNA isolated from one strain of *Pneumococcus* bacteria can confer properties of that strain upon another strain. In addition, DNA is the only component that is inherited by progeny phages from the parental phages. DNA can be used to transfect new properties into eukaryotic cells.

DNA is a double helix consisting of antiparallel strands in which the nucleotide units are linked by 5′to–3′ phosphodiester bonds. The backbone provides the exterior; purine and pyrimidine bases are stacked in the interior in pairs in which A is complementary to T and G is complementary to C. The strands separate and use complementary base pairing to assemble daughter strands in semiconservative replication. Complementary base pairing is also used to transcribe an RNA representing one strand of a DNA duplex.

A stretch of DNA may code for protein. The genetic code describes the relationship between the sequence of DNA and the sequence of the protein. Only one of the two strands of DNA codes for protein. A codon consists of three nucleotides that represent a single amino acid. A coding sequence of DNA consists of a series of codons, which are read from a fixed starting point. Usually only one of the three possible reading frames can be translated into protein.

A mutation consists of a change in the sequence of A-T and G-C base pairs in DNA. A mutation in a coding sequence may change the sequence of amino acids in the corresponding protein. A frameshift mutation alters the subsequent reading frame by inserting or deleting a base; this causes an entirely new series of amino acids to be coded after the site of mutation. A point mutation changes only the amino acid represented by the codon in which the mutation occurs. Point mutations may be reverted by back mutation of the original mutation. Insertions may revert by loss of the inserted material, but deletions cannot revert. Mutations may also be suppressed indirectly when a mutation in a different gene counters the original defect.

The natural incidence of mutations is increased by mutagens. Mutations may be concentrated at hotspots. A type of hotspot responsible for some point mutations is caused by deamination of the modified base 5-methylcytosine.

Forward mutations occur at a rate of ~10^{-6} per locus per generation; back mutations are rarer. Not all mutations have an effect on the phenotype.

Although all genetic information in cells is carried by DNA, viruses have genomes of double-stranded or single-stranded DNA or RNA. Viroids are subviral pathogens that consist solely of small circular molecules of RNA, with no protective packaging. The RNA does not code for protein and its mode of perpetuation and of pathogenesis is unknown. Scrapie consists of a proteinaceous infectious agent.

References

1.1 Introduction

Reviews

Cairns, J., Stent, G., and Watson, J. D. (1966). Phage and the Origins of Molecular Biology. *Cold Spring Harbor Symp. Quant. Biol.*

Judson, H. (1978). *The Eighth Day of Creation.* Knopf, New York.

Olby, R. (1974). *The Path to the Double Helix.* MacMillan, London.

1.2 DNA Is the Genetic Material of Bacteria

Research

Avery, O. T., MacLeod, C. M., and McCarty, M. (1944). Studies on the chemical nature of the substance inducing transformation of pneumococcal types. *J. Exp. Med.* 98, 451–460.

Griffith, F. (1928). The significance of pneumococcal types. *J. Hyg.* 27, 113–159.

1.3 DNA Is the Genetic Material of Viruses

Research

Hershey, A. D. and Chase, M. (1952). Independent functions of viral protein and nucleic acid in growth of bacteriophage *J. Gen. Physiol.* 36, 39–56.

1.4 DNA Is the Genetic Material of Animal Cells

Research

Pellicer, A., Wigler, M., Axel, R., and Silverstein, S. (1978). The transfer and stable integration of the HSV thymidine kinase gene into mouse cells. *Cell* 14, 133–141.

1.6 DNA Is a Double Helix

Research

Watson, J. D., and Crick, F. H. C. (1953). A structure for DNA. *Nature* 171, 737–738.

Watson, J. D., and Crick, F. H. C. (1953). Genetic implications of the structure of DNA. *Nature* 171, 964–967.

Wilkins, M. F. H., Stokes, A. R., and Wilson, H. R. (1953). Molecular structure of DNA. *Nature* 171, 738–740.

1.7 DNA Replication Is Semiconservative

Review

Holmes, F. (2001). *Meselson, Stahl, and the Replication of DNA: A History of the Most Beautiful Experiment in Biology.* Yale University Press, New Haven, CT.

Research

Meselson, M. and Stahl, F. W. (1958). The replication of DNA in *E. coli*. *Proc. Natl. Acad. Sci. USA* 44, 671–682.

1.11 Mutations Change the Sequence of DNA

Reviews

Drake, J. W., Charlesworth, B., Charlesworth, D., and Crow, J. F. (1998). Rates of spontaneous mutation. *Genetics* 148, 1667–1686.

Drake, J. W. and Balz, R. H. (1976). The biochemistry of mutagenesis. *Annu. Rev. Biochem.* 45, 11–37.

Research

Drake, J. W. (1991). A constant rate of spontaneous mutation in DNA-based microbes. *Proc. Natl. Acad. Sci. USA* 88, 7160–7164.

Grogan, D. W., Carver, G. T., and Drake, J. W. (2001). Genetic fidelity under harsh conditions: analysis of spontaneous mutation in the thermoacidophilic archaeon *Sulfolobus acidocaldarius*. *Proc. Natl. Acad. Sci. USA* 98, 7928–7933.

1.12 Mutations May Affect Single Base Pairs or Longer Sequences

Review

Maki, H. (2002). Origins of spontaneous mutations: specificity and directionality of base-substitution, frameshift, and sequence-substitution mutageneses. *Annu. Rev. Genet.* 36, 279–303.

1.15 Many Hotspots Result from Modified Bases

Research

Coulondre, C. et al. (1978). Molecular basis of base substitution hotspots in *E. coli*. *Nature* 274, 775–780.

Millar, C. B., Guy, J., Sansom, O. J., Selfridge, J., MacDougall, E., Hendrich, B., Keightley, P. D., Bishop, S. M., Clarke, A. R., and Bird, A. (2002). Enhanced CpG mutability and tumorigenesis in MBD4-deficient mice. *Science* 297, 403–405.

1.16 Some Hereditary Agents Are Extremely Small

Reviews

Diener, T. O. (1986). Viroid processing: a model involving the central conserved region and hairpin. *Proc. Natl. Acad. Sci. USA* 83, 58–62.

Diener, T. O. (1999). Viroids and the nature of viroid diseases. *Arch. Virol. Suppl.* 15, 203–220.

Prusiner, S. B. (1998). Prions. *Proc. Natl. Acad. Sci. USA* 95, 13363–13383.

Research

Bueler, H. et al. (1993). Mice devoid of PrP are resistant to scrapie. *Cell* 73, 1339–1347.

McKinley, M. P., Bolton, D. C., and Prusiner, S. B. (1983). A protease-resistant protein is a structural component of the scrapie prion. *Cell* 35, 57–62.

2

Genes Code for Proteins

CHAPTER OUTLINE

2.1 Introduction

The gene is the functional unit of heredity. Each gene is a sequence within the genome that functions by giving rise to a discrete product (which may be a protein or an RNA). The basic behavior of the gene was defined by Mendel more than a century ago. Summarized in his two laws, the gene was recognized as a "particulate factor" that passes unchanged from parent to progeny. A gene may exist in alternative forms. These forms are called alleles.

In diploid organisms, which have two sets of chromosomes, one copy of each chromosome is inherited from each parent. This is the same behavior that is displayed by genes. One of the two copies of each gene is the paternal allele (inherited from the father), the other is the maternal allele (inherited from the mother). The equivalence led to the discovery that chromosomes in fact carry the genes.

Each chromosome consists of a linear array of genes. Each gene resides at a particular location on the chromosome. The location is more formally called a genetic locus. The alleles of a gene are the different forms that are found at its locus.

The key to understanding the organization of genes into chromosomes was the discovery of genetic linkage—the tendency for genes on the same chromosome to remain together in the progeny instead of assorting independently as predicted by Mendel's laws. Once the unit of recombination (reassortment) was introduced as the measure of linkage, the construction of genetic maps became possible.

The resolution of the recombination map of a higher eukaryote is restricted by the small number of progeny that can be obtained from each mating. Recombination occurs so infrequently between nearby points that it is rarely observed between different mutations in the same gene. As a result, classical linkage maps of eukaryotes can place the genes in order, but cannot determine relationships within a gene. By moving to a microbial system in which a very large number of progeny can be obtained from each genetic cross, researchers could demonstrate that recombination occurs within genes. It follows the same rules that were previously deduced for recombination between genes.

Mutations within a gene can be arranged into a linear order, showing that the gene itself has the same linear construction as the array of genes on a chromosome. So the genetic map is linear within as well as between loci: it consists of an unbroken sequence within which the genes reside. This conclusion leads naturally into the modern view summarized in FIGURE 2.1 that the genetic material of a chromosome consists of an uninterrupted length of DNA representing many genes.

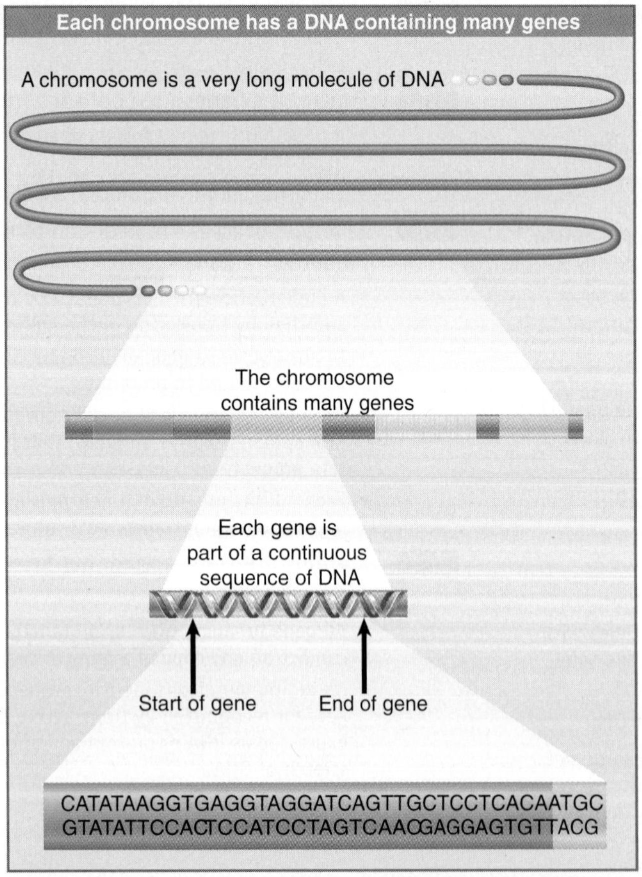

FIGURE 2.1 Each chromosome has a single long molecule of DNA within which are the sequences of individual genes.

2.2 A Gene Codes for a Single Polypeptide

Key concepts

- The one gene : one enzyme hypothesis summarizes the basis of modern genetics: that a gene is a stretch of DNA coding for a single polypeptide chain.
- Most mutations damage gene function.

The first systematic attempt to associate genes with enzymes showed that each stage in a metabolic pathway is catalyzed by a single enzyme and can be blocked by mutation in a different gene. This led to the *one gene : one enzyme hypothesis*. Each metabolic step is catalyzed by a particular enzyme, whose production is the responsibility of a single gene. A mutation in the gene alters the activity of the protein for which it is responsible.

A modification in the hypothesis is needed to accommodate proteins that consist of more than one subunit. If the subunits are all the same, the protein is a **homomultimer,** represented by a single gene. If the subunits are different, the protein is a **heteromultimer.** Stated as a more general rule applicable to any heteromultimeric protein, the one gene : one enzyme hypothesis becomes more precisely expressed as *one gene : one polypeptide chain.*

Identifying which protein represents a particular gene can be a protracted task. The mutation responsible for creating Mendel's wrinkled-pea mutant was identified only in 1990 as an alteration that inactivates the gene for a starch branching enzyme!

It is important to remember that a gene does not directly generate a protein. As shown previously in Figure 1.2, a gene codes for an RNA, which may in turn code for a protein. Most genes code for proteins, but some genes code for RNAs that do not give rise to proteins. These RNAs may be structural components of the apparatus responsible for synthesizing proteins or may have roles in regulating gene expression. The basic principle is that the gene is a sequence of DNA that specifies the sequence of an independent product. The process of gene expression may terminate in a product that is either RNA or protein.

A mutation is a random event with regard to the structure of the gene, so the greatest probability is that it will damage or even abolish gene function. Most mutations that affect gene function are recessive: *they represent an absence of function, because the mutant gene has been prevented from producing its usual protein.* FIGURE 2.2 illustrates the relationship between recessive and wild-type alleles. When a heterozygote contains one wild-type allele and one mutant allele, the wild-type allele is able to direct production of the enzyme. The wild-type allele is therefore dominant. (This assumes that an adequate *amount* of protein is made by the single wild-type allele. When this is not true, the smaller amount made by one allele as compared to two alleles results in the intermediate phenotype of a partially dominant allele in a heterozygote.)

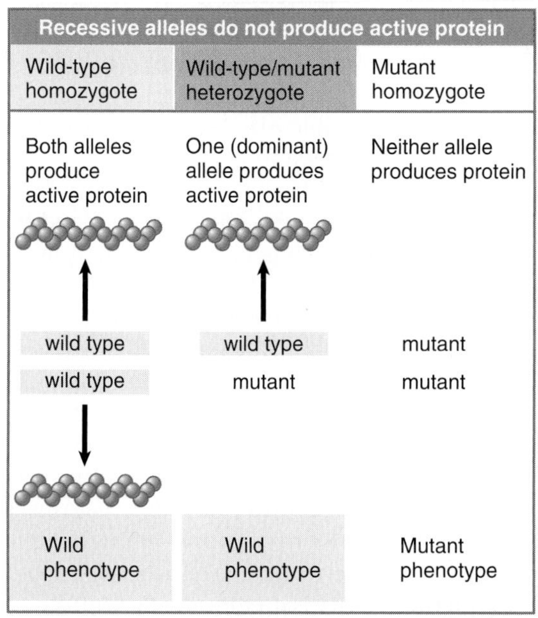

FIGURE 2.2 Genes code for proteins; dominance is explained by the properties of mutant proteins. A recessive allele does not contribute to the phenotype because it produces no protein (or protein that is nonfunctional).

2.3 Mutations in the Same Gene Cannot Complement

Key concepts

- A mutation in a gene affects only the protein coded by the mutant copy of the gene and does not affect the protein coded by any other allele.
- Failure of two mutations to complement (produce wild phenotype) when they are present in *trans* configuration in a heterozygote means that they are part of the same gene.

How do we determine whether two mutations that cause a similar phenotype lie in the same gene? If they map close together, they may be alleles. However, they could also represent mutations in two *different* genes whose proteins are involved in the same function. The **complementation test** is used to determine whether two mutations lie in the same gene or in different genes. The test consists of making a heterozygote for the two mutations (by mating parents homozygous for each mutation).

If the mutations lie in the same gene, the parental genotypes can be represented as:

$$\frac{m_1}{m_1} \text{ and } \frac{m_2}{m_2}$$

The first parent provides an m_1 mutant allele and the second parent provides an m_2 allele, so that the heterozygote has the constitution:

$$\frac{m_1}{m_2}$$

No wild-type gene is present, so the heterozygote has mutant phenotype.

If the mutations lie in different genes, the parental genotypes can be represented as:

$$\frac{m_1 +}{m_1 +} \text{ and } \frac{+ m_2}{+ m_2}$$

Each chromosome has a wild-type copy of one gene (represented by the plus sign) and a mutant copy of the other. Then the heterozygote has the constitution:

$$\frac{m_1 +}{+ m_2}$$

in which the two parents between them have provided a wild-type copy of each gene. The heterozygote has wild phenotype, and thus the two genes are said to **complement.**

The complementation test is shown in more detail in FIGURE 2.3. The basic test consists of the comparison shown in the top part of the figure. If two mutations lie in the same gene, we see a difference in the phenotypes of the *trans* configuration and the *cis* configuration. The *trans* configuration is mutant, because each allele has a (different) mutation. However, the *cis* configuration is wild-type, because one allele has two mutations and the other allele has no mutations. The lower part of the figure shows that if the two mutations lie in different genes, we always see a wild phenotype. There is always one wild-type and one mutant allele of each gene, and the configuration is irrelevant. Failure to complement means that two mutations are part of the *same* genetic unit. Mutations that do not complement one another are said to comprise part of the same **complementation group.** Another term that is used to describe the unit defined by the complementation test is the **cistron.** This is the same as the gene. Basically these three terms all describe a stretch of DNA that functions as a unit to give rise to an RNA or protein product. The properties of the gene with regard to complementation are explained by the fact that this product is a single molecule that behaves as a functional unit.

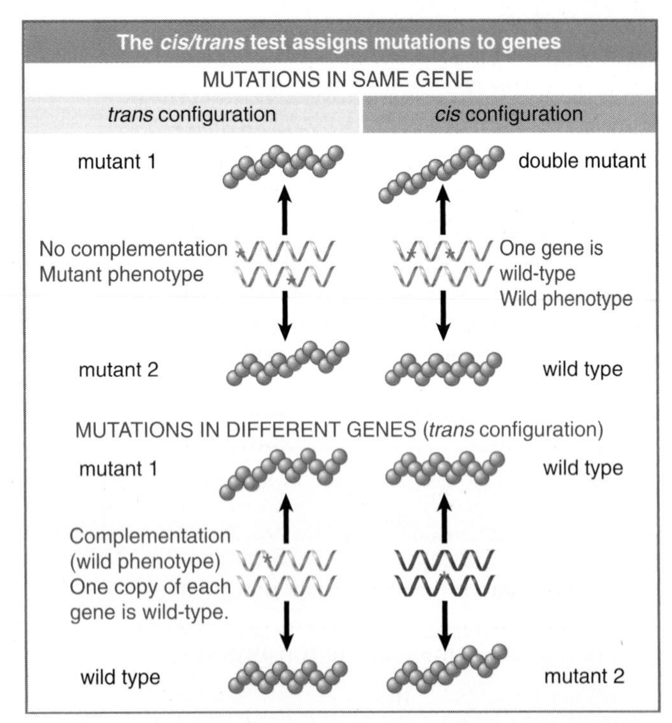

FIGURE 2.3 The cistron is defined by the complementation test. Genes are represented by spirals; red stars identify sites of mutation.

2.4 Mutations May Cause Loss-of-Function or Gain-of-Function

Key concepts

- Recessive mutations are due to loss-of-function by the protein product.
- Dominant mutations result from a gain-of-function.
- Testing whether a gene is essential requires a null mutation (one that completely eliminates its function).
- Silent mutations have no effect, either because the base change does not change the sequence or amount of protein, or because the change in protein sequence has no effect.
- Leaky mutations do affect the function of the gene product, but are not revealed in the phenotype because sufficient activity remains.

The various possible effects of mutation in a gene are summarized in FIGURE 2.4.

When a gene has been identified, insight into its function in principle can be gained by generating a mutant organism that entirely lacks the gene. A mutation that completely eliminates gene function—usually because the gene has been deleted—is called a **null mutation.** If a gene is essential, a null mutation is lethal.

To determine what effect a gene has upon the phenotype, it is essential to characterize a null mutant. When a mutation fails to affect the phenotype, it is always possible that this is because it is a **leaky mutation**—enough active product is made to fulfill its function, even though the activity is quantitatively reduced or qualitatively different from the wild type. However, if a null mutant fails to affect a phenotype, we may safely conclude that the gene function is not necessary.

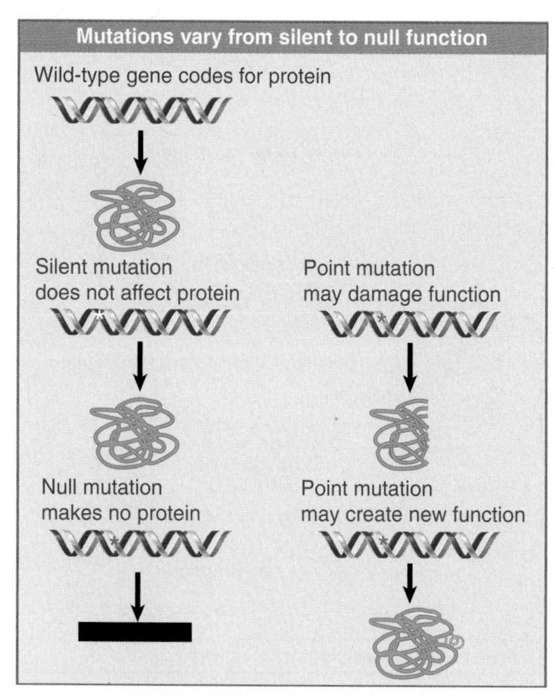

FIGURE 2.4 Mutations that do not affect protein sequence or function are silent. Mutations that abolish all protein activity are null. Point mutations that cause loss-of-function are recessive; those that cause gain-of-function are dominant.

Null mutations, or other mutations that impede gene function (but do not necessarily abolish it entirely), are called **loss-of-function mutations.** A loss-of-function mutation is recessive (as in the example of Figure 2.2). Sometimes a mutation has the opposite effect and causes a protein to acquire a new function; such a change is called a **gain-of-function mutation.** A gain-of-function mutation is dominant.

Not all mutations in DNA lead to a detectable change in the phenotype. Mutations without apparent effect are called **silent mutations.** They fall into two types: One type involves base changes in DNA that do not cause any change in the amino acid present in the corresponding protein. The second type changes the amino acid, but the replacement in the protein does not affect its activity; these are called **neutral substitutions.**

2.5 A Locus May Have Many Different Mutant Alleles

Key concept

- The existence of multiple alleles allows heterozygotes to occur representing any pairwise combination of alleles.

If a recessive mutation is produced by every change in a gene that prevents the production of an active protein, there should be a large number of such mutations in any one gene. Many amino acid replacements may change the structure of the protein sufficiently to impede its function.

Different variants of the same gene are called **multiple alleles,** and their existence makes it possible to create a heterozygote between mutant alleles. The relationship between these multiple alleles takes various forms.

In the simplest case, a wild-type gene codes for a protein product that is functional. Mutant allele(s) code for proteins that are nonfunctional.

But there are often cases in which a series of mutant alleles have different phenotypes. For example, wild-type function of the *white* locus of *Drosophila melanogaster* is required for development of the normal red color of the eye. The locus is named for the effect of extreme (null) mutations, which cause the fly to have a white eye in mutant homozygotes.

To describe wild-type and mutant alleles, wild genotype is indicated by a plus superscript after the name of the locus (w^+ is the wild-type allele for [red] eye color in *D. melanogaster*). Sometimes + is used by itself to describe the wild-type allele, and only the mutant alleles are indicated by the name of the locus.

An entirely defective form of the gene (or absence of phenotype) may be indicated by a minus superscript. To distinguish among a variety of mutant alleles with different effects, other superscripts may be introduced, such as w^i or w^a.

The w^+ allele is dominant over any other allele in heterozygotes. There are many different mutant alleles. FIGURE 2.5 shows a (small) sample. Although some alleles have no eye color, many alleles produce some color. Each of these mutant alleles must therefore represent a different mutation of the gene, which does not eliminate its function entirely, but leaves a residual activity that produces a characteristic phenotype. These alleles are named for the color of the eye in a homozygote. (Most *w* alleles affect the quantity of pigment in the eye. The examples in the figure are arranged in [roughly] declining amount of color, but others, such as w^{sp}, affect the pattern in which it is deposited.)

When multiple alleles exist, an animal may be a heterozygote that carries two different mutant alleles. The phenotype of such a heterozygote depends on the nature of the residual activity of each allele. The relationship

Each allele has a different phenotype	
Allele	Phenotype of homozygote
w^{+}	red eye (wild type)
w^{bl}	blood
w^{ch}	cherry
w^{bf}	buff
w^{h}	honey
w^{a}	apricot
w^{e}	eosin
w^{i}	ivory
w^{z}	zeste (lemon-yellow)
w^{sp}	mottled, color varies
w^{1}	white (no color)

FIGURE 2.5 The *w* locus has an extensive series of alleles whose phenotypes extend from wild-type (red) color to complete lack of pigment.

between two mutant alleles is in principle no different from that between wild-type and mutant alleles: one allele may be dominant, there may be partial dominance, or there may be codominance.

2.6 A Locus May Have More than One Wild-type Allele

Key concept

- A locus may have a polymorphic distribution of alleles with no individual allele that can be considered to be the sole wild-type.

There is not necessarily a unique wild-type allele at any particular locus. Control of the human blood group system provides an example. Lack of function is represented by the null type, *O* group. However, the functional alleles *A* and *B* provide activities that are codominant with one another and dominant over *O* group. The basis for this relationship is illustrated in FIGURE 2.6.

The O (or H) antigen is generated in all individuals and consists of a particular carbohydrate group that is added to proteins. The *ABO* locus codes for a galactosyltransferase enzyme that adds a further sugar group to the O antigen. The specificity of this enzyme determines the blood group. The *A* allele produces an enzyme that uses the cofactor UDP-N-acetylgalactose, creating the A antigen. The *B* allele produces an enzyme that uses the cofactor UDP-galactose, creating the B antigen. The A and B versions of the transferase protein differ in four amino acids that presumably affect its recognition of the type of cofactor. The *O* allele has a mutation (a small deletion) that eliminates activity, so no modification of the O antigen occurs.

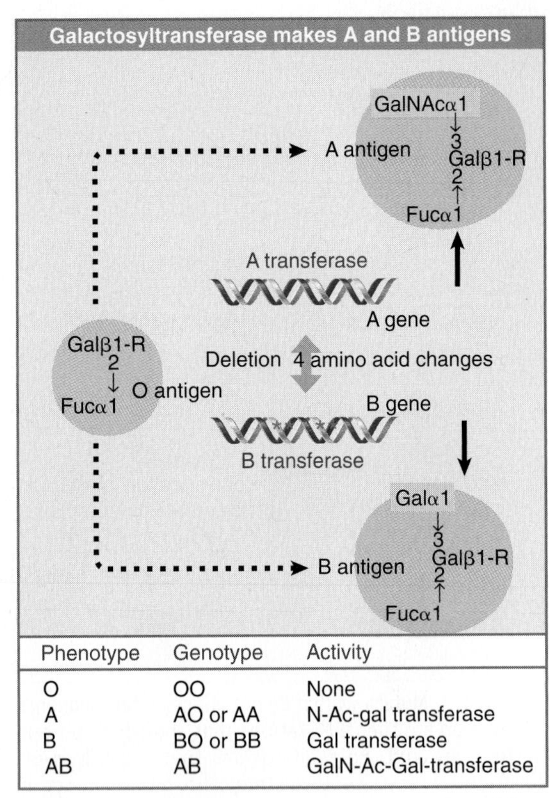

Phenotype	Genotype	Activity
O	OO	None
A	AO or AA	N-Ac-gal transferase
B	BO or BB	Gal transferase
AB	AB	GalN-Ac-Gal-transferase

FIGURE 2.6 The ABO blood group locus codes for a galactosyltransferase whose specificity determines the blood group.

This explains why *A* and *B* alleles are dominant in the *AO* and *BO* heterozygotes: the corresponding transferase activity creates the A or B antigen. The *A* and *B* alleles are codominant in *AB* heterozygotes, because both transferase activities are expressed. The *OO* homozygote is a null that has neither activity and therefore lacks both antigens.

Neither *A* nor *B* can be regarded as uniquely wild type, because they represent alternative activities rather than loss or gain of function. A situation such as this, in which there are multiple functional alleles in a population, is described as a **polymorphism** (see Section 4.3, Individual Genomes Show Extensive Variation).

2.7 Recombination Occurs by Physical Exchange of DNA

Key concepts

- Recombination is the result of crossing-over that occurs at chiasmata and involves two of the four chromatids.
- Recombination occurs by a breakage and reunion that proceeds via an intermediate of hybrid DNA.

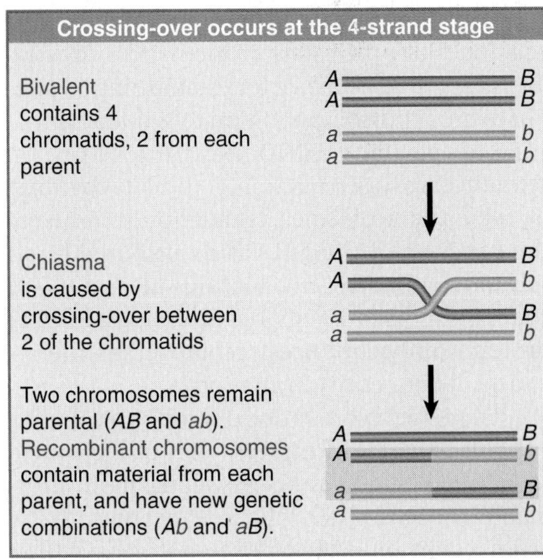

FIGURE 2.7 Chiasma formation is responsible for generating recombinants.

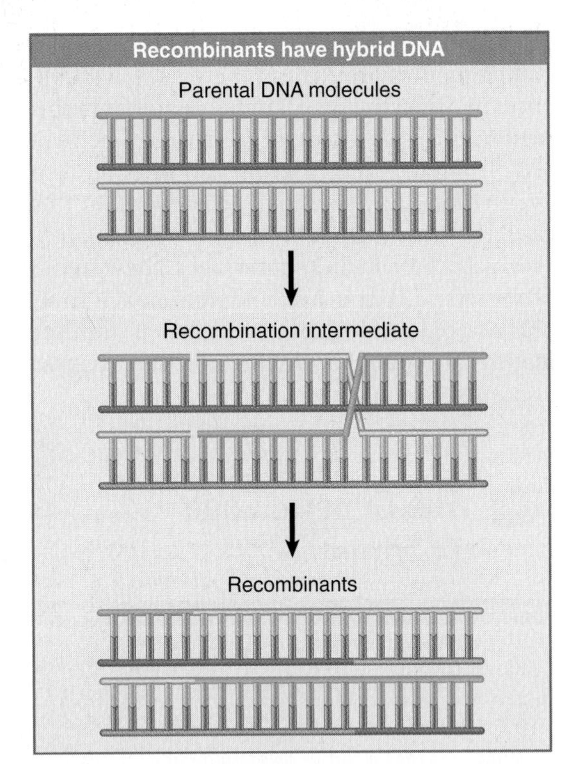

FIGURE 2.8 Recombination involves pairing between complementary strands of the two parental duplex DNAs.

Genetic recombination describes the generation of new combinations of alleles that occurs at each generation in diploid organisms. The two copies of each chromosome may have different alleles at some loci. By exchanging corresponding parts between the chromosomes, recombinant chromosomes can be generated that are different from the parental chromosomes.

Recombination results from a physical exchange of chromosomal material. This is visible in the form of the **crossing-over** that occurs during meiosis (the specialized division that produces haploid germ cells). Meiosis starts with a cell that has duplicated its chromosomes, so that it has four copies of each chromosome. Early in meiosis, all four copies are closely associated (synapsed) in a structure called a **bivalent.** Each individual chromosomal unit is called a **chromatid** at this stage. Pairwise exchanges of material occur between the chromatids.

The visible result of a crossing-over event is called a **chiasma** and is illustrated diagrammatically in FIGURE 2.7. A chiasma represents a site at which two of the chromatids in a bivalent have been broken at corresponding points. The broken ends have been rejoined crosswise, generating new chromatids. Each new chromatid consists of material derived from one chromatid on one side of the junction point, with material from the other chromatid on the opposite side. The two recombinant chromatids have reciprocal structures. The event is described as a **breakage and reunion.** Its nature explains why a single recombination event can produce only 50% recombinants: each individual recombination event involves only two of the four associated chromatids.

The complementarity of the two strands of DNA is essential for the recombination process. Each of the chromatids shown in Figure 2.7 consists of a very long duplex of DNA. For them to be broken and reconnected without any loss of material requires a mechanism to recognize exactly corresponding positions. This is provided by complementary base pairing.

Recombination involves a process in which the single strands in the region of the crossover exchange their partners. FIGURE 2.8 shows that this creates a stretch of **hybrid DNA,** in which the single strand of one duplex is paired with its complement from the other duplex. The mechanism, of course, involves other stages (strands must be broken and resealed), which we discuss in more detail in Chapter 20, Repair Systems, but the crucial feature that makes precise recombination possible is the complementarity of DNA strands. The figure shows only some stages of the reaction, but we see that a stretch of hybrid DNA forms in the recombination intermediate when a single strand crosses over from one duplex to the other. Each recombinant consists of one parental duplex DNA at the left, which is connected by a stretch of hybrid DNA to the other parental duplex at the right. Each duplex DNA corresponds to one of the chromatids involved in recombination in Figure 2.7.

The formation of hybrid DNA requires the sequences of the two recombining duplexes to be close enough to allow pairing between the complementary strands. If there are no differences between the two parental genomes in this region, formation of hybrid DNA will be perfect. However, the reaction can be tolerated even when there are small differences. In this case, the hybrid DNA has points of mismatch, at which a base in one strand faces a base in the other strand that is not complementary to it. The correction of such mismatches is another feature of genetic recombination (see Chapter 20 Repair Systems).

2.8 The Genetic Code Is Triplet

Key concepts

- The genetic code is read in triplet nucleotides called codons.
- The triplets are nonoverlapping and are read from a fixed starting point.
- Mutations that insert or delete individual bases cause a shift in the triplet sets after the site of mutation.
- Combinations of mutations that together insert or delete three bases (or multiples of three) insert or delete amino acids, but do not change the reading of the triplets beyond the last site of mutation.

Each gene represents a particular protein chain. The concept that each protein consists of a particular series of amino acids dates from Sanger's characterization of insulin in the 1950s. The discovery that a gene consists of DNA presents us with the issue of how a sequence of nucleotides in DNA represents a sequence of amino acids in protein.

A crucial feature of the general structure of DNA is that *it is independent of the particular sequence of its component nucleotides.* The sequence of nucleotides in DNA is important not because of its structure *per se,* but because it *codes* for the sequence of amino acids that constitutes the corresponding polypeptide. The relationship between a sequence of DNA and the sequence of the corresponding protein is called the **genetic code.**

The structure and/or enzymatic activity of each protein follows from its primary sequence of amino acids. By determining the sequence of amino acids in each protein, the gene is able to carry all the information needed to specify an active polypeptide chain. In this way, a single type of structure—the gene—is able to represent itself in innumerable polypeptide forms.

Together the various protein products of a cell undertake the catalytic and structural activities that are responsible for establishing its phenotype. Of course, in addition to sequences that code for proteins, DNA also contains certain sequences whose function is to be recognized by regulator molecules, usually proteins. Here the function of the DNA is determined by its sequence directly, not via any intermediary code. Both types of region—genes expressed as proteins and sequences recognized as such—constitute genetic information.

The genetic code is deciphered by a complex apparatus that interprets the nucleic acid sequence. This apparatus is essential if the information carried in DNA is to have meaning. In any given region, only one of the two strands of DNA codes for protein, so we write the genetic code as a sequence of bases (rather than base pairs).

The genetic code is read in groups of three nucleotides, each group representing one amino acid. Each trinucleotide sequence is called a **codon.** A gene includes a series of codons that is read sequentially from a starting point at one end to a termination point at the other end. Written in the conventional 5′ to 3′ direction, the nucleotide sequence of the DNA strand that codes for protein corresponds to the amino acid sequence of the protein written in the direction from N-terminus to C-terminus.

The genetic code is read in *nonoverlapping triplets from a fixed starting point:*

- Nonoverlapping implies that each codon consists of three nucleotides and that successive codons are represented by successive trinucleotides.
- The use of a *fixed starting point* means that assembly of a protein must start at one end and work to the other, so that different parts of the coding sequence cannot be read independently.

The nature of the code predicts that two types of mutations will have different effects. If a particular sequence is read sequentially, such as:

UUU AAA GGG CCC (codons)
aa1 aa2 aa3 aa4 (amino acids)

then a point mutation will affect only one amino acid. For example, the substitution of an A by some other base (X) causes aa2 to be replaced by aa5:

UUU AAX GGG CCC
aa1 aa5 aa3 aa4

because only the second codon has been changed.

But a mutation that inserts or deletes a single base will change the triplet sets for the entire subsequent sequence. A change of this sort is called a **frameshift.** An insertion might take the form:

UUU AAX AGG GCC C
aa1 aa5 aa6 aa7

Because the new sequence of triplets is completely different from the old one, the entire amino acid sequence of the protein is altered beyond the site of mutation. So the function of the protein is likely to be lost completely.

Frameshift mutations are induced by the **acridines,** compounds that bind to DNA and distort the structure of the double helix, causing additional bases to be incorporated or omitted during replication. Each mutagenic event sponsored by an acridine results in the addition or removal of a single base pair.

If an acridine mutant is produced by, say, addition of a nucleotide, it should revert to wild type by deletion of the nucleotide. However, reversion also can be caused by deletion of a different base at a site close to the first. Combinations of such mutations provided revealing evidence about the nature of the genetic code.

FIGURE 2.9 illustrates the properties of frameshift mutations. An insertion or deletion changes the entire protein sequence following the site of mutation. However, the combination of an insertion *and* a deletion causes the code to be read incorrectly only between the two sites of mutation; correct reading resumes after the second site.

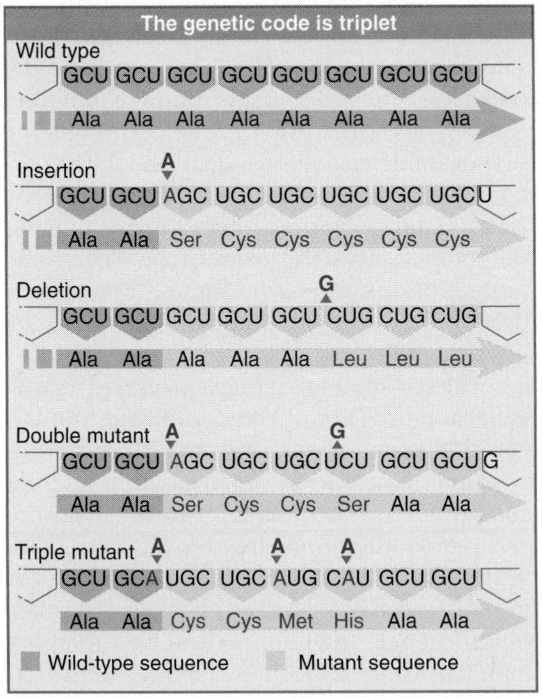

In 1961, genetic analysis of acridine mutations in the *rII* region of the phage T6 showed that all the mutations could be classified into one of two sets, described as (+) and (–). Either type of mutation by itself causes a frameshift: the (+) type by virtue of a base addition, and the (–) type by virtue of a base deletion. Double mutant combinations of the types (+ +) and (– –) continue to show mutant behavior. However, combinations of the types (+ –) or (– +) suppress one another, giving rise to a description in which one mutation is described as a **frameshift supressor** of the other. (In the context of this work, "suppressor" is used in an unusual sense because the second mutation is in the same gene as the first.)

These results show that the genetic code must be read as a sequence that is fixed by the starting point. Thus additions or deletions compensate for each other, whereas double additions or double deletions remain mutant. However, this does not reveal how many nucleotides make up each codon.

When triple mutants are constructed, only (+ + +) and (– – –) combinations show the wild phenotype, whereas other combinations remain mutant. If we take three additions or three deletions to correspond respectively to the addition or omission overall of a single amino acid, this implies that the code is read in triplets. An incorrect amino acid sequence is found between the two outside sites of mutation and the sequence on either side remains wild type, as indicated in Figure 2.9.

FIGURE 2.9 Frameshift mutations show that the genetic code is read in triplets from a fixed starting point.

2.9 Every Sequence Has Three Possible Reading Frames

Key concept

- Usually only one reading frame is translated and the other two are blocked by frequent termination signals.

If the genetic code is read in nonoverlapping triplets, there are three possible ways of translating any nucleotide sequence into protein, depending on the starting point. These are called **reading frames.** For the sequence

A C G A C G A C G A C G A C G A C G

the three possible reading frames are

ACG ACG ACG ACG ACG ACG ACG
CGA CGA CGA CGA CGA CGA CGA
GAC GAC GAC GAC GAC GAC GAC

A reading frame that consists exclusively of triplets representing amino acids is called an **open reading frame** or **ORF.** A sequence that is translated into protein has a reading frame that starts with a special **initiation codon** (AUG) and then extends through a series of triplets representing amino acids until it ends at one of three types of **termination codon** (see Chapter 7, Messenger RNA).

A reading frame that cannot be read into protein because termination codons occur frequently is said to be **blocked.** If a sequence is blocked in all three reading frames, it cannot have the function of coding for protein.

When the sequence of a DNA region of unknown function is obtained, each possible reading frame is analyzed to determine whether it is open or blocked. Usually no more than one of the three possible frames of reading is open in any single stretch of DNA. FIGURE 2.10 shows an example of a sequence that can be read in only one reading frame because the alternative reading frames are blocked by frequent termination codons. A long open reading frame is unlikely to exist by chance; if it were not translated into protein, there would have been no selective pressure to prevent the accumulation of termination codons. So the identification of a lengthy open reading frame is taken to be *prima facie* evidence that the sequence is translated into protein in that frame. An ORF for which no protein product has been identified is sometimes called an unidentified reading frame (URF).

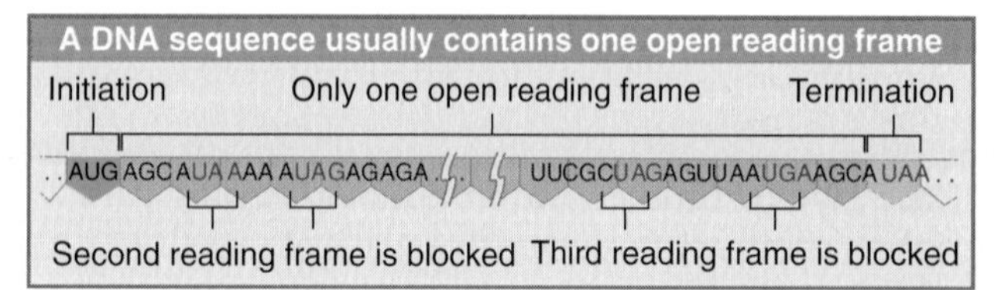

FIGURE 2.10 An open reading frame starts with AUG and continues in triplets to a termination codon. Blocked reading frames may be interrupted frequently by termination codons.

2.10 Prokaryotic Genes Are Colinear with Their Proteins

Key concepts

- A prokaryotic gene consists of a continuous length of 3*N* nucleotides that codes for *N* amino acids.
- The gene, mRNA, and protein are all colinear.

By comparing the nucleotide sequence of a gene with the amino acid sequence of a protein, we can determine directly whether the gene and the protein are **colinear;** that is, whether the sequence of nucleotides in the gene corresponds exactly with the sequence of amino acids in the protein. In bacteria and their viruses, there is an exact equivalence. Each gene contains a continuous stretch of DNA whose length is directly related to the number of amino acids in the protein that it represents. A gene of 3*N* bp is required to code for a protein of *N* amino acids, according to the genetic code.

The equivalence of the bacterial gene and its product means that a physical map of DNA will exactly match an amino acid map of the protein. How well do these maps fit with the recombination map?

The colinearity of gene and protein was originally investigated in the tryptophan synthetase gene of *E. coli.* Genetic distance was measured by the percent recombination between mutations; protein distance was measured by the number of amino acids separating sites of replacement. FIGURE 2.11 compares the two maps. The order of seven sites of mutation is the same as the order of the corresponding sites of amino acid replacement, and the recombination distances are relatively similar to the actual distances in the protein. The recombination map expands the distances between some mutations, but otherwise there is little distortion of the recombination map relative to the physical map.

The recombination map makes two further general points about the organization of the gene. Different mutations may cause a wild-type amino acid to be replaced with different substituents. If two such mutations cannot recombine, they must involve different point mutations at the same position in DNA. If the mutations can be separated on the genetic map, but affect the same amino acid on the upper map (the connecting lines converge in the figure), they must involve point mutations at dif-

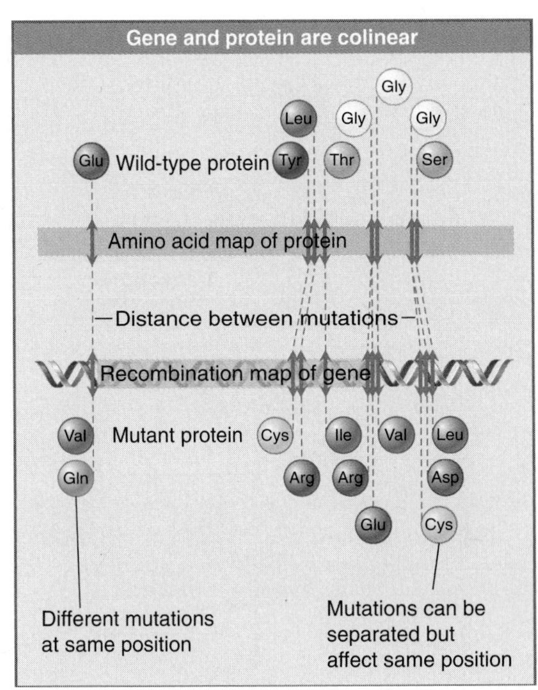

FIGURE 2.11 The recombination map of the tryptophan synthetase gene corresponds with the amino acid sequence of the protein.

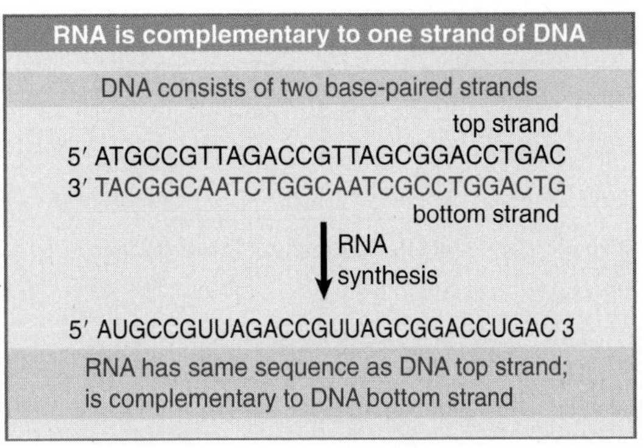

FIGURE 2.12 RNA is synthesized by using one strand of DNA as a template for complementary base pairing.

ferent positions that affect the same amino acid. This happens because the unit of genetic recombination (actually 1 bp) is smaller than the unit coding for the amino acid (actually 3 bp).

2.11 Several Processes Are Required to Express the Protein Product of a Gene

Key concepts

- A prokaryotic gene is expressed by transcription into mRNA and then by translation of the mRNA into protein.
- In eukaryotes, a gene may contain internal regions that are not represented in protein.
- Internal regions are removed from the RNA transcript by RNA splicing to give an mRNA that is colinear with the protein product.
- Each mRNA consists of a nontranslated 5′ leader, a coding region, and a nontranslated 3′ trailer.

In comparing gene and protein, we are restricted to dealing with the sequence of DNA stretching between the points corresponding to the ends of the protein. However, a gene is not directly translated into protein, but is expressed via the production of a **messenger RNA** (abbreviated to **mRNA**), a nucleic acid intermediate actually used to synthesize a protein (as we see in detail in Chapter 7, Messenger RNA).

Messenger RNA is synthesized by the same process of complementary base pairing used to replicate DNA, with the important difference that it corresponds to only one strand of the DNA double helix. FIGURE 2.12 shows that the sequence of mRNA is complementary with the sequence of one strand of DNA and is identical (apart from the replacement of T with U) with the other strand of DNA. The convention for writing DNA sequences is that the top strand runs 5′→3′, with the sequence that is the same as RNA.

The process by which a gene gives rise to a protein is called gene expression. In bacteria, it consists of two stages. The first stage is **transcription,** when an mRNA copy of one strand of the DNA is produced. The second stage is **translation** of the mRNA into protein. This is the process by which the sequence of an mRNA is read in triplets to give the series of amino acids that make the corresponding protein.

A mRNA includes a sequence of nucleotides that corresponds with the sequence of amino acids in the protein. This part of the nucleic acid is called the **coding region.** However, the mRNA includes additional sequences on either end; these sequences do not directly represent protein. The 5′ nontranslated region is called the **leader,** and the 3′ nontranslated region is called the **trailer.**

The *gene* includes the entire sequence represented in messenger RNA. Sometimes mutations impeding gene function are found in the additional, noncoding regions, confirming the view that these comprise a legitimate part of the genetic unit.

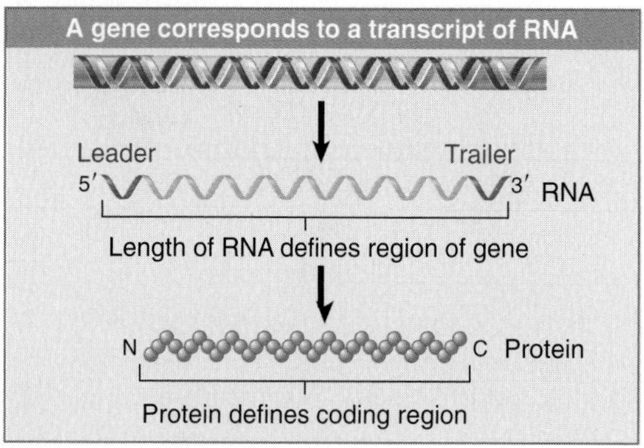

FIGURE 2.13 The gene may be longer than the sequence coding for protein.

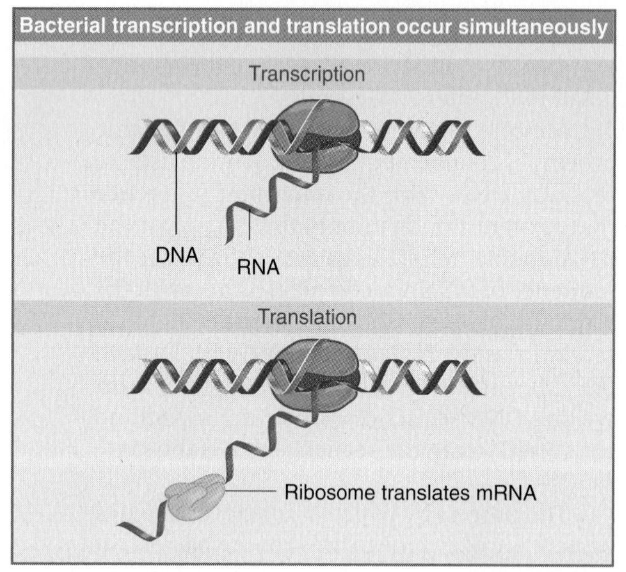

FIGURE 2.14 Transcription and translation take place in the same compartment in bacteria.

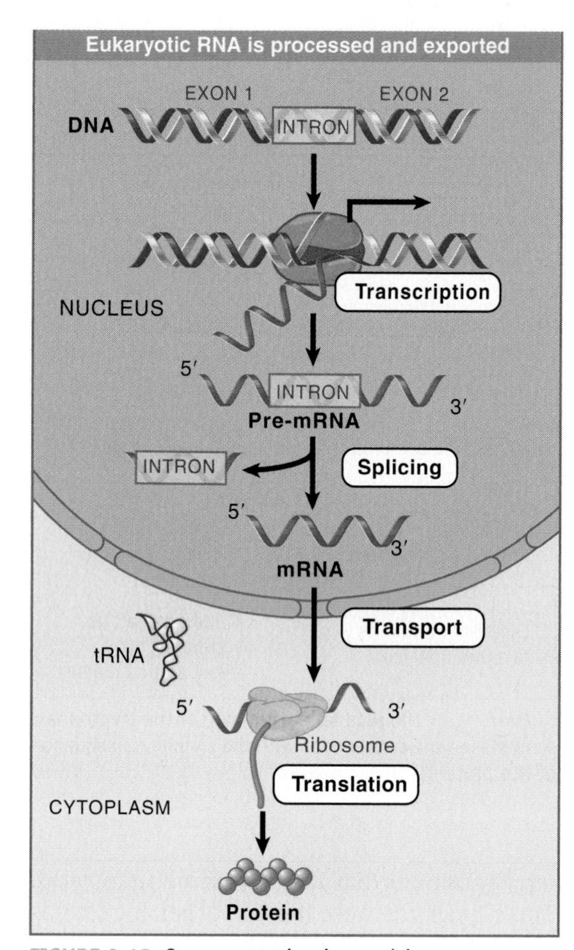

FIGURE 2.15 Gene expression is a multistage process.

FIGURE 2.13 illustrates this situation, in which the gene is considered to comprise a continuous stretch of DNA, needed to produce a particular protein. It includes the sequence coding for that protein, but also includes sequences on either side of the coding region.

A bacterium consists of only a single compartment, so transcription and translation occur in the same place, as illustrated in FIGURE 2.14.

In eukaryotes transcription occurs in the nucleus, but the RNA product must be *transported* to the cytoplasm in order to be translated. For the simplest eukaryotic genes (just like in bacteria) the transcript RNA is in fact the mRNA. However, for more complex genes, the immediate transcript of the gene is a **pre-mRNA** that requires **processing** to generate the mature mRNA. The basic stages of gene expression in a eukaryote are outlined in FIGURE 2.15. This results in a spatial separation between transcription (in the nucleus) and translation (in the cytoplasm).

The most important stage in processing is **RNA splicing.** Many genes in eukaryotes (and a majority in higher eukaryotes) contain internal regions that do not code for protein. The process of splicing removes these regions from the pre-mRNA to generate an RNA that has a continuous open reading frame (see Figure 3.1). Other processing events that occur at this stage involve the modification of the 5′ and 3′ ends of the pre-mRNA (see Figure 7.16).

Translation is accomplished by a complex apparatus that includes both protein and RNA components. The actual "machine" that undertakes the process is the *ribosome,* a large complex that includes some large RNAs (*ribosomal RNAs*, abbreviated to *rRNAs*) and many small proteins. The process of recognizing which amino acid corresponds to a particular nucleotide triplet requires an intermediate *transfer RNA* (abbreviated to *tRNA*); there is at least one tRNA species for every amino acid. Many ancillary proteins are involved. We describe translation in Chapter 7, Messenger RNA, but note for now that the ribosomes are the large structures in Figure 2.14 that move along the mRNA.

The important point to note at this stage is that the process of gene expression involves RNA not only as the essential substrate, but also in providing components of the apparatus. The rRNA and tRNA components are coded by genes and are generated by the process of transcription (just like mRNA, except that there is no subsequent stage of translation).

2.12 Proteins Are *Trans*-acting, but Sites on DNA Are *Cis*-acting

Key concepts

- All gene products (RNA or proteins) are *trans*-acting. They can act on any copy of a gene in the cell.
- *cis*-acting mutations identify sequences of DNA that are targets for recognition by *trans*-acting products. They are not expressed as RNA or protein and affect only the contiguous stretch of DNA.

A crucial step in the definition of the gene was the realization that all its parts must be present on one contiguous stretch of DNA. In genetic terminology, sites that are located on the same DNA are said to be in ***cis***. Sites that are located on two different molecules of DNA are described as being in ***trans***. So two mutations may be in *cis* (on the same DNA) or in *trans* (on different DNAs). The complementation test uses this concept to determine whether two mutations are in the same gene (see Figure 2.3 in Section 2.3, Mutations in the Same Gene Cannot Complement). We may now extend the concept of the difference between *cis* and *trans* effects from defining the coding region of a gene to describing the interaction between regulatory elements and a gene.

Suppose that the ability of a gene to be expressed is controlled by a protein that binds to the DNA close to the coding region. In the example depicted in FIGURE 2.16, mRNA can be synthesized only when the protein is bound to the DNA. Now suppose that a mutation occurs in the DNA sequence to which this protein binds, so that the protein can no longer recognize the DNA. As a result, the DNA can no longer be expressed.

So a gene can be inactivated either by a mutation in a control site or by a mutation in a coding region. The mutations cannot be distinguished genetically, because both have the property of acting only on the DNA sequence of the single allele in which they occur. They have identical properties in the complementation test, and a mutation in a control region is therefore defined as comprising part of the gene in the same way as a mutation in the coding region.

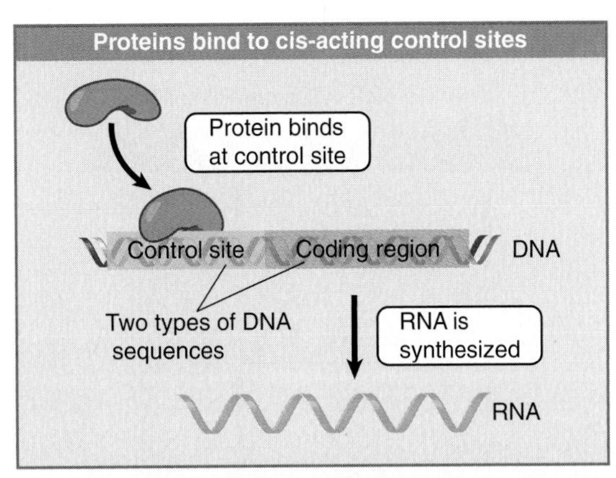

FIGURE 2.16 Control sites in DNA provide binding sites for proteins; coding regions are expressed via the synthesis of RNA.

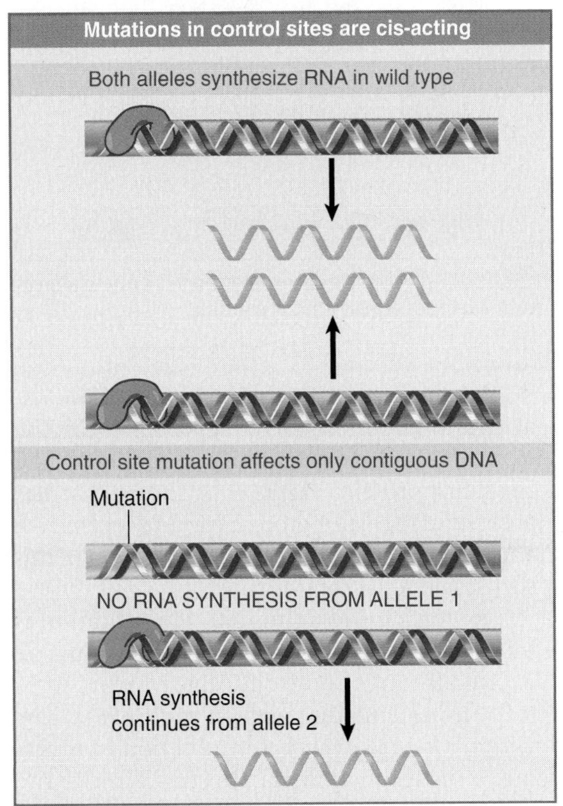

FIGURE 2.17 A *cis*-acting site controls the adjacent DNA, but does not influence the other allele.

FIGURE 2.17 shows that a deficiency in the control site *affects only the coding region to which it is connected; it does not affect the ability of the other allele to be expressed.* A mutation that acts solely by affecting the properties of the contiguous sequence of DNA is called ***cis*-acting.**

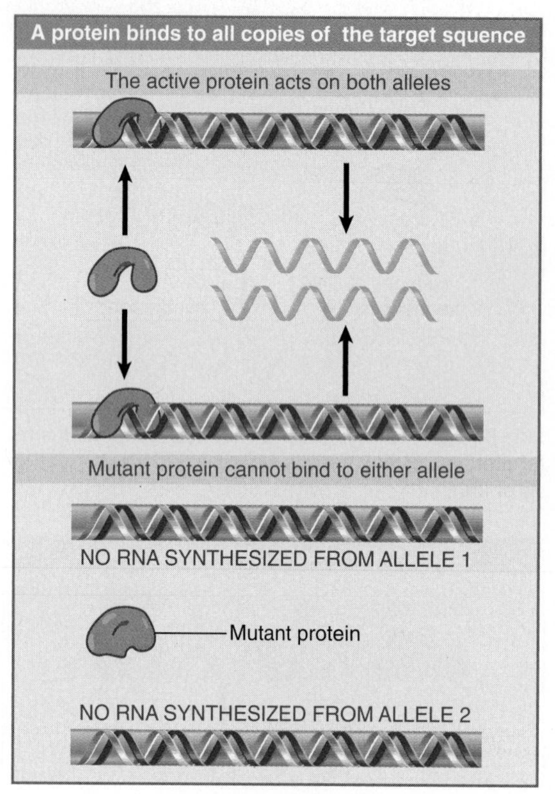

FIGURE 2.18 A *trans*-acting mutation in a protein affects both alleles of a gene that it controls.

We may contrast the behavior of the *cis*-acting mutation shown in Figure 2.17 with the result of a mutation in the gene coding for the regulator protein. FIGURE 2.18 shows that the absence of regulator protein would prevent *both* alleles from being expressed. A mutation of this sort is said to be *trans*-acting.

Reversing the argument, if a mutation is *trans*-acting, we know that its effects must be exerted through some diffusible product (typically a protein) that acts on multiple targets within a cell. However, if a mutation is *cis*-acting, it must function via affecting directly the properties of the contiguous DNA, which means that it is *not expressed in the form of RNA or protein.*

2.13 Summary

A chromosome consists of an uninterrupted length of duplex DNA that contains many genes. Each gene (or cistron) is transcribed into an RNA product, which in turn is translated into a polypeptide sequence if the gene codes for protein. An RNA or protein product of a gene is said to be *trans*-acting. A gene is defined as a unit of a single stretch of DNA by the complementation test. A site on DNA that regulates the activity of an adjacent gene is said to be *cis*-acting.

When a gene codes for protein, the relationship between the sequence of DNA and sequence of the protein is given by the genetic code. Only one of the two strands of DNA codes for protein. A codon consists of three nucleotides that represent a single amino acid. A coding sequence of DNA consists of a series of codons, read from a fixed starting point. Usually one of the three possible reading frames can be translated into protein.

A gene may have multiple alleles. Recessive alleles are caused by loss-of-function mutations that interfere with the function of the protein. A null allele has total loss-of-function. Dominant alleles are caused by gain-of-function mutations that create a new property in the protein.

References

2.8 The Genetic Code Is Triplet

Review

Roth, J. R. (1974). Frameshift mutations. *Annu. Rev. Genet.* 8, 319–346.

Research

Benzer, S. and Champe, S. P. (1961). Ambivalent rII mutants of phage T4. *Proc. Natl. Acad. Sci. USA* 47, 403–416.

Crick, F. H. C., Barnett, L., Brenner, S., and Watts-Tobin, R. J. (1961). General nature of the genetic code for proteins. *Nature* 192, 1227–1232.

2.10 Prokaryotic Genes Are Colinear with Their Proteins

Research

Yanofsky, C., Drapeau, G. R., Guest, J. R., and Carlton, B. C. (1967). The complete amino acid sequence of the tryptophan synthetase A protein (μ subunit) and its colinear relationship with the genetic map of the A gene. *Proc. Natl. Acad. Sci. USA* 57, 2966–2968.

Yanofsky, C. et al. (1964). On the colinearity of gene structure and protein structure. *Proc. Natl. Acad. Sci. USA,* 51, 266–272.

3

The Interrupted Gene

CHAPTER OUTLINE

3.1 Introduction

The simplest form of a gene is a length of DNA that is colinear with its protein product. Bacterial genes are almost always of this type, in which a continuous coding sequence of 3*N* base pairs represents a protein of *N* amino acids. In eukaryotes, however, a gene may include additional sequences that lie within the coding region, which interrupts the sequence that represents the protein. These sequences are removed from the RNA product during gene expression, generating an mRNA that includes a nucleotide sequence exactly corresponding with the protein product according to the rules of the genetic code.

The sequences of DNA comprising an interrupted gene are divided into the two categories depicted in FIGURE 3.1:

- **Exons** are the sequences represented in the mature RNA. By definition, a gene starts and ends with exons that correspond to the 5′ and 3′ ends of the RNA.
- **Introns** are the intervening sequences that are removed when the primary transcript is processed to give the mature RNA.

The exon sequences are in the same order in the gene and in the RNA, but an interrupted gene is longer than its final RNA product because of the presence of the introns.

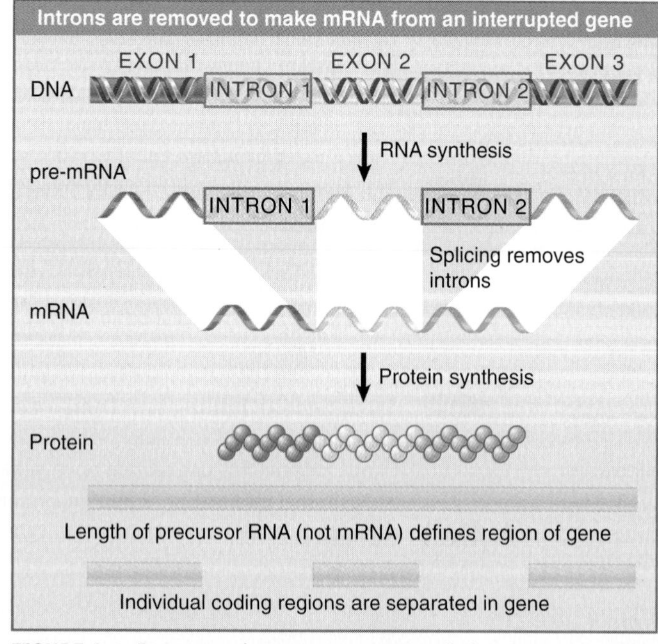

FIGURE 3.1 Interrupted genes are expressed via a precursor RNA. Introns are removed when the exons are spliced together. The mRNA has only the sequences of the exons.

The expression of interrupted genes requires an additional step that is not needed for uninterrupted genes. The DNA of an interrupted gene gives rise to an RNA copy (a **transcript**) that exactly represents the genome sequence. This RNA is only a precursor, though; it cannot be used for producing protein. First the introns must be removed from the RNA to give a messenger RNA that consists only of the series of exons. This process is called RNA splicing (see Section 2.11. Several Processes Are Required to Express the Protein Product of a Gene) and involves precisely deleting an intron from the primary transcript and then joining the ends of the RNA on either side to form a covalently intact molecule (see Chapter 26, RNA Splicing and Processing).

The gene comprises the region in the genome between points corresponding to the 5′ and 3′ terminal bases of mature mRNA. We know that transcription starts at the 5′ end of the mRNA and usually extends beyond the 3′ end, which is generated by cleavage of the RNA (see Section 26.19, The 3′ Ends of mRNAs Are Generated by Cleavage and Polyadenylation). The gene is considered to include the regulatory regions on both sides of the gene that are required for initiating and (sometimes) terminating gene expression.

3.2 An Interrupted Gene Consists of Exons and Introns

Key concepts

- Introns are removed by the process of RNA splicing, which occurs only in *cis* on an individual RNA molecule.
- Only mutations in exons can affect protein sequence; however, mutations in introns can affect processing of the RNA and therefore prevent production of protein.

How does the existence of introns change our view of the gene? Following splicing, the exons are always joined together in the same order in which they lie in DNA. So the colinearity of gene and protein is maintained between the individual exons and the corresponding parts of the protein chain. FIGURE 3.2 shows that the *order* of mutations in the gene remains the same as the order of amino acid replacements in the protein. But the *distances* in the gene do not correspond at all with the distances in the protein. Genetic distances, as seen on a recombination map, have no relationship to the distances

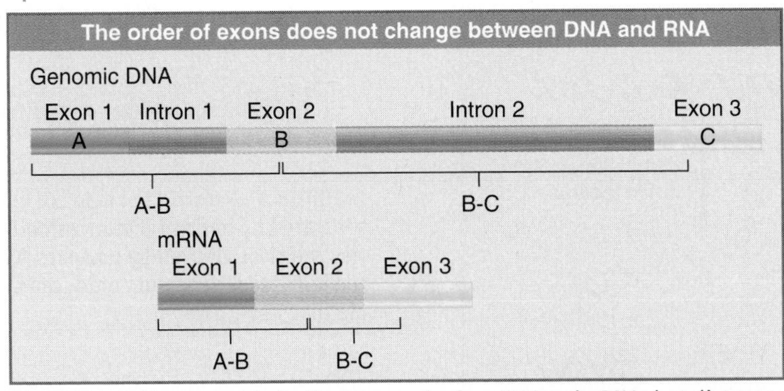

FIGURE 3.2 Exons remain in the same order in mRNA as in DNA, but distances along the gene do not correspond to distances along the mRNA or protein products. The distance from A–B in the gene is smaller than the distance from B–C; but the distance from A–B in the mRNA (and protein) is greater than the distance from B–C.

between the corresponding points in the protein. The length of the gene is defined by the length of the initial (precursor) RNA instead of by the length of the messenger RNA(mRNA).

All of the exons are represented on the same molecule of RNA, and their splicing together occurs only as an *intra*molecular reaction. There is usually no joining of exons carried by *different* RNA molecules, so the mechanism excludes any splicing together of sequences representing different alleles. Mutations located in different exons of a gene cannot complement one another; thus they continue to be defined as members of the same complementation group.

Mutations that directly affect the sequence of a protein must lie in exons. What are the effects of mutations in the introns? The introns are not part of the messenger RNA, thus mutations in them cannot directly affect protein structure. However, they can prevent the production of the messenger RNA—for example, by inhibiting the splicing together of exons. A mutation of this sort acts only on the allele that carries it. As a result, it fails to complement any other mutation in that allele and constitutes part of the same complementation group as the exons.

Mutations that affect splicing are usually deleterious. The majority are single-base substitutions at the junctions between introns and exons. They may cause an exon to be left out of the product, cause an intron to be included, or make splicing occur at an aberrant site. The most common result is to introduce a termination codon that results in truncation of the protein sequence. About 15% of the point mutations that cause human diseases are caused by disruption of splicing.

Eukaryotic genes are not necessarily interrupted. Some correspond directly with the protein product in the same manner as prokaryotic genes. In yeast, most genes are uninterrupted. In higher eukaryotes most genes are interrupted, and the introns are usually much longer than exons. This creates genes that are noticeably larger than their coding regions.

3.3 Restriction Endonucleases Are a Key Tool in Mapping DNA

Key concepts

- Restriction endonucleases can be used to cleave DNA into defined fragments.
- A map can be generated by using the overlaps between the fragments generated by different restriction enzymes.

The characterization of eukaryotic genes was made possible by the development of techniques for physically mapping DNA. The techniques can be extended to (single-stranded) RNA by making a (double-stranded) DNA copy of the RNA. A physical map of any DNA molecule can be obtained by breaking it at defined points whose distance apart can be accurately determined. Specific breaks are made possible by the ability of **restriction endonucleases** to recognize rather short sequences of double-stranded DNA as targets for cleavage.

Each restriction enzyme has a particular target in duplex DNA, usually a specific sequence of four to six base pairs. The enzyme cuts the DNA at every point at which its target sequence occurs. Different restriction enzymes have different target sequences, and a large range of these activities (obtained from a wide variety of bacteria) now is available.

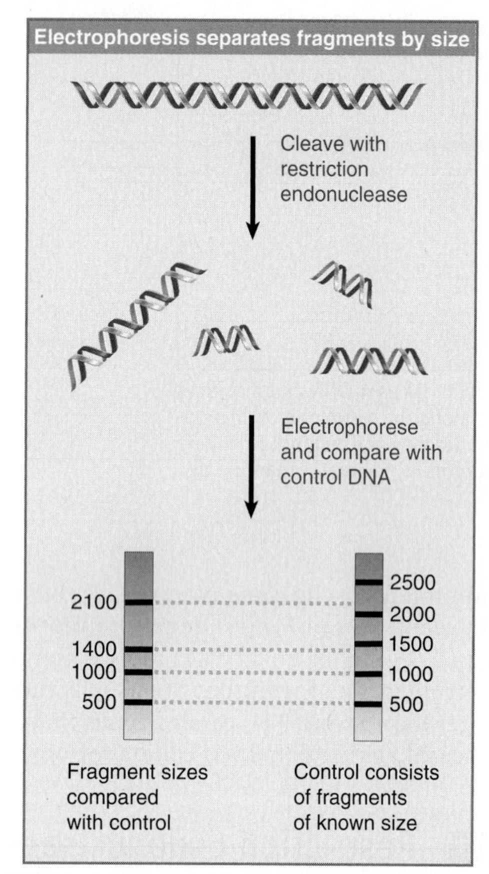

FIGURE 3.3 Fragments generated by cleaving DNA with a restriction endonuclease can be separated according to their sizes.

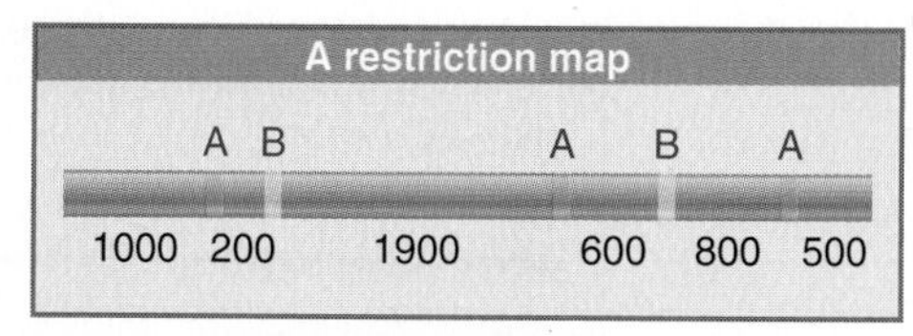

FIGURE 3.4 A restriction map is a linear sequence of sites separated by defined distances on DNA. The map identifies the sites cleaved by enzymes A and B, as defined by the individual fragments produced by the single and double digests.

A **restriction map** represents a linear sequence of the sites at which particular restriction enzymes find their targets. For short distances, the distance along such maps is measured directly in **base pairs (bp).** Longer distances are given in **kilobases (kb),** which correspond to kilobase (10^3) pairs in DNA or to kilobases in RNA. At the level of the chromosome, a map is described in **megabase pairs** (1 **Mb** = 10^6 bp).

When a DNA molecule is cut with a suitable restriction enzyme, it is cleaved into distinct fragments. These fragments can be separated on the basis of their size by gel electrophoresis, as shown in FIGURE 3.3. The cleaved DNA is placed on top of a gel made of agarose or polyacrylamide. When an electric current is passed through the gel, each fragment moves down at a rate that is inversely related to the log of its molecular weight. This movement produces a series of bands. Each band corresponds to a fragment of particular size, decreasing down the gel.

By analyzing the restriction fragments of DNA, we can generate a map of the original molecule in the form shown in FIGURE 3.4. The map shows the positions at which particular restriction enzymes cut DNA; the distances between the sites of cutting are measured in base pairs. *So the DNA is divided into a series of regions of defined lengths that lie between sites recognized by the restriction enzymes.* An important feature is that a restriction map can be obtained for any sequence of DNA, *irrespective of whether mutations have been identified in it,* or, indeed, whether we have any knowledge of its function.

3.4 Organization of Interrupted Genes May Be Conserved

Key concepts

- Introns can be detected by the presence of additional regions when genes are compared with their RNA products by restriction mapping or electron microscopy. The ultimate definition, though, is based on comparison of sequences.
- The positions of introns are usually conserved when homologous genes are compared between different organisms. The lengths of the corresponding introns may vary greatly.
- Introns usually do not code for proteins.

When a gene is uninterrupted, the restriction map of its DNA corresponds exactly with the map of its mRNA.

When a gene possesses an intron, the map at each end of the gene corresponds with the map at each end of the message sequence. Within the gene, though, the maps diverge because additional regions that are found in the gene are not represented in the message. Each such region corresponds to an intron. The example of FIGURE 3.5 compares the restriction maps of a β-globin gene and mRNA. There are two introns, each of which contains a series of restriction sites that are absent from the cDNA. The pattern of restriction sites in the exons is the same in both the cDNA and the gene.

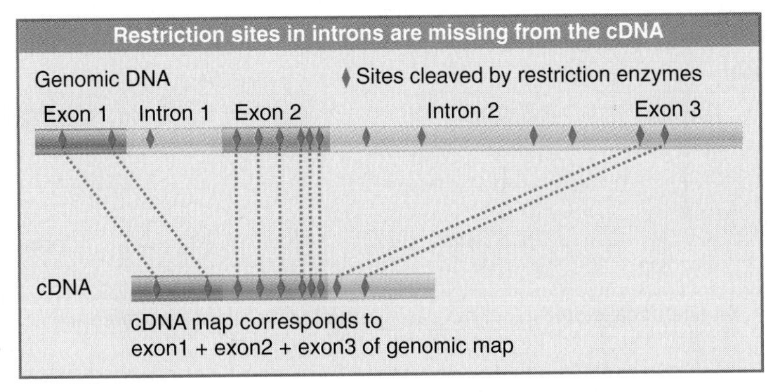

FIGURE 3.5 Comparison of the restriction maps of cDNA and genomic DNA for mouse β globin shows that the gene has two introns that are not present in the cDNA. The exons can be aligned exactly between cDNA and gene.

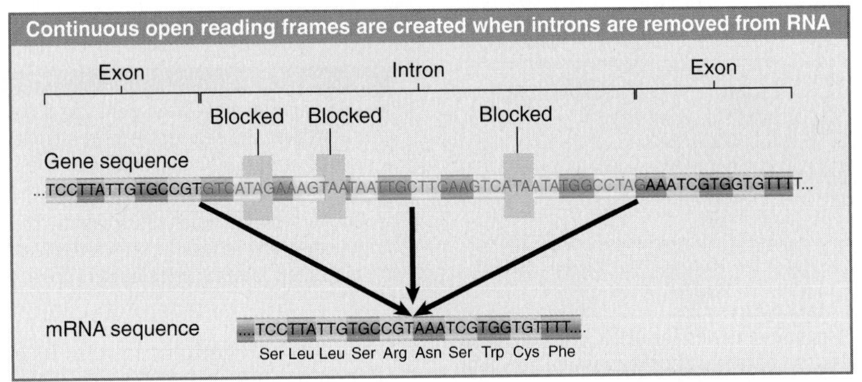

FIGURE 3.6 An intron is a sequence present in the gene but absent from the mRNA (here shown in terms of the cDNA sequence). The reading frame is indicated by the alternating open and shaded blocks; note that all three possible reading frames are blocked by termination codons in the intron.

Ultimately, a comparison of the nucleotide sequences of the genomic and mRNA sequences precisely defines the introns. As indicated in FIGURE 3.6, an intron usually has no open reading frame. An intact reading frame is created in the mRNA sequence by the removal of the introns.

The structures of eukaryotic genes show extensive variation. Some genes are uninterrupted, so that the genomic sequence is colinear with that of the mRNA. Most higher eukaryotic genes are interrupted, but the introns vary enormously in both number and size.

All classes of genes may be interrupted: nuclear genes coding for proteins, nucleolar genes coding for rRNA, and genes coding for tRNA. Interruptions also are found in mitochondrial genes in lower eukaryotes and in chloroplast genes. Interrupted genes do not appear to be excluded from any class of eukaryotes and have been found in bacteria and bacteriophages. They are, however, extremely rare in prokaryotic genomes.

Some interrupted genes possess only one or a few introns. The globin genes provide an extensively studied example (see Section 3.10, The Members of a Gene Family Have a Common Organization). The two general types of globin gene, α and β, share a common type of structure. The consistency of the organization of mammalian globin genes is evident from the structure of the "generic" globin gene summarized in FIGURE 3.7.

Interruptions occur at homologous positions (relative to the coding sequence) in all known active globin genes, including those of mammals, birds, and frogs. The first intron is always fairly short, and the second usually is longer, but the actual lengths can vary. Most of the variation in overall lengths between different globin genes results from the variation in the second intron. In the mouse, the second intron in the α-globin gene is only 150 bp long, so the overall length of the gene is 850 bp, compared with the major β-globin gene for which the intron length of 585 bp gives the gene a

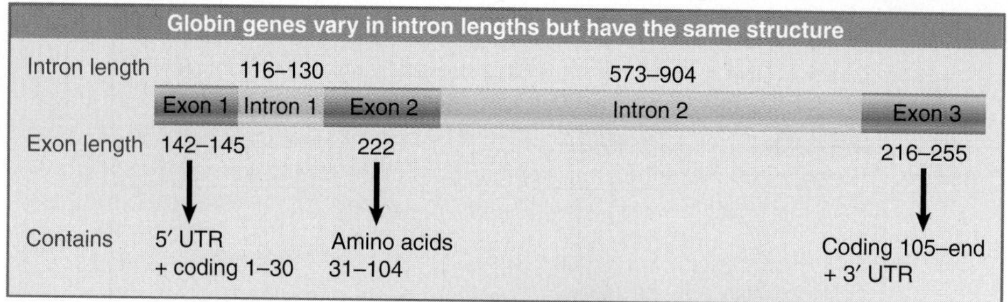

FIGURE 3.7 All functional globin genes have an interrupted structure with three exons. The lengths indicated in the figure apply to the mammalian β-globin genes.

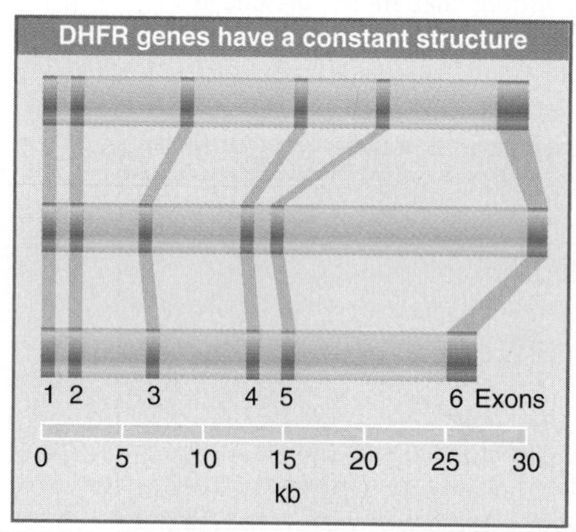

FIGURE 3.8 Mammalian genes for DHFR have the same relative organization of rather short exons and very long introns, but vary extensively in the lengths of introns.

total length of 1382 bp. The variation in length of the genes is much greater than the range of lengths of the mRNAs (α-globin mRNA = 585 bases; β-globin mRNA = 620 bases).

The example of DHFR, a somewhat larger gene, is shown in FIGURE 3.8. The mammalian DHFR (dihydrofolate reductase) gene is organized into six exons that correspond to the 2000-base mRNA. They extend over a much greater length of DNA because the introns are very long. In three mammals the exons remain essentially the same, and the relative positions of the introns are unaltered. The lengths of individual introns vary extensively, though, resulting in a variation in the length of the gene from 25 to 31 kb.

The globin and DHFR genes present examples of a general phenomenon: *Genes that are related by evolution have related organizations with conservation of the positions of (at least some) of the introns. Variations in the lengths of the genes are primarily determined by the lengths of the introns.*

3.5 Exon Sequences Are Conserved but Introns Vary

Key concepts

- Comparisons of related genes in different species show that the sequences of the corresponding exons are usually conserved but the sequences of the introns are much less well related.
- Introns evolve much more rapidly than exons because of the lack of selective pressure to produce a protein with a useful sequence.

Is a structural gene unique in its genome? The answer can be ambiguous. The entire length of the gene is unique as such, but its exons often are related to those of other genes. As a general rule, when two genes are related, the relationship between their exons is closer than the relationship between their introns. In an extreme case, the exons of two genes may code for the same protein sequence, whereas the introns may be different. This implies that the two genes originated by a duplication of some common ancestral gene. Then differences accumulated between the copies, but they were restricted in the exons by the need to code for protein functions.

As we will see later when we consider the evolution of the gene, exons can be considered basic building blocks that are assembled in various combinations. A gene may have some exons that are related to exons of another gene, but the other exons may be unrelated. Usually the introns are not related at all in such cases. Such genes may arise by duplication and translocation of individual exons.

The relationship between two genes can be plotted in the form of the dot matrix comparison of FIGURE 3.9. A dot is placed to indicate each position at which the same sequence is found

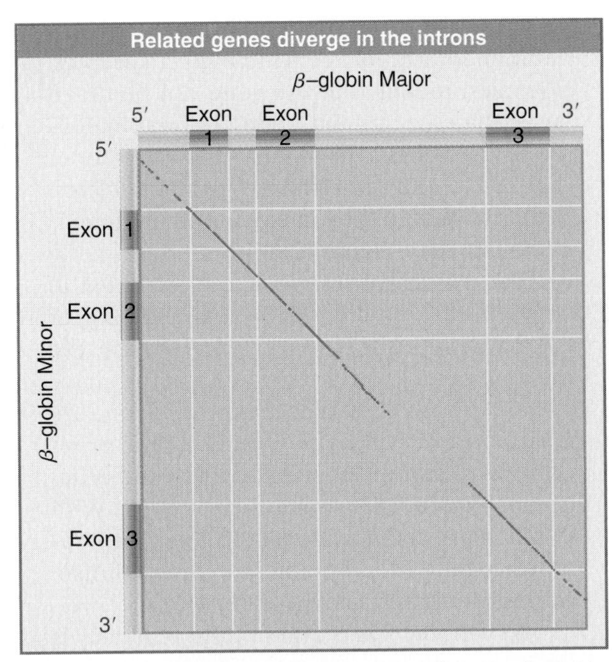

FIGURE 3.9 The sequences of the mouse α^{maj}- and α^{min}-globin genes are closely related in coding regions but differ in the flanking regions and long intron. Data provided by Philip Leder, Harvard Medical School.

in each gene. The dots form a line at an angle of 45° if two sequences are identical. The line is broken by regions that lack similarity and is displaced laterally or vertically by deletions or insertions in one sequence relative to the other.

When the two β-globin genes of the mouse are compared, such a line extends through the three exons and through the small intron. The line peters out in the flanking regions and in the large intron. This is a typical pattern, in which coding sequences are well related and the relationship can extend beyond the boundaries of the exons. The pattern is lost, though, in longer introns and the regions on either side of the gene.

The overall degree of divergence between two exons is related to the differences between the proteins. It is caused mostly by base substitutions. In the translated regions, the exons are under the constraint of needing to code for amino acid sequences, so they are limited in their potential to change sequence. Many of the changes do not affect codon meanings, because they change one codon into another that represents the same amino acid. Changes occur more freely in nontranslated regions (corresponding to the 5′ leader and 3′ trailer of the mRNA).

In corresponding introns, the pattern of divergence involves both changes in size (due to deletions and insertions) and base substitutions. Introns evolve much more rapidly than exons. When a gene is compared in different species, there are times when its exons are homologous but its introns have diverged so much that corresponding sequences cannot be recognized.

Mutations occur at the same rate in both exons and introns, but are removed more effectively from the exons by adverse selection. However, in the absence of the constraints imposed by a coding function, an intron is able quite freely to accumulate point substitutions and other changes. These changes imply that the intron does not have a sequence-specific function. Whether its presence is at all necessary for gene function is not clear.

3.6 Genes Show a Wide Distribution of Sizes

Key concepts

- Most genes are uninterrupted in yeasts, but are interrupted in higher eukaryotes.
- Exons are usually short, typically coding for <100 amino acids.
- Introns are short in lower eukaryotes, but range up to several 10s of kb in length in higher eukaryotes.
- The overall length of a gene is determined largely by its introns.

FIGURE 3.10 shows the overall organization of genes in yeasts, insects, and mammals. In *Saccharomyces cerevisiae*, the great majority of genes (>96%) are not interrupted, and those that have exons usually remain reasonably compact. There are virtually no *S. cerevisiae* genes with more than four exons.

In insects and mammals the situation is reversed. Only a few genes have uninterrupted coding sequences (6% in mammals). Insect genes tend to have a fairly small number of exons—typically fewer than 10. Mammalian genes are split into more pieces, and some have several 10s of exons. Approximately 50% of mammalian genes have >10 introns.

Examining the consequences of this type of organization for the overall size of the gene, we see in FIGURE 3.11 that there is a striking difference between yeast and the higher eukaryotes. The average yeast gene is 1.4 kb long, and

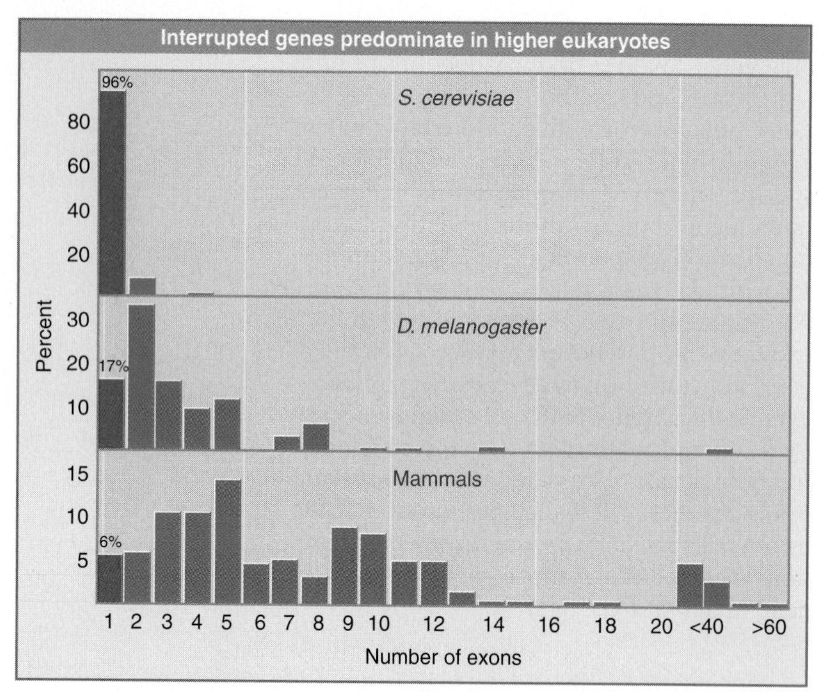

FIGURE 3.10 Most genes are uninterrupted in yeast, but most genes are interrupted in flies and mammals. (Uninterrupted genes have only one exon and are totaled in the leftmost column.)

very few are longer than 5 kb. The predominance of interrupted genes in high eukaryotes, however, means that the gene can be much larger than the unit that codes for protein. Relatively few genes in flies or mammals are shorter than 2 kb, and many have lengths between 5 kb and 100 kb. The average human gene is 27 kb long (see Figure 5.11).

The switch from largely uninterrupted to largely interrupted genes occurs in the lower eukaryotes. In fungi (except the yeasts), the majority of genes are interrupted, but they have a relatively small number of exons (<6) and are fairly short (<5 kb). The switch to long genes occurs within the higher eukaryotes, and genes become significantly larger in the insects. With this increase in the length of the gene, the relationship between genome size and organism complexity is lost (see Figure 4.5).

As genome size increases, the tendency is for introns to become rather large, whereas exons remain quite small.

FIGURE 3.12 shows that the exons coding for stretches of protein tend to be fairly small. In higher eukaryotes, the average exon codes for ~50 amino acids, and the general distribution fits well with the idea that genes have evolved

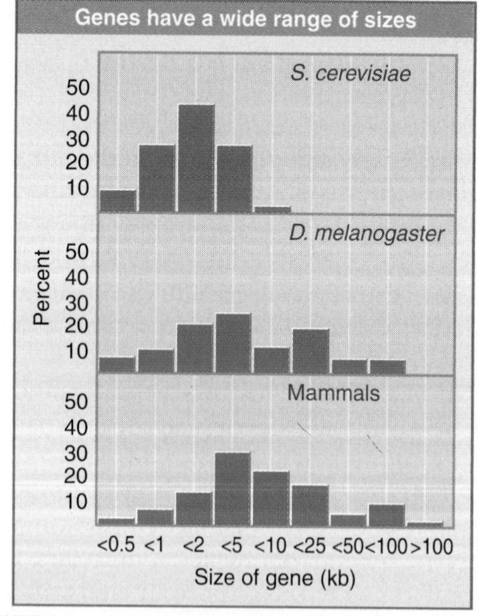

FIGURE 3.11 Yeast genes are short, but genes in flies and mammals have a dispersed distribution extending to very long sizes.

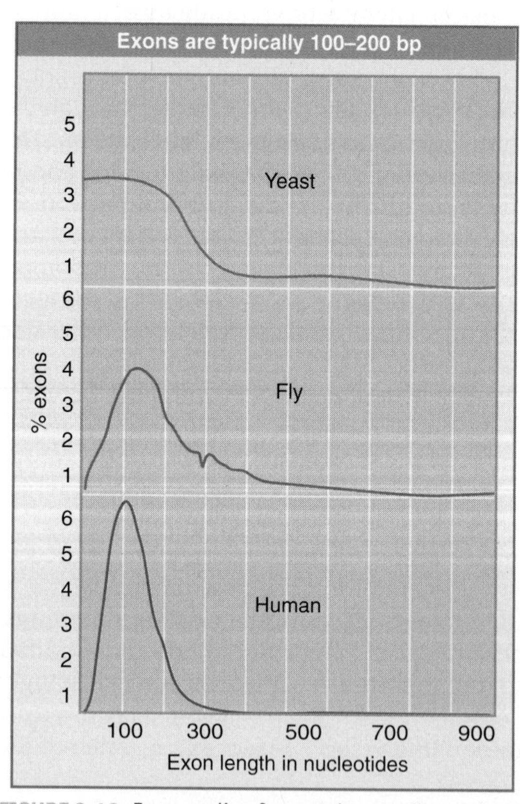

FIGURE 3.12 Exons coding for proteins usually are short.

by the slow addition of units that code for small, individual domains of proteins (see Section 3.8, How Did Interrupted Genes Evolve?). There is no significant difference in the sizes of exons in different types of higher eukaryotes, although the distribution is more compact in vertebrates for which there are few exons longer than 200 bp. In yeast, there are some longer exons that represent uninterrupted genes for which the coding sequence is intact. There is a tendency for exons coding for untranslated 5′ and 3′ regions to be longer than those that code for proteins.

FIGURE 3.13 shows that introns vary widely in size. In worms and flies, the average intron is not much longer than the exons. There are no very long introns in worms, but flies contain a significant proportion. In vertebrates, the size distribution is much wider, extending from approximately the same length as the exons (<200 bp) to lengths measured in 10s of kbs and extending up to 50 to 60 kb in extreme cases.

Very long genes are the result of very long introns, not the result of coding for longer products. There is no correlation between gene size and mRNA size in higher eukaryotes, nor is there a good correlation between gene size and number of exons. The size of a gene therefore depends primarily on the lengths of its individual introns. In mammals, insects, and birds, the "average" gene is approximately 5× the length of its mRNA.

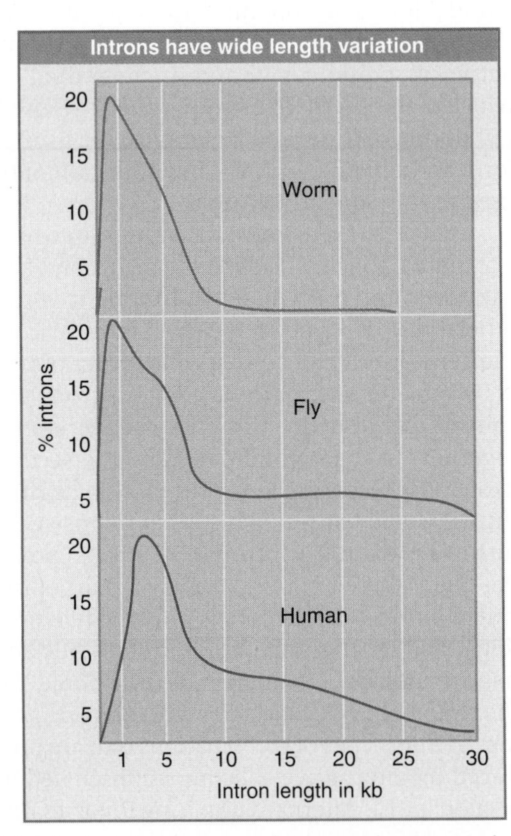

FIGURE 3.13 Introns range from very short to very long.

3.7 Some DNA Sequences Code for More Than One Protein

Key concepts

- The use of alternative initiation or termination codons allows two proteins to be generated where one is equivalent to a fragment of the other.
- Nonhomologous protein sequences can be produced from the same sequence of DNA when it is read in different reading frames by two (overlapping) genes.
- Homologous proteins that differ by the presence or absence of certain regions can be generated by differential (alternative) splicing when certain exons are included or excluded. This may take the form of including or excluding individual exons or of choosing between alternative exons.

Most genes consist of a sequence of DNA that is devoted solely to the purpose of coding for one protein (although the gene may include noncoding regions at either end and introns within the coding region). However, there are some cases in which a single sequence of DNA codes for more than one protein.

Overlapping genes occur in the relatively simple situation in which one gene is part of the other. The first half (or second half) of a gene is used independently to specify a protein that represents the first (or second) half of the protein specified by the full gene. This relationship is illustrated in FIGURE 3.14. The end result is much the same as though a partial cleavage took place in the protein product to generate part-length as well as full-length forms.

Two genes overlap in a more subtle manner when the same sequence of DNA is shared between two *nonhomologous* proteins. This situation arises when the same sequence of DNA is translated in more than one reading frame. In cellular genes, a DNA sequence usually is read in only one of the three potential reading frames. In some viral and mitochondrial genes, however, there is an overlap between two adjacent genes that are read in different reading frames. This situation is illustrated in FIGURE 3.15. The distance of overlap is usually relatively short, so that most of the sequence representing the protein retains a unique coding function.

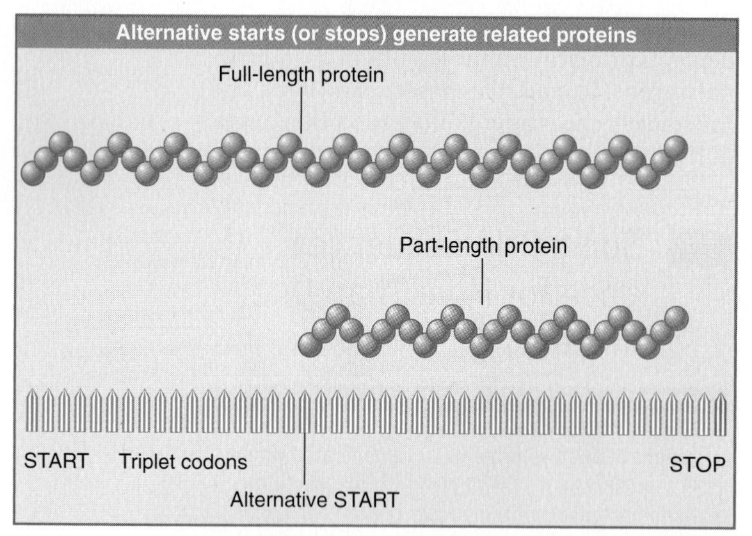

FIGURE 3.14 Two proteins can be generated from a single gene by starting (or terminating) expression at different points.

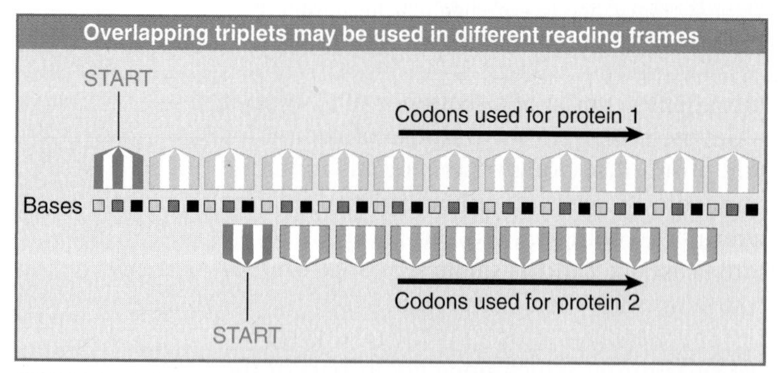

FIGURE 3.15 Two genes may share the same sequence by reading the DNA in different frames.

In some genes, *alternative* patterns of gene expression create switches in the pathway for connecting the exons. A single gene may generate a variety of mRNA products that differ in their content of exons. The difference may be that certain exons are optional; i.e., they may be included or spliced out. There also may be exons that are treated as mutually exclusive—one or the other is included, but not both. The alternative forms produce proteins in which one part is common and the other part is different.

In some cases, the alternative means of expression do not affect the sequence of the protein. For example, changes that affect the 5′ nontranslated leader or the 3′ nontranslated trailer may have regulatory consequences, but the same protein is made. In other cases, one exon is substituted for another, as indicated in FIGURE 3.16.

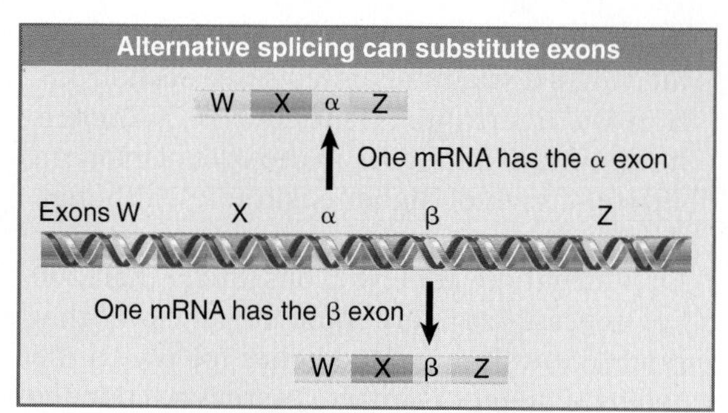

FIGURE 3.16 Alternative splicing generates the α and β variants of troponin T.

In this example, the proteins produced by the two mRNAs contain sequences that overlap extensively but are different within the alternatively spliced region. The 3′ half of the troponin T gene of rat muscle contains five exons, but only four are used to construct an individual mRNA. Three exons, *WXZ*, are the same in both expression patterns. However, in one pattern the α exon is spliced between *X* and *Z;* in the other pattern, the β exon is used. The α and β forms of troponin T therefore differ in the sequence of the amino acids present between sequences W and Z, depending on which of the alternative exons—α or β—is used. Either one of the α and β exons can be used to form an individual mRNA, but both cannot be used in the same mRNA.

FIGURE 3.17 illustrates an example in which alternative splicing leads to the inclusion of an exon in some mRNAs while leaving it out of others. A single type of transcript is made from the gene, but it can be spliced in either of two ways. In the first pathway, two introns are spliced out and the three exons are joined together. In the second pathway, the second exon is not recognized. As a result, a single large intron is spliced out. This intron consists of intron 1 + exon 2 + intron 2. In effect, exon 2 has been treated in this pathway as part of the single intron. The pathways produce two proteins that are the same at their ends, but one has an additional sequence in the middle. So the region of DNA codes for more than one protein. (Other types of combinations that are produced by alternative splicing are discussed in Section 26.12, Alternative Splicing Involves Differential Use of Splice Junctions).

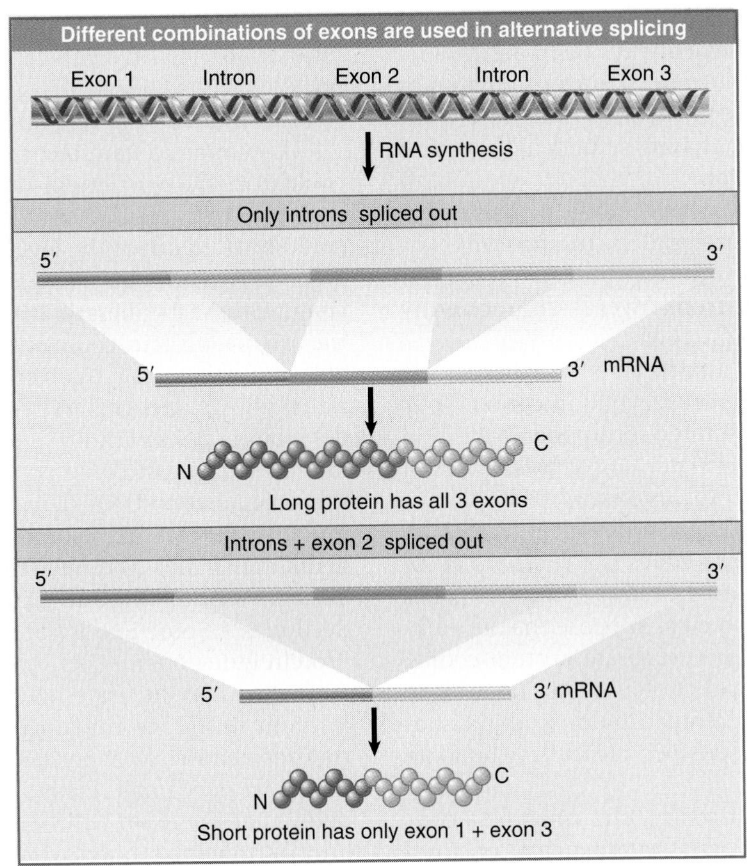

FIGURE 3.17 Alternative splicing uses the same pre-mRNA to generate mRNAs that have different combinations of exons.

Sometimes two pathways operate simultaneously, with a certain proportion of the RNA being spliced in each way. Sometimes the pathways are alternatives that are expressed under different conditions—one in one cell type and one in another cell type.

So alternative (or differential) splicing can generate proteins with overlapping sequences from a single stretch of DNA. It is curious that the higher eukaryotic genome is extremely spacious and has large genes that are often quite dispersed, but at the same time it may make multiple products from an individual locus. Alternative splicing expands the number of proteins relative to the number of genes by ~15% in flies and worms, but has much bigger effects in man, for which ~60% of genes may have alternative modes of expression (see Section 5.5, The Human Genome Has Fewer Genes Than Expected). About 80% of the alternative splicing events result in a change in the protein sequence.

3.8 How Did Interrupted Genes Evolve?

Key concepts

- The major evolutionary question is whether genes originated as sequences interrupted by introns or whether they were originally uninterrupted.
- Most protein-coding genes probably originated in an interrupted form, but interrupted genes that code for RNA may have originally been uninterrupted.
- A special class of introns is mobile and can insert itself into genes.

The highly interrupted structure of eukaryotic genes suggests a picture of the eukaryotic genome as a sea of introns (mostly but not exclusively unique in sequence), in which islands of exons (sometimes very short) are strung out in individual archipelagoes that represent genes.

What was the original form of genes that today are interrupted?

- The "introns early" model proposes that introns have always been an integral part of the gene. Genes originated as interrupted structures, and those without introns have lost them in the course of evolution.
- The "introns late" model proposes that the ancestral protein-coding units consisted of uninterrupted sequences of DNA. Introns were subsequently inserted into them.

A test of the models is to ask whether the difference between eukaryotic and prokaryotic genes can be accounted for by the acquisition of introns in the eukaryotes or by the loss of introns from the prokaryotes.

The "introns early" model suggests that the mosaic structure of genes is a remnant of an ancient approach to the reconstruction of genes to make novel proteins. Suppose that an early cell had a number of separate protein-coding sequences: One aspect of its evolution is likely to have been the reorganization and juxtaposition of different polypeptide units to build up new proteins.

If the protein-coding unit must be a continuous series of codons, every such reconstruction would require a precise recombination of DNA to place the two protein-coding units in register, end to end in the same reading frame. Furthermore, if this combination is not successful, the cell has been damaged because it has lost the original protein-coding units.

If an approximate recombination of DNA could place the two protein-coding units within the same transcription unit, splicing patterns could be tried out at the level of RNA to combine the two proteins into a single polypeptide chain. If these combinations are not successful, the original protein-coding units remain available for further trials. Such an approach essentially allows the cell to try out controlled deletions in RNA without suffering the damaging instability that could occur from applying this procedure to DNA. This argument is supported by the fact that we can find related exons in different genes, as though the gene had been assembled by mixing and matching exons (see Section 3.9, Some Exons Can Be Equated with Protein Functions).

FIGURE 3.18 illustrates the outcome when a random sequence that includes an exon is translocated to a new position in the genome. Exons are very small relative to introns, so it is likely that the exon will find itself within an intron. Only the sequences at the exon–intron junctions are required for splicing, and as a result the exon is likely to be flanked by functional 3′

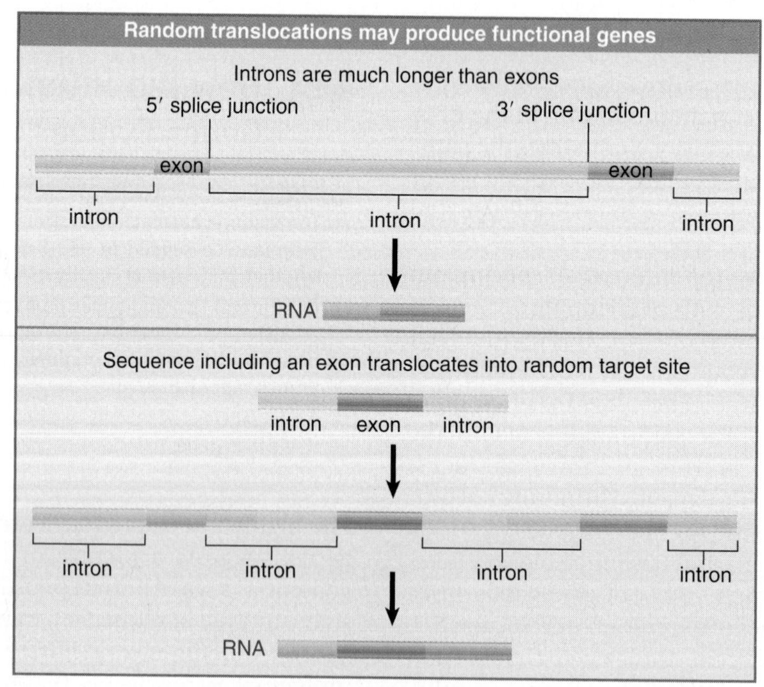

FIGURE 3.18 An exon surrounded by flanking sequences that is translocated into an intron may be spliced into the RNA product.

and 5′ splice junctions, respectively. Splicing junctions are recognized in pairs; thus the 5′ splicing junction of the original intron is likely to interact with the 3′ splicing junction introduced by the new exon, instead of with its original partner. Similarly, the 5′ splicing junction of the new exon will interact with the 3′ splicing junction of the original intron. The result is to insert the new exon into the RNA product between the original two exons. As long as the new exon is in the same coding frame as the original exons, a new protein sequence will be produced. This type of event could have been responsible for generating new combinations of exons during evolution. Note that the principle of this type of event is mimicked by the technique of exon trapping that is used to screen for functional exons (see Figure 4.11).

Alternative forms of genes for rRNA and tRNA are sometimes found, both with and without introns. In the case of the tRNAs, for which all the molecules conform to the same general structure, it seems unlikely that evolution brought together the two regions of the gene. After all, the different regions are involved in the base pairing that gives significance to the structure. So here it must be that the introns were inserted into continuous genes.

Organelle genomes provide some striking connections between the prokaryotic and eukaryotic worlds. There are many general similarities between mitochondria or chloroplasts and bacteria, and because of this it seems likely that the organelles originated by an *endosymbiosis* in which an early bacterial prototype was inserted into eukaryotic cytoplasm. Yet in contrast with the resemblances with bacteria—for example, as seen in protein or RNA synthesis—some organelle genes possess introns and therefore resemble eukaryotic nuclear genes.

Introns are found in several chloroplast genes, including some that have homologies with genes of *E. coli*. This suggests that the endosymbiotic event occurred before introns were lost from the prokaryotic line. If a suitable gene can be found, it may be possible to trace gene lineage back to the period when endosymbiosis occurred.

The mitochondrial genome presents a particularly striking case. The genes of yeast and mammalian mitochondria code for virtually identical mitochondrial proteins in spite of a considerable difference in gene organization. Vertebrate mitochondrial genomes are very small, with an extremely compact organization of continuous genes, whereas yeast mitochondrial genomes are larger and have some complex interrupted genes. Which is the ancestral form? The yeast mitochondrial introns (and certain other introns) can have the property of mobility—they are self-contained sequences that can splice out of the RNA and insert DNA copies elsewhere—which suggests that they may have arisen by insertions into the genome (see Section 27.5, Some Group I Introns Code for Endonucleases That Sponsor Mobility and Section 27.6, Group II Introns May Code for Multifunction Proteins).

3.9 Some Exons Can Be Equated with Protein Functions

Key concepts

- Facts suggesting that exons were the building blocks of evolution and the first genes were interrupted are:
  - Gene structure is conserved between genes in very distant species.
  - Many exons can be equated with coding for protein sequences that have particular functions.
  - Related exons are found in different genes.

If current proteins evolved by combining ancestral proteins that were originally separate, the accretion of units is likely to have occurred sequentially over some period of time, with one exon added at a time. Can the different functions from which these genes were pieced together be seen in their present structures? In other words, can we equate particular functions of current proteins with individual exons?

In some cases, there is a clear relationship between the structures of the gene and the protein. The example *par excellence* is provided by the immunoglobulin proteins, which are coded by genes in which every exon corresponds exactly with a known functional domain of the protein. FIGURE 3.19 compares the structure of an immunoglobulin with its gene.

An immunoglobulin is a tetramer of two light chains and two heavy chains, which aggregate to generate a protein with several distinct domains. Light chains and heavy chains differ in structure, and there are several types of heavy chain. Each type of chain is expressed from a gene that has a series of exons corresponding with the structural domains of the protein.

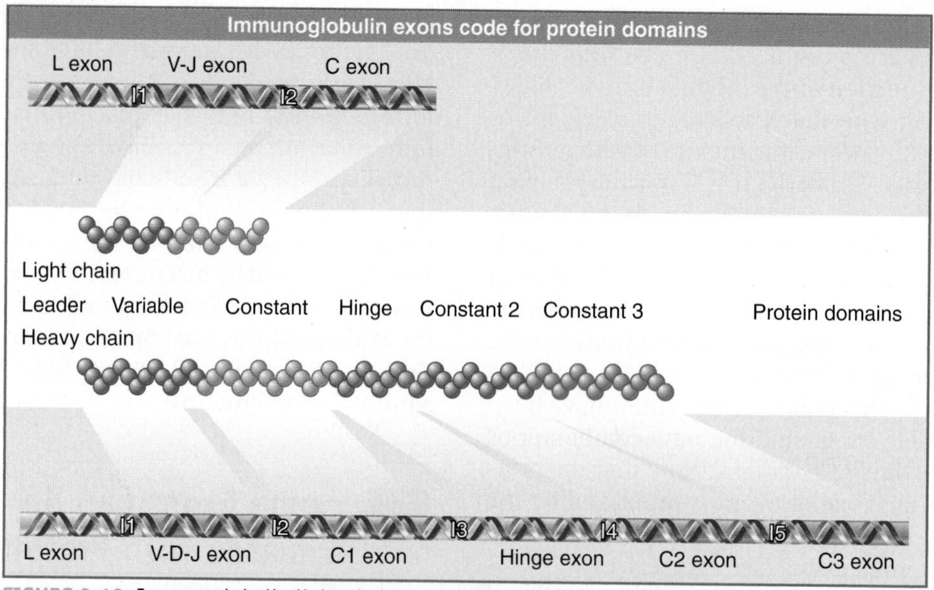

FIGURE 3.19 Immunoglobulin light chains and heavy chains are coded by genes whose structures (in their expressed forms) correspond with the distinct domains in the protein. Each protein domain corresponds to an exon; introns are numbered 1 to 5.

In many instances, some of the exons of a gene can be identified with particular functions. In secretory proteins, the first exon, coding for the N-terminal region of the polypeptide, often specifies the signal sequence involved in membrane secretion. An example is insulin.

The view that exons are the functional building blocks of genes is supported by cases in which two genes may have some exons that are related to one another, whereas other exons are found only in one of the genes. FIGURE 3.20 summarizes the relationship between the receptor for human LDL (plasma low density lipoprotein) and other proteins. In the center of the LDL receptor gene is a series of exons related to the exons of the gene for the precursor for EGF (epidermal growth factor). In the N-terminal part of the protein, a series of exons codes for a sequence related to the blood protein complement factor C9. So the LDL receptor gene was created by assembling *modules* for its various functions. These modules also are used in different combinations in other proteins.

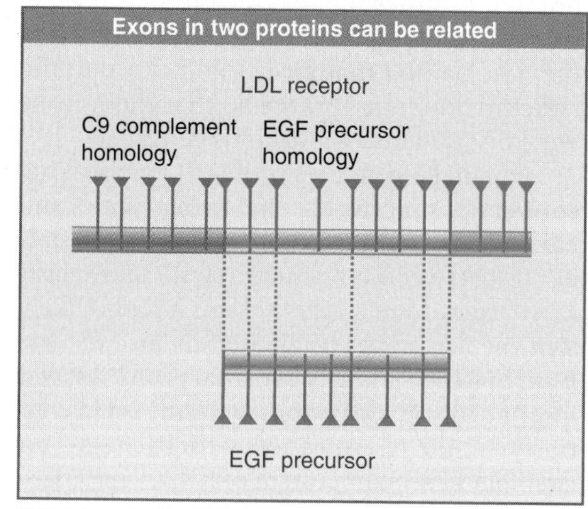

FIGURE 3.20 The LDL receptor gene consists of 18 exons, some of which are related to EGF precursor exons and some of which are related to the C9 blood complement gene. Triangles mark the positions of introns.

Exons tend to be fairly small (see Figure 4.11), around the size of the smallest polypeptide that can assume a stable folded structure (~20 to 40 residues). Perhaps proteins were originally assembled from rather small modules. Each module need not necessarily correspond to a current function; several modules could have combined to generate a function. The number of exons in a gene tends to increase with the length of its protein, which is consistent with the view that proteins acquire multiple functions by successively adding appropriate modules.

This idea might explain another feature of protein structure. It appears that the sites represented at exon–intron boundaries often are located at the surface of a protein. As modules are added to a protein, the connections—at least of the most recently added modules—could tend to lie at the surface.

3.10 The Members of a Gene Family Have a Common Organization

Key concepts

- A common feature in a set of genes is assumed to identify a property that preceded their separation in evolution.
- All globin genes have a common form of organization with three exons and two introns, suggesting that they are descended from a single ancestral gene.

Many genes in a higher eukaryotic genome are related to others in the same genome. A **gene family** can be defined as a group of genes that code for related or identical proteins. A family originates when a gene is duplicated. Initially the two copies are identical, but then they diverge as mutations accumulate in them. Further duplications and divergence extend the family further. The globin genes are an example of a family that can be divided into two subfamilies (α globin and β globin), but all its members have the same basic structure and function. The concept can be extended further when we find genes that are more distantly related, but still can be recognized as having common ancestry; in this case, a group of gene families can be considered to make up a **superfamily.**

A fascinating case of evolutionary conservation is presented by the α and β globins and two other proteins related to them. Myoglobin is a monomeric oxygen-binding protein of animals whose amino acid sequence suggests a common (though ancient) origin with the globin subunits. Leghemoglobins are oxygen-binding proteins present in the legume class of plants; like myoglobin, they are monomeric. They, too, share a common origin with the other heme-binding proteins. Together, the globins, myoglobin, and leghemoglobin constitute the globin superfamily—a set of gene families all descended from some (distant) common ancestor.

Both α- and β-globin genes have three exons (see Figure 3.7). The two introns are located at constant positions relative to the coding sequence. The central exon represents the heme-binding domain of the globin chain.

Myoglobin is represented by a single gene in the human genome whose structure is essentially the same as that of the globin genes. The three-exon structure therefore predates the evolution of separate myoglobin and globin functions.

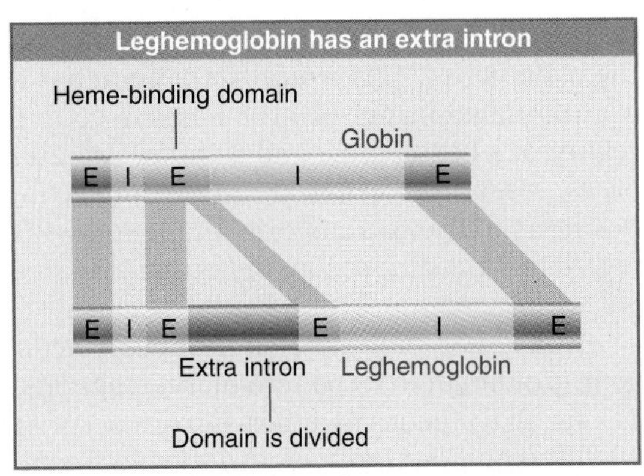

FIGURE 3.21 The exon structure of globin genes corresponds with protein function, but leghemoglobin has an extra intron in the central domain.

Leghemoglobin genes contain three introns, the first and last of which occur at points in the coding sequence that are homologous to the locations of the two introns in the globin genes. This remarkable similarity suggests an exceedingly ancient origin for the heme-binding proteins in the form of a split gene, as illustrated in FIGURE 3.21.

The central intron of leghemoglobin separates two exons that together code for the sequence corresponding to the single central exon in globin. Could the central exon of the globin gene have been derived by a fusion of two central exons in the ancestral gene? Or is the single central exon the ancestral form—in this case, an intron that must have been inserted into it at the start of plant evolution?

Cases in which homologous genes differ in structure may provide information about their evolution. An example is insulin. Mammals and birds have only one gene for insulin, except for rodents, which have two. FIGURE 3.22 illustrates the structures of these genes.

The principle we use in comparing the organization of related genes in different species is that *a common feature identifies a structure that predated the evolutionary separation of the two species.* In chickens, the single insulin gene has two introns; one of the two rat genes has the same structure. The common structure implies that the ancestral insulin gene had two introns. However, the second rat gene has only one intron. It must have evolved by a gene duplication in rodents that was followed by the precise removal of one intron from one of the copies.

The organization of some genes shows extensive discrepancies between species. In these cases, there must have been extensive removal or insertion of introns during evolution.

A well-characterized case is represented by the actin genes. The typical actin gene has a nontranslated leader of <100 bases, a coding region of ~1200 bases, and a trailer of ~200 bases. Most actin genes are interrupted; the positions of the introns can be aligned with regard to the coding sequence (except for a single intron sometimes found in the leader).

FIGURE 3.23 shows that almost every actin gene is different in its pattern of interruptions. Taking all the genes together, introns occur at 19 different sites. However, no individual gene has more than six introns; some genes have only one intron, and one is uninterrupted altogether. How did this situation arise? If we suppose that the primordial actin gene was interrupted, and that all current actin genes are related to it by loss of introns, different introns have been lost in each evolutionary branch. Probably some introns have been lost entirely, so the primordial gene could well have had 20 introns or more. The alternative is to suppose that a process of intron insertion continued independently in the different lines of evolution. The relationships between the intron locations found in different species may be used ultimately to construct a tree for the evolution of the gene.

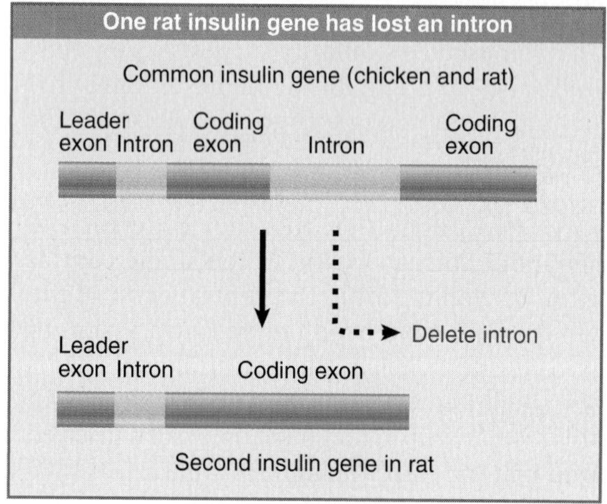

FIGURE 3.22 The rat insulin gene with one intron evolved by loss of an intron from an ancestor with two introns.

The relationship between exons and protein domains is somewhat erratic. In some cases there is a clear 1:1 relationship; in others no pattern can be discerned. One possibility is that the removal of introns has fused the adjacent exons. This means that the intron must have been precisely removed, without changing the integrity of the coding region. An alternative is that some introns arose by insertion into a coherent domain. Together with the variations that we see in exon placement in cases such as the actin genes, this argues that intron positions can be adjusted in the course of evolution.

The equation of at least some exons with protein domains, and the appearance of related exons in different proteins, leaves no doubt that the duplication and juxtaposition of exons has played an important role in evolution. It is possible that the number of ancestral exons—from which all proteins have been derived by duplication, variation, and recombination—could be relatively small (a few thousand or tens of thousands). By taking exons as the building blocks of evolution, this view implicitly accepts the introns early model for the origin of genes coding for proteins.

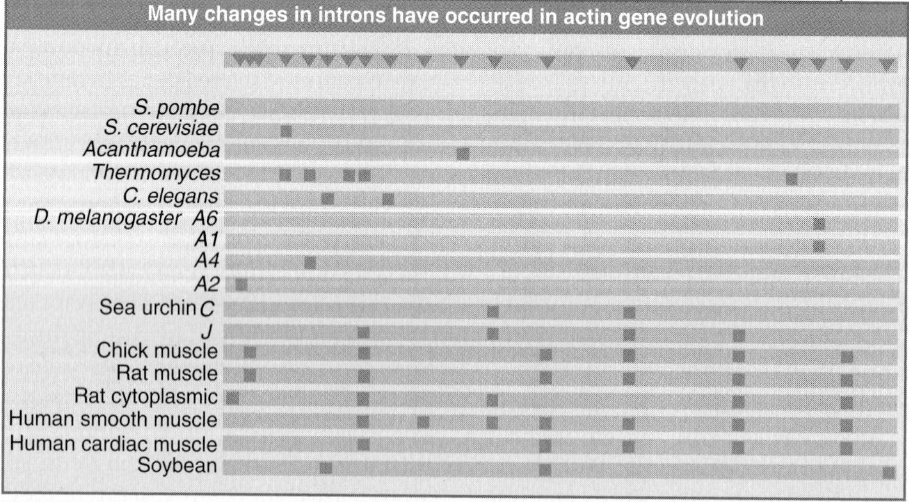

FIGURE 3.23 Actin genes vary widely in their organization. The sites of introns are indicated in purple.

3.11 Is All Genetic Information Contained in DNA?

Key concepts

- The definition of the gene has reversed from "one gene : one protein" to "one protein : one gene."
- Positional information is also important in development.

The concept of the gene has evolved significantly in the past several years. The question of what's in a name is especially appropriate for the gene. We can no longer say that a gene is a sequence of DNA that continuously and uniquely codes for a particular protein. In situations in which a stretch of DNA is responsible for production of one particular protein, current usage regards the entire sequence of DNA—from the first point represented in the messenger RNA to the last point corresponding to its end—as comprising the "gene," exons, introns, and all.

When the sequences representing proteins overlap or have alternative forms of expression, we may reverse the usual description of the gene. Instead of saying "one gene–one polypeptide," we may describe the relationship as "one polypeptide–one gene." So we regard the sequence actually responsible for production of the polypeptide (including introns as well as exons) as constituting the gene while recognizing that from the perspective of another protein, part of this same sequence also belongs to *its* gene. This allows the use of descriptions such as "overlapping" or "alternative" genes.

We can now see how far we have come from the original one gene : one enzyme hypothesis. Up to that time, the driving question was the nature of the gene. Once it was discovered that genes represent proteins, the paradigm became fixed in the form of the concept that every genetic unit functions through the synthesis of a particular protein.

This view remains the central paradigm of molecular biology: A sequence of DNA functions either by directly coding for a particular protein or by being necessary for the use of an adjacent segment that actually codes for the protein. How far does this paradigm take us beyond explaining the basic relationship between genes and proteins?

The development of multicellular organisms rests on the use of different genes to generate the different cell phenotypes of each tissue. The expression of genes is determined by a regulatory network that takes the form of a cascade. Expression of the first set of genes at the start of embryonic development leads to expression of the genes involved in the next stage of development, which in turn leads to a further stage, and so on until all the tissues of the adult are functioning. The molecular nature of this regulatory network is largely unknown, but we assume that it consists of genes that code for products (probably protein, perhaps sometimes RNA) that act on other genes.

Although such a series of interactions is almost certainly the means by which the developmental program is executed, we can ask whether it is entirely sufficient. One specific question concerns the nature and role of **positional information.** We know that all parts of a fertilized egg are not equal; one of the features responsible for development of different tissue parts from different regions of the egg is location of information (presumably specific macromolecules) within the cell.

We do not know how these particular regions are formed. We can, however, speculate that the existence of positional information in the egg leads to the differential expression of genes in the cells subsequently formed in these regions. This leads to the development of the adult organism, which in turn leads to the development of an egg with the appropriate positional information.

This possibility prompts us to ask whether some information needed for development of the organism is contained in a form that we cannot directly attribute to a sequence of DNA (although the expression of particular sequences may be needed to perpetuate the positional information). Put in a more general way, we might ask the following: When we read out the entire sequence of DNA comprising the genome of some organism and interpret it in terms of proteins and regulatory regions, could we in principle construct an organism (or even a single living cell) by controlled expression of the proper genes?

3.12 Summary

All types of eukaryotic genomes contain interrupted genes. The proportion of interrupted genes is low in yeasts and increases in the lower eukaryotes; few genes are uninterrupted in higher eukaryotes.

Introns are found in all classes of eukaryotic genes. The structure of the interrupted gene is the same in all tissues: exons are joined together in RNA in the same order as their organization in DNA, and the introns usually have no coding function. Introns are removed from RNA by splicing. Some genes are expressed by alternative splicing patterns, in which a particular sequence is removed as an intron in some situations, but retained as an exon in others.

Positions of introns often are conserved when the organization of homologous genes is compared between species. Intron sequences vary—and may even be unrelated—although exon sequences remain well related. The conservation of exons can be used to isolate related genes in different species.

The size of a gene is determined primarily by the lengths of its introns. Introns become larger early in the higher eukaryotes, when gene sizes therefore increase significantly. The range of gene sizes in mammals is generally from 1 to 100 kb, but it is possible to have even larger genes: the longest known case is dystrophin, at 2000 kb.

Some genes share only some of their exons with other genes, suggesting that they have been assembled by addition of exons representing individual modules of the protein. Such modules may have been incorporated into a variety of different proteins. The idea that genes have been assembled by accretion of exons implies that introns were present in genes of primitive organisms. Some of the relationships between homologous genes can be explained by loss of introns from the primordial genes, with different introns being lost in different lines of descent.

References

3.2 An Interrupted Gene Consists of Exons and Introns

Reviews

Breathnach, R. and Chambon, P. (1981). Organization and expression of eukaryotic split genes coding for proteins. *Annu. Rev. Biochem.* 50, 349–383.

Faustino, N. A. and Cooper, T. A. (2003). Pre-mRNA splicing and human disease. *Genes Dev.* 17, 419–437.

3.3 Restriction Endonucleases Are a Key Tool in Mapping DNA

Reviews

Nathans, D. and Smith, H. O. (1975). Restriction endonucleases in the analysis and restructuring of DNA molecules. *Annu. Rev. Biochem.* 44, 273–293.

Wu, R. (1978). DNA sequence analysis. *Annu. Rev. Biochem.* 47, 607–734.

Research

Danna, K. J., Sack, G. H., and Nathans, D. (1973). Studies of SV40 DNA VII A cleavage map of the SV40 genome. *J. Mol. Biol.* 78, 363–376.

3.4 Organization of Interrupted Genes May Be Conserved

Research

Berget, S. M., Moore, C., and Sharp, P. (1977). Spliced segments at the 5′ terminus of adenovirus 2 late mRNA. *Proc. Natl. Acad. Sci. USA* 74, 3171–3175.

Chow, L. T., Gelinas, R. E., Broker, T. R., and Roberts, R. J. (1977). An amazing sequence arrangement at the 5′ ends of adenovirus 2 mRNA. *Cell* 12, 1–8.

Glover, D. M. and Hogness, D. S. (1977). A novel arrangement of the 8S and 28S sequences in a repeating unit of *D. melanogaster* rDNA. *Cell* 10, 167–176.

Jeffreys, A. J. and Flavell, R. A. (1977). The rabbit β-globin gene contains a large insert in the coding sequence. *Cell* 12, 1097–1108.

Wenskink, P. et al. (1974). A system for mapping DNA sequences in the chromosomes of *D. melanogaster*. *Cell* 3, 315–325.

3.9 Some Exons Can Be Equated with Protein Functions

Review

Blake, C. C. (1985). Exons and the evolution of proteins. *Int. Rev. Cytol.* 93, 149–185.

4

The Content of the Genome

CHAPTER OUTLINE

4.1 Introduction

The key question about the genome is how many genes it contains. We can think about the total number of genes at four levels, which correspond to successive stages in gene expression:

- The **genome** is the complete set of genes of an organism. Ultimately it is defined by the complete DNA sequence, although as a practical matter it may not be possible to identify every gene unequivocally solely on the basis of sequence.
- The **transcriptome** is the complete set of genes expressed under particular conditions. It is defined in terms of the set of RNA molecules that is present and can refer to a single cell type, or to any more complex assembly of cells, up to the complete organism. Because some genes generate multiple mRNAs, the transcriptome is likely to be larger than the number of genes defined directly in the genome. The transcriptome includes noncoding RNAs as well as mRNAs.
- The **proteome** is the complete set of proteins. It should correspond to the mRNAs in the transcriptome, although there can be differences of detail reflecting changes in the relative abundance or stabilities of mRNAs and proteins. It can be used to refer to the set of proteins coded by the whole genome or produced in any particular cell or tissue.
- Proteins may function independently or as part of multiprotein assemblies. If we could identify all protein–protein interactions, we could define the total number of independent assemblies of proteins.

The number of genes in the genome can be identified directly by defining open reading frames. Large-scale mapping of this nature is complicated by the fact that interrupted genes may consist of many separated open reading frames. We do not necessarily have information about the functions of the protein products—or indeed proof that they are expressed at all—so this approach is restricted to defining the *potential* of the genome. However, a strong presumption exists that any conserved open reading frame is likely to be expressed.

Another approach is to define the number of genes directly in terms of the transcriptome (by directly identifying all the mRNAs) or proteome (by directly identifying all the proteins). This gives an assurance that we are dealing with *bona fide* genes that are expressed under known circumstances. It allows us to ask how many genes are expressed in a particular tissue or cell type, what variation exists in the relative levels of expression, and how many of the genes expressed in one particular cell are unique to that cell or are also expressed elsewhere.

Concerning the types of genes, we may ask whether a particular gene is essential: what happens to a null mutant? If a null mutation is lethal, or the organism has a visible defect, we may conclude that the gene is essential or at least conveys a selective advantage. But some genes can be deleted without apparent effect on the phenotype. Are these genes really dispensable, or does a selective disadvantage result from the absence of the gene, perhaps in other circumstances, or over longer periods of time?

4.2 Genomes Can Be Mapped by Linkage, Restriction Cleavage, or DNA Sequence

Defining the contents of a genome essentially means making a map. We can think about mapping genes and genomes at several levels of resolution:

- A genetic (or linkage) map identifies the distance between mutations in terms of recombination frequencies. It is limited by its reliance on the occurrence of mutations that affect the phenotype. Because recombination frequencies can be distorted relative to the physical distance between sites, it does not accurately represent physical distances along the genetic material.
- A linkage map also can be constructed by measuring recombination between sites in genomic DNA. These sites have sequence variations that generate differences in the susceptibility to cleavage by certain (restriction) enzymes. Because such variations are common, such a map can be prepared for any organism irrespective of the occurrence of mutants. It has the same disadvantage as any linkage map in that the relative distances are based on recombination.
- A restriction map is constructed by cleaving DNA into fragments with restriction enzymes and measuring the distances between the sites of cleavage. This represents distances in terms of the length of DNA, so it provides a physical map of the genetic material. A restric-

tion map does not intrinsically identify sites of genetic interest. For it to be related to the genetic map, mutations have to be characterized in terms of their effects upon the restriction sites. Large changes in the genome can be recognized because they affect the sizes or numbers of restriction fragments. Point mutations are more difficult to detect.

- The ultimate map is to determine the sequence of the DNA. From the sequence, we can identify genes and the distances between them. By analyzing the protein-coding potential of a sequence of the DNA, we can deduce whether it represents a protein. The basic assumption here is that natural selection prevents the accumulation of damaging mutations in sequences that code for proteins. Reversing the argument, we may assume that an intact coding sequence is likely to be used to generate a protein.

By comparing the sequence of a wild-type DNA with that of a mutant allele, we can determine the nature of a mutation and its exact site of occurrence. This defines the relationship between the genetic map (based entirely on sites of mutation) and the physical map (based on, or even comprising, the sequence of DNA).

Similar techniques are used to identify and sequence genes and to map the genome, although there is of course a difference of scale. In each case, the principle is to obtain a series of overlapping fragments of DNA that can be connected into a continuous map. The crucial feature is that each segment is related to the next segment on the map by characterizing the overlap between them, so that we can be sure no segments are missing. This principle is applied both at the level of ordering large fragments into a map and in connecting the sequences that make up the fragments.

4.3 Individual Genomes Show Extensive Variation

Key concepts

- Polymorphism may be detected at the phenotypic level when a sequence affects gene function, at the restriction fragment level when it affects a restriction enzyme target site, and at the sequence level by direct analysis of DNA.
- The alleles of a gene show extensive polymorphism at the sequence level, but many sequence changes do not affect function.

The original Mendelian view of the genome classified alleles as either wild-type or mutant. Subsequently we recognized the existence of multiple alleles, each with a different effect on the phenotype. In some cases it may not even be appropriate to define any one allele as "wild-type."

The coexistence of multiple alleles at a locus is called genetic **polymorphism.** Any site at which multiple alleles exist as stable components of the population is by definition polymorphic. An allele is usually defined as polymorphic if it is present at a frequency of >1% in the population.

What is the basis for the polymorphism among the mutant alleles? They possess different mutations that alter the protein function, thus producing changes in phenotype. If we compare the restriction maps or the DNA sequences of these alleles they, too, will be polymorphic in the sense that each map or sequence will be different from the others.

Although not evident from the phenotype, the wild type may itself be polymorphic. Multiple versions of the wild-type allele may be distinguished by differences in sequence that do not affect their function, and which therefore do not produce phenotypic variants. A population may have extensive polymorphism at the level of genotype. Many different sequence variants may exist at a given locus; some of them are evident because they affect the phenotype, but others are hidden because they have no visible effect.

So there may be a continuum of changes at a locus, including those that change DNA sequence but do not change protein sequence, those that change protein sequence without changing function, those that create proteins with different activities, and those that create mutant proteins that are nonfunctional.

A change in a single nucleotide when alleles are compared is called a **single nucleotide polymorphism (SNP).** One occurs every ~1330 bases in the human genome. Defined by their SNPs, every human being is unique. SNPs can be detected by various means, ranging from direct comparisons of sequence to mass spectroscopy or biochemical methods that produce differences based on sequence variations in a defined region.

One aim of genetic mapping is to obtain a catalog of common variants. The observed frequency of SNPs per genome predicts that, over the human population as a whole (taking the sum of all human genomes of all living individuals), there should be >10 million SNPs that

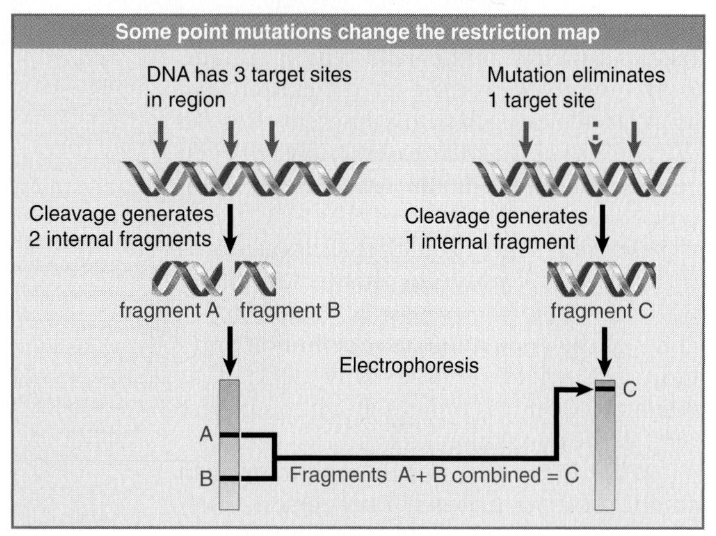

FIGURE 4.1 A point mutation that affects a restriction site is detected by a difference in restriction fragments.

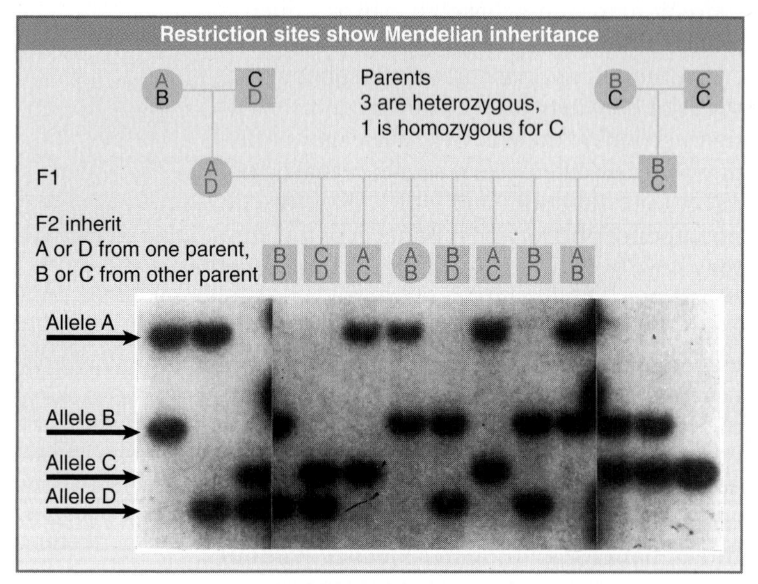

FIGURE 4.2 Restriction site polymorphisms are inherited according to Mendelian rules. Four alleles for a restriction marker are found in all possible pairwise combinations and segregate independently at each generation. Photo courtesy of Ray White, Ernest Gallo Clinic and Research Center, University of California, San Francisco.

occur at a frequency of >1%. Already >1 million have been identified.

Some polymorphisms in the genome can be detected by comparing the restriction maps of different individuals. The criterion is a change in the pattern of fragments produced by cleavage with a restriction enzyme. FIGURE 4.1 shows that when a target site is present in the genome of one individual and absent from another, the extra cleavage in the first genome will generate two fragments corresponding to the single fragment in the second genome.

The restriction map is independent of gene function; as a result, a polymorphism at this level can be detected *irrespective of whether the sequence change affects the phenotype.* Probably very few of the restriction site polymorphisms in a genome actually affect the phenotype. Most involve sequence changes that have no effect on the production of proteins (for example, because they lie between genes).

A difference in restriction maps between two individuals is called a **restriction fragment length polymorphism (RFLP).** Basically, an RFLP is an SNP that is located in the target site for a restriction enzyme. It can be used as a genetic marker in exactly the same way as any other marker. Instead of examining some feature of the phenotype, we directly assess the genotype, as revealed by the restriction map. FIGURE 4.2 shows a pedigree of a restriction polymorphism followed through three generations. It displays Mendelian segregation at the level of DNA marker fragments.

4.4 RFLPs and SNPs Can Be Used for Genetic Mapping

Key concept

- RFLPs and SNPs can be the basis for linkage maps and are useful for establishing parent–progeny relationships.

Recombination frequency can be measured between a restriction marker and a visible phenotypic marker, as illustrated in FIGURE 4.3. Thus a genetic map can include both genotypic and phenotypic markers.

Restriction markers are not restricted to those genome changes that affect the phenotype; as a result, they provide the basis for an extremely powerful technique for identifying genetic loci at the molecular level. A typical problem concerns a mutation with known effects on the phenotype, where the relevant genetic locus can be placed on a genetic map, but for which we have no knowledge about the corresponding gene or protein. Many damaging or fatal human diseases fall into this category. For example, cystic fibrosis shows Mendelian inheritance, but the molecular nature of the mutant function was unknown until it could be identified as a result of characterizing the gene.

If restriction polymorphisms occur at random in the genome, some should occur near any particular target gene. We can identify such restriction markers by virtue of their tight link-

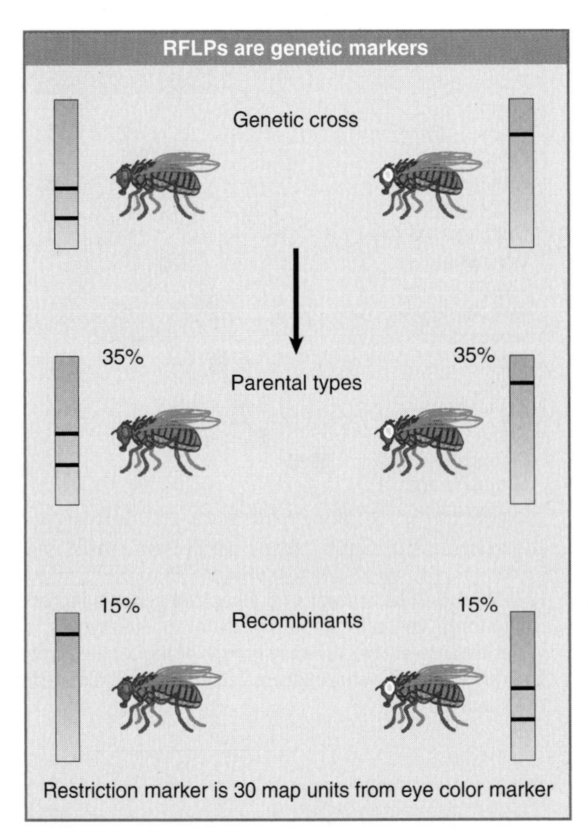

FIGURE 4.3 A restriction polymorphism can be used as a genetic marker to measure recombination distance from a phenotypic marker (such as eye color). The figure simplifies the situation by showing only the DNA bands corresponding to the allele of one genome in a diploid.

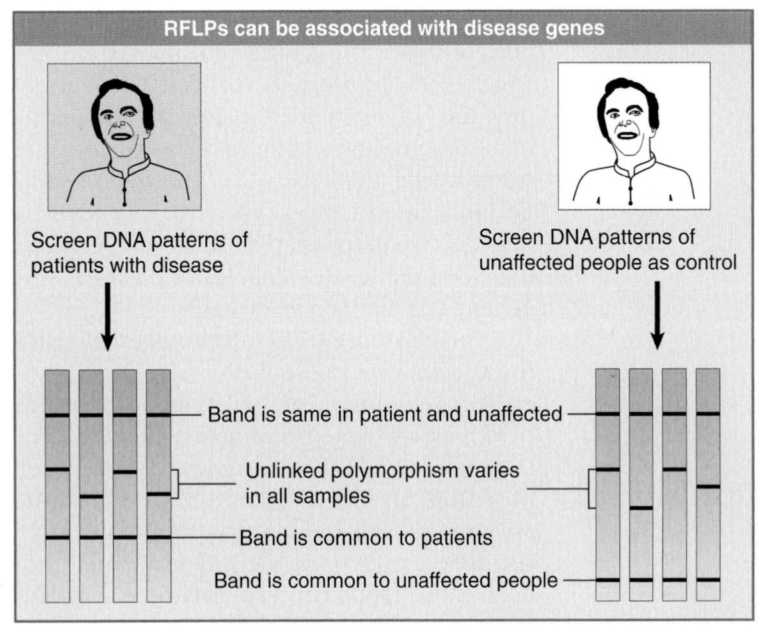

FIGURE 4.4 If a restriction marker is associated with a phenotypic characteristic, the restriction site must be located near the gene responsible for the phenotype. The mutation changing the band that is common in healthy people into the band that is common in patients is very closely linked to the disease gene.

age to the mutant phenotype. If we compare the restriction map of DNA from patients suffering from a disease with the DNA of healthy people, we may find that a particular restriction site is always present (or always absent) from the patients.

A hypothetical example is shown in FIGURE 4.4. This situation corresponds to finding 100% linkage between the restriction marker and the phenotype. It would imply that the restriction marker lies so close to the mutant gene that it is never separated from it by recombination.

The identification of such a marker has two important consequences:

- It may offer a diagnostic procedure for detecting the disease. Some of the human diseases that are genetically well characterized but ill defined in molecular terms cannot be easily diagnosed. If a restriction marker is reliably linked to the phenotype, then its presence can be used to diagnose the disease.
- It may lead to isolation of the gene. The restriction marker must lie relatively near the gene on the genetic map if the two loci rarely or never recombine. "Relatively near" in genetic terms can be a substantial distance in terms of base pairs of DNA; nonetheless, it provides a starting point from which we can proceed along the DNA to the gene itself.

The frequent occurrence of SNPs in the human genome makes them useful for genetic mapping. From the 1.5×10^6 SNPs that have already been identified, there is on average an SNP every 1 to 2 kb. This should allow rapid localization of new disease genes by locating them between the nearest SNPs.

On the same principle, RFLP mapping has been in use for some time. Once an RFLP has been assigned to a linkage group, it can be placed on the genetic map. RFLP mapping in man and mouse has led to the construction of linkage maps for both genomes. Any unknown site can be tested for linkage to these sites, and by this means can be rapidly placed on the map. There are fewer RFLPs than SNPs; thus the resolution of the RFLP map is in principle more limited.

The frequency of polymorphism means that every individual has a unique constellation of SNPs or RFLPs. The particular combination of sites found in a specific region is called a **haplotype,** a genotype in miniature. Haplotype was originally introduced as a concept to describe the genetic constitution of the major histocompatibility locus, a region specifying proteins of importance in the immune system (see Chapter 23, Immune Diversity). The concept now has been extended to describe the particular combination of alleles or restriction sites (or any other genetic marker) present in some defined area of the genome. Using SNPs, a detailed haplotype map of the human genome has been made; this enables disease-causing genes to be mapped more easily.

The existence of RFLPs provides the basis for a technique to establish unequivocal parent–progeny relationships. In cases for which parentage is in doubt, a comparison of the RFLP map in a suitable chromosome region between potential parents and child allows absolute assignment of the relationship. The use of DNA restriction analysis to identify individuals has been called **DNA fingerprinting.** Analysis of especially variable "minisatellite" sequences is used in mapping the human genome (see Section 6.14, Minisatellites Are Useful for Genetic Mapping).

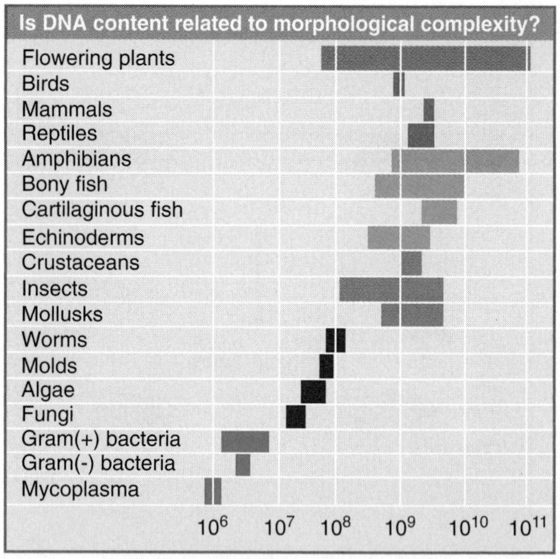

Is DNA content related to morphological complexity?
Flowering plants
Birds
Mammals
Reptiles
Amphibians
Bony fish
Cartilaginous fish
Echinoderms
Crustaceans
Insects
Mollusks
Worms
Molds
Algae
Fungi
Gram(+) bacteria
Gram(-) bacteria
Mycoplasma
10^6 10^7 10^8 10^9 10^10 10^11

FIGURE 4.5 DNA content of the haploid genome increases with morphological complexity of lower eukaryotes, but varies extensively within some groups of higher eukaryotes. The range of DNA values within each group is indicated by the shaded area.

4.5 Why Are Genomes So Large?

Key concepts

- There is no good correlation between genome size and genetic complexity.
- There is an increase in the minimum genome size required to make organisms of increasing complexity.
- There are wide variations in the genome sizes of organisms within many phyla.

The total amount of DNA in the (haploid) genome is a characteristic of each living species known as its **C-value.** There is enormous variation in the range of C-values, from $<10^6$ bp for a mycoplasma to $>10^{11}$ bp for some plants and amphibians.

FIGURE 4.5 summarizes the range of C-values found in different evolutionary phyla. There is an increase in the minimum genome size found in each group as the complexity increases. As absolute amounts of DNA increase in the higher eukaryotes, though, we see some wide variations in the genome sizes within some phyla.

Plotting the *minimum* amount of DNA required for a member of each group suggests in FIGURE 4.6 that an increase in genome size is required to make more complex prokaryotes and lower eukaryotes.

Minimum genome size increases with the phylum
10^10
10^9
10^8
10^7
10^6
Genome size
Mycoplasma
Bacteria
Yeast
Molds
Worms
Insects
Birds
Amphibians
Mammals

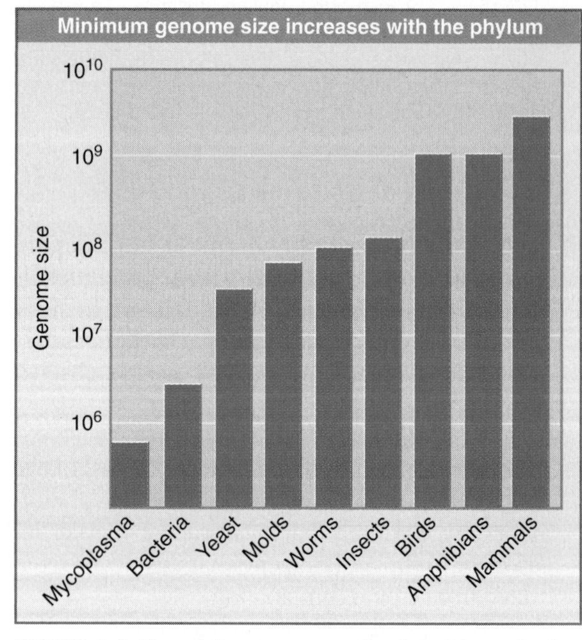

FIGURE 4.6 The minimum genome size found in each phylum increases from prokaryotes to mammals.

Mycoplasma are the smallest prokaryotes and have genomes only ~3× the size of a large bacteriophage. Bacteria start at $\sim 2 \times 10^6$ bp. Unicellular eukaryotes (whose lifestyles may resemble the prokaryotic) get by with genomes that are small, too, although they are larger than those of the bacteria. Being eukaryotic *per se*

Useful genome sizes		
Phylum	Species	Genome (bp)
Algae	*Pyrenomas salina*	6.6×10^5
Mycoplasma	*M. pneumoniae*	1.0×10^6
Bacterium	*E. coli*	4.2×10^6
Yeast	*S. cerevisiae*	1.3×10^7
Slime mold	*D. discoideum*	5.4×10^7
Nematode	*C. elegans*	8.0×10^7
Insect	*D. melanogaster*	1.4×10^8
Bird	*G. domesticus*	1.2×10^9
Amphibian	*X. laevis*	3.1×10^9
Mammal	*H. sapiens*	3.3×10^9

FIGURE 4.7 The genome sizes of some common experimental organisms.

does not imply a vast increase in genome size; a yeast may have a genome size of $\sim 1.3 \times 10^7$ bp, which is only about twice the size of an average bacterial genome.

A further twofold increase in genome size is adequate to support the slime mold *Dictyostelium discoideum,* which is able to live in either unicellular or multicellular modes. Another increase in complexity is necessary to produce the first fully multicellular organisms; the nematode worm *Caenorhabditis elegans* has a DNA content of 8×10^7 bp.

We also can see the steady increase in genome size with complexity in the listing in FIGURE 4.7 of some of the most commonly analyzed organisms. It is necessary to increase the genome size in order to make insects, birds or amphibians, and mammals. After this point, though, there is no good relationship between genome size and morphological complexity of the organism.

We know that genes are much larger than the sequences needed to code for proteins, because exons (coding regions) may comprise only a small part of the total length of a gene. This explains why there is much more DNA than is needed to provide reading frames for all the proteins of the organism. Large parts of an interrupted gene may not be concerned with coding for protein. In addition, there also may be significant lengths of DNA between genes. So it is not possible to deduce from the overall size of the genome anything about the number of genes.

The **C-value paradox** refers to the lack of correlation between genome size and genetic complexity. There are some extremely curious variations in relative genome size. The toad *Xenopus* and man have genomes of essentially the same size. We assume, though, that man is more complex in terms of genetic development! In some phyla there are extremely large variations in DNA content between organisms that do not vary much in complexity (see Figure 4.5). (This is especially marked in insects, amphibians, and plants, but does not occur in birds, reptiles, and mammals, which all show little variation within the group, with an ~2× range of genome sizes.) A cricket has a genome 11× the size of a fruit fly. In amphibians, the smallest genomes are $<10^9$ bp, whereas the largest are $\sim 10^{11}$ bp. There is unlikely to be a large difference in the number of genes needed to specify these amphibians. It is not understood why natural selection allows this variation and whether it has evolutionary consequences.

4.6 Eukaryotic Genomes Contain Both Nonrepetitive and Repetitive DNA Sequences

Key concepts

- The kinetics of DNA reassociation after a genome has been denatured distinguish sequences by their frequency of repetition in the genome.
- Genes are generally coded by sequences in nonrepetitive DNA.
- Larger genomes within a phylum do not contain more genes, but have large amounts of repetitive DNA.
- A large part of repetitive DNA may be made up of transposons.

The general nature of the eukaryotic genome can be assessed by the kinetics of reassociation of denatured DNA. This technique was used extensively before large scale DNA sequencing became possible.

Reassociation kinetics identifies two general types of genomic sequences:

- **Nonrepetitive DNA** consists of sequences that are unique: there is only one copy in a haploid genome.
- **Repetitive DNA** describes sequences that are present in more than one copy in each genome.

Repetitive DNA often is divided into two general types:

- Moderately repetitive DNA consists of relatively short sequences that are repeated typically 10–1000× in the

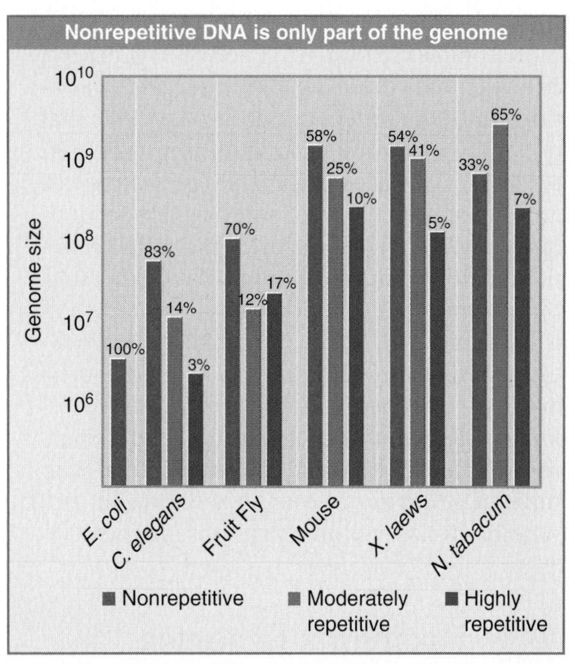

FIGURE 4.8 The proportions of different sequence components vary in eukaryotic genomes. The absolute content of nonrepetitive DNA increases with genome size but reaches a plateau at ~2×10^9 bp.

genome. The sequences are dispersed throughout the genome and are responsible for the high degree of secondary structure formation in pre-mRNA, when (inverted) repeats in the introns pair to form duplex regions.

- Highly repetitive DNA consists of very short sequences (typically <100 bp) that are present many thousands of times in the genome, often organized as long tandem repeats (see Section 6.11, Satellite DNAs Often Lie in Heterochromatin). Neither class represents protein.

The proportion of the genome occupied by nonrepetitive DNA varies widely. FIGURE 4.8 summarizes the genome organization of some representative organisms. Prokaryotes contain only nonrepetitive DNA. For lower eukaryotes, most of the DNA is nonrepetitive; <20% falls into one or more moderately repetitive components. In animal cells, up to half of the DNA often is occupied by moderately and highly repetitive components. In plants and amphibians, the moderately and highly repetitive components may account for up to 80% of the genome, so that the nonrepetitive DNA is reduced to a minority component.

A significant part of the moderately repetitive DNA consists of **transposons,** short sequences of DNA (~1 kb) that have the ability to move to new locations in the genome and/or to make additional copies of themselves (see Chapter 21, Transposons, and Chapter 22, Retroviruses and Retroposons). In some higher eukaryotic genomes they may even occupy more than half of the genome (see Section 5.5, The Human Genome Has Fewer Genes Than Expected).

Transposons are sometimes viewed as fitting the concept of **selfish DNA,** which is defined as sequences that propagate themselves within a genome without contributing to the development of the organism. Transposons may sponsor genome rearrangements, and these could confer selective advantages. It is fair to say, though, that we do not really understand why selective forces do not act against transposons becoming such a large proportion of the genome. Another term that is used to describe the apparent excess of DNA is *junk DNA,* meaning genomic sequences without any apparent function. Of course, it is likely that there is a balance in the genome between the generation of new sequences and the elimination of unwanted sequences, and some proportion of DNA that apparently lacks function may be in the process of being eliminated.

The length of the nonrepetitive DNA component tends to increase with overall genome size as we proceed up to a total genome size ~3×10^9 (characteristic of mammals). Further increases in genome size, however, generally reflect an increase in the amount and proportion of the repetitive components, so that it is rare for an organism to have a nonrepetitive DNA component >2×10^9. The nonrepetitive DNA content of genomes therefore accords better with our sense of the relative complexity of the organism. *E. coli* has 4.2×10^6 bp, *C. elegans* increases an order of magnitude to 6.6×10^7 bp, *D. melanogaster* increases further to ~10^8 bp, and mammals increase another order of magnitude to ~2×10^9 bp.

What type of DNA corresponds to protein-coding genes? Reassociation kinetics typically shows that mRNA is derived from nonrepetitive DNA. The amount of nonrepetitive DNA is therefore a better indication than the total DNA of the coding potential. (However, more detailed analysis based on genomic sequences shows that many exons have related sequences in other exons [see Section 3.5, Exon Sequences Are Conserved but Introns Vary]. Such exons evolve by a duplication to give copies that initially are identical, but which then diverge in sequence during evolution.)

4.7 Genes Can Be Isolated by the Conservation of Exons

Key concept

- Conservation of exons can be used as the basis for identifying coding regions by identifying fragments whose sequences are present in multiple organisms.

Some major approaches to identifying genes are based on the contrast between the conservation of exons and the variation of introns. In a region containing a gene whose function has been conserved among a range of species, the sequence representing the protein should have two distinctive properties:

- It must have an open reading frame,
- and it is likely to have a related sequence in other species.

These features can be used to isolate genes.

Suppose we know by genetic data that a particular genetic trait is located in a given chromosomal region. If we lack knowledge about the nature of the gene product, how are we to identify the gene in a region that may be, for example, >1 Mb?

A heroic approach that has proved successful with some genes of medical importance is to screen relatively short fragments from the region for the two properties expected of a conserved gene. First we seek to identify fragments that cross-hybridize with the genomes of other species, then we examine these fragments for open reading frames.

The first criterion is applied by performing a **zoo blot.** We use short fragments from the region as (radioactive) probes to test for related DNA from a variety of species by Southern blotting. If we find hybridizing fragments in several species related to that of the probe—the probe is usually human—the probe becomes a candidate for an exon of the gene.

The candidates are sequenced, and if they contain open reading frames they are used to isolate surrounding genomic regions. If these appear to be part of an exon, we may then use them to identify the entire gene, to isolate the corresponding cDNA or mRNA, and ultimately to identify the protein.

This approach is especially important when the target gene is spread out because it has many large introns. This proved to be the case with Duchenne muscular dystrophy (DMD), a degenerative disorder of muscle that is X-linked and affects 1 in 3500 human male births. The steps in identifying the gene are summarized in FIGURE 4.9.

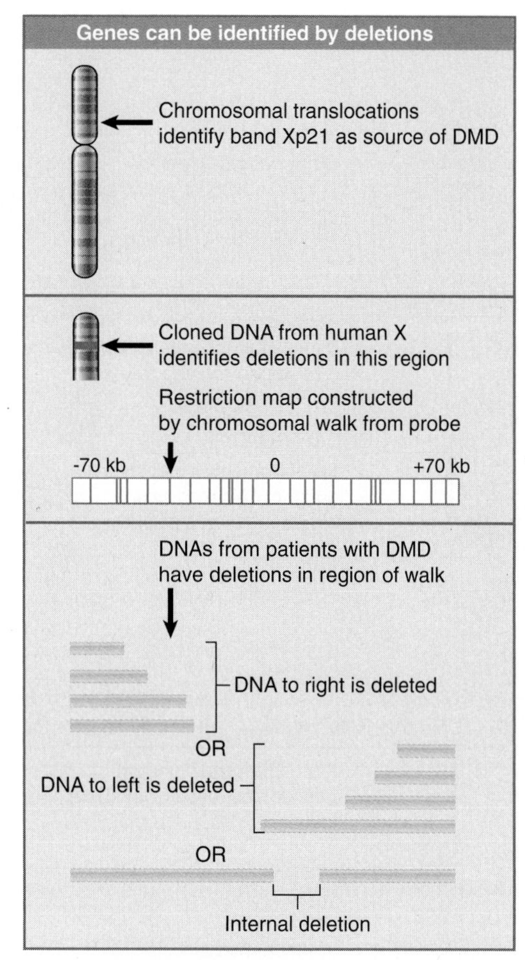

FIGURE 4.9 The gene involved in Duchenne muscular dystrophy was tracked down by chromosome mapping and "walking" to a region in which deletions can be identified with the occurrence of the disease.

Linkage analysis localized the DMD locus to chromosomal band Xp21. Patients with the disease often have chromosomal rearrangements involving this band. By comparing the ability of X-linked DNA probes to hybridize with DNA from patients with normal DNA, cloned fragments were obtained that correspond to the region that was rearranged or deleted in patients' DNA.

Once some DNA in the general vicinity of the target gene has been obtained, it is possible to "walk" along the chromosome until the gene is reached. A chromosomal walk was used to construct a restriction map of the region on either side of the probe, which covered a region

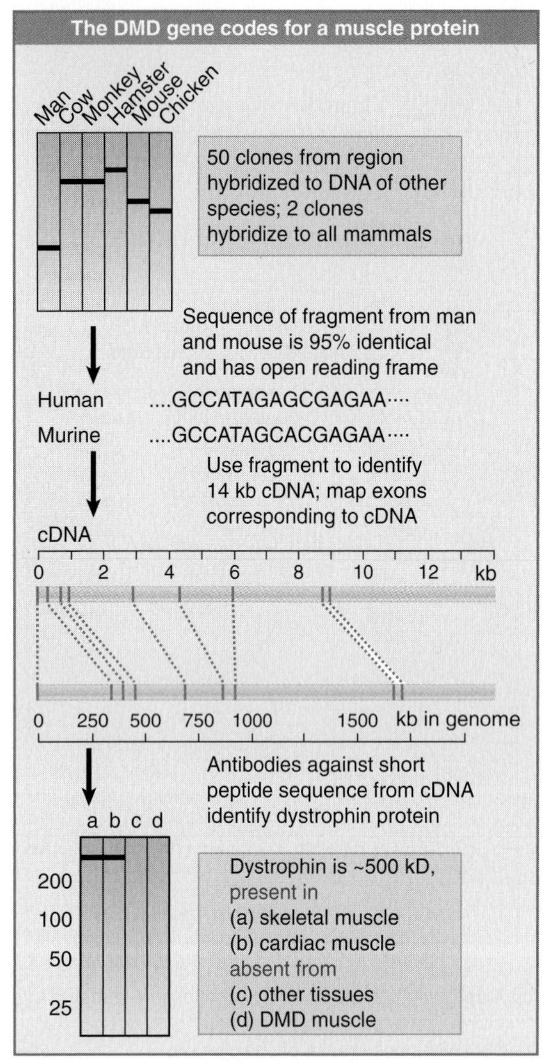

FIGURE 4.10 The Duchenne muscular dystrophy gene was characterized by zoo blotting, cDNA hybridization, genomic hybridization, and identification of the protein.

of >100 kb. Analysis of the DNA from a series of patients identified large deletions in this region that extended in either direction. The most telling deletion is one that is contained entirely within the region, because this delineates a segment that must be important in gene function and indicates that the gene—or at least part of it—lies in this region.

Having now come into the region of the gene, we need to identify its exons and introns. A zoo blot identified fragments that cross-hybridize with the mouse X chromosome and with other mammalian DNAs. As summarized in FIGURE 4.10, these were scrutinized for open reading frames and the sequences typical of exon–intron junctions. Fragments that met these criteria were used as probes to identify homologous sequences in a cDNA library prepared from muscle mRNA.

The cDNA corresponding to the gene identifies an unusually large mRNA of approximately 14 kb. Hybridization back to the genome shows that the mRNA is represented in >60 exons, which are spread over ~2000 kb of DNA. This makes DMD the longest gene identified.

The gene codes for a protein of ~500 kD called dystrophin, which is a component of muscle and is present in rather low amounts. All patients with the disease have deletions at this locus and lack (or have defective) dystrophin.

Muscle also has the distinction of having the largest known protein, titin, with almost 27,000 amino acids. Its gene has the largest number of exons (178) and the longest single exon in the human genome (17,000 bp).

Another technique that allows genomic fragments to be scanned rapidly for the presence of exons is called **exon trapping.** FIGURE 4.11 shows that it starts with a vector that contains a strong promoter and has a single intron between two exons. When this vector is transfected into cells, its transcription generates large amounts of an RNA containing the sequences of the two exons. A restriction-cloning site lies within the intron and is used to insert genomic fragments from a region of interest. If a fragment does not contain an exon, there is no change in the splicing pattern, and the RNA contains only the same sequences as the parental vector. If the genomic fragment contains an exon flanked by two partial intron sequences, though, the splicing sites on either side of this exon are recognized and the sequence of the exon is inserted into the RNA between the two exons of the vector. This can be detected readily by reverse transcribing the cytoplasmic RNA into cDNA and using PCR to amplify the sequences between the two exons of the vector. So the appearance in the amplified population of sequences from the genomic fragment indicates that an exon has been trapped. Because introns are usually large and exons are small in animal cells, there is a high probability that a random piece of genomic DNA will contain the required structure of an exon surrounded by partial introns. In fact, exon trapping may mimic the events that have occurred naturally during evolution of genes (see Section 3.8, How Did Interrupted Genes Evolve?).

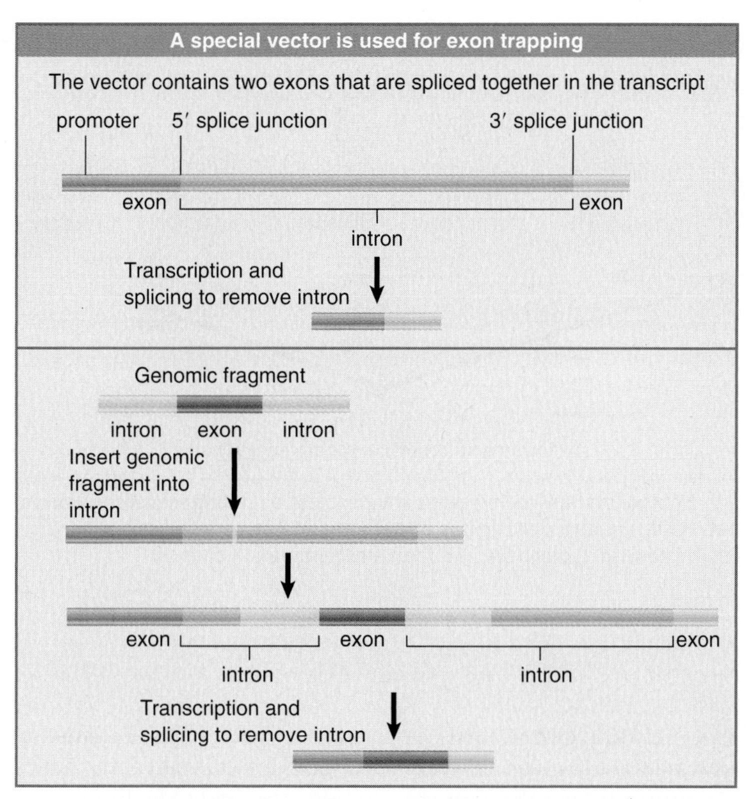

FIGURE 4.11 A special splicing vector is used for exon trapping. If an exon is present in the genomic fragment, its sequence will be recovered in the cytoplasmic RNA. If the genomic fragment consists solely of sequences from within an intron, though, splicing does not occur, and the mRNA is not exported to the cytoplasm.

4.8 The Conservation of Genome Organization Helps to Identify Genes

Key concepts

- Algorithms for identifying genes are not perfect and many corrections must be made to the initial data set.
- Pseudogenes must be distinguished from active genes.
- Syntenic relationships are extensive between mouse and human genomes, and most active genes are in a syntenic region.

Once we have assembled the sequence of a genome, we still have to identify the genes within it. Coding sequences represent a very small fraction. Exons can be identified as uninterrupted open reading frames flanked by appropriate sequences. What criteria need to be satisfied to identify an active gene from a series of exons?

FIGURE 4.12 shows that an active gene should consist of a series of exons for which the first exon immediately follows a promoter, the internal exons are flanked by appropriate splicing junctions, the last exon is followed by 3′ processing signals, and a single open reading frame starting with an initiation codon and ending with a termination codon can be deduced by joining the exons together. Internal exons can be identified as open reading frames flanked by splicing junctions. In the simplest cases, the first and last exons contain the start and end of the coding region, respectively (as well as the 5′ and 3′ untranslated regions). In more complex cases, the first or last exons may have only untranslated regions and may therefore be more difficult to identify.

The algorithms that are used to connect exons are not completely effective when the genome is very large and the exons may be separated by very large distances. For example, the initial analysis of the human genome mapped

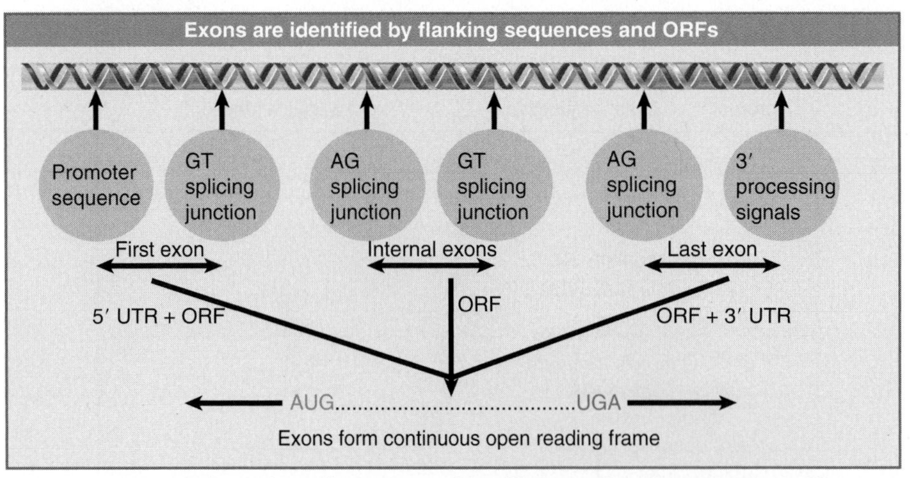

FIGURE 4.12 Exons of protein-coding genes are identified as coding sequences flanked by appropriate signals (with untranslated regions at both ends). The series of exons must generate an open reading frame with appropriate initiation and termination codons.

170,000 exons into 32,000 genes. This is unlikely to be correct because it gives an average of 5.3 exons per gene, whereas the average of individual genes that have been fully characterized is 10.2. Either we have missed many exons, or they should be connected differently into a smaller number of genes in the whole genome sequence.

Even when the organization of a gene is correctly identified, there is the problem of distinguishing active genes from pseudogenes. Many pseudogenes can be recognized by obvious defects in the form of multiple mutations that create an inactive coding sequence. Pseudogenes that have arisen more recently have not accumulated so many mutations and thus may be more difficult to recognize. In an extreme example, the mouse has only one active *Gapdh* gene (coding for glyceraldehyde phosphate dehydrogenase), but has ~400 pseudogenes. Approximately 100 of these pseudogenes initially appeared to be active in the mouse genome sequence, and individual examination was necessary to exclude them from the list of active genes.

Confidence that a gene is active can be increased by comparing regions of the genomes of different species. There has been extensive overall reorganization of sequences between the mouse and human genomes, as seen in the simple fact that there are 23 chromosomes in the human haploid genome and 20 chromosomes in the mouse haploid genome. However, at the local level the order of genes is generally the same: When pairs of human and mouse homologues are compared, the genes located on either side also tend to be homologues. This relationship is called **synteny.**

FIGURE 4.13 shows the relationship between mouse chromosome 1 and the human chromosomal set. We can recognize 21 segments in this mouse chromosome that have syntenic counterparts in human chromosomes. The extent of reshuffling that has occurred between the genomes is shown by the fact that the segments are spread among six different human chromosomes. The same types of relationships are found in all mouse chromosomes except for the X chromosome, which is syntenic only with the human X chromosome. This is explained by the fact that the X is a special case, subject to dosage compensation to adjust for the difference between males (one copy) and females (two copies) (see Section 31.5, X Chromosomes Undergo Global Changes). This may apply selective pressure against the translocation of genes to and from the X chromosome.

Comparison of the mouse and human genome sequences shows that >90% of each genome lies in syntenic blocks that range widely in size (from 300 kb to 65 Mb). There is a total of 342 syntenic segments, with an average length of 7 Mb (0.3% of the genome). Ninety-nine percent of mouse genes have a homologue in the human genome; for 96% that homologue is in a syntenic region.

Comparing the genomes provides interesting information about the evolution of species.

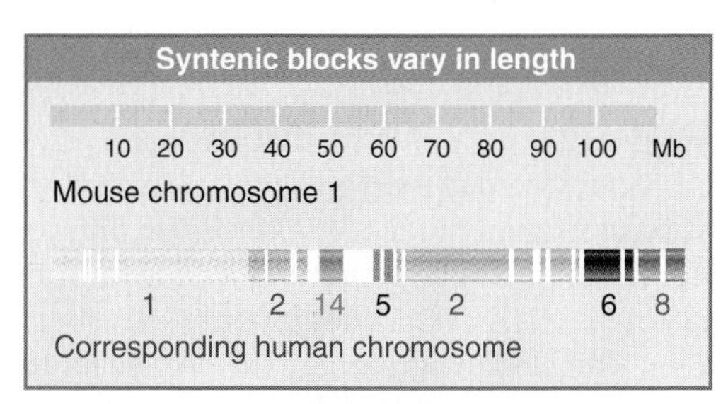

FIGURE 4.13 Mouse chromosome 1 has 21 segments of 1 to 25 Mb that are syntenic with regions corresponding to parts of six human chromosomes.

The number of gene families in the mouse and human genomes is the same, and a major difference between the species is the differential expansion of particular families in one of the genomes. This is especially noticeable in genes that affect phenotypic features that are unique to the species. Of 25 families for which the size has been expanded in mouse, 14 contain genes specifically involved in rodent reproduction, and 5 contain genes specific to the immune system.

A validation of the importance of syntenic blocks comes from pairwise comparisons of the genes within them. Looking for likely pseudogenes on the basis of sequence comparisons, a gene that is not in a syntenic location (that is, its context is different in the two species) is twice as likely to be a pseudogene. Put another way, translocation away from the original locus tends to be associated with the creation of pseudogenes. The lack of a related gene in a syntenic position is therefore grounds for suspecting that an apparent gene may really be a pseudogene. Overall, >10% of the genes that are initially identified by analysis of the genome are likely to turn out to be pseudogenes.

As a general rule, comparisons between genomes add significantly to the effectiveness of gene prediction. When sequence features indicating active genes are conserved—for example, between Man and mouse—there is an increased probability that they identify active homologues.

Identifying genes coding for RNA is more difficult because we cannot use the criterion of the open reading frame. It also is true that comparative genome analysis increased the rigor of the analysis. For example, analysis of either the human or the mouse genome alone identifies ~500 genes coding for tRNA, but comparison of features suggests that <350 of these genes are in fact active in each genome.

4.9 Organelles Have DNA

Key concepts

- Mitochondria and chloroplasts have genomes that show non-Mendelian inheritance. Typically they are maternally inherited.
- Organelle genomes may undergo somatic segregation in plants.
- Comparisons of mitochondrial DNA suggest that humans are descended from a single female who lived 200,000 years ago in Africa.

The first evidence for the presence of genes outside the nucleus was provided by non-Mendelian inheritance in plants (observed in the early years of the twentieth century, just after the rediscovery of Mendelian inheritance). Non-Mendelian inheritance sometimes is associated with the phenomenon of somatic segregation. Both have a similar cause:

- Non-Mendelian inheritance is defined by the failure of the progeny of a mating to display Mendelian segregation for parental characters. It reflects lack of association between the segregating character and the meiotic spindle.
- Somatic segregation describes a phenomenon in which parental characters segregate in somatic cells and therefore display heterogeneity in the organism. This is a notable feature of plant development, as it reflects lack of association between the segregating character and the mitotic spindle.

Non-Mendelian inheritance and somatic segregation are therefore taken to indicate the presence of genes that reside outside the nucleus and do not utilize segregation on the meiotic and mitotic spindles to distribute replicas to gametes or to daughter cells, respectively. FIGURE 4.14 shows that this happens when the mitochondria inherited from the male and female parents have different alleles, and by chance a daughter cell receives an unbalanced distribution of mitochondria that represents only one parent (see Section 17.12, How Do Mitochondria Replicate and Segregate?).

The extreme form of non-Mendelian inheritance is uniparental inheritance, which occurs when the genotype of only one parent is inherited and that of the other parent is permanently lost. In less extreme examples, the progeny of one parental genotype exceed those of the other genotype. Usually it is the mother whose genotype is preferentially (or solely) inherited. This effect is sometimes described as **maternal inheritance.** The important point is that the genotype contributed by the parent of one

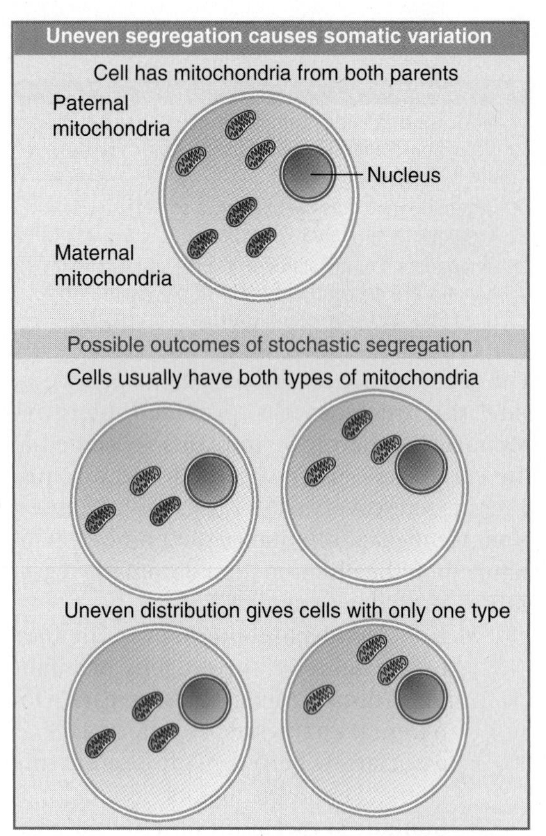

FIGURE 4.14 When paternal and maternal mitochondrial alleles differ, a cell has two sets of mitochondrial DNAs. Mitosis usually generates daughter cells with both sets. Somatic variation may result if unequal segregation generates daughter cells with only one set.

particular sex predominates, as seen in abnormal segregation ratios when a cross is made between mutant and wild type. This contrasts with the behavior of Mendelian genetics, which occurs when reciprocal crosses show the contributions of both parents to be equally inherited.

The bias in parental genotypes is established at or soon after the formation of a zygote. There are various possible causes. The contribution of maternal or paternal information to the organelles of the zygote may be unequal; in the most extreme case, only one parent contributes. In other cases the contributions are equal, but the information provided by one parent does not survive. Combinations of both effects are possible. Whatever the cause, the unequal representation of the information from the two parents contrasts with nuclear genetic information, which derives equally from each parent.

Non-Mendelian inheritance results from the presence in mitochondria and chloroplasts of DNA genomes that are inherited independently of nuclear genes. In effect, the organelle genome comprises a length of DNA that has been physically sequestered in a defined part of the cell and is subject to its own form of expression and regulation. An organelle genome can code for some or all of the RNAs, but codes for only some of the proteins needed to perpetuate the organelle. The other proteins are coded in the nucleus, expressed via the cytoplasmic protein synthetic apparatus, and imported into the organelle.

Genes not residing within the nucleus are generally described as **extranuclear genes;** they are transcribed and translated in the *same* organelle compartment (mitochondrion or chloroplast) in which they reside. By contrast, *nuclear* genes are expressed by means of *cytoplasmic* protein synthesis. (The term **cytoplasmic inheritance** sometimes is used to describe the behavior of genes in organelles. We shall not use this description, though, because it is important to be able to distinguish between events in the general cytosol and those in specific organelles.)

Higher animals show maternal inheritance, which can be explained if the mitochondria are contributed entirely by the ovum and not at all by the sperm. FIGURE 4.15 shows that the sperm contributes only a copy of the nuclear DNA. Thus the mitochondrial genes are derived exclusively from the mother, and in males they are discarded each generation.

Conditions in the organelle are different from those in the nucleus, and organelle DNA therefore evolves at its own distinct rate. If inheritance is uniparental, there can be no recombination between parental genomes. In fact, recombination usually does not occur in those cases for which organelle genomes are inherited from both parents. Organelle DNA has a different replication system from that of the nucleus; as a result, the error rate during replication may be different. Mitochondrial DNA accumulates mutations more rapidly than nuclear DNA in mammals, but in plants the accumulation in the mitochondrion is slower than in the nucleus (the chloroplast is intermediate).

One consequence of maternal inheritance is that the sequence of mitochondrial DNA is more sensitive than nuclear DNA to reductions in the size of the breeding population. Comparisons of mitochondrial DNA sequences in a range of human populations allow an evolutionary tree to be constructed. The divergence among human mitochondrial DNAs spans 0.57%. A tree can be constructed in which the

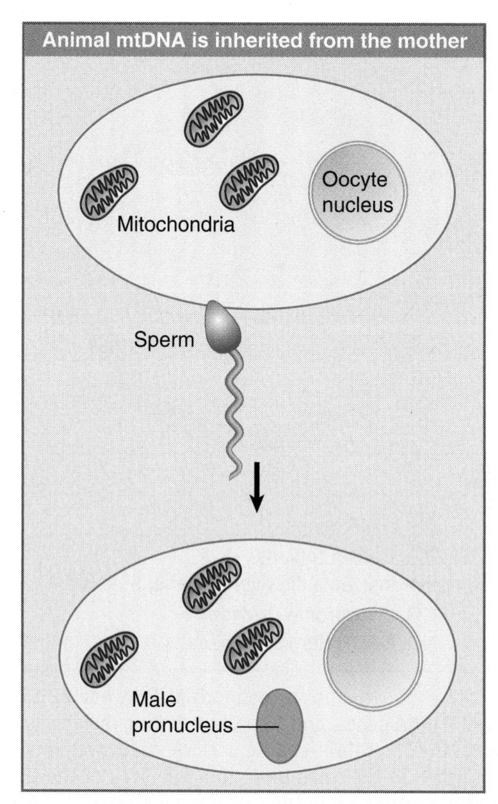

FIGURE 4.15 DNA from the sperm enters the oocyte to form the male pronucleus in the fertilized egg, but all the mitochondria are provided by the oocyte.

mitochondrial variants diverged from a common (African) ancestor. The rate at which mammalian mitochondrial DNA accumulates mutations is 2% to 4% per million years, which is >10× faster than the rate for globin. Such a rate would generate the observed divergence over an evolutionary period of 140,000 to 280,000 years. This implies that the human race is descended from a single female who lived in Africa ~200,000 years ago.

4.10 Organelle Genomes Are Circular DNAs That Code for Organelle Proteins

Key concepts

- Organelle genomes are usually (but not always) circular molecules of DNA.
- Organelle genomes code for some, but not all, of the proteins found in the organelle.

Most organelle genomes take the form of a single circular molecule of DNA of unique sequence (denoted **mtDNA** in the mitochondrion and **ctDNA** in the chloroplast). There are a few exceptions for which mitochondrial DNA is a linear molecule; these generally occur in lower eukaryotes.

Usually there are several copies of the genome in the individual organelle. There are multiple organelles per cell, therefore there are many organelle genomes per cell. Although the organelle genome itself is unique, it constitutes a repetitive sequence relative to any nonrepetitive nuclear sequence.

Chloroplast genomes are relatively large, usually ~140 kb in higher plants and <200 kb in lower eukaryotes. This is comparable to the size of a large bacteriophage, for example, T4 at ~165 kb. There are multiple copies of the genome per organelle, typically 20 to 40 in a higher plant, and multiple copies of the organelle per cell, typically 20 to 40.

Mitochondrial genomes vary in total size by more than an order of magnitude. Animal cells have small mitochondrial genomes—approximately 16.5 kb in mammals. There are several hundred mitochondria per cell. Each mitochondrion has multiple copies of the DNA. The total amount of mitochondrial DNA relative to nuclear DNA is small; it is estimated to be <1%.

In yeast, the mitochondrial genome is much larger. In *Saccharomyces cerevisiae*, the exact size varies among different strains but averages ~80 kb. There are ~22 mitochondria per cell, which corresponds to ~4 genomes per organelle. In growing cells, the proportion of mitochondrial DNA can be as high as 18%.

Plants show an extremely wide range of variation in mitochondrial DNA size, with a minimum of ~100 kb. The size of the genome makes it difficult to isolate intact, but restriction mapping in several plants suggests that the mitochondrial genome is usually a single sequence that is organized as a circle. Within this circle there are short homologous sequences. Recombination between these elements generates smaller, subgenomic circular molecules that coexist with the complete, "master" genome—a good example of the apparent complexity of plant mitochondrial DNAs.

With mitochondrial genomes sequenced from many organisms, we can now see some general patterns in the representation of functions in mitochondrial DNA. FIGURE 4.16 summarizes the distribution of genes in mitochondrial genomes. The total number of protein-coding genes is rather small and does not correlate with the size of the genome. Mammalian

Mitochondria code for RNAs and proteins			
Species	Size (kb)	Protein-coding genes	RNA-coding genes
Fungi	19–100	8–14	10–28
Protists	6–100	3–62	2–29
Plants	186–366	27–34	21–30
Animals	16–17	13	4–24

FIGURE 4.16 Mitochondrial genomes have genes coding for (mostly complex I–IV) proteins, rRNAs, and tRNAs.

mitochondria use their 16 kb genomes to code for 13 proteins, whereas yeast mitochondria use their 60 to 80 kb genomes to code for as few as eight proteins. Plants, which have much larger mitochondrial genomes, code for more proteins. Introns are found in most mitochondrial genomes, although not in the very small mammalian genomes.

The two major rRNAs are always coded by the mitochondrial genome. The number of tRNAs coded by the mitochondrial genome varies from none to the full complement (25 to 26 in mitochondria). This accounts for the variation in Figure 4.16.

The major part of the protein-coding activity is devoted to the components of the multisubunit assemblies of respiration complexes I–IV. Many ribosomal proteins are coded in protist and plant mitochondrial genomes, but there are few or none in fungi and animal genomes. There are genes coding for proteins involved in import in many protist mitochondrial genomes.

4.11 Mitochondrial DNA Organization Is Variable

Key concepts

- Animal cell mitochondrial DNA is extremely compact and typically codes for 13 proteins, 2 rRNAs, and 22 tRNAs.
- Yeast mitochondrial DNA is 5× longer than animal cell mtDNA because of the presence of long introns.

Animal mitochondrial DNA is extremely compact. There are extensive differences in the detailed gene organization found in different animal phyla, but the general principle is maintained of a small genome coding for a restricted number of functions. In mammalian mitochondrial genomes, the organization is extremely compact. There are no introns, some genes actually overlap, and almost every single base pair can be

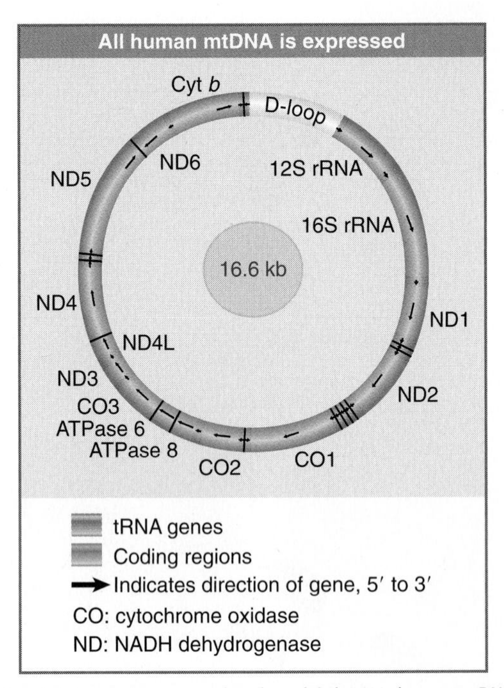

FIGURE 4.17 Human mitochondrial DNA has 22 tRNA genes, 2 rRNA genes, and 13 protein-coding regions. Fourteen of the 15 protein-coding or rRNA-coding regions are transcribed in the same direction. Fourteen of the tRNA genes are expressed in the clockwise direction and 8 are read counter-clockwise.

assigned to a gene. With the exception of the D loop, a region concerned with the initiation of DNA replication, no more than 87 of the 16,569 bp of the human mitochondrial genome can be regarded as lying in intercistronic regions.

The complete nucleotide sequences of mitochondrial genomes in animal cells show extensive homology in organization. The map of the human mitochondrial genome is summarized in FIGURE 4.17. There are 13 protein-coding regions. All of the proteins are components of the apparatus concerned with respiration. These include cytochrome *b*, three subunits of cytochrome oxidase, one of the subunits of ATPase, and seven subunits (or associated proteins) of NADH dehydrogenase.

The fivefold discrepancy in size between the *S. cerevisiae* (84 kb) and mammalian (16 kb) mitochondrial genomes alone alerts us to the fact that there must be a great difference in their genetic organization in spite of their common function. The number of endogenously synthesized products concerned with mitochondrial enzymatic functions appears to be similar. Does the additional genetic material in yeast mitochondria represent other proteins, perhaps concerned with regulation, or is it unexpressed?

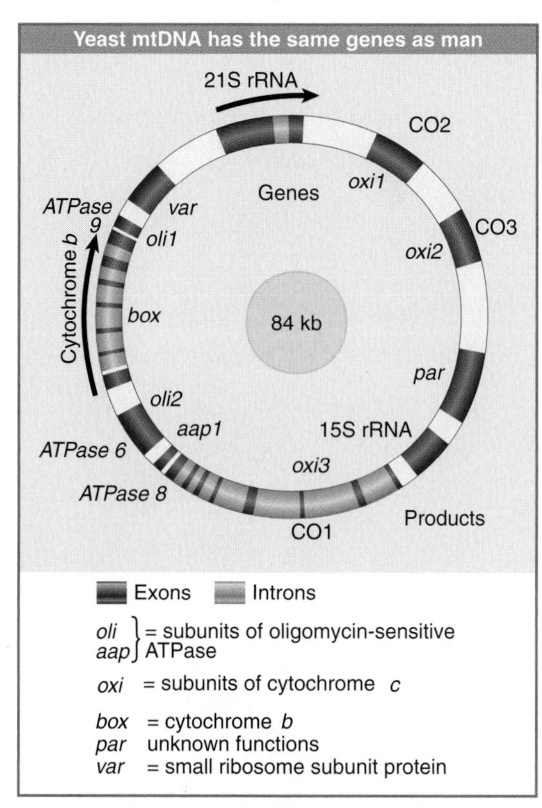

FIGURE 4.18 The mitochondrial genome of *S. cerevisiae* contains both interrupted and uninterrupted protein-coding genes, rRNA genes, and tRNA genes (positions not indicated). Arrows indicate direction of transcription.

The map shown in FIGURE 4.18 accounts for the major RNA and protein products of the yeast mitochondrion. The most notable feature is the dispersion of loci on the map.

The two most prominent loci are the interrupted genes *box* (coding for cytochrome *b*) and *oxi3* (coding for subunit 1 of cytochrome oxidase). Together these two genes are almost as long as the entire mitochondrial genome in mammals! Many of the long introns in these genes have open reading frames in register with the preceding exon (see Section 27.5, Some Group I Introns Code for Endonucleases That Sponsor Mobility). This adds several proteins, all synthesized in low amounts, to the complement of the yeast mitochondrion.

The remaining genes are uninterrupted. They correspond to the other two subunits of cytochrome oxidase coded by the mitochondrion, to the subunit(s) of the ATPase, and (in the case of *var1*) to a mitochondrial ribosomal protein. The total number of yeast mitochondrial genes is unlikely to exceed ~25.

Chloroplasts have >100 genes

Genes	Types
RNA-coding	
16S rRNA	1
23S rRNA	1
4.5S rRNA	1
5S rRNA	1
tRNA	30–32
Gene Expression	
r-proteins	20–21
RNA polymerase	3
Others	2
Chloroplast functions	
Rubisco and thylakoids	31–32
NADH dehydrogenase	11
Total	105–113

FIGURE 4.19 The chloroplast genome in land plants codes for 4 rRNAs, 30 tRNAs, and ~60 proteins.

4.12 The Chloroplast Genome Codes for Many Proteins and RNAs

Key concept

- Chloroplast genomes vary in size, but are large enough to code for 50 to 100 proteins as well as the rRNAs and tRNAs.

What genes are carried by chloroplasts? Chloroplast DNAs vary in length from 120 to 190 kb. The sequenced chloroplast genomes (>10 in total) have 87 to 183 genes. FIGURE 4.19 summarizes the functions coded by the chloroplast genome in land plants. There is more variation in the chloroplast genomes of algae.

The situation is generally similar to that of mitochondria, except that more genes are involved. The chloroplast genome codes for all the rRNA and tRNA species needed for protein synthesis. The ribosome includes two small rRNAs in addition to the major species. The tRNA set may include all of the necessary genes. The chloroplast genome codes for ~50 proteins, including RNA polymerase and ribosomal proteins. Again, the rule is that organelle genes are transcribed and translated by the apparatus of the organelle.

About half of the chloroplast genes code for proteins involved in protein synthesis. The endosymbiotic origin of the chloroplast is emphasized by the relationships between these genes and their counterparts in bacteria. The organization of the rRNA genes in particular is closely related to that of a cyanobacterium, which pins down more precisely the last common ancestor between chloroplasts and bacteria.

Introns in chloroplasts fall into two general classes. Those in tRNA genes are usually (although not inevitably) located in the anticodon loop, like the introns found in yeast nuclear tRNA genes (see Section 26.14, Yeast tRNA Splicing Involves Cutting and Rejoining). Those in protein-coding genes resemble the introns of mitochondrial genes (see Chapter 27, Catalytic RNA). This places the endosymbiotic event at a time in evolution before the separation of prokaryotes with uninterrupted genes.

The role of the chloroplast is to undertake photosynthesis. Many of its genes code for proteins of complexes located in the thylakoid membranes. The constitution of these complexes shows a different balance from that of mitochondrial complexes. Although some complexes are like mitochondrial complexes in having some subunits coded by the organelle genome and some by the nuclear genome, other chloroplast complexes are coded entirely by one genome.

4.13 Mitochondria Evolved by Endosymbiosis

How did a situation evolve in which an organelle contains genetic information for some of its functions, whereas others are coded in the nucleus? FIGURE 4.20 shows the endosymbiosis model for mitochondrial evolution, in which primitive cells captured bacteria that provided the functions that evolved into mitochondria and chloroplasts. At this point, the proto-organelle must have contained all of the genes needed to specify its functions.

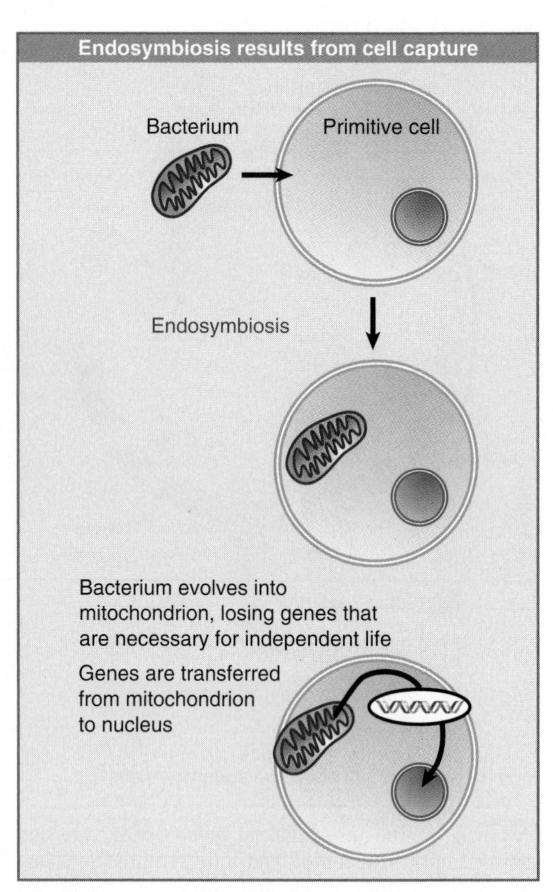

FIGURE 4.20 Mitochondria originated by a endosymbiotic event when a bacterium was captured by a eukaryotic cell.

Sequence homologies suggest that mitochondria and chloroplasts evolved separately, from lineages that are common with eubacteria, with mitochondria sharing an origin with α-purple bacteria and chloroplasts sharing an origin with cyanobacteria. The closest known relative of mitochondria among the bacteria is *Rickettsia* (the causative agent of typhus), which is an obligate intracellular parasite that is probably descended from free-living bacteria. This reinforces the idea that mitochondria originated in an endosymbiotic event involving an ancestor that is also common to *Rickettsia.*

Two changes must have occurred as the bacterium became integrated into the recipient cell and evolved into the mitochondrion (or chloroplast). The organelles have far fewer genes than an independent bacterium and have lost many of the gene functions that are necessary for independent life (such as metabolic pathways). The majority of genes coding for organelle functions are in fact now located in the nucleus, thus these genes must have been transferred there from the organelle.

Transfer of DNA between organelle and nucleus has occurred over evolutionary time periods and still continues. The rate of transfer can be measured directly by introducing into an organelle a gene that can function only in the nucleus, for example, because it contains a nuclear intron, or because the protein must function in the cytosol. In terms of providing the material for evolution, the transfer rates from organelle to nucleus are roughly equivalent to the rate of single gene mutation. DNA introduced into mitochondria is transferred to the nucleus at a rate of 2×10^{-5} per generation. Experiments to measure transfer in the reverse direction, from nucleus to mitochondrion, suggest that the rate is a much lower $<10^{-10}$. When a nuclear-specific antibiotic resistance gene is introduced into chloroplasts, its transfer to the nucleus and successful expression can be followed by screening seedlings for resistance to the antibiotic. This shows that transfer occurs at a rate of 1 in 16,000 seedlings, or 6×10^{-5}.

Transfer of a gene from an organelle to the nucleus requires physical movement of the DNA, of course, but successful expression also requires changes in the coding sequence. Organelle proteins that are coded by nuclear genes have special sequences that allow them to be imported into the organelle after they have been synthesized in the cytoplasm (see Section 10.16, Posttranslational Membrane Insertion Depends on Leader Sequences). These sequences are not required by proteins that are synthesized within the organelle. Perhaps the process of effective gene transfer occurred at a period when compartments were less rigidly defined, so that it was easier both for the DNA to be relocated and for the proteins to be incorporated into the organelle irrespective of the site of synthesis.

Phylogenetic maps show that gene transfers have occurred independently in many different lineages. It appears that transfers of mitochondrial genes to the nucleus occurred only early in animal cell evolution, but it is possible that the process is still continuing in plant cells. The number of transfers can be large; there are >800 nuclear genes in *Arabidopsis* whose sequences are related to genes in the chloroplasts of other plants. These genes are candidates for evolution from genes that originated in the chloroplast.

4.14 Summary

The DNA sequences composing a eukaryotic genome can be classified in three groups:

- nonrepetitive sequences are unique;
- moderately repetitive sequences are dispersed and repeated a small number of times as related by not identical copies;
- and highly repetitive sequences are short and usually repeated as tandem arrays.

The proportions of the types of sequences are characteristic for each genome, although larger genomes tend to have a smaller proportion of nonrepetitive DNA. Almost 50% of the human genome consists of repetitive sequences, the vast majority corresponding to transposons sequences. Most structural genes are located in nonrepetitive DNA. The amount of nonrepetitive DNA is a better reflection of the complexity of the organism than the total genome size; the greatest amount of nonrepetitive DNA is genomes is $\sim 2 \times 10^9$ bp.

Non-Mendelian inheritance is explained by the presence of DNA in organelles in the cytoplasm. Mitochondria and chloroplasts are membrane-bounded systems in which some proteins are synthesized within the organelle, while others are imported. The organelle genome is usually a circular DNA that codes for all the RNAs and some of the proteins required by the organelle.

Mitochondrial genomes vary greatly in size from the 16 kb minimalist mammalian genome to the 570 kb genome of higher plants. The larger genomes may code for additional functions. Chloroplast genomes range from 120–200 kb. Those that have been sequenced have similar organization and coding functions. In both mitochondria and chloroplasts, many of the major proteins contain some subunits synthesized in the organelle and some subunits imported from the cytosol.

Rearrangements occur in mitochondrial DNA rather frequently in yeast, and recombination between mitochondrial or between chloroplast genomes has been found. Transfers of DNA have occurred between chloroplasts or mitochondria and nuclear genomes.

References

4.3 Individual Genomes Show Extensive Variation

Research

Altshuler, D., Brooks, L. D., Chakravarti, A., Collins, F. S., Daly, M. J., and Donnelly, P. (2005). A haplotype map of the human genome. *Nature* 437, 1299–1320.

Altshuler, D., Pollara, V. J., Cowles, C. R., Van Etten, W. J., Baldwin, J., Linton, L., and Lander, E. S. (2000). An SNP map of the human genome generated by reduced representation shotgun sequencing. *Nature* 407, 513–516.

Mullikin, J. C., Hunt, S. E., Cole, C. G., Mortimore, B. J., Rice, C. M., Burton, J., Matthews, L. H., Pavitt, R., Plumb, R. W., Sims, S. K., Ainscough, R. M., Attwood, J., Bailey, J. M., Barlow, K., Bruskiewich, R. M., Butcher, P. N., Carter, N. P., Chen, Y., and Clee, C. M. (2000). An SNP map of human chromosome 22. *Nature* 407, 516–520.

4.4 RFLPs and SNPs Can Be Used for Genetic Mapping

Reviews

Gusella, J. F. (1986). DNA polymorphism and human disease. *Annu. Rev. Biochem.* 55, 831–854.

White, R., Leppert, M., Bishop, D. T., et al. (1985). Construction of linkage maps with DNA markers for human chromosomes. *Nature* 313, 101–105.

Research

Altshuler, D., Brooks, L. D., Chakravarti, A., Collins, F. S., Daly, M. J., and Donnelly, P. (2005). A haplotype map of the human genome. *Nature* 437, 1299–1320.

Dib, C., Faure, S., Fizames, C., et al. (1996). A comprehensive genetic map of the human genome based on 5,264 microsatellites. *Nature* 380, 152–154.

Dietrich, W. F., Miller, J., Steen, R., et al. (1996). A comprehensive genetic map of the mouse genome. *Nature* 380, 149–152.

Donis-Keller, J., Green, P., Helms, C., et al. (1987). A genetic linkage map of the human genome. *Cell* 51, 319–337.

Hinds, D. A., Stuve, L. L., Nilsen, G. B., Halperin, E., Eskin, E., Ballinger, D. G., Frazer, K. A., and Cox, D. R. (2005). Whole-genome patterns of common DNA variation in three human populations. *Science* 307, 1072–1079.

Sachidanandam, R., Weissman, D., Schmidt, S., et al. (2001). A map of human genome sequence variation containing 1.42 million single nucleotide polymorphisms. The International SNP Map Working Group. *Nature* 409, 928–933.

4.5 Why Are Genomes So Large?

Reviews

Gall, J. G. (1981). Chromosome structure and the C-value paradox. *J. Cell Biol.* 91, 3s–14s.

Gregory, T. R. (2001). Coincidence, coevolution, or causation? DNA content, cell size, and the C-value enigma. *Biol. Rev. Camb. Philos. Soc.* 76, 65–101.

4.6 Eukaryotic Genomes Contain Both Nonrepetitive and Repetitive DNA Sequences

Reviews

Britten, R. J. and Davidson, E. H. (1971). Repetitive and nonrepetitive DNA sequences and a speculation on the origins of evolutionary novelty. *Q. Rev. Biol.* 46, 111–133.

Davidson, E. H. and Britten, R. J. (1973). Organization, transcription, and regulation in the animal genome. *Q. Rev. Biol.* 48, 565–613.

4.7 Genes Can Be Isolated by the Conservation of Exons

Research

Buckler, A. J., Chang, D. D., Graw, S. L., Brook, J. D., Haber, D. A., Sharp, P. A., and Housman, D. E. (1991). Exon amplification: a strategy to isolate mammalian genes based on RNA splicing. *Proc. Natl. Acad. Sci. USA* 88, 4005–4009.

Kunkel, L. M., Monaco, A. P., Middlesworth, W., Ochs, H. D., and Latt, S. A. (1985). Specific cloning of DNA fragments absent from the DNA of a male patient with an X chromosome deletion. *Proc. Natl. Acad. Sci. USA* 82, 4778–4782.

Monaco, A. P., Bertelson, C. J., Middlesworth, W., Colletti, C. A., Aldridge, J., Fischbeck, K. H., Bartlett, R., Pericak-Vance, M. A., Roses, A. D., and Kunkel, L. M. (1985). Detection of deletions spanning the Duchenne muscular dystrophy locus using a tightly linked DNA segment. *Nature* 316, 842–845.

4.9 Organelles Have DNA

Research

Cann, R. L., Stoneking, M., and Wilson, A. C. (1987). Mitochondrial DNA and human evolution. *Nature* 325, 31–36.

4.10 Organelle Genomes Are Circular DNAs That Code for Organelle Proteins

Review

Lang, B. F., Gray, M. W., and Burger, G. (1999). Mitochondrial genome evolution and the origin of eukaryotes. *Annu. Rev. Genet.* 33, 351–397.

4.11 Mitochondrial DNA Organization Is Variable

Reviews

Attardi, G. (1985). Animal mitochondrial DNA: an extreme example of economy. *Int. Rev. Cytol.* 93, 93–146.

Boore, J. L. (1999). Animal mitochondrial genomes. *Nucleic Acids Res.* 27, 1767–1780.

Clayton, D. A. (1984). Transcription of the mammalian mitochondrial genome. *Annu. Rev. Biochem.* 53, 573–594.

Gray, M. W. (1989). Origin and evolution of mitochondrial DNA. *Annu. Rev. Cell Biol.* 5, 25–50.

Research

Anderson, S., Bankier, A. T., Barrell, B. G., et al. (1981). Sequence and organization of the human mitochondrial genome. *Nature* 290, 457–465.

4.12 The Chloroplast Genome Codes for Many Proteins and RNAs

Reviews

Palmer, J. D. (1985). Comparative organization of chloroplast genomes. *Annu. Rev. Genet.* 19, 325–354.

Shimada, H. and Sugiura, M. (1991). Fine structural features of the chloroplast genome: comparison of the sequenced chloroplast genomes. *Nucleic Acids Res.* 11, 983–995.

Sugiura, M., Hirose, T., and Sugita, M. (1998). Evolution and mechanism of translation in chloroplasts. *Annu. Rev. Genet.* 32, 437–459.

4.13 Mitochondria Evolved by Endosymbiosis

Review

Lang, B. F., Gray, M. W., and Burger, G. (1999). Mitochondrial genome evolution and the origin of eukaryotes. *Annu. Rev. Genet.* 33, 351–397.

Research

Adams, K. L., Daley, D. O., Qiu, Y. L., Whelan, J., and Palmer, J. D. (2000). Repeated, recent and diverse transfers of a mitochondrial gene to the nucleus in flowering plants. *Nature* 408, 354–357.

Arabidopsis Initiative (2000). Analysis of the genome sequence of the flowering plant *Arabidopsis thaliana*. *Nature* 408, 796–815.

Huang, C. Y., Ayliffe, M. A., and Timmis, J. N. (2003). Direct measurement of the transfer rate of chloroplast DNA into the nucleus. *Nature* 422, 72–76.

Thorsness, P. E. and Fox, T. D. (1990). Escape of DNA from mitochondria to the nucleus in *S. cerevisiae*. *Nature* 346, 376–379.

5

Genome Sequences and Gene Numbers

CHAPTER OUTLINE

5.1 Introduction

Since the first genomes were sequenced in 1995, both the speed and range of sequencing have improved greatly. The first genomes to be sequenced were small bacterial genomes, < 2Mb. By 2002, the human genome of 3000 Mb had been sequenced. Genomes have now been sequenced from a wide range of organisms, including bacteria, archaea, yeasts, lower eukaryotes, plants, and animals, including worms, flies, rodents, and mammals.

Perhaps the most important single piece of information provided by a genome sequence is the number of genes. *Mycoplasma genitalium*, an obligate intracellular bacterium, has the smallest known genome of any organism, with only ~470 genes. Free-living bacteria range from 1700 genes to 7500 genes. Archaea are in a similar range. Single-celled eukaryotes start with about 5300 genes. Worms and flies have roughly 18,500 and 13,500 genes, respectively, but the number rises only to ~ 25,000 for mouse and man.

FIGURE 5.1 summarizes the minimum number of genes found in six groups of organisms. It takes ~500 genes to make a cell, ~1500 to make a free-living cell, >5000 to make a cell with a nucleus, >10,000 to make a multicellular organism, and >13,000 to make an organism with a nervous system. Of course, many species may have more than the minimum number required for their type, so the number of genes can vary widely even among closely related species.

Within bacteria and the lower eukaryotes, most genes are unique. Within higher eukaryotic genomes, however, genes can be divided into families of related members. Of course, some genes are unique (formally the family has only one member), but many belong to families with ten or more members. The number of different families may be better related to the overall complexity of the organism than the number of genes.

Some of the most insightful information comes from comparing genome sequences. With the sequences now available for both the human and chimpanzee genomes, it may be possible to begin to address some of the questions as to what makes humans unique.

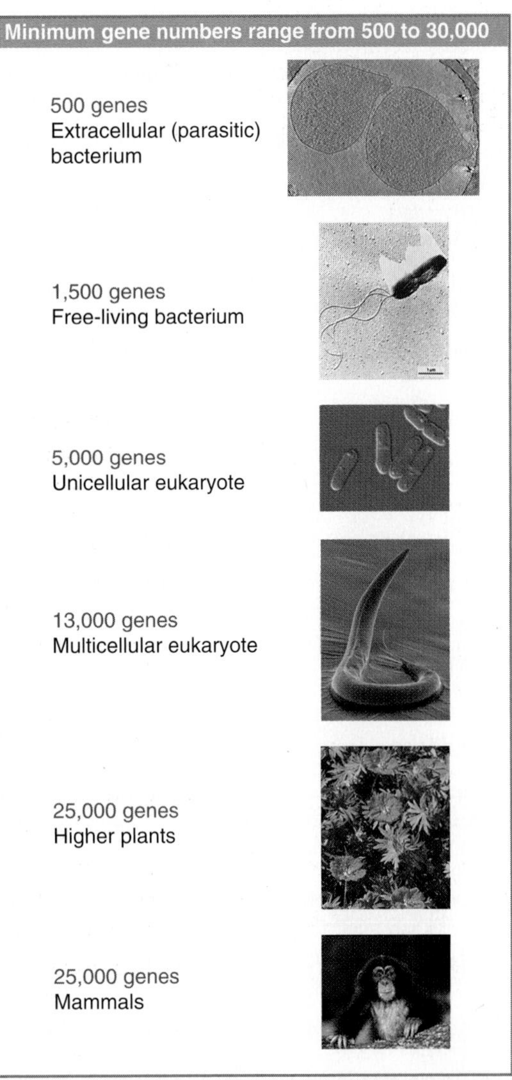

FIGURE 5.1 The minimum gene number required for any type of organism increases with its complexity. Photo of extracellular bacterium courtesy of Gregory P. Henderson and Grant J. Jensen, California Institute of Technology. Photo of free-living bacterium courtesy of Karl O. Stetter, Universität Regensburg. Photo of unicellular eukaryote courtesy of Eishi Noguchi, Drexel University College of Medicine. Photo of multicellular eukaryote courtesy of Carolyn B. Marks and David H. Hall, Albert Einstein College of Medicine, Bronx, NY. Photo of higher plant courtesy of Keith Weller/USDA. Mammal photo © Photodisc.

5.2 Bacterial Gene Numbers Range Over an Order of Magnitude

Large-scale efforts have now led to the sequencing of many genomes. A range is summarized in FIGURE 5.2. They extend from the 0.6×10^6 bp of a mycoplasma to the 3.3×10^9 bp of the human genome and include several important experimental animals, such as yeasts, the fruit fly, and a nematode worm.

Species	Genomes (Mb)	Genes	Lethal loci
Mycoplasma genitalium	0.58	470	~300
Rickettsia prowazekii	1.11	834	
Haemophilus influenzae	1.83	1,743	
Methanococcus jannaschi	1.66	1,738	
B. subtilis	4.2	4,100	
E. coli	4.6	4,288	1,800
S. cerevisiae	13.5	6,034	1,090
S. pombe	12.5	4,929	
A. thaliana	119	25,498	
O. sativa (rice)	466	~30,000	
D. melanogaster	165	13,601	3,100
C. elegans	97	18,424	
H. sapiens	3,300	~25,000	

Sequenced genomes vary from 470 to 30,000 genes

FIGURE 5.2 Genome sizes and gene numbers are known from complete sequences for several organisms. Lethal loci are estimated from genetic data.

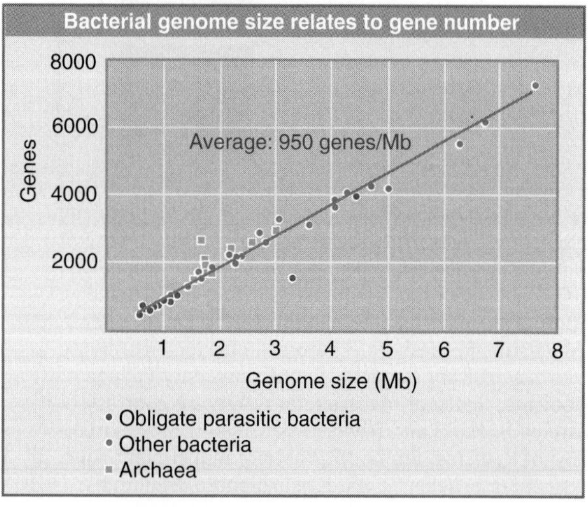

FIGURE 5.3 The number of genes in bacterial and archaeal genomes is proportional to genome size.

The sequences of the genomes of bacteria and archaea show that virtually all of the DNA (typically 85%–90%) codes for RNA or protein. FIGURE 5.3 shows that the range of genome sizes is about an order of magnitude, and that the genome size is proportional to the number of genes. The typical gene is about 1000 bp in length.

All of the bacteria with genome sizes below 1.5 Mb are obligate intracellular parasites—they live within a eukaryotic host that provides them with small molecules. Their genomes identify the minimum number of functions required to construct a cell. All classes of genes are reduced in number compared with bacteria with larger genomes, but the most significant reduction is in loci coding for enzymes concerned with metabolic functions (which are largely provided by the host cell) and with regulation of gene expression. *Mycoplasma genitalium* has the smallest genome, ~470 genes.

The archaea have biological properties that are intermediate between the prokaryotes and eukaryotes, but their genome sizes and gene numbers fall in the same range as bacteria. Their genome sizes vary from 1.5 to 3 Mb, corresponding to 1500–2700 genes. *Methanococcus jannaschii* is a methane-producing species that lives under high pressure and temperature. Its total gene number is similar to that of *Haemophilus influenzae,* but fewer of its genes can be identified on the basis of comparison with genes known in other organisms. Its apparatus for gene expression resembles eukaryotes more than prokaryotes, but its apparatus for cell division better resembles prokaryotes.

The archaea and the smallest free-living bacteria identify the minimum number of genes required to make a cell able to function independently in the environment. The smallest archaeal genome has ~1500 genes. The free-living bacterium with the smallest known genome is the thermophile *Aquifex aeolicus,* with 1.5 Mb and 1512 genes. A "typical" gram-negative bacterium, *H. influenzae,* has 1743 genes, each of ~900 bp. So we can conclude that ~1500 genes are required to make a free-living organism.

Bacterial genome sizes extend over about an order of magnitude, from 0.6 Mb to <8 Mb. The larger genomes have more genes. The bacteria with the largest genomes, *Sinorhizobium meliloti* and *Mesorhizobium loti,* are nitrogen-fixing bacteria that live on plant roots. Their genome sizes (~7 Mb) and total gene numbers (>7500) are similar to those of yeasts.

The size of the genome of *E. coli* is in the middle of the range. The common laboratory strain has 4288 genes, with an average length ~950 bp, and an average separation between genes of 118 bp. There can be quite significant differences between strains, though. The known extremes of *E. coli* are from the smallest strain, which has 4.6 Mb with 4249 genes, to the largest strain, which has 5.5 Mb bp with 5361 genes.

We still do not know the functions of all the genes. In most of these genomes, ~60% of the genes can be identified on the basis of homology with known genes in other species. These genes fall approximately equally into classes whose products are concerned with metabolism, cell structure or transport of components, and gene expression and its regulation. In virtually every genome, >25% of the genes cannot be ascribed any function. Many of these genes can be found in related organisms, which implies that they have a conserved function.

There has been some emphasis on sequencing the genomes of pathogenic bacteria, given their medical importance. An important insight into the nature of pathogenicity has been provided by the demonstration that "pathogenicity islands" are a characteristic feature of their genomes. These are large regions, ~10 to 200 kb, which are present in the genome of a pathogenic species but absent from the genomes of nonpathogenic variants of the same or related species. Their G-C content often differs from that of the rest of the genome, and it is likely that they migrate between bacteria by a process of horizontal transfer. For example, the bacterium that causes anthrax (*Bacillus anthracis*) has two large plasmids (extrachromosomal DNA), one of which has a pathogenicity island that includes the gene coding for the anthrax toxin.

5.3 Total Gene Number Is Known for Several Eukaryotes

Key concept

- There are 6000 genes in yeast; 18,500 in a worm; 13,600 in a fly; 25,000 in the small plant *Arabidopsis;* and probably ~25,000 in mouse and man.

As soon as we look at eukaryotic genomes, the relationship between genome size and gene number is lost. The genomes of unicellular eukaryotes fall in the same size range as the largest bacterial genomes. Higher eukaryotes have more genes, but the number does not correlate with genome size, as can be seen from FIGURE 5.4.

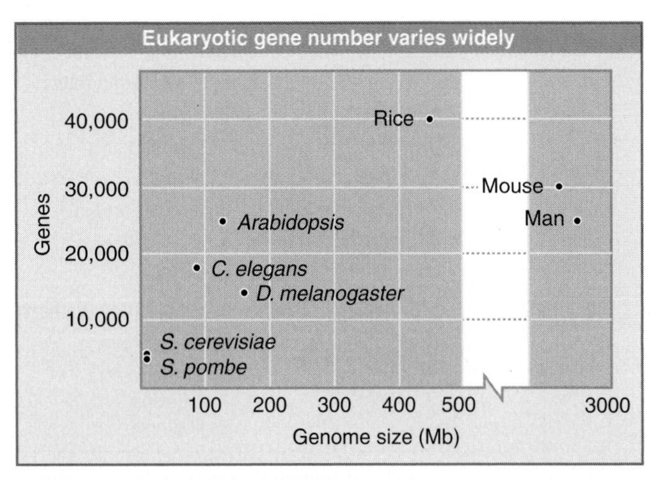

FIGURE 5.4 The number of genes in a eukaryote varies from 6000 to 40,000 but does not correlate with the genome size or the complexity of the organism.

The most extensive data for lower eukaryotes are available from the sequences of the genomes of the yeasts *Saccharomyces cerevisiae* and *Schizosaccharomyces pombe*. FIGURE 5.5 summarizes the most important features. The yeast genomes of 12.5 Mb and 13.5 Mb have ~6000 and ~5000 genes, respectively. The average open reading frame (ORF) is ~1.4 kb, so that ~70% of the genome is occupied by coding regions.

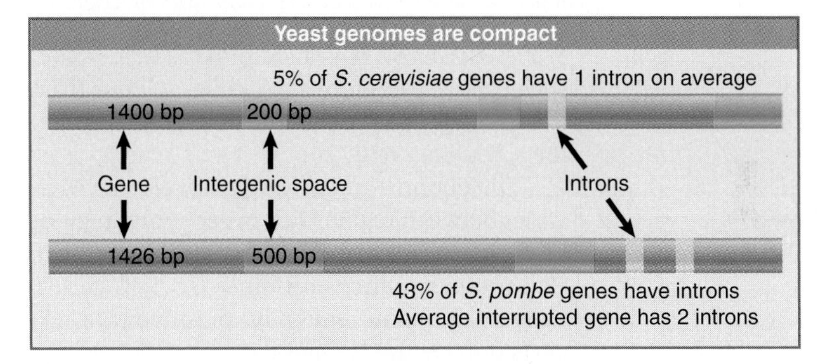

FIGURE 5.5 The *S. cerevisiae* genome of 13.5 Mb has 6000 genes, almost all uninterrupted. The *S. pombe* genome of 12.5 Mb has 5000 genes, almost half having introns. Gene sizes and spacing are fairly similar.

The major difference between them is that only 5% of *S. cerevisiae* genes have introns, compared to 43% in *S. pombe*. The density of genes is high; organization is generally similar, although the spaces between genes are a bit shorter in *S. cerevisiae*. About half of the genes identified by sequence were either known previously or related to known genes. The remainder are new, which gives some indication of the number of new types of genes that may be discovered.

The identification of long reading frames on the basis of sequence is quite accurate. However, ORFs coding for <100 amino acids cannot be identified solely by sequence because of the high occurrence of false positives. Analysis of gene expression suggests that ~300 of 600 such ORFs in *S. cerevisiae* are likely to be genuine genes.

A powerful way to validate gene structure is to compare sequences in closely related species—if a gene is active, it is likely to be conserved. Comparisons between the sequences of four closely related yeast species suggest that 503 of the genes originally identified in

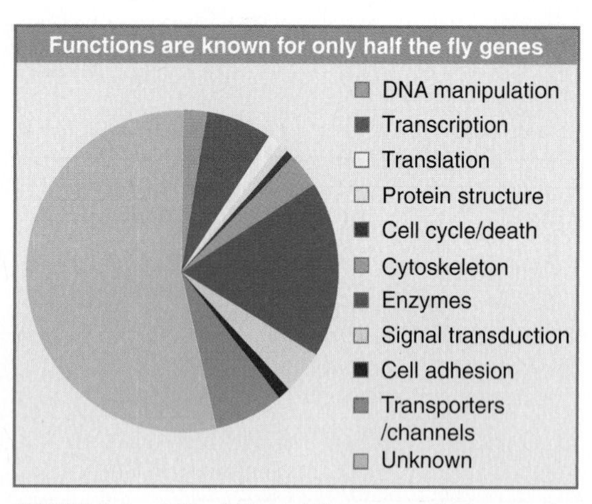

FIGURE 5.6 ~20% of *Drosophila* genes code for proteins concerned with maintaining or expressing genes, ~20% code for enzymes, and <10% code for proteins concerned with the cell cycle or signal transduction. Half of the genes of *Drosophila* code for products of unknown function.

S. cerevisiae do not have counterparts in the other species and therefore should be deleted from the catalog. This reduces the total gene number for *S. cerevisiae* to 5726.

The genome of *Caenorhabditis elegans* DNA varies between regions rich in genes and regions in which genes are more sparsely organized. The total sequence contains ~18,500 genes. Only ~42% of the genes have putative counterparts outside the Nematoda.

Although the fly genome is larger than the worm genome, there are fewer genes (13,600) in *D. melanogaster.* The number of different transcripts is slightly larger (14,100) as the result of alternative splicing. We do not understand why the fly—a much more complex organism—has only 70% of the number of genes in the worm. This emphasizes forcefully the lack of an exact relationship between gene number and complexity of the organism.

The plant *Arabidopsis thaliana* has a genome size intermediate between the worm and the fly, but has a larger gene number (25,000) than either. This again shows the lack of a clear relationship and also emphasizes the special quality of plants, which may have more genes (due to ancestral duplications) than animal cells. A majority of the *Arabidopsis* genome is found in duplicated segments, suggesting that there was an ancient doubling of the genome (to give a tetraploid). Only 35% of *Arabidopsis* genes are present as single copies.

The genome of rice *(Oryza sativa)* is ~4 larger than *Arabidopsis,* but the number of genes is only ~50% larger, probably ~40,000. Repetitive DNA occupies 42%–45% of the genome. More than 80% of the genes found in *Arabidopsis* are represented in rice. Of these common genes, ~8000 are found in *Arabidopsis* and rice but not in any of the bacterial or animal genomes that have been sequenced. This is probably the set of genes that codes for plant-specific functions, such as photosynthesis.

From the fly genome, we can form an impression of how many genes are devoted to each type of function. FIGURE 5.6 breaks down the functions into different categories. Among the genes that are identified, we find 2500 enzymes, ~750 transcription factors, ~700 transporters and ion channels, and ~700 proteins involved with signal transduction. Just over half of the genes code for products of unknown function. Approximately 20% of the proteins reside in membranes.

Protein size increases from prokaryotes and archaea to eukaryotes. The archaea *M. jannaschi* and bacterium *E. coli* have average protein lengths of 287 and 317 amino acids, respectively, whereas *S. cerevisiae* and *C. elegans* have average lengths of 484 and 442 amino acids, respectively. Large proteins (500 amino acids) are rare in bacteria, but comprise a significant component (~1/3) in eukaryotes. The increase in length is due to the addition of extra domains, with each domain typically constituting 100–300 amino acids. The increase in protein size, however, is responsible for only a very small part of the increase in genome size.

Another insight into gene number is obtained by counting the number of expressed genes. If we rely upon the estimates of the number of different mRNA species that can be counted in a cell, we would conclude that the average vertebrate cell expresses ~10,000 to 20,000 genes. The existence of significant overlaps between the messenger populations in different cell types would suggest that the total expressed gene number for the organism should be within a few fold of this amount. The estimate for the total human genome number of 20 to 25,000 (see Section 5.5, The Human Genome Has Fewer Genes Than Expected) would imply that a significant proportion of the total gene number is actually expressed in any given cell.

Eukaryotic genes are transcribed individually, with each gene producing a monocistronic messenger. There is only one general exception to this rule: in the genome of *C. elegans,* ~15% of the genes are organized into polycistronic units (which are associated with the use of *trans*-splicing to allow expression of the downstream genes in these units; see Section 26.13, *trans*-splicing Reactions Use Small RNAs).

5.4 How Many Different Types of Genes Are There?

Some genes are unique; others belong to families in which the other members are related (but not usually identical). The proportion of unique genes declines with genome size and the proportion of genes in families increases. The minimum number of gene families required to code a bacterium is >1000, a yeast is >4000, and a higher eukaryote 11,000 to 14,000.

Some genes are present in more than one copy or are related to one another; thus the number of different types of genes is less than the total number of genes. We can divide the total number of genes into sets that have related members, as defined by comparing their exons. (A gene family arises by duplication of an ancestral gene followed by accumulation of changes in sequence between the copies. Most often the members of a family are related but not identical.) The number of types of genes is calculated by adding the number of unique genes (for which there is no other related gene at all) to the numbers of families that have two or more members.

FIGURE 5.7 compares the total number of genes with the number of distinct families in each of six genomes. In bacteria most genes are unique, so the number of distinct families is close to the total gene number. The situation is different even in the lower eukaryote *S. cerevisiae*, for which there is a significant proportion of repeated genes. The most striking effect is that the number of genes increases quite sharply in the higher eukaryotes, but the number of gene families does not change much.

FIGURE 5.8 shows that the proportion of unique genes drops sharply with genome size. When genes are present in families, the number of members in a family is small in bacteria and lower eukaryotes, but is large in higher eukaryotes. Much of the extra genome size of *Arabidopsis* is accounted for by families with >4 members.

If every gene is expressed, the total number of genes will account for the total number of proteins required to make the organism (the proteome). Two effects mean, however, that the proteome is different from the total gene number. Genes are duplicated, and as a result some of them code for the same protein (although it may be expressed in a different time or place) and others may code for related proteins that again play the same role in different times or places. The proteome can be larger than the number of genes because some genes can produce more than one protein by means of alternative splicing.

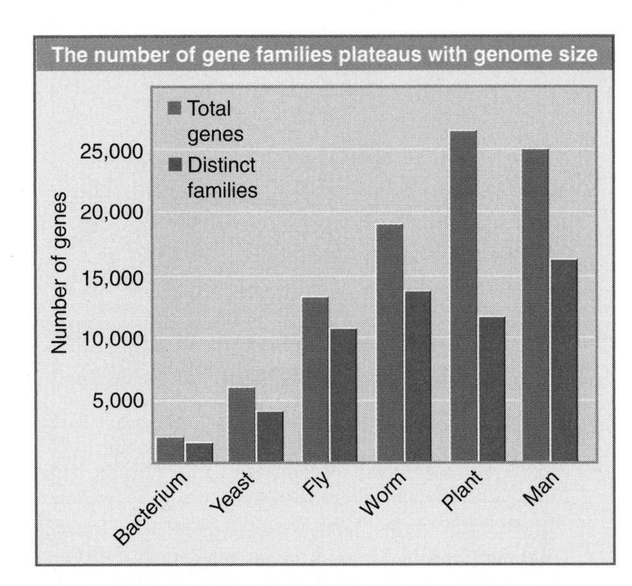

FIGURE 5.7 Many genes are duplicated, and as a result the number of different gene families is much less than the total number of genes. The histogram compares the total number of genes with the number of distinct gene families.

Family size increases with genome size

	Unique genes	Families with 2–4 members	Families with >4 members
H. influenzae	89%	10%	1%
S. cerevisiae	72%	19%	9%
D. melanogaster	72%	14%	14%
C. elegans	55%	20%	26%
A. thaliana	35%	24%	41%

FIGURE 5.8 The proportion of genes that are present in multiple copies increases with genome size in higher eukaryotes.

What is the core proteome—the basic number of the different types of proteins in the organism? A minimum estimate is given by the number of gene families, ranging from 1400 in the bacterium, >4000 in the yeast, and 11,000 to 14,000 for the fly and the worm.

What is the distribution of the proteome among types of proteins? The 6000 proteins of the yeast proteome include 5000 soluble proteins and 1000 transmembrane proteins. About half of the proteins are cytoplasmic, a quarter

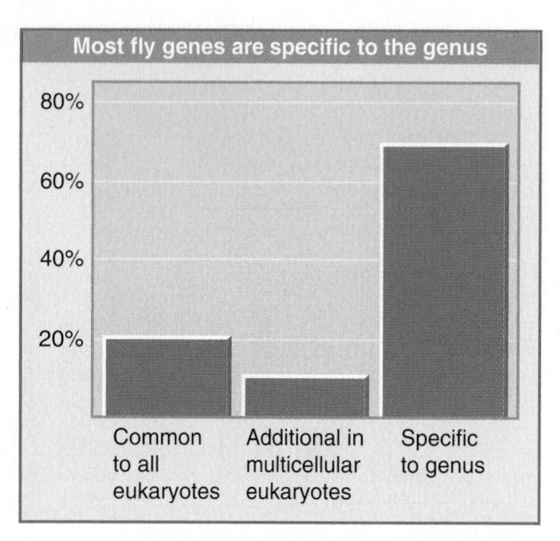

FIGURE 5.9 The fly genome can be divided into genes that are (probably) present in all eukaryotes, additional genes that are (probably) present in all multicellular eukaryotes, and genes that are more specific to subgroups of species that include flies.

are in the nucleolus, and the remainder are split between the mitochondrion and the endoplasmic reticulum (ER)/Golgi system.

How many genes are common to all organisms (or to groups such as bacteria or higher eukaryotes), and how many are specific for the individual type of organism? FIGURE 5.9 summarizes the comparison between yeast, worm, and fly. Genes that code for corresponding proteins in different organisms are called **orthologs.** Operationally, we usually reckon that two genes in different organisms can be considered to provide corresponding functions if their sequences are similar over >80% of the length. By this criterion, ~20% of the fly genes have orthologs in both yeast and the worm. These genes are probably required by all eukaryotes. The proportion increases to 30% when fly and worm are compared, probably representing the addition of gene functions that are common to multicellular eukaryotes. This still leaves a major proportion of genes as coding for proteins that are required specifically by either flies or worms, respectively.

The proteome can be deduced from the number and structures of genes, and can also be directly measured by analyzing the total protein content of a cell or organism. By such approaches, some proteins have been identified that were not suspected on the basis of genome analysis; this has led to the identification of new genes. Several methods are used for large scale analysis of proteins. Mass spectrometry can be used for separating and identifying proteins in a mixture obtained directly from cells or tissues. Hybrid proteins bearing tags can be obtained by expression of cDNAs made by linking the sequences of ORFs to appropriate expression vectors that incorporate the sequences for affinity tags. This allows array analysis to be used to analyze the products. These methods also can be effective in comparing the proteins of two tissues—for example, a tissue from a healthy individual and one from a patient with disease—to pinpoint the differences.

Once we know the total number of proteins, we can ask how they interact. By definition, proteins in structural multiprotein assemblies must form stable interactions with one another. Proteins in signaling pathways interact with one another transiently. In both cases, such interactions can be detected in test systems where essentially a readout system magnifies the effect of the interaction. One popular such system is the two hybrid assay discussed in Section 25.3, Independent Domains Bind DNA and Activate Transcription. Such assays cannot detect all interactions: for example, if one enzyme in a metabolic pathway releases a soluble metabolite that then interacts with the next enzyme, the proteins may not interact directly.

As a practical matter, assays of pairwise interactions can give us an indication of the minimum number of independent structures or pathways. An analysis of the ability of all 6000 (predicted) yeast proteins to interact in pairwise combinations shows that ~1000 proteins can bind to at least one other protein. Direct analyses of complex formation have identified 1440 different proteins in 232 multiprotein complexes. This is the beginning of an analysis that will lead to definition of the number of functional assemblies or pathways. A comparable analysis of 8100 human proteins identified 2800 interactions, but is more difficult to interpret in the context of the larger proteome.

In addition to functional genes, there are also copies of genes that have become nonfunctional (identified as such by interruptions in their protein-coding sequences). These are called pseudogenes (see Section 6.6, Pseudogenes Are Dead Ends of Evolution). The number of pseudogenes can be large. In the mouse and human genomes, the number of pseudogenes is ~10% of the number of (potentially) active genes (see Section 4.8, The Conservation of Genome Organization Helps to Identify Genes).

Besides needing to know the density of genes to estimate the total gene number, we must also ask: is it important in itself? Are there structural constraints that make it necessary for genes to have a certain spacing, and does this contribute to the large size of eukaryotic genomes?

5.5 The Human Genome Has Fewer Genes Than Expected

Key concepts

- Only 1% of the human genome consists of coding regions.
- The exons comprise ~5% of each gene, so genes (exons plus introns) comprise ~25% of the genome.
- The human genome has 20,000 to 25,000 genes.
- ~60% of human genes are alternatively spliced.
- Up to 80% of the alternative splices change protein sequence, so the proteome has ~50,000 to 60,000 members.

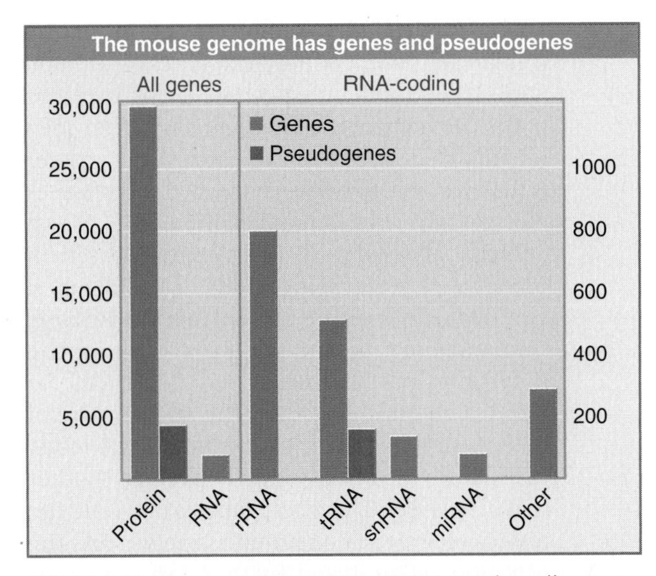

FIGURE 5.10 The mouse genome has ~30,000 protein-coding genes, which have ~4000 pseudogenes. There are ~1600 RNA-coding genes. The data for RNA-coding genes are replotted on the right at an expanded scale to show that there are ~800 rRNA genes, ~350 tRNA genes and 150 pseudogenes, and ~450 other noncoding RNA genes, including snRNAs and miRNAs.

The human genome was the first vertebrate genome to be sequenced. This massive task has revealed a wealth of information about the genetic makeup of our species and about the evolution of the genome in general. Our understanding is deepened further by the ability to compare the human genome sequence with the more recently sequenced mouse genome.

Mammal and rodent genomes generally fall into a narrow size range, ~3×10^9 bp (see Section 4.5, Why Are Genomes So Large?). The mouse genome is ~14% smaller than the human genome, probably because it has had a higher rate of deletion. The genomes contain similar gene families and genes, with most genes having an ortholog in the other genome, but with differences in the number of members of a family, especially in those cases for which the functions are specific to the species (see Section 4.8, The Conservation of Genome Organization Helps to Identify Genes). Originally estimated to have ~30,000 genes, the mouse genome is now thought to have about the same number as the human genome, 20 to 25,000. FIGURE 5.10 plots the distribution of the mouse genes. The 30,000 protein-coding genes are accompanied by ~4000 pseudogenes. There are ~800 genes representing RNAs that do not code for proteins; these are generally small (aside from the ribormal RNAs). Almost half of these genes code for transfer RNAs, for which a large number of pseudogenes also have been identified.

The human (haploid) genome contains 22 autosomes plus the X or Y. The chromosomes range in size from 45 to 279 Mb of DNA, making a total genome content of 3,286 Mb (~3.3×10^9 bp). On the basis of chromosome structure, the overall genome can be divided into regions of euchromatin (potentially containing active genes) and heterochromatin (see Section 28.7, Chromatin Is Divided into Euchromatin and Heterochromatin). The euchromatin comprises the majority of the genome, ~2.9×10^9 bp. The identified genome sequence represents ~90% of the euchromatin. In addition to providing information on the genetic content of the genome, the sequence also identifies features that may be of structural importance (see Section 28.8, Chromosomes Have Banding Patterns).

FIGURE 5.11 shows that a tiny proportion (~1%) of the human genome is accounted for by the exons that actually code for proteins. The introns that constitute the remaining sequences in the genes bring the total of DNA concerned with producing proteins to ~25%. As shown in FIGURE 5.12, the average human gene is 27 kb long, with nine exons that include a total coding sequence of 1,340 bp. The average coding sequence is therefore only 5% of the length of the gene.

Two independent sequencing efforts for the human genome produced estimates of ~30,000 and ~40,000 genes, respectively. One measure of the accuracy of the analyses is whether they identify the same genes. The surprising answer is that the overlap between the two sets of genes is only ~50%, as summarized in FIGURE 5.13. An earlier analysis of the human gene set based on RNA transcripts had identified ~11,000 genes, almost all of which are present in both the large human gene sets, and which account for the major part of the overlap between them. So there is no question about the authenticity of half of each human gene set, but we have yet to establish the relationship between the other half of each set. The discrepancies illustrate the pitfalls of large scale sequence analysis! As the sequence is analyzed further (and as other genomes are sequenced with which it can be compared), the number of valid genes seems to decline, and is now generally thought to be ~20,000 to 25,000.

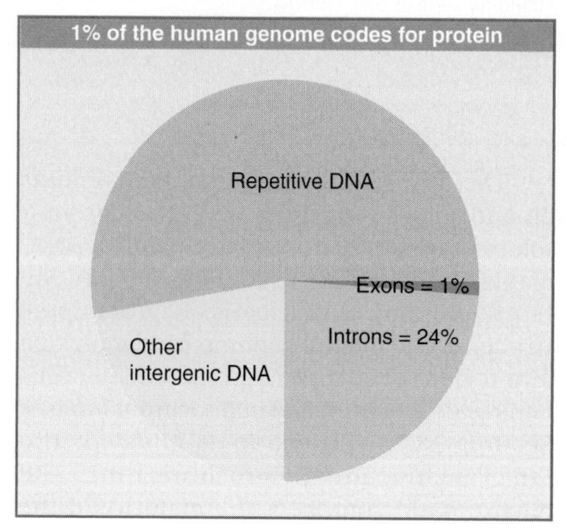

FIGURE 5.11 Genes occupy 25% of the human genome, but protein-coding sequences are only a tiny part of this fraction.

By any measure, the total human gene number is much less than we had expected—most estimates before the genome was sequenced were ~100,000. It shows a relatively small increase over flies and worms (13,600 and 18,500, respectively), not to mention the plant *Arabidopsis* (25,000) (see Figure 5.2). However, we should not be particularly surprised by the notion that it does not take a great number of additional genes to make a more complex organism. The difference in DNA sequences between man and chimpanzee is extremely small (there is >99% similarity), so it is clear that the functions and interactions between a similar set of genes can produce very different results. The functions of specific groups of genes may be especially important, because detailed comparisons of orthologous genes in man and chimpanzee suggest that there has been accelerated evolution of certain classes of genes, including some involved in early development, olfaction, hearing—all functions that are relatively specific for the species.

The number of genes is less than the number of potential proteins because of alternative splicing. The extent of alternative splicing is greater in man than in fly or worms; it may affect as many as 60% of the genes, so the increase in size of the human proteome relative to the other eukaryotes may be larger than the increase in the number of genes. A sample of genes from two chromosomes suggests that the proportion of the alternative splices that actually result in changes in the protein sequence may be as high as 80%. This could increase the size of the proteome to 50,000 to 60,000 members.

In terms of the diversity of the number of gene families, however, the discrepancy

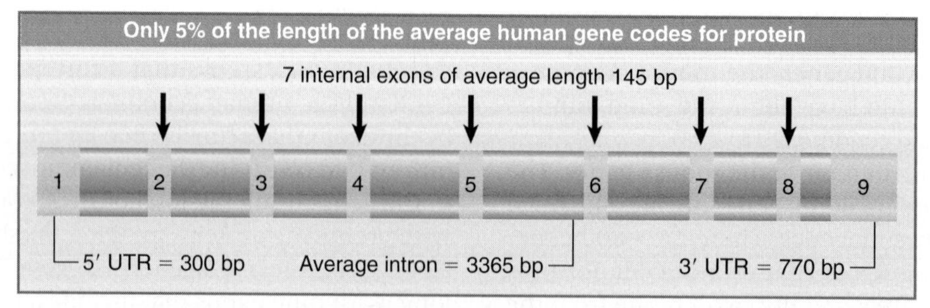

FIGURE 5.12 The average human gene is 27 kb long and has nine exons, usually comprising two longer exons at each end and seven internal exons. The UTRs in the terminal exons are the untranslated (noncoding) regions at each end of the gene. (This is based on the average. Some genes are extremely long, which makes the median length 14 kb with seven exons.)

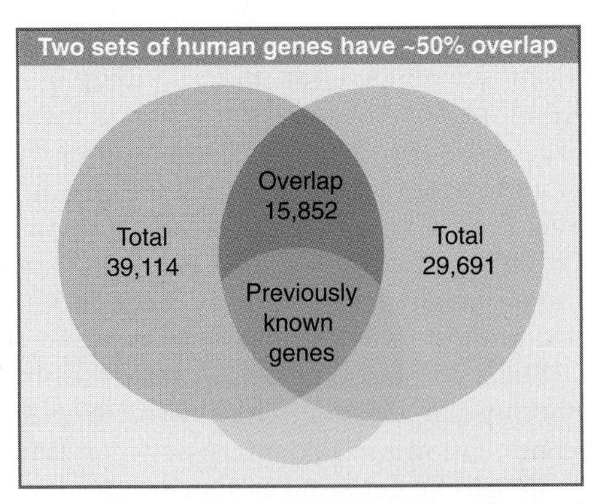

FIGURE 5.13 The two sets of genes identified in the human genome overlap only partially, as shown in the two large upper circles. They include, however, almost all previously known genes, as shown by the overlap with the smaller, lower circle.

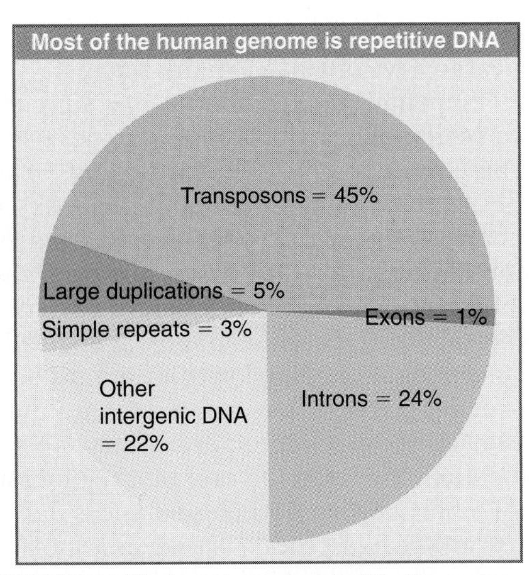

FIGURE 5.14 The largest component of the human genome consists of transposons. Other repetitive sequences include large duplications and simple repeats.

between man and the other eukaryotes may not be so great. Many of the human genes belong to families. An analysis of ~25,000 genes identified 3500 unique genes and 10,300 gene pairs. As can be seen from Figure 5.7, this extrapolates to a number of gene families only slightly larger than worm or fly.

5.6 How Are Genes and Other Sequences Distributed in the Genome?

Key concepts

- Repeated sequences (present in more than one copy) account for >50% of the human genome.
- The great bulk of repeated sequences consist of copies of nonfunctional transposons.
- There are many duplications of large chromosome regions.

Are genes uniformly distributed in the genome? Some chromosomes are relatively poor in genes and have >25% of their sequences as "deserts"—regions longer than 500 kb where there are no genes. Even the most gene-rich chromosomes have >10% of their sequences as deserts. So overall, ~20% of the human genome consists of deserts that have no genes.

Repetitive sequences account for >50% of the human genome, as seen in FIGURE 5.14. The repetitive sequences fall into five classes:

- Transposons (either active or inactive) account for the vast majority (45% of the genome). All transposons are found in multiple copies.
- Processed pseudogenes (~3000 in all, account for ~0.1% of total DNA). (These are sequences that arise by insertion of a copy of an mRNA sequence into the genome; see Section 6.6, Pseudogenes Are Dead Ends of Evolution.)
- Simple sequence repeats (highly repetitive DNA such as (CA)n account for ~3%).
- Segmental duplications (blocks of 10 to 300 kb that have been duplicated into a new region) account for ~5%. Only a minority of these duplications are found on the same chromosome; in the other cases, the duplicates are on different chromosomes.
- Tandem repeats form blocks of one type of sequence (especially found at centromeres and telomeres).

The sequence of the human genome emphasizes the importance of transposons. (Transposons have the capacity to replicate themselves and insert into new locations. They may function exclusively as DNA elements [see Chapter 21, Transposons] or may have an active form that is RNA [see Chapter 22, Retroviruses and Retroposons]. Their distribution in the human genome is summarized in Figure 22.18.) Most of the transposons in the human genome are nonfunctional; very few are currently active. However, the high proportion of the genome occupied by these elements indicates that they have played an active role in shaping the genome. One interesting feature is that some present genes originated as transposons and evolved into their present condition after losing the ability to transpose. Almost 50 genes appear to have originated in this manner.

Segmental duplication at its simplest involves the tandem duplication of some region within a chromosome (typically because of an aberrant recombination event at meiosis; see Section 6.7, Unequal Crossing-over Rearranges Gene Clusters). In many cases, however, the duplicated regions are on different chromosomes, implying that either there was originally a tandem duplication followed by a translocation of one copy to a new site, or that the duplication arose by some different mechanism altogether. The extreme case of a segmental duplication is when a whole genome is duplicated, in which case the diploid genome initially becomes tetraploid. As the duplicated copies develop differences from one another, the genome may gradually become effectively a diploid again, although homologies between the diverged copies leave evidence of the event. This is especially common in plant genomes. The present state of analysis of the human genome identifies many individual duplicated regions, but does not indicate whether there was a whole genome duplication in the vertebrate lineage.

One curious feature of the human genome is the presence of sequences that do not appear to have coding functions, but that nonetheless show an evolutionary conservation higher than the background level. As detected by comparison with other genomes (initially the mouse genome), these represent about 5% of the total genome. Are these sequences connected with protein-coding sequences in some functional way? Their density on chromosome 18 is the same as elsewhere in the genome, although chromosome 18 has a significantly lower concentration of protein-coding genes. This suggests indirectly that their function is not connected with structure or expression of protein-coding genes.

5.7 The Y Chromosome Has Several Male-Specific Genes

Key concepts

- The Y chromosome has ~60 genes that are expressed specifically in the testis.
- The male-specific genes are present in multiple copies in repeated chromosomal segments.
- Gene conversion between multiple copies allows the active genes to be maintained during evolution.

The sequence of the human genome has significantly extended our understanding of the role of the sex chromosomes. It is generally thought that the X and Y chromosomes have descended from a common (very ancient) autosome. Their development has involved a process in which the X chromosome has retained most of the original genes, whereas the Y chromosome has lost most of them.

The X chromosome behaves like the autosomes insofar as females have two copies and recombination can take place between them. The density of genes on the X chromosome is comparable to the density of genes on other chromosomes.

The Y chromosome is much smaller than the X chromosome and has many fewer genes. Its unique role results from the fact that only males have the Y chromosome, of which there is only one copy, so Y-linked loci are effectively haploid instead of diploid like all other human genes.

For many years, the Y chromosome was thought to carry almost no genes except for one (or more) sex-determining genes that determine maleness. The vast majority of the Y chromosome (>95% of its sequence) does not undergo crossing-over with the X chromosome, which led to the view that it could not contain active genes because there would be no means to prevent the accumulation of deleterious mutations. This region is flanked by short pseudoautosomal regions that exchange frequently with the X chromosome during male meiosis. It was originally called the nonrecombining region, but now has been renamed as the male-specific region.

Detailed sequencing of the Y chromosome shows that the male-specific region contains three types of regions, as illustrated in FIGURE 5.15:

- The *X-transposed sequences* consist of a total of 3.4 Mb comprising some large blocks resulting from a transposition from band q21 in the X chromosome about 3 or 4 million years ago. This is specific to the human lineage. These sequences do not recombine with the X chromosome and have become largely inactive. They now contain only two active genes.
- The *X-degenerate segments* of the Y are sequences that have a common origin with the X chromosome (going back to the common autosome from which both X and Y have descended) and contain

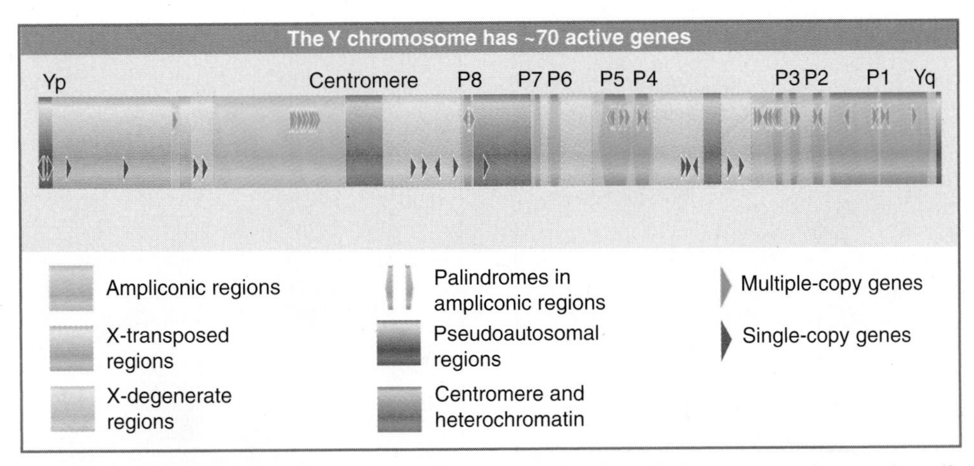

FIGURE 5.15 The Y chromosome consists of X-transposed regions, X-degenerate regions, and amplicons. The X-transposed X-degenerate regions have two and fourteen single-copy genes, respectively. The amplicons have eight large palindromes (P1–P8), which contain nine gene families. Each family contains at least two copies.

genes or pseudogenes related to X-linked genes. There are 14 active genes and 13 pseudogenes. The active genes have, in a sense, thus far defied the trend for genes to be eliminated from chromosomal regions that cannot recombine at meiosis.

- The *ampliconic segments* have a total length of 10.2 Mb and are internally repeated on the Y chromosome. There are eight large palindromic blocks. They include nine protein-coding gene families, with copy numbers per family ranging from 2 to 35. The name "amplicon" reflects the fact that the sequences have been internally amplified on the Y chromosome.

Totaling the genes in these three regions, the Y chromosome contains many more genes than had been expected. There are 156 transcription units, of which half represent protein-coding genes and half represent pseudogenes.

The presence of the active genes is explained by the fact that the existence of closely related genes copies in the ampliconic segments allows gene conversion between multiple copies of a gene to be used to regenerate active copies. The most common needs for multiple copies of a gene are quantitative (to provide more protein product) or qualitative (to code for proteins with slightly different properties or that are expressed in different times or places). In this case, though, the essential function is evolutionary. In effect, the existence of multiple copies allows recombination within the Y chromosome itself to substitute for the evolutionary diversity that is usually provided by recombination between allelic chromosomes.

Most of the protein-coding genes in the ampliconic segments are expressed specifically in testis and are likely to be involved in male development. If there are ~60 such genes out of a total human gene set of ~25,000, then the genetic difference between man and woman is ~0.2%.

5.8 More Complex Species Evolve by Adding New Gene Functions

Key concepts

- Comparisons of different genomes show a steady increase in gene number as additional genes are added to make eukaryotes, multicellular organisms, animals, and vertebrates.
- Most of the genes that are unique to vertebrates are concerned with the immune or nervous systems.

Comparison of the human genome sequence with sequences found in other species is revealing about the process of evolution. FIGURE 5.16 analyzes human genes according to the breadth of their distribution in nature. Starting with the most generally distributed (top right corner of the figure), 21% of genes are common to eukaryotes and prokaryotes. These tend to code

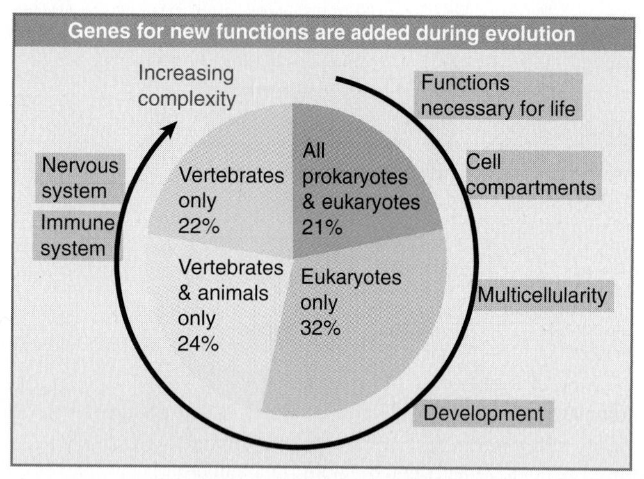

FIGURE 5.16 Human genes can be classified according to how widely their homologues are distributed in other species.

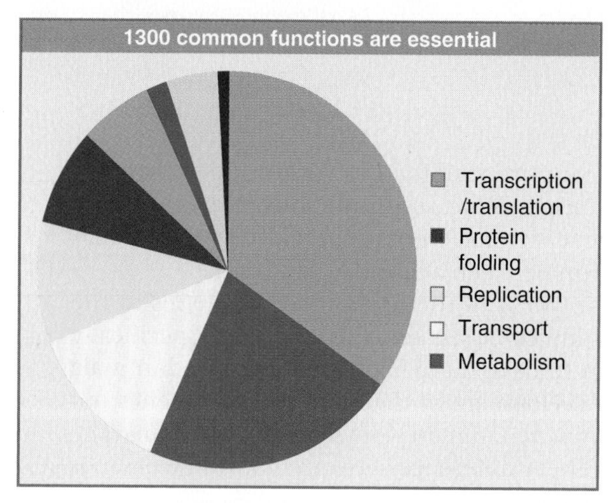

FIGURE 5.17 Common eukaryotic proteins are concerned with essential cellular functions.

for proteins that are essential for all living forms—typically basic metabolism, replication, transcription, and translation. Moving clockwise, another 32% of genes are added in eukaryotes in general—for example, they may be found in yeast. These tend to code for proteins involved in functions that are general to eukaryotic cells but not to bacteria—for example, they may be concerned with specifying organelles or cytoskeletal components. Another 24% of genes are needed to specify animals. These include genes necessary for multicellularity and for development of different tissue types. Twenty-two percent of genes are unique to vertebrates. These mostly code for proteins of the immune and nervous systems; they code for very few enzymes, consistent with the idea that enzymes have ancient origins, and that metabolic pathways originated early in evolution. We see, therefore, that the progression from bacteria to

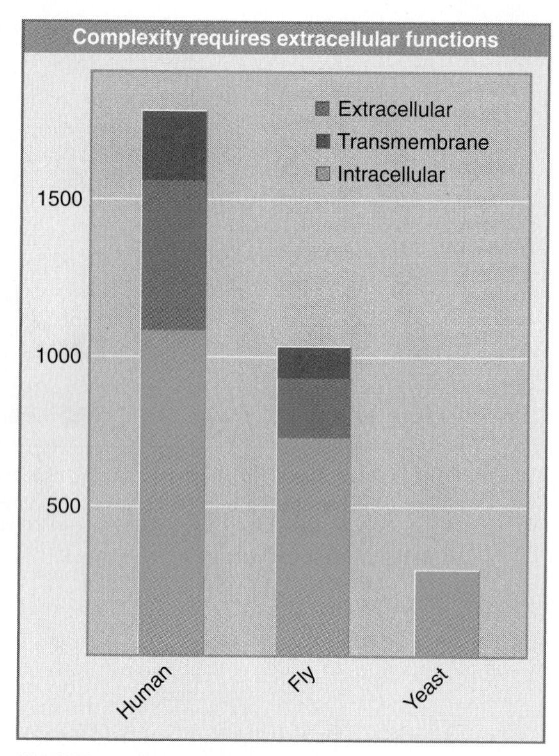

FIGURE 5.18 Increasing complexity in eukaryotes is accompanied by accumulation of new proteins for transmembrane and extracellular functions.

vertebrates requires addition of groups of genes representing the necessary new functions at each stage.

One way to define commonly needed proteins is to identify the proteins present in all proteomes. Comparing the human proteome in more detail with the proteomes of other organisms, 46% of the yeast proteome, 43% of the worm proteome, and 61% of the fly proteome is represented in the human proteome. A key group of ~1300 proteins is present in all four proteomes. The common proteins are basic housekeeping proteins required for essential functions, falling into the types summarized in FIGURE 5.17. The main functions are concerned with transcription and translation (35%), metabolism (22%), transport (12%), DNA replication and modification (10%), protein folding and degradation (8%), and cellular processes (6%).

One of the striking features of the human proteome is that it has many new proteins compared with other eukaryotes, but has relatively few new protein domains. Most protein domains appear to be common to the animal kingdom. There are many new protein architectures, however, defined as new combinations of domains. FIGURE 5.18 shows that the greatest increase occurs in transmembrane and extracellular proteins. In yeast, the vast majority of architectures are concerned with intracellular proteins. About twice as many intracellular architectures are found in fly (or worm), but there is a very striking increase in transmembrane and extracellular proteins, as might be expected from the addition of functions required for the interactions between the cells of a multicellular organism. The increase in intracellular architectures required to make a vertebrate (man) is relatively small, but there is again a large increase in transmembrane and extracellular architectures.

It has long been known that the genetic difference between man and chimpanzee (our nearest relative) is very small, with ~99% identity between genomes. The sequence of the chimpanzee genome now allows us to investigate the 1% of differences in more detail to see whether features responsible for "humanness" can be identified. The comparison shows 35×10^6 nucleotide substitutions (1.2% sequence difference overall), 5×10^6 deletions or insertions (making ~1.5% of the euchromatic sequence specific to each species), and many chromosomal rearrangements. Corresponding proteins are usually very similar; 29% are identical, and in most cases there are only one or two amino acid changes in the protein between the species. In fact, nucleotide substitutions occur less often in genes coding for proteins than are likely to be involved in specifically human traits, suggesting that protein evolution is not a major effect in human–chimpanzee differences. This leaves larger-scale changes in gene structure and/or changes in gene regulation as the major candidates. Some 25% of nucleotide substitutions occur in CpG dinucleotides (among which are many potential regulator sites).

5.9 How Many Genes Are Essential?

Key concepts

- Not all genes are essential. In yeast and fly, deletions of <50% of the genes have detectable effects.
- When two or more genes are redundant, a mutation in any one of them may not have detectable effects.
- We do not fully understand the survival in the genome of genes that are apparently dispensable.

Natural selection is the force that ensures that useful genes are retained in the genome. Mutations occur at random, and their most common effect in an ORF will be to damage the protein product. An organism with a damaging mutation will be at a disadvantage in evolution, and ultimately the mutation will be eliminated by the competitive failure of organisms carrying it. The frequency of a disadvantageous allele in the population is balanced between the generation of new mutations and the elimination of old mutations. Reversing this argument, whenever we see an intact ORF in the genome, we assume that its product plays a useful role in the organism. Natural selection must have prevented mutations from accumulating in the gene. The ultimate fate of a gene that ceases to be useful is to accumulate mutations until it is no longer recognizable.

The maintenance of a gene implies that it confers a selective advantage on the organism. In the course of evolution, though, even a small relative advantage may be the subject of natural selection, and a phenotypic defect may not necessarily be immediately detectable as the result of a mutation. However, we should like to know how many genes are actually *essential.* This means that their absence is lethal to the organism. In the case of diploid organisms, it means of course that the homozygous null mutation is lethal.

We might assume that the proportion of essential genes will decline with increase in genome size, given that larger genomes may have multiple related copies of particular gene functions. So far this expectation has not been borne out by the data (see Figure 5.2).

One approach to the issue of gene number is to determine the number of essential genes by mutational analysis. If we saturate some specified region of the chromosome with mutations that are lethal, the mutations should map into a number of complementation groups that correspond to the number of lethal loci in that region. By extrapolating to the genome as a whole, we may calculate the total essential gene number.

In the organism with the smallest known genome (*M. genitalium*), random insertions have detectable effects only in about two thirds of the genes. Similarly, fewer than half of the genes of *E. coli* appear to be essential. The proportion is even lower in the yeast *S. cerevisiae.* When insertions were introduced at random into the genome in one early analysis only 12% were lethal, and another 14% impeded growth. The majority (70%) of the insertions had no effect. A more systematic survey based on completely deleting each of 5,916 genes (>96% of the identified genes) shows that only 18.7% are essential for growth on a rich medium (that is, when nutrients are fully provided). FIGURE 5.19 shows that these include genes in all categories. The only notable concentration of defects is in genes coding for products involved in protein synthesis, where ~50% are essential. Of course, this approach underestimates the number of genes that are essential for the yeast to live in the wild, when it is not so well provided with nutrients.

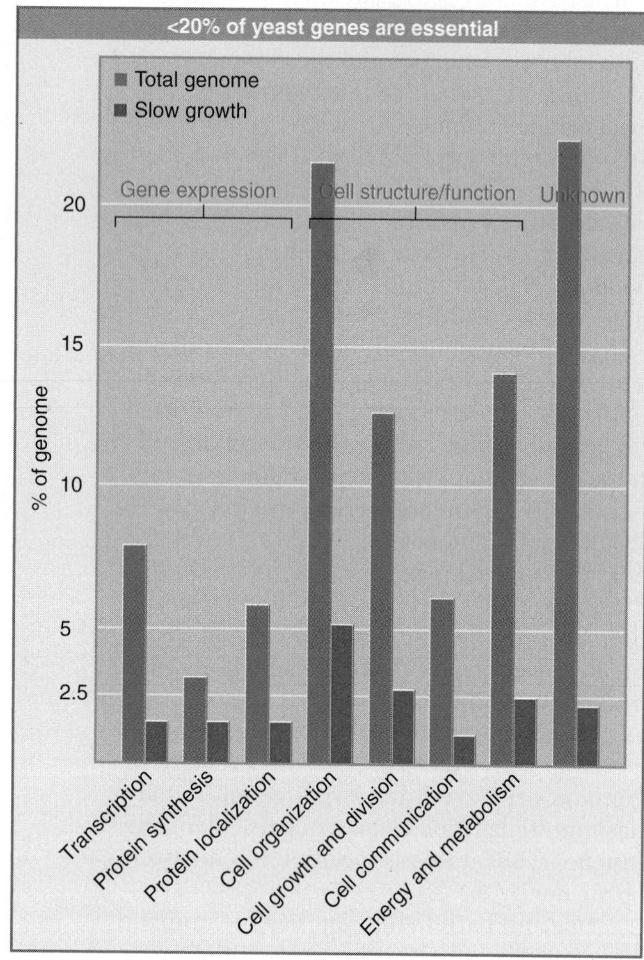

FIGURE 5.19 Essential yeast genes are found in all classes. Blue bars show total proportion of each class of genes, red bars shows those that are essential.

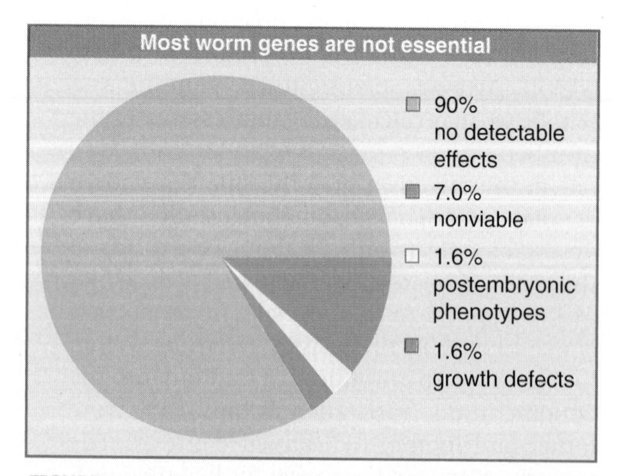

FIGURE 5.20 A systematic analysis of loss of function for 86% of worm genes shows that only 10% have detectable effects on the phenotype.

FIGURE 5.20 summarizes the results of a systematic analysis of the effects of loss of gene function in the worm *C. elegans*. The sequences of individual genes were predicted from the genome sequence, and by targeting an inhibitory RNA against these sequences (see Section 13.10, RNA Interference Is Related to Gene Silencing) a large set of worms were made in which one (predicted) gene was prevented from functioning in each worm. Detectable effects on the phenotype were only observed for 10% of these knockouts, suggesting that most genes do not play essential roles.

There is a greater proportion of essential genes (21%) among those worm genes that have counterparts in other eukaryotes, suggesting that widely conserved genes tend to play more basic functions. There is also an increased proportion of essential genes among those that are present in only one copy per haploid genome, compared with those where there are multiple copies of related or identical genes. This suggests that many of the multiple genes might be relatively recent duplications that can substitute for one another's functions.

Extensive analyses of essential gene number in a higher eukaryote have been made in *Drosophila* through attempts to correlate visible aspects of chromosome structure with the number of functional genetic units. The notion that this might be possible arose originally from the presence of bands in the polytene chromosomes of *D. melanogaster*. (These chromosomes are found at certain developmental stages and represent an unusually extended physical form, in which a series of bands [more formally called chromomeres] are evident; see Section 28.10, Polytene Chromosomes Form Bands.) From the early concept that the bands might represent a linear order of genes, we have come to the attempt to correlate the organization of genes with the organization of bands. There are ~5000 bands in the *D. melanogaster* haploid set; they vary in size over an order of magnitude, but on average there is ~20 kb of DNA per band.

The basic approach is to saturate a chromosomal region with mutations. Usually the mutations are simply collected as lethals, without analyzing the cause of the lethality. Any mutation that is lethal is taken to identify a locus that is essential for the organism. Sometimes mutations cause visible deleterious effects short of lethality, in which case we also count them as identifying an essential locus. When the mutations are placed into complementation groups, the number can be compared with the number of bands in the region, or individual complementation groups may even be assigned to individual bands. The purpose of these experiments has been to determine whether there is a consistent relationship between bands and genes. For example, does every band contain a single gene?

Totaling the analyses that have been carried out over the past 30 years, the number of lethal complementation groups is ~70% of the number of bands. It is an open question whether there is any functional significance to this relationship. Irrespective of the cause, the equivalence gives us a reasonable estimate for the lethal gene number of ~3600. By any measure, the number of lethal loci in *Drosophila* is significantly less than the total number of genes.

If the proportion of essential human genes is similar to other eukaryotes, we would predict a range of 4000 to 8000 genes, in which mutations would be lethal or produce evidently damaging effects. At present, 1300 genes have been identified in which mutations cause evident defects. This is a substantial proportion of the expected total, especially in view of the fact that many lethal genes may act so early that we never see their effects. This sort of bias may also explain the results in FIGURE 5.21, which show that the majority of known genetic defects are due to point mutations (where there is more likely to be at least some residual function of the gene).

How do we explain the survival of genes whose deletion appears to have no effect? The most likely explanation is that the organism has alternative ways of fulfilling the same function. The simplest possibility is that there is **redundancy**, and that some genes are present in multiple copies. This is certainly true in some cases, in which multiple (related) genes must be knocked out in order to produce an effect. In a slightly more complex scenario, an organism might have two separate pathways capable of providing some activity. Inactivation of either pathway by itself would not be damaging, but the simultaneous occurrence of mutations in genes from both pathways would be deleterious.

Such situations can be tested by combining mutations. In principle, deletions in two genes, neither of which is lethal by itself, are introduced into the same strain. If the double mutant dies, the strain is called a **synthetic lethal**. This technique has been used to great effect with yeast, where the isolation of double mutants can be automated. The procedure is called **synthetic genetic array analysis (SGA)**. FIGURE 5.22 summarizes the results of an analysis in which an SGA screen was made for each of 132 viable deletions by testing whether it could survive in combination with any one of 4,700 viable deletions. Every one of the test genes had at least one partner with which the combination was lethal, and most of the test genes had many such partners; the median is ~25 partners, and the greatest number is shown by one test gene that had 146 lethal partners. A small proportion (~10%) of the interacting mutant pairs code for proteins that interact physically.

This result goes some way toward explaining the apparent lack of effect of so many deletions. Natural selection will act against these deletions when they find themselves in lethal pairwise combinations. To some degree, the

Most human mutations causing defects are small	
Missense/nonsense	58%
Splicing	10%
Regulatory	<1%
Small deletions	16%
Small insertions	6%
Large deletions	5%
Large rearrangements	2%

FIGURE 5.21 Most known genetic defects in human genes are due to point mutations. The majority directly affect the protein sequence. The remainder are due to insertions, deletions, or rearrangements of varying sizes.

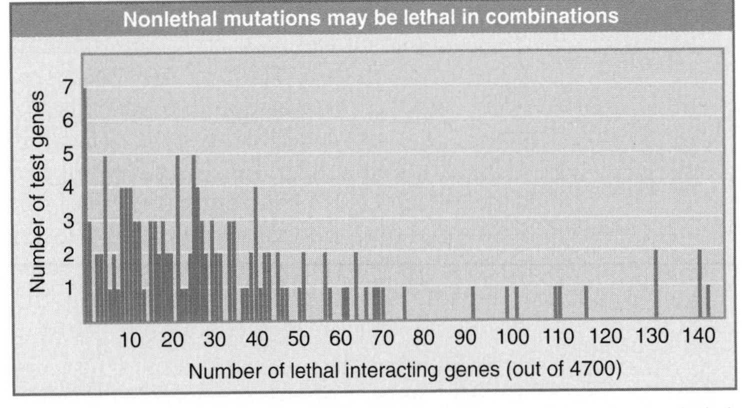

FIGURE 5.22 All 132 mutant test genes have some combinations that are lethal when they are combined with each of 4700 nonlethal mutations. The chart shows how many lethal interacting genes there are for each test gene.

organism has protected itself against the damaging effects of mutations by building in redundancy. However, it pays a price in the form of accumulating the "genetic load" of mutations that are not deleterious in themselves, but that may cause serious problems when combined with other such mutations in future generations. The theory of natural selection would suggest that the loss of the individual genes in such circumstances produces a sufficient disadvantage to maintain the active gene during the course of evolution.

5.10 Genes Are Expressed at Widely Differing Levels

Key concepts

- In any given cell, most genes are expressed at a low level.
- Only a small number of genes, whose products are specialized for the cell type, are highly expressed.

The proportion of DNA represented in an mRNA population can be determined by the amount of the DNA that can hybridize with the RNA. Such a saturation analysis typically identifies ~1% of the DNA as providing a template for mRNA. From this we can calculate the number of genes, so long as we know the average length of an mRNA. For a lower eukaryote such as yeast, the total number of expressed genes is ~4000. For somatic tissues of higher eukaryotes, the number usually is 10,000 to 15,000. The value is similar for plants and for vertebrates. (The only consistent exception to this type of value is presented by mammalian brain, for which much larger numbers of genes appear to be expressed, although the exact quantitation is not certain.)

Kinetic analysis of the reassociation of an RNA population can be used to determine its sequence complexity. This type of analysis typically identifies three components in a eukaryotic cell. Just as with a DNA reassociation curve, a single component hybridizes over about two decades of Rot (RNA concentration × time) values, and a reaction extending over a greater range must be resolved by computer curve-fitting into individual components. Again, this represents what is really a continuous spectrum of sequences.

An example of an excess mRNA × cDNA reaction that generates three components is given in FIGURE 5.23:

- The first component has the same characteristics as a control reaction of ovalbumin mRNA with its DNA copy. This suggests that the first component is in fact just ovalbumin mRNA (which indeed occupies about half of the messenger mass in oviduct tissue).
- The next component provides 15% of the reaction, with a total complexity of 15 kb. This corresponds to 7 to 8 mRNA species of average length 2000 bases.
- The last component provides 35% of the reaction, which corresponds to a complexity of 26 Mb. This corresponds to ~13,000 mRNA species of average length 2000 bases.

From this analysis, we can see that about half of the mass of mRNA in the cell represents a single mRNA, ~15% of the mass is provided by a mere 7 to 8 mRNAs, and ~35% of the mass is divided into the large number of 13,000 mRNA species. It is therefore obvious that the mRNAs comprising each component must be present in very different amounts.

The average number of molecules of each mRNA per cell is called its **abundance.** It can be calculated quite simply if the total mass of RNA in the cell is known. In the example shown

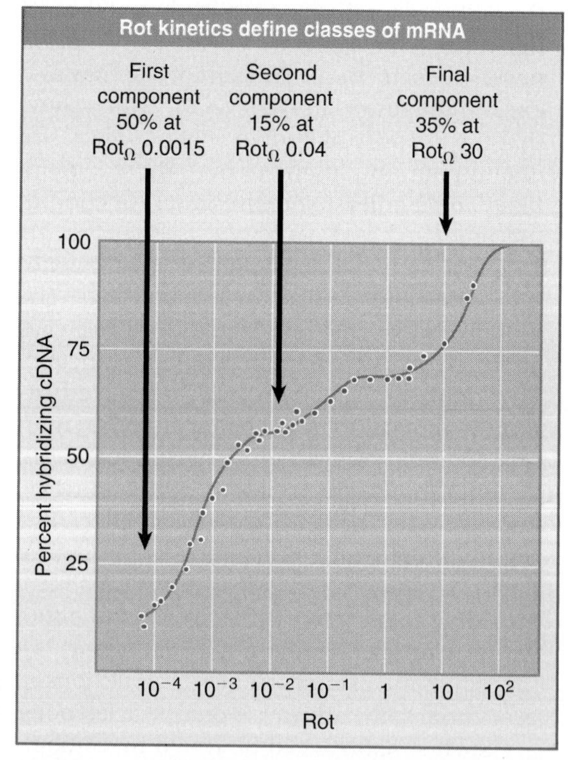

FIGURE 5.23 Hybridization between excess mRNA and cDNA identifies several components in chick oviduct cells, each characterized by the $Rot_{1/2}$ of reaction.

in Figure 5.23, the total mRNA can be accounted for as 100,000 copies of the first component (ovalbumin mRNA), 4000 copies of each of the 7 to 8 mRNAs in the second component, but only ~5 copies of each of the 13,000 mRNAs that constitute the last component.

We can divide the mRNA population into two general classes, according to their abundance:

- The oviduct is an extreme case, with so much of the mRNA represented in only one species, but most cells do contain a small number of RNAs present in many copies each. This **abundant mRNA** component typically consists of <100 different mRNAs present in 1000 to 10,000 copies per cell. It often corresponds to a major part of the mass, approaching 50% of the total mRNA.
- About half of the mass of the mRNA consists of a large number of sequences, of the order of 10,000, each represented by only a small number of copies in the mRNA—say, <10. This is the **scarce mRNA** or **complex mRNA** class. It is this class that drives a saturation reaction.

5.11 How Many Genes Are Expressed?

Key concepts

- mRNAs expressed at low levels overlap extensively when different cell types are compared.
- The abundantly expressed mRNAs are usually specific for the cell type.
- ~10,000 expressed genes may be common to most cell types of a higher eukaryote.

Many somatic tissues of higher eukaryotes have an expressed gene number in the range of 10,000 to 20,000. How much overlap is there between the genes expressed in different tissues? For example, the expressed gene number of chick liver is ~11,000 to 17,000, compared with the value for oviduct of ~13,000 to 15,000. How many of these two sets of genes are identical? How many are specific for each tissue? These questions are usually addressed by analyzing the transcriptome—the set of sequences represented in RNA.

We see immediately that there are likely to be substantial differences among the genes expressed in the abundant class. Ovalbumin, for example, is synthesized only in the oviduct, and not at all in the liver. This means that 50% of the mass of mRNA in the oviduct is specific to that tissue.

The abundant mRNAs represent only a small proportion of the number of expressed genes, though. In terms of the total number of genes of the organism, and of the number of changes in transcription that must be made between different cell types, we need to know the extent of overlap between the genes represented in the scarce mRNA classes of different cell phenotypes.

Comparisons between different tissues show that, for example, ~75% of the sequences expressed in liver and oviduct are the same. In other words, ~12,000 genes are expressed in both liver and oviduct, ~5000 additional genes are expressed only in liver, and ~3000 additional genes are expressed only in oviduct.

The scarce mRNAs overlap extensively. Between mouse liver and kidney, ~90% of the scarce mRNAs are identical, leaving a difference between the tissues of only 1000 to 2000 in terms of the number of expressed genes. The general result obtained in several comparisons of this sort is that only ~10% of the mRNA sequences of a cell are unique to it. The majority of sequences are common to many—perhaps even all—cell types.

This suggests that the common set of expressed gene functions, numbering perhaps ~10,000 in mammals, comprise functions that are needed in all cell types. Sometimes this type of function is referred to as a **housekeeping gene** or **constitutive gene.** It contrasts with the activities represented by specialized functions (such as ovalbumin or globin) needed only for particular cell phenotypes. These are sometimes called **luxury genes.**

5.12 Expressed Gene Number Can Be Measured *En Masse*

Key concepts

- "Chip" technology allows a snapshot to be taken of the expression of the entire genome in a yeast cell.
- ~75% (~4500 genes) of the yeast genome is expressed under normal growth conditions.
- Chip technology allows detailed comparisons of related animal cells to determine (for example) the differences in expression between a normal cell and a cancer cell.

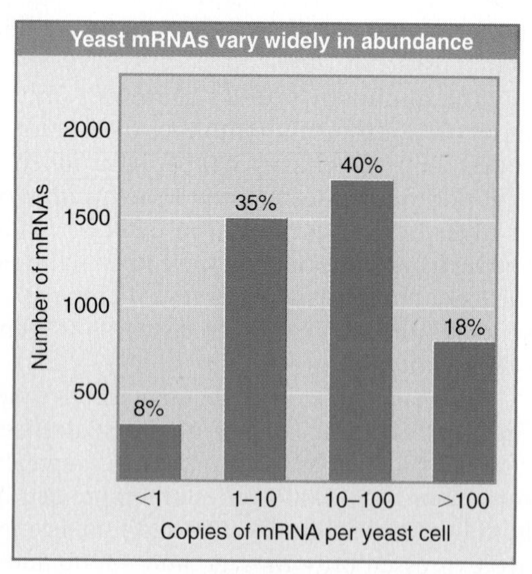

FIGURE 5.24 The abundances of yeast mRNAs vary from <1 per cell (meaning that not every cell has a copy of the mRNA) to >100 per cell (coding for the more abundant proteins).

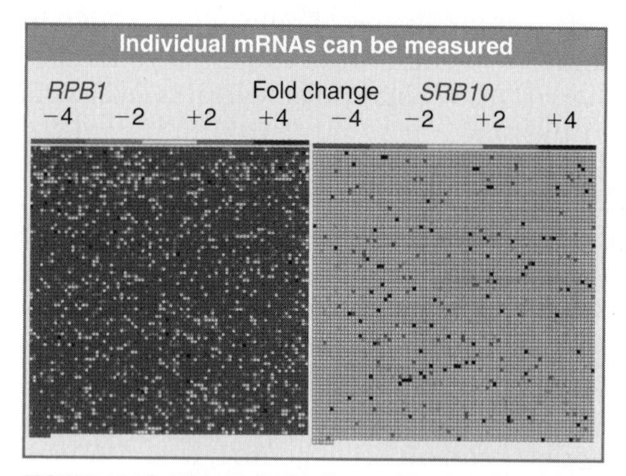

FIGURE 5.25 HDA analysis allows change in expression of each gene to be measured. Each square represents one gene (top left is first gene on chromosome I, bottom right is last gene on chromosome XVI). Change in expression relative to wild type is indicated by red (reduction), white (no change), or blue (increase). Photos courtesy of Rick A. Young, Whitehead Institute, Massachusetts Institute of Technology.

Recent technology allows more systematic and accurate estimates of the number of expressed genes. One approach (serial analysis of gene expression, or SAGE) allows a unique sequence tag to be used to identify each mRNA. The technology then allows the abundance of each tag to be measured. This approach identifies 4,665 expressed genes in *S. cerevisiae* growing under normal conditions, with abundances varying from 0.3 to >200 transcripts/cell. This means that ~75% of the total gene number (~6000) is expressed under these conditions. FIGURE 5.24 summarizes the number of different mRNAs that is found at each different abundance level.

The most powerful new technology uses chips that contain high-density oligonucleotide arrays (HDAs). Their construction is made possible by knowledge of the sequence of the entire genome. In the case of *S. cerevisiae,* each of 6181 ORFs is represented on the HDA by 20 25-mer oligonucleotides that perfectly match the sequence of the message and 20 mismatch oligonucleotides that differ at one base position. The expression level of any gene is calculated by subtracting the average signal of a mismatch from its perfect match partner. The entire yeast genome can be represented on four chips. This technology is sensitive enough to detect transcripts of 5460 genes (~90% of the genome), and shows that many genes are expressed at low levels, with abundances of 0.1 to 0.2 transcripts/cell. An abundance of <1 transcript/cell means that not all cells have a copy of the transcript at any given moment.

The technology allows not only measurement of levels of gene expression, but also detection of differences in expression in mutant cells compared with wild-type cells growing under different growth conditions, and so on. The results of comparing two states are expressed in the form of a grid, in which each square represents a particular gene and the relative change in expression is indicated by color. The left part of FIGURE 5.25 shows the effect of a mutation in RNA polymerase II, the enzyme that produces mRNA, which as might be expected causes the expression of most genes to be heavily reduced. By contrast, the right part shows that a mutation in an ancillary component of the transcription apparatus (*SRB10*) has much more restricted effects, causing increases in expression of some genes.

The extension of this technology to animal cells will allow the general descriptions based on RNA hybridization analysis to be replaced by exact descriptions of the genes that are expressed, and the abundances of their products, in any given cell type. A gene expression map of *D. melanogaster* detects transcriptional activity in some stage of the life cycle in almost all (93%) of predicted genes and shows that 40% have alternatively spliced forms.

5.13 Summary

Genomes that have been sequenced include many bacteria and archaea, yeasts, and a worm, a fly, a mouse, and man. The minimum number of genes required to make a living cell (an obligatory intracellular parasite) is ~470. The minimum number required to make a free-living cell is ~1700. A typical gram-negative bacterium has ~1500 genes. Strains of *E. coli* vary from 4300 to 5400 genes. The average bacterial gene is ~1000 bp long and is separated from the next gene by a space of ~100 bp. The yeasts *S. pombe* and *S. cerevisiae* have 5000 and 6000 genes, respectively.

Although the fly *D. melanogaster* is a more complex organism and has a larger genome than the worm *C. elegans*, the fly has fewer genes (13,600) than the worm (18,500). The plant *Arabidopsis* has 25,000 genes, and the lack of a clear relationship between genome size and gene number is shown by the fact that the rice genome is 4× larger but contains only a 50% increase in gene number, to ~40,000. Mouse and man each have 20,000 to 30,000 genes,which is much less than had been expected. The complexity of development of an organism may depend on the nature of the interactions between genes as well as their total number.

About 8000 genes are common to prokaryotes and eukaryotes and are likely to be involved in basic functions. A further 12,000 genes are found in multicellular organisms. Another 8000 genes are added to make an animal, and an additional 8000 (largely involved with the immune and nervous systems) are found in vertebrates. In each organism that has been sequenced, only ~50% of the genes have defined functions. Analysis of lethal genes suggests that only a minority of genes is essential in each organism.

The sequences comprising a eukaryotic genome can be classified in three groups: nonrepetitive sequences are unique; moderately repetitive sequences are dispersed and repeated a small number of times in the form of related, but not identical, copies; and highly repetitive sequences are short and usually repeated as tandem arrays. The proportions of the types of sequence are characteristic for each genome, although larger genomes tend to have a smaller proportion of nonrepetitive DNA. Almost 50% of the human genome consists of repetitive sequences, the vast majority corresponding to transposon sequences. Most structural genes are located in nonrepetitive DNA. The complexity of nonrepetitive DNA is a better reflection of the complexity of the organism than the total genome complexity; nonrepetitive DNA reaches a maximum complexity of $\sim 2 \times 10^9$ bp.

Genes are expressed at widely varying levels. There may be 10^5 copies of mRNA for an abundant gene whose protein is the principal product of the cell, 10^3 copies of each mRNA for <10 moderately abundant messages, and <10 copies of each mRNA for >10,000 scarcely expressed genes. Overlaps between the mRNA populations of cells of different phenotypes are extensive; the majority of mRNAs are present in most cells.

Non-Mendelian inheritance is explained by the presence of DNA in organelles in the cytoplasm. Mitochondria and chloroplasts both represent membrane-bounded systems in which some proteins are synthesized within the organelle, whereas others are imported. The organelle genome is usually a circular DNA that codes for all of the RNAs and for some of the proteins that are required.

Mitochondrial genomes vary greatly in size, from the 16 kb minimalist mammalian genome to the 570 kb genome of higher plants. It is assumed that the larger genomes code for additional functions. Chloroplast genomes range from 120 to 200 kb. Those that have been sequenced have a similar organization and coding functions. In both mitochondria and chloroplasts, many of the major proteins contain some subunits synthesized in the organelle and some subunits imported from the cytosol.

Mammalian mtDNAs are transcribed into a single transcript from the major coding strand, and individual products are generated by RNA processing. Rearrangements occur in mitochondrial DNA rather frequently in yeast, and recombination between mitochondrial or between chloroplast genomes has been found. Transfers of DNA have occurred from chloroplasts or mitochondria to nuclear genomes.

References

5.2 Bacterial Gene Numbers Range Over an Order of Magnitude

Reviews

Bentley, S. D. and Parkhill, J. (2004). Comparative genomic structure of prokaryotes. *Annu. Rev. Genet.* 38, 771–792.

Hacker, J. and Kaper, J. B. (2000). Pathogenicity islands and the evolution of microbes. *Annu. Rev. Microbiol.* 54, 641–679.

Research

Blattner, F. R. et al. (1997). The complete genome sequence of *Escherichia coli* K-12. *Science* 277, 1453–1474.

Deckert, G. et al. (1998). The complete genome of the hyperthermophilic bacterium *Aquifex aeolicus*. *Nature* 392, 353–358.

Galibert, F. et al. (2001). The composite genome of the legume symbiont *Sinorhizobium meliloti*. *Science* 293, 668–672.

5.3 Total Gene Number Is Known for Several Eukaryotes

Research

Adams, M. D. et al. (2000). The genome sequence of *D. melanogaster. Science* 287, 2185–2195.

Arabidopsis Initiative (2000). Analysis of the genome sequence of the flowering plant *Arabidopsis thaliana. Nature* 408, 796–815.

C. elegans Sequencing Consortium (1998). Genome sequence of the nematode *C. elegans:* a platform for investigating biology. *Science* 282, 2012–2022.

Duffy, A., and Grof, P. (2001). Psychiatric diagnoses in the context of genetic studies of bipolar disorder. *Bipolar Disord.* 3, 270–275.

Dujon, B. et al. (1994). Complete DNA sequence of yeast chromosome XI. *Nature* 369, 371–378.

Goff, S. A. et al. (2002). A draft sequence of the rice genome *(Oryza sativa L. ssp. japonica). Science* 296, 92–114.

Johnston, M. et al. (1994). Complete nucleotide sequence of *S. cerevisiae* chromosome VIII. *Science* 265, 2077–2082.

Kellis, M., Patterson, N., Endrizzi, M., Birren, B., and Lander, E. S. (2003). Sequencing and comparison of yeast species to identify genes and regulatory elements. *Nature* 423, 241–254.

Oliver, S. G. et al. (1992). The complete DNA sequence of yeast chromosome III. *Nature* 357, 38–46.

Wilson, R. et al. (1994). 22 Mb of contiguous nucleotide sequence from chromosome III of *C. elegans. Nature* 368, 32–38.

Wood, V. et al. (2002). The genome sequence of *S. pombe. Nature* 415, 871–880.

5.4 How Many Different Types of Genes Are There?

Reference

Rual, J. F., Venkatesan, K., Hao, T., Hirozane-Kishikawa, T., Dricot, A., Li, N., Berriz, G. F., Gibbons, F. D., Dreze, M., Ayivi-Guedehoussou, N., et al. (2005). Towards a proteome-scale map of the human protein–protein interaction network. *Nature* 437, 1173–1178.

Reviews

Aebersold, R. and Mann, M. (2003). Mass spectrometry-based proteomics. *Nature* 422, 198–207.

Hanash, S. (2003). Disease proteomics. *Nature* 422, 226–232.

Phizicky, E., Bastiaens, P. I., Zhu, H., Snyder, M., and Fields, S. (2003). Protein analysis on a proteomic scale. *Nature* 422, 208–215.

Sali, A., Glaeser, R., Earnest, T., and Baumeister, W. (2003). From words to literature in structural proteomics. *Nature* 422, 216–225.

Research

Agarwal, S., Heyman, J. A., Matson, S., Heidtman, M., Piccirillo, S., Umansky, L., Drawid, A., Jansen, R., Liu, Y., Miller, P., Gerstein, M., Roeder, G. S., and Snyder, M. (2002). Subcellular localization of the yeast proteome. *Genes Dev.* 16, 707–719.

Arabidopsis Initiative (2000). Analysis of the genome sequence of the flowering plant *Arabidopsis thaliana. Nature* 408, 796–815.

Gavin, A. C. et al. (2002). Functional organization of the yeast proteome by systematic analysis of protein complexes. *Nature* 415, 141–147.

Ho, Y. et al. (2002). Systematic identification of protein complexes in *S. cerevisiae* by mass spectrometry. *Nature* 415, 180–183.

Rubin, G. M. et al. (2000). Comparative genomics of the eukaryotes. *Science* 287, 2204–2215.

Uetz, P. et al. (2000). A comprehensive analysis of protein–protein interactions in *S. cerevisiae. Nature* 403, 623–630.

Venter, J. C. et al. (2001). The sequence of the human genome. *Science* 291, 1304–1350.

5.5 The Human Genome Has Fewer Genes Than Expected

Research

Clark, A. G. et al. (2003). Inferring nonneutral evolution from human–chimp–mouse orthologous gene trios. *Science* 302, 1960–1963.

Hogenesch, J. B., Ching, K. A., Batalov, S., Su, A. I., Walker, J. R., Zhou, Y., Kay, S. A., Schultz, P. G., and Cooke, M. P. (2001). A comparison of the Celera and Ensembl predicted gene sets reveals little overlap in novel genes. *Cell* 106, 413–415.

International Human Genome Sequencing Consortium (2001). Initial sequencing and analysis of the human genome. *Nature* 409, 860–921.

International Human Genome Sequencing Consortium (2004). Finishing the euchromatic sequence of the human genome. *Nature* 431, 931–945.

Venter, J. C. et al. (2001). The sequence of the human genome. *Science* 291, 1304–1350.

Waterston et al. (2002). Initial sequencing and comparative analysis of the mouse genome. *Nature* 420, 520–562.

5.6 How Are Genes and Other Sequences Distributed in the Genome?

Reference

Nusbaum, C., Cody, M. C., Borowsky, M. L., Kamal, M., Kodira, C. D., Taylor, T. D., Whittaker, C. A., Chang, J. L., Cuomo, C. A., Dewar, K., et al. (2005). DNA sequence and analysis of human chromosome 18. *Nature* 437, 551–555.

5.7 The Y Chromosome Has Several Male-Specific Genes

Research

Skaletsky, H. et al. (2003). The male-specific region of the human Y chromosome is a mosaic of discrete sequence classes. *Nature* 423, 825–837.

5.8 More Complex Species Evolve by Adding New Gene Functions

Reference

The Chimpanzee Sequencing and Analysis Consortium (2005). Initial sequence of the chimpanzee genome and comparison with the human genome. *Nature* 437, 69–87.

5.9 How Many Genes Are Essential?

Research

Giaever et al. (2002). Functional profiling of the *S. cerevisiae* genome. *Nature* 418, 387–391.

Goebl, M. G. and Petes, T. D. (1986). Most of the yeast genomic sequences are not essential for cell growth and division. *Cell* 46, 983–992.

Hutchison, C. A. et al. (1999). Global transposon mutagenesis and a minimal mycoplasma genome. *Science* 286, 2165–2169.

Kamath, R. S., Fraser, A. G., Dong, Y., Poulin, G., Durbin, R., Gotta, M., Kanapin, A., Le Bot, N., Moreno, S., Sohrmann, M., Welchman, D. P., Zipperlen, P., and Ahringer, J. (2003). Systematic functional analysis of the *C. elegans* genome using RNAi. *Nature* 421, 231–237.

Tong, A. H. et al. (2004). Global mapping of the yeast genetic interaction network. *Science* 303, 808–813.

5.10 Genes Are Expressed at Widely Differing Levels

Research

Hastie, N. B. and Bishop, J. O. (1976). The expression of three abundance classes of mRNA in mouse tissues. *Cell* 9, 761–774.

5.12 Expressed Gene Number Can Be Measured *En Masse*

Reviews

Mikos, G. L. G. and Rubin, G. M. (1996). The role of the genome project in determining gene function: insights from model organisms. *Cell* 86, 521–529.

Young, R. A. (2000). Biomedical discovery with DNA arrays. *Cell* 102, 9–15.

Research

Holstege, F. C. P. et al. (1998). Dissecting the regulatory circuitry of a eukaryotic genome. *Cell* 95, 717–728.

Hughes, T. R., Marton, M. J., Jones, A. R., Roberts, C. J., Stoughton, R., Armour, C. D., Bennett, H. A., Coffey, E., Dai, H., He, Y. D., Kidd, M. J., King, A. M., Meyer, M. R., Slade, D., Lum, P. Y., Stepaniants, S. B., Shoemaker, D. D. et al. (2000). Functional discovery via a compendium of expression profiles. *Cell* 102, 109–126.

Stolc, V. et al. (2004). A gene expression map for the euchromatic genome of *Drosophila melanogaster. Science* 306, 655–660.

Velculescu, V. E. et al. (1997). Characterization of the yeast transcriptosome. *Cell* 88, 243–251.

6

Clusters and Repeats

CHAPTER OUTLINE

6.1 Introduction

A set of genes descended by duplication and variation from some ancestral gene is called a **gene family.** Its members may be clustered together or dispersed on different chromosomes (or a combination of both). Genome analysis shows that many genes belong to families; the 25,000 genes identified in the human genome fall into ~15,000 families, so the average gene has a couple of relatives in the genome (see Figure 5.7). Gene families vary enormously in the degree of relatedness between members, from those consisting of multiple identical members to those for which the relationship is quite distant. Genes are usually related only by their exons, with introns having diverged (see Section 3.5, Exon Sequences Are Conserved but Introns Vary). Genes may also be related by only some of their exons, whereas others are unique (see Section 3.9, Some Exons Can Be Equated with Protein Functions).

The initial event that allows related exons or genes to develop is a duplication, when a copy is generated of some sequence within the genome. Tandem duplication (when the duplicates remain together) may arise through errors in replication or recombination. Separation of the duplicates can occur by a **translocation** that transfers material from one chromosome to another. A duplicate at a new location may also be produced directly by a transposition event that is associated with copying a region of DNA from the vicinity of the transposon. Duplications may apply either to intact genes or to collections of exons or even individual exons. When an intact gene is involved, the act of duplication generates two copies of a gene whose activities are indistinguishable, but then usually the copies diverge as each accumulates different mutations.

The members of a well-related structural gene family usually have related or even identical functions, although they may be expressed at different times or in different cell types. As a result, different globin proteins are expressed in embryonic and adult red blood cells, whereas different actins are utilized in muscle and nonmuscle cells. When genes have diverged significantly, or when only some exons are related, the proteins may have different functions.

Some gene families consist of identical members. Clustering is a prerequisite for maintaining identity between genes, although clustered genes are not necessarily identical. **Gene clusters** range from extremes in which a duplication has generated two adjacent related genes to cases where hundreds of identical genes lie in a tandem array. Extensive tandem repetition of a gene may occur when the product is needed in unusually large amounts. Examples are the genes for rRNA or histone proteins. This creates a special situation with regard to the maintenance of identity and the effects of selective pressure.

Gene clusters offer us an opportunity to examine the forces involved in evolution of the genome over larger regions than single genes. Duplicated sequences, especially those that remain in the same vicinity, provide the substrate for further evolution by recombination. A population evolves by the classical recombination illustrated in FIGURES 6.1 and 6.2, in which an exact crossing-over occurs. The recombinant chromosomes have the same organization as the parental chromosome. They contain precisely the same loci in the same order, but contain different combinations of alleles, providing the raw material for natural selection. However, the existence of duplicated sequences allows aberrant events to occur occasionally, which changes the content of genes and not just the combination of alleles.

Unequal crossing-over (also known as **nonreciprocal recombination**) describes a recombination event occurring between two sites that are not homologous. The feature that makes such events possible is the existence of repeated sequences. FIGURE 6.3 shows that this allows one copy of a repeat in one chromosome to misalign for recombination with a different

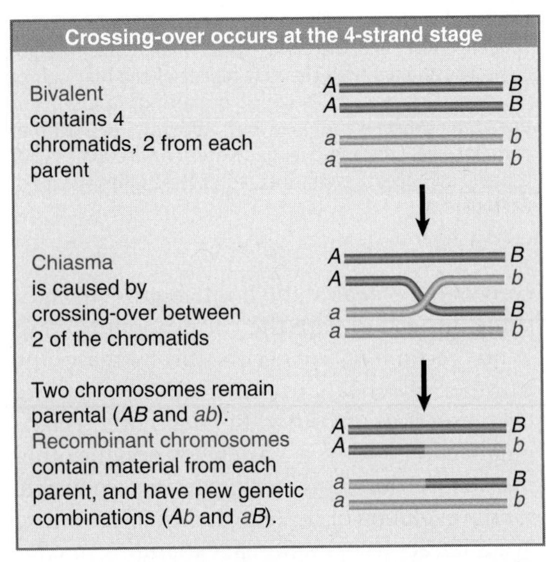

FIGURE 6.1 Chiasma formation represents the generation of recombinants.

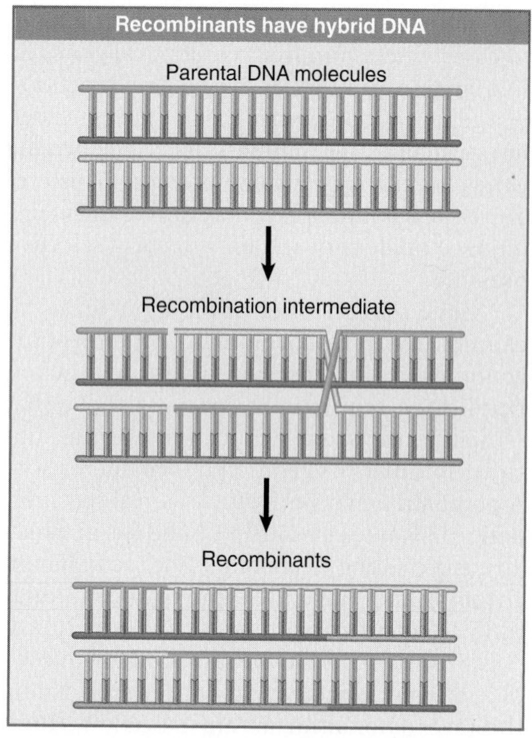

FIGURE 6.2 Recombination involves pairing between complementary strands of the two parental duplex DNAs.

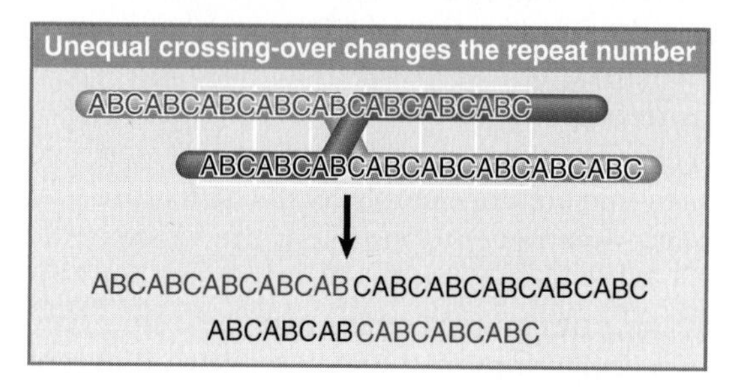

FIGURE 6.3 Unequal crossing-over results from pairing between nonequivalent repeats in regions of DNA consisting of repeating units. Here the repeating unit is the sequence ABC, and the third repeat of the blue chromosome has aligned with the first repeat of the black chromosome. Throughout the region of pairing, ABC units of one chromosome are aligned with ABC units of the other chromosome. Crossing-over generates chromosomes with ten and six repeats each, instead of the eight repeats of each parent.

copy of the repeat in the homologous chromosome, instead of with the corresponding copy. When recombination occurs, this increases the number of repeats in one chromosome and decreases it in the other. In effect, one recombinant chromosome has a deletion and the other has an insertion. This mechanism is responsible for the evolution of clusters of related sequences. We can trace its operation in expanding or contracting the size of an array in both gene clusters and regions of highly repeated DNA.

The highly repetitive fraction of the genome consists of multiple tandem copies of very short repeating units. These often have unusual properties. One is that they may be identified as a separate peak on a density gradient analysis of DNA, which gave rise to the name **satellite DNA.** They often are associated with inert regions of the chromosomes and in particular with centromeres (which contain the points of attachment for segregation on a mitotic or meiotic spindle). As a result of their repetitive organization, they show some of the same behavior with regard to evolution as the tandem gene clusters. In addition to the satellite sequences, there are shorter stretches of DNA called **minisatellites** that show similar behavior. They are useful in showing a high degree of divergence between individual genomes that can be used for mapping purposes.

All of these events that change the constitution of the genome are rare, but they are significant over the course of evolution.

6.2 Gene Duplication Is a Major Force in Evolution

Key concept

- Duplicated genes may diverge to generate different genes or one copy may become inactive.

Exons behave like modules for building genes that are tried out in the course of evolution in various combinations. At one extreme, an individual exon from one gene may be copied and used in another gene. At the other extreme, an entire gene, including both exons and introns, may be duplicated. In such a case, mutations can accumulate in one copy without attracting the adverse attention of natural selection. This copy may then evolve to a new function; become expressed in a different time or place from the first copy, or acquire different activities.

FIGURE 6.4 summarizes our present view of the rates at which these processes occur. There is ~1% probability that a given gene will be included in a duplication in a period of one million years. After the gene has duplicated, differences develop as the result of the occurrence of different mutations in each copy. These accumulate at a rate of ~0.1% per million years (see Section 6.4, Sequence Divergence Is the Basis for the Evolutionary Clock).

The organism is not likely to need to retain two identical copies of the gene. As differences

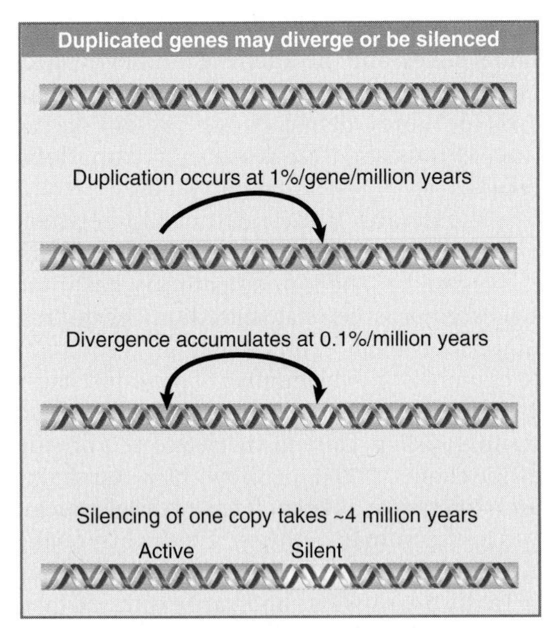

FIGURE 6.4 After a gene has been duplicated, differences may accumulate between the copies. The genes may acquire different functions or one of the copies may become inactive.

develop between the duplicated genes, one of two types of event is likely to occur:

- Both of the genes become necessary. This can happen either because the differences between them generate proteins with different functions, or because they are expressed specifically in different times or places.
- If this does not happen, one of the genes is likely to be eliminated because it will by chance gain a deleterious mutation, and there will be no adverse selection to eliminate this copy. Typically this takes ~ 4 million years. In such a situation, it is purely a matter of chance in terms of which of the two copies becomes inactive. (This can contribute to incompatibility between different individuals, and ultimately to speciation, if different copies become inactive in different populations.)

Analysis of the human genome sequence shows that ~5% comprises duplications of identifiable segments ranging in length from 10 to 300 kb. These duplications have arisen relatively recently, that is, there has not been sufficient time for divergence between them to eliminate their relationship. They include a proportional share (~6%) of the expressed exons, which shows that the duplications are occurring more or less irrespective of genetic content. The genes in these duplications may be especially interesting because of the implication that they have evolved recently and therefore could be important for recent evolutionary developments (such as the separation of man from monkey).

6.3 Globin Clusters Are Formed by Duplication and Divergence

Key concepts

- All globin genes are descended by duplication and mutation from an ancestral gene that had three exons.
- The ancestral gene gave rise to myoglobin, leghemoglobin, and α and β globins.
- The α- and β-globin genes separated in the period of early vertebrate evolution, after which duplications generated the individual clusters of separate α- and β-like genes.
- Once a gene has been inactivated by mutation, it may accumulate further mutations and become a pseudogene, which is homologous to the active gene(s) but has no functional role.

The most common type of duplication generates a second copy of the gene close to the first copy. In some cases, the copies remain associated and further duplication may generate a cluster of related genes. The best characterized example of a gene cluster is presented by the globin genes, which constitute an ancient gene family concerned with a function that is central to the animal kingdom: the transport of oxygen through the bloodstream.

The major constituent of the red blood cell is the globin tetramer, which is associated with its heme (iron-binding) group in the form of hemoglobin. Functional globin genes in all species have the same general structure: they are divided into three exons, as shown previously in Figure 3.7. We conclude that all globin genes are derived from a single ancestral gene, and by tracing the development of individual globin genes within and between species we may learn about the mechanisms involved in the evolution of gene families.

In adult cells, the globin tetramer consists of two identical α chains and two identical β chains. Embryonic blood cells contain hemoglobin tetramers that are different from the adult form. Each tetramer contains two identical α-like chains and two identical β-like chains, each of which is related to the adult polypeptide and is later replaced by it. This is an example of

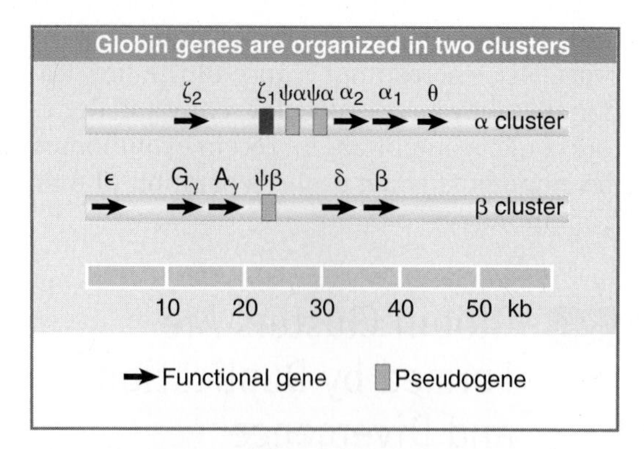

FIGURE 6.5 Each of the α-like and β-like globin gene families is organized into a single cluster that includes functional genes and pseudogenes (ψ).

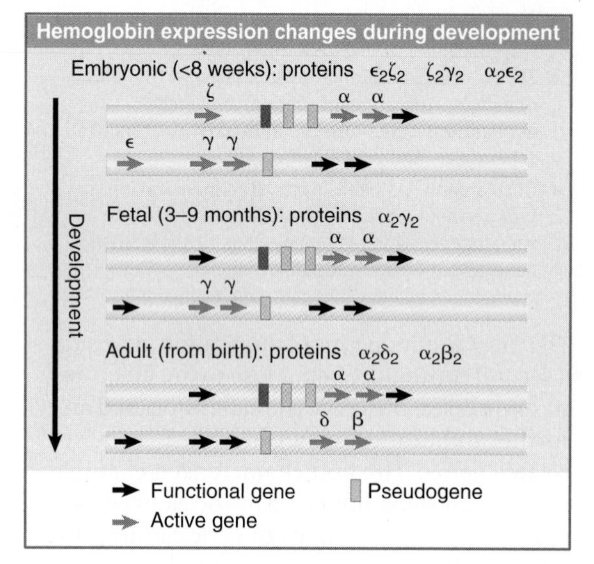

FIGURE 6.6 Different hemoglobin genes are expressed during embryonic, fetal, and adult periods of human development.

developmental control, in which different genes are successively switched on and off to provide alternative products that fulfill the same function at different times.

The division of globin chains into α-like and β-like reflects the organization of the genes. Each type of globin is coded by genes organized into a single cluster. The structures of the two clusters in the higher primate genome are illustrated in FIGURE 6.5.

Stretching over 50 kb, the β cluster contains five functional genes (ε, two γ, δ, and β) and one nonfunctional gene (ψβ). The two γ genes differ in their coding sequence in only one amino acid; the G variant has glycine at position 136, whereas the A variant has alanine.

The more compact α cluster extends over 28 kb and includes one active ζ gene, one nonfunctional ζ gene, two α genes, two α nonfunctional genes, and the θ gene of unknown function. The two α genes code for the same protein. Two (or more) identical genes present on the same chromosome are described as **nonallelic** copies.

The details of the relationship between embryonic and adult hemoglobins vary with the organism. The human pathway has three stages: embryonic, fetal, and adult. The distinction between embryonic and adult is common to mammals, but the number of pre-adult stages varies. In man, zeta and alpha are the two α-like chains. Epsilon, gamma, delta, and beta are the β-like chains. FIGURE 6.6 shows how the chains are expressed at different stages of development.

In the human pathway, ζ is the first α-like chain to be expressed, but it is soon replaced by α. In the β-pathway, ε and γ are expressed first, with δ and β replacing them later. In adults, the $\alpha_2\beta_2$ form provides 97% of the hemoglobin, $\alpha_2\delta_2$ provides ~2%, and ~1% is provided by persistence of the fetal form $\alpha_2\gamma_2$.

What is the significance of the differences between embryonic and adult globins? The embryonic and fetal forms have a higher affinity for oxygen. This is necessary in order to obtain oxygen from the mother's blood. This explains why there is no equivalent in (for example) chicken, for which the embryonic stages occur outside the body (that is, within the egg).

Functional genes are defined by their expression in RNA and ultimately by the proteins for which they code. Nonfunctional genes are defined as such by their inability to code for proteins; the reasons for their inactivity vary, and the deficiencies may be in transcription or translation (or both). They are called **pseudogenes** and are indicated by the symbol ψ. A similar general organization is found in other vertebrate globin gene clusters, but details of the types, numbers, and order of genes all vary, as illustrated in FIGURE 6.7. Each cluster contains both embryonic and adult genes. The total lengths of the clusters vary widely. The longest is found in the goat, where a basic cluster of four genes has been duplicated twice. The distribution of active genes and pseudogenes differs in each case, illustrating the random nature of the conversion of one copy of a duplicated gene into the inactive state.

The characterization of these gene clusters makes an important general point. *There may be more members of a gene family, both functional and nonfunctional, than we would suspect on the basis*

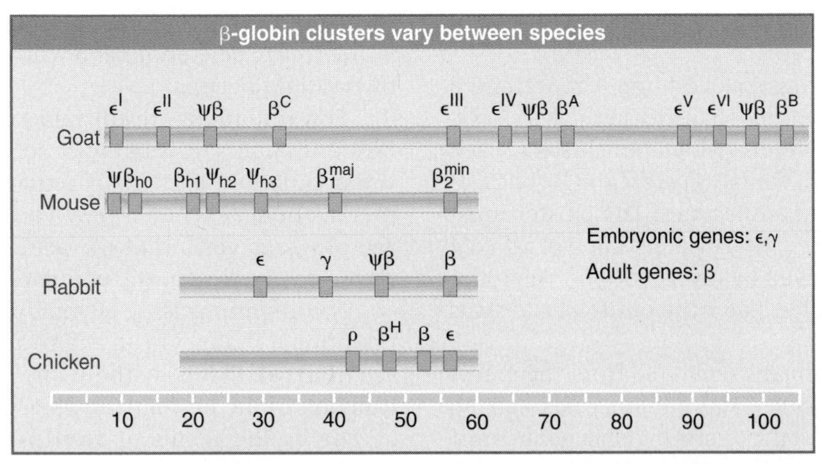

FIGURE 6.7 Clusters of β-globin genes and pseudogenes are found in vertebrates. Seven mouse genes include two early embryonic genes, one late embryonic gene, two adult genes, and two pseudogenes. Rabbit and chick each have four genes.

of protein analysis. The extra functional genes may represent duplicates that code for identical polypeptides, or they may be related to—but different from—known proteins (and presumably expressed only briefly or in low amounts).

With regard to the question of how much DNA is needed to code for a particular function, we see that coding for the β-like globins requires a range of 20 to 120 kb in different mammals. This is much greater than we would expect just from scrutinizing the known β-globin proteins or from even considering the individual genes. However, clusters of this type are not common; most genes are found as individual loci.

From the organization of globin genes in a variety of species, we should be able to trace the evolution of present globin gene clusters from a single ancestral globin gene. Our present view of the evolutionary descent is pictured in FIGURE 6.8.

The leghemoglobin gene of plants, which is related to the globin genes, may represent the ancestral form. The furthest back that we can trace a globin gene in modern form is provided by the sequence of the single chain of mammalian myoglobin, which diverged from the globin line of descent ~800 million years ago. The myoglobin gene has the same organization as globin genes, so we may take the three-exon structure to represent their common ancestor.

Some "primitive fish" have only a single type of globin chain, so they must have diverged from the line of evolution before the ancestral globin gene was duplicated to give rise to the α and β variants. This appears to have occurred

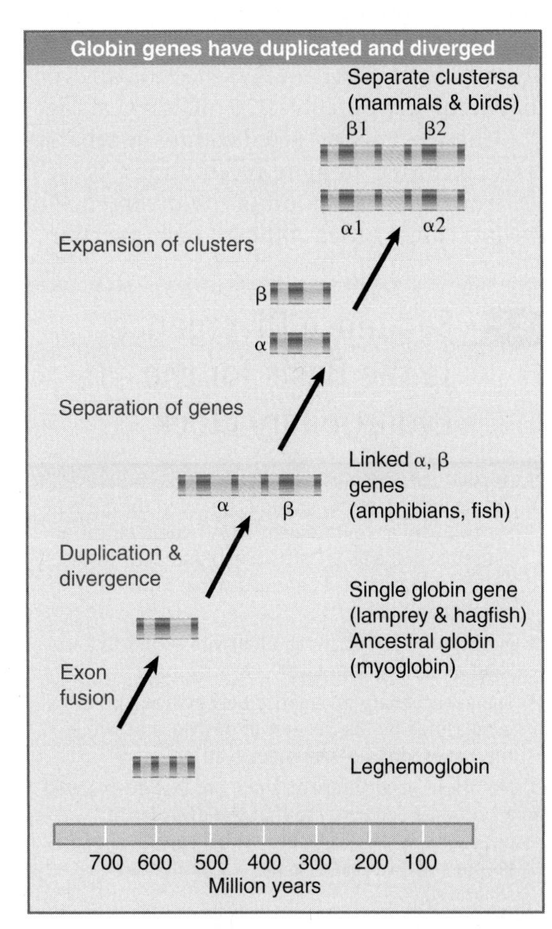

FIGURE 6.8 All globin genes have evolved by a series of duplications, transpositions, and mutations from a single ancestral gene.

~500 million years ago, during the evolution of the bony fish.

The next stage of evolution is represented by the state of the globin genes in the frog *Xenopus laevis,* which has two globin clusters. Each cluster, though, contains *both* α and β genes, of both larval and adult types. The cluster must therefore have evolved by duplication of a linked α–β pair, followed by divergence between the individual copies. Later the entire cluster was duplicated.

The amphibians separated from the mammalian/avian line ~350 million years ago, so the separation of the α- and β-globin genes must have resulted from a transposition in the mammalian/avian forerunner after this time. This probably occurred in the period of early vertebrate evolution. There are separate clusters for α and β globins in both birds and mammals, thus the α and β genes must have been physically separated before the mammals and birds diverged from their common ancestor, an event that occurred probably ~270 million years ago.

Changes have occurred within the separate α and β clusters in more recent times, as we'll see from the description of the divergence of the individual genes in the following section.

6.4 Sequence Divergence Is the Basis for the Evolutionary Clock

Key concepts

- The sequences of homologous genes in different species vary at replacement sites (where mutation causes amino acid substitutions) and silent sites (where mutation does not affect the protein sequence).
- Mutations accumulate at silent sites ~10× faster than at replacement sites.
- The evolutionary divergence between two proteins is measured by the percent of positions at which the corresponding amino acids differ.
- Mutations accumulate at a more or less even speed after genes separate, so that the divergence between any pair of globin sequences is proportional to the time since their genes separated.

Most changes in protein sequences occur by small mutations that accumulate slowly over time. Point mutations and small insertions and deletions occur by chance, probably with more or less equal probability in all regions of the genome. The exceptions to this are hotspots, where mutations occur much more frequently. Most mutations that change the amino acid sequence are deleterious and will be eliminated by natural selection.

Few mutations are advantageous. When a rare one occurs, it is likely to spread through the population, eventually replacing the former sequence. When a new variant replaces the previous version of the gene, it is said to have become *fixed* in the population.

A contentious issue is what proportion of mutational changes in an amino acid sequence are **neutral,** that is, without any effect on the function of the protein and able, therefore, to accrue as the result of **random drift and fixation.**

The rate at which mutational changes accumulate is a characteristic of each protein, presumably depending at least in part on its flexibility with regard to change. Within a species, a protein evolves by mutational substitution, followed by elimination or fixation within the single breeding pool. Remember that when we scrutinize the gene pool of a species, we see only the variants that have survived. When multiple variants are present they may be stable (because none has any selective advantage), or one may in fact be transient because it is in the process of being displaced.

When a species separates into two new species, each of the resulting species now constitutes an independent pool for evolution. By comparing the corresponding proteins in two species, we see the differences that have accumulated between them *since the time when their ancestors ceased to interbreed.* Some proteins are highly conserved, showing little or no change from species to species. This indicates that almost any change is deleterious and therefore not selected.

The difference between two proteins is expressed as their **divergence,** the percent of positions at which the amino acids are different. The divergence between proteins can be different from the divergence between the corresponding nucleic acid sequences. The source of this difference is the representation of each amino acid in a three-base codon, in which the third base often has no effect on the meaning.

We may divide the nucleotide sequence of a coding region into potential **replacement sites** and **silent sites:**

- At replacement sites, a mutation alters the amino acid that is coded. The effect of the mutation (deleterious, neutral, or advantageous) depends on the result of the amino acid replacement.

- At silent sites, mutation only substitutes one synonym codon for another, so there is no change in the protein. Usually the replacement sites account for 75% of a coding sequence and the silent sites provide 25%.

In addition to the coding sequence, a gene contains nontranslated regions. Here again, mutations are potentially neutral, apart from their effects on either secondary structure or (usually rather short) regulatory signals.

Although silent mutations are neutral with regard to the protein, they could affect gene expression via the sequence change in RNA. For example, a change in secondary structure might influence transcription, processing, or translation. Another possibility is that a change in synonym codons calls for a different tRNA to respond, influencing the efficiency of translation.

The mutations in replacement sites should correspond with the amino acid divergence (determined by the percent of changes in the protein sequence). A nucleic acid divergence of 0.45% at replacement sites corresponds to an amino acid divergence of 1% (assuming that the average number of replacement sites per codon is 2.25). Actually, the measured divergence underestimates the differences that have occurred during evolution, because of the occurrence of multiple events at one codon. Usually a correction is made for this.

To take the example of the human β- and δ-globin chains, there are 10 differences in 146 residues, a divergence of 6.9%. The DNA sequence has 31 changes in 441 residues. However, these changes are distributed very differently in the replacement and silent sites. There are 11 changes in the 330 replacement sites, but 20 changes in only 111 silent sites. This gives (corrected) rates of divergence of 3.7% in the replacement sites and 32% in the silent sites, almost an order of magnitude in difference.

The striking difference in the divergence of replacement and silent sites demonstrates the existence of much greater constraints on nucleotide positions that influence protein constitution relative to those that do not. So probably very few of the amino acid changes are neutral.

Suppose we take the rate of mutation at silent sites to indicate the underlying rate of mutational fixation (this assumes that there is no selection at all at the silent sites). Then over the period since the β and δ genes diverged, there should have been changes at 32% of the 330 replacement sites, for a total of 105. All but 11 of them have been eliminated, which means that ~90% of the mutations did not survive.

The divergence between any pair of globin sequences is (more or less) proportional to the time since they separated. This provides an **evolutionary clock** that measures the accumulation of mutations at an apparently even rate during the evolution of a given protein.

The rate of divergence can be measured as the percent difference per million years, or as its reciprocal, the unit evolutionary period (UEP), the time in millions of years that it takes for 1% divergence to develop. Once the clock has been established by pairwise comparisons between species (remembering the practical difficulties in establishing the actual time of speciation), it can be applied to related genes *within* a species. From their divergence, we can calculate how much time has passed since the duplication that generated them.

By comparing the sequences of homologous genes in different species, the rate of divergence at both replacement and silent sites can be determined, as plotted in FIGURE 6.9.

In pairwise comparisons, there is an average divergence of 10% in the replacement sites of either the α- or β-globin genes of mammals that have been separated since the mammalian radiation occurred ~85 million years ago. This corresponds to a replacement divergence rate of 0.12% per million years.

The rate is steady when the comparison is extended to genes that diverged in the more

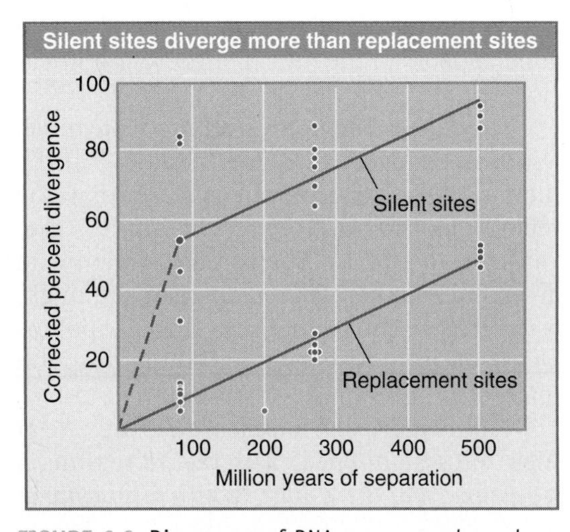

FIGURE 6.9 Divergence of DNA sequences depends on evolutionary separation. Each point on the graph represents a pairwise comparison.

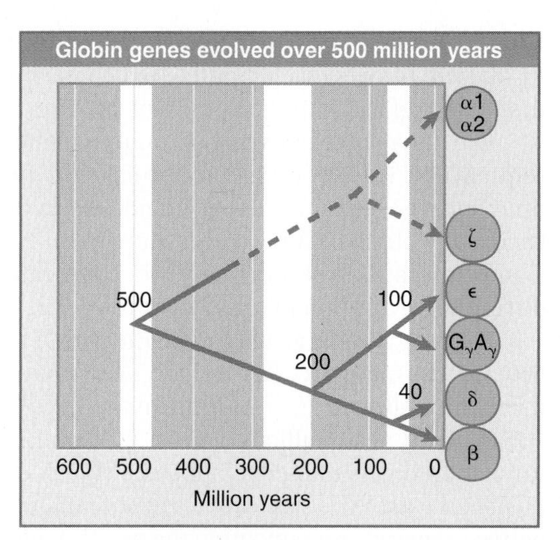

FIGURE 6.10 Replacement site divergences between pairs of β-globin genes allow the history of the human cluster to be reconstructed. This tree accounts for the separation of classes of globin genes.

distant past. For example, the average replacement divergence between corresponding mammalian and chicken globin genes is 23%. Relative to a separation ~270 million years ago, this gives a rate of 0.09% per million years.

Going further back, we can compare the α- with the β-globin genes within a species. They have been diverging since the individual gene types separated 500 million years ago (see Figure 6.8). They have an average replacement divergence of ~50%, which gives a rate of 0.1% per million years.

The summary of these data in Figure 6.9 shows that replacement divergence in the globin genes has an average rate of ~0.096% per million years (or a UEP of 10.4). Considering the uncertainties in estimating the times at which the species diverged, the results lend good support to the idea that there is a linear clock.

The data on silent site divergence are much less clear. In every case, it is evident that the silent site divergence is much greater than the replacement site divergence, by a factor that varies from 2 to 10. The spread of silent site divergences in pairwise comparisons, though, is too great to show whether a clock is applicable (so we must base temporal comparisons on the replacement sites).

From Figure 6.9, it is clear that the rate at silent sites is not linear with regard to time. *If we assume that there must be zero divergence at zero years of separation*, we see that the rate of silent site divergence is much greater for the first ~100 million years of separation. One interpretation is that a fraction of roughly half of the silent sites is rapidly (within 100 million years) saturated by mutations; this fraction behaves as neutral sites. The other fraction accumulates mutations more slowly, at a rate approximately the same as that of the replacement sites; this fraction identifies sites that are silent with regard to the protein, but that come under selective pressure for some other reason.

Now we can reverse the calculation of divergence rates to estimate the times since genes within a species have been apart. The difference between the human β and δ genes is 3.7% for replacement sites. At a UEP of 10.4, these genes must have diverged $10.4 \times 3.7 = 40$ million years ago—about the time of the separation of the lines leading to New World monkeys, Old World monkeys, great apes, and man. All of these higher primates have both β and δ genes, which suggests that the gene divergence commenced just before this point in evolution.

Proceeding further back, the divergence between the replacement sites of γ and ε genes is 10%, which corresponds to a time of separation ~100 million years ago. The separation between embryonic and fetal globin genes therefore may have just preceded or accompanied the mammalian radiation.

An evolutionary tree for the human globin genes is constructed in FIGURE 6.10. Features that evolved before the mammalian radiation—such as the separation of β/δ from γ—should be found in all mammals. Features that evolved afterward—such as the separation of β- and δ-globin genes—should be found in individual lines of mammals.

In each species, there have been comparatively recent changes in the structures of the clusters. We know this because we see differences in gene number (one adult β-globin gene in man, two in mouse) or in type (most often concerning whether there are separate embryonic and fetal genes).

When sufficient data have been collected on the sequences of a particular gene, the arguments can be reversed, and comparisons between genes in different species can be used to assess taxonomic relationships.

6.5 The Rate of Neutral Substitution Can Be Measured from Divergence of Repeated Sequences

Key concept

- The rate of substitution per year at neutral sites is greater in the mouse than in the human genome.

We can make the best estimate of the rate of substitution at neutral sites by examining sequences that do not code for protein. (We use the term neutral here rather than silent, because there is no coding potential). An informative comparison can be made by comparing the members of a common repetitive family in the human and mouse genomes.

The principle of the analysis is summarized in FIGURE 6.11. We start with a family of related sequences that have evolved by duplication and substitution from an original family member. We assume that the common ancestral sequence can be deduced by taking the base that is most common at each position. Then we can calculate the divergence of each individual family member as the proportion of bases that differ from the deduced ancestral sequence. In this example, individual members vary from 0.13 to 0.18 divergence and the average is 0.16.

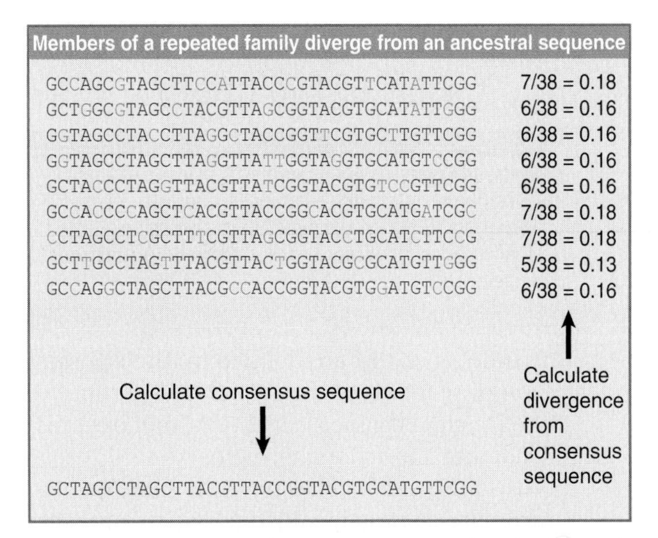

FIGURE 6.11 An ancestral consensus sequence for a family is calculated by taking the most common base at each position. The divergence of each existing current member of the family is calculated as the proportion of bases at which it differs from the ancestral sequence.

One family used for this analysis in the human and mouse genomes derives from a sequence that is thought to have ceased to be active at about the time of the divergence between man and rodents (the LINES family; see Section 22.9, Retroposons Fall into Three Classes). This means that it has been diverging without any selective pressure for the same length of time in both species. Its average divergence in man is ~0.17 substitutions per site, corresponding to a rate of 2.2×10^{-9} substitutions per base per year over the 75 million years since the separation. In the mouse genome, however, neutral substitutions have occurred at twice this rate, corresponding to 0.34 substitutions per site in the family, or a rate of 4.5×10^{-9}. Note, however, that if we calculated the rate per generation instead of per year, it would be greater in man than in mouse (~2.2×10^{-8} as opposed to ~10^{-9}).

These figures probably underestimate the rate of substitution in the mouse; at the time of divergence the rates in both species would have been the same, and the difference must have evolved since then. The current rate of neutral substitution per year in the mouse is probably 2–3× greater than the historical average. These rates reflect the balance between the occurrence of mutations and the ability of the genetic system of the organism to correct them. The difference between the species demonstrates that each species has systems that operate with a characteristic efficiency.

Comparing the mouse and human genomes allows us to assess whether syntenic (corresponding) sequences show signs of conservation or have differed at the rate expected from accumulation of neutral substitutions. The proportion of sites that show signs of selection is ~5%. This is much higher than the proportion that codes for protein or RNA (~1%). It implies that the genome includes many more stretches whose sequence is important for noncoding functions than for coding functions. Known regulatory elements are likely to comprise only a small part of this proportion. This number also suggests that most (i.e., the rest) of the genome sequences do not have any function that depends on the exact sequence.

6.6 Pseudogenes Are Dead Ends of Evolution

Key concept

- Pseudogenes have no coding function, but they can be recognized by sequence similarities with existing functional genes. They arise by the accumulation of mutations in (formerly) functional genes.

Pseudogenes (Ψ) are defined by their possession of sequences that are related to those of the functional genes, but that cannot be translated into a functional protein.

Some pseudogenes have the same general structure as functional genes, with sequences corresponding to exons and introns in the usual locations. They may have been rendered inactive by mutations that prevent any or all of the stages of gene expression. The changes can take the form of abolishing the signals for initiating transcription, preventing splicing at the exon–intron junctions, or prematurely terminating translation.

Usually a pseudogene has several deleterious mutations. Presumably once it ceased to be active, there was no impediment to the accumulation of further mutations. Pseudogenes that represent inactive versions of currently active genes have been found in many systems, including globin, immunoglobulins, and histocompatibility antigens, where they are located in the vicinity of the gene cluster, often interspersed with the active genes.

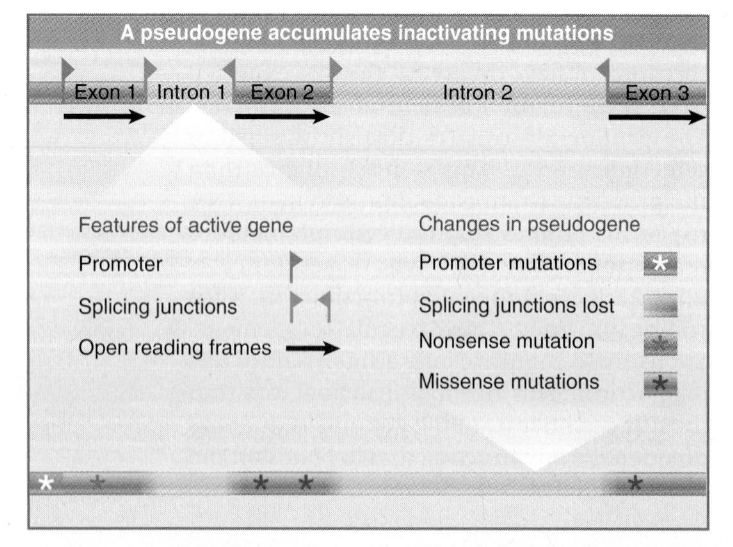

FIGURE 6.12 Many changes have occurred in a β globin gene since it became a pseudogene.

A typical example is the rabbit pseudogene, Ψβ2, which has the usual organization of exons and introns and is related most closely to the functional globin gene β1. The rabbit pseudogene is not functional, though. FIGURE 6.12 summarizes the many changes that have occurred in the pseudogene. The deletion of a base pair at codon 20 of Ψβ2 has caused a frameshift that would lead to termination shortly after. Several point mutations have changed later codons representing amino acids that are highly conserved in the β globins. Neither of the two introns any longer possesses recognizable boundaries with the exons, so probably the introns could not be spliced out even if the gene were transcribed. However, there are no transcripts corresponding to the gene, possibly because there have been changes in the 5′ flanking region.

This list of defects includes mutations potentially preventing each stage of gene expression, thus we have no means of telling which event originally inactivated this gene. If we measure the divergence between the pseudogene and the functional gene, though, we can estimate when the pseudogene originated and when its mutations started to accumulate.

If the pseudogene had become inactive as soon as it was generated by duplication from β1, we should expect both replacement site and silent site divergence rates to be the same. (They will be different only if the gene is translated to create selective pressure on the replacement sites.) In fact, there are fewer replacement site substitutions than silent site substitutions. This suggests that at first (while the gene was expressed) there was selection against replacement site substitution. From the relative extents of substitution in the two types of site, we can calculate that Ψβ2 diverged from β1 ~55 million years ago, remained a functional gene for 22 million years, but has been a pseudogene for the last 33 million years.

Similar calculations can be made for other pseudogenes. Some appear to have been active for some time before becoming pseudogenes; others appear to have been inactive from the very time of their original generation. The general point made by the structures of these pseudogenes is that each has evolved independently during the development of the globin gene cluster in each species. This reinforces the conclusion that the creation of new genes—followed by their acceptance as functional duplicates, variation to become new functional genes, or inactivation as pseudogenes—is a continuing process in the gene cluster. Most gene families have members that are pseudogenes. Usually the pseudogenes represent a small minority of the total gene number.

The mouse Ψα3-globin gene has an interesting property: it precisely lacks both introns. Its sequence can be aligned (allowing for accumulated mutations) with the α-globin mRNA. The apparent time of inactivation coincides with the original duplication, which suggests that the original inactivating event was associated with the loss of introns.

Inactive genomic sequences that resemble the RNA transcript are called **processed pseudogenes.** Following a retrotransposition event, processed pseudogenes originate by insertion at some random site of a product derived from the RNA, as discussed in Chapter 22, Retroviruses and Retroposons. Their characteristic features are summarized in Figure 22.19.

If pseudogenes are evolutionary dead ends—simply unwanted accompaniments to the rearrangement of functional genes—why are they still present in the genome? Do they fulfill any function or are they entirely without purpose, in which case there should be no selective pressure for their retention?

We should remember that we see those genes that have survived in present populations. In past times, any number of other pseudogenes may have been eliminated. This elimination could occur by deletion of the sequence as a sudden event or by the accretion of mutations to the point where the pseudogene can no longer be recognized as a member of its original sequence family (probably the ultimate fate of any pseudogene that is not suddenly eliminated).

Even relics of evolution can be duplicated. In the β-globin genes of the goat, there are three adult species: βA, βB, and βC (see Figure 6.7). Each of these has a pseudogene a few kilobases upstream of it. The pseudogenes are better related to each other than to the adult β-globin genes; in particular, they share several inactivating mutations. In addition, the adult β-globin genes are better related to each other than to the pseudogenes. This implies that an original Ψβ–β structure was itself duplicated, giving functional β genes (which diverged further) and two nonfunctional genes (which diverged into the current pseudogenes).

The mechanisms responsible for gene duplication, deletion, and rearrangement act on all sequences that are recognized as members of the cluster, whether or not they are functional. It is left to selection to discriminate among the products.

By definition, pseudogenes do not code for proteins, and typically they have no function at all. In at least one exceptional case, though, a pseudogene has a regulatory function. Transcription of a pseudogene inhibits degradation of the mRNA produced by its homologous active gene. Most likely there is a protein responsible for this degradation that binds a specific sequence in the mRNA. If this sequence is also present in the RNA transcribed from the pseudogene, the effect of the protein will be diluted when the pseudogene is transcribed. It is not clear how common such effects may be, but as a general rule we might expect dilution effects of this type to be possible whenever pseudogenes are transcribed.

6.7 Unequal Crossing-over Rearranges Gene Clusters

Key concepts

- When a genome contains a cluster of genes with related sequences, mispairing between nonallelic genes can cause unequal crossing-over. This produces a deletion in one recombinant chromosome and a corresponding duplication in the other.
- Different thalassemias are caused by various deletions that eliminate α- or β-globin genes. The severity of the disease depends on the individual deletion.

There are frequent opportunities for rearrangement in a cluster of related or identical genes. We can see the results by comparing the mammalian β clusters included in Figure 6.7. Although the clusters serve the same function, and all have the same general organization, each is different in size, there is variation in the total number and types of β-globin genes, and the numbers and structures of pseudogenes are different. All of these changes must have occurred since the mammalian radiation ~85 million years ago (the last point in evolution common to all the mammals).

The comparison makes the general point that gene duplication, rearrangement, and variation is as important a factor in evolution as the slow accumulation of point mutations in individual genes. What types of mechanisms are responsible for gene reorganization?

As described in the introduction to this chapter, unequal crossing-over can occur as the result

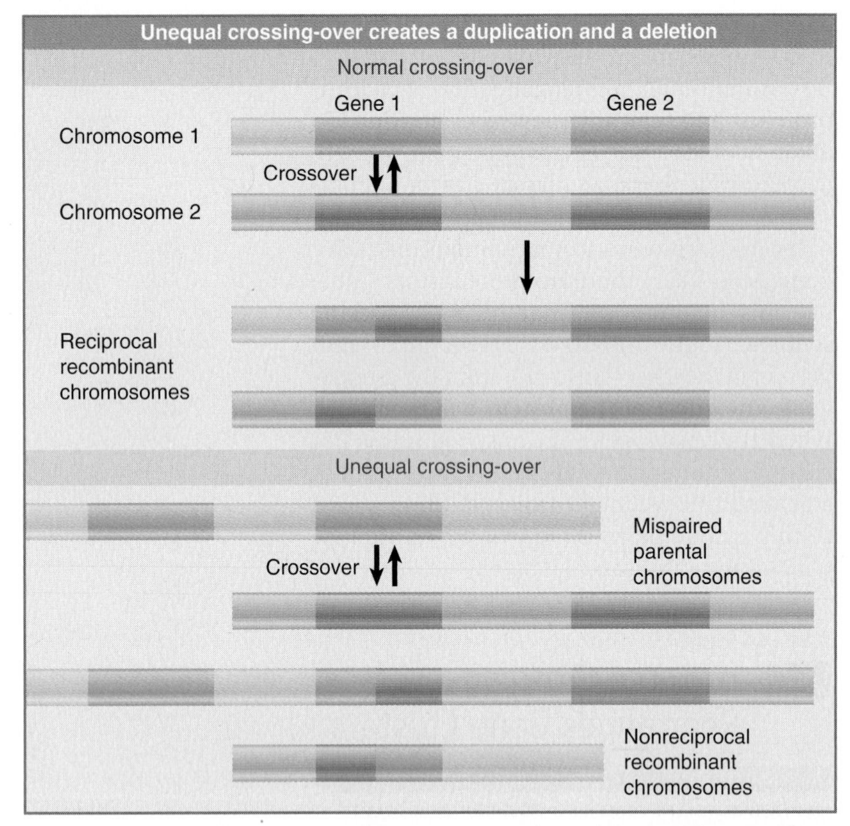

FIGURE 6.13 Gene number can be changed by unequal crossing-over. If gene 1 of one chromosome pairs with gene 2 of the other chromosome, the other gene copies are excluded from pairing. Recombination between the mispaired genes produces one chromosome with a single (recombinant) copy of the gene and one chromosome with three copies of the gene (one from each parent and one recombinant).

of pairing between two sites that are *not* homologous. Usually, recombination involves corresponding sequences of DNA held in exact alignment between the two homologous chromosomes. However, when there are two copies of a gene on each chromosome, an occasional misalignment allows pairing between them. (This requires some of the adjacent regions to go unpaired.) This can happen in a region of short repeats (see Figure 6.3) or in a gene cluster. FIGURE 6.13 shows that unequal crossing-over in a gene cluster can have two consequences—quantitative and qualitative:

- The number of repeats increases in one chromosome and decreases in the other. In effect, one recombinant chromosome has a deletion and the other has an insertion. This happens irrespective of the exact location of the crossover. In the figure, the first recombinant has an increase in the number of gene copies from two to three, whereas the second has a decrease from two to one.
- If the recombination event occurs within a gene (as opposed to between genes), the result depends on whether the recombining genes are identical or only related. If the noncorresponding gene copies one and two are entirely homologous, there is no change in the sequence of either gene. However, unequal crossing-over also can occur when the adjacent genes are well related (although the probability is less than when they are identical). In this case, each of the recombinant genes has a sequence that is different from either parent.

Whether the chromosome has a selective advantage or disadvantage will depend on the consequence of any change in the sequence of the gene product, as well as on the change in the number of gene copies.

An obstacle to unequal crossing-over is presented by the interrupted structure of the genes. In a case such as the globins, the corresponding exons of adjacent gene copies are likely to be well enough related to support pairing; however, the sequences of the introns have diverged appreciably. The restriction of pairing to the exons considerably reduces the continuous length of DNA that can be involved. This lowers the chance of unequal crossing-over. So divergence between introns could enhance the stability of gene clusters by hindering the occurrence of unequal crossing-over.

Thalassemias result from mutations that reduce or prevent synthesis of either α or β globin. The occurrence of unequal crossing-over in the human globin gene clusters is revealed by the nature of certain thalassemias.

Many of the most severe thalassemias result from deletions of part of a cluster. In at least some cases, the ends of the deletion lie in regions that are homologous, which is exactly what would be expected if it had been generated by unequal crossing-over.

FIGURE 6.14 summarizes the deletions that cause the α-thalassemias. α-thal-1 deletions are long, varying in the location of the left end, with the positions of the right ends located beyond the known genes. They eliminate both the α genes. The α-thal-2 deletions are short and eliminate only one of the two α genes. The L deletion removes 4.2 kb of DNA, including the $\alpha 2$ gene. It probably results from unequal crossing-over, because the ends of the deletion lie in homologous regions, just to the right of the $\psi\alpha$ and $\alpha 2$ genes, respectively. The R deletion results from the removal of exactly 3.7 kb of DNA, the precise distance between the α1 and α2 genes. It appears to have been generated by unequal crossing-over between the α1 and α2 genes themselves. This is precisely the situation depicted in Figure 6.13.

Depending on the diploid combination of thalassemic chromosomes, an affected individual may have any number of α chains from zero to three. There are few differences from the wild type (four α genes) in individuals with three or two α genes. If an individual has only one α gene, though, the excess β chains form the unusual tetramer β_4, which causes hemoglobin H (**HbH**) disease. The complete absence of α genes results in **hydrops fetalis,** which is fatal at or before birth.

The same unequal crossing-over that generated the thalassemic chromosome should also have generated a chromosome with three α genes. Individuals with such chromosomes have been identified in several populations. In some populations, the frequency of the triple α locus is about the same as that of the single α locus; in others, the triple α genes are much *less* common than single α genes. This suggests that (unknown) selective factors operate in different populations to adjust the gene levels.

Variations in the number of α genes are found relatively frequently, which argues that unequal crossing-over in the cluster must be fairly common. It occurs more often in the α cluster than in the β cluster, possibly because the introns in α genes are much shorter and therefore present less impediment to mispairing between nonhomologous genes.

The deletions that cause β-thalassemias are summarized in FIGURE 6.15. In some (rare) cases, only the β gene is affected. These have a deletion of 600 bp, extending from the second intron through the 3′ flanking regions. In the other cases, more than one gene of the cluster is affected. Many of the deletions are very long, extending from the 5′ end indicated on the map for >50 kb toward the right.

The **Hb Lepore** type provided the classic evidence that deletion can result from unequal crossing-over between linked genes. The β and δ genes differ only ~7% in sequence. Unequal recombination deletes the material between the genes, thus fusing them together (see Figure 6.13). The fused gene produces a single β-like chain that consists of the N-terminal sequence of δ joined to the C-terminal sequence of β.

Several types of Hb Lepore now are known, the difference between them lying in the point of transition from δ to β sequences. Thus when the δ and β genes pair for unequal crossing-over, the exact point of recombination determines the position at which the switch from δ to β sequence occurs in the amino acid chain.

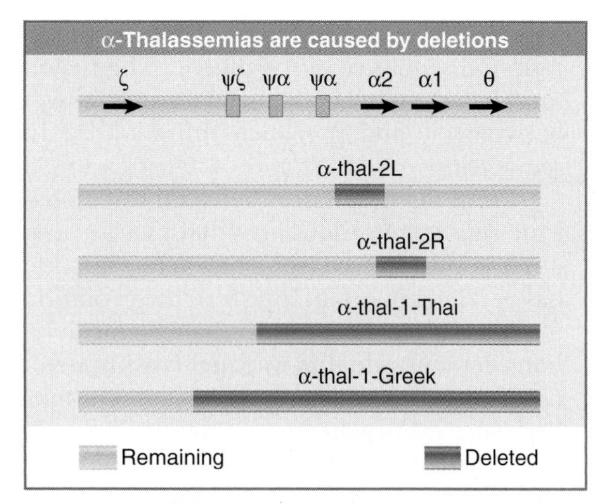

FIGURE 6.14 α thalassemias result from various deletions in the α-globin gene cluster.

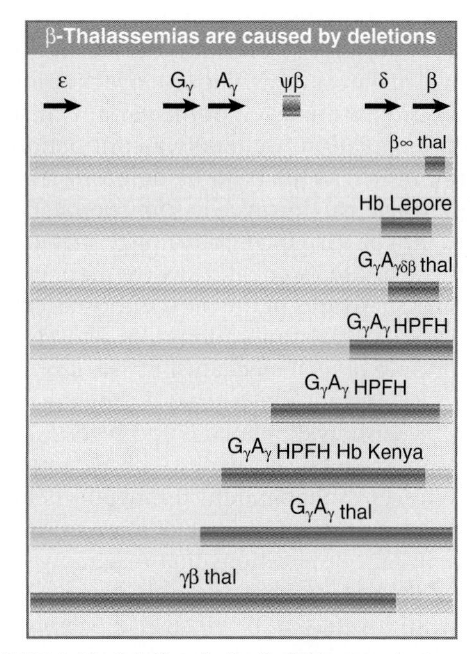

FIGURE 6.15 Deletions in the β-globin gene cluster cause several types of thalassemia.

The reciprocal of this event has been found in the form of **Hb anti-Lepore,** which is produced by a gene that has the N-terminal part of β and the C-terminal part of δ. The fusion gene lies between normal δ and β genes.

Evidence that unequal crossing-over can occur between more distantly related genes is provided by the identification of **Hb Kenya,** another fused hemoglobin. This contains the

N-terminal sequence of the $^{A}\gamma$ gene and the C-terminal sequence of the β gene. The fusion must have resulted from unequal crossing-over between $^{A}\gamma$ and β, which differ ~20% in sequence.

From the differences between the globin gene clusters of various mammals, we see that duplication followed (sometimes) by variation has been an important feature in the evolution of each cluster. The human thalassemic deletions demonstrate that unequal crossing-over continues to occur in both globin gene clusters. Each such event generates a duplication as well as the deletion, and we must account for the fate of both recombinant loci in the population. Deletions can also occur (in principle) by recombination between homologous sequences lying on the *same* chromosome. This does not generate a corresponding duplication.

It is difficult to estimate the natural frequency of these events, because selective forces rapidly adjust the levels of the variant clusters in the population. Generally a contraction in gene number is likely to be deleterious and selected against. However, in some populations, there may be a balancing advantage that maintains the deleted form at a low frequency.

The structures of the present human clusters show several duplications that attest to the importance of such mechanisms. The *functional* sequences include two α genes coding the same protein, fairly well related β and δ genes, and two almost identical γ genes. These comparatively recent independent duplications have survived in the population, not to mention the more distant duplications that originally generated the various types of globin genes. Other duplications may have given rise to pseudogenes or have been lost. We expect continual duplication and deletion to be a feature of all gene clusters.

6.8 Genes for rRNA Form Tandem Repeats

Key concepts

- Ribosomal RNA is coded by a large number of identical genes that are tandemly repeated to form one or more clusters.
- Each rDNA cluster is organized so that transcription units giving a joint precursor to the major rRNAs alternate with nontranscribed spacers.

In the cases we have discussed thus far, there are differences between the individual members of a gene cluster that allow selective pressure to act independently upon each gene. A contrast is provided by two cases of large gene clusters that contain many identical copies of the same gene or genes. Most organisms contain multiple copies of the genes for the histone proteins that are a major component of the chromosomes, and there are almost always multiple copies of the genes that code for the ribosomal RNAs. These situations pose some interesting evolutionary questions.

Ribosomal RNA is the predominant product of transcription, constituting some 80%–90% of the total mass of cellular RNA in both eukaryotes and prokaryotes. The number of major rRNA genes varies from seven in *E. coli*, 100 to 200 in lower eukaryotes, to several hundred in higher eukaryotes. The genes for the large and small rRNA (found in the large and small subunits of the ribosome, respectively) usually form a tandem pair. (The sole exception is the yeast mitochondrion.)

The lack of any detectable variation in the sequences of the rRNA molecules implies that all the copies of each gene must be identical, or at least must have differences below the level of detection in rRNA (~1%). A point of major interest is what mechanism(s) are used to prevent variations from accruing in the individual sequences.

In bacteria, the multiple rRNA gene pairs are dispersed. In most eukaryotic nuclei, the rRNA genes are contained in a tandem cluster or clusters. Sometimes these regions are called **rDNA.** (In some cases, the proportion of rDNA in the total DNA, together with its atypical base composition, is great enough to allow its isolation as a separate fraction directly from sheared genomic DNA.) An important diagnostic feature of a tandem cluster is that it generates a circular restriction map, as shown in FIGURE 6.16.

Suppose that each repeat unit has three restriction sites. When we map these fragments by conventional means, we find that A is next to B, which is next to C, which is next to A, generating the circular map. If the cluster is large, the internal fragments (A, B, and C) will be present in much greater quantities than the terminal fragments (X and Y), which connect the cluster to adjacent DNA. In a cluster of 100 repeats, X and Y would be present at 1% of the level of A, B, and C. This can make it difficult to obtain the ends of a gene cluster for mapping purposes.

The region of the nucleus where rRNA synthesis occurs has a characteristic appearance,

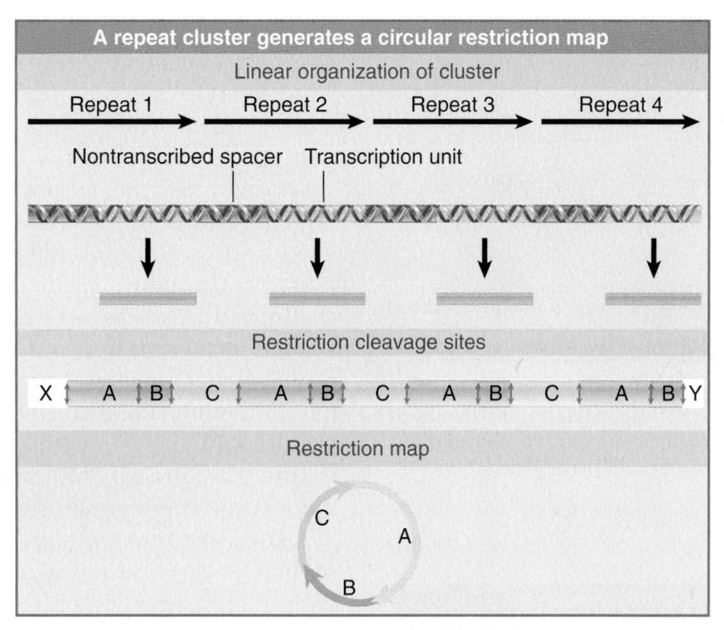

FIGURE 6.16 A tandem gene cluster has an alternation of transcription unit and nontranscribed spacer and generates a circular restriction map.

with a core of fibrillar nature surrounded by a granular cortex. The fibrillar core is where the rRNA is transcribed from the DNA template, and the granular cortex is formed by the ribonucleoprotein particles into which the rRNA is assembled. The whole area is called the **nucleolus.** Its characteristic morphology is evident in FIGURE 6.17.

The particular chromosomal regions associated with a nucleolus are called **nucleolar organizers.** Each nucleolar organizer corresponds to a cluster of tandemly repeated rRNA genes on one chromosome. The concentration of the tandemly repeated rRNA genes, together with their very intensive transcription, is responsible for creating the characteristic morphology of the nucleoli.

The pair of major rRNAs is transcribed as a single precursor in both bacteria and eukaryotic nuclei. Following transcription, the precursor is cleaved to release the individual rRNA molecules. The transcription unit is shortest in bacteria and is longest in mammals (where it is known as 45S RNA, according to its rate of sedimentation). An rDNA cluster contains many transcription units, each separated from the next by a **nontranscribed spacer.** The alternation of transcription unit and nontranscribed spacer can be seen directly in electron micrographs. The example shown in FIGURE 6.18 is taken from the newt *Notopthalmus viridescens*, in which each transcription unit is intensively expressed, so that many RNA polymerases are simultaneously engaged in transcription on one repeating unit. The polymerases are so closely packed that the RNA transcripts form a characteristic matrix displaying increasing length along the transcription unit.

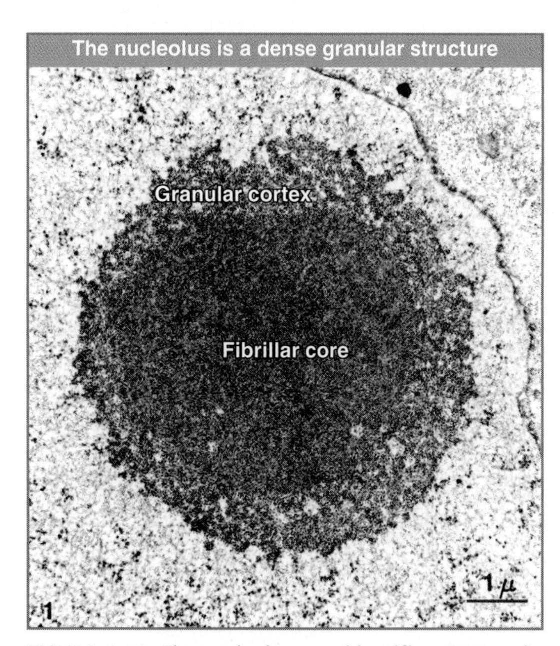

FIGURE 6.17 The nucleolar core identifies rDNA under transcription, and the surrounding granular cortex consists of assembling ribosomal subunits. This thin section shows the nucleolus of the newt *Notopthalmus viridescens*. Photo courtesy of Oscar Miller.

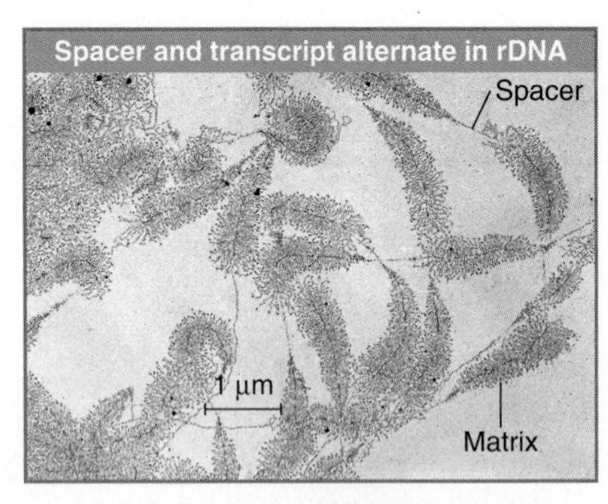

FIGURE 6.18 Transcription of rDNA clusters generates a series of matrices, each corresponding to one transcription unit and separated from the next by the nontranscribed spacer. Photo courtesy of Oscar Miller.

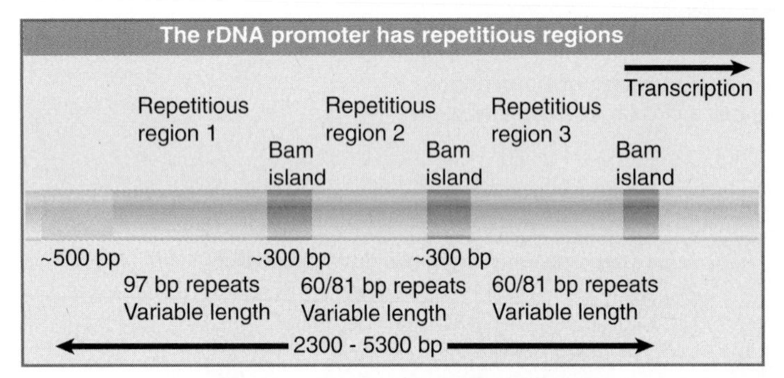

FIGURE 6.19 The nontranscribed spacer of *X. laevis* rDNA has an internally repetitious structure that is responsible for its variation in length. The Bam islands are short constant sequences that separate the repetitious regions.

6.9 The Repeated Genes for rRNA Maintain Constant Sequence

Key concepts

- The genes in an rDNA cluster all have an identical sequence.
- The nontranscribed spacers consist of shorter repeating units whose number varies so that the lengths of individual spacers are different.

The nontranscribed spacer varies widely in length between and (sometimes) within species. In yeast there is a short nontranscribed spacer that is relatively constant in length. In *D. melanogaster* there is almost a twofold variation in the length of the nontranscribed spacer between different copies of the repeating unit. A similar situation is seen in *X. laevis*. In each of these cases, all of the repeating units are present as a single tandem cluster on one particular chromosome. (In the example of *D. melanogaster*, this happens to be the sex chromosome. The cluster on the X chromosome is larger than the one on the Y chromosome, so female flies have more copies of the rRNA genes than male flies.)

In mammals the repeating unit is very much larger, comprising the transcription unit of ~13 kb and a nontranscribed spacer of ~30 kb. Usually, the genes lie in several dispersed clusters—in the case of man and mouse the clusters reside on five and six chromosomes, respectively. One interesting (but unanswered) question is how the corrective mechanisms that presumably function within a single cluster to ensure constancy of rRNA sequence are able to work when there are several clusters.

The variation in length of the nontranscribed spacer in a single gene cluster contrasts with the conservation of sequence of the transcription unit. In spite of this variation, the sequences of longer nontranscribed spacers remain homologous with those of the shorter nontranscribed spacers. This implies that each nontranscribed spacer is *internally repetitious*, so that the variation in length results from changes in the number of repeats of some subunit.

The general nature of the nontranscribed spacer is illustrated by the example of *X. laevis*. FIGURE 6.19 illustrates the situation. Regions that are fixed in length alternate with regions that vary. Each of the three repetitious regions comprises a variable number of repeats of a rather short sequence. One type of repetitious region has repeats of a 97 bp sequence; the other, which occurs in two locations, has a repeating unit found in two forms, 60 bp and 81 bp long. The variation in the number of repeating units in the repetitious regions accounts for the overall variation in spacer length. The repetitious regions are separated by shorter constant sequences called **Bam islands.** (This description takes its name from their isolation via the use of the BamHI restriction enzyme.) From this type of organization, we see that the cluster has evolved by duplications involving the promoter region.

We need to explain the lack of variation in the expressed copies of the repeated genes. One model would suppose that there is a quantitative demand for a certain number of "good" sequences. This would, however, enable mutated sequences to accumulate up to a point at which their proportion of the cluster is great enough for selective pressure to be exerted. We can exclude such models because of the lack of such variation in the cluster.

The lack of variation implies the existence of selective pressure in some form that is sensitive to individual variations. One model would suppose that the entire cluster is regenerated periodically from one or from a very few members. As a practical matter, any mechanism would need to involve regeneration every generation. We can exclude such models because a regenerated cluster would not show variation in the nontranscribed regions of the individual repeats.

We are left with a dilemma. Variation in the nontranscribed regions suggests that there is frequent unequal crossing-over. This will change the size of the cluster, but will not otherwise change the properties of the individual repeats. So how are mutations prevented from accumulating? We'll see in the next section that continuous contraction and expansion of a cluster may provide a mechanism for homogenizing its copies.

6.10 Crossover Fixation Could Maintain Identical Repeats

Key concepts

- Unequal crossing-over changes the size of a cluster of tandem repeats.
- Individual repeating units can be eliminated or can spread through the cluster.

The same problem is encountered whenever a gene has been duplicated. How can selection be imposed to prevent the accumulation of deleterious mutations?

The duplication of a gene is likely to result in an immediate relaxation of the evolutionary pressure on its sequence. Now that there are two identical copies, a change in the sequence of either one will not deprive the organism of a functional protein, because the original amino acid sequence continues to be coded by the other copy. Then the selective pressure on the two genes is diffused, until one of them mutates sufficiently away from its original function to refocus all the selective pressure on the other.

Immediately following a gene duplication, changes might accumulate more rapidly in one of the copies, leading eventually to a new function (or to its disuse in the form of a pseudogene). If a new function develops, the gene then evolves at the same, slower rate characteristic of the original function. Probably this is the sort of mechanism responsible for the separation of functions between embryonic and adult globin genes.

Yet there are instances in which duplicated genes retain the same function, coding for the identical or nearly identical proteins. Identical proteins are coded by the two human α-globin genes, and there is only a single amino acid difference between the two γ-globin proteins. How is selective pressure exerted to maintain their sequence identity?

The most obvious possibility is that the two genes do not actually have identical functions, but instead differ in some (undetected) property, such as time or place of expression. Another possibility is that the need for two copies is quantitative, because neither by itself produces a sufficient amount of protein.

In more extreme cases of repetition, however, it is impossible to avoid the conclusion that no single copy of the gene is essential. When there are many copies of a gene, the immediate effects of mutation in any one copy must be very slight. The consequences of an individual mutation are diluted by the large number of copies of the gene that retain the wild-type sequence. Many mutant copies could accumulate before a lethal effect is generated.

Lethality becomes quantitative, a conclusion reinforced by the observation that half of the units of the rDNA cluster of *X. laevis* or *D. melanogaster* can be deleted without ill effect. So how are these units prevented from gradually accumulating deleterious mutations? What chance is there for the rare favorable mutation to display its advantages in the cluster?

The basic principle of models to explain the maintenance of identity among repeated copies is to suppose that nonallelic genes are not independently inherited, but rather must be continually regenerated from *one* of the copies of a preceding generation. In the simplest case of two identical genes, when a mutation occurs in one copy, either it is by chance eliminated (because the sequence of the other copy takes over), or it is spread to both duplicates (because the mutant copy becomes the dominant version). Spreading exposes a mutation to selection. The result is that the two genes evolve together as though only a single locus existed. This is called **coincidental evolution** or **concerted evolution** (occasionally **coevolution**). It can be applied to a pair of identical genes or (with further assumptions) to a cluster containing many genes.

One mechanism supposes that the sequences of the nonallelic genes are directly compared with one another and homogenized by enzymes that recognize any differences. This can be done by exchanging single strands between them to form genes, one of whose strands derives from one copy, and one from the other copy. Any differences are revealed as improperly paired bases, which attract attention from enzymes able to excise and replace a base, so that only A-T and G-C pairs survive. This type of event is called **gene conversion** and is associated with genetic recombination. We should be able to ascertain the scope of such events by comparing the sequences of duplicate genes. If they are subject to concerted evolution, we should not see the accumulation of silent site substitutions between them (because the homogenization process applies to these as well as to the replacement sites). We know that the extent of the maintenance mechanism need not extend beyond the gene itself, as there are cases of duplicate genes whose flanking sequences are entirely different. Indeed, we may see abrupt boundaries that mark the ends of the sequences that were homogenized.

We must remember that the existence of such mechanisms can invalidate the determination of the history of such genes via their divergence, *because the divergence reflects only the time since the last homogenization/regeneration event, not the original duplication.*

The **crossover fixation** model supposes that an entire cluster is subject to continual rearrangement by the mechanism of unequal crossing-over. Such events can explain the concerted evolution of multiple genes if unequal crossing-over causes all the copies to be regenerated physically from one copy.

Following the sort of event depicted in Figure 6.13, for example, the chromosome carrying a triple locus could suffer deletion of one of the genes. Of the two remaining genes, 1½ represent the sequence of one of the original copies; only ½ of the sequence of the other original copy has survived. Any mutation in the first region now exists in both genes and is subject to selective pressure.

Tandem clustering provides frequent opportunities for "mispairing" of genes whose sequences are the same, but that lie in different positions in their clusters. By continually expanding and contracting the number of units via unequal crossing-over, it is possible for all the units in one cluster to be derived from rather a small proportion of those in an ancestral cluster. The variable lengths of the spacers are consistent with the idea that unequal crossing-over events take place in spacers that are internally mispaired. This can explain the homogeneity of the genes compared with the variability of the spacers. The genes are exposed to selection when individual repeating units are amplified within the cluster; however, the spacers are irrelevant and can accumulate changes.

In a region of nonrepetitive DNA, recombination occurs between precisely matching points on the two homologous chromosomes, thus generating reciprocal recombinants. The basis for this precision is the ability of two duplex DNA sequences to align exactly. We know that unequal recombination can occur when there are multiple copies of genes whose exons are related, even though their flanking and intervening sequences may differ. This happens because of the mispairing between corresponding exons in *nonallelic* genes.

Imagine how much more frequently misalignment must occur in a tandem cluster of identical or nearly identical repeats. Except at the very ends of the cluster, the close relationship between successive repeats makes it impossible even to define the exactly corresponding repeats! This has two consequences: there is continual adjustment of the size of the cluster; and there is homogenization of the repeating unit.

Consider a sequence consisting of a repeating unit "ab" with ends "x" and "y." If we represent one chromosome in black and the other in color, the exact alignment between "allelic" sequences would be:

xabababababababababababababababy
xabababababababababababababababy

It is likely, however, that *any* sequence *ab* in one chromosome could pair with *any* sequence *ab* in the other chromosome. In a misalignment such as:

xabababababababababababababababy
      xabababababababababababababababy

the region of pairing is no less stable than in the perfectly aligned pair, although it is shorter. We do not know very much about how pairing is initiated prior to recombination, but very likely it starts between short corresponding regions and then spreads. If it starts within satellite DNA, it is more likely than not to involve repeating units that do not have exactly corresponding locations in their clusters.

Now suppose that a recombination event occurs within the unevenly paired region. The recombinants will have different numbers of repeating units. In one case, the cluster has become longer; in the other, it has become shorter,

xabababababababababababababababy
x
xabababababababababababababababy
↓
xababababababababababababababababy
+
xabababababababababababababy

where "x" indicates the site of the crossover.

If this type of event is common, clusters of tandem repeats will undergo continual expansion and contraction. This can cause a particular repeating unit to spread through the cluster, as illustrated in FIGURE 6.20. Suppose that the cluster consists initially of a sequence *abcde,* where each letter represents a repeating unit. The different repeating units are closely enough related to one another to mispair for recombination. Then by a series of unequal recombination events, the size of the repetitive region increases or decreases, and one unit spreads to replace all the others.

The crossover fixation model predicts that *any sequence of DNA that is not under selective pressure will be taken over by a series of identical tandem repeats generated in this way.* The critical assumption is that the process of crossover fixation is fairly rapid relative to mutation, so that new mutations either are eliminated (their repeats are lost) or come to take over the entire cluster. In the case of the rDNA cluster, of course, a further factor is imposed by selection for an effective transcribed sequence.

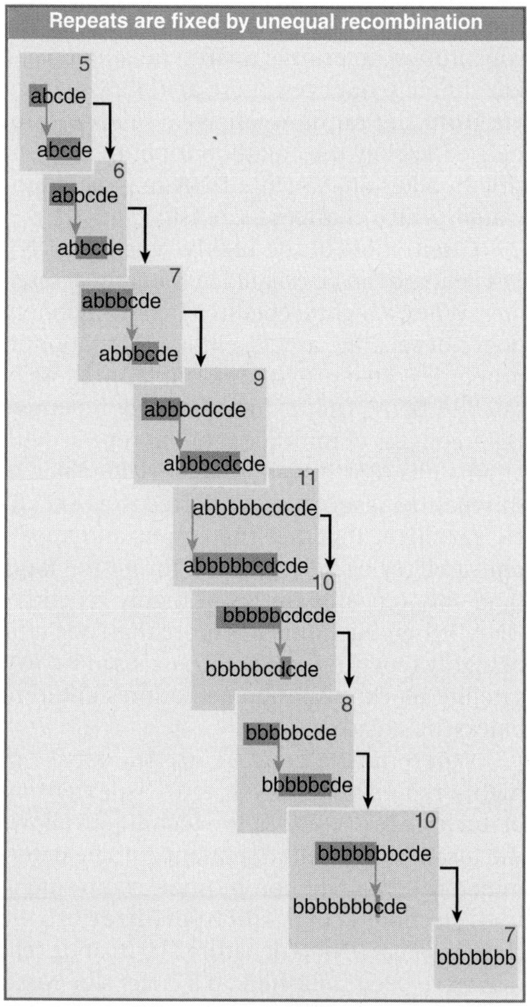

FIGURE 6.20 Unequal recombination allows one particular repeating unit to occupy the entire cluster. The numbers indicate the length of the repeating unit at each stage.

6.11 Satellite DNAs Often Lie in Heterochromatin

Key concepts

- Highly repetitive DNA has a very short repeating sequence and no coding function.
- It occurs in large blocks that can have distinct physical properties.
- It is often the major constituent of centromeric heterochromatin.

Repetitive DNA is defined by its (relatively) rapid rate of renaturation. The component that renatures most rapidly in a eukaryotic genome is called *highly repetitive* DNA and consists of very short sequences repeated many times in tandem in large clusters. As a result of its short repeating unit, it is sometimes described as **simple sequence DNA.** This type of component is present in almost all higher eukaryotic genomes, but its overall amount is extremely variable. In mammalian genomes it is typically <10%, but in (for example) *Drosophila virilis,* it amounts to ~50%. In addition to the large clusters in which this type of sequence was originally discovered, there are smaller clusters interspersed with nonrepetitive DNA. It typically consists of short sequences that are repeated in identical or related copies in the genome.

The tandem repetition of a short sequence often creates a fraction with distinctive physical properties that can be used to isolate it. In some cases, the repetitive sequence has a base composition distinct from the genome average, which allows it to form a separate fraction by virtue of its distinct buoyant density. A fraction of this sort

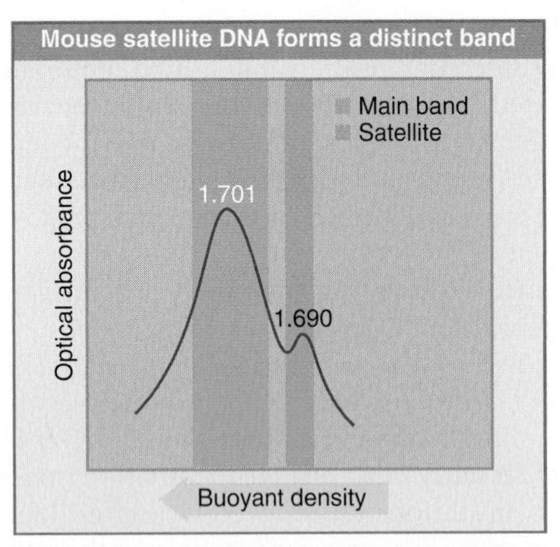

FIGURE 6.21 Mouse DNA is separated into a main band and a satellite by centrifugation through a density gradient of CsCl.

is called satellite DNA. The term satellite DNA is essentially synonymous with simple sequence DNA. Consistent with its simple sequence, this DNA is not transcribed or translated.

Tandemly repeated sequences are especially liable to undergo misalignments during chromosome pairing, and thus the sizes of tandem clusters tend to be highly polymorphic, with wide variations between individuals. In fact, the smaller clusters of such sequences can be used to characterize individual genomes in the technique of "DNA fingerprinting" (see Section 6.14, Minisatellites Are Useful for Genetic Mapping).

The buoyant density of a duplex DNA depends on its G-C content according to the empirical formula

$$\rho = 1.660 + 0.00098\ (\%\text{G-C})\ \text{g-cm}^{-3}$$

Buoyant density usually is determined by centrifuging DNA through a **density gradient** of CsCl. The DNA forms a band at the position corresponding to its own density. Fractions of DNA differing in G-C content by >5% can usually be separated on a density gradient.

When eukaryotic DNA is centrifuged on a density gradient, two types of material may be distinguished:

- Most of the genome forms a continuum of fragments that appear as a rather broad peak centered on the buoyant density corresponding to the average G-C content of the genome. This is called the main band.
- Sometimes an additional, smaller peak (or peaks) is seen at a different value. This material is the satellite DNA.

Satellites are present in many eukaryotic genomes. They may be either heavier or lighter than the main band, but it is uncommon for them to represent >5% of the total DNA. A clear example is provided by mouse DNA, shown in FIGURE 6.21. The graph is a quantitative scan of the bands formed when mouse DNA is centrifuged through a CsCl density gradient. The main band contains 92% of the genome and is centered on a buoyant density of 1.701 g-cm^{-3} (corresponding to its average G-C of 42%, typical for a mammal). The smaller peak represents 8% of the genome and has a distinct buoyant density of 1.690 g-cm^{-3}. It contains the mouse satellite DNA, whose G-C content (30%) is much lower than any other part of the genome.

The behavior of satellite DNA on density gradients is often anomalous. When the actual base composition of a satellite is determined, it is different from the prediction based on its buoyant density. The reason is that ρ is a function not just of base composition, but of the constitution in terms of nearest neighbor pairs. For simple sequences, these are likely to deviate from the random pairwise relationships needed to obey the equation for buoyant density. In addition, satellite DNA may be methylated, which changes its density.

Often, most of the highly repetitive DNA of a genome can be isolated in the form of satellites. When a highly repetitive DNA component does not separate as a satellite, on isolation its properties often prove to be similar to those of satellite DNA. That is to say, highly repetitive DNA consists of multiple tandem repeats with anomalous centrifugation. Material isolated in this manner is sometimes referred to as a **cryptic satellite.** Together the cryptic and apparent satellites usually account for all the large tandemly repeated blocks of highly repetitive DNA. When a genome has more than one type of highly repetitive DNA, each exists in its own satellite block (although sometimes different blocks are adjacent).

Where in the genome are the blocks of highly repetitive DNA located? An extension of nucleic acid hybridization techniques allows the location of satellite sequences to be determined directly in the chromosome complement. In the technique of ***in situ* hybridization,** the chromosomal DNA is denatured by treating cells that have been squashed on a cover slip. Next, a solution containing a radioactively labeled DNA or RNA probe is added. The probe hybridizes with its complements in the denatured genome. The location of the sites of

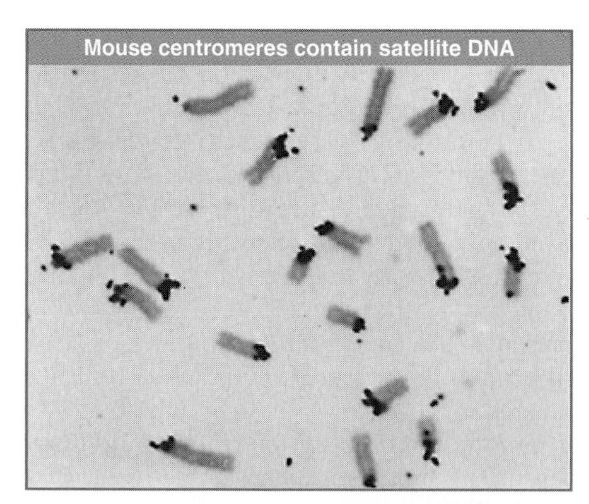

FIGURE 6.22 Cytological hybridization shows that mouse satellite DNA is located at the centromeres. Photo courtesy of Mary Lou Pardue and Joseph G. Gall, Carnegie Institution.

D. virilis has four related satellites			
Satellite	Predominant Sequence	Total Length	Genome Proportion
I	ACAAACT TGTTTGA	1.1×10^7	25%
II	ATAAACT TATTTGA	3.6×10^6	8%
III	ACAAATT TGTTTAA	3.6×10^6	8%
Cryptic	AATATAG TTATATC		

FIGURE 6.23 Satellite DNAs of *D. virilis* are related. More than 95% of each satellite consists of a tandem repetition of the predominant sequence.

hybridization can be determined by autoradiography (see Figure 28.19).

Satellite DNAs are found in regions of **heterochromatin.** Heterochromatin is the term used to describe regions of chromosomes that are permanently tightly coiled up and inert, in contrast with the **euchromatin** that represents most of the genome (see Section 28.7, Chromatin Is Divided into Euchromatin and Heterochromatin). Heterochromatin is commonly found at centromeres (the regions where the kinetochores are formed at mitosis and meiosis for controlling chromosome movement). The centromeric location of satellite DNA suggests that it has some structural function in the chromosome. This function could be connected with the process of chromosome segregation.

An example of the localization of satellite DNA for the mouse chromosomal complement is shown in FIGURE 6.22. In this case, one end of each chromosome is labeled, because this is where the centromeres are located in *Mus musculus* chromosomes.

6.12 Arthropod Satellites Have Very Short Identical Repeats

Key concept

- The repeating units of arthropod satellite DNAs are only a few nucleotides long. Most of the copies of the sequence are identical.

In the arthropods, as typified by insects and crabs, each satellite DNA appears to be rather homogeneous. Usually, a single, very short repeating unit accounts for >90% of the satellite. This makes it relatively straightforward to determine the sequence.

Drosophila virilis has three major satellites and a cryptic satellite; together they represent >40% of the genome. The sequences of the satellites are summarized in FIGURE 6.23. The three major satellites have closely related sequences. A single base substitution is sufficient to generate either satellite II or III from the sequence of satellite I.

The satellite I sequence is present in other species of *Drosophila* related to *virilis* and so may have preceded speciation. The sequences of satellites II and III seem to be specific to *D. virilis,* and so may have evolved from satellite I after speciation.

The main feature of these satellites is their very short repeating unit: only 7 bp. Similar satellites are found in other species. *D. melanogaster* has a variety of satellites, several of which have very short repeating units (5, 7, 10, or 12 bp). Comparable satellites are found in the crabs.

The close sequence relationship found among the *D. virilis* satellites is not necessarily a feature of other genomes, for which the satellites may have unrelated sequences. *Each satellite has arisen by a lateral amplification of a very short sequence.* This sequence may represent a variant of a previously existing satellite (as in *D. virilis*), or could have some other origin.

Satellites are continually generated and lost from genomes. This makes it difficult to ascertain evolutionary relationships, because a current satellite could have evolved from some previous satellite that has since been lost. The important feature of these satellites is that *they represent very long stretches of DNA of very low sequence complexity, within which constancy of sequence can be maintained*

One feature of many of these satellites is a pronounced asymmetry in the orientation of base pairs on the two strands. In the example of the *D. virilis* satellites shown in Figure 6.22, in each of the major satellites one of the strands is much richer in T and G bases. This increases its buoyant density, so that upon denaturation this **heavy strand** (H) can be separated from the complementary light strand (L). This can be useful in sequencing the satellite.

6.13 Mammalian Satellites Consist of Hierarchical Repeats

Key concept

- Mouse satellite DNA has evolved by duplication and mutation of a short repeating unit to give a basic repeating unit of 234 bp in which the original half, quarter, and eighth repeats can be recognized.

In the mammals, as typified by various rodents, the sequences comprising each satellite show appreciable divergence between tandem repeats. Common short sequences can be recognized by their preponderance among the oligonucleotide fragments released by chemical or enzymatic treatment. However, the predominant short sequence usually accounts for only a small minority of the copies. The other short sequences are related to the predominant sequence by a variety of substitutions, deletions, and insertions.

A series of these variants of the short unit can constitute a longer repeating unit, though, that is itself repeated in tandem with some variation. Thus mammalian satellite DNAs are constructed from a hierarchy of repeating units. These longer repeating units constitute the sequences that renature in reassociation analysis. They also can be recognized by digestion with restriction enzymes.

When any satellite DNA is digested with an enzyme that has a recognition site in its repeating unit, one fragment will be obtained for every repeating unit in which the site occurs. In fact, when the DNA of a eukaryotic genome is digested with a restriction enzyme, most of it gives a general smear due to the random distribution of cleavage sites. Satellite DNA generates sharp bands, though, because a large number of fragments of identical or almost identical size are created by cleavage at restriction sites that lie a regular distance apart.

Determining the sequence of satellite DNA can be difficult. Using the discrete bands generated by restriction cleavage, we can attempt to obtain a sequence directly. However, if there is appreciable divergence between individual repeating units, different nucleotides will be present at the same position in different repeats, so the sequencing gels will be obscure. If the divergence is not too great—say, within ~2%—it may be possible to determine an average repeating sequence.

Individual segments of the satellite can be inserted into plasmids for cloning. A difficulty is that the satellite sequences tend to be excised from the chimeric plasmid by recombination in the bacterial host. However, when the cloning succeeds it is possible to determine the sequence of the cloned segment unambiguously. Although this gives the actual sequence of a repeating unit or units, we should need to have many individual such sequences to reconstruct the type of divergence typical of the satellite as a whole.

Using either sequencing approach, the information we can gain is limited to the distance that can be analyzed on one set of sequence gels. The repetition of divergent tandem copies makes it impossible to reconstruct longer sequences by obtaining overlaps between individual restriction fragments.

The satellite DNA of the mouse *M. musculus* is cleaved by the enzyme EcoRII into a series of bands, including a predominant monomeric fragment of 234 bp. This sequence must be repeated with few variations throughout the 60%–70% of the satellite that is cleaved into the monomeric band. We may analyze this sequence in terms of its successively smaller constituent repeating units.

FIGURE 6.24 depicts the sequence in terms of two half-repeats. By writing the 234 bp sequence so that the first 117 bp are aligned with the second 117 bp, we see that the two halves are quite well related. They differ at 22 positions, corresponding to 19% divergence. This means that the current 234 bp repeating unit must have been generated at some time in the past by duplicating a 117 bp repeating unit, after which differences accumulated between the duplicates.

Within the 117 bp unit we can recognize two further subunits. Each of these is a quarter-repeat relative to the whole satellite. The four

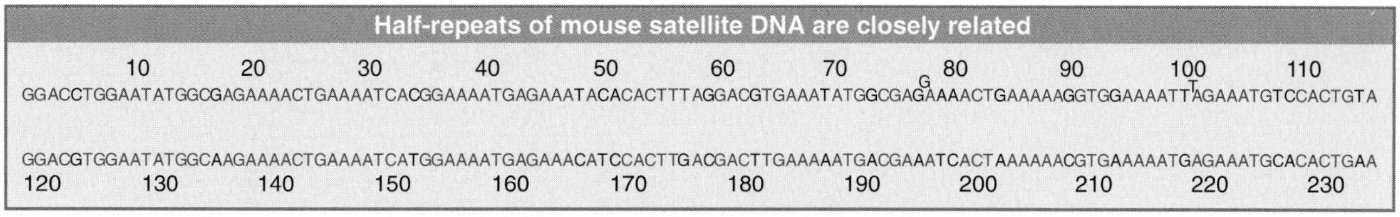

Half-repeats of mouse satellite DNA are closely related

10 20 30 40 50 60 70 80 90 100 110

G T

GGACCTGGAATATGGCGAGAAAACTGAAAATCACGGAAAATGAGAAATACACACTTTAGGACGTGAAATATGGCGAGAAAACTGAAAAAGGTGGAAAATTAGAAATGTCCACTGTA

GGACGTGGAATATGGCAAGAAAACTGAAAATCATGGAAAATGAGAAACATCCACTTGACGACTTGAAAAATGACGAAATCACTAAAAAACGTGAAAAATGAGAAATGCACACTGAA

120 130 140 150 160 170 180 190 200 210 220 230

FIGURE 6.24 The repeating unit of mouse satellite DNA contains two half-repeats, which are aligned to show the identities (in blue).

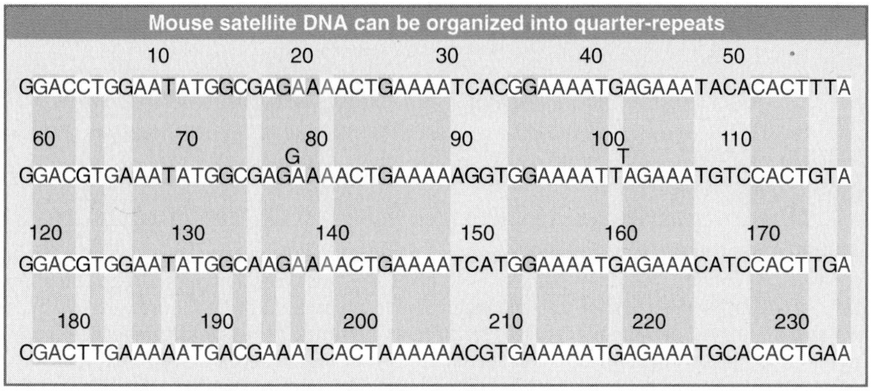

Mouse satellite DNA can be organized into quarter-repeats

10 20 30 40 50

GGACCTGGAATATGGCGAGAAAACTGAAAATCACGGAAAATGAGAAATACACACTTTA

60 70 80 90 100 110

G T

GGACGTGAAATATGGCGAGAAAACTGAAAAAGGTGGAAAATTAGAAATGTCCACTGTA

120 130 140 150 160 170

GGACGTGGAATATGGCAAGAAAACTGAAAATCATGGAAAATGAGAAACATCCACTTGA

180 190 200 210 220 230

CGACTTGAAAAATGACGAAATCACTAAAAAACGTGAAAAATGAGAAATGCACACTGAA

FIGURE 6.25 The alignment of quarter-repeats identifies homologies between the first and second half of each half-repeat. Positions that are the same in all four quarter-repeats are shown in gray; identities that extend only through three quarter-repeats are indicated by black letters in the green area.

quarter-repeats are aligned in FIGURE 6.25. The upper two lines represent the first half-repeat of Figure 6.24; the lower two lines represent the second half-repeat. We see that the divergence between the four quarter-repeats has increased to 23 out of 58 positions, or 40%. The first three quarter-repeats are somewhat better related, and a large proportion of the divergence is due to changes in the fourth quarter-repeat.

Looking within the quarter-repeats, we find that each consists of two related subunits (one-eighth-repeats), shown as the α and β sequences in FIGURE 6.26. The α sequences all have an insertion of a C, and the β sequences all have an insertion of a trinucleotide, relative to a common consensus sequence. This suggests that the quarter-repeat originated by the duplication of a sequence like the consensus sequence, after which changes occurred to generate the components we now see as α and β. Further changes then took place between tandemly repeated αβ sequences to generate the individual quarter- and half-repeats that exist today. Among the one-eighth-repeats, the present divergence is 19/31 = 61%.

The consensus sequence is analyzed directly in FIGURE 6.27, which demonstrates that the current satellite sequence can be treated as derivatives of a 9 bp sequence. We can recognize three variants of this sequence in the satellite, as indicated at the bottom of the figure. If in one of the repeats we take the next most frequent base at two positions instead of the most frequent, we obtain three well-related 9 bp sequences:

G A A A A A C G T
G A A A A A T G A
G A A A A A A C T

The origin of the satellite could well lie in an amplification of one of these three nonamers. The overall consensus sequence of the present satellite is GAAAAA$^{AG}_{TC}$T, which is effectively an amalgam of the three 9 bp repeats.

The average sequence of the monomeric fragment of the mouse satellite DNA explains its properties. The longest repeating unit of 234 bp is identified by the restriction cleavage. The unit of reassociation between single strands of denatured satellite DNA is probably the 117 bp half-repeat, because the 234 bp fragments can anneal both in register and in half-register (in the latter case, the first half-repeat of one strand renatures with the second half-repeat of the other).

So far, we have treated the present satellite as though it consisted of identical copies of

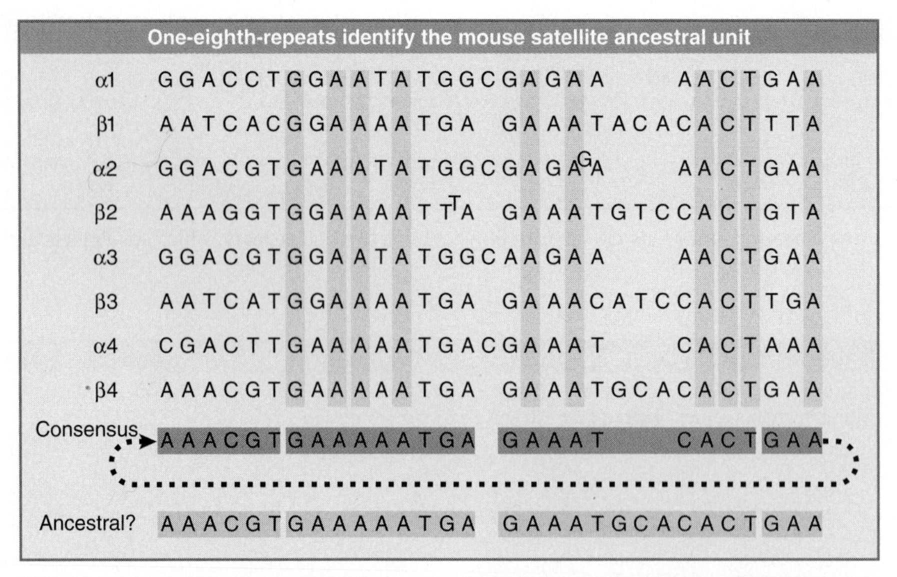

FIGURE 6.26 The alignment of eighth-repeats shows that each quarter-repeat consists of an α half and a β half. The consensus sequence gives the most common base at each position. The "ancestral" sequence shows a sequence very closely related to the consensus sequence, which could have been the predecessor to the α and β units. (The satellite sequence is continuous, so that for the purposes of deducing the consensus sequence we can treat it as a circular permutation, as indicated by joining the last GAA triplet to the first 6 bp.)

the 234 bp repeating unit. Although this unit accounts for the majority of the satellite, variants of it also are present. Some of them are scattered at random throughout the satellite; others are clustered.

The existence of variants is implied by our description of the starting material for the sequence analysis as the "monomeric" fragment. When the satellite is digested by an enzyme that has one cleavage site in the 234 bp sequence, it also generates dimers, trimers, and tetramers relative to the 234 bp length. They arise when a repeating unit has lost the enzyme cleavage site as the result of mutation.

The monomeric 234 bp unit is generated when two adjacent repeats each have the recognition site. A dimer occurs when one unit has lost the site, a trimer is generated when two adjacent units have lost the site, and so on. With some restriction enzymes, most of the satellite is cleaved into a member of this repeating series, as shown in the example of FIGURE 6.28. The declining number of dimers, trimers, and so forth shows that there is a random distribution of the repeats in which the enzyme's recognition site has been eliminated by mutation.

Other restriction enzymes show a different type of behavior with the satellite DNA. They continue to generate the same series of bands. They cleave, however, only a small proportion of the DNA, say 5%–10%. This implies that a certain region of the satellite contains a concentration of the repeating units with this particular restriction site. Presumably the series of repeats in this domain all are derived from an ancestral variant that possessed this recognition site (although in the usual way, some members since have lost it by mutation).

A satellite DNA suffers unequal recombination. This has additional consequences when there is internal repetition in the repeating unit. Let us return to our cluster consisting of "ab" repeats. Suppose that the "a" and "b" components of the repeating unit are themselves sufficiently well related to pair. Then the two clusters can align in *half-register,* with the "a" sequence of one aligned with the "b" sequence of the other. How frequently this occurs will depend on the closeness of the relationship between the two halves of the repeating unit. In mouse satellite DNA, reassociation between the denatured satellite DNA strands *in vitro* commonly occurs in the half-register.

When a recombination event occurs out of register, it changes the length of the repeating units that are involved in the reaction:

xababababababababababababababababy

×

xababababababababababababababababy

↓

xabababababababababaabababababababy

\+

xababababababababbabababababababy

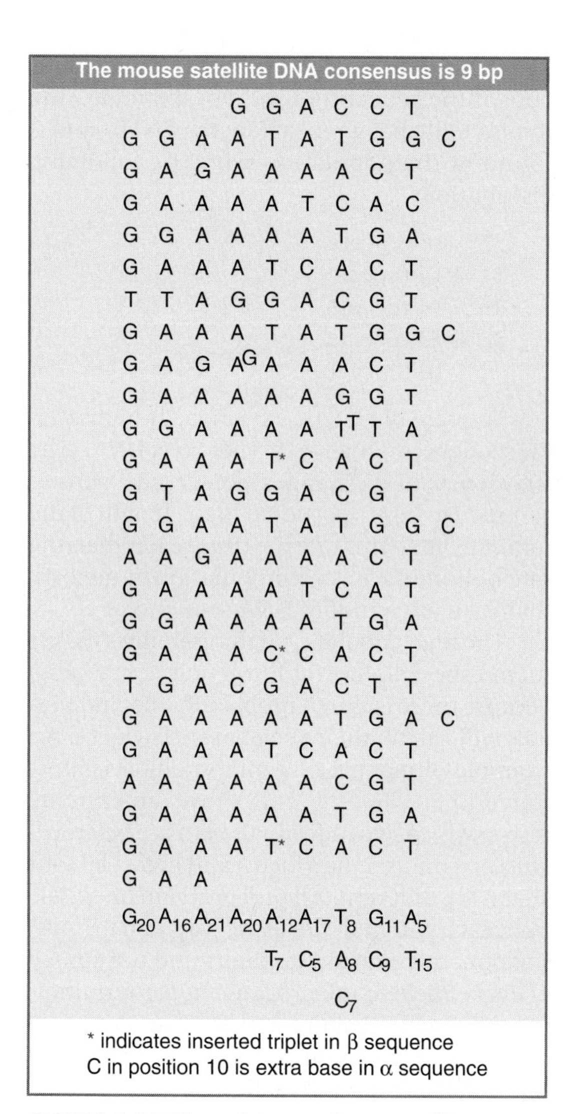

FIGURE 6.27 The existence of an overall consensus sequence is shown by writing the satellite sequence in terms of a 9 bp repeat.

In the upper recombinant cluster, an "ab" unit has been replaced by an "aab" unit. In the lower cluster, the "ab" unit has been replaced by a "b" unit.

This type of event explains a feature of the restriction digest of mouse satellite DNA. Figure 6.28 shows a fainter series of bands at lengths of ½, 1½, 2½, and 3½ repeating units, in addition to the stronger integral length repeats. Suppose that in the preceding example, "ab" represents the 234 bp repeat of mouse satellite DNA, generated by cleavage at a site in the "b" segment. The "a" and "b" segments correspond to the 117 bp half-repeats.

Then, in the upper recombinant cluster, the "aab" unit generates a fragment of 1½ times the usual repeating length. In the lower recombinant cluster, the "b" unit generates a fragment of half of the usual length. (The multiple fragments in the half-repeat series are generated in the same way as longer fragments in the integral series, when some repeating units have lost the restriction site by mutation.)

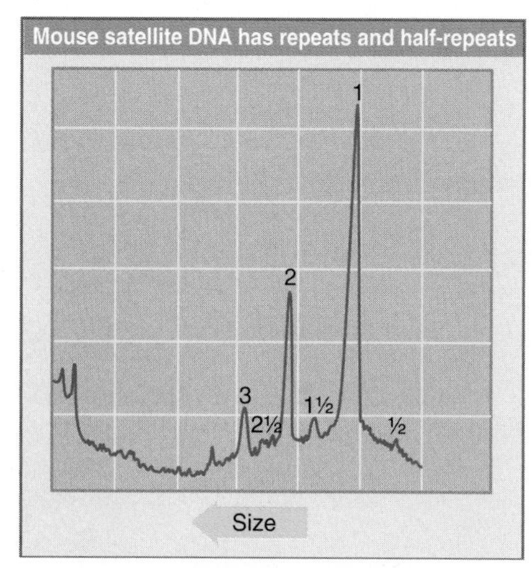

FIGURE 6.28 Digestion of mouse satellite DNA with the restriction enzyme EcoRII identifies a series of repeating units (1, 2, 3) that are multimers of 234 bp and also a minor series (½, 1½, 2½) that includes half-repeats (see text this page). The band at the far left is a fraction resistant to digestion.

Turning the argument the other way around, the identification of the half-repeat series on the gel shows that the 234 bp repeating unit consists of two half-repeats well enough related to pair sometimes for recombination. Also visible in Figure 6.28 are some rather faint bands corresponding to ¼- and ¾-spacings. These will be generated in the same way as the ½-spacings, when recombination occurs between clusters aligned in a quarter-register. The decreased relationship between quarter-repeats compared with half-repeats explains the reduction in frequency of the ¼- and ¾-bands compared with the ½-bands.

6.14 Minisatellites Are Useful for Genetic Mapping

Key concept

- The variation between microsatellites or minisatellites in individual genomes can be used to identify heredity unequivocally by showing that 50% of the bands in an individual are derived from a particular parent.

Sequences that resemble satellites in consisting of tandem repeats of a short unit, but that

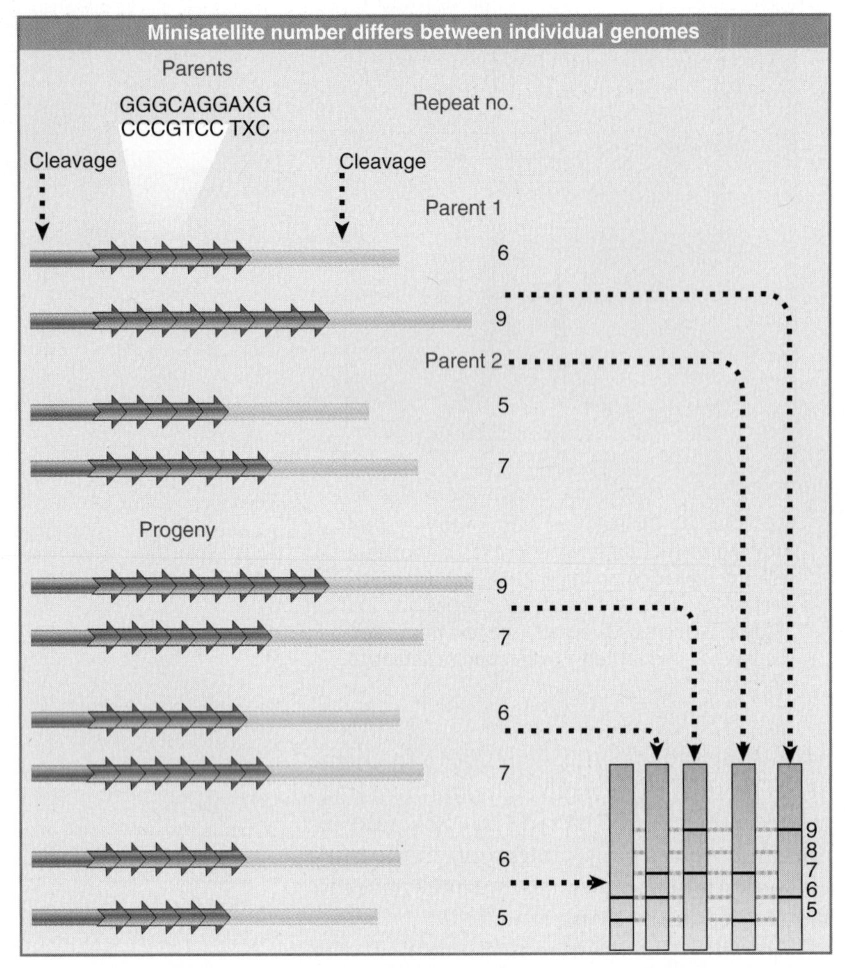

FIGURE 6.29 Alleles may differ in the number of repeats at a minisatellite locus, so that cleavage on either side generates restriction fragments that differ in length. By using a minisatellite with alleles that differ between parents, the pattern of inheritance can be followed.

overall are much shorter—consisting of (for example) 5 to 50 repeats—are common in mammalian genomes. They were discovered by chance as fragments whose size is extremely variable in genomic libraries of human DNA. The variability is seen when a population contains fragments of many different sizes that represent the same genomic region; when individuals are examined, it turns out that there is extensive polymorphism, and that many different alleles can be found.

The name **microsatellite** is usually used when the length of the repeating unit is <10 bp, and the name minisatellite is used when the length of the repeating unit is ~10 to 100 bp. The terminology is not, however, precisely defined. These types of sequences are also called variable number tandem repeat **(VNTR)** regions.

The cause of the variation between individual genomes at microsatellites or minisatellites is that individual alleles have different numbers of the repeating unit. For example, one minisatellite has a repeat length of 64 bp and is found in the population with the following distribution:

7% 18 repeats
11% 16 repeats
43% 14 repeats
36% 13 repeats
4% 10 repeats

The rate of genetic exchange at minisatellite sequences is high, ~10^{-4} per kb of DNA. (The frequency of exchanges per actual locus is assumed to be proportional to the length of the minisatellite.) This rate is ~10× greater than the rate of homologous recombination at meiosis, that is, in any random DNA sequence.

The high variability of minisatellites makes them especially useful for genomic mapping, because there is a high probability that individuals will vary in their alleles at such a locus. An example of mapping by minisatellites is illustrated in FIGURE 6.29. This shows an extreme case in which two individuals both are heterozygous at a minisatellite locus, and in fact all four alleles are different. All progeny gain one allele from each parent in the usual way, and it is possible unambiguously to determine the source of every allele in the progeny. In the terminology of human genetics, the meioses described in this figure are highly informative because of the variation between alleles.

One family of minisatellites in the human genome share a common "core" sequence. The core is a G-C-rich sequence of 10 to 15 bp, showing an asymmetry of purine/pyrimidine distribution on the two strands. Each individual minisatellite has a variant of the core sequence, but ~1000 minisatellites can be detected on Southern blot by a probe consisting of the core sequence.

Consider the situation shown in Figure 6.29, but multiplied many times by the existence of many such sequences. The effect of the variation at individual loci is to create a unique pattern for every individual. This makes it possible to assign heredity unambiguously between parents and progeny by showing that 50% of the bands in any individual are derived from a particular parent. This is the basis of the technique known as DNA fingerprinting.

Both microsatellites and minisatellites are unstable, although for different reasons. Microsatellites undergo intrastrand mispairing, when slippage during replication leads to expansion of the repeat, as shown in FIGURE 6.30. Systems that repair damage to DNA—in particular those that recognize mismatched base pairs—are important in reversing such changes, as shown by a large increase in frequency when repair genes are inactivated. Mutations in repair systems are an important contributory factor in the development of cancer, thus tumor cells often display variations in microsatellite sequences. Minisatellites undergo the same sort of unequal crossing-over between repeats that we have discussed for satellites (see Figure 6.3). One telling case is that increased variation is associated with a meiotic hotspot. The recombination event is not usually associated with recombination between flanking markers, but has a complex form in which the new mutant allele gains information from both the sister chromatid and the other (homologous) chromosome.

It is not clear at what repeating length the cause of the variation shifts from replication slippage to recombination.

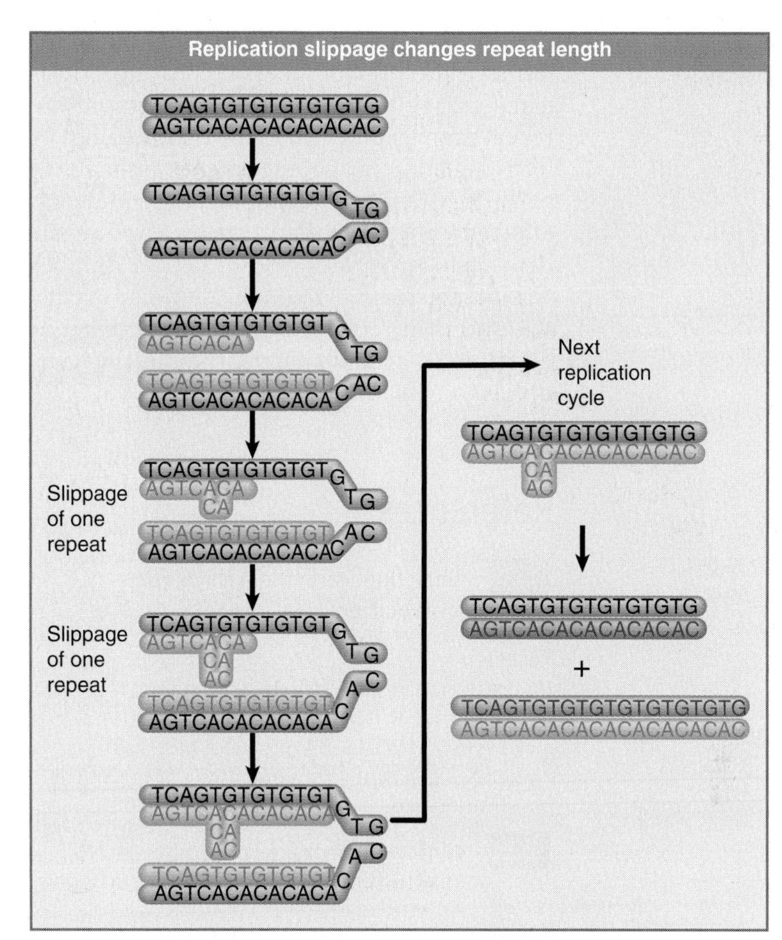

FIGURE 6.30 Replication slippage occurs when the daughter strand slips back one repeating unit in pairing with the template strand. Each slippage event adds one repeating unit to the daughter strand. The extra repeats are extruded as a single-strand loop. Replication of this daughter strand in the next cycle generates a duplex DNA with an increased number of repeats.

6.15 Summary

Almost all genes belong to families, which are defined by the possession of related sequences in the exons of individual members. Families evolve by the duplication of a gene (or genes), followed by divergence between the copies. Some copies suffer inactivating mutations and become pseudogenes that no longer have any function. Pseudogenes also may be generated as DNA copies of the mRNA sequences.

An evolving set of genes may remain together in a cluster or may be dispersed to new locations by chromosomal rearrangement. The organization of existing clusters can sometimes be used to infer the series of events that has occurred. These events act with regard to sequence rather than function, and therefore include pseudogenes as well as active genes.

Mutations accumulate more rapidly in silent sites than in replacement sites (which affect the amino acid sequence). The rate of divergence at replacement sites can be used to establish a clock, which can be calibrated in percent divergence per million years. The clock can then be used to calculate the time of divergence between any two members of the family.

A tandem cluster consists of many copies of a repeating unit that includes the transcribed sequence(s) and a nontranscribed spacer(s). rRNA gene clusters code only for a single rRNA precursor. Maintenance of active genes in clusters depends on mechanisms such as gene conversion or unequal crossing-over, which cause mutations to spread through the cluster so that they become exposed to evolutionary pressure.

Satellite DNA consists of very short sequences repeated many times in tandem. Its distinct centrifugation properties reflect its biased base composition. Satellite DNA is concentrated in centromeric heterochromatin, but its function (if any) is unknown. The individual repeating units of arthropod satellites are identical. Those of mammalian satellites are related and can be organized into a hierarchy reflecting the evolution of the satellite by the amplification and divergence of randomly chosen sequences.

Unequal crossing-over appears to have been a major determinant of satellite DNA organization. Crossover fixation explains the ability of variants to spread through a cluster.

Minisatellites and microsatellites consist of even shorter repeating sequences than satellites:

<10 bp for microsatellites and 10 to 50 bp for minisatellites. The number of repeating units is usually 5 to 50. There is high variation in the repeat number between individual genomes. A microsatellite repeat number varies as the result of slippage during replication; the frequency is affected by systems that recognize and repair damage in DNA. Minisatellite repeat number varies as the result of recombination-like events. Variations in repeat number can be used to determine hereditary relationships by the technique known as DNA fingerprinting.

References

6.2 Gene Duplication Is a Major Force in Evolution

Research

Bailey, J. A., Gu, Z., Clark, R. A., Reinert, K., Samonte, R. V., Schwartz, S., Adams, M. D., Myers, E. W., Li, P. W., and Eichler, E. E. (2002). Recent segmental duplications in the human genome. *Science* 297, 1003–1007.

6.3 Globin Clusters Are Formed by Duplication and Divergence

Review

Hardison, R. (1998). Hemoglobins from bacteria to man: evolution of different patterns of gene expression. *J. Exp. Biol.* 201, 1099–1117.

6.5 The Rate of Neutral Substitution Can Be Measured from Divergence of Repeated Sequences

Research

Waterston, R. H. et al. (2002). Initial sequencing and comparative analysis of the mouse genome. *Nature* 420, 520–562.

6.6 Pseudogenes Are Dead Ends of Evolution

Research

Hirotsune, S., Yoshida, N., Chen, A., Garrett, L., Sugiyama, F., Takahashi, S., Yagami, K., Wynshaw-Boris, A., and Yoshiki, A. (2003). An expressed pseudogene regulates the messenger-RNA stability of its homologous coding gene. *Nature* 423, 91–96.

6.10 Crossover Fixation Could Maintain Identical Repeats

Review

Charlesworth, B., Sniegowski, P., and Stephan, W. (1994). The evolutionary dynamics of repetitive DNA in eukaryotes. *Nature* 371, 215–220.

6.14 Minisatellites Are Useful for Genetic Mapping

Research

Jeffreys, A. J., Murray, J., and Neumann, R. (1998). High-resolution mapping of crossovers in human sperm defines a minisatellite-associated recombination hotspot. *Mol. Cell* 2, 267–273.

Jeffreys, A. J., Royle, N. J., Wilson, V., and Wong, Z. (1988). Spontaneous mutation rates to new length alleles at tandem-repetitive hypervariable loci in human DNA. *Nature* 332, 278–281.

Jeffreys, A. J., Tamaki, K., MacLeod, A., Monckton, D. G., Neil, D. L., and Armour, J. A. (1994). Complex gene conversion events in germline mutation at human minisatellites. *Nat. Genet.* 6, 136–145.

Jeffreys, A. J., Wilson, V., and Thein, S. L. (1985). Hypervariable minisatellite regions in human DNA. *Nature* 314, 67–73.

Strand, M., Prolla, T. A., Liskay, R. M., and Petes, T. D. (1993). Destabilization of tracts of simple repetitive DNA in yeast by mutations affecting DNA mismatch repair. *Nature* 365, 274–276.

7

Messenger RNA

CHAPTER OUTLINE

Continued on next page

7.14 Nonsense Mutations Trigger a Surveillance System

- Nonsense mutations cause mRNA to be degraded.
- Genes coding for the degradation system have been found in yeast and worms.

7.15 Eukaryotic RNAs Are Transported

- RNA is transported through a membrane as a ribonucleoprotein particle.
- All eukaryotic RNAs that function in the cytoplasm must be exported from the nucleus.
- tRNAs and the RNA component of a ribonuclease are imported into mitochondria.
- mRNAs can travel long distances between plant cells.

7.16 mRNA Can Be Specifically Localized

- Yeast Ash1 mRNA forms a ribonucleoprotein that binds to a myosin motor.
- A motor transports it along actin filaments into the daughter bud.
- It is anchored and translated in the bud, so that the protein is found only in the bud.

7.17 Summary

7.1 Introduction

RNA is a central player in gene expression. It was first characterized as an intermediate in protein synthesis, but since then many other RNAs have been discovered that play structural or functional roles at other stages of gene expression. The involvement of RNA in many functions concerned with gene expression supports the general view that the entire process may have evolved in an "RNA world" in which RNA was originally the active component in maintaining and expressing genetic information. Many of these functions were subsequently assisted or taken over by proteins, with a consequent increase in versatility and probably efficiency.

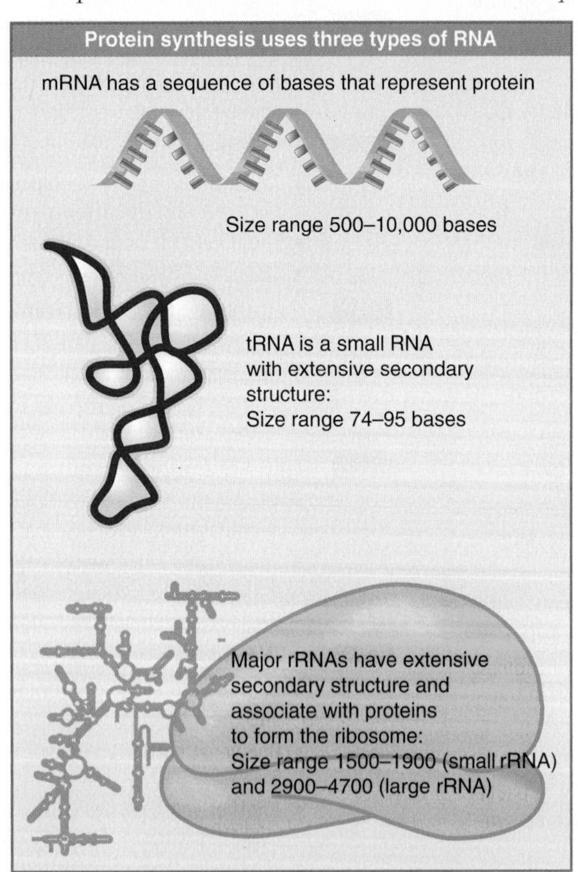

FIGURE 7.1 The three types of RNA universally required for gene expression are mRNA (carries the coding sequence), tRNA (provides the amino acid corresponding to each codon), and rRNA (a major component of the ribosome that provides the environment for protein synthesis).

As summarized in FIGURE 7.1, three major classes of RNA are directly involved in the production of proteins:

- Messenger RNA (mRNA) provides an intermediate that carries the copy of a DNA sequence that represents protein.
- Transfer RNAs **(tRNA)** are small RNAs that are used to provide amino acids corresponding to each particular codon in mRNA.
- Ribosomal RNAs **(rRNA)** are components of the ribosome, a large ribonucleoprotein complex that contains many proteins as well as its RNA components, and which provides the apparatus for actually polymerizing amino acids into a polypeptide chain.

The type of role that RNA plays in each of these cases is distinct. For messenger RNA, its sequence is the important feature: each nucleotide triplet within the coding region of the mRNA represents an amino acid in the corresponding protein. However, the structure of the mRNA—in particular the sequences on either side of the coding region—can play an important role in controlling its activity, and therefore the amount of protein that is produced from it.

In tRNA, we see two of the common themes governing the use of RNA: its three-dimensional structure is important, and it has the ability to base pair with another RNA (mRNA). The three-dimensional structure is recognized first by an enzyme as providing a target that is appropriate for linkage to a specific amino acid. The linkage creates an aminoacyl-tRNA, which is recognized as the structure that is used for protein synthesis. The specificity with which an aminoacyl-tRNA is used is controlled by base pairing, when a short triplet sequence (the anticodon) pairs with the nucleotide triplet representing its amino acid.

With rRNA, we see another type of activity. One role of RNA is structural, in providing a framework to which ribosomal proteins attach. It also participates directly in the activities of the ribosome. One of the crucial activities of the ribosome is the ability to catalyze the formation of a peptide bond by which an amino acid is incorporated into protein. This activity resides in one of the rRNAs.

The important thing about this background is that, as we consider the role of RNA in protein synthesis, we have to view it as a component that plays an active role and that can be a target for regulation by either proteins or by other RNAs, and we should remember that the RNAs may have been the basis for the original apparatus. The theme that runs through all of the activities of RNA, in both protein synthesis and elsewhere, is that its functions depend critically upon base pairing, both to form its secondary structure and to interact specifically with other RNA molecules. The coding function of mRNA is unique, but tRNA and rRNA are examples of a much broader class of noncoding RNAs with a variety of functions in gene expression.

7.2 mRNA Is Produced by Transcription and Is Translated

Key concept

- Only one of the two strands of DNA is transcribed into RNA.

Gene expression occurs by a two-stage process.

- Transcription generates a single-stranded RNA identical in sequence with one of the strands of the duplex DNA.
- Translation converts the nucleotide sequence of mRNA into the sequence of amino acids comprising a protein. The entire length of an mRNA is not translated, but each mRNA contains at least one coding region that is related to a protein sequence by the genetic code: each nucleotide triplet (codon) of the coding region represents one amino acid.

Only one strand of a DNA duplex is transcribed into a messenger RNA. We distinguish the two strands of DNA as depicted in FIGURE 7.2:

- The strand of DNA that directs synthesis of the mRNA via complementary base pairing is called the template strand or **antisense strand**. (*Antisense* is used as a general term to describe a sequence of DNA or RNA that is complementary to mRNA.)
- The other DNA strand bears the *same* sequence as the mRNA (except for possessing T instead of U) and is called the **coding strand** or **sense strand**.

In this chapter we discuss mRNA and its use as a template for protein synthesis. In Chapter 8, Protein Synthesis, we discuss the process by which a protein is synthesized. In Chapter 9, Using the Genetic Code, we discuss the way the genetic code is used to interpret the meaning of a sequence of mRNA. In Chapter 10, Protein Localization, we turn to the question of how a protein finds its proper location in the cell when or after it is synthesized.

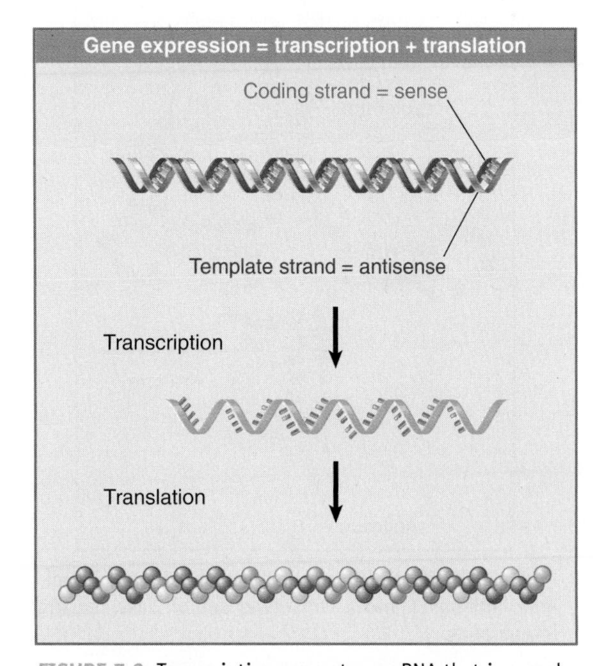

FIGURE 7.2 Transcription generates an RNA that is complementary to the DNA template strand and has the same sequence as the DNA coding strand. Translation reads each triplet of bases into one amino acid. Three turns of the DNA double helix contain 30 bp, which code for ten amino acids.

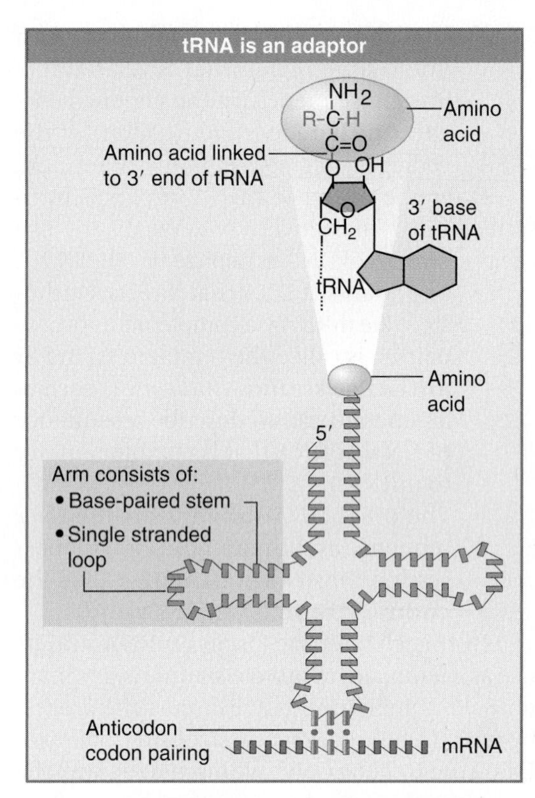

FIGURE 7.3 A tRNA has the dual properties of an adaptor that recognizes both the amino acid and codon. The 3′ adenosine is covalently linked to an amino acid. The anticodon pairs with the codon on mRNA.

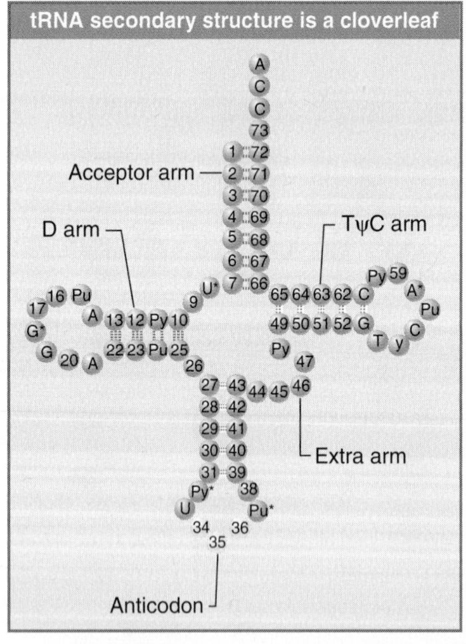

FIGURE 7.4 The tRNA cloverleaf has invariant and semi-invariant bases and a conserved set of base pairing interactions.

7.3 Transfer RNA Forms a Cloverleaf

Key concepts

- A tRNA has a sequence of 74 to 95 bases that folds into a cloverleaf secondary structure with four constant arms (and an additional arm in the longer tRNAs).
- tRNA is charged to form aminoacyl-tRNA by forming an ester link from the 2′ or 3′ OH group of the adenylic acid at the end of the acceptor arm to the COOH group of the amino acid.
- The sequence of the anticodon is solely responsible for the specificity of the aminoacyl-tRNA.

Messenger RNA can be distinguished from the apparatus responsible for its translation by the use of *in vitro* cell-free systems to synthesize proteins. *A protein-synthesizing system from one cell type can translate the mRNA from another, demonstrating that both the genetic code and the translation apparatus are universal.*

Each nucleotide triplet in the mRNA represents an amino acid. The incongruity of structure between trinucleotide and amino acid immediately raises the question of how each codon is matched to its particular amino acid. The "adaptor" is transfer RNA (tRNA). A tRNA has two crucial properties:

- It represents a single amino acid, to which it is *covalently linked.*
- It contains a trinucleotide sequence, the **anticodon,** which is *complementary to the codon representing its amino acid.* The anticodon enables the tRNA to recognize the codon via complementary base pairing.

All tRNAs have common secondary and tertiary structures. The tRNA secondary structure can be written in the form of a **cloverleaf,** illustrated in FIGURE 7.3, in which complementary base pairing forms **stems** for single-stranded **loops.** The stem–loop structures are called the **arms** of tRNA. Their sequences include "unusual" bases that are generated by modification of the four standard bases after synthesis of the polynucleotide chain.

The construction of the cloverleaf is illustrated in more detail in FIGURE 7.4. The four major arms are named for their structure or function:

- The **acceptor arm** consists of a base-paired stem that ends in an unpaired sequence whose free 2′– or 3′–OH group can be linked to an amino acid.

- The *TψC arm* is named for the presence of this triplet sequence. (ψ stands for pseudouridine, a modified base.)
- The **anticodon arm** always contains the anticodon triplet in the center of the loop.
- The **D arm** is named for its content of the base dihydrouridine (another of the modified bases in tRNA).
- The **extra arm** lies between the TψC and anticodon arms and varies from 3 to 21 bases.

The numbering system for tRNA illustrates the constancy of the structure. Positions are numbered from 5′ to 3′ according to the most common tRNA structure, which has 76 residues. The overall range of tRNA lengths is 74 to 95 bases. The variation in length is caused by differences in the D arm and extra arm.

The base pairing that maintains the secondary structure is shown in Figure 7.4. Within a given tRNA, most of the base pairings are conventional partnerships of A-U and G-C, but occasional G-U, G-ψ, or A-ψ pairs are found. The additional types of base pairs are less stable than the regular pairs, but still allow a double-helical structure to form in RNA.

When the sequences of tRNAs are compared, the bases found at some positions are **invariant** (or **conserved**); almost always a particular base is found at the position. Some positions are described as **semi-invariant** (or **semiconserved**) because they are restricted to one type of base (purine versus pyrimidine), but either base of that type may be present.

When a tRNA is *charged* with the amino acid corresponding to its anticodon, it is called **aminoacyl-tRNA.** The amino acid is linked by an ester bond from its carboxyl group to the 2′ or 3′ hydroxyl group of the ribose of the 3′ terminal base of the tRNA (which is always adenine). The process of charging a tRNA is catalyzed by a specific enzyme, **aminoacyl-tRNA synthetase.** There are (at least) 20 aminoacyl-tRNA synthetases. Each recognizes a single amino acid and all the tRNAs on to which it can legitimately be placed.

There is at least one tRNA (but usually more) for each amino acid. A tRNA is named by using the three-letter abbreviation for the amino acid as a superscript. If there is more than one tRNA for the same amino acid, subscript numerals are used to distinguish them. So two tRNAs for tyrosine would be described as $tRNA_1^{Tyr}$ and $tRNA_2^{Tyr}$. A tRNA carrying an amino acid—that is, an aminoacyl-tRNA—is indicated by a prefix that identifies the amino acid. Ala-tRNA describes $tRNA^{Ala}$ carrying its amino acid.

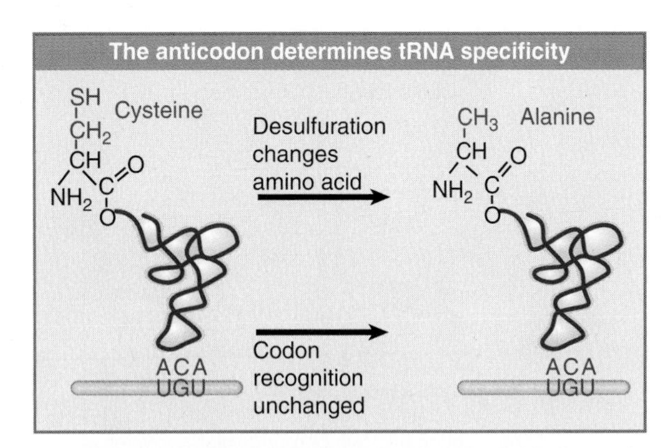

FIGURE 7.5 The meaning of tRNA is determined by its anticodon and not by its amino acid.

Does the anticodon sequence alone allow aminoacyl-tRNA to recognize the correct codon? A classic experiment to test this question is illustrated in FIGURE 7.5. Reductive desulfuration converts the amino acid of cysteinyl-tRNA into alanine, generating alanyl-$tRNA^{Cys}$. The tRNA has an anticodon that responds to the codon UGU. Modification of the amino acid does not influence the specificity of the anticodon–codon interaction, so the alanine residue is incorporated into protein in place of cysteine. *Once a tRNA has been charged, the amino acid plays no further role in its specificity, which is determined exclusively by the anticodon.*

7.4 The Acceptor Stem and Anticodon Are at Ends of the Tertiary Structure

Key concept

- The cloverleaf forms an L-shaped tertiary structure with the acceptor arm at one end and the anticodon arm at the other end.

The secondary structure of each tRNA folds into a compact L-shaped tertiary structure in which the 3′ end that binds the amino acid is distant from the anticodon that binds the mRNA. All tRNAs have the same general tertiary structure, although they are distinguished by individual variations.

The base paired double-helical stems of the secondary structure are maintained in the tertiary structure, but their arrangement in three dimensions essentially creates two double helices at right angles to each other, as

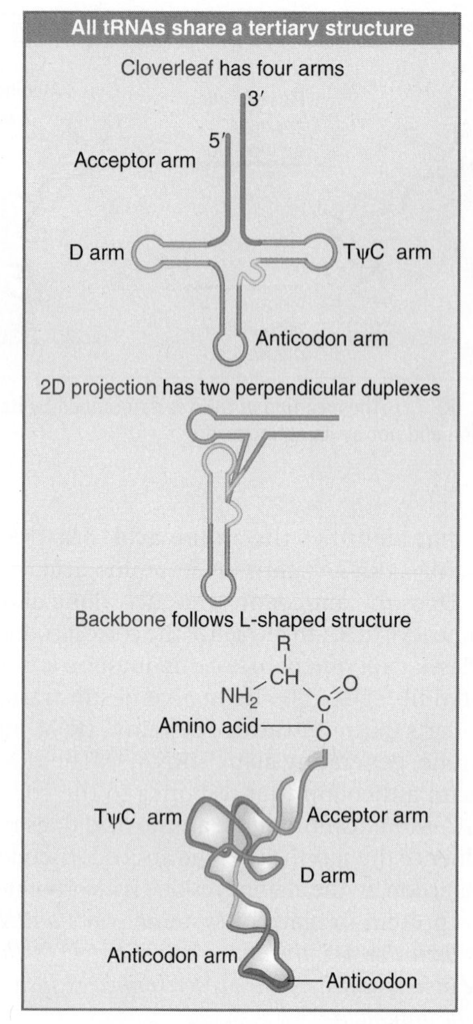

FIGURE 7.6 Transfer RNA folds into a compact L-shaped tertiary structure with the amino acid at one end and the anticodon at the other end.

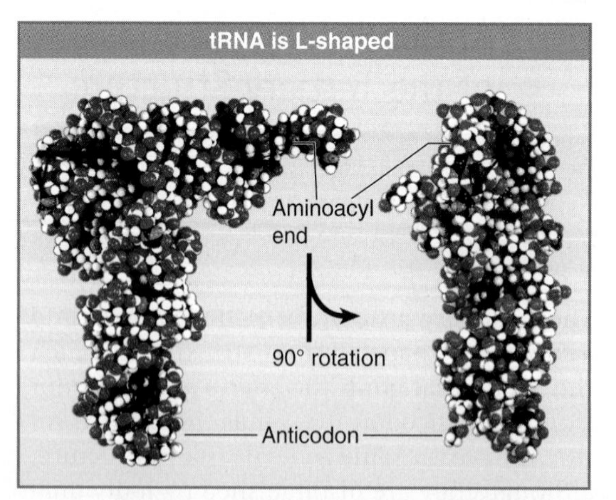

FIGURE 7.7 A space-filling model shows that the $tRNA^{Phe}$ tertiary structure is compact. The two views of tRNA are rotated by 90°. Photo courtesy of Sung-Hou Kim, University of California, Berkeley.

illustrated in FIGURE 7.6. The acceptor stem and the TψC stem form one continuous double helix with a single gap; the D stem and anticodon stem form another continuous double helix, also with a gap. The region between the double helices, where the turn in the L-shape is made, contains the TψC loop and the D loop. So the amino acid resides at the extremity of one arm of the L-shape and the anticodon loop forms the other end.

The tertiary structure is created by hydrogen bonding, mostly involving bases that are unpaired in the secondary structure. Many of the invariant and semi-invariant bases are involved in these H-bonds, which explains their conservation. Not every one of these interactions is universal, but probably they identify the *general* pattern for establishing tRNA structure.

A molecular model of the structure of yeast $tRNA^{Phe}$ is shown in FIGURE 7.7. The left view corresponds with the bottom panel in Figure 7.6. Differences in the structure are found in other tRNAs, thus accommodating the dilemma that all tRNAs must have a similar shape, yet it must be possible to recognize differences between them. For example, in $tRNA^{Asp}$, the angle between the two axes is slightly greater, so the molecule has a slightly more open conformation.

The structure suggests a general conclusion about the function of tRNA: *Its sites for exercising particular functions are maximally separated.* The amino acid is as far distant from the anticodon as possible, which is consistent with their roles in protein synthesis.

7.5 Messenger RNA Is Translated by Ribosomes

Key concepts

- Ribosomes are characterized by their rate of sedimentation (70S for bacterial ribosomes and 80S for eukaryotic ribosomes).
- A ribosome consists of a large subunit (50S or 60S for bacteria and eukaryotes) and a small subunit (30S or 40S).
- The ribosome provides the environment in which aminoacyl-tRNAs add amino acids to the growing polypeptide chain in response to the corresponding triplet codons.
- A ribosome moves along an mRNA from 5′ to 3′.

Translation of an mRNA into a polypeptide chain is catalyzed by the **ribosome**. Ribosomes are traditionally described in terms of their (approximate) rate of sedimentation (measured in Svedbergs, in which a higher S value indicates a greater rate of sedimentation and a larger mass). Bacterial ribosomes generally sediment at ~70S. The ribosomes of the cytoplasm of higher eukaryotic cells are larger, usually sedimenting at ~80S.

The ribosome is a compact **ribonucleoprotein** particle consisting of two subunits. Each subunit has an RNA component, including one very large RNA molecule, and many proteins. The relationship between a ribosome and its subunits is depicted in FIGURE 7.8. The two subunits dissociate *in vitro* when the concentration of Mg^{2+} ions is reduced. In each case, the **large subunit** is about twice the mass of the **small subunit.** Bacterial (70S) ribosomes have subunits that sediment at 50S and 30S. The subunits of eukaryotic cytoplasmic (80S) ribosomes sediment at 60S and 40S. The two subunits work together as part of the complete ribosome, but each undertakes distinct reactions in protein synthesis.

All the ribosomes of a given cell compartment are identical. *They undertake the synthesis of different proteins by associating with the different mRNAs that provide the actual coding sequences.*

The ribosome provides the environment that controls the recognition between a codon of mRNA and the anticodon of tRNA. Reading the genetic code as a series of adjacent triplets, protein synthesis proceeds from the start of a coding region to the end. *A protein is assembled by the sequential addition of amino acids in the direction from the N-terminus to the C-terminus as a ribosome moves along the mRNA.*

A ribosome begins translation at the 5′ end of a coding region; it translates each triplet codon into an amino acid as it proceeds toward the 3′ end. At each codon, the appropriate aminoacyl-tRNA associates with the ribosome, donating its amino acid to the polypeptide chain. At any given moment, the ribosome can accommodate the two aminoacyl-tRNAs corresponding to successive codons, making it possible for a peptide bond to form between the two corresponding amino acids. At each step, the growing polypeptide chain becomes longer by one amino acid.

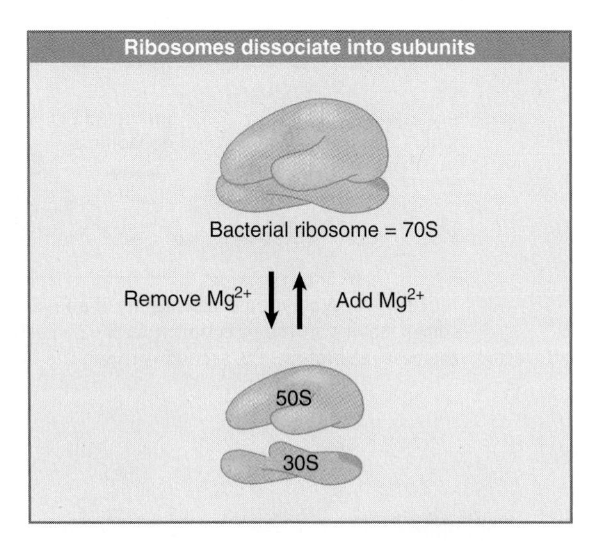

FIGURE 7.8 A ribosome consists of two subunits.

7.6 Many Ribosomes Bind to One mRNA

Key concept

- An mRNA is simultaneously translated by several ribosomes. Each ribosome is at a different stage of progression along the mRNA.

When active ribosomes are isolated in the form of the fraction associated with newly synthesized proteins, they are found in the form of a complex consisting of an mRNA associated with several ribosomes. This is the **polyribosome** or **polysome.** The 30S subunit of each ribosome is associated with the mRNA, and the 50S subunit carries the newly synthesized protein. The tRNA spans both subunits.

Each ribosome in the polysome independently synthesizes a single polypeptide during its traverse of the messenger sequence. Essentially the mRNA is pulled through the ribosome, and each triplet nucleotide is translated into an amino acid. Thus the mRNA has a series of ribosomes that carry increasing lengths of the protein product, moving from the 5′ to the 3′ end, as illustrated in FIGURE 7.9. A polypeptide chain in the process of synthesis is sometimes called a **nascent protein.**

Roughly the most recent 30 to 35 amino acids added to a growing polypeptide chain are protected from the environment by the structure of the ribosome. Probably all of the preceding part of the polypeptide protrudes and is free to start folding into its proper conformation. Thus proteins can display parts of the mature conformation even before synthesis has been completed.

A classic characterization of polysomes is shown in the electron micrograph of FIGURE 7.10. Globin protein is synthesized by a set of five ribosomes attached to each mRNA (pentasomes). The ribosomes appear as squashed spherical objects of ~7 nm (70 Å) in diameter, connected by a thread of mRNA. The ribosomes are located at various positions along the messenger. Those at one end have just started

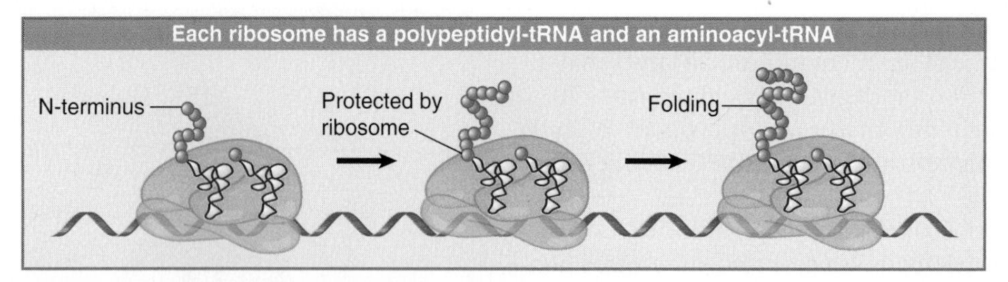

FIGURE 7.9 A polyribosome consists of an mRNA being translated simultaneously by several ribosomes moving in the direction from 5′–3′. Each ribosome has two tRNA molecules, one carrying the nascent protein, the second carrying the next amino acid to be added.

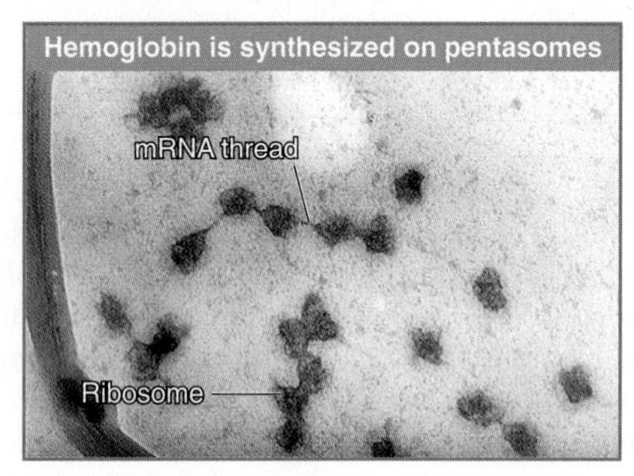

FIGURE 7.10 Protein synthesis occurs on polysomes. Photo courtesy of Alexander Rich, Massachusetts Institute of Technology.

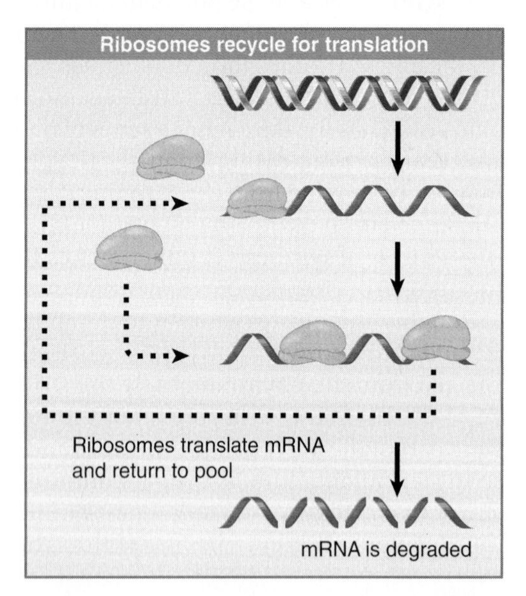

FIGURE 7.11 Messenger RNA is translated by ribosomes that cycle through a pool.

protein synthesis; those at the other end are about to complete production of a polypeptide chain.

The size of the polysome depends on several variables. In bacteria it is very large, with tens of ribosomes simultaneously engaged in translation. The size is due in part to the length of the mRNA (which usually codes for several proteins); it is also due in part to the high efficiency with which the ribosomes attach to the mRNA.

Polysomes in the cytoplasm of a eukaryotic cell are likely to be smaller than those in bacteria; again, their size is a function both of the length of the mRNA (usually representing only a single protein in eukaryotes) and of the characteristic frequency with which ribosomes attach. An average eukaryotic mRNA probably has ~8 ribosomes attached at any one time.

FIGURE 7.11 illustrates the life cycle of the ribosome. Ribosomes are drawn from a pool (actually the pool consists of ribosomal subunits), used to translate an mRNA, and then returned to the pool for further cycles. The number of ribosomes on each mRNA molecule synthesizing a particular protein is not precisely determined, in either bacteria or eukaryotes, but is a matter of statistical fluctuation, determined by the variables of mRNA size and efficiency.

An overall view of the attention devoted to protein synthesis in the intact bacterium is given in FIGURE 7.12. The 20,000 or so ribosomes account for a quarter of the cell mass. There are >3000 copies of each tRNA, and altogether the tRNA molecules outnumber the ribosomes by almost tenfold; most of them are present as aminoacyl-tRNAs, that is, ready to be used at once in protein synthesis. As a result of their instability, it is difficult to calculate the number of mRNA molecules, but a reasonable guess would be ~1500, in varying states of synthesis and decomposition. There are ~600 different types of mRNA in a bacterium. This suggests that there are usually only two to three copies of each mRNA per bacterium. On average, each probably codes for ~3 proteins. If there are 1850 different soluble proteins, there must be on average >1000 copies of each protein in a bacterium.

7.7 The Life Cycle of Bacterial Messenger RNA

Key concepts

- Transcription and translation occur simultaneously in bacteria, as ribosomes begin translating an mRNA before its synthesis has been completed.
- Bacterial mRNA is unstable and has a half-life of only a few minutes.
- A bacterial mRNA may be polycistronic in having several coding regions that represent different genes.

Messenger RNA has the same function in all cells, but there are important differences in the details of the synthesis and structure of prokaryotic and eukaryotic mRNA.

A major difference in the production of mRNA depends on the locations where transcription and translation occur:

- In bacteria, mRNA is transcribed and translated in the single cellular compartment; the two processes are so closely linked that they occur simultaneously. Ribosomes attach to bacterial mRNA even before its transcription has been completed, thus the polysome is likely still to be attached to DNA. Bacterial mRNA usually is unstable, and is therefore translated into proteins for only a few minutes.
- In a eukaryotic cell, synthesis and maturation of mRNA occur exclusively in the nucleus. Only after these events are completed is the mRNA exported to the cytoplasm, where it is translated by ribosomes. Eukaryotic mRNA is relatively stable and continues to be translated for several hours.

FIGURE 7.13 shows that transcription and translation are intimately related in bacteria. Transcription begins when the enzyme RNA polymerase binds to DNA and then moves along making a copy of one strand. As soon as transcription begins, ribosomes attach to the 5′ end of the mRNA and start translation, even before the rest of the message has been synthesized. A bunch of ribosomes moves along the mRNA while it is being synthesized. The 3′ end of the mRNA is generated when transcription terminates. Ribosomes continue to translate the mRNA while it survives, but it is degraded in

~30% of bacterial dry mass is concerned with gene expression

Component	Dry cell mass (%)	Molecules /cell	Different types	Copies of each type
Wall	10	1	1	1
Membrane	10	2	2	1
DNA	1.5	1	1	1
mRNA	1	1,500	600	2–3
tRNA	3	200,000	60	>3,000
rRNA	16	38,000	2	19,000
Ribosomal proteins	9	10^6	52	19,000
Soluble proteins	46	2.0×10^6	1,850	>1,000
Small molecules	3	7.5×10^6	800	

FIGURE 7.12 Considering *E. coli* in terms of its macromolecular components.

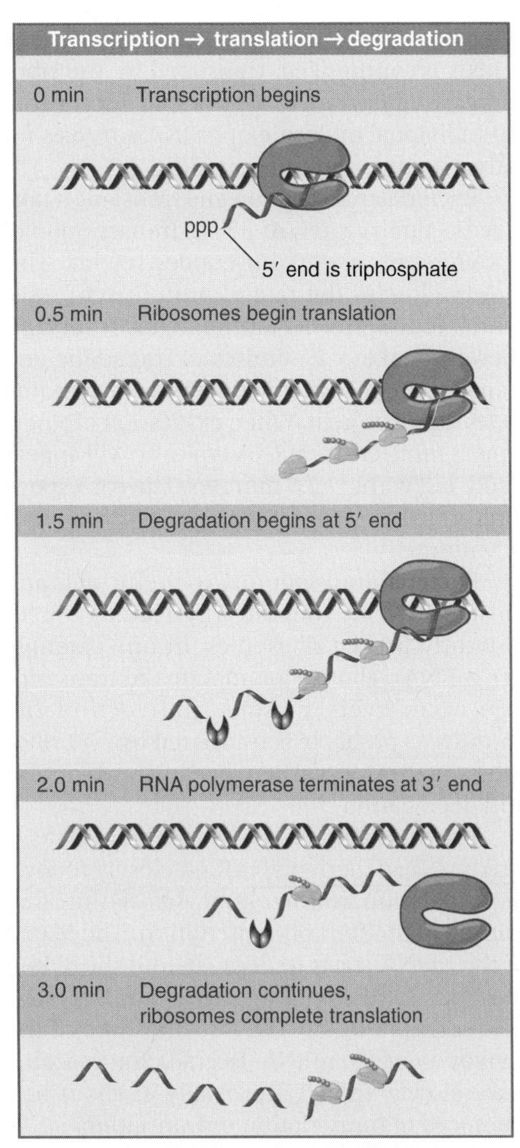

FIGURE 7.13 Overview: mRNA is transcribed, translated, and degraded simultaneously in bacteria.

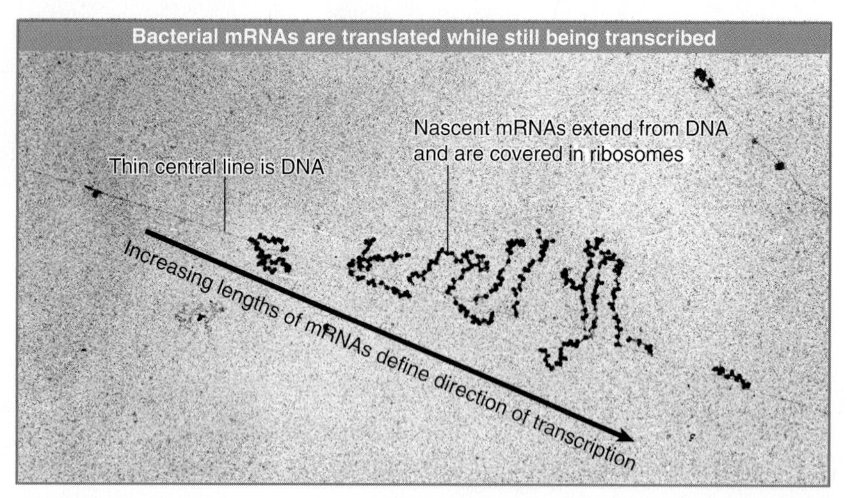

FIGURE 7.14 Transcription units can be visualized in bacteria. Photo courtesy of Oscar Miller.

the overall 5′→3′ direction quite rapidly. The mRNA is synthesized, translated by the ribosomes, and degraded, all in rapid succession. An individual molecule of mRNA survives for only a matter of minutes at most.

Bacterial transcription and translation take place at similar rates. At 37° C, transcription of mRNA occurs at ~40 nucleotides/second. This is very close to the rate of protein synthesis, which is roughly 15 amino acids/second. It therefore takes ~2 minutes to transcribe and translate an mRNA of 5000 bp, corresponding to 180 kD of protein. When expression of a new gene is initiated, its mRNA typically will appear in the cell within ~2.5 minutes. The corresponding protein will appear within perhaps another 0.5 minute.

Bacterial translation is very efficient, and most mRNAs are translated by a large number of tightly packed ribosomes. In one example (*trp* mRNA), about 15 initiations of transcription occur every minute, and each of the 15 mRNAs probably is translated by ~30 ribosomes in the interval between its transcription and degradation.

The instability of most bacterial mRNAs is striking. Degradation of mRNA closely follows its translation and likely begins within one minute of the start of transcription. The 5′ end of the mRNA starts to decay before the 3′ end has been synthesized or translated. Degradation seems to follow the last ribosome of the convoy along the mRNA. Degradation proceeds more slowly, though—probably at about half the speed of transcription or translation.

The stability of mRNA has a major influence on the amount of protein that is produced. It is usually expressed in terms of the half-life. The mRNA representing any particular gene has a characteristic half-life, but the average is ~2 minutes in bacteria.

This series of events is only possible, of course, because transcription, translation, and degradation all occur in the same direction. The dynamics of gene expression have been caught *in flagrante delicto* in the electron micrograph of FIGURE 7.14. In these (unknown) transcription units, several mRNAs are under synthesis simultaneously, and each carries many ribosomes engaged in translation. (This corresponds to the stage shown in the second panel in Figure 7.13.) An RNA whose synthesis has not yet been completed is often called a **nascent RNA.**

Bacterial mRNAs vary greatly in the number of proteins for which they code. Some mRNAs represent only a single gene: they are **monocistronic.** Others (the majority) carry sequences coding for several proteins: they are **polycistronic.** In these cases, a single mRNA is transcribed from a group of adjacent genes. (Such a cluster of genes constitutes an operon that is controlled as a single genetic unit; see Chapter 12, The Operon.)

All mRNAs contain two types of region. The coding region consists of a series of codons representing the amino acid sequence of the protein, starting (usually) with AUG and ending with a termination codon. The mRNA is always longer than the coding region, though, as extra regions are present at both ends. An additional sequence at the 5′ end, preceding the start of the coding region, is described as the leader or **5′ UTR** (untranslated region). An additional sequence following the termination signal, forming the 3′ end, is called the trailer or **3′ UTR**. Although part of the transcription unit, these sequences are not used to code for protein.

A polycistronic mRNA also contains **intercistronic regions,** as illustrated in FIGURE 7.15. They vary greatly in size: They may be as long as 30 nucleotides in bacterial mRNAs (and even longer in phage RNAs), or they may be very short, with as few as one or two nucleotides separating the termination codon for one protein from the initiation codon for the next. In an extreme case, two genes actually overlap, so that the last base of one coding region is also the first base of the next coding region.

The number of ribosomes engaged in translating a particular cistron depends on the efficiency of its initiation site. The initiation site for the first cistron becomes available as soon as the 5′ end of the mRNA is synthesized. How are subsequent cistrons translated? Are the several coding regions in a polycistronic mRNA translated independently or is their expression connected? Is the mechanism of initiation the same for all cistrons, or is it different for the first cistron and the internal cistrons?

Translation of a bacterial mRNA proceeds sequentially through its cistrons. At the time when ribosomes attach to the first coding region, the subsequent coding regions have not yet even been transcribed. By the time the second ribosome site is available, translation is well under way through the first cistron. Typically ribosomes terminate translation at the end of the first cistron (and dissociate into subunits), and a new ribosome assembles independently at the start of the next coding region. (We discuss the processes of initiation and termination in Chapter 8, Protein Synthesis.)

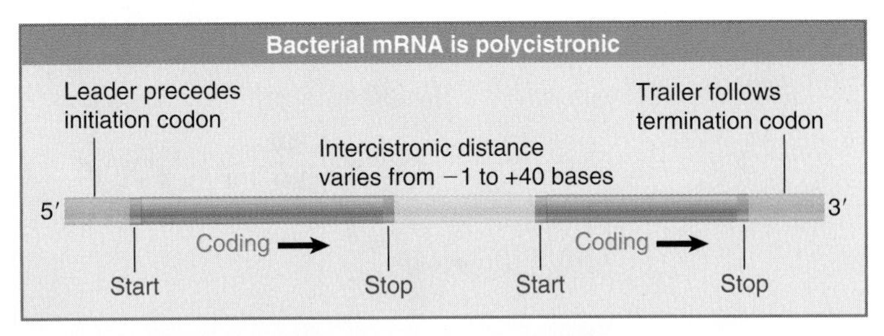

FIGURE 7.15 Bacterial mRNA includes untranslated as well as translated regions. Each coding region has its own initiation and termination signals. A typical mRNA may have several coding regions.

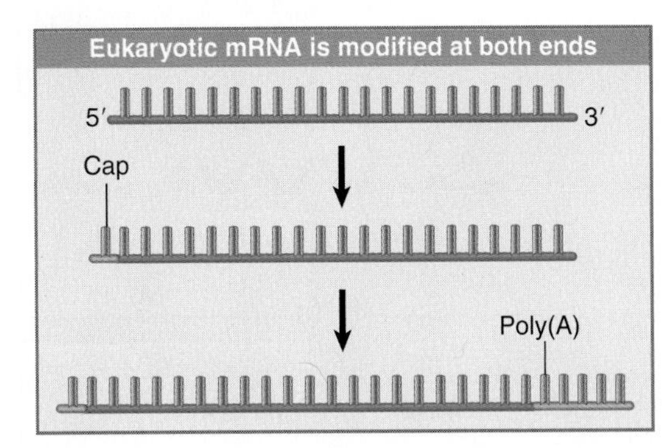

FIGURE 7.16 Eukaryotic mRNA is modified by the addition of a cap to the 5′ end and poly(A) to the 3′ end.

7.8 Eukaryotic mRNA Is Modified During or after Its Transcription

Key concepts

- A eukaryotic mRNA transcript is modified in the nucleus during or shortly after transcription.
- The modifications include the addition of a methylated cap at the 5′ end and a sequence of poly(A) at the 3′ end.
- The mRNA is exported from the nucleus to the cytoplasm only after all modifications have been completed.

The production of eukaryotic mRNA involves additional stages after transcription. Transcription occurs in the usual way, initiating a transcript with a 5′ triphosphate end. However, the 3′ end is generated by cleaving the transcript, rather than by terminating transcription at a fixed site. Those RNAs that are derived from interrupted genes require splicing to remove the introns, generating a smaller mRNA that contains an intact coding sequence.

FIGURE 7.16 shows that both ends of the transcript are modified by additions of further nucleotides (involving additional enzyme systems). The 5′ end of the RNA is modified by the addition of a "cap" virtually as soon as it appears. This replaces the triphosphate of the initial transcript with a nucleotide in reverse (3′→5′) orientation, thus "sealing" the end. The 3′ end is modified by addition of a series of adenylic acid nucleotides [polyadenylic acid or **poly(A)**] immediately after its cleavage. Only after the completion of all modification and processing events can the mRNA be exported from the nucleus to the cytoplasm. The average delay in leaving for the cytoplasm is ~20 minutes. Once the mRNA has entered the cytoplasm, it is recognized by ribosomes and translated.

FIGURE 7.17 shows that the life cycle of eukaryotic mRNA is more protracted than that of bacterial mRNA. Transcription in animal cells occurs at about the same speed as in bacteria, ~40 nucleotides per second. Many eukaryotic genes are large; a gene of 10,000 bp takes ~5 minutes to transcribe. Transcription of mRNA is not terminated by the release of enzyme from the DNA; instead the enzyme continues past the end of the gene. A coordinated series of events generates the 3′ end of the mRNA by cleavage, and adds a length of poly(A) to the newly generated 3′ end.

Eukaryotic mRNA constitutes only a small proportion of the total cellular RNA (~3% of the mass). Half-lives are relatively short in yeast, ranging from 1–60 minutes. There is a substantial increase in stability in higher eukaryotes; animal cell mRNA is relatively stable, with half-lives ranging from 4–24 hours.

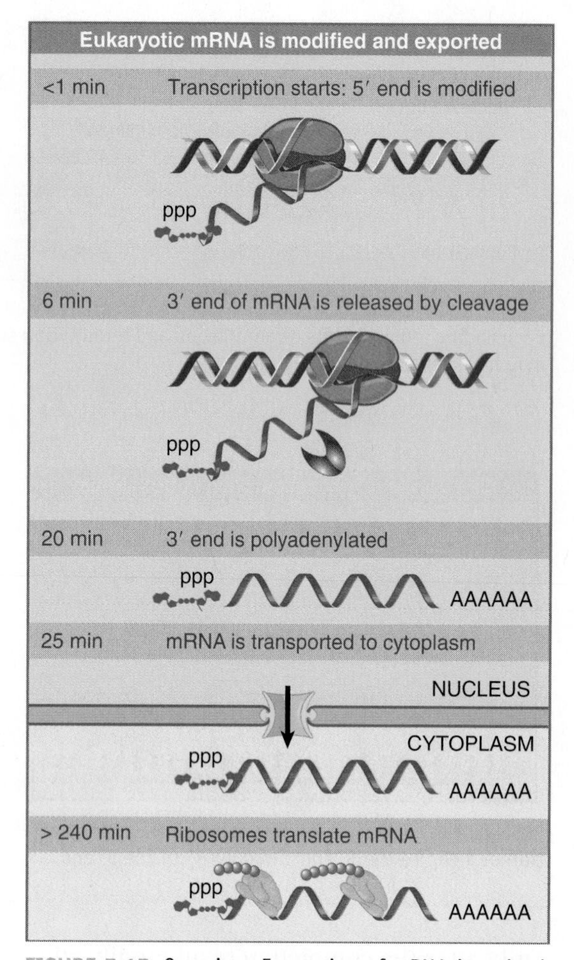

FIGURE 7.17 Overview: Expression of mRNA in animal cells requires transcription, modification, processing, nucleo-cytoplasmic transport, and translation.

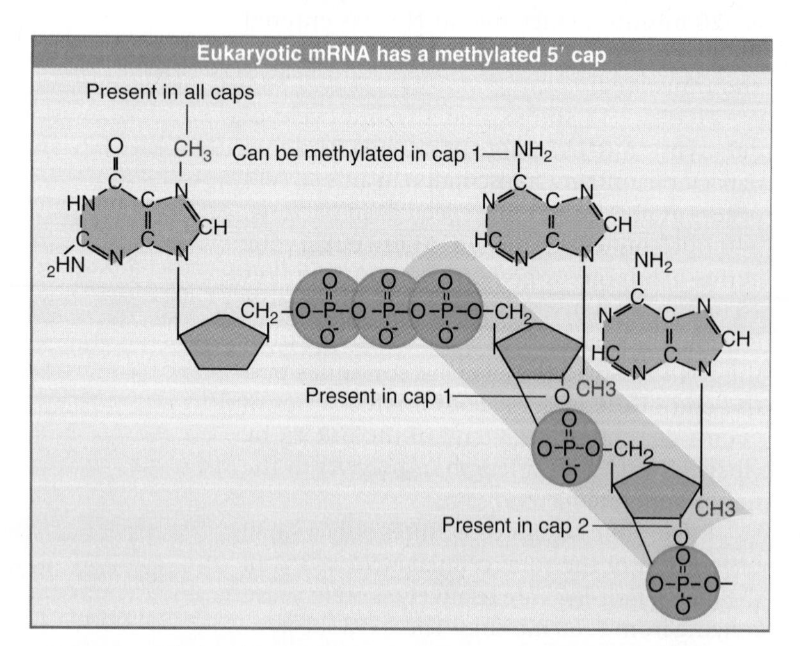

FIGURE 7.18 The cap blocks the 5′ end of mRNA and may be methylated at several positions.

Eukaryotic polysomes are reasonably stable. The modifications at both ends of the mRNA contribute to their stability.

7.9 The 5′ End of Eukaryotic mRNA Is Capped

Key concept

- A 5′ cap is formed by adding a G to the terminal base of the transcript via a 5′–5′ link. One to three methyl groups are added to the base or ribose of the new terminal guanosine.

Transcription starts with a nucleoside triphosphate (usually a purine, A or G). The first nucleotide retains its 5′ triphosphate group and makes the usual phosphodiester bond from its 3′ position to the 5′ position of the next nucleotide. The initial sequence of the transcript can be represented as:

$$5'ppp^{A}/_{G}pNpNpNp\ldots$$

When the mature mRNA is treated *in vitro* with enzymes that should degrade it into individual nucleotides, however, the 5′ end does not give rise to the expected nucleoside triphosphate. Instead it contains two nucleotides, connected by a 5′–5′ triphosphate linkage and also bearing methyl groups. The terminal base is always a guanine that is added to the original RNA molecule after transcription.

Addition of the 5′ terminal G is catalyzed by a nuclear enzyme, guanylyl transferase. The reaction occurs so soon after transcription has started that it is not possible to detect more than trace amounts of the original 5′ triphosphate end in the nuclear RNA. The overall reaction can be represented as a condensation between GTP and the original 5′ triphosphate terminus of the RNA. Thus

5′ 5′
Gppp + pppApNpNp . . .
↓
5′—5′
GpppApNpNp . . . + pp + p

The new G residue added to the end of the RNA is in the reverse orientation from all the other nucleotides.

This structure is called a **cap.** It is a substrate for several methylation events. FIGURE 7.18 shows the full structure of a cap after all possible methyl groups have been added. Types of caps are distinguished by how many of these methylations have occurred:

- The first methylation occurs in all eukaryotes, and consists of the addition of a methyl group to the 7 position of the terminal guanine. A cap that possesses this single methyl group is known as a **cap 0.** This is as far as the reaction proceeds in unicellular eukaryotes. The enzyme responsible for this modification is called guanine-7-methyltransferase.
- The next step is to add another methyl group, to the 2′–O position of the penultimate base (which was actually the original first base of the transcript before any modifications were made). This reaction is catalyzed by another enzyme (2′–O-methyl-transferase). A cap with the two methyl groups is called **cap 1.** This is the predominant type of cap in all eukaryotes except unicellular organisms.
- In a small minority of cases in higher eukaryotes, another methyl group is added to the second base. This happens only when the position is occupied by adenine; the reaction involves addition of a methyl group at the N^6 position. The enzyme responsible acts only on an adenosine substrate that already has the methyl group in the 2′–O position.
- In some species, a methyl group is added to the third base of the capped mRNA. The substrate for this reaction is the cap 1 mRNA that already possesses two methyl groups. The third-base modification is always a 2′–O ribose methylation. This creates the **cap 2** type. This cap usually represents less than 10%–15% of the total capped population.

In a population of eukaryotic mRNAs, every molecule is capped. The proportions of the different types of cap are characteristic for a particular organism. We do not know whether the structure of a particular mRNA is invariant or can have more than one type of cap.

In addition to the methylation involved in capping, a low frequency of internal methylation occurs in the mRNA only of higher eukaryotes. This is accomplished by the generation of N^6 methyladenine residues at a frequency of about one modification per 1000 bases. There are one or two methyladenines in a typical higher eukaryotic mRNA, although their presence is not obligatory; some mRNAs do not have any.

7.10 The 3′ Terminus Is Polyadenylated

Key concepts

- A length of poly(A) ~200 nucleotides long is added to a nuclear transcript after transcription.
- The poly(A) is bound by a specific protein (PABP).
- The poly(A) stabilizes the mRNA against degradation.

The 3′ terminal stretch of A residues is often described as the poly(A) tail; mRNA with this feature is denoted **poly(A)$^+$**.

The poly(A) sequence is not coded in the DNA, but rather is added to the RNA in the nucleus after transcription. The addition of poly(A) is catalyzed by the enzyme **poly(A) polymerase,** which adds ~200 A residues to the free 3′-OH end of the mRNA. The poly(A) tract of both nuclear RNA and mRNA is associated with a protein, the **poly(A)-binding protein (PABP)**. Related forms of this protein are found in many eukaryotes. One PABP monomer of ~70 kD is bound every 10 to 20 bases of the poly(A) tail. Thus a common feature in many or most eukaryotes is that the 3′ end of the mRNA consists of a stretch of poly(A) bound to a large mass of protein. Addition of poly(A) occurs as part of a reaction in which the 3′ end of the mRNA is generated and modified by a complex of enzymes (see Section 26.19, The 3′ Ends of mRNAs Are Generated by Cleavage and Polyadenylation).

Binding of the PABP to the initiation factor eIF4G generates a closed loop, in which the 5′ and 3′ ends of the mRNA find themselves held in the same protein complex (see Figure 8.20 in Section 8.9, Eukaryotes Use a Complex of Many Initiation Factors). The formation of this complex may be responsible for some of the effects of poly(A) on the properties of mRNA. Poly(A) usually stabilizes mRNA. The ability of the poly(A) to protect mRNA against degradation requires binding of the PABP.

Removal of poly(A) inhibits the initiation of translation *in vitro,* and depletion of PABP has the same effect in yeast *in vivo.* These effects could depend on the binding of PABP to the initiation complex at the 5′ end of mRNA. There are many examples in early embryonic development where polyadenylation of a particular mRNA is correlated with its translation. In some cases, mRNAs are stored in a nonpolyadenylated form, and poly(A) is added when their translation is required; in other cases, poly(A)$^+$

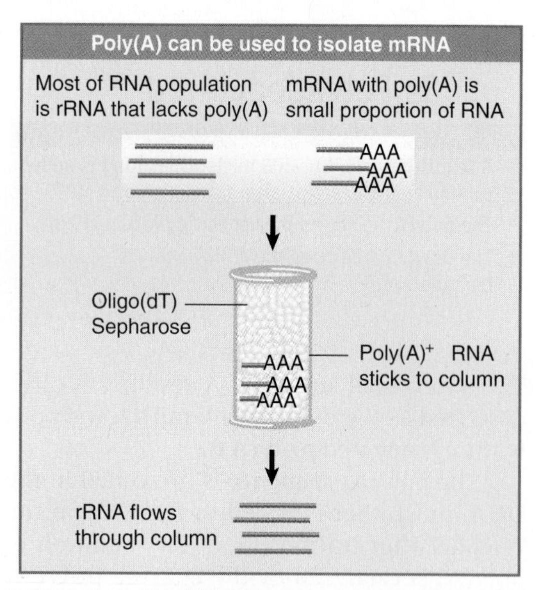

FIGURE 7.19 Poly(A)$^+$ RNA can be separated from other RNAs by fractionation on Sepharose-oligo(dT).

mRNAs are de-adenylated, and their translation is reduced.

The presence of poly(A) has an important practical consequence. The poly(A) region of mRNA can base pair with oligo(U) or oligo(dT), and this reaction can be used to isolate poly(A)$^+$ mRNA. The most convenient technique is to immobilize the oligo (U or dT) on solid support material. Then, when an RNA population is applied to the column, as illustrated in FIGURE 7.19, only the poly(A)$^+$ RNA is retained. It can be retrieved by treating the column with a solution that breaks the bonding to release the RNA.

The only drawback to this procedure is that it isolates all the RNA that contains poly(A). If RNA of the whole cell is used, for example, both nuclear and cytoplasmic poly(A)$^+$ RNA will be retained. If preparations of polysomes are used (a common procedure), most of the isolated poly(A)$^+$ RNA will be active mRNA. However, in addition to mRNA in polysomes, there are also ribonucleoprotein particles in the cytosol that contain poly(A)$^+$ mRNA, but which are not translated. This RNA may be "stored" for use at some other time. Isolation of total poly(A)$^+$ mRNA therefore does not correspond exactly with the active mRNA population.

The "cloning" approach for purifying mRNA uses a procedure in which the mRNA is copied to make a complementary DNA strand (known as **cDNA**). Then the cDNA can be used as a template to synthesize a DNA strand that is identical with the original mRNA sequence. The product of these reactions is a double-stranded DNA corresponding to the sequence of the mRNA. This DNA can be reproduced in large amounts.

The availability of a cloned DNA makes it easy to isolate the corresponding mRNA by hybridization techniques. Even mRNAs that are present in only very few copies per cell can be isolated by this approach. Indeed, only mRNAs that are present in relatively large amounts can be isolated directly without using a cloning step.

Almost all cellular mRNAs possess poly(A). A significant exception is provided by the mRNAs that code for the histone proteins (a major structural component of chromosomal material). These mRNAs comprise most or all of the **poly(A)$^-$** fraction. The significance of the absence of poly(A) from histone mRNAs is not clear, and there is no particular aspect of their function for which this appears to be necessary.

7.11 Bacterial mRNA Degradation Involves Multiple Enzymes

Key concepts

- The overall direction of degradation of bacterial mRNA is 5′→3′.
- Degradation results from the combination of endonucleolytic cleavages followed by exonucleolytic degradation of the fragment from 3′→5′.

Bacterial mRNA is constantly degraded by a combination of endonucleases and exonucleases. Endonucleases cleave an RNA at an internal site. Exonucleases are involved in trimming reactions in which the extra residues are whittled away, base by base, from the end. Bacterial exonucleases that act on single-stranded RNA proceed along the nucleic acid chain from the 3′ end.

The way the two types of enzymes work together to degrade an mRNA is shown in FIGURE 7.20. Degradation of a bacterial mRNA is initiated by an endonucleolytic attack. Several 3′ ends may be generated by endonucleolytic cleavages within the mRNA. The overall direction of degradation (as measured by loss of ability to synthesize proteins) is 5′→3′. This probably results from a succession of endonucleolytic cleavages following the last ribosome. Degradation of the released fragments of mRNA into nucleotides then proceeds by exonucleolytic attack from the free 3′–OH end toward the 5′ terminus (that is, in the opposite direction from transcription). Endonucleolytic attack releases fragments that may have different susceptibilities to exonucleases. A region of secondary structure within the mRNA may provide an obstacle to the exonuclease, thus protecting the regions on its 5′ side. The stability of each mRNA is therefore determined by the susceptibility of its particular sequence to both endo- and exonucleolytic cleavages.

There are ~12 ribonucleases in *E. coli.* Mutants in the endoribonucleases (except ribonuclease I, which is without effect) accumulate unprocessed precursors to rRNA and tRNA, but are viable. Mutants in the exonucleases often have apparently unaltered phenotypes, which suggests that one enzyme can substitute for the absence of another. Mutants lacking multiple enzymes sometimes are inviable.

RNAase E is the key enzyme in initiating cleavage of mRNA. It may be the enzyme that makes the first cleavage for many mRNAs. Bacterial mutants that have a defective ribonuclease E have increased stability (two- to threefold) of mRNA. However, this is not its only function. RNAase E was originally discovered as the enzyme that is responsible for processing 5′ rRNA from the primary transcript by a specific endonucleolytic processing event.

The process of degradation may be catalyzed by a multienzyme complex (sometimes called the **degradosome**) that includes ribonuclease E, PNPase, and a helicase. RNAase E plays dual roles. Its N-terminal domain provides an endonuclease activity. The C-terminal domain provides a scaffold that holds together the other components. The helicase unwinds the substrate RNA to make it available to PNPase. According to this model, RNAase E makes the initial cut and then passes the fragments to the other components of the complex for processing.

Polyadenylation may play a role in initiating degradation of some mRNAs in bacteria. Poly(A) polymerase is associated with ribosomes in *E. coli,* and short (10 to 40 nucleotide) stretches of poly(A) are added to at least some mRNAs. Triple mutations that remove poly(A) polymerase, ribonuclease E, and polynucleotide phosphorylase (PNPase is a 3′–5′ exonuclease) have a strong effect on stability. (Mutations in individual genes or pairs of genes have only a weak effect.) Poly(A) polymerase may create a poly(A) tail that acts as a binding site for the nucleases. The role of poly(A) in bacteria would therefore be different from that in eukaryotic cells.

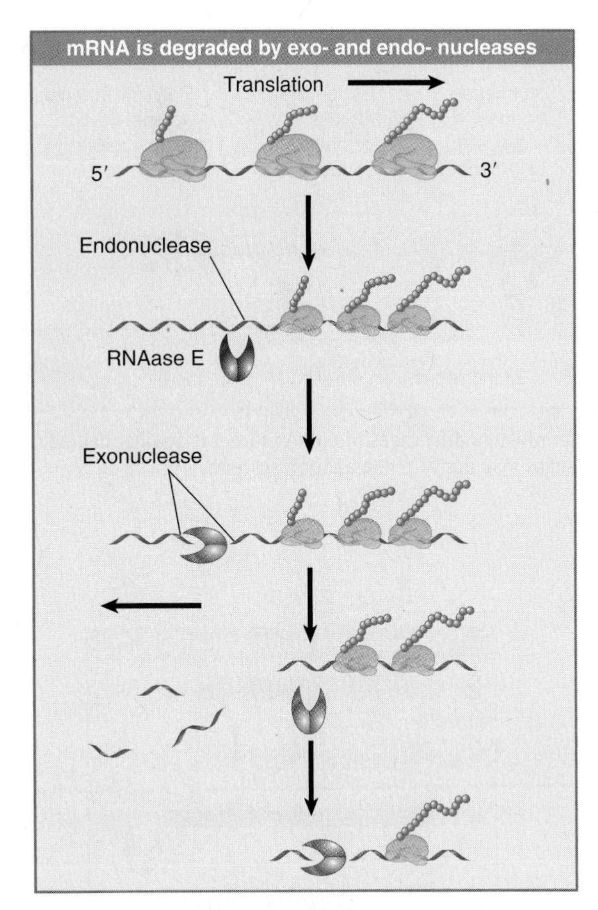

FIGURE 7.20 Degradation of bacterial mRNA is a two-stage process. Endonucleolytic cleavages proceed 5′–3′ behind the ribosomes. The released fragments are degraded by exonucleases that move 3′–5′.

7.12 mRNA Stability Depends on Its Structure and Sequence

Key concepts

- The modifications at both ends of mRNA protect it against degradation by exonucleases.
- Specific sequences within an mRNA may have stabilizing or destabilizing effects.
- Destabilization may be triggered by loss of poly(A).

The major features of mRNA that affect its stability are summarized in FIGURE 7.21. Both structure and sequence are important. The 5′ and 3′ terminal structures protect against degradation, and specific sequences within the mRNA may either serve as targets to trigger degradation or may protect against degradation:

- The modifications at the 5′ and 3′ ends of mRNA play an important role in

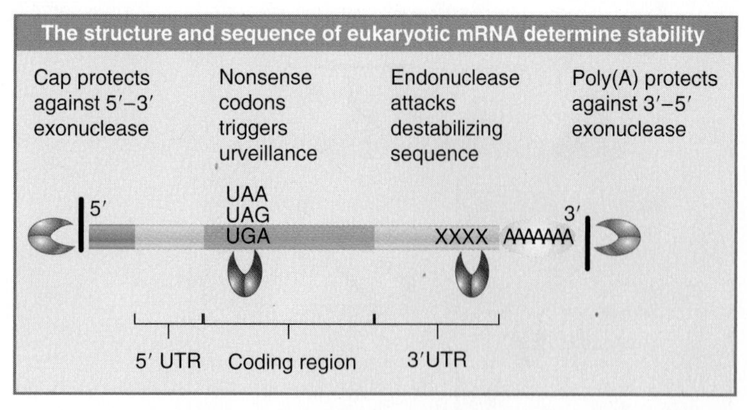

FIGURE 7.21 The terminal modifications of mRNA protect it against degradation. Internal sequences may activate degradation systems.

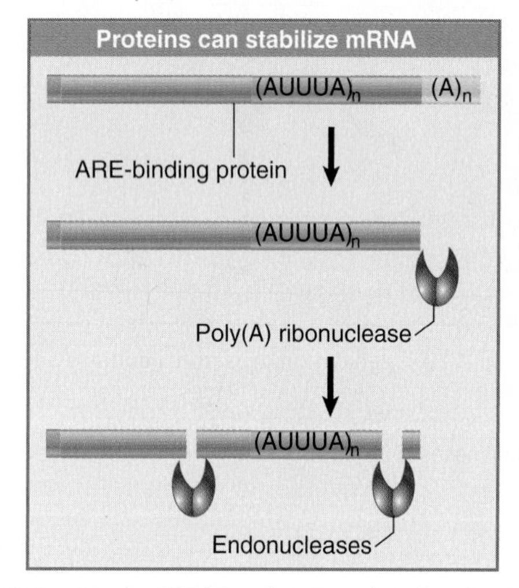

FIGURE 7.22 An ARE in a 3′ nontranslated region initiates degradation of mRNA.

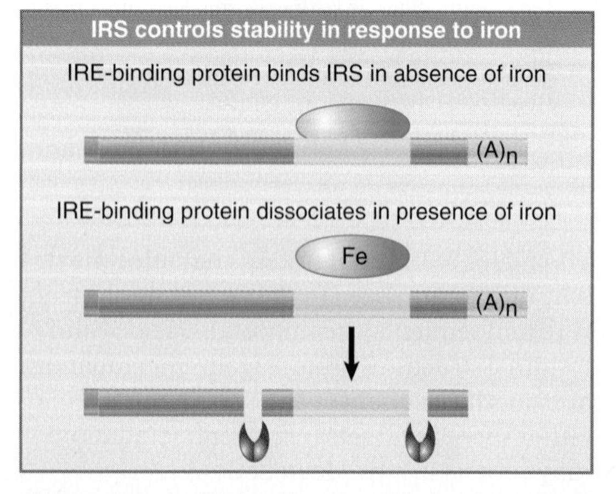

FIGURE 7.23 An IRE in a 3′ nontranslated region controls mRNA stability.

preventing exonuclease attack. The cap prevents 5′–3′ exonucleases from attacking the 5′ end, and the poly(A) prevents 3′–5′ exonucleases from attacking the 3′ end.

- Specific sequence elements within the mRNA may stabilize or destabilize it. The most common location for destabilizing elements is within the 3′ untranslated region. The presence of such an element shortens the lifetime of the mRNA.
- Within the coding region, mutations that create termination codons trigger a surveillance system that degrades the mRNA (see Section 7.14, Nonsense Mutations Trigger a Surveillance System).

Destabilizing elements have been found in several yeast mRNAs, although as yet we do not see any common sequences or know how they destabilize the mRNA. They do not necessarily act directly (by providing targets for endonucleases), but may function indirectly, perhaps by encouraging deadenylation. The criterion for defining a destabilizing sequence element is that its introduction into a new mRNA may cause it to be degraded. The removal of an element from an mRNA does not necessarily stabilize it, suggesting that an individual mRNA can have more than one destabilizing element.

A common feature in some unstable mRNAs is the presence of an AU-rich sequence of ~50 bases (called the ARE) that is found in the 3′ trailer region. The consensus sequence in the ARE is the pentanucleotide AUUUA, repeated several times. FIGURE 7.22 shows that the ARE triggers destabilization by a two-stage process: first the mRNA is deadenylated, and then it decays. The deadenylation is probably needed because it causes loss of the poly(A)-binding protein, whose presence stabilizes the 3′ region (see Section 7.13, mRNA Degradation Involves Multiple Activities).

In some cases, an mRNA can be stabilized by specifically inhibiting the function of a destabilizing element. Transferrin mRNA contains a sequence called the iron-responsive element (IRE), which controls the response of the mRNA to changes in iron concentration. The IRE is located in the 3′ nontranslated region, and contains stem–loop structures that bind a protein whose affinity for the mRNA is controlled by iron. FIGURE 7.23 shows that binding of the protein to the IRE stabilizes the mRNA by inhibiting the function of (unidentified) destabilizing sequences in the vicinity. This is a general model for the stabilization of mRNA, that is, stability is conferred by inhibiting the function of destabilizing sequences.

Decapping and degradation occur in large protein complexes, which may be localized in the form of discrete cytoplasmic units called P-bodies, to which other enzymes involved in mRNA production or metabolism may be localized. In fact, P-bodies may provide a means for sequestering any mRNAs that are not actively involved in translation.

7.13 mRNA Degradation Involves Multiple Activities

Key concepts

- Degradation of yeast mRNA requires removal of the 5′ cap and the 3′ poly(A).
- One yeast pathway involves exonucleolytic degradation from 5′→3′.
- Another yeast pathway uses a complex of several exonucleases that work in the 3′→5′ direction.
- The deadenylase of animal cells may bind directly to the 5′ cap.

We know the most about the degradation of mRNA in yeast. There are basically two pathways. Both start with removal of the poly(A) tail. This is catalyzed by a specific deadenylase, which probably functions as part of a large protein complex. (The catalytic subunit is the exonuclease Ccr4 in yeast; it is also the exonuclease PARN in vertebrates, which is related to RNAase D.) The enzyme action is processive—once it has started to degrade a particular mRNA substrate, it continues to whittle away at that mRNA, base by base.

The major degradation pathway is summarized in FIGURE 7.24. Deadenylation at the 3′ end triggers decapping at the 5′ end. The basis for this relationship is that the presence of the PABP (poly(A)-binding protein) on the poly(A) prevents the decapping enzyme from binding to the 5′ end. PABP is released when the length of poly(A) falls below 10 to 15 residues. Decapping occurs by cleaving the methylated base off the 5′ end to leave a monophosphate in the reaction: $m^7GpppX \ldots \rightarrow m^7GDP + pX \ldots$ The enzyme requires the 7-methyl group.

Each end of the mRNA influences events that occur at the other end. This is explained by the fact that the two ends of the mRNA are held together by the factors involved in protein

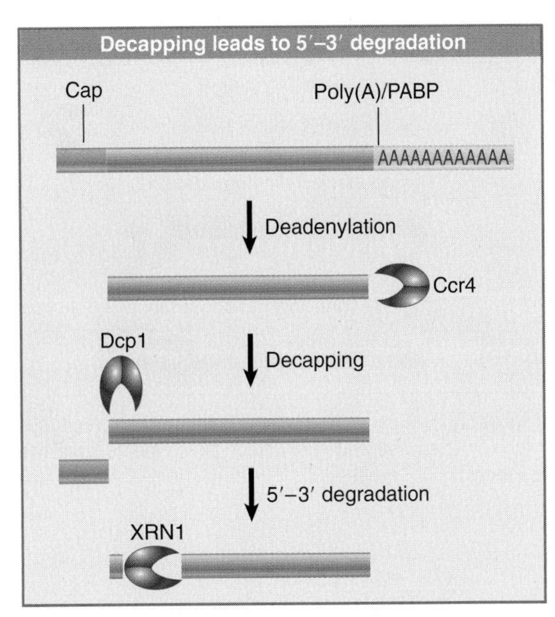

FIGURE 7.24 Deadenylation allows decapping to occur, which leads to endonucleolytic cleavage from the 5′ end.

synthesis (see Section 8.9, Eukaryotes Use a Complex of Many Initiation Factors). The effect of PABP on decapping allows the 3′ end to have an effect in stabilizing the 5′ end. There is also a connection between the structure at the 5′ end and degradation at the 3′ end. The deadenylase directly binds to the 5′ cap, and this interaction is in fact needed for its exonucleolytic attack on the poly(A).

What is the rationale for the connection between events occurring at both ends of an mRNA? Perhaps it is necessary to ensure that the mRNA is not left in a state (having the structure of one end but not the other) that might compete with active mRNA for the proteins that bind to the ends.

Removal of the cap triggers the 5′–3′ degradation pathway in which the mRNA is degraded rapidly from the 5′ end, by the 5′–3′ exonuclease XRN1. The decapping enzyme is concentrated in discrete cytoplasmic foci, which may be "processing bodies" where the mRNA is deadenylated and then degraded after it has been decapped.

In the second pathway, deadenylated yeast mRNAs can be degraded by the 3′–5′ exonuclease activity of the exosome , a complex of >9 exonucleases. The exosome is also involved in processing precursors for rRNAs. The aggregation of the individual exonucleases into the exosome complex may enable 3′–5′ exonucleolytic activities to be coordinately controlled. The exosome may also degrade fragments of mRNA

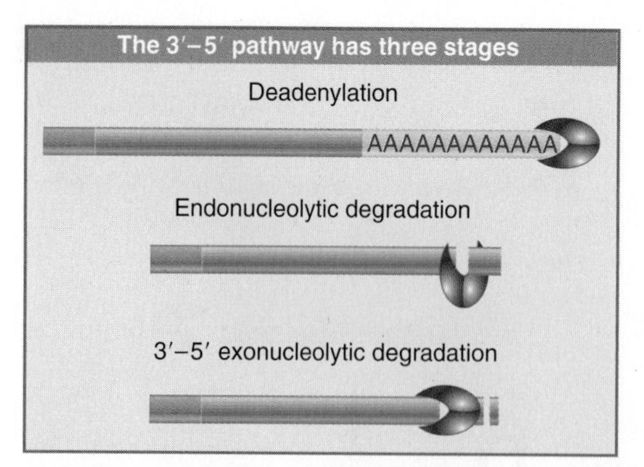

FIGURE 7.25 Deadenylation may lead directly to endonucleolytic cleavage and exonucleolytic cleavage from the 3′ end(s).

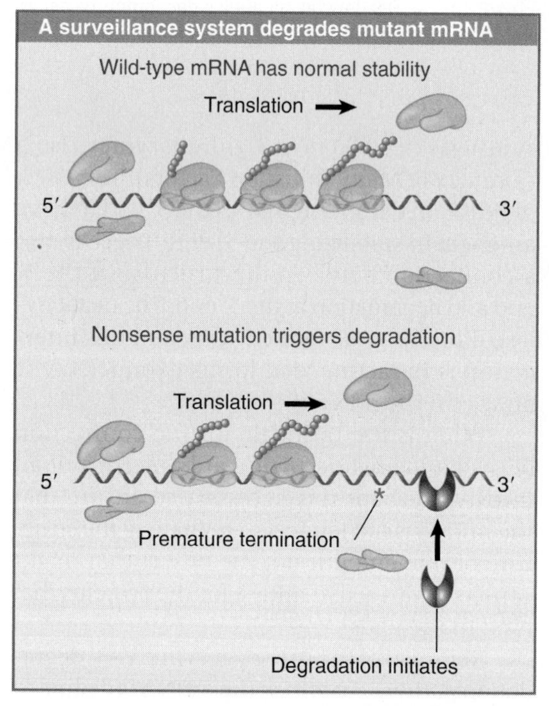

FIGURE 7.26 Nonsense mutations may cause mRNA to be degraded.

released by endonucleolytic cleavage. FIGURE 7.25 shows that the 3′–5′ degradation pathway may actually involve combinations of endonucleolytic and exonucleolytic action. The exosome is also found in the nucleus, where it degrades unspliced precursors to mRNA.

Yeast mutants lacking either exonucleolytic pathway degrade their mRNAs more slowly, but the loss of both pathways is lethal.

7.14 Nonsense Mutations Trigger a Surveillance System

Key concepts

- Nonsense mutations cause mRNA to be degraded.
- Genes coding for the degradation system have been found in yeast and worms.

Another pathway for degradation is identified by **nonsense-mediated mRNA decay.** FIGURE 7.26 shows that the introduction of a nonsense mutation often leads to increased degradation of the mRNA. As may be expected from dependence on a termination codon, the degradation occurs in the cytoplasm. It may represent a quality control or **surveillance** system for removing nonfunctional mRNAs.

The surveillance system has been studied best in yeast and *C. elegans,* but may also be important in animal cells. For example, during the formation of immunoglobulins and T cell receptors in cells of the immune system, genes are modified by somatic recombination and mutation (see Chapter 23, Immune Diversity). This generates a significant number of nonfunctional genes whose RNA products are disposed of by a surveillance system.

In yeast, the degradation requires sequence elements (called *DSE*) that are downstream of the nonsense mutation. The simplest possibility would be that these are destabilizing elements and that translation suppresses their use. When translation is blocked, however, the mRNA is stabilized. This suggests that the process of degradation is linked to translation of the mRNA or to the termination event in some direct way.

Genes that are required for the process have been identified in *S. cerevisiae* (*upf* loci) and *C. elegans* (*smg* loci) by identifying suppressors of nonsense-mediated degradation. Mutations in these genes stabilize aberrant mRNAs, but do not affect the stability of most wild-type transcripts. One of these genes is conserved in eukaryotes *(upf1/smg2).* It codes for an ATP-dependent helicase (an enzyme that unwinds double-stranded nucleic acids into single strands). This implies that recognition of the mRNA as an appropriate target for degradation requires a change in its structure.

Upf1 interacts with the release factors (eRF1 and eRF3) that catalyze termination, which is probably how it recognizes the termination event. It may then "scan" the mRNA by moving toward the 3′ end to look for the downstream sequence elements.

In mammalian cells, the surveillance system appears to work only on mutations located prior to the last exon—in other words, there must be an intron after the site of mutation. This suggests that the system requires some event to occur in the nucleus, before the introns are removed by splicing. One possibility is that proteins attach to the mRNA in the nucleus at the exon–exon boundary when a splicing event occurs. FIGURE 7.27 shows a general model for the operation of such a system. This is similar to the way in which an mRNA may be marked for export from the nucleus (see Section 26.10, Splicing Is Connected to Export of mRNA). Attachment of a protein to the exon–exon junction creates a mark of the event that persists into the cytoplasm. Human homologues of the yeast Upf2,3 proteins may be involved in such a system. They bind specifically to mRNA that has been spliced.

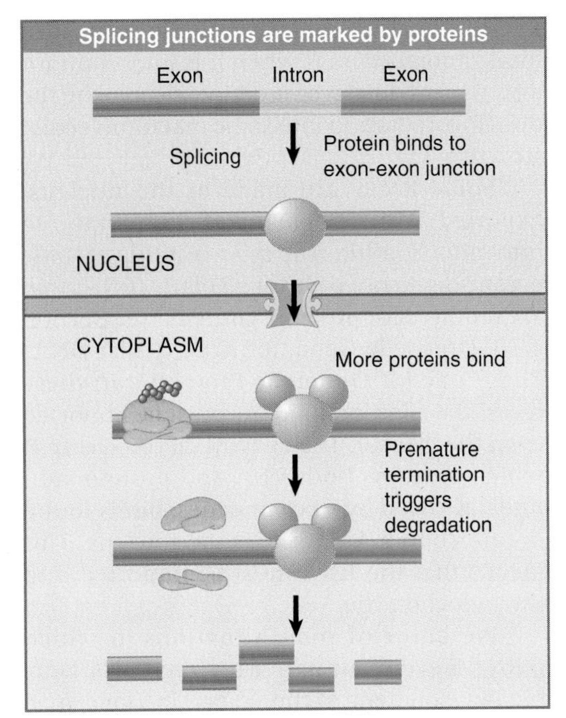

FIGURE 7.27 A surveillance system could have two types of components. Protein(s) must bind in the nucleus to mark the result of a splicing event. Other proteins could bind to the mark either in the nucleus or cytoplasm. They are triggered to act to degrade the mRNA when ribosomes terminate prematurely.

7.15 Eukaryotic RNAs Are Transported

Key concepts

- RNA is transported through a membrane as a ribonucleoprotein particle.
- All eukaryotic RNAs that function in the cytoplasm must be exported from the nucleus.
- tRNAs and the RNA component of a ribonuclease are imported into mitochondria.
- mRNAs can travel long distances between plant cells.

A bacterium consists of only a single compartment, so all the RNAs function in the same environment in which they are synthesized. This is most striking in the case of mRNA, where translation occurs simultaneously with transcription (see Section 7.7, The Life Cycle of Bacterial Messenger RNA).

RNA is transported through membranes in the variety of instances summarized in FIGURE 7.28. It poses a significant thermodynamic problem to transport a highly negative RNA through a hydrophobic membrane, and the solution is to transport the RNA packaged with proteins.

In eukaryotic cells, RNAs are transcribed in the nucleus, but translation occurs in the cytoplasm. Each type of RNA must be transported into the cytoplasm to assemble the apparatus for translation. The rRNA assembles with ribosomal proteins into immature ribosome subunits that are the substrates for the transport system. tRNA is transported by a specific protein system. mRNA is transported as a ribonucleoprotein, which forms on the RNA transcript in the nucleus (see Chapter 26, RNA Splicing and Processing). These processes are common to all eukaryotic cells. Many mRNAs are translated in the cytosol, but some are localized within the cell by means of attachment to

Eukaryotic RNA can be transported between cell compartments

RNA	Transport	Location
All RNA	Nucleus → cytoplasm	All cells
tRNA	Nucleus → mitochondrion	Many cells
mRNA	Nurse cell → oocyte	Fly embryogenesis
mRNA	Anterior → posterior oocyte	Fly embryogenesis
mRNA	Cell → cell	Plant phloem

FIGURE 7.28 RNAs are transported through membranes in a variety of systems.

a cytoskeletal element. One situation in which localization occurs is when it is important for a protein product to be produced near to the site of its incorporation into some macromolecular structure.

Some RNAs are made in the nucleus, exported to the cytosol, and then imported into mitochondria. The mitochondria of some organisms do not code for all of the tRNAs that are required for protein synthesis (see Section 4.10, Organelle Genomes Are Circular DNAs That Code for Organelle Proteins). In these cases, the additional tRNAs must be imported from the cytosol. The enzyme ribonuclease P, which contains both RNA and protein subunits, is coded by nuclear genes, but is found in mitochondria as well as the nucleus. This means that the RNA must be imported into the mitochondria.

We know of some situations in which mRNA is even transported between cells. During development of the oocyte in *Drosophila*, certain mRNAs are transported into the egg from the nurse cells that surround it. The nurse cells have specialized junctions with the oocyte that allow passage of material needed for early development. This material includes certain mRNAs. Once in the egg, these mRNAs take up specific locations. Some simply diffuse from the anterior end where they enter, but others are transported the full length of the egg to the posterior end by a motor attached to microtubules.

The most striking case of transport of mRNA has been found in plants. Movement of individual nucleic acids over long distances was first discovered in plants, where viral movement proteins help propagate the viral infection by transporting an RNA virus genome through the plasmodesmata (connections between cells). Plants also have a defense system, which causes cells to silence an infecting virus. This, too, may involve the spread of components including RNA over long distance between cells. Now it has turned out that similar systems may transport mRNAs between plant cells. Although the existence of the systems has been known for some time, it is only recently that their functional importance has been demonstrated. This was shown by grafting wild-type tomato plants onto plants that had the dominant mutation *Me* (which causes a change in the shape of the leaf). mRNA from the mutant stock was transported into the leaves of the wild-type graft, where it changed their shape.

7.16 mRNA Can Be Specifically Localized

Key concepts

- Yeast Ash1 mRNA forms a ribonucleoprotein that binds to a myosin motor.
- A motor transports it along actin filaments into the daughter bud.
- It is anchored and translated in the bud, so that the protein is found only in the bud.

An mRNA is synthesized in the nucleus but translated in the cytoplasm of a eukaryotic cell. It passes into the cytoplasm in the form of a ribonucleoprotein particle that is transported through the nuclear pore. Once in the cytosol, the mRNA may associate with ribosomes and be translated. The cytosol is a crowded place occupied by a high concentration of proteins. It is not clear how freely a polysome can diffuse within the cytosol, and most mRNAs are probably translated in random locations, determined by their point of entry into the cytosol, and the distance that they may have moved away from it. However, some mRNAs are translated at specific sites. This may be accomplished by several mechanisms:

- An mRNA may be specifically transported to a site where it is translated.
- It may be universally distributed but degraded at all sites except the site of translation.
- It may be freely diffusible but become trapped at the site of translation.

One of the best characterized cases of localization within a cell is that of Ash1 in yeast. Ash1 represses expression of the HO endonuclease in the budding daughter cell, with the result that HO is expressed only in the mother cell. The consequence is that mating type is changed only in the mother cell (see Section 19.24, Regulation of HO Expression Controls Switching). The cause of the restriction to the daughter cell is that all the Ash1 mRNA is transported from the mother cell, where it is made, into the budding daughter cell.

Mutations in any one of five genes, called *SHE1-5*, prevent the specific localization and cause Ash1 mRNA to be symmetrically distributed in both mother and daughter compartments. The proteins She1, -2, and -3 bind Ash1 mRNA into a ribonucleoprotein particle that transports the mRNA into the daughter cell. FIGURE 7.29 shows the functions of the proteins. She1p is a myosin (previously identified as Myo4), and She3 and

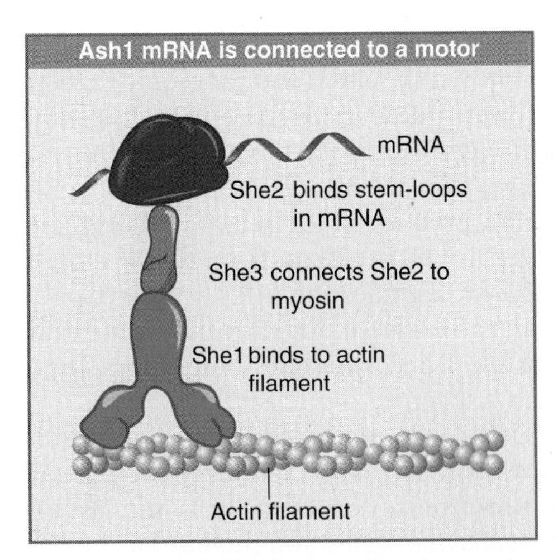

FIGURE 7.29 Ash1 mRNA forms a ribo-nucleoprotein containing a myosin motor that moves it along an actin filament.

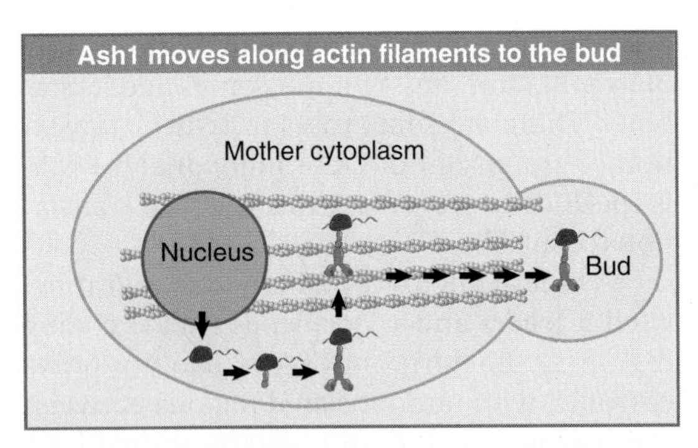

FIGURE 7.30 Ash1 mRNA is exported from the nucleus into the cytoplasm, where it is assembled into a complex with the She proteins. The complex transports it along actin filaments to the bud.

She2 are proteins that connect the myosin to the mRNA. The myosin is a motor that moves the mRNA along actin filaments.

FIGURE 7.30 summarizes the overall process. Ash1 mRNA is exported from the nucleus in the form of a ribonucleoprotein. In the cytoplasm it is first bound by She2, which recognizes some stem–loop secondary structures within the mRNA. Then She3 binds to She2, after which the myosin She1 binds. Next, the particle hooks on to an actin filament and moves to the bud. When Ash1 mRNA reaches the bud, it is anchored there, probably by proteins that bind specifically to the mRNA.

Similar principles govern other cases where mRNAs are transported to specific sites. The mRNA is recognized by means of *cis*-acting sequences, which usually are regions of secondary structure in the 3′ untranslated region. (Ash1 mRNA is unusual in that the *cis*-acting regions are in the coding frame.) The mRNA is packaged into a ribonucleoprotein particle. In some cases, the transported mRNA can be visualized in very large particles called mRNA granules. The particles are large enough (several times the size of a ribosome) to contain many protein and RNA components.

A transported mRNP must be connected to a motor that moves it along a system of tracks. The tracks can be either actin filaments or microtubules. Whereas Ash1 uses a myosin motor on actin tracks, *oscar* mRNA in the *Drosophila* egg uses a kinesin motor to move along microtubules. Once the mRNA reaches its destination, it needs to be anchored in order to prevent it from diffusing away. Less is known about this, but the process appears to be independent of transport. An mRNA that is transported along microtubules may anchor to actin filaments at its destination.

7.17 Summary

Genetic information carried by DNA is expressed in two stages: transcription of DNA into mRNA and translation of the mRNA into protein. Messenger RNA is transcribed from one strand of DNA. It is complementary to this (noncoding) strand and identical with the other (coding) strand. The sequence of mRNA, in triplet codons 5′–3′, is related to the amino acid sequence of protein, N- to C-terminal.

The adaptor that interprets the meaning of a codon is transfer RNA, which has a compact L-shaped tertiary structure; one end of the tRNA has an anticodon that is complementary to the codon, and the other end can be covalently linked to the specific amino acid that corresponds to the target codon. A tRNA carrying an amino acid is called an aminoacyl-tRNA.

The ribosome provides the apparatus that allows aminoacyl-tRNAs to bind to their codons on mRNA. The small subunit of the ribosome is bound to mRNA; the large subunit carries the nascent polypeptide. A ribosome moves along mRNA from an initiation site in the 5′ region to a termination site in the 3′ region, and the appropriate aminoacyl-tRNAs respond to their codons, unloading their amino acids, so that the growing polypeptide chain extends by one residue for each codon traversed.

The translational apparatus is not specific for tissue or organism; an mRNA from one source can be translated by the ribosomes and tRNAs from another source. The number of times any mRNA is translated is a function of the affinity of its initiation site(s) for ribosomes and its stability. There are some cases in which translation of groups of mRNA or individual mRNAs is specifically prevented: this is called translational control.

A typical mRNA contains both a nontranslated 5′ leader and a 3′ trailer as well as coding region(s). Bacterial mRNA is usually polycistronic, with nontranslated regions between the cistrons. Each cistron is represented by a coding region that starts with a specific initiation site and ends with a termination site. Ribosome subunits associate at the initiation site and dissociate at the termination site of each coding region.

A growing *E. coli* bacterium has ~20,000 ribosomes and ~200,000 tRNAs, mostly in the form of aminoacyl-tRNA. There are ~1500 mRNA molecules, representing two to three copies of each of 600 different messengers.

A single mRNA can be translated by many ribosomes simultaneously, generating a polyribosome (or polysome). Bacterial polysomes are large, typically with tens of ribosomes bound to a single mRNA. Eukaryotic polysomes are smaller, typically with fewer than ten ribosomes; each mRNA carries only a single coding sequence.

Bacterial mRNA has an extremely short half-life of only a few minutes. The 5′ end starts translation even while the downstream sequences are being transcribed. Degradation is initiated by endonucleases that cut at discrete sites, following the ribosomes in the 5′–3′ direction, after which exonucleases reduce the fragments to nucleotides by degrading them from the released 3′ end toward the 5′ end. Individual sequences may promote or retard degradation in bacterial mRNAs.

Eukaryotic mRNA must be processed in the nucleus before it is transported to the cytoplasm for translation. A methylated cap is added to the 5′ end. It consists of a nucleotide added to the original end by a 5′–5′ bond, after which methyl groups are added. Most eukaryotic mRNA has an ~200 base sequence of poly(A) added to its 3′ terminus in the nucleus after transcription, but poly(A)$^-$ mRNAs appear to be translated and degraded with the same kinetics as poly(A)$^+$ mRNAs. Eukaryotic mRNA exists as a ribonucleoprotein particle; in some cases mRNPs are stored that fail to be translated. Eukaryotic mRNAs are usually stable for several hours. They may have multiple sequences that initiate degradation; examples are known in which the process is regulated.

Yeast mRNA is degraded by (at least) two pathways. Both start with the removal of poly(A) from the 3′ end, causing loss of poly(A)-binding protein, which in turn leads to removal of the methylated cap from the 5′ end. One pathway degrades the mRNA from the 5′ end by an exonuclease. Another pathway degrades from the 3′ end by the exosome, a complex containing several exonucleases.

Nonsense-mediated degradation leads to the destruction of mRNAs that have a termination (nonsense) codon prior to the last exon. The *upf* loci in yeast and the *smg* loci in worms are required for the process. They include a helicase activity to unwind mRNA and a protein that interacts with the factors that terminate protein synthesis. The features of the process in mammalian cells suggest that some of the proteins attach to the mRNA in the nucleus when RNA splicing occurs to remove introns.

mRNAs can be transported to specific locations within a cell (especially in embryonic development). In the Ash1 system in yeast, mRNA is transported from the mother cell into the daughter cell by a myosin motor that moves on actin filaments. In plants, mRNAs can be transported long distances between cells.

References

7.3 Transfer RNA Forms a Cloverleaf

Review

Soll, D. and RajBhandary, U. L. (1995). *tRNA Structure, Biosynthesis, and Function.* Washington, DC: American Society for Microbiology.

Research

Chapeville, F. et al. (1962). On the role of soluble RNA in coding for amino acids. *Proc. Natl. Acad. Sci. USA* 48, 1086–1092.

Hoagland, M. B. et al. (1958). A soluble RNA intermediate in protein synthesis. *J. Biol. Chem.* 231, 241–257.

Holley, R. W. et al. (1965). Structure of an RNA. *Science* 147, 1462–1465

7.5 Messenger RNA Is Translated by Ribosomes

Research

Dintzis, H. M. (1961). Assembly of the peptide chain of hemoglobin. *Proc. Natl. Acad. Sci. USA* 47, 247–261.

7.6 Many Ribosomes Bind to One mRNA

Research

Slayter, H. S. et al. (1963). The visualization of polyribosome structure. *J. Mol. Biol.* 7, 652–657.

7.7 The Life Cycle of Bacterial Messenger RNA

Research

Brenner, S. Jacob, F., and Meselson, M. (1961). An unstable intermediate carrying information from genes to ribosomes for protein synthesis. *Nature* 190, 576–581.

7.9 The 5′ End of Eukaryotic mRNA Is Capped

Review

Bannerjee, A. K. (1980). 5′-Terminal cap structure in eukaryotic mRNAs. *Microbiol. Rev.* 44, 175–205.

7.10 The 3′ Terminus Is Polyadenylated

Review

Jackson, R. J. and Standart, N. (1990). Do the poly(A) tail and 3′ untranslated region control mRNA translation? *Cell* 62, 15–24.

Research

Darnell, J. et al. (1971). Poly(A) sequences: role in conversion of nuclear RNA into mRNA. *Science* 174, 507–510.

7.11 Bacterial mRNA Degradation Involves Multiple Enzymes

Reviews

Caponigro, G. and Parker, R. (1996). Mechanisms and control of mRNA turnover in *S. cerevisiae*. *Microbiol. Rev.* 60, 233–249.

Grunberg-Manago, M. (1999). mRNA stability and its role in control of gene expression in bacteria and phages. *Annu. Rev. Genet.* 33, 193–227.

Research

Miczak, A., Kaberdin, V. R., Wei, C.-L., and Lin-chao, S. (1996). Proteins associated with RNAase E in a multicomponent ribonucleolytic complex. *Proc. Natl. Acad. Sci. USA* 93, 3865–3869.

O'Hara, E. B. et al. (1995). Polyadenylation helps regulate mRNA decay in *E. coli*. *Proc. Natl. Acad. Sci. USA* 92, 1807–1811.

Vanzo, N. F. et al. (1998). RNAase E organizes the protein interactions in the *E. coli* RNA degradaosome. *Genes Dev.* 12, 2770–2781.

7.12 mRNA Stability Depends on Its Structure and Sequence

Reviews

Coller, J., and Parker, R. (2004). Eukaryotic mRNA decapping. *Annu. Rev. Biochem.* 73, 861–890.

Moore, M. J. (2005). From birth to death: the complex lives of eukaryotic mRNAs. *Science* 309, 1514–1518.

Ross, J. (1995). mRNA stability in mammalian cells. *Microbiol. Rev.* 59, 423–450.

Sachs, A. (1993). Messenger RNA degradation in eukaryotes. *Cell* 74, 413–421.

7.13 mRNA Degradation Involves Multiple Activities

Reviews

Beelman, C. A. and Parker, R. (1995). Degradation of mRNA in eukaryotes. *Cell* 81, 179–183.

Coller, J. and Parker, R. (2004). Eukaryotic mRNA decapping. *Annu. Rev. Biochem.* 73, 861–890.

Jacobson, A. and Peltz, S. W. (1996). Interrelationships of the pathways of mRNA decay and translation in eukaryotic cells. *Annu. Rev. Biochem.* 65, 693–739.

Research

Allmang, C., Petfalski, E., Podtelejnikov, A., Mann, M., Tollervey, D., and Mitchell, P. (1999). The yeast exosome and human PM-Scl are related complexes of 3′–5′ exonucleases. *Genes Dev.* 13, 2148–2158.

Bousquet-Antonelli, C., Presutti, C., and Tollervey, D. (2000). Identification of a regulated pathway for nuclear pre-mRNA turnover. *Cell* 102, 765–775.

Gao, M., Fritz, D. T., Ford, L. P., and Wilusz, J. (2000). Interaction between a poly(A)-specific ribonuclease and the 5′ cap influences mRNA deadenylation rates *in vitro*. *Mol. Cell* 5, 479–488.

Mitchell, P. et al. (1997). The exosome: a conserved eukaryotic RNA processing complex containing multiple 3′–5′ exoribonuclease activities. *Cell* 91, 457–466.

Muhlrad, D., Decker, C. J., and Parker, R. (1994). Deadenylation of the unstable mRNA encoded by the yeast MFA2 gene leads to decapping followed by 5′–3′ digestion of the transcript. *Genes Dev.* 8, 855–866.

Sheth, U., and Parker, R. (2003). Decapping and decay of messenger RNA occur in cytoplasmic processing bodies. *Science* 300, 805–808.

Tucker, M., Valencia-Sanchez, M. A., Staples, R. R., Chen, J., Denis, C. L., and Parker, R. (2001). The transcription factor associated Ccr4 and Caf1 proteins are components of the major cytoplasmic mRNA deadenylase in *S. cerevisiae*. *Cell* 104, 377–386.

7.14 Nonsense Mutations Trigger a Surveillance System

Review

Hilleren, P. and Parker, R. (1999). Mechanisms of mRNA surveillance in eukaryotes. *Annu. Rev. Genet.* 33, 229–260.

Research

Cui, Y., Hagan, K. W., Zhang, S., and Peltz, S. W. (1995). Identification and characterization of genes that are required for the accelerated degradation of mRNAs containing a premature translational termination codon. *Genes Dev.* 9, 423–436.

Czaplinski, K., Ruiz-Echevarria, M. J., Paushkin, S. V., Han, X., Weng, Y., Perlick, H. A., Dietz, H. C., Ter-Avanesyan, M. D., and Peltz, S. W. (1998). The surveillance complex interacts with the translation release factors to enhance termination and degrade aberrant mRNAs. *Genes Dev.* 12, 1665–1677.

Le Hir, H., Moore, M. J., and Maquat, L. E. (2000). Pre-mRNA splicing alters mRNP composition: evidence for stable association of proteins at exon–exon junctions. *Genes Dev.* 14, 1098–1108.

Lykke-Andersen, J., Shu, M. D., and Steitz, J. A. (2000). Human Upf proteins target an mRNA for nonsense-mediated decay when bound downstream of a termination codon. *Cell* 103, 1121–1131.

Peltz, S. W., Brown, A. H., and Jacobson, A. (1993). mRNA destabilization triggered by premature translational termination depends on at least three *cis*-acting sequence elements and one *trans*-acting factor. *Genes Dev.* 7, 1737–1754.

Pulak, R. and Anderson, P. (1993). mRNA surveillance by the *C. elegans smg* genes. *Genes Dev.* 7, 1885–1897.

Ruiz-Echevarria, M. J. et al. (1998). Identifying the right stop: determining how the surveillance complex recognizes and degrades an aberrant mRNA. *EMBO J.* 15, 2810–2819.

Weng, Y., Czaplinski, K., and Peltz, S. (1996). Genetic and biochemical characterization of mutants in the ATPase and helicase regions of the Upf1 protein. *Mol. Cell Biol.* 16, 5477–5490.

Weng, Y., Czaplinski, K., and Peltz, S. (1996). Identification and characterization of mutations in the *upf1* gene that affect the Upf protein complex, nonsense suppression, but not mRNA turnover. *Mol. Cell Biol.* 16, 5491–5506.

7.15 Eukaryotic RNAs Are Transported

Reviews

Ghoshroy, S., Lartey, R., Sheng, J., and Citovsky, V. (1997). Transport of proteins and nucleic acids through plasmodesmata. *Ann. Rev. Plant. Physiol. Plant. Mol. Biol.* 48, 27–50.

Jansen, R. P. (2001). mRNA localization: message on the move. *Nat. Rev. Mol. Cell Biol.* 2, 247–256.

Lucas, W. J. and Gilbertson, R. L. (1994). Plasmodesmata in relation to viral movement within leaf tissues. *Ann. Rev. Phytopathol.* 32, 387–411.

Vance, V. and Vaucheret, H. (2001). RNA silencing in plants—defense and counterdefense. *Science* 292, 2277–2280.

Research

Kim, M., Canio, W., Kessler, S., and Sinha, N. (2001). Developmental changes due to long-distance movement of a homeobox fusion transcript in tomato. *Science* 293, 287–289.

Puranam, R. S. and Attardi, G. (2001). The RNase P associated with HeLa cell mitochondria contains an essential RNA component identical in sequence to that of the nuclear RNase P. *Mol. Cell Biol.* 21, 548–561.

7.16 mRNA Can Be Specifically Localized

Reviews

Chartrand, P., Singer, R. H., and Long, R. M. (2001). RNP localization and transport in yeast. *Annu. Rev. Cell Dev. Biol.* 17, 297–310.

Jansen, R. P. (2001). mRNA localization: message on the move. *Nat. Rev. Mol. Cell Biol.* 2, 247–256.

Kloc, M., Zearfoss, N. R., and Etkin, L. D. (2002). Mechanisms of subcellular mRNA localization. *Cell* 108, 533–544.

Palacios, I. M. and Johnston, D. S. (2001). Getting the message across: the intracellular localization of mRNAs in higher eukaryotes. *Annu. Rev. Cell Dev. Biol.* 17, 569–614.

Research

Bertrand, E., Chartrand, P., Schaefer, M., Shenoy, S. M., Singer, R. H., and Long, R. M. (1998). Localization of ASH1 mRNA particles in living yeast. *Mol. Cell* 2, 437–445.

Long, R. M., Singer, R. H., Meng, X., Gonzalez, I., Nasmyth, K., and Jansen, R. P. (1997). Mating type switching in yeast controlled by asymmetric localization of ASH1 mRNA. *Science* 277, 383–387.

8

Protein Synthesis

CHAPTER OUTLINE

8.1 Introduction

8.2 Protein Synthesis Occurs by Initiation, Elongation, and Termination
- The ribosome has three tRNA-binding sites.
- An aminoacyl-tRNA enters the A site.
- Peptidyl-tRNA is bound in the P site.
- Deacylated tRNA exits via the E site.
- An amino acid is added to the polypeptide chain by transferring the polypeptide from peptidyl-tRNA in the P site to aminoacyl-tRNA in the A site.

8.3 Special Mechanisms Control the Accuracy of Protein Synthesis
- The accuracy of protein synthesis is controlled by specific mechanisms at each stage.

8.4 Initiation in Bacteria Needs 30S Subunits and Accessory Factors
- Initiation of protein synthesis requires separate 30S and 50S ribosome subunits.
- Initiation factors (IF-1, -2, and -3), which bind to 30S subunits, are also required.
- A 30S subunit carrying initiation factors binds to an initiation site on mRNA to form an initiation complex.
- IF-3 must be released to allow 50S subunits to join the 30S-mRNA complex.

8.5 A Special Initiator tRNA Starts the Polypeptide Chain
- Protein synthesis starts with a methionine amino acid usually coded by AUG.
- Different methionine tRNAs are involved in initiation and elongation.
- The initiator tRNA has unique structural features that distinguish it from all other tRNAs.
- The NH_2 group of the methionine bound to bacterial initiator tRNA is formylated.

8.6 Use of fMet-$tRNA_f$ Is Controlled by IF-2 and the Ribosome
- IF-2 binds the initiator fMet-$tRNA_f$ and allows it to enter the partial P site on the 30S subunit.

8.7 Initiation Involves Base Pairing Between mRNA and rRNA
- An initiation site on bacterial mRNA consists of the AUG initiation codon preceded with a gap of ~10 bases by the Shine–Dalgarno polypurine hexamer.
- The rRNA of the 30S bacterial ribosomal subunit has a complementary sequence that base pairs with the Shine–Dalgarno sequence during initiation.

8.8 Small Subunits Scan for Initiation Sites on Eukaryotic mRNA
- Eukaryotic 40S ribosomal subunits bind to the 5′ end of mRNA and scan the mRNA until they reach an initiation site.
- A eukaryotic initiation site consists of a ten-nucleotide sequence that includes an AUG codon.
- 60S ribosomal subunits join the complex at the initiation site.

8.9 Eukaryotes Use a Complex of Many Initiation Factors
- Initiation factors are required for all stages of initiation, including binding the initiator tRNA, 40S subunit attachment to mRNA, movement along the mRNA, and joining of the 60S subunit.
- Eukaryotic initiator tRNA is a Met-tRNA that is different from the Met-tRNA used in elongation, but the methionine is not formulated.
- eIF2 binds the initiator Met-$tRNA_i$ and GTP, and the complex binds to the 40S subunit before it associates with mRNA.

8.10 Elongation Factor Tu Loads Aminoacyl-tRNA into the A Site
- EF-Tu is a monomeric G protein whose active form (bound to GTP) binds aminoacyl-tRNA.
- The EF-Tu-GTP-aminoacyl-tRNA complex binds to the ribosome A site.

8.11 The Polypeptide Chain Is Transferred to Aminoacyl-tRNA
- The 50S subunit has peptidyl transferase activity.
- The nascent polypeptide chain is transferred from peptidyl-tRNA in the P site to aminoacyl-tRNA in the A site.
- Peptide bond synthesis generates deacylated tRNA in the P site and peptidyl-tRNA in the A site.

8.12 Translocation Moves the Ribosome
- Ribosomal translocation moves the mRNA through the ribosome by three bases.
- Translocation moves deacylated tRNA into the E site and peptidyl-tRNA into the P site, and empties the A site.
- The hybrid state model proposes that translocation occurs in two stages, in which the 50S moves relative to the 30S, and then the 30S moves along mRNA to restore the original conformation.

Continued on next page

8.1 Introduction

An mRNA contains a series of codons that interact with the anticodons of aminoacyl-tRNAs so that a corresponding series of amino acids is incorporated into a polypeptide chain. The ribosome provides the environment for controlling the interaction between mRNA and aminoacyl-tRNA. The ribosome behaves like a small migrating factory that travels along the template engaging in rapid cycles of peptide bond synthesis. Aminoacyl-tRNAs shoot in and out of the particle at a fearsome rate while depositing amino acids, and elongation factors cyclically associate with and dissociate from the ribosome. Together with its accessory factors, the ribosome provides the full range of activities required for all the steps of protein synthesis.

FIGURE 8.1 shows the relative dimensions of the components of the protein synthetic apparatus. The ribosome consists of two subunits that have specific roles in protein synthesis. Messenger RNA is associated with the small subunit; ~30 bases of the mRNA are bound at any time. The mRNA threads its way along the surface close to the junction of the subunits. Two tRNA molecules are active in protein synthesis at any moment, so polypeptide elongation involves reactions taking place at just two of the (roughly) ten codons covered by the ribosome. The two tRNAs are inserted into internal sites that stretch across the subunits. A third tRNA may remain on the ribosome after it has been used in protein synthesis before being recycled.

The basic form of the ribosome has been conserved in evolution, but there are appreciable variations in the overall size and proportions of RNA and protein in the ribosomes of bacteria, eukaryotic cytoplasm, and organelles. FIGURE 8.2 compares the components of bacterial and mammalian ribosomes. Both are ribonucleoprotein particles that contain more RNA than protein. The ribosomal proteins are known as *r-proteins*.

Each of the ribosome subunits contains a major rRNA and a number of small proteins. The large subunit may also contain smaller RNA(s). In *E. coli*, the small (30S) subunit consists of the 16S rRNA and 21 r-proteins. The large (50S) subunit contains 23S rRNA, the small 5S RNA, and 31 proteins. With the exception of one protein present at four copies per ribosome, there is one copy of each protein. The major RNAs constitute the major part of the mass of the bacterial ribosome. Their presence is pervasive, and probably most or all of the ribosomal proteins actually contact rRNA. So the major rRNAs form what is sometimes thought of as the backbone of each subunit—a continuous thread whose presence dominates the structure and which determines the positions of the ribosomal proteins.

The ribosomes of higher eukaryotic cytoplasm are larger than those of bacteria. The total content of both RNA and protein is greater; the major RNA molecules are longer (called 18S and 28S rRNAs), and there are more proteins. Probably most or all of the proteins are present in stoichiometric amounts. RNA is still the predominant component by mass.

Organelle ribosomes are distinct from the ribosomes of the cytosol and take varied forms. In some cases, they are almost the size of bacterial ribosomes and have 70% RNA; in other cases, they are only 60S and have <30% RNA.

The ribosome possesses several active centers, each of which is constructed from a group of proteins associated with a region of ribosomal RNA. The active centers require the direct participation of rRNA in a structural or even catalytic role. Some catalytic functions require individual proteins, but none of the activities can be reproduced by isolated proteins or groups of proteins; they function only in the context of the ribosome.

Two types of information are important in analyzing the ribosome. Mutations implicate particular ribosomal proteins or bases in rRNA in participating in particular reactions. Structural analysis, including direct modification of components of the ribosome and comparisons to identify conserved features in rRNA, identifies the physical locations of components involved in particular functions.

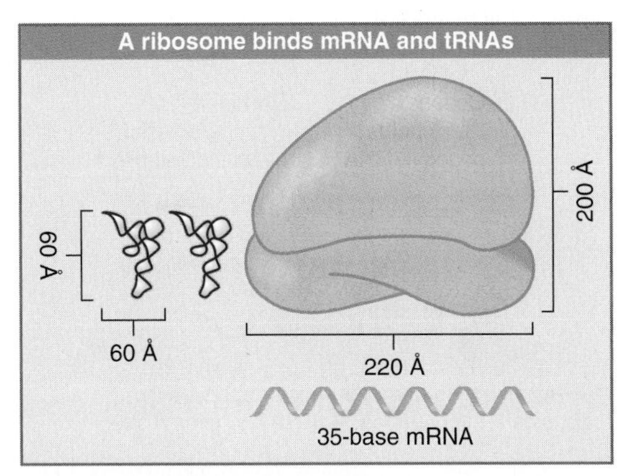

FIGURE 8.1 Size comparisons show that the ribosome is large enough to bind tRNAs and mRNA.

Ribosomes are ribonucleoprotein particles

Ribosomes		rRNAs	r-proteins
Bacterial (70S) mass: 2.5 MDa 66% RNA	50S	23S = 2904 bases 5S = 120 bases	31
	30S	16S = 1542 bases	21
Mammalian (80S) mass: 4.2 MDa 60% RNA	60S	28S = 4718 bases 5.8S = 160 bases 5S = 120 bases	49
	40S	18S = 1874 bases	33

FIGURE 8.2 Ribosomes are large ribonucleoprotein particles that contain more RNA than protein and dissociate into large and small subunits.

8.2 Protein Synthesis Occurs by Initiation, Elongation, and Termination

Key concepts

- The ribosome has three tRNA-binding sites.
- An aminoacyl-tRNA enters the A site.
- Peptidyl-tRNA is bound in the P site.
- Deacylated tRNA exits via the E site.
- An amino acid is added to the polypeptide chain by transferring the polypeptide from peptidyl-tRNA in the P site to aminoacyl-tRNA in the A site.

An amino acid is brought to the ribosome by an aminoacyl-tRNA. Its addition to the growing protein chain occurs by an interaction with the tRNA that brought the previous amino acid.

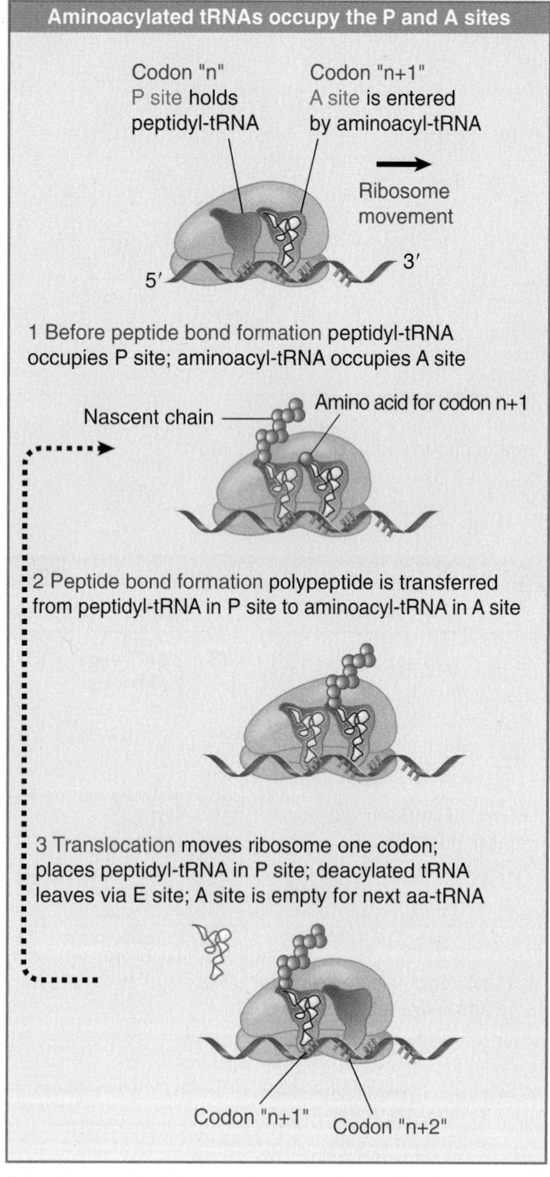

FIGURE 8.3 The ribosome has two sites for binding charged tRNA.

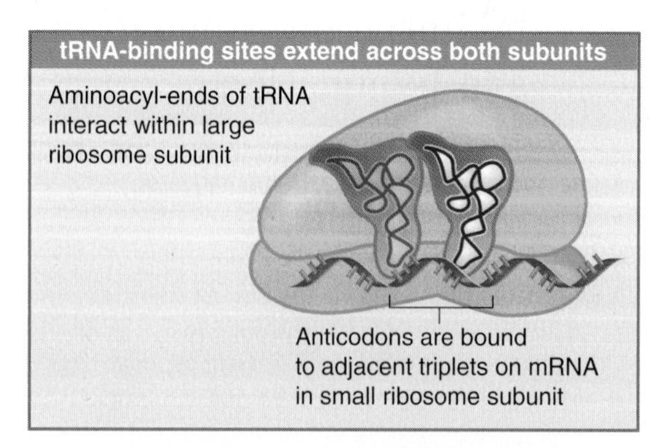

FIGURE 8.4 The P and A sites position the two interacting tRNAs across both ribosome subunits.

Each of these tRNA lies in a distinct site on the ribosome. FIGURE 8.3 shows that the two sites have different features:

- An incoming aminoacyl-tRNA binds to the **A site.** Prior to the entry of aminoacyl-tRNA, the site exposes the codon representing the next amino acid due to be added to the chain.
- The codon representing the most recent amino acid to have been added to the nascent polypeptide chain lies in the **P site.** This site is occupied by **peptidyl-tRNA,** a tRNA carrying the nascent polypeptide chain.

FIGURE 8.4 shows that the aminoacyl end of the tRNA is located on the large subunit, whereas the anticodon at the other end interacts with the mRNA bound by the small subunit. So the P and A sites each extend across both ribosomal subunits.

For a ribosome to synthesize a peptide bond, it must be in the state shown in step 1 in Figure 8.3, when peptidyl-tRNA is in the P site and aminoacyl-tRNA is in the A site. Peptide bond formation occurs when the polypeptide carried by the peptidyl-tRNA is transferred to the amino acid carried by the aminoacyl-tRNA. This reaction is catalyzed by the large subunit of the ribosome.

Transfer of the polypeptide generates the ribosome shown in step 2, in which the **deacylated tRNA,** lacking any amino acid, lies in the P site and a new peptidyl-tRNA has been created in the A site. This peptidyl-tRNA is one amino acid residue longer than the peptidyl-tRNA that had been in the P site in step 1.

The ribosome now moves one triplet along the messenger. This stage is called **translocation.** The movement transfers the deacylated tRNA out of the P site and moves the peptidyl-tRNA into the P site (see step 3 in the figure). The next codon to be translated now lies in the A site, ready for a new aminoacyl-tRNA to enter, when the cycle will be repeated. FIGURE 8.5 summarizes the interaction between tRNAs and the ribosome.

The deacylated tRNA leaves the ribosome via another tRNA-binding site, the E site. This site is transiently occupied by the tRNA en route between leaving the P site and being released from the ribosome into the cytosol. Thus the flow of tRNA is into the A site, through the P site, and out through the E site (see also Figure 8.28 in Section 8.12). FIGURE 8.6 compares the movement of tRNA and mRNA, which may be thought of as a sort of ratchet in which the reaction is driven by the codon–anticodon interaction.

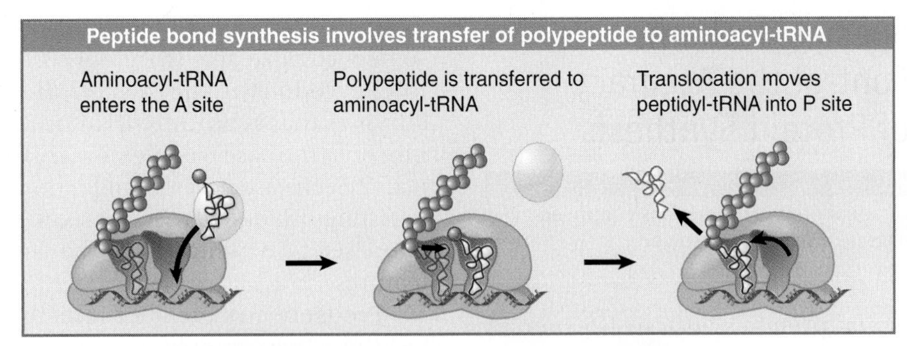

FIGURE 8.5 Aminoacyl-tRNA enters the A site, receives the polypeptide chain from peptidyl-tRNA, and is transferred into the P site for the next cycle of elongation.

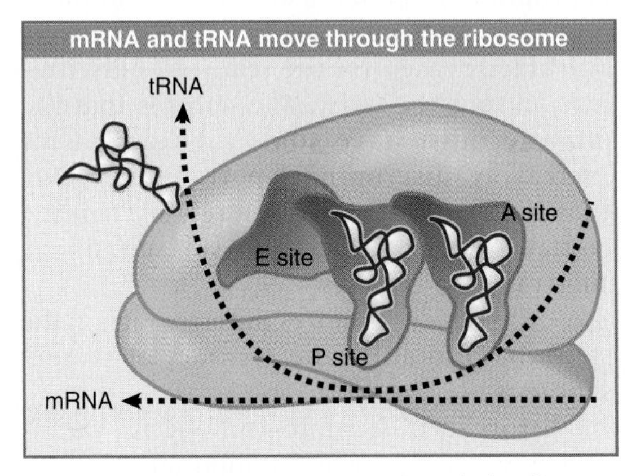

FIGURE 8.6 tRNA and mRNA move through the ribosome in the same direction.

Protein synthesis falls into the three stages shown in FIGURE 8.7:

- Initiation involves the reactions that precede formation of the peptide bond between the first two amino acids of the protein. It requires the ribosome to bind to the mRNA, which forms an initiation complex that contains the first aminoacyl-tRNA. This is a relatively slow step in protein synthesis and usually determines the rate at which an mRNA is translated.
- **Elongation** includes all the reactions from synthesis of the first peptide bond to addition of the last amino acid. Amino acids are added to the chain one at a time; the addition of an amino acid is the most rapid step in protein synthesis.
- **Termination** encompasses the steps that are needed to release the completed polypeptide chain; at the same time, the ribosome dissociates from the mRNA.

Different sets of accessory factors assist the ribosome at each stage. Energy is provided at various stages by the hydrolysis of guanine triphosphate (GTP).

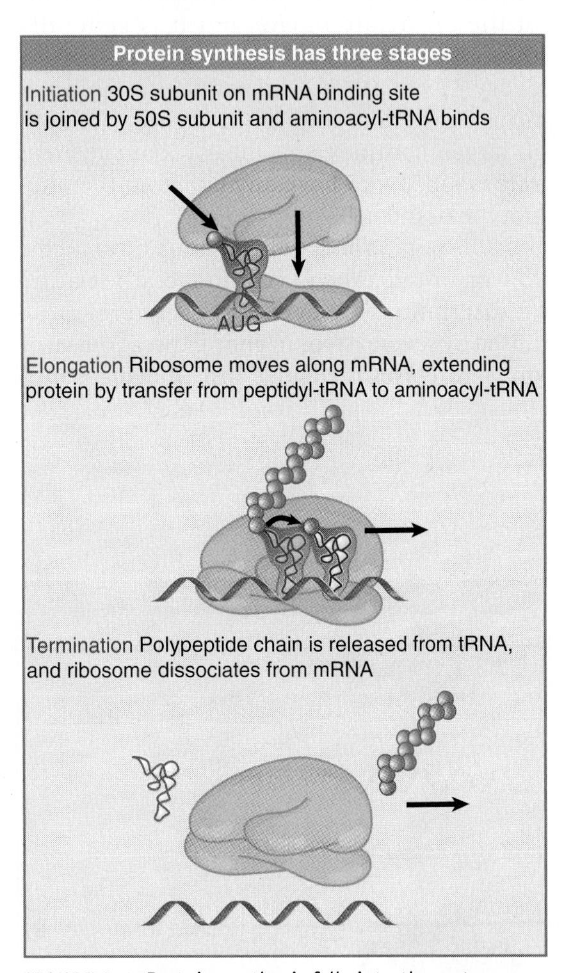

FIGURE 8.7 Protein synthesis falls into three stages.

During initiation, the small ribosomal subunit binds to mRNA and then is joined by the 50S subunit. During elongation, the mRNA moves through the ribosome and is translated in triplets. (Although we usually talk about the ribosome moving along mRNA, it is more realistic to think in terms of the mRNA being pulled through the ribosome.) At termination the protein is released, mRNA is released, and the individual ribosomal subunits dissociate in order to be used again.

8.3 Special Mechanisms Control the Accuracy of Protein Synthesis

Key concept

- The accuracy of protein synthesis is controlled by specific mechanisms at each stage.

We know that protein synthesis is generally accurate, because of the consistency that is found when we determine the sequence of a protein. There are few detailed measurements of the error rate *in vivo*, but it is generally thought to lie in the range of one error for every 10^4 to 10^5 amino acids incorporated. Considering that most proteins are produced in large quantities, this means that the error rate is too low to have any effect on the phenotype of the cell.

It is not immediately obvious how such a low error rate is achieved. In fact, the nature of discriminatory events is a general issue raised by several steps in gene expression. How do synthetases recognize just the corresponding tRNAs and amino acids? How does a ribosome recognize only the tRNA corresponding to the codon in the A site? How do the enzymes that synthesize DNA or RNA recognize only the base complementary to the template? Each case poses a similar problem: how to distinguish one particular member from the entire set, all of which share the same general features.

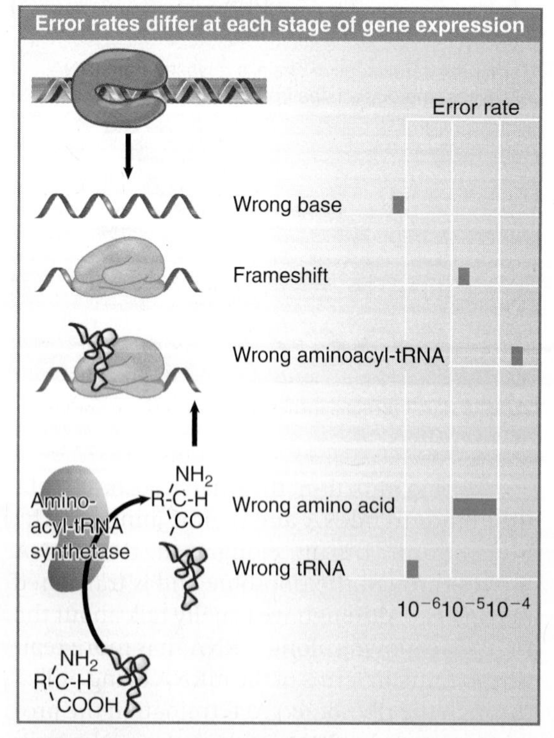

FIGURE 8.8 Errors occur at rates from 10^{-6} to 5×10^{-4} at different stages of protein synthesis.

Probably any member initially can contact the active center by a random-hit process, but then the wrong members are rejected and only the appropriate one is accepted. The appropriate member is always in a minority (one of twenty amino acids, one of ~40 tRNAs, one of four bases), so the criteria for discrimination must be strict. The point is that the enzyme must have some mechanism for increasing discrimination from the level that would be achieved merely by making contacts with the available surfaces of the substrates.

FIGURE 8.8 summarizes the error rates at the steps that can affect the accuracy of protein synthesis.

Errors in transcribing mRNA are rare—probably $<10^{-6}$. This is an important stage to control, because a single mRNA molecule is translated into many protein copies. We do not know very much about the mechanisms.

The ribosome can make two types of errors in protein synthesis. It may cause a frameshift by skipping a base when it reads the mRNA (or in the reverse direction by reading a base twice—once as the last base of one codon and then again as the first base of the next codon). These errors are rare, occurring at $\sim10^{-5}$. Or it may allow an incorrect aminoacyl-tRNA to (mis)pair with a codon, so that the wrong amino acid is incorporated. This is probably the most common error in protein synthesis, occurring at $\sim 5 \times 10^{-4}$. It is controlled by ribosome structure and velocity (see Section 9.15, The Ribosome Influences the Accuracy of Translation).

A tRNA synthetase can make two types of error: It can place the wrong amino acid on its tRNA, or it can charge its amino acid with the wrong tRNA. The incorporation of the wrong amino acid is more common, probably because the tRNA offers a larger surface with which the enzyme can make many more contacts to ensure specificity. Aminoacyl-tRNA synthetases have specific mechanisms to correct errors before a mischarged tRNA is released (see Section 9.11, Synthetases Use Proofreading to Improve Accuracy).

8.4 Initiation in Bacteria Needs 30S Subunits and Accessory Factors

Key concepts

- Initiation of protein synthesis requires separate 30S and 50S ribosome subunits.
- Initiation factors (IF-1, -2, and -3), which bind to 30S subunits, are also required.
- A 30S subunit carrying initiation factors binds to an initiation site on mRNA to form an initiation complex.
- IF-3 must be released to allow 50S subunits to join the 30S-mRNA complex.

Bacterial ribosomes engaged in elongating a polypeptide chain exist as 70S particles. At termination, they are released from the mRNA as free ribosomes. In growing bacteria, the majority of ribosomes are synthesizing proteins; the free pool is likely to contain ~20% of the ribosomes.

Ribosomes in the free pool can dissociate into separate subunits; this means that 70S ribosomes are in dynamic equilibrium with 30S and 50S subunits. *Initiation of protein synthesis is not a function of intact ribosomes, but is undertaken by the separate subunits,* which reassociate during the initiation reaction. FIGURE 8.9 summarizes the ribosomal subunit cycle during protein synthesis in bacteria.

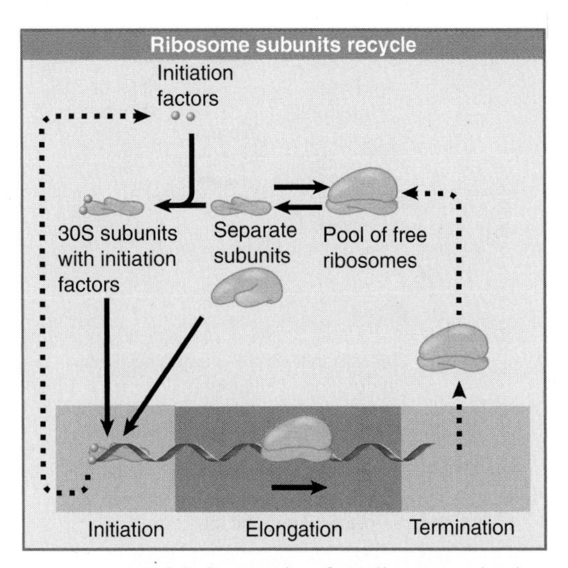

FIGURE 8.9 Initiation requires free ribosome subunits. When ribosomes are released at termination, the 30S subunits bind initiation factors and dissociate to generate free subunits. When subunits reassociate to give a functional ribosome at initiation, they release the factors.

Initiation occurs at a special sequence on mRNA called the **ribosome-binding site.** This is a short sequence of bases that precedes the coding region (see Figure 8.1). The small and large subunits associate at the ribosome-binding site to form an intact ribosome. The reaction occurs in two steps:

- Recognition of mRNA occurs when a small subunit binds to form an **initiation complex** at the ribosome-binding site.
- A large subunit then joins the complex to generate a complete ribosome.

Although the 30S subunit is involved in initiation, it is not by itself competent to undertake the reactions of binding mRNA and tRNA. It requires additional proteins called **initiation factors (IF).** These factors are found only on 30S subunits, and they are released when the 30S subunits associate with 50S subunits to generate 70S ribosomes. This behavior distinguishes initiation factors from the structural proteins of the ribosome. The initiation factors are concerned solely with formation of the initiation complex, they are absent from 70S ribosomes, and they play no part in the stages of elongation. FIGURE 8.10 summarizes the stages of initiation.

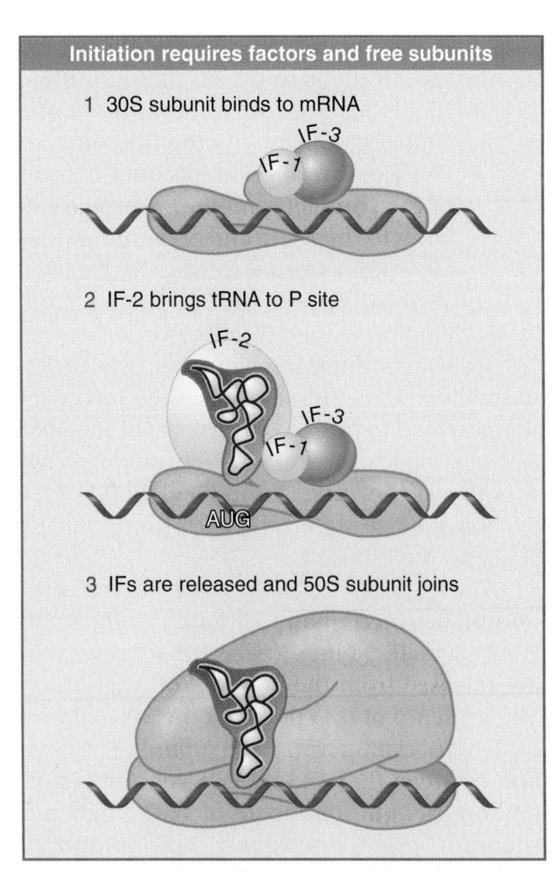

FIGURE 8.10 Initiation factors stabilize free 30S subunits and bind initiator tRNA to the 30S-mRNA complex.

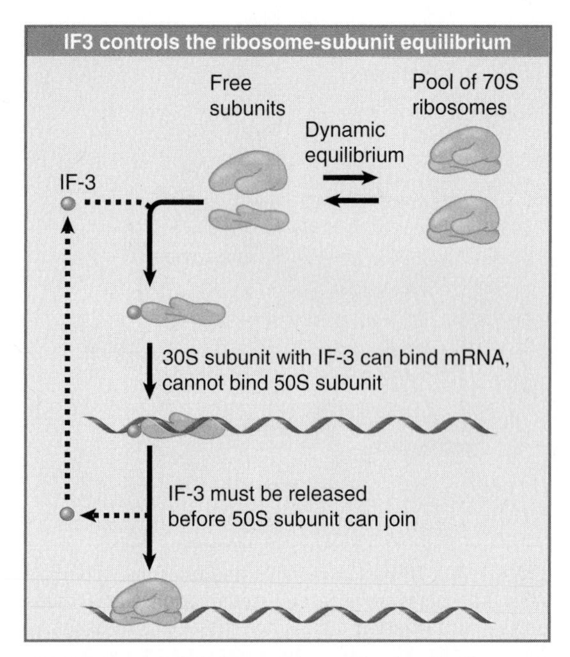

FIGURE 8.11 Initiation requires 30S subunits that carry IF-3.

Bacteria use three initiation factors, numbered **IF-1, IF-2,** and **IF-3.** They are needed for both mRNA and tRNA to enter the initiation complex:

- IF-3 is needed for 30S subunits to bind specifically to initiation sites in mRNA.
- IF-2 binds a special initiator tRNA and controls its entry into the ribosome.
- IF-1 binds to 30S subunits only as a part of the complete initiation complex. It binds to the A site and prevents aminoacyl-tRNA from entering. Its location also may impede the 30S subunit from binding to the 50S subunit.

IF-3 has multiple functions: it is needed first to stabilize (free) 30S subunits; then it enables them to bind to mRNA; and as part of the 30S-mRNA complex, it checks the accuracy of recognition of the first aminoacyl-tRNA (see Section 8.6, Use of fMet-tRNA$_f$ Is Controlled by IF-2 and the Ribosome).

The first function of IF-3 controls the equilibrium between ribosomal states, as shown in FIGURE 8.11. IF-3 binds to free 30S subunits that are released from the pool of 70S ribosomes. The presence of IF-3 prevents the 30S subunit from reassociating with a 50S subunit. The reaction between IF-3 and the 30S subunit is stoichiometric: one molecule of IF-3 binds per subunit. There is a relatively small amount of IF-3, so its availability determines the number of free 30S subunits.

IF-3 binds to the surface of the 30S subunit in the vicinity of the A site. There is significant overlap between the bases in 16S rRNA protected by IF-3 and those protected by binding of the 50S subunit, suggesting that it physically prevents junction of the subunits. IF-3 therefore behaves as an anti-association factor that causes a 30S subunit to remain in the pool of free subunits.

The second function of IF-3 controls the ability of 30S subunits to bind to mRNA. Small subunits must have IF-3 in order to form initiation complexes with mRNA. IF-3 must be released from the 30S-mRNA complex in order to enable the 50S subunit to join. On its release, IF-3 immediately recycles by finding another 30S subunit.

IF-2 has a ribosome-dependent GTPase activity: It sponsors the hydrolysis of GTP in the presence of ribosomes, releasing the energy stored in the high-energy bond. The GTP is hydrolyzed when the 50S subunit joins to generate a complete ribosome. The GTP cleavage could be involved in changing the conformation of the ribosome, so that the joined subunits are converted into an active 70S ribosome.

8.5 A Special Initiator tRNA Starts the Polypeptide Chain

Key concepts

- Protein synthesis starts with a methionine amino acid usually coded by AUG.
- Different methionine tRNAs are involved in initiation and elongation.
- The initiator tRNA has unique structural features that distinguish it from all other tRNAs.
- The NH_2 group of the methionine bound to bacterial initiator tRNA is formylated.

Synthesis of all proteins starts with the same amino acid: methionine. The signal for initiating a polypeptide chain is a special initiation codon that marks the start of the reading frame. Usually the initiation codon is the triplet AUG, but in bacteria GUG or UUG are also used.

The AUG codon represents methionine, and two types of tRNA can carry this amino acid. One is used for initiation, the other for recognizing AUG codons during elongation.

In bacteria and in eukaryotic organelles, the initiator tRNA carries a methionine residue that has been formylated on its amino group, forming a molecule of **N-formyl-methionyl-**

tRNA. The tRNA is known as **$tRNA_f^{Met}$**. The name of the aminoacyl-tRNA is usually abbreviated to fMet-$tRNA_f$.

The initiator tRNA gains its modified amino acid in a two-stage reaction. First, it is charged with the amino acid to generate Met-$tRNA_f$; and then the formylation reaction shown in FIGURE 8.12 blocks the free NH_2 group. Although the blocked amino acid group would prevent the initiator from participating in chain elongation, it does not interfere with the ability to initiate a protein.

This tRNA is used only for initiation. It recognizes the codons AUG or GUG (occasionally UUG). The codons are not recognized equally well: the extent of initiation declines by about half when AUG is replaced by GUG, and declines by about half again when UUG is employed.

The species responsible for recognizing AUG codons in internal locations is **$tRNA_m^{Met}$**. This tRNA responds only to internal AUG codons. Its methionine cannot be formylated.

What features distinguish the fMet-$tRNA_f$ initiator and the Met-$tRNA_m$ elongator? Some characteristic features of the tRNA sequence are important, as summarized in FIGURE 8.13. Some of these features are needed to prevent the initiator from being used in elongation, whereas others are necessary for it to function in initiation:

- Formylation is not strictly necessary, because nonformylated Met-$tRNA_f$ can function as an initiator. Formylation improves the efficiency with which the Met-$tRNA_f$ is used, though, because it is one of the features recognized by the factor IF-2 that binds the initiator tRNA.
- The bases that face one another at the last position of the stem to which the amino acid is connected are paired in all tRNAs except $tRNA_f^{Met}$. Mutations that create a base pair in this position of $tRNA_f^{Met}$ allow it to function in elongation. The absence of this pair is therefore important in preventing $tRNA_f^{Met}$ from being used in elongation. It is also needed for the formylation reaction.
- A series of 3 G-C pairs in the stem that precedes the loop containing the anticodon is unique to $tRNA_f^{Met}$. These base pairs are required to allow the fMet-$tRNA_f$ to be inserted directly into the P site.

In bacteria and mitochondria, the formyl residue on the initiator methionine is removed by a specific deformylase enzyme to generate a normal NH_2 terminus. If methionine is to be the N-terminal amino acid of the protein, this is the only necessary step. In about half the proteins, the methionine at the terminus is removed by an aminopeptidase, which creates a new terminus from R_2 (originally the second amino acid incorporated into the chain). When both steps are necessary, they occur sequentially. The removal reaction(s) occur rather rapidly, probably when the nascent polypeptide chain has reached a length of 15 amino acids.

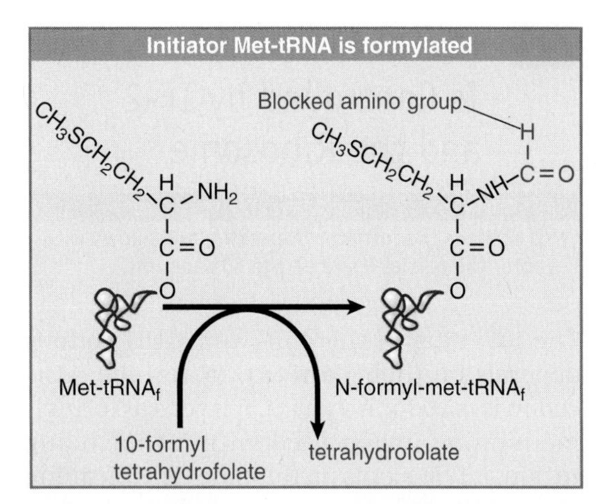

FIGURE 8.12 The initiator N-formyl-methionyl-tRNA (fMet-$tRNA_f$) is generated by formylation of methionyl-tRNA, using formyl-tetrahydrofolate as cofactor.

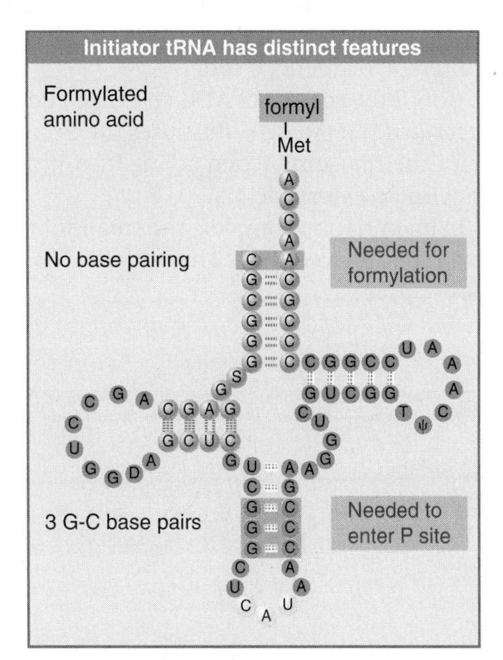

FIGURE 8.13 fMet-$tRNA_f$ has unique features that distinguish it as the initiator tRNA.

8.6 Use of fMet-$tRNA_f$ Is Controlled by IF-2 and the Ribosome

Key concept

- IF-2 binds the initiator fMet-$tRNA_f$ and allows it to enter the partial P site on the 30S subunit.

The meaning of the AUG and GUG codons depends on their **context.** When the AUG codon is used for initiation, it is read as formyl-methionine; when used within the coding region, it represents methionine. The meaning of the GUG codon is even more dependent on its location. When present as the first codon, it is read via the initiation reaction as formyl-methionine. Yet when present within a gene, it is read by Val-tRNA, one of the regular members of the tRNA set, to provide valine as required by the genetic code.

How is the context of AUG and GUG codons interpreted? FIGURE 8.14 illustrates the decisive role of the ribosome when acting in conjunction with accessory factors.

In an initiation complex, the small subunit alone is bound to mRNA. The initiation codon lies within the part of the P site carried by the small subunit. The only aminoacyl-tRNA that can become part of the initiation complex is the initiator, which has the unique property of being able to enter directly into the partial P site to recognize its codon.

When the large subunit joins the complex, the partial tRNA-binding sites are converted into the intact P and A sites. The initiator fMet-$tRNA_f$ occupies the P site, and the A site is available for entry of the aminoacyl-tRNA complementary to the second codon of the gene. The first peptide bond forms between the initiator and the next aminoacyl-tRNA.

Initiation prevails when an AUG (or GUG) codon lies within a ribosome-binding site, because only the initiator tRNA can enter the partial P site generated when the 30S subunit binds *de novo* to the mRNA. Internal reading prevails subsequently, when the codons are encountered by a ribosome that is continuing to translate an mRNA, because only the regular aminoacyl-tRNAs can enter the (complete) A site.

Accessory factors are critical in controlling the usage of aminoacyl-tRNAs. All aminoacyl-tRNAs associate with the ribosome by binding to an accessory factor. The factor used in initiation is IF-2 (see Section 8.4, Initiation in Bacteria Needs 30S Subunits and Accessory Factors), and the corresponding factor used at elongation is EF-Tu (see Section 8.10, Elongation Factor Tu Loads Aminoacyl-tRNA into the A Site).

The initiation factor IF-2 places the initiator tRNA into the P site. By forming a complex specifically with fMet-$tRNA_f$, IF-2 ensures that only the initiator tRNA, and none of the regular aminoacyl-tRNAs, participates in the initiation reaction. Conversely, EF-Tu, which places aminoacyl-tRNAs in the A site, cannot bind fMet-$tRNA_f$, which is therefore excluded from use during elongation.

An additional check on accuracy is made by IF-3, which stabilizes binding of the initiator tRNA by recognizing correct base pairing with the second and third bases of the AUG initiation codon.

FIGURE 8.15 details the series of events by which IF-2 places the fMet-$tRNA_f$ initiator in the P site. IF-2, bound to GTP, associates with the P site of the 30S subunit. At this point, the 30S subunit carries all the initiation factors. fMet-$tRNA_f$ binds to the IF-2 on the 30S subunit, and then IF-2 transfers the tRNA into the partial P site.

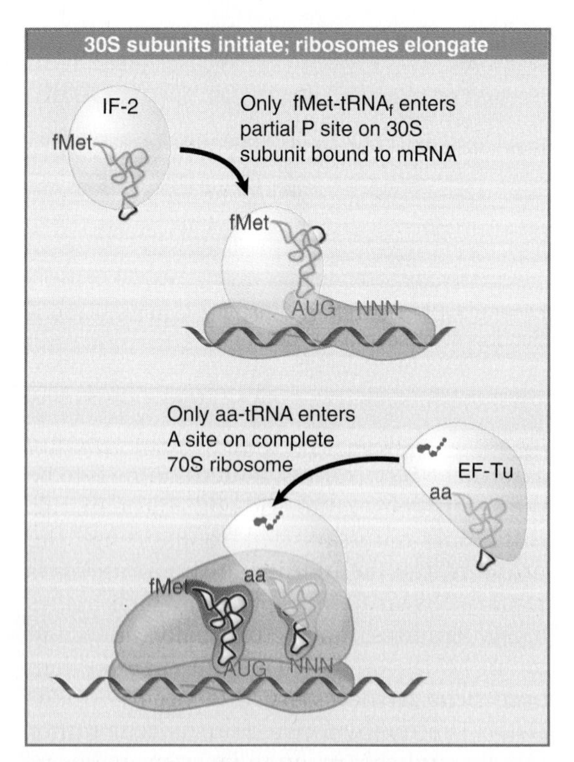

FIGURE 8.14 Only fMet-$tRNA_f$ can be used for initiation by 30S subunits; other aminoacyl-tRNAs ($\alpha\alpha$-tRNA) must be used for elongation by 70S ribosomes.

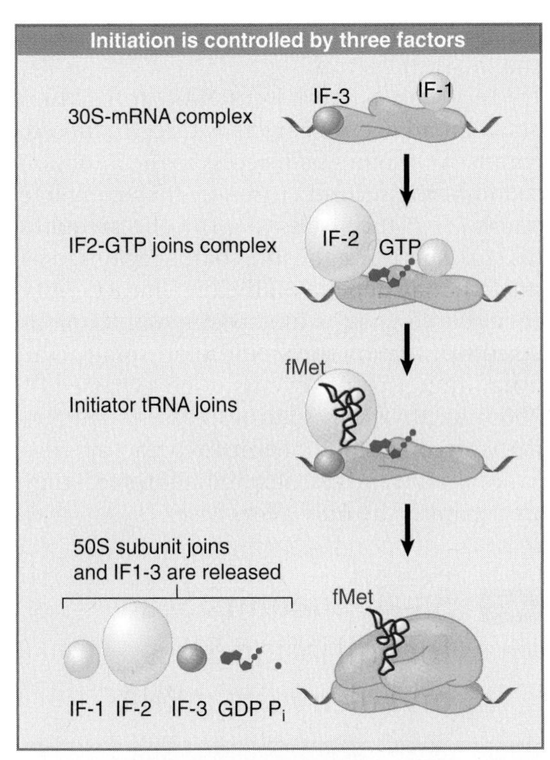

FIGURE 8.15 IF-2 is needed to bind fMet-$tRNA_f$ to the 30S-mRNA complex. After 50S binding, all IF factors are released and GTP is cleaved.

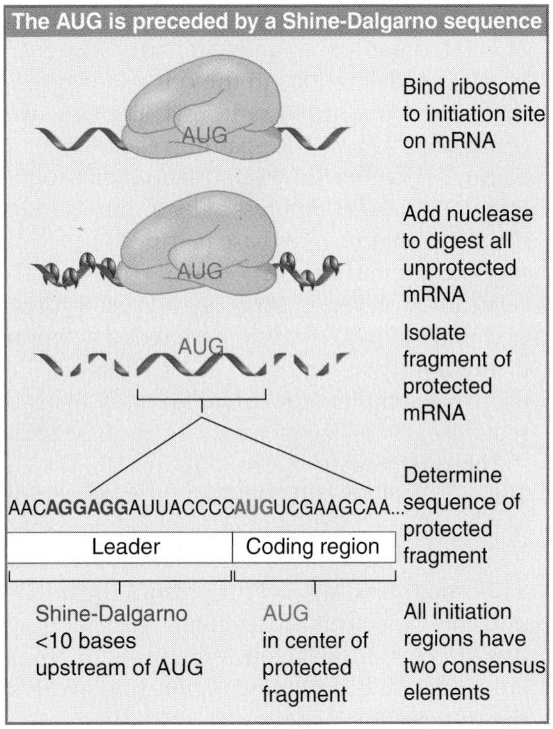

The AUG is preceded by a Shine-Dalgarno sequence

Bind ribosome to initiation site on mRNA

AUG

Add nuclease to digest all unprotected mRNA

AUG

Isolate fragment of protected mRNA

AUG

Determine sequence of protected fragment

AACAGGAGGAUUACCCCAUGUCGAAGCAA...

Leader | Coding region

Shine-Dalgarno <10 bases upstream of AUG

AUG in center of protected fragment

All initiation regions have two consensus elements

FIGURE 8.16 Ribosome-binding sites on mRNA can be recovered from initiation complexes. They include the upstream Shine–Dalgarno sequence and the initiation codon.

8.7 Initiation Involves Base Pairing Between mRNA and rRNA

Key concepts

- An initiation site on bacterial mRNA consists of the AUG initiation codon preceded with a gap of ~10 bases by the Shine–Dalgarno polypurine hexamer.
- The rRNA of the 30S bacterial ribosomal subunit has a complementary sequence that base pairs with the Shine–Dalgarno sequence during initiation.

An mRNA contains many AUG triplets: How is the initiation codon recognized as providing the starting point for translation? The sites on mRNA where protein synthesis is initiated can be identified by binding the ribosome to mRNA under conditions that block elongation. Then the ribosome remains at the initiation site. When ribonuclease is added to the blocked initiation complex, all the regions of mRNA outside the ribosome are degraded.Those actually bound to it are protected, though, as illustrated in FIGURE 8.16. The protected fragments can be recovered and characterized.

The initiation sequences protected by bacterial ribosomes are ~30 bases long. The ribosome-binding sites of different bacterial mRNAs display two common features:

- The AUG (or less often, GUG or UUG) initiation codon is always included within the protected sequence.
- Within ten bases upstream of the AUG is a sequence that corresponds to part or all of the hexamer.

5′. . . A G G A G G . . . 3′

This polypurine stretch is known as the **Shine–Dalgarno** sequence. It is complementary to a highly conserved sequence close to the 3′ end of 16S rRNA. (The extent of complementarity differs with individual mRNAs, and may extend from a four-base core sequence GAGG to a nine-base sequence extending beyond each end of the hexamer.) Written in reverse direction, the rRNA sequence is the hexamer:

3′. . . U C C U C C . . . 5′

Does the Shine–Dalgarno sequence pair with its complement in rRNA during mRNA-ribosome binding? Mutations of both partners in this reaction demonstrate its importance in initiation. Point mutations in the Shine–Dalgarno sequence can prevent an mRNA from being translated. In addition, the introduction of mutations into the complementary sequence in rRNA is deleterious to the cell and changes the pattern of protein synthesis. The decisive confirmation of the base-pairing reaction is that a mutation in the Shine–Dalgarno sequence of an mRNA can be suppressed by a mutation in the rRNA that restores base pairing.

The sequence at the 3′ end of rRNA is conserved between prokaryotes and eukaryotes, except that in all eukaryotes there is a deletion of the five-base sequence CCUCC that is the principal complement to the Shine–Dalgarno sequence. There does not appear to be base pairing between eukaryotic mRNA and 18S rRNA. This is a significant difference in the mechanism of initiation.

In bacteria, a 30S subunit binds directly to a ribosome-binding site. As a result, the initiation complex forms at a sequence surrounding the AUG initiation codon. When the mRNA is polycistronic, each coding region starts with a ribosome-binding site.

The nature of bacterial gene expression means that translation of a bacterial mRNA proceeds sequentially through its cistrons. At the time when ribosomes attach to the first coding region, the subsequent coding regions have not yet even been transcribed. By the time the second ribosome site is available, translation is well under way through the first cistron.

What happens between the coding regions depends on the individual mRNA. In most cases, the ribosomes probably bind independently at the beginning of each cistron. The most common series of events is illustrated in FIGURE 8.17. When synthesis of the first protein terminates, the ribosomes leave the mRNA and dissociate into subunits. Then a new ribosome must assemble at the next coding region and set out to translate the next cistron.

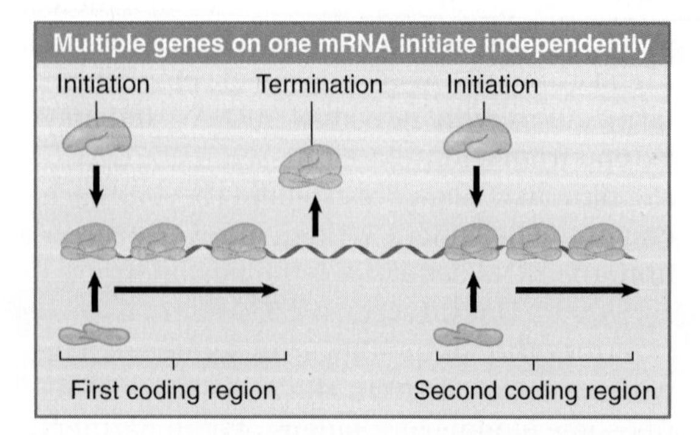

FIGURE 8.17 Initiation occurs independently at each cistron in a polycistronic mRNA. When the intercistronic region is longer than the span of the ribosome, dissociation at the termination site is followed by independent reinitiation at the next cistron.

In some bacterial mRNAs, translation between adjacent cistrons is directly linked, because ribosomes gain access to the initiation codon of the second cistron as they complete translation of the first cistron. This effect requires the space between the two coding regions to be small. It may depend on the high local density of ribosomes, or the juxtaposition of termination and initiation sites could allow some of the usual intercistronic events to be bypassed. A ribosome physically spans ~30 bases of mRNA, so that it could simultaneously contact a termination codon and the next initiation site if they are separated by only a few bases.

8.8 Small Subunits Scan for Initiation Sites on Eukaryotic mRNA

Key concepts

- Eukaryotic 40S ribosomal subunits bind to the 5′ end of mRNA and scan the mRNA until they reach an initiation site.
- A eukaryotic initiation site consists of a ten-nucleotide sequence that includes an AUG codon.
- 60S ribosomal subunits join the complex at the initiation site.

Initiation of protein synthesis in eukaryotic cytoplasm resembles the process in bacteria, but the order of events is different and the number of accessory factors is greater. Some of the differences in initiation are related to a difference in the way that bacterial 30S and eukaryotic 40S subunits find their binding sites for initiating protein synthesis on mRNA. In eukaryotes, small subunits first recognize the 5′ end of the mRNA and then move to the initiation site, where they are joined by large subunits. (In prokaryotes, small subunits bind directly to the initiation site.)

Virtually all eukaryotic mRNAs are monocistronic, but each mRNA usually is substantially longer than necessary just to code for its protein. The average mRNA in eukaryotic cytoplasm is 1000 to 2000 bases long, has a methylated cap at the 5′ terminus, and carries 100 to 200 bases of poly(A) at the 3′ terminus.

The nontranslated 5′ leader is relatively short, usually <100 bases. The length of the coding region is determined by the size of the protein. The nontranslated 3′ trailer is often rather long, at times reaching lengths of up to ~1000 bases.

The first feature to be recognized during translation of a eukaryotic mRNA is the methylated cap that marks the 5′ end. Messengers whose caps have been removed are not translated efficiently *in vitro*. Binding of 40S subunits to mRNA requires several initiation factors, including proteins that recognize the structure of the cap.

Modification at the 5′ end occurs to almost all cellular or viral mRNAs and is essential for their translation in eukaryotic cytoplasm (although it is not needed in organelles). The sole exception to this rule is provided by a few viral mRNAs (such as poliovirus) that are not capped; only these exceptional viral mRNAs can be translated *in vitro* without caps. They use an alternative pathway that bypasses the need for the cap.

Some viruses take advantage of this difference. Poliovirus infection inhibits the translation of host mRNAs. This is accomplished by interfering with the cap-binding proteins that are needed for initiation of cellular mRNAs, but that are superfluous for the noncapped poliovirus mRNA.

We have dealt with the process of initiation as though the ribosome-binding site is always freely available. However, its availability may be impeded by secondary structure. The recognition of mRNA requires several additional factors; an important part of their function is to remove any secondary structure in the mRNA (see Figure 8.20).

Sometimes the AUG initiation codon lies within 40 bases of the 5′ terminus of the mRNA, so that both the cap and AUG lie within the span of ribosome binding. In many mRNAs, however, the cap and AUG are farther apart—in extreme cases, they can be as much as 1000 bases away from each other. Yet the presence of the cap still is necessary for a stable complex to be formed at the initiation codon. How can the ribosome rely on two sites so far apart?

FIGURE 8.18 illustrates the "scanning" model, which supposes that the 40S subunit initially recognizes the 5′ cap and then "migrates" along the mRNA. Scanning from the 5′ end is a linear process. When 40S subunits scan the leader region, they can melt secondary structure hairpins with stabilities <–30 kcal, but hairpins of greater stability impede or prevent migration.

Migration stops when the 40S subunit encounters the AUG initiation codon. Usually, although not always, the first AUG triplet sequence to be encountered will be the initiation codon. However, the AUG triplet by itself is not sufficient to halt migration; it is recognized efficiently as an initiation codon only when it is in the right context. The most important determinants of context are the bases in positions –4 and +1. An initiation codon may be recognized in the sequence NNNPuNN*AUG*G. The purine (A or G) 3 bases before the AUG codon, and the G immediately following it, can influence the efficiency of translation by 10×. When the leader sequence is long, further 40S subunits can recognize the 5′ end before the first has left the initiation site, creating a queue of subunits proceeding along the leader to the initiation site.

It is probably true that the initiation codon is the first AUG to be encountered in the most efficiently translated mRNAs. What happens, though, when there is an AUG triplet in the 5′ nontranslated region? There are two possible escape mechanisms for a ribosome that starts scanning at the 5′ end. The most common is that scanning is leaky, that is, a ribosome may continue past a noninitiation AUG because it is not in the right context. In the rare case that it does recognize the AUG, it may initiate translation but terminate before the proper initiation codon, after which it resumes scanning.

The vast majority of eukaryotic initiation events involve scanning from the 5′ cap, but

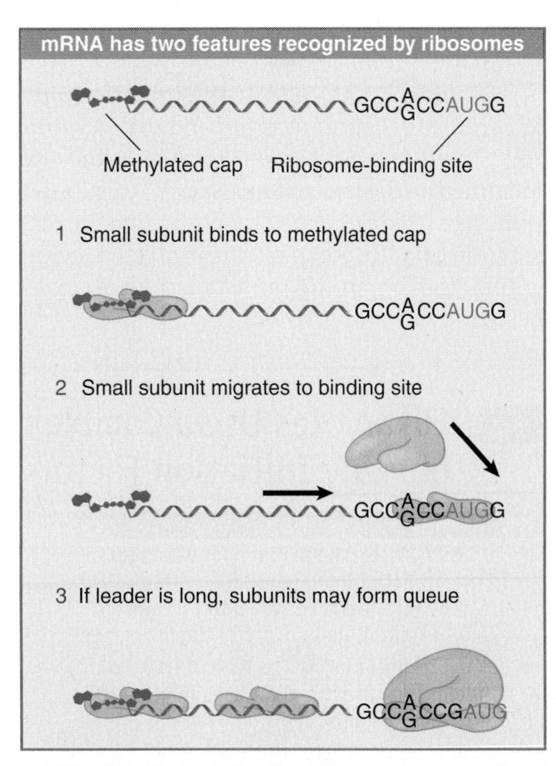

FIGURE 8.18 Eukaryotic ribosomes migrate from the 5′ end of mRNA to the ribosome binding site, which includes an AUG initiation codon.

there is an alternative means of initiation, used especially by certain viral RNAs, in which a 40S subunit associates directly with an internal site called an IRES. (This entirely bypasses any AUG codons that may be in the 5′ nontranslated region.) There are few sequence homologies between known IRES elements. We can distinguish three types on the basis of their interaction with the 40S subunit:

- One type of IRES includes the AUG initiation codon at its upstream boundary. The 40S subunit binds directly to it, using a subset of the same factors that are required for initiation at 5′ ends.
- Another is located as much as 100 nucleotides upstream of the AUG, requiring a 40S subunit to migrate, again probably by a scanning mechanism.
- An exceptional type of IRES in hepatitis C virus can bind a 40S subunit directly, without requiring any initiation factors. The order of events is different from all other eukaryotic initiation. Following 40S-mRNA binding, a complex containing initiator factors and the initiator tRNA binds.

Use of the IRES is especially important in picornavirus infection, where it was first discovered, because the virus inhibits host protein synthesis by destroying cap structures and inhibiting the initiation factors that bind them (see Section 8.9, Eukaryotes Use a Complex of Many Initiation Factors).

Binding is stabilized at the initiation site. When the 40S subunit is joined by a 60S subunit, the intact ribosome is located at the site identified by the protection assay. A 40S subunit protects a region of up to 60 bases; when the 60S subunits join the complex, the protected region contracts to about the same length of 30 to 40 bases seen in prokaryotes.

8.9 Eukaryotes Use a Complex of Many Initiation Factors

Key concepts

- Initiation factors are required for all stages of initiation, including binding the initiator tRNA, 40S subunit attachment to mRNA, movement along the mRNA, and joining of the 60S subunit.
- Eukaryotic initiator tRNA is a Met-tRNA that is different from the Met-tRNA used in elongation, but the methionine is not formulated.
- eIF2 binds the initiator Met-$tRNA_i$ and GTP, and the complex binds to the 40S subunit before it associates with mRNA.

Initiation in eukaryotes has the same general features as in bacteria in using a specific initiation codon and initiator tRNA. Initiation in eukaryotic cytoplasm uses AUG as the initiator. The initiator tRNA is a distinct species, but its methionine does not become formylated. It is called **$tRNA_i^{Met}$**. Thus the difference between the initiating and elongating Met-tRNAs lies solely in the tRNA moiety, with Met-$tRNA_i$ used for initiation and Met-$tRNA_m$ used for elongation.

At least two features are unique to the initiator $tRNA_i^{Met}$ in yeast: it has an unusual tertiary structure, and it is modified by phosphorylation of the 2′ ribose position on base 64 (if this modification is prevented, the initiator can be used in elongation). Thus the principle of a distinction between initiator and elongator Met-tRNAs is maintained in eukaryotes, but its structural basis is different from that in bacteria (for comparison see Figure 8.13).

Eukaryotic cells have more initiation factors than bacteria—the current list includes 12 factors that are directly or indirectly required for initiation. The factors are named similarly to those in bacteria, sometimes by analogy with the bacterial factors, and are given the prefix "e" to indicate their eukaryotic origin. They act at all stages of the process, including:

- forming an initiation complex with the 5′ end of mRNA;
- forming a complex with Met-$tRNA_i$;
- binding the mRNA-factor complex to the Met-$tRNA_i$-factor complex;
- enabling the ribosome to scan mRNA from the 5′ end to the first AUG;
- detecting binding of initiator tRNA to AUG at the start site; and
- mediating joining of the 60S subunit.

FIGURE 8.19 summarizes the stages of initiation and shows which initiation factors are involved at each stage: eIF2 and eIF3 bind to the 40S ribosome subunit; eIF4A, eIF4B, and eIF4F bind to the mRNA; and eIF1 and eIF1A bind to the ribosome subunit-mRNA complex.

FIGURE 8.20 shows the group of factors that bind to the 5′ end of mRNA. The factor eIF4F is a protein complex that contains three of the initiation factors. It is not clear whether it preassembles as a complex before binding to mRNA or whether the individual subunits are added individually to form the complex on mRNA. It includes the cap-binding subunit eIF4E, the helicase eIF4A, and the "scaffolding" subunit eIF4G. After eIF4E binds the cap, eIF4A unwinds any secondary structure that exists in the first 15 bases of the mRNA. Energy for the unwinding is provided by hydrolysis of ATP. Unwind-

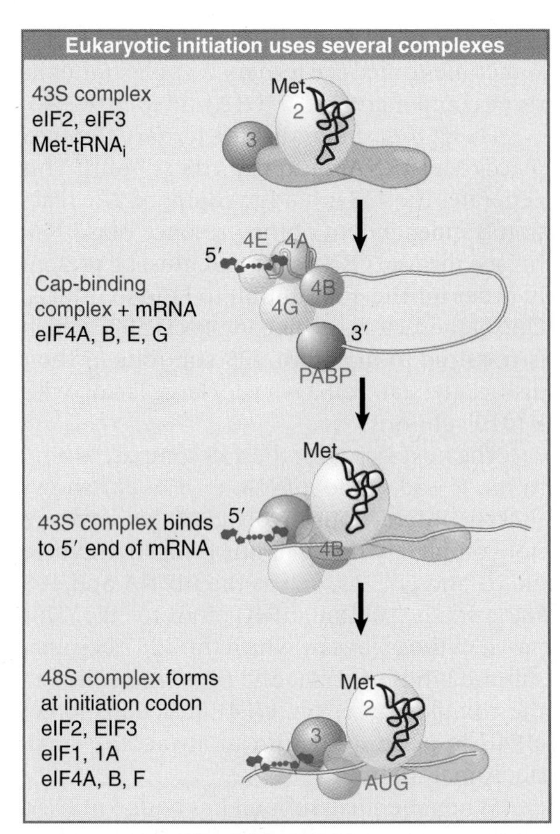

FIGURE 8.19 Some initiation factors bind to the 40S ribosome subunit to form the 43S complex; others bind to mRNA. When the 43S complex binds to mRNA, it scans for the initiation codon and can be isolated as the 48S complex.

ing of structure farther along the mRNA is accomplished by eIF4A together with another factor, eIF4B. The main role of eIF4G is to link other components of the initiation complex.

The subunit eIF4E is a focus for regulation. Its activity is increased by phosphorylation, which is triggered by stimuli that increase protein synthesis and reversed by stimuli that repress protein synthesis. The subunit eIF4F has a kinase activity that phosphorylates eIF4E. The availability of eIF4E is also controlled by proteins that bind to it (called 4E-BP1, -2, and -3), to prevent it from functioning in initiation. The subunit eIF4G is also a target for degradation during picornavirus infection, as part of the destruction of the capacity to initiate at 5′ cap structures (see Section 8.8, Small Subunits Scan for Initiation Sites on Eukaryotic mRNA).

The presence of poly(A) on the 3′ tail of an mRNA stimulates the formation of an initiation complex at the 5′ end. The poly(A)-binding protein (Pab1p in yeast) is required for this effect. Pab1p binds to the eIF4G scaffolding protein. This implies that the mRNA will have a circular organization so long as eIFG is bound, with both the 5′ and 3′ ends held in this complex

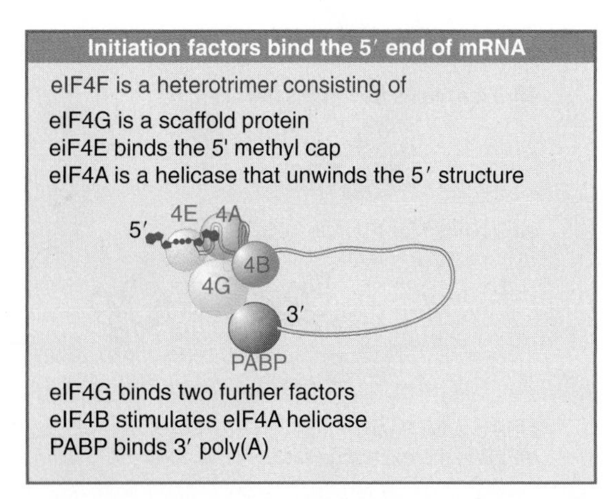

FIGURE 8.20 The heterotrimer eIF4F binds the 5′ end of mRNA and also binds further factors.

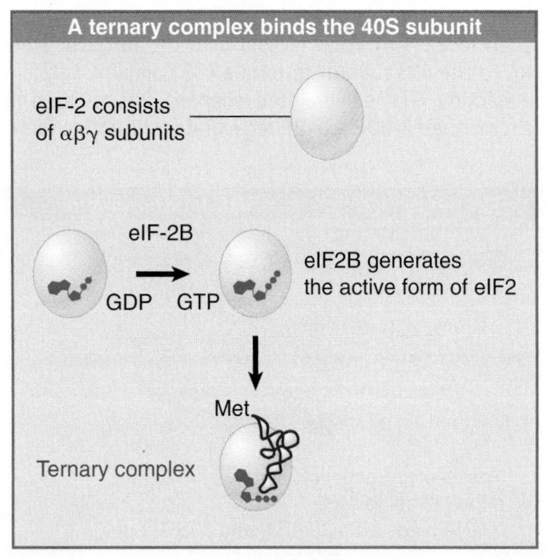

FIGURE 8.21 In eukaryotic initiation, eIF-2 forms a ternary complex with Met-tRNA$_i$. The ternary complex binds to free 40S subunits, which attach to the 5′ end of mRNA. Later in the reaction, GTP is hydrolyzed when eIF-2 is released in the form of eIF2-GDP. eIF-2B regenerates the active form.

(see Figure 8.20). The significance of the formation of this closed loop is not clear, although it could have several effects, such as:

- stimulating initiation of translation;
- promoting reinitiation of ribosomes, so that when they terminate at the 3′ end, the released subunits are already in the vicinity of the 5′ end;
- stabilizing the mRNA against degradation; and
- allowing factors that bind to the 3′ end to regulate the initiation of translation.

The subunit eIF2 is the key factor in binding Met-tRNA$_i$. It is a typical monomeric GTP-binding protein that is active when bound to GTP and inactive when bound to guanine diphosphate (GDP). FIGURE 8.21 shows that the

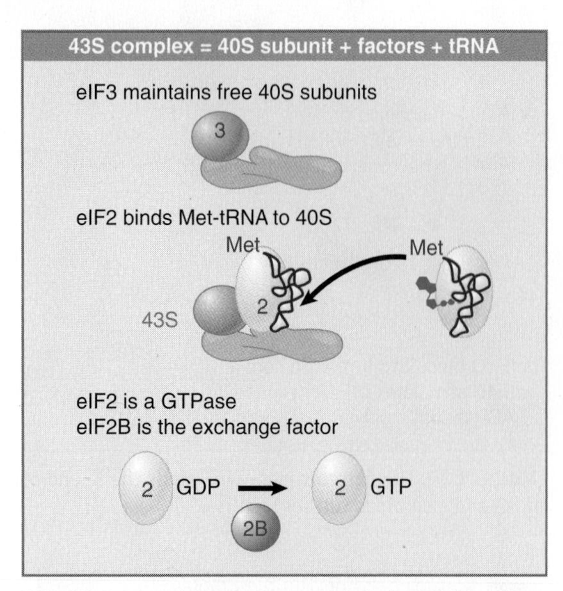

FIGURE 8.22 Initiation factors bind the initiator Met-tRNA to the 40S subunit to form a 43S complex. Later in the reaction, GTP is hydrolyzed when eIF-2 is released in the form of eIF2-GDP. eIF-2B regenerates the active form.

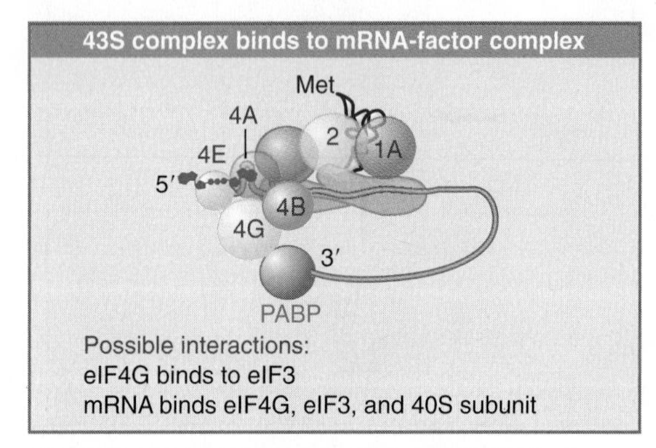

FIGURE 8.23 Interactions involving initiation factors are important when mRNA binds to the 43S complex.

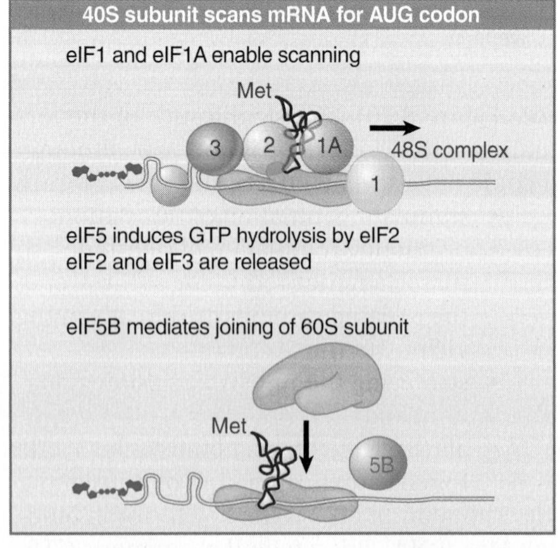

FIGURE 8.24 eIF1 and eIF1A help the 43S initiation complex to scan the mRNA until it reaches an AUG codon. eIF2 hydrolyzes its GTP to enable its release together with IF3. eIF5B mediates 60S–40S joining.

eIF2-GTP binds to Met-$tRNA_i$. The product is sometimes called the ternary complex (after its three components, eIF2, GTP, and Met-$tRNA_i$).

FIGURE 8.22 shows that the ternary complex places Met-$tRNA_i$ onto the 40S subunit. This generates the 43S initiation complex. The reaction is independent of the presence of mRNA. In fact, the Met-$tRNA_i$ initiator must be present in order for the 40S subunit to bind to mRNA. One of the factors in this complex is eIF3, which is required to maintain 40S subunits in their dissociated state. eIF3 is a very large factor, with 8 to10 subunits.

The next step is for the 43S complex to bind to the 5′ end of the mRNA. FIGURE 8.23 shows that the interactions involved at this stage are not completely defined, but probably involve eIF4G and eIF3 as well as the mRNA and 40S subunit. The subunit eIF4G binds to eIF3. This provides the means by which the 40S ribosomal subunit binds to eIF4F, and thus is recruited to the complex. In effect, eIF4F functions to get eIF4G in place so that it can attract the small ribosomal subunit.

When the small subunit has bound mRNA, it migrates to (usually) the first AUG codon. This requires expenditure of energy in the form of ATP. It is assisted by the factors eIF1 and eIF1A. FIGURE 8.24 shows that the small subunit stops when it reaches the initiation site, forming a 48S complex.

Junction of the 60S subunits with the initiation complex cannot occur until eIF2 and eIF3 have been released from the initiation complex. This is mediated by eIF5 and causes eIF2 to hydrolyze its GTP. The reaction occurs on the small ribosome subunit and requires the initiator tRNA to be base-paired with the AUG initiation codon. All of the remaining factors likely are released when the complete 80S ribosome is formed.

Finally, the factor eIF5B enables the 60S subunit to join the complex, forming an intact ribosome that is ready to start elongation. The subunit eIF5B has a similar sequence to the prokaryotic factor IF2, which has a similar role in hydrolyzing GTP (in addition to its role in binding the initiator tRNA).

Once the factors have been released, they can associate with the initiator tRNA and ribosomal subunits in another initiation cycle. The subunit eIF2 has hydrolyzed its GTP; as a result, the active form must be regenerated. This is accomplished by another factor, eIF2B, which displaces the GDP so that it can be replaced by GTP.

The subunit eIF2 is a target for regulation. Several regulatory kinases act on the α subunit of eIF2. Phosphorylation prevents eIF2B from regenerating the active form. This limits the action of eIF2B to one cycle of initiation, and thereby inhibits protein synthesis.

8.10 Elongation Factor Tu Loads Aminoacyl-tRNA into the A Site

Key concepts

- EF-Tu is a monomeric G protein whose active form (bound to GTP) binds aminoacyl-tRNA.
- The EF-Tu-GTP-aminoacyl-tRNA complex binds to the ribosome A site.

Once the complete ribosome is formed at the initiation codon, the stage is set for a cycle in which aminoacyl-tRNA enters the A site of a ribosome whose P site is occupied by peptidyl-tRNA. Any aminoacyl-tRNA except the initiator can enter the A site. Its entry is mediated by an **elongation factor** (**EF-Tu** in bacteria). The process is similar in eukaryotes. EF-Tu is a highly conserved protein throughout bacteria and mitochondria and is homologous to its eukaryotic counterpart.

Just like its counterpart in initiation (IF-2), EF-Tu is associated with the ribosome only during the process of aminoacyl-tRNA entry. Once the aminoacyl-tRNA is in place, EF-Tu leaves the ribosome, to work again with another aminoacyl-tRNA. Thus it displays the cyclic association with, and dissociation from, the ribosome that is the hallmark of the accessory factors.

The pathway for aminoacyl-tRNA entry to the A site is illustrated in FIGURE 8.25. EF-Tu carries a guanine nucleotide. The factor is a monomeric G protein whose activity is controlled by the state of the guanine nucleotide:

- When GTP is present, the factor is in its active state.
- When the GTP is hydrolyzed to GDP, the factor becomes inactive.
- Activity is restored when the GDP is replaced by GTP.

The binary complex of EF-Tu-GTP binds aminoacyl-tRNA to form a ternary complex of aminoacyl-tRNA-EF-Tu-GTP. The ternary complex binds only to the A site of ribosomes whose P site is already occupied by peptidyl-tRNA. This is the critical reaction in ensuring that the aminoacyl-tRNA and peptidyl-tRNA are correctly positioned for peptide bond formation.

Aminoacyl-tRNA is loaded into the A site in two stages. First, the anticodon end binds to the A site of the 30S subunit. Then, codon–anticodon recognition triggers a change in the conformation of the ribosome. This stabilizes tRNA binding and causes EF-Tu to hydrolyze its GTP. The CCA end of the tRNA now moves into the A site on the 50S subunit. The binary complex EF-Tu-GDP is released. This form of EF-Tu is inactive and does not bind aminoacyl-tRNA effectively.

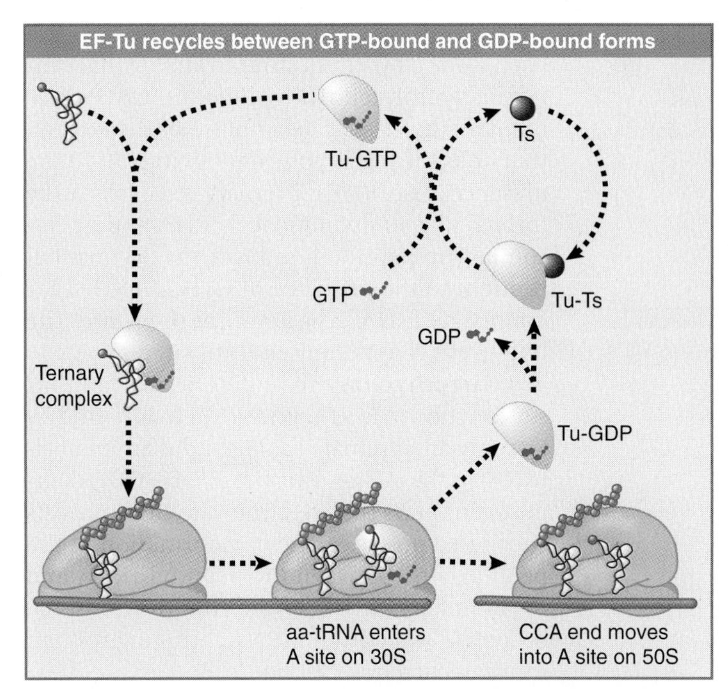

FIGURE 8.25 EF-Tu-GTP places aminoacyl-tRNA on the ribosome and then is released as EF-Tu-GDP. EF-Ts is required to mediate the replacement of GDP by GTP. The reaction consumes GTP and releases GDP. The only aminoacyl-tRNA that cannot be recognized by EF-Tu-GTP is fMet-$tRNA_f$, whose failure to bind prevents it from responding to internal AUG or GUG codons.

Another factor, EF-Ts, mediates the regeneration of the used form, EF-Tu-GDP, into the active form EF-Tu-GTP. First, EF-Ts displaces the GDP from EF-Tu, forming the combined factor EF-Tu-EF-Ts. Then the EF-Ts is in turn displaced by GTP, reforming EF-Tu-GTP. The active binary complex binds aminoacyl-tRNA, and the released EF-Ts can recycle.

There are ~70,000 molecules of EF-Tu per bacterium (~5% of the total bacterial protein), which approaches the number of aminoacyl-tRNA molecules. This implies that most aminoacyl-tRNAs are likely to be present in ternary complexes. There are only ~10,000 molecules of EF-Ts per cell (about the same as the number of ribosomes). The kinetics of the interaction between EF-Tu and EF-Ts suggest that the EF-Tu-EF-Ts complex exists only transiently, so that the EF-Tu is very rapidly converted to the GTP-bound form, and then to a ternary complex.

The role of GTP in the ternary complex has been studied by substituting an analog that cannot be hydrolyzed. The compound **GMP-PCP** has a methylene bridge in place of the oxygen that links the β and γ phosphates in GTP. In the presence of GMP-PCP, a ternary complex can be formed that binds aminoacyl-tRNA to the ribosome. The peptide bond cannot be formed, though, so the presence of GTP is needed for aminoacyl-tRNA to be bound at the A site. The hydrolysis is not required until later.

Kirromycin is an antibiotic that inhibits the function of EF-Tu. When EF-Tu is bound by kirromycin, it remains able to bind aminoacyl-tRNA to the A site. But the EF-Tu-GDP complex cannot be released from the ribosome. Its continued presence prevents formation of the peptide bond between the peptidyl-tRNA and the aminoacyl-tRNA. As a result, the ribosome becomes "stalled" on mRNA, bringing protein synthesis to a halt.

This effect of kirromycin demonstrates that inhibiting one step in protein synthesis blocks the next step. The reason is that the continued presence of EF-Tu prevents the aminoacyl end of aminoacyl-tRNA from entering the A site on the 50S subunit (see Figure 8.31). Thus the release of EF-Tu-GDP is needed for the ribosome to undertake peptide bond formation. The same principle is seen at other stages of protein synthesis: one reaction must be completed properly before the next can occur.

The interaction with EF-Tu also plays a role in quality control. Aminoacyl-tRNAs are brought into the A site without knowing whether their anticodons will fit the codon. The hydrolysis of EF-Tu-GTP is relatively slow: it takes longer than the time required for an incorrect aminoacyl-tRNA to dissociate from the A site, therefore most incorrect species are removed at this stage. The release of EF-Tu-GDP after hydrolysis also is slow, so any surviving incorrect aminoacyl-tRNAs may dissociate at this stage. The basic principle is that the reactions involving EF-Tu occur slowly enough to allow incorrect aminoacyl-tRNAs to dissociate before they become trapped in protein synthesis.

In eukaryotes, the factor eEF1α is responsible for bringing aminoacyl-tRNA to the ribosome, again in a reaction that involves cleavage of a high-energy bond in GTP. Like its prokaryotic homolog (EF-Tu), it is an abundant protein. After hydrolysis of GTP, the active form is regenerated by the factor eEF1βγ, a counterpart to EF-Ts.

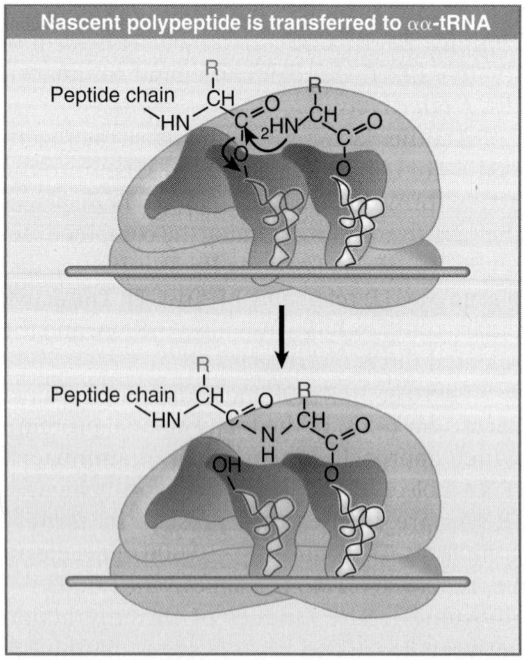

Nascent polypeptide is transferred to αα-tRNA

Peptide chain

Peptide chain

FIGURE 8.26 Peptide bond formation takes place by reaction between the polypeptide of peptidyl-tRNA in the P site and the amino acid of aminoacyl-tRNA in the A site.

8.11 The Polypeptide Chain Is Transferred to Aminoacyl-tRNA

Key concepts

- The 50S subunit has peptidyl transferase activity.
- The nascent polypeptide chain is transferred from peptidyl-tRNA in the P site to aminoacyl-tRNA in the A site.
- Peptide bond synthesis generates deacylated tRNA in the P site and peptidyl-tRNA in the A site.

The ribosome remains in place while the polypeptide chain is elongated by transferring the polypeptide attached to the tRNA in the P site to the aminoacyl-tRNA in the A site. The reaction is shown in FIGURE 8.26. The activity responsible for synthesis of the peptide bond is called **peptidyl transferase.**

The nature of the transfer reaction is revealed by the ability of the antibiotic **puromycin** to inhibit protein synthesis. Puromycin resembles an amino acid attached to the terminal adenosine of tRNA. FIGURE 8.27 shows that puromycin has an N instead of the O that joins an amino acid to tRNA. The antibiotic is treated by the ribosome as though it were an incoming aminoacyl-tRNA, after which the polypeptide attached to peptidyl-tRNA is transferred to the NH_2 group of the puromycin.

The puromycin moiety is not anchored to the A site of the ribosome, and as a result the polypeptidyl-puromycin adduct is released from the ribosome in the form of polypeptidyl-puromycin. This premature termination of protein synthesis is responsible for the lethal action of the antibiotic.

Peptidyl transferase is a function of the large (50S or 60S) ribosomal subunit. The reaction is triggered when EF-Tu releases the aminoacyl end of its tRNA. The aminoacyl end then swings into a location close to the end of the peptidyl-tRNA. This site has a peptidyl transferase activity that essentially ensures a rapid transfer of the peptide chain to the aminoacyl-tRNA. Both rRNA and 50S subunit proteins are necessary for this activity, but the actual act of catalysis is a property of the ribosomal RNA of the 50S subunit (see Section 8.19, 23S rRNA Has Peptidyl Transferase Activity).

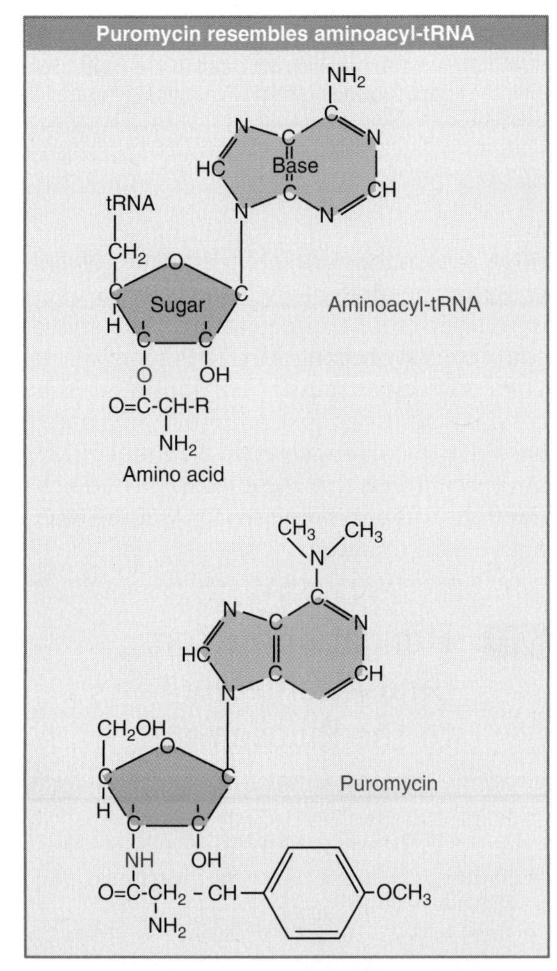

FIGURE 8.27 Puromycin mimics aminoacyl-tRNA because it resembles an aromatic amino acid linked to a sugar-base moiety.

8.12 Translocation Moves the Ribosome

Key concepts

- Ribosomal translocation moves the mRNA through the ribosome by three bases.
- Translocation moves deacylated tRNA into the E site and peptidyl-tRNA into the P site, and empties the A site.
- The hybrid state model proposes that translocation occurs in two stages, in which the 50S moves relative to the 30S, and then the 30S moves along mRNA to restore the original conformation.

The cycle of addition of amino acids to the growing polypeptide chain is completed by translocation, when the ribosome advances three nucleotides along the mRNA. FIGURE 8.28 shows that translocation expels the uncharged tRNA from the P site, so that the new peptidyl-tRNA can enter. The ribosome then has an empty A site ready for entry of the aminoacyl-tRNA corresponding to the next codon. As the figure shows, in bacteria the discharged tRNA is transferred from the P site to the E site (from which it is then expelled into the cytoplasm). In eukaryotes it is expelled directly into the cytosol. The A and P sites straddle both the large and small subunits; the E site (in bacteria) is located largely on the 50S subunit, but has some contacts in the 30S subunit.

Most thinking about translocation follows the hybrid state model, which proposes that translocation occurs in two stages. FIGURE 8.29 shows that first there is a shift of the 50S subunit relative to the 30S subunit, followed by a second shift that occurs when the 30S subunit moves along mRNA to restore the original conformation. The basis for this model was the observation that the pattern of contacts that tRNA makes with the ribosome (measured by chemical footprinting) changes in two stages. When puromycin is added to a ribosome that has an aminoacylated tRNA in the P site, the contacts of tRNA on the 50S subunit change from the P site to the E site, but the contacts on the 30S subunit do not change. This suggests that the 50S subunit has moved to a posttransfer state, but the 30S subunit has not changed.

The interpretation of these results is that first the aminoacyl ends of the tRNAs (located in the 50S subunit) move into the new sites (while the anticodon ends remain bound to their anticodons in the 30S subunit). At this stage, the tRNAs are effectively bound in

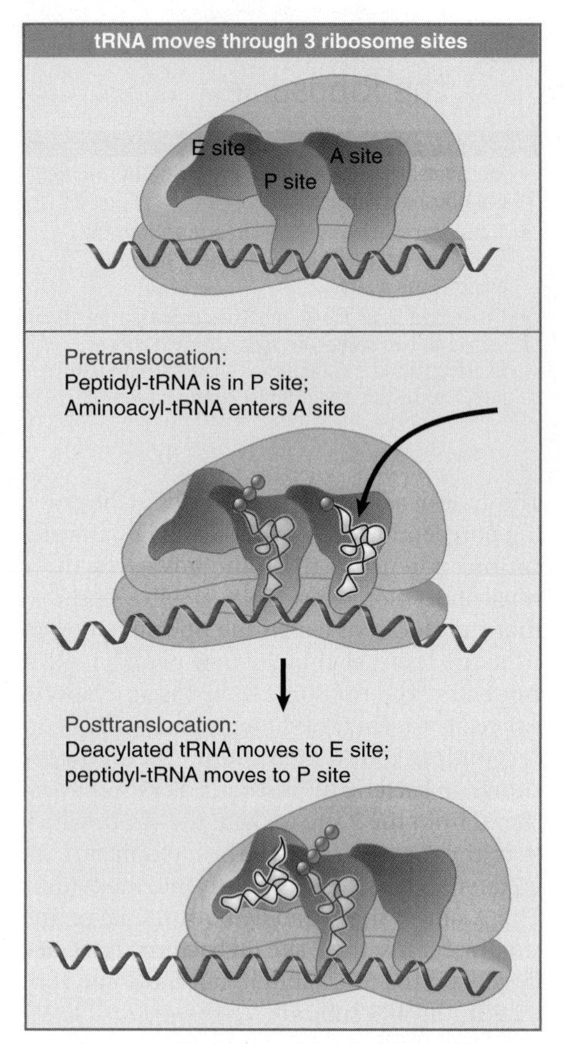

FIGURE 8.28 A bacterial ribosome has three tRNA-binding sites. Aminoacyl-tRNA enters the A site of a ribosome that has peptidyl-tRNA in the P site. Peptide bond synthesis deacylates the P site tRNA and generates peptidyl-tRNA in the A site. Translocation moves the deacylated tRNA into the E site and moves peptidyl-tRNA into the P site.

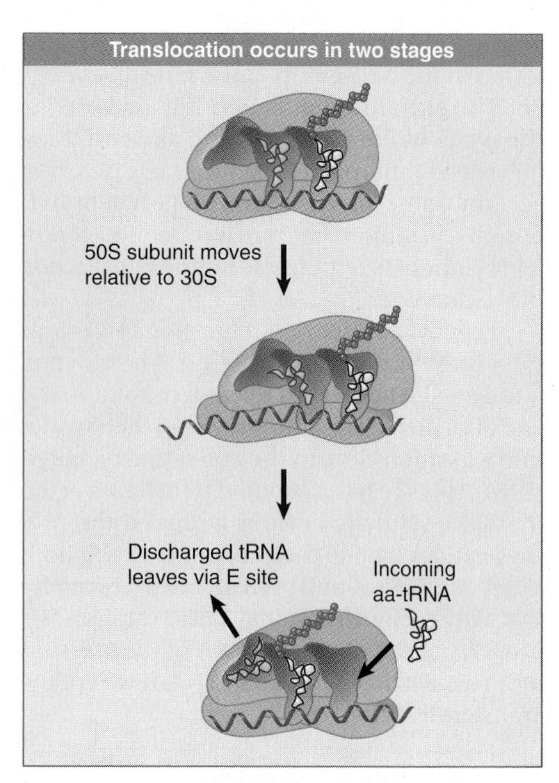

FIGURE 8.29 Models for translocation involve two stages. First, at peptide bond formation the aminoacyl end of the tRNA in the A site becomes relocated in the P site. Second, the anticodon end of the tRNA becomes relocated in the P site.

hybrid sites, consisting of the 50SE/30S P and the 50SP/30S A sites. Then movement is extended to the 30S subunits, so that the anticodon–codon pairing region finds itself in the right site. The most likely means of creating the hybrid state is by a movement of one ribosomal subunit relative to the other, so that translocation in effect involves two stages, with the normal structure of the ribosome being restored by the second stage.

The ribosome faces an interesting dilemma at translocation. It needs to break many of its contacts with tRNA in order to allow movement. At the same time, however, it must maintain pairing between tRNA and the anticodon (breaking the pairing of the deacylated tRNA only at the right moment). One possibility is that the ribosome switches between alternative, discrete conformations. The switch could consist of changes in rRNA base pairing. The accuracy of translation is influenced by certain mutations that influence alternative base pairing arrangements. The most likely interpretation is that the effect is mediated by the tightness of binding to tRNA of the alternative conformations.

8.13 Elongation Factors Bind Alternately to the Ribosome

Key concepts

- Translocation requires EF-G, whose structure resembles the aminoacyl-tRNA-EF-Tu-GTP complex.
- Binding of EF-Tu and EF-G to the ribosome is mutually exclusive.
- Translocation requires GTP hydrolysis, which triggers a change in EF-G, which in turn triggers a change in ribosome structure.

Translocation requires GTP and another elongation factor, EF-G. This factor is a major constituent of the cell: it is present at a level of ~1 copy per ribosome (20,000 molecules per cell).

Ribosomes cannot bind EF-Tu and EF-G simultaneously, so protein synthesis follows the cycle illustrated in Figure 8.31, in which the factors are alternately bound to, and released from, the ribosome. Thus EF-Tu-GDP must be released before EF-G can bind; and then EF-G must be released before aminoacyl-tRNA-EF-Tu-GTP can bind.

Does the ability of each elongation factor to exclude the other rely on an allosteric effect on the overall conformation of the ribosome or on direct competition for overlapping binding sites? FIGURE 8.30 shows an extraordinary similarity between the structures of the ternary complex of aminoacyl-tRNA-EF-Tu-GDP and EF-G. The structure of EF-G mimics the overall structure of EF-Tu bound to the amino acceptor stem of aminoacyl-tRNA. This creates the immediate assumption that they compete for the same binding site (presumably in the vicinity of the A site). The need for each factor to be released before the other can bind ensures that the events of protein synthesis proceed in an orderly manner.

Both elongation factors are monomeric GTP-binding proteins that are active when bound to GTP, but inactive when bound to GDP. The triphosphate form is required for binding to the ribosome, which ensures that each factor obtains access to the ribosome only in the company of the GTP that it needs to fulfill its function.

EF-G binds to the ribosome to sponsor translocation, and then is released following ribosome movement. EF-G can still bind to the ribosome when GMP-PCP is substituted for GTP; thus the presence of a guanine nucleotide is needed for binding, but its hydrolysis is not absolutely essential for translocation (although translocation is much slower in the absence of GTP hydrolysis). The hydrolysis of GTP is needed to release EF-G.

The need for EF-G release was discovered by the effects of the steroid antibiotic fusidic acid, which "jams" the ribosome in its post-translocation state (see FIGURE 8.31). In the presence of fusidic acid, one round of translocation occurs: EF-G binds to the ribosome, GTP is hydrolyzed, and the ribosome moves three nucleotides. Fusidic acid stabilizes the ribosome-EF-G-GDP complex, though, so that EF-G and GDP remain on the ribosome instead of being

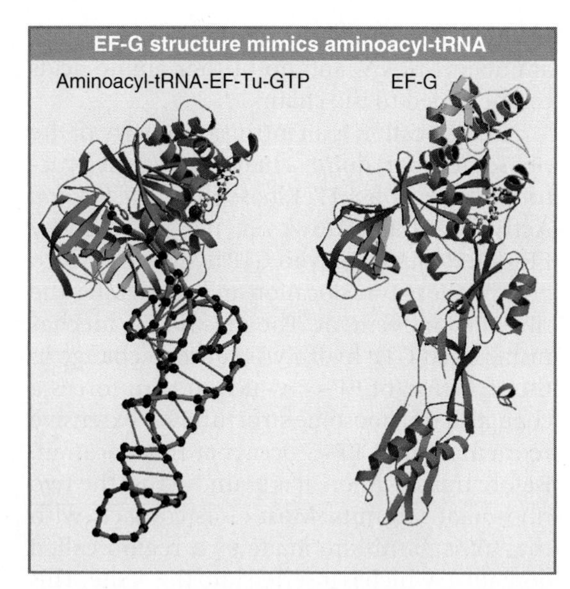

FIGURE 8.30 The structure of the ternary complex of aminoacyl-tRNA-EF-Tu-GTP (left) resembles the structure of EF-G (right). Structurally conserved domains of EF-Tu and EF-G are in red and green; the tRNA and the domain resembling it in EF-G are in purple. Photo courtesy of Poul Nissen, University of Aarhua, Denmark.

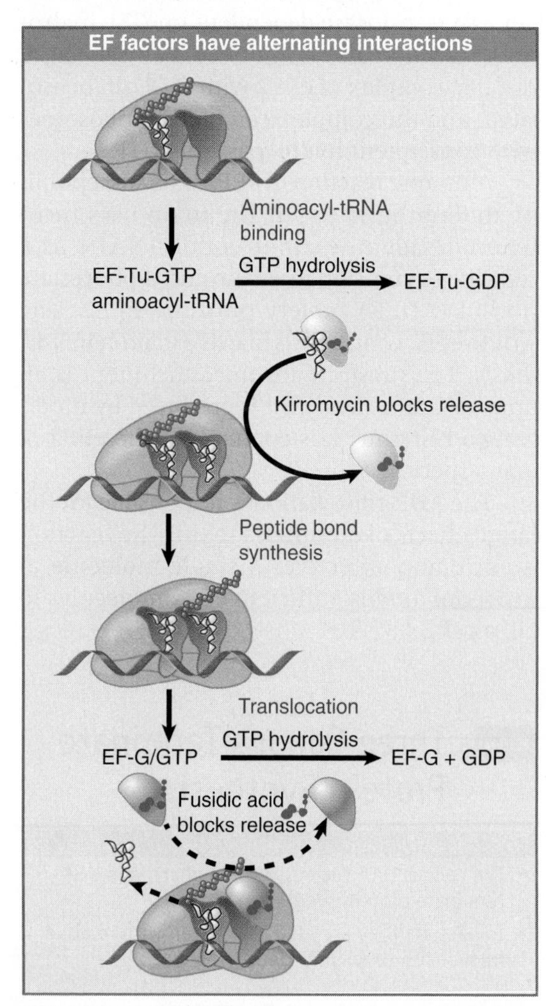

FIGURE 8.31 Binding of factors EF-Tu and EF-G alternates as ribosomes accept new aminoacyl-tRNA, form peptide bonds, and translocate.

released. As a result, the ribosome cannot bind aminoacyl-tRNA, and no further amino acids can be added to the chain.

Translocation is an intrinsic property of the ribosome that requires a major change in structure (see Section 8.17, Ribosomes Have Several Active Centers). However, it's activated by EF-G in conjunction with GTP hydrolysis, which occurs before translocation and accelerates the ribosome movement. The most likely mechanism is that GTP hydrolysis causes a change in the structure of EF-G, which in turn forces a change in the ribosome structure. An extensive reorientation of EF-G occurs at translocation. Before translocation, it is bound across the two ribosomal subunits. Most of its contacts with the 30S subunit are made by a region called domain 4, which is inserted into the A site. This domain could be responsible for displacing the tRNA. After translocation, domain 4 is instead oriented toward the 50S subunit.

The eukaryotic counterpart to EF-G is the protein eEF2, which functions in a similar manner as a translocase dependent on GTP hydrolysis. Its action also is inhibited by fusidic acid. A stable complex of eEF2 with GTP can be isolated, and the complex can bind to ribosomes with consequent hydrolysis of its GTP.

A unique reaction of eEF2 is its susceptibility to diphtheria toxin. The toxin uses nicotinamide adenine dinucleotide (NAD) as a cofactor to transfer an adenosine diphosphate ribosyl (ADPR) moiety onto the eEF2. The ADPR-eEF2 conjugate is inactive in protein synthesis. The substrate for the attachment is an unusual amino acid that is produced by modifying a histidine; it is common to the eEF2 of many species.

The ADP-ribosylation is responsible for the lethal effects of diphtheria toxin. The reaction is extremely effective: A single molecule of toxin can modify sufficient eEF2 molecules to kill a cell.

8.14 Three Codons Terminate Protein Synthesis

Key concepts

- The codons UAA (ochre), UAG (amber), and UGA terminate protein synthesis.
- In bacteria they are used most often with relative frequencies UAA>UGA>UAG.

Only 61 triplets are assigned to amino acids. The other three triplets are termination codons (or **stop codons**), which end protein synthesis. They have casual names from the history of their discovery. The UAG triplet is called the **amber** codon, UAA is the **ochre** codon, and UGA is sometimes called the **opal** codon.

The nature of these triplets was originally shown by a genetic test that distinguished two types of point mutation:

- A point mutation that changes a codon to represent a different amino acid is called a **missense** mutation. One amino acid replaces the other in the protein; the effect on protein function depends on the site of mutation and the nature of the amino acid replacement.
- When a point mutation creates one of the three termination codons, it causes **premature termination** of protein synthesis at the mutant codon. Only the first part of the protein is made in the mutant cell. This is likely to abolish protein function (depending, of course, on how far along the protein the mutant site is located). A change of this sort is called a **nonsense mutation.**

(Sometimes the term *nonsense codon* is used to describe the termination triplets. "Nonsense" is really a misnomer, given that the codons do have meaning—albeit a disruptive one in a mutant gene. A better term is stop codon.)

In every gene that has been sequenced, one of the termination codons lies immediately after the codon representing the C-terminal amino acid of the wild-type sequence. Nonsense mutations show that any one of the three codons is sufficient to terminate protein synthesis within a gene. The UAG, UAA, and UGA triplet sequences are therefore necessary and sufficient to end protein synthesis, whether occurring naturally at the end of a gene or created by mutation within a coding sequence.

In bacterial genes, UAA is the most commonly used termination codon. UGA is used more heavily than UAG, although there appear to be more errors reading UGA. (An error in reading a termination codon, when an aminoacyl-tRNA improperly responds to it, results in the continuation of protein synthesis until another termination codon is encountered.)

8.15 Termination Codons Are Recognized by Protein Factors

Key concepts

- Termination codons are recognized by protein release factors, not by aminoacyl-tRNAs.
- The structures of the class 1 release factors resemble aminoacyl-tRNA-EF-Tu and EF-G.
- The class 1 release factors respond to specific termination codons and hydrolyze the polypeptide-tRNA linkage.
- The class 1 release factors are assisted by class 2 release factors that depend on GTP.
- The mechanism is similar in bacteria (which have two types of class 1 release factors) and eukaryotes (which have only one class 1 release factor).

Two stages are involved in ending translation. The *termination reaction* itself involves release of the protein chain from the last tRNA. The *post-termination reaction* involves release of the tRNA and mRNA and dissociation of the ribosome into its subunits.

None of the termination codons is represented by a tRNA. They function in an entirely different manner from other codons and are recognized directly by protein factors. (The reaction does not depend on codon-anticodon recognition, so there seems to be no particular reason why it should require a triplet sequence. Presumably this reflects the evolution of the genetic code.)

Termination codons are recognized by class 1 **release factors (RF).** In *E. coli,* two class 1 release factors are specific for different sequences. **RF1** recognizes UAA and UAG; **RF2** recognizes UGA and UAA. The factors act at the ribosomal A site and require polypeptidyl-tRNA in the P site. The RF are present at much lower levels than initiation or elongation factors; there are ~600 molecules of each per cell, equivalent to one RF per ten ribosomes. At one time there probably was only a single release factor that recognized all termination codons, which later evolved into two factors with specificities for particular codons. In eukaryotes, there is only a single class 1 release factor, called eRF. The efficiency with which the bacterial factors recognize their target codons is influenced by the bases on the 3′ side.

The class 1 release factors are assisted by class 2 release factors, which are not codon-specific. The class 2 factors are GTP-binding proteins. In *E. coli,* the role of the class 2 factor is to release the class 1 factor from the ribosome.

Although the general mechanism of termination is similar in prokaryotes and eukaryotes, the interactions between the class 1 and class 2 factors have some differences.

The class 1 factors RF1 and RF2 recognize the termination codons and activate the ribosome to hydrolyze the peptidyl tRNA. Cleavage of polypeptide from tRNA takes place by a reaction analogous to the usual peptidyl transfer, except that the acceptor is H_2O instead of aminoacyl-tRNA (see Figure 8.34).

At this point RF1 or RF2 is released from the ribosome by the class 2 factor **RF3,** which is related to EF-G. RF3-GDP binds to the ribosome before the termination reaction occurs, and the GDP is replaced by GTP. This enables RF3 to contact the ribosome GTPase center, where it causes RF1/2 to be released when the polypeptide chain is terminated.

RF3 resembles the GTP-binding domains of EF-Tu and EF-G, and RF1 and -2 resemble the C-terminal domain of EF-G, which mimics tRNA. This suggests that the release factors utilize the same site that is used by the elongation factors. FIGURE 8.32 illustrates the basic idea that

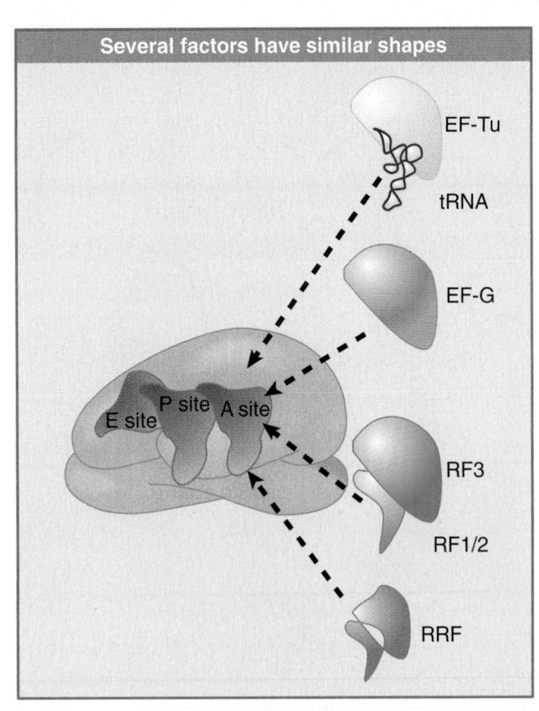

FIGURE 8.32 Molecular mimicry enables the elongation factor Tu-tRNA complex, the translocation factor EF-G, and the release factors RF1/2-RF3 to bind to the same ribosomal site. RRF is the ribosome recycling factor (see Figure 8.35)

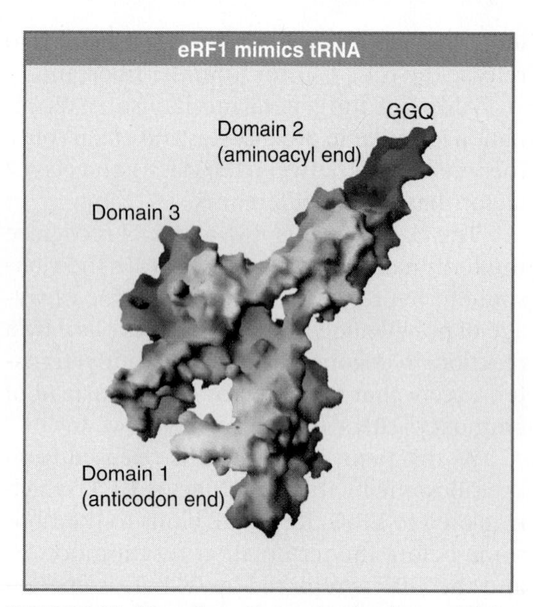

FIGURE 8.33 The eukaryotic termination factor eRF1 has a structure that mimics tRNA. The motif GGQ at the tip of domain 2 is essential for hydrolyzing the polypeptide chain from tRNA. Photo courtesy of David Barford, The Institute of Cancer Research.

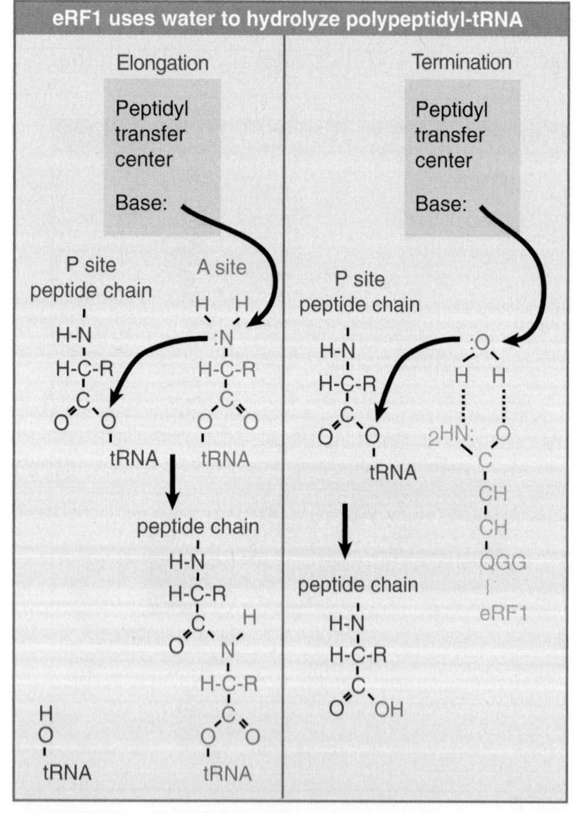

FIGURE 8.34 Peptide transfer and termination are similar reactions in which a base in the peptidyl transfer center triggers a transesterification reaction by attacking an N–H or O–H bond, releasing the N or O to attack the link to tRNA.

these factors all have the same general shape and bind to the ribosome successively at the same site (basically the A site or a region extensively overlapping with it).

The eukaryotic class 1 release factor, eRF1, is a single protein that recognizes all three termination codons. Its sequence is unrelated to the bacterial factors. It can terminate protein synthesis *in vitro* without the class 2 factor, eRF2, although eRF2 is essential in yeast *in vivo*. The structure of eRF1 follows a familiar theme: FIGURE 8.33 shows that it consists of three domains that mimic the structure of tRNA.

An essential motif of three amino acids, GGQ, is exposed at the top of domain 2. Its position in the A site corresponds to the usual location of an amino acid on an aminoacyl-tRNA. This positions it to use the glutamine (Q) to position a water molecule to substitute for the amino acid of aminoacyl-tRNA in the peptidyl transfer reaction. FIGURE 8.34 compares the termination reaction with the usual peptide transfer reaction. Termination transfers a hydroxyl group from the water, thus effectively hydrolyzing the peptide-tRNA bond (see Figure 8.48 for discussion of how the peptidyl transferase center works).

Mutations in the RF genes reduce the efficiency of termination, as seen by an increased ability to continue protein synthesis past the termination codon. Overexpression of RF1 or RF2 increases the efficiency of termination at the codons on which it acts. This suggests that codon recognition by RF1 or RF2 competes with aminoacyl-tRNAs that erroneously recognize the termination codons. The release factors recognize their target sequences very efficiently.

The termination reaction involves release of the completed polypeptide, but leaves a deacylated tRNA and the mRNA still associated with the ribosome. FIGURE 8.35 shows that the dissociation of the remaining components (tRNA, mRNA, 30S, and 50S subunits) requires ribosome recycling factor (RRF). RRF acts together with EF-G in a reaction that uses hydrolysis of GTP. As for the other factors involved in release, RRF has a structure that mimics tRNA, except that it lacks an equivalent for the 3′ amino acid-binding region. IF-3 is also required, which brings the wheel full circle to its original discovery, when it was proposed to be a dissociation factor! RRF acts on the 50S subunit, and IF-3 acts to remove deacylated tRNA from the 30S subunit. Once the subunits have separated, IF-3 remains necessary, of course, to prevent their reassociation.

8.16 Ribosomal RNA Pervades Both Ribosomal Subunits

Key concepts

- Each rRNA has several distinct domains that fold independently.
- Virtually all ribosomal proteins are in contact with rRNA.
- Most of the contacts between ribosomal subunits are made between the 16S and 23S rRNAs.

Two thirds of the mass of the bacterial ribosome is made up of rRNA. The most penetrating approach to analyzing secondary structure of large RNAs is to compare the sequences of corresponding rRNAs in related organisms. Those regions that are important in the secondary structure retain the ability to interact by base pairing. Thus if a base pair is required, it can form at the same relative position in each rRNA. This approach has enabled detailed models to be constructed for both 16S and 23S rRNA.

Each of the major rRNAs can be drawn in a secondary structure with several discrete domains. Four general domains are formed by 16S rRNA, in which just under half of the sequence is base paired (see Figure 8.45). Six general domains are formed by 23S rRNA. The individual double-helical regions tend to be short (<8 bp). Often the duplex regions are not perfect and contain bulges of unpaired bases. Comparable models have been drawn for mitochondrial rRNAs (which are shorter and have fewer domains) and for eukaryotic cytosolic rRNAs (which are longer and have more domains). The increase in length in eukaryotic rRNAs is due largely to the acquisition of sequences representing additional domains. The crystal structure of the ribosome shows that in each subunit the domains of the major rRNA fold independently and have a discrete location in the subunit.

Differences in the ability of 16S rRNA to react with chemical agents are found when 30S subunits are compared with 70S ribosomes; also there are differences between free ribosomes and those engaged in protein synthesis. Changes in the reactivity of the rRNA occur when mRNA is bound, when the subunits associate, or when tRNA is bound. Some changes reflect a direct interaction of the rRNA with mRNA or tRNA, whereas others are caused indirectly by other changes in ribosome structure. The main point is that ribosome conformation is flexible during protein synthesis.

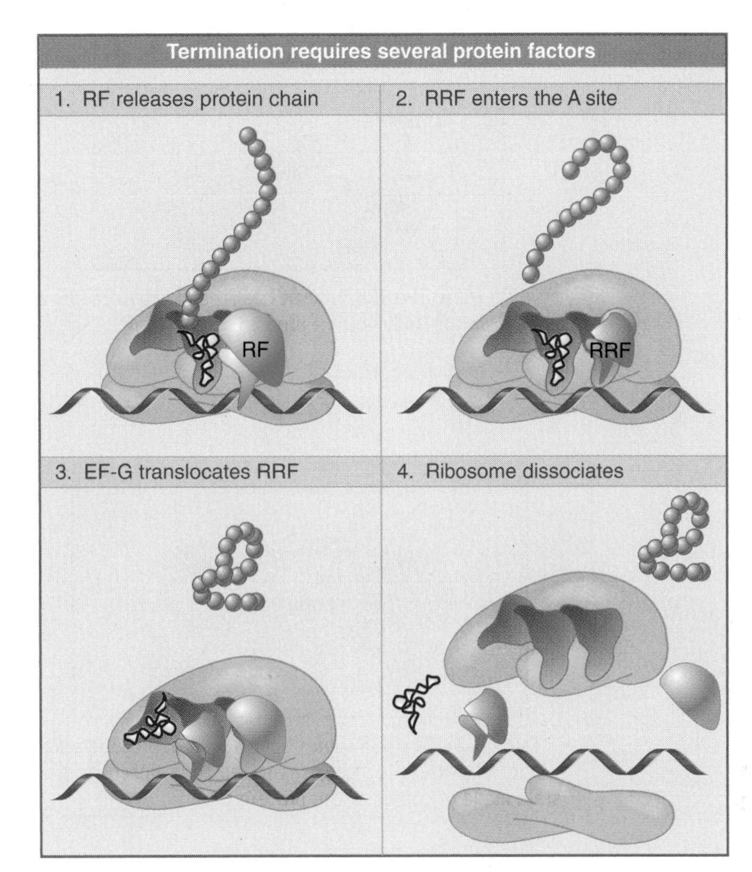

FIGURE 8.35 The RF (release factor) terminates protein synthesis by releasing the protein chain. The RRF (ribosome recycling factor) releases the last tRNA, and EF-G releases RRF, causing the ribosome to dissociate.

A feature of the primary structure of rRNA is the presence of methylated residues. There are ~10 methyl groups in 16S rRNA (located mostly toward the 3′ end of the molecule) and ~20 in 23S rRNA. In mammalian cells, the 18S and 28S rRNAs carry 43 and 74 methyl groups, respectively, so ~2% of the nucleotides are methylated (about three times the proportion methylated in bacteria).

The large ribosomal subunit also contains a molecule of a 120 base **5S RNA** (in all ribosomes except those of mitochondria). The sequence of 5S RNA is less well conserved than those of the major rRNAs. All 5S RNA molecules display a highly base-paired structure.

In eukaryotic cytosolic ribosomes, another small RNA is present in the large subunit. This is the **5.8S RNA**. Its sequence corresponds to the 5′ end of the prokaryotic 23S rRNA.

Some ribosomal proteins bind strongly to isolated rRNA. Others do not bind to free rRNA, but can bind after other proteins have bound. This suggests that the conformation of the rRNA is important in determining whether binding

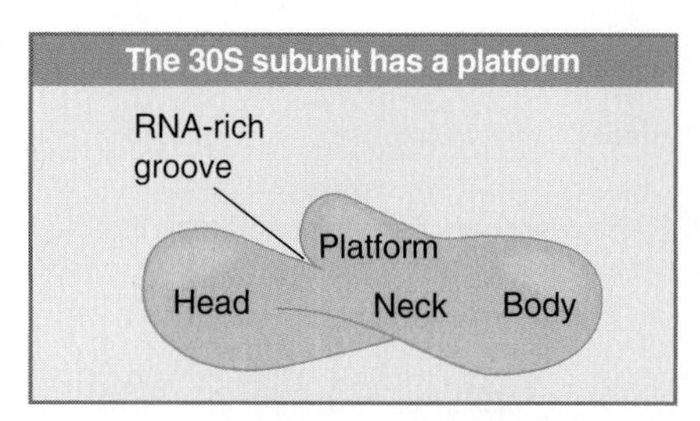

FIGURE 8.36 The 30S subunit has a head separated by a neck from the body, with a protruding platform.

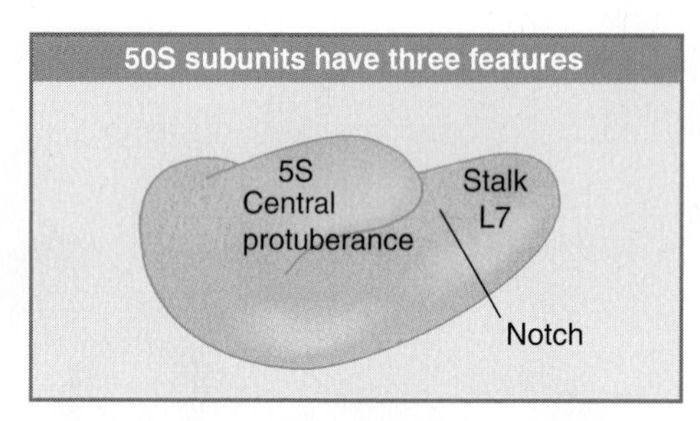

FIGURE 8.37 The 50S subunit has a central protuberance where 5S rRNA is located, separated by a notch from a stalk made of copies of the protein L7.

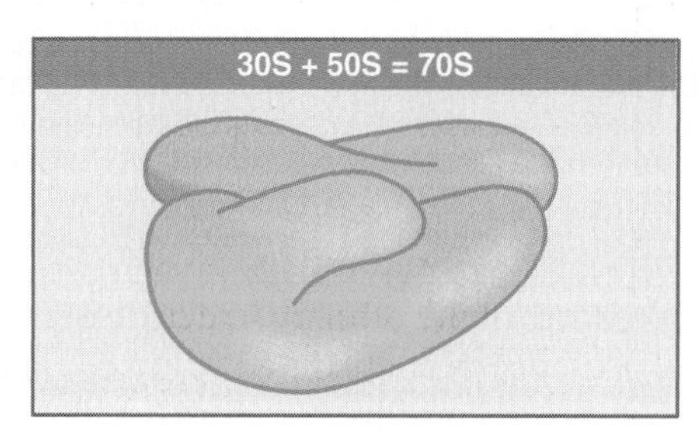

FIGURE 8.38 The platform of the 30S subunit fits into the notch of the 50S subunit to form the 70S ribosome.

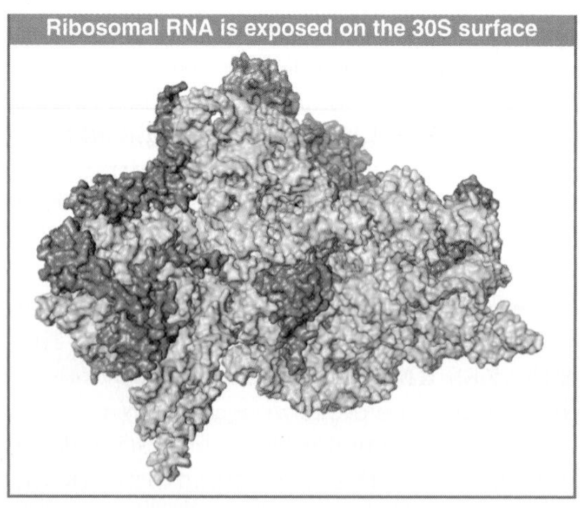

FIGURE 8.39 The 30S ribosomal subunit is a ribonucleoprotein particle. Proteins are in yellow. Photo courtesy of V. Ramakrishnan, Medical Research Council (UK).

sites exist for some proteins. As each protein binds, it induces conformational changes in the rRNA that make it possible for other proteins to bind. In *E. coli,* virtually all the 30S ribosomal proteins interact (albeit to varying degrees) with 16S rRNA. The binding sites on the proteins show a wide variety of structural features, suggesting that protein–RNA recognition mechanisms may be diverse.

The 70S ribosome has an asymmetric construction. FIGURE 8.36 shows a schematic of the structure of the 30S subunit, which is divided into four regions: the head, neck, body, and platform. FIGURE 8.37 shows a similar representation of the 50S subunit, where two prominent features are the central protuberance (where 5S rRNA is located) and the stalk (made of multiple copies of protein L7). FIGURE 8.38 shows that the platform of the small subunit fits into the notch of the large subunit. There is a cavity between the subunits that contains some of the important sites.

The structure of the 30S subunit follows the organization of 16S rRNA, with each structural feature corresponding to a domain of the rRNA. The body is based on the 5′ domain, the platform on the central domain, and the head on the 3′ region. FIGURE 8.39 shows that the 30S subunit has an asymmetrical distribution of RNA and protein. One important feature is that the platform of the 30S subunit that provides the interface with the 50S subunit is composed almost entirely of RNA. Only two proteins (a small part of S7 and possibly part of S12) lie near the interface. This means that the association and dissociation of ribosomal subunits must depend on interactions with the 16S rRNA. Subunit association is affected by a mutation in a loop of 16S rRNA (at position 791) that is located at the subunit interface, and other nucleotides in 16S rRNA have been shown to be involved by modification/interference experiments. This behavior supports the idea that the evolutionary origin of the ribosome may have been as a particle consisting of RNA rather than protein.

The 50S subunit has a more even distribution of components than the 30S, with long rods of double-stranded RNA crisscrossing the structure. The RNA forms a mass of tightly packed helices. The exterior surface largely consists of protein, except for the peptidyl transferase center (see Section 8.19, 23S rRNA Has Peptidyl Transferase Activity). Almost all segments of the 23S rRNA interact with protein, but many of the proteins are relatively unstructured.

The junction of subunits in the 70S ribosome involves contacts between 16S rRNA (many in the platform region) and 23S rRNA. There are also some interactions between rRNA of each subunit with proteins in the other, and a few protein–protein contacts. FIGURE 8.40 identifies the contact points on the rRNA structures. FIGURE 8.41 opens out the structure (imagine the 50S subunit rotated counterclockwise and the 30S subunit rotated clockwise around the axis shown in the figure) to show the locations of the contact points on the face of each subunit.

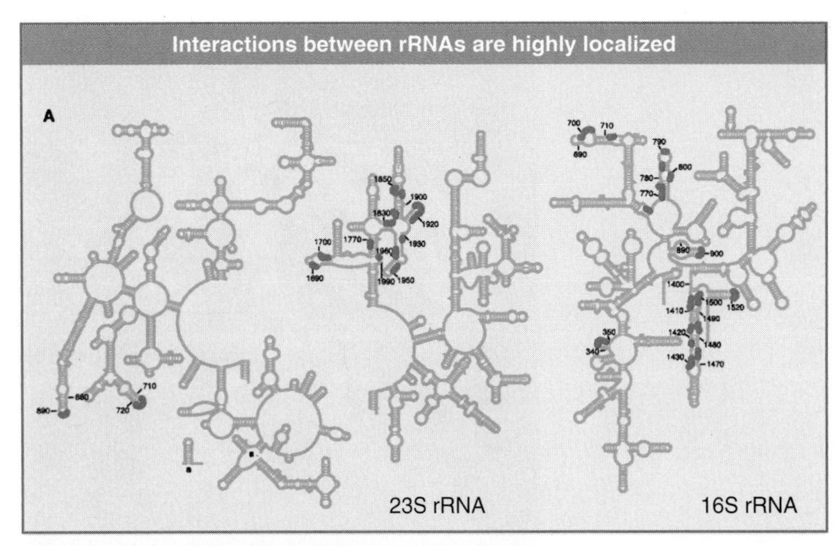

FIGURE 8.40 Contact points between the rRNAs are located in two domains of 16S rRNA and one domain of 23S rRNA. Reproduced from Yusupov, M. M. 2001. *Science*. 292: 883–896. © 2001 AAAS. Photo courtesy of Harry Noller, University of California, Santa Cruz.

8.17 Ribosomes Have Several Active Centers

Key concepts

- Interactions involving rRNA are a key part of ribosome function.
- The environment of the tRNA-binding sites is largely determined by rRNA.

The basic message to remember about the ribosome is that it is a cooperative structure that depends on changes in the relationships among its active sites during protein synthesis. The active sites are not small, discrete regions like the active centers of enzymes. They are large regions whose construction and activities may depend just as much on the rRNA as on the ribosomal proteins. The crystal structures of the individual subunits and bacterial ribosomes give us a good impression of the overall organization and emphasize the role of the rRNA. The most recent structure, at 5.5 Å resolution, clearly identifies the locations of the tRNAs and the functional sites. We can now account for many ribosomal functions in terms of its structure.

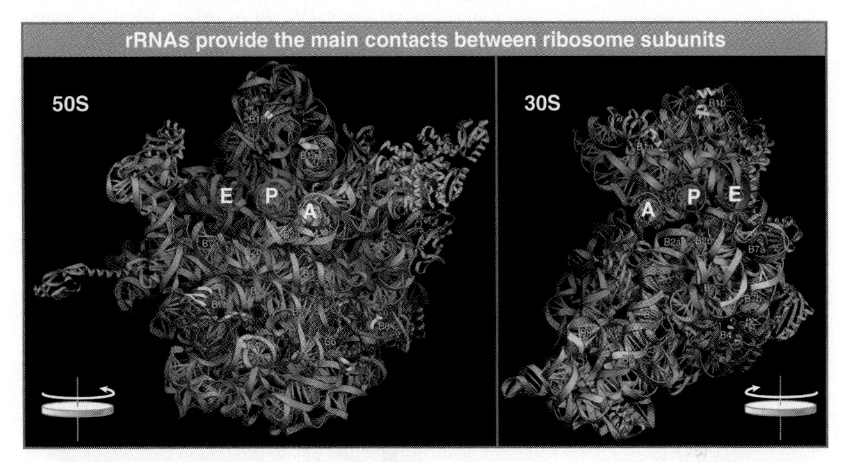

FIGURE 8.41 Contacts between the ribosomal subunits are mostly made by RNA (shown in purple). Contacts involving proteins are shown in yellow. The two subunits are rotated away from one another to show the faces where contacts are made; from a plane of contact perpendicular to the screen, the 50S subunit is rotated 90° counterclockwise, and the 30S is rotated 90° clockwise (this shows it in the reverse of the usual orientation). Photo courtesy of Harry Noller, University of California, Santa Cruz.

Ribosomal functions are centered around the interaction with tRNAs. FIGURE 8.42 shows the 70S ribosome with the positions of tRNAs in the three binding sites. The tRNAs in the A and P sites are nearly parallel to one another. All three tRNAs are aligned with their anticodon loops bound to the mRNA in the groove on the 30S subunit. The rest of each tRNA is bound to the 50S subunit. The environment surrounding each tRNA is mostly provided by rRNA. In each site, the rRNA contacts the tRNA at parts of the structure that are universally conserved.

It has always been a big puzzle to understand how two bulky tRNAs can fit next to one another in reading adjacent codons. The crystal structure shows a 45° kink in the mRNA between the P and A sites, which allows the tRNAs to fit as shown in the expansion of FIGURE 8.43. The tRNAs in the P and A sites are angled at 26° relative to each other at their anticodons. The closest approach between the backbones of the tRNAs occurs at the 3′ ends, where they converge to within 5 Å (perpendicular to the plane of the screen). This allows the peptide chain to be transferred from the peptidyl-tRNA in the A site to the aminoacyl-tRNA in the A site.

Aminoacyl-tRNA is inserted into the A site by EF-Tu, and its pairing with the codon is necessary for EF-Tu to hydrolyze GTP and be released from the ribosome (see Section 8.10, Elongation Factor Tu Loads Aminoacyl-tRNA

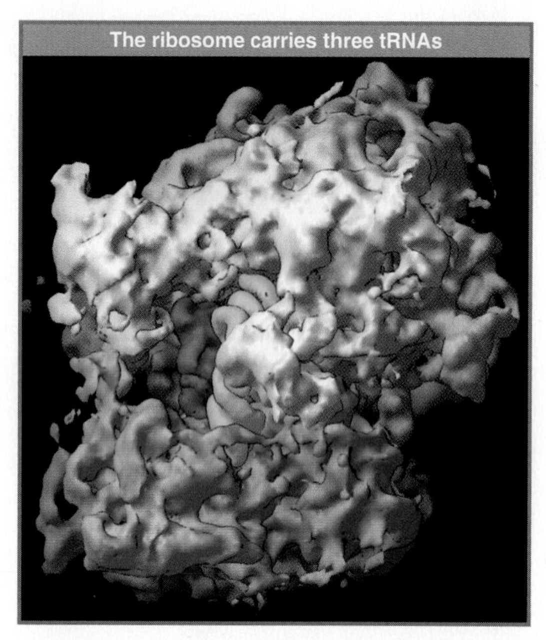

FIGURE 8.42 The 70S ribosome consists of the 50S subunit (white) and the 30S subunit (purple) with three tRNAs located superficially: yellow in the A site, blue in the P site, and green in the E site. Photo courtesy of Harry Noller, University of California, Santa Cruz.

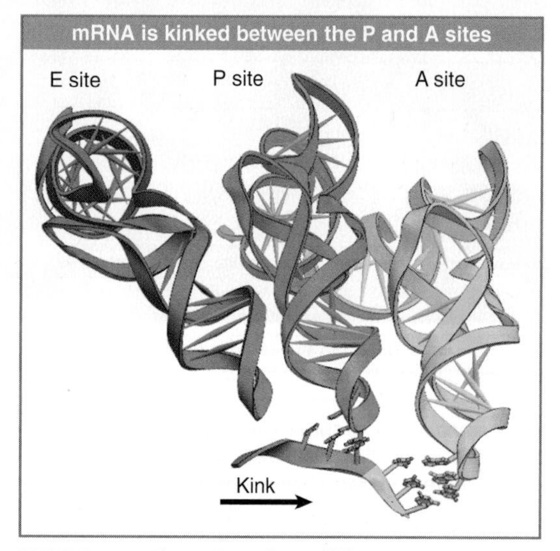

FIGURE 8.43 Three tRNAs have different orientations on the ribosome. mRNA turns between the P and A sites to allow aminoacyl-tRNAs to bind adjacent codons. Photo courtesy of Harry Noller, University of California, Santa Cruz.

into the A Site). EF-Tu initially places the aminoacyl-tRNA into the small subunit, where the anticodon pairs with the codon. Movement of the tRNA is required to bring it fully into the A site, when its 3′ end enters the peptidyl transferase center on the large subunit. There are different models for how this process may occur. One calls for the entire tRNA to swivel, so that the elbow in the L-shaped structure made by the D and TΨC arms moves into the ribosome, enabling the TΨC arm to pair with rRNA. Another calls for the internal structure of the tRNA to change, using the anticodon loop as a hinge, with the rest of the tRNA rotating from a position in which it is stacked on the 3′ side of the anticodon loop to one in which it is stacked on the 5′ side. Following the transition, EF-Tu hydrolyzes GTP, allowing peptide synthesis to proceed.

Translocation involves large movements in the positions of the tRNAs within the ribosome. The anticodon end of tRNA moves ~28 Å from the A site to the P site, and then a further 20 Å from the P site to the E site. As a result of the angle of each tRNA relative to the anticodon, the bulk of the tRNA moves much larger distances: 40 Å from A site to P site and 55 Å from P site to E site. This suggests that translocation requires a major reorganization of structure.

For many years, it was thought that translocation could occur only in the presence of the factor EF-G. However, the antibiotic sparsomycin (which inhibits the peptidyl transferase activity) triggers translocation. This suggests that the energy to drive translocation actually is stored in the ribosome after peptide bond formation has occurred. Usually EF-G acts on the ribosome to release this energy and enable it to drive translocation, but sparsomycin can have the same role. Sparsomycin inhibits peptidyl transferase by binding to the peptidyl-tRNA, blocking its interaction with aminoacyl-tRNA. It probably creates a conformation that resembles the usual posttranslocation conformation, which in turn promotes movement of the peptidyl-tRNA. The important point is that translocation is an intrinsic property of the ribosome.

The hybrid states model suggests that translocation may take place in two stages, with one ribosomal subunit moving relative to the other to create an intermediate stage in which there are hybrid tRNA-binding sites (50S E/30S P and 50SP/30S A) (see Figure 8.29). Comparisons of the ribosome structure between pre- and posttranslocation states, and comparisons in 16S rRNA conformation between free 30S subunits and 70S ribosomes, suggest that mobility of structure is especially marked in the head and platform regions of the 30S subunit. An interesting insight on the hybrid states model is cast by the fact that many bases in rRNA involved in subunit association are close to bases involved in interacting with tRNA. This suggests that tRNA-binding sites are close to the interface between subunits, and carries the implication that changes in subunit interaction could be connected with movement of tRNA.

Much of the structure of the ribosome is occupied by its active centers. The schematic view of the ribosomal sites in FIGURE 8.44 shows they comprise about two thirds of the ribosomal structure. A tRNA enters the A site, is transferred by translocation into the P site, and then leaves the (bacterial) ribosome by the E site. The A and P sites extend across both ribosome subunits; tRNA is paired with mRNA in the 30S subunit, but peptide transfer takes place in the 50S subunit. The A and P sites are adjacent, enabling translocation to move the tRNA from one site into the other. The E site is located near the P site (representing a position en route to the surface of the 50S subunit). The peptidyl transferase center is located on the 50S subunit, close to the aminoacyl ends of the tRNAs in the A and P sites (see Section 8.18, 16S rRNA Plays an Active Role in Protein Synthesis).

All of the GTP-binding proteins that function in protein synthesis (EF-Tu, EF-G, IF-2, and RF1, -2, and -3) bind to the same factor-binding site (sometimes called the GTPase center), which probably triggers their hydrolysis of GTP. This site is located at the base of the stalk of the large subunit, which consists of the proteins L7 and L12. (L7 is a modification of L12 and has an acetyl group on the N-terminus.) In addition to this region, the complex of protein L11 with a 58-base stretch of 23S rRNA provides the binding site for some antibiotics that affect GTPase activity. Neither of these ribosomal structures actually possesses GTPase activity, but they are both necessary for it. The role of the ribosome is to trigger GTP hydrolysis by factors bound in the factor-binding site.

Initial binding of 30S subunits to mRNA requires protein S1, which has a strong affinity for single-stranded nucleic acid. It is responsible for maintaining the single-stranded state in mRNA that is bound to the 30S subunit. This action is necessary to prevent the mRNA from taking up a base-paired conformation that would be unsuitable for translation. S1 has an extremely elongated structure and associates with S18 and S21. The three proteins constitute a domain that is involved in the initial binding of mRNA and in binding initiator tRNA. This locates the mRNA-binding site in the vicinity of the cleft of the small subunit (see Figure 8.3). The 3′ end of rRNA, which pairs with the mRNA initiation site, is located in this region.

The initiation factors bind in the same region of the ribosome. IF-3 can be crosslinked to the 3′ end of the rRNA, as well as to several ribosomal proteins, including those probably involved in binding mRNA. The role of IF-3 could be to stabilize mRNA-30S subunit binding; then it would be displaced when the 50S subunit joins.

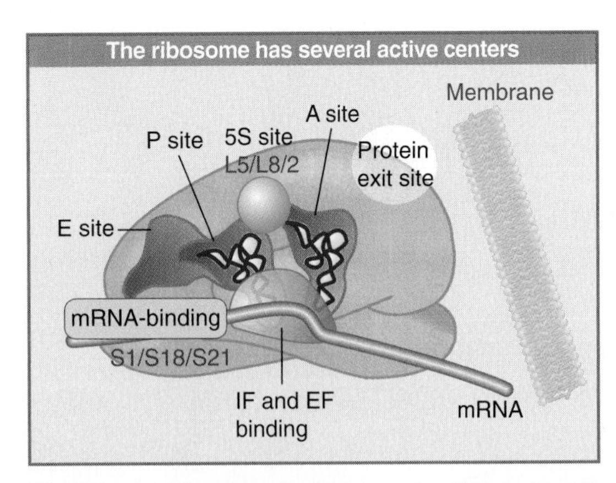

FIGURE 8.44 The ribosome has several active centers. It may be associated with a membrane. mRNA takes a turn as it passes through the A and P sites, which are angled with regard to each other. The E site lies beyond the P site. The peptidyl transferase site (not shown) stretches across the tops of the A and P sites. Part of the site bound by EF-Tu/G lies at the base of the A and P sites.

The incorporation of 5S RNA into 50S subunits that are assembled *in vitro* depends on the ability of three proteins—L5, L8, and L25—to form a stoichiometric complex with it. The complex can bind to 23S rRNA, although none of the isolated components can do so. It lies in the vicinity of the P and A sites.

A nascent protein debouches through the ribosome, away from the active sites, into the region in which ribosomes may be attached to membranes (see Chapter 10, Protein Localization). A polypeptide chain emerges from the ribosome through an exit channel, which leads from the peptidyl transferase site to the surface of the 50S subunit. The tunnel is composed mostly of rRNA. It is quite narrow—only 1 to 2 nm wide—and is ~10 nm long. The nascent polypeptide emerges from the ribosome ~15 Å away from the peptidyl transferase site. The tunnel can hold ~50 amino acids, and probably constrains the polypeptide chain so that it cannot fold until it leaves the exit domain.

8.18 16S rRNA Plays an Active Role in Protein Synthesis

Key concept

- 16S rRNA plays an active role in the functions of the 30S subunit. It interacts directly with mRNA, with the 50S subunit, and with the anticodons of tRNAs in the P and A sites.

The ribosome was originally viewed as a collection of proteins with various catalytic

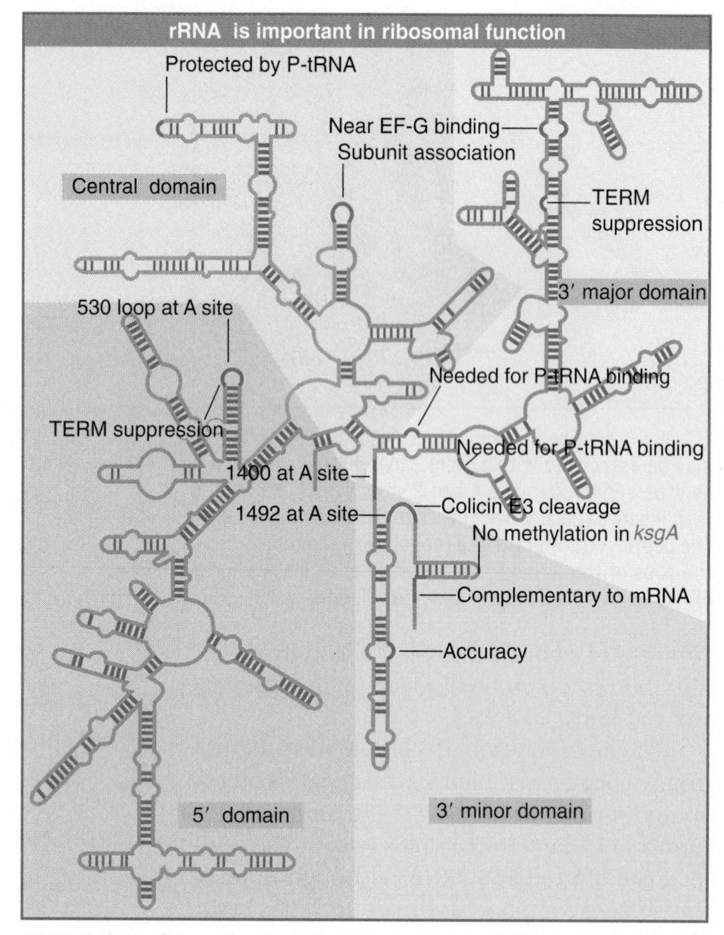

FIGURE 8.45 Some sites in 16S rRNA are protected from chemical probes when 50S subunits join 30S subunits or when aminoacyl-tRNA binds to the A site. Others are the sites of mutations that affect protein synthesis. TERM suppression sites may affect termination at some or several termination codons. The large colored blocks indicate the four domains of the rRNA.

activities held together by protein–protein interactions and by binding to rRNA. The discovery of RNA molecules with catalytic activities (see Chapter 26, RNA Splicing and Processing) immediately suggests, however, that rRNA might play a more active role in ribosome function. There is now evidence that rRNA interacts with mRNA or tRNA at each stage of translation, and that the proteins are necessary to maintain the rRNA in a structure in which it can perform the catalytic functions. Several interactions involve specific regions of rRNA:

- The 3′ terminus of the rRNA interacts directly with mRNA at initiation.
- Specific regions of 16S rRNA interact directly with the anticodon regions of tRNAs in both the A site and the P site. Similarly, 23S rRNA interacts with the CCA terminus of peptidyl-tRNA in both the P site and A site.
- Subunit interaction involves interactions between 16S and 23S rRNAs (see Section 8.16, Ribosomal RNA Pervades Both Ribosomal Subunits).

Much information about the individual steps of bacterial protein synthesis has been obtained by using antibiotics that inhibit the process at particular stages. The target for the antibiotic can be identified by the component in which resistant mutations occur. Some antibiotics act on individual ribosomal proteins, but several act on rRNA, which suggests that the rRNA is involved with many or even all of the functions of the ribosome.

The functions of rRNA have been investigated by two types of approach. Structural studies show that particular regions of rRNA are located in important sites of the ribosome, and that chemical modifications of these bases impede particular ribosomal functions. In addition, mutations identify bases in rRNA that are required for particular ribosomal functions. FIGURE 8.45 summarizes the sites in 16S rRNA that have been identified by these means.

An indication of the importance of the 3′ end of 16S rRNA is given by its susceptibility to the lethal agent colicin E3. Produced by some bacteria, the colicin cleaves ~50 nucleotides from the 3′ end of the 16S rRNA of *E. coli*. The cleavage entirely abolishes initiation of protein synthesis. Several important functions require the region that is cleaved: binding the factor IF-3, recognition of mRNA, and binding of tRNA.

The 3′ end of the 16S rRNA is directly involved in the initiation reaction by pairing with the Shine–Dalgarno sequence in the ribosome-binding site of mRNA (see Figure 8.16). Another direct role for the 3′ end of 16S rRNA in protein synthesis is shown by the properties of kasugamycin-resistant mutants, which lack certain modifications in 16S rRNA. Kasugamycin blocks initiation of protein synthesis. Resistant mutants of the type *ksgA* lack a methylase enzyme that introduces four methyl groups into two adjacent adenines at a site near the 3′ terminus of the 16S rRNA. The methylation generates the highly conserved sequence G–m_2^6A–m_2^6A, found in both prokaryotic and eukaryotic small rRNA. The methylated sequence is involved in the joining of the 30S and 50S subunits, which in turn is connected also with the retention of initiator tRNA in the complete ribosome. Kasugamycin causes fMet-$tRNA_f$ to be released from the sensitive (methylated) ribosomes, but the resistant ribosomes are able to retain the initiator.

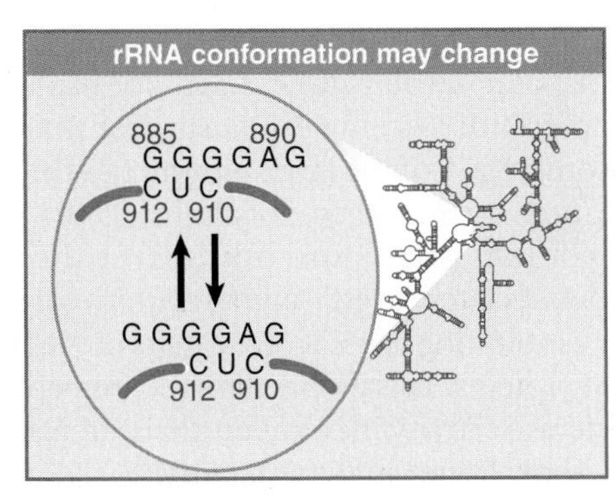

FIGURE 8.46 A change in conformation of 16S rRNA may occur during protein synthesis.

Changes in the structure of 16S rRNA occur when ribosomes are engaged in protein synthesis, as seen by protection of particular bases against chemical attack. The individual sites fall into a few groups that are concentrated in the 3′ minor and central domains. Although the locations are dispersed in the linear sequence of 16S rRNA, it seems likely that base positions involved in the same function are actually close together in the tertiary structure.

Some of the changes in 16S rRNA are triggered by joining with 50S subunits, binding of mRNA, or binding of tRNA. They indicate that these events are associated with changes in ribosome conformation that affect the exposure of rRNA. They do not necessarily indicate direct participation of rRNA in these functions. One change that occurs during protein synthesis is shown in FIGURE 8.46; it involves a local movement to change the nature of a short duplex sequence.

The 16S rRNA is involved in both A site and P site function, and significant changes in its structure occur when these sites are occupied. Certain distinct regions are protected by tRNA bound in the A site (see Figure 8.45). One is the 530 loop (which also is the site of a mutation that prevents termination at the UAA, UAG, and UGA codons). The other is the 1400 to 1500 region (so-called because bases 1399 to 1492 and the adenines at 1492 and 1493 are two single-stranded stretches that are connected by a long hairpin). All of the effects that tRNA binding has on 16S rRNA can be produced by the isolated oligonucleotide of the anticodon stem–loop, so that tRNA-30S subunit binding must involve this region.

The adenines at 1492 and 1493 provide a mechanism for detecting properly paired codon–anticodon complexes. The principle of the interaction is that the structure of the 16S rRNA responds to the structure of the first two bases pairs in the minor groove of the duplex

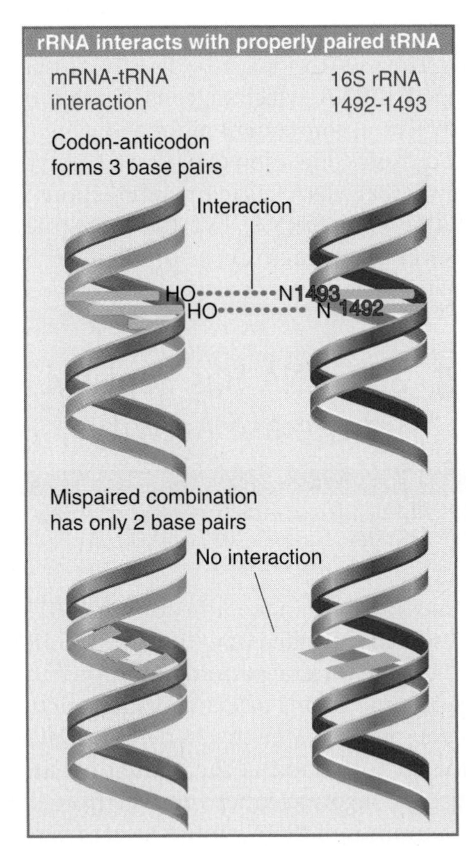

FIGURE 8.47 Codon-anticodon pairing supports interaction with adenines 1492–1493 of 16S rRNA, but mispaired tRNA-mRNA cannot interact.

formed by the codon-anticodon interaction. Modification of the N1 position of either base 1492 or 1493 in rRNA prevents tRNA from binding in the A site. However, mutations at 1492 or 1493 can be suppressed by the introduction of fluorine at the 2′ position of the corresponding bases in mRNA (which restores the interaction). FIGURE 8.47 shows that codon–anticodon pairing allows the N1 of each adenine to interact with the 2′–OH in the mRNA backbone. When an incorrect tRNA enters the A site, the structure of the codon–anticodon complex is distorted and this interaction cannot occur. The interaction stabilizes the association of tRNA with the A site.

A variety of bases in different positions of 16S rRNA are protected by tRNA in the P site; most likely the bases lie near one another in the tertiary structure. In fact, there are more contacts with tRNA when it is in the P site than when it is in the A site. This may be responsible for the increased stability of peptidyl-tRNA compared with aminoacyl-tRNA. This makes sense, because once the tRNA has reached the P site, the ribosome has decided that it is correctly bound, whereas in the A site, the assessment of binding is being made. The 1400 region can be directly crosslinked to peptidyl-tRNA, which suggests that this region is a structural component of the P site.

The basic conclusion to be drawn from these results is that rRNA has many interactions with both tRNA and mRNA, and that these interactions recur in each cycle of peptide bond formation.

8.19 23S rRNA Has Peptidyl Transferase Activity

Key concept

- Peptidyl transferase activity resides exclusively in the 23S rRNA.

The sites involved in the functions of 23S rRNA are less well identified than those of 16S rRNA, but the same general pattern is observed: bases at certain positions affect specific functions. Bases at some positions in 23S rRNA are affected by the conformation of the A site or P site. In particular, oligonucleotides derived from the 3′ CCA terminus of tRNA protect a set of bases in 23S rRNA that essentially are the same as those protected by peptidyl-tRNA. This suggests that the major interaction of 23S rRNA with peptidyl-tRNA in the P site involves the 3′ end of the tRNA.

The tRNA makes contacts with the 23S rRNA in both the P and A sites. At the P site, G2552 of 23S rRNA base pairs with C74 of the peptidyl tRNA. A mutation in the G in the rRNA prevents interaction with tRNA, but interaction is restored by a compensating mutation in the C of the amino acceptor end of the tRNA. At the A site, G2553 of the 23S rRNA base pairs with C75 of the aminoacyl-tRNA. Thus there is a close role for rRNA in both the tRNA-binding sites. Indeed, when we have a clearer structural view of the region, we should be able to understand the movements of tRNA between the A and P sites in terms of making and breaking contacts with rRNA.

Another site that binds tRNA is the E site, which is localized almost exclusively on the 50S subunit. Bases affected by its conformation can be identified in 23S rRNA.

What is the nature of the site on the 50S subunit that provides peptidyl transferase function? The involvement of rRNA was first indicated because a region of the 23S rRNA is the site of mutations that confer resistance to antibiotics that inhibit peptidyl transferase.

A long search for ribosomal proteins that might possess the catalytic activity has been unsuccessful. Recent results suggest that the ribosomal RNA of the large subunit has the catalytic activity. Extraction of almost all the protein content of 50S subunits leaves the 23S rRNA associated largely with fragments of proteins, amounting to <5% of the mass of the ribosomal proteins. This preparation retains peptidyl transferase activity. Treatments that damage the RNA abolish the catalytic activity.

Following from these results, 23S rRNA prepared by transcription *in vitro* can catalyze the formation of a peptide bond between Ac-Phe-tRNA and Phe-tRNA. The yield of Ac-Phe-Phe is very low, suggesting that the 23S rRNA requires proteins in order to function at a high efficiency. Given that the rRNA has the basic catalytic activity, though, the role of the proteins must be indirect, serving to fold the rRNA properly or to present the substrates to it. The reaction also works, although less effectively, if the domains of 23S rRNA are synthesized separately and then combined. In fact, some activity is shown by domain V alone, which has the catalytic center. Activity is abolished by mutations in position 2252 of domain V that lies in the P site.

The crystal structure of an archaeal 50S subunit shows that the peptidyl transferase site basically consists of 23S rRNA. There is no protein within 18 Å of the active site where the transfer reaction occurs between peptidyl-tRNA and aminoacyl-tRNA!

Peptide bond synthesis requires an attack by the amino group of one amino acid on the carboxyl group of another amino acid. Catalysis requires a basic residue to accept the hydrogen atom that is released from the amino group, as shown in FIGURE 8.48. If rRNA is the catalyst it must provide this residue, but we do not know how this happens. The purine and pyrimidine bases are not basic at physiological pH. A highly conserved base (at position 2451 in *E. coli*) had been implicated, but appears now neither to have the right properties nor to be crucial for peptidyl transferase activity.

Proteins that are bound to the 23S rRNA outside of the peptidyl transfer region are almost certainly required to enable the rRNA to form the proper structure *in vivo*. The idea that rRNA is the catalytic component is consistent with the results discussed in Chapter 26, RNA Splicing and Processing, which identify catalytic properties in RNA that are involved with several RNA processing reactions. It fits with the notion

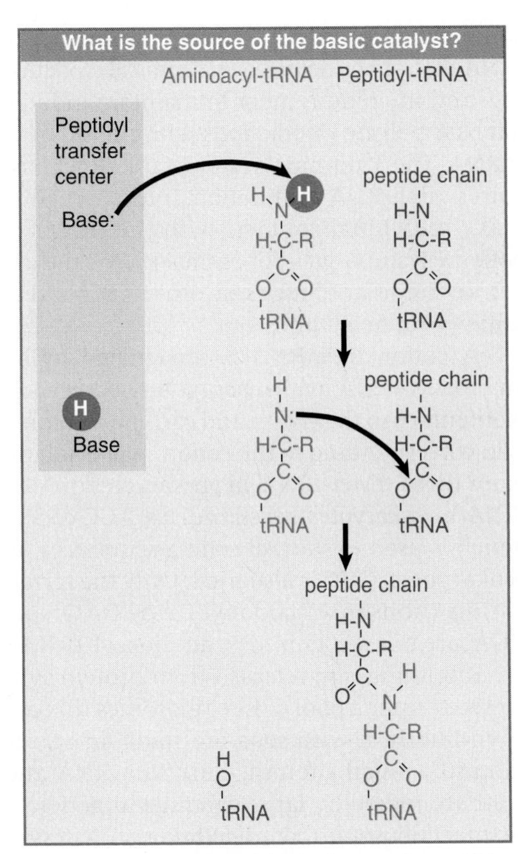

FIGURE 8.48 Peptide bond formation requires acid-base catalysis in which an H atom is transferred to a basic residue.

that the ribosome evolved from functions originally possessed by RNA.

8.20 Ribosomal Structures Change When the Subunits Come Together

Key concepts

- The head of the 30S subunit swivels around the neck when complete ribosomes are formed.
- The peptidyl transferase active site of the 50S subunit is more active in complete ribosomes than in individual 50S subunits.
- The interface between the 30S and 50S subunits is very rich in solvent contacts.

Much indirect evidence suggests that the structures of the individual subunits change significantly when they join together to form a complete ribosome. Differences in the susceptibilities of the rRNAs to outside agents are one of the strongest indicators (see Section 8.18, 16S rRNA Plays an Active Role in Protein Synthesis). More directly, comparisons of the high resolution crystal structures of the individual subunits with the lower resolution structure of the intact ribosome suggests the existence of significant differences. These ideas have been confirmed by a crystal structure of the *E. coli* ribosome at 3.5 Å, which furthermore identifies two different conformations of the ribosome, possibly representing different stages in protein synthesis.

The crystal contains two ribosomes per unit, each with a different conformation. The differences are due to changes in the positioning of domains within each subunit, the most important being that in one conformation the head of the small subunit has swiveled 6° around the neck region toward the E site. Also, a 6° rotation in the opposite direction is seen in the (low resolution) structures of *Thermus thermophilus* ribosomes that are bound to mRNA and have tRNAs in both A and P sites, suggesting that the head may swivel overall by 12° depending on the stage of protein synthesis. The rotation of the head follows the path of tRNAs through the ribosome, raising the possibility that its swiveling controls movement of mRNA and tRNA.

The changes in conformation that occur when subunits join together are much more marked in the 30S subunit than in the 50S subunit. The changes are probably concerned with controlling the position and movement of mRNA. The most significant change in the 50S subunit concerns the peptidyl transferase center. 50S subunits are ~1000× less effective in catalyzing peptide bond synthesis than complete ribosomes; the reason may be a change in structure that positions the substrate more effectively in the active site in the complete ribosome.

One of the main features emerging from the structure of the complete ribosome is the very high density of solvent contacts at their interface; this may help the making and breaking of contacts that is essential for subunit association and dissociation, and may also be involved in structural changes that occur during translocation.

8.21 Summary

Ribosomes are ribonucleoprotein particles in which a majority of the mass is provided by rRNA. The shapes of all ribosomes are generally similar, but only those of bacteria (70S) have been characterized in detail. The small (30S) subunit has a squashed shape, with a "body" containing about two thirds of the mass divided from the "head" by a cleft. The large (50S) subunit is more spherical, with a prominent "stalk" on the right and a "central protuberance." Locations of all proteins are known approximately in the small subunit.

Each subunit contains a single major rRNA, 16S and 23S in prokaryotes, and 18S and 28S in eukaryotic cytosol. There are also minor rRNAs, most notably 5S rRNA in the large subunit. Both major rRNAs have extensive base pairing, mostly in the form of short, imperfectly paired duplex stems with single-stranded loops. Conserved features in the rRNA can be identified by comparing sequences and the secondary structures that can be drawn for rRNA of a variety of organisms. The 16S rRNA has four distinct domains; the 23S rRNA has six distinct domains. Eukaryotic rRNAs have additional domains.

The crystal structure shows that the 30S subunit has an asymmetrical distribution of RNA and protein. RNA is concentrated at the interface with the 50S subunit. The 50S subunit has a surface of protein, with long rods of double-stranded RNA crisscrossing the structure. 30Sto-50S joining involves contacts between 16S rRNA and 23S rRNA. The interface between the subunits is very rich in contacts for solvent. Structural changes occur in both subunits when they join to form a complete ribosome.

Each subunit has several active centers, which are concentrated in the translational domain of the ribosome where proteins are synthesized. Proteins leave the ribosome through the exit domain, which can associate with a membrane. The major active sites are the P and A sites, the E site, the EF-Tu and EF-G binding sites, peptidyl transferase, and the mRNA-binding site. Ribosome conformation may change at stages during protein synthesis; differences in the accessibility of particular regions of the major rRNAs have been detected.

The tRNAs in the A and P sites are parallel to one another. The anticodon loops are bound to mRNA in a groove on the 30S subunit. The rest of each tRNA is bound to the 50S subunit. A conformational shift of tRNA within the A site is required to bring its aminoacyl end into juxtaposition with the end of the peptidyl-tRNA in the P site. The peptidyl transferase site that links the P- and A-binding sites is made of 23S rRNA, which has the peptidyl transferase catalytic activity, although proteins are probably needed to acquire the right structure.

An active role for the rRNAs in protein synthesis is indicated by mutations that affect ribosomal function, interactions with mRNA or tRNA that can be detected by chemical crosslinking, and the requirement to maintain individual base pairing interactions with the tRNA or mRNA. The 3′ terminal region of the rRNA base pairs with mRNA at initiation. Internal regions make individual contacts with the tRNAs in both the P and A sites. Ribosomal RNA is the target for some antibiotics or other agents that inhibit protein synthesis

A codon in mRNA is recognized by an aminoacyl-tRNA, which has an anticodon complementary to the codon and carries the amino acid corresponding to the codon. A special initiator tRNA (fMet-tRNA$_f$ in prokaryotes or Met-tRNA$_i$ in eukaryotes) recognizes the AUG codon, which is used to start all coding sequences. In prokaryotes, GUG is also used. Only the termination (nonsense) codons, UAA, UAG, and UGA, are not recognized by aminoacyl-tRNAs.

Ribosomes are released from protein synthesis to enter a pool of free ribosomes that are in equilibrium with separate small and large subunits. Small subunits bind to mRNA and then are joined by large subunits to generate an intact ribosome that undertakes protein synthesis. Recognition of a prokaryotic initiation site involves binding of a sequence at the 3′ end of rRNA to the Shine–Dalgarno motif, which precedes the AUG (or GUG) codon in the mRNA. Recognition of a eukaryotic mRNA involves binding to the 5′ cap; the small subunit then migrates to the initiation site by scanning for AUG codons. When it recognizes an appropriate AUG codon (usually, but not always, the first it encounters), it is joined by a large subunit.

A ribosome can carry two aminoacyl-tRNAs simultaneously: its P site is occupied by a polypeptidyl-tRNA, which carries the polypeptide chain synthesized so far, whereas the A site is used for entry by an aminoacyl-tRNA carrying the next amino acid to be added to the chain. Bacterial ribosomes also have an E site, through which deacylated tRNA passes before it is released after being used in protein synthesis. The polypeptide chain in the P site is transferred to the aminoacyl-tRNA in the A site, creating a deacylated tRNA in the P site and a peptidyl-tRNA in the A site.

Following peptide bond synthesis, the ribosome translocates one codon along the mRNA, moving deacylated tRNA into the E site and peptidyl tRNA from the A site into the P site. Translocation is catalyzed by the elongation factor EF-G and, like several other stages of ribosome function, requires hydrolysis of GTP. During translocation, the ribosome passes through a hybrid stage in which the 50S subunit moves relative to the 30S subunit.

Protein synthesis is an expensive process. ATP is used to provide energy at several stages, including the charging of tRNA with its amino acid and the unwinding of mRNA. It has been estimated that up to 90% of all the ATP molecules synthesized in a rapidly growing bacterium are consumed in assembling amino acids into protein!

Additional factors are required at each stage of protein synthesis. They are defined by their cyclic association with, and dissociation from, the ribosome. IF factors are involved in prokaryotic initiation. IF-3 is needed for 30S subunits to bind to mRNA, and also is responsible for maintaining the 30S subunit in a free form. IF-2 is needed for fMet-tRNA$_f$ to bind to the 30S subunit and is responsible for excluding other aminoacyl-tRNAs from the initiation reaction. GTP is hydrolyzed after the initiator tRNA has been bound to the initiation complex. The initiation factors must be released in order to allow a large subunit to join the initiation complex.

Eukaryotic initiation involves a greater number of factors. Some of them are involved in the initial binding of the 40S subunit to the capped 5′ end of the mRNA, at which point the initiator tRNA is bound by another group of factors. After this initial binding, the small subunit scans the mRNA until it recognizes the correct AUG codon. At this point, initiation factors are released and the 60S subunit joins the complex.

Prokaryotic EF factors are involved in elongation. EF-Tu binds aminoacyl-tRNA to the 70S ribosome. GTP is hydrolyzed when EF-Tu is released, and EF-Ts is required to regenerate the active form of EF-Tu. EF-G is required for translocation. Binding of the EF-Tu and EF-G factors to ribosomes is mutually exclusive, which ensures that each step must be completed before the next can be started.

Termination occurs at any one of the three special codons, UAA, UAG, and UGA. Class 1 RF factors that specifically recognize the termination codons activate the ribosome to hydrolyze the peptidyl-tRNA. A class 2 RF factor is required to release the class 1 RF factor from the ribosome. The GTP-binding factors IF-2, EF-Tu, EF-G, and RF3 all have similar structures, with the latter two mimicking the RNA-protein structure of the first two when they are bound to tRNA.They all bind to the same ribosomal site, the G-factor binding site.

References

8.4 Initiation in Bacteria Needs 30S Subunits and Accessory Factors

Review

Maitra, U., (1982). Initiation factors in protein biosynthesis. *Annu. Rev. Biochem.* 51, 869–900.

Research

Carter, A. P., Clemons, W. M., Brodersen, D. E., Morgan-Warren, R. J., Hartsch, T., Wimberly, B. T., and Ramakrishnan, V. (2001). Crystal structure of an initiation factor bound to the 30S ribosomal subunit. *Science* 291, 498–501.

Dallas, A. and Noller, H. F. (2001). Interaction of translation initiation factor 3 with the 30S ribosomal subunit. *Mol. Cell* 8, 855–864.

Moazed, D., Samaha, R. R., Gualerzi, C., and Noller, H. F. (1995). Specific protection of 16S rRNA by translational initiation factors. *J. Mol. Biol.* 248, 207–210.

8.5 A Special Initiator tRNA Starts the Polypeptide Chain

Research

Lee, C. P., Seong, B. L., and RajBhandary, U. L. (1991). Structural and sequence elements important for recognition of *E. coli* formylmethionine tRNA by methionyl-tRNA transformylase are clustered in the acceptor stem. *J. Biol. Chem.* 266, 18012–18017.

Marcker, K. and Sanger, F. (1964). N-Formyl-methionyl-S-RNA. *J. Mol. Biol.* 8, 835–840.

Sundari, R. M., Stringer, E. A., Schulman, L. H., and Maitra, U. (1976). Interaction of bacterial initiation factor 2 with initiator tRNA. *J. Biol. Chem.* 251, 3338–3345.

8.8 Small Subunits Scan for Initiation Sites on Eukaryotic mRNA

Reviews

Hellen, C. U. and Sarnow, P. (2001). Internal ribosome entry sites in eukaryotic mRNA molecules. *Genes Dev.* 15, 1593–1612.

Kozak, M. (1978). How do eukaryotic ribosomes select initiation regions in mRNA? *Cell* 15, 1109–1123.

Kozak, M. (1983). Comparison of initiation of protein synthesis in prokaryotes, eukaryotes, and organelles. *Microbiol. Rev.* 47, 1–45.

Research

Kaminski, A., Howell, M. T., and Jackson, R. J. (1990). Initiation of encephalomyocarditis virus RNA translation: the authentic initiation site is not selected by a scanning mechanism. *EMBO J.* 9, 3753–3759.

Pelletier, J. and Sonenberg, N. (1988). Internal initiation of translation of eukaryotic mRNA directed by a sequence derived from poliovirus RNA. *Nature* 334, 320–325.

Pestova, T. V., Hellen, C. U., and Shatsky, I. N. (1996). Canonical eukaryotic initiation factors determine initiation of translation by internal ribosomal entry. *Mol. Cell Biol.* 16, 6859–6869.

Pestova, T. V., Shatsky, I. N., Fletcher, S. P., Jackson, R. J., and Hellen, C. U. (1998). A prokaryotic-like mode of cytoplasmic eukaryotic ribosome binding to the initiation codon during internal translation initiation of hepatitis C and classical swine fever virus RNAs. *Genes Dev.* 12, 67–83.

8.9 Eukaryotes Use a Complex of Many Initiation Factors

Reviews

Dever, T. E. (2002). Gene-specific regulation by general translation factors. *Cell* 108, 545–556.

Gingras, A. C., Raught, B., and Sonenberg, N. (1999). eIF4 initiation factors: effectors of mRNA recruitment to ribosomes and regulators of translation. *Annu. Rev. Biochem.* 68, 913–963.

Hershey, J. W. B. (1991). Translational control in mammalian cells. *Annu. Rev. Biochem.* 60, 717–755.

Merrick, W. C. (1992). Mechanism and regulation of eukaryotic protein synthesis. *Microbiol. Rev.* 56, 291–315.

Pestova, T. V., Kolupaeva, V. G., Lomakin, I. B., Pilipenko, E. V., Shatsky, I. N., Agol, V. I., and Hellen, C. U. (2001). Molecular mechanisms of translation initiation in eukaryotes. *Proc. Natl. Acad. Sci. USA* 98, 7029–7036.

Sachs, A., Sarnow, P., and Hentze, M. W. (1997). Starting at the beginning, middle, and end: translation initiation in eukaryotes. *Cell* 89, 831–838.

Research

Asano, K., Clayton, J., Shalev, A., and Hinnebusch, A. G. (2000). A multifactor complex of eukaryotic initiation factors, eIF1, eIF2, eIF3, eIF5, and initiator tRNA(Met) is an important translation initiation intermediate *in vitro*. *Genes Dev.* 14, 2534–2546.

Huang, H. K., Yoon, H., Hannig, E. M., and Donahue, T. F. (1997). GTP hydrolysis controls stringent selection of the AUG start codon during translation initiation in *S. cerevisiae*. *Genes Dev.* 11, 2396–2413.

Pestova, T. V., Lomakin, I. B., Lee, J. H., Choi, S. K., Dever, T. E., and Hellen, C. U. (2000). The joining of ribosomal subunits in eukaryotes requires eIF5B. *Nature* 403, 332–335.

Tarun, S. Z. and Sachs, A. B. (1996). Association of the yeast poly(A) tail binding protein with translation initiation factor eIF-4G. *EMBO J.* 15, 7168–7177.

8.12 Translocation Moves the Ribosome

Reviews

Ramakrishnan, V. (2002). Ribosome structure and the mechanism of translation. *Cell* 108, 557–572.

Wilson, K. S. and Noller, H. F. (1998). Molecular movement inside the translational engine. *Cell* 92, 337–349.

Research

Moazed, D., and Noller, H. F. (1986). Transfer RNA shields specific nucleotides in 16S ribosomal RNA from attack by chemical probes. *Cell* 47, 985–994.

Moazed, D. and Noller, H. F. (1989). Intermediate states in the movement of tRNA in the ribosome. *Nature* 342, 142–148.

8.13 Elongation Factors Bind Alternately to the Ribosome

Research

Nissen, P., Kjeldgaard, M., Thirup, S., Polekhina, G., Reshetnikova, L., Clark, B. F., and Nyborg, J. (1995). Crystal structure of the ternary complex of Phe-tRNAPhe, EF-Tu, and a GTP analog. *Science* 270, 1464–1472.

Stark, H., Rodnina, M. V., Wieden, H. J., van Heel, M., and Wintermeyer, W. (2000). Large-scale movement of elongation factor G and extensive conformational change of the ribosome during translocation. *Cell* 100, 301–309.

8.15 Termination Codons Are Recognized by Protein Factors

Reviews

Eggertsson, G. and Soll, D. (1988). Transfer RNA-mediated suppression of termination codons in *E. coli*. *Microbiol. Rev.* 52, 354–374.

Frolova, L. et al. (1994). A highly conserved eukaryotic protein family possessing properties of polypeptide chain release factor. *Nature* 372, 701–703.

Nissen, P., Kjeldgaard, M., and Nyborg, J. (2000). Macromolecular mimicry. *EMBO J.* 19, 489–495.

Research

Freistroffer, D. V., Kwiatkowski, M., Buckingham, R. H., and Ehrenberg, M. (2000). The accuracy of codon recognition by polypeptide release factors. *Proc. Natl. Acad. Sci. USA* 97, 2046–2051.

Ito, K., Ebihara, K., Uno, M., and Nakamura, Y. (1996). Conserved motifs in prokaryotic and eukaryotic polypeptide release factors: tRNA-protein mimicry hypothesis. *Proc. Natl. Acad. Sci. USA* 93, 5443–5448.

Klaholz, B. P., Myasnikov, A. G., and Van Heel, M. (2004). Visualization of release factor 3 on the

ribosome during termination of protein synthesis. *Nature* 427, 862–865.
Mikuni, O., Ito, K., Moffat, J., Matsumura, K., McCaughan, K., Nobukuni, T., Tate, W., and Nakamura, Y. (1994). Identification of the *prfC* gene, which encodes peptide-chain-release factor 3 of *E. coli*. *Proc. Natl. Acad. Sci. USA* 91, 5798–5802.
Milman, G., Goldstein, J., Scolnick, E., and Caskey, T. (1969). Peptide chain termination. 3. Stimulation of *in vitro* termination. *Proc. Natl. Acad. Sci. USA* 63, 183–190.
Scolnick, E. et al. (1968). Release factors differing in specificity for terminator codons. *Proc. Natl. Acad. Sci. USA* 61, 768–774.
Selmer, M., Al-Karadaghi, S., Hirokawa, G., Kaji, A., and Liljas, A. (1999). Crystal structure of *Thermotoga maritima* ribosome recycling factor: a tRNA mimic. *Science* 286, 2349–2352.
Song, H., Mugnier, P., Das, A. K., Webb, H. M., Evans, D. R., Tuite, M. F., Hemmings, B. A., and Barford, D. (2000). The crystal structure of human eukaryotic release factor eRF1—mechanism of stop codon recognition and peptidyl-tRNA hydrolysis. *Cell* 100, 311–321.

8.16 Ribosomal RNA Pervades Both Ribosomal Subunits

Reviews

Hill, W. E. et al. (1990). *The Ribosome*. Washington, DC: American Society for Microbiology.
Noller, H. F. (1984). Structure of ribosomal RNA. *Annu. Rev. Biochem.* 53, 119–162.
Noller, H. F. (2005). RNA structure: reading the ribosome. *Science* 309, 1508–1514.
Noller, H. F. and Nomura, M. (1987). *E. coli and S. typhimurium*. Washington, DC: American Society for Microbiology.
Wittman, H. G. (1983). Architecture of prokaryotic ribosomes. *Annu. Rev. Biochem.* 52, 35–65.

Research

Ban, N., Nissen, P., Hansen, J., Capel, M., Moore, P. B., and Steitz, T. A. (1999). Placement of protein and RNA structures into a 5 Å-resolution map of the 50S ribosomal subunit. *Nature* 400, 841–847.
Ban, N., Nissen, P., Hansen, J., Moore, P. B., and Steitz, T. A. (2000). The complete atomic structure of the large ribosomal subunit at 2.4 Å resolution. *Science* 289, 905–920.
Clemons, W. M. et al. (1999). Structure of a bacterial 30S ribosomal subunit at 5.5 Å resolution. *Nature* 400, 833–840.
Wimberly, B. T., Brodersen, D. E., Clemons W. M. Jr., Morgan-Warren, R. J., Carter, A. P., Vonrhein, C., Hartsch, T., and Ramakrishnan, V. (2000). Structure of the 30S ribosomal subunit. *Nature* 407, 327–339.
Yusupov, M. M., Yusupova, G. Z., Baucom, A., Lieberman, A., Earnest, T. N., Cate, J. H. D., and Noller, H. F. (2001). Crystal structure of the ribosome at 5.5 Å resolution. *Science* 292, 883–896.

8.17 Ribosomes Have Several Active Centers

Reviews

Lafontaine, D. L. and Tollervey, D. (2001). The function and synthesis of ribosomes. *Nat. Rev. Mol. Cell Biol.* 2, 514–520.
Moore, P. B. and Steitz, T. A. (2003). The structural basis of large ribosomal subunit function. *Annu. Rev. Biochem.* 72, 813–850.
Ramakrishnan, V. (2002). Ribosome structure and the mechanism of translation. *Cell* 108, 557–572.

Research

Cate, J. H., Yusupov, M. M., Yusupova, G. Z., Earnest, T. N., and Noller, H. F. (1999). X-ray crystal structures of 70S ribosome functional complexes. *Science* 285, 2095–2104.
Fredrick, K., and Noller, H. F. (2003). Catalysis of ribosomal translocation by sparsomycin. *Science* 300, 1159–1162.
Sengupta, J., Agrawal, R. K., and Frank, J. (2001). Visualization of protein S1 within the 30S ribosomal subunit and its interaction with messenger RNA. *Proc. Natl. Acad. Sci. USA* 98, 11991–11996.
Simonson, A. B. and Simonson, J. A. (2002). The transorientation hypothesis for codon recognition during protein synthesis. *Nature* 416, 281–285.
Valle, M., Sengupta, J., Swami, N. K., Grassucci, R. A., Burkhardt, N., Nierhaus, K. H., Agrawal, R. K., and Frank, J. (2002). Cryo-EM reveals an active role for aminoacyl-tRNA in the accommodation process. *EMBO J.* 21, 3557–3567.
Yusupov, M. M., Yusupova, G. Z., Baucom, A., Lieberman, A., Earnest, T. N., Cate, J. H. D., and Noller, H. F. (2001). Crystal structure of the ribosome at 5.5 Å resolution. *Science* 292, 883–896.

8.18 16S rRNA Plays an Active Role in Protein Synthesis

Reviews

Noller, H. F. (1991). Ribosomal RNA and translation. *Annu. Rev. Biochem.* 60, 191–227.
Yonath, A. (2005). Antibiotics targeting ribosomes: resistance, selectivity, synergism and cellular regulation. *Annu. Rev. Biochem.* 74, 649–679.

Research

Lodmell, J. S. and Dahlberg, A. E. (1997). A conformational switch in *E. coli* 16S rRNA during decoding of mRNA. *Science* 277, 1262–1267.

Moazed, D., and Noller, H. F. (1986). Transfer RNA shields specific nucleotides in 16S ribosomal RNA from attack by chemical probes. *Cell* 47, 985–994.

Yoshizawa, S., Fourmy, D., and Puglisi, J. D. (1999). Recognition of the codon-anticodon helix by rRNA. *Science* 285, 1722–1725.

8.19 23S rRNA Has Peptidyl Transferase Activity

Research

Ban, N., Nissen, P., Hansen, J., Moore, P. B., and Steitz, T. A. (2000). The complete atomic structure of the large ribosomal subunit at 2.4 Å resolution. *Science* 289, 905–920.

Bayfield, M. A., Dahlberg, A. E., Schulmeister, U., Dorner, S., and Barta, A. (2001). A conformational change in the ribosomal peptidyl transferase center upon active/inactive transition. *Proc. Natl. Acad. Sci. USA* 98, 10096–10101.

Noller, H. F., Hoffarth, V., and Zimniak, L. (1992). Unusual resistance of peptidyl transferase to protein extraction procedures. *Science* 256, 1416–1419.

Samaha, R. R., Green, R., and Noller, H. F. (1995). A base pair between tRNA and 23S rRNA in the peptidyl transferase center of the ribosome. *Nature* 377, 309–314.

Thompson, J., Thompson, D. F., O'Connor, M., Lieberman, K. R., Bayfield, M. A., Gregory, S. T., Green, R., Noller, H. F., and Dahlberg, A. E. (2001). Analysis of mutations at residues A2451 and G2447 of 23S rRNA in the peptidyltransferase active site of the 50S ribosomal subunit. *Proc. Natl. Acad. Sci. USA* 98, 9002–9007.

8.20 Ribosomal Structures Change When the Subunits Come Together

Reference

Schuwirth, B. S., Borovinskaya, M. A., Hau, C. W., Zhang, W., Vila-Sanjurjo, A., Holton, J. M., and Cate, J. H. (2005). Structures of the bacterial ribosome at 3.5 Å resolution. *Science* 310, 827–834.

9

Using the Genetic Code

CHAPTER OUTLINE

Continued on next page

9.1 Introduction

The sequence of a coding strand of DNA, read in the direction from 5′ to 3′, consists of nucleotide triplets (codons) corresponding to the amino acid sequence of a protein read from N-terminus to C-terminus. Sequencing of DNA and proteins makes it possible to compare corresponding nucleotide and amino acid sequences directly. There are sixty-four codons (each of four possible nucleotides can occupy each of the three positions of the codon, making $4^3 = 64$ possible trinucleotide sequences). Each of these codons has a specific meaning in protein synthesis: sixty-one codons represent amino acids; three codons cause the termination of protein synthesis.

The meaning of a codon that represents an amino acid is determined by the tRNA that corresponds to it; the meaning of the termination codons is determined directly by protein factors.

The breaking of the genetic code originally showed that genetic information is stored in the form of nucleotide triplets, but did not reveal how each codon specifies its corresponding amino acid. Before the advent of sequencing, codon assignments were deduced on the basis of two types of *in vitro* studies. A system involving the translation of synthetic polynucleotides was introduced in 1961, when Nirenberg showed that polyuridylic acid [poly(U)] directs the assembly of phenylalanine into polyphenylalanine. This result means that UUU must be a codon for phenylalanine. A second system was later introduced in which a trinucleotide was used to mimic a codon, thus causing the corresponding aminoacyl-tRNA to bind to a ribosome. By identifying the amino acid component of the aminoacyl-tRNA, the meaning of the codon can be found. The two techniques together assigned meaning to all of the codons that represent amino acids.

Sixty-one of the sixty-four codons represent amino acids. The other three cause termination of protein synthesis. The assignment of amino acids to codons is not random, but shows relationships in which the third base has less effect on codon meaning. In addition, related amino acids are often represented by related codons.

9.2 Related Codons Represent Related Amino Acids

Key concepts

- Sixty-one of the sixty-four possible triplets code for twenty amino acids.
- Three codons do not represent amino acids and cause termination.
- The genetic code was frozen at an early stage of evolution and is universal.
- Most amino acids are represented by more than one codon.
- The multiple codons for an amino acid are usually related.
- Related amino acids often have related codons, minimizing the effects of mutation.

The code is summarized in FIGURE 9.1. There are more codons than there are amino acids, and as a result almost all amino acids are represented by more than one codon. The only exceptions are methionine and tryptophan. Codons that have the same meaning are called **synonyms.**

The genetic code is actually read on the mRNA, and thus it is usually described in terms of the four bases present in RNA: U, C, A, and G.

Codons representing the same or related amino acids tend to be similar in sequence. Often the base in the third position of a codon is not significant, because the four codons differing only in the third base represent the same amino acid. Sometimes a distinction is made only between a purine versus a pyrimidine in this position. The reduced specificity at the last position is known as **third-base degeneracy.**

The interpretation of a codon requires base pairing with the anticodon of the corresponding aminoacyl-tRNA. The reaction occurs within the ribosome: complementary trinucleotides in isolation would usually be too short to pair in a stable manner, but the interaction is stabilized by the environment of the ribosomal A site. Also, base pairing between codon and anticodon is not solely a matter of A-U and G-C base pairing. The ribosome controls the environment in such a way that conventional pairing occurs at the first two positions of the codon, but additional reactions are permitted at the third base. As a result, a single aminoacyl-tRNA may recognize more than one codon, corresponding with the pattern of degeneracy. Furthermore, pairing interactions may also be influenced by the introduction of special bases into tRNA, especially by modification in or close to the anticodon.

The tendency for similar amino acids to be represented by related codons minimizes the effects of mutations. It increases the probability that a single random base change will result in no amino acid substitution or in one involving amino acids of similar character. For example, a mutation of CUC to CUG has no effect, because both codons represent leucine.A mutation of CUU to AUU results in replacement of leucine with isoleucine, a closely related amino acid.

FIGURE 9.2 plots the number of codons representing each amino acid against the frequency with which the amino acid is used in proteins (in *E. coli*). There is only a slight tendency for amino acids that are more common to be represented by more codons, and therefore it does not seem that the genetic code has been optimized with regard to the utilization of amino acids.

The three codons (UAA, UAG, and UGA) that do not represent amino acids are used specifically to terminate protein synthesis. One of these stop codons marks the end of every gene.

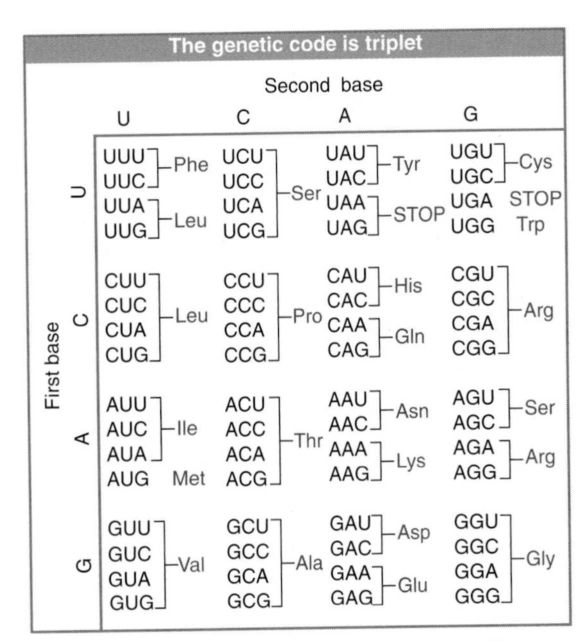

FIGURE 9.1 All the triplet codons have meaning: Sixty-one represent amino acids, and three cause termination (STOP).

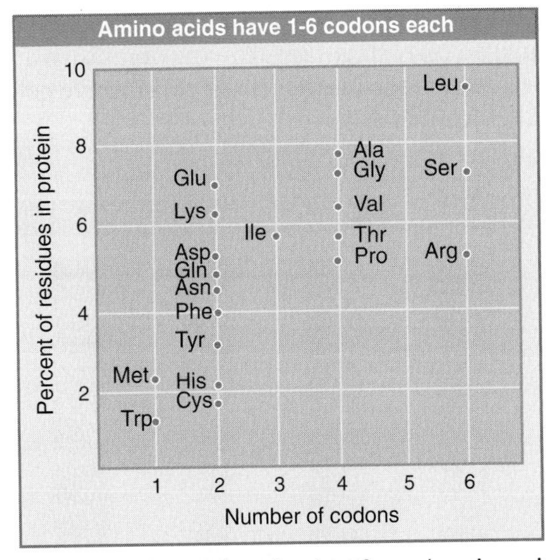

FIGURE 9.2 The number of codons for each amino acid does not correlate closely with its frequency of use in proteins.

Is the genetic code the same in all living organisms?

Comparisons of DNA sequences with the corresponding protein sequences reveal that the identical set of codon assignments is used in bacteria and in eukaryotic cytoplasm. As a result, mRNA from one species usually can be translated correctly *in vitro* or *in vivo* by the protein synthetic apparatus of another species. Thus the codons used in the mRNA of one species have the same meaning for the ribosomes and tRNAs of other species.

The universality of the code argues that it must have been established very early in evolution. Perhaps the code started in a primitive form in which a small number of codons were used to represent comparatively few amino acids, possibly even with one codon corresponding to any member of a group of amino acids. More precise codon meanings and additional amino acids could have been introduced later. One possibility is that at first only two of the three bases in each codon were used; discrimination at the third position could have evolved later. (Originally there might have been a stereochemical relationship between amino acids and the codons representing them. Then a more complex system evolved.)

Evolution of the code could have become "frozen" at a point at which the system had become so complex that any changes in codon meaning would disrupt existing proteins by substituting unacceptable amino acids. Its universality implies that this must have happened at such an early stage that all living organisms are descended from a single pool of primitive cells in which this occurred.

Third bases have least meaning

UUU UUC	UCU UCC	UAU UAC	UGU UGC
UUA UUG	UCA UCG	UAA UAG	UGA UGG
CUU CUC CUA CUG	CCU CCC CCA CCG	CAU CAC CAA CAG	CGU CGC CGA CGG
AUU AUC AUA AUG	ACU ACC ACA ACG	AAU AAC AAA AAG	AGU AGC AGA AGG
GUU GUC GUA GUG	GCU GCC GCA GCG	GAU GAC GAA GAG	GGU GGC GGA GGG

Third-base relationship	Third bases with same meaning	Codon Number
Third base irrelevant	U, C, A, G	32
	U, C, A	3
Purines differ from pyrimidines	A or G	14
	U or C	10
Unique	G only	2

FIGURE 9.3 Third bases have the least influence on codon meanings. Boxes indicate groups of codons within which third-base degeneracy ensures that the meaning is the same.

Exceptions to the universal genetic code are rare. Changes in meaning in the principal genome of a species usually concern the termination codons. For example, in a mycoplasma, UGA codes for tryptophan; in certain species of the ciliates *Tetrahymena* and *Paramecium,* UAA and UAG code for glutamine. Systematic alterations of the code have occurred only in mitochondrial DNA (see Section 9.7, There Are Sporadic Alterations of the Universal Code).

9.3 Codon–Anticodon Recognition Involves Wobbling

Key concepts

- Multiple codons that represent the same amino acid most often differ at the third base position.
- The wobble in pairing between the first base of the anticodon and the third base of the codon results from the structure of the anticodon loop.

The function of tRNA in protein synthesis is fulfilled when it recognizes the codon in the ribosomal A site. The interaction between anticodon and codon takes place by base pairing, but under rules that extend pairing beyond the usual G-C and A-U partnerships.

We can deduce the rules governing the interaction from the sequences of the anticodons that correspond to particular codons. The ability of any tRNA to respond to a given codon can be measured directly by the trinucleotide binding assay or by its use in an *in vitro* protein synthetic system.

The genetic code itself yields some important clues about the process of codon recognition. The pattern of third-base degeneracy is drawn in FIGURE 9.3, which shows that in almost all cases either the third base is irrelevant or a distinction is made only between purines and pyrimidines.

There are eight codon families in which all four codons sharing the same first two bases have the same meaning, so that the third base has no role at all in specifying the amino acid. There are seven codon pairs in which the meaning is the same regardless of which pyrimidine is present at the third position, and there are five codon pairs in which either purine may be present without changing the amino acid that is coded.

There are only three cases in which a unique meaning is conferred by the presence of a particular base at the third position: AUG (for methionine), UGG (for tryptophan), and UGA (termination). So C and U never have a unique meaning in the third position, and A never signifies a unique amino acid.

The anticodon is complementary to the codon; thus it is the first base in the anticodon sequence written conventionally in the direction from 5′ to 3′ that pairs with the third base in the codon sequence written by the same convention. So the combination

Codon	5′ A C G 3′
Anticodon	3′ U G C 5′

is usually written as codon ACG/anticodon CGU, where the anticodon sequence must be read backward for complementarity with the codon.

To avoid confusion, we shall retain the usual convention in which all sequences are written 5′–3′, but indicate anticodon sequences with a backward arrow as a reminder of the relationship with the codon. Thus the codon/anticodon pair shown above will be written as ACG and $\overleftarrow{\text{C}}$GU, respectively.

Does each triplet codon demand its own tRNA with a complementary anticodon? Or can a single tRNA respond to both members of a codon pair and to all (or at least some) of the four members of a codon family?

Often one tRNA can recognize more than one codon. This means that the base in the first position of the anticodon must be able to partner alternative bases in the corresponding third position of the codon. Base pairing at this position cannot be limited to the usual G-C and A-U partnerships.

The rules governing the recognition patterns are summarized in the **wobble hypothesis,** which states that the pairing between codon and anticodon at the first two codon positions always follows the usual rules, but that exceptional "wobbles" occur at the third position. Wobbling occurs because the conformation of the tRNA anticodon loop permits flexibility at the first base of the anticodon. FIGURE 9.4 shows that G-U pairs can form in addition to the usual pairs.

This single change creates a pattern of base pairing in which A can no longer have a unique meaning in the codon (because the U that recognizes it must also recognize G). Similarly, C also no longer has a unique meaning (because the G that recognizes it also must recognize U). FIGURE 9.5 summarizes the pattern of recognition.

It is therefore possible to recognize unique codons only when the third bases are G or U. This option is not used often, as UGG and AUG are the only examples of the first type and there is none of the second type. (G-U pairs are common in RNA duplex structures. When only three base pairs can be formed the formation of stable contacts between codon and anticodon is more constrained, though, and thus G-U pairs can contribute only in the last position of the codon.)

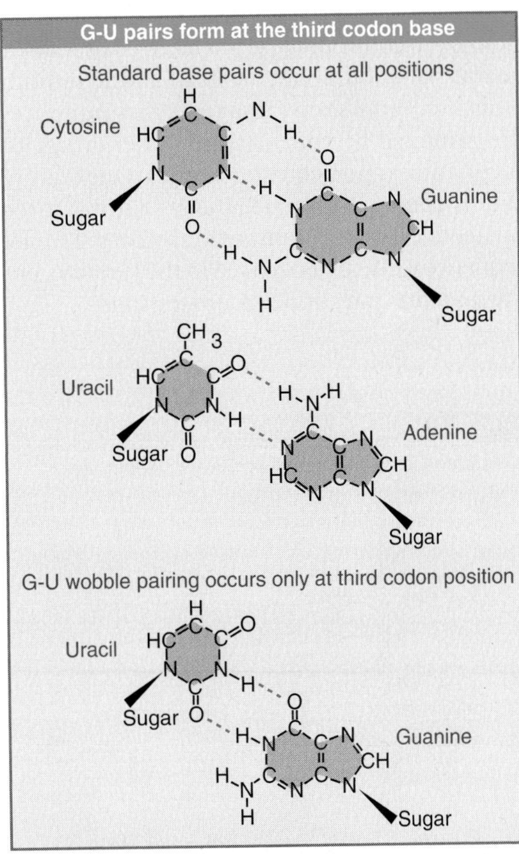

FIGURE 9.4 Wobble in base pairing allows G-U pairs to form between the third base of the codon and the first base of the anticodon.

The third codon base wobbles

Base in first position of anticodon	Base(s) recognized in third position of codon
U	A or G
C	G only
A	U only
G	C or U

FIGURE 9.5 Codon–anticodon pairing involves wobbling at the third position.

9.4 tRNAs Are Processed from Longer Precursors

Key concepts

- A mature tRNA is generated by processing a precursor.
- The 5′ end is generated by cleavage by the endonuclease RNAase P.
- The 3′ end is generated by cleavage followed by trimming of the last few bases, followed by addition of the common terminal trinucleotide sequence CCA.

tRNAs are commonly synthesized as precursor chains with additional material at one or both ends. FIGURE 9.6 shows that the extra sequences are removed by combinations of endonucleolytic and exonucleolytic activities. One feature that is common to all tRNAs is that the three nucleotides at the 3′ terminus, always the triplet sequence CCA, are not coded in the genome, but are added as part of tRNA processing.

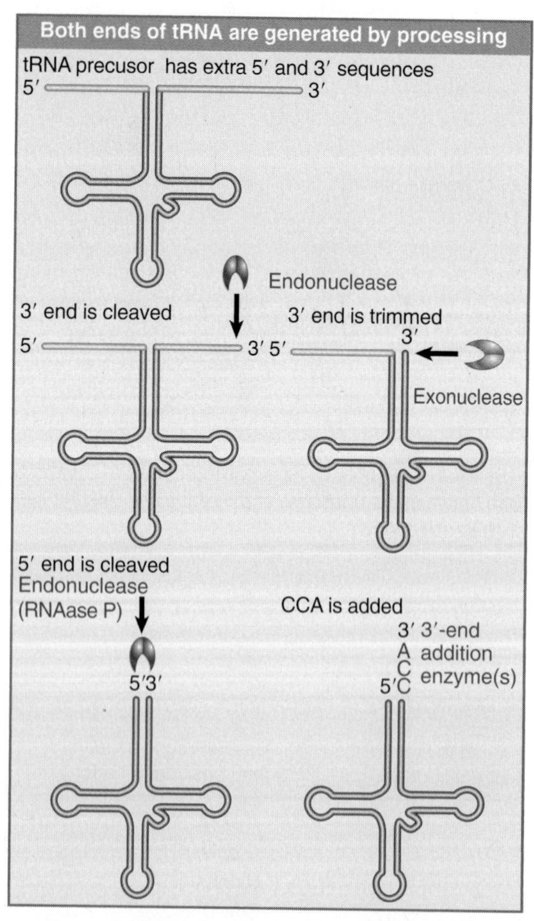

FIGURE 9.6 The tRNA 3′ end is generated by cutting and trimming followed by addition of CCA; the 5′ end is generated by cutting.

The 5′ end of tRNA is generated by a cleavage action catalyzed by the enzyme ribonuclease P.

The enzymes that process the 3′ end are best characterized in *E. coli,* where an endonuclease triggers the reaction by cleaving the precursor downstream, and several exonucleases then trim the end by degradation in the 3′–5′ direction. The reaction also involves several enzymes in eukaryotes. It generates a tRNA that needs the CCA trinucleotide sequence to be added to the 3′ end.

The addition of CCA is the result solely of an enzymatic process, that is, the enzymatic activity carries the specificity for the sequence of the trinucleotide, which is not determined by a template. There are several models for the process, which may be different in different organisms.

In some organisms, the process is catalyzed by a single enzyme. One model for its action proposes that a single enzyme binds to the 3′ end, and sequentially adds C, C, and A, the specificity at each stage being determined by the structure of the 3′ end. Other models propose that the enzyme has different active sites for cytosine triphosphate (CTP) and adenosine triphosphate (ATP).

In other organisms, different enzymes are responsible for adding the C and A residues, and they function sequentially.

When a tRNA is not properly processed, it attracts the attention of a quality control system that degrades it. This ensures that the protein synthesis apparatus does not become blocked by nonfunctional tRNAs.

9.5 tRNA Contains Modified Bases

Key concepts

- tRNAs contain >50 modified bases.
- Modification usually involves direct alteration of the primary bases in tRNA, but there are some exceptions in which a base is removed and replaced by another base.

Transfer RNA is unique among nucleic acids in its content of "unusual" bases. An unusual base is any purine or pyrimidine ring except the usual A, G, C, and U from which all RNAs are synthesized. All other bases are produced by **modification** of one of the four bases after it has been incorporated into the polyribonucleotide chain.

All classes of RNA display some degree of modification, but in all cases except tRNA this is confined to rather simple events, such as the addition of methyl groups. In tRNA, there is a vast range of modifications, ranging from simple methylation to wholesale restructuring of the purine ring. Modifications occur in all parts of the tRNA molecule. There are >50 different types of modified bases in tRNA.

FIGURE 9.7 shows some of the more common modified bases. Modifications of pyrimidines (C and U) are less complex than those of purines (A and G). In addition to the modifications of the bases themselves, methylation at the 2′–O position of the ribose ring also occurs.

The most common modifications of uridine are straightforward. Methylation at position 5 creates ribothymidine (T). The base is the same commonly found in DNA, but here it is attached to ribose rather than deoxyribose. In RNA, thymine constitutes an unusual base that originates by modification of U.

Dihydrouridine (D) is generated by the saturation of a double bond, which changes the ring structure. Pseudouridine (ψ) interchanges the positions of N and C atoms (see Figure 26.40). In 4-thiouridine, sulfur is substituted for oxygen.

The nucleoside inosine is found normally in the cell as an intermediate in the purine biosynthetic pathway. It is not, however, incorporated directly into RNA. Instead, its existence depends on modification of A to create I. Other modifications of A include the addition of complex groups.

Two complex series of nucleotides depend on modification of G. The Q bases, such as queuosine, have an additional pentenyl ring added via an NH linkage to the methyl group of 7-methylguanosine. The pentenyl ring may carry various further groups. The Y bases, such as wyosine, have an additional ring fused with the purine ring itself. This extra ring carries a long carbon chain; again, to which further groups are added in different cases.

The modification reaction usually involves the alteration of, or addition to, existing bases in the tRNA. An exception is the synthesis of Q bases, for which a special enzyme exchanges free queuosine with a guanosine residue in the tRNA. The reaction involves breaking and remaking bonds on either side of the nucleoside.

The modified nucleosides are synthesized by specific tRNA-modifying enzymes. The original nucleoside present at each position can be determined either by comparing the sequence of tRNA with that of its gene or (less efficiently) by isolating precursor molecules that lack some or all of the modifications. The sequences of precursors show that different modifications are introduced at different stages during the maturation of tRNA.

Some modifications are constant features of all tRNA molecules—for example, the D residues that give rise to the name of the D arm and the ψ found in the TψC sequence. On the 3′ side of the anticodon there is always a modified purine, although the modification varies widely.

Other modifications are specific for particular tRNAs or groups of tRNAs. For example, wyosine bases are characteristic of $tRNA^{Phe}$ in bacteria, yeast, and mammals. There are also some species-specific patterns.

The many tRNA-modifying enzymes (~60 in yeast) vary greatly in specificity. In some cases, a single enzyme acts to make a particular modification at a single position. In other cases, an enzyme can modify bases at several different target positions. Some enzymes undertake single reactions with individual tRNAs;

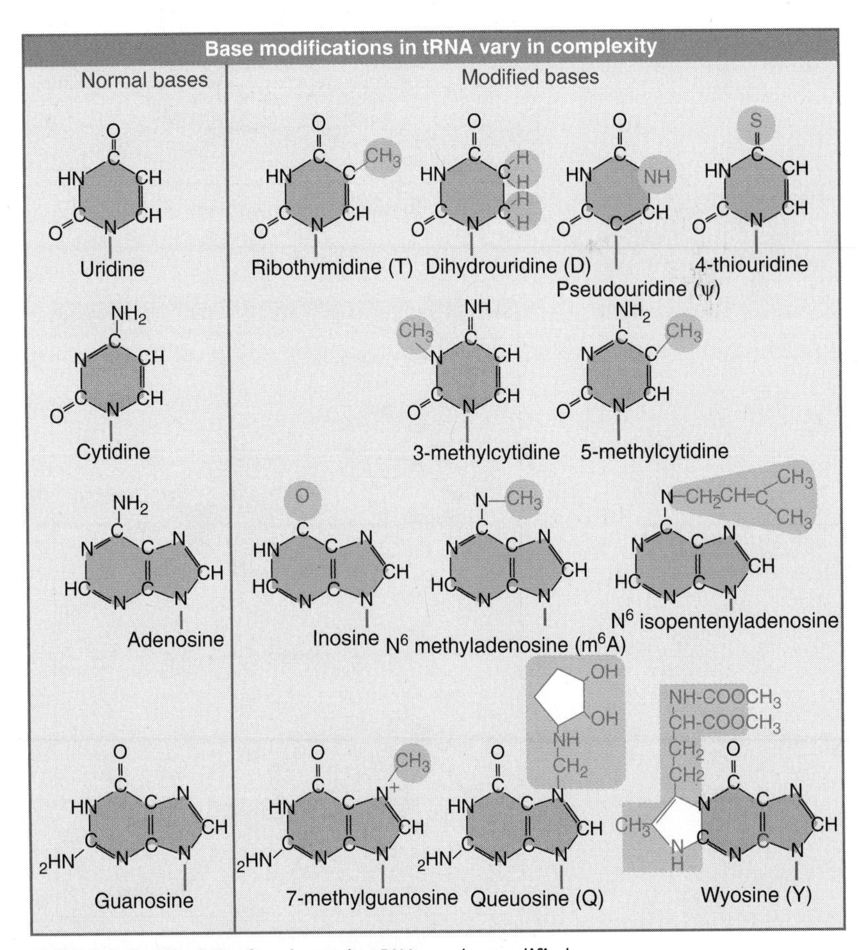

FIGURE 9.7 All of the four bases in tRNA can be modified.

others have a range of substrate molecules. The features recognized by the tRNA-modifying enzymes are unknown, but probably involve recognition of structural features surrounding the site of modification. Some modifications require the successive actions of more than one enzyme.

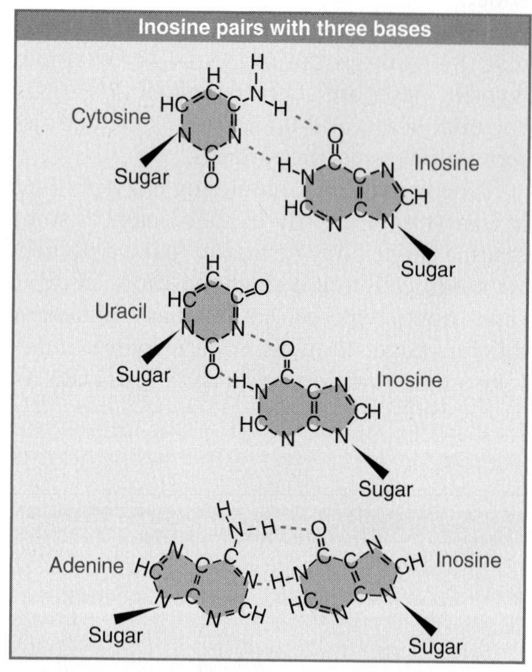

FIGURE 9.8 Inosine can pair with any of U, C, and A.

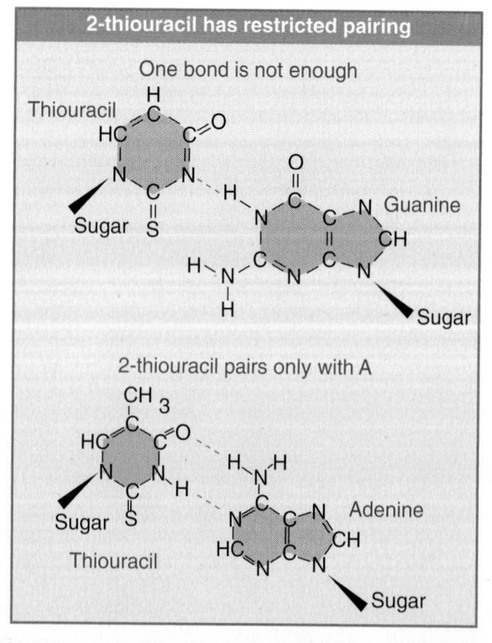

FIGURE 9.9 Modification to 2-thiouridine restricts pairing to A alone because only one H-bond can form with G.

9.6 Modified Bases Affect Anticodon–Codon Pairing

Key concept

- Modifications in the anticodon affect the pattern of wobble pairing and therefore are important in determining tRNA specificity.

The most direct effect of modification is seen in the anticodon, where change of sequence influences the ability to pair with the codon, thus determining the meaning of the tRNA. Modifications elsewhere in the vicinity of the anticodon also influence its pairing.

When bases in the anticodon are modified, further pairing patterns become possible in addition to those predicted by the regular and wobble pairing involving A, C, U, and G. FIGURE 9.8 shows the use of inosine (I), which is often present at the first position of the anticodon. Inosine can pair with any one of three bases, U, C, and A.

This ability is especially important in the isoleucine codons, where AUA codes for isoleucine, whereas AUG codes for methionine. It is not possible with the usual bases to recognize A alone in the third position, and as a result any tRNA with U starting its anticodon would have to recognize AUG as well as AUA. Thus AUA must be read together with AUU and AUC, a problem that is solved by the existence of tRNA with I in the anticodon.

Actually, some of the predicted regular combinations do not occur, because some bases are always modified. There seems to be an absolute ban on the employment of A; usually it is converted to I. In most cases, U at the first position of the anticodon is converted to a modified form that has altered pairing properties.

Some modifications create preferential readings of some codons with respect to others. Anticodons with uridine-5-oxyacetic acid and 5-methoxyuridine in the first position recognize A and G efficiently as third bases of the codon, but recognize U less efficiently. Another case in which multiple pairings can occur—but with some pairings preferred to others—is provided by the series of queuosine and its derivatives. These modified G bases continue to recognize both C and U, but pair with U more readily.

A restriction not allowed by the usual rules can be achieved by the employment of 2-thiouridine in the anticodon. FIGURE 9.9 shows that its modification allows the base to continue to pair with A, but prevents it from indulging in wobble pairing with G.

These and other pairing relationships make the general point that there are multiple ways to construct a set of tRNAs able to recognize all the sixty-one codons representing amino acids. No particular pattern predominates in any given organism, although the absence of a certain pathway for modification can prevent the use of some recognition patterns. Thus a particular codon family is read by tRNAs with different anticodons in different organisms.

Often the tRNAs will have overlapping responses, so that a particular codon is read by more than one tRNA. In such cases there may be differences in the efficiencies of the alternative recognition reactions. (As a general rule, codons that are commonly used tend to be more efficiently read.) In addition to the construction of a set of tRNAs able to recognize all the codons, there may be multiple tRNAs that respond to the same codons.

The predictions of wobble pairing accord very well with the observed abilities of almost all tRNAs. There are, however, exceptions in which the codons recognized by a tRNA differ from those predicted by the wobble rules. Such effects probably result from the influence of neighboring bases and/or the conformation of the anticodon loop in the overall tertiary structure of the tRNA. Indeed, the importance of the structure of the anticodon loop is inherent in the idea of the wobble hypothesis itself. Further support for the influence of the surrounding structure is provided by the isolation of occasional mutants in which a change in a base in some other region of the molecule alters the ability of the anticodon to recognize codons.

Another unexpected pairing reaction is presented by the ability of the bacterial initiator, fMet-tRNA$_f$, to recognize both AUG and GUG. This misbehavior involves the third base of the anticodon.

9.7 There Are Sporadic Alterations of the Universal Code

Key concepts

- Changes in the universal genetic code have occurred in some species.
- These changes are more common in mitochondrial genomes, where a phylogenetic tree can be constructed for the changes.
- In nuclear genomes, the changes are sporadic and usually affect only termination codons.

The universality of the genetic code is striking, but some exceptions exist. They tend to affect the codons involved in initiation or termination and result from the production (or absence) of tRNAs representing certain codons. The changes found in principal (bacterial or nuclear) genomes are summarized in FIGURE 9.10.

Almost all of the changes that allow a codon to represent an amino acid affect termination codons:

- In the prokaryote *Mycoplasma capricolum*, UGA is not used for termination, but instead codes for tryptophan (Trp). In fact, it is the predominant Trp codon, and UGG is used only rarely. Two Trp-tRNA species exist, which have the anticodons UCA$^{\leftarrow}$ (reads UGA and UGG) and CCA$^{\leftarrow}$ (reads only UGG).
- Some ciliates (unicellular protozoa) read UAA and UAG as glutamine instead of termination signals. *Tetrahymena thermophila*, which is one of the ciliates, contains three tRNA^{Glu} species. One recognizes the usual codons CAA and CAG for glutamine, a second recognizes both UAA and UAG (in accordance with the wobble hypothesis), and the third recognizes only UAG. We assume that a further change is that the release factor eRF has a restricted specificity, compared with that of other eukaryotes.
- In another ciliate *(Euplotes octacarinatus)*, UGA codes for cysteine. Only UAA is used as a termination codon, and UAG

Changes in the genetic code usually involve Stop/None signals			
UUU Phe	UCU	UAU Tyr	UGU Cys
UUC	UCC Ser	UAC	UGC
UUA	UCA	UAA STOP→Gln	UGA STOP→Trp, Cys, Sel
UUG Leu	UCG	UAG	UGG Trp
CUU	CCU	CAU His	CGU
CUC Leu	CCC Pro	CAC	CGC
CUA	CCA	CAA Gln	CGA Arg
CUG Leu→Ser	CCG	CAG	CGG Arg→NONE
AUU	ACU	AAU Asn	AGU Ser
AUC Ile	ACC Thr	AAC	AGC
AUA Ile→NONE	ACA	AAA Lys	AGA Arg→NONE
AUG Met	ACG	AAG	AGG Arg
GUU	GCU	GAU Asp	GGU
GUC	GCC Ala	GAC	GGC
GUA Val	GCA	GAA	GGA Gly
GUG	GCG	GAG Glu	GGG

FIGURE 9.10 Changes in the genetic code in bacterial or eukaryotic nuclear genomes usually assign amino acids to stop codons or change a codon so that it no longer specifies an amino acid. A change in meaning from one amino acid to another is unusual.

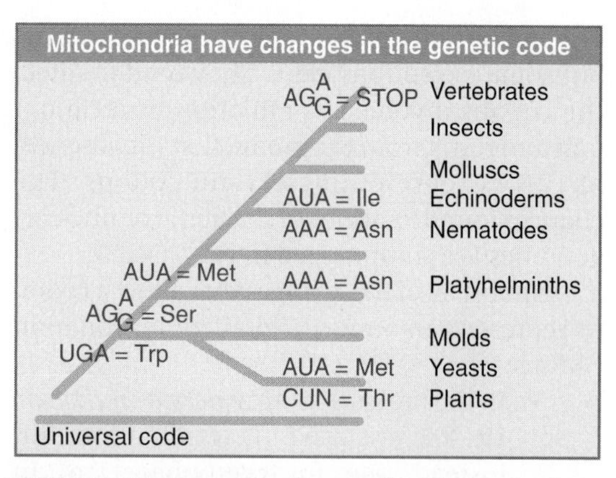

FIGURE 9.11 Changes in the genetic code in mitochondria can be traced in phylogeny. The minimum number of independent changes is generated by supposing that the AUA=Met and the AAA=Asn changes each occurred independently twice, and that the early AUA=Met change was reversed in echinoderms.

is not found. The change in meaning of UGA might be accomplished by a modification in the anticodon of $tRNA^{Cys}$ to allow it to read UGA with the usual codons UGU and UGC.

- The only substitution in coding for amino acids occurs in a yeast *(Candida)*, where CUG means serine instead of leucine (and UAG is used as a sense codon).

Acquisition of a coding function by a termination codon requires two types of change: a tRNA must be mutated so as to recognize the codon, and the class 1 release factor must be mutated so that it does not terminate at this codon.

The other common type of change is loss of the tRNA that responds to a codon, so that the codon no longer specifies any amino acid. What happens at such a codon will depend on whether the termination factor evolves to recognize it.

All of these changes are sporadic, which is to say that they appear to have occurred independently in specific lines of evolution. They may be concentrated on termination codons, because these changes do not involve substitution of one amino acid for another. Once the genetic code was established, early in evolution, any general change in the meaning of a codon would cause a substitution in all the proteins that contain that amino acid. It seems likely that the change would be deleterious in at least some of these proteins, with the result that it would be strongly selected against. The divergent uses of the termination codons could represent their "capture" for normal coding purposes. If some termination codons were used only rarely, they could be recruited to coding purposes by changes that allowed tRNAs to recognize them.

Exceptions to the universal genetic code also occur in the mitochondria from several species. FIGURE 9.11 constructs a phylogeny for the changes. It suggests that there was a universal code that was changed at various points in mitochondrial evolution. The earliest change was the employment of UGA to code for tryptophan, which is common to all (nonplant) mitochondria.

Some of these changes make the code simpler by replacing two codons that had different meanings with a pair that has a single meaning. Pairs treated like this include UGG and UGA (both Trp instead of one Trp and one termination) and AUG and AUA (both Met instead of one Met and the other Ile).

Why have changes been able to evolve in the mitochondrial code? The mitochondrion synthesizes only a small number of proteins (~10), and as a result the problem of disruption by changes in meaning is much less severe. It is likely that the codons that are altered were not used extensively in locations where amino acid substitutions would have been deleterious. The variety of changes found in mitochondria of different species suggests that they have evolved separately rather than by common descent from an ancestral mitochondrial code.

According to the wobble hypothesis, a minimum of 31 tRNAs (excluding the initiator) are required to recognize all sixty-one codons (at least two tRNAs are required for each codon family and one tRNA is needed per codon pair or single codon). An unusual situation exists in (at least) mammalian mitochondria, however, in which there are only twenty-two different tRNAs. How does this limited set of tRNAs accommodate all the codons?

The critical feature lies in a simplification of codon–anticodon pairing, in which one tRNA recognizes all four members of a codon family. This reduces to twenty-three the minimum number of tRNAs required to respond to all usual codons. The use of AG^{A}_{G} for termination reduces the requirement by one further tRNA, to twenty-two.

In all eight codon families, the sequence of the tRNA contains an unmodified U at the first position of the anticodon. The remaining codons are grouped into pairs in which all the codons ending in pyrimidines are read by G in the anticodon, and all the codons ending in purines are read by a modified U in the anticodon, as predicted by the wobble hypothesis. The complication of the single UGG codon is avoided by the change in the code to read UGA with UGG as tryptophan. In mammals, AUA ceases to represent isoleucine and instead is read with AUG as methionine. This allows all the nonfamily codons to be read as fourteen pairs.

The twenty-two identified tRNA genes therefore code for fourteen tRNAs representing pairs and eight tRNAs representing families. This leaves the two usual termination codons UAG and UAA unrecognized by tRNA, together with the codon pair $AG^{A}{}_{G}$. Similar rules are followed in the mitochondria of fungi.

9.8 Novel Amino Acids Can Be Inserted at Certain Stop Codons

Key concepts

- Changes in the reading of specific codons can occur in individual genes.
- The insertion of seleno-Cys-tRNA at certain UGA codons requires several proteins to modify the Cys-tRNA and insert it into the ribosome.
- Pyrrolysine can be inserted at certain UAG codons.

Specific changes in reading the code occur in individual genes. The specificity of such changes implies that the reading of the particular codon must be influenced by the surrounding bases. In two cases, amino acids other than the classical twenty are inserted by special aminoacyl-tRNAs.

A striking example is the incorporation of the modified amino acid seleno-cysteine at certain UGA codons within the genes that code for selenoproteins in both prokaryotes and eukaryotes. Usually these proteins catalyze oxidation-reduction reactions, and contain a single seleno-cysteine residue, which forms part of the active site. The most is known about the use of the UGA codons in three *E. coli* genes coding for formate dehydrogenase isozymes. The internal UGA codon is read by a seleno-Cys-tRNA. This unusual reaction is determined by the local secondary structure of mRNA, in particular by the presence of a hairpin loop downstream of the UGA.

Mutations in 4 *sel* genes create a deficiency in selenoprotein synthesis. *selC* codes for tRNA (with the anticodon ACU←) that is charged with serine. *selA* and *selD* are required to modify the serine to seleno-cysteine. SelB is an alternative elongation factor. It is a guanine nucleotide-binding protein that acts as a specific translation factor for entry of seleno-Cys-tRNA into the A site; it thus provides (for this single tRNA) a replacement for factor EF-Tu. The sequence of SelB is related to both EF-Tu and IF-2.

Why is seleno-Cys-tRNA inserted only at certain UGA codons? These codons are followed by a stem-loop structure in the mRNA. FIGURE 9.12 shows that the stem of this structure is recognized by an additional domain in SelB (one that is not present in EF-Tu or IF-2). A similar mechanism interprets some UGA codons in mammalian cells, except that two proteins are required to identify the appropriate UGA codons. One protein (SBP2) binds a stem-loop structure far downstream from the UGA codon, whereas the counterpart of SelB (called SECIS) binds to SBP2 and simultaneously binds the tRNA to the UGA codon.

Another example of the insertion of a special amino acid is the placement of pyrrolysine at an UAG codon. This happens in both an archaea and a bacterium. The mechanism is probably similar to the insertion of seleno-cysteine. An unusual tRNA is charged with lysine, which is presumably then modified. The tRNA has a CUA anticodon, which responds to UAG. There must be other components of the system that restricts its response to the appropriate UAG codons.

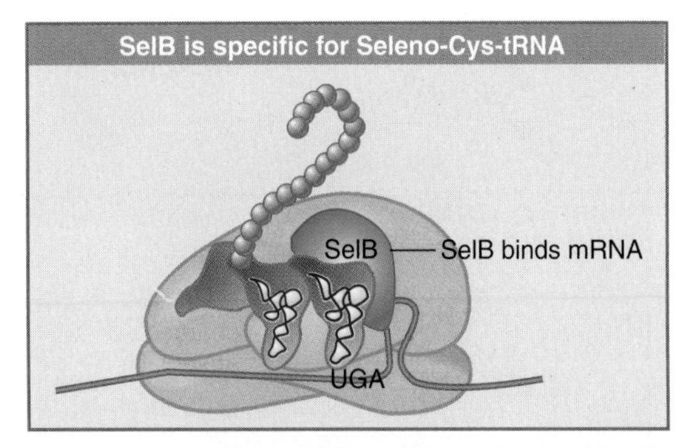

FIGURE 9.12 SelB is an elongation factor that specifically binds Seleno-Cys-tRNA to a UGA codon that is followed by a stem-loop structure in mRNA.

9.9 tRNAs Are Charged with Amino Acids by Synthetases

Key concepts

- Aminoacyl-tRNA synthetases are enzymes that charge tRNA with an amino acid to generate aminoacyl-tRNA in a two-stage reaction that uses energy from ATP.
- There are twenty aminoacyl-tRNA synthetases in each cell. Each charges all the tRNAs that represent a particular amino acid.
- Recognition of a tRNA is based on a small number of points of contact in the tRNA sequence.

It is necessary for tRNAs to have certain characteristics in common, yet be distinguished by others. The crucial feature that confers this capacity is the ability of tRNA to fold into a specific tertiary structure. Changes in the details of this structure, such as the angle of the two arms of the "L" or the protrusion of individual bases, may distinguish the individual tRNAs.

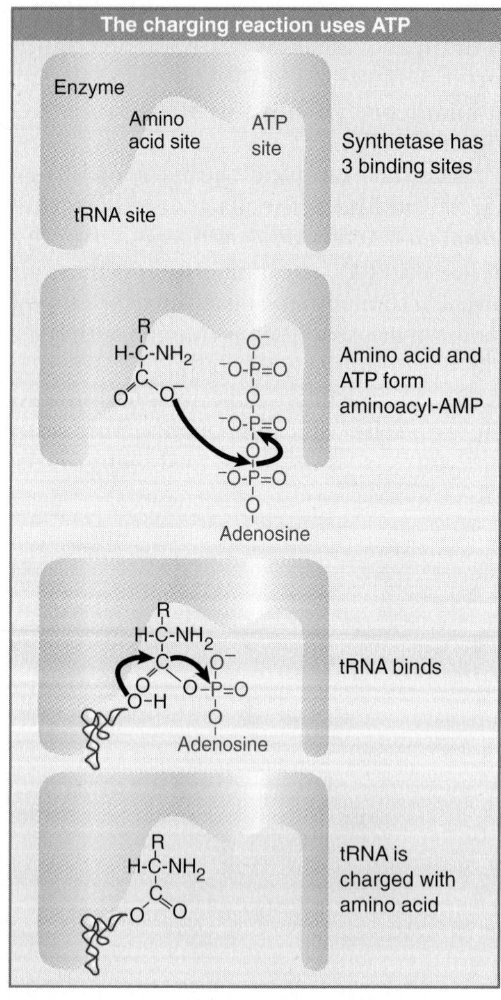

FIGURE 9.13 An aminoacyl-tRNA synthetase charges tRNA with an amino acid.

All tRNAs can fit in the P and A sites of the ribosome. At one end they are associated with mRNA via codon–anticodon pairing, and at the other end the polypeptide is being transferred. Similarly, all tRNAs (except the initiator) share the ability to be recognized by the translation factors (EF-Tu or eEF1) for binding to the ribosome. The initiator tRNA is recognized instead by IF-2 or eIF2. Thus the tRNA set must possess common features for interaction with elongation factors, but the initiator tRNA can be distinguished.

Amino acids enter the protein synthesis pathway through the aminoacyl-tRNA synthetases, which provide the interface for connection with nucleic acid. All synthetases function by the two-step mechanism depicted in FIGURE 9.13:

- First, the amino acid reacts with ATP to form aminoacyl-adenylate, releasing pyrophosphate. Energy for the reaction is provided by cleaving the high energy bond of the ATP.
- Then the activated amino acid is transferred to the tRNA, releasing AMP.

The synthetases sort the tRNAs and amino acids into corresponding sets. Each synthetase recognizes a single amino acid and all the tRNAs that should be charged with it. Usually, each amino acid is represented by more than one tRNA. Several tRNAs may be needed to respond to synonym codons, and sometimes there are multiple species of tRNA reacting with the same codon. Multiple tRNAs representing the same amino acid are called **isoaccepting tRNAs;** because they are all recognized by the same synthetase, they are also described as its **cognate tRNAs.**

Many attempts to deduce similarities in sequence between cognate tRNAs, or to induce chemical alterations that affect their charging, have shown that the basis for recognition is not the same for different tRNAs, and does not necessarily lie in some feature of primary or secondary structure alone. We know from the crystal structure that the acceptor stem and the anticodon stem make tight contacts with the synthetase, and mutations that alter recognition of a tRNA are found in these two regions. (The anticodon itself is not necessarily recognized as such; for example, the "suppressor" mutations discussed later in this chapter change a base in the anticodon, and therefore the codons to which a tRNA responds, without altering its charging with amino acids.)

A group of isoaccepting tRNAs must be charged only by the single aminoacyl-tRNA synthetase specific for their amino acid. So isoaccepting tRNAs must share some common feature(s) enabling the enzyme to distinguish them from the other tRNAs. The entire complement of tRNAs is divided into twenty isoaccepting groups, and each group is able to identify itself to its particular synthetase.

tRNAs are identified by their synthetases by contacts that recognize a small number of bases, typically from one to five. Three types of features commonly are used:

- Usually (but not always), at least one base of the anticodon is recognized. Sometimes all the positions of the anticodon are important.
- Often one of the last three base pairs in the acceptor stem is recognized. An extreme case is represented by alanine tRNA, which is identified by a single unique base pair in the acceptor stem.
- The so-called discriminator base, which lies between the acceptor stem and the CCA terminus, is always invariant among isoacceptor tRNAs.

No one of these features constitutes a unique means of distinguishing twenty sets of tRNAs, or provides sufficient specificity, so it appears that recognition of tRNAs is idiosyncratic, with each following its own rules.

Several synthetases can specifically charge a "minihelix," which consists only of the acceptor and TψC arms (equivalent to one arm of the L-shaped molecule) with the correct amino acid. For certain tRNAs, specificity depends exclusively upon the acceptor stem. However, it is clear that there are significant variations between tRNAs, and in some cases the anticodon region is important. Mutations in the anticodon can affect recognition by the class II Phe-tRNA synthetase. Multiple features may be involved; minihelices from the $tRNA^{Val}$ and $tRNA^{Met}$ (where we know that the anticodon is important *in vivo*) can react specifically with their synthetases.

Thus recognition depends on an interaction between a few points of contact in the tRNA, concentrated at the extremities, and a few amino acids constituting the active site in the protein. The relative importance of the roles played by the acceptor stem and anticodon is different for each tRNA-synthetase interaction.

9.10 Aminoacyl-tRNA Synthetases Fall into Two Groups

Key concept

- Aminoacyl-tRNA synthetases are divided into the class I and class II groups by sequence and structural similarities.

In spite of their common function, synthetases are a rather diverse group of proteins. The individual subunits vary from 40 to 110 kD, and the enzymes may be monomeric, dimeric, or tetrameric. Homologies between them are rare. Of course, the active site that recognizes tRNA comprises a rather small part of the molecule. It is interesting to compare the active sites of different synthetases.

Synthetases have been divided into two general groups, each containing ten enzymes, on the basis of the structure of the domain that contains the active site. A general type of organization that applies to both groups is represented in FIGURE 9.14. The catalytic domain includes the binding sites for ATP and amino acid. It can be recognized as a large region that is interrupted by an insertion of the domain that binds the acceptor helix of the tRNA. This places the terminus of the tRNA in proximity to the catalytic site. A separate domain binds the anticodon region of tRNA. Those synthetases that are multimeric also possess an oligomerization domain.

Class I synthetases have an N-terminal catalytic domain that is identified by the presence of two short, partly conserved sequences of amino acids, sometimes called "signature

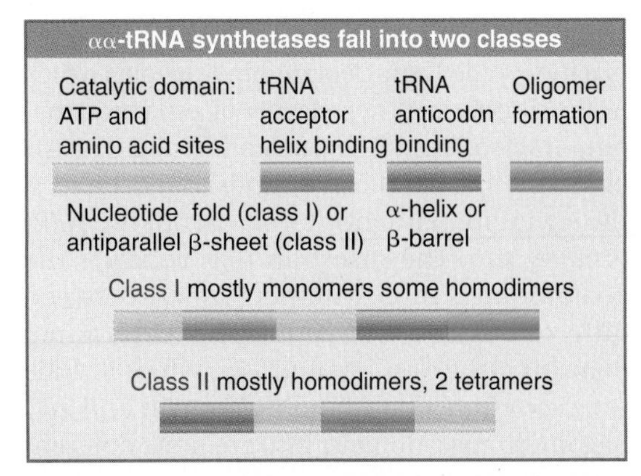

FIGURE 9.14 An aminoacyl-tRNA synthetase contains three or four regions with different functions. (Only multimeric synthetases possess an oligomerization domain.)

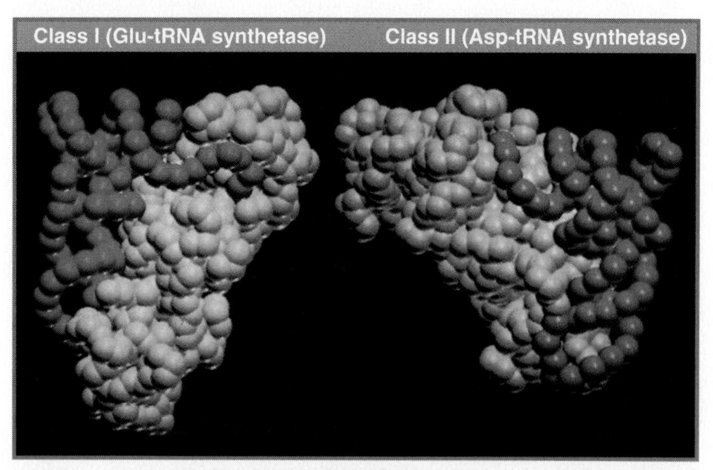

FIGURE 9.15 Crystal structures show that class I and class II aminoacyl-tRNA synthetases bind the opposite faces of their tRNA substrates. The tRNA is shown in red and the protein in blue. Photo courtesy of Dino Moras, Institute of Genetics and Molecular and Cellular Biology.

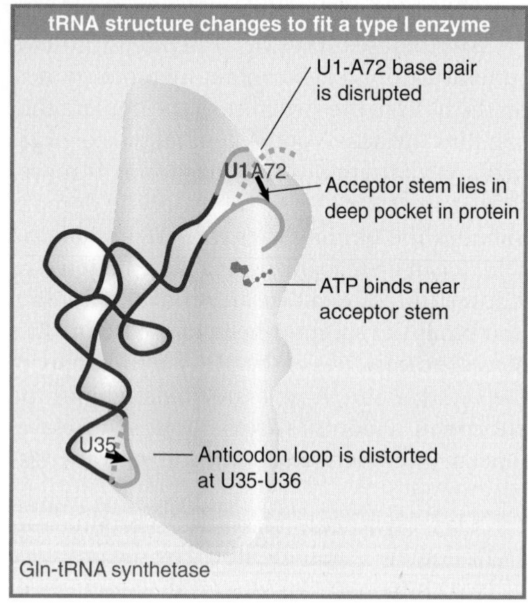

FIGURE 9.16 A class I tRNA synthetase contacts tRNA at the minor groove of the acceptor stem and at the anticodon.

sequences." The catalytic domain takes the form of a motif called a nucleotide-binding fold (which is also found in other classes of enzymes that bind nucleotides). The nucleotide fold consists of alternating parallel β-strands and α-helices; the signature sequence forms part of the ATP-binding site. The insertion that contacts the acceptor helix of tRNA differs widely between different class I enzymes. The C-terminal domains of the class I synthetases, which include the tRNA anticodon-binding domain and any oligomerization domain, also are quite different from one another.

Class II enzymes share three rather general similarities of sequence in their catalytic domains. The active site contains a large antiparallel β-sheet surrounded by α-helices. Again, the acceptor helix-binding domain that interrupts the catalytic domain has a structure that depends on the individual enzyme. The anticodon-binding domain tends to be N-terminal. The location of any oligomerization domain is widely variable.

The lack of any apparent relationship between the two groups of synthetases is a puzzle. Perhaps they evolved independently of one another. This makes it seem possible even that an early form of life could have existed with proteins that were made up of just the ten amino acids coded by one type or the other.

A general model for synthetase-tRNA binding suggests that the protein binds the tRNA along the "side" of the L-shaped molecule. The same general principle applies for all synthetase-tRNA binding: The tRNA is bound principally at its two extremities, and most of the tRNA sequence is not involved in recognition by a synthetase. However, the detailed nature of the interaction is different between class I and class II enzymes, as can be seen from the models of FIGURE 9.15, which are based on crystal structures. The two types of enzyme approach the tRNA from opposite sides, with the result that the tRNA-protein models look almost like mirror images of one another.

A class I enzyme (Gln-tRNA synthetase) approaches the D-loop side of the tRNA. It recognizes the minor groove of the acceptor stem at one end of the binding site, and interacts with the anticodon loop at the other end. FIGURE 9.16 is a diagrammatic representation of the crystal structure of the $tRNA^{Gln}$-synthetase complex. A revealing feature of the structure is that contacts with the enzyme change the structure of the tRNA at two important points. These can be seen by comparing the dotted and solid lines in the anticodon loop and acceptor stem:

- Bases U35 and U36 in the anticodon loop are pulled farther out of the tRNA into the protein.
- The end of the acceptor stem is seriously distorted, with the result that base pairing between U1 and A72 is disrupted. The single-stranded end of the stem pokes into a deep pocket in the synthetase protein, which also contains the binding site for ATP.

This structure explains why changes in U35, G73, or the U1-A72 base pair affect the recognition of the tRNA by its synthetase. At all of these positions, hydrogen bonding occurs between the protein and tRNA.

A class II enzyme (Asp-tRNA synthetase) approaches the tRNA from the other side; it recognizes both the variable loop and the major groove of the acceptor stem, as drawn in FIGURE 9.17. The acceptor stem remains in its regular helical conformation. ATP is probably bound near to the terminal adenine. At the other end of the binding site, there is a tight contact with the anticodon loop, which has a change in conformation that allows the anticodon to be in close contact with the protein.

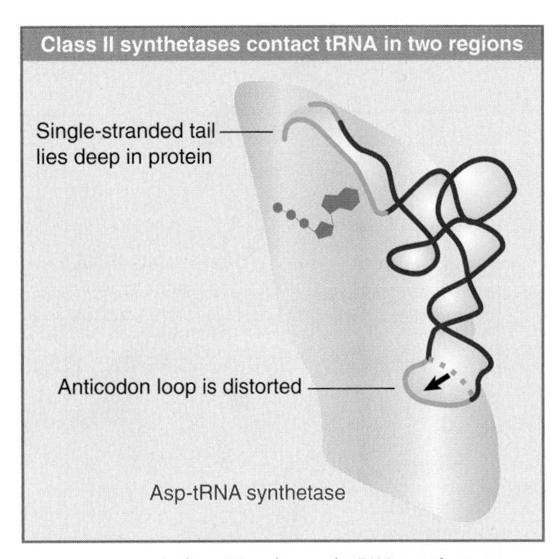

FIGURE 9.17 A class II aminoacyl-tRNA synthetase contacts tRNA at the major groove of the acceptor helix and at the anticodon loop.

9.11 Synthetases Use Proofreading to Improve Accuracy

Key concept

- Specificity of recognition of both amino acid and tRNA is controlled by aminoacyl-tRNA synthetases by proofreading reactions that reverse the catalytic reaction if the wrong component has been incorporated.

Aminoacyl-tRNA synthetases have a difficult job. Each synthetase must distinguish one out of twenty amino acids, and must differentiate cognate tRNAs (typically one to three) from the total set (perhaps 100 in all).

Many amino acids are closely related to one another, and all amino acids are related to the metabolic intermediates in their particular synthetic pathway. It is especially difficult to distinguish between two amino acids that differ only in the length of the carbon backbone (that is, by one CH_2 group). Intrinsic discrimination based on relative energies of binding two such amino acids would be only ~1/5. The synthetase enzymes improve this ratio ~1000-fold.

Intrinsic discrimination between tRNAs is better, because the tRNA offers a larger surface with which to make more contacts. It is still true, however, that all tRNAs conform to the same general structure, and there may be a quite limited set of features that distinguish the cognate tRNAs from the noncognate tRNAs.

We can imagine two general ways in which the enzyme might select its substrate:

- The cycle of admittance, scrutiny, and rejection/acceptance could represent a single binding step that precedes all other stages of whatever reaction is involved. This is tantamount to saying that the affinity of the binding site is sufficient to control the entry of substrate. In the case of synthetases, this would mean that only the correct amino acids and cognate tRNAs could form a stable attachment at the site.
- Alternatively, the reaction proceeds through some of its stages, after which a decision is reached on whether the correct species is present. If it is not present, the reaction is reversed, or a bypass route is taken, and the wrong member is expelled. This sort of postbinding scrutiny is generally described as **proofreading.** In the example of synthetases, it would require that the charging reaction proceeds through certain stages even if the wrong tRNA or amino acid is present.

Synthetases use proofreading mechanisms to control the recognition of both types of substrates. They improve significantly on the intrinsic differences among amino acids or among tRNAs, but, consistent with the intrinsic differences in each group, make more mistakes in selecting amino acids (error rates are 10^{-4} to 10^{-5}) than in selecting tRNAs (for which error rates are ~10^{-6}) (see Figure 8.8). Transfer RNA binds to synthetase by the two-stage reaction depicted in FIGURE 9.18. Cognate tRNAs have a greater intrinsic affinity for the binding site, so they are bound more rapidly and dissociate more slowly. Following binding, the enzyme scrutinizes the tRNA that has been bound. If the correct tRNA is present, binding is stabilized by a conformational change in the enzyme. This allows aminoacylation to occur rapidly. If the

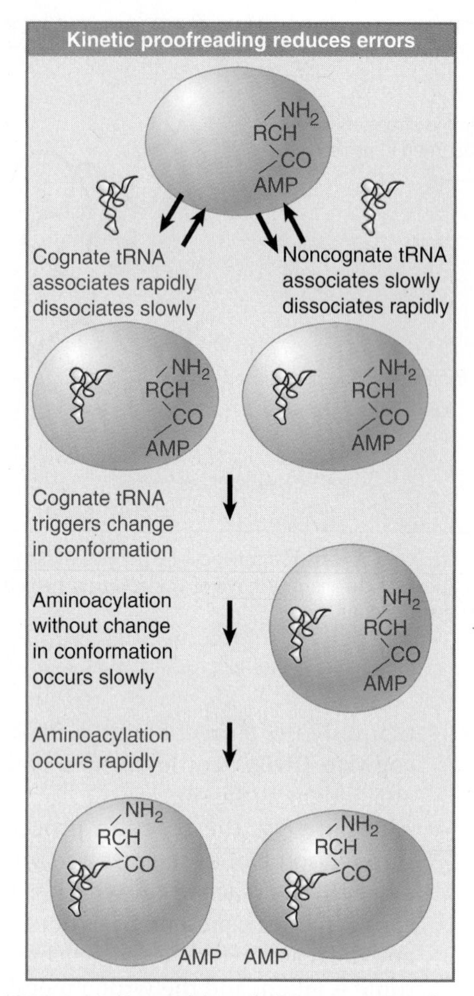

FIGURE 9.18 Recognition of the correct tRNA by synthetase is controlled at two steps. First, the enzyme has a greater affinity for its cognate tRNA. Second, the aminoacylation of the incorrect tRNA is very slow.

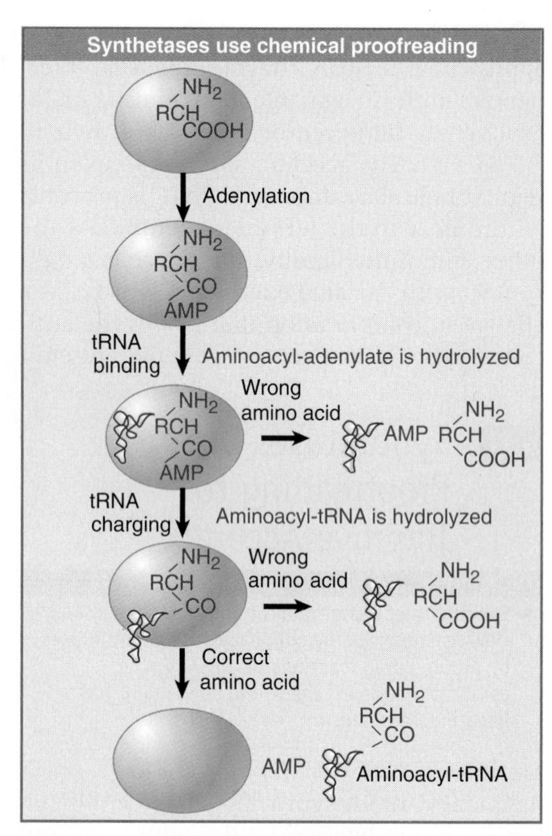

FIGURE 9.19 When a synthetase binds the incorrect amino acid, proofreading requires binding of the cognate tRNA. It may take place either by a conformation change that causes hydrolysis of the incorrect aminoacyl-adenylate or by transfer of the amino acid to tRNA, followed by hydrolysis.

wrong tRNA is present, the conformational change does not occur. As a result, the reaction proceeds much more slowly; this increases the chance that the tRNA will dissociate from the enzyme before it is charged. This type of control is called **kinetic proofreading.**

Specificity for amino acids varies among the synthetases. Some are highly specific for initially binding a single amino acid, whereas others can also activate amino acids closely related to the proper substrate. The analog amino acid can sometimes be converted to the adenylate form, but in none of these cases is an incorrectly activated amino acid actually used to form a stable aminoacyl-tRNA.

The presence of the cognate tRNA usually is needed to trigger proofreading, even if the reaction occurs at the stage before formation of aminoacyl-adenylate. (An exception is provided by Met-tRNA synthetase, which can reject noncognate aminoacyl-adenylate complexes even in the absence of tRNA.)

There are two stages at which proofreading of an incorrect aminoacyl-adenylate may occur during formation of aminoacyl-tRNA. FIGURE 9.19 shows that both use **chemical proofreading,** in which the catalytic reaction is reversed. The extent to which one pathway or the other predominates varies with the individual synthetase:

- The noncognate aminoacyl-adenylate may be hydrolyzed when the cognate tRNA binds. This mechanism is used predominantly by several synthetases, including those for methionine, isoleucine, and valine. (Usually, the reaction cannot be seen *in vivo*, but it can be followed for Met-tRNA synthetase when the incorrectly activated amino acid is homocysteine, which lacks the methyl group of methionine). Proofreading releases the amino acid in an altered form, as homocysteine thiolactone. In fact, homocysteine thiolactone is produced in *E. coli* as a by-product of the charging reaction of Met-tRNA synthetase. This shows that continuous proofreading is part of the process of charging a tRNA with its amino acid.

- Some synthetases use chemical proofreading at a later stage. The wrong amino acid is actually transferred to tRNA, is then recognized as incorrect by its structure in the tRNA binding site, and so is hydrolyzed and released. The process requires a continual cycle of linkage and hydrolysis until the correct amino acid is transferred to the tRNA.

A classic example in which discrimination between amino acids depends on the presence of tRNA is provided by the Ile-tRNA synthetase of *E. coli.* The enzyme can charge valine with AMP, but hydrolyzes the valyl-adenylate when $tRNA^{Ile}$ is added. The overall error rate depends on the specificities of the individual steps, as summarized in FIGURE 9.20. The overall error rate of 1.5×10^{-5} is less than the measured rate at which valine is substituted for isoleucine (in rabbit globin), which ranges from 2 to 5×10^{-4}. So mischarging probably provides only a small fraction of the errors that actually occur in protein synthesis.

Ile-tRNA synthetase uses size as a basis for discrimination among amino acids. FIGURE 9.21 shows that it has two active sites: the synthetic (or activation) site and the editing (or hydrolytic) site. The crystal structure of the enzyme shows that the synthetic site is too small to allow leucine (a close analog of isoleucine) to enter. All amino acids larger than isoleucine are excluded from activation because they cannot enter the synthetic site. An amino acid that can enter the synthetic site is placed on tRNA. Then the enzyme tries to transfer it to the editing site. Isoleucine is safe from editing because it is too large to enter the editing site. However, valine can enter this site, and as a result an incorrect $Val\text{-}tRNA^{Ile}$ is hydrolyzed. Essentially the enzyme provides a double molecular sieve, in which size of the amino acid is used to discriminate between closely related species.

One interesting feature of Ile-tRNA synthetase is that the synthetic and editing sites are a considerable distance apart, ~34 Å. A crystal structure of the enzyme complexed with an edited analog of isoleucine shows that the amino acid is transported from the synthetic site to the editing site. FIGURE 9.22 shows that this involves a change in the conformation of the tRNA. The amino acid acceptor stem of $tRNA^{Ile}$ can exist in alternative conformations. It adopts an unusual

Errors are controlled at each stage	
Step	Frequency of Error
Activation of valine to $Val\text{-}AMP^{Ile}$	1/225
Release of Val–tRNA	1/270
Overall rate of error	1/225 × 1/270 = 1/60,000

FIGURE 9.20 The accuracy of charging $tRNA^{Ile}$ by its synthetase depends on error control at two stages.

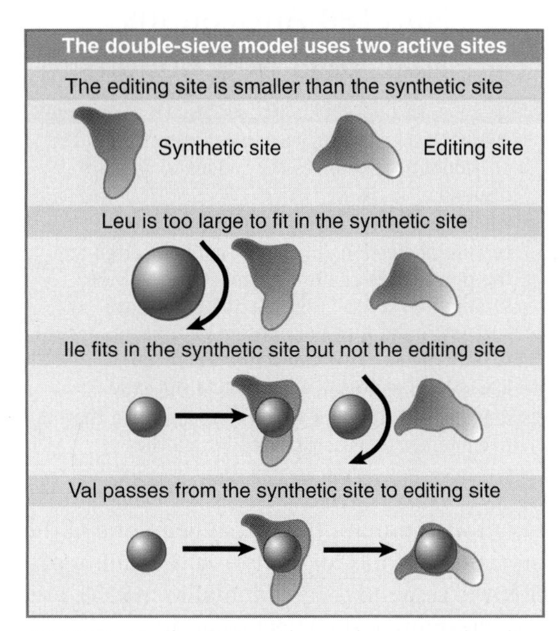

FIGURE 9.21 Ile-tRNA synthetase has two active sites. Amino acids larger than Ile cannot be activated because they do not fit in the synthetic site. Amino acids smaller than Ile are removed because they are able to enter the editing site.

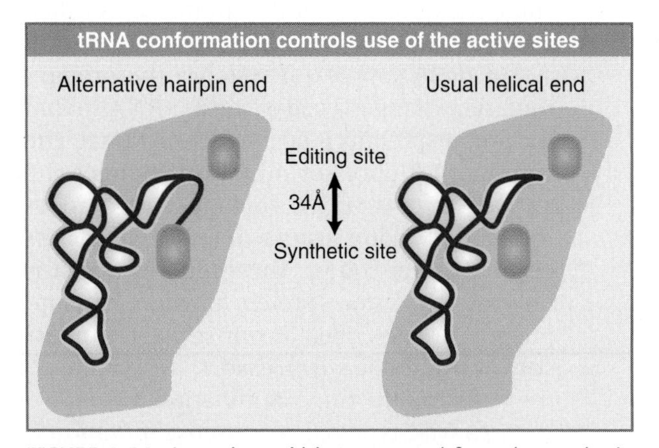

FIGURE 9.22 An amino acid is transported from the synthetic site to the editing site of Ile-tRNA synthetase by a change in the conformation of the amino acceptor stem of tRNA.

hairpin in order to be aminoacylated by an amino acid in the synthetic site. Then it returns to the more common helical structure in order to move the amino acid to the editing site. The translocation between sites is the rate-limiting step in proofreading. Ile-tRNA synthetase is a class I synthetase, but the double sieve mechanism is used also by class II synthetases.

9.12 Suppressor tRNAs Have Mutated Anticodons That Read New Codons

Key concepts

- A suppressor tRNA typically has a mutation in the anticodon that changes the codons to which it responds.
- When the new anticodon corresponds to a termination codon, an amino acid is inserted and the polypeptide chain is extended beyond the termination codon. This results in nonsense suppression at a site of nonsense mutation, or in readthrough at a natural termination codon.
- Missense suppression occurs when the tRNA recognizes a different codon from usual, so that one amino acid is substituted for another.

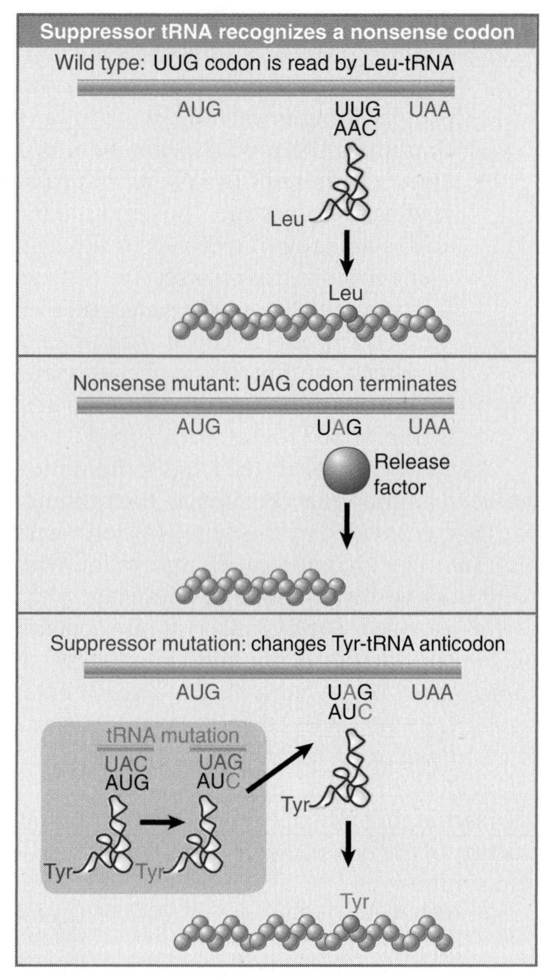

FIGURE 9.23 Nonsense mutations can be suppressed by a tRNA with a mutant anticodon, which inserts an amino acid at the mutant codon, producing a full length protein in which the original Leu residue has been replaced by Tyr.

Isolation of mutant tRNAs has been one of the most potent tools for analyzing the ability of a tRNA to respond to its codon(s) in mRNA, and for determining the effects that different parts of the tRNA molecule have on codon–anticodon recognition.

Mutant tRNAs are isolated by virtue of their ability to overcome the effects of mutations in genes coding for proteins. In general genetic terminology, a mutation that is able to overcome the effects of another mutation is called a suppressor.

In tRNA suppressor systems, the primary mutation changes a codon in an mRNA so that the protein product is no longer functional. The secondary suppressor mutation changes the anticodon of a tRNA, so that it recognizes the mutant codon instead of (or as well as) its original target codon. The amino acid that is now inserted restores protein function. The suppressors are described as **nonsense suppressors** or **missense suppressors,** depending on the nature of the original mutation.

In a wild-type cell, a nonsense mutation is recognized only by a release factor, which terminates protein synthesis. The suppressor mutation creates an aminoacyl-tRNA that can recognize the termination codon; by inserting an amino acid, it allows protein synthesis to continue beyond the site of nonsense mutation. This new capacity of the translation system allows a full-length protein to be synthesized, as illustrated in FIGURE 9.23. If the amino acid inserted by suppression is different from the amino acid that was originally present at this site in the wild-type protein, the activity of the protein may be altered.

Missense mutations change a codon representing one amino acid into a codon representing another amino acid—one that cannot function in the protein in place of the original residue. (Formally, any substitution of amino acids constitutes a missense mutation, but in practice it is detected only if it changes the activity of the protein.) The mutation can be suppressed by the insertion either of the original amino acid or of some other amino acid that is acceptable to the protein.

FIGURE 9.24 demonstrates that missense suppression can be accomplished in the same way as nonsense suppression, by mutating the anticodon of a tRNA carrying an acceptable amino acid so that it responds to the mutant codon. So missense suppression involves a change in the meaning of the codon from one amino acid to another.

9.13 There Are Nonsense Suppressors for Each Termination Codon

Key concepts

- Each type of nonsense codon is suppressed by tRNAs with mutant anticodons.
- Some rare suppressor tRNAs have mutations in other parts of the molecule.

Nonsense suppressors fall into three classes, one for each type of termination codon. FIGURE 9.25 describes the properties of some of the best characterized suppressors.

The easiest to characterize have been amber suppressors. In *E. coli,* at least six tRNAs have been mutated to recognize UAG codons. All of the amber suppressor tRNAs have the anticodon CUA$^{\leftarrow}$, in each case derived from wild type by a single base change. The site of mutation can be any one of the three bases of the anticodon, as seen from *supD, supE,* and *supF.* Each suppressor tRNA recognizes only the UAG codon, instead of its former codon(s). The amino acids inserted are serine, glutamine, or tyrosine, the same as those carried by the corresponding wild-type tRNAs.

Ochre suppressors also arise by mutations in the anticodon. The best known are *supC* and *supG,* which insert tyrosine or lysine in response to both ochre (UAA) and amber (UAG) codons. This conforms with the prediction of the wobble hypothesis that UAA cannot be recognized alone.

A UGA suppressor has an unexpected property. It is derived from $tRNA^{Trp}$, but its only mutation is the substitution of A in place of G at position 24. This change replaces a G-U pair in the D stem with an A-U pair, increasing the stability of the helix. The sequence of the anticodon remains the same as the wild type, CCA$^{\leftarrow}$. So the mutation in the D stem must in some way alter the conformation of the anticodon loop, allowing CCA$^{\leftarrow}$ to pair with UGA in an unusual wobble pairing of C with A. The

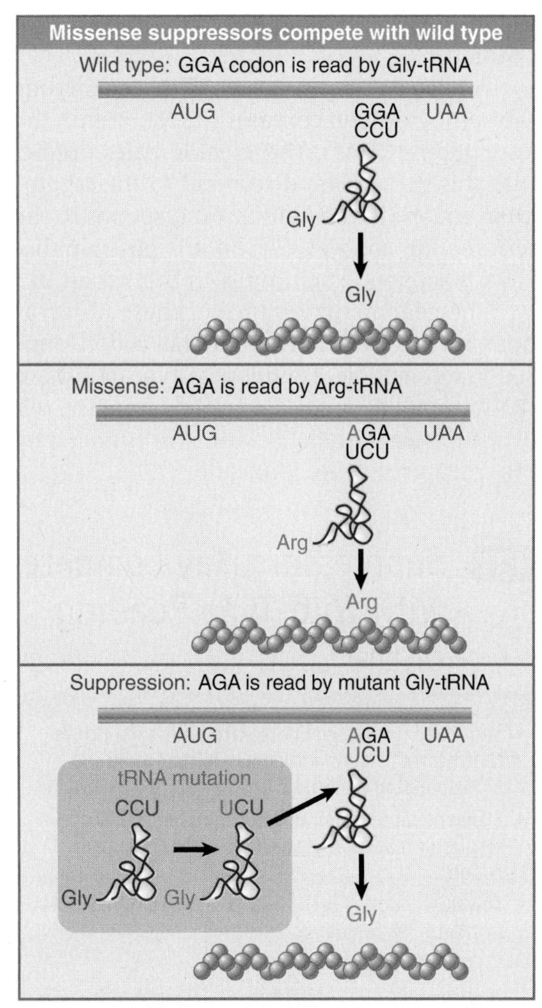

FIGURE 9.24 Missense suppression occurs when the anticodon of tRNA is mutated so that it responds to the wrong codon. The suppression is only partial because both the wild-type tRNA and the suppressor tRNA can respond to AGA.

Suppressors have anticodon mutations

Locus	tRNA	Wild Type		Suppressor	
		Codon	Anti	Anti	Codon
supD (su1)	Ser	UCG	CGA	CUA	UAG
supE (su2)	Gln	CAG	CUG	CUA	UAG
supF (su3)	Tyr	UA$^{C}_{U}$	GUA	CUA	UAG
supC (su4)	Tyr	UA$^{C}_{U}$	GUA	UUA	UA$^{A}_{G}$
supG (su5)	Lys	AA$^{A}_{G}$	UUU	UUA	UA$^{A}_{G}$
supU (su7)	Trp	UGG	CCA	UCA	UG$^{A}_{G}$

FIGURE 9.25 Nonsense suppressor tRNAs are generated by mutations in the anticodon.

suppressor tRNA continues to recognize its usual codon, UGG.

A related response is seen with a eukaryotic tRNA. Bovine liver contains a $tRNA^{Ser}$ with the anticodon $^{m}CCA^{\leftarrow}$. The wobble rules predict that this tRNA should respond to the tryptophan codon UGG, but in fact it responds to the termination codon UGA. So it is possible that UGA is suppressed naturally in this situation.

The general importance of these observations lies in the demonstration that codon–anticodon recognition of either wild-type or mutant tRNA cannot be predicted entirely from the relevant triplet sequences, but is influenced by other features of the molecule.

9.14 Suppressors May Compete with Wild-Type Reading of the Code

Key concepts

- Suppressor tRNAs compete with wild-type tRNAs that have the same anticodon to read the corresponding codon(s).
- Efficient suppression is deleterious because it results in readthrough past normal termination codons.
- The UGA codon is leaky and is misread by Trp-tRNA at 1% to 3% frequency.

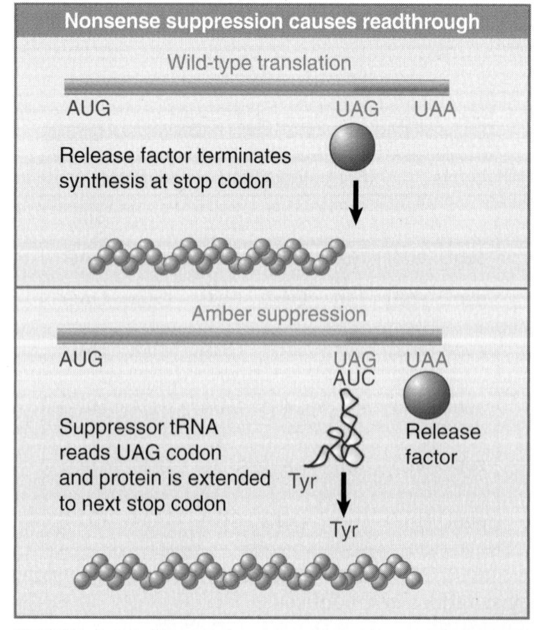

FIGURE 9.26 Nonsense suppressors also read through natural termination codons, synthesizing proteins that are longer than wild-type.

There is an interesting difference between the usual recognition of a codon by its proper aminoacyl-tRNA and the situation in which mutation allows a suppressor tRNA to recognize a new codon. In the wild-type cell, only one meaning can be attributed to a given codon, which represents either a particular amino acid or a signal for termination. In a cell carrying a suppressor mutation, though, the mutant codon has the alternatives of being recognized by the suppressor tRNA or of being read with its usual meaning.

A nonsense suppressor tRNA must compete with the release factors that recognize the termination codon(s). A missense suppressor tRNA must compete with the tRNAs that respond properly to its new codon. The extent of competition influences the efficiency of suppression; thus the effectiveness of a particular suppressor depends not only on the affinity between its anticodon and the target codon, but also on its concentration in the cell, and on the parameters governing the competing termination or insertion reactions.

The efficiency with which any particular codon is read is influenced by its location. Thus the extent of nonsense suppression by a given tRNA can vary quite widely, depending on the context of the codon. We do not understand the effect that neighboring bases in mRNA have on codon–anticodon recognition, but the context can change the frequency with which a codon is recognized by a particular tRNA by more than an order of magnitude. The base on the 3′ side of a codon appears to have a particularly strong effect.

A nonsense suppressor is isolated by its ability to respond to a mutant nonsense codon. The same triplet sequence, however, constitutes one of the normal termination signals of the cell! The mutant tRNA that suppresses the nonsense mutation must in principle be able to suppress natural termination at the end of any gene that uses this codon. FIGURE 9.26 shows that this **readthrough** results in the synthesis of a longer protein, with additional C-terminal material. The extended protein will end at the next termination triplet sequence found in the phase of the reading frame. Any extensive suppression of termination is likely to be deleterious to the cell by producing extended proteins whose functions are thereby altered.

Amber suppressors tend to be relatively efficient, usually in the range of 10% to 50%, depending on the system. This efficiency is possible because amber codons are used relatively infrequently to terminate protein synthesis in *E. coli*.

Ochre suppressors are difficult to isolate. They are always much less efficient, usually with activities below 10%. All ochre suppressors grow rather poorly, which indicates that suppression of both UAA and UAG is damaging to *E. coli*, probably because the ochre codon is used most frequently as a natural termination signal.

UGA is the least efficient of the termination codons in its natural function; it is misread by Trp-tRNA as frequently as 1% to 3% in wild-type situations. In spite of this deficiency, however, it is used more commonly than the amber triplet to terminate bacterial genes.

One gene's missense suppressor is likely to be another gene's mutator. A suppressor corrects a mutation by substituting one amino acid for another at the mutant site. In other locations, though, the same substitution will replace the wild-type amino acid with a new amino acid. The change may inhibit normal protein function.

This poses a dilemma for the cell: it must suppress what is a mutant codon at one location while failing to change too extensively its normal meaning at other locations. The absence of any strong missense suppressors is therefore explained by the damaging effects that would be caused by a general and efficient substitution of amino acids.

A mutation that creates a suppressor tRNA can have two consequences. First, it allows the tRNA to recognize a new codon. Second, it sometimes prevents the tRNA from recognizing the codons to which it previously responded. It is significant that all the high-efficiency amber suppressors are derived by mutation of one copy of a redundant tRNA set. In these cases, the cell has several tRNAs able to respond to the codon originally recognized by the wild-type tRNA. Thus the mutation does not abolish recognition of the old codons, which continue to be served adequately by the tRNAs of the set. In the unusual situation in which there is only a single tRNA that responds to a particular codon, any mutation that prevents the response is lethal.

Suppression is most often considered in the context of a mutation that changes the reading of a codon. There are, however, some situations in which a stop codon is read as an amino acid at a low frequency in the wild-type situation. The first example to be discovered was the coat protein gene of the RNA phage Qβ. The formation of infective Qβ particles requires that the stop codon at the end of this gene is suppressed at a low frequency to generate a small proportion of coat proteins with a C-terminal extension. In effect, this stop codon is leaky. The reason is that Trp-tRNA recognizes the codon at a low frequency.

Readthrough past stop codons also occurs in eukaryotes, where it is employed most often by RNA viruses. This may involve the suppression of UAG/UAA by Tyr-tRNA, Gln-tRNA, or Leu-tRNA, or the suppression of UGA by Trp-tRNA or Arg-tRNA. The extent of partial suppression is dictated by the context surrounding the codon.

9.15 The Ribosome Influences the Accuracy of Translation

Key concept

- The structure of the 16S rRNA at the P and A sites of the ribosome influences the accuracy of translation.

The lack of detectable variation when the sequence of a protein is analyzed demonstrates that protein synthesis must be extremely accurate. Very few mistakes are apparent in the form of substitutions of one amino acid for another. There are two general stages in protein synthesis at which errors might be made (see Figure 8.8 in Section 8.3, Special Mechanisms Control the Accuracy of Protein Synthesis):

- Charging a tRNA only with its correct amino acid clearly is critical. This is a function of the aminoacyl-tRNA synthetase. The error rate probably varies with the particular enzyme, but in general mistakes occur in $<1/10^5$ aminoacylations.
- The specificity of codon–anticodon recognition is crucial, but puzzling. Although binding constants vary with the individual codon–anticodon reaction, the specificity is always much too low to provide an error rate of $<10^{-5}$. When free in solution, tRNAs bind to their trinucleotide codon sequences only rather weakly. Related, but erroneous, triplets (with two correct bases out of three) are recognized 10^{-1} to 10^{-2} times as efficiently as the correct triplets.

Codon–anticodon base pairing therefore seems to be a weak point in the accuracy of translation. The ribosome has an important role in controlling the specificity of this interaction: It functions directly or indirectly as a "proofreader" in order to distinguish correct and incorrect codon–anticodon pairs, thus amplifying the rather modest intrinsic difference by ~1000×. In addition to the role of the ribosome itself, the factors that place initiator- and aminoacyl-tRNAs in the ribosome also may influence the pairing reaction.

There must be some mechanism for stabilizing the correct aminoacyl-tRNA, allowing its amino acid to be accepted as a substrate for receipt of the polypeptide chain; contacts with an incorrect aminoacyl-tRNA must be rapidly broken, so that the complex leaves without reacting. Suppose that there is no specificity in the initial collision between the aminoacyl-tRNA-EF-Tu-GTP complex and the ribosome. If any complex, irrespective of its tRNA, can enter the A site, the number of incorrect entries must far exceed the number of correct entries.

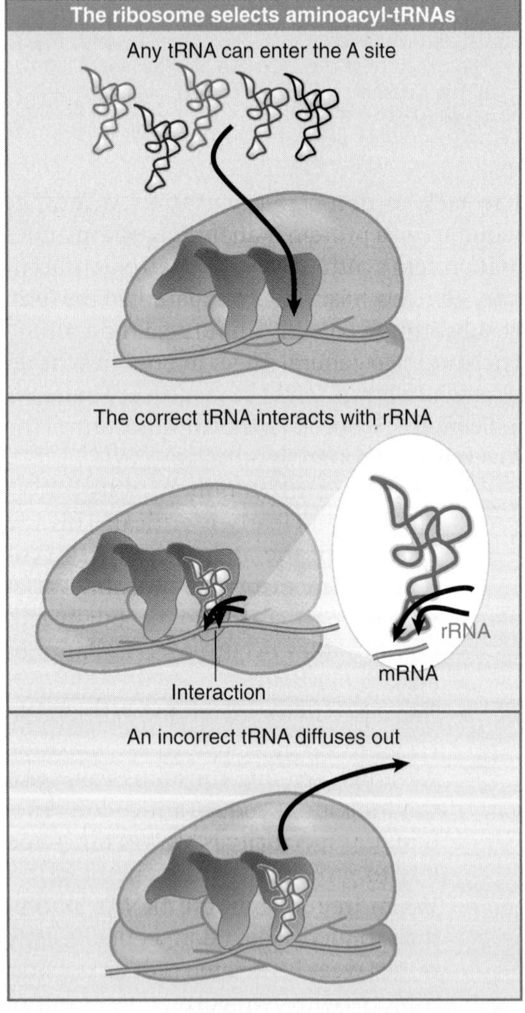

FIGURE 9.27 Any aminoacyl-tRNA can be placed in the A site (by EF-Tu), but only one that pairs with the anticodon can make stabilizing contacts with rRNA. In the absence of these contacts, the aminoacyl-tRNA diffuses out of the A site.

There are two basic models for how the ribosome might discriminate between correctly and incorrectly paired aminoacyl-tRNAs. The actual situation incorporates elements of both models.

- The direct recognition model supposes that the structure of the ribosome is designed to recognize aminoacyl-tRNAs that are correctly paired. This would mean that the correct pairing results in some small change in the conformation of the aminoacyl-tRNA that the ribosome can recognize. Discrimination occurs before any further reaction occurs.
- The kinetic proofreading model proposes that there are at least two stages in the process, so that the aminoacyl-tRNA has multiple opportunities to disengage. An incorrectly paired aminoacyl-tRNA may pass through some stages of the reaction before it is rejected. Overall selectivity can in principle be the product of the selectivities at each stage.

FIGURE 9.27 illustrates diagrammatically what happens to correctly and incorrectly paired aminoacyl-tRNAs. A correctly paired aminoacyl-tRNA is able to make stabilizing contacts with rRNA. An incorrectly paired aminoacyl-tRNA does not make these contacts, and therefore is able to diffuse out of the A site.

The path to discovering these interactions started with investigations of the effects of the antibiotic streptomycin in the 1960s. Streptomycin inhibits protein synthesis by binding to 16S rRNA and inhibiting the ability of EF-G to catalyze translocation. It also increases the level of misreading of the pyrimidines U and C (usually one is mistaken for the other, occasionally for A). The site at which streptomycin acts is influenced by the S12 protein; the sequence of this protein is altered in resistant mutants. Ribosomes with an S12 protein derived from resistant bacteria show a reduction in the level of misreading compared with wild-type ribosomes. In effect, S12 controls the level of misreading. When it is mutated to decrease misreading, it suppresses the effect of streptomycin.

S12 stabilizes the structure of 16S rRNA in the region that is bound by streptomycin. *The important point to note here is that the P/A site region influences the accuracy of translation: translation can be made more or less accurate by changing the structure of 16S rRNA.* The combination of the effects of the S12 protein and streptomycin on the rRNA structure explains the behavior of different mutants in S12, some of which even make the ribosome *dependent* on the presence of streptomycin for correct translation.

We now know from the crystal structure of the ribosome that 16S rRNA is in a position to make contacts with aminoacyl-tRNA. Two bases of 16S rRNA can contact the minor groove of the helix formed by pairing between the anticodon in tRNA with the first two bases of the codon in mRNA. This directly stabilizes the structure when the correct codon-anticodon contacts are made at the first two codon positions, but it does not monitor contacts at the third position.

The stabilization of correctly paired aminoacyl-tRNA may have two effects. By holding the aminoacyl-tRNA in the A site, it prevents it from escaping before the next stage of protein synthesis. The conformational change in the rRNA may help to trigger the next stage of the reaction, which is the hydrolysis of GTP by EF-Tu.

Part of the proofreading effect is determined by timing. An aminoacyl-tRNA in the A site may in effect be trapped if the next stage of protein synthesis occurs while it is there. Thus a delay between entry into the A site and peptidyl transfer may give more opportunity for a mismatched aminoacyl-tRNA to dissociate. Mismatched aminoacyl-tRNA dissociates more rapidly than correctly matched aminoacyl-tRNA, probably by a factor of ~5×. Its chance of escaping is therefore increased when the peptide transfer step is slowed.

The specificity of decoding has been assumed to reside with the ribosome itself, but some recent results suggest that translation factors influence the process at both the P site and A site. An indication that EF-Tu is involved in maintaining the reading frame is provided by mutants of the factor that suppress frameshifting. This implies that EF-Tu does not merely bring aminoacyl-tRNA to the A site, but also is involved in positioning the incoming aminoacyl-tRNA relative to the peptidyl-tRNA in the P site.

A striking case in which factors influence meaning is found at initiation. Mutation of the AUG initiation codon to UUG in the yeast gene *HIS4* prevents initiation. Extragenic suppressor mutations can be found that allow protein synthesis to be initiated at the mutant UUG codon. Two of these suppressors prove to be in genes coding for the α and β subunits of eIF2, the factor that binds Met-$tRNA_i$ to the P site. The mutation in eIFβ2 resides in a part of the protein that is almost certainly involved in binding nucleic acid. It seems likely that its target is either the initiation sequence of mRNA as such or the base-paired association between the mRNA codon and $tRNA_i^{Met}$ anticodon. This suggests that eIF2 participates in the discrimination of initiation codons as well as bringing the initiator tRNA to the P site.

The cost of protein synthesis in terms of high-energy bonds may be increased by proofreading processes. An important question in calculating the cost of protein synthesis is the stage at which the decision is taken on whether to accept a tRNA. If a decision occurs immediately to release an aminoacyl-tRNA-EF-TuGTP complex, there is little extra cost for rejecting the large number of incorrect tRNAs that are likely (statistically) to enter the A site before the correct tRNA is recognized. If, however, GTP is hydrolyzed before the mismatched aminoacyl-tRNA dissociates, the cost will be greater. A mismatched aminoacyl-tRNA can be rejected either before or after the cleavage of GTP, although we do not know yet where on average it is rejected. There is some evidence that the use of GTP *in vivo* is greater than the three high-energy bonds that are used in adding every (correct) amino acid to the chain.

9.16 Recoding Changes Codon Meanings

Key concepts

- Changes in codon meaning can be caused by mutant tRNAs or by tRNAs with special properties.
- The reading frame can be changed by frameshifting or bypassing, both of which depend on properties of the mRNA.

The reading frame of a messenger usually is invariant. Translation starts at an AUG codon and continues in triplets to a termination codon. Reading takes no notice of sense: insertion or deletion of a base causes a frameshift mutation, in which the reading frame is changed beyond the site of mutation. Ribosomes and tRNAs continue ineluctably in triplets, synthesizing an entirely different series of amino acids.

There are some exceptions to the usual pattern of translation that enable a reading frame with an interruption of some sort—such as a

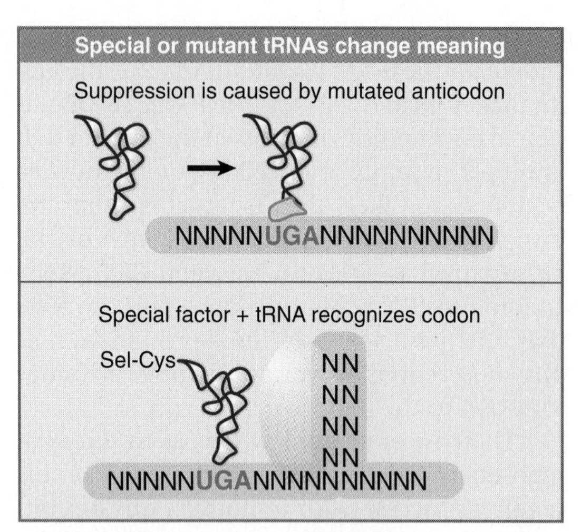

FIGURE 9.28 A mutation in an individual tRNA (usually in the anticodon) can suppress the usual meaning of that codon. In a special case, a specific tRNA is bound by an unusual elongation factor to recognize a termination codon adjacent to a hairpin loop.

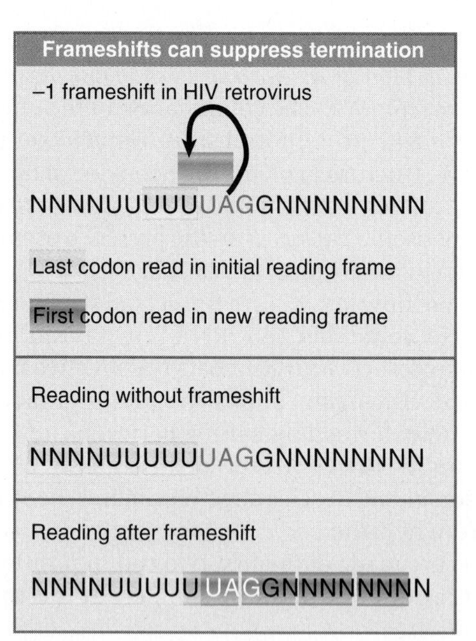

FIGURE 9.29 A tRNA that slips one base in pairing with a codon causes a frameshift that can suppress termination. The efficiency is usually ~5%.

nonsense codon or frameshift—to be translated into a full-length protein. **Recoding** events are responsible for making exceptions to the usual rules, and can involve several types of events.

Changing the meaning of a single codon allows one amino acid to be substituted in place of another, or for an amino acid to be inserted at a termination codon. FIGURE 9.28 shows that these changes rely on the properties of an individual tRNA that responds to the codon:

- Suppression involves recognition of a codon by a (mutant) tRNA that usually would respond to a different codon (see Section 9.12, Suppressor tRNAs Have Mutated Anticodons That Read New Codons).
- Redefinition of the meaning of a codon occurs when an aminoacyl-tRNA is modified (see Section 9.8, Novel Amino Acids Can Be Inserted at Certain Stop Codons).

Changing the reading frame occurs in two types of situations:

- Frameshifting typically involves changing the reading frame when aminoacyl-tRNA slips by one base, either +1 forward or –1 backward (see the following section, Frameshifting Occurs at Slippery Sequences). The result shown in FIGURE 9.29 is that translation continues past a termination codon.
- Bypassing involves a movement of the ribosome to change the codon that is paired with the peptidyl-tRNA in the P site. The sequence between the two codons fails to be represented in protein. As shown in FIGURE 9.30, this allows translation to continue past any termination codons in the intervening region.

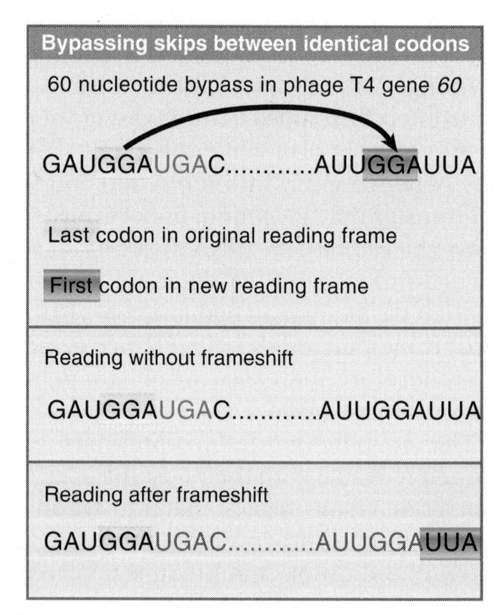

FIGURE 9.30 Bypassing occurs when the ribosome moves along mRNA so that the peptidyl-tRNA in the P site is released from pairing with its codon and then repairs with another codon farther along.

9.17 Frameshifting Occurs at Slippery Sequences

Key concepts

- The reading frame may be influenced by the sequence of mRNA and the ribosomal environment.
- Slippery sequences allow a tRNA to shift by one base after it has paired with its anticodon, thereby changing the reading frame.
- Translation of some genes depends upon the regular occurrence of programmed frameshifting.

Frameshifting is associated with specific tRNAs in two circumstances:

- Some mutant tRNA suppressors recognize a "codon" for four bases instead of the usual three bases.
- Certain "slippery" sequences allow a tRNA to move a base up or down mRNA in the A site.

Frameshift mutants result from the insertion or deletion of a base. They can be suppressed by restoring the original reading frame. This can be achieved by compensating base deletions and insertions within a gene (see Section 2.8, The Genetic Code Is Triplet). However, extragenic frameshift suppressors also can be found in the form of tRNAs with aberrant properties.

The simplest type of external frameshift suppressor corrects the reading frame when a mutation has been caused by inserting an additional base within a stretch of identical residues. For example, a G may be inserted in a run of several contiguous G bases. The frameshift suppressor is a $tRNA^{Gly}$ that has an extra base inserted in its anticodon loop, converting the anticodon from the usual triplet sequence $CCC^{\leftarrow}$ to the quadruplet sequence $CCCC^{\leftarrow}$. The suppressor tRNA recognizes a 4-base "codon."

Some frameshift suppressors can recognize more than one 4-base "codon." For example, a bacterial $tRNA^{Lys}$ suppressor can respond to either AAAA or AAAU, instead of the usual codon AAA. Another suppressor can read any 4-base "codon" with ACC in the first three positions; the next base is irrelevant. In these cases, the alternative bases that are acceptable in the fourth position of the longer "codon" are not related by the usual wobble rules. The suppressor tRNA probably recognizes a 3-base codon, but for some other reason—most likely steric hindrance—the adjacent base is blocked. This forces one base to be skipped before the next tRNA can find a codon.

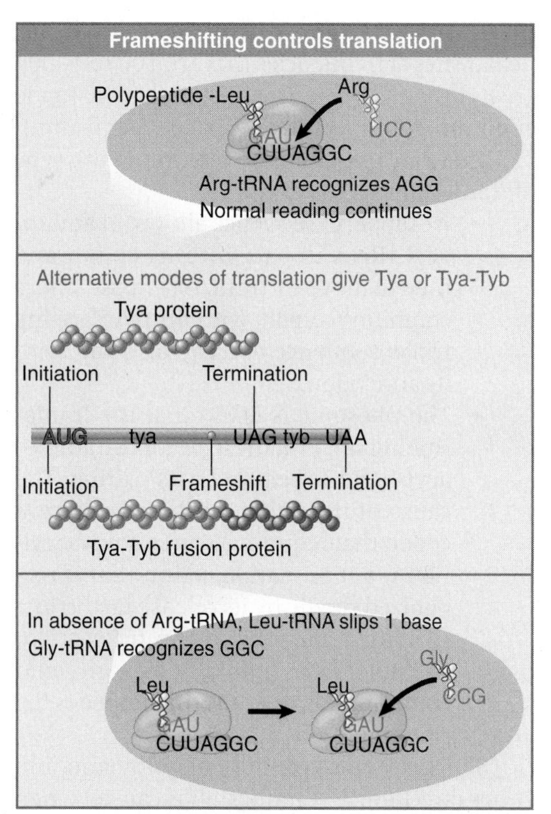

FIGURE 9.31 A +1 frameshift is required for expression of the *tyb* gene of the yeast Ty element. The shift occurs at a 7-base sequence at which two Leu codon(s) are followed by a scarce Arg codon.

Situations in which frameshifting is a normal event are presented by phages and viruses. Such events may affect the continuation or termination of protein synthesis, and result from the intrinsic properties of the mRNA.

In retroviruses, translation of the first gene is terminated by a nonsense codon in phase with the reading frame. The second gene lies in a different reading frame, and (in some viruses) is translated by a frameshift that changes into the second reading frame and therefore bypasses the termination codon (see Figure 9.29 and also Section 22.3, Retroviral Genes Code for Polyproteins). The efficiency of the frameshift is low, typically ~5%. In fact, this is important in the biology of the virus; an increase in efficiency can be damaging. FIGURE 9.31 illustrates the similar situation of the yeast Ty element, in which the termination codon of *tya* must be bypassed by a frameshift in order to read the subsequent *tyb* gene.

Such situations makes the important point that the rare (but predictable) occurrence of "misreading" events can be relied on as a necessary step in natural translation. This is called

programmed frameshifting. It occurs at particular sites at frequencies that are 100 to 1000× greater than the rate at which errors are made at nonprogrammed sites (~3×10^{-5} per codon).

There are two common features in this type of frameshifting:

- A "slippery" sequence allows an aminoacyl-tRNA to pair with its codon and then to move +1 (rare) or –1 base (more common) to pair with an overlapping triplet sequence that can also pair with its anticodon.
- The ribosome is delayed at the frameshifting site to allow time for the aminoacyl-tRNA to rearrange its pairing. The cause of the delay can be an adjacent codon that requires a scarce aminoacyl-tRNA, a termination codon that is recognized slowly by its release factor, or a structural impediment in mRNA (for example, a "pseudoknot," a particular conformation of RNA) that impedes the ribosome.

Slippery events can involve movement in either direction; a –1 frameshift is caused when the tRNA moves backward, and a +1 frameshift is caused when it moves forward. In either case, the result is to expose an out-of-phase triplet in the A site for the next aminoacyl-tRNA. The frameshifting event occurs before peptide bond synthesis. In the most common type of case, when it is triggered by a slippery sequence in conjunction with a downstream hairpin in mRNA, the surrounding sequences influence its efficiency.

The frameshifting in Figure 9.31 shows the behavior of a typical slippery sequence. The seven-nucleotide sequence CUUAGGC is usually recognized by Leu-tRNA at CUU, followed by Arg-tRNA at AGC. However, the Arg-tRNA is scarce, and when its scarcity results in a delay, the Leu-tRNA slips from the CUU codon to the overlapping UUA triplet. This causes a frameshift, because the next triplet in phase with the new pairing (GGC) is read by Gly-tRNA. Slippage usually occurs in the P site (when the Leu-tRNA actually has become peptidyl-tRNA, carrying the nascent chain).

Frameshifting at a stop codon causes readthrough of the protein. The base on the 3′ side of the stop codon influences the relative frequencies of termination and frameshifting, and thus affects the efficiency of the termination signal. This helps to explain the significance of context on termination.

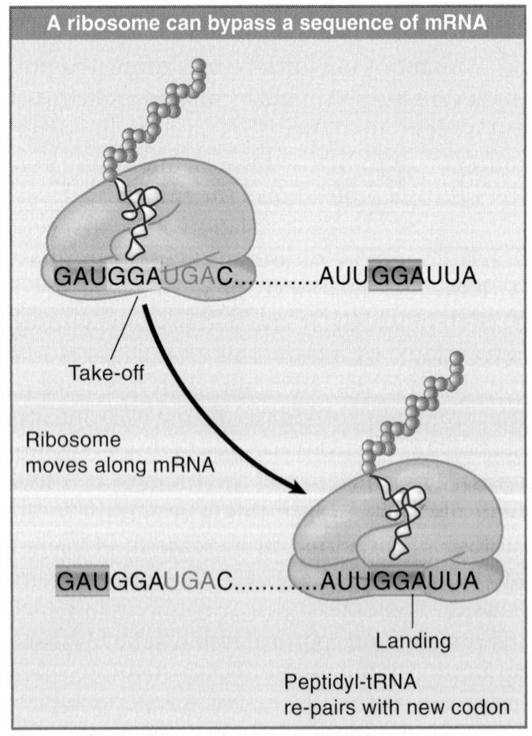

FIGURE 9.32 In bypass mode, a ribosome with its P site occupied can stop translation. It slides along mRNA to a site where peptidyl-tRNA pairs with a new codon in the P site. Then protein synthesis is resumed.

9.18 Bypassing Involves Ribosome Movement

Certain sequences trigger a bypass event, when a ribosome stops translation, slides along mRNA with peptidyl-tRNA remaining in the P site, and then resumes translation (see Figure 9.30). This is a rather rare phenomenon, with only ~3 authenticated examples. The most dramatic example of bypassing is in gene *60* of phage T4, where the ribosome moves sixty nucleotides along the mRNA.

The key to the bypass system is that there are identical (or synonymous) codons at either end of the sequence that is skipped. They are sometimes referred to as the "take-off" and "landing" sites. Before bypass, the ribosome is positioned with a peptidyl-tRNA paired with the take-off codon in the P site, with an empty A site waiting for an aminoacyl-tRNA to enter. FIGURE 9.32 shows that the ribosome slides along mRNA in this condition until the peptidyl-tRNA can become paired with the codon in the landing site. A remarkable feature of the system is its high efficiency, ~50%.

The sequence of the mRNA triggers the bypass. The important features are the two GGA codons for take-off and landing, the spacing between them, a stem-loop structure that includes the take-off codon, and the stop codon adjacent to the take-off codon. The protein under synthesis is also involved.

The take-off stage requires the peptidyl-tRNA to unpair from its codon. This is followed by a movement of the mRNA that prevents it from re-pairing. Then the ribosome scans the mRNA until the peptidyl-tRNA can repair with the codon in the landing reaction. This is followed by the resumption of protein synthesis when aminoacyl-tRNA enters the A site in the usual way.

Like frameshifting, the bypass reaction depends on a pause by the ribosome. The probability that peptidyl-tRNA will dissociate from its codon in the P site is increased by delays in the entry of aminoacyl-tRNA into the A site. Starvation for an amino acid can trigger bypassing in bacterial genes because of the delay that occurs when there is no aminoacyl-tRNA available to enter the A site. In phage T4 gene *60*, one role of mRNA structure may be to reduce the efficiency of termination, thus creating the delay that is needed for the take-off reaction.

9.19 Summary

The sequence of mRNA read in triplets 5′ → 3′ is related by the genetic code to the amino acid sequence of protein read from N- to C-terminus. Of the sixty-four triplets, sixty-one code for amino acids and three provide termination signals. Synonym codons that represent the same amino acids are related, often by a change in the third base of the codon. This third-base degeneracy, coupled with a pattern in which related amino acids tend to be coded by related codons, minimizes the effects of mutations. The genetic code is universal and must have been established very early in evolution. Changes in nuclear genomes are rare, but some changes have occurred during mitochondrial evolution.

Multiple tRNAs may respond to a particular codon. The set of tRNAs responding to the various codons for each amino acid is distinctive for each organism. Codon–anticodon recognition involves wobbling at the first position of the anticodon (third position of the codon), which allows some tRNAs to recognize multiple codons. All tRNAs have modified bases, introduced by enzymes that recognize target bases in the tRNA structure. Codon–anticodon pairing is influenced by modifications of the anticodon itself and also by the context of adjacent bases, especially on the 3′ side of the anticodon. Taking advantage of codon–anticodon wobble allows vertebrate mitochondria to use only twenty-two tRNAs to recognize all codons, compared with the usual minimum of thirty-one tRNAs; this is assisted by the changes in the mitochondrial code.

Each amino acid is recognized by a particular aminoacyl-tRNA synthetase, which also recognizes all of the tRNAs coding for that amino acid. Aminoacyl-tRNA synthetases have a proofreading function that scrutinizes the aminoacyl-tRNA products and hydrolyzes incorrectly joined aminoacyl-tRNAs.

Aminoacyl-tRNA synthetases vary widely, but fall into two general groups according to the structure of the catalytic domain. Synthetases of each group bind the tRNA from the side, making contacts principally with the extremities of the acceptor stem and the anticodon stem-loop; the two types of synthetases bind tRNA from opposite sides. The relative importance attached to the acceptor stem and the anticodon region for specific recognition varies with the individual tRNA.

Mutations may allow a tRNA to read different codons; the most common form of such mutations occurs in the anticodon itself. Alteration of its specificity may allow a tRNA to suppress a mutation in a gene coding for protein. A tRNA that recognizes a termination codon provides a nonsense suppressor; one that changes the amino acid responding to a codon is a missense suppressor. Suppressors of UAG and UGA codons are more efficient than those of UAA codons, which is explained by the fact that UAA is the most commonly used natural termination codon. The efficiency of all suppressors, however, depends on the context of the individual target codon.

Frameshifts of the +1 type may be caused by aberrant tRNAs that read "codons" of four bases. Frameshifts of either +1 or –1 may be caused by slippery sequences in mRNA that allow a peptidyl-tRNA to slip from its codon to an overlapping sequence that can also pair with its anticodon. This frameshifting also requires another sequence that causes the ribosome to delay. Frameshifts determined by the mRNA sequence may be required for expression of natural genes. Bypassing occurs when a ribosome stops translation and moves along mRNA with its peptidyl-tRNA in the P site until the peptidyl-tRNA pairs with an appropriate codon; then translation resumes.

References

9.1 Introduction

Research

Nirenberg, M. W. and Leder, P. (1964). The effect of trinucleotides upon the binding of sRNA to ribosomes. *Science* 145, 1399–1407.

Nirenberg, M. W. and Matthaei, H. J. (1961). The dependence of cell-free protein synthesis in *E. coli* upon naturally occurring or synthetic polyribonucleotides. *Proc. Natl. Acad. Sci. USA* 47, 1588–1602.

9.3 Codon-Anticodon Recognition Involves Wobbling

Research

Crick, F. H. C. (1966). Codon-anticodon pairing: the wobble hypothesis. *J. Mol. Biol.* 19, 548–555.

9.4 tRNAs Are Processed from Longer Precursors

Review

Hopper, A. K. and Phizicky, E. M. (2003). tRNA transfers to the limelight. *Genes Dev.* 17, 162–180.

9.5 tRNA Contains Modified Bases

Reviews

Hopper, A. K. and Phizicky, E. M. (2003). tRNA transfers to the limelight. *Genes Dev.* 17, 162–180.

9.6 Modified Bases Affect Anticodon-Codon Pairing

Review

Bjork, G. R. (1987). Transfer RNA modification. *Annu. Rev. Biochem.* 56, 263–287.

9.7 There Are Sporadic Alterations of the Universal Code

Reviews

Fox, T. D. (1987). Natural variation in the genetic code. *Annu. Rev. Genet.* 21, 67–91.

Osawa, S. et al. (1992). Recent evidence for evolution of the genetic code. *Microbiol. Rev.* 56, 229–264.

9.8 Novel Amino Acids Can Be Inserted at Certain Stop Codons

Reviews

Bock, A. (1991). Selenoprotein synthesis: an expansion of the genetic code. *Trends Biochem. Sci.* 16, 463–467.

Ibba, M. and Söll, D. (2004). Aminoacyl-tRNAs: setting the limits of the genetic code. *Genes Dev.* 18, 731–738.

Research

Fagegaltier, D., Hubert, N., Yamada, K., Mizutani, T., Carbon, P., and Krol, A. (2000). Characterization of mSelB, a novel mammalian elongation factor for selenoprotein translation. *EMBO J.* 19, 4796–4805.

Hao, B., Gong, W., Ferguson, T. K., James, C. M., Krzycki, J. A., and Chan, M. K. (2002). A new UAG-encoded residue in the structure of a methanogen methyltransferase. *Science* 296, 1462–1466.

Srinivasan, G., James, C. M., and Krzycki, J. A. (2002). Pyrrolysine encoded by UAG in Archaea: charging of a UAG-decoding specialized tRNA. *Science* 296, 1459–1462.

9.9 tRNAs Are Charged with Amino Acids by Synthetases

Review

Schimmel, P. (1989). Parameters for the molecular recognition of tRNAs. *Biochemistry* 28, 2747–2759.

9.10 Aminoacyl-tRNA Synthetases Fall into Two Groups

Review

Schimmel, P. (1987). Aminoacyl-tRNA synthetases: general scheme of structure-function relationships on the polypeptides and recognition of tRNAs. *Annu. Rev. Biochem.* 56, 125–158.

Research

Rould, M. A. et al. (1989). Structure of *E. coli* glutaminyl-tRNA synthetase complexed with $tRNA^{Gln}$ and ATP at 28Å resolution. *Science* 246, 1135–1142.

Ruff, M. et al. (1991). Class II aminoacyl tRNA synthetases: crystal structure of yeast aspartyl-tRNA synthetase complexes with $tRNA^{Asp}$. *Science* 252, 1682–1689.

9.11 Synthetases Use Proofreading to Improve Accuracy

Review

Jakubowski, H. and Goldman, E. (1992). Editing of errors in selection of amino acids for protein synthesis. *Microbiol. Rev.* 56, 412–429.

Research

Dock-Bregeon, A., Sankaranarayanan, R., Romby, P., Caillet, J., Springer, M., Rees, B., Francklyn, C. S., Ehresmann, C., and Moras, D. (2000). Transfer RNA-mediated editing in threonyl-tRNA synthetase. The class II solution to the double discrimination problem. *Cell* 103, 877–884.

Hopfield, J. J. (1974). Kinetic proofreading: a new mechanism for reducing errors in biosynthetic processes requiring high specificity. *Proc. Natl. Acad. Sci. USA* 71, 4135–4139.

Jakubowski, H. (1990). Proofreading in vivo: editing of homocysteine by methionyl-tRNA synthetase in *E. coli*. *Proc. Natl. Acad. Sci. USA* 87, 4504–4508.

Nomanbhoy, T. K., Hendrickson, T. L., and Schimmel, P. (1999). Transfer RNA-dependent translocation of misactivated amino acids to prevent errors in protein synthesis. *Mol. Cell* 4, 519–528.

Nureki, O. et al. (1998). Enzyme structure with two catalytic sites for double sieve selection of substrate. *Science* 280, 578–581.

Silvian, L. F., Wang, J., and Steitz, T. A. (1999). Insights into editing from an Ile-tRNA synthetase structure with $tRNA^{Ile}$ and mupirocin. *Science* 285, 1074–1077.

9.14 Suppressors May Compete with Wild-type Reading of the Code

Reviews

Atkins, J. F. (1991). Towards a genetic dissection of the basis of triplet decoding, and its natural subversion: programmed reading frameshifts and hops. *Annu. Rev. Genet.* 25, 201–228.

Beier, H. and Grimm, M. (2001). Misreading of termination codons in eukaryotes by natural nonsense suppressor tRNAs. *Nucleic Acids Res.* 29, 4767–4782.

Eggertsson, G. and Soll, D. (1988). Transfer RNA-mediated suppression of termination codons in *E. coli*. *Microbiol. Rev.* 52, 354–374.

Murgola, E. J. (1985). tRNA, suppression, and the code. *Annu. Rev. Genet.* 19, 57–80.

Normanly, J. and Abelson, J. (1989). Transfer RNA identity. *Annu. Rev. Biochem.* 58, 1029–1049.

Research

Hirsh, D. (1971). Tryptophan transfer RNA as the UGA suppressor. *J. Mol. Biol.* 58, 439–458.

Weiner, A. M. and Weber, K. (1973). A single UGA codon functions as a natural termination signal in the coliphage q beta coat protein cistron. *J. Mol. Biol.* 80, 837–855.

9.15 The Ribosome Influences the Accuracy of Translation

Reviews

Kurland, C. G. (1992). Translational accuracy and the fitness of bacteria. *Annu. Rev. Genet.* 26, 29–50.

Ogle, J. M., and Ramakrishnan, V. (2005). Structural insights into translational fidelity. *Annu. Rev. Biochem.* 74, 129–177.

Ramakrishnan, V. (2002). Ribosome structure and the mechanism of translation. *Cell* 108, 557–572.

Research

Carter, A. P., Clemons, W. M., Brodersen, D. E., Morgan-Warren, R. J., Wimberly, B. T., and Ramakrishnan, V. (2000). Functional insights from the structure of the 30S ribosomal subunit and its interactions with antibiotics. *Nature* 407, 340–348.

Ogle, J. M., Brodersen, D. E., Clemons, W. M., Tarry, M. J., Carter, A. P., and Ramakrishnan, V. (2001). Recognition of cognate transfer RNA by the 30S ribosomal subunit. *Science* 292, 897–902.

9.17 Frameshifting Occurs at Slippery Sequences

Reviews

Farabaugh, P. J. (1995). Programmed translational frameshifting. *Microbiol. Rev.* 60, 103–134.

Farabaugh, P. J. and Bjorkk, G. R. (1999). How translational accuracy influences reading frame maintenance. *EMBO J.* 18, 1427–1434.

Gesteland, R. F. and Atkins, J. F. (1996). Recoding: dynamic reprogramming of translation. *Annu. Rev. Biochem.* 65, 741–768.

Research

Jacks, T., Power, M. D., Masiarz, F. R., Luciw, P. A., Barr, P. J., and Varmus, H. E. (1988). Characterization of ribosomal frameshifting in HIV-1 gag-pol expression. *Nature* 331, 280–283.

9.18 Bypassing Involves Ribosome Movement

Review

Herr, A. J., Atkins, J. F., and Gesteland, R. F. (2000). Coupling of open reading frames by translational bypassing. *Annu. Rev. Biochem.* 69, 343–372.

Research

Gallant, J. A. and Lindsley, D. (1998). Ribosomes can slide over and beyond "hungry" codons, resuming protein chain elongation many nucleotides downstream. *Proc. Natl. Acad. Sci. USA* 95, 13771–13776.

Huang, W. M., Ao, S. Z., Casjens, S., Orlandi, R., Zeikus, R., Weiss, R., Winge, D., and Fang, M. (1988). A persistent untranslated sequence within bacteriophage T4 DNA topoisomerase gene 60. *Science* 239, 1005–1012.

10

Protein Localization

CHAPTER OUTLINE

10.12 Reverse Translocation Sends Proteins to the Cytosol for Degradation

- Sec61 translocons can be used for reverse translocation of proteins from the ER into the cytosol.

10.13 Proteins Reside in Membranes by Means of Hydrophobic Regions

- Group I proteins have the N-terminus on the far side of the membrane; group II proteins have the opposite orientation.
- Some proteins have multiple membrane-spanning domains.

10.14 Anchor Sequences Determine Protein Orientation

- An anchor sequence halts the passage of a protein through the translocon. Typically this is located at the C-terminal end and results in a group I orientation in which the N-terminus has passed through the membrane.
- A combined signal-anchor sequence can be used to insert a protein into the membrane and anchor the site of insertion. Typically this is internal and results in a group II orientation in which the N-terminus is cytosolic.

10.15 How Do Proteins Insert into Membranes?

- Transfer of transmembrane domains from the translocon into the lipid bilayer is triggered by the interaction of the transmembrane region with the translocon.

10.16 Posttranslational Membrane Insertion Depends on Leader Sequences

- N-terminal leader sequences provide the information that allows proteins to associate with mitochondrial or chloroplast membranes.

10.17 A Hierarchy of Sequences Determines Location within Organelles

- The N-terminal part of a leader sequence targets a protein to the mitochondrial matrix or chloroplast lumen.
- An adjacent sequence can control further targeting to a membrane or the intermembrane spaces.
- The sequences are cleaved successively from the protein.

10.18 Inner and Outer Mitochondrial Membranes Have Different Translocons

- Transport through the outer and inner mitochondrial membranes uses different receptor complexes.
- The TOM (outer membrane) complex is a large complex in which substrate proteins are directed to the Tom40 channel by one of two subcomplexes.
- Different TIM (inner membrane) complexes are used depending on whether the substrate protein is targeted to the inner membrane or to the lumen.
- Proteins pass directly from the TOM to the TIM complex.

10.19 Peroxisomes Employ Another Type of Translocation System

- Proteins are imported into peroxisomes in their fully folded state.
- They have either a PTS1 sequence at the C-terminus or a PTS2 sequence at the N-terminus.
- The receptor Pex5p binds the PTS1 sequence and the receptor Pex7p binds the PTS2 sequence.
- The receptors are cytosolic proteins that shuttle into the peroxisome carrying a substrate protein and then return to the cytosol.

10.20 Bacteria Use Both Cotranslational and Posttranslational Translocation

- Bacterial proteins that are exported to or through membranes use both posttranslational and cotranslational mechanisms.

10.21 The Sec System Transports Proteins into and Through the Inner Membrane

- The bacterial SecYEG translocon in the inner membrane is related to the eukaryotic Sec61 translocon.
- Various chaperones are involved in directing secreted proteins to the translocon.

10.22 Sec-Independent Translocation Systems in *E. coli*

- *E. coli* and organelles have related systems for protein translocation.
- One system allows certain proteins to insert into membranes without a translocation apparatus.
- YidC is homologous to a mitochondrial system for transferring proteins into the inner membrane.
- The tat system transfers proteins with a twin arginine motif into the periplasmic space.

10.23 Summary

10.1 Introduction

Proteins are synthesized in two types of location:

- The vast majority of proteins are synthesized by ribosomes in the cytosol.
- A small minority are synthesized by ribosomes within organelles (mitochondria or chloroplasts).

Proteins synthesized in the cytosol can be divided into two general classes with regard to localization: those that are not associated with membranes, and those that are associated with membranes. FIGURE 10.1 maps the cell in terms of the possible ultimate destinations for a newly synthesized protein and the systems that transport it:

- Cytosolic (or "soluble") proteins are not localized in any particular organelle. They are synthesized in the cytosol, and remain there, where they function as individual catalytic centers, acting on metabolites that are in solution in the cytosol.
- Macromolecular structures may be located at particular sites in the cytoplasm; for example, centrioles are associated with the regions that become the poles of the mitotic spindle.
- Nuclear proteins must be transported from their site of synthesis in the cytosol through the nuclear envelope into the nucleus.
- Most of the proteins in cytoplasmic organelles are synthesized in the cytosol and transported specifically to (and through) the organelle membrane, for example, to the mitochondrion or peroxisome or (in plant cells) to the chloroplast. (Those proteins that are synthesized within the organelle remain within it.)
- The cytoplasm contains a series of membranous bodies, including endoplasmic reticulum (ER), Golgi apparatus, endosomes, and lysosomes. This is sometimes referred to as the "reticuloendothelial system." Proteins that reside within these compartments are inserted into ER membranes and then are directed to their particular locations by the transport system of the Golgi apparatus.
- Proteins that are secreted from the cell are transported to the plasma membrane and then must pass through it to the exterior. They start their synthesis in the same way as proteins associated with the reticuloendothelial system, but pass entirely through the system instead of halting at some particular point within it.

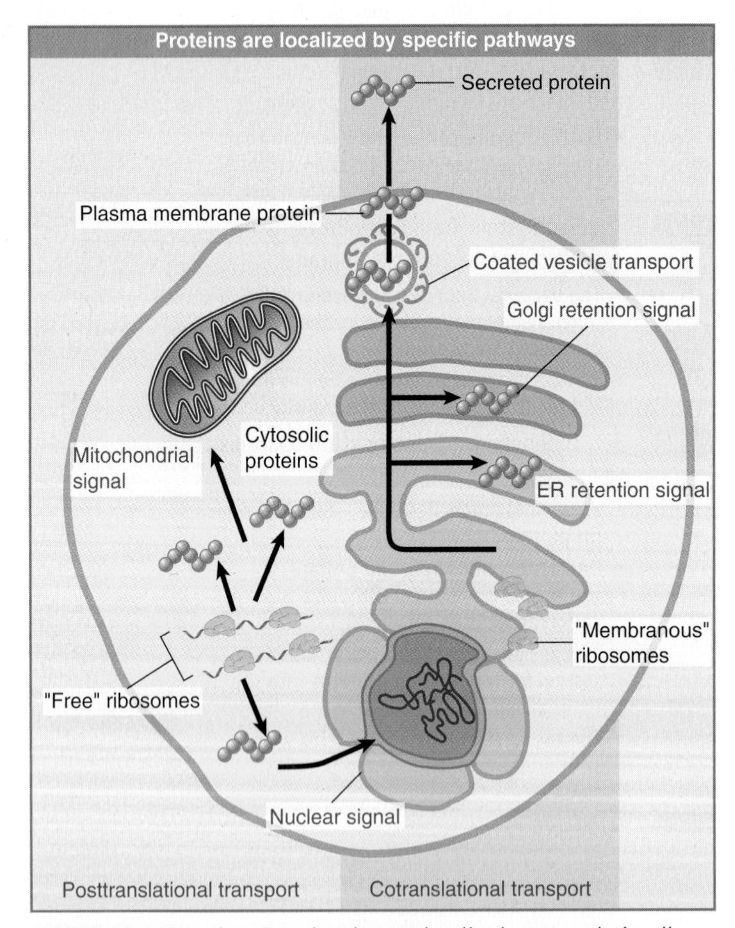

FIGURE 10.1 Overview: Proteins that are localized posttranslationally are released into the cytosol after synthesis on free ribosomes. Some have signals for targeting to organelles such as the nucleus or mitochondria. Proteins that are localized cotranslationally associate with the ER membrane during synthesis, so their ribosomes are "membrane bound." The proteins pass into the endoplasmic reticulum, along to the Golgi apparatus, and then through the plasma membrane, unless they have signals that cause retention at one of the steps on the pathway. They may also be directed to other organelles, such as endosomes or lysosomes.

10.2 Passage Across a Membrane Requires a Special Apparatus

Key concepts

- Proteins pass across membranes through specialized protein structures embedded in the membrane.
- Substrate proteins interact directly with the transport apparatus of the ER, mitochondria, or chloroplasts, but require carrier proteins to interact with peroxisomes.
- A much larger and complex apparatus is required for transport into the nucleus.

The process of inserting into or passing through a membrane is called protein **translocation.** The same dilemma must be solved for every situation in which a protein passes through a membrane. The protein presents a hydrophilic surface, but the membrane is hydrophobic. Like oil and water, the two would prefer not to mix. The solution is to create a special structure in the membrane through which the protein can pass. There are three different types of arrangements for such structures.

The endoplasmic reticulum, mitochondria, and chloroplasts contain proteinaceous structures embedded in their membranes that allow proteins to pass through without contacting the surrounding hydrophobic lipids. FIGURE 10.2 shows that a substrate protein binds directly to the structure, is transported by it to the other side, and then is released.

Peroxisomes also have such structures in their membranes, but the substrate proteins do not bind directly to them. FIGURE 10.3 shows that instead they bind to carrier proteins in the cytosol, the carrier protein is transported through the channel into the peroxisome, and then the substrate protein is released.

For transport into the nucleus, a much larger and more complex structure is employed. This is the nuclear pore. FIGURE 10.4 shows that, although the pore provides the environment that allows a substrate to enter (or to leave) the nucleus, it does not actually provide the apparatus that binds to the substrate proteins and moves them through. Included in this apparatus are carrier proteins that bind to the substrates and transport them through the pore to the other side.

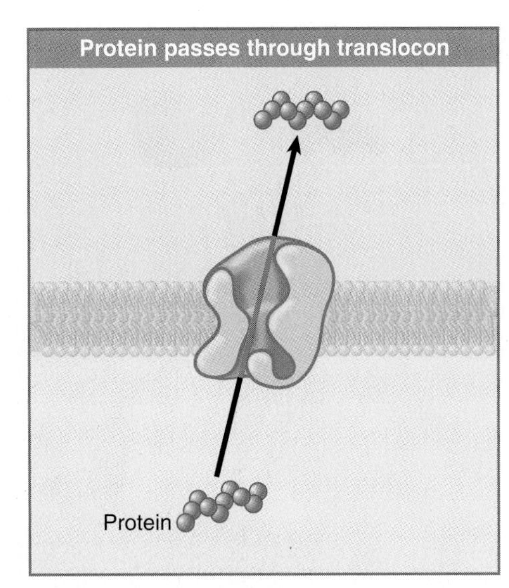

FIGURE 10.2 Proteins enter the ER or a mitochondrion by binding to a translocon that transports them across the membrane.

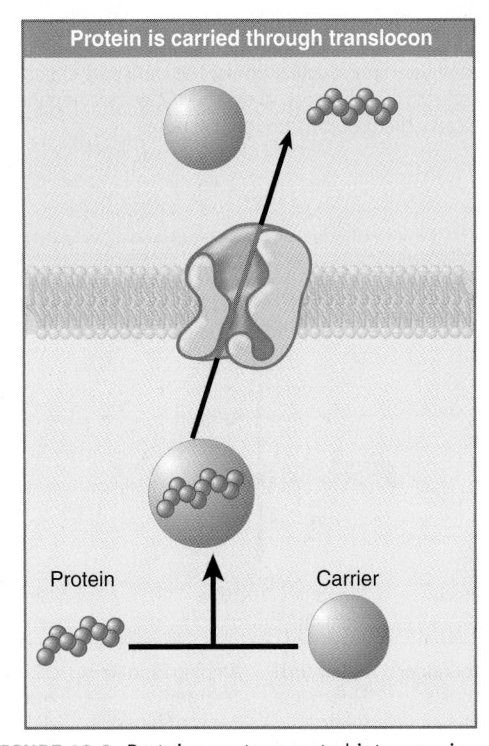

FIGURE 10.3 Proteins are transported into peroxisomes by a carrier protein that binds them in the cytosol, passes with them through the membrane channel, and releases them on the other side.

10.3 Protein Translocation May Be Posttranslational or Cotranslational

Key concepts

- Proteins that are imported into cytoplasmic organelles are synthesized on free ribosomes in the cytosol.
- Proteins that are imported into the ER-Golgi system are synthesized on ribosomes that are associated with the ER.
- Proteins associate with membranes by means of specific amino acid sequences called signal sequences.
- Signal sequences are most often leaders that are located at the N-terminus.
- N-terminal signal sequences are usually cleaved off the protein during the insertion process.

There are two ways for a protein to make its initial contact with a membrane:

- The nascent protein may associate with the translocation apparatus while it is still being synthesized on the ribosome. This is called **cotranslational translocation.**

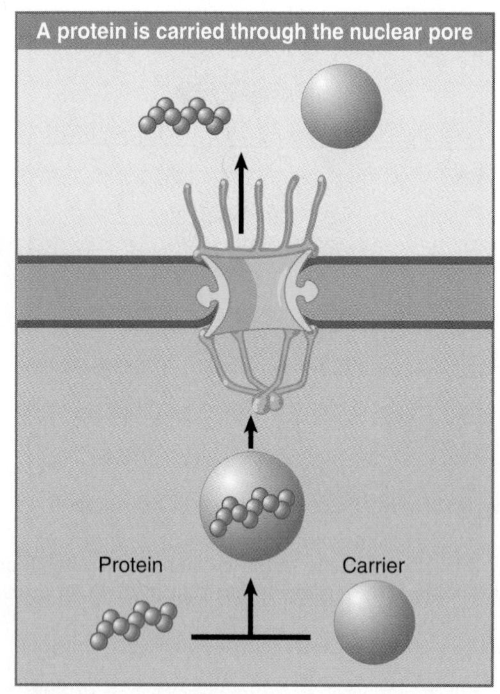

FIGURE 10.4 Proteins enter the nucleus by passage through very large nuclear pores. The transport apparatus is distinct from the pore itself and includes components that carry the protein through the pore.

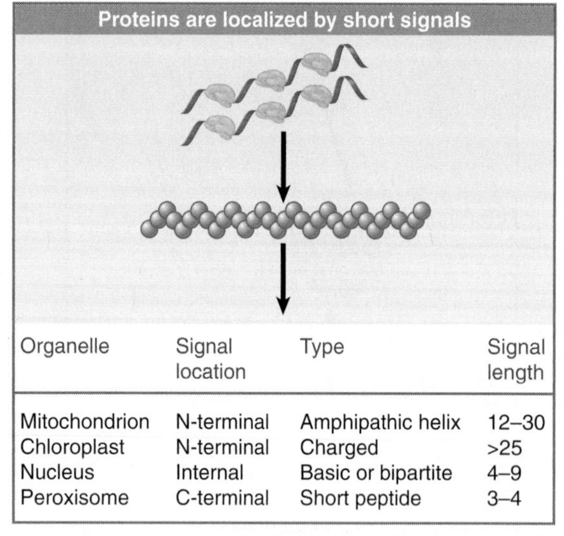

Organelle	Signal location	Type	Signal length
Mitochondrion	N-terminal	Amphipathic helix	12–30
Chloroplast	N-terminal	Charged	>25
Nucleus	Internal	Basic or bipartite	4–9
Peroxisome	C-terminal	Short peptide	3–4

FIGURE 10.5 Proteins synthesized on free ribosomes in the cytosol are directed after their release to specific destinations by short signal motifs.

- The protein may be released from a ribosome after translation has been completed. Then the completed protein diffuses to the appropriate membrane and associates with the translocation apparatus. This is called **posttranslational translocation.**

The location of a ribosome depends on whether the protein under synthesis is associating with a membrane cotranslationally:

- Cotranslational translocation is used for proteins that enter the endoplasmic reticulum (ER). The consequence of this association is that the ribosome is localized to the surface of the ER. The ribosomes are associated with the ER membranes during synthesis of these proteins, and therefore are found in membrane fractions of the cell; thus they are sometimes described as "membrane bound."
- All other ribosomes are located in the cytosol; because they are not associated with any organelle and fractionate separately from membranes, they are sometimes called "free ribosomes." The free ribosomes synthesize all proteins except those that are translocated cotranslationally. The proteins are released into the cytosol when their synthesis is completed. Some of these proteins remain free in the cytosol in quasi-soluble form; others associate with macromolecular cytosolic structures, such as filaments, microtubules, centrioles, etc., or are transported to the nucleus, or associate with membrane-bound organelles by posttranslational translocation.

To associate with a membrane (or any other type of structure), a protein requires an appropriate signal, typically a sequence motif that causes it to be recognized by a translocation system (or to be assembled into a macromolecular structure).

FIGURE 10.5 summarizes some signals used by proteins released from cytosolic ribosomes. Import into the nucleus results from the presence of a variety of rather short sequences within proteins. These "nuclear localization signals" enable the proteins to pass through nuclear pores. One type of signal that determines transport to the peroxisome is a very short C-terminal sequence. Mitochondrial and chloroplast proteins are synthesized on "free" ribosomes; after their release into the cytosol they associate with the organelle membranes by means of N-terminal sequences of ~25 amino acids in length that are recognized by receptors on the organelle envelope.

Proteins that reside within the reticuloendothelial system enter the ER while they are being synthesized. The principle of cotranslational translocation is summarized in

FIGURE 10.6. An important feature of this system is that the nascent protein is responsible for recognizing the translocation apparatus. This requires the signal for cotranslational translocation to be part of the protein that is first synthesized, and, in fact, it is usually located at the N-terminus.

A common feature is found in proteins that use N-terminal sequences to be transported cotranslationally to the ER or posttranslationally to mitochondria or chloroplasts. The N-terminal sequence comprises a **leader** that is not part of the mature protein. The protein carrying this leader is called a **preprotein** and is a transient precursor to the mature protein. The leader is cleaved from the protein during protein translocation.

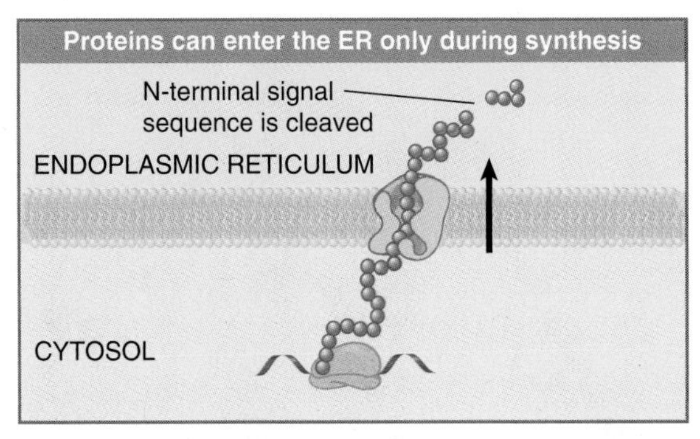

FIGURE 10.6 Proteins can enter the ER-Golgi pathway only by associating with the endoplasmic reticulum while they are being synthesized.

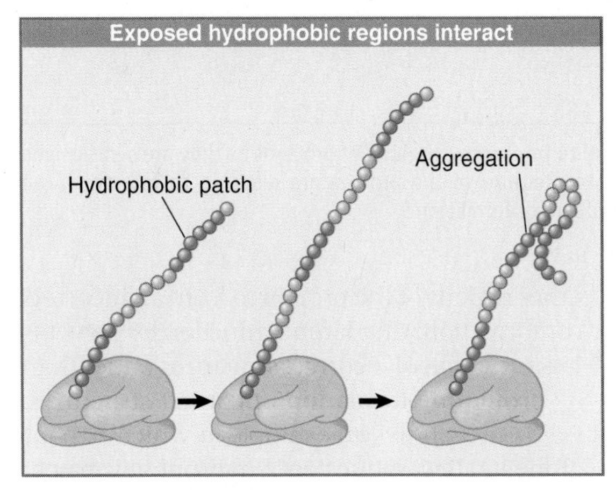

FIGURE 10.7 Hydrophobic regions of proteins are intrinsically interactive, and unless prevented will aggregate with one another when a protein is synthesized (or denatured).

10.4 Chaperones May Be Required for Protein Folding

Key concepts

- Proteins that can acquire their conformation spontaneously are said to self-assemble.
- Proteins can often assemble into alternative structures.
- A chaperone directs a protein into one particular pathway by excluding alternative pathways.
- Chaperones prevent the formation of incorrect structures by interacting with unfolded proteins to prevent them from folding incorrectly.

Some proteins are able to acquire their mature conformation spontaneously. A test for this ability is to denature the protein and determine whether it can then renature into the active form. This capacity is called **self-assembly.** A protein that can self-assemble can fold or refold into the active state from other conformations, including the condition in which it is initially synthesized. This implies that the internal interactions are intrinsically directed toward the right conformation. The classic case is that of ribonuclease; it was shown in the 1970s that, when the enzyme is denatured, it can renature *in vitro* into the correct conformation. More recently the process of intrinsic folding has been described in detail for some small proteins.

When correct folding does not happen, and alternative sets of interactions can occur, a protein may become trapped in a stable conformation that is not the intended final form. Proteins in this category cannot self-assemble. Their acquisition of proper structure requires the assistance of a **chaperone.**

Protein folding takes place by interactions between reactive surfaces. Typically these surfaces consist of exposed hydrophobic side chains. Their interactions form a hydrophobic core. The intrinsic reactivity of these surfaces means that incorrect interactions may occur unless the process is controlled. FIGURE 10.7 illustrates what would happen. As a newly synthesized protein emerges from the ribosome, any hydrophobic patch in the sequence is likely to aggregate with another hydrophobic patch. Such associations are likely to occur at random and therefore will probably not represent the desired conformation of the protein.

Chaperones are proteins that mediate correct assembly by causing a target protein to acquire one possible conformation instead of others. This is accomplished by binding to reactive surfaces in the target protein that are exposed during the assembly process and preventing those surfaces from interacting with

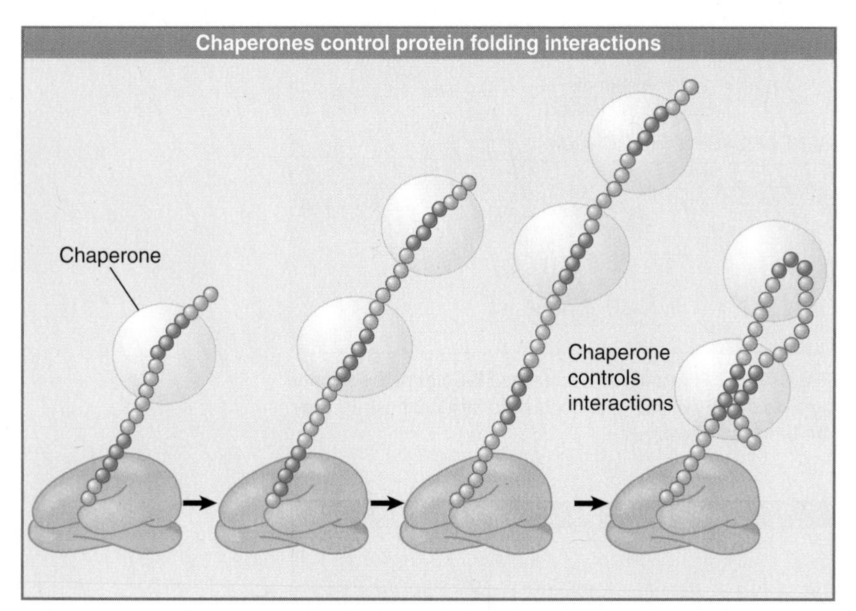

FIGURE 10.8 Chaperones bind to interactive regions of proteins as they are synthesized to prevent random aggregation. Regions of the protein are released to interact in an orderly manner to give the proper conformation.

other regions of the protein to form an incorrect conformation. Chaperones function by preventing formation of incorrect structures rather than by promoting formation of correct structures. FIGURE 10.8 shows an example in which a chaperone in effect sequesters a hydrophobic patch, thus allowing interactions to occur that would not have been possible in its presence, as can be seen by comparing the result with Figure 10.7.

An incorrect structure may be formed either by the misfolding of a single protein or by interactions with another protein. The density of proteins in the cytosol is high, and "macromolecular crowding" can increase the efficiencies of many reactions compared to the rates observed *in vitro*. Crowding can cause folding proteins to aggregate, but chaperones can counteract this effect. Thus one role of chaperones may be to protect a protein so that it can fold without being adversely affected by the crowded conditions in the cytosol.

We do not know what proportion of proteins can self-assemble, as opposed to those that require assistance from a chaperone. (It is not axiomatic that a protein capable of self-assembly *in vitro* actually self-assembles *in vivo*, because there may be rate differences in the two conditions, and chaperones still could be involved *in vivo*. There is, however, a distinction to be drawn between proteins that can in principle self-assemble and those that in principle must have a chaperone to assist in acquisition of the correct structure.)

10.5 Chaperones Are Needed by Newly Synthesized and by Denatured Proteins

Key concepts

- Chaperones act on newly synthesized proteins, proteins that are passing through membranes, or proteins that have been denatured.
- Hsp (heat shock protein) 70 and some associated proteins form a major class of chaperones that act on many target proteins.
- Group I and group II chaperonins are large oligomeric assemblies that act on target proteins they sequester in internal cavities.
- Hsp90 is a specialized chaperone that acts on proteins of signal transduction pathways.

The ability of chaperones to recognize incorrect protein conformations allows them to play two related roles concerned with protein structure:

- When a protein is initially synthesized—that is to say, when it exits the ribosome to enter the cytosol—it appears in an unfolded form. Spontaneous folding then occurs as the emerging sequence interacts with regions of the protein that were synthesized previously. Chaperones influence the folding process by controlling the accessibility of the reactive surfaces. This process is involved in initial acquisition of the correct conformation.
- When a protein is denatured, new regions are exposed and become able to interact. These interactions are similar to those that occur when a protein (transiently) misfolds as it is initially synthesized. They are recognized by chaperones as comprising incorrect folds. This process is involved in recognizing a protein that has been denatured and either assisting renaturation or leading to its removal by degradation.

Chaperones may also be required to assist the formation of oligomeric structures and for the transport of proteins through membranes. A persistent theme in membrane passage is that control (or delay) of protein folding is an important feature. FIGURE 10.9 shows that it may be necessary to maintain a protein in an unfolded state before it enters the membrane because of the geometry of passage: the mature protein could simply be too large to fit into the available channel. Chaperones may prevent a protein

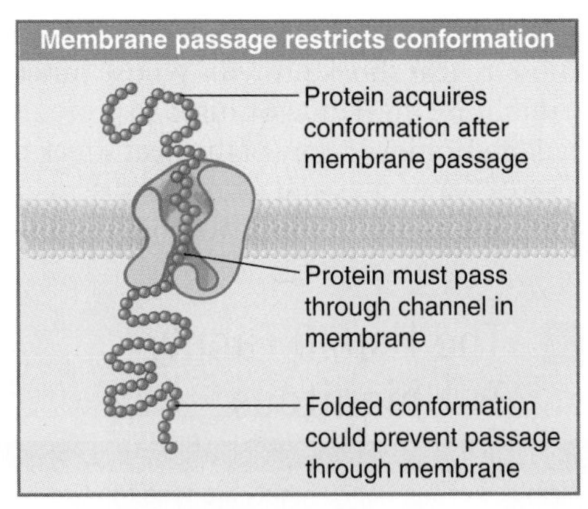

FIGURE 10.9 A protein is constrained to a narrow passage as it crosses a membrane.

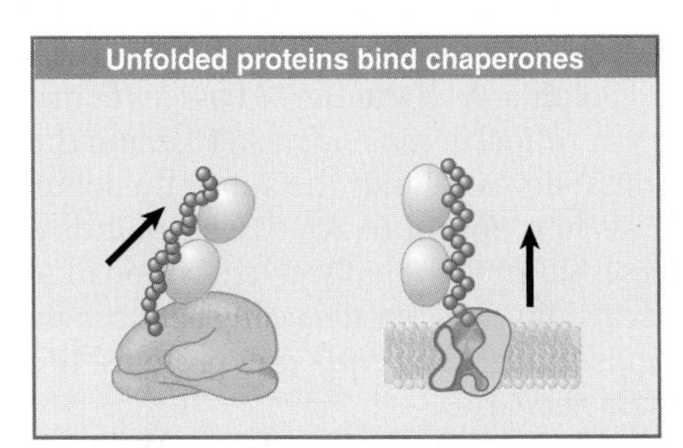

FIGURE 10.10 Proteins emerge from the ribosome or from passage through a membrane in an unfolded state that attracts chaperones to bind and protect them from misfolding.

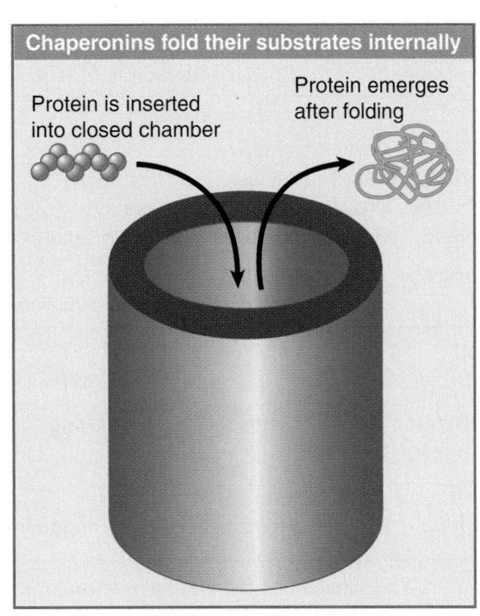

FIGURE 10.11 A chaperonin forms a large oligomeric complex and folds a substrate protein within its interior.

from acquiring a conformation that would prevent passage through the membrane; in this capacity, their role is basically to maintain the protein in an unfolded, flexible state. Once the protein has passed through the membrane, it may require another chaperone to assist with folding to its mature conformation in much the same way that a cytosolic protein requires assistance from a chaperone as it emerges from the ribosome. The state of the protein as it emerges from a membrane is probably similar to that as it emerges from the ribosome—basically extended in a more or less linear condition.

Two major types of chaperones have been well characterized. They affect folding through two different types of mechanism:

- FIGURE 10.10 shows that the *Hsp70 system consists of individual proteins that bind to, and act on, the substrates whose folding is to be controlled.* It recognizes proteins as they are synthesized or emerge from membranes (and also when they are denatured by stress). Basically, it controls the interactions between exposed reactive regions of the protein, enabling it to fold into the correct conformation *in situ*. The components of the system are Hsp70, Hsp40, and GrpE. The name of the system reflects the original identification of Hsp70 as a protein induced by heat shock. The Hsp70 and Hsp40 proteins bind individually to the substrate proteins. They use hydrolysis of ATP to provide the energy for changing the structure of the substrate protein and work in conjunction with an exchange factor that regenerates ATP from ADP.
- FIGURE 10.11 shows that *a chaperonin system consists of a large oligomeric assembly (represented as a cylinder).* This assembly forms a structure into which unfolded proteins are inserted. The protected environment directs their folding. There are two types of chaperonin system: GroEL/GroES is found in all classes of organism and TRiC is found in eukaryotic cytosol.

The components of the systems are summarized in FIGURE 10.12. The Hsp70 system and the two chaperonin systems all act on many different substrate proteins. Another system, the Hsp90 protein, functions in conjunction

There are 2 major types of chaperone systems	
System	Structure/function
Individual chaperones	
Hsp70 system	
Hsp70 (DnaK)	ATPase
Hsp40 (DnaJ)	Stimulates ATPase
GrpE (GrpE)	Nucleotide exchange factor
Hsp90	Functions on proteins involved in signal transduction
Oligomeric structures (chaperonins)	
Group I	
Hsp60 (GroEL)	Forms two heptameric rings;
Hsp10 (GroES)	Forms cap
Group II	
TRiC	Forms two octameric rings

FIGURE 10.12 Chaperone families have eukaryotic and bacterial counterparts (named in parentheses).

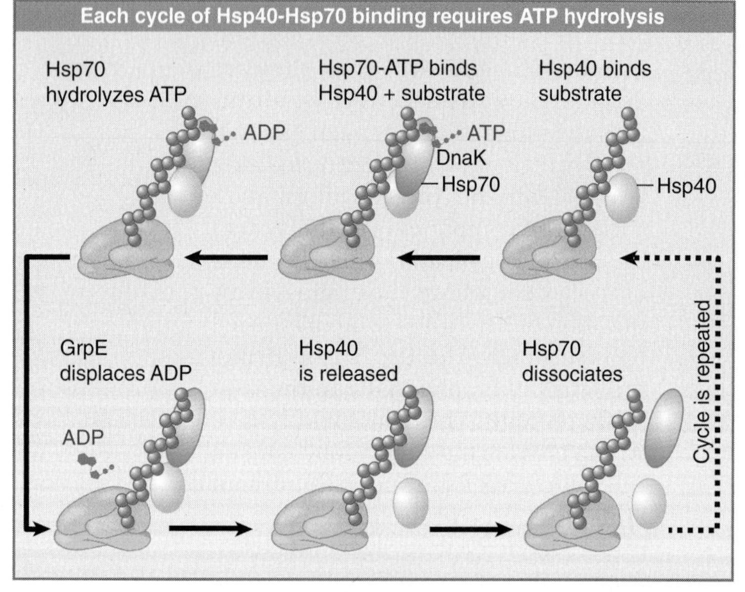

FIGURE 10.13 Hsp40 binds the substrate and then Hsp70. ATP hydrolysis drives conformational change. GrpE displaces the ADP; this causes the chaperones to be released. Multiple cycles of association and dissociation may occur during the folding of a substrate protein.

with Hsp70, but is directed against specific classes of proteins that are involved in signal transduction, especially the steroid hormone receptors and signaling kinases. The Hsp90 system's basic function is to maintain its targets in an appropriate conformation until they are stabilized by interacting with other components of the pathway. (The reason many of these proteins are named "Hsp," which stands for "heat shock protein" is that increase in temperature causes production of heat shock proteins whose function is to minimize the damage caused to proteins by heat denaturation. Many of the heat shock proteins are chaperones and were first discovered, and named, as part of the heat shock response.)

10.6 The Hsp70 Family Is Ubiquitous

Key concepts

- Members of the Hsp70 family are found in the cytosol, in the ER, and in mitochondria and chloroplasts.
- Hsp70 is a chaperone that functions on target proteins in conjunction with DnaJ and GrpE.

The Hsp70 family is found in bacteria, eukaryotic cytosol, in the ER, and in chloroplasts and mitochondria. A typical Hsp70 has two domains: the N-terminal domain is an ATPase and the C-terminal domain binds the substrate polypeptide. When bound to ATP, Hsp70 binds and releases substrates rapidly; when bound to ADP, the reactions are slow. Recycling between these states is regulated by two other proteins, Hsp40 (DnaJ) and GrpE.

FIGURE 10.13 shows that Hsp40 (DnaJ) binds first to a nascent protein as it emerges from the ribosome. Hsp40 contains a region called the J domain (named for DnaJ), which interacts with Hsp70. Hsp70 (DnaK) binds to both Hsp40 and to the unfolded protein. In effect, two interacting chaperones bind to the protein. The J domain accounts for the specificity of the pairwise interaction and drives a particular Hsp40 to select the appropriate partner from the Hsp70 family.

The interaction of Hsp70 (DnaK) with Hsp40 (DnaJ) stimulates the ATPase activity of Hsp70. The ADP-bound form of the complex remains associated with the protein substrate until GrpE displaces the ADP. This causes loss of Hsp40 followed by dissociation of Hsp70. The Hsp70 binds another ATP and the cycle can be repeated. GrpE (or its equivalent) is found only in bacteria, mitochondria, and chloroplasts; in other locations, the dissociation reaction is coupled to ATP hydrolysis in a more complex way.

Protein folding is accomplished by multiple cycles of association and dissociation. As the protein chain lengthens, Hsp70 (DnaK) may dissociate from one binding site and then reassociate with another, thus releasing parts of the substrate protein to fold correctly in an ordered manner. Finally, the intact protein is released from the ribosome folded into its mature conformation.

Different members of the Hsp70 class function on various types of target proteins. Cytosolic proteins (the eponymous Hsp70 and a related protein called Hsc70) act on nascent proteins on ribosomes. Variants in the ER (called BiP or Grp78 in higher eukaryotes and Kar2 in *Saccharomyces cerevisiae*), or in mitochondria or chloroplasts, function in a rather similar manner on proteins as they emerge into the interior of the organelle on passing through the membrane.

What feature does Hsp70 recognize in a target protein? It binds to a linear stretch of amino acids embedded in a hydrophobic context. This is precisely the sort of motif that is buried in the hydrophobic core of a properly folded, mature protein. Its exposure therefore indicates that the protein is nascent or denatured. Motifs of this nature occur about every forty amino acids. Binding to the motif prevents it from misaggregating with another one.

This mode of action explains how the Hsp70 protein Bip can fulfill two functions: to assist in oligomerization and/or folding of newly translocated proteins in the ER, and to remove misfolded proteins. Suppose that BiP recognizes certain peptide sequences that are inaccessible within the conformation of a mature, properly folded protein. These sequences are exposed and attract BiP when the protein enters the ER lumen in an essentially one-dimensional form. If a protein is misfolded or denatured, it may become exposed on its surface instead of being properly buried.

10.7 Signal Sequences Initiate Translocation

Key concepts

- Proteins associate with the ER system only cotranslationally.
- The signal sequence of the substrate protein is responsible for membrane association.

Proteins that associate with membranes via N-terminal leaders use a hierarchy of signals to find their final destination. In the case of the reticuloendothelial system, the ultimate location of a protein depends on how it is directed as it transits the endoplasmic reticulum and Golgi apparatus. The leader sequence itself introduces the protein to the membrane; the intrinsic consequence of the interaction is for the protein to pass through the membrane into the compartment on the other side. For a protein to reside within the membrane, a further signal is required to stop passage through the membrane. Other types of signals are required for a protein to be sorted to a particular destination, that is, to remain within the membrane or lumen of some particular compartment. The general process of finding its ultimate destination by transport through successive membrane systems is called **protein sorting** or **targeting,** and is discussed in protein trafficking.

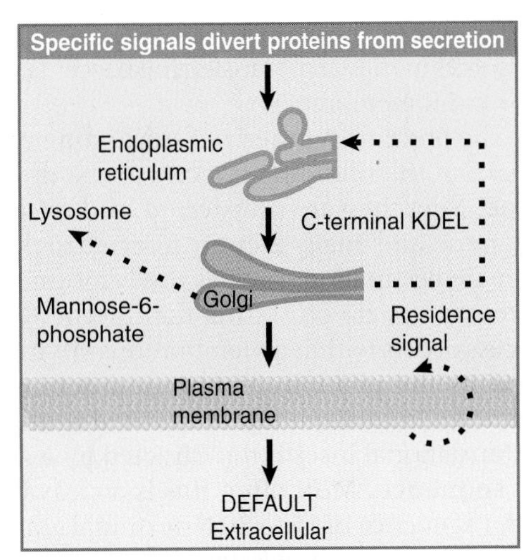

FIGURE 10.14 Proteins that enter the ER-Golgi pathway may flow through to the plasma membrane or may be diverted to other destinations by specific signals.

The overall nature of the pathway is summarized in FIGURE 10.14. The "default pathway" takes a protein through the ER, into the Golgi apparatus, and on to the plasma membrane. Proteins that reside in the ER possess a C-terminal tetrapeptide (KDEL, which actually provides a signal for them to return to the ER from the Golgi apparatus). The signal that diverts a protein to the lysosome is a covalent modification: the addition of a particular sugar residue. Other signals are required for a protein to become a permanent constituent of the Golgi apparatus or the plasma membrane. There is a common starting point for proteins that associate with, or pass through, the reticuloendothelial system of membranes. *These proteins can associate with the membrane only while they are being synthesized.* The ribosomes synthesizing these proteins become associated with the ER, enabling the nascent protein to be cotranslationally transferred to the membrane. Regions in which ribosomes are associated with the ER are sometimes called the "rough ER," in contrast with the "smooth ER" regions that lack associated polysomes and which have a tubular rather than sheetlike appearance. FIGURE 10.15 shows ribosomes in the act of transferring nascent proteins to ER membranes.

The proteins synthesized at the rough ER pass from the ribosome directly to the membrane. Next they are transferred to the Golgi apparatus, and finally they are directed to their ultimate destination, such as the lysosome or secretory vesicle or plasma membrane. The process occurs within a membranous environment as the proteins are carried between organelles in small membrane-coated vesicles. Cotranslational insertion is directed by a **signal sequence.** Most often this is a cleavable leader sequence of 15 to 30 N-terminal amino acids. At or close to the N-terminus are several polar residues, and within the leader is a hydrophobic core consisting exclusively or very largely of hydrophobic amino acids. There is no other conservation of sequence. FIGURE 10.16 gives an example.

The signal sequence is both necessary and sufficient to sponsor transfer of any attached polypeptide into the target membrane. A signal sequence added to the N-terminus of a globin protein, for example, causes it to be secreted through cellular membranes instead of remaining in the cytosol.

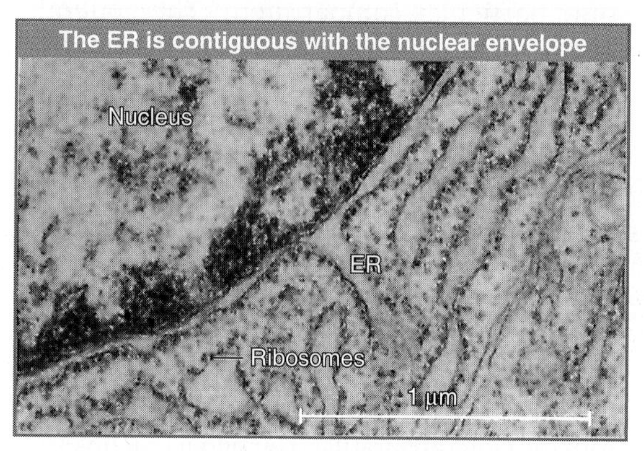

FIGURE 10.15 The endoplasmic reticulum consists of a highly folded sheet of membranes that extends from the nucleus. The small objects attached to the outer surface of the membranes are ribosomes. Photo courtesy of Lelio Orci, University of Geneva, Switzerland.

The signal sequence provides the connection that enables the ribosomes to attach to the membrane. There is no intrinsic difference between free ribosomes (synthesizing proteins in the cytosol) and ribosomes that are attached to the ER. A ribosome starts synthesis of a protein without knowing whether the protein will be synthesized in the cytosol or transferred to a membrane. It is the synthesis of a signal sequence that causes the ribosome to associate with a membrane.

10.8 The Signal Sequence Interacts with the SRP

Key concepts

- The signal sequence binds to the SRP (signal recognition particle).
- Signal-SRP binding causes protein synthesis to pause.
- Protein synthesis resumes when the SRP binds to the SRP receptor in the membrane.
- The signal sequence is cleaved from the translocating protein by the signal peptidase located on the "inside" face of the membrane.

Protein translocation can be divided into two general stages: first, ribosomes carrying nascent polypeptides associate with the membranes, and then the nascent chain is transferred to the channel and translocates through it.

The attachment of ribosomes to membranes requires the **signal recognition particle (SRP).** The SRP has two important abilities:

- It can bind to the signal sequence of a nascent secretory protein.
- It can bind to a protein (the SRP receptor) located in the membrane.

The SRP and SRP receptor function catalytically to transfer a ribosome carrying a nascent

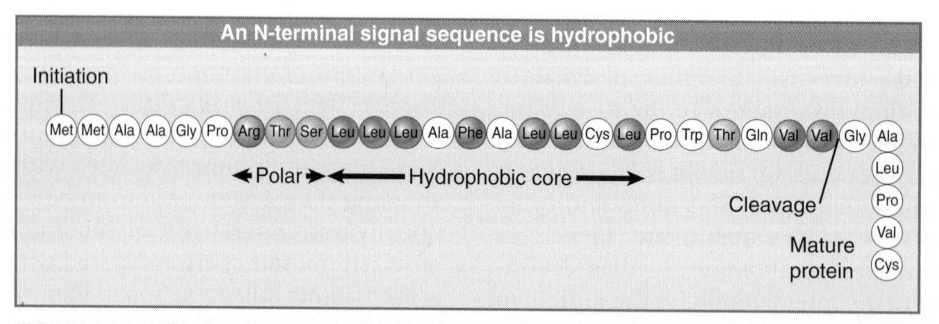

FIGURE 10.16 The signal sequence of bovine growth hormone consists of the N-terminal 29 amino acids and has a central highly hydrophobic region, preceded or flanked by regions containing polar amino acids.

protein to the membrane. The first step is the recognition of the signal sequence by the SRP. The SRP then binds to the SRP receptor and the ribosome binds to the membrane. The stages of translation of membrane proteins are summarized in FIGURE 10.17.

The role of the SRP receptor in protein translocation is transient. When the SRP binds to the signal sequence, it arrests translation. This usually happens when ~70 amino acids have been incorporated into the polypeptide chain (at this point the 25-residue leader has become exposed, with the next ~40 amino acids still buried in the ribosome).

When the SRP binds to the SRP receptor, the SRP releases the signal sequence. The ribosome becomes bound by another component of the membrane. At this point, translation can resume. When the ribosome has been passed on to the membrane, the combined role of SRP and SRP receptor has been played. They now recycle and are free to sponsor the association of another nascent polypeptide with the membrane.

This process may be needed to control the conformation of the protein. If the nascent protein were released into the cytoplasm, it could take up a conformation in which it might be unable to traverse the membrane. The ability of the SRP to inhibit translation while the ribosome is being handed over to the membrane is therefore important in preventing the protein from being released into the aqueous environment.

The signal peptide is cleaved from a translocating protein by a complex of five proteins called the **signal peptidase.** The complex is several times more abundant than the SRP and SRP receptor. Its amount is equivalent roughly to the amount of bound ribosomes, suggesting that it functions in a structural capacity. It is located on the lumenal face of the ER membrane, which implies that the entire signal sequence must cross the membrane before cleavage occurs. Homologous signal peptidases can be recognized in eubacteria, archaea, and eukaryotes.

10.9 The SRP Interacts with the SRP Receptor

Key concepts

- The SRP is a complex of 7S RNA with six proteins.
- The bacterial equivalent to the SRP is a complex of 4.5S RNA with two proteins.
- The SRP receptor is a dimer.
- GTP hydrolysis releases the SRP from the SRP receptor after their interaction.

The interaction between the SRP and the SRP receptor is the key event in eukaryotic translation in transferring a ribosome carrying a nascent protein to the membrane. An analogous interacting system exists in bacteria, although its role is more restricted.

The SRP is an 11S ribonucleoprotein complex, containing six proteins (total mass 240 kD) and a small (305 base, 100 kD) 7S RNA. FIGURE 10.18 shows that the 7S RNA provides the structural backbone of the particle; the individual proteins do not assemble in its absence.

The 7S RNA of the SRP particle is divided into two parts. The 100 bases at the 5′ end and the 45 bases at the 3′ end are closely related to

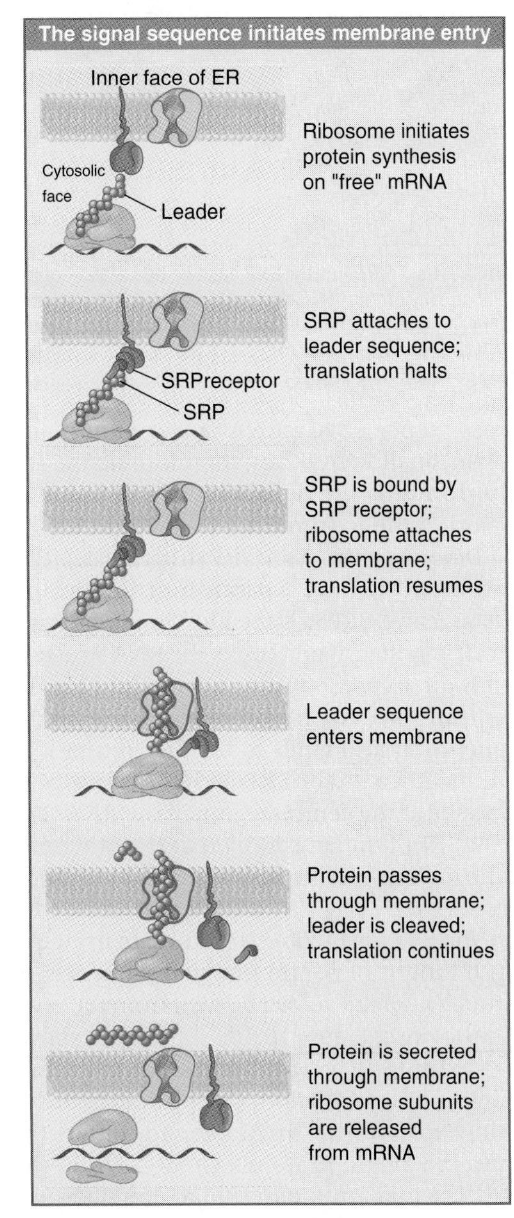

FIGURE 10.17 Ribosomes synthesizing secretory proteins are attached to the membrane via the signal sequence on the nascent polypeptide.

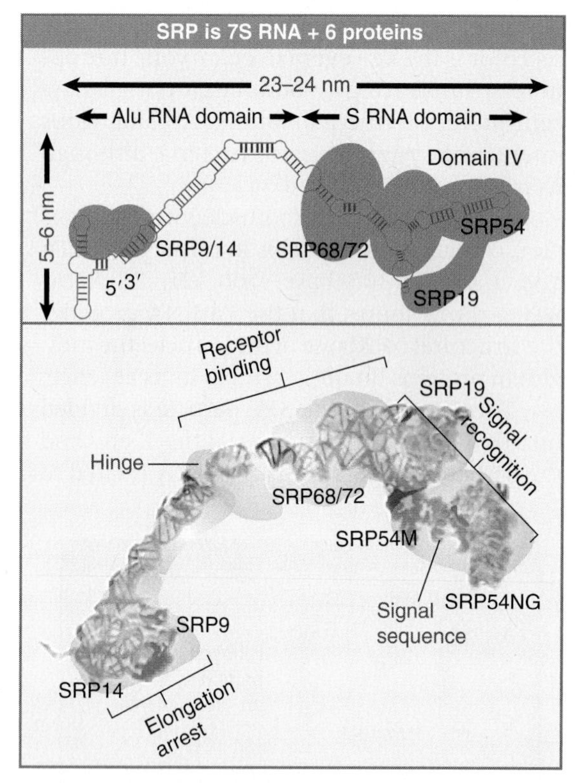

FIGURE 10.18 7S RNA of the SRP has two domains. Proteins bind as shown on the two-dimensional diagram above to form the crystal structure shown below. Each function of the SRP is associated with a discrete part of the structure.

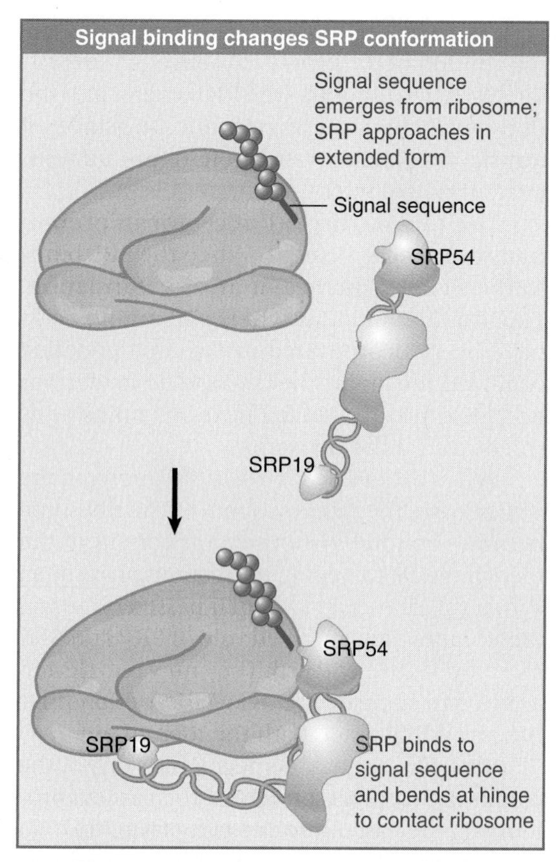

FIGURE 10.19 SRP binds to a signal sequence as it emerges from the ribosome. The binding causes the SRP to change conformation by bending at a "hinge," allowing SRP54 to contact the ribosome at the protein exit site while SRP19 makes a second set of contacts.

the sequence of Alu RNA, a common mammalian small RNA. They therefore define the **Alu domain.** The remaining part of the RNA comprises the **S domain.**

Different parts of the SRP structure depicted in Figure 10.18 have separate functions in protein targeting. SRP54 is the most important subunit. It is located at one end of the RNA structure and is directly responsible for recognizing the substrate protein by binding to the signal sequence. It also binds to the SRP receptor in conjunction with the SRP68-SRP72 dimer that is located at the central region of the RNA. The SRP9-SRP14 dimer is located at the other end of the molecule; it is responsible for elongation arrest.

The SRP is a flexible structure. In its unengaged form (not bound to signal sequence), it is quite extended, as can be seen from the crystal structure of Figure 10.18. FIGURE 10.19 shows that binding to a signal sequence triggers a change of conformation. The protein bends at a hinge to allow the SRP54 end to contact the ribosome at the protein exit site, while the SRP19 swings around to contact the ribosome at the elongation factor binding site. This enables it to cause the elongation arrest that gives time for targeting to the translocation site on the membrane.

The SRP receptor is a dimer containing subunits SRα (72 kD) and SRβ (30 kD). The β subunit is an integral membrane protein. The amino-terminal end of the large α subunit is anchored by the β subunit. The bulk of the α protein protrudes into the cytosol. A large part of the sequence of the cytoplasmic region of the protein resembles a nucleic acid-binding protein with many positive residues. This suggests the possibility that the SRP receptor recognizes the 7S RNA in the SRP.

There is a counterpart to SRP in bacteria, although it contains fewer components. *E. coli* contains a 4.5S RNA that associates with ribosomes and is homologous to the 7S RNA of the SRP. It associates with two proteins: Ffh is homologous to SRP54 and FtsY is homologous to the α subunit of the SRP receptor. In fact, FtsY replaces the functions of both the α and β SRP subunits; its N-terminal domain substitutes for SRPβ in membrane targeting, and the

C-terminal domain interacts with the target protein. The role of this complex is more limited than that of SRP–SRP receptor. It is probably required to keep some (but not all) secreted proteins in a conformation that enables them to interact with the secretory apparatus. This could be the original connection between protein synthesis and secretion; in eukaryotes the SRP has acquired the additional roles of causing translational arrest and targeting to the membrane.

Why should the SRP have an RNA component? The answer must lie in the evolution of the SRP: It must have originated very early in evolution, in an RNA-dominated world, presumably in conjunction with a ribosome whose functions were mostly carried out by RNA. The crystal structure of the complex between the protein-binding domain of 4.5S RNA and the RNA-binding domain of Ffh suggests that RNA continues to play a role in the function of SRP.

The 4.5S RNA has a region (domain IV) that is very similar to domain IV in 7S RNA (see Figure 10.18). Ffh consists of three domains (N, G, and M). The M domain (named for a high content of methionines) performs the key binding functions. It has a hydrophobic pocket that binds the signal sequence of a target protein. The hydrophobic side chains of the methionine residues create the pocket by projecting into a cleft in the protein structure. Next to the pocket is a helix-turn-helix motif that is typical of DNA-binding proteins (see Section 14.11, Repressor Uses a Helix-Turn-Helix Motif to Bind DNA).

The crystal structure shows that the helix-loop-helix of the M domain binds to a duplex region of the 4.5S RNA in domain IV. The negatively charged backbone of the RNA is adjacent to the hydrophobic pocket. This raises the possibility that a signal sequence actually binds to both the protein and RNA components of the SRP. The positively charged sequences that start the signal sequence (see Figure 10.16) could interact with the RNA, while the hydrophobic region of the signal sequence could sit in the pocket.

GTP hydrolysis plays an important role in inserting the signal sequence into the membrane. Both the SRP and the SRP receptor have GTPase capability. The signal-binding subunit of the SRP, SRP54, is a GTPase. Both subunits of the SRP receptor are GTPases. All of the GTPase activities are necessary for a nascent protein to be transferred to the membrane. FIGURE 10.20 shows that the SRP starts out with GDP when it binds to the signal sequence. The ribosome then stimulates replacement of the GDP with GTP. The signal sequence inhibits hydrolysis of the GTP. This ensures that the complex has GTP bound when it encounters the SRP receptor.

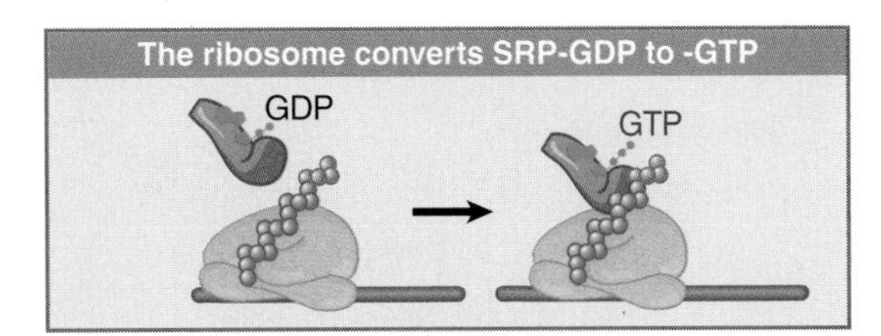

FIGURE 10.20 The SRP carries GDP when it binds the signal sequence. The ribosome causes the GDP to be replaced with GTP.

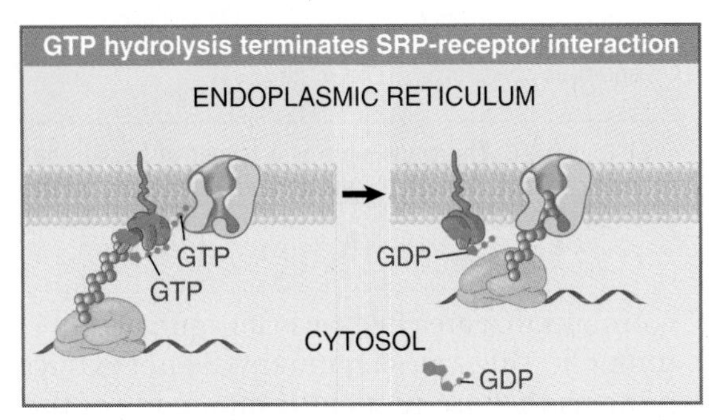

FIGURE 10.21 The SRP and SRP receptor both hydrolyze GTP when the signal sequence is transferred to the membrane.

For the nascent protein to be transferred from the SRP to the membrane, the SRP must be released from the SRP receptor. FIGURE 10.21 shows that this requires hydrolysis of the GTPs of both the SRP and the SRP receptor. The reaction has been characterized in the bacterial system, where it has the unusual feature that Ffh activates hydrolysis by FtsY, and FtsY reciprocally activates hydrolysis by Ffh.

10.10 The Translocon Forms a Pore

Key concepts

- The Sec61 trimeric complex provides the channel for proteins to pass through a membrane.
- A translocating protein passes directly from the ribosome to the translocon without exposure to the cytosol.

There is a basic problem in passing a (largely) hydrophilic protein through a hydrophobic membrane. The energetics of the interaction between the charged protein and the hydrophobic lipids are highly unfavorable. However, a protein in the process of translocation across the ER membrane can be extracted by de-

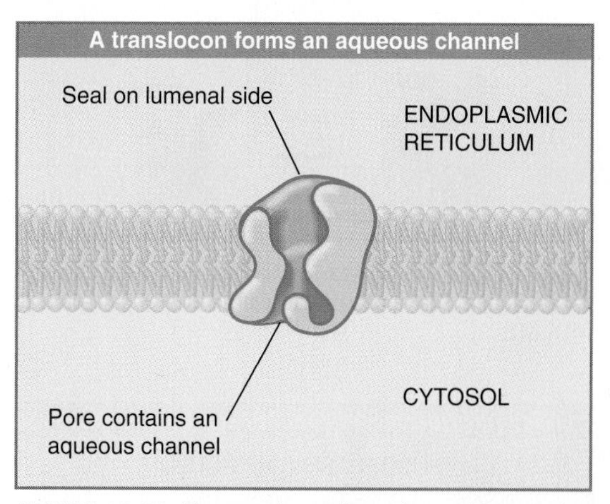

FIGURE 10.22 The translocon is a trimer of Sec61 that forms a channel through the membrane. It is sealed on the lumenal (ER) side.

naturants that are effective in an aqueous environment. The same denaturants do not extract proteins that are resident components of the membrane. This suggests the model for translocation illustrated in FIGURE 10.22, in which proteins that are part of the ER membrane form an aqueous channel through the bilayer. A translocating protein moves through this channel, interacting with the resident proteins rather than with the lipid bilayer. The channel is sealed on the lumenal side to stop free transfer of ions between the ER and the cytosol.

The channel through the membrane is called the **translocon.** Its components have been identified in two ways: Resident ER membrane proteins that are crosslinked to translocating proteins are potential subunits of the channel, and *sec* mutants in yeast (named because they fail to secrete proteins) include a class that cause precursors of secreted or membrane proteins to accumulate in the cytosol. These approaches together identified the *Sec61 complex,* which consists of three transmembrane proteins: Sec61α, β, and γ. Sec61 is the major component of the translocon. In detergent (which provides a hydrophobic milieu that mimics the effect of a surrounding membrane), Sec61 forms cylindrical oligomers with a diameter of ~85 Å and a central pore of ~20 Å. Each oligomer consists of four heterotrimers.

A similar trimeric structure for the channel is found in all organisms. In bacteria and archaea it is called the SecY complex. The Sec61α subunit (or the corresponding SecY subunit in bacteria/archaea) provides the pore through which the protein passes and is the best conserved in sequence. The pore is created from dimers of the trimeric complex, organized back-to-back so that the α subunits are fused into a single channel. A channel complex in the membrane may contain two dimers, probably organized side-by-side, although only one can be accessed by the ribosome at any time.

Is the channel a preexisting structure (as implied in the figure), or might it be assembled in response to the association of a hydrophobic signal sequence with the lipid bilayer? Channels can be detected by their ability to allow the passage of ions (measured as a localized change in electrical conductance). Ion-conducting channels can be detected in the ER membrane, and their state depends on protein translocation. This demonstrates that the channel is a permanent feature of the membrane.

A channel opens when a nascent polypeptide is transferred from a ribosome to the ER membrane. The translocating protein fills the channel completely, so ions cannot pass through during translocation. If, however, the protein is released by treatment with puromycin, then the channel becomes freely permeable. If the ribosomes are removed from the membrane, the channel closes, suggesting that the open state requires the presence of the ribosome. This suggests that the channel is controlled in response to the presence of a translocating protein.

Measurements of the abilities of fluorescence-quenching agents of different sizes to enter the channel suggest that it is large, with an internal diameter of 40 Å to –60 Å. This is much larger than the diameter of an extended α-helical stretch of protein. It is also larger than the pore seen in direct views of the channel; this discrepancy remains to be explained.

The aqueous environment of an amino acid in a protein can be measured by incorporating variant amino acids that have photoreactive residues. The fluorescence of these residues indicates whether they are in an aqueous or hydrophobic environment. Experiments with such probes show that when the signal sequence is first synthesized in the ribosome, it is in an aqueous state, but is not accessible to ions in the cytosol. It remains in the aqueous state throughout its interaction with a membrane. This suggests that the translocating protein travels directly from an enclosed tunnel in the ribosome into an aqueous channel in the membrane.

In fact, access to the pore is controlled (or "gated") on *both* sides of the membrane. Before attachment of the ribosome, the pore is closed on the lumenal side. FIGURE 10.23 shows that when the ribosome attaches, it seals the pore on the cytosolic side. When the nascent protein reaches a length of ~70 amino acids—most likely, when it extends fully across the channel—the pore opens on the lumenal side. Thus at all times, the pore is closed on one side or the other, maintaining the ionic integrities of the separate compartments.

The translocon is versatile and can be used by translocating proteins in several ways:

- It is the means by which nascent proteins are transferred from cytosolic ribosomes to the lumen of ER (see Section 10.11, Translocation Requires Insertion into the Translocon and (Sometimes) a Ratchet in the ER).
- It is also the route by which integral membrane proteins of the ER system are transferred to the membrane; this requires the channel to open or disaggregate in some unknown way so that the protein can move laterally into the lipid bilayer (see Section 10.15, How Do Proteins Insert into Membranes?).
- Proteins can also be transferred from the ER back to the cytosol; this is known as reverse translocation (see Section 10.12, Reverse Translocation Sends Proteins to the Cytosol for Degradation).

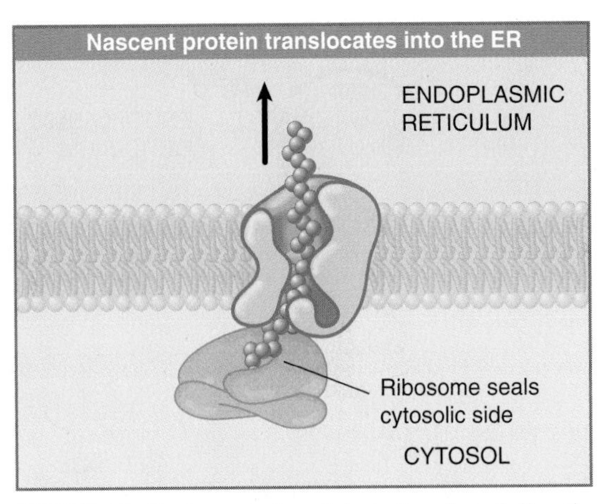

FIGURE 10.23 A nascent protein is transferred directly from the ribosome to the translocon. The ribosome seals the channel on the cytosolic side.

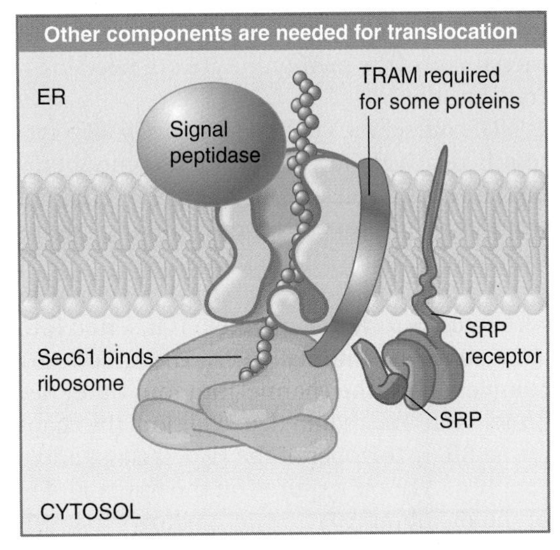

FIGURE 10.24 Translocation requires the translocon, SRP, SRP receptor, Sec61, TRAM, and signal peptidase.

10.11 Translocation Requires Insertion into the Translocon and (Sometimes) a Ratchet in the ER

Key concepts

- The ribosome, SRP, and SRP receptor are sufficient to insert a nascent protein into a translocon.
- Proteins that are inserted posttranslationally require additional components in the cytosol and BiP in the ER.
- BiP is a ratchet that prevents a protein from slipping backward.

The translocon and the SRP receptor are the basic components required for cotranslational translocation. When the Sec61 complex is incorporated into artificial membranes together with the SRP receptor, it can support translocation of some nascent proteins. Other nascent proteins require the presence of an additional component, the translocating chain-associating membrane (TRAM), which is a major protein that becomes crosslinked to a translocating nascent chain. TRAM stimulates the translocation of all proteins.

The components of the translocon and their functions are summarized in FIGURE 10.24. The simplicity of this system makes several important points. We visualize Sec61 as forming the channel and also as interacting with the ribosome. The initial targeting is made when the SRP recognizes the signal sequence as the newly synthesized protein begins to emerge from the ribosome. The SRP binds to the SRP receptor, and the signal sequence is transferred to the translocon. When the signal sequence enters the translocon, the ribosome attaches to Sec61; this forms a seal, which prevents the pore from being exposed to the cytosol. Cleavage of the signal peptide does not occur in this system and therefore cannot be necessary for translocation

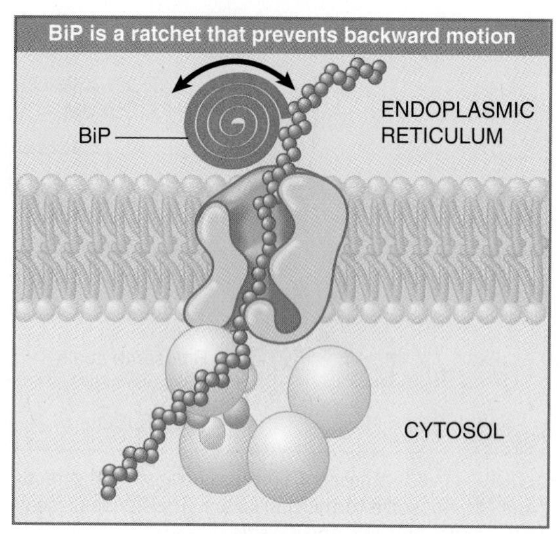

FIGURE 10.25 BiP acts as a ratchet to prevent backward diffusion of a translocating protein.

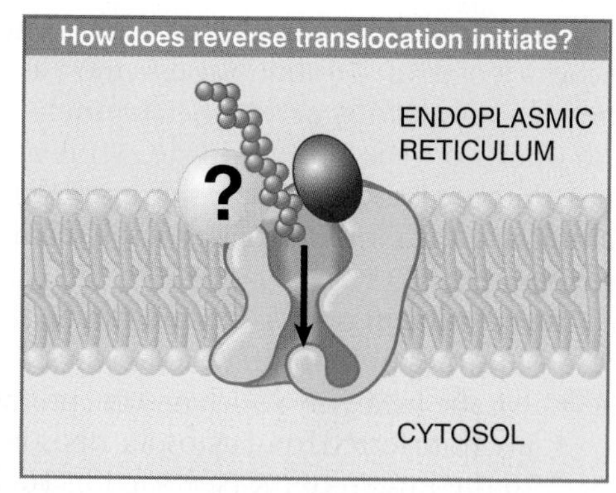

FIGURE 10.26 Reverse translocation uses the translocon to send an unfolded protein from the ER to the cytosol, where it is degraded. The mechanism of putting the translocon into reverse is not known.

per se. In this system, components on the lumenal side of the membrane are not needed for translocation.

Of course, the efficiency of the *in vitro* system is relatively low. Additional components could be required *in vivo* to achieve efficient transfer or to prevent other cellular proteins from interfering with the process.

A more complex apparatus is required in certain cases in which a protein is inserted into a membrane posttranslationally. The same Sec61 complex forms the channel, but four other Sec proteins are also required. In addition, the chaperone BiP (a member of the Hsp70 class) and a supply of ATP are required on the lumenal side of the membrane. FIGURE 10.25 shows that BiP behaves as a ratchet. In the absence of BiP, Brownian motion allows the protein to slip back into the cytosol. When BiP is present, though, it grabs the protein as it exits the pore into the ER. This stops the protein from moving backward. BiP does not pull the protein through; it just stops it from sliding back. (The reason why BiP is required for posttranslational translocation, but not for cotranslational translocation, may be that a newly synthesized protein is continuously extruded from the ribosome and therefore cannot slip backward.)

10.12 Reverse Translocation Sends Proteins to the Cytosol for Degradation

Key concept

- Sec61 translocons can be used for reverse translocation of proteins from the ER into the cytosol.

Several important activities occur within the ER. Proteins move through the ER en route to a variety of destinations. They are glycosylated and folded into their final conformations. The ER provides a "quality control" system in which misfolded proteins are identified and degraded. The degradation itself, however, does not occur in the ER, but may require the protein to be exported back to the cytosol.

The first indication that ER proteins are degraded in the cytosol and not in the ER itself was provided by evidence for the involvement of the proteasome, a large protein aggregate with several proteolytic activities. Inhibitors of the proteasome prevent the degradation of aberrant ER proteins. Proteins are marked for cleavage by the proteasome when they are modified by the addition of ubiquitin, a small polypeptide chain. The important point to note now is that ubiquitination and proteasomal degradation both occur in the cytosol (with a minor proportion in the nucleus).

Transport from the ER back into the cytosol occurs by a reversal of the usual process of import. This is called **reverse translocation** (also sometimes called retrotranslocation or dislocation). The Sec61 translocon is used. The conditions are different; for example, the translocon is not associated with a ribosome. Some mutations in Sec61 prevent reverse translocation, but do not prevent forward translocation. This could be either because there is some difference in the process or (more likely) because these regions interact with other components that are necessary for reverse translocation.

FIGURE 10.26 points out that we do not know how the channel is opened to allow insertion of the protein on the ER side. Special components are presumably involved. One model is that misfolded or misassembled proteins are recognized by chaperones, which transfer them to the translocon. In one particular case, human cytomegalovirus (CMV) codes for cytosolic proteins that destroy newly synthesized MHC class I (cellular major histocompatibility complex) proteins. This requires a viral protein product (US11), which is a membrane protein that functions in the ER. It interacts with the MHC proteins and probably conveys them into the translocon for reverse translocation.

The system involved in the degradation of aberrant ER proteins can be identified by mutations (in yeast) that lead to accumulation of aberrant proteins. In most cases a protein that misfolds (produced by a mutated gene) is degraded instead of being transported through the ER. Yeast mutants that cannot degrade the substrate fall into two classes: some identify components of the proteolytic apparatus, such as the enzymes involved in ubiquitination; others identify components of the transport apparatus, including Sec61, BiP, and Sec63.

Proteins involved in the system have also been identified by their interactions with the CMV system. The CMV protein US11 passes the MHC substrates to a protein called Derlin that is localized in the ER membrane. Derlin in turn passes the substrates to a cytosolic ATPase called p97, which is probably responsible for pulling them out of the channel into the cytosol. Derlin is a homolog of yeast Der1p, one of the proteins identified by mutations as part of the reverse translocation system.

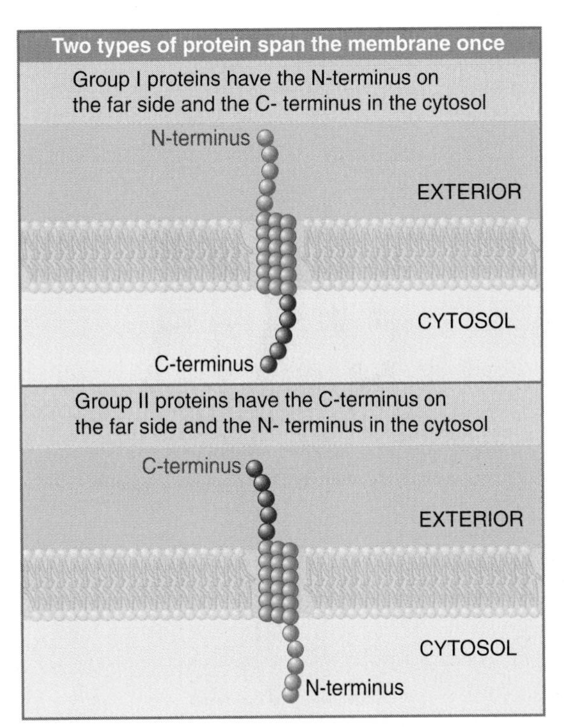

FIGURE 10.27 Group I and group II transmembrane proteins have opposite orientations with regard to the membrane.

10.13 Proteins Reside in Membranes by Means of Hydrophobic Regions

Key concepts

- Group I proteins have the N-terminus on the far side of the membrane; group II proteins have the opposite orientation.
- Some proteins have multiple membrane-spanning domains.

All biological membranes contain proteins, which are held in the lipid bilayer by noncovalent interactions. The operational definition of an **integral membrane protein** is that it requires disruption of the lipid bilayer in order to be released from the membrane. A common feature in such proteins is the presence of at least one **transmembrane domain,** consisting of an α-helical stretch of 21 to 26 hydrophobic amino acids. A sequence that fits the criteria for membrane insertion can be identified by a hydropathy plot, which measures the cumulative hydrophobicity of a stretch of amino acids. A protein that has domains exposed on both sides of the membrane is called a **transmembrane protein.** The association of a protein with a membrane takes several forms. The topography of a membrane protein depends on the number and arrangement of transmembrane regions.

When a protein has a single transmembrane region, its position determines how much of the protein is exposed on either side of the membrane. A protein may have extensive domains exposed on both sides of the membrane or may have a site of insertion close to one end, so that little or no material is exposed on one side. The length of the N-terminal or C-terminal tail that protrudes from the membrane near the site of insertion varies from insignificant to quite bulky.

FIGURE 10.27 shows that proteins with a single transmembrane domain fall into two classes. Group I proteins, in which the N-terminus faces the extracellular space, are more common than group II proteins, in which the orientation has

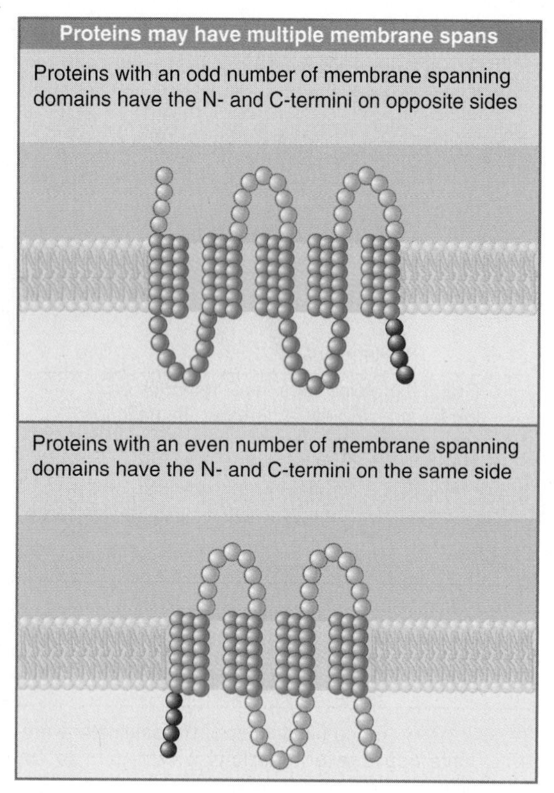

FIGURE 10.28 The orientations of the termini of multiple membrane-spanning proteins depends on whether there is an odd or even number of transmembrane segments.

been reversed so that the N-terminus faces the cytoplasm. Orientation is determined during the insertion of the protein into the ER.

FIGURE 10.28 shows orientations for proteins that have multiple membrane-spanning domains. An odd number means that both termini of the protein are on opposite sides of the membrane, whereas an even number implies that the termini are on the same face. The extent of the domains exposed on one or both sides is determined by the locations of the transmembrane domains. Domains at either terminus may be exposed, and internal sequences between the domains "loop out" into the extracellular space or cytoplasm. One common type of structure is the seven-membrane passage or "serpentine" receptor; another is the twelve-membrane passage component of an ion channel.

Does a transmembrane domain itself play any role in protein function besides allowing the protein to insert into the lipid bilayer? In the simple group I or II proteins, it has little or no additional function; often it can be replaced by any other transmembrane domain. However, transmembrane domains play an important role in the function of proteins that make multiple passes through the membrane or that have subunits that oligomerize within the membrane. The transmembrane domains in such cases often contain polar residues, which are not found in the single membrane-spanning domains of group I and group II proteins. Polar regions in the membrane-spanning domains do not interact with the lipid bilayer, but instead interact with one another. This enables them to form a polar pore or channel within the lipid bilayer. Interaction between such transmembrane domains can create a hydrophilic passage through the hydrophobic interior of the membrane. This can allow highly charged ions or molecules to pass through the membrane and is important for the function of ion channels and transport of ligands. Another case in which conformation of the transmembrane domains is important is provided by certain receptors that bind lipophilic ligands. In such cases, the transmembrane domains (rather than the extracellular domains) bind the ligand within the plane of the membrane.

10.14 Anchor Sequences Determine Protein Orientation

Key concepts

- An anchor sequence halts the passage of a protein through the translocon. Typically this is located at the C-terminal end and results in a group I orientation in which the N-terminus has passed through the membrane.
- A combined signal-anchor sequence can be used to insert a protein into the membrane and anchor the site of insertion. Typically this is internal and results in a group II orientation in which the N-terminus is cytosolic.

Proteins that are secreted from the cell pass through a membrane while remaining in the aqueous channel of the translocon. By contrast, proteins that reside in membranes start the process in the same way, but then transfer from the aqueous channel into the hydrophobic environment. The challenge in accounting for insertion of proteins into membranes is to explain what distinguishes transmembrane proteins from secreted proteins and causes this transfer. The pathway by which proteins of either type I or type II are inserted into the membrane follows the same initial route as that of secretory proteins, relying on a signal sequence that functions cotranslationally. Proteins that are to remain within the membrane, however, possess a second, **stop-transfer** signal. This takes the form of a cluster of hydrophobic amino acids adjacent to some ionic residues. The cluster serves as an **anchor** that latches on to the membrane and stops the protein from passing right through.

A surprising property of anchor sequences is that they can function as signal sequences when engineered into a different location. When placed into a protein lacking other signals, such a sequence may sponsor membrane translocation. One possible explanation for these results is that the signal sequence and anchor sequence interact with some common component of the apparatus for translocation. Binding of the signal sequence initiates translocation, but the appearance of the anchor sequence displaces the signal sequence and halts transfer.

Membrane insertion starts by the insertion of a signal sequence in the form of a hairpin loop, in which the N-terminus remains on the cytoplasmic side. Two features determine the position and orientation of a protein in the membrane: whether the signal sequence is cleaved and the location of the anchor sequence.

The insertion of type I proteins is illustrated in FIGURE 10.29. The signal sequence is N-terminal. The location of the anchor signal determines when transfer of the protein is halted. When the anchor sequence takes root in the membrane, domains on the N-terminal side will be located in the lumen, whereas domains on the C-terminal side are located facing the cytosol.

A common location for a stop-transfer sequence of this type is at the C-terminus. As shown in the figure, transfer is halted only as the last sequences of the protein enter the membrane. This type of arrangement is responsible for the location in the membrane of many proteins, including cell surface proteins. Most of the protein sequence is exposed on the lumenal side of the membrane, with a small or negligible tail facing the cytosol.

Type II proteins do not have a cleavable leader sequence at the N-terminus. The signal sequence is instead combined with an anchor sequence. We imagine that the general pathway for the integration of type I proteins into the membrane involves the steps illustrated in FIGURE 10.30. The signal sequence enters the membrane, but the joint signal-anchor sequence does not pass through. It stays, instead, in the membrane (perhaps interacting directly with the lipid bilayer), while the rest of the growing polypeptide continues to loop into the ER.

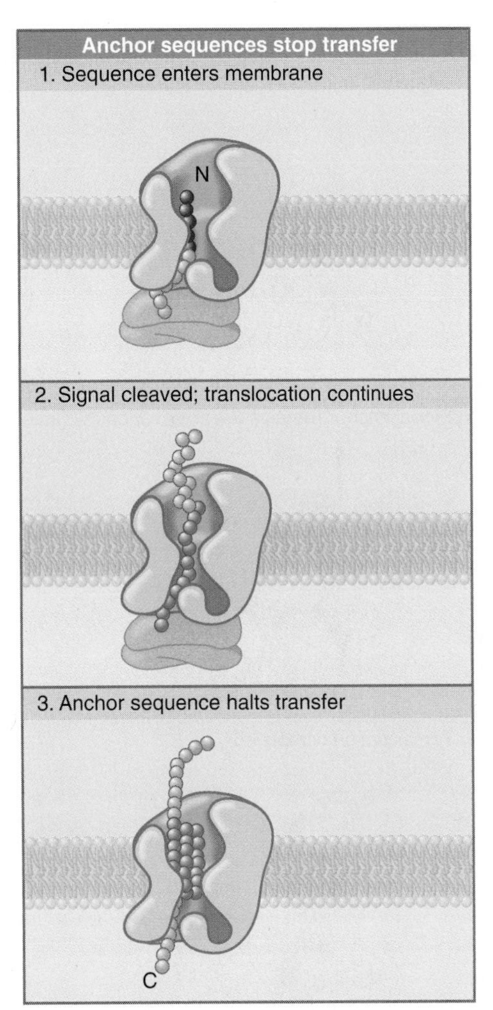

FIGURE 10.29 Proteins that reside in membranes enter by the same route as secreted proteins, but transfer is halted when an anchor sequence passes into the membrane. If the anchor is at the C-terminus, the bulk of the protein passes through the membrane and is exposed on the far surface.

The signal-anchor sequence is usually internal, and its location determines which parts of the protein remain in the cytosol and which are extracellular. Essentially all the N-terminal sequences that precede the signal-anchor are exposed to the cytosol. Most often the cytosolic tail is short, ~6 to 30 amino acids. In effect the N-terminus remains constrained while the rest of the protein passes through the membrane. This reverses the orientation of the protein with regard to the membrane.

The combined signal-anchor sequences of type II proteins resemble cleavable signal sequences. FIGURE 10.31 gives an example. As with cleavable leader sequences, the amino acid composition is more important than the actual sequence. The regions at the extremities of the signal-anchor carry positive charges; the central region is uncharged and resembles

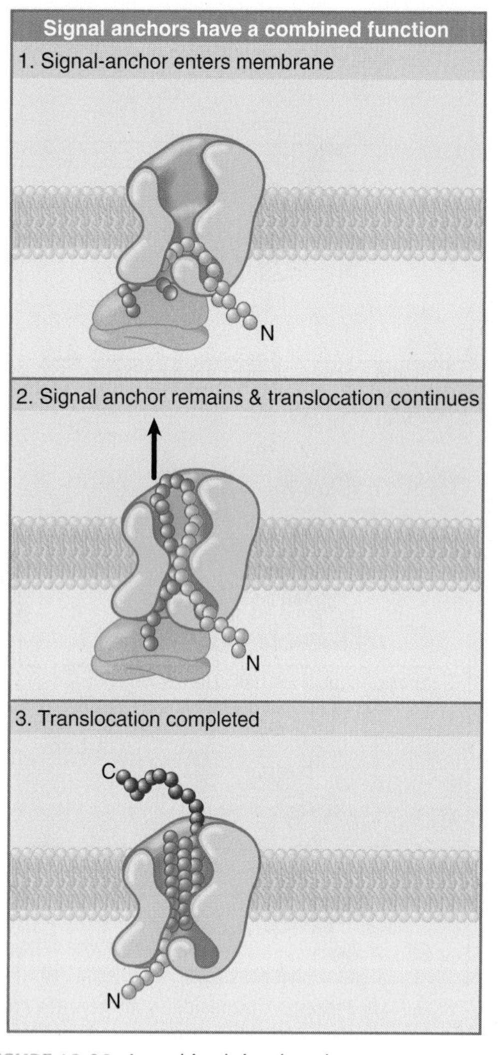

FIGURE 10.30 A combined signal-anchor sequence causes a protein to reverse its orientation, so that the N-terminus remains on the inner face and the C-terminus is exposed on the outer face of the membrane.

a hydrophobic core of a cleavable leader. Mutations to introduce charged amino acids in the core region prevent membrane insertion; mutations on either side prevent the anchor from working, so the protein is secreted or located in an incorrect compartment.

The distribution of charges around the anchor sequence has an important effect on the orientation of the protein. More positive charges are usually found on the cytoplasmic side (N-terminal side in type II proteins). If the positive charges are removed by mutation, the orientation of the protein can be reversed. The effect of charges on orientation is summarized by the "positive inside" rule, which states that the side of the anchor with the most positive charges will be located in the cytoplasm. The positive charges in effect provide a hook that latches on to the cytoplasmic side of the membrane, which in turn controls the direction in which the hydrophobic region is inserted, thus determining the orientation of the protein.

10.15 How Do Proteins Insert into Membranes?

Key concept

- Transfer of transmembrane domains from the translocon into the lipid bilayer is triggered by the interaction of the transmembrane region with the translocon.

We have a reasonable understanding of the processes by which secreted proteins pass through membranes and of how this relates to the insertion of the single-membrane spanning

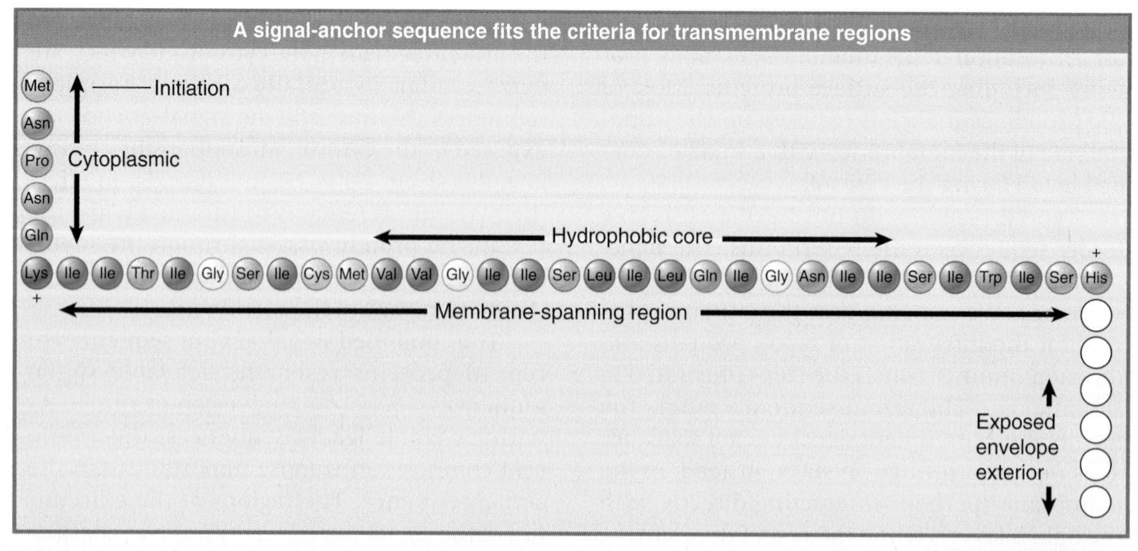

FIGURE 10.31 The signal-anchor of influenza neuraminidase is located close to the N-terminus and has a hydrophobic core.

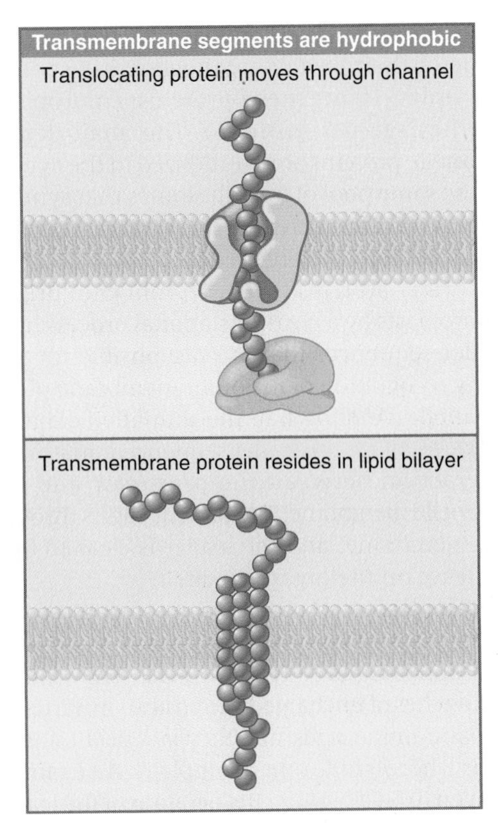

FIGURE 10.32 How does a transmembrane protein make the transition from moving through a proteinaceous channel to interacting directly with the lipid bilayer?

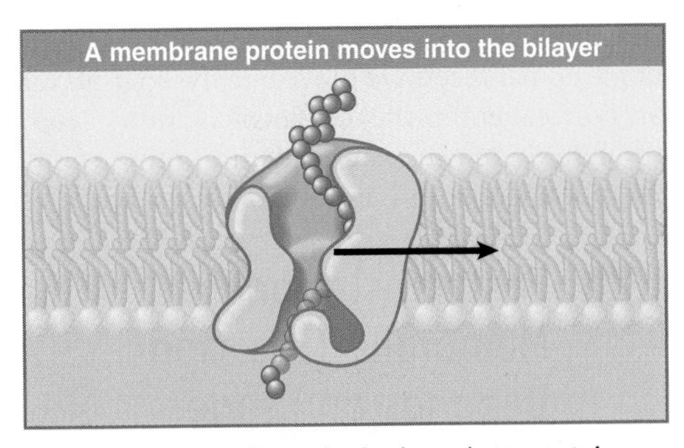

FIGURE 10.33 Newly synthesized membrane proteins are able to transfer laterally from the translocon into the lipid bilayer. The mechanism of transfer is not known.

group I and group II proteins. We cannot yet explain the details of insertion of proteins with multiple membrane-spanning domains.

We understand how a secreted protein passes through a membrane without any conflict, but it is more difficult to apply the same model to a protein that resides in the membrane. FIGURE 10.32 illustrates the difference between the organization of a translocating protein, which is protected from the lipid bilayer by the aqueous channel, and a transmembrane protein, which has a hydrophobic segment directly in contact with the membrane. The critical question is how a protein is transferred from the proteinaceous channel into the lipid bilayer.

The possibility that there is a mechanism for transferring hydrophobic transmembrane domains directly from the channel into the membrane is suggested in FIGURE 10.33. This idea is supported by observations of an *in vitro* system that measured transfer into a lipid environment for proteins with different transmembrane domains. When the domain passed a threshold of hydrophobicity, the protein could pass from a channel consisting of Sec61 and TRAM into the lipid bilayer. In addition to overall hydrophobicity, the locations of polar residues within the transmembrane segments have an important effect. The simplest explanation is that the structure of the channel allows the translocating protein to contact the lipid bilayer, so that a sufficiently hydrophobic segment can simply partition directly into the lipid. The structure of the translocon suggests that there could be a gate located between two helices that is used for the transfer.

It has always been a common assumption that, whatever the exact mechanism for transferring the transmembrane segment into the membrane, it is triggered by the presence of the transmembrane sequence in the pore. However, changes in the pore occur earlier in response to the synthesis of the transmembrane sequence in the ribosome. When a secreted protein passes through the pore, the channel remains sealed on the cytosolic side but opens on the lumenal side after synthesis of the first seventy residues. As soon as a transmembrane sequence has been fully synthesized, though—that is, while it is still entirely within the ribosome—the pore closes on the lumenal side. How this change relates to the transfer of the transmembrane sequence into the membrane is not clear.

The process of insertion into a membrane has been characterized for both type I and type II proteins, in which there is a single transmembrane domain. How is a protein with multiple membrane-spanning regions inserted into a membrane? Much less is known about this process, but we assume that it relies on sequences that provide signal and/or anchor capabilities. One model is to suppose that there is an alternating series of signal and anchor sequences. Translocation is initiated at the first signal sequence and continues until stopped by the first anchor. It then is reinitiated by a subsequent signal sequence until stopped by the next anchor. It is possible that there are multiple pathways for integration into the membrane, because in some cases a transmembrane domain seems to move into the lipid bilayer as soon as it enters the translocon. In other cases, though, there can be a delay until other transmembrane regions have been synthesized.

10.16 Posttranslational Membrane Insertion Depends on Leader Sequences

Key concept

- N-terminal leader sequences provide the information that allows proteins to associate with mitochondrial or chloroplast membranes.

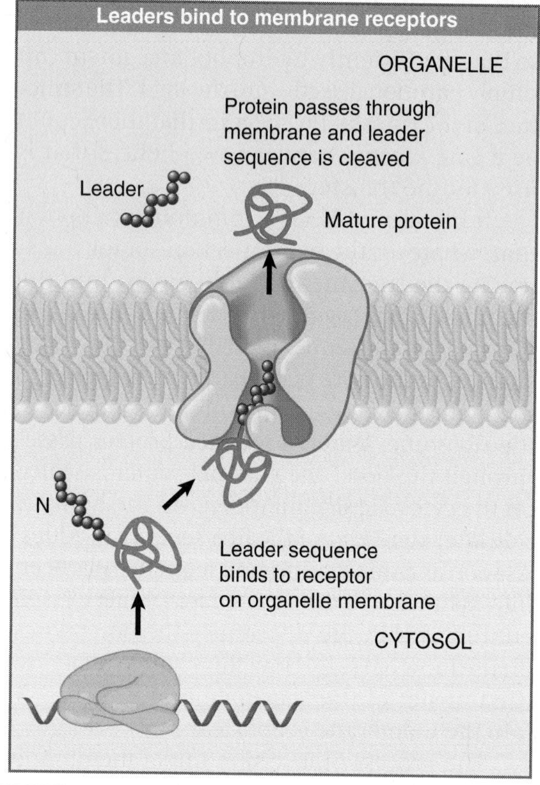

FIGURE 10.34 Leader sequences allow proteins to recognize mitochondrial or chloroplast surfaces by a posttranslational process.

Mitochondria and chloroplasts synthesize only some of their proteins. Mitochondria synthesize only ~10 organelle proteins; chloroplasts synthesize ~50 proteins. The majority of organelle proteins are synthesized in the cytosol by the same pool of free ribosomes that synthesize cytosolic proteins. They must then be imported into the organelle.

Many proteins that enter mitochondria or chloroplasts by a posttranslational process have leader sequences that are responsible for primary recognition of the outer membrane of the organelle. As shown in the simplified diagram of FIGURE 10.34, the leader sequence initiates the interaction between the precursor and the organelle membrane. The protein passes through the membrane, and the leader is cleaved by a protease on the organelle side.

The leaders of proteins imported into mitochondria and chloroplasts usually have both hydrophobic and basic amino acids. They consist of stretches of uncharged amino acids interrupted by basic amino acids, and they lack acidic amino acids. There is little other homology. An example is given in FIGURE 10.35. Recognition of the leader does not depend on its exact sequence, but rather on its ability to form an amphipathic helix, in which one face has hydrophobic amino acids and the other face presents the basic amino acids.

The leader sequence contains all the information needed to localize an organelle protein. The ability of a leader sequence can be tested by constructing an artificial protein in which a leader from an organelle protein is joined to a cytosolic protein. The experiment is performed by constructing a hybrid gene, which is then translated into the hybrid protein.

Several leader sequences have been shown by such experiments to function independently to target any attached sequence to the mitochondrion or chloroplast. For example, if the leader sequence given in Figure 10.35 is attached to the cytosolic protein DHFR (dihydrofolate reductase), the DHFR becomes localized in the mitochondrion.

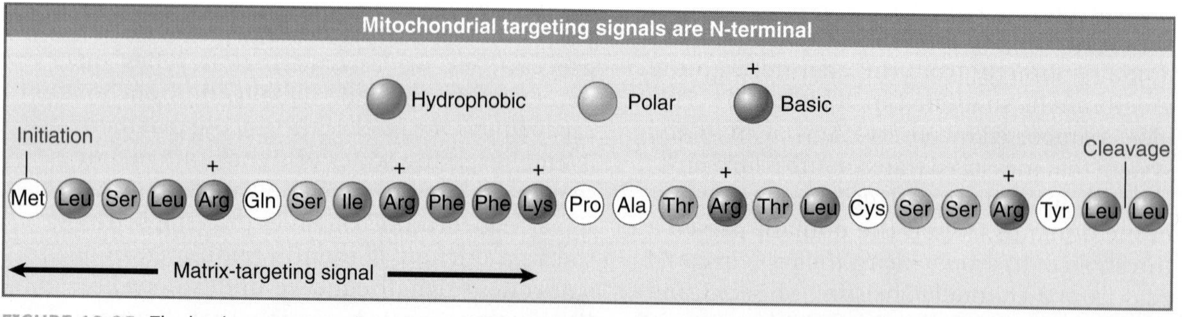

FIGURE 10.35 The leader sequence of yeast cytochrome *c* oxidase subunit IV consists of twenty-five neutral and basic amino acids. The first twelve amino acids are sufficient to transport any attached polypeptide into the mitochondrial matrix.

The leader sequence and the transported protein represent domains that fold independently. Irrespective of the sequence to which it is attached, the leader must be able to fold into an appropriate structure to be recognized by receptors on the organelle envelope. The attached polypeptide sequence plays no part in recognition of the envelope.

What restrictions are there on transporting a hydrophilic protein through the hydrophobic membrane? An insight into this question is given by the observation that methotrexate, a ligand for the enzyme DHFR, blocks transport into mitochondria of DHFR fused to a mitochondrial leader. The tight binding of methotrexate prevents the enzyme from unfolding when it is translocated through the membrane. The sequence of the transported protein is irrelevant for targeting purposes; however, in order to follow its leader through the membrane, the protein requires the flexibility to assume an unfolded conformation.

Hydrolysis of ATP is required both outside and inside for translocation across the membrane. It may be involved with pushing the protein from outside and pulling from inside. In the cases of mitochondrial import and bacterial export, there is also a requirement for an electrochemical potential across the inner membrane to transfer the amino terminal part of the leader.

10.17 A Hierarchy of Sequences Determines Location within Organelles

Key concepts

- The N-terminal part of a leader sequence targets a protein to the mitochondrial matrix or chloroplast lumen.
- An adjacent sequence can control further targeting to a membrane or the intermembrane spaces.
- The sequences are cleaved successively from the protein.

The mitochondrion is surrounded by an envelope consisting of two membranes. Proteins imported into mitochondria may be located in the outer membrane, the intermembrane space, the inner membrane, or the matrix. A protein that is a component of one of the membranes may be oriented so that it faces one side or the other.

What is responsible for directing a mitochondrial protein to the appropriate compartment? The "default" pathway for a protein imported into a mitochondrion is to move through both membranes into the matrix. This property is conferred by the N-terminal part of the leader sequence. A protein that is localized within the intermembrane space or in the inner membrane itself requires an additional signal, which specifies its destination within the organelle. A multipart leader contains signals that function in a hierarchical manner, as summarized in FIGURE 10.36. The first part of the leader targets the protein to the organelle, and the second part is required if its destination is elsewhere than the matrix. The two parts of the leader are removed by successive cleavages.

Cytochrome *c1* is an example. It is bound to the inner membrane and faces the intermembrane space. Its leader sequence consists of sixty-one amino acids and can be divided into regions with different functions. The sequence of the first thirty-two amino acids alone, or even the N-terminal half of this region, can transport DHFR all the way into the matrix. Thus the first part of the leader sequence (thirty-two N-terminal amino acids) comprises a matrix-targeting signal. The

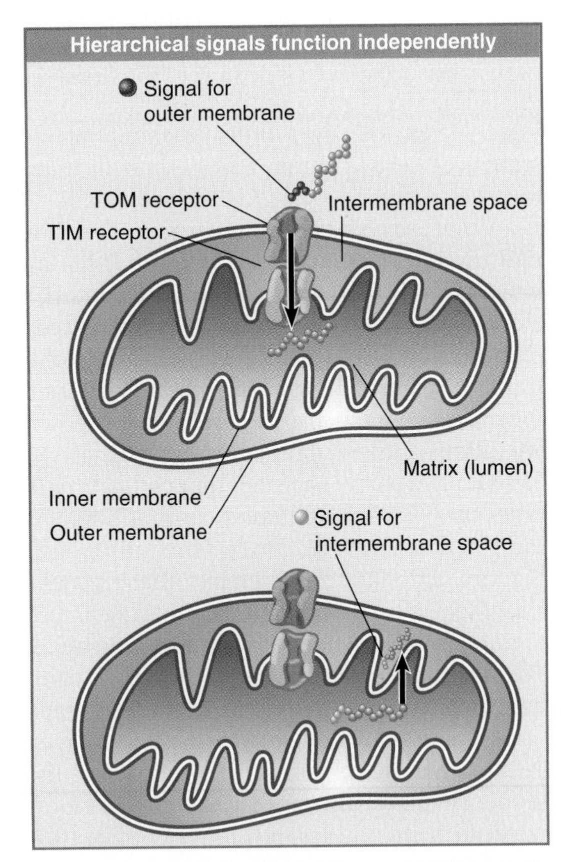

FIGURE 10.36 Mitochondria have receptors for protein transport in the outer and inner membranes. Recognition at the outer membrane may lead to transport through both receptors into the matrix, where the leader is cleaved. If it has a membrane-targeting signal, it may be reexported.

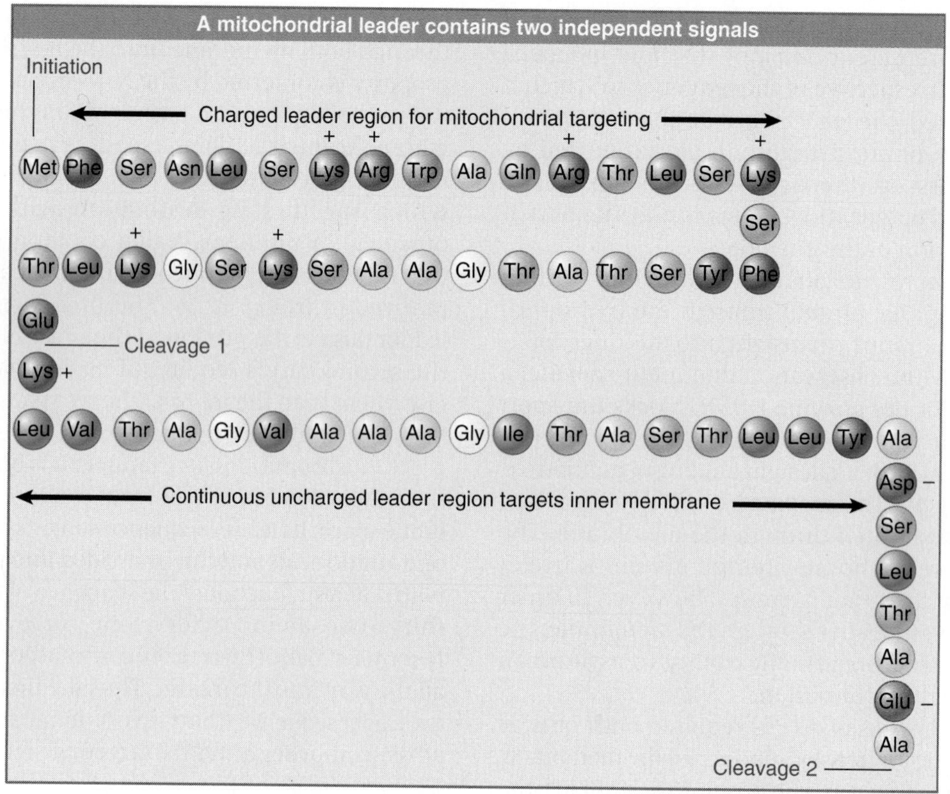

FIGURE 10.37 The leader of yeast cytochrome *c*1 contains an N-terminal region that targets the protein to the mitochondrion, followed by a region that targets the (cleaved) protein to the inner membrane. The leader is removed by two cleavage events.

intact leader, however, transports an attached sequence—such as murine DHFR—into the intermembrane space.

What prevents the protein from proceeding past the intermembrane space when it has an intact leader? The region following the matrix-targeting signal (comprising nineteen amino acids of the leader) provides another signal that localizes the protein at the inner membrane or within the intermembrane space. For working purposes, we call this the membrane-targeting signal.

The two parts of a leader that contains both types of signal have different compositions. As indicated in FIGURE 10.37, the thirty-five N-terminal amino acids resemble other organelle leader sequences in the high content of uncharged amino acids, punctuated by basic amino acids. The next nineteen amino acids, however, comprise an uninterrupted stretch of uncharged amino acids that is long enough to span a lipid bilayer. This membrane-targeting signal resembles the sequences that are involved in protein translocation into membranes of the ER (see Section 10.7, Signal Sequences Initiate Translocation).

Cleavage of the matrix-targeting signal is the sole processing event required for proteins that reside in the matrix. This signal must also be cleaved from proteins that reside in the intermembrane space; however, following this cleavage, the membrane-targeting signal (which is now the N-terminal sequence of the protein) directs the protein to its destination in the outer membrane, intermembrane space, or inner membrane. The signal then, in turn, is cleaved.

The N-terminal matrix-targeting signal functions in the same manner for all mitochondrial proteins. Its recognition by a receptor on the outer membrane leads to transport through the two membranes. Note that the same protease is involved in cleaving the matrix-targeting signal, irrespective of the final destination of the protein. This protease is a water soluble, Mg^{2+}-dependent enzyme that is located in the matrix. Thus the N-terminal sequence must reach the matrix, even if the protein ultimately will reside in the intermembrane space.

Residence in the matrix occurs in the absence of any other signal. If there is a membrane-targeting signal, however, it is activated by cleavage of the matrix-targeting signal. The remaining part of the leader (which is now N-terminal) then causes the protein to take up its final destination.

The nature of the membrane-targeting signal is controversial. One model holds that the entire protein enters the matrix, after which the membrane-targeting signal causes it to be reexported into or through the inner membrane. An alternative model proposes that the membrane-targeting sequence simply prevents the rest of the protein from following the leader through the inner membrane into the matrix. Whichever model applies, another protease (located within the intermembrane space) completes the removal of leader sequences.

Passage through chloroplast membranes is achieved in a similar manner. FIGURE 10.38 illustrates the variety of locations for chloroplast proteins. They pass the outer and inner membranes of the envelope into the stroma, a process involving the same types of passage as into the mitochondrial matrix. Some proteins are transported yet further, though, across the stacks of the thylakoid membrane into the lumen. Proteins destined for the thylakoid membrane or lumen must cross the stroma en route.

Chloroplast targeting signals resemble mitochondrial targeting signals. The leader consists of ~50 amino acids, and the N-terminal half is needed to recognize the chloroplast envelope. A cleavage between positions 20 and 25 occurs during or following passage across the envelope, and proteins destined for the thylakoid membrane or lumen have a new N-terminal leader that guides recognition of the thylakoid membrane. There are several (at least four) different systems in the chloroplast that catalyze import of proteins into the thylakoid membrane.

The general principle governing protein transport into mitochondria and chloroplasts, therefore, is that the N-terminal part of the leader targets a protein to the organelle matrix, and an additional sequence (within the leader) is needed to localize the protein at the outer membrane, intermembrane space, or inner membrane.

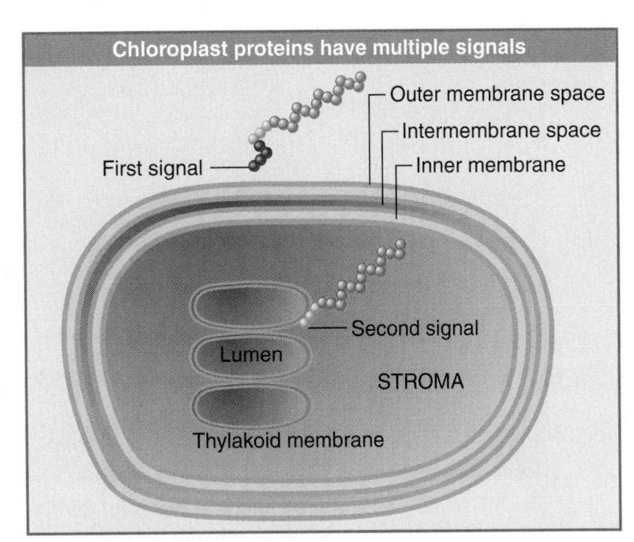

FIGURE 10.38 A protein approaches the chloroplast from the cytosol with a ~50 residue leader. The N-terminal half of the leader sponsors passage into the envelope or through it into the stroma. Cleavage occurs during envelope passage.

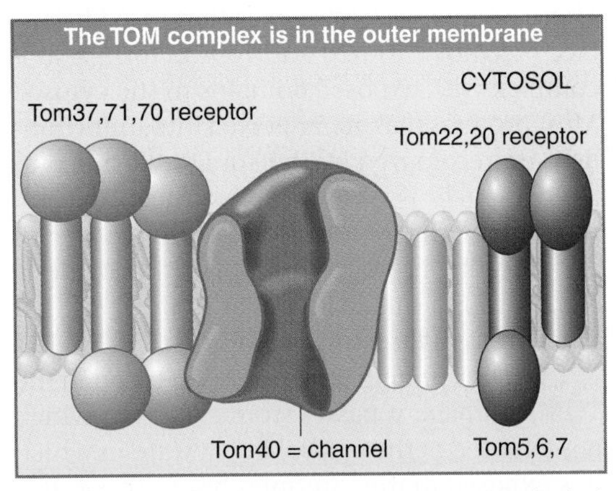

FIGURE 10.39 TOM proteins form receptor complex(es) that are needed for translocation across the mitochondrial outer membrane.

10.18 Inner and Outer Mitochondrial Membranes Have Different Translocons

Key concepts

- Transport through the outer and inner mitochondrial membranes uses different receptor complexes.
- The TOM (outer membrane) complex is a large complex in which substrate proteins are directed to the Tom40 channel by one of two subcomplexes.
- Different TIM (inner membrane) complexes are used depending on whether the substrate protein is targeted to the inner membrane or to the lumen.
- Proteins pass directly from the TOM to the TIM complex.

There are different receptors for transport through each membrane in the chloroplast and mitochondrion. In the chloroplast they are called TOC and TIC, and in the mitochondrion they are called **TOM** and **TIM,** referring to the outer and inner membranes, respectively.

The TOM complex consists of ~9 proteins, many of which are integral membrane proteins. A general model for the complex is shown in FIGURE 10.39. The TOM aggregate has a size of >500 kD, with a diameter of ~138 Å, and forms an ion-conducting channel. A complex contains 2 to 3 individual rings of diameter 75 Å, each with a pore of diameter 20 Å.

Tom40 is deeply imbedded in the membrane and provides the channel for translocation. It contacts preproteins as they pass through the outer membrane. It binds to three smaller

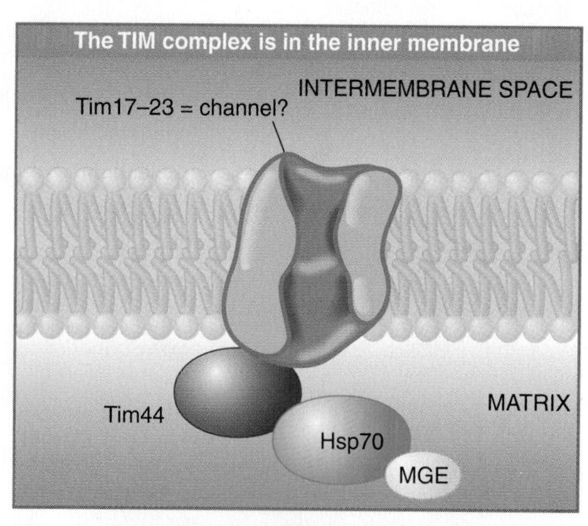

FIGURE 10.40 Tim proteins form the complex for translocation across the mitochondrial inner membrane.

proteins, Tom5, -6, and -7, which may be components of the channel or assembly factors. There are two subcomplexes that provide surface receptors. Tom20 and Tom22 form a subcomplex with exposed domains in the cytosol. Most proteins that are imported into mitochondria are recognized by the Tom20,22 subcomplex, which is the primary receptor and recognizes the N-terminal sequence of the translocating protein. Tom37,70,71 provides a receptor for a smaller number of proteins that have internal targeting sequences.

When a protein is translocated through the TOM complex, it passes from a state in which it is exposed to the cytosol into a state in which it is exposed to the intermembrane space. It is not, however, usually released, but instead is transferred directly to the TIM complex. It is possible to trap intermediates in which the leader is cleaved by the matrix protease, leaving a major part of the precursor exposed on the cytosolic surface of the envelope. This suggests that a protein spans the two membranes during passage. The TOM and TIM complexes do not appear to interact directly (or at least do not form a detectable stable complex), and they may therefore be linked simply by a protein in transit. When a translocating protein reaches the intermembrane space, the exposed residues may immediately bind to a TIM complex, whereas the rest of the protein continues to translocate through the TOM complex.

There are two TIM complexes in the inner membrane. The Tim17-23 complex translocates proteins to the lumen. Substrates are recognized by their possession of a positively charged N-terminal signal. Transmembrane proteins Tim17 through Tim23 comprise the channel. FIGURE 10.40 shows that they are associated with Tim44 on the matrix side of the membrane. Tim 44 in turn binds the chaperone Hsp70. This is also associated with another chaperone, Mge, the counterpart to bacterial GrpE. This association ensures that when the imported protein reaches the matrix, it is bound by the Hsp70 chaperone. The high affinity of Hsp70 for the unfolded conformation of the protein as it emerges from the inner membrane helps to "pull" the protein through the channel.

A major chaperone activity in the mitochondrial matrix is provided by Hsp60 (which forms the same sort of structure as its counterpart GroEL). Association with Hsp60 is necessary for joining of the subunits of imported proteins that form oligomeric complexes. An imported protein may be "passed on" from Hsp70 to Hsp60 in the process of acquiring its proper conformation.

The Tim22-54 complex translocates proteins that reside in the inner membrane.

How does a translocating protein finds its way from the TOM complex to the appropriate TIM complex? Two protein complexes in the intermembrane space escort a translocating protein from TIM to TOM. The Tim9-10 and Tim8-13 complexes act as escorts for different sets of substrate proteins. Tim9-10 may direct its substrates to either Tim22-54 or Tim23-17, whereas Tim8-13 directs substrates only to Tim22-54. Some substrates do not use either Tim9-10 or Tim8-13, so other pathways must also exist. The pathways are summarized in FIGURE 10.41.

What is the role of the escorting complexes? They may be needed to help the protein exit from the TOM complex as well as for recognizing the TIM complex. FIGURE 10.42 shows that a translocating protein may pass directly from the TOM channel to the Tim9,10 complex, and then into the Tim22-54 channel.

A mitochondrial protein folds under different conditions before and after its passage through the membrane. Ionic conditions and the chaperones that are present are different in the cytosol and in the mitochondrial matrix. It is possible that a mitochondrial protein can attain its mature conformation *only* in the mitochondrion.

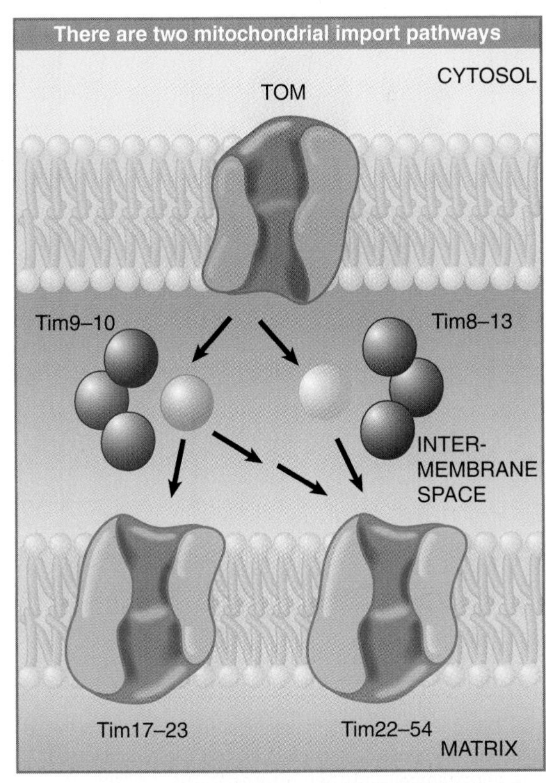

FIGURE 10.41 Tim9-10 takes proteins from TOM to either TIM complex, and Tim8-13 takes proteins to Tim22-54.

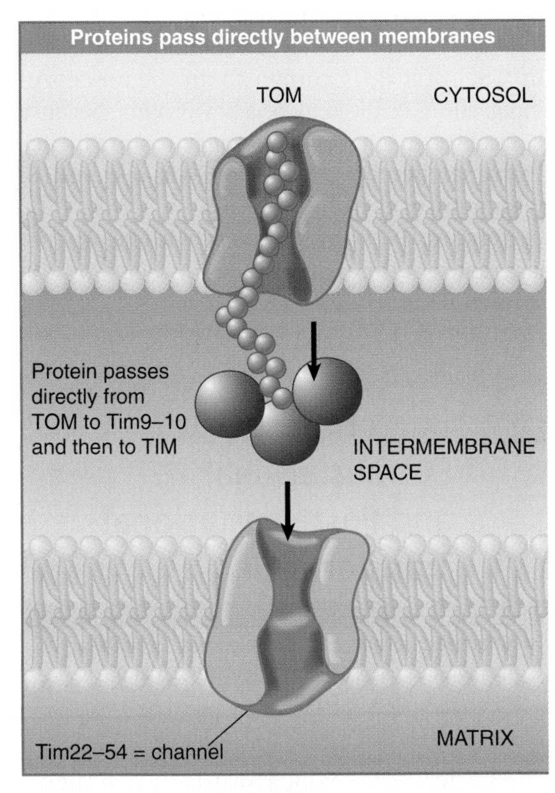

FIGURE 10.42 A translocating protein may be transferred directly from TOM to Tim22-54.

10.19 Peroxisomes Employ Another Type of Translocation System

Key concepts

- Proteins are imported into peroxisomes in their fully folded state.
- They have either a PTS1 sequence at the C-terminus or a PTS2 sequence at the N-terminus.
- The receptor Pex5p binds the PTS1 sequence and the receptor Pex7p binds the PTS2 sequence.
- The receptors are cytosolic proteins that shuttle into the peroxisome carrying a substrate protein and then return to the cytosol.

Peroxisomes are small bodies (0.5 to 1.5 μm diameter) enclosed by a single membrane. They contain enzymes concerned with oxygen utilization, which convert oxygen to hydrogen peroxide by removing hydrogen atoms from substrates. Catalase then uses the hydrogen peroxide to oxidize a variety of other substrates. Their activities are crucial for the cell. Since the fatal disease of Zellweger syndrome was found to be caused by lack of peroxisomes, >15 human diseases have been linked to disorders in peroxisome function.

All of the components of the peroxisome are imported from the cytosol. Proteins that are required for peroxisome formation are called **peroxins.** Twenty-three genes coding for peroxins have been identified, and human peroxisomal diseases have been mapped to twelve complementation groups, most of which are identified with specific genes. Peroxisomes appear to be absent from cells that have null mutations in some of these genes. In some of these cases, introduction of a wild-type gene leads to the reappearance of peroxisomes. It has generally been assumed that, like other membrane-bounded organelles, peroxisomes can arise only by duplication of preexisting peroxisomes. These results, however, raised the question of whether it might be possible to assemble them *de novo* from their components. In at least some cases, the absence of peroxins leaves the cells with peroxisomal ghosts—empty membrane bodies. Even when they cannot be easily seen, it is hard to exclude the possibility that there is some remnant that serves to regenerate the peroxisomes.

Transport of proteins to peroxisomes occurs posttranslationally. Proteins that are imported into the matrix have either of two short sequences, called peroximal targeting signal (PTS)1 and PTS2. The PTS1 signal is a tri- or tetrapeptide at the C-terminus. It was originally characterized as the sequence SKL (Ser-Lys-Leu), but now a large variety of sequences have been shown to act as a PTS1 signal. The addition of a suitable sequence to the C-terminus of cytosolic proteins is sufficient to ensure their import into the organelle. The PTS2 signal is a sequence of nine amino acids, again with much diversity, and this can be located near the N-terminus or internally. It is possible there may be a third type of sequence called PTS3.

Several peroxisomal proteins are necessary for the import of proteins from the cytosol. The peroxisomal receptors that bind the two types of signals are called Pex5p and Pex7p, respectively. The other proteins are part of membrane-associated complexes concerned with the translocation reaction.

Transport into the peroxisome has unusual features that mark important differences from the system used for transport into other organelles.

Proteins can be imported into the peroxisome in their mature, fully folded state. This contrasts with the requirement to unfold a protein for passage into the ER or mitochondrion, where it passes through a channel in the membrane into the organelle in something akin to an unfolded thread of amino acids. It is not clear how the structure of a preexisting channel could expand to permit this. One possibility is to resurrect an old idea and to suppose that the channel assembles around the substrate protein when it associates with the membrane.

The Pex5p and Pex7p receptors are not integral membrane proteins, but rather are largely cytosolic, with only a small proportion associated with peroxisomes. They behave in the same way, cycling between the peroxisome and the cytosol. FIGURE 10.43 shows that the receptor binds a substrate protein in the cytosol, takes it to the peroxisome, moves with it through the membrane into the interior, and then returns to the cytosol to undertake another cycle. This shuttling behavior resembles the carrier system for import into the nucleus.

The import pathways converge at the perxoisomal membrane, where Pex5p and Pex7p both interact with the same membrane protein complex, which consists of Pex14p and Pex13p. The receptors dock with this complex, and then several other peroxins are involved with the process of transport into the lumen. The details of the transport process are not yet clear.

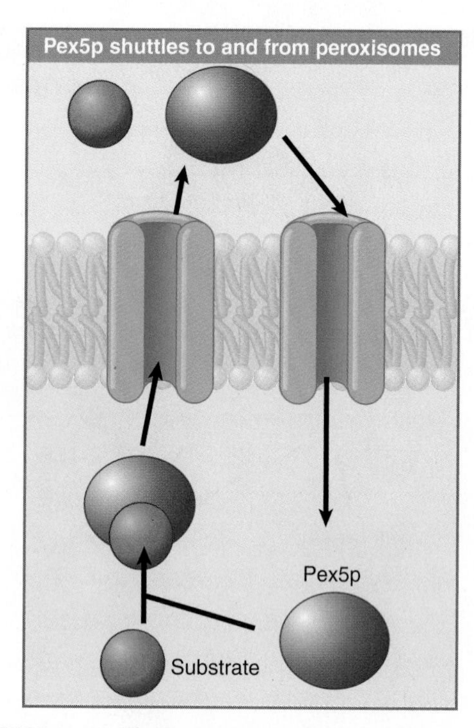

FIGURE 10.43 The Pex5p receptor binds a substrate protein in the cytosol, carries it across the membrane into the peroxisome, and then returns to the cytosol.

Proteins that are incorporated into the peroxisomal membrane have a sequence called the mPTS, but little is known about the process of integration. Pex3p may be a key protein, because in its absence other proteins are not found in peroxisomal membranes. Pex3p has its own mPTS, which raises the question of how it enters the membrane. Perhaps it interacts with Pex3p that is already in the membrane. This bears on the question of whether peroxisomes can ever assemble *de novo*.

10.20 Bacteria Use Both Cotranslational and Posttranslational Translocation

Key concept

- Bacterial proteins that are exported to or through membranes use both posttranslational and cotranslational mechanisms.

The bacterial envelope consists of two membrane layers. The space between them is called the **periplasm.** Proteins are exported from the cytoplasm to reside in the envelope or to be secreted from the cell. The mechanisms of secretion from bacteria are similar to those characterized for eukaryotic cells, and we can recognize some related components. FIGURE 10.44 shows that proteins that are exported from the cytoplasm have one of four fates:

- to be inserted into the inner membrane,
- to be translocated through the inner membrane to rest in the periplasm,
- to be inserted into the outer membrane, or
- to be translocated through the outer membrane into the medium.

Different protein complexes in the inner membrane are responsible for transport of proteins depending on whether their fate is to pass through or stay within the inner membrane. This resembles the situation in mitochondria, where different complexes in each of the inner and outer membranes handle different subsets of protein substrates depending on their destinations (see Section 10.16, Posttranslational Membrane Insertion Depends on Leader Sequences) A difference from import into organelles is that transfer in *E. coli* may be either co- or posttranslational. Some proteins are secreted both cotranslationally and posttranslationally, and the relative kinetics of translation versus secretion through the membrane could determine the balance.

Exported bacterial proteins have N-terminal leader sequences with a hydrophilic N-terminus and an adjacent hydrophobic core. The leader is cleaved by a signal peptidase that recognizes precursor forms of several exported proteins. The signal peptidase is an integral membrane protein located in the inner membrane. Mutations in N-terminal leaders prevent secretion; they are suppressed by mutations in other genes, which are thus defined as components of the protein export apparatus. Several genes given the general description *sec* are implicated in coding for components of the secretory apparatus by the occurrence of mutations that block secretion of many or all exported proteins.

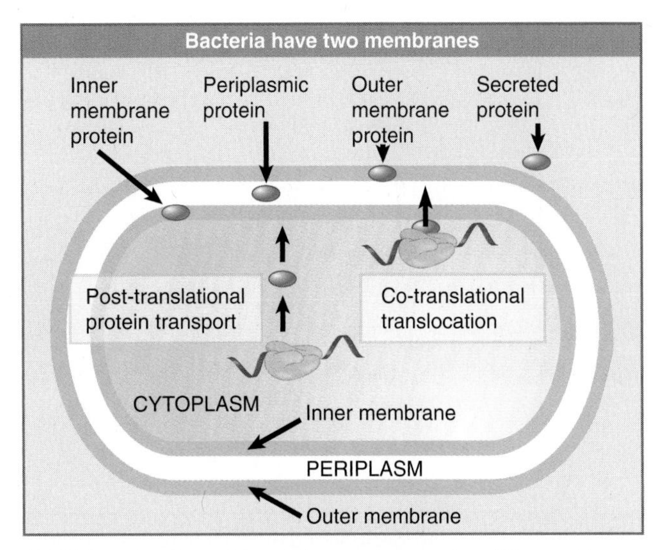

FIGURE 10.44 Bacterial proteins may be exported either posttranslationally or cotranslationally, and may be located within either membrane or the periplasmic space, or may be secreted.

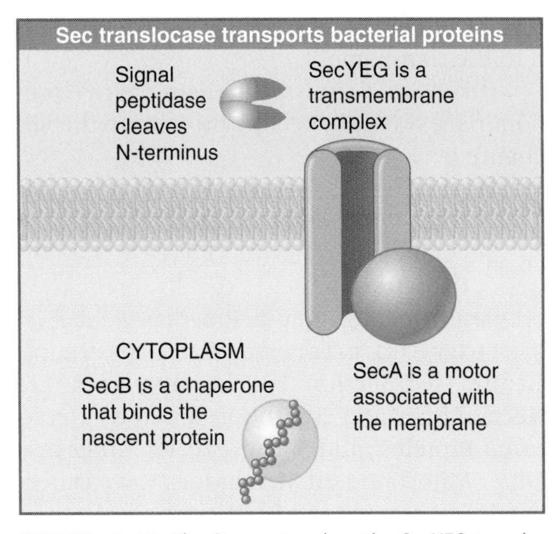

FIGURE 10.45 The Sec system has the SecYEG translocon embedded in the membrane, the SecA-associated protein that pushes proteins through the channel, the SecB chaperone that transfers nascent proteins to SecA, and the signal peptidase that cleaves the N-terminal signal from the translocated protein.

10.21 The Sec System Transports Proteins into and Through the Inner Membrane

Key concepts

- The bacterial SecYEG translocon in the inner membrane is related to the eukaryotic Sec61 translocon.
- Various chaperones are involved in directing secreted proteins to the translocon.

There are several systems for transport through the inner membrane. The best characterized is the Sec system, whose components are shown in FIGURE 10.45. The translocon that is embedded in the membrane consists of three subunits that are related to the components of mammalian/yeast Sec61. Each of the subunits is an integral transmembrane protein. (SecY has ten transmembrane segments; SecE has three.) The functional translocon is a trimer with one copy of each subunit. The major pathway for directing proteins to the translocon consists of SecB and SecA. SecB is a chaperone that binds to the

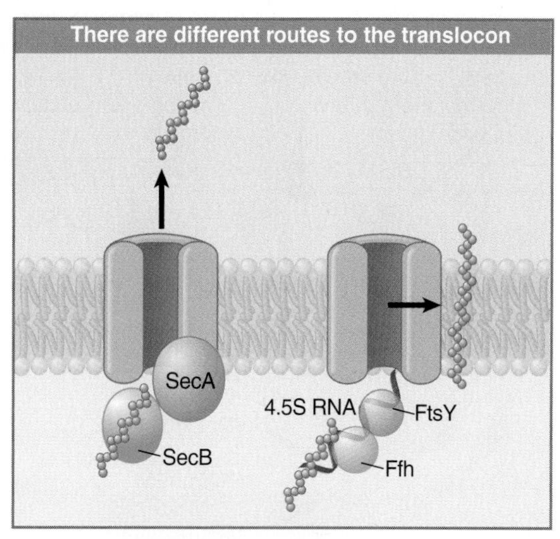

FIGURE 10.46 SecB/SecA transfer proteins to the translocon that pass through the membrane. 4.4S RNA transfers proteins that enter the membrane.

nascent protein to control its folding. It transfers the protein to SecA, which in turn transfers it to the translocon.

FIGURE 10.46 shows that there are two predominant ways of directing proteins to the Sec channel:

- the SecB chaperone, and
- the 4.5S RNA-based SRP.

Several chaperones can increase the efficiency of bacterial protein export by preventing premature folding; they include "trigger factor" (characterized as a chaperone that assists export), GroEL (see Section 10.5, Chaperones Are Needed by Newly Synthesized and by Denatured Proteins, and Section 10.18, Inner and Outer Mitochondrial Membranes Have Different Translocons), and SecB (identified as the product of one of the *sec* mutants). SecB is the least abundant of these proteins; however, it has the major role in promoting export. This role comprises two functions: first, SecB behaves as a chaperone and binds to a nascent protein to retard folding. It cannot reverse the change in structure of a folded protein, so it does not function as an unfolding factor. Its role is therefore to inhibit improper folding of the newly synthesized protein. Second, SecB has an affinity for the protein SecA. This allows it to target a precursor protein to the membrane. The SecB-SecYEG pathway is used for translocation of proteins that are secreted into the periplasm and is summarized in FIGURE 10.47.

SecA is a large peripheral membrane protein that has alternative ways to associate with the membrane. As a peripheral membrane protein, it associates with the membrane by virtue

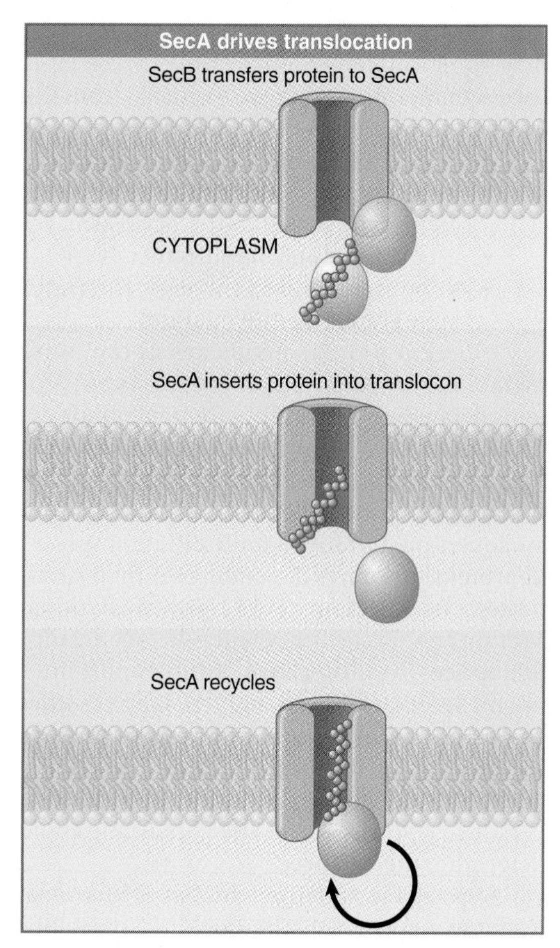

FIGURE 10.47 SecB transfers a nascent protein to SecA, which inserts the protein into the channel. Translocation requires hydrolysis of ATP and a protonmotive force. SecA undergoes cycles of association and dissociation with the channel and provides the motive force to push the protein through.

of its affinity for acidic lipids and for the SecY component of the translocon, which are part of a multisubunit complex that provides the translocase function. In the presence of other proteins (SecD and SecF), however, SecA can be found as a membrane-spanning protein. It probably provides the motor that pushes the substrate protein through the SecYEG translocon.

SecA recognizes both SecB and the precursor protein that it chaperones; most likely, features of the mature protein sequence as well as its leader are required for recognition. SecA has an ATPase activity that depends upon binding to lipids, SecY, and a precursor protein. The ATPase functions in a cyclical manner during translocation. After SecA binds a precursor protein it binds ATP, and ~20 amino acids are translocated through the membrane. Hydrolysis of ATP is required to release the precursor from SecA. The cycle may then be repeated. Precursor protein is bound again to provide the spur to bind more ATP, translocate another segment of protein, and release the precursor. SecA may alternate between the peripheral and integral membrane forms during translocation; with each cycle, a 30 kD domain of SecA may insert into the membrane and then retract.

Another process can also undertake translocation. When a precursor is released by SecA, it can be driven through the membrane by a protonmotive force (that is, an electrical potential across the membrane). This process cannot initiate transfer through the membrane, but it can continue the process initiated by a cycle of SecA ATPase action. Thus after or between cycles of the SecA–ATP driven reaction, the protonmotive force can drive translocation of the precursor.

The *E. coli* ribonucleoprotein complex of 4.5S RNA with Ffh and FtsY proteins is a counterpart to the eukaryotic SRP (see Section 10.9, The SRP Interacts with the SRP Receptor). It probably plays the role of keeping the nascent protein in an appropriate conformation until it interacts with other components of the secretory apparatus. It is needed for the secretion of some, but not all, proteins. As we see in Figure 10.46, its substrates are integral membrane proteins. The basis for differential selection of substrates is that the *E. coli* SRP recognizes an anchor sequence in the protein (anchor sequences by definition are present only in integral membrane proteins). Chloroplasts have counterparts to the Ffh and FtsY proteins, but do not require an RNA component.

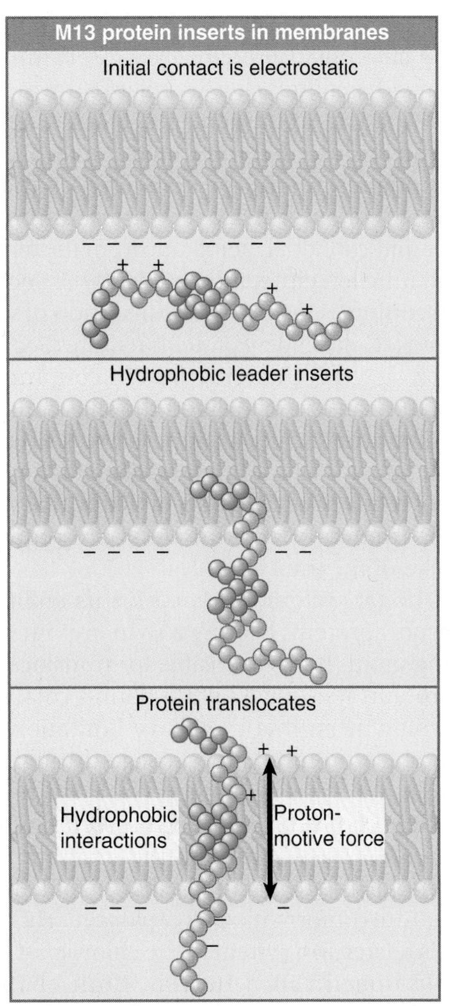

FIGURE 10.48 M13 coat protein inserts into the inner membrane by making an initial electrostatic contact, followed by insertion of hydrophobic sequences. Translocation is driven by hydrophobic interactions and a protonmotive force until the anchor sequence enters the membrane.

10.22 Sec-Independent Translocation Systems in *E. coli*

Key concepts

- *E. coli* and organelles have related systems for protein translocation.
- One system allows certain proteins to insert into membranes without a translocation apparatus.
- YidC is homologous to a mitochondrial system for transferring proteins into the inner membrane.
- The tat system transfers proteins with a twin arginine motif into the periplasmic space.

The most striking alternative system for protein translocation in *E. coli* is revealed by the coat protein of phage M13. FIGURE 10.48 shows that this does not appear to require any translocation apparatus! It can insert posttranslationally into protein-free liposomes. Targeting the protein to the membrane requires specific sequences (comprising basic residues) in the N- and C-terminal regions of the protein. They may interact with negatively charged heads of phospholipids. The protein then enters the membrane by using hydrophobic groups in its N-terminal leader sequence and an internal anchor sequence. Hydrophobicity is the main driving force for translocation, but it can be assisted by a protonmotive force that is generated between the positively charged periplasmic side of the membrane and an acidic region in the protein. This drives the protein through the membrane, and leader peptidase can then cleave the N-terminal sequence. The generality of this mechanism in bacteria is unclear; it may apply only to the special case of bacteriophage coat proteins. Some chloroplast proteins may insert into the thylakoid membrane by a similar pathway.

Mutations in the gene *yidC* block insertion of proteins into the inner membrane. YidC is homologous to the protein Oxa1p that is required when proteins are inserted into the inner mitochondrial membrane from the matrix. It can function either independently of SecYEG or in conjunction with it. The insertion of some of the YidC-dependent proteins requires SecYEG, which suggests that YidC acts in conjunction with the translocon to divert the substrate into membrane insertion as opposed to secretion. Other proteins whose insertion depends on YidC do not require SecYEG: It seems likely that some other (unidentified) functions are required instead of the translocon.

The tat system is named for its ability to transport proteins bearing a twin arginine targeting motif. It is responsible for translocation of proteins that have tightly bound cofactors. This may mean that they have limitations on their ability to unfold for passage through the membrane. This would be contrary to the principle of most translocation systems, where the protein passes through the membrane in an unfolded state and then must be folded into its mature conformation after passage. This system is related to a system in the chloroplast thylakoid lumen called Hcf106. Both of these systems transport proteins into the periplasm.

10.23 Summary

A protein that is inserted into, or passes through, a membrane has a signal sequence that is recognized by a receptor that is part of the membrane or that can associate with it. The protein passes through an aqueous channel that is created by transmembrane protein(s) that reside in the membrane. In almost all cases, the protein passes through the channel in an unfolded form, and association with chaperones when it emerges is necessary in order to acquire the correct conformation. The major exception is the peroxisome, where an imported protein in its mature conformation binds to a cytosolic protein that carries it through the channel in the membrane.

Synthesis of proteins in the cytosol starts on "free" ribosomes. Proteins that are secreted from the cell or that are inserted into membranes of the reticuloendothelial system start with an N-terminal signal sequence that causes the ribosome to become attached to the membrane of the endoplasmic reticulum. The protein is translocated through the membrane by cotranslational transfer. The process starts when the signal sequence is recognized by the SRP (a ribonucleoprotein particle), which interrupts translation. The SRP binds to the SRP receptor in the ER membrane and transfers the signal sequence to the Sec61/TRAM receptor in the membrane. Synthesis resumes, and the protein is translocated through the membrane while it is being synthesized, although there is no energetic connection between the processes. The channel through the membrane provides a hydrophilic environment and is largely made of the protein Sec61.

A secreted protein passes completely through the membrane into the ER lumen. Proteins that are integrated into membranes can be divided into two general types based on their orientation. For type I integral membrane proteins, the N-terminal signal sequence is cleaved, and transfer through the membrane is halted later by an anchor sequence. The protein becomes oriented in the membrane with its N-terminus on the far side and its C-terminus in the cytosol. Type II proteins do not have a cleavable N-terminal signal, but instead have a combined signal-anchor sequence, which enters the membrane and becomes embedded in it. This causes the C-terminus to be located on the far side, whereas the N-terminus remains in the cytosol. The orientation of the signal-anchor is determined by the "positive inside" rule, which states that the side of the anchor with more positive charges will be located in the cytoplasm. Proteins that have single transmembrane spanning regions move laterally from the channel into the lipid bilayer. Proteins may have multiple membrane-spanning regions, with loops between them protruding on either side of the membrane. The mechanism of insertion of multiple segments is unknown.

In the absence of any particular signal, a protein is released into the cytosol when its synthesis is completed. Proteins are imported posttranslationally into mitochondria or chloroplasts. They possess N-terminal leader sequences that target them to the outer membrane of the organelle envelope; they then are transported through the outer and inner membranes into the matrix. Translocation requires ATP and a potential across the inner membrane. The N-terminal leader is cleaved by a protease within the organelle. Proteins that reside within the membranes or intermembrane space possess a signal (which becomes N-terminal when the first part of the leader is removed) that either causes export from the matrix to the appropriate location or which halts transfer before all of the protein has entered the matrix. Control of folding, by Hsp70 and Hsp60 in the mitochondrial matrix, is an important feature of the process.

Mitochondria and chloroplasts have separate receptor complexes that create channels through each of the outer and inner membranes. All imported proteins pass directly from the TOM complex in the outer membrane to a TIM complex in the inner membrane. Proteins that reside in the intermembrane space or in the outer membrane are reexported from the TIM complex after entering the matrix. The TOM complex uses different receptors for imported proteins depending on whether they have N-terminal or internal signal sequences and directs both types into the Tom40 channel. There are two TIM receptors in the inner membrane: one is used for proteins whose ultimate destination is the inner matrix; the other is used for proteins that are reexported to the intermembrane space or the outer membrane.

Bacteria have components for membrane translocation that are related to those of the cotranslational eukaryotic system, but translocation often occurs by a posttranslational mechanism. SecY/E provide the translocase, and SecA associates with the channel and is involved in inserting and propelling the substrate protein. SecB is a chaperone that brings the protein to the channel. Some integral membrane proteins are inserted into the channel by an interaction with an apparatus resembling the SRP, which consists of 4.5S RNA and the Ffh and FtsY proteins. The protein YidC is homologous to a mitochondrial protein and is required for insertion of some membrane proteins.

References

10.4 Chaperones May Be Required for Protein Folding

Reviews

Ellis, R. J. and van der Vies, S. M. (1991). Molecular chaperones. *Annu. Rev. Biochem.* 60, 321–347.

Fersht, A. R. and Daggett, V. (2002). Protein folding and unfolding at atomic resolution. *Cell* 108, 573–582.

Hartl, F. U. and Hayer-Hartl, M. (2002). Molecular chaperones in the cytosol: from nascent chain to folded protein. *Science* 295, 1852–1858.

Research

Anfinsen, C. B. (1973). Principles that govern the folding of protein chains. *Science* 181, 223–230.

van den Berg, B., Ellis, R. J., and Dobson, C. M. (1999). Effects of macromolecular crowding on protein folding and aggregation. *EMBO J.* 18, 6927–6933.

10.5 Chaperones Are Needed by Newly Synthesized and by Denatured Proteins

Reviews

Frydman, J. (2001). Folding of newly translated proteins *in vitro:* the role of molecular chaperones. *Annu. Rev. Biochem.* 70, 603–647.

Moarefi, I. and Hartl, F. U. (2001). Hsp90: a specialized but essential protein-folding tool. *J. Cell Biol.* 154, 267–273.

Research

Queitsch, C., Sangster, T. A., and Lindquist, S. (2002). Hsp90 as a capacitor of phenotypic variation. *Nature* 417, 618–624.

Rutherford, S. L. and Lindquist, S. (1998). Hsp90 as a capacitor for morphological evolution. *Nature* 396, 336–342.

10.6 The Hsp70 Family Is Ubiquitous

Reviews

Bukau, B. and Horwich, A. L. (1998). The Hsp70 and Hsp60 chaperone machines. *Cell* 92, 351–366.

Frydman, J. (2001). Folding of newly translated proteins *in vitro:* the role of molecular chaperones. *Annu. Rev. Biochem.* 70, 603–647.

Georgopoulos, C. and Welch, W. J. (1993). Role of the major heat shock proteins as molecular chaperones. *Annu. Rev. Cell Biol.* 9, 601–634.

Hartl, F. U. (1966). Molecular chaperones in cellular protein folding. *Nature* 381, 571–580.

Research

Blond-Elguindi, S., Cwirla, S. E., Dower, W. J., Lipshutz, R. J., Sprang, S. R., Sambrook, J. F., and Gething, M. J. (1993). Affinity panning of a library of peptides displayed on bacteriophages reveals the binding specificity of BiP. *Cell* 75, 717–728.

Flaherty, K. M., DeLuca-Flaherty, C., and McKay, D. B. (1990). Three-dimensional structure of the ATPase fragment of a 70K heat-shock cognate protein. *Nature* 346, 623–628.

Flynn, G. C., Pohl, J., Flocco, M. T., and Rothman, J. E. (1991). Peptide-binding specificity of the molecular chaperone BiP. *Nature* 353, 726–730.

Zhu, X., Zhao, X., Burkholder, W. F., Gragerov, A., Ogata, C. M., Gottesman, M. E., and Hendrickson, W. A. (1996). Structural analysis of substrate binding by the molecular chaperone DnaK. *Science* 272, 1606–1614.

10.7 Signal Sequences Initiate Translocation

Reviews

Lee, C. and Beckwith, J. (1986). Cotranslational and posttranslational protein translocation in prokaryotic systems. *Annu. Rev. Cell Biol.* 2, 315–336.

Palade, G. (1975). Intracellular aspects of the process of protein synthesis. *Science* 189, 347–358.

Research

Blobel, G. and Dobberstein, B. (1975). Transfer of proteins across membranes. I. Presence of proteolytically processed and unprocessed nascent immunoglobulin light chains on membrane-bound ribosomes of murine myeloma. *J. Cell Biol.* 67, 835–851.

Lingappa, V. R., Chaidez, J., Yost, C. S., and Hedgpeth, J. (1984). Determinants for protein localization: beta-lactamase signal sequence directs globin across microsomal membranes. *Proc. Natl. Acad. Sci. USA* 81, 456–460.

von Heijne, G. (1985). Signal sequences. The limits of variation. *J. Mol. Biol.* 184, 99–105.

10.8 The Signal Sequence Interacts with the SRP

Review

Walter, P. and Johnson, A. E. (1994). Signal sequence recognition and protein targeting to the endoplasmic reticulum membrane. *Annu. Rev. Cell Biol.* 10, 87–119.

Research

Tjalsma, H., Bolhuis, A., van Roosmalen, M. L., Wiegert, T., Schumann, W., Broekhuizen, C. P., Quax, W. J., Venema, G., Bron, S., and van Dijl, M. (1998). Functional analysis of the secretory precursor processing machinery of *Bacillus subtilis:* identification of a eubacterial homolog of archaeal and eukaryotic signal peptidases. *Genes Dev.* 12, 2318–2331.

Walter, P. and Blobel, G. (1981). Translocation of proteins across the ER III SRP causes signal sequence and site specific arrest of chain elongation that is released by microsomal membranes. *J. Cell Biol.* 91, 557–561.

10.9 The SRP Interacts with the SRP Receptor

Reviews

Doudna, J. A. and Batey, R. T. (2004). Structural insights into the signal recognition particle. *Annu. Rev. Biochem.* 73, 539–557.

Keenan, R. J., Freymann, D. M., Stroud, R. M., and Walter, P. (2001). The signal recognition particle. *Annu. Rev. Biochem.* 70, 755–775.

Research

Batey, R. T., Rambo, R. P., Lucast, L., Rha, B., and Doudna, J. A. (2000). Crystal structure of the ribonucleoprotein core of the signal recognition particle. *Science* 287, 1232–1239.

Halic, M., Becker, T., Pool, M. R., Spahn, C. M., Grassucci, R. A., Frank, J., and Beckmann, R. (2004). Structure of the signal recognition particle interacting with the elongation-arrested ribosome. *Nature* 427, 808–814.

Keenan, R. J., Freymann, D. M., Walter, P., and Stroud, R. M. (1998). Crystal structure of the signal sequence-binding subunit of the signal recognition particle. *Cell* 94, 181–191.

Powers, T. and Walter, P. (1995). Reciprocal stimulation of GTP hydrolysis by two directly interacting GTPases. *Science* 269, 1422–1424.

Siegel, V. and Walter, P. (1988). Each of the activities of SRP is contained within a distinct domain: analysis of biochemical mutants of SRP. *Cell* 52, 39–49.

Tajima, S., Lauffer, L., Rath, V. L., and Walter, P. (1986). The signal recognition particle receptor is a complex that contains two distinct polypeptide chains. *J. Cell Biol.* 103, 1167–1178.

Walter, P. and Blobel, G. (1981). Translocation of proteins across the ER III SRP causes signal sequence and site specific arrest of chain elongation that is released by microsomal membranes. *J. Cell Biol.* 91, 557–561.

Walter, P. and Blobel, G. (1982). Signal recognition particle contains a 7S RNA essential for protein translocation across the ER. *Nature* 299, 691–698.

Zopf, D., Bernstein, H. D., Johnson, A. E., and Walter, P. (1990). The methionine-rich domain of the 54 kd protein subunit of the signal recognition particle contains an RNA binding site and can be crosslinked to a signal sequence. *EMBO J.* 9, 4511–4517.

10.10 The Translocon Forms a Pore

Research

Crowley, K. S. (1994). Secretory proteins move through the ER membrane via an aqueous, gated pore. *Cell* 78, 461–471.

Deshaies, R. J. and Schekman, R. (1987). A yeast mutant defective at an early stage in import of secretory protein precursors into the endoplasmic reticulum. *J. Cell Biol.* 105, 633–645.

Esnault, Y., Blondel, M. O., Deshaies, R. J., Scheckman, R., and Kepes, F. (1993). The yeast SSS1 gene is essential for secretory protein translocation and encodes a conserved protein of the endoplasmic reticulum. *EMBO J.* 12, 4083–4093.

Hanein, D., Matlack, K. E., Jungnickel, B., Plath, K., Kalies, K. U., Miller, K. R., Rapoport, T. A., and Akey, C. W. (1996). Oligomeric rings of the Sec61p complex induced by ligands required for protein translocation. *Cell* 87, 721–732.

Liao, S., Lin, J., Do, H., and Johnson, A. E. (1997). Both lumenal and cytosolic gating of the aqueous ER translocon pore are regulated from inside the ribosome during membrane protein integration. *Cell* 90, 31–41.

Mothes, W., Prehn, S., and Rapoport, T. A. (1994). Systematic probing of the environment of a translocating secretory protein during translocation through the ER membrane. *EMBO J.* 13, 3973–3982.

Simon, S. M. and Blobel, G. (1991). A protein-conducting channel in the endoplasmic reticulum. *Cell* 65, 371–380.

van den Berg, B., Clemons, W. M., Collinson, I., Modis, Y., Hartmann, E., Harrison, S. C., and Rapoport, T. A. (2004). X-ray structure of a protein-conducting channel. *Nature* 427, 36–44.

10.11 Translocation Requires Insertion into the Translocon and (Sometimes) a Ratchet in the ER

Reviews

Rapoport, T. A., Jungnickel, B., and Kutay, U. (1996). Protein transport across the eukaryotic endoplasmic reticulum and bacterial inner membranes. *Annu. Rev. Biochem.* 65, 271–303.

Walter, P. and Lingappa, V. (1986). Mechanism of protein translocation across the endoplasmic reticulum membrane. *Annu. Rev. Cell Biol.* 2, 499–516.

Research

Gorlich, D. and Rapoport, T. A. (1993). Protein translocation into proteoliposomes reconstituted from purified components of the endoplasmic reticulum membrane. *Cell* 75, 615–630.

Matlack, K. E., Misselwitz, B., Plath, K., and Rapoport, T. A. (1999). BiP acts as a molecular ratchet during posttranslational transport of prepro-alpha factor across the ER membrane. *Cell* 97, 553–564.

10.12 Reverse Translocation Sends Proteins to the Cytosol for Degradation

Reviews

Johnson, A. E. and Haigh, N. G. (2000). The ER translocon and retrotranslocation: is the shift into reverse manual or automatic? *Cell* 102, 709–712.

Tsai, B., Ye, Y., and Rapoport, T. A. (2002). Retrotranslocation of proteins from the endoplasmic reticulum into the cytosol. *Nat. Rev. Mol. Cell Biol.* 3, 246–255.

Research

Lilley, B. N. and Ploegh, H. L. (2004). A membrane protein required for dislocation of misfolded proteins from the ER. *Nature* 429, 834–840.

Wiertz, E. J. H. J. et al. (1996). Sec61-mediated transfer of a membrane protein from the endoplasmic reticulum to the proteasome for destruction. *Nature* 384, 432–438.

Wilkinson, B. M., Tyson, J. R., Reid, P. J., and Stirling, C. J. (2000). Distinct domains within yeast Sec61p involved in post-translational translocation and protein dislocation. *J. Biol. Chem.* 275, 521–529.

Ye, Y., Shibata, Y., Yun, C., Ron, D., and Rapoport, T. A. (2004). A membrane protein complex mediates retro-translocation from the ER lumen into the cytosol. *Nature* 429, 841–847.

Zhou, M. and Schekman, R. (1999). The engagement of Sec61p in the ER dislocation process. *Mol. Cell* 4, 925–934.

10.15 How Do Proteins Insert into Membranes?

Reviews

Hegde, R. S. and Lingappa, V. R. (1997). Membrane protein biogenesis: regulated complexity at the endoplasmic reticulum. *Cell* 91, 575–582.

Wickner, W. T. and Lodish, H. (1985). Multiple mechanisms of protein insertion into and across membranes. *Science* 230, 400–407.

Research

Borel, A. C., and Simon, S. M. (1996). Biogenesis of polytopic membrane proteins: membrane segments assemble within translocation channels prior to membrane integration. *Cell* 85, 379–389.

Do, H., Falcone, D., Lin, J., Andrews, D. W., and Johnson, A. E. (1996). The cotranslational integration of membrane proteins into the phospholipid bilayer is a multistep process. *Cell* 85, 369–378.

Heinrich, S. U., Mothes, W., Brunner, J., and Rapoport, T. A. (2000). The Sec61p complex mediates the integration of a membrane protein by allowing lipid partitioning of the transmembrane domain. *Cell* 102, 233–244.

Hessa, T., Kim, H., Bihlmaier, K., Lundin, C., Boekel, J., Andersson, H., Nilsson, I., White, S. H., and von Heijne, G. (2005). Recognition of transmembrane helices by the endoplasmic reticulum translocon. *Nature* 433, 377–381.

Kim, P. K., Janiak-Spens, F., Trimble, W. S., Leber, B., and Andrews, D. W. (1997). Evidence for multiple mechanisms for membrane binding and integration via carboxyl-terminal insertion sequences. *Biochemistry* 36, 8873–8882.

Liao, S., Lin, J., Do, H., and Johnson, A. E. (1997). Both lumenal and cytosolic gating of the aqueous ER translocon pore are regulated from inside the ribosome during membrane protein integration. *Cell* 90, 31–41.

Mothes, W., Heinrich, S. U., Graf, R., Nilsson, I., von Heijne, G., Brunner, J., and Rapoport, T. A. (1997). Molecular mechanism of membrane protein integration into the endoplasmic reticulum. *Cell* 89, 523–533.

van den Berg, B., Clemons, W. M., Collinson, I., Modis, Y., Hartmann, E., Harrison, S. C., and Rapoport, T. A. (2004). X-ray structure of a protein-conducting channel. *Nature* 427, 36–44.

10.16 Posttranslational Membrane Insertion Depends on Leader Sequences

Reviews

Baker, K. P. and Schatz, G. (1991). Mitochondrial proteins essential for viability mediate protein import into yeast mitochondria. *Nature* 349, 205–208.

Schatz, G. and Dobberstein, B. (1996). Common principles of protein translocation across membranes. *Science* 271, 1519–1526.

Research

Eilers, M. and Schatz, G. (1986). Binding of a specific ligand inhibits import of a purified precursor protein into mitochondria. *Nature* 322, 228–232.

10.17 A Hierarchy of Sequences Determines Location within Organelles

Review

Cline, K. and Henry, R. (1996). Import and routing of nucleus-encoded chloroplast proteins. *Annu. Rev. Cell Dev. Biol.* 12, 1–26.

Research

Hartl, F. U., Ostermann, J., Guiard, B., and Neupert, W. (1987). Successive translocation into and out of the mitochondrial matrix: targeting of proteins to the intermembrane space by a bipartite signal peptide. *Cell* 51, 1027–1037.

van Loon, A. P. G. M. et al. (1986). The presequences of two imported mitochondrial proteins contain information for intracellular and intramitochondrial sorting. *Cell* 44, 801–812.

10.18 Inner and Outer Mitochondrial Membranes Have Different Translocons

Reviews

Dalbey, R. E. and Kuhn, A. (2000). Evolutionarily related insertion pathways of bacterial, mitochondrial, and thylakoid membrane proteins. *Annu. Rev. Cell Dev. Biol.* 16, 51–87.

Neupert, W. (1997). Protein import into mitochondria. *Annu. Rev. Biochem.* 66, 863–917.

Neupert, W. and Brunner, M. (2002). The protein import motor of mitochondria. *Nat. Rev. Mol. Cell Biol.* 3, 555–565.

Research

Leuenberger, D., Bally, N. A., Schatz, G., and Koehler, C. M. (1999). Different import pathways through the mitochondrial intermembrane space for inner membrane proteins. *EMBO J.* 18, 4816–4822.

Ostermann, J., Horwich, A. L., Neupert, W., and Hartl, F. U. (1989). Protein folding in mitochondria requires complex formation with hsp60 and ATP hydrolysis. *Nature* 341, 125–130.

10.19 Peroxisomes Employ Another Type of Translocation System

Reviews

Purdue, P. E. and Lazarow, P. B. (2001). Peroxisome biogenesis. *Annu. Rev. Cell Dev. Biol.* 17, 701–752.

Subramani, S., Koller, A., and Snyder, W. B. (2000). Import of peroxisomal matrix and membrane proteins. *Annu. Rev. Biochem.* 69, 399–418.

Research

Dodt, G. and Gould, S. J. (1996). Multiple PEX genes are required for proper subcellular distribution and stability of Pex5p, the PTS1 receptor: evidence that PTS1 protein import is mediated by a cycling receptor. *J. Cell Biol.* 135, 1763–1774.

Elgersma, Y., Elgersma-Hooisma, M., Wenzel, T., McCaffery, J. M., Farquhar, M. G., and Subramani, S. (1998). A mobile PTS2 receptor for peroxisomal protein import in Pichia pastoris. *J. Cell Biol.* 140, 807–820.

Elgersma, Y., Vos, A., van den Berg, M., van Roermund, C. W., van der Sluijs, P., Distel, B., and Tabak, H. F. (1996). Analysis of the carboxyl-terminal peroxisomal targeting signal 1 in a homologous context in *S. cerevisiae*. *J. Biol. Chem.* 271, 26375–26382.

Goldfischer, S., Moore, C. L., Johnson, A. B., Spiro, A. J., Valsamis, M. P., Wisniewski, H. K., Ritch, R. H., Norton, W. T., Rapin, I., and Gartner, L. M. (1973). Peroxisomal and mitochondrial defects in the cerebro-hepato-renal syndrome. *Science* 182, 62–64.

Gould, S. J., Keller, G. A., Hosken, N., Wilkinson, J., and Subramani, S. (1989). A conserved tripeptide sorts proteins to peroxisomes. *J. Cell Biol.* 108, 1657–1664.

Matsuzono, Y., Kinoshita, N., Tamura, S., Shimozawa, N., Hamasaki, M., Ghaedi, K., Wanders, R. J., Suzuki, Y., Kondo, N., and Fujiki, Y. (1999). Human PEX19: cDNA cloning by functional complementation, mutation analysis in a patient with Zellweger syndrome, and potential role in peroxisomal membrane assembly. *Proc. Natl. Acad. Sci. USA* 96, 2116–2121.

South, S. T. and Gould, S. J. (1999). Peroxisome synthesis in the absence of preexisting peroxisomes. *J. Cell Biol.* 144, 255–266.

Walton, P. A., Hill, P. E., and Hill, S. (1995). Import of stably folded proteins into peroxisomes. *Mol. Biol. Cell* 6, 675–683.

10.21 The Sec System Transports Proteins into and Through the Inner Membrane

Reviews

Lee, C. and Beckwith, J. (1986). Cotranslational and posttranslational protein translocation in prokaryotic systems. *Annu. Rev. Cell Biol.* 2, 315–336.

Oliver, D. (1985). Protein secretion in *E. coli*. *Annu. Rev. Immunol.* 39, 615–648.

Research

Beck, K., Wu, L. F., Brunner, J., and Muller, M. (2000). Discrimination between SRP- and SecA/SecB-dependent substrates involves selective recognition of nascent chains by SRP and trigger factor. *EMBO J.* 19, 134–143.

Brundage, L. et al. (1990). The purified *E. coli* integral membrane protein SecY/E is sufficient for reconstitution of SecA-dependent precursor protein translocation. *Cell* 62, 649–657.

Collier, D. N. et al. (1988). The antifolding activity of SecB promotes the export of the *E. coli* maltose-binding protein. *Cell* 53, 273–283.

Crooke, E. et al. (1988). ProOmpA is stabilized for membrane translocation by either purified *E. coli* trigger factor or canine signal recognition particle. *Cell* 54, 1003–1011.

Valent, Q. A., Scotti, P. A., High, S., von Heijne, G., Lentzen, G., Wintermeyer, W., Oudega, B., and Luirink, J. (1998). The *E. coli* SRP and SecB targeting pathways converge at the translocon. *EMBO J.* 17, 2504–2512.

Yahr, T. L. and Wickner, W. T. (2000). Evaluating the oligomeric state of SecYEG in preprotein translocase. *EMBO J.* 19, 4393–4401.

10.22 Sec-Independent Translocation Systems in *E. coli*

Reviews

Dalbey, R. E. and Kuhn, A. (2000). Evolutionarily related insertion pathways of bacterial, mitochondrial, and thylakoid membrane proteins. *Annu. Rev. Cell Dev. Biol.* 16, 51–87.

Dalbey, R. E. and Robinson, C. (1999). Protein translocation into and across the bacterial plasma membrane and the plant thylakoid membrane. *Trends Biochem. Sci.* 24, 17–22.

Research

Beck, K., Wu, L. F., Brunner, J., and Muller, M. (2000). Discrimination between SRP- and SecA/SecB-dependent substrates involves selective recognition of nascent chains by SRP and trigger factor. *EMBO J.* 19, 134–143.

Samuelson, J. C., Chen, M., Jiang, F., Moller, I., Wiedmann, M., Kuhn, A., Phillips, G. J., and Dalbey, R. E. (2000). YidC mediates membrane protein insertion in bacteria. *Nature* 406, 637–641.

Scotti, P. A., Urbanus, M. L., Brunner, J., de Gier, J. W., von Heijne, G., van der Does, C., Driessen, A. J., Oudega, B., and Luirink, J. (2000). YidC, the *E. coli* homologue of mitochondrial Oxa1p, is a component of the Sec translocase. *EMBO J.* 19, 542–549.

Soekarjo, M., Eisenhawer, M., Kuhn, A., and Vogel, H. (1996). Thermodynamics of the membrane insertion process of the M13 procoat protein, a lipid bilayer traversing protein containing a leader sequence. *Biochemistry* 35, 1232–1241.

11

Transcription

CHAPTER OUTLINE

11.1 Introduction

11.2 Transcription Occurs by Base Pairing in a "Bubble" of Unpaired DNA

- RNA polymerase separates the two strands of DNA in a transient "bubble" and uses one strand as a template to direct synthesis of a complementary sequence of RNA.
- The length of the bubble is ~12 to 14 bp, and the length of RNA-DNA hybrid within it is ~8 to 9 bp.

11.3 The Transcription Reaction Has Three Stages

- RNA polymerase initiates transcription after binding to a promoter site on DNA.
- During elongation the transcription bubble moves along DNA and the RNA chain is extended in the 5′–3′ direction.
- When transcription stops, the DNA duplex reforms and RNA polymerase dissociates at a terminator site.

11.4 Phage T7 RNA Polymerase Is a Useful Model System

- T3 and T7 phage RNA polymerases are single polypeptides with minimal activities in recognizing a small number of phage promoters.
- Crystal structures of T7 RNA polymerase with DNA identify the DNA-binding region and the active site.

11.5 A Model for Enzyme Movement Is Suggested by the Crystal Structure

- DNA moves through a groove in yeast RNA polymerase that makes a sharp turn at the active site.
- A protein bridge changes conformation to control the entry of nucleotides to the active site.

11.6 Bacterial RNA Polymerase Consists of Multiple Subunits

- Bacterial RNA core polymerases are ~500 kD multisubunit complexes with the general structure $\alpha_2\beta\beta'$.
- DNA is bound in a channel and is contacted by both the β and β' subunits.

11.7 RNA Polymerase Consists of the Core Enzyme and Sigma Factor

- Bacterial RNA polymerase can be divided into the $\alpha_2\beta\beta'$ core enzyme that catalyzes transcription and the sigma subunit that is required only for initiation.
- Sigma factor changes the DNA-binding properties of RNA polymerase so that its affinity for general DNA is reduced and its affinity for promoters is increased.
- Binding constants of RNA polymerase for different promoters vary over six orders of magnitude, corresponding to the frequency with which transcription is initiated at each promoter.

11.8 The Association with Sigma Factor Changes at Initiation

- When RNA polymerase binds to a promoter, it separates the DNA strands to form a transcription bubble and incorporates up to nine nucleotides into RNA.
- There may be a cycle of abortive initiations before the enzyme moves to the next phase.
- Sigma factor may be released from RNA polymerase when the nascent RNA chain reaches eight to nine bases in length.

11.9 A Stalled RNA Polymerase Can Restart

- An arrested RNA polymerase can restart transcription by cleaving the RNA transcript to generate a new 3′ end.

11.10 How Does RNA Polymerase Find Promoter Sequences?

- The rate at which RNA polymerase binds to promoters is too fast to be accounted for by random diffusion.
- RNA polymerase probably binds to random sites on DNA and exchanges them with other sequences very rapidly until a promoter is found.

11.11 Sigma Factor Controls Binding to DNA

- A change in association between sigma factor and holoenzyme changes binding affinity for DNA so that core enzyme can move along DNA.

11.12 Promoter Recognition Depends on Consensus Sequences

- A promoter is defined by the presence of short consensus sequences at specific locations.
- The promoter consensus sequences consist of a purine at the startpoint, the hexamer TATAAT centered at –10, and another hexamer centered at –35.
- Individual promoters usually differ from the consensus at one or more positions.

11.13 Promoter Efficiencies Can Be Increased or Decreased by Mutation

- Down mutations to decrease promoter efficiency usually decrease conformance to the consensus sequences, whereas up mutations have the opposite effect.
- Mutations in the –35 sequence usually affect initial binding of RNA polymerase.

- Mutations in the −10 sequence usually affect the melting reaction that converts a closed to an open complex.

11.14 RNA Polymerase Binds to One Face of DNA

- The consensus sequences at −35 and −10 provide most of the contact points for RNA polymerase in the promoter.
- The points of contact lie on one face of the DNA.

11.15 Supercoiling Is an Important Feature of Transcription

- Negative supercoiling increases the efficiency of some promoters by assisting the melting reaction.
- Transcription generates positive supercoils ahead of the enzyme and negative supercoils behind it, and these must be removed by gyrase and topoisomerase.

11.16 Substitution of Sigma Factors May Control Initiation

- *E. coli* has several sigma factors, each of which causes RNA polymerase to initiate at a set of promoters defined by specific −35 and −10 sequences.
- σ^{70} is used for general transcription, and the other sigma factors are activated by special conditions.

11.17 Sigma Factors Directly Contact DNA

- σ^{70} changes its structure to release its DNA-binding regions when it associates with core enzyme.
- σ^{70} binds both the −35 and −10 sequences.

11.18 Sigma Factors May Be Organized into Cascades

- A cascade of sigma factors is created when one sigma factor is required to transcribe the gene coding for the next sigma factor.
- The early genes of phage SP01 are transcribed by host RNA polymerase.
- One of the early genes codes for a sigma factor that causes RNA polymerase to transcribe the middle genes.
- Two of the middle genes code for subunits of a sigma factor that causes RNA polymerase to transcribe the late genes.

11.19 Sporulation Is Controlled by Sigma Factors

- Sporulation divides a bacterium into a mother cell that is lysed and a spore that is released.
- Each compartment advances to the next stage of development by synthesizing a new sigma factor that displaces the previous sigma factor.
- Communication between the two compartments coordinates the timing of sigma factor substitutions.

11.20 Bacterial RNA Polymerase Terminates at Discrete Sites

- Termination may require both recognition of the terminator sequence in DNA and the formation of a hairpin structure in the RNA product.

11.21 There Are Two Types of Terminators in *E. coli*

- Intrinsic terminators consist of a G-C-rich hairpin in the RNA product followed by a U-rich region in which termination occurs.

11.22 How Does Rho Factor Work?

- Rho factor is a terminator protein that binds to a *rut* site on nascent RNA and tracks along the RNA to release it from the RNA–DNA hybrid structure at the RNA polymerase.

11.23 Antitermination Is a Regulatory Event

- Termination is prevented when antitermination proteins act on RNA polymerase to cause it to read through a specific terminator or terminators.
- Phage lambda has two antitermination proteins, pN and pQ, that act on different transcription units.

11.24 Antitermination Requires Sites That Are Independent of the Terminators

- The site where an antiterminator protein acts is upstream of the terminator site in the transcription unit.
- The location of the antiterminator site varies in different cases and can be in the promoter or within the transcription unit.

11.25 Termination and AntiTermination Factors Interact with RNA Polymerase

- Several bacterial proteins are required for lambda pN to interact with RNA polymerase.
- These proteins are also involved in antitermination in the *rrn* operons of the host bacterium.
- The lambda antiterminator pQ has a different mode of interaction that involves binding to DNA at the promoter.

11.26 Summary

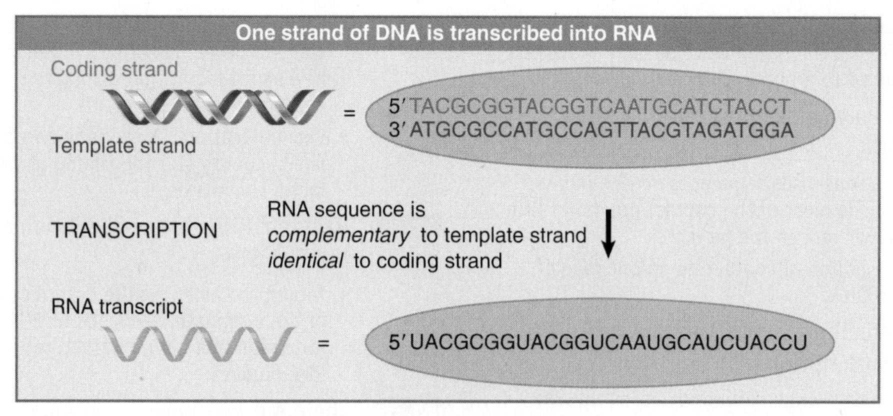

FIGURE 11.1 The function of RNA polymerase is to copy one strand of duplex DNA into RNA.

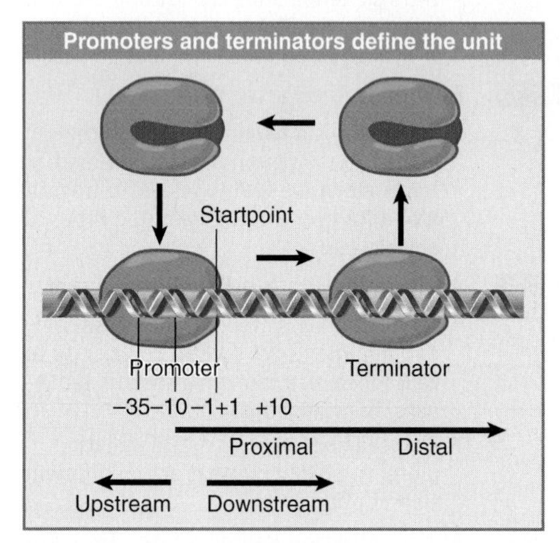

FIGURE 11.2 A transcription unit is a sequence of DNA transcribed into a single RNA, starting at the promoter and ending at the terminator.

11.1 Introduction

Transcription involves synthesis of an RNA chain representing one strand of a DNA duplex. When we say "representing," we mean that the RNA is *identical in sequence* with one strand of the DNA, which is called the **coding strand.** It is *complementary* to the other strand, which provides the **template strand** for its synthesis. FIGURE 11.1 recapitulates the relationship between double-stranded DNA and its single-stranded RNA transcript.

RNA synthesis is catalyzed by the enzyme **RNA polymerase.** Transcription starts when RNA polymerase binds to a special region, the **promoter,** at the start of the gene. The promoter surrounds the first base pair that is transcribed into RNA, the **startpoint.** From this point, RNA polymerase moves along the template, synthesizing RNA, until it reaches a **terminator** *(t)* sequence. This action defines a **transcription unit** that extends from the promoter to the terminator. The critical feature of the transcription unit, depicted in FIGURE 11.2, is that it constitutes a stretch of DNA *expressed via the production of a single RNA molecule.* A transcription unit may include more than one gene.

Sequences prior to the startpoint are described as **upstream** of it; those after the startpoint (within the transcribed sequence) are **downstream** of it. Sequences are conventionally written so that transcription proceeds from left (upstream) to right (downstream). This corresponds to writing the mRNA in the usual $5' \rightarrow 3'$ direction.

The DNA sequence often is written to show only the coding strand, which has the same sequence as the RNA. Base positions are numbered in both directions away from the startpoint, which is assigned the value +1; numbers increase as they go downstream. The base before the startpoint is numbered −1, and the negative numbers increase going upstream. (There is no base assigned the number 0.)

The immediate product of transcription is called the **primary transcript.** It consists of an RNA extending from the promoter to the terminator and possesses the original 5′ and 3′ ends. The primary transcript is, however, almost always unstable. In prokaryotes, it is rapidly degraded (mRNA) or cleaved to give mature products (rRNA and tRNA). In eukaryotes, it is modified at the ends (mRNA) and/or cleaved to give mature products (all RNA).

Transcription is the first stage in gene expression and the principal step at which it is controlled. Regulatory proteins determine whether a particular gene is available to be transcribed by RNA polymerase. The initial (and often the only) step in regulation is the decision on whether or not to transcribe a gene. Most regulatory events occur at the initiation of transcription, although subsequent stages in transcription (or other stages of gene expression) are sometimes regulated.

Within this context, there are two basic questions in gene expression:

- How does RNA polymerase find promoters on DNA? This is a particular example of a more general question: How do proteins distinguish their specific binding sites in DNA from other sequences?
- How do regulatory proteins interact with RNA polymerase (and with one another) to activate or to repress specific steps in the initiation, elongation, or termination of transcription?

In this chapter, we analyze the interactions of bacterial RNA polymerase with DNA from its initial contact with a gene, through the act of transcription, and then finally its release when the transcript has been completed. Chapter 12, The Operon, describes the various means by which regulatory proteins can assist or prevent bacterial RNA polymerase from recognizing a particular gene for transcription. Chapter 13, Regulatory RNA, discusses other means of regulation, including the use of small RNAs, and considers how these interactions can be connected into larger regulatory networks. In Chapter 14, Phage Strategies, we consider how individual regulatory interactions can be connected into more complex networks. In Chapter 24, Promoters and Enhancers, and Chapter 25, Activating Transcription, we consider the analogous reactions between eukaryotic RNA polymerases and their templates.

11.2 Transcription Occurs by Base Pairing in a "Bubble" of Unpaired DNA

Key concepts

- RNA polymerase separates the two strands of DNA in a transient "bubble" and uses one strand as a template to direct synthesis of a complementary sequence of RNA.
- The length of the bubble is ~12 to 14 bp, and the length of RNA-DNA hybrid within it is ~8 to 9 bp.

Transcription takes place by the usual process of complementary base pairing. FIGURE 11.3 illustrates the general principle of transcription. RNA synthesis takes place within a "transcription bubble," in which DNA is transiently separated into its single strands and the template strand is used to direct synthesis of the RNA strand.

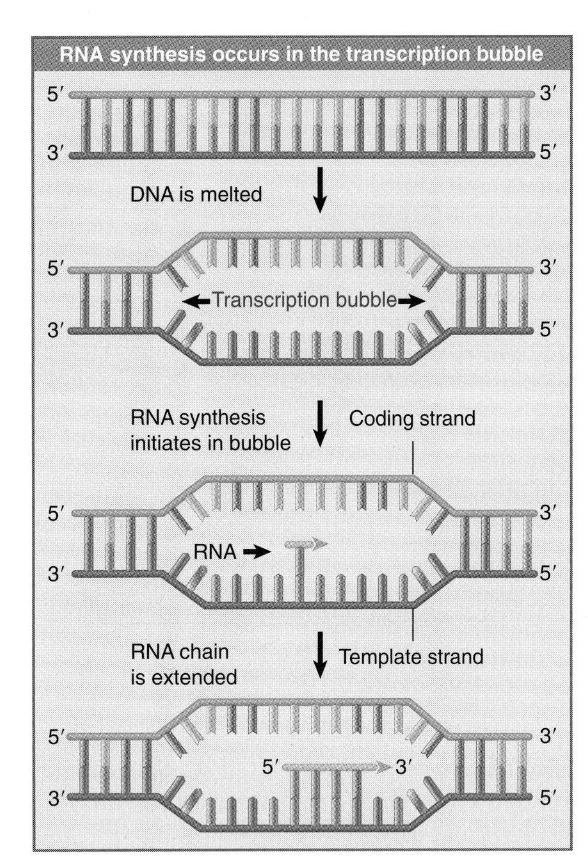

FIGURE 11.3 DNA strands separate to form a transcription bubble. RNA is synthesized by complementary base pairing with one of the DNA strands.

The RNA chain is synthesized from the 5′ end toward the 3′ end. The 3′–OH group of the last nucleotide added to the chain reacts with an incoming nucleoside 5′ triphosphate. The incoming nucleotide loses its terminal two phosphate groups (γ and β); its α group is used in the phosphodiester bond linking it to the chain. The overall reaction rate is ~40 nucleotides/second at 37° C (for the bacterial RNA polymerase); this is about the same as the rate of translation (15 amino acids/sec), but much slower than the rate of DNA replication (800 bp/sec).

RNA polymerase creates the transcription bubble when it binds to a promoter. FIGURE 11.4 shows that as RNA polymerase moves along the DNA, the bubble moves with it and the RNA chain grows longer. The process of base pairing and base addition within the bubble is catalyzed and scrutinized by the enzyme.

The structure of the bubble within RNA polymerase is shown in the expanded view of FIGURE 11.5. As RNA polymerase moves along the DNA template, it unwinds the duplex at the

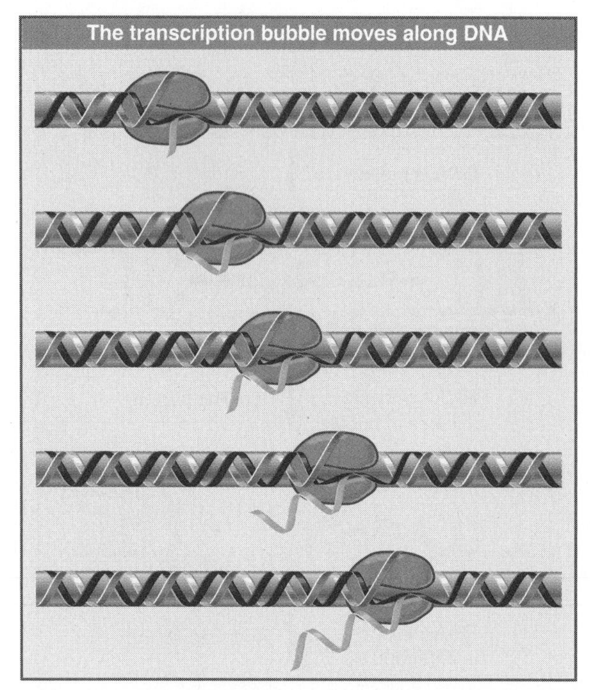

FIGURE 11.4 Transcription takes place in a bubble, in which RNA is synthesized by base pairing with one strand of DNA in the transiently unwound region. As the bubble progresses, the DNA duplex reforms behind it, displacing the RNA in the form of a single polynucleotide chain.

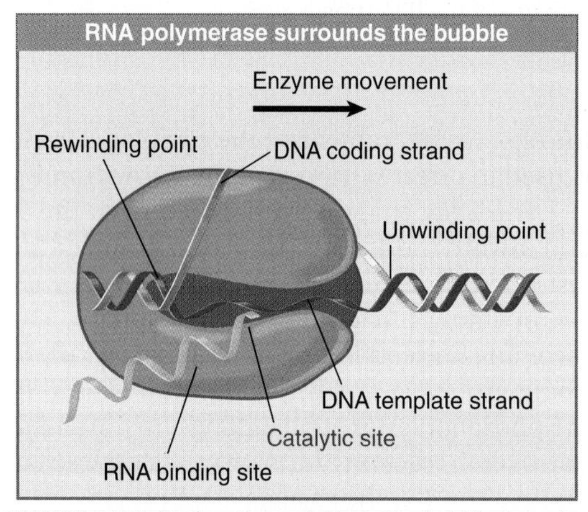

FIGURE 11.5 During transcription, the bubble is maintained within bacterial RNA polymerase, which unwinds and rewinds DNA and synthesizes RNA.

front of the bubble (the unwinding point), and rewinds the DNA at the back (the rewinding point). The length of the transcription bubble is ~12 to 14 bp, but the length of the RNA-DNA hybrid region within it is shorter.

There is a major change in the topology of DNA extending over ~1 turn, but it is not clear how much of this region is actually base paired with RNA at any given moment. Certainly the RNA-DNA hybrid is short and transient. As the enzyme moves on, the DNA duplex reforms, and the RNA is displaced as a free polynucleotide chain. Roughly the last twenty-five ribonucleotides added to a growing chain are complexed with DNA and/or enzyme at any moment.

11.3 The Transcription Reaction Has Three Stages

Key concepts

- RNA polymerase initiates transcription after binding to a promoter site on DNA.
- During elongation the transcription bubble moves along DNA and the RNA chain is extended in the 5′–3′ direction.
- When transcription stops, the DNA duplex reforms and RNA polymerase dissociates at a terminator site.

The transcription reaction can be divided into the stages illustrated in FIGURE 11.6, in which a bubble is created, RNA synthesis begins, the bubble moves along the DNA, and finally the bubble is terminated:

- *Template recognition* begins with the binding of RNA polymerase to the double-stranded DNA at a promoter to form a "closed complex." The strands of DNA are then separated to form the "open complex" that makes the template strand available for base pairing with ribonucleotides. The transcription bubble is created by a local unwinding that begins at the site bound by RNA polymerase.
- **Initiation** describes the synthesis of the first nucleotide bonds in RNA. The enzyme remains at the promoter while it synthesizes the first ~9 nucleotide bonds. The initiation phase is protracted by the occurrence of abortive events, in which the enzyme makes short transcripts, releases them, and then starts synthesis of RNA again. The initiation phase ends when the enzyme succeeds in extending the chain and clears the promoter. *The sequence of DNA needed for RNA polymerase to bind to the template and accomplish the initiation reaction defines the promoter.* Abortive initiation probably involves synthesizing an RNA chain that

fills the active site. If the RNA is released, the initiation is aborted and must start again. Initiation is accomplished if and when the enzyme manages to move along the template to move the next region of the DNA into the active site.

- During **elongation** the enzyme moves along the DNA and extends the growing RNA chain. As the enzyme moves, it unwinds the DNA helix to expose a new segment of the template in single-stranded condition. Nucleotides are covalently added to the 3′ end of the growing RNA chain, forming an RNA-DNA hybrid in the unwound region. Behind the unwound region, the DNA template strand pairs with its original partner to reform the double helix. The RNA emerges as a free single strand. *Elongation involves the movement of the transcription bubble by a disruption of DNA structure, in which the template strand of the transiently unwound region is paired with the nascent RNA at the growing point.*
- **Termination** involves recognition of the point at which no further bases should be added to the chain. To terminate transcription, the formation of phosphodiester bonds must cease, and the transcription complex must come apart. When the last base is added to the RNA chain, the transcription bubble collapses as the RNA-DNA hybrid is disrupted, the DNA reforms in duplex state, and the enzyme and RNA are both released. *The sequence of DNA required for these reactions defines the terminator.*

The traditional view of elongation has been that it is a monotonic process, in which the enzyme moves forward 1 bp along DNA for every nucleotide added to the RNA chain. Changes in this pattern occur in certain circumstances, in particular when RNA polymerase pauses. One type of pattern is for the "front end" of the enzyme to remain stationary while the "back end" continues to move, thus compressing the footprint on DNA. After movement of several base pairs, the "front end" is released, restoring a footprint of full length. This gave rise to the "inchworm" model of transcription, in which the enzyme proceeds discontinuously, alternatively compressing and releasing the footprint on DNA. It may, however, be the case that these events describe an aberrant situation rather than normal transcription.

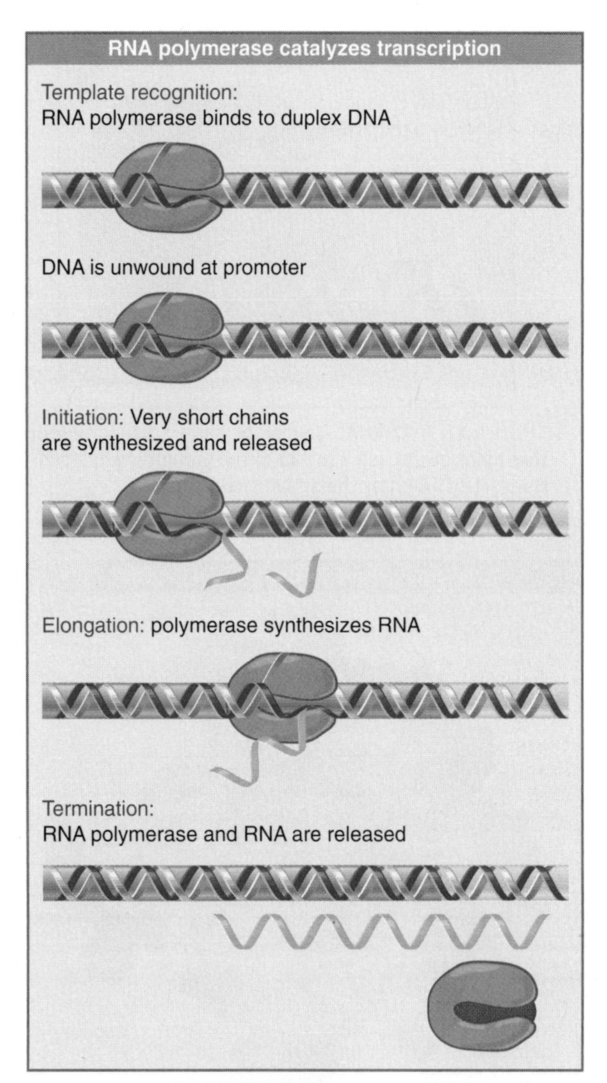

FIGURE 11.6 Transcription has four stages: The enzyme binds to the promoter and melts DNA, remains stationary during initiation, moves along the template during elongation, and dissociates at termination.

11.4 Phage T7 RNA Polymerase Is a Useful Model System

Key concepts

- T3 and T7 phage RNA polymerases are single polypeptides with minimal activities in recognizing a small number of phage promoters.
- Crystal structures of T7 RNA polymerase with DNA identify the DNA-binding region and the active site.

The existence of very small RNA polymerases, comprising single polypeptide chains coded by certain phages, gives some idea of the "minimum" apparatus necessary for transcription. These RNA polymerases recognize just a few promoters on the phage DNA, and they have no ability to change the set of promoters to which

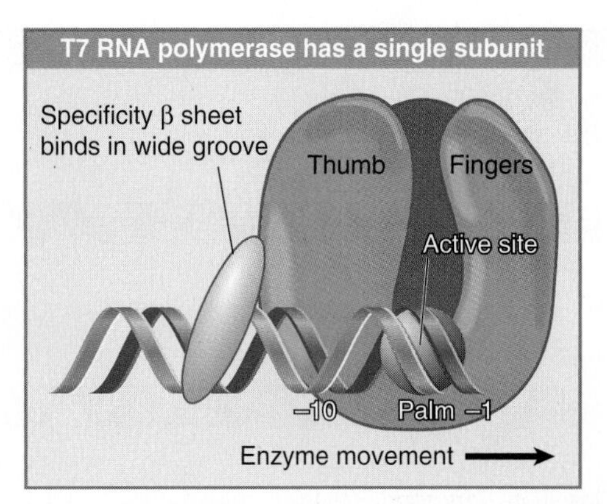

FIGURE 11.7 T7 RNA polymerase has a specificity loop that binds positions –7 to –11 of the promoter while positions –1 to –4 enter the active site.

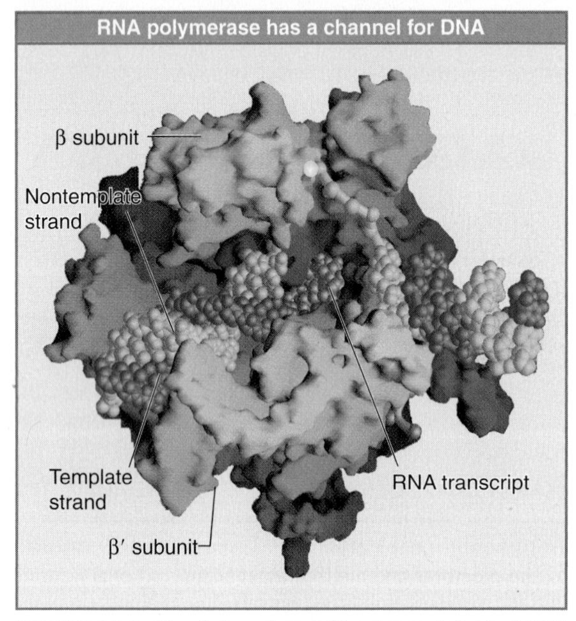

FIGURE 11.8 The β (cyan) and β′ subunit (pink) of RNA polymerase have a channel for the DNA template. Synthesis of an RNA transcript (copper) has just begun; the DNA template (red) and coding (yellow) strands are separated in a transcription bubble. Photo courtesy of Seth Darst, Rockefeller University.

they respond. They provide simple model systems for characterizing the binding of RNA polymerase to DNA and the initiation reaction.

The RNA polymerases coded by the related phages T3 and T7 are single polypeptide chains of <100 kD each. They synthesize RNA at rates of ~200 nucleotides/second at 37°C, a rate that is more rapid than that of bacterial RNA polymerase.

The T7 RNA polymerase is homologous to DNA polymerases and has a similar structure, in which DNA lies in a "palm" surrounded by "fingers" and a "thumb" (see Figure 18.7). We now have a direct view of the active site from a crystal structure of a phage T7 RNA polymerase engaged in transcription.

The T7 RNA polymerase recognizes its target sequence in DNA by binding to bases in the major groove at a position upstream from the startpoint, as shown in FIGURE 11.7. The enzyme uses a *specificity loop* that is formed by a β ribbon. This feature is unique to the RNA polymerase (it is not found in DNA polymerases). The common point with all RNA polymerases is that the enzyme recognizes specific bases in DNA that are upstream of the sequence that is transcribed.

When transcription initiates, the conformation of the enzyme remains essentially the same while several nucleotides are added, and the transcribed template strand is "scrunched" in the active site. The active site can hold a transcript of six to nine nucleotides. The transition from initiation to elongation is defined as the point when the enzyme begins to move along DNA. This occurs when the nascent transcript extends beyond the active site and interacts with the specificity loop. The RNA emerges to the surface of the enzyme when twelve to fourteen nucleotides have been synthesized. These features are similar to those displayed by bacterial RNA polymerase.

11.5 A Model for Enzyme Movement Is Suggested by the Crystal Structure

Key concepts

- DNA moves through a groove in yeast RNA polymerase that makes a sharp turn at the active site.
- A protein bridge changes conformation to control the entry of nucleotides to the active site.

We now have much information about the structure and function of RNA polymerase as the result of the crystal structures of the bacterial and yeast enzymes. Bacterial RNA polymerase has overall dimensions of ~90 × 95 × 160 Å; eukaryotic RNA polymerase is larger but less elongated. Structural analysis shows that they share a common type of structure, in which there is a "channel" or groove on the surface ~25 Å wide that could be the path for DNA. An example of this channel in bacterial RNA polymerase is illustrated in FIGURE 11.8. The length of the groove could hold 16 bp in the bacterial enzyme, and ~25 bp in the eukaryotic enzyme, but this represents only part of the total length of DNA bound during transcription. The enzyme surface is largely negatively charged, but the groove is lined with positive charges, enabling it to interact with the negatively charged phosphate groups of DNA.

The yeast enzyme is a large structure with twelve subunits (see Section 24.2, Eukaryotic RNA Polymerases Consist of Many Subunits). Ten subunits of the yeast RNA polymerase II have been located on the crystal structure, as shown in FIGURE 11.9. The catalytic site is formed by a cleft between the two large subunits (#1 and #2), which grasp DNA downstream in "jaws" as it enters the RNA polymerase. Subunits 4 and 7 are missing from this structure; they form a subcomplex that dissociates from the complete enzyme. The structure is generally similar to that of bacterial RNA polymerase. This can be seen more clearly in the crystal structure of FIGURE 11.10. RNA polymerase surrounds the DNA, as seen in the view of FIGURE 11.11. A catalytic Mg^{2+} ion is found at the active site. The DNA is clamped in position at the active site by subunits 1, 2, and 6. FIGURE 11.12 shows that DNA is forced to take a turn at the entrance to the site because of an adjacent wall of protein. The length of the RNA hybrid is limited by another protein obstruction, called the rudder. Nucleotides probably enter the active site from below, via pores through the structure.

The expanded view of the active site in FIGURE 11.13 shows that the transcription bubble includes 9 bp of DNA-RNA hybrid. Where the DNA takes its turn, the bases downstream are flipped out of the DNA helix. As the enzyme moves along DNA, the base in the template strand at the start of the turn will be flipped to face the nucleotide entry site. The 5′ end of the RNA is forced to leave the DNA when it hits the protein rudder (see Figure 11.12).

Once DNA has been melted, the individual strands have a flexible structure in the transcription bubble. This enables DNA to take its turn in the active site. Before transcription starts, though, the DNA double helix is a relatively rigid straight structure. How does this structure enter the polymerase without being blocked by the wall? The answer is that a large conformational shift must occur in the enzyme. Adjacent to the wall is a clamp. In the free form of RNA polymerase, this clamp swings away from the wall to allow DNA to follow a straight path through the enzyme. After DNA has been melted to create the transcription bubble, the clamp must swing back into position against the wall.

One of the dilemmas of any nucleic acid polymerase is that the enzyme must make tight

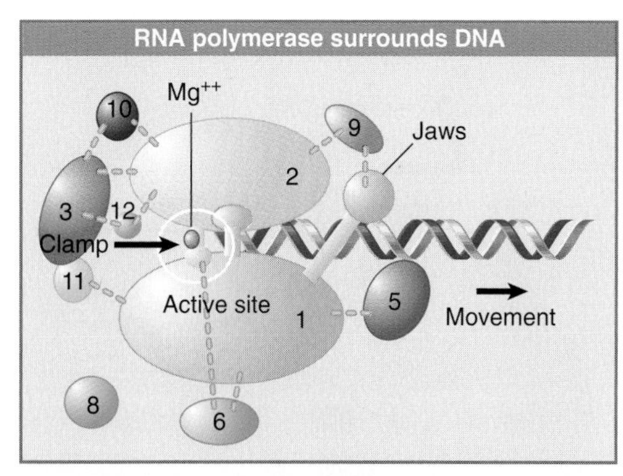

FIGURE 11.9 Ten subunits of RNA polymerase are placed in position from the crystal structure. The colors of the subunits are the same as in the crystal structures of the following figures.

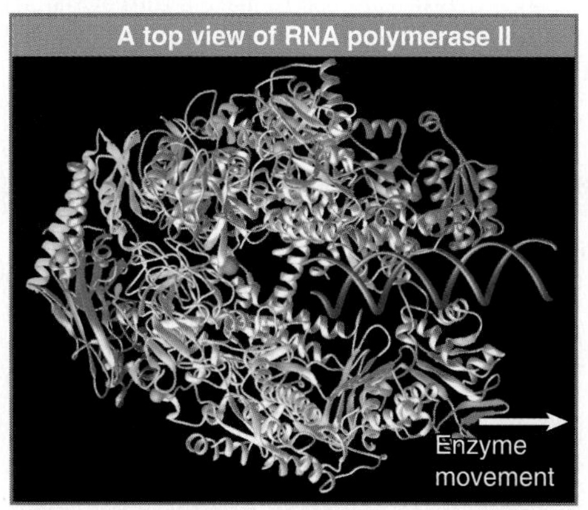

FIGURE 11.10 The top view of the crystal structure of RNA polymerase II from yeast shows that DNA is held downstream by a pair of jaws and is clamped in position in the active site, which contains an Mg^{++} ion. Photo courtesy of Roger Kornberg, Stanford University School of Medicine.

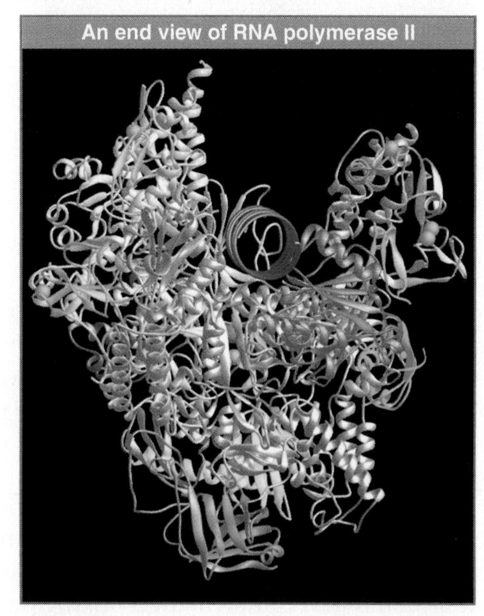

FIGURE 11.11 The end view of the crystal structure of RNA polymerase II from yeast shows that DNA is surrounded by ~270° of protein. Photo courtesy of Roger Kornberg, Stanford University School of Medicine.

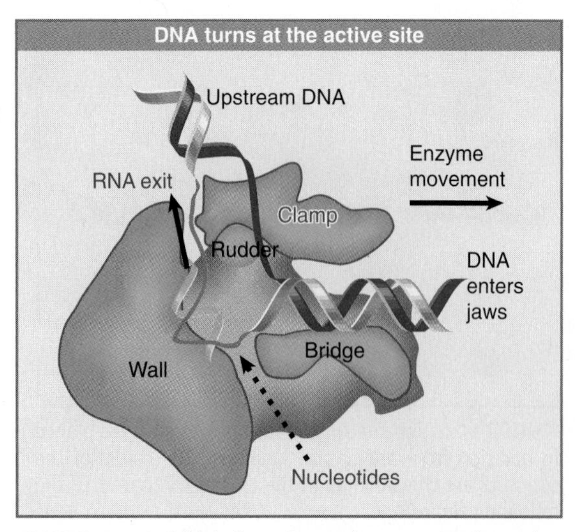

FIGURE 11.12 DNA is forced to make a turn at the active site by a wall of protein. Nucleotides may enter the active site through a pore in the protein.

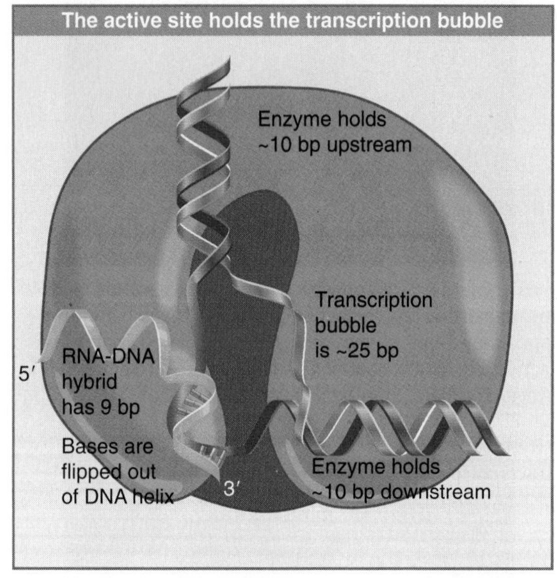

FIGURE 11.13 An expanded view of the active site shows the sharp turn in the path of DNA.

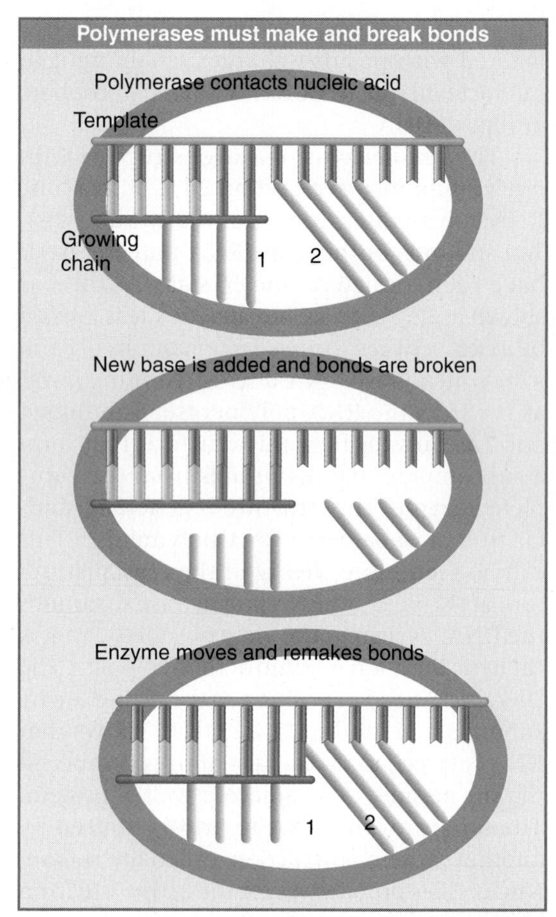

FIGURE 11.14 Movement of a nucleic acid polymerase requires breaking and remaking bonds to the nucleotides at fixed positions relative to the enzyme structure. The nucleotides in these positions change each time the enzyme moves a base along the template.

contacts with the nucleic acid substrate and product, but must break these contacts and remake them with each cycle of nucleotide addition. Consider the situation illustrated in FIGURE 11.14. A polymerase makes a series of specific contacts with the bases at particular positions. For example, contact "1" is made with the base at the end of the growing chain and contact "2" is made with the base in the template strand that is complementary to the next base to be added. Note, however, that the bases that occupy these locations in the nucleic acid chains change every time a nucleotide is added!

The top and bottom panels of the figure show the same situation: a base is about to be added to the growing chain. The difference is that the growing chain has been extended by one base in the bottom panel. The geometry of both complexes is exactly the same, but contacts "1" and "2" in the bottom panel are made to bases in the nucleic acid chains that are located one position farther along the chain. The middle panel shows that this must mean that, after the base is added, and before the enzyme moves relative to the nucleic acid, the contacts made to specific positions must be broken so that they can be remade to bases that occupy those positions after the movement.

The RNA polymerase structure suggests an insight into how the enzyme retains contact with its substrate while breaking and remaking bonds. A structure in the protein called the bridge is adjacent to the active site (see Figure 11.12). This feature is found in both the bacterial and yeast enzymes, but it has different shapes in the different crystal structures. In one it is bent, and in the other it is straight. FIGURE 11.15 suggests that the change in conformation of the bridge structure is closely related to translocation of the enzyme along the nucleic acid.

At the start of the cycle of translocation, the bridge has a straight conformation adjacent to the nucleotide entry site. This allows the next nucleotide to bind at the nucleotide entry site. The bridge is in contact with the newly added nucleotide. The protein then moves one base pair along the substrate. The bridge changes its conformation, bending to keep contact with the newly added nucleotide. In this conformation, the bridge obscures the nucleotide entry site. To end the cycle, the bridge returns to its straight conformation, allowing access again to the nucleotide entry site. The bridge acts as a ratchet that releases the DNA and RNA strands for translocation while holding on to the end of the growing chain.

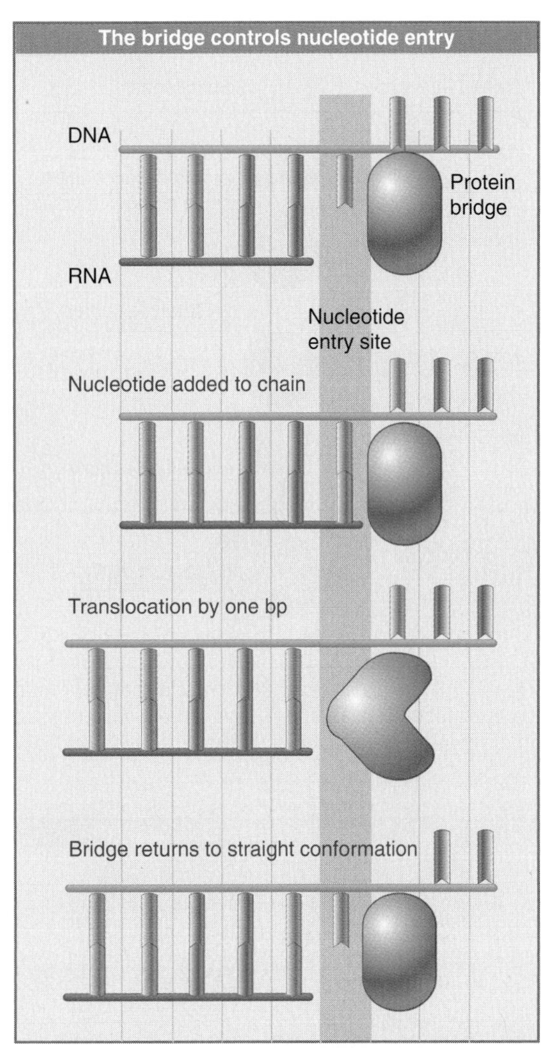

FIGURE 11.15 The RNA polymerase elongation cycle starts with a straight bridge adjacent to the nucleotide entry site. After nucleotide addition, the enzyme moves one base pair and the bridge bends as it retains contact with the newly added nucleotide. When the bridge is released, the cycle can start again.

11.6 Bacterial RNA Polymerase Consists of Multiple Subunits

Key concepts

- Bacterial RNA core polymerases are ~500 kD multisubunit complexes with the general structure $\alpha_2\beta\beta'$.
- DNA is bound in a channel and is contacted by both the β and β' subunits.

The best characterized RNA polymerases are those of eubacteria, for which *Escherichia coli* is a typical case. *A single type of RNA polymerase appears to be responsible for almost all synthesis of mRNA, and all rRNA and tRNA, in a eubacterium.* About 7000 RNA polymerase molecules are present in an *E. coli* cell. Many of them are engaged in transcription; probably 2000 to 5000 enzymes are synthesizing RNA at any one time, with the number depending on the growth conditions.

The **complete enzyme** or **holoenzyme** in *E. coli* has a molecular weight of ~465 kD. Its subunit composition is summarized in FIGURE 11.16.

The β and β' subunits together make up the catalytic center. Their sequences are related to those of the largest subunits of eukaryotic RNA polymerases (see Section 24.2, Eukaryotic RNA Polymerases Consist of Many Subunits), suggesting that there are common features to the actions of all RNA polymerases. The β subunit can be crosslinked to the template DNA, the product RNA, and the substrate ribonucleotides; mutations in *rpoB* affect all stages of transcription. Mutations in *rpoC* show that β' also is involved at all stages.

The α subunit is required for assembly of the core enzyme. When phage T4 infects *E. coli*, the α subunit is modified by ADP-ribosylation of an arginine. The modification is associated with a reduced affinity for the promoters formerly recognized by the holoenzyme, suggesting that the α subunit plays a role in promoter recognition. The α subunit also plays a role in the interaction of RNA polymerase with some regulatory factors.

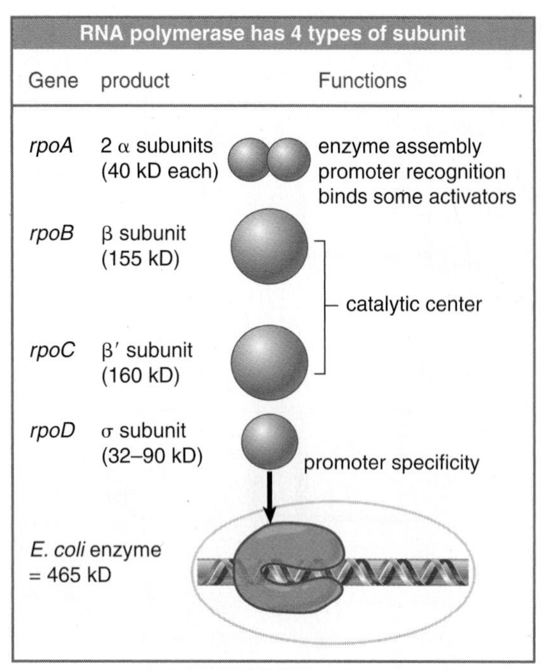

FIGURE 11.16 Eubacterial RNA polymerases have four types of subunit; α, β, and β′ have rather constant sizes in different bacterial species, but σ varies more widely.

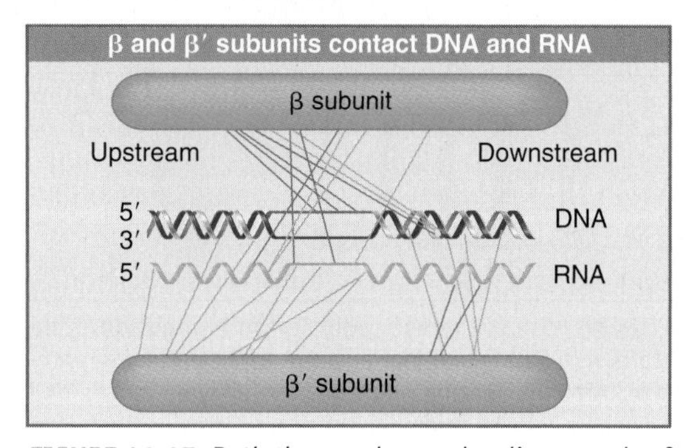

FIGURE 11.17 Both the template and coding strands of DNA are contacted by the β and β′ subunits largely in the region of the transcription bubble and downstream. The RNA is contacted mostly in the transcription bubble. There is no downstream RNA, except in the special case when the enzyme backtracks.

The σ subunit is concerned specifically with promoter recognition, and we have more information about its functions than on any other subunit (see Section 11.7, RNA Polymerase Consists of the Core Enzyme and Sigma Factor).

The crystal structure of the bacterial enzyme (see Figure 11.8) shows that the channel for DNA lies at the interface of the β and β′ subunits. (The α subunits are not visible in this view.) The DNA is unwound at the active site, where an RNA chain is being synthesized. Crosslinking experiments identify the points at which the RNA polymerase subunits contact DNA. These are summarized in FIGURE 11.17. The β and β′ subunits contact DNA at many points downstream of the active site. They make several contacts with the coding strand in the region of the transcription bubble, thus stabilizing the separated single strands. The RNA is contacted largely in the region of the transcription bubble.

The drug rifampicin (a member of the rifamycin antibiotic family) blocks transcription by bacterial RNA polymerase. It is a major drug used against tuberculosis. The crystal structure of RNA polymerase bound to rifampicin explains its action: it binds in a pocket of the β subunit, >12 Å away from the active site, but in a position where it blocks the path of the elongating RNA. By preventing the RNA chain from extending beyond two to three nucleotides, this blocks transcription.

Originally defined simply by its ability to incorporate nucleotides into RNA under the direction of a DNA template, the enzyme RNA polymerase now is seen as part of a more complex apparatus involved in transcription. *The ability to catalyze RNA synthesis defines the minimum component that can be described as RNA polymerase.* It supervises the base pairing of the substrate ribonucleotides with DNA and catalyzes the formation of phosphodiester bonds between them.

All of the subunits of the basic polymerase that participate in elongation are necessary for initiation and termination. Transcription units differ, however, in their dependence on additional polypeptides at the initiation and termination stages. Some of these additional polypeptides are needed at all genes, whereas others may be needed specifically for initiation or termination at particular genes. The analogy with the division of labors between the ribosome and the protein synthesis factors is obvious.

E. coli RNA polymerase can transcribe any one of many (>1000) transcription units. The enzyme therefore requires the ability to interact with a variety of host and phage functions that modify its intrinsic transcriptional activities. The complexity of the enzyme therefore, at least in part, reflects its need to interact with regulatory factors, rather than any demand inherent in its catalytic activity.

11.7 RNA Polymerase Consists of the Core Enzyme and Sigma Factor

Key concepts

- Bacterial RNA polymerase can be divided into the $\alpha_2\beta\beta'$ core enzyme that catalyzes transcription and the sigma subunit that is required only for initiation.
- Sigma factor changes the DNA-binding properties of RNA polymerase so that its affinity for general DNA is reduced and its affinity for promoters is increased.
- Binding constants of RNA polymerase for different promoters vary over six orders of magnitude, corresponding to the frequency with which transcription is initiated at each promoter.

The holoenzyme ($\alpha_2\beta\beta'\sigma$) can be separated into two components, the **core enzyme** ($\alpha_2\beta\beta'$) and the **sigma factor** (the σ polypeptide). *Only the holoenzyme can initiate transcription. Sigma factor ensures that bacterial RNA polymerase binds in a stable manner to DNA only at promoters.* The sigma "factor" is usually released when the RNA chain reaches eight to nine bases, leaving the core enzyme to undertake elongation. *Core enzyme has the ability to synthesize RNA on a DNA template, but cannot initiate transcription at the proper sites.*

The core enzyme has a general affinity for DNA, in which electrostatic attraction between the basic protein and the acidic nucleic acid plays a major role. Any (random) sequence of DNA that is bound by core polymerase in this general binding reaction is described as a **loose binding site.** No change occurs in the DNA, which remains duplex. The complex at such a site is stable, with a half-life for dissociation of the enzyme from DNA ~60 minutes. *Core enzyme does not distinguish between promoters and other sequences of DNA.*

FIGURE 11.18 shows that the sigma factor introduces a major change in the affinity of RNA polymerase for DNA. *The holoenzyme has a drastically reduced ability to recognize loose binding sites*—that is, to bind to any general sequence of DNA. The association constant for the reaction is reduced by a factor of $\sim 10^4$, and the half-life of the complex is <1 second. So sigma factor destabilizes the general binding ability very considerably.

Note that sigma factor also *confers the ability to recognize specific binding sites.* The holoenzyme binds to promoters very tightly, with an association constant increased from that of core enzyme by (on average) 1000 times and with a half-life of several hours.

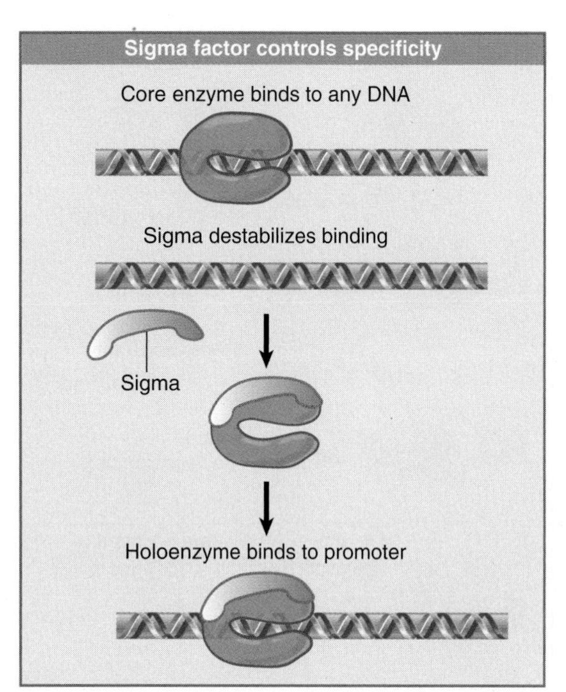

FIGURE 11.18 Core enzyme binds indiscriminately to any DNA. Sigma factor reduces the affinity for sequence-independent binding and confers specificity for promoters.

The specificity of holoenzyme for promoters compared to other sequences is $\sim 10^7$, but this is only an average because there is wide variation in the rate at which the holoenzyme binds to different promoter sequences. This is an important parameter in determining the efficiency of an individual promoter in initiating transcription. The binding constants range from $\sim 10^{12}$ to $\sim 10^6$. Other factors also affect the frequency of initiation, which varies from ~1/sec (rRNA genes) to ~1/30 min (the *lacI* promoter).

11.8 The Association with Sigma Factor Changes at Initiation

Key concepts

- When RNA polymerase binds to a promoter, it separates the DNA strands to form a transcription bubble and incorporates up to nine nucleotides into RNA.
- There may be a cycle of abortive initiations before the enzyme moves to the next phase.
- Sigma factor may be released from RNA polymerase when the nascent RNA chain reaches eight to nine bases in length.

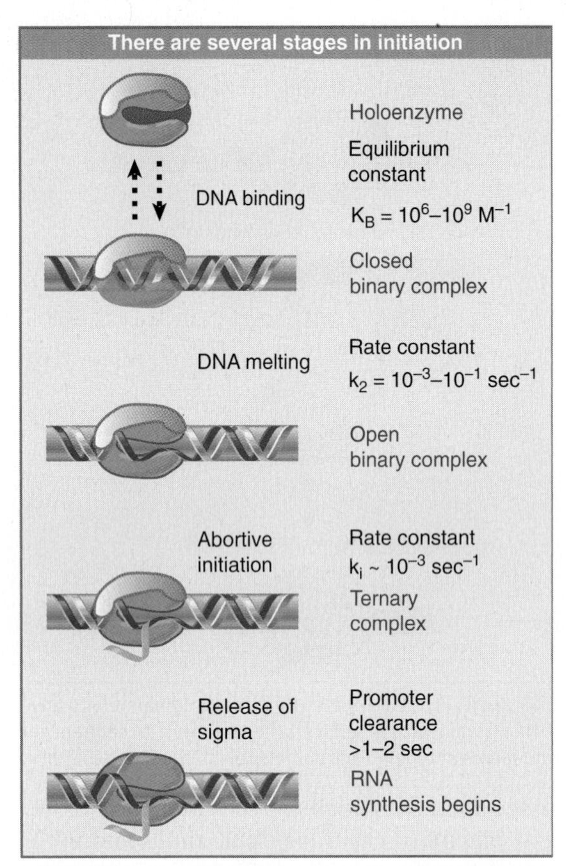

FIGURE 11.19 RNA polymerase passes through several steps prior to elongation. A closed binary complex is converted to an open form and then into a ternary complex.

We can now describe the stages of transcription in terms of the interactions between different forms of RNA polymerase and the DNA template. The initiation reaction can be described by the parameters that are summarized in FIGURE 11.19:

- The holoenzyme-promoter reaction starts by forming a closed binary complex. "Closed" means that the DNA remains duplex. The formation of the closed binary complex is reversible; thus it is usually described by an equilibrium constant (K_B). There is a wide range in values of the equilibrium constant for forming the closed complex.
- The closed complex is converted into an **open complex** by "melting" of a short region of DNA within the sequence bound by the enzyme. The series of events leading to formation of an open complex is called **tight binding.** For strong promoters, conversion into an open binary complex is irreversible, so this reaction is described by a rate constant (k_2). This reaction is fast. Sigma factor is involved in the melting reaction (see Section 11.16, Substitution of Sigma Factors May Control Initiation).
- The next step is to incorporate the first two nucleotides; a phosphodiester bond then forms between them. This generates a **ternary complex** that contains RNA as well as DNA and enzyme. Formation of the ternary complex is described by the rate constant k_i; this is even faster than the rate constant k_2. Further nucleotides can be added without any enzyme movement to generate an RNA chain of up to nine bases. After each base is added, there is a certain probability that the enzyme will release the chain. This comprises an **abortive initiation,** after which the enzyme begins again with the first base. A cycle of abortive initiations usually occurs to generate a series of very short oligonucleotides.
- When initiation succeeds, sigma is no longer necessary, and the enzyme makes the transition to the elongation ternary complex of core polymerase-DNA-nascent RNA. The critical parameter here is *how long it takes for the polymerase to leave the promoter so another polymerase can initiate.* This parameter is the promoter clearance time; its minimum value of one to two seconds establishes the maximum frequency of initiation as <1 event per second. The enzyme then moves along the template, and the RNA chain extends beyond ten bases.

When RNA polymerase binds to DNA, the elongated dimension of the protein extends along the DNA, but some interesting changes in shape occur during transcription. Transitions in shape and size identify three forms of the complex, as illustrated in FIGURE 11.20:

- When RNA polymerase holoenzyme initially binds to DNA, it covers some 75 to 80 bp, extending from –55 to +20. (The long dimension of RNA polymerase (160 Å) could cover ~50 bp of DNA in extended form, which implies that binding of a longer stretch of DNA must involve some bending of the nucleic acid.)
- The shape of the RNA polymerase changes at the transition from initiation to elongation. This is associated with the loss of contacts in the –55 to –35 region,

leaving only ~60 bp of DNA covered by the enzyme. This corresponds with the concept that the more upstream part of the promoter is involved in initial recognition by RNA polymerase, but is not required for the later stages of initiation (see Section 11.13, Promoter Efficiencies Can Be Increased or Decreased by Mutation).

- When the RNA chain extends to 15 to 20 bases, the enzyme makes a further transition, to form the complex that undertakes elongation; now it covers 30 to 40 bp (depending on the stage in the elongation cycle).

It has been a tenet of transcription since soon after the discovery of sigma factor that it is released after initiation. This may not be strictly true. Direct measurements of elongating RNA polymerase complexes show that ~70% of them retain sigma factor. A third of elongating polymerases lack sigma; hence the original conclusion is certainly correct that it is not necessary for elongation. In those cases where it remains associated with core enzyme, the nature of the association has almost certainly changed (see Section 11.11, Sigma Factor Controls Binding to DNA).

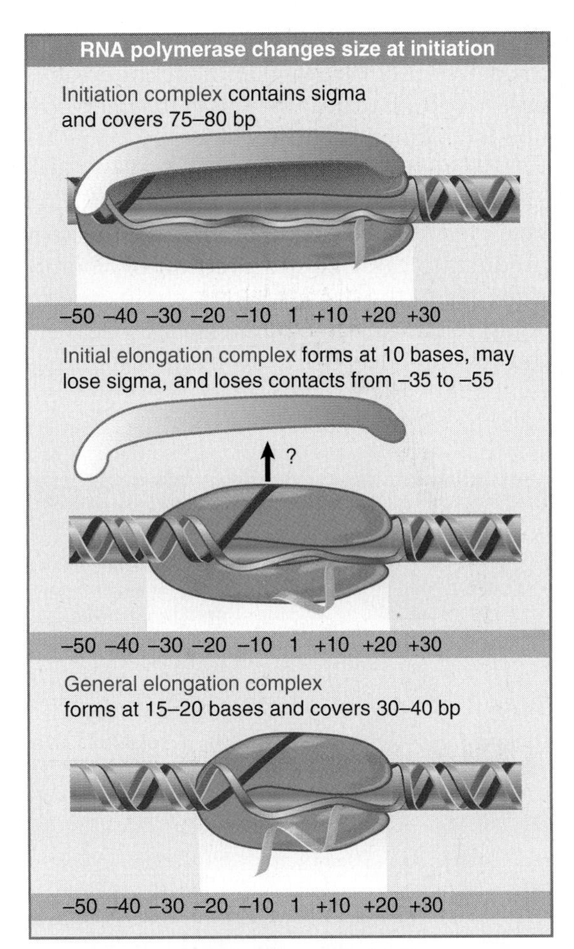

FIGURE 11.20 RNA polymerase initially contacts the region from –55 to +20. When sigma dissociates, the core enzyme contracts to –30; when the enzyme moves a few base pairs, it becomes more compactly organized into the general elongation complex.

11.9 A Stalled RNA Polymerase Can Restart

Key concept

- An arrested RNA polymerase can restart transcription by cleaving the RNA transcript to generate a new 3′ end.

RNA polymerase must be able to handle situations when transcription is blocked. This can happen, for example, when DNA is damaged. A model system for such situations is provided by arresting elongation *in vitro* by omitting one of the necessary precursor nucleotides. When the missing nucleotide is restored, the enzyme can overcome the block by cleaving the 3′ end of the RNA, to create a new 3′ terminus for chain elongation. The cleavage involves accessory factors in addition to the enzyme itself. In the case of *E. coli* RNA polymerase, the proteins GreA and GreB release the RNA polymerase from elongation arrest. In eukaryotic cells, RNA polymerase II requires an accessory factor ($TF_{II}S$), which enables the polymerase to cleave a few ribonucleotides from the 3′ terminus of the RNA product.

The catalytic site of RNA polymerase undertakes the actual cleavage in each case. The roles of GreB and $TF_{II}S$ are to convert the enzyme's catalytic site into a ribonucleolytic site. Even though there is no sequence homology between the factors, crystal structures of their complexes with the respective RNA polymerases suggest that they function in a similar way. Each of the factors inserts a narrow protein domain (in one case a zinc ribbon, in the other a coiled coil) deep into RNA polymerase, where it terminates within the catalytic site. The inserted domain positions two acidic amino acids close to the primary catalytic magnesium ion of the active site; this allows the introduction of a second magnesium ion, which converts the catalytic site to a ribonucleolytic site.

The reason for this reaction may be that stalling causes the template to be mispositioned, so that the 3′ terminus is no longer located in the active site. Cleavage and backtracking is necessary to place the terminus in the right location for addition of further bases.

We see, therefore, that RNA polymerase has the facility to unwind and rewind DNA, to hold the separated strands of DNA and the RNA product, to catalyze the addition of ribonucleotides to the growing RNA chain, and to adjust to difficulties in progressing by cleaving the RNA product and restarting RNA synthesis (with the assistance of some accessory factors).

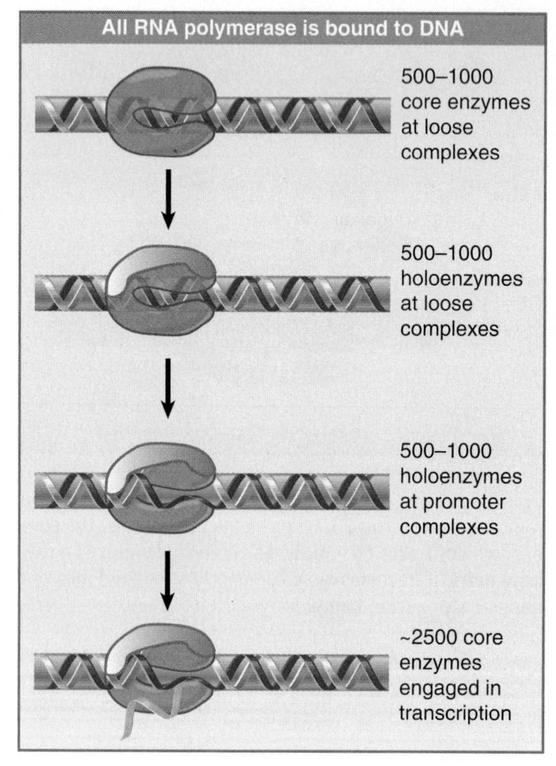

FIGURE 11.21 Core enzyme and holoenzyme are distributed on DNA, and very little RNA polymerase is free.

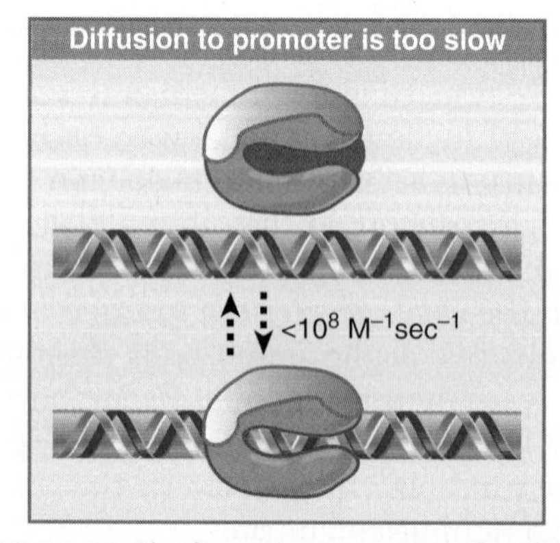

FIGURE 11.22 The forward rate constant for RNA polymerase binding to promoters is faster than random diffusion.

11.10 How Does RNA Polymerase Find Promoter Sequences?

Key concepts

- The rate at which RNA polymerase binds to promoters is too fast to be accounted for by random diffusion.
- RNA polymerase probably binds to random sites on DNA and exchanges them with other sequences very rapidly until a promoter is found.

How is RNA polymerase distributed in the cell? A (somewhat speculative) picture of the enzyme's situation is depicted in FIGURE 11.21:

- Excess core enzyme exists largely as closed loose complexes because the enzyme enters into them rapidly and leaves them slowly. There is very little, if any, free core enzyme.
- There is enough sigma factor for about one third of the polymerases to exist as holoenzymes, and they are distributed between loose complexes at nonspecific sites and binary complexes (mostly closed) at promoters.
- About half of the RNA polymerases consist of core enzymes engaged in transcription.
- How much holoenzyme is free? We do not know, but we suspect that the amount is very small.

RNA polymerase must find promoters within the context of the genome. Suppose that a promoter is a stretch of ~60 bp. How is it distinguished from the 4×10^6 bp that comprise the *E. coli* genome? The next three figures illustrate the principle of some possible models.

FIGURE 11.22 shows the simplest model for promoter binding, in which RNA polymerase moves by random diffusion. Holoenzyme very rapidly associates with, and dissociates from, loose binding sites. Thus it could continue to make and break a series of closed complexes until (by chance) it encounters a promoter, and its recognition of the specific sequence would allow tight binding to occur by formation of an open complex.

For RNA polymerase to move from one binding site on DNA to another, it must dissociate from the first site, find the second site, and then associate with it. Movement from one site to another is limited by the speed of diffusion through the medium. Diffusion sets an upper limit for the rate constant for associating with a 60 bp target of $<10^8$ M^{-1} sec^{-1}. The actual forward rate constant for some promoters *in vitro*, however, appears to be $\sim 10^8$ M^{-1} sec^{-1}, at or above the diffusion limit. If this value applies *in vivo*, the time required for random cycles of successive association and dissociation at loose binding sites is too great to account for the way RNA polymerase finds its promoter.

RNA polymerase must therefore use some other means to seek its binding sites. FIGURE 11.23 shows that the process could be speeded up if the initial target for RNA polymerase is the whole genome, not just a specific promoter sequence. By increasing the target size, the rate constant for diffusion to DNA is correspondingly increased and is no longer limiting.

If this idea is correct, a free RNA polymerase binds DNA and then remains in contact with it. How does the enzyme move from a random (loose) binding site on DNA to a promoter? The most likely model is to suppose that the bound sequence is directly displaced by another sequence. Having taken hold of DNA, the enzyme exchanges this sequence with another sequence very rapidly and continues to exchange sequences until a promoter is found. The enzyme then forms a stable, open complex, after which initiation occurs. The search process becomes much faster because association and dissociation are virtually simultaneous and time is not spent commuting between sites. Direct displacement can give a "directed walk," in which the enzyme moves preferentially from a weak site to a stronger site.

Another idea supposes that the enzyme slides along the DNA by a one-dimensional random walk, as shown in FIGURE 11.24, being halted only when it encounters a promoter. There is, however, no evidence that RNA polymerase (or other DNA-binding proteins) can function in this manner.

11.11 Sigma Factor Controls Binding to DNA

Key concept

- A change in association between sigma factor and holoenzyme changes binding affinity for DNA so that core enzyme can move along DNA.

RNA polymerase encounters a dilemma in reconciling its needs for initiation with those for elongation. Initiation requires tight binding *only* to particular sequences (promoters), whereas elongation requires close association with *all* sequences that the enzyme encounters during transcription. FIGURE 11.25 illustrates how the dilemma is solved by the reversible association between sigma factor and core enzyme. As mentioned previously (see Section 11.8, The Association with Sigma Factor Changes at Initiation), sigma factor is either released following initiation or changes its association with core enzyme so that it no longer participates in DNA binding. There are fewer molecules of sigma than of core enzyme; thus the utilization of core enzyme requires that sigma recycles. This occurs immediately after initiation (as shown in the figure) in about one third of cases; presumably sigma and core dissociate at some later point in the other cases.

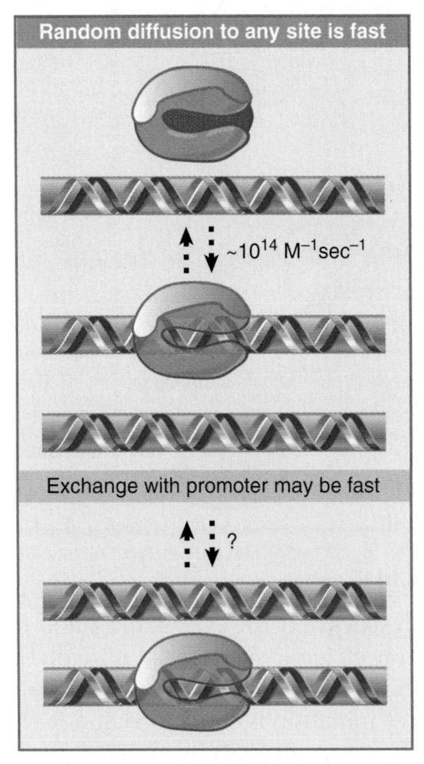

FIGURE 11.23 RNA polymerase binds very rapidly to random DNA sequences and could find a promoter by direct displacement of the bound DNA sequence.

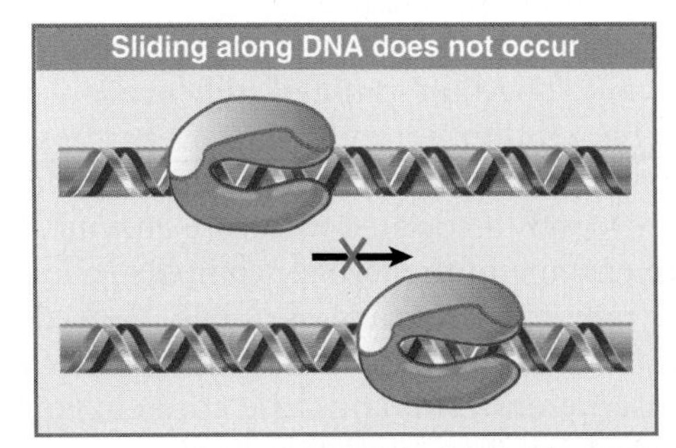

FIGURE 11.24 RNA polymerase does not slide along DNA.

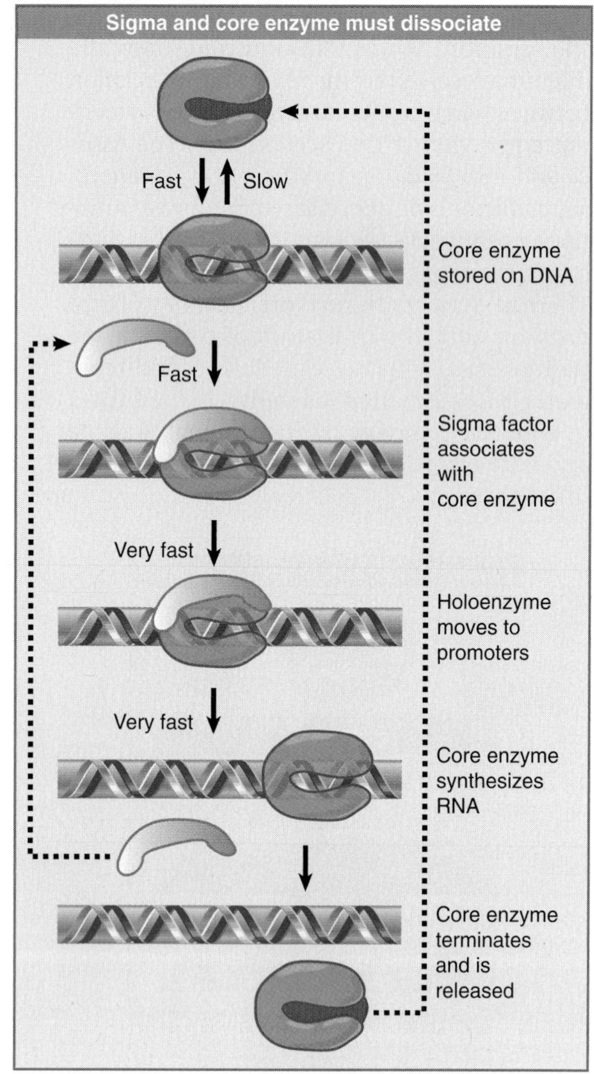

FIGURE 11.25 Sigma factor and core enzyme recycle at different points in transcription.

Irrespective of the exact timing of its release from core enzyme, sigma factor is involved only in initiation. It becomes unnecessary when abortive initiation is concluded and RNA synthesis has been successfully initiated. We do not know whether the state of polymerase changes as a consequence of overcoming abortive initiation, or whether instead it is the change in state that ends abortive initiation and allows elongation to commence.

When sigma factor is released from core enzyme, it becomes immediately available for use by another core enzyme. Regardless of whether sigma factor is released or remains more loosely associated with core enzyme, the core enzyme in the ternary complex is bound very tightly to DNA. It is essentially "locked in" until elongation has been completed. When transcription terminates, the core enzyme is released. It is then "stored" by binding to a loose site on DNA. If it has lost its sigma factor, it must find another sigma factor in order to undertake a further cycle of transcription.

Core enzyme has a high intrinsic affinity for DNA, which is increased by the presence of nascent RNA. Its affinity for loose binding sites is, however, too high to allow the enzyme to distinguish promoters efficiently from other sequences. By reducing the stability of the loose complexes, sigma factor allows the process to occur much more rapidly; and by stabilizing the association at tight binding sites, the factor drives the reaction irreversibly into the formation of open complexes. When the enzyme releases sigma factor (or changes its association with it), it reverts to a general affinity for all DNA, irrespective of sequence, that suits it to continue transcription.

What is responsible for the ability of holoenzyme to bind specifically to promoters? Sigma factor has domains that recognize the promoter DNA. As an independent polypeptide, sigma factor does not bind to DNA, but when holoenzyme forms a tight binding complex, σ contacts the DNA in the region upstream of the startpoint. This difference is due to a change in the conformation of sigma factor when it binds to core enzyme. The N-terminal region of free sigma factor suppresses the activity of the DNA-binding region; when sigma binds to core, this inhibition is released, and it becomes able to bind specifically to promoter sequences (see also Figure 11.35 in Section 11.17). The inability of free sigma factor to recognize promoter sequences may be important: if σ could freely bind to promoters, it might block holoenzyme from initiating transcription.

11.12 Promoter Recognition Depends on Consensus Sequences

Key concepts

- A promoter is defined by the presence of short consensus sequences at specific locations.
- The promoter consensus sequences consist of a purine at the startpoint, the hexamer TATAAT centered at −10, and another hexamer centered at −35.
- Individual promoters usually differ from the consensus at one or more positions.

As a sequence of DNA whose function is to be *recognized by proteins,* a promoter differs from sequences whose role is to be transcribed or translated. The information for promoter function is provided directly by the DNA sequence: its structure is the signal. This is a classic example of a *cis*-acting site, as defined previously in Figure 2.16 and Figure 2.17. By contrast, expressed regions gain their meaning only after the information is transferred into the form of some other nucleic acid or protein.

A key question in examining the interaction between an RNA polymerase and its promoter is how the protein recognizes a specific promoter sequence. Does the enzyme have an active site that distinguishes the chemical structure of a particular sequence of bases in the DNA double helix? How specific are its requirements?

One way to design a promoter would be for a particular sequence of DNA to be recognized by RNA polymerase. Every promoter would consist of, or at least include, this sequence. In the bacterial genome, the minimum length that could provide an adequate signal is 12 bp. (Any shorter sequence is likely to occur—just by chance—a sufficient number of additional times to provide false signals. The minimum length required for unique recognition increases with the size of genome.) The 12 bp sequence need not be contiguous. If a specific number of base pairs separates two constant shorter sequences, their combined length could be less than 12 bp, because the *distance* of separation itself provides a part of the signal (even if the intermediate *sequence* is itself irrelevant).

Attempts to identify the features in DNA that are necessary for RNA polymerase binding started by comparing the sequences of different promoters. Any essential nucleotide sequence should be present in all the promoters. Such a sequence is said to be **conserved.** However, a conserved sequence need not necessarily be conserved at every single position; some variation is permitted. How do we analyze a sequence of DNA to determine whether it is sufficiently conserved to constitute a recognizable signal?

Putative DNA recognition sites can be defined in terms of an idealized sequence that represents the base most often present at each position. A **consensus sequence** is defined by aligning all known examples so as to maximize their homology. For a sequence to be accepted as a consensus, each particular base must be reasonably predominant at its position, and most of the actual examples must be related to the consensus by only one or two substitutions.

The striking feature in the sequence of promoters in *E. coli* is the *lack of any extensive conservation of sequence* over the 60 bp associated with RNA polymerase. The sequence of much of the binding site is irrelevant. Some short stretches within the promoter are conserved, however, and they are critical for its function. *Conservation of only very short consensus sequences is a typical feature of regulatory sites (such as promoters) in both prokaryotic and eukaryotic genomes.*

There are four (perhaps five) conserved features in a bacterial promoter: the startpoint, the −10 sequence, the −35 sequence, the separation between the −10 and −35 sequences, and (sometimes) the UP element:

- The startpoint is usually (>90% of the time) a purine. It is common for the startpoint to be the central base in the sequence CAT, but the conservation of this triplet is not great enough to regard it as an obligatory signal.
- Just upstream of the startpoint, a 6-bp region is recognizable in almost all promoters. The center of the hexamer generally is close to 10 bp upstream of the startpoint; the distance varies in known promoters from position −18 to −9. Named for its location, the hexamer is often called the **−10 sequence.** Its consensus is *TATAAT* and can be summarized in the form

  $$T_{80}\ A_{95}\ T_{45}\ A_{60}\ A_{50}\ T_{96}$$

  where the subscript denotes the percent occurrence of the most frequently found base, which varies from 45% to 96%. (A position at which there is no discernible preference for any base would be indicated by N.) If the frequency of occurrence indicates likely importance in binding RNA polymerase, we would expect the initial highly conserved TA and the final almost completely conserved T in the −10 sequence to be the most important bases.
- Another conserved hexamer is centered ~35 bp upstream of the startpoint. This is called the **−35 sequence.** The consensus is *TTGACA;* in more detailed form, the conservation is

  $$T_{82}\ T_{84}\ G_{78}\ A_{65}\ C_{54}\ A_{45}.$$

- The distance separating the −35 and −10 sites is between 16 and 18 bp in 90% of promoters; in the exceptions, it is as little as 15 bp or as great as 20 bp. *Although the actual sequence in the intervening region is unimportant, the distance is critical in*

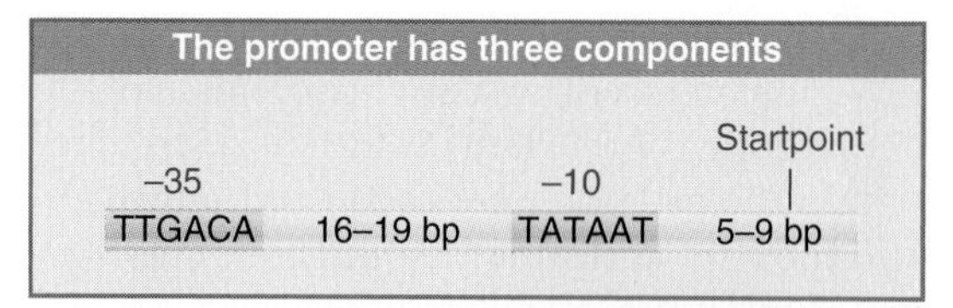

FIGURE 11.26 A typical promoter has three components, consisting of consensus sequences at −35, −10, and the startpoint.

holding the two sites at the appropriate separation for the geometry of RNA polymerase.

- Some promoters have an A-T-rich sequence located farther upstream. This is called the UP element. It interacts with the α subunit of the RNA polymerase. It is typically found in promoters that are highly expressed, such as the promoters for rRNA genes.

The optimal promoter is a sequence consisting of the −35 hexamer, separated by 17 bp from the −10 hexamer, lying 7 bp upstream of the startpoint. The structure of a promoter, showing the permitted range of variation from this optimum, is illustrated in FIGURE 11.26.

11.13 Promoter Efficiencies Can Be Increased or Decreased by Mutation

Key concepts

- Down mutations to decrease promoter efficiency usually decrease conformance to the consensus sequences, whereas up mutations have the opposite effect.
- Mutations in the −35 sequence usually affect initial binding of RNA polymerase.
- Mutations in the −10 sequence usually affect the melting reaction that converts a closed to an open complex.

Mutations are a major source of information about promoter function. Mutations in promoters affect the level of expression of the gene(s) they control without altering the gene products themselves. Most are identified as bacterial mutants that have lost, or have very much reduced, transcription of the adjacent genes. They are known as **down mutations.** Less often, mutants are found in which there is increased transcription from the promoter. They have **up mutations.**

It is important to remember that "up" and "down" mutations are defined relative to the *usual* efficiency with which a particular promoter functions. This varies widely. Thus a change that is recognized as a down mutation in one promoter might never have been isolated in another (which in its wild-type state could be even less efficient than the mutant form of the first promoter). Information gained from studies *in vivo* simply identifies the overall direction of the change caused by mutation.

Is the most effective promoter one that has the actual consensus sequences? This expectation is borne out by the simple rule that up mutations usually increase homology with one of the consensus sequences or bring the distance between them closer to 17 bp. Down mutations usually decrease the resemblance of either site with the consensus or make the distance between them more distant from 17 bp. Down mutations tend to be concentrated in the most highly conserved positions, which confirms their particular importance as the main determinant of promoter efficiency. There are, however, occasional exceptions to these rules.

To determine the absolute effects of promoter mutations, we must measure the affinity of RNA polymerase for wild-type and mutant promoters *in vitro*. There is ~100-fold variation in the rate at which RNA polymerase binds to different promoters *in vitro*, which correlates well with the frequencies of transcription when their genes are expressed *in vivo*. Taking this analysis further, we can investigate the stage at which a mutation influences the capacity of the promoter. Does it change the affinity of the promoter for binding RNA polymerase? Does it leave the enzyme able to bind but unable to initiate? Is the influence of an ancillary factor altered?

By measuring the kinetic constants for formation of a closed complex and its conversion to an open complex, as defined in Figure 11.19, we can dissect the two stages of the initiation reaction:

- Down mutations in the −35 sequence reduce the rate of closed complex formation (they reduce K_B), but they do not inhibit the conversion to an open complex.
- Down mutations in the −10 sequence do not affect the initial formation of a closed complex, but they slow its conversion to the open form (they reduce k_2).

These results suggest the model shown in FIGURE 11.27. The function of the −35 sequence is to provide the signal for recognition by RNA polymerase, whereas the −10 sequence allows the complex to convert from closed to open form. We might view the −35 sequence as comprising a "recognition domain," whereas the −10 sequence comprises an "unwinding domain" of the promoter.

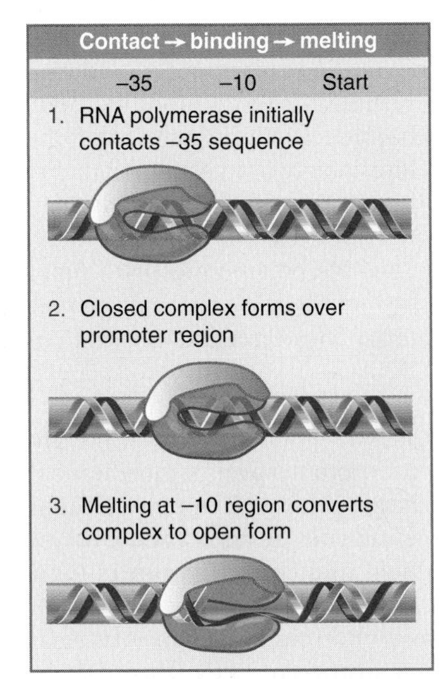

FIGURE 11.27 The −35 sequence is used for initial recognition and the −10 sequence is used for the melting reaction that converts a closed complex to an open complex.

The consensus sequence of the −10 site consists exclusively of A-T base pairs, a configuration that assists the initial melting of DNA into single strands. The lower energy needed to disrupt A-T pairs compared with G-C pairs means that a stretch of A-T pairs demands the minimum amount of energy for strand separation.

The sequence immediately around the startpoint influences the initiation event. The initial transcribed region (from +1 to +30) influences the rate at which RNA polymerase clears the promoter and therefore has an effect upon promoter strength. Thus the overall strength of a promoter cannot be predicted entirely from its −35 and −10 consensus sequences.

A "typical" promoter relies upon its −35 and −10 sequences to be recognized by RNA polymerase, but one or the other of these sequences can be absent from some (exceptional) promoters. In at least some of these cases, the promoter cannot be recognized by RNA polymerase alone; the reaction requires ancillary proteins, which overcome the deficiency in intrinsic interaction between RNA polymerase and the promoter.

11.14 RNA Polymerase Binds to One Face of DNA

Key concepts

- The consensus sequences at −35 and −10 provide most of the contact points for RNA polymerase in the promoter.
- The points of contact lie on one face of the DNA.

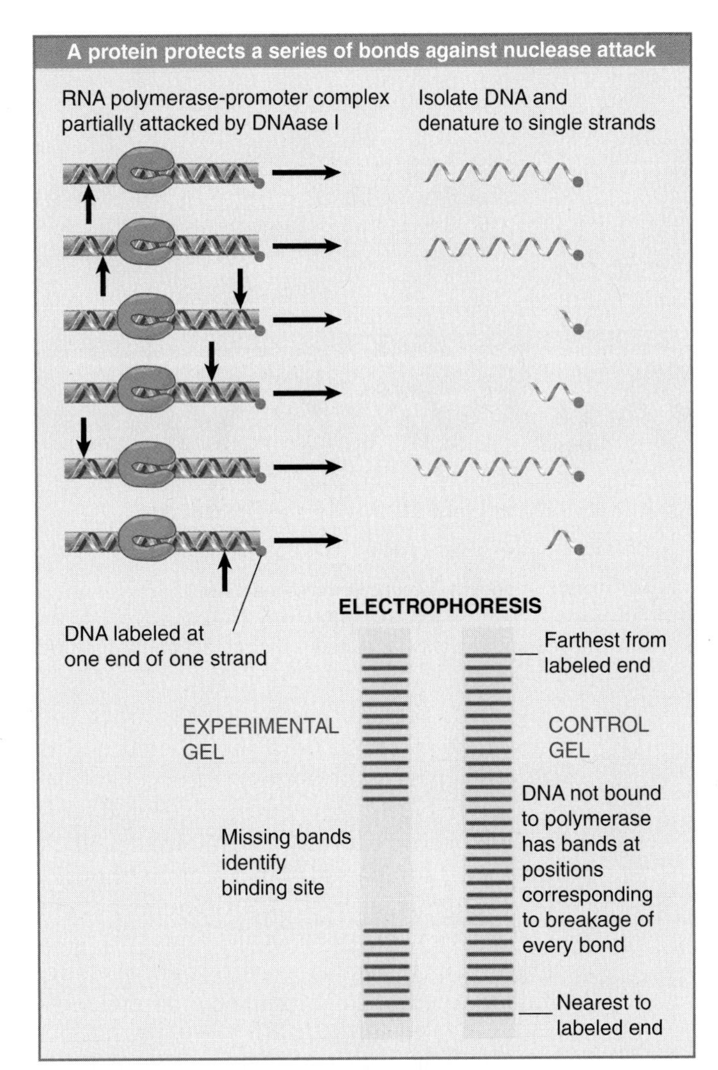

FIGURE 11.28 Footprinting identifies DNA-binding sites for proteins by their protection against nicking.

The ability of RNA polymerase (or indeed any protein) to recognize DNA can be characterized by **footprinting.** A sequence of DNA bound to the protein is *partially* digested with an endonuclease to attack individual phosphodiester bonds within the nucleic acid. Under appropriate conditions, any particular phosphodiester bond is broken in some, but not in all, DNA molecules. The positions that are cleaved are recognized by using DNA labeled on one strand at one end only. The principle is the same as that involved in DNA sequencing: partial cleavage of an end-labeled molecule at a susceptible site creates a fragment of unique length.

As FIGURE 11.28 shows, following the nuclease treatment, the broken DNA fragments are recovered and electrophoresed on a gel that separates them according to length. Each fragment that retains a labeled end produces a radioactive band. The position of the band corresponds to the number of bases in the

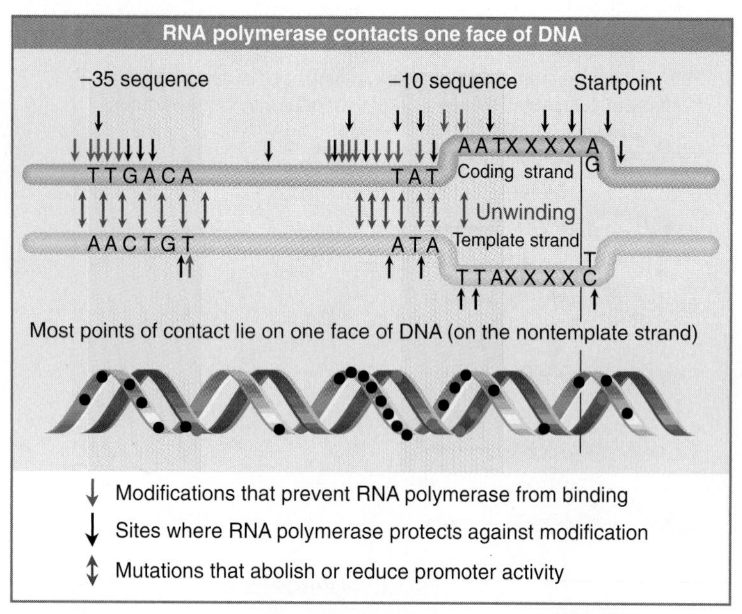

FIGURE 11.29 One face of the promoter contains the contact points for RNA.

fragment. The shortest fragments move the fastest, so distance from the labeled end is counted up from the bottom of the gel.

In a free DNA, *every* susceptible bond position is broken in one or another molecule. When the DNA is complexed with a protein, though, the region covered by the DNA-binding protein is protected in every molecule. Thus two reactions are run in parallel: a control of DNA alone, and an experimental mixture containing molecules of DNA bound to the protein. *When a bound protein blocks access of the nuclease to DNA, the bonds in the bound sequence fail to be broken in the experimental mixture.*

In the control, every bond is broken. This generates a series of bands, with one band representing each base. There are thirty-one bands in the figure. In the protected fragment, bonds cannot be broken in the region bound by the protein, so bands representing fragments of the corresponding sizes are not generated. The absence of bands 9–18 in the figure identifies a protein-binding site covering the region located 9–18 bases from the labeled end of the DNA. By comparing the control and experimental lanes with a sequencing reaction that is run in parallel, it becomes possible to "read off" the corresponding sequence directly, thus identifying the nucleotide sequence of the binding site.

As described previously (see Figure 11.20), RNA polymerase initially binds the region from –50 to +20. The points at which RNA polymerase actually contacts the promoter can be identified by modifying the footprinting technique to treat RNA polymerase-promoter complexes with reagents that modify particular bases. We can perform the experiment in two ways:

- The DNA can be modified before it is bound to RNA polymerase. If the modification prevents RNA polymerase from binding, we have identified a base position where contact is essential.
- The RNA polymerase-DNA complex can be modified. We then compare the pattern of protected bands with that of free DNA and the unmodified complex. Some bands disappear, thus identifying sites at which the enzyme has protected the promoter against modification. Other bands increase in intensity, thus identifying sites at which the DNA must be held in a conformation in which it is more exposed.

These changes in sensitivity reveal the geometry of the complex, as summarized in FIGURE 11.29 for a typical promoter. The regions at –35 and –10 contain most of the contact points for the enzyme. Within these regions, the same sets of positions tend both to prevent binding if previously modified, and to show increased or decreased susceptibility to modification after binding. The points of contact do not coincide completely with sites of mutation; however, they occur in the same limited region.

It is noteworthy that the same *positions* in different promoters provide the contact points, even though a different base is present. This indicates that there is a common mechanism for RNA polymerase binding, although the reaction does not depend on the presence of particular bases at some of the points of contact. This model explains why some of the points of contact are not sites of mutation. In addition, not every mutation lies in a point of contact; the mutations may influence the neighborhood without actually being touched by the enzyme.

It is especially significant that the experiments with prior modification identify *only* sites in the same region that is protected by the enzyme against subsequent modification. These two experiments measure different things. Prior modification identifies all those sites that the enzyme must recognize in order to bind to DNA. Protection experiments recognize all those sites that actually make contact in the binary complex. The protected sites include all the recognition sites and also some additional positions, which suggests that the enzyme first recognizes a set of bases necessary for it to "touch down" and then extends its points of contact to additional bases.

The region of DNA that is unwound in the binary complex can be identified directly by chemical changes in its availability. When the strands of DNA are separated, the unpaired bases become susceptible to reagents that cannot reach them in the double helix. Such experiments implicate positions between –9 and +3 in the initial melting reaction. The region unwound during initiation therefore includes the right end of the –10 sequence and extends just past the startpoint.

Viewed in three dimensions, the points of contact upstream of the –10 sequence all lie on one face of DNA. This can be seen in the lower drawing in Figure 11.29, in which the contact points are marked on a double helix viewed from one side. Most lie on the coding strand. These bases are probably recognized in the initial formation of a closed binary complex. This would make it possible for RNA polymerase to approach DNA from one side and recognize that face of the DNA. As DNA unwinding commences, further sites that originally lay on the other face of DNA can be recognized and bound.

11.15 Supercoiling Is an Important Feature of Transcription

Key concepts

- Negative supercoiling increases the efficiency of some promoters by assisting the melting reaction.
- Transcription generates positive supercoils ahead of the enzyme and negative supercoils behind it, and these must be removed by gyrase and topoisomerase.

The importance of strand separation in the initiation reaction is emphasized by the effects of supercoiling. Both prokaryotic and eukaryotic RNA polymerases can initiate transcription more efficiently *in vitro* when the template is supercoiled, presumably because the supercoiled structure requires less free energy for the initial melting of DNA in the initiation complex.

The efficiency of some promoters is influenced by the degree of supercoiling. The most common relationship is for transcription to be aided by negative supercoiling. We understand in principle how this assists the initiation reaction. Why, though, should some promoters be influenced by the extent of supercoiling whereas others are not? One possibility is that the dependence of a promoter on supercoiling is determined by its sequence. This would predict

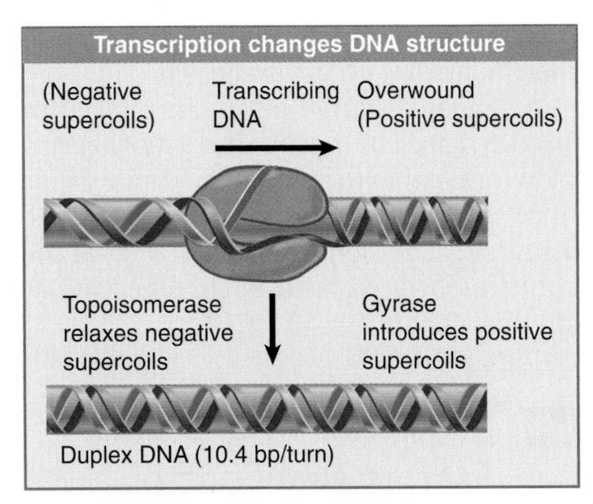

FIGURE 11.30 Transcription generates more tightly wound (positively supercoiled) DNA ahead of RNA polymerase, while the DNA behind becomes less tightly wound (negatively supercoiled).

that some promoters have sequences that are easier to melt (and are therefore less dependent on supercoiling), whereas others have more difficult sequences (and have a greater need to be supercoiled). An alternative is that the location of the promoter might be important if different regions of the bacterial chromosome have different degrees of supercoiling.

Supercoiling also has a continuing involvement with transcription. As RNA polymerase transcribes DNA, unwinding and rewinding occurs, as illustrated in Figure 11.4. This requires that either the entire transcription complex rotates about the DNA or the DNA itself must rotate about its helical axis. The consequences of the rotation of DNA are illustrated in FIGURE 11.30 in the *twin domain* model for transcription. As RNA polymerase pushes forward along the double helix, it generates positive supercoils (more tightly wound DNA) ahead and leaves negative supercoils (partially unwound DNA) behind. For each helical turn traversed by RNA polymerase, +1 turn is generated ahead and –1 turn behind.

Transcription therefore has a significant effect on the (local) structure of DNA. As a result, the enzymes gyrase (introduces negative supercoils) and topoisomerase I (removes negative supercoils) are required to rectify the situation in front of and behind the polymerase, respectively. Blocking the activities of gyrase and topoisomerase causes major changes in the supercoiling of DNA. For example, in yeast lacking an enzyme that relaxes negative supercoils, the density of negative supercoiling doubles in a transcribed region. A possible implication of these results is that transcription is responsible for generating a significant proportion of the supercoiling that occurs in the cell.

A similar situation occurs in replication, when DNA must be unwound at a moving replication fork so that the individual single strands can be used as templates to synthesize daughter strands. (Solutions for the topological constraints associated with such reactions are indicated later, in Figure 19.20.)

11.16 Substitution of Sigma Factors May Control Initiation

Key concepts

- *E. coli* has several sigma factors, each of which causes RNA polymerase to initiate at a set of promoters defined by specific −35 and −10 sequences.
- σ^{70} is used for general transcription, and the other sigma factors are activated by special conditions.

The division of labors between a core enzyme that undertakes chain elongation and a sigma factor involved in site selection immediately raises the question of whether there is more than one type of sigma factor, each specific for a different class of promoters. FIGURE 11.31 shows the principle of a system in which a substitution of the sigma factor changes the choice of promoter.

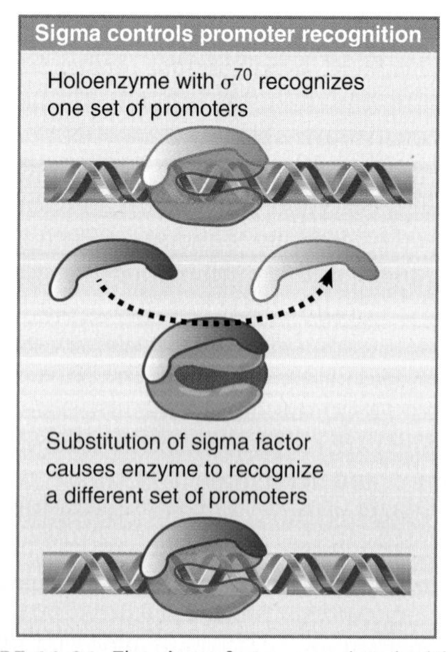

FIGURE 11.31 The sigma factor associated with core enzyme determines the set of promoters at which transcription is initiated.

E. coli uses alternative sigma factors to respond to general environmental changes; they are listed in FIGURE 11.32. (They are named either by molecular weight of the product or for the gene.) The general factor, which is responsible for transcription of most genes under normal conditions, is σ^{70}. The alternative sigma factors σ^{S}, σ^{32}, σ^{E}, and σ^{54} are activated in response to environmental changes; σ^{28} is used for expression of flagellar genes during normal growth, but its level of expression responds to changes in the environment. All the sigma factors except σ^{54} belong to the same protein family and function in the same general manner.

Temperature fluctuation is a common type of environmental challenge. Many organisms, both prokaryotic and eukaryotic, respond in a similar way. Upon an increase in temperature, synthesis of the proteins currently being made is turned off or down, and a new set of proteins is synthesized. The new proteins are the products of the **heat shock genes,** which play a role in protecting the cell against environmental stress. Heat shock genes are synthesized in response to conditions other than heat shock as well. Several of the heat shock proteins are chaperones. In *E. coli,* the expression of seventeen heat shock proteins is triggered by changes at transcription. The gene *rpoH* is a regulator needed to switch on the heat shock response. Its product is σ^{32}, which functions as an alternative sigma factor that causes transcription of the heat shock genes.

The heat shock response is accomplished by increasing the amount of σ^{32} when the temperature increases and decreasing its activity when the temperature change is reversed. The basic signal that induces production of σ^{32} is the accumulation of unfolded (partially denatured) proteins that results from increase in temperature. The σ^{32} protein is unstable, which is important in allowing its quantity to be increased or decreased rapidly. The proteins σ^{70} and σ^{32} can compete for the available core enzyme, so that the set of genes transcribed during heat shock depends on the balance between them.Changing sigma factors is a serious matter that has widespread implications for gene expression in the bacterium. It is not surprising, therefore, that the production of new sigma factors can be the target of many regulatory circuits. The factor σ^{S} is induced when bacteria make the transition from growth phase to stationary phase and also in other stress conditions. It is controlled at two levels. Translation of the *rpoS* mRNA is increased by low temperature or high osmolarity. Proteolysis of the protein product is inhibited by carbon starvation (the typical signal of stationary phase) and by high temperature.

Another group of heat-regulated genes is controlled by the factor σ^E. It responds to more extreme temperature shifts than σ^{32} and is induced by accumulation of unfolded proteins in the periplasmic space or outer membrane. It is controlled by the intricate circuit summarized in FIGURE 11.33. The factor σ^E binds to a protein (RseA) that is located in the inner membrane. As a result, it cannot activate transcription. The accumulation of unfolded proteins activates a protease (DegS) in the periplasmic space, which cleaves off the C-terminal end of the RseA protein. This cleavage activates another protein in the inner membrane (YaeL), which cleaves the N-terminal region of RseA. When this happens, the σ^E factor is released and can then activate transcription. The net result is that the accumulation of unfolded proteins at the periphery of the bacterium is responsible for activating the set of genes controlled by the sigma factor.

This circuit has two interesting parallels with other regulatory circuits. The response to unfolded proteins in eukaryotic cells also uses a pathway in which an unfolded protein (within the endoplasmic reticulum) activates a membrane protein. In this case, the membrane protein is an endonuclease that cleaves an RNA, leading ultimately to a change in splicing that causes the production of a transcription factor (see Section 26.17, The Unfolded Protein Response Is Related to tRNA Splicing). A more direct parallel is with the first case to be discovered, in which cleavage of a membrane protein activates a transcription factor. In this case, the transcription factor itself is synthesized as a membrane protein, and the level of sterols in the membrane controls the activation of proteases that release the transcription factor from the cytosolic domain of the protein.

Another sigma factor is used under conditions of nitrogen starvation. *E. coli* cells contain a small amount of σ^{54}, which is activated when ammonia is absent from the medium. In these conditions, genes are turned on to allow utilization of alternative nitrogen sources. Counterparts to this sigma factor have been found in a wide range of bacteria, so it represents a response mechanism that has been conserved in evolution.

Another case of evolutionary conservation of sigma factors is presented by the factor σ^F, which is present in small amounts and causes RNA polymerase to transcribe genes involved in chemotaxis and flagellar structure. Its counterpart in *Bacillus subtilis* is σ^D, which controls flagellar and motility genes; factors with the same promoter specificity are present in many species of bacteria.

Each sigma factor causes RNA polymerase to initiate at a particular set of promoters. By analyzing the sequences of these promoters, we can show that each set is identified by unique sequence elements. Indeed, the sequence of each type of promoter ensures that it is recognized only by RNA polymerase directed by the appropriate sigma factor. We can deduce the general rules for promoter recognition from the identification of the genes responding to the sigma factors found in *E. coli* and those involved

E. coli has several sigma factors

Gene	Factor	Use
rpoD	σ^{70}	general
rpoS	σ^{S}	stress
rpoH	σ^{32}	heat shock
rpoE	σ^{E}	heat shock
rpoN	σ^{54}	nitrogen starvation
fliA	σ^{28} (σ^{F})	flagellar synthesis

FIGURE 11.32 In addition to σ^{70}, *E. coli* has several sigma factors that are induced by particular environmental conditions. (A number in the name of a factor indicates its mass.)

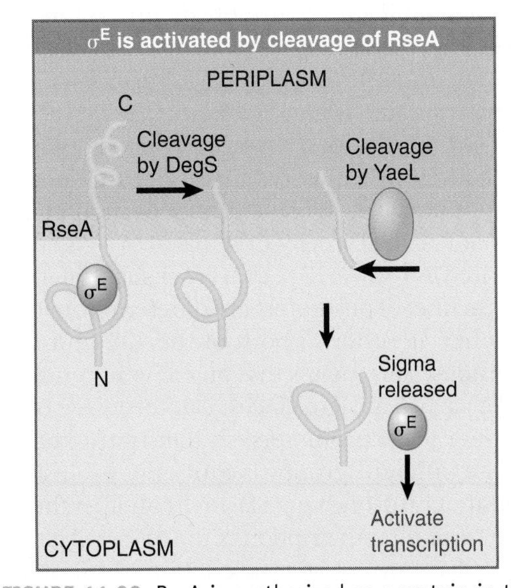

FIGURE 11.33 RseA is synthesized as a protein in the inner membrane. Its cytoplasmic domain binds the σ^E factor. RseA is cleaved sequentially in the periplasmic space and then in the cytoplasm. The cytoplasmic cleavage releases σ^E.

Sigma factors recognize promoters by consensus sequences				
Gene	Factor	−35 Sequence	Separation	−10 Sequence
rpoD	σ^{70}	TTGACA	16–18 bp	TATAAT
rpoH	σ^{32}	CCCTTGAA	13–15 bp	CCCGATNT
rpoN	σ^{54}	CTGGNA	6 bp	TTGCA
fliA	σ^{28} (σ^{F})	CTAAA	15 bp	GCCGATAA
sigH	σ^{H}	AGGANPuPu	11–12 bp	GCTGAATCA

FIGURE 11.34 *E. coli* sigma factors recognize promoters with different consensus sequences.

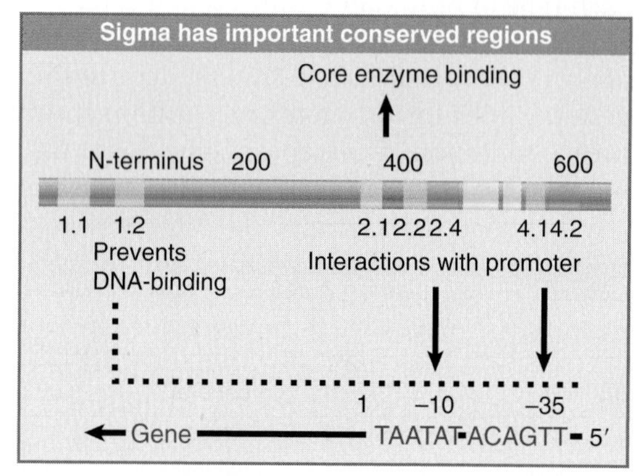

FIGURE 11.35 A map of the *E. coli* σ^{70} factor identifies conserved regions. Regions 2.1 and 2.2 contact core polymerase, 2.3 is required for melting, and 2.4 and 4.2 contact the −10 and −35 promoter elements. The N-terminal region prevents 2.4 and 4.2 from binding to DNA in the absence of core enzyme.

in sporulation in *B. subtilis* (see Section 11.19, Sporulation Is Controlled by Sigma Factors).

A significant feature of the promoters for each enzyme is that *they have the same size and location relative to the startpoint, and they show conserved sequences only around the usual centers of −35 and −10.* (The factor σ^{54} is an exception for which the consensus sequences are closer together and are positioned at −24 and −12; see Section 11.17, Sigma Factors Directly Contact DNA.) As summarized in FIGURE 11.34, the consensus sequences for each set of promoters are different from one another at either or both of the −35 and −10 positions. This means that an enzyme containing a particular sigma factor can recognize only its own set of promoters, so that transcription of the different groups is mutually exclusive. Substitution of one sigma factor by another therefore turns off transcription of the old set of genes as well as turning on transcription of a new set of genes. (Some genes are expressed by RNA polymerases with different sigma factors because they have more than one promoter, each with a different set of consensus sequences.)

11.17 Sigma Factors Directly Contact DNA

Key concepts

- σ^{70} changes its structure to release its DNA-binding regions when it associates with core enzyme.
- σ^{70} binds both the −35 and −10 sequences.

The definition of a series of different consensus sequences recognized at −35 and −10 by holoenzymes containing different sigma factors (see Figure 11.34) carries the immediate implication that the sigma factor subunit must itself contact DNA in these regions. This suggests the general principle that there is a common type of relationship between sigma factor and core enzyme, in which the sigma factor is positioned in such a way as to make critical contacts with the promoter sequences in the vicinity of −35 and −10.

Direct evidence that sigma factor contacts the promoter directly at both the −35 and −10 consensus sequences is provided by mutations in sigma factor that suppress mutations in the consensus sequences. When a mutation at a particular position in the promoter prevents recognition by RNA polymerase, and a compensating mutation in sigma factor allows the polymerase to use the mutant promoter, the most likely explanation is that the relevant base pair in DNA is contacted by the amino acid that has been substituted.

Comparisons of the sequences of several bacterial sigma factors identify regions that have been conserved. Their locations in *E. coli* σ^{70} are summarized in FIGURE 11.35. The crystal structure of a sigma factor fragment from the bacterium *Thermus aquaticus* shows that these regions fold into three independent domains in the protein: domain σ_2 contains 1.2–2.4, σ_3 contains 3.0–1.3, and σ_4 contains 4.1–4.2.

Figure 11.35 shows that two short parts of regions 2 and 4 (named 2.4 and 4.2) are involved in contacting bases in the −10 and −35 elements, respectively. Both of these regions form short stretches of α-helix in the protein. Experiments with heteroduplexes show that σ^{70} makes contacts with bases principally on the coding strand, and it continues to hold these contacts after the DNA has been unwound in this region. This suggests that sigma factor could be important in the melting reaction.

The use of α-helical motifs in proteins to recognize duplex DNA sequences is common (see Section 14.11, Repressor Uses a Helix-Turn-Helix Motif to Bind DNA). Amino acids separated by three to four positions lie on the same face of an α-helix and are therefore in a position to contact adjacent base pairs. FIGURE 11.36 shows that amino acids lying along one face of the 2.4 region α-helix contact the bases at positions –12 to –10 of the –10 promoter sequence.

Region 2.3 resembles proteins that bind single-stranded nucleic acids and is involved in the melting reaction. Regions 2.1 and 2.2 (which comprise the most highly conserved part of sigma) are involved in the interaction with core enzyme. It is assumed that all sigma factors bind the same regions of the core polymerase, which ensures that the reactions are competitive.

The N-terminal region of σ^{70} has important regulatory functions. If it is removed, the shortened protein becomes able to bind specifically to promoter sequences. This suggests that the N-terminal region behaves as an autoinhibition domain. It occludes the DNA-binding domains when σ^{70} is free. Association with core enzyme changes the conformation of sigma so that the inhibition is released, and the DNA-binding domains can contact DNA.

FIGURE 11.37 schematizes the conformational change in sigma factor at open complex formation. When sigma factor binds to the core polymerase, the N-terminal domain swings ~20 Å away from the DNA-binding domains, and the DNA-binding domains separate from one another by ~15 Å, presumably to acquire a more elongated conformation appropriate for contacting DNA. Mutations in either the –10 or –35 sequences prevent an (N-terminal-deleted) σ^{70} from binding to DNA, which suggests that σ^{70} contacts both sequences simultaneously. This implies that the sigma factor must have a rather elongated structure, extending over the ~68 Å of two turns of DNA.

In the free holoenzyme, the N-terminal domain is located in the active site of the core enzyme components, essentially mimicking the location that DNA will occupy when a transcription complex is formed. When the holoenzyme forms an open complex on DNA, the N-terminal sigma domain is displaced from the active site. Its relationship with the rest of the protein is therefore very flexible; the relationship changes when sigma factor binds to core enzyme and again when the holoenzyme binds to DNA.

Comparisons of the crystal structures of the core enzyme and holoenzyme show that

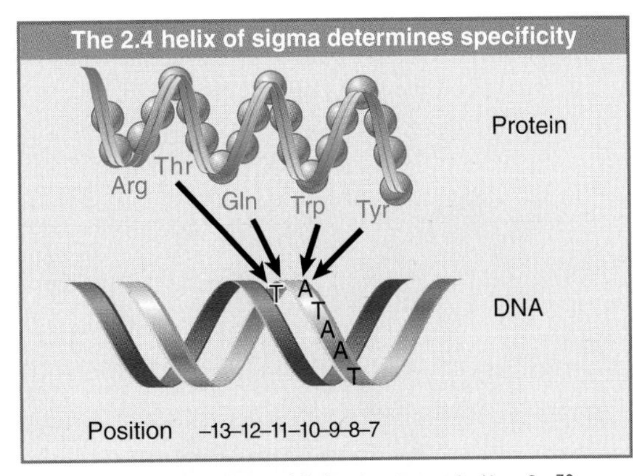

FIGURE 11.36 Amino acids in the 2.4 α-helix of σ^{70} contact specific bases in the coding strand of the –10 promoter sequence.

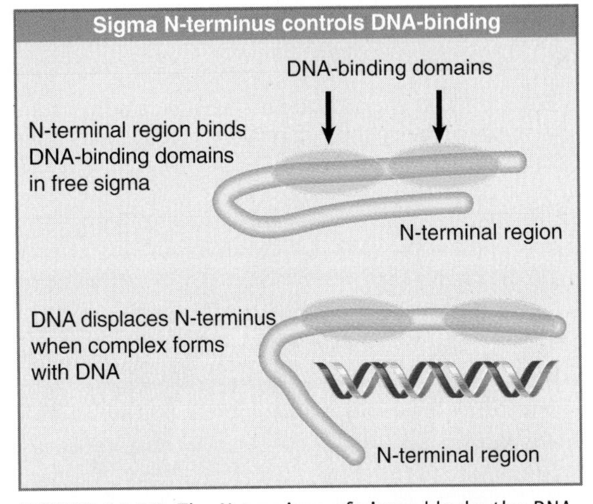

FIGURE 11.37 The N-terminus of sigma blocks the DNA-binding regions from binding to DNA. When an open complex forms, the N-terminus swings 20 Å away, and the two DNA-binding regions separate by 15 Å.

sigma factor lies largely on the surface of the core enzyme. FIGURE 11.38 shows that it has an elongated structure that extends past the DNA-binding site. This places it in a position to contact DNA during the initial binding. The DNA helix has to move some 16 Å from the initial position in order to enter the active site. FIGURE 11.39 illustrates this movement, looking in cross-section down the helical axis of the DNA.

An interesting difference in behavior is found with the σ^{54} factor. This causes RNA polymerase to recognize promoters that have a distinct consensus sequence, with a conserved element at –12 and another close by at –24 (given in the "–35" column of Figure 11.32). Thus the geometry of the polymerase-promoter complex is different under the direction of this sigma factor. Another difference in the

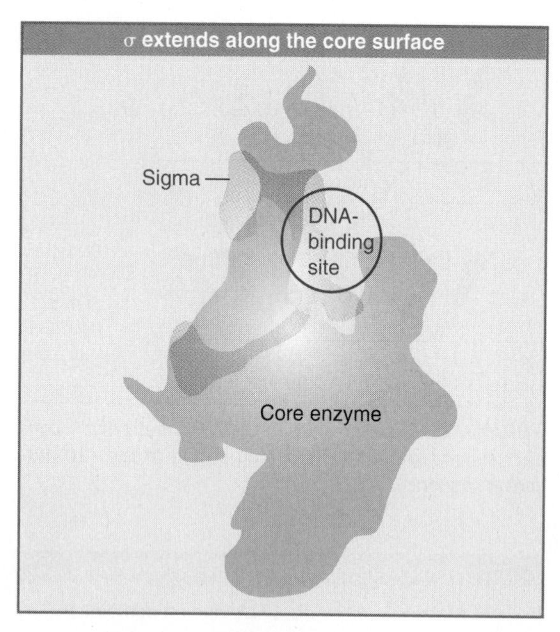

FIGURE 11.38 Sigma factor has an elongated structure that extends along the surface of the core subunits when the holoenzyme is formed.

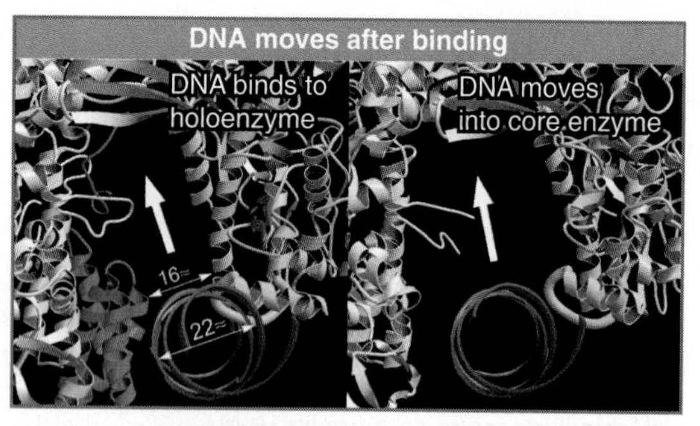

FIGURE 11.39 DNA initially contacts sigma factor (pink) and core enzyme (gray). It moves deeper into the core enzyme to make contacts at the –10 sequence. When sigma is released, the width of the passage containing DNA increases. Reproduced from Vassylyev, D. G., et al. *Nature*. 2002. 417: 712–719. Photo courtesy of Shigeyuki Yokoyama, The University of Tokyo.

mechanism of regulation is that high-level transcription directed by σ^{54} requires other activators to bind to sites that are quite distant from the promoter. This contrasts with the other types of bacterial promoter, for which the regulator sites are always in close proximity to the promoter. The behavior of σ^{54} itself is different from other sigma factors, most notably in its ability to bind to DNA independently of core polymerase. In this regard, σ^{54} is more like the eukaryotic regulators we discuss in Chapter 24, Promoters and Enhancers, than the typical prokaryotic regulators discussed in Chapter 12, The Operon.

11.18 Sigma Factors May Be Organized into Cascades

Key concepts

- A cascade of sigma factors is created when one sigma factor is required to transcribe the gene coding for the next sigma factor.
- The early genes of phage SP01 are transcribed by host RNA polymerase.
- One of the early genes codes for a sigma factor that causes RNA polymerase to transcribe the middle genes.
- Two of the middle genes code for subunits of a sigma factor that causes RNA polymerase to transcribe the late genes.

Sigma factors are used extensively to control initiation of transcription in the bacterium *B. subtilis*, for which ~10 different σ factors are known. Some are present in vegetative cells; others are produced only in the special circumstances of phage infection or the change from vegetative growth to sporulation.

The major RNA polymerase found in *B. subtilis* cells engaged in normal vegetative growth has the same structure as that of *E. coli*, $\alpha_2\beta\beta'\sigma$. Its sigma factor (described as σ^{43} or σ^{A}) recognizes promoters with the same consensus sequences used by the *E. coli* enzyme under direction from σ^{70}. Several variants of the RNA polymerase that contain other sigma factors are found in much smaller amounts. The variant enzymes recognize different promoters on the basis of consensus sequences at –35 and –10.

Transitions from expression of one set of genes to expression of another set are a common feature of bacteriophage infection. In all but the very simplest cases, the development of the phage involves shifts in the pattern of transcription during the infective cycle. These shifts may be accomplished by the synthesis of a phage-encoded RNA polymerase or by the efforts of phage-encoded ancillary factors that control the bacterial RNA polymerase. A well-characterized example of control via the production of new sigma factors occurs during infection of *B. subtilis* by phage SPO1.

The infective cycle of SPO1 passes through three stages of gene expression. Immediately on infection, the **early genes** of the phage are transcribed. After four to five minutes, the early genes cease transcription and the **middle genes** are transcribed. At eight to twelve minutes, middle gene transcription is replaced by transcription of **late genes.**

The early genes are transcribed by the holoenzyme of the host bacterium. They are essentially indistinguishable from host genes whose promoters have the intrinsic ability to be recognized by the RNA polymerase $\alpha_2\beta\beta'\sigma^{43}$.

Expression of phage genes is required for the transitions to middle and late gene transcription. Three regulatory genes, *28*, *33*, and *34*, control the course of transcription. Their functions are summarized in FIGURE 11.40. The pattern of regulation creates a **cascade,** in which the host enzyme transcribes an early gene whose product is needed to transcribe the middle genes. After this transcription, two of the middle genes code for products that are needed to transcribe the late genes.

Mutants in the early gene *28* cannot transcribe the middle genes. The product of gene *28* (called gp28) is a protein of 26 kD that replaces the host sigma factor on the core enzyme. *This substitution is the sole event required to make the transition from early to middle gene expression.* It creates a holoenzyme that can no longer transcribe the host genes but instead specifically transcribes the middle genes. We do not know how gp28 displaces σ^{43} or what happens to the host sigma polypeptide.

Two of the middle genes are involved in the next transition. Mutations in either gene *33* or *34* prevent transcription of the late genes. The products of these genes form a dimer that replaces gp28 on the core polymerase. Again, we do not know how gp33 and gp34 exclude gp28 (or any residual host σ^{43}), *but once they have bound to the core enzyme, it is able to initiate transcription only at the promoters for late genes.*

The successive replacements of sigma factor have dual consequences. Each time the subunit is changed, the RNA polymerase becomes able to recognize a new class of genes *and* it no longer recognizes the previous class. These switches therefore constitute global changes in the activity of RNA polymerase. Probably all, or virtually all, of the core enzyme becomes associated with the sigma factor of the moment, and the change is irreversible.

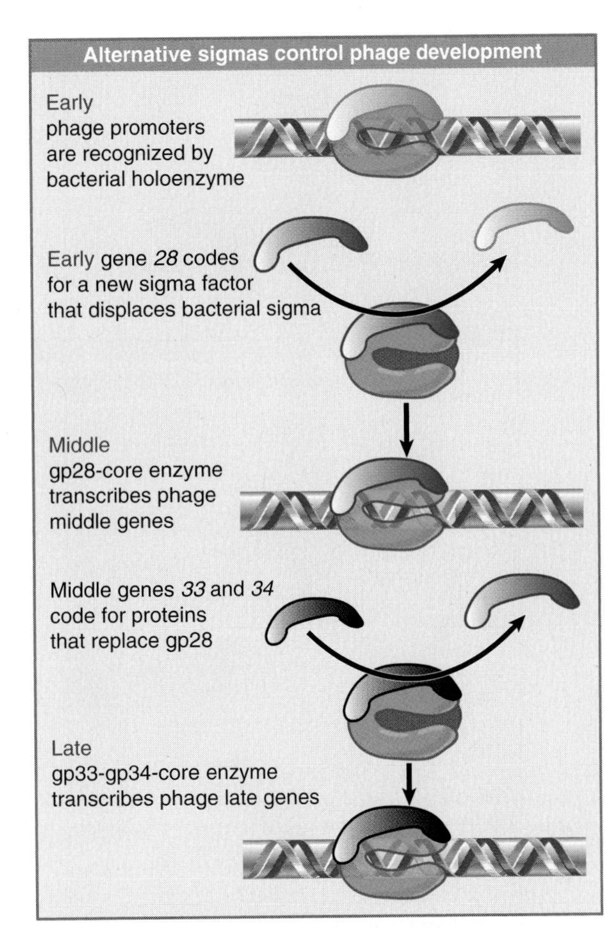

FIGURE 11.40 Transcription of phage SP01 genes is controlled by two successive substitutions of the sigma factor that change the initiation specificity.

11.19 Sporulation Is Controlled by Sigma Factors

Key concepts

- Sporulation divides a bacterium into a mother cell that is lysed and a spore that is released.
- Each compartment advances to the next stage of development by synthesizing a new sigma factor that displaces the previous sigma factor.
- Communication between the two compartments coordinates the timing of sigma factor substitutions.

Perhaps the most extensive example of switches in sigma factors is provided by **sporulation,** an alternative lifestyle available to some bacteria. At the end of the **vegetative phase** in a bacterial culture, logarithmic growth ceases because nutrients in the medium become depleted. This triggers sporulation, as illustrated in FIGURE 11.41. DNA is replicated, a genome is segregated at one end of the cell, and eventually the genome is surrounded by the tough spore coat. When the septum forms, it generates two independent compartments, the mother cell and the forespore. At the start of the process, one chromosome is attached to each pole of the cell. The growing septum traps part of one chromosome in the forespore, and then a translocase (SpoIIIE) pumps the rest of the chromosome into the forespore.

Sporulation takes approximately eight hours. It can be viewed as a primitive sort of differentiation, in which a parent cell (the vegetative bacterium) gives rise to two different daughter cells with distinct fates: the mother

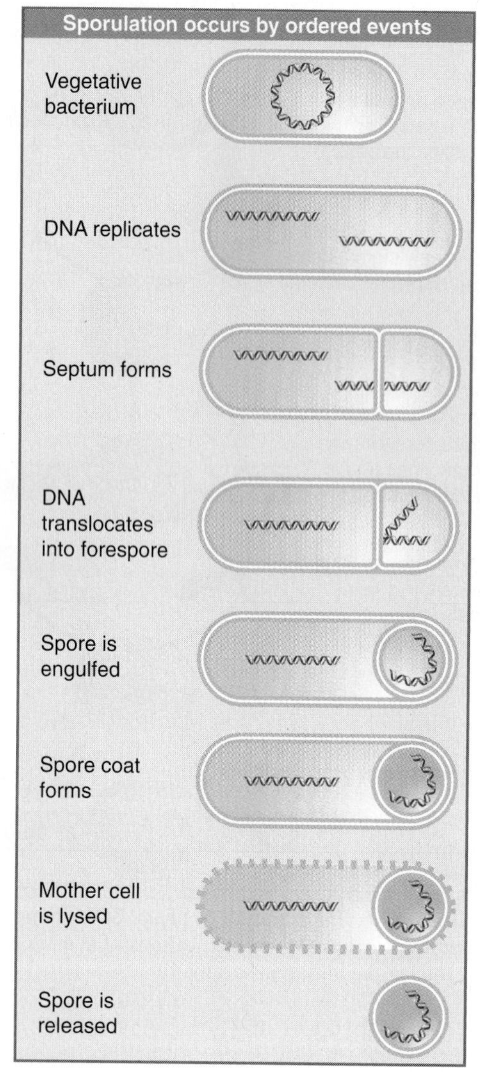

FIGURE 11.41 Sporulation involves the differentiation of a vegetative bacterium into a mother cell that is lysed and a spore that is released.

cell is eventually lysed, and the spore that is released has an entirely different structure from the original bacterium.

Sporulation involves a drastic change in the biosynthetic activities of the bacterium, in which many genes are involved. The basic level of control lies at transcription. Some of the genes that functioned in the vegetative phase are turned off during sporulation, but most continue to be expressed. In addition, the genes specific for sporulation are expressed only during this period. At the end of sporulation, ~40% of the bacterial mRNA is sporulation specific.

New forms of the RNA polymerase become active in sporulating cells; they contain the same core enzyme as vegetative cells, but have different proteins in place of the vegetative σ^{43}. The changes in transcriptional specificity are summarized in FIGURE 11.42. The principle is that in each compartment the existing sigma factor is successively displaced by a new factor that causes transcription of a different set of genes. Communication between the compartments occurs in order to coordinate the timing of the changes in the forespore and mother cell.

The sporulation cascade is initiated when environmental conditions trigger a **phosphorelay,** in which a phosphate group is passed along a series of proteins until it reaches SpoOA. (Several gene products are involved in this process, whose complexity may reflect the need to avoid mistakes in triggering sporulation unnecessarily.) SpoOA is a transcriptional regulator whose activity is affected by phosphorylation. In the phosphorylated form, it activates transcription of two operons, each of which is transcribed by a different form of the host RNA polymerase. Under the direction of phosphorylated SpoOA, host enzyme utilizing the general σ^{43} transcribes the gene coding for the factor σ^F, and host enzyme under the direction of a minor factor, σ^H, transcribes the gene coding for the factor pro-σ^E. Both of these new sigma factors are produced before septum formation, but become active later.

Factor σ^F is the first one to become active in the forespore compartment. It is inhibited by an antisigma factor that binds to it; in the forespore, an anti-antisigma factor removes the inhibitor. This reaction is controlled by a series of phosphorylation/dephosphorylation events. The initial determinant is a phosphatase (SpoIIE) that is an integral membrane protein that accumulates at the pole, with the result that its phosphatase domain becomes more concentrated in the forespore. It dephosphorylates, and thereby activates, SpoIIAA, which in turn displaces the antisigma factor SpoIIAB from the complex of SpoIIAB-σ^F. Release of σ^F activates it.

Activation of σ^F is the start of sporulation. Under the direction of σ^F, RNA polymerase transcribes the first set of sporulation genes instead of the vegetative genes it was previously transcribing. The replacement reaction probably affects only part of the RNA polymerase population, because σ^F is produced only in small amounts. Some vegetative enzyme remains present during sporulation. The displaced σ^{43} is not destroyed; it can be recovered from extracts of sporulating cells.

Two regulatory events follow from the activity of σ^F, as detailed in FIGURE 11.43. In the forespore itself, another factor, σ^G, is the product of one of the early sporulation genes. Factor σ^G

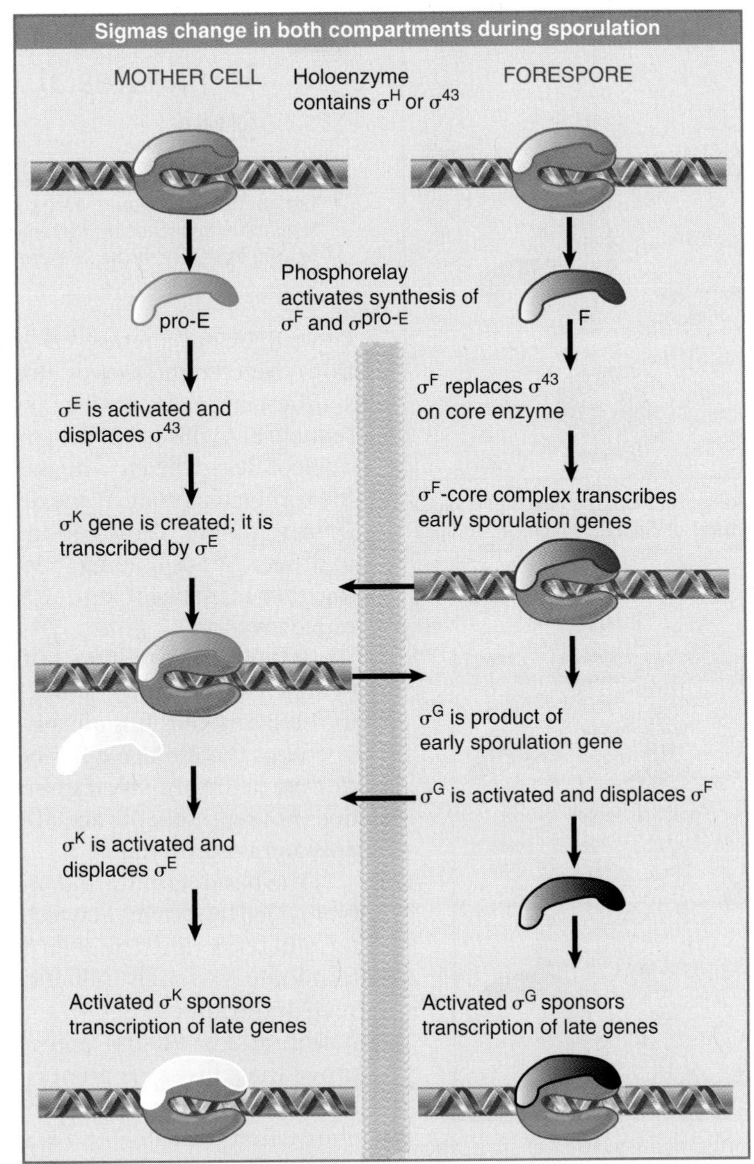

FIGURE 11.42 Sporulation involves successive changes in the sigma factors that control the initiation specificity of RNA polymerase. The cascades in the mother cell (left) and the forespore (right) are related by signals passed across the septum (indicated by horizontal arrows).

causes RNA polymerase to transcribe the late sporulation genes in the forespore. Another early sporulation gene product is responsible for communicating with the mother cell compartment. Factor σ^F activates SpoIIR, which is secreted from the forespore. It then activates the membrane-bound protein SpoIIGA to cleave the inactive precursor pro-σ^E into the active factor σ^E in the mother cell. (Any σ^E that is produced in the forespore is degraded by forespore-specific functions.)

The cascade continues when σ^E in turn is replaced by σ^K. (Actually, the production of σ^K is quite complex, because first its gene must be created by a recombination event!) This factor also is synthesized as an inactive precursor (pro-σ^K) that is activated by a protease. Once σ^K has been activated, it displaces σ^E and causes transcription of the late genes in the mother cell. The timing of these events in the two compartments are coordinated by further signals. The activity of σ^E in the mother cell is necessary for activation of σ^G in the forespore; in turn the activity of σ^G is required to generate a signal that is transmitted across the septum to activate σ^K.

Sporulation is thus controlled by two cascades, in which sigma factors in each

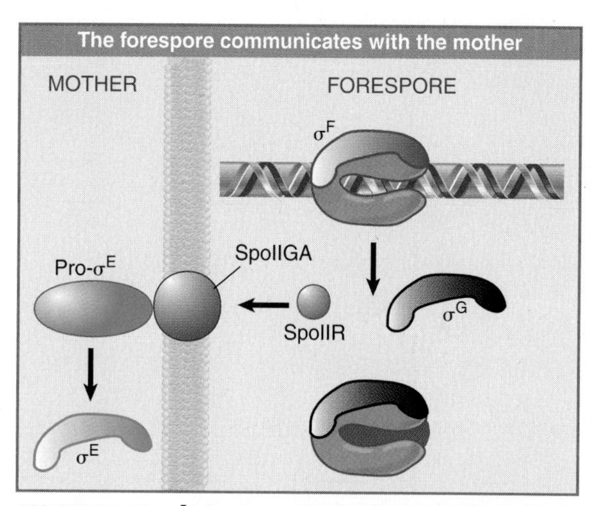

FIGURE 11.43 σ^F triggers synthesis of the next sigma factor in the forespore (σ^G) and turns on SpoIIR, which causes SpoIIGA to cleave pro-σ^E.

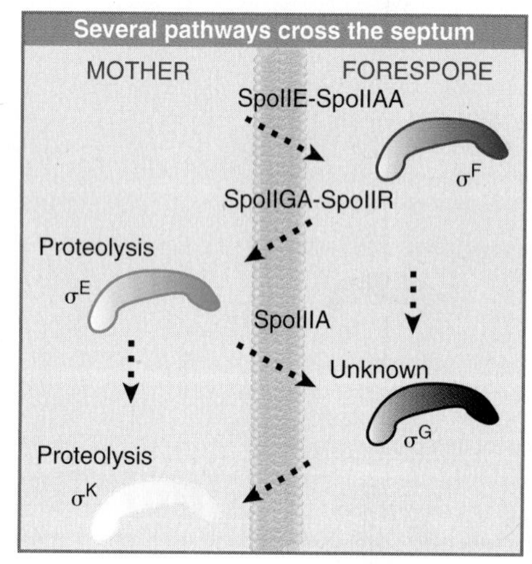

FIGURE 11.44 The crisscross regulation of sporulation coordinates timing of events in the mother cell and forespore.

compartment are successively activated, each directing the synthesis of a particular set of genes. FIGURE 11.44 outlines how the two cascades are connected by the transmission of signals from one compartment to the other. As new sigma factors become active, old sigma factors are displaced, so that transitions in sigma factors turn genes off as well as on. The incorporation of each factor into RNA polymerase dictates when its set of target genes is expressed, and the amount of factor available influences the level of gene expression. More than one sigma factor may be active at any time, and the specificities of some of the sigma factors overlap. We do not know what is responsible for the ability of each sigma factor to replace its predecessor.

11.20 Bacterial RNA Polymerase Terminates at Discrete Sites

Key concept

- Termination may require both recognition of the terminator sequence in DNA and the formation of a hairpin structure in the RNA product.

Once RNA polymerase has started transcription, the enzyme moves along the template, synthesizing RNA, until it meets a **terminator** sequence. At this point, the enzyme stops adding nucleotides to the growing RNA chain, releases the completed product, and dissociates from the DNA template. Termination requires that all hydrogen bonds holding the RNA–DNA hybrid together must be broken, after which the DNA duplex reforms.

It is difficult to define the termination point of an RNA molecule that has been synthesized in the living cell. It is always possible that the 3′ end of the molecule has been generated by *cleavage* of the primary transcript, and therefore does not represent the actual site at which RNA polymerase terminated.

The best identification of termination sites is provided by systems in which RNA polymerase terminates *in vitro*. The ability of the enzyme to terminate is strongly influenced by parameters such as the ionic strength; as a result, its termination at a particular point *in vitro* does not prove that this same point is a natural terminator. We can, however, identify authentic 3′ ends when the same end is generated *in vitro* and *in vivo*.

FIGURE 11.45 summarizes the two types of features found in bacterial terminators.

- Terminators in bacteria and their phages have been identified as sequences that are needed for the termination reaction (*in vitro* or *in vivo*). The sequences at prokaryotic terminators show no similarities beyond the point at which the last base is added to the RNA. The responsibility for termination lies with the *sequences already transcribed* by RNA polymerase. Thus termination relies on scrutiny of the template or product that the polymerase is currently transcribing.
- Many terminators require a hairpin to form in the secondary structure of the RNA being transcribed. *This indicates that termination depends on the RNA product and is not determined simply by scrutiny of the DNA sequence during transcription.*

Terminators vary widely in their efficiencies of termination. At some terminators, the termination event can be *prevented* by specific ancillary factors that interact with RNA polymerase. **Antitermination** causes the enzyme to continue transcription past the terminator sequence, an event called **readthrough** (the same term used in Section 9.14, Suppressors May Compete with Wild-Type Reading of the Code, to describe a ribosome's suppression of termination codons).

In approaching the termination event, we must regard it not simply as a mechanism for generating the 3′ end of the RNA molecule, but as an opportunity to control gene expression. Thus the stages when RNA polymerase associates with DNA (initiation) or dissociates from it (termination) both are subject to specific control. There are interesting parallels between the systems employed in initiation and termination. Both require breaking of hydrogen bonds (initial melting of DNA at initiation and RNA–DNA dissociation at termination), and both require additional proteins to interact with the core enzyme. In fact, they are accomplished by alternative forms of the polymerase. Whereas initiation relies solely upon the interaction between RNA polymerase and duplex DNA, though, the termination event involves recognition of signals in the transcript by RNA polymerase or by ancillary factors as well as the recognition of sequences in DNA.

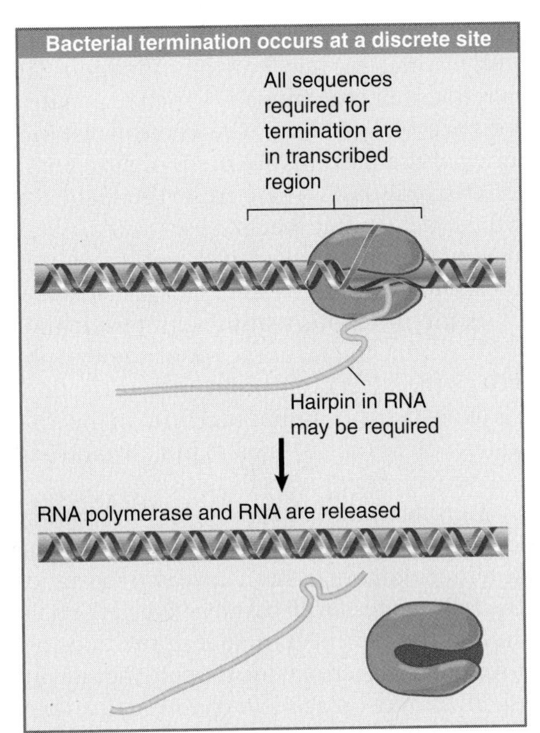

FIGURE 11.45 The DNA sequences required for termination are located upstream of the terminator sequence. Formation of a hairpin in the RNA may be necessary.

11.21 There Are Two Types of Terminators in *E. coli*

Key concept

- Intrinsic terminators consist of a G-C-rich hairpin in the RNA product followed by a U-rich region in which termination occurs.

Terminators are distinguished in *E. coli* according to whether RNA polymerase requires any additional factors to terminate *in vitro:*

- Core enzyme can terminate *in vitro* at certain sites in the absence of any other factor. These sites are called **intrinsic terminators.**
- **Rho-dependent** terminators are defined by the need for addition of **rho factor** (ρ) *in vitro,* and mutations show that the factor is involved in termination *in vivo.*

Intrinsic terminators have the two structural features evident in FIGURE 11.46: a hairpin in the secondary structure, and a region at the

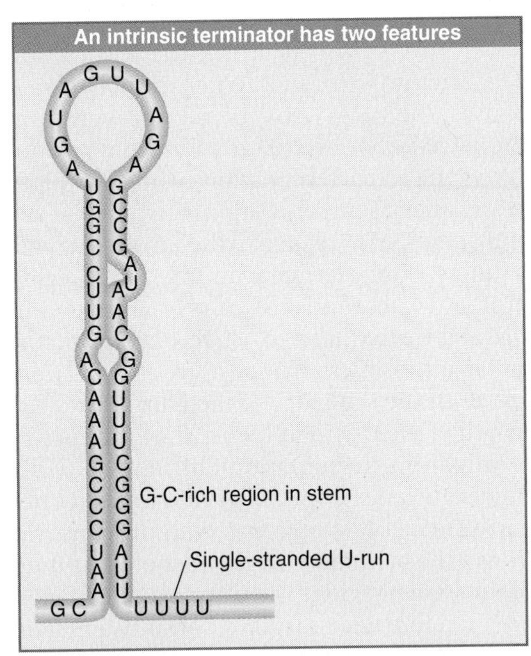

FIGURE 11.46 Intrinsic terminators include palindromic regions that form hairpins varying in length from 7 to 20 bp. The stem-loop structure includes a G-C-rich region and is followed by a run of U residues.

very end of the unit that is rich in U residues. Both features are needed for termination. The hairpin usually contains a G-C-rich region near the base of the stem. The typical distance between the hairpin and the U-rich region is seven to nine bases. There are ~1100 sequences in the *E. coli* genome that fit these criteria, which suggests that about half of the genes have intrinsic terminators.

Point mutations that prevent termination occur within the stem region of the hairpin. What is the effect of a hairpin on transcription? It is likely that all hairpins that form in the RNA product cause the polymerase to slow (and perhaps to pause) in RNA synthesis.

Pausing creates an opportunity for termination to occur. Pausing occurs at sites that resemble terminators but have an increased separation (typically ten to eleven bases) between the hairpin and the U-run. If the pause site does not correspond to a terminator, though, the enzyme usually moves on again to continue transcription. The length of the pause varies, but at a typical terminator lasts ~60 seconds.

A downstream U-rich region destabilizes the RNA–DNA hybrid when RNA polymerase pauses at the hairpin. The rU-dA RNA–DNA hybrid has an unusually weak base-paired structure; it requires the least energy of any RNA–DNA hybrid to break the association between the two strands. When the polymerase pauses, the RNA–DNA hybrid unravels from the weakly bonded rU-dA terminal region. Often, the actual termination event takes place at any one of several positions toward or at the end of the U-rich region, as though the enzyme "stutters" during termination. The U-rich region in RNA corresponds to an A-T-rich region in DNA, so we see that A-T-rich regions are important in intrinsic termination as well as initiation.

Both the sequence of the hairpin and the length of the U-run influence the efficiency of termination. Termination efficiency *in vitro*, however, varies from 2% to 90% and does not correlate in any simple way with the constitution of the hairpin or the number of U residues in the U-rich region. The hairpin and U-region are therefore necessary, but not sufficient, and additional parameters influence the interaction with RNA polymerase. In particular, the sequences both upstream and downstream of the intrinsic terminator influence its efficiency.

Less is known about the signals and ancillary factors involved in termination for eukaryotic polymerases. Each class of polymerase uses a different mechanism (see Chapter 26, RNA Splicing and Processing).

11.22 How Does Rho Factor Work?

Key concept

- Rho factor is a terminator protein that binds to a *rut* site on nascent RNA and tracks along the RNA to release it from the RNA–DNA hybrid structure at the RNA polymerase.

Rho factor is an essential protein in *E. coli* that functions solely at the stage of termination. It acts at rho-dependent terminators, which account for about half of *E. coli* terminators.

FIGURE 11.47 shows how rho functions. First it binds to a sequence within the transcript upstream of the site of termination. This sequence is called a ***rut*** site (an acronym for rho utilization). The rho then tracks along the RNA until it catches up to RNA polymerase. When the RNA polymerase reaches the termination site, rho acts on the RNA–DNA hybrid in the enzyme to cause release of the RNA. Pausing by the polymerase at the site of termination allows time for rho factor to translocate to the hybrid stretch and is an important feature of termination.

We see an important general principle here. When we know the site on DNA at which some protein exercises its effect, we cannot assume that this coincides with the DNA sequence that it initially recognizes. They can be separate, and there need not be a fixed relationship between them. In fact, *rut* sites in different transcription units are found at varying distances preceding the sites of termination. A similar distinction is made by antitermination factors (see Section 11.24, Antitermination Requires Sites That Are Independent of the Terminators).

The common feature of *rut* sites is that the sequence is rich in C residues and poor in G residues and has no secondary structure. An example is given in FIGURE 11.48. C is by far the most common base (41%) and G is the least common base (14%). *rut* sites vary in length. As a general rule, the efficiency of a *rut* site increases with the length of the C-rich/G-poor region.

Rho is a member of the family of hexameric ATP-dependent helicases. The subunit has an RNA-binding domain and an ATP hydrolysis domain. The hexamer functions by passing nucleic acid through the hole in the middle of the assembly formed from the RNA-binding domains of the subunits. FIGURE 11.49 shows that the structure of rho gives some indications of how it functions. It winds RNA from the 3′ end around the exterior of the N-terminal domains, and pushes the 5′ end of the bound region into the interior, where it is bound by a secondary RNA-binding domain in the C-terminal domains. The initial form of rho is a gapped ring, but binding of the RNA converts it to a closed ring.

After binding to the *rut* site, rho uses its helicase activity, driven by ATP hydrolysis, to translocate along RNA until it reaches the RNA–DNA hybrid stretch in RNA polymerase. It then unwinds the duplex structure. We do not know whether this action is sufficient to release the transcript or whether rho also interacts with RNA polymerase to help release RNA. Rho functions as an ancillary factor for RNA polymerase; typically its maximum activity *in vitro* is displayed when it is present at ~10% of the concentration of the RNA polymerase.

Rho needs to translocate along RNA from the *rut* site to the actual point of termination. This requires the factor to move faster than RNA polymerase. The enzyme pauses when it reaches a terminator, and termination occurs if rho catches it there. Pausing is therefore important in rho-dependent termination, just as in intrinsic termination, because it gives time for the other necessary events to occur.

The idea that rho moves along RNA leads to an important prediction about the relationship between transcription and translation. Rho must first have access to a binding sequence on RNA and then must be able to move along the RNA. Either or both of these conditions may be prevented if ribosomes are translating an RNA. Thus the ability of rho factor to reach RNA polymerase at a terminator depends on what is happening in translation.

This model explains a puzzling phenomenon. In some cases, a nonsense mutation in one gene of a transcription unit prevents the expression of subsequent genes in the unit. This effect is called **polarity.** A common cause is the

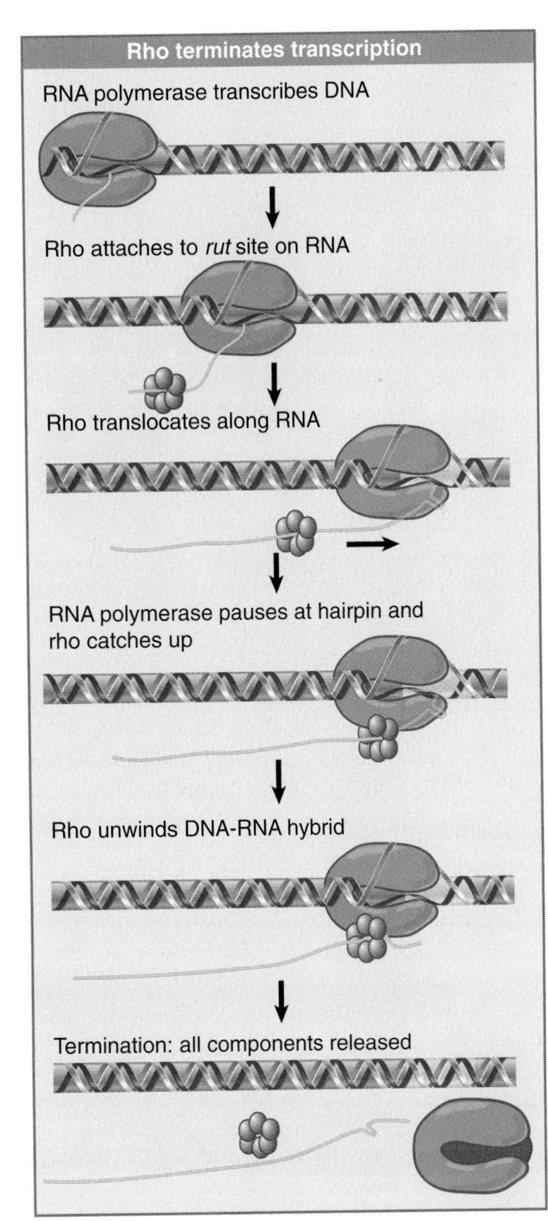

FIGURE 11.47 Rho factor binds to RNA at a *rut* site and translocates along RNA until it reaches the RNA–DNA hybrid in RNA polymerase, where it releases the RNA from the DNA.

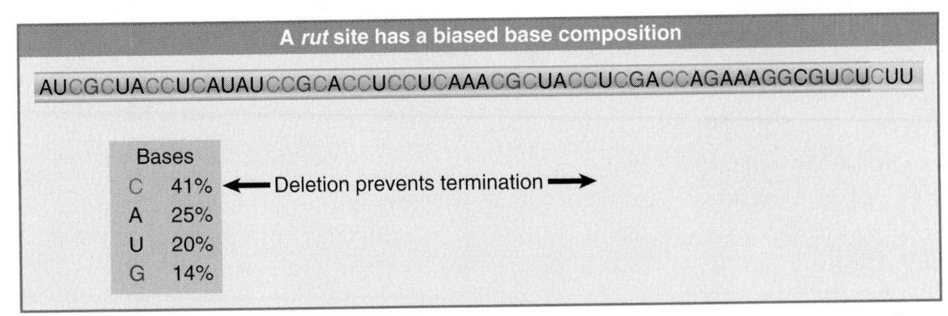

FIGURE 11.48 A *rut* site has a sequence rich in C and poor in G preceding the actual site(s) of termination. The sequence corresponds to the 3′ end of the RNA.

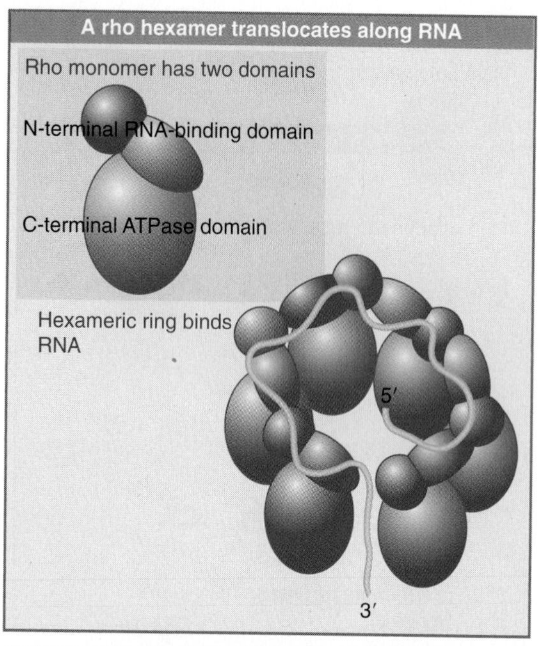

FIGURE 11.49 Rho has an N-terminal RNA-binding domain and a C-terminal ATPase domain. A hexamer in the form of a gapped ring binds RNA along the exterior of the N-terminal domains. The 5′ end of the RNA is bound by a secondary binding site in the interior of the hexamer.

absence of the mRNA corresponding to the subsequent (distal) parts of the unit.

Suppose that there are rho-dependent terminators *within* the transcription unit, that is, before the terminator that *usually* is used. The consequences are illustrated in FIGURE 11.50. Normally these earlier terminators are not used, because the ribosomes prevent rho from reaching RNA polymerase. A nonsense mutation, however, releases the ribosomes, so that rho is free to attach to and/or move along the mRNA, enabling it to act on RNA polymerase at the terminator. As a result, the enzyme is released, and the distal regions of the transcription unit are never expressed. (Why should there be internal terminators? Perhaps they are simply sequences that by coincidence mimic the usual rho-dependent terminator.) Some stable RNAs that have extensive secondary structure are preserved from polar effects, presumably because the structure impedes rho attachment or movement.

rho mutations show wide variations in their influence on termination. The basic nature of the effect is a failure to terminate. The magnitude of the failure, however, as seen in the percent of readthrough *in vivo,* depends on the particular target locus. Similarly, the need for rho

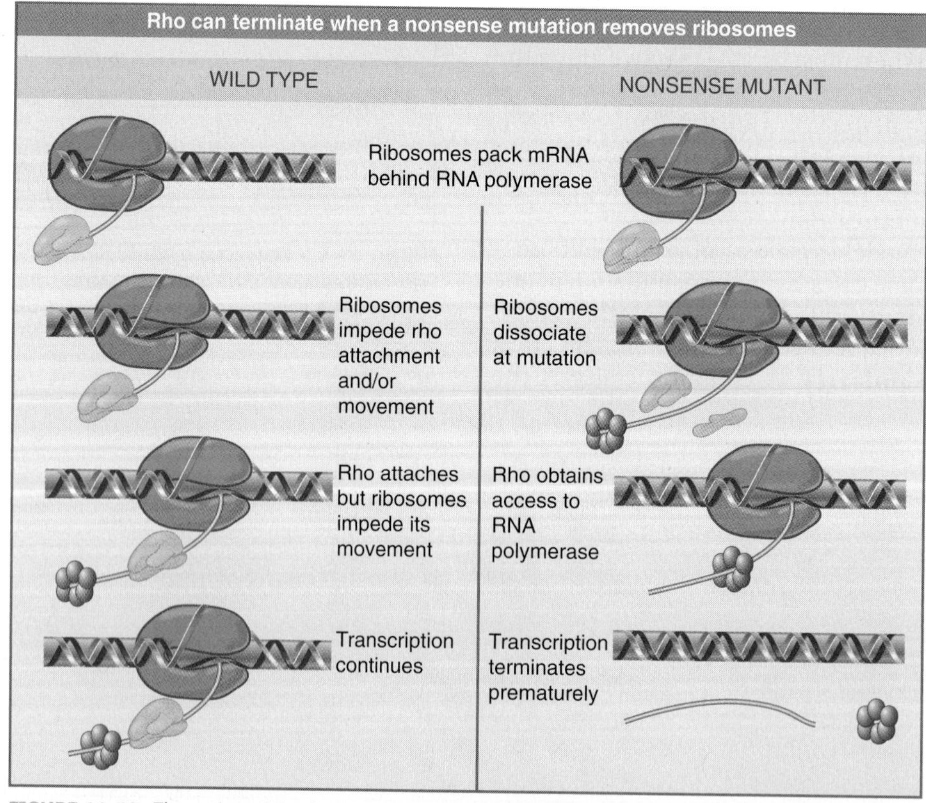

FIGURE 11.50 The action of rho factor may create a link between transcription and translation when a rho-dependent terminator lies soon after a nonsense mutation.

factor *in vitro* is variable. Some (rho-dependent) terminators require relatively high concentrations of rho, whereas others function just as well at lower levels. This suggests that different terminators require different levels of rho factor for termination and therefore respond differently to the residual levels of rho factor in the mutants (*rho* mutants are usually leaky).

Some *rho* mutations can be suppressed by mutations in other genes. This approach provides an excellent way to identify proteins that interact with rho. The β subunit of RNA polymerase is implicated by two types of mutation. First, mutations in the *rpoB* gene can reduce termination at a rho-dependent site. Second, mutations in *rpoB* can restore the ability to terminate transcription at rho-dependent sites in *rho* mutant bacteria. We do not, however, know what function the interaction plays.

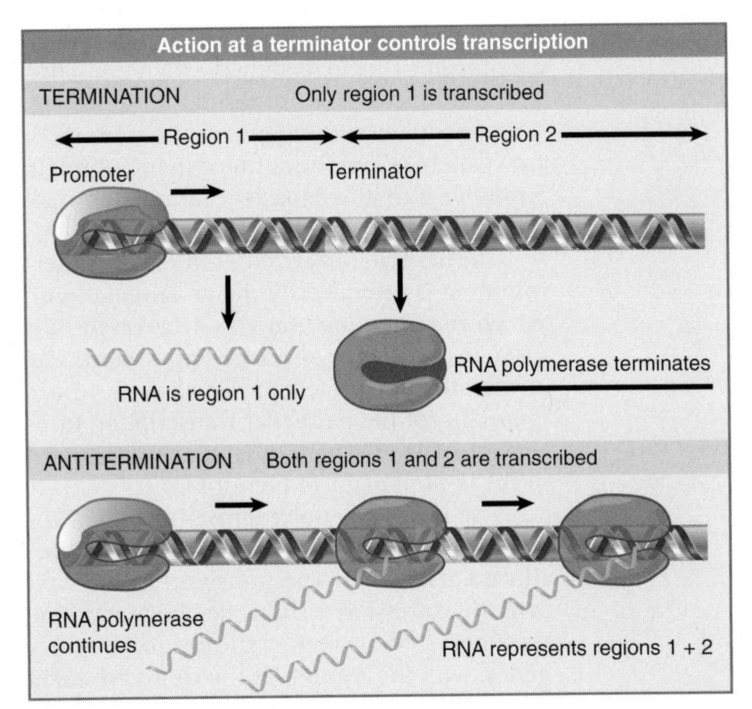

FIGURE 11.51 Antitermination can control transcription by determining whether RNA polymerase terminates or reads through a particular terminator into the following region.

11.23 Antitermination Is a Regulatory Event

Key concepts

- Termination is prevented when antitermination proteins act on RNA polymerase to cause it to read through a specific terminator or terminators.
- Phage lambda has two antitermination proteins, pN and pQ, that act on different transcription units.

Antitermination is used as a control mechanism in both phage regulatory circuits and bacterial operons. FIGURE 11.51 shows that antitermination controls the ability of the enzyme to read past a terminator into genes lying beyond. In the example shown in the figure, the default pathway is for RNA polymerase to terminate at the end of region 1; however, antitermination allows it to continue transcription through region 2. The promoter does not change, so as a result both situations produce an RNA with the same 5′ sequences; the difference is that after antitermination the RNA is extended to include new sequences at the 3′ end.

Antitermination was discovered in bacteriophage infections. A common feature in the control of phage infection is that very few of the phage genes (the "early" genes) can be transcribed by the bacterial host RNA polymerase. Among these genes, however, are regulator(s) whose product(s) allow the next set of phage genes to be expressed (see Section 14.4, Two Types of Regulatory Event Control the Lytic Cascade). One of these types of regulator is an **antitermination protein.** FIGURE 11.52 shows

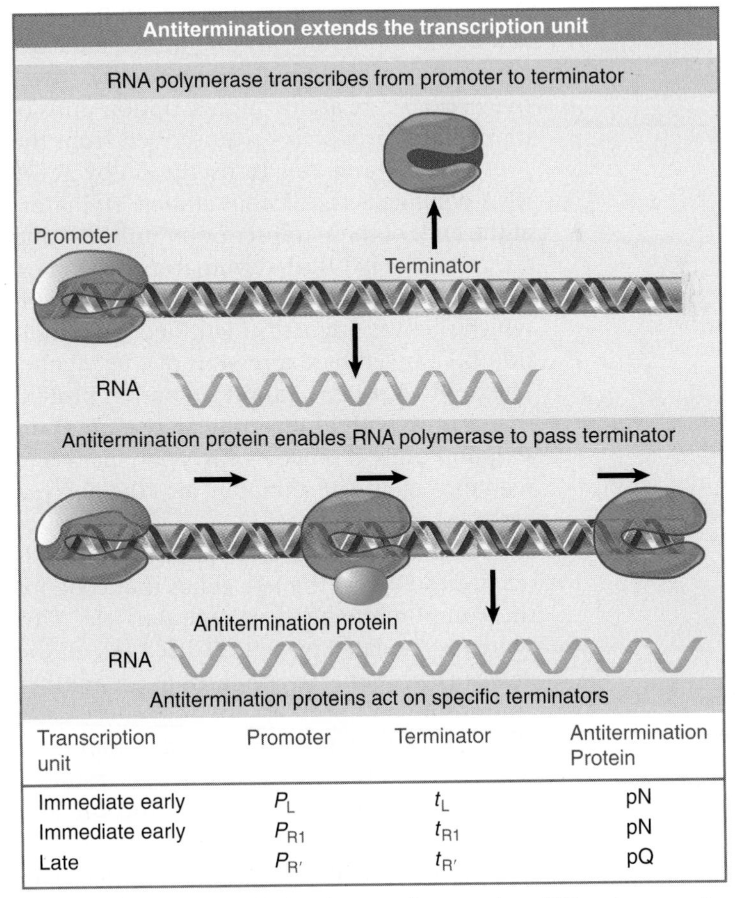

Transcription unit	Promoter	Terminator	Antitermination Protein
Immediate early	P_L	t_L	pN
Immediate early	P_{R1}	t_{R1}	pN
Late	$P_{R'}$	$t_{R'}$	pQ

FIGURE 11.52 An antitermination protein can act on RNA polymerase to enable it to read through a specific terminator.

that it enables RNA polymerase to read through a terminator, thus extending the RNA transcript. In the absence of the antitermination protein, RNA polymerase terminates at the terminator (top panel). When the antitermination protein is present, it continues past the terminator (middle panel).

The best-characterized example of antitermination is provided by phage lambda, with which the phenomenon was discovered. It is used at two stages of phage expression. The antitermination protein produced at each stage is specific for the particular transcription units that are expressed at that stage, as summarized in the bottom panel of Figure 11.52.

The host RNA polymerase initially transcribes two genes, which are called the **immediate early** genes. The transition to the next stage of expression is controlled by preventing termination at the ends of the immediate early genes, with the result that the **delayed early** genes are expressed. (We discuss the overall regulation of lambda development in Chapter 14, Phage Strategies.)

The regulator gene that controls the switch from immediate early to delayed early expression is identified by mutations in lambda gene *N* that can transcribe *only* the immediate early genes; they proceed no further into the infective cycle. There are two transcription units of immediate early genes (transcribed from the promoters P_L and P_R). Transcription by *E. coli* RNA polymerase itself stops at the terminators at the ends of these transcription units (t_{L1} and t_{R1}, respectively.) Both terminators depend on rho; in fact, these were the terminators with which rho was originally identified. The situation is changed by expression of the *N* gene. The product pN is an antitermination protein that acts on both of the immediate early transcription units and allows RNA polymerase to read through the terminators into the delayed early genes beyond them.

As for other phages, still another control is needed to express the late genes that code for the components of the phage particle. This switch is regulated by gene *Q*, itself one of the delayed early genes. Its product, pQ, is another antitermination protein, one that specifically allows RNA polymerase initiating at another site, the late promoter $P_{R'}$, to read through a terminator that lies between it and the late genes.

The different specificities of pN and pQ establish an important general principle: *RNA polymerase interacts with transcription units in such a way that an ancillary factor can sponsor antitermination specifically for some transcripts.* Termination can be controlled with the same sort of precision as initiation.

11.24 Antitermination Requires Sites That Are Independent of the Terminators

Key concepts

- The site where an antiterminator protein acts is upstream of the terminator site in the transcription unit.
- The location of the antiterminator site varies in different cases and can be in the promoter or within the transcription unit.

Which sites are involved in controlling the specificity of antitermination? The antitermination activity of pN is highly specific, but *the antitermination event is not determined by the terminators* t_{L1} *and* t_{R1}*; the recognition site needed for antitermination lies upstream in the transcription unit, that is, at a different place from the terminator site at which the action eventually is accomplished.* FIGURE 11.53 shows the locations of the sites required for antitermination in phage lambda.

The recognition sites required for pN action are called *nut* (for *N utilization*). The sites responsible for determining leftward and rightward antitermination are described as *nutL* and *nutR*, respectively. Mapping of *nut* mutations locates *nutL* between the startpoint of P_L and the beginning of the *N* coding region. By contrast, *nutR* lies between the end of the *cro* gene and t_{L1}. This means that the two *nut* sites lie in different positions relative to the organization of their transcription units. Whereas *nutL* is near the promoter, *nutR* is near to the terminator. (*qut* is different yet again, and lies within the promoter.)

How does antitermination occur? When pN recognizes the *nut* site, it must act on RNA polymerase to ensure that the enzyme can no longer respond to the terminator. The variable locations of the *nut* sites indicate that this event is linked neither to initiation nor to termination, but can occur to RNA polymerase as it elongates the RNA chain past the *nut* site. As illustrated in FIGURE 11.54, the polymerase then becomes a juggernaut that continues past the terminator, heedless of its signal. (This reaction involves antitermination at rho-dependent terminators, but pN also suppresses termination at intrinsic terminators.)

Is the ability of pN to recognize a short sequence within the transcription unit an example of a more widely used mechanism for antitermination? Phages that are related to lambda have different *N* genes and different antitermination specificities. The region of the phage genome in which the *nut* sites lie has a different sequence in each of these phages, and each phage must therefore have characteristic *nut* sites recognized specifically by its own pN. Each of these pN products must have the same general ability to interact with the transcription apparatus in an antitermination capacity, but also have a different specificity for the sequence of DNA that activates the mechanism.

11.25 Termination and Antitermination Factors Interact with RNA Polymerase

Key concepts

- Several bacterial proteins are required for lambda pN to interact with RNA polymerase.
- These proteins are also involved in antitermination in the *rrn* operons of the host bacterium.
- The lambda antiterminator pQ has a different mode of interaction that involves binding to DNA at the promoter.

Termination and antitermination are closely connected and involve bacterial and phage proteins that interact with RNA polymerase. Several proteins concerned with termination have been identified by isolating mutants of *E. coli* in which pN is ineffective. Several of these mutations lie in the *rpoB* gene. This argues that pN (like rho factor) interacts with the β subunit of the core enzyme. Other *E. coli* mutations that prevent pN function identify the *nus* loci: *nusA, nusB, nusE,* and *nusG.* (The term "*nus*" is an acronym for *N* *u*tilization *s*ubstance.)

A lambda *nut* site consists of two sequence elements called *boxA* and *boxB.* Sequence elements related to *boxA* are also found in bacterial operons. *boxA* is required for binding bacterial proteins that are necessary for antitermination in both phage and bacterial operons. *boxB* is specific to the phage genome, and mutations in *boxB* abolish the ability of pN to cause antitermination.

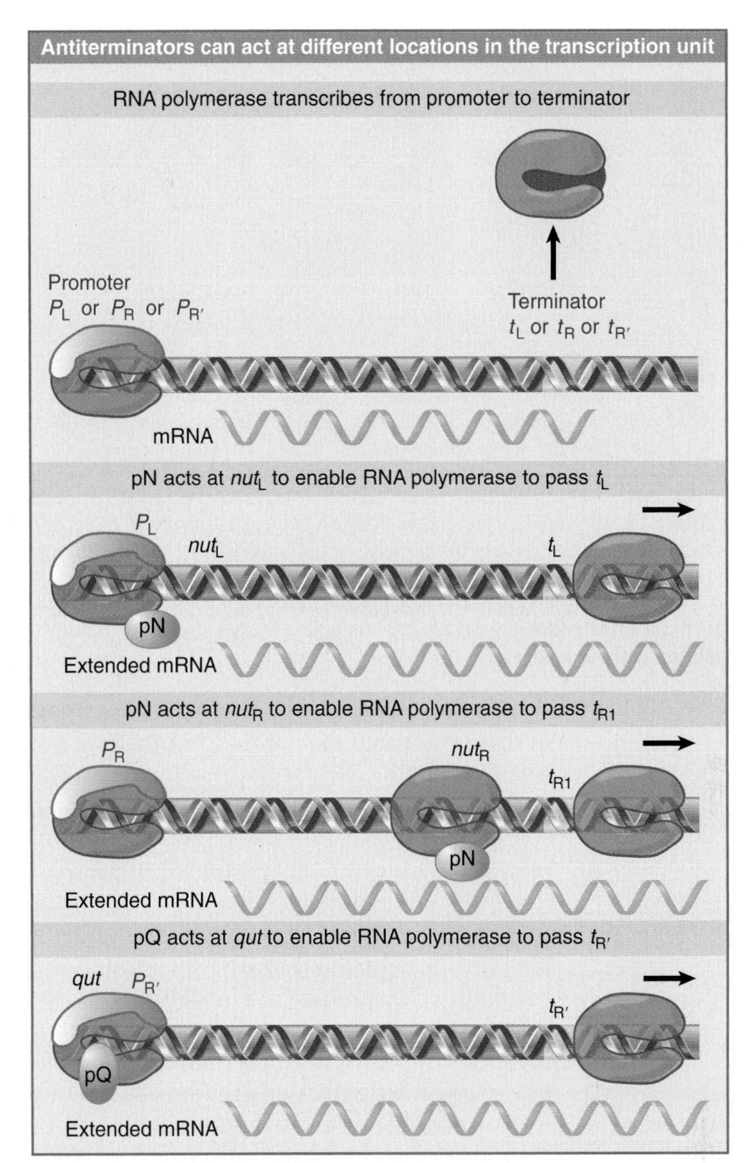

FIGURE 11.53 Host RNA polymerase transcribes lambda genes and terminates at *t* sites. pN allows it to read through terminators in the L and R1 units; pQ allows it to read through the R′ terminator. The sites at which pN acts *(nut)* and at which pQ acts *(qut)* are located at different relative positions in the transcription units.

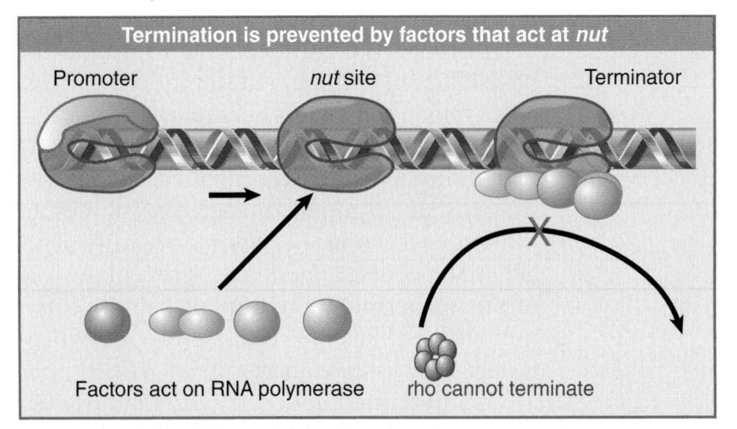

FIGURE 11.54 Ancillary factors bind to RNA polymerase as it passes the *nut* site. They prevent rho from causing termination when the polymerase reaches the terminator.

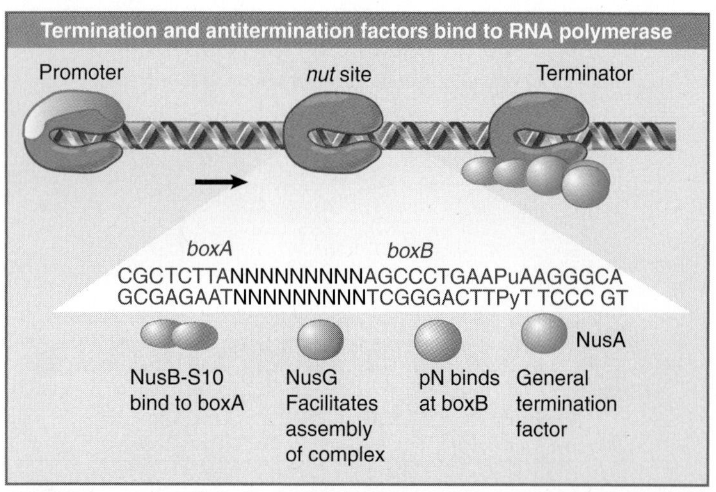

FIGURE 11.55 Ancillary factors bind to RNA polymerase as it passes certain sites. The *nut* site consists of two sequences. NusB-S10 joins core enzyme as it passes *boxA*. Then NusA and pN protein bind as polymerase passes *boxB*. The presence of pN allows the enzyme to read through the terminator, producing a joint mRNA that contains immediate early sequences joined to delayed early sequences.

The *nus* loci code for proteins that form part of the transcription apparatus, but that are not isolated with the RNA polymerase enzyme. The *nusA, nusB,* and *nusG* functions are concerned solely with the termination of transcription. *nusE* codes for ribosomal protein S10; the relationship between its location in the 30S subunit and its function in termination is not clear. The Nus proteins bind to RNA polymerase at the *nut* site, as summarized in FIGURE 11.55.

NusA is a general transcription factor that increases the efficiency of termination, probably by enhancing RNA polymerase's tendency to pause at terminators (and indeed at other regions of secondary structure; see below). NusB and S10 form a dimer that binds specifically to RNA containing a *boxA* sequence. NusG may be concerned with the general assembly of all the Nus factors into a complex with RNA polymerase.

Intrinsic and rho-dependent terminators have different requirements for the Nus factors. NusA is required for termination at intrinsic terminators, and the reaction can be prevented by pN. At rho-dependent terminators, all four Nus proteins are required, and again pN alone can inhibit the reaction. The common feature of pN at both types of terminator is to prevent the role of NusA in termination. Binding of pN to NusA inhibits the ability of NusA to bind RNA, which is necessary for termination.

Antitermination occurs in the *rrn* (rRNA) operons of *E. coli* and involves the same *nus* functions. The leader regions of the *rrn* operons contain *boxA* sequences; NusB-S10 dimers recognize these sequences and bind to RNA polymerase as it elongates past *boxA*. This changes the properties of RNA polymerase in such a way that it can now read through rho-dependent terminators that are present within the transcription unit.

The *boxA* sequence of lambda RNA does not bind NusB-S10 and is probably enabled to do so by the presence of NusA and pN; the *boxB* sequence could be required to stabilize the reaction. Thus variations in *boxA* sequences may determine which particular set of factors is required for antitermination. The consequences are the same: when RNA polymerase passes the *nut* site, it is modified by addition of appropriate factors and fails to terminate when it subsequently encounters the terminator sites.

Antitermination in lambda requires pN to bind to RNA polymerase in a manner that depends on the sequence of the transcription unit. Does pN recognize the *boxB* site in DNA or in the RNA transcript? It does not bind independently to either type of sequence, but does bind to a transcription complex when core enzyme passes the *boxB* site. pN has separate domains that recognize the *boxB* RNA sequence and the NusA protein. After joining the transcription complex, pN remains associated with the core enzyme, in effect becoming an additional subunit whose presence changes recognition of terminators. It is possible that pN in fact continues to bind to both the *boxB* RNA sequence and to RNA polymerase, which maintains a loop in the RNA; thus the role of *boxB* RNA would partly be to tether pN in the vicinity, effectively increasing its local concentration.

pQ, which prevents termination later in phage infection by acting at *qut*, has a different mode of action. The *qut* sequence lies at the start of the late transcription unit. The upstream part of *qut* lies within the promoter, whereas the downstream part lies at the beginning of the transcribed region. This implies that pQ action involves recognition of DNA; it also implies that its mechanism of action, at least concerning the initial binding to the complex, must be different from that of pN. pQ interacts with the holoenzyme during the initiation phase. In fact, σ^{70} is required for the interaction with pQ. This reinforces the view of RNA polymerase as an interactive structure in which conformational changes induced at one phase may affect its activity at a later phase.

The basic action of pQ is to interfere with pausing. Once pQ has acted upon RNA polymerase, the enzyme shows much reduced pausing at all sites, including rho-dependent and intrinsic terminators. This means that pQ does not act directly on termination *per se*, but instead allows the enzyme to pass the terminator more quickly, thus depriving the core polymerase and/or accessory factor of the opportunity to cause termination.

The general principle is that RNA polymerase may exist in forms that are competent to undertake particular stages of transcription, and its activities at these stages can be changed only by modifying the appropriate form. Thus substitutions of sigma factors may change one initiation-competent form into another; and additions of Nus factors may change the properties of termination-competent forms.

Termination seems to be closely connected with the mode of elongation. In its basic transcription mode, core polymerase is subject to many pauses during elongation, and pausing at a terminator site is the prerequisite for termination to occur. Under the influence of factors such as NusA, pausing becomes extended, increasing the efficiency of termination; while under the influence of pN or pQ, pausing is abbreviated, decreasing the efficiency of termination. Recognition sites for these factors are found only in certain transcription units, and as a result pausing and consequently termination are altered only in those units.

11.26 Summary

A transcription unit comprises the DNA between a promoter, where transcription initiates, and a terminator, where it ends. One strand of the DNA in this region serves as a template for synthesis of a complementary strand of RNA. The RNA–DNA hybrid region is short and transient, as the transcription "bubble" moves along DNA. The RNA polymerase holoenzyme that synthesizes bacterial RNA can be separated into two components. Core enzyme is a multimer of structure $\alpha_2\beta\beta'$ that is responsible for elongating the RNA chain. Sigma factor (σ) is a single subunit that is required at the stage of initiation for recognizing the promoter.

Core enzyme has a general affinity for DNA. The addition of sigma factor reduces the affinity of the enzyme for nonspecific binding to DNA, but increases its affinity for promoters. The rate at which RNA polymerase finds its promoters is too great to be accounted for by diffusion and random contacts with DNA; direct exchange of DNA sequences held by the enzyme may be involved.

Bacterial promoters are identified by two short conserved sequences centered at −35 and −10 relative to the startpoint. Most promoters have sequences that are well related to the consensus sequences at these sites. The distance separating the consensus sequences is 16 to 18 bp. RNA polymerase initially "touches down" at the −35 sequence and then extends its contacts over the −10 region. The enzyme covers ~77 bp of DNA. The initial "closed" binary complex is converted to an "open" binary complex by melting of a sequence of ~12 bp that extends from the −10 region to the startpoint. The A-T-rich base pair composition of the −10 sequence may be important for the melting reaction.

The binary complex is converted to a ternary complex by the incorporation of ribonucleotide precursors. There are multiple cycles of abortive initiation, during which RNA polymerase synthesizes and releases very short RNA chains without moving from the promoter. At the end of this stage, there is a change in structure, and the core enzyme contracts to cover ~50 bp. Sigma factor is either released (30% of cases) or changes its form of association with the core enzyme. The core enzyme then moves along DNA, synthesizing RNA. A locally unwound region of DNA moves with the enzyme. The enzyme contracts further in size to cover only 30 to 40 bp when the nascent chain has reached 15 to 20 nucleotides; then it continues to the end of the transcription unit.

The "strength" of a promoter describes the frequency at which RNA polymerase initiates transcription; it is related to the closeness with which its −35 and −10 sequences conform to the ideal consensus sequences, but is influenced also by the sequences immediately downstream of the startpoint. Negative supercoiling increases the strength of certain promoters. Transcription generates positive supercoils ahead of RNA polymerase and leaves negative supercoils behind the enzyme.

The core enzyme can be directed to recognize promoters with different consensus sequences by alternative sigma factors. In *E. coli*, these sigma factors are activated by adverse conditions, such as heat shock or nitrogen starvation. *B. subtilis* contains a single major sigma factor with the same specificity as the *E. coli* sigma factor and also contains a variety of minor sigma factors. Another series of factors is activated when sporulation is initiated; sporulation is regulated by two cascades in which sigma factor replacements occur in the forespore and mother cell. A cascade for regulating transcription by substitution of sigma factors is also used by phage SPO1.

The geometry of RNA polymerase-promoter recognition is similar for holoenzymes containing all sigma factors (except σ^{54}). Each sigma factor causes RNA polymerase to initiate transcription at a promoter that conforms to a particular consensus at −35 and −10. Direct contacts between sigma and DNA at these sites have been demonstrated for *E. coli* σ^{70}. The σ^{70} factor of *E. coli* has an N-terminal autoinhibitory domain that prevents the DNA-binding regions from recognizing DNA. The autoinhibitory region is displaced by DNA when the holoenzyme forms an open complex.

Bacterial RNA polymerase terminates transcription at two types of sites. Intrinsic terminators contain a G-C-rich hairpin followed by a U-rich region. They are recognized *in vitro* by core enzyme alone. Rho-dependent terminators require rho factor both *in vitro* and *in vivo;* rho binds to *rut* sites that are rich in C and poor in G residues and that precede the actual site of termination. Rho is a hexameric ATP-dependent helicase activity that translocates along the RNA until it reaches the RNA–DNA hybrid region in the transcription bubble of RNA polymerase, where it dissociates the RNA from DNA. In both types of termination, pausing by RNA polymerase is important in order to allow time for the actual termination event to occur.

The Nus factors are required for termination. NusA is required for intrinsic terminators, and in addition NusB-S10 is required for rho-dependent terminators. The NusB-S10 dimer recognizes the *boxA* sequence of a *nut* site in the elongating RNA; NusA joins subsequently.

Antitermination is used by some phages to regulate progression from one stage of gene expression to the next. The lambda gene *N* codes for an antitermination protein (pN) that is necessary to allow RNA polymerase to read through the terminators located at the ends of the immediate early genes. Another antitermination protein, pQ, is required later in phage infection. pN and pQ act on RNA polymerase as it passes specific sites (*nut* and *qut,* respectively). These sites are located at different relative positions in their respective transcription units. pN recognizes RNA polymerase carrying NusA when the enzyme passes the sequence *boxB.* pN then binds to the complex and prevents termination by antagonizing the action of NusA when the polymerase reaches the rho-dependent terminator.

References

11.2 Transcription Occurs by Base Pairing in a "Bubble" of Unpaired DNA

Review

Losick, R. and Chamberlin, M. (1976). RNA Polymerase. *Cold Spring Harbor Symp. Quant. Biol.*

Research

Korzheva, N., Mustaev, A., Kozlov, M., Malhotra, A., Nikiforov, V., Goldfarb, A., and Darst, S. A. (2000). A structural model of transcription elongation. *Science* 289, 619–625.

11.3 The Transcription Reaction Has Three Stages

Research

Rice, G. A., Kane, C. M., and Chamberlin, M. (1991). Footprinting analysis of mammalian RNA polymerase II along its transcript: an alternative view of transcription elongation. *Proc. Natl. Acad. Sci. USA* 88, 4245–4281.

Wang, D. et al. (1995). Discontinuous movements of DNA and RNA in RNA polymerase accompany formation of a paused transcription complex. *Cell* 81, 341–350.

11.4 Phage T7 RNA Polymerase Is a Useful Model System

Research

Cheetham, G. M., Jeruzalmi, D., and Steitz, T. A. (1999). Structural basis for initiation of transcription from an RNA polymerase-promoter complex. *Nature* 399, 80–83.

Cheetham, G. M. T. and Steitz, T. A. (1999). Structure of a transcribing T7 RNA polymerase initiation complex. *Science* 286, 2305–2309.

Temiakov, D., Mentesana, D., Temiakov, D., Ma, K., Mustaev, A., Borukhov, S., and McAllister, W. T. (2000). The specificity loop of T7 RNA polymerase interacts first with the promoter and then with the elongating transcript, suggesting a mechanism for promoter clearance. *Proc. Natl. Acad. Sci. USA* 97, 14109–14114.

11.5 A Model for Enzyme Movement Is Suggested by the Crystal Structure

Review

Shilatifard, A., Conaway, R. C., and Conaway, J. W. (2003). The RNA polymerase II elongation complex. *Annu. Rev. Biochem.* 72, 693–715.

Research

Cramer, P., Bushnell, D. A., Fu, J., Gnatt, A. L., Maier-Davis, B., Thompson, N. E., Burgess, R. R., Edwards, A. M., David, P. R., and Kornberg, R. D. (2000). Architecture of RNA polymerase II and implications for

the transcription mechanism. *Science* 288, 640–649.
Cramer, P., Bushnell, P., and Kornberg, R. D. (2001). Structural basis of transcription: RNA polymerase II at 2.8 Å resolution. *Science* 292, 1863–1876.
Gnatt, A. L., Cramer, P., Fu, J., Bushnell, D. A., and Kornberg, R. D. (2001). Structural basis of transcription: an RNA polymerase II elongation complex at 3.3 Å resolution. *Science* 292, 1876–1882.

11.6 Bacterial RNA Polymerase Consists of Multiple Subunits

Review

Helmann, J. D. and Chamberlin, M. (1988). Structure and function of bacterial sigma factors. *Annu. Rev. Biochem.* 57, 839–872.

Research

Campbell, E. A., Korzheva, N., Mustaev, A., Murakami, K., Nair, S., Goldfarb, A., and Darst, S. A. (2001). Structural mechanism for rifampicin inhibition of bacterial RNA polymerase. *Cell* 104, 901–912.
Korzheva, N., Mustaev, A., Kozlov, M., Malhotra, A., Nikiforov, V., Goldfarb, A., and Darst, S. A. (2000). A structural model of transcription elongation. *Science* 289, 619–625.
Zhang, G., Campbell, E. A., Zhang, E. A., Minakhin, L., Richter, C., Severinov, K., and Darst, S. A. (1999). Crystal structure of *Thermus aquaticus* core RNA polymerase at 3.3 Å resolution. *Cell* 98, 811–824.

11.7 RNA Polymerase Consists of the Core Enzyme and Sigma Factor

Research

Travers, A. A. and Burgess, R. R. (1969). Cyclic reuse of the RNA polymerase sigma factor. *Nature* 222, 537–540.

11.8 The Association with Sigma Factor Changes at Initiation

Research

Bar-Nahum, G. and Nudler, E. (2001). Isolation and characterization of sigma(70)-retaining transcription elongation complexes from *E. coli*. *Cell* 106, 443–451.
Krummel, B. and Chamberlin, M. J. (1989). RNA chain initiation by *E. coli* RNA polymerase. Structural transitions of the enzyme in early ternary complexes. *Biochemistry* 28, 7829–7842.
Mukhopadhyay, J., Kapanidis, A. N., Mekler, V., Kortkhonjia, E., Ebright, Y. W., and Ebright, R. H. (2001). Translocation of sigma(70) with RNA polymerase during transcription. Fluorescence resonance energy transfer assay for movement relative to DNA. *Cell* 106, 453–463.

11.9 A Stalled RNA Polymerase Can Restart

Research

Kettenberger, H., Armache, K. J., and Cramer, P. (2003). Architecture of the RNA polymerase II-TFIIS complex and implications for mRNA cleavage. *Cell* 114, 347–357.
Opalka, N., Chlenov, M., Chacon, P., Rice, W. J., Wriggers, W., and Darst, S. A. (2003). Structure and function of the transcription elongation factor GreB bound to bacterial RNA polymerase. *Cell* 114, 335–345.

11.11 Sigma Factor Controls Binding to DNA

Research

Bar-Nahum, G. and Nudler, E. (2001). Isolation and characterization of sigma(70)-retaining transcription elongation complexes from *E. coli*. *Cell* 106, 443–451.
Mukhopadhyay, J., Kapanidis, A. N., Mekler, V., Kortkhonjia, E., Ebright, Y. W., and Ebright, R. H. (2001). Translocation of sigma(70) with RNA polymerase during transcription. Fluorescence resonance energy transfer assay for movement relative to DNA. *Cell* 106, 453–463.

11.12 Promoter Recognition Depends On Consensus Sequences

Review

McClure, W. R. (1985). Mechanism and control of transcription initiation in prokaryotes. *Annu. Rev. Biochem.* 54, 171–204.

Research

Ross, W., Gosink, K. K., Salomon, J., Igarashi, K., Zou, C., Ishihama, A., Severinov, K., and Gourse, R. L. (1993). A third recognition element in bacterial promoters: DNA binding by the alpha subunit of RNA polymerase. *Science* 262, 1407–1413.

11.13 Promoter Efficiencies Can Be Increased or Decreased by Mutation

Review

McClure, W. R. (1985). Mechanism and control of transcription initiation in prokaryotes. *Annu. Rev. Biochem.* 54, 171–204.

11.14 RNA Polymerase Binds to One Face of DNA

Review

Siebenlist, U., Simpson, R. B., and Gilbert, W. (1980). *E. coli* RNA polymerase interacts homologously with two different promoters. *Cell* 20, 269–281.

11.15 Supercoiling Is an Important Feature of Transcription

Research

Wu, H.-Y. et al. (1988). Transcription generates positively and negatively supercoiled domains in the template. *Cell* 53, 433–440.

11.16 Substitution of Sigma Factors May Control Initiation

Review

Hengge-Aronis, R. (2002). Signal transduction and regulatory mechanisms involved in control of the sigma(S) (RpoS) subunit of RNA polymerase. *Microbiol. Mol. Biol. Rev.* 66, 373–393.

Research

Alba, B. M., Onufryk, C., Lu, C. Z., and Gross, C. A. (2002). DegS and YaeL participate sequentially in the cleavage of RseA to activate the sigma(E)-dependent extracytoplasmic stress response. *Genes Dev.* 16, 2156–2168.

Grossman, A. D., Erickson, J. W., and Gross, C. A. (1984). The htpR gene product of *E. coli* is a sigma factor for heat-shock promoters. *Cell* 38, 383–390.

Kanehara, K., Ito, K., and Akiyama, Y. (2002). YaeL (EcfE) activates the sigma(E) pathway of stress response through a site-2 cleavage of anti-sigma(E), RseA. *Genes Dev.* 16, 2147–2155.

Sakai, J., Duncan, E. A., Rawson, R. B., Hua, X., Brown, M. S., and Goldstein, J. L. (1996). Sterol-regulated release of SREBP-2 from cell membranes requires two sequential cleavages, one within a transmembrane segment. *Cell* 85, 1037–1046.

11.17 Sigma Factors Directly Contact DNA

Research

Campbell, E. A., Muzzin, O., Chlenov, M., Sun, J. L., Olson, C. A., Weinman, O., Trester-Zedlitz, M. L., and Darst, S. A. (2002). Structure of the bacterial RNA polymerase promoter specificity sigma subunit. *Mol. Cell* 9, 527–539.

Dombrowski, A. J. et al. (1992). Polypeptides containing highly conserved regions of transcription initiation factor σ^{70} exhibit specificity of binding to promoter DNA. *Cell* 70, 501–512.

Mekler, V., Kortkhonjia, E., Mukhopadhyay, J., Knight, J., Revyakin, A., Kapanidis, A. N., Niu, W., Ebright, Y. W., Levy, R., and Ebright, R. H. (2002). Structural organization of bacterial RNA polymerase holoenzyme and the RNA polymerase-promoter open complex. *Cell* 108, 599–614.

Vassylyev, D. G, Sekine, S., Laptenko, O., Lee, J., Vassylyeva, M. N., Borukhov, S., Yokoyama S. (2002). Crystal structure of a bacterial RNA polymerase holoenzyme at 2.6 Å resolution. *Nature* 417, 712–719.

11.19 Sporulation Is Controlled by Sigma Factors

Reviews

Errington, J. (1993). *B. subtilis* sporulation: regulation of gene expression and control of morphogenesis. *Microbiol. Rev.* 57, 1–33.

Haldenwang, W. G. (1995). The sigma factors of *B. subtilis*. *Microbiol. Rev.* 59, 1–30.

Losick, R. and Stragier, P. (1992). Crisscross regulation of cell-type specific gene expression during development in *B. subtilis*. *Nature* 355, 601–604.

Losick, R. et al. (1986). Genetics of endospore formation in *B. subtilis*. *Annu. Rev. Genet.* 20, 625–669.

Stragier, P and Losick, R. (1996). Molecular genetics of sporulation in *B. subtilis*. *Annu. Rev. Genet.* 30, 297–341.

Research

Haldenwang, W. G., Lang, N., and Losick, R. (1981). A sporulation-induced sigma-like regulatory protein from *B. subtilis*. *Cell* 23, 615–624.

Haldenwang, W. G. and Losick, R. (1980). A novel RNA polymerase sigma factor from *B. subtilis*. *Proc. Natl. Acad. Sci. USA* 77, 7000–7004.

11.20 Bacterial RNA Polymerase Terminates at Discrete Sites

Reviews

Adhya, S. and Gottesman, M. (1978). Control of transcription termination. *Annu. Rev. Biochem.* 47, 967–996.

Friedman, D. I., Imperiale, M. J., and Adhya, S. L. (1987). RNA 3′ end formation in the control of gene expression. *Annu. Rev. Genet.* 21, 453–488.

Platt, T. (1986). Transcription termination and the regulation of gene expression. *Annu. Rev. Biochem.* 55, 339–372.

11.21 There Are Two Types of Terminators in *E. coli*

Review

von Hippel, P. H. (1998). An integrated model of the transcription complex in elongation, termination, and editing. *Science* 281, 660–665.

Research

Lee, D. N., Phung, L., Stewart, J., and Landick, R. (1990). Transcription pausing by *E. coli* RNA polymerase is modulated by downstream DNA sequences. *J. Biol. Chem.* 265, 15145–15153.

Lesnik, E. A., Sampath, R., Levene, H. B., Henderson, T. J., McNeil, J. A., and Ecker, D. J.

(2001). Prediction of rho-independent transcriptional terminators in *E. coli. Nucleic Acids Res.* 29, 3583–3594.

Reynolds, R., Bermadez-Cruz, R. M., and Chamberlin, M. J. (1992). Parameters affecting transcription termination by *E. coli* RNA polymerase. I. Analysis of 13 rho-independent terminators. *J. Mol. Biol.* 224, 31–51.

11.22 How Does Rho Factor Work?

Reviews

Das, A. (1993). Control of transcription termination by RNA-binding proteins. *Annu. Rev. Biochem.* 62, 893–930.

Richardson, J. P. (1996). Structural organization of transcription termination factor Rho. *J. Biol. Chem.* 271, 1251–1254.

von Hippel, P. H. (1998). An integrated model of the transcription complex in elongation, termination, and editing. *Science* 281, 660–665.

Research

Brennan, C. A., Dombroski, A. J., and Platt, T. (1987). Transcription termination factor rho is an RNA-DNA helicase. *Cell* 48, 945–952.

Geiselmann, J., Wang, Y., Seifried, S. E., and von Hippel, P. H. (1993). A physical model for the translocation and helicase activities of *E. coli* transcription termination protein Rho. *Proc. Natl. Acad. Sci. USA* 90, 7754–7758.

Roberts, J. W. (1969). Termination factor for RNA synthesis. *Nature* 224, 1168–1174.

Skordalakes, E. and Berger, J. M. (2003). Structure of the Rho transcription terminator: mechanism of mRNA recognition and helicase loading. *Cell* 114, 135–146.

11.25 Termination and Antitermination Factors Interact with RNA Polymerase

Review

Greenblatt, J., Nodwell, J. R., and Mason, S. W. (1993). Transcriptional antitermination. *Nature* 364, 401–406.

Research

Legault, P., Li, J., Mogridge, J., Kay, L. E., and Greenblatt, J. (1998). NMR structure of the bacteriophage lambda N peptide/boxB RNA complex: recognition of a GNRA fold by an arginine-rich motif. *Cell* 93, 289–299.

Mah, T. F., Kuznedelov, K., Mushegian, A., Severinov, K., and Greenblatt, J. (2000). The alpha subunit of *E. coli* RNA polymerase activates RNA binding by NusA. *Genes Dev.* 14, 2664–2675.

Mogridge, J., Mah, J., and Greenblatt, J. (1995). A protein-RNA interaction network facilitates the template-independent cooperative assembly on RNA polymerase of a stable antitermination complex containing the lambda N protein. *Genes Dev.* 9, 2831–2845.

Olson, E. R., Flamm, E. L., and Friedman, D. I. (1982). Analysis of nutR: a region of phage lambda required for antitermination of transcription. *Cell* 31, 61–70.

12

The Operon

CHAPTER OUTLINE

- The hinge helix inserts into the minor groove of operator DNA.
- Active repressor has a conformation in which the two DNA-binding domains of a dimer can insert into successive turns of the double helix.
- Inducer binding disrupts the hinge helix and changes the conformation so that the two DNA-binding sites are not in the right geometry to make simultaneous contacts.

12.13 Mutant Phenotypes Correlate with the Domain Structure

- Different types of mutations occur in different domains of the repressor subunit.

12.14 Repressor Protein Binds to the Operator

- Repressor protein binds to the double-stranded DNA sequence of the operator.
- The operator is a palindromic sequence of 26 bp.
- Each inverted repeat of the operator binds to the DNA-binding site of one repressor subunit.

12.15 Binding of Inducer Releases Repressor from the Operator

- Inducer binding causes a change in repressor conformation that reduces its affinity for DNA and releases it from the operator.

12.16 Repressor Binds to Three Operators and Interacts with RNA Polymerase

- Each dimer in a repressor tetramer can bind an operator, so that the tetramer can bind two operators simultaneously.
- Full repression requires the repressor to bind to an additional operator downstream or upstream as well as to the operator at the *lacZ* promoter.
- Binding of repressor at the operator stimulates binding of RNA polymerase at the promoter but precludes transcription.

12.17 Repressor Is Always Bound to DNA

- Proteins that have a high affinity for a specific DNA sequence also have a low affinity for other DNA sequences.
- Every base pair in the bacterial genome is the start of a low-affinity binding-site for repressor.
- The large number of low-affinity sites ensures that all repressor protein is bound to DNA.
- Repressor binds to the operator by moving from a low-affinity site rather than by equilibrating from solution.

12.18 The Operator Competes with Low-Affinity Sites to Bind Repressor

- In the absence of inducer, the operator has an affinity for repressor that is $10^7\times$ that of a low affinity site.
- The level of ten repressor tetramers per cell ensures that the operator is bound by repressor 96% of the time.
- Induction reduces the affinity for the operator to $10^4\times$ that of low-affinity sites, so that only 3% of operators are bound.
- Induction causes repressor to move from the operator to a low-affinity site by direct displacement.
- These parameters could be changed by an increase or reduction in the effective concentration of DNA *in vivo*.

12.19 Repression Can Occur at Multiple Loci

- A repressor will act on all loci that have a copy of its target operator sequence.

12.20 Cyclic AMP Is an Effector That Activates CRP to Act at Many Operons

- CRP is an activator protein that binds to a target sequence at a promoter.
- A dimer of CRP is activated by a single molecule of cyclic AMP.

12.21 CRP Functions in Different Ways in Different Target Operons

- CRP introduces a 90° bend into DNA at its binding site.
- CRP-binding sites lie at highly variable locations relative to the promoter.
- CRP interacts with RNA polymerase, but the details of the interaction depend on the relative locations of the CRP-binding site and the promoter.

12.22 Translation Can Be Regulated

- A repressor protein can regulate translation by preventing a ribosome from binding to an initiation codon.
- Accessibility of initiation codons in a polycistronic mRNA can be controlled by changes in the structure of the mRNA that occur as the result of translation.

12.23 r-Protein Synthesis Is Controlled by Autogenous Regulation

- Translation of an r-protein operon can be controlled by a product of the operon that binds to a site on the polycistronic mRNA.

12.24 Phage T4 p32 Is Controlled by an Autogenous Circuit

- p32 binds to its own mRNA to prevent initiation of translation.

12.25 Autogenous Regulation Is Often Used to Control Synthesis of Macromolecular Assemblies

- The precursor to microtubules, free tubulin protein, inhibits translation of tubulin mRNA.

12.26 Summary

12.1 Introduction

Gene expression can be controlled at any of several stages, which we divide broadly into transcription, processing, and translation:

- Transcription often is controlled at the stage of initiation. Transcription is not usually controlled at elongation, but may be controlled at termination to determine whether RNA polymerase is allowed to proceed past a terminator to the gene(s) beyond.
- In eukaryotic cells, processing of the RNA product may be regulated at the stages of modification, splicing, transport, or stability. In bacteria, an mRNA is in principle available for translation as soon as (or even while) it is being synthesized, but during this time the stages of control are not available.
- Translation may be regulated, usually at the stages of initiation and termination (like transcription). Regulation of initiation is formally analogous to the regulation of transcription: the circuitry can be drawn in similar terms for regulating initiation of transcription on DNA or initiation of translation on RNA.

The basic concept for how transcription is controlled in bacteria was provided by the classic formulation of the model for control of gene expression by Jacob and Monod in 1961. They distinguished between two types of sequences in DNA: sequences that code for ***trans*-acting** products and ***cis*-acting** sequences that function exclusively within the DNA. Gene activity is regulated by the specific interactions of the *trans*-acting products (usually proteins) with the *cis*-acting sequences (usually sites in DNA). In more formal terms:

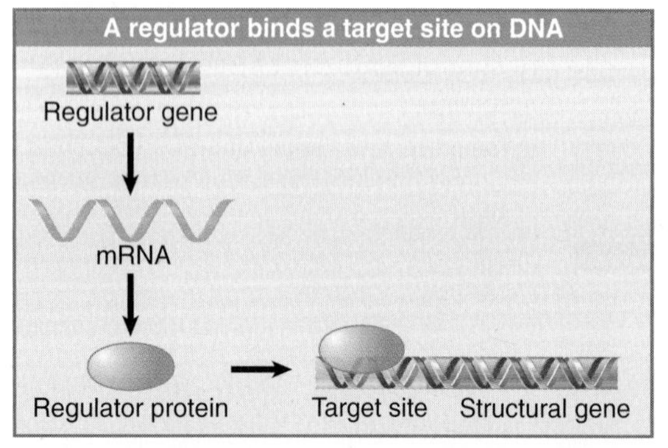

FIGURE 12.1 A regulator gene codes for a protein that acts at a target site on DNA.

- A gene is a sequence of DNA that codes for a diffusible product. This product may be protein (as in the case of the majority of genes) or may be RNA (as in the case of genes that code for tRNA and rRNA). *The crucial feature is that the product diffuses away from its site of synthesis to act elsewhere.* Any gene product that is free to diffuse to find its target is described as *trans*-acting.
- The description *cis*-acting applies to any sequence of DNA that is not converted into any other form, but that functions exclusively as a DNA sequence *in situ*, affecting only the DNA to which it is physically linked. (In some cases, a *cis*-acting sequence functions in an RNA rather than in a DNA molecule.)

To help distinguish between the components of regulatory circuits and the genes that they regulate, we sometimes use the terms structural gene and regulator gene. A **structural gene** is simply any gene that codes for a protein (or RNA) product. Structural genes represent an enormous variety of protein structures and functions, including structural proteins, enzymes with catalytic activities, and regulatory proteins. A **regulator gene** simply describes a gene that codes for a protein (or an RNA) involved in regulating the expression of other genes.

The simplest form of the regulatory model is illustrated in FIGURE 12.1: *a regulator gene codes for a protein that controls transcription by binding to particular site(s) on DNA.* This interaction can regulate a target gene in either a positive manner (the interaction turns the gene on) or in a negative manner (the interaction turns the gene off). The sites on DNA are usually (but not exclusively) located just upstream of the target gene.

The sequences that mark the beginning and end of the transcription unit, the promoter and terminator, are examples of *cis*-acting sites. *A promoter serves to initiate transcription only of the gene or genes physically connected to it on the same stretch of DNA.* In the same way, a terminator can terminate transcription only by an RNA polymerase that has traversed the preceding gene(s). In their simplest forms, promoters and terminators are *cis*-acting elements that are recognized by the same *trans*-acting species, that is, by RNA polymerase (although other factors also participate at each site).

Additional *cis*-acting regulatory sites are often juxtaposed to, or interspersed with, the promoter. A bacterial promoter may have one or more such sites located close by, that is, in the immediate vicinity of the startpoint. A eukaryotic promoter is likely to have a greater number of sites that are spread out over a longer distance.

12.2 Regulation Can Be Negative or Positive

Key concepts

- In negative regulation, a repressor protein binds to an operator to prevent a gene from being expressed.
- In positive regulation, a transcription factor is required to bind at the promoter in order to enable RNA polymerase to initiate transcription.

A classic mode of control in bacteria is *negative:* a **repressor** protein prevents a gene from being expressed. FIGURE 12.2 shows that the "default state" for such a gene is to be expressed via the recognition of its promoter by RNA polymerase. Close to the promoter is another *cis*-acting site called the **operator,** which is the target for the repressor protein. When the repressor binds to the operator, RNA polymerase is prevented from initiating transcription and *gene expression is therefore turned off.*

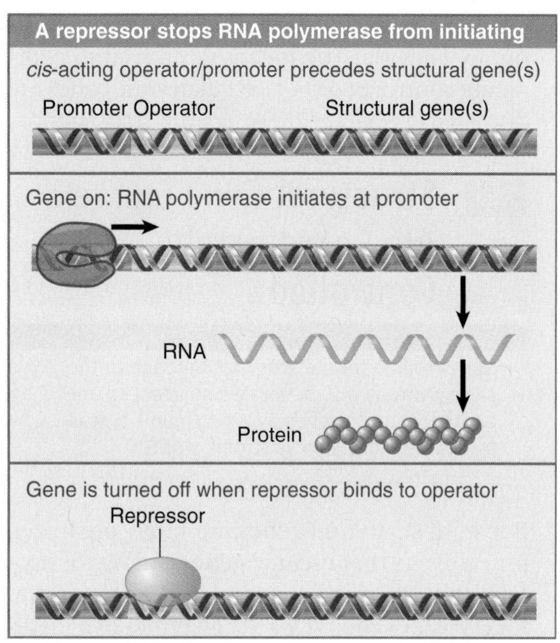

FIGURE 12.2 In negative control, a *trans*-acting repressor binds to the *cis*-acting operator to turn off transcription.

An alternative mode of control is *positive.* This is used in bacteria (probably) with about equal frequency to negative control, and it is the most common mode of control in eukaryotes. A **transcription factor** is required to assist RNA polymerase in initiating at the promoter. FIGURE 12.3 shows that the typical default state of a eukaryotic gene is inactive: RNA polymerase cannot by itself initiate transcription at the promoter. Several *trans*-acting factors have target sites in the vicinity of the promoter, and *binding of some or all of these factors enables RNA polymerase to initiate transcription.*

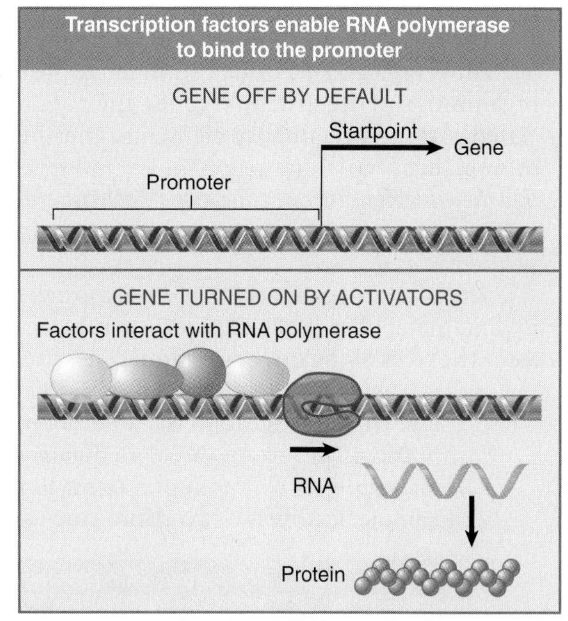

FIGURE 12.3 In positive control, *trans*-acting factors must bind to *cis*-acting sites in order for RNA polymerase to initiate transcription at the promoter.

The unifying theme is that regulatory proteins are *trans*-acting factors that recognize *cis*-acting elements (usually) upstream of the gene. The consequences of this recognition are to activate or to repress the gene, depending on the individual type of regulatory protein. A typical feature is that the protein functions by recognizing a very short sequence in DNA, usually <10 bp in length, although the protein actually binds over a somewhat greater distance of DNA. The bacterial promoter is an example: RNA polymerase covers >70 bp of DNA at initiation, but the crucial sequences that it recognizes are the hexamers centered at –35 and –10.

A significant difference in gene organization between prokaryotes and eukaryotes is that structural genes in bacteria are organized in clusters, whereas those in eukaryotes occur individually. Clustering of structural genes allows them to be coordinately controlled by means of interactions at a single promoter: as a result of these interactions, the entire set of genes is either transcribed or not transcribed. In this chapter, we discuss this mode of control and its use by bacteria. The means employed to coordinate control of dispersed eukaryotic genes are discussed in Chapter 25, Activating Transcription.

12.3 Structural Gene Clusters Are Coordinately Controlled

Key concept

- Genes coding for proteins that function in the same pathway may be located adjacent to one another and controlled as a single unit that is transcribed into a polycistronic mRNA.

Bacterial structural genes are often organized into clusters that include genes coding for proteins whose functions are related. It is common for the genes coding for the enzymes of a metabolic pathway to be organized into such a cluster. In addition to the enzymes actually involved in the pathway, other related activities may be included in the unit of coordinate control; for example, the protein responsible for transporting the small molecule substrate into the cell.

The cluster of the three *lac* structural genes, *lacZYA*, is typical. FIGURE 12.4 summarizes the organization of the structural genes, their associated *cis*-acting regulatory elements, and the *trans*-acting regulatory gene. *The key feature is that the cluster is transcribed into a single polycistronic mRNA from a promoter where initiation of transcription is regulated.*

The protein products enable cells to take up and metabolize β-galactosides, such as lactose. The roles of the three structural genes are:

- *lacZ* codes for the enzyme β-galactosidase, whose active form is a tetramer of ~500 kD. The enzyme breaks a β-galactoside into its component sugars. For example, lactose is cleaved into glucose and galactose (which are then further metabolized). This enzyme also produces an important by-product, β-1, 6-allolactase, which has a role in regulation.
- *lacY* codes for the β-galactoside permease, a 30-kD membrane-bound protein constituent of the transport system. This transports β-galactosides into the cell.
- *lacA* codes for β-galactoside transacetylase, an enzyme that transfers an acetyl group from acetyl-CoA to β-galactosides.

Mutations in either *lacZ* or *lacY* can create the *lac* genotype, in which cells cannot utilize lactose. (The genotypic description "*lac*" without a qualifier indicates loss-of-function.) The *lacZ* mutations abolish enzyme activity, directly preventing metabolism of lactose. The *lacY* mutants cannot take up lactose from the medium. (No defect is identifiable in *lacA* cells, which is puzzling. It is possible that the acetylation reaction gives an advantage when the bacteria grow in the presence of certain analogs of β-galactosides that cannot be metabolized, because the modification results in detoxification and excretion.)

The entire system, including structural genes and the elements that control their expression, forms a common unit of regulation called an **operon.** The activity of the operon is controlled by regulator gene(s) whose protein products interact with the *cis*-acting control elements.

12.4 The *lac* Genes Are Controlled by a Repressor

Key concepts

- Transcription of the *lacZYA* gene cluster is controlled by a repressor protein that binds to an operator that overlaps the promoter at the start of the cluster.
- The repressor protein is a tetramer of identical subunits coded by the gene *lacI*.

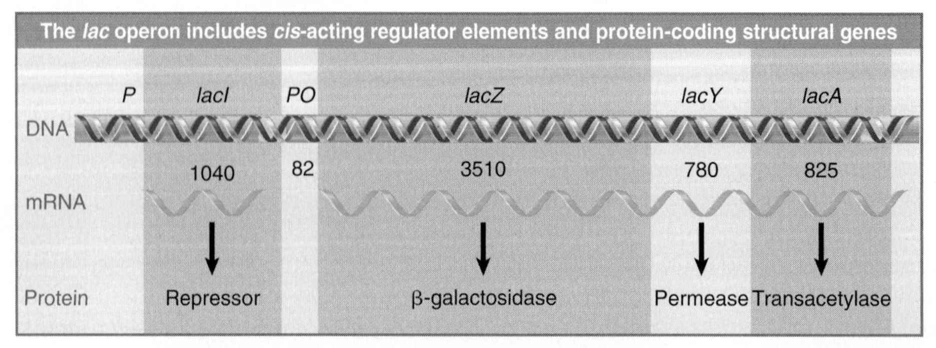

FIGURE 12.4 The *lac* operon occupies ~6000 bp of DNA. At the left the *lacI* gene has its own promoter and terminator. The end of the *lacI* region is adjacent to the promoter, *P*. The operator, *O*, occupies the first 26 bp of the transcription unit. The long *lacZ* gene starts at base 39, and is followed by the *lacY* and *lacA* genes and a terminator.

We can distinguish between structural genes and regulator genes by the effects of mutations. A mutation in a structural gene deprives the cell of the particular protein for which the gene codes. A mutation in a regulator gene, however, influences the expression of all the structural genes that it controls. The consequences of a regulatory mutation reveal the type of regulation.

Transcription of the *lacZYA* genes is controlled by a regulator protein synthesized by the *lacI* gene. It happens that *lacI* is located adjacent to the structural genes, but it comprises an independent transcription unit with its own promoter and terminator. In principle, *lacI* need not be located near the structural genes because it specifies a diffusible product. It can function equally well if moved elsewhere, or can be carried on a separate DNA molecule (the classic test for a *trans*-acting regulator).

The *lac* genes are controlled by **negative regulation:** *they are transcribed unless turned off by the regulator protein.* A mutation that inactivates the regulator causes the structural genes to remain in the expressed condition. The product of *lacI* is called the *Lac repressor,* because its function is to prevent the expression of the structural genes.

The repressor is a tetramer of identical subunits of 38 kD each. There are ~10 tetramers in a wild-type cell. The regulator gene is not controlled by the availability of lactose. It is transcribed into a monocistronic mRNA at a rate that appears to be governed simply by the affinity of its promoter for RNA polymerase.

The repressor functions by binding to an operator (formally denoted O_{lac}) at the start of the *lacZYA* cluster. The operator lies between the promoter (P_{lac}) and the structural genes *(lacZYA). When the repressor binds at the operator, it prevents RNA polymerase from initiating transcription at the promoter.* FIGURE 12.5 expands our view of the region at the start of the *lac* structural genes. The operator extends from position –5 just upstream of the mRNA startpoint to position +21 within the transcription unit; thus it overlaps the right end of the promoter. We discuss the relationship between repressor and RNA polymerase in more detail in Section 12.14, Repressor Protein Binds to the Operator and Section 12.16, Repressor Binds to Three Operators and Interacts with RNA Polymerase.

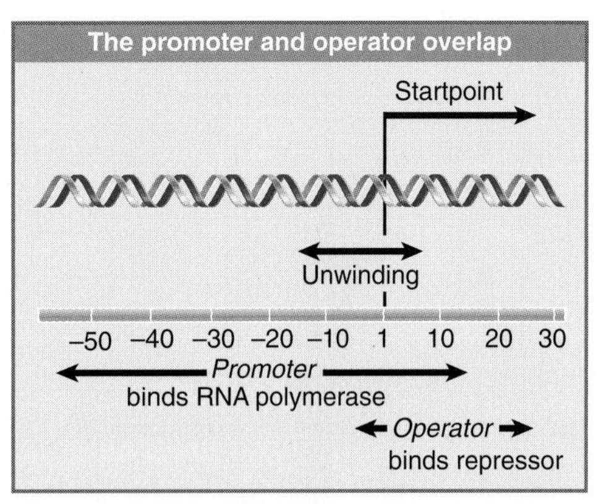

FIGURE 12.5 Repressor and RNA polymerase bind at sites that overlap around the transcription startpoint of the *lac* operon.

12.5 The *lac* Operon Can Be Induced

Key concepts

- Small molecules that induce an operon are identical with or related to the substrate for its enzymes.
- β-galactosides are the substrates for the enzymes coded by *lacZYA*.
- In the absence of β-galactosides, the *lac* operon is expressed only at a very low (basal) level.
- Addition of specific β-galactosides induces transcription of all three genes of the operon.
- The *lac* mRNA is extremely unstable; as a result, induction can be rapidly reversed.
- The same types of systems that allow substrates to induce operons coding for metabolic enzymes can be used to allow end-products to repress the operons that code for biosynthetic enzymes.

Bacteria need to respond swiftly to changes in their environment. Fluctuations in the supply of nutrients can occur at any time, and survival depends on the ability to switch from metabolizing one substrate to another. Yet economy is important, too: a bacterium that indulges in energetically expensive ways to meet the demands of the environment is likely to be at a disadvantage. Thus a bacterium avoids synthesizing the enzymes of a pathway in the absence of the substrate, but is ready to produce the enzymes if the substrate should appear.

The synthesis of enzymes in response to the appearance of a specific substrate is called **induction.** This type of regulation is widespread in bacteria and occurs in unicellular eukaryotes (such as yeasts) as well. The lactose system of *E. coli* provides the paradigm for this sort of control mechanism.

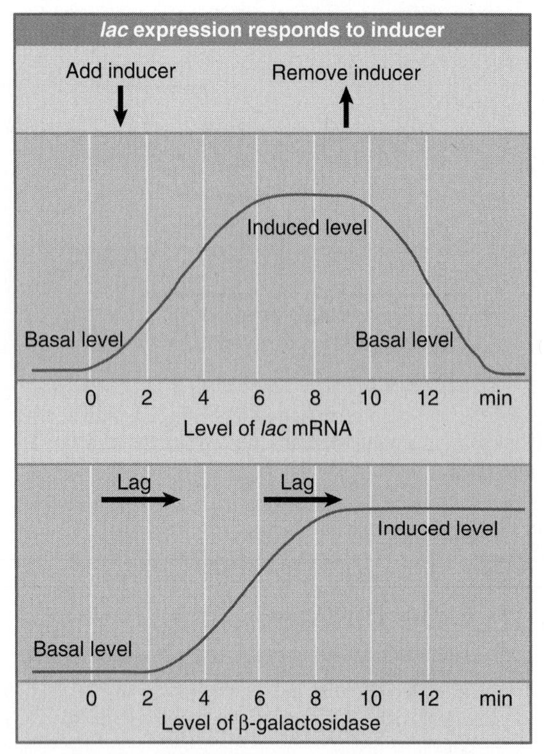

FIGURE 12.6 Addition of inducer results in rapid induction of *lac* mRNA, and is followed after a short lag by synthesis of the enzymes; removal of inducer is followed by rapid cessation of synthesis.

When cells of *E. coli* are grown in the absence of a β-galactoside there is no need for β-galactosidase, and they contain very few molecules of the enzyme—say, <5. When a suitable substrate is added, the enzyme activity appears very rapidly in the bacteria. Within two to three minutes some enzyme is present, and soon there are ~5000 molecules of enzyme per bacterium. (Under suitable conditions, β-galactosidase can account for 5% to 10% of the total soluble protein of the bacterium.) If the substrate is removed from the medium, the synthesis of enzyme stops as rapidly as it started.

FIGURE 12.6 summarizes the essential features of induction. Control of transcription of the *lac* genes responds very rapidly to the inducer, as shown in the upper part of the figure. In the absence of inducer, the operon is transcribed at a very low **basal level.** Transcription is stimulated as soon as inducer is added; the amount of *lac* mRNA increases rapidly to an induced level that reflects a balance between synthesis and degradation of the mRNA.

The *lac* mRNA is extremely unstable and decays with a half-life of only ~3 minutes. This feature allows induction to be reversed rapidly. Transcription ceases as soon as the inducer is removed, and in a very short time all the *lac* mRNA has been destroyed and the cellular content has returned to the basal level.

The production of protein is followed in the lower part of the figure. Translation of the *lac* mRNA produces β-galactosidase (and the products of the other *lac* genes). There is a short lag between the appearance of *lac* mRNA and appearance of the first completed enzyme molecules (it is ~2 minutes after the rise of mRNA from basal level before protein begins to increase). There is a similar lag between reaching maximal induced levels of mRNA and protein. When inducer is removed, synthesis of enzyme ceases almost immediately (as the mRNA is degraded), but the β-galactosidase in the cell is more stable than the mRNA, so the enzyme activity remains at the induced level for longer.

This type of rapid response to changes in nutrient supply not only provides the ability to metabolize new substrates, but also is used to shut off endogenous synthesis of compounds that suddenly appear in the medium. For example, *E. coli* synthesizes the amino acid tryptophan through the action of the enzyme tryptophan synthetase. If, however, tryptophan is provided in the medium on which the bacteria are growing, the production of the enzyme is immediately halted. This effect is called **repression.** It allows the bacterium to avoid devoting its resources to unnecessary synthetic activities.

Induction and repression represent the same phenomenon. In one case the bacterium adjusts its ability to use a given substrate (such as lactose) for growth; in the other it adjusts its ability to synthesize a particular metabolic intermediate (such as an essential amino acid). The trigger for either type of adjustment is a small molecule that is the substrate or related to the substrate for the enzyme or the product of the enzyme activity, respectively. Small molecules that cause the production of enzymes able to metabolize them are called **inducers.** Those that prevent the production of enzymes able to synthesize them are called **corepressors.**

12.6 Repressor Is Controlled by a Small-Molecule Inducer

Key concepts

- An inducer functions by converting the repressor protein into a form with lower operator affinity.
- Repressor has two binding sites, one for the operator and another for the inducer.
- Repressor is inactivated by an allosteric interaction in which binding of inducer at its site changes the properties of the DNA-binding site.

The ability to act as inducer or corepressor is highly specific. Only the substrate/product or a closely related molecule can serve. In most cases, *the activity of the small molecule, however, does not depend on its interaction with the target enzyme.* For the *lac* system, however, the natural inducer is a by-product of the lacZ enzyme, β-1, 6-allolactose, produced by transglucasylation of the β-1, 4-linkage in lactose. Allolactose is also a substrate of the lacZ enzyme, so it does not persist in the cell. Some inducers resemble the natural inducers of the *lac* operon but cannot be metabolized by the enzyme. The example *par excellence* is isopropylthiogalactoside (IPTG), one of several thiogalactosides with this property. IPTG is not recognized by β-galactosidase; even so, it is a very efficient inducer of the *lac* genes.

Molecules that induce enzyme synthesis but are not metabolized are called **gratuitous inducers.** They are extremely useful because they remain in the cell in their original form. (A real inducer would be metabolized, interfering with study of the system.) The existence of gratuitous inducers reveals an important point. *The system must possess some component, distinct from the target enzyme, that recognizes the appropriate substrate, and its ability to recognize related potential substrates is different from that of the enzyme.*

The component that responds to the inducer is the repressor protein coded by *lacI.* The *lacZYA* structural genes are transcribed into a single mRNA from a promoter just upstream of *lacZ.* The state of the repressor determines whether this promoter is turned off or on:

- FIGURE 12.7 shows that in the absence of an inducer the genes are not transcribed, because repressor protein is in an active form that is bound to the operator.
- FIGURE 12.8 shows that when an inducer is added, the repressor is converted into a form without lower affinity for operator or converted into a lower affinity form that leaves the operator. Transcription then starts at the promoter and proceeds through the genes to a terminator located beyond the 3′ end of *lacA.*

The crucial features of the control circuit reside in the dual properties of the repressor: it can prevent transcription, and it can recognize the small-molecule inducer. The repressor has two binding sites; one for the operator and one for the inducer. When the inducer binds at its site, it changes the conformation of the protein in such a way as to influence the activity of the operator-binding site. The ability of one site in the protein to control the activity of another is called **allosteric** control.

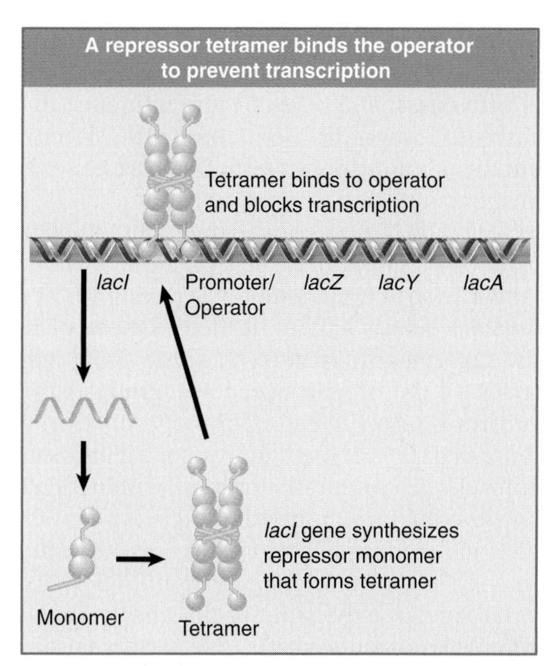

FIGURE 12.7 Repressor maintains the *lac* operon in the inactive condition by binding to the operator. The shape of the repressor is represented as a series of connected domains as revealed by its crystal structure (see later).

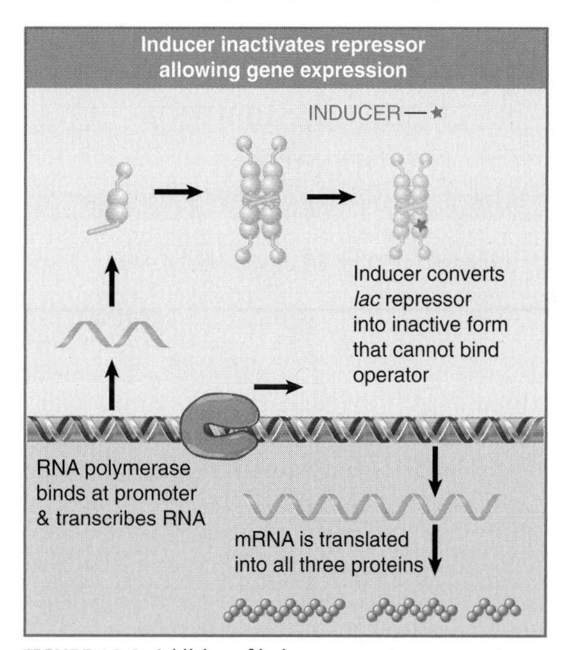

FIGURE 12.8 Addition of inducer converts repressor to an inactive form that cannot bind the operator. This allows RNA polymerase to initiate transcription.

Induction accomplishes a **coordinate regulation:** *all the genes are expressed (or not expressed) in unison.* The mRNA is translated sequentially from its 5′ end, which explains why induction always causes the appearance of β-galactosidase, β-galactoside permease, and β-galactoside transacetylase, in that order. Translation of a common mRNA explains why the relative amounts of the three enzymes always remain the same under varying conditions of induction.

Induction throws a switch that causes the genes to be transcribed. Inducers vary in their effectiveness, and other factors influence the absolute level of transcription or translation, but the relationship between the three genes is predetermined by their organization.

We notice a potential paradox in the constitution of the operon. The lactose operon contains the structural gene *(lacZ)* coding for the β-galactosidase activity needed to metabolize the sugar; it also includes the gene *(lacY)* that codes for the protein needed to transport the substrate into the cell. If the operon is in a repressed state, though, how does the inducer enter the cell to start the process of induction?

Two features ensure that there is always a minimal amount of the protein present in the cell—enough to start the process off. There is a basal level of expression of the operon: even when it is not induced, it is expressed at a residual level (0.1% of the induced level). In addition, some inducer enters anyway via another uptake system.

12.7 *cis*-Acting Constitutive Mutations Identify the Operon

Key concepts

- Mutations in the operator cause constitutive expression of all three *lac* structural genes.
- These mutations are *cis*-acting and affect only those genes on the contiguous stretch of DNA.

Mutations in the regulatory circuit may either abolish expression of the operon or cause unregulated expression. Mutants that cannot be expressed at all are called **uninducible.** The continued expression of a gene that does not respond to regulation is called **constitutive** gene expression, and mutants with this property are called constitutive mutants.

Components of the regulatory circuit of the operon can be identified by mutations that *affect the expression of all the structural genes and map outside them.* They fall into two classes. The promoter and the operator are identified as targets for the regulatory proteins (RNA polymerase and repressor, respectively) by *cis*-acting mutations. The locus *lacI* is identified as the gene that codes for the repressor protein by mutations that eliminate the *trans*-acting product.

The operator was originally identified by constitutive mutations, denoted O^c, whose distinctive properties provided the first evidence for *an element that functions without being represented in a diffusible product.*

The structural genes contiguous with an O^c mutation are expressed constitutively because the mutation changes the operator so that the repressor no longer binds to it. Thus the repressor cannot prevent RNA polymerase from initiating transcription. The operon is transcribed constitutively, as illustrated in FIGURE 12.9.

The operator can control only the lac *genes that are adjacent to it.* If a second *lac* operon is introduced into the bacterium on an independent molecule of DNA, it has its own operator. Neither operator is influenced by the other. Thus if one operon has a wild-type operator it will be repressed under the usual conditions, whereas a second operon with an O^c mutation will be expressed in its characteristic fashion.

These properties define the operator as a typical *cis*-acting site, whose function depends upon recognition of its DNA sequence by some *trans*-acting factor. The operator controls the adjacent genes irrespective of the presence in the cell of other alleles of the site. A mutation in such a site—for example, the O^c mutation—is formally described as ***cis*-dominant.**

A mutation in a *cis*-acting site cannot be assigned to a complementation group. (The ability to complement defines genes that are expressed as diffusible products.) When two *cis*-acting sites lie close together—for example, a promoter and an operator—we cannot classify

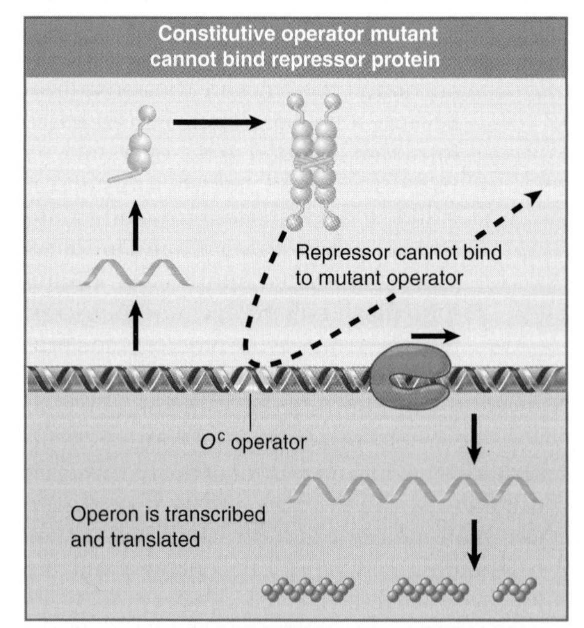

FIGURE 12.9 Operator mutations are constitutive because the operator is unable to bind repressor protein; this allows RNA polymerase to have unrestrained access to the promoter. The O^c mutations are *cis*-acting, because they affect only the contiguous set of structural genes.

the mutations by a complementation test. We are restricted to distinguishing them by their effects on the phenotype.

cis-dominance is a characteristic of any site that is *physically contiguous with the sequences it controls*. If a control site functions as part of a polycistronic mRNA, mutations in it will display *exactly the same pattern* of *cis*-dominance as they would if functioning in DNA. The critical feature is that the control site cannot be physically separated from the genes that it regulates. From the genetic point of view, it does not matter whether the site and genes are together on DNA or on RNA.

12.8 *trans*-Acting Mutations Identify the Regulator Gene

Key concepts

- Mutations in the *lacI* gene are *trans*-acting and affect expression of all *lacZYA* clusters in the bacterium.
- Mutations that eliminate *lacI* function cause constitutive expression and are recessive.
- Mutations in the DNA-binding site of the repressor are constitutive because the repressor cannot bind the operator.
- Mutations in the inducer-binding site of the repressor prevent it from being inactivated and cause uninducibility.
- Mutations in the promoter are uninducible and *cis*-acting.

Constitutive transcription is also caused by mutations of the *lacI*$^-$ type, which are caused by loss of function (including deletions of the gene). When the repressor is inactive or absent, transcription can initiate at the promoter. FIGURE 12.10 shows that the *lacI*$^-$ mutants express the structural genes all the time (constitutively), *irrespective of whether the inducer is present or absent*, because the repressor is inactive.

The two types of constitutive mutations can be distinguished genetically. O^c mutants are *cis*-dominant, whereas *lacI*$^-$ mutants are recessive. This means that the introduction of a normal, *lacI*$^+$ gene restores control, irrespective of the presence of the defective *lacI*$^-$ gene.

Mutants of the operon that are uninducible fall into the same two types of genetic classes as the constitutive mutants:

- Promoter mutations are *cis*-acting. If they prevent RNA polymerase from binding at P_{lac}, they render the operon nonfunctional because it cannot be transcribed.
- Mutations that abolish the ability of repressor to bind the inducer are described as *lacI*^{s}. They are *trans*-acting. The repressor is "locked in" to the active form that recognizes the operator and prevents transcription. The addition of inducer has no effect because its binding is impeded or cannot elicit the allosteric change, and therefore it is impossible to convert the repressor to the form with lower affinity for operator. The mutant repressor binds to all *lac* operators in the cell to prevent their transcription and cannot be pried off, irrespective of the properties of any wild-type repressor protein that is present, so it is genetically dominant.

The two types of mutations in *lacI* can be used to identify the individual active sites in the repressor protein. The *DNA-binding site* recognizes the sequence of the operator. It is identified by constitutive point mutations that prevent repressor from binding to DNA to block RNA polymerase. The *inducer-binding site* is identified by point mutations that cause uninducibility, because inducer cannot bind to trigger the allosteric change in the DNA-binding site.

12.9 Multimeric Proteins Have Special Genetic Properties

Key concepts

- Active repressor is a tetramer of identical subunits.
- When mutant and wild-type subunits are present, a single *lacI*$^{-d}$ mutant subunit can inactivate a tetramer whose other subunits are wild-type.
- *lacI*$^{-d}$ mutations occur in the DNA-binding site. Their effect is explained by the fact that repressor activity requires all DNA-binding sites in the tetramer to be active.

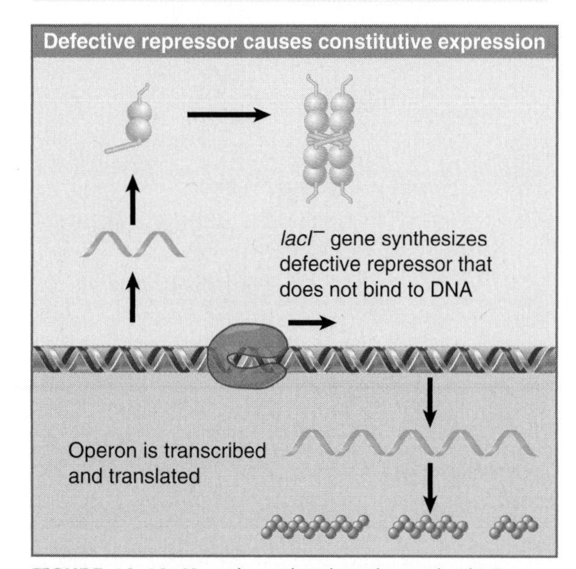

FIGURE 12.10 Mutations that inactivate the *lacI* gene cause the operon to be constitutively expressed, because the mutant repressor protein cannot bind to the operator.

An important feature of the repressor is that it is multimeric. Repressor subunits associate at random in the cell to form the active protein tetramer. When two different alleles of the *lacI* gene are present, the subunits made by each can associate to form a heterotetramer, whose properties differ from those of either homotetramer. This type of interaction between subunits is a characteristic feature of multimeric proteins and is described as **interallelic complementation.**

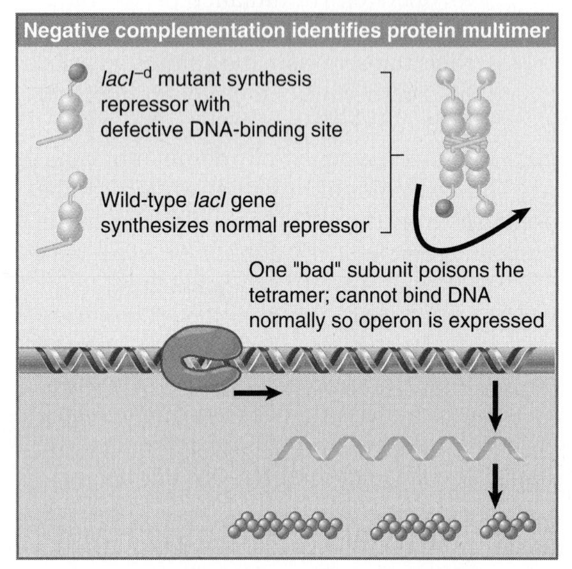

FIGURE 12.11 A $lacI^{-d}$ mutant gene makes a monomer that has a damaged DNA binding site (shown by the red circle). When it is present in the same cell as a wild-type gene, multimeric repressors are assembled at random from both types of subunits. It only requires one of the subunits of the multimer to be of the $lacI^{-d}$ type to block repressor function. This explains the dominant negative behavior of the $lacI^{-d}$ mutation.

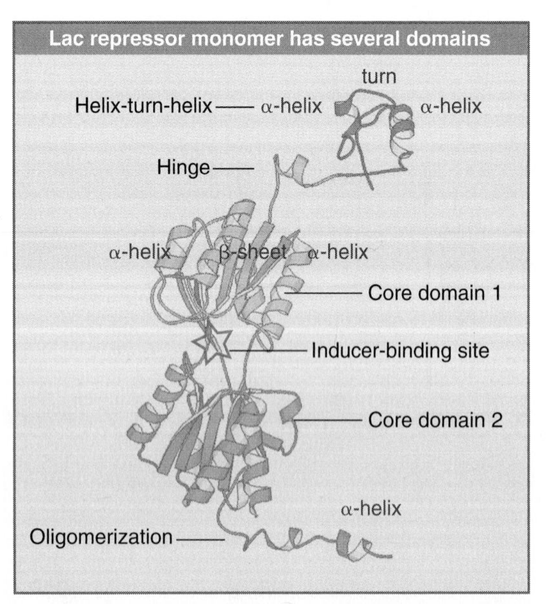

FIGURE 12.12 The structure of a monomer of Lac repressor identifies several independent domains. Structure rendered from PDB file 1lbg by Hongli Zhan and provided by Kathleen S. Matthews, Rice University.

Negative complementation occurs between some repressor mutants, as seen in the combination of $lacI^{-d}$ with $lacI^{+}$ genes. The $lacI^{-d}$ mutation alone results in the production of a repressor that cannot bind the operator, and it is therefore constitutive like the $lacI^{-}$ alleles. The $lacI^{-}$ type of mutation inactivates the repressor; as a result, it is usually recessive to the wild type. The *–d* notation, however, indicates that this variant of the negative type is dominant when paired with a wild-type allele. Such mutations are called **dominant negative.**

FIGURE 12.11 explains this phenomenon. The reason for the dominance is that the $lacI^{-d}$ allele produces a "bad" subunit, which is not only itself unable to bind to operator DNA, but, because it is required as part of the DNA binding site, is also able as part of a tetramer to impede "good" subunits binding. This demonstrates that the repressor tetramer as a whole, rather than the individual monomer, is needed to achieve repression. In fact, we may reverse the argument to say that, whenever a protein has a dominant negative form, this must mean it functions as part of a multimer. The production of dominant negative proteins has become an important technique in eukaryotic genetics.

12.10 The Repressor Monomer Has Several Domains

Key concepts

- A single repressor subunit can be divided into the N-terminal DNA-binding domain, a hinge, and the core of the protein.
- The DNA-binding domain contains two short α-helical regions that bind the major groove of DNA.
- The hinge region inserts into the minor groove as a short, folded α helix.
- The inducer-binding site and the regions responsible for multimerization are located in the core.

The repressor has several domains. The DNA-binding domain occupies residues 1–59 and is known as the **headpiece.** It can be cleaved from the remainder of the monomer, which is known as the *core,* by trypsin. The crystal structure illustrated in FIGURE 12.12 offers a more detailed account of these regions.

The N-terminus of the monomer consists of two α-helices separated by a turn. This is a common DNA-binding motif called the HTH (helix-turn-helix); the two α-helices fit into the major groove of DNA, where they make contacts with specific bases (see Section 14.11, Repressor Uses a Helix-Turn-Helix Motif to Bind DNA). This region is connected by a *hinge* to the main body of the protein. In the DNA-binding form of repressor, the hinge forms a small α-helix (as shown in the figure), but when the repressor is not bound to DNA, this region is disordered. The HTH and hinge together correspond to the DNA binding domain.

The bulk of the core consists of two regions with similar structures (core domains 1 and 2). Each has a six-stranded parallel β-sheet sandwiched between two α-helices on either side. The inducer binds in a cleft between the two regions.

At the C-terminus, there is an α-helix that contains two leucine heptad repeats. This is the oligomerization domain. The oligomerization helices of four monomers associate to maintain the tetrameric structure.

12.11 Repressor Is a Tetramer Made of Two Dimers

Key concepts

- Monomers form a dimer by making contacts between core domains 1 and 2 and potentially between the oligomerization helices.
- Dimers form a tetramer by interactions between the oligomerization helices.

FIGURE 12.13 shows the structure of the tetrameric core (using a different modeling system from Figure 12.12). It consists, in effect, of two dimers. The body of the dimer contains a loose interface between the N-terminal regions of the core monomers, a cleft at which inducer binds, and a hydrophobic core (top). The C-terminal regions of each monomer protrude as parallel helices. (The headpiece would join with the N-terminal regions at the top.) Together the dimers interact to form a tetramer (center) that is held together by a C-terminal bundle of four helices.

Sites of mutations are shown by beads on the structure at the bottom. *lacI*^{s} mutations make the repressor unresponsive to the inducer, so that the operon is uninducible. They map in two groups: gray shows those in the inducer-binding cleft, and yellow shows those that affect the core domain 1 dimer interface. The first group abolishes the inducer binding site; the second group prevents the effects of inducer binding from being transmitted to the DNA-binding site. *lacI*$^{-d}$ mutations that affect oligomerization map in two groups. White shows mutations in core domain 2 that prevent dimer formation. Purple shows those in the oligomerization helix that prevent tetramer formation from dimers.

From these data we can derive the schematic of FIGURE 12.14, which shows how the monomers are organized into the tetramer. Two monomers form a dimer by means of contacts at core domain 2 and in the oligomerization helix. The dimer has two DNA-binding domains at one end of the structure and the oligomerization helices at the other end. Two dimers then form a tetramer by interactions at the oligomerization interface.

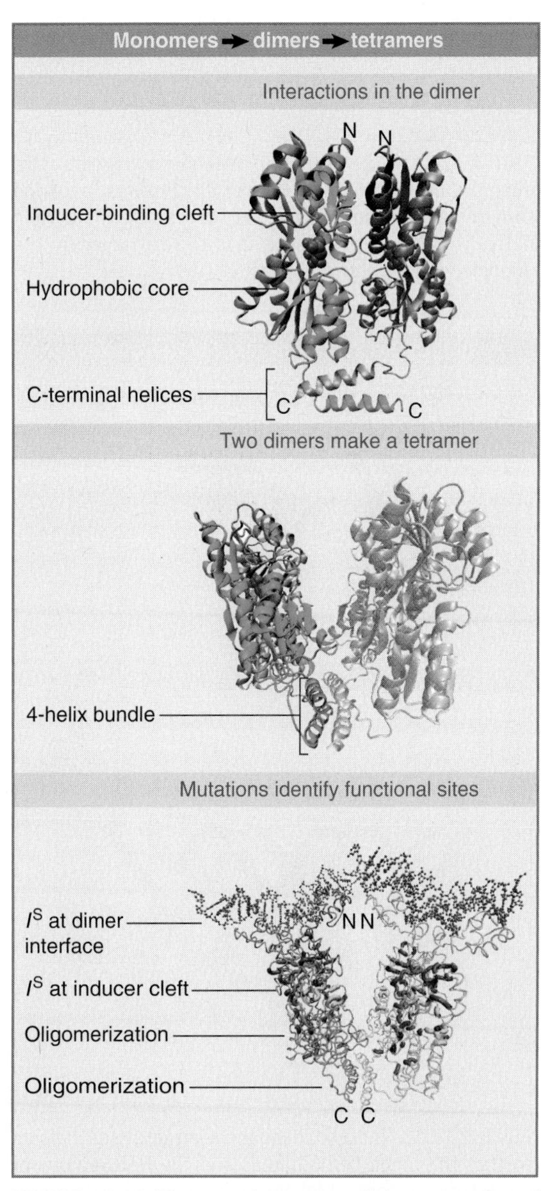

FIGURE 12.13 The crystal structure of the core region of Lac repressor identifies the interactions between monomers in the tetramer. Each monomer is identified by a different color. Mutations are colored as: dimer interface = yellow; inducer-binding = blue; oligomerization = white and purple. Photos courtesy of Benjamin Wieder and Ponzy Lu, University of Pennsylvania.

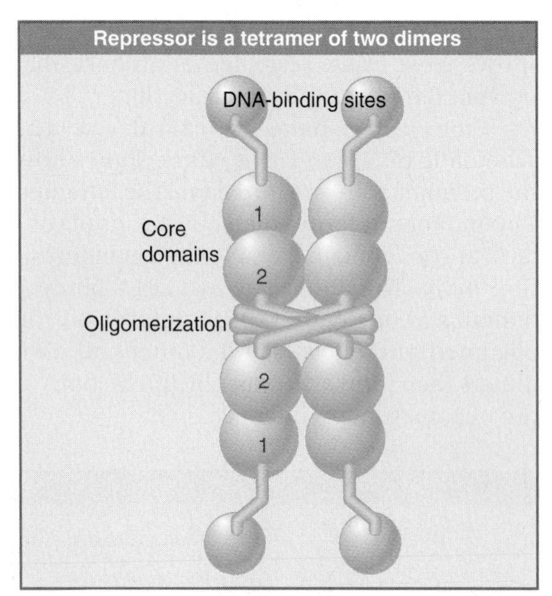

FIGURE 12.14 The repressor tetramer consists of two dimers. Dimers are held together by contacts involving core domains 1 and 2 as well as by the oligomerization helix. The dimers are linked into the tetramer by the oligomerization interface.

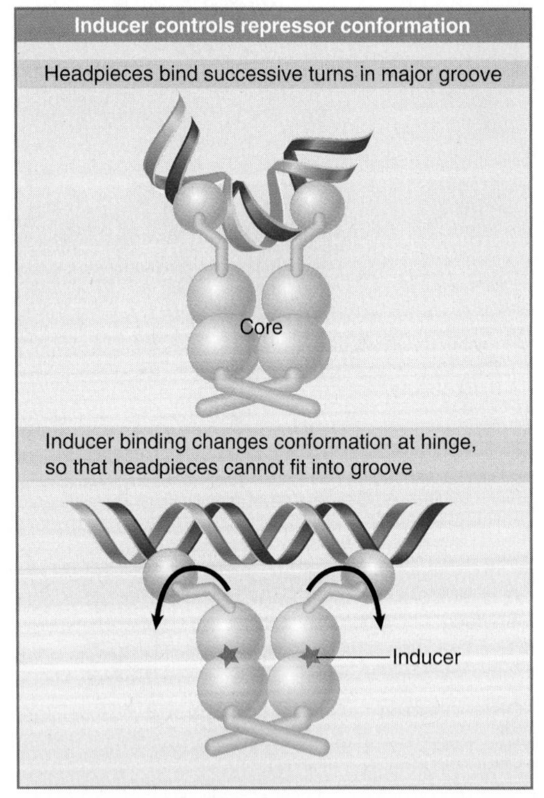

FIGURE 12.15 Inducer changes the structure of the core so that the hinge helix unfolds and the HTH regions of repressor dimer are no longer in an orientation that permits binding to DNA.

12.12 DNA-Binding Is Regulated by an Allosteric Change in Conformation

Key concepts

- The DNA-binding domain of each monomer within a dimer inserts into the major groove of DNA.
- The hinge helix inserts into the minor groove of operator DNA.
- Active repressor has a conformation in which the two DNA-binding domains of a dimer can insert into successive turns of the double helix.
- Inducer binding disrupts the hinge helix and changes the conformation so that the two DNA-binding sites are not in the right geometry to make simultaneous contacts.

Early work suggested a model in which the headpiece is relatively independent of the core. It can bind to operator DNA by making the same pattern of contacts with a half-site as intact repressor. Its affinity for DNA, however, is many orders of magnitude less than that of intact repressor. The reason for the difference is that the dimeric form of intact repressor allows two headpieces to contact the operator simultaneously, each binding to one half-site. FIGURE 12.15 shows that the two DNA-binding domains in a dimeric unit contact DNA by inserting into successive turns of the major groove. This enormously increases affinity for the operator. A key element of binding is the insertion of the short hinge helix into the minor groove of operator DNA, binding the DNA by ~45°. This bend orients the major groove for HTH binding.

Binding of inducer causes an immediate conformational change in the repressor protein. Binding of two molecules of inducer to the repressor tetramer is adequate to release repression. Binding of inducer disrupts the hinge helices and changes the orientation of the headpieces relative to the core, with the result that the two headpieces in a dimer can no longer bind DNA simultaneously. This eliminates the advantage of the multimeric repressor and reduces the affinity for the operator.

12.13 Mutant Phenotypes Correlate with the Domain Structure

Key concept

- Different types of mutations occur in different domains of the repressor subunit.

Mutations in the Lac repressor identified the existence of different domains even before the structure was known. We can now explain the nature of the mutations more fully by reference to the structure, as summarized in FIGURE 12.16.

Recessive mutations of the *lacI*⁻ type can occur anywhere in the bulk of the protein. Basically, any mutation that inactivates the protein will have this phenotype. The more detailed mapping of mutations on to the crystal structure in Figure 12.13 identifies specific impairments for some of these mutations, for example, those that affect oligomerization.

The special class of dominant-negative *lacI*$^{-d}$ mutations lie in the DNA-binding site of the repressor subunit (see Section 12.9, Multimeric Proteins Have Special Genetic Properties). This explains their ability to prevent mixed tetramers from binding to the operator; a reduction in the number of binding sites reduces the specific affinity for the operator. The role of the N-terminal region in specifically binding DNA is shown also by its location as the site of occurrence of "tight binding" mutations. These increase the affinity of the repressor for the operator, sometimes so much that it cannot be released by inducer. They are rare.

Uninducible *lacI*^{s} mutations map largely in a region of the core domain 1 extending from the inducer-binding site to the hinge. One group lies in amino acids that contact the inducer, and these mutations function by preventing binding of inducer. The remaining mutations lie at sites that must be involved in transmitting the allosteric change in conformation to the hinge when inducer binds.

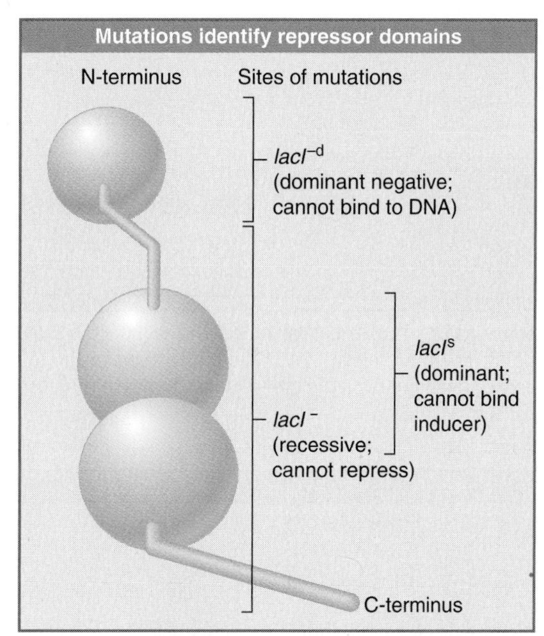

FIGURE 12.16 The locations of three type of mutations in lactose repressor are mapped on the domain structure of the protein. Recessive *lacI*⁻ mutants that cannot repress can map anywhere in the protein. Dominant negative *lacI*$^{-d}$ mutants that cannot repress map to the DNA-binding domain. Dominant *lacI*^{s} mutants that cannot induce because they do not bind inducer or cannot undergo the allosteric change map to core domain 1.

12.14 Repressor Protein Binds to the Operator

Key concepts

- Repressor protein binds to the double-stranded DNA sequence of the operator.
- The operator is a palindromic sequence of 26 bp.
- Each inverted repeat of the operator binds to the DNA-binding site of one repressor subunit.

The repressor was isolated originally by purifying the component able to bind the gratuitous inducer IPTG. (The amount of repressor in the cell is so small that in order to obtain enough material, it was necessary to use a promoter-up mutation to increase *lacI* transcription and to place this *lacI* locus on a DNA molecule present in many copies per cell. This results in an overall overproduction of 100 to 1000-fold.)

The repressor binds to double-stranded DNA containing the sequence of the wild-type *lac* operator. The repressor does not bind DNA from an O^c mutant. The addition of IPTG releases the repressor from operator DNA *in vitro*. The *in vitro* reaction between repressor protein and operator DNA therefore displays the characteristics of control inferred *in vivo;* thus it can be used to establish the basis for repression.

How does the repressor recognize the specific sequence of operator DNA? The operator has a feature common to many recognition sites for bacterial regulator proteins: it is a **palindrome.** The inverted repeats are highlighted in FIGURE 12.17. Each repeat can be regarded as a half-site of the operator.

We can use the same approaches to define the points that the repressor contacts in the operator that we used for analyzing the polymerase-promoter interaction (see Section 11.14, RNA Polymerase Binds to One Face of DNA). Deletions of material on either side define the end points of the region; constitutive point mutations identify individual base pairs that must be crucial. Experiments in which DNA bound to repressor is compared with unbound

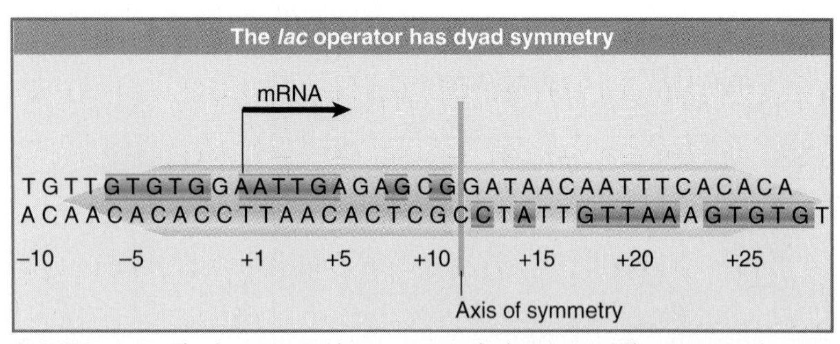

FIGURE 12.17 The *lac* operator has a symmetrical sequence. The sequence is numbered relative to the startpoint for transcription at +1. The pink arrows to the left and to the right identify the two dyad repeats. The green blocks indicate the positions of identity.

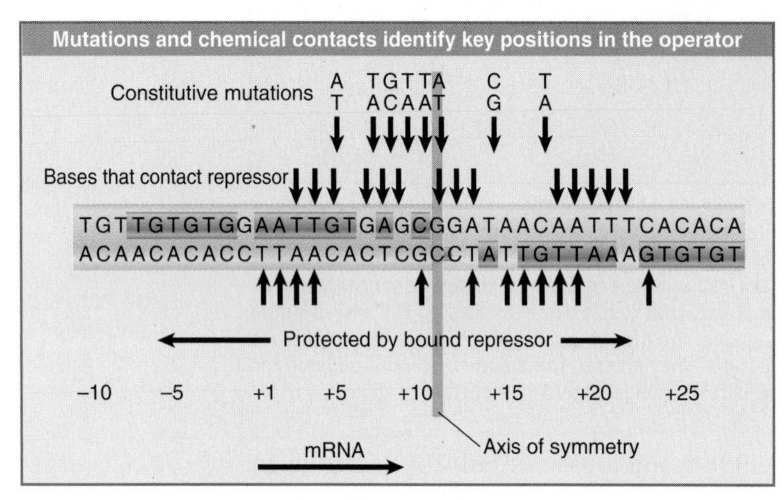

FIGURE 12.18 Bases that contact the repressor can be identified by chemical crosslinking or by experiments to see whether modification prevents binding. They identify positions on both strands of DNA extending from +1 to +23. Constitutive mutations occur at eight positions in the operator between +5 and +17.

DNA for its susceptibility to methylation or UV crosslinking identify bases that are either protected or more susceptible when associated with the protein.

FIGURE 12.18 shows that the region of DNA protected from nucleases by bound repressor lies within the region of symmetry, comprising the 26 bp region from –5 to +21. The area identified by constitutive mutations is even smaller. Within a central region extending over the 13 bp from +5 to +17, there are eight sites at which single base-pair substitutions cause constitutivity. This emphasizes the same point made by the promoter mutations summarized earlier in Figure 11.29. *A small number of essential specific contacts within a larger region can be responsible for sequence-specific association of DNA with protein.*

The symmetry of the DNA sequence reflects the symmetry in the protein. Each of the identical subunits in a repressor tetramer has a DNA-binding site. Two of these sites contact the operator in such a way that each inverted repeat of the operator makes the same pattern of contacts with a repressor monomer. This is shown by symmetry in the contacts that repressor makes with the operator (the pattern between +1 and +6 is identical with that between +21 and +16) and by matching constitutive mutations in each inverted repeat. (The operator is not perfectly symmetrical, though; the left side binds more strongly than the right side to the repressor. A stronger operator is created by a perfect inverted duplication of the left side and eliminated of the central base pair.)

12.15 Binding of Inducer Releases Repressor from the Operator

Key concept

- Inducer binding causes a change in repressor conformation that reduces its affinity for DNA and releases it from the operator.

Various inducers cause characteristic reductions in the affinity of the repressor for the operator *in vitro*. These changes correlate with the effectiveness of the inducers *in vivo*. This suggests that induction results from a reduction in the attraction between operator and repressor. Thus when inducer enters the cell, it binds to free repressors and in effect prevents them from finding their operators. Consider, though, a repressor tetramer that is already bound tightly to the operator. How does inducer cause this repressor to be released?

Two models for repressor action are illustrated in FIGURE 12.19:

- The equilibrium model (left) calls for repressor bound to DNA to be in rapid equilibrium with free repressor. Inducer would bind to the free form of repressor and thus unbalance the equilibrium by preventing reassociation with DNA.
- The rate of dissociation of the repressor from the operator, however, is much too slow to be compatible with this model (the half-life *in vitro* in the absence of inducer is >15 min). This means that instead the *inducer must bind directly to repressor protein complexed with the operator.* As indicated in the model on the right, inducer binding must produce a change in the repressor that makes it release the operator. Indeed, addition of IPTG causes an immediate destabilization of the repressor-operator complex *in vitro*.

Binding of the repressor-IPTG complex to the operator can be studied by using greater concentrations of the protein in the methylation protection/enhancement assay. The large amount compensates for the low affinity of the repressor-IPTG complex for the operator. The complex makes exactly the same pattern of contacts with DNA as the free repressor. An analogous result is obtained with mutant repressors whose affinity for operator DNA is increased; they, too, make the same pattern of contacts.

Overall, a range of repressor variants whose affinities for the operator span seven orders of magnitude all make the same contacts with DNA. *Changes in the affinity of the repressor for DNA must therefore occur by influencing the general conformation of the protein in binding DNA, not by making or breaking one or a few individual bonds.*

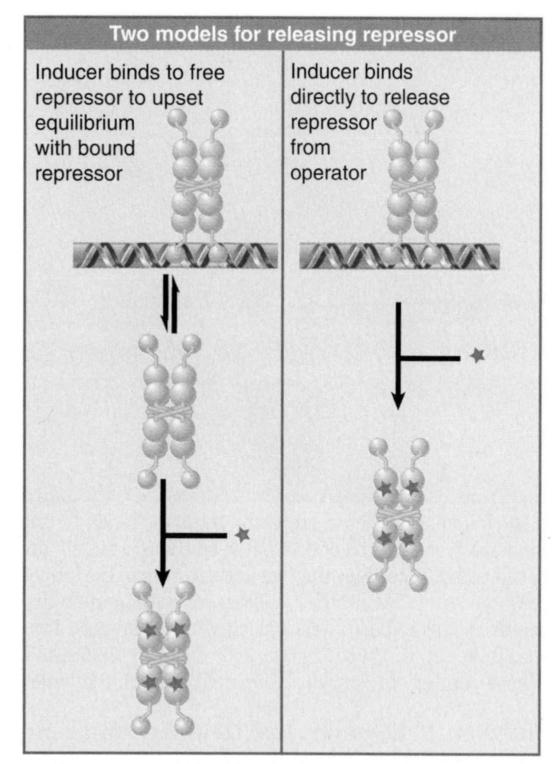

FIGURE 12.19 Does the inducer bind to free repressor to upset an equilibrium (left) or directly to repressor bound at the operator (right)?

12.16 Repressor Binds to Three Operators and Interacts with RNA Polymerase

Key concepts

- Each dimer in a repressor tetramer can bind an operator, so that the tetramer can bind two operators simultaneously.
- Full repression requires the repressor to bind to an additional operator downstream or upstream as well as to the operator at the *lacZ* promoter.
- Binding of repressor at the operator stimulates binding of RNA polymerase at the promoter but precludes transcription.

The allosteric transition that results from binding of inducer occurs in the repressor dimer. Why, then, is a tetramer required to establish full repression?

Each dimer can bind an operator sequence. This enables the intact repressor to bind to two operator sites simultaneously. In fact, there are two further operator sites in the initial region of the *lac* operon. The original operator, *O1*, is located just at the start of the *lacZ* gene. It has the strongest affinity for repressor. Weaker operator sequences (sometimes called pseudo-operators) are located on either side; *O2* is 410 bp downstream of the startpoint in lacZ and *O3* is 88 bp upstream of it in lacI.

FIGURE 12.20 shows what happens when a DNA-binding protein can bind simultaneously to two separated sites on DNA. The DNA between the two sites forms a loop from a base where the protein has bound the two sites. The length of the loop depends on the distance between the two binding sites. When Lac repressor binds simultaneously to *O1* and to one of the other operators, it causes the DNA between them to form a rather short loop, significantly constraining the DNA structure. A scale model for binding of tetrameric repressor to two operators is shown in FIGURE 12.21.

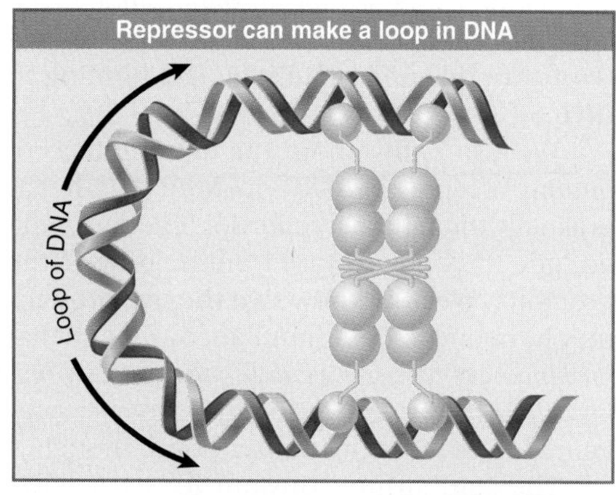

FIGURE 12.20 If both dimers in a repressor tetramer bind to DNA, the DNA between the two binding sites is held in a loop.

Binding at the additional operators affects the level of repression. Elimination of either the downstream operator *(O2)* or the upstream operator *(O3)* reduces the efficiency of repression by

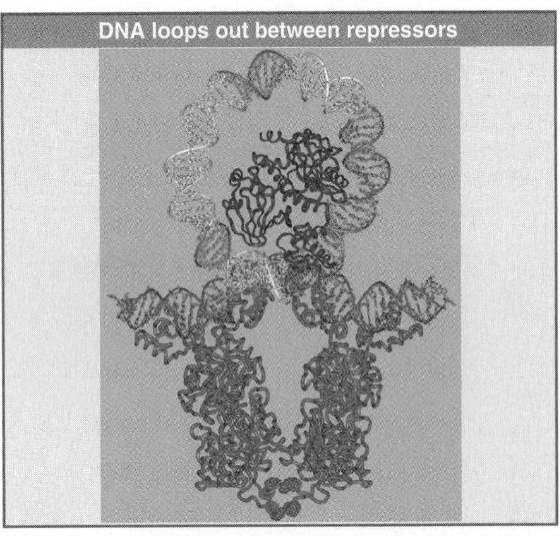

FIGURE 12.21 When a repressor tetramer binds to two operators, the stretch of DNA between them is forced into a tight loop. (The blue structure in the center of the looped DNA represents CAP, which is another regulator protein that binds in this region). Reproduced with permission from Lewis, M., et al. 1996. *Science*. 271: cover. © 1996 AAAS. Photo courtesy of Ponzy Lu, University of Pennsylvania.

2× to 4×. If, however, *both O2* and *O3* are eliminated, repression is reduced 100×. *This suggests that the ability of the repressor to bind to one of the two other operators as well as to O1 is important for establishing repression.* In vitro experiments with supercoiled plasmids containing multiple operators demonstrate significant stabilization of the lacI-DNA complex. Nonetheless, these looped DNAs are released rapidly by lacI binding to IPTE.

We also know about the direct effects of binding of repressor to the operator *(O1)*. It was originally thought that repressor binding would occlude RNA polymerase from binding to the promoter. We now know that the two proteins may be bound to DNA simultaneously, and that *the binding of repressor actually enhances the binding of RNA polymerase!* The bound enzyme is prevented from initiating transcription, though.

The equilibrium constant for RNA polymerase binding alone to the *lac* promoter is $1.9 \times 10^7\ M^{-1}$. The presence of repressor increases this constant by two orders of magnitude to $2.5 \times 10^9\ M^{-1}$. In terms of the range of values for the equilibrium constant K_B given in Figure 11.20, repressor protein effectively converts the formation of closed complex by RNA polymerase at the *lac* promoter from a weak to a strong interaction.

What does this mean for induction of the operon? The higher value for K_B means that, when occupied by repressor, the promoter is 100× more likely to be bound by an RNA polymerase. In addition, by allowing RNA polymerase to be bound at the same time as repressor, it becomes possible for transcription to begin immediately upon induction instead of waiting for an RNA polymerase to be captured.

The repressor in effect causes RNA polymerase to be stored at the promoter. The complex of RNA polymerase-repressor-DNA is blocked at the closed stage. When inducer is added, the repressor is released, and the closed complex is converted to an open complex that initiates transcription. The overall effect of repressor has been to speed up the induction process.

Does this model apply to other systems? The interaction between RNA polymerase, repressor, and the promoter/operator region is distinct in each system, because the operator does not always overlap with the same region of the promoter (see Figure 12.26). For example, in phage lambda, the operator lies in the upstream region of the promoter, and binding of repressor occludes the binding of RNA polymerase (see Chapter 14, Phage Strategies). Thus a bound repressor does not interact with RNA polymerase in the same way in all systems.

12.17 Repressor Is Always Bound to DNA

Key concepts

- Proteins that have a high affinity for a specific DNA sequence also have a low affinity for other DNA sequences.
- Every base pair in the bacterial genome is the start of a low-affinity binding-site for repressor.
- The large number of low-affinity sites ensures that all repressor protein is bound to DNA.
- Repressor binds to the operator by moving from a low-affinity site rather than by equilibrating from solution.

It is likely that all proteins with a high affinity for a specific sequence also possess a low affinity for any (random) DNA sequence. A large number of low-affinity sites will compete just as well for a repressor tetramer as a small number of high-affinity sites. There is only one high-affinity site in the *E. coli* genome: the operator. The remainder of the DNA provides low-affinity binding sites. Every base pair in the genome starts a new low-affinity site. (Just moving one

Repressor + DNA ⇌ Repressor-DNA

The equilibrium for repressor binding to (random) DNA is described by the equation:

$$K_A = \frac{[\text{Repressor-DNA}]}{[\text{Free repressor}]\,[\text{DNA}]}$$

The proportion of free repressor is given by rearranging the equation:

$$\frac{[\text{Free repressor}]}{[\text{Repressor-DNA}]} = \frac{1}{K_A \times [\text{DNA}]}$$

FIGURE 12.22 Repressor binding to random sites is governed by an equilibrium equation.

base pair along the genome, out of phase with the operator itself, creates a low-affinity site!) Thus there are 4.2×10^6 low-affinity sites.

The large number of low-affinity sites means that, even in the absence of a specific binding site, all or virtually all repressor is bound to DNA and none remains free in solution. FIGURE 12.22 shows how the equation of describing the equilibrium between free repressor and DNA-bound repressor can be rearranged to give the proportion of free repressor.

Applying the parameters for the *lac* system, we find that:

- The nonspecific equilibrium binding constant is $K_A = 2 \times 10^6\ M^{-1}$.
- The concentration of nonspecific binding sites is 4×10^6 in a bacterial volume of 10^{-15} liter, which corresponds to $[DNA] = 7 \times 10^{-3}$ M (a very high concentration).

Substituting these values gives: Free/Bound repressor = 10^{-4}.

Thus all but 0.01% of repressor is bound to (random) DNA. Since there are ~10 molecules of repressor per cell, this is tantamount to saying that there is no free repressor protein. This has an important implication for the interaction of repressor with the operator: it means that we are concerned with the *partitioning* of the repressor on DNA, in which the single high-affinity site of the operator *competes* with the large number of low-affinity sites.

In this competition, the absolute values of the association constants for operator and random DNA are not important; what is important is the ratio of K_{sp} (the constant for binding a specific site) to K_{nsp} (the constant for binding any random DNA sequence), that is, the specificity.

12.18 The Operator Competes with Low-Affinity Sites to Bind Repressor

Key concepts

- In the absence of inducer, the operator has an affinity for repressor that is $10^7\times$ that of a low affinity site.
- The level of ten repressor tetramers per cell ensures that the operator is bound by repressor 96% of the time.
- Induction reduces the affinity for the operator to $10^4\times$ that of low-affinity sites, so that only 3% of operators are bound.
- Induction causes repressor to move from the operator to a low-affinity site by direct displacement.
- These parameters could be changed by an increase or reduction in the effective concentration of DNA *in vivo*.

We can define the parameters that influence the ability of a regulator protein to saturate its target site by comparing the equilibrium equations for specific and nonspecific binding. As might be expected intuitively, the important parameters are:

- The size of the genome dilutes the ability of a protein to bind specific target sites.
- The specificity of the protein counters the effect of the mass of DNA.
- The amount of protein that is required increases with the total amount of DNA in the genome and decreases with the specificity.
- The amount of protein also must be in reasonable excess of the total number of specific target sites, so we expect regulators with many targets to be found in greater quantities than regulators with fewer targets.

FIGURE 12.23 compares the equilibrium constants for *lac* repressor/operator binding with repressor/general DNA binding. From these constants, we can deduce how repressor is partitioned between the operator and the rest of DNA and what happens to the repressor when inducer causes it to dissociate from the operator.

Repressor binds ~10^7 times better to operator DNA than to any random DNA sequence of the same length. Thus the operator comprises a single high-affinity site that will compete for the repressor 10^7 better than any low-affinity (random) site. How does this ensure that the

Repressor specifically binds operator DNA		
DNA	Repressor	Repressor + inducer
Operator	2×10^{13}	2×10^{10}
Other DNA	2×10^{6}	2×10^{6}
Specificity	10^{7}	10^{4}
Operators bound	96%	3%
Operon is:	repressed	induced

FIGURE 12.23 Lac repressor binds strongly and specifically to its operator, but is released by inducer. All equilibrium constants are in M^{-1}.

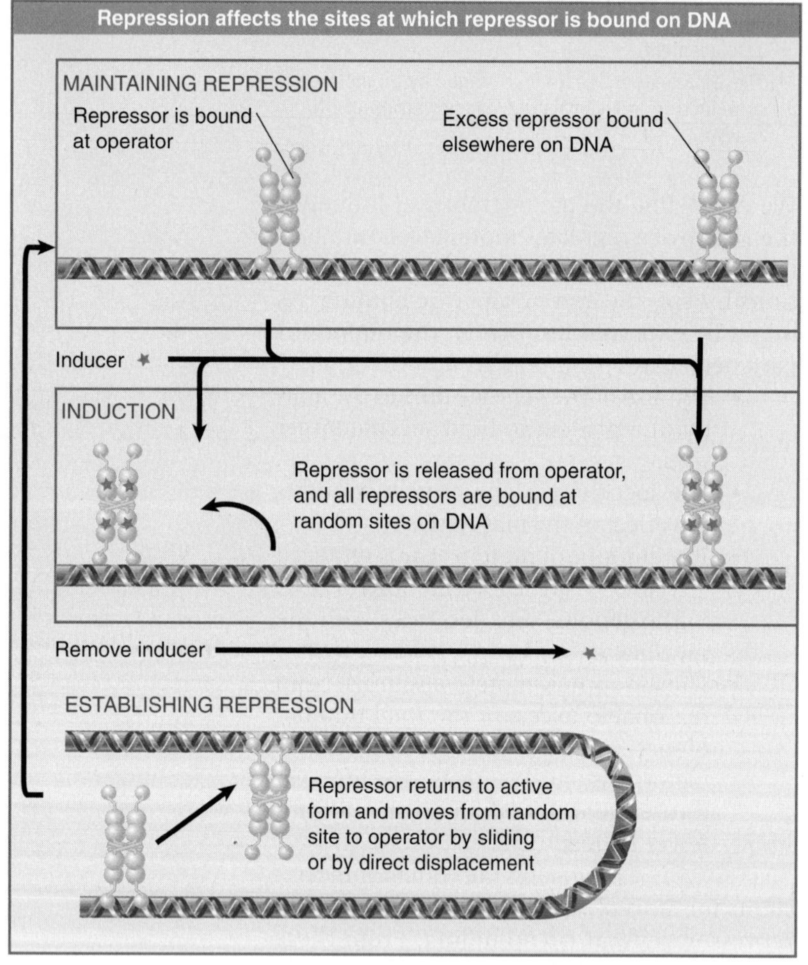

FIGURE 12.24 Virtually all the repressor in the cell is bound to DNA.

repressor can maintain effective control of the operon?

Using the specificity, we can calculate the distribution between random sites and the operator and can express this in terms of occupancy of the operator. If there are ten molecules of *lac* repressor per cell with a specificity for the operator of 10^7, the operator will be bound by repressor 96% of the time. The role of specificity explains two features of the *lac* repressor-operator interaction:

- When inducer binds to the repressor, the affinity for the operator is reduced by ~10^3-fold. The affinity for general DNA sequences remains unaltered. Thus the specificity is now only 10^4, which is insufficient to capture the repressor against competition from the excess of 4.2×10^6 low-affinity sites. Only 3% of operators would be bound under these conditions.
- Mutations that reduce the affinity of the operator for the repressor by as little as 20× to 30× have sufficient effect to be constitutive. Within the genome, the mutant operators can be overwhelmed by the preponderance of random sites. The occupancy of the operator is reduced to ~50% if the repressor's specificity is reduced just 10×.

The consequence of these affinities is that in an uninduced cell, one tetramer of repressor usually is bound to the operator. All, or almost all, of the remaining tetramers are bound at random to other regions of DNA, as illustrated in FIGURE 12.24. There are likely to be very few or no repressor tetramers free within the cell.

The addition of inducer abolishes the ability of repressor to bind specifically at the operator. Those repressors bound at the operator are released and bind to random (low-affinity) sites. Thus in an induced cell, the repressor tetramers are "stored" on random DNA sites. In a noninduced cell, a tetramer is bound at the operator, whereas the remaining repressor molecules are bound to nonspecific sites. *The effect of induction is therefore to change the distribution of repressor on DNA, rather than to generate free repressor.*

When inducer is removed, repressor recovers its ability to bind specifically to the operator and does so very rapidly. This must involve its movement from a nonspecific "storage" site on DNA. What mechanism is used for this rapid movement? The ability to bind to the operator very rapidly is not consistent with the time that would be required for multiple cycles of dissociation and reassociation with nonspecific sites on DNA. The discrepancy excludes random-hit mechanisms for finding the operator, suggesting that the repressor can move directly from a random site on DNA to the operator. This is the same issue that we encountered previously with the ability of RNA polymerase to find its promoters (see Figure 11.22 and Figure 11.23). The same solution is likely: movement could be accomplished by reducing the dimensionality of the search by sliding along DNA or by direct displacement from site to site (as indicated in Figure 12.24). A displacement reaction might be aided by the presence of more binding sites per tetramer (four) than are actually needed to contact DNA at any one time (two).

The parameters involved in finding a high-affinity operator in the face of competition from many low-affinity sites pose a dilemma for repressor. Under conditions of repression, there must be high specificity for the operator. Under conditions of induction, however, this specificity must be relieved. Suppose, for example, that there were 1000 molecules of repressor per cell: only 0.04% of operators would be free under conditions of repression. Upon induction, though, only 40% of operators would become free. We therefore see an inverse correlation between the ability to achieve complete repression and the ability to relieve repression effectively. We assume that the number of repressors synthesized *in vivo* has been subject to selective forces that balance these demands.

The difference in expression of the lactose operon between its induced and repressed states *in vivo* is actually $10^3\times$. In other words, even when inducer is absent, there is a basal level of expression of ~0.1% of the induced level. This would be reduced if there were more repressor protein present and increased if there were less. Thus it could be impossible to establish tight repression if there were fewer repressors than the ten found per cell, and it might become difficult to induce the operon if there were too many. It is possible to introduce the *lac* operator-repressor system into the mouse. When the *lac* operator is connected to a tyrosinase reporter gene, the enzyme is induced by the addition of IPTG. This means that the repressor is finding its target in a genome 10^3 times larger than that of *E. coli.* Induction occurs at approximately the same concentration of IPTG as in bacteria. We do not, however, know the concentration of Lac repressor and how effectively the target is induced.

In order to extrapolate *in vivo* from the affinity of a DNA-protein interaction *in vitro,* we need to know the effective concentration of DNA *in vivo.* The "effective concentration" differs from the mass/volume because of several factors. The effective concentration is increased, for example, by molecular crowding, which occurs when polyvalent cations neutralize ~90% of the charges on DNA, and the nucleic acid collapses into condensed structures. The major force that decreases the effective concentration is the inaccessibility of DNA that results from occlusion or sequestration by DNA-binding proteins.

One way to determine the effective concentration is to compare the rate of a reaction *in vitro* and *in vivo* that depends on DNA concentration. This has been done using intermolecular recombination between two DNA molecules. To provide a control, the same reaction is followed as an intramolecular recombination, that is, the two recombining sites are presented on the same DNA molecule. We assume that concentration is the same *in vivo* and *in vitro* for the *intramolecular* reaction, and therefore any difference in the ratio of intermolecular/ intramolecular recombination rates can be attributed to a change in the effective concentration *in vivo.* The results of such a comparison suggest that the effective concentration of DNA is reduced >10-fold *in vivo.*

This could affect the rates of reactions that depend on DNA concentration, including DNA recombination and protein–DNA binding. It emphasizes the problem encountered by all DNA-binding proteins in finding their targets with sufficient speed and reinforces the conclusion that diffusion is not adequate (see Figure 11.22).

12.19 Repression Can Occur at Multiple Loci

Key concept

- A repressor will act on all loci that have a copy of its target operator sequence.

The *lac* repressor acts only on the operator of the *lacZYA* cluster. Some repressors, however, control dispersed structural genes by binding at more than one operator. An example is the *trp* repressor, which controls three unlinked sets of genes:

- An operator at the cluster of structural genes *trpEDBCA* controls coordinate synthesis of the enzymes that synthesize tryptophan from chorismic acid.
- An operator at another locus controls the *aroH* gene, which codes for one of the three enzymes that catalyze the initial reaction in the common pathway of aromatic amino acid biosynthesis.
- The *trpR* regulator gene is repressed by its own product, the *trp* repressor. Thus

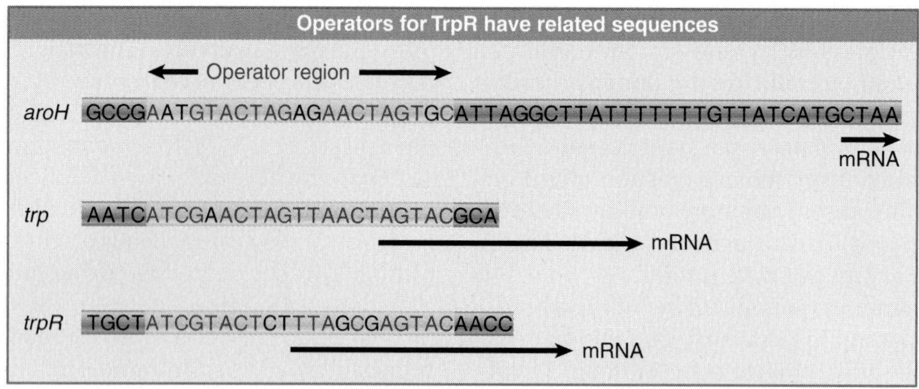

FIGURE 12.25 The *trp* repressor recognizes operators at three loci. Conserved bases are shown in red. The location of the startpoint and mRNA varies, as indicated by the white arrows.

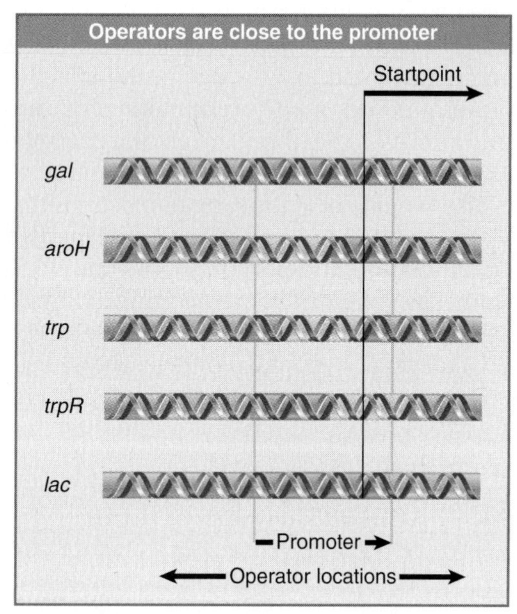

FIGURE 12.26 Operators may lie at various positions relative to the promoter.

the repressor protein acts to reduce its own synthesis. This circuit is an example of **autogenous** control. Such circuits are quite common in regulatory genes and may be either negative or positive (see Section 12.23, r-Protein Synthesis Is Controlled by Autogenous Regulation and Section 14.15, Repressor Maintains an Autogenous Circuit).

A related 21 bp operator sequence is present at each of the three loci at which the *trp* repressor acts. The conservation of sequence is indicated in FIGURE 12.25. Each operator contains appreciable (but not identical) dyad symmetry. The features conserved at all three operators include the important points of contact for *trp* repressor. This explains how one repressor protein acts on several loci: *each locus has a copy of a specific DNA-binding sequence recognized by the repressor* (just as each promoter shares consensus sequences with other promoters).

FIGURE 12.26 summarizes the variety of relationships between operators and promoters. A notable feature of the dispersed operators recognized by TrpR is their presence at different locations within the promoter in each locus. In *trpR* the operator lies between positions –12 and +9, whereas in the *trp* operon it occupies positions –23 to –3. In the *aroH* locus it lies farther upstream, between –49 and –29. In other cases, the operator lies downstream from the promoter (as in *lac*), or apparently just upstream of the promoter (as in *gal*, for which the nature of the repressive effect is not quite clear). The ability of the repressors to act at operators whose positions are different in each target promoter suggests that there could be differences in the exact mode of repression, the common feature being that RNA polymerase is prevented from initiating transcription at the promoter.

12.20 Cyclic AMP Is an Effector That Activates CRP to Act at Many Operons

Key concepts

- CRP is an activator protein that binds to a target sequence at a promoter.
- A dimer of CRP is activated by a single molecule of cyclic AMP.

Thus far we have dealt with the promoter as a DNA sequence that is competent to bind RNA polymerase, which then initiates transcription. There are, however, some promoters at which

RNA polymerase cannot initiate transcription without assistance from an ancillary protein. Such proteins are positive regulators, because their presence is necessary to switch on the transcription unit. Typically, the activator overcomes a deficiency in the promoter, for example, a poor consensus sequence at –35 or –10.

One of the most widely acting activators is a protein called **CRP activator** that controls the activity of a large set of operons in *E. coli*. The protein is a positive control factor whose presence is necessary to initiate transcription at dependent promoters. CRP is active *only in the presence of cyclic AMP*, which behaves as a classic small-molecule inducer for positive control (see FIGURE 12.27; upper right).

Cyclic AMP is synthesized by the enzyme **adenylate cyclase.** The reaction uses ATP as substrate and introduces a 3′–5′ link via phosphodiester bonds, which generates the structure drawn in FIGURE 12.28. Mutations in the gene coding for adenylate cyclase (*cya*⁻) do not respond to changes in glucose levels.

The level of cyclic AMP is inversely related to the level of glucose. The basis for this effect lies with the same component of the Pts system that is responsible for controlling lactose uptake. The phosphorylated form of protein IIA^{Glc} stimulates adenylate cyclase. When glucose is imported, the dephosphorylation of IIA^{Glc} leads to a fall in adenylate cyclase activity.

FIGURE 12.29 shows that reducing the level of cyclic AMP renders the (wild-type) protein unable to bind to the control region, which in turn prevents RNA polymerase from initiating transcription. Thus the effect of glucose in reducing cyclic AMP levels is to deprive the relevant operons of a control factor necessary for their expression.

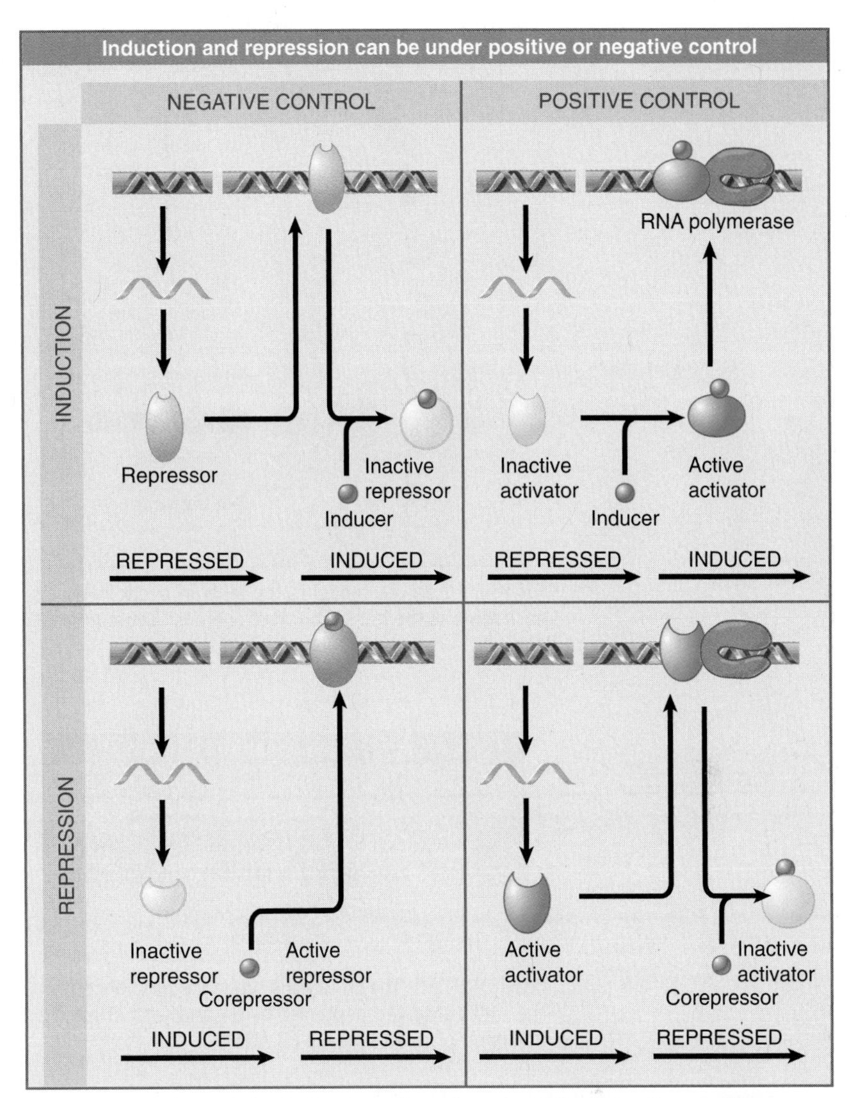

FIGURE 12.27 Control circuits are versatile and can be designed to allow positive or negative control of induction or repression.

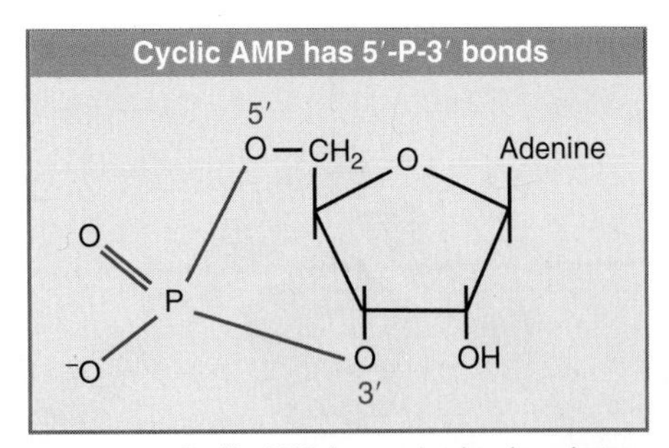

FIGURE 12.28 Cyclic AMP has a single phosphate group connected to both the 3′ and 5′ positions of the sugar ring.

12.21 CRP Functions in Different Ways in Different Target Operons

Key concepts

- CRP introduces a 90° bend into DNA at its binding site.
- CRP-binding sites lie at highly variable locations relative to the promoter.
- CRP interacts with RNA polymerase, but the details of the interaction depend on the relative locations of the CRP-binding site and the promoter.

The CRP factor binds to DNA, and complexes of cyclic AMP·CRP·DNA can be isolated at each promoter at which it functions. The factor is a dimer of two identical subunits of 22.5 kD, which can be activated by a single molecule of cyclic AMP. A CRP monomer contains a DNA-binding region and a transcription-activating region.

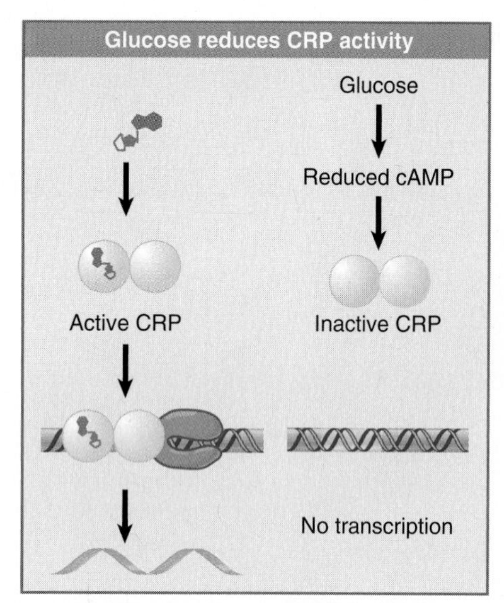

FIGURE 12.29 By reducing the level of cyclic AMP, glucose inhibits the transcription of operons that require CRP activity.

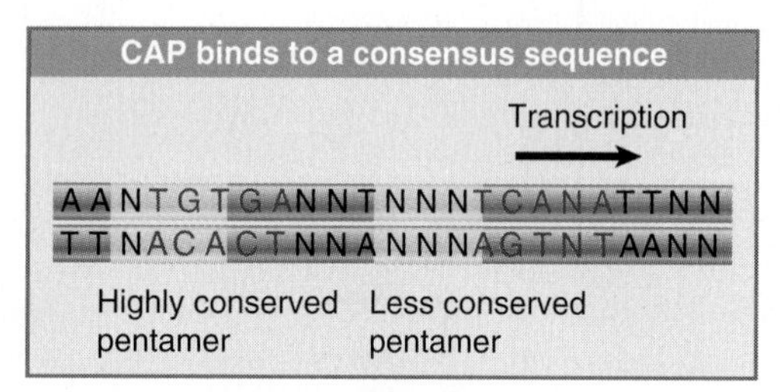

FIGURE 12.30 The consensus sequence for CRP contains the well conserved pentamer TGTGA and (sometimes) an inversion of this sequence (TCANA).

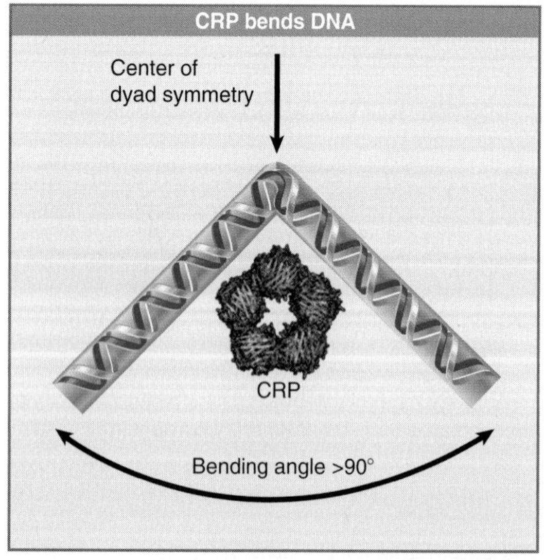

FIGURE 12.31 CRP bends DNA >90° around the center of symmetry.

A CRP dimer binds to a site of ~22 bp at a responsive promoter. The binding sites include variations of the consensus sequence given in FIGURE 12.30. Mutations preventing CRP action usually are located within the well-conserved pentamer $\frac{\text{TGTGA}}{\text{ACACT}}$, which appears to be the essential element in recognition. CRP binds most strongly to sites that contain two (inverted) versions of the pentamer, because this enables both subunits of the dimer to bind to the DNA. Many binding sites lack the second pentamer, however, and in these the second subunit must bind a different sequence (if it binds to DNA). The hierarchy of binding affinities for CRP helps to explain why different genes are activated by different levels of cyclic AMP *in vivo*.

CRP introduces a large bend when it binds DNA. In the *lac* promoter, this point lies at the center of dyad symmetry. The bend is quite severe, >90°, as illustrated in the model of FIGURE 12.31. There is, therefore, a dramatic change in the organization of the DNA double helix when CRP protein binds. The mechanism of bending is to introduce a sharp kink within the TGTGA consensus sequence. When there are inverted repeats of the consensus, the two kinks in each copy present in a palindrome cause the overall 90° bend. It is possible that the bend has some direct effect upon transcription, but it could be the case that it is needed simply to allow CRP to contact RNA polymerase at the promoter.

The action of CRP has the curious feature that its binding sites lie at different locations relative to the startpoint in the various operons that it regulates. The TGTGA pentamer may lie in either orientation. The three examples summarized in FIGURE 12.32 encompass the range of locations:

- The CRP-binding site is adjacent to the promoter, as in the *lac* operon, in which the region of DNA protected by CRP is centered on –61. It is possible that two dimers of CRP are bound. The binding pattern is consistent with the presence of CRP largely on one face of DNA, which is the same face that is bound by RNA polymerase. This location would place the two proteins just about in reach of each other.
- Sometimes the CRP-binding site lies within the promoter, as in the *gal* locus, where the CRP-binding site is centered on –41. It is likely that only a single CRP dimer is bound, probably in quite intimate contact with RNA polymerase,

because the CRP-binding site extends well into the region generally protected by the RNA polymerase.

- In other operons, the CRP-binding site lies well upstream of the promoter. In the *ara* region, the binding site for a single CRP is the farthest from the startpoint, centered at –92.

Dependence on CRP is related to the intrinsic efficiency of the promoter. No CRP-dependent promoter has a good –35 sequence and some also lack good –10 sequences. In fact, we might argue that effective control by CRP would be difficult if the promoter had effective –35 and –10 regions that interacted independently with RNA polymerase.

There are in principle two ways in which CRP might activate transcription: it could interact directly with RNA polymerase, or it could act upon DNA to change its structure in some way that assists RNA polymerase to bind. In fact, CRP has effects upon both RNA polymerase and DNA.

Binding sites for CRP at most promoters resemble either *lac* (centered at –61) or *gal* (centered at –41 bp). The basic difference between them is that in the first type (called class I) the CRP-binding site is entirely upstream of the promoter, whereas in the second type (called class II) the CRP-binding site overlaps the binding site for RNA polymerase. (The interactions at the *ara* promoter may be different.)

In both types of promoter, the CRP binding site is centered an integral number of turns of the double helix from the startpoint. This suggests that CRP is bound to the same face of DNA as RNA polymerase. The nature of the interaction between CRP and RNA polymerase is, however, different at the two types of promoter.

When the α subunit of RNA polymerase has a deletion in the C-terminal end, transcription appears normal except for the loss of ability to be activated by CRP. CRP has an "activating region" that is required for activating both types of its promoters. This activating region, which consists of an exposed loop of ~10 amino acids, is a small patch that interacts directly with the α subunit of RNA polymerase to stimulate the enzyme. At class I promoters, this interaction is sufficient. At class II promoters, a second interaction is required, which involves another region of CRP and the N-terminal region of the RNA polymerase α subunit.

Experiments using CRP dimers in which only one of the subunits has a functional transcription-activating region shows that, when CRP is bound at the *lac* promoter, only the activating region of the subunit nearer the startpoint is required, presumably because it touches RNA polymerase. This offers an explanation for the lack of dependence on the orientation of the binding site: the dimeric structure of CRP ensures that one of the subunits is available to contact RNA polymerase, no matter which subunit binds to DNA and in which orientation.

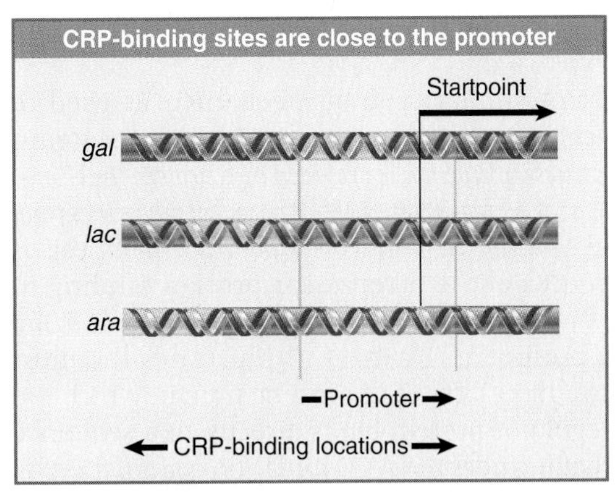

FIGURE 12.32 The CRP protein can bind at different sites relative to RNA polymerase.

The effect upon RNA polymerase binding depends on the relative locations of the two proteins. At class I promoters, where CRP binds adjacent to the promoter, it increases the rate of initial binding to form a closed complex. At class II promoters, where CRP binds within the promoter, it increases the rate of transition from the closed to open complex.

12.22 Translation Can Be Regulated

Key concepts

- A repressor protein can regulate translation by preventing a ribosome from binding to an initiation codon.
- Accessibility of initiation codons in a polycistronic mRNA can be controlled by changes in the structure of the mRNA that occur as the result of translation.

Translational control is a notable feature of operons coding for components of the protein synthetic apparatus. The operon provides an arrangement for *coordinate* regulation of a group of structural genes. Further controls superimposed on the operon, though, such as those at the level of translation, may create *differences* in the extent to which individual genes are expressed.

A similar type of mechanism is used to achieve translational control in several systems. *Repressor function is provided by a protein that binds to a target region on mRNA to prevent ribosomes from recognizing the initiation region.* Formally this is equivalent to a repressor protein binding to DNA to prevent RNA polymerase from utilizing a promoter. FIGURE 12.33 illustrates the most common form of this interaction, in which the regulator protein binds directly to a sequence that includes the AUG initiation codon, thereby preventing the ribosome from binding.

Some examples of translational repressors and their targets are summarized in FIGURE 12.34. A classic example is the coat protein of the RNA phage R17; it binds to a hairpin that encompasses the ribosome-binding site in the phage mRNA. Similarly, the T4 RegA protein binds to a consensus sequence that includes the AUG initiation codon in several T4 early mRNAs, and T4 DNA polymerase binds to a sequence in its own mRNA that includes the Shine–Dalgarno element needed for ribosome binding.

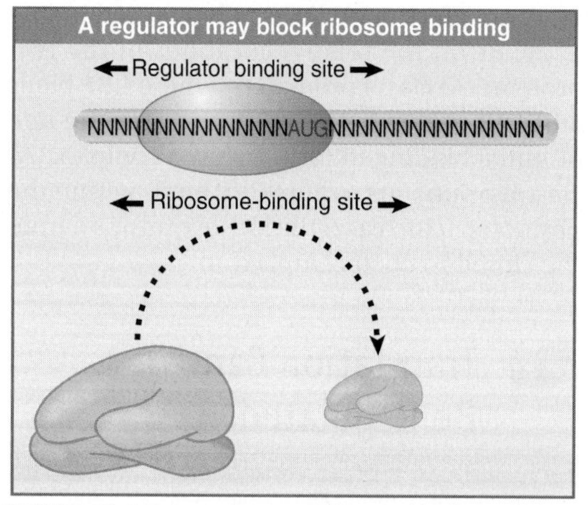

FIGURE 12.33 A regulator protein may block translation by binding to a site on mRNA that overlaps the ribosome-binding site at the initiation codon.

Another form of translational control occurs when translation of one cistron requires changes in secondary structure that depend on translation of a preceding cistron. This happens during translation of the RNA phages, whose cistrons always are expressed in a set order. FIGURE 12.35 shows that the phage RNA takes up a secondary structure in which only one initiation sequence is accessible; the second cannot be recognized by ribosomes because it is

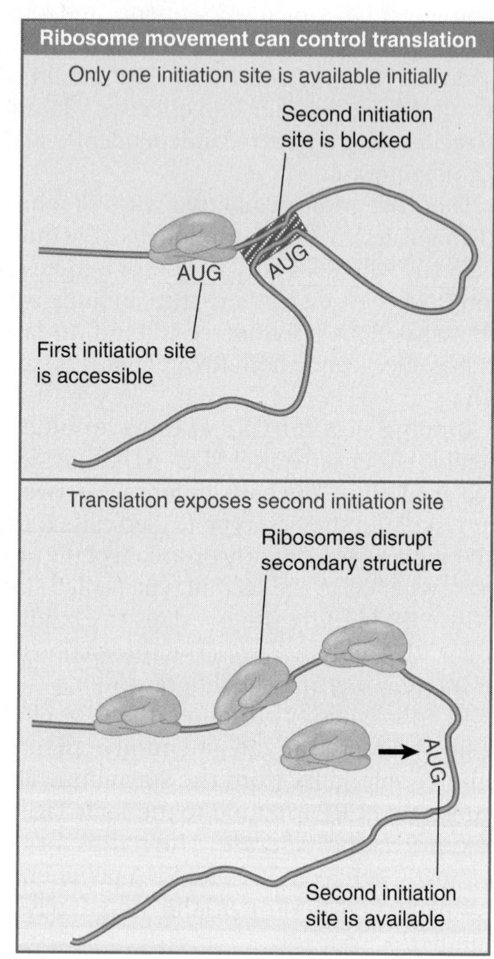

FIGURE 12.35 Secondary structure can control initiation. Only one initiation site is available in the RNA phage, but translation of the first cistron changes the conformation of the RNA so that other initiation site(s) become available.

Translational repressors bind to mRNA

Repressor	Target Gene	Site of Action
R17 coat protein	R17 replicase	hairpin that includes ribosome binding site
T4 RegA	early T4 mRNAs	various sequences including initiation codon
T4 DNA polymerase	T4 DNA polymerase	Shine-Dalgarno sequence
T4 p32	gene 32	single-stranded 5′ leader

FIGURE 12.34 Proteins that bind to sequences within the initiation regions of mRNAs may function as translational repressors.

base-paired with other regions of the RNA. Translation of the first cistron, however, disrupts the secondary structure, allowing ribosomes to bind to the initiation site of the next cistron. In this mRNA, secondary structure controls translatability.

12.23 r-Protein Synthesis Is Controlled by Autogenous Regulation

Key concept

- Translation of an r-protein operon can be controlled by a product of the operon that binds to a site on the polycistronic mRNA.

About seventy or so proteins constitute the apparatus for bacterial gene expression. The ribosomal proteins are the major component, together with the ancillary proteins involved in protein synthesis. The subunits of RNA polymerase and its accessory factors make up the remainder. The genes coding for ribosomal proteins, protein-synthesis factors, and RNA polymerase subunits all are intermingled and organized into a small number of operons. Most of these proteins are represented only by single genes in *E. coli.*

Coordinate controls ensure that these proteins are synthesized in amounts appropriate for the growth conditions: when bacteria grow more rapidly, they devote a greater proportion of their efforts to the production of the apparatus for gene expression. An array of mechanisms is used to control the expression of the genes coding for this apparatus and to ensure that the proteins are synthesized at comparable levels that are related to the levels of the rRNAs.

The organization of six operons is summarized in FIGURE 12.36. About half of the genes for ribosomal proteins **(r-proteins)** map in four operons that lie close together (named *str, spc, S10,* and α simply for the first one of the functions to have been identified in each case). The *rif* and *L11* operons lie together at another location.

Each operon codes for a variety of functions. The *str* operon has genes for small subunit ribosomal proteins as well as for EF-Tu and EF-G. The *spc* and *S10* operons have genes interspersed for both small and large ribosomal subunit proteins. The α operon has genes for proteins of both ribosomal subunit, as well as for the α subunit of RNA polymerase. The *rif* locus has genes for large subunit ribosomal proteins and for the β and β′ subunits of RNA polymerase.

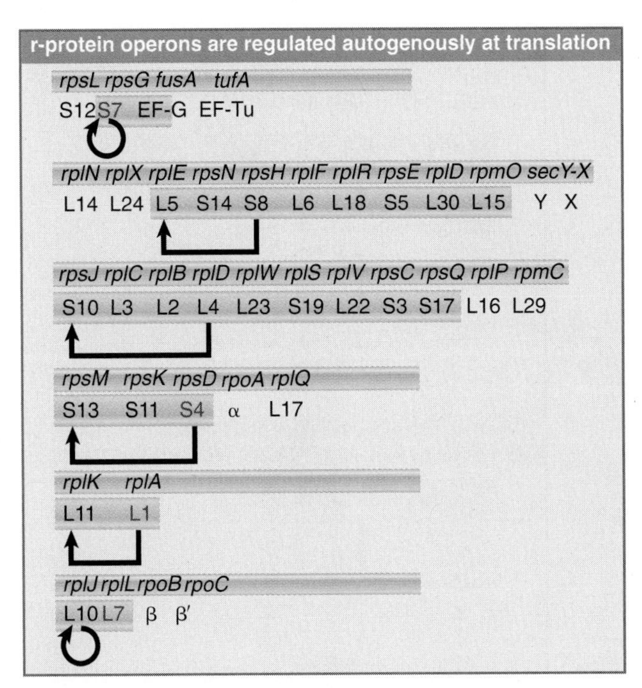

FIGURE 12.36 Genes for ribosomal proteins, protein synthesis factors, and RNA polymerase subunits are interspersed in a small number of operons that are autonomously regulated. The regulator is named in red; the proteins that are regulated are shaded in pink.

All except one of the ribosomal proteins are needed in equimolar amounts, which must be coordinated with the level of rRNA. The dispersion of genes whose products must be equimolar, and their intermingling with genes whose products are needed in different amounts, pose some interesting problems for coordinate regulation.

A feature common to all of the operons described in Figure 12.36 is regulation of some of the genes by one of the products. In each case, the gene coding for the regulatory product is itself one of the targets for regulation. **Autogenous** regulation occurs whenever a protein (or RNA) regulates its own production. In the case of the r-protein operons, the regulatory protein inhibits expression of a contiguous set of genes within the operon, so this is an example of negative autogenous regulation.

In each case, *accumulation of the protein inhibits further synthesis of itself and of some other gene products.* The effect often is exercised at the level of translation of the polycistronic mRNA. Each of the regulators is a ribosomal protein that binds directly to rRNA. *Its effect on translation is a result of its ability also to bind to its own mRNA.* The sites on mRNA at which these proteins bind either overlap the sequence where translation is initiated or lie nearby and

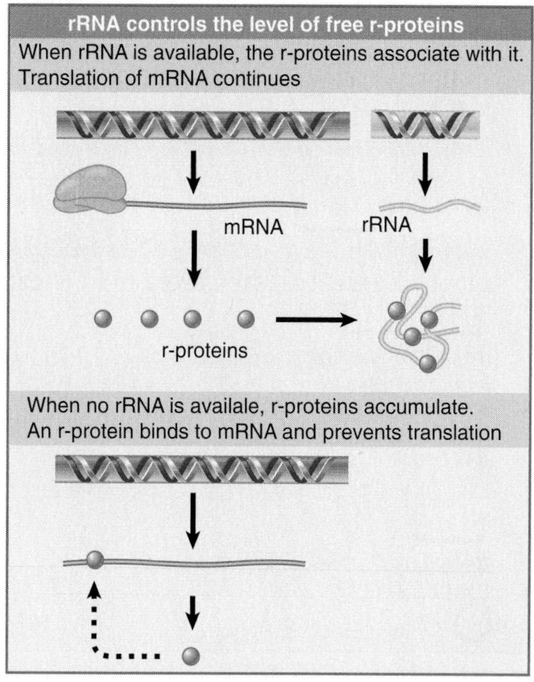

FIGURE 12.37 Translation of the r-protein operons is autogenously controlled and responds to the level of rRNA.

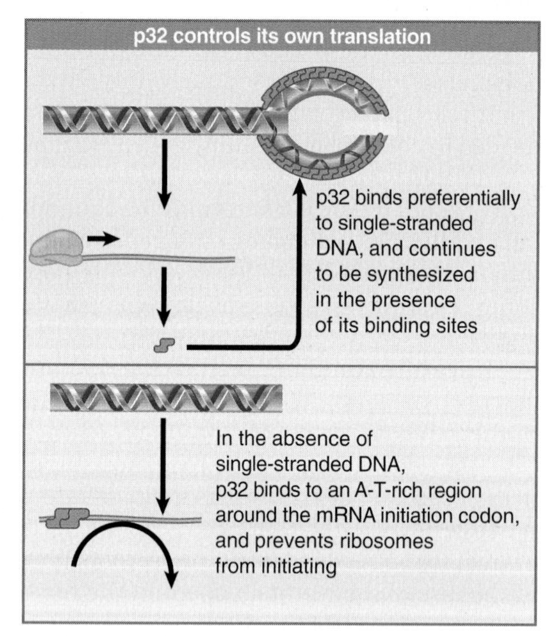

FIGURE 12.38 Excess gene 32 protein (p32) binds to its own mRNA to prevent ribosomes from initiating translation.

probably influence the accessibility of the initiation site by inducing conformational changes. For example, in the S10 operon, protein L4 acts at the very start of the mRNA to inhibit translation of S10 and the subsequent genes. The inhibition may result from a simple block to ribosome access, as illustrated previously in Figure 12.34, or it may prevent a subsequent stage of translation. In two cases (including S4 in the α operon), the regulatory protein stabilizes a particular secondary structure in the mRNA that prevents the initiation reaction from continuing after the 30S subunit has bound.

The use of r-proteins that bind rRNA to establish autogenous regulation immediately suggests that this provides a mechanism to link r-protein synthesis to rRNA synthesis. A generalized model is depicted in FIGURE 12.37. Suppose that the binding sites for the autogenous regulator r-proteins on rRNA are much stronger than those on the mRNAs. As long as any free rRNA is available, the newly synthesized r-proteins will associate with it to start ribosome assembly. There will be no free r-protein available to bind to the mRNA, so its translation will continue. As soon as the synthesis of rRNA slows or stops, though, free r-proteins begin to accumulate. They are then available to bind their mRNAs and thus repress further translation. This circuit ensures that each r-protein operon responds in the same way to the level of rRNA: as soon as there is an excess of r-protein relative to rRNA, synthesis of the protein is repressed.

12.24 Phage T4 p32 Is Controlled by an Autogenous Circuit

Key concept

- p32 binds to its own mRNA to prevent initiation of translation.

Autogenous regulation has been placed on a quantitative basis for gene *32* of phage T4. The protein (p32) plays a central role in genetic recombination, DNA repair, and replication, in which its function is exercised by virtue of its ability to bind to single-stranded DNA. Nonsense mutations cause the inactive protein to be overproduced. *Thus when the function of the protein is prevented, more of it is made.* This effect occurs at the level of translation; the gene *32* mRNA is stable and remains so irrespective of the behavior of the protein product.

FIGURE 12.38 presents a model for the gene *32* control circuit. When single-stranded DNA is present in the phage-infected cell, it sequesters p32. In the absence of single-stranded DNA, however, or at least in conditions in which there is a surplus of p32, the protein prevents trans-

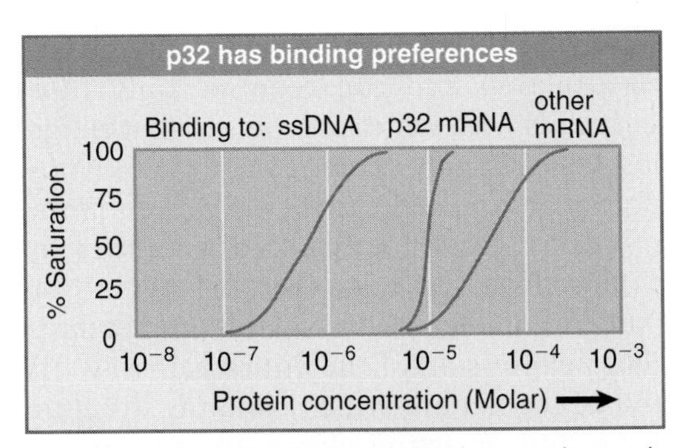

FIGURE 12.39 Gene 32 protein binds to various substrates with different affinities, in the order single-stranded DNA, its own mRNA, and other mRNAs. Binding to its own mRNA prevents the level of p32 from rising $>10^{-6}$ M.

lation of its own mRNA. The effect is mediated directly by p32 binding to mRNA to prevent initiation of translation. In all likelihood this occurs at an A-T-rich region that surrounds the ribosome binding site.

Two features of the binding of p32 to the site on mRNA are required to make the control loop work effectively:

- The affinity of p32 for the site on gene *32* mRNA must be significantly lower than its affinity for single-stranded DNA. The equilibrium constant for binding RNA is in fact almost two orders of magnitude below that for single-stranded DNA.
- The affinity of p32 for the mRNA, however, must be significantly greater than the affinity for other RNA sequences. It is influenced by base composition and by secondary structure; an important aspect of the binding to gene *32* mRNA is that the regulatory region has an extended sequence lacking secondary structure.

Using the known equilibrium constants, we can plot the binding of p32 to its target sites as a function of protein concentration. FIGURE 12.39 shows that at concentrations below 10^{-6} M, p32 binds to single-stranded DNA. At concentrations $>10^{-6}$ M, it binds to gene *32* mRNA. At yet greater concentrations, it binds to other mRNA sequences, with a range of affinities.

These results imply that the level of p32 should be autoregulated to be $<10^{-6}$ M, which corresponds to ~2000 molecules per bacterium. This fits well with the measured level of 1000 to 2000 molecules/cell.

A feature of autogenous control is that each regulatory interaction is unique: a protein acts only on the mRNA responsible for its own synthesis. Phage T4 provides an example of a more general translational regulator, coded by the gene *regA*, which represses the expression of several genes that are transcribed during early infection. RegA protein prevents the translation of mRNAs for these genes by competing with 30S subunits for the initiation sites on the mRNA. Its action is a direct counterpart to the function of a repressor protein that binds multiple operators.

12.25 Autogenous Regulation Is Often Used to Control Synthesis of Macromolecular Assemblies

Key concept

- The precursor to microtubules, free tubulin protein, inhibits translation of tubulin mRNA.

Autogenous regulation is a common type of control among proteins that are incorporated into macromolecular assemblies. The assembled particle itself may be unsuitable as a regulator because it is too large, too numerous, or too restricted in its location. The need for synthesis of its components, though, may be reflected in the pool of free precursor subunits. If the assembly pathway is blocked for any reason, free subunits accumulate and shut off the unnecessary synthesis of further components.

Eukaryotic cells have a common system in which autogenous regulation of this type occurs. Tubulin is the monomer from which microtubules, a major filamentous system of all eukaryotic cells, are synthesized. The production of tubulin mRNA is controlled by the free tubulin pool. When this pool reaches a certain concentration, the production of further tubulin mRNA is prevented. Again, the principle is the same: tubulin sequestered into its macromolecular assembly plays no part in regulation, but the level of the free precursor pool determines whether further monomers are added to it.

The target site for regulation is a short sequence at the start of the coding region. We do not know yet what role this sequence plays, but two models are illustrated in FIGURE 12.40. Tubulin may bind directly to the mRNA, or it

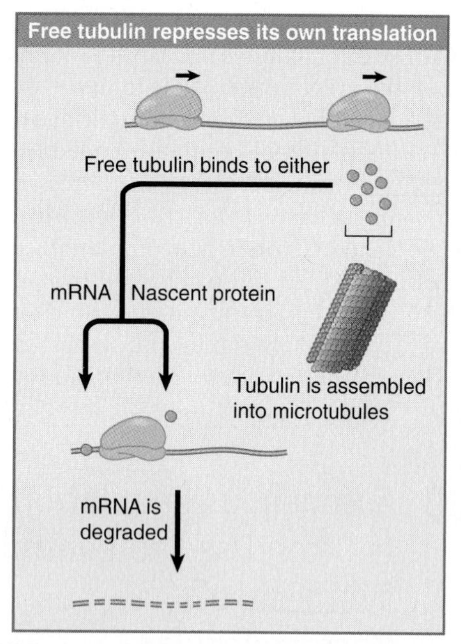

FIGURE 12.40 Tubulin is assembled into microtubules when it is synthesized. Accumulation of excess free tubulin induces instability in the tubulin mRNA by acting at a site at the start of the reading frame in mRNA or at the corresponding position in the nascent protein.

may bind to the nascent polypeptide representing this region. Whichever model applies, excess tubulin causes tubulin mRNA that is located on polysomes to be degraded, so the consequence of the reaction is to make the tubulin mRNA unstable.

Autogenous control is an *intrinsically* self-limiting system, by contrast with the *extrinsic* control that we discussed previously. A repressor protein's ability to bind an operator may be controlled by the level of an extraneous small molecule, which activates or inhibits its activity. In the case of autogenous regulation, though, the critical parameter is the concentration of the protein itself.

12.26 Summary

Transcription is regulated by the interaction between *trans*-acting factors and *cis*-acting sites. A *trans*-acting factor is the product of a regulator gene. It is usually protein but also can be RNA. It diffuses in the cell, and as a result it can act on any appropriate target gene. A *cis*-acting site in DNA (or RNA) is a sequence that functions by being recognized *in situ*. It has no coding function and can regulate only those sequences with which it is physically contiguous. Bacterial genes coding for proteins whose functions are related, such as successive enzymes in a pathway, may be organized in a cluster that is transcribed into a polycistronic mRNA from a single promoter. Control of this promoter regulates expression of the entire pathway. The unit of regulation, which contains structural genes and *cis*-acting elements, is called the operon.

Initiation of transcription is regulated by interactions that occur in the vicinity of the promoter. The ability of RNA polymerase to initiate at the promoter is prevented or activated by other proteins. Genes that are active unless they are turned off are said to be under negative control. Genes that are active only when specifically turned on are said to be under positive control. The type of control can be determined by the dominance relationships between wild type and mutants that are constitutive/derepressed (permanently on) or uninducible/super-repressed (permanently off).

A repressor protein prevents RNA polymerase either from binding to the promoter or from activating transcription. The repressor binds to a target sequence, the operator, that usually is located around or upstream of the startpoint. Operator sequences are short and often are palindromic. The repressor is often a homomultimer whose symmetry reflects that of its target.

The ability of the repressor protein to bind to its operator is regulated by a small molecule. An inducer prevents a repressor from binding; a corepressor activates it. Binding of the inducer or corepressor to its site produces a change in the structure of the DNA-binding site of the repressor. This allosteric reaction occurs in both free repressor proteins and directly in repressor proteins already bound to DNA.

The lactose pathway operates by induction, when an inducer β-galactoside prevents the repressor from binding its operator; transcription and translation of the *lacZ* gene then produce β-galactosidase, the enzyme that metabolizes β-galactosides. The tryptophan pathway operates by repression; the corepressor (tryptophan) activates the repressor protein, so that it binds to the operator and prevents expression of the genes that code for the enzymes that biosynthesize tryptophan. A repressor can control multiple targets that have copies of an operator consensus sequence.

A protein with a high affinity for a particular target sequence in DNA has a lower affinity for all DNA. The ratio defines the specificity of the protein. There are many more nonspecific sites (any DNA sequence) than specific target sites in a genome; as a result, a DNA-binding protein such as a repressor or RNA polymerase is "stored" on DNA. (It is likely that none, or very little, is free.) The specificity for the target sequence must be great enough to counterbalance the excess of nonspecific sites over specific sites. The balance for bacterial proteins is adjusted so that the amount of protein and its specificity allow specific recognition of the target in "on" conditions, but allow almost complete release of the target in "off" conditions.

References

12.1 Introduction

Research

Jacob, F. and Monod, J. (1961). Genetic regulatory mechanisms in the synthesis of proteins. *J. Mol. Biol.* 3, 318–389.

12.2 Regulation Can Be Negative or Positive

Review

Miller, J. and Reznikoff, W., eds. (1980). The Operon, 2nd ed., Woodbury, NY: Cold Spring Harbor Laboratory Press.

12.4 The *lac* Genes Are Controlled by a Repressor

Reviews

Barkley, M. D. and Bourgeois, S. (1978). Repressor recognition of operator and effectors. In *The Operon*, eds. Miller, J. and Reznikoff, W. New York: Cold Spring Harbor Laboratory, 177–220.

Beckwith, J. (1978). *lac:* the genetic system. In *The Operon*, eds. Miller, J. and Reznikoff, W. New York: Cold Spring Harbor Laboratory, 11–30.

Beyreuther, K. (1978). Chemical structure and functional organization of the lac repressor from *E. coli*. In *The Operon*, eds. Miller, J. and Reznikoff, W. New York: Cold Spring Harbor Laboratory, 123–154.

Miller, J. H. (1978). The *lacI* gene: its role in lac operon control and its use as a genetic system. In *The Operon*, eds. Miller, J. and Reznikoff, W. New York: Cold Spring Harbor Laboratory, 31–88.

Weber, K. and Geisler, N. (1978). Lac repressor fragments produced in vivo and in vitro: an approach to the understanding of the interaction of repressor and DNA. In *The Operon*, eds. Miller, J. and Reznikoff, W. New York: Cold Spring Harbor Laboratory, 155–176.

Wilson, C. J., Zahn, H., Swint-Kruse, L., and Matthews, K. S. (2006). The lactose repressor system: paradigms for regulation, allosteric behavior and protein folding. *Cell. Mol. Life Sci.* November 13.

Research

Jacob, F. and Monod, J. (1961). Genetic regulatory mechanisms in the synthesis of proteins. *J. Mol. Biol.* 3, 318–389.

12.10 The Repressor Monomer Has Several Domains

Research

Friedman, A. M., Fischmann, T. O., and Steitz, T. A. (1995). Crystal structure of lac repressor core tetramer and its implications for DNA looping. *Science* 268, 1721–1727.

Lewis, M. et al. (1996). Crystal structure of the lactose operon repressor and its complexes with DNA and inducer. *Science* 271, 1247–1254.

12.13 Mutant Phenotypes Correlate with the Domain Structure

Reviews

Pace, H. C., Kercher, M. A., Lu, P., Markiewicz, P., Miller, J. H., Chang, G., and Lewis, M. (1997). Lac repressor genetic map in real space. *Trends Biochem. Sci.* 22, 334–339.

Markiewicz, P., Kleina, L. G., Cruz, C., Ehret, S., and Miller, J. H. (1994). Genetic studies of the lac repressor. XIV. Analysis of 4000 altered *E. coli lac* repressors reveals essential and nonessential residues, as well as spacers which do not require a specific sequence. *J. Mol. Biol.* 240, 421–433.

Suckow, J., Markiewicz, P., Kleina, L. G., Miller, J., Kisters-Woike, B., and Müller-Hill, B. (1996). Genetic studies of the Lac repressor. XV: 4000 single amino acid substitutions and analysis of the resulting phenotypes on the basis of the protein structure. *J. Mol. Biol.* 261, 509–523.

12.14 Repressor Protein Binds to the Operator

Research

Gilbert, W. and Müller-Hill, B. (1966). Isolation of the lac repressor. *Proc. Natl. Acad. Sci. USA* 56, 1891–1898.

Gilbert, W. and Müller-Hill, B. (1967). The lac operator is DNA. *Proc. Natl. Acad. Sci. USA* 58, 2415–2421.

12.16 Repressor Binds to Three Operators and Interacts with RNA Polymerase

Research

Oehler, S. et al. (1990). The three operators of the lac operon cooperate in repression. *EMBO J.* 9, 973–979.

12.18 The Operator Competes with Low-Affinity Sites to Bind Repressor

Research

Cronin, C. A., Gluba, W., and Scrable, H. (2001). The lac operator-repressor system is functional in the mouse. *Genes Dev.* 15, 1506–1517.
Hildebrandt, E. R. et al. (1995). Comparison of recombination *in vitro* and in *E. coli* cells: measure of the effective concentration of DNA *in vivo*. *Cell* 81, 331–340.
Lin, S.-Y. and Riggs, A. D. (1975). The general affinity of lac repressor for *E. coli* DNA: implications for gene regulation in prokaryotes and eukaryotes. *Cell* 4, 107–111.

12.21 CRP Functions in Different Ways in Different Target Operons

Reviews

Botsford, J. L. and Harman, J. G. (1992). Cyclic AMP in prokaryotes. *Microbiol. Rev.* 56, 100–122.
Kolb, A. (1993). Transcriptional regulation by cAMP and its receptor protein. *Annu. Rev. Biochem.* 62, 749–795.

Research

Niu, W., Kim, Y., Tau, G., Heyduk, T., and Ebright, R. H. (1996). Transcription activation at class II CAP-dependent promoters: two interactions between CAP and RNA polymerase. *Cell* 87, 1123–1134.
Zhou, Y., Busby, S., and Ebright, R. H. (1993). Identification of the functional subunit of a dimeric transcription activator protein by use of oriented heterodimers. *Cell* 73, 375–379.
Zhou, Y., Merkel, T. J., and Ebright, R. H. (1994). Characterization of the activating region of *E. coli* catabolite gene activator protein (CAP). II. Role at Class I and class II CAP-dependent promoters. *J. Mol. Biol.* 243, 603–610.

12.23 r-Protein Synthesis Is Controlled by Autogenous Regulation

Review

Nomura, M. et al. (1984). Regulation of the synthesis of ribosomes and ribosomal components. *Annu. Rev. Biochem.* 53, 75–117.

Research

Baughman, G. and Nomura, M. (1983). Localization of the target site for translational regulation of the L11 operon and direct evidence for translational coupling in *E. coli*. *Cell* 34, 979–988.

12.25 Autogenous Regulation Is Often Used to Control Synthesis of Macromolecular Assemblies

Review

Gold, L. (1988). Posttranscriptional regulatory mechanisms in *E. coli*. *Annu. Rev. Biochem.* 57, 199–223.

13

Regulatory RNA

CHAPTER OUTLINE

13.1 Introduction

Key concept

- RNA functions as a regulator by forming a region of secondary structure (either inter- or intramolecular) that changes the properties of a target sequence.

The basic principle of regulation in bacteria is that gene expression is controlled by a regulator that interacts with a specific sequence or structure in DNA or mRNA at some stage prior to the synthesis of protein. The stage of expression that is controlled can be transcription, when the target for regulation is DNA, or it can be at translation, when the target for regulation is RNA. When control is during transcription, it can be at initiation or at termination. The regulator can be a protein or an RNA. "Controlled" can mean that the regulator turns off (represses) the target or that it turns on (activates) the target. Expression of many genes can be coordinately controlled by a single regulator gene on the principle that each target contains a copy of the sequence or structure that the regulator recognizes. Regulators may themselves be regulated, most typically in response to small molecules whose supply responds to environmental conditions. Regulators may be controlled by other regulators to make complex circuits.

Let's compare the ways that different types of regulators work.

Protein regulators work on the principle of allostery. The protein has two binding sites—one for a nucleic acid target, the other for a small molecule. Binding of the small molecule to its site changes the conformation in such a way as to alter the affinity of the other site for the nucleic acid. The way in which this happens is known in detail for the Lac repressor. Protein regulators are often multimeric, with a symmetrical organization that allows two subunits to contact a palindromic target on DNA. This can generate cooperative binding effects that create a more sensitive response to regulation.

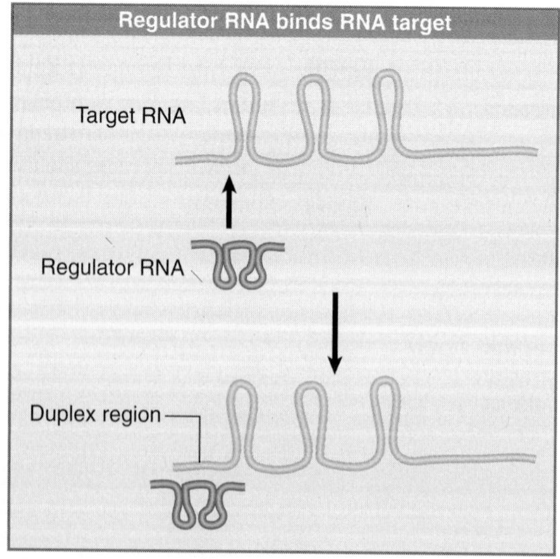

FIGURE 13.1 A regulator RNA is a small RNA with a single-stranded region that can pair with a single-stranded region in a target RNA.

Regulation via RNA uses changes in secondary structure as the guiding principle. The ability of an RNA to shift between different conformations with regulatory consequences is the nucleic acid's alternative to the allosteric changes of protein conformation. The changes in structure may result from either intramolecular or intermolecular interactions.

The most common role for intramolecular changes is for an RNA molecule to assume alternative secondary structures by utilizing different schemes for base pairing. The properties of the alternative conformations may be different. Changes in secondary structure of an mRNA can result in a change in its ability to be translated. Secondary structure also is used to regulate the termination of transcription, when the alternative structures differ in whether they permit termination.

In intermolecular interactions, an RNA regulator recognizes its target by the familiar principle of complementary base pairing. FIGURE 13.1 shows that the regulator is usually a small RNA molecule with extensive secondary structure, but with a single-stranded region(s) that is complementary to a single-stranded region in its target. The formation of a double helical region between regulator and target can have two types of consequence:

- Formation of the double helical structure may itself be sufficient. In some cases, protein(s) can bind only to the single-stranded form of the target sequence and are therefore prevented from acting by duplex formation. In other cases, the duplex region becomes a target for binding; for example, by nucleases that degrade the RNA and therefore prevent its expression.
- Duplex formation may be important because it sequesters a region of the target RNA that would otherwise participate in some alternative secondary structure.

13.2 Alternative Secondary Structures Control Attenuation

Key concepts

- Termination of transcription can be attenuated by controlling formation of the necessary hairpin structure in RNA.
- The most direct mechanisms for attenuation involve proteins that either stabilize or destabilize the hairpin.

RNA structure provides an opportunity for regulation in both prokaryotes and eukaryotes. Its most common role occurs when an RNA molecule can take up alternative secondary structures by utilizing different schemes for intramolecular base pairing. The properties of the alternative conformations may be different. This type of mechanism can be used to regulate the termination of transcription, when the alternative structures differ in whether they permit termination. Another means of controlling conformation (and thereby function) is provided by the cleavage of an RNA; by removing one segment of an RNA, the conformation of the rest may be altered. It is possible also for a (small) RNA molecule to control the activity of a target RNA by base pairing with it; the role of the small RNA is directly analogous to that of a regulator protein (see Section 13.7, Small RNA Molecules Can Regulate Translation). The ability of an RNA to shift between different conformations with regulatory consequences is the nucleic acid's alternative to the allosteric changes of conformation that regulate protein function. Both these mechanisms allow an interaction at one site in the molecule to affect the structure of another site.

Several operons are regulated by **attenuation,** a mechanism that controls the ability of RNA polymerase to read through an **attenuator,** which is an intrinsic terminator located at the beginning of a transcription unit. *The principle of attenuation is that some external event controls the formation of the hairpin needed for intrinsic termination.* If the hairpin is allowed to form, termination prevents RNA polymerase from transcribing the structural genes. If the hairpin is prevented from forming, RNA polymerase elongates through the terminator and the genes are expressed. Different types of mechanisms are used in different systems for controlling the structure of the RNA.

Attenuation may be regulated by proteins that bind to RNA, either to stabilize or to destabilize formation of the hairpin required for termination. FIGURE 13.2 shows an example in which a protein prevents formation of the terminator hairpin. The activity of such a protein may be intrinsic or may respond to a small molecule in the same manner as a repressor protein responds to corepressor.

13.3 Termination of *Bacillus subtilis trp* Genes Is Controlled by Tryptophan and by tRNA^{Trp}

Key concepts

- A terminator protein called TRAP is activated by tryptophan to prevent transcription of *trp* genes.
- Activity of TRAP is (indirectly) inhibited by uncharged tRNA^{Trp}.

The circuitry that controls transcription via termination can use both direct and indirect means to respond to the level of small molecule products or substrates.

In *B. subtilis,* a protein called tryptophan RNA-binding attenuation protein (TRAP) (formerly called MtrB) is activated by tryptophan (Trp) to bind to a sequence in the leader of the

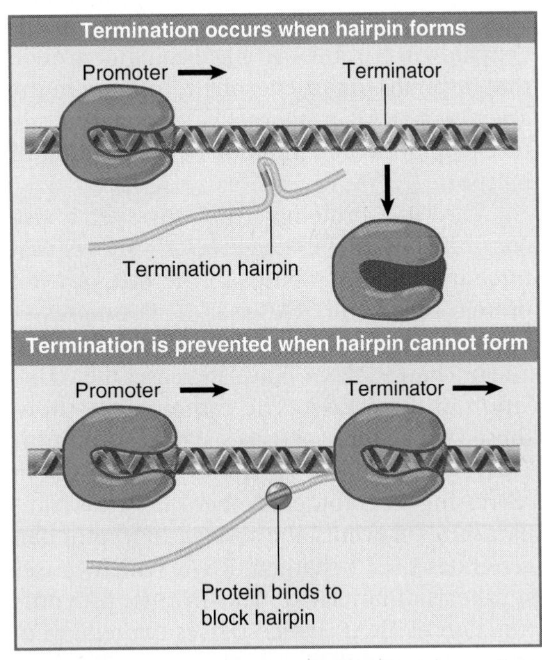

FIGURE 13.2 Attenuation occurs when a terminator hairpin in RNA is prevented from forming.

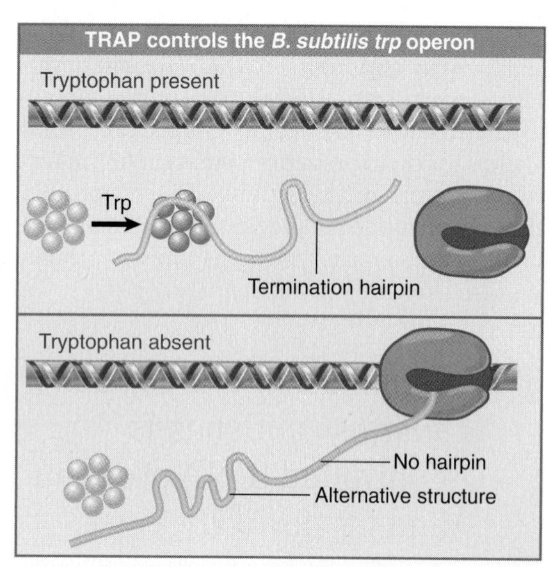

FIGURE 13.3 TRAP is activated by tryptophan and binds to *trp* mRNA. This allows the termination hairpin to form, with the result that RNA polymerase terminates and the genes are not expressed. In the absence of tryptophan, TRAP does not bind, and the mRNA adopts a structure that prevents the terminator hairpin from forming.

nascent transcript. TRAP forms a multimer of eleven subunits. Each subunit binds a single tryptophan amino acid and a trinucleotide (GAG or UAG) of RNA. The RNA is wound in a circle around the protein. FIGURE 13.3 shows that the result is to ensure the availability of the regions that are required to form the terminator hairpin. The termination of transcription then prevents production of the tryptophan biosynthetic enzymes. In effect, TRAP is a terminator protein that responds to the level of tryptophan. In the absence of TRAP, an alternative secondary structure precludes the formation of the terminator hairpin.

The TRAP protein in turn, however, is also controlled by $tRNA^{Trp}$. FIGURE 13.4 shows that uncharged $tRNA^{Trp}$ binds to the mRNA for a protein called antiTRAP (AT). This suppresses formation of a termination hairpin in the mRNA. The uncharged $tRNA^{Trp}$ also increases the translation of the mRNA. The combined result of these two actions increases synthesis of anti-TRAP, which binds to TRAP and prevents it from repressing the tryptophan operon. By this complex series of events, the absence of tryptophan generates the uncharged tRNA, which causes synthesis of antiTRAP. This in turn prevents function of TRAP, which causes expression of tryptophan genes.

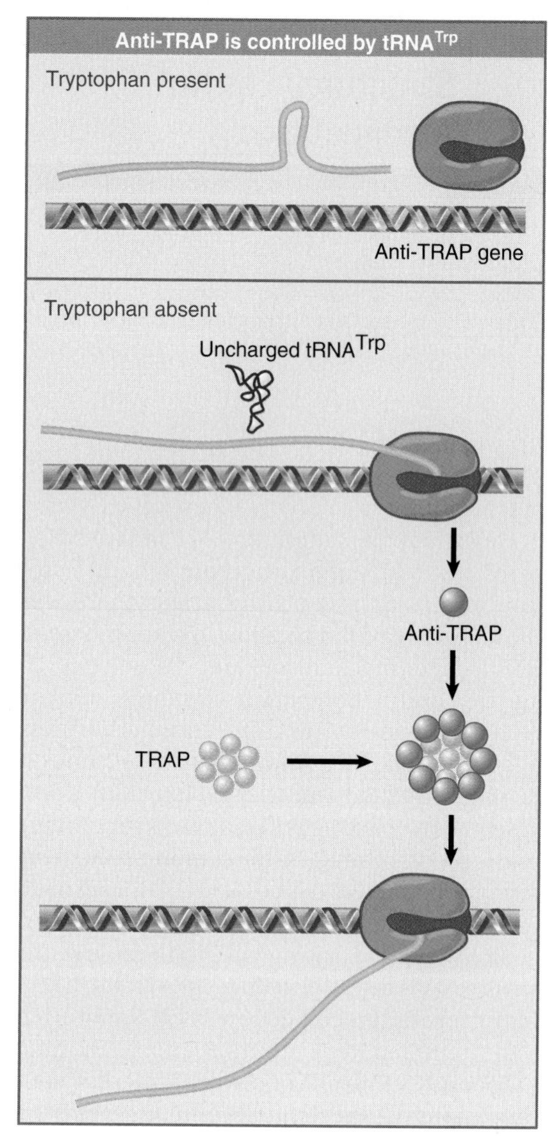

FIGURE 13.4 Under normal conditions (in the presence of tryptophan) transcription terminates before the anti-TRAP gene. When tryptophan is absent, uncharged $tRNA^{Trp}$ base pairs with the anti-TRAP mRNA, preventing formation of the terminator hairpin, thus causing expression of anti-TRAP.

Expression of the *B. subtilis trp* genes is therefore controlled by both tryptophan and $tRNA^{Trp}$. When tryptophan is present, there is no need for it to be synthesized. This is accomplished when tryptophan activates TRAP and therefore inhibits expression of the enzymes that synthesize tryptophan. The presence of uncharged $tRNA^{Trp}$ indicates that there is a shortage of tryptophan. The uncharged tRNA activates the antiTRAP and thereby activates transcription of the *trp* genes.

13.4 The *Escherichia coli tryptophan* Operon Is Controlled by Attenuation

Key concepts

- An attenuator (intrinsic terminator) is located between the promoter and the first gene of the *trp* cluster.
- The absence of tryptophan suppresses termination and results in a 10× increase in transcription.

A complex regulatory system is used in *E. coli* (where attenuation was originally discovered). The changes in secondary structure that control attenuation are determined by the position of the ribosome on mRNA. FIGURE 13.5 shows that termination requires that *the ribosome can translate a leader segment that precedes the* trp *genes in the mRNA.* When the ribosome translates the leader region, a termination hairpin forms at terminator 1. When the ribosome is prevented from translating the leader, though, the termination hairpin does not form, and RNA polymerase transcribes the coding region. *This mechanism of antitermination therefore depends upon the ability of external circumstances to influence ribosome movement in the leader region.*

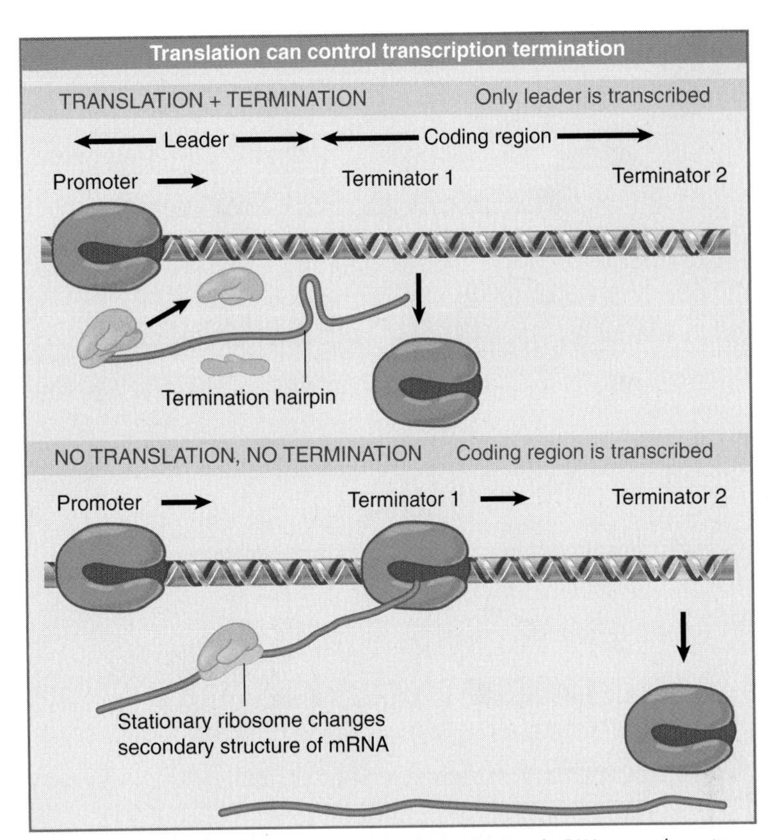

FIGURE 13.5 Termination can be controlled via changes in RNA secondary structure that are determined by ribosome movement.

The *trp* operon consists of five structural genes arranged in a contiguous series, which code for the three enzymes that convert chorismic acid to tryptophan. FIGURE 13.6 shows that transcription starts at a promoter at the left end of the cluster. *trp* operon expression is controlled by two separate mechanisms. Repression of expression is exercised by a repressor protein (coded by the unlinked gene *trpR*) that binds to an operator that is adjacent to the promoter. Attenuation controls the progress of RNA polymerase into the operon by regulating whether termination occurs at a site preceding the first structural gene.

Attenuation was first revealed by the observation that deleting a sequence between the operator and the *trpE* coding region can increase the expression of the structural genes. This effect is independent of repression: both the basal and derepressed levels of transcription are increased. Thus this site influences events that occur *after* RNA polymerase has set out from the promoter (irrespective of the conditions prevailing at initiation).

An attenuator (intrinsic terminator) is located between the promoter and the *trpE* gene.

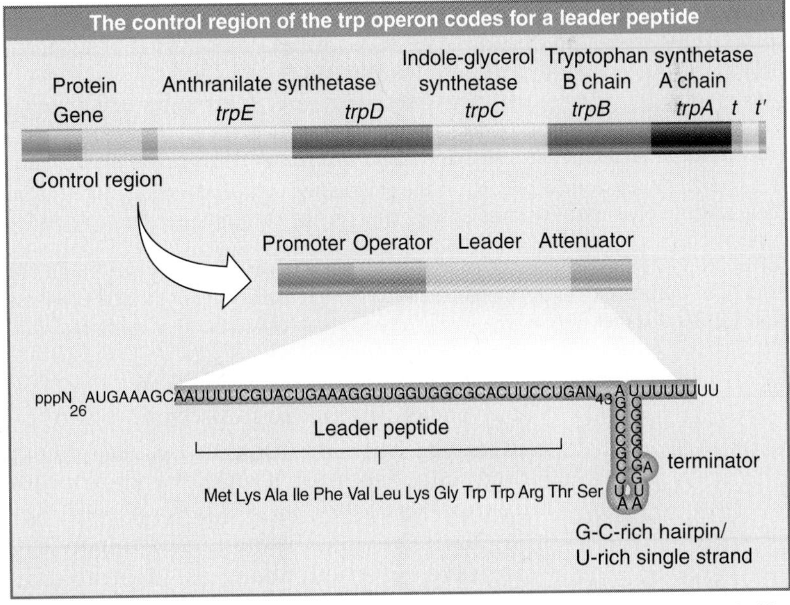

FIGURE 13.6 The *trp* operon consists of five contiguous structural genes preceded by a control region that includes a promoter, operator, leader peptide coding region, and attenuator.

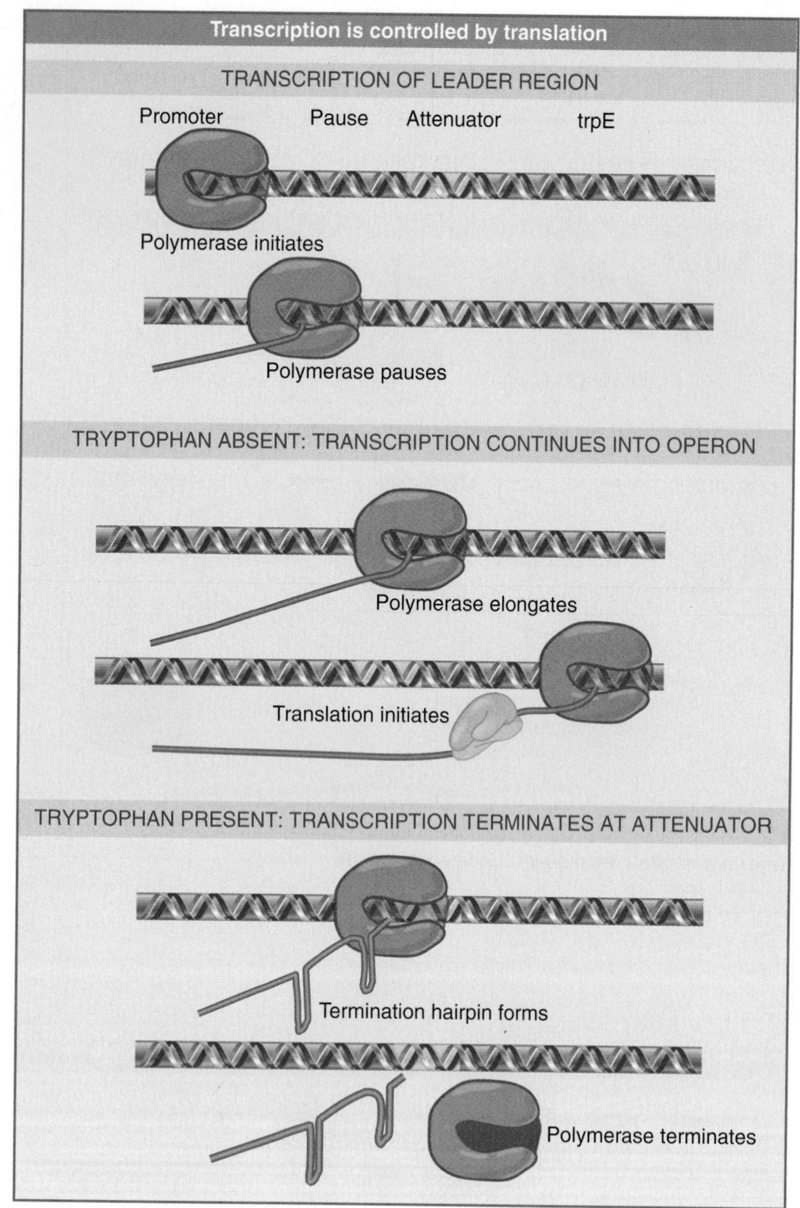

FIGURE 13.7 An attenuator controls the progression of RNA polymerase into the *trp* genes. RNA polymerase initiates at the promoter and then proceeds to position 90, where it pauses before proceeding to the attenuator at position 140. In the absence of tryptophan, the polymerase continues into the structural genes (*trpE* starts at +163). In the presence of tryptophan there is ~90% probability of termination to release the 140-base leader RNA.

It provides a barrier to transcription into the structural genes. RNA polymerase terminates there, either *in vivo* or *in vitro*, to produce a 140-base transcript.

Termination at the attenuator responds to the level of tryptophan, as illustrated in FIGURE 13.7. In the presence of adequate amounts of tryptophan, termination is efficient. In the absence of tryptophan, however, RNA polymerase can continue into the structural genes.

Repression and attenuation respond in the same way to the level of tryptophan. When tryptophan is present the operon is repressed, and most of the RNA polymerases that escape from the promoter then terminate at the attenuator. When tryptophan is removed, RNA polymerase has free access to the promoter, and also is no longer compelled to terminate prematurely.

Attenuation has ~10× effect on transcription. When tryptophan is present termination is effective, and the attenuator allows only ~10% of the RNA polymerases to proceed. In the absence of tryptophan, attenuation allows virtually all of the polymerases to proceed. Together with the ~70× increase in initiation of transcription that results from the release of repression, this allows an ~700-fold range of regulation of the operon.

13.5 Attenuation Can Be Controlled by Translation

Key concepts

- The leader region of the *trp* operon has a fourteen-codon open reading frame that includes two codons for tryptophan.
- The structure of RNA at the attenuator depends on whether this reading frame is translated.
- In the presence of tryptophan, the leader is translated, and the attenuator is able to form the hairpin that causes termination.
- In the absence of tryptophan, the ribosome stalls at the tryptophan codons and an alternative secondary structure prevents formation of the hairpin, so that transcription continues.

How can termination of transcription at the attenuator respond to the level of tryptophan? The sequence of the leader region suggests a mechanism. It has a short coding sequence that could represent a **leader peptide** of fourteen amino acids. Figure 13.6 shows that it contains a ribosome binding site whose AUG codon is followed by a short coding region that contains two successive codons for tryptophan. When the cell runs out of tryptophan, ribosomes initiate translation of the leader peptide but stop when they reach the Trp codons. The sequence of the mRNA suggests that this **ribosome stalling** influences termination at the attenuator.

The leader sequence can be written in alternative base-paired structures. The ability of the ribosome to proceed through the leader region controls transitions between these structures. The structure determines whether the mRNA can provide the features needed for termination.

FIGURE 13.8 draws these structures. In the first, region 1 pairs with region 2 and region 3 pairs with region 4. The pairing of regions 3 and

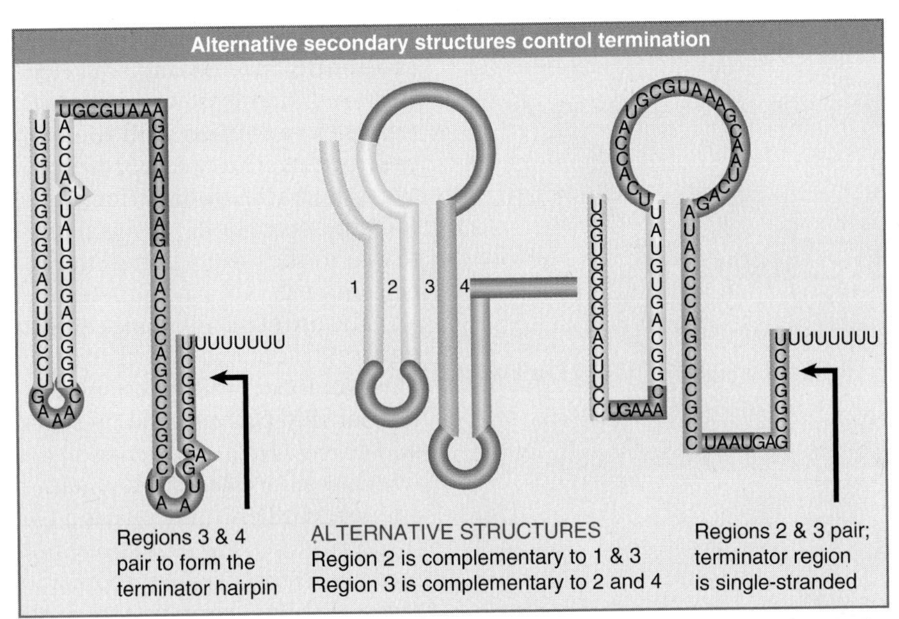

FIGURE 13.8 The *trp* leader region can exist in alternative base-paired conformations. The center shows the four regions that can base pair. Region 1 is complementary to region 2, which is complementary to region 3, which is complementary to region 4. On the left is the conformation produced when region 1 pairs with region 2 and region 3 pairs with region 4. On the right is the conformation when region 2 pairs with region 3, leaving regions 1 and 4 unpaired.

4 generates the hairpin that precedes the U_8 sequence: this is the essential signal for intrinsic termination. It is likely that the RNA would take up this structure in lieu of any outside intervention.

A different structure is formed if region 1 is prevented from pairing with region 2. In this case, region 2 is free to pair with region 3. Region 4 then has no available pairing partner, so it is compelled to remain single-stranded. Thus the terminator hairpin cannot be formed.

FIGURE 13.9 shows that the position of the ribosome can determine which structure is formed in such a way that termination is attenuated only in the absence of tryptophan. The crucial feature is the position of the Trp codons in the leader peptide coding sequence.

When tryptophan is present, ribosomes are able to synthesize the leader peptide. They continue along the leader section of the mRNA to the UGA codon, which lies between regions 1 and 2. As shown in the lower part of the figure, by progressing to this point, the ribosomes extend over region 2 and prevent it from base pairing. The result is that region 3 is available to base pair with region 4, which generates the terminator hairpin. Under these conditions, therefore, RNA polymerase terminates at the attenuator.

When there is no tryptophan ribosomes stall at the Trp codons, which are part of region 1, as shown in the upper part of the figure. Thus

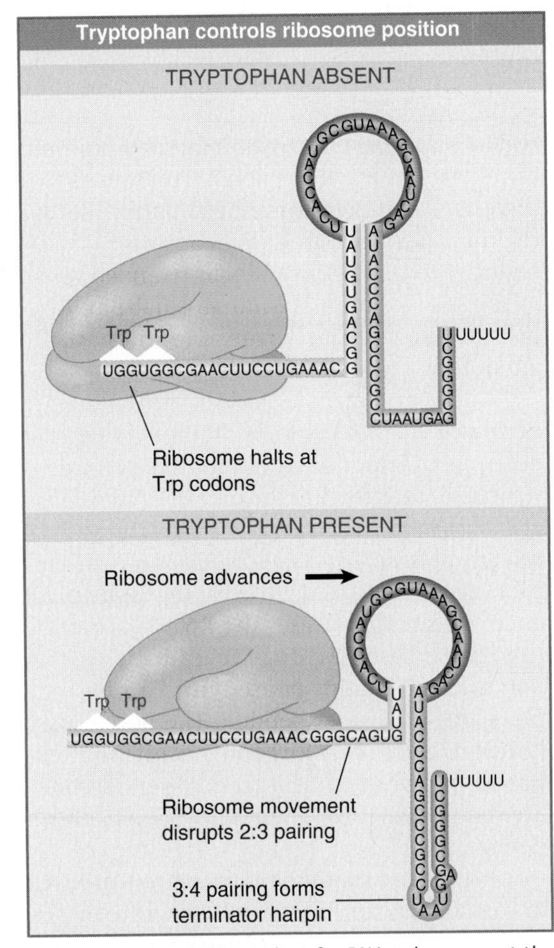

FIGURE 13.9 The alternatives for RNA polymerase at the attenuator depend on the location of the ribosome, which determines whether regions 3 and 4 can pair to form the terminator hairpin.

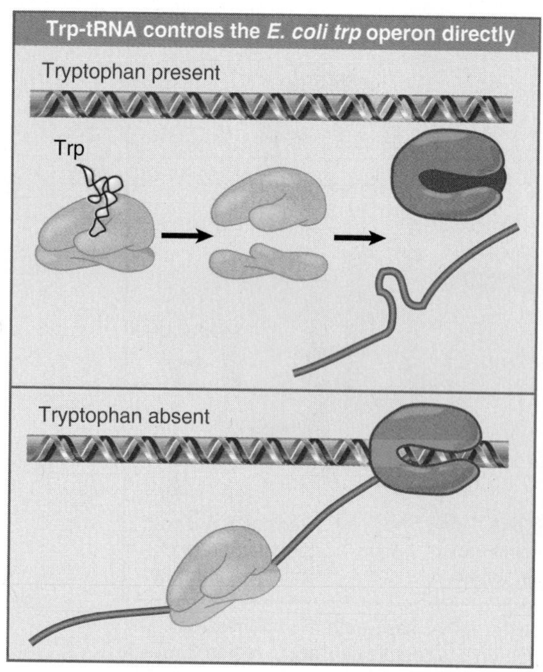

FIGURE 13.10 In the presence of tryptophan tRNA, ribosomes translate the leader peptide and are released. This allows hairpin formation, so that RNA polymerase terminates. In the absence of tryptophan tRNA, the ribosome is blocked, the termination hairpin cannot form, and RNA polymerase continues.

region 1 is sequestered within the ribosome and cannot base pair with region 2. This means that regions 2 and 3 become base paired before region 4 has been transcribed. This compels region 4 to remain in a single-stranded form. In the absence of the terminator hairpin, RNA polymerase continues transcription past the attenuator.

Control by attenuation requires a precise timing of events. For ribosome movement to determine formation of alternative secondary structures that control termination, *translation of the leader must occur at the same time when RNA polymerase approaches the terminator site.* A critical event in controlling the timing is the presence of a site that causes the RNA polymerase to pause at base 90 along the leader. The RNA polymerase remains paused until a ribosome translates the leader peptide. The polymerase is then released and moves off toward the attenuation site. By the time it arrives there, secondary structure of the attenuation region has been determined.

FIGURE 13.10 summarizes the role of Trp-tRNA in controlling expression of the operon. *By providing a mechanism to sense the inadequacy of the supply of Trp-tRNA, attenuation responds directly to the need of the cell for tryptophan in protein synthesis.*

How widespread is the use of attenuation as a control mechanism for bacterial operons? It is used in at least six operons that code for enzymes concerned with the biosynthesis of amino acids. Thus a feedback from the level of the amino acid available for protein synthesis (as represented by the availability of aminoacyl-tRNA) to the production of the enzymes may be common.

The use of the ribosome to control RNA secondary structure in response to the availability of an aminoacyl-tRNA establishes an inverse relationship between the presence of aminoacyl-tRNA and the transcription of the operon, which is equivalent to a situation in which aminoacyl-tRNA functions as a corepressor of transcription. The regulatory mechanism is mediated by changes in the formation of duplex regions; thus attenuation provides a striking example of the importance of secondary structure in the termination event and of its use in regulation.

E. coli and *B. subtilis,* therefore, use the same types of mechanisms, which involve control of mRNA structure in response to the presence or absence of a tRNA, but they have combined the individual interactions in different ways. The end result is the same: to inhibit production of the enzymes when there is an excess supply of the amino acid, and to activate production when a shortage is indicated by the accumulation of uncharged $tRNA^{Trp}$.

13.6 Antisense RNA Can Be Used to Inactivate Gene Expression

Key concept

- Antisense genes block expression of their targets when introduced into eukaryotic cells.

Base pairing offers a powerful means for one RNA to control the activity of another. There are many cases in both prokaryotes and eukaryotes where a (usually rather short) single-stranded RNA base pairs with a complementary region of an mRNA, and as a result it prevents expression of the mRNA. One of the early illustrations of this effect was provided by an artificial situation in which **antisense genes** were introduced into eukaryotic cells.

Antisense genes are constructed by reversing the orientation of a gene with regard to its promoter, so that the "antisense" strand is transcribed, as illustrated in FIGURE 13.11. Synthesis of antisense RNA can inactivate a target RNA in either prokaryotic or eukaryotic cells. An antisense RNA is in effect a synthetic RNA regulator. An antisense thymidine kinase gene inhibits synthesis of thymidine kinase from the endogenous gene. Quantitation of the effect is not entirely reliable, but it seems that an excess (perhaps a considerable excess) of the antisense RNA may be necessary.

At what level does the antisense RNA inhibit expression? It could in principle prevent transcription of the authentic gene, processing of its RNA product, or translation of the messenger. Results with different systems show that the inhibition depends on formation of RNA–RNA duplex molecules, but this can occur either in the nucleus or in the cytoplasm. In the case of an antisense gene stably carried by a cultured cell, sense–antisense RNA duplexes form in the nucleus, preventing normal processing and/or transport of the sense RNA. In another case, injection of antisense RNA into the cytoplasm inhibits translation by forming duplex RNA in the 5′ region of the mRNA.

This technique offers a powerful approach for turning off genes at will; for example, the function of a regulatory gene can be investigated by introducing an antisense version. An extension of this technique is to place the antisense gene under the control of a promoter that is itself subject to regulation. The target gene can then be turned off and on by regulating the production of antisense RNA. This technique allows investigation of the importance of the timing of expression of the target gene.

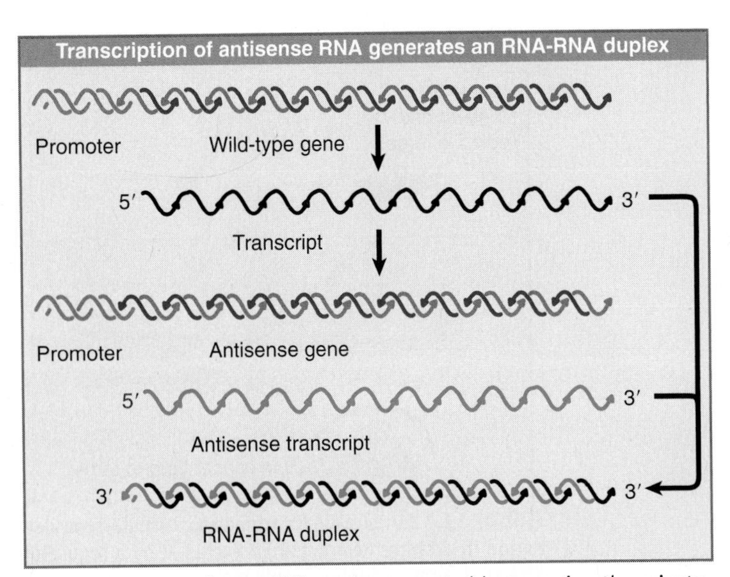

FIGURE 13.11 Antisense RNA can be generated by reversing the orientation of a gene with respect to its promoter and can anneal with the wild-type transcript to form duplex RNA.

13.7 Small RNA Molecules Can Regulate Translation

Key concepts

- A regulator RNA functions by forming a duplex region with a target RNA.
- The duplex may block initiation of translation, cause termination of transcription, or create a target for an endonuclease.

Repressors and activators are *trans*-acting proteins, yet the formal circuitry of a regulatory network could equally well be constructed by using an RNA as regulator. In fact, the original model for the operon left open the question of whether the regulator might be RNA or protein. Indeed, the construction of synthetic antisense RNAs turns out to mimic a class of RNA regulators that is becoming increasingly important.

Like a protein regulator, a small regulator RNA is an independently synthesized molecule that diffuses to a target site consisting of a specific nucleotide sequence. The target for a regulator RNA is a single-stranded nucleic acid sequence. The regulator RNA functions by complementarity with its target, at which it can form a double-stranded region.

We can imagine two general mechanisms for the action of a regulator RNA:

- Formation of a duplex region with the target nucleic acid directly prevents its ability to function by forming or sequestering a specific site. FIGURE 13.12 illustrates the situation in which a protein that binds to single-stranded RNA is prevented from acting by formation of a duplex. FIGURE 13.13 shows the opposite type of relationship, in which the formation of a double-stranded region creates a target site for an endonuclease that destroys the RNA target.
- Formation of a duplex region in one part of the target molecule changes the conformation of another region, thus indirectly affecting its function. FIGURE 13.14 shows an example. The mechanism is essentially similar to the use of secondary structure in attenuation (see

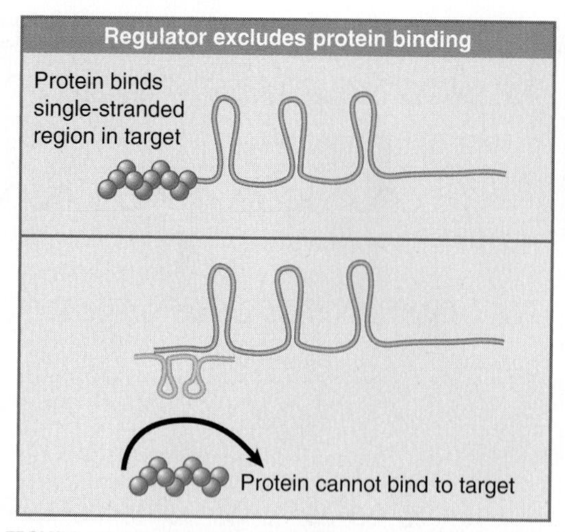

FIGURE 13.12 A protein that binds to a single-stranded region in a target RNA could be excluded by a regulator RNA that forms a duplex in this region.

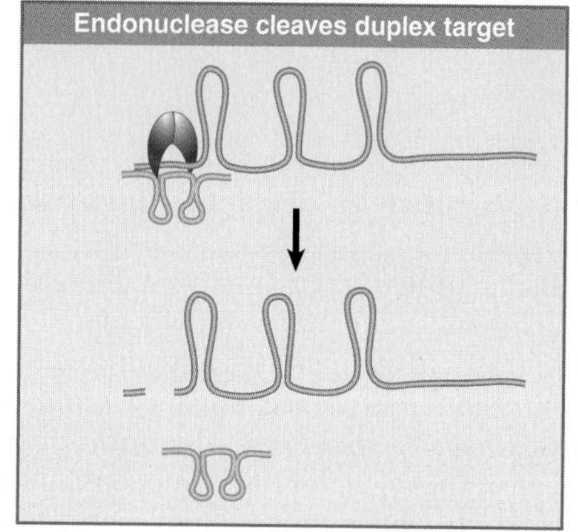

FIGURE 13.13 By binding to a target RNA to form a duplex region, a regulator RNA may create a site that is attacked by a nuclease.

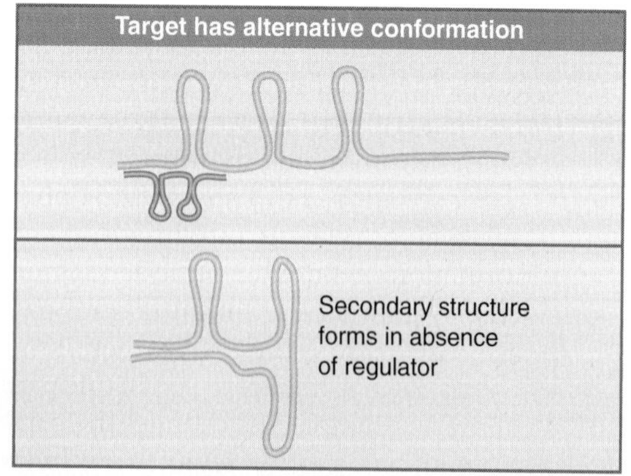

FIGURE 13.14 The secondary structure formed by base pairing between two regions of the target RNA may be prevented from forming by base pairing with a regulator RNA. In this example, the ability of the 3′ end of the RNA to pair with the 5′ end is prevented by the regulator.

Section 13.2, Alternative Secondary Structures Control Attenuation), except that the interacting regions are on different RNA molecules instead of being part of the same RNA molecule.

The feature common to both types of RNA-mediated regulation is that changes in secondary structure of the target control its activity.

A small RNA regulator typically can be turned on by controlling transcription of its gene or turned off by an enzyme that degrades the RNA regulator product. Usually it is not possible otherwise to regulate the activity of an RNA regulator. In fact, it used to be thought that it would not be possible for an RNA to have allosteric properties; unlike repressor proteins that control operons, an RNA usually cannot respond to small molecules by changing its ability to recognize its target.

The discovery of the **riboswitch** provides an exception to this rule. FIGURE 13.15 summarizes the regulation of the system that produces the metabolite GlcN6P. The gene *glmS* codes for an enzyme that synthesizes GlcN6P (Glucosamine-6-phosphate) from fructose-6-phosphate and glutamine. The mRNA contains a long, 5′ untranslated region (UTR) before the coding frame. Within the UTR is a ribozyme—a sequence of RNA that has a catalytic activity (see Section 27.4, Ribozymes Have Various Catalytic Activities). In this case, the catalytic activity is an endonuclease that cleaves its own RNA. It is activated by binding of the metabolite product, GlcN6P, to the ribozyme. The consequence is that accumulation of GlcN6P activates the ribozyme, which cleaves the mRNA, which in turn prevents further translation. This is an exact parallel to allosteric control of a repressor protein by the end product of a metabolic pathway. There are several examples of such riboswitches in bacteria.

Another regulatory mechanism that involves transcription of a noncoding RNA works indirectly. Initiation at a target promoter can be suppressed by transcription from another promoter upstream from it, as shown in FIGURE 13.16. The cause of the inhibition is that the RNA polymerase initiating at the upstream promoter reads through the downstream promoter, which prevents transcription factors and RNA polymerase from binding to it. This type of effect has been demonstrated in eukaryotic systems and may depend on the disruption of chromosomal structure at the target promoter. The RNA that is transcribed from the upstream (regulatory) promoter has no coding function.

This may explain the presence of some of the noncoding RNAs in eukaryotic nuclei; they are not regulator RNAs as such, but are the indirect products of a regulatory system.

13.8 Bacteria Contain Regulator RNAs

Key concepts

- Bacterial regulator RNAs are called sRNAs.
- Several of the sRNAs are bound by the protein Hfq, which increases their effectiveness.
- The OxyS sRNA activates or represses expression of >10 loci at the posttranscriptional level.

In bacteria, regulator RNAs are short molecules that are collectively known as **sRNAs;** *E. coli* contains at least seventeen different sRNAs. Some of the sRNAs are general regulators that affect many target genes. They function by base pairing with target RNAs (typically mRNAs) to control either their stability or function.

An example of stability control is provided by the small antisense regulator RyhB, which regulates six mRNAs coding for proteins concerned with iron storage in *E. coli.* RyhB base pairs with each of the target mRNAs to form double-stranded regions that are substrates for RNAase E. An interesting feature of the circuit is that the ribonuclease destroys the regulator RNA as well as the mRNA.

Oxidative stress provides an interesting example of a general control system in which RNA is the regulator. When exposed to reactive oxygen species, bacteria respond by inducing antioxidant defense genes. Hydrogen peroxide activates the transcription activator OxyR, which controls the expression of several inducible genes. One of these genes is *oxyS,* which codes for a small RNA.

FIGURE 13.17 shows two salient features of the control of *oxyS* expression. In a wild-type bacterium under normal conditions, it is not expressed. The pair of gels on the left side of the figure show that it is expressed at high levels in a mutant bacterium with a constitutively active *oxyR* gene. This identifies *oxyS* as a target for activation by *oxyR.* The pair of gels on the right side of the figure show that OxyS RNA is transcribed within one minute of exposure to hydrogen peroxide.

The OxyS RNA is a short sequence (109 nucleotides) that does not code for protein. It

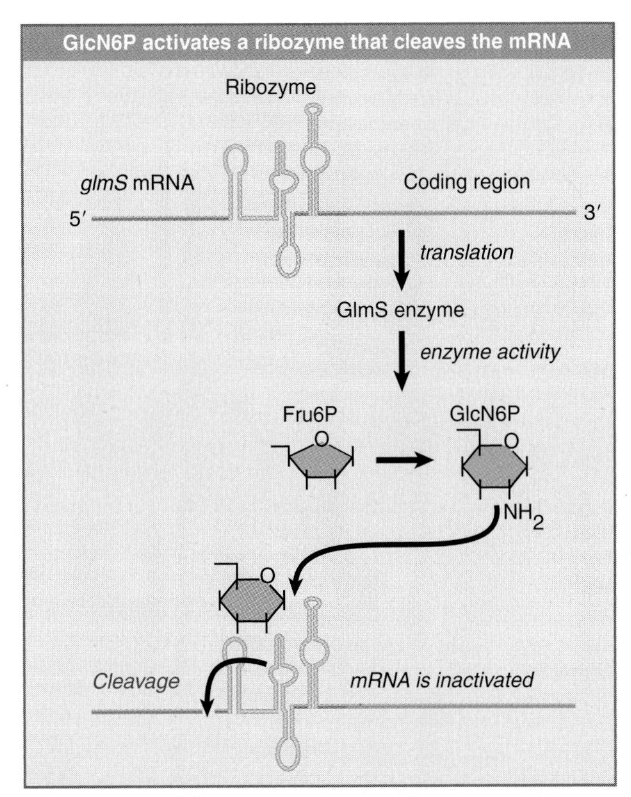

FIGURE 13.15 The 5′ untranslated region of the mRNA for the enzyme that synthesizes GlcN6P contains a ribozyme that is activated by the metabolic product. The ribozyme inactivates the mRNA by cleaving it.

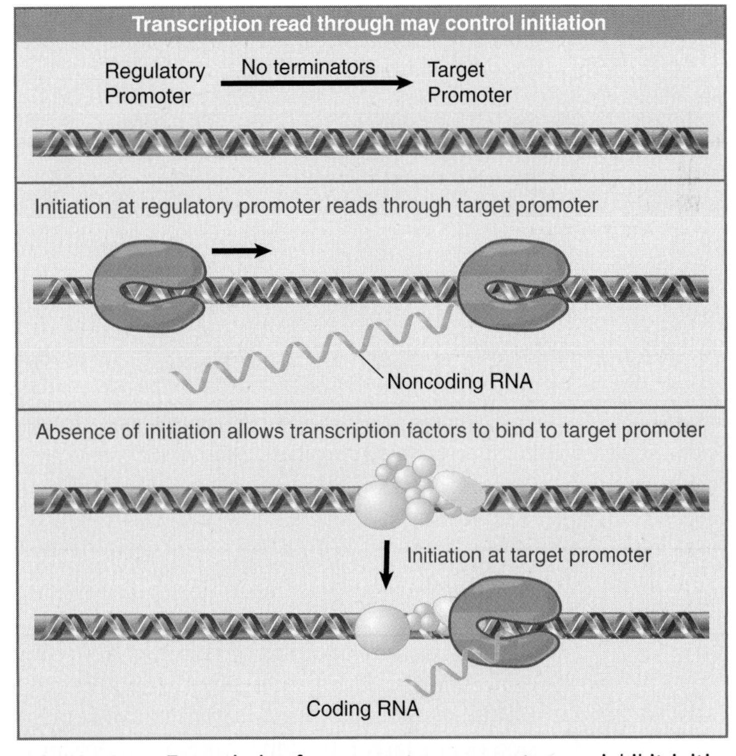

FIGURE 13.16 Transcription from an upstream promoter may inhibit initiation at a promoter downstream from it, because reading through the downstream promoter prevents the necessary transcription factors from binding to it. The RNA transcribed from the upstream promoter has no coding function.

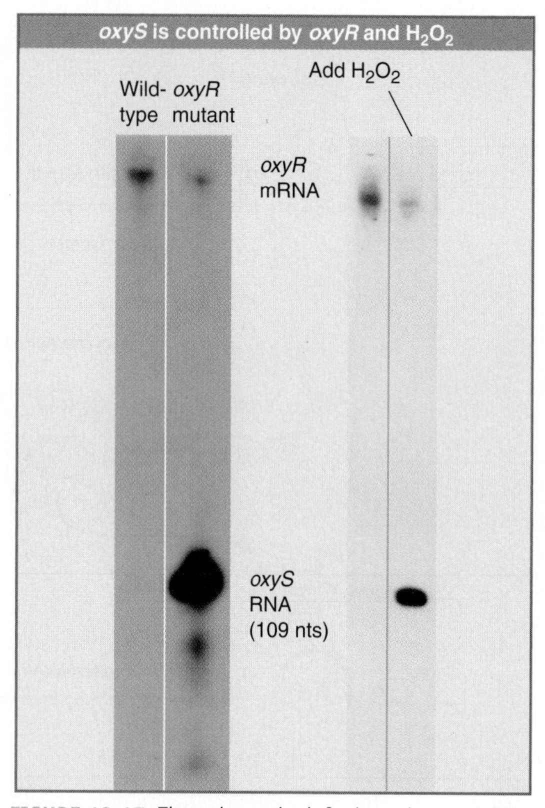

FIGURE 13.17 The gels on the left show that *oxyS* RNA is induced in an *oxyR* constitutive mutant. The gels on the right show that *oxyS* RNA is induced within one minute of adding hydrogen peroxide to a wild-type culture. Reproduced from *Cell*, vol. 90, Altuvia, S., et al., *A small stable RNA . . .* , pp. 44–53. Copyright 1997, with permission from Elsevier. Photo courtesy of Gisela Storz, National Institutes of Health.

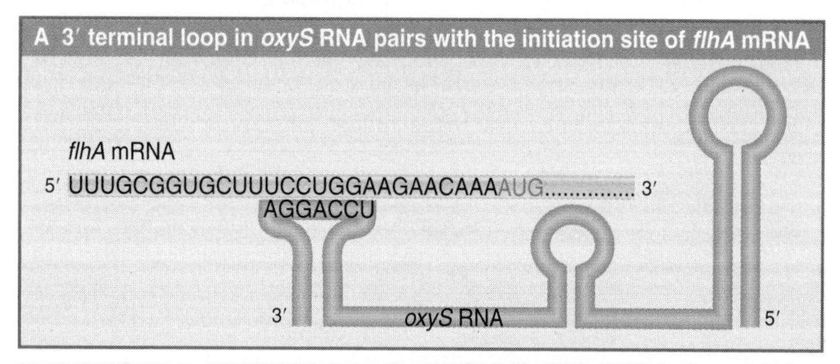

FIGURE 13.18 *oxyS* RNA inhibits translation of *flhA* mRNA by base pairing with a sequence just upstream of the AUG initiation codon.

is a *trans*-acting regulator that affects gene expression at posttranscriptional levels. It has >10 target loci; at some of them, it activates expression; at others it represses expression. FIGURE 13.18 shows the mechanism of repression of one target, the FlhA mRNA. Three stem-loop structures protrude in the secondary structure of OxyS mRNA, and the loop close to the 3′ terminus is complementary to a sequence just preceding the initiation codon of FlhA mRNA. Base pairing between OxyS RNA and FlhA RNA prevents the ribosome from binding to the initiation codon and therefore represses translation. There is also a second pairing interaction that involves a sequence within the coding region of FlhA.

Another target for *oxyS* is *rpoS*, the gene coding for an alternative sigma factor (which activates a general stress response). By inhibiting production of the sigma factor, *oxyS* ensures that the specific response to oxidative stress does not trigger the response that is appropriate for other stress conditions. The *rpoS* gene is also regulated by two other sRNAs (DsrA and RprA), which activate it. These three sRNAs appear to be global regulators that coordinate responses to various environmental conditions.

The actions of all three sRNAs are assisted by an RNA-binding protein called Hfq. The Hfq protein was originally identified as a bacterial host factor needed for replication of the RNA bacteriophage Qβ. It is related to the Sm proteins of eukaryotes that bind to many of the snRNAs (small nuclear RNAs) that have regulatory roles in gene expression (see Section 26.5, snRNAs Are Required for Splicing). Mutations in its gene have many effects, which identifies it as a pleiotropic protein. Hfq binds to many of the sRNAs of *E. coli*, and it increases the effectiveness of OxyS RNA by enhancing its ability to bind to its target mRNAs. The effect of Hfq is probably mediated by causing a small change in the secondary structure of OxyS RNA that improves the exposure of the single-stranded sequences that pair with the target mRNAs.

13.9 MicroRNAs Are Regulators in Many Eukaryotes

Key concepts

- Animal and plant genomes code for many short (~22 base) RNA molecules called microRNAs.
- MicroRNAs regulate gene expression by base pairing with complementary sequences in target mRNAs.

Very small RNAs are gene regulators in many eukaryotes. The first example was discovered in the nematode *Caenorhabditis elegans* as the result of the interaction between the regulator gene *lin4* and its target gene, *lin14*. FIGURE 13.19

illustrates the behavior of this regulatory system. The *lin14* target gene regulates larval development. Expression of *lin14* is controlled by *lin4*, which codes for a small transcript of twenty-two nucleotides. The *lin4* transcripts are complementary to a ten-base sequence that is repeated seven times in the 3′ nontranslated region of *lin14*. Expression of *lin4* represses expression of *lin14* posttranscriptionally, most likely because the base pairing reaction between the two RNAs leads to degradation of the mRNA. This system is especially interesting in implicating the 3′ end as a site for regulation.

The *lin4* RNA is an example of a **microRNA** (miRNA). There are ~80 genes in the *C. elegans* genome coding for miRNAs that are twenty-one to twenty-four nucleotides long. They have varying patterns of expression during development and are likely to be regulators of gene expression. Many of the miRNAs of *C. elegans* are contained in a large (15S) ribonucleoprotein particle.

Many of the *C. elegans* miRNAs have homologs in mammals, so the mechanism may be widespread. They are also found in plants. Of sixteen miRNAs in *Arabidopsis*, eight are completely conserved in rice, suggesting widespread conservation of this regulatory mechanism.

The virus SV40 codes for miRNAs that are complementary to the mRNAs produced during the early period of viral infection. The miRNAs are transcribed later in the viral cycle, base pair with the early mRNAs, and cause them to be degraded at this point in the life cycle, when they are no longer needed.

The mechanism of production of the miRNAs is also widely conserved. In the example of *lin4*, the gene is transcribed into a transcript that forms a double-stranded region that becomes a target for a nuclease called Dicer. This has an N-terminal helicase activity, which enables it to unwind the double-stranded region, and two nuclease domains that are related to the bacterial ribonuclease III. Related enzymes are found in flies, worms, and plants. Cleavage of the initial transcript generates the active miRNA. Interfering with the enzyme activity blocks the production of miRNAs and causes developmental defects.

Another step in the formation of miRNAs has been characterized in plants, in which the 3′ terminal nucleotide is methylated on its ribose by the methyltransferase enzyme HEN1. The methylation stabilizes the miRNA.

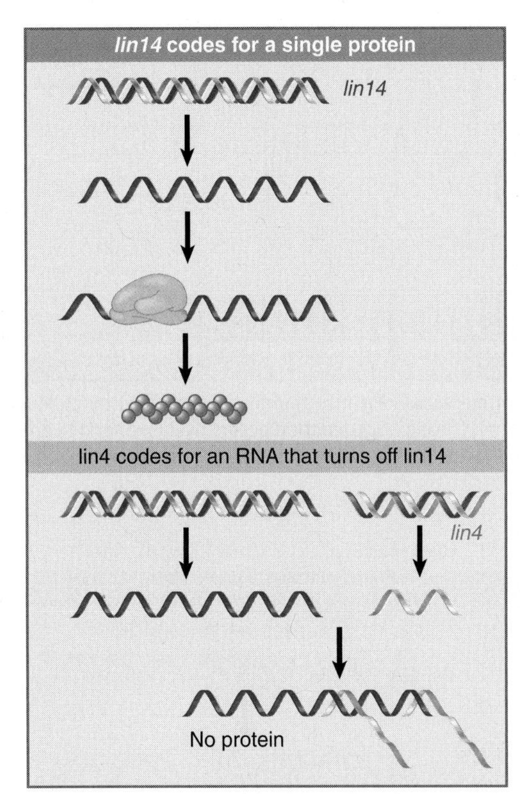

FIGURE 13.19 *lin4* RNA regulates expression of *lin14* by binding to the 3′ nontranslated region.

13.10 RNA Interference Is Related to Gene Silencing

Key concepts

- RNA interference triggers degradation of mRNAs complementary to either strand of a short dsRNA.
- dsRNA may cause silencing of host genes.

The regulation of mRNAs by miRNAs is mimicked by the phenomenon of **RNA interference** (RNAi). This was discovered when it was observed that antisense and sense RNAs can be equally effective in inhibiting gene expression. The reason is that preparations of either type of (supposedly) single-stranded RNA are actually contaminated by small amounts of double-stranded RNA (dsRNA).

Work with an *in vitro* system shows that the dsRNA is degraded by ATP-dependent cleavage to give oligonucleotides of twenty-one to twenty-three bases. The short RNA is sometimes called siRNA (short interfering RNA). FIGURE 13.20 shows that the mechanism of cleavage involves making breaks relative to each 3′

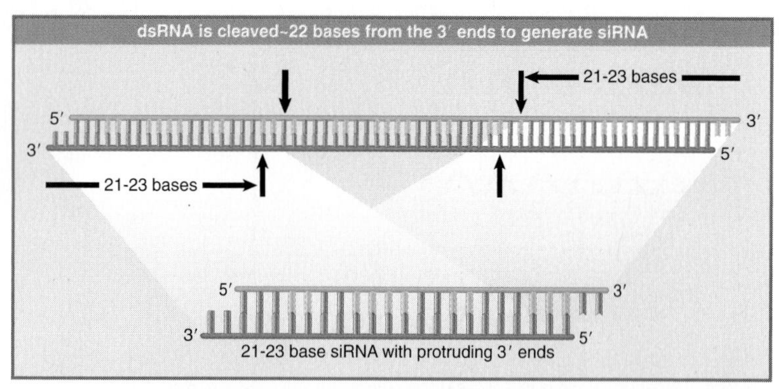

FIGURE 13.20 siRNA that mediates RNA interference is generated by cleaving dsRNA into smaller fragments. The cleavage reaction occurs twenty-one to twenty-three nucleotides from a 3′ end. The siRNA product has protruding bases on its 3′ ends.

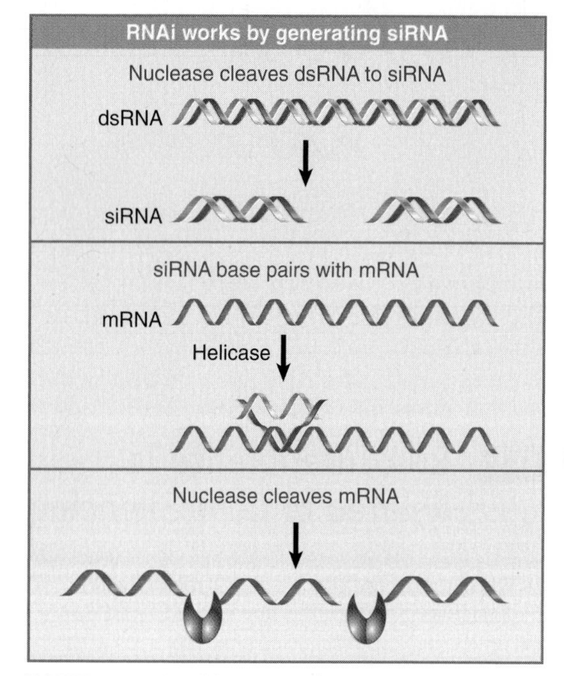

FIGURE 13.21 RNAi is generated when a dsRNA is cleaved into fragments that direct cleavage of the corresponding mRNA.

end of a long dsRNA to generate siRNA fragments with short (two base) protruding 3′ ends. The same enzyme (Dicer) that generates miRNAs is responsible for the cleavage.

RNAi occurs posttranscriptionally when an siRNA induces degradation of a complementary mRNA. FIGURE 13.21 suggests that the siRNA may provide a template that directs a nuclease to degrade mRNAs that are complementary to one or both strands, perhaps by a process in which the mRNA pairs with the fragments. It is likely that a helicase is required to assist the pairing reaction. The siRNA directs cleavage of the mRNA in the middle of the paired segment. These reactions occur within a ribonucleoprotein complex called RISC (RNA-induced silencing complex). Proteins in the Argonaute family are components of this complex and are required for the cleavage reaction. Methylation of the 3′ ribose may be required for the miRNA to be incorporated into the complex.

There is still uncertainty as to how the RISC complex silences gene expression (see Section 31.1, Heterochromatin Depends on Interactions with Histones). The activity of RNA polymerase is required for RNA interference, but it is not clear whether it is direct (because it makes transcripts that are necessary) or indirect (either because it is involved in binding RNAi to the transcripts or because it separates the strands of DNA during transcription, allowing an interaction with RNAi to occur).

RNAi has become a powerful technique for ablating the expression of a specific target gene in invertebrate cells, especially in *C. elegans* and *Drosophila melanogaster.* The technique, however, has been limited in mammalian cells, which have a more generalized response to dsRNA of shutting down protein synthesis and degrading mRNA. FIGURE 13.22 shows that this happens because of two reactions. The dsRNA activates the enzyme PKR, which inactivates the translation initiation factor eIF2a by phosphorylating it. It also activates 2′5′ oligoadenylate synthetase, whose product activates RNAase L, which degrades all mRNAs. It turns out, however, that these reactions require dsRNA that is longer than twenty-six nucleotides. If shorter dsRNA (twenty-one to twenty-three nucleotides) is introduced into mammalian cells, it triggers the specific degradation of complementary RNAs just as with the RNAi technique in worms and flies. With this advance, it seems likely that RNAi will become the universal mechanism of choice for turning off the expression of a specific gene.

As an example of the progress being made with the technique, it has been possible to use RNAi for a systematic analysis of gene expression in *C. elegans.* Loss of function phenotypes can be generated by feeding worms with bacteria expressing a dsRNA that is homologous to a target gene. By making a library of bacteria in which each bacterium expresses a dsRNA corresponding to a different gene, worms have been screened for the effects of knocking out most (86%) of the genes.

RNA interference is related to natural processes in which gene expression is silenced. Plants and fungi show **RNA silencing** (sometimes called posttranscriptional gene silencing), in which dsRNA inhibits expression of a gene. The most common source of the RNA is a replicating virus. This mechanism may have evolved as a defense against viral infection. When a virus infects a plant cell, the formation of dsRNA triggers the suppression of expression from the plant genome. RNA silencing has the further remarkable feature that it is not limited to the cell in which the viral infection occurs: It can spread throughout the plant systemically. Presumably the propagation of the signal involves passage of RNA or fragments of RNA. It may require some of the same features that are involved in movement of the virus itself. It is possible that RNA silencing involves an amplification of the signal by an RNA-dependent RNA synthesis process in which a novel polymerase uses the siRNA as a primer to synthesize more RNA on a template of complementary RNA.

A related process is the phenomenon of **cosuppression,** in which introduction of a transgene causes the corresponding endogenous gene to be silenced. This has been largely characterized in plants. The implication is that the transgene must make both antisense and sense RNA copies, which inhibits expression of the endogenous gene.

Silencing takes place by RNA–RNA interactions. It is also possible that dsRNA may inhibit gene expression by interacting with the DNA. If a DNA copy of a viroid RNA sequence is inserted into a plant genome, it becomes methylated when the viroid RNA replicates. This suggests that the RNA sequence could be inducing methylation of the DNA sequence. Similar targeting of methylation of DNA corresponding to sequences represented in dsRNA has been detected in plant cells. Methylation of DNA is associated with repression of transcription, so this could be another means of silencing genes represented in dsRNA (see Section 24.18, Gene Expression Is Associated with Demethylation). Nothing is known about the mechanism.

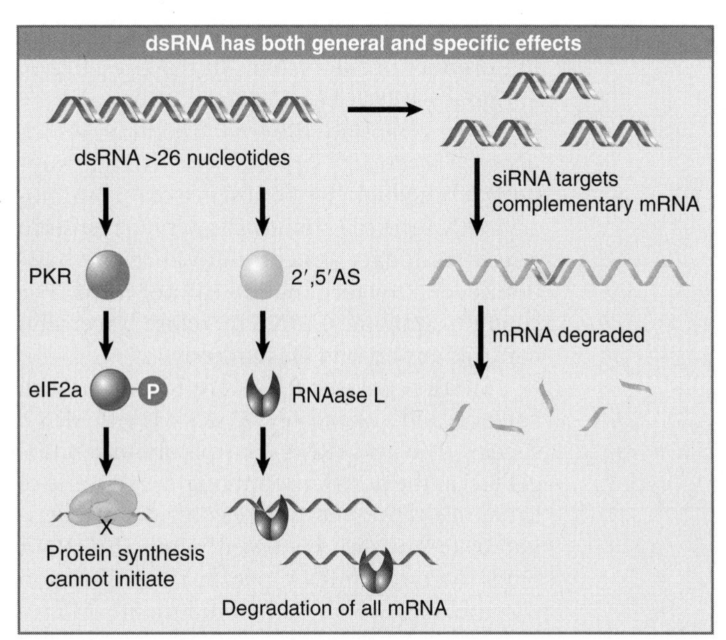

FIGURE 13.22 dsRNA inhibits protein synthesis and triggers degradation of all mRNA in mammalian cells, as well as having sequence-specific effects.

13.11 Summary

Gene expression can be regulated positively by factors that activate a gene or negatively by factors that repress a gene. The first and most common level of control is at the initiation of transcription, but termination of transcription may also be controlled. Translation may be controlled by regulators that interact with mRNA. The regulatory products may be proteins, which often are controlled by allosteric interactions in response to the environment, or RNAs, which function by base pairing with the target RNA to change its secondary structure. Regulatory networks can be created by linking regulators so that the production or activity of one regulator is controlled by another.

Attenuation is a mechanism that relies on regulation of termination to control transcription through bacterial operons. It is commonly used in operons that code for enzymes involved in biosynthesis of an amino acid. The polycistronic mRNA of the operon starts with a sequence that can form alternative secondary structures. One of the structures has a hairpin loop that provides an intrinsic terminator upstream of the structural genes; the alternative structure lacks the hairpin. Various types of interaction determine whether the hairpin forms. One interaction is when a protein binds to the mRNA to prevent formation of the alternative structure. In the *trp* operon of *B. subtilis,* the TRAP protein has this function; it is controlled by the antiTRAP protein, whose production in turn is controlled by the level of uncharged aminoacyl-$tRNA^{Trp}$. In the *trp* (and other) operons of *E. coli,* the choice of which structure forms is controlled by the progress of translation through a short leader sequence that includes codons for the amino acid(s) that are the product of the system. In the presence of aminoacyl-tRNA bearing such amino acid(s), ribosomes translate the leader peptide, which allows a secondary structure to form that supports termination. In the absence of this aminoacyl-tRNA the ribosome stalls, which results in a new secondary structure in which the hairpin needed for termination cannot form. The supply of aminoacyl-tRNA therefore (inversely) controls amino acid biosynthesis.

Small regulator RNAs are found in both bacteria and eukaryotes. *E. coli* has ~17 sRNA species. The *oxyS* sRNA controls about ten target loci at the posttranscriptional level; some of them are repressed whereas others are activated. Repression is caused when the sRNA binds to a target mRNA to form a duplex region that includes the ribosome-binding site. MicroRNAs are ~22 bases long and are produced in many eukaryotes by cleavage of a longer transcript. They function by base pairing with target mRNAs to form duplex regions that are susceptible to cleavage by endonucleases. The degradation of the mRNA prevents its expression. The technique of RNA interference is becoming the method of choice for inactivating eukaryotic genes. It uses the introduction of short dsRNA sequences with one strand complementary to the target RNA, and it works by inducing degradation of the targets. This may be related to a natural defense system in plants called RNA silencing.

References

13.3 Termination of *Bacillus subtilis trp* Genes Is Controlled by Tryptophan and by tRNA^{Trp}

Review

Gollnick, P. (1994). Regulation of the *B. subtilis trp* operon by an RNA-binding protein. *Mol. Microbiol.* 11, 991–997.

Research

Antson, A. A. et al. (1999). Structure of the *trp* RNA-binding attenuation protein, TRAP, bound to RNA. *Nature* 401, 235–242.
Babitzke, P. and Yanoksy, C. (1993). Reconstitution of *B. subtilis trp* attenuation *in vitro* with TRAP, the *trp* RNA-binding attenuation protein. *Proc. Natl. Acad. Sci. USA* 90, 133–137.
Otridge, J. and Gollnick, P. (1993). MtrB from *B. subtilis* binds specifically to *trp* leader RNA in a tryptophan-dependent manner. *Proc. Natl. Acad. Sci. USA* 90, 128–132.
Valbuzzi, A. and Yanofsky, C. (2001). Inhibition of the *B. subtilis* regulatory protein TRAP by the TRAP-inhibitory protein, AT. *Science* 293, 2057–2059.

13.4 The *Escherichia coli tryptophan* Operon Is Controlled by Attenuation

Review

Yanofsky, C. (1981). Attenuation in the control of expression of bacterial operons. *Nature* 289, 751–758.

13.5 Attenuation Can Be Controlled by Translation

Reviews

Bauer, C. E., Carey, J., Kasper, L. M., Lynn, S. P., Waechter, D. A., and Gardner, J. F. (1983). Attenuation in bacterial operons. In Beckwith, J., Davies, J., and Gallant, J. A., eds. *Gene Function in Prokaryotes.* Cold Springs Harbor, NY: Cold Spring Harbor Press. pp. 65–89.
Landick, R. and Yanofsky, C. (1987). In Neidhardt, F. C., ed., E. coli *and* S. typhimurium *Cellular and Molecular Biology.* Washington, D.C.: American Society for Microbiology. pp. 1276–1301.
Yanofsky, C. and Crawford, I. P. (1987). In Ingraham, J. L., et al., eds., Escherichia coli *and* Salmonella typhimurium, Washington, D.C.: American Society for Microbiology. pp. 1453–1472.

Research

Lee, F. and Yanofsky, C. (1977). Transcription termination at the *trp* operon attenuators of *E. coli* and *S. typhimurium:* RNA secondary structure and regulation of termination. *Proc. Natl. Acad. Sci. USA* 74, 4365–4368.
Zurawski, G. et al. (1978). Translational control of transcription termination at the attenuator of the *E. coli tryptophan* operon. *Proc. Natl. Acad. Sci. USA* 75, 5988–5991.

13.6 Antisense RNA Can Be Used to Inactivate Gene Expression

Research

Izant, J. G. and Weintraub, H. (1984). Inhibition of thymidine kinase gene expression by antisense RNA: a molecular approach to genetic analysis. *Cell* 36, 1007–1015.

13.7 Small RNA Molecules Can Regulate Translation

Research

Johnson, A., and O'Donnell, M. (2005). Cellular DNA replicases: components and dynamics at the replication fork. *Annu. Rev. Biochem.* 74, 283–315.

Martens, J. A., Laprade, L., and Winston, F. (2004). Intergenic transcription is required to repress the *Saccharomyces cerevisiae* SER3 gene. *Nature* 429, 571–574.

Winkler, W. C., Nahvi, A., Roth, A., Collins, J. A., and Breaker, R. R. (2004). Control of gene expression by a natural metabolite-responsive ribozyme. *Nature* 428, 281–286.

13.8 Bacteria Contain Regulator RNAs

Review

Gottesman, S. (2002). Stealth regulation: biological circuits with small RNA switches. *Genes Dev.* 16, 2829–2842.

Research

Altuvia, S., Weinstein-Fischer, D., Zhang, A., Postow, L., and Storz, G. (1997). A small, stable RNA induced by oxidative stress: role as a pleiotropic regulator and antimutator. *Cell* 90, 43–53.

Altuvia, S., Zhang, A., Argaman, L., Tiwari, A., and Storz, G. (1998). The *E. coli* OxyS regulatory RNA represses *fhlA* translation by blocking ribosome binding. *EMBO J.* 17, 6069–6075.

Massé, E., Escorcia, F. E., and Gottesman, S. (2003). Coupled degradation of a small regulatory RNA and its mRNA targets in *Escherichia coli*. *Genes Dev.* 17, 2374–2383.

Moller, T., Franch, T., Hojrup, P., Keene, D. R., Bachinger, H. P., Brennan, R. G., and Valentin-Hansen, P. (2002). Hfq: a bacterial Sm-like protein that mediates RNA-RNA interaction. *Mol. Cell* 9, 23–30.

Wassarman, K. M., Repoila, F., Rosenow, C., Storz, G., and Gottesman, S. (2001). Identification of novel small RNAs using comparative genomics and microarrays. *Genes Dev.* 15, 1637–1651.

Zhang, A., Wassarman, K. M., Ortega, J., Steven, A. C., and Storz, G. (2002). The Sm-like Hfq protein increases OxyS RNA interaction with target mRNAs. *Mol. Cell* 9, 11–22.

13.9 MicroRNAs Are Regulators in Many Eukaryotes

Research

Bernstein, E., Caudy, A. A., Hammond, S. M., and Hannon, G. J. (2001). Role for a bidentate ribonuclease in the initiation step of RNA interference. *Nature* 409, 363–366.

Ketting, R. F., Fischer, S. E., Bernstein, E., Sijen, T., Hannon, G. J., and Plasterk, R. H. (2001). Dicer functions in RNA interference and in synthesis of small RNA involved in developmental timing in *C. elegans*. *Genes Dev.* 15, 2654–2659.

Lau, N. C., Lim, l.e. E. P., Weinstein, E. G., and Bartel, d.a. V. P. (2001). An abundant class of tiny RNAs with probable regulatory roles in *C. elegans*. *Science* 294, 858–862.

Lee, R. C. and Ambros, V. (2001). An extensive class of small RNAs in *C. elegans*. *Science* 294, 862–864.

Lee, R. C., Feinbaum, R. L., and Ambros, V. (1993). The *C. elegans* heterochronic gene *lin-4* encodes small RNAs with antisense complementarity to *lin-14*. *Cell* 75, 843–854.

Mourelatos, Z., Dostie, J., Paushkin, S., Sharma, A., Charroux, B., Abel, L., Rappsilber, J., Mann, M., and Dreyfuss, G. (2002). miRNPs: a novel class of ribonucleoproteins containing numerous microRNAs. *Genes Dev.* 16, 720–728.

Reinhart, B. J., Weinstein, E. G., Rhoades, M. W., Bartel, B., and Bartel, D. P. (2002). MicroRNAs in plants. *Genes Dev.* 16, 1616–1626.

Sullivan, C. S., Grundhoff, A. T., Tevethia, S., Pipas, J. M., and Ganem, D. (2005). SV40-encoded microRNAs regulate viral gene expression and reduce susceptibility to cytotoxic T cells. *Nature* 435, 682–686.

Wightman, B., Ha, I., and Ruvkun, G. (1993). Posttranscriptional regulation of the heterochronic gene lin-14 by lin-4 mediates temporal pattern formation in *C. elegans*. *Cell* 75, 855–862.

Yu, B., Yang, Z., Li, J., Minakhina, S., Yang, M., Padgett, R. W., Steward, R., and Chen, X. (2005). Methylation as a crucial step in plant microRNA biogenesis. *Science* 307, 932–935.

Zamore, P. D., and Haley, B. (2005). Ribo-gnome: the big world of small RNAs. *Science* 309, 1519–1524.

13.10 RNA Interference Is Related to Gene Silencing

Reviews

Ahlquist, P. (2002). RNA-dependent RNA polymerases, viruses, and RNA silencing. *Science* 296, 1270–1273.

Matzke, M., Matzke, A. J., and Kooter, J. M. (2001). RNA: guiding gene silencing. *Science* 293, 1080–1083.

Schwartz, D. S. and Zamore, P. D. (2002). Why do miRNAs live in the miRNP? *Genes Dev.* 16, 1025–1031.

Sharp, P. A. (2001). RNA interference—2001. *Genes Dev.* 15, 485–490.

Tijsterman, M., Ketting, R. F., and Plasterk, R. H. (2002). The genetics of RNA silencing. *Annu. Rev. Genet.* 36, 489–519.

Research

Elbashir, S. M., Harborth, J., Lendeckel, W., Yalcin, A., Weber, K., and Tuschl, T. (2001). Duplexes of 21-nucleotide RNAs mediate RNA interference in cultured mammalian cells. *Nature* 411, 494–498.

Fire, A., Xu, S., Montgomery, M. K., Kostas, S. A., Driver, and Mello, C. C. (1998). Potent and specific genetic interference by double-

stranded RNA in Caenorhabditis elegans. *Nature* 391, 806–811.
Hamilton, A. J. and Baulcombe, D. C. (1999). A species of small antisense RNA in posttranscriptional gene silencing in plants. *Science* 286, 950–952.
Kamath, R. S., Fraser, A. G., Dong, Y., Poulin, G., Durbin, R., Gotta, M., Kanapin, A., Le Bot, N., Moreno, S., Sohrmann, M., Welchman, D. P., Zipperlen, P., and Ahringer, J. (2003). Systematic functional analysis of the *C. elegans* genome using RNAi. *Nature* 421, 231–237.
Meister, G., Landthaler, M., Patkaniowska, A., Dorsett, Y., Teng, G., and Tuschl, T. (2004). Human argonaute2 mediates RNA cleavage targeted by miRNAs and siRNAs. *Mol. Cell* 15, 185–197.
Mette, M. F., Aufsatz, W., van der Winden, J., Matzke, M. A., and Matzke, A. J. (2000). Transcriptional silencing and promoter methylation triggered by double-stranded RNA. *EMBO J.* 19, 5194–5201.
Montgomery, M. K., Xu, S., and Fire, A. (1998). RNA as a target of double-stranded RNA-mediated genetic interference in *C. elegans. Proc. Natl. Acad. Sci. USA* 95, 15502–15507.
Ngo, H., Tschudi, C., Gull, K., and Ullu, E. (1998). Double-stranded RNA induces mRNA degradation in *Trypanosoma brucei. Proc. Natl. Acad. Sci. USA* 95, 14687–14692.
Schramke, V., Sheedy, D. M., Denli, A. M., Bonila, C., Ekwall, K., Hannon, G. J., and Allshire, R. C. (2005). RNA-interference-directed chromatin modification coupled to RNA polymerase II transcription. *Nature* 435, 1275–1279.
Voinnet, O., Pinto, Y. M., and Baulcombe, D. C. (1999). Suppression of gene silencing: a general strategy used by diverse DNA and RNA viruses of plants. *Proc. Natl. Acad. Sci. USA* 96, 14147–14152.
Wassenegger, M., Heimes, S., Riedel, L., and Sanger, H. L. (1994). RNA-directed de novo methylation of genomic sequences in plants. *Cell* 76, 567–576.
Waterhouse, P. M., Graham, M. W., and Wang, M. B. (1998). Virus resistance and gene silencing in plants can be induced by simultaneous expression of sense and antisense RNA. *Proc. Natl. Acad. Sci. USA* 95, 13959–13964.
Yu, B., Yang, Z., Li, J., Minakhina, S., Yang, M., Padgett, R. W., Steward, R., and Chen, X. (2005). Methylation as a crucial step in plant microRNA biogenesis. *Science* 307, 932–935.
Zamore, P. D., and Haley, B. (2005). Ribo-gnome: the big world of small RNAs. *Science* 309, 1519–1524.
Zamore, P. D., Tuschl, T., Sharp, P. A., and Bartel, D. P. (2000). RNAi: double-stranded RNA directs the ATP-dependent cleavage of mRNA at 21 to 23 nucleotide intervals. *Cell* 101, 25–33.

14

Phage Strategies

CHAPTER OUTLINE

Continued on next page

14.1 Introduction

Some phages have only a single strategy for survival. On infecting a susceptible host, they subvert its functions to the purpose of producing a large number of progeny phage particles. As the result of this **lytic infection,** the host bacterium dies. In the typical lytic cycle, the phage DNA (or RNA) enters the host bacterium, its genes are transcribed in a set order, the phage genetic material is replicated, and the protein components of the phage particle are produced. Finally, the host bacterium is broken open *(lysed)* to release the assembled progeny particles by the process of **lysis.**

Other phages have a dual existence. They are able to perpetuate themselves via the same sort of lytic cycle in what amounts to an open strategy for producing as many copies of the phage as rapidly as possible. They also have an alternative form of existence, though, in which the phage genome is present in the bacterium in a latent form known as **prophage.** This form of propagation is called **lysogeny.**

In a lysogenic bacterium, the prophage is inserted into the bacterial genome and is inherited in the same way as bacterial genes. The process by which it is converted from an independent phage genome into a prophage that is a linear part of the bacterial genome is described as **integration.** By virtue of its possession of a prophage, a lysogenic bacterium has **immunity** against infection by further phage particles of the same type. Immunity is established by a single integrated prophage, so in general a bacterial genome contains only one copy of a prophage of any particular type.

Transitions occur between the lysogenic and lytic modes of existence. FIGURE 14.1 shows that when a phage produced by a lytic cycle enters a new bacterial host cell, it either repeats the lytic cycle or enters the lysogenic state. The outcome depends on the conditions of infection and the genotypes of phage and bacterium.

A prophage is freed from the restrictions of lysogeny by a process called **induction.** First the phage DNA is released from the bacterial chromosome by **excision;** then the free DNA proceeds through the lytic pathway.

The alternative forms in which these phages are propagated are determined by the regulation of transcription. Lysogeny is maintained by the interaction of a phage repressor with an operator. The lytic cycle requires a cascade of transcriptional controls. The transition between the two lifestyles is accomplished by the establishment of repression (lytic cycle to lysogeny) or by the relief of repression (induction of lysogen to lytic phage).

Another type of existence within bacteria is represented by **plasmids.** These are autonomous units that exist in the cell as **extrachromosomal genomes.** Plasmids are self-replicating circular molecules of DNA that are maintained in the cell in a stable and characteristic number of copies, that is, the number remains constant from generation to generation.

Some plasmids also have alternative lifestyles. They can exist in the autonomous extrachromosomal state or can be inserted into the bacterial chromosome. If inserted into the bacterial chromosome, they are carried as part of it in the same manner as any other sequence would be. Such units are formally called **episomes,** but the terms "plasmid" and "episome" are sometimes used loosely as though interchangeable.

As for lysogenic phages, plasmids and episomes maintain a selfish possession of their bacterium and often make it impossible for another element of the same type to become established. This effect also is called **immunity,** although the basis for plasmid immunity is different from lysogenic immunity. (We discuss the control of plasmid perpetuation in Chapter 15, The Replicon.)

FIGURE 14.2 summarizes the types of genetic units that can be propagated in bacteria as

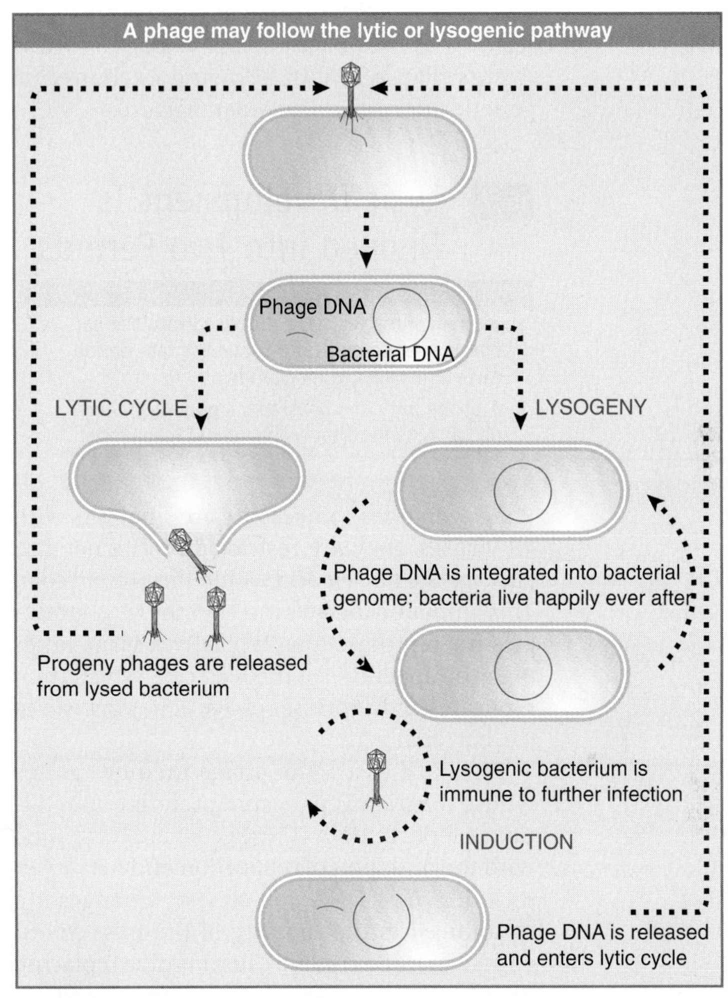

FIGURE 14.1 Lytic development involves the reproduction of phage particles with destruction of the host bacterium, but lysogenic existence allows the phage genome to be carried as part of the bacterial genetic information.

Phages and plasmids live in bacteria

Type of Unit	Genome Structure	Mode of Propagation	Consequences
Lytic phage	ds- or ss-DNA or RNA linear or circular	Infects susceptible host	Usually kills host
Lysogenic phage	ds-DNA	Linear sequence in host chromosome	Immunity to infection
Plasmid	ds-DNA circle	Replicates at defined copy number May be transmissible	Immunity to plasmids in same group
Episome	ds-DNA circle	Free circle or linear integrated	May transfer host DNA

FIGURE 14.2 Several types of independent genetic units exist in bacteria.

independent genomes. Lytic phages may have genomes of any type of nucleic acid; they transfer between cells by release of infective particles. Lysogenic phages have double-stranded DNA genomes, as do plasmids and episomes. Some plasmids and episomes transfer between cells by a conjugative process (involving direct contact between donor and recipient cells). A feature of the transfer process in both cases is that on occasion some bacterial host genes are transferred with the phage or plasmid DNA, so these events play a role in allowing exchange of genetic information between bacteria.

14.2 Lytic Development Is Divided into Two Periods

Key concepts

- A phage infective cycle is divided into the early period (before replication) and the late period (after the onset of replication).
- A phage infection generates a pool of progeny phage genomes that replicate and recombine.

Phage genomes by necessity are small. As with all viruses, they are restricted by the need to package the nucleic acid within the protein coat. This limitation dictates many of the viral strategies for reproduction. Typically a virus takes over the apparatus of the host cell, which then replicates and expresses phage genes instead of the bacterial genes.

In most cases, the phage includes genes whose function is to ensure preferential replication of phage DNA. These genes are concerned with the initiation of replication and may even include a new DNA polymerase. Changes are introduced in the capacity of the host cell to engage in transcription. They involve replacing the RNA polymerase or modifying its capacity for initiation or termination. The result is always the same: phage mRNAs are preferentially transcribed. As far as protein synthesis is concerned, the phage is, for the most part, content to use the host apparatus, redirecting its activities principally by replacing bacterial mRNA with phage mRNA.

Lytic development is accomplished by a pathway in which the phage genes are expressed in a particular order. This ensures that the right amount of each component is present at the appropriate time. The cycle can be divided into the two general parts illustrated in FIGURE 14.3:

- **Early infection** describes the period from entry of the DNA to the start of its replication.

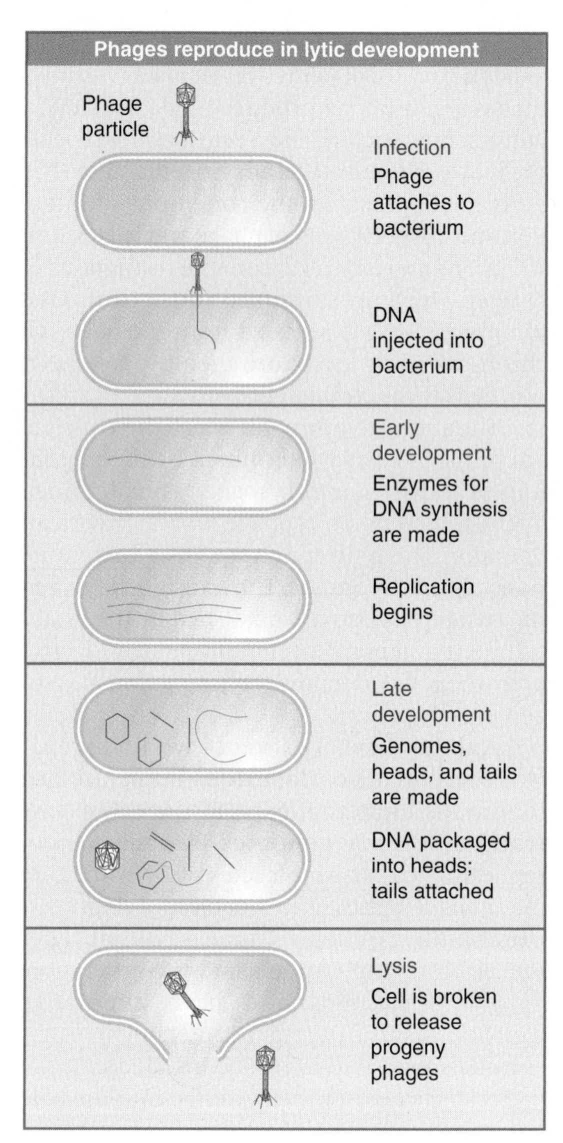

FIGURE 14.3 Lytic development takes place by producing phage genomes and protein particles that are assembled into progeny phages.

- **Late infection** defines the period from the start of replication to the final step of lysing the bacterial cell to release progeny phage particles.

The early phase is devoted to the production of enzymes involved in the reproduction of DNA. These include the enzymes concerned with DNA synthesis, recombination, and sometimes modification. Their activities cause a *pool* of phage genomes to accumulate. In this pool, genomes are continually replicating and recombining, so that *the events of a single lytic cycle concern a population of phage genomes.*

During the late phase, the protein components of the phage particle are synthesized. Often many different proteins are needed to make up head and tail structures, so the largest part of the phage genome consists of late functions. In addition to the structural proteins, "assembly proteins" are needed to help construct the particle, although they are not themselves incorporated into it. By the time the structural components are assembling into heads and tails, replication of DNA has reached its maximum rate. The genomes then are inserted into the empty protein heads, tails are added, and the host cell is lysed to allow release of new viral particles.

14.3 Lytic Development Is Controlled by a Cascade

Key concepts

- The early genes transcribed by host RNA polymerase following infection include, or comprise, regulators required for expression of the middle set of phage genes.
- The middle group of genes includes regulators to transcribe the late genes.
- This results in the ordered expression of groups of genes during phage infection.

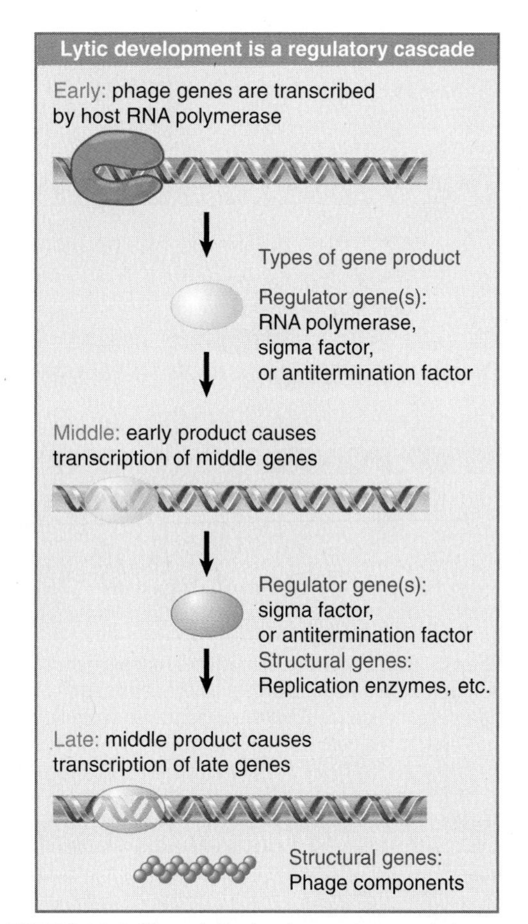

FIGURE 14.4 Phage lytic development proceeds by a regulatory cascade, in which a gene product at each stage is needed for expression of the genes at the next stage.

The organization of the phage genetic map often reflects the sequence of lytic development. The concept of the operon is taken to somewhat of an extreme, in which the genes coding for proteins with related functions are clustered to allow their control with the maximum economy. This allows the pathway of lytic development to be controlled with a small number of regulatory switches.

The lytic cycle is under positive control, so that each group of phage genes can be expressed only when an appropriate signal is given. FIGURE 14.4 shows that the regulatory genes function in a **cascade,** in which a gene expressed at one stage is necessary for synthesis of the genes that are expressed at the next stage.

The first stage of gene expression necessarily relies on the transcription apparatus of the host cell. In general, only a few genes are expressed at this stage. Their promoters are indistinguishable from those of host genes. The name of this class of genes depends on the phage. In most cases, they are known as the **early genes.** In phage lambda, they are given the evocative description of **immediate early.** Irrespective of the name, they constitute only a preliminary set of genes, representing just the initial part of the early period. They are at times exclusively occupied with the transition to the next period. At all events, *one of these genes always codes for a protein that is necessary for transcription of the next class of genes.*

This second class of genes is known variously as the **delayed early** or **middle gene** group. Its expression typically starts as soon as the regulator protein coded by the early gene(s) is available. Depending on the nature of the control circuit, the initial set of early genes may or may not continue to be expressed at this stage. If control is at initiation, the two events are independent (see FIGURE 14.5) and early genes can be switched off when middle genes are transcribed. If control is at termination, the early genes must continue to be expressed (see FIGURE 14.6). Often, the expression of host genes is reduced. Together the two sets of early genes account for all necessary phage functions except those needed to assemble the particle coat itself and to lyse the cell.

When the replication of phage DNA begins, it is time for the **late genes** to be expressed. Their transcription at this stage usually is

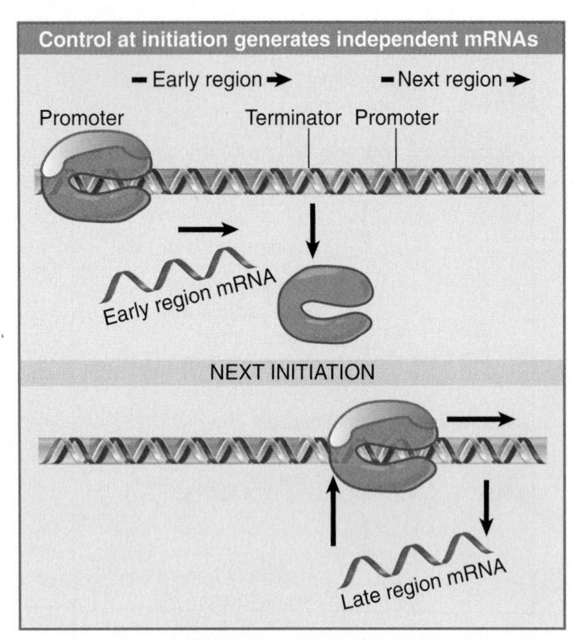

FIGURE 14.5 Control at initiation utilizes independent transcription units, each with its own promoter and terminator, which produce independent mRNAs. The transcription units need not be located near one another.

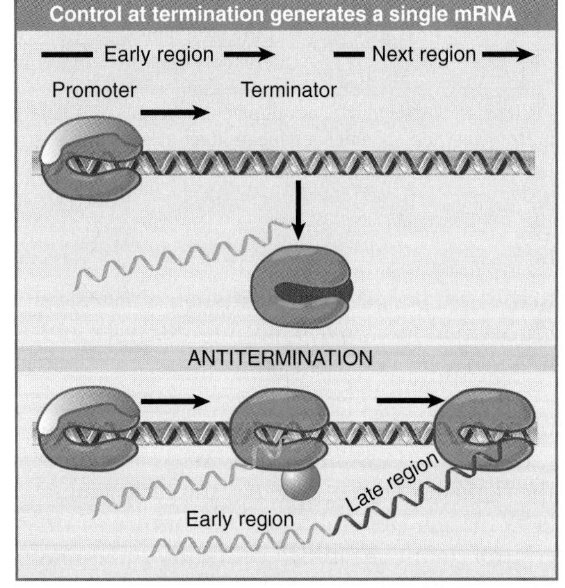

FIGURE 14.6 Control at termination requires adjacent units, so that transcription can read from the first gene into the next gene. This produces a single mRNA that contains both sets of genes.

arranged by embedding a further regulator gene within the previous (delayed early or middle) set of genes. This regulator may be another antitermination factor (as in lambda) or it may be another sigma factor (as in SPO1).

A lytic infection often falls into three stages, as shown in Figure 14.4. The first stage consists of early genes transcribed by host RNA polymerase (sometimes the regulators are the only products at this stage). The second stage consists of genes transcribed under direction of the regulator produced in the first stage (most of these genes code for enzymes needed for replication of phage DNA). The final stage consists of genes for phage components, which are transcribed under direction of a regulator synthesized in the second stage.

The use of these successive controls, in which each set of genes contains a regulator that is necessary for expression of the next set, creates a cascade in which groups of genes are turned on (and sometimes off) at particular times. The means used to construct each phage cascade are different but the results are similar, as the following sections show.

14.4 Two Types of Regulatory Event Control the Lytic Cascade

Key concept

- Regulator proteins used in phage cascades may sponsor initiation at new (phage) promoters or cause the host polymerase to read through transcription terminators.

At every stage of phage expression, one or more of the active genes is a regulator that is needed for the subsequent stage. The regulator may take the form of a new RNA polymerase, a sigma factor that redirects the specificity of the host RNA polymerase (see Section 11.18, Sigma Factors May Be Organized into Cascades), or an antitermination factor that allows it to read a new group of genes (see Section 11.23, Antitermination Is a Regulatory Event). The next two figures compare the use of switching at initiation or termination to control gene expression.

One mechanism for recognizing new phage promoters is to replace the sigma factor of the host enzyme with another factor that redirects its specificity in initiation (see Figure 11.31). An alternative mechanism is to synthesize a new phage RNA polymerase. In either case, the critical feature that distinguishes the new set of genes is their possession of *different promoters from those originally recognized by host RNA polymerase.* Figure 14.5 shows that the two sets of transcripts are independent; as a consequence, early gene expression can cease after the new sigma factor or polymerase has been produced.

Antitermination provides an alternative mechanism for phages to control the switch from early genes to the next stage of expression. The use of antitermination depends on a particular arrangement of genes. Figure 14.6 shows that the early genes lie adjacent to the genes that are to be expressed next, but are separated from them by terminator sites. *If termination is prevented at these sites, the polymerase reads through into the genes on the other side.* Thus in antitermination, the *same promoters* continue to be recognized by RNA polymerase. As a result, the new genes are expressed only by extending the RNA chain to form molecules that contain the early gene sequences at the 5′ end and the new gene sequences at the 3′ end. The two types of sequence remain linked, and so early gene expression inevitably continues.

The regulator gene that controls the switch from immediate early to delayed early expression in phage lambda is identified by mutations in gene *N* that can transcribe *only* the immediate early genes; they proceed no further into the infective cycle (see Figure 11.53). The same effect is seen when gene *28* of phage SPO1 is mutated to prevent the production of σ^{gp28} (see Figure 11.40). From the genetic point of view, the mechanisms of new initiation and antitermination are similar. *Both are positive controls in which an early gene product must be made by the phage in order to express the next set of genes.* By employing either sigma factors or antitermination proteins with different specificities, a cascade for gene expression can be constructed.

14.5 The T7 and T4 Genomes Show Functional Clustering

Key concepts

- Genes concerned with related functions are often clustered.
- Phages T7 and T4 are examples of regulatory cascades in which phage infection is divided into three periods.

The genome of phage T7 has three classes of genes, each of which constitutes a group of adjacent loci. As FIGURE 14.7 shows, the class I genes are the immediate early type and are expressed by host RNA polymerase as soon as the phage DNA enters the cell. Among the products of these genes are a phage RNA polymerase and enzymes that interfere with host gene expression. The phage RNA polymerase is responsible for expressing the class II genes (which are concerned principally with DNA synthesis functions) and the class III genes (which are concerned with assembling the mature phage particle).

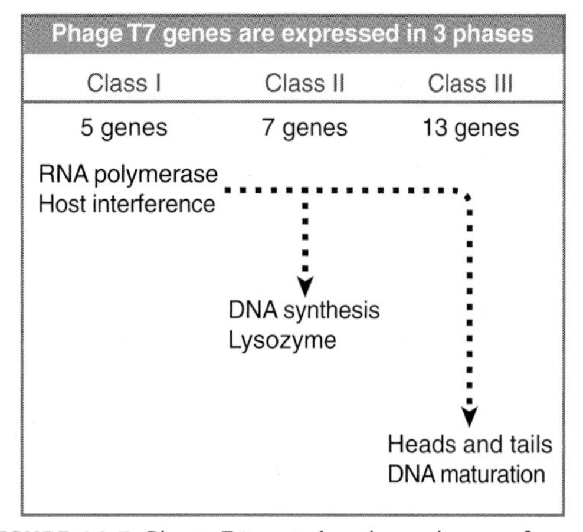

FIGURE 14.7 Phage T7 contains three classes of genes that are expressed sequentially. The genome is ~38 kb.

T4 has one of the larger phage genomes (165 kb), which is organized with extensive functional grouping of genes. FIGURE 14.8 presents the genetic map. *Essential genes* are numbered: A mutation in any one of these loci prevents successful completion of the lytic cycle. *Nonessential genes* are indicated by three-letter abbreviations. (They are defined as nonessential under the usual conditions of infection. We do not really understand the inclusion of many nonessential genes, but presumably they confer a selective advantage in some of T4's habitats. In smaller phage genomes, most or all of the genes are essential.)

There are three phases of gene expression. A summary of the functions of the genes expressed at each stage is given in FIGURE 14.9. The early genes are transcribed by host RNA polymerase. The middle genes are also transcribed by host RNA polymerase, but two phage-encoded products, MotA and AsiA, are also required. The middle promoters lack a consensus –30 sequence and instead have a binding sequence for MotA. The phage protein is an activator that compensates for the deficiency in the promoter by assisting host RNA polymerase to bind. (This is similar to a mechanism employed by phage lambda, which is illustrated later in Figure 14.30.) The early and middle genes account for virtually all of the phage functions concerned with the synthesis of DNA, modifying cell structure, and transcribing and translating phage genes.

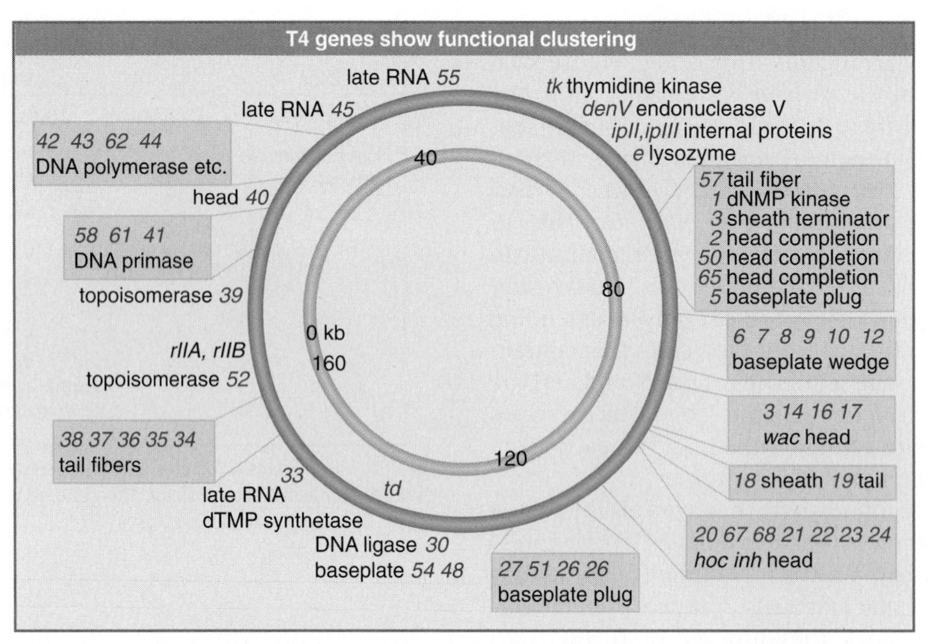

FIGURE 14.8 The map of T4 is circular. There is extensive clustering of genes coding for components of the phage and processes such as DNA replication, but there is also dispersion of genes coding for a variety of enzymatic and other functions. Essential genes are indicated by numbers. Nonessential genes are identified by letters. Only some representative T4 genes are shown on the map.

The two essential genes in the "transcription" category fulfill a regulatory function: their products are necessary for late gene expression. Phage T4 infection depends on a mechanical link between replication and late gene expression. Only actively replicating DNA can be used as a template for late gene transcription. The connection is generated by introducing a new sigma factor and also by making other modifications in the host RNA polymerase so that it is active only with a template of replicating DNA. This link establishes a correlation between the synthesis of phage protein components and the number of genomes available for packaging.

14.6 Lambda Immediate Early and Delayed Early Genes Are Needed for Both Lysogeny and the Lytic Cycle

Key concepts

- Lambda has two immediate early genes, *N* and *cro*, which are transcribed by host RNA polymerase.
- *N* is required to express the delayed early genes.
- Three of the delayed early genes are regulators.
- Lysogeny requires the delayed early genes *cII-cIII*.
- The lytic cycle requires the immediate early gene *cro* and the delayed early gene *Q*.

One of the most intricate cascade circuits is provided by phage lambda. The cascade for lytic development itself actually is straightforward, with two regulators controlling the successive stages of development. The circuit for the lytic cycle, however, is interlocked with the circuit for establishing lysogeny, as summarized in FIGURE 14.10.

When lambda DNA enters a new host cell, the lytic and lysogenic pathways start off the same way. Both require expression of the immediate early and delayed early genes, but then they diverge: lytic development follows if the late genes are expressed, and lysogeny ensues if synthesis of the repressor is established.

Lambda has only two immediate early genes, transcribed independently by host RNA polymerase:

- *N* codes for an antitermination factor whose action at the *nut* sites allows transcription to proceed into the delayed early genes (see Section 11.24, Antitermination Requires Sites That Are Independent of the Terminators).
- *cro* has dual functions: it prevents synthesis of the repressor (a necessary action if the lytic cycle is to proceed), and it turns off expression of the immediate early genes (which are not needed later in the lytic cycle).

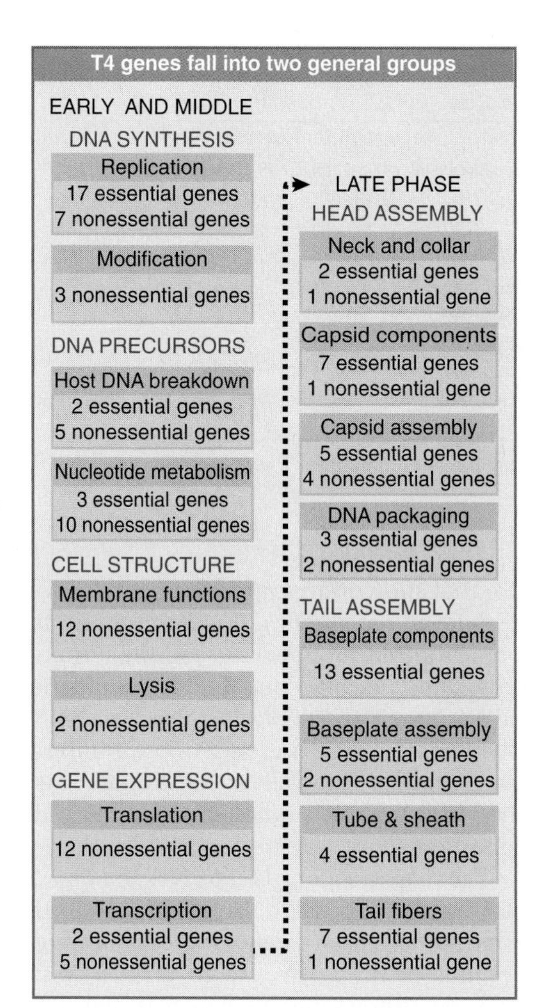

FIGURE 14.9 The phage T4 lytic cascade falls into two parts: early functions are concerned with DNA synthesis and gene expression; late functions are concerned with particle assembly.

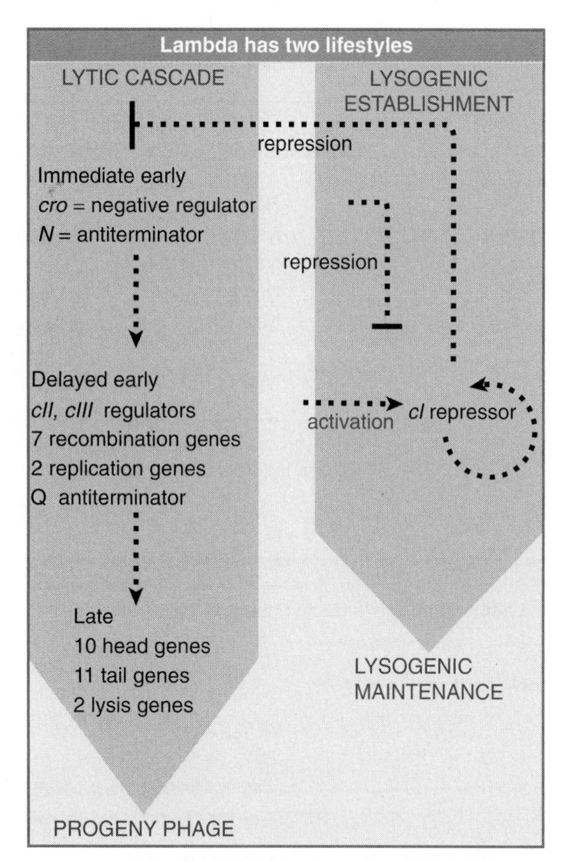

FIGURE 14.10 The lambda lytic cascade is interlocked with the circuitry for lysogeny.

The delayed early genes include two replication genes (needed for lytic infection), seven recombination genes (some involved in recombination during lytic infection, and two necessary to integrate lambda DNA into the bacterial chromosome for lysogeny), and three regulators. The regulators have opposing functions:

- The *cII-cIII* pair of regulators is needed to establish the synthesis of repressor.
- The *Q* regulator is an antitermination factor that allows host RNA polymerase to transcribe the late genes.

Thus the delayed early genes serve two masters: some are needed for the phage to enter lysogeny, and the others are concerned with controlling the order of the lytic cycle.

14.7 The Lytic Cycle Depends on Antitermination

Key concepts

- pN is an antitermination factor that allows RNA polymerase to continue transcription past the ends of the two immediate early genes.
- pQ is the product of a delayed early gene and is an antiterminator that allows RNA polymerase to transcribe the late genes.
- Lambda DNA circularizes after infection; as a result, the late genes form a single transcription unit.

To disentangle the two pathways, let's first consider just the lytic cycle. FIGURE 14.11 gives the map of lambda phage DNA. A group of genes concerned with regulation is surrounded by genes needed for recombination and replication. The genes coding for structural components of the phage are clustered. All of the genes necessary for the lytic cycle are expressed in polycistronic transcripts from three promoters.

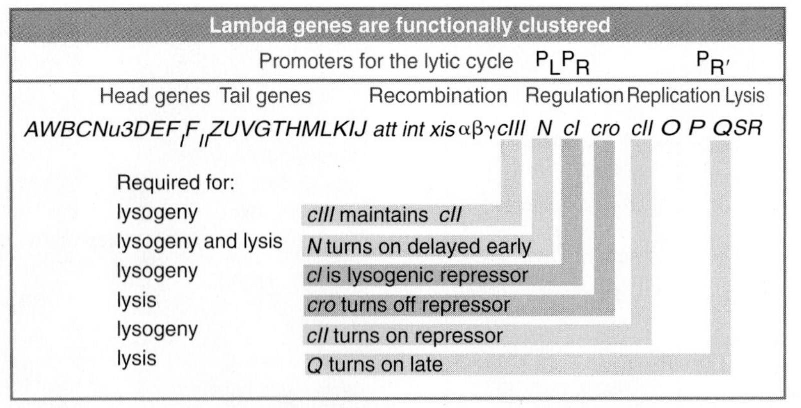

FIGURE 14.11 The lambda map shows clustering of related functions. The genome is 48,514 bp.

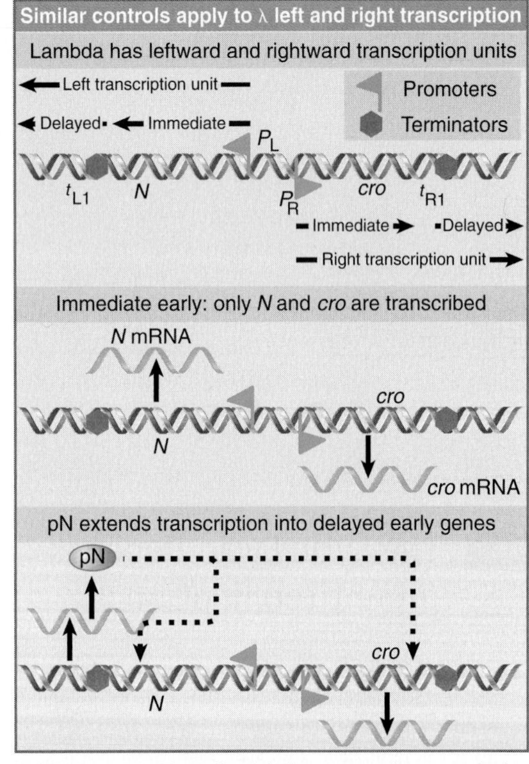

FIGURE 14.12 Phage lambda has two early transcription units. In the "leftward" unit, the "upper" strand is transcribed toward the left; in the "rightward" unit, the "lower" strand is transcribed toward the right. Genes *N* and *cro* are the immediate early functions and are separated from the delayed early genes by the terminators. Synthesis of N protein allows RNA polymerase to pass the terminators t_{L1} to the left and t_{R1} to the right.

FIGURE 14.12 shows that the two immediate early genes, *N* and *cro,* are transcribed by host RNA polymerase. *N* is transcribed toward the left and *cro* toward the right. Each transcript is terminated at the end of the gene. pN is the regulator that allows transcription to continue into the delayed early genes. It is an antitermination factor that suppresses use of the terminators t_L and t_R (see Section 11.25, Termination and Antitermination Factors Interact with RNA Polymerase). In the presence of pN, transcription continues to the left of *N* into the recombination genes and to the right of *cro* into the replication genes.

The map in Figure 14.11 gives the organization of the lambda DNA as it exists in the phage particle. Shortly after infection, though, the ends of the DNA join to form a circle. FIGURE 14.13 shows the true state of lambda DNA during infection. The late genes are welded into a single group, which contain the lysis genes *S-R* from the right end of the linear DNA and the head and tail genes *A-J* from the left end.

The late genes are expressed as a single transcription unit, starting from a promoter $P_{R'}$ that lies between *Q* and *S*. The late promoter is used constitutively. In the absence of the product of gene *Q* (which is the last gene in the rightward delayed early unit), though, late transcription terminates at a site t_{R3}. The transcript resulting from this termination event is 194 bases long; it is known as 6S RNA. When pQ becomes available, it suppresses termination at t_{R3} and the 6S RNA is extended, with the result that the late genes are expressed.

Late gene transcription does not seem to terminate at any specific point, but rather continues through all the late genes into the region beyond. A similar event happens with the leftward delayed early transcription, which continues past the recombination functions. Transcription in each direction is probably terminated before the polymerases can crash into each other.

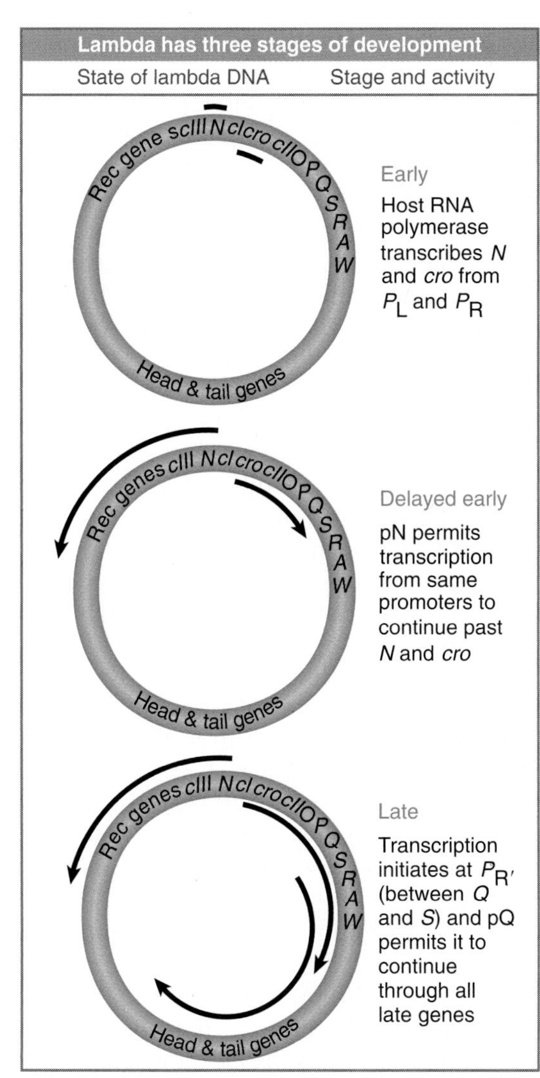

FIGURE 14.13 Lambda DNA circularizes during infection, so that the late gene cluster is intact in one transcription unit.

14.8 Lysogeny Is Maintained by Repressor Protein

Key concepts

- Mutants in the *cI* gene cannot maintain lysogeny.
- *cI* codes for a repressor protein that acts at the O_L and O_R operators to block transcription of the immediate early genes.
- The immediate early genes trigger a regulatory cascade; as a result, their repression prevents the lytic cycle from proceeding.

Looking at the lambda lytic cascade, we see that the entire program is set in motion by the initiation of transcription at the two promoters P_L and P_R for the immediate early genes *N* and *cro*. Lambda uses antitermination to proceed to the next stage of (delayed early) expression; therefore, the same two promoters continue to be used throughout the early period.

The expanded map of the regulatory region drawn in FIGURE 14.14 shows that the promoters P_L and P_R lie on either side of the *cI* gene. Associated with each promoter is an operator (O_L, O_R) at which repressor protein binds to prevent RNA polymerase from initiating transcription. The sequence of each operator overlaps with the promoter that it controls, and because this occurs so often these sequences are described as the P_L/O_L and P_R/O_R control regions.

As a result of the sequential nature of the lytic cascade, the control regions provide a pressure point at which entry to the entire cycle can be controlled. *By denying RNA polymerase access to these promoters, a repressor protein prevents the phage genome from entering the lytic cycle.* The repressor functions in the same way as repressors of bacterial operons: it binds to specific operators.

The repressor protein is coded by the *cI* gene. Mutants in this gene cannot maintain lysogeny,

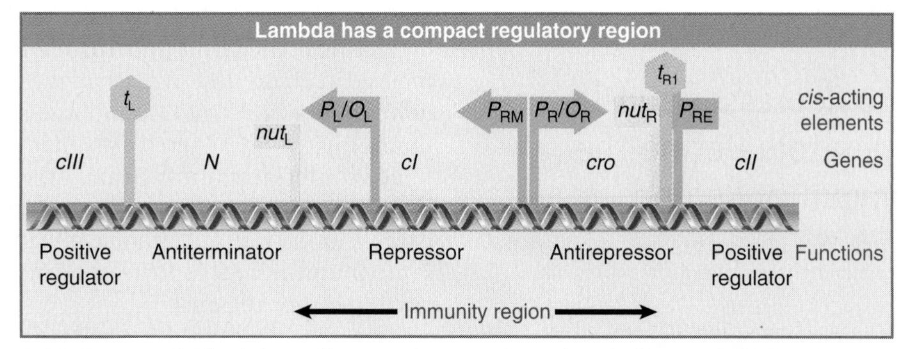

FIGURE 14.14 The lambda regulatory region contains a cluster of *trans*-acting functions and *cis*-acting elements.

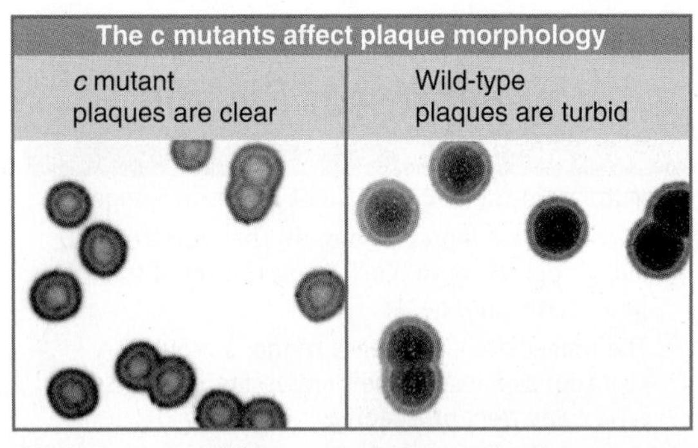

FIGURE 14.15 Wild-type and virulent lambda mutants can be distinguished by their plaque types. Reproduced from *Virology*, vol. 1, Kariser, D., pp. 423–443. Copyright 1955, with permission from Elsevier. Photos courtesy of Dale Kaiser, Stanford University School of Medicine.

but always enter the lytic cycle. In the time since the original isolation of the repressor protein, the characterization of the repressor protein has shown how it both maintains the lysogenic state and provides immunity for a lysogen against superinfection by new phage lambda genomes.

When a bacterial culture is infected with a phage, the cells are lysed to generate regions that can be seen on a culture plate as small areas of clearing called **plaques.** With wild-type phages the plaques are turbid or cloudy, because they contain some cells that have established lysogeny instead of being lysed. The effect of a *cI* mutation is to prevent lysogeny, so that the plaques contain only lysed cells. As a result, such an infection generates only **clear plaques,** and three genes (*cI, cII,* and *cIII*) were named for their involvement in this phenotype. FIGURE 14.15 compares wild-type and mutant plaques.

The *cI* gene is transcribed from a promoter P_{RM} that lies at its right end. (The subscript "RM" stands for repressor maintenance.) Transcription is terminated at the left end of the gene. The mRNA starts with the AUG initiation codon; because of the absence of the usual ribosome binding site, the mRNA is translated somewhat inefficiently, producing only a low level of repressor protein.

14.9 The Repressor and Its Operators Define the Immunity Region

Key concepts

- Several lambdoid phages have different immunity regions.
- A lysogenic phage confers immunity to further infection by any other phage with the same immunity region.

The presence of repressor explains the phenomenon of **immunity.** If a second lambda phage DNA enters a lysogenic cell, repressor protein synthesized from the resident prophage genome will immediately bind to O_L and O_R in the new genome. This prevents the second phage from entering the lytic cycle.

The operators were originally identified as the targets for repressor action by **virulent** mutations (λ*vir*). These mutations prevent the repressor from binding at O_L or O_R, with the result that the phage inevitably proceeds into the lytic pathway when it infects a new host bacterium. Note that λ*vir* mutants can grow on lysogens because the virulent mutations in O_L and O_R allow the incoming phage to ignore the resident repressor and thus enter the lytic cycle. Virulent mutations in phages are the equivalent of operator-constitutive mutations in bacterial operons.

Prophage is induced to enter the lytic cycle when the lysogenic circuit is broken. This happens when the repressor is inactivated (see Section 14.10, The DNA-Binding Form of Repressor Is a Dimer). The absence of repressor allows RNA polymerase to bind at P_L and P_R, starting the lytic cycle as shown in the lower part of Figure 14.26.

The autogenous nature of the repressor-maintenance circuit creates a sensitive response. The presence of repressor is necessary for its own synthesis; therefore, expression of the *cI* gene stops as soon as the existing repressor is destroyed. Thus no repressor is synthesized to replace the molecules that have been damaged. This enables the lytic cycle to start without interference from the circuit that maintains lysogeny.

The region including the left and right operators, the *cI* gene, and the *cro* gene determines the immunity of the phage. Any phage that possesses this region has the same type of immunity, because *it specifies both the repressor protein and the sites on which the repressor acts.* Accordingly, this is called the **immunity region** (as marked in Figure 14.14). Each of the four lambdoid phages φ80, *21, 434,* and λ has a unique immunity region. When we say that a lysogenic phage confers immunity to any other phage of the same type, we mean more precisely that the immunity is to any other phage that has the same immunity region (irrespective of differences in other regions).

14.10 The DNA-Binding Form of Repressor Is a Dimer

Key concepts

- A repressor monomer has two distinct domains.
- The N-terminal domain contains the DNA-binding site.
- The C-terminal domain dimerizes.
- Binding to the operator requires the dimeric form so that two DNA-binding domains can contact the operator simultaneously.
- Cleavage of the repressor between the two domains reduces the affinity for the operator and induces a lytic cycle.

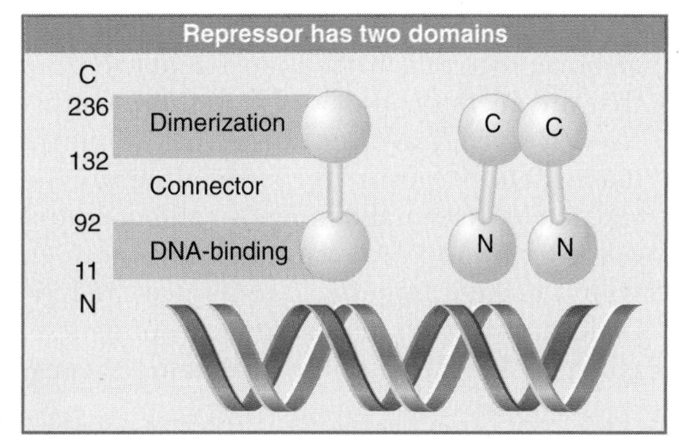

FIGURE 14.16 The N-terminal and C-terminal regions of repressor form separate domains. The C-terminal domains associate to form dimers; the N-terminal domains bind DNA.

The repressor subunit is a polypeptide of 27 kD with the two distinct domains summarized in FIGURE 14.16.

- The N-terminal domain, residues 1–92, provides the operator-binding site.
- The C-terminal domain, residues 132–236, is responsible for dimerization.

The two domains are joined by a connector of forty residues. When repressor is digested by a protease, each domain is released as a separate fragment.

Each domain can exercise its function independently of the other. The C-terminal fragment can form oligomers. The N-terminal fragment can bind the operators, although with a lower affinity than the intact repressor. Thus the information for specifically contacting DNA is contained within the N-terminal domain, but the efficiency of the process is enhanced by the attachment of the C-terminal domain.

The dimeric structure of the repressor is crucial in maintaining lysogeny. The induction of a lysogenic prophage to enter the lytic cycle is caused by cleavage of the repressor subunit in the connector region, between residues 111 and 113. (This is a counterpart to the allosteric change in conformation that results when a small-molecule inducer inactivates the repressor of a bacterial operon, a capacity that the lysogenic repressor does not have.) Induction occurs under certain adverse conditions, such as exposure of lysogenic bacteria to UV irradiation, which leads to proteolytic inactivation of the repressor.

In the intact state, dimerization of the C-terminal domains ensures that when the repressor binds to DNA, its two N-terminal domains each contact DNA simultaneously. Note, however, that cleavage releases the C-terminal domains from the N-terminal domains. As illustrated in FIGURE 14.17, this means that the N-terminal domains can no longer dimerize, which upsets the equilibrium between monomers and dimers. As a result, repressor dissociates from DNA, which allows lytic infection to start. (Another relevant parameter is the loss of cooperative effects between adjacent dimers.)

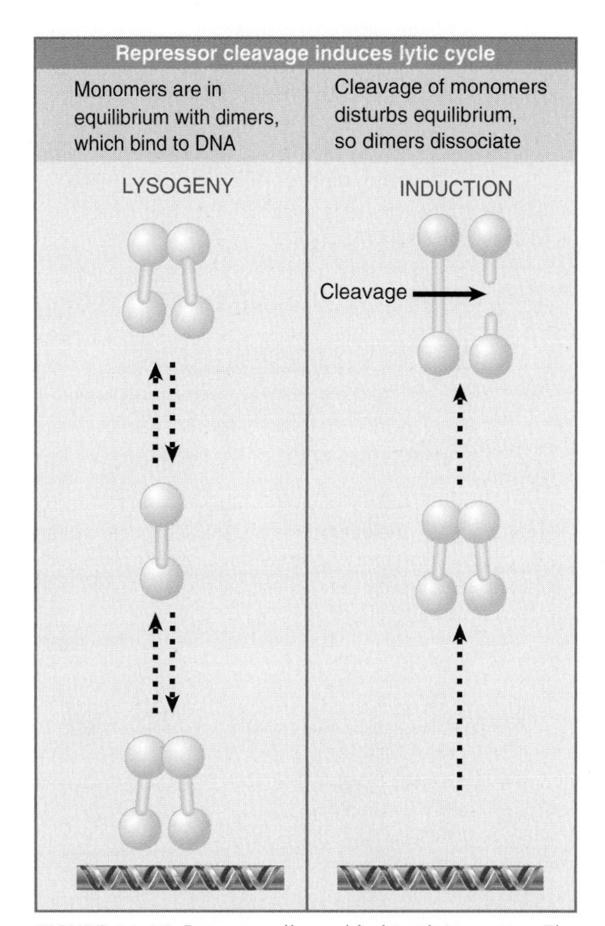

FIGURE 14.17 Repressor dimers bind to the operator. The affinity of the N-terminal domains for DNA is controlled by the dimerization of the C-terminal domains.

The balance between lysogeny and the lytic cycle depends on the concentration of repressor. Intact repressor is present in a lysogenic cell at a concentration sufficient to ensure that the operators are occupied. If the repressor is cleaved, however, this concentration is inadequate, because of the lower affinity of the separate N-terminal domain for the operator. A concentration of repressor that is too high would make it impossible to induce the lytic cycle in this way; a level too low, of course, would make it impossible to maintain lysogeny.

14.11 Repressor Uses a Helix-Turn-Helix Motif to Bind DNA

Key concepts

- Each DNA-binding region in the repressor contacts a half-site in the DNA.
- The DNA-binding site of the repressor includes two short α-helical regions that fit into the successive turns of the major groove of DNA.
- A DNA-binding site is a (partially) palindromic sequence of 17 bp.

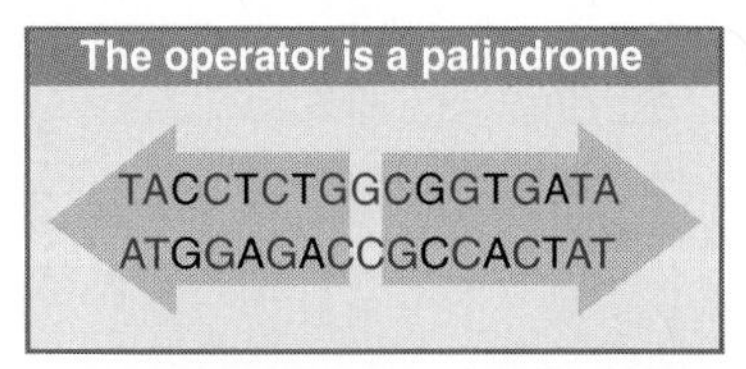

FIGURE 14.18 The operator is a 17-bp sequence with an axis of symmetry through the central base pair. Each half-site is marked in light blue. Base pairs that are identical in each operator half are in dark blue.

A repressor dimer is the unit that binds to DNA. It recognizes a sequence of 17 bp displaying partial symmetry about an axis through the central base pair. FIGURE 14.18 shows an example of a binding site. The sequence on each side of the central base pair is sometimes called a "half-site." Each individual N-terminal region contacts a half-site. Several DNA-binding proteins that regulate bacterial transcription share a similar mode of holding DNA, in which the active domain contains two short regions of α-helix that contact DNA. (Some transcription factors in eukaryotic cells use a similar motif; see Section 25.14, Homeodomains Bind Related Targets in DNA.)

The N-terminal domain of lambda repressor contains several stretches of α-helix, which are arranged as illustrated diagrammatically in FIGURE 14.19. Two of the helical regions are responsible for binding DNA. The **helix-turn-helix** model for contact is illustrated in FIGURE 14.20. Looking at a single monomer, α-helix-3 consists of nine amino acids, each of which lies at an angle to the preceding region of seven amino acids that forms α-helix-2. In the dimer, the two apposed helix-3 regions lie 34 Å apart, enabling them to fit into successive major grooves of DNA. The helix-2 regions lie at an angle that would place them across the groove. The symmetrical binding of dimer to the site means that each N-terminal domain of the dimer contacts a similar set of bases in its half-site.

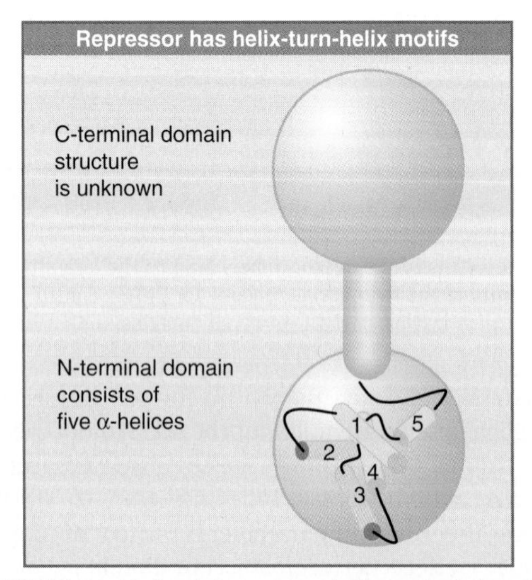

FIGURE 14.19 Lambda repressor's N-terminal domain contains five stretches of α-helix; helices 2 and 3 bind DNA.

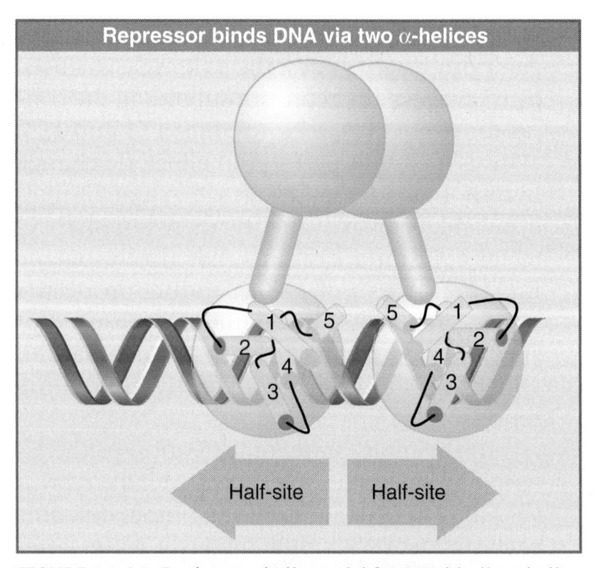

FIGURE 14.20 In the two-helix model for DNA binding, helix-3 of each monomer lies in the wide groove on the same face of DNA, and helix-2 lies across the groove.

14.12 The Recognition Helix Determines Specificity for DNA

Key concept

- The amino acid sequence of the recognition helix makes contacts with particular bases in the operator sequence that it recognizes.

Related forms of the α-helical motifs employed in the helix-turn-helix of the lambda repressor are found in several DNA-binding proteins, including cyclic AMP receptor protein (CRP), the *lac* repressor, and several other phage repressors. By comparing the abilities of these proteins to bind DNA, we can define the roles of each helix:

- Contacts between helix-3 and DNA rely on hydrogen bonds between the amino acid side chains and the exposed positions of the base pairs. This helix is responsible for recognizing the specific target DNA sequence and is therefore also known as the **recognition helix.**
- Contacts from helix-2 to the DNA take the form of hydrogen bonds connecting with the phosphate backbone. These interactions are necessary for binding, but do not control the specificity of target recognition. In addition to these contacts, a large part of the overall energy of interaction with DNA is provided by ionic interactions with the phosphate backbone.

What happens if we manipulate the coding sequence to construct a new protein by substituting the recognition helix in one repressor with the corresponding sequence from a closely related repressor? The specificity of the hybrid protein is that of its new recognition helix. *The amino acid sequence of this short region determines the sequence specificities of the individual proteins and is able to act in conjunction with the rest of the polypeptide chain.*

FIGURE 14.21 shows the details of the binding to DNA of two proteins that bind similar DNA sequences. Both lambda repressor and Cro protein have a similar organization of the helix-turn-helix motif, although their individual specificities for DNA are not identical:

- Each protein uses similar interactions between hydrophobic amino acids to maintain the relationship between helix-2 and helix-3: repressor has an Ala-Val connection, whereas Cro has an Ala-Ile association.
- Amino acids in helix-3 of the repressor make contacts with specific bases in the operator. Three amino acids in repressor recognize three bases in DNA; the amino acids at these positions, and also at additional positions in Cro, recognize five (or possibly six) bases in DNA.

Two of the amino acids involved in specific recognition are identical in repressor and Cro (Gln and Ser at the N-terminal end of the helix), whereas the other contacts are different (Ala in repressor versus Lys and the additional Asn in Cro). In addition, a Thr in helix-2 of Cro directly contacts DNA.

The interactions shown in Figure 14.21 represent binding to the DNA sequence that each protein recognizes most tightly. The sequences shown at the bottom of the figure with the contact points in color differ at three of the nine base pairs. The use of overlapping, but not identical, contacts between amino acids and bases shows how related recognition helices confer recognition of related DNA sequences. This enables repressor and Cro to recognize the same set of sequences, but with different relative affinities for particular members of the group.

The bases contacted by helix-3 of repressor or Cro lie on one face of DNA, as can be seen from the positions indicated on the helical diagram in Figure 14.21. Note, however, that repressor makes an additional contact with the other face of DNA. Removing the last six N-terminal amino acids (which protrude from

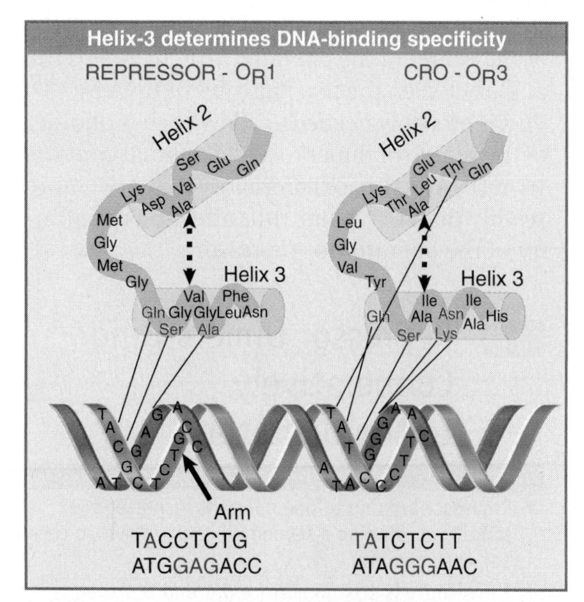

FIGURE 14.21 Two proteins that use the two-helix arrangement to contact DNA recognize lambda operators with affinities determined by the amino acid sequence of helix-3.

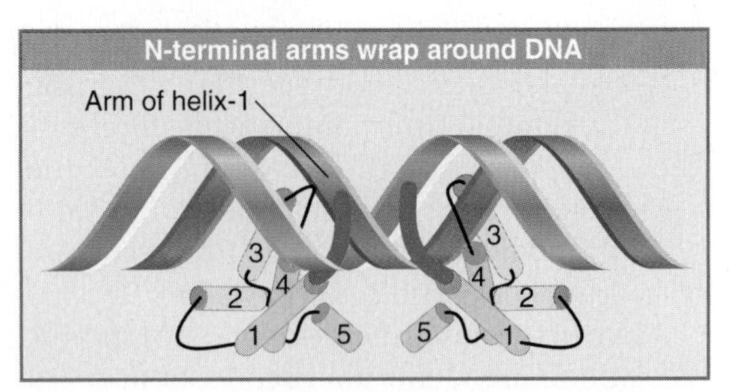

FIGURE 14.22 A view from the back shows that the bulk of the repressor contacts one face of DNA, but its N-terminal arms reach around to the other face.

helix-1) eliminates some of the contacts. This observation provides the basis for the idea that the bulk of the N-terminal domain contacts one face of DNA, whereas the last six N-terminal amino acids form an "arm" extending around the back. FIGURE 14.22 shows the view from the back. Lysine residues in the arm make contacts with G residues in the major groove and also with the phosphate backbone. The interaction between the arm and DNA contributes heavily to DNA binding; the affinity of the armless repressor for DNA is reduced by ~1000-fold.

Bases that are not contacted directly by repressor protein may have an important effect on binding. The related phage *434* repressor binds DNA via a helix-turn-helix motif, and the crystal structure shows that helix-3 is positioned at each half-site so that it contacts the five outermost base pairs, but not the inner two. Operators with A-T base pairs at the inner positions, however, bind *434* repressor more strongly than operators with G-C base pairs. The reason is that *434* repressor binding slightly twists DNA at the center of the operator, which widens the angle between the two half-sites of DNA by ~3°. This is probably needed to allow each monomer of the repressor dimer to make optimal contacts with DNA. A-T base pairs allow this twist more readily than G-C pairs, thus affecting the affinity of the operator for repressor.

14.13 Repressor Dimers Bind Cooperatively to the Operator

Key concepts

- Repressor binding to one operator increases the affinity for binding a second repressor dimer to the adjacent operator.
- The affinity is 10× greater for O_L1 and O_R1 than other operators, so they are bound first.
- Cooperativity allows repressor to bind the *01/02* sites at lower concentrations.

Each operator contains three repressor-binding sites. As can be seen from FIGURE 14.23, no two of the six individual repressor-binding sites are identical, but they all conform with a consensus sequence. The binding sites within each operator are separated by spacers of 3 to 7 bp that are rich in A-T base pairs. The sites at each operator are numbered so that O_R consists of the series of binding sites O_R1-O_R2-O_R3, whereas O_L consists of the series O_L1-O_L2-O_L3. In each case, site 1 lies closest to the startpoint for transcription in the promoter, and sites 2 and 3 lie farther upstream.

Faced with the triplication of binding sites at each operator, how does repressor decide where to start binding? At each operator, site 1 has a greater affinity (roughly tenfold) than the other sites for the repressor. Thus the repressor always binds first to O_L1 and O_R1.

Lambda repressor binds to subsequent sites within each operator in a cooperative manner. The presence of a dimer at site 1 greatly increases the affinity with which a second dimer can bind to site 2. When both sites 1 and 2 are occupied, this interaction does *not* extend farther, to site 3. At the concentrations of repressor usually found in a lysogen, both sites 1 and 2 are filled at each operator, but site 3 is not occupied.

If site 1 is inactive (because of mutation), then repressor binds cooperatively to sites 2 and 3. That is, binding at site 2 assists another dimer to bind at site 3. This interaction occurs directly between repressor dimers and not via conformational change in DNA. The C-terminal domain is responsible for the cooperative interaction between dimers, as well as for the dimer formation between subunits. FIGURE 14.24 shows that it involves both subunits of each dimer, that is, each subunit contacts its counterpart in the other dimer, forming a tetrameric structure.

A result of cooperative binding is to increase the effective affinity of repressor for the operator at physiological concentrations. This enables a lower concentration of repressor to achieve occupancy of the operator. This is an important consideration in a system in which release of repression has irreversible consequences. In an operon coding for metabolic enzymes, after all, failure of repression will merely allow unnecessary synthesis of enzymes. Failure to repress lambda prophage, however, will lead to induction of phage and lysis of the cell.

From the sequences shown in Figure 14.23, we see that O_L1 and O_R1 lie more or less in the center of the RNA polymerase binding sites of P_L and P_R, respectively. Occupancy of O_L1-O_L2 and O_R1-O_R2 thus physically blocks access of RNA polymerase to the corresponding promoters.

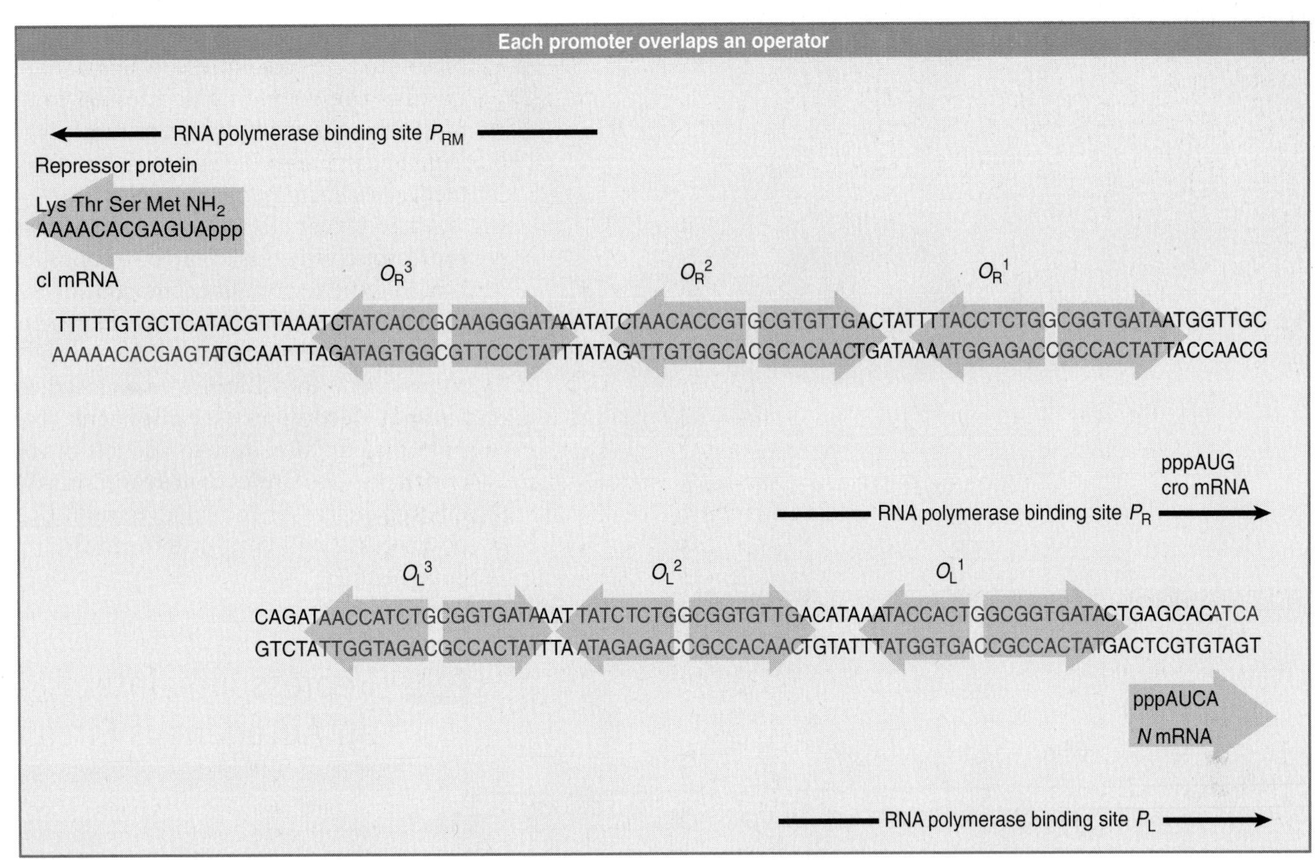

FIGURE 14.23 Each operator contains three repressor-binding sites and overlaps with the promoter at which RNA polymerase binds. The orientation of O_L has been reversed from usual to facilitate comparison with O_R.

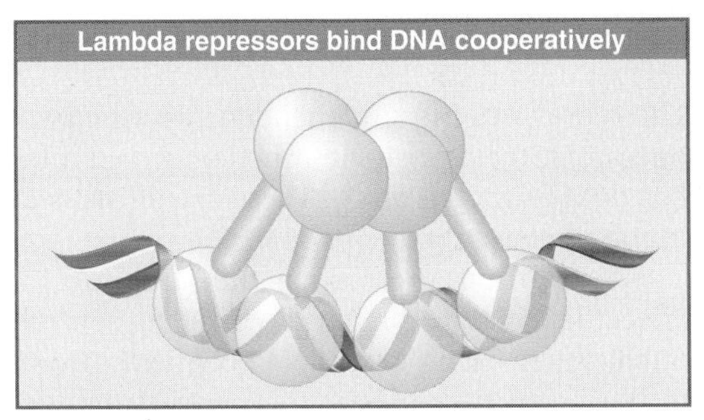

FIGURE 14.24 When two lambda repressor dimers bind cooperatively, each of the subunits of one dimer contacts a subunit in the other dimer.

14.14 Repressor at O_R2 Interacts with RNA Polymerase at P_{RM}

Key concepts

- The DNA-binding region of repressor at O_R2 contacts RNA polymerase and stabilizes its binding to P_{RM}.
- This is the basis for the autogenous control of repressor maintenance.

A different relationship is shown between O_R and the promoter P_{RM} for transcription of *cI*. The RNA polymerase binding site is adjacent to O_R2. This explains how repressor autogenously regulates its own synthesis. When two dimers are bound at O_R1-O_R2, the dimer at O_R2 interacts with RNA polymerase (see Figure 14.26 in Section 14.15, Repressor Maintains an Autogenous Circuit). This effect resides in the amino terminal domain of repressor.

Mutations that abolish positive control map in the *cI* gene. The members of one interesting class of mutants remain able to bind the operator to repress transcription, but they cannot stimulate RNA polymerase to transcribe from P_{RM}. They map within a small group of amino acids that are located on the outside of helix-2 or in the turn between helix-2 and helix-3. The mutations reduce the negative charge of the region; conversely, mutations that increase the negative charge enhance the activation of RNA polymerase. This suggests that the group of amino acids constitutes an "acidic patch" that functions by an electrostatic interaction with a basic region on RNA polymerase.

The location of these "positive control mutations" in the repressor is indicated on FIGURE 14.25. They lie at a site on repressor that is close to a phosphate group on DNA, which is

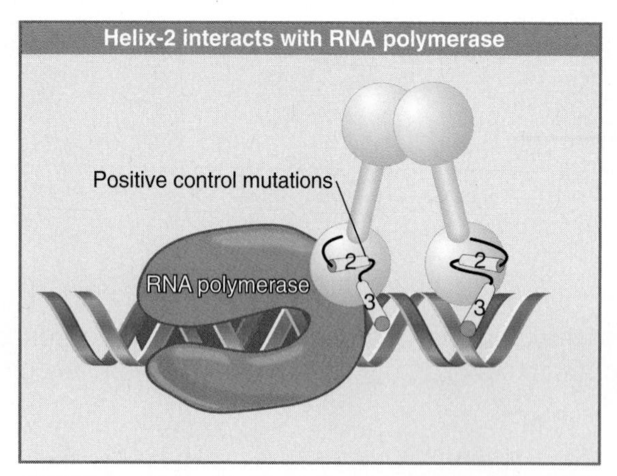

FIGURE 14.25 Positive control mutations identify a small region at helix-2 that interacts directly with RNA polymerase.

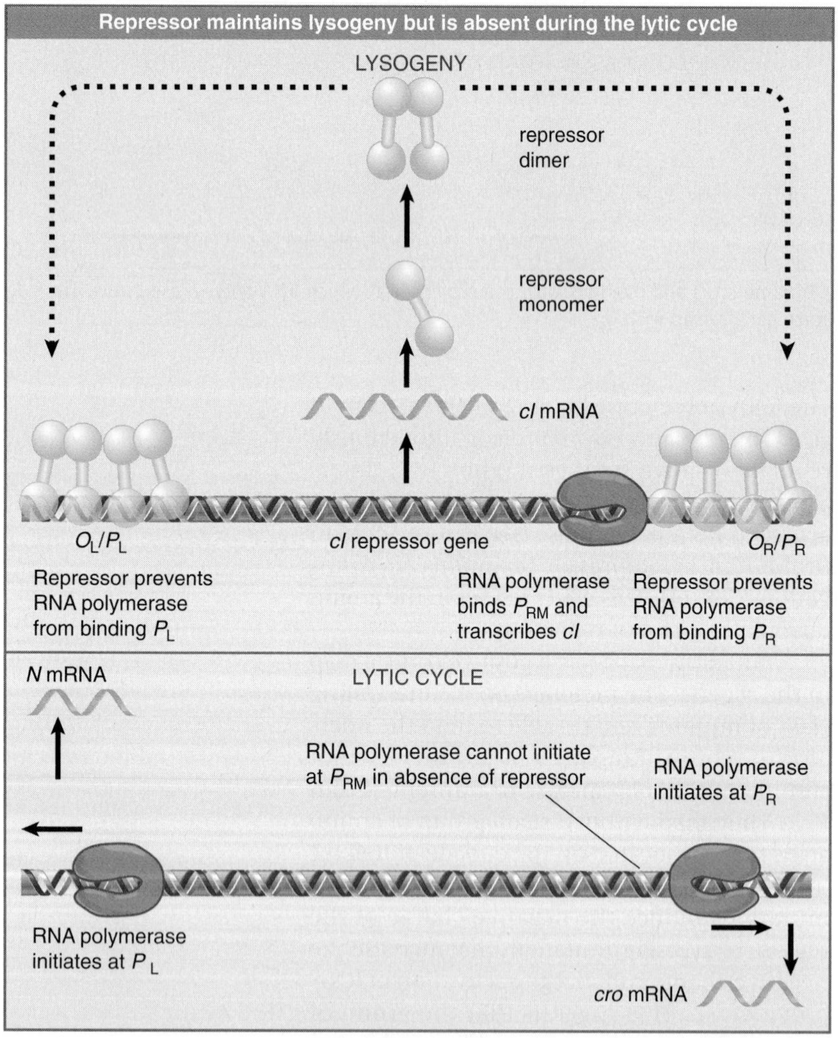

also close to RNA polymerase. Thus the group of amino acids on repressor that is involved in positive control is in a position to contact the polymerase. The important principle is that *protein–protein interactions can release energy that is used to help to initiate transcription.*

The target site on RNA polymerase that repressor contacts is in the σ70 subunit, which is within the region that contacts the –35 region of the promoter. The interaction between repressor and polymerase is needed for the polymerase to make the transition from a closed complex to an open complex (see also Figure 14.31).

This explains how low levels of repressor positively regulate its own synthesis. As long as enough repressor is available to fill O_R2, RNA polymerase will continue to transcribe the *cI* gene from P_{RM}.

Repressor maintains lysogeny but is absent during the lytic cycle

LYSOGENY

repressor dimer

repressor monomer

cI mRNA

O_L/P_L *cI* repressor gene O_R/P_R

Repressor prevents RNA polymerase from binding P_L

RNA polymerase binds P_{RM} and transcribes *cI*

Repressor prevents RNA polymerase from binding P_R

N mRNA

LYTIC CYCLE

RNA polymerase cannot initiate at P_{RM} in absence of repressor

RNA polymerase initiates at P_R

RNA polymerase initiates at P_L

cro mRNA

FIGURE 14.26 Lysogeny is maintained by an autogenous circuit (upper). If this circuit is interrupted, the lytic cycle starts (lower).

14.15 Repressor Maintains an Autogenous Circuit

Key concepts

- Repressor binding at O_L blocks transcription of gene *N* from P_L.
- Repressor binding at O_R blocks transcription of *cro*, but also is required for transcription of *cI*.
- Repressor binding to the operators therefore simultaneously blocks entry to the lytic cycle and promotes its own synthesis.

The repressor binds independently to the two operators. It has a single function at O_L2, but has dual functions at O_R. These are illustrated in the upper part of FIGURE 14.26.

At O_L, the repressor has the same sort of effect that we have already discussed for several other systems: It prevents RNA polymerase from initiating transcription at P_L. This stops the expression of gene *N*. P_L is used for all leftward early gene transcription; as a result, this action prevents expression of the entire leftward early transcription unit. *Thus the lytic cycle is blocked before it can proceed beyond the early stages.*

At O_R1, repressor binding prevents the use of P_R. Thus *cro* and the other rightward early genes cannot be expressed. (We see later why it is important to prevent the expression of *cro* when lysogeny is being maintained.)

The presence of repressor at O_R has another effect, though. The promoter for repressor synthesis, P_{RM}, is adjacent to the rightward operator O_R. It turns out that *RNA polymerase can initiate efficiently at P_{RM} only when repressor is bound at O_R.* The repressor behaves as a positive regulator protein that is necessary for transcription of the *cI* gene (see Section 14.14, Repressor at O_R2 Interacts with RNA Polymerase at P_{RM}). *The repressor is the product of* cI; thus *this interaction creates a positive autogenous circuit, in which the presence of repressor is necessary to support its own continued synthesis.*

The nature of this control circuit explains the biological features of lysogenic existence. Lysogeny is stable because the control circuit ensures that as long as the level of repressor is adequate, there is continued expression of the *cI* gene. The result is that O_L and O_R remain occupied indefinitely. By repressing the entire lytic cascade, this action maintains the prophage in its inert form.

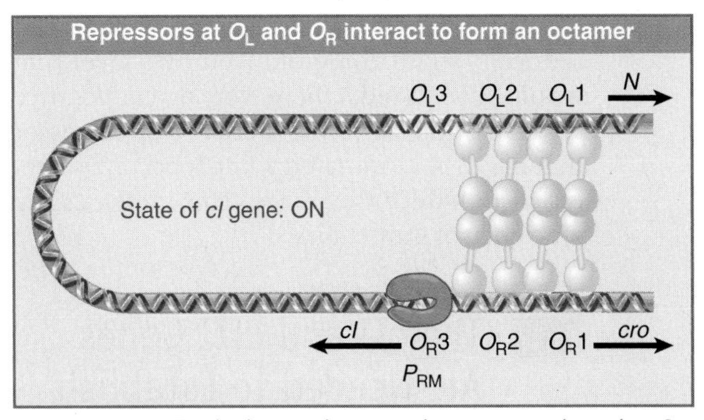

FIGURE 14.27 In the lysogenic state, the repressors bound at O_L1 and O_L2 interact with those bound at O_R1 and O_R2. RNA polymerase is bound at P_{RM} (which overlaps with O_R3) and interacts with the repressor bound at O_R2.

14.16 Cooperative Interactions Increase the Sensitivity of Regulation

Key concepts

- Repressor dimers bound at O_L1 and O_L2 interact with dimers bound at O_R1 and O_R2 to form octamers.
- Octamer formation brings O_L3 close to O_R3, allowing interactions between dimers bound there.
- These cooperative interactions increase the sensitivity of regulation.

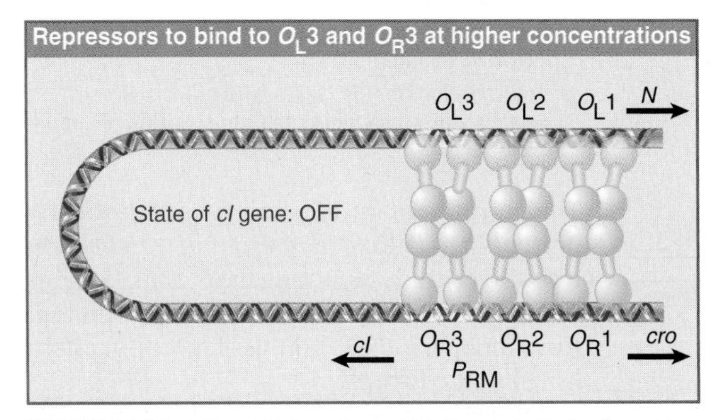

FIGURE 14.28 O_L3 and O_R3 are brought into proximity by formation of the repressor octamer, and an increase in repressor concentration allows dimers to bind at these sites and to interact.

Cooperative interactions between repressor dimers occur at both the left and right operators, so that their normal condition when occupied by repressor is to have dimers at both the 1 and 2 binding sites. In effect, each operator has a tetramer of repressor. This is not the end of the story, though. The two dimers interact with one another to form an octamer. The interaction occurs between the C-terminal domains, which can form an octamer as a crystal structure.

FIGURE 14.27 shows the distribution of repressors at the operator sites that are occupied in a lysogen. Repressors are occupying O_L1, O_L2, O_R1, and O_R2, and the repressor at the last of these sites is interacting with RNA polymerase, which is initiating transcription at P_{RM}.

The interaction between the two operators has several consequences. It stabilizes repressor binding, thereby making it possible for repressor to occupy operators at lower concentrations. Binding at O_R2 stabilizes RNA polymerase binding at P_{RM}, which enables low concentrations of repressor to autogenously stimulate their own production.

The DNA between the O_L and O_R sites (that is, the gene *cI*) forms a large loop, which is held together by the repressor octamer. The octamer brings the sites O_L3 and O_R3 into proximity. As a result, two repressor dimers can bind to these sites and interact with one another, as shown in FIGURE 14.28. The occupation of O_R3 prevents RNA polymerase from binding to P_{RM} and therefore turns off expression of repressor.

This shows us how the expression of the *cI* gene becomes exquisitely sensitive to repressor concentration. At the lowest concentrations, it forms the octamer and activates RNA polymerase in a positive autogenous regulation. An increase in concentration allows binding to O_L3 and O_R3 and turns off transcription in a negative autogenous regulation. The threshold levels of repressor that are required for each of these events is reduced by the cooperative interactions, which makes the overall regulatory system much more sensitive. Any change in repressor level triggers the appropriate regulatory response to restore the lysogenic level.

The overall level of repressor has been reduced (about threefold from the level that would be required if there were no cooperative effects), and as a result there is less repressor that has to be eliminated when it becomes necessary to induce the phage. This increases the efficiency of induction.

14.17 The cII and cIII Genes Are Needed to Establish Lysogeny

Key concepts

- The delayed early gene products cII and cIII are necessary for RNA polymerase to initiate transcription at the promoter P_{RE}.
- cII acts direct at the promoter and cIII protects cII from degradation.
- Transcription from P_{RE} leads to synthesis of repressor and also blocks the transcription of *cro*.

The control circuit for maintaining lysogeny presents a paradox. *The presence of repressor protein is necessary for its own synthesis.* This explains how the lysogenic condition is perpetuated. How, though, is the synthesis of repressor established in the first place?

When a lambda DNA enters a new host cell, RNA polymerase cannot transcribe *cI* because there is no repressor present to aid its binding at P_{RM}. This same absence of repressor, however, means that P_R and P_L are available. Thus the first event after lambda DNA infects a bacterium is when genes *N* and *cro* are transcribed. After this, pN allows transcription to be extended farther. This allows *cIII* (and other genes) to be transcribed on the left, whereas *cII* (and other genes) are transcribed on the right (see Figure 14.14).

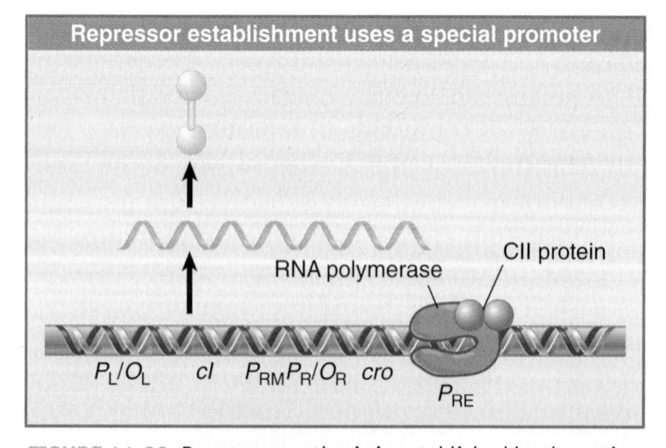

FIGURE 14.29 Repressor synthesis is established by the action of cII and RNA polymerase at P_{RE} to initiate transcription that extends from the antisense strand of *cro* through the *cI* gene.

The *cII* and *cIII* genes share with *cI* the property that mutations in them cause clear plaques. There is, however, a difference. The *cI* mutants can neither establish nor maintain lysogeny. The *cII* or *cIII* mutants have some difficulty in establishing lysogeny, but once it is established they are able to maintain it by the *cI* autogenous circuit.

This implicates the *cII* and *cIII* genes as positive regulators whose products are needed for an alternative system for repressor synthesis. The system is needed only to *initiate* the expression of *cI* in order to circumvent the inability of the autogenous circuit to engage in *de novo* synthesis. They are not needed for continued expression.

The cII protein acts directly on gene expression. Between the *cro* and *cII* genes is another promoter, called P_{RE}. (The subscript "RE" stands for repressor establishment.) This promoter can be recognized by RNA polymerase only in the presence of cII, whose action is illustrated in FIGURE 14.29.

The cII protein is extremely unstable *in vivo*, because it is degraded as the result of the activity of a host protein called HflA. The role of cIII is to protect cII against this degradation.

Transcription from P_{RE} promotes lysogeny in two ways. Its direct effect is that *cI* is translated into repressor protein. An indirect effect is that transcription proceeds through the *cro* gene in the "wrong" direction. Thus the 5′ part of the RNA corresponds to an antisense transcript of *cro;* in fact, it hybridizes to authentic *cro* mRNA, which inhibits its translation. This is important because *cro* expression is needed to enter the lytic cycle (see Section 14.20, The cro Repressor Is Needed for Lytic Infection).

The *cI* coding region on the P_{RE} transcript is very efficiently translated, in contrast with the weak translation of the P_{RM} transcript. In fact, repressor is synthesized approximately seven to eight times more effectively via expression from P_{RE} than from P_{RM}. This reflects the fact that the P_{RE} transcript has an efficient ribosome-binding site, whereas the P_{RM} transcript has no ribosome-binding site and actually starts with the AUG initiation codon.

14.18 A Poor Promoter Requires cII Protein

Key concepts

- P_{RE} has atypical sequences at –10 and –35.
- RNA polymerase binds the promoter only in the presence of cII.
- cII binds to sequences close to the –35 region.

The P_{RE} promoter has a poor fit with the consensus at –10 and lacks a consensus sequence at –35. This deficiency explains its dependence on *cII*. The promoter cannot be transcribed by RNA polymerase alone *in vitro*, but can be transcribed when cII is added. The regulator binds to a region extending from about –25 to –45. When RNA polymerase is added, an additional region is protected, which extends from –12 to +13. As summarized in FIGURE 14.30, the two proteins bind to overlapping sites.

The importance of the –35 and –10 regions for promoter function, in spite of their lack of resemblance with the consensus, is indicated by the existence of *cy* mutations. These have effects similar to those of *cII* and *cIII* mutations in preventing the establishment of lysogeny; but they are *cis*-acting instead of *trans*-acting. They fall into two groups, *cyL* and *cyR*, which are localized at the consensus operator positions of –10 and –35.

The *cyL* mutations are located around –10 and probably prevent RNA polymerase from recognizing the promoter.

The *cyR* mutations are located around –35 and fall into two types, which affect either RNA polymerase or cII binding. Mutations in the center of the region do not affect cII binding; presumably they prevent RNA polymerase binding. On either side of this region, mutations in short tetrameric repeats, TTGC, prevent cII from binding. Each base in the tetramer is 10 bp (one helical turn) separated from its homolog in the other tetramer, so that when cII recognizes the two tetramers, it lies on one face of the double helix.

Positive control of a promoter implies that an accessory protein has increased the efficiency with which RNA polymerase initiates transcription. FIGURE 14.31 reports that either or both stages of the interaction between promoter and polymerase can be the target for regulation. Initial binding to form a closed complex or its conversion into an open complex can be enhanced.

14.19 Lysogeny Requires Several Events

Key concepts

- cII and cIII cause repressor synthesis to be established and also trigger inhibition of late gene transcription.
- Establishment of repressor turns off immediate and delayed early gene expression.
- Repressor turns on the maintenance circuit for its own synthesis.
- Lambda DNA is integrated into the bacterial genome at the final stage in establishing lysogeny.

Now we can see how lysogeny is established during an infection. FIGURE 14.32 recapitulates

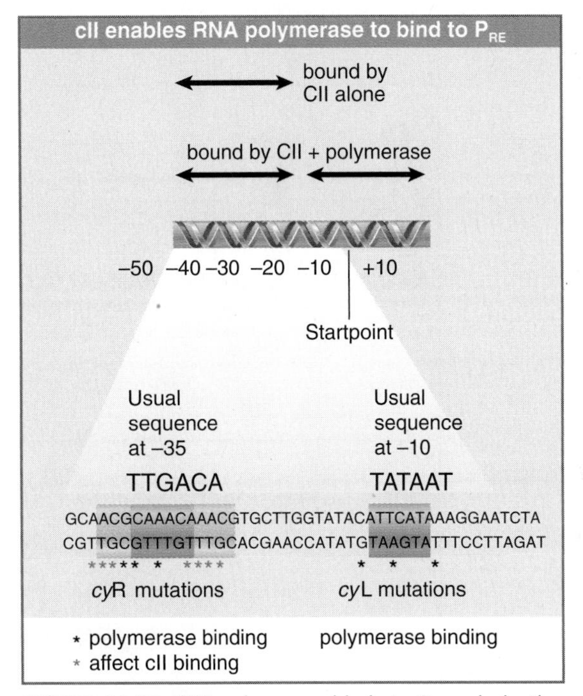

FIGURE 14.30 RNA polymerase binds to P_{RE} only in the presence of cII, which contacts the region around –35.

Positive regulation influences initiation

Promoter	Regulator	Polymerase Binding (equilibrium constant, K_B)	Closed-Open Conversion (rate constant, k_2)
P_{RM}	repressor	no effect	11χ
P_{RE}	CII	100χ	100χ

FIGURE 14.31 Positive regulation can influence RNA polymerase at either stage of initiating transcription.

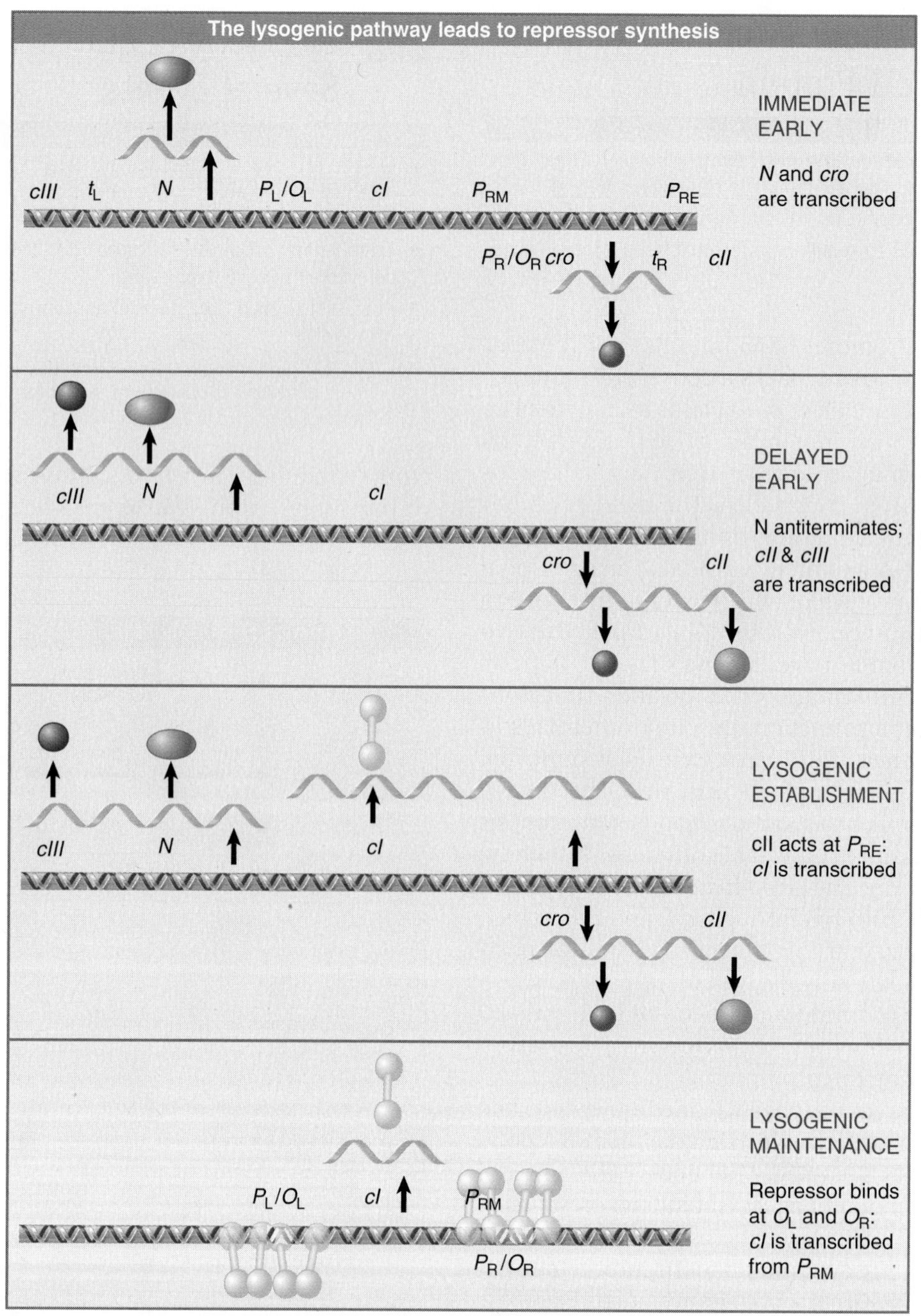

FIGURE 14.32 A cascade is needed to establish lysogeny, but then this circuit is switched off and replaced by the autogenous repressor-maintenance circuit.

the early stages and shows what happens as the result of expression of *cIII* and *cII*. The presence of cII allows P_{RE} to be used for transcription extending through *cI*. Repressor protein is synthesized in high amounts from this transcript and immediately binds to O_L and O_R.

By directly inhibiting any further transcription from P_L and P_R, repressor binding turns off the expression of all phage genes. This halts the synthesis of cII and cIII, which are unstable; they decay rapidly, with the result that P_{RE} can no longer be used. Thus the synthesis of repressor via the establishment circuit is brought to a halt.

Repressor is now present at O_R, though. It switches on the maintenance circuit for expression from P_{RM}. Repressor continues to be synthesized, although at the lower level typical of P_{RM} function. Thus the establishment circuit starts off repressor synthesis at a high level. The repressor then turns off all other functions, while at the same time turning on the maintenance circuit, which functions at the low level adequate to sustain lysogeny.

We shall not at this point deal in detail with the other functions needed to establish lysogeny, but we can just briefly remark that the infecting lambda DNA must be inserted into the bacterial genome (see Section 19.17, Specialized Recombination Involves Specific Sites). The insertion requires the product of gene *int*, which is expressed from its own promoter P_I, at which cII also is necessary. The sequence of P_I shows homology with P_{RE} in the cII binding site (although not in the –10 region). The functions necessary for establishing the lysogenic control circuit are therefore under the same control as the function needed to integrate the phage DNA into the bacterial genome. Thus the establishment of lysogeny is under a control that ensures all the necessary events occur with the same timing.

Emphasizing the tricky quality of lambda's intricate cascade, we now know that cII promotes lysogeny in another, indirect manner. It sponsors transcription from a promoter called $P_{anti\text{-}Q}$, which is located within the *Q* gene. This transcript is an antisense version of the *Q* region, and it hybridizes with *Q* mRNA to prevent translation of Q protein, whose synthesis is essential for lytic development. Thus the same mechanisms that directly promote lysogeny by causing transcription of the *cI* repressor gene also indirectly help lysogeny by inhibiting the expression of *cro* (see above) and *Q*, the regulator genes needed for the antagonistic lytic pathway.

14.20 The cro Repressor Is Needed for Lytic Infection

Key concepts

- Cro binds to the same operators as repressor, but with different affinities.
- When Cro binds to O_R3, it prevents RNA polymerase from binding to P_{RM} and blocks maintenance of repressor.
- When Cro binds to other operators at O_R or O_L, it prevents RNA polymerase from expressing immediate early genes, which (indirectly) blocks repressor establishment.

Lambda has the alternatives of entering lysogeny or starting a lytic infection. Lysogeny is initiated by establishing an autogenous maintenance circuit that inhibits the entire lytic cascade through applying pressure at two points. The program for establishing lysogeny proceeds through some of the same events that are required for the lytic cascade (expression of delayed early genes via expression of *N* is needed). We now face a problem. How does the phage enter the lytic cycle?

The key influence on the lytic cycle is the role of gene *cro*, which codes for another repressor. *Cro is responsible for preventing the synthesis of the repressor protein;* this action shuts off the possibility of establishing lysogeny. *cro* mutants usually establish lysogeny rather than entering the lytic pathway, because they lack the ability to switch events away from the expression of repressor.

Cro forms a small dimer (the subunit is 9 kD) that acts within the immunity region. It has two effects:

- It prevents the synthesis of repressor via the maintenance circuit; that is, it prevents transcription via P_{RM}.
- It also inhibits the expression of early genes from both P_L and P_R.

This means that, when a phage enters the lytic pathway, Cro has responsibility both for preventing the synthesis of repressor and (subsequently) for turning down the expression of the early genes.

Cro achieves its function by binding to the same operators as *(cI)* repressor protein. Cro includes a region with the same general structure as the repressor; a helix-2 is offset at an angle from recognition helix-3. (The remainder of the structure is different, which demonstrates that the helix-turn-helix motif can operate within various contexts.) As for repressor, Cro binds symmetrically at the operators.

The sequences of Cro and repressor in the helix-turn-helix region are related, which explains their ability to contact the same DNA sequences (see Figure 14.21). Cro makes similar contacts to those made by repressor, but binds to only one face of DNA; it lacks the N-terminal arms by which repressor reaches around to the other side.

How can two proteins have the same sites of action, yet have such opposite effects? The answer lies in the different affinities that each protein has for the individual binding sites within the operators. Let us just consider O_R, for which more is known, and where Cro exerts both its effects. The series of events is illustrated in FIGURE 14.33. (Note that the first two stages are identical to those of the lysogenic circuit shown in Figure 14.32.)

The affinity of Cro for O_R3 is greater than its affinity for O_R2 or O_R1. Thus it binds first to

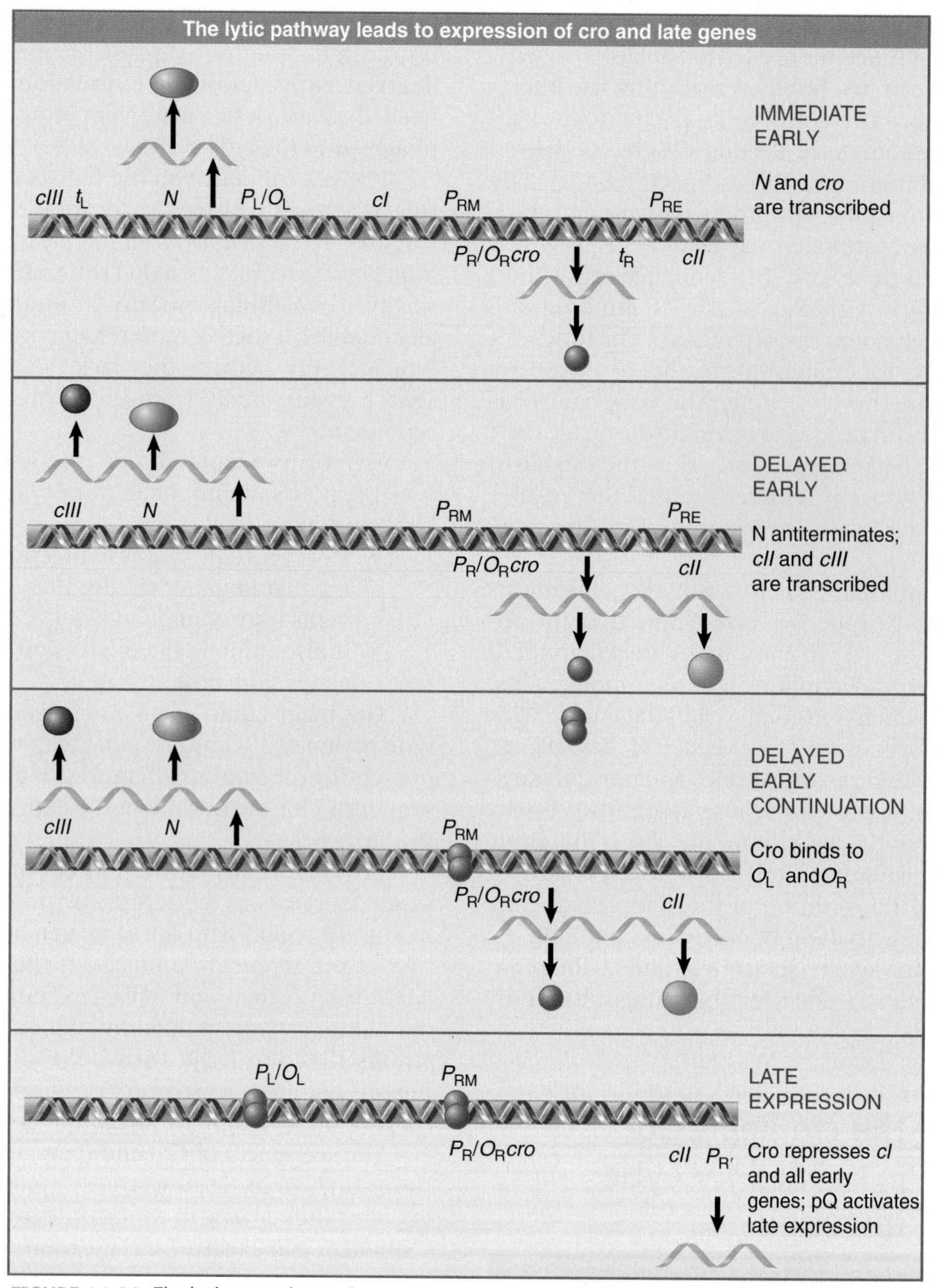

FIGURE 14.33 The lytic cascade requires Cro protein, which directly prevents repressor maintenance via P_{RM}, as well as turning off delayed early gene expression, indirectly preventing repressor establishment.

O_R3. This inhibits RNA polymerase from binding to P_{RM}. As a result, Cro's first action is to prevent the maintenance circuit for lysogeny from coming into play.

Cro then binds to O_R2 or O_R1. Its affinity for these sites is similar, and there is no cooperative effect. Its presence at either site is sufficient to prevent RNA polymerase from using P_R. This in turn stops the production of the early functions (including Cro itself). As a result of cII's instability, any use of P_{RE} is brought to a halt. Thus the two actions of Cro together block *all* production of repressor.

As far as the lytic cycle is concerned, Cro turns down (although it does not completely eliminate) the expression of the early genes. Its incomplete effect is explained by its affinity for O_R1 and O_R2, which is about eight times lower than that of repressor. This effect of Cro does not occur until the early genes have become more or less superfluous, because pQ is present; by this time, the phage has started late gene expression and is concentrating on the production of progeny phage particles.

14.21 What Determines the Balance Between Lysogeny and the Lytic Cycle?

Key concepts

- The delayed early stage when both Cro and repressor are being expressed is common to lysogeny and the lytic cycle.
- The critical event is whether cII causes sufficient synthesis of repressor to overcome the action of Cro.

The programs for the lysogenic and lytic pathways are so intimately related that it is impossible to predict the fate of an individual phage genome when it enters a new host bacterium. Will the antagonism between repressor and Cro be resolved by establishing the autogenous maintenance circuit shown in Figure 14.32, or by turning off repressor synthesis and entering the late stage of development shown in Figure 14.33?

The same pathway is followed in both cases right up to the brink of decision. Both involve the expression of the immediate early genes and extension into the delayed early genes. The difference between them comes down to the question of whether repressor or Cro will obtain occupancy of the two operators.

The early phase during which the decision is made is limited in duration in either case. No matter which pathway the phage follows, expression of all early genes will be prevented as P_L and P_R are repressed and, as a consequence of the disappearance of cII and cIII, production of repressor via P_{RE} will cease.

The critical question comes down to whether the cessation of transcription from P_{RE} is followed by activation of P_{RM} and the establishment of lysogeny, or whether P_{RM} fails to become active and the pQ regulator commits the phage to lytic development. FIGURE 14.34 shows the critical stage, at which both repressor and Cro are being synthesized.

The initial event in establishing lysogeny is the binding of repressor at O_L1 and O_R1. Binding at the first sites is rapidly succeeded by cooperative binding of further repressor dimers at O_L2 and O_R2. This shuts off the synthesis of Cro and starts up the synthesis of repressor via P_{RM}.

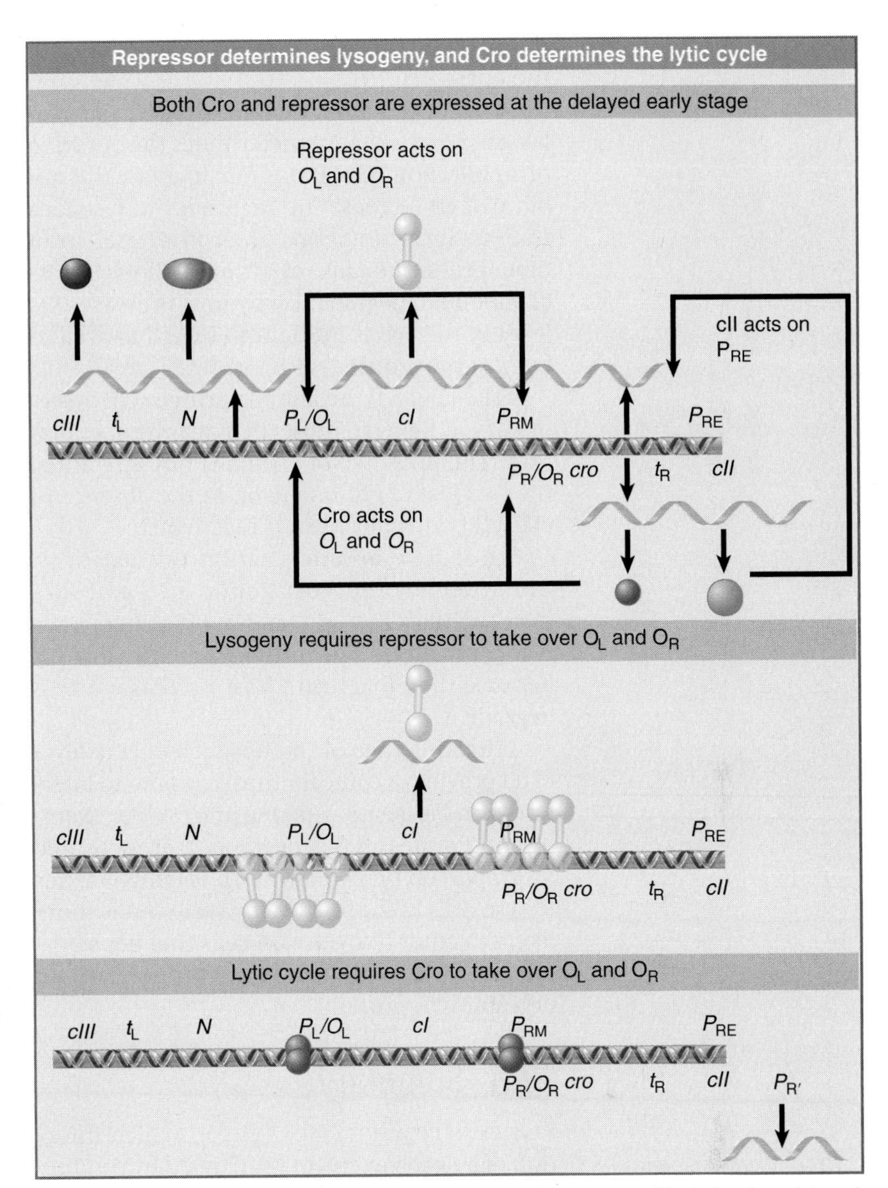

FIGURE 14.34 The critical stage in deciding between lysogeny and lysis is when delayed early genes are being expressed. If cII causes sufficient synthesis of repressor, lysogeny will result because repressor occupies the operators. Otherwise Cro occupies the operators, resulting in a lytic cycle.

The initial event in entering the lytic cycle is the binding of Cro at O_R3. This stops the lysogenic-maintenance circuit from starting up at P_{RM}. Cro must then bind to O_R1 or O_R2, and to O_L1 or O_L2, to turn down early gene expression. By halting production of cII and cIII, this action leads to the cessation of repressor synthesis via P_{RE}. The shutoff of repressor establishment occurs when the unstable cII and cIII proteins decay.

The critical influence over the switch between lysogeny and lysis is cII. If cII is active, synthesis of repressor via the establishment promoter is effective, and as a result, repressor gains occupancy of the operators. If cII is not active repressor establishment fails, and Cro binds to the operators.

The level of cII protein under any particular set of circumstances determines the outcome of an infection. Mutations that increase the stability of cII increase the frequency of lysogenization. Such mutations occur in *cII* itself or in other genes. The cause of *cII*'s instability is its susceptibility to degradation by host proteases. Its level in the cell is influenced by *cIII* as well as by host functions.

The effect of the lambda protein cIII is secondary: it helps to protect cII against degradation. The presence of cIII does not guarantee the survival of cII; however, in the absence of cIII, cII is virtually always inactivated.

Host gene products act on this pathway. Mutations in the host genes *hflA* and *hflB* increase lysogeny—*hfl* stands for *h*igh *f*requency *l*ysogenization. The mutations stabilize cII because they inactivate host protease(s) that degrade it.

The influence of the host cell on the level of cII provides a route for the bacterium to interfere with the decision-taking process. For example, host proteases that degrade cII are activated by growth on rich medium. Thus lambda tends to lyse cells that are growing well, but is more likely to enter lysogeny on cells that are starving (and which lack components necessary for efficient lytic growth).

14.22 Summary

Phages have a lytic life cycle, in which infection of a host bacterium is followed by production of a large number of phage particles, lysis of the cell, and release of the viruses. Some phages also can exist in lysogenic form, in which the phage genome is integrated into the bacterial chromosome and is inherited in this inert, latent form like any other bacterial gene.

In general, lytic infection falls into three phases. In the first phase a small number of phage genes are transcribed by the host RNA polymerase. One or more of these genes is a regulator that controls expression of the group of genes expressed in the second phase. The pattern is repeated in the second phase, when one or more genes is a regulator needed for expression of the genes of the third phase. Genes of the first two phases code for enzymes needed to reproduce phage DNA; genes of the final phase code for structural components of the phage particle. It is common for the very early genes to be turned off during the later phases.

In phage lambda, the genes are organized into groups whose expression is controlled by individual regulatory events. The immediate early gene *N* codes for an antiterminator that allows transcription of the leftward and rightward groups of delayed early genes from the early promoters P_R and P_L. The delayed early gene *Q* has a similar antitermination function that allows transcription of all late genes from the promoter $P_{R'}$. The lytic cycle is repressed, and the lysogenic state maintained, by expression of the *cI* gene, whose product is a repressor protein that acts at the operators O_R and O_L to prevent use of the promoters P_R and P_L, respectively. A lysogenic phage genome expresses only the *cI* gene from its promoter, P_{RM}. Transcription from this promoter involves positive autogenous regulation, in which repressor bound at O_R activates RNA polymerase at P_{RM}.

Each operator consists of three binding sites for repressor. Each site is palindromic, consisting of symmetrical half-sites. Repressor functions as a dimer. Each half binding site is contacted by a repressor monomer. The N-terminal domain of repressor contains a helix-turn-helix motif that contacts DNA. Helix-3 is the recognition helix and responsible for making specific contacts with base pairs in the operator. Helix-2 is involved in positioning helix-3; it is also involved in contacting RNA polymerase at P_{RM}. The C-terminal domain is required for dimerization. Induction is caused by cleavage between the N- and C-terminal domains, which prevents the DNA-binding regions from functioning in dimeric form, thereby reducing their affinity for DNA and making it impossible to maintain lysogeny. Repressor-operator binding is cooperative, so that once one dimer has bound to the first site, a second dimer binds more readily to the adjacent site.

The helix-turn-helix motif is used by other DNA-binding proteins, including lambda Cro. Lambda Cro binds to the same operators but has a different affinity for the individual operator sites, which are determined by the sequence of helix-3. Cro binds individually to operator sites, starting with O_R3, in a noncooperative manner. It is needed for progression through the lytic cycle. Its binding to O_R3 first prevents synthesis of repressor from P_{RM}, and then its binding to O_R2 and O_R1 prevents continued expression of early genes, an effect also seen in its binding to O_L1 and O_L2.

Establishment of repressor synthesis requires use of the promoter P_{RE}, which is activated by the product of the *cII* gene. The product of *cIII* is required to stabilize the *cII* product against degradation. By turning off *cII* and *cIII* expression, Cro acts to prevent lysogeny. By turning off all transcription except that of its own gene, repressor acts to prevent the lytic cycle. The choice between lysis and lysogeny depends on whether repressor or Cro gains occupancy of the operators in a particular infection. The stability of cII protein in the infected cell is a primary determinant of the outcome.

References

14.4 Two Types of Regulatory Event Control the Lytic Cascade

Review

Greenblatt, J., Nodwell, J. R., and Mason, S. W. (1993). Transcriptional antitermination. *Nature* 364, 401–406.

14.6 Lambda Immediate Early and Delayed Early Genes Are Needed for Both Lysogeny and the Lytic Cycle

Review

Ptashne, M. (2004). *The Genetic Switch: Phage Lambda Revisited.* Cold Spring Harbor, NY: Cold Spring Harbor Press.

14.8 Lysogeny Is Maintained by Repressor Protein

Research

Pirrotta, V., Chadwick, P., and Ptashne, M. (1970). Active form of two coliphage repressors. *Nature* 227, 41–44.

Ptashne, M. (1967). Isolation of the lambda phage repressor. *Proc. Natl. Acad. Sci. USA* 57, 306–313.

Ptashne, M. (1967). Specific binding of the lambda phage repressor to lambda DNA. *Nature* 214, 232–234.

14.9 The Repressor and Its Operators Define the Immunity Region

Review

Friedman, D. I. and Gottesman, M. (1982). *Lambda II.* Cambridge, MA: Cell Press.

14.10 The DNA-Binding Form of Repressor Is a Dimer

Research

Pabo, C. O. and Lewis, M. (1982). The operator-binding domain of lambda repressor: structure and DNA recognition. *Nature* 298, 443–447.

14.11 Repressor Uses a Helix-Turn-Helix Motif to Bind DNA

Research

Sauer, R. T. et al. (1982). Homology among DNA-binding proteins suggests use of a conserved super-secondary structure. *Nature* 298, 447–451.

14.12 The Recognition Helix Determines Specificity for DNA

Research

Brennan, R. G. et al. (1990). Protein-DNA conformational changes in the crystal structure of a lambda Cro-operator complex. *Proc. Natl. Acad. Sci. USA* 87, 8165–8169.

Wharton, R. L., Brown, E. L., and Ptashne, M. (1984). Substituting an α-helix switches the sequence specific DNA interactions of a repressor. *Cell* 38, 361–369.

14.13 Repressor Dimers Bind Cooperatively to the Operator

Research

Bell, C. E., Frescura, P., Hochschild, A., and Lewis, M. (2000). Crystal structure of the lambda repressor C-terminal domain provides a model for cooperative operator binding. *Cell* 101, 801–811.

Johnson, A. D., Meyer, B. J., and Ptashne, M. (1979). Interactions between DNA-bound repressors govern regulation by the phage lambda repressor. *Proc. Natl. Acad. Sci. USA* 76, 5061–5065.

14.14 Repressor at O_R2 Interacts with RNA Polymerase at P_{RM}

Research

Hochschild, A., Irwin, N., and Ptashne, M. (1983). Repressor structure and the mechanism of positive control. *Cell* 32, 319–325.

Li, M., Moyle, H., and Susskind, M. M. (1994). Target of the transcriptional activation function of phage lambda cI protein. *Science* 263, 75–77.

14.16 Cooperative Interactions Increase the Sensitivity of Regulation

Review

Ptashne, M. (2004). *The Genetic Switch: Phage Lambda Revisited.* Cold Spring Harbor, NY: Cold Spring Harbor Press.

Research

Bell, C. E. and Lewis, M. (2001). Crystal structure of the lambda repressor C-terminal domain octamer. *J. Mol. Biol.* 314, 1127–1136.

Dodd, I. B., Perkins, A. J., Tsemitsidis, D., and Egan, J. B. (2001). Octamerization of lambda CI repressor is needed for effective repression of P(RM) and efficient switching from lysogeny. *Genes Dev.* 15, 3013–3022.

15

The Replicon

CHAPTER OUTLINE

15.1 Introduction

Whether a cell has only one chromosome (as in prokaryotes) or has many chromosomes (as in eukaryotes), the entire genome must be replicated precisely once for every cell division. How is the act of replication linked to the cell cycle?

Two general principles are used to compare the state of replication with the condition of the cell cycle:

- *Initiation of DNA replication commits the cell (prokaryotic or eukaryotic) to a further division.* From this standpoint, the number of descendants that a cell generates is determined by a series of decisions on whether or not to initiate DNA replication. Replication is controlled at the stage of initiation. *Once replication has started, it continues until the entire genome has been duplicated.*
- If replication proceeds, the consequent division cannot be permitted to occur until the replication event has been completed. Indeed, the completion of replication may provide a trigger for cell division. The duplicate genomes are then segregated one to each daughter cell. The unit of segregation is the chromosome.

In prokaryotes, the initiation of replication is a single event involving a unique site on the bacterial chromosome, and the process of division is accomplished by the development of a septum that grows from the cell wall and divides the cell into two. In eukaryotic cells, initiation of replication is identified by the start of S phase, a protracted period during which DNA synthesis occurs, and which involves many individual initiation events. The act of division is accomplished by the reorganization of the cell at mitosis. In this chapter, we are concerned with the regulation of DNA replication. How is a cycle of replication initiated? What controls its progress and how is its termination signaled?

The unit of DNA in which an individual act of replication occurs is called the **replicon.** Each replicon "fires" once, and only once, in each cell cycle. The replicon is defined by its possession of the control elements needed for replication. It has an **origin** at which replication is initiated. It may also have a **terminus** at which replication stops.

Any sequence attached to an origin—or, more precisely, not separated from an origin by a terminus—is replicated as part of that replicon. The origin is a *cis*-acting site, able to affect only that molecule of DNA on which it resides.

(The original formulation of the replicon [in prokaryotes] viewed it as a unit possessing both the origin *and* the gene coding for the regulator protein. Now, however, "replicon" is usually applied to eukaryotic chromosomes to describe a unit of replication that contains an origin; *trans*-acting regulator protein(s) may be coded elsewhere.)

A genome in a prokaryotic cell constitutes a single replicon; thus the units of replication and segregation coincide. Initiation at a single origin sponsors replication of the entire genome, once for every cell division. Each haploid bacterium has a single chromosome, so this type of replication control is called **single copy.**

Bacteria may contain additional genetic information in the form of plasmids. *A plasmid is an autonomous circular DNA genome that constitutes a separate replicon* (see Figure 14.2). A plasmid replicon may show single copy control, which means that it replicates once every time the bacterial chromosome replicates, or it may be under **multicopy control,** when it is present in a greater number of copies than the bacterial chromosome. Each phage or virus DNA also constitutes a replicon and, thus, is able to initiate many times during an infectious cycle. Perhaps a better way to view the prokaryotic replicon, therefore, is to reverse the definition: *Any DNA molecule that contains an origin can be replicated autonomously in the cell.*

A major difference in the organization of bacterial and eukaryotic genomes is seen in their replication. Each eukaryotic chromosome contains a large number of replicons; thus the unit of segregation includes many units of replication. This adds another dimension to the problem of control: All of the replicons on a chromosome must be fired during one cell cycle. They are not, however, active simultaneously. Each replicon must be activated over a fairly protracted period, *and each must be activated no more than once in each cell cycle.*

Some signal must distinguish replicated from nonreplicated replicons to ensure that replicons do not fire a second time. Many replicons are activated independently, so another signal must exist to indicate when the entire process of replicating all replicons has been completed.

We have begun to collect information about the construction of individual replicons, but we still have little information about the relationship between replicons. We do not know whether the pattern of replication is the same in every cell cycle. Are all origins always used, or are some origins sometimes silent? Do origins always fire in the same order? If there are different classes of origins, what distinguishes them?

In contrast with nuclear chromosomes, which have a single-copy type of control, the DNA of mitochondria and chloroplasts may be regulated more like plasmids that exist in multiple copies per bacterium. There are multiple copies of each organelle DNA per cell, and the control of organelle DNA replication must be related to the cell cycle.

In all these systems, the key question is to define the sequences that function as origins and to determine how they are recognized by the appropriate proteins of the apparatus for replication. We start by considering the basic construction of replicons and the various forms that they take; following the consideration of the origin, we turn to the question of how replication of the genome is coordinated with bacterial division and what is responsible for segregating the genomes to daughter bacteria.

15.2 Replicons Can Be Linear or Circular

Key concepts

- A replicated region appears as an eye within nonreplicated DNA.
- A replication fork is initiated at the origin and then moves sequentially along DNA.
- Replication is unidirectional when a single replication fork is created at an origin.
- Replication is bidirectional when an origin creates two replication forks that move in opposite directions.

Replication eyes form bubbles

Nonreplicated DNA | Replication eye | Nonreplicated DNA

Appearance

Molecular structure

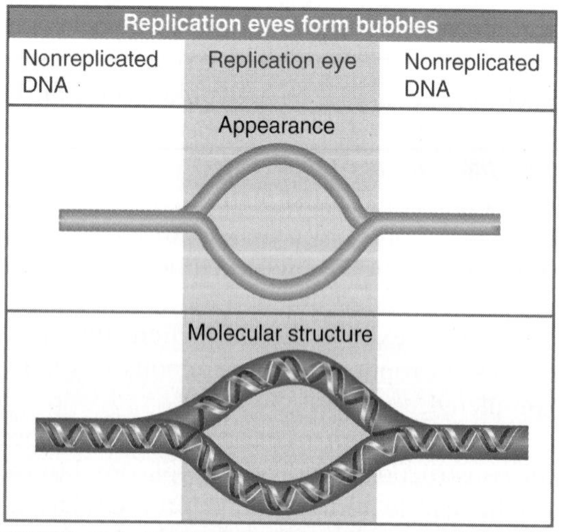

FIGURE 15.1 Replicated DNA is seen as a replication eye flanked by nonreplicated DNA.

A molecule of DNA engaged in replication has two types of regions. FIGURE 15.1 shows that when replicating DNA is viewed by electron microscopy, the replicated region appears as a **replication eye** within the nonreplicated DNA. The nonreplicated region consists of the parental duplex; this opens into the replicated region where the two daughter duplexes have formed.

The point at which replication occurs is called the **replication fork** (sometimes also known as the **growing point**). *A replication fork moves sequentially along the DNA from its starting point at the origin.* The origin may be used to start either **unidirectional replication** or **bidirectional replication.** The type of event is determined by whether one or two replication forks set out from the origin. In unidirectional replication, one replication fork leaves the origin and proceeds along the DNA. In bidirectional replication, two replication forks are formed; they proceed away from the origin in opposite directions.

The appearance of a replication eye does not distinguish between unidirectional and bidirectional replication. As depicted in FIGURE 15.2, the eye can represent either of two structures. If generated by unidirectional replication, the

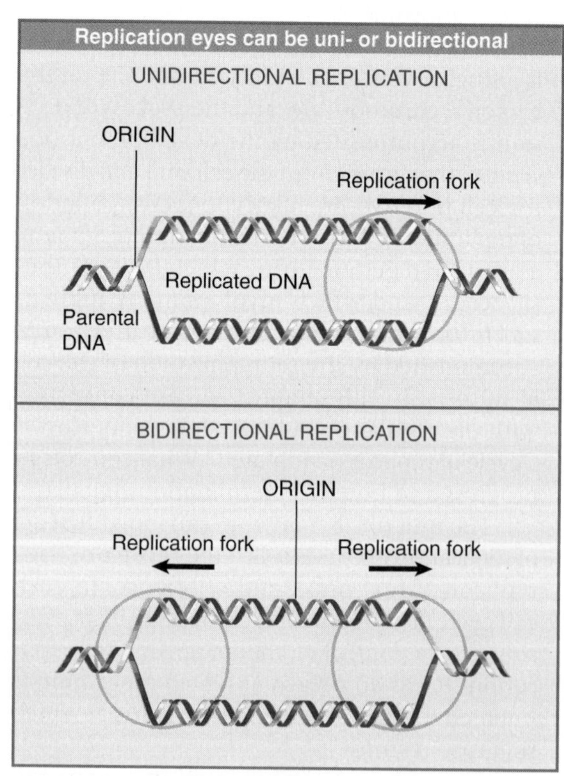

FIGURE 15.2 Replicons may be unidirectional or bidirectional, depending on whether one or two replication forks are formed at the origin.

eye represents one fixed origin and one moving replication fork. If generated by bidirectional replication, the eye represents a pair of replication forks. In either case, the progress of replication expands the eye until ultimately it encompasses the whole replicon.

When a replicon is circular, the presence of an eye forms the θ structure drawn in FIGURE 15.3. The successive stages of replication of the circular DNA of polyoma virus are visualized by electron microscopy in FIGURE 15.4.

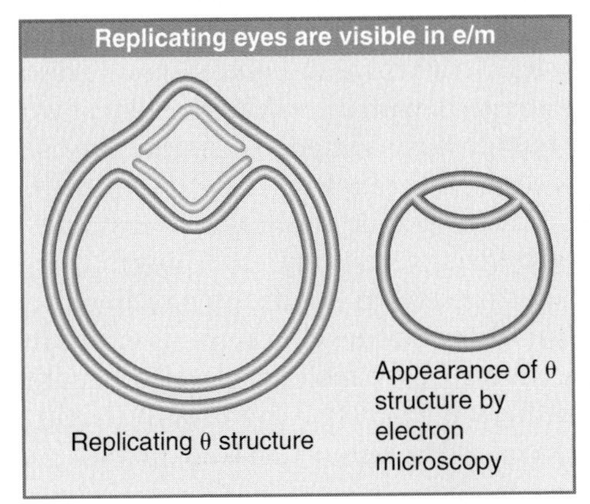

FIGURE 15.3 A replication eye forms a θ structure in circular DNA.

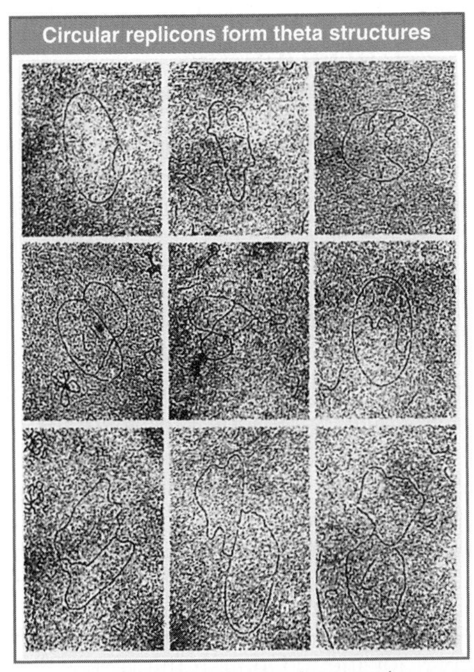

FIGURE 15.4 The replication eye becomes larger as the replication forks proceed along the replicon. Note that the "eye" becomes larger than the nonreplicated segment. The two sides of the eye can be defined because they are both the same length. Photo courtesy of Bernard Hirt, Swiss Institute for Experimental Cancer Research (ISREC).

15.3 Origins Can Be Mapped by Autoradiography and Electrophoresis

Key concepts

- Replication fork movement can be detected by autoradiography using radioactive pulses.
- Replication forks create Y-shaped structures that change the electrophoretic migration of DNA fragments.

Whether a replicating eye has one or two replication forks can be determined in two ways. The choice of method depends on whether the DNA is a defined molecule or an unidentified region of a cellular genome.

With a defined linear molecule, we can use electron microscopy to measure the distance of each end of the eye from the end of the DNA, and then compare the positions of the ends of the eyes in molecules that have eyes of different sizes. If replication is unidirectional, only one of the ends will move; the other is the fixed origin. If replication is bidirectional, both will move; the origin is the point midway between them.

With undefined regions of large genomes, two successive pulses of radioactivity can be used to label the movement of the replication forks. If one pulse has a more intense label than the other, they can be distinguished by the relative intensities of labeling. These can be visualized by autoradiography. FIGURE 15.5 shows that unidirectional replication causes one type

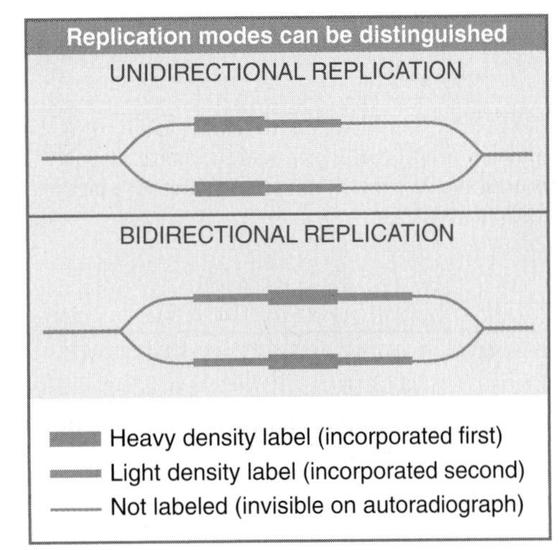

FIGURE 15.5 Different densities of radioactive labeling can be used to distinguish unidirectional and bidirectional replication.

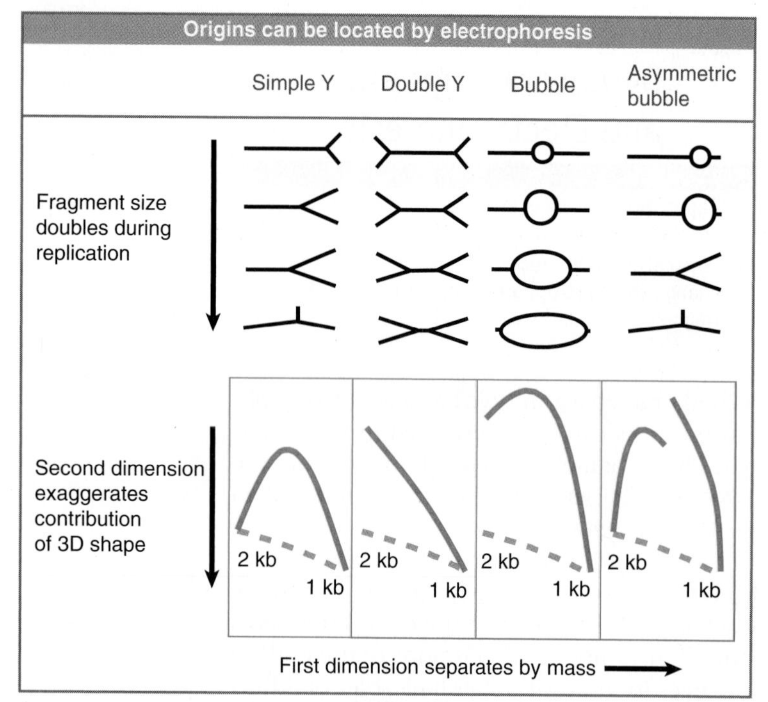

FIGURE 15.6 The position of the origin and the number of replicating forks determine the shape of a replicating restriction fragment, which can be followed by its electrophoretic path (solid line). The dashed line shows the path for a linear DNA.

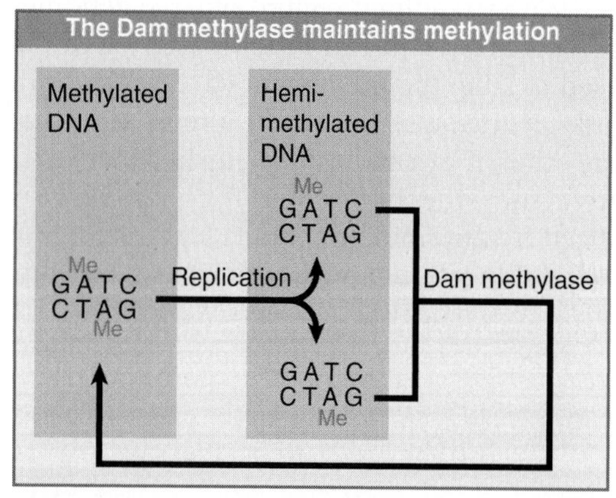

FIGURE 15.7 Replication of methylated DNA gives hemimethylated DNA, which maintains its state at GATC sites until the Dam methylase restores the fully methylated condition.

of label to be followed by the other at *one* end of the eye. Bidirectional replication produces a (symmetrical) pattern at *both* ends of the eye. This is the pattern usually observed in replicons of eukaryotic chromosomes.

A more recent method for mapping origins with greater resolution takes advantage of the effects that changes in shape have upon electrophoretic migration of DNA. FIGURE 15.6 illustrates the two-dimensional mapping technique, in which restriction fragments of replicating DNA are electrophoresed in a first dimension that separates by mass and a second dimension where movement is determined more by shape. Different types of replicating molecules follow characteristic paths, measured by their deviation from the line that would be followed by a linear molecule of DNA that doubled in size.

A simple Y-structure, in which one fork moves along a linear fragment, follows a continuous path. An inflection point occurs when all three branches are the same length, and the structure therefore deviates most extensively from linear DNA. Analogous considerations determine the paths of double Y-structures or bubbles. An asymmetric bubble follows a discontinuous path, with a break at the point at which the bubble is converted to a Y-structure as one fork runs off the end.

Taken together, the various techniques for characterizing replicating DNA show that origins are used most often to initiate bidirectional replication. From this level of resolution, we must now proceed to the molecular level to identify the *cis*-acting sequences that comprise the origin and the *trans*-acting factors that recognize it.

15.4 Does Methylation at the Origin Regulate Initiation?

Key concepts

- *oriC* contains eleven $\frac{\text{GATC}}{\text{CTAG}}$ repeats that are methylated on adenine on both strands.
- Replication generates hemimethylated DNA, which cannot initiate replication.
- There is a 13-minute delay before the $\frac{\text{GATC}}{\text{CTAG}}$ repeats are remethylated.

What feature of a bacterial (or plasmid) origin ensures that it is used to initiate replication only once per cycle? Is initiation associated with some change that marks the origin so that a replicated origin can be distinguished from a nonreplicated origin?

Some sequences that are used for this purpose are included in the origin. *oriC* contains eleven copies of the sequence $\frac{\text{GATC}}{\text{CTAG}}$, which is a target for methylation at the N^6 position of adenine by the Dam methylase. The reaction is illustrated in FIGURE 15.7.

Before replication, the palindromic target site is methylated on the adenines of each strand. Replication inserts the normal (nonmodified) bases into the daughter strands. This generates

hemimethylated DNA, in which one strand is methylated and one strand is unmethylated. Thus the replication event converts Dam target sites from fully methylated to hemimethylated condition.

What is the consequence for replication? The ability of a plasmid relying upon *oriC* to replicate in *dam*⁻*E. coli* depends on its state of methylation. If the plasmid is methylated it undergoes a single round of replication, and then the hemimethylated products accumulate, as described in FIGURE 15.8. Thus a hemimethylated origin cannot be used to initiate a replication cycle.

This suggests two explanations: Initiation may require full methylation of the Dam target sites in the origin, or it may be inhibited by hemimethylation of these sites. The latter seems to be the case, because an origin of nonmethylated DNA can function effectively.

Thus hemimethylated origins cannot initiate again until the Dam methylase has converted them into fully methylated origins. The GATC sites at the origin remain hemimethylated for ~13 minutes after replication. This long period is unusual because at typical GATC sites elsewhere in the genome, remethylation begins immediately (<1.5 minutes) following replication. One other region behaves like *oriC:* The promoter of the *dnaA* gene also shows a delay before remethylation begins.

While it is hemimethylated the *dnaA* promoter is repressed, which causes a reduction in the level of DnaA protein. Thus the origin itself is inert, and production of the crucial initiator protein is repressed during this period.

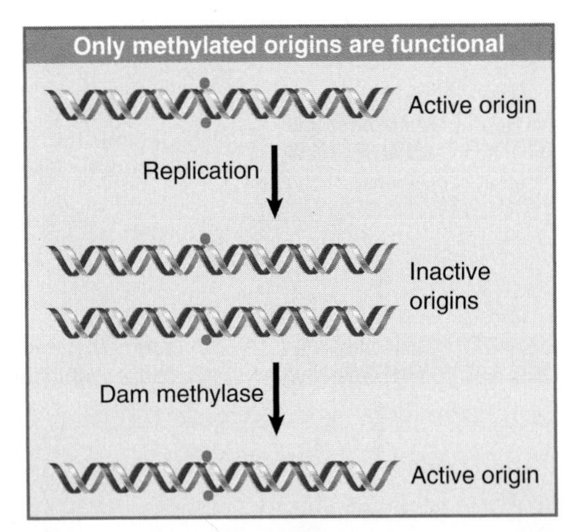

FIGURE 15.8 Only fully methylated origins can initiate replication; hemimethylated daughter origins cannot be used again until they have been restored to the fully methylated state.

15.5 Origins May Be Sequestered after Replication

Key concepts

- SeqA binds to hemimethylated DNA and is required for delaying rereplication.
- SeqA may interact with DnaA.
- As the origins are hemimethylated they bind to the cell membrane and may be unavailable to methylases.
- The nature of the connection between the origin and the membrane is still unclear.

What is responsible for the delay in remethylation at *oriC* and *dnaA?* The most likely explanation is that these regions are sequestered in a form in which they are inaccessible to the Dam methylase.

A circuit responsible for controlling reuse of origins is identified by mutations in the gene *seqA*. The mutants reduce the delay in remethylation at both *oriC* and *dnaA*. As a result, they initiate DNA replication too soon, thereby accumulating an excessive number of origins. This suggests that *seqA* is part of a negative regulatory circuit that prevents origins from being remethylated. SeqA binds to hemimethylated DNA more strongly than to fully methylated DNA. It may initiate binding when the DNA becomes hemimethylated, at which point its continued presence prevents formation of an open complex at the origin. SeqA does not have specificity for the *oriC* sequence, and it seems likely that this is conferred by DnaA protein. This would explain genetic interactions between *seqA* and *dnaA*.

Hemimethylation of the GATC sequences in the origin is required for its association with the cell membrane *in vitro*. Hemimethylated *oriC* DNA binds to the membranes, but DNA that is fully methylated does not bind. One possibility is that membrane association is involved in controlling the activity of the origin. This function could be separate from any role that the membrane plays in segregation (see Figure 17.4). Association with the membrane could prevent reinitiation from occurring prematurely, either indirectly because the origins are sequestered, or directly because some component at the membrane inhibits the reaction.

The properties of the membrane fraction suggest that it includes components that regulate replication. An inhibitor found in this

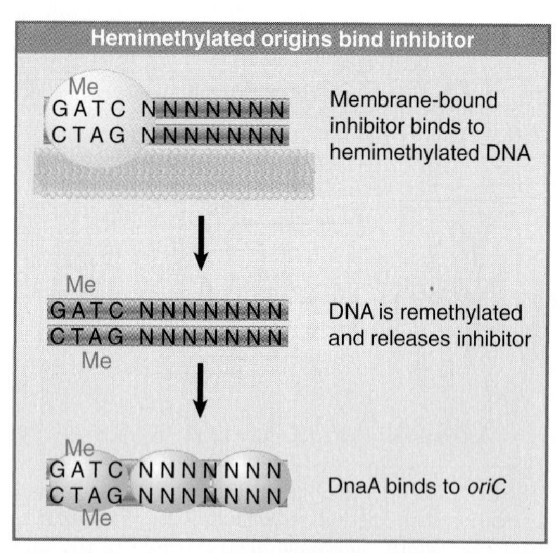

FIGURE 15.9 A membrane-bound inhibitor binds to hemimethylated DNA at the origin and may function by preventing the binding of DnaA. It is released when the DNA is remethylated.

fraction competes with DnaA protein. This inhibitor can prevent initiation of replication only if it is added to an *in vitro* system before DnaA protein. This suggests the model shown in FIGURE 15.9, in which the inhibitor specifically recognizes hemimethylated DNA and prevents DnaA from binding. When the DNA is remethylated, the inhibitor is released and DnaA is free to initiate replication. If the inhibitor is associated with the membrane, association and dissociation of DNA with the membrane may be involved in the control of replication.

The full scope of the system used to control reinitiation is not clear, but several mechanisms may be involved: physical sequestration of the origin, delay in remethylation, inhibition of DnaA binding, and repression of *dnaA* transcription. It is not immediately obvious which of these events cause the others, and whether their effects on initiation are direct or indirect.

We still have to come to grips with the central issue of which feature has the basic responsibility for timing. One possibility is that attachment to the membrane occurs at initiation, and that assembly of some large structure is required to release the DNA. The period of sequestration appears to increase with the length of the cell cycle, which suggests that it directly reflects the clock that controls reinitiation.

As the only member of the replication apparatus uniquely required at the origin, DnaA has attracted much attention. DnaA is a target for several regulatory systems. It may be that no one of these systems alone is adequate to control frequency of initiation, but that when combined they achieve the desired result. Some mutations in *dnaA* render replication asynchronous, which suggests that DnaA could be the "titrator" or "clock" that measures the number of origins relative to cell mass. Overproduction of DnaA yields conflicting results, which vary from no effect to causing initiation to take place at reduced mass.

It has been difficult to identify the protein component(s) that mediate membrane attachment. A hint that this is a function of DnaA is provided by its response to phospholipids. Phospholipids promote the exchange of ATP with ADP bound to DnaA. We do not know what role this plays in controlling the activity of DnaA (which requires ATP), but the reaction implies that DnaA is likely to interact with the membrane. This would imply that more than one event is involved in associating with the membrane. Perhaps a hemimethylated origin is bound by the membrane-associated inhibitor, but when the origin becomes fully methylated, the inhibitor is displaced by DnaA associated with the membrane.

If DnaA is the initiator that triggers a replication cycle, the key event will be its accumulation at the origin to a critical level. There are no cyclic variations in the overall concentration or expression of DnaA, which suggests that local events must be responsible. To be active in initiating replication, DnaA must be in the ATP-bound form. DnaA has a weak intrinsic activity that converts the ATP to ADP. This activity is enhanced by the β subunit of DNA polymerase III. When the replicase is incorporated into the replication complex, this interaction causes hydrolysis of the ATP bound to DnaA, thereby inactivating DnaA, and preventing it from starting another replication cycle. This reaction has been called *RIDA* (regulatory inactivation of DnaA). It is enhanced by a protein called Hda. We do not yet know what controls the timing of the reactivation of DnaA.

Another factor that controls availability of DnaA at the origin is the competition for binding it to other sites on DNA. In particular, a locus called *dat* has a large concentration of DnaA-binding sites. It binds about 8× more DnaA than the origin. Deletion of *dat* causes initiation to occur more frequently. This significantly reduces the amount of DnaA available to the origin, but we do not yet understand exactly what role this may play in controlling the timing of initiation.

15.6 Each Eukaryotic Chromosome Contains Many Replicons

Key concepts

- Eukaryotic replicons are 40 to 100 kb in length.
- A chromosome is divided into many replicons.
- Individual replicons are activated at characteristic times during S phase.
- Regional activation patterns suggest that replicons near one another are activated at the same time.

In eukaryotic cells, the replication of DNA is confined to part of the cell cycle. **S phase** usually lasts a few hours in a higher eukaryotic cell. Replication of the large amount of DNA contained in an eukaryotic chromosome is accomplished by dividing it into many individual replicons. Only some of these replicons are engaged in replication at any point in S phase. Presumably each replicon is activated at a specific time during S phase, although the evidence on this issue is not decisive.

The start of S phase is signaled by the activation of the first replicons. Over the next few hours, initiation events occur at other replicons in an ordered manner. Much of our knowledge about the properties of the individual replicons is derived from autoradiographic studies, generally using the types of protocols illustrated in Figure 15.5 and Figure 15.6. Chromosomal replicons usually display bidirectional replication.

How large is the average replicon, and how many are there in the genome? A difficulty in characterizing the individual unit is that adjacent replicons may fuse to give large replicated eyes, as illustrated in FIGURE 15.10. The approach usually used to distinguish individual replicons from fused eyes is to rely on stretches of DNA in which several replicons can be seen to be active, presumably captured at a stage when all have initiated around the same time, but before the forks of adjacent units have met.

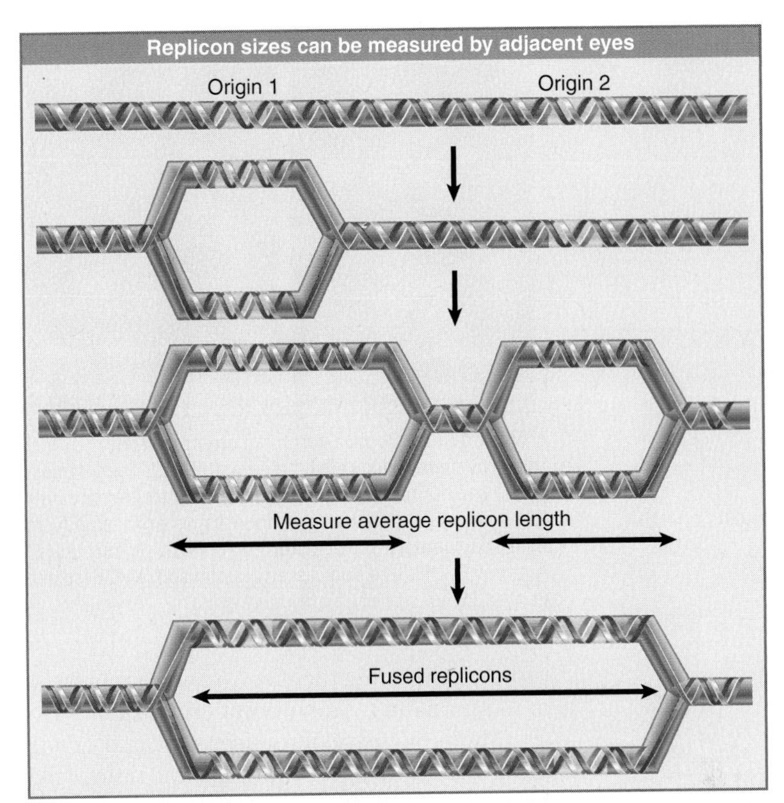

FIGURE 15.10 Measuring the size of the replicon requires a stretch of DNA in which adjacent replicons are active.

In groups of active replicons, the average size of the unit is measured by the distance between the origins (that is, between the midpoints of adjacent replicons). The rate at which the replication fork moves can be estimated from the maximum distance that the autoradiographic tracks travel during a given time.

Individual replicons in eukaryotic genomes are relatively small, typically ~40 kb in yeast or fly and ~100 kb in animals cells. However, they can vary greater than tenfold in length within a genome. The rate of replication is ~2000 bp/min, which is much slower than the 50,000 bp/min of bacterial replication fork movement.

From the speed of replication, it is evident that a mammalian genome could be replicated in ~1 hour if all replicons functioned simultaneously. S phase actually lasts for >6 hours in a typical somatic cell, though, which implies that no more than 15% of the replicons are likely to be active at any given moment. There are some exceptional cases, such as the early embryonic divisions of *Drosophila* embryos, where the duration of S phase is compressed by the simultaneous functioning of a large number of replicons.

How are origins selected for initiation at different times during S phase? In *Saccharomyces cerevisiae*, the default appears to be for origins to replicate early, but *cis*-acting sequences can cause origins linked to them to replicate at late times.

Available evidence suggests that chromosomal replicons do not have termini at which the replication forks cease movement and (presumably) dissociate from the DNA. It seems more likely that a replication fork continues from its origin until it meets a fork proceeding toward it from the adjacent replicon. We have already mentioned the potential topological

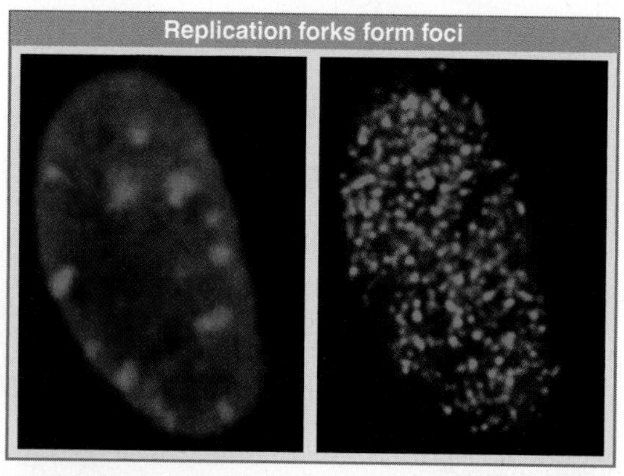

FIGURE 15.11 Replication forks are organized into foci in the nucleus. Cells were labeled with BrdU. The leftmost panel was stained with propidium iodide to identify bulk DNA. The right panel was stained using an antibody to BrdU to identify replicating DNA. Photo courtesy of Anthony D. Mills and Ron Laskey, Hutchinson/MRC Research Center, University of Cambridge.

problem of joining the newly synthesized DNA at the junction of the replication forks.

The propensity of replicons located in the same vicinity to be active at the same time could be explained by "regional" controls, in which groups of replicons are initiated more or less coordinately, as opposed to a mechanism in which individual replicons are activated one by one in dispersed areas of the genome. Two structural features suggest the possibility of large-scale organization. Quite large regions of the chromosome can be characterized as "early replicating" or "late replicating," implying that there is little interspersion of replicons that fire at early or late times. Visualization of replicating forks by labeling with DNA precursors identifies 100 to 300 "foci" instead of uniform staining; each focus shown in FIGURE 15.11 probably contains >300 replication forks. The foci could represent fixed structures through which replicating DNA must move.

15.7 Replication Origins Can Be Isolated in Yeast

Key concepts

- Origins in *S. cerevisiae* are short A-T-rich sequences that have an essential 11-bp sequence.
- The ORC is a complex of six proteins that binds to an *ARS*.

Any segment of DNA that has an origin should be able to replicate, so although plasmids are rare in eukaryotes, it may be possible to construct them by suitable manipulation *in vitro*. This has been accomplished in yeast, although not in higher eukaryotes.

S. cerevisiae mutants can be "transformed" to the wild phenotype by addition of DNA that carries a wild-type copy of the gene. The discovery of yeast origins resulted from the observation that some yeast DNA fragments (when circularized) are able to transform defective cells very efficiently. These fragments can survive in the cell in the unintegrated (autonomous) state, that is, as self-replicating plasmids.

A high-frequency transforming fragment possesses a sequence that confers the ability to replicate efficiently in yeast. This segment is called an **ARS** (for autonomously replicating sequence). *ARS* elements are derived from origins of replication.

Where *ARS* elements have been systematically mapped over extended chromosomal regions, it seems that only some of them are actually used to initiate replication. The others are silent, or possibly used only occasionally. If it is true that some origins have varying probabilities of being used, it follows that there can be no fixed termini between replicons. In this case, a given region of a chromosome could be replicated from different origins in different cell cycles.

An *ARS* element consists of an A-T-rich region that contains discrete sites in which mutations affect origin function. Base composition rather than sequence may be important in the rest of the region. FIGURE 15.12 shows a systematic mutational analysis along the length of an origin. Origin function is abolished completely by mutations in a 14-bp "core" region, called the **A domain,** which contains an 11-bp consensus sequence consisting of A-T base pairs. This consensus sequence (sometimes called the *ACS* for ARS Consensus Sequence) is the only homology between known *ARS* elements.

Mutations in three adjacent elements, numbered B1 to B3, reduce origin function. An origin can function effectively with any two of the B elements, so long as a functional A element is present. (Imperfect copies of the core consensus, typically conforming at 9/11 positions, are found close to, or overlapping with, each B element, but they do not appear to be necessary for origin function.)

The ORC (origin recognition complex) is a complex of six proteins with a mass of ~400 kD. ORC binds to the A and B1 elements on the A-T-rich strand and is associated with *ARS* elements throughout the cell cycle. This means that initiation depends on changes in its condition rather than *de novo* association with an origin (see Section 15.9, Licensing Factor Consists of MCM Proteins). By counting the number of sites to which ORC binds, we can estimate that there are about 400 origins of replication in the yeast genome. This means that the average length of a replicon is ~35,000 bp. Counterparts to ORC are found in higher eukaryotic cells.

ORC was first found in *S. cerevisiae* (where it is called scORC), but similar complexes have now been characterized in *Schizosaccharomyces pombe* (spORC), *Drosophila* (DmORC), and *Xenopus* (XlORC). All of the ORC complexes bind to DNA. Although none of the binding sites have been characterized in the same detail as in *S. cerevisiae,* in several cases they are at locations associated with the initiation of replication. It seems clear that ORC is an initiation complex whose binding identifies an origin of replication. Details of the interaction, however, are clear only in *S. cerevisiae;* it is possible that additional components are required to recognize the origin in the other cases.

ARS elements satisfy the classic definition of an origin as a *cis*-acting sequence that causes DNA replication to initiate. Are similar elements to be found in higher eukaryotes? The conservation of the ORC suggests that origins are likely to take the same sort of form in other eukaryotes, but in spite of this, there is little conservation of sequence among putative origins in different organisms.

Difficulties in finding consensus origin sequences suggest the possibility that origins may be more complex (or determined by features other than discrete *cis*-acting sequences). There are suggestions that some animal cell replicons may have complex patterns of initiation: In some cases, many small replication bubbles are found in one region, posing the question of whether there are alternative or multiple starts to replication, and whether there is a small discrete origin.

A reconciliation between this phenomenon and the use of ORCs is suggested by the discovery that environmental effects can influence the use of origins. At one location where multiple bubbles are found, there is a primary origin that is used predominantly when the nucleotide supply is high. When the nucleotide supply is limiting, though, many secondary origins are also used, giving rise to a pattern of multiple bubbles. One possible molecular explanation is that ORCs dissociate from the primary origin and initiate elsewhere in the vicinity if the supply of nucleotides is insufficient for the initiation reaction to occur quickly. At all events, it now seems likely that we will be able in due course to characterize discrete sequences that function as origins of replication in higher eukaryotes.

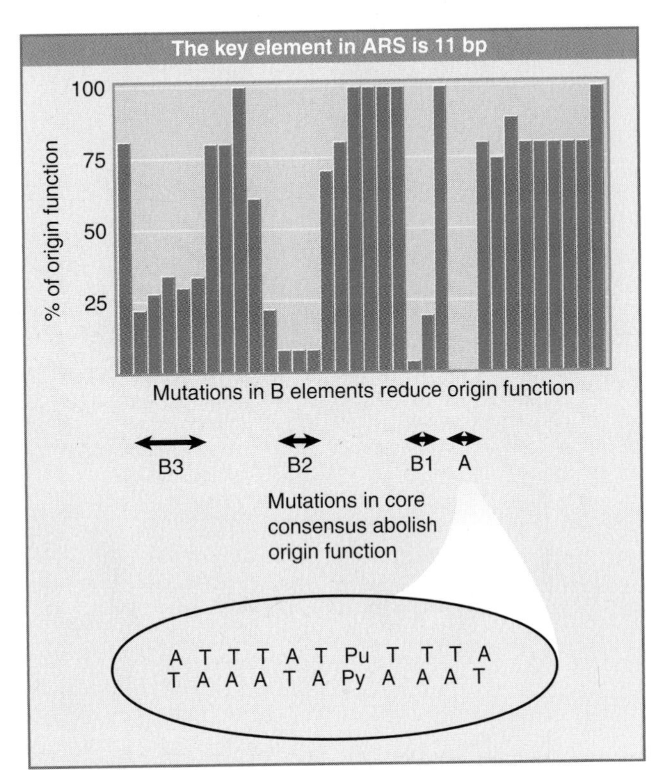

FIGURE 15.12 An *ARS* extends for ~50 bp and includes a consensus sequence (A) and additional elements (B1–B3).

15.8 Licensing Factor Controls Eukaryotic Rereplication

Key concepts

- Licensing factor is necessary for initiation of replication at each origin.
- It is present in the nucleus prior to replication, but is inactivated or destroyed by replication.
- Initiation of another replication cycle becomes possible only after licensing factor reenters the nucleus after mitosis.

A eukaryotic genome is divided into multiple replicons, and the origin in each replicon is activated once and only once in a single division cycle. This could be achieved by providing some rate-limiting component that functions only once at an origin or by the presence of a repressor that prevents rereplication at origins that have been used. The critical questions about

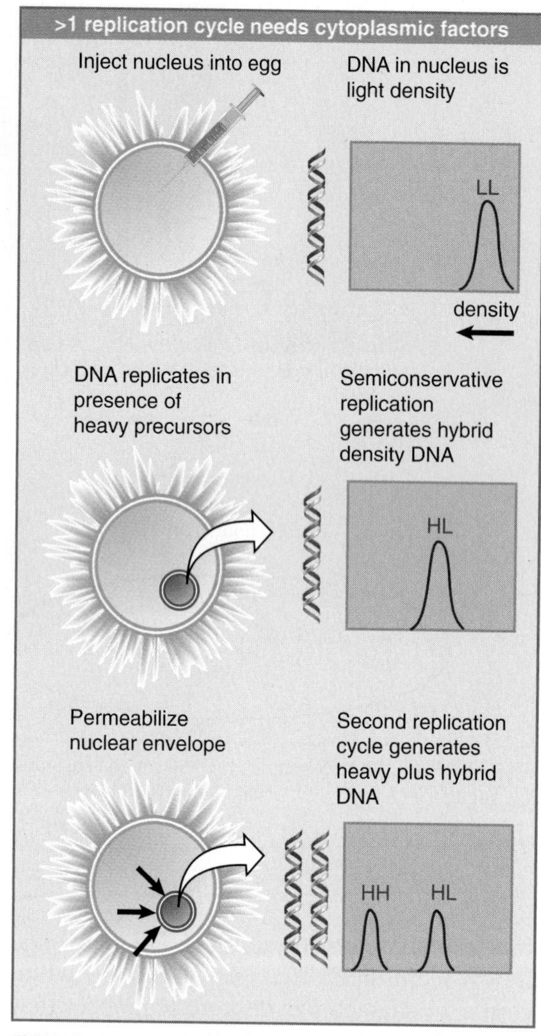

FIGURE 15.13 A nucleus injected into a *Xenopus* egg can replicate only once unless the nuclear membrane is permeabilized to allow subsequent replication cycles.

the nature of this regulatory system are how the system determines whether any particular origin has been replicated and what protein components are involved.

Insights into the nature of the protein components have been provided by using a system in which a substrate DNA undergoes only one cycle of replication. *Xenopus* eggs have all the components needed to replicate DNA—in the first few hours after fertilization they undertake eleven division cycles without new gene expression—and they can replicate the DNA in a nucleus that is injected into the egg. FIGURE 15.13 summarizes the features of this system.

When a sperm or interphase nucleus is injected into the egg, its DNA is replicated only once (this can be followed by use of a density label, just like the original experiment that characterized semiconservative replication, shown previously in Figure 1.12). If protein synthesis is blocked in the egg, the membrane around the injected material remains intact and the DNA cannot replicate again. In the presence of protein synthesis, however, the nuclear membrane breaks down just as it would for a normal cell division, and in this case subsequent replication cycles can occur. The same result can be achieved by using agents that permeabilize the nuclear membrane. This suggests that the nucleus contains a protein(s) needed for replication that is used up in some way by a replication cycle, so even though more of the protein is present in the egg cytoplasm, it can only enter the nucleus if the nuclear membrane breaks down. The system can in principle be taken further by developing an *in vitro* extract that supports nuclear replication, thus allowing the components of the extract to be isolated and the relevant factors identified.

FIGURE 15.14 explains the control of reinitiation by proposing that this protein is a **licensing factor.** It is present in the nucleus prior to replication. One round of replication either inactivates or destroys the factor, and another round cannot occur until further factor is provided. Factor in the cytoplasm can gain access to the nuclear material only at the subsequent mitosis when the nuclear envelope breaks down. This regulatory system achieves two purposes. By removing a necessary component after replication, it prevents more than one cycle of replication from occurring. It also provides a feedback loop that makes the initiation of replication dependent on passing through cell division.

15.9 Licensing Factor Consists of MCM Proteins

Key concepts

- The ORC is a protein complex that is associated with yeast origins throughout the cell cycle.
- Cdc6 protein is an unstable protein that is synthesized only in G1.
- Cdc6 binds to ORC and allows MCM proteins to bind.
- When replication is initiated, Cdc6 and MCM proteins are displaced. The degradation of Cdc6 prevents reinitiation.
- Some MCM proteins are in the nucleus throughout the cycle, but others may enter only after mitosis.

The key event in controlling replication is the behavior of the ORC complex at the origin. Recall that ORC is a 400-kD complex that binds to the *S. cerevisiae ARS* sequence (see Section 15.7, Replication Origins Can Be Isolated

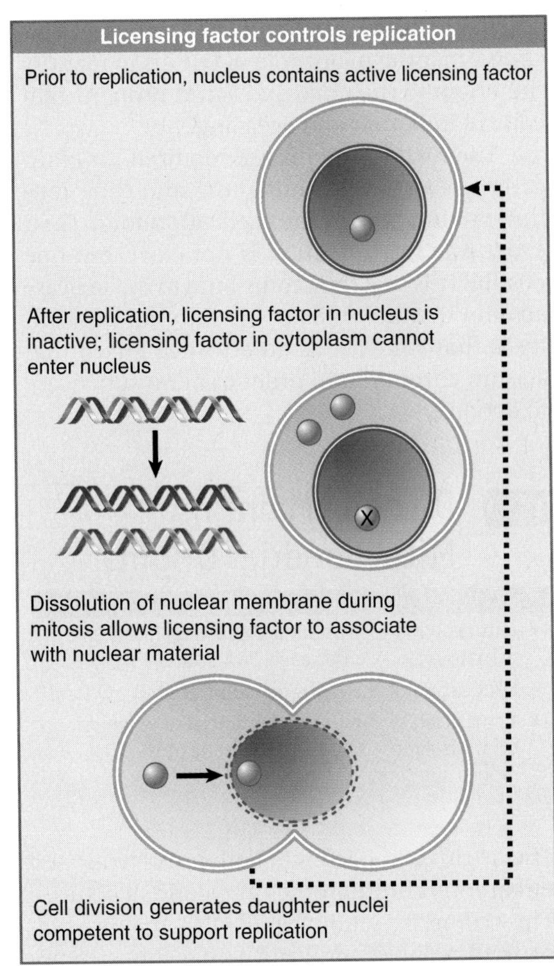

FIGURE 15.14 Licensing factor in the nucleus is inactivated after replication. A new supply of licensing factor can enter only when the nuclear membrane breaks down at mitosis.

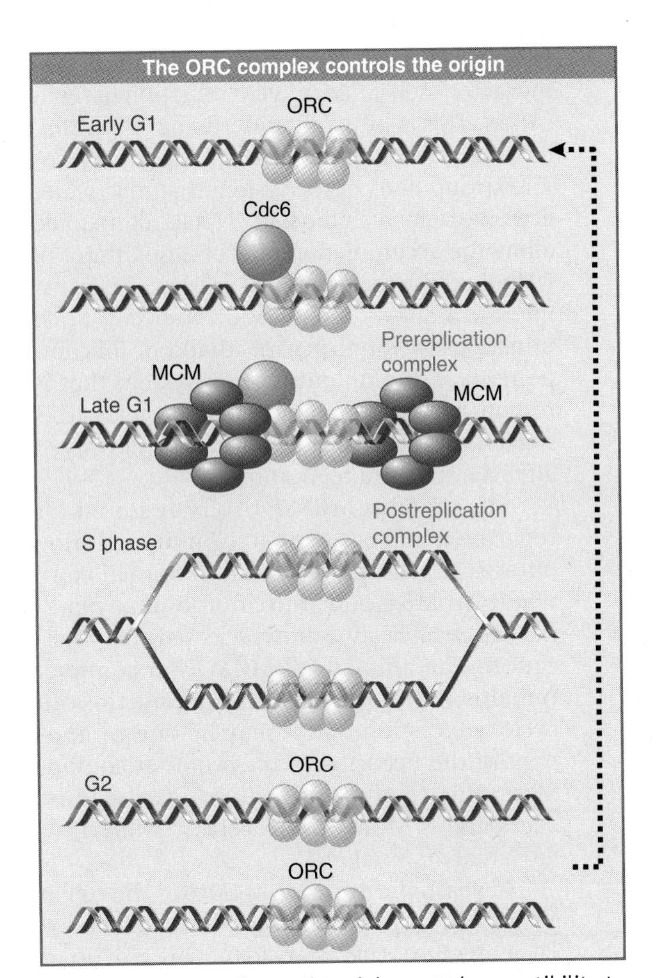

FIGURE 15.15 Proteins at the origin control susceptibility to initiation.

in Yeast). The origin *(ARS)* consists of the A consensus sequence and three B elements (see Figure 15.12). The ORC complex of six proteins (all of which are coded by essential genes) binds to the A and adjacent B1 element. ATP is required for the binding, but is not hydrolyzed until some later stage. The transcription factor ABF1 binds to the B3 element; this assists initiation, but it is the events that occur at the A and B1 elements that actually cause initiation. Most origins are localized in regions between genes, which suggests that it may be important for the local chromatin structure to be in a nontranscribed condition.

The striking feature is that ORC remains bound at the origin through the entire cell cycle. Changes occur, however, in the pattern of protection of DNA as a result of the binding of other proteins to the ORC-origin complex. FIGURE 15.15 summarizes the cycle of events at the origin.

At the end of the cell cycle, ORC is bound to A-B1 and generates a pattern of protection *in vivo* that is similar to that found when it binds to free DNA *in vitro*. Basically, the region across A-B1 is protected against DNAase, but there is a hypersensitive site in the center of B1.

During G1 this pattern changes, most strikingly by the loss of the hypersensitive site. This is due to the binding of Cdc6 protein to the ORC. In yeast, Cdc6 is a highly unstable protein, with a half-life of <5 minutes. It is synthesized during G1 and typically binds to the ORC between the exit from mitosis and late G1. Its rapid degradation means that no protein is available later in the cycle. In mammalian cells it is controlled differently; it is phosphorylated during S phase, and as a result it is exported from the nucleus. Cdc6 provides the connection between ORC and a complex of proteins that is involved in licensing and initiation. Cdc6 has an ATPase activity that is required for it to support initiation.

An insight into the system that controls availability of licensing factor is provided by certain mutants in yeast. Mutations in the licensing factor itself could prevent initiation of replication. This is how mutations behave in minichromosome maintenance (MCM) proteins 2,3,5. Mutations in the system that inactivates licensing factor after the start of replication should allow the accumulation of excess quantities of DNA, because the continued presence of licensing factor allows rereplication to occur. Such mutations are found in genes that code for components of the ubiquitination system that is responsible for degrading certain proteins. This suggests that licensing factor may be destroyed after the start of a replication cycle.

The proteins MCM2,3,5 are required for replication and enter the nucleus only during mitosis. Homologs are found in animal cells, where MCM3 is bound to chromosomal material before replication, but is released after replication. The animal cell MCM2,3,5 complex remains in the nucleus throughout the cell cycle, suggesting that it may be one component of the licensing factor. Another component, able to enter only at mitosis, may be necessary for MCM2,3,5 to associate with chromosomal material.

In yeast, the presence of Cdc6 at the origin allows MCM proteins to bind to the complex. Their presence is necessary for initiation to occur at the origin. The origin therefore enters S phase in the condition of a **prereplication complex,** which contains ORC, Cdc6, and MCM proteins. When initiation occurs, Cdc6 and MCM are displaced, returning the origin to the state of the **postreplication complex,** which contains only ORC. Cdc6 is rapidly degraded during S phase; thus it is not available to support reloading of MCM proteins, and so the origin cannot be used for a second cycle of initiation during the S phase.

If Cdc6 is made available to bind to the origin during G2 (by ectopic expression), MCM proteins do not bind until the following G1, which suggests that there is a secondary mechanism to ensure that they associate with origins only at the right time. This could be another part of licensing control. At least in *S. cerevisiae,* this control does not seem to be exercised at the level of nuclear entry, but this could be a difference between yeasts and animal cells. The MCM2-7 proteins form a six-member ring-shaped complex around DNA. Some of the ORC proteins have similarities to replication proteins that load DNA polymerase on to DNA. It is possible that ORC uses hydrolysis of ATP to load the MCM ring on to DNA. In *Xenopus* extracts, replication can be initiated if ORC is removed after it has loaded Cdc6 and MCM proteins. This shows that the major role of ORC is to identify the origin to the Cdc6 and MCM proteins that control initiation and licensing.

The MCM proteins are required for elongation as well as for initiation, and they continue to function at the replication fork. Their exact role in elongation is not clear, but one possibility is that they contribute to the helicase activity that unwinds DNA. Another possibility is that they act as an advance guard that acts on chromatin in order to allow a helicase to act on DNA.

15.10 D Loops Maintain Mitochondrial Origins

Key concepts

- Mitochondria use different origin sequences to initiate replication of each DNA strand.
- Replication of the H strand is initiated in a D loop.
- Replication of the L strand is initiated when its origin is exposed by the movement of the first replication fork.

The origins of replicons in both prokaryotic and eukaryotic chromosomes are static structures: They comprise sequences of DNA that are recognized in duplex form and used to initiate replication at the appropriate time. Initiation requires separating the DNA strands and commencing bidirectional DNA synthesis. A different type of arrangement is found in mitochondria.

Replication starts at a specific origin in the circular duplex DNA. Initially, though, only one of the two parental strands (the H strand in mammalian mitochondrial DNA) is used as a template for synthesis of a new strand. Synthesis proceeds for only a short distance, displacing the original partner (L) strand, which remains single-stranded, as illustrated in FIGURE 15.16. The condition of this region gives rise to its name as the *displacement loop,* or **D loop.**

DNA polymerases cannot initiate synthesis, but require a priming 3′ end (see Section 18.8, Priming Is Required to Start DNA Synthesis). Replication at the H-strand origin is initiated when RNA polymerase transcribes a primer. The 3′ ends are generated in the primer by an endonuclease that cleaves the DNA–RNA hybrid at several discrete sites. The endonuclease is specific for the triple structure of DNA–RNA hybrid plus the displaced DNA single strand. The 3′ end is then extended into DNA by the DNA polymerase.

A single D loop is found as an opening of 500 to 600 bases in mammalian mitochondria. The short strand that maintains the D loop is unstable and turns over; it is frequently degraded and resynthesized to maintain the opening of the duplex at this site. Some mitochondrial DNAs possess several D loops, reflecting the use of multiple origins. The same mechanism is employed in chloroplast DNA, where (in higher plants) there are two D loops.

To replicate mammalian mitochondrial DNA, the short strand in the D loop is extended. The displaced region of the original L strand becomes longer, expanding the D loop. This expansion continues until it reaches a point about two-thirds of the way around the circle. Replication of this region exposes an origin in the displaced L strand. Synthesis of an H strand initiates at this site, which is used by a special primase that synthesizes a short RNA. The RNA is then extended by DNA polymerase, proceeding around the displaced single-stranded L template in the opposite direction from L-strand synthesis.

As a result of the lag in its start, H-strand synthesis has proceeded only a third of the way around the circle when L-strand synthesis finishes. This releases one completed duplex circle and one gapped circle, the latter of which remains partially single-stranded until synthesis of the H strand is completed. Finally, the new strands are sealed to become covalently intact.

The existence of D loops exposes a general principle: *An origin can be a sequence of DNA that serves to initiate DNA synthesis using one strand as template.* The opening of the duplex does not necessarily lead to the initiation of replication on the other strand. In the case of mitochondrial DNA replication, the origins for replicating the complementary strands lie at different locations. Origins that sponsor replication of only one strand are also found in the rolling circle mode of replication (see Section 16.4, Rolling Circles Produce Multimers of a Replicon).

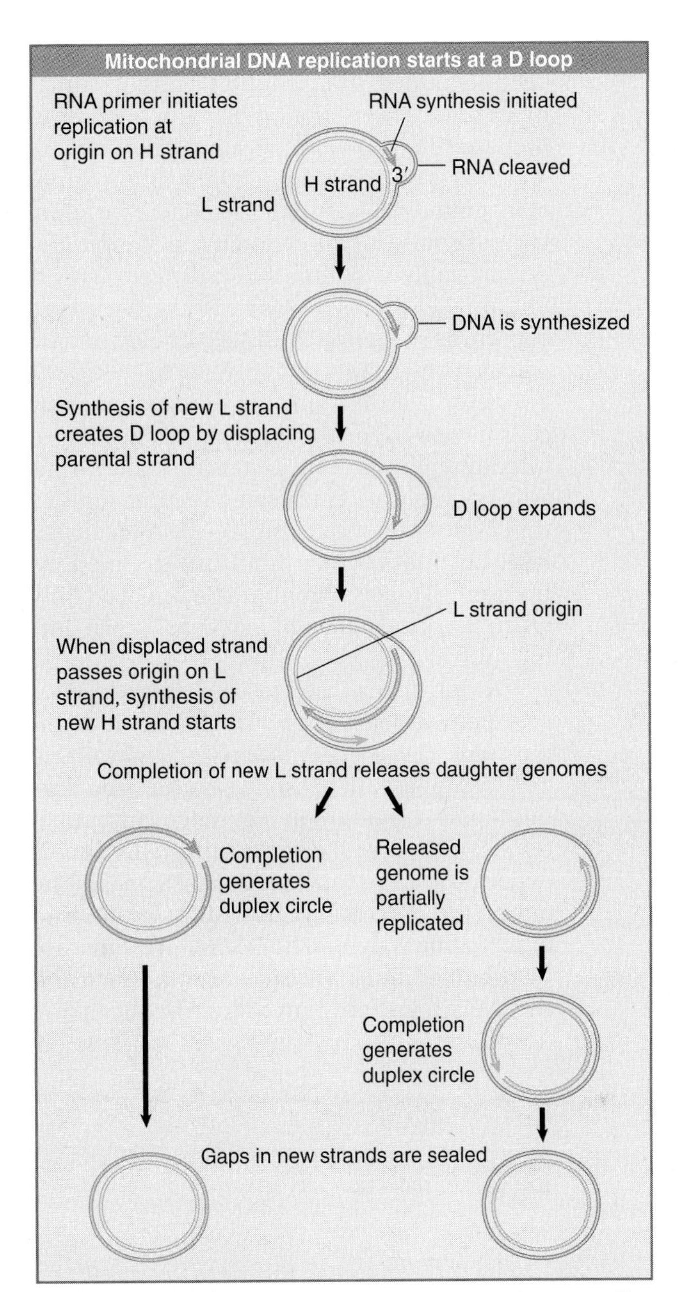

FIGURE 15.16 The D loop maintains an opening in mammalian mitochondrial DNA, which has separate origins for the replication of each strand.

15.11 Summary

The entire chromosome is replicated once for every cell division cycle. Initiation of replication commits the cell to a cycle of division; completion of replication may provide a trigger for the actual division process. The bacterial chromosome consists of a single replicon, but a eukaryotic chromosome is divided into many replicons that function over the protracted period of S phase. The problem of replicating the ends of a linear replicon is solved in a variety of ways, most often by converting the replicon to a circular form. Some viruses have special proteins that recognize ends. Eukaryotic chromosomes encounter the problem at their terminal replicons.

The minimal *E. coli* origin consists of ~245 bp and initiates bidirectional replication. Any DNA molecule with this sequence can replicate in *E. coli.* Two replication forks leave the origin and move around the chromosome, apparently until they meet, although *ter* sequences that would cause the forks to terminate after meeting have been identified. Transcription units are organized so that transcription usually proceeds in the same direction as replication.

Eukaryotic replication is (at least) an order of magnitude slower than bacterial replication. Origins sponsor bidirectional replication and are probably used in a fixed order during S phase. The origins of *S. cerevisiae* have a core consensus sequence consisting of eleven base pairs, mostly A-T.

After cell division, nuclei of eukaryotic cells have a licensing factor that is needed to initiate replication. Its destruction after initiation of replication prevents further replication cycles from occurring in yeast. Licensing factor cannot be imported into the nucleus from the cytoplasm, and can be replaced only when the nuclear membrane breaks down during mitosis.

The origin in yeast is recognized by the ORC proteins, which in yeast remain bound throughout the cell cycle. The protein Cdc6 is available only at S phase. In yeast it is synthesized during S phase and rapidly degraded. In animal cells it is synthesized continuously, but is exported from the nucleus during S phase. The presence of Cdc6 allows the MCM proteins to bind to the origin. The MCM proteins are required for initiation. The action of Cdc6 and the MCM proteins provides the licensing function.

References

15.1 Introduction

Research

Jacob, F., Brenner, S., and Cuzin, F. (1963). On the regulation of DNA replication in bacteria. *Cold Spring Harbor Symp. Quant. Biol.* 28, 329–348.

15.2 Replicons Can Be Linear or Circular

Review

Brewer, B. J. (1988). When polymerases collide: replication and transcriptional organization of the *E. coli* chromosome. *Cell* 53, 679–686.

Research

Cairns, J. (1963). The bacterial chromosome and its manner of replication as seen by autoradiography. *J. Mol. Biol.* 6, 208–213.

Iismaa, T. P. and Wake, R. G. (1987). The normal replication terminus of the *B. subtilis* chromosome, *terC*, is dispensable for vegetative growth and sporulation. *J. Mol. Biol.* 195, 299–310.

Liu, B., Wong, M. L., and Alberts, B. (1994). A transcribing RNA polymerase molecule survives DNA replication without aborting its growing RNA chain. *Proc. Natl. Acad. Sci. USA* 91, 10660–10664.

Steck, T. R. and Drlica, K. (1984). Bacterial chromosome segregation: evidence for DNA gyrase involvement in decatenation. *Cell* 36, 1081–1088.

Zyskind, J. W. and Smith, D. W. (1980). Nucleotide sequence of the *S. typhimurium* origin of DNA replication. *Proc. Natl. Acad. Sci. USA* 77, 2460–2464.

15.3 Origins Can Be Mapped by Autoradiography and Electrophoresis

Research

Huberman, J. and Riggs, A. D. (1968). On the mechanism of DNA replication in mammalian chromosomes. *J. Mol. Biol.* 32, 327–341.

15.6 Each Eukaryotic Chromosome Contains Many Replicons

Review

Fangman, W. L. and Brewer, B. J. (1991). Activation of replication origins within yeast chromosomes. *Annu. Rev. Cell Biol.* 7, 375–402.

Research

Blumenthal, A. B., Kriegstein, H. J., and Hogness, D. S. (1974). The units of DNA replication in *D. melanogaster* chromosomes. *Cold Spring Harbor Symp. Quant. Biol.* 38, 205–223.

15.7 Replication Origins Can Be Isolated in Yeast

Reviews

Bell, S. P. and Dutta, A. (2002). DNA replication in eukaryotic cells. *Annu. Rev. Biochem.* 71, 333–374.

DePamphlis, M. L. (1993). Eukaryotic DNA replication: anatomy of an origin. *Annu. Rev. Biochem.* 62, 29–63.

Gilbert, D. M. (2001). Making sense of eukaryotic DNA replication origins. *Science* 294, 96–100.

Kelly, T. J. and Brown, G. W. (2000). Regulation of chromosome replication. *Annu. Rev. Biochem.* 69, 829–880.

Research

Anglana, M., Apiou, F., Bensimon, A., and Debatisse, M. (2003). Dynamics of DNA replication in mammalian somatic cells: nucleotide pool modulates origin choice and interorigin spacing. *Cell* 114, 385–394.

Chesnokov, I., Remus, D., and Botchan, M. (2001). Functional analysis of mutant and wild-type *Drosophila* origin recognition complex. *Proc. Natl. Acad. Sci. USA* 98, 11997–12002.

Marahrens, Y. and Stillman, B. (1992). A yeast chromosomal origin of DNA replication defined by multiple functional elements. *Science* 255, 817–823.

Wyrick, J. J., Aparicio, J. G., Chen, T., Barnett, J. D., Jennings, E. G., Young, R. A., Bell, S. P., and Aparicio, O. M. (2001). Genome-wide distribution of ORC and MCM proteins in *S. cerevisiae:* high-resolution mapping of replication origins. *Science* 294, 2357–2360.

15.10 D Loops Maintain Mitochondrial Origins

Reviews

Clayton, D. (1982). Replication of animal mitochondrial DNA. *Cell* 28, 693–705.

Clayton, D., A. (1991). Replication and transcription of vertebrate mitochondrial DNA. *Annu. Rev. Cell Biol.* 7, 453–478.

Shadel, G. S. and Clayton, D. A. (1997). Mitochondrial DNA maintenance in vertebrates. *Annu. Rev. Biochem.* 66, 409–435.

16

Extrachromosomal Replicons

CHAPTER OUTLINE

16.1 Introduction

A bacterium may be a host for independently replicating genetic units in addition to its chromosome. These extrachromosomal genomes fall into two general types: plasmids and bacteriophages (phages). Some plasmids, and all phages, have the ability to transfer from a donor bacterium to a recipient by an infective process. An important distinction between them is that plasmids exist only as free DNA genomes, whereas bacteriophages are viruses that package a nucleic acid genome into a protein coat and are released from the bacterium at the end of an infective cycle.

Plasmids are self-replicating circular molecules of DNA that are maintained in the cell in a stable and characteristic number of copies, that is, the number remains constant from generation to generation. Single-copy plasmids are maintained at the same relative quantity as the bacterial host chromosome, at one per unit bacterium. As with the host chromosome, they rely on a specific apparatus to be segregated equally at each bacterial division. Multicopy plasmids exist in many copies per unit bacterium and may be segregated to daughter bacteria stochastically (meaning that there are enough copies that each daughter cell always gains some by a random distribution).

Plasmids and phages are defined by their ability to reside in a bacterium as independent genetic units. Certain plasmids, and some phages, can also exist as sequences within the bacterial genome, though. In this case, the same sequence that constitutes the independent plasmid or phage genome is found within the chromosome, and is inherited like any other bacterial gene. Phages that are found as part of the bacterial chromosome are said to show **lysogeny;** plasmids that have the ability to behave like this are called **episomes.** Related processes are used by phages and episomes to insert into and excise from the bacterial chromosome.

A parallel between lysogenic phages and plasmids and episomes is that they maintain a selfish possession of their bacterium and often make it impossible for another element of the same type to become established. This effect is called **immunity,** although the molecular basis for plasmid immunity is different from lysogenic immunity, and is a consequence of the replication control system.

Figure 14.2 summarizes the types of genetic units that can be propagated in bacteria as independent genomes. Lytic phages may have genomes of any type of nucleic acid; they transfer between cells by release of infective particles. Lysogenic phages have double-stranded DNA genomes, as do plasmids and episomes. Some plasmids transfer between cells by a conjugative process (with direct contact between donor and recipient cells). A feature of the transfer process in both cases is that on occasion some bacterial host genes are transferred with the phage or plasmid DNA, so these events play a role in allowing exchange of genetic information between bacteria.

The key feature in determining the behavior of each type of unit is how its origin is used. An origin in a bacterial or eukaryotic chromosome is used to initiate a single replication event that extends across the replicon. Replicons, however, can also be used to sponsor other forms of replication. The most common alternative is used by the small, independently replicating units of viruses. The objective of a viral replication cycle is to produce many copies of the viral genome before the host cell is lysed to release them. Some viruses replicate in the same way as a host genome, with an initiation event leading to production of duplicate copies, each of which then replicates again, and so on. Others use a mode of replication in which many copies are produced as a tandem array following a single initiation event. A similar type of event is triggered by episomes when an integrated plasmid DNA ceases to be inert and initiates a replication cycle.

Many prokaryotic replicons are circular, and this indeed is a necessary feature for replication modes that produce multiple tandem copies. Some extrachromosomal replicons are linear, though, and in such cases we have to account for the ability to replicate the end of the replicon. (Of course, eukaryotic chromosomes are linear, so the same problem applies to the replicons at each end. These replicons, however, have a special system for resolving the problem.)

16.2 The Ends of Linear DNA Are a Problem for Replication

Key concept

- Special arrangements must be made to replicate the DNA strand with a 5′ end.

None of the replicons that we have considered so far have a linear end: either they are

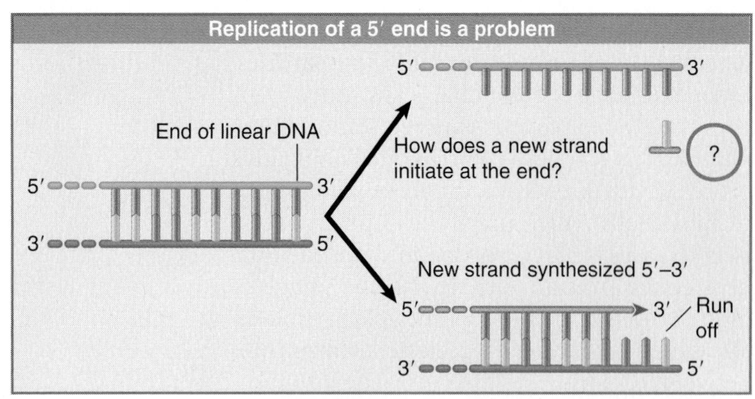

FIGURE 16.1 Replication could run off the 3′ end of a newly synthesized linear strand, but could it initiate at a 5′ end?

circular (as in the *E. coli* or mitochondrial genomes), or they are part of longer segregation units (as in eukaryotic chromosomes). Linear replicons do occur, though—in some cases as single extrachromosomal units, and of course at the ends of eukaryotic chromosomes.

The ability of all known nucleic acid polymerases, DNA or RNA, to proceed only in the 5′– 3′ direction poses a problem for synthesizing DNA at the end of a linear replicon. Consider the two parental strands depicted in FIGURE 16.1. The lower strand presents no problem: It can act as template to synthesize a daughter strand that runs right up to the end, where presumably the polymerase falls off. To synthesize a complement at the end of the upper strand, however, synthesis must start right at the very last base, or else this strand would become shorter in successive cycles of replication.

We do not know whether initiation right at the end of a linear DNA is feasible. We usually think of a polymerase as binding at a site *surrounding* the position at which a base is to be incorporated. Thus a special mechanism must be employed for replication at the ends of linear replicons. Several types of solution may be imagined to accommodate the need to copy a terminus:

- The problem may be circumvented by converting a linear replicon into a circular or multimeric molecule. Phages such as T4 or lambda use such mechanisms (see Section 16.4, Rolling Circles Produce Multimers of a Replicon).
- The DNA may form an unusual structure—for example, by creating a hairpin at the terminus, so that there is no free end. Formation of a crosslink is involved in replication of the linear mitochondrial DNA of *Paramecium.*
- Instead of being precisely determined, the end may be variable. Eukaryotic chromosomes may adopt this solution, in which the number of copies of a short repeating unit at the end of the DNA changes (see Section 28.18, Telomeres Are Synthesized by a Ribonucleoprotein Enzyme). A mechanism to add or remove units makes it unnecessary to replicate right up to the very end.
- A protein may intervene to make initiation possible at the actual terminus. Several linear viral nucleic acids have proteins that are *covalently linked to the 5′ terminal base.* The best characterized examples are adenovirus DNA, phage φ29 DNA, and poliovirus RNA.

16.3 Terminal Proteins Enable Initiation at the Ends of Viral DNAs

Key concept

- A terminal protein binds to the 5′ end of DNA and provides a cytidine nucleotide with a 3′-OH end that primes replication.

An example of initiation at a linear end is provided by adenovirus and φ29 DNAs, which actually replicate from both ends using the mechanism of **strand displacement** illustrated in FIGURE 16.2. The same events can occur independently at either end. Synthesis of a new strand starts at one end, displacing the homologous strand that was previously paired in the duplex. When the replication fork reaches the other end of the molecule, the displaced strand is released as a free single strand. It is then replicated independently; this requires the formation of a duplex origin by base pairing between some short complementary sequences at the ends of the molecule.

In several viruses that use such mechanisms, a protein is found covalently attached to each 5′ end. In the case of adenovirus, a **terminal protein** is linked to the mature viral DNA via a phosphodiester bond to serine, as indicated in FIGURE 16.3.

How does the attachment of the protein overcome the initiation problem? The terminal protein has a dual role: It carries a cytidine nucleotide that provides the primer, and it is associated with DNA polymerase. In fact, linkage of terminal protein to a nucleotide is undertaken by DNA polymerase in the presence of

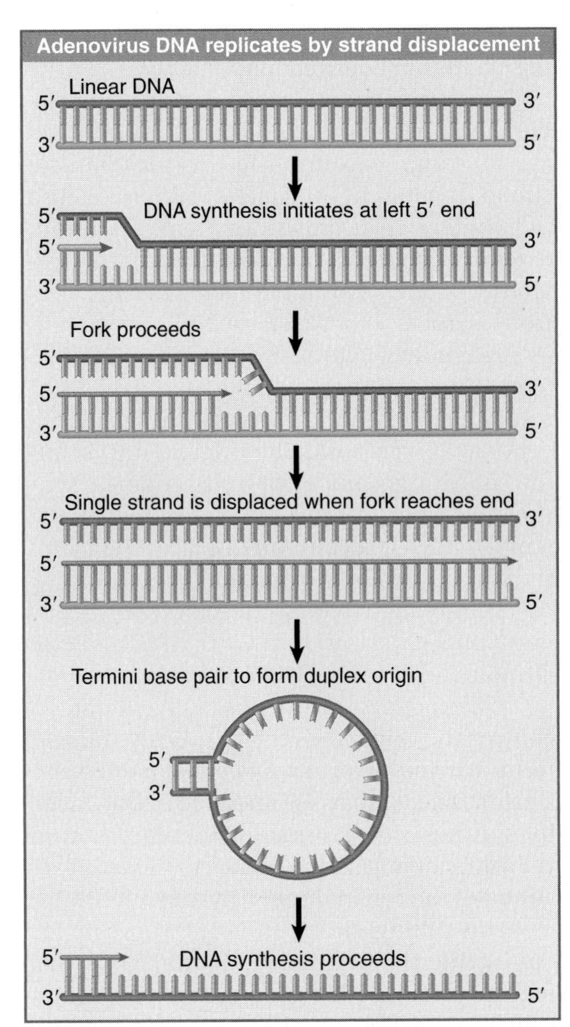

FIGURE 16.2 Adenovirus DNA replication is initiated separately at the two ends of the molecule and proceeds by strand displacement.

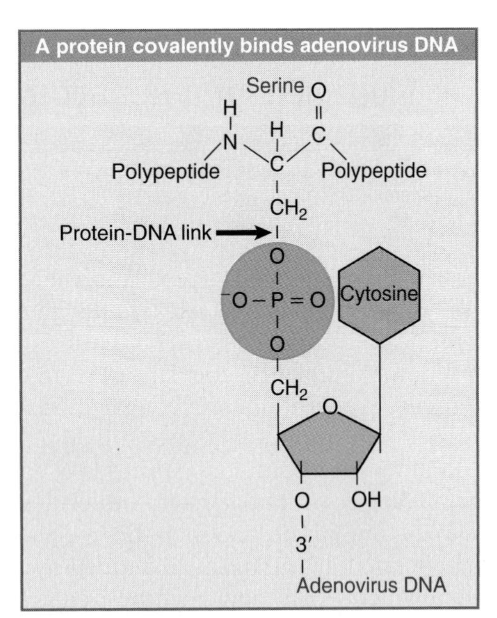

FIGURE 16.3 The 5′ terminal phosphate at each end of adenovirus DNA is covalently linked to serine in the 55 kD Ad-binding protein.

adenovirus DNA. This suggests the model illustrated in FIGURE 16.4. The complex of polymerase and terminal protein, bearing the priming C nucleotide, binds to the end of the adenovirus DNA. The free 3′–OH end of the C nucleotide is used to prime the elongation reaction by the DNA polymerase. This generates a new strand whose 5′ end is covalently linked to the initiating C nucleotide. (The reaction actually involves displacement of protein from DNA rather than binding *de novo*. The 5′ end of adenovirus DNA is bound to the terminal protein that was used in the previous replication cycle. The old terminal protein is displaced by the new terminal protein for each new replication cycle.)

Terminal protein binds to the region located between 9 and 18 bp from the end of the DNA. The adjacent region, between positions 17 and 48, is essential for the binding of a host protein, nuclear factor I, which is also required for the

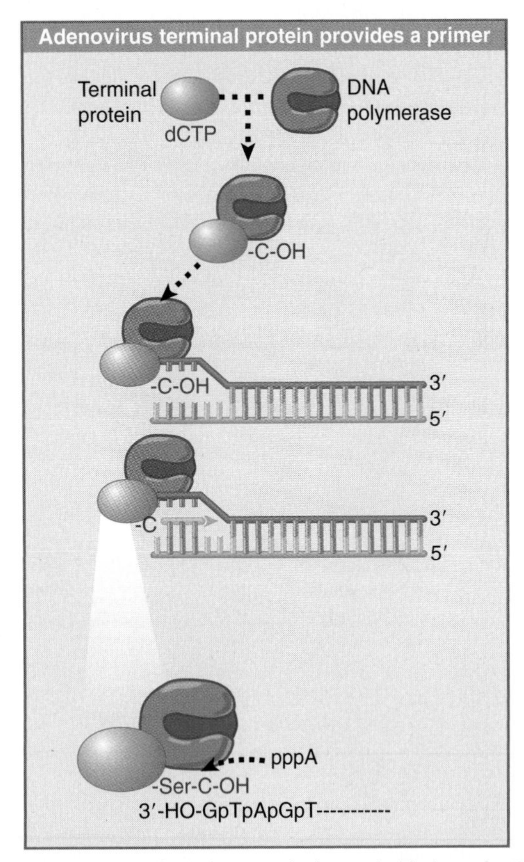

FIGURE 16.4 Adenovirus terminal protein binds to the 5′ end of DNA and provides a C-OH end to prime synthesis of a new DNA strand.

initiation reaction. The initiation complex may therefore form between positions 9 and 48, a fixed distance from the actual end of the DNA.

16.4 Rolling Circles Produce Multimers of a Replicon

Key concept

- A rolling circle generates single-stranded multimers of the original sequence.

The structures generated by replication depend on the relationship between the template and the replication fork. The critical features are whether the template is circular or linear, and whether the replication fork is engaged in synthesizing both strands of DNA or only one.

Replication of only one strand is used to generate copies of some circular molecules. A nick opens one strand, and then the free 3′–OH end generated by the nick is extended by the DNA polymerase. The newly synthesized strand displaces the original parental strand. The ensuing events are depicted in FIGURE 16.5.

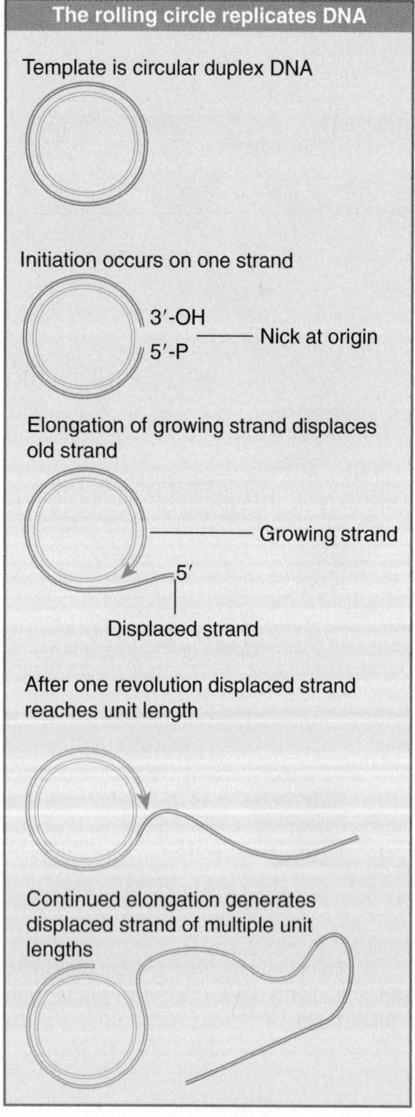

FIGURE 16.5 The rolling circle generates a multimeric single-stranded tail.

This type of structure is called a **rolling circle,** because the growing point can be envisaged as rolling around the circular template strand. It could in principle continue to do so indefinitely. As it moves, the replication fork extends the outer strand and displaces the previous partner. An example is shown in the electron micrograph of FIGURE 16.6.

The newly synthesized material is covalently linked to the original material, and as a result the displaced strand has the original unit genome at its 5′ end. The original unit is followed by any number of unit genomes, synthesized by continuing revolutions of the template. Each revolution displaces the material synthesized in the previous cycle.

The rolling circle is put to several uses *in vivo.* Some pathways that are used to replicate DNA are depicted in FIGURE 16.7.

Cleavage of a unit length tail generates a copy of the original circular replicon in linear form. The linear form may be maintained as a single strand, or may be converted into a duplex by synthesis of the complementary strand (which is identical in sequence to the template strand of the original rolling circle).

The rolling circle provides a means for amplifying the original (unit) replicon. This mechanism is used to generate amplified ribosomal DNA (rDNA) in the *Xenopus* oocyte. The genes for ribosomal RNA (rRNA) are organized as a large number of contiguous repeats in the genome. A single repeating unit from the genome is converted into a rolling circle. The displaced tail, which contains many units, is converted into duplex DNA; later it is cleaved from

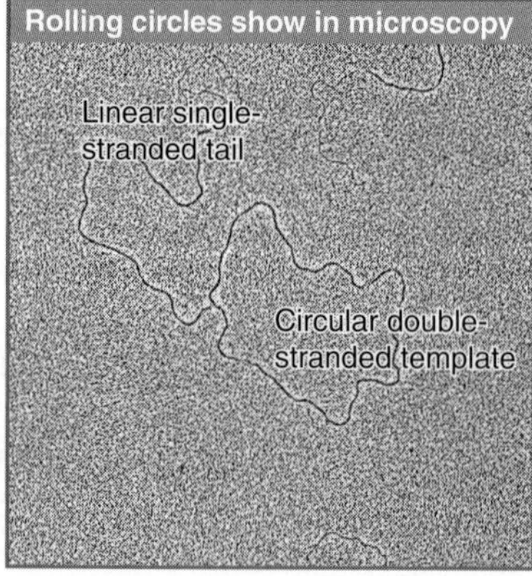

FIGURE 16.6 A rolling circle appears as a circular molecule with a linear tail by electron microscopy. Courtesy of Ross B. Inman, Institute of Molecular Virology, Bock Laboratory and Department of Biochemistry, University of Wisconsin, Madison, Wisconsin, USA.

the circle so that the two ends can be joined together to generate a large circle of amplified rDNA. The amplified material therefore consists of a large number of identical repeating units.

16.5 Rolling Circles Are Used to Replicate Phage Genomes

Key concept

- The φX A protein is a *cis*-acting relaxase that generates single-stranded circles from the tail produced by rolling circle replication.

Replication by rolling circles is common among bacteriophages. Unit genomes can be cleaved from the displaced tail, generating monomers that can be packaged into phage particles or used for further replication cycles. A more detailed view of a phage replication cycle that is centered on the rolling circle is given in FIGURE 16.8.

Phage φX174 consists of a single-stranded circular DNA known as the plus (+) strand. A complementary strand, called the minus (–) strand, is synthesized. This action generates the duplex circle shown at the top of the figure, which is then replicated by a rolling circle mechanism.

The duplex circle is converted to a covalently closed form, which becomes supercoiled. A protein coded by the phage genome, the A protein, nicks the (+) strand of the duplex DNA at a specific site that defines the origin for replication. After nicking the origin, the A protein remains connected to the 5′ end that it

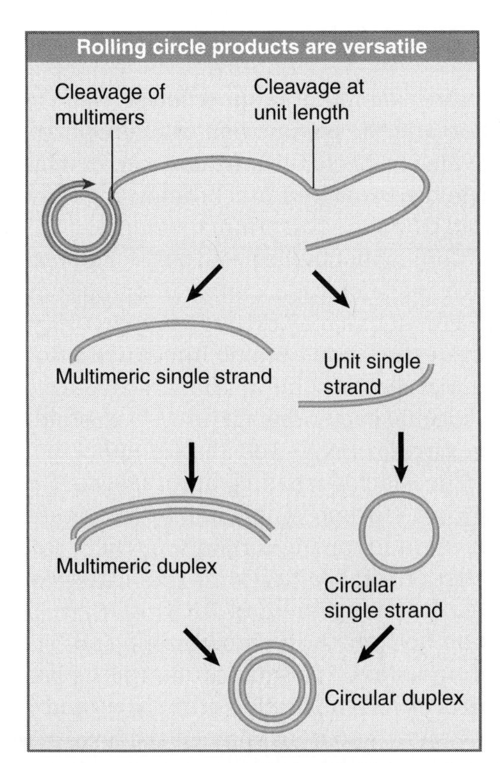

FIGURE 16.7 The fate of the displaced tail determines the types of products generated by rolling circles. Cleavage at unit length generates monomers, which can be converted to duplex and circular forms. Cleavage of multimers generates a series of tandemly repeated copies of the original unit. Note that the conversion to double-stranded form could occur earlier, before the tail is cleaved from the rolling circle.

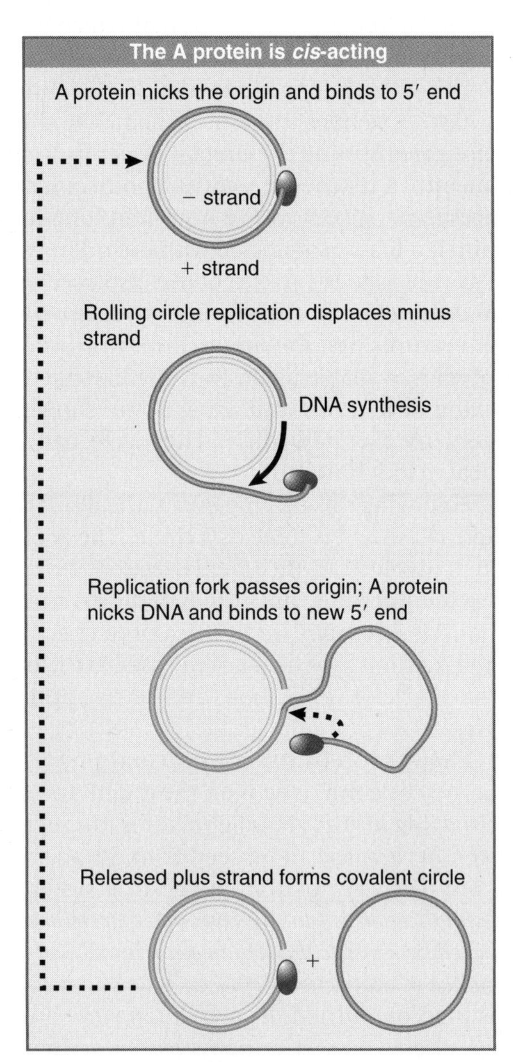

FIGURE 16.8 φX174 RF DNA is a template for synthesizing single-stranded viral circles. The A protein remains attached to the same genome through indefinite revolutions, each time nicking the origin on the viral (+) strand and transferring to the new 5′ end. At the same time, the released viral strand is circularized.

generates, while the 3′ end is extended by DNA polymerase.

The structure of the DNA plays an important role in this reaction, for the DNA can be nicked *only when it is negatively supercoiled* (i.e., wound about its axis in space in the opposite sense from the handedness of the double helix; see Section 19.12, Supercoiling Affects the Structure of DNA). The A protein is able to bind to a single-stranded decamer fragment of DNA that surrounds the site of the nick. This suggests that the supercoiling is needed to assist the formation of a single-stranded region that provides the A protein with its binding site. (An enzymatic activity in which a protein cleaves duplex DNA and binds to a released 5′ end is sometimes called a **relaxase.**) The nick generates a 3′–OH end and a 5′–phosphate end (covalently attached to the A protein), both of which have roles to play in φX174 replication.

Using the rolling circle, the 3′–OH end of the nick is extended into a new chain. The chain is elongated around the circular (–) strand template until it reaches the starting point and displaces the origin. Now the A protein functions again. It remains connected with the rolling circle as well as to the 5′ end of the displaced tail, and is therefore in the vicinity as the growing point returns past the origin. Thus the same A protein is available again to recognize the origin and nick it, now attaching to the end generated by the new nick. The cycle can be repeated indefinitely.

Following this nicking event, the displaced single (+) strand is freed as a circle. The A protein is involved in the circularization. In fact, the joining of the 3′ and 5′ ends of the (+) strand product is accomplished by the A protein as part of the reaction by which it is released at the end of one cycle of replication, and starts another cycle.

The A protein has an unusual property that may be connected with these activities. It is *cis*-acting *in vivo.* (This behavior is not reproduced *in vitro,* as can be seen from its activity on any DNA template in a cell-free system.) *The implication is that* in vivo *the A protein synthesized by a particular genome can attach only to the DNA of that genome.* We do not know how this is accomplished. Its activity *in vitro,* however, shows how it remains associated with the same parental (–) strand template. The A protein has two active sites; this may allow it to cleave the "new" origin while still retaining the "old" origin; it then ligates the displaced strand into a circle.

The displaced (+) strand may follow either of two fates after circularization. During the replication phase of viral infection, it may be used as a template to synthesize the complementary (–) strand. The duplex circle may then be used as a rolling circle to generate more progeny. During phage morphogenesis, the displaced (+) strand is packaged into the phage virion.

16.6 The F Plasmid Is Transferred by Conjugation between Bacteria

Key concepts

- A free F factor is a replicon that is maintained at the level of one plasmid per bacterial chromosome.
- An F factor can integrate into the bacterial chromosome, in which case its own replication system is suppressed.
- The F factor codes for specific pili that form on the surface of the bacterium.
- An F-pilus enables an F-positive bacterium to contact an F-negative bacterium and to initiate conjugation.

Another example of a connection between replication and the propagation of a genetic unit is provided by bacterial **conjugation,** in which a plasmid genome or host chromosome is transferred from one bacterium to another.

Conjugation is mediated by the **F plasmid,** which is the classic example of an episome—an element that may exist as a free circular plasmid, or that may become integrated into the bacterial chromosome as a linear sequence (like a lysogenic bacteriophage). The F plasmid is a large circular DNA ~100 kb in length.

The F factor can integrate at several sites in the *E. coli* chromosome, often by a recombination event involving certain sequences (called IS sequences; see Section 21.5, Transposons Cause Rearrangement of DNA) that are present on both the host chromosome and F plasmid. In its free (plasmid) form, the F plasmid utilizes its own replication origin *(oriV)* and control system, and is maintained at a level of one copy per bacterial chromosome. When it is integrated into the bacterial chromosome, this system is suppressed, and F DNA is replicated as a part of the chromosome.

The presence of the F plasmid, whether free or integrated, has important consequences for the host bacterium. Bacteria that are F-positive are able to conjugate (or mate) with bacteria that are F-negative. Conjugation involves a contact between donor (F-positive) and recipient (F-negative) bacteria; contact is followed by transfer of the F factor. If the F factor exists as a free plasmid in the donor bacterium, it is transferred as a plasmid, and the infective process converts the F-negative recipient into an F-positive state. If the F factor is present in an integrated form in the donor, the transfer process may also cause some or all of the bacterial chromosome to be transferred. Many plasmids have conjugation systems that operate in a generally similar manner, but the F factor was the first to be discovered and remains the paradigm for this type of genetic transfer.

A large (~33 kb) region of the F plasmid called the **transfer region** is required for conjugation. It contains ~40 genes that are required for the transmission of DNA; their organization is summarized in FIGURE 16.9. The genes are named *tra* and *trb* loci. Most of them are expressed coordinately as part of a single 32-kb transcription unit (the *traY-I* unit). *traM* and *traJ* are expressed separately. *traJ* is a regulator that turns on both *traM* and *traY-I*. On the opposite strand, *finP* is a regulator that codes for a small antisense RNA that turns off *traJ*. Its activity requires expression of another gene, *finO*. Only four of the *tra* genes in the major transcription unit are concerned directly with the transfer of DNA; most are concerned with the properties of the bacterial cell surface and with maintaining contacts between mating bacteria.

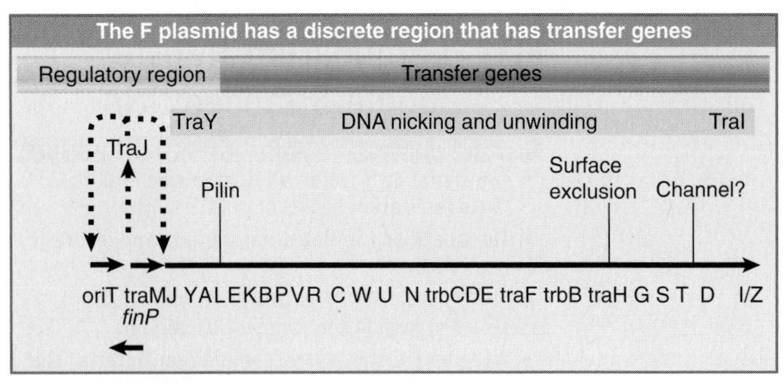

FIGURE 16.9 The *tra* region of the F plasmid contains the genes needed for bacterial conjugation.

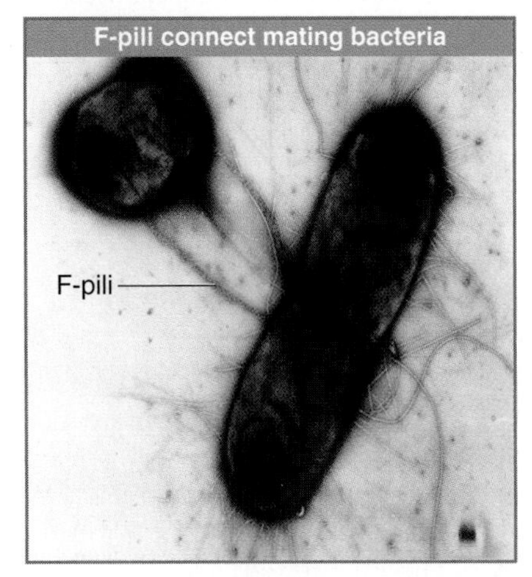

FIGURE 16.10 Mating bacteria are initially connected when donor F pili contact the recipient bacterium. Photo courtesy of Ron Skurray, School of Biological Sciences, University of Sydney.

F-positive bacteria possess surface appendages called pili (singular **pilus**) that are coded by the F factor. The gene *traA* codes for the single subunit protein, **pilin,** that is polymerized into the pilus. At least 12 *tra* genes are required for the modification and assembly of pilin into the pilus. The F-pili are hairlike structures, 2 to 3 μm long, that protrude from the bacterial surface. A typical F-positive cell has two to three pili. The pilin subunits are polymerized into a hollow cylinder, ~8 nm in diameter, with a 2 nm axial hole.

Mating is initiated when the tip of the F-pilus contacts the surface of the recipient cell. FIGURE 16.10 shows an example of *E. coli* cells beginning to mate. A donor cell does not contact other cells carrying the F factor, because the genes *traS* and *traT* code for "surface exclusion" proteins that make the cell a poor recipient in such contacts. This effectively restricts donor cells to mating with F-negative cells. (The presence of F-pili has secondary consequences; they provide the sites to which RNA phages and some single-stranded DNA phages attach, so F-positive bacteria are susceptible to infection by these phages, whereas F-negative bacteria are resistant.)

The initial contact between donor and recipient cells is easily broken, but other *tra* genes act to stabilize the association; this brings the mating cells closer together. The F-pili are essential for initiating pairing, but retract or disassemble as part of the process by which the mating cells are brought into close contact. There must be a channel through which DNA is transferred, but the pilus itself does not appear to provide it. TraD is an inner membrane protein in F^+ bacteria that is necessary for transport of DNA, and it may provide or be part of the channel.

16.7 Conjugation Transfers Single-Stranded DNA

Key concepts

- Transfer of an F factor is initiated when rolling circle replication begins at *oriT*.
- The free 5′ end initiates transfer into the recipient bacterium.
- The transferred DNA is converted into double-stranded form in the recipient bacterium.
- When an F factor is free, conjugation "infects" the recipient bacterium with a copy of the F factor.
- When an F factor is integrated, conjugation causes transfer of the bacterial chromosome until the process is interrupted by (random) breakage of the contact between donor and recipient bacteria.

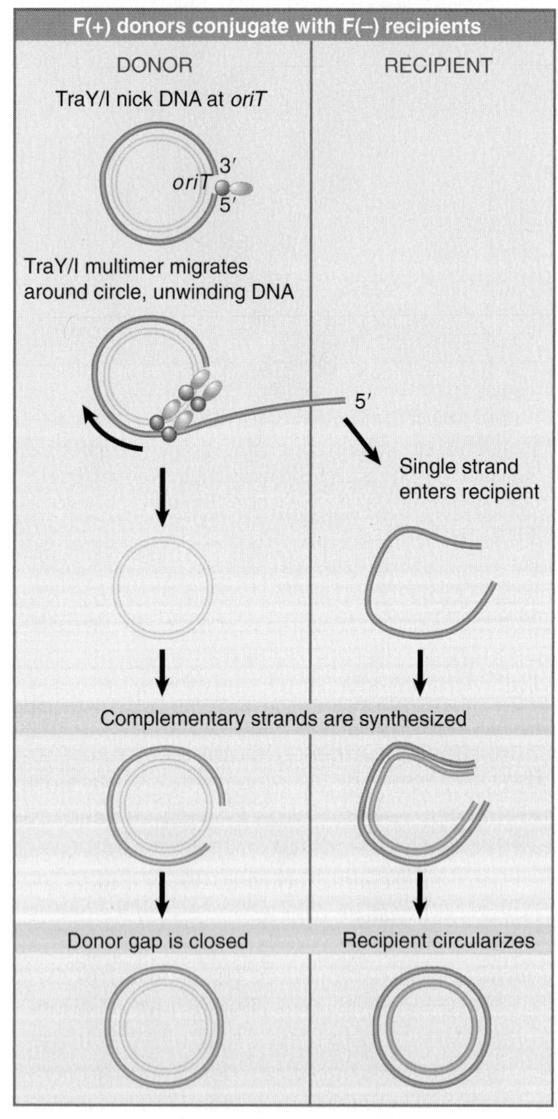

FIGURE 16.11 Transfer of DNA occurs when the F factor is nicked at *oriT* and a single strand is led by the 5′ end into the recipient. Only one unit length is transferred. Complementary strands are synthesized to the single strand remaining in the donor and to the strand transferred into the recipient.

Transfer of the F factor is initiated at a site called *oriT*, the origin of transfer, which is located at one end of the transfer region. The transfer process may be initiated when TraM recognizes that a mating pair has formed. TraY then binds near *oriT* and causes TraI to bind. TraI is a relaxase, like φX174 A protein. TraI nicks *oriT* at a unique site (called *nic*), and then forms a covalent link to the 5′ end that has been generated. TraI also catalyzes the unwinding of ~200 bp of DNA (this is a helicase activity; see Section 18.7, The φX Model System Shows How Single-Stranded DNA Is Generated for Replication). FIGURE 16.11 shows that the freed 5′ end leads the way into the recipient bacterium. A complement for the transferred single strand is synthesized in the recipient bacterium, which as a result is converted to the F-positive state.

A complementary strand must be synthesized in the donor bacterium to replace the strand that has been transferred. If this happens concomitantly with the transfer process, the state of the F plasmid will resemble the rolling circle of Figure 16.5 (and will not generate the extensive single-stranded regions shown in Figure 16.11). Conjugating DNA usually appears like a rolling circle, but replication as such is not necessary to provide the driving energy, and single-strand transfer is independent of DNA synthesis. Only a single unit length of the F factor is transferred to the recipient bacterium. This implies that some (unidentified) feature terminates the process after one revolution, after which the covalent integrity of the F plasmid is restored.

When an integrated F plasmid initiates conjugation, the orientation of transfer is directed away from the transfer region and into the bacterial chromosome. FIGURE 16.12 shows that, following a short leading sequence of F DNA, bacterial DNA is transferred. The process continues until it is interrupted by the breaking of contacts between the mating bacteria. It takes ~100 minutes to transfer the entire bacterial chromosome, and under standard conditions contact is often broken before the completion of transfer.

Donor DNA that enters a recipient bacterium is converted to double-stranded form and may recombine with the recipient chromosome. (Note that two recombination events are required to insert the donor DNA.) Thus conjugation affords a means to exchange genetic material between bacteria (a contrast with their usual asexual growth). A strain of *E. coli* with an integrated F factor supports such recombination at relatively high frequencies (compared to strains that lack integrated F factors); such strains are described as **Hfr** (for high frequency recombination). Each position of integration for the F factor gives rise to a different Hfr strain, with a characteristic pattern of transferring bacterial markers to a recipient chromosome.

Contact between conjugating bacteria is usually broken before transfer of DNA is complete. As a result, the probability that a region of the bacterial chromosome will be transferred depends upon its distance from *oriT.* Bacterial genes located close to the site of F integration (in the direction of transfer) enter recipient bacteria first, and are therefore found at greater frequencies than those that are located farther away and enter later. This gives rise to a gradient of transfer frequencies around the chromosome, declining from the position of F integration. Marker positions on the donor chromosome can be assayed in terms of the time at which transfer occurs; this gave rise to the standard description of the *E. coli* chromosome as a map divided into 100 minutes. The map refers to transfer times from a particular Hfr strain; the starting point for the gradient of transfer is different for each Hfr strain because it is determined by the site where the F factor has integrated into the bacterial genome.

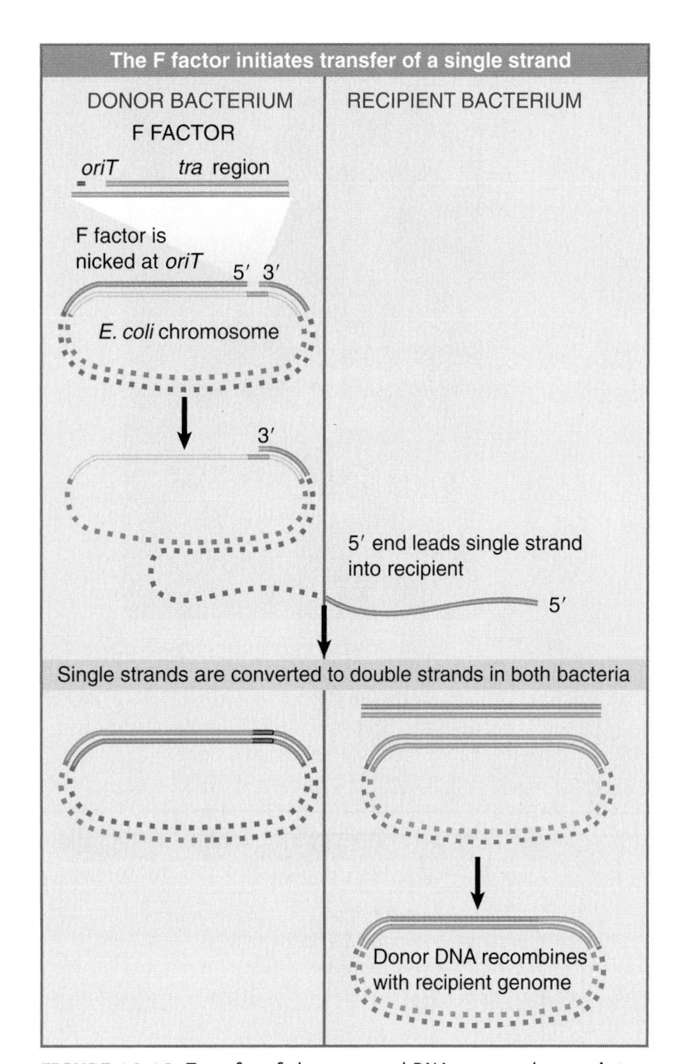

FIGURE 16.12 Transfer of chromosomal DNA occurs when an integrated F factor is nicked at *oriT.* Transfer of DNA starts with a short sequence of F DNA and continues until prevented by loss of contact between the bacteria.

16.8 The Bacterial Ti Plasmid Causes Crown Gall Disease in Plants

Key concepts

- Infection with the bacterium *A. tumefaciens* can transform plant cells into tumors.
- The infectious agent is a plasmid carried by the bacterium.
- The plasmid also carries genes for synthesizing and metabolizing opines (arginine derivatives) that are used by the tumor cell.

Most events in which DNA is rearranged or amplified occur within a genome, but the interaction between bacteria and certain plants involves the transfer of DNA from the bacterial genome to the plant genome. **Crown gall disease,** shown in FIGURE 16.13, can be induced in most dicotyledonous plants by the soil bacterium *Agrobacterium tumefaciens.* The bacterium is a parasite that effects a genetic change in the eukaryotic host cell, with consequences for both parasite and host: It improves conditions for survival of the parasite and causes the plant cell to grow as a tumor.

Agrobacteria are required to induce tumor formation, but the tumor cells do not require the continued presence of bacteria. As with animal tumors, the plant cells have been transformed into a state in which new mechanisms govern growth and differentiation. Transformation is caused by the expression within the plant cell of genetic information transferred from the bacterium.

The tumor-inducing principle of *Agrobacterium* resides in the **Ti plasmid,** which is perpetuated as an independent replicon within the bacterium. The plasmid carries genes involved in various bacterial and plant cell activities, including those required to generate the transformed state, and a set of genes concerned with synthesis or utilization of **opines** (novel derivatives of arginine).

FIGURE 16.13 An *Agrobacterium* carrying a Ti plasmid of the nopaline type induces a teratoma, in which differentiated structures develop. Photo courtesy of the estate of Jeff Schell. Used with permission of the Max Planck Institute for Plant Breeding Research, Cologne.

Ti genes function in bacteria and in plants

Locus	Function	Ti Plasmid
vir	DNA transfer into plant	all
shi	shoot induction	all
roi	root induction	all
nos	nopaline synthesis	nopaline
noc	nopaline catabolism	nopaline
ocs	octopine synthesis	octopine
occ	octopine catabolism	octopine
tra	bacterial transfer genes	all
Inc	incompatibility genes	all
oriV	origin for replication	all

FIGURE 16.14 Ti plasmids carry genes involved in both plant and bacterial functions.

Ti plasmids (and thus the *Agrobacteria* in which they reside) can be divided into four groups, according to the types of opine that are made:

- **Nopaline** plasmids carry genes for synthesizing nopaline in tumors and for utilizing it in bacteria. Nopaline tumors can differentiate into shoots with abnormal structures. They have been called **teratomas** by analogy with certain mammalian tumors that retain the ability to differentiate into early embryonic structures.
- **Octopine** plasmids are similar to nopaline plasmids, but the relevant opine is different. Octopine tumors are usually undifferentiated, however, and do not form teratoma shoots.
- **Agropine** plasmids carry genes for agropine metabolism; the tumors do not differentiate, and they develop poorly and die early.
- **Ri plasmids** can induce hairy root disease on some plants and crown gall on others. They have agropine type genes, and may have segments derived from both nopaline and octopine plasmids.

The types of genes carried by a Ti plasmid are summarized in FIGURE 16.14. Genes utilized in the bacterium code for plasmid replication and incompatibility, transfer between bacteria, sensitivity to phages, and synthesis of other compounds, some of which are toxic to other soil bacteria. Genes used in the plant cell code for transfer of DNA into the plant, induction of the transformed state, and shoot and root induction.

The specificity of the opine genes depends on the type of plasmid. Genes needed for opine synthesis are linked to genes whose products catabolize the same opine; thus each strain of *Agrobacterium* causes crown gall tumor cells to synthesize opines that are useful for survival of the parasite. The opines can be used as the sole carbon and/or nitrogen source for the inducing *Agrobacterium* strain. The principle is that the transformed plant cell synthesizes those opines that the bacterium can use.

16.9 T-DNA Carries Genes Required for Infection

Key concepts

- Part of the DNA of the Ti plasmid is transferred to the plant cell nucleus.
- The *vir* genes of the Ti plasmid are located outside the transferred region and are required for the transfer process.
- The *vir* genes are induced by phenolic compounds released by plants in response to wounding.
- The membrane protein VirA is autophosphorylated on histidine when it binds an inducer.
- VirA activates VirG by transferring the phosphate group to it.
- The VirA-VirG is one of several bacterial two component systems that use a phosphohistidine relay.

The interaction between *Agrobacterium* and a plant cell is illustrated in FIGURE 16.15. The bacterium does not enter the plant cell, but rather transfers part of the Ti plasmid to the plant nucleus. The transferred part of the Ti genome is called **T-DNA.** It becomes integrated into the plant genome, where it expresses the functions needed to synthesize opines and to transform the plant cell.

Transformation of plant cells requires three types of function carried in the *Agrobacterium:*

- Three loci on the *Agrobacterium* chromosome, *chvA, chvB,* and *pscA,* are required for the initial stage of binding the bacterium to the plant cell. They are responsible for synthesizing a polysaccharide on the bacterial cell surface.
- The *vir* region carried by the Ti plasmid outside the T-DNA region is required to release and initiate transfer of the T-DNA.
- The T-DNA is required to transform the plant cell.

The organization of the major two types of Ti plasmid is illustrated in FIGURE 16.16. About 30% of the ~200 kb Ti genome is common to nopaline and octopine plasmids. The common regions include genes involved in all stages of the interaction between *Agrobacterium* and a plant host, but considerable rearrangement of the sequences has occurred between the plasmids.

The T-region occupies ~23 kb. Some 9 kb is the same in the two types of plasmid. The Ti plasmids carry genes for opine synthesis (*Nos* or *Ocs*) within the T-region; corresponding genes for opine catabolism (*Noc* or *Occ*) reside elsewhere on the plasmid. The plasmids code for similar, but not identical, morphogenetic functions, as seen in the induction of characteristic types of tumors.

Functions affecting oncogenicity—the ability to form tumors—are not confined to the T-region. Those genes located outside the T-region must be concerned with establishing the tumorigenic state, but their products are not needed to perpetuate it. They may be concerned with transfer of T-DNA into the plant nucleus or perhaps with subsidiary functions such as the balance of plant hormones in the infected tissue. Some of the mutations are host-specific, preventing tumor formation by some plant species but not by others.

The virulence genes code for the functions required for the transfer process. Six loci (*virA, -B, -C, -D, -E,* and *-G*) reside in a 40-kb region outside the T-DNA. Their organization is summarized in FIGURE 16.17. Each locus is transcribed as an individual unit; some contain more than one open reading frame.

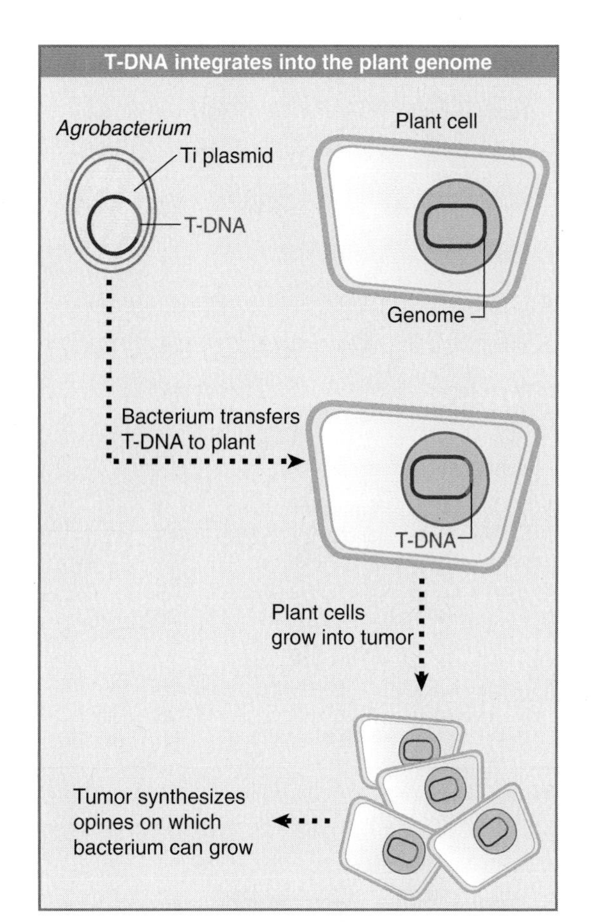

FIGURE 16.15 T-DNA is transferred from *Agrobacterium* carrying a Ti plasmid into a plant cell, where it becomes integrated into the nuclear genome and expresses functions that transform the host cell.

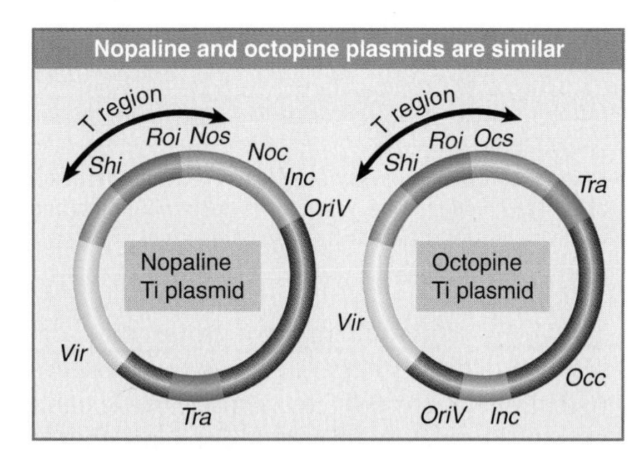

FIGURE 16.16 Nopaline and octopine Ti plasmids carry a variety of genes, including T-regions that have overlapping functions.

We may divide the transforming process into (at least) two stages:

- Agrobacterium contacts a plant cell and the *vir* genes are induced.
- *vir* gene products cause T-DNA to be transferred to the plant cell nucleus, where it is integrated into the genome.

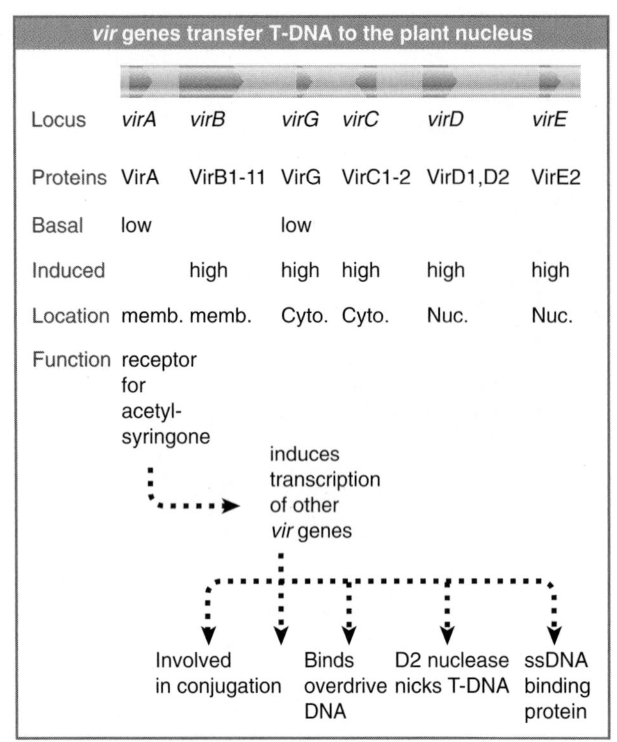

FIGURE 16.17 The *vir* region of the Ti plasmid has six loci that are responsible for transferring T-DNA to an infected plant.

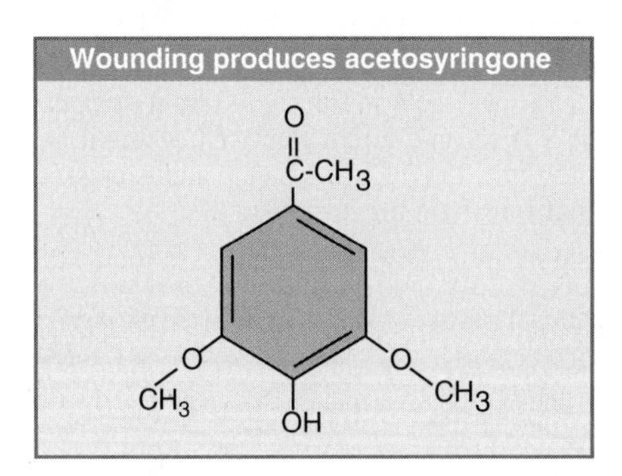

FIGURE 16.18 Acetosyringone (4-acetyl-2,6-dimethoxy-phenol) is produced by *N. tabacum* upon wounding and induces transfer of T-DNA from *Agrobacterium*.

The *vir* genes fall into two groups that correspond to these stages. Genes *virA* and *virG* are regulators that respond to a change in the plant by inducing the other genes. Thus mutants in *virA* and *virG* are avirulent and cannot express the remaining *vir* genes. Genes *virB, -C, -D,* and *-E* code for proteins involved in the transfer of DNA. Mutants in *virB* and *virD* are avirulent in all plants, but the effects of mutations in *virC* and *virE* vary with the type of host plant.

virA and *virG* are expressed constitutively (at a rather low level). The signal to which they respond is provided by phenolic compounds generated by plants as a response to wounding. FIGURE 16.18 presents an example. *Nicotiana tabacum* (tobacco) generates the molecules acetosyringone and α-hydroxyacetosyringone. Exposure to these compounds activates *virA,* which acts on *virG,* which in turn induces the expression *de novo* of *virB, -C, -D, and -E.* This reaction explains why *Agrobacterium* infection succeeds only on wounded plants.

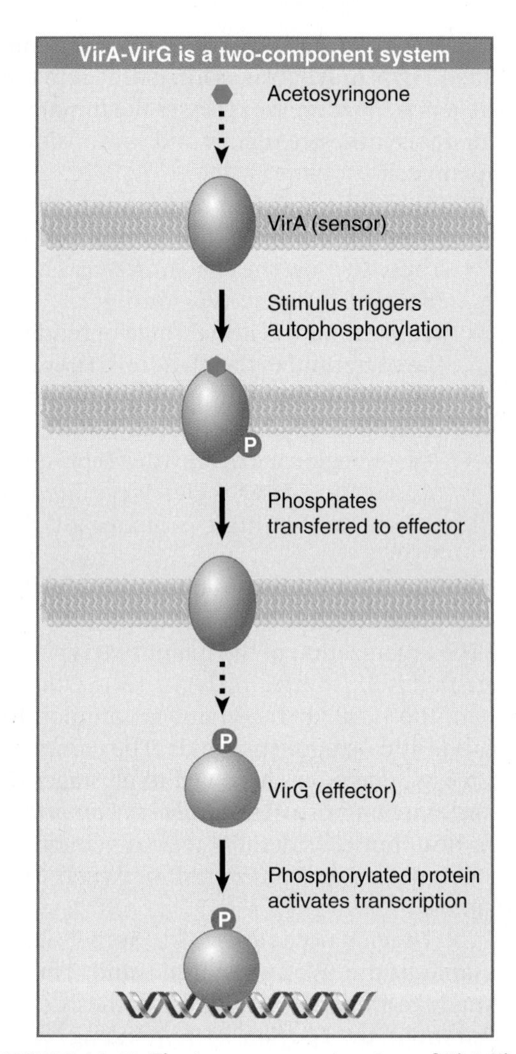

FIGURE 16.19 The two-component system of VirA-VirG responds to phenolic signals by activating transcription of target genes.

VirA and VirG are an example of a classic type of bacterial system in which stimulation of a sensor protein causes autophosphorylation and transfer of the phosphate to the second protein. The relationship is illustrated in FIGURE 16.19. The VirA-VirG system resembles the EnvZ-OmpR system that responds to osmolarity. The sequence of *virA* is related to *envZ,* and the sequences of *virG* and *ompR* are closely related, which suggests that the effector proteins function in a similar manner.

VirA forms a homodimer that is located in the inner membrane; it may respond to the presence of the phenolic compounds in the periplasmic space. Exposure to these compounds causes VirA to become autophosphorylated on histidine. The phosphate group is then transferred to an Asp residue in VirG. The phosphorylated VirG binds to promoters of the *virB, -C, -D, and -E* genes to activate transcription. When *virG* is activated, its transcription is induced from a new startpoint—a different one from the one used for constitutive expression—with the result that the amount of VirG protein is increased.

Of the other *vir* loci, *virD* is the best characterized. The *virD* locus has four open reading frames. Two of the proteins coded at *virD,* VirD1 and VirD2, provide an endonuclease that initiates the transfer process by nicking T-DNA at a specific site.

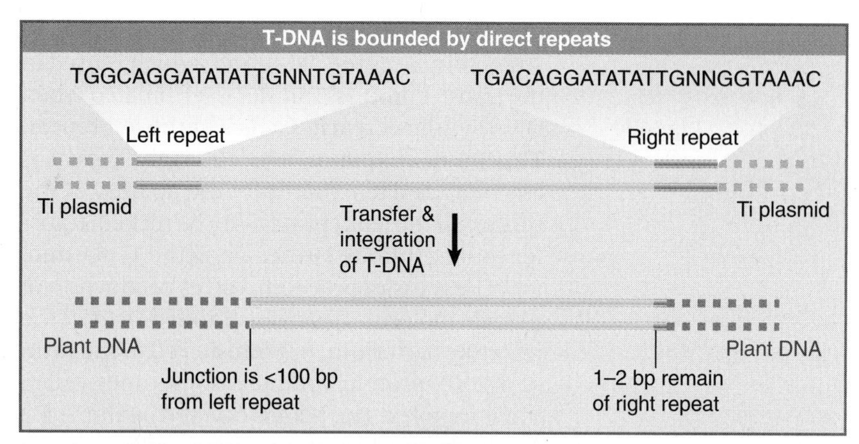

FIGURE 16.20 T-DNA has almost identical repeats of 25 bp at each end in the Ti plasmid. The right repeat is necessary for transfer and integration to a plant genome. T-DNA that is integrated in a plant genome has a precise junction that retains 1 to 2 bp of the right repeat, but the left junction varies and may be up to 100 bp short of the left repeat.

16.10 Transfer of T-DNA Resembles Bacterial Conjugation

Key concepts

- T-DNA is generated when a nick at the right boundary creates a primer for synthesis of a new DNA strand.
- The preexisting single strand that is displaced by the new synthesis is transferred to the plant cell nucleus.
- Transfer is terminated when DNA synthesis reaches a nick at the left boundary.
- The T-DNA is transferred as a complex of single-stranded DNA with the VirE2 single strand-binding protein.
- The single stranded T-DNA is converted into double-stranded DNA and integrated into the plant genome.
- The mechanism of integration is not known. T-DNA can be used to transfer genes into a plant nucleus.

The transfer process actually selects the T-region for entry into the plant. FIGURE 16.20 shows that the T-DNA of a nopaline plasmid is demarcated from the flanking regions in the Ti plasmid by repeats of 25 bp, which differ at only two positions between the left and right ends. When T-DNA is integrated into a plant genome, it has a well-defined right junction, which retains 1 to 2 bp of the right repeat. The left junction is variable; the boundary of T-DNA in the plant genome may be located at the 25 bp repeat or at one of a series of sites extending over ~100 bp within the T-DNA. At times multiple tandem copies of T-DNA are integrated at a single site.

A model for transfer is illustrated in FIGURE 16.21. A nick is made at the right 25 bp repeat. It provides a priming end for synthesis

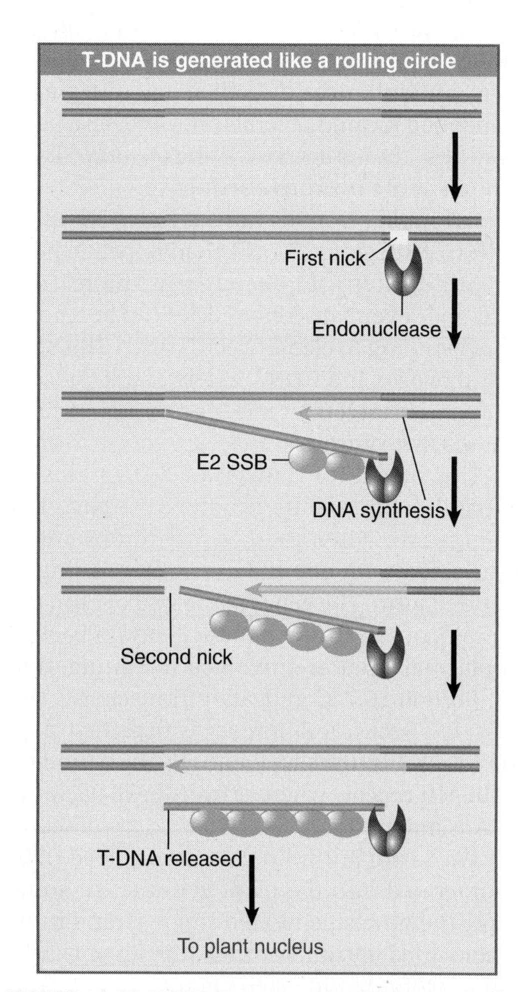

FIGURE 16.21 T-DNA is generated by displacement when DNA synthesis starts at a nick made at the right repeat. The reaction is terminated by a nick at the left repeat.

of a DNA single strand. Synthesis of the new strand displaces the old strand, which is used in the transfer process. Transfer is terminated when DNA synthesis reaches a nick at the left repeat. This model explains why the right repeat is essential, and it accounts for the polarity of the process. If the left repeat fails to be nicked, transfer could continue farther along the Ti plasmid.

The transfer process involves production of a single molecule of single-stranded DNA in the infecting bacterium. It is transferred in the form of a DNA-protein complex, sometimes called the T-complex. The DNA is covered by the VirE2 single-strand binding protein, which has a nuclear localization signal and is responsible for transporting T-DNA into the plant cell nucleus. A single molecule of the D2 subunit of the endonuclease remains bound at the 5′ end. The *virB* operon codes for eleven products that are involved in the transfer reaction.

Outside T-DNA, but immediately adjacent to the right border, is another short sequence, called *overdrive*, which greatly stimulates the transfer process. Overdrive functions like an enhancer: It must lie on the same molecule of DNA, but enhances the efficiency of transfer even when located several thousand base pairs away from the border. VirC1, and possibly VirC2, may act at the overdrive sequence.

Octopine plasmids have a more complex pattern of integrated T-DNA than nopaline plasmids. The pattern of T-strands is also more complex, and several discrete species can be found, corresponding to elements of T-DNA. This suggests that octopine T-DNA has several sequences that provide targets for nicking and/or termination of DNA synthesis.

This model for transfer of T-DNA closely resembles the events involved in bacterial conjugation, when the *E. coli* chromosome is transferred from one cell to another in single-stranded form. The genes of the *virB* operon are homologous to the *tra* genes of certain bacterial plasmids that are involved in conjugation (see Section 16.7, Conjugation Transfers Single-Stranded DNA). A difference is that the transfer of T-DNA is (usually) limited by the boundary of the left repeat, whereas transfer of bacterial DNA is indefinite.

We do not know how the transferred DNA is integrated into the plant genome. At some stage, the newly generated single strand must be converted into duplex DNA. Circles of T-DNA that are found in infected plant cells appear to be generated by recombination between the left and right 25-bp repeats, but we do not know if they are intermediates. The actual event is likely to involve a nonhomologous recombination, because there is no homology between the T-DNA and the sites of integration.

Is T-DNA integrated into the plant genome as an integral unit? How many copies are integrated? What sites in plant DNA are available for integration? Are genes in T-DNA regulated exclusively by functions on the integrated segment? These questions are central to defining the process by which the Ti plasmid transforms a plant cell into a tumor.

What is the structure of the target site? Sequences flanking the integrated T-DNA tend to be rich in A-T base pairs (a feature displayed in target sites for some transposable elements). The sequence rearrangements that occur at the ends of the integrated T-DNA make it difficult to analyze the structure. We do not know whether the integration process generates new sequences in the target DNA comparable to the target repeats created in transposition.

T-DNA is expressed at its site of integration. The region contains several transcription units, each of which probably contains a gene expressed from an individual promoter. Their functions are concerned with the state of the plant cell, maintaining its tumorigenic properties, controlling shoot and root formation, and suppressing differentiation into other tissues. None of these genes is needed for T-DNA transfer.

The Ti plasmid presents an interesting organization of functions. Outside the T-region, it carries genes needed to initiate oncogenesis; at least some are concerned with the transfer of T-DNA, and we would like to know whether others function in the plant cell to affect its behavior at this stage. Also outside the T-region are the genes that enable the *Agrobacterium* to catabolize the opine that the transformed plant cell will produce. Within the T-region are the genes that control the transformed state of the plant, as well as the genes that cause it to synthesize the opines that will benefit the *Agrobacterium* that originally provided the T-DNA.

As a practical matter, the ability of *Agrobacterium* to transfer T-DNA to the plant genome makes it possible to introduce new genes into plants. The transfer/integration and oncogenic functions are separate; thus it is possible to engineer new Ti plasmids in which the oncogenic functions have been replaced by other genes whose effect on the plant we wish to test. The existence of a natural system for delivering genes to the plant genome should greatly facilitate genetic engineering of plants.

16.11 Summary

The rolling circle is an alternative form of replication for circular DNA molecules in which an origin is nicked to provide a priming end. One strand of DNA is synthesized from this end; this displaces the original partner strand, which is extruded as a tail. Multiple genomes can be produced by continuing revolutions of the circle.

Rolling circles are used to replicate some phages. The A protein that nicks the φX174 origin has the unusual property of *cis*-action. It acts only on the DNA from which it was synthesized. It remains attached to the displaced strand until an entire strand has been synthesized, and then nicks the origin again; this releases the displaced strand and starts another cycle of replication.

Rolling circles also characterize bacterial conjugation, which occurs when an F plasmid is transferred from a donor to a recipient cell following the initiation of contact between the cells by means of the F-pili. A free F plasmid infects new cells by this means; an integrated F factor creates an Hfr strain that may transfer chromosomal DNA. In conjugation, replication is used to synthesize complements to the single strand remaining in the donor and to the single strand transferred to the recipient, but does not provide the motive power.

Agrobacteria induce tumor formation in wounded plant cells. The wounded cells secrete phenolic compounds that activate *vir* genes carried by the Ti plasmid of the bacterium. The *vir* gene products cause a single strand of DNA from the T-DNA region of the plasmid to be transferred to the plant cell nucleus. Transfer is initiated at one boundary of T-DNA, but ends at variable sites. The single strand is converted into a double strand and integrated into the plant genome. Genes within the T-DNA transform the plant cell and cause it to produce particular opines (derivatives of arginine). Genes in the Ti plasmid allow *Agrobacteria* to metabolize the opines produced by the transformed plant cell. T-DNA has been used to develop vectors for transferring genes into plant cells.

References

16.4 Rolling Circles Produce Multimers of a Replicon

Research

Gilbert, W. and Dressler, D. (1968). DNA replication: the rolling circle model. *Cold Spring Harbor Symp. Quant. Biol.* 33, 473–484.

16.6 The F Plasmid Is Transferred by Conjugation between Bacteria

Research

Ihler, G. and Rupp, W. D. (1969). Strand-specific transfer of donor DNA during conjugation in *E. coli. Proc. Natl. Acad. Sci. USA* 63, 138–143.

16.7 Conjugation Transfers Single-Stranded DNA

Reviews

Frost, L. S., Ippen-Ihler, K., and Skurray, R. A. (1994). Analysis of the sequence and gene products of the transfer region of the F sex factor. *Microbiol. Rev.* 58, 162–210.

Ippen-Ihler, K. A. and Minkley, E. G. (1986). The conjugation system of F, the fertility factor of *E. coli. Annu. Rev. Genet.* 20, 593–624.

Lanka, E. and Wilkins, B. M. (1995). DNA processing reactions in bacterial conjugation. *Annu. Rev. Biochem.* 64, 141–169.

Willetts, N. and Skurray, R. (1987). Structure and function of the F factor and mechanism of conjugation. In Neidhardt, F. C., ed. Escherichia coli *and* Salmonella typhimurium. Washington, DC: American Society for Microbiology, pp. 1110–1133.

17

Bacterial Replication Is Connected to the Cell Cycle

CHAPTER OUTLINE

17.1 Introduction

17.2 Replication Is Connected to the Cell Cycle

- The doubling time of *E. coli* can vary over a $10\times$ range, depending on growth conditions.
- It requires 40 minutes to replicate the bacterial chromosome (at normal temperature).
- Completion of a replication cycle triggers a bacterial division 20 minutes later.
- If the doubling time is <60 minutes, a replication cycle is initiated before the division resulting from the previous replication cycle.
- Fast rates of growth therefore produce multiforked chromosomes.
- A replication cycle is initiated at a constant ratio of mass/number of chromosome origins.
- There is one origin per unit cell of 1.7 μm in length.

17.3 The Septum Divides a Bacterium into Progeny That Each Contain a Chromosome

- Septum formation is initiated at the annulus, which is a ring around the cell where the structure of the envelope is altered.
- New annuli are initiated at 50% of the distance from the septum to each end of the bacterium.
- When the bacterium divides, each daughter has an annulus at the mid-center position.
- Septation starts when the cell reaches a fixed length.
- The septum consists of the same peptidoglycans that comprise the bacterial envelope.

17.4 Mutations in Division or Segregation Affect Cell Shape

- *fts* mutants form long filaments because the septum fails to form to divide the daughter bacteria.
- Minicells form in mutants that produce too many septa; they are small and lack DNA.
- Anucleate cells of normal size are generated by partition mutants in which the duplicate chromosomes fail to separate.

17.5 FtsZ Is Necessary for Septum Formation

- The product of *ftsZ* is required for septum formation at preexisting sites.
- FtsZ is a GTPase that forms a ring on the inside of the bacterial envelope. It is connected to other cytoskeletal components.

17.6 *min* Genes Regulate the Location of the Septum

- The location of the septum is controlled by *minC, -D,* and *-E.*
- The number and location of septa is determined by the ratio of MinE/MinC,D.
- The septum forms where MinE is able to form a ring.
- At normal concentrations, MinC/D allows a mid-center ring, but prevents additional rings of MinE from forming at the poles.

17.7 Chromosomal Segregation May Require Site-Specific Recombination

- The Xer site-specific recombination system acts on a target sequence near the chromosome terminus to recreate monomers if a generalized recombination event has converted the bacterial chromosome to a dimer.

17.8 Partitioning Involves Separation of the Chromosomes

- Replicon origins may be attached to the inner bacterial membrane.
- Chromosomes make abrupt movements from the mid-center to the 1/4 and 3/4 positions.

17.9 Single-Copy Plasmids Have a Partitioning System

- Single-copy plasmids exist at one plasmid copy per bacterial chromosome origin.
- Multicopy plasmids exist at >1 plasmid copy per bacterial chromosome origin.
- Homologous recombination between circular plasmids generates dimers and higher multimers.
- Plasmids have site-specific recombination systems that undertake intramolecular recombination to regenerate monomers.
- Partition systems ensure that duplicate plasmids are segregated to different daughter cells produced by a division.

17.10 Plasmid Incompatibility Is Determined by the Replicon

- Plasmids in a single compatibility group have origins that are regulated by a common control system.

17.11 The ColE1 Compatibility System Is Controlled by an RNA Regulator

- Replication of ColE1 requires transcription to pass through the origin, where the transcript is cleaved by RNAase H to generate a primer end.

- The regulator RNA I is a short antisense RNA that pairs with the transcript and prevents the cleavage that generates the priming end.
- The Rom protein enhances pairing between RNA I and the transcript.

17.1 Introduction

The way in which replication is controlled and linked to the cell cycle is a major difference between prokaryotes and eukaryotes.

In eukaryotes, the following are true:

- chromosomes reside in the nucleus,
- each chromosome consists of many replicons,
- replication requires coordination of these replicons to reproduce DNA during a discrete period of the cell cycle,
- the decision on whether to replicate is taken by a complex pathway that regulates the cell cycle, and
- the duplicated chromosomes are segregated to daughter cells during mitosis by means of a special apparatus.

FIGURE 17.1 shows that in bacteria, replication is triggered at a single origin when the cell mass increases past a threshold level, and the segregation of the daughter chromosomes is accomplished by ensuring that they find themselves on opposite sides of the septum that grows to divide the bacterium into two.

How does the cell know when to initiate the replication cycle? The initiation event occurs at a constant ratio of cell mass to the number of chromosome origins. Cells growing more rapidly are larger and possess a greater number of origins. The growth of *E. coli* can be described in terms of the **unit cell,** an entity 1.7 μm long. A bacterium contains one origin per unit cell; a rapidly growing cell with two origins is 1.7 to 3.4 μm long.

How is cell mass titrated? An initiator protein could be synthesized continuously throughout the cell cycle; accumulation of a critical amount would trigger initiation. This explains why protein synthesis is needed for the initiation event. Another possibility is that an inhibitor protein might be synthesized at a fixed point and diluted below an effective level by the increase in cell volume.

Some of the events in partitioning the daughter chromosomes are consequences of the circularity of the bacterial chromosome. Circular chromosomes are said to be *catenated* when one passes through another, connecting them. Topoisomerases are required to separate them. An alternative type of structure is formed when a recombination event occurs: A single recombination between two monomers converts them into a single dimer. This is resolved by a specialized recombination system that recreates the independent monomers. Essentially the partitioning process is handled by enzyme systems that act directly on discrete DNA sequences.

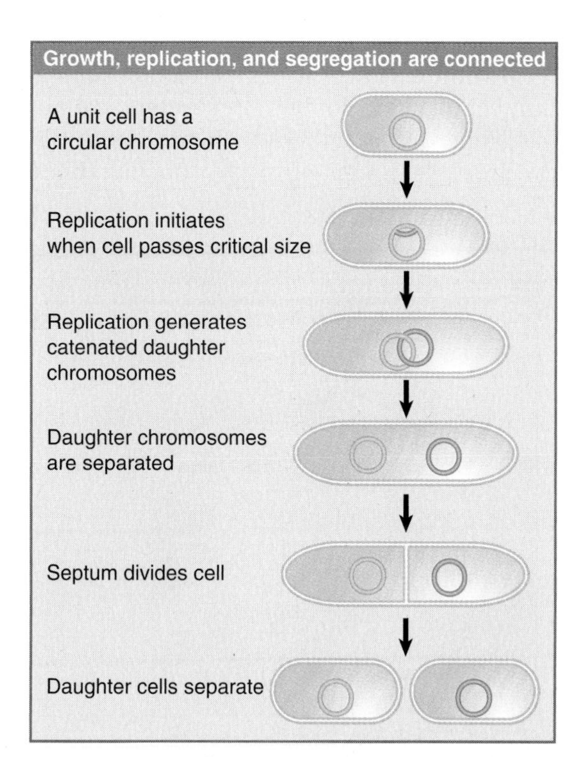

FIGURE 17.1 Replication initiates at the bacterial origin when a cell passes a critical threshold of size. Completion of replication produces daughter chromosomes that may be linked by recombination or that may be catenated. They are separated and moved to opposite sides of the septum before the bacterium is divided into two.

17.2 Replication Is Connected to the Cell Cycle

Key concepts

- The doubling time of *E. coli* can vary over a 10× range, depending on growth conditions.
- It requires 40 minutes to replicate the bacterial chromosome (at normal temperature).
- Completion of a replication cycle triggers a bacterial division 20 minutes later.
- If the doubling time is <60 minutes, a replication cycle is initiated before the division resulting from the previous replication cycle.
- Fast rates of growth therefore produce multiforked chromosomes.
- A replication cycle is initiated at a constant ratio of mass/number of chromosome origins.
- There is one origin per unit cell of 1.7 μm in length.

Bacteria have two links between replication and cell growth:

- The frequency of initiation of cycles of replication is adjusted to fit the rate at which the cell is growing.
- The completion of a replication cycle is connected with division of the cell.

The rate of bacterial growth is assessed by the **doubling time,** the period required for the number of cells to double. The shorter the doubling time, the faster the growth rate. *E. coli* cells can grow at rates ranging from doubling times as fast as 18 minutes to slower than 180 minutes. The bacterial chromosome is a single replicon; thus the frequency of replication cycles is controlled by the number of initiation events at the single origin. The replication cycle can be defined in terms of two constants:

- *C* is the fixed time of ~40 minutes required to replicate the entire bacterial chromosome. Its duration corresponds to a rate of replication fork movement of ~50,000 bp/minute. (The rate of DNA synthesis is more or less invariant at a constant temperature; it proceeds at the same speed unless and until the supply of precursors becomes limiting.)
- *D* is the fixed time of ~20 minutes that elapses between the completion of a round of replication and the cell division with which it is connected. This period may represent the time required to assemble the components needed for division.

(The constants *C* and *D* can be viewed as representing the maximum speed with which the bacterium is capable of completing these processes. They apply for all growth rates between doubling times of 18 and 60 minutes, but both constant phases become longer when the cell cycle occupies >60 minutes.)

A cycle of chromosome replication must be initiated at a fixed time of $C + D = 60$ minutes before a cell division. For bacteria dividing more frequently than every 60 minutes, a cycle of replication must be initiated before the end of the preceding division cycle. You might say that a cell is born already pregnant with the next generation.

Consider the example of cells dividing every 35 minutes. The cycle of replication connected with a division must have been initiated 25 minutes before the preceding division. This situation is illustrated in FIGURE 17.2, which shows the chromosomal complement of a bacterial cell at 5-minute intervals throughout the cycle.

At division (35/0 minutes), the cell receives a partially replicated chromosome. The replication fork continues to advance. At 10 minutes, when this "old" replication fork has not yet reached the terminus, initiation occurs at both origins on the partially replicated chromosome. The start of these "new" replication forks creates a **multiforked chromosome.**

At 15 minutes—that is, at 20 minutes before the next division—the old replication fork reaches the terminus. Its arrival allows the two

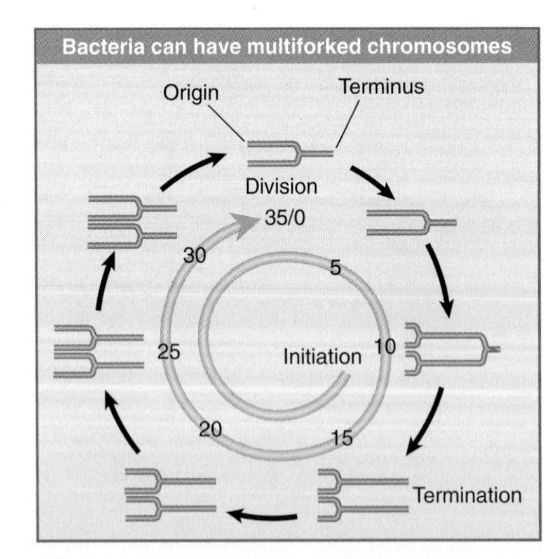

FIGURE 17.2 The fixed interval of 60 minutes between initiation of replication and cell division produces multiforked chromosomes in rapidly growing cells. Note that only the replication forks moving in one direction are shown; the chromosome actually is replicated symmetrically by two sets of forks moving in opposite directions on circular chromosomes.

daughter chromosomes to separate; each of them has already been partially replicated by the new replication forks (which now are the only replication forks). These forks continue to advance.

At the point of division, the two partially replicated chromosomes segregate. This recreates the point at which we started. The single replication fork becomes "old," it terminates at 15 minutes, and 20 minutes later there is a division. We see that the initiation event occurs $1\frac{25}{35}$ cell cycles before the division event with which it is associated.

The general principle of the link between initiation and the cell cycle is that, as cells grow more rapidly (the cycle is shorter), the initiation event occurs an increasing number of cycles before the related division. There are correspondingly more chromosomes in the individual bacterium. This relationship can be viewed as the cell's response to its inability to reduce the periods of *C* and *D* to keep pace with the shorter cycle.

17.3 The Septum Divides a Bacterium into Progeny That Each Contain a Chromosome

Key concepts

- Septum formation is initiated at the annulus, which is a ring around the cell where the structure of the envelope is altered.
- New annuli are initiated at 50% of the distance from the septum to each end of the bacterium.
- When the bacterium divides, each daughter has an annulus at the mid-center position.
- Septation starts when the cell reaches a fixed length.
- The septum consists of the same peptidoglycans that comprise the bacterial envelope.

Chromosome segregation in bacteria is especially interesting because the DNA itself is involved in the mechanism for partition. (This contrasts with eukaryotic cells, in which segregation is achieved by the complex apparatus of mitosis.) The bacterial apparatus is quite accurate; however, anucleate cells form <0.03% of a bacterial population.

The division of a bacterium into two daughter cells is accomplished by the formation of a **septum,** a structure that forms in the center of the cell as an invagination from the surrounding envelope. The septum forms an impenetrable barrier between the two parts of the cell and provides the site at which the two daughter cells eventually separate entirely. Two related questions address the role of the septum in division: What determines the location at which it forms, and what ensures that the daughter chromosomes lie on opposite sides of it?

The formation of the septum is preceded by the organization of the **periseptal annulus.** This is observed as a zone in *E. coli* or *Salmonella typhimurium,* for which the structure of the envelope is altered so that the inner membrane is connected more closely to the cell wall and outer membrane layer. As its name suggests, the annulus extends around the cell. FIGURE 17.3 summarizes its development.

The annulus starts at a central position in a new cell. As the cell grows, two events occur: A septum forms at the mid-cell position defined by the annulus, and new annuli form on either side of the initial annulus. These new annuli are displaced from the center and move along the cell to positions at one quarter and three quarters of the cell length. These will become the mid-cell positions after the next division. The displacement of the periseptal annulus to

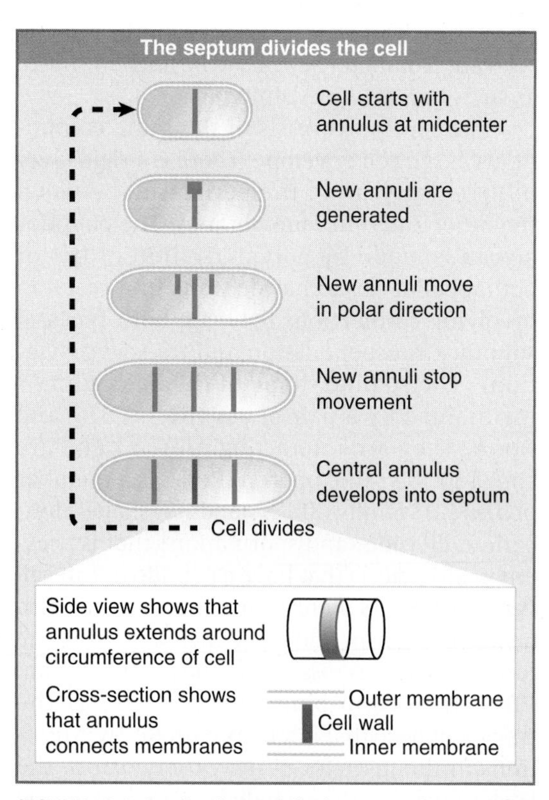

FIGURE 17.3 Duplication and displacement of the periseptal annulus give rise to the formation of a septum that divides the cell.

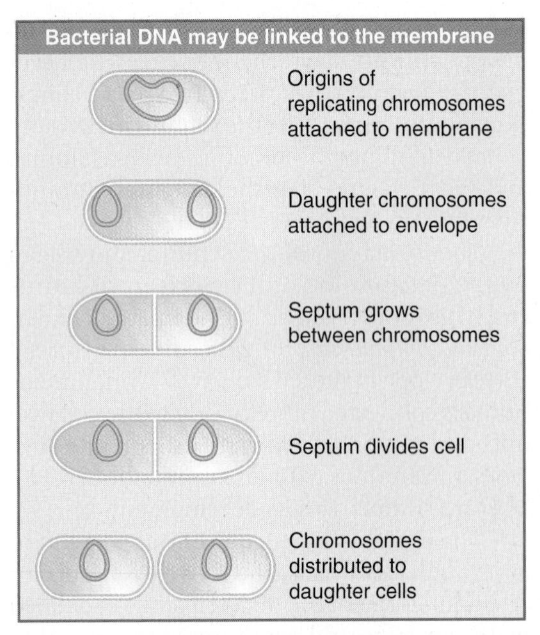

FIGURE 17.4 Attachment of bacterial DNA to the membrane could provide a mechanism for segregation.

the correct position may be the crucial event that ensures the division of the cell into daughters of equal size. (The mechanism of movement is unknown.) Septation begins when the cell reaches a fixed length (2L), and the distance between the new annuli is always L. We do not know how the cell measures length, but the relevant parameter appears to be linear distance as such (not area or volume).

The septum consists of the same components as the cell envelope: There is a rigid layer of peptidoglycan in the periplasm, between the inner and outer membranes. The peptidoglycan is made by polymerization of tri- or pentapeptide-disaccharide units in a reaction involving connections between both types of subunit (transpeptidation and transglycosylation). The rodlike shape of the bacterium is maintained by a pair of activities, PBP2 and RodA. They are interacting proteins and are coded by the same operon. RodA is a member of the SEDS family (SEDS stands for shape, elongation, division, and sporulation) that is present in all bacteria that have a peptidoglycan cell wall. Each SEDS protein functions together with a specific transpeptidase, which catalyzes the formation of the crosslinks in the peptidoglycan. PBP2 (penicillin-binding protein 2) is the transpeptidase that interacts with RodA. Mutations in the gene for either protein cause the bacterium to lose its extended shape and become round. This demonstrates the important principle that shape and rigidity can be determined by the simple extension of a polymeric structure. Another enzyme is responsible for generating the peptidoglycan in the septum (see Section 17.5, FtsZ Is Necessary for Septum Formation). The septum initially forms as a double layer of peptidoglycan, and the protein EnvA is required to split the covalent links between the layers so that the daughter cells may separate.

The behavior of the periseptal annulus suggests that the mechanism for measuring position is associated with the cell envelope. It is plausible to suppose that the envelope could also be used to ensure segregation of the chromosomes. A direct link between DNA and the membrane could account for segregation. If daughter chromosomes are attached to the membrane, they could be physically separated when the septum forms. FIGURE 17.4 shows that the formation of a septum could segregate the chromosomes into the different daughter cells if the origins are connected to sites that lie on either side of the periseptal annulus.

17.4 Mutations in Division or Segregation Affect Cell Shape

Key concepts

- *fts* mutants form long filaments because the septum fails to form to divide the daughter bacteria.
- Minicells form in mutants that produce too many septa; they are small and lack DNA.
- Anucleate cells of normal size are generated by partition mutants in which the duplicate chromosomes fail to separate.

A difficulty in isolating mutants that affect cell division is that mutations in the critical functions may be lethal and/or pleiotropic. For example, if formation of the annulus occurs at a site that is essential for overall growth of the envelope, it would be difficult to distinguish mutations that specifically interfere with annulus formation from those that inhibit envelope growth generally. Most mutations in the division apparatus have been identified as conditional mutants (whose division is affected under nonpermissive conditions; typically they are temperature sensitive). Mutations that affect cell division or chromosome segregation cause striking phenotypic changes. FIGURE 17.5 and FIGURE 17.6 illustrate the opposite consequences of failure in the division process and failure in segregation:

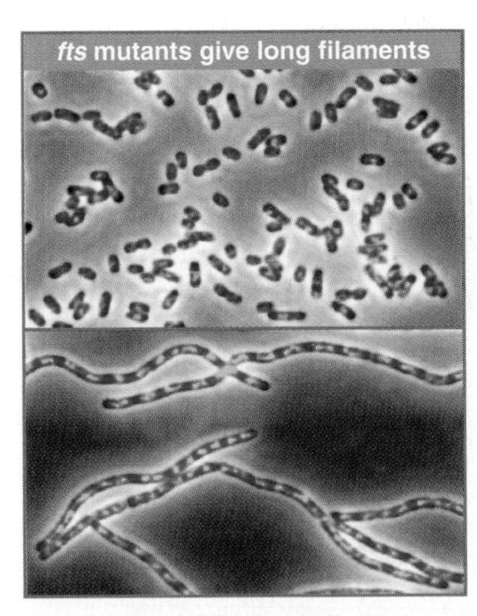

FIGURE 17.5 *Top panel:* Wild-type cells. *Bottom panel:* Failure of cell division under nonpermissive temperatures generates multinucleated filaments. Photos courtesy of Sota Hiraga, Kyoto University.

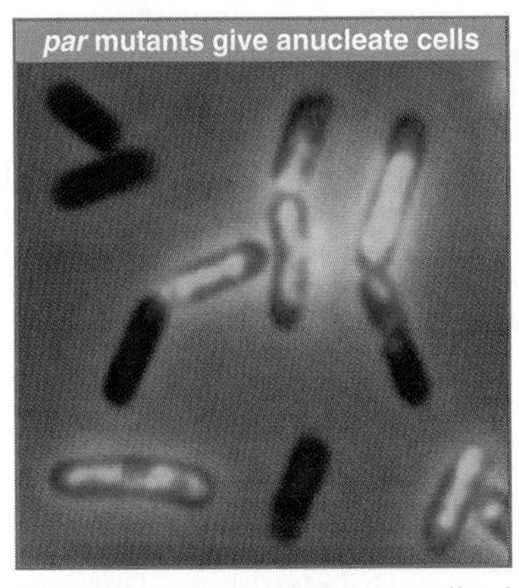

FIGURE 17.6 *E. coli* generate anucleate cells when chromosome segregation fails. Cells with chromosomes stain blue; daughter cells lacking chromosomes have no blue stain. This field shows cells of the *mukB* mutant; both normal and abnormal divisions can be seen. Photo courtesy of Sota Hiraga, Kyoto University.

- Long *filaments* form when septum formation is inhibited, but chromosome replication is unaffected. The bacteria continue to grow—and even continue to segregate their daughter chromosomes—but septa do not form. Thus the cell consists of a very long filamentous structure, with the nucleoids (bacterial chromosomes) regularly distributed along the length of the cell. This phenotype is displayed by *fts* mutants (named for temperature-sensitive filamentation), which identify a defect or multiple defects that lie in the division process itself.
- **Minicells** form when septum formation occurs too frequently or in the wrong place, with the result that one of the new daughter cells lacks a chromosome. The minicell has a rather small size and lacks DNA, but otherwise appears morphologically normal. **Anucleate** cells form when segregation is aberrant; like minicells, they lack a chromosome, but because septum formation is normal, their size is unaltered. This phenotype is caused by *par* (partition) mutants (named because they are defective in chromosome segregation).

17.5 FtsZ Is Necessary for Septum Formation

Key concepts

- The product of *ftsZ* is required for septum formation at preexisting sites.
- FtsZ is a GTPase that forms a ring on the inside of the bacterial envelope. It is connected to other cytoskeletal components.

The gene *ftsZ* plays a central role in division. Mutations in *ftsZ* block septum formation and generate filaments. Overexpression induces minicells by causing an increased number of septation events per unit cell mass. *ftsZ* mutants act at stages varying from the displacement of the periseptal annuli to septal morphogenesis. FtsZ is therefore required for usage of preexisting sites for septum formation, but does not itself affect the formation of the periseptal annuli or their localization.

FtsZ functions at an early stage of septum formation. Early in the division cycle, FtsZ is localized throughout the cytoplasm. As the cell elongates and begins to constrict in the middle, FtsZ becomes localized in a ring around the circumference. The structure is sometimes called the **Z-ring.** FIGURE 17.7 shows that it lies in the position of the mid-center annulus of Figure 17.3. The formation of the Z-ring is the rate-limiting step in septum formation. In a typical division cycle, it forms in the center of the cell 1 to 5 minutes after division, remains for 15 minutes, and then quickly constricts to pinch the cell into two.

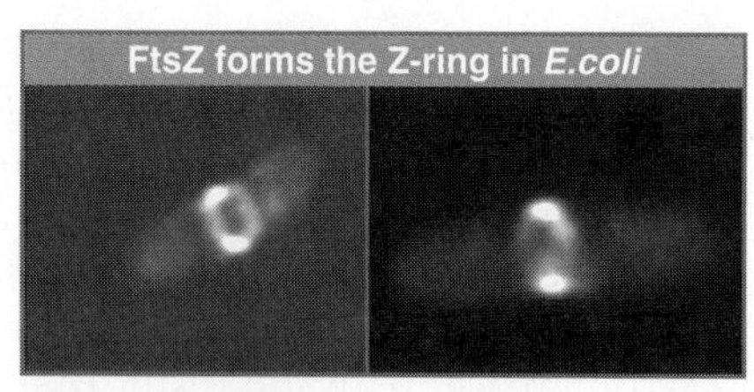

FIGURE 17.7 Immunofluorescence with an antibody against FtsZ shows that it is localized at the mid-cell. Photo courtesy of William Margolin, University of Texas Medical School at Houston.

The structure of FtsZ resembles tubulin, suggesting that assembly of the ring could resemble the formation of microtubules in eukaryotic cells. FtsZ has GTPase activity, and GTP cleavage is used to support the oligomerization of FtsZ monomers into the ring structure. The Z-ring is a dynamic structure, in which there is continuous exchange of subunits with a cytoplasmic pool.

Two other proteins needed for division, ZipA and FtsA, interact directly and independently with FtsZ. ZipA is an integral membrane protein that is located in the inner bacterial membrane. It provides the means for linking FtsZ to the membrane. FtsA is a cytosolic protein, but is often found associated with the membrane. The Z-ring can form in the absence of either ZipA or FtsA, but cannot form if both are absent. This suggests that they have overlapping roles in stabilizing the Z-ring and perhaps in linking it to the membrane.

The products of several other *fts* genes join the Z-ring in a defined order after FtsA has been incorporated. They are all transmembrane proteins. The final structure is sometimes called the **septal ring.** It consists of a multiprotein complex that is presumed to have the ability to constrict the membrane. One of the last components to be incorporated into the septal ring is FtsW, which is a protein belonging to the SEDS family. *ftsW* is expressed as part of an operon with *ftsI*, which codes for a transpeptidase (also called PBP3 for penicillin-binding protein 3), a membrane-bound protein that has its catalytic site in the periplasm. FtsW is responsible for incorporating FtsI into the septal ring. This suggests a model for septum formation in which the transpeptidase activity then causes the peptidoglycan to grow inward, thus pushing the inner membrane and pulling the outer membrane.

FtsZ is the major cytoskeletal component of septation. It is common in bacteria, and also is found in chloroplasts. FIGURE 17.8 shows the localization of the plant homologs to a ring at the midpoint of the chloroplast. Chloroplasts also have other genes related to the bacterial division genes. Consistent with the common evolutionary origins of bacteria and chloroplasts, the apparatus for division seems generally to have been conserved.

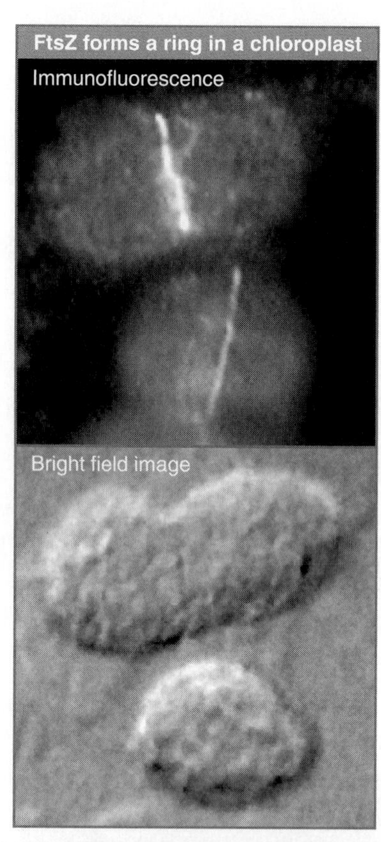

FIGURE 17.8 Immunofluorescence with antibodies against the *Arabidopsis* proteins FtsZ1 and FtsZ2 show that they are localized at the midpoint of the chloroplast (top panel). The bright field image (lower panel) shows the outline of the chloroplast more clearly. Photos courtesy of Katherine Osteryoung, Michigan State University.

Mitochondria, which also share an evolutionary origin with bacteria, usually do not have FtsZ. Instead, they use a variant of the protein dynamin, which is involved in pinching off vesicles from membranes of eukaryotic cytoplasm. This functions from the outside of the organelle, squeezing the membrane to generate a constriction.

The common feature, then, in the division of bacteria, chloroplasts, and mitochondria is the use of a cytoskeletal protein that forms a ring around the organelle and either pulls or pushes the membrane to form a constriction.

17.6 *min* Genes Regulate the Location of the Septum

Key concepts

- The location of the septum is controlled by *minC,-D,* and *-E*.
- The number and location of septa is determined by the ratio of MinE/MinC,D.
- The septum forms where MinE is able to form a ring.
- At normal concentrations, MinC/D allows a mid-center ring, but prevents additional rings of MinE from forming at the poles.

Information about the localization of the septum is provided by minicell mutants. The original minicell mutation lies in the locus *minB;* deletion of *minB* generates minicells by allowing septation to occur at the poles as well as (or instead of) at mid-cell. This suggests that the cell possesses the ability to initiate septum formation either at mid-cell or at the poles, and that the role of the wild-type *minB* locus is to suppress septation at the poles. In terms of the events depicted in Figure 17.3, this implies that a newborn cell has potential septation sites associated both with the annulus at mid-center and with the poles. One pole was formed from the septum of the previous division; the other pole represents the septum from the division before that. Perhaps the poles retain remnants of the annuli from which they were derived and these remnants can nucleate septation.

The *minB* locus consists of three genes, *minC, -D,* and *-E*. Their roles are summarized in FIGURE 17.9. The products of *minC* and *minD* form a division inhibitor. MinD is required to activate MinC, which prevents FtsZ from polymerizing into the Z-ring).

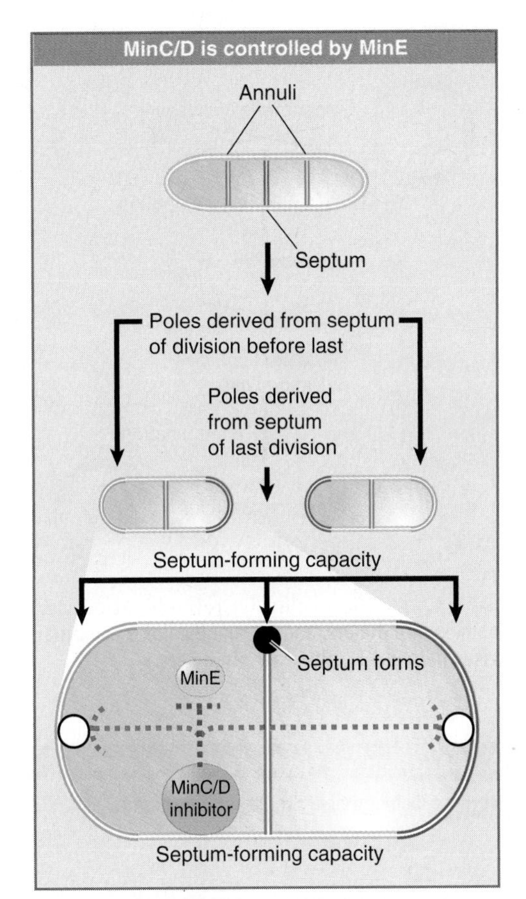

FIGURE 17.9 MinC/D is a division inhibitor whose action is confined to the polar sites by MinE.

Expression of MinCD in the absence of MinE, or overexpression even in the presence of MinE, causes a generalized inhibition of division. The resulting cells grow as long filaments without septa. Expression of MinE at levels comparable to MinCD confines the inhibition to the polar regions, so restoring normal growth. MinE protects the mid-cell sites from inhibition. Overexpression of MinE induces minicells, because the presence of excess MinE counteracts the inhibition at the poles as well as at mid-cell, allowing septa to form at both locations.

The determinant of septation at the proper (mid-cell) site is, therefore, the ratio of MinCD to MinE. The wild-type level prevents polar septation but permits mid-cell septation. The effects of MinC/D and MinE are inversely related; absence of MinCD, or too much MinE, causes indiscriminate septation, forming minicells; too much MinCD or absence of MinE inhibits mid-cell as well as polar sites, resulting in filamentation.

MinE forms a ring at the septal position. Its accumulation suppresses the action of MinCD in the vicinity, thus allowing formation of the septal ring (which includes FtsZ and ZipA). Curiously, MinD is required for formation of the MinE ring.

17.7 Chromosomal Segregation May Require Site-Specific Recombination

Key concept

- The Xer site-specific recombination system acts on a target sequence near the chromosome terminus to recreate monomers if a generalized recombination event has converted the bacterial chromosome to a dimer.

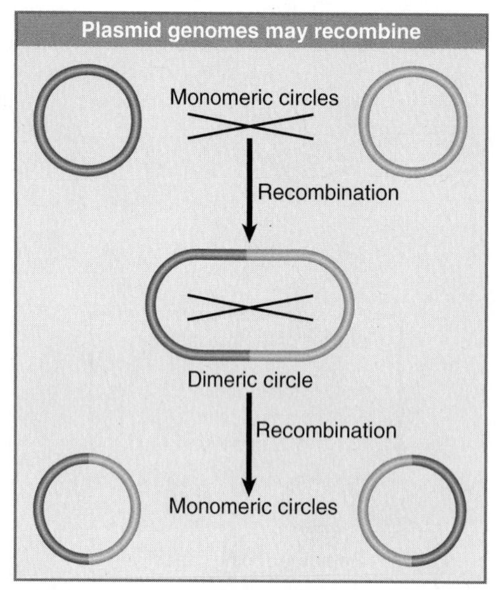

FIGURE 17.10 Intermolecular recombination merges monomers into dimers, and intramolecular recombination releases individual units from oligomers.

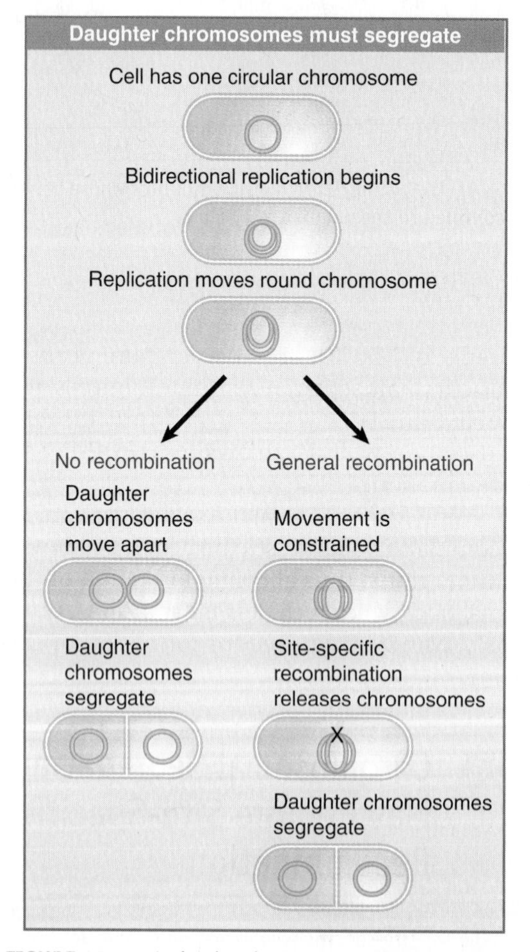

FIGURE 17.11 A circular chromosome replicates to produce two monomeric daughters that segregate to daughter cells. A generalized recombination event, however, generates a single dimeric molecule. This can be resolved into two monomers by a site-specific recombination.

The multiple copies of a plasmid in a bacterium consist of the same DNA sequences, and as a result they are able to recombine. FIGURE 17.10 demonstrates the consequences. A single intermolecular recombination event between two circles generates a dimeric circle; further recombination can generate higher multimeric forms. Such an event reduces the number of physically segregating units. In the extreme case of a single-copy plasmid that has just replicated, formation of a dimer by recombination means that the cell only has one unit to segregate, and the plasmid therefore must inevitably be lost from one daughter cell. To counteract this effect, plasmids often have **site-specific recombination** systems that act upon particular sequences to sponsor an intramolecular recombination that restores the monomeric condition.

The same types of event can occur with the bacterial chromosome; FIGURE 17.11 shows how they affect its segregation. If no recombination occurs there is no problem, and the separate daughter chromosomes can segregate to the daughter cells. A dimer will be produced, however, if homologous recombination occurs between the daughter chromosomes produced by a replication cycle. If there has been such a recombination event, the daughter chromosomes cannot separate. In this case, a second recombination is required to achieve resolution in the same way as a plasmid dimer.

Most bacteria with circular chromosomes possess the Xer site-specific recombination system. In *E. coli,* this consists of two recombinases, XerC and XerD, which act on a 28-bp target site, called *dif,* that is located in the terminus region of the chromosome. The use of the Xer system is related to cell division in an interesting way. The relevant events are summarized in FIGURE 17.12. XerC can bind to a pair of *dif* sequences and form a Holliday junction between them. The complex may form soon after the replication fork passes over the *dif* sequence, which explains how the two copies of the target sequence can find one another consistently. Resolution of the junction to give recombinants, however, occurs only in the presence of FtsK, a protein located in the septum that is required for chromosome segregation and cell division. In addition, the *dif* target sequence must be located in a region of ~30 kb; if it is moved outside of this region, it cannot support the reaction.

Thus there is a site-specific recombination available when the terminus sequence of the chromosome is close to the septum. The bacterium, however, wants to have a recombination only when there has already been a general recombination event to generate a dimer. (Otherwise the site-specific recombination would create the dimer!) How does the system know whether the daughter chromosomes exist as independent monomers or have been recombined into a dimer?

The answer may be that segregation of chromosomes starts soon after replication. If there has been no recombination, the two chromosomes move apart from one another. The ability of the relevant sequences to move apart from one another, however, may be constrained if a dimer has been formed. This forces them to remain in the vicinity of the septum, where they are exposed to the Xer system.

Bacteria that have the Xer system always have an FtsK homolog and vice-versa, which suggests that the system has evolved so that resolution is connected to the septum. FtsK is a large transmembrane protein. Its N-terminal domain is associated with the membrane and causes it to be localized to the septum. Its C-terminal domain has two functions. One is to cause Xer to resolve a dimer into two monomers. It also has an ATPase activity, which it can use to translocate along DNA *in vitro*. This could be used to pump DNA through the septum, in the same way that SpoIIIE transports DNA from the mother compartment into the prespore during sporulation. (See Section 17.8, Partitioning Involves Separation of the Chromosomes.)

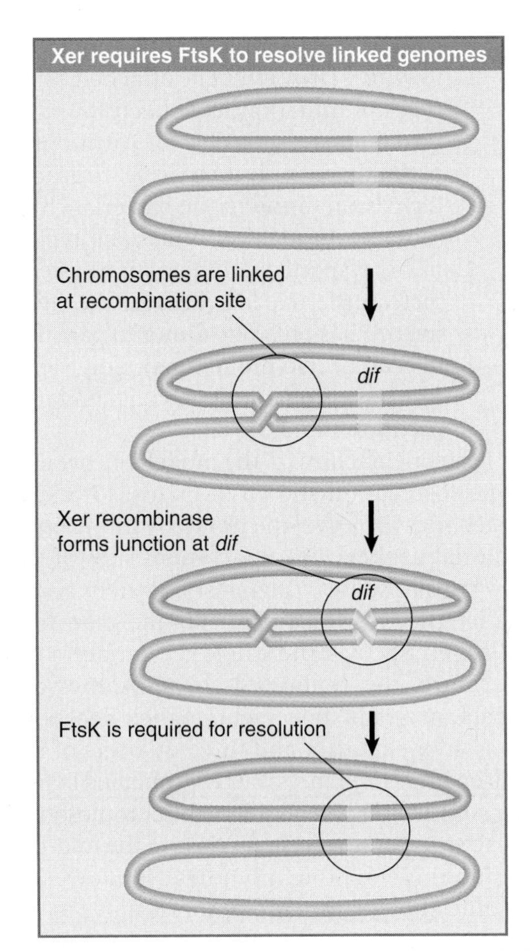

FIGURE 17.12 A recombination event creates two linked chromosomes. Xer creates a Holliday junction at the *dif* site, but can resolve it only in the presence of FtsK.

17.8 Partitioning Involves Separation of the Chromosomes

Key concepts

- Replicon origins may be attached to the inner bacterial membrane.
- Chromosomes make abrupt movements from the mid-center to the 1/4 and 3/4 positions.

Partitioning is the process by which the two daughter chromosomes find themselves on either side of the position at which the septum forms. Two types of event are required for proper partitioning:

- The two daughter chromosomes must be released from one another so that they can segregate following termination. This requires disentangling of DNA regions that are coiled around each other in the vicinity of the terminus. Most mutations affecting partitioning map in genes coding for topoisomerases-enzymes with the ability to pass DNA strands through one another. The mutations prevent the daughter chromosomes from segregating, with the result that the DNA is located in a single large mass at mid-cell. Septum formation then releases an anucleate cell and a cell containing both daughter chromosomes. This tells us that the bacterium must be able to disentangle its chromosomes topologically in order to be able to segregate them into different daughter cells.
- Mutations that affect the partition process itself are rare. We expect to find two classes: *cis*-acting mutations should occur in DNA sequences that are the targets for the partition process; *trans*-acting mutations should occur in genes that code for the protein(s) that cause segregation, which could include proteins that bind to DNA or activities that

control the locations on the envelope to which DNA might be attached. Both types of mutation have been found in the systems responsible for partitioning plasmids, but only *trans*-acting functions have been found in the bacterial chromosome. In addition, mutations in plasmid site-specific recombination systems increase plasmid loss (because the dividing cell has only one dimer to partition instead of two monomers), and therefore have a phenotype that is similar to partition mutants.

The original form of the model for chromosome segregation shown in Figure 17.8 suggested that the envelope grows by insertion of material between the attachment sites of the two chromosomes, thus pushing them apart. In fact, the cell wall and membrane grow heterogeneously over the whole cell surface. Furthermore, the replicated chromosomes are capable of abrupt movements to their final positions at one quarter and three quarters of the cell length. If protein synthesis is inhibited before the termination of replication, the chromosomes fail to segregate and remain close to the mid-cell position. When protein synthesis is allowed to resume, though, the chromosomes move to the quarter positions in the absence of any further envelope elongation. This suggests that an active process—one that requires protein synthesis—may move the chromosomes to specific locations.

Segregation is interrupted by mutations of the *muk* class, which give rise to anucleate progeny at a much increased frequency: Both daughter chromosomes remain on the same side of the septum instead of segregating. Mutations in the *muk* genes are not lethal, and they may identify components of the apparatus that segregates the chromosomes. The gene *mukA* is identical with the gene for a known outer membrane protein *(tolC)* whose product could be involved with attaching the chromosome to the envelope. The gene *mukB* codes for a large (180 kD) globular protein, which has the same general type of organization as the two groups of structural maintenance of chromosomes (SMC) proteins that are involved in condensing and in holding together eukaryotic chromosomes (see Section 31.6, Chromosome Condensation Is Caused by Condensins). SMC-like proteins have also been found in other bacteria.

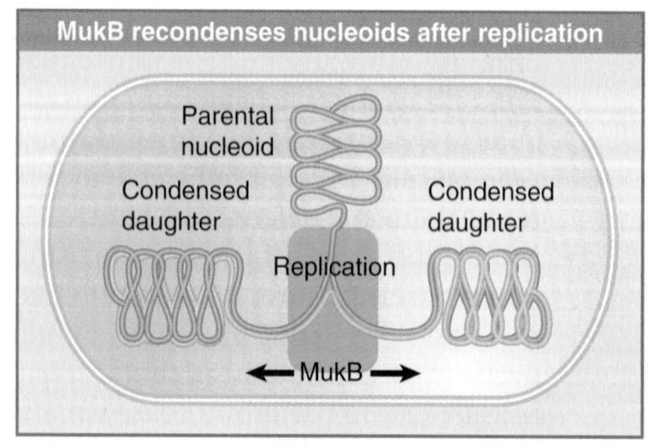

FIGURE 17.13 The DNA of a single parental nucleoid becomes decondensed during replication. MukB is an essential component of the apparatus that recondenses the daughter nucleoids.

The insight into the role of MukB was the discovery that some mutations in *mukB* can be suppressed by mutations in *topA*, the gene that codes for topoisomerase I. MukB forms a complex with two other proteins, MukE and MukF, and the MukBEF complex is considered to be a condensin analogous to eukaryotic condensins. It uses a supercoiling mechanism to condense the chromosome. A defect in this function is the cause of failure to segregate properly. The defect can be compensated by preventing topoisomerases from relaxing negative supercoils; the resulting increase in supercoil density helps to restore the proper state of condensation and thus allows segregation.

We still do not understand how genomes are positioned in the cell, but the process may be connected with condensation. FIGURE 17.13 shows a current model. The parental genome is centrally positioned. It must be decondensed in order to pass through the replication apparatus. The daughter chromosomes emerge from replication, are disentangled by topoisomerases, and then passed in an uncondensed state to MukBEF, which causes them to form condensed masses at the positions that will become the centers of the daughter cells.

It has been suspected for years that a physical link exists between bacterial DNA and the membrane, but the evidence remains indirect. Bacterial DNA can be found in membrane fractions, which tend to be enriched in genetic markers near the origin, the replication fork, and the terminus. The proteins present in these membrane fractions may be affected by mutations that interfere with the initiation of replication. The growth site could be a structure on the membrane to which the origin must be attached for initiation.

During sporulation in *Bacillus subtilis*, one daughter chromosome must be segregated into the small forespore compartment (see Figure 11.41). This is an unusual process that involves transfer of the chromosome across the nascent septum. One of the sporulation genes, *spoIIIE*, is required for this process. The SpoIIIE protein is located at the septum and probably has a translocation function that pumps DNA through to the forespore compartment.

17.9 Single-Copy Plasmids Have a Partitioning System

Key concepts

- Single-copy plasmids exist at one plasmid copy per bacterial chromosome origin.
- Multicopy plasmids exist at >1 plasmid copy per bacterial chromosome origin.
- Homologous recombination between circular plasmids generates dimers and higher multimers.
- Plasmids have site-specific recombination systems that undertake intramolecular recombination to regenerate monomers.
- Partition systems ensure that duplicate plasmids are segregated to different daughter cells produced by a division.

The type of system that a plasmid uses to ensure that it is distributed to both daughter cells at division depends upon its type of replication system. Each type of plasmid is maintained in its bacterial host at a characteristic **copy number:**

- Single-copy control systems resemble that of the bacterial chromosome and result in one replication per cell division. A single-copy plasmid effectively maintains parity with the bacterial chromosome.
- Multicopy control systems allow multiple initiation events per cell cycle, with the result that there are several copies of the plasmid per bacterium. Multicopy plasmids exist in a characteristic number (typically 10 to 20) per bacterial chromosome.

Copy number is primarily a consequence of the type of replication control mechanism. The system responsible for initiating replication determines how many origins can be present in the bacterium. Each plasmid consists of a single replicon, and as a result the number of origins is the same as the number of plasmid molecules.

Single-copy plasmids have a system for replication control whose consequences are similar to those of the system for replication governing the bacterial chromosome. A single origin can be replicated once, and then the daughter origins are segregated to the different daughter cells.

Multicopy plasmids have a replication system that allows a pool of origins to exist. If the number is great enough (in practice, >10 per bacterium), an active segregation system becomes unnecessary, because even a statistical distribution of plasmids to daughter cells will result in the loss of plasmids at frequencies of $<10^{-6}$.

Plasmids are maintained in bacterial populations with very low rates of loss ($<10^{-7}$ per cell division is typical, even for a single-copy plasmid). The systems that control plasmid segregation can be identified by mutations that increase the frequency of loss, but that do not act upon replication itself. Several types of mechanism are used to ensure the survival of a plasmid in a bacterial population. It is common for a plasmid to carry several systems, often of different types, all acting independently to ensure its survival. Some of these systems act indirectly, whereas others are concerned directly with regulating the partition event. In terms of evolution, however, all serve the same purpose: to help ensure perpetuation of the plasmid to the maximum number of progeny bacteria.

Single-copy plasmids require partitioning systems to ensure that the duplicate copies find themselves on opposite sides of the septum at cell division, and are therefore segregated to a different daughter cell. In fact, functions involved in partitioning were first identified in plasmids. The components of a common system are summarized in FIGURE 17.14. Typically there are two *trans*-acting loci (*parA* and *parB*) and a *cis*-acting element (*parS*) located just downstream of the two genes. ParA is an ATPase. It binds to ParB, which binds to the *parS* site on DNA. Deletions of any of the three loci

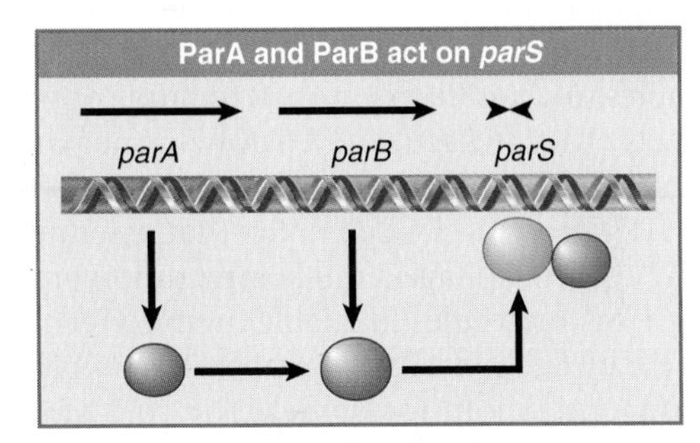

FIGURE 17.14 A common segregation system consists of genes *parA* and *parB* and the target site *parS*.

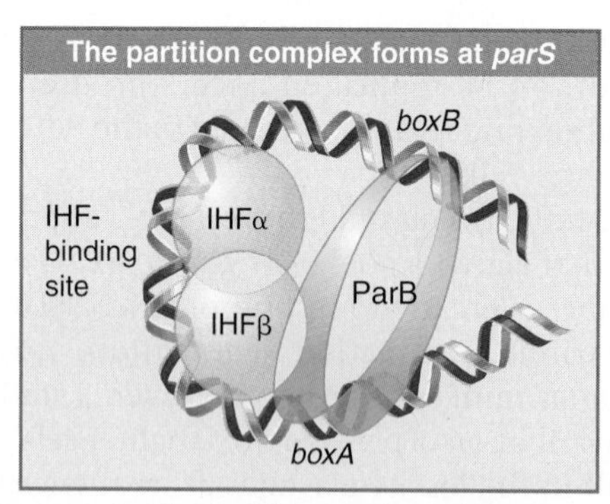

FIGURE 17.15 The partition complex is formed when IHF binds to DNA at *parS* and bends it so that ParB can bind to sites on either side. The complex is initiated by a heterodimer of IHF and a dimer of ParB, and then more ParB dimers bind.

prevent proper partition of the plasmid. Systems of this type have been characterized for the plasmids F, P1, and R1. In spite of their overall similarities, there are no significant sequence homologies between the corresponding genes or *cis*-acting sites.

parS plays a role for the plasmid that is equivalent to the centromere in a eukaryotic cell. Binding of the ParB protein to it creates a structure that segregates the plasmid copies to opposite daughter cells. A bacterial protein, IHF, also binds at this site to form part of the structure. The complex of ParB and IHF with *parS* is called the partition complex. *parS* is a 34-bp sequence containing the IHF-binding site and is flanked on either side by sequences called *boxA* and *boxB* that are bound by ParB.

IHF is the integration host factor. It is named for the role in which it was first discovered (forming a structure that is involved in the integration of phage lambda DNA into the host chromosome). IHF is a heterodimer that has the capacity to form a large structure in which DNA is wrapped on the surface. The role of IHF is to bend the DNA so that ParB can bind simultaneously to the separated *boxA* and *boxB* sites, as indicated in FIGURE 17.15. Complex formation is initiated when *parS* is bound by a heterodimer of IHF together with a dimer of ParB. This enables further dimers of ParB to bind cooperatively. The interaction of ParA with the partition complex structure is essential but transient.

The protein-DNA complex that assembles on IHF during phage lambda integration binds two DNA molecules to enable them to recombine (see Section 19.19, Lambda Recombination Occurs in an Intasome). The role of the partition complex is different: to ensure that two DNA molecules segregate apart from one another. We do not know yet how the formation of the individual complex accomplishes this task. One possibility is that it attaches the DNA to some physical site—for example, on the membrane—and then the sites of attachment are segregated by growth of the septum.

Proteins related to ParA and ParB are found in several bacteria. In *B. subtilis,* they are called Soj and SpoOJ, respectively. Mutations in these loci prevent sporulation because of a failure to segregate one daughter chromosome into the forespore (see Figure 11.42). In sporulating cells, SpoOJ localizes at the pole and may be responsible for localizing the origin there. SpoOJ binds to a sequence that is present in multiple copies that are dispersed over ~20% of the chromosome in the vicinity of the origin. It is possible that SpoOJ binds both old and newly synthesized origins, maintaining a status equivalent to chromosome pairing until the chromosomes are segregated to the opposite poles. In *Caulobacter crescentus,* ParA and ParB localize to the poles of the bacterium and ParB binds sequences close to the origin, thus localizing the origin to the pole. These results suggest that a specific apparatus is responsible for localizing the origin to the pole. The next stage of the analysis will be to identify the cellular components with which this apparatus interacts.

The importance to the plasmid of ensuring that all daughter cells gain replica plasmids is emphasized by the existence of multiple, independent systems in individual plasmids that ensure proper partition. **Addiction systems,** which operate on the basis that "we hang together or we hang separately," ensure that a bacterium carrying a plasmid can survive only as long as it retains the plasmid. There are several ways to ensure that a cell dies if it is "cured" of a plasmid, all of which share the principle illustrated in FIGURE 17.16 that the plasmid produces both a poison and an antidote. The poison is a killer substance that is relatively stable, whereas the antidote consists of a substance that blocks killer action but is relatively short lived. When the plasmid is lost the antidote decays, and then the killer substance causes the death of the cell. Thus bacteria that lose the plasmid inevitably die, and the population is condemned to retain the plasmid indefinitely. These systems take various forms. One specified by the F plasmid consists of killer and blocking proteins. The plasmid R1 has a killer that is the mRNA for a toxic protein; the antidote is a small antisense RNA that prevents expression of the mRNA.

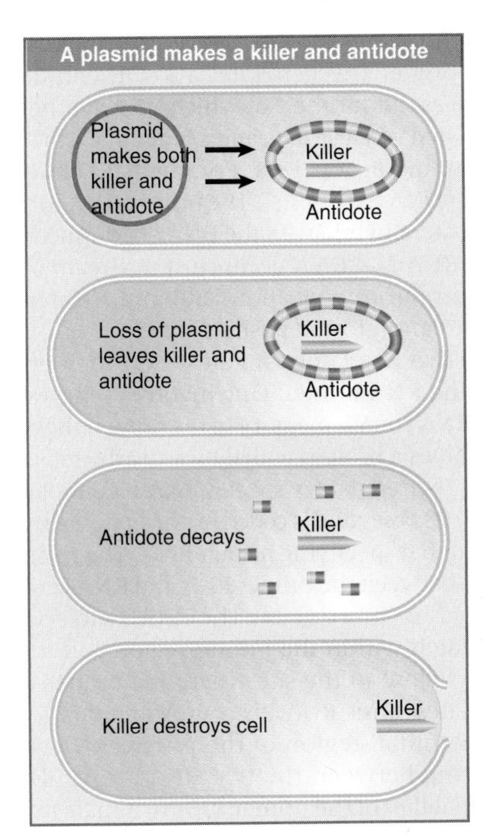

FIGURE 17.16 Plasmids may ensure that bacteria cannot live without them by synthesizing a long-lived killer and a short-lived antidote.

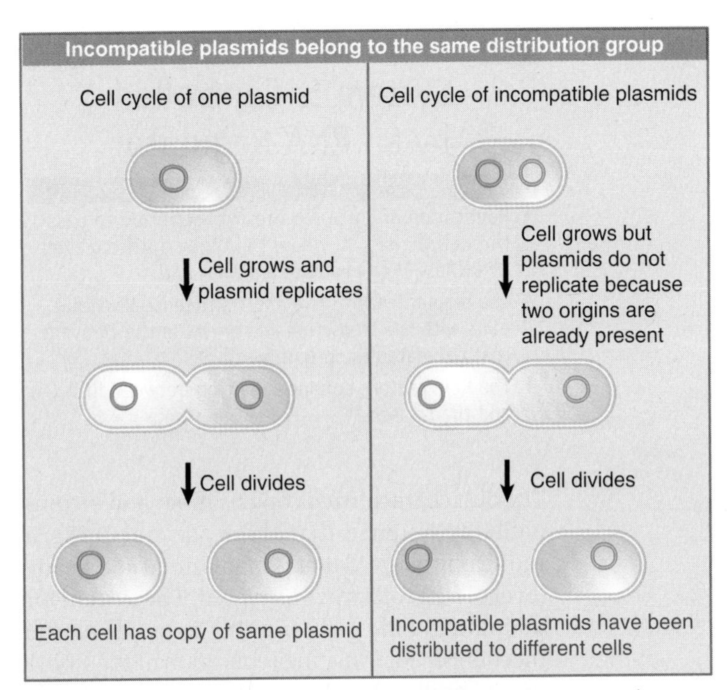

FIGURE 17.17 Two plasmids are incompatible (they belong to the same compatibility group) if their origins cannot be distinguished at the stage of initiation. The same model could apply to segregation.

17.10 Plasmid Incompatibility Is Determined by the Replicon

Key concept

- Plasmids in a single compatibility group have origins that are regulated by a common control system.

The phenomenon of plasmid incompatibility is related to the regulation of plasmid copy number and segregation. A **compatibility group** is defined as a set of plasmids whose members are unable to coexist in the same bacterial cell. The reason for their incompatibility is that they cannot be distinguished from one another at some stage that is essential for plasmid maintenance. DNA replication and segregation are stages at which this may apply.

The negative control model for plasmid incompatibility follows the idea that copy number control is achieved by synthesizing a repressor that measures the concentration of origins. (Formally, this is the same as the titration model for regulating replication of the bacterial chromosome.)

The introduction of a new origin in the form of a second plasmid of the same compatibility group mimics the result of replication of the resident plasmid; two origins now are present. Thus any further replication is prevented until after the two plasmids have been segregated to different cells to create the correct prereplication copy number, as illustrated in FIGURE 17.17.

A similar effect would be produced if the system for segregating the products to daughter cells could not distinguish between two plasmids. For example, if two plasmids have the same *cis*-acting partition sites, competition between them would ensure that they would be segregated to different cells, and therefore could not survive in the same line.

The presence of a member of one compatibility group does not directly affect the survival of a plasmid belonging to a different group. Only one replicon of a given compatibility group (of a single-copy plasmid) can be maintained in the bacterium, but it does not interact with replicons of other compatibility groups.

17.11 The ColE1 Compatibility System Is Controlled by an RNA Regulator

Key concepts

- Replication of ColE1 requires transcription to pass through the origin, where the transcript is cleaved by RNAase H to generate a primer end.
- The regulator RNA I is a short antisense RNA that pairs with the transcript and prevents the cleavage that generates the priming end.
- The Rom protein enhances pairing between RNA I and the transcript.

The best characterized copy number and incompatibility system is that of the plasmid ColE1, a multicopy plasmid that is maintained at a steady level of ~20 copies per *E. coli* cell. The system for maintaining the copy number depends on the mechanism for initiating replication at the ColE1 origin, as illustrated in FIGURE 17.18.

Replication starts with the transcription of an RNA that initiates 555 bp upstream of the origin. Transcription continues through the origin. The enzyme RNAase H (whose name reflects its specificity for a substrate of RNA hybridized with DNA) cleaves the transcript at the origin. This generates a 3′–OH end that is used as the "primer" at which DNA synthesis is initiated (the use of primers is discussed in more detail in Section 18.8, Priming Is Required to Start DNA Synthesis). The primer RNA forms a persistent hybrid with the DNA. Pairing between the RNA and DNA occurs just upstream of the origin (around position –20) and also farther upstream (around position –265).

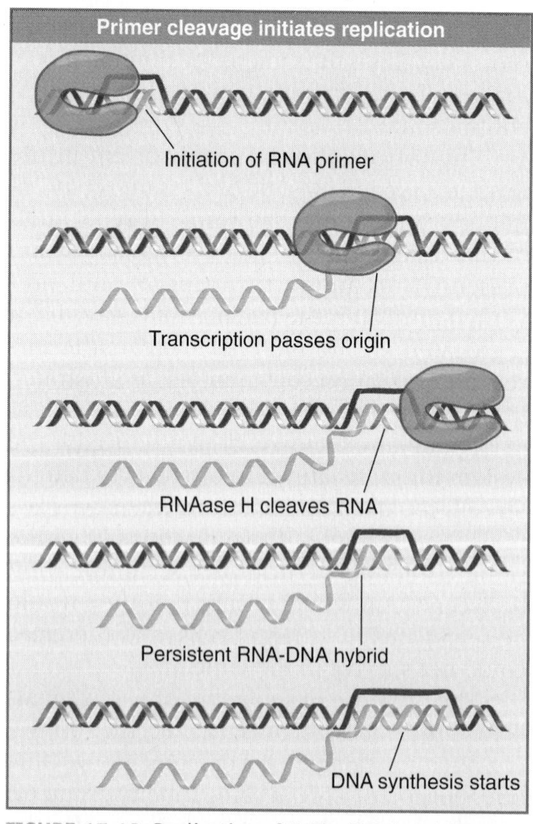

FIGURE 17.18 Replication of ColE1 DNA is initiated by cleaving the primer RNA to generate a 3′–OH end. The primer forms a persistent hybrid in the origin region.

Two regulatory systems exert their effects on the RNA primer. One involves synthesis of an RNA complementary to the primer; the other involves a protein coded by a nearby locus.

The regulatory species RNA I is a molecule of ~108 bases and is coded by the opposite strand from that specifying primer RNA. The relationship between the primer RNA and RNA I is illustrated in FIGURE 17.19. The RNA I molecule is initiated within the primer region and terminates close to the site where the primer RNA initiates. Thus RNA I is complementary to the 5′–terminal region of the primer RNA. Base pairing between the two RNAs controls the availability of the primer RNA to initiate a cycle of replication.

An RNA molecule such as RNA I that functions by virtue of its complementarity with another RNA coded in the same region is called a **countertranscript.** This type of mechanism, of course, is another example of the use of antisense RNA (see Section 13.7, Small RNA Molecules Can Regulate Translation).

Mutations that reduce or eliminate incompatibility between plasmids can be obtained by selecting plasmids of the same group for their ability to coexist. Incompatibility mutations in ColE1 map in the region of overlap between RNA I and primer RNA. This region is represented in two different RNAs, so either or both might be involved in the effect.

When RNA I is added to a system for replicating ColE1 DNA *in vitro,* it inhibits the formation of active primer RNA. The presence of RNA I, however, does not inhibit the initiation

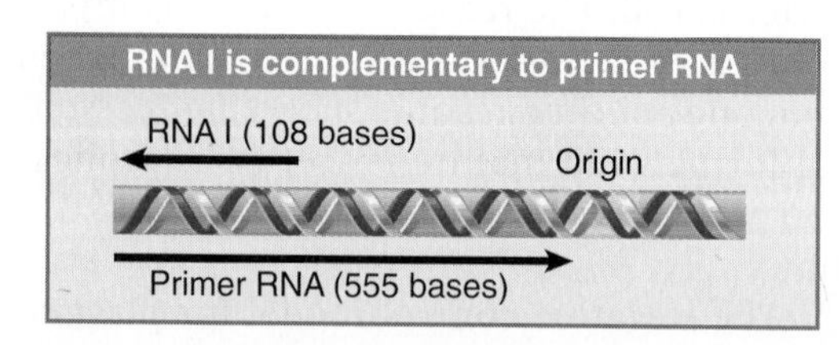

FIGURE 17.19 The sequence of RNA I is complementary to the 5′ region of primer RNA.

or elongation of primer RNA synthesis. This suggests that RNA I prevents RNAase H from generating the 3′ end of the primer RNA. The basis for this effect lies in base pairing between RNA I and primer RNA.

Both RNA molecules have the same potential secondary structure in this region, with three duplex hairpins terminating in single-stranded loops. Mutations reducing incompatibility are located in these loops, which suggests that the initial step in base pairing between RNA I and primer RNA is contact between the unpaired loops.

How does pairing with RNA I prevent cleavage to form primer RNA? A model is illustrated in FIGURE 17.20. In the absence of RNA I, the primer RNA forms its own secondary structure (involving loops and stems). When RNA I is present, though, the two molecules pair and become completely double-stranded for the entire length of RNA I. The new secondary structure prevents the formation of the primer, probably by affecting the ability of the RNA to form the persistent hybrid.

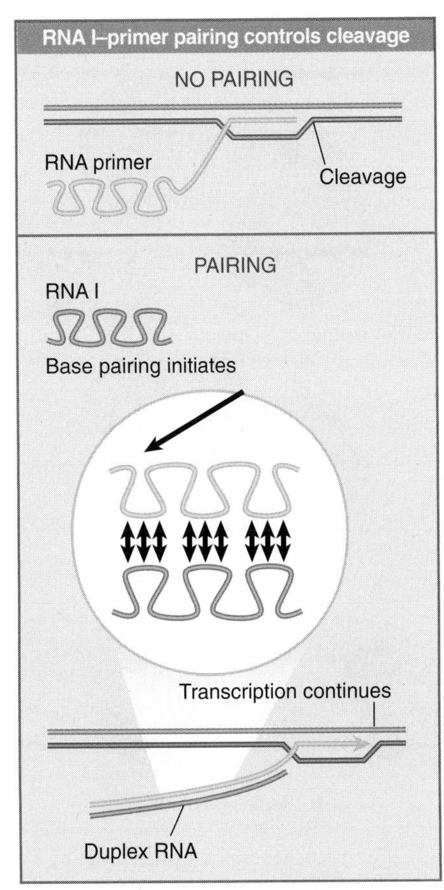

FIGURE 17.20 Base pairing with RNA I may change the secondary structure of the primer RNA sequence and thus prevent cleavage from generating a 3′–OH end.

The model resembles the mechanism involved in attenuation of transcription, in which the alternative pairings of an RNA sequence permit or prevent formation of the secondary structure needed for termination by RNA polymerase (see Section 13.2, Alternative Secondary Structures Control Attenuation). The action of RNA I is exercised by its ability to affect distant regions of the primer precursor.

Formally, the model is equivalent to postulating a control circuit involving two RNA species. A large RNA primer precursor is a positive regulator and is needed to initiate replication. The small RNA I is a negative regulator that is able to inhibit the action of the positive regulator.

In its ability to act on any plasmid present in the cell, RNA I provides a repressor that prevents newly introduced DNA from functioning. This is analogous to the role of the lambda lysogenic repressor (see Section 14.9, The Repressor and Its Operators Define the Immunity Region). Instead of a repressor protein that binds the new DNA, an RNA binds the newly synthesized precursor to the RNA primer.

Binding between RNA I and primer RNA can be influenced by the Rom protein, which is coded by a gene located downstream of the origin. Rom enhances binding between RNA I and primer RNA transcripts of >200 bases. The result is to inhibit formation of the primer.

How do mutations in the RNAs affect incompatibility? FIGURE 17.21 shows the situation when a cell contains two types of

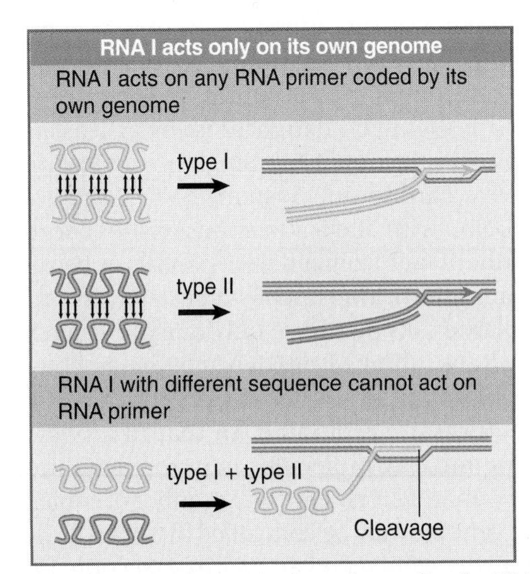

FIGURE 17.21 Mutations in the region coding for RNA I and the primer precursor need not affect their ability to pair; but they may prevent pairing with the complementary RNA coded by a different plasmid.

RNA I/primer RNA sequence. The RNA I and primer RNA made from each type of genome can interact, but RNA I from one genome does not interact with primer RNA from the other genome. This situation would arise when a mutation in the region that is common to RNA I and primer RNA occurred at a location involved in the base pairing between them. Each RNA I would continue to pair with the primer RNA coded by the same plasmid, but might be unable to pair with the primer RNA coded by the other plasmid. This would cause the original and the mutant plasmids to behave as members of different compatibility groups.

17.12 How Do Mitochondria Replicate and Segregate?

Key concepts

- mtDNA replication and segregation to daughter mitochondria is stochastic.
- Mitochondrial segregation to daughter cells is also stochastic.

Mitochondria must be duplicated during the cell cycle and segregated to the daughter cells. We understand some of the mechanics of this process, but not its regulation.

At each stage in the duplication of mitochondria—DNA replication, DNA segregation to duplicate mitochondria, and organelle segregation to daughter cells—the process appears to be stochastic, governed by a random distribution of each copy. The theory of distribution in this case is analogous to that of multicopy bacterial plasmids, with the same conclusion that >10 copies are required to ensure that each daughter gains at least one copy (see Section 17.9, Single-Copy Plasmids Have a Partitioning System). When there are mtDNAs with allelic variations (either because of inheritance from different parents or because of mutation), the stochastic distribution may generate cells that have only one of the alleles.

Replication of mtDNA may be stochastic because there is no control over which particular copies are replicated, so that in any cycle some mtDNA molecules may replicate more times than others. The total number of copies of the genome may be controlled by titrating mass in a way similar to bacteria (see Section 17.2, Replication Is Connected to the Cell Cycle).

A mitochondrion divides by developing a ring around the organelle that constricts to pinch it into two halves. The mechanism is similar in principle to that involved in bacterial division. The apparatus that is used in plant cell mitochondria is similar to bacteria and uses a homolog of the bacterial protein FtsZ (see Section 17.5, FtsZ Is Necessary for Septum Formation). The molecular apparatus is different in animal cell mitochondria and uses the protein dynamin that is involved in formation of membranous vesicles. An individual organelle may have more than one copy of its genome.

We do not know whether there is a partitioning mechanism for segregating mtDNA molecules within the mitochondrion, or whether they are simply inherited by daughter mitochondria according to which half of the mitochondrion they happen to lie in. FIGURE 17.22 shows that the combination of replication and segregation mechanisms can result in a stochastic assignment of DNA to each of the copies,

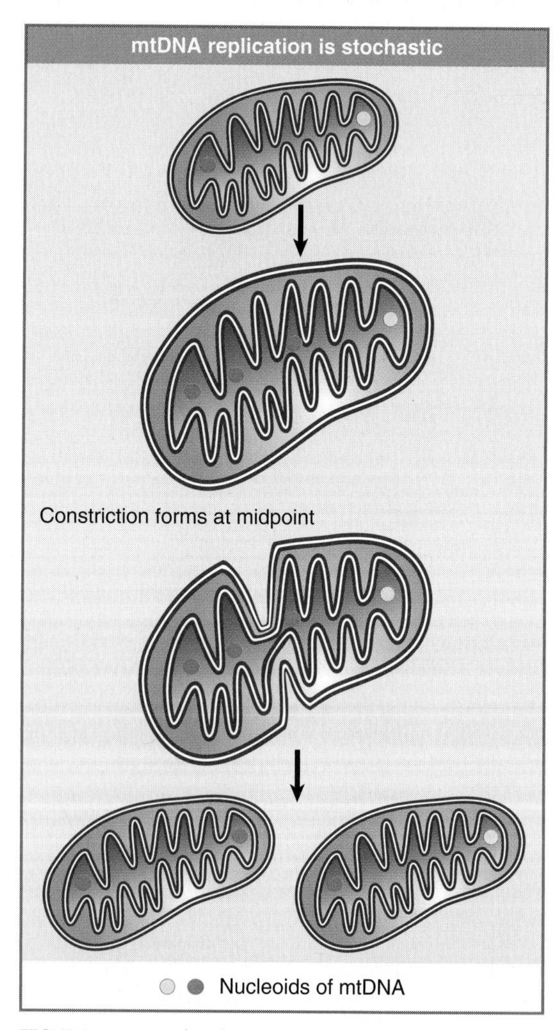

FIGURE 17.22 Mitochondrial DNA replicates by increasing the number of genomes in proportion to mitochondrial mass but without ensuring that each genome replicates the same number of times. This can lead to changes in the representation of alleles in the daughter mitochondria.

that is, so that the distribution of mitochondrial genomes to daughter mitochondria does not depend on their parental origins.

The assignment of mitochondria to daughter cells at mitosis also appears to be random. Indeed, it was the observation of somatic variation in plants that first suggested the existence of genes that could be lost from one of the daughter cells because they were not inherited according to Mendel's laws (see Figure 4.15).

In some situations a mitochondrion has both paternal and maternal alleles. This has two requirements: that both parents provide alleles to the zygote (which of course is not the case when there is maternal inheritance; see Section 4.9, Organelles Have DNA); and that the parental alleles are found in the same mitochondrion. For this to happen, parental mitochondria must have fused.

The size of the individual mitochondrion may not be precisely defined. Indeed, there is a continuing question as to whether an individual mitochondrion represents a unique and discrete copy of the organelle or whether it is in a dynamic flux in which it can fuse with other mitochondria. We know that mitochondria can fuse in yeast, because recombination between mtDNAs can occur after two haploid yeast strains have mated to produce a diploid strain. This implies that the two mtDNAs must have been exposed to one another in the same mitochondrial compartment. Attempts have been made to test for the occurrence of similar events in animal cells by looking for complementation between alleles after two cells have been fused, but the results are not clear.

17.13 Summary

A fixed time of 40 minutes is required to replicate the *E. coli* chromosome and a further 20 minutes is required before the cell can divide. When cells divide more rapidly than every 60 minutes, a replication cycle is initiated before the end of the preceding division cycle. This generates multiforked chromosomes. The initiation event depends on titration of cell mass, probably by accumulating an initiator protein. Initiation may occur at the cell membrane because the origin is associated with the membrane for a short period after initiation.

The septum that divides the cell grows at a location defined by the preexisting periseptal annulus; a locus of three genes (min*C*, -*D*, and -*E*) codes for products that regulate whether the mid-cell periseptal annulus or the polar sites derived from previous annuli are used for septum formation. Absence of septum formation generates multinucleated filaments; an excess of septum formation generates anucleate minicells.

Many transmembrane proteins interact to form the septum. ZipA is located in the inner bacterial membrane and binds to FtsZ, which is a tubulin-like protein that can polymerize into a filamentous structure called a Z-ring. FtsA is a cytosolic protein that binds to FtsZ. Several other *fts* products, all transmembrane proteins, join the Z-ring in an ordered process that generates a septal ring. The last proteins to bind are the SEDS protein FtsW and the transpeptidase ftsI (PBP3), which together function to produce the peptidoglycans of the septum. Chloroplasts use a related division mechanism that has an FtsZ-like protein, but mitochondria use a different process in which the membrane is constricted by a dynamin-like protein.

Plasmids and bacteria have site-specific recombination systems that regenerate pairs of monomers by resolving dimers created by general recombination. The Xer system acts on a target sequence located in the terminus region of the chromosome. The system is active only in the presence of the FtsK protein of the septum, which may ensure that it acts only when a dimer needs to be resolved.

Partitioning involves the interaction of the ParB protein with the *parS* target site to build a structure that includes the IHF protein. This partition complex ensures that replica chromosomes segregate into different daughter cells. The mechanism of segregation may involve movement of DNA, possibly by the action of MukB in condensing chromosomes into masses at different locations as they emerge from replication.

Plasmids have a variety of systems that ensure or assist partition, and an individual plasmid may carry systems of several types. The copy number of a plasmid describes whether it is present at the same level as the bacterial chromosome (one per unit cell) or in greater numbers. Plasmid incompatibility can be a consequence of the mechanisms involved in either replication or partition (for single-copy plasmids). Two plasmids that share the same control system for replication are incompatible because the number of replication events ensures that there is only one plasmid for each bacterial genome.

References

17.2 Replication Is Connected to the Cell Cycle

Review

Donachie, W. D. (1993). The cell cycle of *E. coli*. *Annu. Rev. Immunol.* 47, 199–230.

Research

Donachie, W. D. and Begg, K. J. (1970). Growth of the bacterial cell. *Nature* 227, 1220–1224.

Donachie, W. D., Begg, K. J., and Vicente, M. (1976). Cell length, cell growth and cell division. *Nature* 264, 328–333.

17.3 The Septum Divides a Bacterium into Progeny That Each Contain a Chromosome

Review

de Boer, P. A. J., Cook, W. R., and Rothfield, L. I. (1990). Bacterial cell division. *Annu. Rev. Genet.* 24, 249–274.

Research

Spratt, B. G. (1975). Distinct penicillin binding proteins involved in the division, elongation, and shape of *E. coli* K12. *Proc. Natl. Acad. Sci. USA* 72, 2999–3003.

17.4 Mutations in Division or Segregation Affect Cell Shape

Research

Adler, H. I. et al. (1967). Miniature *E. coli* cells deficient in DNA. *Proc. Natl. Acad. Sci. USA* 57, 321–326.

17.5 FtsZ Is Necessary for Septum Formation

Reviews

Lutkenhaus, J. and Addinall, S. G. (1997). Bacterial cell division and the Z ring. *Annu. Rev. Biochem.* 66, 93–116.

Rothfield, L., Justice, S. and Garcia-Lara, J. (1999). Bacterial cell division. *Annu. Rev. Genet.* 33, 423–438.

Research

Bi, E. F. and Lutkenhaus, J. (1991). FtsZ ring structure associated with division in *Escherichia coli*. *Nature* 354, 161–164.

Hale, C. A. and de Boer, P. A. (1997). Direct binding of FtsZ to ZipA, an essential component of the septal ring structure that mediates cell division in *E. coli*. *Cell* 88, 175–185.

Mercer, K. L. and Weiss, D. S. (2002). The *E. coli* cell division protein FtsW is required to recruit its cognate transpeptidase, FtsI (PBP3), to the division site. *J. Bacteriol.* 184, 904–912.

Pichoff, S. and Lutkenhaus, J. (2002). Unique and overlapping roles for ZipA and FtsA in septal ring assembly in *Escherichia coli*. *EMBO J.* 21, 685–693.

Romberg, L. and Levin, P. A. (2003). Assembly dynamics of the bacterial cell division protein FTSZ: poised at the edge of stability. *Annu. Rev. Microbiol.* 57, 125–154.

Stricker, J., Maddox, P., Salmon, E. D., and Erickson, H. P. (2002). Rapid assembly dynamics of the *Escherichia coli* FtsZ-ring demonstrated by fluorescence recovery after photobleaching. *Proc. Natl. Acad. Sci. USA* 99, 3171–3175.

17.6 *min* Genes Regulate the Location of the Septum

Research

de Boer, P. A. J. et al. (1989). A division inhibitor and a topological specificity factor coded for by the minicell locus determine proper placement of the division septum in *E. coli*. *Cell* 56, 641–649.

Hu, Z., Mukherjee, A., Pichoff, S., and Lutkenhaus, J. (1999). The MinC component of the division site selection system in *E. coli* interacts with FtsZ to prevent polymerization. *Proc. Natl. Acad. Sci. USA* 96, 14819–14824.

Pichoff, S. and Lutkenhaus, J. (2001). *Escherichia coli* division inhibitor MinCD blocks septation by preventing Z-ring formation. *J. Bacteriol.* 183, 6630–6635.

Raskin, D. M. and de Boer, P. A. J. (1997). The MinE ring: an FtsZ-independent cell structure requires for selection of the correct division site in *E. coli*. *Cell* 91, 685–694.

17.7 Chromosomal Segregation May Require Site-Specific Recombination

Research

Aussel, L., Barre, F. X., Aroyo, M., Stasiak, A., Stasiak, A. Z., and Sherratt, D. (2002). FtsK is a DNA motor protein that activates chromosome dimer resolution by switching the catalytic state of the XerC and XerD recombinases. *Cell* 108, 195–205.

Barre, F. X., Aroyo, M., Colloms, S. D., Helfrich, A., Cornet, F., and Sherratt, D. J. (2000). FtsK functions in the processing of a Holliday junction intermediate during bacterial chromosome segregation. *Genes Dev.* 14, 2976–2988.

Blakely, G., May, G., McCulloch, R., Arciszewska, L. K., Burke, M., Lovett, S. T., and Sherratt, D. J. (1993). Two related recombinases are required for site-specific recombination at dif and cer in *E. coli* K12. *Cell* 75, 351–361.

17.8 Partitioning Involves Separation of the Chromosomes

Reviews

Draper, G. C. and Gober, J. W. (2002). Bacterial chromosome segregation. *Annu. Rev. Microbiol.* 56, 567–597.

Errington, J., Bath, J., and Wu, L. J. (2001). DNA transport in bacteria. *Nat. Rev. Mol. Cell Biol.* 2, 538–545.

Gordon, G. S. and Wright, A. (2000). DNA segregation in bacteria. *Annu. Rev. Microbiol.* 54, 681–708.

Hiraga, S. (1992). Chromosome and plasmid partition in *E. coli. Annu. Rev. Biochem.* 61, 283–306.

Wake, R. G. and Errington, J. (1995). Chromosome partitioning in bacteria. *Annu. Rev. Genet.* 29, 41–67.

Research

Case, R. B., Chang, Y. P., Smith, S. B., Gore, J., Cozzarelli, N. R., and Bustamante, C. (2004). The bacterial condensin MukBEF compacts DNA into a repetitive, stable structure. *Science* 305, 222–227.

Jacob, F., Ryter, A., and Cuzin, F. (1966). On the association between DNA and the membrane in bacteria. *Proc. Roy. Soc. Lond. B Biol. Sci.* 164, 267–348.

Sawitzke, J. A. and Austin, S. (2000). Suppression of chromosome segregation defects of *E. coli muk* mutants by mutations in topoisomerase I. *Proc. Natl. Acad. Sci. USA* 97, 1671–1676.

Wu, L. J. and Errington, J. (1997). Septal localization of the SpoIIIE chromosome partitioning protein in *B. subtilis. EMBO J.* 16, 2161–2169.

17.9 Single-Copy Plasmids Have a Partitioning System

Reviews

Draper, G. C. and Gober, J. W. (2002). Bacterial chromosome segregation. *Annu. Rev. Microbiol.* 56, 567–597.

Gordon, G. S. and Wright, A. (2000). DNA segregation in bacteria. *Annu. Rev. Microbiol.* 54, 681–708.

Hiraga, S. (1992). Chromosome and plasmid partition in *E. coli. Annu. Rev. Biochem.* 61, 283–306.

Research

Davis, M. A. and Austin, S. J. (1988). Recognition of the P1 plasmid centromere analog involves binding of the ParB protein and is modified by a specific host factor. *EMBO J.* 7, 1881–1888.

Funnell, B. E. (1988). Participation of *E. coli* integration host factor in the P1 plasmid partition system. *Proc. Natl. Acad. Sci. USA* 85, 6657–6661.

Mohl, D. A. and Gober, J. W. (1997). Cell cycle-dependent polar localization of chromosome partitioning proteins of *C. crescentus. Cell* 88, 675–684.

Surtees, J. A. and Funnell, B. E. (2001). The DNA binding domains of P1 ParB and the architecture of the P1 plasmid partition complex. *J. Biol. Chem.* 276, 12385–12394.

17.10 Plasmid Incompatibility Is Determined by the Replicon

Reviews

Nordstrom, K. and Austin, S. J. (1989). Mechanisms that contribute to the stable segregation of plasmids. *Annu. Rev. Genet.* 23, 37–69.

Scott, J. R. (1984). Regulation of plasmid replication. *Microbiol. Rev.* 48, 1–23.

17.11 The ColE1 Compatibility System Is Controlled by an RNA Regulator

Research

Masukata, H. and Tomizawa, J. (1990). A mechanism of formation of a persistent hybrid between elongating RNA and template DNA. *Cell* 62, 331–338.

Tomizawa, J.-I. and Itoh, T. (1981). Plasmid ColE1 incompatibility determined by interaction of RNA with primer transcript. *Proc. Natl. Acad. Sci. USA* 78, 6096–6100.

17.12 How Do Mitochondria Replicate and Segregate?

Review

Birky, C. W. (2001). The inheritance of genes in mitochondria and chloroplasts: laws, mechanisms, and models. *Annu. Rev. Genet.* 35, 125–148.

18

DNA Replication

CHAPTER OUTLINE

18.1 Introduction

Replication of duplex DNA is a complex endeavor involving a conglomerate of enzyme activities. Different activities are involved in the stages of initiation, elongation, and termination.

- Initiation involves recognition of an origin by a complex of proteins. Before DNA synthesis begins, the parental strands must be separated and (transiently) stabilized in the single-stranded state. After this stage, synthesis of daughter strands can be initiated at the replication fork.
- Elongation is undertaken by another complex of proteins. The **replisome** exists only as a protein complex associated with the particular structure that DNA takes at the replication fork. It does not exist as an independent unit (for example, analogous to the ribosome). As the replisome moves along DNA, the parental strands unwind and daughter strands are synthesized.
- At the end of the replicon, joining and/or termination reactions are necessary. Following termination, the duplicate chromosomes must be separated from one another, which requires manipulation of higher-order DNA structure.

Inability to replicate DNA is fatal for a growing cell. Mutants in replication must therefore be obtained as conditional lethals. These are able to accomplish replication under permissive conditions (provided by the normal temperature of incubation), but they are defective under nonpermissive conditions (provided by the higher temperature of 42°C). A comprehensive series of such temperature-sensitive mutants in *E. coli* identifies a set of loci called the *dna* genes. The **dna mutants** distinguish two stages of replication by their behavior when the temperature is raised:

- The major class of **quick-stop mutants** cease replication immediately on a temperature rise. They are defective in the components of the replication apparatus, typically in the enzymes needed for elongation (but also include defects in the supply of essential precursors).
- The smaller class of **slow-stop mutants** complete the current round of replication,

but cannot start another. They are defective in the events involved in initiating a cycle of replication at the origin.

An important assay used to identify the components of the replication apparatus is called ***in vitro* complementation.** An *in vitro* system for replication is prepared from a *dna* mutant and is operated under conditions in which the mutant gene product is inactive. Extracts from wild-type cells are tested for their ability to restore activity. The protein coded by the *dna* locus can be purified by identifying the active component in the extract.

Each component of the bacterial replication apparatus is now available for study *in vitro* as a biochemically pure product, and is implicated *in vivo* by mutations in its gene. Eukaryotic replication systems are highly purified, and usually have components analogous to the bacterial proteins, but have not necessarily reached the stage of identification of every single component.

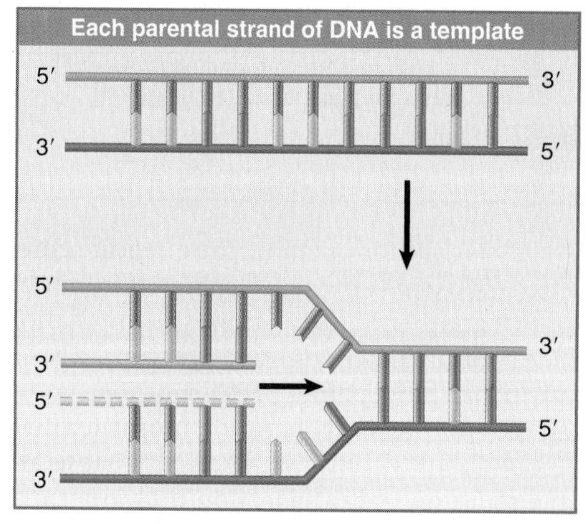

FIGURE 18.1 Semi-conservative replication synthesizes two new strands of DNA.

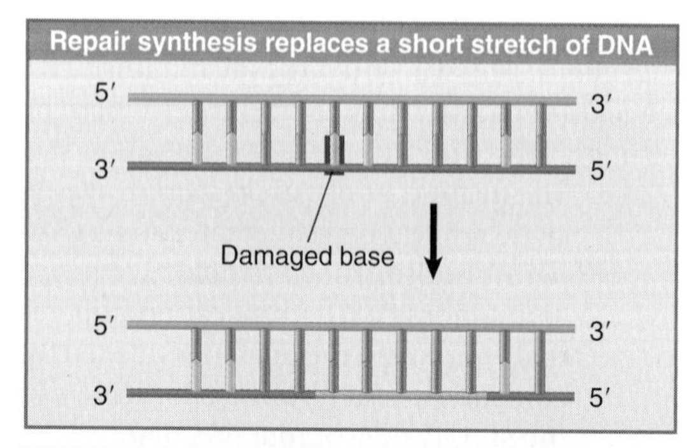

FIGURE 18.2 Repair synthesis replaces a short stretch of one strand of DNA containing a damaged base.

18.2 DNA Polymerases Are the Enzymes That Make DNA

Key concepts

- DNA is synthesized in both semiconservative replication and repair reactions.
- A bacterium or eukaryotic cell has several different DNA polymerase enzymes.
- One bacterial DNA polymerase undertakes semiconservative replication; the others are involved in repair reactions.
- Eukaryotic nuclei, mitochondria, and chloroplasts each have a single unique DNA polymerase required for replication, and other DNA polymerases involved in ancillary or repair activities.

There are two basic types of DNA synthesis.

FIGURE 18.1 shows the result of semiconservative **replication.** The two strands of the parental duplex are separated, and each serves as a template for synthesis of a new strand. The parental duplex is replaced with two daughter duplexes, each of which has one parental strand and one newly synthesized strand.

FIGURE 18.2 shows the consequences of a **repair** reaction. One strand of DNA has been damaged. It is excised and new material is synthesized to replace it. An enzyme that can synthesize a new DNA strand on a template strand is called a **DNA polymerase.** Both prokaryotic and eukaryotic cells contain multiple DNA polymerase activities. Only some of these enzymes actually undertake replication; those that do sometimes are called **DNA replicases.** The remaining enzymes are involved in subsidiary roles in replication and/or participate in repair synthesis.

All prokaryotic and eukaryotic DNA polymerases share the same fundamental type of synthetic activity. Each can extend a DNA chain by adding nucleotides one at a time to a 3′–OH end, as illustrated diagrammatically in FIGURE 18.3. The choice of the nucleotide to add to the chain is dictated by base pairing with the template strand.

Some DNA polymerases function as independent enzymes, but others (most notably the replicases) are incorporated into large protein assemblies. The DNA-synthesizing subunit is only one of several functions of the replicase, which typically contains many other activities concerned with unwinding DNA, initiating new strand synthesis, and so on.

FIGURE 18.4 summarizes the DNA polymerases that have been characterized in *E. coli.* DNA polymerase III, a multisubunit protein, is

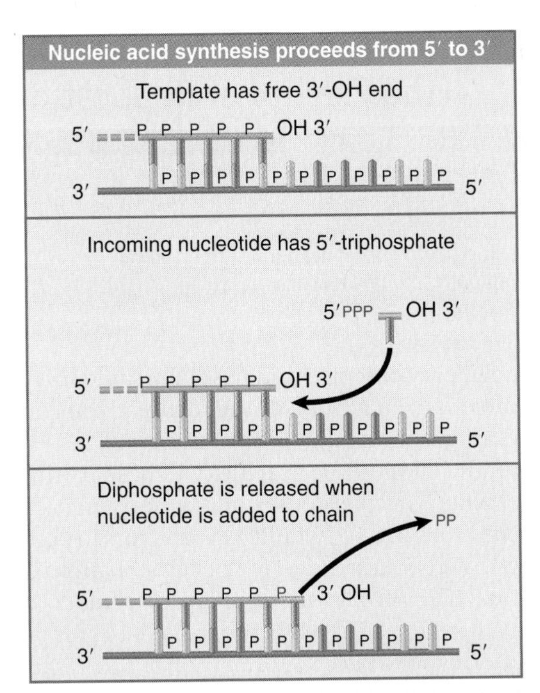

FIGURE 18.3 DNA is synthesized by adding nucleotides to the 3′–OH end of the growing chain, so that the new chain grows in the 5′ → 3′ direction. The precursor for DNA synthesis is a nucleoside triphosphate, which loses the terminal two phosphate groups in the reaction.

E. coli has five DNA polymerases

Enzyme	Gene	Function
I	*polA*	major repair enzyme
II	*polB*	replication restart
III	*polC*	replicase
IV	*dinB*	translesion replication
V	*umuD′$_2$C*	translesion replication

FIGURE 18.4 Only one DNA polymerase is the replicase. The others participate in repair of damaged DNA, restarting stalled replication forks, or bypassing damage in DNA.

the replicase responsible for *de novo* synthesis of new strands of DNA. DNA polymerase I (coded by *polA*) is involved in the repair of damaged DNA and, in a subsidiary role, in semiconservative replication. DNA polymerase II is required to restart a replication fork when its progress is blocked by damage in DNA. DNA polymerases IV and V are involved in allowing replication to bypass certain types of damage.

When extracts of *E. coli* are assayed for their ability to synthesize DNA, the predominant enzyme activity is DNA polymerase I. Its activity is so great that it makes it impossible to detect the activities of the enzymes actually responsible for DNA replication! To develop *in vitro* systems in which replication can be followed, extracts are therefore prepared from *polA* mutant cells.

Some phages code for DNA polymerases. They include T4, T5, T7, and SPO1. The enzymes all possess 5′–3′ synthetic activities and 3′–5′ exonuclease proofreading activities (see Section 18.3, DNA Polymerases Have Various Nuclease Activities). In each case, a mutation in the gene that codes for a single phage polypeptide prevents phage development. Each phage polymerase polypeptide associates with other proteins, of either phage or host origin, to make the intact enzyme.

Several classes of eukaryotic DNA polymerases have been identified. DNA polymerases δ and ε are required for nuclear replication; DNA polymerase α is concerned with "priming" (initiating) replication. Other DNA polymerases are involved in repairing damaged nuclear DNA (β and also ε) or with mitochondrial DNA replication (γ).

18.3 DNA Polymerases Have Various Nuclease Activities

Key concept

- DNA polymerase I has a unique 5′–3′ exonuclease activity that can be combined with DNA synthesis to perform nick translation.

Replicases often have nuclease activities as well as the ability to synthesize DNA. A 3′–5′ exonuclease activity is typically used to excise bases that have been added to DNA incorrectly. This provides a "proofreading" error-control system (see Section 18.4, DNA Polymerases Control the Fidelity of Replication).

The first DNA-synthesizing enzyme to be characterized was DNA polymerase I, which is a single polypeptide of 103 kD. The chain can be cleaved into two parts by proteolytic treatment. Two-thirds of the protein is in the C-terminal and contains the polymerase active site; the remaining third of the protein is in the N-terminal and contains the proofreading exonuclease.

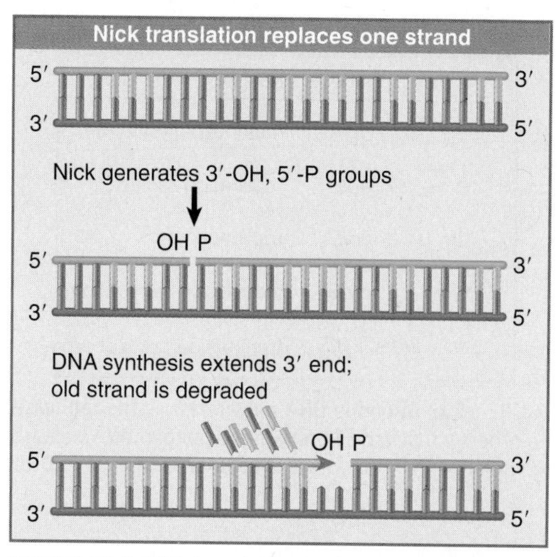

FIGURE 18.5 Nick translation replaces part of a preexisting strand of duplex DNA with newly synthesized material.

The larger cleavage product (68 kD) is called the Klenow fragment. It is used in synthetic reactions *in vitro*. It contains the polymerase and the 3′–5′ exonuclease activities. The active sites are ~30 Å apart in the protein, which indicates that there is spatial separation between adding a base and removing one.

The small fragment (35 kD) possesses a 5′–3′ exonucleolytic activity, which excises small groups of nucleotides, up to ~10 bases at a time. This activity is coordinated with the synthetic/proofreading activity. It provides DNA polymerase I with a unique ability to start replication *in vitro* at a nick in DNA. (No other DNA polymerase has this ability.) At a point where a phosphodiester bond has been broken in a double-stranded DNA, the enzyme extends the 3′–OH end. As the new segment of DNA is synthesized, it displaces the existing homologous strand in the duplex.

This process of **nick translation** is illustrated in FIGURE 18.5. The displaced strand is degraded by the 5′–3′ exonucleolytic activity of the enzyme. The properties of the DNA are unaltered, except that a segment of one strand has been replaced with newly synthesized material, and the position of the nick has been moved along the duplex. This is of great practical use; nick translation has been a major technique for introducing radioactively labeled nucleotides into DNA *in vitro*.

The 5′–3′ synthetic/3′–5′ exonucleolytic action is probably used *in vivo* mostly for filling in short single-stranded regions in double-stranded DNA. These regions arise during replication, as well as when bases that have been damaged are removed from DNA.

18.4 DNA Polymerases Control the Fidelity of Replication

Key concepts

- DNA polymerases often have a 3′–5′ exonuclease activity that is used to excise incorrectly paired bases.
- The fidelity of replication is improved by proofreading by a factor of ~100.

The fidelity of replication poses the same sort of problem we have encountered already in considering (for example) the accuracy of translation. It relies on the specificity of base pairing. Yet when we consider the interactions involved in base pairing, we would expect errors to occur with a frequency of ~10^{-3} per base pair replicated. The actual rate in bacteria seems to be ~10^{-8} to 10^{-10}. This corresponds to ~1 error per genome per 1000 bacterial replication cycles, or ~10^{-6} per gene per generation.

We can divide the errors that DNA polymerase makes during replication into two classes:

- Frameshifts occur when an extra nucleotide is inserted or omitted. Fidelity with regard to frameshifts is affected by the **processivity** of the enzyme: the tendency to remain on a single template rather than to dissociate and reassociate. This is particularly important for the replication of a homopolymeric stretch—for example, a long sequence of dT_n:dA_n, in which "replication slippage" can change the length of the homopolymeric run. As a general rule, increased processivity reduces the likelihood of such events. In multimeric DNA polymerases, processivity is usually increased by a particular subunit that is not needed for catalytic activity *per se*.
- Substitutions occur when the wrong (improperly paired) nucleotide is incorporated. The error level is determined by the efficiency of **proofreading,** in which the enzyme scrutinizes the newly formed base pair and removes the nucleotide if it is mispaired.

All of the bacterial enzymes possess a 3′–5′ exonucleolytic activity that proceeds in the reverse direction from DNA synthesis. This provides the proofreading function illustrated diagrammatically in FIGURE 18.6. In the chain elongation step, a precursor nucleotide enters the position at the end of the growing chain. A bond is formed. The enzyme moves one base pair farther, and then is ready for the next precursor nucleotide to enter. If a mistake has been made, however, the enzyme uses the exonucleolytic activity to excise the last base that was added.

Different DNA polymerases handle the relationship between the polymerizing and proofreading activities in different ways. In some cases, the activities are part of the same protein subunit, but in others they are contained in different subunits. Each DNA polymerase has a characteristic error rate that is reduced by its proofreading activity. Proofreading typically decreases the error rate in replication from $\sim 10^{-5}$ to $\sim 10^{-7}$ per base pair replicated. Systems that recognize errors and correct them following replication then eliminate some of the errors, bringing the overall rate to $<10^{-9}$ per base pair replicated (see Section 20.7, Controlling the Direction of Mismatch Repair).

The replicase activity of DNA polymerase III was originally discovered by a lethal mutation in the *dnaE* locus, which codes for the 130 kD α subunit that possesses the DNA synthetic activity. The 3′–5′ exonucleolytic proofreading activity is found in another subunit, ε, coded by *dnaQ*. The basic role of the ε subunit in controlling the fidelity of replication *in vivo* is demonstrated by the effect of mutations in *dnaQ*: The frequency with which mutations occur in the bacterial strain is increased by $>10^3$-fold.

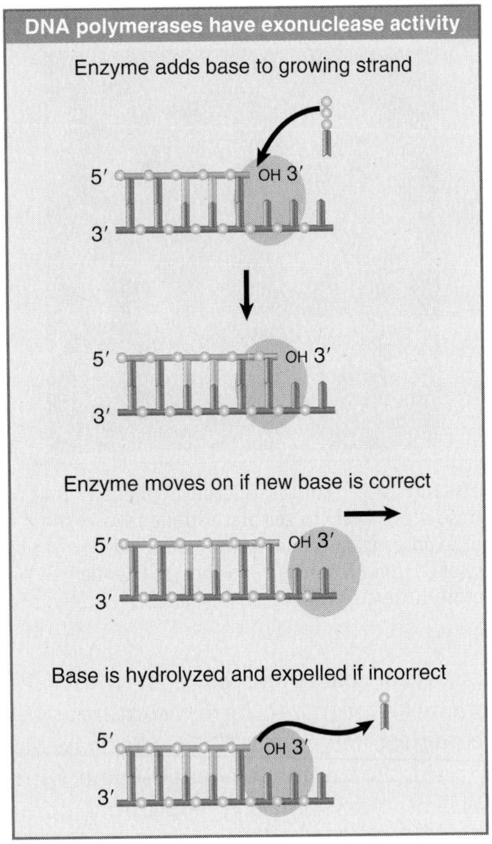

DNA polymerases have exonuclease activity

Enzyme adds base to growing strand

5′ OH 3′

3′

5′ OH 3′

3′

Enzyme moves on if new base is correct

5′ OH 3′

3′

Base is hydrolyzed and expelled if incorrect

5′ OH 3′

3′

FIGURE 18.6 Bacterial DNA polymerases scrutinize the base pair at the end of the growing chain and excise the nucleotide added in the case of a misfit.

18.5 DNA Polymerases Have a Common Structure

Key concepts

- Many DNA polymerases have a large cleft composed of three domains that resemble a hand.
- DNA lies across the "palm" in a groove created by the "fingers" and "thumb."

FIGURE 18.7 shows that all DNA polymerases share some common structural features. The enzyme structure can be divided into several independent domains, which are described by analogy with a human right hand. DNA binds in a large cleft composed of three domains. The "palm" domain has important conserved sequence motifs that provide the catalytic active site. The "fingers" are involved in positioning the template correctly at the active site. The "thumb" binds the DNA as it exits the enzyme, and is important in processivity. The most important

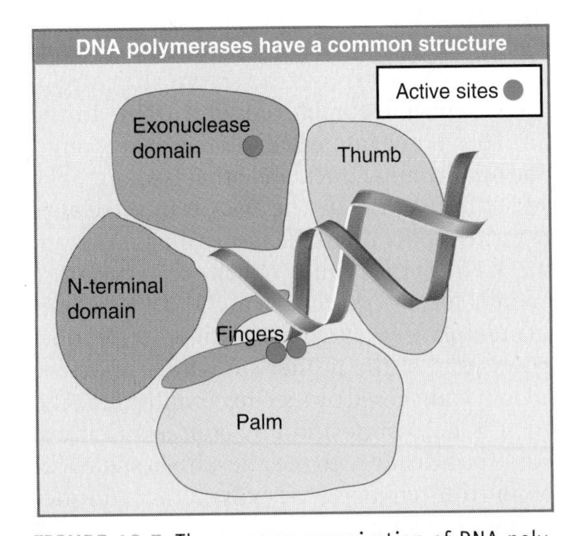

FIGURE 18.7 The common organization of DNA polymerases has a palm that contains the catalytic site, fingers that position the template, a thumb that binds DNA and is important in processivity, an exonuclease domain with its own active site, and an N-terminal domain.

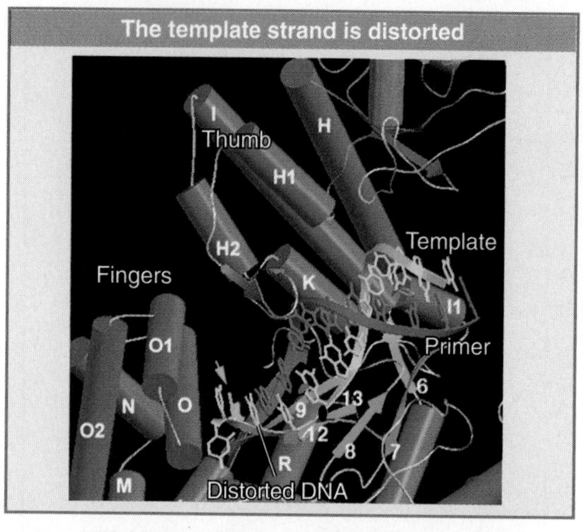

FIGURE 18.8 The crystal structure of phage T7 DNA polymerase shows that the template strand takes a sharp turn that exposes it to the incoming nucleotide. Photo courtesy of Charles Richardson and Thomas Ellenberger, Washington University School of Medicine.

conserved regions of each of these three domains converge to form a continuous surface at the catalytic site. The exonuclease activity resides in an independent domain with its own catalytic site. The N-terminal domain extends into the nuclease domain. DNA polymerases fall into five families based on sequence homologies; the palm is well conserved among them, but the thumb and fingers provide analogous secondary structure elements from different sequences.

The catalytic reaction in a DNA polymerase occurs at an active site in which a nucleotide triphosphate pairs with an (unpaired) single strand of DNA. The DNA lies across the palm in a groove that is created by the thumb and fingers. FIGURE 18.8 shows the crystal structure of the T7 enzyme complexed with DNA (in the form of a primer annealed to a template strand) and an incoming nucleotide that is about to be added to the primer. The DNA is in the classic B-form duplex up to the last two base pairs at the 3′ end of the primer, which are in the more open A-form. A sharp turn in the DNA exposes the template base to the incoming nucleotide. The 3′ end of the primer (to which bases are added) is anchored by the fingers and palm. The DNA is held in position by contacts that are made principally with the phosphodiester backbone (thus enabling the polymerase to function with DNA of any sequence).

In structures of DNA polymerases of this family complexed only with DNA (that is, lacking the incoming nucleotide), the orientation of the fingers and thumb relative to the palm is more open, with the O helix (O, O1, O2; see Figure 18.8) rotated away from the palm. This suggests that an inward rotation of the O helix occurs to grasp the incoming nucleotide and create the active catalytic site. When a nucleotide binds, the fingers domain rotates 60° toward the palm, with the tops of the fingers moving by 30 Å. The thumb domain also rotates toward the palm by 8°. These changes are cyclical: They are reversed when the nucleotide is incorporated into the DNA chain, which then translocates through the enzyme to re-create an empty site.

The exonuclease activity is responsible for removing mispaired bases. The catalytic site of the exonuclease domain is distant from the active site of the catalytic domain, though. The enzyme alternates between polymerizing and editing modes, as determined by a competition between the two active sites for the 3′ primer end of the DNA. Amino acids in the active site contact the incoming base in such a way that the enzyme structure is affected by a mismatched base. When a mismatched base pair occupies the catalytic site, the fingers cannot rotate toward the palm to bind the incoming nucleotide. This leaves the 3′ end free to bind to the active site in the exonuclease domain, which is accomplished by a rotation of the DNA in the enzyme structure.

18.6 DNA Synthesis Is Semidiscontinuous

Key concept

- The DNA replicase advances continuously when it synthesizes the leading strand (5′–3′), but synthesizes the lagging strand by making short fragments that are subsequently joined together.

The antiparallel structure of the two strands of duplex DNA poses a problem for replication. As the replication fork advances, daughter strands must be synthesized on both of the exposed parental single strands. The fork moves in the direction from 5′–3′ on one strand, and in the direction from 3′–5′ on the other strand. Yet nucleic acids are synthesized only from a 5′ end toward a 3′ end. The problem is solved by synthesizing the strand that grows overall from 3′–5′ in a series of short fragments, each actually synthesized in the "backward" direction, that is, with the customary 5′–3′ polarity.

Consider the region immediately behind the replication fork, as illustrated in FIGURE 18.9.

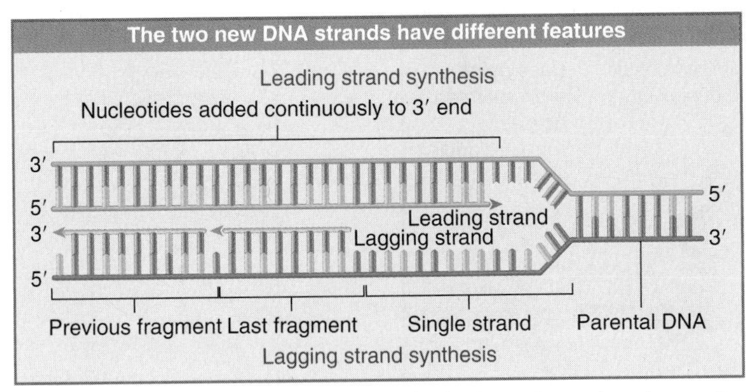

FIGURE 18.9 The leading strand is synthesized continuously, whereas the lagging strand is synthesized discontinuously.

We describe events in terms of the different properties of each of the newly synthesized strands:

- On the **leading strand** DNA synthesis can proceed continuously in the 5′ to 3′ direction as the parental duplex is unwound.
- On the **lagging strand** a stretch of single-stranded parental DNA must be exposed, and then a segment is synthesized in the reverse direction (relative to fork movement). A series of these fragments are synthesized, each 5′–3′; they then are joined together to create an intact lagging strand.

Discontinuous replication can be followed by the fate of a very brief label of radioactivity. The label enters newly synthesized DNA in the form of short fragments, sedimenting in the range of 7S to 11S, corresponding to ~1000 to 2000 bases in length. These **Okazaki fragments** are found in replicating DNA in both prokaryotes and eukaryotes. After longer periods of incubation, the label enters larger segments of DNA. The transition results from covalent linkages between Okazaki fragments.

The lagging strand *must* be synthesized in the form of Okazaki fragments. For a long time it was unclear whether the leading strand is synthesized in the same way or is synthesized continuously. All newly synthesized DNA is found as short fragments in *E. coli*. Superficially, this suggests that both strands are synthesized discontinuously. It turns out, however, that not all of the fragment population represents *bona fide* Okazaki fragments; some are pseudofragments that have been generated by breakage in a DNA strand that actually was synthesized as a continuous chain. The source of this breakage is the incorporation of some uracil into DNA in place of thymine. When the uracil is removed by a repair system, the leading strand has breaks until a thymine is inserted.

Thus the lagging strand is synthesized discontinuously and the leading strand is synthesized continuously. This is called **semi-discontinuous replication.**

18.7 The φX Model System Shows How Single-Stranded DNA Is Generated for Replication

Key concepts

- Replication requires a helicase to separate the strands of DNA using energy provided by hydrolysis of ATP.
- A single-strand binding protein is required to maintain the separated strands.
- The combination of helicase, SSB, and A protein separates a φX174 duplex into a single-stranded circle and a single-stranded linear strand.

As the replication fork advances, it unwinds the duplex DNA. One of the template strands is rapidly converted to duplex DNA as the leading daughter strand is synthesized. The other remains single-stranded until a sufficient length has been exposed to initiate synthesis of an Okazaki fragment of the lagging strand in the backward direction. The generation and maintenance of single-stranded DNA is therefore a crucial aspect of replication. Two types of function are needed to convert double-stranded DNA to the single-stranded state:

- A **helicase** is an enzyme that separates the strands of DNA, usually using the hydrolysis of ATP to provide the necessary energy.

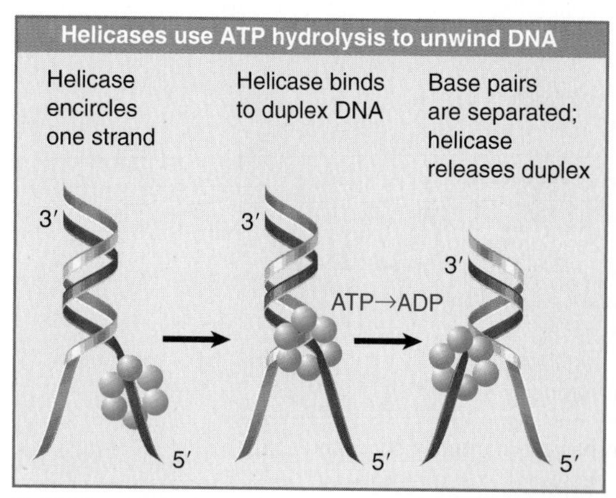

FIGURE 18.10 A hexameric helicase moves along one strand of DNA. It probably changes conformation when it binds to the duplex, uses ATP hydrolysis to separate the strands, and then returns to the conformation it has when bound only to a single strand.

- A **single-strand binding protein (SSB)** binds to the single-stranded DNA, preventing it from reforming the duplex state. The SSB binds as a monomer, but typically in a cooperative manner in which the binding of additional monomers to the existing complex is enhanced.

Helicases separate the strands of a duplex nucleic acid in a variety of situations, ranging from strand separation at the growing point of a replication fork to catalyzing migration of Holliday (recombination) junctions along DNA. There are twelve different helicases in *E. coli*. A helicase is generally multimeric. A common form of helicase is a hexamer. This typically translocates along DNA by using its multimeric structure to provide multiple DNA-binding sites.

FIGURE 18.10 shows a generalized schematic model for the action of a hexameric helicase. It is likely to have one conformation that binds to duplex DNA and another that binds to single-stranded DNA. Alternation between them drives the motor that melts the duplex and requires ATP hydrolysis—typically 1 ATP is hydrolyzed for each base pair that is unwound. A helicase usually initiates unwinding at a single-stranded region adjacent to a duplex. It may function with a particular polarity, preferring single-stranded DNA with a 3′ end (3′–5′ helicase) or with a 5′ end (5′–3′ helicase).

The conversion of φX174 double-stranded DNA into individual single strands illustrates the features of the strand separation process. FIGURE 18.11 shows that a single strand is peeled off the circular strand, resembling the rolling

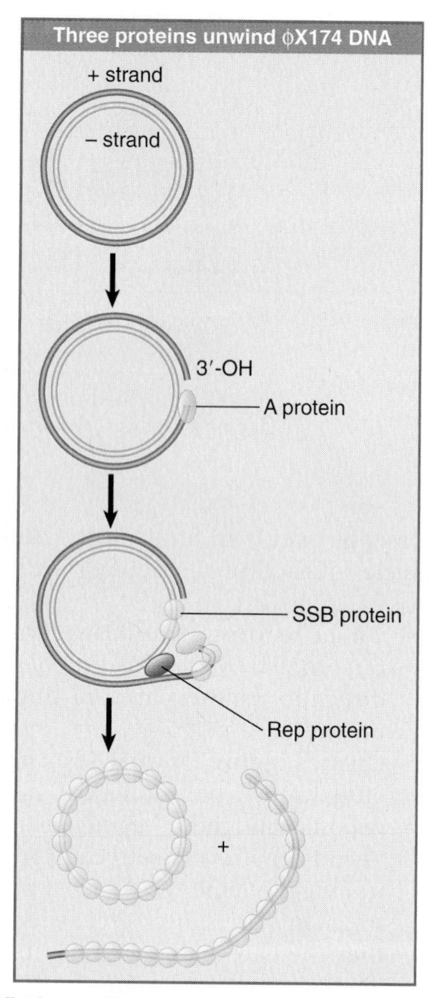

FIGURE 18.11 φX174 DNA can be separated into single strands by the combined effects of three functions: nicking with A protein, unwinding by Rep, and single-strand stabilization by SSB.

circle described previously in Figure 16.6. The reaction can occur in the absence of DNA synthesis when the appropriate three proteins are provided *in vitro*.

The phage A protein nicks the viral (+) strand at the origin of replication. In the presence of two host proteins (Rep and SSB) and ATP, the nicked DNA unwinds. The Rep protein provides a helicase that separates the strands; the SSB traps them in single-stranded form. The *E. coli* SSB is a tetramer of 74 kD that binds cooperatively to single-stranded DNA.

The significance of the cooperative mode of binding is that the binding of one protein molecule makes it much easier for another to bind. Thus once the binding reaction has started on a particular DNA molecule, it is rapidly extended until all of the single-stranded DNA is covered with the SSB protein. Note that this protein is not a DNA-unwinding protein; its function is to stabilize DNA that is already in the single-stranded condition.

Under normal circumstances *in vivo*, the unwinding, coating, and replication reactions proceed in tandem. The SSB binds to DNA as the replication fork advances, keeping the two parental strands separate so that they are in the appropriate condition to act as templates. SSB is needed in stoichiometric amounts at the replication fork. It is required for more than one stage of replication; *ssb* mutants have a quick-stop phenotype, and are defective in repair and recombination as well as in replication. (Some phages use different SSB proteins, notably T4; this shows that there may be specific interactions between components of the replication apparatus and the SSB; see Section 18.14, Phage T4 Provides Its Own Replication Apparatus).

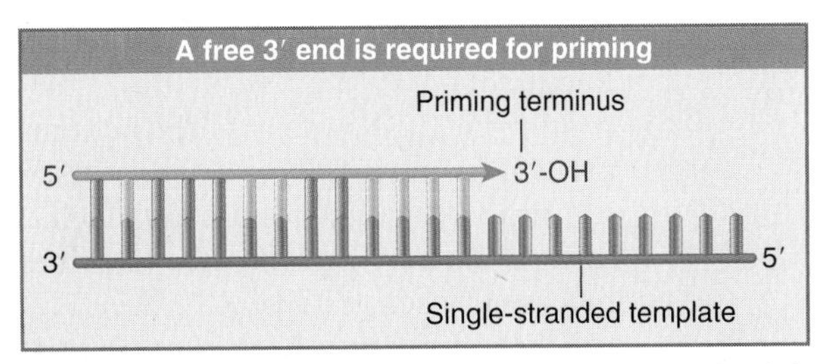

FIGURE 18.12 A DNA polymerase requires a 3′–OH end to initiate replication.

18.8 Priming Is Required to Start DNA Synthesis

Key concepts

- All DNA polymerases require a 3′–OH priming end to initiate DNA synthesis.
- The priming end can be provided by an RNA primer, a nick in DNA, or a priming protein.
- For DNA replication, a special RNA polymerase called a primase synthesizes an RNA chain that provides the priming end.
- *E. coli* has two types of priming reaction, which occur at the bacterial origin *(oriC)* and the φX174 origin.
- Priming of replication on double-stranded DNA always requires a replicase, SSB, and primase.
- DnaB is the helicase that unwinds DNA for replication in *E. coli*.

A common feature of all DNA polymerases is that they cannot initiate synthesis of a chain of DNA *de novo*. FIGURE 18.12 shows the features required for initiation. Synthesis of the new strand can only start from a preexisting 3′–OH end, and the template strand must be converted to a single-stranded condition.

The 3′–OH end is called a **primer.** The primer can take various forms. Types of priming reaction are summarized in FIGURE 18.13:

- A sequence of RNA is synthesized on the template, so that the free 3′–OH end of the RNA chain is extended by the DNA polymerase. This is commonly used in replication of cellular DNA and by some viruses (see Figure 17.18 in Section 17.11, The ColE1 Compatibility

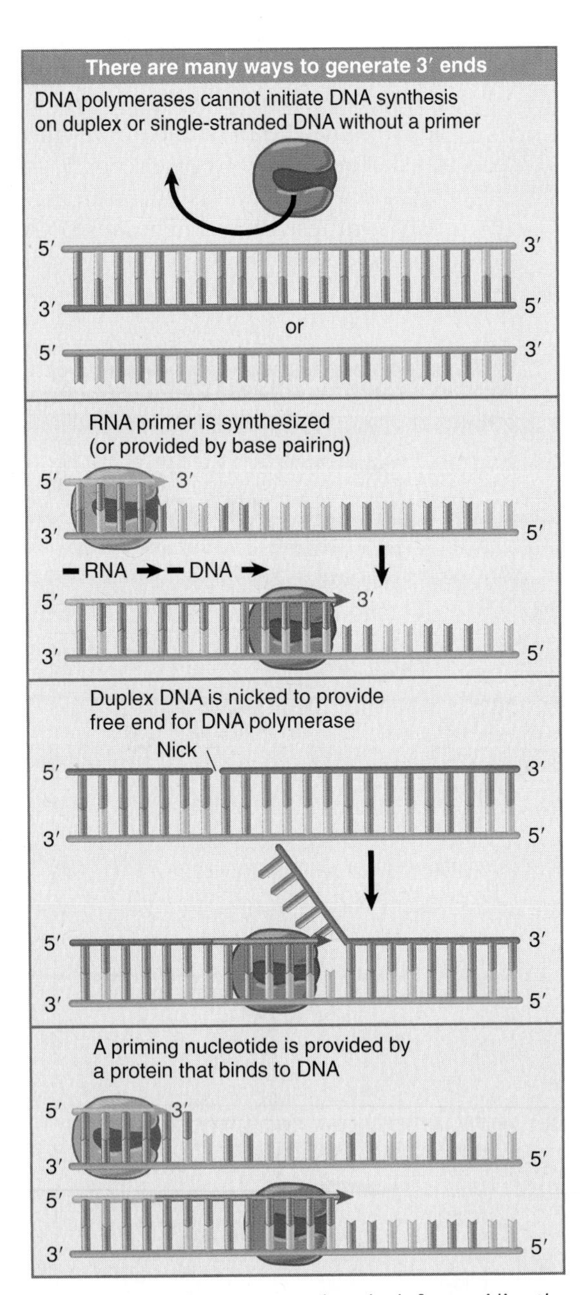

FIGURE 18.13 There are several methods for providing the free 3′–OH end that DNA polymerases require to initiate DNA synthesis.

System Is Controlled by an RNA Regulator).

- A preformed RNA pairs with the template, allowing its 3′–OH end to be used to prime DNA synthesis. This mechanism is used by retroviruses to prime reverse transcription of RNA (see Figure 22.6 in Section 22.4, Viral DNA Is Generated by Reverse Transcription).
- A primer terminus is generated within duplex DNA. The most common mechanism is the introduction of a nick, as used to initiate rolling circle replication (see Figure 16.5). In this case, the preexisting strand is displaced by new synthesis. (Note the difference from nick translation shown in Figure 18.5, in which DNA polymerase I simultaneously synthesizes and degrades DNA from a nick.)

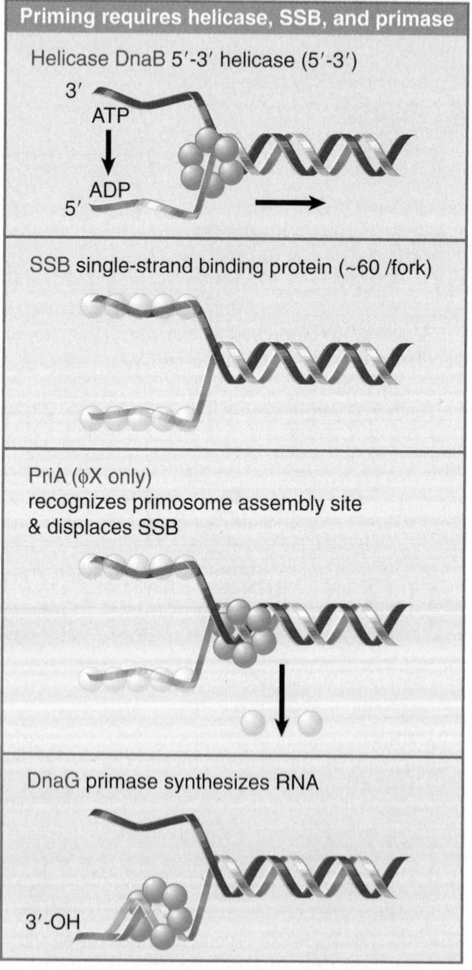

FIGURE 18.14 Initiation requires several enzymatic activities, including helicases, single-strand binding proteins, and synthesis of the primer.

- A protein primes the reaction directly by presenting a nucleotide to the DNA polymerase. This reaction is used by certain viruses (see Figure 16.4 in Section 16.2, The Ends of Linear DNA Are a Problem for Replication).

Priming activity is required to provide 3′–OH ends to start off the DNA chains on both the leading and lagging strands. The leading strand requires only one such initiation event, which occurs at the origin. There must be a series of initiation events on the lagging strand, though, because each Okazaki fragment requires its own start *de novo*. Each Okazaki fragment starts with a primer sequence of RNA ~10 bases long that provides the 3′–OH end for extension by DNA polymerase.

A **primase** is required to catalyze the actual priming reaction. This is provided by a special RNA polymerase activity, the product of the *dnaG* gene. The enzyme is a single polypeptide of 60 kD (much smaller than RNA polymerase). The primase is an RNA polymerase that is used only under specific circumstances, that is, to synthesize short stretches of RNA that are used as primers for DNA synthesis. DnaG primase associates transiently with the replication complex, and typically synthesizes an 11- to 12-base primer. Primers start with the sequence pppAG positioned opposite the sequence 3′–GTC-5′ in the template.

Some systems use alternatives to the DnaG primase. In the examples of the two phages M13 and G4, which were used for early work on replication, an interesting difference emerged. G4 priming uses DnaG, whereas M13 priming uses bacterial RNA polymerase. These phages have another unusual feature: the site of priming is indicated by a region of secondary structure.

There are two types of priming reaction in *E. coli:*

- The *oriC* system, named for the bacterial origin, basically involves the association of the DnaG primase with the protein complex at the replication fork.
- The φX system, named for phage φX174, requires an initiation complex consisting of additional components, called the primosome (see Section 18.17, The Primosome Is Needed to Restart Replication).

At times replicons are referred to as being of the φX or *oriC* type.

The types of activities involved in the initiation reaction are summarized in FIGURE 18.14.

Although other replicons in *E. coli* may have alternatives for some of these particular proteins, the same general types of activity are required in every case. A helicase is required to generate single strands, a single-strand binding protein is required to maintain the single-stranded state, and the primase synthesizes the RNA primer.

DnaB is the central component in both φX and *oriC* replicons. It provides the 5′–3′ helicase activity that unwinds DNA. Energy for the reaction is provided by cleavage of ATP. Basically DnaB is the active component of the growing point. In *oriC* replicons, DnaB is initially loaded at the origin as part of a large complex (see Section 18.15, Creating the Replication Forks at an Origin). It forms the growing point at which the DNA strands are separated as the replication fork advances. It is part of the DNA polymerase complex and interacts with the DnaG primase to initiate synthesis of each Okazaki fragment on the lagging strand.

18.9 DNA Polymerase Holoenzyme Has Three Subcomplexes

Key concepts

- The *E. coli* replicase DNA polymerase III is a 900-kD complex with a dimeric structure.
- Each monomeric unit has a catalytic core, a dimerization subunit, and a processivity component.
- A clamp loader places the processivity subunits on DNA, where they form a circular clamp around the nucleic acid.
- One catalytic core is associated with each template strand.

We can now relate the subunit structure of *E. coli* DNA polymerase III to the activities required for DNA synthesis and propose a model for its action. The holoenzyme is a complex of 900 kD that contains ten proteins organized into four types of subcomplex:

- There are two copies of the catalytic core. Each catalytic core contains the α subunit (the DNA polymerase activity), the ε subunit (the 3′–5′ proofreading exonuclease), and the θ subunit (which stimulates exonuclease).
- There are two copies of the dimerizing subunit, τ, which link the two catalytic cores together.
- There are two copies of the *clamp*, which is responsible for holding catalytic cores on to their template strands. Each clamp consists of a homodimer of β subunits that binds around the DNA and ensures processivity.
- The γ complex is a group of five proteins that comprise the **Clamp loader;** the clamp loader places the clamp on DNA.

A model for the assembly of DNA polymerase III is shown in FIGURE 18.15. The holoenzyme assembles on DNA in three stages:

- First the clamp loader uses hydrolysis of ATP to bind β subunits to a template-primer complex.
- Binding to DNA changes the conformation of the site on β that binds to the clamp loader, and as a result it now has a high affinity for the core polymerase. This enables core polymerase to bind, and this is the means by which the core polymerase is brought to DNA.

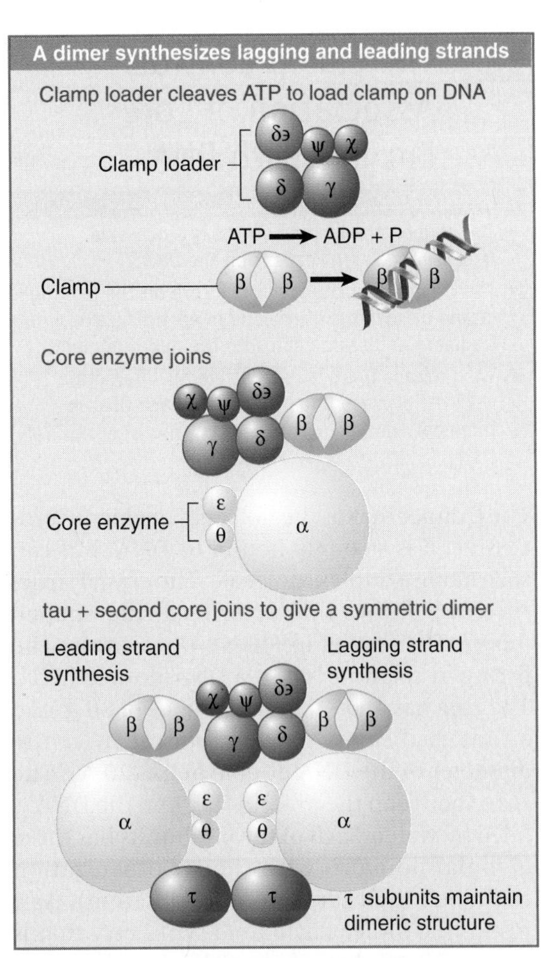

FIGURE 18.15 DNA polymerase III holoenzyme assembles in stages, generating an enzyme complex that synthesizes the DNA of both new strands.

- A τ dimer binds to the core polymerase, and provides a dimerization function that binds a second core polymerase (associated with another β clamp). The holoenzyme is asymmetric because it has only one clamp loader. The clamp loader is responsible for adding a pair of β dimers to each parental strand of DNA.

Each of the core complexes of the holoenzyme synthesizes one of the new strands of DNA. The clamp loader is also needed for unloading the β complex from DNA; as a result, the two cores have different abilities to dissociate from DNA. This corresponds to the need to synthesize a continuous leading strand (where polymerase remains associated with the template) and a discontinuous lagging strand (where polymerase repetitively dissociates and reassociates). The clamp loader is associated with the core polymerase that synthesizes the lagging strand, and plays a key role in the ability to synthesize individual Okazaki fragments.

18.10 The Clamp Controls Association of Core Enzyme with DNA

Key concepts

- The core on the leading strand is processive because its clamp keeps it on the DNA.
- The clamp associated with the core on the lagging strand dissociates at the end of each Okazaki fragment and reassembles for the next fragment.
- The helicase DnaB is responsible for interacting with the primase DnaG to initiate each Okazaki fragment.

The β dimer makes the holoenzyme highly processive. β is strongly bound to DNA, but can slide along a duplex molecule. The crystal structure of β shows that it forms a ring-shaped dimer. The model in FIGURE 18.16 shows the β-ring in relationship to a DNA double helix. The ring has an external diameter of 80 Å and an internal cavity of 35 Å, almost twice the diameter of the DNA double helix (20 Å). The space between the protein ring and the DNA is filled by water. Each of the β subunits has three globular domains with similar organization (although their sequences are different). As a result, the dimer has sixfold symmetry that is reflected in twelve α-helices that line the inside of the ring.

The dimer surrounds the duplex, providing the "sliding clamp" that allows the holoenzyme to slide along DNA. The structure explains the high processivity—there is no way for the enzyme to fall off! The α-helices on the inside have some positive charges that may interact with the DNA via the intermediate water molecules. The protein clamp does not directly contact the DNA, so it may be able to "ice-skate" along the DNA, making and breaking contacts via the water molecules.

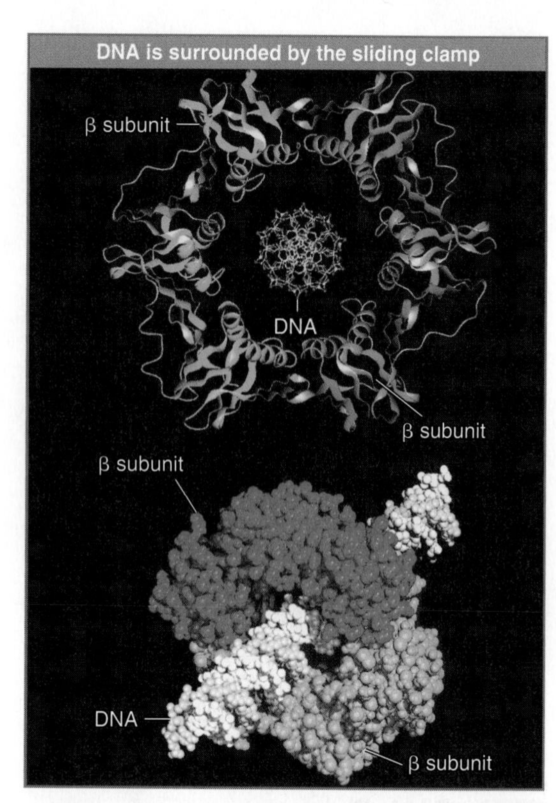

FIGURE 18.16 The β subunit of DNA polymerase III holoenzyme consists of a head-to-tail dimer (the two subunits are shown in red and orange) that forms a ring completely surrounding a DNA duplex (shown in the center). Reprinted from *Cell*, vol. 69, Kong, X. P., et al., pp. 425–437. Copyright 1992, with permission from Elsevier. Photos courtesy of John Kuriyan, University of California, Berkeley.

How does the clamp get on to the DNA? The clamp is a circle of subunits surrounding DNA; thus its assembly or removal requires the use of an energy-dependent process by the clamp loader. The γ clamp loader is a pentameric circular structure that binds an open form of the β ring preparatory to loading it on to DNA. In effect, the ring is opened at one of the interfaces between the two β subunits by the δ subunit of the clamp loader. The clamp loader binds on top of a closed circular clamp, with its ATPase site juxtaposed to the clamp, and uses hydrolysis of ATP to provide the energy to open the ring of the clamp and insert DNA into the central cavity.

The relationship between the β clamp and the γ clamp loader is a paradigm for similar systems used by DNA replicases ranging from bacteriophages to animal cells. The clamp is a heteromer (or possibly a dimer or trimer) that forms a ring around DNA with a set of twelve α-helices forming sixfold symmetry for the structure as a whole. The clamp loader has some subunits that hydrolyze ATP to provide energy for the reaction.

The basic principle that is established by the dimeric polymerase model is that, while one polymerase subunit synthesizes the leading strand continuously, the other cyclically initiates and terminates the Okazaki fragments of the lagging strand within a large single-stranded loop formed by its template strand. FIGURE 18.17 draws a generic model for the operation of such a replicase. The replication fork is created by a helicase—which typically forms a hexameric ring—that translocates in the 5′–3′ direction on the template for the lagging strand. The helicase is connected to two DNA polymerase catalytic subunits, each of which is associated with a sliding clamp.

We can describe this model for DNA polymerase III in terms of the individual components of the enzyme complex, as illustrated in FIGURE 18.18. A catalytic core is associated with each template strand of DNA. The holoenzyme moves continuously along the template for the leading strand; the template for the lagging strand is "pulled through," thus creating a loop in the DNA. DnaB creates the unwinding point, and translocates along the DNA in the "forward" direction.

DnaB contacts the τ subunit(s) of the clamp loader. This establishes a direct connection between the helicase–primase complex and the catalytic cores. This link has two effects. One is to increase the speed of DNA synthesis by increasing the rate of movement by DNA polymerase core by tenfold. The second is to prevent the leading strand polymerase from falling off, that is, to increase its processivity.

Synthesis of the leading strand creates a loop of single-stranded DNA that provides the template for lagging strand synthesis, and this loop becomes larger as the unwinding point advances. After initiation of an Okazaki fragment, the lagging strand core complex pulls the single-stranded template through the β clamp while synthesizing the new strand. The single-stranded template must extend for the length of at least one Okazaki fragment before the lagging polymerase completes one fragment and is ready to begin the next.

What happens when the Okazaki fragment is completed? All of the components of the replication apparatus function processively (that is, they remain associated with the DNA), except for the primase and the β clamp. FIGURE 18.19 shows that they dissociate when the synthesis of each fragment is completed, releasing the loop. A new β clamp is then recruited by the clamp loader to initiate the next Okazaki fragment. The lagging strand polymerase transfers from one β clamp to the next in each cycle, without dissociating from the replicating complex.

What is responsible for recognizing the sites for initiating synthesis of Okazaki fragments? In *oriC* replicons, the connection between priming and the replication fork is provided by the dual properties of DnaB: It is the helicase that propels the replication fork, and it interacts with the DnaG primase at an appropriate site. Following primer synthesis, the primase is released. The length of the priming RNA is limited to eight to fourteen bases. Apparently DNA polymerase III is responsible for displacing the primase.

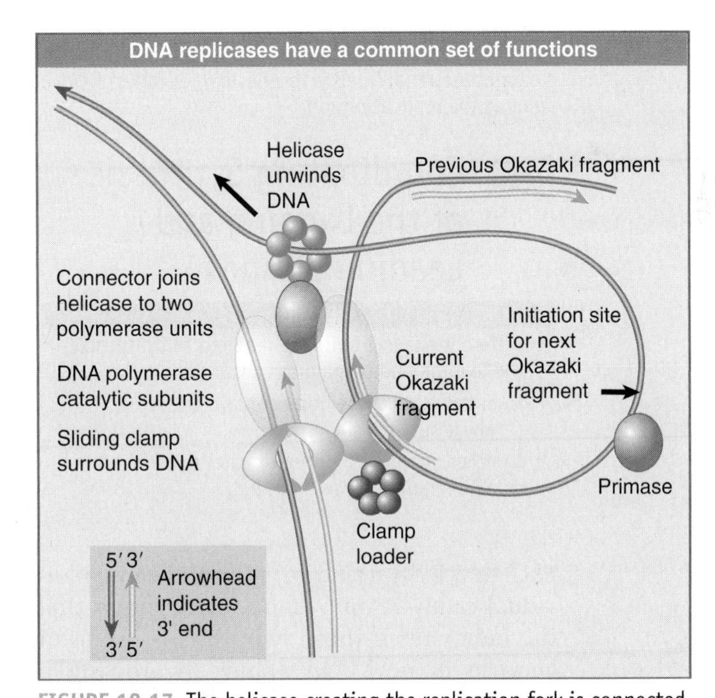

FIGURE 18.17 The helicase creating the replication fork is connected to two DNA polymerase catalytic subunits, each of which is held on to DNA by a sliding clamp. The polymerase that synthesizes the leading strand moves continuously. The polymerase that synthesizes the lagging strand dissociates at the end of an Okazaki fragment and then reassociates with a primer in the single-stranded template loop to synthesize the next fragment.

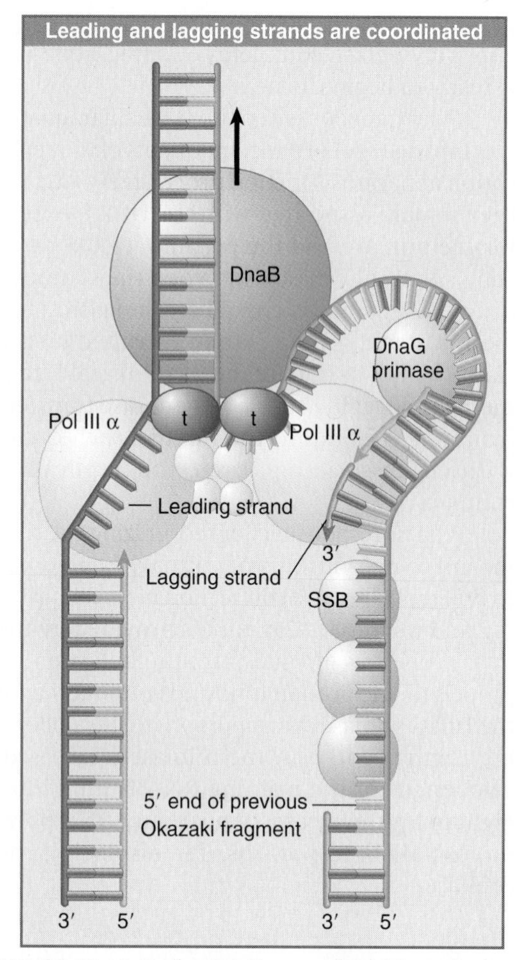

FIGURE 18.18 Each catalytic core of Pol III synthesizes a daughter strand. DnaB is responsible for forward movement at the replication fork.

18.11 Coordinating Synthesis of the Lagging and Leading Strands

Key concepts

- Different enzyme units are required to synthesize the leading and lagging strands.
- In *E. coli* both these units contain the same catalytic subunit (DnaE).
- In other organisms, different catalytic subunits may be required for each strand.

Each new DNA strand is synthesized by an individual catalytic unit. FIGURE 18.20 shows that the behavior of these two units is different because the new DNA strands are growing in opposite directions. One enzyme unit is moving with the unwinding point and synthesizing the leading strand continuously. The other unit is moving "backward" relative to the DNA, along the exposed single strand. Only short segments of template are exposed at any one time. When synthesis of one Okazaki fragment is completed, synthesis of the next Okazaki fragment is required to start at a new location approximately in the vicinity of the growing point for the leading strand. This requires a translocation relative to the DNA of the enzyme unit that is synthesizing the lagging strand.

The term "enzyme unit" avoids the issue of whether the DNA polymerase that synthesizes the leading strand is the same type of enzyme as the DNA polymerase that synthesizes the lagging strand. In the case that we know best, *E. coli,* there is only a single type of DNA polymerase catalytic subunit used in replication, the DnaE protein. The active replicase is a dimer, and each half of the dimer contains DnaE as the catalytic subunit. The DnaE is supported by other proteins (which differ between the leading and lagging strands).

The use of a single type of catalytic subunit, however, may be atypical. In the bacterium *Bacillus subtilis,* there are two different catalytic subunits. PolC is the homolog to *E. coli*'s DnaE, and is responsible for synthesizing the leading strand. A related protein, $DnaE_{BS}$, is the catalytic subunit that synthesizes the lagging strand. Eukaryotic DNA polymerases have the same general structure, with different enzyme units synthesizing the leading and lagging strands, but it is not clear whether the same or different types of catalytic subunits are used (see Section 18.13, Separate Eukaryotic DNA Polymerases Undertake Initiation and Elongation).

A major problem of the semidiscontinuous mode of replication follows from the use of different enzyme units to synthesize each new DNA strand: How is synthesis of the lagging strand coordinated with synthesis of the leading strand? As the replisome moves along DNA, unwinding the parental strands, one enzyme unit elongates the leading strand. Periodically the primosome activity initiates an Okazaki fragment on the lagging strand, and the other enzyme unit must then move in the reverse direction to synthesize DNA. FIGURE 18.21 proposes two types of model for what happens to this enzyme unit when it completes synthesis of an Okazaki fragment. The same complex may be reutilized for synthesis of successive Okazaki fragments. Another possibility is that the complex might dissociate from the template, so that a new complex must be assembled to elongate the next Okazaki fragment. We see in Section 18.10, The Clamp Controls Association of Core Enzyme with DNA that the first model applies.

18.12 Okazaki Fragments Are Linked by Ligase

Key concepts

- Each Okazaki fragment starts with a primer and stops before the next fragment.
- DNA polymerase I removes the primer and replaces it with DNA in an action that resembles nick translation.
- DNA ligase makes the bond that connects the 3′ end of one Okazaki fragment to the 5′ beginning of the next fragment.

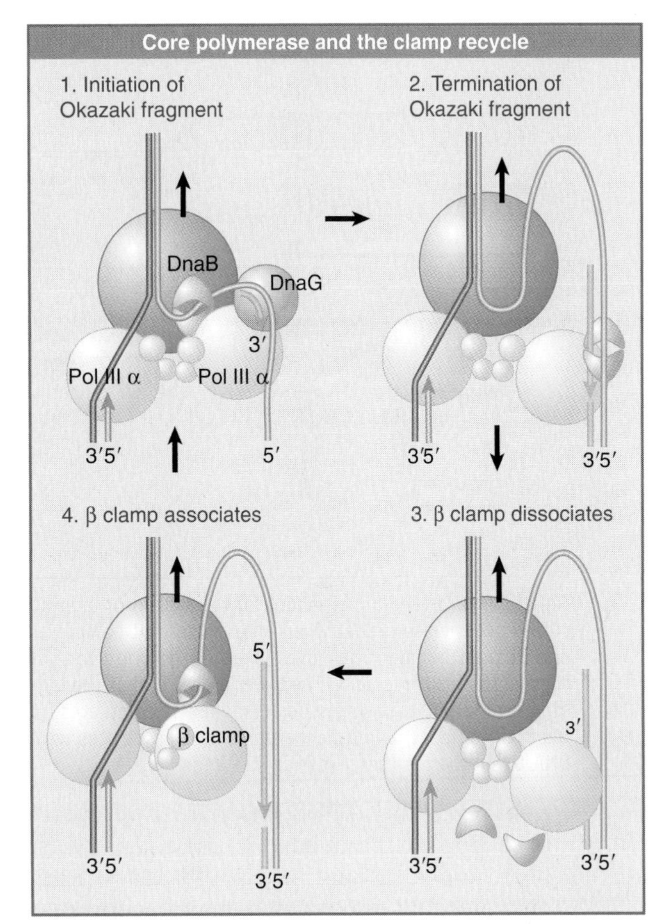

FIGURE 18.19 Core polymerase and the β clamp dissociate at completion of Okazaki fragment synthesis and reassociate at the beginning.

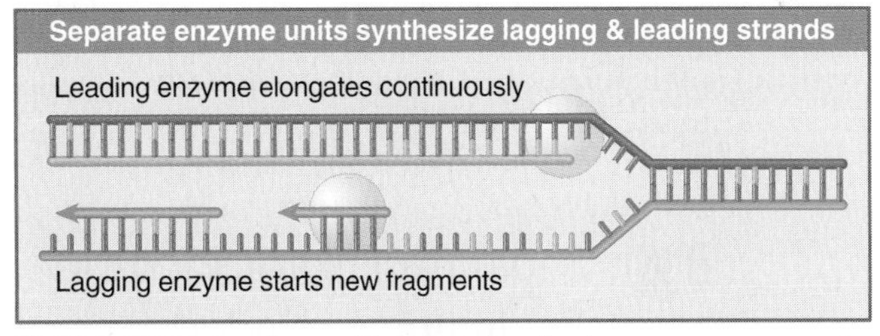

FIGURE 18.20 Leading and lagging strand polymerases move apart.

We can now expand our view of the actions involved in joining Okazaki fragments, as illustrated in FIGURE 18.22. The complete order of events is uncertain, but it must involve synthesis of RNA primer, its extension with DNA, removal of the RNA primer, its replacement by a stretch of DNA, and the covalent linking of adjacent Okazaki fragments.

The figure suggests that synthesis of an Okazaki fragment terminates just before the start of the RNA primer of the preceding fragment. When the primer is removed, there will be a gap. The gap is filled by DNA polymerase I; *polA* mutants fail to join their Okazaki fragments properly. The 5′–3′ exonuclease activity removes the RNA primer while simultaneously replacing it with a DNA sequence extended from the 3′–OH end of the next Okazaki fragment. This is equivalent to nick translation, except that the new DNA replaces a stretch of RNA rather than a segment of DNA.

In mammalian systems (where the DNA polymerase does not have a 5′–3′ exonuclease activity), Okazaki fragments are connected by a two-step process. Synthesis of an Okazaki fragment displaces the RNA primer of the preceding fragment in the form of a "flap." FIGURE 18.23 shows that the base of the flap is cleaved by the enzyme FEN1. In this reaction, FEN1 functions as an endonuclease, but it also has a 5′–3′ exonuclease activity. In DNA repair reactions, FEN1 may cleave next to a displaced nucleotide and then use its exonuclease activity to remove adjacent material.

Failure to remove a flap rapidly can have important consequences in regions of repeated sequences. Direct repeats can be displaced and misaligned with the template; palindromic sequences can form hairpins. These structures may change the number of repeats (see Figure 6.28). The general importance of FEN1 is that it prevents flaps of DNA from generating structures that may cause deletions or duplications in the genome.

Once the RNA has been removed and replaced, the adjacent Okazaki fragments must be linked together. The 3′–OH end of one fragment is adjacent to the 5′–phosphate end of the previous fragment. The responsibility for sealing this nick lies with the enzyme **DNA ligase.** Ligases are present in both prokaryotes and eukaryotes. Unconnected fragments persist in lig^- mutants because they fail to join Okazaki fragments together.

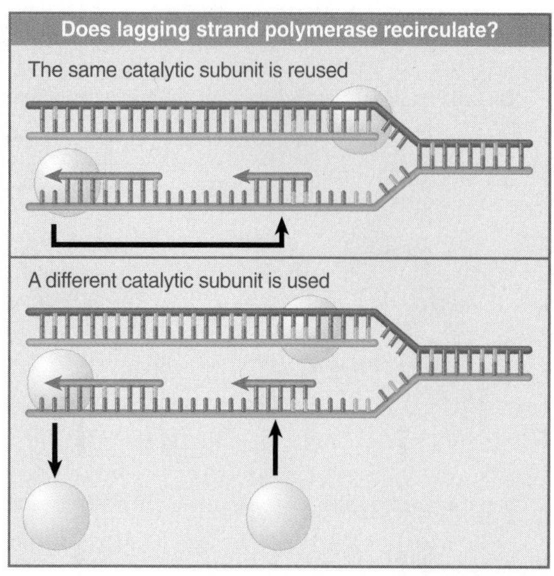

FIGURE 18.21 The upper model for the action of lagging strand polymerase is that when an enzyme unit completes one Okazaki fragment, it moves to a new position to synthesize the next fragment. The lower model is that the lagging strand polymerase dissociates when it completes an Okazaki fragment, and a new enzyme unit associates with DNA to synthesize the next Okazaki fragment.

The *E. coli* and T4 ligases share the property of sealing nicks that have 3′–OH and 5′–phosphate termini, as illustrated in FIGURE 18.24. Both enzymes undertake a two-step reaction that involves an enzyme–AMP complex. (The *E. coli* and T4 enzymes use different cofactors. The *E. coli* enzyme uses NAD [nicotinamide adenine dinucleotide] as a cofactor, whereas the T4 enzyme uses ATP.) The AMP of the enzyme complex becomes attached to the 5′–phosphate of the nick, and then a phosphodiester bond is formed with the 3′–OH terminus of the nick, releasing the enzyme and the AMP.

18.13 Separate Eukaryotic DNA Polymerases Undertake Initiation and Elongation

Key concepts

- A replication fork has one complex of DNA polymerase α/primase and two complexes of DNA polymerase δ and/or ε.
- The DNA polymerase α/primase complex initiates the synthesis of both DNA strands.
- DNA polymerase δ elongates the leading strand and a second DNA polymerase δ or DNA polymerase ε elongates the lagging strand.

Eukaryotic cells have a large number of DNA polymerases. They can be broadly divided into those required for semiconservative replication and those involved in synthesizing material to repair damaged DNA. Nuclear DNA replication requires DNA polymerases α, δ, and ε. All the other nuclear DNA polymerases are concerned with synthesizing stretches of new DNA to replace damaged material. FIGURE 18.25 shows that all of the nuclear replicases are large heterotetrameric enzymes. In each case, one of the subunits has the responsibility for catalysis, and the others are concerned with ancillary functions, such as priming, processivity, or proofreading. These enzymes all replicate DNA with high fidelity, as does the slightly less complex mitochondrial enzyme. The repair polymerases have much simpler structures, which often consist of a single monomeric subunit (although it may function in the context of a complex of other repair enzymes). Of the enzymes involved in repair, only DNA polymerase β has a fidelity approaching the replicases: all of the others have much greater error rates. All mitochondrial DNA synthesis, including repair and recombination reactions as well as replication, is undertaken by DNA polymerase γ.

Each of the three nuclear DNA replicases has a different function:

- DNA polymerase α initiates the synthesis of new strands.
- DNA polymerase δ elongates the leading strand.
- DNA polymerase ε may be involved in lagging strand synthesis, but also has other roles.

DNA polymerase α is unusual because it has the ability to initiate a new strand. It is used to initiate both the leading and lagging strands. The enzyme exists as a complex consisting of a 180-kD catalytic subunit, which is associated with the B subunit that appears necessary for assembly, and two smaller proteins that provide a primase activity. Reflecting its dual capacity to prime and extend chains, it is sometimes called pol α/primase.

The pol α/primase enzyme binds to the initiation complex at the origin and synthesizes a short strand consisting of ~10 bases of RNA followed by 20 to 30 bases of DNA (sometimes called iDNA). It is then replaced by an enzyme that will extend the chain. On the leading strand, this is DNA polymerase δ. This event is called the pol switch. It involves interactions among several components of the initiation complex.

DNA polymerase δ is a highly processive enzyme that continuously synthesizes the leading strand. Its processivity results from its interaction with two other proteins, RF-C and PCNA.

The roles of RF-C and PCNA are analogous to the *E. coli* γ clamp loader and β processivity unit (see Section 18.10, The Clamp Controls Association of Core Enzyme with DNA). RF-C is a clamp loader that catalyzes the loading of PCNA on to DNA. It binds to the 3′ end of the iDNA and uses ATP-hydrolysis to open the ring of PCNA so that it can encircle the DNA. The processivity of DNA polymerase δ is maintained by PCNA, which tethers DNA polymerase δ to the template. (PCNA is called proliferating cell nuclear antigen for historical reasons.) The crystal structure of PCNA closely resembles the *E. coli* β subunit: A trimer forms a ring that surrounds the DNA. The sequence and subunit organization are different from the dimeric β clamp; however, the function is likely to be similar.

We are less certain about events on the lagging strand. One possibility is that DNA polymerase δ also elongates the lagging strand. It has the capability to dimerize, which suggests a model analogous to the behavior of *E. coli* replicase (see Section 18.9, DNA Polymerase Holoenzyme Has Three Subcomplexes). There are, however, some indications that DNA polymerase ε may elongate the lagging strand, although it also has been identified with other roles.

A general model suggests that a replication fork contains one complex of DNA polymerase α/primase and two other DNA polymerase complexes. One is DNA polymerase δ and the other is either a second DNA polymerase δ or may possibly be a DNA polymerase ε. The two complexes of DNA polymerase δ/ε behave in the same way as the two complexes of DNA polymerase III in the *E. coli* replisome: one synthesizes the leading strand, and the other synthesizes Okazaki fragments on the lagging strand. The exonuclease MF1 removes the RNA primers of Okazaki fragments. The enzyme DNA ligase I is specifically required to seal the nicks between the completed Okazaki fragments.

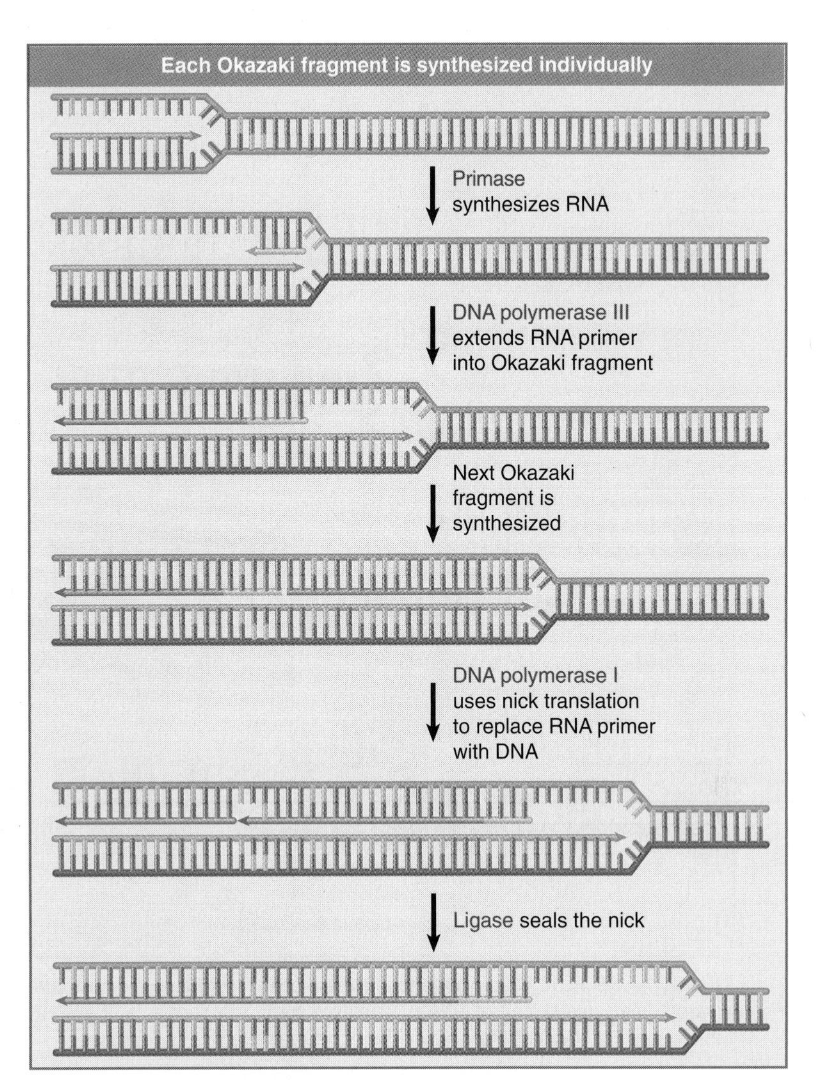

FIGURE 18.22 Synthesis of Okazaki fragments requires priming, extension, removal of RNA, gap filling, and nick ligation.

18.14 Phage T4 Provides Its Own Replication Apparatus

Key concept

- Phage T4 provides its own replication apparatus, which consists of DNA polymerase, the gene *32* SSB, a helicase, a primase, and accessory proteins that increase speed and processivity.

When phage T4 takes over an *E. coli* cell, it provides several functions of its own that either replace or augment the host functions. The phage places little reliance on expression of host functions. The degradation of host DNA is important in releasing nucleotides that are reused in the synthesis of phage DNA. (The phage DNA differs in base composition from cellular DNA in using hydroxymethylcytosine instead of the customary cytosine.)

The phage-coded functions concerned with DNA synthesis in the infected cell can be identified by mutations that impede the production of mature phages. Essential phage functions are identified by conditional lethal mutations, which fall into three phenotypic classes:

- Those in which there is no DNA synthesis at all identify genes whose products either are components of the replication apparatus or are involved in

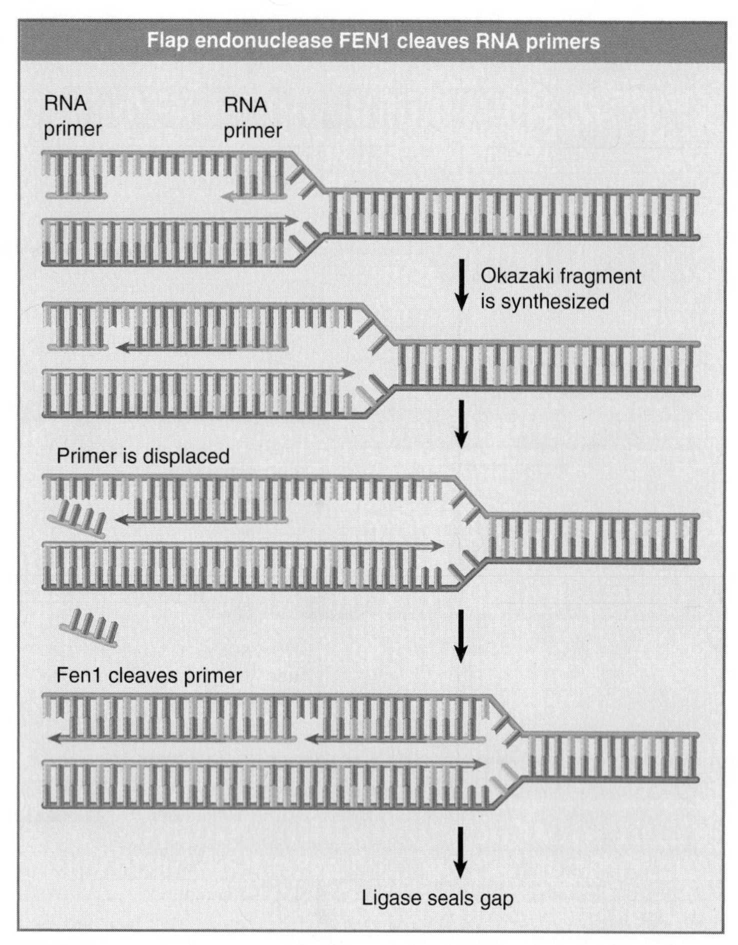

FIGURE 18.23 FEN1 is an exo-/endonuclease that recognizes the structure created when one strand of DNA is displaced from a duplex as a "flap." In replication it cleaves at the base of the flap to remove the RNA primer.

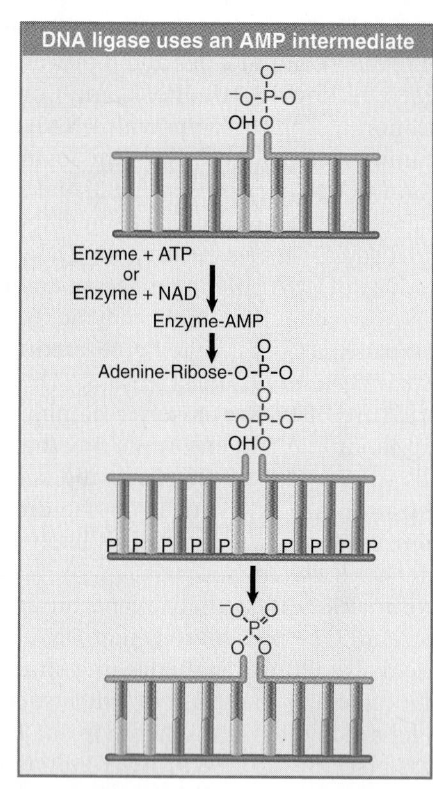

FIGURE 18.24 DNA ligase seals nicks between adjacent nucleotides by employing an enzyme-AMP intermediate.

the provision of precursors (especially the hydroxymethylcytosine).

- Those in which the onset of DNA synthesis is delayed are concerned with the initiation of replication.
- Those in which DNA synthesis starts but then is arrested include regulatory functions, the DNA ligase, and some of the enzymes concerned with host DNA degradation.
- There are also nonessential genes concerned with replication, including those involved in glucosylating the hydroxymethylcytosine in the DNA.

Synthesis of T4 DNA is catalyzed by a multienzyme aggregate assembled from the products of a small group of essential genes.

The gene *32* protein (gp32) is a highly cooperative single-strand binding protein, which is needed in stoichiometric amounts. It was the first example of its type to be characterized. The geometry of the T4 replication fork may specifically require the phage-coded protein, because the *E. coli* SSB cannot substitute. The gp32 forms a complex with the T4 DNA polymerase; this interaction could be important in constructing the replication fork.

The T4 system uses an RNA priming event that is similar to that of its host. With single-stranded T4 DNA as template, the gene *41* and *61* products act together to synthesize short primers. Their behavior is analogous to that of DnaB and DnaG in *E. coli*. The gene *41* protein is the counterpart to DnaB. It is a hexameric helicase that uses hydrolysis of GTP to provide the energy to unwind DNA. The p41/p61 complex moves processively in the 5′–3′ direction in lagging strand synthesis, periodically initiating Okazaki fragments. Another protein, the product of gene *59*, loads the p41/p61 complex onto DNA; it is required to displace the p32 protein in order to allow the helicase to assemble on DNA.

The gene *61* protein is needed in much smaller amounts than most of the T4 replication proteins. There are as few as ten copies of gp61 per cell. (This impeded its characterization. It is required in such small amounts that originally it was missed as a necessary component, because enough was present as a contaminant of the gp32 preparation!) Gene *61* protein has the pri-

DNA polymerases undertake replication or repair		
DNA polymerase	Function	Structure
	High fidelity replicases	
α	Nuclear replication	350 kD tetramer
δ	"	250 kD tetramer
ϵ	"	350 kD tetramer
γ	Mitochondrial replication	200 kD dimer
	High fidelity repair	
β	Base excision repair	39 kD monomer
	Low fidelity repair	
ζ	Thymine dimer bypass	heteromer
η	Base damage repair	monomer
ι	Required in meiosis	monomer
κ	Deletion and base substitution	monomer

FIGURE 18.25 Eukaryotic cells have many DNA polymerases. The replicative enzymes operate with high fidelity. Except for the β enzyme, the repair enzymes all have low fidelity. Replicative enzymes have large structures, with separate subunits for different activities. Repair enzymes have much simpler structures.

Replication requires a common set of functions			
Function	*E.coli*	HeLa/SV40	Phage T4
Helicase	DnaB	T antigen	41
Loading helicase/primase	DnaC	T antigen	59
Single strand maintenance	SSB	RPA	32
Priming	DnaG	Polα/primase	61
Sliding clamp	β	PCNA	45
Clamp loading (ATPase)	γδ complex	RFC	44/62
Catalysis	*Pol III core*	Polδ	43
Holoenzyme dimerization	τ	?	43
RNA removal	*Pol I*	MF1	43
Ligation	*Ligase*	Ligase 1	T4 ligase

FIGURE 18.26 Similar functions are required at all replication forks.

mase activity, which is analogous to DnaG of *E. coli*. The primase recognizes the template sequence 3′–TTG-5′ and synthesizes pentaribonucleotide primers that have the general sequence pppApCpNpNpNp. If the complete replication apparatus is present, these primers are extended into DNA chains.

The gene *43* DNA polymerase has the usual 5′–3′ synthetic activity, which is associated with a 3′–5′ exonuclease proofreading activity. It catalyzes DNA synthesis and removes the primers. When T4 DNA polymerase uses a single-stranded DNA as template, its rate of progress is uneven. The enzyme moves rapidly through single-stranded regions, but proceeds much more slowly through regions that have a base-paired intrastrand secondary structure. The accessory proteins assist the DNA polymerase in passing these roadblocks and maintaining its speed.

The remaining three proteins are referred to as "polymerase accessory proteins." They increase the affinity of the DNA polymerase for the DNA, as well as increase its processivity and speed. The gene *45* product is a trimer that acts as a sliding clamp. The structure of the trimer is similar to that of the *E. coli* β dimer, in that it forms a circle around DNA that holds the DNA polymerase subunit more tightly on the template.

The products of genes *44* and *62* form a tight complex that has ATPase activity. They are the equivalent of the γδ clamp loader complex, and their role is to load p45 onto DNA. Four molecules of ATP are hydrolyzed in loading the p45 clamp and the p43 DNA polymerase on to DNA.

The overall structure of the replisome is similar to that of *E. coli*. It consists of two coupled holoenzyme complexes, one synthesizing the leading strand and the other synthesizing the lagging strand. In this case, the dimerization involves a direct interaction between the p43 DNA polymerase subunits, and p32 plays a role in coordinating the actions of the two DNA polymerase units.

Thus far we have dealt with DNA replication solely in terms of the progression of the replication fork. The need for other functions is shown by the DNA-delay and DNA-arrest mutants. Three of the four genes of the DNA-delay mutants are *39, 52,* and *60,* which code for the three subunits of T4 topoisomerase II, an activity needed for removing supercoils in the template (see Section 19.13, Topoisomerases Relax or Introduce Supercoils in DNA). The essential role of this enzyme suggests that T4 DNA does not remain in a linear form, but rather becomes topologically constrained during some stage of replication. The topoisomerase could be needed to allow rotation of DNA ahead of the replication fork.

Comparison of the T4 apparatus with the *E. coli* apparatus suggests that DNA replication poses a set of problems that are solved in analogous ways in different systems. We may now compare the enzymatic and structural activities found at the replication fork in *E. coli*, T4, and HeLa (human) cells. FIGURE 18.26 summarizes the functions and assigns them to individual proteins. We can interpret the known properties of replication complex proteins in terms of similar functions that involve the unwinding, priming, catalytic, and sealing reactions. The components of each system interact in restricted ways, as shown by the fact that phage T4 requires its own helicase, primase, clamp, and so on, and by the fact that bacterial proteins cannot substitute for their phage counterparts.

18.15 Creating the Replication Forks at an Origin

Key concepts

- Initiation at *oriC* requires the sequential assembly of a large protein complex.
- DnaA binds to short repeated sequences and forms an oligomeric complex that melts DNA.
- Six DnaC monomers bind each hexamer of DnaB, and this complex binds to the origin.
- A hexamer of DnaB forms the replication fork. Gyrase and SSB are also required.

Starting a cycle of replication of duplex DNA requires several successive activities:

- The two strands of DNA must suffer their initial separation. This is, in effect, a melting reaction over a short region.
- An unwinding point begins to move along the DNA; this marks the generation of the replication fork, which continues to move during elongation.
- The first nucleotides of the new chain must be synthesized into the primer. This action is required once for the leading strand, but is repeated at the start of each Okazaki fragment on the lagging strand.

Some events that are required for initiation therefore occur uniquely at the origin; others recur with the initiation of each Okazaki fragment during the elongation phase.

Plasmids carrying the *E. coli oriC* sequence have been used to develop a cell-free system for replication. Initiation of replication at *oriC in vitro* starts with formation of a complex that requires six proteins: DnaA, DnaB, DnaC, HU, Gyrase, and SSB. Of the six proteins involved in prepriming, DnaA draws our attention as the only one uniquely involved in initiation vis-à-vis elongation. DnaB/DnaC provides the "engine" of initiation at the origin.

The first stage in complex formation is binding to *oriC* by DnaA protein. The reaction involves action at two types of sequences: 9 bp and 13 bp repeats. Together the 9 bp and 13 bp repeats define the limits of the 245 bp minimal origin, as indicated in FIGURE 18.27. An origin is activated by the sequence of events summarized in FIGURE 18.28, in which binding of DnaA is succeeded by association with the other proteins.

The four 9 bp consensus sequences on the right side of *oriC* provide the initial binding sites for DnaA. It binds cooperatively to form a central core around which *oriC* DNA is wrapped.

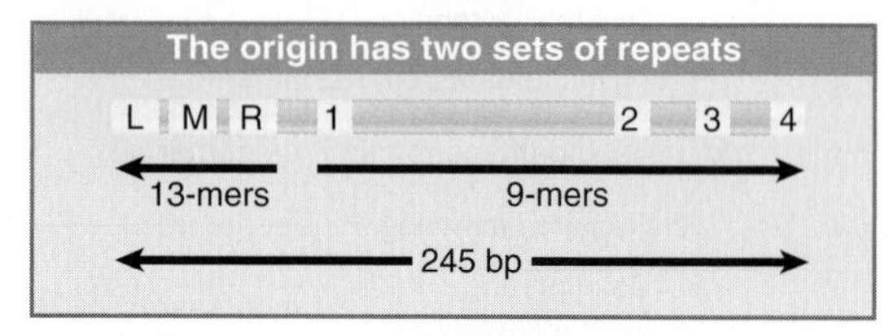

FIGURE 18.27 The minimal origin is defined by the distance between the outside members of the 13-mer and 9-mer repeats.

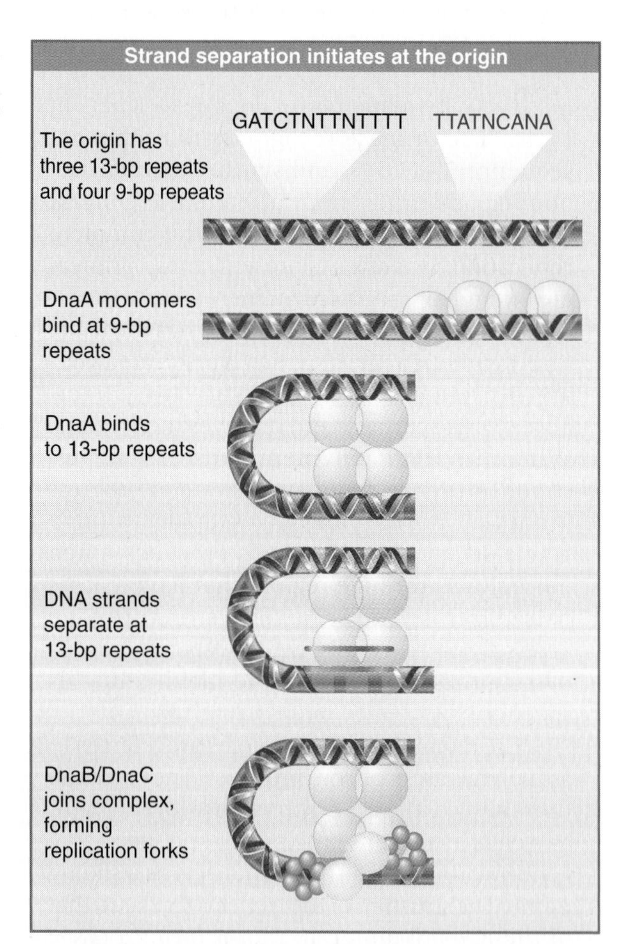

FIGURE 18.28 Prepriming involves formation of a complex by sequential association of proteins, which leads to the separation of DNA strands.

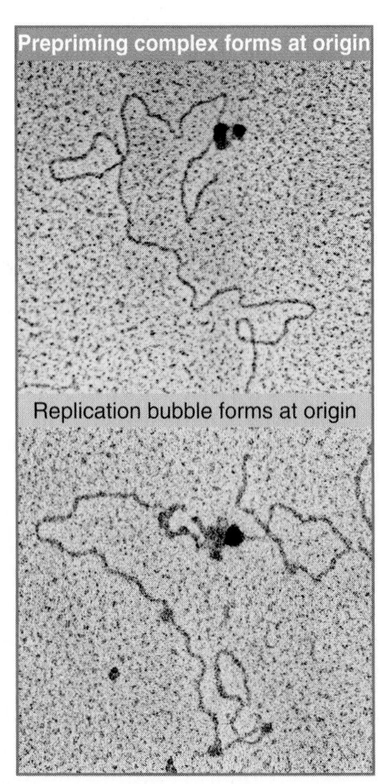

FIGURE 18.29 The complex at *oriC* can be detected by electron microscopy. Both complexes were visualized with antibodies against DnaB protein. Top photo reproduced from Funnel, B. E., et al. *J. Biol. Chem.* 1987. 262: 10327–10334. Copyright 1987 by American Society for Biochemistry & Molecular Biology. Photo courtesy of Barbara E. Funnell, University of Toronto. Bottom photo reproduced from Barker, T. A., et al. *J. Biol. Chem.* 1987. 262: 6877–6885. Copyright 1987 by American Society for Biochemistry & Molecular Biology. Photo courtesy of Barbara E. Funnell, University of Toronto.

DnaA then acts at three A-T-rich 13 bp tandem repeats located in the left side of *oriC*. In the presence of ATP, DnaA melts the DNA strands at each of these sites to form an open complex. All three 13 bp repeats must be opened for the reaction to proceed to the next stage.

Altogether, two to four monomers of DnaA bind at the origin, and they recruit two "prepriming" complexes of DnaB–DnaC to bind, so that there is one for each of the two (bidirectional) replication forks. Each DnaB–DnaC complex consists of six DnaC monomers bound to a hexamer of DnaB. Each DnaB–DnaC complex transfers a hexamer of DnaB to an opposite strand of DNA. DnaC hydrolyzes ATP in order to release DnaB.

The prepriming complex generates a protein aggregate of 480 kD, which corresponds to a sphere of radius 6 nm. The formation of a complex at *oriC* is detectable in the form of the large protein blob visualized in FIGURE 18.29. When replication begins, a replication bubble becomes visible next to the blob.

The region of strand separation in the open complex is large enough for both DnaB hexamers to bind, which initiates the two replication forks. As DnaB binds, it displaces DnaA from the 13 bp repeats and extends the length of the open region. It then uses its helicase activity to extend the region of unwinding. Each DnaB activates a DnaG primase—in one case to initiate the leading strand, and in the other to initiate the first Okazaki fragment of the lagging strand.

Two further proteins are required to support the unwinding reaction. Gyrase provides a swivel that allows one strand to rotate around the other (a reaction discussed in more detail in Section 19.15, Gyrase Functions by Coil Inversion); without this reaction, unwinding would generate torsional strain in the DNA. The protein SSB stabilizes the single-stranded DNA as it is formed. The length of duplex DNA that usually is unwound to initiate replication is probably <60 bp.

The protein HU is a general DNA-binding protein in *E. coli*. Its presence is not absolutely required to initiate replication *in vitro*, but it stimulates the reaction. HU has the capacity to bend DNA, and is involved in building the structure that leads to formation of the open complex.

Input of energy in the form of ATP is required at several stages for the prepriming reaction, and it is required for unwinding DNA. The helicase action of DnaB depends on ATP hydrolysis, and the swivel action of gyrase requires ATP hydrolysis. ATP also is needed for the action of primase and to activate DNA polymerase III.

Following generation of a replication fork as indicated in Figure 18.28, the priming reaction occurs to generate a leading strand. We know that synthesis of RNA is used for the priming event, but the details of the reaction are not known. Some mutations in *dnaA* can be suppressed by mutations in RNA polymerase, which suggests that DnaA could be involved in an initiation step requiring RNA synthesis *in vivo*.

RNA polymerase could be required to read into the origin from adjacent transcription units; by terminating at sites in the origin, it could provide the 3′–OH ends that prime DNA polymerase III. (An example is provided by the use of D loops at mitochondrial origins, as discussed in Section 15.11, D Loops Maintain Mitochondrial Origins.) Alternatively, the act of transcription could be associated with a structural change that assists initiation. This latter idea is supported by observations that transcription does not have to proceed into the origin; it is effective up to 200 bp away from the origin, and can use either strand of DNA as template *in vitro*. The transcriptional event is inversely related to the requirement for supercoiling *in vitro*, which suggests that it acts by changing the local DNA structure so as to aid melting of DNA.

18.16 Common Events in Priming Replication at the Origin

Key concepts

- The general principle of bacterial initiation is that the origin is initially recognized by a protein that forms a large complex with DNA.
- A short region of A-T-rich DNA is melted.
- DnaB is bound to the complex and creates the replication fork.

Another system for investigating interactions at the origin is provided by phage lambda. A map of the region is shown in FIGURE 18.30. Initiation of replication at the lambda origin requires "activation" by transcription starting from P_R. As with the events at *oriC*, this does not necessarily imply that the RNA provides a primer for the leading strand. Analogies between the systems suggest that RNA synthesis could be involved in promoting some structural change in the region.

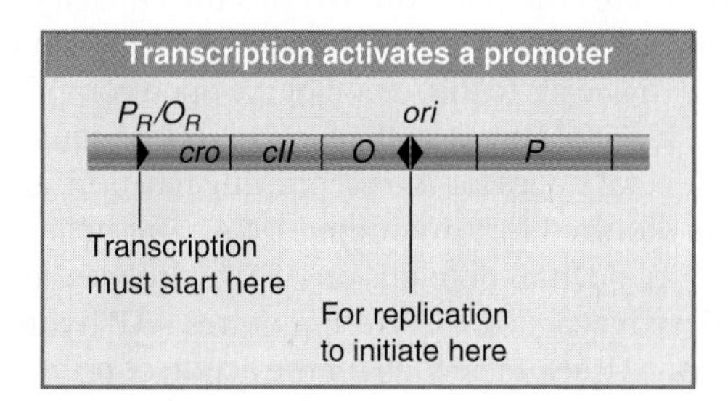

FIGURE 18.30 Transcription initiating at P_R is required to activate the origin of lambda DNA.

Initiation requires the products of phage genes *O* and *P*, as well as several host functions. The phage O protein binds to the lambda origin; the phage P protein interacts with the O protein and with the bacterial proteins. The origin lies within gene *O*, so the protein acts close to its site of synthesis.

Variants of the phage called λ*dv* consist of shorter genomes that carry all the information needed to replicate, but lack infective functions. λ*dv* DNA survives in the bacterium as a plasmid, and can be replicated *in vitro* by a system consisting of the phage-coded proteins O and P together with bacterial replication functions.

Lambda proteins O and P form a complex together with DnaB at the lambda origin, *ori*λ. The origin consists of two regions, as illustrated in FIGURE 18.31: A region comprising a series of four binding sites for the O protein is adjacent to an A-T-rich region.

The first stage in initiation is the binding of O to generate a roughly spherical structure of diameter ~11 nm, a structure sometimes called the O-some. The O-some contains ~100 bp or 60 kD of DNA. There are four 18 bp binding sites for O protein, which is ~34 kD. Each site is palindromic, and probably binds a symmetrical O dimer. The DNA sequences of the O-binding sites appear to be bent, and binding of O protein induces further bending.

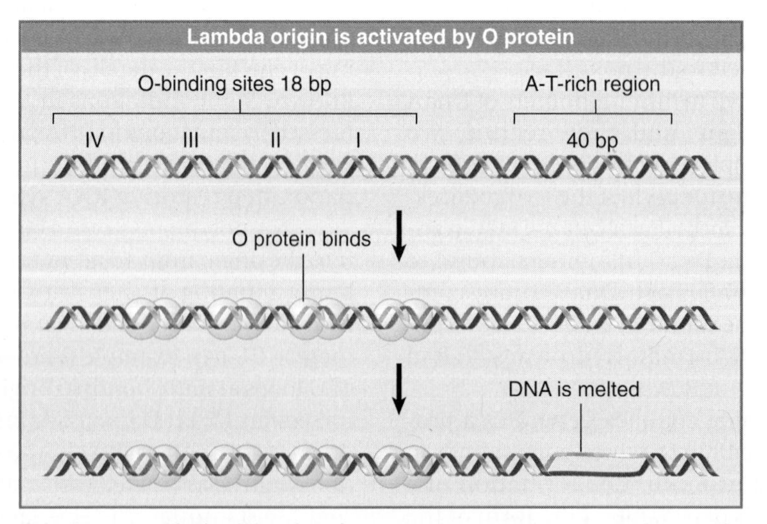

FIGURE 18.31 The lambda origin for replication comprises two regions. Early events are catalyzed by O protein, which binds to a series of four sites, and then DNA is melted in the adjacent A-T-rich region. The DNA is drawn as a straight duplex; it is, however, actually bent at the origin.

If the DNA is supercoiled, binding of O protein causes a structural change in the origin. The A-T-rich region immediately adjacent to the O-binding sites becomes susceptible to S1 nuclease, an enzyme that specifically recognizes unpaired DNA. This suggests that a melting reaction occurs next to the complex of O proteins.

The role of the O protein is analogous to that of DnaA at *oriC:* It prepares the origin for binding of DnaB. Lambda provides its own protein, P, which substitutes for DnaC and brings DnaB to the origin. When lambda P protein and bacterial DnaB proteins are added, the complex becomes larger and asymmetrical. It includes more DNA (a total of ~160 bp) as well as extra proteins. The λ P protein has a special role: It inhibits the helicase action of DnaB. Replication fork movement is triggered when P protein is released from the complex. Priming and DNA synthesis follow.

Some proteins are essential for replication without being directly involved in DNA synthesis as such. Interesting examples are provided by the DnaK and DnaJ proteins. DnaK is a chaperone, and is related to a common stress protein of eukaryotes. Its ability to interact with other proteins in a conformation-dependent manner plays a role in many cellular activities, including replication. The role of DnaK and DnaJ may be to disassemble the prepriming complex: by causing the release of P protein, they allow replication to begin.

The initiation reactions at *oriC* and oriλ are similar. The same stages are involved and rely upon overlapping components. The first step is recognition of the origin by a protein that binds to form a complex with the DNA—DnaA for *oriC* and O protein for *oriλ*. A short region of A-T-rich DNA is melted. DnaB is then loaded; this requires different functions at *oriC* and *oriλ* (still other proteins are required for this stage at other origins). When the helicase DnaB joins the complex, a replication fork is created. Finally an RNA primer is synthesized, after which replication begins.

The use of *oriC* and *oriλ* provides a general model for activation of origins. A similar series of events occurs at the origin of the virus SV40 in mammalian cells. Two hexamers of T antigen, a protein coded by the virus, bind to a series of repeated sites in DNA. In the presence of ATP, changes in DNA structure occur, which culminate in a melting reaction. In the case of SV40, the melted region is rather short and is not A-T-rich, but it has an unusual composition in which one strand consists almost exclusively of pyrimidines and the other of purines. Near this site is another essential region. It consists of A-T base pairs, at which the DNA is bent; the DNA is then underwound by the binding of T antigen. An interesting difference from the prokaryotic systems is that T antigen itself possesses the helicase activity needed to extend unwinding, so that an equivalent for DnaB is not needed.

18.17 The Primosome Is Needed to Restart Replication

Key concepts

- Initiation of φX replication requires the primosome complex to displace SSB from the origin.
- A replication fork stalls when it arrives at damaged DNA.
- After the damage has been repaired, the primosome is required to reinitiate replication.
- The Tus protein binds to *ter* sites and stops DnaB from unwinding DNA, which causes replication to terminate.

Early work on replication made extensive use of phage φX174, and led to the discovery of a complex system for priming. φX174 DNA is not by itself a substrate for the replication apparatus, because the naked DNA does not provide a suitable template. Once the single-stranded form has been coated with SSB, though, replication can proceed. A **primosome** assembles at a unique site on the single-stranded DNA called the assembly site *(pas)*. The *pas* is the equivalent of an origin for synthesis of the complementary strand of φX174. The primosome consists of six proteins: PriA, PriB, PriC, DnaT, DnaB, and DnaC. The key event in localizing the primosome is the ability of PriA to displace SSB from single-stranded DNA.

The primosome forms initially at the *pas* on φX174 DNA; however, primers are initiated at a variety of sites. PriA translocates along the DNA, displacing SSB, to reach additional sites at which priming occurs. As in *oriC* replicons, DnaB plays a key role in unwinding and priming in φX replicons. The role of PriA is to load DnaB to form a replication fork.

It has always been puzzling that φX origins should use a complex structure that is not required to replicate the bacterial chromosome. Why does the bacterium provide this complex?

The answer is provided by the fate of stalled replication forks. FIGURE 18.32 compares an advancing replication fork with what happens when there is damage to a base in the DNA or

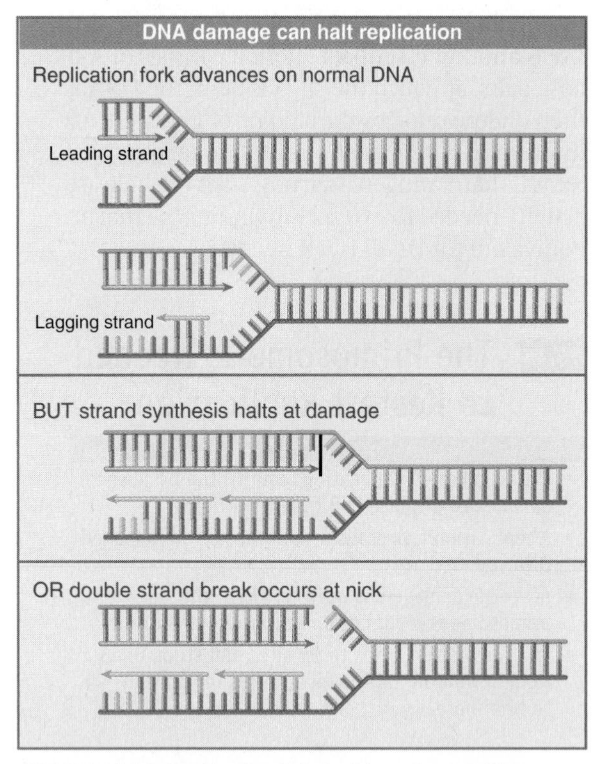

FIGURE 18.32 Replication is halted by a damaged base or a nick in DNA.

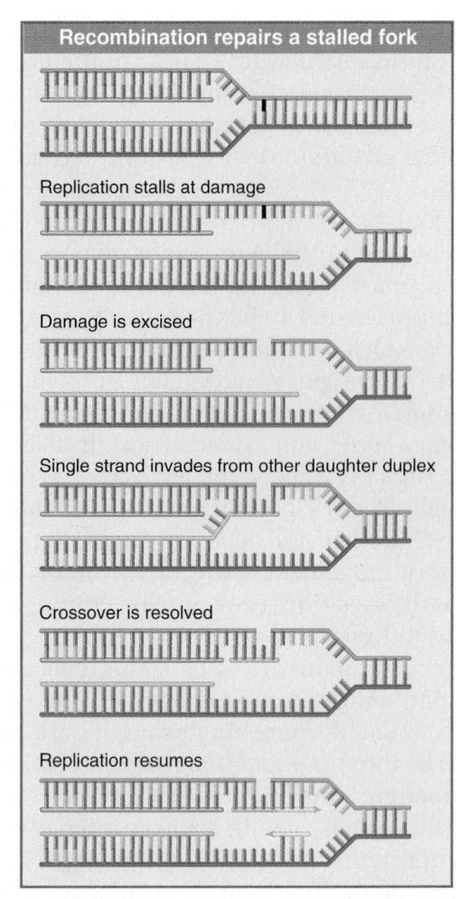

FIGURE 18.33 When replication halts at damaged DNA, the damaged sequence is excised and the complementary (newly synthesized) strand of the other daughter duplex crosses over to repair the gap. Replication can now resume, and the gaps are filled in.

a nick in one strand. In either case, DNA synthesis is halted, and the replication fork is either stalled or disrupted. It is not clear whether the components of the fork remain associated with the DNA or disassemble. Replication-fork stalling appears to be quite common; estimates for the frequency in *E. coli* suggest that 18% to 50% of bacteria encounter a problem during a replication cycle.

The situation is rescued by a recombination event that excises and replaces the damage or provides a new duplex to replace the region containing the double-strand break (see Figure 20.20 in Section 20.8, Recombination-Repair Systems in *E. coli*). The principle of the repair event is to use the built-in redundancy of information between the two DNA strands. FIGURE 18.33 shows the key events in such a repair event. Basically, information from the undamaged DNA daughter duplex is used to repair the damaged sequence. This creates a typical recombination junction that is resolved by the same systems that perform homologous recombination. In fact, one view is that the major importance of these systems for the cell is in repairing damaged DNA at stalled replication forks.

After the damage has been repaired, the replication fork must be restarted. FIGURE 18.34 shows that this may be accomplished by assembly of the primosome, which in effect reloads DnaB so that helicase action can continue.

Replication fork reactivation is a common (and therefore important) reaction. It may be required in most chromosomal replication cycles. It is impeded by mutations in either the retrieval systems that replace the damaged DNA or in the components of the primosome.

Replication forks must stop and disassemble at the termination of replication. How is this accomplished?

Sequences that stop movement of replication forks have been identified in the form of the *ter* elements of the *E. coli* chromosome (see FIGURE 18.35) or equivalent sequences in some plasmids. The common feature of these elements is a 23 bp consensus sequence that provides the binding site for the product of the *tus* gene, a 36 kD protein that is necessary for termination. Tus binds to the consensus sequence, where it provides a contra-helicase activity and stops DnaB from unwinding DNA. The leading strand continues to be synthesized right up to the *ter* ele-

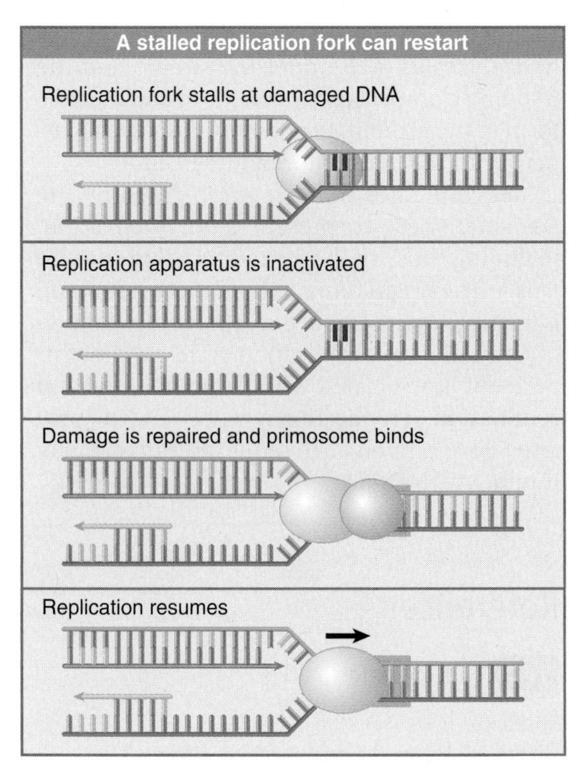

FIGURE 18.34 The primosome is required to restart a stalled replication fork after the DNA has been repaired.

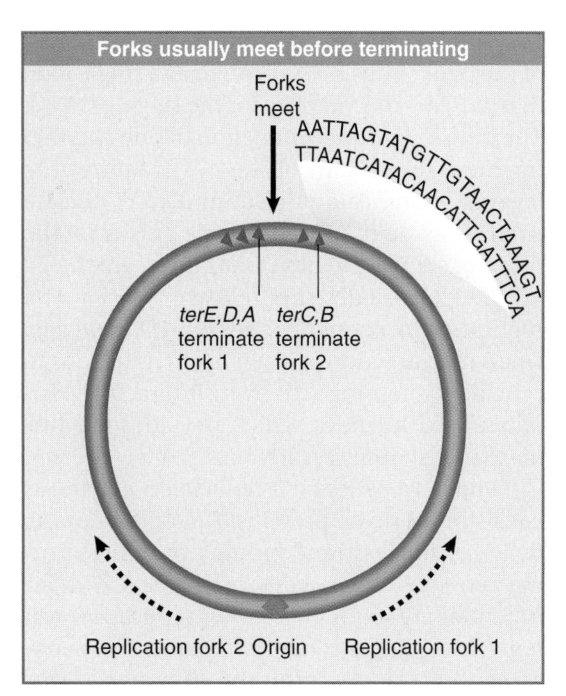

FIGURE 18.35 Replication termini in *E. coli* are located beyond the point at which the replication forks actually meet.

ment, whereas the nearest lagging strand is initiated 50 to100 bp before reaching *ter.*

The result of this inhibition is to halt movement of the replication fork and (presumably) to cause disassembly of the replication apparatus. FIGURE 18.36 reminds us that Tus stops the movement of a replication fork in only one direction. The crystal structure of a Tus–ter complex shows that the Tus protein binds to DNA asymmetrically; α-helices of the protein protrude around the double helix at the end that blocks the replication fork. Presumably a fork proceeding in the opposite direction can displace Tus and thus continue. A difficulty in understanding the function of the system *in vivo* is that it appears to be dispensable, because mutations in the *ter* sites or in *tus* are not lethal.

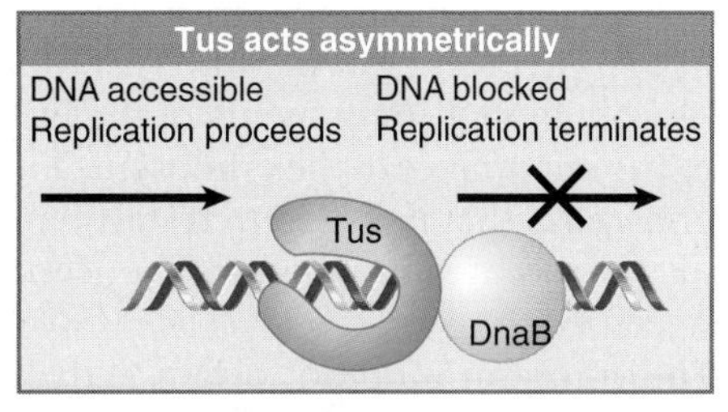

FIGURE 18.36 Tus binds to *ter* asymmetrically and blocks replication in only one direction.

18.18 Summary

DNA synthesis occurs by semidiscontinuous replication, in which the leading strand of DNA growing 5′–3′ is extended continuously, but the lagging strand that grows overall in the opposite 3′–5′ direction is made as short Okazaki fragments, each synthesized 5′–3′. The leading strand and each Okazaki fragment of the lagging strand initiate with an RNA primer that is extended by DNA polymerase. Bacteria and eukaryotes each possess more than one DNA polymerase activity. DNA polymerase III synthesizes both lagging and leading strands in *E. coli.* Many proteins are required for DNA polymerase III action and several constitute part of the replisome within which it functions.

The replisome contains an asymmetric dimer of DNA polymerase III; each new DNA strand is synthesized by a different core complex containing a catalytic (α) subunit. Processivity of the core complex is maintained by the β clamp, which forms a ring around DNA. The clamp is loaded onto DNA by the clamp loader complex. Clamp/clamp loader pairs with similar structural features are widely found in both prokaryotic and eukaryotic replication systems.

The looping model for the replication fork proposes that, as one half of the dimer advances to synthesize the leading strand, the other half of the dimer pulls DNA through as a single loop that provides the template for the lagging strand. The transition from completion of one Okazaki fragment to the start of the next requires the lagging strand catalytic subunit to dissociate from DNA and then reattach to a β clamp at the priming site for the next Okazaki fragment.

DnaB provides the helicase activity at a replication fork; this depends on ATP cleavage. DnaB may function by itself in *oriC* replicons to provide primosome activity by interacting periodically with DnaG, which provides the primase that synthesizes RNA.

Phage T4 codes for a replication apparatus consisting of seven proteins: DNA polymerase, helicase, single-strand binding protein, priming activities, and accessory proteins. Similar functions are required in other replication systems, including a HeLa cell system that replicates SV40 DNA. Different enzymes—DNA polymerase α and DNA polymerase δ—initiate and elongate the new strands of DNA.

The φX priming event also requires DnaB, DnaC, and DnaT. PriA is the component that defines the primosome assembly site *(pas)* for φX replicons; it displaces SSB from DNA in an action that involves cleavage of ATP. PriB and PriC are additional components of the primosome. The importance of the primosome for the bacterial cell is that it is used to restart replication at forks that stall when they encounter damaged DNA.

The common mode of origin activation involves an initial limited melting of the double helix, followed by more general unwinding to create single strands. Several proteins act sequentially at the *E. coli* origin. Replication is initiated at *oriC* in *E. coli* when DnaA binds to a series of 9 bp repeats. This is followed by binding to a series of 13 bp repeats, where it uses hydrolysis of ATP to generate the energy to separate the DNA strands. The prepriming complex of DnaC–DnaB displaces DnaA. DnaC is released in a reaction that depends on ATP hydrolysis; DnaB is joined by the replicase enzyme, and replication is initiated by two forks that set out in opposite directions. Similar events occur at the lambda origin, where phage proteins O and P are the counterparts of bacterial proteins DnaA and DnaC, respectively. In SV40 replication, several of these activities are combined in the functions of T antigen.

The availability of DnaA at the origin is an important component of the system that determines when replication cycles should initiate. Following initiation of replication, DnaA hydrolyzes its ATP under the stimulus of the β sliding clamp, thereby generating an inactive form of the protein. In addition, *oriC* must compete with the *dat* site for binding DnaA.

Several sites that are methylated by the Dam methylase are present in the *E. coli* origin, including those of the 13-mer binding sites for DnaA. The origin remains hemimethylated and is in a sequestered state for ~10 minutes following initiation of a replication cycle. During this period it is associated with the membrane and reinitiation of replication is repressed. The protein SeqA is involved in sequestration and may interact with DnaA.

References

18.1 Introduction

Research

Hirota, Y., Ryter, A., and Jacob, F. (1968). Thermosensitive mutants of *E. coli* affected in the processes of DNA synthesis and cellular division. *Cold Spring Harbor Symp. Quant. Biol.* 33, 677–693.

18.5 DNA Polymerases Have a Common Structure

Reviews

Hubscher, U., Maga, G., and Spadari, S. (2002). Eukaryotic DNA polymerases. *Annu. Rev. Biochem.* 71, 133–163.

Johnson, K. A. (1993). Conformational coupling in DNA polymerase fidelity. *Annu. Rev. Biochem.* 62, 685–713.

Joyce, C. M. and Steitz, T. A. (1994). Function and structure relationships in DNA polymerases. *Annu. Rev. Biochem.* 63, 777–822.

Research

Shamoo, Y. and Steitz, T. A. (1999). Building a replisome from interacting pieces: sliding clamp complexed to a peptide from DNA polymerase and a polymerase editing complex. *Cell* 99, 155–166.

18.7 The φX Model System Shows How Single-Stranded DNA Is Generated for Replication

Research

Dillingham, M. S., Wigley, D. B., and Webb, M. R. (2000). Demonstration of unidirectional single-stranded DNA translocation by PcrA helicase: measurement of step size and translocation speed. *Biochemistry* 39, 205–212.

Singleton, M. R., Sawaya, M. R., Ellenberger, T., and Wigley, D. B. (2000). Crystal structure of T7 gene 4 ring helicase indicates a mechanism for sequential hydrolysis of nucleotides. *Cell* 101, 589–600.

18.9 DNA Polymerase Holoenzyme Has Three Subcomplexes

Reference

Johnson, A., and O'Donnell, M. (2005). Cellular DNA replicases: components and dynamics at the replication fork. *Annu Rev Biochem.* 74, 283–315.

Research

Studwell-Vaughan, P. S. and O'Donnell, M. (1991). Constitution of the twin polymerase of DNA polymerase III holoenzyme. *J. Biol. Chem.* 266, 19833–19841.

Stukenberg, P. T., Studwell-Vaughan, P. S., and O'Donnell, M. (1991). Mechanism of the sliding beta-clamp of DNA polymerase III holoenzyme. *J. Biol. Chem.* 266, 11328–11334.

18.10 The Clamp Controls Association of Core Enzyme with DNA

Reviews

Benkovic, S. J., Valentine, A. M., and Salinas, F. (2001). Replisome-mediated DNA replication. *Annu. Rev. Biochem.* 70, 181–208.

Davey, M. J., Jeruzalmi, D., Kuriyan, J., and O'Donnell, M. (2002). Motors and switches: AAA+ machines within the replisome. *Nat. Rev. Mol. Cell Biol.* 3, 826–835.

Research

Bowman, G. D., O'Donnell, M., and Kuriyan, J. (2004). Structural analysis of a eukaryotic sliding DNA clamp-clamp loader complex. *Nature* 429, 724–730.

Jeruzalmi, D., O'Donnell, M., and Kuriyan, J. (2001). Crystal structure of the processivity clamp loader gamma (gamma) complex of *E. coli* DNA polymerase III. *Cell* 106, 429–441.

Kong, X. P., Onrust, R., O'Donnell, M., and Kuriyan, J. (1992). Three-dimensional structure of the beta subunit of *E. coli* DNA polymerase III holoenzyme: a sliding DNA clamp. *Cell* 69, 425–437.

18.11 Coordinating Synthesis of the Lagging and Leading Strands

Research

Dervyn, E., Suski, C., Daniel, R., Bruand, C., Chapuis, J., Errington, J., Janniere, L., and Ehrlich, S. D. (2001). Two essential DNA polymerases at the bacterial replication fork. *Science* 294, 1716–1719.

18.12 Okazaki Fragments Are Linked by Ligase

Review

Liu, Y., Kao, H. I., and Bambara, R. A. (2004). Flap endonuclease 1: a central component of DNA metabolism. *Annu. Rev. Biochem.* 73, 589–615.

18.13 Separate Eukaryotic DNA Polymerases Undertake Initiation and Elongation

Reviews

Goodman, M. F. (2002). Error-prone repair DNA polymerases in prokaryotes and eukaryotes. *Annu. Rev. Biochem.* 71, 17–50.

Hubscher, U., Maga, G., and Spadari, S. (2002). Eukaryotic DNA polymerases. *Annu. Rev. Biochem.* 71, 133–163.

Kaguni, L. S. (2004). DNA polymerase gamma, the mitochondrial replicase. *Annu. Rev. Biochem.* 73, 293–320.

Research

Karthikeyan, R., Vonarx, E. J., Straffon, A. F., Simon, M., Faye, G., and Kunz, B. A. (2000). Evidence from mutational specificity studies that yeast DNA polymerases delta and epsilon replicate different DNA strands at an intracellular replication fork. *J. Mol. Biol.* 299, 405–419.

Shiomi, Y., Usukura, J., Masamura, Y., Takeyasu, K., Nakayama, Y., Obuse, C., Yoshikawa, H., and Tsurimoto, T. (2000). ATP-dependent structural change of the eukaryotic clamp-loader protein, replication factor C. *Proc. Natl. Acad. Sci. USA* 97, 14127–14132.

Waga, S., Masuda, T., Takisawa, H., and Sugino, A. (2001). DNA polymerase epsilon is required for coordinated and efficient chromosomal DNA replication in *Xenopus* egg extracts. *Proc. Natl. Acad. Sci. USA* 98, 4978–4983.

Zuo, S., Bermudez, V., Zhang, G., Kelman, Z., and Hurwitz, J. (2000). Structure and activity associated with multiple forms of *S. pombe* DNA polymerase delta. *J. Biol. Chem.* 275, 5153–5162.

18.14 Phage T4 Provides Its Own Replication Apparatus

Research

Ishmael, F. T., Alley, S. C., and Benkovic, S. J. (2002). Assembly of the bacteriophage T4 helicase: architecture and stoichiometry of the gp41-gp59 complex. *J. Biol. Chem.* 277, 20555–20562.

Salinas, F., and Benkovic, S. J. (2000). Characterization of bacteriophage T4-coordinated leading- and lagging-strand synthesis on a minicircle substrate. *Proc. Natl. Acad. Sci. USA* 97, 7196–7201.

Schrock, R. D. and Alberts, B. (1996). Processivity of the gene 41 DNA helicase at the bacteriophage T4 DNA replication fork. *J. Biol. Chem.* 271, 16678–16682.

18.15 Creating the Replication Forks at an Origin

Research

Bramhill, D. and Kornberg, A. (1988). Duplex opening by dnaA protein at novel sequences in initiation of replication at the origin of the *E. coli* chromosome. *Cell* 52, 743–755.

Fuller, R. S., Funnell, B. E., and Kornberg, A. (1984). The dnaA protein complex with the *E. coli* chromosomal replication origin (oriC) and other DNA sites. *Cell* 38, 889–900.

Funnell, B. E. and Baker, T. A. (1987). In vitro assembly of a prepriming complex at the origin of the *E. coli* chromosome. *J. Biol. Chem.* 262, 10327–10334.

Sekimizu, K, Bramhill, D., and Kornberg, A. (1987). ATP activates dnaA protein in initiating replication of plasmids bearing the origin of the *E. coli* chromosome. *Cell* 50, 259–265.

Wahle, E., Lasken, R. S., and Kornberg, A. (1989). The dnaB-dnaC replication protein complex of *Escherichia coli*. II. Role of the complex in mobilizing dnaB functions. *J. Biol. Chem.* 264, 2469–2475.

18.17 The Primosome Is Needed to Restart Replication

Reviews

Cox, M. M. (2001). Recombinational DNA repair of damaged replication forks in *E. coli:* questions. *Annu. Rev. Genet.* 35, 53–82.

Cox, M. M., Goodman, M. F., Kreuzer, K. N., Sherratt, D. J., Sandler, S. J., and Marians, K. J. (2000). The importance of repairing stalled replication forks. *Nature* 404, 37–41.

Kuzminov, A. (1995). Collapse and repair of replication forks in *E. coli. Mol. Microbiol.* 16, 373–384.

McGlynn, P. and Lloyd, R. G. (2002). Recombinational repair and restart of damaged replication forks. *Nat. Rev. Mol. Cell Biol.* 3, 859–870.

Research

Seigneur, M., Bidnenko, V., Ehrlich, S. D., and Michel, B. (1998). RuvAB acts at arrested replication forks. *Cell* 95, 419–430.

19

Homologous and Site-Specific Recombination

CHAPTER OUTLINE

Continued on next page

19.13 Topoisomerases Relax or Introduce Supercoils in DNA

- Topoisomerases change the linking number by breaking bonds in DNA, changing the conformation of the double helix in space, and remaking the bonds.
- Type I enzymes act by breaking a single strand of DNA; type II enzymes act by making double-strand breaks.

19.14 Topoisomerases Break and Reseal Strands

- Type I topoisomerases function by forming a covalent bond to one of the broken ends, moving one strand around the other, and then transferring the bound end to the other broken end. Bonds are conserved, and as a result no input of energy is required.

19.15 Gyrase Functions by Coil Inversion

- *E. coli* gyrase is a type II topoisomerase that uses hydrolysis of ATP to provide energy to introduce negative supercoils into DNA.

19.16 Specialized Recombination Involves Specific Sites

- Specialized recombination involves reaction between specific sites that are not necessarily homologous.
- Phage lambda integrates into the bacterial chromosome by recombination between a site on the phage and the *att* site on the *E. coli* chromosome.
- The phage is excised from the chromosome by recombination between the sites at the end of the linear prophage.
- Phage lambda *int* codes for an integrase that catalyzes the integration reaction.

19.17 Site-Specific Recombination Involves Breakage and Reunion

- Cleavages staggered by 7 bp are made in both *attB* and *attP* and the ends are joined crosswise.

19.18 Site-Specific Recombination Resembles Topoisomerase Activity

- Integrases are related to topoisomerases, and the recombination reaction resembles topoisomerase action except that nicked strands from *different* duplexes are sealed together.
- The reaction conserves energy by using a catalytic tyrosine in the enzyme to break a phosphodiester bond and link to the broken 3′ end.
- Two enzyme units bind to each recombination site and the two dimers synapse to form a complex in which the transfer reactions occur.

19.19 Lambda Recombination Occurs in an Intasome

- Lambda integration takes place in a large complex that also includes the host protein IHF.
- The excision reaction requires Int and Xis and recognizes the ends of the prophage DNA as substrates.

19.20 Yeast Can Switch Silent and Active Loci for Mating Type

- The yeast mating type locus *MAT* has either the *MAT***a** or *MAT*α genotype.
- Yeast with the dominant allele *HO* switch their mating type at a frequency ~10^{-6}.
- The allele at *MAT* is called the active cassette.
- There are also two silent cassettes, *HML*α and *HMR***a**.
- Switching occurs if *MAT***a** is replaced by *HML*α or *MAT*α is replaced by *HMR***a**.

19.21 The *MAT* Locus Codes for Regulator Proteins

- In α-type haploids, *MAT*α turns on genes required to specify α-specific functions required for mating and turns off genes required for **a**-mating type.
- In **a**-type cells, *MAT***a** is not required.
- In diploids, **a***1* and α2 products cooperate to repress haploid-specific genes.

19.22 Silent Cassettes at *HML* and *HMR* Are Repressed

- *HML* and *HMR* are repressed by silencer elements.
- Loci required to maintain silencing include *SIR1-4, RAP1,* and genes for histone H4.
- Binding of ORC (origin recognition complex) at the silencers is necessary for inactivation.

19.23 Unidirectional Transposition Is Initiated by the Recipient *MAT* Locus

- Mating type switching is initiated by a double-strand break made at the *MAT* locus by the HO endonuclease.

19.24 Regulation of HO Expression Controls Switching

- HO endonuclease is synthesized in haploid mother cells, so that a switching event causes both daughters to have the new mating type.

19.25 Summary

19.1 Introduction

Evolution could not happen without genetic recombination. If it were not possible to exchange material between (homologous) chromosomes, the content of each individual chromosome would be irretrievably fixed in its particular alleles. When mutations occurred, it would not be possible to separate favorable and unfavorable changes. The length of the target for mutation damage would effectively be increased from the gene to the chromosome. Ultimately a chromosome would accumulate so many deleterious mutations that it would fail to function.

By shuffling the genes, recombination allows favorable and unfavorable mutations to be separated and tested as individual units in new assortments. It provides a means of escape and spreading for favorable alleles, and a means to eliminate an unfavorable allele without bringing down all the other genes with which this allele is associated. This is the basis for natural selection.

Recombination occurs between precisely corresponding sequences, so that not a single base pair is added to or lost from the recombinant chromosomes. Three types of recombination share the feature that the process involves physical exchange of material between duplex DNAs:

- Recombination involving reaction between homologous sequences of DNA is called *generalized* or **homologous recombination.** In eukaryotes, it occurs at meiosis, usually both in males (during spermatogenesis) and females (during oogenesis). We recall that it happens at the "four strand" stage of meiosis and involves only two of the four strands (see Section 2.7, Recombination Occurs by Physical Exchange of DNA).
- Another type of event sponsors recombination between specific pairs of sequences. This was first characterized in prokaryotes where **specialized recombination,** also known as **site-specific recombination,** is responsible for the integration of phage genomes into the bacterial chromosome. The recombination event involves specific sequences of the phage DNA and bacterial DNA, which include a short stretch of homology. The enzymes involved in this event act only on the particular pair of target sequences in an intermolecular reaction. Some related intramolecular reactions are responsible during bacterial division for regenerating two monomeric circular chromosomes when a dimer has been generated by generalized recombination. This latter class also includes recombination events that invert specific regions of the bacterial chromosome.
- A different type of event allows one DNA sequence to be inserted into another without relying on sequence homology. **Transposition** provides a means by which certain elements move from one chromosomal location to another. The mechanisms involved in transposition depend upon breakage and reunion of DNA strands, and thus are related to the processes of recombination (see Chapter 21, Transposons, and Chapter 22, Retroviruses and Retroposons).
- In special circumstances, gene rearrangement is used to control expression. Rearrangement may create new genes, which are needed for expression in particular circumstances, as in the case of the immunoglobulins. Rearrangement also may be responsible for switching expression from one preexisting gene to another, as in the example of yeast mating type, where the sequence at an active locus can be replaced by a sequence from a silent locus. Both these types of rearrangement share mechanistic similarities with transposition; in fact, they can be viewed as specially directed cases of transposition.
- Another type of recombination is used by RNA viruses, in which the polymerase switches from one template to another while it is synthesizing RNA. As a result, the newly synthesized molecule joins sequence information from two different parents. This type of mechanism for recombination is called **copy choice** and is discussed briefly in Section 22.4, Viral DNA Is Generated by Reverse Transcription.

Let's consider the nature and consequences of the generalized and specialized recombination reactions.

FIGURE 19.1 makes the point that generalized recombination occurs between two homologous DNA duplexes and can occur at any point along their length. The two chromosomes are cut at equivalent points, and then each is joined to the other to generate reciprocal recombinants.

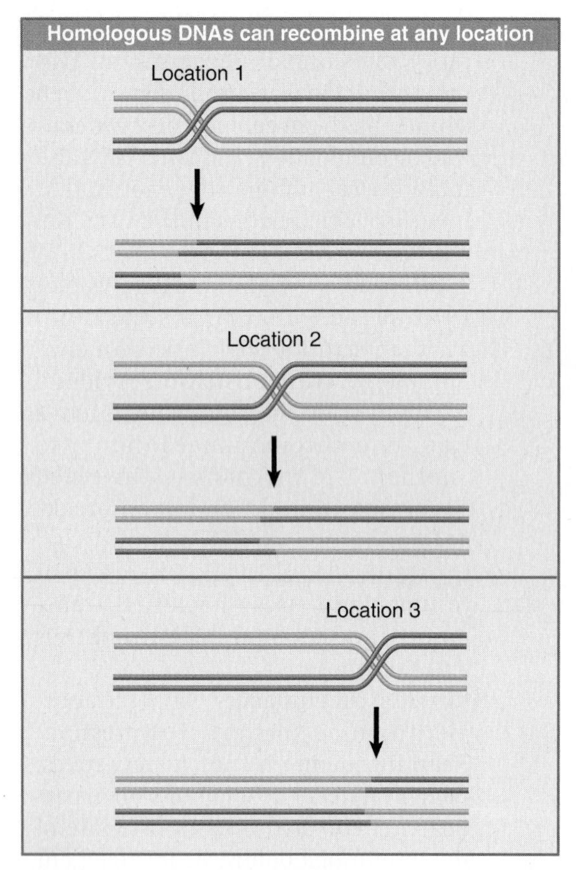

FIGURE 19.1 Generalized recombination can occur at any point along the lengths of two homologous DNAs.

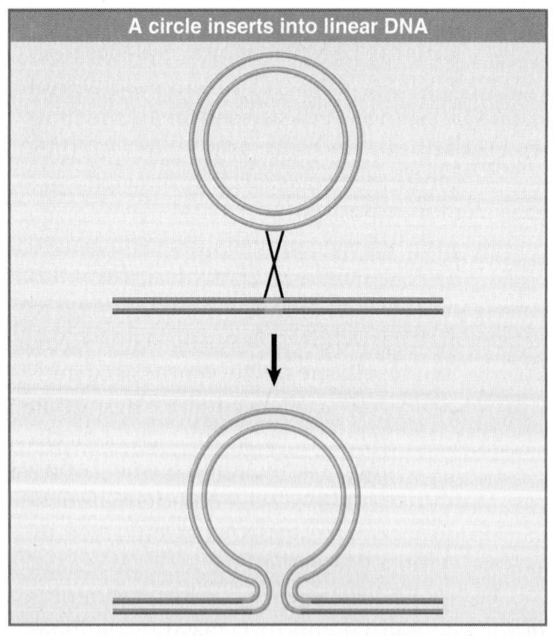

FIGURE 19.2 Site-specific recombination occurs between two specific sequences (identified in green). The other sequences in the two recombining DNAs are not homologous.

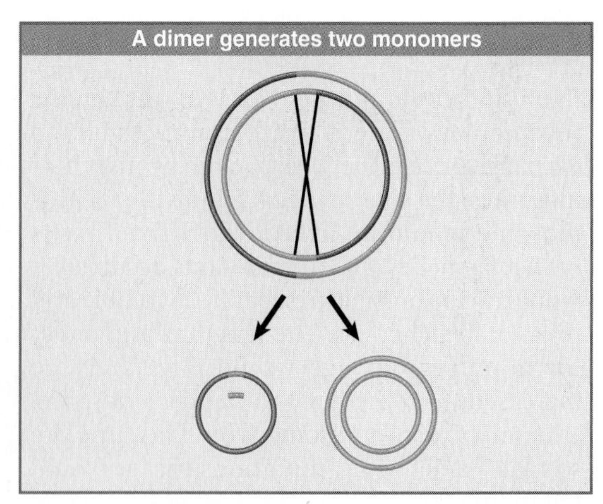

FIGURE 19.3 Site-specific recombination can be used to generate two monomeric circles from a dimeric circle.

The crossover (marked by the X) is the point at which each becomes joined to the other. There is no change in the overall organization of DNA; the products have the same structure as the parents, and both parents and products are homologous.

Specialized recombination occurs only between specific sites. The results depend on the locations of the two recombining sites. FIGURE 19.2 shows that an intermolecular recombination between a circular DNA and a linear DNA inserts the circular DNA into the linear DNA. FIGURE 19.3 shows that an intramolecular recombination between two sites on a circular DNA releases two smaller circular DNAs. Specialized recombination is often used to make changes such as these in the organization of DNA. The change in organization is a consequence of the locations of the recombining sites. We have a large amount of information about the enzymes that undertake specialized recombination, which are related to the topoisomerases that act to change the supercoiling of DNA in space.

19.2 Homologous Recombination Occurs between Synapsed Chromosomes

Key concepts

- Chromosomes must synapse (pair) in order for chiasmata to form where crossing over occurs.
- We can correlate the stages of meiosis with the molecular events that happen to DNA.

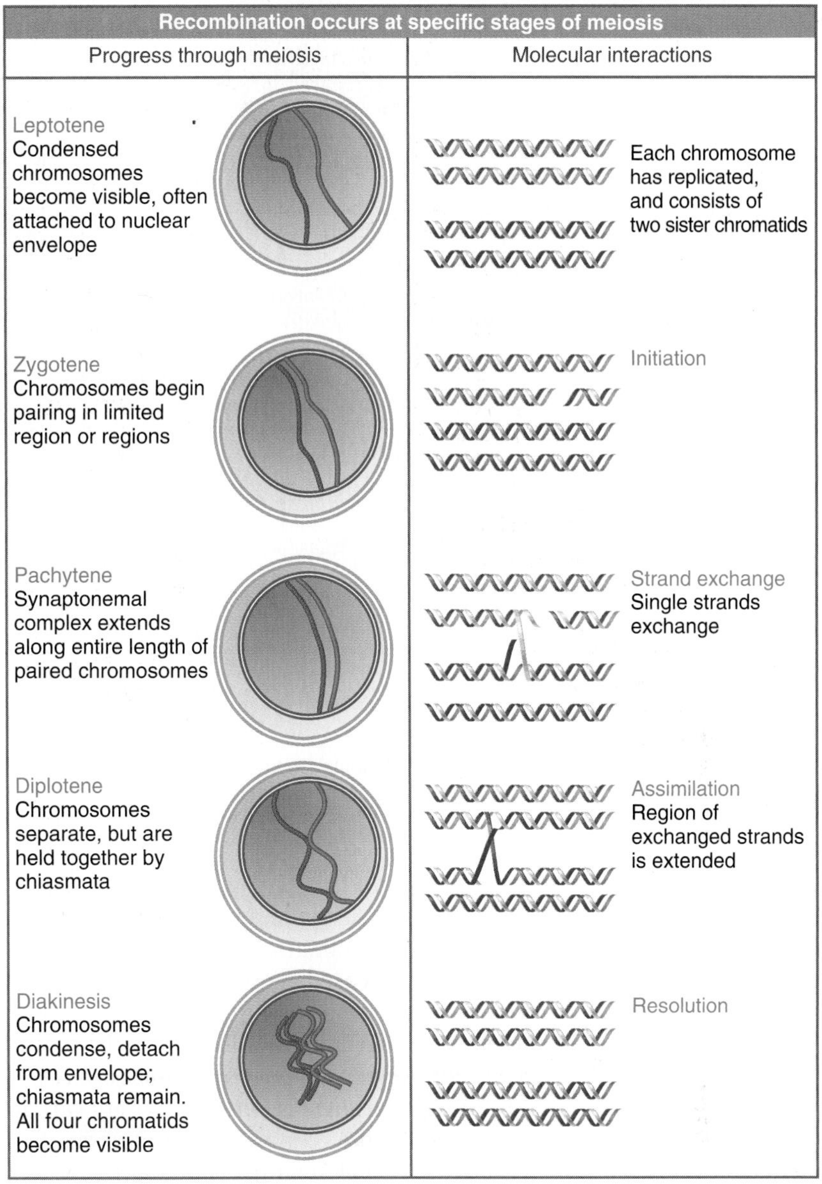

Homologous recombination is a reaction between two duplexes of DNA. Its critical feature is that the enzymes responsible can use any pair of homologous sequences as substrates (although some types of sequences may be favored over others). The frequency of recombination is not constant throughout the genome, but is influenced by both global and local effects. The overall frequency may be different in oocytes and in sperm; recombination occurs twice as frequently in female as in male humans. Within the genome, its frequency depends upon chromosome structure; for example, crossing-over is suppressed in the vicinity of the condensed and inactive regions of heterochromatin.

Recombination occurs during the protracted prophase of meiosis. FIGURE 19.4 compares the visible progress of chromosomes through the five stages of meiotic prophase with the molecular interactions that are involved in exchanging material between duplexes of DNA.

The beginning of meiosis is marked by the point at which individual chromosomes become visible. Each of these chromosomes has replicated previously and consists of two sister chromatids, each of which contains a duplex DNA. The homologous chromosomes approach one another and begin to pair in one or more regions, forming **bivalents.** Pairing extends until the entire length of each chromosome is apposed with its homolog. The process is called **synapsis** or **chromosome pairing.** When the process is completed, the chromosomes are laterally associated in the form of a **synaptonemal complex,** which has a characteristic structure in each species, although there is wide variation in the details between species.

Recombination between chromosomes involves a physical exchange of parts, usually represented as a **breakage and reunion,** in which two nonsister chromatids (each containing a duplex of DNA) have been broken and then linked with each other. When the chromosomes begin to separate, they can be seen to be held together at discrete sites called **chiasmata.** The number and distribution of chiasmata parallel the features of genetic crossing over. Traditional analysis holds that a chiasma represents the crossing-over event (FIGURE 19.5). The chiasmata remain visible when the chromosomes condense and all four chromatids become evident.

What is the molecular basis for these events? Each sister chromatid contains a single DNA duplex, so each bivalent contains four duplex molecules of DNA. Recombination requires a mechanism that allows the duplex DNA of one sister chromatid to interact with the duplex DNA of a sister chromatid from the other chromosome. It must be possible for this reaction to occur between any pair of corresponding sequences in the two molecules in a highly specific manner that allows material to be exchanged with precision at the level of the individual base pair.

Recombination occurs at specific stages of meiosis

Progress through meiosis	Molecular interactions
Leptotene Condensed chromosomes become visible, often attached to nuclear envelope	Each chromosome has replicated, and consists of two sister chromatids
Zygotene Chromosomes begin pairing in limited region or regions	Initiation
Pachytene Synaptonemal complex extends along entire length of paired chromosomes	Strand exchange Single strands exchange
Diplotene Chromosomes separate, but are held together by chiasmata	Assimilation Region of exchanged strands is extended
Diakinesis Chromosomes condense, detach from envelope; chiasmata remain. All four chromatids become visible	Resolution

FIGURE 19.4 Recombination occurs during the first meiotic prophase. The stages of prophase are defined by the appearance of the chromosomes, each of which consists of two replicas (sister chromatids), although the duplicated state becomes visible only at the end. The molecular interactions of any individual crossing-over event involve two of the four duplex DNAs.

We know of only one mechanism for nucleic acids to recognize one another on the basis of sequence: complementarity between single strands. Figure 19.5 shows a general model for the involvement of single strands in

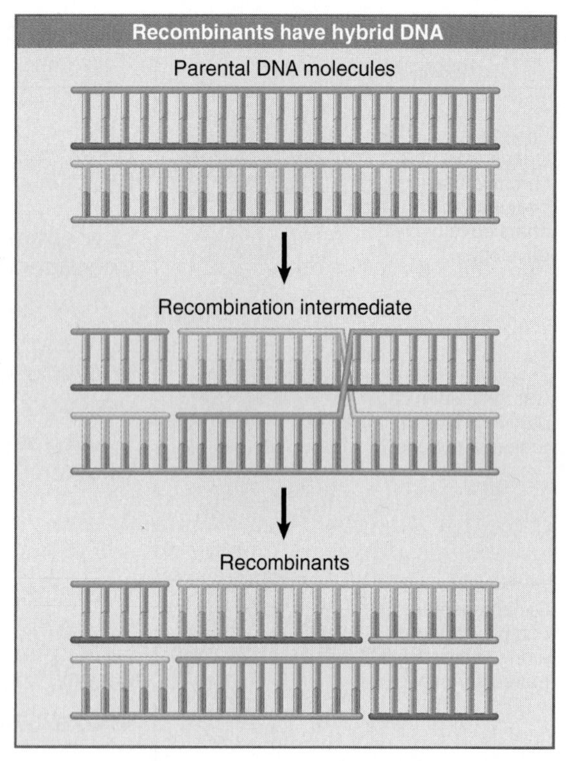

FIGURE 19.5 Recombination involves pairing between complementary strands of the two parental DNAs.

recombination. The first step in providing single strands is to make a break in each DNA duplex. One or both of the strands of that duplex can then be released. If (at least) one strand displaces the corresponding strand in the other duplex, the two duplex molecules will be specifically connected at corresponding sequences. If the strand exchange is extended, there can be more extensive connection between the duplexes. By exchanging both strands and later cutting them, it is possible to connect the parental duplex molecules by means of a crossover that corresponds to the demands of a breakage and reunion.

We cannot at this juncture relate these molecular events rigorously with the changes that are observed at the level of the chromosomes. There is no detailed information about the molecular events involved in recombination in higher eukaryotic cells (in which meiosis has been most closely observed). Recently, however, the isolation of mutants in yeast has made it possible to correlate some of the molecular steps with approximate stages of meiosis. Detailed information about the recombination process is available in bacteria, in which molecular activities are known that cause genetic exchange between duplex molecules. The bacterial reaction involves interaction between restricted regions of the genome, though, rather than an entire pairing of genomes. The synapsis of eukaryotic chromosomes remains the most difficult stage to explain at the molecular level.

19.3 Breakage and Reunion Involves Heteroduplex DNA

Key concepts

- The key event in recombination between two duplex DNA molecules is exchange of single strands.
- When a single strand from one duplex displaces its counterpart in the other duplex, it creates a branched structure.
- The exchange generates a stretch of heteroduplex DNA consisting of one strand from each parent.
- Two (reciprocal) exchanges are necessary to generate a joint molecule.
- The joint molecule is resolved into two separate duplex molecules by nicking two of the connecting strands.
- Whether recombinants are formed depends on whether the strands involved in the original exchange or the other pair of strands are nicked during resolution.

The act of connecting two duplex molecules of DNA is at the heart of the recombination process. Our molecular analysis of recombination therefore starts by expanding our view of the use of base pairing between complementary single strands in recombination. It is useful to imagine the recombination reaction in terms of single-strand exchanges (although we shall see that this is not necessarily how it is actually initiated), because the properties of the molecules created in this way are central to understanding the processes involved in recombination.

FIGURE 19.6 illustrates a process that starts with breakage at the corresponding points of the homologous strands of two paired DNA duplexes. The breakage allows movement of the free ends created by the nicks. Each strand leaves its partner and crosses over to pair with its complement in the other duplex.

The reciprocal exchange creates a connection between the two DNA duplexes. The connected pair of duplexes is called a **joint molecule.** The point at which an individual strand of DNA crosses from one duplex to the other is called the **recombinant joint.**

At the site of recombination, each duplex has a region consisting of one strand from each

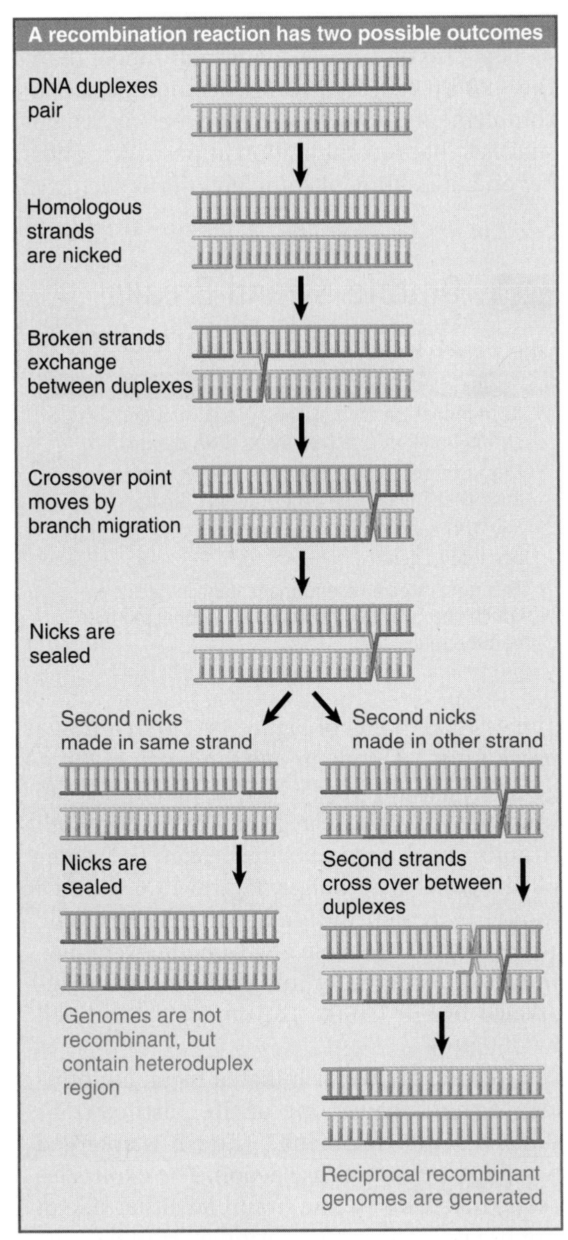

FIGURE 19.6 Recombination between two paired duplex DNAs could involve reciprocal single-strand exchange, branch migration, and nicking.

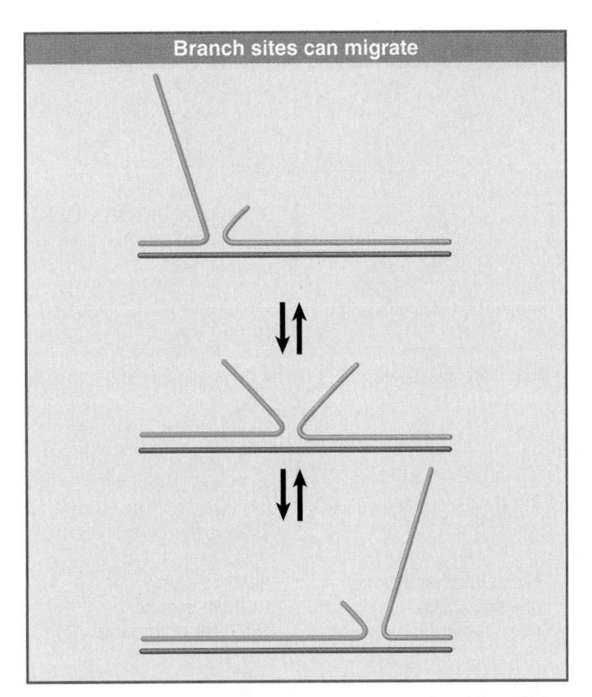

FIGURE 19.7 Branch migration can occur in either direction when an unpaired single strand displaces a paired strand.

of the parental DNA molecules. This region is called **hybrid DNA** or **heteroduplex DNA.**

An important feature of a recombinant joint is its ability to move along the duplex. Such mobility is called **branch migration.** FIGURE 19.7 illustrates the migration of a single strand in a duplex. The branching point can migrate in either direction as one strand is displaced by the other.

Branch migration is important for both theoretical and practical reasons. As a matter of principle, it confers a dynamic property on recombining structures. As a practical feature, its existence means that the point of branching cannot be established by examining a molecule *in vitro* (because the branch may have migrated since the molecule was isolated).

Branch migration could allow the point of crossover in the recombination intermediate to move in either direction. The rate of branch migration is uncertain, but as seen *in vitro* is probably inadequate to support the formation of extensive regions of heteroduplex DNA in natural conditions. Any extensive branch migration *in vivo* must therefore be catalyzed by a recombination enzyme.

The joint molecule formed by strand exchange must be *resolved* into two separate duplex molecules. **Resolution** requires a further pair of nicks. We can most easily visualize the outcome by viewing the joint molecule in one plane as a **Holliday** junction. This is illustrated in FIGURE 19.8, which represents the structure of Figure 19.6 with one duplex rotated relative to the other. The outcome of the reaction depends on which pair of strands is nicked.

If the nicks are made in the pair of strands that were not originally nicked (the pair that did not initiate the strand exchange), all four of the original strands have been nicked. This releases **splice recombinant** DNA molecules. The duplex of one DNA parent is covalently linked to the duplex of the other DNA parent via a stretch of heteroduplex DNA. There has

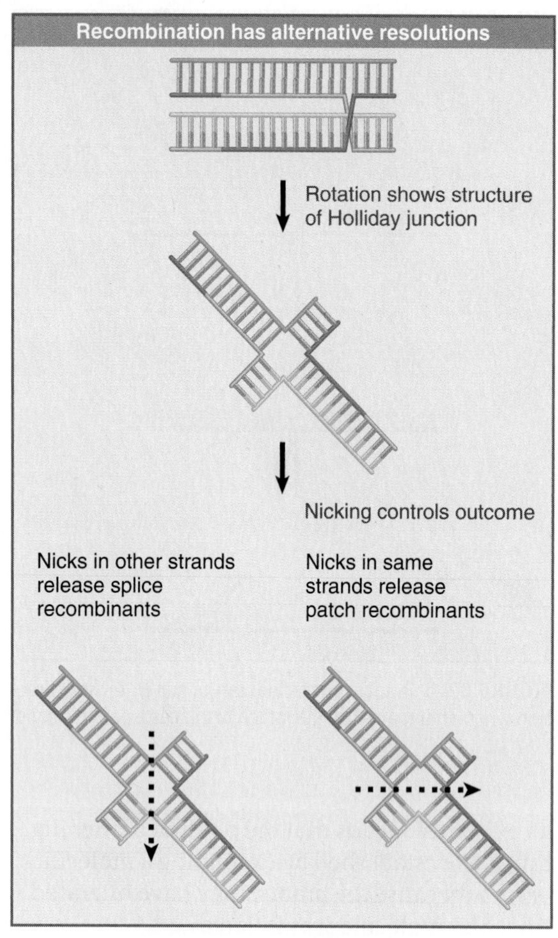

FIGURE 19.8 Resolution of a Holliday junction can generate parental or recombinant duplexes, depending on which strands are nicked. Both types of product have a region of heteroduplex DNA.

been a conventional recombination event between markers located on either side of the heteroduplex region.

If the same two strands involved in the original nicking are nicked again, the other two strands remain intact. The nicking releases the original parental duplexes, which remain intact with the exception that each has a residuum of the event in the form of a length of heteroduplex DNA. These are called **patch recombinants.**

These alternative resolutions of the joint molecule establish the principle that *a strand exchange between duplex DNAs always leaves behind a region of heteroduplex DNA, but the exchange may or may not be accompanied by recombination of the flanking regions.*

What is the minimum length of the region required to establish the connection between the recombining duplexes? Experiments in which short homologous sequences carried by plasmids or phages are introduced into bacteria suggest that the rate of recombination is substantially reduced if the homologous region is <75 bp. This distance is appreciably longer than the ~10 bp required for association between complementary single-stranded regions, which suggests that recombination imposes demands beyond annealing of complements as such.

19.4 Double-Strand Breaks Initiate Recombination

Key concepts

- Recombination is initiated by making a double-strand break in one (recipient) DNA duplex.
- Exonuclease action generates 3′-single-stranded ends that invade the other (donor) duplex.
- New DNA synthesis replaces the material that has been degraded.
- This generates a recombinant joint molecule in which the two DNA duplexes are connected by heteroduplex DNA.

The general model of Figure 19.4 shows that a break must be made in one duplex in order to generate a point from which single strands can unwind to participate in genetic exchange. Both strands of a duplex must be broken to accomplish a genetic exchange. Figure 19.6 shows a model in which individual breaks in single strands occur successively. Genetic exchange, however, is actually initiated by a **double-strand break** (DSB). The model is illustrated in FIGURE 19.9.

Recombination is initiated by an endonuclease that cleaves one of the partner DNA duplexes, the "recipient." The cut is enlarged to a gap by exonuclease action. The exonuclease(s) nibble away one strand on either side of the break, generating 3′ single-stranded termini. One of the free 3′ ends then invades a homologous region in the other ("donor") duplex. This is called single-strand invasion. The formation of heteroduplex DNA generates a D loop, in which one strand of the donor duplex is displaced. The D loop is extended by repair DNA synthesis, using the free 3′ end as a primer to generate double-stranded DNA.

Eventually the D loop becomes large enough to correspond to the entire length of the gap on the recipient chromatid. When the extruded single strand reaches the far side of the gap, the complementary single-stranded sequences anneal. Now there is heteroduplex DNA on either side of the gap, and the gap itself is represented by the single-stranded D loop.

The duplex integrity of the gapped region can be restored by repair synthesis using the 3′ end on the left side of the gap as a primer. Overall, the gap has been repaired by two individual rounds of single-strand DNA synthesis.

Branch migration converts this structure into a molecule with two recombinant joints. The joints must be resolved by cutting.

If both joints are resolved in the same way, the original noncrossover molecules will be released, each with a region of altered genetic information that is a footprint of the exchange event. If the two joints are resolved in opposite ways, a genetic crossover is produced.

The structure of the two-jointed molecule before it is resolved illustrates a critical difference between the double-strand break model and models that invoke only single-strand exchanges.

- Following the double-strand break, heteroduplex DNA has been formed at each end of the region involved in the exchange. Between the two heteroduplex segments is the region corresponding to the gap, which now has the sequence of the donor DNA in both molecules (Figure 19.9). Thus the arrangement of heteroduplex sequences is asymmetric, and part of one molecule has been converted to the sequence of the other (which is why the initiating chromatid is called the recipient).
- Following reciprocal single-strand exchange, each DNA duplex has heteroduplex material covering the region from the initial site of exchange to the migrating branch (Figure 19.6). In variants of the single-strand exchange model in which some DNA is degraded and resynthesized, the initiating chromatid is the donor of genetic information.

The double-strand break model does not reduce the importance of the formation of heteroduplex DNA, which remains the only plausible means by which two duplex molecules can interact. By shifting the responsibility for initiating recombination from single-strand to double-strand breaks, though, it influences our perspective about the ability of the cell to manipulate DNA.

The involvement of double-strand breaks at first seems surprising. Once a break has been made right across a DNA molecule, there is no going back. Compare the events of Figure 19.6 and Figure 19.9. At no point in the single-strand exchange model has any information been lost.

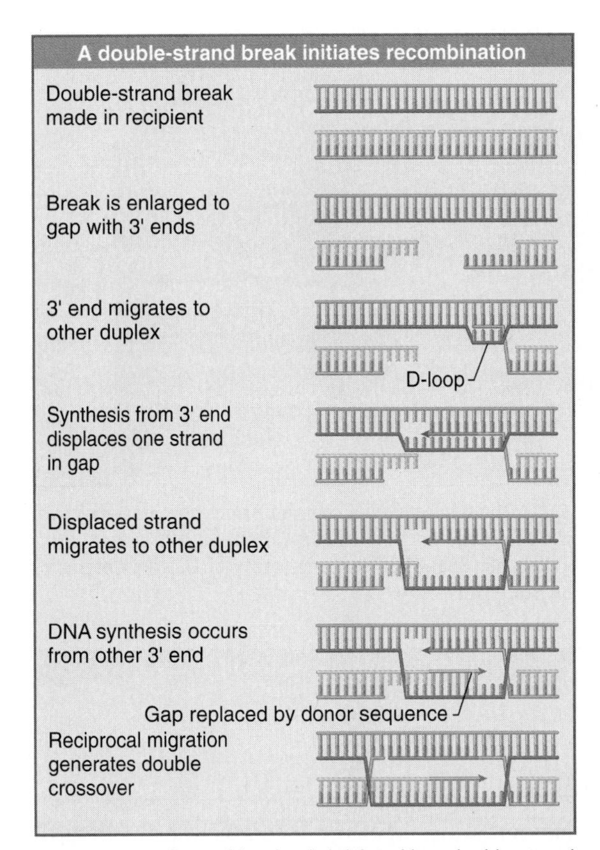

FIGURE 19.9 Recombination is initiated by a double-strand break, followed by formation of single-stranded 3′ ends, one of which migrates to a homologous duplex.

In the double-strand break model, though, the initial cleavage is immediately followed by loss of information. Any error in retrieving the information could be fatal. On the other hand, the very ability to retrieve lost information by resynthesizing it from another duplex provides a major safety net for the cell.

19.5 Recombining Chromosomes Are Connected by the Synaptonemal Complex

Key concepts

- During the early part of meiosis, homologous chromosomes are paired in the synaptonemal complex.
- The mass of chromatin of each homolog is separated from the other by a proteinaceous complex.

A basic paradox in recombination is that the parental chromosomes never seem to be in close enough contact for recombination of DNA to

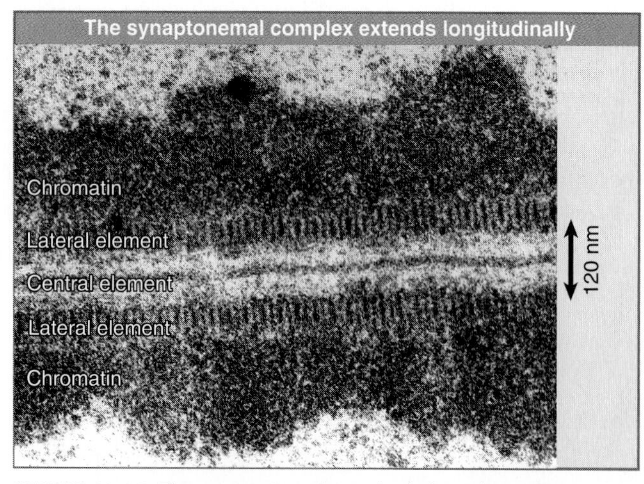

FIGURE 19.10 The synaptonemal complex brings chromosomes into juxtaposition. Reproduced from D. von Wettstein. 1971. *Proc. Natl. Acad. Sci. USA*. 68: 851–855. Photo courtesy of D. von Wettstein, Washington State University.

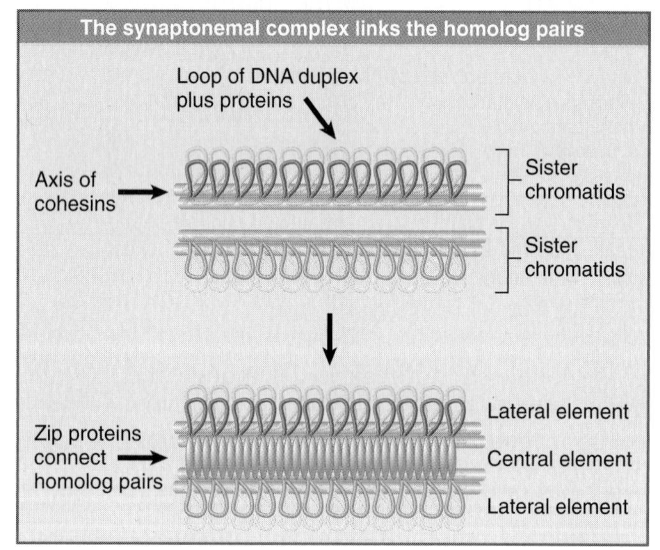

FIGURE 19.11 Each pair of sister chromatids has an axis made of cohesins. Loops of chromatin project from the axis. The synaptonemal complex is formed by linking together the axes via zip proteins.

occur. The chromosomes enter meiosis in the form of replicated (sister chromatid) pairs, which are visible as a mass of chromatin. They pair to form the **synaptonemal complex,** and it has been assumed for many years that this represents some stage involved with recombination—possibly a necessary preliminary to exchange of DNA. A more recent view is that the synaptonemal complex is a consequence rather than a cause of recombination, but, we have yet to define how the structure of the synaptonemal complex relates to molecular contacts between DNA molecules.

Synapsis begins when each chromosome (sister chromatid pair) condenses around a structure called the **axial element,** which is apparently proteinaceous. The axial elements of corresponding chromosomes then become aligned, and the synaptonemal complex forms as a tripartite structure, in which the axial elements, now called **lateral elements,** are separated from each other by a **central element.** FIGURE 19.10 shows an example.

Each chromosome at this stage appears as a mass of chromatin bounded by a lateral element. The two lateral elements are separated from each other by a fine, but dense, central element. The triplet of parallel dense strands lies in a single plane that curves and twists along its axis. The distance between the homologous chromosomes is considerable in molecular terms, at more than 200 nm (the diameter of DNA is 2 nm). Thus a major problem in understanding the role of the complex is that, although it aligns homologous chromosomes, it is far from bringing homologous DNA molecules into contact.

The only visible link between the two sides of the synaptonemal complex is provided by spherical or cylindrical structures observed in fungi and insects. They lie across the complex and are called **nodes** or **recombination nodules;** they occur with the same frequency and distribution as the chiasmata. Their name reflects the hope that they may prove to be the sites of recombination.

From mutations that affect synaptonemal complex formation, we can relate the types of proteins that are involved to its structure. FIGURE 19.11 presents a molecular view of the synaptonemal complex. Its distinctive structural features are due to two groups of proteins:

- The cohesins form a single linear axis for each pair of sister chromatids from which loops of chromatin extend. This is equivalent to the lateral element of Figure 19.10. (The cohesins belong to a general group of proteins involved in connecting sister chromatids so that they segregate properly at mitosis of meiosis.)
- The lateral elements are connected by transverse filaments that are equivalent to the central element of Figure 19.10. These are formed from Zip proteins.

Mutations in proteins that are needed for lateral elements to form are found in the genes coding for cohesins. The cohesins that are used in meiosis include Smc3p (which is also used in mitosis) and Rec8p (which is specific to meiosis and is related to the mitotic cohesin Scc1p). The cohesins appear to bind to specific sites along the chromosomes in both mitosis and meiosis. They are likely to play a structural role in chromosome segregation. At meiosis, the formation of the lateral elements may be necessary for the later stages of recombination, because although these mutations do not prevent the formation of double-strand breaks, they do block formation of recombinants.

The *zip1* mutation allows lateral elements to form and to become aligned, but they do not become closely synapsed. The N-terminal domain of Zip1 protein is localized in the central element, but the C-terminal domain is localized in the lateral elements. Two other proteins, Zip2 and Zip3, are also localized with Zip1. The group of Zip proteins form transverse filaments that connect the lateral elements of the sister chromatid pairs.

19.6 The Synaptonemal Complex Forms after Double-Strand Breaks

Key concepts

- Double-strand breaks that initiate recombination occur before the synaptonemal complex forms.
- If recombination is blocked, the synaptonemal complex cannot form.

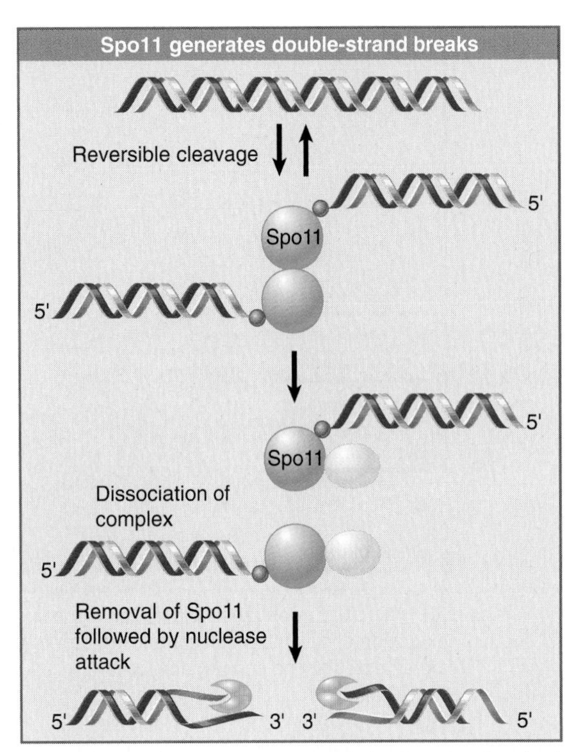

FIGURE 19.12 Spo11 is covalently joined to the 5′ ends of double-strand breaks.

There is good evidence in yeast that double-strand breaks initiate recombination in both homologous and site-specific recombination. Double-strand breaks were initially implicated in the change of mating type, which involves the replacement of one sequence by another (see Section 19.23, Unidirectional Transposition Is Initiated by the Recipient *MAT* Locus). Double-strand breaks also occur early in meiosis at sites that provide hotspots for recombination. Their locations are not sequence specific. They tend to occur in promoter regions and in general to coincide with more accessible regions of chromatin. The frequency of recombination declines in a gradient on one or both sides of the hotspot. The hotspot identifies the site at which recombination is initiated, and the gradient reflects the probability that the recombination events will spread from it.

We may now interpret the role of double-strand breaks in molecular terms. The blunt ends created by the double-strand break are rapidly converted on both sides into long 3′ single-stranded ends, as shown in the model of Figure 19.9. A yeast mutation *(rad50)* that blocks the conversion of the flush end into the single-stranded protrusion is defective in recombination. This suggests that double-strand breaks are necessary for recombination. The gradient is determined by the declining probability that a single-stranded region will be generated as distance increases from the site of the double-strand break.

In *rad50* mutants, the 5′ ends of the double-strand breaks are connected to the protein Spo11, which is homologous to the catalytic subunits of a family of type II topoisomerases. This suggests that Spo11 may be a topoisomerase-like enzyme that generates the double-strand breaks. The model for this reaction shown in FIGURE 19.12 suggests that Spo11 interacts reversibly with DNA; the break is converted into a permanent structure by an interaction with another protein that dissociates the Spo11 complex. Removal of Spo11 is then followed by nuclease action. At least nine other proteins are required to process the double-strand breaks. One group of proteins is required to convert the double-strand breaks into protruding 3′–OH single-stranded ends. Another group then enables the single-stranded ends to invade homologous duplex DNA.

The correlation between recombination and synaptonemal complex formation is well established, and recent work has shown that all mutations that abolish chromosome pairing in

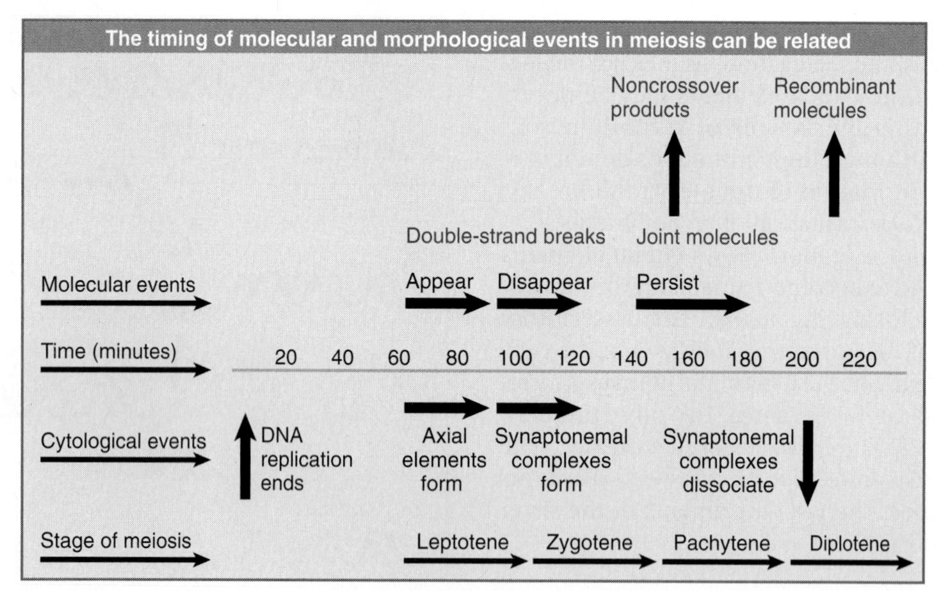

FIGURE 19.13 Double-strand breaks appear when axial elements form and disappear during the extension of synaptonemal complexes. Joint molecules appear and persist until DNA recombinants are detected at the end of pachytene.

Drosophila or in yeast also prevent recombination. The system for generating the double-strand breaks that initiate recombination is generally conserved. Spo11 homologs have been identified in several higher eukaryotes, and a mutation in the *Drosophila* gene blocks all meiotic recombination.

There are few systems in which it is possible to compare molecular and cytological events at recombination, but recently there has been progress in analyzing meiosis in *Saccharomyces cerevisiae.* The relative timing of events is summarized in FIGURE 19.13.

Double-strand breaks appear and then disappear over a 60-minute period. The first joint molecules, which are putative recombination intermediates, appear soon after the double-strand breaks disappear. The sequence of events suggests that double-strand breaks, individual pairing reactions, and formation of recombinant structures occur in succession at the same chromosomal site.

Double-strand breaks appear during the period when axial elements form. They disappear during the conversion of the paired chromosomes into synaptonemal complexes. This relative timing of events suggests that formation of the synaptonemal complex results from the initiation of recombination via the introduction of double-strand breaks and their conversion into later intermediates of recombination. This idea is supported by the observation that the *rad50* mutant cannot convert axial elements into synaptonemal complexes. This refutes the traditional view of meiosis that the synaptonemal complex represents the need for chromosome pairing to precede the molecular events of recombination.

It has been difficult to determine whether recombination occurs at the stage of synapsis, because recombination is assessed by the appearance of recombinants after the completion of meiosis. By assessing the appearance of recombinants in yeast directly in terms of the production of DNA molecules containing diagnostic restriction sites, though, it has been possible to show that recombinants appear at the end of pachytene. This clearly places the completion of the recombination event after the formation of synaptonemal complexes.

Thus the synaptonemal complex forms after the double-strand breaks that initiate recombination, and it persists until the formation of recombinant molecules. It does not appear to be necessary for recombination as such, because some mutants that lack a normal synaptonemal complex can generate recombinants. Mutations that abolish recombination, however, also fail to develop a synaptonemal complex. This suggests that the synaptonemal complex forms as a consequence of recombination, following chromosome pairing, and is required for later stages of meiosis.

Ever since the model for recombination via a Holliday structure was proposed, it has been assumed that the resolution of this structure gives rise to either noncrossover products (with a residual stretch of hybrid DNA) or to crossovers (recombinants), depending on which strands are involved in resolution (see Figure 19.8). Recent measurements of the times of production of noncrossover and crossover molecules, however, suggest that this may not be true. Crossovers do not appear until well after the first appearance of joint molecules, whereas noncrossovers appear almost simultaneously with the joint molecules (see Figure 19.13). If both types of product were produced by the same resolution process, however, we would expect them to appear at the same time. The discrepancy in timing suggests that crossovers are produced as previously thought—by resolution of joint molecules—but that there may be some other route for the production of noncrossovers.

19.7 Pairing and Synaptonemal Complex Formation Are Independent

Key concept

- Mutations can occur in either chromosome pairing or synaptonemal complex formation without affecting the other process.

We can distinguish the processes of pairing and synaptonemal complex formation by the effects of two mutations, each of which blocks one of the processes without affecting the other.

The *zip2* mutation allows chromosomes to pair, but they do not form synaptonemal complexes. Thus recognition between homologs is independent of recombination or synaptonemal complex formation.

The specificity of association between homologous chromosomes is controlled by the gene *hop2* in *S. cerevisiae*. In *hop2* mutants, normal amounts of synaptonemal complex form at meiosis, but the individual complexes contain nonhomologous chromosomes. This suggests that the formation of synaptonemal complexes as such is independent of homology (and therefore cannot be based on any extensive comparison of DNA sequences). The usual role of Hop2 is to prevent nonhomologous chromosomes from interacting.

Double-strand breaks form in the mispaired chromosomes in the synaptonemal complexes of *hop2* mutants, but they are not repaired. This suggests that, if formation of the synaptonemal complex requires double-strand breaks, it does not require any extensive reaction of these breaks with homologous DNA.

It is not clear what usually happens during pachytene, before DNA recombinants are observed. It may be that this period is occupied by the subsequent steps of recombination, which involve the extension of strand exchange, DNA synthesis, and resolution.

At the next stage of meiosis (diplotene), the chromosomes shed the synaptonemal complex; the chiasmata then become visible as points at which the chromosomes are connected. This has been presumed to indicate the occurrence of a genetic exchange, but the molecular nature of a chiasma is unknown. It is possible that it represents the residuum of a completed exchange, or that it represents a connection between homologous chromosomes where a genetic exchange has not yet been resolved. Later in meiosis, the chiasmata move toward the ends of the chromosomes. This flexibility suggests that they represent some remnant of the recombination event rather than providing the actual intermediate.

Recombination events occur at discrete points on meiotic chromosomes, but we cannot as yet correlate their occurrence with the discrete structures that have been observed, that is, recombination nodules and chiasmata. Insights into the molecular basis for the formation of discontinuous structures, however, are provided by the identification of proteins involved in yeast recombination that can be localized to discrete sites. These include MSH4 (which is related to bacterial proteins involved in mismatch repair) and Dmc1 and Rad51 (which are homologs of the *E. coli* RecA protein). The exact roles of these proteins in recombination remain to be established.

Recombination events are subject to a general control. Only a minority of interactions actually mature as crossovers, but these are distributed in such a way that, in general, each pair of homologs acquires only one to two crossovers, yet the probability of zero crossovers for a homolog pair is very low (<0.1%). This process is probably the result of a single **crossover control,** because the nonrandomness of crossovers is generally disrupted in certain mutants. Furthermore, the occurrence of recombination is necessary for progress through meiosis, and a "checkpoint" system exists to block meiosis if recombination has not occurred. (The block is lifted when recombination has been successfully completed; this system provides a safeguard to ensure that cells do not try to segregate their chromosomes until recombination has occurred.)

19.8 The Bacterial RecBCD System Is Stimulated by *chi* Sequences

Key concepts

- The RecBCD complex has nuclease and helicase activities.
- It binds to DNA downstream of a *chi* sequence, unwinds the duplex, and degrades one strand from 3′–5′ as it moves to the *chi* site.
- The *chi* site triggers loss of the RecD subunit and nuclease activity.

The nature of the events involved in exchange of sequences between DNA molecules was first described in bacterial systems. Here the recognition reaction is part and parcel of the recombination mechanism and involves restricted regions of DNA molecules rather than intact chromosomes. The general order of molecular events is similar, though: A single strand from a broken molecule interacts with a partner duplex; the region of pairing is extended; and an endonuclease resolves the partner duplexes. Enzymes involved in each stage are known, although they probably represent only some of the components required for recombination.

Bacterial enzymes implicated in recombination have been identified by the occurrence of ***rec***$^-$ mutations in their genes. The phenotype of *rec*$^-$ mutants is the inability to undertake generalized recombination. Some ten to twenty loci have been identified.

Bacteria do not usually exchange large amounts of duplex DNA, but there may be various routes to initiate recombination in prokaryotes. In some cases, DNA may be available with free single-stranded 3′ ends: DNA may be provided in single-stranded form (as in conjugation; see Section 16.10, Conjugation Transfers Single-Stranded DNA); single-stranded gaps may be generated by irradiation damage; or single-stranded tails may be generated by phage genomes undergoing replication by a rolling circle. In circumstances involving two duplex molecules (as in recombination at meiosis in eukaryotes), however, single-stranded regions and 3′ ends must be generated.

One mechanism for generating suitable ends has been discovered as a result of the existence of certain hotspots that stimulate recombination. These hotspots, which were discovered in phage lambda in the form of mutants called *chi,* have single base-pair changes that create sequences that stimulate recombination. These sites lead us to the role of other proteins involved in recombination.

These sites share a constant nonsymmetrical sequence of 8 bp:

5′ GCTGGTGG 3′
3′ CGACCACC 5′

The *chi* sequence occurs naturally in *E. coli* DNA about once every 5 to 10 kb. Its absence from wild-type lambda DNA, and also from other genetic elements, shows that it is not essential for recombination.

A *chi* sequence stimulates recombination in its general vicinity, say within a distance of up to 10 kb from the site. A *chi* site can be activated by a double-strand break made several kb away on one particular side (to the right of the sequence shown above). This dependence on orientation suggests that the recombination apparatus must associate with DNA at a broken end, and then can move along the duplex only in one direction.

chi sites are targets for the action of an enzyme coded by the genes *recBCD.* This complex exercises several activities. It is a potent nuclease that degrades DNA, which was originally identified as the activity exonuclease V. It has helicase activities that can unwind duplex DNA in the presence of a single-strand binding protein (SSB), and it has an ATPase activity. Its role in recombination may be to provide a single-stranded region with a free 3′ end.

FIGURE 19.14 shows how these reactions are coordinated on a substrate DNA that has a *chi* site. RecBCD binds to DNA at a double-stranded end. Two of its subunits have helicase activities: RecD functions with 5′–3′ polarity, and RecB functions with 3′–5′ polarity. Translocation along DNA and unwinding the double helix is initially driven by the RecD subunit. As RecBCD advances, it degrades the released single strand with the 3′ end. When it reaches the *chi* site, it recognizes the top strand of the *chi* site in single-stranded form. This causes the enzyme to pause. It then cleaves the top strand of the DNA at a position between four and six bases to the right of *chi.* Recognition of the *chi* site causes the RecD subunit to dissociate or become inactivated, at which point the enzyme loses its nuclease activity. It continues, however, to function as a helicase—now using only the RecB subunit to drive translocation—at about half the previous speed. The overall result of this interaction is to generate single-stranded DNA with a 3′ end at the *chi* sequence. This is a substrate for recombination.

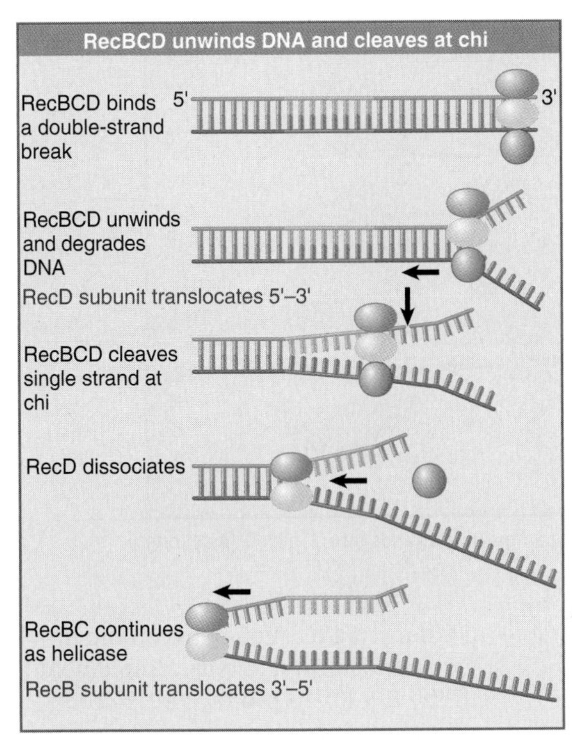

FIGURE 19.14 RecBCD nuclease approaches a *chi* sequence from one side, degrading DNA as it proceeds; at the *chi* site, it makes an endonucleolytic cut, loses RecD, and retains only the helicase activity.

19.9 Strand-Transfer Proteins Catalyze Single-Strand Assimilation

Key concept

- RecA forms filaments with single-stranded or duplex DNA and catalyzes the ability of a single-stranded DNA with a free 3′ to displace its counterpart in a DNA duplex.

The *E. coli* protein RecA was the first example to be discovered of a DNA strand-transfer protein. It is the paradigm for a group that includes several other bacterial and archaeal proteins (Rad51 in *S. cerevisiae*) and the higher eukaryotic protein Dmc1. Analysis of yeast *rad51* mutants shows that this class of protein plays a central role in recombination. They accumulate double-strand breaks and fail to form normal synaptonemal complexes. This reinforces the idea that exchange of strands between DNA duplexes is involved in formation of the synaptonemal complex, and raises the possibility that chromosome synapsis is related to the bacterial strand assimilation reaction.

RecA in bacteria has two quite different types of activity: It can stimulate protease activity in the SOS response (see Section 20.10, RecA Triggers the SOS System), and can promote base pairing between a single strand of DNA and its complement in a duplex molecule. Both activities are activated by single-stranded DNA in the presence of ATP.

The DNA-handling activity of RecA enables a single strand to displace its homolog in a duplex in a reaction that is called **single-strand uptake** or **single-strand assimilation.** The displacement reaction can occur between DNA molecules in several configurations and has three general conditions:

- One of the DNA molecules must have a single-stranded region.
- One of the molecules must have a free 3′ end.
- The single-stranded region and the 3′ end must be located within a region that is complementary between the molecules.

The reaction is illustrated in FIGURE 19.15. When a linear single strand invades a duplex, it displaces the original partner to its complement. The reaction can be followed most easily by making either the donor or recipient a circular molecule. The reaction proceeds 5′–3′ along the strand whose partner is being displaced and replaced, that is, the reaction involves an exchange in which (at least) one of the exchanging strands has a free 3′ end.

Single-strand assimilation is potentially related to the initiation of recombination. All models call for an intermediate in which one or both single strands cross over from one duplex to the other (see Figure 19.6 and Figure 19.9). RecA could catalyze this stage of the reaction. In the bacterial context, RecA acts on substrates generated by RecBCD. RecBCD-mediated unwinding and cleavage can be used to generate ends that initiate the formation of heteroduplex joints. RecA can take the single strand with the 3′ end that is released when RecBCD cuts at *chi*, and can use it to react with a homologous duplex sequence, thus creating a joint molecule.

All of the bacterial and archaeal proteins in the RecA family can aggregate into long filaments with single-stranded or duplex DNA. There are six RecA monomers per turn of the filament, which has a helical structure with a deep groove that contains the DNA. The stoichiometry of binding is three nucleotides (or base pairs) per RecA monomer. The DNA is held in a form that is extended 1.5 times relative to duplex B DNA, making a turn every 18.6 nucleotides (or base pairs). When duplex DNA

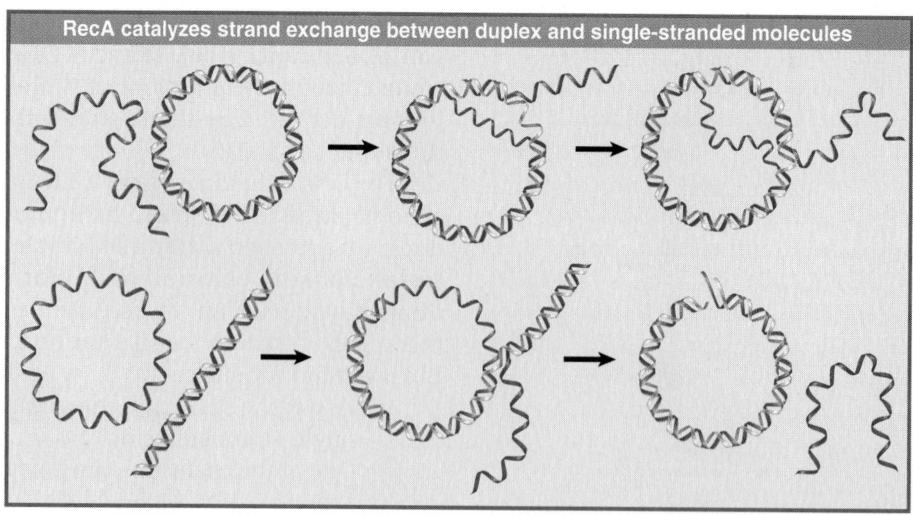

FIGURE 19.15 RecA promotes the assimilation of invading single strands into duplex DNA so long as one of the reacting strands has a free end.

is bound, it contacts RecA via its minor groove, leaving the major groove accessible for possible reaction with a second DNA molecule.

The interaction between two DNA molecules occurs within these filaments. When a single strand is assimilated into a duplex, the first step is for RecA to bind the single strand into a filament. The duplex is then incorporated, probably forming some sort of triple-stranded structure. In this system, synapsis precedes physical exchange of material, because the pairing reaction can take place even in the absence of free ends, when strand exchange is impossible. A free 3′ end is required for strand exchange. The reaction occurs within the filament, and RecA remains bound to the strand that was originally single, so that at the end of the reaction RecA is bound to the duplex molecule.

All of the proteins in this family can promote the basic process of strand exchange without a requirement for energy input. RecA, however, augments this activity by using ATP hydrolysis. Large amounts of ATP are hydrolyzed during the reaction. The ATP may act through an allosteric effect on RecA conformation. When bound to ATP, the DNA-binding site of RecA has a high affinity for DNA; this is needed to bind DNA and for the pairing reaction. Hydrolysis of ATP converts the binding site to low affinity, which is needed to release the heteroduplex DNA.

We can divide the reaction that RecA catalyzes between single-stranded and duplex DNA into three phases:

- a slow presynaptic phase in which RecA polymerizes on single-stranded DNA;
- a fast pairing reaction between the single-stranded DNA and its complement in the duplex to produce a heteroduplex joint; and
- a slow displacement of one strand from the duplex to produce a long region of heteroduplex DNA.

The presence of SSB stimulates the reaction, by ensuring that the substrate lacks secondary structure. It is not clear yet how SSB and RecA both can act on the same stretch of DNA. Like SSB, RecA is required in stoichiometric amounts, which suggests that its action in strand assimilation involves binding cooperatively to DNA to form a structure related to the filament.

When a single-stranded molecule reacts with a duplex DNA, the duplex molecule becomes unwound in the region of the recombinant joint. The initial region of heteroduplex DNA may not even lie in the conventional double helical form, but could consist of the two strands associated side by side. A region of this type is called a **paranemic joint** (compared with the classical intertwined **plectonemic** relationship of strands in a double helix). A paranemic joint is unstable; further progress of the reaction requires its conversion to the double-helical form. This reaction is equivalent to removing negative supercoils and may require an enzyme that solves the unwinding/rewinding problem by making transient breaks that allow the strands to rotate about each other.

All of the reactions we have discussed so far represent only a part of the potential recombination event: the invasion of one duplex by a single strand. Two duplex molecules can inter-

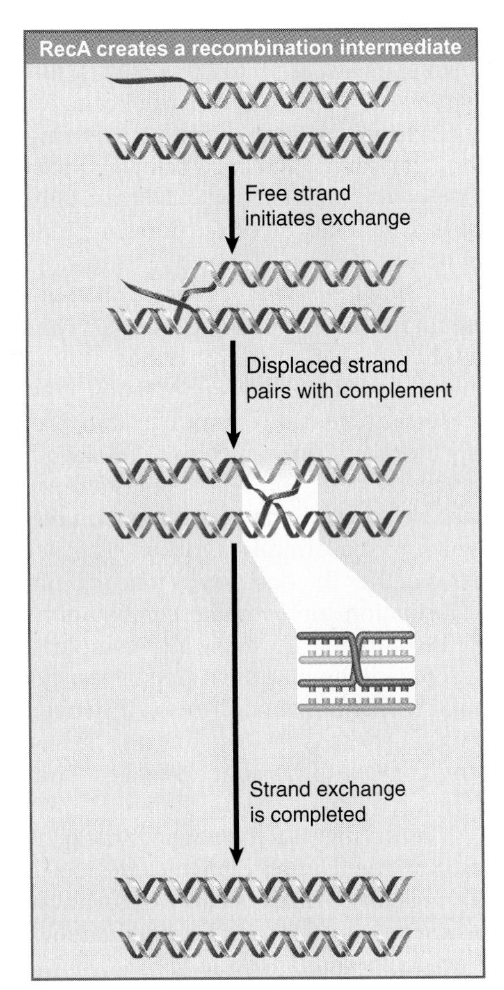

FIGURE 19.16 RecA-mediated strand exchange between partially duplex and entirely duplex DNA generates a joint molecule with the same structure as a recombination intermediate.

act with each other under the sponsorship of RecA, provided that one of them has a single-stranded region of at least fifty bases. The single-stranded region can take the form of a tail on a linear molecule or of a gap in a circular molecule.

The reaction between a partially duplex molecule and an entirely duplex molecule leads to the exchange of strands. An example is illustrated in FIGURE 19.16. Assimilation starts at one end of the linear molecule, where the invading single strand displaces its homolog in the duplex in the customary way. When the reaction reaches the region that is duplex in both molecules, though, the invading strand unpairs from its partner, which then pairs with the other displaced strand.

At this stage, the molecule has a structure indistinguishable from the recombinant joint in Figure 19.8. The reaction sponsored *in vitro* by RecA can generate Holliday junctions, which suggests that the enzyme can mediate reciprocal strand transfer. We know less about the geometry of four-strand intermediates bound by RecA, but presumably two duplex molecules can lie side by side in a way consistent with the requirements of the exchange reaction.

The biochemical reactions characterized *in vitro* leave open many possibilities for the functions of strand-transfer proteins *in vivo*. Their involvement is triggered by the availability of a single-stranded 3′ end. In bacteria, this is most likely generated when RecBCD processes a double-strand break to generate a single-stranded end. One of the main circumstances in which this is invoked may be when a replication fork stalls at a site of DNA damage (see Section 20.9, Recombination Is an Important Mechanism to Recover from Replication Errors). The introduction of DNA during conjugation, when RecA is required for recombination with the host chromosome, is more closely related to conventional recombination. In yeast, double-strand breaks may be generated by DNA damage or as part of the normal process of recombination. In either case, processing of the break to generate a 3′ single-stranded end is followed by loading the single strand into a filament with Rad51, followed by a search for matching duplex sequences. This can be used in both repair and recombination reactions.

19.10 The Ruv System Resolves Holliday Junctions

Key concepts

- The Ruv complex acts on recombinant junctions.
- RuvA recognizes the structure of the junction and RuvB is a helicase that catalyzes branch migration.
- RuvC cleaves junctions to generate recombination intermediates.

One of the most critical steps in recombination is the resolution of the Holliday junction, which determines whether there is a reciprocal recombination or a reversal of the structure that leaves only a short stretch of hybrid DNA (see Figure 19.6 and Figure 19.8). Branch migration from the exchange site (see Figure 19.7) determines the length of the region of hybrid DNA (with or without recombination).The proteins involved in stabilizing and resolving Holliday junctions have been identified as the products of the *ruv* genes in *E. coli*. RuvA and RuvB increase the formation of heteroduplex structures. RuvA recognizes the structure of the

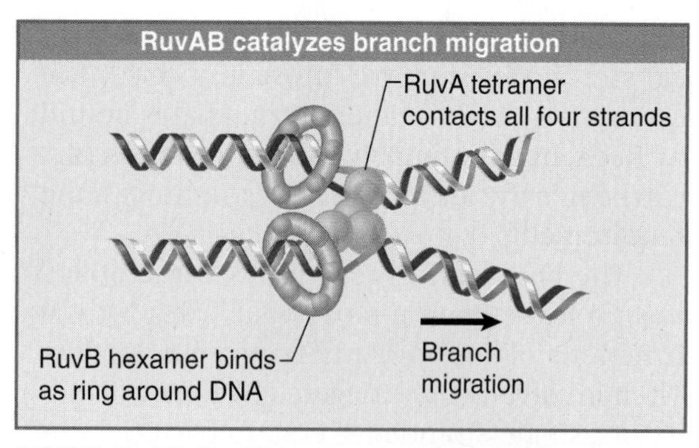

FIGURE 19.17 RuvAB is an asymmetric complex that promotes branch migration of a Holliday junction.

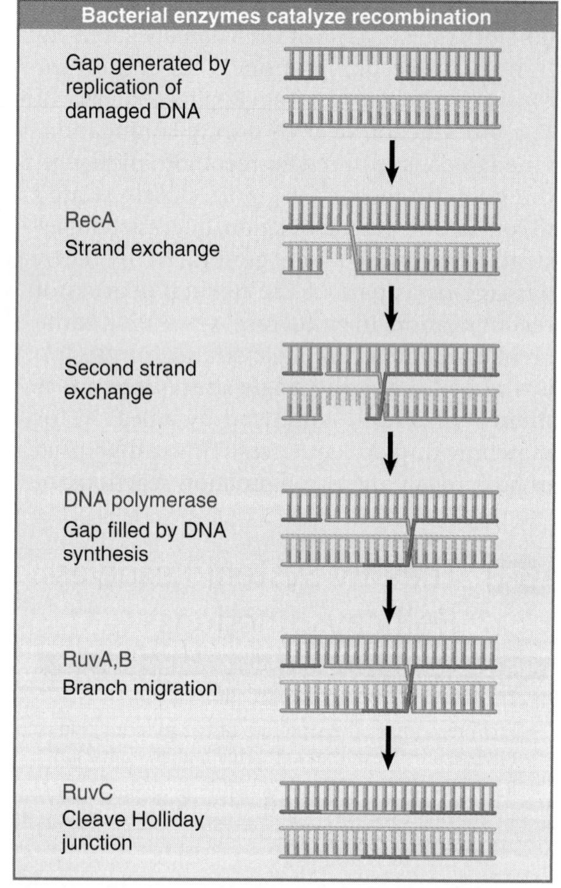

FIGURE 19.18 Bacterial enzymes can catalyze all stages of recombination in the repair pathway following the production of suitable substrate DNA molecules.

Holliday junction. RuvA binds to all four strands of DNA at the crossover point and forms two tetramers that sandwich the DNA. RuvB is a hexameric helicase with an ATPase activity that provides the motor for branch migration. Hexameric rings of RuvB bind around each duplex of DNA upstream of the crossover point. A diagram of the complex is shown in FIGURE 19.17.

The RuvAB complex can cause the branch to migrate as fast as 10 to 20 bp/sec. A similar activity is provided by another helicase, RecG. RuvAB displaces RecA from DNA during its action. The RuvAB and RecG activities both can act on Holliday junctions, but if both are mutant, *E. coli* is completely defective in recombination activity.

The third gene, *ruvC*, codes for an endonuclease that specifically recognizes Holliday junctions. It can cleave the junctions *in vitro* to resolve recombination intermediates. A common tetranucleotide sequence provides a hotspot for RuvC to resolve the Holliday junction. The tetranucleotide (ATTG) is asymmetric, and thus may direct resolution with regard to which pair of strands is nicked. This determines whether the outcome is patch recombinant formation (no overall recombination) or splice recombinant formation (recombination between flanking markers). Crystal structures of RuvC and other junction-resolving enzymes show that there is little structural similarity among the group, in spite of their common function.

All of this suggests that recombination uses a "resolvasome" complex that includes enzymes catalyzing branch migration as well as junction-resolving activity. It is possible that mammalian cells contain a similar complex.

We may now account for the stages of recombination in *E. coli* in terms of individual proteins. FIGURE 19.18 shows the events that are involved in using recombination to repair a gap in one duplex by retrieving material from the other duplex. The major caveat in applying these conclusions to recombination in eukaryotes is that bacterial recombination generally involves interaction between a fragment of DNA and a whole chromosome. It occurs as a repair reaction that is stimulated by damage to DNA, but this is not entirely equivalent to recombination between genomes at meiosis. Nonetheless, similar molecular activities are involved in manipulating DNA.

Another system of resolvases has been characterized in yeast and mammals. Mutants in *S. cerevisiae mus81* are defective in recombination. Mus81 is a component of an endonuclease that resolves Holliday junctions into duplex structures. The resolvase is important both in meiosis and for restarting stalled replication forks (see Section 20.9, Recombination Is an Important Mechanism to Recover from Replication Errors).

19.11 Gene Conversion Accounts for Interallelic Recombination

Key concepts

- Heteroduplex DNA that is created by recombination can have mismatched sequences where the recombining alleles are not identical.
- Repair systems may remove mismatches by changing one of the strands so its sequence is complementary to the other.

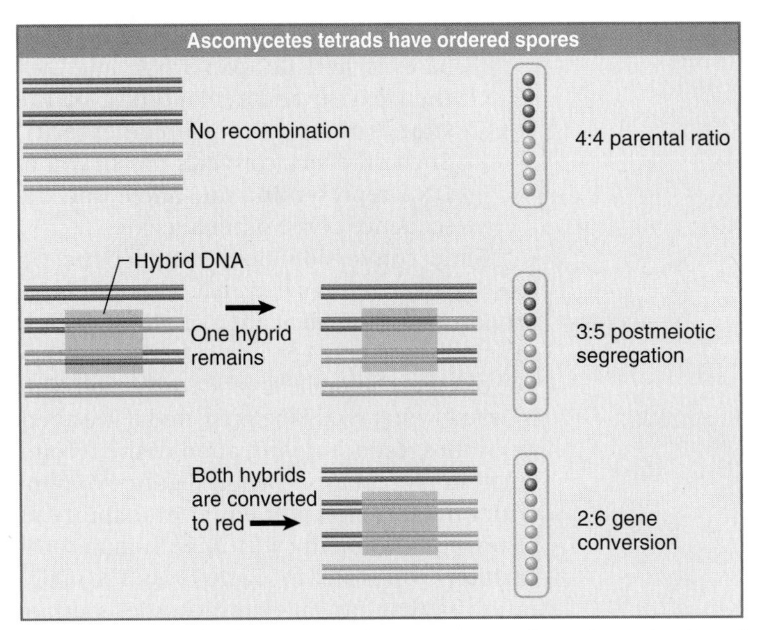

FIGURE 19.19 Spore formation in the ascomycetes allows determination of the genetic constitution of each of the DNA strands involved in meiosis.

The involvement of heteroduplex DNA explains the characteristics of recombination between alleles; indeed, allelic recombination provided the impetus for the development of the heteroduplex model. When recombination between alleles was discovered, the natural assumption was that it takes place by the same mechanism of reciprocal recombination that applies to more distant loci. That is to say, an individual breakage and reunion event occurs within the locus to generate a reciprocal pair of recombinant chromosomes. In the close quarters of a single gene, however, the formation of heteroduplex DNA itself is usually responsible for the recombination event.

Individual recombination events can be studied in the ascomycetes fungi, because the products of a single meiosis are held together in a large cell called the **ascus** (or less commonly, the **tetrad**). Even better is that in some fungi, the four haploid nuclei produced by meiosis are arranged in a linear order. (Actually, a mitosis occurs after the production of these four nuclei, giving a linear series of eight haploid nuclei.) FIGURE 19.19 shows that each of these nuclei effectively represents the genetic character of one of the eight strands of the four chromosomes produced by the meiosis.

Meiosis in a heterozygote should generate four copies of each allele. This is seen in the majority of spores. There are some spores, though, with abnormal ratios. They are explained by the formation and correction of heteroduplex DNA in the region in which the alleles differ. The figure illustrates a recombination event in which a length of hybrid DNA occurs on one of the four meiotic chromosomes, a possible outcome of recombination initiated by a double-strand break.

Suppose that two alleles differ by a single point mutation. When a strand exchange occurs to generate heteroduplex DNA, the two strands of the heteroduplex will be mispaired at the site of mutation. Thus each strand of DNA carries different genetic information. If no change is made in the sequence, the strands separate at the ensuing replication, each giving rise to a duplex that perpetuates its information. This event is called **postmeiotic segregation,** because it reflects the separation of DNA strands after meiosis. Its importance is that it demonstrates directly the existence of heteroduplex DNA in recombining alleles.

Another effect is seen when examining recombination between alleles: the proportions of the alleles differ from the initial 4:4 ratio. This effect is called **gene conversion.** It describes a nonreciprocal transfer of information from one chromatid to another.

Gene conversion results from exchange of strands between DNA molecules, and the change in sequence may have either of two causes at the molecular level:

- As indicated by the double-strand break model in Figure 19.9, one DNA duplex may act as a donor of genetic information that directly replaces the corresponding sequences in the recipient duplex by a process of gap generation, strand exchange, and gap filling.
- As part of the exchange process, heteroduplex DNA is generated when a single strand from one duplex pairs with its complement in the other duplex.

Repair systems recognize mispaired bases in heteroduplex DNA, and may then excise and replace one of the strands to restore complementarity. Such an event converts the strand of DNA representing one allele into the sequence of the other allele.

Gene conversion does not depend on crossing over, but is correlated with it. A large proportion of the aberrant asci show genetic recombination between two markers on either side of a site of interallelic gene conversion. This is exactly what would be predicted if the aberrant ratios result from initiation of the recombination process as shown in Figure 19.6, but with an approximately equal probability of resolving the structure with or without recombination (as indicated in Figure 19.8). The implication is that fungal chromosomes initiate crossing over about twice as often as would be expected from the measured frequency of recombination between distant genes.

Various biases are seen when recombination is examined at the molecular level. Either direction of gene conversion may be equally likely, or allele-specific effects may create a preference for one direction. Gradients of recombination may fall away from hotspots. We now know that hotspots represent sites at which double-strand breaks are initiated, and that the gradient is correlated with the extent to which the gap at the hotspot is enlarged and converted to long single-stranded ends (see Section 19.6, The Synaptonemal Complex Forms after Double-Strand Breaks).

Some information about the extent of gene conversion is provided by the sequences of members of gene clusters. Usually, the products of a recombination event will separate and become unavailable for analysis at the level of DNA sequence. When a chromosome carries two (nonallelic) genes that are related, though, they may recombine by an "unequal crossing-over" event (see Section 6.7, Unequal Crossing-Over Rearranges Gene Clusters). All we need to note for now is that a heteroduplex may be formed between the two nonallelic genes. Gene conversion effectively converts one of the nonallelic genes to the sequence of the other.

The presence of more than one gene copy on the same chromosome provides a footprint to trace these events. For example, if heteroduplex formation and gene conversion occurred over part of one gene, this part may have a sequence identical with, or very closely related to, the other gene, whereas the remaining part shows more divergence. Available sequences suggest that gene conversion events may extend for considerable distances, up to a few thousand bases.

19.12 Supercoiling Affects the Structure of DNA

Key concepts

- Supercoiling occurs only in a closed DNA with no free ends.
- A closed DNA can be a circular DNA molecule or a linear molecule where both ends are anchored in a protein structure.
- Any closed DNA molecule has a linking number, which is the sum of the twisting number and writhing number.
- Turns can be repartitioned between the twisting number and writhing number, so that a change in the structure of the double helix can compensate for a change in its coiling in space.
- The linking number can be changed only by breaking and making bonds in DNA.

The winding of the two strands of DNA around each other in the double helical structure makes it possible to change the structure by influenc-

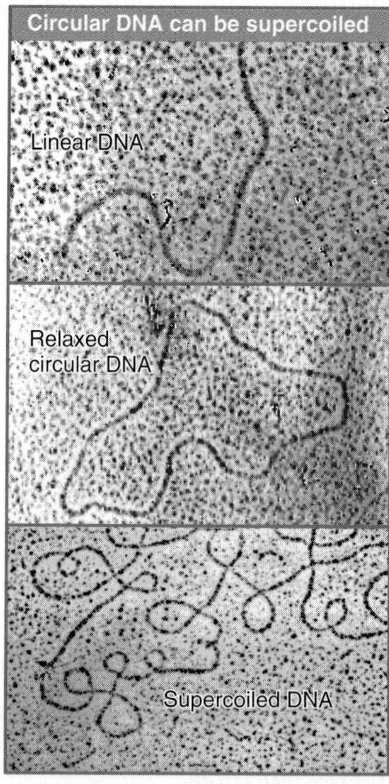

FIGURE 19.20 Linear DNA is extended (a), a circular DNA remains extended if it is relaxed (nonsupercoiled) (b), but a supercoiled DNA has a twisted and condensed form (c). Photos courtesy of Nirupam Roy Choudhury, International Centre for Genetic Engineering and Biotechnology (ICGEB).

ing its conformation in space. If the two ends of a DNA molecule are fixed, the double helix can be wound around itself in space. This is called **supercoiling.** The effect can be imagined like a rubber band twisted around itself. The simplest example of a DNA with no free ends is a circular molecule. The effect of supercoiling can be seen by comparing the nonsupercoiled circular DNA lying flat in FIGURE 19.20 with the supercoiled circular molecule that forms a twisted—and therefore more condensed—shape.

The consequences of supercoiling depend on whether the DNA is twisted around itself in the same sense as the two strands within the double helix (clockwise) or in the opposite sense. Twisting in the same sense produces *positive supercoiling.* This has the effect of causing the DNA strands to wind around one another more tightly, so that there are more base pairs per turn. Twisting in the opposite sense produces *negative supercoiling.* This causes the DNA strands to be twisted around one another less tightly, so there are fewer base pairs per turn. Negative supercoiling can be thought of as creating tension in the DNA that is relieved by unwinding the double helix. The ultimate effect of negative supercoiling is to generate a region in which the two strands of DNA have separated—formally there are zero base pairs per turn.

Topological manipulation of DNA is a central aspect of all its functional activities—recombination, replication, and transcription—as well as of the organization of higher-order structure. All synthetic activities involving double-stranded DNA require the strands to separate. The strands do not simply lie side by side, though; they are intertwined. Their separation therefore requires the strands to rotate about each other in space. Some possibilities for the unwinding reaction are illustrated in FIGURE 19.21.

We might envisage the structure of DNA in terms of a free end that would allow the strands to rotate about the axis of the double helix for unwinding. Given the length of the double helix, however, this would involve the separating strands in a considerable amount of flailing about, which seems unlikely in the confines of the cell.

A similar result is achieved by placing an apparatus to control the rotation at the free end. The effect, however, must be transmitted over a considerable distance, again involving the rotation of an unreasonable length of material.

Consider the effects of separating the two strands in a molecule whose ends are not free to rotate. When two intertwined strands are pulled apart from one end, the result is to increase their winding about each other farther along the molecule. The problem can be overcome by introducing a transient nick in one strand. An internal free end allows the nicked strand to rotate about the intact strand, after which the nick can be sealed. Each repetition of the nicking and sealing reaction releases one superhelical turn.

A closed molecule of DNA can be characterized by its **linking number,** the number of times one strand crosses over the other in space. Closed DNA molecules of identical sequence may have different linking numbers, reflecting different degrees of supercoiling. Molecules of DNA that are the same except for their linking numbers are called **topological isomers.**

The linking number is made up of two components: the writhing number (W) and the twisting number (T).

The **twisting number,** T, is a property of the double helical structure itself, representing the rotation of one strand about the other. It represents the total number of turns of the duplex and is determined by the number of base

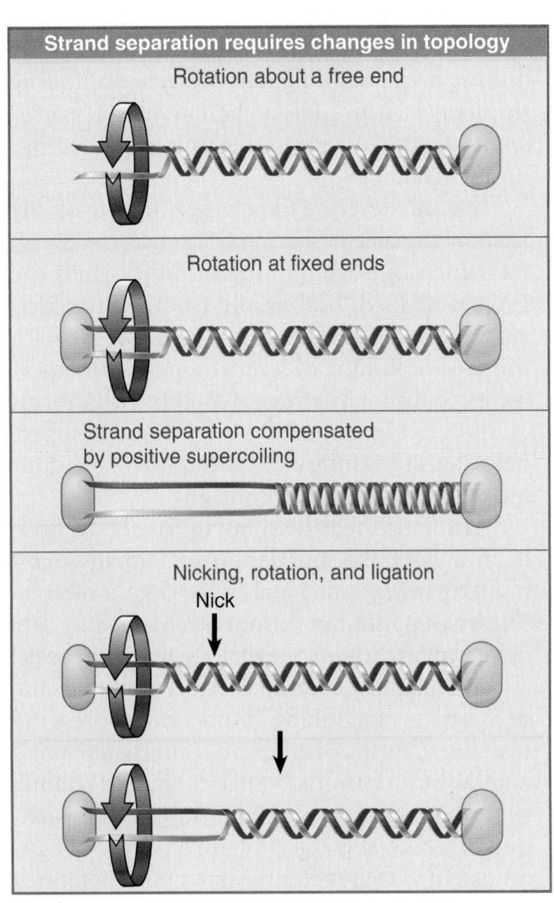

FIGURE 19.21 Separation of the strands of a DNA double helix could be achieved in several ways.

pairs per turn. For a relaxed closed circular DNA lying flat in a plane, the twist is the total number of base pairs divided by the number of base pairs per turn.

The **writhing number,** W, represents the turning of the axis of the duplex in space. It corresponds to the intuitive concept of supercoiling, but does not have exactly the same quantitative definition or measurement. For a relaxed molecule W = 0, and the linking number equals the twist.

We are often concerned with the change in linking number, ΔL, given by the equation

$$\Delta L = \Delta W + \Delta T.$$

The equation states that any change in the total number of revolutions of one DNA strand about the other can be expressed as the sum of the changes of the coiling of the duplex axis in space (ΔW) and changes in the screwing of the double helix itself (ΔT). In a free DNA molecule W and T are freely adjustable, and any ΔL (change in linking number) is likely to be expressed by a change in W, that is, by a change in supercoiling.

A decrease in linking number, that is, a change of –ΔL, corresponds to the introduction of some combination of negative supercoiling and/or underwinding. An increase in linking number, measured as a change of +ΔL, corresponds to a decrease in negative supercoiling/underwinding.

We can describe the change in state of any DNA by the specific linking difference, $\sigma = \Delta L/L_0$, for which L_0 is the linking number when the DNA is relaxed. If all of the change in linking number is due to change in W (that is, ΔT = 0), the specific linking difference equals the supercoiling density. In effect, σ as defined in terms of $\Delta L/L_0$ can be assumed to correspond to superhelix density so long as the structure of the double helix itself remains constant.

The critical feature about the use of the linking number is that this parameter is an invariant property of any individual closed DNA molecule. The linking number cannot be changed by any deformation short of one that involves the breaking and rejoining of strands. A circular molecule with a particular linking number can express the number in terms of different combinations of T and W, but it cannot change their sum so long as the strands are unbroken. (In fact, the partition of L between T and W prevents the assignment of fixed values for the latter parameters for a DNA molecule in solution.)

The linking number is related to the actual enzymatic events by which changes are made in the topology of DNA. The linking number of a particular closed molecule can be changed only by breaking a strand or strands, using the free end to rotate one strand about the other, and rejoining the broken ends. When an enzyme performs such an action, it must change the linking number by an integer; this value can be determined as a characteristic of the reaction. We can then consider the effects of this change in terms of ΔW and ΔT.

19.13 Topoisomerases Relax or Introduce Supercoils in DNA

Key concepts

- Topoisomerases change the linking number by breaking bonds in DNA, changing the conformation of the double helix in space, and remaking the bonds.
- Type I enzymes act by breaking a single strand of DNA; type II enzymes act by making double-strand breaks.

Changes in the topology of DNA can be caused in several ways. FIGURE 19.22 shows some examples. In order to start replication or transcription, the two strands of DNA must be unwound. In the case of replication, the two strands separate permanently, and each reforms a duplex with the newly-synthesized daughter strand. In the case of transcription, the movement of RNA polymerase creates a region of positive supercoiling in front and a region of negative supercoiling behind the enzyme. This must be resolved before the positive supercoils impede the movement of the enzyme (see Section 11.15, Supercoiling Is an Important Feature of Transcription). When a circular DNA molecule is replicated, the circular products may be catenated, with one passed through the other. They must be separated in order for the daughter molecules to segregate to separate daughter cells. Yet another situation in which supercoiling is important is the folding of the DNA thread into a chain of nucleosomes in the eukaryotic nucleus (see Section 29.6, The Periodicity of DNA Changes on the Nucleosome). All of the situations are resolved by the actions of topoisomerases.

DNA **topoisomerases** are enzymes that catalyze changes in the topology of DNA by

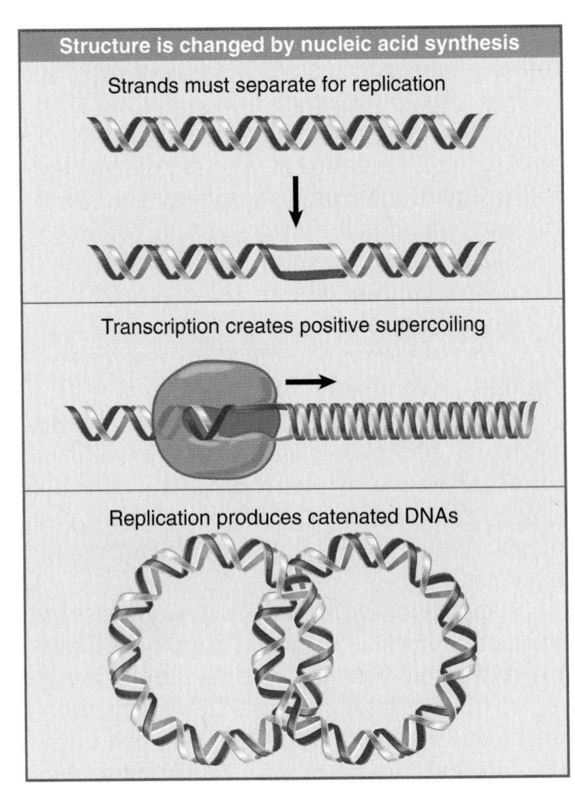

FIGURE 19.22 The topological structure of DNA is changed during replication and transcription. Strand separation requires a base turn of DNA to be unwound. Transcription creates positive supercoils ahead of the RNA polymerase. Replication of a circular template produces two catenated daughter templates.

transiently breaking one or both strands of DNA, passing the unbroken strand(s) through the gap, and then resealing the gap. The ends that are generated by the break are never free, but instead are manipulated exclusively within the confines of the enzyme—in fact, they are covalently linked to the enzyme. Topoisomerases act on DNA irrespective of its sequence, but some enzymes involved in site-specific recombination function in the same way and also fit the definition of topoisomerases (see Section 19.18, Site-Specific Recombination Resembles Topoisomerase Activity).

Topoisomerases are divided into two classes according to the nature of the mechanisms they employ. **Type I topoisomerases** act by making a transient break in one strand of DNA. **Type II topoisomerases** act by introducing a transient double-strand break. Topoisomerases in general vary with regard to the types of topological change they introduce. Some topoisomerases can relax (remove) only negative supercoils from DNA; others can relax both negative and positive supercoils. Enzymes that can introduce negative supercoils are called gyrases; those that can introduce positive supercoils are called reverse gyrases.

There are four topoisomerase enzymes in *E. coli:* topoisomerases I, III, and IV and DNA gyrase. DNA topoisomerases I and III are type I enzymes. Gyrase and DNA topoisomerase IV are type II enzymes. Each of the four enzymes is important in one or more of the situations described in Figure 19.22:

- The overall level of negative supercoiling in the bacterial nucleoid is the result of a balance between the introduction of supercoils by gyrase and their relaxation by topoisomerases I and IV. This is a crucial aspect of nucleoid structure (see Section 28.4, The Bacterial Genome Is Supercoiled), and it affects initiation of transcription at certain promoters (see Section 11.15, Supercoiling Is an Important Feature of Transcription).
- The same enzymes are involved in resolving the problems created by transcription; gyrase converts the positive supercoils that are generated ahead of RNA polymerase into negative supercoils, and topoisomerases I and IV remove the negative supercoils that are left behind the enzyme. Similar, but more complicated, effects occur during replication, and the enzymes have similar roles in dealing with them.
- As replication proceeds, the daughter duplexes can become twisted around one another in a stage known as precatenation. The precatenanes are removed by topoisomerase IV, which also decatenates any catenated genomes that are left at the end of replication. The functions of topoisomerase III partially overlap those of topoisomerase IV.

The enzymes in eukaryotes follow the same principles, although the detailed division of responsibilities may be different. They do not show sequence or structural similarity with the prokaryotic enzymes. Most eukaryotes contain a single topoisomerase I enzyme that is required both for replication fork movement and for relaxing supercoils generated by transcription. A topoisomerase II enzyme(s) is required to unlink chromosomes following replication. Other individual topoisomerases have been implicated in recombination and repair activities.

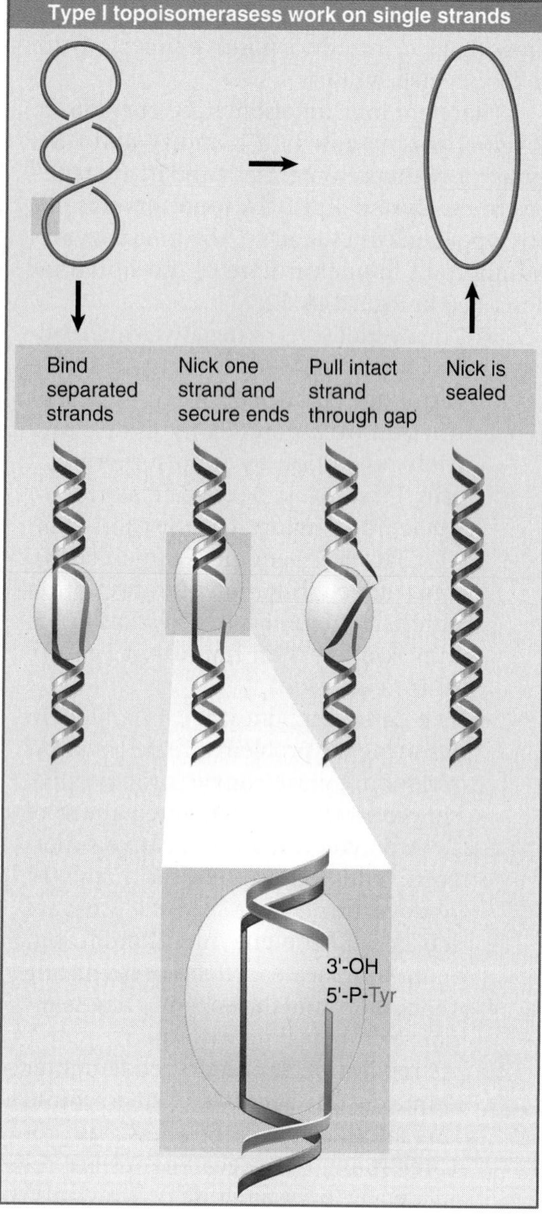

FIGURE 19.23 Bacterial type I topoisomerases recognize partially unwound segments of DNA and pass one strand through a break made in the other.

19.14 Topoisomerases Break and Reseal Strands

Key concept

- Type I topoisomerases function by forming a covalent bond to one of the broken ends, moving one strand around the other, and then transferring the bound end to the other broken end. Bonds are conserved, and as a result no input of energy is required.

The common action for all topoisomerases is to link one end of each broken strand to a tyrosine residue in the enzyme. A type I enzyme links to the single broken strand; a type II enzyme links to one end of each broken strand. The topoisomerases are further divided into the A and B groups according to whether the linkage is to a 5′ phosphate or 3′ phosphate. The use of the transient phosphodiester–tyrosine bond suggests a mechanism for the action of the enzyme; it transfers a phosphodiester bond(s) in DNA to the protein, manipulates the structure of one or both DNA strands, and then rejoins the bond(s) in the original strand.

The *E. coli* enzymes are all of type A and use links to 5′ phosphate. This is the general pattern for bacteria, where there are almost no type B topoisomerases. All four possible types of topoisomerase (IA, IB, IIA, and IIB) are found in eukaryotes.

A model for the action of topoisomerase IA is illustrated in FIGURE 19.23. The enzyme binds to a region in which duplex DNA becomes separated into its single strands. The enzyme then breaks one strand, pulls the other strand through the gap, and finally seals the gap. The transfer of bonds from nucleic acid to protein explains how the enzyme can function without requiring any input of energy. There has been no irreversible hydrolysis of bonds; their energy has been conserved through the transfer reactions. The model is supported by the crystal structure of the enzyme.

The reaction changes the linking number in steps of one. Each time one strand is passed through the break in the other, there is a ΔL of +1. The figure illustrates the enzyme activity in terms of moving the individual strands. In a free supercoiled molecule, the interchangeability of W and T should let the change in linking number be taken up by a change of $\Delta W = +1$, that is, by one less turn of negative supercoiling.

The reaction is equivalent to the rotation illustrated in the bottom part of Figure 19.21, with the restriction that the enzyme limits the reaction to a single-strand passage per event. (By contrast, the introduction of a nick in a supercoiled molecule allows free strand rotation to relieve all the tension by multiple rotations.)

The type I topoisomerase also can pass one segment of a single-stranded DNA through another. This **single-strand passage** reaction can introduce **knots** in DNA and can **catenate** two circular molecules so that they are connected like links on a chain. We do not understand the uses (if any) to which these reactions are put *in vivo*.

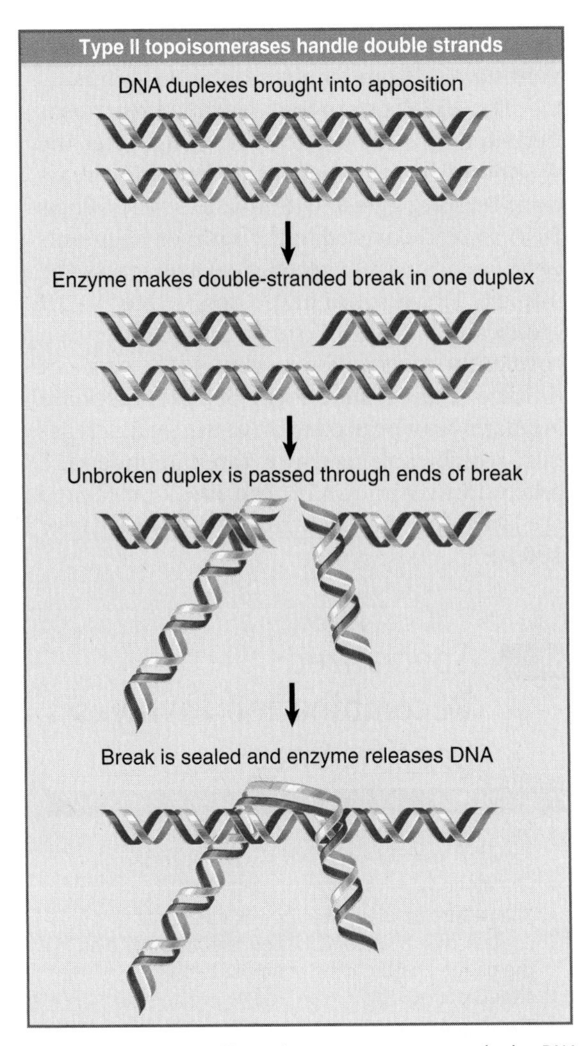

FIGURE 19.24 Type II topoisomerases can pass a duplex DNA through a double-strand break in another duplex.

Type II topoisomerases generally relax both negative and positive supercoils. The reaction requires ATP, with one ATP hydrolyzed for each catalytic event. As illustrated in FIGURE 19.24, the reaction is mediated by making a double-stranded break in one DNA duplex. The double-strand is cleaved with a 4-base stagger between the ends, and each subunit of the dimeric enzyme attaches to a protruding broken end. Another duplex region is then passed through the break. The ATP is used in the following religation/release step, when the ends are rejoined and the DNA duplexes are released. This is why inhibiting the ATPase activity of the enzyme results in a "cleavable complex" that contains broken DNA.

A formal consequence of two-strand transfer is that the linking number is always changed in multiples of two. The topoisomerase II activity also can be used to introduce or resolve catenated duplex circles and knotted molecules.

The reaction probably represents a nonspecific recognition of duplex DNA in which the enzyme binds any two double-stranded segments that cross each other. The hydrolysis of ATP may be used to drive the enzyme through conformational changes that provide the force needed to push one DNA duplex through the break made in the other. As a result of the topology of supercoiled DNA, the relationship of the crossing segments allows supercoils to be removed from either positively or negatively supercoiled circles.

19.15 Gyrase Functions by Coil Inversion

Key concept

- *E. coli* gyrase is a type II topoisomerase that uses hydrolysis of ATP to provide energy to introduce negative supercoils into DNA.

Bacterial DNA gyrase is a topoisomerase of type II that is able to introduce negative supercoils into a relaxed closed circular molecule. DNA gyrase binds to a circular DNA duplex and supercoils it processively and catalytically while continuing to introduce supercoils into the same DNA molecule. One molecule of DNA gyrase can introduce ~100 supercoils per minute.

The supercoiled form of DNA has a higher free energy than the relaxed form, and the energy needed to accomplish the conversion is supplied by the hydrolysis of ATP. In the absence of ATP, the gyrase can relax negative but not positive supercoils, although the rate is more than 10× slower than the rate of introducing supercoils.

The *E. coli* DNA gyrase is a tetramer consisting of two types of subunit, each of which is a target for antibiotics (the most often used being nalidixic acid, which acts on GyrA, and novobiocin, which acts on GyrB). The drugs inhibit replication, which suggests that DNA gyrase is necessary for DNA synthesis to proceed. Mutations that confer resistance to the antibiotics identify the loci that code for the subunits.

Gyrase binds its DNA substrate around the outside of the protein tetramer. Gyrase protects ~140 bp of DNA from digestion by micrococcal nuclease. The **sign inversion** model for gyrase action is illustrated in FIGURE 19.25. The enzyme binds the DNA in a crossover configuration that is equivalent to a positive supercoil. This induces a compensating negative supercoil in the unbound DNA. The enzyme then breaks the double strand at the crossover of the positive

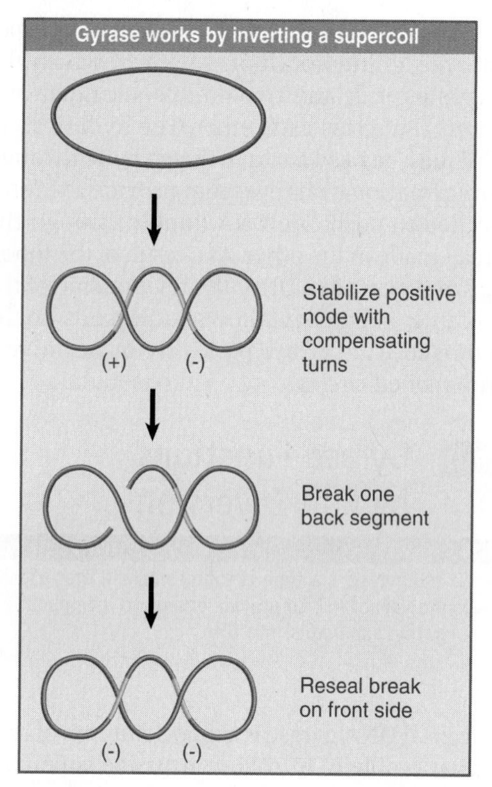

FIGURE 19.25 DNA gyrase may introduce negative supercoils in duplex DNA by inverting a positive supercoil.

supercoil, passes the other duplex through, and seals the break.

The reaction directly inverts the sign of the supercoil: It has been converted from a +1 turn to a –1 turn. Thus the linking number has changed by $\Delta L = -2$, which conforms with the demand that all events involving double-strand passage must change the linking number by a multiple of two.

Gyrase then releases one of the crossing segments of the (now negative) bound supercoil; this allows the negative turns to redistribute along DNA (as a change in either T or W or both), and the cycle begins again. The same type of topological manipulation is responsible for catenation and knotting.

On releasing the inverted supercoil, the conformation of gyrase changes. For the enzyme to undertake another cycle of supercoiling, its original conformation must be restored. This process is called **enzyme turnover.** It is thought to be driven by the hydrolysis of ATP, because the replacement of ATP by an analog that cannot be hydrolyzed allows gyrase to introduce only one inversion (–2 supercoils) per substrate. Thus the enzyme does not need ATP for the supercoiling reaction, but does need it to undertake a second cycle. Novobiocin interferes with the ATP-dependent reactions of gyrase by preventing ATP from binding to the B subunit.

The ATP-independent relaxation reaction is inhibited by nalidixic acid. This implicates the A subunit in the breakage and reunion reaction. Treating gyrase with nalidixic acid allows DNA to be recovered in the form of fragments generated by a staggered cleavage across the duplex. The termini all possess a free 3′–OH group and a 4-base 5′ single-strand extension covalently linked to the A subunit. The covalent linkage retains the energy of the phosphate bond; this can be used to drive the sealing reaction, which explains why gyrase can undertake relaxation without ATP. The sites of cleavage are fairly specific, occurring about once every 100 bp.

19.16 Specialized Recombination Involves Specific Sites

Key concepts

- Specialized recombination involves reaction between specific sites that are not necessarily homologous.
- Phage lambda integrates into the bacterial chromosome by recombination between a site on the phage and the *att* site on the *E. coli* chromosome.
- The phage is excised from the chromosome by recombination between the sites at the end of the linear prophage.
- Phage lambda *int* codes for an integrase that catalyzes the integration reaction.

Specialized recombination involves a reaction between two specific sites. The lengths of target sites are short, and are typically in a range of 14 to 50 bp. In some cases the two sites have the same sequence, but in other cases they are nonhomologous. The reaction is used to insert a free phage DNA into the bacterial chromosome or to excise an integrated phage DNA from the chromosome, and in this case the two recombining sequences are different from one another. It is also used before division to regenerate monomeric circular chromosomes from a dimer that has been created by a generalized recombination event (see Section 17.7, Chromosomal Segregation May Require Site-Specific Recombination). In this case the recombining sequences are identical.

The enzymes that catalyze site-specific recombination are generally called recombinases, and more than one hundred of them are now known. Those involved in phage integration or related to these enzymes are also known as the integrase family. Prominent members of the integrase family are the prototype Int from phage lambda, Cre from phage P1, and the yeast FLP enzyme (which catalyzes a chromosomal inversion).

The classic model for **site-specific recombination** is illustrated by phage lambda. The conversion of lambda DNA between its different life forms involves two types of event. The pattern of gene expression is regulated as described in Chapter 14, Phage Strategies. The physical condition of the DNA is different in the lysogenic and lytic states:

- In the lytic lifestyle, lambda DNA exists as an independent, circular molecule in the infected bacterium.
- In the lysogenic state, the phage DNA is an integral part of the bacterial chromosome (called **prophage**).

Transition between these states involves site-specific recombination:

- To enter the lysogenic condition, free lambda DNA must be inserted into the host DNA. This is called **integration.**
- To be released from lysogeny into the lytic cycle, prophage DNA must be released from the chromosome. This is called **excision.**

Integration and excision occur by recombination at specific loci on the bacterial and phage DNAs called attachment ***(att)*** sites. The attachment site on the bacterial chromosome is called *att*$^{\lambda}$ in bacterial genetics. The locus is defined by mutations that prevent integration of lambda; it is occupied by prophage λ in lysogenic strains. When the *att*$^{\lambda}$ site is deleted from the *E. coli* chromosome, an infecting lambda phage can establish lysogeny by integrating elsewhere, although the efficiency of the reaction is <0.1% of the frequency of integration at *att*$^{\lambda}$. This inefficient integration occurs at **secondary attachment sites,** which resemble the authentic *att* sequences.

For describing the integration/excision reactions, the bacterial attachment site (*att*$^{\lambda}$) is called *attB*, consisting of the sequence components *BOB′*. The attachment site on the phage, *attP*, consists of the components *POP′*. FIGURE 19.26 outlines the recombination reaction between these sites. The sequence O is common to *attB* and *attP*. It is called the **core** sequence; and the recombination event occurs within it. The flanking regions *B*, *B′* and *P*, *P′* are referred to as the **arms;** each is distinct in sequence. The phage DNA is circular, so the recombination event inserts it into the bacterial chromosome as a linear sequence. The prophage is bounded by two new *att* sites—the products of the recombination—called *attL* and *attR*.

An important consequence of the constitution of the *att* sites is that the integration and excision reactions do not involve the same pair of reacting sequences. Integration requires recognition between *attP* and *attB*, whereas excision requires recognition between *attL* and *attR*. The directional character of site-specific recombination is controlled by the identity of the recombining sites.

The recombination event is reversible, but different conditions prevail for each direction of the reaction. This is an important feature in the life of the phage, because it offers a means to ensure that an integration event is not immediately reversed by an excision, and vice versa.

The difference in the pairs of sites reacting at integration and excision is reflected by a difference in the proteins that mediate the two reactions:

- Integration (*attB* × *attP*) requires the product of the phage gene *int*, which codes for an integrase enzyme, and a bacterial protein called integration host factor (IHF).

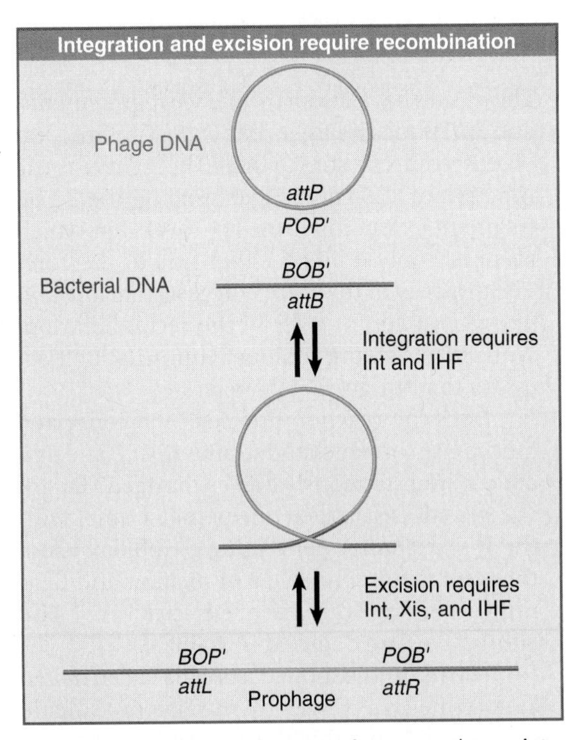

FIGURE 19.26 Circular phage DNA is converted to an integrated prophage by a reciprocal recombination between *attP* and *attB*; the prophage is excised by reciprocal recombination between *attL* and *attR*.

- Excision (*attL* × *attR*) requires the product of phage gene *xis*, in addition to Int and IHF.

Thus Int and IHF are required for both reactions. Xis plays an important role in controlling the direction; it is required for excision, but inhibits integration.

A similar system, but with somewhat simpler requirements for both sequence and protein components, is found in the bacteriophage P1. The Cre recombinase coded by the phage catalyzes a recombination between two target sequences. Unlike phage lambda, for which the recombining sequences are different, in phage P1 they are identical. Each consists of a 34 bp-long sequence called *loxP*. The Cre recombinase is sufficient for the reaction; no accessory proteins are required. As a result of its simplicity and its efficiency, what is now known as the Cre/*lox* system has been adapted for use in eukaryotic cells, where it has become one of the standard techniques for undertaking site-specific recombination.

19.17 Site-Specific Recombination Involves Breakage and Reunion

Key concept

- Cleavages staggered by 7 bp are made in both *attB* and *attP* and the ends are joined crosswise.

The *att* sites have distinct sequence requirements, and *attP* is much larger than *attB*. The function of *attP* requires a stretch of 240 bp, whereas the function of *attB* can be exercised by the 23 bp fragment extending from −11 to +11, in which there are only 4 bp on either side of the core. The disparity in their sizes suggests that *attP* and *attB* play different roles in the recombination, with *attP* providing additional information necessary to distinguish it from *attB*.

Does the reaction proceed by a concerted mechanism in which the strands in *attP* and *attB* are cut simultaneously and exchanged? Or are the strands exchanged one pair at a time, with the first exchange generating a Holliday junction and the second cycle of nicking and ligation occurring to release the structure? The alternatives are depicted in FIGURE 19.27.

The recombination reaction has been halted at intermediate stages by the use of "suicide substrates," in which the core sequence is nicked. The presence of the nick interferes with the recombination process. This makes it possible to identify molecules in which recombination has commenced but has not been completed. The structures of these intermediates suggest that exchanges of single strands take place sequentially.

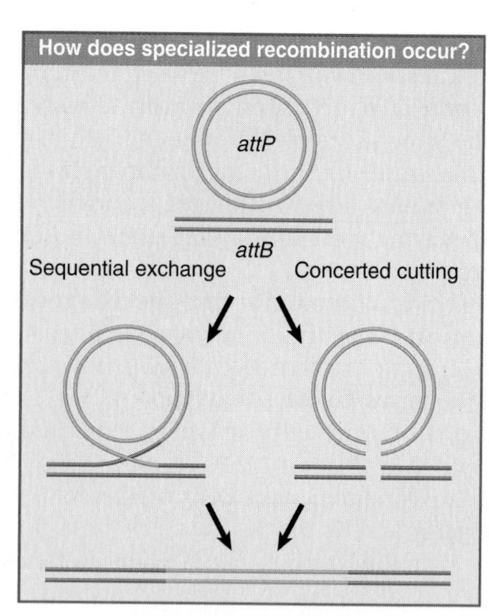

FIGURE 19.27 Does recombination between *attP* and *attB* proceed by sequential exchange or concerted cutting?

The model illustrated in FIGURE 19.28 shows that if *attP* and *attB* sites each suffer the same staggered cleavage, complementary single-stranded ends could be available for crosswise hybridization. The distance between the lambda crossover points is 7 bp, and the reaction generates 3′–phosphate and 5′–OH ends. The reaction is shown for simplicity as generating overlapping single-stranded ends that anneal, but actually occurs by a process akin to the recombination event of Figure 19.6. The corresponding strands on each duplex are cut at the same position, the free 3′ ends exchange between duplexes, the branch migrates for a distance of 7 bp along the region of homology, and then the structure is resolved by cutting the other pair of corresponding strands.

19.18 Site-Specific Recombination Resembles Topoisomerase Activity

Key concepts

- Integrases are related to topoisomerases, and the recombination reaction resembles topoisomerase action except that nicked strands from *different* duplexes are sealed together.
- The reaction conserves energy by using a catalytic tyrosine in the enzyme to break a phosphodiester bond and link to the broken 3′ end.
- Two enzyme units bind to each recombination site and the two dimers synapse to form a complex in which the transfer reactions occur.

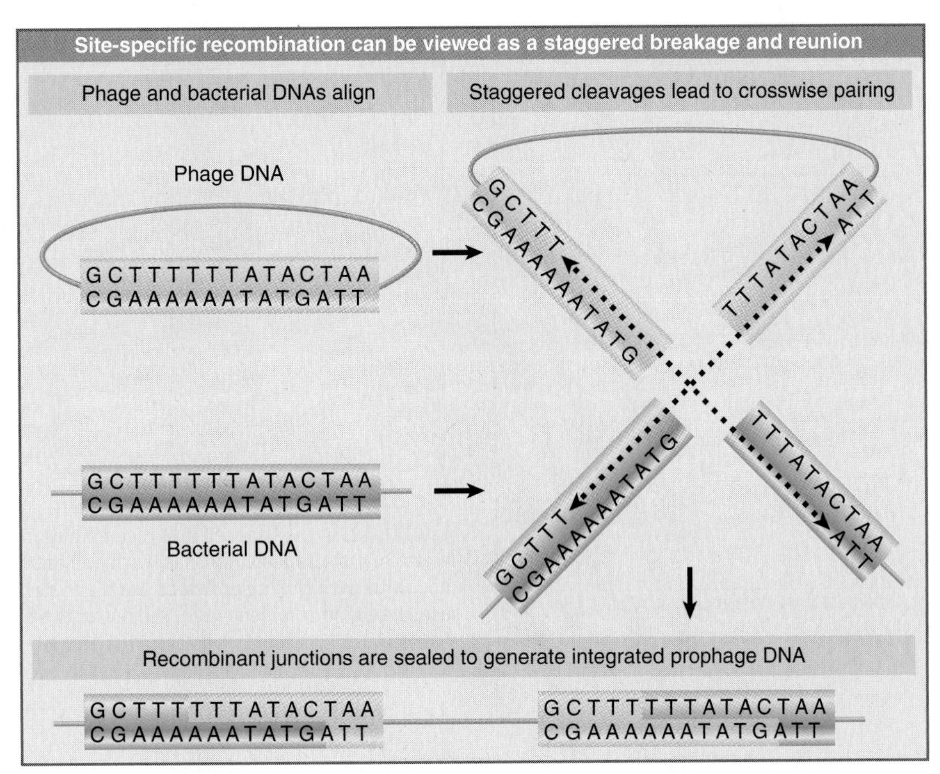

FIGURE 19.28 Staggered cleavages in the common core sequence of *attP* and *attB* allow crosswise reunion to generate reciprocal recombinant junctions.

Integrases use a mechanism similar to that of type I topoisomerases, in which a break is made in one DNA strand at a time. The difference is that a recombinase reconnects the ends crosswise, whereas a topoisomerase makes a break, manipulates the ends, and then rejoins the original ends. The basic principle of the system is that four molecules of the recombinase are required, one to cut each of the four strands of the two duplexes that are recombining.

FIGURE 19.29 shows the nature of the reaction catalyzed by an integrase. The enzyme is a monomeric protein that has an active site capable of cutting and ligating DNA. The reaction involves an attack by a tyrosine on a phosphodiester bond. The 3′ end of the DNA chain is linked through a phosphodiester bond to a tyrosine in the enzyme. This releases a free 5′ hydroxyl end.

Two enzyme units are bound to each of the recombination sites. At each site, only one of the units attacks the DNA. The symmetry of the system ensures that complementary strands are broken in each recombination site. The free 5′–OH end in each site attacks the 3′–phosphotyrosine link in the other site. This generates a Holliday junction.

The structure is resolved when the other two enzyme units (which had not been involved in the first cycle of breakage and reunion) act on the other pair of complementary strands.

The successive interactions accomplish a conservative strand exchange, in which there are no deletions or additions of nucleotides at the exchange site, and there is no need for input of energy. The transient 3′–phosphotyrosine link between protein and DNA conserves the energy of the cleaved phosphodiester bond.

FIGURE 19.30 shows the reaction intermediate, based on the crystal structure. (Trapping the intermediate was made possible by using a "suicide substrate," which consists of a synthetic DNA duplex with a missing phosphodiester bond, so that the attack by the enzyme does not generate a free 5′–OH end.) The structure of the Cre-*lox* complex shows two Cre molecules, each of which is bound to a 15 bp length of DNA. The DNA is bent by ~100° at the center of symmetry. Two of these complexes assemble in an antiparallel way to form a tetrameric protein structure bound to two synapsed DNA molecules. Strand exchange takes place in a central cavity of the protein structure that contains the central six bases of the crossover region.

The tyrosine that is responsible for cleaving DNA in any particular half site is provided by the enzyme subunit that is bound to that half site. This is called *cis* cleavage. This is true also for the Int integrase and XerD recombinase. The

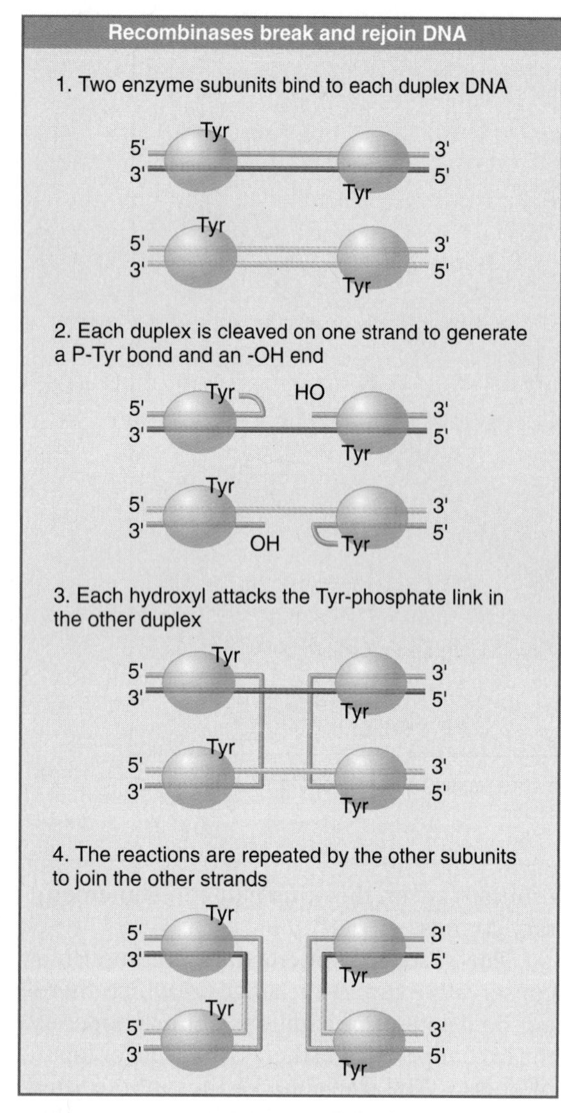

FIGURE 19.29 Integrases catalyze recombination by a mechanism similar to topoisomerases. Staggered cuts are made in DNA and the 3′-phosphate end is covalently linked to a tyrosine in the enzyme. The free hydroxyl group of each strand then attacks the P-Tyr link of the other strand. The first exchange shown in the figure generates a Holliday structure. The structure is resolved by repeating the process with the other pair of strands.

FLP recombinase cleaves in *trans,* however, which involves a mechanism in which the enzyme subunit that provides the tyrosine is *not* the subunit bound to that half site, but rather is one of the other subunits.

19.19 Lambda Recombination Occurs in an Intasome

Key concepts

- Lambda integration takes place in a large complex that also includes the host protein IHF.
- The excision reaction requires Int and Xis and recognizes the ends of the prophage DNA as substrates.

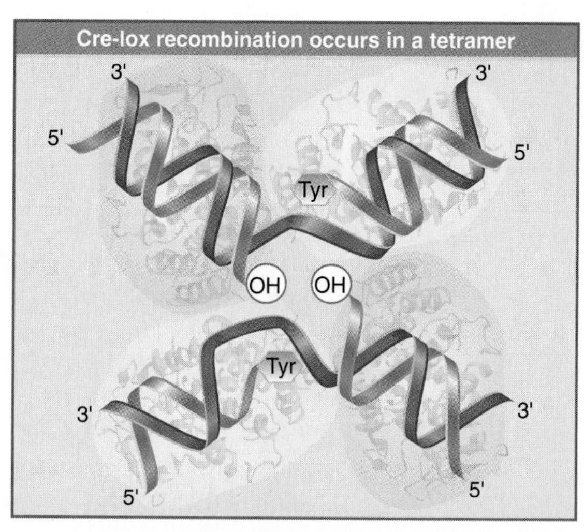

FIGURE 19.30 A synapsed *loxA* recombination complex has a tetramer of Cre recombinases, with one enzyme monomer bound to each half site. Two of the four active sites are in use, acting on complementary strands of the two DNA sites.

Unlike the Cre/*lox* recombination system, which requires only the enzyme and the two recombining sites, phage lambda recombination occurs in a large structure and has different components for each direction of the reaction (integration versus excision).

A host protein called IHF is required for both integration and excision. IHF is a 20 kD protein of two different subunits, which are coded by the genes *himA* and *himD.* IHF is not an essential protein in *E. coli,* and is not required for homologous bacterial recombination. It is one of several proteins with the ability to wrap DNA on a surface. Mutations in the *him* genes prevent lambda site-specific recombination and can be suppressed by mutations in λ*int,* which suggests that IHF and Int interact. Site-specific recombination can be performed *in vitro* by Int and IHF.

The *in vitro* reaction requires supercoiling in *attP,* but not in *attB.* When the reaction is performed *in vitro* between two supercoiled DNA molecules, almost all of the supercoiling is retained by the products. Thus there cannot be any free intermediates in which strand rotation could occur. This was one of the early hints that the reaction proceeds through a Holliday junction. We now know that the reaction proceeds by the mechanism typical of this class of enzymes, which is related to the topoisomerase I mechanism (see Section 19.18, Site-Specific Recombination Resembles Topoisomerase Activity).

Int has two different modes of binding. The C-terminal domain behaves like the Cre recombinase. It binds to inverted sites at the core sequence, positioning itself to make the

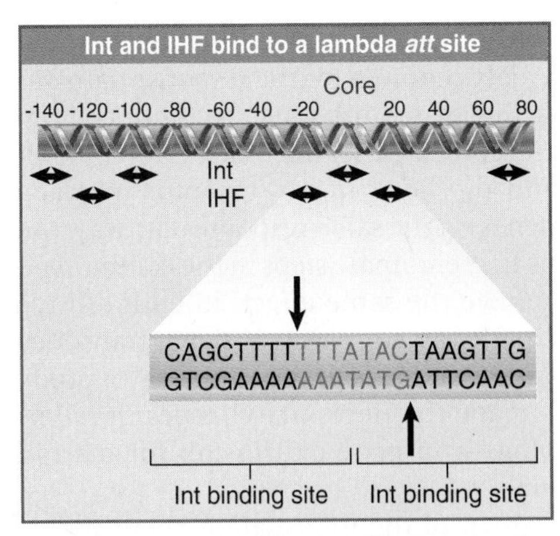

FIGURE 19.31 Int and IHF bind to different sites in *attP*. The Int recognition sequences in the core region include the sites of cutting.

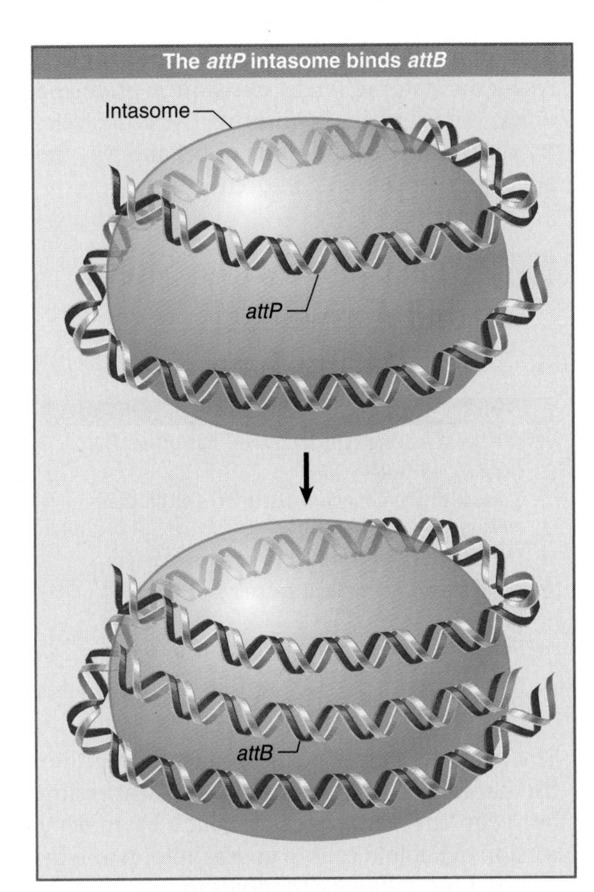

FIGURE 19.32 Multiple copies of Int protein may organize *attP* into an intasome, which initiates site-specific recombination by recognizing *attB* on free DNA.

cleavage and ligation reactions on each strand at the positions illustrated in FIGURE 19.31. The N-terminal domain binds to sites in the arms of *attP* that have a different consensus sequence. This binding is responsible for the aggregation of subunits into the intasome. The two domains probably bind DNA simultaneously, thus bringing the arms of *attP* close to the core.

IHF binds to sequences of ~20 bp in *attP*. The IHF binding sites are approximately adjacent to sites where Int binds. Xis binds to two sites located close to one another in *attP*, so that the protected region extends over 30 to 40 bp. Together, Int, Xis, and IHF cover virtually all of *attP*. The binding of Xis changes the organization of the DNA so that it becomes inert as a substrate for the integration reaction.

When Int and IHF bind to *attP*, they generate a complex in which all the binding sites are pulled together on the surface of a protein. Supercoiling of *attP* is needed for the formation of this **intasome.** The only binding sites in *attB* are the two Int sites in the core. Int does not bind directly to *attB* in the form of free DNA, though. The intasome is the intermediate that "captures" *attB*, as indicated schematically in FIGURE 19.32.

According to this model, the initial recognition between *attP* and *attB* does not depend directly on DNA homology, but instead is determined by the ability of Int proteins to recognize both *att* sequences. The two *att* sites then are brought together in an orientation predetermined by the structure of the intasome. Sequence homology becomes important at this stage, when it is required for the strand-exchange reaction.

The asymmetry of the integration and excision reactions is shown by the fact that Int can form a similar complex with *attR* only if Xis is added. This complex can pair with a condensed complex that Int forms at *attL*. IHF is not needed for this reaction. A significant difference between lambda integration/excision and the recombination reactions catalyzed by Cre or Flp is that Int-catalyzed reactions bind the regulatory sequences in the arms of the target sites, bending the DNA and allowing interactions between arm and core sites that drive each reaction to its conclusion. This is why each lambda reaction is irreversible, whereas recombination catalyzed by Cre or Flp is reversible. Crystal structures of λ-Int tetramers show that, like other recombinases, the tetramer has two active and two inactive subunits that switch roles during recombination. Allosteric interactions triggered by arm-binding control structural transitions in the tetramer that drive the reaction.

Much of the complexity of site-specific recombination may be caused by the need to regulate the reaction so that integration occurs preferentially when the virus is entering the lysogenic state, whereas excision is preferred when the prophage is entering the lytic cycle. By controlling the amounts of Int and Xis, the appropriate reaction will occur.

19.20 Yeast Can Switch Silent and Active Loci for Mating Type

Key concepts

- The yeast mating type locus *MAT* has either the *MAT***a** or *MAT*α genotype.
- Yeast with the dominant allele *HO* switch their mating type at a frequency ~10^{-6}.
- The allele at *MAT* is called the active cassette.
- There are also two silent cassettes, *HML*α and *HMR***a**.
- Switching occurs if *MAT***a** is replaced by *HML*α or *MAT*α is replaced by *HMR***a**.

The yeast *S. cerevisiae* can propagate in either the haploid or diploid condition. Conversion between these states takes place by mating (fusion of haploid cells to give a diploid) and by sporulation (meiosis of diploids to give haploid spores). The ability to engage in these activities is determined by the mating type of the strain, which can be either **a** or α. Haploid cells of type **a** can mate only with haploid cells of type α to generate diploid cells of type **a**/α. The diploid cells can sporulate to regenerate haploid spores of either type.

Mating behavior is determined by the genetic information present at the *MAT* locus. Cells that carry the *MAT***a** allele at this locus are type **a**; likewise, cells that carry the *MAT*α allele are type α. Recognition between cells of opposite mating type is accomplished by the secretion of pheromones: α cells secrete the small polypeptide α-factor; **a** cells secrete **a**-factor. A cell of one mating type carries a surface receptor for the pheromone of the opposite type. When an **a** cell and an α cell encounter one another, their pheromones act on their receptors to arrest the cells in the G1 phase of the cell cycle, and various morphological changes occur. In a successful mating, the cell cycle arrest is followed by cell and nuclear fusion to produce an **a**/αdiploid cell.

Mating is a symmetrical process that is initiated by the interaction of pheromone secreted by one cell type with the receptor carried by the other cell type. The only genes that are uniquely required for the response pathway in a particular mating type are those coding for the receptors. Either the **a** factor–receptor interaction or the α factor–receptor interaction switches on the same response pathway. Mutations that eliminate steps in the common pathway have the same effects in both cell types. The pathway consists of a signal transduction cascade that leads to the synthesis of products that make the necessary changes in cell morphology and gene expression for mating to occur.

Much of the information about the yeast mating-type pathway was deduced from the properties of mutations that eliminate the ability of **a** and/or α cells to mate. The genes identified by such mutations are called *STE* (for sterile). Mutations in the genes for the pheromones or receptors are specific for individual mating types, whereas mutations in the other *STE* genes eliminate mating in both **a** and α cells. This situation is explained by the fact that the events that follow the interaction of factor with receptor are identical for both types.

Some yeast strains have the remarkable ability to switch their mating types. These strains carry a dominant allele *HO* and change their mating type frequently—as often as once every generation. Strains with the recessive allele *ho* have a stable mating type, which is subject to change with a frequency ~10^{-6}.

The presence of *HO* causes the genotype of a yeast population to change. Irrespective of the initial mating type, within a very few generations there are large numbers of cells of both mating types, leading to the formation of *MAT***a**/*MAT*α diploids that take over the population. The production of stable diploids from a haploid population can be viewed as the raison d'être for switching.

The existence of switching suggests that all cells contain the potential information needed to be either *MAT***a** or *MAT*α but express only one type. Where does the information to change mating types come from? Two additional loci are needed for switching. *HML*α is needed for switching to give a *MAT*α type; *HMR***a** is needed for switching to give a *MAT***a** type. These loci lie on the same chromosome that carries *MAT*. *HML* is far to the left, *HMR* is far to the right.

The **cassette** model for mating type is illustrated in FIGURE 19.33. It proposes that *MAT* has an *active cassette* of either type α or type **a**. *HML* and *HMR* have *silent cassettes*. In general, *HML* carries an α cassette, whereas *HMR* carries an *a* cassette. All cassettes carry information that codes for mating type, but only the active cassette at *MAT* is expressed. Mating-type switching occurs when the active cassette is replaced by information from a silent cassette. The newly installed cassette is then expressed.

Switching is nonreciprocal; the copy at *HML* or *HMR* replaces the allele at *MAT*. We know this because a mutation at *MAT* is lost permanently when it is replaced by switching—it does not exchange with the copy that replaces it.

If the silent copy present at *HML* or *HMR* is mutated, switching introduces a mutant allele into the *MAT* locus. The mutant copy at *HML* or *HMR* remains there through an indefinite number of switches. Like replicative transposition, the donor element generates a new copy at the recipient site while itself remaining inviolate.

Mating-type switching is a directed event, in which there is only one recipient *(MAT)*, but two potential donors (*HML* and *HMR*). Switching usually involves replacement of *MAT***a** by the copy at *HML*α or replacement of *MAT*α by the copy at *HMR***a**. In 80%–90% of switches, the *MAT* allele is replaced by one of opposite type. This is determined by the phenotype of the cell. Cells of **a** phenotype preferentially choose *HML* as donor; cells of α phenotype preferentially choose *HMR*.

Several groups of genes are involved in establishing and switching mating type. In addition to the genes that directly determine mating type, they include genes needed to repress the silent cassettes, to switch mating type, or to execute the functions involved in mating.

By comparing the sequences of the two silent cassettes (*HML*α and *HMR***a**) with the sequences of the two types of active cassette (*MAT***a** and *MAT*α), we can delineate the sequences that determine mating type. The organization of the mating type loci is summarized in FIGURE 19.34. Each cassette contains common sequences that flank a central region that differs in the **a** and α types of cassette (called *Y***a** or *Y*α). On either side of this region, the flanking sequences are virtually identical, although they are shorter at *HMR*. The active cassette at *MAT* is transcribed from a promoter within the *Y* region.

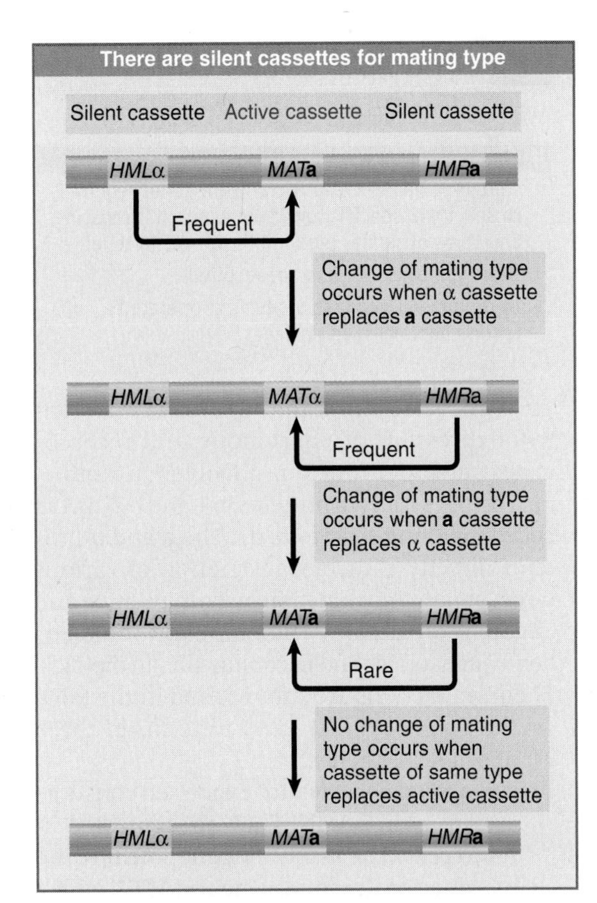

FIGURE 19.33 Changes of mating type occur when silent cassettes replace active cassettes of opposite genotype; when transpositions occur between cassettes of the same type, the mating type remains unaltered.

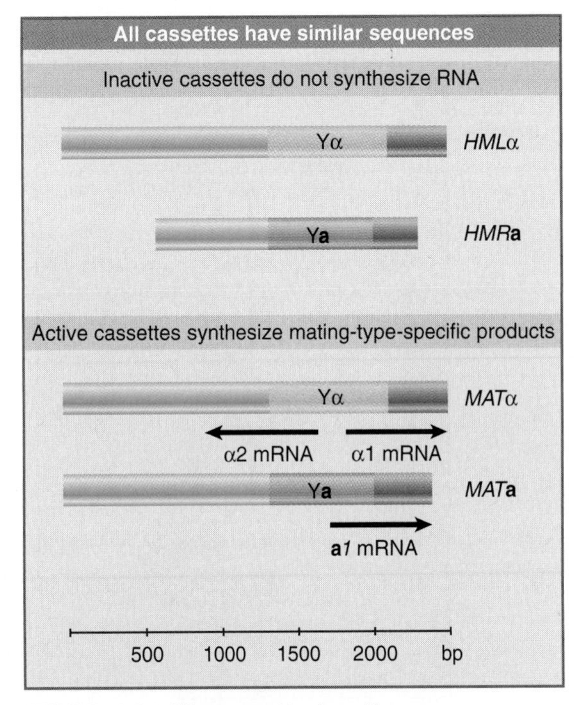

FIGURE 19.34 Silent cassettes have the same sequences as the corresponding active cassettes, except for the absence of the extreme flanking sequences in *HMR***a**. Only the Y region changes between **a** and α types.

19.21 The *MAT* Locus Codes for Regulator Proteins

Key concepts

- In α-type haploids, *MAT*α turns on genes required to specify α-specific functions required for mating and turns off genes required for **a**-mating type.
- In **a**-type cells, *MAT***a** is not required.
- In diploids, **a***1* and α2 products cooperate to repress haploid-specific genes.

The basic function of the *MAT* locus is to control expression of pheromone and receptor genes and other functions involved in mating. *MAT*α codes for two proteins, α1 and α2. *MAT***a** codes for a single protein, **a***1*. The **a** and α proteins directly control transcription of various target genes; they function by both positive and negative regulation. They function independently in haploids and in conjunction in diploids. Their interactions are summarized in the table on the right of FIGURE 19.35 in terms of three groups of target genes:

- **a**-specific genes are expressed constitutively in **a** cells. They are repressed in α cells. The **a**-specific genes include the **a**-factor structural gene and *STE2*, which codes for the α-factor receptor. Thus the **a** phenotype is associated with readiness to recognize the pheromone produced by the opposite mating type.
- α-specific functions are induced in α cells, but are not expressed in **a** cells. They include the α-factor structural gene and the **a**-factor receptor gene, *STE3*. Again, the expression of pheromone of one type is associated with expression of receptor for the pheromone of the opposite type.
- Haploid-specific functions include genes that are needed for transcription of pheromone and receptor genes, the *HO* gene involved in switching, and *RME*, a repressor of sporulation. They are expressed constitutively in both types of haploid, but are repressed in **a**/α diploids. As a result, the **a**-specific and α-specific functions also remain unexpressed in diploids.

We may now view the functions of the regulators and their targets from the perspective of the *MAT* functions expressed in haploid and diploid yeast cells, as outlined in the diagram on the left of Figure 19.35. The **a** and α mating types are regulated by different mechanisms:

- In **a** haploids, mating functions are expressed constitutively. The functions of the products of *MAT***a** in the **a** cell (if any) are unknown. It may be required only to repress haploid functions in diploid cells.
- In α haploids, the α1 product turns on α-specific genes whose products are

Transcription is controlled by mating type

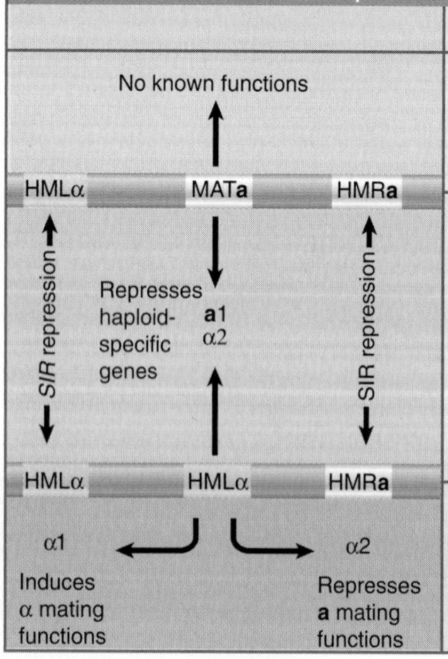

	a genes	**α genes**	**haploid functions**
a haploid	constitutive	not expressed	constitutive
diploid	not expressed	not expressed	repressed
α haploid	repressed	induced	constitutive

No known functions
HMLα MATa HMRa
SIR repression
Represses haploid-specific genes a1 α2
HMLα HMLα HMRa
α1 Induces α mating functions
α2 Represses a mating functions

FIGURE 19.35 In diploids the α1 and α2 proteins cooperate to repress haploid-specific functions. In **a** haploids, mating functions are constitutive. In α haploids, the α2 protein represses **a** mating functions, whereas the α1 protein induces α mating functions.

needed for α mating type. The α2 product represses the genes responsible for producing **a** mating type by binding to an operator sequence located upstream of target genes.

- In diploids, the **a***1* and α2 products cooperate to repress haploid-specific genes. They combine to recognize an operator sequence different from the target for α2 alone.

The abilities of the α2, **a***1*, and α1 proteins to regulate transcription rely upon some interesting protein–protein interactions between themselves and with other protein(s). The pattern of gene control in **a** cells, α cells, and diploids is summarized in FIGURE 19.36.

A protein called pheromone receptor transcription factor (PRTF) (which is not specific for mating type) is involved in many of these interactions. PRTF binds to a short consensus sequence called the P box. The role of PRTF in gene regulation may be quite extensive, as P boxes are found in a variety of locations. In some of these sites, the P box is required for activation of the gene; at other loci, PRTF is needed for repression. Its effects may therefore depend on the other proteins that bind at sites adjacent to the P box.

Genes that are **a**-specific may be activated by PRTF alone. This is adequate to ensure their expression in an **a** haploid.

The **a**-specific genes are repressed in an α haploid by the combined action of the α2 protein and PRTF. The α2 protein contains two domains. The C-terminal domain binds to short palindromic elements at the ends of an operator consensus sequence of 32 bp. Binding of this fragment to DNA, however, does not cause repression. The N-terminal domain is needed for repression and is responsible for making contacts with PRTF. The binding site for PRTF is a P box in the center of the operator. In fact, α and PRTF bind to the operator cooperatively.

Expression of α-specific genes requires another small protein called the α1 activator, which is 175 amino acids long. *Cis*-acting sequences that confer α-specific transcription are 26 bp long, and can be divided into two parts. The first 16 bp form the P box, where PRTF binds; the adjacent 10 bp sequence forms the binding site for α1. The α1 factor binds only when PRTF is present to bind to the P box. Neither protein alone can bind to its target box, but together they can bind to DNA, presumably as a result of protein–protein interactions.

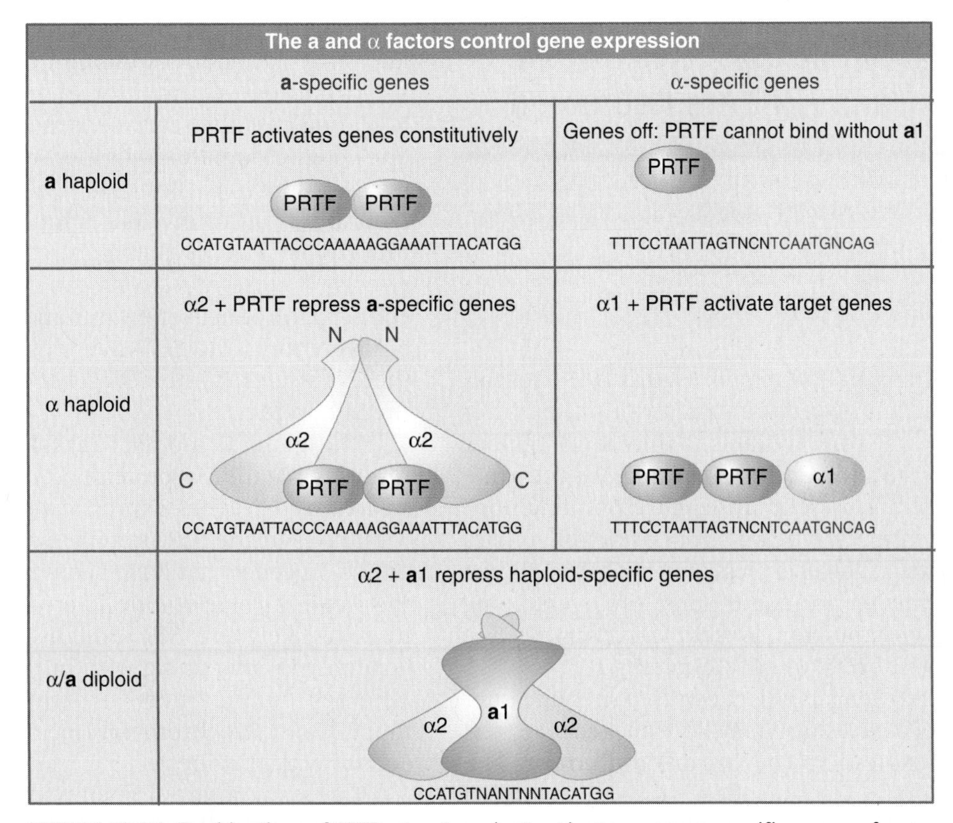

FIGURE 19.36 Combinations of PRTF, **a**1, α1, and α2 activate or repress specific groups of genes to correspond with the mating type of the cell.

The α-specific genes are turned off by default in **a** haploids, because in the absence of α1 protein, PRTF is unable to bind to activate them.

The α2 protein can also cooperate with the α1 protein. The combination of these proteins recognizes a different operator. The operator shares the outlying palindromic sequences with the sequence recognized by α2 alone, but is shorter because the sequence between them is different. The α1/α2 combination represses genes with this motif in diploid cells.

The major point to be made from these results is that the phenotype of each type of cell (**a** or α haploid or **a**/α diploid) is determined by the combination of **a** and α proteins that are expressed. One aspect is the distinction between the haploid and diploid phenotypes; another is the distinction between **a** and α haploid phenotypes. The latter extends to expression of genes corresponding to the appropriate mating type and to the determination of the direction of switching of mating type (see Figure 19.33). *MAT***a** cells activate a recombination enhancer on the left arm of chromosome III, which increases recombination over a 40 kb region that includes *HML*. *MAT*α cells inactivate the left end of chromosome III.

19.22 Silent Cassettes at *HML* and *HMR* Are Repressed

Key concepts

- *HML* and *HMR* are repressed by silencer elements.
- Loci required to maintain silencing include *SIR1-4*, *RAP1*, and genes for histone H4.
- Binding of ORC (origin recognition complex) at the silencers is necessary for inactivation.

The transcription map in Figure 19.34 reveals an intriguing feature: transcription of either *MATa* or *MAT*α initiates within the *Y* region. Only the *MAT* locus is expressed, yet the same *Y* region is present in the corresponding non-transcribed cassette (*HML* or *HMR*). This implies that regulation of expression is not accomplished by direct recognition of some site overlapping with the promoter. *A site outside the cassettes must distinguish HML and HMR from MAT.*

Deletion analysis shows that sites on either side of both *HML* and *HMR* are needed to repress their expression. They are called **silencers.** The sites on the left side of each cassette are called the E silencers, and the sites on the right side are called the I silencers. These control sites can function at a distance (up to 2.5 kb away from a promoter) and in either orientation. They behave like negative enhancers (enhancers are elements distant from the promoter that activate transcription; see Section 24.15, Enhancers Contain Bidirectional Elements That Assist Initiation).

Can we find the basis for the control of cassette activity by identifying genes that are responsible for keeping the cassettes silent? We would expect the products of these genes to act on the silencers. A convenient assay for mutation in such genes is provided by the fact that, when a mutation allows the usually silent cassettes at *HML* and *HMR* to be expressed, both *a* and α functions are produced, so the cells behave like *MATa*/*MAT*α diploids.

Mutations in several loci abolish silencing and lead to expression of *HML* and *HMR*. The first to be discovered were the four *SIR* loci (silent information regulators). All four wild-type *SIR* loci are needed to maintain *HML* and *HMR* in the repressed state; mutation in any one of these loci to give a *sir*⁻ allele has two effects. Both *HML* and *HMR* can be transcribed, and both the silent cassettes become targets for replacement by switching. Thus the same regulatory event is involved in repressing a silent cassette and in preventing it from being a recipient for replacement by another cassette.

Other loci required for silencing include *RAP1* (which is also required to maintain telomeric heterochromatin in its inert state) and the genes coding for histone H4. Deletions of the N-terminus of histone H4 or individual point mutations activate the silent cassettes. The effects of these mutations can be overcome either by introducing new mutations in *SIR3* or by overexpressing *SIR1*, which suggests that there is a specific interaction between H4 and the SIR proteins.

The general model suggested by these results is that the SIR proteins act on chromatin structure to prevent expression of the genes. Mutations in the SIR proteins have the same effects on genes that have been inactivated by the proximity of telomeric heterochromatin, so it seems likely that SIR proteins are involved generally in interacting with histones to form heterochromatic (inert) structures (see Section 31.3, Heterochromatin Depends on Interactions with Histones).

There is an interesting connection between repression at the silencers and DNA replication.

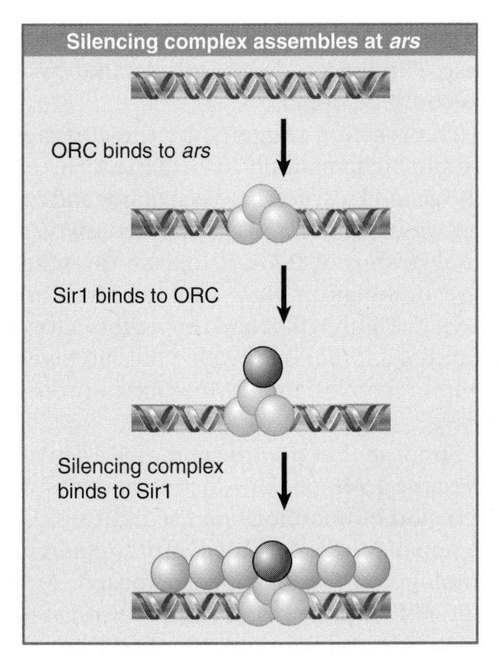

FIGURE 19.37 Silencing at HMR depends on recruiting Sir1.

Each silencer contains an *ARS* sequence (an origin of replication). The *ARS* is bound by the ORC (origin recognition complex) that is involved in initiating replication. Mutations in *ORC* genes prevent silencing, indicating that the binding of ORC protein at the silencer is required for silencing.

There are two separate types of connection between silencing and the replication apparatus:

- the presence of Sir1 is necessary, and
- replication is required.

If a Sir1 protein is localized at the silencer (by linkage to another protein that is bound there), the binding of ORC is no longer necessary. This means that the role of ORC is solely to bring in Sir1; it is not required to initiate replication. As illustrated in FIGURE 19.37, the role of ORC could, therefore, be to provide an initiating center from which the silencing effect can spread. ORC provides the structure to which Sir1 binds, and Sir1 then recruits the other SIR proteins. This is different from its role in replication.

Passage through S phase, however, is necessary for silencing to be established. This does not require initiation to occur at the *ARS* in the silencer. The effect could depend on the passage of a replication fork through the silencer, perhaps in order to allow the chromatin structure to be changed.

19.23 Unidirectional Transposition Is Initiated by the Recipient *MAT* Locus

Key concept

- Mating type switching is initiated by a double-strand break made at the *MAT* locus by the HO endonuclease.

A switch in mating type is accomplished by a gene conversion in which the recipient site *(MAT)* acquires the sequence of the donor type (*HML* or *HMR*). Sites needed for transposition have been identified by mutations at *MAT* that prevent switching. The unidirectional nature of the process is indicated by lack of mutations in *HML* or *HMR*.

The mutations identify a site at the right boundary of *Y* at *MAT* that is crucial for the switching event. The nature of the boundary is shown by analyzing the locations of these point mutations relative to the site of switching (this is done by examining the results of rare switches that occur in spite of the mutation). Some mutations lie within the region that is replaced (and thus disappear from *MAT* after a switch), whereas others lie just outside the replaced region (and therefore continue to impede switching). Thus sequences both within and outside the replaced region are needed for the switching event.

Switching is initiated by a double-strand break close to the *Y-Z* boundary that coincides with a site that is sensitive to attack by DNAase. (This is a common feature of chromosomal sites that are involved in initiating transcription or recombination.) It is recognized by an endonuclease coded by the *HO* locus. The HO endonuclease makes a staggered double-strand break just to the right of the *Y* boundary. Cleavage generates the single-stranded ends of four bases drawn in FIGURE 19.38. The nuclease does not attack mutant *MAT* loci that cannot switch. Deletion analysis shows that most or all of the sequence of 24 bp surrounding the *Y* junction is required for cleavage *in vitro*. The recognition site is relatively large for a nuclease, and it occurs only at the three mating-type cassettes.

Only the *MAT* locus, and not the *HML* or *HMR* loci, is a target for the endonuclease. It seems plausible that the same mechanisms that keep the silent cassettes from being transcribed

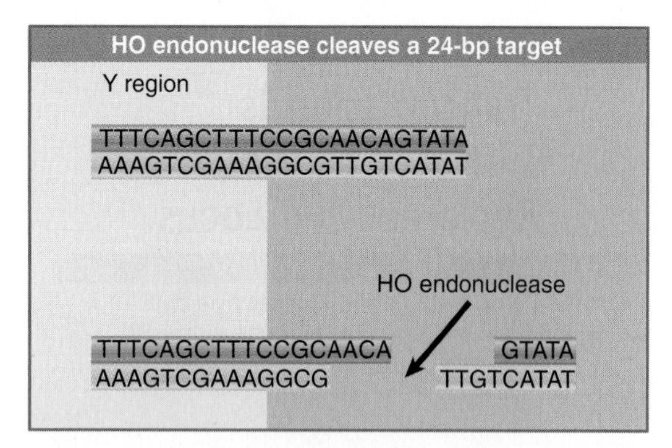

FIGURE 19.38 HO endonuclease cleaves *MAT* just to the right of the Y region, which generates sticky ends with a 4-base overhang.

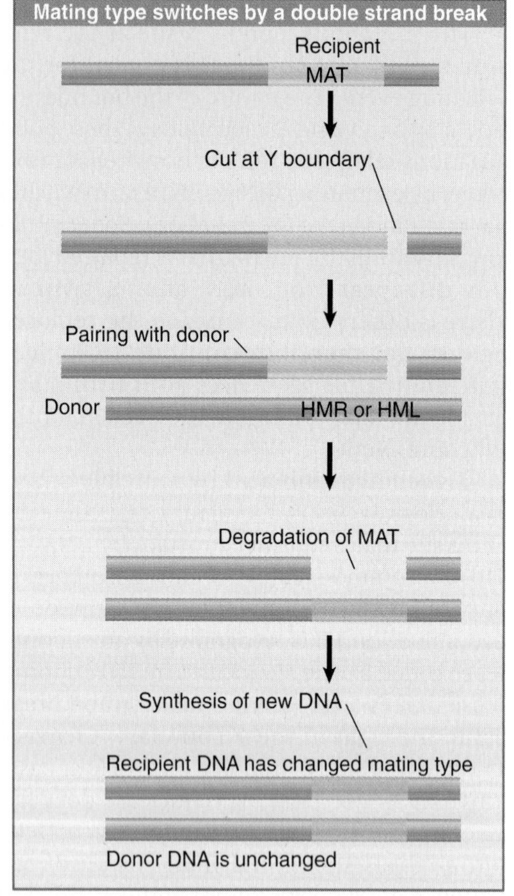

FIGURE 19.39 Cassette substitution is initiated by a double-strand break in the recipient *(MAT)* locus, and may involve pairing on either side of the Y region with the donor (*HMR* or *HML*) locus.

also keep them inaccessible to the HO endonuclease. This inaccessibility ensures that switching is unidirectional.

The reaction triggered by the cleavage is illustrated schematically in FIGURE 19.39 in terms of the general reaction between donor and recipient regions. In terms of the interactions of individual strands of DNA, it follows the scheme for recombination via a double-strand break drawn in Figure 19.9, and the stages following the initial cut require the enzymes involved in general recombination. Mutations in some of these genes prevent switching.

Suppose that the free end of *MAT* invades either the *HML* or *HMR* locus and pairs with the region of homology on the right side. The *Y* region of *MAT* is degraded until a region with homology on the left side is exposed. At this point, *MAT* is paired with *HML* or *HMR* at both the left side and the right side. The *Y* region of *HML* or *HMR* is copied to replace the region lost from *MAT* (which might extend beyond the limits of *Y* itself). The paired loci separate. (The order of events could be different.)

Like the double-strand break model for recombination, the process is initiated by *MAT*, the locus that is to be replaced. In this sense, the description of *HML* and *HMR* as donor loci refers to their ultimate role, but not to the mechanism of the process. Like replicative transposition, the donor site is unaffected, but a change in sequence occurs at the recipient. Unlike transposition, the recipient locus suffers a substitution rather than addition of material.

19.24 Regulation of HO Expression Controls Switching

Key concept

- HO endonuclease is synthesized in haploid mother cells, so that a switching event causes both daughters to have the new mating type.

Production of the HO endonuclease is regulated at the level of gene transcription. There are three separate control systems:

- *HO* is under mating-type control. It is not synthesized in *MAT***a**/*MAT*α diploids. The reason could be that there is no need for switching when both *MAT* alleles are expressed anyway.
- *HO* is transcribed in mother cells but not in daughter cells.

- *HO* transcription also responds to the cell cycle. The gene is expressed only at the end of the G1 phase of a mother cell.

The timing of nuclease production explains the relationship between switching and cell lineage. FIGURE 19.40 shows that switching is detected only in the products of a division; both daughter cells have the same mating type, which is switched from that of the parent. The reason is that the restriction of *HO* expression to G1 phase ensures that the mating type is switched before the *MAT* locus is replicated, with the result that both progeny have the new mating type.

cis-acting sites that control *HO* transcription reside in the 1500 bp upstream of the gene. The general pattern of control is that repression at any one of many sites, responding to several regulatory circuits, may prevent transcription of *HO*. FIGURE 19.41 summarizes the types of sites that are involved.

Mating type control resembles that of other haploid-specific genes. Transcription is prevented (in diploids) by the *a1/α2* repressor. There are ten binding sites for the repressor in the upstream region. These sites vary in their conformity to the consensus sequence; we do not know which and how many of them are required for haploid-specific repression.

The control of *HO* transcription involves interplay between a series of activating and repressing events. The genes *SWI1-5* are required for *HO* transcription. They function by preventing products of the genes *SIN1-6* from repressing *HO*. The *SWI* genes were discovered first, as mutants unable to switch; the *SIN* genes were then discovered for their ability to release the blocks caused by particular *SWI* mutations. *SWI-SIN* interactions are involved in both cell-cycle control and the restriction of expression to mother cells.

Some of the *SWI* and *SIN* genes are not specifically concerned with mating type, but are global regulators of transcription whose functions are needed for expression of many loci. They include the *SWI/SNF* chromatin remodeling complex containing Swi1,2,3 and the loci *SIN1-4* that code for chromosomal proteins or chromatin in regulators. Their role in mating type expression is incidental. The "real" regulator is therefore the Swi protein that counteracts the general repression system specifically at the *HO* locus.

Cell-cycle control is conferred by nine copies of an octanucleotide sequence called *URS2*. A copy of the consensus sequence can confer cell-cycle control on a gene to which it is attached. A gene linked to this sequence is repressed except during a transient period toward the end of G1 phase. Swi4 and Swi6 are the activators that release repression at *URS2*. Their activity depends on the function of the cell-cycle regulator Cdc28, which executes the decision that commits the cell to divide.

The target for restricting expression to alternate generations is the activator Swi5 (which antagonizes a general repression system exercised by Sin3,4). In mutants that lack these functions, *HO* is transcribed equally well in mother

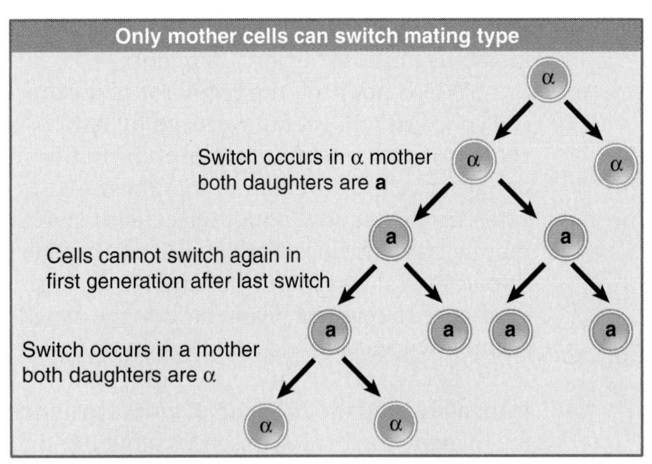

FIGURE 19.40 Switching occurs only in mother cells; both daughter cells have the new mating type. A daughter cell must pass through an entire cycle before it becomes a mother cell that is able to switch again.

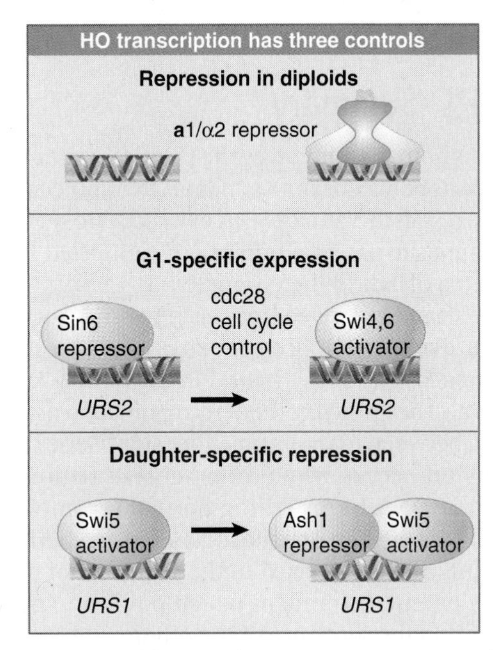

FIGURE 19.41 Three regulator systems act on transcription of the *HO* gene. Transcription occurs only when all repression is lifted.

and daughter cells. This system acts on *URS1* elements in the far upstream region.

SWI5 is not itself the regulator of mother-cell specificity, but is antagonized by Ash1p, a repressor that accumulates preferentially in daughter cells at the end of anaphase. Mutations in *ASH1* allow daughter cells to switch mating type. The localization of Ash1p is determined by the transport of its mRNA from the mother cell along actin filaments into the daughter bud (see Section 7.16, mRNA Can Be Specifically Localized). Its presence prevents SWI5 from activating the *HO* gene. It works by binding to many copies of a consensus sequence that are distributed throughout the regulatory regions *URS1* and *URS2*. When the daughter cell grows to become a mother cell, the concentration of Ash1p is diluted, and it becomes possible to express the *HO* gene again.

19.25 Summary

Recombination involves the physical exchange of parts between corresponding DNA molecules. This results in a duplex DNA in which two regions of opposite parental origins are connected by a stretch of hybrid (heteroduplex) DNA, in which one strand is derived from each parent. Correction events may occur at sites that are mismatched within the hybrid DNA. Hybrid DNA can also be formed without recombination occurring between markers on either side. Gene conversion occurs when an extensive region of hybrid DNA forms during normal recombination (or between nonallelic genes in an aberrant event) and is corrected to the sequence of only one parental strand, at which point one gene takes on the sequence of the other.

Recombination is initiated by a double-strand break in DNA. The break is enlarged to a gap with a single-stranded end; the free single-stranded end then forms a heteroduplex with the allelic sequence. The DNA in which the break occurs actually incorporates the sequence of the chromosome that it invades, so the initiating DNA is called the recipient. Hotspots for recombination are sites where double-strand breaks are initiated. A gradient of gene conversion is determined by the likelihood that a sequence near the free end will be converted to a single strand; this decreases with distance from the break.

Recombination is initiated in yeast by Spo11, a topoisomerase-like enzyme that becomes linked to the free 5′ ends of DNA. The DSB is then processed by generating single-stranded DNA that can anneal with its complement in the other chromosome. Yeast mutations that block synaptonemal complex formation show that recombination is required for its formation. Formation of the synaptonemal complex may be initiated by double-strand breaks, and it may persist until recombination is completed. Mutations in components of the synaptonemal complex block its formation but do not prevent chromosome pairing, so homolog recognition is independent of recombination and synaptonemal complex formation.

The full set of reactions required for recombination can be undertaken by the Rec and Ruv proteins of *E. coli*. A single-stranded region with a free end is generated by the RecBCD nuclease. The enzyme binds to DNA on one side of a *chi* sequence and then moves to the *chi* sequence, unwinding DNA as it progresses. A single-strand break is made at the *chi* sequence. *chi* sequences provide hotspots for recombination. The single-strand provides a substrate for RecA, which has the ability to synapse homologous DNA molecules by sponsoring a reaction in which a single strand from one molecule invades a duplex of the other molecule. Heteroduplex DNA is formed by displacing one of the original strands of the duplex. These actions create a recombination junction, which is resolved by the Ruv proteins. RuvA and RuvB act at a heteroduplex, and RuvC cleaves Holliday junctions.

Recombination, like replication and (probably) transcription, requires topological manipulation of DNA. Topoisomerases may relax (or introduce) supercoils in DNA, and are required to disentangle DNA molecules that have become catenated by recombination or by replication. Type I topoisomerases introduce a break in one strand of a DNA duplex; type II topoisomerases make double-stranded breaks. The enzyme becomes linked to the DNA by a bond from tyrosine to either 5′ phosphate (type A enzymes) or 3′ phosphate (type B enzymes).

The enzymes involved in site-specific recombination have actions related to those of topoisomerases. Among this general class of recombinases, those concerned with phage integration form the subclass of integrases. The Cre/*lox* system uses two molecules of Cre to bind to each *lox* site, so that the recombining complex is a tetramer. This is one of the standard systems for inserting DNA into a foreign genome. Phage lambda integration requires the phage Int protein and host IHF protein and involves a precise breakage and reunion in the absence of any synthesis of DNA. The reaction involves wrapping of the *attP* sequence of phage DNA into the nucleoprotein structure of the intasome, which contains several copies of Int and IHF; the host *attB* sequence is then bound and recombination occurs. Reaction in the reverse direction requires the phage protein Xis. Some integrases function by *cis*-cleavage, where the tyrosine that reacts with DNA in a half site is provided by the enzyme subunit bound to that half site; others function by *trans*-cleavage, for which a different protein subunit provides the tyrosine.

The yeast *S. cerevisiae* can propagate in either the haploid or diploid condition. Conversion between these states takes place by mating (fusion of haploid cells to give a diploid) and by sporulation (meiosis of diploids to give haploid spores). The ability to engage in these activities is determined by the mating type of the strain. The mating type is determined by the sequence of the *MAT* locus, and can be changed by a recombination event that substitutes a different sequence at this locus. The recombination event is initiated by a double-strand break—such as a homologous recombination event—but then the subsequent events ensure a unidirectional replacement of the sequence at the *MAT* locus.

Replacement is regulated so that *MATa* is usually replaced by the sequence from *HMLα*, whereas *MATα* is usually replaced by the sequence from *HMRa*. The endonuclease *HO* triggers the reaction by recognizing a unique target site at *MAT*. *HO* is regulated at the level of transcription by a system that ensures its expression in mother cells but not daughter cells, with the consequence that both progeny have the same (new) mating type.

References

19.4 Double-Strand Breaks Initiate Recombination

Research

Hunter, N. and Kleckner, N. (2001). The single-end invasion: an asymmetric intermediate at the double-strand break to double-Holliday junction transition of meiotic recombination. *Cell* 106, 59–70.

Reviews

Lichten, M. and Goldman, A. S. (1995). Meiotic recombination hotspots. *Annu. Rev. Genet.* 29, 423–444.

Szostak, J. W., Orr-Weaver, T. L., Rothstein, R. J., and Stahl, F. W. (1983). The double-strand-break repair model for recombination. *Cell* 33, 25–35.

19.5 Recombining Chromosomes Are Connected by the Synaptonemal Complex

Research

Blat, Y, and Kleckner, N. (1999). Cohesins bind to preferential sites along yeast chromosome III, with differential regulation along arms versus the central region. *Cell* 98, 249–259.

Dong, H. and Roeder, G. S. (2000). Organization of the yeast Zip1 protein within the central region of the synaptonemal complex. *J. Cell Biol.* 148, 417–426.

Klein, F. et al. (1999). A central role for cohesins in sister chromatid cohesion, formation of axial elements, and recombination during yeast meiosis. *Cell* 98, 91–103.

Sym, M., Engebrecht, J. A., and Roeder, G. S. (1993). ZIP1 is a synaptonemal complex protein required for meiotic chromosome synapsis. *Cell* 72, 365–378.

Reviews

Roeder, G. S. (1997). Meiotic chromosomes: it takes two to tango. *Genes Dev.* 11, 2600–2621.

Zickler, D. and Kleckner, N. (1999). Meiotic chromosomes: integrating structure and function. *Annu. Rev. Genet.* 33, 603–754.

19.6 The Synaptonemal Complex Forms After Double-Strand Breaks

Research

Allers, T. and Lichten, M. (2001). Differential timing and control of noncrossover and crossover recombination during meiosis. *Cell* 106, 47–57.

Weiner, B. M. and Kleckner, N. (1994). Chromosome pairing via multiple interstitial interactions before and during meiosis in yeast. *Cell* 77, 977–991.

Reviews

McKim, K. S., Jang, J. K., and Manheim, E. A. (2002). Meiotic recombination and chromosome segregation in *Drosophila* females. *Annu. Rev. Genet.* 36, 205–232.

Petes, T. D. (2001). Meiotic recombination hot spots and cold spots. *Nat. Rev. Genet.* 2, 360–369.

19.8 The Bacterial RecBCD System Is Stimulated by *chi* Sequences

Research

Dillingham, M. S., Spies, M., and Kowalczykowski, S. C. (2003). RecBCD enzyme is a bipolar DNA helicase. *Nature* 423, 893–897.

Spies, M., Bianco, P. R., Dillingham, M. S., Handa, N., Baskin, R. J., and Kowalczykowski, S. C. (2003). A molecular throttle: the recombination hotspot

chi controls DNA translocation by the RecBCD helicase. *Cell* 114, 647–654.
Taylor, A. F. and Smith, G. R. (2003). RecBCD enzyme is a DNA helicase with fast and slow motors of opposite polarity. *Nature* 423, 889–893.

19.9 Strand-Transfer Proteins Catalyze Single-Strand Assimilation

Reviews

Kowalczykowski, S. C., Dixon, D. A., Eggleston, A. K., Lauder, S. D., and Rehrauer, W. M. (1994). Biochemistry of homologous recombination in *Escherichia coli*. *Microbiol. Rev.* 58, 401–465.
Kowalczykowski, S. C. and Eggleston, A. K. (1994). Homologous pairing and DNA strand-exchange proteins. *Annu. Rev. Biochem.* 63, 991–1043.
Lusetti, S. L. and Cox, M. M. (2002). The bacterial RecA protein and the recombinational DNA repair of stalled replication forks. *Annu. Rev. Biochem.* 71, 71–100.

19.10 The Ruv System Resolves Holliday Junctions

Research

Boddy, M. N., Gaillard, P. H., McDonald, W. H., Shanahan, P., Yates, J. R., and Russell, P. (2001). Mus81-Eme1 are essential components of a Holliday junction resolvase. *Cell* 107, 537–548.
Chen, X. B., Melchionna, R., Denis, C. M., Gaillard, P. H., Blasina, A., Van de Weyer, I., Boddy, M. N., Russell, P., Vialard, J., and McGowan, C. H. (2001). Human Mus81-associated endonuclease cleaves Holliday junctions *in vitro*. *Mol. Cell* 8, 1117–1127.
Constantinou, A., Davies, A. A., and West, S. C. (2001). Branch migration and Holliday junction resolution catalyzed by activities from mammalian cells. *Cell* 104, 259–268.
Kaliraman, V., Mullen, J. R., Fricke, W. M., Bastin-Shanower, S. A., and Brill, S. J. (2001). Functional overlap between Sgs1-Top3 and the Mms4-Mus81 endonuclease. *Genes Dev.* 15, 2730–2740.

Reviews

Lilley, D. M. and White, M. F. (2001). The junction-resolving enzymes. *Nat. Rev. Mol. Cell Biol.* 2, 433–443.
West, S. C. (1997). Processing of recombination intermediates by the RuvABC proteins. *Annu. Rev. Genet.* 31, 213–244.

19.13 Topoisomerases Relax or Introduce Supercoils in DNA

Reviews

Champoux, J. J. (2001). DNA topoisomerases: structure, function, and mechanism. *Annu. Rev. Biochem.* 70, 369–413.
Wang, J. C. (2002). Cellular roles of DNA topoisomerases: a molecular perspective. *Nat. Rev. Mol. Cell Biol.* 3, 430–440.

19.14 Topoisomerases Break and Reseal Strands

Research

Lima, C. D., Wang, J. C., and Mondragon, A. (1994). Three-dimensional structure of the 67K N-terminal fragment of *E. coli* DNA topoisomerase I. *Nature* 367, 138–146.

Review

Champoux, J. J. (2001). DNA topoisomerases: structure, function, and mechanism. *Annu. Rev. Biochem.* 70, 369–413.

19.16 Specialized Recombination Involves Specific Sites

Research

Metzger, D., Clifford, J., Chiba, H., and Chambon, P. (1995). Conditional site-specific recombination in mammalian cells using a ligand-dependent chimeric Cre recombinase. *Proc. Natl. Acad. Sci. USA* 92, 6991–6995.
Nunes-Duby, S. E., Kwon, H. J., Tirumalai, R. S., Ellenberger, T., and Landy, A. (1998). Similarities and differences among 105 members of the Int family of site-specific recombinases. *Nucleic Acids Res.* 26, 391–406.

Review

Craig, N. L. (1988). The mechanism of conservative site-specific recombination. *Annu. Rev. Genet.* 22, 77–105.

19.18 Site-Specific Recombination Resembles Topoisomerase Activity

Research

Guo, F., Gopaul, D. N., and van Duyne, G. D. (1997). Structure of Cre recombinase complexed with DNA in a site-specific recombination synapse. *Nature* 389, 40–46.

19.19 Lambda Recombination Occurs in an Intasome

Research

Biswas, T., Aihara, H., Radman-Livaja, M., Filman, D., Landy, A., and Ellenberger, T. (2005). A structural basis for allosteric control of DNA recombination by lambda integrase. *Nature* 435, 1059–1066.
Wojciak, J. M., Sarkar, D., Landy, A., and Clubb, R. T. (2002). Arm-site binding by lambda integrase: solution structure and functional characterization of its amino-terminal domain. *Proc. Natl. Acad. Sci. USA* 99, 3434–3439.

20

Repair Systems

CHAPTER OUTLINE

20.1 Introduction

20.2 Repair Systems Correct Damage to DNA

- Repair systems recognize DNA sequences that do not conform to standard base pairs.
- Excision systems remove one strand of DNA at the site of damage and then replace it.
- Recombination-repair systems use recombination to replace the double-stranded region that has been damaged.
- All these systems are prone to introducing errors during the repair process.
- Photoreactivation is a nonmutagenic repair system that acts specifically on pyrimidine dimers.

20.3 Excision Repair Systems in *E. coli*

- The Uvr system makes incisions ~12 bases apart on both sides of damaged DNA, removes the DNA between them, and resynthesizes new DNA.

20.4 Excision-Repair Pathways in Mammalian Cells

- Mammalian excision repair is triggered by directly removing a damaged base from DNA.
- Base removal triggers the removal and replacement of a stretch of polynucleotides.
- The nature of the base removal reaction determines which of two pathways for excision repair is activated.
- The polδ/ε pathway replaces a long polynucleotide stretch; the polβ pathway replaces a short stretch.

20.5 Base Flipping Is Used by Methylases and Glycosylases

- Uracil and alkylated bases are recognized by glycosylases and removed directly from DNA.
- Pyrimidine dimers are reversed by breaking the covalent bonds between them.
- Methylases add a methyl group to cytosine.
- All these types of enzyme act by flipping the base out of the double helix where, depending on the reaction, it is either removed or is modified and returned to the helix.

20.6 Error-Prone Repair and Mutator Phenotypes

- Damaged DNA that has not been repaired causes DNA polymerase III to stall during replication.
- DNA polymerase V (coded by *umuCD*) or DNA polymerase IV (coded by *dinB*) can synthesize a complement to the damaged strand.
- The DNA synthesized by the repair DNA polymerase often has errors in its sequence.
- Proteins that affect the fidelity of replication may be identified by mutator genes, in which mutation causes an increased rate of spontaneous mutation.

20.7 Controlling the Direction of Mismatch Repair

- The *mut* genes code for a mismatch-repair system that deals with mismatched base pairs.
- There is a bias in the selection of which strand to replace at mismatches.
- The strand lacking methylation at a hemimethylated $\genfrac{}{}{0pt}{}{\text{GATC}}{\text{CTAG}}$ is usually replaced.
- This repair system is used to remove errors in a newly synthesized strand of DNA. At G-T and C-T mismatches, the T is preferentially removed.

20.8 Recombination-Repair Systems in *E. coli*

- The *rec* genes of *E. coli* code for the principal retrieval system.
- The principal retrieval system functions when replication leaves a gap in a newly synthesized strand that is opposite a damaged sequence.
- The single strand of another duplex is used to replace the gap.
- The damaged sequence is then removed and resynthesized.

20.9 Recombination Is an Important Mechanism to Recover from Replication Errors

- A replication fork may stall when it encounters a damaged site or a nick in DNA.
- A stalled fork may reverse by pairing between the two newly synthesized strands.
- A stalled fork may restart after repairing the damage and use a helicase to move the fork forward.
- The structure of the stalled fork is the same as a Holliday junction and may be converted to a duplex and DSB by resolvases.

20.10 RecA Triggers the SOS System

- Damage to DNA causes RecA to trigger the SOS response, which consists of genes coding for many repair enzymes.
- RecA activates the autocleavage activity of LexA.
- LexA represses the SOS system; its autocleavage activates those genes.

Continued on next page

20.11 Eukaryotic Cells Have Conserved Repair Systems

- The yeast *RAD* mutations, identified by radiation sensitive phenotypes, are in genes that code for repair systems.
- Xeroderma pigmentosum (XP) is a human disease caused by mutations in any one of several repair genes.
- A complex of proteins including XP products and the transcription factor TF_{IIH} provides a human excision-repair mechanism.
- Transcriptionally active genes are preferentially repaired.

20.12 A Common System Repairs Double-Strand Breaks

- The NHEJ pathway can ligate blunt ends of duplex DNA.
- Mutations in the NHEJ pathway cause human diseases.

20.13 Summary

20.1 Introduction

Any event that introduces a deviation from the usual double-helical structure of DNA is a threat to the genetic constitution of the cell. Injury to DNA is minimized by systems that recognize and correct the damage. The repair systems are as complex as the replication apparatus itself, which indicates their importance for the survival of the cell. When a repair system reverses a change to DNA, there is no consequence. A mutation may result, though, when it fails to do so. The measured rate of mutation reflects a balance between the number of damaging events occurring in DNA and the number that have been corrected (or miscorrected).

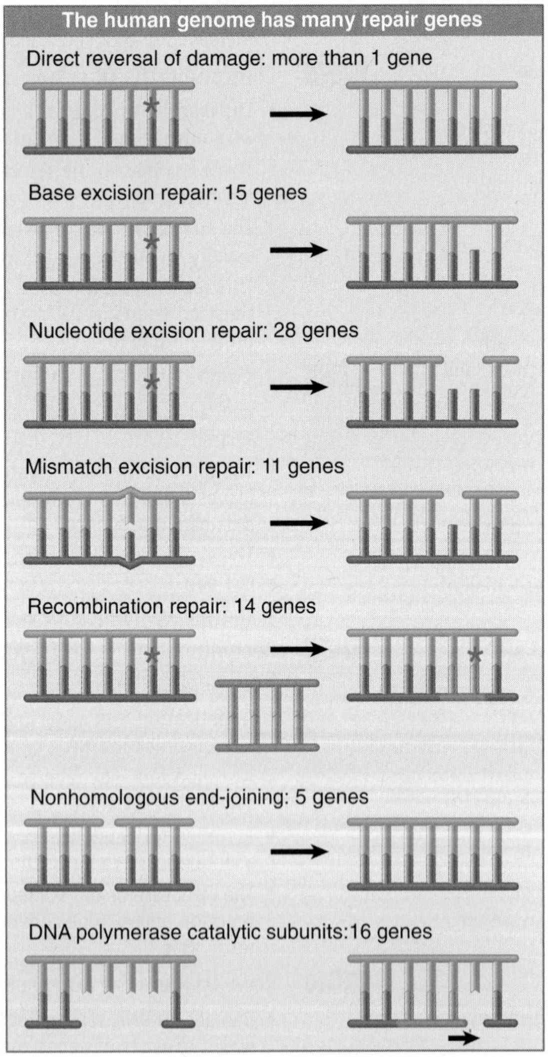

The human genome has many repair genes

Direct reversal of damage: more than 1 gene

Base excision repair: 15 genes

Nucleotide excision repair: 28 genes

Mismatch excision repair: 11 genes

Recombination repair: 14 genes

Nonhomologous end-joining: 5 genes

DNA polymerase catalytic subunits:16 genes

FIGURE 20.1 Repair genes can be classified into pathways that use different mechanisms to reverse or bypass damage to DNA.

Repair systems often can recognize a range of distortions in DNA as signals for action, and a cell is likely to have several systems able to deal with DNA damage. The importance of DNA repair in eukaryotes is indicated by the identification of >130 repair genes in the human genome. We may divide the repair systems into several general types, as summarized in FIGURE 20.1

- Some enzymes directly reverse specific sorts of damage to DNA.
- There are pathways for base excision repair, nucleotide excision repair, and mismatch repair, all of which function by removing and replacing material.
- There are systems that function by using recombination to retrieve an undamaged copy that is then used to replace a damaged duplex sequence.
- The nonhomologous end-joining pathway rejoins broken double-stranded ends.
- Several different DNA polymerases can resynthesize stretches of replacement DNA.

Direct repair is rare and involves the reversal or simple removal of the damage. *Photoreactivation* of pyrimidine dimers, in which the offending covalent bonds are reversed by a light-dependent enzyme, is a good example.

This system is widespread in nature, ocurring in all but placental mammals, and appears to be especially important in plants. In *Escherichia coli* it depends on the product of a single gene *(phr)* that codes for an enzyme called photolyase.

Mismatches between the strands of DNA are one of the major targets for repair systems. *Mismatch repair* is accomplished by scrutinizing DNA for apposed bases that do not pair properly. Mismatches that arise during replication are corrected by distinguishing between the "new" and "old" strands and preferentially correcting the sequence of the newly synthesized strand. Other systems deal with mismatches generated by base conversions, such as the result of deamination. The importance of these systems is emphasized by the fact that cancer is caused in human populations by mutation of genes related to those involved in mismatch repair in yeast.

Mismatches are usually corrected by *excision repair,* which is initiated by a recognition enzyme that sees an actual damaged base or a change in the spatial path of DNA. There are two types of excision repair system.

- *Base excision repair* systems directly remove the damaged base and replace it in DNA. A good example is DNA uracil glycolase, which removes uracils that are mispaired with guanines (see Section 20.5, Base Flipping Is Used by Methylases and Glycosylases).
- *Nucleotide excision repair* systems excise a sequence that includes the damaged base(s); a new stretch of DNA is then synthesized to replace the excised material. FIGURE 20.2 summarizes the main events in the operation of such a system. Such systems are common. Some recognize general damage to DNA; others act upon specific types of base damage. There are often multiple excision repair systems in a single cell type.

Recombination-repair systems handle situations in which damage remains in a daughter molecule and replication has been forced to bypass the site, which typically creates a gap in the daughter strand. A retrieval system uses recombination to obtain another copy of the sequence from an undamaged source; the copy is then used to repair the gap.

A major feature in recombination and repair is the need to handle double-strand breaks (DSBs). DSBs initiate crossovers in homologous recombination. They can also be created by problems in replication, when they may trigger the use of recombination-repair systems. When DSBs are created by environmental damage (for example, by radiation damage) or are the result of the shortening of telomeres, they can cause mutations. One system for handling DSBs can join together nonhomologous DNA ends.

Mutations that affect the ability of *E. coli* cells to engage in DNA repair fall into groups that correspond to several repair pathways (not necessarily all independent). The major known pathways are the *uvr* excision repair system, the methyl-directed mismatch-repair system, and the *recB* and *recF* recombination and recombination-repair pathways. The enzyme activities associated with these systems are endonucleases and exonucleases (important in removing damaged DNA), resolvases (endonucleases that act specifically on recombinant junctions), helicases to unwind DNA, and DNA polymerases to synthesize new DNA. Some of these enzyme activities are unique to particular repair pathways, whereas others participate in multiple pathways.

The replication apparatus devotes a lot of attention to quality control. DNA polymerases use proofreading to check the daughter strand sequence and to remove errors. Some of the repair systems are less accurate when they synthesize DNA to replace damaged material. For this reason, these systems have been known historically as *error-prone* systems.

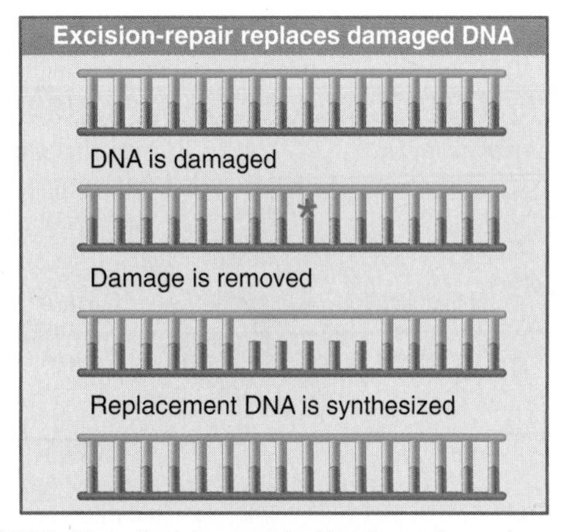

FIGURE 20.2 Excision repair directly replaces damaged DNA and then resynthesizes a replacement stretch for the damaged strand.

20.2 Repair Systems Correct Damage to DNA

Key concepts

- Repair systems recognize DNA sequences that do not conform to standard base pairs.
- Excision systems remove one strand of DNA at the site of damage and then replace it.
- Recombination-repair systems use recombination to replace the double-stranded region that has been damaged.
- All these systems are prone to introducing errors during the repair process.
- Photoreactivation is a nonmutagenic repair system that acts specifically on pyrimidine dimers.

The types of damage that trigger repair systems can be divided into two general classes:

- *Single-base changes* affect the sequence of DNA but not its overall structure. They do not affect transcription or replication, when the strands of the DNA duplex are separated. Thus these changes exert their damaging effects on future generations through the consequences of the change in DNA sequence. The cause of this type of effect is the conversion of one base into another that is not properly paired with the partner base. Single-base changes may happen as the result of mutation of a base *in situ* or by replication errors. FIGURE 20.3 shows that deamination of cytosine to uracil (spontaneously or by chemical mutagen) creates a mismatched U-G pair. FIGURE 20.4 shows that a replication error might insert adenine instead of cytosine to create an A-G pair. Similar consequences could result from covalent addition of a small group to a base that modifies its ability to base pair. These changes may result in very minor structural distortion (as in the case of a U-G pair) or quite significant change (as in the case of an A-G pair), but the common feature is that the mismatch persists only until the next replication. Thus only limited time is available to repair the damage before it is fixed by replication.
- **Structural distortions** may provide a physical impediment to replication or transcription. Introduction of covalent links between bases on one strand of DNA or between bases on opposite strands inhibits replication and transcription. FIGURE 20.5 shows the exam-

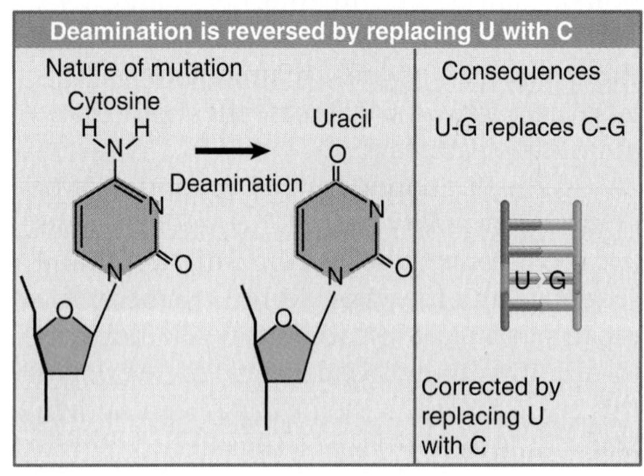

FIGURE 20.3 Deamination of cytosine creates a U-G base pair. Uracil is preferentially removed from the mismatched pair.

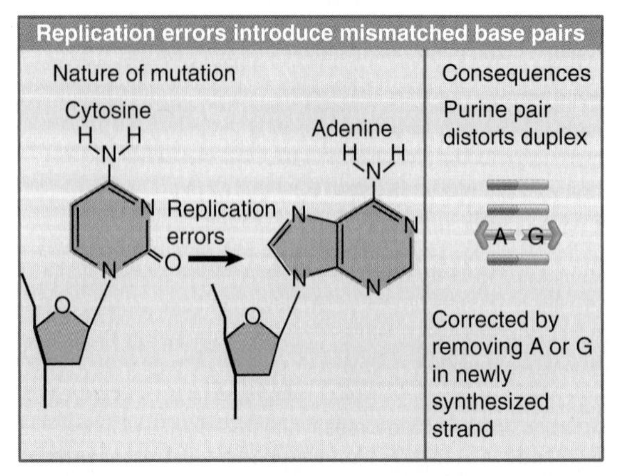

FIGURE 20.4 A replication error creates a mismatched pair that may be corrected by replacing one base; if uncorrected, a mutation is fixed in one daughter duplex.

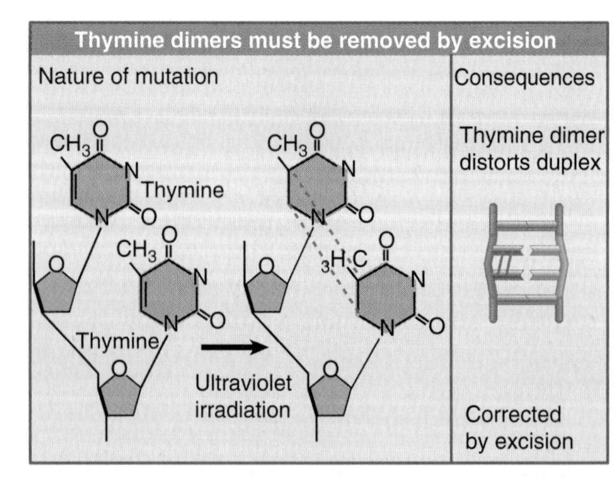

FIGURE 20.5 Ultraviolet irradiation causes dimer formation between adjacent thymines. The dimer blocks replication and transcription.

ple of ultraviolet (UV) irradiation, which introduces covalent bonds between two adjacent thymine bases and results in an intrastrand **pyrimidine dimer.** FIGURE 20.6 shows that similar consequences could result from addition of a bulky adduct to a base that distorts the structure of the double helix. A single-strand nick or the removal of a base, as shown in FIGURE 20.7, prevents a strand from serving as a proper template for synthesis of RNA or DNA. The common feature in all these changes is that the damaged adduct remains in the DNA and continues to cause structural problems and/or induce mutations until it is removed.

When the repair systems are eliminated, cells become exceedingly sensitive to ultraviolet irradiation. The introduction of UV-induced damage has been a major test for repair systems, and so in assessing their activities and relative efficiencies, we should remember that the emphasis might be different if another damaged adduct were studied.

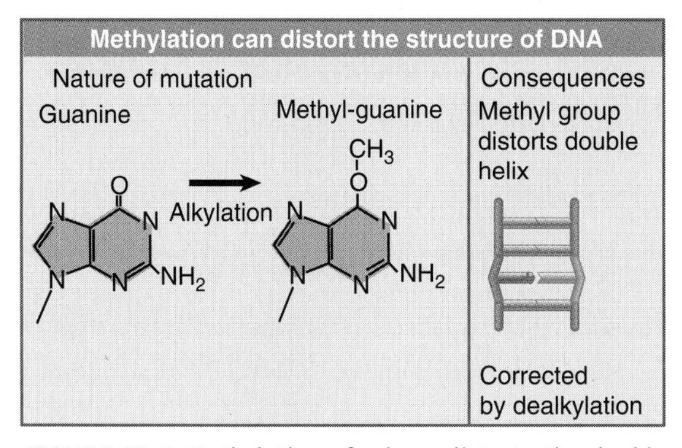

FIGURE 20.6 Methylation of a base distorts the double helix and causes mispairing at replication. Star indicates the methyl group.

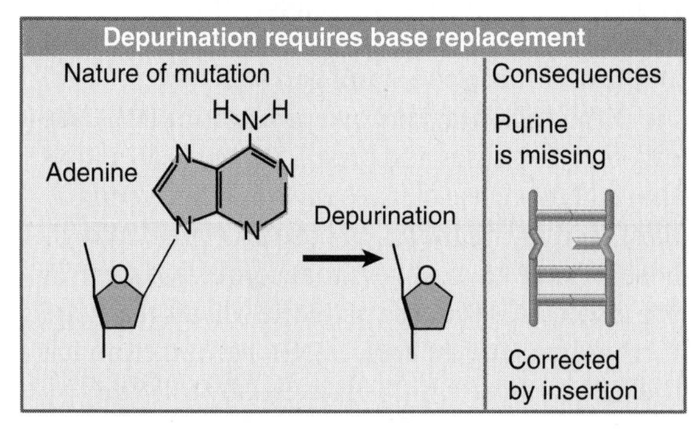

FIGURE 20.7 Depurination removes a base from DNA, blocking replication and transcription.

20.3 Excision Repair Systems in *E. coli*

Key concept

- The Uvr system makes incisions ~12 bases apart on both sides of damaged DNA, removes the DNA between them, and resynthesizes new DNA.

Excision repair systems vary in their specificity, but share the same general features. Each system removes mispaired or damaged bases from DNA and then synthesizes a new stretch of DNA to replace them. The main type of pathway for excision repair is illustrated in FIGURE 20.8.

In the **incision** step, the damaged structure is recognized by an endonuclease that cleaves the DNA strand on both sides of the damage.

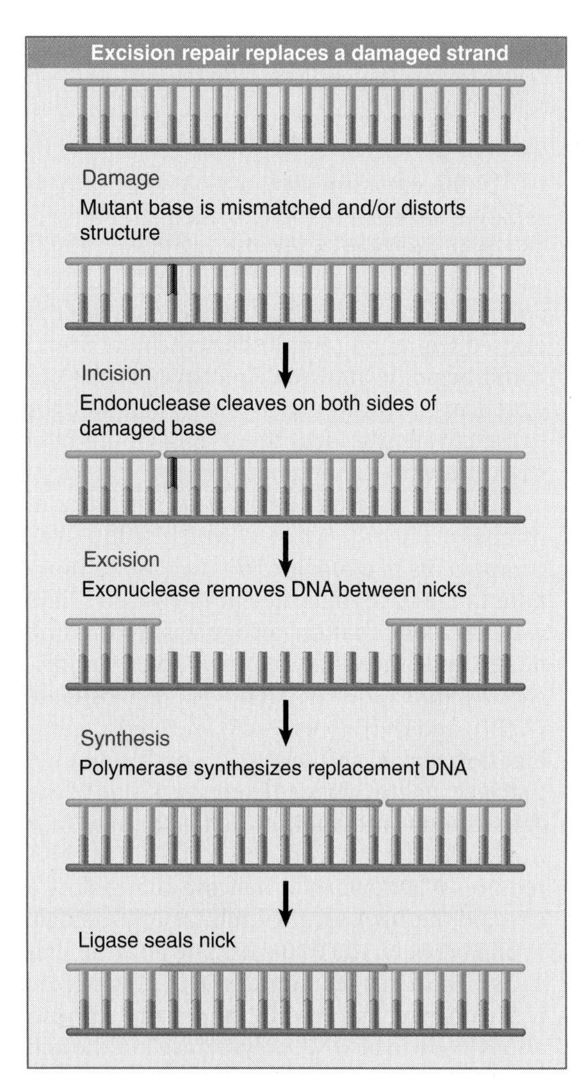

FIGURE 20.8 Excision repair removes and replaces a stretch of DNA that includes the damaged base(s).

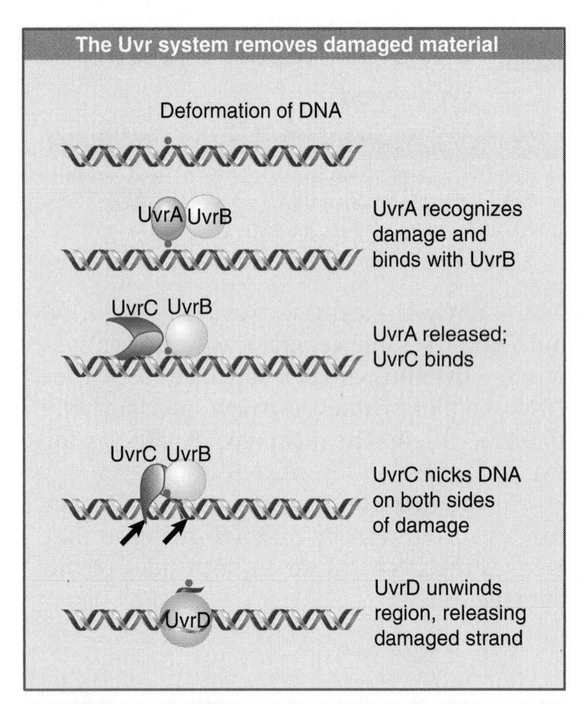

FIGURE 20.9 The Uvr system operates in stages in which UvrAB recognizes damage, UvrBC nicks the DNA, and UvrD unwinds the marked region.

In the **excision** step, a 5′–3′ exonuclease removes a stretch of the damaged strand.

In the *synthesis* step, the resulting single-stranded region serves as a template for a DNA polymerase to synthesize a replacement for the excised sequence. (Synthesis of the new strand could be associated with removal of the old strand, in one coordinated action.) Finally, DNA ligase covalently links the 3′ end of the new DNA strand to the original DNA.

The *uvr* system of excision repair includes three genes, *uvrA, -B, and -C,* which code for the components of a repair endonuclease. It functions in the stages indicated in FIGURE 20.9. First, a UvrAB combination recognizes pyrimidine dimers and other bulky lesions. Next, UvrA dissociates (this requires adenosine triphosphate [ATP]), and UvrC joins UvrB. The UvrBC combination makes an incision on each side, one that is seven nucleotides from the 5′ side of the damaged site and another that is three to four nucleotides away from the 3′ side. This also requires ATP. UvrD is a helicase that helps to unwind the DNA to allow release of the single strand between the two cuts. The enzyme that excises the damaged strand is DNA polymerase I. The enzyme involved in the repair synthesis also is likely to be DNA polymerase I (although DNA polymerases II and III can substitute for it). The events are basically the same, although their order is different in terms of the eukaryotic repair pathway shown in Figure 20.23.

UvrABC repair accounts for virtually all of the excision repair events in *E. coli.* In almost all (99%) of cases, the average length of replaced DNA is ~12 nucleotides. (For this reason, this is sometimes described as short-patch repair.) The remaining 1% of cases involve the replacement of stretches of DNA mostly ~1500 nucleotides long, but extending as much as >9000 nucleotides (sometimes called long-patch repair). We do not know why some events trigger the long-patch rather than short-patch mode.

The Uvr complex can be directed to damaged sites by other proteins. Damage to DNA may prevent transcription, but the situation is handled by a protein called Mfd that displaces the RNA polymerase and recruits the Uvr complex (see Figure 24.19 in Section 24.12, A Connection between Transcription and Repair).

20.4 Excision-Repair Pathways in Mammalian Cells

Key concepts

- Mammalian excision repair is triggered by directly removing a damaged base from DNA.
- Base removal triggers the removal and replacement of a stretch of polynucleotides.
- The nature of the base removal reaction determines which of two pathways for excision repair is activated.
- The polδ/ε pathway replaces a long polynucleotide stretch; the polβ pathway replaces a short stretch.

The general principle of excision-repair in mammalian cells is similar to that of bacteria. The process usually starts in a different way, however, with the removal of an individual damaged base. This serves as the trigger to activate the enzymes that excise and replace a stretch of DNA, including the damaged site.

Enzymes that remove bases from DNA are called glycosylases and lyases. FIGURE 20.10 shows that a glycosylase cleaves the bond between the damaged or mismatched base and the deoxyribose. FIGURE 20.11 shows that some glycosylases are also lyases that can take the reaction a stage further by using an amino (NH_2) group to attack the deoxyribose ring. This is usually followed by

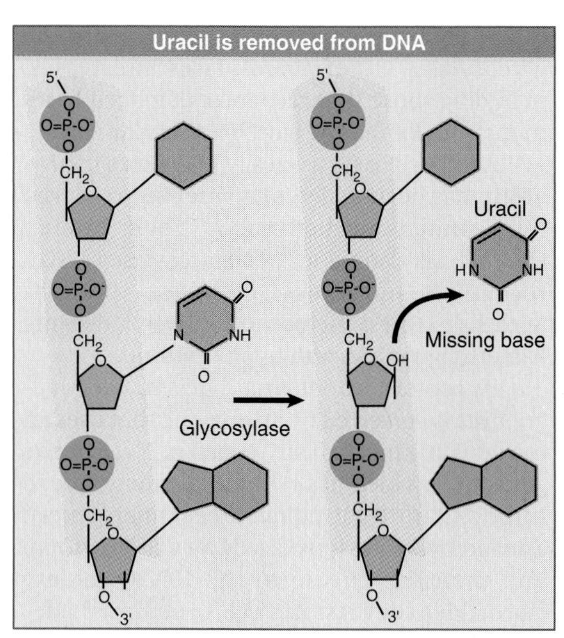

FIGURE 20.10 A glycosylase removes a base from DNA by cleaving the bond to the deoxyribose.

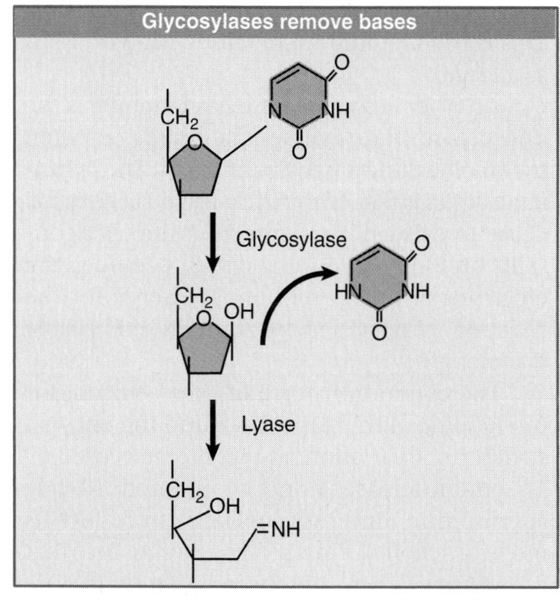

FIGURE 20.11 A glycosylase hydrolyzes the bond between base and deoxyribose (using H_2O), but a lyase takes the reaction further by opening the sugar ring (using NH_2).

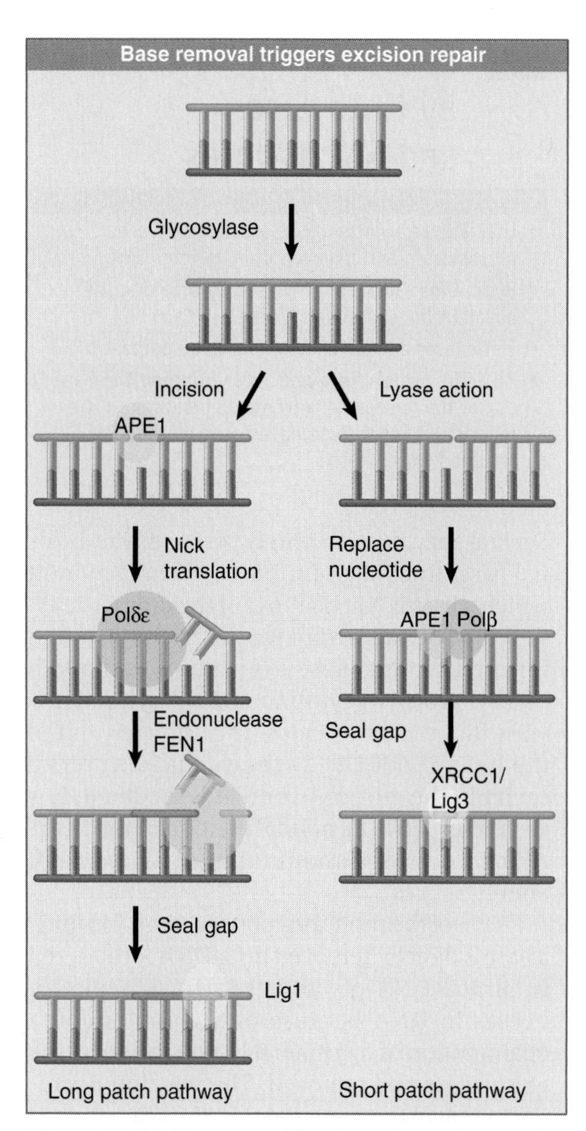

FIGURE 20.12 Base removal by glycosylase or lyase action triggers mammalian excision-repair pathways.

a reaction that introduces a nick into the polynucleotide chain.

FIGURE 20.12 shows that the exact form of the pathway depends on whether the damaged base is removed by a glycosylase or lyase.

Glycosylase action is followed by the endonuclease APE1, which cleaves the polynucleotide chain on the 5′ side. This in turn attracts a replication complex including the DNA polymerase δ/ε and ancillary components, which undertakes a process of nick translation extending for two to ten nucleotides. The displaced material is removed by the endonuclease FEN1. The enzyme ligase-1 seals the chain. This is called the long patch pathway.

When the initial removal involves a lyase action, endonuclease APE1 combines with DNA polymerase β to replace a single nucleotide. The nick is then sealed by the ligase action of XRCC1/ligase-3. This is called the short patch pathway.

20.5 Base Flipping Is Used by Methylases and Glycosylases

Key concepts

- Uracil and alkylated bases are recognized by glycosylases and removed directly from DNA.
- Pyrimidine dimers are reversed by breaking the covalent bonds between them.
- Methylases add a methyl group to cytosine.
- All these types of enzyme act by flipping the base out of the double helix where, depending on the reaction, it is either removed or is modified and returned to the helix.

Several enzymes that remove or modify individual bases in DNA use a remarkable reaction in which a base is "flipped" out of the double helix. This type of interaction was first demonstrated for methyltransferases—enzymes that add a methyl group to cytosine in DNA. The methylase flips the target cytosine completely out of the helix. FIGURE 20.13 shows that it enters a cavity in the enzyme where it is modified. It is then returned to its normal position in the helix. All this occurs without input of an external energy source.

One of the most common reactions in which a base is directly removed from DNA is catalyzed by uracil-DNA glycosylase. Uracil typically occurs in DNA because of a (spontaneous) deamination of cytosine. It is recognized by the glycosylase and removed. The reaction is similar to that of the methylase: The uracil is flipped out of the helix and into the active site in the glycosylase. Other glycosylases and lyases, including those that recognize damaged bases in mammalian DNA, work in a similar way.

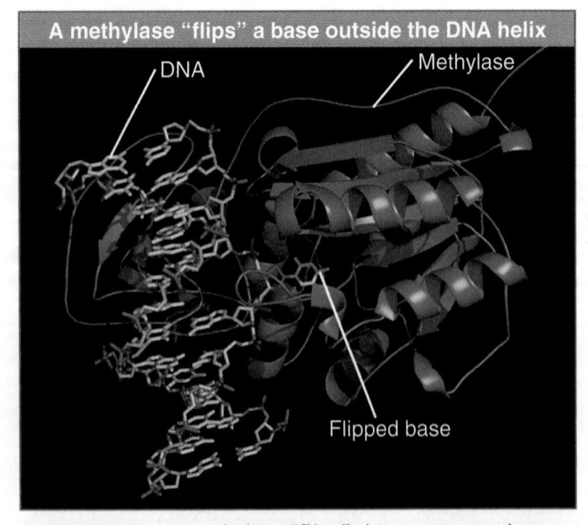

FIGURE 20.13 A methylase "flips" the target cytosine out of the double helix in order to modify it. Photo courtesy of Sanjay Kumar, New England Biolabs.

Alkylated bases (typically in which a methyl group has been added to a base) are removed by a similar mechanism. A single human enzyme, alkyladenine DNA glycosylase (AAG), recognizes and removes a variety of alkylated substrates, including 3-methyladenine, 7-methylguanine, and hypoxanthine.

By contrast with this mechanism, 1-methyladenine is corrected by an enzyme that uses an oxygenating mechanism (coded in *E. coli* by the gene *alkB*, which has homologs widely spread through nature, including three human genes). The methyl group is oxidized to a CH_2OH group, and then the release of the HCHO moiety (formaldehyde) restores the structure of adenine. A very interesting development is the discovery that the bacterial enzyme, and one of the human enzymes, can also repair the same damaged base in RNA. In the case of the human enzyme, the main target may be ribosomal RNA. This is the first known repair event with RNA as a target.

Another enzyme to use base flipping is the photolyase that reverses the bonds between pyrimidine dimers (see Figure 20.5). The pyrimidine dimer is flipped into a cavity in the enzyme. Close to this cavity is an active site that contains an electron donor, which provides the electrons to break the bonds. Energy for the reaction is provided by light in the visible wavelength.

The common feature of these enzymes is the flipping of the target base into the enzyme structure. A variation on this theme is used by T4 endonuclease V, now renamed T4-pdg (pyrimidine dimer glycosylase) to reflect its mode of action. It flips out the adenine base that is *complementary* to the thymine on the 5′ side of the pyrimidine dimer. Thus in this case, the target for the catalytic action of the enzyme remains in the DNA duplex, and the enzyme uses flipping as an indirect mechanism to get access to its target.

When a base is removed from DNA, the reaction is followed by excision of the phosphodiester backbone by an endonuclease, DNA synthesis by a DNA polymerase to fill the gap, and ligation by a ligase to restore the integrity of the polynucleotide chain, as described for the mammalian excision-repair pathway (see Section 20.4, Excision-Repair Pathways in Mammalian Cells).

20.6 Error-Prone Repair and Mutator Phenotypes

Key concepts

- Damaged DNA that has not been repaired causes DNA polymerase III to stall during replication.
- DNA polymerase V (coded by *umuCD*) or DNA polymerase IV (coded by *dinB*) can synthesize a complement to the damaged strand.
- The DNA synthesized by the repair DNA polymerase often has errors in its sequence.
- Proteins that affect the fidelity of replication may be identified by mutator genes, in which mutation causes an increased rate of spontaneous mutation.

The existence of repair systems that engage in DNA synthesis raises the question of whether their quality control is comparable with that of DNA replication. As far as we know, most systems, including *uvr*-controlled excision repair, do not differ significantly from DNA replication in the frequency of mistakes. **Error-prone** synthesis of DNA, however, occurs in *E. coli* under certain circumstances.

The error-prone feature was first observed when it was found that the repair of damaged λ phage DNA is accompanied by the induction of mutations if the phage is introduced into cells that had previously been irradiated with UV. This suggests that the UV irradiation of the host has activated functions that generate mutations. The mutagenic response also operates on the bacterial host DNA.

What is the actual error-prone activity? It is a DNA polymerase that inserts incorrect bases, which represent mutations, when it passes any site at which it cannot insert complementary base pairs in the daughter strand. Functions involved in this error-prone pathway are identified by mutations in the genes *umuD* and *umuC,* which abolish UV-induced mutagenesis. This implies that the UmuC and UmuD proteins cause mutations to occur after UV irradiation. The genes constitute the *umuDC* operon, whose expression is induced by DNA damage. Their products form a complex $UmuD'_2C$, which consists of two subunits of a truncated UmuD protein and one subunit of UmuC. UmuD is cleaved by RecA, which is activated by DNA damage (see Section 20.10, RecA Triggers the SOS System).

The $UmuD'_2C$ complex has DNA polymerase activity. It is called DNA polymerase V, and is responsible for synthesizing new DNA to replace sequences that have been damaged by UV. This is the only enzyme in *E. coli* that can bypass the classic pyrimidine dimers produced by UV (or other bulky adducts). The polymerase activity is error-prone. Mutation of *umuC* or *umuD* inactivate the enzyme, which makes UV irradiation lethal. Some plasmids carry genes called *mucA* and *mucB,* which are homologs of *umuD* and *umuC,* and whose introduction into a bacterium increases resistance to UV killing and susceptibility to mutagenesis.

How does an alternative DNA polymerase get access to the DNA? When the replicase (DNA polymerase III) encounters a block, such as a thymidine dimer, it stalls. It is then displaced from the replication fork and replaced by DNA polymerase V. In fact, DNA polymerase V uses some of the same ancillary proteins as DNA polymerase III. The same situation is true for DNA polymerase IV, the product of *dinB,* which is another enzyme that acts on damaged DNA. DNA polymerases IV and V are part of a larger family, which includes eukaryotic DNA polymerases, whose members are involved in repairing damaged DNA (see Section 20.11, Eukaryotic Cells Have Conserved Repair Systems).

20.7 Controlling the Direction of Mismatch Repair

Key concepts

- The *mut* genes code for a mismatch-repair system that deals with mismatched base pairs.
- There is a bias in the selection of which strand to replace at mismatches.
- The strand lacking methylation at a hemimethylated $\frac{\text{GATC}}{\text{CTAG}}$ is usually replaced.
- This repair system is used to remove errors in a newly synthesized strand of DNA. At G-T and C-T mismatches, the T is preferentially removed.

Genes whose products are involved in controlling the fidelity of DNA synthesis during either replication or repair may be identified by mutations that have a **mutator** phenotype. A mutator mutant has an increased frequency of spontaneous mutation. If identified originally by the mutator phenotype, a gene is described as *mut;* often, though, a *mut* gene is later found to be equivalent with a known replication or repair activity.

The general types of activities identified by *mut* genes fall into two groups:

- The major group consists of components of mismatch-repair systems. Failure to

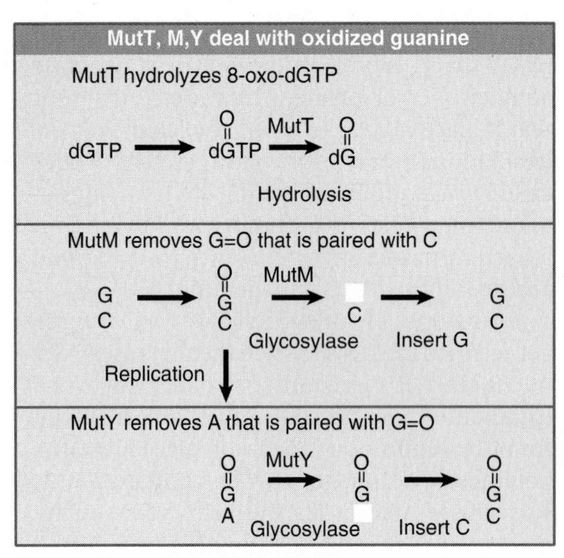

FIGURE 20.14 Preferential removal of bases in pairs that have oxidized guanine is designed to minimize mutations.

remove a damaged or mispaired base before replication allows it to induce a mutation. Functions in this group include the *dam* methylase that identifies the target for repair, and enzymes that participate directly or indirectly in the removal of particular types of damage *(mutH, -S, -L, and -Y)*.

- A smaller group, typified by *dnaQ* (which codes for a subunit of DNA polymerase III), is concerned with the accuracy of synthesizing new DNA.

When a structural distortion is removed from DNA, the wild-type sequence is restored. In most cases, the distortion is due to the creation of a base that is not naturally found in DNA, and which is therefore recognized and removed by the repair system.

A problem arises if the target for repair is a mispaired partnership of (normal) bases created when one was mutated. The repair system has no intrinsic means of knowing which is the wild-type base and which is the mutant! All it sees are two improperly paired bases, either of which can provide the target for excision repair.

If the mutated base is excised, the wild-type sequence is restored. If it happens to be the original (wild-type) base that is excised, though, the new (mutant) sequence becomes fixed. Often, however, the direction of excision repair is not random, but instead is biased in a way that is likely to lead to restoration of the wild-type sequence.

Some precautions are taken to direct repair in the right direction. For example, for cases such as the deamination of 5-methylcytosine to thymine, there is a special system to restore the proper sequence (see also Section 1.15, Many Hotspots Result from Modified Bases). The deamination generates a G-T pair, and the system that acts on such pairs has a bias to correct them to G-C pairs (rather than to A-T pairs). The system that undertakes this reaction includes the MutL, S products that remove T from both G-T and C-T mismatches.

The *mutT, M, Y* system handles the consequences of oxidative damage. A major type of chemical damage is caused by oxidation of G to 8-oxo-G. FIGURE 20.14 shows that the system operates at three levels. MutT hydrolyzes the damaged precursor (8-oxo-dGTP), which prevents it from being incorporated into DNA. When guanine is oxidized in DNA its partner is cytosine, and MutM preferentially removes the 8-oxo-G from 8-oxo-G-C pairs. Oxidized guanine mispairs with A, and so when 8-oxo-G survives and is replicated, it generates an 8-oxo-G-A pair. MutY removes A from these pairs. MutM and MutY are glycosylases that directly remove a base from DNA. This creates an apurinic site that is recognized by an endonuclease whose action triggers the involvement of the excision repair system.

When mismatch errors occur during replication in *E. coli*, it is possible to distinguish the original strand of DNA. Immediately after replication of methylated DNA, only the original parental strand carries the methyl groups. In the period while the newly synthesized strand awaits the introduction of methyl groups, the two strands can be distinguished.

This provides the basis for a system to correct replication errors. The *dam* gene codes for a methylase whose target is the adenine in the sequence $\genfrac{}{}{0pt}{}{\text{GATC}}{\text{CTAG}}$ (see Figure 15.7). The hemimethylated state is used to distinguish replicated origins from nonreplicated origins. The same target sites are used by a replication-related repair system.

FIGURE 20.15 shows that DNA containing mismatched base partners is repaired preferentially by excising the strand that lacks the methylation. The excision is quite extensive; mismatches can be repaired preferentially for >1 kb around a GATC site. The result is that the newly synthesized strand is corrected to the sequence of the parental strand.

E. coli dam$^-$ mutants show an increased rate of spontaneous mutation. This repair system therefore helps reduce the number of mutations caused by errors in replication. It consists

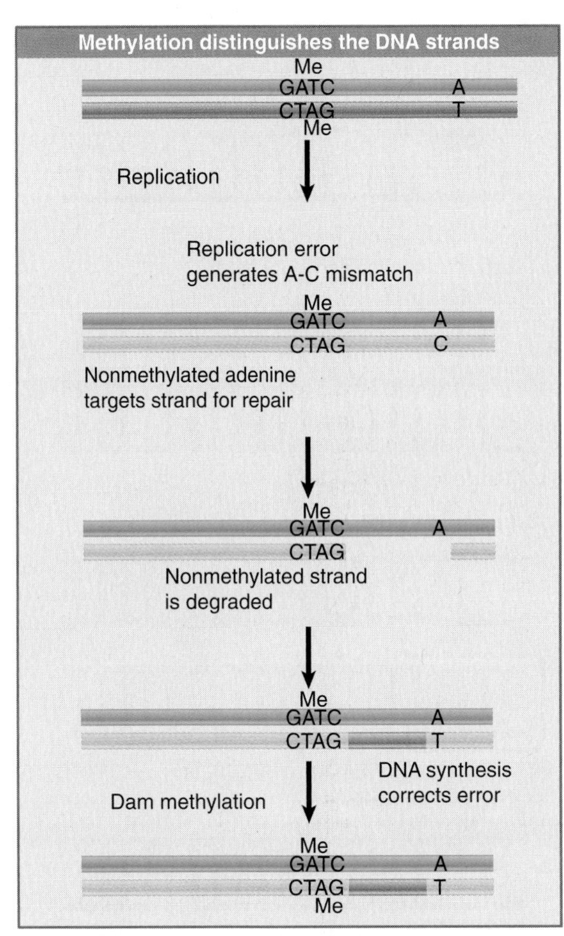

FIGURE 20.15 GATC sequences are targets for the Dam methylase after replication. During the period before this methylation occurs, the nonmethylated strand is the target for repair of mismatched bases.

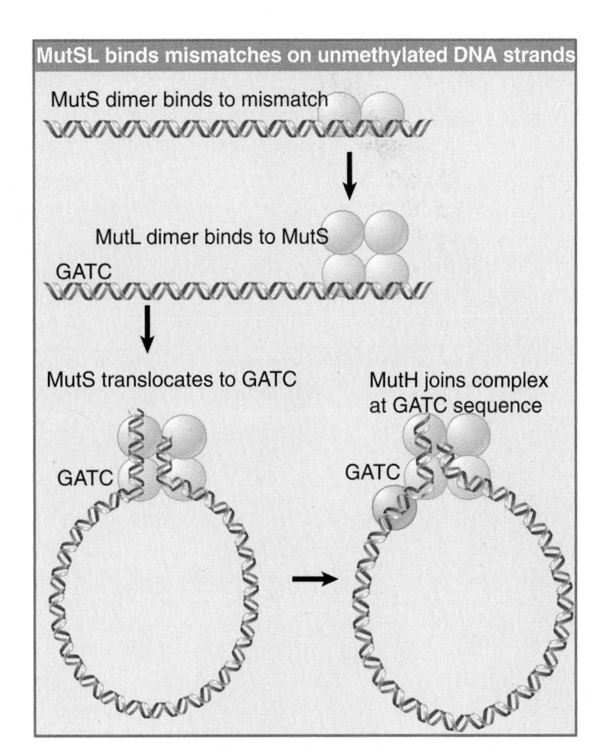

FIGURE 20.16 MutS recognizes a mismatch and translocates to a GATC site. MutH cleaves the unmethylated strand at the GATC. Endonucleases degrade the strand from the GATC to the mismatch site.

of several proteins, coded by the *mut* genes. MutS binds to the mismatch and is joined by MutL. MutS can use two DNA-binding sites, as illustrated in FIGURE 20.16. The first specifically recognizes mismatches. The second is not specific for sequence or structure, and is used to translocate along DNA until a GATC sequence is encountered. Hydrolysis of ATP is used to drive the translocation. MutS is bound to both the mismatch site and to DNA as it translocates, and as a result it creates a loop in the DNA.

Recognition of the GATC sequence causes the MutH endonuclease to bind to MutSL. The endonuclease then cleaves the unmethylated strand. This strand is then excised from the GATC site to the mismatch site. The excision can occur in either the 5′–3′ direction (using RecJ or exonuclease VII) or in the 3′–5′ direction (using exonuclease I), and is assisted by the helicase UvrD. The new DNA strand is synthesized by DNA polymerase III.

The *MSH* repair system of *Sacharomyces cerevisiae* is homologous to the *E. coli mut* system. Msh2 provides a scaffold for the apparatus that recognizes mismatches. Msh3 and Msh6 provide specificity factors. The Msh2-Msh3 complex binds mismatched loops of two to four nucleotides, and the Msh2-Msh6 complex binds to single-base mismatches or insertions or deletions. Other proteins are then required for the repair process itself.

Homologs of the MutSL system also are found in higher eukaryotic cells. In addition to repairing single-base mismatches, they are responsible for repairing mismatches that arise as the result of replication slippage. In a region such as a microsatellite, where a very short sequence is repeated several times, realignment between the newly synthesized daughter strand and its template can lead to a stuttering in which the DNA polymerase slips backward and synthesizes extra repeating units. These units in the daughter strand are extruded as a single-stranded loop from the double helix (see Figure 6.28). They are repaired by homologs of the MutSL system, as shown in FIGURE 20.17.

The importance of the MutSL system for mismatch repair is indicated by the high rate at which it is found to be defective in human cancers. Loss of this system leads to an increased mutation rate.

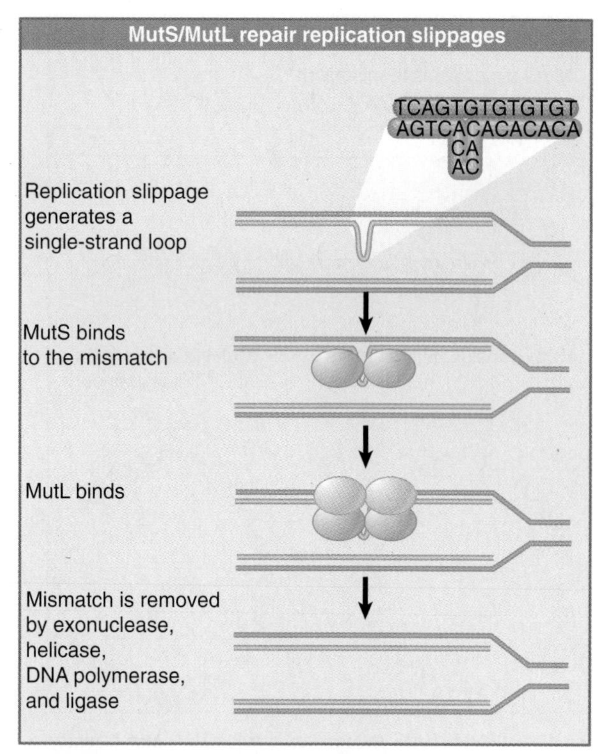

FIGURE 20.17 The MutS/MutL system initiates repair of mismatches produced by replication slippage.

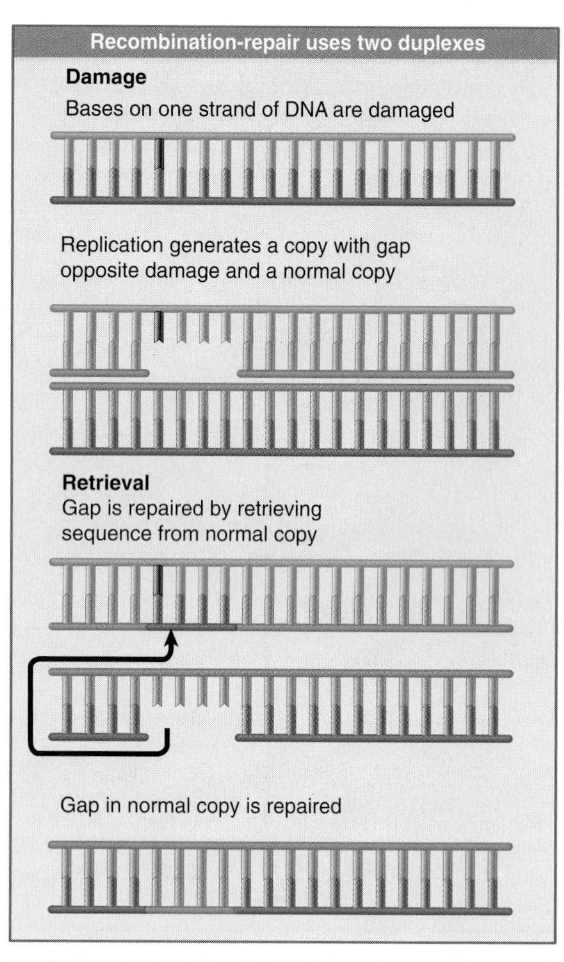

FIGURE 20.18 An *E. coli* retrieval system uses a normal strand of DNA to replace the gap left in a newly synthesized strand opposite a site of unrepaired damage.

20.8 Recombination-Repair Systems in *E. coli*

Key concepts

- The *rec* genes of *E. coli* code for the principal retrieval system.
- The principal retrieval system functions when replication leaves a gap in a newly synthesized strand that is opposite a damaged sequence.
- The single strand of another duplex is used to replace the gap.
- The damaged sequence is then removed and resynthesized.

Recombination-repair systems use activities that overlap with those involved in genetic recombination. They are also sometimes called "post-replication repair," because they function after replication. Such systems are effective in dealing with the defects produced in daughter duplexes by replication of a template that contains damaged bases. An example is illustrated in FIGURE 20.18. Restarting stalled replication forks could be the major role of the recombination-repair systems (see Section 18.17, The Primosome Is Needed to Restart Replication).

Consider a structural distortion, such as a pyrimidine dimer, on one strand of a double helix. When the DNA is replicated, the dimer prevents the damaged site from acting as a template. Replication is forced to skip past it.

DNA polymerase probably proceeds up to or close to the pyrimidine dimer. The polymerase then ceases synthesis of the corresponding daughter strand. Replication restarts some distance farther along. A substantial gap is left in the newly synthesized strand.

The resulting daughter duplexes are different in nature. One has the parental strand containing the damaged adduct, which faces a newly synthesized strand with a lengthy gap. The other duplicate has the undamaged parental strand, which has been copied into a normal complementary strand. The retrieval system takes advantage of the normal daughter.

The gap opposite the damaged site in the first duplex is filled by stealing the homologous single strand of DNA from the normal duplex. Following this **single-strand exchange,** the recipient duplex has a parental (damaged) strand facing a wild-type strand. The donor duplex has a normal parental strand facing a gap; the gap can be filled by repair synthesis in the usual way, generating a normal duplex. Thus the damage is confined to the original distortion (although the same recombination-repair events must be repeated after every replication cycle unless and until the damage is removed by an excision repair system).

The principal pathway for recombination-repair in *E. coli* is identified by the *rec* genes (see Figures 19.13–19.15). In *E. coli* deficient in excision repair, mutation in the *recA* gene essentially abolishes all the remaining repair and recovery facilities. Attempts to replicate DNA in *uvr⁻recA⁻* cells produce fragments of DNA whose size corresponds with the expected distance between thymine dimers. This result implies that the dimers provide a lethal obstacle to replication in the absence of RecA function. It explains why the double mutant cannot tolerate >1 to 2 dimers in its genome (compared with the ability of a wild-type bacterium to handle as many as 50).

One *rec* pathway involves the *recBC* genes and is well characterized; the other involves *recF* and is not so well defined. They fulfill different functions *in vivo*. The RecBC pathway is involved in restarting stalled replication forks (see Section 20.9, Recombination Is an Important Mechanism to Recover from Replication Errors). The RecF pathway is involved in repairing the gaps in a daughter strand that are left after replicating past a pyrimidine dimer.

The RecBC and RecF pathways both function prior to the action of RecA (although in different ways). They lead to the association of RecA with a single-stranded DNA. The ability of RecA to exchange single strands allows it to perform the retrieval step in Figure 20.18. Nuclease and polymerase activities then complete the repair action.

The RecF pathway contains a group of three genes: *recF, recO,* and *recR.* The proteins form two types of complex, RecOR and RecOF. They promote the formation of RecA filaments on single-stranded DNA. One of their functions is to make it possible for the filaments to assemble in spite of the presence of the SSB, which is inhibitory. They are thought to function at gaps; however, the reaction *in vitro* requires a free 5′ end.

The designations of repair and recombination genes are based on the phenotypes of the mutants, but sometimes a mutation isolated in one set of conditions and named as a *uvr* locus turns out to have been isolated in another set of conditions as a *rec* locus. This uncertainty makes an important point. We cannot yet define how many functions belong to each pathway or how the pathways interact. The *uvr* and *rec* pathways are not entirely independent, because *uvr* mutants show reduced efficiency in recombination-repair. We must expect to find a network of nuclease, polymerase, and other activities, which constitute repair systems that are partially overlapping (or in which an enzyme usually used to provide some function can be substituted by another from a different pathway).

20.9 Recombination Is an Important Mechanism to Recover from Replication Errors

Key concepts

- A replication fork may stall when it encounters a damaged site or a nick in DNA.
- A stalled fork may reverse by pairing between the two newly synthesized strands.
- A stalled fork may restart repairing the damage and use a helicase to move the fork forward.
- The structure of the stalled fork is the same as a Holliday junction and may be converted to a duplex and DSB by resolvases.

All cells have many pathways to repair damage in DNA. Which pathway is used will depend upon the type of damage and the situation. Excision-repair pathways can in principle be used at any time, but recombination-repair can be used only when there is a second duplex with a copy of the damaged sequence, that is, postreplication. A special situation is presented when damaged DNA is replicated, because the replication fork may stall at the site of damage. Recombination-repair pathways are involved in allowing the fork to be restored after the damage has been repaired or to allow it to bypass the damage.

FIGURE 20.19 shows one possible outcome when a replication fork stalls. The fork stops moving forward when it encounters the damage. The replication apparatus disassembles, at least partially. This allows branch migration to occur, when the fork effectively moves backward, and the new daughter strands pair to form a duplex structure. After the damage has been repaired, a helicase rolls the fork forward to restore its structure. Then the replication

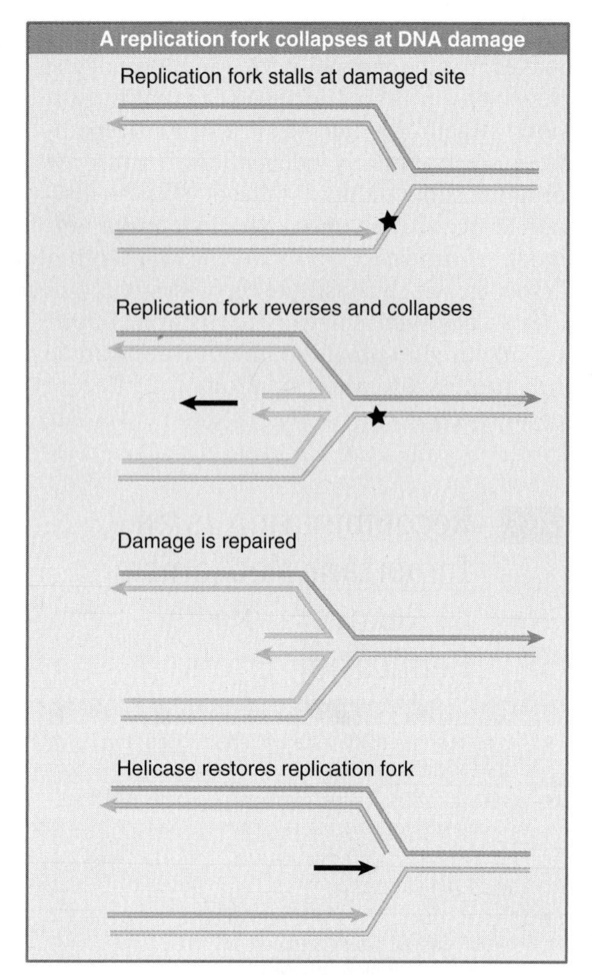

FIGURE 20.19 A replication fork stalls when it reaches a damaged site in DNA. Reversing the fork allows the two daughter strands to pair. After the damage has been repaired, the fork is restored by forward-branch migration catalyzed by a helicase. Arrowheads indicate 3′ ends.

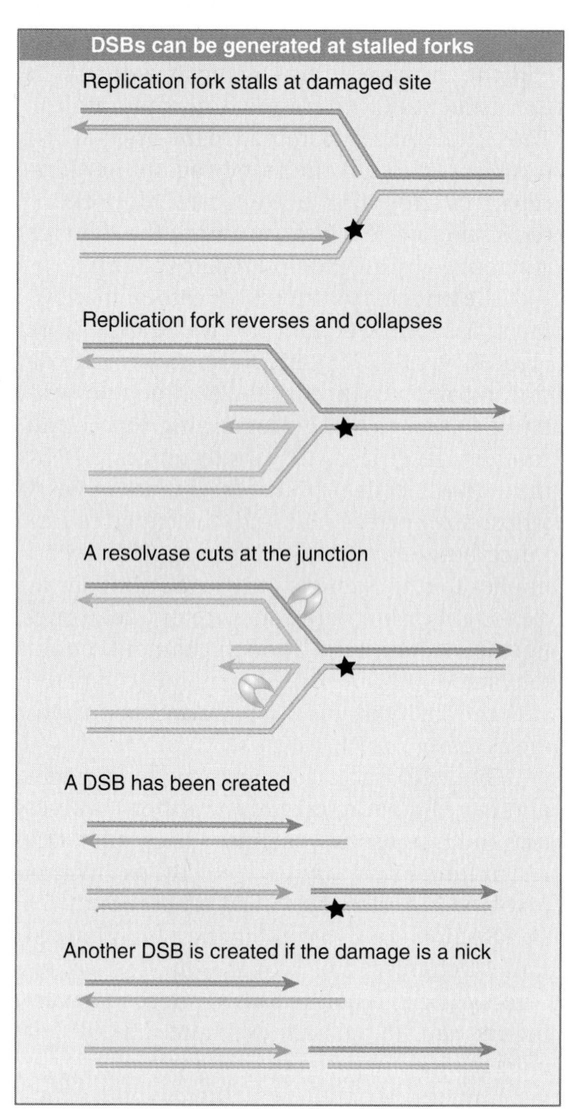

FIGURE 20.20 The structure of a stalled replication fork resembles a Holliday junction and can be resolved in the same way by resolvases. The results depend on whether the site of damage contains a nick. Result 1 shows that a double-strand break is generated by cutting a pair of strands at the junction. Result 2 shows a second DSB is generated at the site of damage if it contains a nick. Arrowheads indicate 3′ ends.

apparatus can reassemble, and replication is restarted (see Section 18.17, The Primosome Is Needed to Restart Replication). DNA polymerase II is required for the replication restart, and is later replaced by DNA polymerase III.

The pathway for handling a stalled replication fork requires repair enzymes. In *E. coli,* the RecA and RecBC systems have an important role in this reaction (in fact, this may be their major function in the bacterium). One possible pathway is for RecA to stabilize single-stranded DNA by binding to it at the stalled replication fork and possibly acting as the sensor that detects the stalling event. RecBC is involved in excision-repair of the damage. After the damage has been repaired, replication can resume.

Another pathway may use recombination-repair—possibly the strand-exchange reactions of RecA. FIGURE 20.20 shows that the structure of the stalled fork is essentially the same as a Holliday junction created by recombination between two duplex DNAs. This makes it a target for resolvases. A double-strand break is generated if a resolvase cleaves either pair of complementary strands. In addition, if the damage is in fact a nick, another double-strand break is created at this site.

Stalled replication forks can be rescued by recombination-repair. We don't know the exact

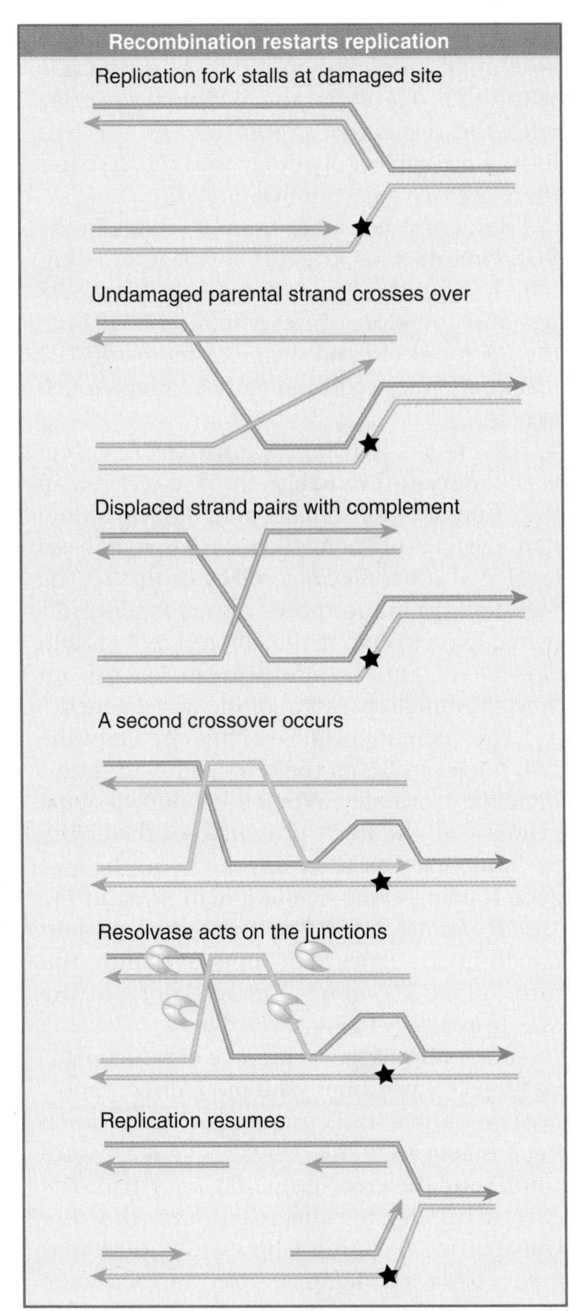

FIGURE 20.21 When a replication fork stalls, recombination-repair can place an undamaged strand opposite the damaged site. This allows replication to continue.

sequence of events, but one possible scenario is outlined in FIGURE 20.21. The principle is that a recombination event occurs on either side of the damaged site, allowing an undamaged single strand to pair with the damaged strand. This allows the replication fork to be reconstructed so that replication can continue, effectively bypassing the damaged site.

20.10 RecA Triggers the SOS System

Key concepts

- Damage to DNA causes RecA to trigger the SOS response, which consists of genes coding for many repair enzymes.
- RecA activates the autocleavage activity of LexA.
- LexA represses the SOS system; its autocleavage activates those genes.

The direct involvement of RecA protein in recombination-repair is only one of its activities. This extraordinary protein also has another, quite distinct function. It can be activated by many treatments that damage DNA or inhibit replication in *E. coli*. This causes it to trigger a complex series of phenotypic changes called the **SOS response,** which involves the expression of many genes whose products include repair functions. These dual activities of the RecA protein make it difficult to know whether a deficiency in repair in *recA* mutant cells is due to loss of the DNA strand-exchange function of RecA or to some other function whose induction depends on the protease activity.

The inducing damage can take the form of ultraviolet irradiation (the most studied case) or can be caused by crosslinking or alkylating agents. Inhibition of replication by any of several means—including deprivation of thymine, addition of drugs, or mutations in several of the *dna* genes—has the same effect.

The response takes the form of increased capacity to repair damaged DNA, which is achieved by inducing synthesis of the components of both the long-patch excision repair system and the Rec recombination-repair pathways. In addition, cell division is inhibited. Lysogenic prophages may be induced.

The initial event in the response is the activation of RecA by the damaging treatment. We do not know very much about the relationship between the damaging event and the sudden change in RecA activity. A variety of damaging events can induce the SOS response; thus current work focuses on the idea that RecA is activated by some common intermediate in DNA metabolism.

The inducing signal could consist of a small molecule released from DNA, or it might be some structure formed in the DNA itself. *In vitro,* the activation of RecA requires the presence of single-stranded DNA and ATP. Thus the

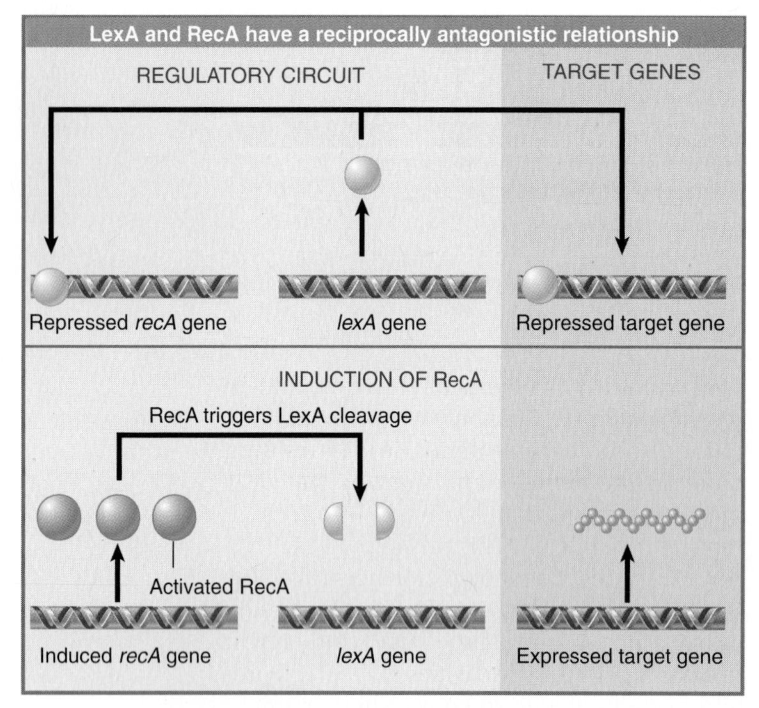

FIGURE 20.22 The LexA protein represses many genes, including repair functions, *recA* and *lexA*. Activation of RecA leads to proteolytic cleavage of LexA and induces all of these genes.

activating signal could be the presence of a single-stranded region at a site of damage. Whatever form the signal takes, its interaction with RecA is rapid: The SOS response occurs within a few minutes of the damaging treatment.

Activation of RecA causes proteolytic cleavage of the product of the *lexA* gene. LexA is a small (22 kD) protein that is relatively stable in untreated cells, where it functions as a repressor at many operons. The cleavage reaction is unusual; LexA has a latent protease activity that is activated by RecA. When RecA is activated, it causes LexA to undertake an autocatalytic cleavage; this inactivates the LexA repressor function, and coordinately induces all the operons to which it was bound. The pathway is illustrated in FIGURE 20.22.

The target genes for LexA repression include many repair functions. Some of these SOS genes are active only in treated cells; others are active in untreated cells, but the level of expression is increased by cleavage of LexA. In the case of *uvrB,* which is a component of the excision repair system, the gene has two promoters; one functions independently of LexA, the other is subject to its control. Thus after cleavage of LexA, the gene can be expressed from the second promoter as well as from the first.

LexA represses its target genes by binding to a 20 bp stretch of DNA called an **SOS box,** which includes a consensus sequence with eight absolutely conserved positions. As is common with other operators, the SOS boxes overlap with the respective promoters. At the *lexA* locus—the subject of autogenous repression—there are two adjacent SOS boxes.

RecA and LexA are mutual targets in the SOS circuit: RecA triggers cleavage of LexA, which represses *recA* and itself. The SOS response therefore causes amplification of both the RecA protein and the LexA repressor. The results are not so contradictory as might at first appear.

The increase in expression of RecA protein is necessary (presumably) for its direct role in the recombination-repair pathways. On induction, the level of RecA is increased from its basal level of ~1200 molecules/cell by up to 50×. The high level in induced cells means there is sufficient RecA to ensure that all the LexA protein is cleaved. This should prevent LexA from reestablishing repression of the target genes.

The main importance of this circuit for the cell, however, lies in the cell's ability to return rapidly to normalcy. When the inducing signal is removed, the RecA protein loses the ability to destabilize LexA. At this moment, the *lexA* gene is being expressed at a high level; in the absence of activated RecA, the LexA protein rapidly accumulates in the uncleaved form and turns off the SOS genes. This explains why the SOS response is freely reversible.

RecA also triggers cleavage of other cellular targets, sometimes with more direct consequences. The UmuD protein is cleaved when RecA is activated; the cleavage event activates UmuD and the error-prone repair system. The current model for the reaction is that the $UmuD_2UmuC$ complex binds to a RecA filament near a site of damage, RecA activates the complex by cleaving UmuD to generate UmuD′, and the complex then synthesizes a stretch of DNA to replace the damaged material.

Activation of RecA also causes cleavage of some other repressor proteins, including those of several prophages. Among these is the lambda repressor (with which the protease activity was discovered). This explains why lambda is induced by ultraviolet irradiation; the lysogenic repressor is cleaved, releasing the phage to enter the lytic cycle.

This reaction is not a cellular SOS response, but instead represents a recognition by the prophage that the cell is in trouble. Survival is then best assured by entering the lytic cycle to generate progeny phages. In this sense, prophage induction is piggybacking onto the cellular system by responding to the same indicator (activation of RecA).

The two activities of RecA are relatively independent. The *recA441* mutation allows the SOS response to occur without inducing treatment, probably because RecA remains spontaneously in the activated state. Other mutations abolish the ability to be activated. Neither type of mutation affects the ability of RecA to handle DNA. The reverse type of mutation, inactivating the recombination function but leaving intact the ability to induce the SOS response, would be useful in disentangling the direct and indirect effects of RecA in the repair pathways.

20.11 Eukaryotic Cells Have Conserved Repair Systems

Key concepts

- The yeast *RAD* mutations, identified by radiation sensitive phenotypes, are in genes that code for repair systems.
- Xeroderma pigmentosum (XP) is a human disease caused by mutations in any one of several repair genes.
- A complex of proteins including XP products and the transcription factor $TF_{II}H$ provides a human excision-repair mechanism.
- Transcriptionally active genes are preferentially repaired.

The types of repair functions recognized in *E. coli* are common to a wide range of organisms. The best characterized eukaryotic systems are in yeast, where Rad51 is the counterpart to RecA. In yeast, the main function of the strand-transfer protein is homologous recombination. Many of the repair systems found in yeast have direct counterparts in higher eukaryotic cells, and in several cases these systems are involved with human diseases.

RAD genes are genes that are involved in repair functions; they have been characterized genetically in yeast by virtue of their sensitivity to radiation. There are three general groups of repair genes in the yeast *S. cerevisiae*, identified by the *RAD3* group (involved in excision repair), the *RAD6* group (required for postreplication repair), and the *RAD52* group (concerned with recombination-like mechanisms). The RAD52 group is divided into two subgroups by a difference in mutant phenotypes. One subgroup affects homologous recombination, as seen by a reduction in mitotic recombination in *RAD50*, *RAD51*, *RAD54*, *RAD55*, and *RAD57*. These Rad proteins form a multiprotein complex at a double-strand break. After an exonuclease has acted on the free ends to generate single-stranded tails, Rad51 initiates the process by binding to the single-stranded DNA to form a nucleoprotein filament. Rad52, Rad55, and Rad54 then bind sequentially to the filament. By contrast, recombination rates are increased in *RAD59*, *MRE11*, and *XRS2* mutants; this subgroup is not deficient in homologous recombination, but is deficient in nonhomologous DNA joining reactions.

A superfamily of DNA polymerases involved in synthesizing DNA to replace material at damaged sites is identified by the *dinB* and *umuCD* genes that code for DNA polymerases IV and V in *E. coli*, the *RAD30* gene coding for DNA polymerase η of *S. cerevisiae*, and the gene *XPV* that codes for the human homolog. They are sometimes called translesion DNA polymerases. A difference between the bacterial and eukaryotic enzymes is that the latter are not error-prone at thymine dimers: They accurately introduce an A-A pair opposite a T-T dimer. When they replicate through other sites of damage, however, they are more prone to introduce errors.

An interesting feature of repair that has been best characterized in yeast is its connection with transcription. Transcriptionally active genes are preferentially repaired. The consequence is that the transcribed strand is preferentially repaired (removing the impediment to transcription). The cause appears to be a mechanistic connection between the repair apparatus and RNA polymerase. The Rad3 protein, which is a helicase required for the incision step, is a component of a transcription factor associated with RNA polymerase (see Section 24.12, A Connection between Transcription and Repair).

Mammalian cells show heterogeneity in the amount of DNA resynthesized at each lesion after damage. The patches are always relatively short, though, at <10 bases.

An indication of the existence and importance of the mammalian repair systems is given by certain human hereditary disorders. The best investigated of these is xeroderma pigmentosum (XP), a recessive disease resulting in hypersensitivity to sunlight, and in particular, ultraviolet. The deficiency results in skin disorders (and sometimes more severe defects).

The disease is caused by a deficiency in excision repair. Fibroblasts from XP patients cannot excise pyrimidine dimers and other bulky adducts. Mutations fall into eight genes called

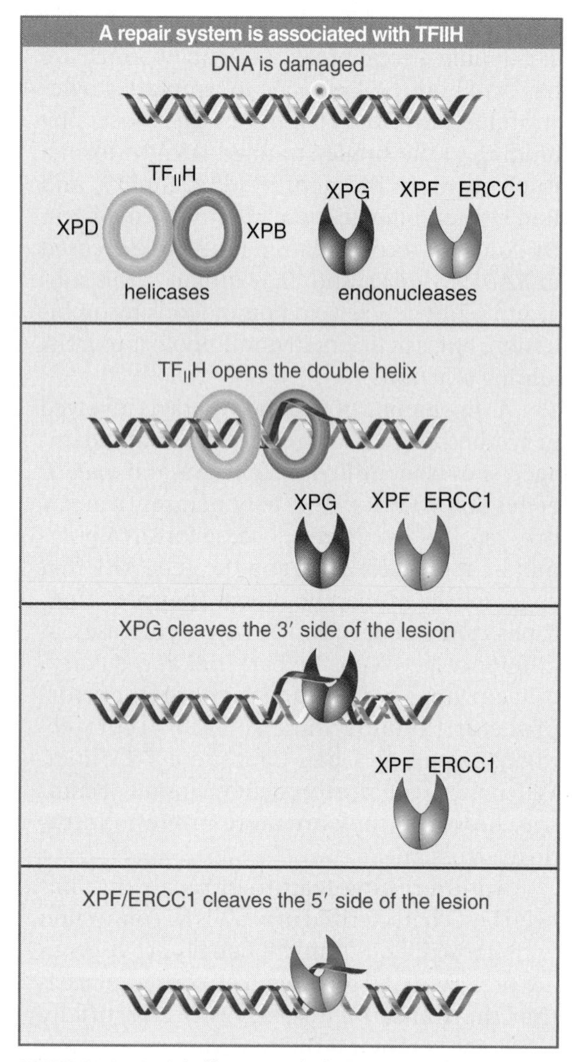

FIGURE 20.23 A helicase unwinds DNA at a damaged site, endonucleases cut on either side of the lesion, and new DNA is synthesized to replace the excised stretch.

XP-A to *XP-G*. They have homologs in the *RAD* genes of yeast, showing that this pathway is widely used in eukaryotes.

A protein complex that includes products of several of the *XP* genes is responsible for excision of thymine dimers. FIGURE 20.23 shows its role in the repair pathway. The complex binds to DNA at a site of damage, perhaps by a mechanism involving the cooperative acion of several of its components. The strands of DNA are then unwound for ~20 bp around the damaged site. This action is undertaken by the helicase activity of the transcription factor $TF_{II}H$, itself a large complex, which includes the products of several *XP* genes, and which is involved with the repair of damaged DNA that is encountered by RNA polymerase during transcription. Then cleavages are made on either side of the lesion by endonucleases coded by *XP* genes. The single-stranded stretch including the damaged bases can then be replaced by synthesis of a replacement.

In cases where replication encounters a thymine dimer that has not been removed, replication requires the DNA polymerase η activity in order to proceed past the dimer. This is coded by *XPV*. Skin cancers that occur in *XPV* mutants are presumably due to loss of the DNA polymerase.

20.12 A Common System Repairs Double-Strand Breaks

Key concepts

- The NHEJ pathway can ligate blunt ends of duplex DNA.
- Mutations in the NHEJ pathway cause human diseases.

Double-strand breaks occur in cells in various circumstances. They initiate the process of homologous recombination and are an intermediate in the recombination of immunoglobulin genes (see Section 23.10, The RAG Proteins Catalyze Breakage and Reunion). They also occur as the result of damage to DNA—for example, by irradiation. The major mechanism to repair these breaks is called **nonhomologous end-joining (NHEJ),** and consists of ligating the blunt ends together.

The steps involved in NHEJ are summarized in FIGURE 20.24. The same enzyme complex undertakes the process in both NHEJ and immune recombination. The first stage is recognition of the broken ends by a heterodimer consisting of the proteins Ku70 and Ku80. They form a scaffold that holds the ends together and allows other enzymes to act on them. A key component is the DNA-dependent protein kinase (DNA-PKcs), which is activated by DNA to phosphorylate protein targets. One of these targets is the protein Artemis, which in its activated form has both exonuclease and endonuclease activities, and can both trim overhanging ends and cleave the hairpins generated by recombination of immunoglobulin genes. The DNA polymerase activity that fills in any remaining single-stranded protrusions is not known. The actual joining of the double-stranded ends is undertaken by the DNA ligase IV, which functions in conjunction with the protein XRCC4. Mutations in any of these components may render eukaryotic cells more sensitive to radiation.

Some of the genes for these proteins are mutated in patients who have diseases due to deficiencies in DNA repair.

The Ku heterodimer is the sensor that detects DNA damage by binding to the broken ends. The crystal structure in FIGURE 20.25 shows why it binds only to ends. The bulk of the protein extends for about two turns along one face of DNA (lower), but a narrow bridge between the subunits, located in the center of the structure, completely encircles DNA. This means that the heterodimer needs to slip onto a free end.

Ku can bring broken ends together by binding two DNA molecules. The ability of Ku heterodimers to associate with one another suggests that the reaction might take place as illustrated in FIGURE 20.26. This would predict that the ligase would act by binding in the region between the bridges on the individual heterodimers. Presumably Ku must change its structure in order to be released from DNA.

Deficiency in DNA repair causes several human diseases. The common feature is that an inability to repair double-strand breaks in DNA leads to chromosomal instability. The instability is revealed by chromosomal aberrations, which are associated with an increased rate of mutation, which in turn leads to an increased susceptibility to cancer in patients with the disease. The basic cause can be mutation in pathways that control DNA repair or in the genes that code for enzymes of the repair complexes. The phenotypes can be very similar, as in the case of Ataxia telangiectasia (AT), which is caused by failure of a cell cycle checkpoint pathway, and Nijmegan breakage syndrome (NBS), which is caused by a mutation of a repair enzyme. One of the lessons that we learned from characterizing the repair pathways is that they are conserved in mammals, yeast, and bacteria.

The recessive human disorder of Bloom's syndrome is caused by mutations in a helicase gene (called BLM) that is homologous to *recQ* of *E. coli*. The mutation results in an increased frequency of chromosomal breaks and sister chromatid exchanges. BLM associates with other repair proteins as part of a large complex. One of the proteins with which it interacts is hMLH1, a mismatch-repair protein that is the human homolog of bacterial *mutL*. The yeast homologs of these two proteins, Sgs1 and MLH1, also associate, identifying these genes as parts of a well-conserved repair pathway.

Nijmegan breakage syndrome results from mutations in a gene coding for a protein

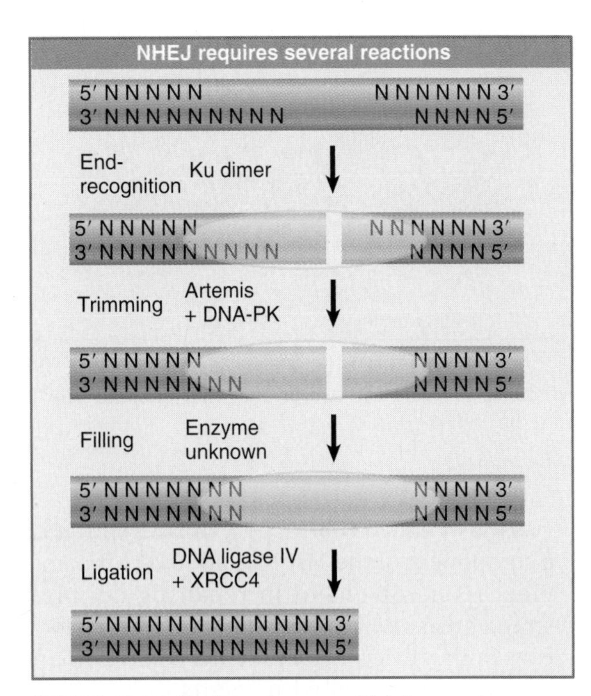

FIGURE 20.24 Nonhomologous end joining requires recognition of the broken ends, trimming of overhanging ends and/or filling, followed by ligation.

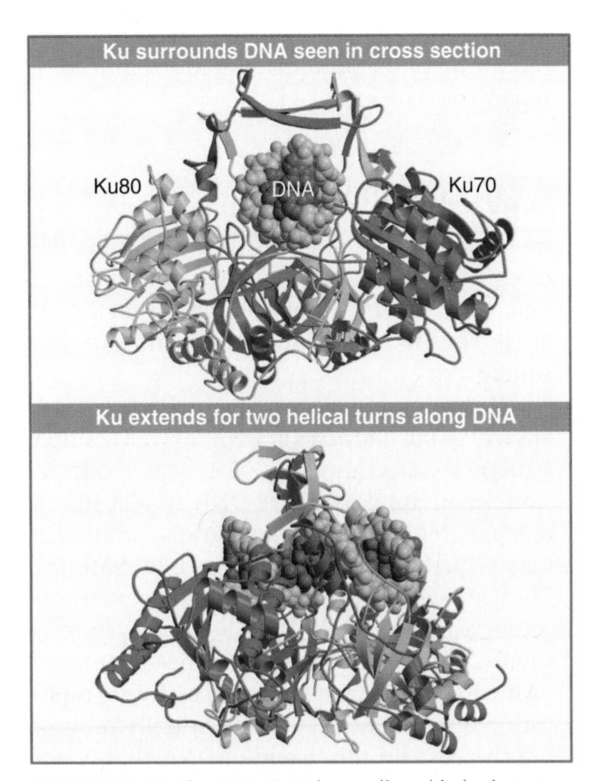

FIGURE 20.25 The Ku70-Ku80 heterodimer binds along two turns of the DNA double helix and surrounds the helix at the center of the binding site. Reproduced from Walker, J.R., et al. *Nature*. 2001. 412:607–614. Photos courtesy of Jonathan Goldberg, Memorial Sloan-Kettering Cancer Center.

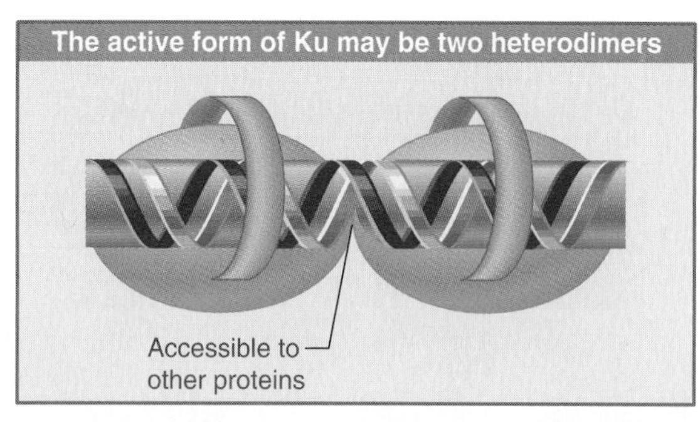

FIGURE 20.26 If two heterodimers of Ku bind to DNA, the distance between the two bridges that encircle DNA is ~12 bp.

(variously called Nibrin, p95, or NBS1) that is a component of the Mre11/Rad50 repair complex. Its involvement in repairing double-strand breaks is shown by the formation of foci containing the group of proteins when human cells are irradiated with agents that induce double-strand breaks. After irradiation, the kinase ATMP (coded by the *AT* gene) phosphorylates NBS1; this activates the complex, which localizes at sites of DNA damage. Subsequent steps involve triggering a checkpoint (a mechanism that prevents the cell cycle from proceeding until the damage is repaired) and recruiting other proteins that are required to repair the damage

20.13 Summary

Bacteria contain systems that maintain the integrity of their DNA sequences in the face of damage or errors of replication and that distinguish the DNA from sequences of a foreign source.

Repair systems can recognize mispaired, altered, or missing bases in DNA, as well as other structural distortions of the double helix. Excision repair systems cleave DNA near a site of damage, remove one strand, and synthesize a new sequence to replace the excised material. The Uvr system provides the main excision-repair pathway in *E. coli*. The *dam* system is involved in correcting mismatches generated by incorporation of incorrect bases during replication and functions by preferentially removing the base on the strand of DNA that is not methylated at the *dam* target sequence. Eukaryotic homologs of the *E. coli* MutSL system are involved in repairing mismatches that result from replication slippage; mutations in this pathway are common in certain types of cancer.

Recombination-repair systems retrieve information from a DNA duplex and use it to repair a sequence that has been damaged on both strands. The RecBC and RecF pathways both act prior to RecA, whose strand-transfer function is involved in all bacterial recombination. A major use of recombination-repair may be to recover from the situation created when a replication fork stalls.

The other capacity of recA is the ability to induce the SOS response. RecA is activated by damaged DNA in an unknown manner. It triggers cleavage of the LexA repressor protein, thus releasing repression of many loci, and inducing synthesis of the enzymes of both excision repair and recombination-repair pathways. Genes under LexA control possess an operator SOS box. RecA also directly activates some repair activities. Cleavage of repressors of lysogenic phages may induce the phages to enter the lytic cycle.

Repair systems can be connected with transcription in both prokaryotes and eukaryotes. Human diseases are caused by mutations in genes coding for repair activities that are associated with the transcription factor TFIIH. They have homologs in the *RAD* genes of yeast, which suggests that this repair system is widespread.

Nonhomologous end-joining (NHEJ) is a general reaction for repairing broken ends in (eukaryotic) DNA. The Ku heterodimer brings the broken ends together so they can be ligated. Several human diseases are caused by mutations in enzymes of this pathway.

References

20.2 Repair Systems Correct Damage to DNA

Reviews

Sancar, A., Lindsey-Boltz, L. A., Unsal-Kaçmaz, K., and Linn, S. (2004). Molecular mechanisms of mammalian DNA repair and the DNA damage checkpoints. *Annu. Rev. Biochem.* 73, 39–85.

Wood, R. D., Mitchell, M., Sgouros, J., and Lindahl, T. (2001). Human DNA repair genes. *Science* 291, 1284–1289.

20.4 Excision-Repair Pathways in Mammalian Cells

Reviews

Barnes, D. E. and Lindahl, T. (2004). Repair and genetic consequences of endogenous DNA base damage in mammalian cells. *Annu. Rev. Genet.* 38, 445–476.

McCullough, A. K., Dodson, M. L., and Lloyd, R. S. (1999). Initiation of base excision repair: glycosylase mechanisms and structures. *Annu. Rev. Biochem.* 68, 255–285.

Sancar, A., Lindsey-Boltz, L. A., Unsal-Kaçmaz, K., and Linn, S. (2004). Molecular mechanisms of mammalian DNA repair and the DNA damage checkpoints. *Annu. Rev. Biochem.* 73, 39–85.

Research

Klungland, A. and Lindahl, T. (1997). Second pathway for completion of human DNA base excision-repair: reconstitution with purified proteins and requirement for DNase IV (FEN1). *EMBO J.* 16, 3341–3348.

Matsumoto, Y. and Kim, K. (1995). Excision of deoxyribose phosphate residues by DNA polymerase beta during DNA repair. *Science* 269, 699–702.

20.5 Base Flipping Is Used by Methylases and Glycosylases

Research

Aas, P. A., Otterlei, M., Falnes, P. A., Vagbe, C. B., Skorpen, F., Akbari, M., Sundheim, O., Bjoras, M., Slupphaug, G., Seeberg, E., and Krokan, H. E.,(2003). Human and bacterial oxidative demethylases repair alkylation damage in both RNA and DNA. *Nature* 421, 859–863.

Falnes, P. A., Johansen, R. F., and Seeberg, E. (2002). AlkB-mediated oxidative demethylation reverses DNA damage in *E. coli*. *Nature* 419, 178–182.

Klimasauskas, S., Kumar, S., Roberts, R. J. and Cheng, X. (1994). HhaI methyltransferase flips its target base out of the DNA helix. *Cell* 76, 357–369.

Lau, A. Y., Glassner, B. J., Samson, L. D., and Ellenberger, T. (2000). Molecular basis for discriminating between normal and damaged bases by the human alkyladenine glycosylase, AAG. *Proc. Natl. Acad. Sci. USA* 97, 13573–13578.

Lau, A. Y., Scherer, O. D., Samson, L., Verdine, G. L., and Ellenberger, T. (1998). Crystal structure of a human alkylbase-DNA repair enzyme complexed to DNA: mechanisms for nucleotide flipping and base excision. *Cell* 95, 249–258.

Mol, D. D. et al. (1995). Crystal structure and mutational analysis of human uracil-DNA glycosylase: structural basis for specificity and catalysis. *Cell* 80, 869–878.

Park, H. W., Kim, S. T., Sancar, A., and Deisenhofer, J. (1995). Crystal structure of DNA photolyase from *E. coli*. *Science* 268, 1866–1872.

Savva, R. et al. (1995). The structural basis of specific base-excision repair by uracil-DNA glycosylase. *Nature* 373, 487–493.

Trewick, S. C., Henshaw, T. F., Hausinger, R. P., Lindahl, T., and Sedgwick, B. (2002). Oxidative demethylation by *E. coli* AlkB directly reverts DNA base damage. *Nature* 419, 174–178.

Vassylyev, D. G. et al. (1995). Atomic model of a pyrimidine dimer excision repair enzyme complexed with a DNA substrate: structural basis for damaged DNA recognition. *Cell* 83, 773–782.

20.6 Error-Prone Repair and Mutator Phenotypes

Research

Friedberg, E. C., Feaver, W. J., and Gerlach, V. L. (2000). The many faces of DNA polymerases: strategies for mutagenesis and for mutational avoidance. *Proc. Natl. Acad. Sci. USA* 97, 5681–5683.

Goldsmith, M., Sarov-Blat, L., and Livneh, Z. (2000). Plasmid-encoded MucB protein is a DNA polymerase (pol RI) specialized for lesion bypass in the presence of MucA, RecA, and SSB. *Proc. Natl. Acad. Sci. USA* 97, 11227–11231.

Maor-Shoshani, A., Reuven, N. B., Tomer, G., and Livneh, Z. (2000). Highly mutagenic replication by DNA polymerase V (UmuC) provides a mechanistic basis for SOS untargeted mutagenesis. *Proc. Natl. Acad. Sci. USA* 97, 565–570.

Wagner, J., Gruz, P., Kim, S. R., Yamada, M., Matsui, K., Fuchs, R. P., and Nohmi, T. (1999). The *dinB* gene encodes a novel *E. coli* DNA polymerase, DNA pol IV, involved in mutagenesis. *Mol. Cell* 4, 281–286.

20.7 Controlling the Direction of Mismatch Repair

Review

Kunkel, T. A., and Erie, D. A. (2005). DNA mismatch repair. *Annu. Rev. Biochem.* 74, 681–710.

Research

Strand, M., Prolla, T. A., Liskay, R. M., and Petes, T. D. (1993). Destabilization of tracts of simple repetitive DNA in yeast by mutations affecting DNA mismatch repair. *Nature* 365, 274–276.

20.8 Recombination-Repair Systems in *E. coli*

Review

West, S. C. (1997). Processing of recombination intermediates by the RuvABC proteins. *Annu. Rev. Genet.* 31, 213–244.

Research

Bork, J. M. and Inman, R. B. (2001). The RecOR proteins modulate RecA protein function at 5′ ends of single-stranded DNA. *EMBO J.* 20, 7313–7322.

20.9 Recombination Is an Important Mechanism to Recover from Replication Errors

Reviews

Cox, M. M., Goodman, M. F., Kreuzer, K. N., Sherratt, D. J., Sandler, S. J., and Marians, K. J. (2000). The importance of repairing stalled replication forks. *Nature* 404, 37–41.

McGlynn, P. and Lloyd, R. G. (2002). Recombinational repair and restart of damaged replication forks. *Nat. Rev. Mol. Cell Biol.* 3, 859–870.

Michel, B., Viguera, E., Grompone, G., Seigneur, M., and Bidnenko, V. (2001). Rescue of arrested replication forks by homologous recombination. *Proc. Natl. Acad. Sci. USA* 98, 8181–8188.

Research

Courcelle, J. and Hanawalt, P. C. (2003). RecA-dependent recovery of arrested DNA replication forks. *Annu. Rev. Genet.* 37, 611–646.

Kuzminov, A. (2001). Single-strand interruptions in replicating chromosomes cause double-strand breaks. *Proc. Natl. Acad. Sci. USA* 98, 8241–8246.

Rangarajan, S., Woodgate, R., and Goodman, M. F. (1999). A phenotype for enigmatic DNA polymerase II: a pivotal role for pol II in replication restart in UV-irradiated *Escherichia coli. Proc. Natl. Acad. Sci. USA* 96, 9224–9229.

20.10 RecA Triggers the SOS System

Research

Tang, M. et al. (1999). UmuD$'_2$C is an error-prone DNA polymerase, *E. coli* pol V. *Proc. Natl. Acad. Sci. USA* 96, 8919–8924.

20.11 Eukaryotic Cells Have Conserved Repair Systems

Reviews

Krogh, B. O. and Symington, L. S. (2004). Recombination proteins in yeast. *Annu. Rev. Genet.* 38, 233–271.

Prakash, S. and Prakash, L. (2002). Translesion DNA synthesis in eukaryotes: a one- or two-polymerase affair. *Genes Dev.* 16, 1872–1883.

Sancar, A., Lindsey-Boltz, L. A., Unsal-Kaçmaz, K., and Linn, S. (2004). Molecular mechanisms of mammalian DNA repair and the DNA damage checkpoints. *Annu. Rev. Biochem.* 73, 39–85.

Research

Friedberg, E. C., Feaver, W. J., and Gerlach, V. L. (2000). The many faces of DNA polymerases: strategies for mutagenesis and for mutational avoidance. *Proc. Natl. Acad. Sci. USA* 97, 5681–5683.

Johnson, R. E., Prakash, S., and Prakash, L. (1999). Efficient bypass of a thymine-thymine dimer by yeast DNA polymerase, Pol eta. *Science* 283, 1001–1004.

Rattray, A. J. and Strathern, J. N. (2003). Error-prone DNA polymerases: when making a mistake is the only way to get ahead. *Annu. Rev. Genet.* 37, 31–66.

Reardon, J. T. and Sancar, A. (2003). Recognition and repair of the cyclobutane thymine dimer, a major cause of skin cancers, by the human excision nuclease. *Genes Dev.* 17, 2539–2551.

Wolner, B., van Komen, S., Sung, P., and Peterson, C. L. (2003). Recruitment of the recombinational repair machinery to a DNA double-strand break in yeast. *Mol. Cell* 12, 221–232.

20.12 A Common System Repairs Double-Strand Breaks

Review

D'Amours, D. and Jackson, S. P. (2002). The Mre11 complex: at the crossroads of DNA repair and checkpoint signalling. *Nat. Rev. Mol. Cell Biol.* 3, 317–327.

Research

Carney, J. P., Maser, R. S., Olivares, H., Davis, E. M., Le Beau, M., Yates, J. R., Hays, L., Morgan, W. F., and Petrini, J. H. (1998). The hMre11/hRad50 protein complex and Nijmegen breakage syndrome: linkage of double-strand break repair to the cellular DNA damage response. *Cell* 93, 477–486.

Cary, R. B., Peterson, S. R., Wang, J., Bear, D. G., Bradbury, E. M., and Chen, D. J. (1997). DNA looping by Ku and the DNA-dependent protein kinase. *Proc. Natl. Acad. Sci. USA* 94, 4267–4272.

Ellis, N. A., Groden, J., Ye, T. Z., Straughen, J., Lennon, D. J., Ciocci, S., Proytcheva, M., and German, J. (1995). The Bloom's syndrome gene product is homologous to RecQ helicases. *Cell* 83, 655–666.

Ma, Y., Pannicke, U., Schwarz, K., and Lieber, M. R. (2002). Hairpin opening and overhang processing by an Artemis/DNA-Dependent protein kinase complex in nonhomologous end joining and V(D)J Recombination. *Cell* 108, 781–794.

Ramsden, D. A. and Gellert, M. (1998). Ku protein stimulates DNA end joining by mammalian DNA ligases: a direct role for Ku in repair of DNA double-strand breaks. *EMBO J.* 17, 609–614.

Varon, R. et al. (1998). Nibrin, a novel DNA double-strand break repair protein, is mutated in Nijmegen breakage syndrome. *Cell* 93, 467–476.

Walker, J. R., Corpina, R. A., and Goldberg, J. (2001). Structure of the Ku heterodimer bound to DNA and its implications for double-strand break repair. *Nature* 412, 607–614.

21

Transposons

CHAPTER OUTLINE

Continued on next page

- The acentric fragment is lost during mitosis; this can be detected by the disappearance of dominant alleles in a heterozygote.
- Fusion between the broken ends of the chromosome generates dicentric chromosomes, which undergo further cycles of breakage and fusion.
- The fusion-breakage-bridge cycle is responsible for the occurrence of somatic variegation.

21.12 Controlling Elements Form Families of Transposons

- Each family of transposons in maize has both autonomous and nonautonomous controlling elements.
- Autonomous controlling elements code for proteins that enable them to transpose.
- Nonautonomous controlling elements have mutations that eliminate their capacity to catalyze transposition, but they can transpose when an autonomous element provides the necessary proteins.
- Autonomous controlling elements have changes of phase, when their properties alter as a result of changes in the state of methylation.

21.13 Spm Elements Influence Gene Expression

- Spm elements affect gene expression at their sites of insertion, when the TnpA protein binds to its target sites at the ends of the transposon.
- Spm elements are inactivated by methylation.

21.14 The Role of Transposable Elements in Hybrid Dysgenesis

- P elements are transposons that are carried in P strains of *Drosophila melanogaster,* but not in M strains.
- When a P male is crossed with an M female, transposition is activated.
- The insertion of P elements at new sites in these crosses inactivates many genes and makes the cross infertile.

21.15 P Elements Are Activated in the Germline

- P elements are activated in the germline of P male × M female crosses because a tissue-specific splicing event removes one intron, which generates the coding sequence for the transposase.
- The P element also produces a repressor of transposition, which is inherited maternally in the cytoplasm.
- The presence of the repressor explains why M male × P female crosses remain fertile.

21.16 Summary

21.1 Introduction

Genomes evolve both by acquiring new sequences and by rearranging existing sequences.

The sudden introduction of new sequences results from the ability of vectors to carry information between genomes. Extrachromosomal elements move information horizontally by mediating the transfer of (usually rather short) lengths of genetic material. In bacteria, plasmids move by conjugation (see Section 16.10, Conjugation Transfers Single-Stranded DNA), whereas phages spread by infection (see Chapter 14, Phage Strategies). Both plasmids and phages occasionally transfer host genes along with their own replicon. Direct transfer of DNA occurs between some bacteria by means of transformation (see Section 1.2, DNA Is the Genetic Material of Bacteria). In eukaryotes, some viruses, notably the retroviruses, can transfer genetic information during an infective cycle (see Section 22.6, Retroviruses May Transduce Cellular Sequences).

Rearrangements are sponsored by processes internal to the genome. Two of the major causes are summarized in FIGURE 21.1.

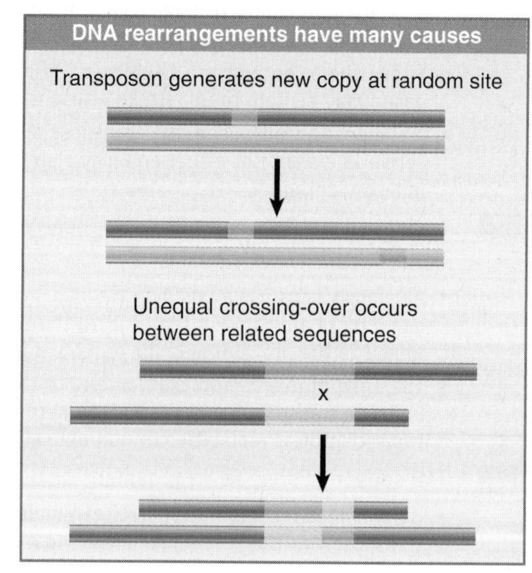

FIGURE 21.1 A major cause of sequence change within a genome is the movement of a transposon to a new site. This may have direct consequences on gene expression. Unequal crossing over between related sequences causes rearrangements. Copies of transposons can provide targets for such events.

Unequal recombination results from mispairing by the cellular systems for homologous recombination. Nonreciprocal recombination results in duplication or rearrangement of loci (see Section 6.7, Unequal Crossing-Over Rearranges Gene Clusters). Duplication of sequences within a genome provides a major source of new sequences. One copy of the sequence can retain its original function, whereas the other may evolve into a new function. Furthermore, significant differences between individual genomes are found at the molecular level because of polymorphic variations caused by recombination. We saw in Section 6.14, Minisatellites Are Useful for Genetic Mapping, that recombination between minisatellites adjusts their lengths so that every individual genome is distinct.

Another major cause of variation is provided by **transposable elements** or **transposons:** these are discrete sequences in the genome that are mobile—they are able to transport themselves to other locations within the genome. The mark of a transposon is that it does not utilize an independent form of the element (such as phage or plasmid DNA), but moves directly from one site in the genome to another. Unlike most other processes involved in genome restructuring, transposition does not rely on any relationship between the sequences at the donor and recipient sites. Transposons are restricted to moving themselves, and sometimes additional sequences, to new sites elsewhere within the same genome; they are, therefore, an internal counterpart to the vectors that can transport sequences from one genome to another. They may provide the major source of mutations in the genome.

Transposons fall into two general classes. The groups of transposons reviewed in this chapter exist as sequences of DNA coding for proteins that are able directly to manipulate DNA so as to propagate themselves within the genome. The transposons reviewed in Chapter 22, Retroviruses and Retroposons, are related to retroviruses, and the source of their mobility is the ability to make DNA copies of their RNA transcripts; the DNA copies then become integrated at new sites in the genome.

Transposons that mobilize via DNA are found in both prokaryotes and eukaryotes. Each bacterial transposon carries gene(s) that code for the enzyme activities required for its own transposition, although it may also require ancillary functions of the genome in which it resides (such as DNA polymerase or DNA gyrase). Comparable systems exist in eukaryotes, although their enzymatic functions are not so well characterized. A genome may contain both functional and nonfunctional (defective) elements. Often the majority of elements in a eukaryotic genome are defective, and have lost the ability to transpose independently, although they may still be recognized as substrates for transposition by the enzymes produced by functional transposons. A eukaryotic genome contains a large number and variety of transposons. The fly genome has >50 types of transposons, with a total of several hundred individual elements.

Transposable elements can promote rearrangements of the genome directly or indirectly:

- The transposition event itself may cause deletions or inversions or lead to the movement of a host sequence to a new location.
- Transposons serve as substrates for cellular recombination systems by functioning as "portable regions of homology"; two copies of a transposon at different locations (even on different chromosomes) may provide sites for reciprocal recombination. Such exchanges result in deletions, insertions, inversions, or translocations.

The intermittent activities of a transposon seem to provide a somewhat nebulous target for natural selection. This concern has prompted suggestions that (at least some) transposable elements confer neither advantage nor disadvantage on the phenotype, but could constitute "selfish DNA"—DNA concerned only with their own propagation. Indeed, in considering transposition as an event that is distinct from other cellular recombination systems, we tacitly accept the view that the transposon is an independent entity that resides in the genome.

Such a relationship of the transposon to the genome would resemble that of a parasite with its host. Presumably the propagation of an element by transposition is balanced by the harm done if a transposition event inactivates a necessary gene, or if the number of transposons becomes a burden on cellular systems. Yet we must remember that any transposition event conferring a selective advantage—for example, a genetic rearrangement—will lead to preferential survival of the genome carrying the active transposon.

21.2 Insertion Sequences Are Simple Transposition Modules

Key concepts

- An insertion sequence is a transposon that codes for the enzyme(s) needed for transposition flanked by short inverted terminal repeats.
- The target site at which a transposon is inserted is duplicated during the insertion process to form two repeats in direct orientation at the ends of the transposon.
- The length of the direct repeat is 5 to 9 bp and is characteristic for any particular transposon.

Transposable elements were first identified at the molecular level in the form of spontaneous insertions in bacterial operons. Such an insertion prevents transcription and/or translation of the gene in which it is inserted. Many different types of transposable elements have now been characterized.

The simplest transposons are called **insertion sequences** (reflecting the way in which they were detected). Each type is given the prefix **IS,** followed by a number that identifies the type. (The original classes were numbered IS1 to IS4; later classes have numbers reflecting the history of their isolation, but not corresponding to the total number of elements so far isolated!)

The IS elements are normal constituents of bacterial chromosomes and plasmids. A standard strain of *E. coli* is likely to contain several (<10) copies of any one of the more common IS elements. To describe an insertion into a particular site, a double colon is used; so λ::IS1 describes an IS1 element inserted into phage lambda.

The IS elements are autonomous units, each of which codes only for the proteins needed to sponsor its own transposition. Each IS element is different in sequence, but there are some common features in organization. The structure of a generic transposon before and after insertion at a target site is illustrated in FIGURE 21.2, which also summarizes the details of some common IS elements.

An IS element ends in short **inverted terminal repeats;** usually the two copies of the repeat are closely related rather than identical. As illustrated in the figure, the presence of the inverted terminal repeats means that the same sequence is encountered proceeding toward the element from the flanking DNA on either side of it.

When an IS element transposes, a sequence of host DNA at the site of insertion is duplicated. The nature of the duplication is revealed by comparing the sequence of the target site before and after an insertion has occurred. Figure 21.2 shows that at the site of insertion, the IS DNA is always flanked by very short **direct repeats.** (In this context, "direct" indicates that two copies of a sequence are repeated in the same orientation, not that the repeats are adjacent.) In the original gene (prior to insertion), however, the target site has the sequence of only one of these repeats. In the figure, the target site consists of the sequence $\frac{\text{ATGCA}}{\text{TACGT}}$. After transposition, one copy of this sequence is present on either side of the transposon.

The sequence of the direct repeat varies among individual transposition events undertaken by a transposon, but the length is constant for any particular IS element (a reflection of the mechanism of transposition). The most common length for the direct repeats is 9 bp.

An IS element therefore displays a characteristic structure in which its ends are identified by the inverted terminal repeats, whereas the adjacent ends of the flanking host DNA are

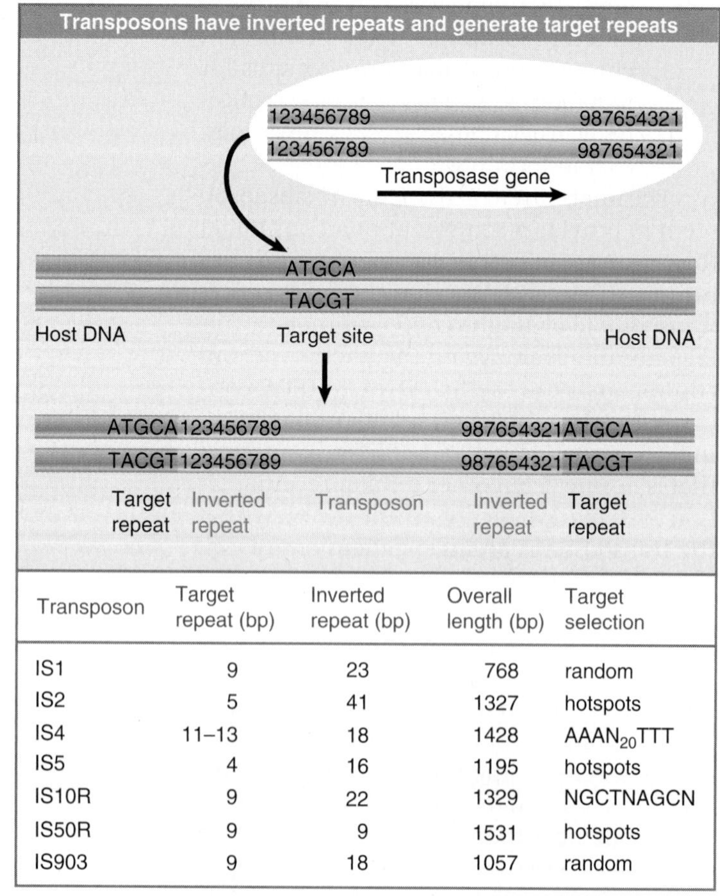

Transposon	Target repeat (bp)	Inverted repeat (bp)	Overall length (bp)	Target selection
IS1	9	23	768	random
IS2	5	41	1327	hotspots
IS4	11–13	18	1428	$AAAN_{20}TTT$
IS5	4	16	1195	hotspots
IS10R	9	22	1329	NGCTNAGCN
IS50R	9	9	1531	hotspots
IS903	9	18	1057	random

FIGURE 21.2 Transposons have inverted terminal repeats and generate direct repeats of flanking DNA at the target site. In this example, the target is a 5 bp sequence. The ends of the transposon consist of inverted repeats of 9 bp, where the numbers 1 through 9 indicate a sequence of base pairs.

identified by the short direct repeats. When observed in a sequence of DNA, this type of organization is taken to be diagnostic of a transposon, and suggests that the sequence originated in a transposition event.

Most IS elements insert at a variety of sites within host DNA. Some, though, show (varying degrees of) preference for particular hotspots.

The inverted repeats define the ends of a transposon. Recognition of the ends is common to transposition events sponsored by all types of transposon. *Cis*-acting mutations that prevent transposition are located in the ends, which are recognized by a protein(s) responsible for transposition. The protein is called a **transposase.**

All the IS elements except IS1 contain a single long coding region, which starts just inside the inverted repeat at one end and terminates just before or within the inverted repeat at the other end. This codes for the transposase. IS1 has a more complex organization, with two separate reading frames; the transposase is produced by making a frameshift during translation to allow both reading frames to be used.

The frequency of transposition varies among different elements. The overall rate of transposition is ~10^{-3} to 10^{-4} per element per generation. Insertions in individual targets occur at a level comparable with the spontaneous mutation rate, usually ~10^{-5} to 10^{-7} per generation. Reversion (by precise excision of the IS element) is usually infrequent, with a range of rates of 10^{-6} to 10^{-10} per generation, which is ~10^{3} times less frequent than insertion.

21.3 Composite Transposons Have IS Modules

Key concepts

- Transposons can carry other genes in addition to those coding for transposition.
- Composite transposons have a central region flanked by an IS element at each end.
- Either one or both of the IS elements of a composite transposon may be able to undertake transposition.
- A composite transposon may transpose as a unit, but an active IS element at either end may also transpose independently.

Some transposons carry drug resistance (or other) markers in addition to the functions concerned with transposition. These transposons are named **Tn** followed by a number. One class of larger transposons are called **composite elements,** because a central region carrying the drug marker(s) is flanked on either side by "arms" that consist of IS elements.

The arms may be in either the same or (more commonly) inverted orientation. Thus a composite transposon with arms that are direct repeats has the structure

If the arms are inverted repeats, the structure is

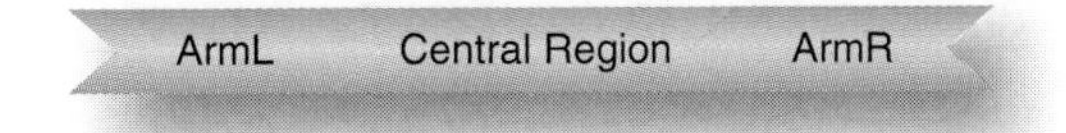

The arrows indicate the orientation of the arms, which are identified as L and R according to an (arbitrary) orientation of the genetic map of the transposon from left to right. The structure of a composite transposon is illustrated in more detail in FIGURE 21.3, which also summarizes

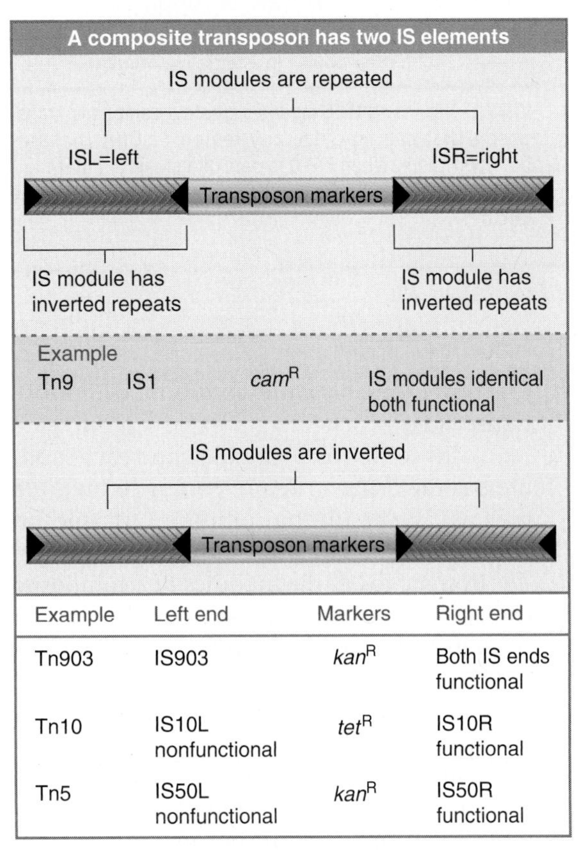

Example	Left end	Markers	Right end
Tn903	IS903	*kan*R	Both IS ends functional
Tn10	IS10L nonfunctional	*tet*R	IS10R functional
Tn5	IS50L nonfunctional	*kan*R	IS50R functional

FIGURE 21.3 A composite transposon has a central region carrying markers (such as drug resistance) flanked by IS modules. The modules have short inverted terminal repeats. If the modules themselves are in inverted orientation, the short inverted terminal repeats at the ends of the transposon are identical.

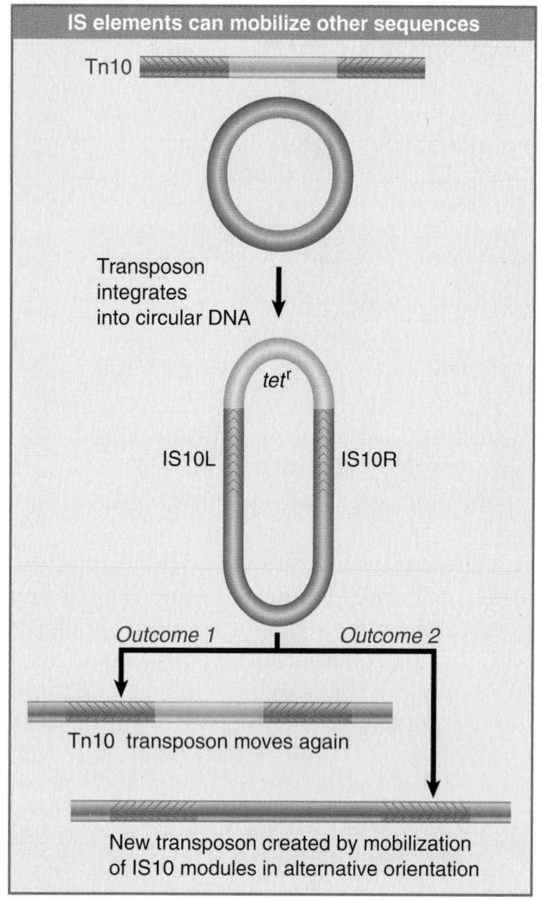

FIGURE 21.4 Two IS10 modules create a composite transposon that can mobilize any region of DNA that lies between them. When Tn10 is part of a small circular molecule, the IS10 repeats can transpose either side of the circle.

the properties of some common composite transposons.

Arms consist of IS modules, and each module has the usual structure ending in inverted repeats; as a result the composite transposon also ends in the same short inverted repeats.

In some cases, the modules of a composite transposon are identical, such as Tn9 (direct repeats of IS1) or Tn903 (inverted repeats of IS903). In other cases, the modules are closely related, but not identical. Thus we can distinguish the L and R modules in Tn10 or in Tn5.

A functional IS module can transpose either itself or the entire transposon. When the modules of a composite transposon are identical, presumably either module can sponsor movement of the transposon, as in the case of Tn9 or Tn903. When the modules are different, they may differ in functional ability, so transposition can depend entirely or principally on one of the modules, as in the case of Tn10 or Tn5.

We assume that composite transposons evolved when two originally independent modules associated with the central region. Such a situation could arise when an IS element transposes to a recipient site close to the donor site. The two identical modules may remain identical or diverge. The ability of a single module to transpose the entire composite element explains the lack of selective pressure for both modules to remain active.

What is responsible for transposing a composite transposon instead of just the individual module? This question is especially pressing in cases where both the modules are functional. In the example of Tn9, where the modules are IS1 elements, presumably each is active in its own right as well as on behalf of the composite transposon. Why is the transposon preserved as a whole, instead of each insertion sequence looking out for itself?

Two IS elements in fact can transpose any sequence residing between them, as well as themselves. FIGURE 21.4 shows that if Tn10 resides on a circular replicon, its two modules can be considered to flank either the *tet*^{R} gene of the original Tn10 or the sequence in the other part of the circle. Thus a transposition event can involve either the original Tn10 transposon (marked by the movement of *tet*^{R}) or the creation of the new "inside-out" transposon with the alternative central region.

Note that both the original and "inside-out" transposons have inverted modules, but these modules evidently can function in either orientation relative to the central region. The frequency of transposition for composite transposons declines with the distance between the modules. Thus length dependence is a factor in determining the sizes of the common composite transposons.

A major force supporting the transposition of composite transposons is selection for the marker(s) carried in the central region. An IS10 module is free to move around on its own, and mobilizes an order of magnitude more frequently than Tn10. Tn10 is held together by selection for *tet*^{R}, though, so that under selective conditions, the relative frequency of intact Tn10 transposition is much increased.

The IS elements code for transposase activities that are responsible both for creating a target site and for recognizing the ends of the transposon. Only the ends are needed for a transposon to serve as a substrate for transposition.

21.4 Transposition Occurs by Both Replicative and Nonreplicative Mechanisms

Key concepts

- All transposons use a common mechanism in which staggered nicks are made in target DNA, the transposon is joined to the protruding ends, and the gaps are filled.
- The order of events and exact nature of the connections between transposon and target DNA determine whether transposition is replicative or nonreplicative.

The insertion of a transposon into a new site is illustrated in FIGURE 21.5. It consists of making staggered breaks in the target DNA, joining the transposon to the protruding single-stranded ends, and filling in the gaps. The generation and filling of the staggered ends explain the occurrence of the direct repeats of target DNA at the site of insertion. The stagger between the cuts on the two strands determines the length of the direct repeats; thus the target repeat characteristic of each transposon reflects the geometry of the enzyme involved in cutting target DNA.

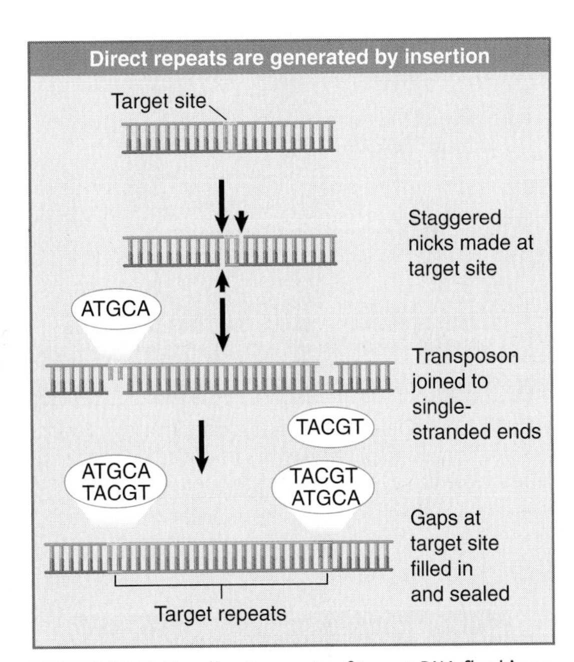

FIGURE 21.5 The direct repeats of target DNA flanking a transposon are generated by the introduction of staggered cuts whose protruding ends are linked to the transposon.

The use of staggered ends is common to all means of transposition, but we can distinguish three different types of mechanism by which a transposon moves:

- In **replicative transposition,** the element is duplicated during the reaction, so that the transposing entity is a copy of the original element. FIGURE 21.6 summarizes the results of such a transposition. The transposon is copied as part of its movement. One copy remains at the original site, whereas the other inserts at the new site. Thus transposition is accompanied by an increase in the number of copies of the transposon. Replicative transposition involves two types of enzymatic activity: a transposase that acts on the ends of the original transposon, and a **resolvase** that acts on the duplicated copies. A group of transposons related to TnA move only by replicative transposition (see Section 21.7, Replicative Transposition Proceeds through a Cointegrate).
- In **nonreplicative transposition,** the transposing element moves as a physical entity directly from one site to another and is conserved. The insertion sequences and composite transposons Tn10 and Tn5 use the mechanism shown in FIGURE 21.7, which involves the release of the transposon from the flanking donor DNA during transfer. This type of mechanism requires only a transposase. Another mechanism utilizes the connection of donor and target DNA sequences and shares some steps with replicative transposition (see Section 21.6, Common Intermediates for Transposition). Both mechanisms of nonreplicative transposition cause the element to be inserted at the target site and lost from the donor site. What happens to the donor molecule after a

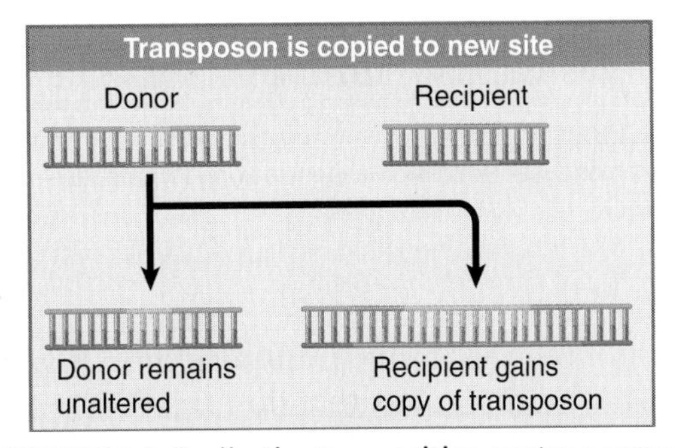

FIGURE 21.6 Replicative transposition creates a copy of the transposon, which inserts at a recipient site. The donor site remains unchanged, so both donor and recipient have a copy of the transposon.

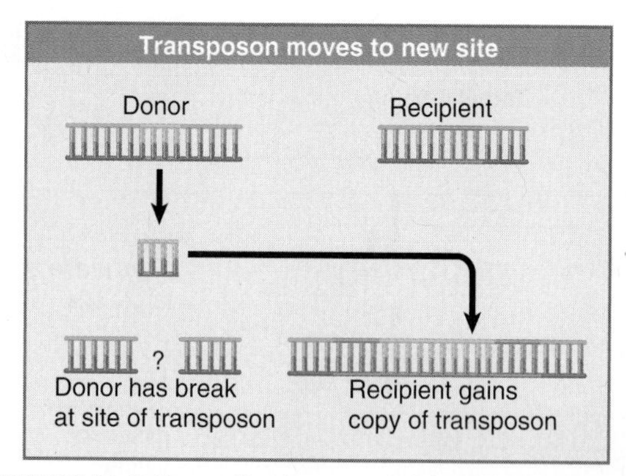

FIGURE 21.7 Nonreplicative transposition allows a transposon to move as a physical entity from a donor to a recipient site. This leaves a break at the donor site, which is lethal unless it can be repaired.

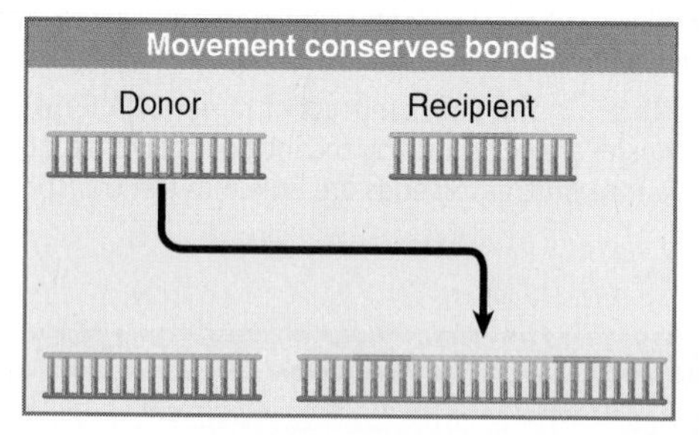

FIGURE 21.8 Conservative transposition involves direct movement with no loss of nucleotide bonds; compare with lambda integration and excision.

nonreplicative transposition? Its survival requires that host repair systems recognize the double-strand break and repair it.

- **Conservative transposition** describes another sort of nonreplicative event, in which the element is excised from the donor site and inserted into a target site by a series of events in which every nucleotide bond is conserved. FIGURE 21.8 summarizes the result of a conservative event. This exactly resembles the mechanism of lambda integration discussed in Section 19.16, Site-Specific Recombination Involves Breakage and Reunion, and the transposases of such elements are related to the λ integrase family. The elements that use this mechanism are large, and can mediate transfer not only of the element itself but also of donor DNA from one bacterium to another. Such elements were originally classified as transposons, but may more properly be regarded as episomes.

Some transposons use only one type of pathway for transposition, whereas others may be able to use multiple pathways. The elements IS1 and IS903 use both nonreplicative and replicative pathways, and the ability of phage Mu to turn to either type of pathway from a common intermediate has been well characterized (see Section 21.6, Common Intermediates for Transposition).

The same basic types of reaction are involved in all classes of transposition event. The ends of the transposon are disconnected from the donor DNA by cleavage reactions that generate 3′–OH ends. The exposed ends are then joined to the target DNA by transfer reactions, involving transesterification in which the 3′–OH end directly attacks the target DNA. These reactions take place within a nucleoprotein complex that contains the necessary enzymes and both ends of the transposon. Transposons differ as to whether the target DNA is recognized before or after the cleavage of the transposon itself.

The choice of target site is in effect made by the transposase. In some cases, the target is chosen virtually at random. In others, there is specificity for a consensus sequence or for some other feature in DNA. The feature can take the form of a structure in DNA, such as bent DNA, or for a protein-DNA complex. In the latter case, the nature of the target complex can cause the transposon to insert at specific promoters (such as Ty1 or Ty3, which select pol III promoters in yeast), inactive regions of the chromosome, or replicating DNA.

21.5 Transposons Cause Rearrangement of DNA

Key concepts

- Homologous recombination between multiple copies of a transposon causes rearrangement of host DNA.
- Homologous recombination between the repeats of a transposon may lead to precise or imprecise excision.

In addition to the "simple" intermolecular transposition that results in insertion at a new site, transposons promote other types of DNA rearrangements. Some of these events are consequences of the relationship between the multiple copies of the transposon. Others represent alternative outcomes of the transposition mechanism, and they leave clues about the nature of the underlying events.

Rearrangements of host DNA may result when a transposon inserts a copy at a second site near its original location. Host systems may undertake reciprocal recombination between the two copies of the transposon; the consequences are determined by whether the repeats are the same or in inverted orientation.

FIGURE 21.9 illustrates the general rule that recombination between any pair of direct repeats will delete the material between them. The intervening region is excised as a circle of DNA (which is lost from the cell); the chromosome retains a single copy of the direct repeat. A recombination between the directly repeated IS1 modules of the composite transposon Tn9 would replace the transposon with a single IS1 module.

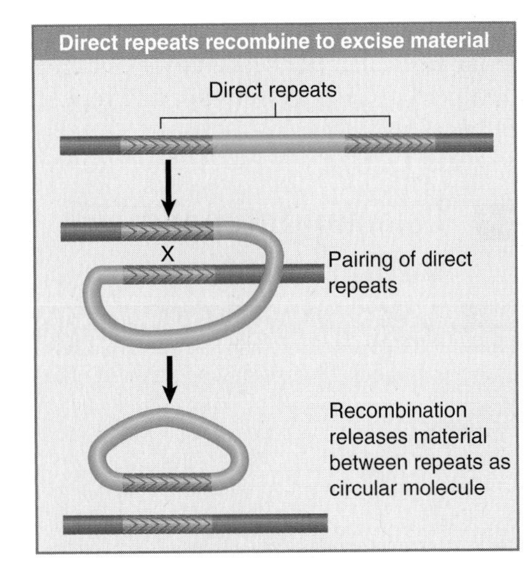

FIGURE 21.9 Reciprocal recombination between direct repeats excises the material between them; each product of recombination has one copy of the direct repeat.

Deletion of sequences adjacent to a transposon could therefore result from a two-stage process; transposition generates a direct repeat of a transposon, and recombination occurs between the repeats. The majority of deletions that arise in the vicinity of transposons, however, probably result from a variation in the pathway followed in the transposition event itself.

FIGURE 21.10 depicts the consequences of a reciprocal recombination between a pair of inverted repeats. The region between the repeats becomes inverted; the repeats themselves remain available to sponsor further inversions. A composite transposon whose modules are inverted is a stable component of the genome, although the direction of the central region with regard to the modules could be inverted by recombination.

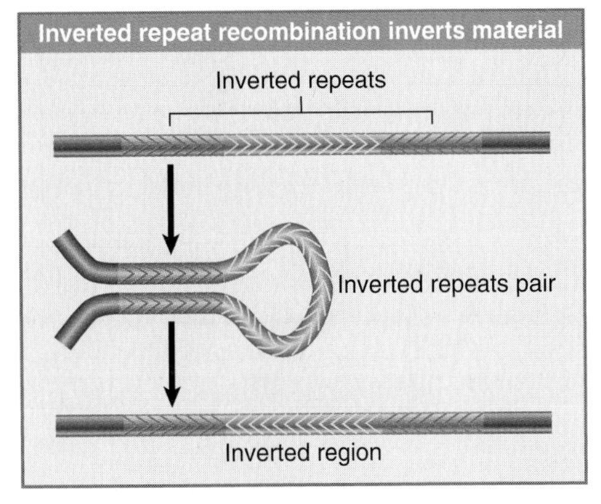

FIGURE 21.10 Reciprocal recombination between inverted repeats inverts the region between them.

Excision is not supported by transposons themselves, but may occur when bacterial enzymes recognize homologous regions in the transposons. This is important because the loss of a transposon may restore function at the site of insertion. **Precise excision** requires removal of the transposon plus one copy of the duplicated sequence. This is rare; it occurs at a frequency of ~10^{-6} for Tn5 and ~10^{-9} for Tn10. It probably involves a recombination between the 9 bp duplicated target sites.

Imprecise excision leaves a remnant of the transposon. The remnant may be sufficient to prevent reactivation of the target gene, but it may be insufficient to cause polar effects in adjacent genes so that a change of phenotype occurs. Imprecise excision occurs at a frequency of ~10^{-6} for Tn10. It involves recombination between sequences of 24 bp in the IS10 modules; these sequences are inverted repeats, but since the IS10 modules themselves are inverted, they form direct repeats in Tn10.

The greater frequency of imprecise excision compared with precise excision probably reflects the increase in the length of the direct repeats (24 bp as opposed to 9 bp). Neither type of excision relies on transposon-coded functions, but the mechanism is not known. Excision is

RecA-independent and could occur by some cellular mechanism that generates spontaneous deletions between closely spaced repeated sequences.

21.6 Common Intermediates for Transposition

Key concepts

- Transposition starts by forming a strand transfer complex in which the transposon is connected to the target site through one strand at each end.
- The Mu transposase forms the complex by synapsing the ends of Mu DNA, followed by nicking, and then a strand transfer reaction.
- Replicative transposition follows if the complex is replicated and nonreplicative transposition follows if it is repaired.

Many mobile DNA elements transpose from one chromosomal location to another by a fundamentally similar mechanism. They include IS elements, prokaryotic and eukaryotic transposons, and bacteriophage Mu. Insertion of the DNA copy of retroviral RNA uses a similar mechanism (see Section 22.2, The Retrovirus Life Cycle Involves Transposition-Like Events). The first stages of immunoglobulin recombination also are similar (see Section 23.10, The RAG Proteins Catalyze Breakage and Reunion).

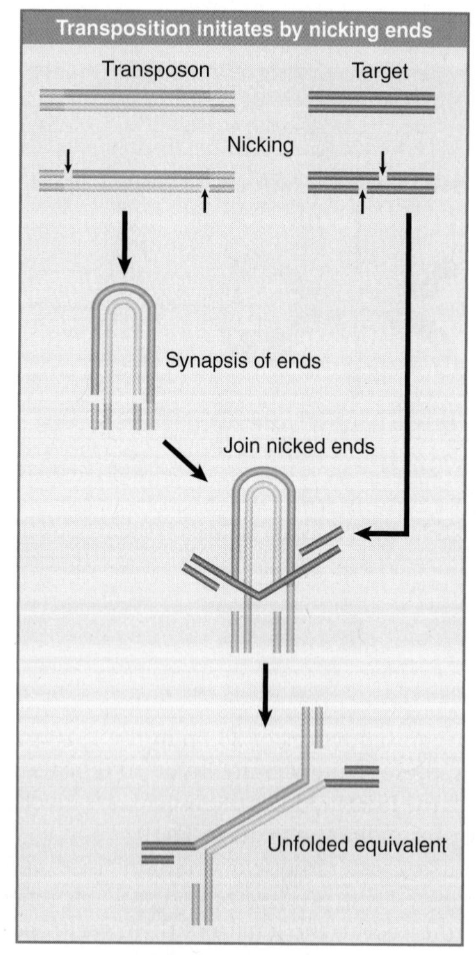

FIGURE 21.11 Transposition is initiated by nicking the transposon ends and target site and joining the nicked ends into a strand transfer complex.

Transposition starts with a common mechanism for joining the transposon to its target. FIGURE 21.11 shows that the transposon is nicked at both ends and the target site is nicked on both strands. The nicked ends are joined crosswise to generate a covalent connection between the transposon and the target. The two ends of the transposon are brought together in this process; for simplicity in following the cleavages, the synapsis stage is shown after cleavage, but actually occurs previously.

Much of this pathway was first revealed with phage Mu, which uses the process of transposition in two ways. Upon infecting a host cell, Mu integrates into the genome by nonreplicative transposition; during the ensuing lytic cycle, the number of copies is amplified by replicative transposition. Both types of transposition involve the same type of reaction between the transposon and its target, but the subsequent reactions are different.

The initial manipulations of the phage DNA are performed by the MuA transposase. Three MuA-binding sites with a 22 bp consensus are located at each end of Mu DNA. L1, L2, and L3 are at the left end; R1, R2, and R3 are at the right end. A monomer of MuA can bind to each site. MuA also binds to an internal site in the phage genome. Binding of MuA at both the left and right ends and at the internal site forms a complex. The role of the internal site is not clear; it appears to be necessary for formation of the complex, but not for strand cleavage and subsequent steps.

Joining the Mu transposon DNA to a target site passes through the three stages illustrated in FIGURE 21.12. This involves only the two sites closest to each end of the transposon. MuA subunits bound to these sites form a tetramer. This achieves synapsis of the two ends of the transposon. The tetramer now functions in a way that ensures a coordinated reaction on both ends of Mu DNA. MuA has two sites for manipulating DNA, and their mode of action compels subunits of the transposase to act in *trans*. The consensus-binding site binds to the 22 bp sequences that constitute the L1, L2, R1, and

R2 sites. The active site cleaves the Mu DNA strands at positions adjacent to the MuA-binding sites L1 and R1. The active site cannot cleave the DNA sequence that is adjacent to the consensus sequence in the consensus-binding site. It can, however, cleave the appropriate sequence on a different stretch of DNA.

The ends of the transposon are thus cleaved by MuA subunits acting in *trans*. The *trans* mode of action means that the monomers actually bound to L1 and R1 do not cleave the adjacent sites. One of the monomers bound to the left end nicks the site at the right end, and vice versa. (We do not know which monomer is active at this stage of the reaction.) The strand transfer reaction also occurs in *trans;* the monomer at L1 transfers the strand at R1, and vice versa. It could be the case that different monomers catalyze the cleavage and strand transfer reactions for a given end.

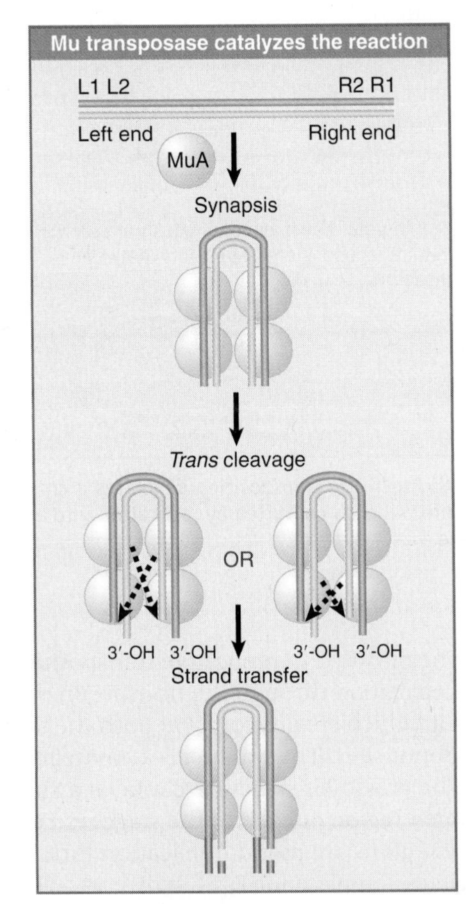

FIGURE 21.12 Mu transposition passes through three stable stages. MuA transposase forms a tetramer that synapses the ends of phage Mu. Transposase subunits act in *trans* to nick each end of the DNA, and then a second *trans* action joins the nicked ends to the target DNA.

A second protein, MuB, assists the reaction. It has an influence on the choice of target sites. Mu has a preference for transposing to a target site >10 to 15 kb away from the original insertion. This is called "target immunity." It is demonstrated in an *in vitro* reaction containing donor (Mu-containing) and target (Mu-deficient) plasmids, MuA and MuB proteins, *E. coli* HU protein, and Mg^{2+} and ATP. The presence of MuB and ATP restricts transposition exclusively to the target plasmid. The reason is that when MuB binds to the MuA-Mu DNA complex, MuA causes MuB to hydrolyze ATP, after which MuB is released. MuB binds (nonspecifically) to the target DNA, though, where it stimulates the recombination activity of MuA when a transposition complex forms. In effect, the prior presence of MuA "clears" MuB from the donor, thus giving a preference for transposition to the target.

The product of these reactions is a strand transfer complex in which the transposon is connected to the target site through one strand at each end. The next step of the reaction differs and determines the type of transposition. We see in the next two sections how the common structure can be a substrate for replication (leading to replicative transposition) or used directly for breakage and reunion (leading to nonreplicative transposition).

21.7 Replicative Transposition Proceeds through a Cointegrate

Key concepts

- Replication of a strand transfer complex generates a cointegrate, which is a fusion of the donor and target replicons.
- The cointegrate has two copies of the transposon, which lie between the original replicons.
- Recombination between the transposon copies regenerates the original replicons, but the recipient has gained a copy of the transposon.
- The recombination reaction is catalyzed by a resolvase coded by the transposon.

The basic structures involved in replicative transposition are illustrated in FIGURE 21.13:

- The 3′ ends of the strand transfer complex are used as primers for replication. This generates a structure called a **cointegrate,** which represents a fusion of the two original molecules. The cointegrate has two copies of the transposon,

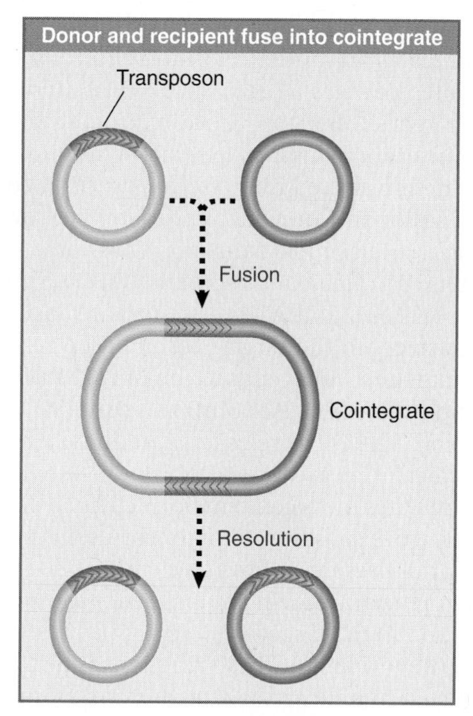

FIGURE 21.13 Transposition may fuse a donor and recipient replicon into a cointegrate. Resolution releases two replicons, each containing a copy of the transposon.

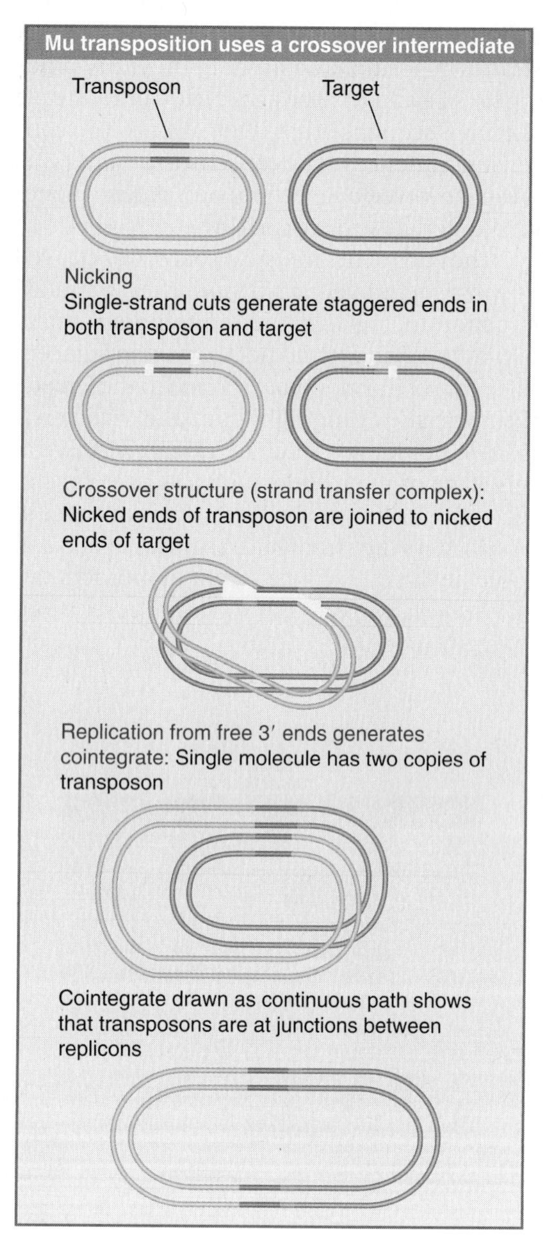

FIGURE 21.14 Mu transposition generates a crossover structure, which is converted by replication into a cointegrate.

one at each junction between the original replicons, oriented as direct repeats. The crossover is formed by the transposase, as described in the previous section. Its conversion into the cointegrate requires host replication functions.

- A homologous recombination between the two copies of the transposon releases two individual replicons, each of which has a copy of the transposon. One of the replicons is the original donor replicon. The other is a target replicon that has gained a transposon flanked by short direct repeats of the host target sequence. The recombination reaction is called **resolution;** the enzyme activity responsible is called the **resolvase.**

The reactions involved in generating a cointegrate have been defined in detail for phage Mu and are illustrated in FIGURE 21.14. The process starts with the formation of the strand transfer complex (sometimes also called a crossover complex). The donor and target strands are ligated so that each end of the transposon sequence is joined to one of the protruding single strands generated at the target site. The strand transfer complex generates a crossover-shaped structure held together at the duplex transposon. The fate of the crossover structure determines the mode of transposition.

The principle of replicative transposition is that replication through the transposon duplicates it, which creates copies at both the target and donor sites. The product is a cointegrate.

The crossover structure contains a single-stranded region at each of the staggered ends. These regions are pseudoreplication forks that provide a template for DNA synthesis. (Use of the ends as primers for replication implies that the strand breakage must occur with a polarity that generates a 3′–OH terminus at this point.)

If replication continues from both the pseudoreplication forks, it will proceed through the transposon, separating its strands and terminating at its ends. Replication is probably accomplished by host-coded functions. At this juncture, the structure has become a cointegrate, possessing direct repeats of the transposon at the junctions between the replicons (as can be seen by tracing the path around the cointegrate).

21.8 Nonreplicative Transposition Proceeds by Breakage and Reunion

Key concepts

- Nonreplicative transposition results if a crossover structure is nicked on the unbroken pair of donor strands and the target strands on either side of the transposon are ligated.
- Two pathways for nonreplicative transposition differ according to whether the first pair of transposon strands are joined to the target before the second pair are cut (Tn5), or whether all four strands are cut before joining to the target (Tn10).

The crossover structure can also be used in nonreplicative transposition. The principle of nonreplicative transposition by this mechanism is that a breakage and reunion reaction allows the target to be reconstructed with the insertion of the transposon; the donor remains broken. No cointegrate is formed.

FIGURE 21.15 shows the cleavage events that generate nonreplicative transposition of phage Mu. Once the unbroken donor strands have been nicked, the target strands on either side of the transposon can be ligated. The single-stranded regions generated by the staggered cuts must be filled in by repair synthesis. The product of this reaction is a target replicon in which the transposon has been inserted between repeats of the sequence created by the original single-strand nicks. The donor replicon has a double-strand break across the site where the transposon was originally located.

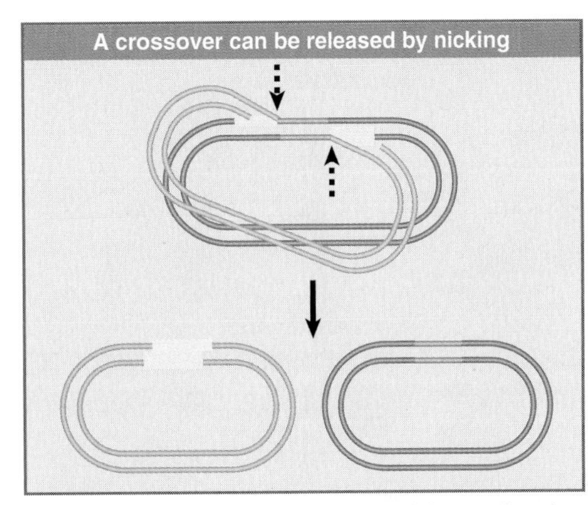

FIGURE 21.15 Nonreplicative transposition results when a crossover structure is released by nicking. This inserts the transposon into the target DNA, flanked by the direct repeats of the target, and the donor is left with a double-strand break.

Nonreplicative transposition can also occur by an alternative pathway in which nicks are made in target DNA, but a double-strand break is made on either side of the transposon, releasing it entirely from flanking donor sequences (as envisaged in Figure 21.7). This "cut and paste" pathway is used by Tn10, as illustrated in FIGURE 21.16.

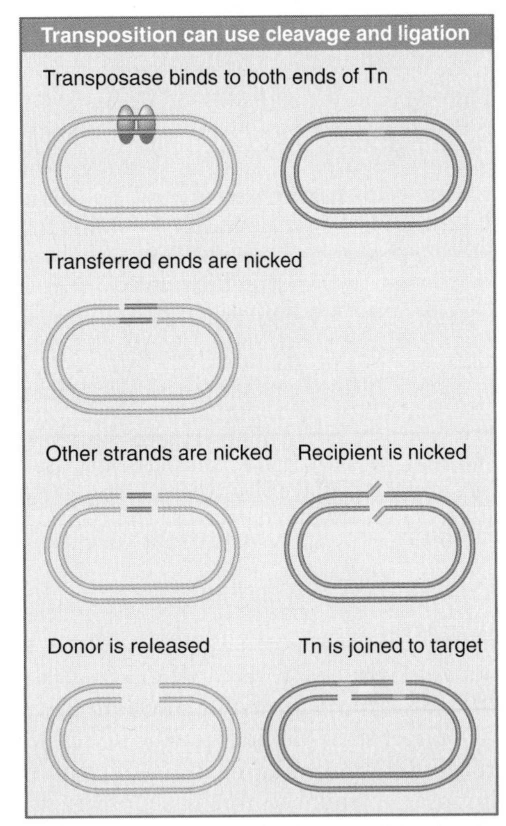

FIGURE 21.16 Both strands of Tn10 are cleaved sequentially, and then the transposon is joined to the nicked target site.

A neat experiment to prove that Tn10 transposes nonreplicatively made use of an artificially constructed heteroduplex of Tn10 that contained single base mismatches. If transposition involves replication, the transposon at the new site will contain information from only one of the parent Tn10 strands. If, however, transposition takes place by physical movement of the existing transposon, the mismatches will be conserved at the new site. This proved to be the case.

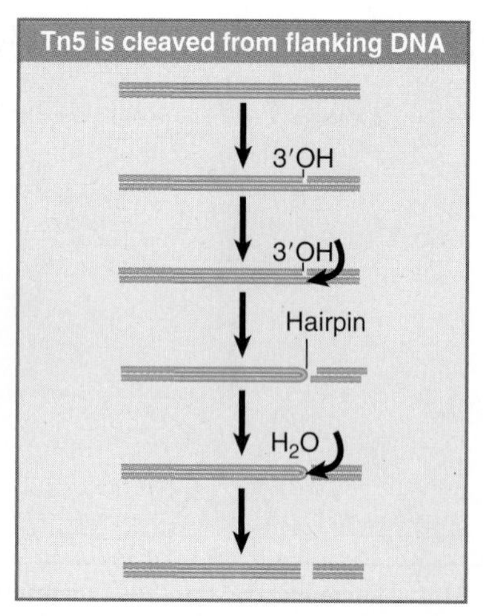

FIGURE 21.17 Cleavage of Tn5 from flanking DNA involves nicking, interstrand reaction, and hairpin cleavage.

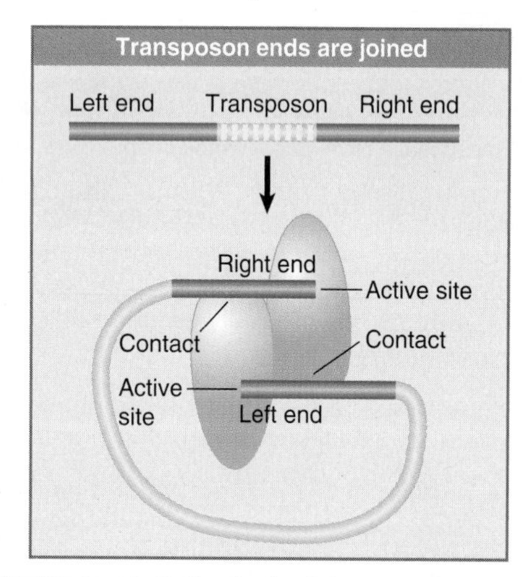

FIGURE 21.18 Each subunit of the Tn5 transposase has one end of the transposon located in its active site and also makes contact at a different site with the other end of the transposon.

The basic difference in Figure 21.16 from the model of Figure 21.15 is that both strands of Tn10 are cleaved before any connection is made to the target site. The first step in the reaction is recognition of the transposon ends by the transposase, forming a proteinaceous structure within which the reaction occurs. At each end of the transposon, the strands are cleaved in a specific order: the transferred strand (the one to be connected to the target site) is cleaved first, followed by the other strand. (This is the same order as in the Mu transposition of Figure 21.14 and Figure 21.15.)

Tn5 also transposes by nonreplicative transposition, and FIGURE 21.17 shows the interesting cleavage reaction that separates the transposon from the flanking sequences. First one DNA strand is nicked. The 3'–OH end that is released then attacks the other strand of DNA. This releases the flanking sequence and joins the two strands of the transposon in a hairpin. An activated water molecule then attacks the hairpin to generate free ends for each strand of the transposon.

In the next step, the cleaved donor DNA is released, and the transposon is joined to the nicked ends at the target site. The transposon and the target site remain constrained in the proteinaceous structure created by the transposase (and other proteins). The double-strand cleavage at each end of the transposon precludes any replicative-type transposition and forces the reaction to proceed by nonreplicative transposition, thus giving the same outcome as in Figure 21.14, but with the individual cleavage and joining steps occurring in a different order.

The Tn5 and Tn10 transposases both function as dimers. Each subunit in the dimer has an active site that successively catalyzes the double-strand breakage of the two strands at one end of the transposon, and then catalyzes staggered cleavage of the target site. FIGURE 21.18 illustrates the structure of the Tn5 transposase bound to the cleaved transposon. Each end of the transposon is located in the active site of one subunit. One end of the subunit also contacts the other end of the transposon. This controls the geometry of the transposition reaction. Each of the active sites will cleave one strand of the target DNA. It is the geometry of the complex that determines the distance between these sites on the two target strands (nine base pairs in the case of Tn5).

21.9 TnA Transposition Requires Transposase and Resolvase

Key concepts

- Replicative transposition of TnA requires a transposase to form the cointegrate structure and a resolvase to release the two replicons.
- The action of the resolvase resembles lambda Int protein and belongs to the general family of topoisomerase-like site-specific recombination reactions, which pass through an intermediate in which the protein is covalently bound to the DNA.

Replicative transposition is the only mode of mobility of the TnA family, which consists of large (~5 kb) transposons. They are not composites relying on IS-type transposition modules, but rather comprise independent units carrying genes for transposition as well as for features such as drug resistance. The TnA family includes several related transposons, of which Tn3 and Tn1000 (also known as γδ) are the best characterized. They have the usual terminal feature of closely related inverted repeats, generally ~38 bp in length. *Cis*-acting deletions in either repeat prevent transposition of an element. A 5 bp direct repeat is generated at the target site. They carry resistance markers such as *amp*^{r}.

The two stages of TnA-mediated transposition are accomplished by the transposase and the resolvase, whose genes (*tnpA* and *tnpR*) are identified by recessive mutations. The transposition stage involves the ends of the element, as it does in IS-type elements. Resolution requires a specific internal site. This feature is unique to the TnA family.

Mutants in *tnpA* cannot transpose. The gene product is a transposase that binds to a sequence of ~25 bp located within the 38 bp of the inverted terminal repeat. A binding site for the *E. coli* protein IHF exists adjacent to the transposase binding site, and transposase and IHF bind cooperatively. The transposase recognizes the ends of the element and also makes the staggered 5 bp breaks in target DNA where the transposon is to be inserted. IHF is a DNA-binding protein that often is involved in assembling large structures in *E. coli;* its role in the transposition reaction may not be essential.

The *tnpR* gene product has dual functions. It acts as a repressor of gene expression and it provides the resolvase function.

Mutations in *tnpR* increase the transposition frequency. The reason is that TnpR represses the transcription of both *tnpA* and its own gene. Thus inactivation of TnpR protein allows increased synthesis of TnpA, which results in an increased frequency of transposition. This implies that the amount of the TnpA transposase must be a limiting factor in transposition.

The *tnpA* and *tnpR* genes are expressed divergently from an A-T-rich intercistronic control region, which is indicated in the map of Tn3 given in FIGURE 21.19. Both effects of TnpR are mediated by its binding in this region.

In its capacity as the resolvase, TnpR is involved in recombination between the direct repeats of Tn3 in a cointegrate structure. A cointegrate can in principle be resolved by a homologous recombination between any corresponding pair of points in the two copies of the transposon. The Tn3 resolution reaction, however, occurs only at a specific site.

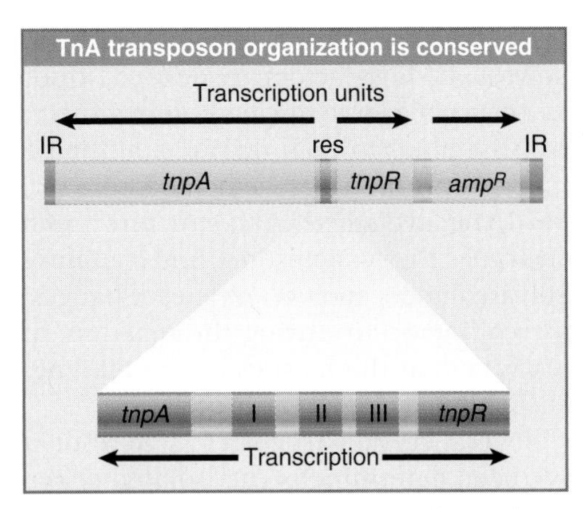

FIGURE 21.19 Transposons of the TnA family have inverted terminal repeats, an internal *res* site, and three known genes.

The site of resolution is called *res*. It is identified by *cis*-acting deletions that block completion of transposition, causing the accumulation of cointegrates. In the absence of *res*, the resolution reaction can be substituted by RecA-mediated general recombination, but this is much less efficient.

The sites bound by the TnpR resolvase are summarized in the lower part of Figure 21.19. Binding occurs independently at each of three sites, each of which is 30 to 40 bp long. The three binding sites share a sequence homology that defines a consensus sequence with dyad symmetry.

Site I includes the region genetically defined as the *res* site; in its absence, the resolution reaction does not proceed at all. Resolution also involves binding at sites II and III, though, because the reaction proceeds poorly if either of these sites is deleted. Site I overlaps with the startpoint for *tnpA* transcription. Site II overlaps with the startpoint for *tnpR* transcription; an operator mutation maps just at the left end of the site.

Do the sites interact? One possibility is that binding at all three sites is required to hold the DNA in an appropriate topology. Binding at a single set of sites may repress *tnpA* and *tnpR* transcription without introducing any change in the DNA.

An *in vitro* resolution assay uses a cointegrate-like DNA molecule as substrate. The substrate must be supercoiled; its resolution produces two catenated circles, each containing one *res* site.

The reaction requires large amounts of the TnpR resolvase; no host factors are needed. Resolution occurs in a large nucleoprotein structure. Resolvase binds to each *res* site, and then the bound sites are brought together to form a structure ~10 nm in diameter. This structure is sometimes called the synaptosome, and contains six resolvase dimers and two *res* sites. Changes in supercoiling occur during the reaction, and DNA is bent at the *res* sites by the binding of transposase.

Several crystal structures of γδ resolvase have been determined. The isolated enzyme can form a dimer of dimers, that is, a subunit has an interface that promotes dimer formation, and the dimer then uses a second interface to form a tetramer. The dimer can bind to site I, forming a structure in which the catalytic residues are distant from the cleavage target sites on DNA. The crystal structure of the synaptosome in which a resolvase tetramer is linked via Ser^{10} to two cleaved site I duplex DNAs shows that the target sites in the duplexes that are to be recombined lie on opposite sides of the tetramer. This implies that one dimer must rotate by 180° relative to the other dimer to reposition the DNAs in order to accomplish the recombination reaction. This part of the mechanism is different from that employed by the tyrosine recombinases (such as Cre or Flp), where the enzyme basically brings together the recombining DNAs in the form of a Holliday junction (see Section 19.18, Site-Specific Recombination Resembles Topoisomerase Activity).

Resolution occurs by breaking and rejoining bonds without input of energy. Cleavage is accomplished by a transesterification reaction that covalently links Ser^{10} of a resolvase subunit to a 5′ phosphate at the target bond in site I. The product consists of resolvase covalently attached to both 5′ ends of double-stranded cuts made at the *res* site. The cleavage occurs symmetrically at a short palindromic region to generate two base extensions. Expanding the view of the crossover region located in site I, we can describe the cutting reaction as:

5′ TTATAA3′
3′ AATATT5′
⬇
5′ TTAT + protein-AA3′
3′ AA- protein TATT5′

The reaction resembles the action of lambda Int at the *att* sites (see Section 19.16, Site-Specific Recombination Involves Breakage and Reunion). Indeed, 15 of the 20 bp of the *res* site are identical to the bases at corresponding positions in *att*. This suggests that the site-specific recombination of lambda and resolution of TnA have evolved from a common type of recombination reaction. Indeed, we see in Section 23.10, The RAG Proteins Catalyze Breakage and Reunion, that recombination involving immunoglobulin genes has the same basis. The common feature in all these reactions is the transfer of the broken end to the catalytic protein as an intermediate stage before it is rejoined to another broken end (see Section 19.18, Site-Specific Recombination Resembles Topoisomerase Activity).

The reactions themselves are analogous in terms of manipulation of DNA, although resolution occurs only between intramolecular sites, whereas the recombination between *att* sites is intermolecular and directional (as seen by the differences in *attB* and *attP* sites). The mechanism of protein action, however, is different in each case. Resolvase functions in a manner in which four subunits bind to the recombining *res* sites. Each subunit makes a single-strand cleavage. A reorganization of the subunits relative to one another then physically moves the DNA strands, placing them in a recombined conformation. This allows the nicks to be sealed, along with the release of resolvase.

21.10 Transposition of Tn10 Has Multiple Controls

Key concepts

- Multicopy inhibition reduces the rate of transposition of any one copy of a transposon when other copies of the same transposon are introduced into the genome.
- Multiple mechanisms affect the rate of transposition.

Control of the frequency of transposition is important for the cell. A transposon must be able to maintain a certain minimum frequency of movement in order to survive, but too great a frequency could be damaging to the host cell. Every transposon appears to have mechanisms that control its frequency of transposition. A variety of mechanisms have been characterized for Tn10.

Tn10 is a composite transposon in which the element IS10R provides the active module. The organization of IS10R is summarized in FIGURE 21.20. Two promoters are found close to

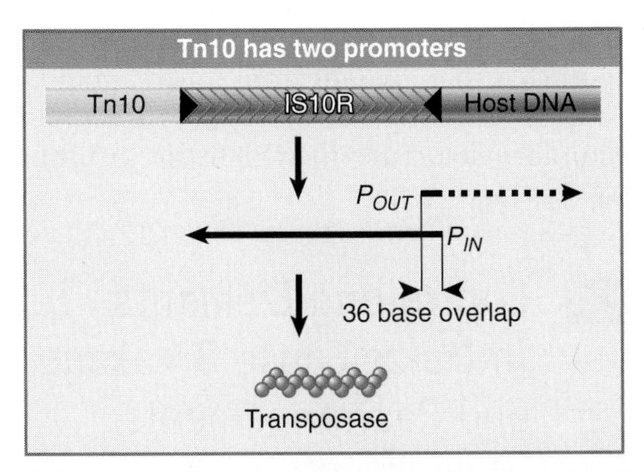

FIGURE 21.20 Two promoters in opposite orientation lie near the outside boundary of IS10R. The strong promoter P_{OUT} sponsors transcription toward the flanking host DNA. The weaker promoter P_{IN} causes transcription of an RNA that extends the length of IS10R and is translated into the transposase.

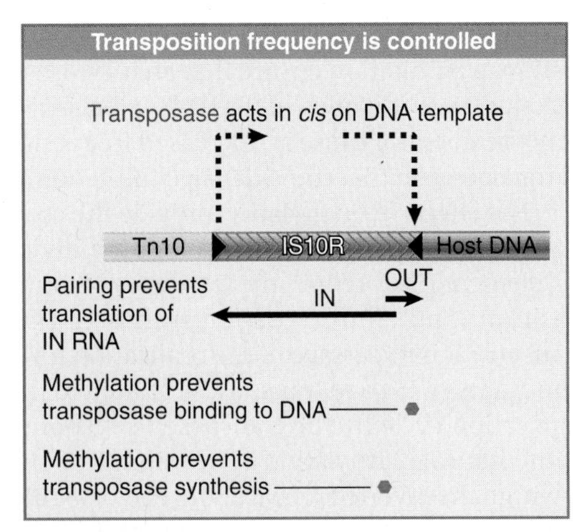

FIGURE 21.21 Several mechanisms restrain the frequency of Tn10 transposition by affecting either the synthesis or function of transposase protein. Transposition of an individual transposon is restricted by methylation to occur only after replication. In multicopy situations, *cis*-preference restricts the choice of target, and OUT/IN RNA pairing inhibits synthesis of transposase.

the outside boundary. The promoter P_{IN} is responsible for transcription of IS10R. The promoter P_{OUT} causes transcription to proceed toward the adjacent flanking DNA. Transcription usually terminates within the transposon, but occasionally continues into the host DNA; sometimes this readthrough transcription is responsible for activating adjacent bacterial genes.

The phenomenon of "multicopy inhibition" reveals that expression of the IS10R transposase gene is regulated. Transposition of a Tn10 element on the bacterial chromosome is reduced when additional copies of IS10R are introduced via a multicopy plasmid. The inhibition requires the P_{OUT} promoter, and is exercised at the level of translation. The basis for the effect lies with the overlap in the 5′ terminal regions of the transcripts from P_{IN} and P_{OUT}. OUT RNA is a transcript of 69 bases. It is present at >100× the level of IN RNA for two reasons: P_{OUT} is a much stronger promoter than P_{IN}; and OUT RNA is more stable than IN RNA.

OUT RNA functions as an antisense RNA (see Section 13.7, Small RNA Molecules Can Regulate Translation). The level of OUT RNA has no effect in a single-copy situation, but has a significant effect when >5 copies are present. There are usually ~5 copies of OUT RNA per copy of IS10 (which corresponds to ~150 copies of OUT RNA in a typical multicopy situation). OUT RNA base pairs with IN RNA, and the excess of OUT RNA ensures that IN RNA is bound rapidly before a ribosome can attach. Thus the paired IN RNA cannot be translated.

The quantity of transposase protein is often a critical feature. Tn10, whose transposase is synthesized at the low level of 0.15 molecules per cell per generation, displays several interesting mechanisms. FIGURE 21.21 summarizes the various effects that influence transposition frequency.

A continuous reading frame on one strand of IS10R codes for the transposase. The level of the transposase limits the rate of transposition. Mutants in this gene can be complemented in *trans* by another, wild-type IS10 element, but only with some difficulty. This reflects a strong preference of the transposase for *cis*-action; the enzyme functions efficiently only with the DNA template from which it was transcribed and translated. *Cis*-preference is a common feature of transposases coded by IS elements. (Other proteins that display *cis*-preference include the A protein involved in φX174 replication; see Section 16.5, Rolling Circles Are Used to Replicate Phage Genomes.)

Does *cis*-preference reflect an ability of the transposase to recognize more efficiently those DNA target sequences that lie nearer to the site where the enzyme is synthesized? One possible explanation is that the transposase binds to DNA so tightly after (or even during) protein synthesis that it has a very low probability of diffusing elsewhere. Another possibility is that the enzyme may be unstable when it is not bound to DNA, so that protein molecules failing to bind quickly (and therefore nearby) never have a chance to become active.

Together the results of *cis*-preference and multicopy inhibition ensure that an increase in the number of copies of Tn10 in a bacterial genome does not cause an increased frequency of transposition that could damage the genome.

The effects of methylation provide the most important system of regulation for an individual element. They reduce the frequency of transposition and (more importantly) couple transposition to passage of the replication fork. The ability of IS10 to transpose is related to the replication cycle by the transposon's response to the state of methylation at two sites. One site is within the inverted repeat at the end of IS10R, where the transposase binds. The other site is in the promoter P_{IN}, from which the transposase gene is transcribed.

Both of these sites are methylated by the *dam* system described in Section 15.5, Does Methylation at the Origin Regulate Initiation? The Dam methylase modifies the adenine in the sequence GATC on a newly synthesized strand generated by replication. The frequency of Tn10 transposition is increased 1000-fold in *dam*$^-$ strains in which the two target sites lack methyl groups.

Passage of a replication fork over these sites generates hemimethylated sequences; this activates the transposon by a combination of transcribing the transposase gene more frequently from P_{IN} and enhancing binding of transposase to the end of IS10R. In a wild-type bacterium, the sites remain hemimethylated for a short period after replication.

Why should it be desirable for transposition to occur soon after replication? The nonreplicative mechanism of Tn10 transposition places the donor DNA at risk of being destroyed (see Figure 21.7). The cell's chances of survival may be increased if replication has just occurred to generate a second copy of the donor sequence. The mechanism is effective because only one of the two newly replicated copies gives rise to a transposition event (determined by which strand of the transposon is unmethylated at the *dam* sites).

A transposon selects its target site at random; as a result there is a reasonable probability that it may land in an active operon. Will transcription from the outside continue through the transposon and thus activate the transposase, whose overproduction may in turn lead to high (perhaps lethal) levels of transposition? Tn10 protects itself against such events by two mechanisms. Transcription across the IS10R terminus decreases its activity, presumably by inhibiting its ability to bind transposase, and the mRNA that extends from upstream of the promoter is poorly translated, because it has a secondary structure in which the initiation codon is inaccessible.

21.11 Controlling Elements in Maize Cause Breakage and Rearrangements

Key concepts

- Transposition in maize was discovered because of the effects of the chromosome breaks generated by transposition of "controlling elements."
- The break generates one chromosome that has a centromere and a broken end and one acentric fragment.
- The acentric fragment is lost during mitosis; this can be detected by the disappearance of dominant alleles in a heterozygote.
- Fusion between the broken ends of the chromosome generates dicentric chromosomes, which undergo further cycles of breakage and fusion.
- The fusion-breakage-bridge cycle is responsible for the occurrence of somatic variegation.

One of the most visible consequences of the existence and mobility of transposons occurs during plant development, when somatic variation occurs. This is due to changes in the location or behavior of **controlling elements** (the name that transposons were given in maize before their molecular nature was discovered).

Two features of maize have helped to follow transposition events. Controlling elements often insert near genes that have visible but nonlethal effects on the phenotype. Maize displays clonal development, which means that the occurrence and timing of a transposition event can be visualized as depicted diagrammatically in FIGURE 21.22.

The nature of the event does not matter: It may be a point mutation, insertion, excision, or chromosome break. What is important is that it occurs in a heterozygote to alter the expression of one allele. The descendants of a cell that has suffered the event then display a new phenotype, whereas the descendants of cells not affected by the event continue to display the original phenotype.

Mitotic descendants of a given cell remain in the same location and give rise to a **sector** of tissue. A change in phenotype during somatic development is called **variegation;** it is revealed by a sector of the new phenotype residing within the tissue of the original phenotype. The size of the sector depends on the number of divisions in the lineage giving rise to it, so the size of the area of the new phenotype is determined by the timing of the change in genotype. The earlier its occurrence in the cell lineage, the greater the number of descendants and thus the size of patch in the mature tissue. This is seen most vividly in the variation in kernel color, when patches of one color appear within another color.

Insertion of a controlling element may affect the activity of adjacent genes. Deletions, duplications, inversions, and translocations all occur at the sites where controlling elements are present. Chromosome breakage is a common consequence of the presence of some elements. A unique feature of the maize system is that the activities of the controlling elements are regulated during development. The elements transpose and promote genetic rearrangements at characteristic times and frequencies during plant development.

The characteristic behavior of controlling elements in maize is typified by the Ds element, which was originally identified by its ability to provide a site for chromosome breakage. The consequences are illustrated in FIGURE 21.23. Consider a heterozygote in which Ds lies on one homolog between the centromere and a series of dominant markers. The other homolog lacks Ds and has recessive markers (*C, bz,* and *wx*). Breakage at Ds generates an **acentric fragment** carrying the dominant markers. As a result of its lack of a centromere, this fragment is lost at mitosis. Thus the descendant cells have only the recessive markers carried by the intact chromosome. This gives the type of situation whose results are depicted in Figure 21.22.

FIGURE 21.24 shows that breakage at Ds leads to the formation of two unusual chromosomes. These are generated by joining the broken ends of the products of replication. One is a U-shaped acentric fragment consisting of the joined sister chromatids for the region distal to Ds (on the left as drawn in the figure). The other is a U-shaped **dicentric chromosome** comprising the sister chromatids proximal to Ds (on its right in the figure). The latter structure leads to the classic **breakage-fusion-bridge** cycle illustrated in the figure.

Follow the fate of the dicentric chromosome when it attempts to segregate on the mitotic spindle. Each of its two centromeres pulls toward an opposite pole. The tension

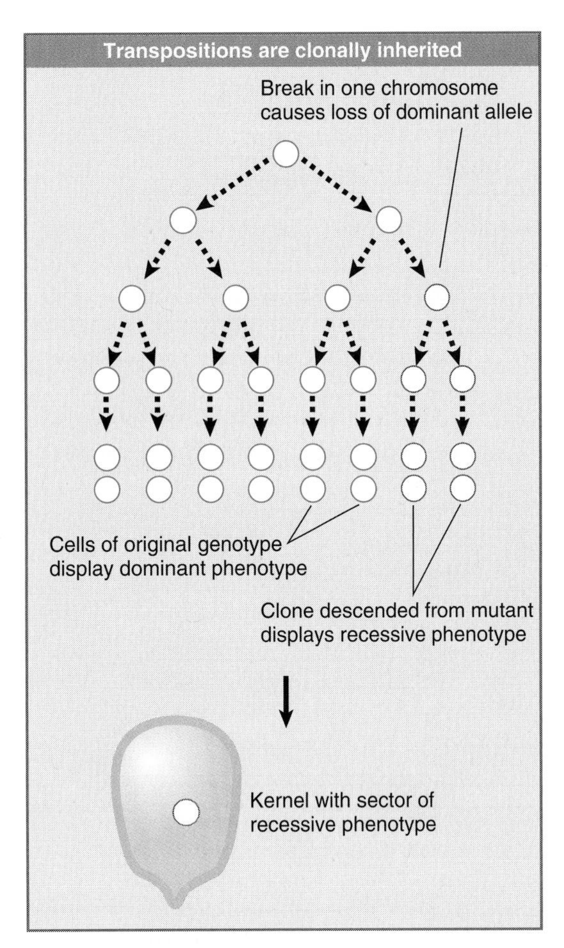

FIGURE 21.22 Clonal analysis identifies a group of cells descended from a single ancestor in which a transposition-mediated event altered the phenotype. Timing of the event during development is indicated by the number of cells; tissue specificity of the event may be indicated by the location of the cells.

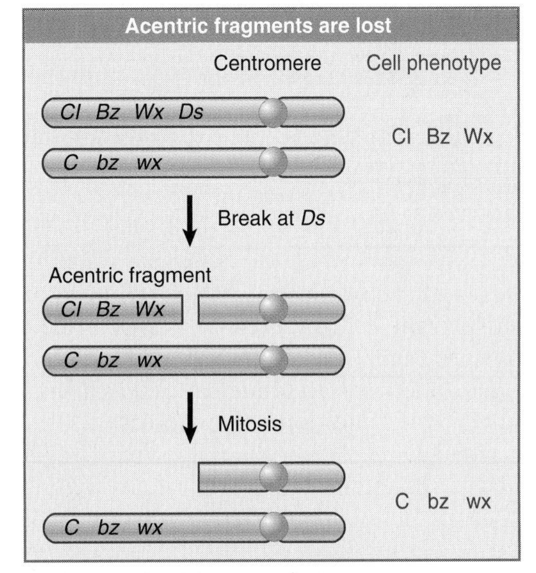

FIGURE 21.23 A break at a controlling element causes loss of an acentric fragment; if the fragment carries the dominant markers of a heterozygote, its loss changes the phenotype. The effects of the dominant markers, *CI, Bz,* and *Wx,* can be visualized by the color of the cells or by appropriate staining.

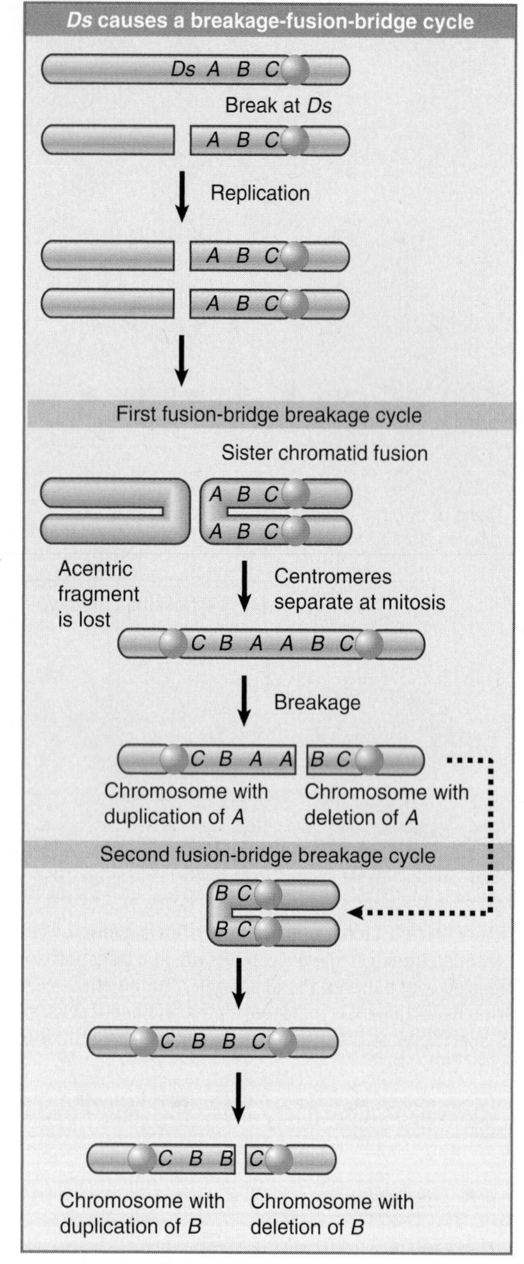

FIGURE 21.24 *Ds* provides a site to initiate the chromatid breakage-fusion-bridge cycle. The products can be followed by clonal analysis.

breaks the chromosome at a random site between the centromeres. In the example of the figure, breakage occurs between loci A and B, with the result that one daughter chromosome has a duplication of A, whereas the other has a deletion. If A is a dominant marker, the cells with the duplication will retain *a* phenotype, but cells with the deletion will display the recessive *a* phenotype.

The breakage-fusion-bridge cycle continues through further cell generations, allowing genetic changes to continue in the descendants. For example, consider the deletion chromosome that has lost A. In the next cycle, a break occurs between B and C, so that the descendants are divided into those with a duplication of B and those with a deletion. Successive losses of dominant markers are revealed by subsectors within sectors.

21.12 Controlling Elements Form Families of Transposons

Key concepts

- Each family of transposons in maize has both autonomous and nonautonomous controlling elements.
- Autonomous controlling elements code for proteins that enable them to transpose.
- Nonautonomous controlling elements have mutations that eliminate their capacity to catalyze transposition, but they can transpose when an autonomous element provides the necessary proteins.
- Autonomous controlling elements have changes of phase, when their properties alter as a result of changes in the state of methylation.

The maize genome contains several families of controlling elements. The numbers, types, and locations of the elements are characteristic for each individual maize strain. They may occupy a significant part of the genome. The members of each family are divided into two classes:

- **Autonomous controlling elements** have the ability to excise and transpose. As a result of the continuing activity of an autonomous element, its insertion at any locus creates an unstable or "mutable" allele. Loss of the autonomous element itself, or of its ability to transpose, converts a mutable allele to a stable allele.
- **Nonautonomous controlling elements** are stable; they do not transpose or suffer other spontaneous changes in condition. They become unstable only when an autonomous member of the same family is present elsewhere in the genome. When complemented in *trans* by an autonomous element, a nonautonomous element displays the usual range of activities associated with autonomous elements, including the ability to transpose to new sites. Nonautonomous elements are derived from autonomous elements by loss of *trans*-acting functions needed for transposition.

Families of controlling elements are defined by the interactions between autonomous and nonautonomous elements. A family consists of a single type of autonomous element accompanied by many varieties of nonautonomous elements. A nonautonomous element is placed in a family by its ability to be activated in *trans* by the autonomous elements. The major families of controlling elements in maize are summarized in FIGURE 21.25.

Characterized at the molecular level, the maize transposons share the usual form of organization—inverted repeats at the ends and short direct repeats in the adjacent target DNA—but otherwise vary in size and coding capacity. All families of transposons share the same type of relationship between the autonomous and nonautonomous elements. The autonomous elements have open reading frames between the terminal repeats, whereas the nonautonomous elements do not code for functional proteins. Sometimes the internal sequences are related to those of autonomous elements; at other times they have diverged completely.

The Mutator transposon is one of the simplest elements. The autonomous element MuDR codes for the genes *mudrA* (which codes for the MURA transposase) and *mudrB* (which codes for a nonessential accessory protein). The ends of the elements are marked by 200 bp inverted repeats. Nonautonomous elements—basically any unit that has the inverted repeats, which may not have any internal sequence relationship to MuDR—are also mobilized by MURA.

There are typically several members (~10) of each transposon family in a plant genome. By analyzing autonomous and nonautonomous elements of the Ac/Ds family, we have molecular information about many individual examples of these elements. FIGURE 21.26 summarizes their structures.

Most of the length of the autonomous Ac element is occupied by a single gene consisting of five exons. The product is the transposase. The element itself ends in inverted repeats of 11 bp, and a target sequence of 8 bp is duplicated at the site of insertion.

Ds elements vary in both length and sequence, but are related to Ac. They end in the same 11 bp inverted repeats. They are shorter than Ac, and the length of deletion varies. At one extreme, the element Ds9 has a deletion of only 194 bp. In a more extensive deletion, the Ds6 element retains a length of only 2 kb, representing 1 kb from each end of Ac. A complex double Ds element has one Ds6 sequence inserted in reverse orientation into another.

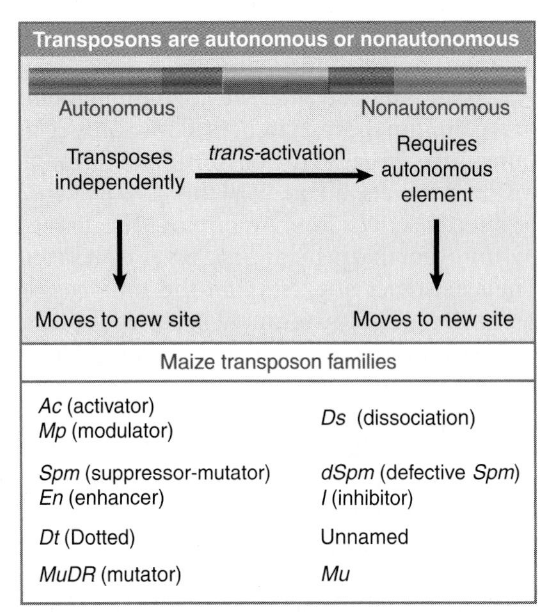

Maize transposon families	
Ac (activator) *Mp* (modulator)	*Ds* (dissociation)
Spm (suppressor-mutator) *En* (enhancer)	*dSpm* (defective *Spm*) *I* (inhibitor)
Dt (Dotted)	Unnamed
MuDR (mutator)	*Mu*

FIGURE 21.25 Each controlling element family has both autonomous and nonautonomous members. Autonomous elements are capable of transposition. Nonautonomous elements are deficient in transposition. Pairs of autonomous and nonautonomous elements can be classified in >4 families.

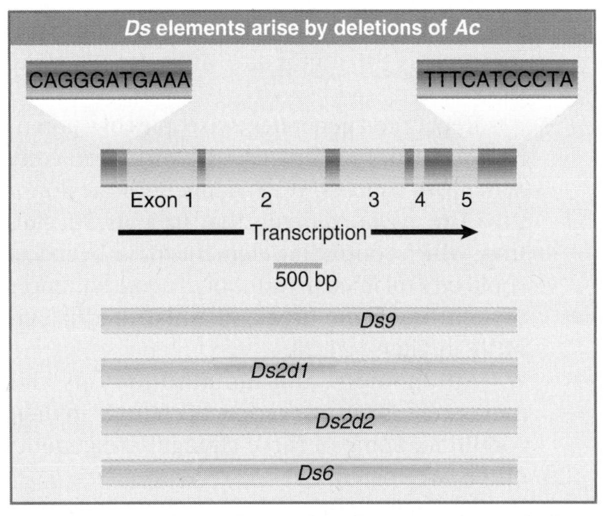

FIGURE 21.26 The *Ac* element has five exons that code for a transposase; *Ds* elements have internal deletions.

Nonautonomous elements lack internal sequences, but possess the terminal inverted repeats (and possibly other sequence features). Nonautonomous elements are derived from autonomous elements by deletions (or other changes) that inactivate the *trans*-acting transposase, but leave intact the sites (including the termini) on which the transposase acts. Their structures range from minor (but inactivating) mutations of Ac to sequences that have major deletions or rearrangements.

At another extreme, the Ds1 family members comprise short sequences whose only relationship to Ac lies in the possession of terminal inverted repeats. Elements of this class need not be directly derived from Ac, but could be derived by any event that generates the inverted repeats. Their existence suggests that the transposase recognizes only the terminal inverted repeats, or possibly the terminal repeats in conjunction with some short internal sequence.

Transposition of Ac/Ds occurs by a nonreplicative mechanism, and is accompanied by its disappearance from the donor location. Clonal analysis suggests that transposition of Ac/Ds almost always occurs soon after the donor element has been replicated. These features resemble transposition of the bacterial element Tn10 (see Section 21.10, Transposition of Tn10 Has Multiple Controls). The cause is the same: transposition does not occur when the DNA of the transposon is methylated on both strands (the typical state before methylation), and is activated when the DNA is hemimethylated (the typical state immediately after replication). The recipient site is frequently on the same chromosome as the donor site, and often is quite close to it.

Replication generates two copies of a potential Ac/Ds donor, but usually only one copy actually transposes. What happens to the donor site? The rearrangements that are found at sites from which controlling elements have been lost could be explained in terms of the consequences of a chromosome break, as illustrated previously in Figure 21.23.

Autonomous and nonautonomous elements are subject to a variety of changes in their condition. Some of these changes are genetic; others are epigenetic.

The major change is (of course) the conversion of an autonomous element into a nonautonomous element, but further changes may occur in the nonautonomous element. *Cis*-acting defects may render a nonautonomous element impervious to autonomous elements. Thus a nonautonomous element may become permanently stable because it can no longer be activated to transpose.

Autonomous elements are subject to "changes of phase," which are heritable but relatively unstable alterations in their properties. These take the form of a reversible inactivation in which the element cycles between an active and inactive condition during plant development.

Phase changes in both the Ac and Mu types of autonomous element result from changes in the methylation of DNA. Comparisons of the susceptibilities of active and inactive elements to restriction enzymes suggest that the inactive form of the element is methylated in the target sequence $\frac{\text{CAG}}{\text{GTC}}$. There are several target sites in each element, and we do not know which sites control the effect. In the case of MuDR, demethylation of the terminal repeats increases transposase expression, suggesting that the effect may be mediated through control of the promoter for the transposase gene. We should like to know what controls the methylation and demethylation of the elements.

The effect of methylation is common generally among transposons in plants. The best demonstration of the effect of methylation on activity comes from observations made with the *Arabidopsis* mutant *ddm1,* which causes a loss of methylation in heterochromatin. Among the targets that lose methyl groups is a family of transposons related to MuDR. Direct analysis of genome sequences shows that the demethylation causes transposition events to occur. Methylation is probably the major mechanism that is used to prevent transposons from damaging the genome by transposing too frequently.

There may be self-regulating controls of transposition, analogous to the immunity effects displayed by bacterial transposons. An increase in the number of Ac elements in the genome decreases the frequency of transposition. The Ac element may code for a repressor of transposition; the activity could be carried by the same protein that provides transposase function.

21.13 Spm Elements Influence Gene Expression

Key concepts

- Spm elements affect gene expression at their sites of insertion, when the TnpA protein binds to its target sites at the ends of the transposon.
- Spm elements are inactivated by methylation.

The Spm and En autonomous elements are virtually identical; they differ at <10 positions. FIGURE 21.27 summarizes the structure. The 13 bp inverted terminal repeats are essential for transposition, as indicated by the transposition-defective phenotype of deletions at the termini. Transposons related to Spm are found in other plants, and are defined as members of the same family by their generally similar organization. They all share nearly identical inverted terminal repeats, and generate 3 bp duplications of target DNA upon transposition. Named for the terminal similarities, they are known as the CACTA group of transposons.

A sequence of 8300 bp is transcribed from a promoter in the left end of the element. The 11 exons contained in the transcript are spliced into a 2500-base messenger. The mRNA codes for a protein of 621 amino acids. The gene is called *tnpA*, and the protein binds to a 12 bp consensus sequence present in multiple copies in the terminal regions of the element. Function of *tnpA* is required for excision, but may not be sufficient.

All of the nonautonomous elements of this family (denoted dSpm for defective Spm) are closely related in structure to the Spm element itself. They have deletions that affect the exons of *tnpA*.

Two additional open reading frames (ORF1 and ORF2) are located within the first, long intron of *tnpA*. They are contained in an alternatively spliced 6000-base RNA, which is present at 1% of the level of the *tnpA* mRNA. The function containing ORFs 1 and 2 is called *tnpB*. It may provide the protein that binds to the 13 bp terminal inverted repeats to cleave the termini for transposition.

In addition to the fully active Spm element, there are Spm-w derivatives that show weaker activity in transposition. The example given in Figure 21.27 has a deletion that eliminates both ORF1 and ORF2. This suggests that the need for TnpB in transposition can be bypassed or substituted.

Spm insertions can control the expression of a gene at the site of insertion. A recipient locus may be brought under either negative or positive control. An Spm-suppressible locus suffers inhibition of expression. An Spm-dependent locus is expressed only with the aid of Spm. When the inserted element is a dSpm, suppression or dependence responds to the *trans*-acting function supplied by an autonomous Spm. What is the basis for these opposite effects?

A dSpm-suppressible allele contains an insertion of dSpm within an exon of the gene. This structure raises the immediate question of how a gene with a dSpm insertion in an exon can ever be expressed! The dSpm sequence can be excised from the transcript by using sequences at its termini. The splicing event may leave a change in the sequence of the mRNA, thus explaining a change in the properties of the protein for which it codes. A similar ability to be excised from a transcript has been found for some Ds insertions.

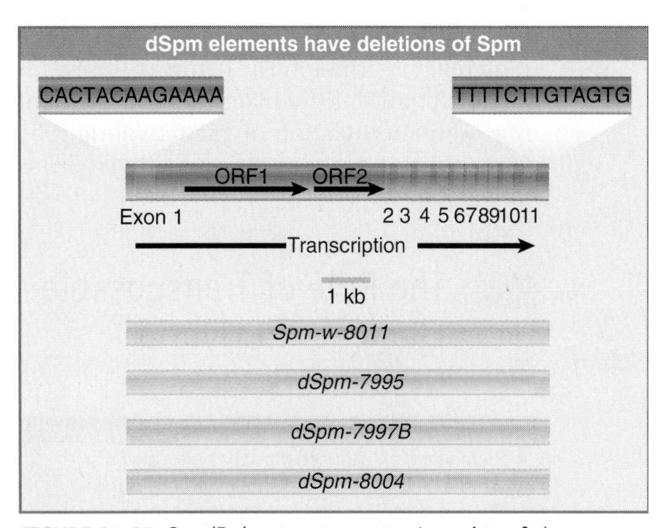

FIGURE 21.27 *Spm/En* has two genes. *tnpA* consists of eleven exons that are transcribed into a spliced 2500-base mRNA. *tnpB* may consist of a 6000-base mRNA containing ORF1 + ORF2.

tnpA provides the suppressor function for which the Spm element was originally named. The presence of a defective element may reduce, but not eliminate, expression of a gene in which it resides. The introduction of an autonomous element that possesses a functional *tnpA* gene, however, may suppress expression of the target gene entirely. Suppression is caused by the ability of TnpA to bind to its target sites in the defective element, which blocks transcription from proceeding.

A dSpm-dependent allele contains an insertion near but not within a gene. The insertion appears to provide an enhancer that activates the promoter of the gene at the recipient locus.

Suppression and dependence at dSpm elements appear to rely on the same interaction between the *trans*-acting product of the *tnpA* gene of an autonomous Spm element and the *cis*-acting sites at the ends of the element. Thus a single interaction between the protein and the ends of the element either suppresses or activates a target locus depending on whether the element is located upstream of or within the recipient gene.

Spm elements exist in a variety of states ranging from fully active to cryptic. A cryptic element is silent and neither transposes itself nor activates dSpm elements. A cryptic element may be reactivated transiently or converted to the active state by interaction with a fully active Spm element. Inactivation is caused by methylation of sequences in the vicinity of the transcription startpoint. The nature of the events that are responsible for inactivating an element by *de novo* methylation or for activating it by demethylation (or preventing methylation) is not yet known.

21.14 The Role of Transposable Elements in Hybrid Dysgenesis

Key concepts

- P elements are transposons that are carried in P strains of *Drosophila melanogaster,* but not in M strains.
- When a P male is crossed with an M female, transposition is activated.
- The insertion of P elements at new sites in these crosses inactivates many genes and makes the cross infertile.

Certain strains of *D. melanogaster* encounter difficulties in interbreeding. When flies from two of these strains are crossed, the progeny display "dysgenic traits"—a series of defects including mutations, chromosomal aberrations, distorted segregation at meiosis, and sterility. The appearance of these correlated defects is called **hybrid dysgenesis.**

Two systems responsible for hybrid dysgenesis have been identified in *D. melanogaster.* In the first, flies are divided into the types I (inducer) and R (reactive). Reduced fertility is seen in crosses of I males with R females, but not in the reverse direction. In the second system, flies are divided into the two types P (paternal contributing) and M (maternal contributing). FIGURE 21.28 illustrates the asymmetry of the system; a cross between a P male and an M female causes dysgenesis, but the reverse cross does not.

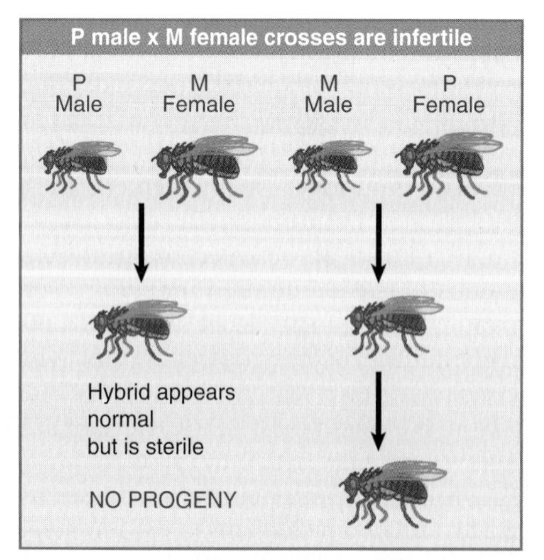

FIGURE 21.28 Hybrid dysgenesis is asymmetrical; it is induced by P male × M female crosses, but not by M male × P female crosses.

Dysgenesis is principally a phenomenon of the germ cells. In crosses involving the P-M system, the F1 hybrid flies have normal somatic tissues. Their gonads, however, do not develop. The morphological defect in gamete development dates from the stage at which rapid cell divisions commence in the germline.

Any one of the chromosomes of a P male can induce dysgenesis in a cross with an M female. The construction of recombinant chromosomes shows that several regions within each P chromosome are able to cause dysgenesis. This suggests that a P male has sequences at many different chromosomal locations that can induce dysgenesis. The locations differ between individual P strains. The P-specific sequences are absent from chromosomes of M flies.

The nature of the P-specific sequences was first identified by mapping the DNA of *w* mutants found among the dysgenic hybrids. All the mutations result from the insertion of DNA into the *w* locus. (The insertion inactivates the gene, causing the white-eye phenotype for which the locus is named.) The inserted sequence is called the **P element.**

The P element insertions form a classic transposable system. Individual elements vary in length but are homologous in sequence. All P elements possess inverted terminal repeats of 31 bp, and generate direct repeats of target DNA of 8 bp upon transposition. The longest P elements are ~2.9 kb long and have four open reading frames. The shorter elements arise, apparently rather frequently, by internal deletions of a full-length P factor. At least some of the shorter P elements have lost the capacity to produce the transposase, but may be activated in *trans* by the enzyme coded by a complete P element.

A P strain carries 30 to 50 copies of the P element, about a third of them full length. The elements are absent from M strains. In a P strain the elements are carried as inert components of the genome, but they become activated to transpose when a P male is crossed with an M female.

Chromosomes from P-M hybrid dysgenic flies have P elements inserted at many new sites. The insertions inactivate the genes in which they are located and often cause chromosomal breaks. The result of the transpositions is therefore to inactivate the genome.

21.15 P Elements Are Activated in the Germline

Key concepts

- P elements are activated in the germline of P male × M female crosses because a tissue-specific splicing event removes one intron, which generates the coding sequence for the transposase.
- The P element also produces a repressor of transposition, which is inherited maternally in the cytoplasm.
- The presence of the repressor explains why M male × P female crosses remain fertile.

Activation of P elements is tissue-specific: it occurs only in the germline. P elements are transcribed, though, in both germline and somatic tissues. Tissue-specificity is conferred by a change in the splicing pattern.

FIGURE 21.29 depicts the organization of the element and its transcripts. The primary transcript extends for 2.5 kb or 3.0 kb, the difference probably reflecting merely the leakiness of the termination site. Two protein products can be produced:

- In somatic tissues, only the first two introns are excised, creating a coding region of ORF0-ORF1-ORF2. Translation of this RNA yields a protein of 66 kD. This protein is a repressor of transposon activity.
- In germline tissues, an additional splicing event occurs to remove intron 3. This connects all four open reading frames into an mRNA that is translated to generate a protein of 87 kD. This protein is the transposase.

Two types of experiment have demonstrated that splicing of the third intron is needed for transposition. First, if the splicing junctions are mutated *in vitro* and the P element is reintroduced into flies, its transposition activity is abolished. Second, if the third intron is deleted, so that ORF3 is constitutively included in the mRNA in all tissues, transposition occurs in somatic tissues as well as the germline. Thus whenever ORF3 is spliced to the preceding reading frame, the P element becomes active. This is the crucial regulatory event, and usually it occurs only in the germline.

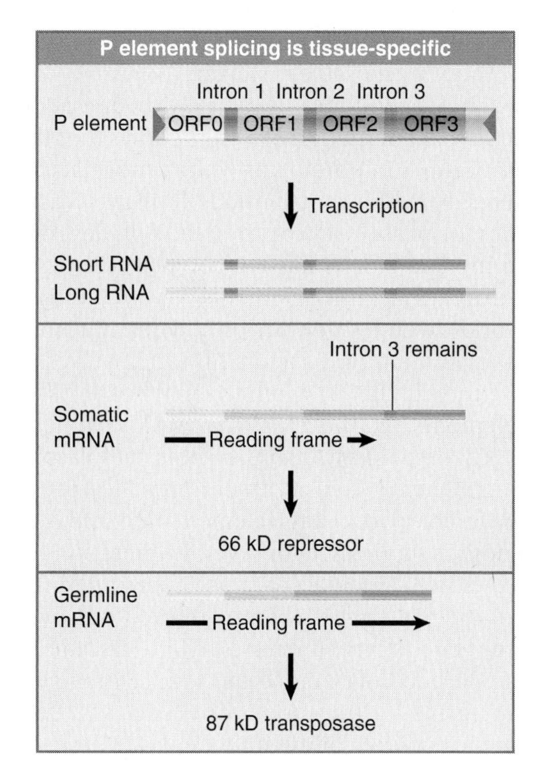

FIGURE 21.29 The P element has four exons. The first three are spliced together in somatic expression; all four are spliced together in germline expression.

What is responsible for the tissue-specific splicing? Somatic cells contain a protein that binds to sequences in exon 3 to prevent splicing of the last intron (see Section 26.12, Alternative Splicing Involves Differential Use of Splice Junctions). The absence of this protein in germline cells allows splicing to generate the mRNA that codes for the transposase.

Transposition of a P element requires ~150 bp of terminal DNA. The transposase binds to 10 bp sequences that are adjacent to the 31 bp inverted repeats. Transposition occurs by a nonreplicative "cut and paste" mechanism resembling that of Tn10. (It contributes to hybrid dysgenesis in two ways: Insertion of the transposed element at a new site may cause mutations, and the break that is left at the donor site—see Figure 21.7—has a deleterious effect.)

It is interesting that, in a significant proportion of cases, the break in donor DNA is repaired by using the sequence of the homologous chromosome. If the homolog has a P element, the presence of a P element at the donor site may be restored (so the event resembles the result of a replicative transposition). If the homolog lacks a P element, repair may generate a sequence lacking the P element, thus apparently providing a precise excision (an unusual event in other transposable systems).

The dependence of hybrid dysgenesis on the sexual orientation of a cross shows that the cytoplasm is important, as well as the P factors themselves. The contribution of the cytoplasm is described as the **cytotype;** a line of flies containing P elements has P cytotype, whereas a line of flies lacking P elements has M cytotype. Hybrid dysgenesis occurs only when chromosomes containing P factors find themselves in M cytotype, that is, when the male parent has P elements and the female parent does not.

Cytotype shows an inheritable cytoplasmic effect; when a cross occurs through P cytotype (the female parent has P elements), hybrid dysgenesis is suppressed for several generations of crosses with M female parents. Thus something in P cytotype, which can be diluted out over some generations, suppresses hybrid dysgenesis.

The effect of cytotype is explained in molecular terms by the model of FIGURE 21.30. It depends on the ability of the 66 kD protein to repress transposition. The protein is provided as a maternal factor in the egg. In a P line, there must be sufficient protein to prevent transposition from occurring, even though the P elements are present. In any cross involving a P female, its presence prevents either synthesis or activity of the transposase. When the female parent is M type, though, there is no repressor in the egg, and the introduction of a P element from the male parent results in activity of transposase in the germline. The ability of P cytotype to exert an effect through more than one generation suggests that there must be enough repressor protein in the egg, and it must be stable enough, to be passed on through the adult to be present in the eggs of the next generation.

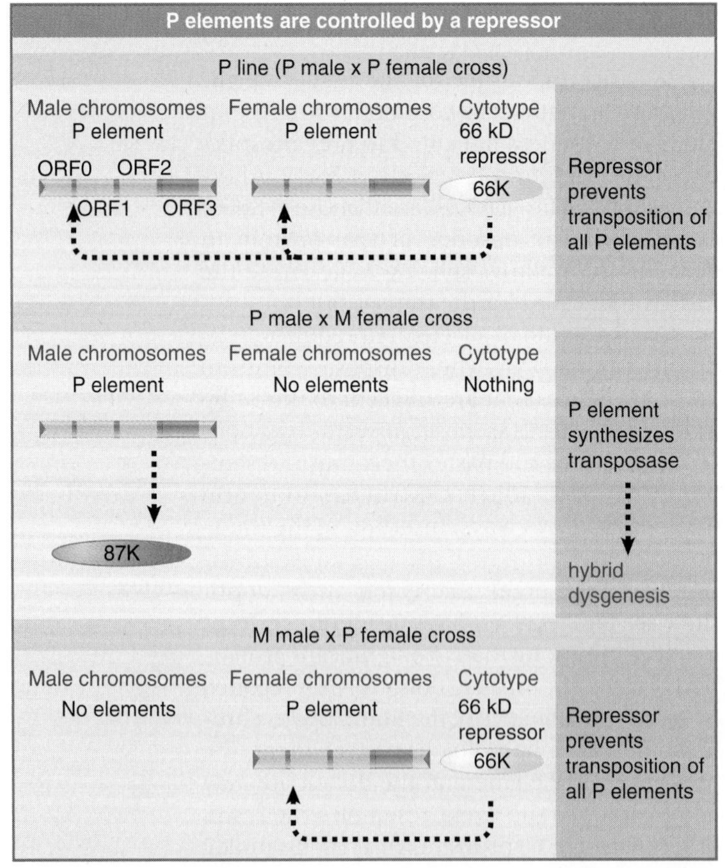

FIGURE 21.30 Hybrid dysgenesis is determined by the interactions between P elements in the genome and 66 kD repressor in the cytotype.

Strains of *D. melanogaster* descended from flies caught in the wild more than 30 years ago are always M. Strains descended from flies caught in the past 10 years are almost always P. Does this mean that the P element family has invaded wild populations of *D. melanogaster* in recent years? P elements are indeed highly invasive when introduced into a new population; the source of the invading element would have to be another species.

Hybrid dysgenesis reduces interbreeding, and thus it is a step on the path to speciation. Suppose that a dysgenic system is created by a transposable element in some geographic location. Another element may create a different system in some other location. Flies in the two areas will be dysgenic for two (or possibly more) systems. If this renders them intersterile and the populations become genetically isolated, further separation may occur. Multiple dysgenic systems therefore lead to inability to mate—and to speciation.

21.16 Summary

Prokaryotic and eukaryotic cells contain a variety of transposons that mobilize by moving or copying DNA sequences. The transposon can be identified only as an entity within the genome; its mobility does not involve an independent form. The transposon could be selfish DNA, concerned only with perpetuating itself within the resident genome; if it conveys any selective advantage upon the genome, this must be indirect. All transposons have systems to limit the extent of transposition, because unbridled transposition is presumably damaging, but the molecular mechanisms are different in each case.

The archetypal transposon has inverted repeats at its termini and generates direct repeats of a short sequence at the site of insertion. The simplest types are the bacterial insertion sequences (IS), which consist essentially of the inverted terminal repeats flanking a coding frame(s) whose product(s) provide transposition activity. Composite transposons have terminal modules that consist of IS elements; one or both of the IS modules provides transposase activity, and the sequences between them (often carrying antibiotic resistance) are treated as passengers.

The generation of target repeats flanking a transposon reflects a common feature of transposition. The target site is cleaved at points that are staggered on each DNA strand by a fixed distance (often five or nine base pairs). The transposon is in effect inserted between protruding single-stranded ends generated by the staggered cuts. Target repeats are generated by filling in the single-stranded regions.

IS elements, composite transposons, and P elements mobilize by nonreplicative transposition, in which the element moves directly from a donor site to a recipient site. A single transposase enzyme undertakes the reaction. It occurs by a "cut and paste" mechanism in which the transposon is separated from flanking DNA. Cleavage of the transposon ends, nicking of the target site, and connection of the transposon ends to the staggered nicks all occur in a nucleoprotein complex containing the transposase. Loss of the transposon from the donor creates a double-strand break whose fate is not clear. In the case of Tn10, transposition becomes possible immediately after DNA replication, when sites recognized by the *dam* methylation system are transiently hemimethylated. This imposes a demand for the existence of two copies of the donor site, which may enhance the cell's chances for survival.

The TnA family of transposons mobilize by replicative transposition. After the transposon at the donor site becomes connected to the target site, replication generates a cointegrate molecule that has two copies of the transposon. A resolution reaction that involves recombination between two particular sites then frees the two copies of the transposon, so that one remains at the donor site and one appears at the target site. Two enzymes coded by the transposon are required: Transposase recognizes the ends of the transposon and connects them to the target site, and resolvase provides a site-specific recombination function.

Phage Mu undergoes replicative transposition by the same mechanism as TnA. It also can use its cointegrate intermediate to transpose by a nonreplicative mechanism. The difference between this reaction and the nonreplicative transposition of IS elements is that the cleavage events occur in a different order.

The best characterized transposons in plants are the controlling elements of maize, which fall into several families. Each family contains a single type of autonomous element that is analogous to bacterial transposons in its ability to mobilize. A family also contains many different nonautonomous elements that are derived by mutations (usually deletions) of the autonomous element. The nonautonomous elements lack the ability to transpose, but display transposition activity and other abilities of the autonomous element when an autonomous element is present to provide the necessary *trans*-acting functions.

In addition to the direct consequences of insertion and excision, the maize elements may also control the activities of genes at or near the sites where they are inserted; this control may be subject to developmental regulation. Maize elements inserted into genes may be excised from the transcripts, which explains why they do not simply impede gene activity. Control of target gene expression involves a variety of molecular effects, including activation by provision of an enhancer and suppression by interference with posttranscriptional events.

Transposition of maize elements (in particular Ac) is nonreplicative, and probably requires only a single transposase enzyme coded by the element. Transposition occurs preferentially after replication of the element. It is likely that there are mechanisms to limit the frequency of transposition. Advantageous rearrangements of the maize genome may have been connected with the presence of the elements.

P elements in *D. melanogaster* are responsible for hybrid dysgenesis, which could be a forerunner of speciation. A cross between a male carrying P elements and a female lacking them generates hybrids that are sterile. A P element has four open reading frames, which are separated by introns. Splicing of the first three ORFs generates a 66 kD repressor and occurs in all cells. Splicing of all four ORFs to generate the 87 kD transposase occurs only in the germline by a tissue-specific splicing event. P elements mobilize when exposed to cytoplasm lacking the repressor. The burst of transposition events inactivates the genome by random insertions. Only a complete P element can generate transposase, but defective elements can be mobilized in *trans* by the enzyme.

References

21.1 Introduction

Reviews

Campbell, A. (1981). Evolutionary significance of accessory DNA elements in bacteria. *Annu. Rev. Immunol.* 35, 55–83.

Finnegan, D. J. (1985). Transposable elements in eukaryotes. *Int. Rev. Cytol.* 93, 281–326.

21.2 Insertion Sequences Are Simple Transposition Modules

Reviews

Berg, D. E. and Howe, M., eds. (1989). *Mobile DNA.* Washington, DC: American Society for Microbiology Press.

Calos, M. and Miller, J. H. (1980). Transposable elements. *Cell* 20, 579–595.

Craig, N. L. (1997). Target site selection in transposition. *Annu. Rev. Biochem.* 66, 437–474.

Galas, D. J. and Chandler, M. (1989). Bacterial insertion sequence. In Berg, D. E. and Howe, M., eds., *Mobile DNA.* Washington, DC: American Society for Microbiology Press, pp. 109–162.

Kleckner, N.(1977). Translocatable elements in prokaryotes. *Cell* 11, 11–23.

Kleckner, N. (1981). Transposable elements in prokaryotes. *Annu. Rev. Genet.* 15, 341–404.

Research

Grindley, N. D. (1978). IS1 insertion generates duplication of a 9 bp sequence at its target site. *Cell* 13, 419–426.

Johnsrud, L., Calos, M. P., and Miller, J. H. (1978). The transposon Tn9 generates a 9 bp repeated sequence during integration. *Cell* 15, 1209–1219.

21.3 Composite Transposons Have IS Modules

Review

Kleckner, N. (1989). Tn10 transposon. In Berg, D. E. and Howe, M. M., eds., *Mobile DNA.* Washington, DC: American Society for Microbiology Press, pp. 227–268.

21.4 Transposition Occurs by Both Replicative and Nonreplicative Mechanisms

Reviews

Craig, N. L. (1997). Target site selection in transposition. *Annu. Rev. Biochem.* 66, 437–474.

Grindley, N. D. and Reed, R. R. (1985). Transpositional recombination in prokaryotes. *Annu. Rev. Biochem.* 54, 863–896.

Haren, L., Ton-Hoang, B., and Chandler, M. (1999). Integrating DNA: transposases and retroviral integrases. *Annu. Rev. Microbiol.* 53, 245–281.

Scott, J. R. and Churchward, G. G. (1995). Conjugative transposition. *Annu. Rev. Immunol.* 49, 367–397.

21.6 Common Intermediates for Transposition

Reviews

Mizuuchi, K. (1992). Transpositional recombination: mechanistic insights from studies of Mu and other elements. *Annu. Rev. Biochem.* 61, 1011–1051.

Pato, M. L. (1989). Bacteriophage mu. In Berg, D. E. and Howe, M., eds., *Mobile DNA.* Washington, DC: American Society for Microbiology Press, pp. 23–52.

Research

Aldaz, H., Schuster, E., and Baker, T. A. (1996). The interwoven architecture of the Mu transposase couples DNA synthesis to catalysis. *Cell* 85, 257–269.

Savilahti, H. and Mizuuchi, K. (1996). Mu transpositional recombination: donor DNA cleavage and strand transfer in *trans* by the Mu transpose. *Cell* 85, 271–280.

21.8 Nonreplicative Transposition Proceeds by Breakage and Reunion

Research

Bender, J. and Kleckner, N. (1986). Genetic evidence that Tn10 transposes by a nonreplicative mechanism. *Cell* 45, 801–815.

Bolland, S. and Kleckner, N. (1996). The three chemical steps of Tn10/IS10 transposition involve repeated utilization of a single active site. *Cell* 84, 223–233.

Davies, D. R., Goryshin, I. Y., Reznikoff, W. S., and Rayment, I. (2000). Three-dimensional structure of the Tn5 synaptic complex transposition intermediate. *Science* 289, 77–85.

Haniford, D. B., Benjamin, H. W., and Kleckner, N. (1991). Kinetic and structural analysis of a cleaved donor intermediate and a strand transfer intermediate in Tn10 transposition. *Cell* 64, 171–179.

Kennedy, A. K., Guhathakurta, A., Kleckner, N., and Haniford, D. B. (1998). Tn10 transposition via a DNA hairpin intermediate. *Cell* 95, 125–134.

21.9 TnA Transposition Requires Transposase and Resolvase

Review

Sherratt, D. (1989). Tn3 and related transposable elements: site-specific recombination and transposition. In Berg, D. E. and Howe, M. M., eds., *Mobile DNA.* Washington, DC: American Society for Microbiology Press, pp. 163–184.

Research

Droge, P. et al. (1990). The two functional domains of gamma delta resolvase act on the same recombination site: implications for the mechanism of strand exchange. *Proc. Natl. Acad. Sci. USA* 87, 5336–5340.

Grindley, N. D. et al. (1982). Transposon-mediated site-specific recombination: identification of three binding sites for resolvase at the *res* sites of $\gamma\delta$ and Tn3. *Cell* 30, 19–27.

Li, W., Kamtekar, S., Xiong, Y., Sarkis, G. J., Grindley, N. D., and Steitz, T. A. (2005). Structure of a synaptic gammadelta resolvase tetramer covalently linked to two cleaved DNAs. *Science* 309, 1210–1215.

21.10 Transposition of Tn10 Has Multiple Controls

Reviews

Kleckner, N. (1989). Tn10 transposon. In Berg, D. E. and Howe, M. M., eds., *Mobile DNA*. Washington, DC: American Society for Microbiology Press, pp. 227–268.

Kleckner, N. (1990). Regulation of transposition in bacteria. *Annu. Rev. Cell Biol.* 6, 297–327.

Research

Roberts, D. et al. (1985). IS10 transposition is regulated by DNA adenine methylation. *Cell* 43, 117–130.

21.12 Controlling Elements Form Families of Transposons

Reviews

Fedoroff, N. (1989). Maize transposable elements. In Berg, D. E. and Howe, M. M., eds., *Mobile DNA*. Washington, DC: American Society for Microbiology Press, pp. 375–411.

Fedoroff, N. (2000). Transposons and genome evolution in plants. *Proc. Natl. Acad. Sci. USA* 97, 7002–7007.

Gierl, A., Saedler, H., and Peterson, P. A. (1989). Maize transposable elements. *Annu. Rev. Genet.* 23, 71–85.

Research

Benito, M. I. and Walbot, V. (1997). Characterization of the maize Mutator transposable element MURA transposase as a DNA-binding protein. *Mol. Cell Biol.* 17, 5165–5175.

Chandler, V. L. and Walbot, V. (1986). DNA modification of a maize transposable element correlates with loss of activity. *Proc. Natl. Acad. Sci. USA* 83, 1767–1771.

Ros, F. and Kunze, R. (2001). Regulation of activator/dissociation transposition by replication and DNA methylation. *Genetics* 157, 1723–1733.

Singer, T., Yordan, C., and Martienssen, R. A. (2001). Robertson's Mutator transposons in A. thaliana are regulated by the chromatin-remodeling gene decrease in DNA Methylation (DDM1). *Genes Dev.* 15, 591–602.

21.14 The Role of Transposable Elements in Hybrid Dysgenesis

Reviews

Engels, W. R. (1983). The P family of transposable elements in *Drosophila*. *Annu. Rev. Genet.* 17, 315–344.

Engels, W. R. (1989). P elements in *Drosophila*. In Berg, D. E. and Howe, M. M., eds., *Mobile DNA*. Washington, DC: American Society for Microbiology Press, pp. 437–484.

21.15 P Elements Are Activated in the Germline

Research

Laski, F. A., Rio, D. C., and Rubin, G. M. (1986). Tissue specificity of *Drosophila* P element transposition is regulated at the level of mRNA splicing. *Cell* 44, 7–19.

22

Retroviruses and Retroposons

CHAPTER OUTLINE

22.1 Introduction

Transposition that involves an obligatory intermediate of RNA is unique to eukaryotes, and is provided by the ability of **retroviruses** to insert DNA copies (proviruses) of an RNA viral genome into the chromosomes of a host cell. Some eukaryotic transposons are related to retroviral proviruses in their general organization, and they transpose through RNA intermediates. As a class, these elements are called **retroposons** (or sometimes **retrotransposons**). The very simplest such elements do not themselves have transposition activity, but have sequences that are recognized as substrates for transposition by active elements. Thus elements that use RNA-dependent transposition range from the retroviruses themselves (which are able freely to infect host cells), to sequences that transpose via RNA, to those that do not themselves possess the ability to transpose. They share with all transposons the diagnostic feature of generating short direct repeats of target DNA at the site of an insertion.

Even in genomes where active transposons have not been detected, footprints of ancient transposition events are found in the form of direct target repeats flanking dispersed repetitive sequences. The features of these sequences sometimes implicate an RNA sequence as the progenitor of the genomic (DNA) sequence. This suggests that the RNA must have been converted into a duplex DNA copy that was inserted into the genome by a transposition-like event.

Like any other reproductive cycle, the cycle of a retrovirus or retroposon is continuous; it is arbitrary to consider the point at which we interrupt it a "beginning." Our perspectives of these elements are biased, though, by the forms in which we usually observe them (FIGURE 22.1). Retroviruses were first observed as infectious virus particles that were capable of transmission between cells, and so the intracellular cycle (involving duplex DNA) is thought of as the means of reproducing the RNA virus. Retroposons were discovered as components of the genome, and the RNA forms have been mostly characterized for their functions as mRNAs. Thus we think of retroposons as genomic (duplex DNA) sequences that may transpose within a genome; they do not migrate between cells.

22.2 The Retrovirus Life Cycle Involves Transposition-Like Events

Key concepts

- A retrovirus has two copies of its genome of single-stranded RNA.
- An integrated provirus is a double-stranded DNA sequence.
- A retrovirus generates a provirus by reverse transcription of the retroviral genome.

Retroviruses have genomes of single-stranded RNA that are replicated through a double-stranded DNA intermediate. The life cycle of the virus involves an obligatory stage in which the double-stranded DNA is inserted into the host genome by a transposition-like event that generates short direct repeats of target DNA.

The significance of this reaction extends beyond the perpetuation of the virus. Some of its consequences are that:

- a retroviral sequence that is integrated in the germline remains in the cellular genome as an endogenous **provirus.**

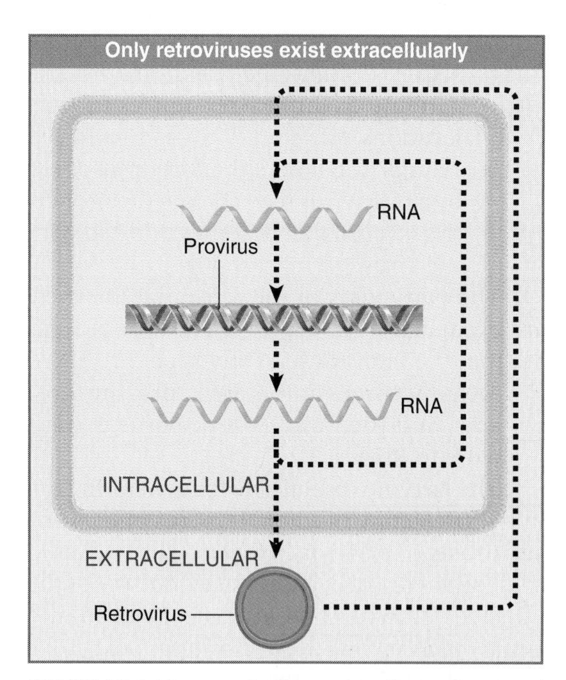

FIGURE 22.1 The reproductive cycles of retroviruses and retroposons alternate reverse transcription from RNA to DNA with transcription from DNA to RNA. Only retroviruses can generate infectious particles. Retroposons are confined to an intracellular cycle.

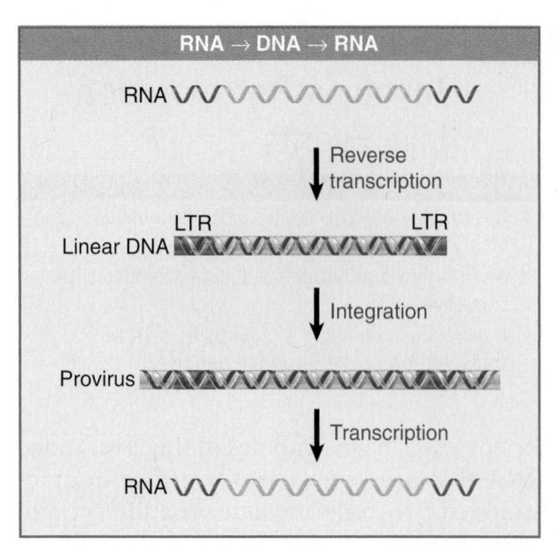

FIGURE 22.2 The retroviral life cycle proceeds by reverse transcribing the RNA genome into duplex DNA, which is inserted into the host genome, in order to be transcribed into RNA.

Like a lysogenic bacteriophage, a provirus behaves as part of the genetic material of the organism.

- cellular sequences occasionally recombine with the retroviral sequence and then are transposed with it; these sequences may be inserted into the genome as duplex sequences in new locations.
- cellular sequences that are transposed by a retrovirus may change the properties of a cell that becomes infected with the virus.

The particulars of the retroviral life cycle are expanded in FIGURE 22.2. The crucial steps are that the viral RNA is converted into DNA, the DNA becomes integrated into the host genome, and then the DNA provirus is transcribed into RNA.

The enzyme responsible for generating the initial DNA copy of the RNA is **reverse transcriptase.** The enzyme converts the RNA into a linear duplex of DNA in the cytoplasm of the infected cell. The DNA also is converted into circular forms, but these do not appear to be involved in reproduction.

The linear DNA makes its way to the nucleus. One or more DNA copies become integrated into the host genome. A single enzyme called **integrase** is responsible for integration. The provirus is transcribed by the host machinery to produce viral RNAs, which serve both as mRNAs and as genomes for packaging into virions. Integration is a normal part of the life cycle and is necessary for transcription.

Two copies of the RNA genome are packaged into each virion, making the individual virus particle effectively diploid. When a cell is simultaneously infected by two different but related viruses, it is possible to generate heterozygous virus particles carrying one genome of each type. The diploidy may be important in allowing the virus to acquire cellular sequences. The enzymes reverse transcriptase and integrase are carried with the genome in the viral particle.

22.3 Retroviral Genes Codes for Polyproteins

Key concepts

- A typical retrovirus has three genes: *gag, pol,* and *env.*
- Gag and Pol proteins are translated from a full-length transcript of the genome.
- Translation of Pol requires a frameshift by the ribosome.
- Env is translated from a separate mRNA that is generated by splicing.
- Each of the three protein products is processed by proteases to give multiple proteins.

A typical retroviral sequence contains three or four "genes." (In this context, the term *genes* is used to identify coding regions, each of which actually gives rise to multiple proteins by processing reactions.) A typical retrovirus genome with three genes is organized in the sequence *gag-pol-env,* as indicated in FIGURE 22.3.

Retroviral mRNA has a conventional structure; it is capped at the 5′ end and polyadenylated at the 3′ end. It is represented in two mRNAs. The full-length mRNA is translated to give the Gag and Pol polyproteins. The Gag product is translated by reading from the initiation codon to the first termination codon. This termination codon must be bypassed to express Pol.

Different mechanisms are used in different viruses to proceed beyond the *gag* termination codon, depending on the relationship between the *gag* and *pol* reading frames. When *gag* and *pol* follow continuously, suppression by a glutamyl-tRNA that recognizes the termination codon allows a single protein to be generated. When *gag* and *pol* are in different reading frames,

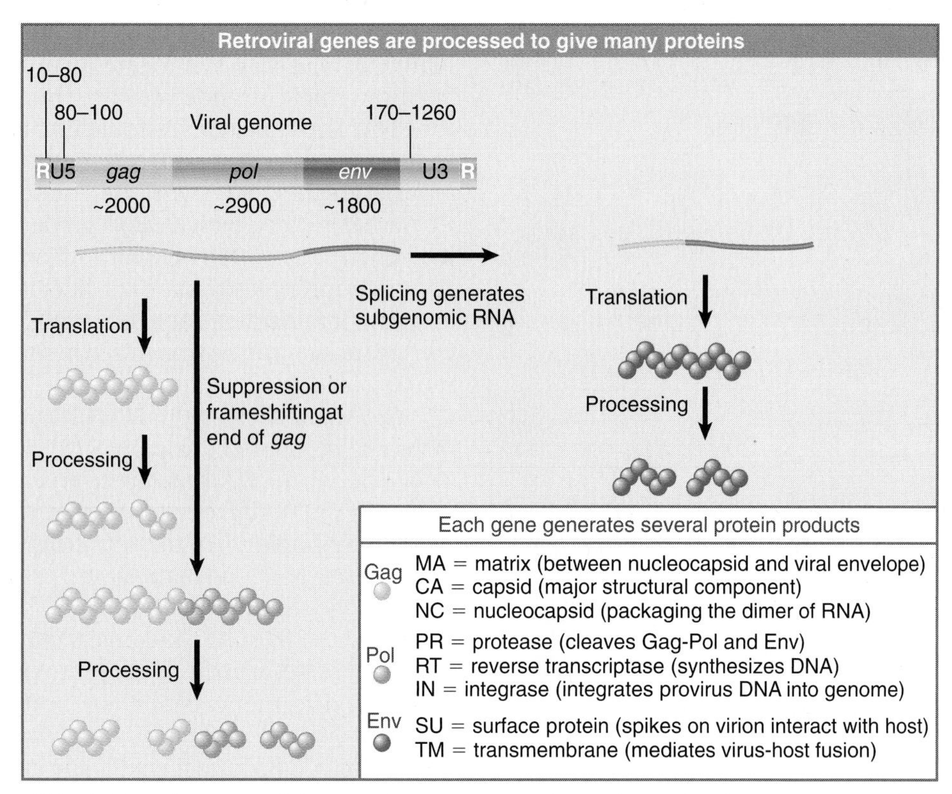

FIGURE 22.3 The genes of the retrovirus are expressed as polyproteins that are processed into individual products.

a ribosomal frameshift occurs to generate a single protein. Usually the readthrough is ~5% efficient, so Gag protein outnumbers Gag-Pol protein about 20-fold.

The Env polyprotein is expressed by another means: Splicing generates a shorter *subgenomic* messenger that is translated into the Env product.

The *gag* gene gives rise to the protein components of the nucleoprotein core of the virion. The *pol* gene codes for functions concerned with nucleic acid synthesis and recombination. The *env* gene codes for components of the envelope of the particle, which also sequesters components from the cellular cytoplasmic membrane.

Both the Gag or Gag-Pol and the Env products are polyproteins that are cleaved by a protease to release the individual proteins that are found in mature virions. The protease activity is coded by the virus in various forms: It may be part of Gag or Pol, and at times it takes the form of an additional independent reading frame.

The production of a retroviral particle involves packaging the RNA into a core, surrounding it with capsid proteins, and pinching off a segment of membrane from the host cell.

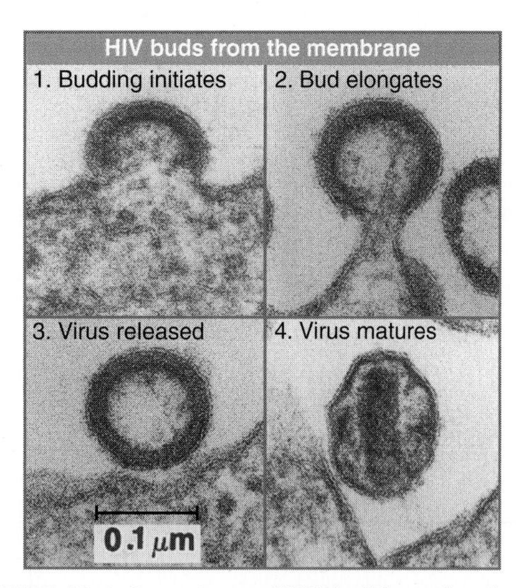

FIGURE 22.4 Retroviruses (HIV) bud from the plasma membrane of an infected cell. Photos courtesy of Matthew A. Gonda, Ph.D., Chief Executive Officer, International Medical Innovations, Inc.

The release of infective particles by such means is shown in FIGURE 22.4. The process is reversed during infection: A virus infects a new host cell by fusing with the plasma membrane and then releasing the contents of the virion.

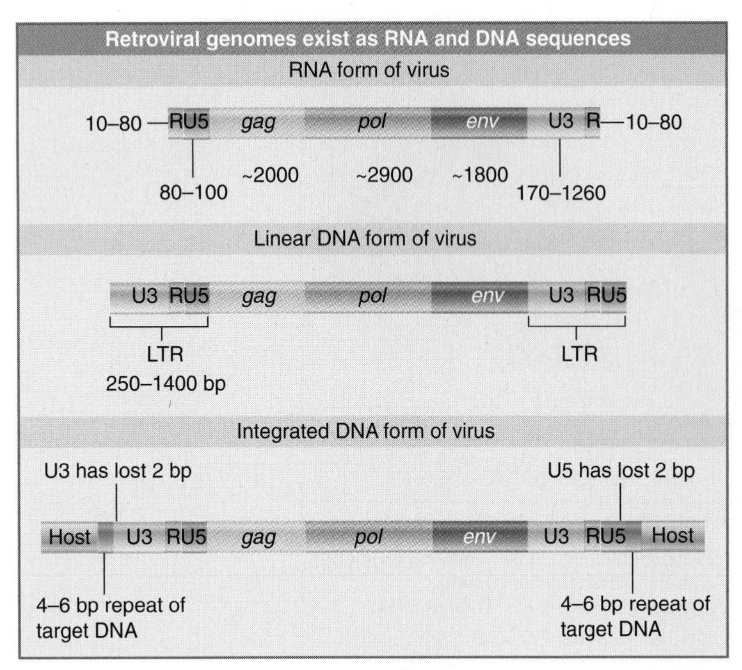

FIGURE 22.5 Retroviral RNA ends in direct repeats (R), the free linear DNA ends in LTRs, and the provirus ends in LTRs that are shortened by two bases each.

22.4 Viral DNA Is Generated by Reverse Transcription

Key concepts

- A short sequence (R) is repeated at each end of the viral RNA, so the 5′ and 3′ ends are R-U5 and U3-R, respectively.
- Reverse transcriptase starts synthesis when a tRNA primer binds to a site 100 to 200 bases from the 5′ end.
- When the enzyme reaches the end, the 5′-terminal bases of RNA are degraded, exposing the 3′ end of the DNA product.
- The exposed 3′ end base pairs with the 3′ terminus of another RNA genome.
- Synthesis continues, generating a product in which the 5′ and 3′ regions are repeated, giving each end the structure U3-R-U5.
- Similar strand switching events occur when reverse transcriptase uses the DNA product to generate a complementary strand.
- Strand switching is an example of the copy choice mechanism of recombination.

Retroviruses are called **plus strand viruses,** because the viral RNA itself codes for the protein products. As its name implies, reverse transcriptase is responsible for converting the genome (plus strand RNA) into a complementary DNA strand, which is called the **minus strand DNA.** Reverse transcriptase also catalyzes subsequent stages in the production of duplex DNA. It has a DNA polymerase activity, which enables it to synthesize a duplex DNA from the single-stranded reverse transcript of the RNA. The second DNA strand in this duplex is called the **plus strand DNA.** As a necessary adjunct to this activity, the enzyme has an RNAase H activity, which can degrade the RNA part of the RNA–DNA hybrid. All retroviral reverse transcriptases share considerable similarities of amino acid sequence, and homologous sequences can be recognized in some other retroposons.

The structures of the DNA forms of the virus are compared with the RNA in FIGURE 22.5. The viral RNA has direct repeats at its ends. These **R segments** vary in different strains of virus from 10 to 80 nucleotides. The sequence at the 5′ end of the virus is R-**U5,** and the sequence at the 3′ end is **U3**-R. The R segments are used during the conversion from the RNA to the DNA form to generate the more extensive direct repeats that are found in linear DNA (FIGURE 22.6 and FIGURE 22.7). The shortening of 2 bp at each end in the integrated form is a consequence of the mechanism of integration (see Figure 22.9).

Like other DNA polymerases, reverse transcriptase requires a primer. The native primer is tRNA. An uncharged host tRNA is present in the virion. A sequence of 18 bases at the 3′ end of the tRNA is base paired to a site 100 to 200 bases from the 5′ end of one of the viral RNA molecules. The tRNA may also be base paired to another site near the 5′ end of the other viral RNA, thus assisting in dimer formation between the viral RNAs.

Here is a dilemma: Reverse transcriptase starts to synthesize DNA at a site only 100 to 200 bases downstream from the 5′ end. How can DNA be generated to represent the intact RNA genome? (This is an extreme variant of the general problem in replicating the ends of any linear nucleic acid; see Section 16.2, The Ends of Linear DNA Are a Problem for Replication.)

Synthesis *in vitro* proceeds to the end, generating a short DNA sequence called minus strong-stop DNA. This molecule is not found *in vivo* because synthesis continues by the reaction illustrated in Figure 22.6. Reverse transcriptase switches templates, carrying the nascent DNA with it to the new template. This is the first of two jumps between templates.

In this reaction, the R region at the 5′ terminus of the RNA template is degraded by the RNAase H activity of reverse transcriptase. Its

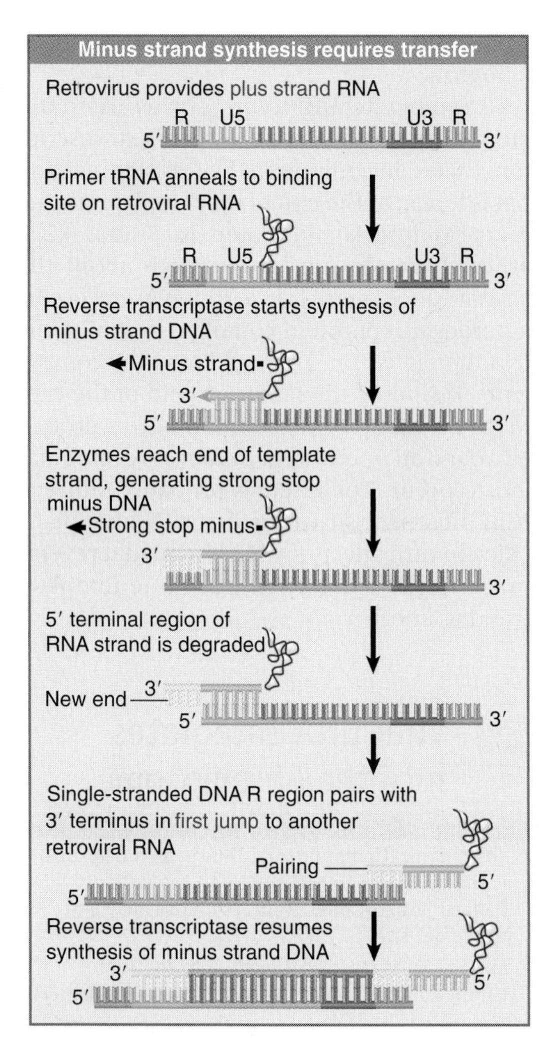

FIGURE 22.6 Minus strand DNA is generated by switching templates during reverse transcription.

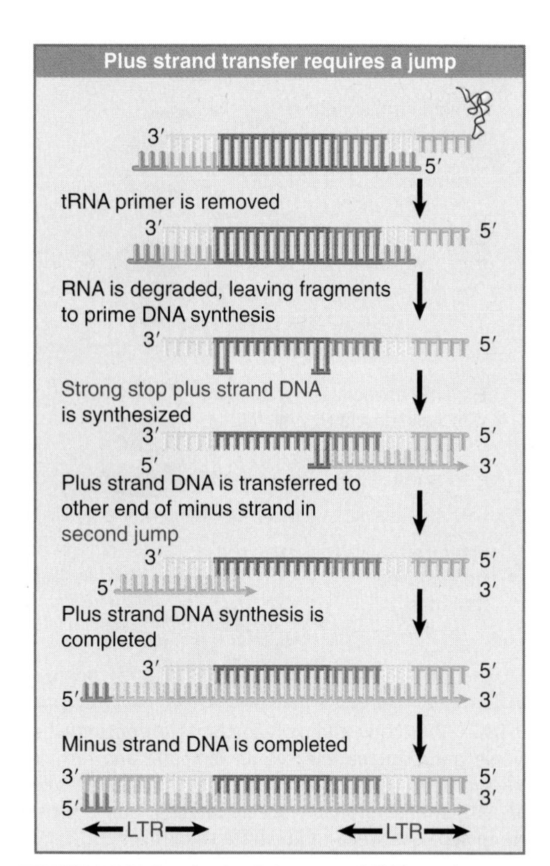

FIGURE 22.7 Synthesis of plus-strand DNA requires a second jump.

removal allows the R region at a 3′ end to base pair with the newly synthesized DNA. Reverse transcription then continues through the U3 region into the body of the RNA.

The source of the R region that pairs with the strong-stop minus DNA can be either the 3′ end of the same RNA molecule (intramolecular pairing) or the 3′ end of a different RNA molecule (intermolecular pairing). The switch to a different RNA template is used in the figure because there is evidence that the sequence of the tRNA primer is not inherited in a retroposon life cycle. (If intramolecular pairing occurred, we would expect the sequence to be inherited, because it would provide the only source for the primer binding sequence in the next cycle. Intermolecular pairing allows another retroviral RNA to provide this sequence.)

The result of the switch and extension is to add a U3 segment to the 5′ end. The stretch of sequence U3-R-U5 is called the **long terminal repeat (LTR)** because a similar series of events adds a U5 segment to the 3′ end, giving it the same structure of U5-R-U3. Its length varies from 250 to 1400 bp (see Figure 22.5).

We now need to generate the plus strand of DNA and to generate the LTR at the other end. The reaction is shown in Figure 22.7. Reverse transcriptase primes synthesis of plus strand DNA from a fragment of RNA that is left after degrading the original RNA molecule. A strong-stop plus strand DNA is generated when the enzyme reaches the end of the template. This DNA is then transferred to the other end of a minus strand, where it is probably released by a displacement reaction when a second round of DNA synthesis occurs from a primer fragment farther upstream (to its left in the figure). It uses the R region to pair with the 3′ end of a minus strand DNA. This double-stranded DNA then requires completion of both strands to generate a duplex LTR at each end.

Each retroviral particle carries two RNA genomes. This makes it possible for

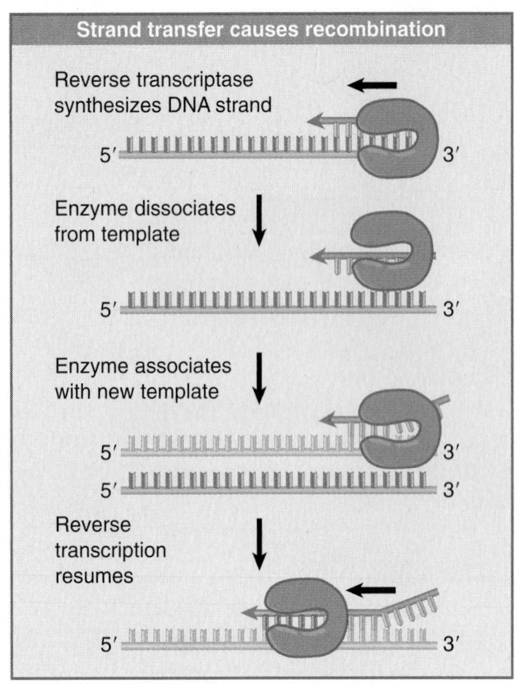

FIGURE 22.8 Copy choice recombination occurs when reverse transcriptase releases its template and resumes DNA synthesis using a new template. Transfer between template strands probably occurs directly, but is shown here in separate steps to illustrate the process.

recombination to occur during a viral life cycle. In principle this could occur during minus strand synthesis and/or during plus strand synthesis:

- The intermolecular pairing shown in Figure 22.6 allows a recombination to occur between sequences of the two successive RNA templates when minus strand DNA is synthesized. Retroviral recombination is mostly due to strand transfer at this stage, when the nascent DNA strand is transferred from one RNA template to another during reverse transcription.
- Plus strand DNA may be synthesized discontinuously, in a reaction that involves several internal initiations. Strand transfer during this reaction can also occur, but is less common.

The common feature of both events is that recombination results from a change in the template during the act of DNA synthesis. This is a general example of a mechanism for recombination called **copy choice.** For many years this was regarded as a possible mechanism for general recombination. It is unlikely to be employed by cellular systems, but is a common basis for recombination during infection by RNA viruses, including those that replicate exclusively through RNA forms, such as poliovirus.

Strand switching occurs with a certain frequency during each cycle of reverse transcription, that is, in addition to the transfer reaction that is forced at the end of the template strand. The principle is illustrated in FIGURE 22.8, although we do not know much about the mechanism. Reverse transcription *in vivo* occurs in a ribonucleoprotein complex, in which the RNA template strand is bound to virion components, including the major protein of the capsid. In the case of HIV, addition of this protein (NCp7) to an *in vitro* system causes recombination to occur. The effect is probably indirect: NCp7 affects the structure of the RNA template, which in turn affects the likelihood that reverse transcriptase will switch from one template strand to another.

22.5 Viral DNA Integrates into the Chromosome

Key concepts

- The organization of proviral DNA in a chromosome is the same as a transposon, with the provirus flanked by short direct repeats of a sequence at the target site.
- Linear DNA is inserted directly into the host chromosome by the retroviral integrase enzyme.
- Two base pairs of DNA are lost from each end of the retroviral sequence during the integration reaction.

The organization of the integrated provirus resembles that of the linear DNA. The LTRs at each end of the provirus are identical. The 3′ end of U5 consists of a short inverted repeat relative to the 5′ end of U3, so the LTR itself ends in short inverted repeats. The integrated proviral DNA is like a transposon: The proviral sequence ends in inverted repeats and is flanked by short direct repeats of target DNA.

The provirus is generated by directly inserting a linear DNA into a target site. (In addition to linear DNA, there are circular forms of the viral sequences. One has two adjacent LTR sequences generated by joining the linear ends. The other has only one LTR—presumably generated by a recombination event and actually comprising the majority of circles. For a long time it appeared that the circle might be an integration intermediate (by analogy with the integration of lambda DNA; we now know that the linear form is used for integration).

Integration of linear DNA is catalyzed by a single viral product, the integrase. Integrase acts on both the retroviral linear DNA and the target DNA. The reaction is illustrated in FIGURE 22.9.

The ends of the viral DNA are important. For example, mutations in the ends of transposons prevent integration. The most conserved feature is the presence of the dinucleotide sequence CA close to the end of each inverted repeat. The integrase brings the ends of the linear DNA together in a ribonucleoprotein complex, and then converts the blunt ends into recessed ends by removing the bases beyond the conserved CA. In general, this involves a loss of 2 bases.

Target sites are chosen at random with respect to sequence. The integrase makes staggered cuts at a target site. In the example of Figure 22.9, the cuts are separated by 4 bp. The length of the target repeat depends on the particular virus; it may be 4, 5, or 6 bp. Presumably it is determined by the geometry of the reaction of integrase with target DNA.

The 5′ ends generated by the cleavage of target DNA are covalently joined to the 3′ recessed ends of the viral DNA. At this point, both termini of the viral DNA are joined by one strand to the target DNA. The single-stranded region is repaired by enzymes of the host cell, and in the course of this reaction the protruding two bases at each 5′ end of the viral DNA are removed. The result is that the integrated viral DNA has lost 2 bp at each LTR; this corresponds to the loss of 2 bp from the left end of the 5′ terminal U3 and to the loss of 2 bp from the right end of the 3′ terminal U5. There is a characteristic short direct repeat of target DNA at each end of the integrated retroviral genome.

The viral DNA integrates into the host genome at randomly selected sites. A successfully infected cell gains one to ten copies of the provirus. (An infectious virus enters the cytoplasm, of course, but the DNA form becomes integrated into the genome in the nucleus. Retroviruses can replicate only in proliferating cells, because entry into the nucleus requires the cell to pass through mitosis, when the viral genome gains access to the nuclear material.)

The U3 region of each LTR carries a promoter. The promoter in the left LTR is responsible for initiating transcription of the provirus. Recall that the generation of proviral DNA is required to place the U3 sequence at the left LTR; thus we see that the promoter is in fact generated by the conversion of the RNA into duplex DNA.

Sometimes (probably rather rarely), the promoter in the right LTR sponsors transcription of the host sequences that are adjacent to the site of integration. The LTR also carries an enhancer (a sequence that activates promoters in the vicinity) that can act on cellular as well as viral sequences. Integration of a retrovirus can be responsible for converting a host cell into a tumorigenic state when certain types of genes are activated in this way. Can integrated proviruses be excised from the genome? Homologous recombination could take place between the LTRs of a provirus; solitary LTRs that could be relics of an excision event are present in some cellular genomes.

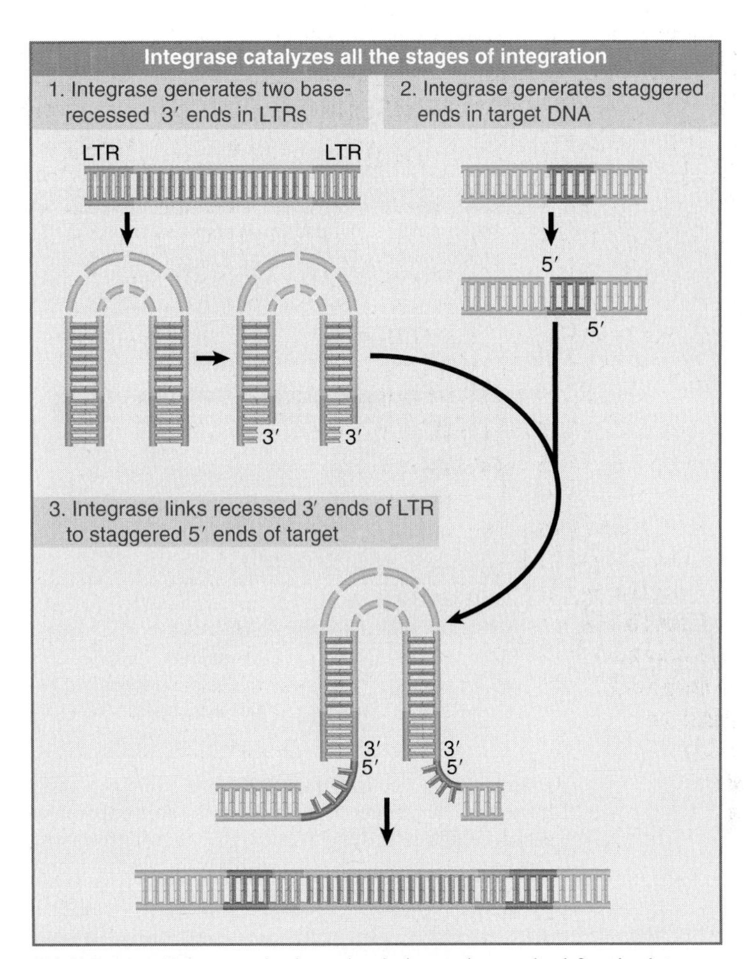

FIGURE 22.9 Integrase is the only viral protein required for the integration reaction, in which each LTR loses 2 bp and is inserted between 4 bp repeats of target DNA.

We have dealt thus far with retroviruses in terms of the infective cycle, in which integration is necessary for the production of further copies of the RNA. When a viral DNA integrates in a germline cell, though, it becomes an inherited "endogenous provirus" of the organism. Endogenous viruses usually are not expressed, but sometimes they are activated by external events, such as infection with another virus.

22.6 Retroviruses May Transduce Cellular Sequences

Key concept

- Transforming retroviruses are generated by a recombination event in which a cellular RNA sequence replaces part of the retroviral RNA.

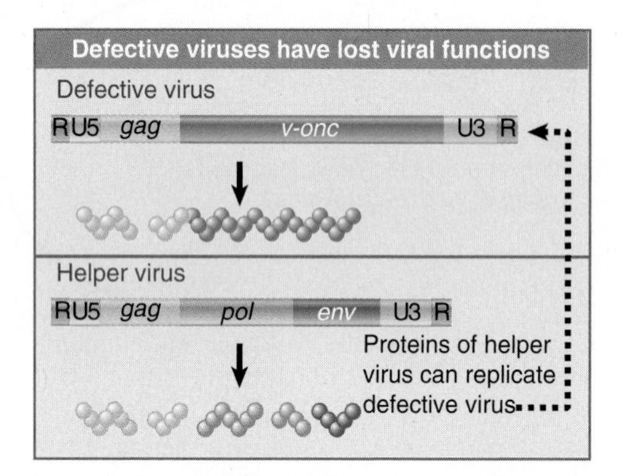

FIGURE 22.10 Replication-defective transforming viruses have a cellular sequence substituted for part of the viral sequence. The defective virus may replicate with the assistance of a helper virus that carries the wild-type functions.

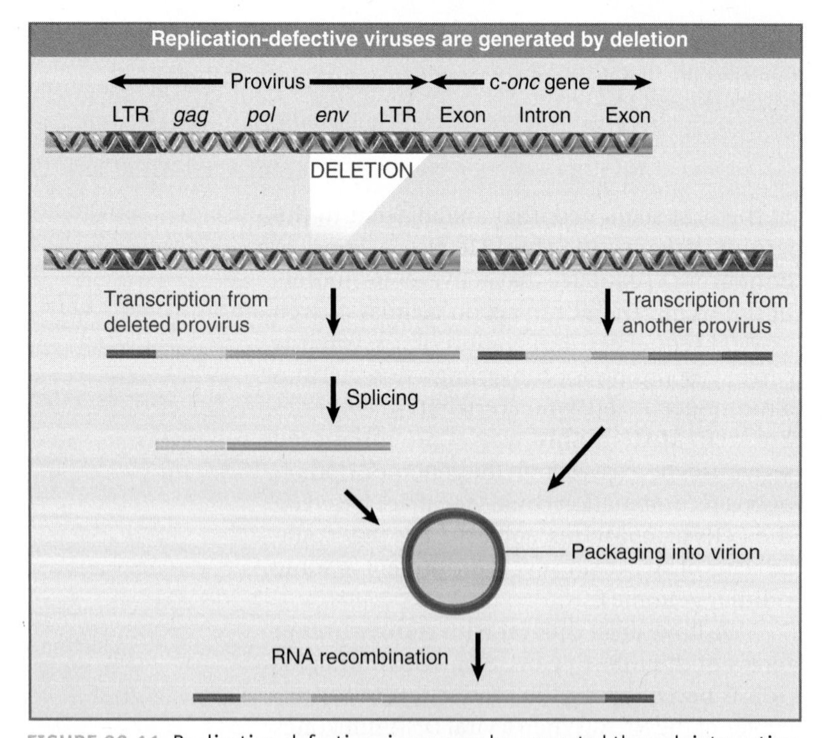

FIGURE 22.11 Replication-defective viruses may be generated through integration and deletion of a viral genome to generate a fused viral-cellular transcript that is packaged with a normal RNA genome. Nonhomologous recombination is necessary to generate the replication-defective transforming genome.

An interesting light on the viral life cycle is cast by the occurrence of **transducing viruses,** which are variants that have acquired cellular sequences in the form illustrated in FIGURE 22.10. Part of the viral sequence has been replaced by the *v-onc* gene. Protein synthesis generates a Gag-v-Onc protein instead of the usual Gag, Pol, and Env proteins. The resulting virus is **replication-defective;** it cannot sustain an infective cycle by itself. It can, however, be perpetuated in the company of a **helper virus** that provides the missing viral functions.

Onc is an abbreviation for **oncogenesis,** the ability to *transform* cultured cells so that the usual regulation of growth is released to allow unrestricted division. Both viral and cellular *onc* genes may be responsible for creating tumorigenic cells.

A *v-onc* gene confers upon a virus the ability to transform a certain type of host cell. Loci with homologous sequences found in the host genome are called *c-onc* genes. How are the *onc* genes acquired by the retroviruses? A revealing feature is the discrepancy in the structures of *c-onc* and *v-onc* genes. The *c-onc* genes usually are interrupted by introns, whereas the *v-onc* genes are uninterrupted. This suggests that the *v-onc* genes originate from spliced RNA copies of the *c-onc* genes.

A model for the formation of transforming viruses is illustrated in FIGURE 22.11. A retrovirus has integrated near a *c-onc* gene. A deletion occurs to fuse the provirus to the *c-onc* gene; transcription then generates a joint RNA, which contains viral sequences at one end and cellular *onc* sequences at the other end. Splicing removes the introns in both the viral and cellular parts of the RNA. The RNA has the appropriate signals for packaging into the virion; virions will be generated if the cell also contains another, intact copy of the provirus. At this point, some of the diploid virus particles may contain one fused RNA and one viral RNA.

A recombination between these sequences could generate the transforming genome, in which the viral repeats are present at both ends. (Recombination occurs by various means at a high frequency during the retroviral infective cycle. We do not know anything about its demands for homology in the substrates, but we assume that the nonhomologous reaction between a viral genome and the cellular part of the fused RNA proceeds by the same mechanisms responsible for viral recombination.)

The common features of the entire retroviral class suggest that it may be derived from a single ancestor. Primordial insertion sequences (IS) elements could have surrounded a host gene for a nucleic acid polymerase; the resulting unit would have the form LTR-pol-LTR. It might evolve into an infectious virus by acquiring more sophisticated abilities to manipulate both DNA and RNA substrates, including the incorporation of genes whose products allowed packaging of the RNA. Other functions, such as transforming genes, might be incorporated later. (There is no reason to suppose that the mechanism involved in acquisition of cellular functions is unique for *onc* genes; viruses carrying these genes may have a selective advantage, though, because of their stimulatory effect on cell growth.)

22.7 Yeast *Ty* Elements Resemble Retroviruses

Key concepts

- *Ty* transposons have a similar organization to endogenous retroviruses.
- *Ty* transposons are retroposons, with a reverse transcriptase activity, that transpose via an RNA intermediate.

Ty elements comprise a family of dispersed repetitive DNA sequences that are found at different sites in different strains of yeast. ***Ty*** is an abbreviation for "transposon yeast." A transposition event creates a characteristic footprint: 5 bp of target DNA are repeated on either side of the inserted *Ty* element. *Ty* elements are **retroposons** that transpose by the same mechanism as retroviruses. The frequency of *Ty* transposition is lower than that of most bacterial transposons, $\sim10^{-7}$–10^{-8}.

There is considerable divergence between individual *Ty* elements. Most elements fall into one of two major classes, called *Ty1* and *Ty917*. They have the same general organization illustrated in FIGURE 22.12. Each element is 6.3 kb long; the last 330 bp at each end constitute direct repeats, called δ. Individual *Ty* elements of each type have many changes from the prototype of their class, including base pair substitutions, insertions, and deletions. There are ~30 copies of the *Ty1* type and ~6 of the *Ty917* type in a typical yeast genome. In addition, there are ~100 independent *delta* elements, called solo δs.

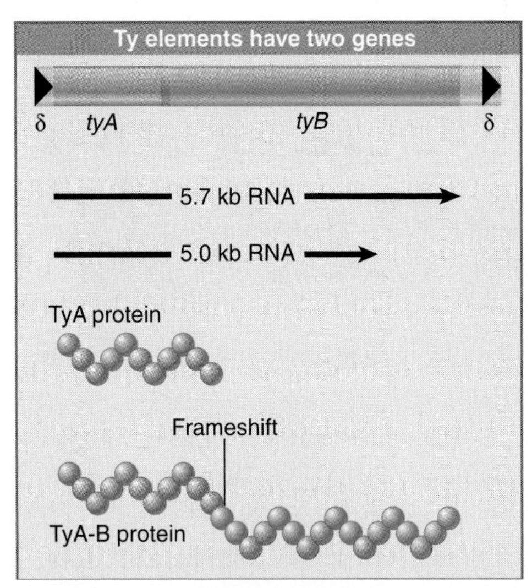

FIGURE 22.12 Ty elements terminate in short direct repeats and are transcribed into two overlapping RNAs. They have two reading frames, with sequences related to the retroviral *gag* and *pol* genes.

The *delta* sequences also show considerable heterogeneity, although the two repeats of an individual *Ty* element are likely to be identical or at least very closely related. The *delta* sequences associated with *Ty* elements show greater conservation of sequence than the solo *delta* elements, which suggests that recognition of the repeats is involved in transposition.

The *Ty* element is transcribed into two poly$(A)^+$ RNA species, which constitute >5% of the total mRNA of a haploid yeast cell. Both species initiate within a promoter in the δ element at the left end. One terminates after 5 kb; the other terminates after 5.7 kb, within the delta sequence at the right end.

The sequence of the *Ty* element has two open reading frames. These frames are expressed in the same direction, but are read in different phases and overlap by 13 amino acids. The sequence of *TyA* suggests that it codes for a DNA-binding protein. The sequence of *TyB* contains regions that have homologies with reverse transcriptase, protease, and integrase sequences of retroviruses.

The organization and functions of *TyA* and *TyB* are analogous to the behavior of the retroviral *gag* and *pol* functions. The reading frames *TyA* and *TyB* are expressed in two forms. The TyA protein represents the *TyA* reading frame and terminates at its end. The *TyB* reading frame, however, is expressed only as part of a joint protein, in which the *TyA* region is fused to the *TyB* region by a specific frameshift event that allows

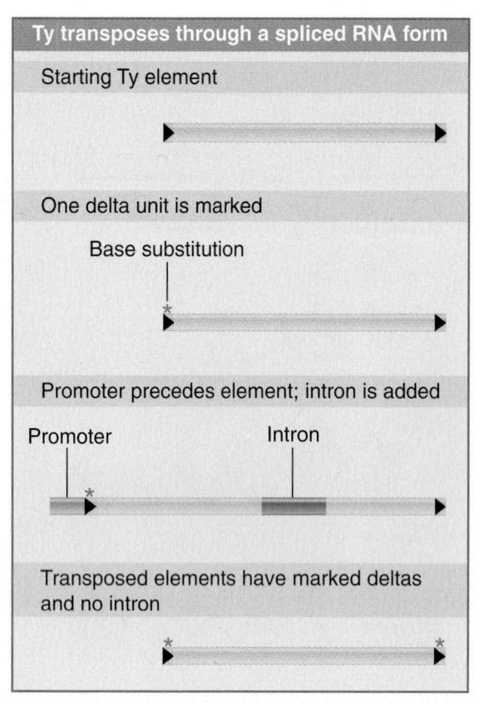

FIGURE 22.13 A unique Ty element, engineered to contain an intron, transposes to give copies that lack the intron. The copies possess identical terminal repeats, which are generated from one of the termini of the original Ty element.

the termination codon to be bypassed. (This is analogous to *gag-pol* translation in retroviruses.)

Recombination between *Ty* elements seems to occur in bursts; when one event is detected, there is an increased probability of finding others. Gene conversion occurs between *Ty* elements at different locations, with the result that one element is "replaced" by the sequence of the other.

Ty elements can excise by homologous recombination between the directly repeated *delta* sequences. The large number of solo *delta* elements may be footprints of such events. An excision of this nature may be associated with reversion of a mutation caused by the insertion of *Ty;* the level of reversion may depend on the exact *delta* sequences left behind.

A paradox is that both *delta* elements have the same sequence, yet a promoter is active in the *delta* at one end and a terminator is active in the *delta* at the other end. (A similar feature is found in other transposable elements, including the retroviruses.)

Ty elements are classic retroposons, in that they transpose through an RNA intermediate. An ingenious protocol used to detect this event is illustrated in FIGURE 22.13. An intron was inserted into an element to generate a unique *Ty* sequence. This sequence was placed under the control of a *GAL* promoter on a plasmid and introduced into yeast cells. Transposition results in the appearance of multiple copies of the transposon in the yeast genome, but the copies all lack the intron.

We know of only one way to remove introns: RNA splicing. This suggests that transposition occurs by the same mechanism as with retroviruses. The *Ty* element is transcribed into an RNA that is recognized by the splicing apparatus. The spliced RNA is recognized by a reverse transcriptase and regenerates a duplex DNA copy.

The analogy with retroviruses extends further. The original *Ty* element has a difference in sequence between its two *delta* elements. The transposed elements possess identical *delta* sequences, however, which are derived from the 5′ delta of the original element. If we consider the *delta* sequence to be exactly like an LTR, consisting of the regions U3-R-U5, the *Ty* RNA extends from R region to R region. Just as shown for retroviruses in Figures 22.3–22.6, the complete LTR is regenerated by adding a U5 to the 3′ end and a U3 to the 5′ end.

Transposition is controlled by genes within the *Ty* element. The *GAL* promoter used to control transcription of the marked *Ty* element is inducible: It is turned on by the addition of galactose. Induction of the promoter has two effects. It is necessary to activate transposition of the marked element, and its activation also increases the frequency of transposition of the other *Ty* elements on the yeast chromosome. This implies that the products of the *Ty* element can act in *trans* on other elements (actually on their RNAs).

The *Ty* element does not give rise to infectious particles, but virus-like particles (VLPs) accumulate within the cells in which transposition has been induced. The particles, which can be seen in FIGURE 22.14, contain full-length RNA, double-stranded DNA, reverse transcriptase activity, and a TyB product with integrase activity. The TyA product is cleaved like a *gag* precursor to produce the mature core proteins of the VLP. This takes the analogy between the *Ty* transposon and the retrovirus even further. The *Ty* element behaves, in short, like a retrovirus that has lost its *env* gene and therefore cannot properly package its genome.

Not all of the *Ty* elements in any yeast genome are active: Most have lost the ability to transpose (and are analogous to inert endogenous proviruses). These "dead" elements retain the δ repeats, though, and as a result they pro-

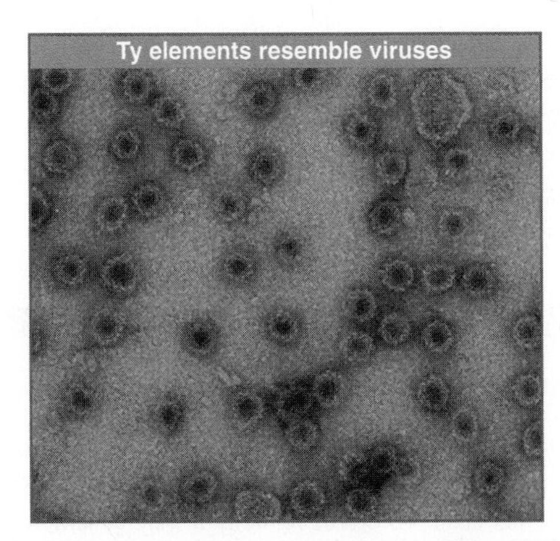

FIGURE 22.14 Ty elements generate virus-like particles. Reproduced from *J. Mol. Biol.*, vol. 292, AL-Khayat, H. A., et al., Yeast Ty retrotransposons . . . , pp. 65–73. Copyright 1999, with permission from Elsevier. Photo courtesy of Dr. Hind A. AL-Khayat, Imperial College London, United Kingdom.

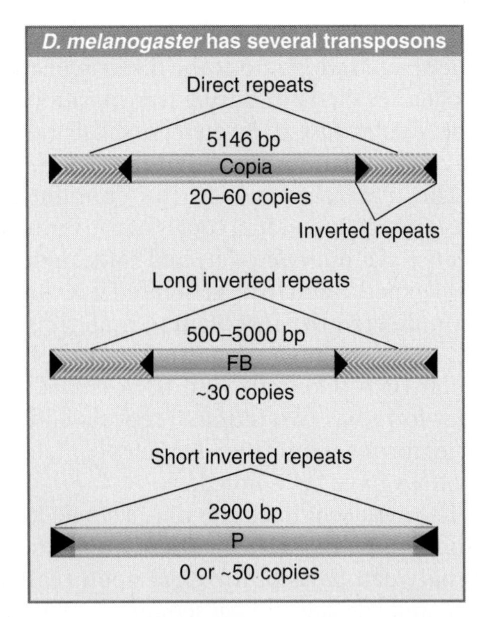

FIGURE 22.15 Three types of transposable element in *D. melanogaster* have different structures.

vide targets for transposition in response to the proteins synthesized by an active element.

22.8 Many Transposable Elements Reside in *Drosophila melanogaster*

Key concept

- *copia* is a retroposon that is abundant in *D. melanogaster.*

The presence of transposable elements in *D. melanogaster* was first inferred from observations analogous to those that identified the first insertion sequences in *E. coli.* Unstable mutations are found that revert to wild type by deletion, or that generate deletions of the flanking material with an endpoint at the original site of mutation. They are caused by several types of transposable sequence, which are illustrated in FIGURE 22.15. These sequences include the *copia* retroposon, the FB family, and the P elements discussed previously in Section 21.14, The Role of Transposable Elements in Hybrid Dysgenesis.

The best-characterized family of retroposons is *copia.* Its name reflects the presence of a large number of closely related sequences that code for abundant mRNAs. The *copia* family is taken as a paradigm for several other types of elements whose sequences are unrelated, but whose structure and general behavior appear to be similar.

The number of copies of the *copia* element depends on the strain of fly; usually it is 20 to 60. The members of the family are widely dispersed. The locations of *copia* elements show a different (although overlapping) spectrum in each strain of *D. melanogaster.*

These differences have developed over evolutionary periods. Comparisons of strains that have diverged recently (over the past 40 years or so) as the result of their propagation in the laboratory reveal few changes. We cannot estimate the rate of change, but the nature of the underlying events is indicated by the result of growing cells in culture. The number of *copia* elements per genome then increases substantially, by as much as threefold. The additional elements represent insertions of *copia* sequences at new sites. Adaptation to culture in some unknown way transiently increases the rate of transposition to a range of 10^{-3} to 10^{-4} events per generation.

The *copia* element is ~5000 bp long, with identical direct terminal repeats of 276 bp. Each of the direct repeats itself ends in related inverted repeats. A direct repeat of 5 bp of target DNA is generated at the site of insertion. The divergence between individual members of the *copia* family is slight, at <5%; variants often contain small deletions. All of these features are common to the other *copia*-like families, although their individual members display greater divergence.

The identity of the two direct repeats of each *copia* element implies either that they interact to permit correction events, or that both are generated from one of the direct repeats of a progenitor element during transposition. As in the similar case of *Ty* elements, this is suggestive of a relationship with retroviruses.

The *copia* elements in the genome are always intact; individual copies of the terminal repeats have not been detected (although we would expect them to be generated if recombination deleted the intervening material). At times *copia* elements are found in the form of free circular DNA; like retroviral DNA circles, the longer form has two terminal repeats and the shorter form has only one. Particles containing *copia* RNA have been noted.

The *copia* sequence contains a single long-reading frame of 4227 bp. There are homologies between parts of the *copia* open reading frame and the *gag* and *pol* sequences of retroviruses. A notable absence from the homologies is any relationship with retroviral *env* sequences required for the envelope of the virus, which means that *copia* is unlikely to be able to generate virus-like particles.

Transcripts of *copia* are found as abundant poly(A)$^+$ mRNAs, representing both full-length and part-length transcripts. The mRNAs have a common 5′ terminus, which results from initiation in the middle of one of the terminal repeats. Several proteins are produced, probably involving events such as splicing of RNA and cleavage of polyproteins.

We lack direct evidence for *copia*'s mode of transposition; as a result, there are so many resemblances with retroviral organization that it seems inevitable that *copia* must have an origin related to the retroviruses. It is hard to say how many retroviral functions it possesses. We know, of course, that it transposes, but (as is the case with *Ty* elements) there is no evidence for any infectious capacity.

22.9 Retroposons Fall into Three Classes

Key concepts

- Retroposons of the viral superfamily are transposons that mobilize via an RNA that does not form an infectious particle.
- Some retroposons directly resemble retroviruses in their use of LTRs, whereas others do not have LTRs.
- Other elements can be found that were generated by an RNA-mediated transposition event, but they do not themselves code for enzymes that can catalyze transposition.
- Transposons and retroposons constitute almost half of the human genome.

Retroposons are defined by their use of mechanisms for transposition that involve reverse transcription of RNA into DNA. Three classes of retroposons are distinguished in FIGURE 22.16:

- Members of the **viral superfamily** code for reverse transcriptase and/or integrase activities. Like other retroposons, they reproduce in the same manner as retroviruses but differ from them in not passing through an independent infectious form. They are best characterized in the *Ty* and *copia* elements of yeast and flies.
- The long interspersed repeated sequences (LINES) also have reverse transcriptase activity (and may therefore be considered to comprise more distant members of the viral superfamily), but they lack LTRs and use a different mech-

Eukaryotic genomes have three types of retroposons			
	Viral Superfamily	LINES	Nonviral Superfamily
Common types	Ty (*S. cerevisiae*) copia (*D.melanogaster*)	L1 (human) B1, B2 ID, B4 (mouse)	SINES (mammals) Pseudogenes of pol III transcripts
Termini	Long terminal repeats	No repeats	No repeats
Target repeats	4–6 bp	7–21 bp	7–21 bp
Enzyme activities	Reverse transcriptase and/or integrase	Reverse transcriptase /endonuclease	None (or none coding for transposon products)
Organization	May contain introns (removed in subgenomic mRNA)	One or two uninterrupted ORFs	No introns

FIGURE 22.16 Retroposons can be divided into the viral superfamilies that are retrovirus-like or LINES and the nonviral superfamilies that do not have coding functions.

anism from retroviruses to prime the reverse transcription reaction. They are derived from RNA polymerase II transcripts. A minority of the elements in the genome are fully functional and can transpose autonomously; others have mutations, and thus can only transpose as the result of the action of a *trans*-acting autonomous element.

- Members of the **nonviral superfamily** are identified by external and internal features that suggest that they originated in RNA sequences. In these cases, though, we can only speculate on how a DNA copy was generated. We assume that they were targets for a transposition event by an enzyme system coded elsewhere, that is, they are always nonautonomous. They originated in cellular transcripts. They do not code for proteins that have transposition functions. The most prominent component of this family is called short interspersed repeated sequences (SINES). These components are derived from RNA polymerase III transcripts.

FIGURE 22.17 shows the organization and sequence relationships of elements that code for reverse transcriptase. Like retroviruses, the LTR-containing retroposons can be classified into groups according to the number of independent reading frames for *gag, pol,* and *int,* and the order of the genes. In spite of these superficial differences of organization, the common feature is the presence of reverse transcriptase and integrase activities. Typical mammalian LINES elements have two reading frames; one codes for a nucleic acid-binding protein and the other codes for reverse transcriptase and endonuclease activity.

LTR-containing elements can vary from integrated retroviruses to retroposons that have lost the capacity to generate infectious particles. Yeast and fly genomes have the *Ty* and *copia* elements that cannot generate infectious particles. Mammalian genomes have endogenous retroviruses that, when active, can generate infectious particles. The mouse genome has several active endogenous retroviruses, which are able to generate particles that propagate horizontal infections. By contrast, almost all endogenous retroviruses lost their activity some 50 million years ago in the human lineage, and the genome now has mostly inactive remnants of the endogenous retroviruses.

LINES and SINES comprise a major part of the animal genome. They were defined originally by the existence of a large number of relatively short sequences that are related to one another (comprising the moderately repetitive DNA described in Section 4.6, Eukaryotic Genomes Contain Both Nonrepetitive and Repetitive DNA Sequences). The LINES comprise long interspersed sequences, and the SINES comprise short interspersed sequences. (They are described as interspersed sequences or **interspersed repeats** because of their common occurrence and widespread distribution.)

Plants contain another type of small mobile element, called MITE (for miniature inverted-repeat transposable element). Such elements terminate in inverted repeats, have a 2 or 3 bp target sequence, do not have coding sequences, and are 200 to 500 bp long. At least nine such families exist in (for example) the rice genome. They are often found in the regions flanking protein-coding genes. They have no relationship to SINES or LINES.

LINES and SINES comprise a significant part of the repetitive DNA of animal genomes. In many higher eukaryotic genomes, they occupy ~50% of the total DNA. FIGURE 22.18 summarizes the distribution of the different types of transposons that constitute almost half of the human genome. Except for the SINES, which are always nonfunctional, the other types of elements all consist of both functional elements and elements that have suffered deletions that eliminated parts of the reading frames that code for the protein(s) needed for transposition. The relative proportions of these types of transposons are generally similar in the mouse genome.

A common LINES in mammalian genomes is called L1. The typical member is ~6500 bp long and terminates in an A-rich tract. The two open reading frames of a full-length element

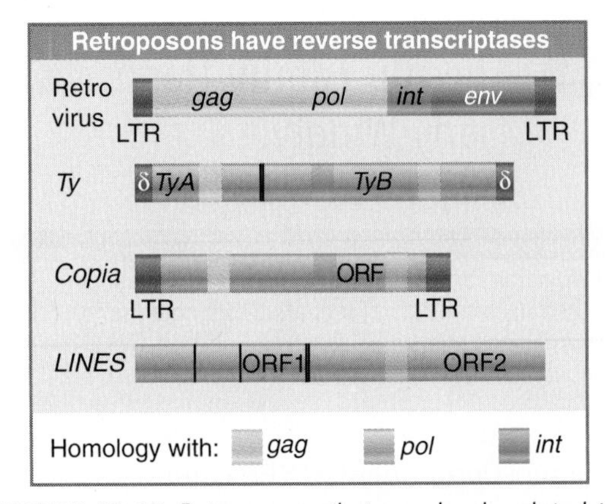

FIGURE 22.17 Retroposons that are closely related to retroviruses have a similar organization, but LINES share only the reverse transcriptase activity.

Retroviruses and transposons constitute half the human genome				
Element	Organization	Length (Kb)	Human genome	
			Number	Fraction
Retrovirus/retroposon	LTR *gag pol (env)* LTR	1–11	450,000	8%
LINES (autonomous), e.g., L1	*ORF1 (pol)* $(A)_n$	6–8	850,000	17%
SINES (nonautonomous), e.g., Alu	$(A)_n$	<0.3	1,500,000	15%
DNA transposon	Transposase	2–3	300,000	3%

FIGURE 22.18 Four types of transposable elements constitute almost half of the human genome.

are called ORF1 and ORF2. The number of full-length elements is usually small (~50), and the remainder of the copies are truncated. Transcripts can be found. As implied by its presence in repetitive DNA, the LINES family shows sequence variation among individual members. The members of the family within a species, however, are relatively homogeneous compared to the variation shown between species. L1 is the only member of the LINES family that has been active in either the mouse or human lineages. It seems to have remained highly active in the mouse, but has declined in the human lineage.

Only one SINES has been active in the human lineage: the common Alu element. The mouse genome has a counterpart to this element (B1), and also other SINES (B2, ID, B4) that have been active. Human Alu and mouse B1 SINES are probably derived from the 7SL RNA (see Section 22.10, The Alu Family Has Many Widely Dispersed Members). The other mouse SINES appear to have originated from reverse transcripts of tRNAs. The transposition of the SINES probably results from their recognition as substrates by an active L1 element.

22.10 The Alu Family Has Many Widely Dispersed Members

Key concept

- A major part of repetitive DNA in mammalian genomes consists of repeats of a single family organized like transposons and derived from RNA polymerase III transcripts.

The most prominent SINES comprises members of a single family. Its short length and high degree of repetition make it comparable to simple sequence (satellite) DNA, except that the individual members of the family are dispersed around the genome instead of being confined to tandem clusters. Again, there is significant similarity between the members within a species compared with variation between species.

In the human genome, a large part of the moderately repetitive DNA exists as sequences of ~300 bp that are interspersed with nonrepetitive DNA. At least half of the renatured duplex material is cleaved by the restriction enzyme AluI at a single site located 170 bp along the sequence. The cleaved sequences all are members of a single family known as the **Alu family,** after the means of its identification. There are ~300,000 members in the haploid genome (equivalent to one member per 6 kb of DNA). The individual Alu sequences are widely dispersed. A related sequence family is present in the mouse (where the 50,000 members are called the B1 family), in the Chinese hamster (where it is called the Alu-equivalent family), and in other mammals.

The individual members of the Alu family are related rather than identical. The human family seems to have originated by means of a 130 bp tandem duplication, with an unrelated sequence of 31 bp inserted in the right half of the dimer. The two repeats are sometimes called the "left half" and the "right half" of the Alu sequence. The individual members of the Alu family have an average identity with the consensus sequence of 87%. The mouse B1 repeating unit is 130 bp long and corresponds to a monomer of the human unit. It has 70%–80% homology with the human sequence.

The Alu sequence is related to 7SL RNA, a component of the signal recognition particle (see Section 10.9, The SRP Interacts with the SRP Receptor). The 7SL RNA corresponds to the left half of an Alu sequence with an insertion in the middle. Thus the ninety 5′ terminal bases of 7SL RNA are homologous to the left end of Alu, the central 160 bases of 7SL RNA have no homology to Alu, and the forty 3′ terminal bases of 7SL RNA are homologous to the right end of Alu. The 7SL RNA is coded by genes that are actively transcribed by RNA polymerase III. It is possible that these genes (or genes related to them) gave rise to the inactive Alu sequences.

The members of the Alu family resemble transposons in being flanked by short direct repeats. They display, however, the curious feature that the lengths of the repeats are different for individual members of the family. They derive from RNA polymerase III transcripts, and as a result it is possible that individual members carry internal active promoters.

A variety of properties have been found for the Alu family, and its ubiquity has prompted many suggestions for its function. It is not yet possible, though, to discern its true role.

At least some members of the family can be transcribed into independent RNAs. In the Chinese hamster, some (though not all) members of the Alu-equivalent family appear to be transcribed *in vivo*. Transcription units of this sort are found in the vicinity of other transcription units.

Members of the Alu family may be included within structural gene transcription units, as seen by their presence in long nuclear RNA. The presence of multiple copies of the Alu sequence in a single nuclear molecule can generate secondary structure. In fact, the presence of Alu family members in the form of inverted repeats is responsible for most of the secondary structure found in mammalian nuclear RNA.

22.11 Processed Pseudogenes Originated as Substrates for Transposition

Key concept

- A processed pseudogene is derived from an mRNA sequence by reverse transcription.

When a sequence generated by reverse transcription of an mRNA is inserted into the genome, we can recognize its relationship to the gene from which the mRNA was transcribed. Such a sequence is called a **processed pseudogene** to reflect the fact that it was processed from RNA and is not active. In FIGURE 22.19, the characteristic features of a processed pseudogene are compared with the features of the original gene and the mRNA. The figure shows all the relevant diagnostic features, only some of which are found in any individual example. Any transcript of RNA polymerase II could in principle give rise to such a pseudogene, and there are many examples, including the processed globin pseudogenes that were the first to be discovered (see Section 3.11, Pseudogenes Are Dead Ends of Evolution).

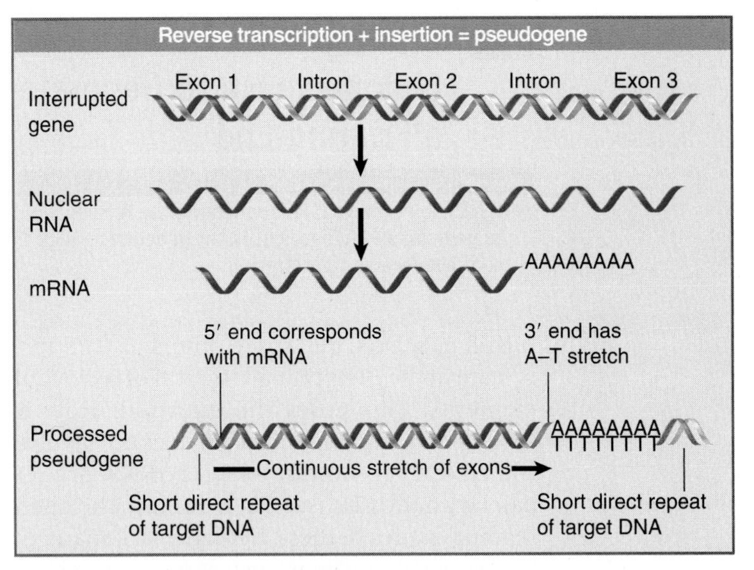

FIGURE 22.19 Pseudogenes could arise by reverse transcription of RNA to give duplex DNAs that become integrated into the genome.

The pseudogene may start at the point equivalent to the 5′ terminus of the RNA, which would be expected only if the DNA had originated from the RNA. Several pseudogenes consist of precisely joined exon sequences; we know of no mechanism to recognize introns in DNA, so this feature argues for an RNA-mediated stage. The pseudogene may end in a short stretch of A-T base pairs, presumably derived from the poly(A) tail of the RNA. On either side of the pseudogene is a short direct repeat, presumed to have been generated by a transposition-like event. Processed pseudogenes reside at locations unrelated to their presumed sites of origin.

The processed pseudogenes do not carry any information that might be used to sponsor a transposition event (or to carry out the preceding reverse transcription of the RNA). This suggests that the RNA was a substrate for another system and coded by a retroposon. In fact, it seems likely that the active LINES elements provide most of the reverse transcriptase activity, and they are responsible not only for their own transposition, but also for acting on the SINES and for generating processed pseudogenes.

22.12 LINES Use an Endonuclease to Generate a Priming End

Key concept

- LINES do not have LTRs and require the retroposon to code for an endonuclease that generates a nick to prime reverse transcription.

LINES elements, and some others, do not terminate in the LTRs that are typical of retroviral elements. This poses the question: How is reverse transcription primed? It does not involve the typical reaction in which a tRNA primer pairs with the LTR (see Figure 22.6). The open reading frames in these elements lack many of the retroviral functions, such as protease or integrase domains, but typically have reverse transcriptase-like sequences and code for an endonuclease activity. In the human LINES L1, ORF1 is a DNA-binding protein and ORF2 has both reverse transcriptase and endonuclease activities; both products are required for transposition.

FIGURE 22.20 shows how these activities support transposition. A nick is made in the DNA target site by an endonuclease activity coded by the retroposon. The RNA product of the element associates with the protein bound at the nick. The nick provides a 3′–OH end that primes synthesis of cDNA on the RNA template. A second cleavage event is required to open the other strand of DNA, and the RNA/DNA hybrid is linked to the other end of the gap either at this stage or after it has been converted into a DNA duplex. A similar mechanism is used by some mobile introns (see Figure 27.12).

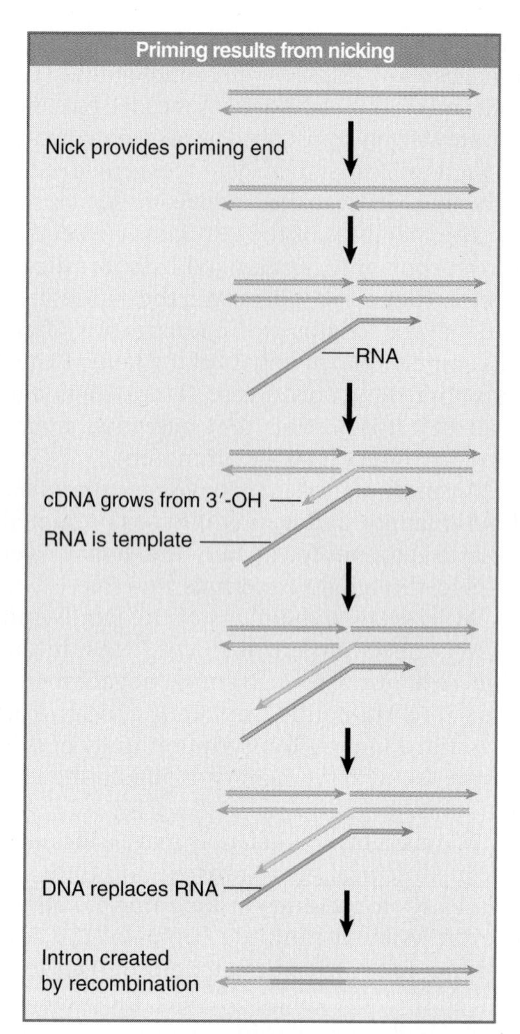

FIGURE 22.20 Retrotransposition of non-LTR elements occurs by nicking the target to provide a primer for cDNA synthesis on an RNA template. The arrowheads indicate 3′ ends.

When elements originate from RNA polymerase II transcripts, the genomic sequences are necessarily inactive: They lack the promoter that was upstream of the original startpoint for transcription. They usually possess the features of the mature transcript; as a result they are called **processed pseudogenes.**

One of the reasons why LINES elements are so effective lies with their method of propagation. When a LINES mRNA is translated, the protein products show a *cis*-preference for binding to the mRNA from which they were translated. FIGURE 22.21 shows that the ribonucleoprotein complex then moves to the nucleus, where the proteins insert a DNA copy into the genome. Reverse transcription often does not proceed fully to the end, so the copy is inactive. There is, however, the potential for

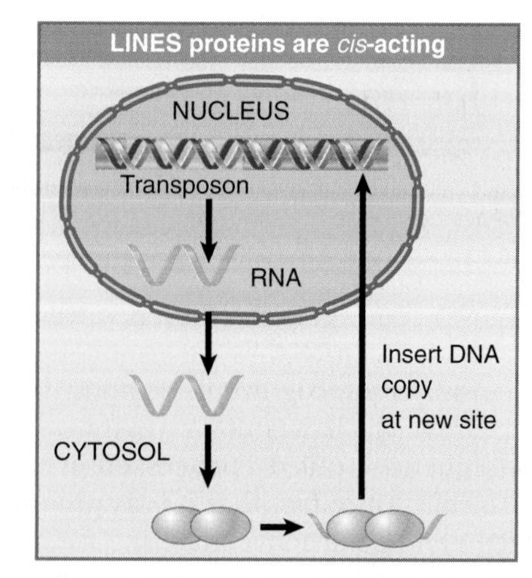

FIGURE 22.21 A LINES is transcribed into an RNA that is translated into proteins that assemble into a complex with the RNA. The complex translocates to the nucleus, where it inserts a DNA copy into the genome.

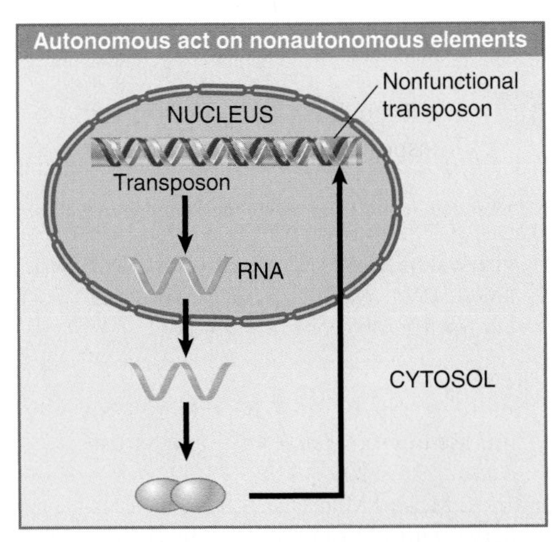

FIGURE 22.22 A transposon is transcribed into an RNA that is translated into proteins that move independently to the nucleus, where they act on any pair of inverted repeats with the same sequence as the original transposon.

insertion of an active copy, because the proteins are acting on a transcript of the original active element.

By contrast, the proteins produced by the DNA transposons must be imported into the nucleus after being synthesized in the cytoplasm, but they have no means of distinguishing full-length transposons from inactive deleted transposons. FIGURE 22.22 shows that instead of distinguishing these two types of transposons, the proteins will indiscriminately recognize any element by virtue of the repeats that mark the ends. This greatly reduces their chance of acting on a full-length element as opposed to one that has been deleted. The consequence is that inactive elements accumulate, and eventually the family dies out because a transposase has such a small chance of finding a target that is a fully functional transposon.

Are transposition events currently occurring in these genomes, or are we seeing only the footprints of ancient systems? This varies with the species. There are only a few currently active transposons in the human genome, but by contrast several active transposons are known in the mouse genome. This explains the fact that spontaneous mutations caused by LINES insertions occur at a rate of ~3% in mouse, but only 0.1% in man. There appear to be ~10 to 50 active LINES elements in the human genome. Some human diseases can be pinpointed as the result of transposition of L1 into genes, and others result from unequal crossing-over events involving repeated copies of L1. A model system in which LINES transposition occurs in tissue culture cells suggests that a transposition event can introduce several types of collateral damage as well as inserting into a new site; the damage includes chromosomal rearrangements and deletions. Such events may be viewed as agents of genetic change. Neither DNA transposons nor retroviral-like retroposons seem to have been active in the human genome for 40 to 50 million years, but several active examples of both are found in the mouse.

Note that for transpositions to survive, they must occur in the germline. Presumably similar events occur in somatic cells, but do not survive beyond one generation.

22.13 Summary

Reverse transcription is the unifying mechanism for reproduction of retroviruses and perpetuation of retroposons. The cycle of each type of element is in principle similar, although retroviruses are usually regarded from the perspective of the free viral (RNA) form, whereas retroposons are regarded from the stance of the genomic (duplex DNA) form.

Retroviruses have genomes of single-stranded RNA that are replicated through a double-stranded DNA intermediate. An individual retrovirus contains two copies of its genome. The genome contains the *gag*, *pol*, and *env* genes that are translated into polyproteins, each of which is cleaved into smaller functional proteins. The Gag and Env components are concerned with packing RNA and generating the virion; the Pol components are concerned with nucleic acid synthesis.

Reverse transcriptase is the major component of Pol, and is responsible for synthesizing a DNA (minus strand) copy of the viral (plus strand) RNA. The DNA product is longer than the RNA template; by switching template strands, reverse transcriptase copies the 3′ sequence of the RNA to the 5′ end of the DNA, and copies the 5′ sequence of the RNA to the 3′ end of the DNA. This generates the characteristic LTRs (long terminal repeats) of the DNA. A similar switch of templates occurs when the plus strand of DNA is synthesized using the minus strand as a template. Linear duplex DNA is inserted into a host genome by the integrase enzyme. Transcription of the integrated DNA from a promoter in the left LTR generates further copies of the RNA sequence.

Switches in template during nucleic acid synthesis allow recombination to occur by copy choice. During an infective cycle, a retrovirus may exchange part of its usual sequence for a cellular sequence; the resulting virus is usually replication-defective, but can be perpetuated in the course of a joint infection with a helper virus. Many of the defective viruses have gained an RNA version *(v-onc)* of a cellular gene *(c-onc)*. The *onc* sequence may be any one of a number of genes whose expression in *v-onc* form causes the cell to be transformed into a tumorigenic phenotype.

The integration event generates direct target repeats (like transposons that mobilize via DNA). An inserted provirus therefore has direct terminal repeats of the LTRs, flanked by short repeats of target DNA. Mammalian and avian genomes have endogenous (inactive) proviruses with such structures. Other elements with this organization have been found in a variety of genomes, most notably in *S. cerevisiae* and *D. melanogaster*. *Ty* elements of yeast and *copia* elements of flies have coding sequences with homology to reverse transcriptase and mobilize via an RNA form. They may generate particles resembling viruses, but do not have infectious capability. The LINES sequences of mammalian genomes are further removed from the retroviruses, but retain enough similarities to suggest a common origin. They use a different type of priming event to initiate reverse transcription, in which an endonuclease activity associated with the reverse transcriptase makes a nick that provides a 3′–OH end for priming synthesis on an RNA template. The frequency of LINES transposition is increased because its protein products are *cis*-acting; they associate with the mRNA from which they were translated to form a ribonucleoprotein complex that is transported into the nucleus.

The members of another class of retroposons have the hallmarks of transposition via RNA, but have no coding sequences (or at least none resembling retroviral functions). They may have originated as passengers in a retroviral-like transposition event, in which an RNA was a target for a reverse transcriptase. Processed pseudogenes arise by such events. A particularly prominent family that appears to have originated from a processing event is the mammalian SINES; it includes the human Alu family. Some snRNAs, including 7SL snRNA (a component of the SRP), are related to this family.

References

22.2 The Retrovirus Life Cycle Involves Transposition-Like Events

Review

Varmus, H. E. and Brown, P. O. (1989). Retroviruses. In Howe, M. M. and Berg, D. E., eds., *Mobile DNA*. Washington, DC: American Society for Microbiology, pp. 53–108.

Research

Baltimore, D. (1970). RNA-dependent DNA polymerase in virions of RNA tumor viruses. *Nature* 226, 1209–1211.

Temin, H. M. and Mizutani, S. (1970). RNA-dependent DNA polymerase in virions of Rous sarcoma virus. *Nature* 226, 1211–1213.

22.4 Viral DNA Is Generated by Reverse Transcription

Reviews

Katz, R. A. and Skalka, A. M. (1994). The retroviral enzymes. *Annu. Rev. Biochem.* 63, 133–173.

Lai, M. M. C. (1992). RNA recombination in animal and plant viruses. *Microbiol. Rev.* 56, 61–79.

Negroni, M. and Buc, H. (2001). Mechanisms of retroviral recombination. *Annu. Rev. Genet.* 35, 275–302.

Research

Hu, W. S. and Temin, H. M. (1990). Retroviral recombination and reverse transcription. *Science* 250, 1227–1233.

Negroni, M. and Buc, H. (2000). Copy-choice recombination by reverse transcriptases: reshuffling of genetic markers mediated by RNA chaperones. *Proc. Natl. Acad. Sci. USA* 97, 6385–6390.

22.5 Viral DNA Integrates into the Chromosome

Review

Goff, S. P. (1992). Genetics of retroviral integration. *Annu. Rev. Genet.* 26, 527–544.

Research

Craigie, R., Fujiwara, T., and Bushman, F. (1990). The IN protein of Moloney murine leukemia virus processes the viral DNA ends and accomplishes their integration *in vitro*. *Cell* 62, 829–837.

22.7 Yeast *Ty* Elements Resemble Retroviruses

Research

Boeke, J. D. et al. (1985). Ty elements transpose through an RNA intermediate. *Cell* 40, 491–500.

22.8 Many Transposable Elements Reside in *D. melanogaster*

Research

Mount, S. M. and Rubin, G. M. (1985). Complete nucleotide sequence of the *Drosophila* transposable element copia: homology between copia and retroviral proteins. *Mol. Cell Biol.* 5, 1630–1638.

22.9 Retroposons Fall into Three Classes

Reviews

Deininger, P. L. (1989). SINEs: short interspersed repeated DNA elements in higher eukaryotes. In Howe, M. M. and Berg, D. E., eds., *Mobile DNA*. Washington, DC: American Society for Microbiology, pp. 619–636.

Hutchison, C. L. A., Hardies, S. C., Loer, D. D., Shehee, W. R., and Edgell, M. H. (1989). LINES and related retroposons: Long interspersed repeated sequences in the eukaryotic genome. In Howe, M. M. and Berg, D. E., eds., *Mobile DNA*. Washington, DC: American Society for Microbiology, pp. 593–617.

Ostertag, E. M. and Kazazian, H. H. (2001). Biology of mammalian L1 retrotransposons. *Annu. Rev. Genet.* 35, 501–538.

Weiner, A. M., Deininger, P. L., and Efstratiadis, A. (1986). Nonviral retroposons: genes, pseudogenes, and transposable elements generated by the reverse flow of genetic information. *Annu. Rev. Biochem.* 55, 631–661.

Research

Bureau, T. E., Ronald, P. C., and Wessler, S. R. (1996). A computer-based systematic survey reveals the predominance of small inverted-repeat elements in wild-type rice genes. *Proc. Natl. Acad. Sci. USA* 93, 8524–8529.

Bureau, T. E. and Wessler, S. R. (1992). Tourist: a large family of small inverted repeat elements frequently associated with maize genes. *Plant Cell* 4, 1283–1294.

Loeb, D. D. et al. (1986). The sequence of a large L1Md element reveals a tandemly repeated 5′ end and several features found in retrotransposons. *Mol. Cell Biol.* 6, 168–182.

Sachidanandam, R. et al. (2001). A map of human genome sequence variation containing 1.42 million single nucleotide polymorphisms. The International SNP Map Working Group. *Nature* 409, 928–933.

Waterston et al. (2002). Initial sequencing and comparative analysis of the mouse genome. *Nature* 420, 520–562.

22.12 LINES Use an Endonuclease to Generate a Priming End

Review

Ostertag, E. M. and Kazazian, H. H. (2001). Biology of mammalian L1 retrotransposons. *Annu. Rev. Genet.* 35, 501–538.

Research

Feng, Q., Moran, J. V., Kazazian, H. H., and Boeke, J. D. (1996). Human L1 retrotransposon encodes a conserved endonuclease required for retrotransposition. *Cell* 87, 905–916.

Gilbert, N., Lutz-Prigge, S., and Moran, J. V. (2002). Genomic deletions created upon LINE-1 retrotransposition. *Cell* 110, 315–325.

Lauermann, V. and Boeke, J. D. (1994). The primer tRNA sequence is not inherited during Ty1 retrotransposition. *Proc. Natl. Acad. Sci. USA* 91, 9847–9851.

Luan, D. D. et al. (1993). Reverse transcription of R2Bm RNA is primed by a nick at the chromosomal target site: a mechanism for non-LTR retrotransposition. *Cell* 72, 595–605.

Moran, J. V., Holmes, S. E., Naas, T. P., DeBerardinis, R. J., Boeke, J. D., and Kazazian, H. H. (1996). High frequency retrotransposition in cultured mammalian cells. *Cell* 87, 917–927.

Symer, D. E., Connelly, C., Szak, S. T., Caputo, E. M., Cost, G. J., Parmigiani, G., and Boeke, J. D. (2002). Human l1 retrotransposition is associated with genetic instability *in vitro*. *Cell* 110, 327–338.

23

Immune Diversity

CHAPTER OUTLINE

- It proceeds through a hairpin intermediate at the coding end; opening of the hairpin is responsible for insertion of extra bases (P nucleotides) in the recombined gene.
- Deoxynucleoside transferase inserts additional N nucleotides at the coding end.
- The codon at the site of the V-(D)J joining reaction has an extremely variable sequence and codes for amino acid 96 in the antigen-binding site.
- The double-strand breaks at the coding joints are repaired by the same system involved in nonhomologous end-joining of damaged DNA.
- An enhancer in the C gene activates the promoter of the V gene after recombination has generated the intact immunoglobulin gene.

23.11 Early Heavy Chain Expression Can Be Changed by RNA Processing

- All lymphocytes start by synthesizing the membrane-bound form of IgM.
- A change in RNA splicing causes this to be replaced by the secreted form when the B cell differentiates.

23.12 Class Switching Is Caused by DNA Recombination

- Immunoglobulins are divided into five classes according to the type of constant region in the heavy chain.
- Class switching to change the C_H region occurs by a recombination between S regions that deletes the region between the old C_H region and the new C_H region.
- Multiple successive switch recombinations can occur.

23.13 Switching Occurs by a Novel Recombination Reaction

- Switching occurs by a double-strand break followed by the nonhomologous end-joining reaction.
- The important feature of a switch region is the presence of inverted repeats.
- Switching requires activation of promoters that are upstream of the switch sites.

23.14 Somatic Mutation Generates Additional Diversity in Mouse and Human Being

- Active immunoglobulin genes have V regions with sequences that are changed from the germline because of somatic mutation.
- The mutations occur as substitutions of individual bases.
- The sites of mutation are concentrated in the antigen-binding site.
- The process depends on the enhancer that activates transcription at the Ig locus.

23.15 Somatic Mutation Is Induced by Cytidine Deaminase and Uracil Glycosylase

- A cytidine deaminase is required for somatic mutation as well as for class switching.
- Uracil-DNA glycosylase activity influences the pattern of somatic mutations.
- Hypermutation may be initiated by the sequential action of these enzymes.

23.16 Avian Immunoglobulins Are Assembled from Pseudogenes

- An immunoglobulin gene in chicken is generated by copying a sequence from one of 25 pseudogenes into the V gene at a single active locus.

23.17 B Cell Memory Allows a Rapid Secondary Response

- The primary response to an antigen is mounted by B cells that do not survive beyond the response period.
- Memory B cells are produced that have specificity for the same antigen, but that are inactive.
- A reexposure to antigen triggers the secondary response in which the memory cells are rapidly activated.

23.18 T Cell Receptors Are Related to Immunoglobulins

- T cells use a similar mechanism of V(D)J-C joining to B cells to produce either of two types of T cell receptor.
- TCR $\alpha\beta$ is found on >95% of T lymphocytes; TCR $\gamma\delta$ is found on <5%.

23.19 The T Cell Receptor Functions in Conjunction with the MHC

- The TCR recognizes a short peptide that is bound in a groove of an MHC protein on the surface of the presenting cell.

23.20 The Major Histocompatibility Locus Codes for Many Genes of the Immune System

- The MHC locus codes for the class I and class II proteins as well as for other proteins of the immune system.
- Class I proteins are the transplantation antigens that are responsible for distinguishing "self" from "nonself" tissue.
- An MHC class I protein is active as a heterodimer with β_2 microglobulin.
- Class II proteins are involved in interactions between T cells.
- An MHC class II protein is a heterodimer of α and β chains.

23.21 Innate Immunity Utilizes Conserved Signaling Pathways

- Innate immunity is triggered by receptors that recognize motifs (PAMPs) that are highly conserved in bacteria or other infective agents.
- Toll-like receptors are commonly used to activate the response pathway.
- The pathways are highly conserved from invertebrates to vertebrates, and an analogous pathway is found in plants.

23.22 Summary

23.1 Introduction

It is an axiom of genetics that the genetic constitution created in the zygote by the combination of sperm and egg is inherited by all somatic cells of the organism. We look to differential control of gene expression, rather than to changes in DNA content, to explain the different phenotypes of particular somatic cells.

Yet there are exceptional situations in which the reorganization of certain DNA sequences is used to regulate gene expression or to create new genes. The immune system provides a striking and extensive case in which the content of the genome changes, when recombination creates active genes in lymphocytes. Other cases are represented by the substitution of one sequence for another to change the mating type of yeast or to generate new surface antigens by trypanosomes.

The **immune response** of vertebrates provides a protective system that distinguishes foreign proteins from the proteins of the organism itself. Foreign material (or part of the foreign material) is recognized as comprising an **antigen.** Usually the antigen is a protein (or protein-attached moiety) that has entered the bloodstream of the animal—for example, the coat protein of an infecting virus. Exposure to an antigen initiates production of an immune response that *specifically recognizes the antigen and destroys it.*

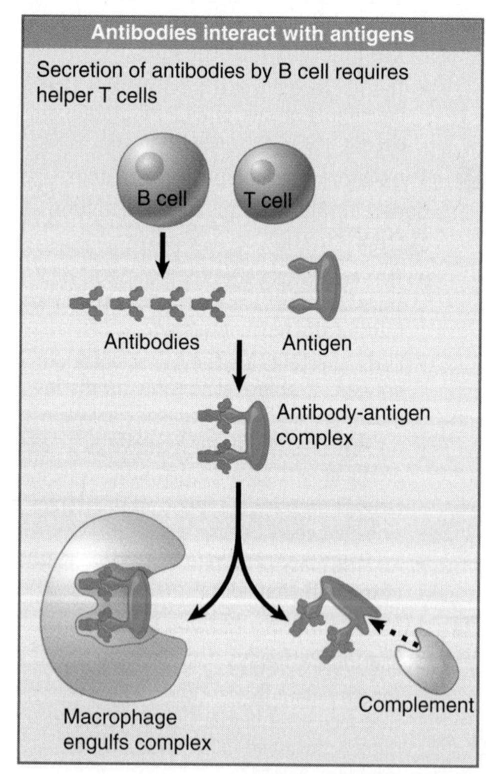

FIGURE 23.1 Humoral immunity is conferred by the binding of free antibodies to antigens to form antigen-antibody complexes that are removed from the bloodstream by macrophages or that are attacked directly by the complement proteins.

Immune reactions are the responsibility of white blood cells—the B and T lymphocytes, and macrophages. The lymphocytes are named after the tissues that produce them. In mammals, **B cells** mature in the bone marrow, whereas **T cells** mature in the thymus. *Each class of lymphocyte uses the rearrangement of DNA as a mechanism for producing the proteins that enable it to participate in the immune response.*

The immune system has many ways to destroy an antigenic invader, but it is useful to consider them in two general classes. Which type of response the immune system mounts when it encounters a foreign structure depends partly on the nature of the antigen. The response is defined according to whether it is executed principally by B cells or T cells.

The **humoral response** depends on B cells. It is mediated by the secretion of antibodies, which are **immunoglobulin** proteins. *Production of an **antibody** specific for a foreign molecule is the primary event responsible for recognition of an antigen.* Recognition requires the antibody to bind to a small region or structure on the antigen.

The function of antibodies is represented in FIGURE 23.1. Foreign material circulating in the bloodstream—for example, a toxin or pathogenic bacterium—has a surface that presents antigens. The antigen(s) are recognized by the antibodies, which form an antigen-antibody complex. This complex then attracts the attention of other components of the immune system.

The humoral response depends on these other components in two ways. First, B cells need signals provided by T cells to enable them to secrete antibodies. These T cells are called **helper T cells,** because they assist the B cells. Second, antigen-antibody formation is a trigger for the antigen to be destroyed. The major pathway is provided by the action of **complement,** a component whose name reflects its ability to "complement" the action of the antibody itself. Complement consists of a set of ~20 proteins that function through a cascade of proteolytic actions. If the target antigen is part of a cell—for example, an infecting bacterium—the action of complement culminates in lysing the target cell. The action of complement also provides a means of attracting macrophages, which scavenge the target cells or their products. Alternatively, the antigen-antibody complex may be taken up directly by macrophages (scavenger cells) and destroyed.

The **cell-mediated response** is executed by a class of T lymphocytes called **cytotoxic T cells** (also called killer T cells). The basic function of the T cell in recognizing a target antigen is indicated in FIGURE 23.2. A cell-mediated response typically is elicited by an intracellular parasite, such as a virus that infects the body's own cells. As a result of the viral infection, fragments of foreign (viral) antigens are displayed on the surface of the cell. These fragments are recognized by the **T cell receptor (TCR),** which is the T cells' equivalent of the antibody produced by a B cell.

A crucial feature of this recognition reaction is that *the antigen must be* presented *by a cellular protein that is a member of the* **MHC (major histocompatibility complex).** The MHC protein has a groove on its surface that binds a peptide fragment derived from the foreign antigen. The combination of peptide fragment and MHC protein is recognized by the T cell receptor. Every individual has a characteristic set of MHC proteins. They are important in graft reactions; a graft of tissue from one individual to another is rejected because of the difference in MHC proteins between the donor and recipient, an issue of major medical importance. The demand that the T lymphocytes recognize both foreign antigen and MHC protein ensures that the cell-mediated response acts only on host cells that have been infected with a foreign antigen. (We discuss the division of MHC proteins into the general types of class I and class II later in Section 23.20, The Major Histocompatibility Locus Codes for Many Genes of the Immune System.)

The purpose of each type of immune response is to attack a foreign target. Target recognition is the prerogative of B-cell immunoglobulins and T cell receptors. A crucial aspect of their function lies in the ability to distinguish "self" from "nonself." Proteins and cells of the body itself must *never* be attacked. Foreign targets must be *destroyed entirely.* The property of failing to attack "self" is called **tolerance.** Loss of this ability results in an **autoimmune disease,** in which the immune system attacks its own body, often with disastrous consequences.

What prevents the lymphocyte pool from responding to "self" proteins? Tolerance probably arises early in lymphocyte cell development when B cells and T cells that recognize "self" antigens are destroyed. This is called **clonal deletion.** In addition to this negative selection, there is also positive selection for T cells carrying certain sets of T cell receptors.

A corollary of tolerance is that it can be difficult to obtain antibodies against proteins that are closely related to those of the organism itself. As a practical matter, therefore, it may be difficult to use (for example) mice or rabbits to obtain antibodies against human proteins that have been highly conserved in mammalian evolution. The tolerance of the mouse or rabbit for its own protein may extend to the human protein in such cases.

Each of the three groups of proteins required for the immune response—immunoglobulins, T cell receptors, and MHC proteins—is diverse. Examining a large number of individuals, we find many variants of each protein. Each protein is coded by a large family of genes; in the case of antibodies and the T cell receptors, the diversity of the

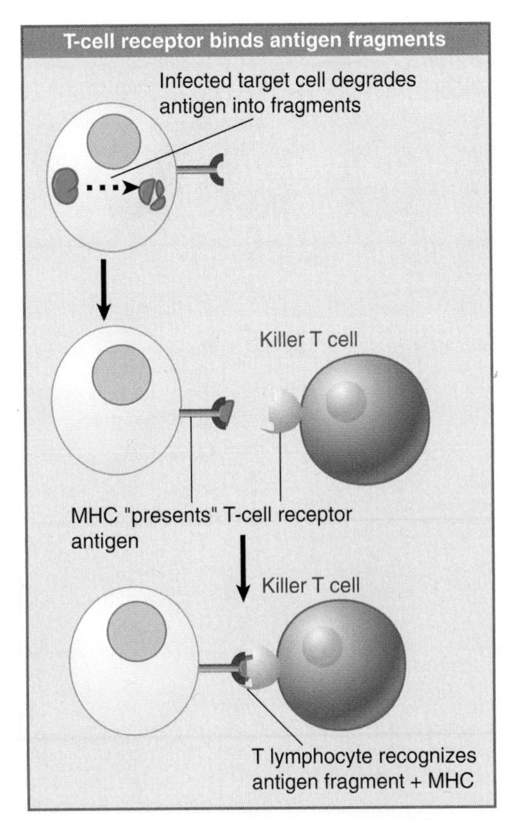

FIGURE 23.2 In cell-mediated immunity, killer T cells use the T cell receptor to recognize a fragment of the foreign antigen that is presented on the surface of the target cell by the MHC protein.

population is increased by DNA rearrangements that occur in the relevant lymphocytes.

Immunoglobulins and T cell receptors are direct counterparts, each produced by its own type of lymphocyte. The proteins are related in structure, and their genes are related in organization. The sources of variability are similar. The MHC proteins also share some common features with the antibodies, as do other lymphocyte-specific proteins. In dealing with the genetic organization of the immune system, we are therefore concerned with a series of related gene families, indeed a **superfamily** that may have evolved from some common ancestor representing a primitive immune response.

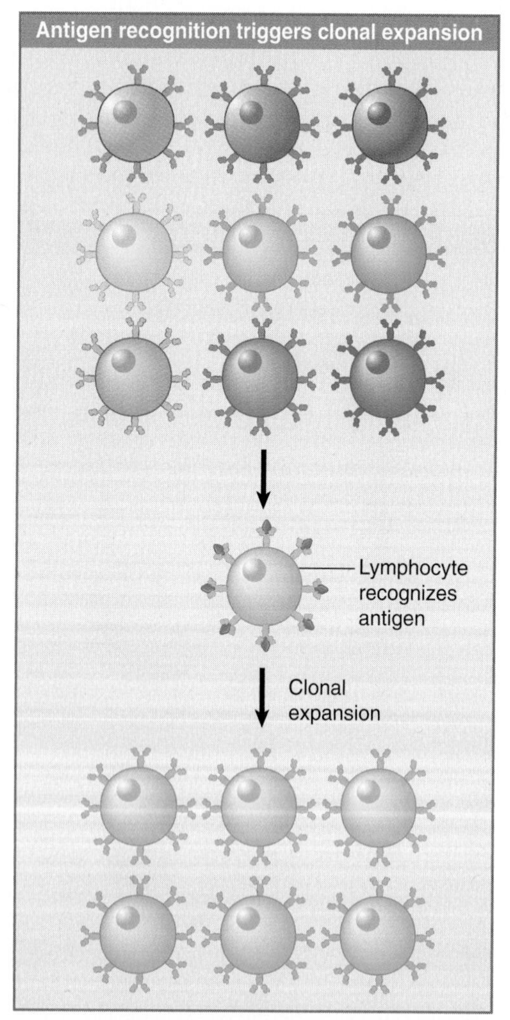

FIGURE 23.3 The pool of immature lymphocytes contains B cells and T cells making antibodies and receptors with a variety of specificities. Reaction with an antigen leads to clonal expansion of the lymphocyte with the antibody (B cell) or receptor (T cell) that can recognize the antigen.

23.2 Clonal Selection Amplifies Lymphocytes That Respond to Individual Antigens

Key concepts

- Each B lymphocyte expresses a single immunoglobulin and each T lymphocyte expresses a single T cell receptor.
- There is a very large variety of immunoglobulins and T cell receptors.
- Antigen binding to an immunoglobulin or T cell receptor triggers clonal multiplication of the cell.

The name of the immune response describes one of its central features. After an organism has been exposed to an antigen, it becomes *immune* to the effects of a new infection. Before exposure to a particular antigen, the organism lacks adequate capacity to deal with any toxic effects. This ability is acquired during the immune response. After the infection has been defeated, the organism retains the ability to respond rapidly in the event of a reinfection.

These features are accommodated by the **clonal selection** theory illustrated in FIGURE 23.3. The pool of lymphocytes contains B cells and T cells carrying a large variety of immunoglobulins or T cell receptors. *Any individual B lymphocyte produces one immunoglobulin, however, which is capable of recognizing only a single antigen; similarly, any individual T lymphocyte produces only one particular T cell receptor.*

In the pool of immature lymphocytes, the unstimulated B cells and T cells are morphologically indistinguishable. On exposure to antigen, though, a B cell whose antibody is able to bind the antigen, or a T cell whose receptor can recognize it, is stimulated to divide, probably by some feedback from the surface of the cell, where the antibody/receptor-antigen reaction occurs. The stimulated cells then develop into mature B or T lymphocytes; this includes morphological changes involving (for example) an increase in cell size (especially pronounced for B cells).

The initial expansion of a specific B or T cell population upon first exposure to an antigen is called the **primary immune response.** Large numbers of B or T lymphocytes with specificity for the target antigen are produced. Each population represents a clone of the original responding cell. Antibody is secreted from the B cells in large quantities, and it may even come to dominate the antibody population.

After a successful primary immune response has been mounted, the organism retains B cells and T cells carrying the corresponding antibody or receptor. These **memory cells** represent an intermediate state between the immature cell and the mature cell. They have not acquired all of the features of the mature cell, but they are long-lived, and can rapidly be converted to mature cells. Their presence allows a **secondary immune response** to be mounted rapidly if the animal is exposed to the same antigen again.

The pool of immature lymphocytes in a mammal contains ~10^{12} cells. This pool contains some lymphocytes that have unique specificities (because a corresponding antigen has never been encountered), whereas others are represented by up to 10^6 cells (because clonal selection has expanded the pool to respond to an antigen).

What features are recognized in an antigen? Antigens are usually macromolecular. Small molecules may have antigenic determinants and can be recognized by antibodies, although they usually are not effective in provoking an immune response (because of their small size). They do, however, provoke a response when conjugated with a larger carrier molecule (usually a protein). A small molecule that is used to provoke a response by such means is called a **hapten.**

Only a small part of the surface of a macromolecular antigen is actually recognized by any one antibody. The binding site consists of only five or six amino acids. Of course, any particular protein may have more than one such binding site, in which case it provokes antibodies with specificities for different regions. The region provoking a response is called an **antigenic determinant** or **epitope.** When an antigen contains several epitopes, some may be more effective than others in provoking the immune response; in fact, they may be so effective that they entirely dominate the response.

How do lymphocytes find target antigens and where does their maturation take place? Lymphocytes are peripatetic cells. They develop from immature stem cells that are located in the adult bone marrow. They migrate to the peripheral lymphoid tissues (spleen, lymph nodes) either directly via the bloodstream (if they are B cells) or via the thymus (where they become T cells). The lymphocytes recirculate between blood and lymph; the process of dispersion ensures that an antigen will be exposed to lymphocytes of all possible specificities. When a lymphocyte encounters an antigen that binds its antibody or receptor, clonal expansion begins the immune response.

23.3 Immunoglobulin Genes Are Assembled from Their Parts in Lymphocytes

Key concepts

- An immunoglobulin is a tetramer of two light chains and two heavy chains.
- Light chains fall into the lambda and kappa families; heavy chains form a single family.
- Each chain has an N-terminal variable region (V) and a C-terminal constant region (C).
- The V domain recognizes antigen and the C domain provides the effector response.
- V domains and C domains are separately coded by V gene segments and C gene segments.
- A gene coding for an intact immunoglobulin chain is generated by somatic recombination to join a V gene segment with a C gene segment.

A remarkable feature of the immune response is an animal's ability to produce an appropriate antibody whenever it is exposed to a new antigen. How can the organism be prepared to produce antibody proteins, each of which is designed specifically to recognize an antigen whose structure cannot be anticipated?

For practical purposes, we usually reckon that a mammal has the ability to produce 10^6 to 10^8 different antibodies. Each antibody is an immunoglobulin tetramer consisting of two identical **light chains** (**L**) and two identical **heavy chains** (**H**). If any light chain can associate with any heavy chain, to produce 10^6 to 10^8 potential antibodies requires 10^3 to 10^4 different light chains and 10^3 to 10^4 different heavy chains.

There are two types of light chain and ~10 types of heavy chain. Different classes of immunoglobulins have different effector functions. The class is determined by the heavy chain constant region, which exercises the effector function (see Figure 23.17).

The structure of the immunoglobulin tetramer is illustrated in FIGURE 23.4. Light chains and heavy chains share the same general type of organization in which each protein chain consists of two principal regions: the N-terminal **variable region (V region);** and the C-terminal **constant region (C region).** They were defined originally by comparing the amino acid sequences of different immunoglobulin chains. As the names

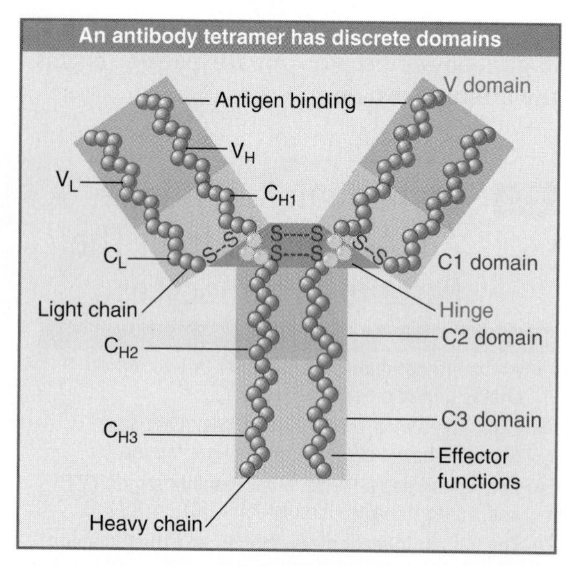

FIGURE 23.4 Heavy and light chains combine to generate an immunoglobulin with several discrete domains.

suggest, the variable regions show considerable changes in sequence from one protein to the next, whereas the constant regions show substantial homology.

Corresponding regions of the light chains and heavy chains associate to generate distinct domains in the immunoglobulin protein.

The variable (V) domain is generated by association between the variable regions of the light chain and heavy chain. *The V domain is responsible for recognizing the antigen.* An immunoglobulin has a Y-shaped structure in which the arms of the Y are identical, and each arm has a copy of the V domain. Production of V domains of different specificities creates the ability to respond to diverse antigens. The total number of variable regions for either light- or heavy-chain proteins is measured in hundreds. *Thus the protein displays the maximum versatility in the region responsible for binding the antigen.*

The number of constant regions is vastly smaller than the number of variable regions—typically there are only one to ten C regions for any particular type of chain. The constant regions in the subunits of the immunoglobulin tetramer associate to generate several individual C domains. The first domain results from association of the single constant region of the light chain (C_L) with the C_{H1} part of the heavy-chain constant region. The two copies of this domain complete the arms of the Y-shaped molecule. Association between the C regions of the heavy chains generates the remaining C domains, which vary in number depending on the type of heavy chain.

Comparing the characteristics of the variable and constant regions, we see the central dilemma in immunoglobulin gene structure. How does the genome code for a set of proteins in which any individual polypeptide chain must have one of <10 possible C regions, but can have any one of several hundred possible V regions? It turns out that the number of coding sequences for each type of region reflects its variability. There are many genes coding for V regions, but only a few genes coding for C regions.

In this context, *"gene" means a sequence of DNA coding for a discrete part of the final immunoglobulin polypeptide* (heavy or light chain). Thus **V genes** code for variable regions and **C genes** code for constant regions, although *neither type of gene is expressed as an independent unit.* To construct a unit that can be expressed in the form of an authentic light or heavy chain, a V gene must be joined physically to a C gene. In this system, two "genes" code for one polypeptide. To avoid confusion, we will refer to these units as "gene segments" rather than "genes."

The sequences coding for light chains and heavy chains are assembled in the same way: *any one of many V gene segments may be joined to any one of a few C gene segments.* This **somatic recombination** occurs *in the B lymphocyte in which the antibody is expressed.* The large number of available V gene segments is responsible for a major part of the diversity of immunoglobulins. Not all diversity is coded in the genome, though; some is generated by changes that occur during the process of constructing a functional gene.

Essentially the same description applies to the formation of functional genes coding for the protein chains of the T cell receptor. Two types of receptor are found on T cells—one consisting of two types of chain called α and β, and the other consisting of γ and δ chains. Like the genes coding for immunoglobulins, the genes coding for the individual chains in T cell receptors consist of separate parts, including V and C regions, that are brought together in an active T cell (see Section 23.18, T Cell Receptors Are Related to Immunoglobulins).

The crucial fact about the synthesis of immunoglobulins, therefore, is that *the arrangement of V gene segments and C gene segments is different in the cells producing the immunoglobulins (or T cell receptors) from all other somatic cells or germ cells.*

The construction of a functional immunoglobulin or T cell receptor gene might seem to be a Lamarckian process, representing a change in the genome that responds to a particular feature of the phenotype (the antigen). At birth, the organism does not possess the functional gene for producing a particular antibody or T cell receptor. It possesses a large number of V gene segments and a smaller number of C gene segments. The subsequent construction of an active gene from these parts allows the antibody/receptor to be synthesized so that it is available to react with the antigen. The clonal selection theory requires that this rearrangement of DNA occur *before the exposure to antigen*, which then results in *selection* for those cells carrying a protein able to bind the antigen. The entire process occurs in somatic cells and does not affect the germline; thus the response to an antigen is not inherited by progeny of the organism.

There are two families of immunoglobulin light chains, κ and λ, and one family containing all the types of heavy chain (H). Each family resides on a different chromosome and consists of its own set of both V gene segments and C gene segments. This is called the *germline pattern*, and is found in the germline and in somatic cells of all lineages other than the immune system.

In a cell expressing an antibody, though, each of its chains—one light type (either κ or λ) and one heavy type—is coded by a single intact gene. The recombination event that brings a V gene segment to partner a C gene segment creates an active gene consisting of exons that correspond precisely with the functional domains of the protein. The introns are removed in the usual way by RNA splicing.

Recombination between V and C gene segments to give functional loci occurs in a population of immature lymphocytes. A B lymphocyte usually has only one productive rearrangement of light-chain gene segments (either κ or λ) and one of heavy-chain gene segments. Similarly, a T lymphocyte productively rearranges an α gene and a β gene, or one δ gene and one γ gene. The antibody or T cell receptor produced by any one cell is determined by the particular configuration of V gene segments and C gene segments that has been joined.

The principles by which functional genes are assembled are the same in each family, but there are differences in the details of the organization of the V gene segments and C gene segments, and correspondingly of the recombination reaction between them. In addition to the V gene segments and C gene segments, other short DNA sequences (including J segments and D segments) are included in the functional somatic loci.

23.4 Light Chains Are Assembled by a Single Recombination

Key concepts

- A lambda light chain is assembled by a single recombination between a V gene and a J-C gene segment.
- The V gene segment has a leader exon, intron, and variable-coding region.
- The J-C gene segment has a short J-coding exon, intron, and C-coding region.
- A kappa light chain is assembled by a single recombination between a V gene segment and one of five J segments preceding the C gene.

A λ light chain is assembled from two parts, as illustrated in FIGURE 23.5. The V gene segment consists of the leader exon (L) separated by a single intron from the variable (V) segment. The C gene segment consists of the J segment separated by a single intron from the constant (C) exon.

The name of the **J segment** is an abbreviation for joining, because it identifies the region to which the V segment becomes connected. Thus the joining reaction does not directly involve V and C gene segments, but occurs via the J segment; when we discuss the joining of "V and C gene segments" for light chains, we really mean V-JC joining.

The J segment is short and codes for the last few amino acids of the variable region, as defined by amino acid sequences. In the intact gene generated by recombination, the V-J segment constitutes a single exon coding for the entire variable region.

The consequences of the κ joining reaction are illustrated in FIGURE 23.6. A κ light chain also is assembled from two parts, but there is a difference in the organization of the C gene segment. A group of five J segments is spread over a region of 500 to 700 bp, separated by an intron of 2 to 3 kb from the C_κ exon. In the mouse, the central J segment is nonfunctional (ψJ3). A V_κ segment may be joined to any one of the J segments.

Whichever J segment is used becomes the terminal part of the intact variable exon. Any J segments on the left of the recombining J segment are lost (J1 has been lost in the figure). Any

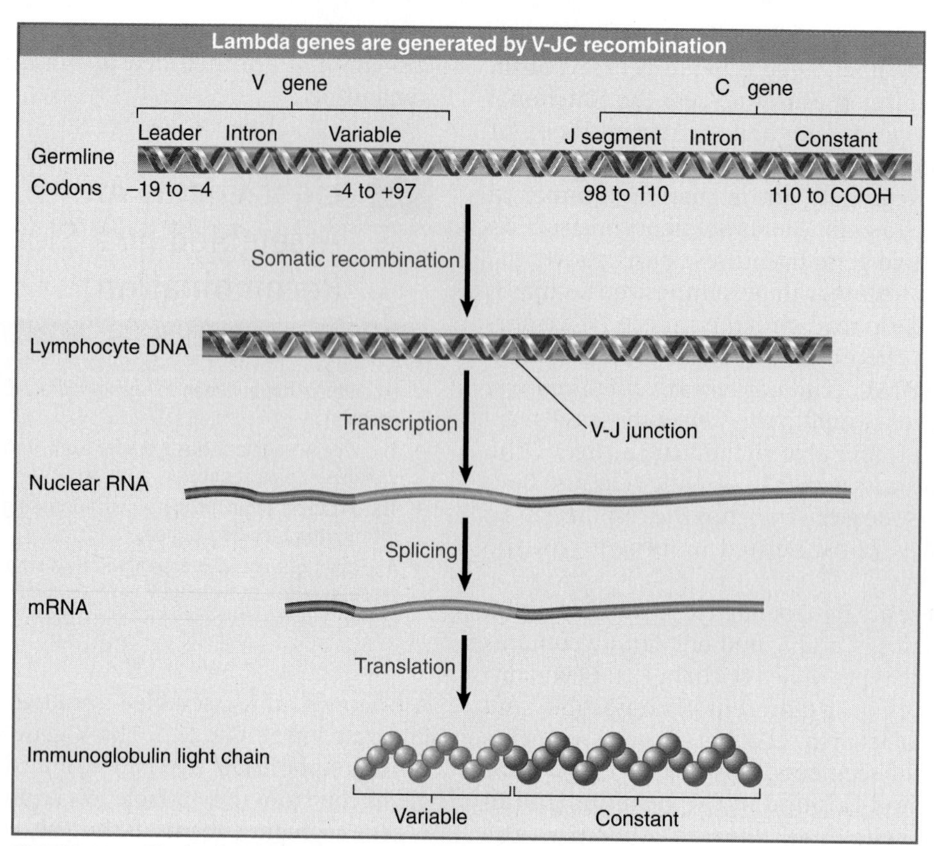

FIGURE 23.5 The lambda C gene segment is preceded by a J segment, so that V-J recombination generates a functional lambda light-chain gene.

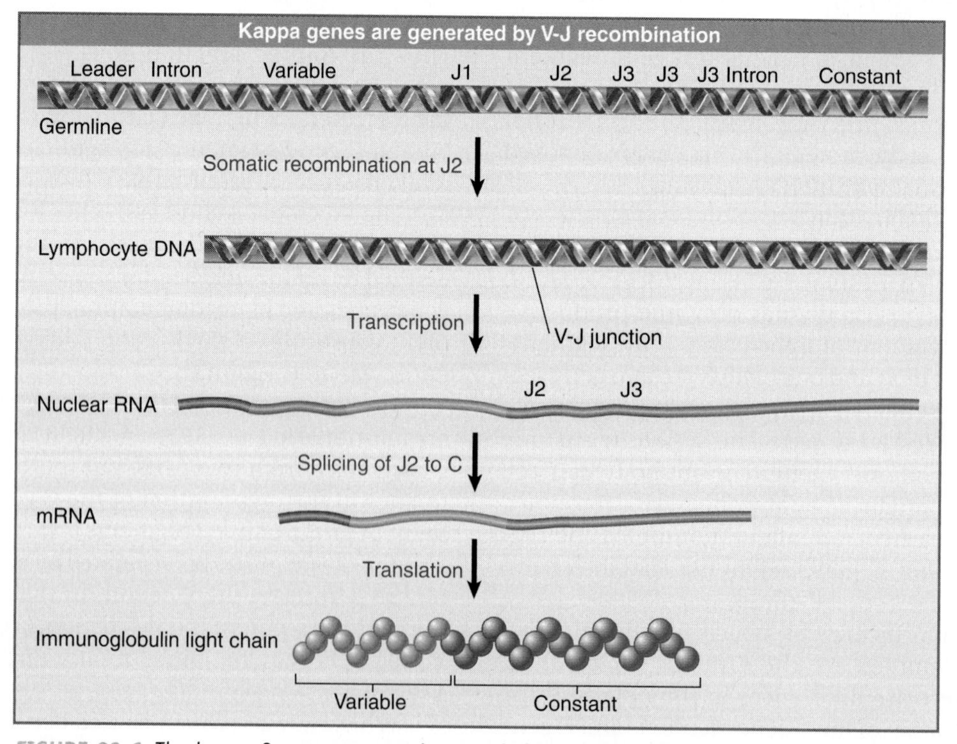

FIGURE 23.6 The kappa C gene segment is preceded by multiple J segments in the germ line. V-J joining may recognize any one of the J segments, which is then spliced to the C gene segment during RNA processing.

J segment on the right of the recombining J segment is treated as part of the intron between the variable and constant exons (J3 is included in the intron that is spliced out in the figure).

All functional J segments possess a signal at the left boundary that makes it possible to recombine with the V segment; they also possess a signal at the right boundary that can be used for splicing to the C exon. Whichever J segment is recognized in DNA V-J joining uses its splicing signal in RNA processing.

23.5 Heavy Chains Are Assembled by Two Recombinations

Key concepts

- The units for heavy chain recombination are a V gene, a D segment, and a J-C gene segment.
- The first recombination joins D to J-C.
- The second recombination joins V to D-J-C.
- The C segment consists of several exons.

Heavy chain construction involves an additional segment. The **D segment** (for diversity) was discovered by the presence in the protein of an extra two to thirteen amino acids between the sequences coded by the V segment and the J segment. An array of >10 D segments lies on the chromosome between the V_H segments and the four J_H segments.

V-D-J joining takes place in two stages, as illustrated in FIGURE 23.7. First one of the D segments recombines with a J_H segment; a V_H segment then recombines with the DJ_H combined segment. The reconstruction leads to expression of the adjacent C_H segment (which consists of several exons). (We discuss the use of different C_H gene segments in Section 23.12, Class Switching Is Caused by DNA Recombination; for now we will just consider the reaction in terms of the connection to one of several J segments that precede a C_H gene segment.)

The D segments are organized in a tandem array. The mouse heavy-chain locus contains twelve D segments of variable length; the human locus has ~30 D segments (not all necessarily active). Some unknown mechanism must ensure that the *same* D segment is involved in the D-J joining and V-D joining reactions. (When we discuss joining of V and C gene segments for heavy chains, we assume the process has been completed by V-D and D-J joining reactions.)

The V gene segments of all three immunoglobulin families are similar in organization. The first exon codes for the signal sequence (involved in membrane attachment), and the second exon codes for the major part of the variable region itself (<100 codons long). The remainder of the variable region is provided by the D segment (in the H family only) and by a J segment (in all three families).

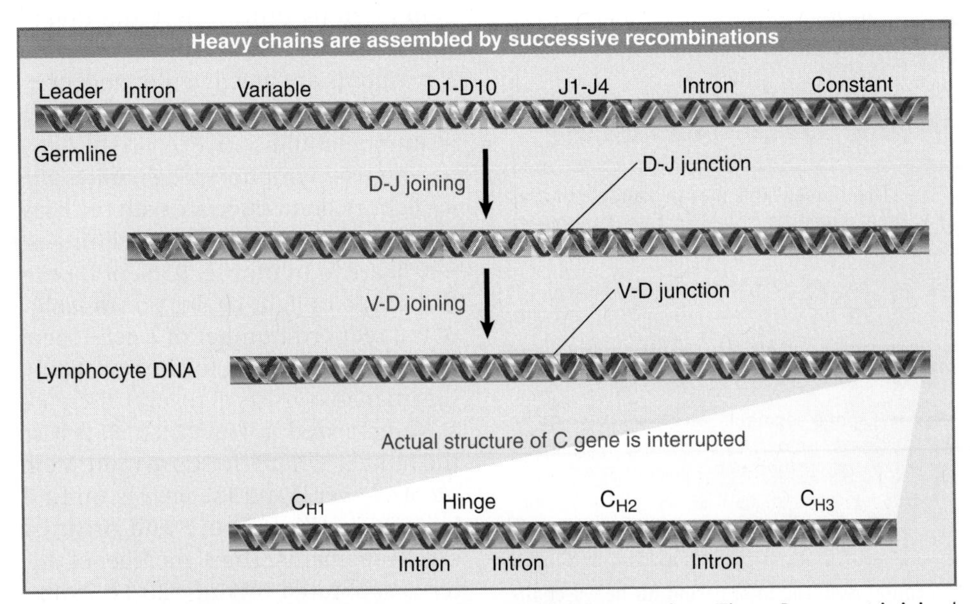

FIGURE 23.7 Heavy genes are assembled by sequential joining reactions. First a D segment is joined to a J segment, and then a V gene segment is joined to the D segment.

The structure of the constant region depends on the type of chain. For both κ and λ light chains, the constant region is coded by a single exon (which becomes the third exon of the reconstructed, active gene). For H chains, the constant region is coded by several exons; corresponding with the protein chain shown in Figure 23.4, separate exons code for the regions C_{H1}, hinge, C_{H2}, and C_{H3}. Each C_H exon is ~100 codons long; the hinge is shorter. The introns usually are relatively small (~300 bp).

23.6 Recombination Generates Extensive Diversity

Key concepts

- A light chain locus can produce >1000 chains by combining 300 V genes with four to five C genes.
- An H locus can produce >4000 chains by combining 300 V genes, 20 D segments, and four J segments.

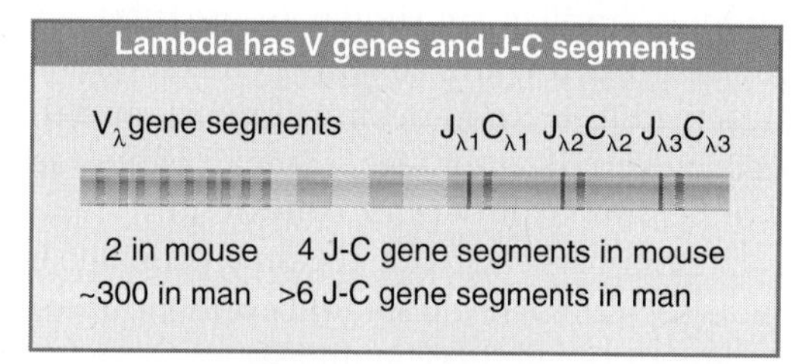

FIGURE 23.8 The lambda family consists of V gene segments linked to a small number of J-C gene segments.

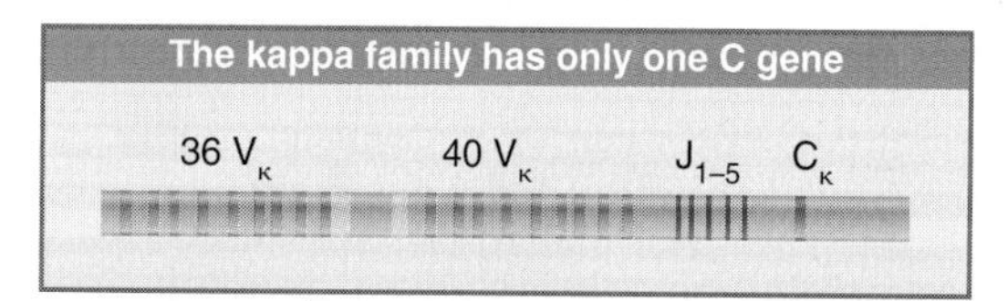

FIGURE 23.9 The human and mouse kappa families consist of V gene segments linked to five J segments connected to a single C gene segment.

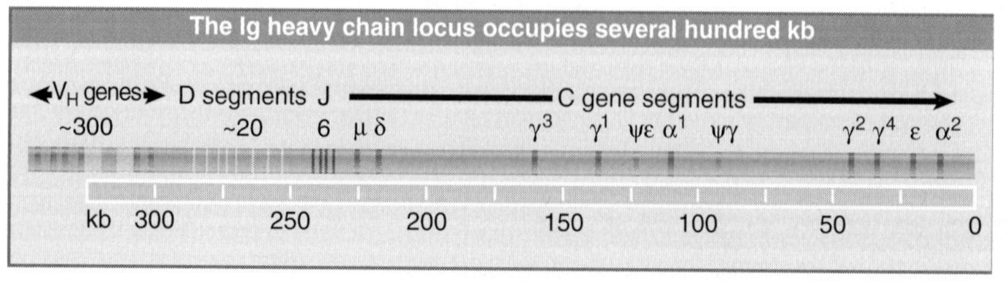

FIGURE 23.10 A single gene cluster in man contains all the information for heavy-chain gene assembly.

Now we must examine the different types of V and C gene segments to see how much diversity can be accommodated by the variety of the coding regions carried in the germline. In each light Ig gene family, many V gene segments are linked to a much smaller number of C gene segments.

FIGURE 23.8 shows that the λ locus has ~6 C gene segments, each preceded by its own J segment. The λ locus in mouse is much less diverse than the human locus. The main difference is that in a mouse there are only two V_λ gene segments; each is linked to two J-C regions. Of the four C_λ gene segments, one is inactive. At some time in the past, the mouse suffered a catastrophic deletion of most of its germline V_λ gene segments.

FIGURE 23.9 shows that the κ locus has only one C gene segment, although it is preceded by five J segments (one of them inactive). The V_κ gene segments occupy a large cluster on the chromosome, upstream of the constant region. The human cluster has two regions. Just preceding the C_κ gene segment, a region of 600 kb contains the five J_κ segments and 40 V_κ gene segments. A gap of 800 kb separates this region from another group of 36 V_κ gene segments.

The V_κ gene segments can be subdivided into families, which are defined by the criterion that members of a family have >80% amino acid identity. The mouse family is unusually large (~1000 genes), and there are ~18 V_κ families that vary in size from 2 to 100 members. Like other families of related genes, related V gene segments form subclusters, which are generated by duplication and divergence of individual ancestral members. Many of the V segments are inactive pseudogenes, though, and <50 are likely to be used to generate immunoglobulins.

A given lymphocyte generates *either* a κ *or* a λ light chain to associate with the heavy chain. In man, ~60% of the light chains are κ and ~40% are λ. In mouse, 95% of B cells express the κ type of light chain, presumably because of the reduced number of λ gene segments.

The single locus for heavy chain production in man consists of several discrete sections, as summarized in FIGURE 23.10. It is similar in the mouse, where there are more V_H gene segments, fewer D and J segments, and a slight difference in the number and organization of C gene segments. The 3′ member of the V_H cluster is separated by only 20 kb from the first D segment. The D segments are spread over

~50 kb, followed by the cluster of J segments. Over the next 220 kb lie all the C_H gene segments. There are nine functional C_H gene segments and two pseudogenes. The organization suggests that a γ gene segment must have been duplicated to give the subcluster of γ-γ-ε -α, after which the entire group was then duplicated.

How far is the diversity of germline information responsible for V region diversity in immunoglobulin proteins? By combining any one of ~50 V gene segments with any one of four to five J segments, a typical light chain locus has the potential to produce some 250 chains. There is even greater diversity in the H chain locus; by combining any one of ~50 V_H gene segments, 20 D segments, and four J segments, the genome potentially can produce 4000 variable regions to accompany any C_H gene segment. In mammals, this is the starting point for diversity, but additional mechanisms introduce further changes. *When closely related variants of immunoglobulins are examined, there often are more proteins than can be accounted for by the number of corresponding V gene segments.* The new members are created by somatic changes in individual genes during or after the recombination process (see Section 23.14, Somatic Mutation Generates Additional Diversity in Mouse and Man).

23.7 Immune Recombination Uses Two Types of Consensus Sequence

Key concepts

- The consensus sequence used for recombination is a heptamer separated by either 12 or 23 base pairs from a nonamer.
- Recombination occurs between two consensus sequences that have different spacings.

Assembly of light- and heavy-chain genes involves the same mechanism (although the number of parts is different). The same consensus sequences are found at the boundaries of all germline segments that participate in joining reactions. Each consensus sequence consists of a heptamer separated by either 12 or 23 bp from a nonamer.

FIGURE 23.11 illustrates the relationship between the consensus sequences at the mouse Ig loci. At the κ locus, each V_κ gene segment is followed by a consensus sequence with a 12 bp spacing. Each J_κ segment is preceded by a consensus sequence with a 23 bp spacing. The V and J consensus sequences are inverted in orientation. At the λ locus, each V_λ gene segment is followed by a consensus sequence with 23 bp spacing; each J_λ gene segment is preceded by a consensus of the 12 bp spacer type.

The rule that governs the joining reaction is that *a consensus sequence with one type of spacing can be joined only to a consensus sequence with the other type of spacing.* The consensus sequences at V and J segments can lie in either order; thus the different spacings do not impart any directional information, but instead serve to prevent one V or J gene segment from recombining with another of the same.

This concept is borne out by the structure of the components of the heavy gene segments. Each V_H gene segment is followed by a consensus sequence of the 23 bp spacer type. The D segments are flanked on either side by consensus sequences of the 12 bp spacer type. The J_H segments are preceded by consensus sequences of the 23 bp spacer type. Thus the V gene segment must be joined to a D segment, and the D segment must be joined to a J segment. A V gene segment cannot be joined directly to a J segment, because both possess the same type of consensus sequence.

The spacing between the components of the consensus sequences corresponds almost to one or two turns of the double helix. This may reflect a geometric relationship in the recombination reaction. For example, the recombination protein(s) may approach the DNA from one side, in the same way that RNA polymerase and repressors approach recognition elements such as promoters and operators.

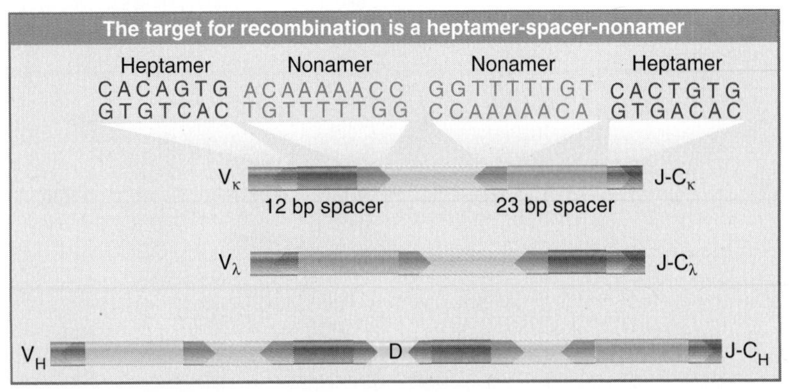

FIGURE 23.11 Consensus sequences are present in inverted orientation at each pair of recombining sites. One member of each pair has a spacing of 12 bp between its components; the other has 23 bp spacing.

23.8 Recombination Generates Deletions or Inversions

Key concepts

- Recombination occurs by double-strand breaks at the heptamers of two consensus sequences.
- The signal ends of the fragment between the breaks usually join to generate an excised circular fragment.
- The coding ends are covalently linked to join V to J-C (L chain), or D to J-C and V to D-J-C (H chain).
- If the recombining genes are inverted instead of direct orientation, there is an inversion instead of deletion of an excised circle.

Recombination of the components of immunoglobulin genes is accomplished by a physical rearrangement of sequences, involving breakage and reunion, but the mechanism is different from homologous recombination. The general nature of the reaction for a λ light chain is illustrated in FIGURE 23.12. (The reaction is similar at a heavy chain locus, with the exception that there are two recombination events: first D-J, then V-DJ.)

Breakage and reunion occur as separate reactions. A double-strand break is made at the heptamers that lie at the ends of the coding units. This releases the entire fragment between the V gene segment and J-C gene segment; the cleaved termini of this fragment are called **signal ends.** The cleaved termini of the V and J-C loci are called **coding end.** The two coding ends are covalently linked to form a coding joint; this is the connection that links the V and J segments. If the two signal ends are also connected, the excised fragment would form a circular molecule.

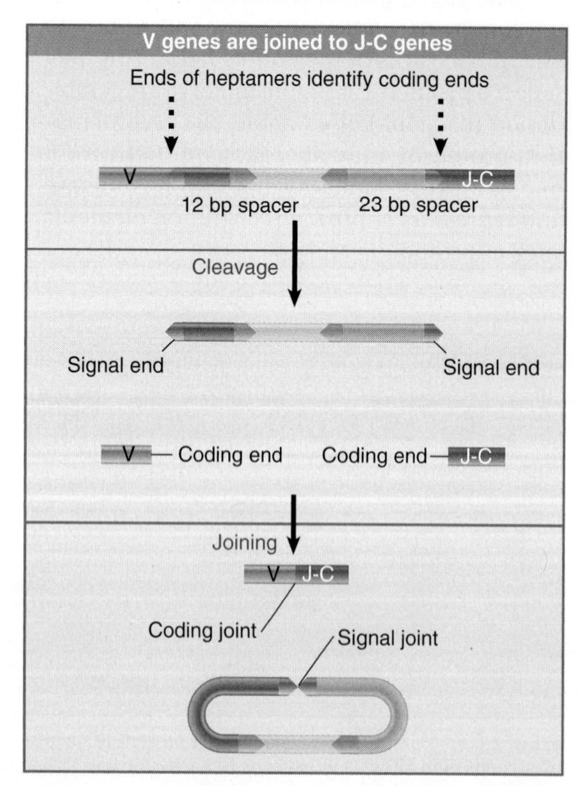

FIGURE 23.12 Breakage and reunion at consensus sequences generates immunoglobulin genes.

We have shown the V and J-C loci as organized in the same orientation. As a result, the cleavage at each consensus sequence releases the region between them as a linear fragment. If the signal ends are joined, it is converted into a circular molecule, as indicated in Figure 23.12. Deletion to release an excised circle is the predominant mode of recombination at the immunoglobulin and TCR loci.

In some exceptional cases the V gene segment is inverted in orientation on the chromosome relative to the J-C loci. In such a case, breakage and reunion inverts the intervening material instead of deleting it. The outcomes of deletion versus inversion are the same as shown previously for homologous recombination between direct or inverted repeats in Figure 21.9 and Figure 21.10. There is one further proviso, however; recombination with an inverted V gene segment makes it *necessary* for the signal ends to be joined, because otherwise there is a break in the locus. Inversion occurs in TCR recombination, and also occurs at times in the κ light chain locus.

23.9 Allelic Exclusion Is Triggered by Productive Rearrangement

Key concepts

- Recombination to generate an intact immunoglobulin gene is productive if it leads to expression of an active protein.
- A productive rearrangement prevents any further rearrangement from occurring, whereas a nonproductive rearrangement does not.
- Allelic exclusion applies separately to light chains (only one kappa *or* lambda may be productively rearranged) and to heavy chains (one heavy chain is productively rearranged).

Each B cell expresses a single type of light chain and a single type of heavy chain, because only a single productive rearrangement of each type occurs in a given lymphocyte in order to produce one light and one heavy chain gene. Each event involves the genes of only *one* of the homologous chromosomes, and as a result *the alleles on the other chromosome are not expressed in the same cell.* This phenomenon is called **allelic exclusion.**

The occurrence of allelic exclusion complicates the analysis of somatic recombination. A probe reacting with a region that has rearranged on one homolog will also detect the allelic sequences on the other homolog. We are therefore compelled to analyze the different fates of the two chromosomes together.

The usual pattern displayed by a rearranged active gene can be interpreted in terms of a deletion of the material between the recombining V and C loci.

Two types of gene organization are seen in active cells:

- Probes to the active gene may reveal one rearranged copy and one germline copy. We assume, then, that joining has occurred on one chromosome, whereas the other chromosome has remained unaltered.
- Two different rearranged patterns may be found, indicating that the chromosomes have suffered independent rearrangements. In some of these instances, material between the recombining V and C gene segments is entirely absent from the cell line. This is most easily explained by the occurrence of independent deletions (resulting from recombination) on each chromosome.

When two chromosomes both lack the germline pattern, usually only one of them has passed through a **productive rearrangement** to generate a functional gene. The other has suffered a **nonproductive rearrangement;** this may take several forms, but in each case the gene sequence cannot be expressed as an immunoglobulin chain. (It may be incomplete, for example because D-J joining has occurred but V-D joining has not followed; or it may be aberrant, with the process completed but failing to generate a gene that codes for a functional protein.)

The coexistence of productive and nonproductive rearrangements suggests the existence of a feedback loop to control the recombination process. A model is outlined in FIGURE 23.13. Suppose that each cell starts with two loci in the unrearranged germline configuration Ig^0. Either of these loci may be rearranged to generate a productive gene Ig^+ or a nonproductive gene Ig^-.

If the rearrangement is productive, the synthesis of an active chain provides a trigger to prevent rearrangement of the other allele. The active cell has the configuration Ig^0/Ig^+.

If the rearrangement is nonproductive, it creates a cell with the configuration Ig^0/Ig^-. There is no impediment to rearrangement of the remaining germline allele. If this rearrangement is productive, the expressing cell has the configuration Ig^+/Ig^-. Again, the presence of an active chain suppresses the possibility of further rearrangements.

Two successive nonproductive rearrangements produce the cell Ig^-/Ig^-. In some cases an Ig^-/Ig^- cell can try yet again. Sometimes the observed patterns of DNA can only have been generated by successive rearrangements.

The crux of the model is that the cell keeps trying to recombine V gene segments and C gene segments until a productive rearrangement is achieved. Allelic exclusion is caused by the suppression of further rearrangement as soon as an active chain is produced. The use of this mechanism *in vivo* is demonstrated by the creation of transgenic mice whose germline has a rearranged immunoglobulin gene.

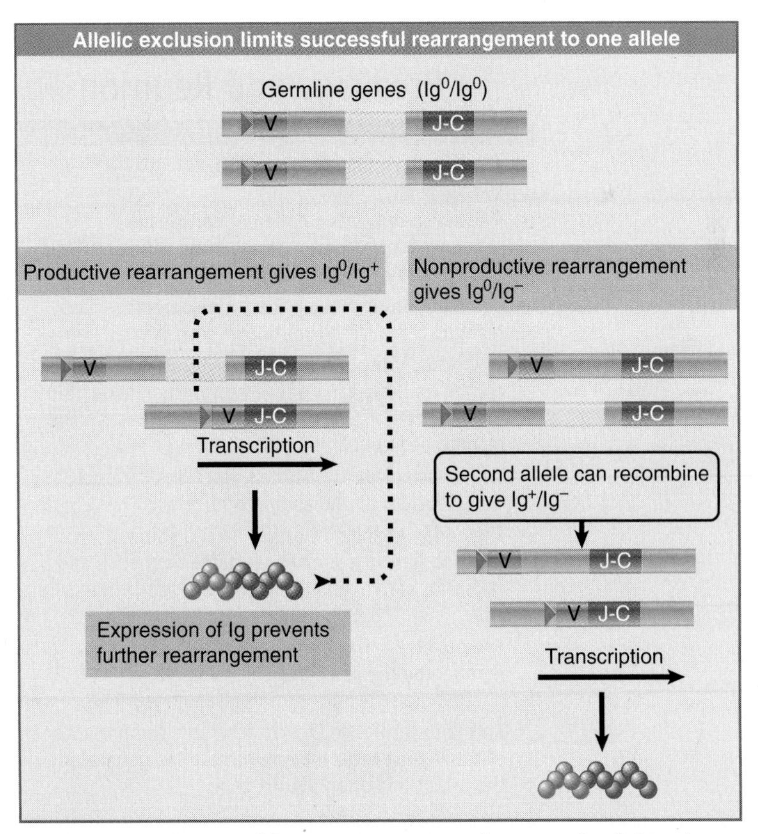

FIGURE 23.13 A successful rearrangement to produce an active light or heavy chain suppresses further rearrangements of the same type, and results in allelic exclusion.

Expression of the transgene in B cells suppresses the rearrangement of endogenous genes.

Allelic exclusion is independent for the heavy- and light-chain loci. Heavy-chain genes usually rearrange first. Allelic exclusion for light chains must apply equally to both families (cells may have *either* active κ or λ light chains). It is likely that the cell rearranges its κ genes first, and tries to rearrange λ only if both κ attempts are unsuccessful.

There is an interesting paradox in this series of events. The same consensus sequences and the same V(D)J recombinase are involved in the recombination reactions at H, κ, and λ loci, and yet the three loci rearrange in a set order. What ensures that heavy rearrangement precedes light rearrangement, and that κ precedes λ? The loci may become accessible to the enzyme at different times, possibly as the result of transcription. Transcription occurs even before rearrangement, although of course the products have no coding function. The transcriptional event may change the structure of chromatin, making the consensus sequences for recombination available to the enzyme.

23.10 The RAG Proteins Catalyze Breakage and Reunion

Key concepts

- The RAG proteins are necessary and sufficient for the cleavage reaction.
- RAG1 recognizes the nonamer consensus sequences for recombination. RAG2 binds to RAG1 and cleaves at the heptamer.
- The reaction resembles the topoisomerase-like resolution reaction that occurs in transposition.
- It proceeds through a hairpin intermediate at the coding end; opening of the hairpin is responsible for insertion of extra bases (P nucleotides) in the recombined gene.
- Deoxynucleoside transferase inserts additional N nucleotides at the coding end.
- The codon at the site of the V-(D)J joining reaction has an extremely variable sequence and codes for amino acid 96 in the antigen-binding site.
- The double-strand breaks at the coding joints are repaired by the same system involved in nonhomologous end-joining of damaged DNA.
- An enhancer in the C gene activates the promoter of the V gene after recombination has generated the intact immunoglobulin gene.

The proteins RAG1 and RAG2 are necessary and sufficient to cleave DNA for V(D)J recombination. They are coded by two genes, separated by <10 kb on the chromosome, whose transfection into fibroblasts causes a suitable substrate DNA to undergo the V(D)J joining reaction. Mice that lack either *RAG1* or *RAG2* are unable to recombine their immunoglobulins or T cell receptors, and as a result have immature B lymphocytes and T lymphocytes. The RAG proteins together undertake the catalytic reactions of cleaving and rejoining DNA, and also provide a structural framework within which the reactions occur.

RAG1 recognizes the heptamer/nonamer signals with the appropriate 12/23 spacing and recruits RAG2 to the complex. The nonamer provides the site for initial recognition, and the heptamer directs the site of cleavage.

The reactions involved in recombination are shown in FIGURE 23.14. The complex nicks one strand at each junction. The nick has 3′–OH and 5′–P ends. The free 3′–OH end then attacks the phosphate bond at the corresponding position *in the other strand of the duplex.* This creates a hairpin at the coding end, in which the 3′ end of one strand is covalently linked to the 5′ end of the other strand; it leaves a blunt double-strand break at the signal end.

This second cleavage is a transesterification reaction in which bond energies are conserved. It resembles the topoisomerase-like reactions catalyzed by the resolvase proteins of bacterial transposons (see Section 21.9, TnA Transposition Requires Transposase and Resolvase). The parallel with these reactions is supported further by a homology between RAG1 and bacterial invertase proteins (which invert specific segments of DNA by similar recombination reactions). In fact, the RAG proteins can insert a donor DNA whose free ends consist of the appropriate signal sequences (heptamer-12/23-spacer nonamer) into an unrelated target DNA in an *in vitro* transposition reaction. This suggests that somatic recombination of immune genes evolved from an ancestral transposon. It also suggests that the RAG proteins are responsible for chromosomal translocations in which Ig or TCR loci are connected to other loci.

The hairpins at the coding ends provide the substrate for the next stage of reaction. If a single-strand break is introduced into one strand close to the hairpin, an unpairing reaction at the end generates a single-stranded protrusion. Synthesis of a complement to the exposed single strand then converts the coding end to an extended duplex. This reaction explains the introduction of **P nucleotides** at coding ends;

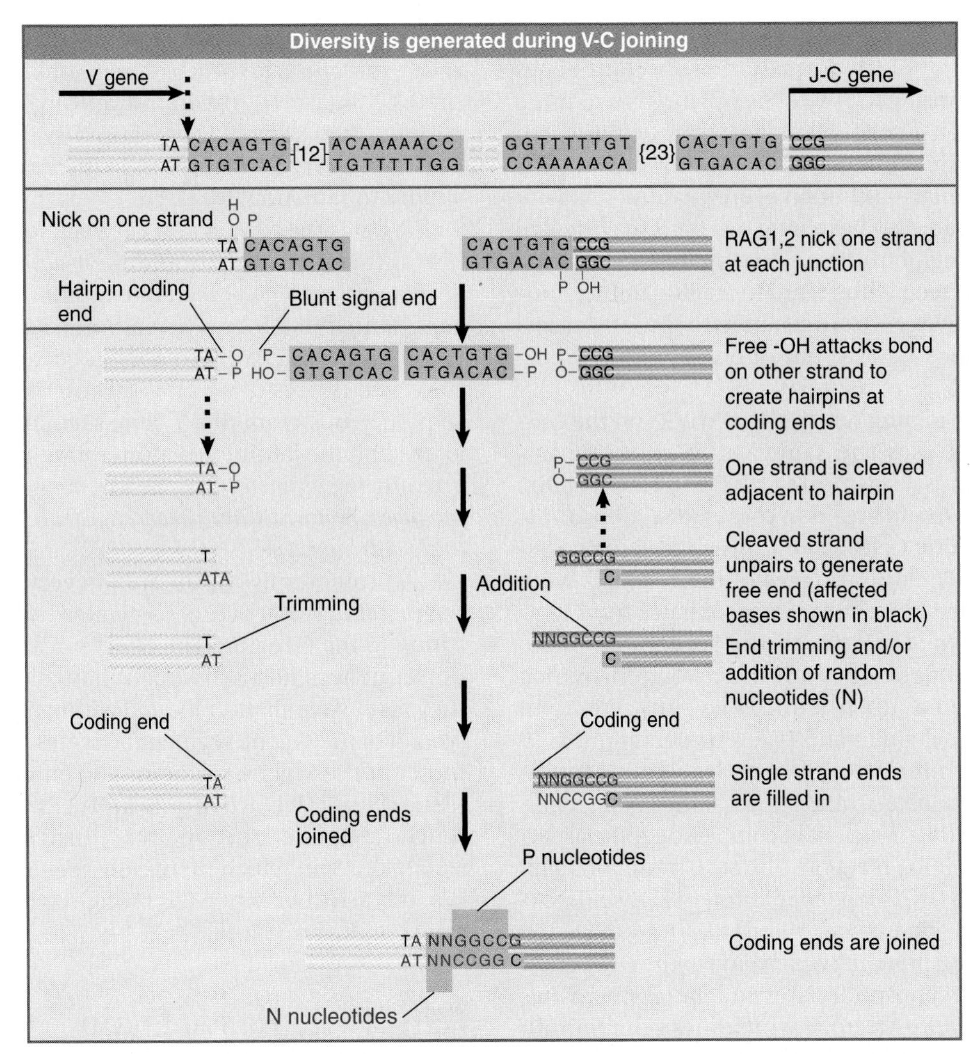

FIGURE 23.14 Processing of coding ends introduces variability at the junction.

they consist of a few extra base pairs related to, but reversed in orientation from, the original coding end.

Some extra bases also may be inserted, apparently with random sequences, between the coding ends. They are called **N nucleotides.** Their insertion occurs via the activity of the enzyme deoxynucleoside transferase (known to be an active component of lymphocytes) at a free 3′ coding end generated during the joining process.

Changes in sequence during recombination are therefore a consequence of the enzymatic mechanisms involved in breaking and rejoining the DNA. In heavy chain recombination, base pairs are lost or inserted at the V_H-D or D-J junctions, or both. Deletion also occurs in V_λ-J_λ joining, but insertion at these joints is unusual. The changes in sequence affect the amino acid coded at V-D junctions and D-J junctions in heavy chains or at the V-J junction in light chains.

These various mechanisms together ensure that a coding joint may have a sequence that is different from what would be predicted by a direct joining of the coding ends of the V, D, and J regions.

Changes in the sequence at the junction make it possible for a great variety of amino acids to be coded at this site. It is interesting that the amino acid at position 96 is created by the V-J joining reaction. It forms part of the antigen-binding site and also is involved in making contacts between the light chains and the heavy chains. Thus the maximum diversity is generated at the site that contacts the target antigen.

Changes in the number of base pairs at the coding joint affect the reading frame. The joining process appears to be random with regard to reading frame, so that probably only one third of the joined sequences retain the proper frame of reading through the junctions. If the V-J region is joined so that the J segment is out of phase, translation is terminated prematurely by a nonsense codon in the incorrect frame. We may think of the formation of aberrant genes as comprising the price the cell must pay for the increased diversity that it gains by being able to adjust the sequence at the joining site.

Similar—although even greater—diversity is generated in the joining reactions that involve the D segment of the heavy chain. The same result is seen with regard to reading frame; nonproductive genes are generated by joining events that place J and C out of phase with the preceding V gene segment.

The joining reaction that works on the coding end uses the same pathway of nonhomologous end-joining (NHEJ) that repairs double-strand breaks in cells (see Section 20.11, Eukaryotic Cells Have Conserved Repair Systems). The initial stages of the reaction were identified by isolating intermediates from lymphocytes of mice with the severe combined immunodeficiency (*SCID*) mutation, which results in a much-reduced level of activity in immunoglobulin and TCR recombination. *SCID* mice accumulate broken molecules that terminate in double-strand breaks at the coding ends, and are thus deficient in completing some aspect of the joining reaction. The *SCID* mutation inactivates a DNA-dependent protein kinase (DNA-PK). The kinase is recruited to DNA by the Ku70 and Ku80 proteins, which bind to the DNA ends. DNA-PK phosphorylates and thereby activates the protein Artemis, which nicks the hairpin ends (it also has exonuclease and endonuclease activities that function in the NHEJ pathway). The actual ligation is undertaken by DNA ligase IV and also requires the protein XRCC4. As a result, mutations in the Ku proteins or in XRCC4 or DNA ligase IV are found among human patients who have diseases caused by deficiencies in DNA repair that result in increased sensitivity to radiation.

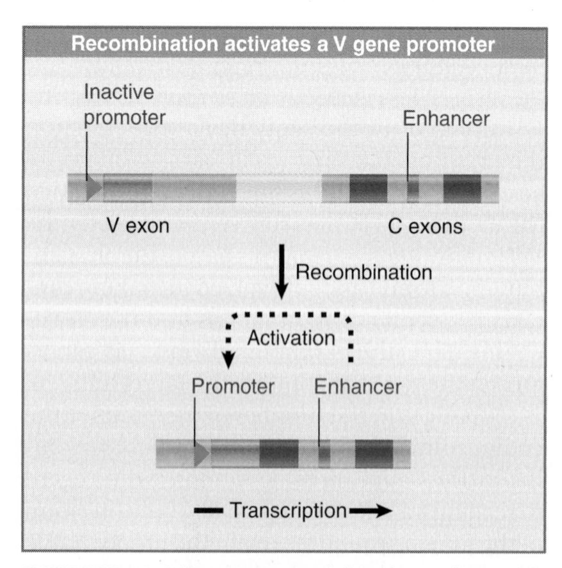

FIGURE 23.15 A V gene promoter is inactive until recombination brings it into the proximity of an enhancer in the C gene segment. The enhancer is active only in B lymphocytes.

What is the connection between joining of V and C gene segments and their activation? Unrearranged V gene segments are not actively represented in RNA. When a V gene segment is joined productively to a C_κ gene segment, however, the resulting unit is transcribed. The sequence upstream of a V gene segment is not altered by the joining reaction, though, and as a result *the promoter must be the same in unrearranged, nonproductively rearranged, and productively rearranged genes.*

A promoter lies upstream of every V gene segment but is inactive. It is activated by its relocation to the C region. The effect must depend on sequences downstream. What role might they play? An enhancer located within or downstream of the C gene segment activates the promoter at the V gene segment. The enhancer is tissue specific; it is active only in B cells. Its existence suggests the model illustrated in FIGURE 23.15, in which the V gene segment promoter is activated when it is brought within the range of the enhancer.

23.11 Early Heavy Chain Expression Can Be Changed by RNA Processing

Key concepts

- All lymphocytes start by synthesizing the membrane-bound form of IgM.
- A change in RNA splicing causes this to be replaced by the secreted form when the B cell differentiates.

The period of IgM synthesis that begins lymphocyte development falls into two parts, during which different versions of the μ constant region are synthesized:

- As a stem cell differentiates to a pre-B lymphocyte, an accompanying light chain is synthesized, and the IgM molecule ($L_2\mu_2$) appears at the surface of the cell. This form of IgM contains the μ_m version of the constant region (*m* indicates that IgM is located in the membrane). The membrane location may be related to the need to initiate cell proliferation in response to the initial recognition of an antigen.

- When the B lymphocyte differentiates further into a plasma cell, the μ_s version of the constant region is expressed. The IgM actually is secreted as a pentamer IgM_5J, in which J is a joining polypeptide (no connection with the J region) that forms disulfide linkages with μ chains. Secretion of the protein is followed by the humoral response depicted in Figure 23.1.

The μ_m and μ_s versions of the μ heavy chain differ only at the C-terminal end. The μ_m chain ends in a hydrophobic sequence that probably secures it in the membrane. This sequence is replaced by a shorter hydrophilic sequence in μ_s; the substitution allows the μ heavy chain to pass through the membrane. The change of C-terminus is accomplished by an alternative splicing event, which is controlled by the 3′ end of the nuclear RNA, as illustrated in FIGURE 23.16.

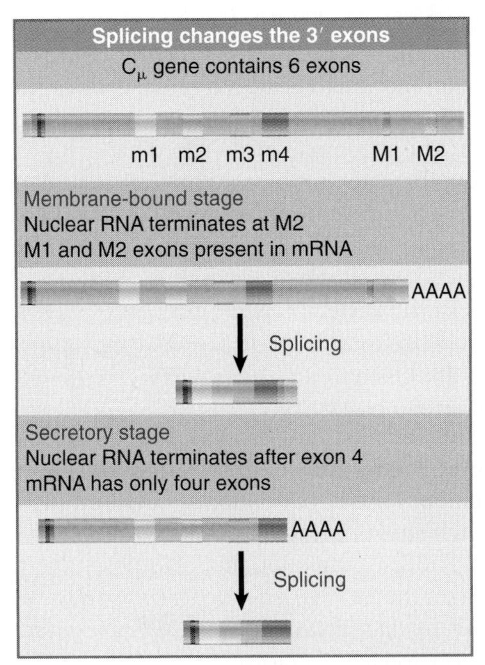

FIGURE 23.16 The 3′ end controls the use of splicing junctions so that alternative forms of the heavy gene are expressed.

At the membrane-bound stage, the RNA terminates after exon M2, and the constant region is produced by splicing together six exons. The first four exons code for the four domains of the constant region. The last two exons, M1 and M2, code for the 41-residue hydrophobic C-terminal region and its nontranslated trailer. The 5′ splice junction within exon 4 is connected to the 3′ splice junction at the beginning of M1.

At the secreted stage, the nuclear RNA terminates after exon 4. The 5′ splice junction within this exon that had been linked to M1 in the membrane form is ignored. This allows the exon to extend for an additional 20 codons.

A similar transition from membrane to secreted forms is found with other constant regions. The conservation of exon structures suggests that the mechanism is the same.

23.12 Class Switching Is Caused by DNA Recombination

Key concepts

- Immunoglobulins are divided into five classes according to the type of constant region in the heavy chain.
- Class switching to change the C_H region occurs by a recombination between S regions that deletes the region between the old C_H region and the new C_H region.
- Multiple successive switch recombinations can occur.

The *class* of immunoglobulin is defined by the type of C_H region. FIGURE 23.17 summarizes the five Ig classes. IgM (the first immunoglobulin to

There are five types of heavy chain

Type	IgM	IgD	IgG	IgA	IgE
Heavy chain	μ	δ	γ	α	ε
Structure	$(\mu_2L_2)_5J$	δ_2L_2	γ_2L_2	$(\alpha_2L_2)_2J$	ε_2L_2
Proportion	5%	1%	80%	14%	<1%
Effector function	Activates complement	Development of tolerance (?)	Activates complement	Found in secretions	Allergic response

FIGURE 23.17 Immunoglobulin type and function are determined by the heavy chain. J is a joining protein in IgM; all other Ig types exist as tetramers.

be produced by any B cell) and IgG (the most common immunoglobulin) possess the central ability to activate complement, which leads to destruction of invading cells. IgA is found in secretions (such as saliva), and IgE is associated with the allergic response and defense against parasites.

All lymphocytes start productive life as immature cells engaged in synthesis of IgM. Cells expressing IgM have the germline arrangement of the C_H gene segment cluster shown in Figure 23.10. The V-D-J joining reaction triggers expression of the C_μ gene segment. A lymphocyte generally produces only a single class of immunoglobulin at any one time, but the class may change during the cell lineage. A change in expression is called **class switching.** It is accomplished by a substitution in the type of C_H region that is expressed. Switching can be stimulated by environmental effects; for example, the growth factor TGFβ causes switching from C_μ to C_α.

Switching involves only the C_H gene segment; the same V_H gene segment continues to be expressed. Thus a given V_H gene segment may be expressed successively in combination with more than one C_H gene segment. The same light chain continues to be expressed throughout the lineage of the cell. Class switching therefore allows the type of effector response (mediated by the C_H region) to change while maintaining the same capacity to recognize antigen (mediated by the V regions).

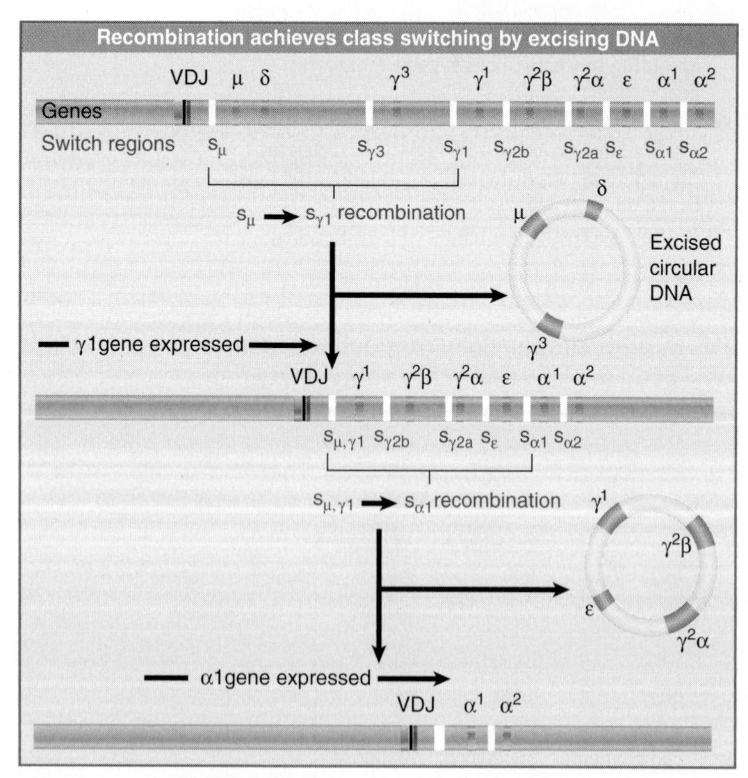

FIGURE 23.18 Class switching of heavy genes may occur by recombination between switch regions (S), deleting the material between the recombining S sites. Successive switches may occur.

Changes in the expression of C_H gene segments are made in two ways. The majority occur via further DNA recombination events, which involve a system different from that concerned with V-D-J joining (and able to operate only later during B cell development). Another type of change occurs at the level of RNA processing, but generally this is involved with changing the C-terminal sequence of the C_H region rather than its class (see Section 23.11, Early Heavy Chain Expression Can Be Changed by RNA Processing).

Cells expressing downstream C_H gene segments have deletions of C_μ and the other gene segments preceding the expressed C_H gene segment. Class switching is accomplished by a recombination to bring a new C_H gene segment into juxtaposition with the expressed V-D-J unit. The sequences of switched V-D-J-C_H units show that the sites of switching lie upstream of the C_H gene segments themselves. The switching sites are called **S regions.** FIGURE 23.18 depicts two successive switches.

In the first switch, expression of C_μ is succeeded by expression of $C_{\gamma 1}$. The $C_{\gamma 1}$ gene segment is brought into the expressed position by recombination between the sites S_μ and $S_{\gamma 1}$. The S_μ site lies between V-D-J and the C_μ gene segment. The $S_{\gamma 1}$ site lies upstream of the $C_{\gamma 1}$ gene segment. The DNA sequence between the two switch sites is excised as a circular molecule.

The linear deletion model imposes a restriction on the heavy-gene locus: *Once a class switch has been made, it becomes impossible to express any C_H gene segment that used to reside between C_μ and the new C_H gene segment.* In the example of Figure 23.18, cells expressing $C_{\gamma 1}$ should be unable to give rise to cells expressing $C_{\gamma 3}$, which has been deleted.

It should be possible, however, to undertake another switch to any C_H gene segment *downstream* of the expressed gene. Figure 23.18 shows a second switch to C_α expression, which is accomplished by recombination between $S_{\alpha 1}$ and the switch region $S_{\mu,\gamma 1}$ that was generated by the original switch.

We assume that all of the C_H gene segments have S regions upstream of the coding sequences. We do not know whether there are any restrictions on the use of S regions. Sequential switches do occur, but we do not know whether they are optional or an obligatory means to proceed to later C_H gene segments. We should like to know whether IgM can switch directly to *any* other class. The S regions lie within the introns that precede the C_H coding regions; as a result, switching does not alter the translational reading frame.

23.13 Switching Occurs by a Novel Recombination Reaction

Key concepts

- Switching occurs by a double-strand break followed by the nonhomologous end-joining reaction.
- The important feature of a switch region is the presence of inverted repeats.
- Switching requires activation of promoters that are upstream of the switch sites.

We know that switch sites are not uniquely defined, because different cells expressing the same C_H gene segment prove to have recombined at different points. Switch regions vary in length (as defined by the limits of the sites involved in recombination) from 1 to 10 kb. They contain groups of short inverted repeats, with repeating units that vary from 20 to 80 nucleotides in length. The primary sequence of the switch region does not seem to be important; what matters is the presence of the inverted repeats.

An S region typically is located ~2 kb upstream of a C_H gene segment. The switching reaction releases the excised material between the switch sites as a circular DNA molecule. Two of the proteins required for the joining phase of VDJ recombination (and also for the general nonhomologous end-joining pathway, NHEJ), Ku and DNA-PKcs, are required, suggesting that the joining reaction may use the NHEJ pathway. Basically, this implies that the reaction occurs by a double-strand break followed by rejoining of the cleaved ends.

We can put together the features of the reaction to propose a model for the generation of the double-strand break. The critical points are:

- transcription through the S region is required;
- the inverted repeats are crucial; and
- the break can occur at many different places within the S region.

FIGURE 23.19 shows the stages of the class-switching reaction. A promoter (I) lies immediately upstream of each switch region. Switching requires transcription from this promoter. The promoter may respond to activators that respond to environmental conditions, such as stimulation by cytokines, thus creating a mechanism to regulate switching. The first stage in switching is therefore to activate the I promoters that are upstream of each of the switch regions that will be involved. When these promoters are activated, they generate sterile transcripts that are spliced to join the I region with the corresponding heavy constant region.

The key insight into the mechanism of switching was the discovery of the requirement for the enzyme AID (activation-induced cytidine deaminase). In the absence of AID, class switching is blocked before the nicking stage. Somatic mutation is also blocked, showing an interesting connection between two important processes in immune diversification (see Section 23.15, Somatic Mutation Is Induced by Cytidine Deaminase and Uracil Glycosylase).

AID is expressed only at a specific stage during the differentiation of B lymphocytes, restricting the processes of class switching and somatic mutation to this stage. AID is a member of a class of enzymes that act on RNA to change a

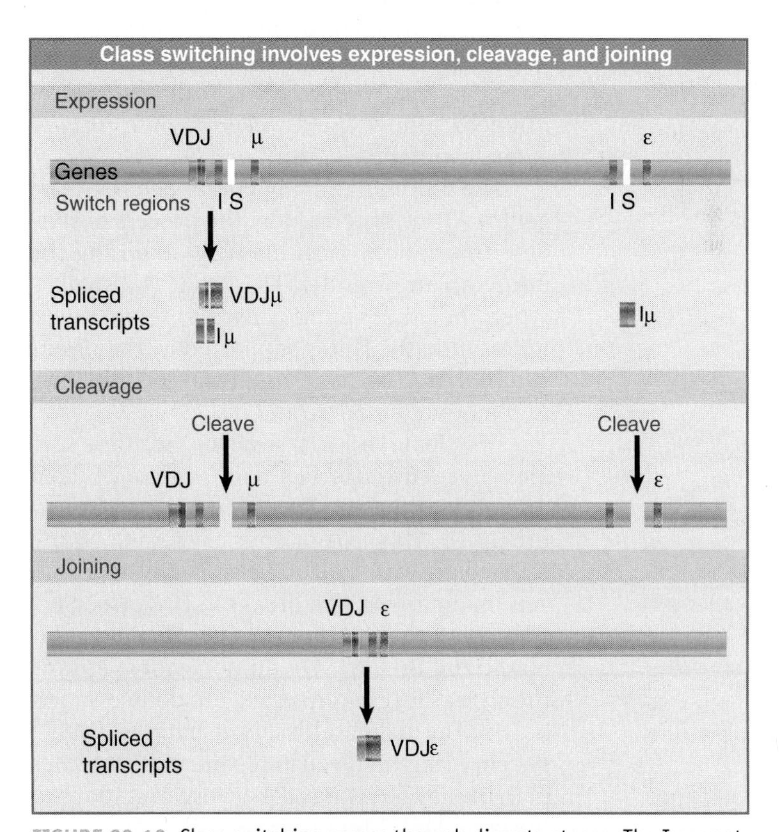

FIGURE 23.19 Class switching passes through discrete stages. The I promoters initiate transcription of sterile transcripts. The switch regions are cleaved. Joining occurs at the cleaved regions.

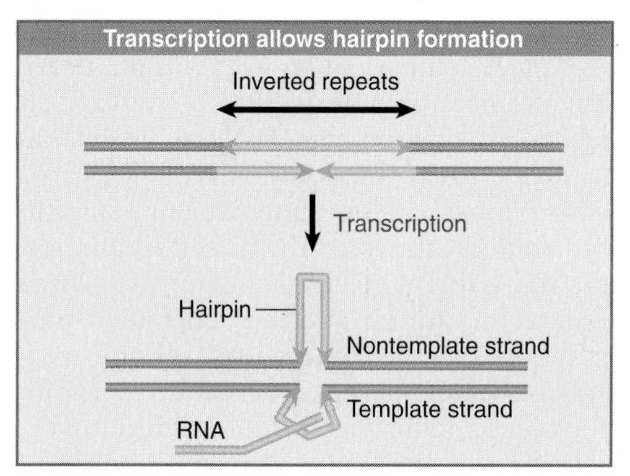

FIGURE 23.20 When transcription separates the strands of DNA, one strand may form an inverted repeat if the sequence is palindromic.

cytidine to a uridine (see Section 27.10, RNA Editing Occurs at Individual Bases). AID has a different specificity, however, and acts on single-stranded DNA.

Another enzyme is also required for both class switching and somatic mutation. This is UNG, a uracil DNA glycosylase, which removes the uracil that AID generates by deaminating cytidine. Mice that are deficient in UNG have a 10-fold reduction in class switching. This suggests a model in which the successive actions of AID and UNG create sites from which a base has been removed in DNA. Different consequences follow in the class switching and somatic mutation systems.

The source of the single-stranded DNA target for AID is generated by the process of sterile transcription, most likely by exposing the nontemplate strand of DNA that is displaced when the other strand is used as template for RNA synthesis. This is supported by the observation that AID preferentially targets cytidines in the nontemplate strand.

In order to cause class switching, these sites are converted into breaks in the nucleotide chain that provide the cleavage events shown in Figure 23.19. The broken ends are joined by the NHEJ pathway, which is a repair system that acts on double-strand breaks in DNA (see Section 20.12, A Common System Repairs Double-Strand Breaks). We do not know yet how the abasic sites are converted into double-strand breaks. The magnesium silicate hydrous (MSH) system that is involved in repairing mismatches in DNA may be required, because mutations in the gene MSH2 reduce class switching.

One unexplained feature is the involvement of inverted repeats. One possibility is that hairpins are formed by an interaction between the inverted repeats on the displaced nontemplate strand, as shown in FIGURE 23.20. In conjunction with the generation of abasic sites on this strand, this might lead to breakage.

Two critical questions that remain unanswered are how the system is targeted to the appropriate regions in the heavy chain locus, and what controls the use of switching sites.

23.14 Somatic Mutation Generates Additional Diversity in Mouse and Human Being

Key concepts

- Active immunoglobulin genes have V regions with sequences that are changed from the germline because of somatic mutation.
- The mutations occur as substitutions of individual bases.
- The sites of mutation are concentrated in the antigen-binding site.
- The process depends on the enhancer that activates transcription at the Ig locus.

Comparisons between the sequences of expressed immunoglobulin genes and the corresponding V gene segments of the germline show that new sequences appear in the expressed population. Some of this additional diversity results from sequence changes at the V-J or V-D-J junctions that occur during the recombination process. Other changes occur upstream, however, at locations within the variable domain.

Two types of mechanism can generate changes in V gene sequences after rearrangement has generated a functional immunoglobulin gene. In mouse and human, the mechanism is the induction of **somatic mutations** at individual locations within the gene specifically in the active lymphocyte. The process is sometimes called **hypermutation.** In chicken, rabbit, and pig, a different mechanism uses gene conversion to change a segment of the expressed V gene into the corresponding sequence from a different V gene (see Section 23.16, Avian Immunoglobulins Are Assembled from Pseudogenes).

A probe representing an expressed V gene segment can be used to identify all the corresponding fragments in the germline. Their sequences should identify the complete repertoire available to the organism. Any expressed gene whose sequence is different must have been generated by somatic changes.

One difficulty is to ensure that every potential contributor in the germline V gene segments actually has been identified. This problem is overcome by the simplicity of the mouse λ chain system. A survey of several myelomas producing λ_1 chains showed that many have the sequence of the single germline gene segment. *Others, however, have new sequences that must have been generated by mutation of the germline gene segment.*

To determine the frequency of somatic mutation in other cases, we need to examine a large number of cells in which the same V gene segment is expressed. A practical procedure for identifying such a group is to characterize the immunoglobulins of a series of cells, all of which express an immune response to a particular antigen.

Epitopes used for this purpose are small molecules—haptens—whose discrete structure is likely to provoke a consistent response, unlike a large protein, different parts of which provoke different antibodies. A hapten is conjugated with a nonreactive protein to form the antigen. The cells are obtained by immunizing mice with the antigen, obtaining the reactive lymphocytes, and sometimes fusing these lymphocytes with a myeloma (immortal tumor) cell to generate a **hybridoma** that continues to express the desired antibody indefinitely.

In one example, 10 out of 19 different cell lines producing antibodies directed against the hapten phosphorylcholine had the same V_H sequence. This sequence was the germline V gene segment T15, one of four related V_H genes. The other nine expressed gene segments differed from each other and from all four germline members of the family. They were more closely related to the T15 germline sequence than to any of the others, and their flanking sequences were the same as those around T15. This suggested that they arose from the T15 member by somatic mutation.

FIGURE 23.21 shows that sequence changes are localized around the V gene segment, extending in a region from ~150 bp downstream of the V gene promoter for ~1.5 kb. They take the form of substitutions of individual nucleotide pairs. Usually there are ~3 to ~15 substitutions, corresponding to <10 amino acid changes in the protein. They are concentrated in the antigen-binding site (thus generating the maximum diversity for recognizing new antigens). Only some of the mutations affect the amino acid

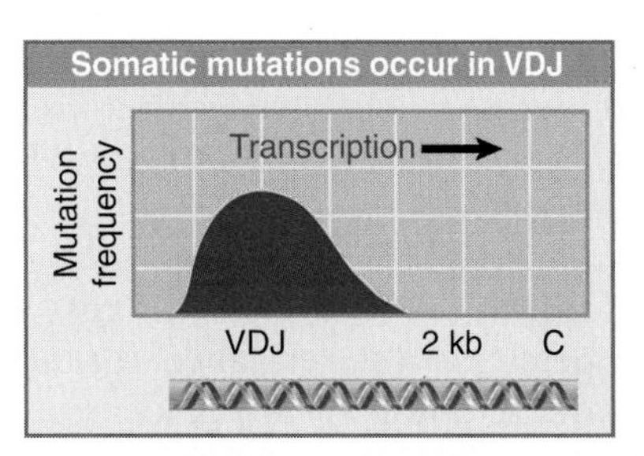

FIGURE 23.21 Somatic mutation occurs in the region surrounding the V segment and extends over the joined VDJ segments.

sequence, because others lie in third-base coding positions as well as in nontranslated regions.

The large proportion of ineffectual mutations suggests that somatic mutation occurs more or less at random in a region including the V gene segment and extending beyond it. There is a tendency for some mutations to recur on multiple occasions. These may represent hotspots as a result of some intrinsic preference in the system.

Somatic mutation occurs during clonal proliferation, apparently at a rate $\sim 10^{-3}$ per bp per cell generation. Approximately half of the progeny cells gain a mutation; as a result, cells expressing mutated antibodies become a high fraction of the clone.

In many cases, a single family of V gene segments is used consistently to respond to a particular antigen. Upon exposure to an antigen, presumably the V region with highest intrinsic affinity provides a starting point. Somatic mutation then increases the repertoire. Random mutations have unpredictable effects on protein function; some inactivate the protein, whereas others confer high specificity for a particular antigen. The proportion and effectiveness of the lymphocytes that respond is increased by selection among the lymphocyte population for those cells bearing antibodies in which mutation has increased the affinity for the antigen.

23.15 Somatic Mutation Is Induced by Cytidine Deaminase and Uracil Glycosylase

Key concepts

- A cytidine deaminase is required for somatic mutation as well as for class switching.
- Uracil-DNA glycosylase activity influences the pattern of somatic mutations.
- Hypermutation may be initiated by the sequential action of these enzymes.

Somatic mutation has many of the same requirements as class switching (see Section 23.13, Switching Occurs by a Novel Recombination Reaction:

- transcription must occur in the target region (as shown in this case by the demand for the enhancer that activates transcription at each Ig locus;
- it requires the enzymes AID and UNG and
- the MSH mismatch-repair system is involved.

The way in which the removal of the deaminated base leads to somatic mutation is suggested by the experiment summarized in FIGURE 23.22. When AID deaminates cytosine, it generates uracil, which then is removed from the DNA by UNG. Normally all possible substitutions occur at the abasic site. If the action of uracil-DNA glycosylase is blocked, though, we see a different result. If uracil is not removed from DNA, it should pair with adenine during replication. The ultimate result is to replace the original C-G pair with a T-A pair. Uracil-DNA glycosylase can be blocked by introducing into cells the gene coding for a protein that inhibits the enzyme. (The gene is a component of the bacteriophage PSB-2, whose genome is unusual in containing uracil, so that the enzyme needs to be blocked during a phage infection.) When the gene is introduced into a lymphocyte cell line, there is a dramatic change in the pattern of mutations, with almost all comprising the predicted transition from C-G to A-T.

The key event in generating a random spectrum of mutations is therefore to create the abasic site. The MSH repair system is then recruited to excise and replace the stretch of DNA containing the damage. The simplest possibility is that if the replacement is performed by an error-prone DNA polymerase, mutations may be introduced. Another possibility is that so many abasic sites are created that the repair systems are overwhelmed. When replication occurs, this could lead to the random insertion of bases opposite the abasic sites. We don't know yet what restricts the action of this system to the target region for hypermutation.

The difference in the systems is at the end of the process, when double-strand breaks are introduced in class switching, but individual point mutations are created during somatic

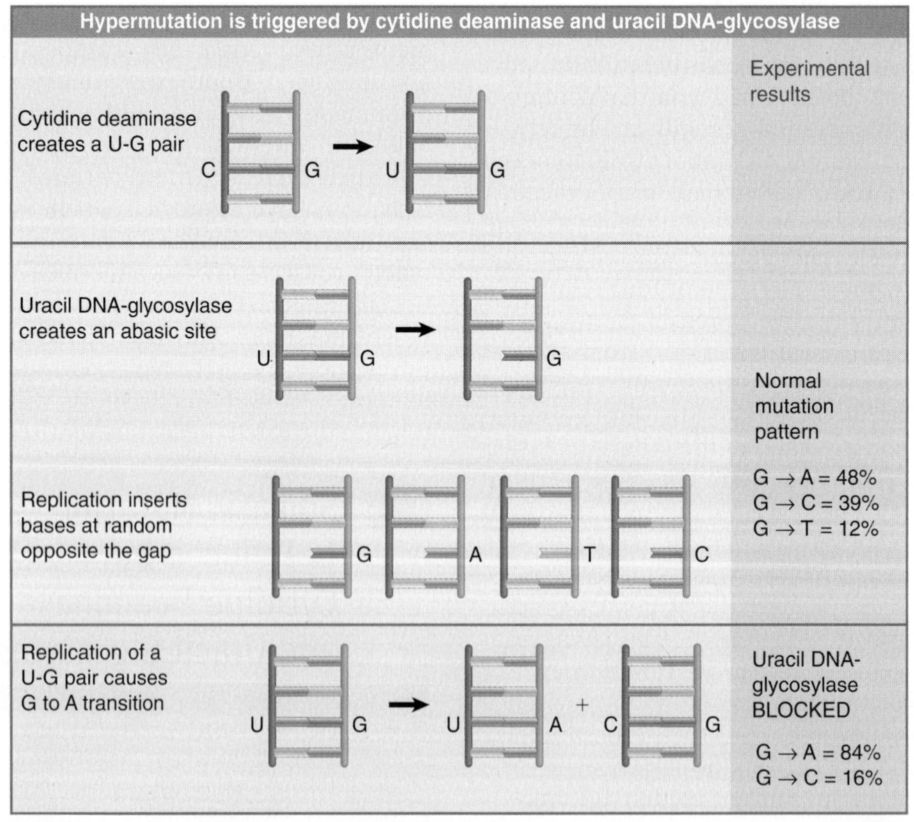

FIGURE 23.22 When the action of cytidine deaminase (top) is followed by that of uracil DNA-glycosylase, an abasic site is created. Replication past this site should insert all four bases at random into the daughter strand (center). If the uracil is not removed from the DNA, its replication causes a C-G to T-A transition.

mutation. We do not yet know exactly where the systems diverge. One possibility is that breaks are introduced at abasic sites in class switching, but the sites are erratically repaired in somatic mutation. Another possibility is that breaks are introduced in both cases, but are repaired in an error-prone manner in somatic mutation.

23.16 Avian Immunoglobulins Are Assembled from Pseudogenes

Key concept

- An immunoglobulin gene in chicken is generated by copying a sequence from one of 25 pseudogenes into the V gene at a single active locus.

The chick immune system is the paradigm for rabbits, cows, and pigs, which rely upon using the diversity that is coded in the genome. A similar mechanism is used by both the single light chain locus (of the λ type) and the H chain locus. The organization of the λ locus is drawn in FIGURE 23.23. It has only one functional V gene segment, J segment, and C gene segment. Upstream of the functional $V_{\lambda 1}$ gene segment lie 25 V_{λ} pseudogenes, organized in either orientation. They are classified as pseudogenes because either the coding segment is deleted at one or both ends, or proper signals for recombination are missing, or both. This assignment is confirmed by the fact that only the $V_{\lambda 1}$ gene segment recombines with the J-C_{λ} gene segment.

Sequences of active rearranged V_{λ}-J-C_{λ} gene segments show considerable diversity, though! A rearranged gene has one or more positions at which a cluster of changes has occurred in the sequence. A sequence identical to the new sequence can almost always be found in one of the pseudogenes (which themselves remain unchanged). The exceptional sequences that are not found in a pseudogene always represent changes at the junction between the original sequence and the altered sequence.

Thus a novel mechanism is employed to generate diversity. Sequences from the pseudogenes, between 10 and 120 bp in length, are substituted into the active $V_{\lambda 1}$ region by gene conversion. The unmodified $V_{\lambda 1}$ sequence is not expressed, even at early times during the immune response. A successful conversion event probably occurs every ten to twenty cell divisions to every rearranged $V_{\lambda 1}$ sequence. At the end of the immune maturation period, a rearranged $V_{\lambda 1}$ sequence has four to six converted segments spanning its entire length, which are derived from different donor pseudogenes. If all pseudogenes participate, this allows 2.5×10^8 possible combinations!

The enzymatic basis for copying pseudogene sequences into the expressed locus depends on enzymes involved in recombination and is related to the mechanism for somatic hypermutation that introduces diversity in mouse and man. Some of the genes involved in recombination are required for the gene conversion process; for example, it is prevented by deletion of *RAD54*. Deletion of other recombination genes (*XRCC2*, *XRCC3*, and *RAD51B*) has another, very interesting effect: Somatic mutation occurs at the V gene in the expressed locus. The frequency of the somatic mutation is ~10× greater than the usual rate of gene conversion.

These results show that the absence of somatic mutation in chick is not due to a deficiency in the enzymatic systems that are responsible in mouse and man. The most likely explanation for a connection between (lack of) recombination and somatic mutation is that unrepaired breaks at the locus trigger the induction of mutations. The reason why somatic mutation occurs in mouse and man but not in chick may therefore lie with the details of the operation of the repair system that operates on

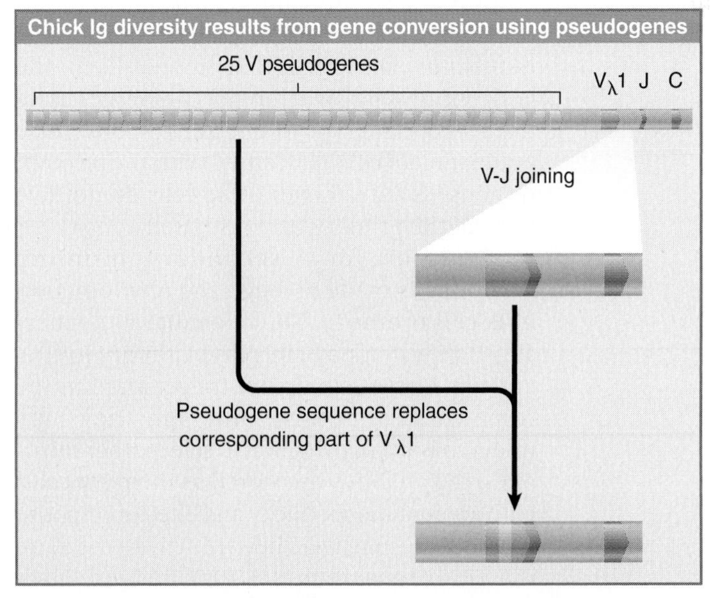

FIGURE 23.23 The chicken lambda light locus has 25 V pseudogenes upstream of the single functional V-J-C region. Sequences derived from the pseudogenes, however, are found in active rearranged V-J-C genes.

breaks at the locus. It is more efficient in chick, so that the gene is repaired by gene conversion before mutations can be induced.

23.17 B Cell Memory Allows a Rapid Secondary Response

Key concepts

- The primary response to an antigen is mounted by B cells that do not survive beyond the response period.
- Memory B cells are produced that have specificity for the same antigen, but that are inactive.
- A reexposure to antigen triggers the secondary response in which the memory cells are rapidly activated.

We are now in a position to summarize the relationship between the generation of high-affinity antibodies and the differentiation of the B cell. FIGURE 23.24 shows that B cells are derived from a self-renewing population of stem cells in the bone marrow. Maturation to give B cells depends upon Ig gene rearrangement, which requires the functions of the *SCID* and *RAG1,2* (and other) genes. If gene rearrangement is blocked, mature B cells are not produced. The antibodies carried by the B cells have specificities determined by the particular combinations of V(D)J regions and any additional nucleotides incorporated during the joining process.

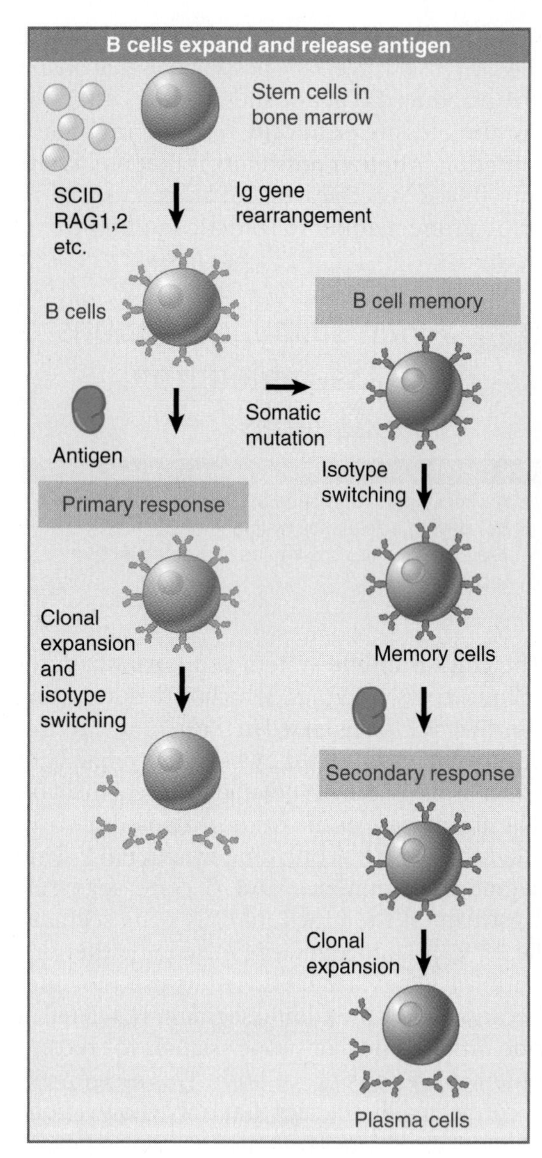

FIGURE 23.24 B cell differentiation is responsible for acquired immunity. Pre-B cells are converted to B cells by Ig gene rearrangement. Initial exposure to antigen provokes both the primary response and storage of memory cells. Subsequent exposure to antigen provokes the secondary response of the memory cells.

Exposure to antigen triggers two aspects of the immune response. The **primary immune response** occurs by clonal expansion of B cells responding to the antigen. This generates a large number of plasma cells that are specific for the antigen; isotype switching occurs to generate the appropriate type of effector response. The population of cells concerned with the primary response is a dead end; these cells do not live beyond the primary response itself.

Provision for a **secondary immune response** is made through the phenomenon of **B cell memory.** Somatic mutation generates B cells that have increased affinity for the antigen. These cells do not trigger an immune response at this time, although they may undergo isotype switching to select other forms of C_H region. They are stored as memory cells, with appropriate specificity and effector response type, but are inactive. They are activated if there is a new exposure to the same antigen. They are preselected for the antigen, and as a result they enable a secondary response to be mounted very rapidly, simply by clonal expansion; no further somatic mutation or isotype switching occurs during the secondary response.

The pathways summarized in Figure 23.24 show the development of acquired immunity, that is, the response to an antigen. In addition to these cells, there is a separate set of B cells, named the Ly-1 cells. These cells have gone through the process of V gene rearrangement, and apparently are selected for expression of a particular repertoire of antibody specificities. They do not undergo somatic mutation or the memory response. They may be involved in natural immunity, that is, an intrinsic ability to respond to certain antigens.

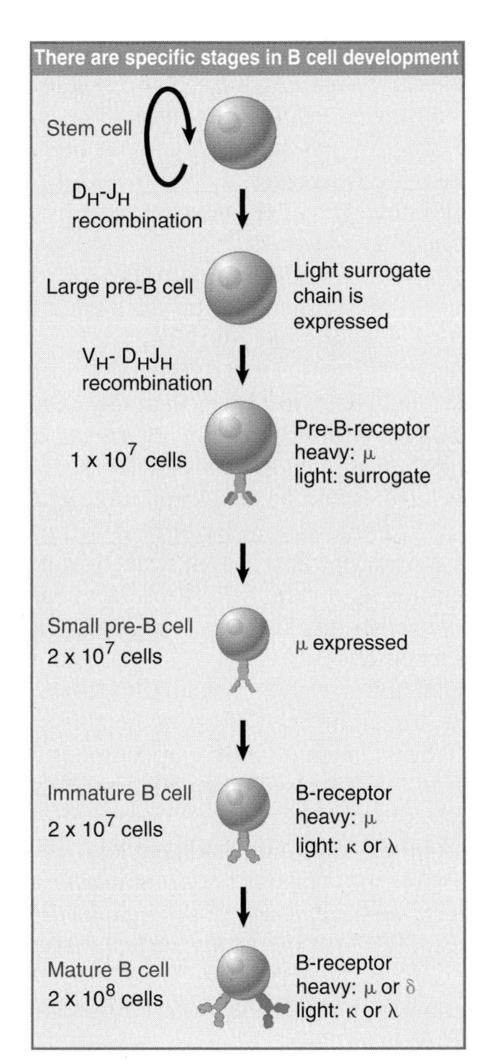

FIGURE 23.25 B cell development proceeds through sequential stages.

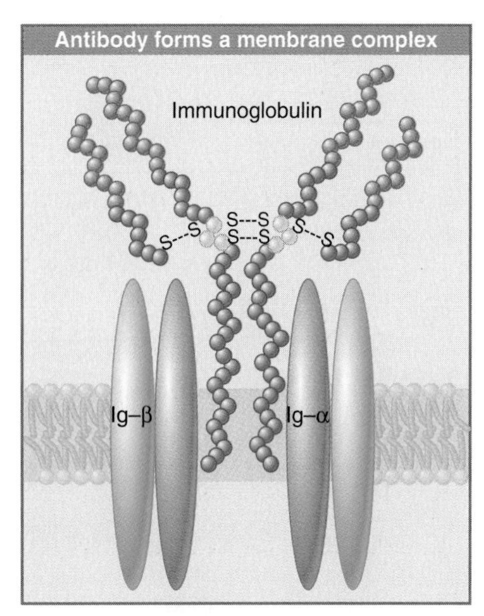

FIGURE 23.26 The B cell antigen receptor consists of an immunoglobulin tetramer (H_2L_2) linked to two copies of the signal-transducing heterodimer (Igαβ).

A more detailed view of B cell development is shown in FIGURE 23.25. The first step is recombination between the D segments and the J segments of the μ heavy chain. This is succeeded by V-D recombination, which generates a μ heavy chain. Several recombination events involving a succession of nonproductive and productive rearrangements may occur, as shown previously in Figure 23.13. These cells express a protein resembling a λ chain called the surrogate light (SL) chain, which is expressed on the surface and associates with the μ heavy chain to form the pre-B-receptor. It resembles an immunoglobulin complex, but does not function as one.

The production of μ chain represses synthesis of SL chain, and the cells divide to become small pre-B cells. Light chain is then expressed and functional immunoglobulin appears on the surface of the immature B cells. Further cell divisions occur, and the expression of δ heavy chain is added to that of μ chain as the cells mature into B cells.

Immunoglobulins function both by secretion from B cells and by surface expression. FIGURE 23.26 shows that the active complex on the cell surface is called the **B cell receptor (BCR),** and consists of an immunoglobulin associated with transmembrane proteins called Igα and Igβ. They provide the signaling components that trigger intracellular pathways in response to antigen-antibody binding. The activation of the BCR is also influenced by interactions with other receptors, for example, to mediate the interaction of antigen-activated B cells with helper T cells.

23.18 T Cell Receptors Are Related to Immunoglobulins

Key concepts

- T cells use a similar mechanism of V(D)J-C joining to B cells to produce either of two types of T cell receptor.
- TCR αβ is found on >95% of T lymphocytes; TCR γδ is found on <5%.

The lymphocyte lineage presents an example of evolutionary opportunism: A similar procedure is used in both B cells and T cells to generate proteins that have a variable region able

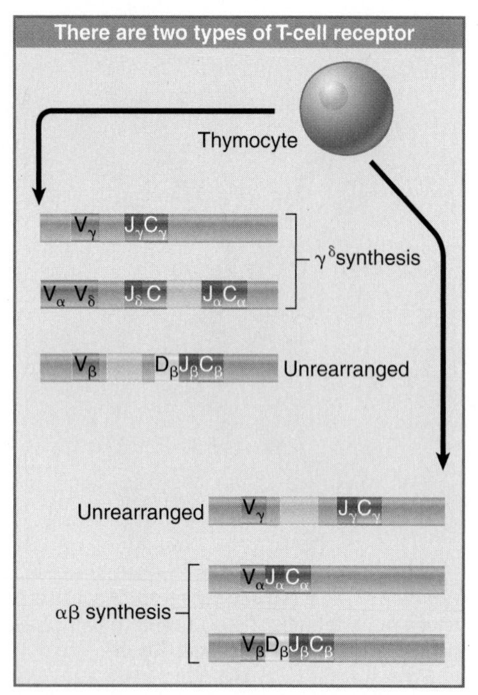

FIGURE 23.27 The γδ receptor is synthesized early in T cell development. TCR αβ is synthesized later and is responsible for "classical" cell-mediated immunity, in which target antigen and host histocompatibility antigen are recognized together.

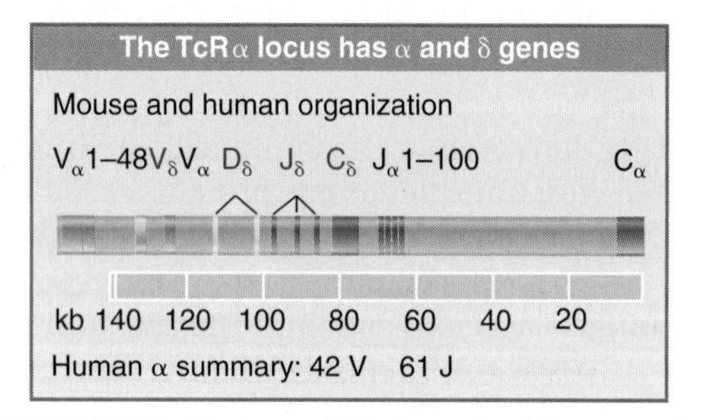

FIGURE 23.28 The human TCRα locus has interspersed α and δ segments. A V_δ segment is located within the V_α cluster. The D-J-C_δ segments lie between the V gene segments and the J-C_α segments. The mouse locus is similar, but has more V_δ segments.

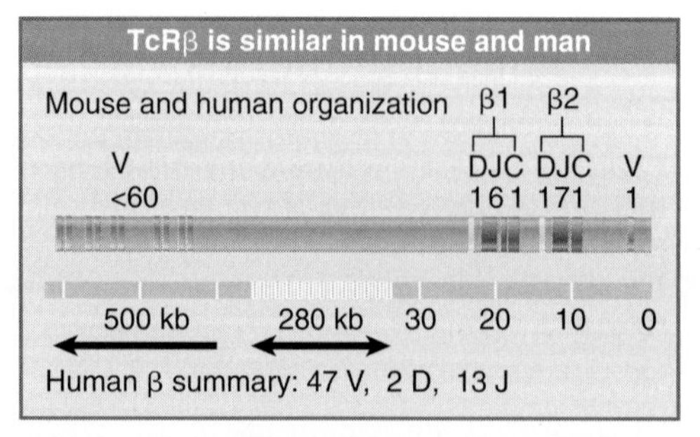

FIGURE 23.29 The TCRβ locus contains many V gene segments spread over ~500 kb that lie ~280 kb upstream of the two D-J-C clusters.

to provide significant diversity, whereas constant regions are more limited and account for a small range of effector functions. T cells produce either of two types of T cell receptor. The different T cell receptors are synthesized at different times during T cell development, as summarized in FIGURE 23.27.

The γδ receptor is found on <5% of T lymphocytes. It is synthesized only at an early stage of T cell development. In mice, it is the only receptor detectable at <15 days of gestation, but has virtually been lost by birth at day 20.

TCR αβ is found on >95% of lymphocytes. It is synthesized later in T cell development than γδ. In mice, it first becomes apparent at 15 to 17 days after gestation. By birth it is the predominant receptor. It is synthesized by a separate lineage of cells from those involved in TCR γδ synthesis, and involves independent rearrangement events.

Like immunoglobulins, a TCR must recognize a foreign antigen of unpredictable structure. The problem of antigen recognition by B cells and T cells is resolved in the same way, and the organization of the T cell receptor genes resembles the immunoglobulin genes in the use of variable and constant regions. *Each locus is organized in the same way as the immunoglobulin genes, with separate segments that are brought together by a recombination reaction specific to the lymphocyte.* The components are the same as those found in the three Ig families.

The organization of the TCR proteins resembles that of the immunoglobulins. The V sequences have the same general internal organization in both Ig and TCR proteins. The TCR C region is related to the constant Ig regions and has a single constant domain followed by transmembrane and cytoplasmic portions. Exon-intron structure is related to protein function.

The resemblance of the organization of TCR genes with the Ig genes is striking. As summarized in FIGURE 23.28, the organization of TCR α resembles that of Ig κ, with V gene segments separated from a cluster of J segments that precedes a single C gene segment. The organization of the locus is similar in both human and mouse, with some differences only in the number of V_α gene segments and J_α segments. (In addition to the α segments, this locus also contains δ segments, which we discuss shortly.)

The components of TCR β resemble those of IgH. FIGURE 23.29 shows that the organization is different, with V gene segments separated from two clusters each containing a

D segment, several J segments, and a C gene segment. Again, the only differences between human being and mouse are in the numbers of the V_β and J_β units.

Diversity is generated by the same mechanisms as in immunoglobulins. Intrinsic diversity results from the combination of a variety of V, D, J, and C segments; some additional diversity results from the introduction of new sequences at the junctions between these components (in the form of P and N nucleotides; see Figure 23.14). Some TCR β chains incorporate two D segments, which are generated by D-D joins (directed by an appropriate organization of the nonamer and heptamer sequences). A difference between TCR and Ig is that somatic mutation does not occur at the TCR loci. Measurements of the extent of diversity show that the 10^{12} T cells in the human being contain 2.5×10^7 different α chains associated with 10^6 different β chains.

The same mechanisms are likely to be involved in the reactions that recombine Ig genes in B cells and TCR genes in T cells. The recombining TCR segments are surrounded by nonamer and heptamer consensus sequences identical to those used by the Ig genes. This argues strongly that the same enzymes are involved. Most rearrangements probably occur by the deletion model (see Figure 23.12). We do not know how the process is controlled so that Ig loci are rearranged in B cells, whereas T cell receptors are rearranged in T cells.

The organization of the γ locus resembles that of Ig λ, with V gene segments separated from a series of J-C segments. FIGURE 23.30 shows that this locus has relatively little diversity, with ~8 functional V segments. The organization is different in human and mouse. Mouse has 3 functional J-C loci, but some segments are inverted in orientation. The human being has multiple J segments for each C gene segment.

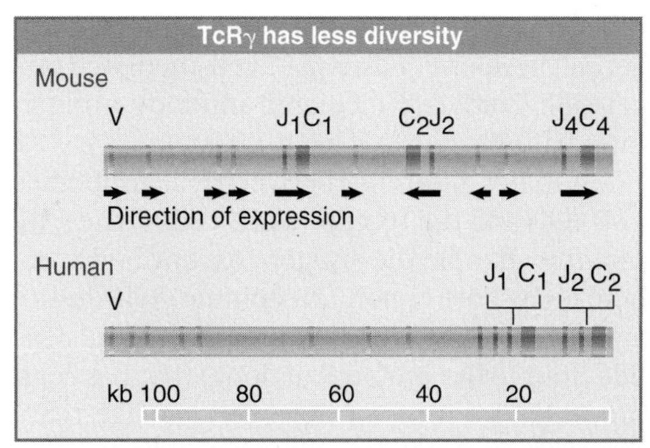

FIGURE 23.30 The TCRγ locus contains a small number of functional V gene segments (and also some pseudogenes not shown) that lie upstream of the J-C loci.

The δ subunit is coded by segments that lie at the TCR α locus, as illustrated previously in Figure 23.28. The segments D_δ-D_δ-J_δ-C_δ lie between the V gene segments and the J_α-C_α segments. Both of the D segments may be incorporated into the δ chain to give the structure VDDJ. The nature of the V gene segments used in the δ rearrangement is an interesting question. Very few V sequences are found in active TCR δ chains. In the human being, only one V gene segment is in general use for δ rearrangement. In mouse, several V_δ segments are found; some are unique for δ rearrangement, but some are also found in α rearrangements. The basis for specificity in choosing V segments in α and δ rearrangement is not known. One possibility is that many of the V_α gene segments can be joined to the DDJ_δ segment, but that only some (therefore defined as V_δ) can give active proteins.

For the present, we have labeled the V segments that are found in δ chains as V_δ gene segments; however, we must reserve judgment on whether they are really unique to δ rearrangement. The interspersed arrangement of genes implies that synthesis of the TCR αβ receptor and the γδ receptor is mutually exclusive at any one allele, because the δ locus is lost entirely when the V_α-J_α rearrangement occurs.

Rearrangements at the TCR loci, like those of immunoglobulin genes, may be productive or nonproductive. The β locus shows allelic exclusion in much the same way as immunoglobulin loci; rearrangement is suppressed once a productive allele has been generated. The α locus may be different; several cases of continued rearrangement suggest the possibility that substitution of V_α sequences may continue after a productive allele has been generated.

23.19 The T Cell Receptor Functions in Conjunction with the MHC

Key concept

- The TCR recognizes a short peptide that is bound in a groove of an MHC protein on the surface of the presenting cell.

T cells with αβ receptors are divided into several subtypes that have a variety of functions connected with interactions between cells involved in the immune response. Cytotoxic T cells possess the capacity to lyse an infected target cell. Helper T cells assist T cell-mediated target killing or B cell-mediated antibody-antigen interaction.

A major difference between the antibodies of B cells and the receptors of T cells is the way that they handle the antigen. An antibody recognizes a short region (an epitope) within the antigen. The T cell receptor binds to a small peptide (four to five amino acids long) that has been processed from the antigen by cleavage reactions. The peptide fragment is "presented" to the T cell by an MHC protein. Thus the T cell simultaneously recognizes the foreign antigen and an MHC protein carried by the presenting cell, as illustrated previously in Figure 23.2. Both helper T cells and killer T cells work in this way, but they have different requirements for the presentation of antigen; different types of MHC protein are used in each case (see Section 23.20, The Major Histocompatibility Locus Codes for Many Genes of the Immune System). Helper T cells require antigen to be presented by an MHC class II protein, whereas killer T cells require antigen to be presented by an MHC class I protein.

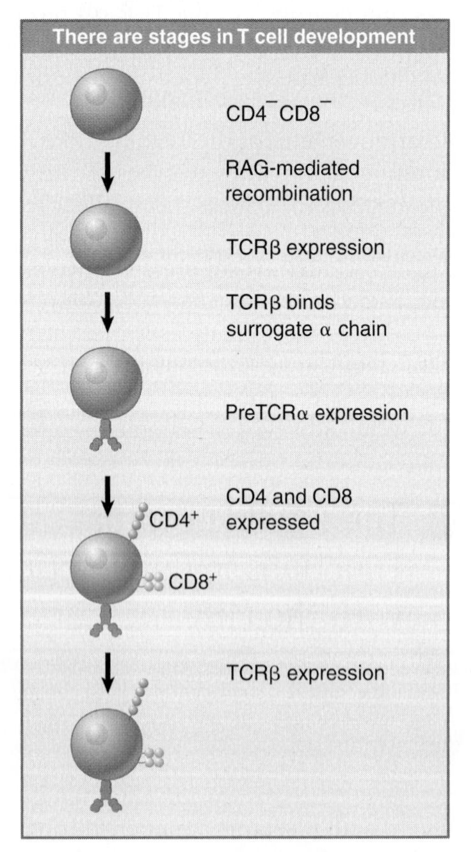

FIGURE 23.31 T cell development proceeds through sequential stages.

The TCR αβ receptor is responsible for helper T cell function in humoral immunity and for killer T cell function in cell-mediated immunity. This places upon it the responsibility of recognizing both the foreign antigen and the host MHC protein. The MHC protein binds a short peptide derived from the foreign antigen, and the TCR then recognizes the peptide in a groove on the surface of the MHC. The MHC is said to *present* the peptide to the TCR. (The peptide is generated when the proteasome degrades the foreign protein to generate fragments of eight to ten residues long.) A given TCR has specificity for a particular MHC as well as for the foreign antigen. The basis for this dual capacity is one of the most interesting issues to be defined about the αβ TCR.

Recombination to generate functional TCR chains is linked to the development of the T lymphocyte, as summarized in FIGURE 23.31. The first stage is rearrangement to form an active TCR β chain. This binds a nonrearranging surrogate α chain called preTCRα. At this stage, the lymphocyte has not expressed either of the surface proteins CD4 or CD8. The preTCR heterodimer then binds to the CD3 signaling complex (see below). Signaling from the complex triggers several rounds of cell division, during which α chains are rearranged, and the CD4 and CD8 genes are turned on, so that the lymphocyte is converted from $CD4^-CD8^-$ to $CD4^+CD8^+$. This point in development is called *β selection.* It generates **DP thymocytes.**

α chain rearrangement continues in the DP thymocytes. The maturation process continues by both positive selection (for mature TCR complexes able to bind a ligand) and by negative selection (against complexes that interact with inappropriate—self—ligands). Both types of selection involve interaction with MHC proteins. The DP thymocytes either die after ~3 days or become mature lymphocytes as the result of the selective processes. The surface TCR αβ heterodimer becomes crosslinked on the surface during positive selection (which rescues the thymocytes from cell death). If the thymocytes survive the subsequent negative selection they give rise to the separate T lymphocyte classes, which are $CD4^+CD8^-$ and $CD4^-CD8^+$.

The T cell receptor is associated with a complex of proteins called **CD3,** which is involved in transmitting a signal from the surface of the cell to the interior when its associated receptor is activated by binding antigen. Our present picture of the components of the receptor complex on a T cell is illustrated in FIGURE 23.32. The important point is that the interaction of the TCR variable regions with antigen causes the ζ subunits of the CD3 complex to activate the T cell response. The activation of CD3 provides the means by which either αβ or γδ TCR signals that it has recognized an antigen. This is comparable to the constitution of the B cell receptor, in which immunoglobulin associates with the Igαβ signaling chains (see Figure 23.26).

A central dilemma about T cell function remains to be resolved. Cell-mediated immunity requires two recognition processes. Recognition of the foreign antigen requires the ability to respond to novel structures. Recognition of the MHC protein is of course restricted to one of those coded by the genome, but even so there are many different MHC proteins. Thus considerable diversity is required in both recognition reactions. Helper and killer T cells rely upon different classes of MHC proteins; however, they use the same pool of α and β gene segments to assemble their receptors. Even allowing for the introduction of additional variation during the TCR recombination process, it is not clear how enough different versions of the T cell receptor are made available to accommodate all these demands.

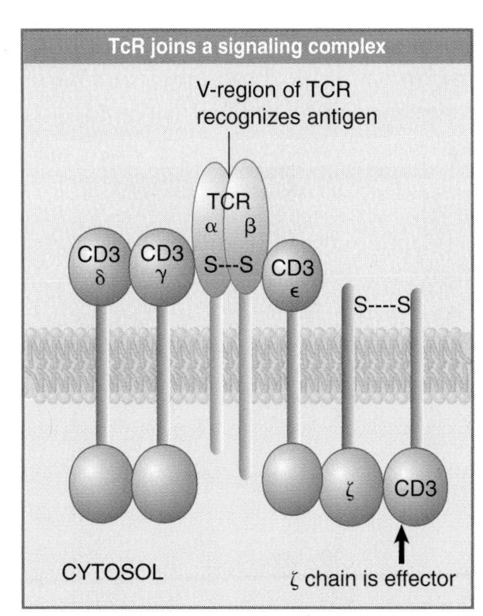

FIGURE 23.32 The two chains of the T cell receptor associate with the polypeptides of the CD3 complex. The variable regions of the TCR are exposed on the cell surface. The cytoplasmic domains of the ζ chains of CD3 provide the effector function.

23.20 The Major Histocompatibility Locus Codes for Many Genes of the Immune System

Key concepts

- The MHC locus codes for the class I and class II proteins as well as for other proteins of the immune system.
- Class I proteins are the transplantation antigens that are responsible for distinguishing "self" from "nonself" tissue.
- An MHC class I protein is active as a heterodimer with β_2 microglobulin.
- Class II proteins are involved in interactions between T cells.
- An MHC class II protein is a heterodimer of α and β chains.

The major histocompatibility locus occupies a small segment of a single chromosome in the mouse (where it is called the **H2 locus**) and in man (called the **HLA** locus). Within this segment are many genes coding for functions concerned with the immune response. At individual gene loci whose products have been identified, many alleles have been found in the population; the locus is described as highly *polymorphic,* meaning that individual genomes are likely to be different from one another. Genes coding for certain other functions also are located in this region.

Histocompatibility antigens are classified into three types by their immunological properties. In addition, other proteins found on lymphocytes and macrophages have a related structure and are important in the function of cells of the immune system:

MHC class I proteins are the **transplantation antigens.** They are present on every cell of the mammal. As their name suggests, these proteins are responsible for the rejection of foreign tissue, which is recognized as such by virtue of its particular array of transplantation antigens. In the immune system, their presence on target cells is required for the cell-mediated response. The types of class I proteins are defined serologically (by their antigenic properties). The murine class I genes code for the H2-K and H2-D/L proteins. Each mouse strain has one of several possible alleles for each of these functions. The human class I functions include the classical transplantation antigens, HLA-A, B, and C.

MHC class II proteins are found on the surfaces of both B lymphocytes and T lymphocytes, as well as on macrophages. These proteins are involved in communications between cells that are necessary to execute the immune response; in particular, they are required for helper T cell function. The murine class II functions are defined genetically as I-A and I-E. The human class II region (also called HLA-D) is arranged into four subregions, DR, DQ, DZ/DO, and DP.

The **complement** proteins are coded by a genetic locus that is also known as the S region; S stands for serum, indicating that the proteins are components of the serum. Their role is to interact with antibody-antigen complexes to cause the lysis of cells in the classical pathway of the humoral response.

The *Qa* and *Tla* loci proteins are found on murine hematopoietic cells. They are known as differentiation antigens, because each is found only on a particular subset of the blood cells, presumably related to their function. They are structurally related to the class I H2 proteins, and like them are polymorphic.

We can now relate the types of proteins to the organization of the genes that code for them. The MHC region was originally defined by genetics in the mouse, where the classical H2 region occupies 0.3 map units. Together with the adjacent region where mutations affecting immune function are also found, this corresponds to a region of ~2000 kb of DNA. The MHC region has been completely sequenced in several mammals, as well as in some birds and fish. By comparing these sequences, we find that the organization has been generally conserved.

The gene organization in mouse and man is summarized in FIGURE 23.33. The genomic regions where the class I and class II genes are located mark the original boundaries of the locus (going in the direction from telomere to centromere; right to left as shown in the figure). The genes in the region that separates the class I and class II genes code for a variety of functions; this is called the class III region. Defining the ends of the locus varies with the species, and the region beyond the class I genes on the telomeric side is called the extended class I region. Similarly, the region beyond the class II genes on the centromeric side is called the extended class II region. The major difference between mouse and human is that the extended class II region contains some class I (H2-K) genes in the mouse.

There are several hundred genes in the MHC regions of mammals, but it is possible for MHC functions to be provided by far fewer genes, as in the case of the chicken, where the MHC region is 92 Kb and has only nine genes.

As in comparisons of other gene families, we find differences in the exact numbers of genes devoted to each function. The MHC locus shows extensive variation between individuals, and as a result the number of genes may be different in different individuals. As a general rule, however, a mouse genome has fewer active H2 genes than a human genome. The class II genes are unique to mammals (except for one subgroup), and as a rule, birds and fish have different genes in their place. There are ~8 functional

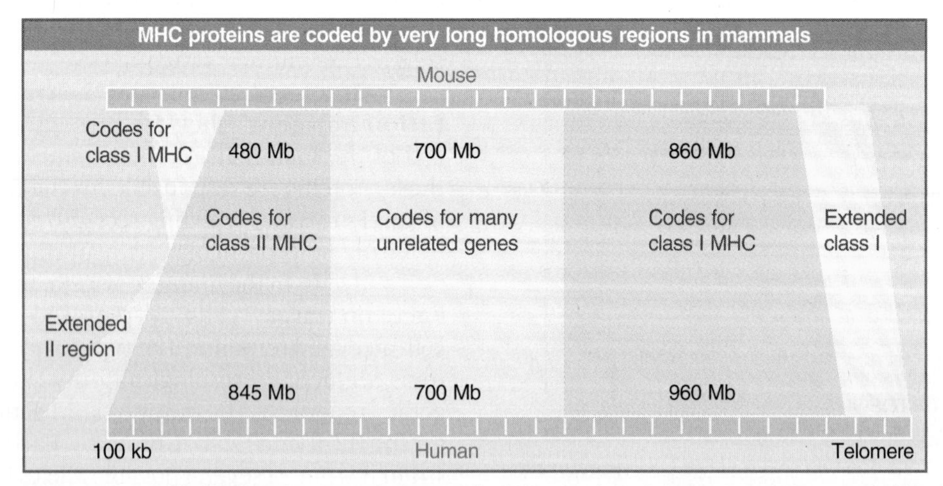

FIGURE 23.33 The MHC region extends for >2 Mb. MHC proteins of classes I and II are coded by two separate regions. The class III region is defined as the segment between them. The extended regions describe segments that are syntenic on either end of the cluster. The major difference between mouse and human is the presence of H2 class I genes in the extended region on the left. The murine locus is located on chromosome 17, and the human locus is located on chromosome 6.

class I genes in the human being and ~30 in the mouse. The class I region also includes many other genes. The class III regions are very similar in human and mouse.

All MHC proteins are dimers located in the plasma membrane, with a major part of the protein protruding on the extracellular side. The structures of class I and class II MHC proteins are related, although they have different components, as summarized in FIGURE 23.34.

Class II antigens consist of two chains, α and β, whose combination generates an overall structure in which there are two extracellular domains.

All class I MHC proteins consist of a dimer between the class I chain itself and the β2-microglobulin protein. The class I chain is a 45 kD transmembrane component that has three **external domains** (each ~90 amino acids long, one of which interacts with β2 microglobulin), a **transmembrane region** of ~40 residues, and a short **cytoplasmic domain** of ~30 residues that resides within the cell.

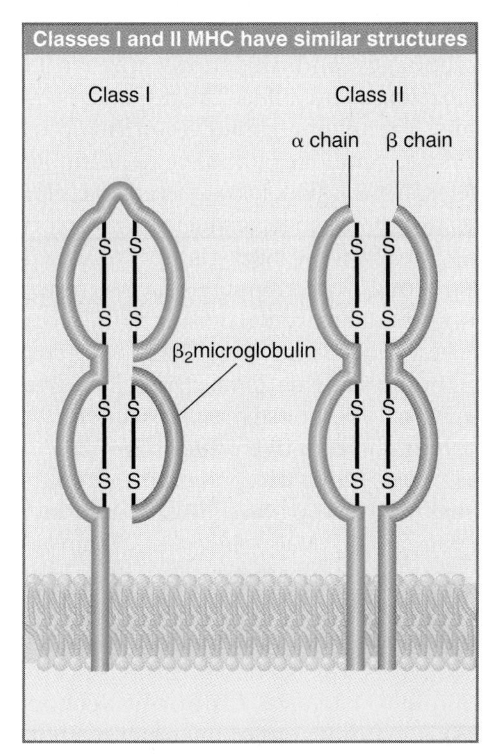

FIGURE 23.34 Class I and class II histocompatibility antigens have a related structure. Class I antigens consist of a single (α) polypeptide with three external domains (α1, α2, and α3) that interacts with β2 microglobulin (β2 m). Class II antigens consist of two (α and β) polypeptides, each with two domains (α1 and α2, and β1 and β2) with a similar overall structure.

The β2 microglobulin is a secreted protein of 12 kD. It is needed for the class I chain to be transported to the cell surface. Mice that lack the β2 microglobulin gene have no MHC class I antigen on the cell surface.

The organization of class I genes summarized in FIGURE 23.35 coincides with the protein structure. The first exon codes for a signal sequence (cleaved from the protein during membrane passage). The next three exons code for each of the external domains. The fifth exon codes for the transmembrane domain. The last three rather small exons together code for the cytoplasmic domain. The only difference in the genes for human transplantation antigens is that their cytoplasmic domain is coded by only two exons.

The exon coding for the third external domain of the class I genes is highly conserved relative to the other exons. The conserved domain probably represents the region that interacts with β2 microglobulin, which explains the need for constancy of structure. This domain also exhibits homologies with the constant region domains of immunoglobulins.

What is responsible for generating the high degree of polymorphism in these genes? Most of the sequence variation between alleles occurs in the first and second external domains, sometimes taking the form of a cluster of base substitutions in a small region. One mechanism involved in their generation is gene conversion between class I genes. Pseudogenes are present, as are functional genes.

The gene for β2 microglobulin is located on a separate chromosome. It has four exons, the first coding for a signal sequence, the second for the bulk of the protein (from amino acids

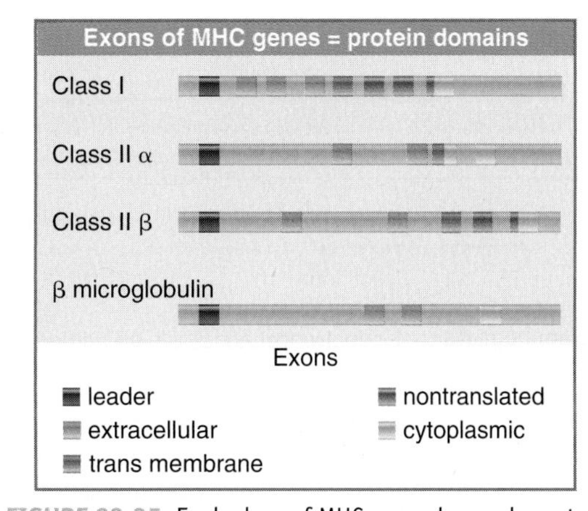

FIGURE 23.35 Each class of MHC genes has a characteristic organization, in which exons represent individual protein domains.

3 to 95), the third for the last four amino acids and some of the nontranslated trailer, and the last for the rest of the trailer.

The length of β2 microglobulin is similar to that of an immunoglobulin V gene; there are certain similarities in amino acid constitution, and there are some (limited) homologies of nucleotide sequence between β2 microglobulin and Ig constant domains or type I gene third external domains. All the groups of genes that we have discussed in this chapter may have descended from a common ancestor that coded for a primitive domain.

23.21 Innate Immunity Utilizes Conserved Signaling Pathways

Key concepts

- Innate immunity is triggered by receptors that recognize motifs (PAMPs) that are highly conserved in bacteria or other infective agents.
- Toll-like receptors are commonly used to activate the response pathway.
- The pathways are highly conserved from invertebrates to vertebrates, and an analogous pathway is found in plants.

The immune response described in this chapter comprises a set of reactions that respond to a pathogen by selecting lymphocytes (B cells or T cells) whose receptors (antibodies or TCR) have a high affinity for the pathogen. The basis for this selective process is the generation of a very large number of receptors, so as to create a high possibility of recognizing a foreign molecule. Receptors that recognize the body's own proteins are screened out early in the process. Activation of the receptors on B cells triggers the pathways of the humoral response; activation of the receptors on T cells triggers the pathways of the cell-mediated response. The overall response to an antigen via selection of receptors on the lymphocytes is called **adaptive immunity** (or **acquired immunity**). The response typically is mounted over several days, following the initial activation of B cells or T cells that recognize the foreign pathogen. The organism retains a memory of the response, which enables it to respond more rapidly if it is exposed again to the same pathogen. The principles of adaptive immunity are similar through the vertebrate kingdoms, although details vary.

Another sort of immune response occurs more quickly and is found in a greater range of animals (including those that do not have adaptive immunity). **Innate immunity** depends on the recognition of certain, predefined patterns in foreign pathogens. These patterns are motifs that are conserved in the pathogens because they have an essential role in their function, but they are not found in higher eukaryotes. The motif is typically recognized by a receptor dedicated to the purpose of triggering the innate response upon an infection. Receptors that trigger the innate response are found on cells such as neutrophils and macrophages, and cause the pathogen to be phagocytosed and killed. The response is rapid, because the set of receptors is already present on the cells and does not have to be amplified by selection. It is widely conserved and is found in organisms ranging from flies to human being. When the innate response is able to deal effectively with an infection, the adaptive response will not be triggered. There is some overlap between the responses in that they activate some of the same pathways, so cells activated by the innate response may subsequently participate in the adaptive response.

The motifs that trigger innate immunity are sometimes called pathogen-associated molecular patterns **(PAMPs).** FIGURE 23.36 shows that they are widely distributed across broad ranges of organisms. Formyl-methionine is used to initiate most bacterial proteins, but is not found in eukaryotes. The peptidoglycan of the cell wall is unique to bacteria. **Lipopolysaccharide (LPS)** is a component of the outer membrane of most gram-negative bacteria; also known as **endotoxin,** it is responsible for septic shock syndrome.

A key insight into the nature of innate immunity was the discovery of the involvement of **Toll-like receptors (TLRs).** The receptor

PAMPs are ubiquitous		
Organism	Pathogen	Location
All bacteria	Formyl-methionine	Most proteins
Most bacteria	Peptidoglycan	Cell wall
Gram-negative bacteria	Lipopolysaccharide	Cell wall
Yeast	Zymosan	Cell wall

FIGURE 23.36 Pathogen-associated molecular patterns (PAMPs) are compounds that are common to large ranges of bacteria or yeasts and are exposed when an infection occurs.

Toll, which is related to mammalian IL1 receptor, triggers the pathway in *Drosophila* that controls dorsal-ventral development. This leads to activation of the transcription factor dorsal, a member of the Rel family, which is related to the mammalian factor NF-κB. The pathway of innate immunity is parallel to the Toll pathway, with similar components. In fact, one of the first indications of the nature of innate immunity in flies was the discovery of the transcription factor Dif (dorsal-related immunity factor), which is activated by one of the pathways.

Flies have no system of adaptive immunity, but are resistant to microbial infections. This is because their innate immune systems trigger synthesis of potent antimicrobial peptides. Seven distinct peptides have been identified in *Drosophila,* where they are synthesized in the fat body (the equivalent organ to the liver). Two of the peptides are antifungal, and five act largely on bacteria. The general mode of action is to kill the target organism by permeabilizing its membrane. All of these peptides are coded by genes whose promoters respond to transcription factors of the Rel family. FIGURE 23.37 summarizes the components of the innate pathways in *Drosophila.*

Two innate response pathways function in *Drosophila;* one responds principally to fungi, whereas the other responds principally to gram-negative bacteria. Gram-positive bacteria may be able to trigger both pathways. FIGURE 23.38 outlines the steps in each pathway. Fungi and gram-positive bacteria activate a proteolytic cascade that generates peptides that activate a TLR. This is the NF-κB-like pathway. The dToll receptor activates the transcription factor Dif (a relative of NF-κB), leading ultimately to activation of the antifungal peptide drosomycin. Gram-negative bacteria trigger a pathway via a different receptor that activates the transcription factor Relish, leading to production of the bactericidal peptide attacin. This pathway is called the Imd pathway after one of its components, a protein that has a "death domain" related to those found in the pathways for apoptosis.

The key agents in responding to the bacteria are proteins called PGRPs because of their high affinities for bacterial peptidoglycans. There are two types of these proteins. PGRP-SAs are short extracellular proteins. They probably function by activating the proteases that trigger the Toll pathway. PGRP-LCs are transmembrane proteins with an extracellular PGRP domain. Their exact role has to be determined.

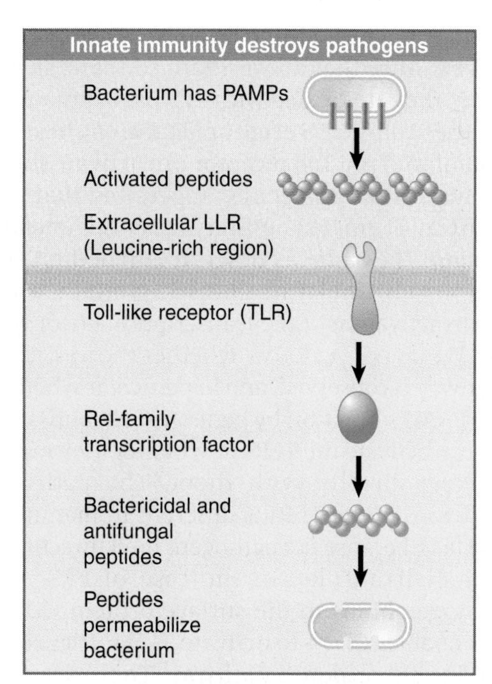

FIGURE 23.37 Innate immunity is triggered by PAMPs. In flies, they cause the production of peptides that activate Toll-like receptors. The receptors lead to a pathway that activates a transcription factor of the Rel family. Target genes for this factor include bactericidal and antifungal peptides. The peptides act by permeabilizing the membrane of the pathogenic organism.

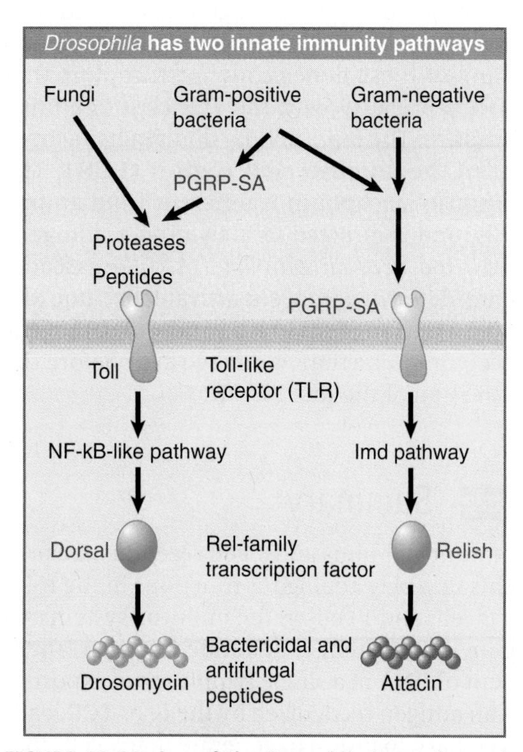

FIGURE 23.38 One of the *Drosophila* innate immunity pathways is closely related to the mammalian pathway for activating NF-κB; the other has components related to those of apoptosis pathways.

The innate immune response is highly conserved. Mice that are resistant to septic shock when they are treated with LPS have mutations in the Toll-like receptor TLR4. A human homolog of the Toll receptor can activate some immune-response genes, suggesting that the pathway of innate immunity may also function in man. The pathway downstream of the TLRs is generally similar in all cases, typically leading to activation of the transcription factor NF-κB. We do not yet know whether the upstream pathway is conserved, and in particular whether the PAMPs function by generating ligands that in turn activate the TLRs or whether they might interact directly with them. The pathway upstream of the TLRs is different in mammals and flies, because the pathogens directly activate mammalian TLRs. In the case of LPS, the pathogen binds to the surface protein CD14; this enables CD14 to activate TLR4, triggering the innate response pathway. There are ~20 receptors in the TLR class in the human genome, which gives some indication of how many pathogens can trigger the innate response.

Plants have extensive defense mechanisms, among which are pathways analogous to the innate response in animals. The same principle applies that PAMPs are the motifs that identify the infecting agent as a pathogen. The proteins that respond to the pathogens are coded by a class of genes called the disease resistance genes. Many of these genes code for receptors that share a property with the TLR class of animal receptors: The extracellular domain has a motif called the **leucine-rich region (LLR).** The response mechanism is different from animal cells, and is directed to activating a mitogen-activated protein kinoset (MAPK) cascade. Many different pathogens activate the same cascade, which suggests that a variety of pathogen-receptor interactions converge at or before the activation of the first MAPK.

23.22 Summary

Immunoglobulins and T cell receptors are proteins that play analogous functions in the roles of B cells and T cells in the immune system. An Ig or TCR protein is generated by rearrangement of DNA in a single lymphocyte; exposure to an antigen recognized by the Ig or TCR leads to clonal expansion to generate many cells that have the same specificity as the original cell. Many different rearrangements occur early in the development of the immune system, thereby creating a large repertoire of cells of different specificities.

Each immunoglobulin protein is a tetramer containing two identical light chains and two identical heavy chains. A TCR is a dimer containing two different chains. Each polypeptide chain is expressed from a gene created by linking one of many V segments via D segments and J segments to one of a few C segments. Ig L chains (either κ or λ) have the general structure V-J-C, Ig H chains have the structure V-D-J-C, TCR α and γ have components like Ig L chains, and TCR δ and β are like Ig H chains.

Each type of chain is coded by a large cluster of V genes separated from the cluster of D, J, and C segments. The numbers of each type of segment and their organization are different for each type of chain, but the principle and mechanism of recombination appear to be the same. The same nonamer and heptamer consensus sequences are involved in each recombination; the reaction always involves joining of a consensus with 23 bp spacing to a consensus with 12 bp spacing. The cleavage reaction is catalyzed by the RAG1 and RAG2 proteins, and the joining reaction is catalyzed by the same NHEJ pathway that repairs double-strand breaks in cells. The mechanism of action of the RAG proteins is related to the action of site-specific recombination catalyzed by resolvases.

Considerable diversity is generated by joining different V, D, and J segments to a C segment; however, additional variations are introduced in the form of changes at the junctions between segments during the recombination process. Changes are also induced in immunoglobulin genes by somatic mutation, which requires the actions of cytidine deaminase and uracil glycosylase. Mutations induced by cytidine deaminase probably lead to removal of uracil by uracil glycosylase, followed by the induction of mutations at the sites where bases are missing.

Allelic exclusion ensures that a given lymphocyte synthesizes only a single Ig or TCR. A productive rearrangement inhibits the occurrence of further rearrangements. The use of the V region is fixed by the first productive rearrangement, but B cells switch use of C_H genes from the initial μ chain to one of the H chains coded farther downstream. This process involves a different type of recombination in which the sequences between the VDJ region and the new C_H gene are deleted. More than one switch occurs in C_H gene usage. Class switching requires the same cytidine deaminase that is required for somatic mutation, but its role is not known. At an earlier stage of Ig production, switches occur from synthesis of a membrane-bound version of the protein to a secreted version. These switches are accomplished by alternative splicing of the transcript.

Innate immunity is a response triggered by receptors whose specificity is predefined for certain common motifs found in bacteria and other infective agents. The receptor that triggers the pathway is typically a member of the Toll-like class, and the pathway resembles the pathway triggered by Toll receptors during embryonic development. The pathway culminates in activation of transcription factors that cause genes to be expressed whose products inactivate the infective agent, typically by permeabilizing its membrane.

References

23.3 Immunoglobulin Genes Are Assembled from Their Parts in Lymphocytes

Reviews

Alt, F. W., Blackwell, T. K., and Yancopoulos, G. D. (1987). Development of the primary antibody repertoire. *Science* 238, 1079–1087.

Blackwell, T. K. and Alt, F. W. (1989). Mechanism and developmental program of immunoglobulin gene rearrangement in mammals. *Annu. Rev. Genet.* 23, 605–636.

Hood, L., Kronenberg, M., and Hunkapiller, T. (1985). T cell antigen receptors and the immunoglobulin supergene family. *Cell* 40, 225–229.

Tonegawa, S. (1983). Somatic generation of antibody diversity. *Nature* 302, 575–581.

Yancopoulos, G. D. and Alt, F. W. (1986). Regulation of the assembly and expression of variable-region genes. *Annu. Rev. Immunol.* 4, 339–368.

Research

Hozumi, N. and Tonegawa, S. (1976). Evidence for somatic rearrangement of immunoglobulin genes coding for variable and constant regions. *Proc. Natl. Acad. Sci. USA* 73, 3628–3632.

23.4 Light Chains Are Assembled by a Single Recombination

Research

Max, E. E., Seidman, J. G., and Leder, P. (1979). Sequences of five potential recombination sites encoded close to an immunoglobulin κ constant region gene. *Proc. Natl. Acad. Sci. USA* 76, 3450–3454.

23.7 Immune Recombination Uses Two Types of Consensus Sequence

Research

Lewis, S., Gifford, A., and Baltimore, D. (1985). DNA elements are asymmetrically joined during the site-specific recombination of kappa immunoglobulin genes. *Science* 228, 677–685.

23.9 Allelic Exclusion Is Triggered by Productive Rearrangement

Review

Storb, U. (1987). Transgenic mice with immunoglobulin genes. *Annu. Rev. Immunol.* 5, 151–174.

23.10 The RAG Proteins Catalyze Breakage and Reunion

Reviews

Gellert, M. (1992). Molecular analysis of VDJ recombination *Annu. Rev. Genet.* 26, 425–446.

Jeggo, P. A. (1998). DNA breakage and repair. *Adv. Genet.* 38, 185–218.

Schatz, D. G., Oettinger, M. A., and Schlissel, M. S. (1992). VDJ recombination: molecular biology and regulation. *Annu. Rev. Immunol.* 10, 359–383.

Research

Agrawal, A., Eastman, Q. M., and Schatz, D. G. (1998). Transposition mediated by RAG1 and RAG2 and its implications for the evolution of the immune system. *Nature* 394, 744–751.

Hiom, K., Melek, M., and Gellert, M. (1998). DNA transposition by the RAG1 and RAG2 proteins: a possible source of oncogenic translocations. *Cell* 94, 463–470.

Ma, Y., Pannicke, U., Schwarz, K., and Lieber, M. R. (2002). Hairpin opening and overhang processing by an Artemis/DNA-dependent protein kinase complex in nonhomologous end joining and V(D)J recombination. *Cell* 108, 781–794.

Melek, M. and Gellert, M. (2000). RAG1/2-mediated resolution of transposition intermediates: two pathways and possible consequences. *Cell* 101, 625–633.

Qiu, J. X., Kale, S. B., Yarnell Schultz, H., and Roth, D. B. (2001). Separation-of-function mutants reveal critical roles for RAG2 in both the cleavage and joining steps of V(D)J recombination. *Mol. Cell* 7, 77–87.

Roth, D. B., Menetski, J. P., Nakajima, P. B., Bosma, M. J., and Gellert, M. (1992). V D J recombination: broken DNA molecules with covalently sealed (hairpin) coding ends in SCID mouse thymocytes. *Cell* 70, 983–991.

Schatz, D. G. and Baltimore, D. (1988). Stable expression of immunoglobulin gene V(D)J recombinase activity by gene transfer into 3T3 fibroblasts. *Cell* 53, 107–115.

Schatz, D. G., Oettinger, M. A., and Baltimore, D. (1989). The V(D)J recombination activating gene, RAG-1. *Cell* 59, 1035–1048.

Tsai, C. L., Drejer, A. H., and Drejer, A. H. (2002). Evidence of a critical architectural function for the RAG proteins in end processing, protection, and joining in V(D)J recombination. *Genes Dev.* 16, 1934–1949.

Yarnell Schultz, H., Landree, M. A., Qiu, J. X., Kale, S. B., and Roth, D. B. (2001). Joining-deficient RAG1 mutants block V(D)J recombination *in vitro* and hairpin opening *in vitro.* *Mol. Cell* 7, 65–75.

23.13 Switching Occurs by a Novel Recombination Reaction

Reviews

Honjo, T., Kinoshita, K., and Muramatsu, M. (2002). Molecular mechanism of class switch recombination: linkage with somatic hypermutation. *Annu. Rev. Immunol.* 20, 165–196.

Li, Z., Woo, C. J., Iglesias-Ussel, M. D., Ronai, D., and Scharff, M. D. (2004). The generation of antibody diversity through somatic hypermutation and class switch recombination. *Genes Dev.* 18, 1–11.

Research

Bransteitter, R., Pham, P., Scharff, M. D., and Goodman, M. F. (2003). Activation-induced cytidine deaminase deaminates deoxycytidine on single-stranded DNA but requires the action of RNase. *Proc. Natl. Acad. Sci. USA* 100, 4102–4107.

Gu, H., Zou, Y. R., and Rajewsky, K. (1993). Independent control of immunoglobulin switch recombination at individual switch regions evidenced through Cre-loxP-mediated gene targeting. *Cell* 73, 1155–1164.

Iwasato, T., Shimizu, A., Honjo, T., and Yamagishi, H. (1990). Circular DNA is excised by immunoglobulin class switch recombination. *Cell* 62, 143–149.

Kinoshita, K., Tashiro, J., Tomita, S., Lee, C. G., and Honjo, T. (1998). Target specificity of immunoglobulin class switch recombination is not determined by nucleotide sequences of S regions. *Immunity* 9, 849–858.

Manis, J. P., Gu, Y., Lansford, R., Sonoda, E., Ferrini, R., Davidson, L., Rajewsky, K., and Alt, F. W. (1998). Ku70 is required for late B cell development and immunoglobulin heavy chain class switching. *J. Exp. Med.* 187, 2081–2089.

Matsuoka, M., Yoshida, K., Maeda, T., Usuda, S., and Sakano, H. (1990). Switch circular DNA formed in cytokine-treated mouse splenocytes: evidence for intramolecular DNA deletion in immunoglobulin class switching. *Cell* 62, 135–142.

Muramatsu, M., Kinoshita, K., Fagarasan, S., Yamada, S., Shinkai, Y., and Honjo, T. (2000). Class switch recombination and hypermutation require activation-induced cytidine deaminase (AID), a potential RNA editing enzyme. *Cell* 102, 553–563.

Pham, P., Bransteitter, R., Petruska, J., and Goodman, M. F. (2003). Processive AID-catalysed cytosine deamination on single-stranded DNA simulates somatic hypermutation. *Nature* 424, 103–107.

Revy, P., Muto, T., Levy, Y., Geissmann, F., Plebani, A., Sanal, O., Catalan, N., Forveille, M., Dufourcq-Labelouse, R., Gennery, A., Tezcan, I., Ersoy, F., Kayserili, H., Ugazio, A. G., Brousse, N., Muramatsu, M., Notarangelo, L. D., Kinoshita, K., and Honjo, T. (2000). Activation-induced cytidine deaminase (AID) deficiency causes the autosomal recessive form of the Hyper-IgM syndrome (HIGM2). *Cell* 102, 565–575.

Rolink, A., Melchers, F., and Andersson, J. (1996). The SCID but not the RAG-2 gene product is required for S mu-S epsilon heavy chain class switching. *Immunity* 5, 319–330.

von Schwedler, U., Jack, H. M., and Wabl, M. (1990). Circular DNA is a product of the immunoglobulin class switch rearrangement. *Nature* 345, 452–456.

Xu, L., Gorham, B., Li, S. C., Bottaro, A., Alt, F. W., and Rothman, P. (1993). Replacement of germ-line epsilon promoter by gene targeting alters control of immunoglobulin heavy chain class switching. *Proc. Natl. Acad. Sci. USA* 90, 3705–3709.

23.14 Somatic Mutation Generates Additional Diversity in Mouse and Human Being

Reviews

French, D. L., Laskov, R., and Scharff, M. D. (1989). The role of somatic hypermutation in the generation of antibody diversity. *Science* 244, 1152–1157.

Kocks, C. and Rajewsky, K. (1989). Stable expression and somatic hypermutation of antibody V regions in B-cell developmental pathways. *Annu. Rev. Immunol.* 7, 537–559.

Research

Kim, S., Davis, M., Sinn, E., Patten, P., and Hood, L. (1981). Antibody diversity: somatic hypermutation of rearranged VH genes. *Cell* 27, 573–581.

23.15 Somatic Mutation Is Induced by Cytidine Deaminase and Uracil Glycosylase

Reviews

Honjo, T., Kinoshita, K., and Muramatsu, M. (2002). Molecular mechanism of class switch

recombination: linkage with somatic hypermutation. *Annu. Rev. Immunol.* 20, 165–196.
Kinoshita, K. and Honjo, T. (2001). Linking class-switch recombination with somatic hypermutation. *Nat. Rev. Mol. Cell Biol.* 2, 493–503.
Li, Z., Woo, C. J., Iglesias-Ussel, M. D., Ronai, D., and Scharff, M. D. (2004). The generation of antibody diversity through somatic hypermutation and class switch recombination. *Genes Dev.* 18, 1–11.

Research

Di Noia, J. and Neuberger, J. (2002). Altering the pathway of immunoglobulin hypermutation by inhibiting uracil-DNA glycosylase. *Nature* 419, 43–48.
Muramatsu, M., Kinoshita, K., Fagarasan, S., Yamada, S., Shinkai, Y., and Honjo, T. (2000). Class switch recombination and hypermutation require activation-induced cytidine deaminase (AID), a potential RNA editing enzyme. *Cell* 102, 553–563.
Peters, A. and Storb, U. (1996). Somatic hypermutation of immunoglobulin genes is linked to transcription initiation. *Immunity* 4, 57–65.
Revy, P., Muto, T., Levy, Y., Geissmann, F., Plebani, A., Sanal, O., Catalan, N., Forveille, M., Dufourcq-Labelouse, R., Gennery, A., Tezcan, I., Ersoy, F., Kayserili, H., Ugazio, A. G., Brousse, N., Muramatsu, M., Notarangelo, L. D., Kinoshita, K., and Honjo, T. (2000). Activation-induced cytidine deaminase (AID) deficiency causes the autosomal recessive form of the Hyper-IgM syndrome (HIGM2). *Cell* 102, 565–575.

23.16 Avian Immunoglobulins Are Assembled from Pseudogenes

Research

Reynaud, C. A., Anquez, V., Grimal, H., and Weill, J. C. (1987). A hyperconversion mechanism generates the chicken light chain preimmune repertoire. *Cell* 48, 379–388.
Sale, J. E., Calandrini, D. M., Takata, M., Takeda, S., and Neuberger, M. S. (2001). Ablation of XRCC2/3 transforms immunoglobulin V gene conversion into somatic hypermutation. *Nature* 412, 921–926.

23.17 B Cell Memory Allows a Rapid Secondary Response

Review

Rajewsky, K. (1996). Clonal selection and learning in the antibody system. *Nature* 381, 751–758.

23.18 T Cell Receptors Are Related to Immunoglobulins

Reviews

Davis, M. M. (1990). T-cell receptor gene diversity and selection. *Annu. Rev. Biochem.* 59, 475–496.
Kronenberg, M., Siu, G., Hood, L. E., and Shastri, N. (1986). The molecular genetics of the T-cell antigen receptor and T-cell antigen recognition. *Annu. Rev. Immunol.* 4, 529–591.
Marrack, P. and Kappler, J. (1987). The T-cell receptor. *Science* 238, 1073–1079.
Raulet, D. H. (1989). The structure, function, and molecular genetics of the gamma/delta T-cell receptor. *Annu. Rev. Immunol.* 7, 175–207.

23.19 The T Cell Receptor Functions in Conjunction with the MHC

Review

Goldrath, A. W. and Bevan, M. J. (1999). Selecting and maintaining a diverse T cell repertoire. *Nature* 402, 255–262.

23.20 The Major Histocompatibility Locus Codes for Many Genes of the Immune System

Reviews

Flavell, R. A., Allen, H., Burkly, L. C., Sherman, D. H., Waneck, G. L., and Widera, G. (1986). Molecular biology of the H-2 histocompatibility complex. *Science* 233, 437–443.
Kumnovics, A., Takada, T., and Lindahl, K. F. (2003). Genomic organization of the mammalian MHC. *Annu. Rev. Immunol.* 21, 629–657.
Steinmetz, M. and Hood, L. (1983). Genes of the MHC complex in mouse and man. *Science* 222, 727–732.

Research

Kaufman, J. et al. (1999). The chicken B locus is a minimal essential major histocompatibility complex. *Nature* 401, 923–925.

23.21 Innate Immunity Utilizes Conserved Signaling Pathways

Reviews

Aderem, A. and Ulevitch, R. J. (2000). Toll-like receptors in the induction of the innate immune response. *Nature* 406, 782–787.
Dangl, J. L. and Jones, J. D. (2001). Plant pathogens and integrated defence responses to infection. *Nature* 411, 826–833.
Hoffmann, J. A., Kafatos, F. C., Janeway, C. A., and Ezekowitz, R. A. (1999). Phylogenetic perspectives in innate immunity. *Science* 284, 1313–1318.
Janeway, C. A. and Medzhitov, R. (2002). Innate immune recognition. *Annu. Rev. Immunol.* 20, 197–216.

Research

Asai, T., Tena, G., Plotnikova, J., Willmann, M. R., Chiu, W. L., Gomez-Gomez, L., Boller, T., Ausubel, F. M., and Sheen, J. (2002). MAP kinase signalling cascade in Arabidopsis innate immunity. *Nature* 415, 977–983.

Hoffmann, J. A. (2003). The immune response of *Drosophila*. *Nature* 426, 33–38.

Ip, Y. T., Reach, M., Engstrom, Y., Kadalayil, L., Cai, H., González-Crespo, S., Tatei, K., and Levine, M. (1993). Dif, a dorsal-related gene that mediates an immune response in *Drosophila*. *Cell* 75, 753–763.

Lemaitre, B., Nicolas, E., Michaut, L., Reichhart, J. M., and Hoffmann, J. A. (1996). The dorsoventral regulatory gene cassette spÃtzle/Toll/cactus controls the potent antifungal response in *Drosophila* adults. *Cell* 86, 973–983.

Medzhitov, R., Preston-Hurlburt, P., and Janeway, C. A. (1997). A human homologue of the *Drosophila* Toll protein signals activation of adaptive immunity. *Nature* 388, 394–397.

Poltorak, A., He, X., Smirnova, I., Liu, M. Y., Huffel, C. V., Du, X., Birdwell, D., Alejos, E., Silva, M., Galanos, C., Freudenberg, M., Ricciardi-Castagnoli, P., Layton, B., and Beutler, B. (1998). Defective LPS signaling in C3H/HeJ and C57BL/10ScCr mice: mutations in Tlr4 gene. *Science* 282, 2085–2088.

Rutschmann, S., Jung, A. C., Hetru, C., Reichhart, J. M., Hoffmann, J. A., and Ferrandon, D. (2000). The Rel protein DIF mediates the antifungal but not the antibacterial host defense in *Drosophila. Immunity* 12, 569–580.

Williams, M. J., Rodriguez, A., Kimbrell, D. A., and Eldon, E. D. (1997). The 18-wheeler mutation reveals complex antibacterial gene regulation in *Drosophila* host defense. *EMBO J.* 16, 6120–6130.

24

Promoters and Enhancers

CHAPTER OUTLINE

Continued on next page

24.1 Introduction

Initiation of transcription requires the enzyme RNA polymerase and transcription factors. Any protein that is needed for the initiation of transcription, but which is not itself part of RNA polymerase, is defined as a transcription factor. Many transcription factors act by recognizing *cis*-acting sites on DNA. Binding to DNA, however, is not the only means of action for a transcription factor. A factor may recognize another factor, or may recognize RNA polymerase, or may be incorporated into an initiation complex only in the presence of several other proteins. The ultimate test for membership of the transcription apparatus is functional: A protein must be needed for transcription to occur at a specific promoter or set of promoters.

A significant difference between the transcription of eukaryotic and prokaryotic mRNAs is that initiation at a eukaryotic promoter involves a large number of factors that bind to a variety of *cis*-acting elements. The promoter is defined as the region containing all these binding sites, that is, all the binding sites that can support transcription at the normal efficiency and with the proper control. Thus the major feature defining the promoter for a eukaryotic mRNA is the location of binding sites for transcription factors. RNA polymerase itself binds around the startpoint, but does not directly contact the extended upstream region of the promoter. By contrast, the bacterial promoters discussed in Chapter 11, Transcription, are largely defined in terms of the binding site for RNA polymerase in the immediate vicinity of the startpoint.

Transcription in eukaryotic cells is divided into three classes. Each class is transcribed by a different RNA polymerase:

- RNA polymerase I transcribes rRNA.
- RNA polymerase II transcribes mRNA.
- RNA polymerase III transcribes tRNA and other small RNAs.

Transcription factors are needed for initiation but are not required subsequently. For the three eukaryotic enzymes, the *factors,* rather than the RNA polymerases themselves, are principally responsible for recognizing the promoter. This is different from bacterial RNA polymerase, where it is the enzyme that recognizes the promoter sequences. For all eukaryotic RNA polymerases, the factors create a structure at the promoter to provide the target that is recognized by the enzyme. For RNA polymerases I and III, these factors are relatively simple, but for RNA polymerase II they form a sizeable group collectively known as the **basal,** or general, **factors.** The basal factors join with RNA polymerase II to form a complex surrounding the startpoint, and they determine the site of initiation. The basal factors together with RNA polymerase constitute the **basal transcription apparatus.**

The promoters for RNA polymerases I and II are (mostly) upstream of the startpoint, but some promoters for RNA polymerase III lie downstream of the startpoint. Each promoter contains characteristic sets of short conserved sequences that are recognized by the appropriate class of factors. RNA polymerases I and III each recognize a relatively restricted set of promoters, and rely upon a small number of accessory factors.

Promoters utilized by RNA polymerase II show more variation in sequence, and have a modular organization. Short sequence elements that are recognized by transcription factors lie upstream of the startpoint. These *cis*-acting sites usually are spread out over a region of >200 bp. Some of these elements and the factors that recognize them are common: They are found in a variety of promoters and are used constitutively. Others are specific: they identify particular classes of genes and their use is regulated. The elements occur in different combinations in individual promoters.

All RNA polymerase II promoters have sequence elements close to the startpoint that are bound by the basal apparatus and that establish the site of initiation. The sequences farther upstream determine whether the promoter is expressed in all cell types or is specifically regulated. Promoters that are constitutively expressed (their genes are sometimes called housekeeping genes) have upstream sequence elements that are recognized by ubiquitous activators. No element/factor combination is an essential component of the promoter, which suggests that initiation by RNA polymerase II may be sponsored in many different ways. Promoters that are expressed only in certain times or places have sequence elements that require activators that are available only at those times or places.

Sequence components of the promoter are defined operationally by the demand that they must be located in the general vicinity of the startpoint and are required for initiation. The **enhancer** is another type of site involved in initiation. It is identified by sequences that stimulate initiation, but that are located a considerable distance from the startpoint. Enhancer elements are often targets for tissue-specific or temporal regulation. FIGURE 24.1 illustrates the general properties of promoters and enhancers.

The components of an enhancer resemble those of the promoter, in that they consist of a variety of modular elements. The elements of an enhancer, however, are organized in a closely packed array and function like those in the promoter. The enhancer does not need to be near the startpoint.

Proteins bound at enhancer elements interact with proteins bound at promoter elements. The distinction between promoters and enhancers is operational, rather than implying a fundamental difference in mechanism. This view is strengthened by the fact that some types of element are found in both promoters and enhancers.

Eukaryotic transcription is most often under positive regulation: A transcription factor is provided under tissue-specific control to activate a promoter or set of promoters that contain a common target sequence. Regulation by specific repression of a target promoter is less common.

A eukaryotic transcription unit generally contains a single gene, and termination occurs beyond the end of the coding region. Termination lacks the regulatory importance that applies in prokaryotic systems. RNA polymerases I and III terminate at discrete sequences in defined reactions, but the mode of termination by RNA polymerase II is not clear. The significant event in generating the 3′ end of an mRNA, however, is not the

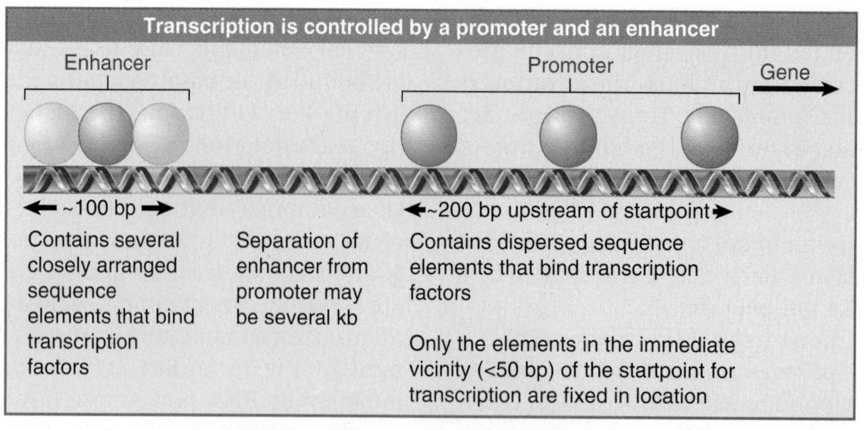

FIGURE 24.1 A typical gene transcribed by RNA polymerase II has a promoter that extends upstream from the site where transcription is initiated. The promoter contains several short (<10 bp) sequence elements that bind transcription factors, dispersed over >200 bp. An enhancer containing a more closely packed array of elements that also bind transcription factors may be located several kb distant. (DNA may be coiled or otherwise rearranged so that transcription factors at the promoter and at the enhancer interact to form a large protein complex.)

termination event itself, but instead results from a cleavage reaction in the primary transcript (see Chapter 26, RNA Splicing and Processing).

24.2 Eukaryotic RNA Polymerases Consist of Many Subunits

Key concepts

- RNA polymerase I synthesizes rRNA in the nucleolus.
- RNA polymerase II synthesizes mRNA in the nucleoplasm.
- RNA polymerase III synthesizes small RNAs in the nucleoplasm.
- All eukaryotic RNA polymerases have ~12 subunits and are aggregates of >500 kD.
- Some subunits are common to all three RNA polymerases.
- The largest subunit in RNA polymerase II has a CTD (carboxy-terminal domain) consisting of multiple repeats of a heptamer.

The three eukaryotic RNA polymerases have different locations in the nucleus that correspond with the genes that they transcribe.

The most prominent activity is the enzyme RNA polymerase I, which resides in the nucleolus and is responsible for transcribing the genes coding for rRNA. It accounts for most cellular RNA synthesis (in terms of quantity).

The other major enzyme is RNA polymerase II, which is located in the nucleoplasm (the part of the nucleus excluding the nucleolus). It represents most of the remaining cellular activity and is responsible for synthesizing heterogeneous nuclear RNA (hnRNA), the precursor for mRNA.

RNA polymerase III is a minor enzyme activity. This nucleoplasmic enzyme synthesizes tRNAs and other small RNAs.

All eukaryotic RNA polymerases are large proteins, appearing as aggregates of >500 kD. They typically have ~12 subunits. The purified enzyme can undertake template-dependent transcription of RNA, but is not able to initiate selectively at promoters. The general constitution of a eukaryotic RNA polymerase II enzyme as typified in *Sacchasomyces cerevisiae* is illustrated in FIGURE 24.2. The two largest subunits are homologous to the β and β′ subunits of bacterial RNA polymerase. Three of the remaining subunits are common to all the RNA polymerases, that is, they are also components of RNA polymerases I and III.

The largest subunit in RNA polymerase II has a **carboxy-terminal domain (CTD),** which consists of multiple repeats of a consensus sequence of seven amino acids. The sequence is unique to RNA polymerase II. There are ~26 repeats in yeast and ~50 in mammals. The number of repeats is important because deletions that remove (typically) more than half of the repeats are lethal (in yeast). The CTD can be highly phosphorylated on serine or threonine residues; this is involved in the initiation reaction (see Section 24.11, Initiation Is Followed by Promoter Clearance).

The RNA polymerases of mitochondria and chloroplasts are smaller, and they resemble bacterial RNA polymerase rather than any of the nuclear enzymes. Of course, the organelle

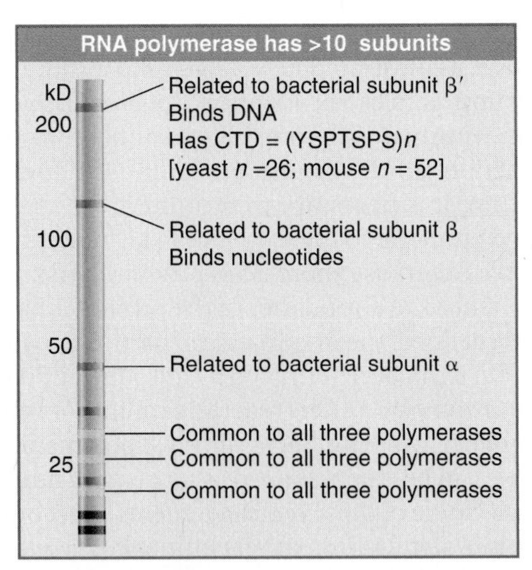

FIGURE 24.2 Some subunits are common to all classes of eukaryotic RNA polymerases and some are related to bacterial RNA polymerase.

genomes are much smaller, the resident polymerase needs to transcribe relatively few genes, and the control of transcription is likely to be very much simpler (if it exists at all). Thus these enzymes are analogous to the phage enzymes that do not need the ability to respond to a more complex environment.

A major practical distinction between the eukaryotic enzymes is drawn from their response to the bicyclic octapeptide α amanitin. In basically all eukaryotic cells, the activity of RNA polymerase II is rapidly inhibited by low concentrations of α amanitin. RNA polymerase I is not inhibited. The response of RNA polymerase III to α amanitin is less well conserved; in animal cells it is inhibited by high levels, but in yeast and insects it is not inhibited.

24.3 Promoter Elements Are Defined by Mutations and Footprinting

Key concept

- Promoters are defined by their ability to cause transcription of an attached sequence in an appropriate test system *in vitro* or *in vivo*.

The first step in characterizing a promoter is to define the overall length of DNA that contains all the necessary sequence elements. To do this, we need a test system in which the promoter is responsible for the production of an easily assayed product. Historically, several types of systems have been used:

- In the *oocyte system,* a DNA template is injected into the nucleus of the *Xenopus laevis* oocyte. The RNA transcript can be recovered and analyzed. The main limitation of this system is that it is restricted to the conditions that prevail in the oocyte. It allows characterization of DNA sequences, but not of the factors that normally bind them.
- *Transfection systems* allow exogenous DNA to be introduced into a cultured cell and expressed. The system is genuinely *in vivo* in the sense that transcription is accomplished by the same apparatus responsible for expressing the cell's own genome. It differs from the natural situation, though, because the template consists of a gene that would not usually be transcribed in the host cell. The usefulness of the system may be extended by using a variety of host cells.
- *Transgenic systems* involve the addition of a gene to the germline of an animal. Expression of the **transgene** can be followed in any or all of the tissues of the animal. Some common limitations apply to transgenic systems and to transfection: The additional gene often is present in multiple copies, and is integrated at a different location from the endogenous gene. Discrepancies between the expression of a gene *in vitro* and its expression as a transgene can yield important information about the role of the genomic context of the gene.
- The *in vitro* system takes the classic approach of purifying all the components and manipulating conditions until faithful initiation is seen. "Faithful" initiation is defined as production of an RNA starting at the site corresponding to the 5′ end of mRNA (or rRNA or tRNA precursors). Ultimately this allows us to characterize the individual sequence elements in the promoter and the transcription factors that bind to them.

When a promoter is analyzed, it is important that *only* the promoter sequence changes. FIGURE 24.3 shows that the same long upstream sequence is always placed next to the promoter to ensure that it is always in the same context. Termination does not occur properly in the *in vitro* systems, and as a result the template is cut at some distance from the promoter (usually ~500 bp downstream). This ensures that all polymerases "run off" at the same point, thus generating an identifiable transcript.

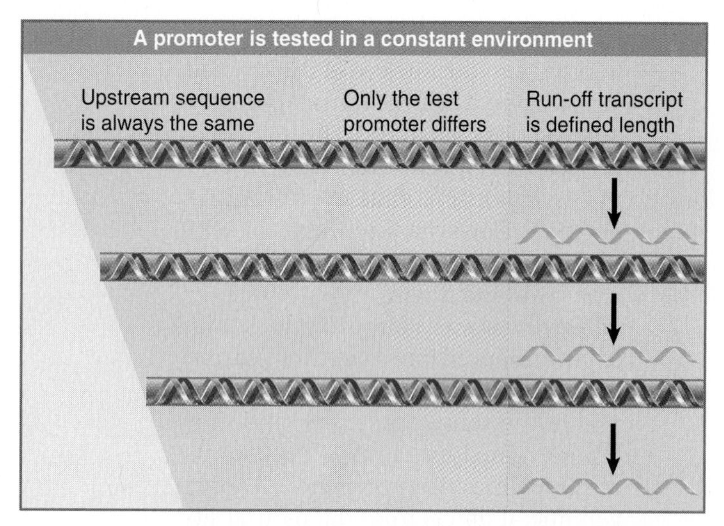

FIGURE 24.3 A promoter is tested by modifying the sequence that is connected to a constant upstream sequence and a constant downstream transcription unit.

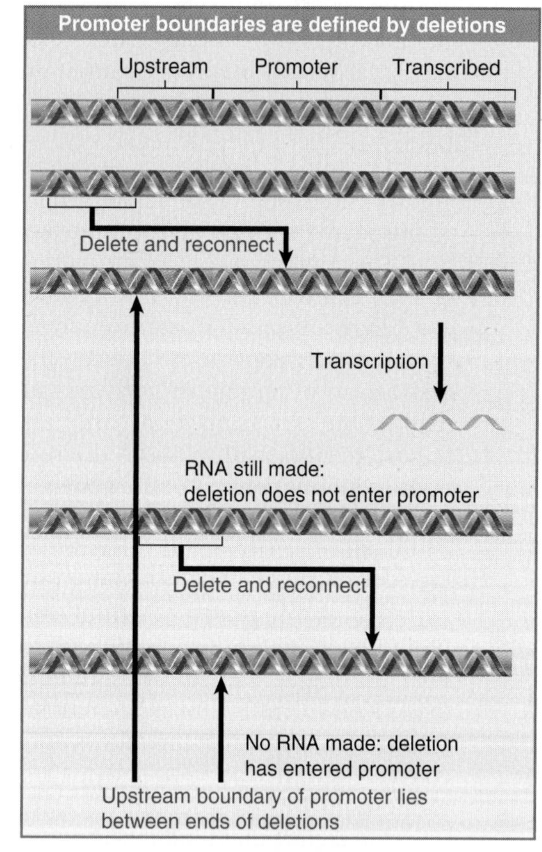

FIGURE 24.4 Promoter boundaries can be determined by making deletions that progressively remove more material from one side. When one deletion fails to prevent RNA synthesis but the next stops transcription, the boundary of the promoter must lie between them.

We start with a particular fragment of DNA that can initiate transcription in one of these systems. The boundaries of the sequence constituting the promoter then can be determined by reducing the length of the fragment from either end, until at some point it ceases to be active, as illustrated in FIGURE 24.4. The boundary upstream can be identified by progressively removing material from this end until promoter function is lost. To test the boundary downstream, it is necessary to reconnect the shortened promoter to the sequence to be transcribed (since otherwise there is no product to assay).

Once the boundaries of the promoter have been defined, the importance of particular bases within it can be determined by introducing point mutations or other rearrangements in the sequence. As with bacterial RNA polymerase, these can be characterized as *up* or *down* mutations. Some of these rearrangements affect only the *rate* of initiation; others influence the *site* at which initiation occurs, as seen in a change of the startpoint. To be sure that we are dealing with comparable products, in each case it is necessary to characterize the 5′ end of the RNA.

We can apply several criteria in identifying the sequence components of a promoter (or any other site in DNA):

- Mutations in the site prevent function *in vitro* or *in vivo*. (Many techniques now exist for introducing point mutations at particular base pairs, and in principle every position in a promoter can be mutated and the mutant sequence tested *in vitro* or *in vivo*.)
- Proteins that act by binding to a site may be footprinted on it. There should be a correlation between the ability of mutations to prevent promoter function and to prevent binding of the factor.
- When a site recognized by a particular factor is present at multiple promoters, it should be possible to derive a consensus sequence that is bound by the factor. A new promoter should become responsive to this factor when an appropriate copy of the element is introduced.

24.4 RNA Polymerase I Has a Bipartite Promoter

Key concepts

- The RNA polymerase I promoter consists of a core promoter and an upstream control element (UPE).
- The factor UBF1 wraps DNA around a protein structure to bring the core and UPE into proximity.
- SL1 includes the factor TBP that is involved in initiation by all three RNA polymerases.
- RNA polymerase binds to the UBF1-SL1 complex at the core promoter.

RNA polymerase I transcribes from a single type of promoter only the genes for ribosomal RNA. The transcript includes the sequences of both large and small rRNAs, which are later released by cleavages and processing. There are many copies of the transcription unit. They alternate with nontranscribed **spacers** and are organized in a cluster as discussed in Section 6.8, Genes for rRNA Form Tandem Repeats. The organization of the promoter, and the events involved in initiation, are illustrated in FIGURE 24.5.

The promoter consists of two separate regions. The **core promoter** surrounds the startpoint, extending from –45 to +20, and is sufficient for transcription to initiate. It is generally G-C-rich (unusual for a promoter), except for the only conserved sequence element, a short A-T-rich sequence around the startpoint called the Inr. The core promoter's efficiency, however, is very much increased by the upstream promoter element (UPE). The UPE is another G-C-rich sequence related to the core promoter sequence, and extends from –180 to –107. This type of organization is common to pol I promoters in many species, although the actual sequences vary widely.

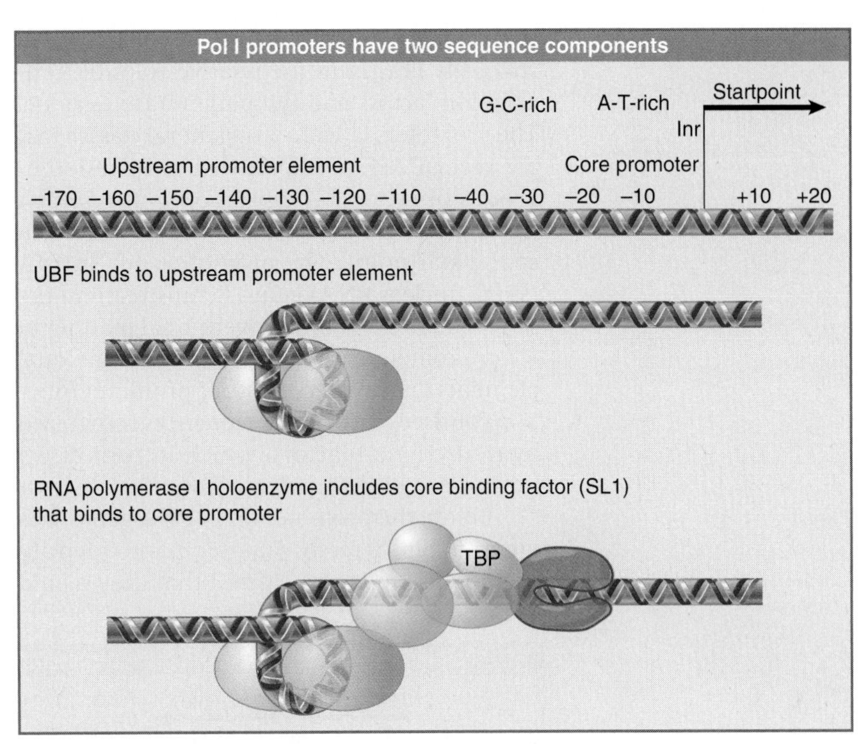

FIGURE 24.5 Transcription units for RNA polymerase I have a core promoter separated by ~70 bp from the upstream promoter element. UBF binding to the UPE increases the ability of core-binding factor to bind to the core promoter. Core-binding factor (SL1) positions RNA polymerase I at the startpoint.

RNA polymerase I requires two ancillary factors. The factor that binds to the core promoter consists of four proteins. (It is called SL1, TIF-IB, and Rib1 in different species). One of its components, TATA-binding protein (TBP), is a factor that also is required for initiation by RNA polymerases II and III (see Section 24.8, TBP Is a Universal Factor). TBP does not bind directly to G-C-rich DNA, and DNA-binding is the responsibility of the other components of the core-binding factor. It is likely that TBP interacts with RNA polymerase, probably with a common subunit or a feature that has been conserved among polymerases. Core-binding factor enables RNA polymerase I to initiate from the promoter at a low basal frequency.

The core-binding factor has primary responsibility for ensuring that the RNA polymerase is properly localized at the startpoint. We see shortly that a comparable function is provided for RNA polymerases II and III by a factor that consists of TBP associated with other proteins. Thus a common feature in initiation by all three polymerases is a reliance on a "positioning" factor that consists of TBP associated with proteins that are specific for each type of promoter.

For high frequency initiation, the factor UBF is required. This is a single polypeptide that binds to a G-C-rich element in the UPE. One indication of how UBF interacts with the core-binding factor is given by the importance of the spacing between the UPE and the core promoter. This can be changed by distances involving integral numbers of turns of DNA, but not by distances that introduce half turns. UBF binds to the minor groove of DNA and wraps the DNA in a loop of almost 360°, thus bringing the core and the UPE into close proximity.

Figure 24.5 shows initiation as a series of sequential interactions. RNA polymerase I, however, exists as a holoenzyme that contains most or all of the factors required for initiation, and which is probably recruited directly to the promoter.

24.5 RNA Polymerase III Uses Both Downstream and Upstream Promoters

Key concepts

- RNA polymerase III has two types of promoters.
- Internal promoters have short consensus sequences located within the transcription unit and cause initiation to occur a fixed distance upstream.
- Upstream promoters can contain three short consensus sequences upstream of the startpoint that are bound by transcription factors.

Recognition of promoters by RNA polymerase III strikingly illustrates the relative roles of transcription factors and the polymerase enzyme. The promoters fall into two general classes that are recognized in different ways by different groups of factors. The promoters for 5S and tRNA genes are *internal;* they lie downstream of the startpoint. The promoters for snRNA (small nuclear RNA) genes lie upstream of the startpoint in the more conventional manner of other promoters. In both cases, the individual elements that are necessary for promoter function consist exclusively of sequences recognized by transcription factors, which in turn direct the binding of RNA polymerase.

Before the promoter of 5S RNA genes was identified in *X. laevis,* all attempts to identify promoter sequences assumed that they would lie upstream of the startpoint. Deletion analysis, however, showed that the 5S RNA product continues to be synthesized when the entire sequence upstream of the gene is removed!

When the deletions continue into the gene, a product very similar in size to the usual 5S RNA continues to be synthesized so long as the deletion ends before base +55. FIGURE 24.6 shows that the first part of the RNA product corresponds to plasmid DNA; the second part represents the segment remaining of the usual 5S RNA sequence. When the deletion extends past +55, though, transcription does not occur. Thus the promoter lies *downstream of position +55,* but causes RNA polymerase III to initiate transcription a more or less fixed distance upstream.

When deletions extend into the gene from its distal end, transcription is unaffected so long as the first 80 bp remain intact. Once the deletion cuts into this region, transcription ceases. This places the downstream boundary position of the promoter at about position +80.

Thus the promoter for 5S RNA transcription lies between positions +55 and +80 within the gene. A fragment containing this region can sponsor initiation of any DNA in which it is placed, from a startpoint ~55 bp farther upstream. (The wild-type startpoint is unique; in deletions that lack it, transcription initiates at the purine base nearest to the position 55 bp upstream of the promoter.)

The structures of three types of promoters for RNA polymerase III are summarized in FIGURE 24.7. There are two types of internal promoter. Each contains a bipartite structure, in which two short sequence elements are separated by a variable sequence. Type 1 consists of a boxA sequence separated from a boxC sequence, and type 2 consists of a boxA sequence separated from a boxB sequence. The distance between boxA and boxB in a type 2 promoter can vary quite extensively, but the boxes usually cannot be brought too close together without abolishing function. A common group of type 3 promoters have three sequence elements that are all located upstream of the startpoint.

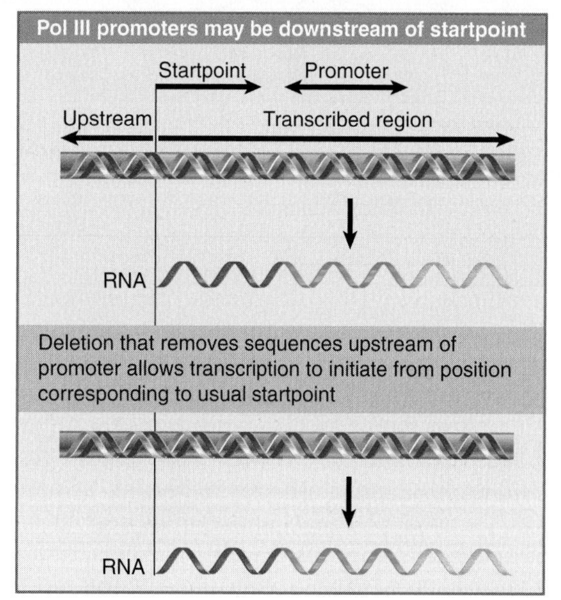

FIGURE 24.6 Deletion analysis shows that the promoter for 5S RNA genes is internal; initiation occurs a fixed distance (~55 bp) upstream of the promoter.

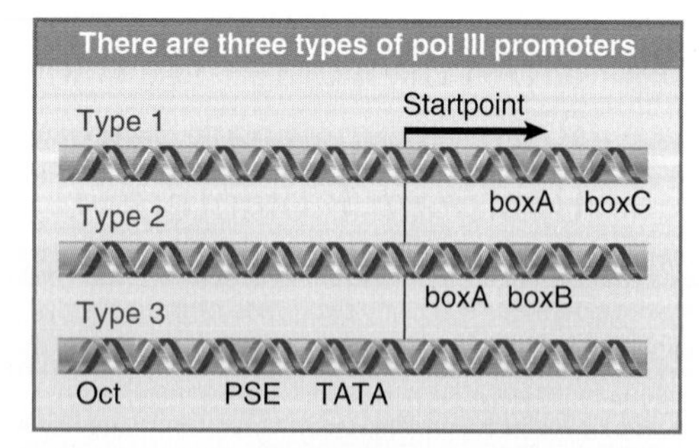

FIGURE 24.7 Promoters for RNA polymerase III may consist of bipartite sequences downstream of the startpoint, with boxA separated from either boxC or boxB, or they may consist of separated sequences upstream of the startpoint (Oct, PSE, TATA).

24.6 $TF_{III}B$ Is the Commitment Factor for Pol III Promoters

Key concepts

- $TF_{III}A$ and $TF_{III}C$ bind to the consensus sequences and enable $TF_{III}B$ to bind at the startpoint.
- $TF_{III}B$ has TBP as one subunit and enables RNA polymerase to bind.

The detailed interactions are different at the two types of internal promoter, but the principle is the same. $TF_{III}C$ binds downstream of the startpoint, either independently (type 2 promoters) or in conjunction with $TF_{III}A$ (type 1 promoters). The presence of $TF_{III}C$ enables the positioning factor $TF_{III}B$ to bind at the startpoint. RNA polymerase is then recruited.

FIGURE 24.8 summarizes the stages of reaction at type 2 internal promoters. $TF_{III}C$ binds to both boxA and boxB. This enables $TF_{III}B$ to bind at the startpoint. At this point RNA polymerase III can bind.

The difference at type 1 internal promoters is that $TF_{III}A$ must bind at boxA to enable $TF_{III}C$ to bind at boxC. FIGURE 24.9 shows that, once $TF_{III}C$ has bound, events follow the same course as at type 2 promoters, with $TF_{III}B$ binding at the startpoint, and RNA polymerase III joining the complex. Type 1 promoters are found only in the genes for 5S rRNA.

$TF_{III}A$ and $TF_{III}C$ are **assembly factors,** whose sole role is to assist the binding of $TF_{III}B$ at the right location. Once $TF_{III}B$ has bound, $TF_{III}A$ and $TF_{III}C$ can be removed from the promoter (by high salt concentration *in vitro*) without affecting the initiation reaction. *$TF_{III}B$ remains bound in the vicinity of the startpoint and its presence is sufficient to allow RNA polymerase III to bind at the startpoint.* Thus $TF_{III}B$ is the only true initiation factor required by RNA polymerase III. This sequence of events explains how the promoter boxes downstream can cause RNA polymerase to bind at the startpoint, farther upstream. Although the ability to transcribe these genes is conferred by the internal promoter, changes in the region immediately upstream of the startpoint can alter the efficiency of transcription.

$TF_{III}C$ is a large protein complex (>500 kD), which is comparable in size to RNA polymerase itself and contains six subunits. $TF_{III}A$ is a member of an interesting class of proteins containing a nucleic acid-binding motif called a zinc finger (see Section 25.9, A Zinc Finger Motif Is a DNA-Binding Domain). The positioning factor, $TF_{III}B$, consists of three subunits. It includes the same protein, TBP, that is present in the core-binding factor for pol I promoters, and also in the corresponding transcription factor ($TF_{II}D$) for RNA polymerase II. It also contains Brf, which is related to the factor $TF_{II}B$ that is used by RNA polymerase II. The third subunit is called B″; it is dispensable if the DNA duplex is partially melted, which suggests that its function is to initiate the transcription bubble. The role of B″

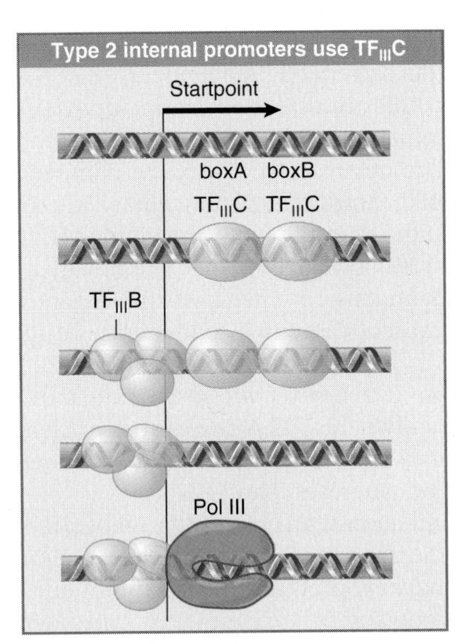

FIGURE 24.8 Internal type 2 pol III promoters use binding of $TF_{III}C$ to boxA and boxB sequences to recruit the positioning factor $TF_{III}B$, which recruits RNA polymerase III.

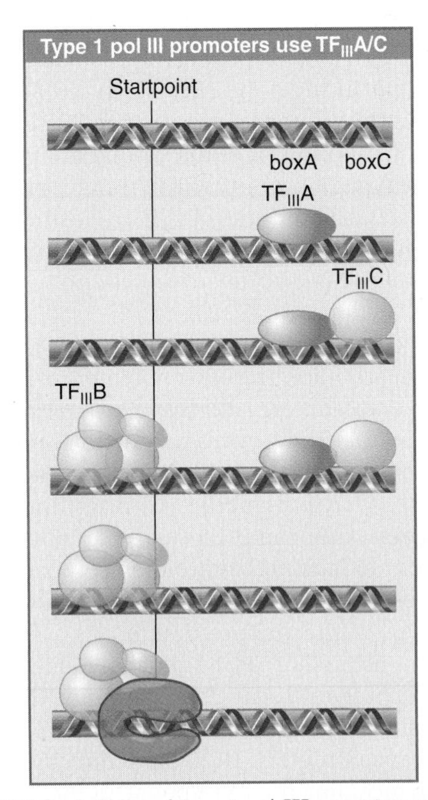

FIGURE 24.9 Internal type 1 pol III promoters use the assembly factors $TF_{III}A$ and $TF_{III}C$, at boxA and boxC, to recruit the positioning factor $TF_{III}B$, which recruits RNA polymerase III.

may be comparable to the role played by sigma factor in bacterial RNA polymerase (see Section 11.16, Substitution of Sigma Factors May Control Initiation).

The upstream region has a conventional role in the third class of polymerase III promoters. In the example shown in Figure 24.8, there are three upstream elements. These elements are also found in promoters for snRNA genes that are transcribed by RNA polymerase II. (Genes for some snRNAs are transcribed by RNA polymerase II, whereas others are transcribed by RNA polymerase III.) The upstream elements function in a similar manner in promoters for both polymerases II and III.

Initiation at an upstream promoter for RNA polymerase III can occur on a short region that immediately precedes the startpoint and contains only the TATA element. Efficiency of transcription, however, is much increased by the presence of the PSE and OCT elements. The factors that bind at these elements interact cooperatively. (The PSE element may be essential at promoters used by RNA polymerase II, whereas it is stimulatory in promoters used by RNA polymerase III; its name stands for proximal sequence element.)

The TATA element confers specificity for the type of polymerase (II or III) that is recognized by an snRNA promoter. It is bound by a factor that includes the TBP, which actually recognizes the sequence in DNA. The TBP is associated with other proteins, which are specific for the type of promoter. The function of TBP and its associated proteins is to position the RNA polymerase correctly at the startpoint. We discuss this in more detail for RNA polymerase II (see Section 24.8, TBP Is a Universal Factor).

The factors work in the same way for both types of promoters for RNA polymerase III. *The factors bind at the promoter before RNA polymerase itself can bind.* They form a **preinitiation complex** that directs binding of the RNA polymerase. RNA polymerase III does not itself recognize the promoter sequence, but binds adjacent to factors that are themselves bound just upstream of the startpoint. For the type 1 and type 2 internal promoters, the assembly factors ensure that $TF_{III}B$ (which includes TBP) is bound just upstream of the startpoint, thereby providing the positioning information. For the upstream promoters, $TF_{III}B$ binds directly to the region including the TATA box. This means that irrespective of the location of the promoter sequences, factor(s) are bound close to the startpoint in order to direct binding of RNA polymerase III.

24.7 The Startpoint for RNA Polymerase II

Key concepts

- RNA polymerase II requires general transcription factors (called **$TF_{II}X$**) to initiate transcription.
- RNA polymerase II promoters commonly have a short conserved sequence Py_2CAPy_5 (the initiator InrR) at the startpoint.
- The TATA box is a common component of RNA polymerase II promoters and consists of an A-T-rich octamer located ~25 bp upstream of the startpoint.
- The DPE is a common component of RNA polymerase II promoters that do not contain a TATA box.
- A core promoter for RNA polymerase II generally includes the InR and either a TATA box or a DPE.

The basic organization of the apparatus for transcribing protein-coding genes was revealed by the discovery that purified RNA polymerase II can catalyze synthesis of mRNA, but cannot initiate transcription unless an additional extract is added. The purification of this extract led to the definition of the general transcription factors—a group of proteins that are needed for initiation by RNA polymerase II at all promoters. RNA polymerase II in conjunction with these factors constitutes the basal transcription apparatus that is needed to transcribe any promoter. The general factors are described as $TF_{II}X$, where "X" is a letter that identifies the individual factor. The subunits of RNA polymerase II and the general transcription factors are conserved among eukaryotes.

Our starting point for considering promoter organization is to define the core promoter as the shortest sequence at which RNA polymerase II can initiate transcription. A core promoter can in principle be expressed in any cell. It comprises the minimum sequence that enables the general transcription factors to assemble at the startpoint. Core promoters are involved in the mechanics of binding to DNA and enable RNA polymerase II to initiate transcription. A core promoter functions at only a low efficiency. Other proteins, called activators, are required for a proper level of function (see Section 24.13, Short Sequence Elements Bind Activators). The activators are not described systematically, but have casual names reflecting their histories of identification.

We may expect any sequence components involved in the binding of RNA polymerase and general transcription factors to be conserved at most or all promoters. As with bacterial pro-

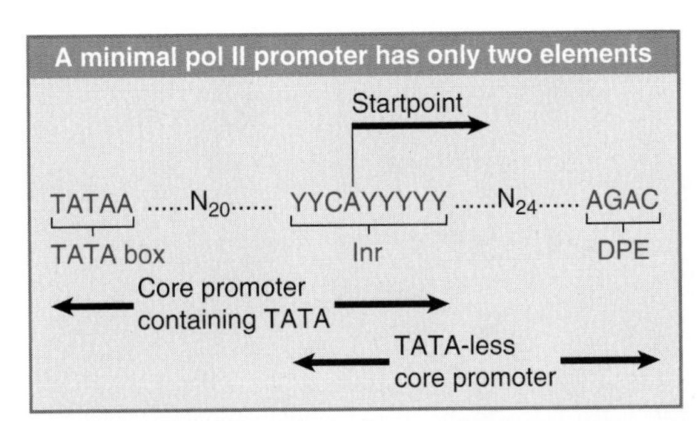

FIGURE 24.10 The minimal pol II promoter may have a TATA box ~25 bp upstream of the InR. The TATA box has the consensus sequence of TATAA. The Inr has pyrimidines (Y) surrounding the CA at the startpoint. The DEP is downstream of the startpoint. The sequence shows the coding strand.

moters, when promoters for RNA polymerase II are compared, homologies in the regions near the startpoint are restricted to rather short sequences. These elements correspond with the sequences implicated in promoter function by mutation. FIGURE 24.10 shows the construction of a typical pol II core promoter.

At the startpoint, there is no extensive homology of sequence, but there is a tendency for the first base of mRNA to be A, flanked on either side by pyrimidines. (This description is also valid for the CAT start sequence of bacterial promoters.) This region is called the *initiator* (**Inr**), and may be described in the general form Py_2CAPy_5. The Inr is contained between positions –3 and +5.

Many promoters have a sequence called the **TATA box**, usually located ~25 bp upstream of the startpoint in higher eukaryotes. It constitutes the only upstream promoter element that has a relatively fixed location with respect to the startpoint. The core sequence is TATAA, usually followed by three more A-T base pairs. The TATA box tends to be surrounded by G-C-rich sequences, which could be a factor in its function. It is almost identical with the –10 sequence found in bacterial promoters; in fact, it could pass for one except for the difference in its location at –25 instead of –10.

Single-base substitutions in the TATA box act as strong down mutations. Some mutations reverse the orientation of an A-T pair, so base composition alone is not sufficient for its function. Thus the TATA box comprises an element whose behavior is analogous to our concept of the bacterial promoter: a short, well-defined sequence just upstream of the startpoint, which is necessary for transcription.

Promoters that do not contain a TATA element are called **TATA-less promoters**. Surveys of promoter sequences suggest that 50% or more of promoters may be TATA-less. When a promoter does not contain a TATA box, it usually contains another element, the DPE (downstream promoter element), which is located at +28 – +32.

Most core promoters consist either of a TATA box plus InR or of an InR plus DPE.

24.8 TBP Is a Universal Factor

Key concepts

- TBP is a component of the positioning factor that is required for each type of RNA polymerase to bind its promoter.
- The factor for RNA polymerase II is $TF_{II}D$, which consists of TBP and 11 TAFs, with a total mass ~800 kD.

The first step in complex formation at a promoter containing a TATA box is binding of the factor **$TF_{II}D$** to a region that extends upstream from the TATA sequence. $TF_{II}D$ contains two types of component. Recognition of the TATA box is conferred by the **TATA-binding protein (TBP)**, a small protein of ~30 kD. The other subunits are called **TAFs** (for TBP-associated factors). Some TAFs are stoichiometric with TBP; others are present in lesser amounts. $TF_{II}Ds$ containing different TAFs could recognize different promoters. Some (substoichiometric) TAFs are tissue-specific. The total mass of $TF_{II}D$ typically is ~800 kD, contains TBP and 11 TAFs, varying in mass from 30 to 250 kD. The TAFs in $TF_{II}D$ are named in the form $TAF_{II}00$, where "00" gives the molecular mass of the subunit.

Positioning factors that consist of TBP associated with a set of TAFs are responsible for identifying all classes of promoters. $TF_{III}B$ (for pol III promoters) and SL1 (for pol I promoters) may both be viewed as consisting of TBP associated with a particular group of proteins that substitute for the TAFs that are found in $TF_{II}D$. TBP is the key component, and is incorporated at each type of promoter by a different mechanism. In the case of promoters for RNA polymerase II, a key feature in positioning is the fixed distance of the TATA box from the startpoint.

FIGURE 24.11 shows that the positioning factor recognizes the promoter in a different way in each case. At promoters for RNA polymerase III, $TF_{III}B$ binds adjacent to $TF_{III}C$. At promoters for RNA polymerase I, SL1 binds in conjunction with UBF. $TF_{II}D$ is solely responsible for recognizing promoters for RNA polymerase II. At a promoter that has a TATA element, TBP binds specifically to DNA, but at other promoters it may be incorporated by association with

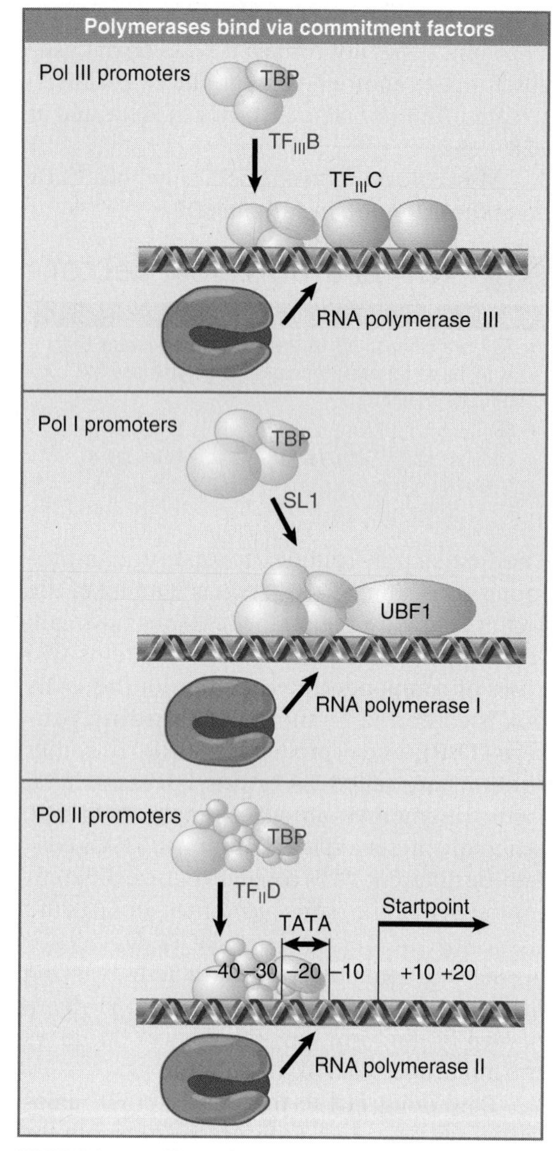

FIGURE 24.11 RNA polymerases are positioned at all promoters by a factor that contains TBP.

FIGURE 24.12 A view in cross-section shows that TBP surrounds DNA from the side of the narrow groove. TBP consists of two related (40% identical) conserved domains, which are shown in light and dark blue. The N-terminal region varies extensively and is shown in green. The two strands of the DNA double helix are in light and dark gray. Photo courtesy of Stephen K. Burley.

other proteins that bind to DNA. Whatever its means of entry into the initiation complex, it has the common purpose of interaction with the RNA polymerase.

$TF_{II}D$ is ubiquitous, but not unique. All multicellular eukaryotes also express an alternative complex, which has TLF (TBP-like factor) instead of TBP. A TLF is typically ~60% similar to TBP. It probably initiates complex formation by the usual set of TF_{II} factors. TLF does not, however, bind to the TATA box, and we do not yet know how it works. *Drosophila* also has a third factor, TRF1, which behaves in the same way as TBP and binds its own set of TAFs to form a complex that functions as an alternative to $TF_{II}D$ at a specific set of promoters.

24.9 TBP Binds DNA in an Unusual Way

Key concepts

- TBP binds to the TATA box in the minor groove of DNA.
- It forms a saddle around the DNA and bends it by ~80°.
- Some of the TAFs resemble histones and may form a structure resembling a histone octamer.

TBP has the unusual property of binding to DNA in the minor groove. (Virtually all known DNA-binding proteins bind in the major groove.) The crystal structure of TBP suggests a detailed model for its binding to DNA. FIGURE 24.12 shows that it surrounds one face of DNA, forming a "saddle" around the double helix. In effect, the inner surface of TBP binds to DNA, and the larger outer surface is available to extend contacts to other proteins. The DNA-binding site consists of a C-terminal domain that is conserved between species, and the variable N-terminal tail is exposed to interact with other proteins. It is a measure of the conservation of mechanism in transcriptional initiation that the DNA-binding sequence of TBP is 80% conserved between yeast and human beings.

Binding of TBP may be inconsistent with the presence of nucleosomes. Nucleosomes form preferentially by placing A-T-rich sequences with the minor grooves facing inward; as a result, they could prevent binding of TBP. This may explain why the presence of nucleosomes prevents initiation of transcription.

TBP binds to the minor groove and bends the DNA by ~80°, as illustrated in FIGURE 24.13. The TATA box bends toward the major groove, widening the minor groove. The distortion is restricted to the 8 bp of the TATA box; at each end of the sequence, the minor groove has its usual width of ~5 Å, but at the center of the sequence the minor groove is >9 Å. This is a deformation of the structure, but does not actually separate the strands of DNA because base pairing is maintained. The extent of the bend can vary with the exact sequence of the TATA box, and is correlated with the efficiency of the promoter.

This structure has several functional implications. By changing the spatial organization of DNA on either side of the TATA box, it allows the transcription factors and RNA polymerase to form a closer association than would be possible on linear DNA. The bending at the TATA box corresponds to unwinding of about one-third of a turn of DNA, and is compensated by a positive writhe.

The presence of TBP in the minor groove, combined with other proteins binding in the major groove, creates a high density of protein–DNA contacts in this region. Binding of purified TBP to DNA *in vitro* protects ~1 turn of the double helix at the TATA box, typically extending from –37 to –25. Binding of the $TF_{II}D$ complex in the initiation reaction, however, regularly protects the region from –45 to –10, and also extends farther upstream beyond the startpoint. TBP is the only general transcription factor that makes sequence-specific contacts with DNA.

Within $TF_{II}D$ as a free protein complex, the factor $TAF_{II}230$ binds to TBP, where it occupies the concave DNA-binding surface. In fact, the structure of the binding site, which lies in the N-terminal domain of $TAF_{II}230$, mimics the surface of the minor groove in DNA. This molecular mimicry allows $TAF_{II}230$ to control the ability of TBP to bind to DNA; the N-terminal domain of $TAF_{II}230$ must be displaced from the DNA-binding surface of TBP in order for $TF_{II}D$ to bind to DNA.

Some TAFs resemble histones; in particular, $TAF_{II}42$ and $TAF_{II}62$ appear to be (distant) homologs of histones H3 and H4, and they form a heterodimer using the same motif (the histone fold) that histones use for the interaction. (Histones H3 and H4 form the kernel of the histone octamer—the basic complex that binds DNA in eukaryotic chromatin; see Section 29.7, Organization of the Histone Octamer.) Together with other TAFs, $TAF_{II}42$ and $TAF_{II}62$ may form the basis for a structure resembling a histone octamer; such a structure may be responsible for the nonsequence-specific interactions of $TF_{II}D$ with DNA. Histone folds are also used in pairwise interactions between other TAF_{II}s.

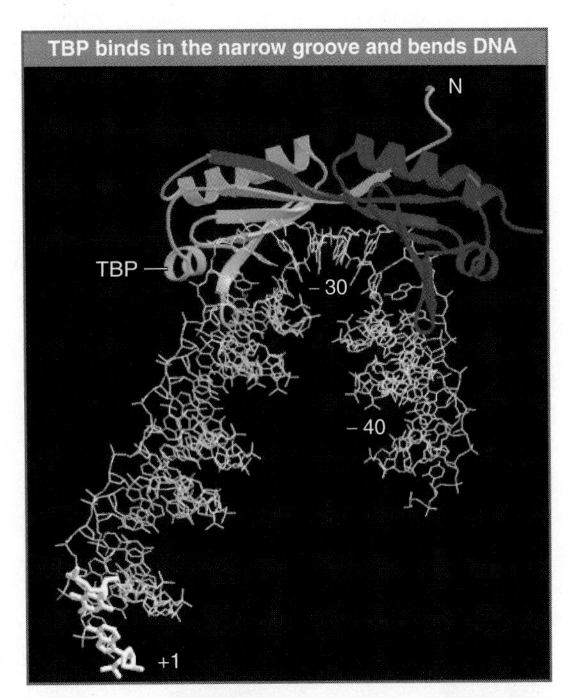

FIGURE 24.13 The cocrystal structure of TBP with DNA from –40 to the startpoint shows a bend at the TATA box that widens the narrow groove where TBP binds. Photo courtesy of Stephen K. Burley.

Some of the TAF_{II}s may be found in other complexes as well as in $TF_{II}D$. In particular, the histone-like TAF_{II}s are found also in protein complexes that modify the structure of chromatin prior to transcription (see Section 30.7, Acetylases Are Associated with Activators).

24.10 The Basal Apparatus Assembles at the Promoter

Key concepts

- Binding of $TF_{II}D$ to the TATA box is the first step in initiation.
- Other transcription factors bind to the complex in a defined order, extending the length of the protected region on DNA.
- When RNA polymerase II binds to the complex, it initiates transcription.

Initiation requires the transcription factors to act in a defined order to build a complex that is joined by RNA polymerase. The series of events was initially defined by following the increasing size

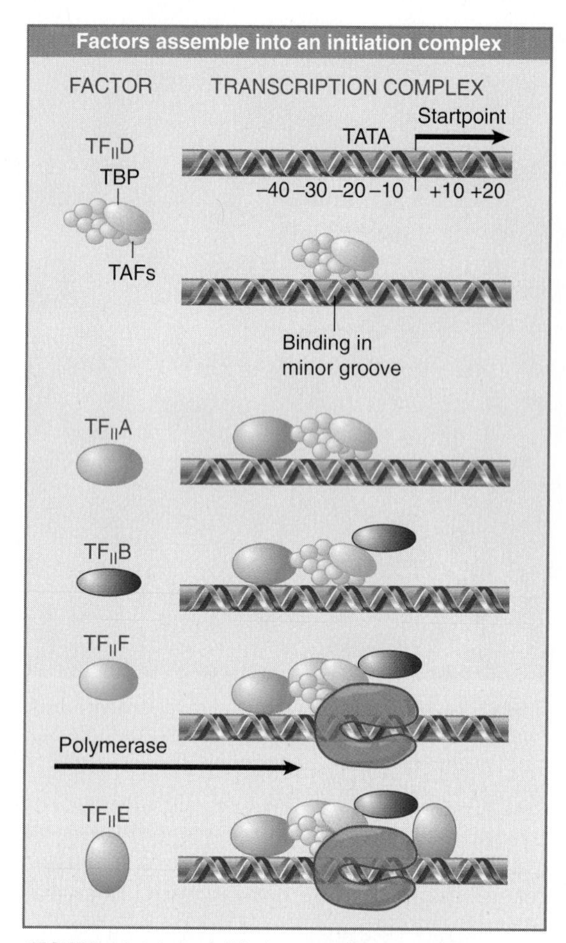

FIGURE 24.14 An initiation complex assembles at promoters for RNA polymerase II by an ordered sequence of association with transcription factors.

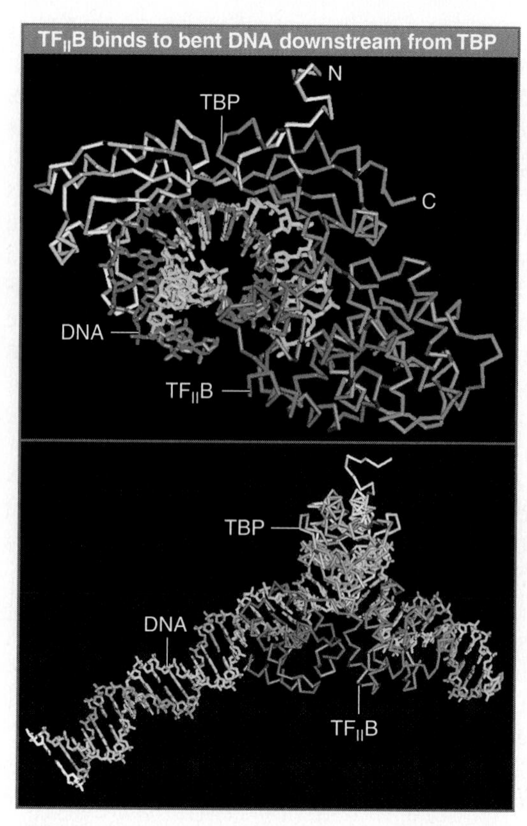

FIGURE 24.15 Two views of the ternary complex of $TF_{II}B$-TBP-DNA show that $TF_{II}B$ binds along the bent face of DNA. The two strands of DNA are green and yellow, TBP is blue, and $TF_{II}B$ is red and purple. Photos courtesy of Stephen K. Burley.

of the protein complex associated with DNA. Now we can define the events in more detail in terms of the interactions revealed by the crystal structures of the various factors and in terms of RNA polymerase bound to DNA.

Footprinting of the DNA regions protected by each complex suggests the model summarized in FIGURE 24.14. As each TF_{II} factor joins the complex, an increasing length of DNA is covered. RNA polymerase is incorporated at a late stage.

Commitment to a promoter is initiated when $TF_{II}D$ binds the TATA box. ($TF_{II}D$ also recognizes the InR sequence at the startpoint.) When $TF_{II}A$ joins the complex, $TF_{II}D$ becomes able to protect a region extending farther upstream. $TF_{II}A$ may activate TBP by relieving the repression that is caused by the $TAF_{II}230$.

Addition of $TF_{II}B$ gives partial protection of the region of the template strand in the vicinity of the startpoint, from –10 to +10. This suggests that $TF_{II}B$ is bound downstream of the TATA box, perhaps loosely associated with DNA and asymmetrically oriented with regard to the two DNA strands. The crystal structure shown in FIGURE 24.15 extends this model. $TF_{II}B$ binds adjacent to TBP, extending contacts along one face of DNA. It makes contacts in the minor groove downstream of the TATA box, and contacts the major groove upstream of the TATA box in a region called the BRE. In archaea, the homolog of $TF_{II}B$ actually makes sequence-specific contacts with the promoter in the BRE region. $TF_{II}B$ may provide the surface that is in turn recognized by RNA polymerase, so that it is responsible for the directionality of the binding of the enzyme.

The crystal structure of $TF_{II}B$ with RNA polymerase shows that three domains of the factor interact with the enzyme. As illustrated schematically in FIGURE 24.16, an N-terminal zinc ribbon from $TF_{II}B$ contacts the enzyme near the site where RNA exits; it is possible that this interferes with the exit of RNA and influences the switch from abortive initiation to promoter escape. An elongated "finger" of $TF_{II}B$ is inserted into the polymerase active center. The C-terminal domain interacts with the RNA polymerase and with $TF_{II}D$ to orient the DNA. It also determines the path of the DNA where it contacts the factors $TF_{II}E$, $TF_{II}F$, and $TF_{II}H$, which may align them in the basal factor complex.

The factor $TF_{II}F$ is a heterotetramer consisting of two types of subunit. The larger subunit (RAP74) has an ATP-dependent DNA helicase activity that could be involved in melting the DNA at initiation. The smaller subunit (RAP38) has some homology to the regions of bacterial sigma factor that contact the core polymerase; it binds tightly to RNA polymerase II. $TF_{II}F$ may bring RNA polymerase II to the assembling transcription complex and provide the means by which it binds. The complex of TBP and TAFs may interact with the CTD tail of RNA polymerase, and interaction with $TF_{II}B$ may also be important when $TF_{II}F$/polymerase joins the complex.

Polymerase binding extends the sites that are protected downstream to +15 on the template strand and +20 on the nontemplate strand. The enzyme extends the full length of the complex because additional protection is seen at the upstream boundary.

What happens at TATA-less promoters? The same general transcription factors, including $TF_{II}D$, are needed. The Inr provides the positioning element; $TF_{II}D$ binds to it via an ability of one or more of the TAFs to recognize the Inr directly. Other TAFs in $TF_{II}D$ also recognize the DPE element downstream from the startpoint. The function of TBP at these promoters is more like that at promoters for RNA polymerase I and at internal promoters for RNA polymerase III.

Assembly of the RNA polymerase II initiation complex provides an interesting contrast with prokaryotic transcription. Bacterial RNA polymerase is essentially a coherent aggregate with intrinsic ability to bind DNA; the sigma factor, needed for initiation but not for elongation, becomes part of the enzyme before DNA is bound, although it is later released. RNA polymerase II can bind to the promoter, but only after separate transcription factors have bound. The factors play a role analogous to that of bacterial sigma factor—to allow the basic polymerase to recognize DNA specifically at promoter sequences—but have evolved more independence. Indeed, the factors are primarily responsible for the specificity of promoter recognition. Only some of the factors participate in protein–DNA contacts (and only TBP makes sequence-specific contacts); thus protein–protein interactions are important in the assembly of the complex.

When a TATA box is present, it determines the location of the startpoint. Its deletion causes the site of initiation to become erratic, although

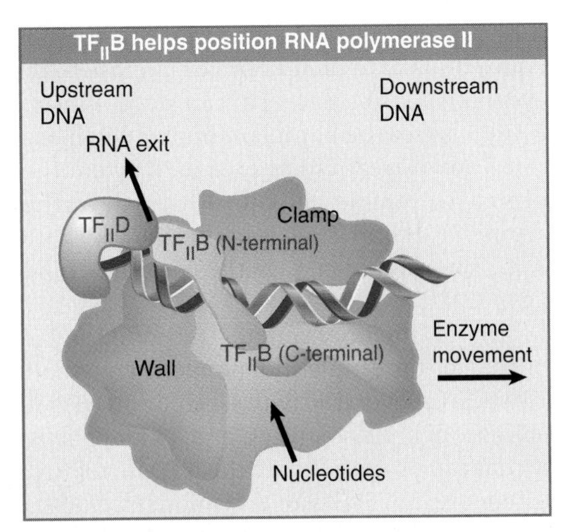

FIGURE 24.16 $TF_{II}B$ binds to DNA and contacts RNA polymerase near the RNA exit site and at the active center, and orients it on DNA. Compare with Figure 24.17, which shows the polymerase structure engaged in transcription.

any overall reduction in transcription is relatively small. Indeed, some TATA-less promoters lack unique startpoints; initiation occurs instead at any one of a cluster of startpoints. The TATA box aligns the RNA polymerase (via the interaction with $TF_{II}D$ and other factors) so that it initiates at the proper site. This explains why its location is fixed with respect to the startpoint. Binding of TBP to TATA is the predominant feature in recognition of the promoter, but two large TAFs ($TAF_{II}250$ and $TAF_{II}150$) also contact DNA in the vicinity of the startpoint and influence the efficiency of the reaction.

Although assembly can take place just at the core promoter *in vitro*, this reaction is not sufficient for transcription *in vivo*, where interactions with activators that recognize the more upstream elements are required. The activators interact with the basal apparatus at various stages during its assembly (see Section 25.5, Activators Interact with the Basal Apparatus).

24.11 Initiation Is Followed by Promoter Clearance

Key concepts

- $TF_{II}E$ and $TF_{II}H$ are required to melt DNA to allow polymerase movement.
- Phosphorylation of the CTD may be required for elongation to begin.
- Further phosphorylation of the CTD is required at some promoters to end abortive initiation.
- The CTD may coordinate processing of RNA with transcription.

Most of the general transcription factors are required solely to bind RNA polymerase to the promoter, but some act at a later stage. Binding of $TF_{II}E$ causes the boundary of the region protected downstream to be extended by another turn of the double helix, to +30. Two further factors, $TF_{II}H$ and $TF_{II}J$, join the complex after $TF_{II}E$. They do not change the pattern of binding to DNA.

$TF_{II}H$ is the only general transcription factor that has multiple independent enzymatic activities. Its several activities include an ATPase, helicases of both polarities, and a kinase activity that can phosphorylate the CTD tail of RNA polymerase II. $TF_{II}H$ is an exceptional factor that may also play a role in elongation. Its interaction with DNA downstream of the startpoint is required for RNA polymerase to escape from the promoter. $TF_{II}H$ is also involved in repair of damage to DNA (see Section 24.12, A Connection between Transcription and Repair).

The initiation reaction, as defined by formation of the first phosphodiester bond, occurs once RNA polymerase has bound. FIGURE 24.17 proposes a model in which phosphorylation of the tail is needed to release RNA polymerase II from the transcription factors so that it can make the transition to the elongating form. Most of the transcription factors are released from the promoter at this stage.

On a linear template, ATP hydrolysis, $TF_{II}E$, and the helicase activity of $TF_{II}H$ (provided by the XPB subunit) are required for polymerase movement. This requirement is bypassed with a supercoiled template. This suggests that $TF_{II}E$ and $TF_{II}H$ are required to melt DNA to allow polymerase movement to begin. The helicase activity of the XPB subunit of $TF_{II}H$ is responsible for the actual melting of DNA.

RNA polymerase II stutters at some genes when it starts transcription. (The result is not dissimilar to the abortive initiation of bacterial RNA polymerase discussed in Section 11.11, Sigma Factor Controls Binding to DNA, although the mechanism is different.) At many genes, RNA polymerase II terminates after a short distance. The short RNA product is degraded rapidly. To extend elongation into the gene, a kinase called P-TEFb is required. This kinase is a member of the cdk family that controls the cell cycle. P-TEFb acts on the CTD to phosphorylate it further. We do not yet understand why this effect is required at some promoters but not others or how it is regulated.

The CTD may also be involved, directly or indirectly, in processing RNA after it has been synthesized by RNA polymerase II. FIGURE 24.18 summarizes processing reactions in which the CTD may be involved. The capping enzyme (guanylyl transferase), which adds the G residue to the 5′ end of newly synthesized mRNA, binds to the phosphorylated CTD: This may be important in enabling it to modify the 5′ end as soon as it is synthesized. A set of proteins called SCAFs bind to the CTD, and they may in turn bind to splicing factors. This may be a means of coordinating transcription and splicing. Some components of the cleavage/polyadenylation apparatus also bind to the CTD. Oddly enough, they do so at the time of initiation, so that RNA polymerase is ready for the 3′ end processing reactions as soon as it sets out! All of this suggests that the CTD may be a general focus for connecting other processes with transcription. In the cases of capping and splicing, the CTD

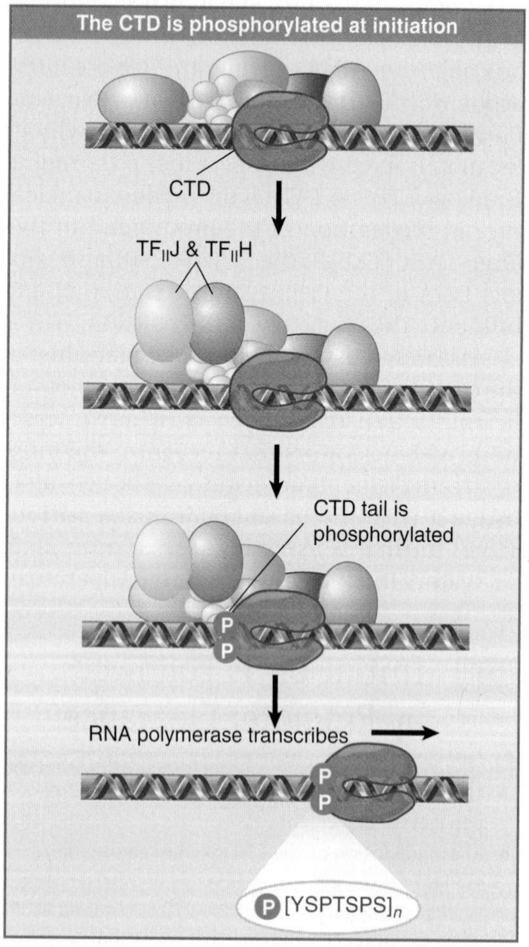

The CTD is phosphorylated at initiation

CTD

$TF_{II}J$ & $TF_{II}H$

CTD tail is phosphorylated

P P

RNA polymerase transcribes

P [YSPTSPS]$_n$

FIGURE 24.17 Phosphorylation of the CTD by the kinase activity of $TF_{II}H$ may be needed to release RNA polymerase to start transcription.

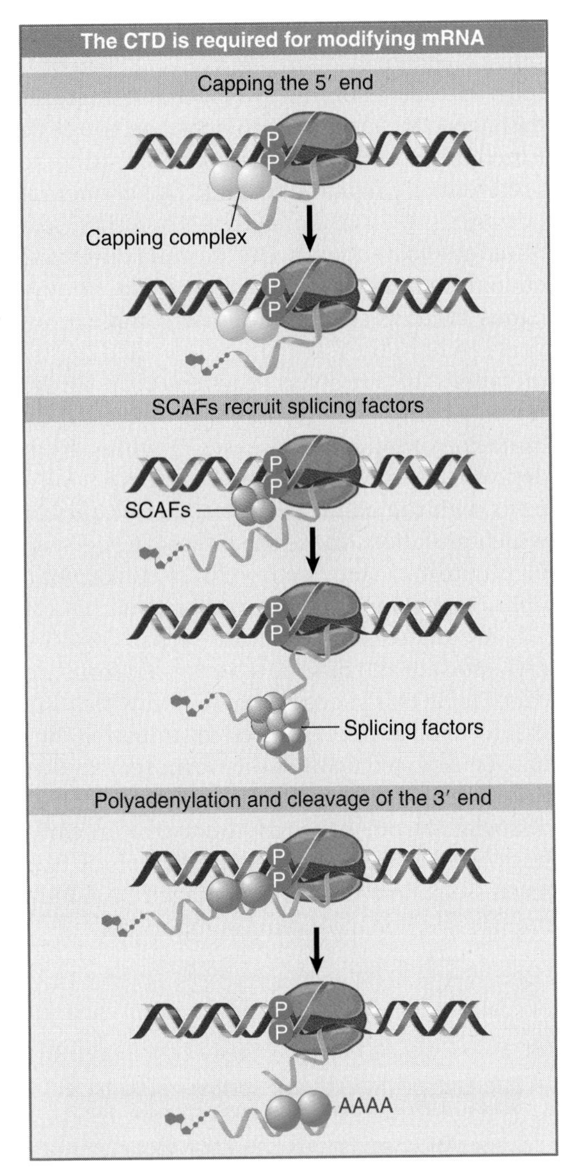

FIGURE 24.18 The CTD is important in recruiting enzymes that modify RNA.

functions indirectly to promote formation of the protein complexes that undertake the reactions. In the case of 3′ end generation, it may participate directly in the reaction.

The general process of initiation is similar to that catalyzed by bacterial RNA polymerase. Binding of RNA polymerase generates a closed complex, which is converted at a later stage to an open complex in which the DNA strands have been separated. In the bacterial reaction, formation of the open complex completes the necessary structural change to DNA; a difference in the eukaryotic reaction is that further unwinding of the template is needed after this stage.

24.12 A Connection between Transcription and Repair

Key concepts

- Transcribed genes are preferentially repaired when DNA damage occurs.
- $TF_{II}H$ provides the link to a complex of repair enzymes.
- Mutations in the XPD component of $TF_{II}H$ cause three types of human diseases

In both bacteria and eukaryotes, there is a direct link from RNA polymerase to the activation of repair. The basic phenomenon was first observed because transcribed genes are preferentially repaired. It was then discovered that it is only the template strand of DNA that is the target—the nontemplate strand is repaired at the same rate as bulk DNA.

In bacteria, the repair activity is provided by the *uvr* excision-repair system (see Section 20.3, Excision Repair Systems in *E. coli*). Preferential repair is abolished by mutations in the gene *mfd,* whose product provides the link from RNA polymerase to the Uvr enzymes.

FIGURE 24.19 shows a model for the link between transcription and repair. When RNA polymerase encounters DNA damage in the template strand, it stalls because it cannot use the damaged sequences as a template to direct complementary base pairing. This explains the specificity of the effect for the template strand (damage in the nontemplate strand does not impede progress of the RNA polymerase).

The Mfd protein has two roles. First, it displaces the ternary complex of RNA polymerase from DNA. Second, it causes the UvrABC enzyme to bind to the damaged DNA. This leads to repair of DNA by the excision-repair mechanism (see Figure 20.11). After the DNA has been repaired, the next RNA polymerase to traverse the gene is able to produce a normal transcript.

A similar mechanism, albeit one that relies on different components, is used in eukaryotes. The template strand of a transcribed gene is preferentially repaired following UV-induced damage. The general transcription factor $TF_{II}H$ is involved. $TF_{II}H$ is found in alternative forms, which consist of a core associated with other subunits.

$TF_{II}H$ has a common function in both initiating transcription and repairing damage. The same helicase subunit (XPD) creates the initial transcription bubble and melts DNA at a damaged site. Its other functions differ between transcription and repair, as provided by the appropriate form of the complex.

FIGURE 24.20 shows that the basic factor involved in transcription consists of a core (of five subunits) associated with other subunits that have a kinase activity; this complex also includes a repair subunit. The kinase catalytic subunit that phosphorylates the CTD of RNA polymerase belongs to a group of kinases that are involved in cell cycle control. It is possible that this connection influences transcription in response to the stage of the cell cycle.

The alternative complex consists of the core associated with a large group of proteins that are coded by repair genes. (The basic model for repair is shown in Figure 20.25.) The repair proteins include a subunit (XPC) that recognizes damaged DNA, which provides the coupling function that enables a template strand to be preferentially repaired when RNA polymerase becomes stalled at damaged DNA. Other proteins associated with the complex include endonucleases (XPG, XPF, and ERCC1). Homologous proteins are found in the complexes in yeast (where they are often identified by *rad* mutations that are defective in repair) and in the human being (where they are identified by mutations that cause diseases resulting from deficiencies in repairing damaged DNA). Subunits with the name XP are coded by genes in which mutations cause the disease xeroderma pigmentosum (see Section 20.11, Eukaryotic Cells Have Conserved Repair Systems).

The kinase complex and the repair complex can associate and dissociate reversibly from the core $TF_{II}H$. This suggests a model in which the first form of $TF_{II}H$ is required for initiation, but may be replaced by the other form (perhaps in response to encountering DNA damage). $TF_{II}H$ dissociates from RNA polymerase at an early stage of elongation (after transcription of ~50 bp); its reassociation at a site of damaged DNA may require additional coupling components.

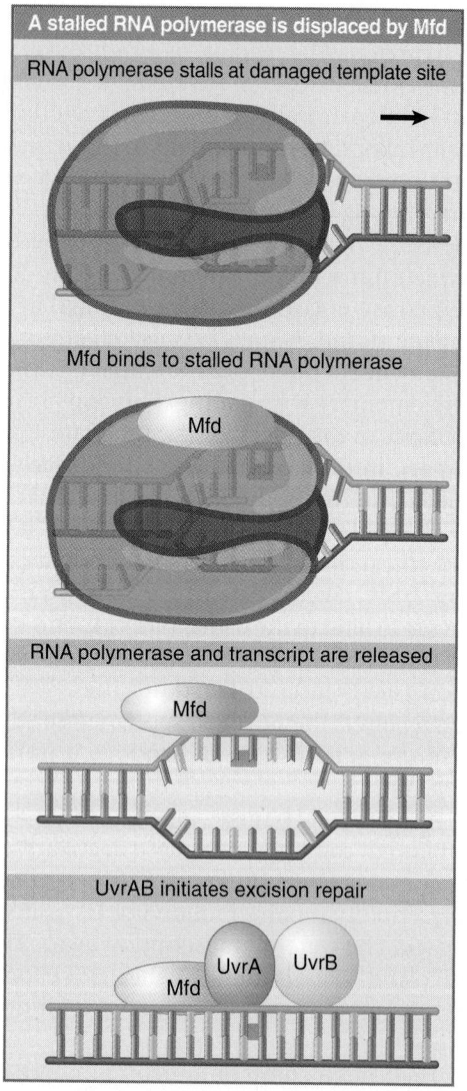

FIGURE 24.19 Mfd recognizes a stalled RNA polymerase and directs DNA repair to the damaged template strand.

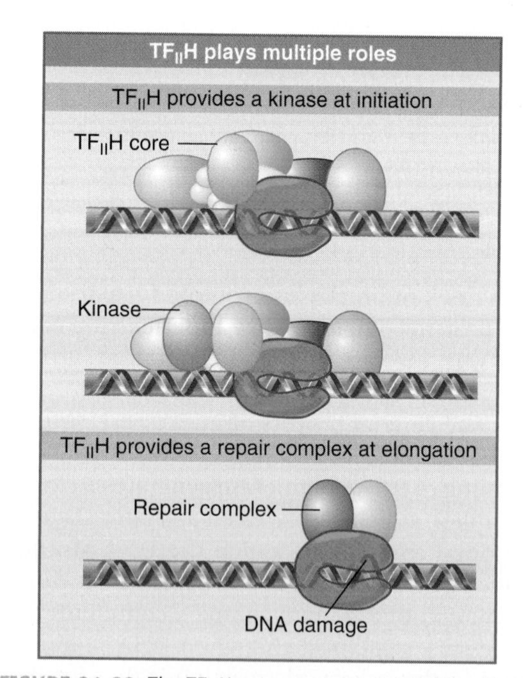

FIGURE 24.20 The $TF_{II}H$ core may associate with a kinase at initiation and associate with a repair complex when damaged DNA is encountered.

The repair function may require modification or degradation of RNA polymerase. The large subunit of RNA polymerase is degraded when the enzyme stalls at sites of UV damage. We do not yet understand the connection between the transcription/repair apparatus as such and the degradation of RNA polymerase. It is possible that removal of the polymerase is necessary once it has become stalled.

This degradation of RNA polymerase is deficient in cells from patients with Cockayne's syndrome (a repair disorder). Cockayne's syndrome is caused by mutations in either of two genes (*CSA* and *CSB*), both of whose products appear to be part of or bound to $TF_{II}H$. Cockayne's syndrome is also occasionally caused by mutations in XPD.

Another disease that can be caused by mutations in XPD is trichothiodystrophy, which has little in common with XP or Cockayne's (it involves mental retardation and is marked by changes in the structure of hair). All of this marks XPD as a pleiotropic protein, in which different mutations can affect different functions. In fact, XPD is required for the stability of the $TF_{II}H$ complex during transcription, but the helicase activity as such is not needed. Mutations that prevent XPD from stabilizing the complex cause trichothiodystrophy. The helicase activity is required for the repair function. Mutations that affect the helicase activity cause the repair deficiency that results in XP or Cockayne's syndrome.

24.13 Short Sequence Elements Bind Activators

Key concepts

- Short conserved sequence elements are dispersed in the region preceding the startpoint.
- The upstream elements increase the frequency of initiation.
- The factors that bind to them to stimulate transcription are called activators.

A promoter for RNA polymerase II consists of two types of region. The startpoint itself is identified by the Inr and/or by the TATA box close by. In conjunction with the general transcription factors, RNA polymerase II forms an initiation complex surrounding the startpoint, as we have just described. The efficiency and specificity with which a promoter is recognized, however, depend upon short sequences farther upstream, which are recognized by a different group of factors, usually called **activators.** In general, the target sequences are ~100 bp upstream of the startpoint, but sometimes they are more distant. Binding of activators at these sites may influence the formation of the initiation complex at (probably) any one of several stages.

An analysis of a typical promoter is summarized in FIGURE 24.21. Individual base substitutions were introduced at almost every position in the 100 bp upstream of the β-globin

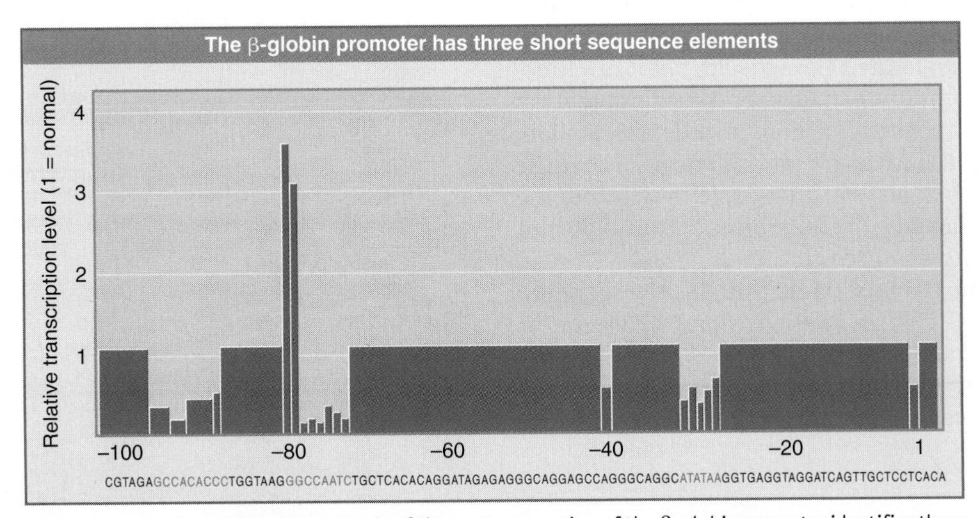

FIGURE 24.21 Saturation mutagenesis of the upstream region of the β-globin promoter identifies three short regions (centered at −30, −75, and −90) that are needed to initiate transcription. These correspond to the TATA, CAAT, and GC boxes.

startpoint. The striking result is that *most mutations do not affect the ability of the promoter to initiate transcription.* Down mutations occur in three locations, corresponding to three short discrete elements. The two upstream elements have a greater effect on the level of transcription than the element closest to the startpoint. Up mutations occur in only one of the elements. We conclude that the three short sequences centered at –30, –75, and –90 constitute the promoter. Each of them corresponds to the consensus sequence for a common type of promoter element.

The TATA box (centered at –30) is the least effective component of the promoter as measured by the reduction in transcription that is caused by mutations. Note that although initiation is not prevented when a TATA box is mutated, the startpoint varies from its usual precise location. This confirms the role of the TATA box as a crucial positioning component of the core promoter.

The basal elements and the elements upstream of them have different types of functions. The basal elements (the TATA box and Inr) primarily determine the location of the startpoint, but can sponsor initiation only at a rather low level. They identify the *location* at which the general transcription factors assemble to form the basal complex. The sequence elements farther upstream influence the *frequency* of initiation, most likely by acting directly on the general transcription factors to enhance the efficiency of assembly into an initiation complex (see Section 25.5, Activators Interact with the Basal Apparatus).

The sequence at –75 is the **CAAT box.** Named for its consensus sequence, it was one of the first common elements to be described. It is often located close to –80, but it can function at distances that vary considerably from the startpoint. It functions in either orientation. Susceptibility to mutations suggests that the CAAT box plays a strong role in determining the efficiency of the promoter, but does not influence its specificity.

The **GC box** at –90 contains the sequence GGGCGG. Often multiple copies are present in the promoter, and they occur in either orientation. The GC box, too, is a relatively common promoter component.

24.14 Promoter Construction Is Flexible but Context Can Be Important

Key concepts

- No individual upstream element is essential for promoter function, although one or more elements must be present for efficient initiation.
- Some elements are recognized by multiple factors, and the factor that is used at any particular promoter may be determined by the context of the other factors that are bound.

Promoters are organized on a principle of "mix and match." A variety of elements can contribute to promoter function, but none is essential for all promoters. Some examples are summarized in FIGURE 24.22. Four types of elements are found altogether in these promoters: TATA, GC boxes, CAAT boxes, and the octamer (an 8 bp element). The elements found in any individual promoter differ in number, location, and orientation. No element is common to all of the promoters. Although the promoter conveys directional information (transcription proceeds only in the downstream direction), the GC and CAAT boxes seem to be able to function in either orientation. This implies that the elements function solely as DNA-binding sites to bring transcription factors into the vicinity of the startpoint; the structure of a factor must be flexible enough to allow it to make protein–protein contacts with the basal apparatus irrespective of the way in which its DNA-binding domain is oriented and its exact distance from the startpoint.

Activators that are more or less ubiquitous are assumed to be available to any promoter that has a copy of the element that they recog-

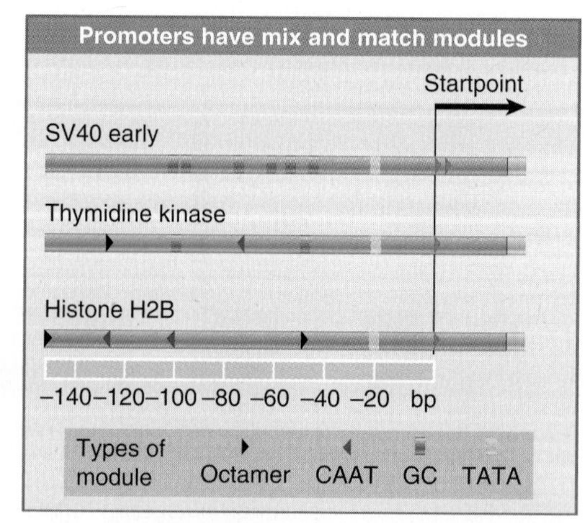

FIGURE 24.22 Promoters contain different combinations of TATA boxes, CAAT boxes, GC boxes, and other elements.

nize. Common elements recognized by ubiquitous activators include the CAAT box, GC box, and the octamer. All promoters probably require one or more of these elements in order to function efficiently. An activator typically binds a consensus sequence of <10 bp, but actually covers a length of ~20 bp of DNA. Given the sizes of the activators, and the length of DNA each covers, we expect that the various proteins will together cover the entire region upstream of the startpoint in which the elements reside.

In general, a particular consensus sequence is recognized by a corresponding activator (or by a member of a family of factors). Sometimes, though, a particular promoter sequence can be recognized by one of several activators. A ubiquitous activator, Oct-1, binds to the octamer to activate the histone H2B (and presumably also other) genes. Oct-1 is the only octamer-binding factor in nonlymphoid cells. In lymphoid cells, however, a different activator, Oct-2, binds to the octamer to activate the immunoglobulin κ light gene. Thus Oct-2 is a tissue-specific activator, whereas Oct-1 is ubiquitous. The exact details of recognition are not so important to know as the fact that a variety of activators recognize CAAT boxes.

The use of the same octamer in the ubiquitously expressed H2B gene and the lymphoid-specific immunoglobulin genes poses a paradox. Why does the ubiquitous Oct-1 fail to activate the immunoglobulin genes in nonlymphoid tissues? The *context* must be important: Oct-2 rather than Oct-1 may be needed to interact with other proteins that bind at the promoter. These results mean that we cannot predict whether a gene will be activated by a particular activator simply on the basis of the presence of particular elements in its promoter.

24.15 Enhancers Contain Bidirectional Elements That Assist Initiation

Key concepts

- An enhancer activates the nearest promoter to it, and can be any distance either upstream or downstream of the promoter.
- A UAS (upstream activator sequence) in yeast behaves like an enhancer but works only upstream of the promoter.
- Similar sequence elements are found in enhancers and promoters.
- Enhancers form complexes of activators that interact directly or indirectly with the promoter.

Until now, we have considered the promoter an isolated region responsible for binding RNA polymerase. Eukaryotic promoters do not necessarily function alone, though. In at least some cases, the activity of a promoter is enormously increased by the presence of an enhancer, which consists of another group of elements but is located at a variable distance from those regarded as comprising part of the promoter itself.

The concept that the enhancer is distinct from the promoter reflects two characteristics. The position of the enhancer relative to the promoter need not be fixed, but can vary substantially. FIGURE 24.23 shows that it can be either upstream or downstream. In addition, it can function in either orientation (that is, it can be inverted) relative to the promoter. Manipulations of DNA show that an enhancer can stimulate any promoter placed in its vicinity. In natural genomes, enhancers can be located within genes (that is, just downstream of the promoter) or tens of kilobases away in either direction.

For operational purposes, it is sometimes useful to define the promoter *as a sequence or sequences of DNA that must be in a (relatively) fixed location with regard to the startpoint.* By this definition, the TATA box and other upstream elements are included, but the enhancer is excluded. This is, however, a working definition rather than a rigid classification.

Elements analogous to enhancers, called **upstream activator sequences (UAS),** are

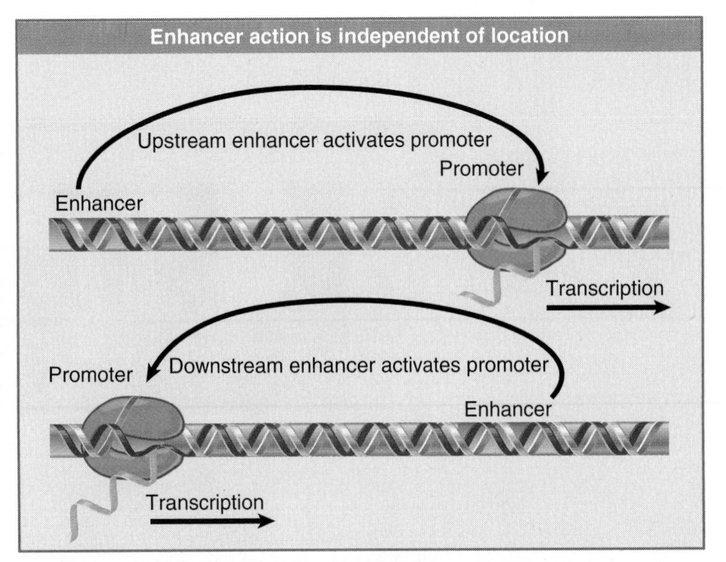

FIGURE 24.23 An enhancer can activate a promoter from upstream or downstream locations, and its sequence can be inverted relative to the promoter.

found in yeast. They can function in either orientation at variable distances upstream of the promoter, but cannot function when located downstream. They have a regulatory role: In several cases the UAS is bound by the regulatory protein(s) that activates the genes downstream.

Reconstruction experiments in which the enhancer sequence is removed from the DNA and then is inserted elsewhere show that normal transcription can be sustained so long as it is present *anywhere* on the DNA molecule. If a β-globin gene is placed on a DNA molecule that contains an enhancer, its transcription is increased *in vivo* more than 200-fold, even when the enhancer is several kb upstream or downstream of the startpoint, in either orientation. We have yet to discover at what distance the enhancer fails to work.

24.16 Enhancers Contain the Same Elements That Are Found at Promoters

Key concepts

- Enhancers are made of the same short sequence elements that are found in promoters.
- The density of sequence components is greater in the enhancer than in the promoter.

A difference between the enhancer and a typical promoter is presented by the density of regulatory elements. FIGURE 24.24 summarizes the susceptibility of the SV40 enhancer to damage by mutation, and we see that a much greater proportion of its sites directly influences its function than is the case with the promoter analyzed in the same way in Figure 24.21. There is a corresponding increase in the density of protein-binding sites. Many of these sites are common elements in promoters; for example, AP1 and the octamer.

The specificity of transcription may be controlled by either a promoter or an enhancer. A promoter may be specifically regulated and a nearby enhancer used to increase the efficiency of initiation, or a promoter may lack specific regulation but become active only when a nearby enhancer is specifically activated. An example is provided by immunoglobulin genes, which carry enhancers *within* the transcription unit. The immunoglobulin enhancers appear to be active only in the B lymphocytes in which the immunoglobulin genes are expressed. Such enhancers provide part of the regulatory network by which gene expression is controlled.

A difference between enhancers and promoters may be that an enhancer shows greater cooperativity between the binding of factors. A complex that assembles at the enhancer that responds to IFN (interferon) γ assembles cooperatively to form a functional structure called the **enhanceosome.** Binding of the nonhistone protein HMGI(Y) bends the DNA into a structure that then binds several activators (NF-κB, IRF, and ATF-Jun). In contrast with the "mix and match" construction of promoters, all of these components are required to create an active structure at the enhancer. These components do not themselves directly bind to RNA polymerase, but they create a surface that binds a *coactivating complex.* The complex helps the preinitiation complex of basal transcription factors that is assembling at the promoter to recruit RNA polymerase. We discuss the function of

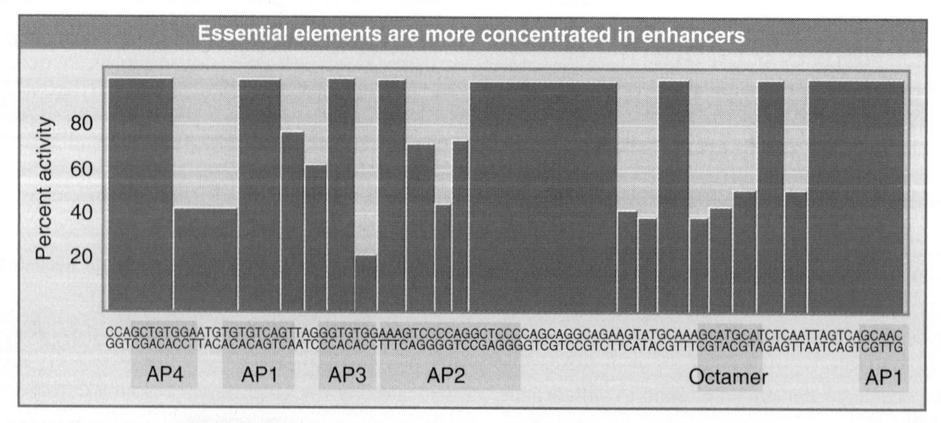

FIGURE 24.24 An enhancer contains several structural motifs. The histogram plots the effect of all mutations that reduce enhancer function to <75% of wild type. Binding sites for proteins are indicated below the histogram.

coactivators in more detail in Section 25.5, Activators Interact with the Basal Apparatus.

24.17 Enhancers Work by Increasing the Concentration of Activators Near the Promoter

Key concepts

- Enhancers usually work only in *cis* configuration with a target promoter.
- Enhancers can be made to work in *trans* configuration by linking the DNA that contains the target promoter to the DNA that contains the enhancer via a protein bridge or by catenating the two molecules.
- The principle is that an enhancer works in any situation in which it is constrained to be in the same proximity as the promoter.

How can an enhancer stimulate initiation at a promoter that can be located any distance away on either side of it? When enhancers were first discovered, several possibilities were considered for their action as elements distinctly different from promoters:

- An enhancer could change the overall structure of the template—for example, by influencing the density of supercoiling.
- It could be responsible for locating the template at a particular place within the cell—for example, attaching it to the nuclear matrix.
- An enhancer could provide an "entry site"—a point at which RNA polymerase (or some other essential protein) initially associates with chromatin.

Now we take the view that enhancer function involves the same sort of interaction with the basal apparatus as the interactions sponsored by upstream promoter elements. Enhancers are modular, like promoters. Some elements are found in both enhancers and promoters. Some individual elements found in promoters share with enhancers the ability to function at variable distance and in either orientation. Thus the distinction between enhancers and promoters is blurred: Enhancers might be viewed as containing promoter elements, which are grouped closely together, and as having the ability to function at increased distances from the startpoint.

The essential role of the enhancer may be to increase the concentration of activator in the vicinity of the promoter (vicinity in this sense being a relative term). Two types of experiment illustrated in FIGURE 24.25 suggest that this is the case.

A fragment of DNA that contains an enhancer at one end and a promoter at the other is not effectively transcribed, but the enhancer can stimulate transcription from the promoter when they are connected by a protein bridge. Structural effects, such as changes in supercoiling, could not be transmitted across such a bridge; this suggests that the critical feature is bringing the enhancer and promoter into close proximity.

A bacterial enhancer provides a binding site for the regulator NtrC, which acts upon RNA polymerase using promoters recognized by σ^{54}. When the enhancer is placed upon a circle of DNA that is catenated (interlocked) with a circle that contains the promoter, initiation is almost as effective as when the enhancer and promoter are on the same circular molecule. There is, however, no initiation when the enhancer and promoter are on separated circles. Again, this suggests that the critical feature is localization of the protein bound at the enhancer, which increases the enhancer's chance of contacting a protein bound at the promoter.

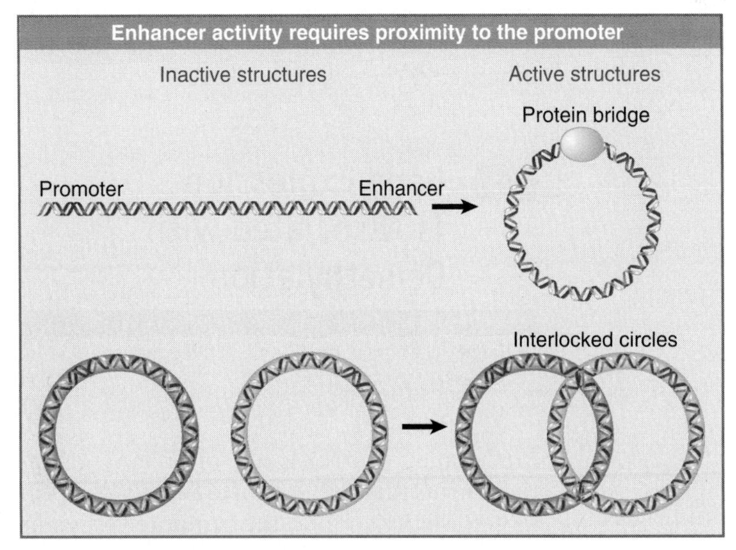

FIGURE 24.25 An enhancer may function by bringing proteins into the vicinity of the promoter. An enhancer does not act on a promoter at the opposite end of a long linear DNA, but becomes effective when the DNA is joined into a circle by a protein bridge. An enhancer and promoter on separate circular DNAs do not interact, but can interact when the two molecules are catenated.

If proteins bound at an enhancer several kb distant from a promoter interact directly with proteins bound in the vicinity of the startpoint, the organization of DNA must be flexible enough to allow the enhancer and promoter to be closely located. This requires the intervening DNA to be extruded as a large "loop." Such loops have been directly observed in the case of the bacterial enhancer.

There is an interesting exception to the rule that enhancers are *cis*-acting in natural situations. This is seen in the phenomenon of transvection. Pairing of somatic chromosomes allows an enhancer on one chromosome to activate a promoter on the partner chromosome. This reinforces the view that enhancers work by proximity.

What limits the activity of an enhancer? Typically it works upon the nearest promoter. There are situations in which an enhancer is located between two promoters, but activates only one of them on the basis of specific protein–protein contacts between the complexes bound at the two elements. The action of an enhancer may be limited by an insulator—an element in DNA that prevents the enhancer from acting on promoters beyond the insulator (see Section 29.14, Insulators Block the Actions of Enhancers and Heterochromatin).

The generality of enhancement is not yet clear. We do not know what proportion of cellular promoters require an enhancer to achieve their usual level of expression, nor do we know how often an enhancer provides a target for regulation. Some enhancers are activated only in the tissues in which their genes function, but others could be active in all cells.

24.18 Gene Expression Is Associated with Demethylation

Key concept

- Demethylation at the 5′ end of the gene is necessary for transcription.

Methylation of DNA is one of several regulatory events that influence the activity of a promoter. Methylation at the promoter prevents transcription, and the methyl groups must be removed in order to activate a promoter. This effect is well characterized at promoters for both RNA polymerase I and RNA polymerase II. In effect, methylation is a reversible regulatory event. It is triggered by modifications to histones that include deacetylation and protein methylation (see Section 30.9, Methylation of Histones and DNA Is Connected).

Methylation also occurs as an epigenetic event. In this case, modification may occur specifically in sperm or oocyte, with the result that there may be a difference between two alleles in the next generation. This can result in differences in the expression of the paternal and maternal alleles (see Section 31.8, DNA Methylation Is Responsible for Imprinting).

In this chapter we are concerned with the means by which methylation influences transcription, which is the same whether the methyl groups were added or removed as a local regulatory event or as an epigenetic event.

Methylation at promoters for RNA polymerase II occurs at CG doublets. The distribution of methyl groups can be examined by taking advantage of restriction enzymes that cleave target sites containing the CG doublet. Two types of restriction activity are compared in FIGURE 24.26. These **isoschizomers** are enzymes that cleave the same target sequence in DNA, but have a different response to its state of methylation.

The enzyme HpaII cleaves the sequence CCGG (writing the sequence of only one strand of DNA). If the second C is methylated, though, the enzyme can no longer recognize the site. The enzyme MspI, however, cleaves the same

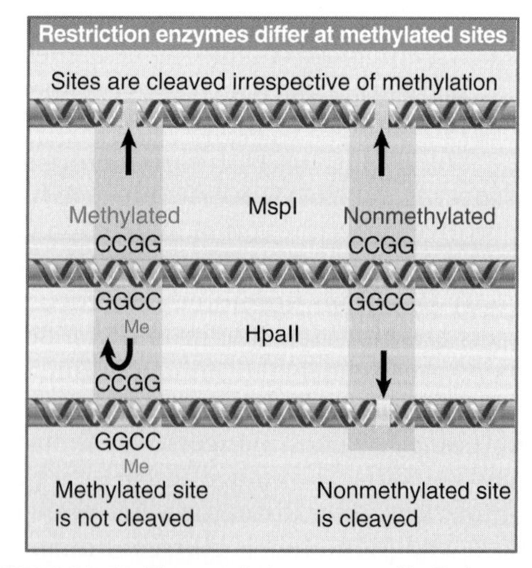

FIGURE 24.26 The restriction enzyme MspI cleaves all CCGG sequences whether or not they are methylated at the second C, but HpaII cleaves only nonmethylated CCGG tetramers.

target site *irrespective* of the state of methylation at this C. Thus MspI can be used to identify all the CCGG sequences, and HpaII can be used to determine whether they are methylated.

With a substrate of nonmethylated DNA, the two enzymes would generate the same restriction bands. In methylated DNA, however, the modified positions are not cleaved by HpaII. For every such position, one larger HpaII fragment replaces two MspI fragments. FIGURE 24.27 gives an example.

Many genes show a pattern in which the state of methylation is constant at most sites but varies at others. Some of the sites are methylated in all tissues examined; some sites are unmethylated in all tissues. *A minority of sites are methylated in tissues in which the gene is not expressed, but are not methylated in tissues in which the gene is active.* Thus an active gene may be described as *undermethylated.*

Experiments with the drug 5-azacytidine produce indirect evidence that demethylation can result in gene expression. The drug is incorporated into DNA in place of cytidine and cannot be methylated, because the 5′ position is blocked. This leads to the appearance of demethylated sites in DNA as the consequence of replication (following the scheme on the right of Figure 15.7).

The phenotypic effects of 5-azacytidine include the induction of changes in the state of cellular differentiation. For example, muscle cells are induced to develop from nonmuscle cell precursors. The drug also activates genes on a silent X chromosome, which raises the possibility that the state of methylation could be connected with chromosomal inactivity.

As well as examining the state of methylation of resident genes, we can compare the results of introducing methylated or nonmethylated DNA into new host cells. Such experiments show a clear correlation: *The methylated gene is inactive, but the nonmethylated gene is active.*

What is the extent of the undermethylated region? In the chicken α-globin gene cluster in adult erythroid cells, the undermethylation is confined to sites that extend from ~500 bp upstream of the first of the two adult α genes to ~500 bp downstream of the second. Sites of undermethylation are present in the entire region, including the spacer between the genes. The region of undermethylation coincides with the region of maximum sensitivity to DNAase I. This argues that undermethylation is a feature of a domain that contains a transcribed gene or genes. As with other changes in chromatin, it seems likely that the absence of methyl groups is associated with the *ability to be transcribed* rather than with the act of transcription itself.

Our problem in interpreting the general association between undermethylation and gene activation is that only a minority (sometimes a small minority) of the methylated sites are involved. It is likely that the state of methylation is critical at specific sites or in a restricted region. It is also possible that a reduction in the level of methylation (or even the complete removal of methyl groups from some stretch of DNA) is part of some structural change needed to permit transcription to proceed.

In particular, demethylation at the promoter may be necessary to make it available for the initiation of transcription. In the γ-globin gene, for example, the presence of methyl groups in the region around the startpoint, between –200 and +90, suppresses transcription. Removal of the three methyl groups located upstream of the startpoint, or of the three methyl groups located downstream, does not relieve the suppression. Removal of all methyl groups, though, allows the promoter to function. Transcription may therefore require a methyl-free region at the promoter (see Section 24.19, CpG Islands Are Regulatory Targets). There are exceptions to this general relationship.

Some genes can be expressed even when they are extensively methylated. Any connection between methylation and expression thus is not universal in an organism, but the general rule is that methylation prevents gene expression and demethylation is required for expression.

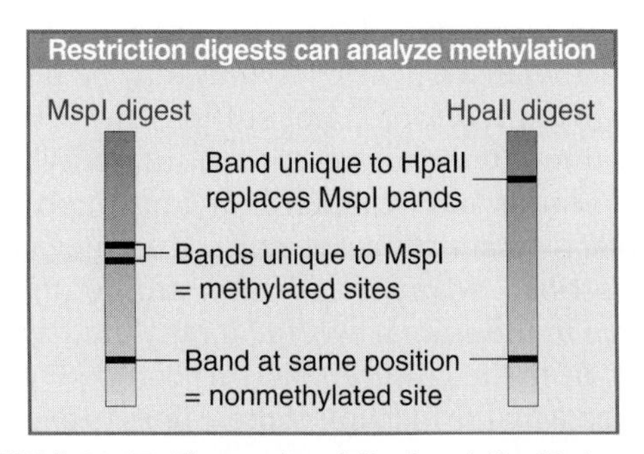

FIGURE 24.27 The results of MspI and HpaII cleavage are compared by gel electrophoresis of the fragments.

24.19 CpG Islands Are Regulatory Targets

Key concepts

- CpG islands surround the promoters of constitutively expressed genes where they are unmethylated.
- CpG islands also are found at the promoters of some tissue-regulated genes.
- There are ~29,000 CpG islands in the human genome.
- Methylation of a CpG island prevents activation of a promoter within it.
- Repression is caused by proteins that bind to methylated CpG doublets.

The presence of **CpG islands** in the 5′ regions of some genes is connected with the effect of methylation on gene expression. These islands are detected by the presence of an increased density of the dinucleotide sequence, CpG. (CpG = 5′-CG-3′)

The CpG doublet occurs in vertebrate DNA at only ~20% of the frequency that would be expected from the proportion of G-C base pairs. (This may be because CpG doublets are methylated on C, and spontaneous deamination of methyl-C converts it to T, which introduces a mutation that removes the doublet.) In certain regions, however, the density of CpG doublets reaches the predicted value; in fact, it is increased by 10× relative to the rest of the genome. The CpG doublets in these regions are generally unmethylated.

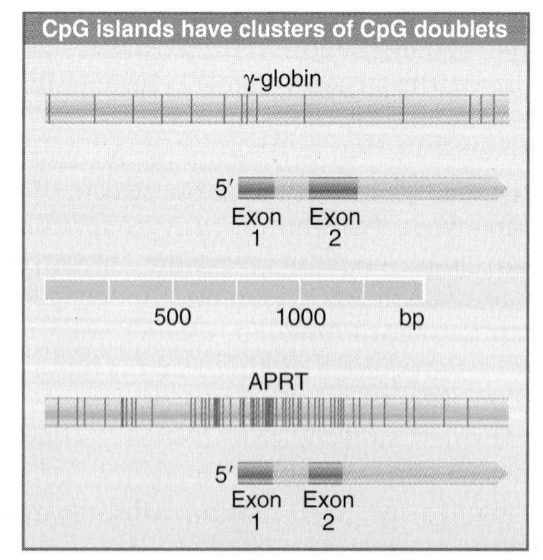

FIGURE 24.28 The typical density of CpG doublets in mammalian DNA is ~1/100 bp, as seen for a γ-globin gene. In a CpG-rich island, the density is increased to >10 doublets/100 bp. The island in the APRT gene starts ~100 bp upstream of the promoter and extends ~400 bp into the gene. Each vertical line represents a CpG doublet.

These CpG-rich islands have an average G-C content of ~60%, compared with the 40% average in bulk DNA. They take the form of stretches of DNA typically 1 to 2 kb long. There are ~45,000 such islands in the human genome. Some of the islands are present in repeated Alu elements, and may just be the consequence of their high G-C-content. The human genome sequence confirms that, excluding these, there are ~29,000 islands. There are fewer in the mouse genome, ~15,500. About 10,000 of the predicted islands in both species appear to reside in a context of sequences that are conserved between the species, suggesting that these may be the islands with regulatory significance. The structure of chromatin in these regions has changes associated with gene expression (see Section 30.11, Promoter Activation Involves an Ordered Series of Events); there is a reduced content of histone H1 (which probably means that the structure is less compact), the other histones are extensively acetylated (a feature that tends to be associated with gene expression), and there are hypersensitive sites (as would be expected of active promoters).

In several cases, CpG-rich islands begin just upstream of a promoter and extend downstream into the transcribed region before petering out. FIGURE 24.28 compares the density of CpG doublets in a "general" region of the genome with a CpG island identified from the DNA sequence. The CpG island surrounds the 5′ region of the APRT gene, which is constitutively expressed.

All of the "housekeeping" genes that are constitutively expressed have CpG islands; this accounts for about half of the islands. The other half of the islands occur at the promoters of tissue-regulated genes; only a minority (<40%) of these genes have islands. In these cases, the islands are unmethylated irrespective of the state of expression of the gene. The presence of unmethylated CpG-rich islands may be necessary, but therefore is not sufficient, for transcription. Thus the presence of unmethylated CpG islands may be taken as an indication that a gene is potentially active rather than inevitably transcribed. Many islands that are nonmethylated in the animal become methylated in cell lines in tissue culture, and this could be connected with the inability of these lines to express all of the functions typical of the tissue from which they were derived.

Methylation of a CpG island can affect transcription. One of two mechanisms can be involved:

- Methylation of a binding site for some factor may prevent it from binding. This happens in a case of binding to a regulatory site other than the promoter (see Section 31.9, Oppositely Imprinted Genes Can Be Controlled by a Single Center).
- Methylation may cause specific repressors to bind to the DNA.

Repression is caused by either of two types of protein that bind to methylated CpG sequences. The protein MeCP1 requires the presence of several methyl groups to bind to DNA, whereas MeCP2 and a family of related proteins can bind to a single methylated CpG base pair. This explains why a methylation-free zone is required for initiation of transcription. Binding of proteins of either type prevents transcription *in vitro* by a nuclear extract.

MeCP2, which directly represses transcription by interacting with complexes at the promoter, is bound also to the Sin3 repressor complex, which contains histone deacetylase activities (see Figure 30.16). This observation provides a direct connection between two types of repressive modifications: methylation of DNA and deacetylation of histones.

The absence of methyl groups is associated with gene expression. There are, however, some difficulties in supposing that the state of methylation provides a general means for controlling gene expression. In the case of *Drosophila melanogaster* (and other Dipteran insects), there is very little methylation of DNA (although there is a gene potentially coding a methyltransferase), and in the nematode *Clostridium elegans* there is no methylation of DNA. The other differences between inactive and active chromatin appear to be the same as in species that display methylation. Thus in these organisms, any role that methylation has in vertebrates is replaced by some other mechanism.

Three changes that occur in active genes are:

- A hypersensitive site(s) is established near the promoter.
- The nucleosomes of a domain including the transcribed region become more sensitive to DNAase I.
- The DNA of the same region is undermethylated.

All of these changes are necessary for transcription.

24.20 Summary

Of the three eukaryotic RNA polymerases, RNA polymerase I transcribes rDNA and accounts for the majority of activity, RNA polymerase II transcribes structural genes for mRNA and has the greatest diversity of products, and RNA polymerase III transcribes small RNAs. The enzymes have similar structures, with two large subunits and many smaller subunits; there are some common subunits among the enzymes.

None of the three RNA polymerases recognize their promoters directly. A unifying principle is that transcription factors have primary responsibility for recognizing the characteristic sequence elements of any particular promoter, and they serve in turn to bind the RNA polymerase and to position it correctly at the startpoint. At each type of promoter, the initiation complex is assembled by a series of reactions in which individual factors join (or leave) the complex. The factor TBP is required for initiation by all three RNA polymerases. In each case it provides one subunit of a transcription factor that binds in the vicinity of the startpoint.

A promoter consists of a number of short sequence elements in the region upstream of the startpoint. Each element is bound by a transcription factor. The basal apparatus, which consists of the TF_{II} factors, assembles at the startpoint and enables RNA polymerase to bind. The TATA box (if there is one) near the startpoint, and the initiator region immediately at the startpoint, are responsible for selection of the exact startpoint at promoters for RNA polymerase II. TBP binds directly to the TATA box when there is one; in TATA-less promoters it is located near the startpoint by binding to the DPE downstream. After binding of $TF_{II}D$, the other general transcription factors for RNA polymerase II assemble the basal transcription apparatus at the promoter. Other elements in the promoter, located upstream of the TATA box, bind activators that interact with the basal apparatus. The activators and basal factors are released when RNA polymerase begins elongation.

The CTD of RNA polymerase II is phosphorylated during the initiation reaction. $TF_{II}D$ and SRB proteins both may interact with the CTD. It may also provide a point of contact for proteins that modify the RNA transcript, including the 5′ capping enzyme, splicing factors, and the 3′ processing complex.

Promoters may be stimulated by enhancers, sequences that can act at great distances and in either orientation on either side of a gene. Enhancers also consist of sets of elements, although they are more compactly organized. Some elements are found in both promoters and enhancers. Enhancers probably function by assembling a protein complex that interacts with the proteins bound at the promoter, requiring that DNA between is "looped out."

CpG islands contain concentrations of CpG doublets and often surround the promoters of constitutively expressed genes, although they are also found at the promoters of regulated genes. The island including a promoter must be unmethylated for that promoter to be able to initiate transcription. A specific protein binds to the methylated CpG doublets and prevents initiation of transcription.

References

24.2 Eukaryotic RNA Polymerases Consist of Many Subunits

Reviews

Doi, R. H. and Wang, L.-F. (1986). Multiple prokaryotic RNA polymerase sigma factors. *Microbiol. Rev.* 50, 227–243.

Young, R. A. (1991). RNA polymerase II. *Annu. Rev. Biochem.* 60, 689–715.

24.4 RNA Polymerase I Has a Bipartite Promoter

Reviews

Grummt, I. (2003). Life on a planet of its own: regulation of RNA polymerase I transcription in the nucleolus. *Genes Dev.* 17, 1691–1702.

Paule, M. R. and White, R. J. (2000). Survey and summary: transcription by RNA polymerases I and III. *Nucleic Acids Res.* 28, 1283–1298.

Research

Bell, S. P., Learned, R. M., Jantzen, H. M., and Tjian, R. (1988). Functional cooperativity between transcription factors UBF1 and SL1 mediates human ribosomal RNA synthesis. *Science* 241, 1192–1197.

24.5 RNA Polymerase III Uses Both Downstream and Upstream Promoters

Research

Bogenhagen, D. F., Sakonju, S., and Brown, D. D. (1980). A control region in the center of the 5S RNA gene directs specific initiation of transcription: II the 3′ border of the region. *Cell* 19, 27–35.

Galli, G., Hofstetter, H., and Birnstiel, M. L. (1981). Two conserved sequence blocks within eukaryotic tRNA genes are major promoter elements. *Nature* 294, 626–631.

Kunkel, G. R. and Pederson, T. (1988). Upstream elements required for efficient transcription of a human U6 RNA gene resemble those of U1 and U2 genes even though a different polymerase is used. *Genes Dev.* 2, 196–204.

Pieler, T., Hamm, J., and Roeder, R. G. (1987). The 5S gene internal control region is composed of three distinct sequence elements, organized as two functional domains with variable spacing. *Cell* 48, 91–100.

Sakonju, S., Bogenhagen, D. F., and Brown, D. D. (1980). A control region in the center of the 5S RNA gene directs specific initiation of transcription: I the 5′ border of the region. *Cell* 19, 13–25.

24.6 $TF_{III}B$ Is the Commitment Factor for Pol III Promoters

Reviews

Geiduschek, E. P. and Tocchini-Valentini, G. P. (1988). Transcription by RNA polymerase III. *Annu. Rev. Biochem.* 57, 873–914.

Schramm, L. and Hernandez, N. (2002). Recruitment of RNA polymerase III to its target promoters. *Genes Dev.* 16, 2593–2620.

Research

Kassavatis, G. A., Braun, B. R., Nguyen, L. H., and Geiduschek, E. P. (1990). *S. cerevisiae* TFIIIB is the transcription initiation factor proper of RNA polymerase III, while TFIIIA and TFIIIC are assembly factors. *Cell* 60, 235–245.

Kassavetis, G. A., Joazeiro, C. A., Pisano, M., Geiduschek, E. P., Colbert, T., Hahn, S., and Blanco, J. A. (1992). The role of the TATA-binding protein in the assembly and function of the multisubunit yeast RNA polymerase III transcription factor, TFIIIB. *Cell* 71, 1055–1064.

Kassavetis, G. A., Letts, G. A., and Geiduschek, E. P. (1999). A minimal RNA polymerase III transcription system. *EMBO J.* 18, 5042–5051.

24.7 The Startpoint for RNA Polymerase II

Reviews

Butler, J. E. and Kadonaga, J. T. (2002). The RNA polymerase II core promoter: a key component in the regulation of gene expression. *Genes Dev.* 16, 2583–2592.

Smale, S. T., Jain, A., Kaufmann, J., Emami, K. H., Lo, K., and Garraway, I. P. (1998). The initiator element: a paradigm for core promoter heterogeneity within metazoan protein-coding genes. *Cold Spring Harb Symp Quant Biol.* 63, 21–31.

Smale, S. T. and Kadonaga, J. T. (2003). The RNA polymerase II core promoter. *Annu. Rev. Biochem.* 72, 449–479.

Woychik, N. A. and Hampsey, M. (2002). The RNA polymerase II machinery: structure illuminates function. *Cell* 108, 453–463.

Research

Burke, T. W. and Kadonaga, J. T. (1996). *Drosophila* TFIID binds to a conserved downstream basal promoter element that is present in many TATA-box-deficient promoters. *Genes Dev.* 10, 711–724.

Singer, V. L., Wobbe, C. R., and Struhl, K. (1990). A wide variety of DNA sequences can functionally replace a yeast TATA element for transcriptional activation. *Genes Dev.* 4, 636–645.

Smale, S. T. and Baltimore, D. (1989). The "initiator" as a transcription control element. *Cell* 57, 103–113.

Weil, P. A., Luse, D. S., Segall, J., and Roeder, R. G. (1979). Selective and accurate initiation of transcription at the Ad2 major late promoter in a soluble system dependent on purified RNA polymerase II and DNA. *Cell* 18, 469–484.

24.8 TBP Is a Universal Factor

Reviews

Berk, A. J. (2000). TBP-like factors come into focus. *Cell* 103, 5–8.

Hernandez, N. (1993). TBP, a universal eukaryotic transcription factor? *Genes Dev.* 7, 1291–1308.

Lee, T. I. and Young, R. A. (1998). Regulation of gene expression by TBP-associated proteins. *Genes Dev.* 12, 1398–1408.

Research

Crowley, T. E., Hoey, T., Liu, J. K., Jan, Y. N., Jan, L. Y., and Tjian, R. (1993). A new factor related to TATA-binding protein has highly restricted expression patterns in *Drosophila*. *Nature* 361, 557–561.

24.9 TBP Binds DNA in an Unusual Way

Reviews

Burley, S. K. and Roeder, R. G. (1996). Biochemistry and structural biology of TFIID. *Annu. Rev. Biochem.* 65, 769–799.

Lee, T. I. and Young, R. A. (1998). Regulation of gene expression by TBP-associated proteins. *Genes Dev.* 12, 1398–1408.

Orphanides, G., Lagrange, T., and Reinberg, D. (1996). The general transcription factors of RNA polymerase II. *Genes Dev.* 10, 2657–2683.

Research

Horikoshi, M. et al. (1988). Transcription factor ATD interacts with a TATA factor to facilitate establishment of a preinitiation complex. *Cell* 54, 1033–1042.

Kim, J. L., Nikolov, D. B., and Burley, S. K. (1993). Cocrystal structure of TBP recognizing the minor groove of a TATA element. *Nature* 365, 520–527.

Kim, Y. et al. (1993). Crystal structure of a yeast TBP/TATA box complex. *Nature* 365, 512–520.

Liu, D. et al. (1998). Solution structure of a TBP-TAFII230 complex: protein mimicry of the minor groove surface of the TATA box unwound by TBP. *Cell* 94, 573–583.

Martinez, E. et al. (1994). TATA-binding protein-associated factors in TFIID function through the initiator to direct basal transcription from a TATA-less class II promoter. *EMBO J.* 13, 3115–3126.

Nikolov, D. B. et al. (1992). Crystal structure of TFIID TATA-box binding protein. *Nature* 360, 40–46.

Ogryzko, V. V. et al. (1998). Histone-like TAFs within the PCAF histone acetylase complex. *Cell* 94, 35–44.

Verrijzer, C. P. et al. (1995). Binding of TAFs to core elements directs promoter selectivity by RNA polymerase II. *Cell* 81, 1115–1125.

Wu, J., Parkhurst, K. M., Powell, R. M., Brenowitz, M., and Parkhurst, L. J. (2001). DNA bends in TATA-binding protein-TATA complexes in solution are DNA sequence-dependent. *J. Biol. Chem.* 276, 14614–14622.

24.10 The Basal Apparatus Assembles at the Promoter

Reviews

Nikolov, D. B. and Burley, S. K. (1997). RNA polymerase II transcription initiation: a structural view. *Proc. Natl. Acad. Sci. USA* 94, 15–22.

Zawel, L. and Reinberg, D. (1993). Initiation of transcription by RNA polymerase II: a multi-step process. *Prog. Nucleic Acid Res. Mol. Biol.* 44, 67–108.

Research

Buratowski, S., Hahn, S., Guarente, L., and Sharp, P. A. (1989). Five intermediate complexes in transcription initiation by RNA polymerase II. *Cell* 56, 549–561.

Burke, T. W. and Kadonaga, J. T. (1996). *Drosophila* TFIID binds to a conserved downstream basal promoter element that is present in many TATA-box-deficient promoters. *Genes Dev.* 10, 711–724.

Bushnell, D. A., Westover, K. D., Davis, R. E., and Kornberg, R. D. (2004). Structural basis of transcription: an RNA polymerase II-TFIIB cocrystal at 4.5 Angstroms. *Science* 303, 983–988.

Littlefield, O., Korkhin, Y., and Sigler, P. B. (1999). The structural basis for the oriented assembly of a TBP/TFB/promoter complex. *Proc. Natl. Acad. Sci. USA* 96, 13668–13673.

Nikolov, D. B. et al. (1995). Crystal structure of a TFIIB-TBP-TATA-element ternary complex. *Nature* 377, 119–128.

24.11 Initiation Is Followed by Promoter Clearance

Reviews

Calvo, O. and Manley, J. L. (2003). Strange bedfellows: polyadenylation factors at the promoter. *Genes Dev.* 17, 1321–1327.

Hirose, Y. and Manley, J. L. (2000). RNA polymerase II and the integration of nuclear events. *Genes Dev.* 14, 1415–1429.

Price, D. H. (2000). P-TEFb, a cyclin dependent kinase controlling elongation by RNA polymerase II. *Mol. Cell Biol.* 20, 2629–2634.

Proudfoot, N. J., Furger, A., and Dye, M. J. (2002). Integrating mRNA processing with transcription. *Cell* 108, 501–512.

Shilatifard, A., Conaway, R. C., and Conaway, J. W. (2003). The RNA polymerase II elongation complex. *Annu. Rev. Biochem.* 72, 693–715.

Woychik, N. A. and Hampsey, M. (2002). The RNA polymerase II machinery: structure illuminates function. *Cell* 108, 453–463.

Research

Douziech, M., Coin, F., Chipoulet, J. M., Arai, Y., Ohkuma, Y., Egly, J. M., and Coulombe, B. (2000). Mechanism of promoter melting by the xeroderma pigmentosum complementation group B helicase of transcription factor IIH revealed by protein-DNA photo-cross-linking. *Mol. Cell Biol.* 20, 8168–8177.

Fong, N. and Bentley, D. L. (2001). Capping, splicing, and 3′ processing are independently stimulated by RNA polymerase II: different functions for different segments of the CTD. *Genes Dev.* 15, 1783–1795.

Goodrich, J. A. and Tjian, R. (1994). Transcription factors IIE and IIH and ATP hydrolysis direct promoter clearance by RNA polymerase II. *Cell* 77, 145–156.

Holstege, F. C., van der Vliet, P. C., and Timmers, H. T. (1996). Opening of an RNA polymerase II promoter occurs in two distinct steps and requires the basal transcription factors IIE and IIH. *EMBO J.* 15, 1666–1677.

Kim, T. K., Ebright, R. H., and Reinberg, D. (2000). Mechanism of ATP-dependent promoter melting by transcription factor IIH. *Science* 288, 1418–1422.

Spangler, L., Wang, X., Conaway, J. W., Conaway, R. C., and Dvir, A. (2001). TFIIH action in transcription initiation and promoter escape requires distinct regions of downstream promoter DNA. *Proc. Natl. Acad. Sci. USA* 98, 5544–5549.

24.12 A Connection between Transcription and Repair

Reviews

Lehmann, A. R. (2001). The xeroderma pigmentosum group D (XPD) gene: one gene, two functions, three diseases. *Genes Dev.* 15, 15–23.

Selby, C. P. and Sancar, A. (1994). Mechanisms of transcription-repair coupling and mutation frequency decline. *Microbiol. Rev.* 58, 317–329.

Research

Bregman, D. et al. (1996). UV-induced ubiquitination of RNA polymerase II: a novel modification deficient in Cockayne syndrome cells. *Proc. Natl. Acad. Sci. USA* 93, 11586–11590.

Schaeffer, L. et al. (1993). DNA repair helicase: a component of BTF2 (TFIIH) basic transcription factor. *Science* 260, 58–63.

Selby, C. P. and Sancar, A. (1993). Molecular mechanism of transcription-repair coupling. *Science* 260, 53–58.

Svejstrup, J. Q. et al. (1995). Different forms of TFIIH for transcription and DNA repair: holo-TFIIH and a nucleotide excision repairosome. *Cell* 80, 21–28.

24.15 Enhancers Contain Bidirectional Elements That Assist Initiation

Review

Muller, M. M., Gerster, T., and Schaffner, W. (1988). Enhancer sequences and the regulation of gene transcription. *Eur. J. Biochem.* 176, 485–495.

Research

Banerji, J., Rusconi, S., and Schaffner, W. (1981). Expression of β-globin gene is enhanced by remote SV40 DNA sequences. *Cell* 27, 299–308.

24.16 Enhancers Contain the Same Elements That Are Found at Promoters

Reviews

Maniatis, T., Falvo, J. V., Kim, T. H., Kim, T. K., Lin, C. H., Parekh, B. S., and Wathelet, M. G. (1998). Structure and function of the interferon-beta enhanceosome. *Cold Spring Harbor Symp. Quant. Biol.* 63, 609–620.

Munshi, N., Yie, Y., Merika, M., Senger, K., Lomvardas, S., Agalioti, T., and Thanos, D. (1999). The IFN-beta enhancer: a paradigm for understanding activation and repression of inducible gene expression. *Cold Spring Harbor Symp. Quant. Biol.* 64, 149–159.

24.17 Enhancers Work by Increasing the Concentration of Activators Near the Promoter

Review

Blackwood, E. M. and Kadonaga, J. T. (1998). Going the distance: a current view of enhancer action. *Science* 281, 60–63.

Research

Mueller-Storm, H. P., Sogo, J. M., and Schaffner, W. (1989). An enhancer stimulates transcription in *trans* when attached to the promoter via a protein bridge. *Cell* 58, 767–777.

Zenke, M. et al. (1986). Multiple sequence motifs are involved in SV40 enhancer function. *EMBO J.* 5, 387–397.

24.19 CpG Islands Are Regulatory Targets

Review

Bird, A. (2002). DNA methylation patterns and epigenetic memory. *Genes Dev.* 16, 6–21.

Research

Antequera, F. and Bird, A. (1993). Number of CpG islands and genes in human and mouse. *Proc. Natl. Acad. Sci. USA* 90, 11995–11999.

Bird, A. et al. (1985). A fraction of the mouse genome that is derived from islands of non-methylated, Cp-G-rich DNA. *Cell* 40, 91–99.

Boyes, J. and Bird, A. (1991). DNA methylation inhibits transcription indirectly via a methyl-CpG binding protein. *Cell* 64, 1123–1134.

25

Activating Transcription

CHAPTER OUTLINE

25.1 Introduction

Key concept

- Eukaryotic gene expression is usually controlled at the level of initiation of transcription.

The phenotypic differences that distinguish the various kinds of cells in a higher eukaryote are largely due to differences in the expression of genes that code for proteins, that is, those transcribed by RNA polymerase II. In principle, the expression of these genes might be regulated at any one of several stages. We can distinguish (at least) five potential control points, which form the following series:

Activation of gene structure
↓
Initiation of transcription
↓
Processing the transcript
↓
Transport to cytoplasm
↓
Translation of mRNA

As we see in FIGURE 25.1, gene expression in eukaryotes is largely controlled at the initiation of transcription. For most genes, this is the major control point in their expression. It involves changes in the structure of chromatin at the promoter (see Section 30.11, Promoter Activation Involves an Ordered Series of Events), accompanied by the binding of the basal transcription apparatus (including RNA polymerase II) to the promoter. (Regulation at subsequent stages of transcription is rare in eukaryotic cells. Premature termination occurs at some genes and is counteracted by a kinase, P-TEFb, but otherwise antitermination does not seem to be employed.)

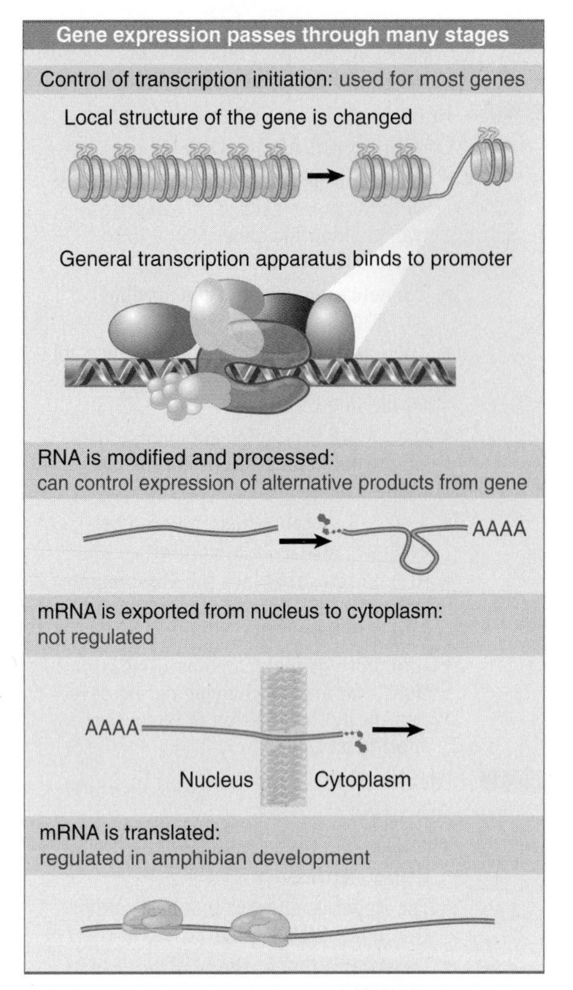

FIGURE 25.1 Gene expression is controlled principally at the initiation of transcription. It is rare for the subsequent stages of gene expression to be used to determine whether a gene is expressed, although control of processing may be used to determine which form of a gene is represented in mRNA.

The primary transcript is modified by capping at the 5′ end, and in general also is modified by polyadenylation at the 3′ end. Introns must be excised from the transcripts of interrupted genes. The mature RNA must be exported from the nucleus to the cytoplasm. Regulation of gene expression by selection of sequences at the level of nuclear RNA might involve any or all of these stages, but the one for which we have most evidence concerns changes in splicing; some genes are expressed by means of alternative splicing patterns whose regulation controls the type of protein product (see Section 26.12, Alternative Splicing Involves Differential Use of Splice Junctions).

Finally, the translation of an mRNA in the cytoplasm can be specifically controlled. While employment of this mechanism is uncommon adult somatic cells, it does occur in some embryonic situations. This can involve localization of the mRNA to specific sites where it is expressed and/or the blocking of initiation of translation by specific protein factors.

Regulation of tissue-specific gene transcription lies at the heart of eukaryotic differentiation. A regulatory transcription factor serves to provide common control of a large number of target genes, and we seek to answer two questions about this mode of regulation: How does the transcription factor identify its group of target genes, and how is the activity of the transcription factor itself regulated in response to intrinsic or extrinsic signals?

25.2 There Are Several Types of Transcription Factors

Key concepts

- The basal apparatus determines the startpoint for transcription.
- Activators determine the frequency of transcription.
- Activators work by making protein–protein contacts with the basal factors.
- Activators may work via coactivators.
- Some components of the transcriptional apparatus work by changing chromatin structure.

Initiation of transcription involves many protein–protein interactions among transcription factors bound at the promoter or at an enhancer, as well as with RNA polymerase. We can divide the factors required for transcription into several classes, which are described in the following list. FIGURE 25.2 summarizes their properties.

- Basal factors, together with RNA polymerase, bind at the startpoint and TATA box (see Section 24.10, The Basal Apparatus Assembles at the Promoter).
- Activators are transcription factors that recognize specific short consensus elements. They bind to sites in the promoter or in enhancers (see Section 24.13, Short Sequence Elements Bind Activators). They act by increasing the efficiency with which the basal apparatus binds to the promoter. They therefore increase the frequency of tran-

scription, and are required for a promoter to function at an adequate level. Some activators act constitutively (they are ubiquitous), whereas others have a regulatory role and are synthesized or activated at specific times or in specific tissues. These factors are therefore responsible for the control of transcription patterns in time and space. The sequences that they bind are called **response elements.**
- Members of another group of factors necessary for efficient transcription do not themselves bind DNA. **Coactivators** provide a connection between activators and the basal apparatus (see Section 25.5, Activators Interact with the Basal Apparatus). They work by protein–protein interactions, forming bridges between activators and the basal transcription apparatus.
- Some coactivators and other regulators act to make changes in chromatin (see Section 30.7, Acetylases Are Associated with Activators).

The diversity of elements from which a functional promoter may be constructed, and the variations in their locations relative to the startpoint, argues that the activators have an ability to interact with one another by protein–protein interactions in multiple ways. There appear to be no constraints on the potential relationships between the elements. The modular nature of the promoter is illustrated by experiments in which equivalent regions of different promoters have been exchanged. Hybrid promoters, for example, between the thymidine kinase and β-globin genes, work well. This suggests that the main purpose of the elements is to bring the activators they bind into the vicinity of the initiation complex, where protein–protein interactions determine the efficiency of the initiation reaction.

The organization of RNA polymerase II promoters contrasts with that of bacterial promoters, where all the transcription factors must interact directly with RNA polymerase. In the eukaryotic system, only the basal factors interact directly with the enzyme. Activators may interact with the basal factors, or they may interact with coactivators that in turn interact with the basal factors. The construction of the apparatus through layers of interactions explains the flexibility with which elements may be arranged and the distance over which they can be dispersed.

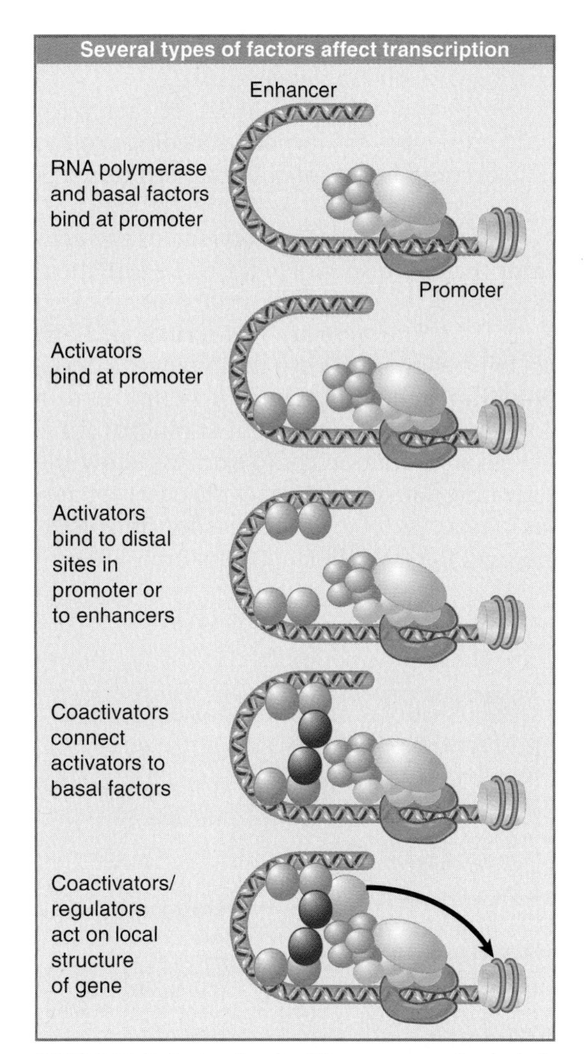

FIGURE 25.2 Factors involved in gene expression include RNA polymerase and the basal apparatus, activators that bind directly to DNA at the promoter or at enhancers, coactivators that bind to both activators and the basal apparatus, and regulators that act on chromatin structure.

25.3 Independent Domains Bind DNA and Activate Transcription

Key concepts

- DNA-binding activity and transcription-activation are carried by independent domains of an activator.
- The role of the DNA-binding domain is to bring the transcription-activation domain into the vicinity of the promoter.

Activators and other regulatory proteins require two types of ability:

- They recognize specific target sequences located in enhancers, promoters, or other regulatory elements that affect a particular target gene.

- Having bound to DNA, an activator exercises its function by binding to other components of the transcription apparatus.

Can we characterize domains in the activator that are responsible for these activities? Often an activator has separate domains that bind DNA and activate transcription. Each domain behaves as a separate module that functions independently when it is linked to a domain of the other type. The geometry of the overall transcription complex must allow the activating domain to contact the basal apparatus irrespective of the exact location and orientation of the DNA-binding domain.

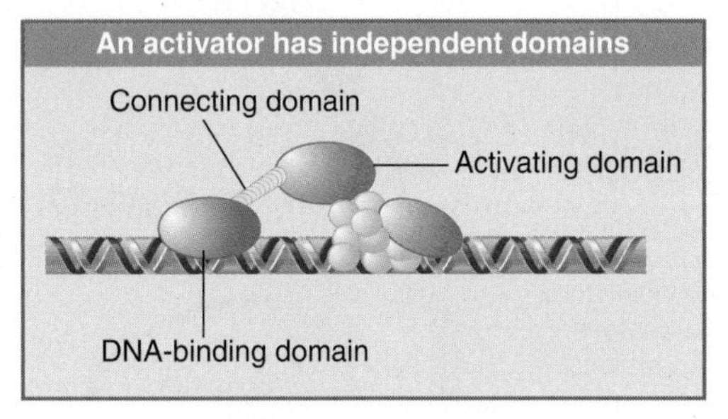

FIGURE 25.3 DNA-binding and activating functions in a transcription factor may comprise independent domains of the protein.

Upstream promoter elements may be an appreciable distance from the startpoint, and in many cases may be oriented in either direction. Enhancers may be even farther away and always show orientation independence. This organization has implications for both the DNA and proteins. The DNA may be looped or condensed in some way to allow the formation of the transcription complex. In addition, the domains of the activator may be connected in a flexible way, as illustrated diagrammatically in FIGURE 25.3. The main point here is that the DNA-binding and activating domains are independent, and are connected in a way that allows the activating domain to interact with the basal apparatus irrespective of the orientation and exact location of the DNA-binding domain.

Binding to DNA is necessary for activating transcription, but does activation depend on the *particular* DNA-binding domain?

FIGURE 25.4 illustrates an experiment to answer this question. The activator Gal4 has a DNA-binding domain that recognizes a *UAS* and an activating domain that stimulates initiation at the target promoter. The bacterial repressor LexA has an N-terminal DNA-binding domain that recognizes a specific operator. When LexA binds to this operator, it represses the adjacent promoter. In a "swap" experiment, the DNA-binding domain of LexA can be substituted for the DNA-binding domain of Gal4. The hybrid gene can then be introduced

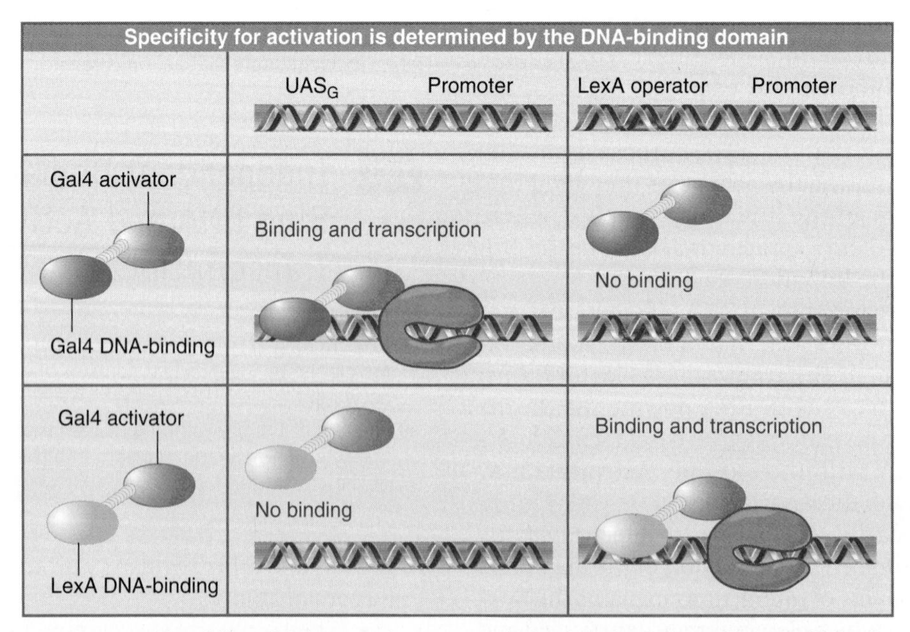

FIGURE 25.4 The ability of Gal4 to activate transcription is independent of its specificity for binding DNA. When the Gal4 DNA-binding domain is replaced by the LexA DNA-binding domain, the hybrid protein can activate transcription when a LexA operator is placed near a promoter.

into yeast together with a target gene that contains either the *UAS* or a LexA operator.

An authentic Gal4 protein can activate a target gene only if it has a *UAS*. The LexA repressor by itself of course lacks the ability to activate either sort of target. The LexA-Gal4 hybrid can no longer activate a gene with a *UAS*, but it can now activate a gene that has a LexA operator!

This result fits the modular view of transcription activators. The DNA-binding domain serves to bring the protein into the right location. Precisely how or where it is bound to DNA is irrelevant, but once it is there, the transcription-activating domain can play its role. According to this view, it does not matter whether the transcription-activating domain is brought to the vicinity of the promoter by recognition of a *UAS* via the DNA-binding domain of Gal4 or by recognition of a LexA operator via the LexA specificity module. The ability of the two types of module to function in hybrid proteins suggests that each domain of the protein folds independently into an active structure that is not influenced by the rest of the protein.

The idea that activators have independent domains that bind DNA and that activate transcription is reinforced by the ability of the tat protein of HIV to stimulate initiation without binding DNA at all. The tat protein binds to a region of secondary structure in the RNA product; the part of the RNA required for tat action is called the *tar* sequence. A model for the role of the tat-*tar* interaction in stimulating transcription is shown in FIGURE 25.5.

The *tar* sequence is located just downstream of the startpoint, so that when tat binds to *tar*, it is brought into the vicinity of the initiation complex. This is sufficient to ensure that its activation domain is in close enough proximity to the initiation complex. The activation domain interacts with one or more of the transcription factors bound at the complex in the same way as an activator. (Of course, the first transcript must be made in the absence of tat in order to provide the binding site.)

An extreme demonstration of the independence of the localizing and activating domains is indicated by some constructs in which tat was engineered so that the activating domain was connected to a DNA-binding domain instead of to the usual *tar*-binding sequence. When an appropriate target site was placed into the promoter, the tat activating-domain could activate transcription. This suggests that we should think of the DNA-binding (or in this case the RNA-binding) domain as providing a "tethering" function, *whose main purpose is to ensure that the activating domain is in the vicinity of the initiation complex.*

The notion of tethering is a more specific example of the general idea that initiation requires a high concentration of transcription factors in the vicinity of the promoter. This may be achieved when activators bind to enhancers in the general vicinity, when they bind to upstream promoter components, or, in an extreme case, when they bind by tethering to the RNA product. The common requirement of all these situations is flexibility in the exact three-dimensional arrangement of DNA and proteins. The principle of independent domains is common in transcriptional activators.

We might view the function of the DNA-binding domain as *bringing the activating domain into the vicinity of the startpoint.* This explains why the exact locations of DNA-binding sites can vary within the promoter.

25.4 The Two Hybrid Assay Detects Protein–Protein Interactions

Key concept

- The two hybrid assay works by requiring an interaction between two proteins, where one has a DNA-binding domain and the other has a transcription-activation domain.

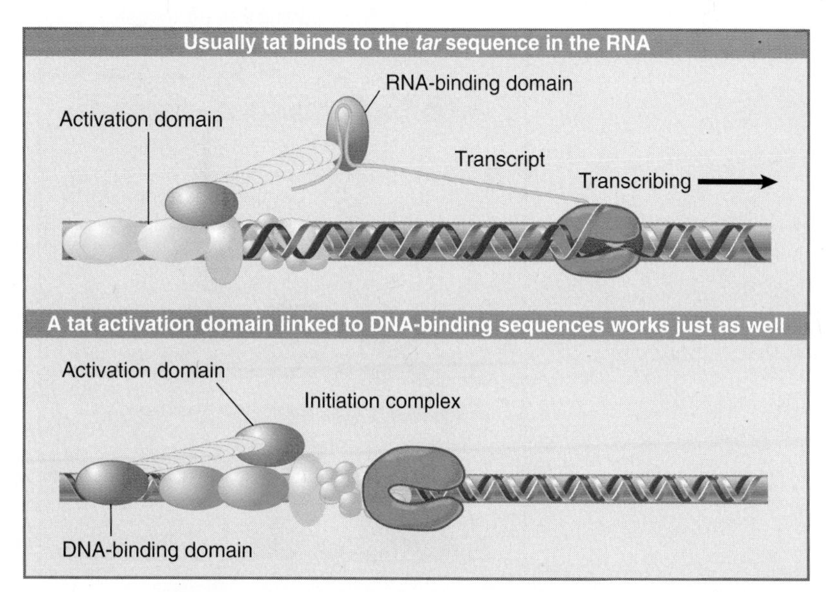

FIGURE 25.5 The activating domain of the tat protein of HIV can stimulate transcription if it is tethered in the vicinity by binding to the RNA product of a previous round of transcription. Activation is independent of the means of tethering, as shown by the substitution of a DNA-binding domain for the RNA-binding domain.

The model of domain independence is the basis for an extremely useful assay for detecting protein interactions. In effect, we replace the connecting domain in Figure 25.3 with a protein–protein interaction. The principle is illustrated in FIGURE 25.6. We fuse one of the proteins to be tested to a DNA-binding domain. We fuse the other protein to a transcription-activating domain. (This is done by linking the appropriate coding sequences in each case and making synthetic proteins by expressing each hybrid gene.)

If the two proteins that are being tested can interact with one another, the two hybrid proteins will interact. This is reflected in the name of the technique: the **two hybrid** assay. The protein with the DNA-binding domain binds to a reporter gene that has a simple promoter containing its target site. It cannot, however, activate the gene by itself. Activation occurs only if the second hybrid binds to the first hybrid to bring the activation domain to the promoter. Any reporter gene can be used where the product is readily assayed, and this technique has given rise to several automated procedures for rapidly testing protein–protein interactions.

The effectiveness of the technique dramatically illustrates the modular nature of proteins. Even when fused to another protein, the DNA-binding domain can bind to DNA and the transcription-activating domain can activate transcription. Correspondingly, the interaction ability of the two proteins being tested is not inhibited by the attachment of the DNA-binding or transcription-activating domains. (Of course, there are some exceptions for which these simple rules do not apply and interference between the domains of the hybrid protein prevents the technique from working.)

The power of this assay is that it requires only that the two proteins being tested can interact with each other. They need not have anything to do with transcription. As a result of the independence of the DNA-binding and transcription-activating domains, all we require is that they are brought together. This will happen so long as the two proteins being tested can interact in the environment of the nucleus.

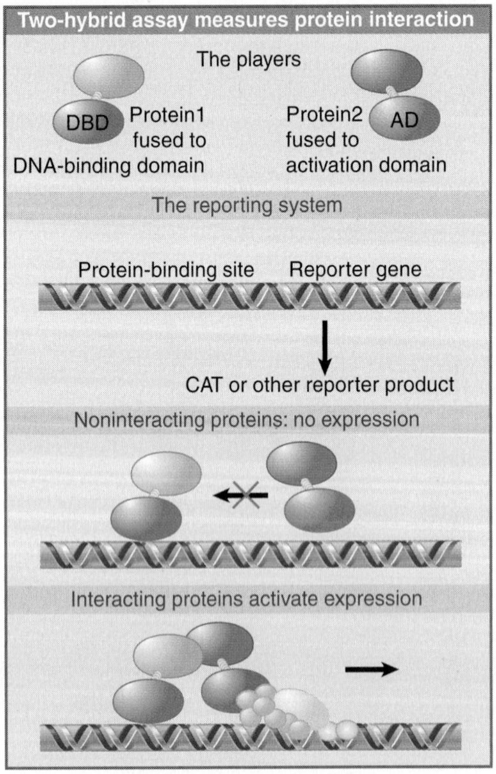

FIGURE 25.6 The two-hybrid technique tests the ability of two proteins to interact by incorporating them into hybrid proteins where one has a DNA-binding domain and the other has a transcription-activating domain.

25.5 Activators Interact with the Basal Apparatus

Key concepts

- The principle that governs the function of all activators is that a DNA-binding domain determines specificity for the target promoter or enhancer.
- The DNA-binding domain is responsible for localizing a transcription-activating domain in the proximity of the basal apparatus.
- An activator that works directly has a DNA-binding domain and an activating domain.
- An activator that does not have an activating domain may work by binding a coactivator that has an activating domain.
- Several factors in the basal apparatus are targets with which activators or coactivators interact.
- RNA polymerase may be associated with various alternative sets of transcription factors in the form of a holoenzyme complex.

An activator may work directly when it consists of a DNA-binding domain linked to a transcription-activating domain, as illustrated in Figure 25.3. In other cases, the activator does not itself have a transcription-activating domain, but binds another protein—a coactivator—that has the transcription-activating domain. FIGURE 25.7 shows the action of such an activator. We may regard coactivators as transcription factors whose specificity is conferred by the ability to bind to DNA-binding transcription factors instead of directly to DNA. A particular activator may require a specific coactivator.

Although the protein components are organized differently, the mechanism is the same. An activator that contacts the basal apparatus directly has an activation domain covalently connected to the DNA-binding domain. When an activator works through a coactivator, the connections involve noncovalent binding between protein subunits (compare Figure 25.3 and Figure 25.7). The same interactions are responsible for activation, irrespective of whether the various domains are present in the same protein subunit or divided into multiple protein subunits.

A transcription-activating domain works by making protein–protein contacts with general transcription factors that promote assembly of the basal apparatus. Contact with the basal apparatus may be made with any one of several basal factors, but typically occurs with $TF_{II}D$, $TF_{II}B$, or $TF_{II}A$. All of these factors participate in early stages of assembly of the basal apparatus (see Figure 24.14). FIGURE 25.8 illustrates the situation when such a contact is made. The major effect of the activators is to influence the assembly of the basal apparatus.

$TF_{II}D$ may be the most common target for activators, which may contact any one of several TAFs. In fact, a major role of the TAFs is to provide the connection from the basal apparatus to activators. This explains why TBP alone can support basal-level transcription, whereas the TAFs of $TF_{II}D$ are required for the higher levels of transcription that are stimulated by activators. Different TAFs in $TF_{II}D$ may provide surfaces that interact with different activators. Some activators interact only with individual TAFs; others interact with multiple TAFs. We assume that the interaction either assists binding of $TF_{II}D$ to the TATA box or assists the binding of other activators around the $TF_{II}D$-TATA box complex. In either case, the interaction stabilizes the basal transcription complex; this speeds the process of initiation, and thereby increases use of the promoter.

The activating domains of the yeast activators Gal4 and Gcn4 have multiple negative charges, giving rise to their description as "acidic activators." Another particularly effective activator of this type is carried by the VP16 protein of the herpes simplex virus. (VP16 does not itself have a DNA-binding domain, but interacts with the transcription apparatus via an intermediary protein.) Experiments to characterize acidic activator functions have often made use of the VP16 activating region linked to a DNA-binding motif.

Acidic activators function by enhancing the ability of $TF_{II}B$ to join the basal initiation complex. Experiments *in vitro* show that binding of $TF_{II}B$ to an initiation complex at an adenovirus promoter is stimulated by the presence of Gal4 or VP16 acid activators, and that the VP16 activator can bind directly to $TF_{II}B$. Assembly of $TF_{II}B$ into the complex at this promoter is therefore a rate-limiting step that is stimulated by the presence of an acidic activator.

The resilience of an RNA polymerase II promoter to the rearrangement of elements, and its indifference even to the particular elements present, suggests that the events by which it is activated are relatively general in nature. Any activators whose activating region is brought within range of the basal initiation complex may be able to stimulate its formation. Some striking illustrations of such versatility have been accomplished by constructing promoters consisting of new combinations of elements. For example, when a yeast UAS_G element is

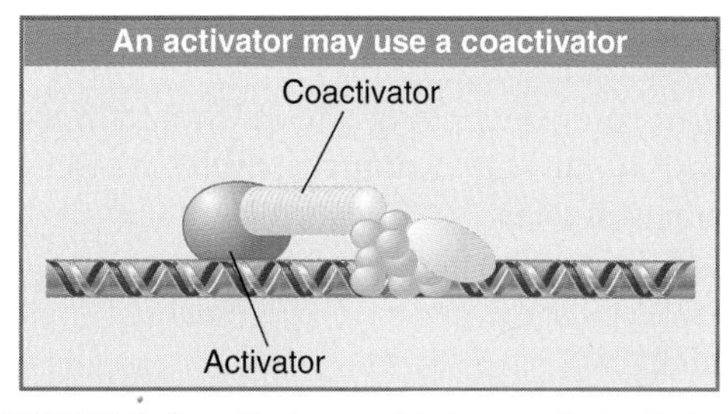

FIGURE 25.7 An activator may bind a coactivator that contacts the basal apparatus.

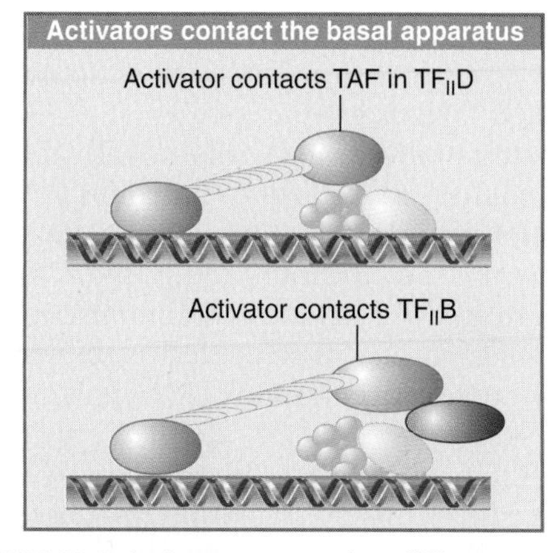

FIGURE 25.8 Activators may work at different stages of initiation by contacting the TAFs of $TF_{II}D$ or by contacting $TF_{II}B$.

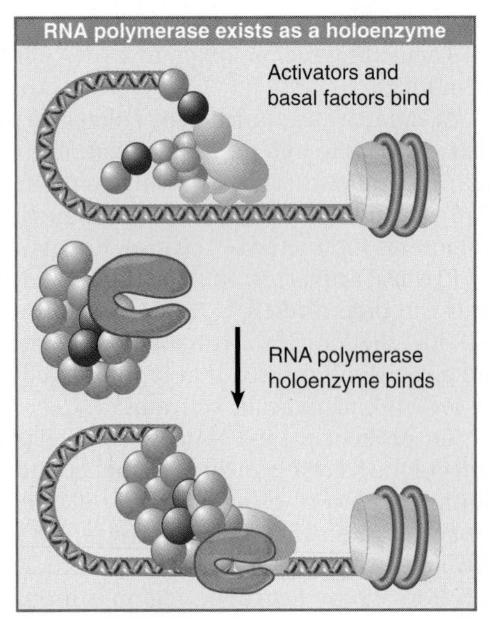

FIGURE 25.9 RNA polymerase exists as a holoenzyme containing many activators.

inserted near the promoter of a higher eukaryotic gene, this gene can be activated by Gal4 in a mammalian cultured cell. Whatever means Gal4 uses to activate the promoter seems therefore to have been conserved between yeast and higher eukaryotes. The Gal4 protein must recognize some feature of the mammalian transcription apparatus that resembles its normal contacts in yeast.

How does an activator stimulate transcription? We can imagine two general types of model:

- The recruitment model argues that its sole effect is to increase the binding of RNA polymerase to the promoter.
- An alternative model is to suppose that it induces some change in the transcriptional complex, for example, in the conformation of the enzyme, which increases its efficiency.

A test of these models in one case in yeast showed that recruitment can account for activation. When the concentration of RNA polymerase was increased sufficiently, the activator failed to produce any increase in transcription. This suggests that its sole effect is to increase the effective concentration of RNA polymerase at the promoter.

When we add up all the components required for efficient transcription—basal factors, RNA polymerase, activators, and coactivators—we get a very large apparatus that consists of >40 proteins. Is it feasible for this apparatus to assemble step-by-step at the promoter? Some activators, coactivators, and basal factors may assemble stepwise at the promoter, but then may be joined by a very large complex consisting of RNA polymerase preassembled with further activators and coactivators, as illustrated in FIGURE 25.9.

Several forms of RNA polymerase have been found in which the enzyme is associated with various transcription factors. The most prominent "holoenzyme complex" in yeast (defined as being capable of initiating transcription without additional components) consists of RNA polymerase associated with a 20-subunit complex called **mediator.** Mediator includes products of several genes in which mutations block transcription, including some *SRB* loci (so named because many of their genes were originally identified as *s*uppressors of mutations in *R*NA polymerase *B*). The name was suggested by its ability to mediate the effects of activators. Mediator is necessary for transcription of most yeast genes. Homologous complexes are required for the transcription of most higher eukaryotic genes. Mediator undergoes a conformational change when it interacts with the C-terminal domain (CTD) of RNA polymerase. It can transmit either activating or repressing effects from upstream components to the RNA polymerase. It is probably released when a polymerase starts elongation. Some transcription factors influence transcription directly by interacting with RNA polymerase or the basal apparatus, but others work by manipulating structure of chromatin (see Section 30.3, Chromatin Remodeling Is an Active Process).

25.6 Some Promoter-Binding Proteins Are Repressors

Key concept

- Repression is usually achieved by affecting chromatin structure, but there are repressors that act by binding to specific promoters.

Repression of transcription in eukaryotes is generally accomplished at the level of influencing chromatin structure; regulator proteins that function like *trans*-acting bacterial repressors to block transcription are relatively rare, but some examples are known. One case is the global repressor NC2/Dr1/DRAP1, a heterodimer that binds to TBP to prevent it from interacting with other components of the basal apparatus. The importance of this interaction is suggested by the

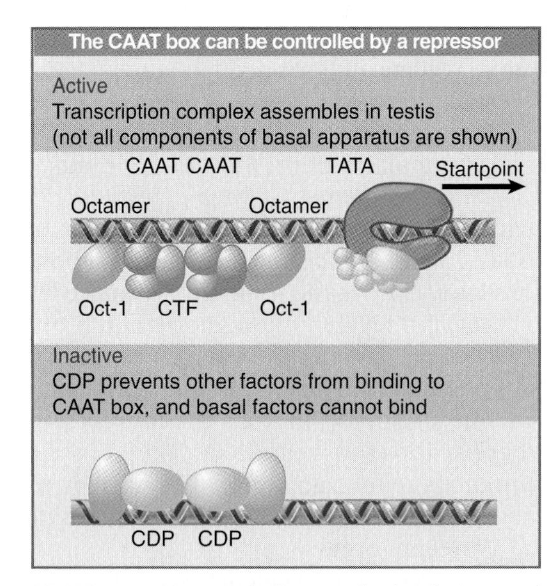

FIGURE 25.10 A transcription complex involves recognition of several elements in the sea urchin H2B promoter in testis. Binding of the CAAT displacement factor in embryo prevents the CAAT-binding factor from binding, so an active complex cannot form.

lethality of null mutations in the genes that code for the repressor in yeast. Repressors that work in this way have an active role in inhibiting basal apparatus function.

In a more specific case, the CAAT sequence is a target for regulation. Two copies of this element are found in the promoter of a gene for histone H2B (see Figure 24.22) that is expressed only during spermatogenesis in a sea urchin. CAAT-binding factors can be extracted from testis tissue and also from embryonic tissues, but only the former can bind to the CAAT box. In the embryonic tissues another protein, called the CAAT-displacement protein (CDP), binds to the CAAT boxes, *thereby preventing the activator from recognizing them.*

FIGURE 25.10 illustrates the consequences for gene expression. In testis, the promoter is bound by transcription factors at the TATA box, CAAT boxes, and octamer sequences. In embryonic tissue, the exclusion of the CAAT-binding factor from the promoter prevents a transcription complex from being assembled. The analogy with the effect of a bacterial repressor in preventing RNA polymerase from initiating at the promoter is obvious. These results also make the point that the function of a protein in binding to a known promoter element cannot be assumed: It may be an activator, a repressor, or even irrelevant to gene transcription.

25.7 Response Elements Are Recognized by Activators

Key concepts

- Response elements may be located in promoters or enhancers.
- Each response element is recognized by a specific activator.
- A promoter may have many response elements, which in turn may activate transcription independently or in certain combinations.

The principle that emerges from characterizing groups of genes under common control is that *they share a promoter (or enhancer) element that is recognized by an activator.* An element that causes a gene to respond to such a factor is called a response element; examples are the **HSE (heat shock response element),** the **GRE (glucocorticoid response element),** and the **SRE (serum response element).** Response elements contain short consensus sequences; copies of the response elements found in different genes are closely related, but not necessarily identical. The region bound by the factor extends for a short distance on either side of the consensus sequence. In promoters, the elements are not present at fixed distances from the startpoint, but are usually <200 bp upstream of it. The presence of a single element usually is sufficient to confer the regulatory response, but sometimes there are multiple copies.

Response elements may be located in promoters or in enhancers. Some types of elements are typically found in one rather than the other: Usually an HSE is found in a promoter, whereas a GRE is found in an enhancer. We assume that all response elements function by the same general principle. Binding of an activator to the response element is required to allow RNA polymerase to initiate transcription. The difference from constitutively active activators is that the protein is either available or is active only under certain conditions. The conditions determine when the gene is to be expressed.

An example of a situation in which many genes are controlled by a single factor is provided by the heat shock response. This is common to a wide range of prokaryotes and eukaryotes and involves multiple controls of gene expression: An increase in temperature turns off transcription of some genes, turns on transcription of the **heat shock genes,** and causes changes in the translation of mRNAs. The control of the heat shock genes illustrates the differences between prokaryotic and eukaryotic modes of control. In bacteria, a new sigma factor is synthesized that directs RNA polymerase holoenzyme to recognize an alternative −10 sequence common to the promoters of heat shock genes (see Section 11.16, Substitution of Sigma Factors May Control Initiation). In eukaryotes, the heat shock genes also possess a common consensus sequence (HSE), but it is located at various positions relative to the startpoint and is recognized by an independent activator, heat shock transcription factor (HSTF). The activation of this factor therefore provides a means to initiate transcription at the specific group of ~20 genes that contains the appropriate target sequence at its promoter.

All the heat shock genes of *Drosophila melanogaster* contain multiple copies of the HSE. The HSTF binds cooperatively to adjacent response elements. Both the HSE and HSTF have been conserved in evolution, and it is striking that a heat shock gene from *D. melanogaster* can be activated in species as distant as mammals or sea urchins. The HSTF proteins of fruit fly and yeast appear similar, and show the same footprint pattern on DNA containing HSE sequences. Yeast HSTF becomes phosphorylated when cells are heat-shocked; this modification is responsible for activating the protein.

The metallothionein (MT) gene provides an example of how a single gene may be regulated by many different circuits. The metallothionein protein protects the cell against excess concentrations of heavy metals by binding the metal and removing it from the cell. The gene is expressed at a basal level, but is induced to greater levels of expression by heavy metal ions (such as cadmium) or by glucocorticoids. The control region combines several different kinds of regulatory elements.

The organization of the promoter for an MT gene is summarized in FIGURE 25.11. A major feature of this map is the high density of elements that can activate transcription. The TATA and GC boxes are located at their usual positions fairly close to the startpoint. Also needed for the basal level of expression are the two basal level elements (BLE), which fit the formal description of enhancers. Although located near the startpoint, they can be moved elsewhere without loss of effect. They contain sequences related to those found in other enhancers, and are bound by proteins that bind the SV40 enhancer.

The TPA response element (TRE) is a consensus sequence that is present in several enhancers, including one BLE of metallothionein and the 72 bp repeats of the virus SV40. The TRE has a binding site for factor AP1; this interaction is part of the mechanism for constitutive expression, for which AP1 is an activator. AP1 binding also has a second function: The TRE confers a response to phorbol esters such as TPA (an agent that promotes tumors), and this response is mediated by the interaction of AP1 with the TRE. This binding reaction is one (but not necessarily the sole) means by which phorbol esters trigger a series of transcriptional changes.

The inductive response to metals is conferred by the multiple metal response element (MRE) sequences, which function as promoter elements. The presence of one MRE confers the ability to respond to heavy metal; a greater level of induction is achieved by the inclusion of mul-

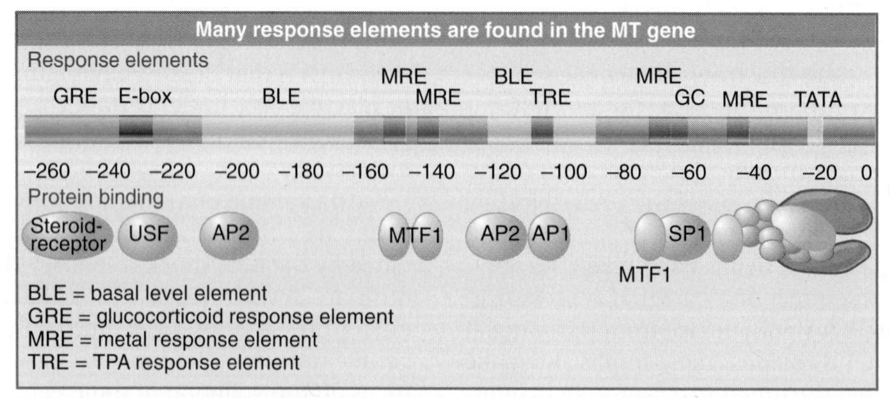

FIGURE 25.11 The regulatory region of a human metallothionein gene contains regulator elements in both its promoter and enhancer. The promoter has elements for metal induction; an enhancer has an element for response to glucocorticoid. Promoter elements are shown above the map, and proteins that bind them are indicated below.

tiple elements. The factor MTF1 binds to the MRE in response to the presence of metal ions.

The response to steroid hormones is governed by a GRE located 250 bp upstream of the startpoint, which behaves as an enhancer. Deletion of this region does not affect the basal level of expression or the level induced by metal ions, but it is absolutely needed for the response to steroids.

The regulation of metallothionein illustrates the general principle that *any one of several different elements, located in either an enhancer or promoter, can independently activate the gene.* The absence of an element needed for one mode of activation does not affect activation in other modes. The variety of elements, their independence of action, and the apparently unlimited flexibility of their relative arrangements suggest that a factor binding to any one element is able independently to increase the efficiency of initiation by the basal transcription apparatus, probably by virtue of protein–protein interactions that stabilize or otherwise assist formation of the initiation complex.

25.8 There Are Many Types of DNA-Binding Domains

Key concepts

- Activators are classified according to the type of DNA-binding domain.
- Members of the same group have sequence variations of a specific motif that confer specificity for individual target sites.

It is common for an activator to have a modular structure in which different domains are responsible for binding to DNA and for activating transcription. Factors are often classified according to the type of DNA-binding domain. In general, a relatively short motif in this domain is responsible for binding to DNA:

- The **zinc finger** motif comprises a DNA-binding domain. It was originally recognized in factor $TF_{III}A$, which is required for RNA polymerase III to transcribe 5S rRNA genes. It has since been identified in several other transcription factors (and presumed transcription factors). A distinct form of the motif is found also in the steroid receptors.
- The **steroid receptors** are defined as a group by a functional relationship: Each receptor is activated by binding a particular steroid. The glucocorticoid receptor is the most fully analyzed. Together with other receptors, such as the thyroid hormone receptor or the retinoic acid receptor, the steroid receptors are members of the superfamily of ligand-activated activators with the same general *modus operandi: The protein factor is inactive until it binds a small ligand.*
- The **helix-turn-helix** motif was originally identified as the DNA-binding domain of phage repressors. One α-helix lies in the major groove of DNA; the other lies at an angle across DNA. A related form of the motif is present in the **homeodomain,** a sequence first characterized in several proteins coded by genes concerned with developmental regulation in *Drosophila.* It is also present in genes for mammalian transcription factors.
- The amphipathic **helix-loop-helix (HLH)** motif has been identified in some developmental regulators and in genes coding for eukaryotic DNA-binding proteins. Each amphipathic helix presents a face of hydrophobic residues on one side and charged residues on the other side. The length of the connecting loop varies from 12 to 28 amino acids. The motif enables proteins to dimerize, and a basic region near this motif contacts DNA.
- **Leucine zippers** consist of a stretch of amino acids with a leucine residue in every seventh position. A leucine zipper in one polypeptide interacts with a zipper in another polypeptide to form a dimer. Adjacent to each zipper is a stretch of positively charged residues that is involved in binding to DNA.

The activity of an inducible activator may be regulated in any one of several ways, as illustrated schematically in FIGURE 25.12:

- A factor is tissue-specific because it is synthesized only in a particular type of cell. This is typical of factors that regulate development, such as homeodomain proteins.
- The activity of a factor may be directly controlled by modification. HSTF is converted to the active form by phosphorylation. AP1 (a heterodimer between the subunits Jun and Fos) is converted to the active form by phosphorylating the Jun subunit.

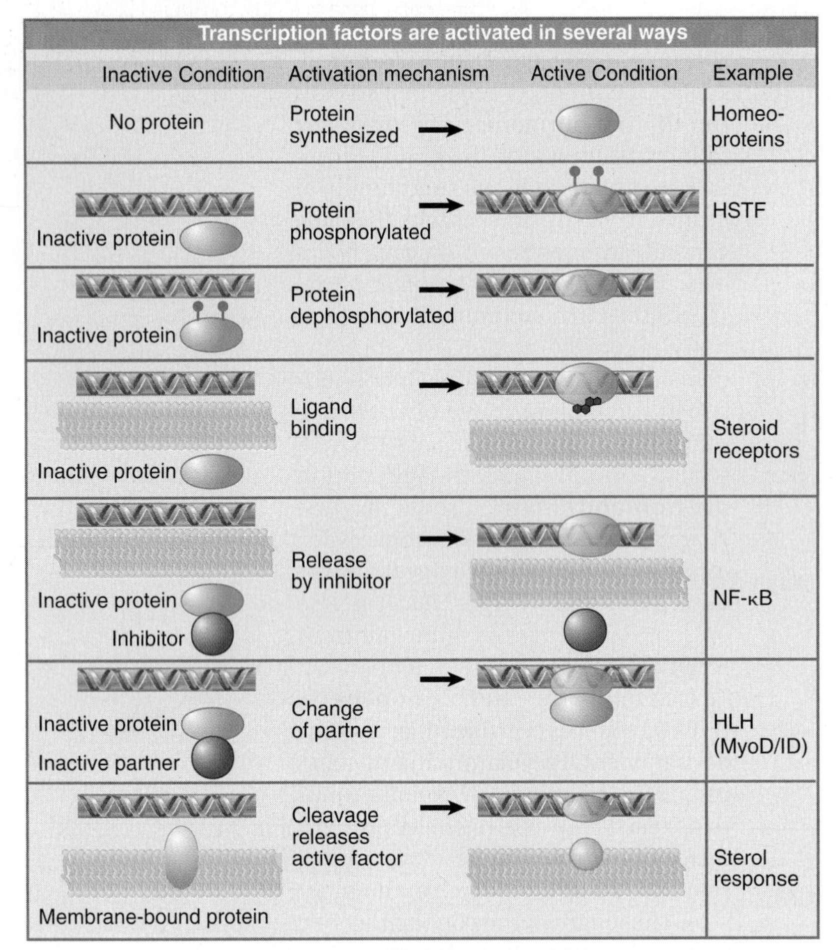

FIGURE 25.12 The activity of a regulatory transcription factor may be controlled by synthesis of protein, covalent modification of protein, ligand binding, or binding of inhibitors that sequester the protein or affect its ability to bind to DNA.

- A factor is activated or inactivated by binding a ligand. The steroid receptors are prime examples. Ligand binding may influence the localization of the protein (causing transport from cytoplasm to nucleus), as well as determining its ability to bind to DNA.
- Availability of a factor may vary; for example, the factor NF-κB (which activates immunoglobulin κ genes in B lymphocytes) is present in many cell types. It is sequestered in the cytoplasm, however, by the inhibitory protein I-κB. In B lymphocytes, NF-κB is released from I-κB and moves to the nucleus, where it activates transcription.
- A dimeric factor may have alternative partners. One partner may cause it to be inactive; synthesis of the active partner may displace the inactive partner. Such situations may be amplified into networks in which various alternative partners pair with one another, especially among the HLH proteins.
- The factor may be cleaved from an inactive precursor. One activator is produced as a protein bound to the nuclear envelope and endoplasmic reticulum. The absence of sterols (such as cholesterol) causes the cytosolic domain to be cleaved; it then translocates to the nucleus and provides the active form of the activator.

We now discuss in more detail the DNA-binding and activation reactions that are sponsored by some of these classes of proteins.

25.9 A Zinc Finger Motif Is a DNA-Binding Domain

Key concepts

- A zinc finger is a loop of ~23 amino acids that protrudes from a zinc-binding site formed by His and Cys amino acids.
- A zinc finger protein usually has multiple zinc fingers.
- The C-terminal part of each finger forms an α-helix that binds one turn of the major groove of DNA.
- Some zinc finger proteins bind RNA instead of, or as well as, DNA.

Zinc fingers take their name from the structure illustrated in FIGURE 25.13, in which a small group of conserved amino acids binds a zinc ion to form an independent domain in the protein. Two types of DNA-binding proteins have structures of this type: The classic "zinc finger" proteins and the steroid receptors.

A "finger protein" typically has a series of zinc fingers, as depicted in the figure. The consensus sequence of a single finger is:

$$Cys\text{-}X_{2\text{-}4}\text{-}Cys\text{-}X_3\text{-}Phe\text{-}X_5\text{-}Leu\text{-}X_2\text{-}His\text{-}X_3\text{-}His$$

The motif takes its name from the loop of amino acids that protrudes from the zinc-binding site and is described as the Cys_2/His_2 finger. The zinc is held in a tetrahedral structure formed by the conserved Cys and His residues. The finger itself comprises ~23 amino acids, and the linker between fingers is usually seven to eight amino acids.

Zinc fingers are a common motif in DNA-binding proteins. The fingers usually are organized as a single series of tandem repeats; occasionally there is more than one group of fingers. The stretch of fingers ranges from nine repeats that occupy almost the entire protein (as in $TF_{III}A$) to providing just one small domain consisting of two fingers (as in the *Drosophila* regulator ADR1). The activator Sp1 has a DNA-binding domain that consists of three zinc fingers.

The crystal structure of DNA bound by a protein with three fingers suggests the structure illustrated schematically in FIGURE 25.14. The C-terminal part of each finger forms α helices that bind DNA; the N-terminal part forms a β sheet. (For simplicity, the β sheet and the location of the zinc ion are not shown in the lower part of the figure.) The three α-helical stretches fit into one turn of the major groove; each α helix (and thus each finger) makes two sequence-specific contacts with DNA (indicated by the arrows). We expect that the nonconserved amino acids in the C-terminal side of each finger are responsible for recognizing specific target sites.

Knowing that zinc fingers are found in authentic activators that assist both RNA polymerases II and III, we may view finger proteins from the reverse perspective. When a protein is found to have multiple zinc fingers, there is at least a *prima facie* case for investigating a possible role as a transcription factor. This type of identification suggests that several loci involved in embryonic development of *D. melanogaster* are regulators of transcription.

It is necessary to be cautious about interpreting the presence of (putative) zinc fingers, though, especially when the protein contains only a single finger motif. Fingers may be involved in binding RNA rather than DNA, or may not even be connected with any nucleic acid binding activity. For example, the prototype zinc finger protein, $TF_{III}A$, binds both to the 5S gene and to the product, 5S rRNA. A translation initiation factor, eIF2β, has a zinc finger, and mutations in the finger influence the recognition of initiation codons. Retroviral capsid proteins have a motif related to the finger that may be involved in binding the viral RNA.

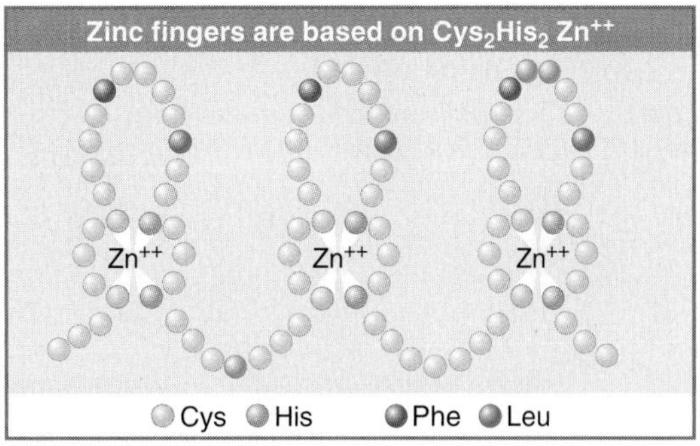

Zinc fingers are based on Cys_2His_2 Zn^{++}

FIGURE 25.13 Transcription factor SP1 has a series of three zinc fingers, each with a characteristic pattern of cysteine and histidine residues that constitute the zinc-binding site.

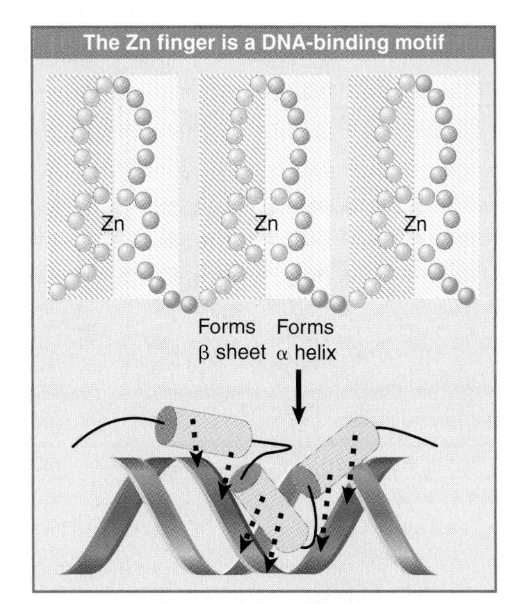

FIGURE 25.14 Zinc fingers may form α helices that insert into the major groove, which is associated with β sheets on the other side.

25.10 Steroid Receptors Are Activators

Key concepts

- Steroid receptors are examples of ligand-responsive activators that are activated by binding a steroid (or other related molecules).
- There are separate DNA-binding and ligand-binding domains.

Steroid hormones are synthesized in response to a variety of neuroendocrine activities and exert major effects on growth, tissue development, and body homeostasis in the animal world. The major groups of steroids and some other compounds with related (molecular) activities are classified in FIGURE 25.15.

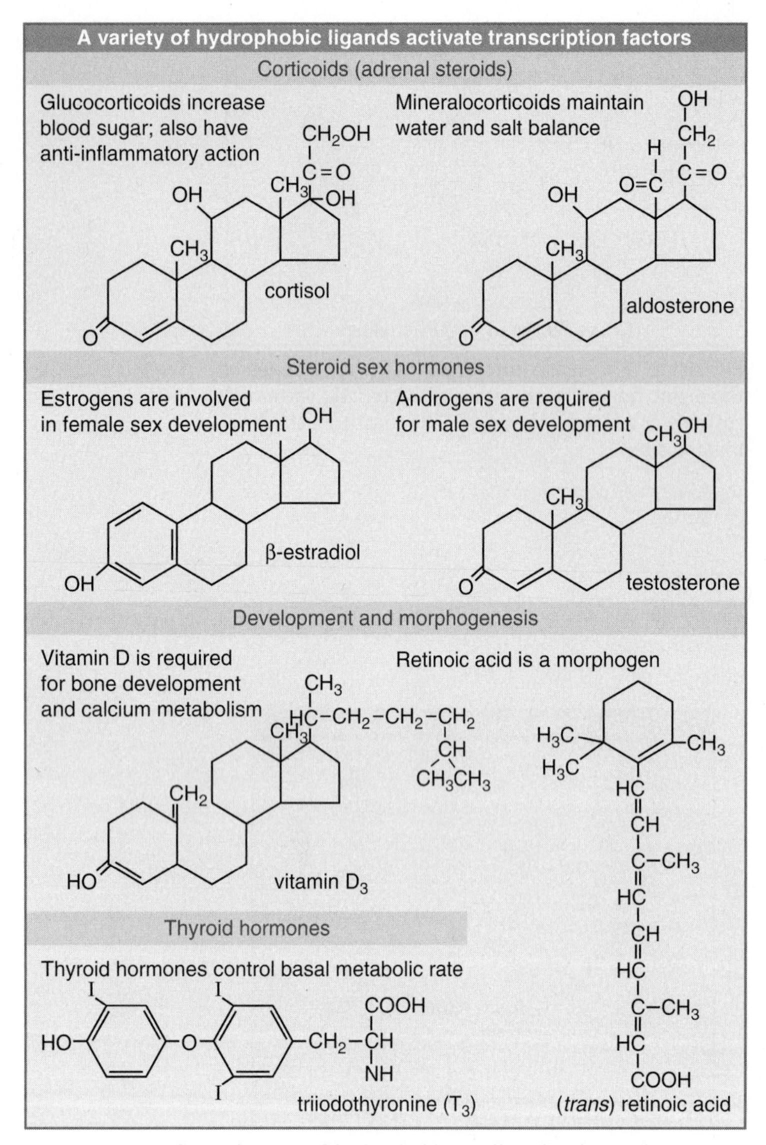

FIGURE 25.15 Several types of hydrophobic small molecules activate transcription factors.

The adrenal gland secretes >30 steroids, the two major groups being the glucocorticoids and mineralocorticoids. Steroids provide the reproductive hormones (androgen male sex hormones and estrogen female sex hormones). Vitamin D is required for bone development.

Other hormones, which have unrelated structures and physiological purposes, function at the molecular level in a similar way to the steroid hormones. Thyroid hormones, which are based on iodinated forms of tyrosine, control basal metabolic rate in animals. Steroid and thyroid hormones also may be important in metamorphosis (ecdysteroids in insects and thyroid hormones in frogs).

Retinoic acid (vitamin A) is a morphogen responsible for development of the anterior–posterior axis in the developing chick limb bud. Its metabolite, 9-*cis* retinoic acid, is found in tissues that are major sites for storage and metabolism of vitamin A.

We may account for these various actions in terms of pathways for regulating gene expression. These diverse compounds share a common mode of action: *Each is a small molecule that binds to a specific receptor that activates gene transcription.* ("Receptor" may be a misnomer: The protein is a receptor for steroid or thyroid hormone in the same sense that *lac* repressor is a receptor for a β galactoside, i.e., it is not a receptor in the sense of comprising a membrane-bound protein that is exposed to the cell surface.)

Receptors for the diverse groups of steroid hormones, thyroid hormones, and retinoic acid represent a new "superfamily" of gene regulators, the ligand-responsive activators. All the receptors have independent domains for DNA-binding and hormone binding that are in the same relative locations. Their general organization is summarized in FIGURE 25.16.

The central part of the protein is the DNA-binding domain. These regions are closely related for the various steroid receptors (from the most closely related pair, with 94% sequence identity, to the least well related pair, at 42% identity). The act of binding DNA cannot be disconnected from the ability to activate transcription, because mutations in this domain affect both activities.

The N-terminal regions of the receptors show the least conservation of sequence. They include other regions that are needed to activate transcription.

The C-terminal domains bind the hormones. Those in the steroid receptor family show identities ranging from 30% to 57%, reflecting specificity for individual hormones. Their relationships with the other receptors are minimal and reflect specificity for a variety of compounds—thyroid hormones, vitamin D, retinoic acid, and so forth. This domain also has the motifs responsible for dimerization and a region involved in transcriptional activation.

Some ligands have multiple receptors that are closely related, such as the three retinoic acid receptors (RARα, β, and γ) and the three receptors for 9-*cis*-retinoic acid (RXRα, β, and γ).

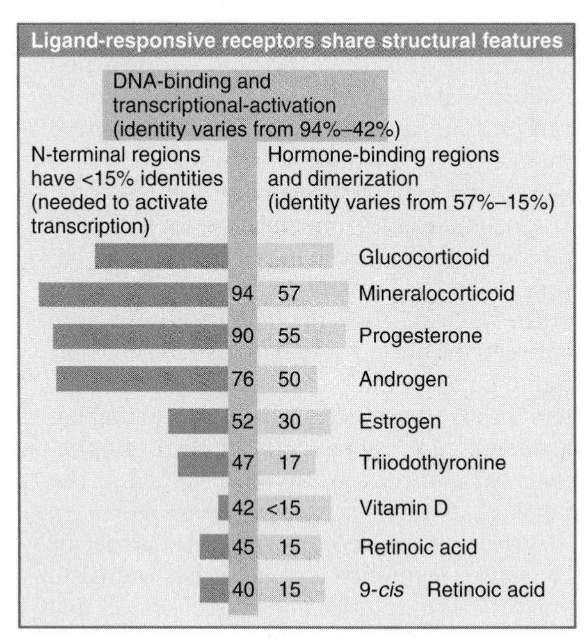

FIGURE 25.16 Receptors for many steroid and thyroid hormones have a similar organization, with an individual N-terminal region, conserved DNA-binding region, and a C-terminal hormone-binding region. Identities are relative to GR.

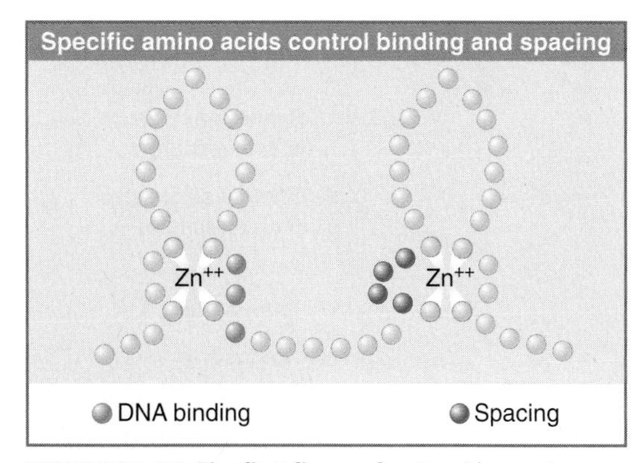

FIGURE 25.17 The first finger of a steroid receptor controls which DNA sequence is bound (positions shown in purple); the second finger controls spacing between the sequences (positions shown in blue).

25.11 Steroid Receptors Have Zinc Fingers

Key concepts

- The DNA binding domain of a steroid receptor is a type of zinc finger that has Cys but not His residues.
- Glucocorticoid and estrogen receptors each have two zinc fingers, the first of which determines the DNA target sequence.
- Steroid receptors bind to DNA as dimers.

Steroid receptors (and some other proteins) have another type of zinc finger that is different from Cys_2/His_2 fingers. The structure is based on a sequence with the zinc-binding consensus:

$$Cys\text{-}X_2\text{-}Cys\text{-}X_{13}\text{-}Cys\text{-}X_2\text{-}Cys$$

These sequences are called Cys_2/Cys_2 fingers. Proteins with Cys_2/Cys_2 fingers often have nonrepetitive fingers, in contrast with the tandem repetition of the Cys_2/His_2 type. Binding sites in DNA (where known) are short and palindromic.

The glucocorticoid and estrogen receptors each have two fingers, each with a zinc atom at the center of a tetrahedron of cysteines. The two fingers form α-helices that fold together to form a large globular domain. The aromatic sides of the α-helices form a hydrophobic center together with a β sheet that connects the two helices. One side of the N-terminal helix makes contacts in the major groove of DNA. Two glucocorticoid receptors dimerize upon binding to DNA, and each engages a successive turn of the major groove. This fits with the palindromic nature of the response element (see Section 25.13, Steroid Receptors Recognize Response Elements by a Combinatorial Code).

Each finger controls one important property of the receptor. FIGURE 25.17 identifies the relevant amino acids. Those on the right side of the first finger determine the sequence of the target in DNA; those on the left side of the second finger control the spacing between the target sites recognized by each subunit in the dimer (see Section 25.13, Steroid Receptors Recognize Response Elements by a Combinatorial Code).

Direct evidence that the first finger binds DNA was obtained by a "specificity swap" experiment. The finger of the estrogen receptor was deleted and replaced by the sequence of the glucocorticoid receptor. The new protein recognized the GRE sequence (the usual target of the glucocorticoid receptor) instead of the ERE (the usual target of the estrogen receptor. This region therefore establishes the specificity with which DNA is recognized.

The differences between the sequences of the glucocorticoid receptor and estrogen receptor fingers lie mostly at the base of the finger. The substitution at two positions shown in

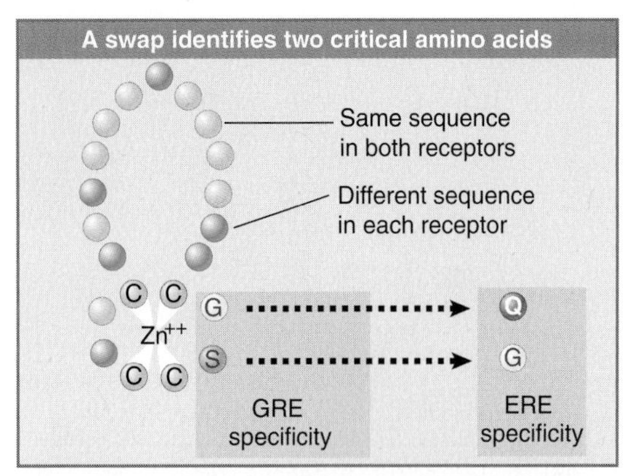

FIGURE 25.18 Discrimination between GRE and ERE target sequences is determined by two amino acids at the base of the first zinc finger in the receptor.

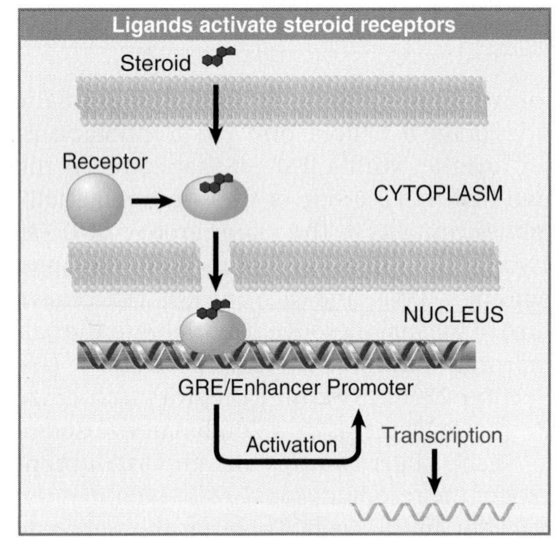

FIGURE 25.19 Glucocorticoids regulate gene transcription by causing their receptor to bind to an enhancer whose action is needed for promoter function.

FIGURE 25.18 allows the glucocorticoid receptor to bind at an ERE instead of a GRE.

25.12 Binding to the Response Element Is Activated by Ligand-Binding

Key concept

- Binding of ligand to the C-terminal domain increases the affinity of the DNA-binding domain for its specific target site in DNA.

We know the most about the interaction of glucocorticoids with their receptor, whose action is illustrated in FIGURE 25.19. A steroid hormone can pass through the cell membrane to enter the cell by simple diffusion. Within the cell, a glucocorticoid binds the glucocorticoid receptor. (Work on the glucocorticoid receptor has relied on the synthetic steroid hormone, dexamethasone.) The localization of free receptors is not entirely clear; they may be in equilibrium between the nucleus and cytoplasm. When hormone binds to the receptor, though, the protein is converted into an activated form that has an increased affinity for DNA, so the hormone-receptor complex is always localized in the nucleus.

The activated receptor recognizes a specific consensus sequence that identifies the GRE. The GRE is typically located in an enhancer that may be several kb upstream or downstream of the promoter. When the steroid-receptor complex binds to the enhancer, the nearby promoter is activated and transcription initiates there. Enhancer activation provides the general mechanism by which steroids regulate a wide set of target genes.

The C-terminal region regulates the activity of the receptor in a way that varies for the individual receptor. If the C-terminal domain of the glucocorticoid receptor is deleted, the remaining N-terminal protein is constitutively active: It no longer requires steroids for activity. This suggests that, in the absence of steroid, the steroid-binding domain prevents the receptor from recognizing the GRE; it functions as an internal negative regulator. The addition of steroid inactivates the inhibition, releasing the receptor's ability to bind the GRE and activate transcription. The basis for the repression could be internal, relying on interactions with another part of the receptor, or it could result from an interaction with some other protein that is displaced when steroid binds.

The interaction between the domains is different in the estrogen receptor. If the hormone-binding domain is deleted, the protein is unable to activate transcription, although it continues to bind to the ERE. This region is therefore required to activate rather than to repress activity.

25.13 Steroid Receptors Recognize Response Elements by a Combinatorial Code

Key concepts

- A steroid response element consists of two short half sites that may be palindromic or directly repeated.
- There are only two types of half sites.
- A receptor recognizes its response element by the orientation and spacing of the half sites.
- The sequence of the half site is recognized by the first zinc finger.
- The second zinc finger is responsible for dimerization, which determines the distance between the subunits.
- Subunit separation in the receptor determines the recognition of spacing in the response element.
- Some steroid receptors function as homodimers, whereas others form heterodimers.
- Homodimers recognize palindromic response elements; heterodimers recognize response elements with directly repeated half sites.

Each receptor recognizes a response element that consists of two short repeats (or half sites). This immediately suggests that the receptor binds as a dimer, so that each half of the consensus is contacted by one subunit (reminiscent of the λ operator-repressor interaction described in Section 14.11, Repressor Uses a Helix-Turn-Helix Motif to Bind DNA).

The half sites may be arranged either as palindromes or as repeats in the same orientation. They are separated by zero to four base pairs whose sequence is irrelevant. Only two types of half site are used by the various receptors. Their orientation and spacing determine which receptor recognizes the response element. This behavior allows response elements that have restricted consensus sequences to be recognized specifically by a variety of receptors. The rules that govern recognition are not absolute, but may be modified by context, and there are also cases in which palindromic response elements are recognized permissively by more than one receptor.

The receptors fall into two groups:

- Glucocorticoid (GR), mineralocorticoid (MR), androgen (AR), and progesterone (PR) receptors all form homodimers. They recognize response elements whose half sites have the consensus sequence TGTTCT. FIGURE 25.20 shows that the half sites are arranged as palindromes, and that the spacing between the sites determines the type of element. The estrogen (ER) receptor functions in the same way, but has the half site sequence TGACCT.
- The 9-*cis*-retinoic acid (RXR) receptor forms homodimers, and also forms heterodimers with ~15 other receptors, including thyroid (T3R), vitamin D (VDR), and retinoic acid (RAR). FIGURE 25.21 shows that the dimers recognize half elements with the sequence TGACCT. The half sites are arranged as direct repeats, and recognition is controlled by spacing between them. Some of the heterodimeric receptors are activated when the ligand binds to the partner for RXR; others can be activated by ligand binding either to this subunit or to the RXR subunit. These receptors can also form homodimers, which recognize palindromic sequences.

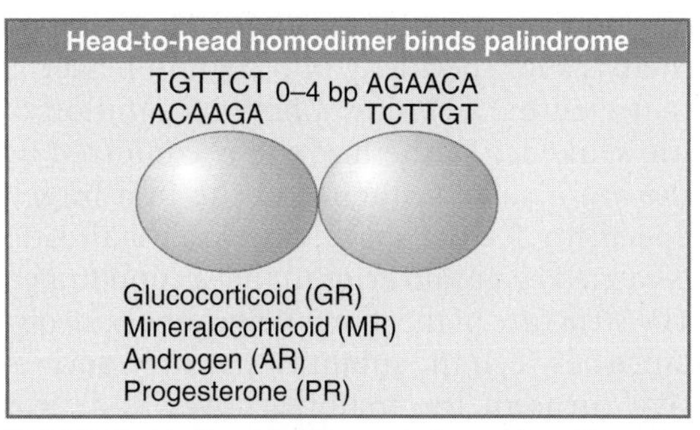

FIGURE 25.20 Response elements formed from the palindromic half site TGTTCT are recognized by several different receptors depending on the spacing between the half sites.

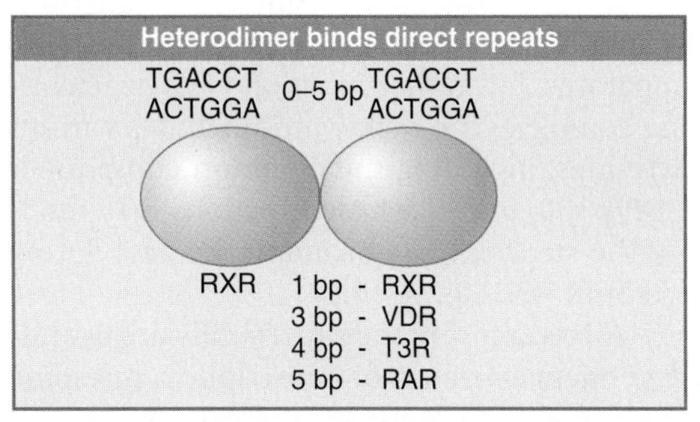

FIGURE 25.21 Response elements with the direct repeat TGACCT are recognized by heterodimers, of which one member is RXR.

Now we are in a position to understand the basis for specificity of recognition. Recall that Figure 25.17 shows how recognition of the sequence of the half site is conferred by the amino acid sequence in the first finger. Specificity for the spacing between half sites is carried by amino acids in the second finger. The structure of the dimer determines the distance between the subunits that sit in successive turns of the major groove, and thus controls the response to the spacing of half sites. The exact positions of the residues responsible for dimerization differ in individual pairwise combinations.

How do the steroid receptors activate transcription? They do not act directly on the basal apparatus, but rather function via a coactivating complex. The coactivator includes various activities, including the common component CBP/p300, one of whose functions is to modify the structure of chromatin by acetylating histones (see Figure 30.14).

All receptors in the superfamily are ligand-dependent activators of transcription, but some are also able to repress transcription. The TR and RAR receptors, in the form of heterodimers with RXR, bind to certain loci in the *absence* of ligand and repress transcription by means of their ability to interact with a corepressor protein. The corepressor functions by the reverse of the mechanism used by coactivators: It inhibits the function of the basal transcription apparatus, one of its actions being the deacetylation of histones (see Figure 30.16). We do not know the relative importance of the repressor activity *vis-à-vis* the ligand-dependent activation in the physiological response to hormone.

The effect of ligand binding on the receptor is to convert it from a repressing complex to an activating complex, as shown in FIGURE 25.22. In the absence of ligand, the receptor is bound to a corepressor complex. The component of the corepressor that binds to the receptor is SMRT. Binding of ligand causes a conformational change that displaces SMRT. This allows the coactivator to bind.

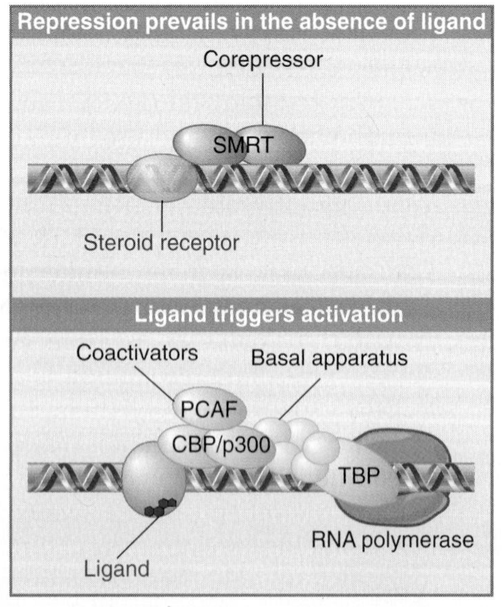

FIGURE 25.22 The steroid receptors TR and RAR bind the SMRT corepressor in the absence of ligand. The promoter is not expressed. When SMRT is displaced by binding of ligand, the receptor binds a coactivator complex. This leads to activation of transcription by the basal apparatus.

25.14 Homeodomains Bind Related Targets in DNA

Key concepts

- The homeodomain is a DNA-binding domain of 60 amino acids that has three α-helices.
- The C-terminal α-helix-3 is 17 amino acids and binds in the major groove of DNA.
- The N-terminal arm of the homeodomain projects into the minor groove of DNA.
- Proteins containing homeodomains may be either activators or repressors of transcription.

The homeobox is a sequence that codes for a domain of 60 amino acids present in proteins of many or even all eukaryotes. Its name derives from its original identification in *Drosophila* homeotic loci (whose genes determine the identity of body structures). It is present in many of the genes that regulate early development in *Drosophila,* and a related motif is found in genes in a wide range of higher eukaryotes. The homeodomain is found in many genes concerned with developmental regulation. Sequences related to the homeodomain are found in several types of animal transcription factors.

In *Drosophila* homeotic genes, the homeodomain often (but not always) occurs close to the C-terminal end. Some examples of genes containing homeoboxes are summarized in FIGURE 25.23. Often the genes have little conservation of sequence except in the homeobox. The conservation of the homeobox sequence varies. A major group of homeobox-containing genes in *Drosophila* has a well conserved sequence, with 80% to 90% similarity in pairwise comparisons. Other genes have less closely related homeoboxes. The homeodomain is sometimes combined with other motifs in animal transcription factors. One example is presented by the Oct (octamer-binding) proteins, in which a conserved stretch of 75 amino acids called the Pou region is located close to a region resembling the homeodomain. The homeoboxes of the Pou group of proteins are the least closely related to the original group, and thus comprise the farthest extension of the family.

The homeodomain is responsible for binding to DNA, and experiments to swap homeodomains between proteins suggest that the specificity of DNA recognition lies within the homeodomain. As with phage repressors, though, no simple code relating protein and DNA sequences can be deduced. The C-terminal region of the homeodomain shows homology with the helix-turn-helix motif of prokaryotic repressors. We recall from Section 14.11, Repressor Uses a Helix-Turn-Helix Motif to Bind DNA, that the λ repressor has a "recognition helix" (α-helix-3) that makes contacts in the major groove of DNA, whereas the other helix (α-helix-2) lies at an angle across the DNA. The homeodomain can be organized into three potential helical regions; the sequences of three examples are compared in FIGURE 25.24. The best conserved part of the sequence lies in the third helix. The difference between these structures and the prokaryotic repressor structures lies in the length of the helix that recognizes DNA, helix-3, which is 17 amino acids long in the homeodomain, compared to nine residues long in the λ repressor.

The structure of the homeodomain of the *D. melanogaster* engrailed protein is represented schematically in FIGURE 25.25. Helix 3 binds in the major groove of DNA and makes the majority of the contacts between protein and nucleic acid. Many of the contacts that orient the helix in the major groove are made with the phosphate backbone, so they are not specific for DNA sequence. They lie largely on one face of the double helix, and flank the bases with which specific contacts are made. The remaining contacts are made by the N-terminal arm of the homeodomain, the sequence that just precedes

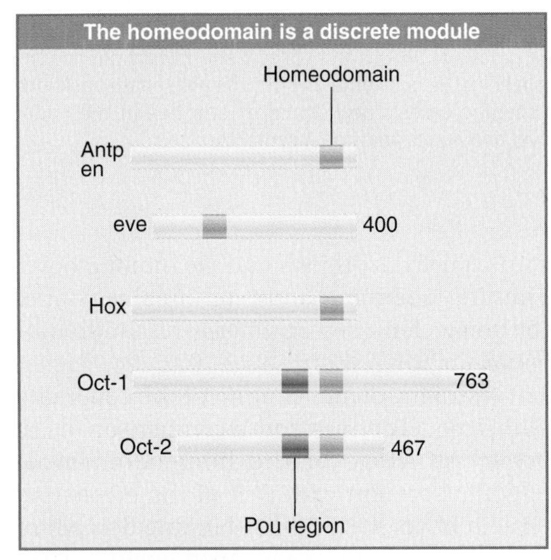

FIGURE 25.23 The homeodomain may be the sole DNA-binding motif in a transcriptional regulator, or may be combined with other motifs. It represents a discrete (60 residue) part of the protein.

The homeodomain is a module of 60 amino acids

N-terminal arm (2–7); Helix 1 (10–22)

	1									10										20		
En	Glu	Lys	Arg	Pro	Arg	Thr	Ala	Phe	Ser	Ser	Glu	Gln	Leu	Ala	Arg	Leu	Lys	Arg	Glu	Phe	Asn	Glu
Antp	Arg	Lys	Arg	Gly	Arg	Gln	Thr	Tyr	Thr	Arg	Tyr	Gln	Thr	Leu	Glu	Leu	Glu	Lys	Glu	Phe	His	Phe
Oct-2	Arg	Arg	Lys	Lys	Arg	Thr	Ser	Ile	Glu	Thr	Asn	Val	Arg	Phe	Ala	Leu	Glu	Lys	Ser	Phe	Leu	Ala

Helix 2

							30										40
En	Asn	Arg	Tyr	Leu	Thr	Glu	Arg	Arg	Arg	Glu	Glu	Leu	Ser	Ser	Glu	Leu	Gly	Leu
Antp	Asn	Arg	Tyr	Leu	Thr	Arg	Arg	Arg	Arg	Ile	Glu	Ile	Ala	His	Ala	Leu	Cys	Leu
Oct-2	Asn	Glu	Lys	Pro	Thr	Ser	Glu	Glu	Ile	Leu	Leu	Ile	Ala	Glu	Gln	Leu	His	Met

Helix 3

	41									50										60
En	Asn	Glu	Ala	Gln	Ile	Lys	Ile	Trp	Phe	Gln	Asn	Lys	Arg	Ala	Lys	Ile	Lys	Lys	Ser	Asn
Antp	Thr	Glu	Arg	Gln	Ile	Lys	Ile	Trp	Phe	Gln	Asn	Arg	Arg	Met	Lys	Trp	Lys	Lys	Glu	Asn
Oct-2	Glu	Lys	Glu	Val	Ile	Arg	Val	Trp	Phe	Cys	Asn	Arg	Arg	Gln	Lys	Glu	Lys	Arg	Ile	Asn

FIGURE 25.24 The homeodomain of the *Antennapedia* gene represents the major group of genes containing homeoboxes in *Drosophila; engrailed (en)* represents another type of homeotic gene; and the mammalian factor Oct-2 represents a distantly related group of transcription factors. The homeodomain is conventionally numbered from 1 to 60. It starts with the N-terminal arm, and the three helical regions occupy residues 10–22, 28–38, and 42–58. Amino acids in blue are conserved in all three examples.

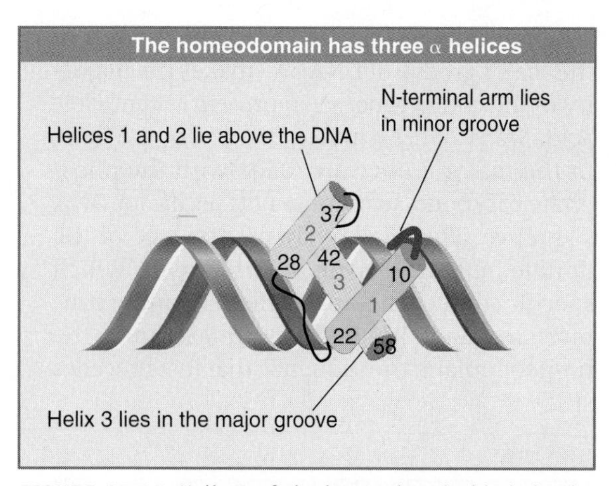

FIGURE 25.25 Helix 3 of the homeodomain binds in the major groove of DNA, with helices 1 and 2 lying outside the double helix. Helix 3 contacts both the phosphate backbone and specific bases. The N-terminal arm lies in the minor groove, and makes additional contacts.

the first helix. It projects into the minor groove. Thus the N-terminal and C-terminal regions of the homeodomain are primarily responsible for contacting DNA.

A striking demonstration of the generality of this model derives from a comparison of the crystal structure of the homeodomain of engrailed protein with that of the α2 mating protein of yeast. The DNA-binding domain of this protein resembles a homeodomain and can form three similar helices: Its structure in the DNA groove can be superimposed almost exactly on that of the engrailed homeodomain. These similarities suggest that all homeodomains bind to DNA in the same manner. This means that a relatively small number of residues in helix-3 and in the N-terminal arm are responsible for specificity of contacts with DNA.

One group of homeodomain-containing proteins is the set of Hox proteins. They bind to DNA with rather low sequence specificity, and it has been puzzling how these proteins can have different specificities. It turns out that Hox proteins often bind to DNA as heterodimers with a partner (called Exd in flies and Pbx in vertebrates). The heterodimer has a more restricted specificity *in vitro* than an individual Hox protein; typically it binds the 10 bp sequence TGATNNATNN. This still is not enough to account for the differences in the specificities of Hox proteins. A third protein, Hth, which is necessary to localize Exd in the nucleus, also forms part of the complex that binds DNA, and may restrict the binding sites further. The same partners (Exd and Hth) are present together with each Hox protein in the trimeric complex, though, so it remains puzzling how each Hox protein has sufficient specificity.

Homeodomain proteins can be either transcriptional activators or repressors. The nature of the factor depends on the other domain(s)—the homeodomain is responsible solely for binding to DNA. The activator or repressor domains both act by influencing the basal apparatus. Activator domains may interact with coactivators that in turn bind to components of the basal apparatus. Repressor domains also interact with the transcription apparatus (that is, they do not act by blocking access to DNA as such). The repressor Eve, for example, interacts directly with $TF_{II}D$.

25.15 Helix-Loop-Helix Proteins Interact by Combinatorial Association

Key concepts

- Helix-loop-helix proteins have a motif of 40 to 50 amino acids that comprises two amphipathic α helices of 15 to 16 residues separated by a loop.
- The helices are responsible for dimer formation.
- bHLH proteins have a basic sequence adjacent to the HLH motif that is responsible for binding to DNA.
- Class A bHLH proteins are ubiquitously expressed. Class B bHLH proteins are tissue-specific.
- A class B protein usually forms a heterodimer with a class A protein.
- HLH proteins that lack the basic region prevent a bHLH partner in a heterodimer from binding to DNA.
- HLH proteins form combinatorial associations that may be changed during development by the addition or removal of specific proteins.

Two common features in DNA-binding proteins are the presence of helical regions that bind DNA and the ability of the protein to dimerize. Both features are represented in the group of helix-loop-helix (HLH) proteins that share a common type of sequence motif: A stretch of 40 to 50 amino acids contains two amphipathic α helices separated by a linker region (the loop) of varying length. (An amphipathic helix forms two faces, one presenting hydrophobic amino acids, and the other presenting charged amino acids.) The proteins in this group form both homodimers and heterodimers by means of interactions between the hydrophobic residues on the corresponding faces of the two helices. The helical regions

HLH proteins have two helical regions		
MyoD	Ala Asp Arg Arg Lys Ala Ala Thr Met Arg Gln Arg Arg Arg	Basic region
Id	Arg Leu Pro Ala Leu Leu Asp Gln Glu Glu Val Asn Val Leu	Six conserved residues are absent from Id
MyoD	Leu Ser Lys Val Asn Gln Ala Phe Gln Thr Leu Lys Arg Cys Thr	Helix 1
Id	Leu Tyr Asp Met Asn Gly Cys Tyr Ser Arg Leu Lys Gln Leu Val	Conserved residues are found in both MyoD and Id
MyoD	Lys Val Gln Ile Leu Arg Asn Ala Ile Arg Tyr Ile Gln Gly Leu Glu	Helix 2
Id	Lys Val Gln Ile Leu Glu His Val Ile Asp Tyr Ile Arg Asp Leu Glu	

FIGURE 25.26 All HLH proteins have regions corresponding to helix 1 and helix 2, which are separated by a loop of 10 to 24 residues. Basic HLH proteins have a region with conserved positive charges immediately adjacent to helix 1.

are 15 to 16 amino acids long, and each contains several conserved residues. Two examples are compared in FIGURE 25.26. The ability to form dimers resides with these amphipathic helices and is common to all HLH proteins. The loop is probably important only for allowing the freedom for the two helical regions to interact independently of one another.

Most HLH proteins contain a region adjacent to the HLH motif itself that is highly basic, and which is needed for binding to DNA. There are ~6 conserved residues in a stretch of 15 amino acids (see Figure 25.26). Members of the group with such a region are called **bHLH proteins.** A dimer in which both subunits have the basic region can bind to DNA. The HLH domains probably correctly orient the two basic regions contributed by the individual subunits.

The bHLH proteins fall into two general groups. Class A consists of proteins that are ubiquitously expressed, including mammalian E12/E47. Class B consists of proteins that are expressed in a tissue-specific manner, including mammalian MyoD, myogenin, and Myf-5 (a group of activators that are involved in myogenesis [muscle formation]). A common *modus operandi* for a tissue-specific bHLH protein is to form a heterodimer with a ubiquitous partner. There is also a group of gene products that specify development of the nervous system in *D. melanogaster* (where *Ac-S* is the tissue-specific component and *da* is the ubiquitous component). The Myc proteins (which are the cellular counterparts of oncogene products and are involved in growth regulation) form a separate class of bHLH proteins, whose partners and targets are different.

Dimers formed from bHLH proteins differ in their abilities to bind to DNA. For example, E47 homodimers, E12-E47 heterodimers, and MyoD-E47 heterodimers all form efficiently and bind strongly to DNA; E12 homodimerizes well but binds DNA poorly, whereas MyoD homodimerizes only poorly. Thus both dimer formation and DNA binding may represent important regulatory points. At this juncture, it is possible to define groups of HLH proteins whose members form various pairwise combinations, but it is not possible to predict from the sequences the strengths of dimer formation or DNA binding. All of the dimers in this group that bind DNA recognize the same consensus sequence, but we do not know yet whether different homodimers and heterodimers have preferences for slightly different target sites that are related to their functions.

Differences in DNA-binding result from properties of the region in or close to the HLH motif; for example, E12 differs from E47 in possessing an inhibitory region just by the basic region, which prevents DNA binding by homodimers. Some HLH proteins lack the basic region and/or contain proline residues that appear to disrupt its function. The example of the protein Id is shown in Figure 25.26. Proteins of this type have the same capacity to dimerize as bHLH proteins, but a dimer that contains one subunit of this type can no longer bind to DNA specifically. This is a forceful demonstration of the importance of doubling the DNA-binding motif in DNA-binding proteins.

The importance of the distinction between the nonbasic HLH and bHLH proteins is suggested by the properties of two pairs of HLH proteins: the *da-Ac-S/emc* pair and the MyoD/Id pair. A model for their functions in forming a regulatory network is illustrated in FIGURE 25.27.

In *D. melanogaster,* the gene *emc (extramacrochaetae)* is required to establish the normal spatial pattern of adult sensory organs. It

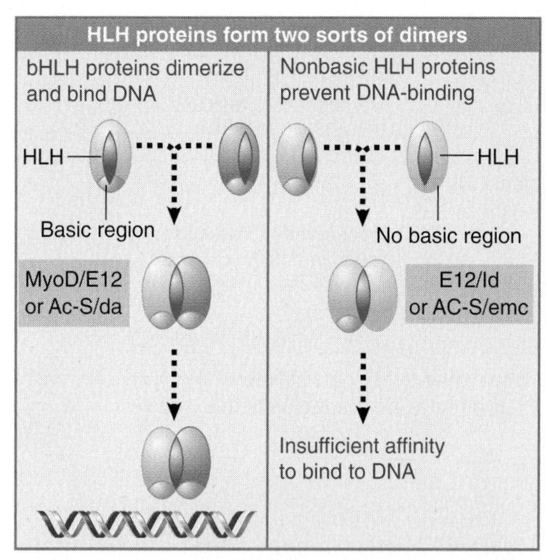

FIGURE 25.27 An HLH dimer in which both subunits are of the bHLH type can bind DNA, but a dimer in which one subunit lacks the basic region cannot bind DNA.

functions by *suppressing* the functions of several genes, including *da (daughterless)* and the *achaete-scute (Ac-S)* complex. *Ac-S* and *da* are genes of the bHLH type. The suppressor *emc* codes for an HLH protein that lacks the basic region. We suppose that, in the absence of *emc* function, the *da* and *Ac-S* proteins form dimers that activate transcription of appropriate target genes, but the production of *emc* protein causes the formation of heterodimers that cannot bind to DNA. Thus production of *emc* protein in the appropriate cells is necessary to suppress the function of *Ac-S/da.*

The formation of muscle cells is triggered by a change in the transcriptional program that requires several bHLH proteins, including MyoD. MyoD is produced specifically in myogenic cells and, indeed, overexpression of MyoD in certain other cells can induce them to commence a myogenic program. The trigger for muscle differentiation is probably a heterodimer consisting of MyoD-E12 or MyoD-E47 rather than a MyoD homodimer. Before myogenesis begins, a member of the nonbasic HLH type, the Id protein, may bind to MyoD and/or E12 and E47 to form heterodimers that cannot bind to DNA. It binds to E12/E47 better than to MyoD, and so might function by sequestering the ubiquitous bHLH partner. Overexpression of Id can prevent myogenesis. Thus the removal of Id could be the trigger that releases MyoD to initiate myogenesis.

A bHLH activator such as MyoD can be controlled in several ways. It is prevented from binding to DNA when it is sequestered by an HLH partner such as Id. It can activate transcription when bound to a bHLH partner such as E12 or E47. It can also act as a site-specific repressor when bound to another partner; the bHLH protein MyoR forms a MyoD-MyoR dimer in proliferating myoblasts that represses transcription (at the same target loci at which MyoD-E12/E47 activate transcription).

The behavior of the HLH proteins therefore illustrates two general principles of transcriptional regulation. A small number of proteins form combinatorial associations. Particular combinations have different functions with regard to DNA binding and transcriptional regulation. Differentiation may depend either on the presence or on the removal of particular partners.

25.16 Leucine Zippers Are Involved in Dimer Formation

Key concepts

- The leucine zipper is an amphipathic helix that dimerizes.
- The zipper is adjacent to a basic region that binds DNA.
- Dimerization forms the bZIP motif in which the two basic regions symmetrically bind inverted repeats in DNA.

Interactions between proteins are a common theme in building a transcription complex, and a motif found in several activators (and other proteins) is involved in both homo- and heteromeric interactions. The leucine zipper is a stretch of amino acids rich in leucine residues that provides a dimerization motif. Dimer formation itself has emerged as a common principle in the action of proteins that recognize specific DNA sequences, and in the case of the leucine zipper, its relationship to DNA binding is especially clear, because we can see how dimerization juxtaposes the DNA-binding regions of each subunit. The reaction is depicted diagrammatically in FIGURE 25.28.

An amphipathic α helix has a structure in which the hydrophobic groups (including leucine) face one side while charged groups face the other side. A leucine zipper forms an amphipathic helix in which the leucines of the zipper on one protein could protrude from the α-helix and interdigitate with the leucines of the zipper of another protein in parallel to form a coiled coil. The two right-handed helices wind around each other, with 3.5 residues per turn, so the pattern repeats integrally every seven residues.

How is this structure related to DNA binding? The region adjacent to the leucine repeats is highly basic in each of the zipper proteins, and could comprise a DNA-binding site. The two leucine zippers in effect form a Y-shaped structure, in which the zippers comprise the stem and the two basic regions stick out to form the arms that bind to DNA. This is known as the **bZIP** structural motif. It explains why the target sequences for such proteins are inverted repeats with no separation.

Zippers may be used to sponsor formation of homodimers or heterodimers. They are lengthy motifs. Leucine (or another hydrophobic amino acid) occupies every seventh residue in the potential zipper. There are four repeats of the zipper (Leu-X_6) in the protein C/EBP (a factor that binds as a dimer to both the CAAT box and the SV40 core enhancer) and five repeats in the factors Jun and Fos (which form the heterodimeric activator, AP1).

AP1 was originally identified by its binding to a DNA sequence in the SV40 enhancer (see Figure 24.24). The active preparation of AP1 includes several polypeptides. A major component is Jun, the product of the gene *c-jun*, which was identified by its relationship with the oncogene *v-jun* carried by an avian sarcoma virus. The mouse genome contains a family of related genes, *c-jun* (the original isolate), *junB*, and *junD* (identified by sequence homology with *jun*). There are considerable sequence similarities in the three Jun proteins; they have leucine zippers that can interact to form homodimers or heterodimers.

The other major component of AP1 is the product of another gene with an oncogenic counterpart. The *c-fos* gene is the cellular homolog to the oncogene *v-fos* carried by a murine sarcoma virus. Expression of *c-fos* activates genes whose promoters or enhancers possess an AP1 target site. The *c-fos* product is a nuclear phosphoprotein that is one of a group of proteins. The others are described as Fos-related antigens (FRA); they constitute a family of Fos-like proteins.

Fos also has a leucine zipper. Fos cannot form homodimers, but can form a heterodimer with Jun. A leucine zipper in each protein is required for the reaction. The ability to form dimers is a crucial part of the interaction of these factors with DNA. Fos cannot by itself bind to DNA, possibly because of its failure to form a dimer. The Jun-Fos heterodimer can, however,

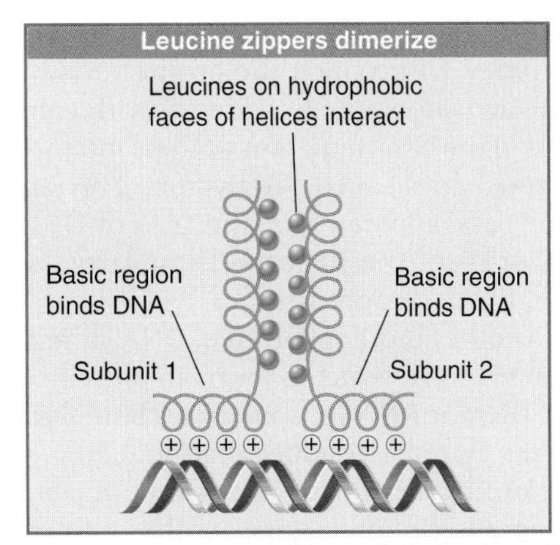

FIGURE 25.28 The basic regions of the bZIP motif are held together by the dimerization at the adjacent zipper region when the hydrophobic faces of two leucine zippers interact in parallel orientation.

bind to DNA with same target specificity as the Jun-Jun dimmer, and this heterodimer binds to the AP1 site with an affinity ~10× that of the Jun homodimer.

25.17 Summary

Transcription factors include basal factors, activators, and coactivators. Basal factors interact with RNA polymerase at the startpoint. Activators bind specific short response elements (REs) located in promoters or enhancers. Activators function by making protein–protein interactions with the basal apparatus. Some activators interact directly with the basal apparatus; others require coactivators to mediate the interaction. Activators often have a modular construction, in which there are independent domains responsible for binding to DNA and for activating transcription. The main function of the DNA-binding domain may be to tether the activating domain in the vicinity of the initiation complex. Some response elements are present in many genes and are recognized by ubiquitous factors; others are present in a few genes and are recognized by tissue-specific factors.

Promoters for RNA polymerase II contain a variety of short *cis*-acting elements, each of which is recognized by a *trans*-acting factor. The *cis*-acting elements are located upstream of the TATA box and may be present in either orientation and at a variety of distances with regard to the startpoint. The upstream elements are recognized by activators that interact with the basal transcription complex to determine the efficiency with which the promoter is used. Some activators interact directly with components of the basal apparatus; others interact via intermediaries called coactivators. The targets in the basal apparatus are the TAFs of $TF_{II}D$, or $TF_{II}B$ or $TF_{II}A$. The interaction stimulates assembly of the basal apparatus.

Another motif involved in DNA-binding is the zinc finger, which is found in proteins that bind DNA or RNA (or sometimes both). A finger has cysteine residues that bind zinc. One type of finger is found in multiple repeats in some transcription factors; another is found in single or double repeats in others.

Several groups of transcription factors have been identified by sequence homologies. The homeodomain is a 60-residue sequence found in genes that regulates development in insects and worms and in mammalian transcription factors. It is related to the prokaryotic helix-turn-helix motif and provides the motif by which the factors bind to DNA.

The leucine zipper contains a stretch of amino acids rich in leucine that are involved in dimerization of transcription factors. An adjacent basic region is responsible for binding to DNA.

Steroid receptors were the first members identified of a group of transcription factors in which the protein is activated by binding a small hydrophobic hormone. The activated factor becomes localized in the nucleus and binds to its specific response element, where it activates transcription. The DNA-binding domain has zinc fingers. The receptors are homodimers or heterodimers. The homodimers all recognize palindromic response elements with the same consensus sequence; the difference between the response elements is the spacing between the inverted repeats. The heterodimers recognize direct repeats, again being distinguished by the spacing between the repeats. The DNA-binding motif of these receptors includes two zinc fingers; the first determines which consensus sequence is recognized, and the second responds to the spacing between the repeats.

HLH (helix-loop-helix) proteins have amphipathic helices that are responsible for dimerization, which are adjacent to basic regions that bind to DNA. bHLH proteins have a basic region that binds to DNA and fall into two groups: ubiquitously expressed and tissue-specific. An active protein is usually a heterodimer between two subunits, one from each group. When a dimer has one subunit that does not have the basic region it fails to bind DNA, so such subunits can prevent gene expression. Combinatorial associations of subunits form regulatory networks.

Many transcription factors function as dimers, and it is common for there to be multiple members of a family that form homodimers and heterodimers. This creates the potential for complex combinations to govern gene expression. In some cases, a family includes inhibitory members, whose participation in dimer formation prevents the partner from activating transcription.

References

25.2 There Are Several Types of Transcription Factors

Reviews

Lee, T. I. and Young, R. A. (2000). Transcription of eukaryotic protein-coding genes. *Annu. Rev. Genet.* 34, 77–137.

Lemon, B. and Tjian, R. (2000). Orchestrated response: a symphony of transcription factors for gene control. *Genes Dev.* 14, 2551–2569.

25.3 Independent Domains Bind DNA and Activate Transcription

Reviews

Guarente, L. (1987). Regulatory proteins in yeast. *Annu. Rev. Genet.* 21, 425–452.

Ptashne, M. (1988). How eukaryotic transcriptional activators work. *Nature* 335, 683–689.

25.4 The Two-Hybrid Assay Detects Protein–Protein Interactions

Research

Fields, S. and Song, O. (1989). A novel genetic system to detect protein-protein interactions. *Nature* 340, 245–246.

25.5 Activators Interact with the Basal Apparatus

Reviews

Lemon, B. and Tjian, R. (2000). Orchestrated response: a symphony of transcription factors for gene control. *Genes Dev.* 14, 2551–2569.

Maniatis, T., Goodbourn, S., and Fischer, J. A. (1987). Regulation of inducible and tissue-specific gene expression. *Science* 236, 1237–1245.

Mitchell, P., and Tjian, R. (1989). Transcriptional regulation in mammalian cells by sequence-specific DNA-binding proteins. *Science* 245, 371–378.

Myers, L. C. and Kornberg, R. D. (2000). Mediator of transcriptional regulation. *Annu. Rev. Biochem.* 69, 729–749.

Research

Asturias, F. J., Jiang, Y. W., Myers, L. C., Gustafsson, C. M., and Kornberg, R. D. (1999). Conserved structures of mediator and RNA polymerase II holoenzyme. *Science* 283, 985–987.

Chen, J.-L. et al. (1994). Assembly of recombinant TFIID reveals differential coactivator requirements for distinct transcriptional activators. *Cell* 79, 93–105.

Dotson, M. R., Yuan, C. X., Roeder, R. G., Myers, L. C., Gustafsson, C. M., Jiang, Y. W., Li, Y., Kornberg, R. D., and Asturias, F. J. (2000). Structural organization of yeast and mammalian mediator complexes. *Proc. Natl. Acad. Sci. USA* 97, 14307–14310.

Dynlacht, B. D., Hoey, T., and Tjian, R. (1991). Isolation of coactivators associated with the TATA-binding protein that mediate transcriptional activation. *Cell* 66, 563–576.

Kim, Y. J., Bjorklund, S., Li, Y., Sayre, M. H., and Kornberg, R. D. (1994). A multiprotein mediator of transcriptional activation and its interaction with the C-terminal repeat domain of RNA polymerase II. *Cell* 77, 599–608.

Ma, J. and Ptashne, M. (1987). A new class of yeast transcriptional activators *Cell* 51, 113–119.

Pugh, B. F. and Tjian, R. (1990). Mechanism of transcriptional activation by Sp1: evidence for coactivators. *Cell* 61, 1187–1197.

25.6 Some Promoter-Binding Proteins Are Repressors

Research

Goppelt, A., Stelzer, G., Lottspeich, F., and Meisterernst, M. (1996). A mechanism for repression of class II gene transcription through specific binding of NC2 to TBP-promoter complexes via heterodimeric histone fold domains. *EMBO J.* 15, 3105–3116.

Inostroza, J. A., Mermelstein, F. H., Ha, I., Lane, W. S., and Reinberg, D. (1992). Dr1, a TATA-binding protein-associated phosphoprotein and inhibitor of class II gene transcription. *Cell* 70, 477–489.

Kim, T. K., Zhao, Y., Ge, H., Bernstein, R., and Roeder, R. G. (1995). TATA-binding protein residues implicated in a functional interplay between negative cofactor NC2 (Dr1) and general factors TFIIA and TFIIB. *J. Biol. Chem.* 270, 10976–10981.

25.8 There Are Many Types of DNA-Binding Domains

Reviews

Harrison, S. C. (1991). A structural taxonomy of DNA-binding proteins. *Nature* 353, 715–719.

Pabo, C. T. and Sauer, R. T. (1992). Transcription factors: structural families and principles of DNA recognition. *Annu. Rev. Biochem.* 61, 1053–1095.

Research

Miller, J. et al. (1985). Repetitive zinc binding domains in the protein transcription factor IIIA from *Xenopus* oocytes. *EMBO J.* 4, 1609–1614.

Murre, C., McCaw, P. S., and Baltimore, D. (1989). A new DNA binding and dimerization motif in immunoglobulin enhancer binding, daughterless, MyoD, and myc proteins. *Cell* 56, 777–783.

25.9 A Zinc Finger Motif Is a DNA-Binding Domain

Research

Kadonaga, J. et al. (1987). Isolation of cDNA encoding transcription factor Sp1 and functional analysis of the DNA binding domain. *Cell* 51, 1079–1090.

Miller, J. et al. (1985). Repetitive zinc binding domains in the protein transcription factor IIIA from *Xenopus* oocytes. *EMBO J.* 4, 1609–1614.

Pavletich, N. P. and Pabo, C. O. (1991). Zinc finger-DNA recognition: crystal structure of a Zif268-DNA complex at 21 Å. *Science* 252, 809–817.

25.10 Steroid Receptors Are Activators

Reviews

Evans, R. M. (1988). The steroid and thyroid hormone receptor superfamily. *Science* 240, 889–895.

Mangelsdorf, D. J. and Evans, R. (1995). The RXR heterodimers and orphan receptors. *Cell* 83, 841–850.

25.11 Steroid Receptors Have Zinc Fingers

Review

Tsai, M J, and O'Malley, B. W. (1994). Molecular mechanisms of action of steroid/thyroid receptor superfamily members. *Annu. Rev. Biochem.* 63, 451–486.

Research

Umesono, K. and Evans, R. M. (1989). Determinants of target gene specificity for steroid/thyroid hormone receptors. *Cell* 57, 1139–1146.

25.13 Steroid Receptors Recognize Response Elements by a Combinatorial Code

Reviews

Mangelsdorf, D. J. and Evans, R. (1995). The RXR heterodimers and orphan receptors. *Cell* 83, 841–850.

Yamamoto, K. R. (1985). Steroid receptor regulated transcription of specific genes and gene networks. *Annu. Rev. Genet.* 19, 209–252.

Research
Hurlein, A. J. et al. (1995). Ligand-independent repression by the thyroid hormone receptor mediated by a nuclear receptor corepressor. *Nature* 377, 397–404.
Rastinejad, F., Perlmann, T., Evans, R. M., and Sigler, P. B. (1995). Structural determinants of nuclear receptor assembly on DNA direct repeats. *Nature* 375, 203–211.
Umesono, K., Murakami, K. K., Thompson, C. C., and Evans, R. M. (1991). Direct repeats as selective response elements for the thyroid hormone, retinoic acid, and vitamin D3 receptors. *Cell* 65, 1255–1266.

25.14 Homeodomains Bind Related Targets in DNA

Review
Gehring, W. J. et al. (1994). Homeodomain-DNA recognition. *Cell* 78, 211–223.

Research
Han, K., Levine, M. S., and Manley, J. L. (1989). Synergistic activation and repression of transcription by *Drosophila* homeobox proteins. *Cell* 56, 573–583.
Wolberger, C. et al. (1991). Crystal structure of a MATα2 homeodomain-operator complex suggests a general model for homeodomain-DNA interactions. *Cell* 67, 517–528.

25.15 Helix-Loop-Helix Proteins Interact by Combinatorial Association

Review
Weintraub, H. (1991). The MyoD gene family: nodal point during specification of the muscle cell lineage. *Science* 251, 761–766.

Research
Benezra, R. et al. (1990). The protein Id: a negative regulator of helix-loop-helix DNA-binding proteins. *Cell* 61, 49–59.
Davis, R. L. et al. (1987). Expression of a single transfected cDNA converts fibroblasts to myoblasts. *Cell* 51, 987–1000.
Davis, R. L. et al. (1990). The MyoD DNA binding domain contains a recognition code for muscle-specific gene activation. *Cell* 60, 733–746.
Lassar, A. B. et al. (1991). Functional activity of myogenic HLH proteins requires hetero-oligomerization with E12/E47-like proteins *in vitro*. *Cell* 66, 305–315.
Murre, C., McCaw, P. S., and Baltimore, D. (1989). A new DNA binding and dimerization motif in immunoglobulin enhancer binding, daughterless, MyoD, and myc proteins. *Cell* 56, 777–783.

25.16 Leucine Zippers Are Involved in Dimer Formation

Review
Vinson, C. R., Sigler, P. B., and McKnight, S. L. (1989). Scissors-grip model for DNA recognition by a family of leucine zipper proteins. *Science* 246, 911–916.

Research
Landschulz, W. H., Johnson, P. F., and McKnight, S. L. (1988). The leucine zipper: a hypothetical structure common to a new class of DNA binding proteins. *Science* 240, 1759–1764.

26

RNA Splicing and Processing

CHAPTER OUTLINE

26.1 Introduction

26.2 Nuclear Splice Junctions Are Short Sequences

- Splice sites are the sequences immediately surrounding the exon–intron boundaries. They are named for their positions relative to the intron.
- The 5′ splice site at the 5′ (left) end of the intron includes the consensus sequence GU.
- The 3′ splice site at the 3′ (right) end of the intron includes the consensus sequence AG.
- The GU-AG rule (originally called the GT-AG rule in terms of DNA sequence) describes the requirement for these constant dinucleotides at the first two and last two positions of introns in pre-mRNAs.

26.3 Splice Junctions Are Read in Pairs

- Splicing depends only on recognition of pairs of splice junctions.
- All 5′ splice sites are functionally equivalent, and all 3′ splice sites are functionally equivalent.

26.4 Pre-mRNA Splicing Proceeds through a Lariat

- Splicing requires the 5′ and 3′ splice sites and a branch site just upstream of the 3′ splice site.
- The branch sequence is conserved in yeast but less well conserved in higher eukaryotes.
- A lariat is formed when the intron is cleaved at the 5′ splice site, and the 5′ end is joined to a 2′ position at an A at the branch site in the intron.
- The intron is released as a lariat when it is cleaved at the 3′ splice site, and the left and right exons are then ligated together.
- The reactions occur by transesterifications, in which a bond is transferred from one location to another.

26.5 snRNAs Are Required for Splicing

- The five snRNPs involved in splicing are U1, U2, U5, U4, and U6.
- Together with some additional proteins, the snRNPs form the spliceosome.
- All the snRNPs except U6 contain a conserved sequence that binds the Sm proteins that are recognized by antibodies generated in autoimmune disease.

26.6 U1 snRNP Initiates Splicing

- U1 snRNP initiates splicing by binding to the 5′ splice site by means of an RNA–RNA pairing reaction.
- The E complex contains U1 snRNP bound at the 5′ splice site, the protein U2AF bound to a pyrimidine tract between the branch site and the 3′ splice site, and SR proteins connecting U1 snRNP to U2AF.

26.7 The E Complex Can be Formed by Intron Definition or Exon Definition

- The direct way of forming an E complex is for U1 snRNP to bind at the 5′ splice site and U2AF to bind at a pyrimidine tract between the branch site and the 3′ splice site.
- Another possibility is for the complex to form between U2AF at the pyrimidine tract and U1 snRNP at a downstream 5′ splice site.
- The E complex is converted to the A complex when U2 snRNP binds at the branch site.
- If an E complex forms using a downstream 5′ splice site, this splice site is replaced by the appropriate upstream 5′ splice site when the E complex is converted to the A complex.
- Weak 3′ splice sites may require a splicing enhancer located in the exon downstream to bind SR proteins directly.

26.8 5 snRNPs Form the Spliceosome

- Binding of U5 and U4/U6 snRNPs converts the A complex to the B1 spliceosome, which contains all the components necessary for splicing.
- The spliceosome passes through a series of further complexes as splicing proceeds.
- Release of U1 snRNP allows U6 snRNA to interact with the 5′ splice site, and converts the B1 spliceosome to the B2 spliceosome.
- When U4 dissociates from U6 snRNP, U6 snRNA can pair with U2 snRNA to form the catalytic active site.

26.9 An Alternative Splicing Apparatus Uses Different snRNPs

- An alternative splicing pathway uses another set of snRNPs that comprise the U12 spliceosome.
- The target introns are defined by longer consensus sequences at the splice junctions, but usually include the same GU-AG junctions.
- Some introns have the splice junctions AU-AC, including some that are U1-dependent and some that are U12-dependent.

Continued on next page

26.10 Splicing Is Connected to Export of mRNA

- The REF proteins bind to splicing junctions by associating with the spliceosome.
- After splicing, they remain attached to the RNA at the exon–exon junction.
- They interact with the transport protein TAP/Mex that exports the RNA through the nuclear pore.

26.11 Group II Introns Autosplice via Lariat Formation

- Group II introns excise themselves from RNA by an autocatalytic splicing event.
- The splice junctions and mechanism of splicing of group II introns are similar to splicing of nuclear introns.
- A group II intron folds into a secondary structure that generates a catalytic site resembling the structure of U6-U2-nuclear intron.

26.12 Alternative Splicing Involves Differential Use of Splice Junctions

- Specific exons may be excluded or included in the RNA product by using or failing to use a pair of splicing junctions.
- Exons may be extended by changing one of the splice junctions to use an alternative junction.
- Sex determination in *Drosophila* involves a series of alternative splicing events in genes coding for successive products of a pathway.
- P elements of *Drosophila* show germline-specific alternative splicing.

26.13 *trans*-Splicing Reactions Use Small RNAs

- Splicing reactions usually occur only in *cis* between splice junctions on the same molecule of RNA.
- *trans*-splicing occurs in trypanosomes and worms where a short sequence (SL RNA) is spliced to the 5′ ends of many precursor mRNAs.
- SL RNA has a structure resembling the Sm-binding site of U snRNAs and may play an analogous role in the reaction.

26.14 Yeast tRNA Splicing Involves Cutting and Rejoining

- tRNA splicing occurs by successive cleavage and ligation reactions.

26.15 The Splicing Endonuclease Recognizes tRNA

- An endonuclease cleaves the tRNA precursors at both ends of the intron.
- The yeast endonuclease is a heterotetramer with two (related) catalytic subunits.
- It uses a measuring mechanism to determine the sites of cleavage by their positions relative to a point in the tRNA structure.
- The archaeal nuclease has a simpler structure and recognizes a bulge-helix-bulge structural motif in the substrate.

26.16 tRNA Cleavage and Ligation Are Separate Reactions

- Release of the intron generates two half-tRNAs that pair to form the mature structure.
- The halves have the unusual ends 5′ hydroxyl and 2′–3′ cyclic phosphate.
- The 5′–OH end is phosphorylated by a polynucleotide kinase, the cyclic phosphate group is opened by phosphodiesterase to generate a 2′–phosphate terminus and 3′–OH group, exon ends are joined by an RNA ligase, and the 2′–phosphate is removed by a phosphatase.

26.17 The Unfolded Protein Response Is Related to tRNA Splicing

- Ire1p is an inner nuclear membrane protein with its N-terminal domain in the ER lumen, and its C-terminal domain in the nucleus.
- Binding of an unfolded protein to the N-terminal domain activates the C-terminal nuclease by autophosphorylation.
- The activated nuclease cleaves Hac1 mRNA to release an intron and generate exons that are ligated by a tRNA ligase.
- The spliced Hac1 mRNA codes for a transcription factor that activates genes coding for chaperones that help to fold unfolded proteins.

26.18 The 3′ Ends of polI and polIII Transcripts Are Generated by Termination

- RNA polymerase I terminates transcription at an 18-base terminator sequence.
- RNA polymerase III terminates transcription in poly$(U)_4$ sequence embedded in a G-C-rich sequence.

26.19 The 3′ Ends of mRNAs Are Generated by Cleavage and Polyadenylation

- The sequence AAUAAA is a signal for cleavage to generate a 3′ end of mRNA that is polyadenylated.
- The reaction requires a protein complex that contains a specificity factor, an endonuclease, and poly(A) polymerase.
- The specificity factor and endonuclease cleave RNA downstream of AAUAAA.
- The specificity factor and poly(A) polymerase add ~200 A residues processively to the 3′ end.

- A-U-rich sequences in the 3′ tail control cytoplasmic polyadenylation or deadenylation during *Xenopus* embryonic development.

- Histone mRNAs are not polyadenylated; their 3′ ends are generated by a cleavage reaction that depends on the structure of the mRNA.
- The cleavage reaction requires the SLBP to bind to a stem-loop structure and the U7 snRNA to pair with an adjacent single-stranded region.

- The large and small rRNAs are released by cleavage from a common precursor RNA.

- The C/D group of snoRNAs is required for modifying the 2′ position of ribose with a methyl group.
- The H/ACA group of snoRNAs is required for converting uridine to pseudouridine.
- In each case the snoRNA base pairs with a sequence of rRNA that contains the target base to generate a typical structure that is the substrate for modification.

26.1 Introduction

Interrupted genes are found in all classes of organisms. They represent a minor proportion of the genes of the very lowest eukaryotes, but the vast majority of genes in higher eukaryotic genomes. Genes vary widely according to the numbers and lengths of introns, but a typical mammalian gene has seven to eight exons spread out over ~16 kb. The exons are relatively short (~100 to 200 bp) and the introns are relatively long (>1 kb) (see Section 3.6, Genes Show a Wide Distribution of Sizes).

The discrepancy between the interrupted organization of the gene and the uninterrupted organization of its mRNA requires processing of the primary transcription product. The primary transcript has the same organization as the gene and is sometimes called the **pre-mRNA.** Removal of the introns from pre-mRNA leaves a typical messenger of ~2.2 kb. The process by which the introns are removed is called **RNA splicing.** Removal of introns is a major part of the production of RNA in all eukaryotes. (Although interrupted genes are relatively rare in lower eukaryotes such as yeast, the overall proportion underestimates the importance of introns because most of the genes that are interrupted code for relatively abundant proteins. Splicing is therefore involved in the production of a greater proportion of total mRNA than would be apparent from analysis of the genome, perhaps as much as 50%.)

One of the first clues about the nature of the discrepancy in size between nuclear genes and their products in higher eukaryotes was provided by the properties of nuclear RNA. Its average size is much larger than mRNA, it is very unstable, and it has a much greater sequence complexity. Taking its name from its

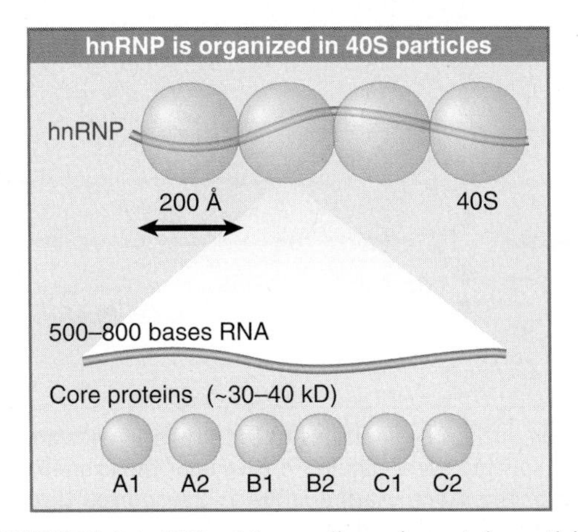

FIGURE 26.1 hnRNA exists as a ribonucleoprotein particle organized as a series of beads.

broad size distribution, it was called **heterogeneous nuclear RNA (hnRNA).** It includes pre-mRNA, but could also include other transcripts (that is, transcripts that are not ultimately processed to mRNA).

The physical form of hnRNA is a ribonucleoprotein particle **(hnRNP),** in which the hnRNA is bound by proteins. As characterized *in vitro,* an hnRNP particle takes the form of beads connected by a fiber. The structure is summarized in FIGURE 26.1. The most abundant proteins in the particle are the core proteins, but other proteins are present at lower stoichiometry, making a total of ~20 proteins. The proteins typically are present at ~10^8 copies per nucleus, compared with ~10^6 molecules of hnRNA. Some of the proteins may have a structural role in packaging the hnRNA; several are known to shuttle between the nucleus and cytoplasm and play roles in exporting the RNA or otherwise controlling its activity.

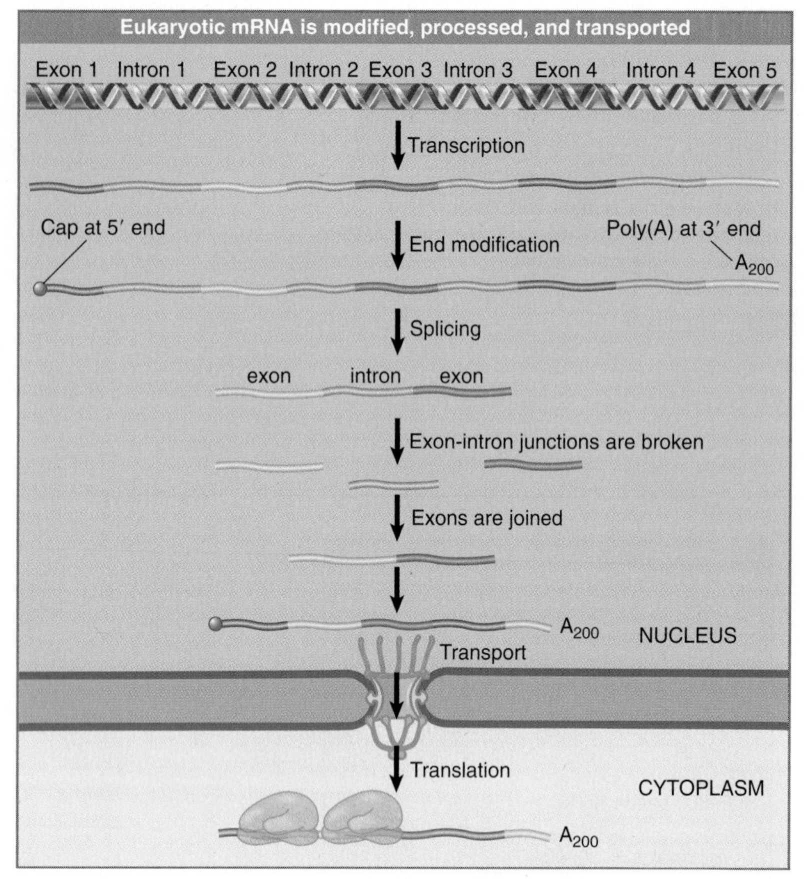

FIGURE 26.2 RNA is modified in the nucleus by additions to the 5′ and 3′ ends and by splicing to remove the introns. The splicing event requires breakage of the exon–intron junctions and joining of the ends of the exons. Mature mRNA is transported through nuclear pores to the cytoplasm, where it is translated.

Splicing occurs in the nucleus, together with the other modifications that are made to newly synthesized RNAs. The process of expressing an interrupted gene is reviewed in FIGURE 26.2. The transcript is capped at the 5′ end (see Section 7.9, The 5′ End of Eukaryotic mRNA Is Capped), has the introns removed, and is polyadenylated at the 3′ end (see Section 7.10, The 3′ Terminus Is Polyadenylated). The RNA is then transported through nuclear pores to the cytoplasm, where it is available to be translated.

With regard to the various processing reactions that occur in the nucleus, we should like to know at what point splicing occurs *vis-à-vis* the other modifications of RNA. Does splicing occur at a particular location in the nucleus, and is it connected with other events—for example, nucleocytoplasmic transport? Does the lack of splicing make an important difference in the expression of uninterrupted genes?

With regard to the splicing reaction itself, one of the main questions is how its specificity is controlled. What ensures that the ends of each intron are recognized in pairs so that the correct sequence is removed from the RNA? Are introns excised from a precursor in a particular order? Is the maturation of RNA used to *regulate* gene expression by discriminating among the available precursors or by changing the pattern of splicing?

We can identify several types of splicing systems:

- Introns are removed from the nuclear pre-mRNAs of higher eukaryotes by a system that recognizes only short consensus sequences conserved at exon–intron boundaries and within the intron. This reaction requires a large splicing apparatus, which takes the form of an array of proteins and ribonucleoproteins that functions as a large particulate complex (the spliceosome). The mechanism of splicing involves transesterifications, and the catalytic center includes RNA as well as proteins.
- Certain RNAs have the ability to excise their introns autonomously. Introns of this type fall into two groups, as distinguished by secondary/tertiary structure. Both groups use transesterification reactions in which the RNA is the catalytic agent (see Chapter 27, Catalytic RNA).
- The removal of introns from yeast nuclear tRNA precursors involves enzymatic activities that handle the substrate in a way resembling the tRNA processing enzymes, in which a critical feature is the conformation of the tRNA precursor. These splicing reactions are accomplished by enzymes that use cleavage and ligation.

26.2 Nuclear Splice Junctions Are Short Sequences

Key concepts

- Splice sites are the sequences immediately surrounding the exon–intron boundaries. They are named for their positions relative to the intron.
- The 5′ splice site at the 5′ (left) end of the intron includes the consensus sequence GU.
- The 3′ splice site at the 3′ (right) end of the intron includes the consensus sequence AG.
- The GU-AG rule (originally called the GT-AG rule in terms of DNA sequence) describes the requirement for these constant dinucleotides at the first two and last two positions of introns in pre-mRNAs.

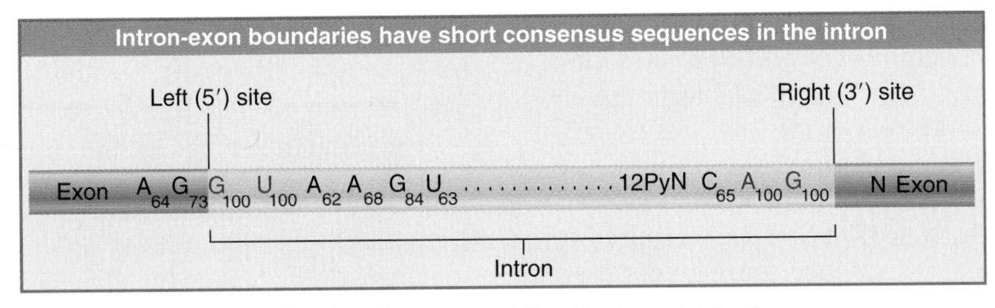

FIGURE 26.3 The ends of nuclear introns are defined by the GU-AG rule.

To focus on the molecular events involved in nuclear intron splicing, we must consider the nature of the **splice sites,** the two exon–intron boundaries that include the sites of breakage and reunion.

By comparing the nucleotide sequence of mRNA with that of the structural gene, the junctions between exons and introns can be assigned. There is no extensive homology or complementarity between the two ends of an intron. The junctions, however, have well conserved, though rather short, consensus sequences.

It is possible to assign a specific end to every intron by relying on the conservation of exon–intron junctions. They can all be aligned to conform to the consensus sequence given in FIGURE 26.3.

The subscripts indicate the percent occurrence of the specified base at each consensus position. High conservation is found only *immediately within the intron* at the presumed junctions. This identifies the sequence of a generic intron as:

GU . . . . . . AG

The intron defined in this way starts with the dinucleotide GU and ends with the dinucleotide AG; as a result, the junctions are often described as conforming to the **GT-AG rule.** (This reflects the fact that the sequences were originally analyzed in terms of DNA, but of course the GT in the coding strand sequence of DNA becomes a GU in the RNA.)

Note that the two sites have different sequences and so they define the ends of the intron *directionally.* They are named proceeding from left to right along the intron as the 5′ splice site (sometimes called the left or donor site) and the 3′ splice site (also called the right or acceptor site). The consensus sequences are implicated as the sites recognized in splicing by point mutations that prevent splicing *in vivo* and *in vitro*.

26.3 Splice Junctions Are Read in Pairs

Key concepts

- Splicing depends only on recognition of pairs of splice junctions.
- All 5′ splice sites are functionally equivalent, and all 3′ splice sites are functionally equivalent.

A typical mammalian mRNA has many introns. The basic problem of pre-mRNA splicing results from the simplicity of the splice sites and is illustrated in FIGURE 26.4: What ensures that the correct pairs of sites are spliced together? The corresponding GU-AG pairs must be connected across great distances (some introns are >10 kb long). We can imagine two types of principle that might be responsible for pairing the appropriate 5′ and 3′ sites:

- It could be an *intrinsic property* of the RNA to connect the sites at the ends of a particular intron. This would require matching of specific sequences or structures.

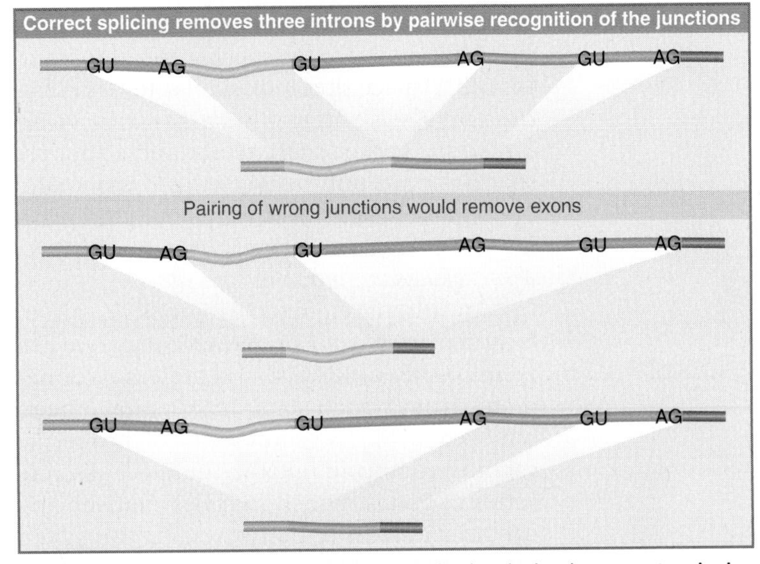

FIGURE 26.4 Splicing junctions are recognized only in the correct pairwise combinations.

- It could be that all 5′ sites may be functionally equivalent and all 3′ sites may be similarly indistinguishable, but splicing could follow *rules* that ensure a 5′ site is always connected to the 3′ site that comes next in the RNA.

Neither the splice sites nor the surrounding regions have any sequence complementarity, which excludes models for complementary base pairing between intron ends. Experiments using hybrid RNA precursors show that any 5′ splice site can in principle be connected to any 3′ splice site. For example, when the first exon of the early SV40 transcription unit is linked to the third exon of mouse β globin, the hybrid intron can be excised to generate a perfect connection between the SV40 exon and the β-globin exon. Indeed, this interchangeability is the basis for the exon-trapping technique described previously in Figure 4.11. Such experiments make two general points:

- *Splice sites are generic:* They do not have specificity for individual RNA precursors, and individual precursors do not convey specific information (such as secondary structure) that is needed for splicing.
- *The apparatus for splicing is not tissue specific:* An RNA can usually be properly spliced by any cell, whether or not it is usually synthesized in that cell. (We discuss exceptions in which there are tissue-specific alternative splicing patterns in Section 26.12, Alternative Splicing Involves Differential Use of Splice Junctions.)

Here is a paradox. It is likely that all 5′ splice sites look similar to the splicing apparatus, and that all 3′ splice sites look similar to it. *In principle, any 5′ splice site may be able to react with any 3′ splice site.* In the usual circumstances, though, splicing occurs only between the 5′ and 3′ sites of the *same* intron. *What rules ensure that recognition of splice sites is restricted so that only the 5′ and 3′ sites of the same intron are spliced?*

Are introns removed in a specific *order* from a particular RNA? Using RNA blotting, we can identify nuclear RNAs that represent intermediates from which some introns have been removed. FIGURE 26.5 shows a blot of the precursors to ovomucoid mRNA. There is a discrete series of bands, which suggests that splicing occurs via definite pathways. (If the seven introns were removed in an entirely random order, there would be more than 300 precursors with different combinations of introns, and we would not see discrete bands.)

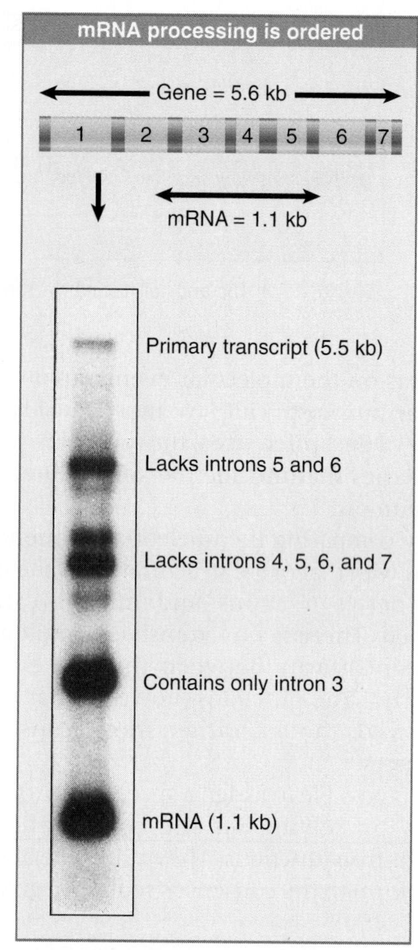

FIGURE 26.5 Northern blotting of nuclear RNA with an ovomucoid probe identifies discrete precursors to mRNA. The contents of the more prominent bands are indicated. Photo courtesy of Bert W. O'Malley, Baylor College of Medicine.

There does not seem to be a *unique* pathway, because intermediates can be found in which different combinations of introns have been removed. There is, however, evidence for a *preferred* pathway or pathways. When only one intron has been lost, it is virtually always 5 or 6. Either can be lost first. When two introns have been lost, 5 and 6 are again the most frequent, but there are other combinations. Intron 3 is never (or at least very rarely) lost at one of the first three splicing steps. From this pattern, we see that there is a preferred pathway in which introns are removed in the order 5/6, 7/4, 2/1, 3. There are other pathways, though, because (for example) there are some molecules in which 4 or 7 is lost last. A caveat in interpreting these results is that we do not have proof that all these intermediates actually lead to mature mRNA.

The general conclusion suggested by this analysis is that the conformation of the RNA influences the accessibility of the splice sites. As particular introns are removed, the conformation changes, and new pairs of splice sites become available. The ability of the precursor to remove its introns in more than one order, though, suggests that alternative conformations are available at each stage. Of course, the longer the molecule, the more structural options become available, and when we consider larger genes, it becomes difficult to see how specific secondary structures could control the reaction. One important conclusion of this analysis is that *the reaction does not proceed sequentially along the precursor.*

A simple model to control recognition of splice sites would be for the splicing apparatus to act in a processive manner. Having recognized a 5′ site, the apparatus might scan the RNA in the appropriate direction until it meets the next 3′ site. This would restrict splicing to adjacent sites. This model, however, is excluded by experiments that show that splicing can occur in *trans* as an intermolecular reaction under special circumstances (see Section 26.13, *trans*-splicing Reactions Use Small RNAs) or in RNA molecules in which part of the nucleotide chain is replaced by a chemical linker. This means that there cannot be a requirement for strict scanning along the RNA from the 5′ splice site to the 3′ splice site. Another problem with the scanning model is that it cannot explain the existence of alternative splicing patterns, where (for example) a common 5′ site is spliced to more than one 3′ site. The basis for proper recognition of correct splice site pairs remains incompletely defined.

26.4 Pre-mRNA Splicing Proceeds through a Lariat

Key concepts

- Splicing requires the 5′ and 3′ splice sites and a branch site just upstream of the 3′ splice site.
- The branch sequence is conserved in yeast but less well conserved in higher eukaryotes.
- A lariat is formed when the intron is cleaved at the 5′ splice site, and the 5′ end is joined to a 2′ position at an A at the branch site in the intron.
- The intron is released as a lariat when it is cleaved at the 3′ splice site, and the left and right exons are then ligated together.
- The reactions occur by transesterifications, in which a bond is transferred from one location to another.

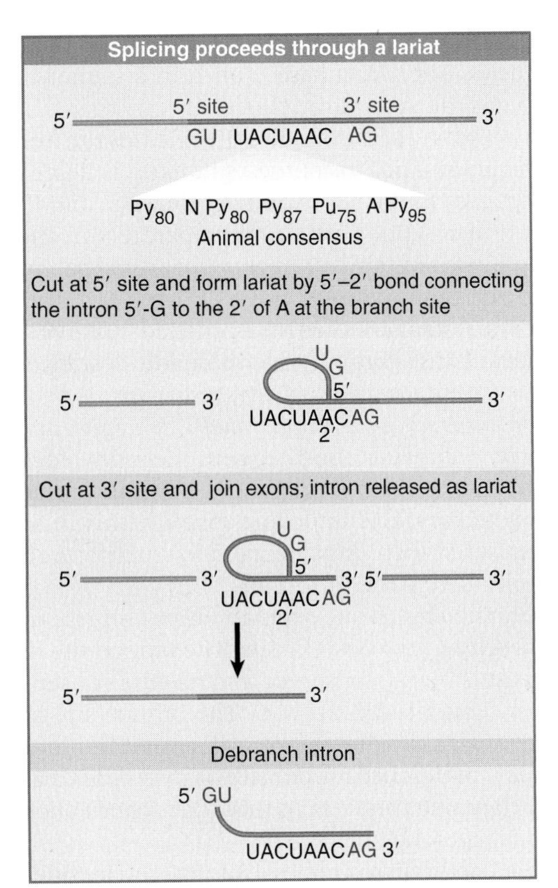

FIGURE 26.6 Splicing occurs in two stages. First the 5′ exon is cleaved off, and then it is joined to the 3′ exon.

The mechanism of splicing has been characterized *in vitro* using systems in which introns can be removed from RNA precursors. Nuclear extracts can splice purified RNA precursors; this shows that the action of splicing is not linked to the process of transcription. Splicing can occur in RNAs that are neither capped nor polyadenylated. The splicing reaction as such is independent of transcription or modification to the RNA; these events, however, normally occur in a coordinated manner, and the efficiency of splicing may be influenced by other processing events.

The stages of splicing *in vitro* are illustrated in the pathway of FIGURE 26.6. We discuss the reaction in terms of the individual RNA species that can be identified, but remember that *in vivo* the species containing exons are not released as free molecules, but remain held together by the splicing apparatus.

The first step is to make a cut at the 5′ splice site, separating the left exon and the right intron–exon molecule. The left exon takes the form of a linear molecule. The right intron–exon molecule forms a **lariat,** in which the 5′ terminus generated at the end of the intron becomes linked by a 5′–2′ bond to a base within the intron. The target base is an A in a sequence that is called the **branch site.**

Cutting at the 3′ splice site releases the free intron in lariat form; the right exon is ligated (spliced) to the left exon. The cleavage and ligation reactions are shown separately in the figure for illustrative purposes, but actually occur as one coordinated transfer.

The lariat is then "debranched" to give a linear excised intron, which is rapidly degraded.

The sequences needed for splicing are the short consensus sequences at the 5′ and 3′ splice sites and at the branch site. Together with the knowledge that most of the sequence of an intron can be deleted without impeding splicing, this indicates that there is no demand for specific conformation in the intron (or exon).

The branch site plays an important role in identifying the 3′ splice site. The branch site in yeast is highly conserved and has the consensus sequence UACUAAC. The branch site in higher eukaryotes is not well conserved, but has a preference for purines or pyrimidines at each position and retains the target A nucleotide (see Figure 26.6).

The branch site lies 18 to 40 nucleotides upstream of the 3′ splice site. Mutations or deletions of the branch site in yeast prevent splicing. In higher eukaryotes, the relaxed constraints in its sequence result in the ability to use related sequences (called cryptic sites) when the authentic branch is deleted. Proximity to the 3′ splice site appears to be important, because the cryptic site is always close to the authentic site. A cryptic site is used only when the branch site has been inactivated. When a cryptic branch sequence is used in this manner, splicing otherwise appears to be normal, and the exons give the same products as wild type. *The role of the branch site therefore is to identify the nearest 3′ splice site as the target for connection to the 5′* splice site. This can be explained by the fact that an interaction occurs between protein complexes that bind to these two sites.

The bond that forms the lariat goes from the 5′ position of the invariant G that was at the 5′ end of the intron to the 2′ position of the invariant A in the branch site. This corresponds to the third A residue in the yeast UACUAAC box.

The chemical reactions proceed by **transesterification:** A bond is in effect *transferred* from one location to another. FIGURE 26.7 shows that the first step is a nucleophilic attack by the 2′–OH of the invariant A of the UACUAAC sequence on the 5′ splice site. In the second step, the free 3′–OH of the exon that was released by the first reaction now attacks the bond at the 3′ splice site. Note that the number of phosphodiester bonds is conserved. There were originally two 5′–3′ bonds at the exon–intron splice sites; one has been replaced by the 5′–3′ bond between the exons, and the other has been replaced by the 5′–2′ bond that forms the lariat.

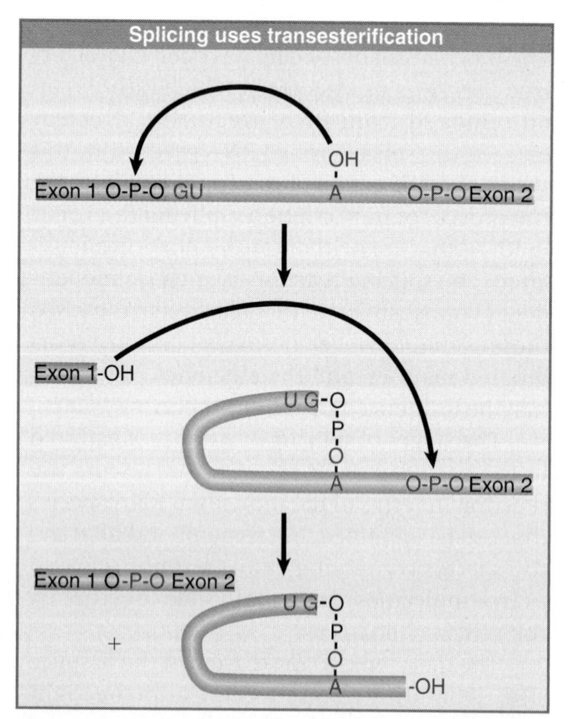

FIGURE 26.7 Nuclear splicing occurs by two transesterification reactions in which an OH group attacks a phosphodiester bond.

26.5 snRNAs Are Required for Splicing

Key concepts

- The five snRNPs involved in splicing are U1, U2, U5, U4, and U6.
- Together with some additional proteins, the snRNPs form the spliceosome.
- All the snRNPs except U6 contain a conserved sequence that binds the Sm proteins that are recognized by antibodies generated in autoimmune disease.

The 5′ and 3′ splice sites and the branch sequence are recognized by components of the splicing apparatus that assemble to form a large complex. This complex brings together the 5′ and 3′ splice sites before any reaction occurs, which explains why a deficiency in any one of the sites may prevent the reaction from initiating. The complex assembles sequentially on the pre-mRNA, and several intermediates can be recognized by fractionating complexes of different sizes. Splicing occurs only after all the components have assembled.

The splicing apparatus contains both proteins and RNAs (in addition to the pre-mRNA). The RNAs take the form of small molecules that exist as ribonucleoprotein particles. Both the nucleus and cytoplasm of eukaryotic cells contain many discrete small RNA species. They range in size from 100 to 300 bases in higher eukaryotes and extend in length to ~1000 bases in yeast. They vary considerably in abundance, from 10^5 to 10^6 molecules per cell to concentrations too low to be detected directly.

Those restricted to the nucleus are called **small nuclear RNAs (snRNAs);** those found in the cytoplasm are called **small cytoplasmic RNAs (scRNAs).** In their natural state, they exist as ribonucleoprotein particles (snRNP and scRNP). Colloquially, they are sometimes known as **snurps** and **scyrps.** There is also a class of small RNAs found in the nucleolus, called snoRNAs, which are involved in processing ribosomal RNA (see Section 26.22, Small RNAs Are Required for rRNA Processing).

The snRNPs involved in splicing, together with many additional proteins, form a large particulate complex called the **spliceosome.** Isolated from the *in vitro* splicing systems, it comprises a 50S to 60S ribonucleoprotein particle. The spliceosome may be formed in stages as the snRNPs join, proceeding through several "presplicing complexes." The spliceosome is a large body, greater in mass than the ribosome.

FIGURE 26.8 summarizes the components of the spliceosome. The 5 snRNAs account for more than a quarter of the mass; together with their 41 associated proteins, they account for almost half of the mass. Some 70 other proteins found in the spliceosome are described as splicing factors. They include proteins required for assembly of the spliceosome, proteins required for it to bind to the RNA substrate, and proteins involved in the catalytic process. In addition to these proteins, another ~30 proteins associated with the spliceosome have been implicated in acting at other stages of gene expression, which suggests that the spliceosome may serve as a coordinating apparatus.

The spliceosome forms on the intact precursor RNA and passes through an intermediate state in which it contains the individual 5′ exon linear molecule and the right lariat-intron–exon. Little spliced product is found in the complex, which suggests that it is usually released immediately following the cleavage of the 3′ site and ligation of the exons.

We may think of the snRNP particles as being involved in building the structure of the spliceosome. Like the ribosome, the spliceosome depends on RNA–RNA interactions as well as protein–RNA and protein–protein interactions. Some of the reactions involving the snRNPs require their RNAs to base pair directly with sequences in the RNA being spliced; other reactions require recognition between snRNPs or between their proteins and other components of the spliceosome.

The importance of snRNA molecules can be tested directly in yeast by making mutations in their genes. Mutations in 5 snRNA genes are lethal and prevent splicing. All of the snRNAs involved in splicing can be recognized in conserved forms in animal, bird, and insect cells. The corresponding RNAs in yeast are often rather larger, but conserved regions include features that are similar to the snRNAs of higher eukaryotes.

The snRNPs involved in splicing are U1, U2, U5, U4, and U6. They are named according to the snRNAs that are present. Each snRNP contains a single snRNA and several (<20) proteins. The U4 and U6 snRNPs are usually found as a single (U4/U6) particle. A common structural core for each snRNP consists of a group of eight proteins, all of which are recognized by an autoimmune antiserum called **anti-Sm;**

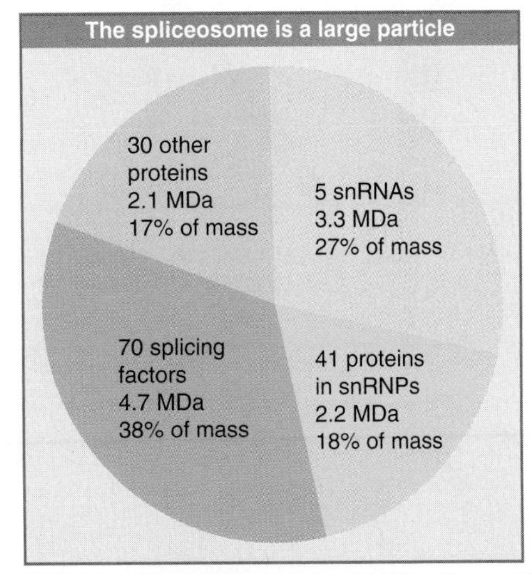

FIGURE 26.8 The spliceosome is ~12 MDa. Five snRNPs account for almost half of the mass. The remaining proteins include known splicing factors, as well as proteins that are involved in other stages of gene expression.

conserved sequences in the proteins form the target for the antibodies. The other proteins in each snRNP are unique to it. The Sm proteins bind to the conserved sequence $PuAU_{3-6}Gpu$, which is present in all snRNAs except U6. The U6 snRNP instead contains a set of Sm-like (Lsm) proteins. The Sm proteins must be involved in the autoimmune reaction, although their relationship to the phenotype of the autoimmune disease is not clear.

Some of the proteins in the snRNPs may be involved directly in splicing; others may be required in structural roles or just for assembly or interactions between the snRNP particles. About one third of the proteins involved in splicing are components of the snRNPs. Increasing evidence for a direct role of RNA in the splicing reaction suggests that relatively few of the splicing factors play a direct role in catalysis; most are involved in structural or assembly roles.

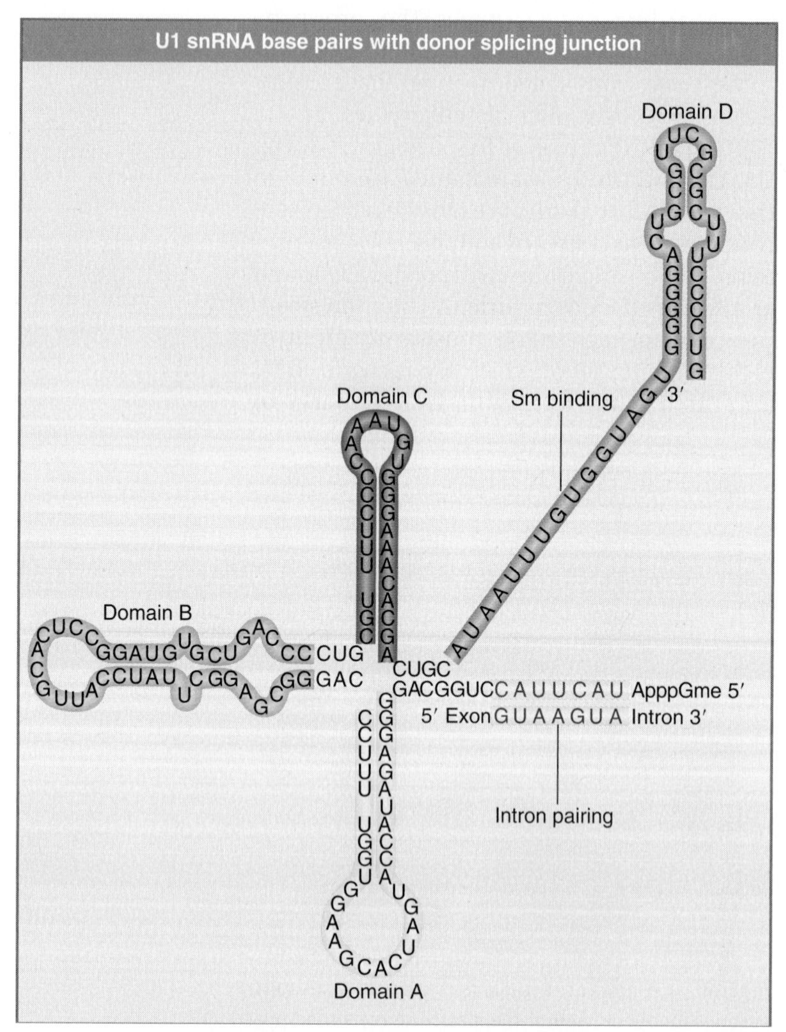

FIGURE 26.9 U1 snRNA has a base-paired structure that creates several domains. The 5′ end remains single stranded and can base pair with the 5′ splicing site.

26.6 U1 snRNP Initiates Splicing

Key concepts

- U1 snRNP initiates splicing by binding to the 5′ splice site by means of an RNA–RNA pairing reaction.
- The E complex contains U1 snRNP bound at the 5′ splice site, the protein U2AF bound to a pyrimidine tract between the branch site and the 3′ splice site, and SR proteins connecting U1 snRNP to U2AF.

Splicing can be broadly divided into two stages:

- First the consensus sequences at the 5′ splice site, branch sequence, and adjacent pyrimidine tract are recognized. A complex assembles that contains all of the splicing components.
- The cleavage and ligation reactions then change the structure of the substrate RNA. Components of the complex are released or reorganized as it proceeds through the splicing reactions.

The important point is that all of the splicing components are assembled and have ensured that the splice sites are available before any irreversible change is made to the RNA.

Recognition of the consensus sequences involves both RNAs and proteins. Certain snRNAs have sequences that are complementary to the consensus sequences or to one another, and base pairing between snRNA and pre-mRNA, or between snRNAs, plays an important role in splicing.

The human U1 snRNP contains eight proteins as well as the RNA. The secondary structure of the U1 snRNA is drawn in FIGURE 26.9. It contains several domains. The Sm-binding site is required for interaction with the common snRNP proteins. Domains identified by the individual stem-loop structures provide binding sites for proteins that are unique to U1 snRNP.

Binding of U1 snRNP to the 5′ splice site is the first step in splicing. The recruitment of U1 snRNP involves an interaction between one of its proteins (U1-70k) and the protein ASF/SF2 (a general splicing factor in the SR class: see below). U1 snRNA base pairs with the 5′ site by means of a single-stranded region at its 5′ terminus, which usually includes a stretch of four to six bases that is complementary with the splice site.

Mutations in the 5′ splice site and U1 snRNA can be used to test directly whether pairing between them is necessary. The results of such an experiment are illustrated in FIGURE 26.10. The wild-type sequence of the splice site of the 12S adenovirus pre-mRNA pairs at five out of six positions with U1 snRNA. A mutant in the 12S RNA that cannot be spliced has two sequence changes; the GG residues at positions 5 to 6 in the intron are changed to AU. The mutation changes the pattern of base pairing between U1 snRNA and the 5′ splice site, although it does not alter the *overall* extent of pairing (because complementarity is lost at one position and gained at the other). The effect on splicing suggests that the base-pairing interaction is important.

When a mutation is introduced into U1 snRNA that restores pairing at position 5, normal splicing is regained. Other cases in which corresponding mutations are made in U1 snRNA to see whether they can suppress the mutation in the splice site suggests the general rule: Complementarity between U1 snRNA and the 5′ splice site is necessary for splicing, but the efficiency of splicing is not determined solely by the number of base pairs that can form. The pairing reaction is stabilized by the proteins of the U1 snRNP.

FIGURE 26.11 shows the early stages of splicing. The first complex formed during splicing is the E (early presplicing) complex, which contains U1 snRNP, the splicing factor U2AF, and members of a family called **SR proteins,** which comprise an important group of splicing factors and regulators. They take their name from the presence of an Arg-Ser-rich region that is variable in length. SR proteins interact with one another via their Arg-Ser-rich regions. They also bind to RNA. They are an essential component of the spliceosome, forming a framework on the RNA substrate. They connect U2AF to U1 (FIGURE 26.12). The E complex is sometimes called the commitment complex, because its formation identifies a pre-mRNA as a substrate for formation of the splicing complex.

In the E complex, U2AF is bound to the region between the branch site and the 3′ splice site. The name of U2AF reflects its original isolation as the U2 auxiliary factor. In most organisms, it has a large subunit (U2AF65) that contacts a pyrimidine tract downstream of the branch site; a small subunit (U2AF35) directly contacts the dinucleotide AG at the 3′ splice site. In *Saccharomyces cerevisiae,* this function is filled by the protein Mud2, which is a counterpart of U2AF65, and binds only to the pyrimidine tract. This marks a difference in the mechanism of splicing between *S. cerevisiae* and other organisms. In yeast, the 3′ splice site is not involved in the early stages of forming the splicing complex, but in all other known cases it is required.

Another splicing factor, called SF1 in mammals and BBP in yeast, connects U2AF/Mud2 to the U1 snRNP bound at the 5′ splice site. Complex formation is enhanced by the cooperative reactions of the two proteins; SF1 and U2AF (or BBP and Mud2) bind together to the RNA substrate ~10× more effectively than either alone. This interaction is probably responsible for making the first connection between the two splice sites across the intron.

The E complex is converted to the A complex when U2 snRNP binds to the branch site. Both U1 snRNP and U2AF/Mud2 are needed for U2 binding. The U2 snRNA includes sequences complementary to the branch site.

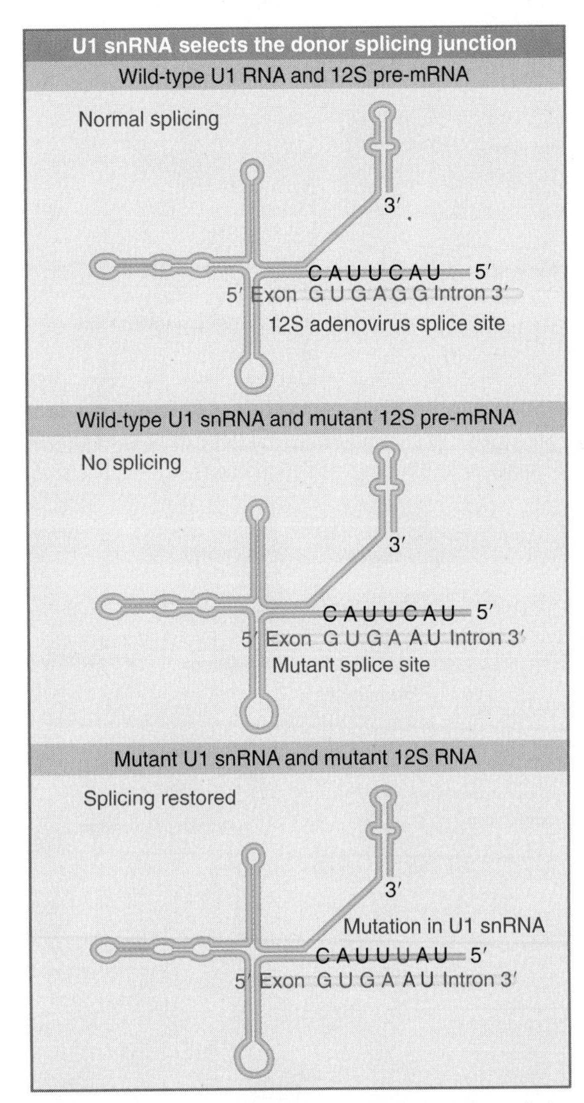

FIGURE 26.10 Mutations that abolish function of the 5′ splicing site can be suppressed by compensating mutations in U1 snRNA that restore base pairing.

A sequence near the 5′ end of the snRNA base pairs with the branch sequence in the intron. In yeast this typically involves formation of a duplex with the UACUAAC box (see Figure 26.14). Several proteins of the U2 snRNP are bound to the substrate RNA just upstream of the branch site. The addition of U2 snRNP to the E complex generates the A presplicing complex. The binding of U2 snRNP requires ATP hydrolysis, and commits a pre-mRNA to the splicing pathway.

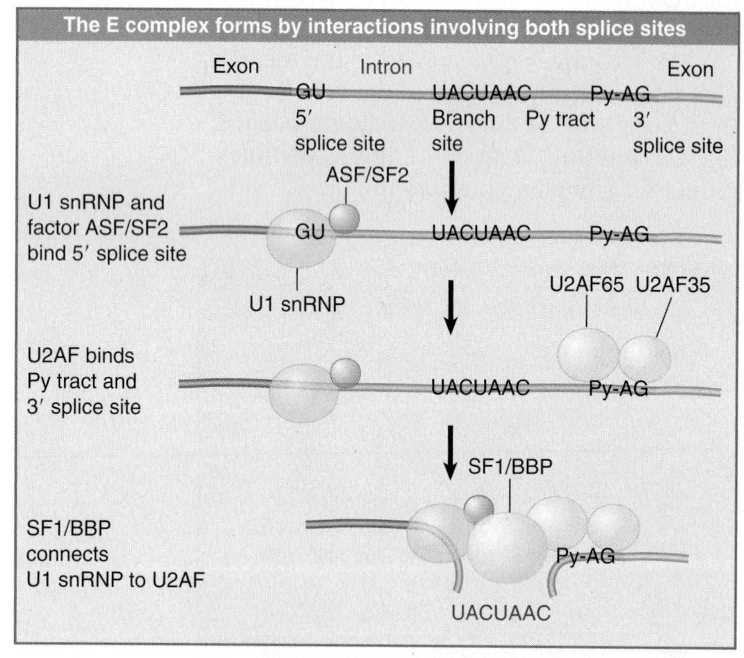

FIGURE 26.11 The commitment (E) complex forms by the successive addition of U1 snRNP to the 5′ splice site, U2AF to the pyrimidine tract/3′ splice site, and the bridging protein SF1/BBP.

26.7 The E Complex Can Be Formed by Intron Definition or Exon Definition

Key concepts

- The direct way of forming an E complex is for U1 snRNP to bind at the 5′ splice site and U2AF to bind at a pyrimidine tract between the branch site and the 3′ splice site.
- Another possibility is for the complex to form between U2AF at the pyrimidine tract and U1 snRNP at a downstream 5′ splice site.
- The E complex is converted to the A complex when U2 snRNP binds at the branch site.
- If an E complex forms using a downstream 5′ splice site, this splice site is replaced by the appropriate upstream 5′ splice site when the E complex is converted to the A complex.
- Weak 3′ splice sites may require a splicing enhancer located in the exon downstream to bind SR proteins directly.

There is more than one way to form the E complex. Figure 26.12 illustrates some possibilities. The most direct reaction is for both splice sites to be recognized across the intron. The presence of U1 snRNP at the 5′ splice site is necessary for U2AF to bind at the pyrimidine tract downstream of the branch site, making it possible that the 5′ and 3′ ends of the intron are brought together in this complex. The E complex is converted to the A complex when U2 snRNP binds at the branch site. *The basic feature of this route for splicing is that the two splice sites are*

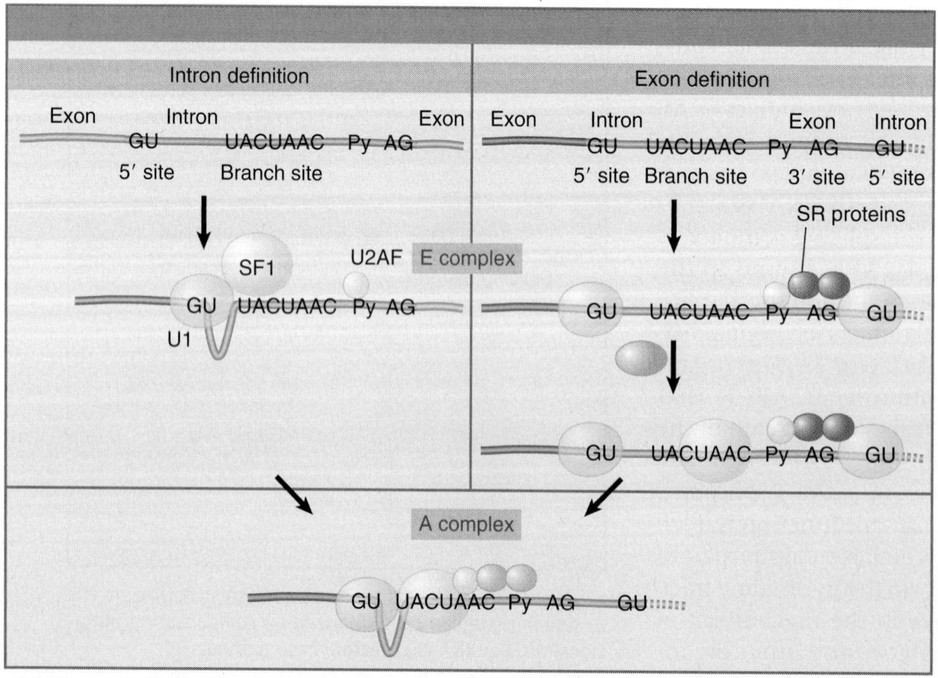

FIGURE 26.12 There may be multiple routes for initial recognition of 5′ and 3′ splice sites.

recognized without requiring any sequences outside of the intron. This process is called **intron definition.**

In an extreme case, the SR proteins may enable U2AF/U2 snRNP to bind *in vitro* in the absence of U1, raising the possibility that there could be a U1-independent pathway for splicing.

An alternative route to form the spliceosome may be followed when the introns are long and the splice sites are weak. As shown on the right of the figure, the 5′ splice site is recognized by U1 snRNA in the usual way. The 3′ splice site is recognized as part of a complex that forms across the *next exon,* though, in which the next 5′ splice site is also bound by U1 snRNA. This U1 snRNA is connected by SR proteins to the U2AF at the pyrimidine tract. When U2 snRNP joins to generate the A complex, there is a rearrangement in which the correct (leftmost) 5′ splice site displaces the downstream 5′ splice site in the complex. The important feature of this route for splicing is that sequences downstream of the intron itself are required. Usually these sequences include the next 5′ splice site. This process is called **exon definition.** This mechanism is not universal: Neither SR proteins nor exon definition are found in *S. cerevisiae.*

"Weak" 3′ splice sites do not bind U2AF and U2 snRNP effectively. Additional sequences are needed to bind the SR proteins, which assist U2AF in binding to the pyrimidine tract. Such sequences are called "splicing enhancers," and they are most commonly found in the exon downstream of the 3′ splice site.

26.8 5 snRNPs Form the Spliceosome

Key concepts

- Binding of U5 and U4/U6 snRNPs converts the A complex to the B1 spliceosome, which contains all the components necessary for splicing.
- The spliceosome passes through a series of further complexes as splicing proceeds.
- Release of U1 snRNP allows U6 snRNA to interact with the 5′ splice site, and converts the B1 spliceosome to the B2 spliceosome.
- When U4 dissociates from U6 snRNP, U6 snRNA can pair with U2 snRNA to form the catalytic active site.

Following formation of the E complex, the other snRNPs and factors involved in splicing associate with the complex in a defined order. FIGURE 26.13 shows the components of the complexes that can be identified as the reaction proceeds.

The B1 complex is formed when a trimer containing the U5 and U4/U6 snRNPs binds to the A complex containing U1 and U2 snRNPs. This complex is regarded as a spliceosome because it contains the components needed for the splicing reaction. It is converted to the B2 complex after U1 is released. The dissociation of U1 is necessary to allow other components to come into juxtaposition with the 5′ splice site, most notably U6 snRNA. At this point U5 snRNA changes its position; initially it is close to exon sequences at the 5′ splice site, but it shifts to the vicinity of the intron sequences.

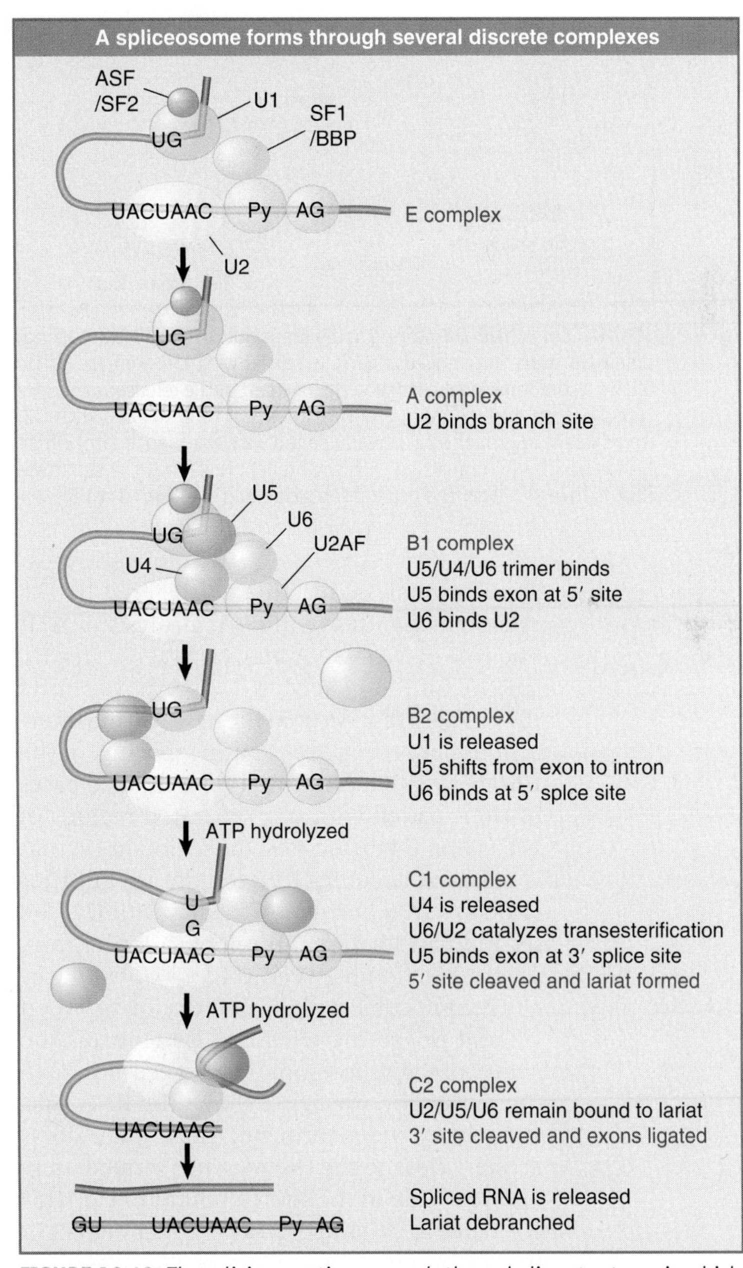

FIGURE 26.13 The splicing reaction proceeds through discrete stages in which spliceosome formation involves the interaction of components that recognize the consensus sequences.

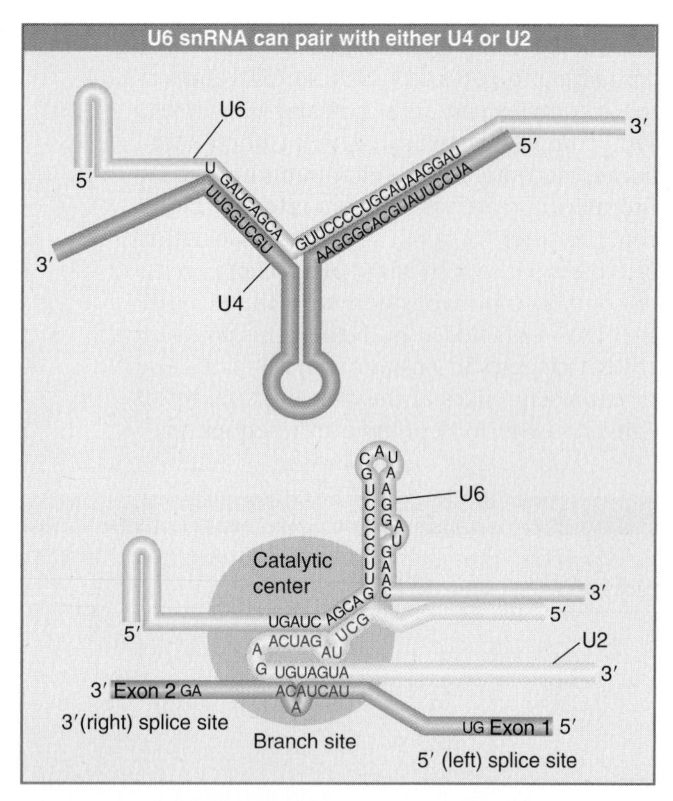

FIGURE 26.14 U6-U4 pairing is incompatible with U6-U2 pairing. When U6 joins the spliceosome it is paired with U4. Release of U4 allows a conformational change in U6; one part of the released sequence forms a hairpin (dark gray), and the other part (black) pairs with U2. An adjacent region of U2 is already paired with the branch site, which brings U6 into juxtaposition with the branch. Note that the substrate RNA is reversed from the usual orientation and is shown 3′ to 5′.

The catalytic reaction is triggered by the release of U4; this requires hydrolysis of ATP. The role of U4 snRNA may be to sequester U6 snRNA until it is needed. FIGURE 26.14 shows the changes that occur in the base pairing interactions between snRNAs during splicing. In the U6/U4 snRNP, a continuous length of 26 bases of U6 is paired with two separated regions of U4. When U4 dissociates, the region in U6 that is released becomes free to take up another structure. The first part of it pairs with U2; the second part forms an intramolecular hairpin. The interaction between U4 and U6 is mutually incompatible with the interaction between U2 and U6, so the release of U4 controls the ability of the spliceosome to proceed.

For clarity, the figure shows the RNA substrate in extended form, but the 5′ splice site is actually close to the U6 sequence immediately on the 5′ side of the stretch bound to U2. This sequence in U6 snRNA pairs with sequences in the intron just downstream of the conserved GU at the 5′ splice site (mutations that enhance such pairing improve the efficiency of splicing).

Thus several pairing reactions between snRNAs and the substrate RNA occur in the course of splicing. They are summarized in FIGURE 26.15. The snRNPs have sequences that pair with the substrate and with one another. They also have single-stranded regions in loops that are in close proximity to sequences in the substrate, and which play an important role, as judged by the ability of mutations in the loops to block splicing.

The base pairing between U2 and the branch point, and between U2 and U6, creates a structure that resembles the active center of group II self-splicing introns (see Figure 26.20). This suggests the possibility that the catalytic component could comprise an RNA structure generated by the U2-U6 interaction. U6 is paired with the 5′ splice site, and crosslinking experiments show that a loop in U5 snRNA is immediately adjacent to the first base positions in both exons. Although we can define the proximities of the substrate (5′ splice site and branch site) and snurps (U2 and U6) at the catalytic center (as shown in Figure 26.14), the components that undertake the transesterifications have not been directly identified.

The formation of the lariat at the branch site is responsible for determining the use of the 3′ splice site, because the 3′ consensus sequence nearest to the 3′ side of the branch becomes the target for the second transesterification. The second splicing reaction follows rapidly. Binding of U5 snRNP to the 3′ splice site is needed for this reaction, but there is no evidence for a base-pairing reaction.

The important conclusion suggested by these results is that *the snRNA components of the splicing apparatus interact both among themselves and with the substrate RNA by means of base pairing interactions, and these interactions allow for changes in structure that may bring reacting groups into apposition and may even create catalytic centers.* Furthermore, the conformational changes in the snRNAs are reversible; for example, U6 snRNA is not used up in a splicing reaction and at completion must be released from U2 so that it can reform the duplex structure with U4 to undertake another cycle of splicing.

We have described individual reactions in which each snRNP participates, but as might be expected from a complex series of reactions, any particular snRNP may play more than one role in splicing. Thus the ability of U1 snRNP to promote binding of U2 snRNP to the branch site is independent of its ability to bind to the 5′ splice site. Similarly, different regions of U2 snRNA can be defined that are needed to bind to the branch site and to interact with other splicing components.

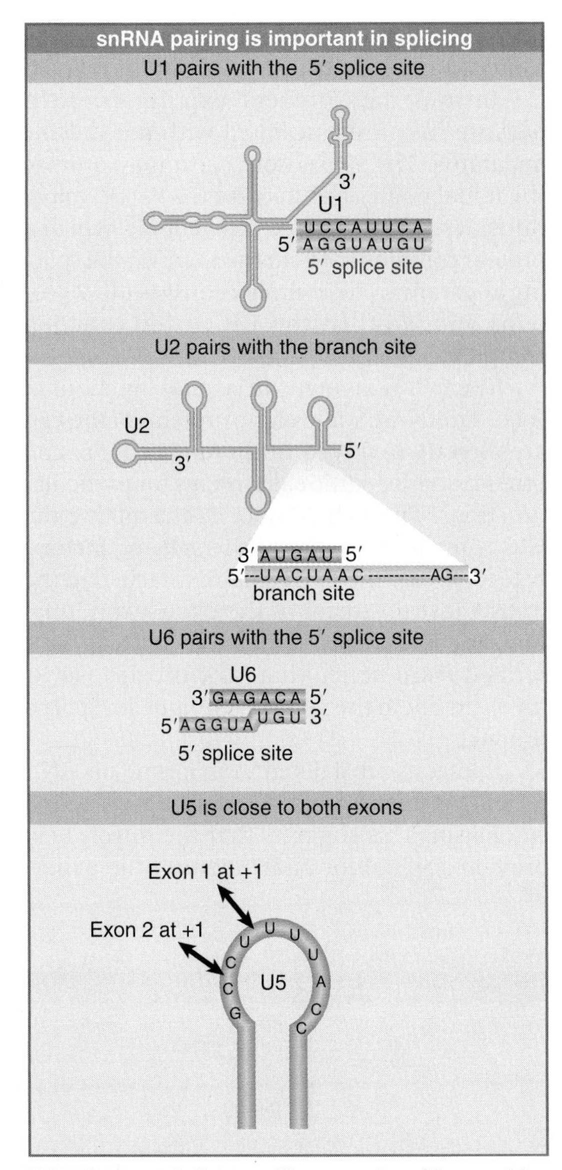

FIGURE 26.15 Splicing utilizes a series of base pairing reactions between snRNAs and splice sites.

An extensive mutational analysis has been undertaken in yeast to identify both the RNA and protein components of the spliceosome. Mutations in genes needed for splicing are identified by the accumulation of unspliced precursors. A series of loci that identify genes potentially coding for proteins involved in splicing were originally called *RNA*, but are now known as *PRP* mutants (for pre-RNA processing). Several of the products of these genes have motifs that identify them as RNA-binding proteins, and some appear to be related to a family of ATP-dependent RNA helicases. We suppose that, in addition to RNA–RNA interactions, protein–RNA interactions are important in creating or releasing structures in the pre-mRNA or snRNA components of the spliceosomes.

Some of the PRP proteins are components of snRNP particles, but others function as independent factors. One interesting example is PRP16, a helicase that hydrolyzes ATP and associates transiently with the spliceosome to participate in the second catalytic step. Another example is PRP22, another ATP-dependent helicase, which is required to release the mature mRNA from the spliceosome. The conservation of bonds during the splicing reaction means that input of energy is not required to drive bond formation *per se*, which implies that the ATP hydrolysis is required for other purposes. The use of ATP by PRP16 and PRP22 may be examples of a more general phenomenon: the use of ATP hydrolysis to drive conformational changes that are needed to proceed through splicing.

26.9 An Alternative Splicing Apparatus Uses Different snRNPs

Key concepts

- An alternative splicing pathway uses another set of snRNPs that comprise the U12 spliceosome.
- The target introns are defined by longer consensus sequences at the splice junctions, but usually include the same GU-AG junctions.
- Some introns have the splice junctions AU-AC, including some that are U1-dependent and some that are U12-dependent.

GU-AG introns comprise the vast majority (>98% of splicing junctions in the human genome). Less than 1% use the related junctions GC-AG, and then there is a minor class of introns marked by the ends AU-AC (comprising 0.1% of introns). The first of these introns to be discovered required an alternative splicing apparatus, called the U12 spliceosome, which consisted of U11 and U12 (related to U1 and U2, respectively), a U5 variant, and the $U4_{atac}$ and $U6_{atac}$ snRNAs. The splicing reaction is essentially similar to that at GU-AG introns, and the snRNAs play analogous roles. Whether there are differences in the protein components of this apparatus is not known.

It now turns out that the dependence on the type of spliceosome is also influenced by sequences in the intron, so that there are some AU-AC introns spliced by U2-type spliceosomes, and some GU-AG introns spliced by U12-type spliceosomes. A strong consensus sequence at the left end defines the U12-dependent type of intron: $5'{}^{G}{}_{A}$UAUCCUUU. . . . . .PyA${}^{G}{}_{C}$ 3′. In fact, most U12-dependent introns have the GU. . . . .AG termini. In addition, they have a highly conserved branch point, UCCUUPuAPy, which pairs with U12. For this reason, the term U12-dependent intron is used instead of AU-AC intron.

The two types of introns coexist in a variety of genomes, and in some cases are found in the same gene. U12-dependent introns tend to be flanked by U2-dependent introns. What is known about the phylogeny of these introns suggests that AU-AC U12-dependent introns may once have been more common, but tend to be converted to GU-AG termini, and to U2-dependence, in the course of evolution. The common evolution of the systems is emphasized by the fact that they use analogous sets of base pairing between the snRNAs and with the substrate pre-mRNA.

The involvement of snRNPs in splicing is only one example of their involvement in RNA processing reactions. snRNPs are required for several reactions in the processing of nuclear RNA to mature rRNAs. Especially in view of the demonstration that group I introns are self-splicing, and that the RNA of ribonuclease P has catalytic activity (as discussed in Chapter 27, Catalytic RNA), it is plausible to think that RNA–RNA reactions are important in many RNA processing events.

26.10 Splicing Is Connected to Export of mRNA

Key concepts

- The REF proteins bind to splicing junctions by associating with the spliceosome.
- After splicing, they remain attached to the RNA at the exon–exon junction.
- They interact with the transport protein TAP/Mex that exports the RNA through the nuclear pore.

After it has been synthesized and processed, mRNA is exported from the nucleus to the cytoplasm in the form of a ribonucleoprotein complex. The proteins that are responsible for transport "shuttle" between the nucleus and cytoplasm, remain in the compartment only briefly. Two important questions are how these proteins recognize their RNA substrates, and what ensures that only fully processed mRNAs are exported. The answers in part may lie in the relative timing of events: Spliceosomes may form to remove introns before transcription has been completed. There may, however, also be a direct connection between splicing and export.

Introns may prevent export of mRNA because they are associated with the splicing apparatus. The spliceosome also may provide the initial point of contact for the export apparatus. FIGURE 26.16 shows a model in which a protein complex binds to the RNA via the splicing apparatus. The complex consists of >9 proteins and is called the EJC (exon junction complex).

The EJC is involved in several functions of spliced mRNAs. Some of the proteins of the EJC are directly involved in these functions, and others recruit additional proteins for particular functions. The first contact in assembling the EJC is made with one of the splicing factors. After splicing, the EJC remains attached to the mRNA just upstream of the exon–exon junction. The EJC is not associated with RNAs transcribed from genes that lack introns, so its involvement in the process is unique for spliced products.

If introns are deleted from a gene, its RNA product is exported much more slowly to the cytoplasm. This suggests that the intron may provide a signal for attachment of the export

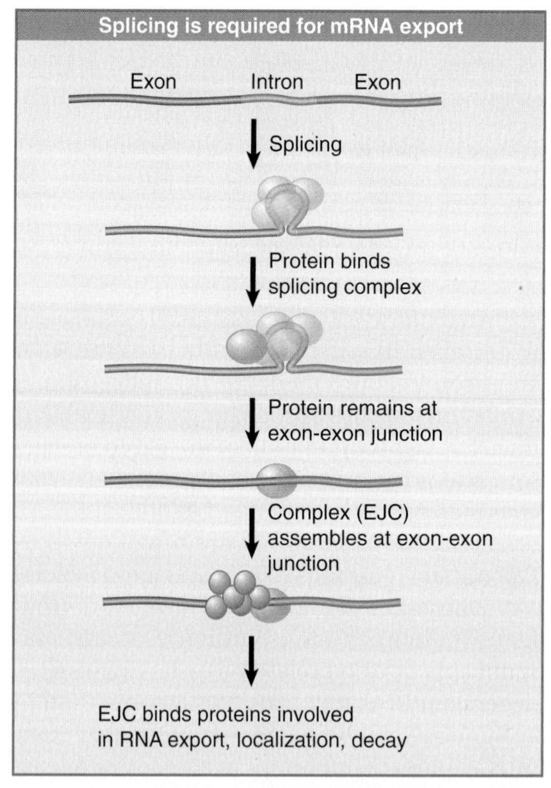

FIGURE 26.16 The EJC (exon junction complex) binds to RNA by recognizing the splicing complex.

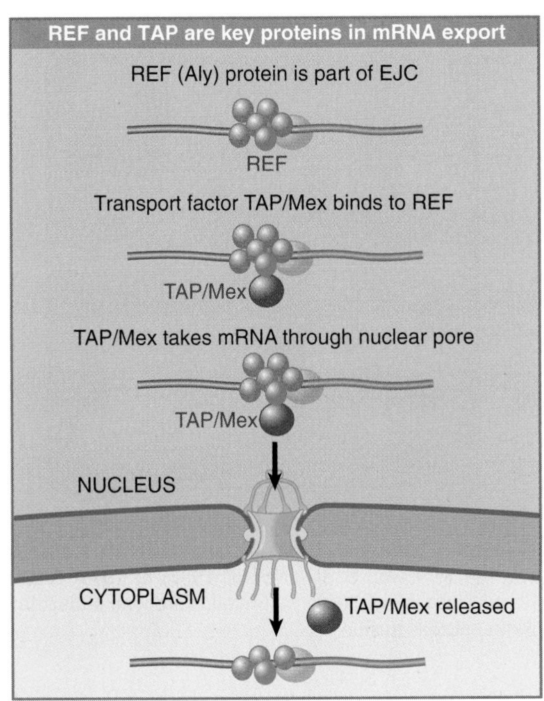

FIGURE 26.17 A REF protein binds to a splicing factor and remains with the spliced RNA product. REF binds to an export factor that binds to the nuclear pore.

apparatus. We can now account for this phenomenon in terms of a series of protein interactions, as shown in FIGURE 26.17. The EJC includes a group of proteins called the REF family (the best characterized member is called Aly). The REF proteins in turn interact with a transport protein (variously called TAP and Mex), which has direct responsibility for interaction with the nuclear pore.

A similar system may be used to identify a spliced RNA so that nonsense mutations prior to the last exon trigger its degradation in the cytoplasm (see Section 7.14, Nonsense Mutations Trigger a Surveillance System).

26.11 Group II Introns Autosplice via Lariat Formation

Key concepts

- Group II introns excise themselves from RNA by an autocatalytic splicing event.
- The splice junctions and mechanism of splicing of group II introns are similar to splicing of nuclear introns.
- A group II intron folds into a secondary structure that generates a catalytic site resembling the structure of U6-U2-nuclear intron.

Introns in protein-coding genes (in fact, in all genes except nuclear tRNA-coding genes) can be divided into three general classes. Nuclear pre-mRNA introns are identified only by the possession of the GU . . . AG dinucleotides at the 5′ and 3′ ends and the branch site/pyrimidine tract near the 3′ end. They do not show any common features of secondary structure. Group I and group II introns are found in organelles and in bacteria. (Group I introns are found also in the nucleus in lower eukaryotes.) Group I and group II introns are classified according to their internal organization. Each can be folded into a typical type of secondary structure.

The group I and group II introns have the remarkable ability to excise themselves from an RNA. This is called **autosplicing.** Group I introns are more common than group II introns. There is little relationship between the two classes, but in each case the RNA can perform the splicing reaction *in vitro* by itself, without requiring enzymatic activities provided by proteins; however, proteins are almost certainly required *in vivo* to assist with folding (see Chapter 27, Catalytic RNA).

FIGURE 26.18 shows that three classes of introns are excised by two successive transesterifications (shown previously for nuclear introns in Figure 26.6). In the first reaction, the 5′ exon–intron junction is attacked by a free hydroxyl group (provided by an internal 2′–OH position in nuclear and group II introns, and by a free guanine nucleotide in group I introns). In the second reaction, the free 3′–OH at the end of the released exon in turn attacks the 3′ intron–exon junction.

There are parallels between group II introns and pre-mRNA splicing. Group II mitochondrial introns are excised by the same mechanism as nuclear pre-mRNAs via a lariat that is held together by a 5′–2′ bond. An example of a lariat produced by splicing a group II intron is shown in FIGURE 26.19. When an isolated group II RNA is incubated *in vitro* in the absence of additional components, it is able to perform the splicing reaction. This means that the two transesterification reactions shown in Figure 26.18 can be performed by the group II intron RNA sequence itself. The number of phosphodiester bonds is conserved in the reaction, and as a result an external supply of energy is not required; this could have been an important feature in the evolution of splicing.

A group II intron forms into a secondary structure that contains several domains formed

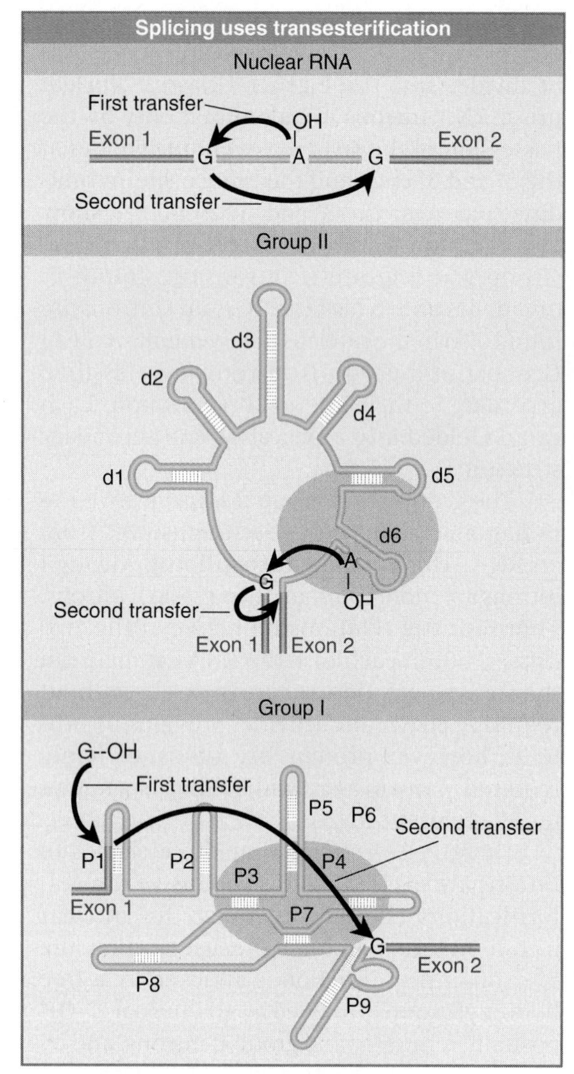

FIGURE 26.18 Three classes of splicing reactions proceed by two transesterifications. First, a free OH group attacks the exon 1–intron junction. Second, the OH created at the end of exon 1 attacks the intron–exon 2 junction.

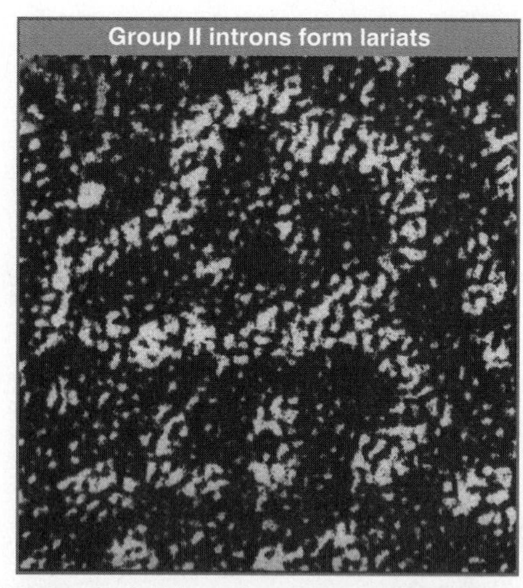

FIGURE 26.19 Splicing releases a mitochondrial group II intron in the form of a stable lariat. Reproduced from Van der Veen, R., et al. *EMBO J.* 1987. 6: 1079–1084. Photo courtesy of Leslie A. Grivell, European Molecular Biology Organisation.

by base-paired stems and single-stranded loops. Domain 5 is separated by two bases from domain 6, which contains an A residue that donates the 2′–OH group for the first transesterification. This constitutes a catalytic domain in the RNA. FIGURE 26.20 compares this secondary structure with the structure formed by the combination of U6 with U2 and of U2 with the branch site. The similarity suggests that U6 may have a catalytic role.

The features of group II splicing suggest that splicing evolved from an autocatalytic reaction undertaken by an individual RNA molecule, in which it accomplished a controlled deletion of an internal sequence. It is likely that such a reaction requires the RNA to fold into a specific conformation, or series of conformations, and would occur exclusively in *cis* conformation.

The ability of group II introns to remove themselves by an autocatalytic splicing event stands in great contrast to the requirement of nuclear introns for a complex splicing apparatus. We may regard the snRNAs of the spliceosome as compensating for the lack of sequence information in the intron, and providing the information required to form particular structures in RNA. The functions of the snRNAs may have evolved from the original autocatalytic system. These snRNAs act in *trans* upon the substrate pre-mRNA; we might imagine that the ability of U1 to pair with the 5′ splice site, or of U2 to pair with the branch sequence, replaced a similar reaction that required the relevant sequence to be carried by the intron. Thus the snRNAs may undergo reactions with the pre-mRNA substrate, and with one another, that have substituted for the series of conformational changes that occur in RNAs that splice by group II mechanisms. In effect, these changes have relieved the substrate pre-mRNA of the obligation to carry the sequences needed to sponsor the reaction. As the splicing apparatus has become more complex (and as the number of potential substrates has increased), proteins have played a more important role.

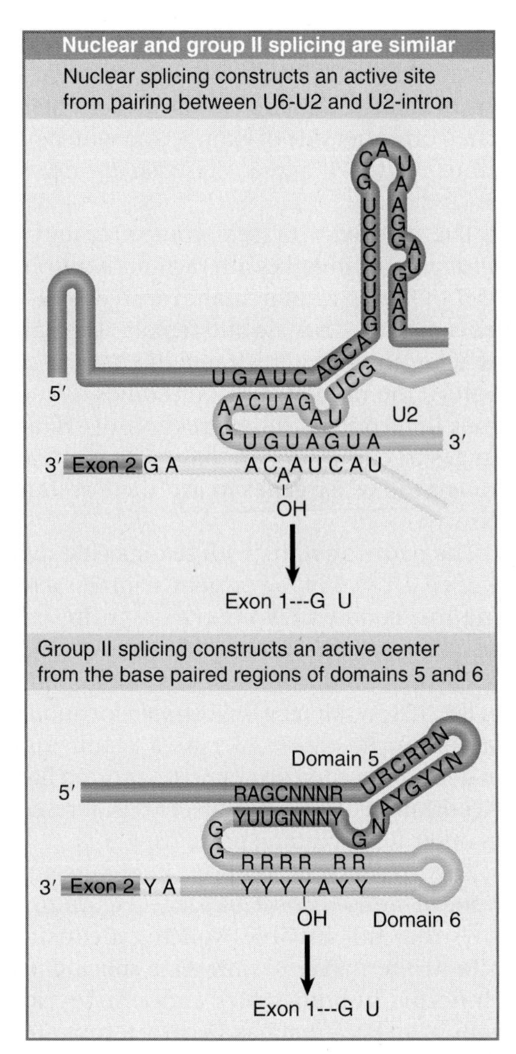

FIGURE 26.20 Nuclear splicing and group II splicing involve the formation of similar secondary structures. The sequences are more specific in nuclear splicing; group II splicing uses positions that may be occupied by either purine (R) or pyrimidine (Y).

26.12 Alternative Splicing Involves Differential Use of Splice Junctions

Key concepts

- Specific exons may be excluded or included in the RNA product by using or failing to use a pair of splicing junctions.
- Exons may be extended by changing one of the splice junctions to use an alternative junction.
- Sex determination in *Drosophila* involves a series of alternative splicing events in genes coding for successive products of a pathway.
- P elements of *Drosophila* show germline-specific alternative splicing.

When an interrupted gene is transcribed into an RNA that gives rise to a single type of spliced mRNA, there is no ambiguity in assignment of exons and introns. The RNAs of some genes, however, follow patterns of **alternative splicing,** which occurs when a single gene gives rise to more than one mRNA sequence. In some cases, the ultimate pattern of expression is dictated by the primary transcript, because the use of different startpoints or the generation of alternative 3′ ends alters the pattern of splicing. In other cases, a single primary transcript is spliced in more than one way, and internal exons are substituted, added, or deleted. In some cases, the multiple products all are made in the same cell, but in others the process is regulated so that particular splicing patterns occur only under particular conditions.

One of the most pressing questions in splicing is to determine what controls the use of such alternative pathways. Proteins that intervene to bias the use of alternative splice sites have been identified in two ways. In some mammalian systems, it has been possible to characterize alternative splicing *in vitro,* and to identify proteins that are required for the process. In *D. melanogaster,* aberrations in alternative splicing may be caused either by mutations in the genes that are alternatively spliced or in the genes whose products are necessary for the reaction.

FIGURE 26.21 shows examples of alternative splicing in which one splice site remains constant, but the other varies. The large T/small t antigens of SV40 and the products of the adenovirus E1A region are generated by connecting a varying 5′ site to a constant 3′ site. In the case of the T/t antigens, the 5′ site used for T antigen removes a termination codon that is present in the t antigen mRNA, so that T antigen is larger than t antigen. In the case of the E1A transcripts, one of the 5′ sites connects to the last exon in a different reading frame, again making a significant change in the C-terminal part of the protein. In these examples, all the relevant splicing events take place in every cell in which the gene is expressed, so that all the protein products are made.

There are differences in the ratios of T/t antigens in different cell types. A protein extracted from cells that produce relatively more small t antigen can cause preferential production of small t RNA in extracts from other cell types. This protein, which was called ASF (alternative splicing factor), turns out to be the same

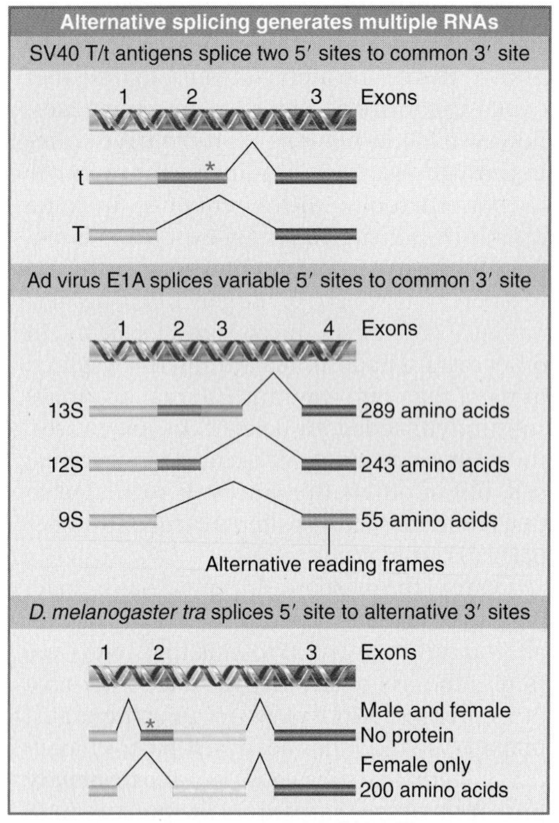

FIGURE 26.21 Alternative forms of splicing may generate a variety of protein products from an individual gene. Changing the splice sites may introduce termination codons (shown by asterisks) or change reading frames.

as the splicing factor SF2, which is required for early steps in spliceosome assembly and for the first cleavage-ligation reaction (see Figure 26.13). ASF/SF2 is an RNA-binding protein in the SR family. When a pre-mRNA has more than one 5′ splice site preceding a single 3′ splice site, increased concentrations of ASF/SF2 promote use of the 5′ site nearest to the 3′ site at the expense of the other site. This effect of ASF/SF2 can be counteracted by another splicing factor, SF5.

The exact molecular roles of the factors in controlling splice utilization are not yet known, but we see in general terms that alternative splicing involving different 5′ sites may be influenced by proteins involved in spliceosome assembly. In the case of T/t antigens, the effect probably rests on increased binding of the SR proteins to the site that is preferentially used. Alternative splicing also may be influenced by repression of one site. Exons 2 and 3 of the mouse troponin T gene are mutually exclusive; exon 2 is used in smooth muscle, whereas exon 3 is used in other tissues. Smooth muscle contains proteins that bind to repeated elements located on either side of exon 3, and which prevent use of the 3′ and 5′ sites that are needed to include it.

The pathway of sex determination in *D. melanogaster* involves interactions between a series of genes in which alternative splicing events distinguish male and female. The pathway takes the form illustrated in FIGURE 26.22, in which the ratio of X chromosomes to autosomes determines the expression of *sxl*, and changes in expression are passed sequentially through the other genes to *dsx*, the last in the pathway.

The pathway starts with sex-specific splicing of *sxl*. Exon 3 of the *sxl* gene contains a termination codon that prevents synthesis of functional protein. This exon is included in the mRNA produced in males, but is skipped in females. (Exon skipping is illustrated for another example in FIGURE 26.23.) As a result, only females produce Sxl protein. The protein has a concentration of basic amino acids that resembles other RNA-binding proteins.

The presence of Sxl protein changes the splicing of the *transformer (tra)* gene. Figure 26.21 shows that this involves splicing a constant 5′ site to alternative 3′ sites. One splicing pattern occurs in both males and females, and results in an RNA that has an early termination codon. The presence of Sxl protein inhibits usage of the normal 3′ splice site by binding to the polypyrimidine tract at its branch site. When this site is skipped, the next 3′ site is used. This generates a female-specific mRNA that codes for a protein.

Thus *tra* produces a protein only in females; this protein is a splicing regulator. *tra2* has a similar function in females (but is also expressed in the male germline). The Tra and Tra2 proteins are SR splicing factors that act directly upon the target transcripts. Tra and Tra2 cooperate (in females) to affect the splicing of *dsx*.

Figure 26.23 shows examples of cases in which splice sites are used to add or to substitute exons or introns, again with the consequence that different protein products are generated. In the *doublesex (dsx)* gene, females splice the 5′ site of intron 3 to the 3′ site of that intron; as a result translation terminates at the end of exon 4. Males splice the 5′ site of intron 3 directly to the 3′ site of intron 4, thus omitting exon 4 from the mRNA and allowing translation to continue through exon 6. The result of the alternative splicing is that different proteins are produced in each sex: The male product blocks female sexual differentiation, whereas the female product represses expression of male-specific genes.

Alternative splicing of *dsx* RNA is controlled by competition between 3′ splice sites. *dsx* RNA has an element downstream of the leftmost 3′ splice site that is bound by Tra2; Tra and SR proteins associate with Tra2 at the site, which becomes an enhancer that assists binding of U2AF at the adjacent pyrimidine tract. This commits the formation of the spliceosome to use this 3′ site in females rather than the alternative 3′ site. The proteins recognize the enhancer cooperatively, possibly relying on formation of some secondary structure as well as sequence *per se*.

Sex determination therefore has a pleasing symmetry: The pathway starts with a female-specific splicing event that causes omission of an exon that has a termination codon, and ends with a female-specific splicing event that causes inclusion of an exon that has a termination codon. The events have different molecular bases. At the first control point, Sxl inhibits the default splicing pattern. At the last control point, Tra and Tra2 cooperate to promote the female-specific splice.

The Tra and Tra2 proteins are not needed for normal splicing, because in their absence flies develop normally (as males). As specific regulators, they need not necessarily participate in the mechanics of the splicing reaction; in this respect they differ from SF2, which is a factor required for general splicing, but can also influence choice of alternative splice sites.

P elements of *D. melanogaster* show a tissue-specific splicing pattern. In somatic cells there are two splicing events, but in germline an additional splicing event removes another intron. A termination codon lies in the germline-specific intron; as a result, a longer protein (with different properties) is produced in germline. We discuss the consequences for control of transposition in Section 21.15, P Elements Are Activated in the Germline, and note for now that the tissue specificity results from differences in the splicing apparatus.

The default splicing pathway of the P element pre-mRNA when the RNA is subjected to a heterologous (human) splicing extract is the germline pattern, in which intron 3 is excised. Extracts of somatic cells of *D. melanogaster*, however, contain a protein that inhibits excision of

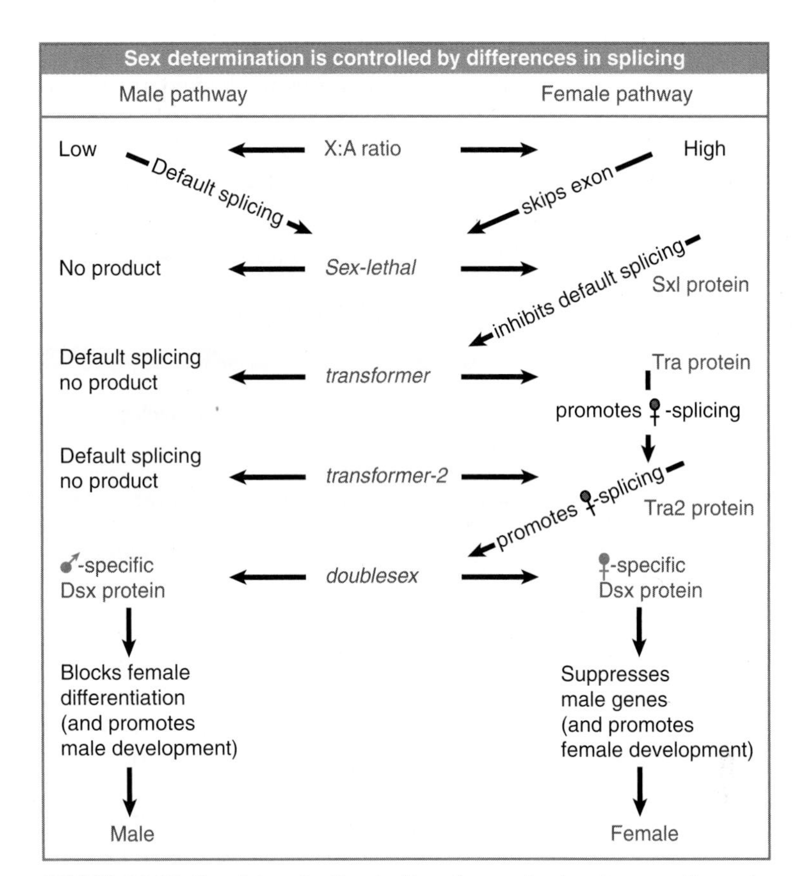

FIGURE 26.22 Sex determination in *D. melanogaster* involves a pathway in which different splicing events occur in females. Blocks at any stage of the pathway result in male development.

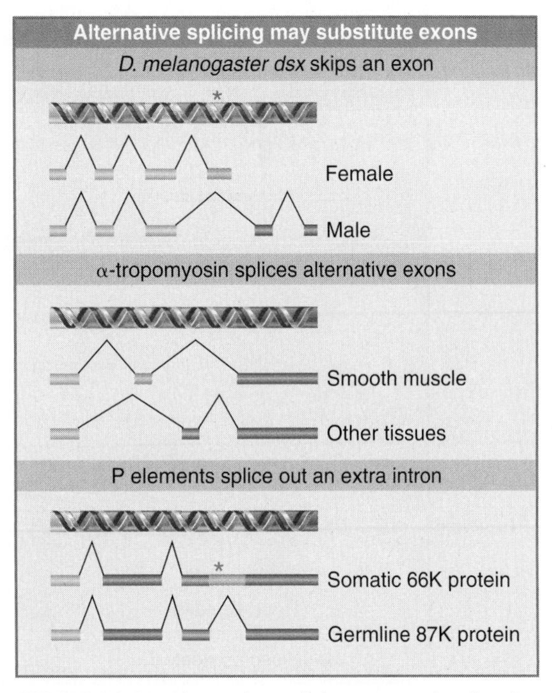

FIGURE 26.23 Alternative splicing events that involve both sites may cause exons to be added or substituted.

this intron. The protein binds to sequences in exon 3; if these sequences are deleted, the intron is excised. The function of the protein is therefore probably to repress association of the spliceosome with the 5′ site of intron 3.

26.13 *trans*-Splicing Reactions Use Small RNAs

Key concepts

- Splicing reactions usually occur only in *cis* between splice junctions on the same molecule of RNA.
- *trans*-splicing occurs in trypanosomes and worms where a short sequence (SL RNA) is spliced to the 5′ ends of many precursor mRNAs.
- SL RNA has a structure resembling the Sm-binding site of U snRNAs and may play an analogous role in the reaction.

In both mechanistic and evolutionary terms, splicing has been viewed as an *intramolecular* reaction, essentially amounting to a controlled deletion of the intron sequences at the level of RNA. In genetic terms, splicing occurs only in *cis*. This means that *only sequences on the same molecule of RNA can be spliced together*. The upper part of FIGURE 26.24 shows the normal situation. The introns can be removed from each RNA molecule, allowing the exons of that RNA molecule to be spliced together, but there is no *intermolecular* splicing of exons between different RNA molecules. We cannot say that *trans* splicing never occurs between pre-mRNA transcripts of the same gene, but we know that it must be exceedingly rare, because if it were prevalent the exons of a gene would be able to complement one another genetically instead of belonging to a single complementation group.

Some manipulations can generate *trans*-splicing. In the example illustrated in the lower part of Figure 26.24, complementary sequences were introduced into the introns of two RNAs. Base pairing between the complements should create an H-shaped molecule. This molecule could be spliced in *cis*, to connect exons that are covalently connected by an intron, or it could be spliced in *trans*, to connect exons of the juxtaposed RNA molecules. Both reactions occur *in vitro*.

Another situation in which *trans*-splicing is possible *in vitro* occurs when substrate RNAs are provided in the form of one containing a 5′ splice site and the other containing a 3′ splice site

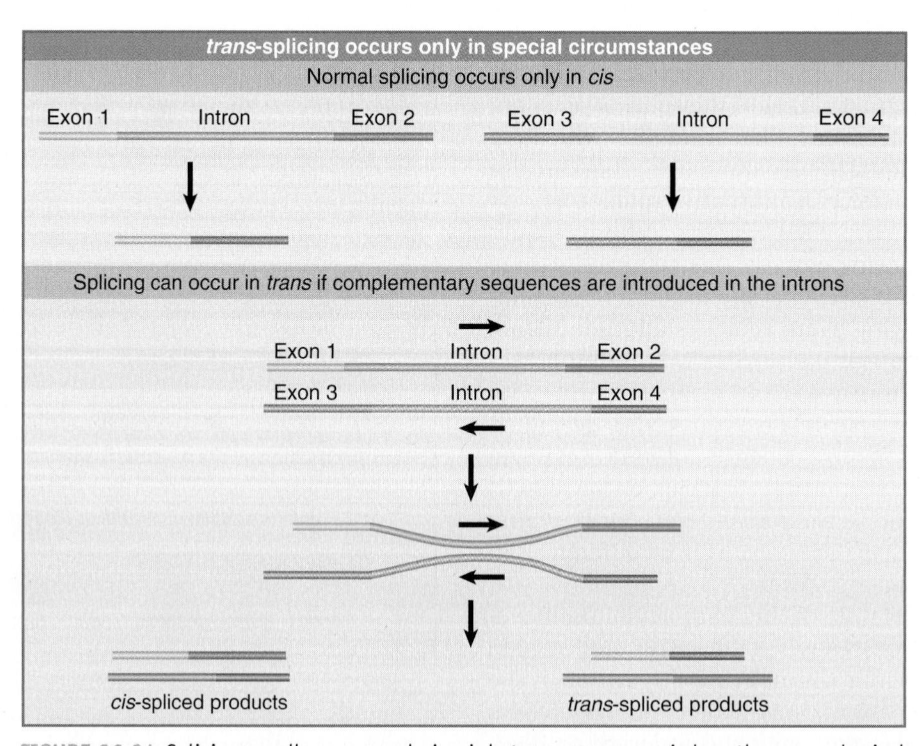

FIGURE 26.24 Splicing usually occurs only in *cis* between exons carried on the same physical RNA molecule, but *trans*-splicing can occur when special constructs are made that support base pairing between introns.

together with appropriate downstream sequences (which may be either the next 5′ splice site or a splicing enhancer). In effect, this mimics splicing by exon definition (see the right side of Figure 26.12), and shows that *in vitro* it is not necessary for the left and right splice sites to be on the same RNA molecule.

These results show that there is no *mechanistic* impediment to *trans*-splicing. They exclude models for splicing that require processive movement of a spliceosome along the RNA. It must be possible for a spliceosome to recognize the 5′ and 3′ splice sites of different RNAs when they are in close proximity.

Although *trans*-splicing is rare, it occurs *in vivo* in some special situations. One is revealed by the presence of a common 35-base leader sequence at the end of numerous mRNAs in the trypanosome. The leader sequence is not coded upstream of the individual transcription units, though. Instead it is transcribed into an independent RNA, carrying additional sequences at its 3′ end, from a repetitive unit located elsewhere in the genome. FIGURE 26.25 shows that this RNA carries the 35-base leader sequence followed by a 5′ splice site sequence. The sequences coding for the mRNAs carry a 3′ splice site just preceding the sequence found in the mature mRNA.

When the leader and the mRNA are connected by a *trans*-splicing reaction, the 3′ region of the leader RNA and the 5′ region of the mRNA in effect comprise the 5′ and 3′ halves of an intron. When splicing occurs, a 5′–2′ link forms by the usual reaction between the GU of the 5′ intron and the branch sequence near the AG of the 3′ intron. The two parts of the intron are not covalently linked, and thus generate a Y-shaped molecule instead of a lariat.

A similar situation is presented by the expression of actin genes in *Clostridium elegans*. Three actin mRNAs (and some other RNAs) share the same 22-base leader sequence at the 5′ terminus. The leader sequence is not coded in the actin gene, but is transcribed independently as part of a 100-base RNA coded by a gene elsewhere. *trans*-splicing also occurs in chloroplasts.

The RNA that donates the 5′ exon for *trans*-splicing is called the **SL RNA (spliced leader RNA).** The SL RNAs found in several species of trypanosomes and also in the nematode *(C. elegans)* have some common features. They fold into a common secondary structure that has three stem-loops and a single-stranded region that resembles the Sm-binding site. The SL RNAs therefore exist as snRNPs that count as members of the Sm snRNP class. Trypanosomes possess the U2, U4, and U6 snRNAs, but do not have U1 or U5 snRNAs. The absence of U1 snRNA can be explained by the properties of the SL RNA, which can carry out the functions that U1 snRNA usually performs at the 5′ splice site; thus SL RNA in effect consists of an snRNA sequence possessing U1 function that is linked to the exon–intron site that it recognizes.

There are two types of SL RNA in *C. elegans.* SL1 RNA (the first to be discovered) is used for splicing to coding sequences that are preceded only by 5′ nontranslated regions (the most common situation). SL2 RNA is used in cases in which a pre-mRNA contains two coding sequences; it is spliced to the second sequence, thus releasing it from the first and allowing it to be used as an independent mRNA. About 15% of all genes in *C. elegans* are organized in transcription units that include more than one gene (most often two to three genes). The significance of this form of organization for control of gene expression is not clear. These transcription units do not generally resemble operons where the genes function coordinately in a pathway.

The *trans*-splicing reaction of the SL RNA may represent a step toward the evolution of the pre-mRNA splicing apparatus. The SL RNA provides in *cis* the ability to recognize the 5′ splice

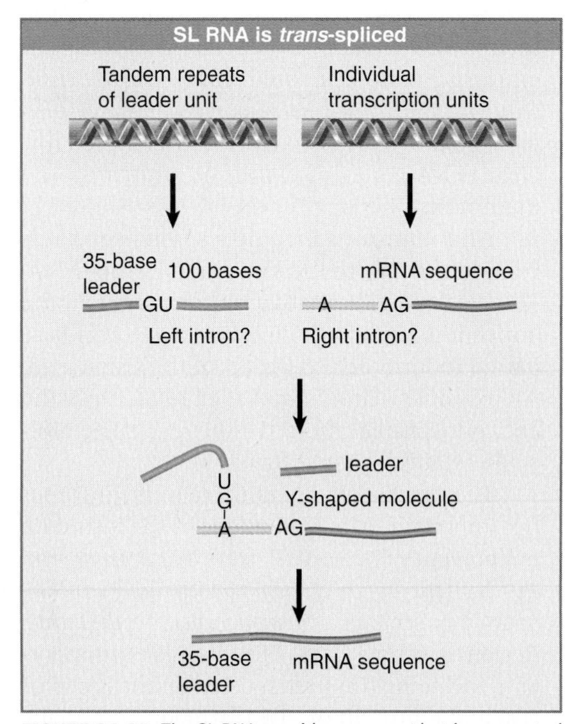

FIGURE 26.25 The SL RNA provides an exon that is connected to the first exon of an mRNA by *trans*-splicing. The reaction involves the same interactions as nuclear *cis*-splicing, but generates a Y-shaped RNA instead of a lariat.

site, and this probably depends upon the specific conformation of the RNA. The remaining functions required for splicing are provided by independent snRNPs. The SL RNA can function without participation of proteins like those in U1 snRNP, which suggests that the recognition of the 5′ splice site depends directly on RNA.

26.14 Yeast tRNA Splicing Involves Cutting and Rejoining

Key concept

- tRNA splicing occurs by successive cleavage and ligation reactions.

Most splicing reactions depend on short consensus sequences and occur by transesterification reactions in which breaking and making of bonds is coordinated. The splicing of tRNA genes is achieved by a different mechanism that relies upon separate cleavage and ligation reactions.

Some 59 of the 272 nuclear tRNA genes in the yeast *S. cerevisiae* are interrupted. Each has a single intron that is located just one nucleotide beyond the 3′ side of the anticodon. The introns vary in length from 14 to 60 bp. Those in related tRNA genes are related in sequence, but the introns in tRNA genes representing different amino acids are unrelated. *There is no consensus sequence that could be recognized by the splicing enzymes.* This is also true of interrupted nuclear tRNA genes of plants, amphibians, and mammals.

All the introns include a sequence that is complementary to the anticodon of the tRNA. This creates an alternative conformation for the anticodon arm in which the anticodon is base paired to form an extension of the usual arm. An example is drawn in FIGURE 26.26. Only the anticodon arm is affected—the rest of the molecule retains its usual structure.

The exact sequence and size of the intron is not important. Most mutations in the intron do not prevent splicing. *Splicing of tRNA depends principally on recognition of a common secondary structure in tRNA rather than a common sequence of the intron.* Regions in various parts of the molecule are important, including the stretch between the acceptor arm and D arm, in the TψC arm, and especially the anticodon arm. This is reminiscent of the structural demands placed on tRNA for protein synthesis (see Chapter 8, Protein Synthesis).

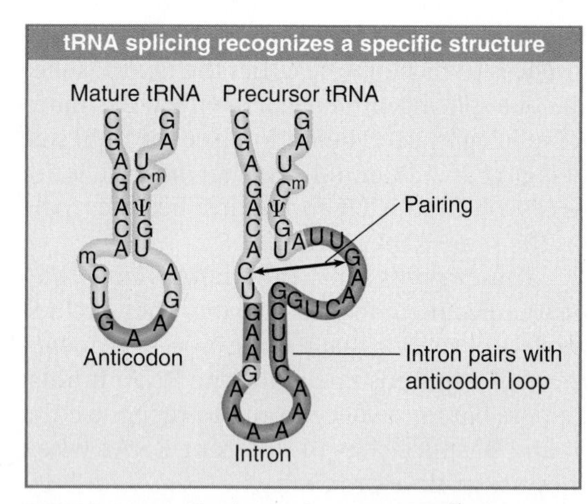

FIGURE 26.26 The intron in yeast $tRNA^{Phe}$ base pairs with the anticodon to change the structure of the anticodon arm. Pairing between an excluded base in the stem and the intron loop in the precursor may be required for splicing.

The intron is not entirely irrelevant, however. Pairing between a base in the intron loop and an unpaired base in the stem is required for splicing. Mutations at other positions that influence this pairing (for example, to generate alternative patterns for pairing) influence splicing. The rules that govern availability of tRNA precursors for splicing resemble the rules that govern recognition by aminoacyl-tRNA synthetases (see Section 9.9, tRNAs Are Charged with Amino Acids by Synthetases).

In a temperature-sensitive mutant of yeast that fails to remove the introns, the interrupted precursors accumulate in the nucleus. The precursors can be used as substrates for a cell-free system extracted from wild-type cells. The splicing of the precursor can be followed by virtue of the resulting size reduction. This is seen by the change in position of the band on gel electrophoresis, as illustrated in FIGURE 26.27. The reduction in size can be accounted for by the appearance of a band representing the intron.

The cell-free extract can be fractionated by assaying the ability to splice the tRNA. The *in vitro* reaction requires ATP. Characterizing the reactions that occur with and without ATP shows that the *two separate stages of the reaction are catalyzed by different enzymes.*

- The first step does not require ATP. It involves phosphodiester bond cleavage by an atypical nuclease reaction. It is catalyzed by an endonuclease.
- The second step requires ATP and involves bond formation; it is a ligation reaction, and the responsible enzyme activity is described as an **RNA ligase.**

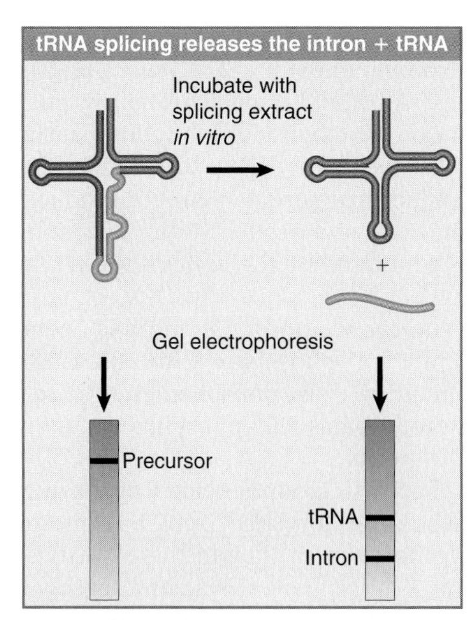

FIGURE 26.27 Splicing of yeast tRNA *in vitro* can be followed by assaying the RNA precursor and products by gel electrophoresis.

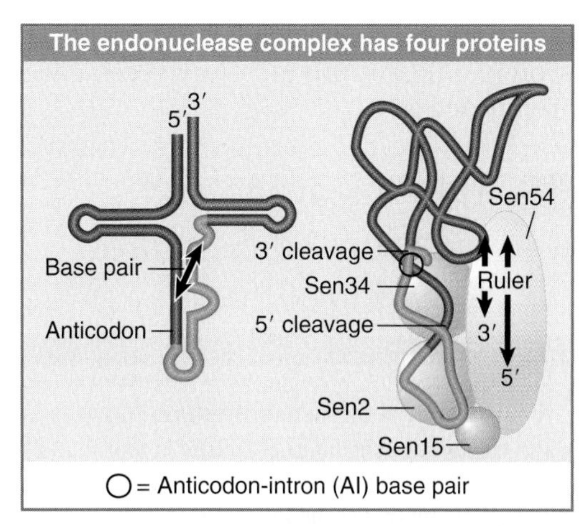

FIGURE 26.28 The 3′ and 5′ cleavages in *S. cerevisiae* pre-tRNA are catalyzed by different subunits of the endonuclease. Another subunit may determine location of the cleavage sites by measuring distance from the mature structure. The AI base pair is also important.

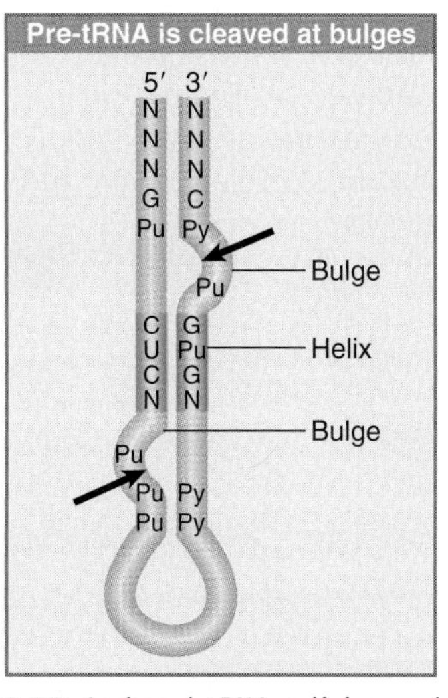

FIGURE 26.29 Archaeal tRNA splicing endonuclease cleaves each strand at a bulge in a bulge-helix-bulge motif.

26.15 The Splicing Endonuclease Recognizes tRNA

Key concepts

- An endonuclease cleaves the tRNA precursors at both ends of the intron.
- The yeast endonuclease is a heterotetramer with two (related) catalytic subunits.
- It uses a measuring mechanism to determine the sites of cleavage by their positions relative to a point in the tRNA structure.
- The archaeal nuclease has a simpler structure and recognizes a bulge-helix-bulge structural motif in the substrate.

The endonuclease is responsible for the specificity of intron recognition. It cleaves the precursor at both ends of the intron. The yeast endonuclease is a heterotetrameric protein. Its activities are illustrated in FIGURE 26.28. The related subunits Sen34 and Sen2 cleave the 3′ and 5′ splice sites, respectively. Subunit Sen54 may determine the sites of cleavage by "measuring" distance from a point in the tRNA structure. This point is in the elbow of the (mature) L-shaped structure. The role of subunit Sen15 is not known, but its gene is essential in yeast. The base pair that forms between the first base in the anticodon loop and the base preceding the 3′ splice site is required for 3′ splice site cleavage.

An interesting insight into the evolution of tRNA splicing is provided by the endonucleases of archaea. These are homodimers or homotetramers, in which each subunit has an active site (although only two of the sites function in the tetramer) that cleaves one of the splice sites. The subunit has sequences related to the sequences of the active sites in the Sen34 and Sen2 subunits of the yeast enzyme. The archaeal enzymes recognize their substrates in a different way, though. Instead of measuring distance from particular sequences, they recognize a structural feature called the bulge-helix-bulge. FIGURE 26.29 shows that cleavage occurs in the two bulges.

Thus the origin of splicing of tRNA precedes the separation of the archaea and the eukaryotes. If it originated by insertion of the intron into tRNAs, this must have been a very ancient event.

26.16 tRNA Cleavage and Ligation Are Separate Reactions

Key concepts

- Release of the intron generates two half-tRNAs that pair to form the mature structure.
- The halves have the unusual ends 5′ hydroxyl and 2′–3′ cyclic phosphate.
- The 5′-OH end is phosphorylated by a polynucleotide kinase, the cyclic phosphate group is opened by phosphodiesterase to generate a 2′-phosphate terminus and 3′-OH group, exon ends are joined by an RNA ligase, and the 2′-phosphate is removed by a phosphatase.

The overall tRNA splicing reaction is summarized in FIGURE 26.30. The products of cleavage are a linear intron and two half-tRNA molecules. These intermediates have unique ends.

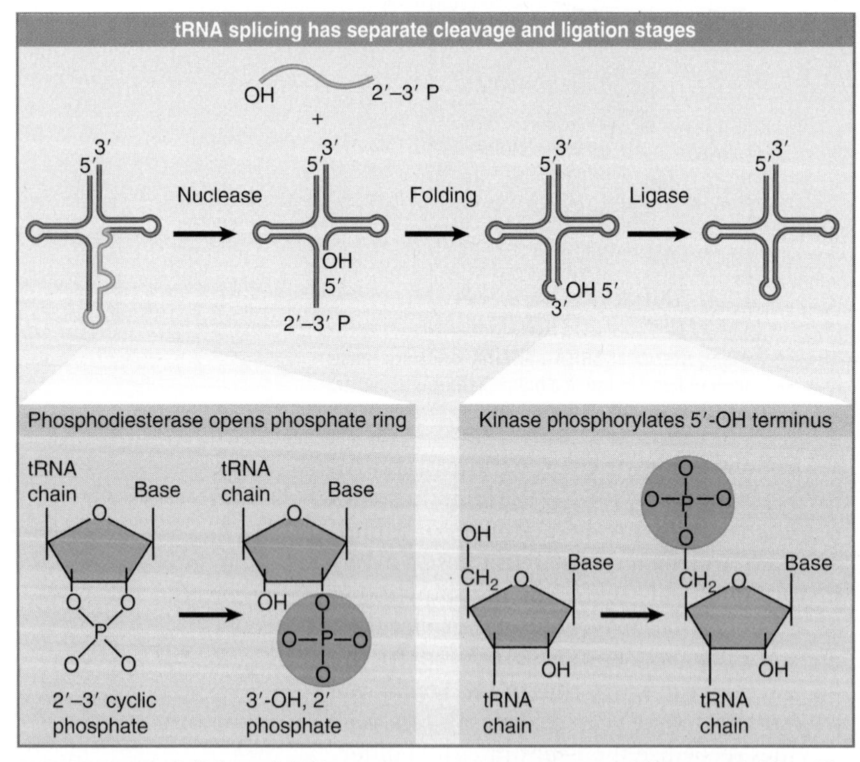

FIGURE 26.30 Splicing of tRNA requires separate nuclease and ligase activities. The exon–intron boundaries are cleaved by the nuclease to generate 2′ to 3′ cyclic phosphate and 5′ OH termini. The cyclic phosphate is opened to generate 3′-OH and 2′ phosphate groups. The 5′-OH is phosphorylated. After releasing the intron, the tRNA half molecules fold into a tRNA-like structure that now has a 3′-OH, 5′-P break. This is sealed by a ligase.

Each 5′ terminus ends in a hydroxyl group; each 3′ terminus ends in a 2′,3′–cyclic phosphate group. (All other known RNA splicing enzymes cleave on the other side of the phosphate bond.)

The two half-tRNAs base pair to form a tRNA-like structure. When ATP is added, the second reaction occurs. Both of the unusual ends generated by the endonuclease must be altered.

The cyclic phosphate group is opened to generate a 2′–phosphate terminus. This reaction requires cyclic phosphodiesterase activity. The product has a 2′–phosphate group and a 3′–OH group.

The 5′–OH group generated by the nuclease must be phosphorylated to give a 5′–phosphate. This generates a site in which the 3′–OH is next to the 5′–phosphate. Covalent integrity of the polynucleotide chain is then restored by ligase activity.

All three activities—phosphodiesterase, polynucleotide kinase, and adenylate synthetase (which provides the ligase function)—are arranged in different functional domains on a single protein. They act sequentially to join the two tRNA halves.

The spliced molecule is now uninterrupted, with a 5′–3′ phosphate linkage at the site of splicing, but it also has a 2′–phosphate group marking the event. The surplus group must be removed by a phosphatase.

Generation of a 2′,3′–cyclic phosphate also occurs during the tRNA-splicing reaction in plants and mammals. The reaction in plants seems to be the same as in yeast, but the detailed chemical reactions are different in mammals.

The yeast tRNA precursors also can be spliced in an extract obtained from the germinal vesicle (nucleus) of *Xenopus* oocytes. This shows that the reaction is not species-specific. *Xenopus* must have enzymes able to recognize the introns in the yeast tRNAs.

The ability to splice the products of tRNA genes is therefore well conserved, but is likely to have a different origin from the other splicing reactions (such as that of nuclear pre-mRNA). The tRNA-splicing reaction uses cleavage and synthesis of bonds and is determined by sequences that are external to the intron. Other splicing reactions use transesterification, in which bonds are transferred directly, and the sequences required for the reaction lie within the intron.

26.17 The Unfolded Protein Response Is Related to tRNA Splicing

Key concepts

- Ire1p is an inner nuclear membrane protein with its N-terminal domain in the ER lumen, and its C-terminal domain in the nucleus.
- Binding of an unfolded protein to the N-terminal domain activates the C-terminal nuclease by autophosphorylation.
- The activated nuclease cleaves Hac1 mRNA to release an intron and generate exons that are ligated by a tRNA ligase.
- The spliced Hac1 mRNA codes for a transcription factor that activates genes coding for chaperones that help to fold unfolded proteins.

An unusual splicing system that is related to tRNA splicing mediates the response to unfolded proteins in yeast. The accumulation of unfolded proteins in the lumen of the endoplasmic reticulum (ER) triggers a response pathway that leads to increased transcription of genes coding for chaperones that assist protein folding in the ER. A signal must therefore be transmitted from the lumen of the ER to the nucleus.

The sensor that activates the pathway is the protein Ire1p. It is an integral membrane protein (Ser/Thr) kinase that has domains on each side of the ER membrane. The N-terminal domain in the lumen of the ER detects the presence of unfolded proteins, presumably by binding to exposed motifs. This causes aggregation of monomers and activates the C-terminal domain on the other side of the membrane by autophosphorylation.

Genes that are activated by this pathway have a common promoter element called the UPRE (unfolded protein response element). The transcription factor Hac1p binds to the UPRE, and is produced in response to accumulation of unfolded proteins. The trigger for production of Hac1p is the action of Ire1p on Hac1 mRNA.

The operation of the pathway is summarized in FIGURE 26.31. Under normal conditions, when the pathway is not activated, Hac1 mRNA is translated into a protein that is rapidly degraded. The activation of Ire1p results in the splicing of the Hac1 mRNA to change the sequence of the protein to a more stable form. This form provides the functional transcription factor that activates genes with the UPRE.

Unusual splicing components are involved in this reaction. Ire1P has an endonuclease activity that acts directly on Hac1 mRNA to cleave the two splicing junctions. The two junctions are ligated by the tRNA ligase that acts in the tRNA splicing pathway. The endonuclease reaction resembles the cleavage of tRNA during splicing.

Where does the modification of Hac1 mRNA occur? Ire1p is probably located in the inner nuclear membrane, with the N-terminal sensor domain in the ER lumen, and the C-terminal kinase/nuclease domain in the nucleus. This would enable it to act directly on Hac1 RNA before it is exported to the cytoplasm. It also would allow easy access by the tRNA ligase. There is no apparent relationship between the Ire1p nuclease activity and the tRNA splicing endonuclease, so it is not obvious how this specialized system would have evolved.

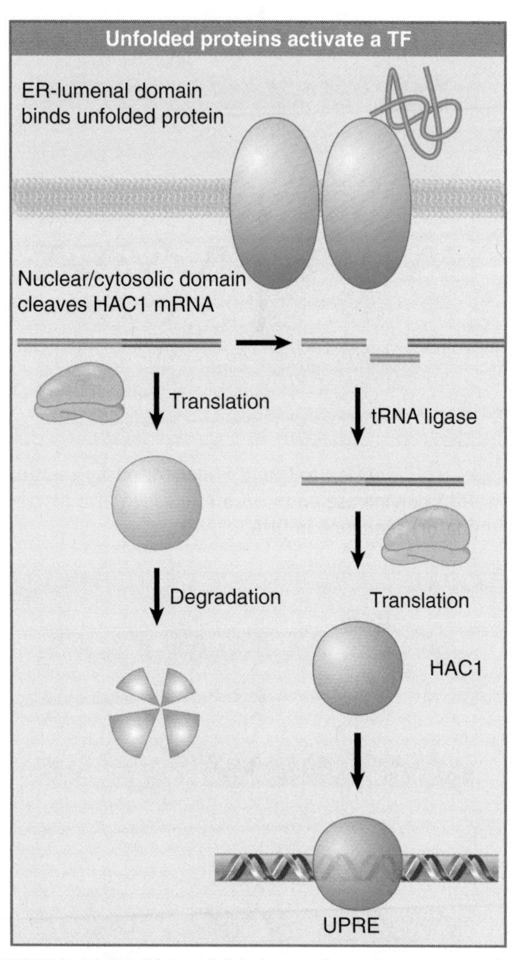

FIGURE 26.31 The unfolded protein response occurs by activating special splicing of HAC1 mRNA to produce a transcription factor that recognizes the UPRE.

26.18 The 3′ Ends of polI and polIII Transcripts Are Generated by Termination

Key concepts

- RNA polymerase I terminates transcription at an 18-base terminator sequence.
- RNA polymerase III terminates transcription in poly(U)$_4$ sequence embedded in a G-C-rich sequence.

3′ ends of RNAs can be generated in two ways. Some RNA polymerases terminate transcription at a defined (terminator) sequence in DNA, as shown in FIGURE 26.32. Other RNA polymerases do not show discrete termination, but continue past the site corresponding to the 3′ end, which is generated by cleavage of the RNA by an endonuclease, as shown in FIGURE 26.33.

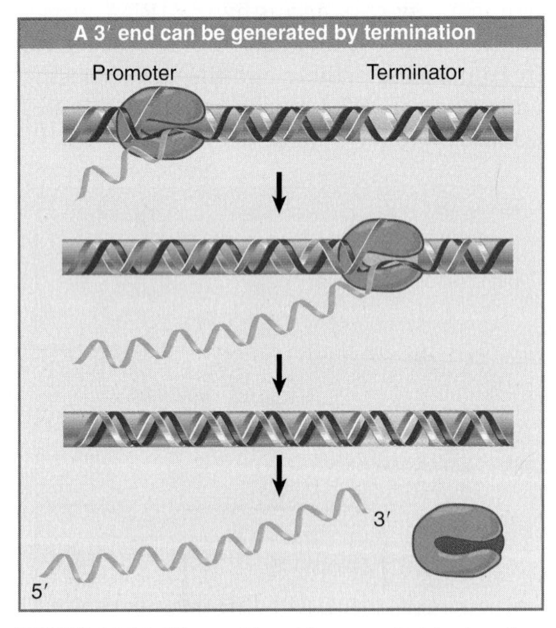

FIGURE 26.32 When a 3′ end is generated by termination, RNA polymerase and RNA are released at a discrete (terminator) sequence in DNA.

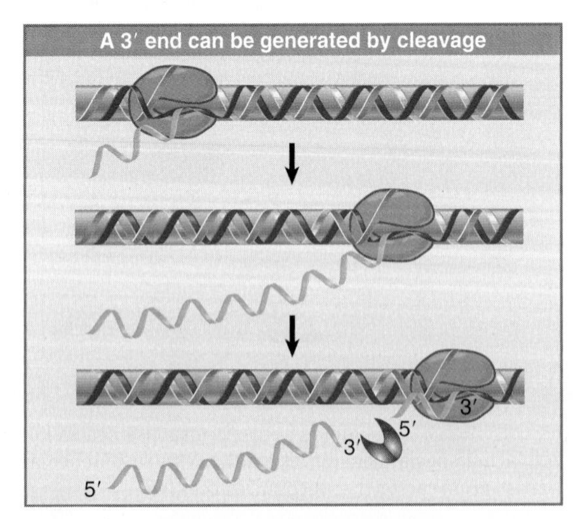

FIGURE 26.33 When a 3′ end is generated by cleavage, RNA polymerase continues transcription while an endonuclease cleaves at a defined sequence in the RNA.

Information about the termination reaction for eukaryotic RNA polymerases is less detailed than our knowledge of initiation. RNA polymerases I and III have discrete termination events (like bacterial RNA polymerase), but it is not clear whether RNA polymerase II usually terminates in this way.

For RNA polymerase I, the sole product of transcription is a large precursor that contains the sequences of the major rRNA. The precursor is subjected to extensive processing. Termination occurs at a discrete site >1000 bp downstream of the mature 3′ end, which is generated by cleavage. Termination involves recognition of an 18-base terminator sequence by an ancillary factor.

With RNA polymerase III, transcription *in vitro* generates molecules with the same 5′ and 3′ ends as those synthesized *in vivo*. The termination reaction resembles intrinsic termination by bacterial RNA polymerase (see Section 11.21, There Are Two Types of Terminators in *E. coli*). Termination usually occurs at the second U within a run of four U bases, but there is heterogeneity, with some molecules ending in three or even four U bases. The same heterogeneity is seen in molecules synthesized *in vivo*, so it seems to be a *bona fide* feature of the termination reaction.

Just like the prokaryotic terminators, the U run is embedded in a G-C-rich region. Although sequences of dyad symmetry are present, they are not needed for termination, because mutations that abolish the symmetry do not prevent the normal completion of RNA synthesis. Nor are any sequences beyond the U run necessary, because all distal sequences can be replaced without any effect on termination.

The U run itself is not sufficient for termination, because regions of four successive U residues exist within transcription units read by RNA polymerase III. (There are no internal U_5 runs, though, which fits with the greater efficiency of termination when the terminator is a U_5 rather than a U_4 sequence.) The critical feature in termination must therefore be the recognition of a U_4 sequence in a context that is rich in G-C base pairs.

How does the termination reaction occur? It cannot rely on the weakness of the rU-dA RNA–DNA hybrid region that lies at the end of the transcript, because often only the first two U residues are transcribed. Perhaps the G-C-rich region plays a role in slowing down the enzyme, but there does not seem to be a counterpart to the hairpin involved in prokaryotic termination. We remain puzzled about how the enzyme can respond so specifically to such a short signal. In contrast with the initiation reaction, which RNA polymerase III cannot accomplish alone, termination seems to be a function of the enzyme itself.

26.19 The 3′ Ends of mRNAs Are Generated by Cleavage and Polyadenylation

Key concepts

- The sequence AAUAAA is a signal for cleavage to generate a 3′ end of mRNA that is polyadenylated.
- The reaction requires a protein complex that contains a specificity factor, an endonuclease, and poly(A) polymerase.
- The specificity factor and endonuclease cleave RNA downstream of AAUAAA.
- The specificity factor and poly(A) polymerase add ~200 A residues processively to the 3′ end.
- A-U-rich sequences in the 3′ tail control cytoplasmic polyadenylation or deadenylation during *Xenopus* embryonic development.

The 3′ ends of mRNAs are generated by cleavage followed by polyadenylation. Addition of poly(A) to nuclear RNA can be prevented by the analog 3′–deoxyadenosine, which is also known as **cordycepin.** Although cordycepin does not stop the transcription of nuclear RNA, its addition prevents the appearance of mRNA in the cytoplasm. This shows that polyadenylation is *necessary* for the maturation of mRNA from nuclear RNA.

Generation of the 3′ end is illustrated in FIGURE 26.34. RNA polymerase transcribes past the site corresponding to the 3′ end, and sequences in the RNA are recognized as targets for an endonucleolytic cut followed by polyadenylation. A single processing complex undertakes both the cutting and polyadenylation. The polyadenylation stabilizes the mRNA against degradation from the 3′ end. Its 5′ end is already stabilized by the cap. RNA polymerase continues transcription after the cleavage, but the 5′ end that is generated by the cleavage is unprotected.

The cleavage event provides an indirect trigger for termination by RNA polymerase II. An exonuclease binds to the 5′ end of the RNA that is continuing to be transcribed after cleavage. It degrades the RNA faster than it is synthesized, so that it catches up with RNA polymerase. It then interacts with ancillary proteins that are bound to the carboxy-terminal domain of the polymerase, and this interaction triggers the release of RNA polymerase from DNA, causing transcription to terminate. The overall model is similar to that for the role of rho in terminating transcription by bacterial RNA polymerase (see Section 11.22, How Does Rho Factor Work?). This explains why the termination sites for RNA polymerase II are not well defined, but may occur at varying locations within a long region downstream of the site corresponding to the 3′ end of the RNA.

A common feature of mRNAs in higher eukaryotes (but not in yeast) is the presence of the highly conserved sequence AAUAAA in the region from 11 to 30 nucleotides upstream of the site of poly(A) addition. Deletion or mutation of the AAUAAA hexamer prevents generation of the polyadenylated 3′ end. The signal is needed for both cleavage and polyadenylation.

The development of a system in which polyadenylation occurs *in vitro* opened the route to analyzing the reactions. The formation and functions of the complex that undertakes 3′ processing are illustrated in FIGURE 26.35. Generation of the proper 3′ terminal structure

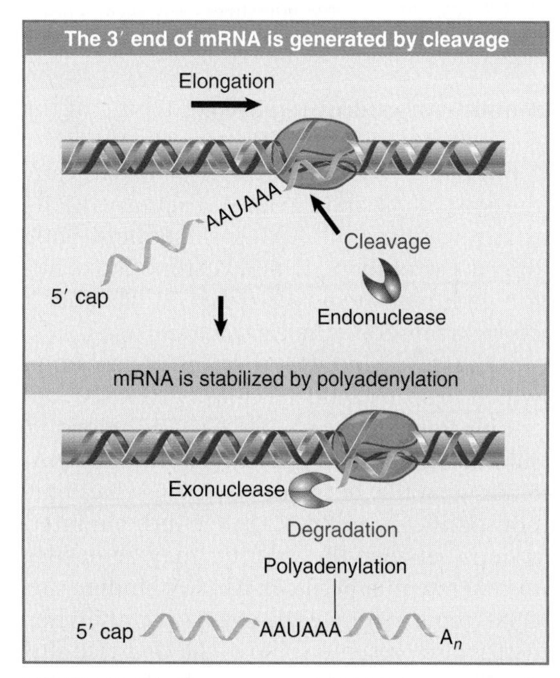

FIGURE 26.34 The sequence AAUAAA is necessary for cleavage to generate a 3′ end for polyadenylation.

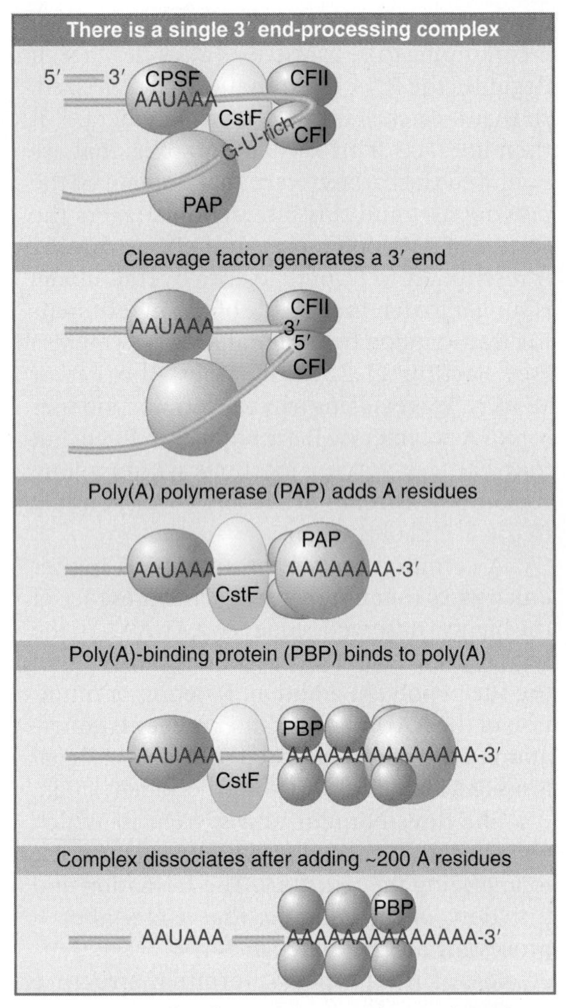

FIGURE 26.35 The 3′ processing complex consists of several activities. CPSF and CstF each consist of several subunits; the other components are monomeric. The total mass is >900 kD.

requires an **endonuclease** consisting of the components CFI and CFII) to cleave the RNA, a **poly(A) polymerase (PAP)** to synthesize the poly(A) tail, and a *specificity component* (CPSF) that recognizes the AAUAAA sequence and directs the other activities. A stimulatory factor, CstF, binds to a G-U-rich sequence that is downstream from the cleavage site itself.

The specificity factor contains four subunits, which together bind specifically to RNA containing the sequence AAUAAA. The individual subunits are proteins that have common RNA-binding motifs, but which by themselves bind nonspecifically to RNA. Protein–protein interactions between the subunits may be needed to generate the specific AAUAAA-binding site. CPSF binds strongly to AAUAAA only when CstF is also present to bind to the G-U-rich site.

The specificity factor is needed for both the cleavage and polyadenylation reactions. It exists in a complex with the endonuclease and poly(A) polymerase, and this complex usually undertakes cleavage followed by polyadenylation in a tightly coupled manner.

The two components CFI and CFII (cleavage factors I and II), together with specificity factor, are necessary and sufficient for the endonucleolytic cleavage.

The poly(A) polymerase has a nonspecific catalytic activity. When it is combined with the other components, the synthetic reaction becomes specific for RNA containing the sequence AAUAAA. The polyadenylation reaction passes through two stages. First, a rather short oligo(A) sequence (~10 residues) is added to the 3′ end. This reaction is absolutely dependent on the AAUAAA sequence, and poly(A) polymerase performs it under the direction of the specificity factor. In the second phase, the oligo(A) tail is extended to the full ~200 residue length. This reaction requires another stimulatory factor that recognizes the oligo(A) tail and directs poly(A) polymerase specifically to extend the 3′ end of a poly(A) sequence.

The poly(A) polymerase by itself adds A residues individually to the 3′ position. Its intrinsic mode of action is distributive; it dissociates after each nucleotide has been added. In the presence of CPSF and PABP (poly(A)-binding protein), however, it functions processively to extend an individual poly(A) chain. The PABP is a 33 kD protein that binds stoichiometrically to the poly(A) stretch. The length of poly(A) is controlled by the PABP, which in some way limits the action of poly(A) polymerase to ~200 additions of A residues. The limit may represent the accumulation of a critical mass of PABP on the poly(A) chain. PABP binds to the translation initiation factor eIF4G, thus generating a closed loop in which a protein complex contains both the 5′ and 3′ ends of the mRNA (see Figure 8.20 in Section 8.9, Eukaryotes Use a Complex of Many Initiation Factors).

Polyadenylation is an important determinant of mRNA function. It may affect both stability and initiation of translation (see Section 7.10, The 3′ Terminus Is Polyadenylated). In embryonic development in some organisms, the presence of poly(A) is used to control translation, and preexisting mRNAs may either be polyadenylated (to stimulate translation) or deadenylated (to terminate translation). During *Xenopus* embryonic development, polyadenylation of mRNA in the cytoplasm in *Xenopus* depends on a specific *cis*-acting element (the CPE) in the 3′ tail. This is another AU-rich sequence, UUUUUAU.

In *Xenopus* embryos at least two type of *cis*-acting sequences found in the 3′ tail can trigger deadenylation. EDEN (embryonic deadenylation element) is a 17-nucleotide sequence. ARE elements are AU-rich and usually contain tandem repeats of AUUUA. There is a poly(A)-specific RNAase (PARN) that could be involved in the degradation. Of course, deadenylation is not always triggered by specific elements; in some situations (including the normal degradation of mRNA as it ages), poly(A) is degraded unless it is specifically stabilized.

26.20 Cleavage of the 3′ End of Histone mRNA May Require a Small RNA

Key concepts

- Histone mRNAs are not polyadenylated; their 3′ ends are generated by a cleavage reaction that depends on the structure of the mRNA.
- The cleavage reaction requires the SLBP to bind to a stem-loop structure and the U7 snRNA to pair with an adjacent single-stranded region.

Some mRNAs are not polyadenylated. The formation of their 3′ ends is therefore different from the coordinated cleavage/polyadenylation reaction. The most prominent members of this mRNA class are the mRNAs coding for histones that are synthesized during DNA replication. Formation of their 3′ ends depends upon secondary structure. The structure at the 3′ terminus is a highly conserved stem-loop structure, with a stem of 6 bp and a loop of four nucleotides. Cleavage occurs four to five bases downstream of the stem-loop. Two factors are required for the cleavage reaction: The stem-loop binding protein (SLBP) recognizes the structure, and the U7 snRNA pairs with a purine-rich sequence (the histone downstream element, or HDE) located ~10 nucleotides downstream of the cleavage site.

Mutations that prevent formation of the duplex stem of the stem-loop prevent formation of the end of the RNA. Secondary mutations that restore duplex structure (though not necessarily the original sequence) behave as revertants. This suggests that *formation of the secondary structure is more important than the exact sequence.* The SLBP binds to the stem-loop and then interacts with U7 snRNP to enhance its interaction with the downstream binding site for U7 snRNA.

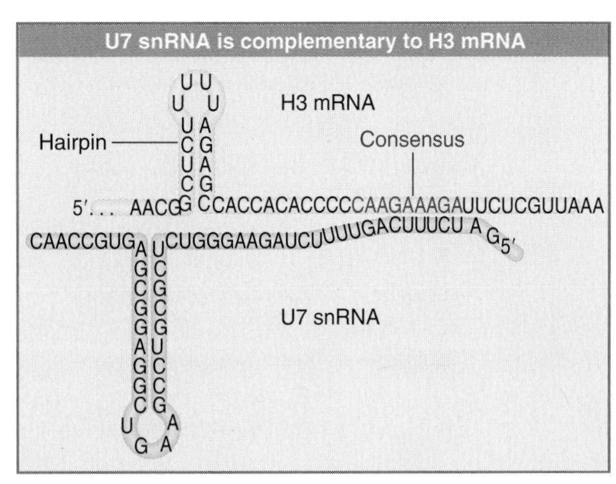

FIGURE 26.36 Generation of the 3′ end of histone H3 mRNA depends on a conserved hairpin and a sequence that base pairs with U7 snRNA.

U7 snRNP is a minor snRNP consisting of the 63 nucleotide U7 snRNA and a set of several proteins (including Sm proteins; see Section 26.5, snRNAs Are Required for Splicing).

The reaction between histone H3 mRNA and U7 snRNA is drawn in FIGURE 26.36. The upstream hairpin and the HDE that pairs with U7 snRNA are conserved in histone H3 mRNAs of several species. The U7 snRNA has sequences toward its 5′ end that pair with the histone mRNA consensus sequences. 3′ processing is inhibited by mutations in the HDE that reduce ability to pair with U7 snRNA. Compensatory mutations in U7 snRNA that restore complementarity also restore 3′ processing. This suggests that U7 snRNA functions by base pairing with the histone mRNA. The sequence of the HDE varies among the various histone mRNAs, with the result that binding of snRNA is not by itself necessarily stable, but requires also the interaction with SLBP.

Cleavage to generate a 3′ terminus occurs at a fixed distance from the site recognized by U7 snRNA, which suggests that the snRNA is involved in defining the cleavage site. The factor(s) actually responsible for cleavage, however, have not yet been identified.

26.21 Production of rRNA Requires Cleavage Events

Key concept

- The large and small rRNAs are released by cleavage from a common precursor RNA.

The major rRNAs are synthesized as part of a single primary transcript that is processed to

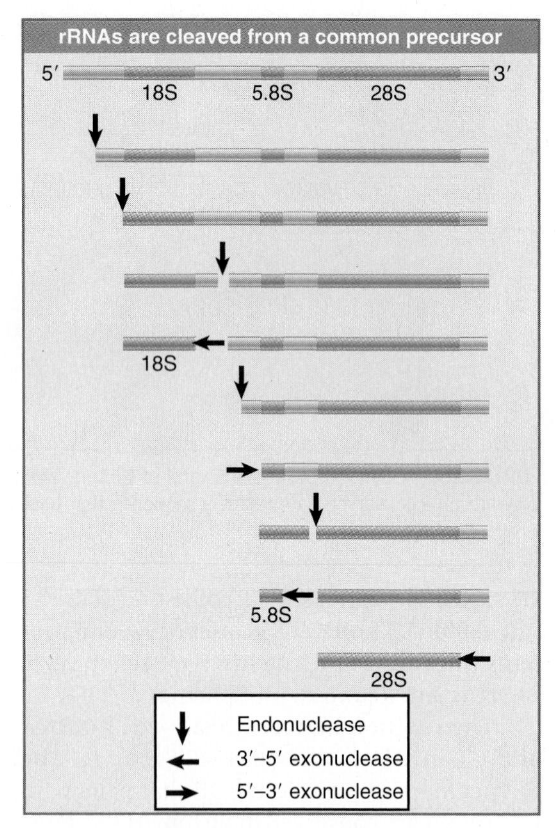

FIGURE 26.37 Mature eukaryotic rRNAs are generated by cleavage and trimming events from a primary transcript.

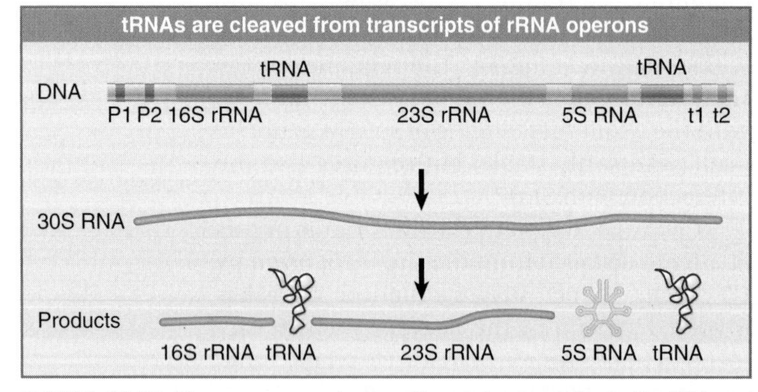

FIGURE 26.38 The *rrn* operons in *E. coli* contain genes for both rRNA and tRNA. The exact lengths of the transcripts depend on which promoters (P) and terminators (t) are used. Each RNA product must be released from the transcript by cuts on either side.

generate the mature products. The precursor contains the sequences of the 18S, 5.8S, and 28S rRNAs. In higher eukaryotes, the precursor is named for its sedimentation rate as **45S RNA.** In lower eukaryotes it is smaller (35S in yeast).

The mature rRNAs are released from the precursor by a combination of cleavage events and trimming reactions. FIGURE 26.37 shows the general pathway in yeast. There can be variations in the order of events, but basically similar reactions are involved in all eukaryotes. Most of the 5′ ends are generated directly by a cleavage event. Most of the 3′ ends are generated by cleavage followed by a 3′–5′ trimming reaction.

Many ribonucleases have been implicated in processing rRNA, including the exosome, which is an assembly of several exonucleases that also participates in mRNA degradation (see Section 7.13, mRNA Degradation Involves Multiple Activities). Mutations in individual enzymes usually do not prevent processing, which suggests that their activities are redundant and that different combinations of cleavages can be used to generate the mature molecules.

There are always multiple copies of the transcription unit for the rRNAs. The copies are organized as tandem repeats (see Section 6.9, The Repeated Genes for rRNA Maintain Constant Sequence).

5S RNA is transcribed from separate genes by RNA polymerase III. In general, the 5S genes are clustered, but are separate from the genes for the major rRNAs. (In the case of yeast, a 5S gene is associated with each major transcription unit, but is transcribed independently.)

There is a difference in the organization of the precursor in bacteria. The sequence corresponding to 5.8S rRNA forms the 5′ end of the large (23S) rRNA, that is, there is no processing between these sequences. FIGURE 26.38 shows that the precursor also contains the 5S rRNA and one or two tRNAs. In *E. coli,* the seven *rrn* operons are dispersed around the genome; four *rrn* loci contain one tRNA gene between the 16S and 23S rRNA sequences, and the other *rrn* loci contain two tRNA genes in this region. Additional tRNA genes may or may not be present between the 5S sequence and the 3′ end. Thus the processing reactions required to release the products depend on the content of the particular *rrn* locus.

In both prokaryotic and eukaryotic rRNA processing, ribosomal proteins (and possibly other proteins) bind to the precursor, so that the substrate for processing is not the free RNA, but rather a ribonucleoprotein complex.

26.22 Small RNAs Are Required for rRNA Processing

Key concepts

- The C/D group of snoRNAs is required for modifying the 2′ position of ribose with a methyl group.
- The H/ACA group of snoRNAs is required for converting uridine to pseudouridine.
- In each case the snoRNA base pairs with a sequence of rRNA that contains the target base to generate a typical structure that is the substrate for modification.

Processing and modification of rRNA requires a class of small RNAs called snoRNAs (small nucleolar RNAs). There are 71 snoRNAs in the yeast *(S. cerevisiae)* genome. They are associated with the protein fibrillarin, which is an abundant component of the nucleolus (the region of the nucleus where the rRNA genes are transcribed). Some snoRNAs are required for cleavage of the precursor to rRNA; one example is U3 snoRNA, which is required for the first cleavage event in both yeast and *Xenopus*. We do not know what role the snoRNA plays in cleavage. It could be required to pair with the rRNA sequence to form a secondary structure that is recognized by an endonuclease.

Two groups of snoRNAs are required for the modifications that are made to bases in the rRNA. The members of each group are identified by very short conserved sequences and common features of secondary structure.

The C/D group of snoRNAs is required for adding a methyl group to the 2′ position of ribose. There are >100 2′–O–methyl groups at conserved locations in vertebrate rRNAs. This group takes its name from two short conserved sequences motifs called boxes C and D. Each snoRNA contains a sequence near the D box that is complementary to a region of the 18S or 28S rRNA that is methylated. Loss of a particular snoRNA prevents methylation in the rRNA region to which it is complementary.

FIGURE 26.39 suggests that the snoRNA base pairs with the rRNA to create the duplex region that is recognized as a substrate for methylation. Methylation occurs within the region of complementarity at a position that is fixed five bases on the 5′ side of the D box. It is likely that each methylation event is specified by a different snoRNA; ~40 snoRNAs have been characterized so far. The methylase(s) has not been characterized; one possibility is that the snoRNA itself provides part of the methylase activity.

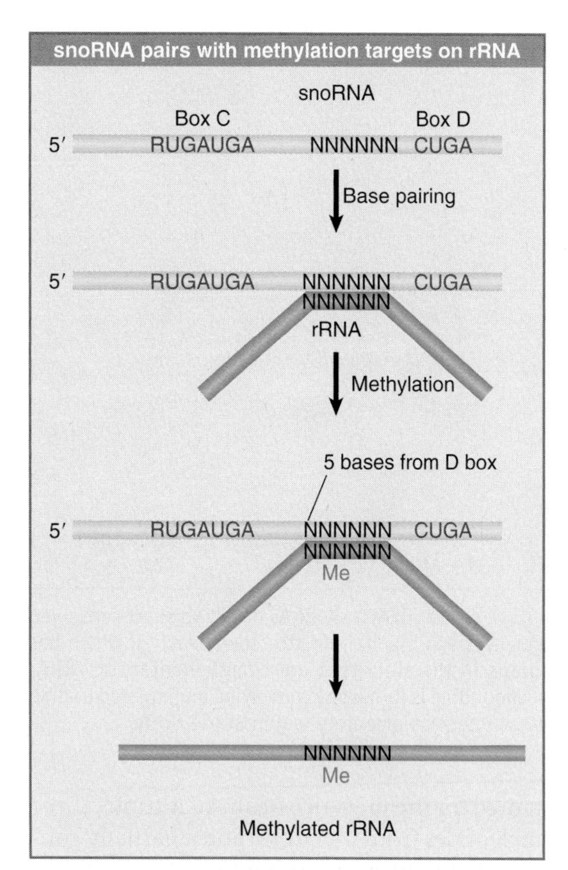

FIGURE 26.39 A snoRNA base pairs with a region of rRNA that is to be methylated.

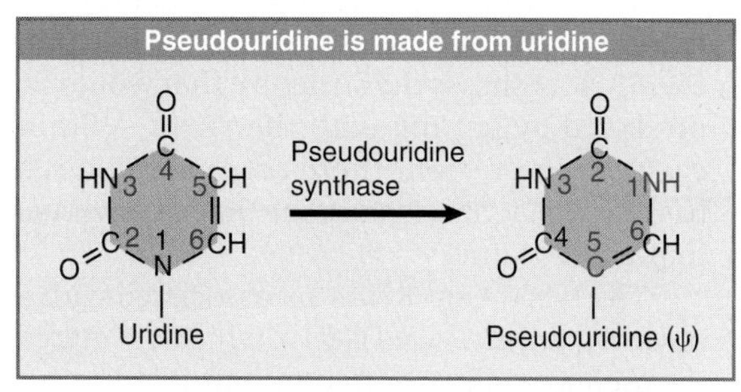

FIGURE 26.40 Uridine is converted to pseudouridine by replacing the N1-sugar bond with a C5-sugar bond and rotating the base relative to the sugar.

Another group of snoRNAs is involved in the synthesis of pseudouridine. There are 43 ψ residues in yeast rRNAs and ~100 in vertebrate rRNAs. The synthesis of pseudouridine involves the reaction shown in FIGURE 26.40, in which the N1 bond from uridylic acid to ribose is broken, the base is rotated, and C5 is rejoined to the sugar.

Pseudouridine formation in rRNA requires the H/ACA group of ~20 snoRNAs. They are

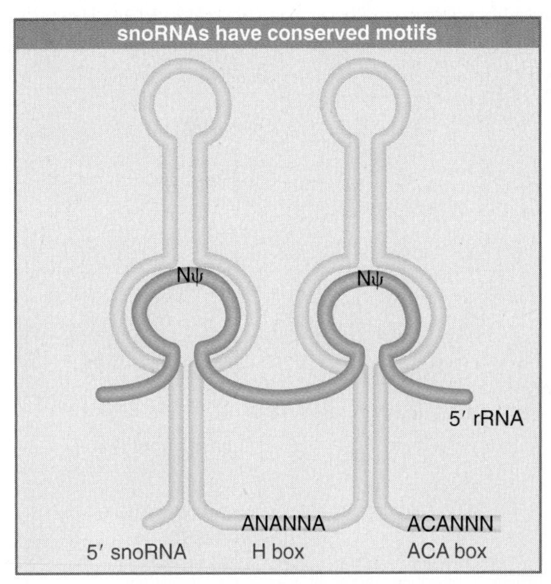

FIGURE 26.41 H/ACA snoRNAs have two short conserved sequences and two hairpin structures, each of which has regions in the stem that are complementary to rRNA. Pseudouridine is formed by converting an unpaired uridine within the complementary region of the rRNA.

named for the presence of an ACA triplet three nucleotides from the 3′ end and a partially conserved sequence (the H box) that lies between two stem-loop hairpin structures. Each of these snoRNAs has a sequence complementary to rRNA within the stem of each hairpin. FIGURE 26.41 shows the structure that would be produced by pairing with the rRNA. Within each pairing region, there are two unpaired bases, one of which is a uridine that is converted to pseudouridine.

The H/ACA snoRNAs are associated with a nucleolar protein called Gar1p, which is required for pseudouridine formation, but its function is unknown. The known pseudouridine synthases are proteins that function without an RNA cofactor. Synthases that could be involved in snoRNA-mediated pseudouridine synthesis have not been identified.

The involvement of the U7 snRNA in 3′ end generation, and the role of snoRNAs in rRNA processing and modification, is consistent with the view we develop in Chapter 27, Catalytic RNA, that many—perhaps all—RNA processing events depend on RNA–RNA interactions. As with splicing reactions, the snRNA probably functions in the form of a ribonucleoprotein particle containing proteins as well as the RNA. It is common (although not the only mechanism of action) for the RNA of the particle to base pair with a short sequence in the substrate RNA.

26.23 Summary

Splicing accomplishes the removal of introns and the joining of exons into the mature sequence of RNA. There are at least four types of reaction, as distinguished by their requirements *in vitro* and the intermediates that they generate. The systems include eukaryotic nuclear introns, group I and group II introns, and tRNA introns. Each reaction involves a change of organization within an individual RNA molecule, and is therefore a *cis*-acting event.

Pre-mRNA splicing follows preferred but not obligatory pathways. Only very short consensus sequences are necessary; the rest of the intron appears irrelevant. All 5′ splice sites are probably equivalent, as are all 3′ splice sites. The required sequences are given by the GU-AG rule, which describes the ends of the intron. The UACUAAC branch site of yeast, or a less well conserved consensus in mammalian introns, is also required. The reaction with the 5′ splice site involves formation of a lariat that joins the GU end of the intron via a 5′–2′ linkage to the A at position 6 of the branch site. The 3′–OH end of the exon then attacks the 3′ splice site, so that the exons are ligated and the intron is released as a lariat. Both reactions are transesterifications in which bonds are conserved. Several stages of the reaction require hydrolysis of ATP, probably to drive conformational changes in the RNA and/or protein components. Lariat formation is responsible for choice of the 3′ splice site. Alternative splicing patterns are caused by protein factors that either stimulate use of a new site or that block use of the default site.

Pre-mRNA splicing requires formation of a spliceosome—a large particle that assembles the consensus sequences into a reactive conformation. The spliceosome most often forms by the process of intron definition, involving recognition of the 5′ splice site, branch site, and 3′ splice site. An alternative pathway involves exon definition, which involves initial recognition of the 5′ splice sites of both the substrate intron and the next intron. Its formation passes through a series of stages from the E (commitment) complex, which contains U1 snRNP and splicing factors, through the A and B complexes as additional components are added.

The spliceosome contains the U1, U2, U4/U6, and U5 snRNPs, as well as some additional splicing factors. The U1, U2, and U5 snRNPs each contain a single snRNA and several proteins; the U4/U6 snRNP contains two snRNAs and several proteins. Some proteins are common to all snRNP particles. The snRNPs recognize consensus sequences. U1 snRNA base pairs with the 5′ splice site, U2 snRNA base pairs with the branch sequence, and U5 snRNP acts at the 5′ splice site. When U4 releases U6, the U6 snRNA base pairs with U2; this may create the catalytic center for splicing. An alternative set of snRNPs provides analogous functions for splicing the U12-dependent subclass of introns. The snRNA molecules may have catalytic-like roles in splicing and other processing reactions.

In the nucleolus, two groups of snoRNAs are responsible for pairing with rRNAs at sites that are modified; group C/D snoRNAs indicate target sites for methylation, and group ACA snoRNAs identify sites where uridine is converted to pseudouridine.

Splicing is usually intramolecular, but *trans*-splicing (intermolecular splicing) occurs in trypanosomes and nematodes. It involves a reaction between a small SL RNA and the pre-mRNA. The SL RNA resembles U1 snRNA and may combine the role of providing the exon and the functions of U1. In worms there are two types of SL RNA: One is used for splicing to the 5′ end of an mRNA, and the other is used for splicing to an internal site.

Group II introns share with nuclear introns the use of a lariat as intermediate, but are able to perform the reaction as a self-catalyzed property of the RNA. These introns follow the GT-AG rule, but form a characteristic secondary structure that holds the reacting splice sites in the appropriate apposition.

Yeast tRNA splicing involves separate endonuclease and ligase reactions. The endonuclease recognizes the secondary (or tertiary) structure of the precursor and cleaves both ends of the intron. The two half-tRNAs released by loss of the intron can be ligated in the presence of ATP.

The termination capacity of RNA polymerase II has not been characterized, and 3′ ends of its transcripts are generated by cleavage. The sequence AAUAAA, located 11 to 30 bases upstream of the cleavage site, provides the signal for both cleavage and polyadenylation. An endonuclease and the poly(A) polymerase are associated in a complex with other factors that confer specificity for the AAUAAA signal. Transcription is terminated when an exonuclease, which binds to the 5′ end of the nascent RNA chain created by the cleavage, catches up to RNA polymerase.

References

26.1 Introduction

Reviews

Dreyfuss, G., Kim, V. N., and Kataoka, N. (2002). Messenger-RNA-binding proteins and the messages they carry. *Nat. Rev. Mol. Cell Biol.* 3, 195–205.

Dreyfuss, G. et al. (1993). hnRNP proteins and the biogenesis of mRNA. *Annu. Rev. Biochem.* 62, 289–321.

Lewin, B. (1975). Units of transcription and translation: sequence components of hnRNA and mRNA. *Cell* 4, 77–93.

26.2 Nuclear Splice Junctions Are Short Sequences

Reviews

Padgett, R. A. (1986). Splicing of messenger RNA precursors. *Annu. Rev. Biochem.* 55, 1119–1150.

Sharp, P. A. (1987). Splicing of mRNA precursors. *Science* 235, 766–771.

26.4 Pre-mRNA Splicing Proceeds through a Lariat

Reviews

Sharp, P. A. (1994). Split genes and RNA splicing. *Cell* 77, 805–815.

Weiner, A. (1993). mRNA splicing and autocatalytic introns: distant cousins or the products of chemical determinism. *Cell* 72, 161–164.

Research

Reed, R. and Maniatis, T. (1985). Intron sequences involved in lariat formation during pre-mRNA splicing. *Cell* 41, 95–105.

Reed, R. and Maniatis, T. (1986). A role for exon sequences and splice-site proximity in splice-site selection. *Cell* 46, 681–690.

Zhuang, Y. A., Goldstein, A. M., and Weiner, A. M. (1989). UACUAAC is the preferred branch site for mammalian mRNA splicing. *Proc. Natl. Acad. Sci. USA* 86, 2752–2756.

26.5 snRNAs Are Required for Splicing

Reviews

Guthrie, C. (1991). Messenger RNA splicing in yeast: clues to why the spliceosome is a ribonucleoprotein. *Science* 253, 157–163.

Guthrie, C. and Patterson, B. (1988). Spliceosomal snRNAs. *Annu. Rev. Genet.* 22, 387–419.

Maniatis, T. and Reed, R. (1987). The role of small nuclear ribonucleoprotein particles in pre-mRNA splicing. *Nature* 325, 673–678.

Research

Grabowski, P. J., Seiler, S. R., and Sharp, P. A. (1985). A multicomponent complex is

involved in the splicing of messenger RNA precursors. *Cell* 42, 345–353.
Zhou, Z., Licklider, L. J., Gygi, S. P., and Reed, R. (2002). Comprehensive proteomic analysis of the human spliceosome. *Nature* 419, 182–185.

26.6 U1 snRNP Initiates Splicing

Review

Brow, D. A. (2002). Allosteric cascade of spliceosome activation. *Annu. Rev. Genet.* 36, 333–360.

Research

Abovich, N. and Rosbash, M. (1997). Cross-intron bridging interactions in the yeast commitment complex are conserved in mammals. *Cell* 89, 403–412.
Berglund, J. A., Chua, K., Abovich, N., Reed, R., and Rosbash, M. (1997). The splicing factor BBP interacts specifically with the pre-mRNA branchpoint sequence UACUAAC. *Cell* 89, 781–787.
Burgess, S., Couto, J. R., and Guthrie, C. (1990). A putative ATP binding protein influences the fidelity of branchpoint recognition in yeast splicing. *Cell* 60, 705–717.
Parker, R., Siliciano, P. G., and Guthrie, C. (1987). Recognition of the TACTAAC box during mRNA splicing in yeast involves base pairing to the U2-like snRNA. *Cell* 49, 229–239.
Singh, R., Valcárcel, J., and Green, M. R. (1995). Distinct binding specificities and functions of higher eukaryotic polypyrimidine tract-binding proteins. *Science* 268, 1173–1176.
Wu, S., Romfo, C. M., Nilsen, T. W., and Green, M. R. (1999). Functional recognition of the 3′ splice site AG by the splicing factor U2AF35. *Nature* 402, 832–835.
Zamore, P. D. and Green, M. R. (1989). Identification, purification, and biochemical characterization of U2 small nuclear ribonucleoprotein auxiliary factor. *Proc. Natl. Acad. Sci. USA* 86, 9243–9247.
Zhang, D. and Rosbash, M. (1999). Identification of eight proteins that cross-link to pre-mRNA in the yeast commitment complex. *Genes Dev.* 13, 581–592.
Zhuang, Y. and Weiner, A. M. (1986). A compensatory base change in U1 snRNA suppresses a 5 splice site mutation. *Cell* 46, 827–835.

26.7 The E Complex Can Be Formed by Intron Definition or Exon Definition

Research

Bruzik, J. P. and Maniatis, T. (1995). Enhancer-dependent interaction between 5′ and 3′ splice sites in *trans*. *Proc. Natl. Acad. Sci. USA* 92, 7056–7059.

26.8 5 snRNPs Form the Spliceosome

Reviews

Krämer, A. (1996). The structure and function of proteins involved in mammalian pre-mRNA splicing. *Annu. Rev. Biochem.* 65, 367–409.
Madhani, H. D. and Guthrie, C. (1994). Dynamic RNA-RNA interactions in the spliceosome. *Annu. Rev. Genet.* 28, 1–26.

Research

Lamond, A. I. (1988). Spliceosome assembly involves the binding and release of U4 small nuclear ribonucleoprotein. *Proc. Natl. Acad. Sci. USA* 85, 411–415.
Lesser, C. F. and Guthrie, C. (1993). Mutations in U6 snRNA that alter splice site specificity: implications for the active site. *Science* 262, 1982–1988.
Madhani, H. D. and Guthrie, C. (1992). A novel base-pairing interaction between U2 and U6 snRNAs suggests a mechanism for the catalytic activation of the spliceosome. *Cell* 71, 803–817.
Newman, A. and Norman, C. (1991). Mutations in yeast U5 snRNA alter the specificity of 5′ splice site cleavage. *Cell* 65, 115–123.
Sontheimer, E. J. and Steitz, J. A. (1993). The U5 and U6 small nuclear RNAs as active site components of the spliceosome. *Science* 262, 1989–1996.

26.9 An Alternative Splicing Apparatus Uses Different snRNPs

Research

Burge, C. B., Padgett, R. A., and Sharp, P. A. (1998). Evolutionary fates and origins of U12-type introns. *Mol. Cell* 2, 773–785.
Dietrich, R. C., Incorvaia, R., and Padgett, R. A. (1997). Terminal intron dinucleotide sequences do not distinguish between U2- and U12-dependent introns. *Mol. Cell* 1, 151–160.
Tarn, W.-Y. and Steitz, J. (1996). A novel spliceosome containing U11, U12, and U5 snRNPs excises a minor class AT-AC intron *in vitro*. *Cell* 84, 801–811.

26.10 Splicing Is Connected to Export of mRNA

Reviews

Dreyfuss, G., Kim, V. N., and Kataoka, N. (2002). Messenger-RNA-binding proteins and the messages they carry. *Nat. Rev. Mol. Cell Biol.* 3, 195–205.
Reed, R. and Hurt, E. (2002). A conserved mRNA export machinery coupled to pre-mRNA splicing. *Cell* 108, 523–531.

Research

Kataoka, N., Yong, J., Kim, V. N., Velazquez, F., Perkinson, R. A., Wang, F., and Dreyfuss, G. (2000). Pre-mRNA splicing imprints mRNA in the nucleus with a novel RNA-binding protein that persists in the cytoplasm. *Mol. Cell* 6, 673–682.

Le Hir, H., Gatfield, D., Izaurralde, E., and Moore, M. J. (2001). The exon-exon junction complex provides a binding platform for factors involved in mRNA export and nonsense-mediated mRNA decay. *EMBO J.* 20, 4987–4997.

Le Hir, H., Izaurralde, E., Maquat, L. E., and Moore, M. J. (2000). The spliceosome deposits multiple proteins 20–24 nucleotides upstream of mRNA exon-exon junctions. *EMBO J.* 19, 6860–6869.

Luo, M. J. and Reed, R. (1999). Splicing is required for rapid and efficient mRNA export in metazoans. *Proc. Natl. Acad. Sci. USA* 96, 14937–14942.

Luo, M. L., Zhou, Z., Magni, K., Christoforides, C., Rappsilber, J., Mann, M., and Reed, R. (2001). Pre-mRNA splicing and mRNA export linked by direct interactions between UAP56 and Aly. *Nature* 413, 644–647.

Reichert, V. L., Le Hir, H., Jurica, M. S., and Moore, M. J. (2002). 5′ exon interactions within the human spliceosome establish a framework for exon junction complex structure and assembly. *Genes Dev.* 16, 2778–2791.

Rodrigues, J. P., Rode, M., Gatfield, D., Blencowe, B., Blencowe, M., and Izaurralde, E. (2001). REF proteins mediate the export of spliced and unspliced mRNAs from the nucleus. *Proc. Natl. Acad. Sci. USA* 98, 1030–1035.

Strasser, K. and Hurt, E. (2001). Splicing factor Sub2p is required for nuclear mRNA export through its interaction with Yra1p. *Nature* 413, 648–652.

Zhou, Z., Luo, M. J., Straesser, K., Katahira, J., Hurt, E., and Reed, R. (2000). The protein Aly links pre-messenger-RNA splicing to nuclear export in metazoans. *Nature* 407, 401–405.

26.11 Group II Introns Autosplice via Lariat Formation

Review

Michel, F. and Ferat, J.-L. (1995). Structure and activities of group II introns. *Annu. Rev. Biochem.* 64, 435–461.

26.12 Alternative Splicing Involves Differential Use of Splice Junctions

Review

Green, M. R. (1991). Biochemical mechanisms of constitutive and regulated pre-mRNA splicing. *Annu. Rev. Cell Biol.* 7, 559–599.

Research

Handa, N., Nureki, O., Kurimoto, K., Kim, I., Sakamoto, H., Shimura, Y., Muto, Y., and Yokoyama, S. (1999). Structural basis for recognition of the tra mRNA precursor by the Sex-lethal protein. *Nature* 398, 579–585.

Lynch, K. W. and Maniatis, T. (1996). Assembly of specific SR protein complexes on distinct regulatory elements of the *Drosophila* doublesex splicing enhancer. *Genes Dev.* 10, 2089–2101.

Sun, Q., Mayeda, A., Hampson, R. K., Krainer, A. R., and Rottman, F. M. (1993). General splicing factor SF2/ASF promotes alternative splicing by binding to an exonic splicing enhancer. *Genes Dev.* 7, 2598–2608.

Tian, M. and Maniatis, T. (1993). A splicing enhancer complex controls alternative splicing of doublesex pre-mRNA. *Cell* 74, 105–114.

Wu, J. Y. and Maniatis, T. (1993). Specific interactions between proteins implicated in splice site selection and regulated alternative splicing. *Cell* 75, 1061–1070.

26.13 *trans*-Splicing Reactions Use Small RNAs

Review

Nilsen, T. (1993). *Trans*-splicing of nematode pre-mRNA. *Annu. Rev. Immunol.* 47, 413–440.

Research

Blumenthal, T., Evans, D., Link, C. D., Guffanti, A., Lawson, D., Thierry-Mieg, J., Thierry-Mieg, D., Chiu, W. L., Duke, K., Kiraly, M., and Kim, S. K. (2002). A global analysis of *C. elegans* operons. *Nature* 417, 851–854.

Hannon, G. J. et al. (1990). *trans*-splicing of nematode pre-mRNA *in vitro*. *Cell* 61, 1247–1255.

Huang, X. Y. and Hirsh, D. (1989). A second *trans*-spliced RNA leader sequence in the nematode *C. elegans*. *Proc. Natl. Acad. Sci. USA* 86, 8640–8644.

Krause, M. and Hirsh, D. (1987). A *trans*-spliced leader sequence on actin mRNA in *C. elegans*. *Cell* 49, 753–761.

Murphy, W. J., Watkins, K. P., and Agabian, N. (1986). Identification of a novel Y branch structure as an intermediate in trypanosome mRNA processing: evidence for *trans*-splicing. *Cell* 47, 517–525.

Sutton, R. and Boothroyd, J. C. (1986). Evidence for *trans*-splicing in trypanosomes. *Cell* 47, 527–535.

26.15 The Splicing Endonuclease Recognizes tRNA

Research

Baldi, I. M. et al. (1992). Participation of the intron in the reaction catalyzed by the *Xenopus* tRNA splicing endonuclease. *Science* 255, 1404–1408.

Diener, J. L. and Moore, P. B. (1998). Solution structure of a substrate for the archaeal pre-tRNA splicing endonucleases: the bulge-helix-bulge motif. *Mol. Cell* 1, 883–894.

Di Nicola Negri, E., Fabbri, S., Bufardeci, E., Baldi, M. I., Mattoccia, E., and Tocchini-Valentini, G. P. (1997). The eucaryal tRNA splicing endonuclease recognizes a tripartite set of RNA elements. *Cell* 89, 859–866.

Kleman-Leyer, K., Armbruster, D. W., and Daniels, C. J. (2000). Properties of H. volcanii tRNA intron endonuclease reveal a relationship between the archaeal and eucaryal tRNA intron processing systems. *Cell* 89, 839–847.

Lykke-Andersen, J. and Garrett, R. A. (1997). RNA-protein interactions of an archaeal homotetrameric splicing endoribonuclease with an exceptional evolutionary history. *EMBO J.* 16, 6290–6300.

Mattoccia, E. et al. (1988). Site selection by the tRNA splicing endonuclease of *X. laevis*. *Cell* 55, 731–738.

Reyes, V. M. and Abelson, J. (1988). Substrate recognition and splice site determination in yeast tRNA splicing. *Cell* 55, 719–730.

Trotta, C. R., Miao, F., Arn, E. A., Stevens, S. W., Ho, C. K., Rauhut, R., and Abelson, J. N. (1997). The yeast tRNA splicing endonuclease: a tetrameric enzyme with two active site subunits homologous to the archaeal tRNA endonucleases. *Cell* 89, 849–858.

26.17 The Unfolded Protein Response Is Related to tRNA Splicing

Research

Gonzalez, T. N., Sidrauski, C., Dorfler, S., and Walter, P. (1999). Mechanism of non-spliceosomal mRNA splicing in the unfolded protein response pathway. *EMBO J.* 18, 3119–3132.

Sidrauski, C., Cox, J. S., and Walter, P. (1996). tRNA ligase is required for regulated mRNA splicing in the unfolded protein response. *Cell* 87, 405–413.

Sidrauski, C. and Walter, P. (1997). The transmembrane kinase Ire1p is a site-specific endonuclease that initiates mRNA splicing in the unfolded protein response. *Cell* 90, 1031–1039.

26.19 The 3′ Ends of mRNAs Are Generated by Cleavage and Polyadenylation

Review

Wahle, E. and Keller, W. (1992). The biochemistry of 3′-end cleavage and polyadenylation of messenger RNA precursors. *Annu. Rev. Biochem.* 61, 419–440.

Research

Bouvet, P., Omilli, F., Arlot-Bonnemains, Y., Legagneux, V., Roghi, C., Bassez, T., and Osborne, H. B. (1994). The deadenylation conferred by the 3′ untranslated region of a developmentally controlled mRNA in *Xenopus* embryos is switched to polyadenylation by deletion of a short sequence element. *Mol. Cell Biol.* 14, 1893–1900.

Conway, L. and Wickens, M. (1985). A sequence downstream of AAUAAA is required for formation of SV40 late mRNA 3′ termini in frog oocytes. *Proc. Natl. Acad. Sci. USA* 82, 3949–3953.

Fox, C. A., Sheets, M. D., and Wickens, M. P. (1989). Poly(A) addition during maturation of frog oocytes: distinct nuclear and cytoplasmic activities and regulation by the sequence UUUUUAU. *Genes Dev.* 3, 2151–2162.

Gil, A. and Proudfoot, N. (1987). Position-dependent sequence elements downstream of AAUAAA are required for efficient rabbit β-globin mRNA 3′ end formation. *Cell* 49, 399–406.

Karner, C. G., Wormington, M., Muckenthaler, M., Schneider, S., Dehlin, E., and Wahle, E. (1998). The deadenylating nuclease (DAN) is involved in poly(A) tail removal during the meiotic maturation of *Xenopus* oocytes. *EMBO J.* 17, 5427–5437.

Kim, M., Krogan, N. J., Vasiljeva, L., Rando, O. J., Nedea, E., Greenblatt, J. F., and Buratowski, S. (2004). The yeast Rat1 exonuclease promotes transcription termination by RNA polymerase II. *Nature* 432, 517–522.

McGrew, L. L., Dworkin-Rastl, E., Dworkin, M. B., and Richter, J. D. (1989). Poly(A) elongation during *Xenopus* oocyte maturation is required for translational recruitment and is mediated by a short sequence element. *Genes Dev.* 3, 803–815.

Takagaki, Y., Ryner, L. C., and Manley, J. L. (1988). Separation and characterization of a poly(A) polymerase and a cleavage/specificity factor required for pre-mRNA polyadenylation. *Cell* 52, 731–742.

Voeltz, G. K. and Steitz, J. A. (1998). AUUUA sequences direct mRNA deadenylation uncoupled from decay during *Xenopus* early development. *Mol. Cell Biol.* 18, 7537–7545.

26.20 Cleavage of the 3′ End of Histone mRNA May Require a Small RNA

Review

Birnstiel, M. L. (1985). Transcription termination and 3′ processing: the end is in site. *Cell* 41, 349–359.

Research

Bond, U. M., Yario, T. A., and Steitz, J. A. (1991). Multiple processing-defective mutations in a mammalian histone pre-mRNA are suppressed by compensatory changes in U7 RNA both *in vitro* and *in vitro*. *Genes Dev.* 5, 1709–1722.

Dominski, Z., Erkmann, J. A., Greenland, J. A., and Marzluff, W. F. (2001). Mutations in the RNA binding domain of stem-loop binding protein define separable requirements for RNA binding and for histone pre-mRNA processing. *Mol. Cell. Biol.* 21, 2008–2017.

Galli, G. et al. (1983). Biochemical complementation with RNA in the *Xenopus* oocyte: a small RNA is required for the generation of 3′ histone mRNA termini. *Cell* 34, 823–828.

Mowry, K. L. and Steitz, J. A. (1987). Identification of the human U7 snRNP as one of several factors involved in the 3′ end maturation of histone premessenger RNA's. *Science* 238, 1682–1687.

Wang, Z. F., Whitfield, M. L., Ingledue, T. C., Dominski, Z., and Marzluff, W. F. (1996). The protein that binds the 3′ end of histone mRNA: a novel RNA-binding protein required for histone pre-mRNA processing. *Genes Dev.* 10, 3028–3040.

26.21 Production of rRNA Requires Cleavage Events

Review

Venema, J. and Tollervey, D. (1999). Ribosome synthesis in *S. cerevisiae*. *Annu. Rev. Genet.* 33, 261–311.

26.22 Small RNAs Are Required for rRNA Processing

Research

Balakin, A. G., Smith, L., and Fournier, M. J. (1996). The RNA world of the nucleolus: two major families of small RNAs defined by different box elements with related functions. *Cell* 86, 823–834.

Bousquet-Antonelli, C., Henry, Y., G'elugne, J. P., Caizergues-Ferrer, M., and Kiss, T. (1997). A small nucleolar RNP protein is required for pseudouridylation of eukaryotic ribosomal RNAs. *EMBO J.* 16, 4770–4776.

Ganot, P., Bortolin, M. L., and Kiss, T. (1997). Site-specific pseudouridine formation in preribosomal RNA is guided by small nucleolar RNAs. *Cell* 89, 799–809.

Ganot, P., Caizergues-Ferrer, M., and Kiss, T. (1997). The family of box ACA small nucleolar RNAs is defined by an evolutionarily conserved secondary structure and ubiquitous sequence elements essential for RNA accumulation. *Genes Dev.* 11, 941–956.

Kass, S. et al. (1990). The U3 small nucleolar ribonucleoprotein functions in the first step of preribosomal RNA processing. *Cell* 60, 897–908.

Kiss-Laszlo, Z., Henry, Y., and Kiss, T. (1998). Sequence and structural elements of methylation guide snoRNAs essential for site-specific ribose methylation of pre-rRNA. *EMBO J.* 17, 797–807.

Kiss-Laszlo, Z. et al. (1996). Site-specific ribose methylation of preribosomal RNA: a novel function for small nucleolar RNAs. *Cell* 85, 1077–1068.

Ni, J., Tien, A. L., and Fournier, M. J. (1997). Small nucleolar RNAs direct site-specific synthesis of pseudouridine in rRNA. *Cell* 89, 565–573.

27

Catalytic RNA

CHAPTER OUTLINE

27.1 Introduction

The idea that only proteins have enzymatic activity was deeply rooted in biochemistry (yet devotées of protein function once thought that only proteins could have the versatility to be the genetic material!). A rationale for the identification of enzymes with proteins lies in the view that only proteins, with their varied three-dimensional structures and variety of side-groups, have the flexibility to create the active sites that catalyze biochemical reactions. The characterization of systems involved in RNA processing, though, has shown this view to be an oversimplification.

Several types of catalytic reactions are now known to reside in RNA. **Ribozyme** has become a general term used to describe an RNA with catalytic activity, and it is possible to characterize the enzymatic activity in the same way as a more conventional enzyme. Some RNA catalytic activities are directed against separate substrates, whereas others are intramolecular (which limits the catalytic action to a single cycle).

Introns of the group I and group II classes possess the ability to splice themselves out of the pre-mRNA that contains them. Engineering of group I introns has generated RNA molecules that have several other catalytic activities related to the original activity.

The enzyme ribonuclease P is a ribonucleoprotein that contains a single RNA molecule bound to a protein. The RNA possesses the ability to catalyze cleavage in a tRNA substrate, whereas the protein component plays an indirect role, probably to maintain the structure of the catalytic RNA.

The common theme of these reactions is that the RNA can perform an intramolecular or intermolecular reaction that involves cleavage or joining of phosphodiester bonds *in vitro*. Although the specificity of the reaction and the basic catalytic activity is provided by RNA, proteins associated with the RNA may be needed for the reaction to occur efficiently *in vivo*.

RNA splicing is not the only means by which changes can be introduced in the informational content of RNA. In the process of **RNA editing,** changes are introduced at individual bases, or bases are added at particular positions within an mRNA. The insertion of bases (most commonly uridine residues) occurs for several genes in the mitochondria of certain lower eukaryotes; like splicing, it involves the breakage and reunion of bonds between nucleotides, but also requires a template for coding the information of the new sequence.

27.2 Group I Introns Undertake Self-Splicing by Transesterification

Key concepts

- The only factors required for autosplicing *in vitro* by group I introns are a monovalent cation, a divalent cation, and a guanine nucleotide.
- Splicing occurs by two transesterifications, without requiring input of energy.
- The 3′-OH end of the guanine cofactor attacks the 5′ end of the intron in the first transesterification.
- The 3′-OH end generated at the end of the first exon attacks the junction between the intron and second exon in the second transesterification.
- The intron is released as a linear molecule that circularizes when its 3′-OH terminus attacks a bond at one of two internal positions.
- The G^{414}–A^{16} internal bond of the intron can also be attacked by other nucleotides in a *trans*-splicing reaction.

Group I introns are found in diverse locations. They occur in the genes coding for rRNA in the nuclei of the lower eukaryotes *Tetrahymena thermophila* (a ciliate) and *Physarum polycephalum* (a slime mold). They are common in the genes of fungal mitochondria. They are present in three genes of phage T4 and also are found in bacteria. Group I introns have an intrinsic ability to splice themselves. This is called **self-splicing** or **autosplicing.** (This property also is found in the group II introns discussed in Section 26.11, Group II Introns Autosplice via Lariat Formation.)

Self-splicing was discovered as a property of the transcripts of the rRNA genes in *T. thermophila.* The genes for the two major rRNAs follow the usual organization, in which both are expressed as part of a common transcription unit. The product is a 35S precursor RNA with the sequence of the small rRNA in the 5′ part, and the sequence of the larger (26S) rRNA toward the 3′ end.

In some strains of *T. thermophila,* the sequence coding for 26S rRNA is interrupted by a single, short intron. When the 35S precursor RNA is incubated *in vitro,* splicing occurs as an autonomous reaction. The intron is excised from the precursor and accumulates as a linear fragment of 400 bases, which is subsequently converted to a circular RNA. These events are summarized in FIGURE 27.1.

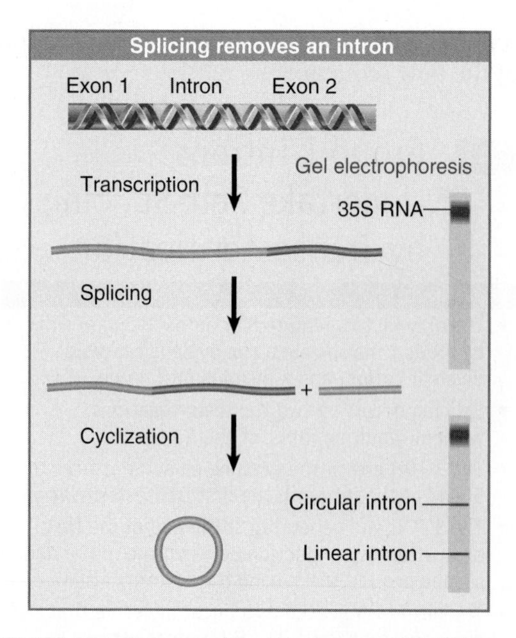

FIGURE 27.1 Splicing of the *Tetrahymena* 35S rRNA precursor can be followed by gel electrophoresis. The removal of the intron is revealed by the appearance of a rapidly moving small band. When the intron becomes circular, it electrophoreses more slowly, as seen by a higher band.

The reaction requires only a monovalent cation, a divalent cation, and a guanine nucleotide cofactor. No other base can be substituted for G, but a triphosphate is not needed: GTP, GDP, GMP, and guanosine itself all can be used, so there is no net energy requirement. The guanine nucleotide must have a 3′–OH group.

The fate of the guanine nucleotide can be followed by using a radioactive label. The radioactivity initially enters the excised linear intron fragment. The G residue becomes linked to the 5′ end of the intron by a normal phosphodiester bond.

FIGURE 27.2 shows that three transfer reactions occur. In the first transfer, the guanine nucleotide behaves as a cofactor that provides a free 3′–OH group that attacks the 5′ end of the intron. This reaction creates the G–intron link and generates a 3′–OH group at the end of the exon. The second transfer involves a similar chemical reaction, in which this 3′–OH then attacks the second exon. The two transfers are connected; no free exons have been observed, so their ligation may occur as part of the same reaction that releases the intron. The intron is released as a linear molecule, but the third transfer reaction converts it to a circle.

Each stage of the self-splicing reaction occurs by a transesterification, in which one phosphate ester is converted directly into another without any intermediary hydrolysis. Bonds are exchanged directly, and energy is conserved, so the reaction does not require input of energy from hydrolysis of ATP or GTP. If each of the consecutive transesterification reactions involves no net change of energy, why does the splicing reaction proceed to completion instead of coming to equilibrium between spliced product and nonspliced precursor? The concentration of GTP is high relative to that of RNA, and therefore drives the reaction forward; a change in secondary structure in the RNA prevents the reverse reaction.

The in vitro *system includes no protein, so the ability to splice is intrinsic to the RNA.* The RNA forms a specific secondary/tertiary structure in which the relevant groups are brought into juxtaposition so that a guanine nucleotide can be bound to a specific site, and then the bond breakage and reunion reactions shown in Figure 27.2 can occur. Although a property of the RNA itself, the reaction is assisted *in vivo* by proteins, which stabilize the RNA structure.

The ability to engage in these transfer reactions resides with the sequence of the intron, which continues to be reactive after its excision as a linear molecule. FIGURE 27.3 summarizes its activities.

The intron can circularize when the 3′ terminal G attacks either of two positions near the 5′ end. The internal bond is broken and the new 5′ end is transferred to the 3′–OH end of the intron. The *primary cyclization* usually involves reaction between the terminal G^{414} and the A^{16}. This is the most common reaction (shown as the third transfer in Figure 27.2). Less frequently, the G^{414} reacts with U^{20}. Each reaction generates a circular intron and a linear fragment that represents the original 5′ region (15 bases long for attack on A^{16}, and 19 bases long for attack on U^{20}). The released 5′ fragment contains the original added guanine nucleotide.

Either type of circle can regenerate a linear molecule *in vitro* by specifically hydrolyzing the bond (G^{414}–A^{16} or G^{414}–U^{20}) that had closed the circle. This is called a *reverse cyclization.* The linear molecule generated by reversing the primary cyclization at A^{16} remains reactive and can perform a secondary cyclization by attacking U^{20}.

The final product of the spontaneous reactions following release of the intron is the L-19 RNA, a linear molecule generated by reversing the shorter circular form. This molecule has an enzymatic activity that allows it to catalyze the extension of short oligonucleotides (not shown in the figure, but see Figure 27.8).

The reactivity of the released intron extends beyond merely reversing the cyclization reaction. Addition of the oligonucleotide UUU reopens the primary circle by reacting with the G^{414}–A^{16} bond. The UUU (which resembles the 3′ end of the 15-mer released by the primary cyclization) becomes the 5′ end of the linear molecule that is formed. This is an *intermolecular* reaction, and thus demonstrates the ability to connect together two different RNA molecules.

This series of reactions demonstrates vividly that the autocatalytic activity reflects a generalized ability of the RNA molecule to form an active center that can bind guanine cofactors, recognize oligonucleotides, and bring together the reacting groups in a conformation that allows bonds to be broken and rejoined. Other group I introns have not been investigated in as much detail as the *Tetrahymena* intron, but their properties are generally similar.

The autosplicing reaction is an intrinsic property of RNA *in vitro*, but to what degree are proteins involved *in vivo?* Some indications for the involvement of proteins are provided by mitochondrial systems, where splicing of group I introns requires the *trans*-acting products of other genes. One striking case is presented by the *cyt18* mutant of *Neurospora crassa*, which is defective in splicing several mitochondrial group I introns. The product of this gene turns out to be the mitochondrial tyrosyl-tRNA synthetase! This is explained by the fact that the intron can take up a tRNA-like tertiary structure that is stabilized by the synthetase and which promotes the catalytic reaction.

This relationship between the synthetase and splicing is consistent with the idea that splicing originated as an RNA-mediated reaction, subsequently assisted by RNA-binding proteins that originally had other functions. The *in vitro* self-splicing ability may represent the basic biochemical interaction. The RNA structure creates the active site, but is able to function efficiently *in vivo* only when assisted by a protein complex.

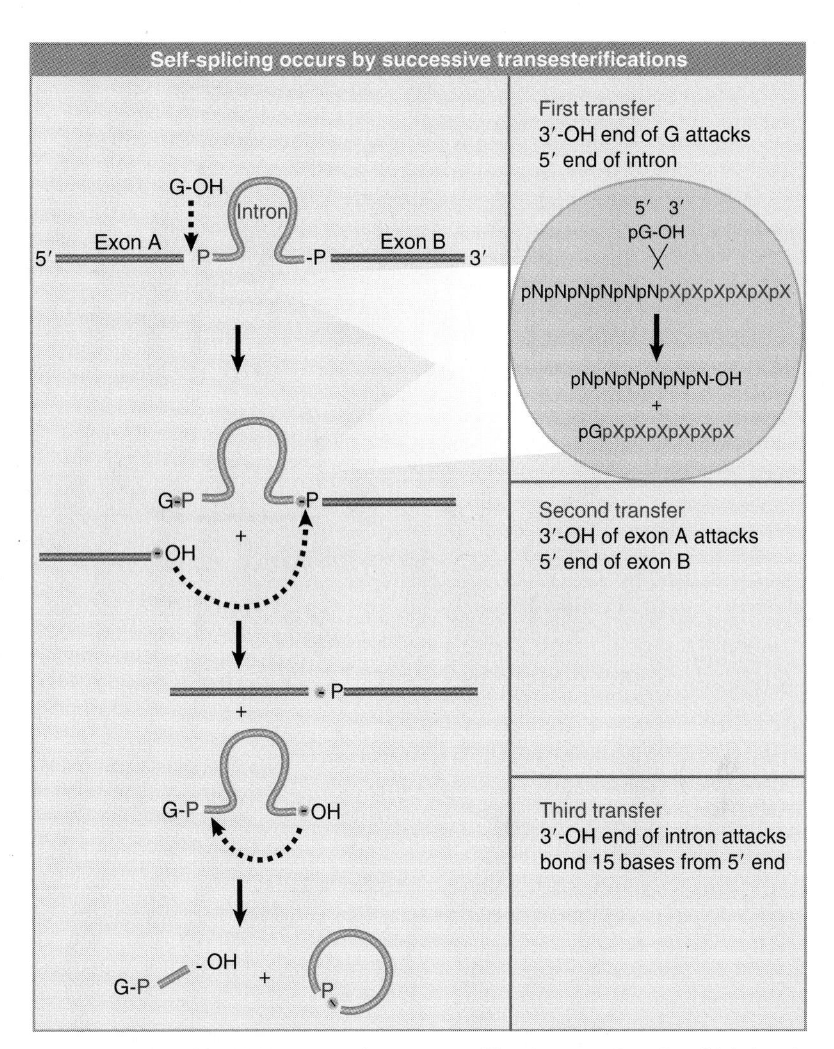

FIGURE 27.2 Self-splicing occurs by transesterification reactions in which bonds are exchanged directly. The bonds that have been generated at each stage are indicated by the shaded boxes.

27.3 Group I Introns Form a Characteristic Secondary Structure

Key concepts

- Group I introns form a secondary structure with nine duplex regions.
- The cores of regions P3, P4, P6, and P7 have catalytic activity.
- Regions P4 and P7 are both formed by pairing between conserved consensus sequences.
- A sequence adjacent to P7 base pairs with the sequence that contains the reactive G.

All group I introns can be organized into a characteristic secondary structure with nine helices (P1–P9). FIGURE 27.4 shows a model for the secondary structure of the *Tetrahymena* intron. Two of the base-paired regions are generated by pairing between conserved sequence elements that are common to group I introns. P4 is constructed from the sequences *P* and *Q;* P7 is formed from sequences *R* and *S*. The other base-paired regions vary in sequence in individual introns. Mutational analysis identifies an intron "core" containing P3, P4, P6, and P7, which provides the minimal region that can undertake a catalytic reaction. The lengths of group I introns vary widely, and the consensus sequences are located a considerable distance from the actual splice junctions.

Some of the pairing reactions are directly involved in bringing the splice junctions into a conformation that supports the enzymatic

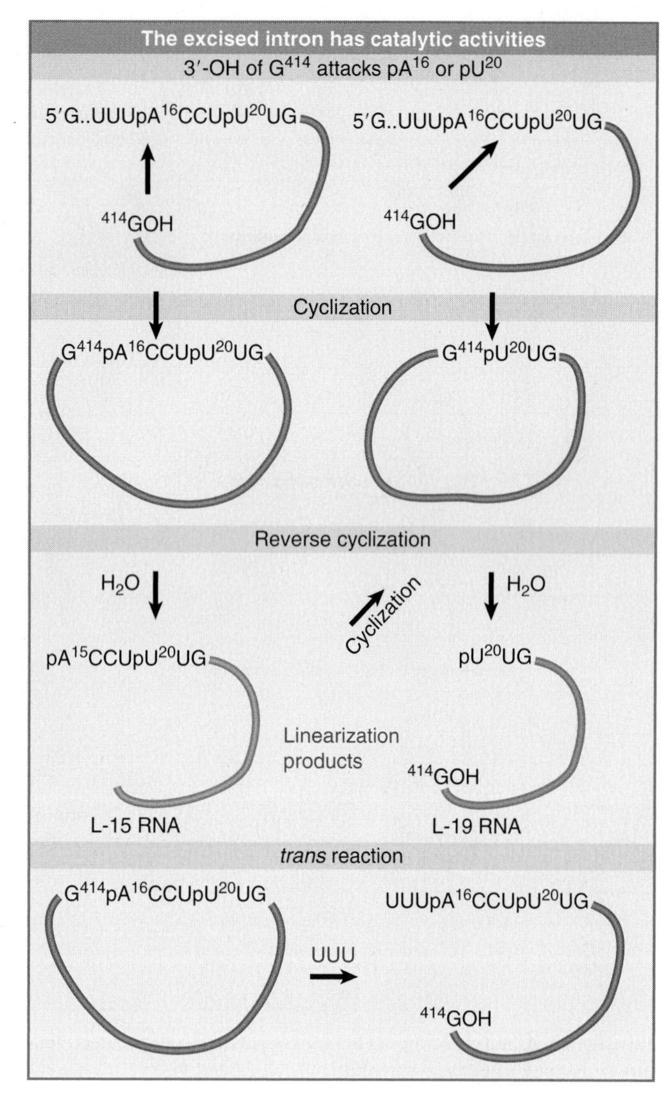

FIGURE 27.3 The excised intron can form circles by using either of two internal sites for reaction with the 5′ end, and can reopen the circles by reaction with water or oligo-nucleotides.

reaction. P1 includes the 3′ end of the left exon. The sequence within the intron that pairs with the exon is called the IGS, or internal guide sequence. (Its name reflects the fact that originally the region immediately 3′ to the IGS sequence shown in the figure was thought to pair with the 3′ splice junction, thus bringing the two junctions together. This interaction may occur, but does not seem to be essential.) A very short sequence—sometimes as short as two bases—between P7 and P9 base pairs with the sequence that immediately precedes the reactive G (position 414 in *Tetrahymena*) at the 3′ end of the intron.

The importance of base pairing in creating the necessary core structure in the RNA is emphasized by the properties of *cis*-acting mutations that prevent splicing of group I introns.

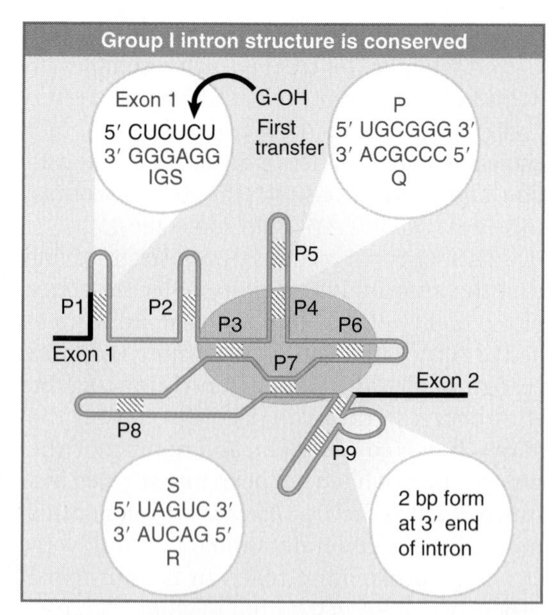

FIGURE 27.4 Group I introns have a common secondary structure that is formed by nine base-paired regions. The sequences of regions P4 and P7 are conserved, and identify the individual sequence elements P, Q, R, and S. P1 is created by pairing between the end of the left exon and the IGS of the intron; a region between P7 and P9 pairs with the 3′ end of the intron.

Such mutations have been isolated for the mitochondrial introns through mutants that cannot remove an intron *in vivo,* and they have been isolated for the *Tetrahymena* intron by transferring the splicing reaction into a bacterial environment. The construct shown in FIGURE 27.5 allows the splicing reaction to be followed in *E. coli.* The self-splicing intron is placed at a location that interrupts the tenth codon of the β galactosidase coding sequence. The protein can therefore be successfully translated from an RNA only after the intron has been removed.

The synthesis of β galactosidase in this system indicates that splicing can occur in conditions quite distant from those prevailing in *Tetrahymena* or even *in vitro.* One interpretation of this result is that self-splicing can occur in the bacterial cell. Another possibility is that there are bacterial proteins that assist the reaction.

Using this assay, we can introduce mutations into the intron to see whether they prevent the reaction. Mutations in the group I consensus sequences that disrupt their base pairing stop splicing. The mutations can be reverted by making compensating changes that restore base pairing.

Mutations in the corresponding consensus sequences in mitochondrial group I introns have

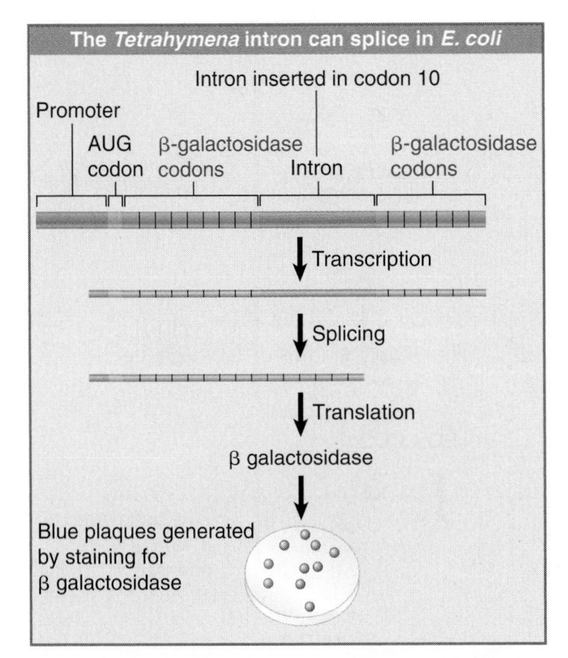

FIGURE 27.5 Placing the *Tetrahymena* intron within the β-galactosidase coding sequence creates an assay for self-splicing in *E. coli*. Synthesis of β galactosidase can be tested by adding a compound that is turned blue by the enzyme. The sequence is carried by a bacteriophage, so the presence of blue plaques (containing infected bacteria) indicates successful splicing.

similar effects. A mutation in one consensus sequence may be reverted by a mutation in the complementary consensus sequence to restore pairing; for example, mutations in the R consensus can be compensated by mutations in the S consensus.

Together these results suggest that the group I splicing reaction depends on the formation of secondary structure between pairs of consensus sequences within the intron. The principle established by this work is that *sequences distant from the splice junctions themselves are required to form the active site that makes self-splicing possible.*

27.4 Ribozymes Have Various Catalytic Activities

Key concepts

- By changing the substrate binding-site of a group I intron, it is possible to introduce alternative sequences that interact with the reactive G.
- The reactions follow classical enzyme kinetics with a low catalytic rate.
- Reactions using 2′–OH bonds could have been the basis for evolving the original catalytic activities in RNA.

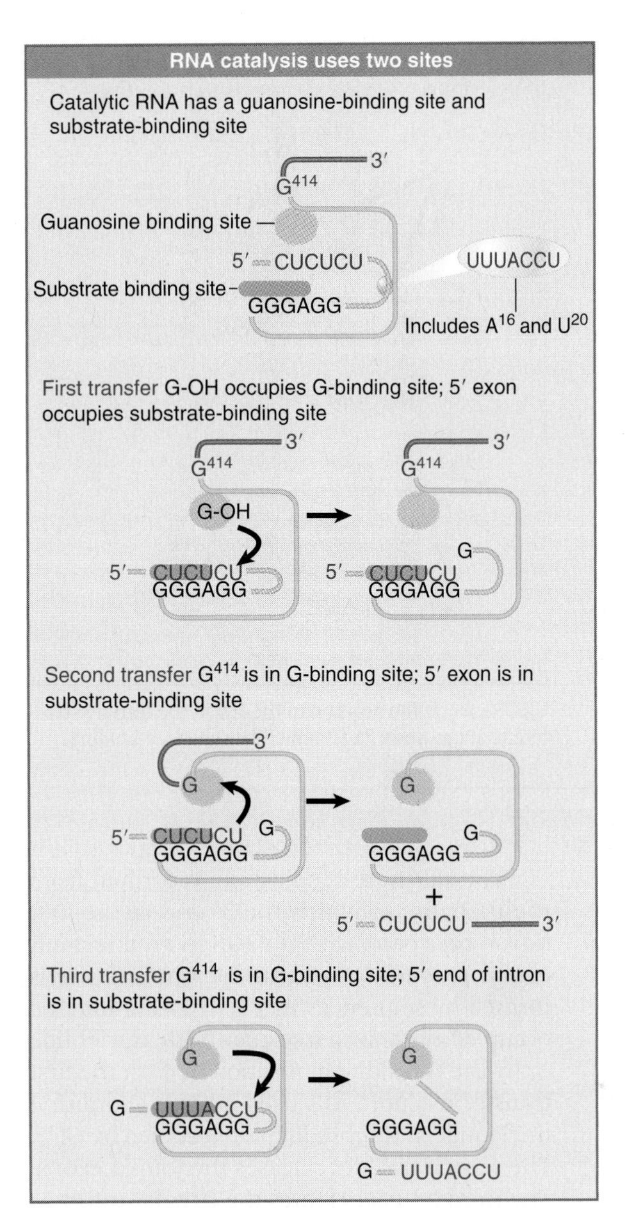

FIGURE 27.6 Excision of the group I intron in *Tetrahymena* rRNA occurs by successive reactions between the occupants of the guanosine-binding site and the substrate-binding site. The left exon is pink, and the right exon is purple.

The catalytic activity of group I introns was discovered by virtue of their ability to autosplice, but they are able to undertake other catalytic reactions *in vitro*. All of these reactions are based on transesterifications. We analyze these reactions in terms of their relationship to the splicing reaction itself.

The catalytic activity of a group I intron is conferred by its ability to generate particular secondary and tertiary structures that create active sites that are equivalent to the active sites of a conventional (proteinaceous) enzyme. FIGURE 27.6 illustrates the splicing reaction in terms of these sites (this is the same series of reactions shown previously in Figure 27.2).

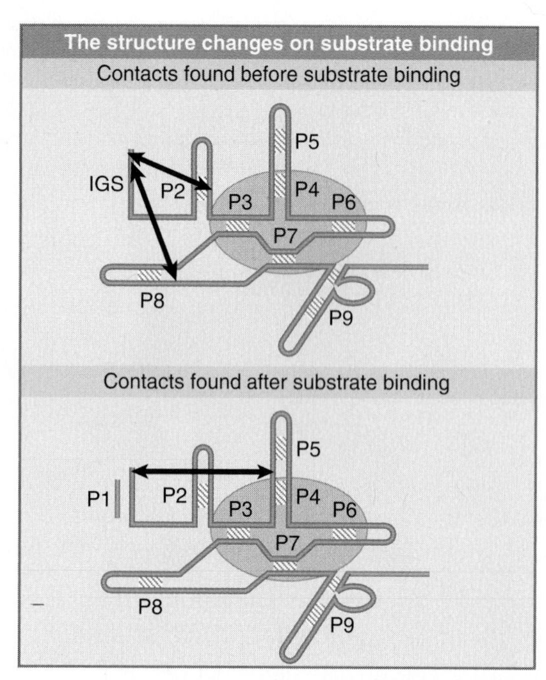

FIGURE 27.7 The position of the IGS in the tertiary structure changes when P1 is formed by substrate binding.

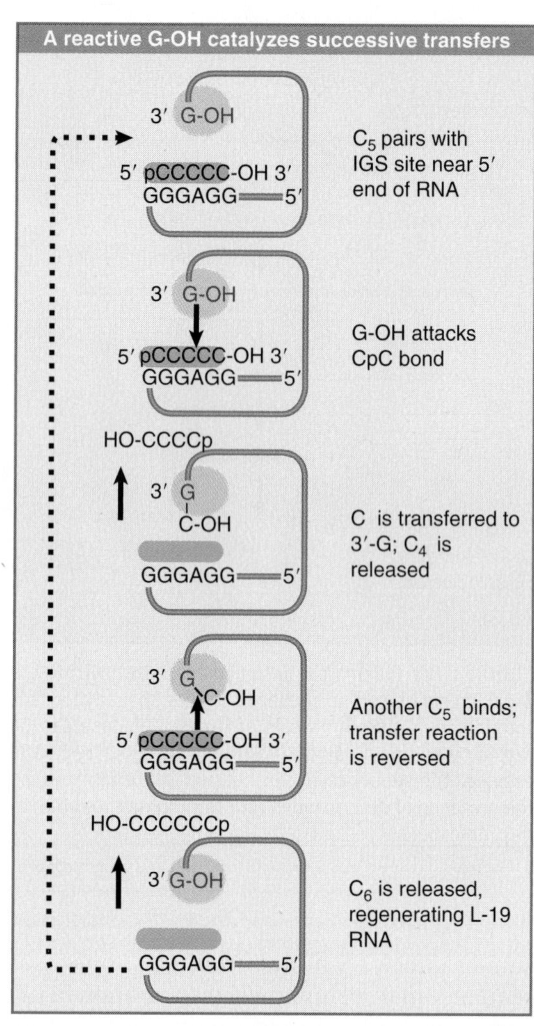

FIGURE 27.8 The L-19 linear RNA can bind C in the substrate-binding site; the reactive G-OH 3′ end is located in the G-binding site, and catalyzes transfer reactions that convert two C_5 oligo-nucleotides into a C_4 and a C_6 oligonucleotide.

The substrate-binding site is formed from the P1 helix, in which the 3′ end of the first intron base pairs with the IGS in an intermolecular reaction. A guanosine-binding site is formed by sequences in P7. This site may be occupied either by a free guanosine nucleotide or by the G residue in position 414. In the first transfer reaction, it is used by free guanosine nucleotide; it is subsequently occupied by G^{414}. The second transfer releases the joined exons. The third transfer creates the circular intron.

Binding to the substrate involves a change of conformation. Before substrate binding, the 5′ end of the IGS is close to P2 and P8; after binding, when it forms the P1 helix, it is close to conserved bases that lie between P4 and P5. The reaction is visualized by contacts that are detected in the secondary structure in FIGURE 27.7. In the tertiary structure, the two sites alternatively contacted by P1 are 37 Å apart, which implies a substantial movement in the position of P1.

The L-19 RNA is generated by opening the circular intron (shown as the last stage of the intramolecular rearrangements in Figure 27.3). It still retains enzymatic abilities, which resemble the activities involved in the original splicing reaction. We may consider ribozyme function in terms of the ability to bind an intramolecular sequence complementary to the IGS in the substrate-binding site while binding either the terminal G^{414} or a free G-nucleotide in the G-binding site.

FIGURE 27.8 illustrates the mechanism by which the oligonucleotide C_5 is extended to generate a C_6 chain. The C_5 oligonucleotide binds in the substrate-binding site, whereas G^{414} occupies the G-binding site. By transesterification reactions, a C is transferred from C_5 to the 3′–terminal G, and then back to a new C_5 molecule. Further transfer reactions lead to the accumulation of longer cytosine oligonucleotides. The reaction is a true catalysis, because the L-19 RNA remains unchanged and is available to catalyze multiple cycles. The ribozyme is behaving as a nucleotidyl transferase.

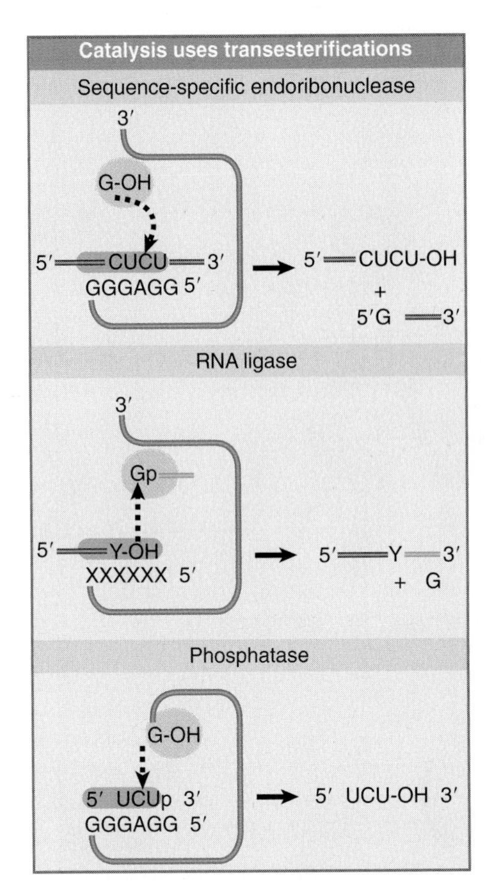

FIGURE 27.9 Catalytic reactions of the ribozyme involve transesterifications between a group in the substrate-binding site and a group in the G-binding site.

RNA catalysis is enzymatic			
Enzyme	Substrate	K_M (mM)	Turnover (/min)
19-base virusoid	24-base RNA	0.0006	0.5
L-19 Intron	CCCCCC	0.04	1.7
RNAase P RNA	pre-tRNA	0.00003	0.4
RNAase P complete	pre-tRNA	0.00003	29
RNAase T1	GpA	0.05	5,700
β galactosidase	lactose	4.0	12,500

FIGURE 27.10 Reactions catalyzed by RNA have the same features as those catalyzed by proteins, although the rate is slower. The K_M gives the concentration of substrate required for half-maximum velocity; this is an inverse measure of the affinity of the enzyme for substrate. The turnover number gives the number of substrate molecules transformed in unit time by a single catalytic site.

Some further enzymatic reactions are characterized in FIGURE 27.9. The ribozyme can function as a sequence-specific endoribonuclease by utilizing the ability of the IGS to bind complementary sequences. In this example, it binds an external substrate containing the sequence CUCU, instead of binding the analogous sequence that is usually contained at the end of the left exon. A guanine-containing nucleotide is present in the G-binding site, and attacks the CUCU sequence in precisely the same way that the exon is usually attacked in the first transfer reaction. This cleaves the target sequence into a 5′ molecule that resembles the left exon, and a 3′ molecule that bears a terminal G residue. By mutating the IGS element, it is possible to change the specificity of the ribozyme so that it recognizes sequences complementary to the new sequence at the IGS region.

Altering the IGS so that the specificity of the substrate-binding site is changed to enable other RNA targets to enter can be used to generate a ligase activity. An RNA terminating in a 3′–OH is bound in the substrate site, and an RNA terminating in a 5′–G residue is bound in the G-binding site. An attack by the hydroxyl on the phosphate bond connects the two RNA molecules, with the loss of the G residue.

The phosphatase reaction is not directly related to the splicing transfer reactions. An oligonucleotide sequence that is complementary to the IGS and terminates in a 3′–phosphate can be attacked by the G^{414}. The phosphate is transferred to the G^{414}, and an oligonucleotide with a free 3′–OH end is then released. The phosphate can then be transferred either to an oligonucleotide terminating in 3′–OH (effectively reversing the reaction) or indeed to water (releasing inorganic phosphate and completing an authentic phosphatase reaction).

The reactions catalyzed by RNA can be characterized in the same way as classical enzymatic reactions in terms of Michaelis–Menten kinetics. FIGURE 27.10 analyzes the reactions catalyzed by RNA. The K_M values for RNA-catalyzed reactions are low, and therefore imply that the RNA can bind its substrate with high specificity. The turnover numbers are low, which reflects a low catalytic rate. In effect, the RNA molecules behave in the same general manner as traditionally defined for enzymes, although they are relatively slow compared to protein catalysts (where a typical range of turnover numbers is 10^3 to 10^6).

A powerful extension of the activities of ribozymes has been made with the discovery that they can be regulated by ligands (see Section 13.7, Small RNA Molecules Can Regulate

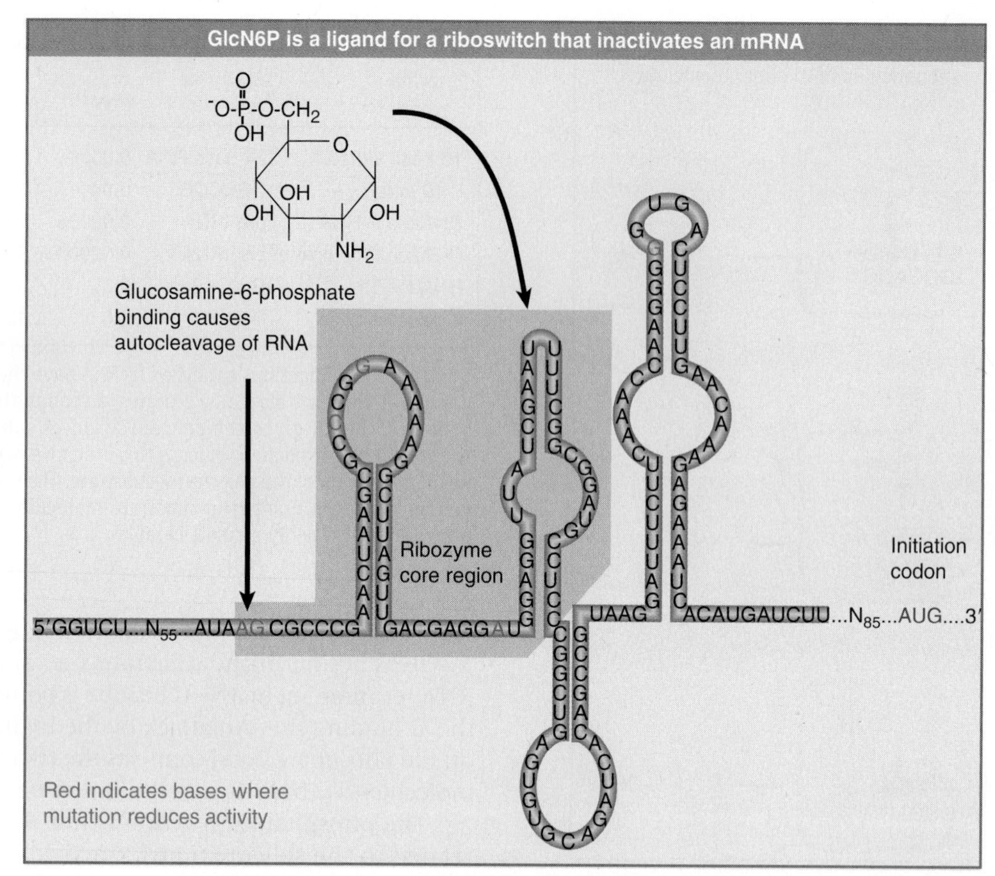

FIGURE 27.11 A ribozyme is contained within the 5′ untranslated region of the mRNA coding for the enzyme that produces glucosamine-6-phosphate. When Glc-6-P binds to the ribozyme, it cleaves off the 5′ end of the mRNA, thereby inactivating it and preventing further production of the enzyme.

Translation). FIGURE 27.11 summarizes the regulation of a **riboswitch.** The small metabolite GlcN6P binds to a ribozyme and activates its ability to cleave the RNA in an intramolecular reaction. The purpose of the system is to regulate production of GlcN6P; the ribozyme is located in the 5′ untranslated region of the mRNA that codes for the enzyme involved in producing GlcN6P, and the cleavage prevents translation.

How does RNA provide a catalytic center? Its ability seems reasonable if we think of an active center as a surface that exposes a series of active groups in a fixed relationship. In a protein, the active groups are provided by the side chains of the amino acids, which have appreciable variety, including positive and negative ionic groups and hydrophobic groups. In an RNA, the available moieties are more restricted, consisting primarily of the exposed groups of bases. Short regions are held in a particular structure by the secondary/tertiary conformation of the molecule, providing a surface of active groups able to maintain an environment in which bonds can be broken and made in another molecule. It seems inevitable that the interaction between the RNA catalyst and the RNA substrate will rely on base pairing to create the environment. Divalent cations (typically Mg^{2+}) play an important role in structure, typically being present at the active site where they coordinate the positions of the various groups. They play a direct role in the endonucleolytic activity of virusoid ribozymes (see Section 27.9, Viroids Have Catalytic Activity).

The evolutionary implications of these discoveries are intriguing. The split personality of the genetic apparatus—in which RNA is present in all components, but proteins undertake catalytic reactions—has always been puzzling. It seems unlikely that the very first replicating systems could have contained both nucleic acid and protein.

Suppose, though, that the first systems contained only a self-replicating nucleic acid with primitive catalytic activities—just those needed to make and break phosphodiester bonds. If we suppose that the involvement of 2′–OH bonds in current splicing reactions is derived from these primitive catalytic activities, we may argue that the original nucleic acid was RNA, because DNA lacks the 2′–OH group and therefore could not undertake such reactions. Proteins could have been added for their ability to stabilize the RNA structure. The greater versatility of proteins then could have allowed them to take over catalytic reactions, leading eventually to the complex and sophisticated apparatus of modern gene expression.

27.5 Some Group I Introns Code for Endonucleases That Sponsor Mobility

Key concepts

- Mobile introns are able to insert themselves into new sites.
- Mobile group I introns code for an endonuclease that makes a double-strand break at a target site.
- The intron transposes into the site of the double-strand break by a DNA-mediated replicative mechanism.

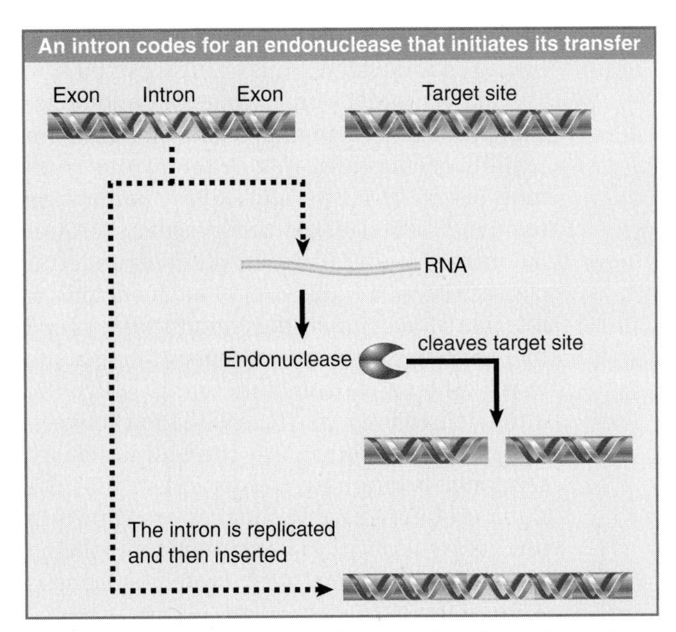

FIGURE 27.12 An intron codes for an endonuclease that makes a double-strand break in DNA. The sequence of the intron is duplicated and then inserted at the break.

Certain introns of both the group I and group II classes contain open reading frames that are translated into proteins. Expression of the proteins allows the intron (either in its original DNA form or as a DNA copy of the RNA) to be *mobile:* It is able to insert itself into a new genomic site. Introns of both groups I and II are extremely widespread, being found in both prokaryotes and eukaryotes. Group I introns migrate by DNA-mediated mechanisms, whereas group II introns migrate by RNA-mediated mechanisms.

Intron mobility was first detected by crosses in which the alleles for the relevant gene differ with regard to their possession of the intron. Polymorphisms for the presence or absence of introns are common in fungal mitochondria. This is consistent with the view that these introns originated by insertion into the gene. Some light on the process that could be involved is cast by an analysis of recombination in crosses involving the large rRNA gene of the yeast mitochondrion.

This gene has a group I intron that contains a coding sequence. The intron is present in some strains of yeast (called ω^+) but absent in others (ω^-). Genetic crosses between ω^+ and ω^- are *polar:* The progeny are usually ω^+.

If we think of the ω^+ strain as a donor and the ω^- strain as a recipient, we form the view that in $\omega^+ \times \omega^-$ crosses a new copy of the intron is generated in the ω^- genome. As a result, the progeny are all ω^+.

Mutations can occur in either parent to abolish the polarity. Mutants show normal segregation, with equal numbers of ω^+ and ω^- progeny. The mutations indicate the nature of the process. Mutations in the ω^- strain occur close to the site where the intron would be inserted. Mutations in the ω^+ strain lie in the reading frame of the intron and prevent production of the protein. This suggests the model of FIGURE 27.12, in which the protein coded by the intron in an ω^+ strain recognizes the site where the intron should be inserted in an ω^- strain and causes it to be preferentially inherited.

What is the action of the protein? The product of the ω intron is an endonuclease *that recognizes the ω^- gene as a target for a double-strand break.* The endonuclease recognizes an 18 bp target sequence that contains the site where the intron is inserted. The target sequence is cleaved on each strand of DNA two bases to the 3′ side of the insertion site. Thus the cleavage sites are 4 bp apart and generate overhanging single strands.

This type of cleavage is related to the cleavage characteristic of transposons when they migrate to new sites (see Chapter 21, Transposons). The double-strand break probably initiates a gene conversion process in which the sequence of the ω^+ gene is copied to replace the sequence of the ω^- gene. The reaction involves transposition by a duplicative mechanism, and occurs solely at the level of DNA. Insertion of the intron interrupts the sequence recognized by the endonuclease, thus ensuring stability.

Many group I introns code for endonucleases that make them mobile. Several different families of endonucleases are found; one common feature is the presence of the amino acid sequence LAGLIDADG near the active site. Similar introns often carry quite different endonucleases. There are differences in the details of insertion; for example, the endonuclease coded by the phage T4 *td* intron cleaves a target site that is 24 bp upstream of the site at which the intron is itself inserted. The dissociation between the intron sequence and the endonuclease sequence is emphasized by the fact that the same endonuclease sequences are found in inteins (sequences that code for self-splicing proteins; see Section 27.12, Protein Splicing Is Autocatalytic).

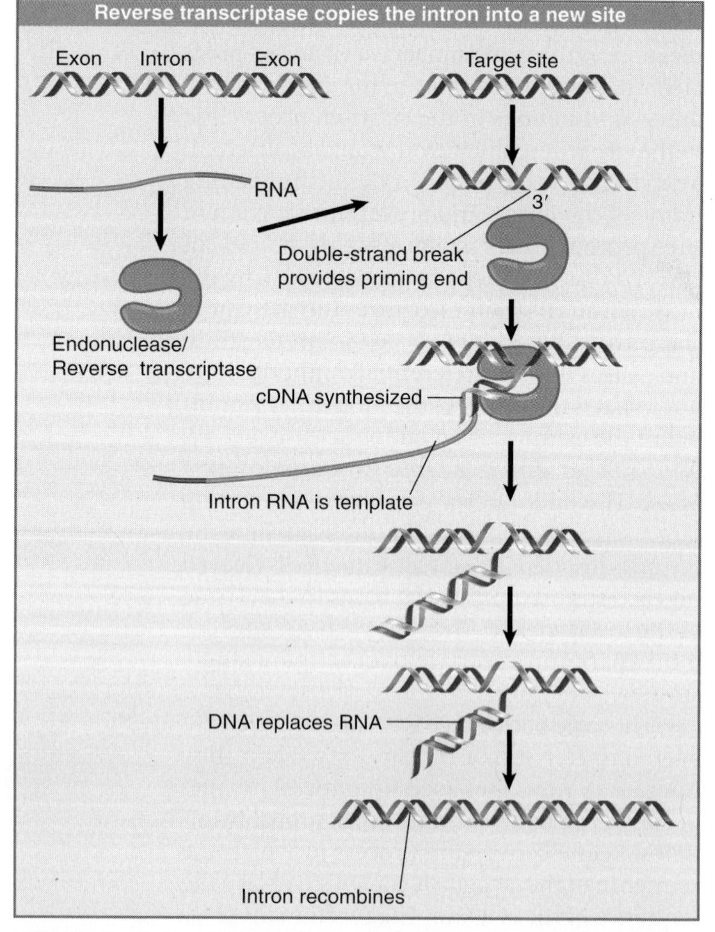

FIGURE 27.13 Reverse transcriptase coded by an intron allows a copy of the RNA to be inserted at a target site generated by a double-strand break.

The variation in the endonucleases means that there is no homology between the sequences of their target sites. The target sites are among the longest and therefore the most specific known for any endonucleases (with a range of 14 to 40 bp). The specificity ensures that the intron perpetuates itself only by insertion into a single target site and not elsewhere in the genome. This is called **intron homing.**

Introns carrying sequences that code for endonucleases are found in a variety of bacteria and lower eukaryotes. These results strengthen the view that introns carrying coding sequences originated as independent elements.

27.6 Group II Introns May Code for Multifunction Proteins

Key concepts

- Group II introns can autosplice *in vitro,* but are usually assisted by protein activities coded within the intron.
- A single coding frame specifies a protein with reverse transcriptase activity, maturase activity, DNA-binding motif, and a DNA endonuclease.
- The reverse transcriptase generates a DNA copy of the RNA sequence that transposes by a retroposon-like mechanism.
- The endonuclease cleaves target DNA to allow insertion of the transposon at a new site.

The best characterized mobile group II introns code for a single protein in a region of the intron beyond its catalytic core. The typical protein contains an N-terminal reverse transcriptase activity, a central domain associated with an ancillary activity that assists folding of the intron into its active structure (called the maturase; see Section 27.7, Some Autosplicing Introns Require Maturases), a DNA-binding domain, and a C-terminal endonuclease domain.

The endonuclease initiates the transposition reaction, and plays the same role in homing as its counterpart in a group I intron. The reverse transcriptase generates a DNA copy of the intron that is inserted at the homing site. The endonuclease also cleaves target sites that resemble, but are not identical to, the homing site at much lower frequency, leading to insertion of the intron at new locations.

FIGURE 27.13 illustrates the transposition reaction for a typical group II intron. The endonuclease makes a double-strand break at the target site. The endonuclease activity requires both the domain of the protein and the intron RNA. The protein domain cleaves the antisense strand of DNA, and the intron RNA actually cleaves the sense strand. This reaction directly inserts the intron into the DNA target site.

The intron RNA provides the template for the synthesis of cDNA. Almost all group II introns have a reverse transcriptase activity that is specific for the intron. The reverse transcriptase generates a DNA copy of the intron, the result being the insertion of the intron into the target site as a duplex DNA. The mechanism resembles the transposition of retroviruses, in which the RNA is an obligatory intermediate (see Section 22.2, The Retrovirus Life Cycle Involves Transposition-Like Events). The type of retrotransposition involved in this case resembles that of a group of retroposons that lack LTRs, and which generate the 3′–OH needed for priming by making a nick in the target (see Figure 22.20 in Section 22.12, LINES Use an Endonuclease to Generate a Priming End).

27.7 Some Autosplicing Introns Require Maturases

Key concept

- Autosplicing introns may require maturase activities encoded within the intron to assist folding into the active catalytic structure.

Although group I and group II introns both have the capacity to autosplice *in vitro*, under physiological conditions they usually require assistance from proteins. Both types of intron may code for **maturase** activities that are required to assist the splicing reaction.

The maturase activity is part of the single open reading frame coded by the intron. In the example of introns that code for homing endonucleases, the single protein product has both endonuclease and maturase activity. Mutational analysis shows that the two activities are independent.

Structural analysis shows that the endonuclease and maturase activities are provided by different active sites in the protein, each coded by a separate domain. The endonuclease site binds to DNA, but the maturase site binds to the intron RNA. FIGURE 27.14 shows the structure of one such protein bound to DNA. A characteristic feature of the endonuclease is the presence of parallel α helices, which contain the hallmark LAGLIDADG sequences, leading to the two catalytic amino acids. The maturase activity is located some distance away on the surface of the protein.

Introns that code for maturases may be unable to splice themselves effectively in the absence of the protein activity. The maturase is in effect a splicing factor that is required specifically for splicing of the sequence that codes for it. It functions to assist the folding of the catalytic core to form an active site.

The coexistence of endonuclease and maturase activities in the same protein suggests a route for the evolution of the intron. FIGURE 27.15 suggests that the intron originated in an independent autosplicing element. The insertion into this element of a sequence coding for an endonuclease gave it mobility. The insertion, however, might well disrupt the ability of the RNA sequence to fold into the active structure. This would create pressure for assistance from proteins that could restore folding ability. The incorporation of such a sequence into the intron would maintain its independence.

Some group II introns that do not code for maturase activities may use comparable proteins that are coded by sequences in the host genome. This suggests a possible route for the evolution of general splicing factors. The factor may have originated as a maturase that specifically assisted the splicing of a particular intron. The coding sequence became isolated from the intron in the host genome, and then it evolved to function with a wider range of substrates that the original intron sequence. The catalytic core of the intron could have evolved into an snRNA.

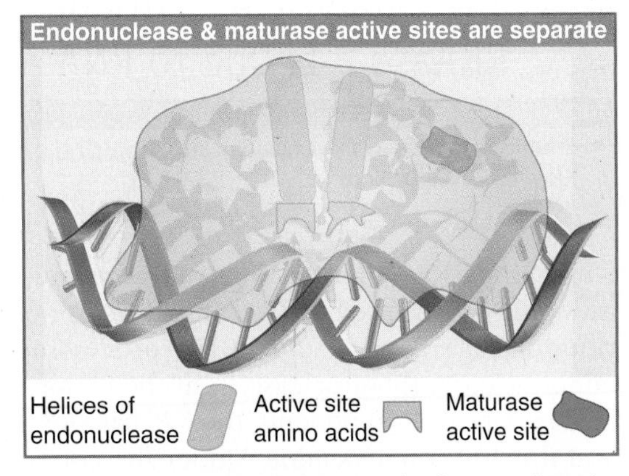

FIGURE 27.14 A homing intron codes for an endonuclease of the LAGLIDADG family that also has maturase activity. The LAGLIDADG sequences are part of the two α helices that terminate in the catalytic amino acids close to the DNA duplex. The maturase active site is identified by an arginine residue elsewhere on the surface of the protein.

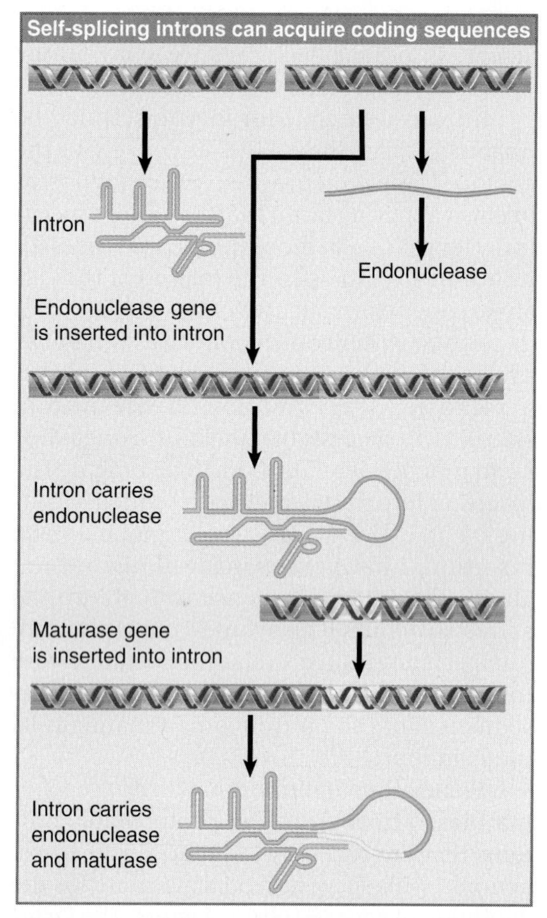

FIGURE 27.15 The intron originated as independent sequence coding for a self-splicing RNA. The insertion of the endonuclease sequence created a homing intron that was mobile. The insertion of the maturase sequence then enhanced the ability of the intron sequences to fold into the active structure for splicing.

27.8 The Catalytic Activity of RNAase P Is Due to RNA

Key concept

- Ribonuclease P is a ribonucleoprotein in which the RNA has catalytic activity.

One of the first demonstrations of the capabilities of RNA was provided by the dissection of ribonuclease P, an *E. coli* tRNA-processing endonuclease. Ribonuclease P can be dissociated into its two components, the 375 base RNA and the 20 kD polypeptide. Under the conditions initially used to characterize the enzyme activity *in vitro,* both components were necessary to cleave the tRNA substrate.

A change in ionic conditions (an increase in the concentration of Mg^{2+}) renders the protein component superfluous, though. *The RNA alone can catalyze the reaction!* Analyzing the results as though the RNA were an enzyme, each "enzyme" catalyzes the cleavage of multiple substrates. Although the catalytic activity resides in the RNA, the protein component greatly increases the speed of the reaction, as seen in the increase in turnover number (see Figure 27.10).

Mutations in either the gene for the RNA or the gene for protein can inactivate RNAase P *in vivo,* so we know that both components are necessary for natural enzyme activity. Originally it had been assumed that the protein provided the catalytic activity, whereas the RNA filled some subsidiary role—for example, assisting in the binding of substrate (it has some short sequences complementary to exposed regions of tRNA). These roles, however, are reversed!

27.9 Viroids Have Catalytic Activity

Key concepts

- Viroids and virusoids form a hammerhead structure that has a self-cleaving activity.
- Similar structures can be generated by pairing a substrate strand that is cleaved by an enzyme strand.
- When an enzyme strand is introduced into a cell, it can pair with a substrate strand target that is then cleaved.

Another example of the ability of RNA to function as an endonuclease is provided by some small plant RNAs (~350 bases) that undertake a self-cleavage reaction. As with the case of the *Tetrahymena* group I intron, however, it is possible to engineer constructs that can function on external substrates.

These small plant RNAs fall into two general groups: viroids and virusoids. The **viroids** are infectious RNA molecules that function independently without encapsidation by any protein coat. The **virusoids** are similar in organization but are encapsidated by plant viruses, being packaged together with a viral genome. The virusoids cannot replicate independently, but require assistance from the virus. The virusoids are sometimes called **satellite RNAs.**

Viroids and virusoids both replicate via rolling circles (see Figure 16.6). The strand of RNA that is packaged into the virus is called the plus strand. The complementary strand, generated during replication of the RNA, is called the minus strand. Multimers of both plus and minus strands are found. Both types of monomer are generated by cleaving the tail of a rolling circle; circular plus-strand monomers are generated by ligating the ends of the linear monomer.

Both plus and minus strands of viroids and virusoids undergo self-cleavage *in vitro*. The cleavage reaction is promoted by divalent metal cations; it generates 5′–OH and 2′–3′–cyclic phosphodiester termini. Some of the RNAs cleave *in vitro* under physiological conditions. Others do so only after a cycle of heating and cooling; this suggests that the isolated RNA has an inappropriate conformation, but can generate an active conformation when it is denatured and renatured.

The viroids and virusoids that undergo self-cleavage form a "hammerhead" secondary structure at the cleavage site, as drawn in the upper part of FIGURE 27.16. The sequence of this structure is sufficient for cleavage. When the surrounding sequences are deleted, the need for a heating–cooling cycle is obviated, and the small RNA self-cleaves spontaneously. This suggests that the sequences beyond the hammerhead usually interfere with its formation.

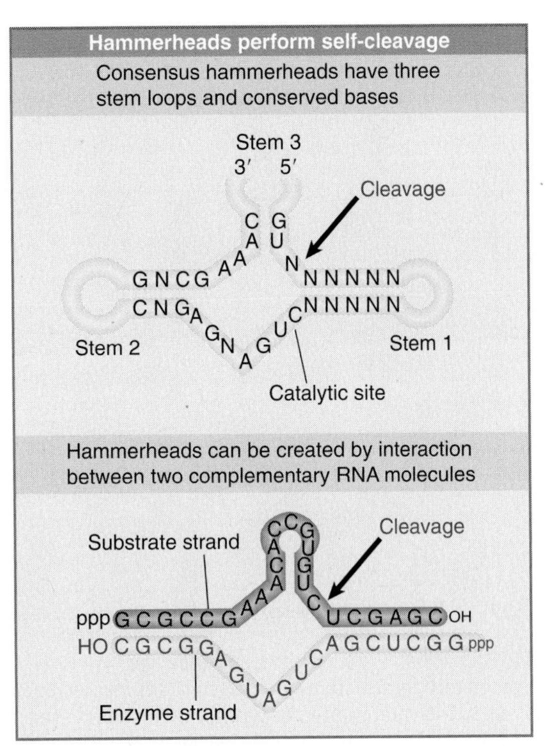

FIGURE 27.16 Self-cleavage sites of viroids and virusoids have a consensus sequence and form a hammerhead secondary structure by intramolecular pairing. Hammerheads can also be generated by pairing between a substrate strand and an "enzyme" strand.

The active site is a sequence of only 58 nucleotides. The hammerhead contains three stem-loop regions whose position and size are constant, and 13 conserved nucleotides, mostly in the regions connecting the center of the structure. The conserved bases and duplex stems generate an RNA with the intrinsic ability to cleave.

An active hammerhead can also be generated by pairing an RNA representing one side of the structure with an RNA representing the other side. The lower part of Figure 27.16 shows an example of a hammerhead generated by hybridizing a 19 base molecule with a 24 base molecule. The hybrid mimics the hammerhead structure, with the omission of loops I and III. When the 19 base RNA is added to the 24 base RNA, cleavage occurs at the appropriate position in the hammerhead.

We may regard the top (24 base) strand of this hybrid as comprising the "substrate" and the bottom (19 base) strand as comprising the "enzyme." When the 19 base RNA is mixed with an excess of the 24 base RNA, multiple copies of the 24 base RNA are cleaved. This suggests that there is a cycle of 19 base–24 base pairing, cleavage, dissociation of the cleaved fragments from the 19 base RNA, and pairing of the 19 base RNA with a new 24 base substrate. The 19 base RNA is therefore a ribozyme with endonuclease activity. The parameters of the reaction are similar to those of other RNA-catalyzed reactions (see Figure 27.10).

The crystal structure of a hammerhead shows that it forms a compact V-shape in which the catalytic center lies in a turn, as indicated diagrammatically in FIGURE 27.17. An Mg^{2+} ion located in the catalytic site plays a crucial role in the reaction. It is positioned by the target cytidine and by the cytidine at the base of stem 1; it may also be connected to the adjacent uridine. It extracts a proton from the 2′–OH of the target cytidine, and then directly attacks the labile phosphodiester bond. Mutations in the hammerhead sequence that affect the transition state of the cleavage reaction occur in both the active site and other locations, suggesting that there may be a substantial rearrangement of structure prior to cleavage.

It is possible to design enzyme-substrate combinations that can form hammerhead structures, and these have been used to demonstrate that introduction of the appropriate RNA molecules into a cell can allow the enzymatic

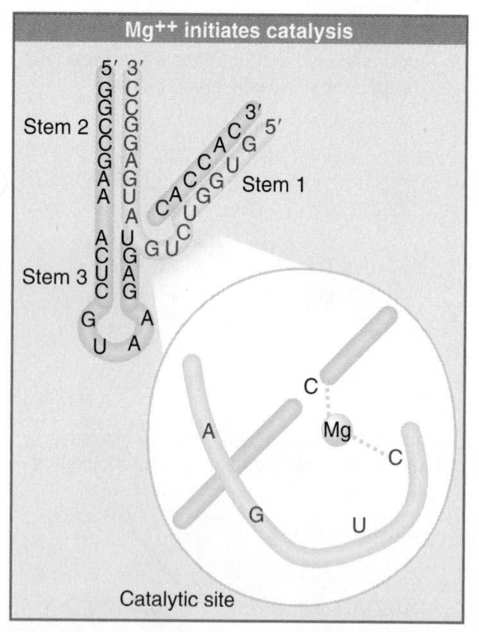

FIGURE 27.17 A hammerhead ribozyme forms a V-shaped tertiary structure in which stem 2 is stacked upon stem 3. The catalytic center lies between stems 2 and 3 and stem 1. It contains a magnesium ion that initiates the hydrolytic reaction.

reaction to occur *in vivo*. A ribozyme designed in this way essentially provides a highly specific restriction-like activity directed against an RNA target. By placing the ribozyme under control of a regulated promoter, it can be used in the same way as (for example) antisense constructs specifically to turn off expression of a target gene under defined circumstances.

27.10 RNA Editing Occurs at Individual Bases

Key concept

- Apolipoprotein-B and glutamate receptors have site-specific deaminations catalyzed by cytidine and adenosine deaminases that change the coding sequence.

A prime axiom of molecular biology is that the sequence of an mRNA can only represent what is coded in the DNA. The central dogma envisaged a linear relationship in which a continuous sequence of DNA is transcribed into a sequence of mRNA that is in turn directly translated into protein. The occurrence of interrupted genes and the removal of introns by RNA splicing introduces an additional step into the process of gene expression: The coding sequences (exons) in DNA must be reconnected in RNA. The process remains one of information transfer, though, in which the actual coding sequence in DNA remains unchanged.

Changes in the information coded by DNA occur in some exceptional circumstances, most notably in the generation of new sequences coding for immunoglobulins in mammals and birds. These changes occur specifically in the somatic cells (B lymphocytes) in which immunoglobulins are synthesized (see Chapter 23, Immune Diversity). New information is generated in the DNA of an individual during the process of reconstructing an immunoglobulin gene, and information coded in the DNA is changed by somatic mutation. The information in DNA continues to be faithfully transcribed into RNA.

RNA editing is a process in which *information changes at the level of mRNA*. It is revealed by situations in which the coding sequence in an RNA differs from the sequence of DNA from which it was transcribed. RNA editing occurs in two different situations, each with different causes. In mammalian cells there are cases in which a substitution occurs in an individual base in mRNA, causing a change in the sequence of the protein that is coded. In trypanosome mitochondria, more widespread changes occur in transcripts of several genes, when bases are systematically added or deleted.

FIGURE 27.18 summarizes the sequences of the apolipoprotein-B gene and mRNA in mam-

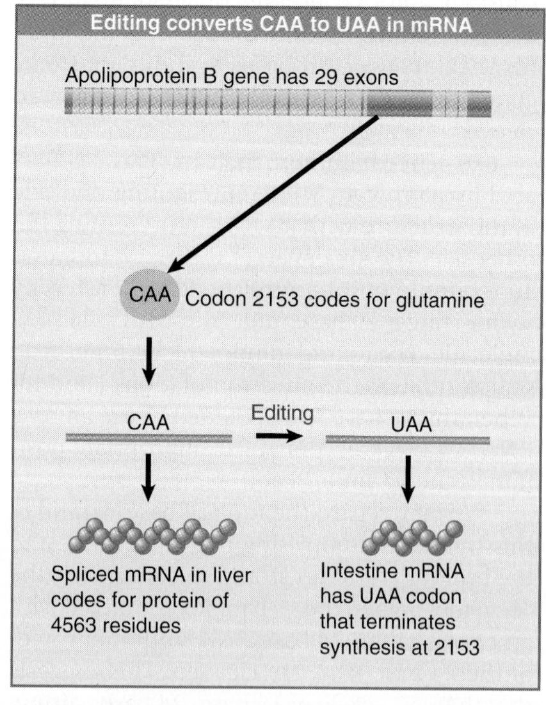

FIGURE 27.18 The sequence of the apo-B gene is the same in intestine and liver, but the sequence of the mRNA is modified by a base change that creates a termination codon in intestine.

malian intestine and liver. The genome contains a single (interrupted) gene whose sequence is identical in all tissues, with a coding region of 4563 codons. This gene is transcribed into an mRNA that is translated into a protein of 512 kD representing the full coding sequence in the liver.

A shorter form of the protein, ~250 kD, is synthesized in the intestine. This protein consists of the N-terminal half of the full-length protein. It is translated from an mRNA whose sequence is identical with that of liver except for a change from C to U at codon 2153. This substitution changes the codon CAA for glutamine into the ochre codon UAA for termination.

What is responsible for this substitution? No alternative gene or exon is available in the genome to code for the new sequence, and no change in the pattern of splicing can be discovered. We are forced to conclude that a change has been made directly in the sequence of the transcript.

Another example is provided by glutamate receptors in rat brain. Editing at one position changes a glutamine codon in DNA into a codon for arginine in RNA; the change affects the conductivity of the channel and therefore has an important effect on controlling ion flow through the neurotransmitter. At another position in the receptor, an arginine codon is converted to a glycine codon.

The editing event in apo-B causes C_{2153} to be changed to U; both changes in the glutamate receptor are from A to I (inosine). These events are *deaminations* in which the amino group on the nucleotide ring is removed. Such events are catalyzed by enzymes called cytidine and adenosine deaminases, respectively. This type of editing appears to occur largely in the nervous system. There are 16 (potential) targets for cytosine deaminase in *Drosophila melanogaster,* and all are genes involved in neurotransmission. In many cases, the editing event changes an amino acid at a functionally important position in the protein.

What controls the specificity of an editing reaction? Enzymes that undertake deamination as such often have broad specificity—for example, the best characterized adenosine deaminase acts on any A residue in a duplex RNA region. Editing enzymes are related to the general deaminases, but have other regions or additional subunits that control their specificity. In the case of apo-B editing, the catalytic subunit of an editing complex is related to bacterial cytidine deaminase, but has an additional RNA-binding region that helps to recognize the specific target site for editing. A special adenosine deaminase enzyme recognizes the target sites in the glutamate receptor RNA, and similar events occur in a serotonin receptor RNA.

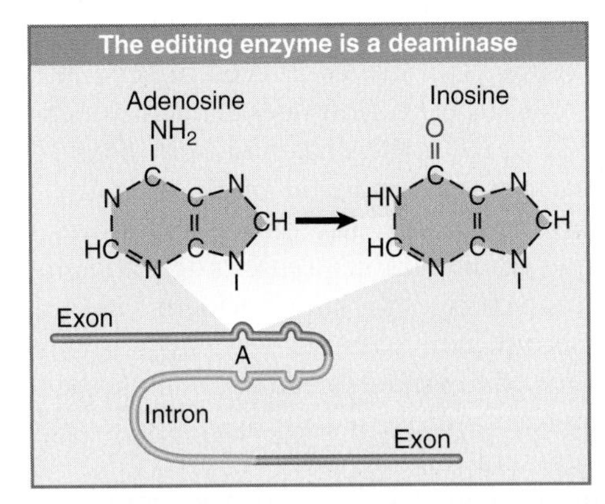

FIGURE 27.19 Editing of mRNA occurs when a deaminase acts on an adenine in an imperfectly paired RNA duplex region.

The complex may recognize a particular region of secondary structure in a manner analogous to tRNA-modifying enzymes, or could directly recognize a nucleotide sequence. The development of an *in vitro* system for the apo-B editing event suggests that a relatively small sequence (~26 bases) surrounding the editing site provides a sufficient target. FIGURE 27.19 shows that in the case of the GluR-B RNA, a base-paired region that is necessary for recognition of the target site is formed between the edited region in the exon and a complementary sequence in the downstream intron. A pattern of mispairing within the duplex region is necessary for specific recognition. Thus different editing systems may have different types of requirement for sequence specificity in their substrates.

27.11 RNA Editing Can Be Directed by Guide RNAs

Key concepts

- Extensive RNA editing in trypanosome mitochondria occurs by insertions or deletions of uridine.
- The substrate RNA base pairs with a guide RNA on both sides of the region to be edited.
- The guide RNA provides the template for addition (or less often, deletion) of uridines.
- Editing is catalyzed by a complex of endonuclease, terminal uridyltransferase activity, and RNA ligase.

Another type of editing is revealed by dramatic changes in sequence in the products of several genes of trypanosome mitochondria. In the first case to be discovered, the sequence of the cytochrome oxidase subunit II protein has a frameshift relative to the sequence of the *coxII* gene. The sequences of the gene and protein given in FIGURE 27.20 are conserved in several trypanosome species. How does this gene function?

The *coxII* mRNA has an insert of an additional four nucleotides (all uridines) around the site of frameshift. The insertion restores the proper reading frame; it inserts an extra amino acid and changes the amino acids on either side. No second gene with this sequence can be discovered, and we are forced to conclude that the extra bases are inserted during or after transcription. A similar discrepancy between mRNA and genomic sequences is found in genes of the SV5 and measles paramyxoviruses, in these cases involving the addition of G residues in the mRNA.

Similar editing of RNA sequences occurs for other genes, and includes deletions as well as additions of uridine. The extraordinary case of the *coxIII* gene of *Trypanosoma brucei* is summarized in FIGURE 27.21.

More than half of the residues in the mRNA consist of uridines that are not coded in the gene. Comparison between the genomic DNA and the mRNA shows that no stretch longer than seven nucleotides is represented in the mRNA without alteration, and runs of uridine up to seven bases long are inserted.

What provides the information for the specific insertion of uridines? A **guide RNA** contains a sequence that is complementary to the correctly edited mRNA. FIGURE 27.22 shows a model for its action in the cytochrome *b* gene of *Leishmania*.

The sequence at the top of the figure shows the original transcript, or preedited RNA. Gaps show where bases will be inserted in the editing process. Eight uridines must be inserted into this region to create the valid mRNA sequence.

The guide RNA is complementary to the mRNA for a significant distance, including and surrounding the edited region. Typically the complementarity is more extensive on the 3′ side of the edited region and is rather short on the 5′ side. Pairing between the guide RNA and the preedited RNA leaves gaps where unpaired A residues in the guide RNA do not find complements in the preedited RNA. The guide RNA provides a template that allows the missing U residues to be inserted at these positions. When the reaction is completed the guide RNA separates from the mRNA, which becomes available for translation.

Specification of the final edited sequence can be quite complex. In this example, a lengthy stretch of the transcript is edited by the insertion of a total of 39 U residues, which appears to require two guide RNAs that act at adjacent sites. The first guide RNA pairs at the 3′–most site, and the edited sequence then becomes a substrate for further editing by the next guide RNA.

The guide RNAs are encoded as independent transcription units. FIGURE 27.23 shows a map of the relevant region of the *Leishmania*

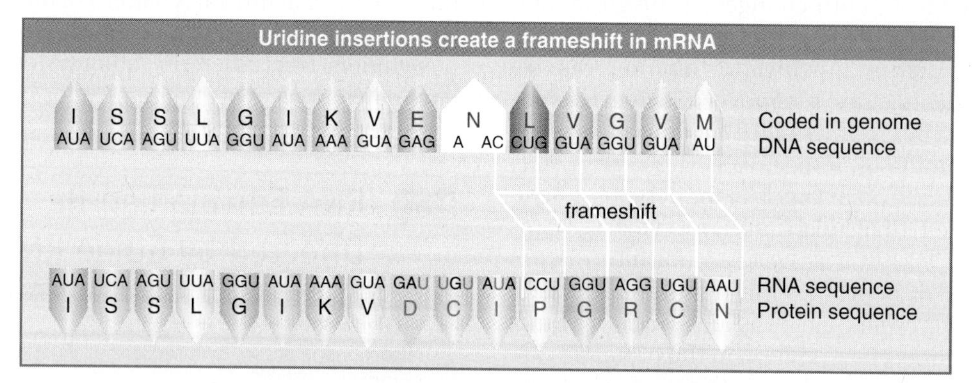

FIGURE 27.20 The mRNA for the trypanosome *coxII* gene has a frameshift relative to the DNA; the correct reading frame is created by the insertion of four uridines.

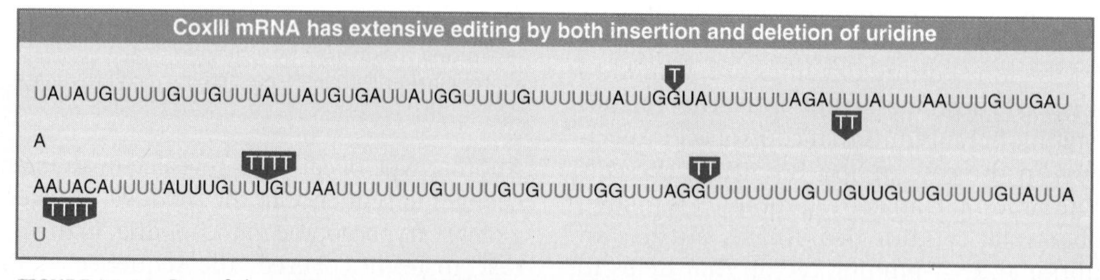

FIGURE 27.21 Part of the mRNA sequence of *T. brucei coxIII* shows many uridines that are not coded in the DNA (shown in red) or that are removed from the RNA (shown as T).

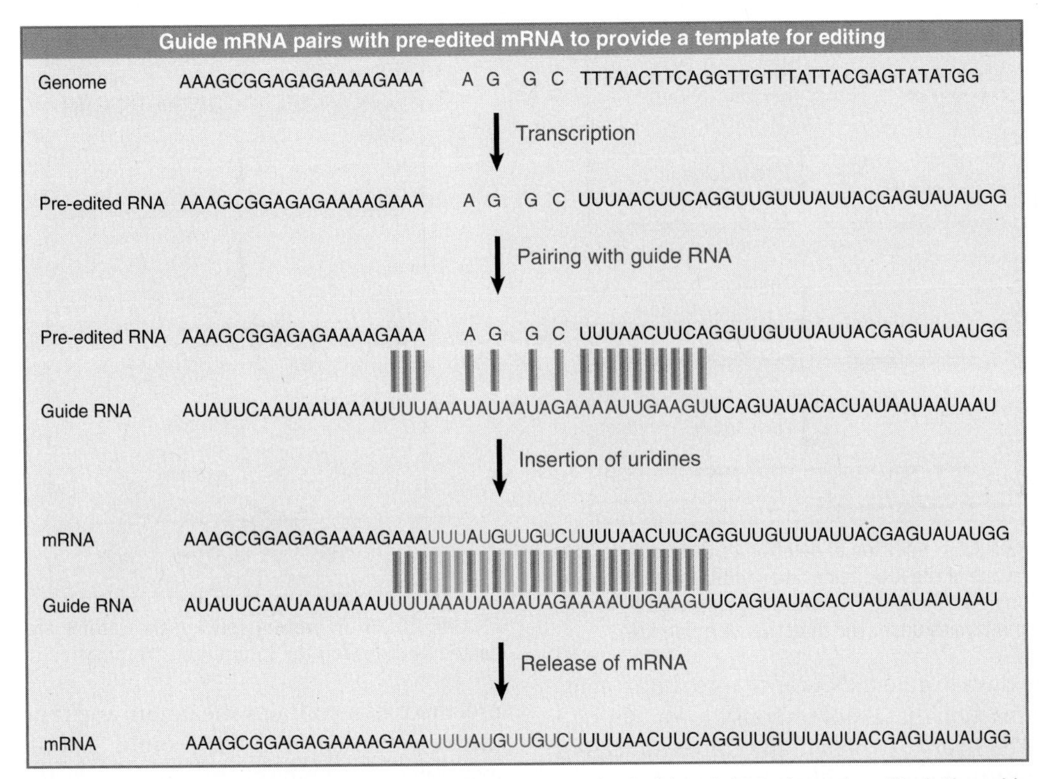

FIGURE 27.22 Preedited RNA base pairs with a guide RNA on both sides of the region to be edited. The guide RNA provides a template for the insertion of uridines. The mRNA produced by the insertions is complementary to the guide RNA.

mitochondrial DNA. It includes the "gene" for cytochrome *b,* which codes for the preedited sequence, and two regions that specify guide RNAs. Genes for the major coding regions and for their guide RNAs are interspersed.

In principle, a mutation in either the "gene" or one of its guide RNAs could change the primary sequence of the mRNA, and thus the primary sequence of the protein. By genetic criteria, each of these units could be considered to comprise part of the "gene." The units are independently expressed, and as a result they should of course complement in *trans.* If mutations were available, we should therefore find that three complementation groups were needed to code for the primary sequence of a single protein.

The characterization of intermediates that are partially edited suggests that the reaction proceeds along the preedited RNA in the 3′–5′ direction. The guide RNA determines the specificity of uridine insertions by its pairing with the preedited RNA.

Editing of uridines is catalyzed by a 20S enzyme complex that contains an endonuclease, a terminal uridyltransferase (TUTase), and an RNA ligase, as illustrated in FIGURE 27.24. It binds the guide RNA and uses it to pair with the preedited mRNA. The substrate RNA is cleaved at a site that is (presumably) identified by the absence of pairing with the guide RNA, a uridine is inserted or deleted to base pair with the guide RNA, and then the substrate RNA is ligated. Uracil triphosphate (UTP) provides the source for the uridyl residue. It is added by the TUTase activity; it is not clear whether this activity, or a separate exonuclease, is responsible for deletion. (At one time it was thought that a stretch of U residues at the end of guide RNA might provide the source for added U residues or a sink for deleted residues, but transfer of

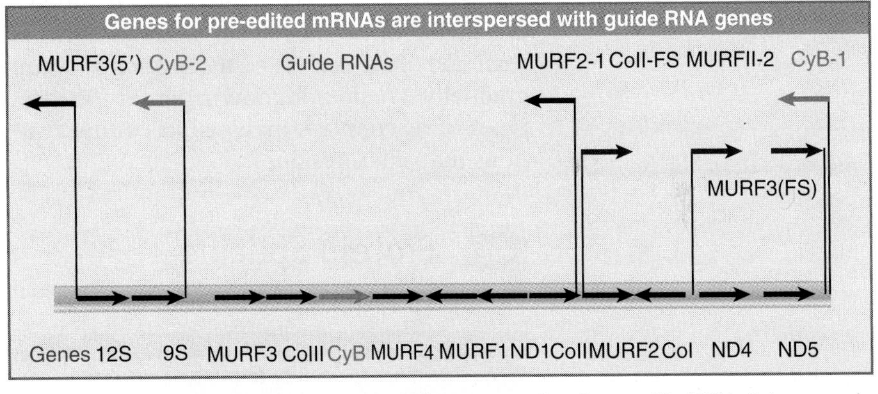

FIGURE 27.23 The *Leishmania* genome contains genes coding for preedited RNAs interspersed with units that code for the guide RNAs required to generate the correct mRNA sequences. Some genes have multiple guide RNAs. *CyB* is the gene for preedited cytochrome b, and *CyB-1* and *CyB-2* are genes for the guide RNAs involved in its editing.

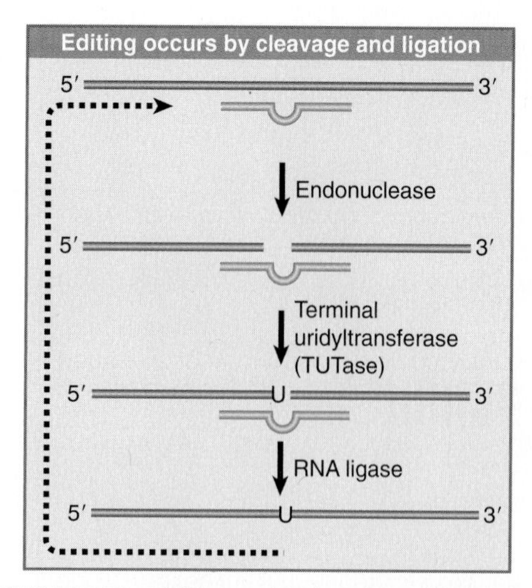

FIGURE 27.24 Addition or deletion of U residues occurs by cleavage of the RNA, removal or addition of the U, and ligation of the ends. The reactions are catalyzed by a complex of enzymes under the direction of guide RNA.

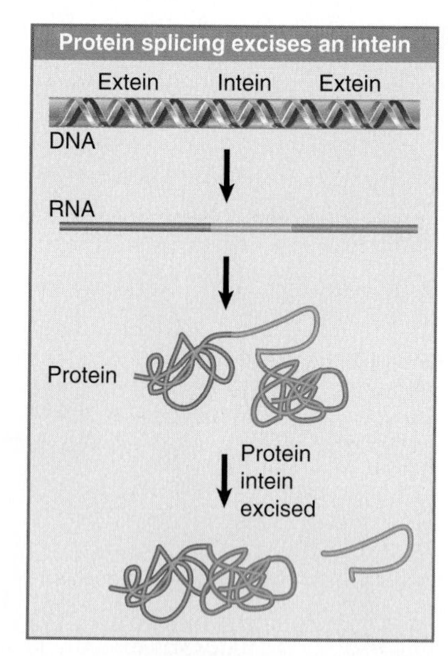

FIGURE 27.25 In protein splicing the exteins are connected by removing the intein from the protein.

U residues to guide RNAs appears to be an aberrant reaction that is not responsible for editing.)

The structures of partially edited molecules suggest that the U residues are added one at a time rather than in groups. It is possible that the reaction proceeds through successive cycles in which U residues are added, tested for complementarity with the guide RNA, retained if acceptable, and removed if not, so that the construction of the correct edited sequence occurs gradually. We do not know whether the same types of reaction are involved in editing reactions that add C residues.

27.12 Protein Splicing Is Autocatalytic

Key concepts

- An intein has the ability to catalyze its own removal from a protein in such a way that the flanking exteins are connected.
- Protein splicing is catalyzed by the intein.
- Most inteins have two independent activities: protein splicing and a homing endonuclease.

Protein splicing has the same effect as RNA splicing: A sequence that is represented within the gene fails to be represented in the protein. The parts of the protein are named by analogy with RNA splicing: **exteins** are the sequences that are represented in the mature protein, and **inteins** are the sequences that are removed. The mechanism of removing the intein is completely different from RNA splicing. FIGURE 27.25 shows that the gene is translated into a protein precursor that contains the intein, and then the intein is excised from the protein. About 100 examples of protein splicing are known and are spread throughout all classes of organisms. The typical gene whose product undergoes protein splicing has a single intein.

The first intein was discovered in an archaeal DNA polymerase gene in the form of an intervening sequence in the gene that does not conform to the rules for introns. It was then demonstrated that the purified protein can splice this sequence out of itself in an autocatalytic reaction. The reaction does not require input of energy and occurs through the series of bond rearrangements shown in FIGURE 27.26. The reaction is a function of the intein, although its efficiency can be influenced by the exteins.

The first reaction is an attack by an –OH or –SH side chain of the first amino acid in the intein on the peptide bond that connects it to the first extein. This transfers the extein from the amino-terminal group of the intein to an N-O or N-S acyl connection. This bond is then attacked by the –OH or –SH side chain of the first amino acid in the second extein. The result is to transfer extein1 to the side chain of the amino-terminal acid of extein2. Finally, the C-terminal asparagine of the intein cyclizes, and the terminal NH of extein2 attacks the acyl bond to replace it with a conventional peptide bond. Each of these reactions can occur spontaneously at very low rates, but their occurrence in a coordinate manner rapidly enough to achieve protein splicing requires catalysis by the intein.

Inteins have characteristic features. They are found as in-frame insertions into coding sequences. They can be recognized as such because of the existence of homologous genes that lack the insertion. They have an N-terminal serine or cysteine (to provide the -XH side chain) and a C-terminal asparagine. A typical intein has a sequence of ~150 amino acids at the N-terminal end and ~50 amino acids at the C-terminal end that are involved in catalyzing the protein splicing reaction. The sequence in the center of the intein can have other functions.

An extraordinary feature of many inteins is that they have homing endonuclease activity. A homing endonuclease cleaves a target DNA to create a site into which the DNA sequence coding for the intein can be inserted (see Figure 27.12 in Section 27.5, Some Group I Introns Code for Endonucleases That Sponsor Mobility). The protein splicing and homing endonuclease activities of an intein are independent.

We do not really understand the connection between the presence of both these activities in an intein, but two types of model have been suggested. One is to suppose that there was originally some sort of connection between the activities, but that they have since become independent and some inteins have lost the homing endonuclease. The other is to suppose that inteins may have originated as protein splicing units, most of which (for unknown reasons) were subsequently invaded by homing endonucleases. This is consistent with the fact that homing endonucleases appear to have invaded other types of units as well, including, most notably, group I introns.

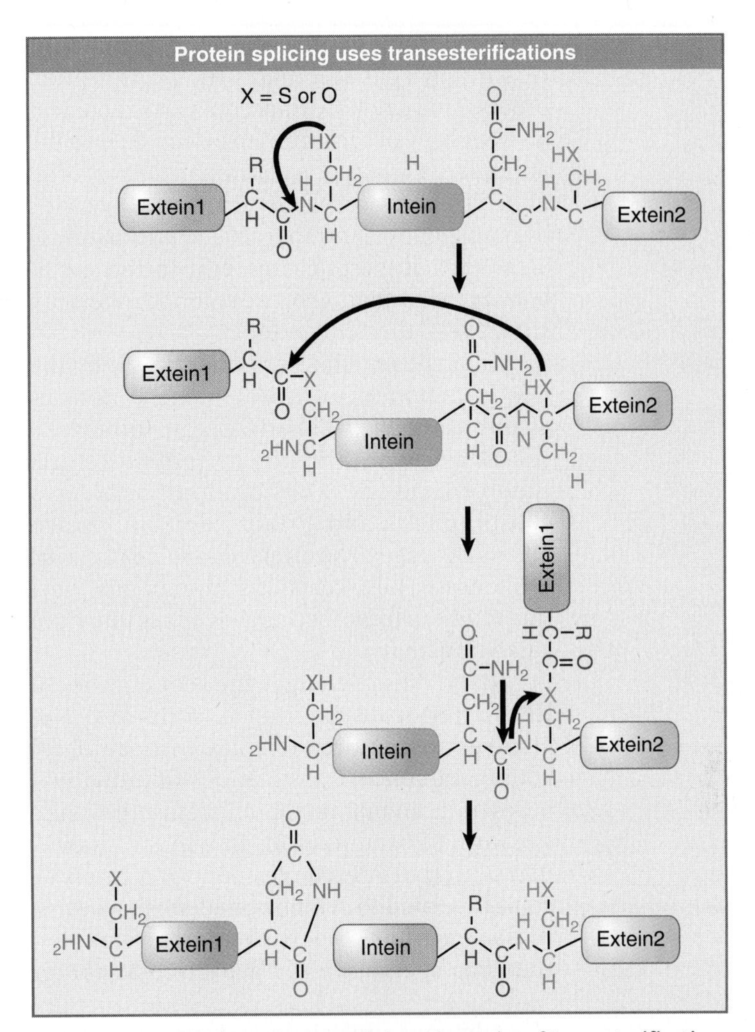

FIGURE 27.26 Bonds are rearranged through a series of transesterifications involving the -OH groups of serine or threonine or the -SH group of cysteine until the exteins are connected by a peptide bond and the intein is released with a circularized C-terminus.

27.13 Summary

Self-splicing is a property of two groups of introns, which are widely dispersed in lower eukaryotes, prokaryotic systems, and mitochondria. The information necessary for the reaction resides in the intron sequence (although the reaction is actually assisted by proteins *in vivo*). For both group I and group II introns, the reaction requires formation of a specific secondary/tertiary structure involving short consensus sequences. Group I intron RNA creates a structure in which the substrate sequence is held by the IGS region of the intron, and other conserved sequences generate a guanine nucleotide binding site. It occurs by a transesterification involving a guanosine residue as cofactor. No input of energy is required. The guanosine breaks the bond at the 5′ exon–intron junction and becomes linked to the intron; the hydroxyl at the free end of the exon then attacks the 3′ exon–intron junction. The intron cyclizes and loses the guanosine and the terminal 15 bases. A series of related reactions can be catalyzed via attacks by the terminal G-OH residue of the intron on internal phosphodiester bonds. By providing appropriate substrates, it has been possible to engineer ribozymes that perform a variety of catalytic reactions, including nucleotidyl transferase activities.

Some group I and group II mitochondrial introns have open reading frames. The proteins coded by group I introns are endonucleases that make double-stranded cleavages in target sites in DNA; the cleavage initiates a gene conversion process in which the sequence of the intron itself is copied into the target site. The proteins coded by group II introns include an endonuclease activity that initiates the transposition process, and a reverse transcriptase that enables an RNA copy of the intron to be copied into the target site. These types of introns probably originated by insertion events. The proteins coded by both groups of introns may include maturase activities that assist splicing of the intron by stabilizing the formation of the secondary/tertiary structure of the active site.

Catalytic reactions are undertaken by the RNA component of the RNAase P ribonucleoprotein. Virusoid RNAs can undertake self-cleavage at a "hammerhead" structure. Hammerhead structures can form between a substrate RNA and a ribozyme RNA, which allows cleavage to be directed at highly specific sequences. These reactions support the view that RNA can form specific active sites that have catalytic activity.

RNA editing changes the sequence of an RNA after or during its transcription. The changes are required to create a meaningful coding sequence. Substitutions of individual bases occur in mammalian systems; they take the form of deaminations in which C is converted to U or A is converted to I. A catalytic subunit related to cytidine or adenosine deaminase functions as part of a larger complex that has specificity for a particular target sequence.

Additions and deletions (most often of uridine) occur in trypanosome mitochondria and in paramyxoviruses. Extensive editing reactions occur in trypanosomes in which as many as half of the bases in an mRNA are derived from editing. The editing reaction uses a template consisting of a guide RNA that is complementary to the mRNA sequence. The reaction is catalyzed by an enzyme complex that includes an endonuclease, terminal uridyltransferase, and RNA ligase, using free nucleotides as the source for additions, or releasing cleaved nucleotides following deletion.

Protein splicing is an autocatalytic reaction that occurs by bond transfer reactions and input of energy is not required. The intein catalyzes its own splicing out of the flanking exteins. Many inteins have a homing endonuclease activity that is independent of the protein splicing activity.

References

27.2 Group I Introns Undertake Self-Splicing by Transesterification

Reviews

Cech, T. R. (1985). Self-splicing RNA: implications for evolution. *Int. Rev. Cytol.* 93, 3–22.

Cech, T. R. (1987). The chemistry of self-splicing RNA and RNA enzymes. *Science* 236, 1532–1539.

Research

Been, M. D. and Cech, T. R. (1986). One binding site determines sequence specificity of *Tetrahymena* pre-rRNA self-splicing, *trans*-splicing, and RNA enzyme activity. *Cell* 47, 207–216.

Belfort, M., Pedersen-Lane, J., West, D., Ehrenman, K., Maley, G., Chu, F., and Maley, F. (1985). Processing of the intron-containing thymidylate synthase (td) gene of phage T4 is at the RNA level. *Cell* 41, 375–382.

Cech, T. R. et al. (1981). *In vitro* splicing of the rRNA precursor of *Tetrahymena*: involvement of a guanosine nucleotide in the excision of the intervening sequence. *Cell* 27, 487–496.

Kruger, K., Grabowski, P. J., Zaug, A. J., Sands, J., Gottschling, D. E., and Cech, T. R. (1982). Self-splicing RNA: autoexcision and autocyclization of the ribosomal RNA intervening sequence of *Tetrahymena*. *Cell* 31, 147–157.

Myers, C. A., Kuhla, B., Cusack, S., and Lambowitz, A. M. (2002). tRNA-like recognition of group I introns by a tyrosyl-tRNA synthetase. *Proc. Natl. Acad. Sci. USA* 99, 2630–2635.

27.3 Group I Introns Form a Characteristic Secondary Structure

Research

Burke, J. M. et al. (1986). Role of conserved sequence elements 9L and 2 in self-splicing of the *Tetrahymena* ribosomal RNA precursor. *Cell* 45, 167–176.

Michel, F. and Wetshof, E. (1990). Modeling of the three-dimensional architecture of group I catalytic introns based on comparative sequence analysis. *J. Mol. Biol.* 216, 585–610.

27.4 Ribozymes Have Various Catalytic Activities

Review

Cech, T. R. (1990). Self-splicing of group I introns. *Annu. Rev. Biochem.* 59, 543–568.

Research

Winkler, W. C., Nahvi, A., Roth, A., Collins, J. A., and Breaker, R. R. (2004). Control of gene expression by a natural metabolite-responsive ribozyme. *Nature* 428, 281–286.

27.5 Some Group I Introns Code for Endonucleases That Sponsor Mobility

Review

Belfort, M. and Roberts, R. J. (1997). Homing endonucleases: keeping the house in order. *Nucleic Acids Res.* 25, 3379–3388.

27.6 Group II Introns May Code for Multifunction Proteins

Reviews

Lambowitz, A. M. and Belfort, M.(1993). Introns as mobile genetic elements. *Annu. Rev. Biochem.* 62, 587–622.

Lambowitz, A. M. and Zimmerly, S. (2004). Mobile group II introns. *Annu. Rev. Genet.* 38, 1–35.

Research

Dickson, L., Huang, H. R., Liu, L., Matsuura, M., Lambowitz, A. M., and Perlman, P. S. (2001). Retrotransposition of a yeast group II intron occurs by reverse splicing directly into ectopic DNA sites. *Proc. Natl. Acad. Sci. USA* 98, 13207–13212.

Zimmerly, S. et al. (1995). Group II intron mobility occurs by target DNA-primed reverse transcription. *Cell* 82, 545–554.

Zimmerly, S. et al. (1995). A group II intron is a catalytic component of a DNA endonuclease involved in intron mobility. *Cell* 83, 529–538.

27.7 Some Autosplicing Introns Require Maturases

Research

Bolduc et al. (2003). Structural and biochemical analyses of DNA and RNA binding by a bifunctional homing endonuclease and group I splicing factor. *Genes. Dev.* 17, 2875–2888.

Carignani, G. et al. (1983). An RNA maturase is encoded by the first intron of the mitochondrial gene for the subunit I of cytochrome oxidase in *S. cerevisiae*. *Cell* 35, 733–742.

Henke, R. M., Butow, R. A., and Perlman, P. S. (1995). Maturase and endonuclease functions depend on separate conserved domains of the bifunctional protein encoded by the group I intron aI4 alpha of yeast mitochondrial DNA. *EMBO J.* 14, 5094–5099.

Matsuura, M., Noah, J. W., and Lambowitz, A. M. (2001). Mechanism of maturase-promoted group II intron splicing. *EMBO J.* 20, 7259–7270.

27.9 Viroids Have Catalytic Activity

Reviews

Doherty, E. A. and Doudna, J. A. (2000). Ribozyme structures and mechanisms. *Annu. Rev. Biochem.* 69, 597–615.

Symons, R. H. (1992). Small catalytic RNAs. *Annu. Rev. Biochem.* 61, 641–671.

Research

Forster, A. C. and Symons, R. H. (1987). Self-cleavage of virusoid RNA is performed by the proposed 55-nucleotide active site. *Cell* 50, 9–16.

Guerrier-Takada, C., Gardiner, K., Marsh, T., Pace, N., and Altman, S. (1983). The RNA moiety of ribonuclease P is the catalytic subunit of the enzyme. *Cell* 35, 849–857.

Scott, W. G., Finch, J. T., and Klug, A. (1995). The crystal structure of an all-RNA hammerhead ribozyme: a proposed mechanism for RNA catalytic cleavage. *Cell* 81, 991–1002.

27.10 RNA Editing Occurs at Individual Bases

Research

Higuchi, M. et al. (1993). RNA editing of AMPA receptor subunit GluR-B: a base-paired intron-exon structure determines position and efficiency. *Cell* 75, 1361–1370.

Navaratnam, N. et al. (1995). Evolutionary origins of apoB mRNA editing: catalysis by a cytidine deaminase that has acquired a novel RNA-binding motif at its active site. *Cell* 81, 187–195.

Powell, L. M., Wallis, S. C., Pease, R. J., Edwards, Y. H., Knott, T. J., and Scott, J. (1987). A novel form of tissue-specific RNA processing produces apolipoprotein-B48 in intestine. *Cell* 50, 831–840.

Sommer, B. et al. (1991). RNA editing in brain controls a determinant of ion flow in glutamate-gated channels. *Cell* 67, 11–19.

27.11 RNA Editing Can Be Directed by Guide RNAs

Research

Aphasizhev, R., Sbicego, S., Peris, M., Jang, S. H., Aphasizheva, I., Simpson, A. M., Rivlin, A., and Simpson, L. (2002). Trypanosome mitochondrial 3′ terminal uridylyl transferase (TUTase): the key enzyme in U-insertion/deletion RNA editing. *Cell* 108, 637–648.

Benne, R., Van den Burg J., Brakenhoff, J. P., Sloof, P., Van Boom, J. H., and Tromp, M. C. (1986). Major transcript of the frameshifted *coxII* gene from trypanosome mitochondria contains four nucleotides that are not encoded in the DNA. *Cell* 46, 819–826.

Blum, B., Bakalara, N., and Simpson, L. (1990). A model for RNA editing in kinetoplastid mitochondria: "guide" RNA molecules transcribed from maxicircle DNA provide the edited information. *Cell* 60, 189–198.

Feagin, J. E., Abraham, J. M., and Stuart, K. (1988). Extensive editing of the cytochrome c oxidase III transcript in *Trypanosoma brucei*. *Cell* 53, 413–422.

Seiwert, S. D., Heidmann, S. and Stuart, K. (1996). Direct visualization of uridylate deletion *in vitro* suggests a mechanism for kinetoplastid editing. *Cell* 84, 831–841.

27.12 Protein Splicing Is Autocatalytic

Review

Paulus, H. (2000). Protein splicing and related forms of protein autoprocessing. *Annu. Rev. Biochem.* 69, 447–496.

Research

Derbyshire, V., Wood, D. W., Wu, W., Dansereau, J. T., Dalgaard, J. Z., and Belfort, M. (1997). Genetic definition of a protein-splicing domain: functional mini-inteins support structure predictions and a model for intein evolution. *Proc. Natl. Acad. Sci. USA* 94, 11466–11471.

Perler, F. B. et al. (1992). Intervening sequences in an Archaea DNA polymerase gene. *Proc. Natl. Acad. Sci. USA* 89, 5577–5581.

Xu, M. Q., Southworth, M. W., Mersha, F. B., Hornstra, L. J., and Perler, F. B. (1993). *in vitro* protein splicing of purified precursor and the identification of a branched intermediate. *Cell* 75, 1371–1377.

28

Chromosomes

CHAPTER OUTLINE

Continued on next page

28.1 Introduction

A general principle is evident in the organization of all cellular genetic material. It exists as a compact mass that is confined to a limited volume, and its various activities, such as replication and transcription, must be accomplished within this space. The organization of this material must accommodate transitions between inactive and active states.

The condensed state of nucleic acid results from its binding to basic proteins. The positive charges of these proteins neutralize the negative charges of the nucleic acid. The structure of the nucleoprotein complex is determined by the interactions of the proteins with the DNA (or RNA).

A common problem is presented by the packaging of DNA into phages, viruses, bacterial cells, and eukaryotic nuclei. The length of the DNA as an extended molecule would vastly exceed the dimensions of the compartment that contains it. The DNA (or in the case of some viruses, the RNA) must be compressed exceedingly tightly to fit into the space available. *Thus in contrast with the customary picture of DNA as an extended double helix, structural deformation of DNA to bend or fold it into a more compact form is the rule rather than exception.*

The magnitude of the discrepancy between the length of the nucleic acid and the size of its compartment is evident from the examples summarized in FIGURE 28.1. For bacteriophages and for eukaryotic viruses, the nucleic acid genome, whether single-stranded or double-stranded DNA or RNA, effectively fills the container (which can be rodlike or spherical).

For bacteria or for eukaryotic cell compartments, the discrepancy is hard to calculate exactly, because the DNA is contained in a compact area that occupies only part of the compartment. The genetic material is seen in the form of the **nucleoid** in bacteria and as the mass of **chromatin** in eukaryotic nuclei at interphase (between divisions).

The density of DNA in these compartments is high. In a bacterium it is ~10 mg/ml, in a eukaryotic nucleus it is ~100 mg/ml, and in the phage T4 head it is >500mg/ml. Such a concentration in solution would be equivalent to a gel

DNA is highly compacted in all types of genomes

Compartment	Shape	Dimensions	Type of Nucleic Acid	Length
TMV	filament	0.008 x 0.3 μm	One single-stranded RNA	2 μm = 6.4 kb
Phage fd	filament	0.006 x 0.85 μm	One single-stranded DNA	2 μm = 6.0 kb
Adenovirus	icosahedron	0.07 μm diameter	One double-stranded DNA	11 μm = 35.0 kb
Phage T4	icosahedron	0.065 x 0.10 μm	One double-stranded DNA	55 μm = 170.0 kb
E. coli	cylinder	1.7 x 0.65 μm	One double-stranded DNA	1.3 mm = 4.2 x 10^3 kb
Mitochondrion (human)	oblate spheroid	3.0 x 0.5 μm	~10 identical double-stranded DNAs	50 μm = 16.0 kb
Nucleus (human)	spheroid	6 μm diameter	46 chromosomes of double-stranded DNA	1.8 m = 6 x 10^6 kb

FIGURE 28.1 The length of nucleic acid is much greater than the dimensions of the surrounding compartment.

of great viscosity. We do not entirely understand the physiological implications, such as the effect this has upon the ability of proteins to find their binding sites on DNA.

The packaging of chromatin is flexible; it changes during the eukaryotic cell cycle. At the time of division (mitosis or meiosis), the genetic material becomes even more tightly packaged, and individual **chromosomes** become recognizable.

The overall compression of the DNA can be described by the **packing ratio,** which is the length of the DNA divided by the length of the unit that contains it. For example, the smallest human chromosome contains ~4.6×10^7 bp of DNA (~10 times the genome size of the bacterium *E. coli*). This is equivalent to 14,000 μm (= 1.4 cm) of extended DNA. At the most condensed moment of mitosis, the chromosome is ~2 μm long. Thus the packing ratio of DNA in the chromosome can be as great as 7000.

Packing ratios cannot be established with such certainty for the more amorphous overall structures of the bacterial nucleoid or eukaryotic chromatin. The usual reckoning, however, is that mitotic chromosomes are likely to be five to ten times more tightly packaged than interphase chromatin, which indicates a typical packing ratio of 1000 to 2000.

A major unanswered question concerns the *specificity* of packaging. Is the DNA folded into a *particular* pattern, or is it different in each individual copy of the genome? How does the pattern of packaging change when a segment of DNA is replicated or transcribed?

28.2 Viral Genomes Are Packaged into Their Coats

Key concepts

- The length of DNA that can be incorporated into a virus is limited by the structure of the headshell.
- Nucleic acid within the headshell is extremely condensed.
- Filamentous RNA viruses condense the RNA genome as they assemble the headshell around it.
- Spherical DNA viruses insert the DNA into a preassembled protein shell.

From the perspective of packaging the *individual* sequence, there is an important difference between a cellular genome and a virus. The cellular genome is essentially indefinite in size; the number and location of individual sequences can be changed by duplication, deletion, and rearrangement. Thus it requires a *generalized* method for packaging its DNA, one that is insensitive to the total content or distribution of sequences. By contrast, two restrictions define the needs of a virus. The amount of nucleic acid to be packaged is *predetermined* by the size of the genome, and it must all fit within a coat assembled from a protein or proteins coded by the viral genes.

A virus particle is deceptively simple in its superficial appearance. The nucleic acid genome is contained within a **capsid,** which is a symmetrical or quasisymmetrical structure assembled from one or only a few proteins. Attached to the capsid (or incorporated into it) are other structures; these structures are assembled from distinct proteins and are necessary for infection of the host cell.

The virus particle is tightly constructed. The internal volume of the capsid is rarely much greater than the volume of the nucleic acid it must hold. The difference is usually less than twofold, and often the internal volume is barely larger than the nucleic acid.

In its most extreme form, the restriction that the capsid must be assembled from proteins coded by the virus means that the entire shell is constructed from a single type of subunit. The rules for assembly of identical subunits into closed structures restrict the capsid to one of two types. For the first type, the protein subunits stack sequentially in a helical array to form a *filamentous* or rodlike shape. For the second type, they form a pseudospherical shell—a type of structure that conforms to a polyhedron with **icosahedral symmetry.** Some viral capsids are assembled from more than a single type of protein subunit, but although this extends the exact types of structures that can be formed, viral capsids still all conform to the general classes of quasicrystalline filaments or icosahedrons.

There are two types of solution to the problem of how to construct a capsid that contains nucleic acid:

- The protein shell can be assembled around the nucleic acid, thereby condensing the DNA or RNA by protein–nucleic acid interactions during the process of assembly.
- The capsid can be constructed from its component(s) in the form of an empty shell, into which the nucleic acid must be inserted, being condensed as it enters.

The capsid is assembled around the genome for single-stranded RNA viruses. The principle

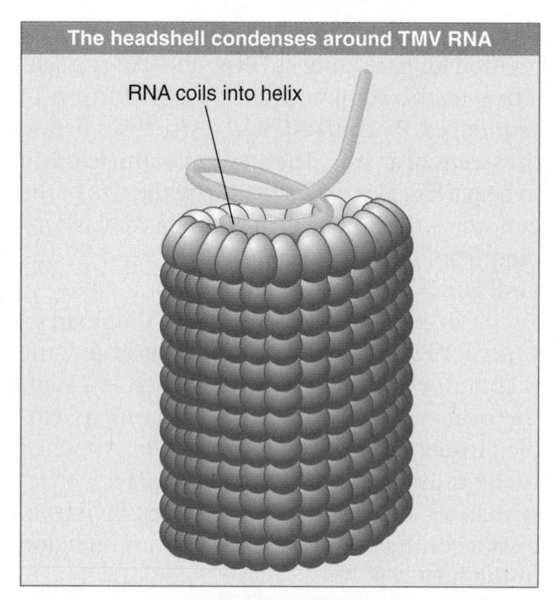

FIGURE 28.2 A helical path for TMV RNA is created by the stacking of protein subunits in the virion.

of assembly is that *the position of the RNA within the capsid is determined directly by its binding to the proteins of the shell.* The best characterized example is TMV (tobacco mosaic virus). Assembly starts at a duplex hairpin that lies within the RNA sequence. From this **nucleation center,** it proceeds bidirectionally along the RNA until it reaches the ends. The unit of the capsid is a two-layer disk, with each layer containing 17 identical protein subunits. The disk is a circular structure, which forms a helix as it interacts with the RNA. At the nucleation center, the RNA hairpin inserts into the central hole in the disk, and the disk changes conformation into a helical structure that surrounds the RNA. Additional disks are added, with each new disk pulling a new stretch of RNA into its central hole. The RNA becomes coiled in a helical array on the inside of the protein shell, as illustrated in FIGURE 28.2.

The spherical capsids of DNA viruses are assembled in a different way, as best characterized for the phages lambda and T4. In each case, an empty headshell is assembled from a small set of proteins. *The duplex genome then is inserted into the head,* accompanied by a structural change in the capsid.

FIGURE 28.3 summarizes the assembly of lambda. It starts with a small headshell that contains a protein "core." This is converted to an empty headshell of more distinct shape. At this point the DNA packaging begins, the headshell expands in size though remaining the same shape, and finally the full head is sealed by the addition of the tail.

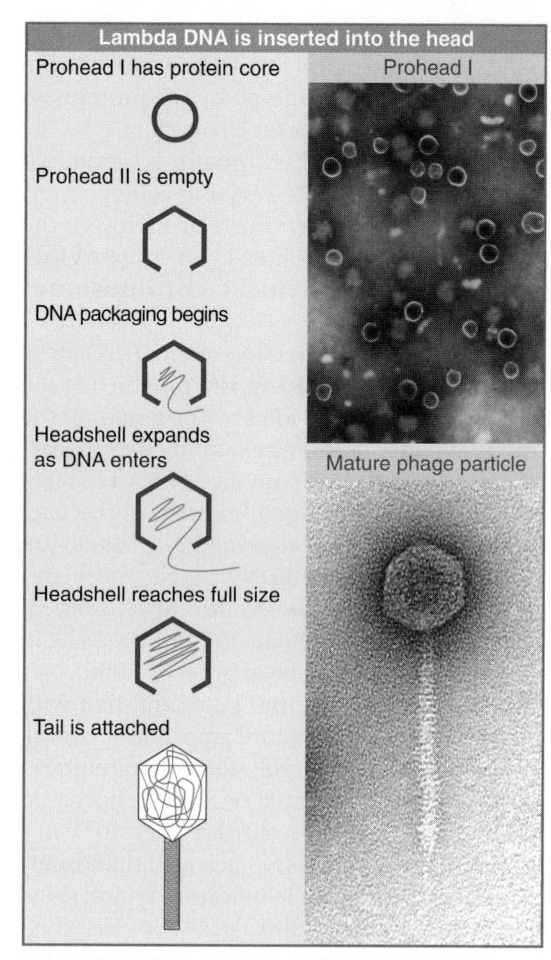

FIGURE 28.3 Maturation of phage lambda passes through several stages. The empty head changes shape and expands when it becomes filled with DNA. The electron micrographs show the particles at the start and the end of the maturation pathway. Top photo reproduced from Cue, D. and Feiss, M. 1993. *Proc. Natl. Acad. Sci. USA.* 90: 9290–9294. Copyright 1993 National Academy of Science, U.S.A. Photo courtesy of Michael G. Feiss, University of Iowa. Bottom photo courtesy of Robert Duda, University of Pittsburgh.

A double-stranded DNA that spans short distances is a fairly rigid rod, yet it must be compressed into a compact structure to fit within the capsid. We should like to know whether packaging involves a smooth coiling of the DNA into the head or whether it requires abrupt bends.

Inserting DNA into a phage head involves two types of reaction: translocation and condensation. Both are energetically unfavorable.

Translocation is an active process in which the DNA is driven into the head by an ATP-dependent mechanism. A common mechanism is used for many viruses that replicate by a rolling circle mechanism to generate long tails that contain multimers of the viral genome. The best characterized example is phage lambda. The genome is packaged into the empty capsid by the **terminase** enzyme. FIGURE 28.4 summarizes the process.

The terminase was first recognized for its role in generating the ends of the linear phage DNA by cleaving at *cos* sites. (The name *cos* reflects the fact that it generates cohesive ends that have complementary single-stranded tails.) The phage genome codes two subunits that make up the terminase. One subunit binds to a *cos* site; at this point it is joined by the other subunit, which cuts the DNA. The terminase assembles into a hetero-oligomer in a complex that also includes IHF (integration host factor, a dimer that is coded by the bacterial genome). It then binds to an empty capsid and uses ATP hydrolysis to power translocation along the DNA. The translocation drives the DNA into the empty capsid.

Another method of packaging uses a structural component of the phage. In the *Bacillus subtilis* phage φ29, the motor that inserts the DNA into the phage head is the structure that connects the head to the tail. It functions as a rotary motor, where the motor action effects the linear translocation of the DAN into the phage head. The same motor is used to eject the DNA from the phage head when it infects a bacterium.

Little is known about the mechanism of condensation into an empty capsid, except that the capsid contains "internal proteins" as well as DNA. One possibility is that they provide some sort of "scaffolding" onto which the DNA condenses. (This would be a counterpart to the use of the proteins of the shell in the plant RNA viruses.)

How specific is the packaging? It cannot depend on particular sequences, because deletions, insertions, and substitutions all fail to interfere with the assembly process. The relationship between DNA and the headshell has been investigated directly by determining which regions of the DNA can be chemically crosslinked to the proteins of the capsid. The surprising answer is that all regions of the DNA are more or less equally susceptible. This probably means that when DNA is inserted into the head it follows a general rule for condensing, but the pattern is not determined by particular sequences.

These varying mechanisms of virus assembly all accomplish the same end: packaging a single DNA or RNA molecule into the capsid. Some viruses, though, have genomes that consist of multiple nucleic acid molecules. Reovirus contains ten double-stranded RNA segments, all of which must be packaged into the capsid. Specific sorting sequences in the segments may be required to ensure that the assembly process selects one copy of each different molecule in order to collect a complete set of genetic information. In the simpler case of phage φ6, which packages three different segments of double-stranded RNA into one capsid, the RNA segments must bind in a specific order: As each is incorporated into the capsid, it triggers a change in the conformation of the capsid that creates binding sites for the next segment.

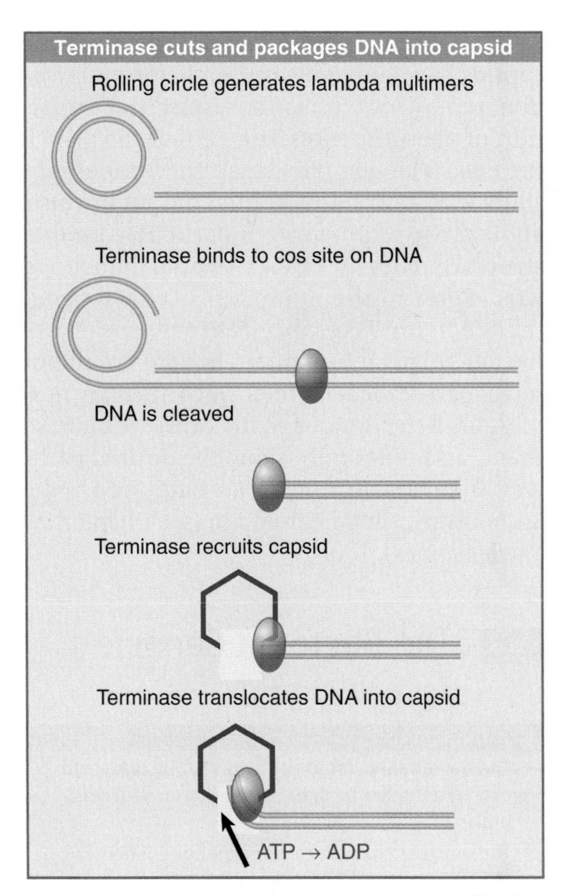

FIGURE 28.4 Terminase protein binds to specific sites on a multimer of virus genomes generated by rolling circle replication. It cuts the DNA and binds to an empty virus capsid, and then uses energy from hydrolysis of ATP to insert the DNA into the capsid.

Some plant viruses are multipartite: Their genomes consist of segments, each of which is packaged into a *different* capsid. An example is alfalfa mosaic virus (AMV), which has four different single-stranded RNAs, each of which is packaged independently into a coat comprising the same protein subunit. A successful infection depends on the entry of one of each type into the cell.

The four components of AMV exist as particles of different sizes. This means that the same capsid protein can package each RNA into its own characteristic particle. This is a departure from the packaging of a unique length of nucleic acid into a capsid of fixed shape.

The assembly pathway of viruses whose capsids have only one authentic form may be diverted by mutations that cause the formation of aberrant **monster** particles in which the head is longer than usual. These mutations show that a capsid protein(s) has an intrinsic ability to assemble into a particular type of structure, but the exact size and shape may vary. Some of the mutations occur in genes that code for **assembly factors,** which are needed for head formation, but are not themselves part of the headshell. Such ancillary proteins limit the options of the capsid protein so that it assembles only along the desired pathway. Comparable proteins are employed in the assembly of cellular chromatin (see Chapter 29, Nucleosomes).

28.3 The Bacterial Genome Is a Nucleoid

Key concepts

- The bacterial nucleoid is ~80% DNA by mass and can be unfolded by agents that act on RNA or protein.
- The proteins that are responsible for condensing the DNA have not been identified.

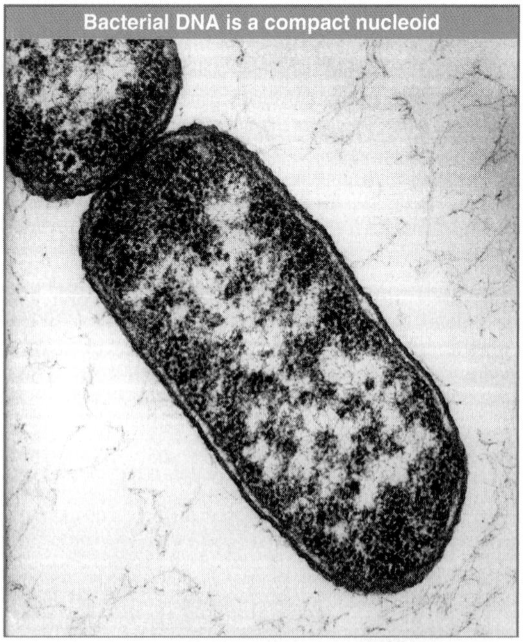

Although bacteria do not display structures with the distinct morphological features of eukaryotic chromosomes, their genomes nonetheless are organized into definite bodies. The genetic material can be seen as a fairly compact clump (or series of clumps) that occupies about a third of the volume of the cell. FIGURE 28.5 displays a thin section through a bacterium in which this nucleoid is evident.

When *E. coli* cells are lysed, fibers are released in the form of loops attached to the broken envelope of the cell. As can be seen from FIGURE 28.6, the DNA of these loops is not found in the extended form of a free duplex, but instead is compacted by association with proteins.

Several DNA-binding proteins with a superficial resemblance to eukaryotic chromosomal proteins have been isolated in *E. coli.* What criteria should we apply for deciding whether a DNA-binding protein plays a structural role in the nucleoid? It should be present in sufficient quantities to bind throughout the genome, and mutations in its gene should cause some disruption of structure or of functions associated with genome survival (for example, segregation to daughter cells). None of the candidate proteins yet satisfies the genetic conditions.

Protein HU is a dimer that condenses DNA, possibly wrapping it into a beadlike structure. It is related to IHF, which has a structural role in building a protein complex in specialized

Bacterial DNA is a compact nucleoid

FIGURE 28.5 A thin section shows the bacterial nucleoid as a compact mass in the center of the cell. Photo courtesy of the Molecular and Cell Biology Instructional Laboratory Program, University of California, Berkeley.

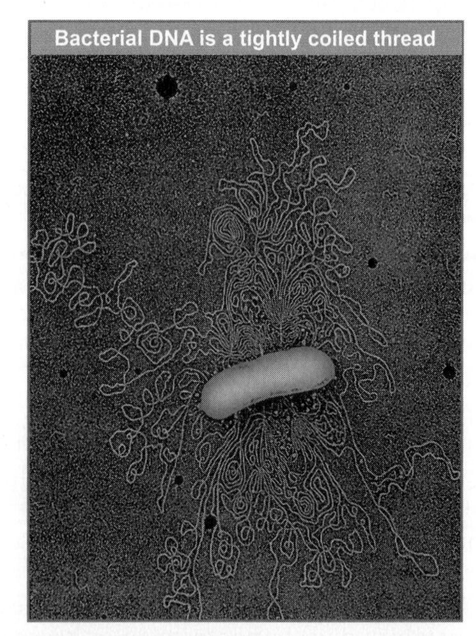

FIGURE 28.6 The nucleoid spills out of a lysed *E. coli* cell in the form of loops of a fiber. Photo © G. Murti/Photo Researchers, Inc.

recombination reactions. Null mutations in either of the genes coding for the subunits of HU (*hupA* and *-B*) have little effect, but loss of both functions causes a cold-sensitive phenotype and some loss of superhelicity in DNA. These results raise the possibility that HU plays some general role in nucleoid condensation.

Protein H1 (also known as H-NS) binds DNA, interacting preferentially with sequences that are bent. Mutations in its gene have turned up in a variety of guises *(osmZ, bglY, pilG)*, each of which is identified as an apparent regulator of a different system. These results probably reflect the effect that H1 has on the local topology of DNA, with effects upon gene expression that depend upon the particular promoter.

We might expect that the absence of a protein required for nucleoid structure would have serious effects upon viability. Why, then, are the effects of deletions in the genes for proteins HU and H1 relatively restricted? One explanation is that these proteins are *redundant*, and that any one can substitute for the others so that deletions of *all* of them would be necessary to interfere seriously with nucleoid structure. Another possibility is that we have yet to identify the proteins responsible for the major features of nucleoid integrity.

The nucleoid can be isolated directly in the form of a very rapidly sedimenting complex, which consists of ~80% DNA by mass. (The analogous complexes in eukaryotes have ~50% DNA by mass; see Section 28.4, The Bacterial Genome Is Supercoiled.) It can be unfolded by treatment with reagents that act on RNA or on protein. The possible role of proteins in stabilizing its structure is evident. The role of RNA has been quite refractory to analysis.

28.4 The Bacterial Genome Is Supercoiled

Key concepts

- The nucleoid has ~100 independent negatively supercoiled domains.
- The average density of supercoiling is ~1 turn/100 bp.

The DNA of the bacterial nucleoid isolated *in vitro* behaves as a closed duplex structure, as judged by its response to ethidium bromide. This small molecule intercalates between base pairs to generate *positive* superhelical turns in "closed" circular DNA molecules, that is, molecules in which both strands have covalent integrity. (In "open" circular molecules, which contain a nick in one strand, or with linear molecules, the DNA can rotate freely in response to the intercalation, thus relieving the tension.)

In a natural closed DNA that is *negatively* supercoiled, the intercalation of ethidium bromide first removes the negative supercoils and then introduces positive supercoils. The amount of ethidium bromide needed to achieve zero supercoiling is a measure of the original density of negative supercoils.

Some nicks occur in the compact nucleoid during its isolation; they can also be generated by limited treatment with DNAase. This does not, however, abolish the ability of ethidium bromide to introduce positive supercoils. This capacity of the genome to retain its response to ethidium bromide in the face of nicking means that it must have many independent chromosomal **domains,** and that *the supercoiling in each domain is not affected by events in the other domains.*

This autonomy suggests that the structure of the bacterial chromosome has the general organization depicted diagrammatically in FIGURE 28.7. Each domain consists of a loop of DNA, the ends of which are secured in some (unknown) way that does not allow rotational events to propagate from one domain to another.

Early data suggested that each domain consists of ~40 kb of DNA, but more recent analysis suggests that the domains may be smaller, at ~10 kb each. This would correspond to ~400

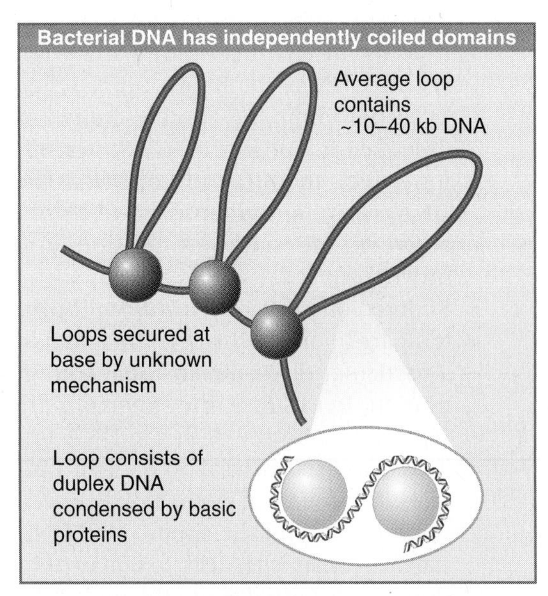

FIGURE 28.7 The bacterial genome consists of a large number of loops of duplex DNA (in the form of a fiber), each of which is secured at the base to form an independent structural domain.

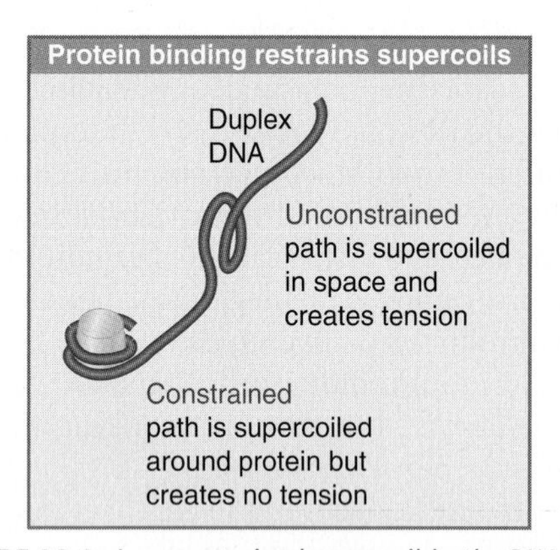

FIGURE 28.8 An unrestrained supercoil in the DNA path creates tension, but no tension is transmitted along DNA when a supercoil is restrained by protein binding.

domains in the *E. coli* genome. The ends of the domains appear to be randomly distributed instead of located at predetermined sites on the chromosome.

The existence of separate domains could permit different degrees of supercoiling to be maintained in different regions of the genome. This could be relevant in considering the different susceptibilities of particular bacterial promoters to supercoiling (see Section 11.15, Supercoiling Is an Important Feature of Transcription).

As shown in FIGURE 28.8, supercoiling in the genome can in principle take either of two forms:

- If a supercoiled DNA is free its path is *unconstrained,* and negative supercoils generate a state of torsional tension that is transmitted freely along the DNA within a domain. It can be relieved by unwinding the double helix, as described in Section 19.12, Supercoiling Affects the Structure of DNA. The DNA is in a dynamic equilibrium between the states of tension and unwinding.
- Supercoiling can be *constrained* if proteins are bound to the DNA to hold it in a particular three-dimensional configuration. In this case, the supercoils are represented by the path the DNA follows in its fixed association with the proteins. The energy of interaction between the proteins and the supercoiled DNA stabilizes the nucleic acid, so that no tension is transmitted along the molecule.

Are the supercoils in *E. coli* DNA constrained *in vivo* or is the double helix subject to the torsional tension characteristic of free DNA? Measurements of supercoiling *in vitro* encounter the difficulty that constraining proteins may have been lost during isolation. Various approaches suggest that DNA is under torsional stress *in vivo.*

One approach is to measure the effect of nicking the DNA. Unconstrained supercoils are released by nicking, whereas constrained supercoils are unaffected. Nicking releases ~50% of the overall supercoiling. This suggests that about half of the supercoiling is transmitted as tension along DNA, with the other half being absorbed by protein binding.

Another approach uses the crosslinking reagent psoralen, which binds more readily to DNA when it is under torsional tension. The reaction of psoralen with *E. coli* DNA *in vivo* corresponds to an average density of one negative superhelical turn/200 bp ($\sigma = -0.05$).

We can also examine the ability of cells to form alternative DNA structures; for example, to generate cruciforms at palindromic sequences. From the change in linking number that is required to drive such reactions, it is possible to calculate the original supercoiling density. This approach suggests an average density of $\sigma = -0.025$, or one negative superhelical turn/100 base pairs.

Thus supercoils *do* create torsional tension *in vivo*. There may be variation about an average level, and the precise range of densities is difficult to measure. It is, however, clear that the level is sufficient to exert significant effects on DNA structure—for example, in assisting melting in particular regions such as origins or promoters.

Many of the important features of the structure of the compact nucleoid remain to be established. What is the specificity with which domains are constructed? Do the same sequences always lie at the same relative locations, or can the contents of individual domains shift? How is the integrity of the domain maintained? Biochemical analysis by itself is unable to answer these questions fully, but if it is possible to devise suitable selective techniques, the properties of structural mutants should lead to a molecular analysis of nucleoid construction.

28.5 Eukaryotic DNA Has Loops and Domains Attached to a Scaffold

Key concepts

- DNA of interphase chromatin is negatively supercoiled into independent domains of ~85 kb.
- Metaphase chromosomes have a protein scaffold to which the loops of supercoiled DNA are attached.

Interphase chromatin is a tangled mass occupying a large part of the nuclear volume, in contrast with the highly organized and reproducible ultrastructure of mitotic chromosomes. What controls the distribution of interphase chromatin within the nucleus?

Some indirect evidence on its nature is provided by the isolation of the genome as a single, compact body. Using the same technique that was developed for isolating the bacterial nucleoid (see Section 28.4, The Bacterial Genome Is Supercoiled), nuclei can be lysed on top of a sucrose gradient. This releases the genome in a form that can be collected by centrifugation. As isolated from *Drosophila melanogaster,* it can be visualized as a compactly folded fiber (10 nm in diameter) consisting of DNA bound to proteins.

Supercoiling measured by the response to ethidium bromide corresponds to about one negative supercoil/200 bp. These supercoils can be removed by nicking with DNAase, although the DNA remains in the form of the 10 nm fiber. This suggests that the supercoiling is caused by the arrangement of the fiber in space, and that it represents the existing torsion.

Full relaxation of the supercoils requires one nick/85 kb, thus identifying the average length of "closed" DNA. This region could comprise a loop or domain similar in nature to those identified in the bacterial genome. Loops can be seen directly when the majority of proteins are extracted from mitotic chromosomes. The resulting complex consists of the DNA associated with ~8% of the original protein content. As seen in FIGURE 28.9, the protein-depleted chromosomes take the form of a central **scaffold** surrounded by a halo of DNA.

The metaphase scaffold consists of a dense network of fibers. Threads of DNA emanate from the scaffold, apparently as loops of average length 10 to 30 μm (30 to 90 kb). The DNA can be digested without affecting the integrity of the scaffold, which consists of a set of specific proteins. This suggests a form of organization in which loops of DNA of ~60 kb are anchored in a central proteinaceous scaffold.

The appearance of the scaffold resembles a mitotic pair of sister chromatids. The sister scaffolds usually are tightly connected (but sometimes are separate), and are joined only by a few fibers. Could this be the structure responsible for maintaining the shape of the mitotic chromosomes? Could it be generated by bringing together the protein components that usually secure the bases of loops in interphase chromatin?

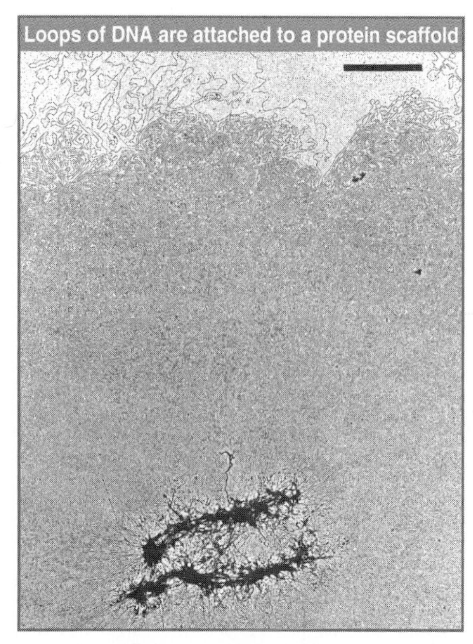

FIGURE 28.9 Histone-depleted chromosomes consist of a protein scaffold to which loops of DNA are anchored. Reproduced from *Cell,* vol. 12, Paulson, J. R., and Laemmli, U. K., *The structure of histone . . .* , pp. 817–828. Copyright 1977, with permission from Elsevier. Photo courtesy of Ulrich K. Laemmli, University of Geneva, Switzerland.

28.6 Specific Sequences Attach DNA to an Interphase Matrix

Key concepts

- DNA is attached to the nuclear matrix at specific sequences called MARs or SARs.
- The MARs are A-T-rich but do not have any specific consensus sequence.

Is DNA attached to the scaffold via specific sequences? DNA sites attached to proteinaceous structures in interphase nuclei are called **MARS (matrix attachment regions);** they are sometimes also called *SAR* (scaffold attachment regions). The nature of the structure in interphase cells to which they are connected is not clear. Chromatin often appears to be attached to a matrix, and there have been many suggestions that this attachment is necessary for transcription or replication. When nuclei are depleted of proteins, the DNA extrudes as loops from a residual proteinaceous structure. Attempts to relate the proteins found in this preparation to structural elements of intact cells have not been successful, though.

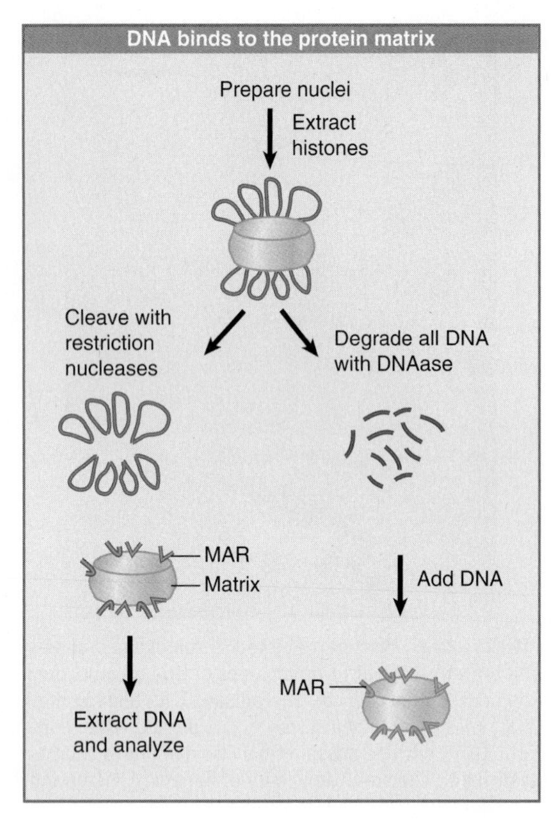

FIGURE 28.10 Matrix-associated regions may be identified by characterizing the DNA retained by the matrix isolated *in vivo* or by identifying the fragments that can bind to the matrix from which all DNA has been removed.

Are particular DNA regions associated with this matrix? *In vivo* and *in vitro* approaches are summarized in FIGURE 28.10. Both start by isolating the matrix as a crude nuclear preparation containing chromatin and nuclear proteins. Different treatments can then be used to characterize DNA in the matrix or to identify DNA able to attach to it.

To analyze the existing MAR, the chromosomal loops can be decondensed by extracting the proteins. Removal of the DNA loops by treatment with restriction nucleases leaves only the (presumptive) *in vivo* MAR sequences attached to the matrix.

The complementary approach is to remove *all* the DNA from the matrix by treatment with DNAase, at which point isolated fragments of DNA can be tested for their ability to bind to the matrix *in vitro*.

The same sequences should be associated with the matrix *in vivo* or *in vitro*. Once a potential MAR has been identified, the size of the minimal region needed for association *in vitro* can be determined by deletions. This enables us to identify proteins that bind to the MAR sequences.

A surprising feature is the lack of conservation of sequence in MAR fragments. They are usually ~70% A-T-rich, but otherwise lack any consensus sequences. Other interesting sequences, however, often are in the DNA stretch containing the MAR. *cis*-acting sites that regulate transcription are common, and a recognition site for topoisomerase II is usually present in the MAR. It is therefore possible that a MAR serves more than one function by providing a site for attachment to the matrix and containing other sites at which topological changes in DNA are effected.

What is the relationship between the chromosome scaffold of dividing cells and the matrix of interphase cells? Are the same DNA sequences attached to both structures? In several cases, the same DNA fragments that are found with the nuclear matrix *in vivo* can be retrieved from the metaphase scaffold. Fragments that contain MAR sequences can bind to a metaphase scaffold, so it therefore seems likely that DNA contains a single type of attachment site. In interphase cells the attachment site is connected to the nuclear matrix, whereas in mitotic cells is connected to the chromosome scaffold.

The nuclear matrix and chromosome scaffold consist of different proteins, although there are some common components. Topoisomerase II is a prominent component of the chromosome scaffold, and is a constituent of the nuclear matrix. This suggests that the control of topology is important in both cases.

28.7 Chromatin Is Divided into Euchromatin and Heterochromatin

Key concepts

- Individual chromosomes can be seen only during mitosis.
- During interphase, the general mass of chromatin is in the form of euchromatin, which is less tightly packed than mitotic chromosomes.
- Regions of heterochromatin remain densely packed throughout interphase.

Each chromosome contains a single, very long duplex of DNA. This explains why chromosome replication is semiconservative like the individual DNA molecule. (This would not necessarily be the case if a chromosome carried many

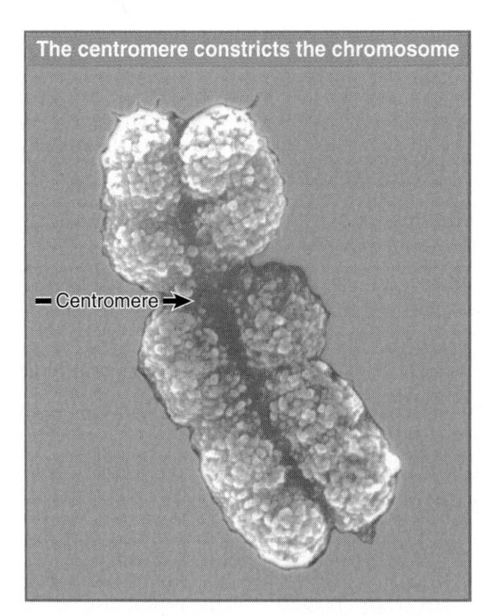

FIGURE 28.11 The sister chromatids of a mitotic pair each consist of a fiber (~30 nm in diameter) compactly folded into the chromosome. Photo courtesy of Daniel L. Hartl, Harvard University.

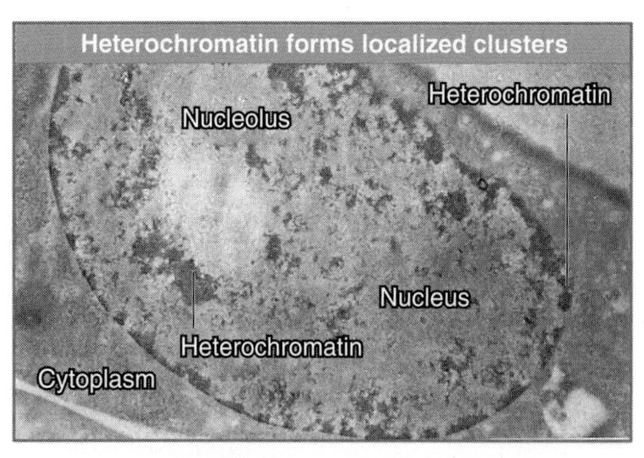

FIGURE 28.12 A thin section through a nucleus stained with Feulgen shows heterochromatin as compact regions clustered near the nucleolus and nuclear membrane. Photo courtesy of Edmund Puvion, Centre National de la Recherche Scientifique.

independent molecules of DNA.) The single duplex of DNA is folded into a fiber that runs continuously throughout the chromosome. *Thus in accounting for interphase chromatin and mitotic chromosome structure, we have to explain the packaging of a single, exceedingly long molecule of DNA into a form in which it can be transcribed and replicated, and can become cyclically more and less compressed.*

Individual eukaryotic chromosomes come into the limelight for a brief period, during the act of cell division. Only then can each be seen as a compact unit. FIGURE 28.11 is an electron micrograph of a sister chromatid pair captured at metaphase. (The sister chromatids are daughter chromosomes produced by the previous replication event, and still joined together at this stage of mitosis.) Each consists of a fiber with a diameter of ~30 nm and a nubbly appearance. The DNA is five to ten times more condensed in chromosomes than in interphase chromatin.

During most of the life cycle of the eukaryotic cell, however, its genetic material occupies an area of the nucleus in which individual chromosomes cannot be distinguished. The structure of the interphase chromatin does not change visibly between divisions. No disruption is evident during the period of replication, when the amount of chromatin doubles. Chromatin is fibrillar, although the overall configuration of the fiber in space is hard to discern in detail. The fiber itself, however, is similar or identical to that of the mitotic chromosomes.

As can be seen in the nuclear section of FIGURE 28.12, chromatin can be divided into two types of material:

- In most regions, the fibers are much less densely packed than in the mitotic chromosome. This material is called **euchromatin.** It has a relatively dispersed appearance in the nucleus and occupies most of the nuclear region in Figure 28.12.
- Some regions of chromatin are very densely packed with fibers, displaying a condition comparable to that of the chromosome at mitosis. This material is called **heterochromatin.** It is typically found at centromeres, but occurs at other locations as well. It passes through the cell cycle with relatively little change in its degree of condensation. It forms a series of discrete clumps in Figure 28.12, but often the various heterochromatic regions aggregate into a densely staining **chromocenter.** (This description applies to regions that are always heterochromatic, which are called constitutive heterochromatin. In addition, there is another sort of heterochromatin, called facultative heterochromatin, in which regions of euchromatin are converted to a heterochromatic state).

The same fibers run continuously between euchromatin and heterochromatin, which implies that these states represent different degrees of condensation of the genetic material. In the same way, euchromatic regions exist in different states of condensation during interphase and during mitosis. Thus the genetic material is organized in a manner that permits alternative states to be maintained side by side in chromatin, and allows cyclical changes to occur in the packaging of euchromatin between interphase and division. We discuss the molecular basis for these states in Chapter 30, Controlling Chromatin Structure.

The structural condition of the genetic material is correlated with its activity. The common features of constitutive heterochromatin are:

- It is permanently condensed.
- It often consists of multiple repeats of a few sequences of DNA that are not transcribed.
- The density of genes in this region is very much reduced compared with euchromatin, and genes that are translocated into or near it are often inactivated.
- It replicates late in S phase and has a reduced frequency of genetic recombination; this is probably a result of its condensed state.

We have some molecular markers for changes in the properties of the DNA and protein components (see Section 31.3, Heterochromatin Depends on Interactions with Histones). They include reduced acetylation of histone proteins, increased methylation of one histone protein, and hypermethylation of cytidine bases in DNA. These molecular changes cause the condensation of the material, which is responsible for its inactivity.

Although active genes are contained within euchromatin, only a small minority of the sequences in euchromatin are transcribed at any time. Thus location in euchromatin is *necessary* for gene expression, but is not *sufficient* for it.

28.8 Chromosomes Have Banding Patterns

Key concepts

- Certain staining techniques cause the chromosomes to have the appearance of a series of striations, which are called G-bands.
- The bands are lower in G-C content than the interbands.
- Genes are concentrated in the G-C-rich interbands.

As a result of the diffuse state of chromatin, we cannot directly determine the specificity of its organization. We can, however, ask whether the structure of the (mitotic) chromosome is ordered. Do particular sequences always lie at particular sites, or is the folding of the fiber into the overall structure a more random event?

At the level of the chromosome, each member of the complement has a different and reproducible ultrastructure. When subjected to certain treatments and then stained with the chemical dye Giemsa, chromosomes generate a series of **G-bands.** FIGURE 28.13 presents an example of the human set.

Until the development of this technique, chromosomes could be distinguished only by their overall size and the relative location of the centromere. G-banding allows each chromosome to be be identified by its characteristic banding pattern. This pattern allows translocations from one chromosome to another to be identified by comparison with the original diploid set. FIGURE 28.14 shows a diagram of the bands of the human X chromosome. The bands are large structures, each ~10^7 bp of DNA, which could include many hundreds of genes.

The banding technique is of enormous practical use, but the mechanism of banding remains a mystery. All that is certain is that the dye stains untreated chromosomes more or less uniformly. Thus the generation of bands depends on a variety of treatments that change the response of the chromosome (presumably by extracting the component that binds the stain from the nonbanded regions). Similar bands can be generated by an assortment of other treatments.

The only known feature that distinguishes bands from interbands is that the bands have a

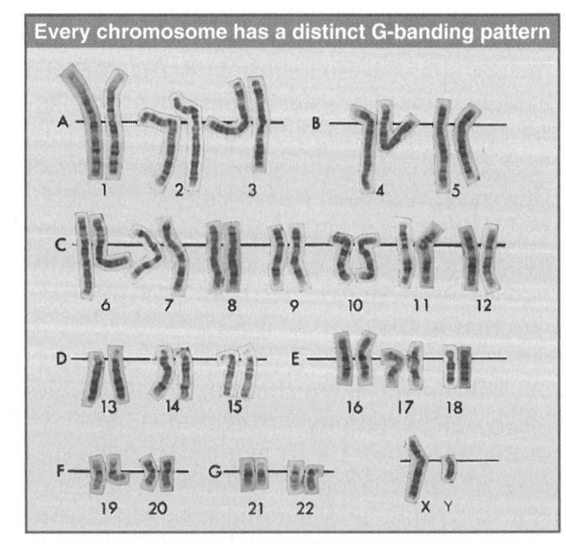

FIGURE 28.13 G-banding generates a characteristic lateral series of bands in each member of the chromosome set. Photo courtesy of Lisa Shaffer, Washington State University–Spokane.

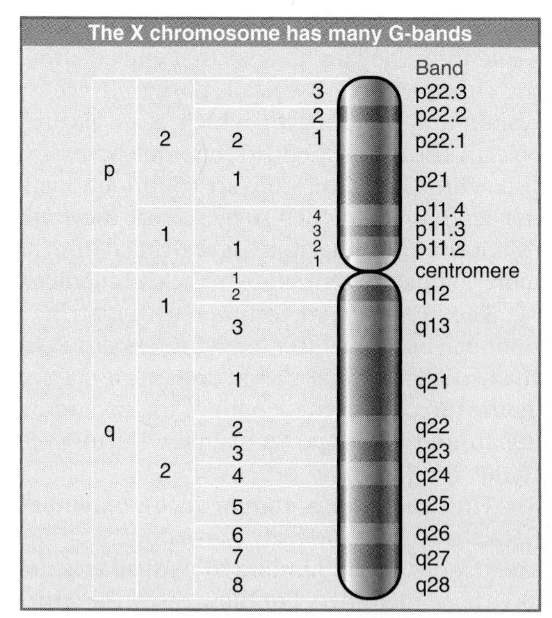

FIGURE 28.14 The human X chromosome can be divided into distinct regions by its banding pattern. The short arm is *p* and the long arm is *q;* each arm is divided into larger regions that are further subdivided. This map shows a low resolution structure; at higher resolution, some bands are further subdivided into smaller bands and interbands, e.g., *p*21 is divided into *p*21.1, *p*21.2, and *p*21.3.

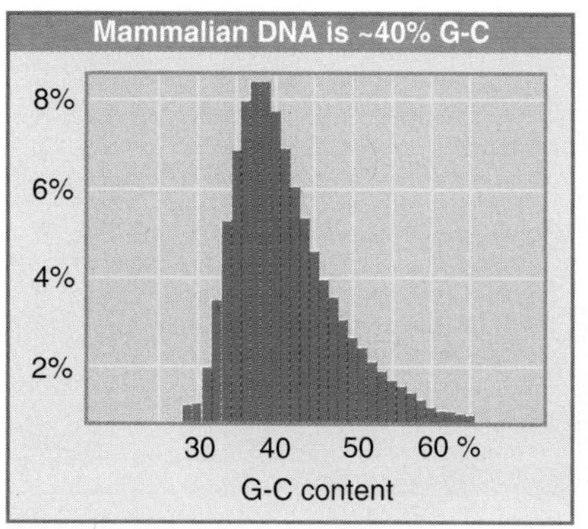

FIGURE 28.15 There are large fluctuations in G-C content over short distances. Each bar shows the percent of 20 kb fragments with the given G-C content.

lower G-C content than the interbands. This is a peculiar result. If there are ~10 bands on a large chromosome with a total content of ~100 Mb, this means that the chromosome is divided into regions of ~5 Mb in length that alternate between low G-C (band) and high G-C (interband) content. There is a tendency for genes (as identified by hybridization with mRNAs) to be located in the interband regions. All of this argues for some long-range sequence-dependent organization.

The human genome sequence confirms the basic observation. FIGURE 28.15 shows that there are distinct fluctuations in G-C content when the genome is divided into small tranches (DNA segments or lengths). The average of 41% G-C is common to mammalian genomes. There are regions as low as 30% or as high as 65%. When longer tranches are examined, there is less variation. The average length of regions with >43% G-C is 200 to 250 kb. This makes it clear that the band/interband structure does not represent homogeneous segments that alternate in G-C content, although the bands do contain a higher content of low G-C segments. Genes are concentrated in regions of higher G-C content. We have yet to understand how the G-C content affects chromosome structure.

28.9 Lampbrush Chromosomes Are Extended

Key concept

- Sites of gene expression on lampbrush chromosomes show loops that are extended from the chromosomal axis.

It would be extremely useful to visualize gene expression in its natural state in order to see what structural changes are associated with transcription. The compression of DNA in chromatin, coupled with the difficulty of identifying particular genes within it, makes it impossible to visualize the transcription of individual active genes.

Gene expression can be visualized directly in certain unusual situations in which the chromosomes are found in a highly extended form that allows individual loci (or groups of loci) to be distinguished. Lateral differentiation of structure is evident in many chromosomes when they first appear for meiosis. At this stage, the chromosomes resemble a series of beads on a string. The beads are densely staining granules, properly known as **chromomeres.** In general, though, there is little gene expression at meiosis, and it is not practical to use this material to identify the activities of individual genes. An exceptional situation that allows the material to be examined is presented by **lampbrush chromosomes,** which have been best characterized in certain amphibians.

Lampbrush chromosomes are formed during an unusually extended meiosis, which can last up to several months! During this period, the chromosomes are held in a stretched-out form in which they can be visualized in the light microscope. At a later point during meiosis, the chromosomes revert to their usual compact size.

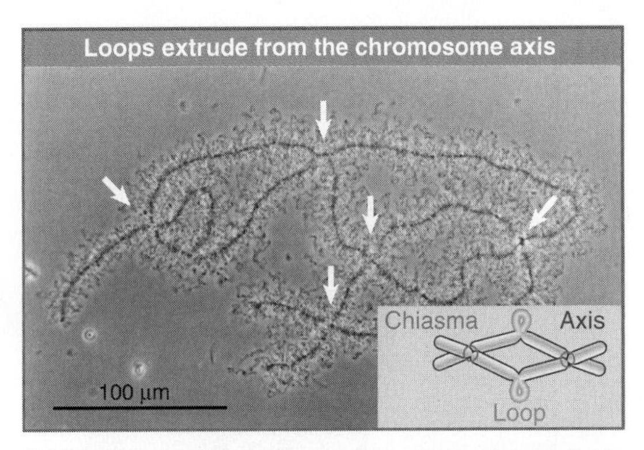

FIGURE 28.16 A lampbrush chromosome is a meiotic bivalent in which the two pairs of sister chromatids are held together at chiasmata (indicated by arrows). Photo courtesy of Joseph G. Gall, Carnegie Institution.

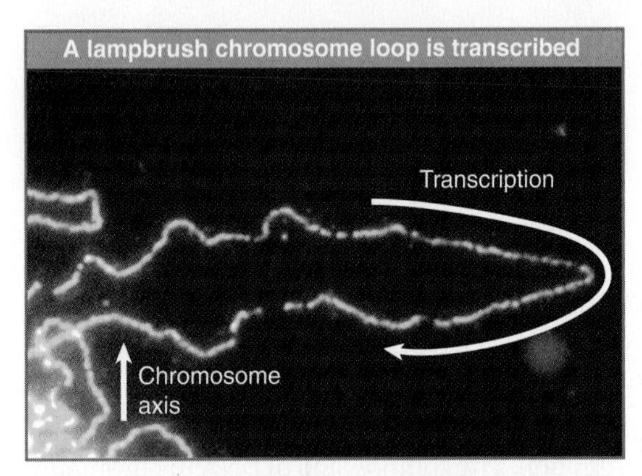

FIGURE 28.17 A lampbrush chromosome loop is surrounded by a matrix of ribonucleoprotein. Reproduced from Gall, J. G., et al. *Mol. Biol. Cell.* December, 1999. 10: 4385–4402. Photo courtesy of Joseph G. Gall, Carnegie Institution.

Thus the extended state essentially proffers an unfolded version of the normal condition of the chromosome.

The lampbrush chromosomes are meiotic bivalents, each consisting of two pairs of sister chromatids. FIGURE 28.16 shows an example in which the sister chromatid pairs have mostly separated so that they are held together only by chiasmata. Each sister chromatid pair forms a series of ellipsoidal chromomeres, ~1 to 2 μm in diameter, which are connected by a very fine thread. This thread contains the two sister duplexes of DNA and runs continuously along the chromosome, through the chromomeres.

The lengths of the individual lampbrush chromosomes in the newt *Notophthalmus viridescens* range from 400 to 800 μm, compared with the range of 15 to 20 μm seen later in meiosis. Thus the lampbrush chromosomes are ~30 times less tightly packed. The total length of the entire lampbrush chromosome set is 5 to 6 mm and is organized into ~5000 chromomeres.

The lampbrush chromosomes take their name from the lateral loops that extrude from the chromomeres at certain positions. (These resemble a lampbrush, which is an extinct object.) The loops extend in pairs, one from each sister chromatid. The loops are continuous with the axial thread, which suggests that they represent chromosomal material extruded from its more compact organization in the chromomere.

The loops are surrounded by a matrix of ribonucleoproteins that contain nascent RNA chains. Often, a transcription unit can be defined by the increase in the length of the RNP moving around the loop. An example is shown in FIGURE 28.17.

Thus the loop is an extruded segment of DNA that is being actively transcribed. In some cases, loops corresponding to particular genes have been identified. For these cases, the structure of the transcribed gene—and the nature of the product—can be scrutinized *in situ*.

28.10 Polytene Chromosomes Form Bands

Key concept

- Polytene chromosomes of dipterans have a series of bands that can be used as a cytological map.

The interphase nuclei of some tissues of the larvae of dipteran flies contain chromosomes that are greatly enlarged relative to their usual condition. They possess both increased diameter and greater length. FIGURE 28.18 shows an example of a chromosome set from the salivary gland of *D. melanogaster.* The members of this set are called **polytene** chromosomes.

Each member of the polytene set consists of a visible series of **bands** (more properly, but rarely, described as chromomeres). The bands range in size from the largest, with a breadth of ~0.5 μm, to the smallest, at ~0.05 μm. (The smallest can be distinguished only under an electron microscope.) The bands contain most of the mass of DNA and stain intensely with appropriate reagents. The regions between them stain more lightly and are called **interbands.** There are ~5000 bands in the *D. melanogaster* set.

The centromeres of all four chromosomes of *D. melanogaster* aggregate to form a chromocenter that consists largely of heterochromatin. (In the male it includes the entire Y chromosome.) Allowing for this, ~75% of the haploid DNA set is organized into alternating bands and interbands. The length of the chromosome

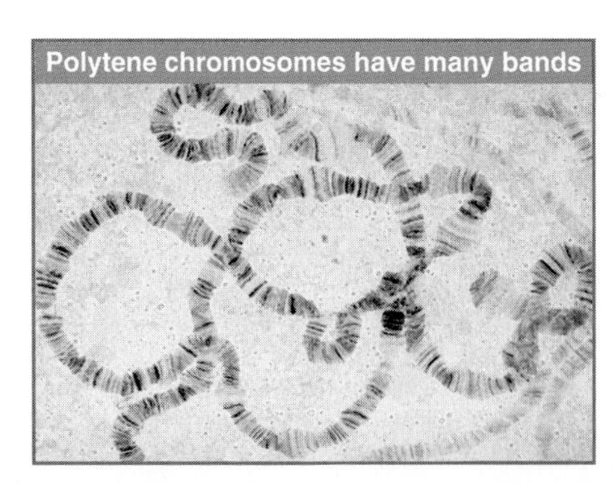

FIGURE 28.18 The polytene chromosomes of *D. melanogaster* form an alternating series of bands and interbands. Photo courtesy of José Bonner, Indiana University.

set is ~2000 μm. The DNA in extended form would stretch for ~40,000 μm, so the packing ratio is ~20. This demonstrates vividly the extension of the genetic material relative to the usual states of interphase chromatin or mitotic chromosomes.

What is the structure of these giant chromosomes? Each is produced by the successive replications of a synapsed diploid pair. The replicas do not separate, but instead remain attached to each other in their extended state. At the start of the process, each synapsed pair has a DNA content of 2C (where C represents the DNA content of the individual chromosome). This amount then doubles up to nine times, at its maximum giving a content of 1024C. The number of doublings is different in the various tissues of the *D. melanogaster* larva.

Each chromosome can be visualized as a large number of parallel fibers running longitudinally that are tightly condensed in the bands and less so in the interbands. It is likely that each fiber represents a single (C) haploid chromosome. This gives rise to the name polytene: The degree of polyteny is the number of haploid chromosomes contained in the giant chromosome.

The banding pattern is characteristic for each strain of *Drosophila.* The constant number and linear arrangement of the bands was first noted in the 1930s, when it was realized that they form a *cytological map* of the chromosomes. Rearrangements—such as deletions, inversions, or duplications—result in alterations of the order of bands.

The linear array of bands can be equated with the linear array of genes. Thus genetic rearrangements, as seen in a linkage map, can be correlated with structural rearrangements of the cytological map. Ultimately, a particular mutation can be located in a particular band. The total number of genes in *D. melanogaster* exceeds the number of bands, so there are probably multiple genes in most or all bands.

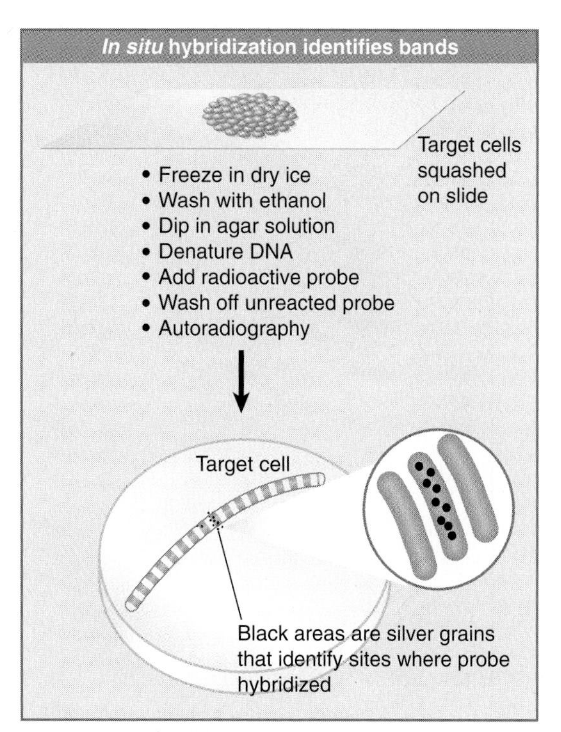

FIGURE 28.19 Individual bands containing particular genes can be identified by *in situ* hybridization.

The positions of particular genes on the cytological map can be determined directly by the technique of ***in situ* hybridization.** The protocol is summarized in FIGURE 28.19. A radioactive probe representing a gene (most often a labeled cDNA clone derived from the mRNA) is hybridized with the denatured DNA of the polytene chromosomes *in situ.* Autoradiography identifies the position or positions of the corresponding genes by the superimposition of grains at a particular band or bands. An example is shown in FIGURE 28.20. More recently, fluorescent probes have replaced radioactive probes. With this type of technique at hand, it is possible to determine directly the band within which a particular sequence lies.

28.11 Polytene Chromosomes Expand at Sites of Gene Expression

Key concept

- Bands that are sites of gene expression on polytene chromosomes expand to give "puffs."

One of the intriguing features of the polytene chromosomes is that active sites can be visualized. Some of the bands pass transiently through

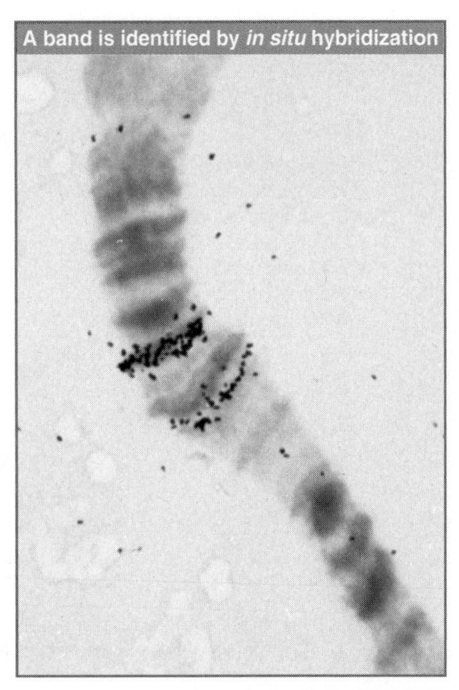

FIGURE 28.20 A magnified view of bands 87A and 87C shows their hybridization *in situ* with labeled RNA extracted from heat-shocked cells. Photo courtesy of José Bonner, Indiana University.

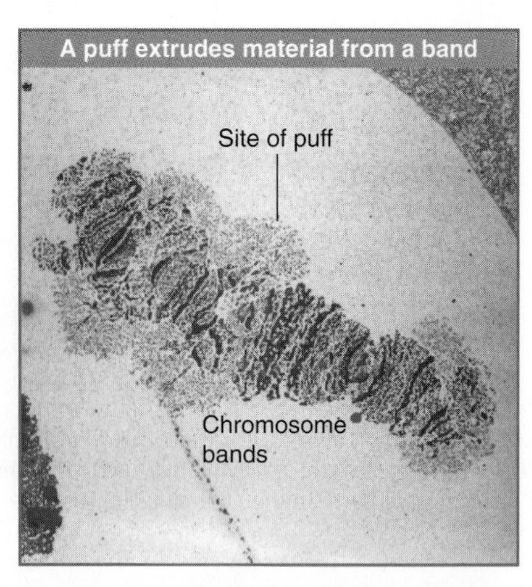

FIGURE 28.21 Chromosome IV of the insect *C. tentans* has three Balbiani rings in the salivary gland. Reproduced from *Cell,* vol. 4, Daneholt, B., et al., *Transcription in polytene chromosomes . . . ,* pp. 1–9. Copyright 1975, with permission from Elsevier. Photo courtesy of Bertil Daneholt, Medical Nobel Institute.

an expanded state in which they appear like a **puff** on the chromosome, when chromosomal material is extruded from the axis. Examples of some very large puffs (called Balbiani rings) are shown in FIGURE 28.21.

What is the nature of the puff? It consists of a region in which the chromosome fibers unwind from their usual state of packing in the band. The fibers remain continuous with those in the chromosome axis. Puffs usually emanate from single bands, although when they are very large, as typified by the Balbiani rings, the swelling may be so extensive as to obscure the underlying array of bands.

The pattern of puffs is related to gene expression. During larval development, puffs appear and regress in a definite, tissue-specific pattern. A characteristic pattern of puffs is found in each tissue at any given time. Puffs are induced by the hormone ecdysone that controls *Drosophila* development. Some puffs are induced directly by the hormone; others are induced indirectly by the products of earlier puffs.

The puffs are *sites where RNA is being synthesized.* The accepted view of puffing has been that expansion of the band is a consequence of the need to relax its structure in order to synthesize RNA. Puffing has therefore been viewed as a consequence of transcription. A puff can be generated by a single active gene. The sites of puffing differ from ordinary bands in accumulating additional proteins, which include RNA polymerase II and other proteins associated with transcription.

The features displayed by lampbrush and polytene chromosomes suggest a general conclusion. In order to be transcribed, the genetic material is dispersed from its usual, more tightly packed state. The question to keep in mind is whether this dispersion at the gross level of the chromosome mimics the events that occur at the molecular level within the mass of ordinary interphase euchromatin.

Do the bands of a polytene chromosome have a functional significance, that is, does each band correspond to some type of genetic unit? You might think that the answer would be immediately evident from the sequence of the fly genome, because by mapping interbands to the sequence it should be possible to determine whether a band has any fixed type of identity. Thus far, though, no pattern has been found that identifies a functional significance for the bands.

28.12 The Eukaryotic Chromosome Is a Segregation Device

Key concepts

- A eukaryotic chromosome is held on the mitotic spindle by the attachment of microtubules to the kinetochore that forms in its centromeric region.
- Centromeres often have heterochromatin that is rich in satellite DNA sequences.

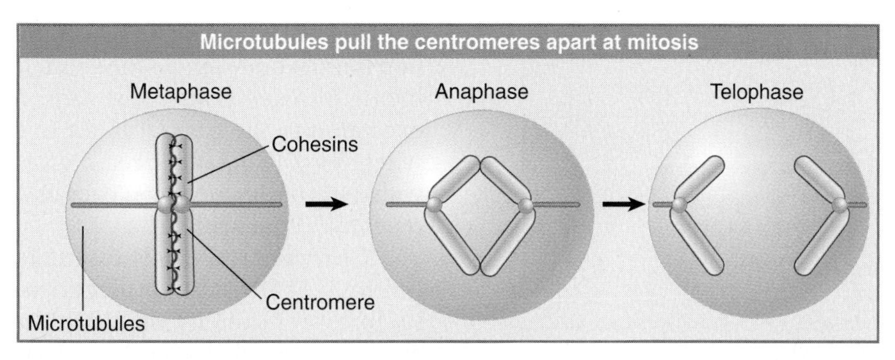

FIGURE 28.22 Chromosomes are pulled to the poles via microtubules that attach at the centromeres. The sister chromatids are held together until anaphase by glue proteins (cohesins). The centromere is shown here in the middle of the chromosome (metacentric), but can be located anywhere along its length, including close to the end (acrocentric) and at the end (telocentric).

During mitosis, the sister chromatids move to opposite poles of the cell. Their movement depends on the attachment of the chromosome to microtubules, which are connected at their other end to the poles. (The microtubules comprise a cellular filamentous system, which is reorganized at mitosis so that they connect the chromosomes to the poles of the cell.) The sites in the two regions where microtubule ends are organized—in the vicinity of the centrioles at the poles and at the chromosomes—are called **MTOCs** (microtubule organizing centers).

FIGURE 28.22 illustrates the separation of sister chromatids as mitosis proceeds from metaphase to telophase. The region of the chromosome that is responsible for its segregation at mitosis and meiosis is called the **centromere.** The centromeric region on each sister chromatid is pulled by microtubules to the opposite pole. Opposing this motive force, "glue" proteins called cohesins hold the sister chromatids together. Initially the sister chromatids separate at their centromeres, and then they are released completely from one another during anaphase when the cohesins are degraded. The centromere is pulled toward the pole during mitosis, and the attached chromosome is 'dragged along' behind it. The chromosome therefore provides a device for attaching a large number of genes to the apparatus for division. It contains the site at which the sister chromatids are held together prior to the separation of the individual chromosomes. This shows as a constricted region connecting all four chromosome arms, as in the photo of Figure 28.11, which shows the sister chromatids at the metaphase stage of mitosis.

The centromere is essential for segregation, as shown by the behavior of chromosomes that have been broken. A single break generates one piece that retains the centromere, and another, an **acentric fragment,** that lacks it. The acentric fragment does not become attached to the mitotic spindle, and as a result it fails to be included in either of the daughter nuclei.

(When chromosome movement relies on discrete centromeres, there can be *only* one centromere per chromosome. When translocations generate chromosomes with more than one centromere, aberrant structures form at mitosis. This is because the two centromeres on the *same* sister chromatid can be pulled toward different poles, thus breaking the chromosome. In some species, though, the centromeres are "diffuse," which creates a different situation. Only discrete centromeres have been analyzed at the molecular level.)

The regions flanking the centromere often are rich in satellite DNA sequences and display a considerable amount of heterochromatin. The entire chromosome is condensed, though, so centromeric heterochromatin is not immediately evident in mitotic chromosomes. It can, however, be visualized by a technique that generates **C-bands.** In the example of FIGURE 28.23, all the centromeres show as darkly staining regions. Although it is common, heterochromatin cannot be identified around *every* known centromere, which suggests that it is unlikely to be essential for the division mechanism.

The region of the chromosome at which the centromere forms is defined by DNA sequences (although the sequences have been defined in only a very small number of cases). The centromeric DNA binds specific proteins that are responsible for establishing the structure that attaches the chromosome to the microtubules. This structure is called the **kinetochore.** It is a darkly staining fibrous object of diameter or length ~400 nm. The kinetochore provides the MTOC on a chromosome. FIGURE 28.24

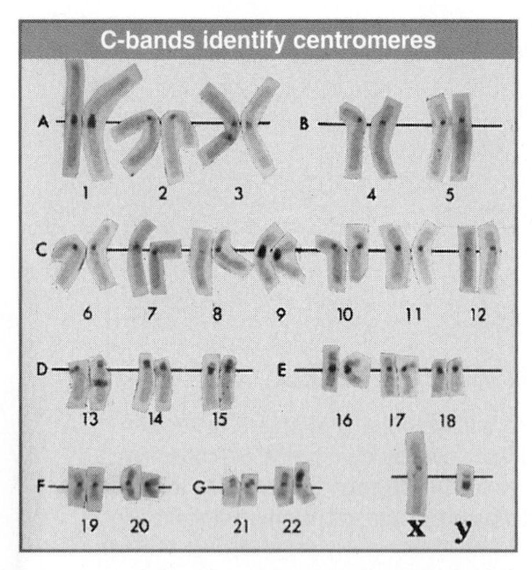

FIGURE 28.23 C-banding generates intense staining at the centromeres of all chromosomes. Photo courtesy of Lisa Shaffer, Washington State University–Spokane.

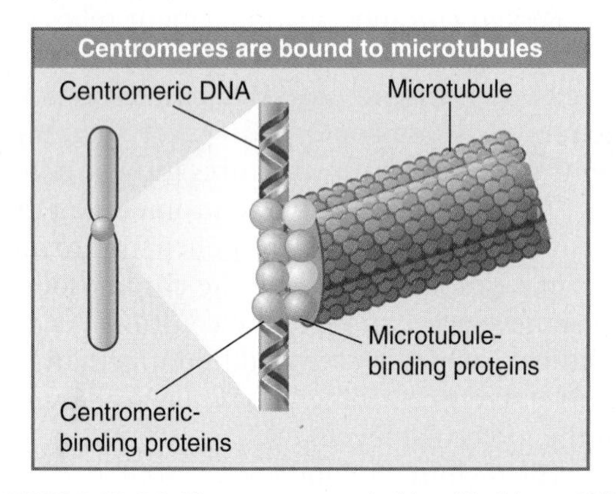

FIGURE 28.24 The centromere is identified by a DNA sequence that binds specific proteins. These proteins do not themselves bind to microtubules, but establish the site at which the microtubule-binding proteins in turn bind.

shows the hierarchy of organization that connects centromeric DNA to the microtubules. Proteins bound to the centromeric DNA bind other proteins that bind to microtubules (see Section 28.15, The Centromere Binds a Protein complex).

28.13 Centromeres May Contain Repetitive DNA

Key concepts

- Centromeres in higher eukaryotic chromosomes contain large amounts of repetitive DNA.
- The function of the repetitive DNA is not known.

The length of DNA required for centromeric function is often quite long. (The short, discrete elements of *Saccharomyces cerevisiae* may be an exception to the general rule.) In those cases where we can equate specific DNA sequences with the centromeric region, they usually include repetitive sequences.

S. cerevisiae is the only case so far in which centromeric DNA can be identified by its ability to confer stability on plasmids. A related approach, though, has been used with the yeast *Schizosaccharomyces pombe*. *S. pombe* has only three chromosomes, and the region containing each centromere has been identified by deleting most of the sequences of each chromosome to create a stable minichromosome. This approach locates the centromeres within regions of 40 to 100 kb that consist largely or entirely of repetitious DNA. It is not clear how much of each of these rather long regions is required for chromosome segregation at mitosis and meiosis.

Attempts to localize centromeric functions in *Drosophila* chromosomes suggest that they are dispersed in a large region, consisting of 200 to 600 kb. The large size of this type of centromere suggests that it is likely to contain several separate specialized functions, including sequences required for kinetochore assembly and sister chromatid pairing.

The size of the centromere in *Arabidopsis* is comparable. Each of the five chromosomes has a centromeric region in which recombination is very largely suppressed. This region occupies >500 kb. Clearly it includes the centromere, but we have no direct information as to how much of it is required. There are expressed genes within these regions, which casts some doubt on whether the entire region is part of the centromere. At the center of the region is a series of 180 bp repeats; this is the type of structure generally associated with centromeres. It is too early to say how these structures relate to centromeric function.

The primary motif comprising the heterochromatin of primate centromeres is the α satellite DNA, which consists of tandem arrays of a 170 bp repeating unit. There is significant variation between individual repeats, although those at any centromere tend to be better related to one another than to members of the family in other locations. It is clear that the sequences required for centromeric function reside within the blocks of α satellite DNA, but it is not clear whether the α satellite sequences themselves provide this function, or whether other sequences are embedded within the α satellite arrays.

The *S. cerevisiae* centromere has short conserved sequences and a long A-T stretch

TCACATGATGATATTTGATTTTATTATATTTTTAAAAAAAGTAAAAAATAAAAAGTAGTTTATTTTTAAAAAATAAAATTTAAAATATTTCACAAAATGATTTCCGAA
AGTGTACTACTATAAACTAAAATAATATAAAAATTTTTTTCATTTTTTATTTTTCATCAAATAAAAATTTTTTATTTTAAATTTTATAAAGTGTTTTACTAAAGGCTT

CDE-I *CDE-II* 80–90 bp, >90% A + T *CDE-III*

FIGURE 28.25 Three conserved regions can be identified by the sequence homologies between yeast *CEN* elements.

28.14 Centromeres Have Short DNA Sequences in *S. cerevisiae*

Key concepts

- CEN elements are identified in *S. cerevisiae* by the ability to allow a plasmid to segregate accurately at mitosis.
- CEN elements consist of the short conserved sequences CDE-I and CDE-III that flank the A-T-rich region CDE-II.

If a centromeric sequence of DNA is responsible for segregation, any molecule of DNA possessing this sequence should move properly at cell division, whereas any DNA lacking it should fail to segregate. This prediction has been used to isolate centromeric DNA in the yeast, *S. cerevisiae*. Yeast chromosomes do not display visible kinetochores comparable to those of higher eukaryotes, but otherwise divide at mitosis and segregate at meiosis by the same mechanisms.

Genetic engineering has produced plasmids of yeast that are replicated like chromosomal sequences (see Section 15.8, Replication Origins Can Be Isolated in Yeast). They are unstable at mitosis and meiosis, though, and disappear from a majority of the cells because they segregate erratically. Fragments of chromosomal DNA containing centromeres have been isolated by their ability to confer mitotic stability on these plasmids.

A centromeric DNA regions (*CEN*) fragment is identified as the minimal sequence that can confer stability upon such a plasmid. Another way to characterize the function of such sequences is to modify them *in vitro* and then reintroduce them into the yeast cell, where they replace the corresponding centromere on the chromosome. This allows the sequences required for *CEN* function to be defined directly in the context of the chromosome.

A *CEN* fragment derived from one chromosome can replace the centromere of another chromosome with no apparent consequence. This result suggests that centromeres are interchangeable. *They are used simply to attach the chromosome to the spindle, and play no role in distinguishing one chromosome from another.*

The sequences required for centromeric function fall within a stretch of ~120 bp. The centromeric region is packaged into a nuclease-resistant structure and binds a single microtubule. We may therefore look to the *S. cerevisiae* centromeric region to identify proteins that bind centromeric DNA and proteins that connect the chromosome to the spindle.

As summarized in FIGURE 28.25, three types of sequence element can be distinguished in the *CEN* region:

- cell cycle-dependent element (CDE)-I is a sequence of 9 bp that is conserved with minor variations at the left boundary of all centromeres.
- CDE-II is a >90% A-T-rich sequence of 80 to 90 bp found in all centromeres; its function could depend on its length rather than exact sequence. Its constitution is reminiscent of some short tandemly repeated (satellite) DNAs (see Section 6.12, Arthropod Satellites Have Very Short Identical Repeats). Its base composition may cause some characteristic distortions of the DNA double helical structure.
- CDE-III is an 11 bp sequence highly conserved at the right boundary of all centromeres. Sequences on either side of the element are less well conserved, and may also be needed for centromeric function. (CDE-III could be longer than 11 bp if it turns out that the flanking sequences are essential.)

Mutations in CDE-I or CDE-II reduce, but do not inactivate, centromere function, but point mutations in the central CCG of CDE-III completely inactivate the centromere.

28.15 The Centromere Binds a Protein Complex

Key concepts

- A specialized protein complex that is an alternative to the usual chromatin structure is formed at CDE-II.
- The CBF3 protein complex that binds to CDE-III is essential for centromeric function.
- The proteins that connect these two complexes may provide the connection to microtubules.

Can we identify proteins that are necessary for the function of CEN sequences? There are several genes in which mutations affect chromosome segregation, and whose proteins are localized at centromeres. The contributions of these proteins to the centromeric structure are summarized in FIGURE 28.26.

A specialized chromatin structure is built by binding the CDE-II region to a protein called Cse4, which resembles one of the histone proteins that comprise the basic subunits of chromatin (see Section 31.3, Heterochromatin Depends on Interactions with Histones). A protein called Mif2 may also be part of this complex, or at least connected to it. Cse4 and Mif2 have counterparts that are localized at higher eukaryotic centromeres, called CENP-A and CENP-C, which suggests that this interaction may be a universal aspect of centromere construction. The basic interaction consists of bending the DNA of the CDE-II region around a protein aggregate; the reaction is probably assisted by the occurrence of intrinsic bending in the CDE-II sequence.

CDE-I is bound by the homodimer CBF1; this interaction is not essential for centromere function, but in its absence the fidelity of chromosome segregation is reduced ~10×. A 240 kD complex of four proteins, called CBF3, binds to CDE-III. This interaction is essential for centromeric function.

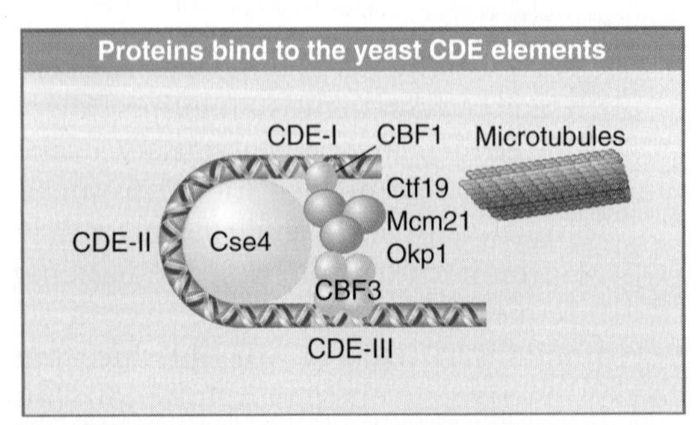

FIGURE 28.26 The DNA at CDE-II is wound around a protein aggregate including Cse4p, CDE-III is bound to CBF3, and CDE-I is bound to CBF1. These proteins are connected by the group of Ctf19, Mcm21, and Okp1.

The proteins bound at CDE-I and CDE-III are connected to each other and also to the protein structure bound at CDE-II by another group of proteins (Ctf19, Mcm21, Okp1). The connection to the microtubule may be made by this complex.

The overall model suggests that the complex is localized at the centromere by a protein structure that resembles the normal building block of chromatin (the nucleosome). The bending of DNA at this structure allows proteins bound to the flanking elements to become part of a single complex. Some components of the complex (possibly not those that bind directly to DNA) link the centromere to the microtubule. The construction of kinetochores probably follows a similar pattern, and uses related components, in a wide variety of organisms.

28.16 Telomeres Have Simple Repeating Sequences

Key concepts

- The telomere is required for the stability of the chromosome end.
- A telomere consists of a simple repeat where a C+A-rich strand has the sequence $C_{>1}(A/T)_{1-4}$.

Another essential feature in all chromosomes is the **telomere,** which "seals" the end. We know that the telomere must be a special structure, because chromosome ends generated by breakage are "sticky" and tend to react with other chromosomes, whereas natural ends are stable.

We can apply two criteria in identifying a telomeric sequence:

- It must lie at the end of a chromosome (or, at least at the end of an authentic linear DNA molecule).
- It must confer stability on a linear molecule.

The problem of finding a system that offers an assay for function again has been brought to the molecular level by using yeast. All the plasmids that survive in yeast (by virtue of possessing *ARS* and *CEN* elements) are circular DNA molecules. Linear plasmids are unstable (because they are degraded). Could an authentic telomeric DNA sequence confer stability on a linear plasmid? Fragments from yeast DNA that prove to be located at chromosome ends can be identified by such an assay, and a region from the end of a known natural linear DNA molecule—the extrachromosomal rDNA of *Tetrahymena*—is able to render a yeast plasmid stable in linear form.

Telomeric sequences have been characterized from a wide range of lower and higher eukaryotes. The same type of sequence is found in plants and human beings, so the construction of the telomere seems to follow a nearly universal principle. Each telomere consists of a long series of short, tandemly repeated sequences. There may be 100 to 1000 repeats, depending on the organism.

All telomeric sequences can be written in the general form $C_n(A/T)_m$, where $n>1$ and m is 1 to 4. FIGURE 28.27 shows a generic example. One unusual property of the telomeric sequence is the extension of the G-T-rich strand, which for 14 to 16 bases is usually a single strand. The G-tail is probably generated because there is a specific limited degradation of the C-A-rich strand.

Some indications about how a telomere functions are given by some unusual properties of the ends of linear DNA molecules. In a trypanosome population, the ends vary in length. When an individual cell clone is followed, the telomere grows longer by 7 to 10 bp (1 to 2 repeats) per generation. Even more revealing is the fate of ciliate telomeres introduced into yeast. After replication in yeast, *yeast telomeric repeats are added onto the ends of the Tetrahymena repeats.*

Addition of telomeric repeats to the end of the chromosome in every replication cycle could solve the difficulty of replicating linear DNA molecules discussed in Section 16.2, The Ends of Linear DNA Are a Problem for Replication. The addition of repeats by *de novo* synthesis would counteract the loss of repeats resulting from failure to replicate up to the end of the chromosome. Extension and shortening would be in dynamic equilibrium.

If telomeres are continually being lengthened (and shortened), their exact sequence may be irrelevant. All that is required is for the end to be recognized as a suitable substrate for addition. This explains how the ciliate telomere functions in yeast.

C-A-strand degradation generates the G-tail

CCCCAACCCCAACCCCAACCCCAACCCCAACCCCAA
GGGGTTGGGGTTGGGGTTGGGGTTGGGGTTGGGGTT

CCCCAACCCCAACCCCAA 5′
GGGGTTGGGGTTGGGGTTGGGGTTGGGGTTGGGGTT3′

FIGURE 28.27 A typical telomere has a simple repeating structure with a G-T-rich strand that extends beyond the C-A-rich strand. The G-tail is generated by a limited degradation of the C-A-rich strand.

Telomere repeats from G quartets

AGGGTTAGGGTTAGGGTTAGGG

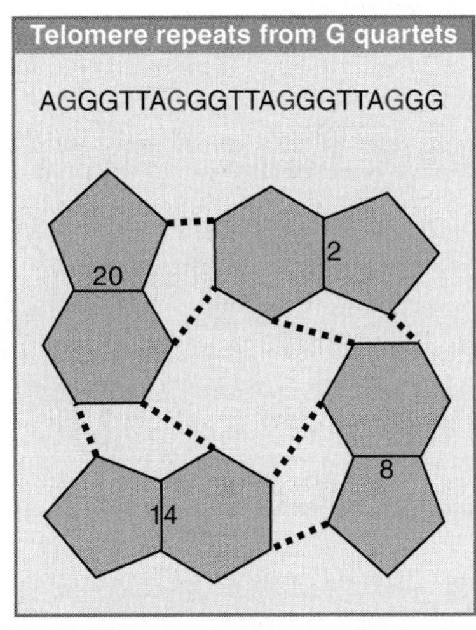

FIGURE 28.28 The crystal structure of a short repeating sequence from the human telomere forms three stacked G quartets. The top quartet contains the first G from each repeating unit. This is stacked above quartets that contains the second G (G3, G9, G15, G21) and the third G (G4, G10, G16, G22).

28.17 Telomeres Seal the Chromosome Ends

Key concept

- The protein TRF2 catalyzes a reaction in which the 3′ repeating unit of the G+T-rich strand forms a loop by displacing its homolog in an upstream region of the telomere.

Isolated telomeric fragments do not behave as though they contain single-stranded DNA; instead, they show aberrant electrophoretic mobility and other properties.

Guanine bases have an unusual capacity to associate with one another. The single-stranded G-rich tail of the telomere can form "quartets" of G residues. Each quartet contains four guanines that hydrogen bond with one another to form a planar structure. Each guanine comes from the corresponding position in a successive TTAGGG repeating unit. FIGURE 28.28 shows an organization based on a recent crystal structure. The quartet that is illustrated represents an association between the first guanine in each repeating unit. It is stacked on top of another quartet that has the same organization, but is

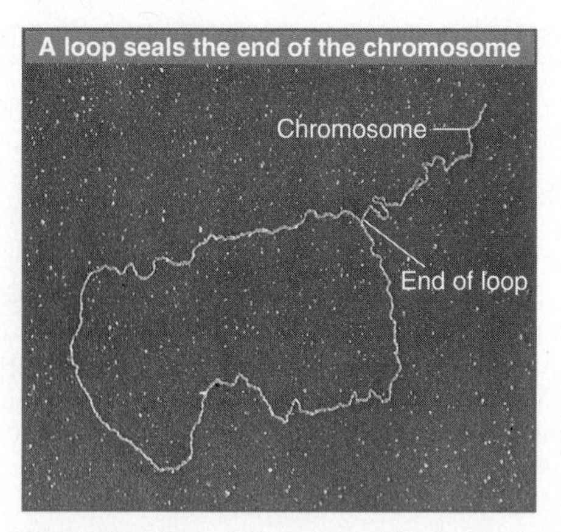

FIGURE 28.29 A loop forms at the end of chromosomal DNA. Photo courtesy of Jack Griffith, University of North Carolina at Chapel Hill.

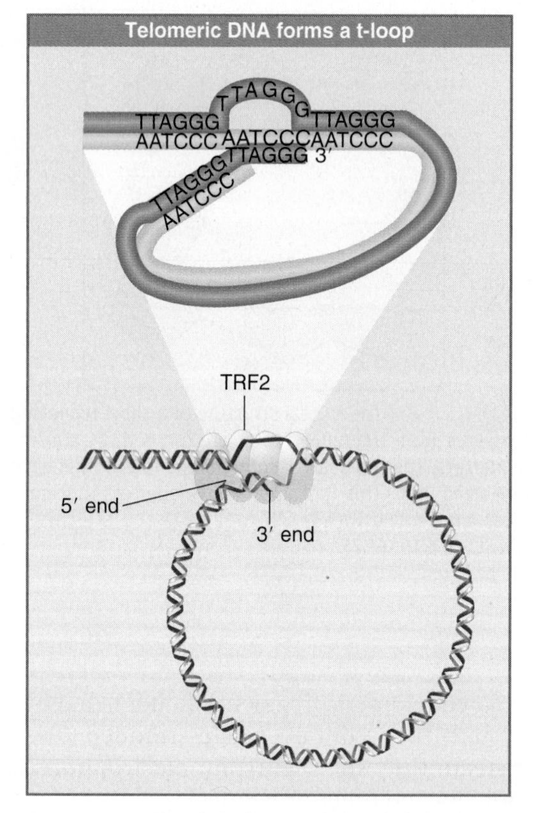

FIGURE 28.30 The 3′ single-stranded end of the telomere $(TTAGGG)_n$ displaces the homologous repeats from duplex DNA to form a t-loop. The reaction is catalyzed by TRF2.

formed from the second guanine in each repeating unit. A series of quartets could be stacked like this in a helical manner. Although the formation of this structure attests to the unusual properties of the G-rich sequence *in vitro*, it does not of course demonstrate whether the quartet forms *in vivo*.

What feature of the telomere is responsible for the stability of the chromosome end? FIGURE 28.29 shows that a loop of DNA forms at the telomere. The absence of any free end may be the crucial feature that stabilizes the end of the chromosome. The average length of the loop in animal cells is 5 to 10 kb.

FIGURE 28.30 shows that the loop is formed when the 3′ single-stranded end of the telomere $(TTAGGG)_n$ displaces the same sequence in an upstream region of the telomere. This converts the duplex region into a structure like a D-loop, where a series of TTAGGG repeats are displaced to form a single-stranded region, and the tail of the telomere is paired with the homologous strand.

The reaction is catalyzed by the telomere-binding protein TRF2, which together with other proteins forms a complex that stabilizes the chromosome ends. Its importance in protecting the ends is indicated by the fact the deletion of TRF2 causes chromosome rearrangements to occur.

28.18 Telomeres Are Synthesized by a Ribonucleoprotein Enzyme

Key concepts

- Telomerase uses the 3′-OH of the G+T telomeric strand to prime synthesis of tandem TTGGGG repeats.
- The RNA component of telomerase has a sequence that pairs with the C+A-rich repeats.
- One of the protein subunits is a reverse transcriptase that uses the RNA as template to synthesis the G+T-rich sequence.

The telomere has two functions:

- One is to protect the chromosome end. Any other DNA end—for example, the end generated by a double-strand break—becomes a target for repair systems. The cell has to be able to distinguish the telomere.
- The second is to allow the telomere to be extended. If it is not extended, it becomes shorter with each replication cycle (because replication cannot start at the very end).

Proteins that bind to the telomere provide the solution for both problems. In yeast, different sets of proteins solve each problem, but both are bound to the telomere via the same protein, Cdc13:

- The Stn1 protein protects against degradation (specifically, against any extension of the degradation of the C-A-strand that generates the G-tail).
- A **telomerase** enzyme extends the C-A-rich strand. Its activity is influenced by two proteins that have ancillary roles, such as controlling the length of the extension.

The telomerase uses the 3′–OH of the G+T telomeric strand as a primer for synthesis of tandem TTGGGG repeats. Only dGTP and dTTP are needed for the activity. The telomerase is a large ribonucleoprotein that consists of a templating RNA (coded by *TLC1*) and a protein with catalytic activity *(EST2)*. The short RNA component (159 bases long in *Tetrahymena,* and 192 bases long in *Euplotes*) includes a sequence of 15 to 22 bases that is identical to two repeats of the C-rich repeating sequence. This RNA provides the template for synthesizing the G-rich repeating sequence. The protein component of the telomerase is a catalytic subunit that can act only upon the RNA template provided by the nucleic acid component.

FIGURE 28.31 shows the action of telomerase. The enzyme progresses discontinuously: The template RNA is positioned on the DNA primer, several nucleotides are added to the primer, and then the enzyme translocates to begin again. The telomerase is a specialized example of a reverse transcriptase, an enzyme that synthesizes a DNA sequence using an RNA template (see Section 22.4, Viral DNA Is Generated by Reverse Transcription). We do not know how the complementary (C-A-rich) strand of the telomere is assembled, but we may speculate that it could be synthesized by using the 3′–OH of a terminal G-T hairpin as a primer for DNA synthesis.

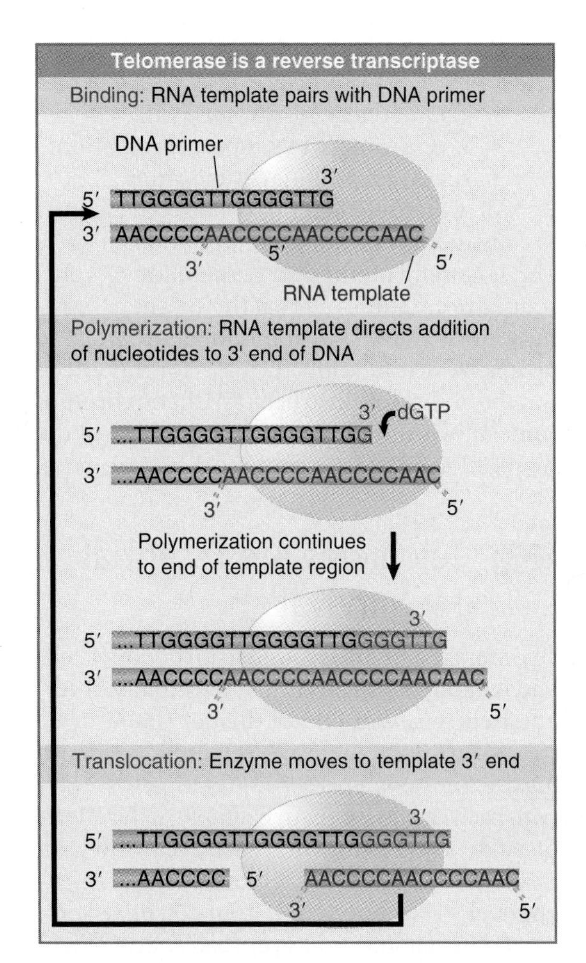

FIGURE 28.31 Telomerase positions itself by base pairing between the RNA template and the protruding single-stranded DNA primer. It adds G and T bases one at a time to the primer, as directed by the template. The cycle starts again when one repeating unit has been added.

Telomerase synthesizes the individual repeats that are added to the chromosome ends, but does not itself control the number of repeats. Other proteins are involved in determining the length of the telomere. They can be identified by the *EST1* and *EST3* mutants in yeast that have altered telomere lengths. These proteins may bind telomerase, and influence the length of the telomere by controlling the access of telomerase to its substrate. Proteins that bind telomeres in mammalian cells have been found, but less is known about their functions.

Each organism has a characterisic range of telomere lengths. They are long in mammals (typically 5 to 15 kb in human beings) and short in yeast (typically ~300 bp in *S. cerevisiae*). The basic control mechanism is that the probability that a telomere will be a substrate for telomerase increases as the length of the telomere shortens; we do not know if this is a continuous effect or if it depends on the length falling below some critical value. When telomerase acts on a telomere, it may add several repeating units. The enzyme's intrinsic mode of action is to dissociate after adding one repeat; addition of several repeating units depends on other proteins that cause telomerase to undertake more than one round of extension. The number of repeats that is added is not influenced by the length of the telomere itself, but instead is controlled by ancillary proteins that associate with telomerase.

The minimum features required for existence as a chromosome are:

- Telomeres to ensure survival.
- A centromere to support segregation.
- An origin to initiate replication.

All of these elements have been put together to construct a yeast artificial chromosome (YAC). This is a useful method for perpetuating foreign sequences. It turns out that the synthetic chromosome is stable only if it is longer than 20 to 50 kb. We do not know the basis for this effect, but the ability to construct a synthetic chromosome allows us to investigate the nature of the segregation device in a controlled environment.

28.19 Telomeres Are Essential for Survival

Telomerase activity is found in all dividing cells and is generally turned off in terminally differentiated cells that do not divide. FIGURE 28.32 shows that if telomerase is mutated in a dividing cell, the telomeres become gradually shorter with each cell division. An example of the effects of such a mutation in yeast are shown in FIGURE 28.33, where the telomere length shortens over ~120 generations from 400 bp to zero.

Loss of telomeres has very bad effects. When the telomere length reaches zero, it becomes difficult for the cells to divide successfully. Attempts to divide typically generate chromosome breaks and translocations. This causes an increased rate of mutation. In yeast this is associated with a loss of viability and the culture becomes predominantly occupied by senescent cells (which are elongated and nondividing, and eventually die).

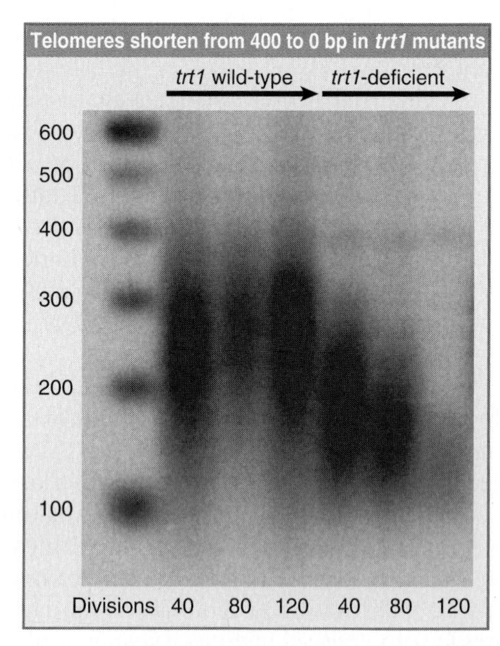

FIGURE 28.33 Telomere length is maintained at ~350 bp in wild-type yeast, but a mutant in the *trt1* gene coding for the RNA component of telomerase rapidly shortens its telomeres to zero length. Reproduced with permission from Nakamura, T. M., et al. 1997. *Science*. 277: 955–959. © 1997 AAAS. Photo courtesy of Thomas R. Cech and Toru Nakamura, University of Colorado.

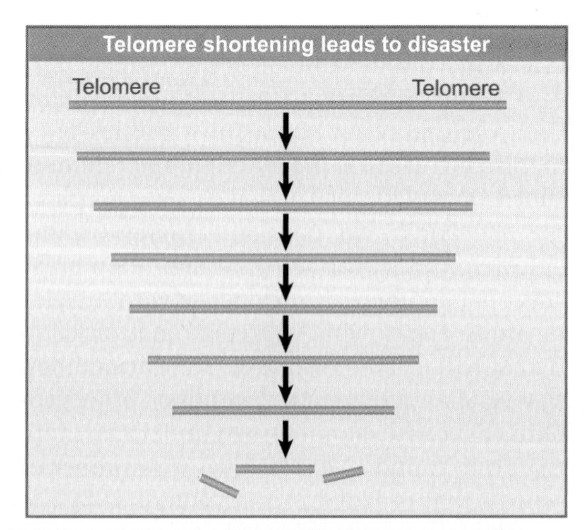

FIGURE 28.32 Mutation in telomerase causes telomeres to shorten in each cell division. Eventual loss of the telomere causes chromosome breaks and rearrangements.

Some cells grow out of the senescing culture. They have acquired the ability to extend their telomeres by an alternative to telomerase activity. The survivors fall into two groups. The members of one group have circularized their chromosomes: They now have no telomeres, and as a result they have become independent of telomerase. The other group uses unequal crossing-over to extend their telomeres (FIGURE 28.34). The telomere is a repeating structure, so it is possible for two telomeres to misalign when chromosomes pair. Recombination between the mispaired regions generates an unequal crossing-over, as shown previously in Figure 6.1: When the length of one recombinant chromosome increases, the length of the other decreases.

Cells usually suppress unequal crossing-over because of its potentially deleterious consequences. Two systems are responsible for suppressing crossing-over between telomeres. One is provided by telomere-binding proteins. In yeast, the frequency of recombination between telomeres is increased by deletion of the gene *taz1*, which codes for a protein that regulates telomerase activity. The second is a general system that undertakes mismatch repair. In addition to correcting mismatched base pairs that may arise in DNA, this system suppresses recombination between mispaired regions. As shown in Figure 28.34, this includes telomeres. When it is mutated, a greater proportion of telomerase-deficient yeast survive the loss of telomeres because recombination between telomeres generates some chromosomes with longer telomeres.

When eukaryotic cells are placed in culture, they usually divide for a fixed number of generations and then enter senescence. The reason appears to be a decline in telomere length because of the absence of telomerase expression. Cells enter a crisis from which some emerge, but typically the cells that emerge have chromosome rearrangements that have resulted from lack of protection of chromosome ends. These rearrangements may cause mutations that contribute to the tumorigenic state. The absence of telomerase expression in this situation is due to failure to express the gene, and reactivation of telomerase is one of the mechanisms by which these cells then survive continued culture. (This of course was not an option in the yeast experiments in which the gene had been deleted.)

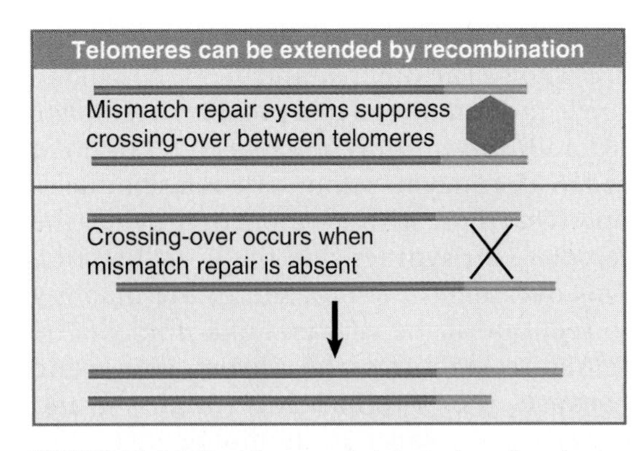

FIGURE 28.34 Crossing-over in telomeric regions is usually suppressed by mismatch-repair systems, but can occur when they are mutated. An unequal crossing-over event extends the telomere of one of the products, allowing the chromosome to survive in the absence of telomerase.

28.20 Summary

The genetic material of all organisms and viruses takes the form of tightly packaged nucleoprotein. Some virus genomes are inserted into preformed virions, whereas others assemble a protein coat around the nucleic acid. The bacterial genome forms a dense nucleoid, with ~20% protein by mass, but details of the interaction of the proteins with DNA are not known. The DNA is organized into ~100 domains that maintain independent supercoiling, with a density of unrestrained supercoils corresponding to ~1/100 to 200 bp. In eukaryotes, interphase chromatin and metaphase chromosomes both appear to be organized into large loops. Each loop may be an independently supercoiled domain. The bases of the loops are connected to a metaphase scaffold or to the nuclear matrix by specific DNA sites.

Transcriptionally active sequences reside within the euchromatin that comprises the majority of interphase chromatin. The regions of heterochromatin are packaged ~5 to 10× more compactly, and are transcriptionally inert. All chromatin becomes densely packaged during cell division, when the individual chromosomes can be distinguished. The existence of a reproducible ultrastructure in chromosomes is indicated by the production of G-bands by treatment with Giemsa stain. The bands are very large regions (~10^7 bp) that can be used to map chromosomal translocations or other large changes in structure.

Lampbrush chromosomes of amphibians and polytene chromosomes of insects have unusually extended structures, with packing ratios <100. Polytene chromosomes of *D. melanogaster* are divided into ~5000 bands. These bands vary in size by an order of magnitude, with an average of ~25 kb. Transcriptionally active regions can be visualized in even more unfolded ("puffed") structures, in which material is extruded from the axis of the chromosome. This may resemble the changes that occur on a smaller scale when a sequence in euchromatin is transcribed.

The centromeric region contains the kinetochore, which is responsible for attaching a chromosome to the mitotic spindle. The centromere often is surrounded by heterochromatin. Centromeric sequences have been identified only in yeast *S. cerevisiae,* where they consist of short conserved elements. These elements, CDE-I and CDE-III, bind CBF1 and the CBF3 complex, respectively, and a long A-T-rich region called CDE-II binds Cse4 to form a specialized structure in chromatin. Another group of proteins that binds to this assembly provides the connection to microtubules.

Telomeres make the ends of chromosomes stable. Almost all known telomeres consist of multiple repeats in which one strand has the general sequence $C_n(A/T)_m$, where $n>1$ and $m = 1$ to 4. The other strand, $G_n(T/A)_m$, has a single protruding end that provides a template for addition of individual bases in defined order. The enzyme telomerase is a ribonucleoprotein whose RNA component provides the template for synthesizing the G-rich strand. This overcomes the problem of the inability to replicate at the very end of a duplex. The telomere stabilizes the chromosome end because the overhanging single strand $G_n(T/A)_m$ displaces its homolog in earlier repeating units in the telomere to form a loop, so there are no free ends.

References

28.2 Viral Genomes Are Packaged into Their Coats

Reviews

Black, L. W. (1989). DNA packaging in dsDNA bacteriophages. *Annu. Rev. Immunol.* 43, 267–292.

Butler, P. J. (1999). Self-assembly of tobacco mosaic virus: the role of an intermediate aggregate in generating both specificity and speed. *Philos. Trans. R. Soc. Lond. B Biol. Sci.* 354, 537–550.

Klug, A. (1999). The tobacco mosaic virus particle: structure and assembly. *Philos. Trans. R. Soc. Lond. B Biol. Sci.* 354, 531–535.

Mindich, L. (2000). Precise packaging of the three genomic segments of the double-stranded-RNA bacteriophage phi6. *Microbiol. Mol. Biol. Rev.* 63, 149–160.

Research

Caspar, D. L. D. and Klug, A. (1962). Physical principles in the construction of regular viruses. *Cold Spring Harbor Symp. Quant. Biol.* 27, 1–24.

de Beer, T., Fang, J., Ortega, M., Yang, Q., Maes, L., Duffy, C., Berton, N., Sippy, J., Overduin, M., Feiss, M., and Catalano, C. E. (2002). Insights into specific DNA recognition during the assembly of a viral genome packaging machine. *Mol. Cell* 9, 981–991.

Dube, P., Tavares, P., Lurz, R., and van Heel, M. (1993). The portal protein of bacteriophage SPP1: a DNA pump with 13-fold symmetry. *EMBO J.* 12, 1303–1309.

Fraenkel-Conrat, H. and Williams, R. C. (1955). Reconstitution of active tobacco mosaic virus from its inactive protein and nucleic acid components. *Proc. Natl. Acad. Sci. USA* 41, 690–698.

Jiang, Y. J., Aerne, B. L., Smithers, L., Haddon, C., Ish-Horowicz, D., and Lewis, J. (2000). Notch signalling and the synchronization of the somite segmentation clock. *Nature* 408, 475–479.

Zimmern, D. (1977). The nucleotide sequence at the origin for assembly on tobacco mosaic virus RNA. *Cell* 11, 463–482.

Zimmern, D. and Butler, P. J. (1977). The isolation of tobacco mosaic virus RNA fragments containing the origin for viral assembly. *Cell* 11, 455–462.

28.3 The Bacterial Genome Is a Nucleoid

Reviews

Brock, T. D. (1988). The bacterial nucleus: a history. *Microbiol. Rev.* 52, 397–411.

Drlica, K. and Rouviere-Yaniv, J. (1987). Histone-like proteins of bacteria. *Microbiol. Rev.* 51, 301–319.

28.4 The Bacterial Genome Is Supercoiled

Review

Hatfield, G. W. and Benham, C. J. (2002). DNA topology-mediated control of global gene expression in *Escherichia coli*. *Annu. Rev. Genet.* 36, 175–203.

Research

Pettijohn, D. E. and Pfenninger, O. (1980). Supercoils in prokaryotic DNA restrained *in vitro*. *Proc. Natl. Acad. Sci. USA* 77, 1331–1335.

Postow, L., Hardy, C. D., Arsuaga, J., and Cozzarelli, N. R. (2004). Topological domain structure of the *Escherichia coli* chromosome. *Genes Dev.* 18, 1766–1779.

28.8 Chromosomes Have Banding Patterns

Research

International Human Genome Sequencing Consortium. (2001). Initial sequencing and analysis of the human genome. *Nature* 409, 860–921.

Saccone, S., De Sario, A., Wiegant, J., Raap, A. K., Della Valle, G., and Bernardi, G. (1993). Correlations between isochores and chromosomal bands in the human genome. *Proc. Natl. Acad. Sci. USA* 90, 11929–11933.

Venter, J. C. et al. (2001). The sequence of the human genome. *Science* 291, 1304–1350.

28.12 The Eukaryotic Chromosome Is a Segregation Device

Review

Hyman, A. A. and Sorger, P. K. (1995). Structure and function of kinetochores in budding yeast. *Annu. Rev. Cell Dev. Biol.* 11, 471–495.

28.13 Centromeres May Contain Repetitive DNA

Review

Wiens, G. R. and Sorger, P. K. (1998). Centromeric chromatin and epigenetic effects in kinetochore assembly. *Cell* 93, 313–316.

Research

Copenhaver, G. P. et al. (1999). Genetic definition and sequence analysis of *Arabidopsis* centromeres. *Science* 286, 2468–2474.

Haaf, T., Warburton, P. E., and Willard, H. F. (1992). Integration of human alpha-satellite DNA into simian chromosomes: centromere protein binding and disruption of normal chromosome segregation. *Cell* 70, 681–696.

Sun, X., Wahlstrom, J., and Karpen, G. (1997). Molecular structure of a functional *Drosophila* centromere. *Cell* 91, 1007–1019.

28.14 Centromeres Have Short DNA Sequences in *S. cerevisiae*

Reviews

Blackburn, E. H. and Szostak, J. W. (1984). The molecular structure of centromeres and telomeres. *Annu. Rev. Biochem.* 53, 163–194.

Clarke, L. and Carbon, J. (1985). The structure and function of yeast centromeres. *Annu. Rev. Genet.* 19, 29–56.

Research

Fitzgerald-Hayes, M., Clarke, L., and Carbon, J. (1982). Nucleotide sequence comparisons and functional analysis of yeast centromere DNAs. *Cell* 29, 235–244.

28.15 The Centromere Binds a Protein Complex

Review

Kitagawa, K. and Hieter, P. (2001). Evolutionary conservation between budding yeast and human kinetochores. *Nat. Rev. Mol. Cell Biol.* 2, 678–687.

Research

Lechner, J. and Carbon, J. (1991). A 240 kd multisubunit protein complex, CBF3, is a major component of the budding yeast centromere. *Cell* 64, 717–725.

Meluh, P. B. and Koshland, D. (1997). Budding yeast centromere composition and assembly as revealed by *in vitro* cross-linking. *Genes Dev.* 11, 3401–3412.

Meluh, P. B. et al. (1998). Cse4p is a component of the core centromere of *S. cerevisiae*. *Cell* 94, 607–613.

Ortiz, J., Stemmann, O., Rank, S., and Lechner, J. (1999). A putative protein complex consisting of Ctf19, Mcm21, and Okp1 represents a missing link in the budding yeast kinetochore. *Genes Dev.* 13, 1140–1155.

28.16 Telomeres Have Simple Repeating Sequences

Reviews

Blackburn, E. H. and Szostak, J. W. (1984). The molecular structure of centromeres and telomeres. *Annu. Rev. Biochem.* 53, 163–194.

Zakian, V. A. (1989). Structure and function of telomeres. *Annu. Rev. Genet.* 23, 579–604.

Research

Wellinger, R. J., Ethier, K., Labrecque, P., and Zakian, V. A. (1996). Evidence for a new step in telomere maintenance. *Cell* 85, 423–433.

28.17 Telomeres Seal the Chromosome Ends

Research

Griffith, J. D. et al. (1999). Mammalian telomeres end in a large duplex loop. *Cell* 97, 503–514.

Henderson, E., Hardin, C. H., Walk, S. K., Tinoco, I., and Blackburn, E. H. (1987). Telomeric oligonucleotides form novel intramolecular structures containing guanine-guanine base pairs. *Cell* 51, 899–908.

Karlseder, J., Broccoli, D., Dai, Y., Hardy, S., and de Lange, T. (1999). p53- and ATM-dependent apoptosis induced by telomeres lacking TRF2. *Science* 283, 1321–1325.

Parkinson, G. N., Lee, M. P., and Neidle, S. (2002). Crystal structure of parallel quadruplexes from human telomeric DNA. *Nature* 417, 876–880.

van Steensel, B., Smogorzewska, A., and de Lange, T. (1998). TRF2 protects human telomeres from end-to-end fusions. *Cell* 92, 401–413.

Williamson, J. R., Raghuraman, K. R., and Cech, T. R. (1989). Monovalent cation-induced structure of telomeric DNA: the G-quartet model. *Cell* 59, 871–880.

28.18 Telomeres Are Synthesized by a Ribonucleoprotein Enzyme

Reviews

Blackburn, E. H. (1991). Structure and function of telomeres. *Nature* 350, 569–573.

Blackburn, E. H. (1992). Telomerases. *Annu. Rev. Biochem.* 61, 113–129.

Collins, K. (1999). Ciliate telomerase biochemistry. *Annu. Rev. Biochem.* 68, 187–218.

Smogorzewska, A. and de Lange, T. (2004). Regulation of telomerase by telomeric proteins. *Annu. Rev. Biochem.* 73, 177–208.

Zakian, V. A. (1995). Telomeres: beginning to understand the end. *Science* 270, 1601–1607.

Zakian, V. A. (1996). Structure, function, and replication of *S. cerevisiae* telomeres. *Annu. Rev. Genet.* 30, 141–172.

Research

Greider, C. and Blackburn, E. H. (1987). The telomere terminal transferase of *Tetrahymena* is a ribonucleoprotein enzyme with two kinds of primer specificity. *Cell* 51, 887–898.

Murray, A., and Szostak, J. W. (1983). Construction of artificial chromosomes in yeast. *Nature* 305, 189–193.

Pennock, E., Buckley, K., and Lundblad, V. (2001). Cdc13 delivers separate complexes to the telomere for end protection and replication. *Cell* 104, 387–396.

Shippen-Lentz, D. and Blackburn, E. H. (1990). Functional evidence for an RNA template in telomerase. *Science* 247, 546–552.

Teixeira, M. T., Arneric, M., Sperisen, P., and Lingner, J. (2004). Telomere length homeostasis is achieved via a switch between telomerase-extendible and -nonextendible states. *Cell* 117, 323–335.

28.19 Telomeres Are Essential for Survival

Research

Hackett, J. A., Feldser, D. M., and Greider, C. W. (2001). Telomere dysfunction increases mutation rate and genomic instability. *Cell* 106, 275–286.

Nakamura, T. M., Cooper, J. P., and Cech, T. R. (1998). Two modes of survival of fission yeast without telomerase. *Science* 282, 493–496.

Nakamura, T. M., Morin, G. B., Chapman, K. B., Weinrich, S. L., Andrews, W. H., Lingner, J., Harley, C. B., and Cech, T. R. (1997). Telomerase catalytic subunit homologs from fission yeast and human. *Science* 277, 955–959.

Rizki, A. and Lundblad, V. (2001). Defects in mismatch repair promote telomerase-independent proliferation. *Nature* 411, 713–716.

29

Nucleosomes

CHAPTER OUTLINE

Continued on next page

29.1 Introduction

Chromatin has a compact organization in which most DNA sequences are structurally inaccessible and functionally inactive. Within this mass is the minority of active sequences. What is the general structure of chromatin, and what is the difference between active and inactive sequences? The high overall packing ratio of the genetic material immediately suggests that DNA cannot be directly packaged into the final structure of chromatin. There must be *hierarchies* of organization.

The fundamental subunit of chromatin has the same type of design in all eukaryotes. The **nucleosome** contains ~200 bp of DNA, organized by an octamer of small, basic proteins into a bead-like structure. The protein components are **histones.** They form an interior core; the DNA lies on the surface of the particle. Nucleosomes are an invariant component of euchromatin and heterochromatin in the interphase nucleus and of mitotic chromosomes. The nucleosome provides the first level of organization, giving a packing ratio of ~6. Its components and structure are well characterized.

The second level of organization is the coiling of the series of nucleosomes into a helical array to constitute the fiber of diameter ~30 nm that is found in both interphase chromatin and mitotic chromosomes (see Figure 28.11). In chromatin this brings the packing ratio of DNA to ~40. The structure of this fiber requires additional proteins, but is not well defined.

The final packing ratio is determined by the third level of organization, the packaging of the 30 nm fiber itself. This gives an overall packing ratio of ~1000 in euchromatin, cyclically interchangeable with packing into mitotic chromosomes to achieve an overall ratio of ~10,000. Heterochromatin generally has a packing ratio of ~10,000 in both interphase and mitosis.

We need to work through these levels of organization to characterize the events involved in cyclical packaging, replication, and transcription. We assume that association with additional proteins, or modifications of existing chromosomal proteins, are involved in changing the structure of chromatin. We do not know the individual targets for controlling cyclical packaging. Both replication and transcription require unwinding of DNA, and thus must involve an unfolding of the structure that allows the relevant enzymes to manipulate the DNA. This is likely to involve changes in all levels of organization.

When chromatin is replicated, the nucleosomes must be reproduced on both daughter duplex molecules. In addition to asking how the nucleosome itself is assembled, we must inquire what happens to other proteins present in chromatin. Replication disrupts the structure of chromatin, which indicates that it both poses a problem for maintaining regions with specific structure and offers an opportunity to change the structure.

The mass of chromatin contains up to twice as much protein as DNA. Approximately half of the protein mass is accounted for by the nucleosomes. The mass of RNA is < 10% of the mass of DNA. Much of the RNA consists of nascent transcripts still associated with the template DNA.

The **nonhistones** include all the proteins of chromatin except the histones. They are more variable between tissues and species, and they comprise a smaller proportion of the mass than the histones. They also comprise a much larger number of proteins, so that any individual protein is present in amounts much smaller than any histone.

The functions of nonhistone proteins include control of gene expression and higher-order structure. Thus RNA polymerase may be considered to be a prominent nonhistone. The HMG (high-mobility group) proteins comprise a discrete and well-defined subclass of nonhistones (at least some of which are transcription factors). A major problem in working with other nonhistones is that they tend to be contaminated with other nuclear proteins, and so far it has proved difficult to obtain those nonhistone proteins responsible for higher-order structures.

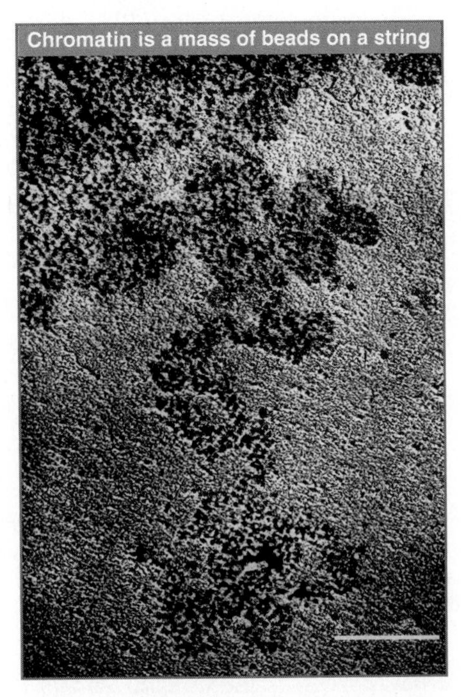

FIGURE 29.1 Chromatin spilling out of lysed nuclei consists of a compactly organized series of particles. The bar is 100 nm. Reproduced from *Cell*, vol. 4, Oudet, P., et al., *Electron microscopic . . .* , pp. 281–300. Copyright 1975, with permission from Elsevier. Photo courtesy of Pierre Chambon.

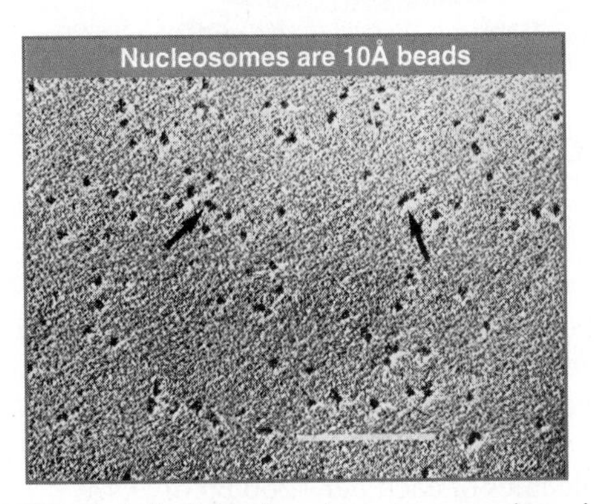

FIGURE 29.2 Individual nucleosomes are released by digestion of chromatin with micrococcal nuclease. The bar is 100 nm. Reproduced from *Cell*, vol. 4, Oudet, P., et al., *Electron microscopic . . .* , pp. 281–300. Copyright 1975, with permission from Elsevier. Photo courtesy of Pierre Chambon.

29.2 The Nucleosome Is the Subunit of All Chromatin

Key concepts

- Micrococcal nuclease releases individual nucleosomes from chromatin as 11S particles.
- A nucleosome contains ~200 bp of DNA, two copies of each core histone (H2A, H2B, H3, and H4).
- DNA is wrapped around the outside surface of the protein octamer.

When interphase nuclei are suspended in a solution of low ionic strength, they swell and rupture to release fibers of chromatin. FIGURE 29.1 shows a lysed nucleus in which fibers are streaming out. In some regions, the fibers consist of tightly packed material, but in regions that have become stretched, they can be seen to consist of discrete particles. These are the nucleosomes. In especially extended regions, individual nucleosomes are connected by a fine thread, which is a free duplex of DNA. *A continuous duplex thread of DNA runs through the series of particles.*

Individual nucleosomes can be obtained by treating chromatin with the endonuclease **micrococcal nuclease,** which cuts the DNA thread at the junction between nucleosomes. First it releases groups of particles, and then, it releases single nucleosomes. Individual nucleosomes can be seen in FIGURE 29.2 as compact particles. They sediment at ~11S.

The nucleosome contains ~200 bp of DNA associated with a histone octamer that consists of two copies each of H2A, H2B, H3, and H4. These are known as the **core histones.** Their association is

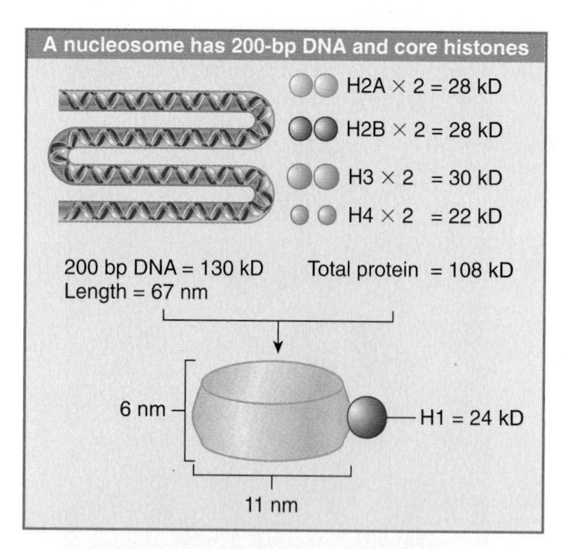

FIGURE 29.3 The nucleosome consists of approximately equal masses of DNA and histones (including H1). The predicted mass of the nucleosome is 262 kD.

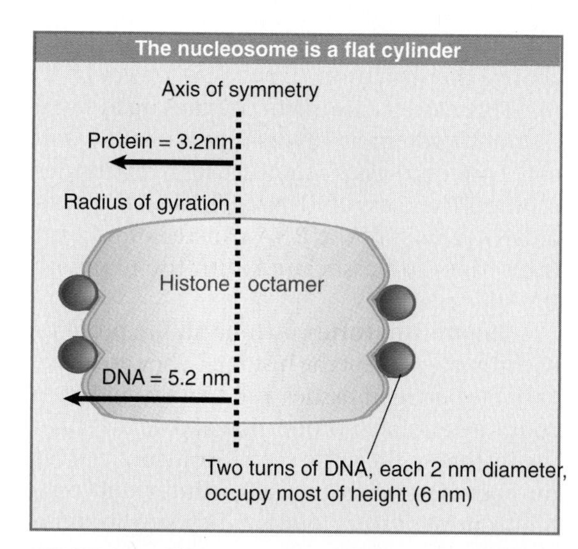

FIGURE 29.5 The two turns of DNA on the nucleosome lie close together.

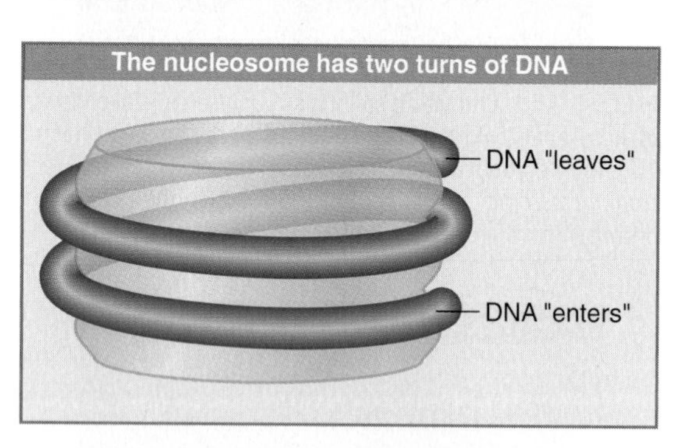

FIGURE 29.4 The nucleosome may be a cylinder with DNA organized into two turns around the surface.

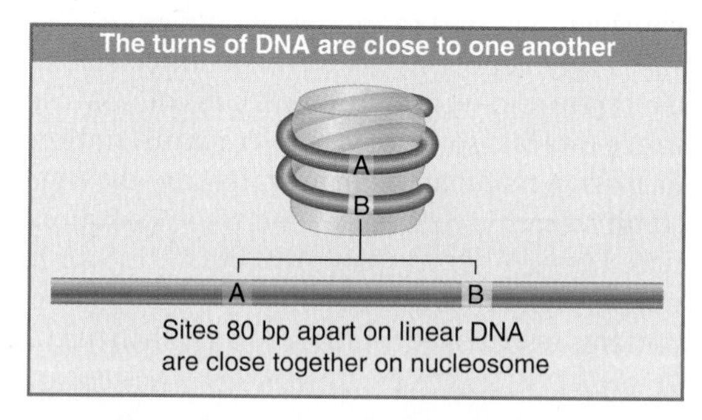

FIGURE 29.6 Sequences on the DNA that lie on different turns around the nucleosome may be close together.

illustrated diagrammatically in FIGURE 29.3. This model explains the stoichiometry of the core histones in chromatin: H2A, H2B, H3, and H4 are present in equimolar amounts, with two molecules of each per ~200 bp of DNA.

Histones H3 and H4 are among the most conserved proteins known. This suggests that their functions are identical in all eukaryotes. The types of H2A and H2B can be recognized in all eukaryotes, but show appreciable species-specific variation in sequence.

Histone H1 comprises a set of closely related proteins that show appreciable variation between tissues and between species. The role of H1 is different from that of the core histones. It is present in half the amount of a core histone and can be extracted more readily from chromatin (typically with dilute salt [0.5 M] solution). *The H1 can be removed without affecting the structure of the nucleosome, which suggests that its location is external to the particle.*

The shape of the nucleosome corresponds to a flat disk or cylinder of diameter 11 nm and height 6 nm. The length of the DNA is roughly twice the ~34 nm circumference of the particle. The DNA follows a symmetrical path around the octamer. FIGURE 29.4 shows the DNA path diagrammatically as a helical coil that makes two turns around the cylindrical octamer. Note that the DNA "enters" and "leaves" the nucleosome at points close to one another. Histone H1 may be located in this region (see Section 29.4, Nucleosomes Have a Common Structure).

Considering this model in terms of a cross-section through the nucleosome, in FIGURE 29.5 we see that the two circumferences made by the DNA lie close to one another. The height of the cylinder is 6 nm, of which 4 nm is occupied by the two turns of DNA (each of diameter 2 nm).

The pattern of the two turns has a possible functional consequence. One turn around the nucleosome takes ~80 bp of DNA, so two points separated by 80 bp in the free double helix may actually be close on the nucleosome surface, as illustrated in FIGURE 29.6.

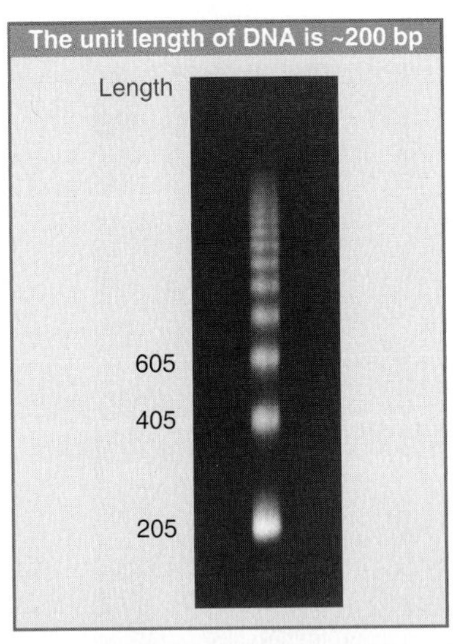

FIGURE 29.7 Micrococcal nuclease digests chromatin in nuclei into a multimeric series of DNA bands that can be separated by gel electrophoresis. Photo courtesy of Markus Noll, Urniversität Zürich.

29.3 DNA Is Coiled in Arrays of Nucleosomes

Key concepts

- >95% of the DNA is recovered in nucleosomes or multimers when micrococcal nuclease cleaves DNA of chromatin.
- The length of DNA per nucleosome varies for individual tissues in a range from 154 to 260 bp.

When chromatin is digested with the enzyme micrococcal nuclease, the DNA is cleaved into integral multiples of a unit length. Fractionation by gel electrophoresis reveals the "ladder" presented in FIGURE 29.7. Such ladders extend for ~10 steps, and the unit length, determined by the increments between successive steps, is ~200 bp.

FIGURE 29.8 shows that the ladder is generated by groups of nucleosomes. When nucleosomes are fractionated on a sucrose gradient, they give a series of discrete peaks that correspond to monomers, dimers, trimers, and so on. When the DNA is extracted from the individual fractions and electrophoresed, each fraction yields a band of DNA whose size corresponds with a step on the micrococcal nuclease ladder. The monomeric nucleosome contains DNA of the unit length, the nucleosome dimer contains DNA of twice the unit length, and so on.

Each step on the ladder represents the DNA derived from a discrete number of nucleosomes. *We therefore take the existence of the 200 bp ladder in any chromatin to indicate that the DNA is organized into nucleosomes.* The micrococcal ladder is generated when only ~2% of the DNA in the nucleus is rendered acid-soluble (degraded to small fragments) by the enzyme. *Thus a small proportion of the DNA is specifically attacked; it must represent especially susceptible regions.*

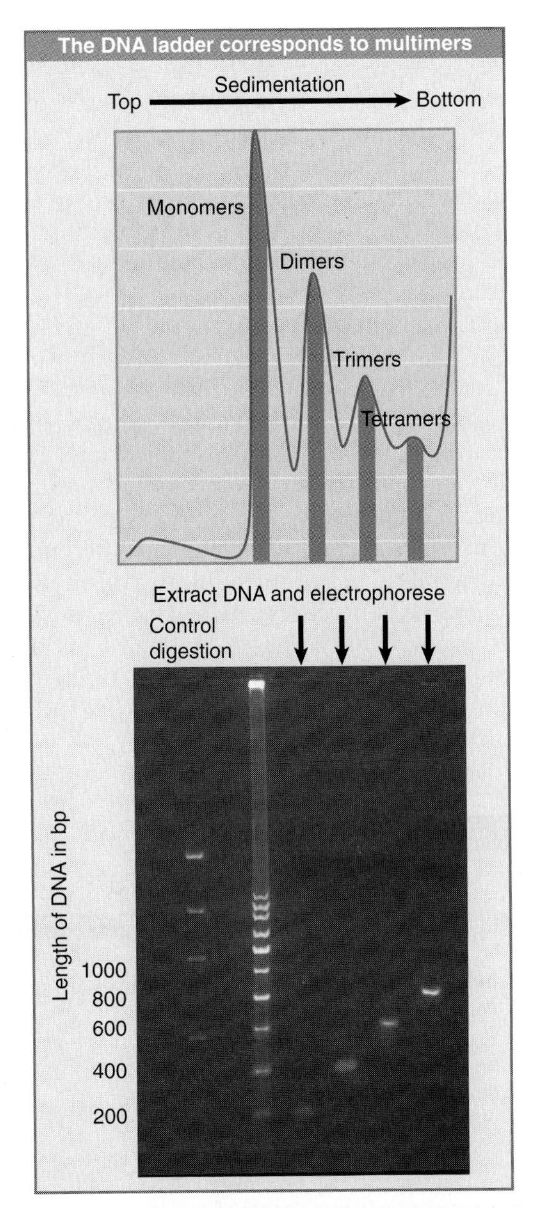

FIGURE 29.8 Each multimer of nucleosomes contains the appropriate number of unit lengths of DNA. In the photo, artificial bands simulate a DNA ladder. The image was constructed using PCR fragments with sizes corresponding to actual band sizes. Photo courtesy of Jan Kieleczawa, Wyeth Research.

When chromatin is spilled out of nuclei, we often see a series of nucleosomes connected by a thread of free DNA (the beads on a string). The need for tight packaging of DNA *in vivo,* however, suggests that probably there is usually little (if any) free DNA.

This view is confirmed by the fact that >95% *of the DNA of chromatin can be recovered in the form of the 200 bp ladder.* Almost all DNA must therefore be organized in nucleosomes. In their natural state, nucleosomes are likely to be closely packed, with DNA passing directly from one to the next. Free DNA is probably generated by the loss of some histone octamers during isolation.

The length of DNA present in the nucleosome varies somewhat from the "typical" value of 200 bp. The chromatin of any particular cell type has a characteristic average value (±5 bp). The average most often is between 180 and 200, but there are extremes as low as 154 bp (in a fungus) or as high as 260 bp (in a sea urchin sperm). The average value may be different in individual tissues of the adult organism, and there can be differences between different parts of the genome in a single cell type. Variations from the genome average include tandemly repeated sequences, such as clusters of 5S RNA genes.

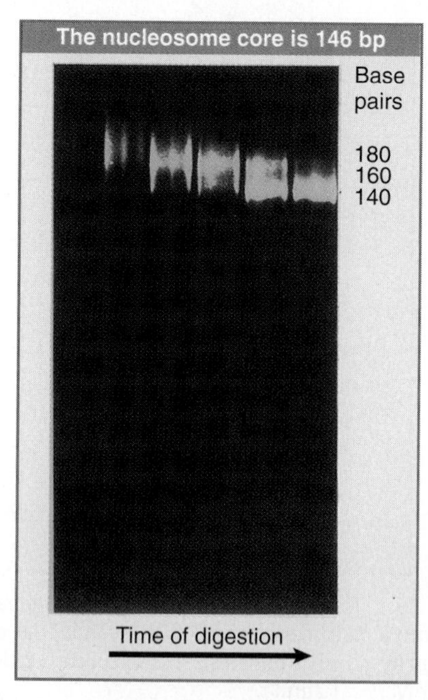

FIGURE 29.9 Micrococcal nuclease reduces the length of nucleosome monomers in discrete steps. Photo courtesy of Roger Kornberg, Stanford University School of Medicine.

29.4 Nucleosomes Have a Common Structure

Key concepts

- Nucleosomal DNA is divided into the core DNA and linker DNA depending on its susceptibility to micrococcal nuclease.
- The core DNA is the length of 146 bp that is found on the core particles produced by prolonged digestion with micrococcal nuclease.
- Linker DNA is the region of 8 to 114 bp that is susceptible to early cleavage by the enzyme.
- Changes in the length of linker DNA account for the variation in total length of nucleosomal DNA.
- H1 is associated with linker DNA and may lie at the point where DNA enters and leaves the nucleosome.

A common structure underlies the varying amount of DNA that is contained in nucleosomes of different sources. The association of DNA with the histone octamer forms a core particle containing 146 bp of DNA, irrespective of the total length of DNA in the nucleosome. The variation in total length of DNA per nucleosome is superimposed on this basic core structure.

The core particle is defined by the effects of micrococcal nuclease on the nucleosome monomer. The initial reaction of the enzyme is to cut between nucleosomes, but if it is allowed to continue after monomers have been generated, then it proceeds to digest some of the DNA of the individual nucleosome. This occurs by a reaction in which DNA is "trimmed" from the ends of the nucleosome.

The length of the DNA is reduced in discrete steps, as shown in FIGURE 29.9. With rat liver nuclei, the nucleosome monomers initially have 205 bp of DNA. After the first step, some monomers are found in which the length of DNA has been reduced to ~165 bp. Finally, this is reduced to the length of the DNA of the core particle, 146 bp. (The core is reasonably stable, but continued digestion generates a "limit digest." In the limit digest, the longest fragments are the 146 bp DNA of the core, whereas the shortest are as small as 20 bp.)

This analysis suggests that the nucleosomal DNA can be divided into two regions:

- Core DNA has an invariant length of 146 bp, and is relatively resistant to digestion by nucleases.
- Linker DNA comprises the rest of the repeating unit. Its length varies from as little as 8 bp to as much as 114 bp per nucleosome.

The sharp size of the band of DNA generated by the initial cleavage with micrococcal nuclease suggests that the region immediately available to the enzyme is restricted. It represents only part of each linker. (If the entire linker DNA were susceptible, the band would range from 146 bp to >200 bp.) Once a cut has been made in the linker DNA, though, the rest of this region becomes susceptible, and it can be removed relatively rapidly by further enzyme action. The connection between nucleosomes is represented in FIGURE 29.10.

Core particles have properties similar to those of the nucleosomes themselves, although they are smaller. Their shape and size are similar to those of nucleosomes; this suggests that the essential geometry of the particle is established by the interactions between DNA and the protein octamer in the core particle. Core particles are more readily obtained as a homogeneous population, and as a result they are often used for structural studies in preference to nucleosome preparations. (Nucleosomes tend to vary because it is difficult to obtain a preparation in which there has been no end-trimming of the DNA.)

What is the physical nature of the core and the linker regions? *These terms were introduced as operational definitions that describe the regions in terms of their relative susceptibility to nuclease treatment.* This description does not make any implication about their actual structure. It turns out, though, that the major part of the core DNA is tightly curved on the nucleosome, whereas the terminal regions of the core and the linker regions are more extended (see Section 29.5, DNA Structure Varies on the Nucleosomal Surface).

The existence of linker DNA depends on factors extraneous to the four core histones. Reconstitution experiments *in vitro* show that histones have an intrinsic ability to organize DNA into core particles, but do not form nucleosomes with the proper unit length. The degree of supercoiling of the DNA is an important factor. Histone H1 and/or nonhistone proteins influence the length of linker DNA associated with the histone octamer in a natural series of nucleosomes. "Assembly proteins" that are not part of the nucleosome structure are involved *in vivo* in constructing nucleosomes from histones and DNA (see Section 29.9, Reproduction of Chromatin Requires Assembly of Nucleosomes).

Where is histone H1 located? The H1 is lost during the degradation of nucleosome monomers. It can be retained on monomers that still have 165 bp of DNA, but is always lost with the final reduction to the 146 bp core particle. This suggests that H1 could be located in the region of the linker DNA immediately adjacent to the core DNA.

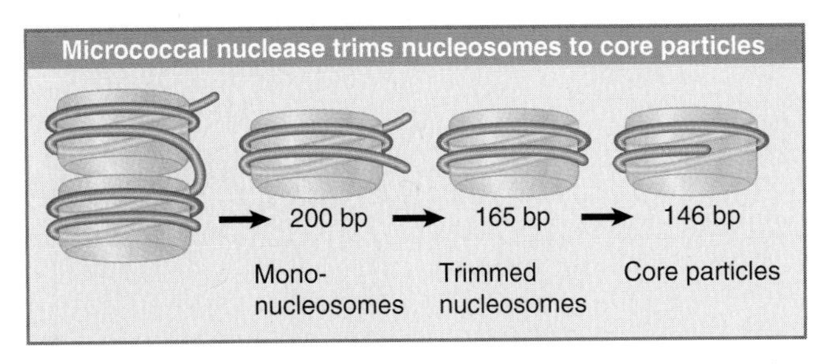

FIGURE 29.10 Micrococcal nuclease initially cleaves between nucleosomes. Mononucleosomes typically have ~200 bp DNA. End-trimming reduces the length of DNA first to ~165 bp, and then generates core particles with 146 bp.

If H1 is located at the linker, it could "seal" the DNA in the nucleosome by binding at the point where the nucleic acid enters and leaves (see Figure 29.4). The idea that H1 lies in the region joining adjacent nucleosomes is consistent with old results that H1 is removed the most readily from chromatin, and that H1-depleted chromatin is more readily "solubilized." In addition, it is easier to obtain a stretched-out fiber of beads on a string when the H1 has been removed.

29.5 DNA Structure Varies on the Nucleosomal Surface

Key concepts

- DNA is wrapped 1.65 times around the histone octamer.
- The structure of the DNA is altered so that it has an increased number of base pairs/turn in the middle, but a decreased number at the ends.

The exposure of DNA on the surface of the nucleosome explains why it is accessible to cleavage by certain nucleases. The reaction with nucleases that attack single strands has been especially informative. The enzymes DNAase I and DNAase II make single-strand nicks in DNA; they cleave a bond in one strand, but the other strand remains intact at this point. Thus no effect is visible in the double-stranded DNA. Upon denaturation, though, short fragments are released instead of full-length single strands. If the DNA has been labeled at its ends, the end fragments can be identified by autoradiography, as summarized in FIGURE 29.11. When DNA is free in solution, it is nicked (relatively) at random. The DNA on nucleosomes also can be nicked by the enzymes, *but only at regular intervals.* When the points of cutting are

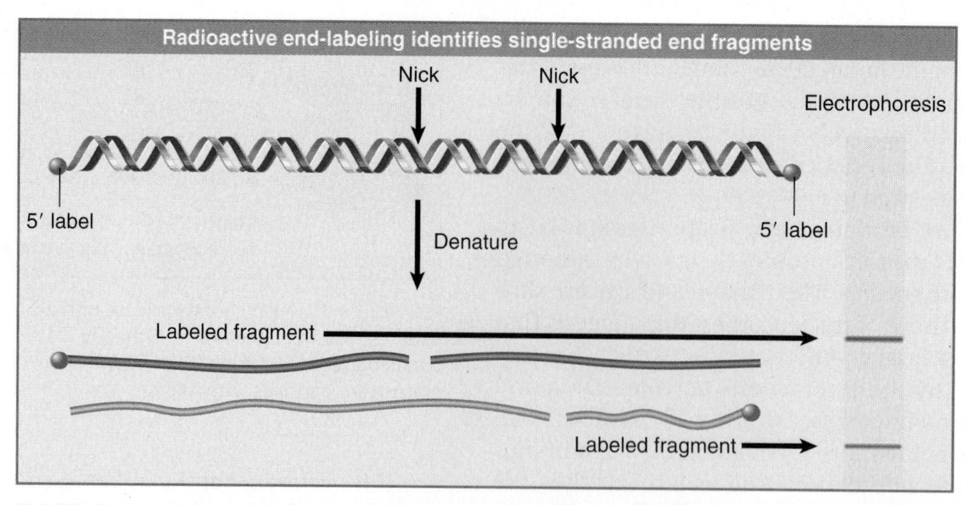

FIGURE 29.11 Nicks in double-stranded DNA are revealed by fragments when the DNA is denatured to give single strands. If the DNA is labeled at (say) 5′ ends, only the 5′ fragments are visible by autoradiography. The size of the fragment identifies the distance of the nick from the labeled end.

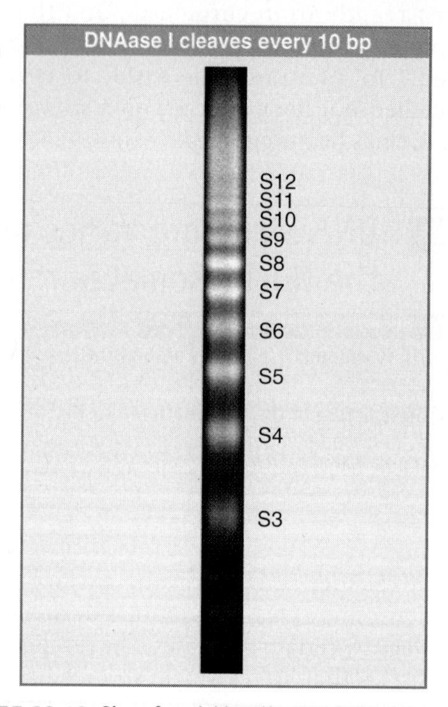

FIGURE 29.12 Sites for nicking lie at regular intervals along core DNA, as seen in a DNAase I digest of nuclei. Photo courtesy of Leonard C. Lutter, Henry Ford Hospital, Detroit, MI.

determined by using radioactively end-labeled DNA and then DNA is denatured and electrophoresed, a ladder of the sort displayed in FIGURE 29.12 is obtained.

The interval between successive steps on the ladder is 10 to 11 bases. The ladder extends for the full distance of core DNA. The cleavage sites are numbered as S1 through S13 (where S1 is ~10 bases from the labeled 5′ end, S2 is ~20 bases from it, and so on). Their positions relative to the DNA superhelix are illustrated in FIGURE 29.13.

Not all sites are cut with equal frequency: Some are cut rather efficiently, whereas others are cut scarcely at all. The enzymes DNAase I and DNAase II generate the same ladder, although with some differences in the intensities of the bands. This shows that the pattern of cutting represents a unique series of targets in DNA, determined by its organization, with only some slight preference for particular sites imposed by the individual enzyme. The same cutting pattern is obtained by cleaving with a hydroxyl radical, which argues that the pattern reflects the structure of the DNA itself rather than any sequence preference.

The sensitivity of nucleosomal DNA to nucleases is analogous to a footprinting experiment. Thus we can assign the lack of reaction at particular target sites to the structure of the nucleosome, in which certain positions on DNA are rendered inaccessible.

There are two strands of DNA in the core particle, so in an end-labeling experiment both of the 5′ (or 3′) ends are labeled, one on each strand. Thus the cutting pattern includes fragments derived from both strands. This is implied in Figure 29.11, where each labeled fragment is derived from a different strand. The corollary is that, in an experiment, each labeled band in fact can represent two fragments that are generated by cutting the *same* distance from *either* of the labeled ends.

How, then, should we interpret discrete preferences at particular sites? One view is that the path of DNA on the particle is symmetrical (about a horizontal axis through the nucleosome, as drawn in Figure 29.4). If, for example, no 80-base fragment is generated by DNAase I, this must mean that the position at 80 bases from the 5′ end of *either* strand is not susceptible to the enzyme. The second numbering scheme used in Figure 29.13 reflects this view, and identifies S7 = site 0 as the center of symmetry.

When DNA is immobilized on a flat surface, sites are cut with a regular separation. FIGURE 29.14 suggests that this reflects the recurrence of the exposed site with the helical periodicity of B-form DNA. The cutting periodicity (the spacing between cleavage points) coincides with—indeed, is a reflection of—the structural periodicity (the number of base pairs per turn of the double helix). Thus the distance between the sites corresponds to the number of base pairs per turn. Measurements of this type suggest that the average value for double-helical B-type DNA is 10.5 bp/turn.

What is the nature of the target sites on the nucleosome? FIGURE 29.15 shows that each site has three to four positions at which cutting occurs; that is, the cutting site is defined ±2 bp. Thus a cutting site represents a short stretch of bonds on both strands that is exposed to nuclease action over three to four base pairs. The relative intensities indicate that some sites are preferred to others.

From this pattern, we can calculate the "average" point that is cut. At the ends of the DNA, pairs of sites from S1 to S4 or from S10 to S13 lie apart a distance of 10.0 bases each. In the center of the particle, the separation from sites S4 to S10 is 10.7 bases. (This analysis deals with *average* positions, so sites need not lie at an integral number of bases apart.)

The variation in cutting periodicity along the core DNA (10.0 at the ends, 10.7 in the middle) means that there is variation in the structural periodicity of core DNA. The DNA has more bp/turn than its solution value in the middle, but has fewer bp/turn at the ends. The average periodicity over the nucleosome is only 10.17 bp/turn, which is significantly less than the 10.5 bp/turn of DNA in solution.

The crystal structure of the core particle suggests that DNA is organized as a flat superhelix, with 1.65 turns wound around the histone octamer. The pitch of the superhelix varies and has a discontinuity in the middle. Regions

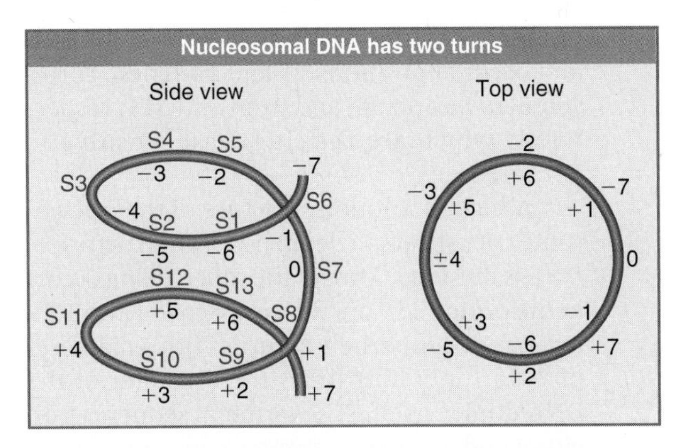

FIGURE 29.13 Two numbering schemes divide core particle DNA into 10 bp segments. Sites may be numbered S1 to S13 from one end; or taking S7 to identify coordinate 0 of the dyad symmetry, they may be numbered −7 to +7.

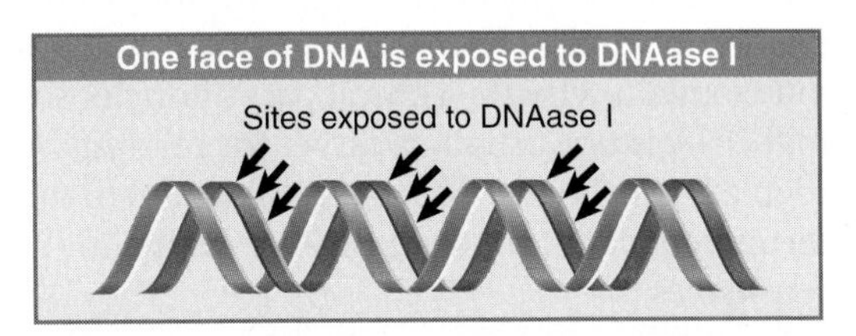

FIGURE 29.14 The most exposed positions on DNA recur with a periodicity that reflects the structure of the double helix. (For clarity, sites are shown for only one strand.)

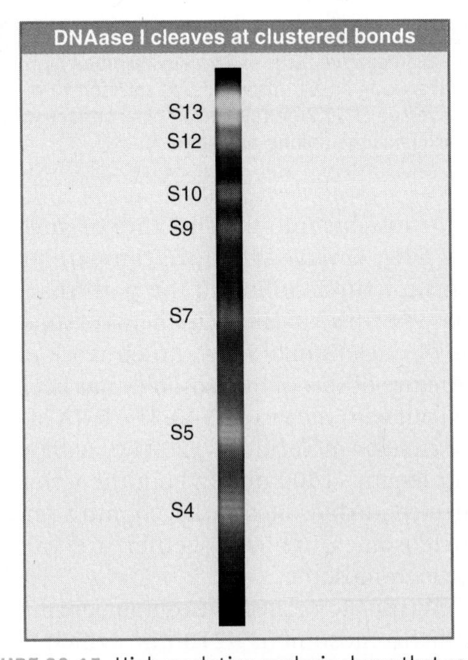

FIGURE 29.15 High resolution analysis shows that each site for DNAase I consists of several adjacent susceptible phosphodiester bonds, as seen in this example of sites S4 and S5 analyzed in end-labeled core particles. Photo courtesy of Leonard C. Lutter, Henry Ford Hospital, Detroit, MI.

of high curvature are arranged symmetrically, and occur at positions ±1 and ±4. These correspond to S6 and S8, and to S3 and S11, respectively, which are the sites least sensitive to DNAase I.

A high-resolution structure of the nucleosome core shows in detail how the structure of DNA is distorted. Most of the supercoiling occurs in the central 129 bp, which are coiled into 1.59 left-handed superhelical turns with a diameter of 80 Å (only four times the diameter of the DNA duplex itself). The terminal sequences on either end make only a very small contribution to the overall curvature.

The central 129 bp are in the form of B-DNA, but with a substantial curvature that is needed to form the superhelix. The major groove is smoothly bent, but the minor groove has abrupt kinks. These conformational changes may explain why the central part of nucleosomal DNA is not usually a target for binding by regulatory proteins, which typically bind to the terminal parts of the core DNA or to the linker sequences.

29.6 The Periodicity of DNA Changes on the Nucleosome

Key concept

- ~0.6 negative turns of DNA are absorbed by the change in bp/turn from 10.5 in solution to an average of 10.2 on the nucleosomal surface, which explains the linking-number paradox.

Some insights into the structure of nucleosomal DNA emerge when we compare predictions for supercoiling in the path that DNA follows with actual measurements of supercoiling of nucleosomal DNA. Much work on the structure of sets of nucleosomes has been carried out with the virus SV40. The DNA of SV40 is a circular molecule of 5200 bp, with a contour length ~1500 nm. In both the virion and infected nucleus, it is packaged into a series of nucleosomes, which together are called a minichromosome.

As usually isolated, the contour length of the minichromosome is ~210 nm, which corresponds to a packing ratio of ~7 (essentially the same as the ~6 of the nucleosome itself). Changes in the salt concentration can convert it to a flexible string of beads with a much lower overall packing ratio. This emphasizes the point that nucleosome strings can take more than one form *in vitro,* depending on the conditions.

The degree of supercoiling on the individual nucleosomes of the minichromosome can be measured as illustrated in FIGURE 29.16. First, the free supercoils of the minichromosome itself are relaxed, so that the nucleosomes form a circular string with a superhelical density of 0. Next, the histone octamers are extracted. This releases the DNA to follow a free path. Every supercoil that was present but constrained in the minichromosome will appear in the deproteinized DNA as –1 turn. Now the total number of supercoils in the SV40 DNA is measured.

The observed value is close to the number of nucleosomes. The reverse result is seen when nucleosomes are assembled *in vitro* onto a supercoiled SV40 DNA: The formation of each nucleosome removes ~1 negative supercoil.

Thus the DNA follows a path on the nucleosomal surface that generates ~1 negative supercoiled turn when the restraining protein is removed. The path that DNA follows on the nucleosome, though, corresponds to –1.67 superhelical turns (see Figure 29.4). This discrepancy is sometimes called the linking number paradox.

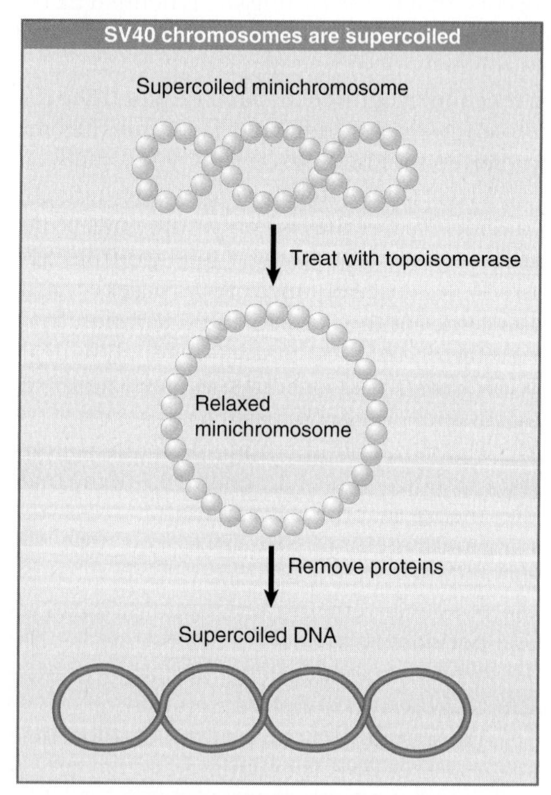

FIGURE 29.16 The supercoils of the SV40 minichromosome can be relaxed to generate a circular structure, whose loss of histones then generates supercoils in the free DNA.

The discrepancy is explained by the difference between the 10.17 average bp/turn of nucleosomal DNA and the 10.5 bp/turn of free DNA. In a nucleosome of 200 bp, there are 200/10.17 = 19.67 turns. When DNA is released from the nucleosome, it now has 200/10.5 = 19.0 turns. The path of the less tightly wound DNA on the nucleosome absorbs −0.67 turns, which explains the discrepancy between the physical path of −1.67 and the measurement of −1.0 superhelical turns. In effect, some of the torsional strain in nucleosomal DNA goes into increasing the number of bp/turn; only the rest is left to be measured as a supercoil.

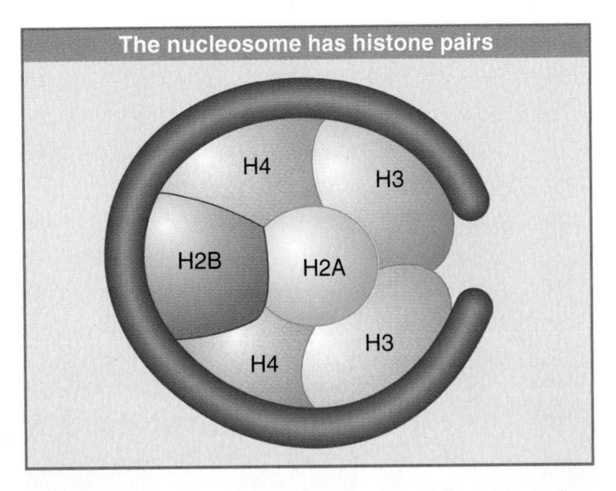

FIGURE 29.17 In a symmetrical model for the nucleosome, the $H3_2$-$H4_2$ tetramer provides a kernel for the shape. One H2A-H2B dimer can be seen in the top view; the other is underneath.

29.7 Organization of the Histone Octamer

Key concepts

- The histone octamer has a kernel of an $H3_2$-$H4_2$ tetramer associated with two H2A-H2B dimers.
- Each histone is extensively interdigitated with its partner.
- All core histones have the structural motif of the histone fold. N-terminal tails extend out of the nucleosome.

Thus far we have considered the construction of the nucleosome from the perspective of how the DNA is organized on the surface. From the perspective of protein, we need to know how the histones interact with each other and with DNA. Do histones react properly only in the presence of DNA, or do they possess an independent ability to form octamers? Most of the evidence about histone–histone interactions is provided by their abilities to form stable complexes, and by crosslinking experiments with the nucleosome.

The core histones form two types of complexes. H3 and H4 form a tetramer ($H3_2$-$H4_2$). Various complexes are formed by H2A and H2B, in particular a dimer (H2A-H2B).

Intact histone octamers can be obtained either by extraction from chromatin or (with more difficulty) by letting histones associate *in vitro* under conditions of high-salt and high-protein concentration. The octamer can dissociate to generate a hexamer of histones that has lost an H2A-H2B dimer. The other H2A-H2B dimer is lost separately at this point, which leaves the $H3_2$-$H4_2$ tetramer. This argues for a form of organization in which the nucleosome has a central "kernel" consisting of the $H3_2$-$H4_2$ tetramer. The tetramer can organize DNA *in vitro* into particles that display some of the properties of the core particle.

Crosslinking studies extend these relationships to show which pairs of histones lie near each other in the nucleosome. (A difficulty with such data is that usually only a small proportion of the proteins becomes crosslinked, so it is necessary to be cautious in deciding whether the results typify the major interactions.) From these data, a model has been constructed for the organization of the nucleosome. It is shown in diagrammatic form in FIGURE 29.17.

Structural studies show that the overall shape of the isolated histone octamer is similar to that of the core particle. This suggests that the histone–histone interactions establish the general structure. The positions of the individual histones have been assigned to regions of the octameric structure on the basis of their interaction behavior and response to crosslinking.

The crystal structure (at 3.1 Å resolution) suggests the model for the histone octamer shown in FIGURE 29.18. Tracing the paths of the individual polypeptide backbones in the crystal structure suggests that the histones are not organized as individual globular proteins, but that each is interdigitated with its partner: H3 with H4, and H2A with H2B. Thus the model distinguishes the $H3_2$-$H4_2$ tetramer (white) from the H2A-H2B dimers (blue) but does not show individual histones.

The top view represents the same perspective that was illustrated schematically in Figure 29.17. The $H3_2$-$H4_2$ tetramer accounts for the diameter of the octamer. It forms the shape

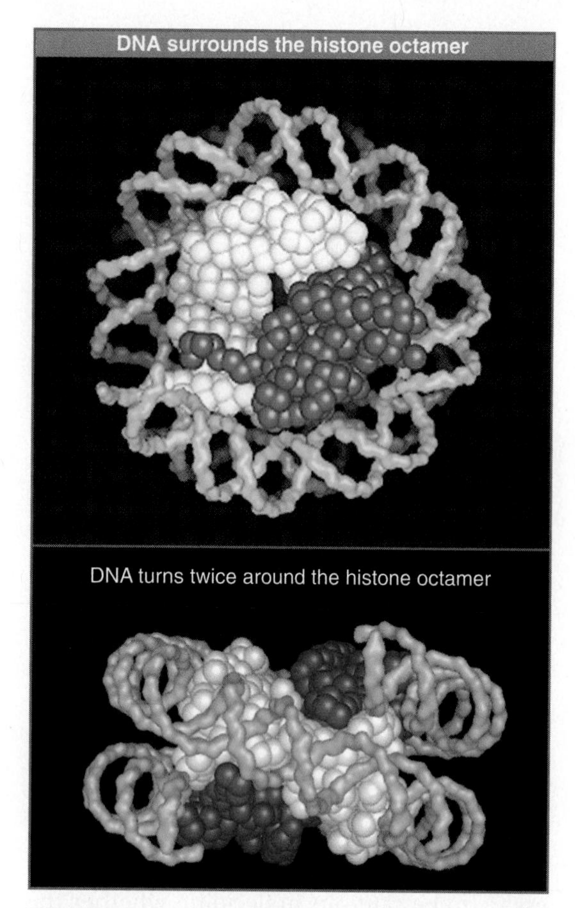

FIGURE 29.18 The crystal structure of the histone core octamer is represented in a space-filling model with the $H3_2$-$H4_2$ tetramer shown in white and the H2A-H2B dimers shown in blue. Only one of the H2A-H2B dimers is visible in the top view, because the other is hidden underneath. The potential path of the DNA is shown in the top view as a narrow tube (one quarter the diameter of DNA), and in the side view is shown by the parallel lines in a 20 Å-wide bundle. Photos courtesy of E. N. Moudrianakis, Johns Hopkins University.

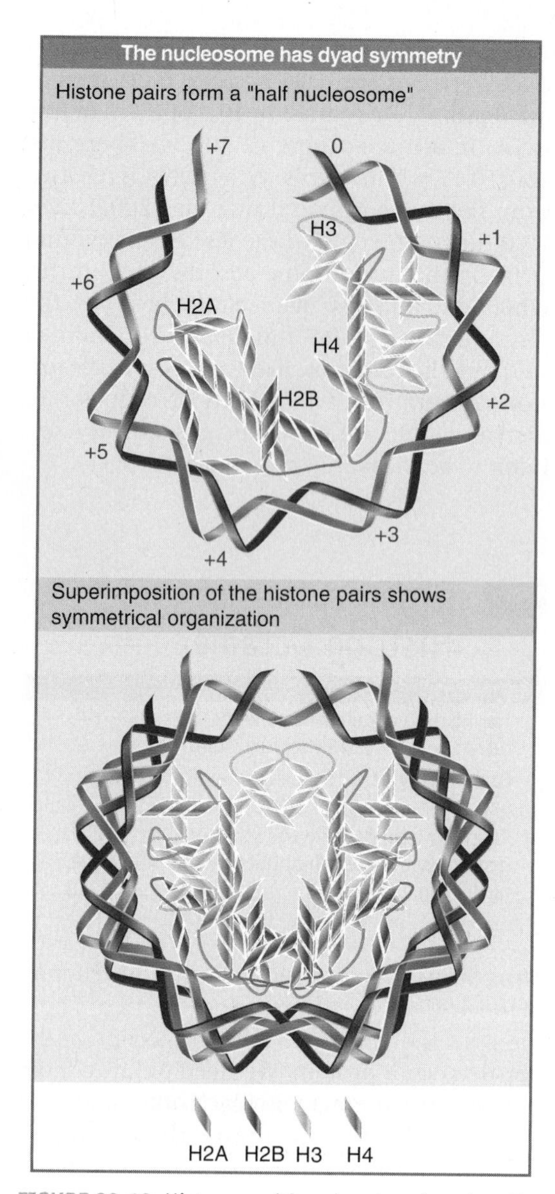

FIGURE 29.19 Histone positions in a top view show H3-H4 and H2A-H2B pairs in a half nucleosome; the symmetrical organization can be seen in the superimposition of both halves in the bottom view.

of a horseshoe. The H2A-H2B pairs fit in as two dimers, but only one can be seen in this view. The side view represents the same perspective that was illustrated in Figure 29.4. Here the responsibilities of the $H3_2$-$H4_2$ tetramer and of the separate H2A-H2B dimers can be distinguished. The protein forms a sort of spool, with a superhelical path that could correspond to the binding site for DNA, which would be wound in almost two full turns in a nucleosome. The model displays twofold symmetry about an axis that would run perpendicular through the side view.

A more detailed view of the positions of the histones (based on a crystal structure at 2.8 Å) is summarized in FIGURE 29.19. The upper view shows the position of one histone of each type relative to one turn around the nucleosome (numbered from 0 to +7). All four core histones show a similar type of structure in which three α helices are connected by two loops: This is called the **histone fold.** These regions interact to form crescent-shaped heterodimers; each heterodimer binds 2.5 turns of the DNA double helix. (H2A-H2B binds at +3.5 – +6; H3-H4 binds at +0.5 – +3 for the circumference that is illustrated.) Binding is mostly to the phosphodiester backbones (consistent with the need to package any DNA irrespective of sequence). The $H3_2$-$H4_2$ tetramer is formed by interactions between the two H3 subunits, as can be seen in the lower part of the figure.

Each of the core histones has a globular body that contributes to the central protein mass of the nucleosome. Each histone also has a flex-

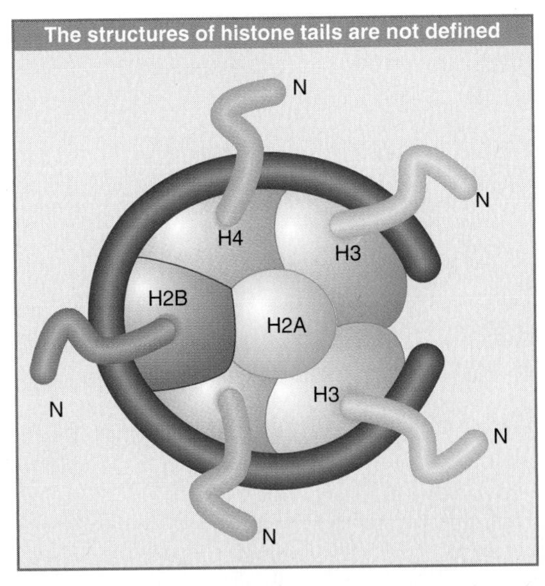

FIGURE 29.20 The globular bodies of the histones are localized in the histone octamer of the core particle. The locations of the N-terminal tails, which carry the sites for modification, are not known, though, and could be more flexible.

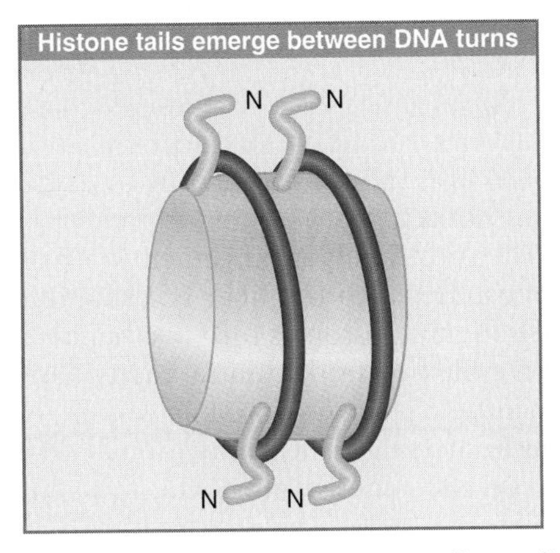

FIGURE 29.21 The N-terminal histone tails are disordered and exit from the nucleosome between turns of the DNA.

ible N-terminal tail, which has sites for modification that may be important in chromatin function. The positions of the tails, which account for about one quarter of the protein mass, are not so well defined, as indicated in FIGURE 29.20. The tails of both H3 and H2B, however, can be seen to pass between the turns of the DNA superhelix and extend out of the nucleosome, as shown in FIGURE 29.21. When histone tails are crosslinked to DNA by UV irradiation, more products are obtained with nucleosomes compared to core particles, which could mean that the tails contact the linker DNA. The tail of H4 appears to contact an H2A-H2B dimer in an adjacent nucleosome; this could be an important feature in the overall structure.

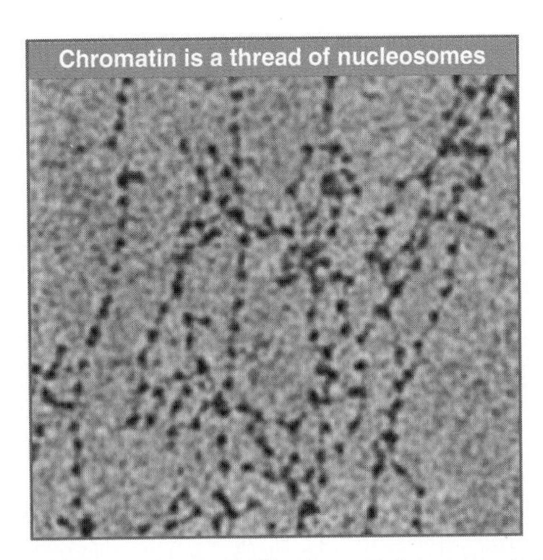

FIGURE 29.22 The 10 nm fiber in partially unwound state can be seen to consist of a string of nucleosomes. Photo courtesy of Barbara Hamkalo, University of California, Irvine.

29.8 The Path of Nucleosomes in the Chromatin Fiber

Key concepts

- 10 nm chromatin fibers are unfolded from 30 nm fibers and consist of a string of nucleosomes.
- 30 nm fibers have six nucleosomes/turn, which are organized into a solenoid.
- Histone H1 is required for formation of the 30 nm fiber.

When chromatin is examined in the electron microscope, two types of fibers are seen: the 10 nm fiber and 30 nm fiber. They are described by the approximate diameter of the thread (that of the 30 nm fiber actually varies from ~25–30 nm).

The **10 nm fiber** is essentially a continuous string of nucleosomes. In fact, at times it runs continuously into a more stretched-out region in which nucleosomes are seen as a string of beads, as indicated in the example of FIGURE 29.22. The 10 nm fibril structure is obtained under conditions of low ionic strength and does not require the presence of histone H1. This means that it is a function strictly of the nucleosomes themselves. It may be visualized essentially as a continuous series of nucleosomes, as shown in FIGURE 29.23. It is not clear whether such a structure exists *in vivo* or is simply a consequence of unfolding during extraction *in vitro*.

When chromatin is visualized in conditions of greater ionic strength, the **30 nm fiber** is

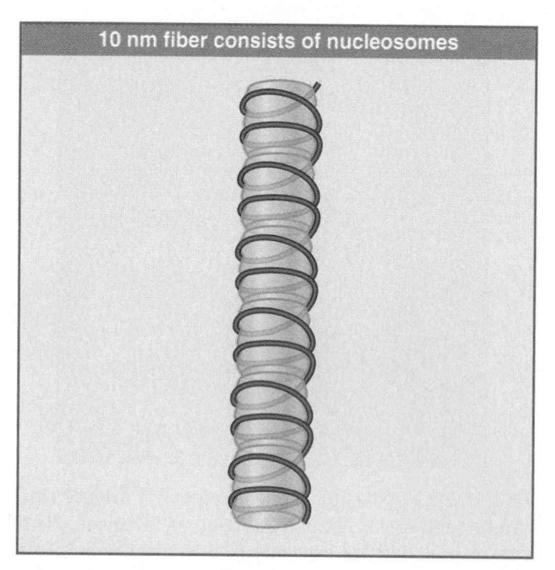

FIGURE 29.23 The 10 nm fiber is a continuous string of nucleosomes.

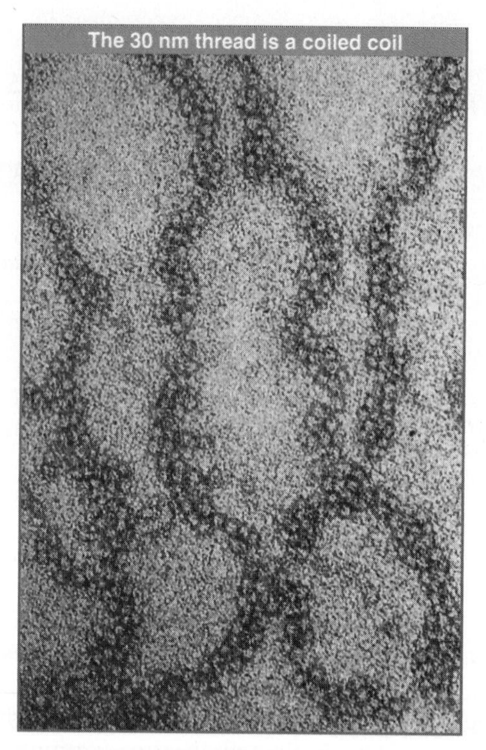

FIGURE 29.24 The 30 nm fiber has a coiled structure. Photo courtesy of Barbara Hamkalo, University of California, Irvine.

obtained. An example is given in FIGURE 29.24. The fiber can be seen to have an underlying coiled structure. It has ~6 nucleosomes for every turn, which corresponds to a packing ratio of 40 (that is, each μm along the axis of the fiber contains 40 μm of DNA). The presence of H1 is required. This fiber is the basic constituent

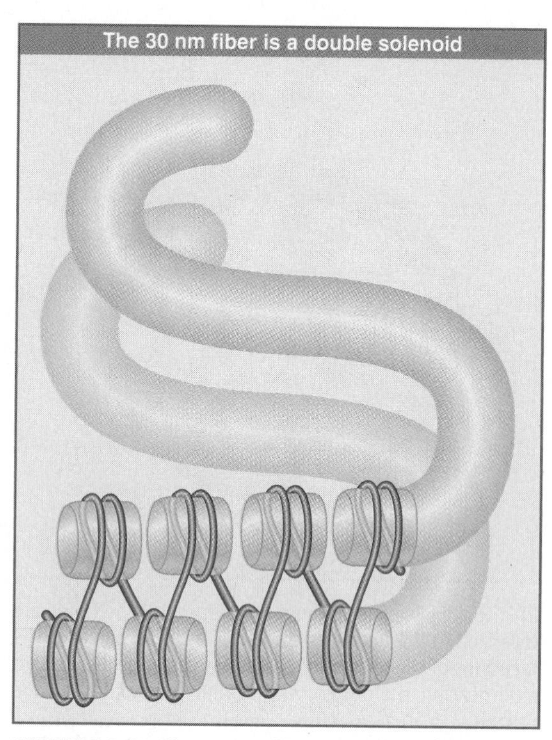

FIGURE 29.25 The 30 nm fiber is a helical ribbon consisting of two parallel rows of nucleosomes coiled into a solenoid.

of both interphase chromatin and mitotic chromosomes.

The most likely arrangement for packing nucleosomes into the fiber is a solenoid, in which the nucleosomes turn in a helical array that is coiled around a central cavity. The two main forms of a solenoid are a single-start, which forms from a single linear array, and a two-start, which in effect consists of a double row of nucleosomes. FIGURE 29.25 shows a two-start model suggested by recent crosslinking data identifying a double stack of nucleosomes in the 30 nm fiber. This is supported by the crystal structure of a tetranucleosome complex.

The 30 nm and 10 nm fibers can be reversibly converted by changing the ionic strength. This suggests that the linear array of nucleosomes in the 10 nm fiber is coiled into the 30 nm structure at higher ionic strength and in the presence of H1.

Although the presence of H1 is necessary for the formation of the 30 nm fiber, information about its location is conflicting. Its relative ease of extraction from chromatin seems to argue that it is present on the outside of the superhelical fiber axis. Diffraction data, though, and the fact that it is harder to find in 30 nm fibers than in 10 nm fibers that retain it, would argue for an interior location.

How do we get from the 30 nm fiber to the specific structures displayed in mitotic chromosomes? Is there any further specificity in the arrangement of interphase chromatin? Do particular regions of 30 nm fibers bear a fixed relationship to one another, or is their arrangement random?

29.9 Reproduction of Chromatin Requires Assembly of Nucleosomes

Key concepts

- Histone octamers are not conserved during replication, but H2A-H2B dimers and $H3_2$-$H4_2$ tetramers are conserved.
- There are different pathways for the assembly of nucleosomes during replication and independently of replication.
- Accessory proteins are required to assist the assembly of nucleosomes.
- CAF-1 is an assembly protein that is linked to the PCNA subunit of the replisome; it is required for deposition of $H3_2$-$H4_2$ tetramers following replication.
- A different assembly protein and a variant of histone H3 may be used for replication-independent assembly.

Replication separates the strands of DNA and therefore must inevitably disrupt the structure of the nucleosome. The transience of the replication event is a major difficulty in analyzing the structure of a particular region while it is being replicated. The structure of the replication fork is distinctive. It is more resistant to micrococcal nuclease and is digested into bands that differ in size from nucleosomal DNA. The region that shows this altered structure is confined to the immediate vicinity of the replication fork. This suggests that a large protein complex is engaged in replicating the DNA, but the nucleosomes re-form more or less immediately behind it as it moves along.

Reproduction of chromatin does not involve any protracted period during which the DNA is free of histones. Once DNA has been replicated, nucleosomes are quickly generated on both the duplicates. This point is illustrated by the electron micrograph of FIGURE 29.26, which shows a recently replicated stretch of DNA that is already packaged into nucleosomes on both daughter duplex segments.

Both biochemical analysis and visualization of the replication fork therefore suggest that the disruption of nucleosome structure is

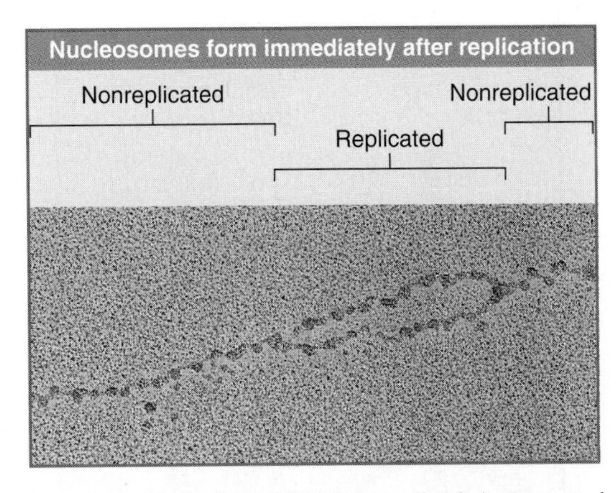

FIGURE 29.26 Replicated DNA is immediately incorporated into nucleosomes. Photo courtesy of Steven L. McKnight, UT Southwestern Medical Center at Dallas.

limited to a short region immediately around the fork. Progress of the fork disrupts nucleosomes, but they form very rapidly on the daughter duplexes as the fork moves forward. In fact, the assembly of nucleosomes is directly linked to the replisome that is replicating DNA.

How do histones associate with DNA to generate nucleosomes? Do the histones *preform* a protein octamer around which the DNA is subsequently wrapped? Or does the histone octamer assemble on DNA from free histones? FIGURE 29.27 shows that two pathways can be used *in vitro* to assemble nucleosomes, depending on the conditions that are employed. In one pathway, a preformed octamer binds to DNA. In the other pathway, a tetramer of $H3_2$-$H4_2$ binds first; and then two H2A-H2B dimers are added. Both these pathways are related to reactions that occur *in vivo*. The first reflects the capacity of chromatin to be remodeled by moving histone octamers along DNA (see Section 30.3, Chromatin Remodeling Is an Active Process). The second represents the pathway that is used in replication.

Accessory proteins are involved in assisting histones to associate with DNA. Candidates for this role can be identified by using extracts that assemble histones and exogenous DNA into nucleosomes. Accessory proteins may act as "molecular chaperones" that bind to the histones in order to release either individual histones or complexes ($H3_2$-$H4_2$ or H2A-H2B) to the DNA in a controlled manner. This could be necessary because the histones, as basic proteins, have a general high affinity for DNA. *Such interactions allow histones to form nucleosomes without becoming trapped in other kinetic intermediates (that is, other complexes resulting from indiscreet binding of histones to DNA).*

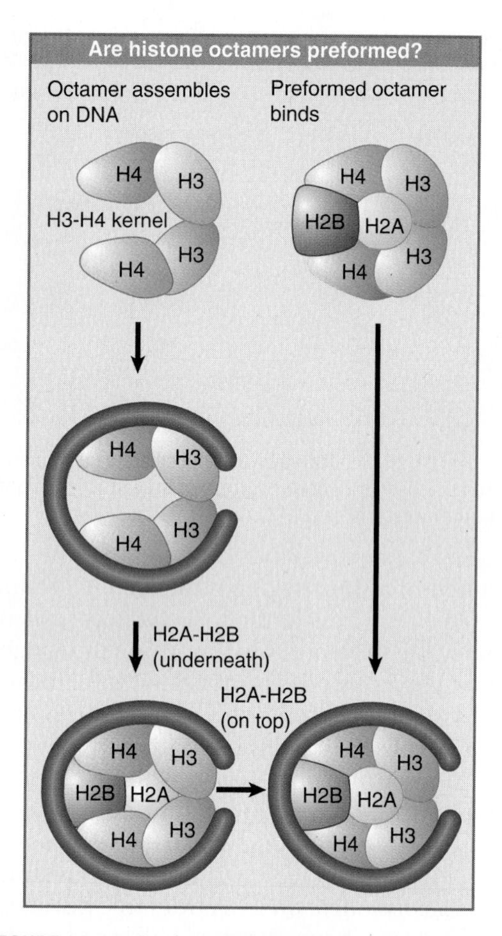

FIGURE 29.27 *In vitro,* DNA can either interact directly with an intact (crosslinked) histone octamer or can assemble with the $H3_2$-$H4_2$ tetramer, after which two H2A-H2B dimers are added.

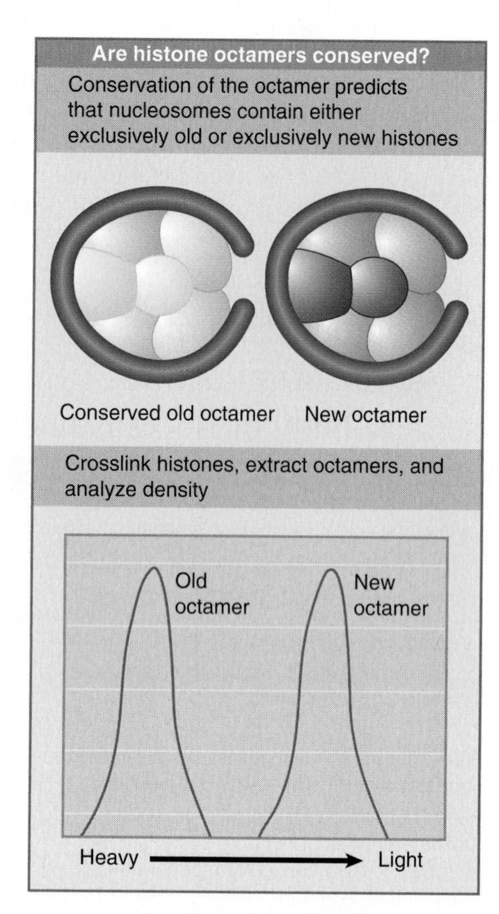

FIGURE 29.28 If histone octamers were conserved, old and new octamers would band at different densities when replication of heavy octamers occurs in light amino acids.

Attempts to produce nucleosomes *in vitro* began by considering a process of assembly between free DNA and histones. Nucleosomes form *in vivo,* though, only when DNA is replicated. A system that mimics this requirement has been developed by using extracts of human cells that replicate SV40 DNA and assemble the products into chromatin. The assembly reaction occurs preferentially on replicating DNA. It requires an ancillary factor, chromatin assembly factor (CAF)-1, that consists of >5 subunits, with a total mass of 238 kD. CAF-1 is recruited to the replication fork by proliferating cell nuclear antigen (PCNA), the processivity factor for DNA polymerase. This provides the link between replication and nucleosome assembly, ensuring that nucleosomes are assembled as soon as DNA has been replicated.

CAF-1 acts stoichiometrically, and functions by binding to newly synthesized H3 and H4. This suggests that new nucleosomes form by assembling first the $H3_2$-$H4_2$ tetramer, and then adding the H2A-H2B dimers. The nucleosomes that are formed *in vitro* have a repeat length of 200 bp. They do not have any H1 histone, though, which suggests that proper spacing can be accomplished without H1.

When chromatin is reproduced, a stretch of DNA *already associated with nucleosomes* is replicated, giving rise to two daughter duplexes. What happens to the preexisting nucleosomes at this point? Are the histone octamers dissociated into free histones for reuse, or do they remain assembled? The integrity of the octamer can be tested by crosslinking the histones. The next two figures compare the possible outcomes from an experiment in which cells are grown in the presence of heavy amino acids to identify the histones before replication. Replication is then allowed to occur in the presence of light amino acids. At this point the histone octamers are crosslinked and centrifuged on a density gradient. FIGURE 29.28 shows that if the original octamers have been conserved, they will be found at a position of high density, and new octamers will occupy a low density position.

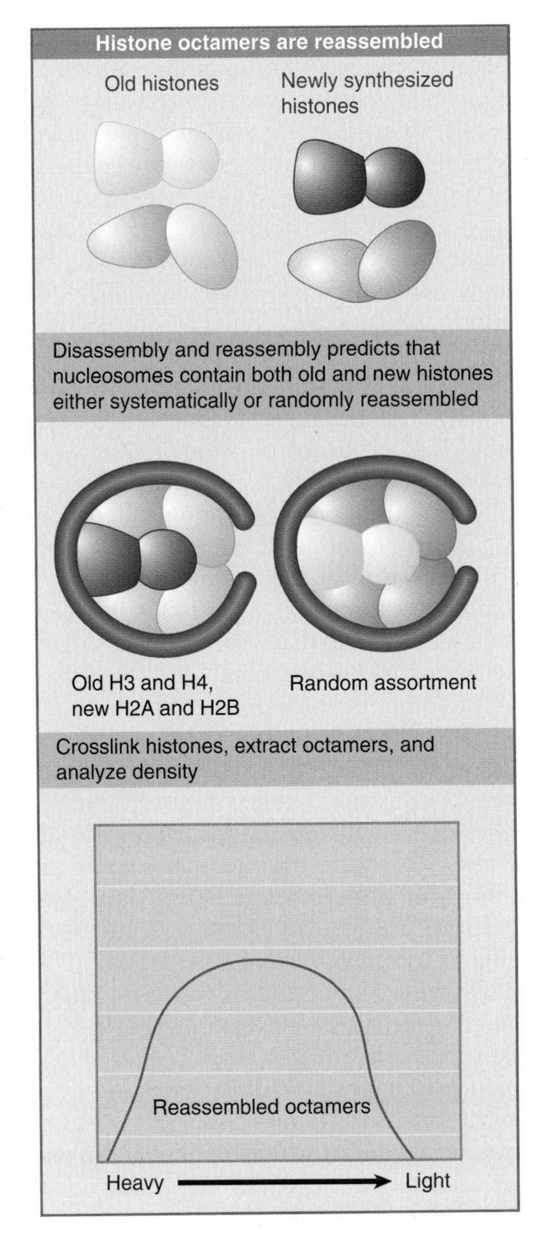

FIGURE 29.29 When heavy octamers are replicated in light amino acids, the new octamers band diffusely between heavy and light densities, which suggests that disassembly and reassembly has occurred.

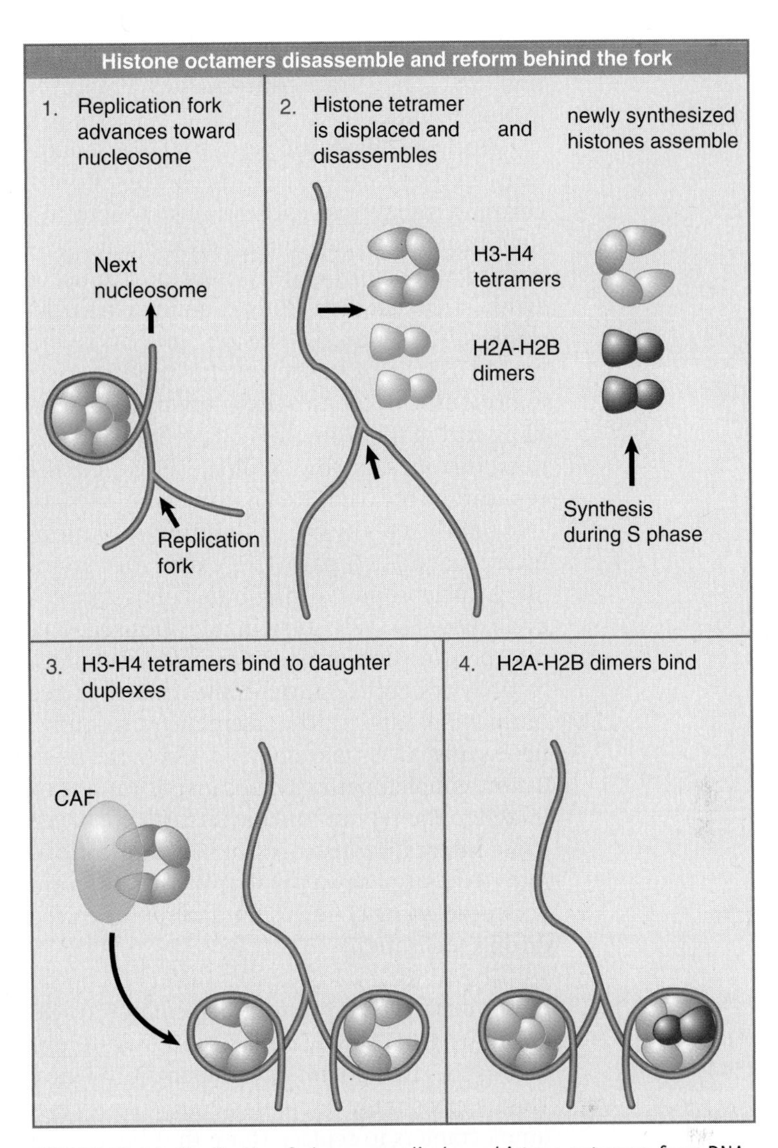

FIGURE 29.30 Replication fork passage displaces histone octamers from DNA. They disassemble into H3-H4 tetramers and H2A-H2B dimers. Newly synthesized histones are assembled into H3-H4 tetramers and H2A-H2B dimers. The old and new tetramers and dimers are assembled with the aid of CAF-1 at random into new nucleosomes immediately behind the replication fork.

This does not happen, though. Little material is found at the high density position, which suggests that histone octamers are not conserved. The octamers have an intermediate density, and FIGURE 29.29 shows that this is the expected result if the old histones have been released and then reassembled with newly synthesized histones.

The pattern of disassembly and reassembly has been difficult to characterize in detail, but a working model is illustrated in FIGURE 29.30. The replication fork displaces histone octamers, which then dissociate into $H3_2$-$H4_2$ tetramers and H2A-H2B dimers. These "old" tetramers and dimers enter a pool that also includes "new" tetramers and dimers, which are assembled from newly synthesized histones. Nucleosomes assemble ~600 bp behind the replication fork. Assembly is initiated when $H3_2$-$H4_2$ tetramers bind to each of the daughter duplexes, assisted by CAF-1. Two H2A-H2B dimers then bind to each $H3_2$-$H4_2$ tetramer to complete the histone octamer. The assembly of tetramers and dimers is random with respect to "old" and "new" subunits, which explains the results of Figure 29.29. The "old" $H3_2$-$H4_2$ tetramer could have an ability to be transiently associated with a single strand of DNA during replication; it may in fact have an increased chance of remaining with the leading strand for reuse. It is possible that nucleosomes are disrupted and reassembled in a similar way during transcription (see Section 29.11, Are Transcribed Genes Organized in Nucleosomes?).

During S phase (the period of DNA replication) in a eukaryotic cell, the duplication of chromatin requires synthesis of sufficient histone proteins to package an entire genome—basically the same quantity of histones must be synthesized that are already contained in nucleosomes. The synthesis of histone mRNAs is controlled as part of the cell cycle, and increases enormously in S phase. The pathway for assembling chromatin from this equal mix of old and new histones during S phase is called the replication-coupled (RC) pathway.

Another pathway, called the replication-independent (RI) pathway, exists for assembling nucleosomes during other phases of cell cycle, when DNA is not being synthesized. This may become necessary as the result of damage to DNA or because nucleosomes are displaced during transcription. The assembly process must necessarily have some differences from the replication-coupled pathway, because it cannot be linked to the replication apparatus. One of the most interesting features of the replication-independent pathway is that it uses different variants of some of the histones from those used during replication.

The histone H3.3 variant differs from the highly conserved H3 histone at four amino acid positions. H3.3 slowly replaces H3 in differentiating cells that do not have replication cycles. This happens as the result of assembly of new histone octamers to replace those that have been displaced from DNA for whatever reason. The mechanism that is used to ensure the use of H3.3 in the replication-independent pathway is different in two cases that have been investigated.

In the protozoan *Tetrahymena*, histone usage is determined exclusively by availability. Histone H3 is synthesized only during the cell cycle; the variant replacement histone is synthesized only in nonreplicating cells. In *Drosophila*, however, there is an active pathway that ensures the usage of H3.3 by the replication-independent pathway. New nucleosomes containing H3.3 assemble at sites of transcription, presumably replacing nucleosomes that were displaced by RNA polymerase. The assembly process discriminates between H3 and H3.3 on the basis of their sequences, specifically excluding H3 from being utilized. By contrast, replication-coupled assembly uses both types of H3 (although H3.3 is available at much lower levels than H3, and therefore enters only a small proportion of nucleosomes).

CAF-1 is probably not involved in replication-independent assembly. (There also are organisms such as yeast and *Arabidopsis* for which its gene is not essential, implying that alternative assembly processes may be used in replication-coupled assembly.) A protein that may be involved in replication-independent assembly is called HIRA. Depletion of HIRA from *in vitro* systems for nucleosome assembly inhibits the formation of nucleosomes on nonreplicated DNA, but not on replicating DNA, which indicates that the pathways do indeed use different assembly mechanisms. HIRA functions as a chaperone to assist the incorporation of histones into nucleosomes. This pathway appears to be generally responsible for replication-independent assembly; for example, HIRA is required for the decondensation of the sperm nucleus, when protamines are replaced by histones, in order to generate chromatin that is competent to be replicated following fertilization.

Assembly of nucleosomes containing an alternative to H3 also occurs at centromeres (see Section 31.3, Heterochromatin Depends on Interactions with Histones). Centromeric DNA replicates early during the replication phase of the cell cycle (in contrast with the surrounding heterochromatic sequences that replicate later; see Section 15.7, Each Eukaryotic Chromosome Contains Many Replicons). The incorporation of H3 at the centromeres is inhibited, and instead a protein called CENP-A is incorporated in higher eukaryotic cells (in *Drosophila* it is called Cid, and in yeast it is called Cse4). This occurs by the replication-independent assembly pathway, apparently because the replication-coupled pathway is inhibited for a brief period while centromeric DNA replicates.

29.10 Do Nucleosomes Lie at Specific Positions?

Key concepts

- Nucleosomes may form at specific positions as the result either of the local structure of DNA or of proteins that interact with specific sequences.
- The most common cause of nucleosome positioning is when proteins binding to DNA establish a boundary.
- Positioning may affect which regions of DNA are in the linker and which face of DNA is exposed on the nucleosome surface.

We know that nucleosomes can be reconstituted *in vitro* without regard to DNA sequence, but this does not mean that their formation *in vivo* is independent of sequence. Does a particular DNA sequence always lie in a certain position *in vivo* with regard to the topography of the nucleosome? Or are nucleosomes arranged randomly on DNA, so that a particular sequence may occur at any location, for example, in the core region in one copy of the genome and in the linker region in another?

To investigate this question, it is necessary to use a defined sequence of DNA; more precisely, we need to determine the position relative to the nucleosome of a defined point in the DNA. FIGURE 29.31 illustrates the principle of a procedure used to achieve this.

Suppose that the DNA sequence is organized into nucleosomes in only one particular configuration, so that each site on the DNA always is located at a particular position on the nucleosome. This type of organization is called **nucleosome positioning** (or sometimes nucleosome phasing). In a series of positioned nucleosomes, the linker regions of DNA comprise unique sites.

Consider the consequences for just a single nucleosome. Cleavage with micrococcal nuclease generates a monomeric fragment that constitutes a *specific sequence.* If the DNA is isolated and cleaved with a restriction enzyme that has only one target site in this fragment, it should be cut at a unique point. This produces two fragments, each of unique size.

The products of the micrococcal/restriction double digest are separated by gel electrophoresis. A probe representing the sequence on one side of the restriction site is used to identify the corresponding fragment in the double digest. This technique is called **indirect end labeling.**

Reversing the argument, the identification of a single sharp band demonstrates that the position of the restriction site is uniquely defined with respect to the end of the nucleosomal DNA (as defined by the micrococcal nuclease cut). Thus the nucleosome has a unique sequence of DNA.

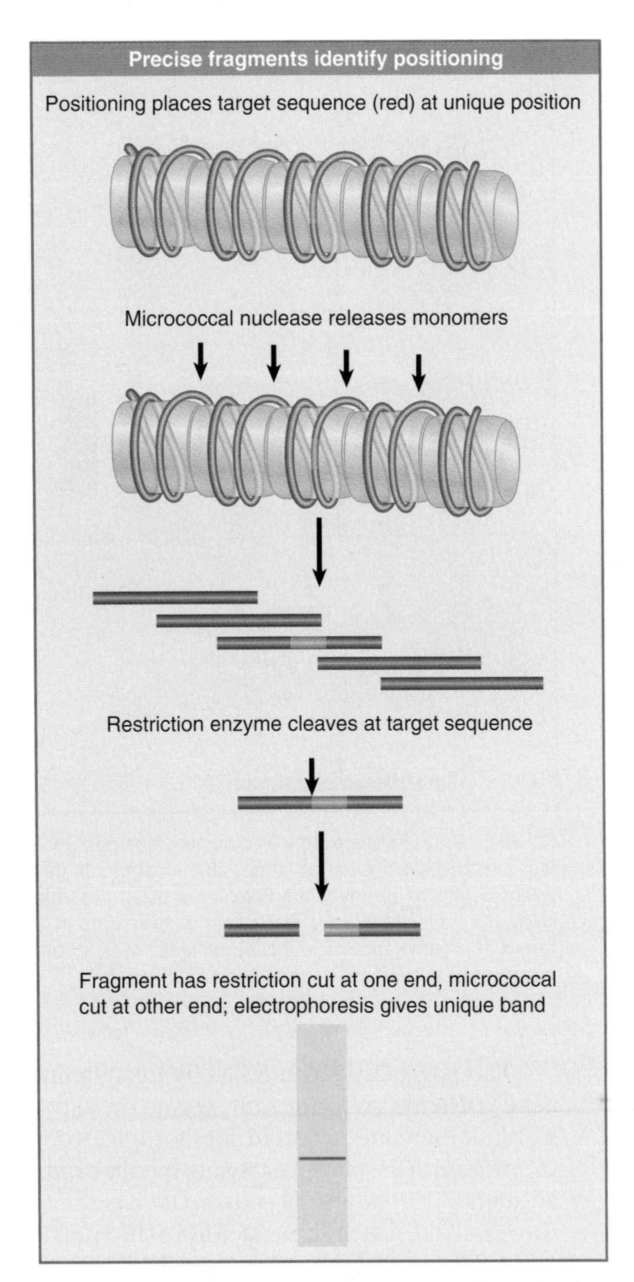

FIGURE 29.31 Nucleosome positioning places restriction sites at unique positions relative to the linker sites cleaved by micrococcal nuclease.

What happens if the nucleosomes do *not* lie at a single position? Now the linkers consist of *different* DNA sequences in each copy of the genome. Thus the restriction site lies at a different position each time; in fact, it lies at all possible locations relative to the ends of the monomeric nucleosomal DNA. FIGURE 29.32 shows that the double cleavage then generates a broad smear, ranging from the smallest detectable fragment (~20 bases) to the length of the monomeric DNA.

In discussing these experiments, we have treated micrococcal nuclease as an enzyme that cleaves DNA at the exposed linker regions without any sort of sequence specificity. The enzyme actually does have some sequence specificity, though, which is biased toward selection of A-T-rich sequences. Thus we cannot assume that the existence of a specific band in the indirect end-labeling technique represents the distance from a restriction cut to the linker region. It could instead represent the distance from the restriction cut to a preferred micrococcal nuclease cleavage site!

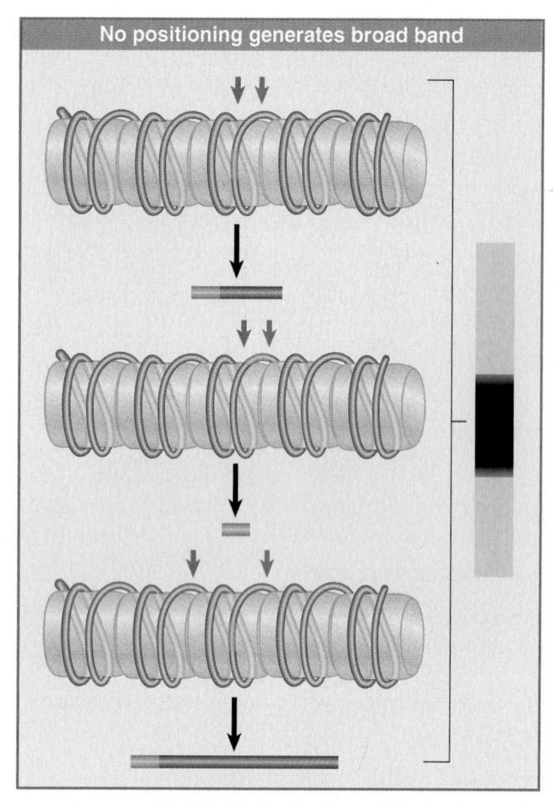

FIGURE 29.32 In the absence of nucleosome positioning, a restriction site lies at all possible locations in different copies of the genome. Fragments of all possible sizes are produced when a restriction enzyme cuts at a target site (red) and micrococcal nuclease cuts at the junctions between nucleosomes (green).

This possibility is controlled by treating the naked DNA in exactly the same way as the chromatin. If there are preferred sites for micrococcal nuclease in the particular region, specific bands are found. This pattern of bands can then be compared with the pattern generated from chromatin.

A *difference* between the control DNA band pattern and the chromatin pattern provides evidence for nucleosome positioning. Some of the bands present in the control DNA digest may disappear from the nucleosome digest, indicating that preferentially cleaved positions are unavailable. New bands may appear in the nucleosome digest when new sites are rendered preferentially accessible by the nucleosomal organization.

Nucleosome positioning might be accomplished in either of two ways:

- It is intrinsic: *Every nucleosome is deposited specifically at a particular DNA sequence.* This modifies our view of the nucleosome as a subunit able to form between any sequence of DNA and a histone octamer.
- It is extrinsic: *The first nucleosome in a region is preferentially assembled at a particular site.* A preferential starting point for nucleosome positioning results from the presence of a region from which nucleosomes are excluded. The excluded region provides a *boundary* that restricts the positions available to the adjacent nucleosome. A series of nucleosomes may then be assembled sequentially, with a defined repeat length.

It is now clear that the deposition of histone octamers on DNA is not random with regard to sequence. The pattern is intrinsic in some cases, in which it is determined by structural features in DNA. It is extrinsic in other cases, in which it results from the interactions of other proteins with the DNA and/or histones.

Certain structural features of DNA affect placement of histone octamers. DNA has intrinsic tendencies to bend in one direction rather than another; thus A-T-rich regions locate so that the minor groove faces in toward the octamer, whereas G-C-rich regions are arranged so that the minor groove points out. Long runs of dA-dT (>8 bp) avoid positioning in the central superhelical turn of the core. It is not yet possible to sum all of the relevant structural effects and thus entirely predict the location of a particular DNA sequence with regard to the nucleosome. Sequences that cause DNA to take up more extreme structures may have effects such as the exclusion of nucleosomes, and thus could cause boundary effects.

Positioning of nucleosomes near boundaries is common. If there is some variability in the construction of nucleosomes—for example, if the length of the linker can vary by, say, 10 bp—the specificity of location would decline proceeding away from the first, defined nucleosome at the boundary. In this case, we might expect the positioning to be maintained rigorously only relatively near the boundary.

The location of DNA on nucleosomes can be described in two ways. FIGURE 29.33 shows that **translational positioning** describes the position of DNA with regard to the boundaries of the nucleosome. In particular, it determines which sequences are found in the linker regions. Shifting the DNA by 10 bp brings the next turn into a linker region. Thus translational positioning determines which regions are more accessible (at least as judged by sensitivity to micrococcal nuclease).

DNA lies on the outside of the histone octamer. As a result, one face of any particular

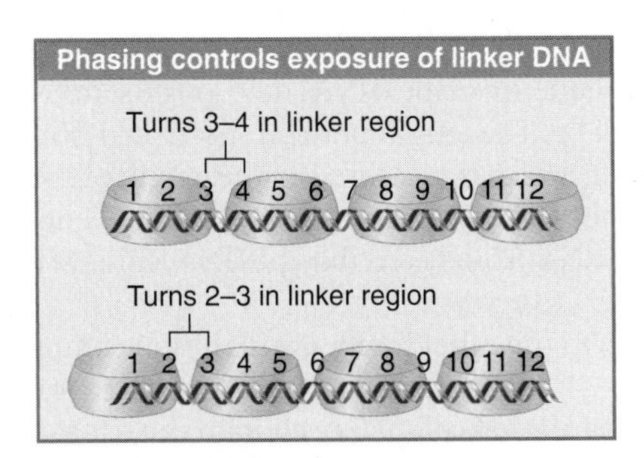

FIGURE 29.33 Translational positioning describes the linear position of DNA relative to the histone octamer. Displacement of the DNA by 10 bp changes the sequences that are in the more exposed linker regions, but does not alter which face of DNA is protected by the histone surface and which is exposed to the exterior. DNA is really coiled around the nucleosomes, and is shown in linear form only for convenience.

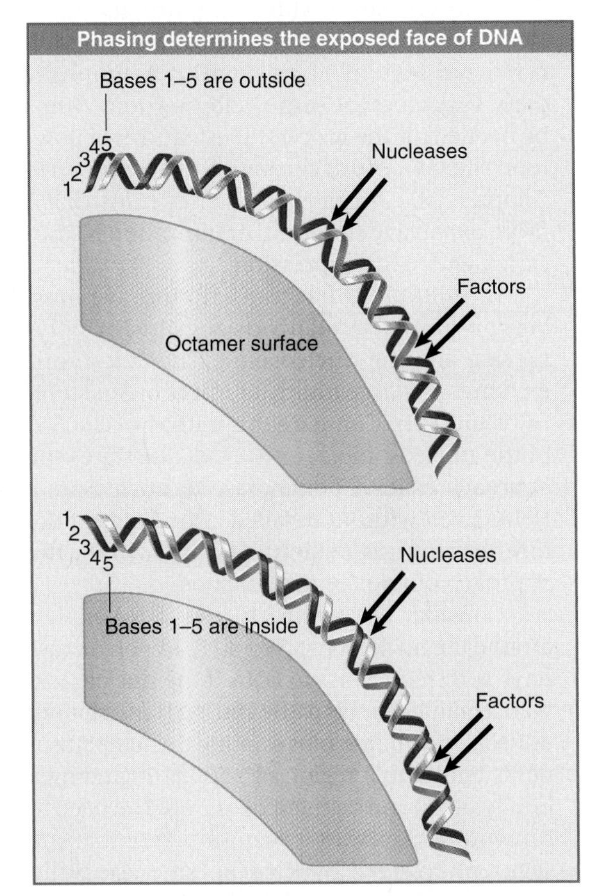

FIGURE 29.34 Rotational positioning describes the exposure of DNA on the surface of the nucleosome. Any movement that differs from the helical repeat (~10.2 bp/turn) displaces DNA with reference to the histone surface. Nucleotides on the inside are more protected against nucleases than nucleotides on the outside.

sequence is obscured by the histones, whereas the other face is accessible. Depending upon its positioning with regard to the nucleosome, a site in DNA that must be recognized by a regulator protein could be inaccessible or available. The exact position of the histone octamer with respect to DNA sequence may therefore be important. FIGURE 29.34 shows the effect of **rotational positioning** of the double helix with regard to the octamer surface. If the DNA is moved by a partial number of turns (imagine the DNA as rotating relative to the protein surface), there is a change in the exposure of sequence to the outside.

Both translational and rotational positioning can be important in controlling access to DNA. The best characterized cases of positioning involve the specific placement of nucleosomes at promoters. Translational positioning and/or the exclusion of nucleosomes from a particular sequence may be necessary to allow a transcription complex to form. Some regulatory factors can bind to DNA only if a nucleosome is excluded to make the DNA freely accessible, and this creates a boundary for translational positioning. In other cases, regulatory factors can bind to DNA on the surface of the nucleosome, but rotational positioning is important to ensure that the face of DNA with the appropriate contact points is exposed.

We discuss the connection between nucleosomal organization and transcription in Section 30.4, Nucleosome Organization May Be Changed at the Promoter, but note for now that promoters (and some other structures) often have short regions that exclude nucleosomes. These regions typically form a boundary next to which nucleosome positions are restricted. A survey of an extensive region in the *Saccharomyces cerevisiae* genome (mapping 2278 nucleosomes over 482 kb of DNA) showed that in fact 60% of the nucleosomes have specific positions as the result of boundary effects, most often from promoters.

29.11 Are Transcribed Genes Organized in Nucleosomes?

Key concepts

- Nucleosomes are found at the same frequency when transcribed genes or nontranscribed genes are digested with micrococcal nuclease.
- Some heavily transcribed genes appear to be exceptional cases that are devoid of nucleosomes.

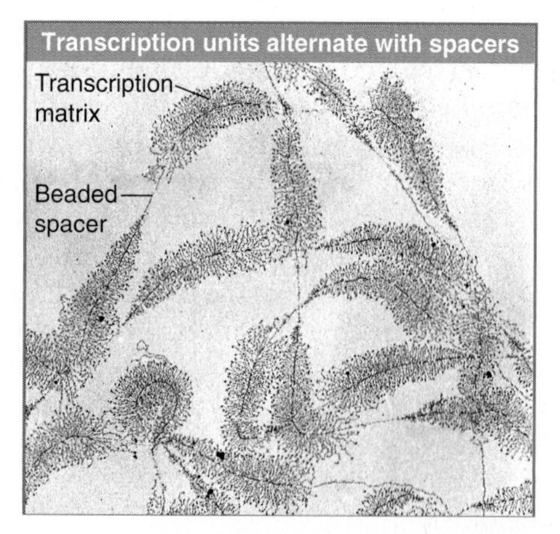

FIGURE 29.35 The isolated nucleolar genes rDNA transcription units alternate with nontranscribed DNA segments. Reproduced from Miller, O. L. and Beatty, B. R. 1969. *Science.* 164: 955–957. Photo courtesy of Oscar Miller.

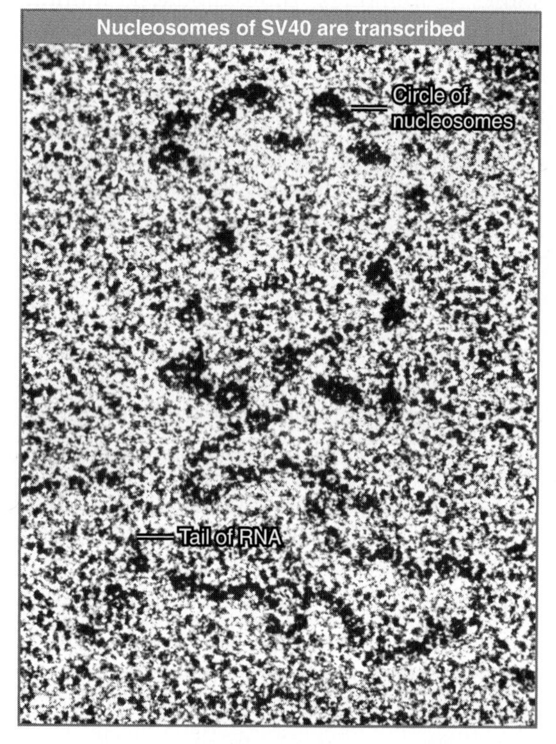

FIGURE 29.36 An SV40 minichromosome can be transcribed. Reproduced from *J. Mol. Bio.*, vol. 131, Gariglio, P., et al., *The template of the isolated . . .*, p. 131. Copyright 1979, with permission from Elsevier. Photo courtesy of Pierre Chambon.

Attempts to visualize genes during transcription have produced conflicting results. The next two figures show each extreme.

Heavily transcribed chromatin can be seen to be rather extended (too extended to be covered in nucleosomes). In the intensively transcribed genes coding for rRNA shown in FIGURE 29.35, the extreme packing of RNA polymerases makes it hard to see the DNA. We cannot directly measure the lengths of the rRNA transcripts because the RNA is compacted by proteins, but we know (from the sequence of the rRNA) how long the transcript must be. The length of the transcribed DNA segment, which is measured by the length of the axis of the "Christmas tree," is ~85% of the length of the rRNA. This means that the DNA is almost completely extended.

On the other hand, transcription complexes of SV40 minichromosomes can be extracted from infected cells. They contain the usual complement of histones and display a beaded structure. Chains of RNA can be seen to extend from the minichromosome, as in the example of FIGURE 29.36. This argues that transcription can proceed while the SV40 DNA is organized into nucleosomes. Of course, the SV40 minichromosome is transcribed less intensively than the rRNA genes.

Transcription involves the unwinding of DNA, and may require the fiber to unfold in restricted regions of chromatin. A simplistic view suggests that some "elbow-room" must be needed for the process. The features of polytene and lampbrush chromosomes described in Chapter 28, Chromosomes, offer hints that a more expansive structural organization is associated with gene expression.

In thinking about transcription, we must bear in mind the relative sizes of RNA polymerase and the nucleosome. The eukaryotic enzymes are large multisubunit proteins, typically >500 kD. Compare this with the ~260 kD of the nucleosome. FIGURE 29.37 illustrates the approach of RNA polymerase to nucleosomal DNA. Even without detailed knowledge of the interaction, it is evident that it involves the approach of two comparable bodies.

Consider the two turns that DNA makes around the nucleosome. Would RNA polymerase have sufficient access to DNA if the nucleic acid were confined to this path? During transcription, as RNA polymerase moves along the template, it binds tightly to a region of ~50 bp, including a locally unwound segment of ~12 bp. The need to unwind DNA makes it seem unlikely that the segment engaged by RNA polymerase could remain on the surface of the histone octamer.

It therefore seems inevitable that transcription must involve a structural change. Thus the first question to ask about the structure of active genes is whether DNA being transcribed remains organized in nucleosomes. If the histone octamers are displaced, do they remain attached in some way to the transcribed DNA?

One experimental approach is to digest chromatin with micrococcal nuclease, and then to use a probe to some specific gene or genes to

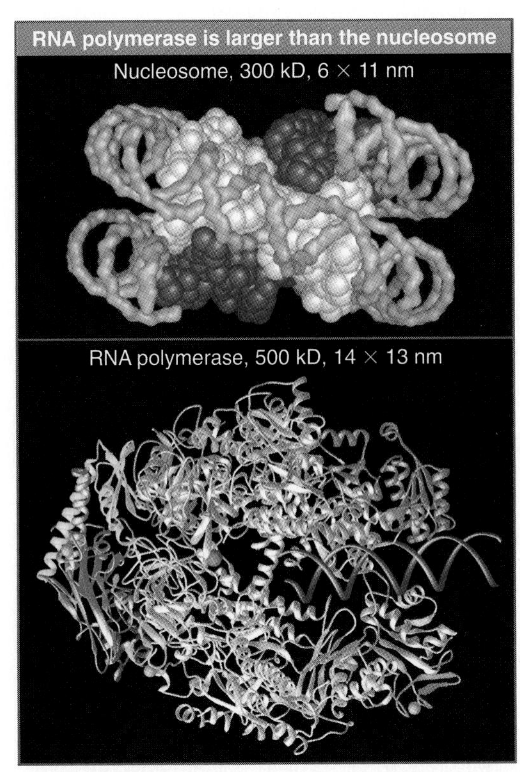

FIGURE 29.37 RNA polymerase is comparable in size to the nucleosome and might encounter difficulties in following the DNA around the histone octamer. Top photo courtesy of E. N. Moudrianakis, Johns Hopkins University. Bottom photo courtesy of Roger Kornberg, Stanford University School of Medicine.

determine whether the corresponding fragments are present in the usual 200 bp ladder at the expected concentration. The conclusions that we can draw from these experiments are limited but important. *Genes that are being transcribed contain nucleosomes at the same frequency as nontranscribed sequences.* Thus genes do not necessarily enter an alternative form of organization in order to be transcribed. *The average transcribed gene probably only has a single RNA polymerase at any given moment, though, so this does not reveal what is happening at sites actually engaged by the enzyme.* Perhaps they retain their nucleosomes; more likely the nucleosomes are temporarily displaced as RNA polymerase passes through, but reform immediately afterward.

29.12 Histone Octamers Are Displaced by Transcription

Key concepts

- RNA polymerase displaces histone octamers during transcription in a model system, but octamers reassociate with DNA as soon as the polymerase has passed.
- Nucleosomes are reorganized when transcription passes through a gene.

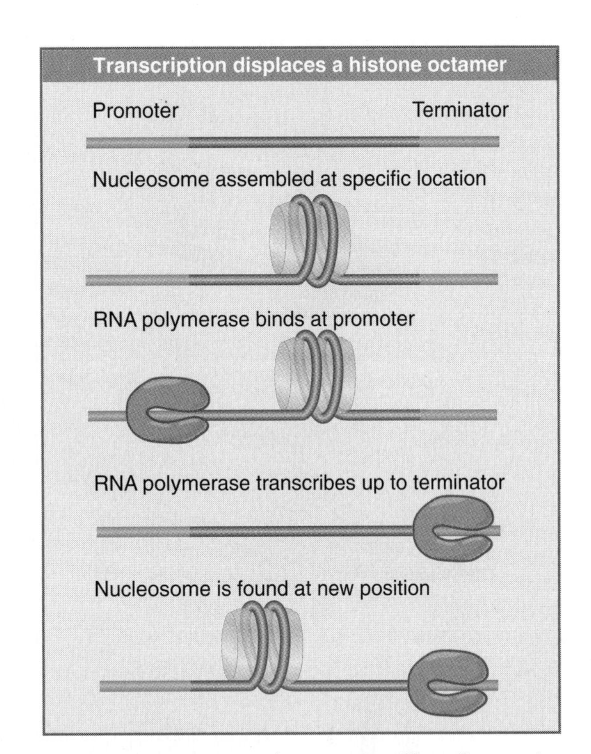

FIGURE 29.38 A protocol to test the effect of transcription on nucleosomes shows that the histone octamer is displaced from DNA and rebinds at a new position.

Experiments to test whether an RNA polymerase can transcribe directly through a nucleosome suggest that the histone octamer is displaced by the act of transcription. FIGURE 29.38 shows what happens when the phage T7 RNA polymerase transcribes a short piece of DNA containing a single octamer core *in vitro*. The core remains associated with the DNA, but is found in a different location. The core is most likely to rebind to the same DNA molecule from which it was displaced.

FIGURE 29.39 shows a model for polymerase progression. DNA is displaced as the polymerase enters the nucleosome, but the polymerase reaches a point at which the DNA loops back and reattaches, thereby forming a closed region. As polymerase advances further, unwinding the DNA, it creates positive supercoils in this loop; the effect could be dramatic, because the closed loop is only ~80 bp, so each base pair through which the polymerase advances makes a significant addition to the supercoiling. In fact, the polymerase progresses easily for the first 30 bp into the nucleosome. It then proceeds more slowly, as though encountering increasing difficulty in progressing. Pauses occur every 10 bp, which suggests that the structure of the loop imposes a constraint related to rotation around each turn of DNA. When the polymerase reaches the midpoint of the

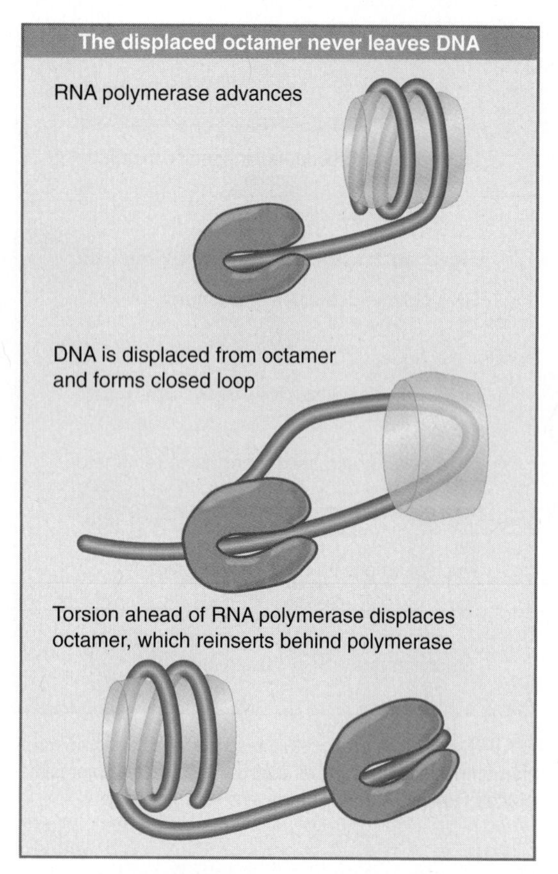

FIGURE 29.39 RNA polymerase displaces DNA from the histone octamer as it advances. The DNA loops back and attaches (to polymerase or to the octamer) to form a closed loop. As the polymerase proceeds, it generates positive supercoiling ahead. This displaces the octamer, which keeps contact with DNA and/or polymerase, and is inserted behind the RNA polymerase.

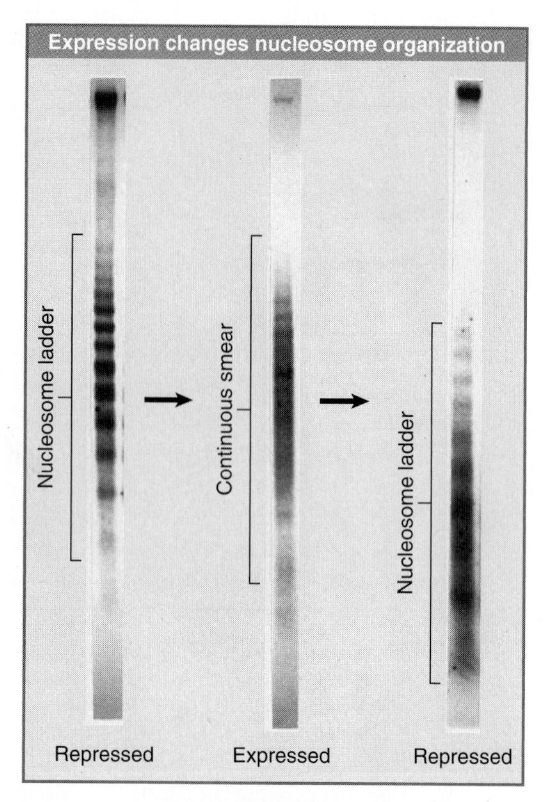

FIGURE 29.40 *URA3* gene sequences are fused to a regulated *GAL1* promoter and to a ribosomal DNA sequence. The *URA3* has transitionally positioned nucleosomes before transcription. When transcription is induced under the control of an inducible promoter, nucleosome positions are randomized. When transcription is repressed, the nucleosomes resume their particular positions. Reproduced from Suter, B., et al. 1997. *EMBO J.* 16: 2150–2160. Copyright © Oxford University Press. Photos courtesy of Fritz Thomas, ETH Zürich.

nucleosome (the next bases to be added are essentially at the axis of dyad symmetry), pausing ceases, and the polymerase advances rapidly. This suggests that the midpoint of the nucleosome marks the point at which the octamer is displaced (possibly because positive supercoiling has reached some critical level that expels the octamer from DNA). This releases tension ahead of the polymerase and allows it to proceed. The octamer then binds to the DNA behind the polymerase and no longer presents an obstacle to progress. It is likely that the octamer changes position without ever completely losing contact with the DNA.

Is the octamer released as an intact unit? Crosslinking the octamer's proteins does not create an obstacle to transcription. Transcription can continue even when crosslinking is extensive enough to ensure that the central regions of the core histones have been linked. This implies that transcription does not require dissociation of the octamer into its component histones, nor is it likely to require any major unfolding of the central structure. The addition of histone H1 to this system, however, causes a rapid decline in transcription. This suggests two conclusions: The histone octamer (whether remaining present or displaced) functions as an intact unit, and it may be necessary to remove H1 from active chromatin or to modify its interactions in some way.

Thus a small RNA polymerase can displace a single nucleosome, which reforms behind it, during transcription. Of course, the situation is more complex in a eukaryotic nucleus. RNA polymerase is much larger, and the impediment to progress is a string of connected nucleosomes. Overcoming this obstacle requires additional factors that act on chromatin (see Chapter 30, Controlling Chromatin Structure).

The organization of nucleosomes may be changed by transcription. FIGURE 29.40 shows what happens to the yeast *URA3* gene when it is transcribed under the control of an inducible promoter. Positioning is examined by using micrococcal nuclease to examine cleavage sites relative to a restriction site at the 5′ end of the gene. Initially the gene displays a pattern of nucleosomes that are organized from the promoter for a significant distance across the gene; positioning is lost in the 3′ regions. When the gene is expressed, a general smear replaces the positioned pattern of nucleosomes. This indicates that nucleosomes are present at the same density but are no longer organized in phase. This in turn suggests that transcription destroys the nucleosomal positioning. When repression is reestablished, positioning appears within ten minutes (although it is not complete). This result makes the interesting point that the positions of the nucleosomes can be adjusted without replication.

The unifying model is to suppose that RNA polymerase displaces histone octamers as it progresses. If the DNA behind the polymerase is available, the octamer reattaches there. (It is possible—or perhaps probable—that the octamer never totally lost contact with the DNA. It remains a puzzle how an octamer could retain contact with DNA, though, without unfolding or losing components as an object of even larger size than itself proceeds along the DNA. Perhaps the octamer is "passed back" by making contacts with RNA polymerase). If the DNA is not available—for example, because another polymerase continues immediately behind the first—then the octamer may be permanently displaced, and the DNA may persist in an extended form.

29.13 Nucleosome Displacement and Reassembly Require Special Factors

Key concept

- Ancillary factors are required both for RNA polymerase to displace octamers during transcription and for the histones to reassemble into nucleosomes after transcription.

Displacing nucleosomes from DNA is a key requirement for all stages of transcription. The process has been characterized best at initiation. Active promoters are marked by sites that are hypersensitive to DNAase, because histone octamers have been displaced from DNA (see Section 29.18, DNAase Hypersensitive Sites Reflect Change in Chromatin Structure). The removal of the octamers requires remodeling complexes that are recruited by transcription factors and which use energy generated by hydrolysis of ATP to change chromatin structure (see Section 30.4, Nucleosome Organization May Be Changed at the Promoter). This means that RNA polymerase starts RNA synthesis on a short stretch of DNA unimpeded by nucleosomes. For it to continue advancing during elongation, the histone octamers ahead of it must be displaced. To avoid leaving naked DNA behind it, the octamers must then reform following transcription.

Transcription *in vitro* by RNA polymerase II requires a protein called facilitates chromatin transcription (FACT), which behaves like a transcription elongation factor. (It is not part of RNA polymerase, but associates with it specifically during the elongation phase of transcription.) FACT consists of two subunits that are well conserved in all eukaryotes. It is associated with the chromatin of active genes.

When FACT is added to isolated nucleosomes, it causes them to lose H2A-H2B dimers. During transcription *in vitro*, it converts nucleosomes to "hexasomes" that have lost H2A-H2B dimers. This suggests that FACT is part of a mechanism for displacing octamers during transcription. FACT may also be involved in the reassembly of nucleosomes after transcription, because it assists formation of nucleosomes from core histones.

This suggests the model shown in FIGURE 29.41, in which FACT detaches H2A-H2B from a nucleosome in front of RNA polymerase and then helps to add it to a nucleosome that is reassembling behind the enzyme. Other factors must be required to complete the process. FACT is also required for other reactions in which nucleosomes may be displaced, including DNA replication and repair.

Other factors are required to maintain the integrity of chromatin in regions that are being transcribed, probably because they are also involved in the disassembly and reassembly of nucleosomes, but we do not yet have detailed information about their functions.

29.14 Insulators Block the Actions of Enhancers and Heterochromatin

Key concepts

- Insulators are able to block passage of any activating or inactivating effects from enhancers, silencers, and LCRs.
- Insulators may provide barriers against the spread of heterochromatin.

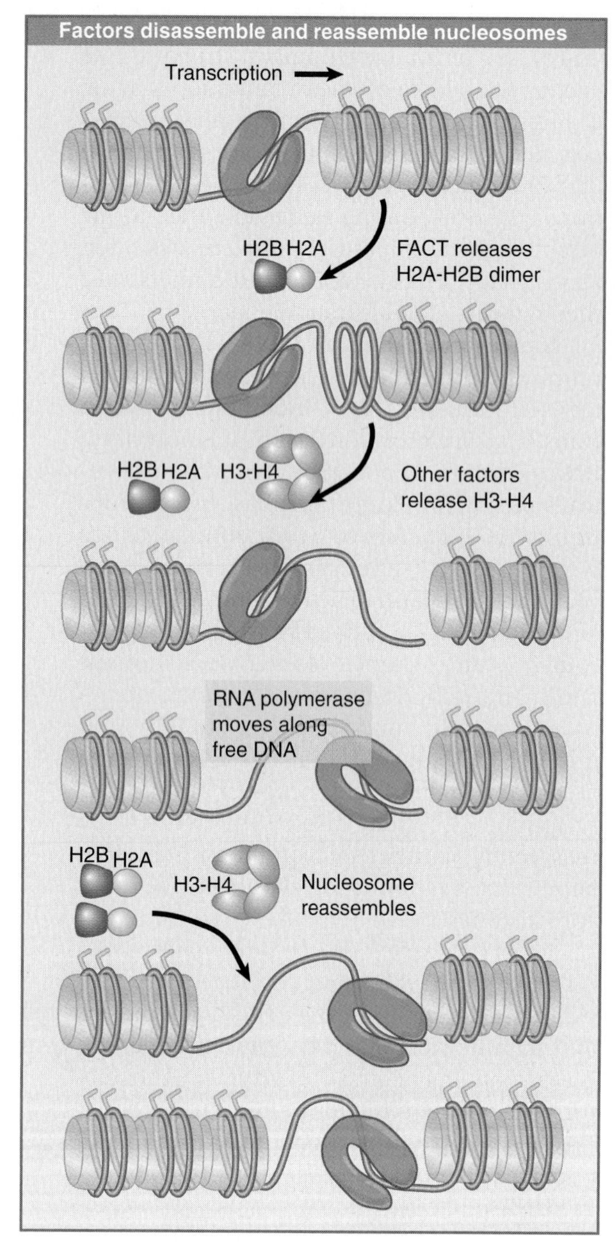

FIGURE 29.41 Histone octamers are disassembled ahead of transcription to remove nucleosomes. They reform following transcription. Release of H2A-H2B dimers probably initiates the disassembly process.

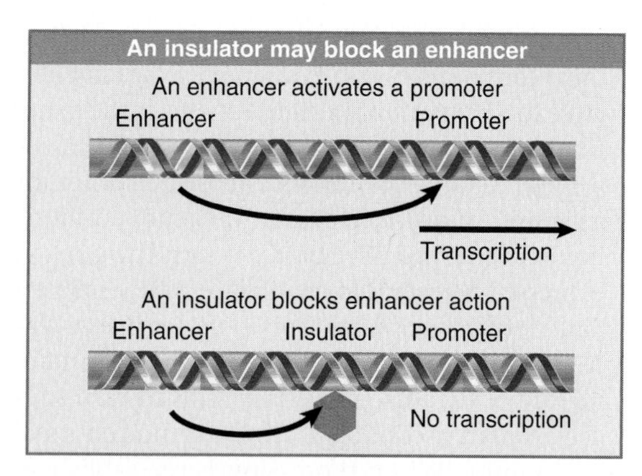

FIGURE 29.42 An enhancer activates a promoter in its vicinity, but may be blocked from doing so by an insulator located between them.

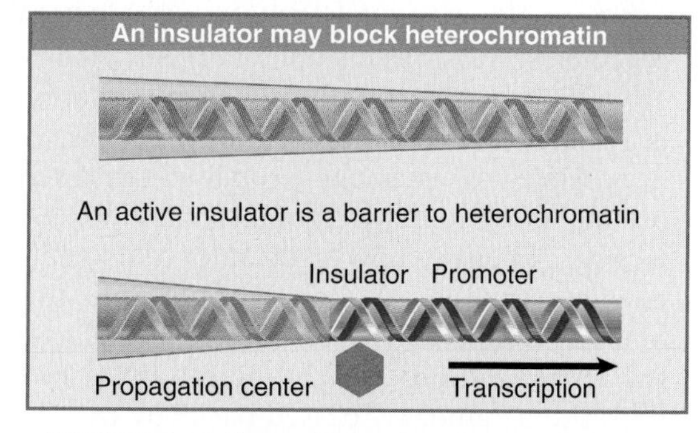

FIGURE 29.43 Heterochromatin may spread from a center and then blocks any promoters that it covers. An insulator may be a barrier to propagation of heterochromatin that allows the promoter to remain active.

Elements that prevent the passage of activating or inactivating effects are called **insulators.** They have either or both of two key properties:

- When an insulator is placed between an enhancer and a promoter, *it prevents the enhancer from activating the promoter.* The blocking effect is shown in FIGURE 29.42. This may explain how the action of an enhancer is limited to a particular promoter.
- When an insulator is placed between an active gene and heterochromatin, *it provides a barrier that protects the gene against the inactivating effect that spreads from the heterochromatin.* (Heterochromatin is a region of chromatin that is inactive as the result of its higher order structure; see Section 31.2, Heterochromatin Propagates from a Nucleation Event.) The barrier effect is shown in FIGURE 29.43.

Some insulators possess both these properties, but others have only one, or the blocking and barrier functions can be separated. Although both actions are likely to be mediated by changing chromatin structure, they may involve different effects. In either case, however, the insulator defines a limit for long-range effects.

What is the purpose of an insulator? A major function may be to counteract the indiscriminate actions of enhancers on promoters. Most enhancers will work with any promoter in the vicinity. An insulator can restrict an enhancer by blocking the effects from passing beyond a certain point, so that it can act only on a specific promoter. Similarly, when a gene is located near heterochromatin, an insulator can prevent it from being inadvertently inactivated by the spread of the heterochromatin. Insulators therefore function as elements for increasing the precision of gene regulation.

29.15 Insulators Can Define a Domain

Key concept

- Insulators are specialized chromatin structures that have hypersensitive sites. Two insulators can protect the region between them from all external effects.

Insulators were discovered during the analysis of the region of the *Drosophila melanogaster* genome summarized in FIGURE 29.44. Two genes for the protein Hsp (heat shock protein) 70 lie within an 18 kb region that constitutes band 87A7. Special structures, called *scs* and *scs'* (specialized chromatin structures), are found at the ends of the band. Each consists of a region that is highly resistant to degradation by DNAase I, and each is flanked on either side by hypersensitive sites that are spaced at about 100 bp. The cleavage pattern at these sites is altered when the genes are turned on by heat shock.

The *scs* elements insulate the *hsp70* genes from the effects of surrounding regions. If we take *scs* units and place them on either side of a *white* gene, the gene can function anywhere it is placed in the genome—even in sites where it would normally be repressed by context, for example, in heterochromatic regions.

The *scs* and *scs'* units do not seem to play either positive or negative roles in controlling gene expression, but just restrict effects from passing from one region to the next. If adjacent regions have repressive effects, however, the *scs* elements might be needed to block the spread of such effects, and therefore could be essential for gene expression. In this case, deletion of such elements could eliminate the expression of the adjacent gene(s).

The *scs* and *scs'* elements have different structures, and each appears to have a different basis for its insulator activity. The key sequence in the *scs* element is a stretch of 24 bp that binds the product of the *zw5* gene. The insulator property of *scs'* resides in a series of CGATA repeats. The repeats bind a group of related proteins called BEAF-32. The protein shows discrete localization within the nucleus, but the most remarkable data derive from its localization on polytene chromosomes. FIGURE 29.45 shows that

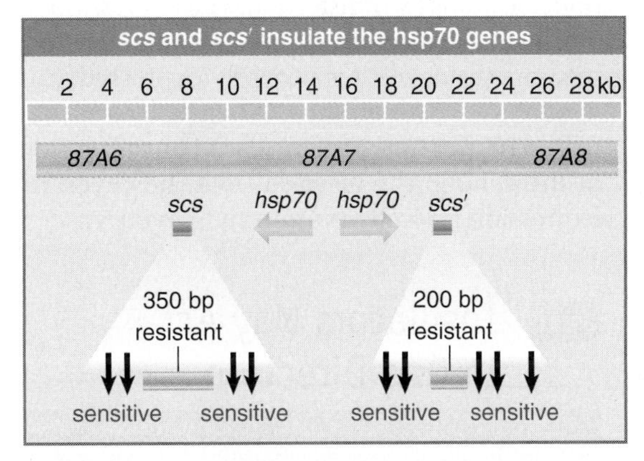

FIGURE 29.44 Specialized chromatin structures that include hypersensitive sites mark the ends of a domain in the *D. melanogaster* genome and insulate genes between them from the effects of surrounding sequences.

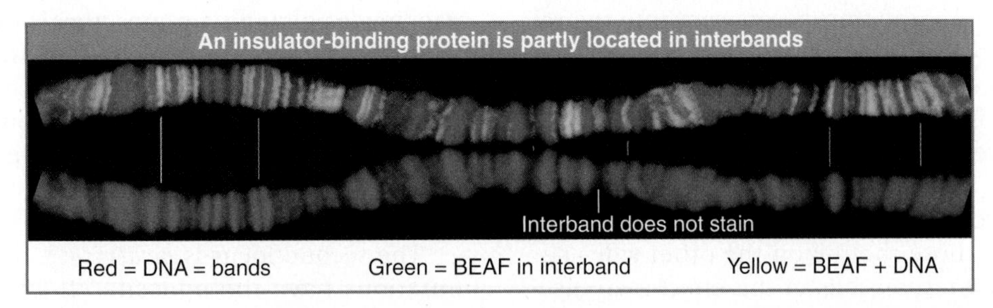

FIGURE 29.45 A protein that binds to the insulator *scs'* is localized at interbands in *Drosophila* polytene chromosomes. Red staining identifies the DNA (the bands) on both the upper and lower samples; green staining identifies BEAF32 (often at interbands) on the upper sample. Yellow shows coincidence of the two labels (meaning that BEAF32 is in a band). Reproduced from *Cell*, vol. 81, Zhao, K., Hart, C. M., and Laemmli, U. K., *Visualization of chromosomal. . . ,* pp. 879–889. Copyright 1995, with permission from Elsevier. Photo courtesy of Ulrich K. Laemmli, University of Geneva, Switzerland.

an anti-BEAF-32 antibody stains ~50% of the interbands of the polytene chromosomes. This suggests that there are many insulators in the genome, and that BEAF-32 is a common part of the insulating apparatus. It would imply that the band is a functional unit, and that interbands often have insulators that block the propagation of activating or inactivating effects.

Another example of an insulator that defines a domain is found in the chick β-globin LCR (the group of hypersensitive sites that controls expression of all β-globin genes; see Section 29.20, An LCR May Control a Domain). The leftmost hypersensitive site of the chick β-globin LCR (HS4) is an insulator that marks the 5′ end of the functional domain. This restricts the LCR to acting only on the globin genes in the domain.

A gene that is surrounded by insulators is usually protected against the propagation of inactivating effects from the surrounding regions. The test is to insert DNA into a genome at random locations by transfection. The expression of a gene in the inserted sequence is often erratic; in some instances it is properly expressed, but in others it is extinguished. When insulators that have a barrier function are placed on either side of the gene in the inserted DNA, however, its expression typically is uniform in every case.

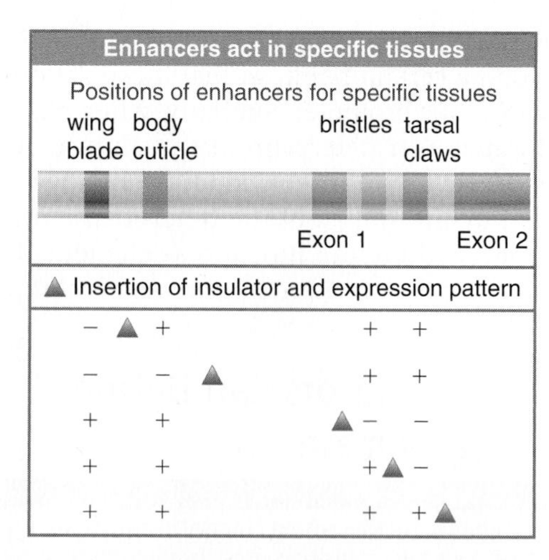

FIGURE 29.46 The insulator of the gypsy transposon blocks the action of an enhancer when it is placed between the enhancer and the promoter.

29.16 Insulators May Act in One Direction

Key concept

- Some insulators have directionality, and may stop passage of effects in one direction but not the other.

Insulators may have directional properties. Insertions of the transposon *gypsy* into the *yellow (y)* locus of *D. melanogaster* cause loss of gene function in some tissues, but not in others. The reason is that the *y* locus is regulated by four enhancers, as shown in FIGURE 29.46. Wherever *gypsy* is inserted, it blocks expression of all enhancers that it separates from the promoter, but not those that lie on the other side. The sequence responsible for this effect is an insulator that lies at one end of the transposon. The insulator works irrespective of its orientation of insertion.

Some of the enhancers are upstream of the promoter and others are downstream, so the effect cannot depend on position with regard to the promoter, nor can it require transcription to occur through the insulator. This is difficult to explain in terms of looping models for enhancer–promoter interaction, which essentially predict the irrelevance of the intervening DNA. The obvious model to invoke is a tracking mechanism, in which some component must move unidirectionally from the enhancer to the promoter, but this is difficult to reconcile with previous characterizations of the independence of enhancers from such effects.

Proteins that act upon the insulator have been identified through the existence of two other loci that affect insulator function in a *trans*-acting manner. Mutations in *su(Hw)* abolish insulation: *y* is expressed in all tissues in spite of the presence of the insulator. This suggests that *su(Hw)* codes for a protein that recognizes the insulator and is necessary for its action. Su(Hw) has a zinc finger DNA-motif; mapping to polytene chromosomes shows that it is bound at a large number of sites. The insulator contains twelve copies of a 26 bp sequence that is bound by Su(Hw). Manipulations show that the strength of the insulator is determined by the number of copies of the binding sequence.

The second locus is *mod(mdg4)*, in which mutations have the opposite effect. This is observed by the loss of directionality. These mutations increase the effectiveness of the insulator by extending its effects so that it blocks utilization of enhancers on both sides. *su(Hw)* is epistatic to *mod(mdg4)*; this means that in a double mutant we see only the effect of *su(Hw)*. This implies that mod(mdg4) acts through

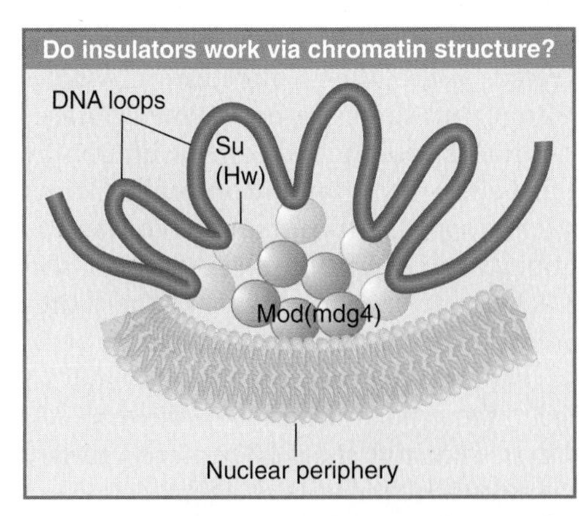

FIGURE 29.47 Su(Hw)/mod(mdg4) complexes are found at the nuclear periphery. They may organize DNA into loops that limit enhancer–promoter interactions.

su(Hw). The basic role of the wild-type protein from the *mod(mdg4)* locus is therefore to impose directionality on the ability of su(Hw) to insulate promoters from the boundary.

Binding of su(Hw) to DNA, followed by binding of mod(mdg4) to su(Hw), therefore creates a unidirectional block to activation of a promoter. This suggests that the insulator bound by su(Hw) can spread inactivity in both directions, but mod(mdg4) stops the effect from spreading in one direction. Perhaps there is some intrinsic directionality to chromatin, which results ultimately in the incorporation of su(Hw), mod(mdg4), or some other component in one orientation, presumably by virtue of an interaction with some component of chromatin that is itself preferentially oriented. Any such directionality would need to reverse at the promoter.

It is likely that insulators act by making changes in chromatin structure. One model is prompted by the observation that Su(Hw) and mod(mdg4) binding sites are present at >500 locations in the *Drosophila* genome. Visualization of the sites where the proteins are bound in the nucleus, however, shows that they are colocalized at ~25 discrete sites around the nuclear periphery. This suggests the model of FIGURE 29.47 in which Su(Hw) proteins bound at different sites on DNA are brought together by binding to mod(mdg4). The Su(Hw)/mod(mdg4) complex is localized at the nuclear periphery. The DNA bound to it is organized into loops. An average complex might have ~20 such loops. Enhancer–promoter actions can occur only within a loop, and cannot propagate between them.

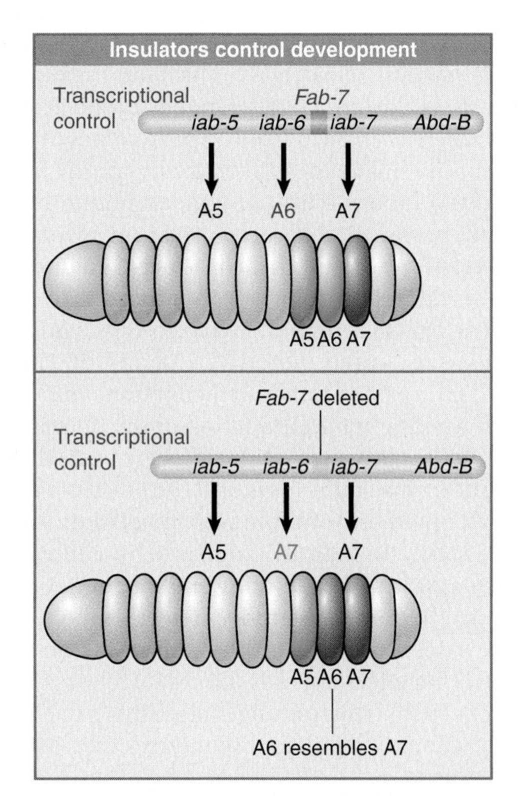

FIGURE 29.48 *Fab-7* is a boundary element that is necessary for the independence of regulatory elements *iab-6* and *iab-7*.

29.17 Insulators Can Vary in Strength

Key concept

- Insulators can differ in how effectively they block passage of an activating signal.

Sometimes elements with different *cis*-acting properties are combined to generate regions with complex regulatory effects. The *Fab-7* region is defined by deletions in the *bithorax* locus of *Drosophila*. This locus contains a series of *cis*-acting regulatory elements that control the activities of three transcription units. The relevant part of the locus is drawn in FIGURE 29.48. The regulatory elements *iab-6* and *iab-7* control expression of the adjacent gene *Abd-B* in successive regions of the embryo (segments A6 and A7). A deletion of *Fab-7* causes A6 to develop like A7, instead of in the usual way. This is a dominant effect, which suggests that *iab-7* has taken over control from *iab-6*. We can interpret this in molecular terms by supposing that *Fab-7* provides a boundary that prevents *iab-7* from acting when *iab-6* is usually active.

Like other boundary elements (insulators), *Fab-7* contains a distinctive chromatin structure that is marked by a series of hypersensitive sites. The region can be divided into two types of elements by smaller deletions and by testing fragments for their ability to provide a boundary. A sequence of ~3.3 kb behaves as an insulator when it is placed in other constructs. A sequence of ~0.8 kb behaves as a repressor that acts on *iab-7*. The presence of these two elements explains the complicated genetic behavior of *Fab-7* (which we have not described in detail).

An insight into the action of the boundary element is provided by the effects of substituting other insulators for *Fab-7*. The effect of *Fab-7* is simply to prevent interaction between *iab-6* and *iab-7*. When *Fab-7* is replaced by a different insulator [in fact a binding site for the protein Su(Hw)], though, a stronger effect is seen: *iab-5* takes over from *iab-7*. When an *scs* element is used, the effect extends to *iab-4*. This suggests a scheme in which stronger elements can block the actions of regulatory sequences that lie farther away.

This conclusion introduces a difficulty for explaining the action of boundary elements. They cannot be functioning in this instance simply by preventing the transmission of effects past the boundary. This argues against models based on simple tracking or inhibiting the linear propagation of structural effects. It suggests that there may be some sort of competitive effect, in which the strength of the element determines how far its effect can stretch.

The situation is further complicated by the existence of **anti-insulator** elements, which allow an enhancer to overcome the blocking effects of an insulator. This again suggests that these effects are mediated by some sort of control over local chromatin structure.

29.18 DNAase Hypersensitive Sites Reflect Changes in Chromatin Structure

Key concepts

- Hypersensitive sites are found at the promoters of expressed genes.
- They are generated by the binding of transcription factors that displace histone octamers.

In addition to the general changes that occur in active or potentially active regions, structural changes occur at specific sites associated with initiation of transcription or with certain structural features in DNA. These changes were first detected by the effects of digestion with very low concentrations of the enzyme DNAase I.

When chromatin is digested with DNAase I, the first effect is the introduction of breaks in the duplex at specific, **hypersensitive sites.** Susceptibility to DNAase I reflects the availability of DNA in chromatin, so we take these sites to represent chromatin regions in which the DNA is particularly exposed because it is not organized in the usual nucleosomal structure. A typical hypersensitive site is 100× more sensitive to enzyme attack than bulk chromatin. These sites are also hypersensitive to other nucleases and to chemical agents.

Hypersensitive sites are created by the (tissue-specific) structure of chromatin. Their locations can be determined by the technique of indirect end labeling that we introduced earlier in the context of nucleosome positioning. This application of the technique is recapitulated in FIGURE 29.49. In this case, cleavage at the hypersensitive site by DNAase I is used to generate one end of the fragment, and its distance is measured from the other end that is generated by cleavage with a restriction enzyme.

Many of the hypersensitive sites are related to gene expression. Every active gene has a site, or sometimes more than one site, in the region of the promoter. *Most hypersensitive sites are found only in chromatin of cells in which the associated gene is being expressed;* they do not occur when the gene is inactive. The 5′ hypersensitive site(s) appear before transcription begins, and the DNA sequences contained within the hypersensitive sites are required for gene expression, as seen by mutational analysis.

A particularly well-characterized nuclease-sensitive region lies on the SV40 minichromosome. A short segment near the origin of replication, just upstream of the promoter for the late transcription unit, is cleaved preferentially by DNAase I, micrococcal nuclease, and other nucleases (including restriction enzymes).

The state of the SV40 minichromosome can be visualized by electron microscopy. In up to 20% of the samples, a "gap" is visible in the nucleosomal organization, as evident in FIGURE 29.50. The gap is a region of ~120 nm in length (about 350 bp), surrounded on either side by nucleosomes. The visible gap corresponds with the nuclease-sensitive region. This shows directly that increased sensitivity to nucleases is associated with the exclusion of nucleosomes.

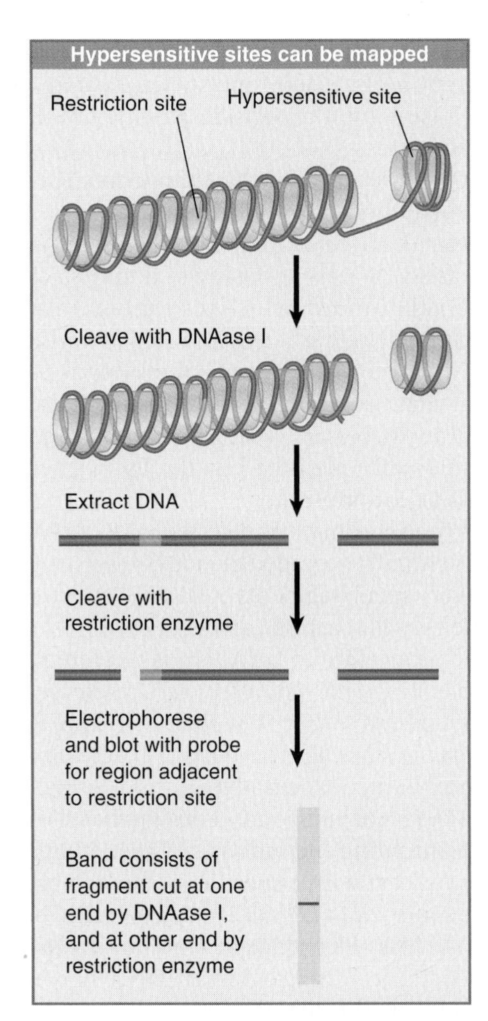

FIGURE 29.49 Indirect end-labeling identifies the distance of a DNAase hypersensitive site from a restriction cleavage site. The existence of a particular cutting site for DNAase I generates a discrete fragment, whose size indicates the distance of the DNAase I hypersensitive site from the restriction site.

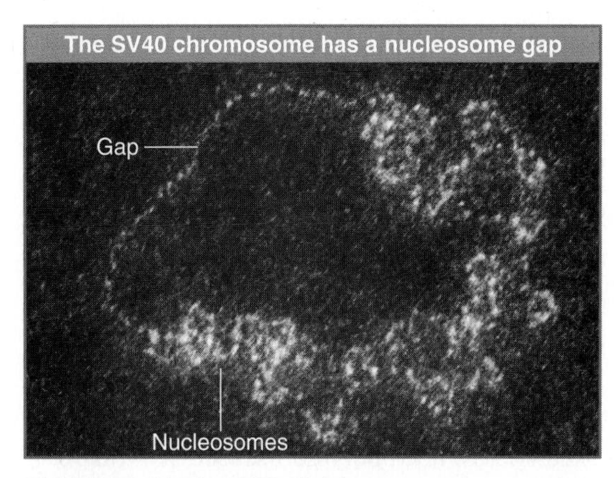

FIGURE 29.50 The SV40 minichromosome has a nucleosome gap. Photo courtesy of Moshe Yaniv, Pasteur Institute.

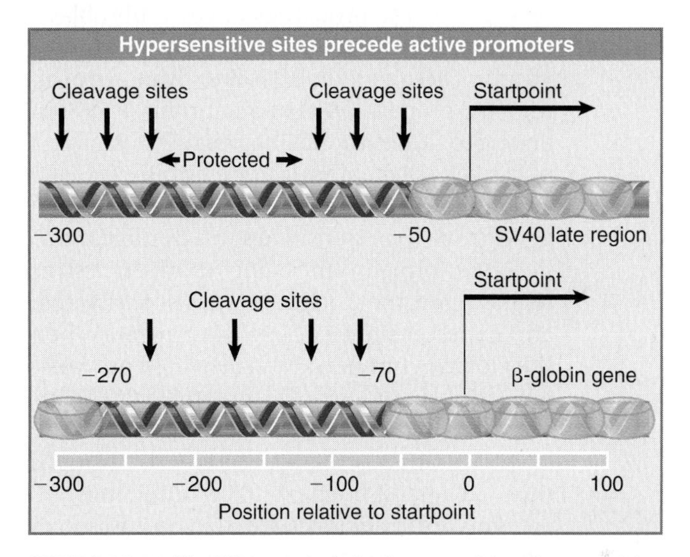

FIGURE 29.51 The SV40 gap includes hypersensitive sites, sensitive regions, and a protected region of DNA. The hypersensitive site of a chicken β-globin gene comprises a region that is susceptible to several nucleases.

A hypersensitive site is not necessarily uniformly sensitive to nucleases. FIGURE 29.51 shows the maps of two hypersensitive sites.

Within the SV40 gap of ~300 bp, there are two hypersensitive DNAase I sites and a "protected" region. The protected region presumably reflects the association of (nonhistone) protein(s) with the DNA. The gap is associated with the DNA sequence elements that are necessary for promoter function.

The hypersensitive site at the β-globin promoter is preferentially digested by several enzymes, including DNAase I, DNAase II, and micrococcal nuclease. The enzymes have preferred cleavage sites that lie at slightly different points in the same general region. Thus a region extending from about –70 to –270 is preferentially accessible to nucleases when the gene is transcribable.

What is the structure of the hypersensitive site? Its preferential accessibility to nucleases indicates that it is not protected by histone octamers, but this does not necessarily imply that it is free of protein. A region of free DNA might be vulnerable to damage, and in any case, how would it be able to exclude nucleosomes? We assume that the hypersensitive site results from the binding of specific regulatory proteins that exclude nucleosomes. Indeed, the binding of such proteins is probably the basis for the existence of the protected region within the hypersensitive site.

The proteins that generate hypersensitive sites are likely to be regulatory factors of various types, because hypersensitive sites are found associated with promoters, other elements that regulate transcription, origins of replication, centromeres, and sites with other structural significance. In some cases, they are associated with more extensive organization of chromatin structure. A hypersensitive site may provide a boundary for a series of positioned nucleosomes. Hypersensitive sites associated with transcription may be generated by transcription factors when they bind to the promoter as part of the process that makes it accessible to RNA polymerase (see Section 30.4, Nucleosome Organization May Be Changed at the Promoter).

The stability of hypersensitive sites is revealed by the properties of chick fibroblasts transformed with temperature-sensitive tumor viruses. These experiments take advantage of an unusual property: Although fibroblasts do not belong to the erythroid lineage, transformation of the cells at the normal temperature leads to activation of the globin genes. The activated genes have hypersensitive sites. If transformation is performed at the higher (nonpermissive) temperature, the globin genes are not activated, and hypersensitive sites do not appear. When the globin genes have been activated by transformation at low temperature, they can be inactivated by raising the temperature. The hypersensitive sites are retained, though, through at least the next 20 cell doublings.

This result demonstrates that acquisition of a hypersensitive site is only one of the features necessary to initiate transcription, and it implies that the events involved in establishing a hypersensitive site are distinct from those concerned with perpetuating it. Once the site has been established, it is perpetuated through replication in the absence of the circumstances needed for induction. Could some specific intervention be needed to abolish a hypersensitive site?

29.19 Domains Define Regions That Contain Active Genes

Key concept

- A domain containing a transcribed gene is defined by increased sensitivity to degradation by DNAase I.

A region of the genome that contains an active gene may have an altered structure. The change in structure precedes, and is different from, the disruption of nucleosome structure that may be caused by the actual passage of RNA polymerase.

One indication of the change in structure of transcribed chromatin is provided by its increased susceptibility to degradation by DNAase I. DNAase I sensitivity defines a chromosomal **domain,** which is a region of altered structure including at least one active transcription unit, and sometimes extending farther. (Note that use of the term "domain" does not imply any necessary connection with the structural domains identified by the loops of chromatin or chromosomes.)

When chromatin is digested with DNAase I, it is eventually degraded into acid-soluble material (very small fragments of DNA). The progress of the overall reaction can be followed in terms of the proportion of DNA that is rendered acid soluble. *When only 10% of the total DNA has become acid soluble, more than 50% of the DNA of an active gene has been lost.* This suggests that active genes are preferentially degraded.

The fate of individual genes can be followed by quantitating the amount of DNA that survives to react with a specific probe. The protocol is outlined in FIGURE 29.52. The principle is that the loss of a particular band indicates that the corresponding region of DNA has been degraded by the enzyme.

FIGURE 29.53 shows what happens to β-globin genes and an ovalbumin gene in chromatin extracted from chicken red blood cells (in which globin genes are expressed and the ovalbumin gene is inactive). The restriction fragments representing the β-globin genes are rapidly lost, whereas those representing the ovalbumin gene show little degradation. (The ovalbumin gene in fact is digested at the same rate as the bulk of DNA.)

Thus the bulk of chromatin is relatively resistant to DNAase I and contains nonexpressed genes (as well as other sequences). *A gene becomes relatively susceptible to the enzyme specifically in the tissue(s) in which it is expressed.*

Is preferential susceptibility a characteristic only of rather actively expressed genes, such as globin, or of all active genes? Experiments using probes representing the entire cellular mRNA population suggest that all active genes, whether coding for abundant or for rare mRNAs, are preferentially susceptible to DNAase I. (There are, however, variations in the degree of susceptibility.) The rarely expressed genes are likely to have very few RNA polymerase molecules

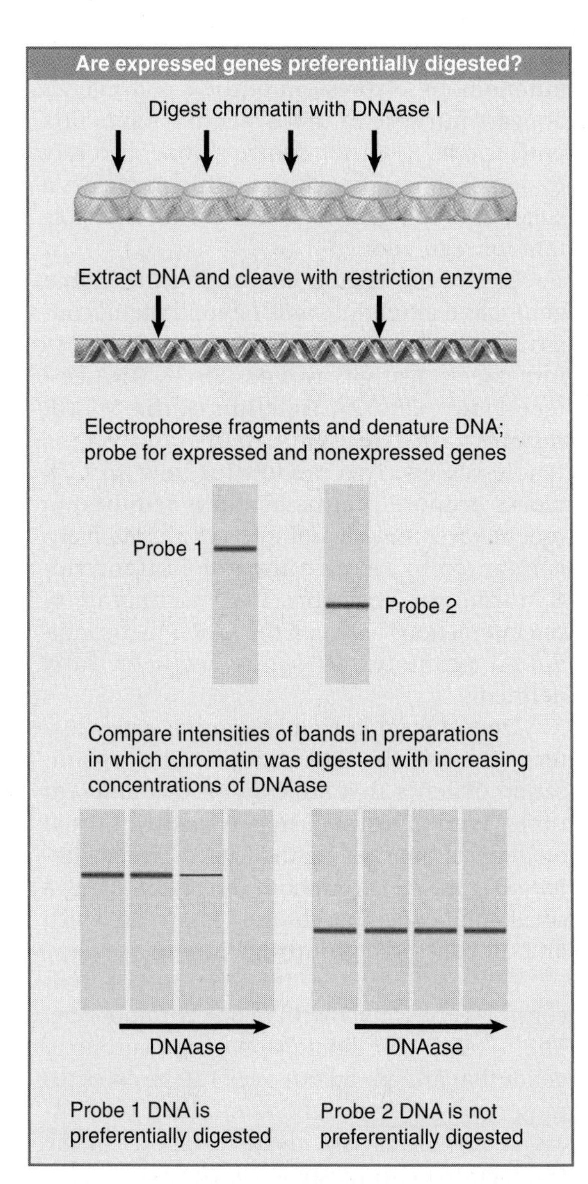

FIGURE 29.52 Sensitivity to DNAase I can be measured by determining the rate of disappearance of the material hybridizing with a particular probe.

actually engaged in transcription at any moment; this implies that the sensitivity to DNAase I does not result from the act of transcription, but instead is a feature of *genes that are able to be transcribed.*

What is the extent of the preferentially sensitive region? This can be determined by using a series of probes representing the flanking regions as well as the transcription unit itself. The sensitive region always extends over the entire transcribed region; an additional region of several kb on either side may show an intermediate level of sensitivity (probably as the result of spreading effects).

The critical concept implicit in the description of the domain is that a region of high sensitivity to DNAase I extends over a considerable distance. Often we think of regulation as residing in events that occur at a discrete site in DNA—for example, in the ability to initiate transcription at the promoter. Even if this is true, such regulation must determine, or must be accompanied by, a more wide-ranging change in structure. This is a difference between eukaryotes and prokaryotes.

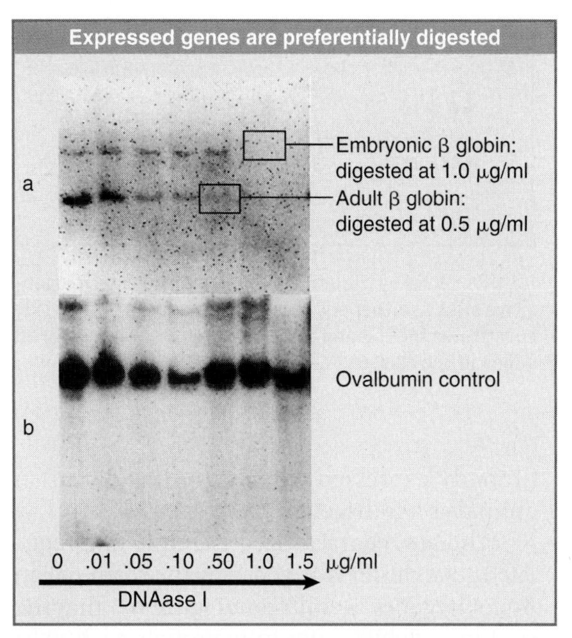

FIGURE 29.53 In adult erythroid cells, the adult β-globin gene is highly sensitive to DNAase I digestion; the embryonic β-globin gene is partially sensitive (probably due to spreading effects), but ovalbumin is not sensitive. Photos courtesy of Harold Weintraub, Fred Hutchinson Cancer Research Center. Used with permission of Mark T. Groudine.

29.20 An LCR May Control a Domain

Key concept

- An LCR is located at the 5′ end of the domain and consists of several hypersensitive sites.

Every gene is controlled by its promoter, and some genes also respond to enhancers (containing similar control elements but located farther away), as discussed in Chapter 24, Promoters and Enhancers. These local controls are not sufficient for all genes, though. In some cases, a gene lies within a domain of several genes, all of which are influenced by regulatory elements that act on the whole domain. The existence of these elements was identified by the inability of a region of DNA including a gene and all its known regulatory elements to be

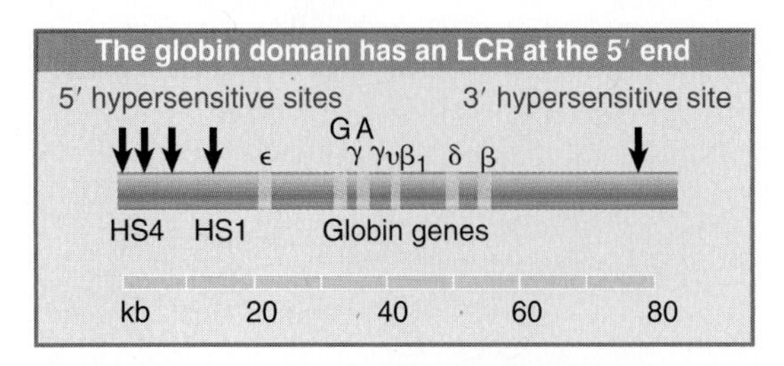

FIGURE 29.54 A globin domain is marked by hypersensitive sites at either end. The group of sites at the 5′ side constitutes the LCR and is essential for the function of all genes in the cluster.

properly expressed when introduced into an animal as a transgene.

The best characterized example of a regulated gene cluster is provided by the mammalian β-globin genes. Recall from Figure 6.3 that the α- and β-globin genes in mammals each exist as clusters of related genes that are expressed at different times during embryonic and adult development. These genes are provided with a large number of regulatory elements, which have been analyzed in detail. In the case of the adult human β-globin gene, regulatory sequences are located both 5′ and 3′ to the gene.The regulatory sequences include both positive and negative elements in the promoter region, as well as additional positive elements within and downstream of the gene.

A human β-globin gene containing all of these control regions, however, is never expressed in a transgenic mouse within an order of magnitude of wild-type levels. Some further regulatory sequence is required. Regions that provide the additional regulatory function are identified by DNAase I hypersensitive sites that are found at the ends of the cluster. The map of FIGURE 29.54 shows that the 20 kb upstream of the ε gene contains a group of five sites, and that there is a single site 30 kb downstream of the β gene. Transfecting various constructs into mouse erythroleukemia cells shows that sequences between the individual hypersensitive sites in the 5′ region can be removed without much effect, but that removal of any of the sites reduces the overall level of expression.

The 5′ regulatory sites are the primary regulators, and the cluster of hypersensitive sites is called the **locus control region (LCR).** We do not know if the 3′ site has any function. The LCR is absolutely required for expression of each of the globin genes in the cluster. Each gene is then further regulated by its own specific controls. Some of these controls are autonomous: Expression of the ε and γ genes appears intrinsic to those loci in conjunction with the LCR. Other controls appear to rely upon position in the cluster, which provides a suggestion that *gene order* in a cluster is important for regulation.

The entire region containing the globin genes, and extending well beyond them, constitutes a chromosomal **domain.** It shows increased sensitivity to digestion by DNAase I (see Figure 29.52). Deletion of the 5′ LCR restores normal resistance to DNAase over the whole region. Two models for how an LCR works propose that its action is required in order to activate the promoter, or alternatively, is required to increase the rate of transcription from the promoter. The exact nature of the interactions between the LCR and the individual promoters has not yet been fully defined.

Does this model apply to other gene clusters? The α-globin locus has a similar organization of genes that are expressed at different times, with a group of hypersensitive sites at one end of the cluster, and increased sensitivity to DNAase I throughout the region. Only a small number of other cases are known in which an LCR controls a group of genes.

One of these cases involves an LCR that controls genes on more than one chromosome. The T_H2 LCR coordinately regulates a group of genes that are spread out over 120 kb on chromosome 11 by interacting with their promoters. It also interacts with the promoter of the IFNγ gene on chromosome 10. The two types of interaction are alternatives that comprise two different cell fates, that is, in one group of cells the LCR causes expression of the genes on chromosome 11, whereas in the other group it causes the gene on chromosome 10 to be expressed.

29.21 What Constitutes a Regulatory Domain?

Key concept

- A domain may have an insulator, an LCR, a matrix attachment site, and transcription unit(s).

If we put together the various types of structures that have been found in different systems, we can think about the possible nature of a chromosomal domain. The basic feature of a regulatory domain is that regulatory elements can act only on transcription units within the same domain. A domain might contain more than one transcription unit and/or enhancer.

FIGURE 29.55 summarizes the structures that might be involved in defining a domain.

An insulator stops activating or repressing effects from passing. In its simplest form, an insulator blocks either type of effect from passing across it, but there can be more complex relationships in which the insulator blocks only one type of effect and/or acts directionally. We assume that insulators act by affecting higher-order chromatin structure, but we do not know the details and varieties of such effects.

A **matrix attachment site (MAR)** may be responsible for attaching chromatin to a site on the nuclear periphery (see Section 28.6, Specific Sequences Attach DNA to an Interphase Matrix). These are likely to be responsible for creating physical domains of DNA that take the form of loops extending out from the attachment sites. This looks very much like one model for insulator action. In fact, some MAR elements behave as insulators in assays *in vitro*, but it seems that their ability to attach DNA to the matrix can be separated from the insulator function, so there is not a simple cause and effect. It would not be surprising if insulator and MAR elements were associated to maintain a relationship between regulatory effects and physical structure.

An LCR functions at a distance and may be required for any and all genes in a domain to be expressed (see Section 29.20, An LCR May Control a Domain). When a domain has an LCR, its function is essential for all genes in the domain, but LCRs do not seem to be common. Several types of *cis*-acting structures could be required for function. As defined originally, the property of the LCR rests with an enhancer-like hypersensitive site that is needed for the full activity of promoter(s) within the domain.

The organization of domains may help to explain the large size of the genome. A certain amount of space could be required for such a structure to operate, for example, to allow chromatin to become decondensed and to become accessible. Although the exact sequences of much of the unit might be irrelevant, there might be selection for the overall amount of DNA within it, or at least selection might prevent the various transcription units from becoming too closely spaced.

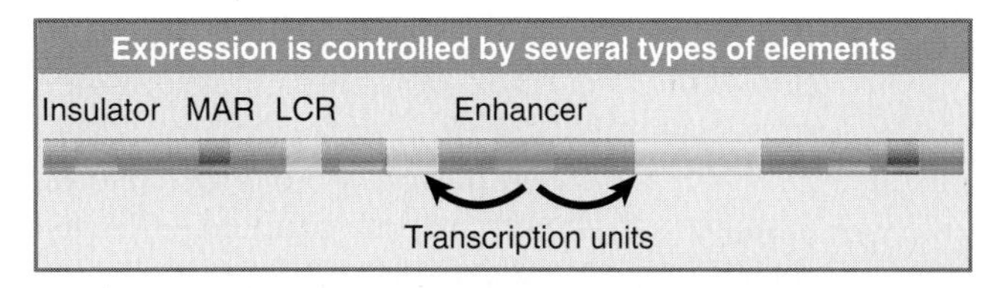

FIGURE 29.55 Domains may possess three types of sites: insulators to prevent effects from spreading between domains, MARs to attach the domain to the nuclear matrix, and LCRs that are required for initiation of transcription. An enhancer may act on more than one promoter within the domain.

29.22 Summary

All eukaryotic chromatin consists of nucleosomes. A nucleosome contains a characteristic length of DNA, usually ~200 bp, which is wrapped around an octamer containing two copies each of histones H2A, H2B, H3, and H4. A single H1 protein is associated with each nucleosome. Virtually all genomic DNA is organized into nucleosomes. Treatment with micrococcal nuclease shows that the DNA packaged into each nucleosome can be divided operationally into two regions. The linker region is digested rapidly by the nuclease; the core region of 146 bp is resistant to digestion. Histones H3 and H4 are the most highly conserved, and an $H3_2$-$H4_2$ tetramer accounts for the diameter of the particle. The H2A and H2B histones are organized as two H2A-H2B dimers. Octamers are assembled by the successive addition of two H2A-H2B dimers to the $H3_2$-$H4_2$ kernel.

The path of DNA around the histone octamer creates –1.65 supercoils. The DNA "enters" and "leaves" the nucleosome in the same vicinity, and could be "sealed" by histone H1. Removal of the core histones releases –1.0 supercoils. The difference can be largely explained by a change in the helical pitch of DNA, from an average of 10.2 bp/turn in nucleosomal form to 10.5 bp/turn when free in solution. There is variation in the structure of DNA from a periodicity of 10.0 bp/turn at the nucleosome ends to 10.7 bp/turn in the center. There are kinks in the path of DNA on the nucleosome.

Nucleosomes are organized into a fiber of 30 nm diameter that has six nucleosomes per turn and a packing ratio of 40. Removal of H1 allows this fiber to unfold into a 10 nm fiber that consists of a linear string of nucleosomes. The 30 nm fiber probably consists of the 10 nm fiber wound into a 2-start solenoid. The 30 nm fiber is the basic constituent of both euchromatin and heterochromatin; nonhistone proteins are responsible for further organization of the fiber into chromatin or chromosome ultrastructure.

There are two pathways for nucleosome assembly. In the replication-coupled pathway, the PCNA processivity subunit of the replisome recruits CAF-1, which is a nucleosome assembly factor. CAF-1 assists the deposition of $H3_2$-$H4_2$ tetramers onto the daughter duplexes resulting from replication. The tetramers may be produced either by disruption of existing nucleosomes by the replication fork or as the result of assembly from newly synthesized histones. Similar sources provide the H2A-H2B dimers that then assemble with the $H3_2$-$H4_2$ tetramer to complete the nucleosome. The $H3_2$-$H4_2$ tetramer and the H2A-H2B dimers assemble at random, so the new nucleosomes may include both preexisting and newly synthesized histones.

RNA polymerase displaces histone octamers during transcription. Nucleosomes reform on DNA after the polymerase has passed, unless transcription is very intensive (such as in rDNA) when they may be displaced completely. The replication-independent pathway for nucleosome assembly is responsible for replacing histone octamers that have been displaced by transcription. It uses the histone variant H3.3 instead of H3. A similar pathway, with another alternative to H3, is used for assembling nucleosomes at centromeric DNA sequences following replication.

An insulator blocks the transmission of activating or inactivating effects in chromatin. An insulator that is located between an enhancer and a promoter prevents the enhancer from activating the promoter. Two insulators define the region between them as a regulatory domain; regulatory interactions within the domain are limited to it, and the domain is insulated from outside effects. Most insulators block regulatory effects from passing in either direction, but some are directional. Insulators usually can block both activating effects (enhancer–promoter interactions) and inactivating effects (mediated by spread of heterochromatin), but some are limited to one or the other. Insulators are thought to act via changing higher-order chromatin structure, but the details are not certain.

Two types of changes in sensitivity to nucleases are associated with gene activity. Chromatin capable of being transcribed has a generally increased sensitivity to DNAase I, reflecting a change in structure over an extensive region that can be defined as a domain containing active or potentially active genes. Hypersensitive sites in DNA occur at discrete locations, and are identified by greatly increased sensitivity to DNAase I. A hypersensitive site consists of a sequence of ~200 bp from which nucleosomes are excluded by the presence of other proteins. A hypersensitive site forms a boundary that may cause adjacent nucleosomes to be restricted in position. Nucleosome positioning may be important in controlling access of regulatory proteins to DNA.

Hypersensitive sites occur at several types of regulators. Those that regulate transcription include promoters, enhancers, and LCRs. Other sites include origins for replication and centromeres. A promoter or enhancer acts on a single gene, but an LCR contains a group of hypersensitive sites and may regulate a domain containing several genes.

References

29.2 The Nucleosome Is the Subunit of All Chromatin

Reviews

Kornberg, R. D. (1977). Structure of chromatin. *Annu. Rev. Biochem.* 46, 931–954.

McGhee, J. D., and Felsenfeld, G. (1980). Nucleosome structure. *Annu. Rev. Biochem.* 49, 1115–1156.

Research

Kornberg, R. D. (1974). Chromatin structure: a repeating unit of histones and DNA. *Science* 184, 868–871.

Richmond, T. J., Finch, J. T., Rushton, B., Rhodes, D., and Klug, A. (1984). Structure of the nucleosome core particle at 7 Å resolution. *Nature* 311, 532–537.

29.3 DNA Is Coiled in Arrays of Nucleosomes

Research

Finch, J. T. et al. (1977). Structure of nucleosome core particles of chromatin. *Nature* 269, 29–36.

29.4 Nucleosomes Have a Common Structure

Research

Shen, X. et al. (1995). Linker histones are not essential and affect chromatin condensation *in vitro*. *Cell* 82, 47–56.

29.5 DNA Structure Varies on the Nucleosomal Surface

Review

Wang, J. (1982). The path of DNA in the nucleosome. *Cell* 29, 724–726.

Research

Richmond, T. J. and Davey, C. A. (2003). The structure of DNA in the nucleosome core. *Nature* 423, 145–150.

29.6 The Periodicity of DNA Changes on the Nucleosome

Review

Travers, A. A. and Klug, A. (1987). The bending of DNA in nucleosomes and its wider implications. *Philos. Trans. R. Soc. Lond. B Biol. Sci.* 317, 537–561.

29.7 Organization of the Histone Octamer

Research

Angelov, D., Vitolo, J. M., Mutskov, V., Dimitrov, S., and Hayes, J. J. (2001). Preferential interaction of the core histone tail domains with linker DNA. *Proc. Natl. Acad. Sci. USA* 98, 6599–6604.

Arents, G., Burlingame, R. W., Wang, B.-C., Love, W. E., and Moudrianakis, E. N. (1991). The nucleosomal core histone octamer at 31 Å resolution: a tripartite protein assembly and a left-handed superhelix. *Proc. Natl. Acad. Sci. USA* 88, 10148–10152.

Luger, K. et al. (1997). Crystal structure of the nucleosome core particle at 28 Å resolution. *Nature* 389, 251–260.

29.8 The Path of Nucleosomes in the Chromatin Fiber

Review

Felsenfeld, G. and McGhee, J. D. (1986). Structure of the 30 nm chromatin fiber. *Cell* 44, 375–377.

Research

Dorigo, B., Schalch, T., Kulangara, A., Duda, S., Schroeder, R. R., and Richmond, T. J. (2004). Nucleosome arrays reveal the two-start organization of the chromatin fiber. *Science* 306, 1571–1573.

Schalch, T., Duda, S., Sargent, D. F., and Richmond, T. J. (2005). X-ray structure of a tetranucleosome and its implications for the chromatin fibre. *Nature* 436, 138–141.

29.9 Reproduction of Chromatin Requires Assembly of Nucleosomes

Reviews

Osley, M. A. (1991). The regulation of histone synthesis in the cell cycle. *Annu. Rev. Biochem.* 60, 827–861.

Verreault, A. (2000). De novo nucleosome assembly: new pieces in an old puzzle. *Genes Dev.* 14, 1430–1438.

Research

Ahmad, K. and Henikoff, S. (2001). Centromeres are specialized replication domains in heterochromatin. *J. Cell Biol.* 153, 101–110.

Ahmad, K. and Henikoff, S. (2002). The histone variant H3.3 marks active chromatin by replication-independent nucleosome assembly. *Mol. Cell* 9, 1191–1200.

Gruss, C., Wu, J., Koller, T., and Sogo, J. M. (1993). Disruption of the nucleosomes at the replication fork. *EMBO J.* 12, 4533–4545.

Loppin, B., Bonnefoy, E., Anselme, C., Laurencon, A., Karr, T. L., and Couble, P. (2005). The histone H3.3 chaperone HIRA is essential for chromatin assembly in the male pronucleus. *Nature* 437, 1386–1390.

Ray-Gallet, D., Quivy, J. P., Scamps, C., Martini, E. M., Lipinski, M., and Almouzni, G. (2002). HIRA is critical for a nucleosome assembly pathway independent of DNA synthesis. *Mol. Cell* 9, 1091–1100.

Shibahara, K., and Stillman, B. (1999). Replication-dependent marking of DNA by PCNA facilitates CAF-1-coupled inheritance of chromatin. *Cell* 96, 575–585.

Smith, S. and Stillman, B. (1989). Purification and characterization of CAF-I, a human cell factor required for chromatin assembly during DNA replication *in vitro*. *Cell* 58, 15–25.

Smith, S. and Stillman, B. (1991). Stepwise assembly of chromatin during DNA replication *in vitro*. *EMBO J.* 10, 971–980.

Tagami, H., Ray-Gallet, D., Almouzni, G., and Nakatani, Y. (2004). Histone H3.1 and H3.3 complexes mediate nucleosome assembly pathways dependent or independent of DNA synthesis. *Cell* 116, 51–61.

Yu, L. and Gorovsky, M. A. (1997). Constitutive expression, not a particular primary sequence, is the important feature of the H3 replacement variant hv2 in *Tetrahymena thermophila*. *Mol. Cell. Biol.* 17, 6303–6310.

29.11 Are Transcribed Genes Organized in Nucleosomes?

Review

Kornberg, R. D. and Lorch, Y. (1992). Chromatin structure and transcription. *Annu. Rev. Cell Biol.* 8, 563–587.

29.12 Histone Octamers Are Displaced by Transcription

Research

Cavalli, G. and Thoma, F. (1993). Chromatin transitions during activation and repression of galactose-regulated genes in yeast. *EMBO J.* 12, 4603–4613.

Studitsky, V. M., Clark, D. J., and Felsenfeld, G. (1994). A histone octamer can step around a transcribing polymerase without leaving the template. *Cell* 76, 371–382.

29.13 Nucleosome Displacement and Reassembly Require Special Factors

Research

Belotserkovskaya, R., Oh, S., Bondarenko, V. A., Orphanides, G., Studitsky, V. M., and Reinberg, D. (2003). FACT facilitates transcription-dependent nucleosome alteration. *Science* 301, 1090–1093.

Saunders, A., Werner, J., Andrulis, E. D., Nakayama, T., Hirose, S., Reinberg, D., and Lis, J. T. (2003). Tracking FACT and the RNA polymerase II elongation complex through chromatin in vivo. *Science* 301, 1094–1096.

29.14 Insulators Block the Actions of Enhancers and Heterochromatin

Reviews

Gerasimova, T. I. and Corces, V. G. (2001). Chromatin insulators and boundaries: effects on transcription and nuclear organization. *Annu. Rev. Genet.* 35, 193–208.

West, A. G., Gaszner, M., and Felsenfeld, G. (2002). Insulators: many functions, many mechanisms. *Genes Dev.* 16, 271–288.

29.15 Insulators Can Define a Domain

Research

Chung, J. H., Whiteley, M., and Felsenfeld, G. (1993). A 5′ element of the chicken β-globin domain serves as an insulator in human erythroid cells and protects against position effect in *Drosophila*. *Cell* 74, 505–514.

Cuvier, O., Hart, C. M., and Laemmli, U. K. (1998). Identification of a class of chromatin boundary elements. *Mol. Cell Biol.* 18, 7478–7486.

Gaszner, M., Vazquez, J., and Schedl, P. (1999). The Zw5 protein, a component of the scs chromatin domain boundary, is able to block enhancer-promoter interaction. *Genes Dev.* 13, 2098–2107.

Kellum, R. and Schedl, P. (1991). A position-effect assay for boundaries of higher order chromosomal domains. *Cell* 64, 941–950.

Pikaart, M. J., Recillas-Targa, F., and Felsenfeld, G. (1998). Loss of transcriptional activity of a transgene is accompanied by DNA methylation and histone deacetylation and is prevented by insulators. *Genes Dev.* 12, 2852–2862.

Zhao, K., Hart, C. M., and Laemmli, U. K. (1995). Visualization of chromosomal domains with boundary element-associated factor BEAF-32. *Cell* 81, 879–889.

29.16 Insulators May Act in One Direction

Research

Gerasimova, T. I., Byrd, K., and Corces, V. G. (2000). A chromatin insulator determines the nuclear localization of DNA. *Mol. Cell* 6, 1025–1035.

Harrison, D. A., Gdula, D. A., Cyne, R. S., and Corces, V. G. (1993). A leucine zipper domain of the suppressor of hairy-wing protein mediates its repressive effect on enhancer function. *Genes Dev.* 7, 1966–1978.

Roseman, R. R., Pirrotta, V., and Geyer, P. K. (1993). The su(Hw) protein insulates expression of the *D. melanogaster white* gene from chromosomal position-effects. *EMBO J.* 12, 435–442.

29.17 Insulators Can Vary in Strength

Research

Hagstrom, K., Muller, M., and Schedl, P. (1996). Fab-7 functions as a chromatin domain boundary to ensure proper segment specification by the *Drosophila* bithorax complex. *Genes Dev.* 10, 3202–3215.

Mihaly, J. et al. (1997). *In situ* dissection of the Fab-7 region of the bithorax complex into a chromatin domain boundary and a polycomb-response element. *Development* 124, 1809–1820.

Zhou, J. and Levine, M. (1999). A novel *cis*-regulatory element, the PTS, mediates an anti-insulator activity in the *Drosophila* embryo. *Cell* 99, 567–575.

29.18 DNAase Hypersensitive Sites Reflect Changes in Chromatin Structure

Review

Gross, D. S. and Garrard, W. T. (1988). Nuclease hypersensitive sites in chromatin. *Annu. Rev. Biochem.* 57, 159–197.

Research

Groudine, M., and Weintraub, H. (1982). Propagation of globin DNAase I-hypersensitive sites in absence of factors required for induction: a possible mechanism for determination. *Cell* 30, 131–139.

Moyne, G., Harper, F., Saragosti, S., and Yaniv, M. (1982). Absence of nucleosomes in a histone-containing nucleoprotein complex obtained by dissociation of purified SV40 virions. *Cell* 30, 123–130.

Scott, W. A. and Wigmore, D. J. (1978). Sites in SV40 chromatin which are preferentially cleaved by endonucleases. *Cell* 15, 1511–1518.

Varshavsky, A. J., Sundin, O., and Bohn, M. J. (1978). SV40 viral minichromosome: preferential exposure of the origin of replication as probed by restriction endonucleases. *Nucleic Acids Res.* 5, 3469–3479.

29.19 Domains Define Regions That Contain Active Genes

Research

Stalder, J. et al. (1980). Tissue-specific DNA cleavage in the globin chromatin domain introduced by DNAase I. *Cell* 20, 451–460.

29.20 An LCR May Control a Domain

Reviews

Bulger, M. and Groudine, M. (1999). Looping versus linking: toward a model for long-distance gene activation. *Genes Dev.* 13, 2465–2477.

Grosveld, F., Antoniou, M., Berry, M., De Boer, E., Dillon, N., Ellis, J., Fraser, P., Hanscombe, O., Hurst, J., and Imam, A. (1993). The regulation of human globin gene switching. *Philos. Trans. R. Soc. Lond. B Biol. Sci.* 339, 183–191.

Research

Gribnau, J., de Boer, E., Trimborn, T., Wijgerde, M., Milot, E., Grosveld, F., and Fraser, P. (1998). Chromatin interaction mechanism of transcriptional control *in vitro*. *EMBO J.* 17, 6020–6027.

Spilianakis, C. G., Lalioti, M. D., Town, T., Lee, G. R., and Flavell, R. A. (2005). Interchromosomal associations between alternatively expressed loci. *Nature* 435, 637–645.

van Assendelft, G. B., Hanscombe, O., Grosveld, F., and Greaves, D. R. (1989). The β-globin dominant control region activates homologous and heterologous promoters in a tissue-specific manner. *Cell* 56, 969–977.

29.21 What Constitutes a Regulatory Domain?

Review

West, A. G., Gaszner, M., and Felsenfeld, G. (2002). Insulators: many functions, many mechanisms. *Genes Dev.* 16, 271–288.

30

Controlling Chromatin Structure

CHAPTER OUTLINE

30.1 Introduction

When transcription is treated in terms of interactions involving DNA and individual transcription factors and RNA polymerases, we get an accurate description of the events that occur *in vitro*, but this lacks an important feature of transcription *in vivo*. The cellular genome is organized as nucleosomes, but initiation of transcription generally is prevented if the promoter region is packaged into nucleosomes. In this sense, histones function as generalized repressors of transcription (a rather old idea), although we see in this chapter that they are also involved in more specific interactions. Activation of a gene requires changes in the state of chromatin: The essential issue is how the transcription factors gain access to the promoter DNA.

Local chromatin structure is an integral part of controlling gene expression. Genes may exist in either of two structural conditions. Genes are found in an "active" state only in the cells in which they are expressed. The change of structure precedes the act of transcription, and indicates that the gene is "transcribable." This suggests that acquisition of the "active" structure must be the first step in gene expression. Active genes are found in domains of euchromatin with a preferential susceptibility to nucleases (see Section 29.19, Domains Define Regions That Contain Active Genes). Hypersensitive sites are created at promoters before a gene is activated (see Section 29.18, DNAase Hypersensitive Sites Change Chromatin Structure).

More recently it has turned out that there is an intimate and continuing connection between initiation of transcription and chromatin structure. Some activators of gene transcription directly modify histones; in particular, acetylation of histones is associated with gene activation. Conversely, some repressors of transcription function by deacetylating histones. Thus a reversible change in histone structure in the vicinity of the promoter is involved in the control of gene expression. This may be part of the mechanism by which a gene is maintained in an active or inactive state.

The mechanisms by which local regions of chromatin are maintained in an inactive (silent) state are related to the means by which an individual promoter is repressed. The proteins involved in the formation of heterochromatin act on chromatin via the histones, and modifications of the histones may be an important feature in the interaction. Once established, such changes in chromatin may persist through cell divisions, creating an **epigenetic** state in which the properties of a gene are determined by the self-perpetuating structure of chromatin. The name epigenetic reflects the fact that a gene may have an inherited condition (it may be active or may be inactive) that does not depend on its sequence. Yet a further insight into epigenetic properties is given by the self-perpetuating structures of **prions** (proteinaceous infectious agents).

30.2 Chromatin Can Have Alternative States

Key concept

- Chromatin structure is stable and cannot be changed by altering the equilibrium of transcription factors and histones.

Two types of models have been proposed to explain how the state of expression of DNA is changed: equilibrium and discontinuous change-of-state.

FIGURE 30.1 shows the equilibrium model. Here the only pertinent factor is the concentration of the repressor or activator protein, which drives an equilibrium between free form and DNA-bound form. When the concentration of the protein is high enough, its DNA-binding site is occupied, and the state of expression of the

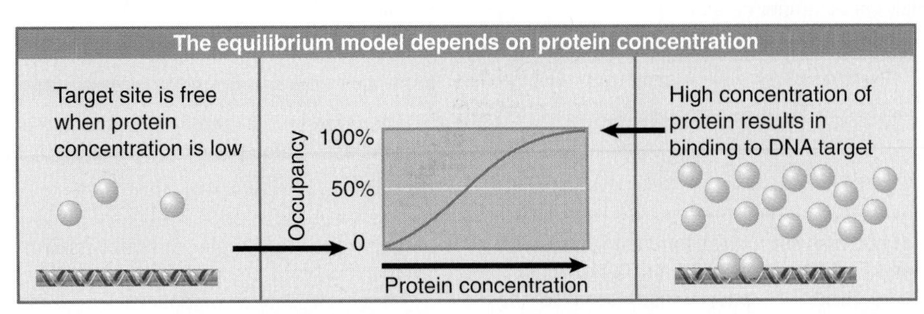

FIGURE 30.1 In an equilibrium model, the state of a binding site on DNA depends on the concentration of the protein that binds to it.

DNA is affected. (Binding might either repress or activate any particular target sequence.) This type of model explains the regulation of transcription in bacterial cells, where gene expression is determined exclusively by the actions of individual repressor and activator proteins (see Chapter 12, The Operon). Whether a bacterial gene is transcribed can be predicted from the sum of the concentrations of the various factors that either activate or repress the individual gene. Changes in these concentrations *at any time* will change the state of expression accordingly. In most cases, the protein binding is cooperative, so that once the concentration becomes high enough, there is a rapid association with DNA, resulting in a switch in gene expression.

A different situation applies with eukaryotic chromatin. Early *in vitro* experiments showed that either an active or inactive state can be established, but this is not affected by the subsequent addition of other components. The transcription factor $TF_{III}A$, which is required for RNA polymerase III to transcribe 5S rRNA genes, cannot activate its target genes *in vitro* if they are complexed with histones. If the factor is presented with free DNA, though, it forms a transcription complex, and then the addition of histones does not prevent the gene from remaining active. Once the factor has bound, it remains at the site; this allows a succession of RNA polymerase molecules to initiate transcription. Whether the factor or histones get to the control site first may be the critical factor.

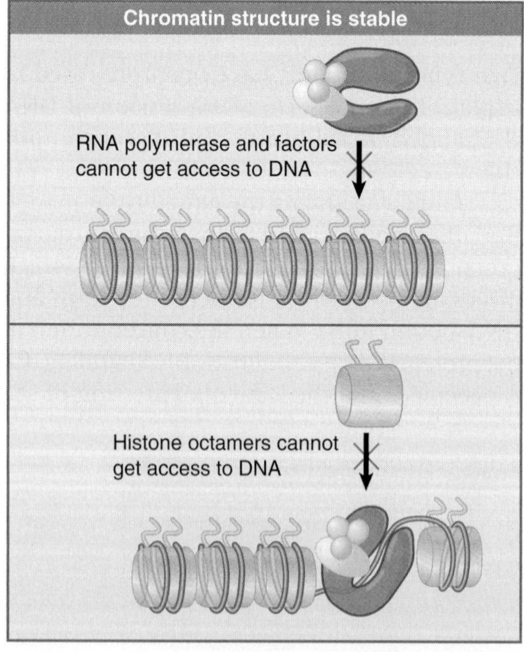

FIGURE 30.2 If nucleosomes form at a promoter, transcription factors (and RNA polymerase) cannot bind. If transcription factors (and RNA polymerase) bind to the promoter to establish a stable complex for initiation, histones are excluded.

FIGURE 30.2 illustrates the two types of condition that can exist at a eukaryotic promoter. In the inactive state, nucleosomes are present, and they prevent basal factors and RNA polymerase from binding. In the active state, the basal apparatus occupies the promoter, and histone octamers cannot bind to it. Each type of state is stable.

A similar situation is seen with the $TF_{II}D$ complex at promoters for RNA polymerase II. A plasmid containing an adenovirus promoter can be transcribed *in vitro* by RNA polymerase II in a reaction that requires $TF_{II}D$ and other transcription factors. The template can be assembled into nucleosomes by the addition of histones. If the histones are added *before* the $TF_{II}D$, transcription cannot be initiated. If the $TF_{II}D$ is added first, though, the template still can be transcribed in its chromatin form. Thus $TF_{II}D$ can recognize free DNA, but either cannot recognize or cannot function on nucleosomal DNA. Only the $TF_{II}D$ must be added before the histones; the other transcription factors and RNA polymerase can be added later. This suggests that binding of $TF_{II}D$ to the promoter creates a structure to which the other components of the transcription apparatus can bind.

It is important to note that these *in vitro* systems use disproportionate quantities of components, which may create unnatural situations. The major importance of these results, therefore, is not that they demonstrate the mechanism used *in vivo*, but that they establish the principle that *transcription factors or nucleosomes may form stable structures that cannot be changed merely by changing the equilibrium with free components.*

30.3 Chromatin Remodeling Is an Active Process

Key concepts

- There are several chromatin remodeling complexes that use energy provided by hydrolysis of ATP.
- The SWI/SNF, RSC, and NURF complexes all are very large, and they share related subunits.
- A remodeling complex does not itself have specificity for any particular target site, but must be recruited by a component of the transcription apparatus.

The general process of inducing changes in chromatin structure is called **chromatin remodeling.** This consists of mechanisms for displacing histones that depend on the input of energy. Many protein–protein and protein–DNA contacts need to be disrupted to release histones from chromatin. There is no free ride: The energy must be provided to disrupt these contacts. FIGURE 30.3 illustrates the principle of a *dynamic model* by a factor that hydrolyzes ATP. When the histone octamer is released from DNA, other proteins (in this case transcription factors and RNA polymerase) can bind.

FIGURE 30.4 summarizes the types of remodeling changes in chromatin that can be characterized *in vitro:*

- Histone octamers may slide along DNA, changing the relationship between the nucleic acid and protein. This alters the position of a particular sequence on the nucleosomal surface.
- The spacing between histone octamers may be changed, again with the result that the positions of individual sequences are altered relative to protein.
- The most extensive change is that an octamer(s) may be displaced entirely from DNA to generate a nucleosome-free gap.

The most common use of chromatin remodeling is to change the organization of nucleosomes at the promoter of a gene that is to be transcribed. This is required to allow the transcription apparatus to gain access to the promoter. Remodeling is also required, however, to enable other manipulations of chromatin, including repair reactions to damaged DNA.

Remodeling most often takes the form of displacing one or more histone octamers. This can be detected by a change in the micrococcal nuclease ladder where protection against cleavage has been lost. It often results in the creation of a site that is hypersensitive to cleavage with DNAase I (see Section 29.18, DNAase Hypersensitive Sites Change Chromatin Structure). Sometimes there are less dramatic changes, for example, involving a change in rotational positioning of a single nucleosome; this may be detected by loss of the DNAase I 10 base ladder. Thus changes in chromatin structure may extend from altering the positions of nucleosomes to removing them altogether.

Chromatin remodeling is undertaken by large complexes that use ATP hydrolysis to provide the energy for remodeling. The heart of the remodeling complex is its ATPase subunit. Remodeling complexes are usually classified according to the type of ATPase subunit—those with related ATPase subunits are considered to belong to the same family (usually some other subunits are common as well). FIGURE 30.5 keeps the names straight. The two major types of complex are SWI/SNF and ISWI (ISWI stands for

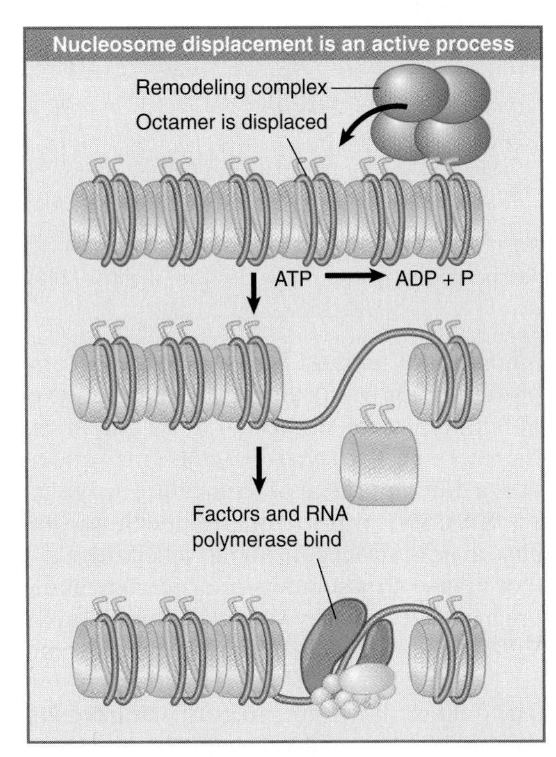

FIGURE 30.3 The dynamic model for transcription of chromatin relies upon factors that can use energy provided by hydrolysis of ATP to displace nucleosomes from specific DNA sequences.

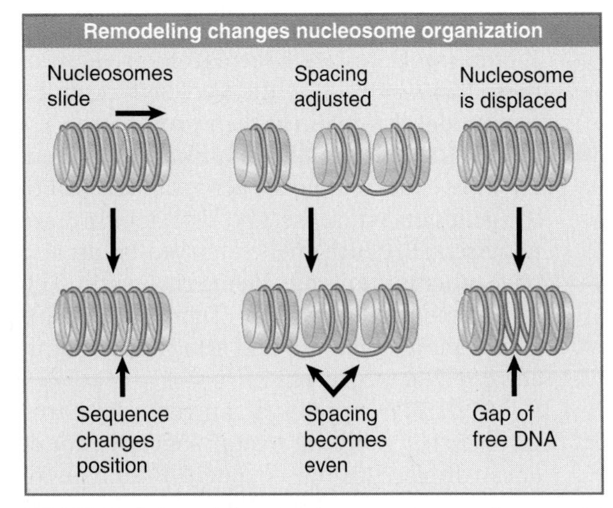

FIGURE 30.4 Remodeling complexes can cause nucleosomes to slide along DNA, can displace nucleosomes from DNA, or can reorganize the spacing between nucleosomes.

There are several types of remodeling complexes			
Type of Complex	SWI/SNF	ISWI	Other
Yeast	SWI/SNF RSC	ISW1 ISW2	INO80 complex SWRI
Fly	dSWI/SNF (brahma)	NURF CHRAC ACF	
Human	hSWI/SNF	RSF hACF/WCFR hCHRAC WICH	NuRD INO80 complex SRCAP
Frog		WICH CHRAC ACF	Mi-2

FIGURE 30.5 Remodeling complexes can be classified by their ATPase subunits.

imitation SWI). Yeast has two SWI/SNF complexes and three ISWI complexes. Complexes of both types are also found in fly and in the human being. Each type of complex may undertake a different range of remodeling activities.

SWI/SNF was the first remodeling complex to be identified. Its name reflects the fact that many of its subunits are coded by genes originally identified by *SWI* or *SNF* mutations in *Saccharomyces cerevisiae.* Mutations in these loci are pleiotropic, and the range of defects is similar to those shown by mutants that have lost the carboxyl-terminaldomain (CTD) tail of RNA polymerase II. These mutations also show genetic interactions with mutations in genes that code for components of chromatin, in particular *SIN1,* which codes for a nonhistone protein, and *SIN2,* which codes for histone H3. The *SWI* and *SNF* genes are required for expression of a variety of individual loci (~120, or 2%, of *S. cerevisiae* genes are affected). Expression of these loci may require the SWI/SNF complex to remodel chromatin at their promoters.

SWI/SNF acts catalytically *in vitro,* and there are only ~150 complexes per yeast cell. All of the genes encoding the SWI/SNF subunits are nonessential, which implies that yeast must also have other ways of remodeling chromatin. The RSC (remodels the structure of chromatin) complex is more abundant and also is essential. It acts at ~ 700 target loci.

SWI/SNF complexes can remodel chromatin *in vitro* without overall loss of histones or can displace histone octamers. Both types of reaction may pass through the same intermediate in which the structure of the target nucleosome is altered, leading either to reformation of a (remodeled) nucleosome on the original DNA or to displacement of the histone octamer to a different DNA molecule. The SWI/SNF complex alters nucleosomal sensitivity to DNAase I at the target site, and induces changes in protein–DNA contacts that persist after it has been released from the nucleosomes. The Swi2 subunit is the ATPase that provides the energy for remodeling by SWI/SNF.

There are many contacts between DNA and a histone octamer; fourteen are identified in the crystal structure. All of these contacts must be broken for an octamer to be released or for it to move to a new position. How is this achieved? Some obvious mechanisms can be excluded because we know that single-stranded DNA is not generated during remodeling (and there are no helicase activities associated with the complexes). Present thinking is that remodeling complexes in the SWI and ISWI classes use the hydrolysis of ATP to twist DNA on the nucleosomal surface. Indirect evidence suggests that this creates a mechanical force that allows a small region of DNA to be released from the surface and then repositioned.

One important reaction catalyzed by remodeling complexes involves nucleosome sliding. It was first observed that the ISWI family affects nucleosome positioning without displacing octamers. This is achieved by a sliding reaction, in which the octamer moves along DNA. Sliding is prevented if the N-terminal tail of histone H4 is removed, but we do not know exactly how the tail functions in this regard. SWI/SNF complexes have the same capacity; the reaction is prevented by the introduction of a barrier in the DNA, which suggests that a sliding reaction is involved, in which the histone octamer moves more or less continuously along DNA without ever losing contact with it.

One puzzle about the action of the SWI/SNF complex is its sheer size. It has eleven subunits with a combined molecular weight ~2×10^6. It dwarfs RNA polymerase and the nucleosome, making it difficult to understand how all of these components could interact with DNA retained on the nucleosomal surface. A transcription complex with full activity, however, called RNA polymerase II holoenzyme, can be found that contains the RNA polymerase itself, all the TF_{II} factors except TBP and $TF_{II}A$, and the SWI/SNF complex, which is associated with the CTD tail of the polymerase. In fact, virtually all of the SWI/SNF complex may be present in holoenzyme preparations. This suggests that the remodeling of chromatin and recognition of promoters is undertaken in a coordinated manner by a single complex.

30.4 Nucleosome Organization May Be Changed at the Promoter

Key concepts

- Remodeling complexes are recruited to promoters by sequence-specific activators.
- The factor may be released once the remodeling complex has bound.
- The MMTV promoter requires a change in rotational positioning of a nucleosome to allow an activator to bind to DNA on the nucleosome.

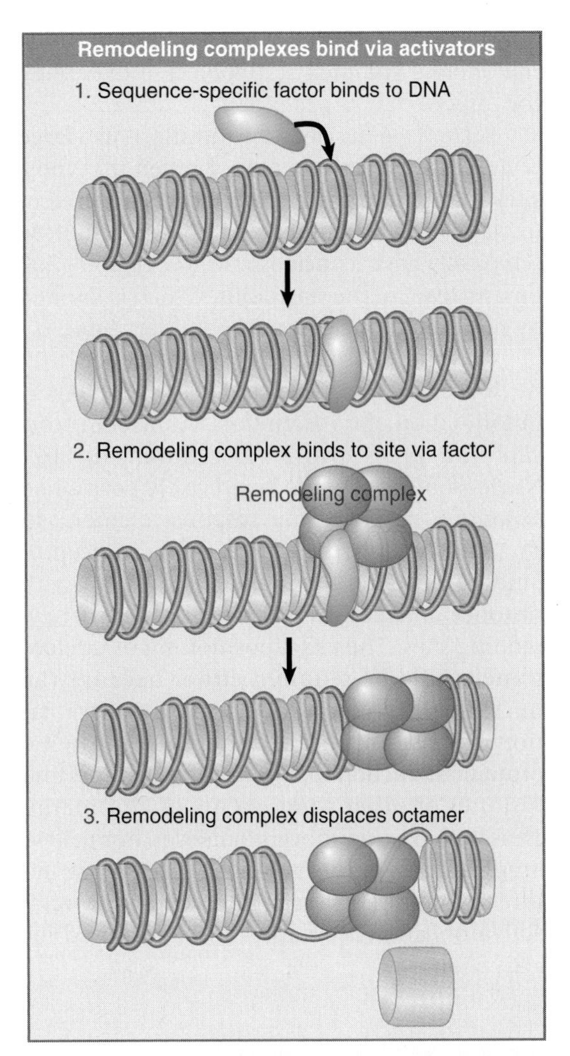

FIGURE 30.6 A remodeling complex binds to chromatin via an activator (or repressor).

How are remodeling complexes targeted to specific sites on chromatin? They do not themselves contain subunits that bind specific DNA sequences. This suggests the model shown in FIGURE 30.6, in which they are recruited by activators or (sometimes) by repressors.

The interaction between transcription factors and remodeling complexes gives a key insight into their modus operandi. The transcription factor Swi5 activates the *HO* locus in yeast. (Note that Swi5 is not a member of the SWI/SNF complex.) Swi5 enters nuclei toward the end of mitosis and binds to the *HO* promoter. It then recruits SWI/SNF to the promoter. Swi5 is then released, leaving SWI/SNF at the promoter. This means that a transcription factor can activate a promoter by a "hit and run" mechanism, in which its function is fulfilled once the remodeling complex has bound.

The involvement of remodeling complexes in gene activation was discovered because the complexes are necessary for the ability of certain transcription factors to activate their target genes. One of the first examples was the GAGA factor, which activates the *hsp70 Drosophila* promoter *in vitro*. Binding of GAGA to four $(CT)_n$-rich sites on the promoter disrupts the nucleosomes, creates a hypersensitive region, and causes the adjacent nucleosomes to be rearranged so that they occupy preferential instead of random positions. Disruption is an energy-dependent process that requires the NURF remodeling complex. The organization of nucleosomes is altered so as to create a boundary that determines the positions of the adjacent nucleosomes. During this process, GAGA binds to its target sites and DNA, and its presence fixes the remodeled state.

The *PHO* system was one of the first in which it was shown that a change in nucleosome organization is involved in gene activation. At the *PHO5* promoter, the bHLH regulator *PHO4* responds to phosphate starvation by inducing the disruption of four precisely positioned nucleosomes. This event is independent of transcription (it occurs in a $TATA^-$ mutant) and independent of replication. There are two binding sites for *PHO4* at the promoter. One is located between nucleosomes, which can be bound by the isolated DNA-binding domain of *PHO4*, and the other lies within a nucleosome, which cannot be recognized. Disruption of the nucleosome to allow DNA binding at the second site is necessary for gene activation. This action requires the presence of the transcription-activating domain. The activator sequence of VP16 can substitute for the *PHO4* activator sequence in nucleosome disruption. This suggests that disruption occurs by protein–protein interactions that involve the same region that makes protein–protein contacts to activate transcription. In this case, it is not known which remodeling complex is involved in executing the effects.

A survey of nucleosome positions in a large region of the yeast genome showed that most sites that bind transcription factors are free of nucleosomes. Promoters for RNA polymerase II typically have a nucleosome-free region ~200 bp upstream of the startpoint, which is flanked by positioned nucleosomes on either side.

It is not always the case, however, that nucleosomes must be excluded in order to permit initiation of transcription. Some activators can bind to DNA on a nucleosomal surface. Nucleosomes appear to be precisely positioned at some steroid hormone response elements in such a way that receptors can bind. Receptor binding may alter the interaction of DNA with histones, and may even lead to exposure of new binding sites. The exact positioning of nucleosomes could be required either because the nucleosome "presents" DNA in a particular rotational phase or because there are protein–protein interactions between the activators and histones or other components of chromatin. Thus we have now moved some way from viewing chromatin exclusively as a repressive structure to considering which interactions between activators and chromatin can be required for activation.

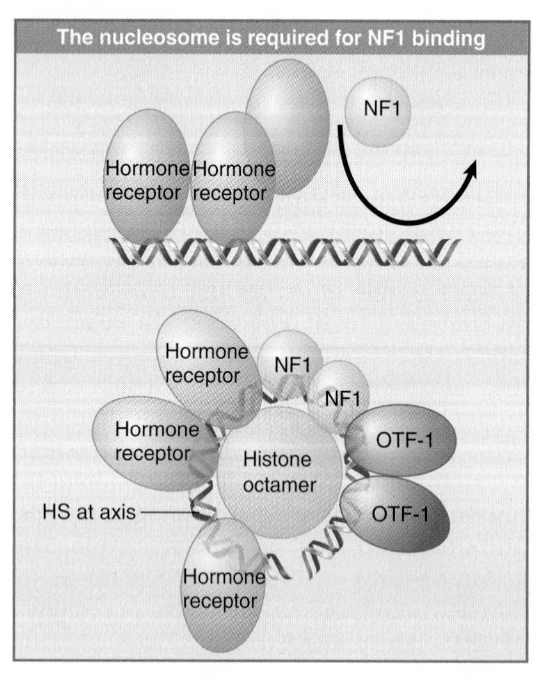

FIGURE 30.7 Hormone receptor and NF1 cannot bind simultaneously to the MMTV promoter in the form of linear DNA, but can bind when the DNA is presented on a nucleosomal surface.

The MMTV promoter presents an example of the need for specific nucleosomal organization. It contains an array of six partly palindromic sites, each bound by one dimer of hormone receptor (HR), which constitute the HRE. It also has a single binding site for the factor NF1, and two adjacent sites for the factor OTF. HR and NF1 cannot bind simultaneously to their sites in free DNA. FIGURE 30.7 shows how the nucleosomal structure controls binding of the factors.

The HR protects its binding sites at the promoter when hormone is added, but does not affect the micrococcal nuclease-sensitive sites that mark either side of the nucleosome. This suggests that HR is binding to the DNA on the nucleosomal surface; however, the rotational positioning of DNA on the nucleosome prior to hormone addition allows access to only two of the four sites. Binding to the other two sites requires a change in rotational positioning on the nucleosome. This can be detected by the appearance of a sensitive site at the axis of dyad symmetry (which is in the center of the binding sites that constitute the HRE). NF1 can be footprinted on the nucleosome after hormone induction, so these structural changes may be necessary to allow NF1 to bind, perhaps because they expose DNA and abolish the steric hindrance by which HR blocks NF1 binding to free DNA.

30.5 Histone Modification Is a Key Event

Key concept

- Histones are modified by methylation, acetylation, and phosphorylation.

Whether a gene is expressed depends on the structure of chromatin both locally (at the promoter) and in the surrounding domain. Chromatin structure correspondingly can be regulated by individual activation events or by changes that affect a wide chromosomal region. The most localized events concern an individual target gene, where changes in nucleosomal structure and organization occur in the immediate vicinity of the promoter. More general changes may affect regions as large as a whole chromosome.

Changes that affect large regions control the potential of a gene to be expressed. The term **silencing** is used to refer to repression of gene activity in a local chromosomal region. The term **heterochromatin** is used to describe chromosomal regions that are large enough to be seen to have a physically more compact structure in the microscope. The basis for both types of change is the same: Additional proteins bind to chromatin and either directly or indirectly prevent transcription factors and RNA polymerase from activating promoters in the region.

Changes at an individual promoter control whether transcription is initiated for a particular gene. These changes may be either activating or repressing.

All of these events depend on interactions with histones. Changes in chromatin structure are initiated by modifying the N-terminal tails of the histones, especially H3 and H4. The histone tails consist of 15–30 amino acids at the N-termini of all four core histones and the C-terminus of H2A. The tails of H2B and H3 pass between the turns of DNA (see Figure 29.21 in Section 29.7, Organization of the Histone Octamer). FIGURE 30.8 shows that they can be modified at several sites by methylation, acetylation, or phosphorylation. Other modifications, such as mono-ubiquitylation or sumoylation, also occur but are less well characterized.

Acetylation and methylation occur on the free (ε) amino group of lysine. As seen in FIGURE 30.9, acetylation removes the positive charge that resides on the NH_3^+ form of the group. Methylation also occurs on arginine. Phosphorylation occurs on the hydroxyl group of serine and also on threonine. This introduces a negative charge in the form of the phosphate group. Lysine can be mono-, di-, or trimethylated (all still positively charged), and arginine can be mono- or dimethylated (symmetrically or asymmetrically).

These modifications are transient. They can change the charge of the protein molecule, and as a result they are potentially able to change the functional properties of the octamers. Modification of histones is associated with structural changes that occur in chromatin at replication and transcription. Phosphorylations on specific positions and on different histones may be required for particular processes, for example, the Ser^{10} position of H3 is phosphorylated when chromosomes condense at mitosis.

In synchronized cells in culture, both the preexisting and newly synthesized core histones appear to be acetylated and methylated during S phase (when DNA is replicated and the histones also are synthesized). During the cell cycle, the modifying groups are later removed.

The coincidence of modification and replication suggests that acetylation (and methylation) could be connected with nucleosome assembly. One speculation has been that the reduction of positive charges on histones might lower their affinity for DNA, thus allowing the reaction to be better controlled. The idea has lost some ground in view of the observation that nucleosomes can be reconstituted, at least *in vitro*, with unmodified histones. Histone acetylation is essential for nucleosome assembly in yeast, and is probably required for some of the protein–protein interactions that occur during later stages of the reaction (see Section 30.6, Histone Acetylation Occurs in Two Circumstances).

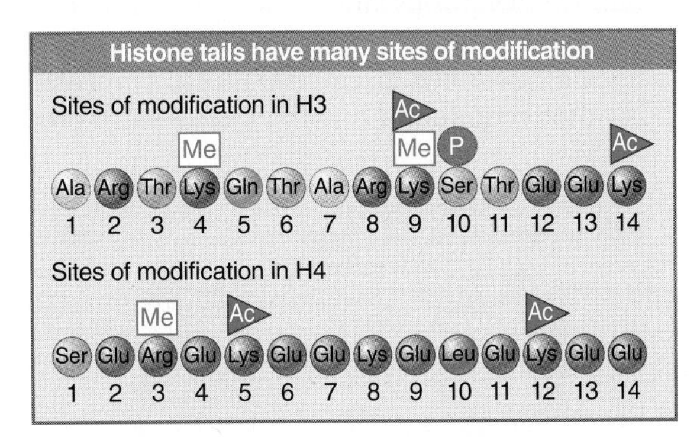

FIGURE 30.8 The N-terminal tails of histones H3 and H4 can be acetylated, methylated, or phosphorylated at several positions.

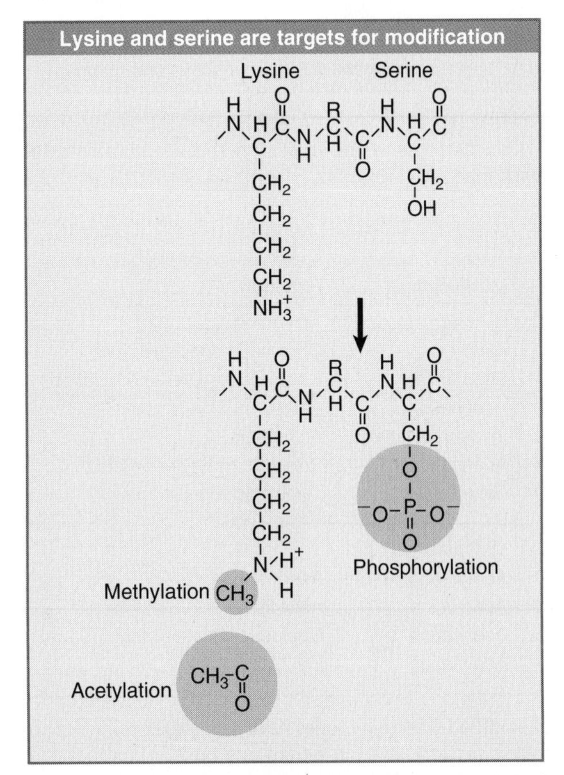

FIGURE 30.9 Acetylation of lysine or phosphorylation of serine reduces the overall positive charge of a protein.

A cycle of phosphorylation and dephosphorylation occurs with H1, but its timing is different from the modification cycle of the other histones. With cultured mammalian cells, one or two phosphate groups are introduced at S phase. The major phosphorylation event is the later addition of more groups at mitosis, though, which brings the total number up to as many as six. All the phosphate groups are removed at the end of the process of division. The phosphorylation of H1 is catalyzed by the M-phase kinase that provides an essential trigger for mitosis. In fact, this enzyme is now often assayed in terms of its H1 kinase activity. Not much is known about phosphatase(s) that remove the groups later.

The timing of the major H1 phosphorylation has prompted speculation that it is involved in mitotic condensation. In *Tetrahymena* (a protozoan), however, it is possible to delete all the genes for H1 without significantly affecting the overall properties of chromatin. There is a relatively small effect on the ability of chromatin to condense at mitosis. Some genes are activated and others are repressed by this change, which suggests that there are alterations in local structure. Mutations that eliminate sites of phosphorylation in H1 have no effect, but mutations that mimic the effects of phosphorylation produce a phenotype that resembles the deletion. This suggests that the effect of phosphorylating H1 is to eliminate its effects on local chromatin structure.

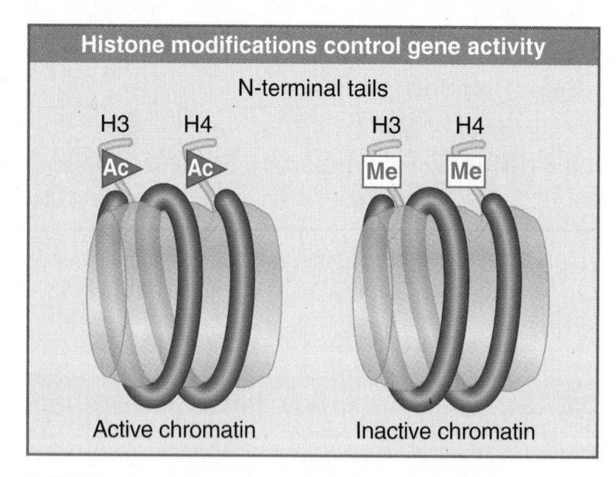

FIGURE 30.10 Acetylation of H3 and H4 is associated with active chromatin, whereas methylation is associated with inactive chromatin.

Histone modification affects the structure and function of chromatin

Histone	Site	Modification	Function
H3	Lys-4	Methylation	Activation
H3	Lys-9	Methylation	Chromatin condensation
	Lys-9	Methylation	Required for DNA methylation
	Lys-9	Acetylation	Activation
H3	Ser-10	Phosphorylation	Activation
H3	Lys-14	Acetylation	Prevents methylation at Lys-9
H3	Lys-79	Methylation	Telomeric silencing
H4	Arg-3	Methylation	
H4	Lys-5	Acetylation	Assembly
H4	Lys-12	Acetylation	Assembly
H4	Lys-16	Acetylation	Nucleosome assembly
	Lys-16	Acetylation	Fly X activation

FIGURE 30.11 Most modified sites in histones have a single, specific type of modification, but some sites can have more than one type of modification. Individual functions can be associated with some of the modifications.

Do histone modifications affect nucleosome structure directly, or is their effect upon chromatin indirect? There is not in fact much evidence for any difference in the properties of nucleosomes depending on the state of the modification of the histones. Several cases have now been characterized, though, in which histone modification creates binding sites for the attachment of nonhistone proteins that change the properties of chromatin.

The range of nucleosomes that is targeted for modification can vary. Modification can be a local event, for example, restricted to nucleosomes at the promoter. It also can be a general event, extending, for example, to an entire chromosome. FIGURE 30.10 shows that there is a general correlation in which acetylation is associated with active chromatin, whereas methylation is associated with inactive chromatin. This is not, however, a simple rule, and the particular sites that are modified (as well as combinations of specific modifications) may be important, so there are certainly exceptions in which (for example) histones methylated at a certain position are found in active chromatin. Mutations in one of the histone acetylase complexes of yeast have the opposite effect from usual (they prevent silencing of some genes); this emphasizes the lack of a uniform effect of acetylation.

The specificity of the modifications is indicated by the fact that many of the modifying enzymes have individual target sites in specific histones. FIGURE 30.11 summarizes the effects of some of the modifications. Most modified sites are subject to only a single type of modification. In some cases, modification of one site may activate or inhibit modification of another site. The idea that combinations of signals may be used to define chromatin types has sometimes been called the *histone code*.

30.6 Histone Acetylation Occurs in Two Circumstances

Key concepts

- Histone acetylation occurs transiently at replication.
- Histone acetylation is associated with activation of gene expression.

All the core histones can be acetylated. The major targets for acetylation are lysines in the N-terminal tails of histones H3 and H4. Acetylation occurs in two different circumstances:

- during DNA replication, and
- when genes are activated.

When chromosomes are replicated, which occurs during the S phase of the cell cycle, histones are transiently acetylated. FIGURE 30.12 shows that this acetylation occurs before the histones are incorporated into nucleosomes. We know that histones H3 and H4 are acetylated at the stage when they are associated with one another in the $H3_2$-$H4_2$ tetramer. The tetramer is then incorporated into nucleosomes. Quite soon after, the acetyl groups are removed.

The importance of the acetylation is indicated by the fact that preventing acetylation of both histones H3 and H4 during replication causes loss of viability in yeast. The two histones are redundant as substrates, because yeast can manage perfectly well so long as they can acetylate either one of these histones during S phase. There are two possible roles for the acetylation: It could be needed for the histones to be recognized by factors that incorporate them into nucleosomes, or it could be required for the assembly and/or structure of the new nucleosome.

The factors that are known to be involved in chromatin assembly do not distinguish between acetylated and nonacetylated histones, which suggests that the modification is more likely to be required for subsequent interactions. It has been thought for a long time that acetylation might be needed to help control protein–protein interactions that occur as histones are incorporated into nucleosomes. Some evidence for such a role is that the yeast SAS histone acetylase complex binds to chromatin assembly complexes at the replication fork, where it acetylates ^{16}Lys of histone H4. This may be part of the system that establishes the histone acetylation patterns after replication.

Outside of S phase, acetylation of histones in chromatin is generally correlated with the state of gene expression. The correlation was first noticed because histone acetylation is increased in a domain containing active genes, and acetylated chromatin is more sensitive to DNAase I and (possibly) to micrococcal nuclease. FIGURE 30.13 shows that this involves the acetylation of histone tails in nucleosomes. We now know that this occurs largely because of acetylation of the nucleosomes in the vicinity of the promoter when a gene is activated.

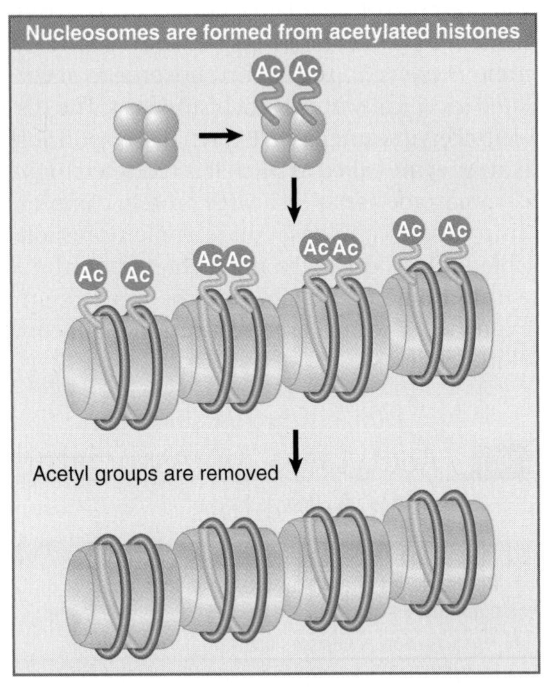

FIGURE 30.12 Acetylation at replication occurs on histones before they are incorporated into nucleosomes.

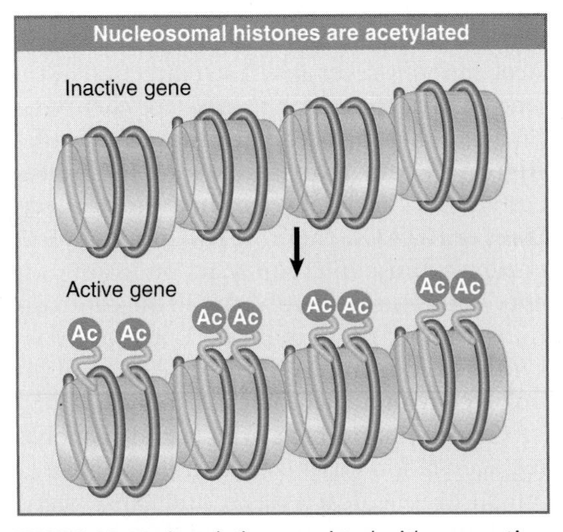

FIGURE 30.13 Acetylation associated with gene activation occurs by directly modifying histones in nucleosomes.

In addition to events at individual promoters, widescale changes in acetylation occur on sex chromosomes. This is part of the mechanism by which the activities of genes on the X chromosome are altered to compensate for the presence of two X chromosomes in one species but only one X chromosome (in addition to the Y chromosome) in other species (see Section 31.5, X Chromosomes Undergo Global Changes). The inactive X chromosome in female mammals has underacetylated H4. The superactive X chromosome in *Drosophila* males has increased acetylation of H4. This suggests that the presence of acetyl groups may be a prerequisite for a less condensed, active structure. In male *Drosophila*, the X chromosome is acetylated specifically at ^{16}Lys of histone H4. The histone acetyltranoferase (HAT) that is responsible is an enzyme called MOF that is recruited to the chromosome as part of a large protein complex. This "dosage compensation" complex is responsible for introducing general changes in the X chromosome that enable it to be more highly expressed. The increased acetylation is only one of its activities.

30.7 Acetylases Are Associated with Activators

Key concepts

- Deacetylated chromatin may have a more condensed structure.
- Transcription activators are associated with histone acetylase activities in large complexes.
- Histone acetylases vary in their target specificity.
- Acetylation could affect transcription in a quantitative or qualitative way.

Acetylation is reversible. Each direction of the reaction is catalyzed by a specific type of enzyme. Enzymes that can acetylate histones are called **histone acetyltransferases** or **HATs;** the acetyl groups are removed by **histone deacetylases** or **HDACs.** There are two groups of HAT enzymes: those in group A act on histones in chromatin and are involved with the control of transcription; those in group B act on newly synthesized histones in the cytosol, and are involved with nucleosome assembly.

Two inhibitors have been useful in analyzing acetylation. Trichostatin and butyric acid inhibit histone deacetylases, and cause acetylated nucleosomes to accumulate. The use of these inhibitors has supported the general view that acetylation is associated with gene expression; in fact, the ability of butyric acid to cause changes in chromatin resembling those found upon gene activation was one of the first indications of the connection between acetylation and gene activity.

The breakthrough in analyzing the role of histone acetylation was provided by the characterization of the acetylating and deacetylating enzymes, and their association with other proteins that are involved in specific events of activation and repression. A basic change in our view of histone acetylation was caused by the discovery that HATs are not necessarily dedicated enzymes associated with chromatin; rather, it turns out that known activators of transcription have HAT activity.

The connection was established when the catalytic subunit of a group A HAT was identified as a homolog of the yeast regulator protein Gcn5. It then was shown that Gcn5 itself has HAT activity (with histones H3 and H4 as substrates). Gcn5 is part of an adaptor complex that is necessary for the interaction between certain enhancers and their target promoters. Its HAT activity is required for activation of the target gene.

This enables us to redraw our picture for the action of coactivators as shown in FIGURE 30.14, where RNA polymerase is bound at a hypersensitive site and coactivators are acetylating histones on the nucleosomes in the vicinity. Many examples are now known of interactions of this type.

Gcn5 leads us into one of the most important acetylase complexes. In yeast, Gcn5 is part

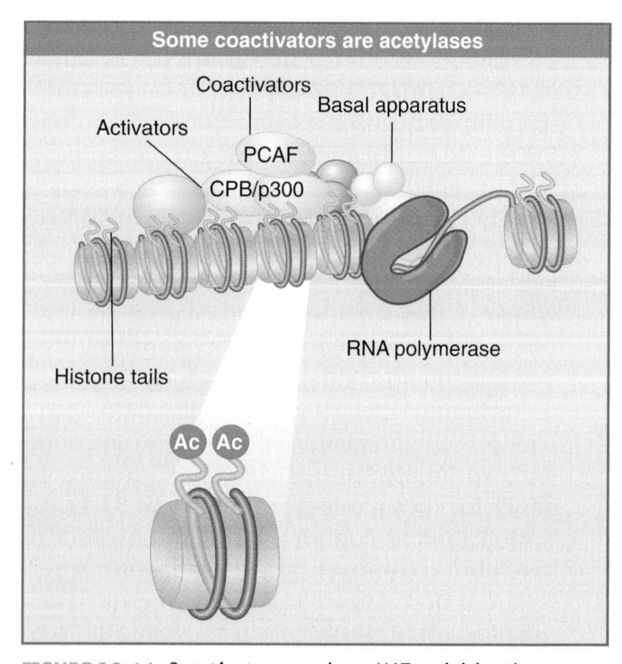

FIGURE 30.14 Coactivators may have HAT activities that acetylate the tails of nucleosomal histones.

of the 1.8 MDa Spt-Ada-Gcn5-acetyltransferase (SAGA) complex, which contains several proteins that are involved in transcription. Among these proteins are several TAF_{II}s. In addition, the TAF_{II}145 subunit of TF_{II}D is an acetylase. (Yeast TAF_{II}145 is the homolog of mammalian TAF_{II}250; both are known as TAF1.) There are some functional overlaps between TF_{II}D and SAGA, most notably that yeast can manage with either TAF_{II}145 or Gcn5, but is damaged by the deletion of both. This suggests that an acetylase activity is essential for gene expression, but can be provided by either TF_{II}D or SAGA. As might be expected from the size of the SAGA complex, acetylation is only one of its functions, although its other functions in gene activation are less well characterized.

One of the first general activators to be characterized as HAT was p300/CREB-binding protein (CBP). (Actually, p300 and CBP are different proteins, but they are so closely related that they are often referred to as a single type of activity.) p300/CBP is a coactivator that links an activator to the basal apparatus (see Figure 25.7). p300/CBP interacts with various activators, including hormone receptors, AP-1 (c-Jun and c-Fos), and MyoD. The interaction is inhibited by the viral regulator proteins adenovirus E1A and SV40 T antigen, which bind to p300/CBP to prevent the interaction with transcription factors; this explains how these viral proteins inhibit cellular transcription. (This inhibition is important for the ability of the viral proteins to contribute to the tumorigenic state.) p300/CBP acetylates the N-terminal tails of H4 in nucleosomes. Another coactivator, PCAF, preferentially acetylates H3 in nucleosomes. p300/CBP and PCAF form a complex that functions in transcriptional activation. In some cases yet another HAT is involved: the coactivator ACTR, which functions with hormone receptors, is itself an HAT that acts on H3 and H4, and also recruits both p300/CBP and PCAF to form a coactivating complex. One explanation for the presence of multiple HAT activities in a coactivating complex is that each HAT has a different specificity, and that multiple different acetylation events are required for activation.

A general feature of acetylation is that a group A HAT is part of a large complex. FIGURE 30.15 shows a simplified model for their behavior. HAT complexes can be targeted to DNA by interactions with DNA-binding factors. This determines the target for the HAT. The complex also contains effector subunits that affect chromatin structure or act directly on transcription. It is likely that at least some of the effectors require the acetylation event in order to act. Deacetylation, catalyzed by an HDAC, may work in a similar way.

Acetylation occurs at both replication (when it is transient) and at transcription (when it is maintained while the gene is active). Is it playing the same role in each case? One possibility is that the important effect is on nucleosome structure. Acetylation may be necessary to "loosen" the nucleosome core. At replication, acetylation of histones could be necessary to allow them to be incorporated into new cores more easily. At transcription, a similar effect could be necessary to allow a related change in structure, possibly even to allow the histone core to be displaced from DNA. Alternatively, acetylation could generate binding sites for other proteins that are required for transcription. In either case, deacetylation would reverse the effect.

Is the effect of acetylation quantitative or qualitative? One possibility is that a certain number of acetyl groups are required to have an effect, and the exact positions at which they occur are largely irrelevant. An alternative is that individual acetylation events have specific effects. We might interpret the existence of complexes containing multiple HAT activities in either way—if individual enzymes have different specificities, we may need multiple activities either to acetylate a sufficient number of different positions or because the individual events are necessary for different effects upon transcription. At replication, it appears (at least with respect to histone H4) that acetylation at any two of three available positions is adequate, favoring a quantitative model in this case. Where chromatin structure is changed to affect transcription, acetylation at specific positions is important (see Section 31.3, Heterochromatin Depends on Interactions with Histones).

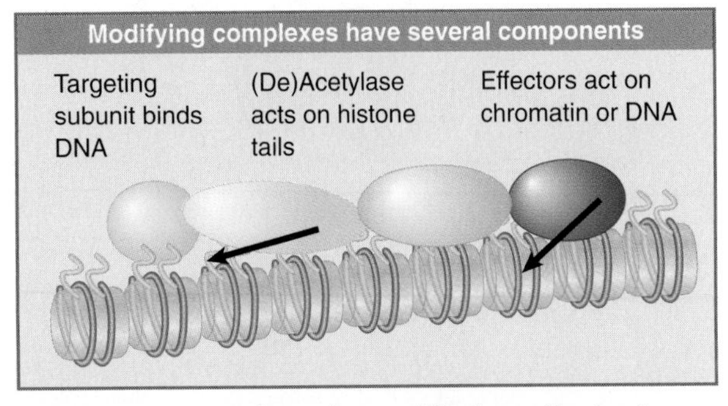

FIGURE 30.15 Complexes that modify chromatin structure or activity have targeting subunits that determine their sites of action, HAT or HDAC enzymes that acetylate or deacetylate histones, and effector subunits that have other actions on chromatin or DNA.

30.8 Deacetylases Are Associated with Repressors

Key concepts

- Deacetylation is associated with repression of gene activity.
- Deacetylases are present in complexes with repressor activity.

In yeast, mutations in *SIN3* and *RPD3* behave as though these loci repress a variety of genes. The proteins form a complex with the DNA-binding protein Ume6, which binds to the *URS1* element. The complex represses transcription at the promoters containing *URS1*, as illustrated in FIGURE 30.16. Rpd3 has histone deacetylase activity; we do not know whether the function of Sin3 is just to bring Rpd3 to the promoter, or whether it has an additional role in repression.

A similar system for repression is found in mammalian cells. The bHLH family of transcription regulators includes activators that function as heterodimers, including MyoD (see Section 25.15, Helix-Loop-Helix Proteins Interact by Combinatorial Association). It also includes repressors, in particular the heterodimer Mad:Max, where Mad can be any one of a group of closely related proteins. The Mad:Max heterodimer (which binds to specific DNA sites) interacts with a homolog of Sin3 (called mSin3 in mouse and hSin3 in human beings). mSin3 is part of a repressive complex that includes histone binding proteins and the histone deacetylases HDAC1 and HDAC2. Deacetylase activity is required for repression. The modular nature of this system is emphasized by other means of employment: A corepressor (SMRT), which enables retinoid hormone receptors to repress certain target genes, functions by binding mSin3, which in turns brings the HDAC activities to the site. Another means of bringing HDAC activities to the site may be a connection with MeCP2, a protein that binds to methylated cytosines (see Section 24.19, CpG Islands Are Regulatory Targets).

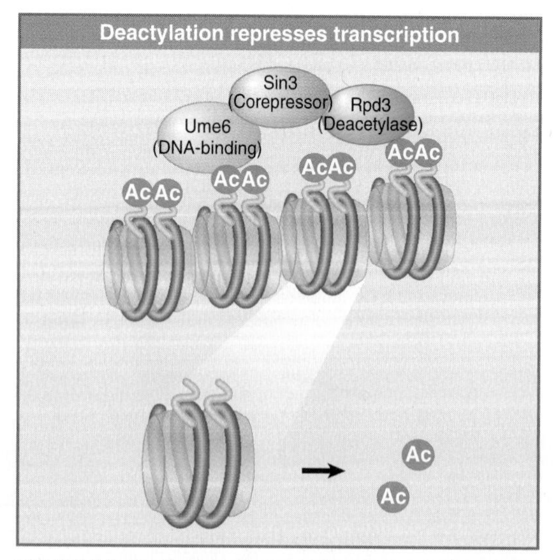

FIGURE 30.16 A repressor complex contains three components: a DNA binding subunit, a corepressor, and a histone deacetylase.

Absence of histone acetylation is also a feature of heterochromatin. This is true of both constitutive heterochromatin (typically involving regions of centromeres or telomeres) and facultative heterochromatin (regions that are inactivated in one cell although they may be active in another). Typically the N-terminal tails of histones H3 and H4 are not acetylated in heterochromatic regions.

30.9 Methylation of Histones and DNA Is Connected

Key concepts

- Methylation of both DNA and histones is a feature of inactive chromatin.
- The two types of methylation event may be connected.

Methylation of both histones and DNA is associated with inactivity. Sites that are methylated in histones include two lysines in the tail of H3 and an arginine in the tail of H4.

Methylation of H3 ^{9}Lys is a feature of condensed regions of chromatin, including heterochromatin as seen in bulk and also smaller regions that are known not to be expressed. The histone methyltransferase enzyme that targets this lysine is called SUV39H1. (We see the origin of this peculiar name in Section 30.13, Some Common Motifs Are Found in Proteins That Modify Chromatin). Its catalytic site has a region called the SET domain. Other histone methyltransferases act on arginine. In addition, methylation may occur on ^{79}Lys in the globular core region of H3; this may be necessary for the formation of heterochromatin at telomeres.

Until recently, it was thought that histone methylation was irreversible. Histone demethylases have now been identified, though, including a lysine-specific demethylase (LSD1) that acts on K4 of histone H3, and an enzyme that demethylates arginine on histones H3 and H4. We do not yet know how demethylation is regulated.

Most of the methylation sites in DNA are CpG islands (see Section 24.19, CpG Islands Are Regulatory Targets). CpG sequences in heterochromatin are usually methylated. Conversely, it is necessary for the CpG islands located in promoter regions to be unmethylated in order for a gene to be expressed (see Section 24.18, Gene Expression Is Associated with Demethylation).

Methylation of DNA and methylation of histones is connected in a mutually reinforcing circuit. Methylation of H3 is the signal that recruits the DNA methylase to chromatin. The order of events is that H3 ^{9}Lys is deacetylated to create the substrate for methylation. H3 is then converted to the Me^{9}Lys or the Me$_2$^{9}Lys condition, which provides a binding site for the DNA methylase. Some histone methyltransferase enzymes contain potential binding sites for the methylated CpG doublet, so the DNA methylation reinforces the circuit by providing a target for the histone methyltransferase to bind. The important point is that one type of modification can be the trigger for another. These systems are widespread, as seen by evidence for these connections in fungi, plants, and animal cells, and for regulating transcription at promoters used by both RNA polymerases I and II, as well as maintaining heterochromatin in an inert state.

30.10 Chromatin States Are Interconverted by Modification

Key concepts

- Acetylation of histones is associated with gene activation.
- Methylation of DNA and of histones is associated with heterochromatin.

FIGURE 30.17 summarizes three types of differences that are found between active chromatin and inactive chromatin:

- Active chromatin is acetylated on the tails of histones H3 and H4.
- Inactive chromatin is methylated on ^{9}Lys of histone H3.
- Inactive chromatin is methylated on cytosines of CpG doublets.

The reverse types of events occur if we compare the activation of a promoter with the generation of heterochromatin. The actions of the enzymes that modify chromatin ensure that activating events are mutually exclusive with inactivating events. Methylation of H3 ^{9}Lys and acetylation of H3 ^{14}Lys are mutually antagonistic.

Acetylases and deacetylases may trigger the initiating events. Deacetylation allows methylation to occur, which causes formation of a heterochromatic complex (see Section 31.3, Heterochromatin Depends on Interactions with Histones). Acetylation marks a region as active (see Section 30.11, Promoter Activation Involves an Ordered Series of Events).

30.11 Promoter Activation Involves an Ordered Series of Events

Key concepts

- The remodeling complex may recruit the acetylating complex.
- Acetylation of histones may be the event that maintains the complex in the activated state.

How are acetylases (or deacetylases) recruited to their specific targets? As we have seen with remodeling complexes, the process is likely to

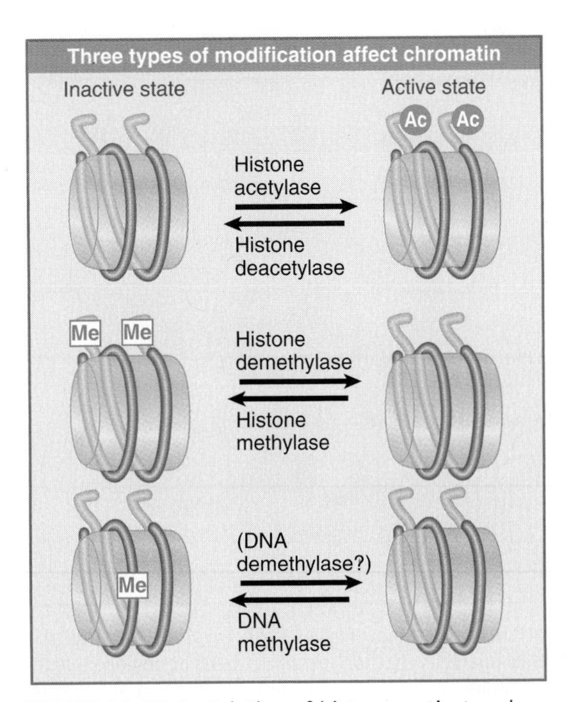

FIGURE 30.17 Acetylation of histones activates chromatin, and methylation of DNA and histones inactivates chromatin.

be indirect. A sequence-specific activator (or repressor) may interact with a component of the acetylase (or deacetylase) complex to recruit it to a promoter.

There may also be direct interactions between remodeling complexes and histone-modifying complexes. Binding by the SWI/SNF remodeling complex may lead in turn to binding by the SAGA acetylase complex. Acetylation of histones may then in fact stabilize the association with the SWI/SNF complex, making a mutual reinforcement of the changes in the components at the promoter.

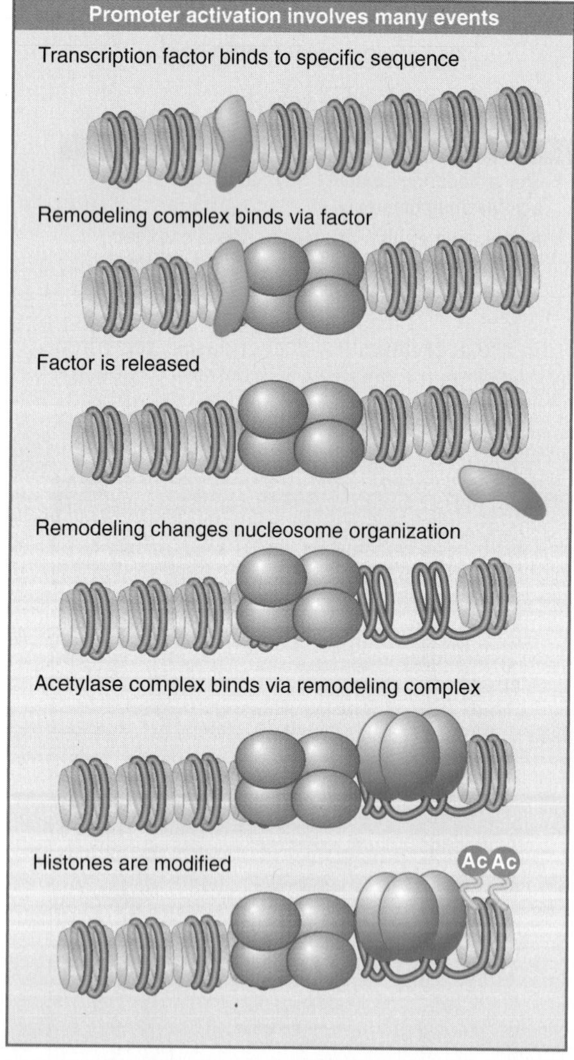

FIGURE 30.18 Promoter activation involves binding of a sequence-specific activator, recruitment and action of a remodeling complex, and recruitment and action of an acetylating complex. The order of events can differ or can even be simultaneous depending on the gene.

We can connect all of the events at the promoter into the series summarized in FIGURE 30.18. The initiating event is binding of a sequence-specific component (which is able to find its target DNA sequence in the context of chromatin). This recruits a remodeling complex. Changes occur in nucleosome structure. An acetylating complex binds, and the acetylation of target histones provides a covalent mark that the locus has been activated.

Modification of DNA also occurs at the promoter. Methylation of cytosine at CpG doublets is associated with gene inactivity (see Section 24.18, Gene Expression Is Associated with Demethylation). The basis for recognition of DNA as a target for methylation is not very well established (see Section 31.8, DNA Methylation Is Responsible for Imprinting).

It is clear that chromatin remodeling at the promoter requires a variety of changes that affect nucleosomes, including acetylation, but what changes are required within the gene to allow an RNA polymerase to traverse it? We know that RNA polymerase can transcribe DNA *in vitro* at rates comparable to the *in vivo* rate (~25 nucleotides per second) only with a template of free DNA. Several proteins have been characterized for their abilities to improve the speed with which RNA polymerase transcribes chromatin *in vivo*. The common feature is that they act on chromatin. A current model for their action is that they associate with RNA polymerase and travel with it along the template, modifying nucleosome structure by acting on histones. Among these factors are histone acetylases. One possibility is that the first RNA polymerase to transcribe a gene is a pioneer polymerase carrying factors that change the structure of the transcription unit so as to make it easier for subsequent polymerases.

30.12 Histone Phosphorylation Affects Chromatin Structure

Key concept

- At least two histones are targets for phosphorylation, possibly with opposing effects.

Histones are phosphorylated in two circumstances:

- cyclically during the cell cycle, and
- in association with chromatin remodeling.

It has been known for a very long time that histone H1 is phosphorylated at mitosis, and more recently it was discovered that H1 is an extremely good substrate for the Cdc2 kinase that controls cell division. This led to speculation that the phosphorylation might be connected with the condensation of chromatin, but so far no direct effect of this phosphorylation event has been demonstrated, and we do not know whether it plays a role in cell division.

Loss of a kinase that phosphorylates histone H3 on ^{10}Ser has devastating effects on chromatin structure. FIGURE 30.19 compares the usual extended structure of the polytene chromosome set of *Drosophila melanogaster* (upper photograph) with the structure that is found in a null mutant that has no JIL-1 kinase (lower photograph). The absence of JIL-1 is lethal, but the chromosomes can be visualized in the larvae before they die.

The cause of the disruption of structure is most likely the failure to phosphorylate histone H3 (of course, JIL-1 may also have other targets). This suggests that H3 phosphorylation is required to generate the more extended chromosome structure of euchromatic regions. Evidence supporting the idea that JIL-1 acts directly on chromatin is that it associates with the complex of proteins that binds to the X chromosome to increase its gene expression in males (see Section 31.5, X Chromosomes Undergo Global Changes).

This leaves us with somewhat conflicting impressions of the roles of histone phosphorylation. If it is important in the cell cycle, it is likely to be as a signal for condensation. Its effect in chromatin remodeling appears to be the opposite. It is of course possible that phosphorylation of different histones, or even of different amino acid residues in one histone, has opposite effects on chromatin structure.

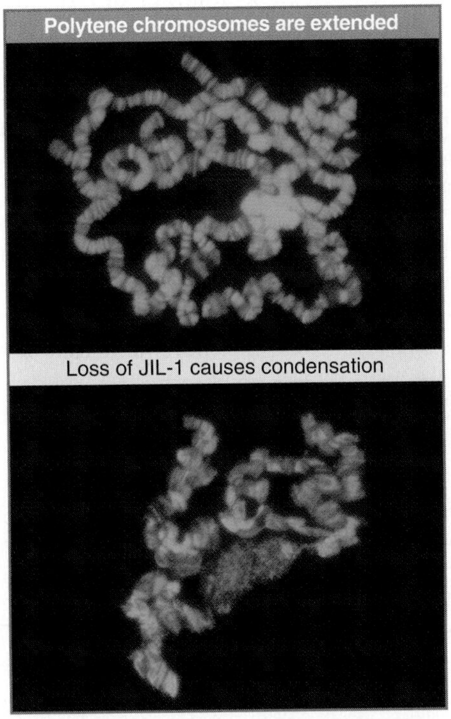

FIGURE 30.19 Flies that have no JIL-1 kinase have abnormal polytene chromosomes that are condensed instead of extended. Photos courtesy of Jorgen Johansen and Kristen M. Johansen, Iowa State University.

30.13 Some Common Motifs Are Found in Proteins That Modify Chromatin

Key concepts

- The chromodomain is found in several chromatin proteins that have either activating or repressing effects on gene expression.
- The SET domain is part of the catalytic site of protein methyltransferases.
- The bromodomain is found in a variety of proteins that interact with chromatin and is used to recognize acetylated sites on histones.

Our insights into the molecular mechanisms for controlling the structure of chromatin start with mutants that affect position effect variegation. Some 30 genes have been identified in *Drosophila*. They are named systematically as *Su(var)* for genes whose products act to suppress variegation and *E(var)* for genes whose products enhance variegation. Remember that the genes were named for the behavior of the mutant loci. *Su(var)* mutations lie in genes whose products are needed for the formation of heterochromatin. They include enzymes that act on chromatin, such as histone deacetylases, and proteins that are localized to heterochromatin. *E(var)* mutations lie in genes whose products are needed to activate gene expression. They include members of the SWI/SNF complex. We see immediately from these properties that modification of chromatin structure is important for controlling the formation of heterochromatin. The universality of these mechanisms is indicated by the fact that many of these loci have homologs in yeast that display analogous properties. Some of the homologs in *Schizosaccharomyces pombe* are *clr* (cryptic loci regulator) genes, in which mutations affect silencing.

Many of the Su(var) and E(var) proteins have a common protein motif of 60 amino acids called the chromodomain. The fact that this domain is found in proteins of both groups suggests that it represents a motif that participates in protein–protein interactions with targets in chromatin. Chromodomain(s) are mostly responsible for targeting proteins to heterochromatin. They function by recognizing methylated lysines in histone tails (see Section 31.3, Heterochromatin Depends on Interactions with Histones and Section 31.4, Polycomb and Trithorax Are Antagonistic Repressors and Activators)

Su(var)3-9 has a chromodomain and also a Su(var)3-9, enhancer-of-zeste, Trithorax (SET) domain, a motif that is found in several Su(var) proteins. Its mammalian homologs localize to centromeric heterochromatin. It is the histone methyltransferase that acts on ^{9}Lys of histone H3 (see Section 30.9, Methylation of Histones and DNA Is Connected). The SET domain is part of the active site, and in fact is a marker for the methylase activity.

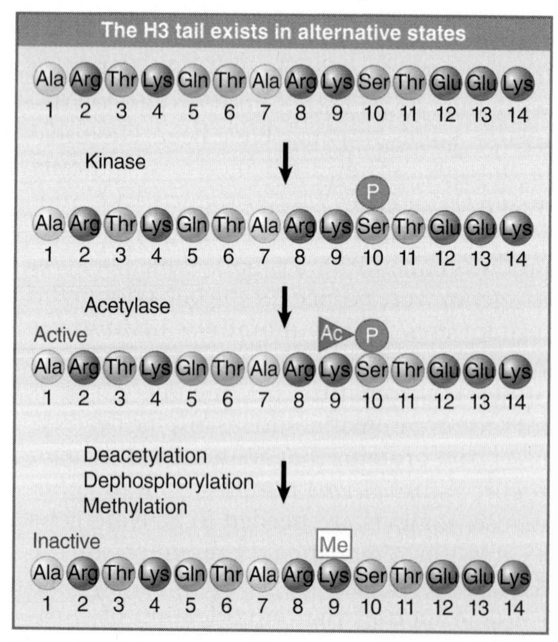

FIGURE 30.20 Multiple modifications in the H3 tail affect chromatin activity.

The bromodomain is found in a variety of proteins that interact with chromatin, including histone acetylases. The crystal structure shows that it has a binding site for acetylated lysine. The bromodomain itself recognizes only a very short sequence of four amino acids, including the acetylated lysine, so specificity for target recognition must depend on interactions involving other regions. Besides the acetylases, the bromodomain is found in a range of proteins that interact with chromatin, including components of the transcription apparatus. This implies that it is used to recognize acetylated histones, which means that it is likely to be found in proteins that are involved with gene activation.

Although there is a general correlation in which active chromatin is acetylated while inactive chromatin is methylated on histones, there are some exceptions to the rule. The best characterized is that acetylation of ^{12}Lys of H4 is associated with heterochromatin.

Multiple modifications may occur on the same histone tail, and one modification may influence another. Phosphorylation of a serine at one position may be necessary for acetylation of a lysine at another position. FIGURE 30.20 shows the situation in the tail of H3, which can exist in either of two alternative states. The inactive state has Methyl-^{9}Lys. The active state has Acetyl-^{9}Lys and Phospho-^{10}Ser. These states can be maintained over extended regions of chromatin. The phosphorylation of ^{10}Ser and the methylation of ^{9}Lys are mutually inhibitory, which suggests the order of events shown in the figure. This situation may cause the tail to flip between the inactive and active states.

30.14 Summary

Genes whose control regions are organized in nucleosomes usually are not expressed. In the absence of specific regulatory proteins, promoters and other regulatory regions are organized by histone octamers into a state in which they cannot be activated. This may explain the need for nucleosomes to be precisely positioned in the vicinity of a promoter, so that essential regulatory sites are appropriately exposed. Some transcription factors have the capacity to recognize DNA on the nucleosomal surface, and a particular positioning of DNA may be required for initiation of transcription.

Active chromatin and inactive chromatin are not in equilibrium. Sudden, disruptive events are needed to convert one to the other. Chromatin remodeling complexes have the ability to displace histone octamers by a mechanism that involves hydrolysis of ATP. Remodeling complexes are large and are classified according to the type of the ATPase subunit. Two common types are SWI/SNF and ISWI. A typical form of this chromatin remodeling is to displace one or more histone octamers from specific sequences of DNA, creating a boundary that results in the precise or preferential positioning of adjacent nucleosomes. Chromatin remodeling may also involve changes in the positions of nucleosomes, sometimes involving sliding of histone octamers along DNA.

Acetylation of histones occurs at both replication and transcription and could be necessary to form a less compact chromatin structure. Some coactivators, which connect transcription factors to the basal apparatus, have histone acetylase activity. Conversely, repressors may be associated with deacetylases. The modifying enzymes are usually specific for particular amino acids in particular histones. The most common sites for modification are located in the N-terminal tails of histones H3 and H4, which extrude from nucleosomes between the turns of DNA. The activating (or repressing) complexes are usually large and often contain several activities that undertake different modifications of chromatin. Some common motifs found in proteins that modify chromatin are the chromodomain (which is concerned with protein–protein interactions), the bromodomain (which targets acetylated lysine), and the SET domain (which is part of the active sites of histone methyltransferases).

Modification of histone tails is a trigger for chromatin reorganization. Acetylation is generally associated with gene activation. Histone acetylases are found in activating complexes, and histone deacetylases are found in inactivating complexes. Histone methylation is associated with gene inactivation. Some histone modifications may be exclusive or synergistic with others.

References

30.2 Chromatin Can Have Alternative States

Reviews

Brown, D. D. (1984). The role of stable complexes that repress and activate eukaryotic genes. *Cell* 37, 359–365.

Weintraub, H. (1985). Assembly and propagation of repressed and derepressed chromosomal states. *Cell* 42, 705–711.

Research

Bogenhagen, D. F., Wormington, W. M., and Brown, D. D. (1982). Stable transcription complexes of *Xenopus* 5S RNA genes: a means to maintain the differentiated state. *Cell* 28, 413–421.

Workman, J. L. and Roeder, R. G. (1987). Binding of transcription factor TFIID to the major late promoter during *in vitro* nucleosome assembly potentiates subsequent initiation by RNA polymerase II. *Cell* 51, 613–622.

30.3 Chromatin Remodeling Is an Active Process

Reviews

Becker, P. B. and Horz, W. (2002). ATP-dependent nucleosome remodeling. *Annu. Rev. Biochem.* 71, 247–273.

Felsenfeld, G. (1992). Chromatin as an essential part of the transcriptional mechanism. *Nature* 355, 219–224.

Grunstein, M. (1990). Histone function in transcription. *Annu. Rev. Cell Biol.* 6, 643–678.

Narlikar, G. J., Fan, H. Y., and Kingston, R. E. (2002). Cooperation between complexes that regulate chromatin structure and transcription. *Cell* 108, 475–487.

Peterson, C. L. and Côté, J. (2004). Cellular machineries for chromosomal DNA repair. *Genes Dev.* 18, 602–616.

Tsukiyama, T. (2002). The *in vivo* functions of ATP-dependent chromatin-remodelling factors. *Nat. Rev. Mol. Cell Biol.* 3, 422–429.

Vignali, M., Hassan, A. H., Neely, K. E., and Workman, J. L. (2000). ATP-dependent chromatin-remodeling complexes. *Mol. Cell Biol.* 20, 1899–1910.

Research

Cairns, B. R., Kim, Y.-J., Sayre, M. H., Laurent, B. C., and Kornberg, R. (1994). A multisubunit complex containing the SWI/ADR6, SWI2/1, SWI3, SNF5, and SNF6 gene products isolated from yeast. *Proc. Natl. Acad. Sci. USA* 91, 1950–1954.

Cote, J., Quinn, J., Workman, J. L., and Peterson, C. L. (1994). Stimulation of GAL4 derivative binding to nucleosomal DNA by the yeast SWI/SNF complex. *Science* 265, 53–60.

Gavin, I., Horn, P. J., and Peterson, C. L. (2001). SWI/SNF chromatin remodeling requires changes in DNA topology. *Mol. Cell* 7, 97–104.

Hamiche, A., Kang, J. G., Dennis, C., Xiao, H., and Wu, C. (2001). Histone tails modulate nucleosome mobility and regulate ATP-dependent nucleosome sliding by NURF. *Proc. Natl. Acad. Sci. USA* 98, 14316–14321.

Kingston, R. E. and Narlikar, G. J. (1999). ATP-dependent remodeling and acetylation as regulators of chromatin fluidity. *Genes Dev.* 13, 2339–2352.

Kwon, H., Imbaizano, A. N., Khavari, P. A., Kingston, R. E., and Green, M. R. (1994). Nucleosome disruption and enhancement of activator binding of human SWI/SNF complex. *Nature* 370, 477–481.

Logie, C. and Peterson, C. L. (1997). Catalytic activity of the yeast SWI/SNF complex on reconstituted nucleosome arrays. *EMBO J.* 16, 6772–6782.

Lorch, Y., Cairns, B. R., Zhang, M., and Kornberg, R. D. (1998). Activated RSC-nucleosome complex and persistently altered form of the nucleosome. *Cell* 94, 29–34.

Lorch, Y., Zhang, M., and Kornberg, R. D. (1999). Histone octamer transfer by a chromatin-remodeling complex. *Cell* 96, 389–392.

Peterson, C. L. and Herskowitz, I. (1992). Characterization of the yeast SWI1, SWI2, and SWI3 genes, which encode a global activator of transcription. *Cell* 68, 573–583.

Robert, F., Young, R. A., and Struhl, K. (2002). Genome-wide location and regulated recruitment of the RSC nucleosome remodeling complex. *Genes Dev.* 16, 806–819.

Schnitzler, G., Sif, S., and Kingston, R. E. (1998). Human SWI/SNF interconverts a nucleosome between its base state and a stable remodeled state. *Cell* 94, 17–27.

Tamkun, J. W., Deuring, R., Scott, M. P., Kissinger, M., Pattatucci, A. M., Kaufman, T. C., and Kennison, J. A. (1992). brahma: a regulator of *Drosophila* homeotic genes structurally related to the yeast transcriptional activator SNF2/SWI2. *Cell* 68, 561–572.

Tsukiyama, T., Daniel, C., Tamkun, J., and Wu, C. (1995). ISWI, a member of the SWI2/SNF2 ATPase family, encodes the 140 kDa subunit of the nucleosome remodeling factor. *Cell* 83, 1021–1026.

Tsukiyama, T., Palmer, J., Landel, C. C., Shiloach, J., and Wu, C. (1999). Characterization of the imitation switch subfamily of ATP-dependent chromatin-remodeling factors in *S. cerevisiae*. *Genes Dev.* 13, 686–697.

Whitehouse, I., Flaus, A., Cairns, B. R., White, M. F., Workman, J. L., and Owen-Hughes, T. (1999). Nucleosome mobilization catalysed by the yeast SWI/SNF complex. *Nature* 400, 784–787.

30.4 Nucleosome Organization May Be Changed at the Promoter

Review

Lohr, D. (1997). Nucleosome transactions on the promoters of the yeast GAL and PHO genes. *J. Biol. Chem.* 272, 26795–26798.

Research

Cosma, M. P., Tanaka, T., and Nasmyth, K. (1999). Ordered recruitment of transcription and chromatin remodeling factors to a cell cycle and developmentally regulated promoter. *Cell* 97, 299–311.

Kadam, S., McAlpine, G. S., Phelan, M. L., Kingston, R. E., Jones, K. A., and Emerson, B. M. (2000). Functional selectivity of recombinant mammalian SWI/SNF subunits. *Genes Dev.* 14, 2441–2451.

McPherson, C. E., Shim, E.-Y., Friedman, D. S., and Zaret, K. S. (1993). An active tissue-specific enhancer and bound transcription factors existing in a precisely positioned nucleosomal array. *Cell* 75, 387–398.

Schmid, V. M., Fascher, K.-D., and Horz, W. (1992). Nucleosome disruption at the yeast PHO5 promoter upon PHO5 induction occurs in the absence of DNA replication. *Cell* 71, 853–864.

Truss, M., Barstch, J., Schelbert, A., Hache, R. J. G., and Beato, M. (1994). Hormone induces binding of receptors and transcription factors to a rearranged nucleosome on the MMTV promoter *in vitro*. *EMBO J.* 14, 1737–1751.

Tsukiyama, T., Becker, P. B., and Wu, C. (1994). ATP-dependent nucleosome disruption at a heat shock promoter mediated by binding of GAGA transcription factor. *Nature* 367, 525–532.

Yudkovsky, N., Logie, C., Hahn, S., and Peterson, C. L. (1999). Recruitment of the SWI/SNF chromatin remodeling complex by transcriptional activators. *Genes Dev.* 13, 2369–2374.

30.5 Histone Modification Is a Key Event

Review

Jenuwein, T. and Allis, C. D. (2001). Translating the histone code. *Science* 293, 1074–1080.

Research

Osada, S., Sutton, A., Muster, N., Brown, C. E., Yates, J. R., Sternglanz, R., and Workman, J. L. (2001). The yeast SAS (something about silencing) protein complex contains a MYST-type putative acetyltransferase and functions with chromatin assembly factor ASF1. *Genes Dev.* 15, 3155–3168.

30.6 Histone Acetylation Occurs in Two Circumstances

Reviews

Hirose, Y. and Manley, J. L. (2000). RNA polymerase II and the integration of nuclear events. *Genes Dev.* 14, 1415–1429.

Verreault, A. (2000). De novo nucleosome assembly: new pieces in an old puzzle. *Genes Dev.* 14, 1430–1438.

Research

Akhtar, A. and Becker, P. B. (2000). Activation of transcription through histone H4 acetylation by MOF, an acetyltransferase essential for dosage compensation in *Drosophila. Mol. Cell* 5, 367–375.

Alwine, J. C., Kemp, D. J., and Stark, G. R. (1977). Method for detection of specific RNAs in agarose gels by transfer to diazobenzyloxymethyl-paper and hybridization with DNA probes. *Proc. Natl. Acad. Sci. USA* 74, 5350–5354.

Jackson, V., Shires, A., Tanphaichitr, N., and Chalkley, R. (1976). Modifications to histones immediately after synthesis. *J. Mol. Biol.* 104, 471–483.

Ling, X., Harkness, T. A., Schultz, M. C., Fisher-Adams, G., and Grunstein, M. (1996). Yeast histone H3 and H4 amino termini are important for nucleosome assembly *in vivo* and *in vitro:* redundant and position-independent functions in assembly but not in gene regulation. *Genes Dev.* 10, 686–699.

Shibahara, K., Verreault, A., and Stillman, B. (2000). The N-terminal domains of histones H3 and H4 are not necessary for chromatin assembly factor-1-mediated nucleosome assembly onto replicated DNA *in vitro. Proc. Natl. Acad. Sci. USA* 97, 7766–7771.

Turner, B. M., Birley, A. J., and Lavender, J. (1992). Histone H4 isoforms acetylated at specific lysine residues define individual chromosomes and chromatin domains in *Drosophila* polytene nuclei. *Cell* 69, 375–384.

30.7 Acetylases Are Associated with Activators

Research

Brownell, J. E. et al. (1996). *Tetrahymena* histone acetyltransferase A: a homologue to yeast Gcn5p linking histone acetylation to gene activation. *Cell* 84, 843–851.

Chen, H. et al. (1997). Nuclear receptor coactivator ACTR is a novel histone acetyltransferase and forms a multimeric activation complex with P/CAF and CP/p300. *Cell* 90, 569–580.

Grant, P. A. et al. (1998). A subset of TAF_{II}s are integral components of the SAGA complex required for nucleosome acetylation and transcriptional stimulation. *Cell* 94, 45–53.

Kingston, R. E. and Narlikar, G. J. (1999). ATP-dependent remodeling and acetylation as regulators of chromatin fluidity. *Genes Dev.* 13, 2339–2352.

Lee, T. I., Causton, H. C., Holstege, F. C., Shen, W. C., Hannett, N., Jennings, E. G., Winston, F., Green, M. R., and Young, R. A. (2000). Redundant roles for the $TF_{II}D$ and SAGA complexes in global transcription. *Nature* 405, 701–704.

30.8 Deacetylases Are Associated with Repressors

Review

Richards, E. J., Elgin, S. C., and Richards, S. C. (2002). Epigenetic codes for heterochromatin formation and silencing: rounding up the usual suspects. *Cell* 108, 489–500.

Research

Ayer, D. E., Lawrence, Q. A., and Eisenman, R. N. (1995). Mad-Max transcriptional repression is mediated by ternary complex formation with mammalian homologs of yeast repressor Sin3. *Cell* 80, 767–776.

Kadosh, D. and Struhl, K. (1997). Repression by Ume6 involves recruitment of a complex containing Sin3 corepressor and Rpd3 histone deacetylase to target promoters. *Cell* 89, 365–371.

Schreiber-Agus, N., Chin, L., Chen, K., Torres, R., Rao, G., Guida, P., Skoultchi, A. I., and DePinho, R. A. (1995). An amino-terminal domain of Mxi1 mediates anti-Myc oncogenic activity and interacts with a homolog of the yeast transcriptional repressor SIN3. *Cell* 80, 777–786.

30.9 Methylation of Histones and DNA Is Connected

Reviews

Bannister, A. J., and Kouzarides, T. (2005). Reversing histone methylation. *Nature* 436, 1103–1106.

Richards, E. J., Elgin, S. C., and Richards, S. C. (2002). Epigenetic codes for heterochromatin formation and silencing: rounding up the usual suspects. *Cell* 108, 489–500.

Zhang, Y. and Reinberg, D. (2001). Transcription regulation by histone methylation: interplay between different covalent modifications of the core histone tails. *Genes Dev.* 15, 2343–2360.

Research

Cuthbert, G. L., Daujat, S., Snowden, A. W., Erdjument-Bromage, H., Hagiwara, T., Yamada, M., Schneider, R., Gregory, P. D., Tempst, P., Bannister, A. J., and Kouzarides, T. (2004). Histone deimination antagonizes arginine methylation. *Cell* 118, 545–553.

Fuks, F., Hurd, P. J., Wolf, D., Nan, X., Bird, A. P., and Kouzarides, T. (2003). The methyl-CpG-binding protein MeCP2 links DNA methylation to histone methylation. *J. Biol. Chem.* 278, 4035–4040.

Gendrel, A. V., Lippman, Z., Yordan, C., Colot, V., and Martienssen, R. A. (2002). Dependence of heterochromatic histone H3 methylation patterns on the *Arabidopsis* gene DDM1. *Science* 297, 1871–1873.

Johnson, L., Cao, X., and Jacobsen, S. (2002). Interplay between two epigenetic marks. DNA methylation and histone H3 lysine 9 methylation. *Curr. Biol.* 12, 1360–1367.

Lawrence, R. J., Earley, K., Pontes, O., Silva, M., Chen, Z. J., Neves, N., Viegas, W., and Pikaard, C. S. (2004). A concerted DNA methylation/histone methylation switch regulates rRNA gene dosage control and nucleolar dominance. *Mol. Cell* 13, 599–609.

Ng, H. H., Feng, Q., Wang, H., Erdjument-Bromage, H., Tempst, P., Zhang, Y., and Struhl, K. (2002). Lysine methylation within the globular domain of histone H3 by Dot1 is important for telomeric silencing and Sir protein association. *Genes Dev.* 16, 1518–1527.

Rea, S., Eisenhaber, F., O'Carroll, D., Strahl, B. D., Sun, Z. W., Sun, M., Opravil, S., Mechtler, K., Ponting, C. P., Allis, C. D., and Jenuwein, T. (2000). Regulation of chromatin structure by site-specific histone H3 methyltransferases. *Nature* 406, 593–599.

Shi, Y., Lan, F., Matson, C., Mulligan, P., Whetstine, J. R., Cole, P. A., and Casero, R. A. (2004). Histone demethylation mediated by the nuclear amine oxidase homolog LSD1. *Cell* 119, 941–953.

Tamaru, H. and Selker, E. U. (2001). A histone H3 methyltransferase controls DNA methylation in *Neurospora crassa*. *Nature* 414, 277–283.

Tamaru, H., Zhang, X., McMillen, D., Singh, P. B., Nakayama, J., Grewal, S. I., Allis, C. D., Cheng, X., and Selker, E. U. (2003). Trimethylated lysine 9 of histone H3 is a mark for DNA methylation in *Neurospora crassa*. *Nat. Genet.* 34, 75–79.

Wang, Y., Wysocka, J., Sayegh, J., Lee, Y. H., Perlin, J. R., Leonelli, L., Sonbuchner, L. S., McDonald, C. H., Cook, R. G., Dou, Y., et al. (2004). Human PAD4 regulates histone arginine methylation levels via demethylimination. *Science* 306, 279–283.

30.11 Promoter Activation Involves an Ordered Series of Events

Review

Orphanides, G. and Reinberg, D. (2000). RNA polymerase II elongation through chromatin. *Nature* 407, 471–475.

Research

Bortvin, A. and Winston, F. (1996). Evidence that Spt6p controls chromatin structure by a direct interaction with histones. *Science* 272, 1473–1476.

Cosma, M. P., Tanaka, T., and Nasmyth, K. (1999). Ordered recruitment of transcription and chromatin remodeling factors to a cell cycle and developmentally regulated promoter. *Cell* 97, 299–311.

Hassan, A. H., Neely, K. E., and Workman, J. L. (2001). Histone acetyltransferase complexes stabilize swi/snf binding to promoter nucleosomes. *Cell* 104, 817–827.

Orphanides, G., LeRoy, G., Chang, C. H., Luse, D. S., and Reinberg, D. (1998). FACT, a factor that facilitates transcript elongation through nucleosomes. *Cell* 92, 105–116.

Wada, T., Takagi, T., Yamaguchi, Y., Ferdous, A., Imai, T., Hirose, S., Sugimoto, S., Yano, K., Hartzog, G. A., Winston, F., Buratowski, S., and Handa, H. (1998). DSIF, a novel transcription elongation factor that regulates RNA polymerase II processivity, is composed of human Spt4 and Spt5 homologs. *Genes Dev.* 12, 343–356.

30.12 Histone Phosphorylation Affects Chromatin Structure

Research

Wang, Y., Zhang, W., Jin, Y., Johansen, J., and Johansen, K. M. (2001). The JIL-1 tandem kinase mediates histone H3 phosphorylation and is required for maintenance of chromatin structure in *Drosophila*. *Cell* 105, 433–443.

30.13 Some Common Motifs Are Found in Proteins That Modify Chromatin

Research

Dhalluin, C., Carlson, J. E., Zeng, L., He, C., Aggarwal, A. K., and Zhou, M. M. (1999). Structure and ligand of a histone acetyltransferase bromo domain. *Nature* 399, 491–496.

Koonin, E. V., Zhou, S., and Lucchesi, J. C. (1995). The chromo superfamily: new members, duplication of the chromo domain and possible role in delivering transcription regulators to chromatin. *Nucleic Acids Res.* 23, 4229–4233.

Litt, M. D., Simpson, M., Gaszner, M., Allis, C. D., and Felsenfeld, G. (2001). Correlation between histone lysine methylation and developmental changes at the chicken beta-globin locus. *Science* 293, 2453–2455.

Owen, D. J., Ornaghi, P., Yang, J. C., Lowe, N., Evans, P. R., Ballario, P., Neuhaus, D., Filetici, P., and Travers, A. A. (2000). The structural basis for the recognition of acetylated histone H4 by the bromo domain of histone acetyltransferase Gcn5p. *EMBO J.* 19, 6141–6149.

Turner, B. M., Birley, A. J., and Lavender, J. (1992). Histone H4 isoforms acetylated at specific lysine residues define individual chromosomes and chromatin domains in *Drosophila* polytene nuclei. *Cell* 69, 375–384.

31

Epigenetic Effects Are Inherited

CHAPTER OUTLINE

31.1 Introduction

Key concept

- Epigenetic effects can result from modification of a nucleic acid after it has been synthesized or by the perpetuation of protein structures.

Epigenetic inheritance describes the ability of different states, which may have different phenotypic consequences, to be inherited without any change in the sequence of DNA. This means that two individuals with the same DNA sequence at the locus that controls the effect may show different phenotypes. The basic cause of this phenomenon is the existence of a self-perpetuating structure in one of the individuals that does not depend on DNA sequence. Several different types of structures have the ability to sustain epigenetic effects:

- A covalent modification of DNA (methylation of a base).
- A proteinaceous structure that assembles on DNA.
- A protein aggregate that controls the conformation of new subunits as they are synthesized.

In each case the epigenetic state results from a difference in function (typically inactivation) that is determined by the structure.

In the case of DNA methylation, a methylated DNA sequence may fail to be transcribed, whereas the nonmethylated sequence will be expressed. FIGURE 31.1 shows how this situation is inherited. One allele has a sequence that is methylated on both strands of DNA, whereas the other allele has an unmethylated sequence. Replication of the methylated allele creates hemimethylated daughters that are restored to the methylated state by a constitutively active methylase enzyme. Replication does not affect the state of the nonmethylated allele. If the state of methylation affects transcription, the two alleles differ in their state of gene

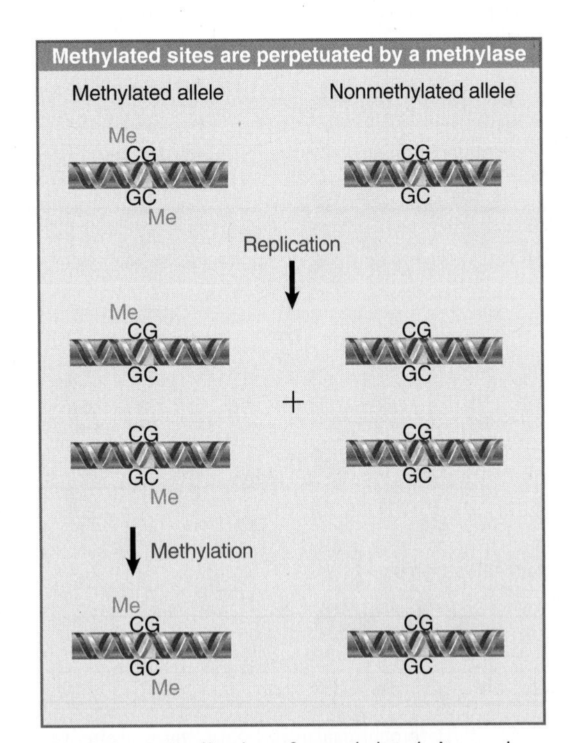

FIGURE 31.1 Replication of a methylated site produces hemimethylated DNA, in which only the parental strand is methylated. A perpetuation methylase recognizes hemimethylated sites and adds a methyl group to the base on the daughter strand. This restores the original situation, in which the site is methylated on both strands. A nonmethylated site remains nonmethylated after replication.

expression, even though their sequences are identical.

Self-perpetuating structures that assemble on DNA usually have a repressive effect by forming heterochromatic regions that prevent the expression of genes within them. Their perpetuation depends on the ability of proteins in a heterochromatic region to remain bound to those regions after replication, and then to recruit more protein subunits to sustain the complex. If individual subunits are distributed at random to each daughter duplex at replication, the two daughters will continue to be marked by the protein, although its density will be reduced to half of the level before replication. FIGURE 31.2 shows that the existence of epigenetic effects forces us to the view that a protein responsible for such a situation must have some sort of self-templating or self-assembling capacity to restore the original complex.

It can be the state of protein modification, rather than the presence of the protein *per se*, that is responsible for an epigenetic effect. Usually the tails of histones H3 and H4 are not acetylated in constitutive heterochromatin. If centromeric heterochromatin is acetylated, though, silenced genes may become active. The effect may be perpetuated through mitosis and meiosis, which suggests that an epigenetic effect has been created by changing the state of histone acetylation.

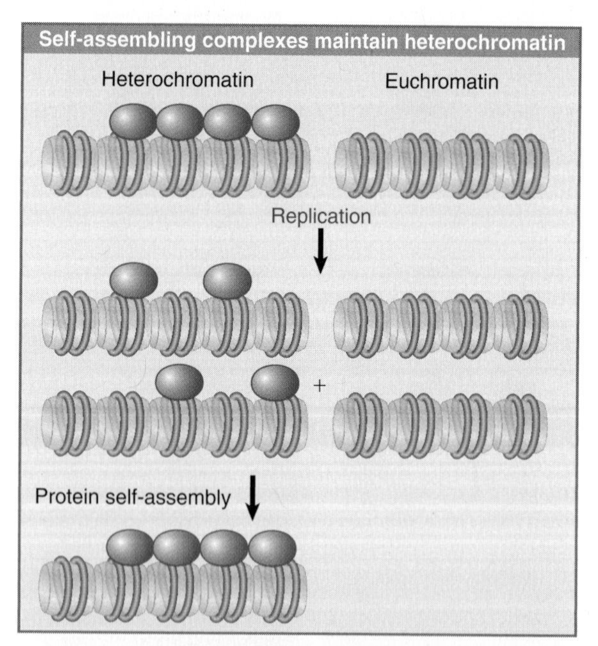

FIGURE 31.2 Heterochromatin is created by proteins that associate with histones. Perpetuation through division requires that the proteins associate with each daughter duplex and then recruit new subunits to reassemble the repressive complexes.

Independent protein aggregates that cause epigenetic effects (called **prions**) work by sequestering the protein in a form in which its normal function cannot be displayed. Once the protein aggregate has formed, it forces newly synthesized protein subunits to join it in the inactive conformation.

31.2 Heterochromatin Propagates from a Nucleation Event

Key concepts

- Heterochromatin is nucleated at a specific sequence and the inactive structure propagates along the chromatin fiber.
- Genes within regions of heterochromatin are inactivated.
- The length of the inactive region varies from cell to cell; as a result, inactivation of genes in this vicinity causes position effect variegation.
- Similar spreading effects occur at telomeres and at the silent cassettes in yeast mating type.

An interphase nucleus contains both euchromatin and heterochromatin. The condensation state of heterochromatin is close to that of mitotic chromosomes. Heterochromatin is inert. It remains condensed in interphase, is transcriptionally repressed, replicates late in S phase, and may be localized to the nuclear periphery. Centromeric heterochromatin typically consists of satellite DNAs; however, the formation of heterochromatin is not rigorously defined by sequence. When a gene is transferred, either by a chromosomal translocation or by transfection and integration, into a position adjacent to heterochromatin, it may become inactive as the result of its new location, implying that it has become heterochromatic.

Such inactivation is the result of an **epigenetic** effect (see Section 31.10, Epigenetic Effects Can Be Inherited). It may differ between individual cells in an animal, and results in the phenomenon of **position effect variegation (PEV),** in which genetically identical cells have different phenotypes. This has been well characterized in *Drosophila*. FIGURE 31.3 shows an example of position effect variegation in the fly eye. Some of the regions in the eye lack color, whereas others are red. This is because the *white* gene was inactivated by adjacent heterochromatin in some cells, but remained active in others.

The explanation for this effect is shown in FIGURE 31.4. Inactivation spreads from heterochromatin into the adjacent region for a variable distance. In some cells it goes far enough to inactivate a nearby gene, but in others it does not. This happens at a certain point in embryonic development, and after that point the state of the gene is inherited by all the progeny cells. Cells descended from an ancestor in which the gene was inactivated form patches corresponding to the phenotype of loss-of-function (in the case of *white,* absence of color).

The closer a gene lies to heterochromatin, the higher the probability that it will be inactivated. This suggests that the formation of heterochromatin may be a two-stage process: A *nucleation* event occurs at a specific sequence, and then the inactive structure *propagates* along the chromatin fiber. The distance for which the inactive structure extends is not precisely determined and may be stochastic, being influenced by parameters such as the quantities of limiting protein components. One factor that may affect the spreading process is the activation of promoters in the region; an active promoter may inhibit spreading.

Genes that are closer to heterochromatin are more likely to be inactivated, and will therefore be inactive in a greater proportion of cells. On this model, the boundaries of a heterochromatic region might be terminated by exhausting the supply of one of the proteins that is required.

The effect of **telomeric silencing** in yeast is analogous to position effect variegation in *Drosophila;* genes translocated to a telomeric location show the same sort of variable loss of activity. This results from a spreading effect that propagates from the telomeres.

A second form of silencing occurs in yeast. Yeast mating type is determined by the activity of a single active locus *(MAT),* but the genome contains two other copies of the mating type sequences (*HML* and *HMR*), which are maintained in an inactive form. The silent loci *HML* and *HMR* share many properties with heterochromatin, and could be regarded as constituting regions of heterochromatin in miniature (see Section 19.22, Silent Cassettes at *HML* and *HMR* Are Repressed).

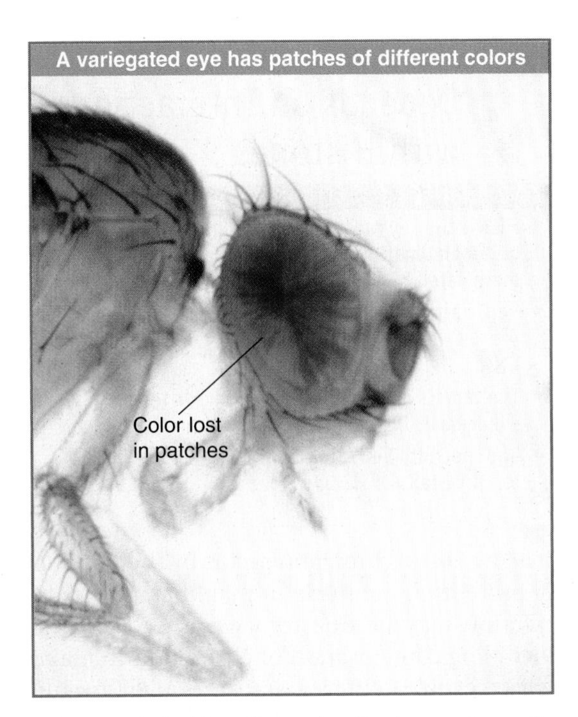

FIGURE 31.3 Position effect variegation in eye color results when the *white* gene is integrated near heterochromatin. Cells in which *white* is inactive give patches of white eye, whereas cells in which *white* is active give red patches. The severity of the effect is determined by the closeness of the integrated gene to heterochromatin. Photo courtesy of Steven Henikoff, Fred Hutchinson Cancer Research Center.

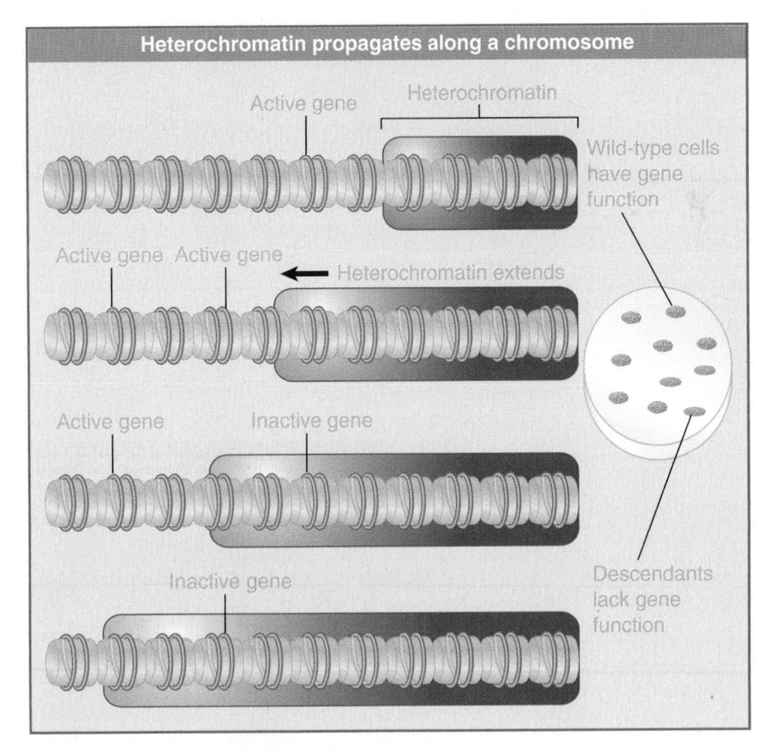

FIGURE 31.4 Extension of heterochromatin inactivates genes. The probability that a gene will be inactivated depends on its distance from the heterochromatin region.

31.3 Heterochromatin Depends on Interactions with Histones

Key concepts

- HP1 is the key protein in forming mammalian heterochromatin, and acts by binding to methylated histone H3.
- Rap1 initiates formation of heterochromatin in yeast by binding to specific target sequences in DNA.
- The targets of Rap1 include telomeric repeats and silencers at *HML* and *HMR*.
- Rap1 recruits Sir3/Sir4, which interact with the N-terminal tails of H3 and H4.

Inactivation of chromatin occurs by the addition of proteins to the nucleosomal fiber. The inactivation may be due to a variety of effects, including condensation of chromatin to make it inaccessible to the apparatus needed for gene expression, addition of proteins that directly block access to regulatory sites, or proteins that directly inhibit transcription.

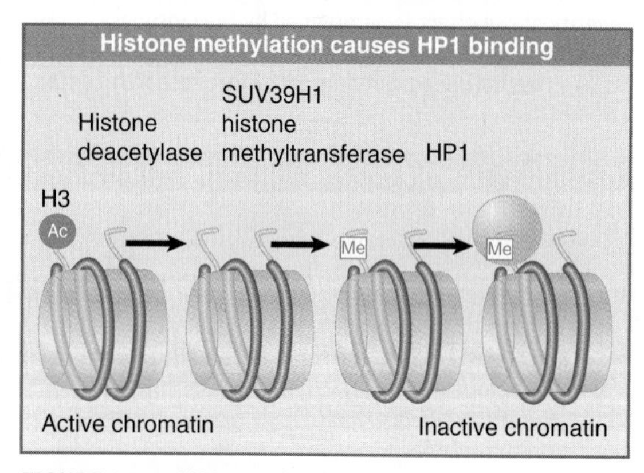

FIGURE 31.5 SUV39H1 is a histone methyltransferase that acts on ^{9}Lys of histone H3. HP1 binds to the methylated histone.

Two systems that have been characterized at the molecular level involve HP1 in mammals and the SIR complex in yeast. Although there are no detailed similarities between the proteins involved in each system, the general mechanism of reaction is similar: The points of contact in chromatin are the N-terminal tails of the histones.

HP1 (heterochromatin protein 1) is one of the most important Su(var) proteins. This was originally identified as a protein that is localized to heterochromatin by staining polytene chromosomes with an antibody directed against the protein. It was later shown to be the product of the gene *Su(var)2-5*. Its homolog in the yeast *Schizosacclaromyces pombe* is coded by *swi6*. The original protein identified as HP1 is now called HP1α because two related proteins, HP1β and HP1γ, have since been found.

HP1 contains a chromodomain near the N-terminus, and another domain that is related to it, called the chromo-shadow domain, at the C-terminus (see Figure 31.6). The importance of the chromodomain is indicated by the fact that it is the location of many of the mutations in HP1.

Mutation of a deacetylase that acts on the H3 Ac-^{14}Lys prevents the methylation at ^{9}Lys. H3 that is methylated at ^{9}Lys binds the protein HP1 via the chromodomain. This suggests the model for initiating formation of heterochromatin shown in FIGURE 31.5. First the deacetylase acts to remove the modification at ^{14}Lys, and then the SUV39H1 methylase acts on the histone H3 tail to create the methylated signal to which HP1 will bind. FIGURE 31.6 expands the reaction to show that the interaction occurs between the chromodomain and the methylated lysine. This is a trigger for forming inactive chromatin. FIGURE 31.7 shows that the

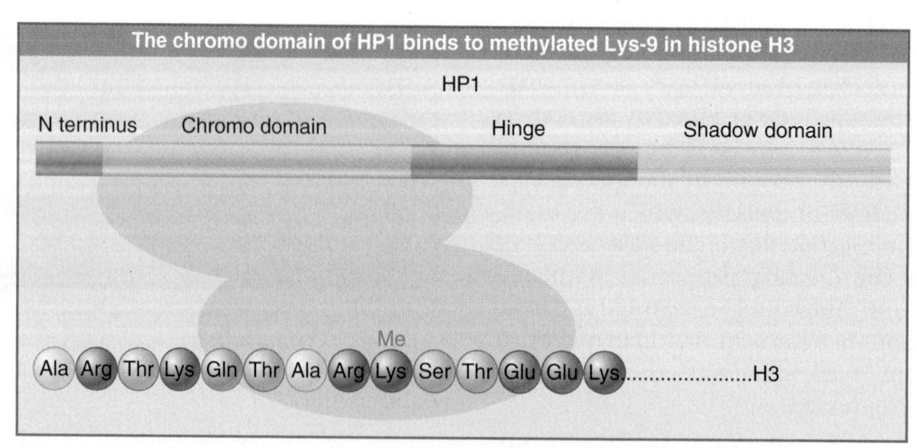

FIGURE 31.6 Methylation of histone H3 creates a binding site for HP1.

inactive region may then be extended by the ability of further HP1 molecules to interact with one another.

The existence of a common basis for silencing in yeast is suggested by its reliance on a common set of genetic loci. Mutations in any one of a number of genes cause *HML* and *HMR* to become activated, and also relieve the inactivation of genes that have been integrated near telomeric heterochromatin. The products of these loci therefore function to maintain the inactive state of both types of heterochromatin.

FIGURE 31.8 proposes a model for actions of these proteins. Only one of them is a sequence-specific DNA-binding protein. This is RAP1, which binds to the $C_{1-3}A$ repeats at the telomeres, and also binds to the *cis*-acting silencer elements that are needed for repression of *HML* and *HMR*. The proteins Sir3 and Sir4 interact with Rap1 and also with one another (they may function as a heteromultimer). Sir3/Sir4 interact with the N-terminal tails of the histones H3 and H4. (In fact, the first evidence that histones might be involved directly in formation of heterochromatin was provided by the discovery that mutations abolishing silencing at *HML/HMR* map to genes coding for H3 and H4.)

Rap1 has the crucial role of identifying the DNA sequences at which heterochromatin forms. It recruits Sir3/Sir4, and they interact directly with the histones H3/H4. Once Sir3/Sir4 have bound to histones H3/H4, the complex may polymerize further and spread along the chromatin fiber. This may inactivate the region, either because coating with Sir3/Sir4 itself has an inhibitory effect, or because binding to histones H3/H4 induces some further change in structure. We do not know what limits the spreading of the complex. The C-terminus of Sir3 has a similarity to nuclear lamin proteins (constituents of the nuclear matrix) and may be responsible for tethering heterochromatin to the nuclear periphery.

A similar series of events forms the silenced regions at *HMR* and *HML* (see also Section 19.22, Silent Cassettes at *HML* and *HMR* Are Repressed). Three sequence-specific factors are involved in triggering formation of the complex: Rap1, Abf1 (a transcription factor), and ORC (the origin replication complex). In this case, Sir1 binds to a sequence-specific factor and recruits Sir2, -3, and -4 to form the repressive structure. Sir2 is a histone deacetylase. The deacetylation reaction is necessary to maintain binding of the Sir complex to chromatin.

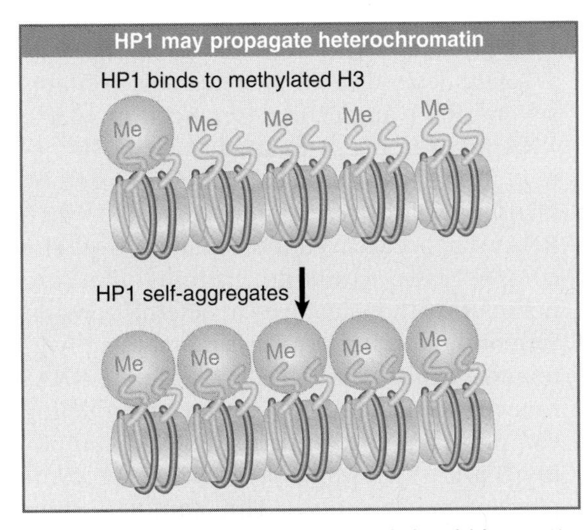

FIGURE 31.7 Binding of HP1 to methylated histone H3 forms a trigger for silencing because further molecules of HP1 aggregate on the nucleosome chain.

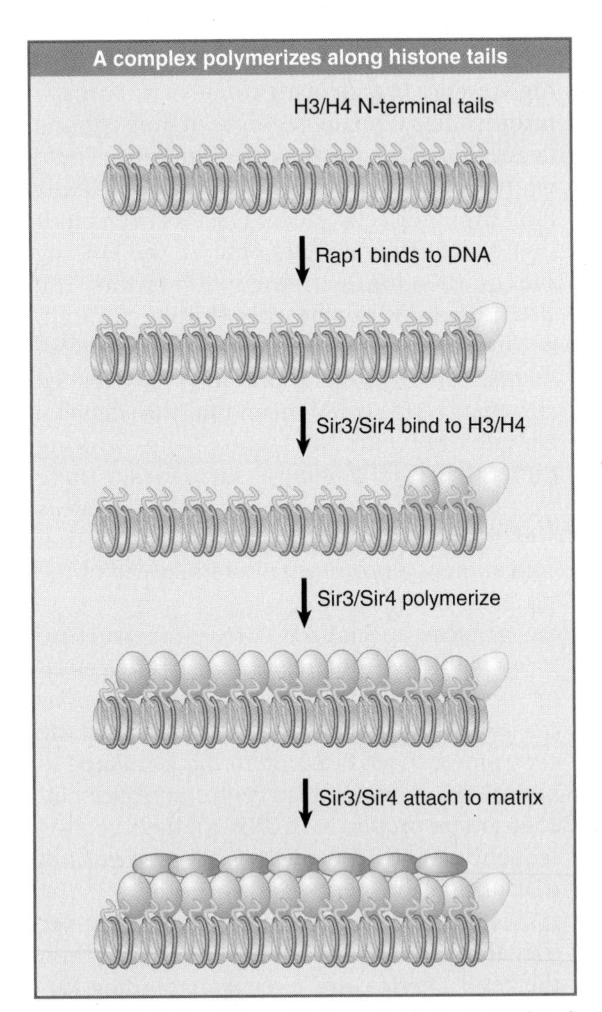

FIGURE 31.8 Formation of heterochromatin is initiated when Rap1 binds to DNA. Sir3/4 bind to Rap1 and also to histones H3/H4. The complex polymerizes along chromatin and may connect telomeres to the nuclear matrix.

Formation of heterochromatin in the yeast *S. pombe* depends on a complex that contains several RNAi molecules (see Section 13.10, RNA Interference Is Related to Gene Silencing). These RNAi molecules are produced by transcription of centromeric repeats to give RNAs that are cleaved into smaller units. The complex also contains proteins that are homologs of those involved in heterochromatin formation in other organisms, including Argonaute, which is involved in targeting RNA-induced silencing complex (RISC) remodeling complexes to chromatin. The RNAi components are responsible for localizing the complex at centromeres. The complex then promotes dimethylation of histone H3 by a histone methyltransferase.

How does a silencing complex repress chromatin activity? It could condense chromatin so that regulator proteins cannot find their targets. The simplest case would be to suppose that the presence of a silencing complex is mutually incompatible with the presence of transcription factors and RNA polymerase. The cause could be that silencing complexes block remodeling (and thus indirectly prevent factors from binding) or that they directly obscure the binding sites on DNA for the transcription factors. The situation may not be this simple, though, because transcription factors and RNA polymerase can be found at promoters in silenced chromatin. This could mean that the silencing complex prevents the factors from working rather than from binding as such. In fact, there may be competition between gene activators and the repressing effects of chromatin, so that activation of a promoter inhibits spread of the silencing complex.

Another specialized chromatin structure forms at the centromere. Its nature is suggested by the properties of an *Saccharomyces cerevisiae* mutation, *cse4*, that disrupts the structure of the centromere. Cse4 is a protein that is related to histone H3. A mammalian centromeric protein, centromere protein A (CENP-A), has a related sequence. Genetic interactions between *cse4* and CDE-II, and between *cse4* and a mutation in the H4 histone gene, suggest that a histone octamer may form around a core of Cse4-H4, and then the centromeric complexes (core binding factors) CBF1 and CBF3 may attach to form the centromere.

The centromere may then be associated with the formation of heterochromatin in the region. In human cells, the centromere-specific protein CENP-B is required to initiate modifications of histone H3 (deacetylation of ^{9}Lys and ^{14}Lys, followed by methylation of ^{9}Lys) that trigger an association with the protein Swi6 that leads to the formation of heterochromatin in the region.

31.4 Polycomb and Trithorax Are Antagonistic Repressors and Activators

Key concepts

- Polycomb group proteins (Pc-G) perpetuate a state of repression through cell divisions.
- The PRE is a DNA sequence that is required for the action of Pc-G.
- The PRE provides a nucleation center from which Pc-G proteins propagate an inactive structure.
- No individual Pc-G protein has yet been found that can bind the PRE.
- Trithorax group proteins antagonize the actions of the Pc-G.

Heterochromatin provides one example of the specific repression of chromatin. Another is provided by the genetics of homeotic genes in *Drosophila*, which have led to the identification of a protein complex that may maintain certain genes in a repressed state. *Pc* mutants show transformations of cell type that are equivalent to gain-of-function mutations in the genes *Antennapedia (Antp)* or *Ultrabithorax*, because these genes are expressed in tissues in which usually they are repressed. This implicates *Pc* in regulating transcription. Furthermore, *Pc* is the prototype for a class of ~15 loci called the *Pc-group (Pc-G)*; mutations in these genes generally have the same result of derepressing homeotic genes, which suggests the possibility that the group of proteins has some common regulatory role.

The Pc proteins function in large complexes. The PRC1 (Polycomb-repressive complex) contains Pc itself, several other Pc-G proteins, and five general transcription factors. The Esc-E(z) complex contains Esc, E(z), other Pc-G proteins, a histone-binding protein, and a histone deacetylase. Pc itself has a chromodomain that binds to methylated H3, and E(z) is a methyltransferase that acts on H3. These properties directly support the connection between chromatin remodeling and repression that was initially suggested by the properties of *brahma*, a fly counterpart to *SWI2*. *brahma* codes for a component of the SWI/SNF remodeling complex, and loss of *brahma* function suppresses mutations in *Polycomb*.

Consistent with the pleiotropy of *Pc* mutations, Pc is a nuclear protein that can be visualized at ~80 sites on polytene chromosomes. These sites include the *Antp* gene. Another member of the *Pc-G, polyhomeotic,* is visualized at a set of polytene chromosome bands that are identical with those bound by Pc. The two proteins coimmunoprecipitate in a complex of ~2.5 × 10^6 D that contains 10 to 15 polypeptides. The relationship between these proteins and the products of the ~30 *Pc-G* genes remains to be established. One possibility is that some of these gene products form a general repressive complex, and then some of the other proteins associate with it to determine its specificity.

The Pc-G proteins are not conventional repressors. They are not responsible for determining the initial pattern of expression of the genes on which they act. In the absence of Pc-G proteins, these genes are initially repressed as usual, but later in development the repression is lost without Pc-G group functions. This suggests that *the Pc-G proteins in some way recognize the state of repression when it is established, and they then act to perpetuate it through cell division of the daughter cells.* FIGURE 31.9 shows a model in which Pc-G proteins bind in conjunction with a repressor, but the Pc-G proteins remain bound after the repressor is no longer available. This is necessary to maintain repression, so that if Pc-G proteins are absent, the gene becomes activated.

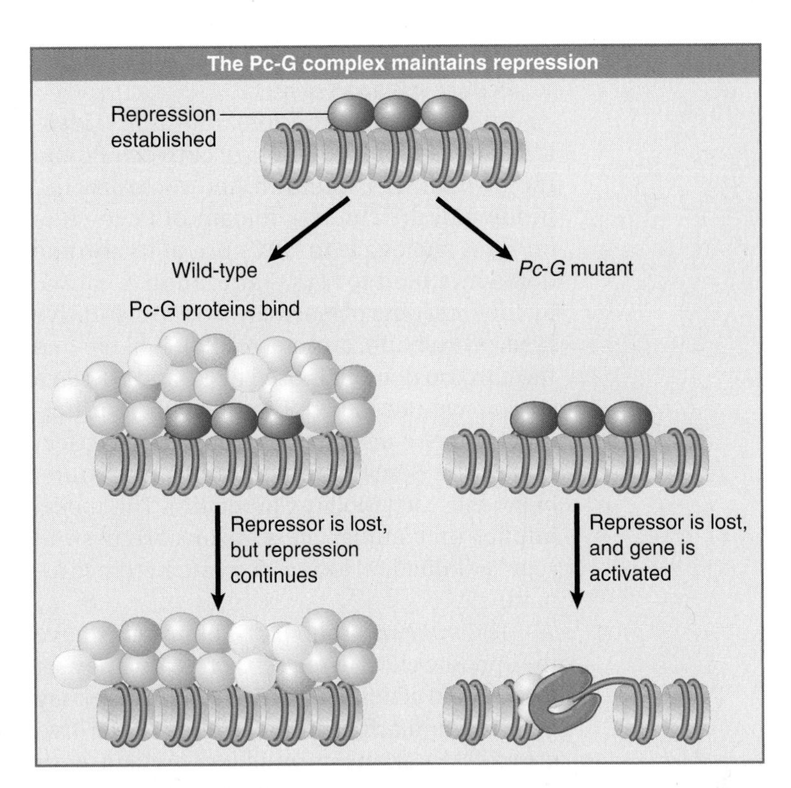

FIGURE 31.9 Pc-G proteins do not initiate repression, but are responsible for maintaining it.

A region of DNA that is sufficient to enable the response to the *Pc-G* genes is called a PRE (*Polycomb* response element). It can be defined operationally by the property that it maintains repression in its vicinity throughout development. The assay for a PRE is to insert it close to a reporter gene that is controlled by an enhancer that is repressed in early development, and then to determine whether the reporter becomes expressed subsequently in the descendants. An effective PRE will prevent such reexpression.

The PRE is a complex structure that measures ~10 kb. Two proteins, Pho and Phol, with DNA-binding activity for sites within the PRE have been identified, but there could be others. When a locus is repressed by Pc-G, however, the Pc-G proteins occupy a much larger length of DNA than the PRE itself. Pc is found locally over a few kilobases of DNA surrounding a PRE.

This suggests that the PRE may provide a nucleation center, from which a structural state depending on Pc-G proteins may propagate. This model is supported by the observation of effects related to position effect variegation (see Figure 31.4), that is, a gene near to a locus whose repression is maintained by Pc-G may become heritably inactivated in some cells but not others. In one typical situation, crosslinking experiments *in vivo* showed that Pc protein is found over large regions of the *bithorax* complex that are inactive, but the protein is excluded from regions that contain active genes. The idea that this could be due to cooperative interactions within a multimeric complex is supported by the existence of mutations in *Pc* that change its nuclear distribution and abolish the ability of other *Pc-G* members to localize in the nucleus. The role of Pc-G proteins in maintaining, as opposed to establishing, repression must mean that the formation of the complex at the PRE also depends on the local state of gene expression.

A working model for Pc-G binding at a PRE is suggested by the properties of the individual proteins. First Pho and Phol bind to specific sequences within the PRE. Esc-E(z) is recruited to Pho/Phol; it then uses its methyltransferase activity to methylate ^{27}Lys of histone H3. This creates the binding site for the PRC, because the chromo domain of Pc binds to the methylated lysine. The Polycomb complex induces a more compact structure in chromatin; each

PRC1 complex causes about three nucleosomes to become less accessible.

In fact, the chromo domain was first identified as a region of homology between Pc and the protein HP1 found in heterochromatin. Binding of the chromo domain of Pc to ^{27}Lys on H3 is analogous to HP1's use of its chromo domain to bind to ^{9}Lys. Variegation is caused by the spreading of inactivity from constitutive heterochromatin, and as a result it is likely that the chromo domain is used by Pc and HP1 in a similar way to induce the formation of heterochromatic or inactive structures (see Section 30.13, Some Common Motifs Are Found in Proteins That Modify Chromatin). This model implies that similar mechanisms are used to repress individual loci or to create heterochromatin.

The *trithorax* group *(trxG)* of proteins have the opposite effect to the Pc-G proteins: They act to maintain genes in an active state. There may be some similarities in the actions of the two groups: Mutations in some loci prevent both Pc-G and trx from functioning, suggesting that they could rely on common components. The GAGA factor, which is coded by the *trithorax-like* gene, has binding sites in the PRE. In fact, the sites where Pc binds to DNA coincide with the sites where GAGA factor binds.

What does this mean? GAGA is probably needed for activating factors, including trxG members, to bind to DNA. Is it also needed for Pc-G proteins to bind and exercise repression? This is not yet clear, but such a model would demand that something other than GAGA determines which of the alternative types of complex subsequently assemble at the site.

31.5 X Chromosomes Undergo Global Changes

Key concepts

- One of the two X chromosomes is inactivated at random in each cell during embryogenesis of eutherian mammals.
- In exceptional cases where there are >2 X chromosomes, all but one are inactivated.
- The *Xic* (X inactivation center) is a *cis*-acting region on the X chromosome that is necessary and sufficient to ensure that only one X chromosome remains active.
- *Xic* includes the *Xist* gene, which codes for an RNA that is found only on inactive X chromosomes.
- The mechanism that is responsible for preventing *Xist* RNA from accumulating on the active chromosome is unknown.

Dosage compensation changes X-expression

	Mammals	Flies	Worms
	Inactivate one ♀ X	Double expression ♂ X	Halve expression two ♀ X
X X			
X Y			

FIGURE 31.10 Different means of dosage compensation are used to equalize X chromosome expression in male and female.

Sex presents an interesting problem for gene regulation, because of the variation in the number of X chromosomes. If X-linked genes were expressed equally well in each sex, females would have twice as much of each product as males. The importance of avoiding this situation is shown by the existence of **dosage compensation,** which equalizes the level of expression of X-linked genes in the two sexes. Mechanisms used in different species are summarized in FIGURE 31.10:

- In mammals, one of the two female X chromosomes is inactivated completely. The result is that females have only one active X chromosome, which is the same situation found in males. The active X chromosome of females and the single X chromosome of males are expressed at the same level.
- In *Drosophila,* the expression of the single male X chromosome is doubled relative to the expression of each female X chromosome.
- In *Caenorhabditis elegans,* the expression of each female X chromosome is halved relative to the expression of the single male X chromosome.

The common feature in all these mechanisms of dosage compensation is that *the entire chromosome is the target for regulation.* A global change occurs that quantitatively affects all of the promoters on the chromosome. We know most about the inactivation of the X chromosome in mammalian females, where the entire chromosome becomes heterochromatic.

The twin properties of heterochromatin are its condensed state and associated inactivity. It can be divided into two types:

- **Constitutive heterochromatin** contains specific sequences that have no coding function. In general these include satellite DNAs, and they are often found at the centromeres. These regions are invariably heterochromatic because of their intrinsic nature.

- **Facultative heterochromatin** takes the form of entire chromosomes that are inactive in one cell lineage, although they can be expressed in other lineages. The example *par excellence* is the mammalian X chromosome. The inactive X chromosome is perpetuated in a heterochromatic state, whereas the active X chromosome is part of the euchromatin. Thus *identical DNA sequences are involved in both states.* Once the inactive state has been established, it is inherited by descendant cells. This is an example of epigenetic inheritance, because it does not depend on the DNA sequence.

Our basic view of the situation of the female mammalian X chromosomes was formed by the **single X hypothesis** in 1961. Female mice that are heterozygous for X-linked coat color mutations have a variegated phenotype in which some areas of the coat are wild-type, but others are mutant. FIGURE 31.11 shows that this can be explained *if one of the two X chromosomes is inactivated at random in each cell of a small precursor population.* Cells in which the X chromosome carrying the wild-type gene is inactivated give rise to progeny that express only the mutant allele on the active chromosome. Cells derived from a precursor where the other chromosome was inactivated have an active wild-type gene. In the case of coat color, cells descended from a particular precursor stay together and thus form a patch of the same color, creating the pattern of visible variegation. In other cases, individual cells in a population will express one or the other of X-linked alleles; for example, in heterozygotes for the X-linked locus G6PD, any particular red blood cell will express only one of the two allelic forms. (Random inactivation of one X chromosome occurs in eutherian mammals. In marsupials, the choice is directed: It is always the X chromosome inherited from the father that is inactivated.)

Inactivation of the X chromosome in females is governed by the **n–1 rule:** However many X chromosomes are present, all but one will be inactivated. In normal females there are of course two X chromosomes, but in rare cases where nondisjunction has generated a 3X or greater genotype, only one X chromosome remains active. This suggests a general model in which a specific event is limited to one X chromosome and protects it from an inactivation mechanism that applies to all the others.

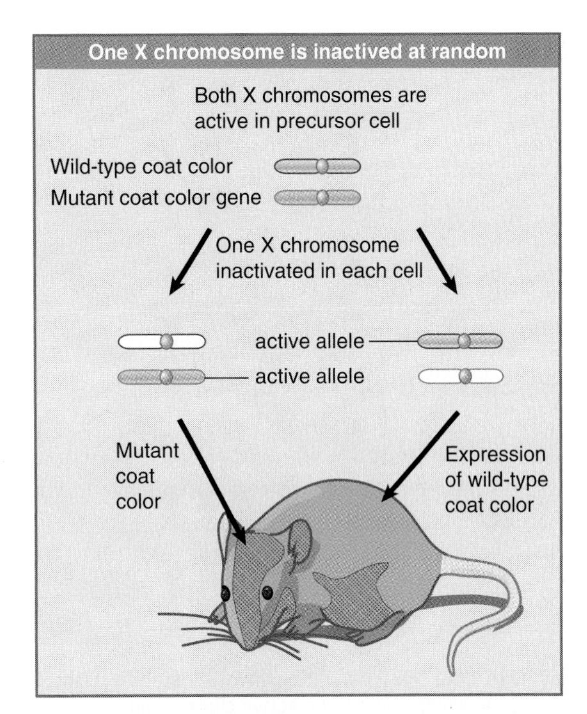

FIGURE 31.11 X-linked variegation is caused by the random inactivation of one X chromosome in each precursor cell. Cells in which the + allele is on the active chromosome have wild phenotype; cells in which the – allele is on the active chromosome have mutant phenotype.

A single locus on the X chromosome is sufficient for inactivation. When a translocation occurs between the X chromosome and an autosome, this locus is present on only one of the reciprocal products, and only that product can be inactivated. By comparing different translocations, it is possible to map this locus, which is called the *Xic* (X-inactivation center). A cloned region of 450 kb contains all the properties of the *Xic.* When this sequence is inserted as a transgene on to an autosome, the autosome becomes subject to inactivation (in a cell culture system).

Xic is a *cis*-acting locus that contains the information necessary to count X chromosomes and inactivate all copies but one. Inactivation spreads from *Xic* along the entire X chromosome. When *Xic* is present on an X chromosome-autosome translocation, inactivation spreads into the autosomal regions (although the effect is not always complete).

Xic contains a gene, called *Xist,* that is expressed only on the *inactive* X chromosome. The behavior of this gene is effectively the opposite from all other loci on the chromosome, which are turned off. Deletion of *Xist* prevents

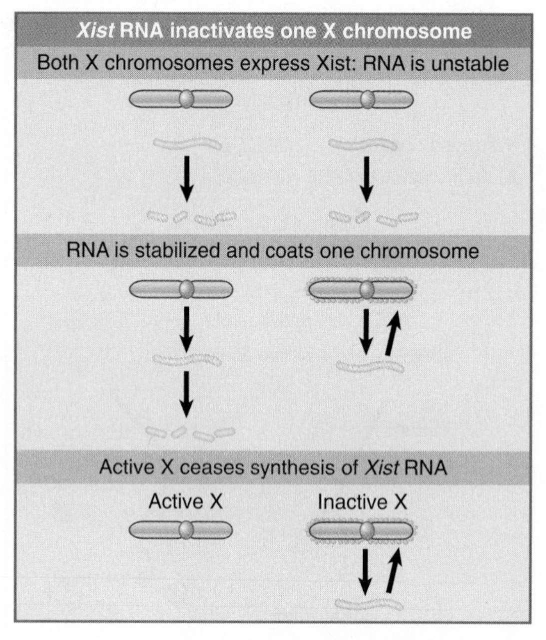

FIGURE 31.12 X-inactivation involves stabilization of *Xist* RNA, which coats the inactive chromosome.

an X chromosome from being inactivated. It does not, however, interfere with the counting mechanism (because other X chromosomes can be inactivated). Thus we can distinguish two features of *Xic:* an unidentified element(s) required for counting, and the *Xist* gene required for inactivation.

FIGURE 31.12 illustrates the role of *Xist* RNA in X-inactivation. *Xist* codes for an RNA that lacks open reading frames. The *Xist* RNA "coats" the X chromosome from which it is synthesized, which suggests that it has a structural role. Prior to X-inactivation, it is synthesized by both female X chromosomes. Following inactivation, the RNA is found only on the inactive X chromosome. The transcription rate remains the same before and after inactivation, so the transition depends on posttranscriptional events.

Prior to X-inactivation, *Xist* RNA decays with a half life of ~2 hours. X-inactivation is mediated by stabilizing the *Xist* RNA on the inactive X chromosome. The *Xist* RNA shows a punctate distribution along the X chromosome, which suggests that association with proteins to form particulate structures may be the means of stabilization. We do not know yet what other factors may be involved in this reaction and how the *Xist* RNA is limited to spreading in *cis* along the chromosome. The characteristic features of the inactive X chromosome, which include a lack of acetylation of histone H4, and methylation of CpG sequences (see Section 24.19, CpG Islands Are Regulatory Targets), presumably occur later as part of the mechanism of inactivation.

The n–1 rule suggests that stabilization of *Xist* RNA is the "default," and that some blocking mechanism prevents stabilization at one X chromosome (which will be the active X). This means that, although *Xic* is necessary and sufficient for a chromosome to be *inactivated,* the products of other loci may be necessary for the establishment of an *active* X chromosome.

Silencing of *Xist* expression is necessary for the active X. Deletion of the gene for DNA methyltransferase prevents silencing of *Xist,* probably because methylation at the *Xist* promoter is necessary for cessation of transcription.

31.6 Chromosome Condensation Is Caused by Condensins

Key concepts

- SMC proteins are ATPases that include the condensins and the cohesins.
- A heterodimer of SMC proteins associates with other subunits.
- The condensins cause chromatin to be more tightly coiled by introducing positive supercoils into DNA.
- Condensins are responsible for condensing chromosomes at mitosis.
- Chromosome-specific condensins are responsible for condensing inactive X chromosomes in *C. elegans.*

The structures of entire chromosomes are influenced by interactions with proteins of the **SMC (structural maintenance of chromosome)** family. They are ATPases that fall into two functional groups. **Condensins** are involved with the control of overall structure, and are responsible for the condensation into compact chromosomes at mitosis. **Cohesins** are concerned with connections between sister chromatids that must be released at mitosis. Both consist of dimers formed by SMC proteins. Condensins form complexes that have a core of the heterodimer SMC2-SMC4 associated with other (non-SMC) proteins. Cohesins have a similar organization based on the heterodimeric core of SMC1-SMC3.

FIGURE 31.13 shows that an SMC protein has a coiled-coil structure in its center that is interrupted by a flexible hinge region. Both the amino and carboxyl termini have ATP- and DNA-binding motifs. Different models have been proposed for the actions of these proteins

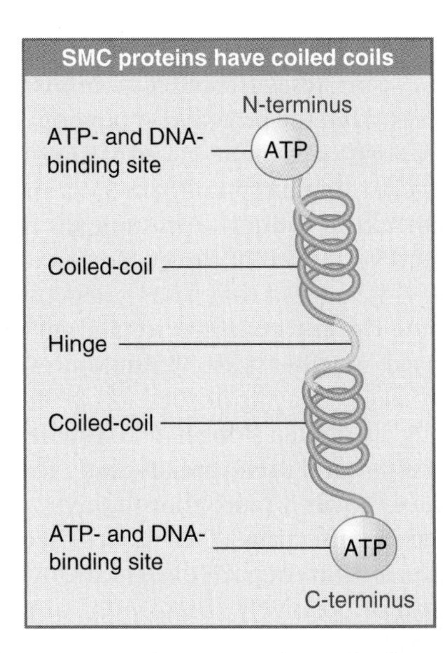

FIGURE 31.13 An SMP protein has a "Walker module" with an ATP-binding motif and DNA-binding site at each end, which are connected by coiled coils that are linked by a hinge region.

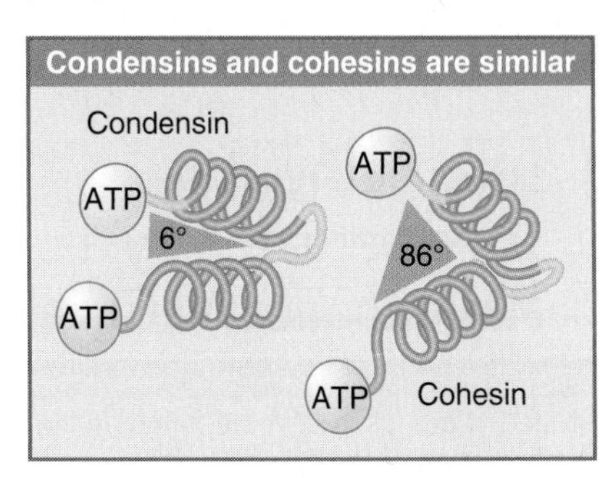

FIGURE 31.14 The two halves of a condensin are folded back at an angle of 6°. Cohesins have a more open conformation with an angle of 86° between the two halves.

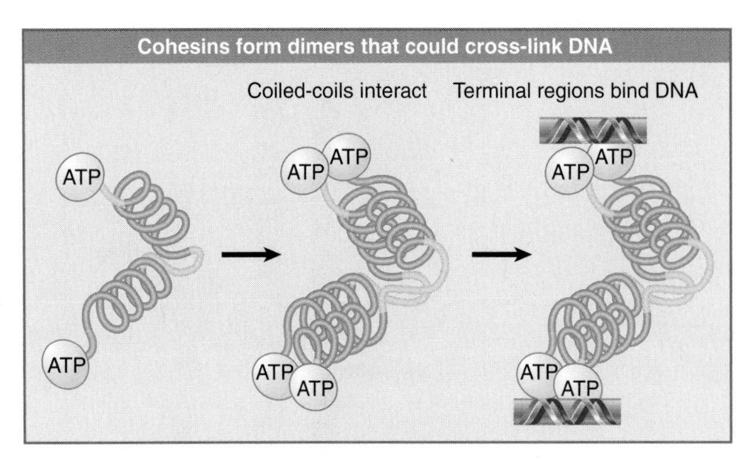

FIGURE 31.15 SMC proteins dimerize by antiparallel interactions between the central coiled coils. Both terminal regions of each subunit have ATP- and DNA-binding motifs. Cohesins may form an extended structure that allows two different DNA molecules to be linked.

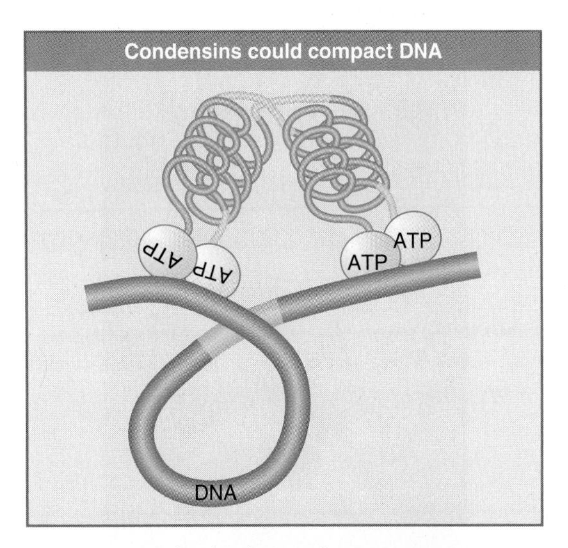

FIGURE 31.16 Condensins may form a compact structure by bending at the hinge, causing DNA to become compacted.

depending on whether they dimerize by intra- or intermolecular interactions.

Experiments with the bacterial homologs of the SMC proteins suggest that a dimer is formed by an antiparallel interaction between the coiled coils, so that the N-terminus of one subunit bonds to the C-terminus of the other subunit. The existence of a flexible hinge region could allow cohesins and condensins to depend on a different mode of action by the dimer. FIGURE 31.14 shows that cohesins have a V-shaped structure, with the arms separated by an 86° angle, whereas condensins are more sharply bent back, with only 6° between the arms. This enables cohesins to hold sister chromatids together, whereas condensins instead condense an individual chromosome. FIGURE 31.15 shows that a cohesin could take the form of an extended dimer that crosslinks two DNA molecules. FIGURE 31.16 shows that a condensin could take the form of a V-shaped dimer—essentially bent at the hinge—that pulls together distant sites on the same DNA molecule, causing it to condense.

An alternative model is suggested by experiments to suggest that the yeast proteins dimerize by intramolecular interactions, that is, a homodimer is formed solely by interaction between two identical subunits. Dimers of two different proteins (in this case, SMC1 and SMC3) may then interact at both their head and hinge regions to form a circular structure as illustrated in FIGURE 31.17. Instead of binding directly to DNA, a structure of this type could hold DNA molecules together by encircling them.

Visualization of mitotic chromosomes shows that condensins are located all along the length of the chromosome, as can be seen in FIGURE 31.18. (By contrast, cohesins are found at discrete locations.) The condensin complex was named for its ability to cause chromatin to

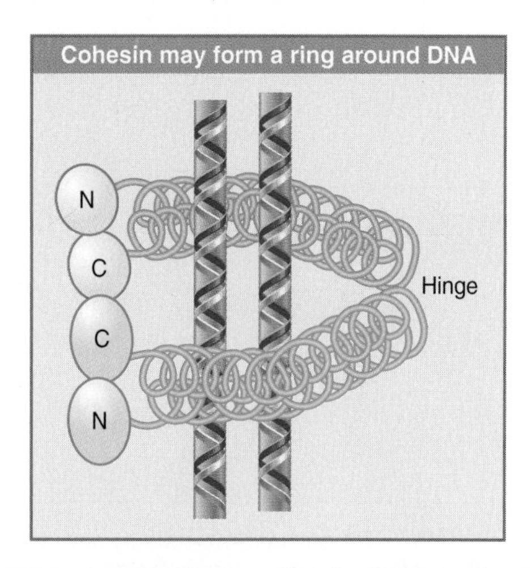

FIGURE 31.17 Cohesins may dimerize by intramolecular connections, and then form multimers that are connected at the heads and at the hinge. Such a structure could hold two molecules of DNA together by surrounding them.

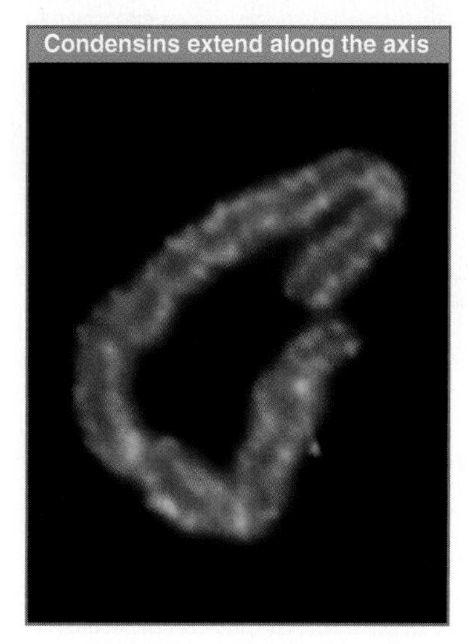

FIGURE 31.18 Condensins are located along the entire length of a mitotic chromosome. DNA is red; condensins are yellow. Photo courtesy of Ana Losada and Tatsuya Hirano.

condense *in vitro*. It has an ability to introduce positive supercoils into DNA in an action that uses hydrolysis of ATP and depends on the presence of topoisomerase I. This ability is controlled by the phosphorylation of the non-SMC subunits, which occurs at mitosis. We do not know yet how this connects with other modifications of chromatin—for example, the phosphorylation of histones. The activation of the condensin complex specifically at mitosis makes it questionable whether it is also involved in the formation of interphase heterochromatin.

Global changes occur in other types of dosage compensation. In *Drosophila,* a complex of proteins is found in males, where it localizes on the X chromosome. In *C. elegans,* a protein complex associates with both X chromosomes in XX embryos, but the protein components remain diffusely distributed in the nuclei of XO embryos. The protein complex contains an SMC core, and is similar to the condensin complexes that are associated with mitotic chromosomes in other species. This suggests that it has a structural role in causing the chromosome to take up a more condensed, inactive state. Multiple sites on the X chromosome may be needed for the complex to be fully distributed along it. The complex binds to these sites, and then spreads along the chromosome to cover it more thoroughly.

Changes affecting all the genes on a chromosome, either negatively (mammals and *C. elegans*) or positively *(Drosophila),* are therefore a common feature of dosage compensation. The components of the dosage compensation apparatus may vary, however, as well as the means by which it is localized to the chromosome, and of course its mechanism of action is different in each case.

31.7 DNA Methylation Is Perpetuated by a Maintenance Methylase

Key concepts

- Most methyl groups in DNA are found on cytosine on both strands of the CpG doublet.
- Replication converts a fully methylated site to a hemimethylated site.
- Hemimethylated sites are converted to fully methylated sites by a maintenance methylase.

Methylation of DNA occurs at specific sites. In bacteria, it is associated with identifying the bacterial restriction-methylation system used for phage defense, and also with distinguishing replicated and nonreplicated DNA (see Section 20.7, Controlling the Direction of Mismatch Repair). In eukaryotes, its principal known function is connected with the control of transcription; methylation is associated with gene inactivation (see Section 24.18, Gene Expression Is Associated with Demethylation).

From 2% to 7% of the cytosines of animal cell DNA are methylated (the value varies with the species). Most of the methyl groups are found in CG "doublets," and, in fact, the majority of the CG sequences are methylated. Usually the C residues on both strands of this short palindromic sequence are methylated, giving the structure.

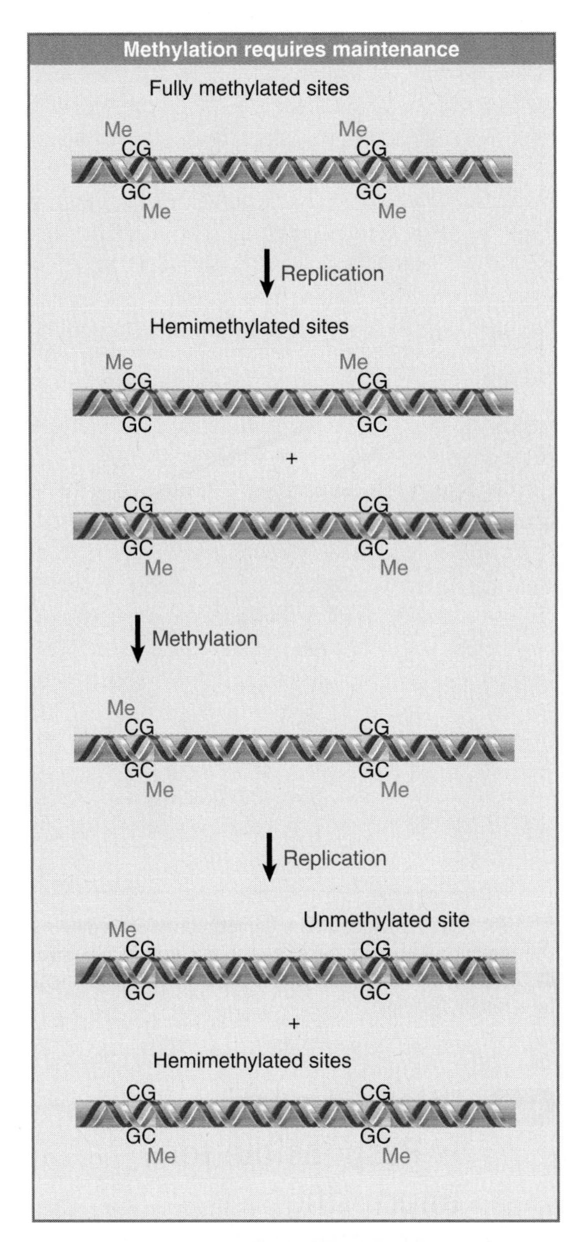

FIGURE 31.19 The state of methylated sites could be perpetuated by an enzyme that recognizes only hemimethylated sites as substrates.

Such a site is described as **fully methylated.** Consider, though, the consequences of replicating this site. FIGURE 31.19 shows that each daughter duplex has one methylated strand and one unmethylated strand. Such a site is called **hemimethylated.**

The perpetuation of the methylated site now depends on what happens to hemimethylated DNA. If methylation of the unmethylated strand occurs, the site is restored to the fully methylated condition. If replication occurs first, though, the hemimethylated condition will be perpetuated on one daughter duplex, but the site will become unmethylated on the other daughter duplex. FIGURE 31.20 shows that the state of methylation of DNA is controlled by **methylases,** which add methyl groups to the 5 position of cytosine, and **demethylases,** which remove the methyl groups. (The more formal name for the enzymes uses **methyltransferase** as the description.)

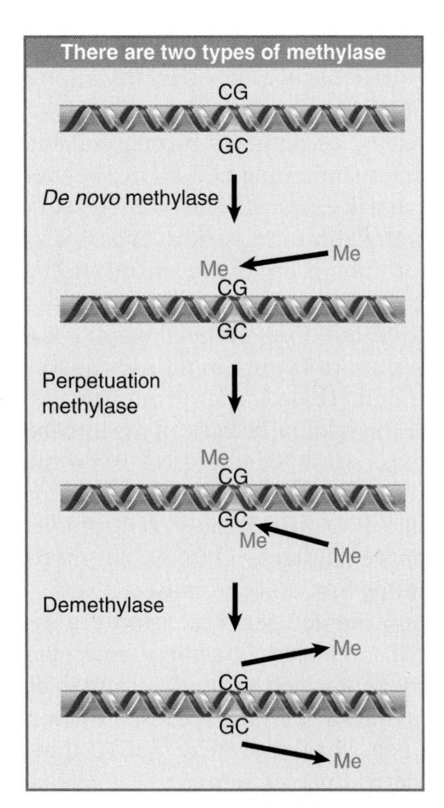

FIGURE 31.20 The state of methylation is controlled by three types of enzyme. *De novo* and perpetuation methylases are known, but demethylases have not been identified.

There are two types of DNA methylase, whose actions are distinguished by the state of the methylated DNA. To modify DNA at a new position requires the action of the ***de novo* methylase,** which recognizes DNA by virtue of a specific sequence. It acts *only* on nonmethylated DNA, to add a methyl group to one strand. There are two *de novo* methylases (Dnmt3A and Dnmt3B) in mouse; they have different target sites, and both are essential for development.

A **maintenance methylase** acts constitutively *only on hemimethylated sites* to convert them to fully methylated sites. Its existence means that any methylated site is perpetuated after replication. There is one maintenance methylase (Dnmt1) in mouse, and it is essential: mouse embryos in which its gene has been disrupted do not survive past early embryogenesis.

Maintenance methylation is virtually 100% efficient, ensuring that the situation shown on the left of Figure 31.19 usually prevails *in vivo*. The result is that, if a *de novo* methylation occurs on one allele but not on the other, this difference will be perpetuated through ensuing cell divisions, maintaining a difference between the alleles that does not depend on their sequences.

Methylation has various types of targets. Gene promoters are the most common target. The promoters are methylated when the gene is inactive, but unmethylated when it is active. The absence of Dnmt1 in mouse causes widespread demethylation at promoters, and we assume this is lethal because of the uncontrolled gene expression. Satellite DNA is another target. Mutations in Dnmt3B prevent methylation of satellite DNA, which causes centromere instability at the cellular level. Mutations in the corresponding human gene cause a disease called ICF (immunodeficiency/centromere instability, facial anomalies). The importance of methylation is emphasized by another human disease, Rhett syndrome, which is caused by mutation of the gene for the protein MeCP2 that binds methylated CpG sequences.

The methylases are conventional enzymes that act on a DNA target. There may, however, also be a methylation system that uses a short RNA sequence to target a corresponding DNA sequence for methylation (see Section 13.6, Antisense RNA Can Be Used to Inactivate Gene Expression) Nothing is known about the mechanism of operation of this system.

How are demethylated regions established and maintained? If a DNA site has not been methylated, a protein that recognizes the unmethylated sequence could protect it against methylation. Once a site has been methylated, there are two possible ways to generate demethylated sites. One is to block the maintenance methylase from acting on the site when it is replicated. After a second replication cycle, one of the daughter duplexes will be unmethylated (as shown on the right side of Figure 31.19). The other is actively to demethylate the site, as shown in FIGURE 31.21, either by removing the methyl group directly from cytosine, or by excising the methylated cytosine or cytidine from DNA for replacement by a repair system. We know that active demethylation can occur to the paternal genome soon after fertilization, but we do not know what mechanism is used. One interesting possibility is that the cytidine deaminase AID may be involved; it can deaminate methylated C residues, creating a mismatched base pair that a repair system might then correct to a standard (unmethylated) C-G pair.

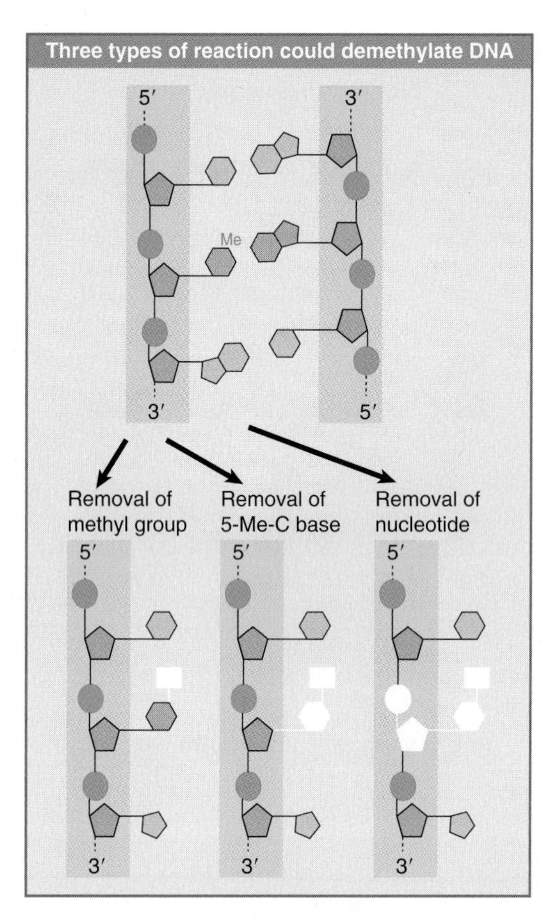

FIGURE 31.21 DNA could be demethylated by removing the methyl group, the base, or the nucleotide. Removal of the base or nucleotide would require its replacement by a repair system.

31.8 DNA Methylation Is Responsible for Imprinting

Key concepts

- Paternal and maternal alleles may have different patterns of methylation at fertilization.
- Methylation is usually associated with inactivation of the gene.
- When genes are differentially imprinted, survival of the embryo may require that the functional allele is provided by the parent with the unmethylated allele.
- Survival of heterozygotes for imprinted genes is different, depending on the direction of the cross.
- Imprinted genes occur in clusters and may depend on a local control site where *de novo* methylation occurs unless specifically prevented.

The pattern of methylation of germ cells is established in each sex during gametogenesis by a two-stage process: First the existing pattern is erased by a genome-wide demethylation, and then the pattern specific for each sex is imposed.

All allelic differences are lost when primordial germ cells develop in the embryo; irrespective of sex, the previous patterns of methylation are erased, and a typical gene is then unmethylated. In males, the pattern develops in two stages. The methylation pattern that is characteristic of mature sperm is established in the spermatocyte, but further changes are made in this pattern after fertilization. In females, the maternal pattern is imposed during oogenesis, when oocytes mature through meiosis after birth.

As may be expected from the inactivity of genes in gametes, the typical state is to be methylated. There are cases of differences between the two sexes, though, for which a locus is unmethylated in one sex. A major question is how the specificity of methylation is determined in the male and female gametes.

Systematic changes occur in early embryogenesis. Some sites will continue to be methylated, whereas others will be specifically unmethylated in cells in which a gene is expressed. From the pattern of changes, we may infer that individual sequence-specific demethylation events occur during somatic development of the organism as particular genes are activated.

The specific pattern of methyl groups in germ cells is responsible for the phenomenon of **imprinting,** which describes a difference in behavior between the alleles inherited from each parent. The expression of certain genes in mouse embryos depends upon the sex of the parent from which they were inherited. For example, the allele coding for IGF-II (insulin-like growth factor II) that is inherited from the father is expressed, but the allele that is inherited from the mother is not expressed. The IGF-II gene of oocytes is methylated, but the IGF-II gene of sperm is not methylated, so that the two alleles behave differently in the zygote. This is the most common pattern, but the dependence on sex is reversed for some genes. In fact, the opposite pattern (expression of maternal copy) is shown for IGF-IIR, the receptor for IGF-II.

This sex-specific mode of inheritance requires that the pattern of methylation is established specifically during each gametogenesis. The fate of a hypothetical locus in a mouse is illustrated in FIGURE 31.22. In the early embryo, the paternal allele is nonmethylated and expressed, and the maternal allele is methylated and silent. What happens when this mouse itself forms gametes? If it is a male, the allele contributed to the sperm must be nonmethylated, irrespective of whether it was originally methylated or not. Thus when the maternal allele finds itself in a sperm, it must be demethylated. If the mouse is a female, the allele contributed to the egg must be methylated; if it was originally the paternal allele, methyl groups must be added.

The consequence of imprinting is that an embryo requires a paternal allele for this gene. Thus in the case of a heterozygous cross where the allele of one parent has an inactivating mutation, the embryo will survive if the wild-type allele comes from the father, but will die if the wild-type allele is from the mother. This type of dependence on the directionality of the cross (in contrast with Mendelian genetics) is an example of epigenetic inheritance, where some factor other than the sequences of the genes themselves influences their effects (see Section 31.10, Epigenetic Effects Can Be Inherited). Although the paternal and maternal alleles have identical sequences, they display different properties, depending on which parent provided them. These properties are inherited through meiosis and the subsequent somatic mitoses.

Imprinted genes are sometimes clustered. More than half of the 17 known imprinted genes in mouse are contained in two particular regions, each containing both maternally and paternally expressed genes. This suggests the possibility that imprinting mechanisms may function over long distances. Some insights into

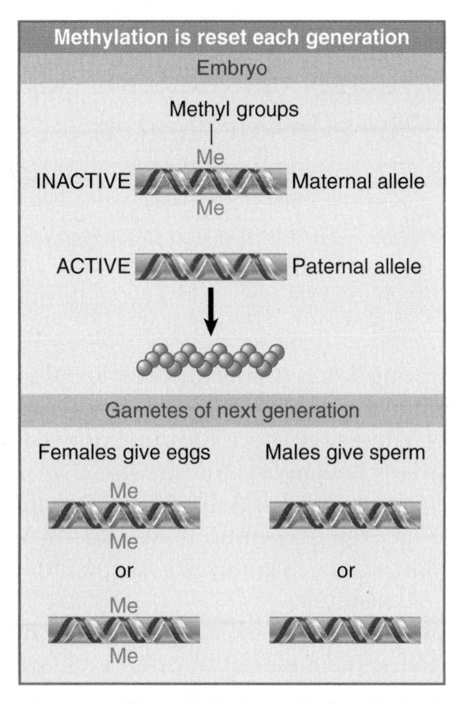

FIGURE 31.22 The typical pattern for imprinting is that a methylated locus is inactive. If this is the maternal allele, only the paternal allele is active, and will be essential for viability. The methylation pattern is reset when gametes are formed, so that all sperm have the paternal type and all oocytes have the maternal type.

this possibility come from deletions in the human population that cause the Prader–Willi and Angelman diseases. Most cases are caused by the same 4 Mb deletion, but the syndromes are different, depending on which parent contributed the deletion. The reason is that the deleted region contains at least one gene that is paternally imprinted and at least one that is maternally imprinted. There are some rare cases, however, with much smaller deletions. Prader–Willi syndrome can be caused by a 20 kb deletion that silences genes that are distant on either side of it. The basic effect of the deletion is to prevent a father from resetting the paternal mode to a chromosome inherited from his mother. The result is that these genes remain in maternal mode, so that the paternal as well as maternal alleles are silent in the offspring. The inverse effect is found in some small deletions that cause Angelman's syndrome. The implication is that this region comprises some sort of "imprint center" that acts at a distance to switch one parental type to the other.

Methylation is also responsible for epigenetic effects that control the expression of rRNA genes. The phenomenon of **nucleolar dominance** describes the transcription of only one set of parental rRNA genes. It results from methylation at cytosines in the promoters for the genes inherited from one parent and not the other.

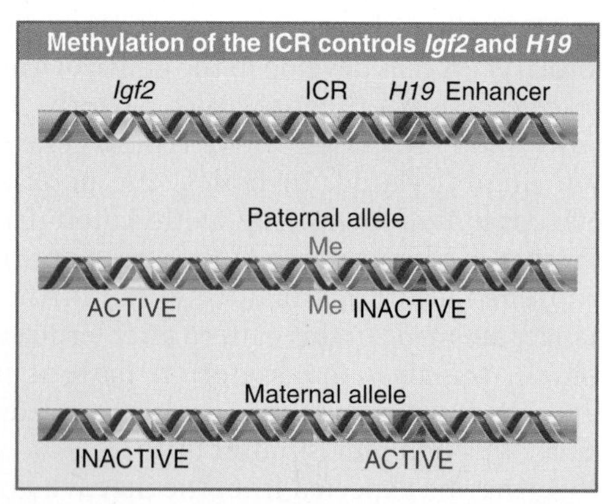

FIGURE 31.23 ICR is methylated on the paternal allele, where *Igf2* is active and *H19* is inactive. ICR is unmethylated on the maternal allele, where *Igf2* is inactive and *H19* is active.

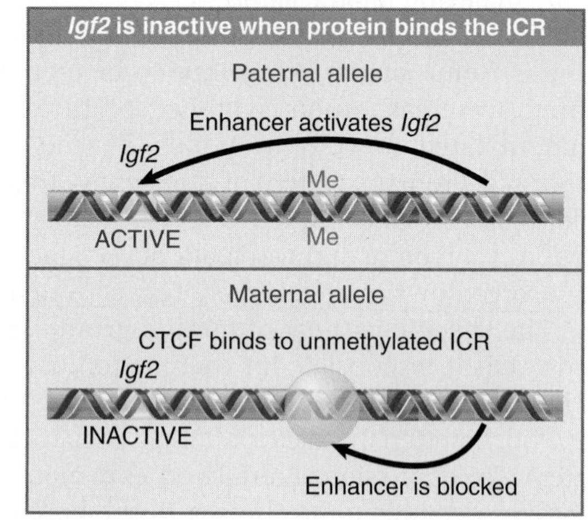

FIGURE 31.24 The ICR is an insulator that prevents an enhancer from activating *Igf2*. The insulator functions only when it binds CTCF to unmethylated DNA.

31.9 Oppositely Imprinted Genes Can Be Controlled by a Single Center

Key concepts

- Imprinted genes are controlled by methylation of *cis*-acting sites.
- Methylation may be responsible for either inactivating or activating a gene.

Imprinting is determined by the state of methylation of a *cis*-acting site near a target gene or genes. These regulatory sites are known as differentially methylated domains (DMDs) or imprinting control regions (ICRs). Deletion of these sites removes imprinting, and the target loci then behave the same in both maternal and paternal genomes.

The behavior of a region containing two genes, *Igf2* and *H19*, illustrates the ways in which methylation can control gene activity. FIGURE 31.23 shows that these two genes react oppositely to the state of methylation at the ICR located between them. The ICR is methylated on the paternal allele. *H19* shows the typical response of inactivation. Note, however, that *Igf2* is expressed. The reverse situation is found on a maternal allele, where the ICR is not methylated. *H19* now becomes expressed, but *Igf2* is inactivated.

The control of *Igf2* is exercised by an insulator function of the ICR. FIGURE 31.24 shows that when the ICR is unmethylated, it binds the protein CTCF. This creates an insulator function that blocks an enhancer from activating the *Igf2* promoter. This is an unusual effect in which methylation indirectly activates a gene by blocking an insulator.

The regulation of *H19* shows the more usual direction of control in which methylation creates an inactive imprinted state. This could reflect a direct effect of methylation on promoter activity.

31.10 Epigenetic Effects Can Be Inherited

Key concept

- Epigenetic effects can result from modification of a nucleic acid after it has been synthesized or by the perpetuation of protein structures.

Epigenetic inheritance describes the ability of different states, which may have different phenotypic consequences, to be inherited without any change in the sequence of DNA. How can this occur? We can divide epigenetic mechanisms into two general classes:

- DNA may be modified by the covalent attachment of a moiety that is then perpetuated. Two alleles with the same sequence may have different states of methylation that confer different properties.
- A self perpetuating protein state may be established. This might involve assembly of a protein complex, modification of specific protein(s), or establishment of an alternative protein conformation.

Methylation establishes epigenetic inheritance so long as the maintenance methylase acts constitutively to restore the methylated state after each cycle of replication, as shown in Figure 31.19. A state of methylation can be perpetuated through an indefinite series of somatic mitoses. This is probably the "default" situation. Methylation can also be perpetuated through meiosis: for example, in the fungus *Ascobolus* there are epigenetic effects that can be transmitted through both mitosis and meiosis by maintaining the state of methylation. In mammalian cells, epigenetic effects are created by resetting the state of methylation differently in male and female meioses.

Situations in which epigenetic effects appear to be maintained by means of protein states are less well understood in molecular terms. Position effect variegation shows that constitutive heterochromatin may extend for a variable distance, and the structure is then perpetuated through somatic divisions. There is no methylation of DNA in *Saccharomyces* and a vanishingly small amount in *Drosophila,* and as a result the inheritance of epigenetic states of position effect variegation or telomeric silencing in these organisms is likely to be due to the perpetuation of protein structures.

FIGURE 31.25 considers two extreme possibilities for the fate of a protein complex at replication.

- A complex could perpetuate itself if it splits symmetrically, so that half complexes associate with each daughter duplex. If the half complexes have the capacity to nucleate formation of full complexes, the original state will be restored. This is basically analogous to the maintenance of methylation. The problem with this model is that there is no evident reason why protein complexes should behave in this way.
- A complex could be maintained as a unit and segregate to one of the two daughter duplexes. The problem with this model is that it requires a new complex to be assembled *de novo* on the other daughter duplex, and it is not evident why this should happen.

Consider now the need to perpetuate a heterochromatic structure consisting of protein complexes. Suppose that a protein is distributed more or less continuously along a stretch of heterochromatin, as implied in Figure 31.4. If individual subunits are distributed at random to each daughter duplex at replication, the two daughters will continue to be marked by the protein, although its density will be reduced to half of the level before replication. If the protein has a self-assembling property that causes

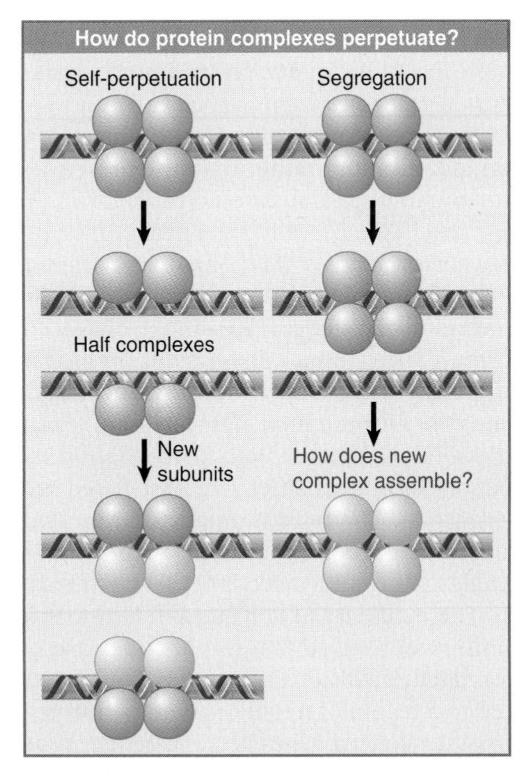

FIGURE 31.25 What happens to protein complexes on chromatin during replication?

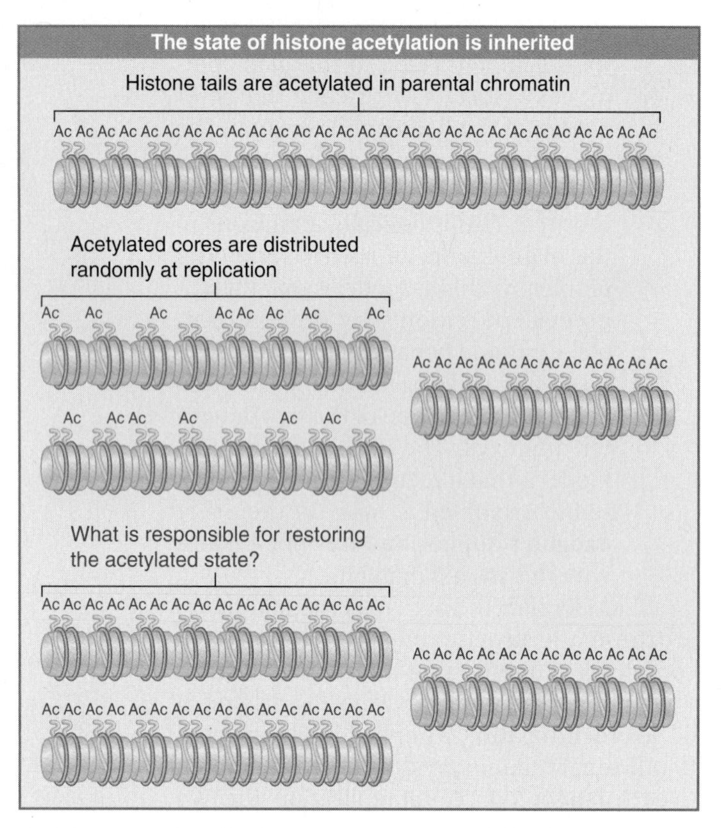

FIGURE 31.26 Acetylated cores are conserved and distributed at random to the daughter chromatin fibers at replication. Each daughter fiber has a mixture of old (acetylated) cores and new (unacetylated) cores.

new subunits to associate with it, the original situation may be restored. *Basically, the existence of epigenetic effects forces us to the view that a protein responsible for such a situation must have some sort of self-templating or self-assembling capacity.*

In some cases, it may be the state of protein modification, rather than the presence of the protein *per se,* that is responsible for an epigenetic effect. There is a general correlation between the activity of chromatin and the state of acetylation of the histones, in particular the acetylation of histones H3 and H4, which occurs on their N-terminal tails. Activation of transcription is associated with acetylation in the vicinity of the promoter; and repression of transcription is associated with deacetylation (see Section 30.7, Acetylases Are Associated with Activators). The most dramatic correlation is that the inactive X chromosome in mammalian female cells is underacetylated on histone H4.

The inactivity of constitutive heterochromatin may require that the histones are not acetylated. If a histone acetyltransferase is tethered to a region of telomeric heterochromatin in yeast, silenced genes become active. When yeast is exposed to trichostatin (an inhibitor of deacetylation), centromeric heterochromatin becomes acetylated, and silenced genes in centromeric regions may become active. *The effect may persist even after trichostatin has been removed.* In fact, it may be perpetuated through mitosis and meiosis. This suggests that an epigenetic effect has been created by changing the state of histone acetylation.

How might the state of acetylation be perpetuated? Suppose that the $H3_2$-$H4_2$ tetramer is distributed at random to the two daughter duplexes. This creates the situation shown in FIGURE 31.26, in which each daughter duplex contains some histone octamers that are fully acetylated on the H3 and H4 tails, whereas others are completely unacetylated. To account for the epigenetic effect, we could suppose that the presence of some fully acetylated histone octamers provides a signal that causes the unacetylated octamers to be acetylated.

(The actual situation is probably more complicated than the one shown in the figure, because transient acetylations occur during replication. If they are simply reversed following deposition of histones into nucleosomes, they may be irrelevant. An alternative possibility is that the usual deacetylation is prevented, instead of, or as well as, inducing acetylation.)

31.11 Yeast Prions Show Unusual Inheritance

Key concepts

- The Sup35 protein in its wild-type soluble form is a termination factor for translation.
- It can also exist in an alternative form of oligomeric aggregates, in which it is not active in protein synthesis.
- The presence of the oligomeric form causes newly synthesized protein to acquire the inactive structure.
- Conversion between the two forms is influenced by chaperones.
- The wild-type form has the recessive genetic state *psi*$^-$ and the mutant form has the dominant genetic state *PSI*$^+$.

One of the clearest cases of the dependence of epigenetic inheritance on the condition of a protein is provided by the behavior of **prions.** They have been characterized in two circumstances: by genetic effects in yeast, and as the causative agents of neurological diseases in mammals, including human beings. A striking epigenetic effect is found in yeast, where two different states can be inherited that map to a single genetic locus, *although the sequence of the*

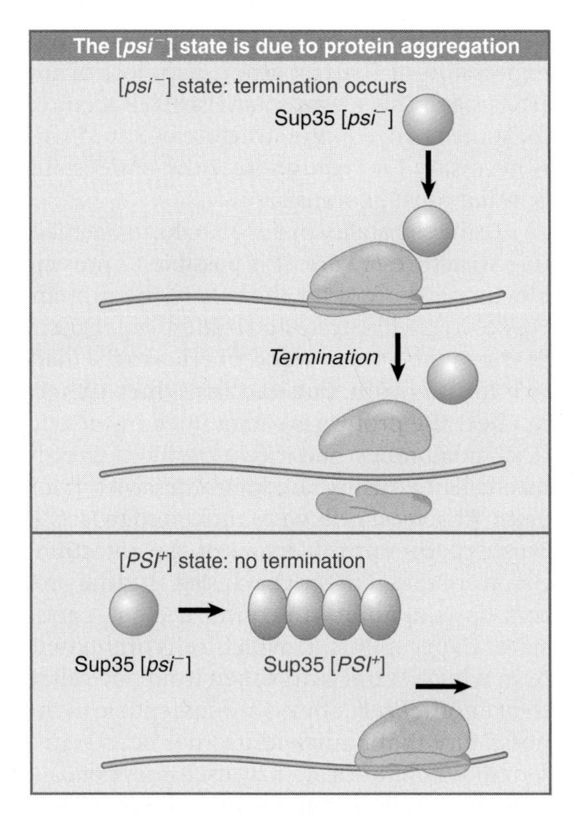

FIGURE 31.27 The state of the Sup35 protein determines whether termination of translation occurs.

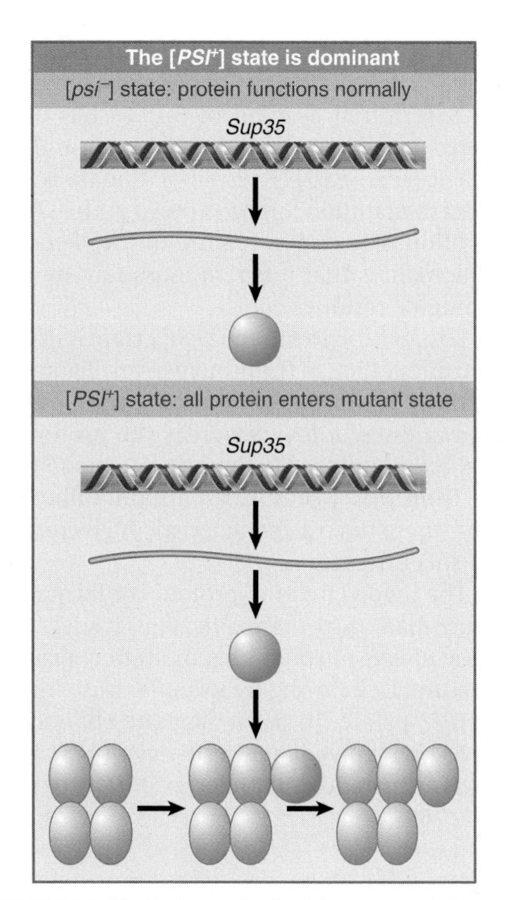

FIGURE 31.28 Newly synthesized Sup35 protein is converted into the [PSI⁺] state by the presence of preexisting [PSI⁺] protein.

gene is the same in both states. The two different states are [*psi*⁻] and [*PSI*⁺]. A switch in condition occurs at a low frequency as the result of a spontaneous transition between the states.

The [*psi*] genotype maps to the locus *SUP35,* which codes for a translation termination factor. FIGURE 31.27 summarizes the effects of the Sup35 protein in yeast. In wild-type cells, which are characterized as [*psi*⁻], the gene is active, and Sup35 protein terminates protein synthesis. In cells of the mutant [*PSI*⁺] type, the factor does not function, which causes a failure to terminate protein synthesis properly. (This was originally detected by the lethal effects of the enhanced efficiency of suppressors of ochre codons in [*PSI*⁺] strains.)

[*PSI*⁺] strains have unusual genetic properties. When a [*psi*⁻] strain is crossed with a [*PSI*⁺] strain, *all of the progeny are* [*PSI*⁺]. This is a pattern of inheritance that would be expected of an extrachromosomal agent, but the [*PSI*⁺] trait cannot be mapped to any such nucleic acid. The [*PSI*⁺] trait is metastable, which means that, although it is inherited by most progeny, it is lost at a higher rate than is consistent with mutation. Similar behavior is shown also by the locus *URE2,* which codes for a protein required for nitrogen-mediated repression of certain catabolic enzymes. When a yeast strain is converted into an alternative state, called [*URE3*], the Ure2 protein is no longer functional.

The [*PSI*⁺] state is determined by the conformation of the Sup35 protein. In a wild-type [*psi*⁻] cell, the protein displays its normal function. In a [*PSI*⁺] cell, though, the protein is present in an alternative conformation in which its normal function has been lost. To explain the unilateral dominance of [*PSI*⁺] over [*psi*⁻] in genetic crosses, we must suppose that *the presence of protein in the* [*PSI*⁺] *state causes all the protein in the cell to enter this state.* This requires an interaction between the [*PSI*⁺] protein and newly synthesized protein, which probably reflects the generation of an oligomeric state in which the [*PSI*⁺] protein has a nucleating role, as illustrated in FIGURE 31.28.

A feature common to both the Sup35 and Ure2 proteins is that each consists of two domains that function independently. The C-terminal domain is sufficient for the activity of the protein. The N-terminal domain is sufficient for formation of the structures that make the protein inactive. Thus yeast in which the N-terminal domain of Sup35 has been deleted cannot acquire the [*PSI*$^+$] state, and the presence of a [*PSI*$^+$] N-terminal domain is sufficient to maintain Sup35 protein in the [*PSI*$^+$] condition. The critical feature of the N-terminal domain is that it is rich in glutamine and asparagine residues.

Loss of function in the [*PSI*$^+$] state is due to the sequestration of the protein in an oligomeric complex. Sup35 protein in [*PSI*$^+$] cells is clustered in discrete foci, whereas the protein in [*psi*$^-$] cells is diffused in the cytosol. Sup35 protein from [*PSI*$^+$] cells forms amyloid fibers *in vitro*—these have a characteristic high content of β-sheet structures.

The involvement of protein conformation (rather than covalent modification) is suggested by the effects of conditions that affect protein structure. Denaturing treatments cause loss of the [*PSI*$^+$] state. In particular, the chaperone Hsp104 is involved in inheritance of [*PSI*$^+$]. Its effects are paradoxical. Deletion of *HSP104* prevents maintenance of the [*PSI*$^+$] state, and overexpression of Hsp104 also causes loss of the [*PSI*$^+$] state. This suggests that Hsp104 is required for some change in the structure of Sup35 that is necessary for acquisition of the [*PSI*$^+$] state, but that must be transitory.

Using the ability of Sup35 to form the inactive structure *in vitro*, it is possible to provide biochemical proof for the role of the protein. FIGURE 31.29 illustrates a striking experiment in which the protein was converted to the inactive form *in vitro*, put into liposomes (where in effect the protein is surrounded by an artificial membrane), and then introduced directly into cells by fusing the liposomes with [*psi*$^-$] yeast. The yeast cells were converted to [*PSI*$^+$]! This experiment refutes all of the objections that were raised to the conclusion that the protein has the ability to confer the epigenetic state. Experiments in which cells are mated, or in which extracts are taken from one cell to treat another cell, always are susceptible to the possibility that a nucleic acid has been transferred. When the protein by itself does not convert target cells, though (even though protein converted to the inactive state can do so), the only difference is the treatment of the protein—which must therefore be responsible for the conversion.

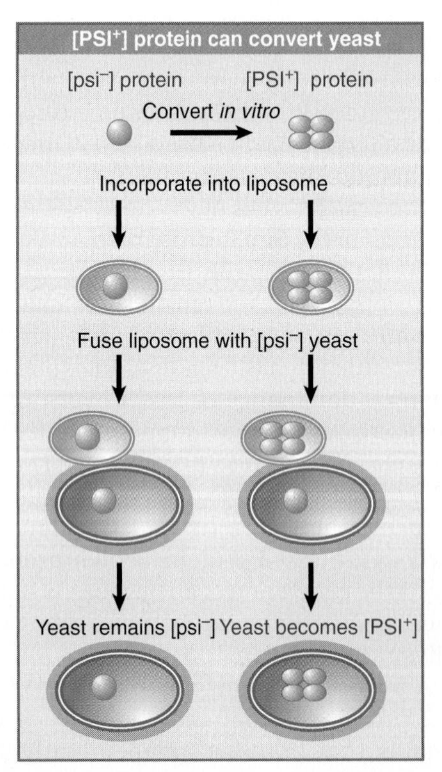

FIGURE 31.29 Purified protein can convert the [*psi*$^-$] state of yeast to [*PSI*$^+$].

The ability of yeast to form the [*PSI*$^+$] prion state depends on the genetic background. The yeast must be [*PIN*$^+$] in order for the [*PSI*$^+$] state to form. The [*PIN*$^+$] condition itself is an epigenetic state. It can be created by the formation of prions from any one of several different proteins. These proteins share the characteristic of Sup35 that they have Gln/Asn-rich domains. Overexpression of these domains in yeast stimulates formation of the [*PSI*$^+$] state. This suggests that there is a common model for the formation of the prion state that involves aggregation of the Gln/Asn domains into self-propagating amyloid structure.

How does the presence of one Gln/Asn protein influence the formation of prions by another? We know that the formation of Sup35 prions is specific to Sup35 protein, that is, it does not occur by cross-aggregation with other proteins. This suggests that the yeast cell may contain soluble proteins that antagonize prion formation. These proteins are not specific for any one prion. As a result, the introduction of any Gln/Asn domain protein that interacts with these proteins will reduce the concentration. This will allow other Gln/Asn proteins to aggregate more easily.

31.12 Prions Cause Diseases in Mammals

Key concepts

- The protein responsible for scrapie exists in two forms: the wild-type noninfectious form PrP^C, which is susceptible to proteases, and the disease-causing form PrP^{Sc}, which is resistant to proteases.
- The neurological disease can be transmitted to mice by injecting the purified PrP^{Sc} protein into mice.
- The recipient mouse must have a copy of the *PrP* gene coding for the mouse protein.
- The PrP^{Sc} protein can perpetuate itself by causing the newly synthesized PrP protein to take up the PrP^{Sc} form instead of the PrP^C form.
- Multiple strains of PrP^{Sc} may have different conformations of the protein.

Prion diseases have been found in sheep, in human beings, and, more recently, in cows. The basic phenotype is an ataxia—a neurodegenerative disorder that is manifested by an inability to remain upright. The name of the disease in sheep, **scrapie,** reflects the phenotype: The sheep rub against walls in order to stay upright. Scrapie can be perpetuated by inoculating sheep with tissue extracts from infected animals. The disease **kuru** was found in New Guinea, where it appeared to be perpetuated by cannibalism, in particular the eating of brains. Related diseases in Western populations with a pattern of genetic transmission include Gerstmann–Straussler syndrome and the related Creutzfeldt–Jakob disease (CJD), which occurs sporadically. Most recently, a disease resembling CJD appears to have been transmitted by consumption of meat from cows suffering from "mad cow" disease.

When tissue from scrapie-infected sheep is inoculated into mice, the disease occurs in a period ranging from 75 to 150 days. The active component is a protease-resistant protein. The protein is coded by a gene that is normally expressed in the brain. The form of the protein in normal brain, called PrP^C, is sensitive to proteases. Its conversion to the resistant form, called Prp^{Sc}, is associated with occurrence of the disease. The infectious preparation has no detectable nucleic acid, is sensitive to UV irradiation at wave lengths that damage protein, and has a low infectivity (1 infectious unit / 10^5 PrP^{Sc} proteins). This corresponds to an epigenetic inheritance in which there is no change in genetic information (because normal and diseased cells have the same *PrP* gene sequence), but the PrP^{Sc} form of the protein is the infectious agent (whereas PrP^C is harmless). The PrP^{Sc} form has a high content of β sheets, which form an amyloid fibrillous structure that is absent from the PrP^C form.

The basis for the difference between the PrP^{Sc} and Prp^C forms appears to lie with a change in conformation rather than with any covalent alteration. Both proteins are glycosylated and linked to the membrane by a GPI-linkage.

The assay for infectivity in mice allows the dependence on protein sequence to be tested. FIGURE 31.30 illustrates the results of some critical experiments. In the normal situation, PrP^{Sc} protein extracted from an infected mouse will induce disease (and ultimately kill) when it is injected into a recipient mouse. If the *PrP* gene is "knocked out," a mouse becomes resistant to infection. This experiment demonstrates two things. First, the endogenous protein is necessary for an infection, presumably because it provides the raw material that is converted into the infectious agent. Second, the cause of disease is not the removal of the PrP^C form of the protein, because a mouse with no PrP^C survives normally: The disease is caused by a gain-of-function in PrP^{Sc}. If the PrP gene is altered to prevent the GPI-linkage from occurring, mice infected with PrP^{Sc} do not develop disease, which suggests that the gain of function involves an altered signalling function for which the GPI-linkage is required.

The existence of species barriers allows hybrid proteins to be constructed to delineate

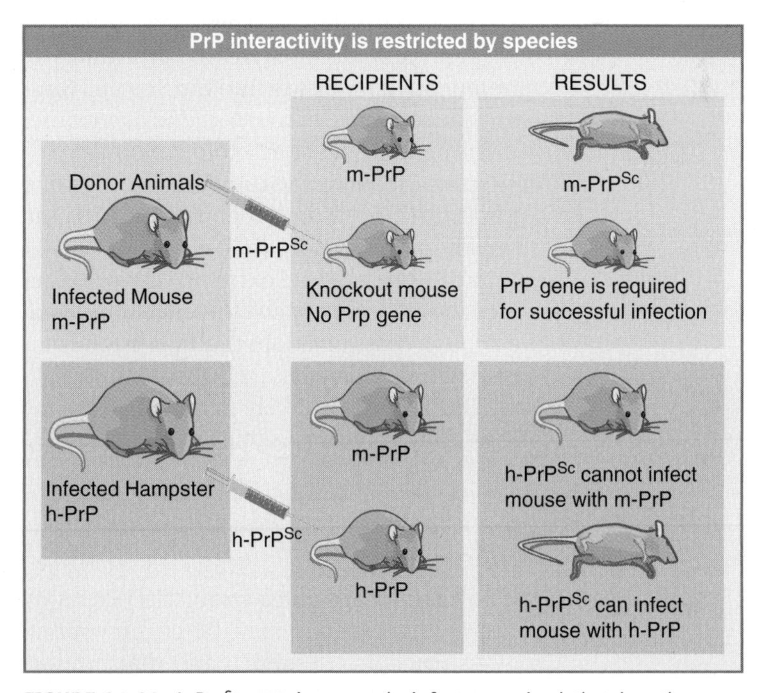

FIGURE 31.30 A Prp^{Sc} protein can only infect an animal that has the same type of endogenous PrP^C protein.

the features required for infectivity. The original preparations of scrapie were perpetuated in several types of animal, but these cannot always be transferred readily. For example, mice are resistant to infection from prions of hamsters. This means that hamster-PrP^{Sc} cannot convert mouse-PrP^{C} to PrP^{Sc}. The situation changes, though, if the mouse *PrP* gene is replaced by a hamster *PrP* gene. (This can be done by introducing the hamster *PrP* gene into the *PrP* knockout mouse.) A mouse with a hamster *PrP* gene is sensitive to infection by hamster PrP^{Sc}. This suggests that the conversion of cellular PrP^{C} protein into the Sc state requires that the PrP^{Sc} and PrP^{C} proteins have matched sequences.

There are different "strains" of PrP^{Sc}, which are distinguished by characteristic incubation periods upon inoculation into mice. This implies that the protein is not restricted solely to alternative states of PrP^{C} and PrP^{Sc}, but rather that there may be multiple Sc states. These differences must depend on some self-propagating property of the protein other than its sequence. If conformation is the feature that distinguishes PrP^{Sc} from PrP^{C}, then there must be multiple conformations, each of which has a self-templating property when it converts PrP^{C}.

The probability of conversion from PrP^{C} to PrP^{Sc} is affected by the sequence of PrP. Gerstmann–Straussler syndrome in human beings is caused by a single amino acid change in PrP. This is inherited as a dominant trait. If the same change is made in the mouse PrP gene, mice develop the disease. This suggests that the mutant protein has an increased probability of spontaneous conversion into the Sc state. Similarly, the sequence of the PrP gene determines the susceptibility of sheep to develop the disease spontaneously; the combination of amino acids at three positions (codons 136, 154, and 171) determines susceptibility.

The prion offers an extreme case of epigenetic inheritance, in which the infectious agent is a protein that can adopt multiple conformations, each of which has a self-templating property. This property is likely to involve the state of aggregation of the protein.

31.13 Summary

The formation of heterochromatin occurs by proteins that bind to specific chromosomal regions (such as telomeres) and that interact with histones. The formation of an inactive structure may propagate along the chromatin thread from an initiation center. Similar events occur in silencing of the inactive yeast mating type loci. Repressive structures that are required to maintain the inactive states of particular genes are formed by the Pc-G protein complex in *Drosophila*. They share with heterochromatin the property of propagating from an initiation center.

Formation of heterochromatin may be initiated at certain sites and then propagated for a distance that is not precisely determined. When a heterochromatic state has been established, it is inherited through subsequent cell divisions. This gives rise to a pattern of epigenetic inheritance, in which two identical sequences of DNA may be associated with different protein structures, and therefore have different abilities to be expressed. This explains the occurrence of position effect variegation in *Drosophila*.

Modification of histone tails is a trigger for chromatin reorganization. Acetylation is generally associated with gene activation. Histone acetylases are found in activating complexes, whereas histone deacetylases are found in inactivating complexes. Histone methylation is associated with gene inactivation. Some histone modifications may be exclusive or synergistic with others.

Inactive chromatin at yeast telomeres and silent mating type loci appears to have a common cause, and involves the interaction of certain proteins with the N-terminal tails of histones H3 and H4. Formation of the inactive complex may be initiated by binding of one protein to a specific sequence of DNA; the other components may then polymerize in a cooperative manner along the chromosome.

Inactivation of one X chromosome in female (eutherian) mammals occurs at random. The *Xic* locus is necessary and sufficient to count the number of X chromosomes. The n–1 rule ensures that all but one X chromosome are inactivated. Xic contains the gene Xist, which codes for an RNA that is expressed only on the inactive X chromosome. Stabilization of Xist RNA is the mechanism by which the inactive X chromosome is distinguished.

Methylation of DNA is inherited epigenetically. Replication of DNA creates hemimethylated products, and a maintenance methylase restores the fully methylated state. Some methylation events depend on parental origin. Sperm and eggs contain specific and different patterns of methylation, with the result that paternal and maternal alleles are differently expressed in the embryo. This is responsible for imprinting, in which the nonmethylated allele inherited from one parent is essential because it is the only active allele; the allele inherited from the other parent is silent. Patterns of methylation are reset during gamete formation in every generation.

Prions are proteinaceous infectious agents that are responsible for the disease of scrapie in sheep and for related diseases in human beings. The infectious agent is a variant of a normal cellular protein. The PrP^{Sc} form has an altered conformation that is self-templating: The normal PrP^{C} form does not usually take up this conformation, but does so in the presence of PrP^{Sc}. A similar effect is responsible for inheritance of the [*PSI*] element in yeast.

References

31.2 Heterochromatin Propagates from a Nucleation Event

Research

Ahmad, K. and Henikoff, S. (2001). Modulation of a transcription factor counteracts heterochromatic gene silencing in *Drosophila*. *Cell* 104, 839–847.

31.3 Heterochromatin Depends on Interactions with Histones

Reviews

Loo, S. and Rine, J. (1995). Silencing and heritable domains of gene expression. *Annu. Rev. Cell Dev. Biol.* 11, 519–548.

Moazed, D. (2001). Common themes in mechanisms of gene silencing. *Mol. Cell* 8, 489–498.

Rusche, L. N., Kirchmaier, A. L., and Rine, J. (2003). The establishment, inheritance, and function of silenced chromatin in *Saccharomyces cerevisiae*. *Annu. Rev. Biochem.* 72, 481–516.

Thompson, J. S., Hecht, A., and Grunstein, M. (1993). Histones and the regulation of heterochromatin in yeast. *Cold Spring Harbor Symp. Quant. Biol.* 58, 247–256.

Zhang, Y. and Reinberg, D. (2001). Transcription regulation by histone methylation: interplay between different covalent modifications of the core histone tails. *Genes Dev.* 15, 2343–2360.

Research

Ahmad, K. and Henikoff, S. (2001). Modulation of a transcription factor counteracts heterochromatic gene silencing in *Drosophila*. *Cell* 104, 839–847.

Bannister, A. J., Zegerman, P., Partridge, J. F., Miska, E. A., Thomas, J. O., Allshire, R. C., and Kouzarides, T. (2001). Selective recognition of methylated lysine 9 on histone H3 by the HP1 chromo domain. *Nature* 410, 120–124.

Bloom, K. S. and Carbon, J. (1982). Yeast centromere DNA is in a unique and highly ordered structure in chromosomes and small circular minichromosomes. *Cell* 29, 305–317.

Cheutin, T., McNairn, A. J., Jenuwein, T., Gilbert, D. M., Singh, P. B., and Misteli, T. (2003). Maintenance of stable heterochromatin domains by dynamic HP1 binding. *Science* 299, 721–725.

Eissenberg, J. C., Morris, G. D., Reuter, G., and Hartnett, T. (1992). The heterochromatin-associated protein HP-1 is an essential protein in *Drosophila* with dosage-dependent effects on position-effect variegation. *Genetics* 131, 345–352.

Hecht, A., Laroche, T., Strahl-Bolsinger, S., Gasser, S. M., and Grunstein, M. (1995). Histone H3 and H4 N-termini interact with the silent information regulators SIR3 and SIR4: a molecular model for the formation of heterochromatin in yeast. *Cell* 80, 583–592.

Imai, S., Armstrong, C. M., Kaeberlein, M., and Guarente, L. (2000). Transcriptional silencing and longevity protein Sir2 is an NAD-dependent histone deacetylase. *Nature* 403, 795–800.

James, T. C. and Elgin, S. C. (1986). Identification of a nonhistone chromosomal protein associated with heterochromatin in *D. melanogaster* and its gene. *Mol. Cell Biol.* 6, 3862–3872.

Kayne, P. S., Kim, U. J., Han, M., Mullen, R. J., Yoshizaki, F., and Grunstein, M. (1988). Extremely conserved histone H4 N terminus is dispensable for growth but essential for repressing the silent mating loci in yeast. *Cell* 55, 27–39.

Lachner, M., O'Carroll, D., Rea, S., Mechtler, K., and Jenuwein, T. (2001). Methylation of histone H3 lysine 9 creates a binding site for HP1 proteins. *Nature* 410, 116–120.

Landry, J., Sutton, A., Tafrov, S. T., Heller, R. C., Stebbins, J., Pillus, L., and Sternglanz, R. (2000). The silencing protein SIR2 and its homologs are NAD-dependent protein deacetylases. *Proc. Natl. Acad. Sci. USA* 97, 5807–5811.

Manis, J. P., Gu, Y., Lansford, R., Sonoda, E., Ferrini, R., Davidson, L., Rajewsky, K., and Alt, F. W. (1998). Ku70 is required for late B cell development and immunoglobulin heavy chain class switching. *J. Exp. Med.* 187, 2081–2089.

Meluh, P. B. et al. (1998). Cse4p is a component of the core centromere of *S. cerevisiae*. *Cell* 94, 607–613.

Moretti, P., Freeman, K., Coodly, L., and Shore, D. (1994). Evidence that a complex of SIR proteins interacts with the silencer and telomere-binding protein RAP1. *Genes Dev.* 8, 2257–2269.

Nakagawa, H., Lee, J. K., Hurwitz, J., Allshire, R. C., Nakayama, J., Grewal, S. I., Tanaka, K., and Murakami, Y. (2002). Fission yeast CENP-B homologs nucleate centromeric heterochromatin by promoting heterochromatin-specific histone tail modifications. *Genes Dev.* 16, 1766–1778.

Nakayama, J., Rice, J. C., Strahl, B. D., Allis, C. D., and Grewal, S. I. (2001). Role of histone H3 lysine 9 methylation in epigenetic control of heterochromatin assembly. *Science* 292, 110–113.

Palladino, F., Laroche, T., Gilson, E., Axelrod, A., Pillus, L., and Gasser, S. M. (1993). SIR3 and SIR4 proteins are required for the positioning and integrity of yeast telomeres. *Cell* 75, 543–555.

Platero, J. S., Hartnett, T., and Eissenberg, J. C. (1995). Functional analysis of the chromo domain of HP1. *EMBO J.* 14, 3977–3986.

Schotta, G., Ebert, A., Krauss, V., Fischer, A., Hoffmann, J., Rea, S., Jenuwein, T., Dorn, R., and Reuter, G. (2002). Central role of *Drosophila* SU(VAR)3-9 in histone H3-K9 methylation and heterochromatic gene silencing. *EMBO J.* 21, 1121–1131.

Sekinger, E. A. and Gross, D. S. (2001). Silenced chromatin is permissive to activator binding and PIC recruitment. *Cell* 105, 403–414.

Shore, D. and Nasmyth, K. (1987). Purification and cloning of a DNA-binding protein from yeast that binds to both silencer and activator elements. *Cell* 51, 721–732.

Smith, J. S., Brachmann, C. B., Celic, I., Kenna, M. A., Muhammad, S., Starai, V. J., Avalos, J. L., Escalante-Semerena, J. C., Grubmeyer, C., Wolberger, C., and Boeke, J. D. (2000). A phylogenetically conserved NAD+-dependent protein deacetylase activity in the Sir2 protein family. *Proc. Natl. Acad. Sci. USA* 97, 6658–6663.

Verdel, A., Jia, S., Gerber, S., Sugiyama, T., Gygi, S., Grewal, S. I., and Moazed, D. (2004). RNAi-mediated targeting of heterochromatin by the RITS complex. *Science* 303, 672–676.

31.4 Polycomb and Trithorax Are Antagonistic Repressors and Activators

Review

Ringrose, L. and Paro, R. (2004). Epigenetic regulation of cellular memory by the Polycomb and Trithorax group proteins. *Annu. Rev. Genet.* 38, 413–443.

Research

Brown, J. L., Fritsch, C., Mueller, J., and Kassis, J. A. (2003). The *Drosophila* pho-like gene encodes a YY1-related DNA binding protein that is redundant with pleiohomeotic in homeotic gene silencing. *Development* 130, 285–294.

Cao, R., Wang, L., Wang, H., Xia, L., Erdjument-Bromage, H., Tempst, P., Jones, R. S., and Zhang, Y. (2002). Role of histone H3 lysine 27 methylation in Polycomb-group silencing. *Science* 298, 1039–1043.

Chan, C. S., Rastelli, L., and Pirrotta, V. (1994). A Polycomb response element in the Ubx gene that determines an epigenetically inherited state of repression. *EMBO J.* 13, 2553–2564.

Czermin, B., Melfi, R., McCabe, D., Seitz, V., Imhof, A., and Pirrotta, V. (2002). *Drosophila* enhancer of Zeste/ESC complexes have a histone H3 methyltransferase activity that marks chromosomal Polycomb sites. *Cell* 111, 185–196.

Eissenberg, J. C., James, T. C., Fister-Hartnett, D. M., Hartnett, T., Ngan, V., and Elgin, S. C. R. (1990). Mutation in a heterochromatin-specific chromosomal protein is associated with suppression of position-effect variegation in *D. melanogaster. Proc. Natl. Acad. Sci. USA* 87, 9923–9927.

Fischle, W., Wang, Y., Jacobs, S. A., Kim, Y., Allis, C. D., and Khorasanizadeh, S. (2003). Molecular basis for the discrimination of repressive methyl-lysine marks in histone H3 by Polycomb and HP1 chromo domains. *Genes Dev.* 17, 1870–1881.

Francis, N. J., Kingston, R. E., and Woodcock, C. L. (2004). Chromatin compaction by a Polycomb group protein complex. *Science* 306, 1574–1577.

Franke, A., DeCamillis, M., Zink, D., Cheng, N., Brock, H. W., and Paro, R. (1992). Polycomb and polyhomeotic are constituents of a multimeric protein complex in chromatin of *D. melanogaster. EMBO J.* 11, 2941–2950.

Geyer, P. K. and Corces, V. G. (1992). DNA position-specific repression of transcription by a *Drosophila* zinc finger protein. *Genes Dev.* 6, 1865–1873.

Orlando, V. and Paro, R. (1993). Mapping Polycomb-repressed domains in the bithorax complex using *in vivo* formaldehyde cross-linked chromatin. *Cell* 75, 1187–1198.

Strutt, H., Cavalli, G., and Paro, R. (1997). Colocalization of Polycomb protein and GAGA factor on regulatory elements responsible for the maintenance of homeotic gene expression. *EMBO J.* 16, 3621–3632.

Wang, L., Brown, J. L., Cao, R., Zhang, Y., Kassis, J. A., and Jones, R. S. (2004). Hierarchical recruitment of Polycomb group silencing complexes. *Mol. Cell* 14, 637–646.

Zink, B. and Paro, R. (1989). *In vivo* binding patterns of a *trans*-regulator of the homeotic genes in *D. melanogaster. Nature* 337, 468–471.

31.5 X Chromosomes Undergo Global Changes

Review

Plath, K., Mlynarczyk-Evans, S., Nusinow, D. A., and Panning, B. (2002). Xist RNA and the mechanism of X chromosome inactivation. *Annu. Rev. Genet.* 36, 233–278.

Research

Jeppesen, P. and Turner, B. M. (1993). The inactive X chromosome in female mammals is distinguished by a lack of histone H4 acetylation, a cytogenetic marker for gene expression. *Cell* 74, 281–289.

Lee, J. T. et al. (1996). A 450 kb transgene displays properties of the mammalian X-inactivation center. *Cell* 86, 83–94.

Lyon, M. F. (1961). Gene action in the X chromosome of the mouse. *Nature* 190, 372–373.

Panning, B., Dausman, J., and Jaenisch, R. (1997). X chromosome inactivation is mediated by Xist RNA stabilization. *Cell* 90, 907–916.

Penny, G. D. et al. (1996). Requirement for Xist in X chromosome inactivation. *Nature* 379, 131–137.

31.6 Chromosome Condensation Is Caused by Condensins

Reviews

Hirano, T. (1999). SMC-mediated chromosome mechanics: a conserved scheme from bacteria to vertebrates? *Genes Dev.* 13, 11–19.

Hirano, T. (2000). Chromosome cohesion, condensation, and separation. *Annu. Rev. Biochem.* 69, 115–144.

Hirano, T. (2002). The ABCs of SMC proteins: two-armed ATPases for chromosome condensation, cohesion, and repair. *Genes Dev.* 16, 399–414.

Jessberger, R. (2002). The many functions of SMC proteins in chromosome dynamics. *Nat. Rev. Mol. Cell Biol.* 3, 767–778.

Nasmyth, K. (2002). Segregating sister genomes: the molecular biology of chromosome separation. *Science* 297, 559–565.

Research

Csankovszki, G., McDonel, P., and Meyer, B. J. (2004). Recruitment and spreading of the *C. elegans* dosage compensation complex along X chromosomes. *Science* 303, 1182–1185.

Haering, C. H., Lowe, J., Hochwage, A., and Nasmyth, K. (2002). Molecular architecture of SMC proteins and the yeast cohesin complex. *Mol. Cell* 9, 773–788.

Kimura, K., Rybenkov, V. V., Crisona, N. J., Hirano, T., and Cozzarelli, N. R. (1999). 13S condensin actively reconfigures DNA by introducing global positive writhe: implications for chromosome condensation. *Cell* 98, 239–248.

31.7 DNA Methylation Is Perpetuated by a Maintenance Methylase

Reviews

Bird, A. (2002). DNA methylation patterns and epigenetic memory. *Genes Dev.* 16, 6–21.

Bird, A. P. (1986). A fraction of the mouse genome that is derived from islands of nonmethylated, CpG-rich DNA. *Nature* 321, 209–213.

Matzke, M., Matzke, A. J., and Kooter, J. M. (2001). RNA: guiding gene silencing. *Science* 293, 1080–1083.

Sharp, P. A. (2001). RNA interference—2001. *Genes Dev.* 15, 485–490.

Research

Amir, R. E., Van den Veyver, I. B., Wan, M., Tran, C. Q., Francke, U., and Zoghbi, H. Y. (1999). Rett syndrome is caused by mutations in X-linked MECP2, encoding methyl-CpG-binding protein 2. *Nat. Genet.* 23, 185–188.

Li, E., Bestor, T. H., and Jaenisch, R. (1992). Targeted mutation of the DNA methyltransferase gene results in embryonic lethality. *Cell* 69, 915–926.

Morgan, H. D., et al. (2004). Activation-induced cytidine deaminase deaminates 5-methylcytosine in DNA and is expressed in pluripotent tissues: implications for epigenetic reprogramming. *J. Biol.Chem.* 279, 52353–52360.

Okano, M., Bell, D. W., Haber, D. A., and Li, E. (1999). DNA methyltransferases Dnmt3a and Dnmt3b are essential for de novo methylation and mammalian development. *Cell* 99, 247–257.

Xu, G. L., Bestor, T. H., Bourc'his, D., Hsieh, C. L., Tommerup, N., Bugge, M., Hulten, M., Qu, X., Russo, J. J., and Viegas-Paquignot, E. (1999). Chromosome instability and immunodeficiency syndrome caused by mutations in a DNA methyltransferase gene. *Nature* 402, 187–191.

31.8 DNA Methylation Is Responsible for Imprinting

Review

Bartolomei, M. S. and Tilghman, S. (1997). Genomic imprinting in mammals. *Annu. Rev. Genet.* 31, 493–525.

Research

Chaillet, J. R., Vogt, T. F., Beier, D. R., and Leder, P. (1991). Parental-specific methylation of an imprinted transgene is established during gametogenesis and progressively changes during embryogenesis. *Cell* 66, 77–83.

Lawrence, R. J., Earley, K., Pontes, O., Silva, M., Chen, Z. J., Neves, N., Viegas, W., and Pikaard, C. S. (2004). A concerted DNA methylation/histone methylation switch regulates rRNA gene dosage control and nucleolar dominance. *Mol. Cell* 13, 599–609.

31.9 Oppositely Imprinted Genes Can Be Controlled by a Single Center

Research

Bell, A. C. and Felsenfeld, G. (2000). Methylation of a CTCF-dependent boundary controls imprinted expression of the Igf2 gene. *Nature* 405, 482–485.

Hark, A. T., Schoenherr, C. J., Katz, D. J., Ingram, R. S., Levorse, J. M., and Tilghman, S. M. (2000). CTCF mediates methylation-sensitive enhancer-blocking activity at the H19/Igf2 locus. *Nature* 405, 486–489.

31.11 Yeast Prions Show Unusual Inheritance

Reviews

Horwich, A. L. and Weissman, J. S. (1997). Deadly conformations: protein misfolding in prion disease. *Cell* 89, 499–510.

Lindquist, S. (1997). Mad cows meet psi-chotic yeast: the expansion of the prion hypothesis. *Cell* 89, 495–498.

Serio, T. R. and Lindquist, S. L. (1999). [PSI$^+$]: an epigenetic modulator of translation termination efficiency. *Annu. Rev. Cell Dev. Biol.* 15, 661–703.

Wickner, R. B. (1996). Prions and RNA viruses of *S. cerevisiae. Annu. Rev. Genet.* 30, 109–139.

Wickner, R. B., Edskes, H. K., Roberts, B. T., Baxa, U., Pierce, M. M., Ross, E. D., and Brachmann, A. (2004). Prions: proteins as genes and infectious entities. *Genes Dev.* 18, 470–485.

Wickner, R. B., Edskes, H. K., Ross, E. D., Pierce, M. M., Baxa, U., Brachmann, A., and Shewmaker, F. (2004). Prion genetics: new rules for a new kind of gene. *Annu. Rev. Genet.* 38, 681–707.

Research

Chernoff, Y. O. et al. (1995). Role of the chaperone protein Hsp104 in propagation of the yeast prion-like factor [PSI$^+$]. *Science* 268, 880–884.

Derkatch, I. L., Bradley, M. E., Hong, J. Y., and Liebman, S. W. (2001). Prions affect the appearance of other prions: the story of [PIN(+)]. *Cell* 106, 171–182.

Derkatch, I. L., Bradley, M. E., Masse, S. V., Zadorsky, S. P., Polozkov, G. V., Inge-Vechtomov, S. G., Liebman S. W. (2000). Dependence and independence of [PSI(+)] and [PIN(+)]: a two-prion system in yeast? *EMBO J.* 19, 1942–1952.

Glover, J. R. et al. (1997). Self-seeded fibers formed by Sup35, the protein determinant of [PSI$^+$], a heritable prion-like factor of *S. cerevisiae. Cell* 89, 811–819.

Masison, D. C. and Wickner, R. B. (1995). Prion-inducing domain of yeast Ure2p and protease resistance of Ure2p in prion-containing cells. *Science* 270, 93–95.

Osherovich, L. Z. and Weissman, J. S. (2001). Multiple gln/asn-rich prion domains confer susceptibility to induction of the yeast. *Cell* 106, 183–194.

Sparrer, H. E., Santoso, A., Szoka, F. C., and Weissman, J. S. (2000). Evidence for the prion hypothesis: induction of the yeast [PSI+] factor by in vitro-converted Sup35 protein. *Science* 289, 595–599.

Wickner, R. B. (1994). [URE3] as an altered URE2 protein: evidence for a prion analog in *S. cerevisiae. Science* 264, 566–569.

31.12 Prions Cause Diseases in Mammals

Reviews

Chien, P., Weissman, J. S., and DePace, A. H. (2004). Emerging principles of conformation-based prion inheritance. *Annu. Rev. Biochem.* 73, 617–656.

Prusiner, S. (1982). Novel proteinaceous infectious particles cause scrapie. *Science* 216, 136–144.

Prusiner, S. B. and Scott, M. R. (1997). Genetics of prions. *Annu. Rev. Genet.* 31, 139–175.

Wickner, R. B., Edskes, H. K., Roberts, B. T., Baxa, U., Pierce, M. M., Ross, E. D., and Brachmann, A. (2004). Prions: proteins as genes and infectious entities. *Genes Dev.* 18, 470–485.

Research

Basler, K., Oesch, B., Scott, M., Westaway, D., Walchli, M., Groth, D. F., McKinley, M. P., Prusiner, S. B., and Weissmann, C. (1986). Scrapie and cellular PrP isoforms are encoded by the same chromosomal gene. *Cell* 46, 417–428.

Bueler, H. et al. (1993). Mice devoid of PrP are resistant to scrapie. *Cell* 73, 1339–1347.

Hsiao, K. et al. (1989). Linkage of a prion protein missense variant to Gerstmann-Straussler syndrome. *Nature* 338, 342–345.

McKinley, M. P., Bolton, D. C., and Prusiner, S. B. (1983). A protease-resistant protein is a structural component of the scrapie prion. *Cell* 35, 57–62.

Oesch, B. et al. (1985). A cellular gene encodes scrapie PrP27-30 protein. *Cell* 40, 735–746.

Scott, M. et al. (1993). Propagation of prions with artificial properties in transgenic mice expressing chimeric PrP genes. *Cell* 73, 979–988.

Glossary

The **10 nm fiber** is a linear array of nucleosomes, generated by unfolding from the natural condition of chromatin.

The **−10 sequence** is the consensus sequence centered about 10 bp before the startpoint of a bacterial gene. It is involved in melting DNA during the initiation reaction.

The **30 nm fiber** is a coiled coil of nucleosomes. It is the basic level of organization of nucleosomes in chromatin.

The **−35 sequence** is the consensus sequence centered about 35 bp before the startpoint of a bacterial gene. It is involved in initial recognition by RNA polymerase.

45S RNA is a precursor that contains the sequences of both major ribosomal RNAs (28S and 18S rRNAs).

5.8S RNA is an independent small RNA present on the large subunit of eukaryotic ribosomes. It is homologous to the 5′ end of bacterial 23S rRNA.

5S RNA is a 120-base RNA that is a component of the large subunit of the ribosome.

The **A domain** is the conserved 11 bp sequence of A-T base pairs in the yeast ARS element that comprises the replication origin.

The **A site** of the ribosome is the site that an aminoacyl-tRNA enters to base pair with the codon.

Abortive initiation describes a process in which RNA polymerase starts transcription but terminates before it has left the promoter. It then reinitiates. Several cycles may occur before the elongation stage begins.

The **abundance** of an mRNA is the average number of molecules per cell.

Abundant mRNAs consist of a small number of individual species, each present in a large number of copies per cell.

The **acceptor arm** of tRNA is a short duplex that terminates in the CCA sequence to which an amino acid is linked.

An **acentric fragment** of a chromosome (generated by breakage) lacks a centromere and is lost at cell division.

Acquired immunity is another term for adaptive immunity.

Acridines are mutagens that act on DNA to cause the insertion or deletion of a single base pair. They were useful in defining the triplet nature of the genetic code.

An **activator** is a protein that stimulates the expression of a gene, typically by acting at a promoter to stimulate RNA polymerase. In eukaryotes, the sequence to which it binds in the promoter is called a response element.

Adaptive immunity is the response mediated by lymphocytes that are activated by their specific interaction with antigen. The adaptive immune response develops over several days as lymphocytes with antigen-specific receptors are stimulated to proliferate and become effector cells. It is responsible for immunological memory.

An **addiction system** is a survival mechanism used by plasmids. The mechanism kills the bacterium upon loss of the plasmid.

Adenylate cyclase is an enzyme that uses ATP as a substrate to generate cyclic AMP, in which 5′ and 3′ positions of the sugar ring are connected via a phosphate group.

Agropine plasmids carry genes coding for the synthesis of opines of the agropine type. The tumors usually die early.

An **allele** is one of several alternative forms of a gene occupying a given locus on a chromosome.

Allelic exclusion describes the expression in any particular lymphocyte of only one allele coding for the expressed immunoglobulin. This is caused by feedback from the first immunoglobulin allele to be expressed that prevents activation of a copy on the other chromosome.

Allosteric regulation describes the ability of a protein to change its conformation (and therefore activity) at one site as the result of binding a small molecule to a second site located elsewhere on the protein.

Alternative splicing describes the production of different RNA products from a single product by changes in the usage of splicing junctions.

The **Alu domain** comprises the parts of the 7S RNA of the SRP that are related to Alu RNA.

The **Alu family** is a set of dispersed, related sequences, each ~300 bp long, in the human genome. The individual members have Alu cleavage sites at each end (hence the name).

Amanitin (more fully α-amanitin) is a bicyclic octapeptide derived from the poisonous mushroom *Amanita phalloides*; it inhibits transcription by certain eukaryotic RNA polymerases, especially RNA polymerase II.

The **amber** codon is the triplet UAG, one of the three termination codons that end protein synthesis.

An **aminoacyl-tRNA** is a tRNA linked to an amino acid. The COOH group of the amino acid is linked to the 3′- or 2′-OH group of the terminal base of the tRNA.

Aminoacyl-tRNA synthetases are enzymes responsible for covalently linking amino acids to the 2′- or 3′-OH position of tRNA.

An **anchor** (often referred to as a "transmembrane anchor") is a segment of a transmembrane protein that resides in the membrane.

Annealing of DNA describes the renaturation of a duplex structure from single strands that were obtained by denaturing duplex DNA.

An **antibody** is a protein that is produced by B lymphocytes and that binds a particular antigen. Antibodies are synthesized in membrane-bound and secreted forms. Antibodies produced during an immune response recruit effector functions to help neutralize and eliminate the pathogen.

The **anticodon** is a trinucleotide sequence in tRNA that is complementary to the codon in mRNA and enables the tRNA to place the appropriate amino acid in response to the codon.

The **anticodon arm** of tRNA is a stem loop structure that exposes the anticodon triplet at one end.

An **antigen** is a molecule that can bind specifically to an antigen receptor, such as an antibody.

An **antigenic determinant** is the portion of an antigen that is recognized by the antigen receptor on lymphocytes. It is also called an epitope.

An **anti-insulator** is a sequence that allows an enhancer to overcome the effect of an insulator.

Antiparallel strands of the double helix are organized in opposite orientation, so that the 5′ end of one strand is aligned with the 3′ end of the other strand.

An **antisense gene** codes for an (antisense) RNA that has a complementary sequence to an RNA that is its target.

The **antisense strand** of DNA is complementary to the sense strand, and is the one that acts as the template for synthesis of mRNA.

Anti-Sm is an autoimmune antiserum that defines the Sm epitope that is common to a group of proteins found in snRNPs that are involved in RNA splicing.

Antitermination is a mechanism of transcriptional control in which termination is prevented at a specific terminator site, allowing RNA polymerase to read into the genes beyond it.

Antitermination proteins allow RNA polymerase to transcribe through certain terminator sites.

Anucleate bacteria lack a nucleoid but are of similar shape to wild-type bacteria.

An **arm** of tRNA is one of the four (or in some cases five) stem-loop structures that make up the secondary structure.

The **arms** of a lambda phage attachment site are the sequences flanking the core region where the recombination event occurs.

ARS is an origin for replication in yeast. The common feature among different *ARS* sequences is a conserved 11 bp sequence called the A domain.

The **ascus** of a fungus contains a tetrad or octad of the (haploid) spores, representing the products of a single meiosis.

An **assembly factor** is a protein that is required for formation of a macromolecular structure but is not itself part of that structure.

att sites are the loci on a lambda phage and the bacterial chromosome at which recombination integrates the phage into, or excises it from, the bacterial chromosome.

Attenuation describes the regulation of bacterial operons by controlling termination of transcription at a site located before the first structural gene.

An **attenuator** is a terminator sequence at which attenuation occurs.

Autogenous control describes the action of a gene product that either inhibits (negative autogenous control) or activates (positive autogenous control) expression of the gene coding for it.

An **autoimmune disease** is a pathological condition in which the immune response is directed to self antigen.

An **autonomous controlling element** in maize is an active transposon with the ability to transpose (*compare with* nonautonomous controlling element).

Autosplicing describes the ability of an intron to excise itself from an RNA by a catalytic action that depends only on the sequence of RNA in the intron.

Avirulent mutants of a bacterium or virus have lost the capacity to infect a host productively, that is, to make more bacterium or virus.

An **axial element** is a proteinaceous structure around which the chromosomes condense at the start of synapsis.

A **B cell** is a lymphocyte that produces antibodies. B cell development occurs primarily in bone marrow.

B cell memory is responsible for rapid antibody production during a secondary immune response and subsequent responses. Memory B cells produce antibodies of higher affinity than naive B cells.

The **B cell receptor** is the antigen receptor complex on the cell surface of B lymphocytes. It consists of membrane-bound immunoglobulin bound noncovalently to Igα and Igβ chains.

A **back mutation** reverses the effect of a mutation that had inactivated a gene; thus it restores wild type.

The **background level** of mutation describes the rate at which sequence changes accumulate in the genome of an organism. It reflects the balance between the occurrence of spontaneous mutations and their removal by repair systems, and is characteristic for any species.

Bam islands are a series of short, repeated sequences found in the nontranscribed spacer of *Xenopus* rDNA genes. The name reflects their isolation by use of the BamI restriction enzyme.

Bands of polytene chromosomes are visible as dense regions that contain the majority of DNA. They include active genes.

A **basal factor** is a transcription factor required by RNA polymerase II to form the initiation complex at all promoters. Factors are identified as $TF_{II}X$, where X is a letter.

The level of response from a system in the absence of a stimulus is its **basal level**. (The basal level of transcription of a gene is the level that occurs in the absence of any specific activation.)

The **basal transcription apparatus** is the complex of transcription factors that assembles at the promoter before RNA polymerase is bound.

Base mispairing is a coupling between two bases that does not conform to the Watson–Crick rule, e.g., adenine with cytosine, thymine with guanine.

Base pairing describes the specific (complementary) interactions of adenine with thymine or of guanine with cytosine in a DNA double helix (thymine is replaced by uracil in double helical RNA).

B-form DNA is a right-handed double helix with ten base pairs per complete turn (360°) of the helix. This is the form found under physiological conditions whose structure was proposed by Crick and Watson.

A **bHLH protein** has a basic DNA-binding region adjacent to the helix-loop-helix motif.

Bidirectional replication describes a system in which an origin generates two replication forks that proceed away from the origin in opposite directions.

A **bivalent** is the structure containing all four chromatids (two representing each homologue) at the start of meiosis.

A **blocked** reading frame cannot be translated into protein because of the occurrence of termination codons.

Branch migration describes the ability of a DNA strand partially paired with its complement in a duplex to extend its pairing by displacing the resident strand with which it is homologous.

The **branch site** is a short sequence just before the end of an intron at which the lariat intermediate is formed in splicing by joining the 5′ nucleotide of the intron to the 2′ position of an adenosine.

Breakage and reunion describes the mode of genetic recombination, in which two DNA duplex molecules are broken at corresponding points and then rejoined crosswise (involving formation of a length of heteroduplex DNA around the site of joining).

The **breakage-fusion-bridge** cycle is a type of chromosomal behavior in which a broken chromatid fuses to its sister, thus forming a "bridge." When the centromeres separate at mitosis, the chromosome breaks again (not necessarily at the bridge), thereby restarting the cycle.

A **bZIP** protein has a basic DNA-binding region adjacent to a leucine zipper dimerization motif.

C genes code for the constant regions of immunoglobulin protein chains.

A **CAAT box** is part of a conserved sequence located upstream of the startpoints of eukaryotic transcription units; it is recognized by a large group of transcription factors.

A **cap** is the structure at the 5′ end of eukaryotic mRNA, and is introduced after transcription by linking the terminal phosphate of 5′ GTP to the terminal base of the mRNA. The added G (and sometimes some other bases) are methylated, giving a structure of the form ^{7}MeG5′ppp5′Np.

A **cap 0** at the 5′ end of mRNA has only a methyl group on 7-guanine.

A **cap 1** at the 5′ end of mRNA has methyl groups on the terminal 7-guanine and the 2′-O position of the next base.

A **cap 2** has three methyl groups (7-guanine, 2′-O position of next base, and N^6 adenine) at the 5′ end of mRNA.

A **capsid** is the external protein coat of a virus particle.

The **carboxy terminal domain (CTD)** of eukaryotic RNA polymerase II is phosphorylated at initiation and is involved in coordinating several activities with transcription.

A **cascade** is a sequence of events, each of which is stimulated by the previous one. In transcriptional regulation, as seen in sporulation and phage lytic development, it means that regulation is divided into stages, and at each stage, one of the genes that are expressed codes for a regulator needed to express the genes of the next stage.

The **cassette** model for yeast mating type describes a single active locus (the active cassette) and two inactive copies of the locus (the silent cassettes). Mating type is changed when an active cassette of one type is replaced by a silent cassette of the other type.

To **catenate** is to link together two circular molecules, as in a chain.

C-bands are generated by staining techniques that react with centromeres. The centromere appears as a darkly staining dot.

CD3 is a complex of proteins that associates with the T cell antigen receptor's α and β chains. Each complex consists of one each of the δ, ε, γ chains and two ζ chains.

cDNA is a single-stranded DNA complementary to an RNA, synthesized from it by reverse transcription *in vitro*.

The **cell-mediated response** is the immune response that is mediated primarily by T lymphocytes. It is defined based on immunity that cannot be transferred from one organism to another by serum antibody.

The **central dogma** describes the basic nature of genetic information: sequences of nucleic acid can be perpetuated and interconverted by replication, transcription, and reverse transcription, but translation from nucleic acid to protein is unidirectional, because nucleic acid sequences cannot be retrieved from protein sequences.

The **central element** is a structure that lies in the middle of the synaptonemal complex, along which the lateral elements of homologous chromosomes align. It is formed from Zip proteins.

The **centromere** is a constricted region of a chromosome that includes the site of attachment (the kinetochore) to the mitotic or meiotic spindle. The centromere consists of unique DNA sequences and proteins not found anywhere else in the chromosome.

Chaperones are a class of proteins that bind to incompletely folded or assembled proteins in order to assist their folding or prevent them from aggregating.

Chemical proofreading is a proofreading mechanism in which the correction event occurs after the addition of an incorrect subunit to a polymeric chain, by means of reversing the addition reaction.

A **chiasma** (*pl.* chiasmata) is a site at which two homologous chromosomes appear to have exchanged material during meiosis.

Chromatids are the copies of a chromosome produced by replication. The name is usually used to describe each of the copies in the period before they separate at the subsequent cell division.

Chromatin describes the state of nuclear DNA and its associated proteins during the interphase (between mitoses) of the eukaryotic cell cycle.

Chromatin remodeling describes the energy-dependent displacement or reorganization of nucleosomes that occurs in conjunction with activation of genes for transcription.

The **chromocenter** is an aggregate of heterochromatin from different chromosomes.

Chromomeres are densely staining granules visible in chromosomes under certain conditions, especially early in meiosis, when a chromosome may appear to consist of a series of chromomeres.

A **chromosome** is a discrete unit of the genome carrying many genes. Each chromosome consists of a very long molecule of duplex DNA and an approximately equal mass of proteins. It is visible as a morphological entity only during cell division.

Chromosome pairing is the coupling of the homologous chromosomes at the start of meiosis.

cis configuration describes two sites on the same molecule of DNA.

A ***cis*-acting** site affects the activity only of sequences on its own molecule of DNA (or RNA); this property usually implies that the site does not code for protein.

A ***cis*-dominant** site or mutation affects the properties only of its own molecule of DNA. *cis*-dominance is taken to indicate that a site does not code for a diffusible product. (A rare exception is that a protein is *cis*-dominant when it is constrained to act only on the DNA or RNA from which it was synthesized.)

A **cistron** is the genetic unit defined by the complementation test; it is equivalent to a gene.

The **clamp loader** is a 5-subunit protein complex that is responsible for loading the β clamp on to DNA at the replication fork.

Class switching describes a change in Ig gene organization in which the C region of the heavy chain is changed but the V region remains the same.

A **clear plaque** is a type of plaque that contains only lysed bacterial cells.

Clonal deletion describes the elimination of a clonal population of lymphocytes. At certain stages of lymphocyte development, clonal deletion can be induced when lymphocyte antigen receptors bind to their cognate antigen.

The **clonal selection** theory proposed that each lymphocyte expresses a single antigen receptor specificity and that only those lymphocytes that bind to a given antigen are stimulated to proliferate and to function in eliminating that antigen. Thus, the antigen "selects" the lymphocytes to be activated. Clonal selection is now an established principle in immunology.

The **cloverleaf** describes the structure of tRNA drawn in two dimensions, forming four distinct arm-loops.

Coactivators are factors required for transcription that do not bind DNA, but are required for (DNA-binding) activators to interact with the basal transcription factors.

A **coding end** is produced during recombination of immunoglobulin and T cell receptor genes. Coding ends are at the termini of the cleaved V and (D)J coding regions. The subsequent joining of the coding ends yields a coding joint.

A **coding region** is a part of the gene that represents a protein sequence.

The **coding strand** of DNA has the same sequence as the mRNA and is related by the genetic code to the protein sequence that it represents.

A **codon** is a triplet of nucleotides that represents an amino acid or a termination signal.

Cognate tRNAs are those recognized by a particular aminoacyl-tRNA synthetase. They all are charged with the same amino acid.

Cohesin proteins form a lateral complex that holds sister chromatids together within the synaptonemal complex. They include some SMC proteins.

Coincidental evolution describes a situation in which two genes evolve together as a single unit.

A **cointegrate** structure is produced by fusion of two replicons, one originally possessing a transposon and the other lacking it; the cointegrate has copies of the transposon present at both junctions of the replicons, oriented as direct repeats.

A **colinear** relationship describes the 1:1 representation of a sequence of triplet nucleotides in a sequence of amino acids.

A **compatibility group** of plasmids contains members unable to coexist in the same bacterial cell.

Two mutants are said to **complement** each other when a diploid that is heterozygous for each mutation produces the wild-type phenotype.

Complementary base pairs are defined by the pairing reactions in double helical nucleic acids (A with T in DNA or with U in RNA, and C with G).

A **complementation group** is a series of mutations unable to complement when tested in pairwise combinations in *trans*; defines a genetic unit (the cistron).

A **complementation test** determines whether two mutations are alleles of the same gene. It is accomplished by crossing two different recessive mutations that have the same phenotype and determining whether the wild-type phenotype can be produced. If so, the mutations are said to complement each other and are probably not mutations in the same gene.

Concerted evolution describes the ability of two related genes to evolve together as though constituting a single locus.

Condensin proteins are components of a complex that binds to chromosomes to cause condensation for meiosis or mitosis. They are members of the SMC family of proteins.

Conjugation is a process in which two cells come in contact and transfer genetic material. In bacteria, DNA is transferred from a donor to a recipient cell. In protozoa, DNA passes from each cell to the other.

A **consensus sequence** is an idealized sequence in which each position represents the base most often found when many actual sequences are compared.

Conservative transposition refers to the movement of large elements, which were originally classified as transposons but now are considered to be episomes. The mechanism of movement resembles that of phage excision and integration.

Conserved sequences are identified when many examples of a particular nucleic acid or protein are compared and the same individual bases or amino acids are always found at particular locations.

A **constant region (C region)** of an immunoglobulin or T cell receptor is the part that varies least in amino acid sequence between different molecules. Constant regions are coded by C gene segments. The heavy chain C regions identify the type of immunoglobulin and recruits effector functions.

A **constitutive** process is one that occurs all the time, unchanged by any form of stimulus or external condition.

Constitutive heterochromatin describes the inert state of permanently nonexpressed sequences, usually satellite DNA.

The **context** of a codon in mRNA refers to the fact that neighboring sequences may change the efficiency with which a codon is recognized by its aminoacyl-tRNA or is used to terminate protein synthesis.

Controlling elements of maize are transposable units originally identified solely by their genetic properties. They may be autonomous (able to transpose independently) or nonautonomous (able to transpose only in the presence of an autonomous element).

Coordinate regulation refers to the common control of a group of genes.

Copy choice is a type of recombination used by RNA viruses, in which the RNA polymerase switches from one template to another during synthesis.

The **copy number** is the number of copies of a plasmid that is maintained in a bacterium (relative to the number of copies of the origin of the bacterial chromosome).

Cordycepin is 3′ deoxyadenosine, an inhibitor of polyadenylation of RNA.

The **core** sequence is the segment of DNA that is common to the attachment sites on both the phage lambda and bacterial genomes. It is the location of the recombination event that allows phage lambda to integrate.

The **core enzyme** is the complex of RNA polymerase subunits needed for elongation. It does not include additional subunits or factors that may be needed for initiation or termination.

A **core histone** is one of the four types of histone (H2A, H2B, H3, and H4) found in the core particle derived from the nucleosome. (This excludes histone H1.)

A **core promoter** is the shortest sequence at which an RNA polymerase can initiate transcription (typically at a much lower level than that displayed by a promoter containing additional elements). For RNA polymerase II it is the minimal sequence at which the basal transcription apparatus can assemble, and it includes two sequence elements: the InR and TATA box. The core promoter is typically ~40 bp long.

A **corepressor** is a small molecule that triggers repression of transcription by binding to a regulator protein.

Cosuppression describes the ability of a transgene (usually in plants) to inhibit expression of the corresponding endogenous gene.

Cotranslational translocation describes the movement of a protein across a membrane as the protein is being synthesized. The term is usually restricted to cases in which the ribosome binds to the channel. This form of translocation may be restricted to the endoplasmic reticulum.

A **countertranscript** is an RNA molecule that prevents an RNA primer from initiating transcription by base pairing with the primer.

A **CpG island** is a stretch of 1 to 2 kb in a mammalian genome that is rich in unmethylated CpG doublets.

Crossing-over is a reciprocal exchange of material between chromosomes that occurs during prophase I of meiosis and is responsible for genetic recombination.

Crossover control limits the number of recombination events between meiotic chromosomes to one to two crossovers per pair of homologs.

Crossover fixation refers to a possible consequence of unequal crossing-over that allows a mutation in one member of a tandem cluster to spread through the whole cluster (or to be eliminated).

Crown gall disease is a tumor that can be induced in many plants by infection with the bacterium *Agrobacterium tumefaciens*.

CRP activator is a positive regulator protein activated by cyclic AMP. It is needed for RNA polymerase to initiate transcription of many operons of *E. coli*.

A **cryptic satellite** is a satellite DNA sequence not identified as such by a separate peak on a density gradient; that is, it remains present in main-band DNA.

The **C-value** is the total amount of DNA in the genome (per haploid set of chromosomes).

The **C-value paradox** describes the lack of relationship between the DNA content (C-value) of an organism and its coding potential.

The **cytoplasmic domain** is the part of a transmembrane protein that is exposed to the cytosol.

Cytoplasmic inheritance is a property of genes located in mitochondria or chloroplasts.

A **cytotoxic T cell** is a T lymphocyte (usually $CD8^+$) that can be stimulated to kill cells containing intracellular pathogens, such as viruses.

Cytotype is a cytoplasmic condition that affects P element activity. The effect of cytotype is due to the presence or absence of a repressor of transposition, which is provided by the mother to the egg.

The **D arm** of tRNA has a high content of the base dihydrouridine.

A **D loop** is a region within mitochondrial DNA in which a short stretch of RNA is paired with one strand of DNA, displacing the original partner DNA strand in this region. The same term is used also to describe the displacement of a region of one strand of duplex DNA by a complementary single-stranded invader.

The **D segment** is an additional sequence that is found between the V and J regions of an immunoglobulin heavy chain.

A **daughter** strand or duplex of DNA refers to the newly synthesized DNA.

A ***de novo* methylase** adds a methyl group to an unmethylated target sequence on DNA.

A **deacetylase** is an enzyme that removes acetyl groups from proteins.

Deacylated tRNA has no amino acid or polypeptide chain attached because it has completed its role in protein synthesis and is ready to be released from the ribosome.

The **degradosome** is a complex of bacterial enzymes, including RNAase and helicase activities, that may be involved in degrading mRNA.

Delayed early genes in phage lambda are equivalent to the middle genes of other phages. They cannot be transcribed until regulator protein(s) coded by the immediate early genes have been synthesized.

A **deletion** is the removal of a sequence of DNA, the regions on either side being joined together except in the case of a terminal deletion at the end of a chromosome.

A **demethylase** is a casual name for an enzyme that removes a methyl group, typically from DNA, RNA, or protein.

Denaturation of protein describes its conversion from the physiological conformation to some other (inactive) conformation.

A **density gradient** is used to separate macromolecules on the basis of differences in their density. It is prepared from a heavy soluble compound such as CsCl.

A **deoxyribonuclease** is an enzyme that attacks bonds in DNA. It may cut only one strand or both strands.

Deoxyribonucleic acid (DNA) is a nucleic acid molecule consisting of long chains of polymerized (deoxyribo)nucleotides. In double-stranded DNA, the two strands are held together by hydrogen bonds between complementary nucleotide base pairs.

A **dicentric chromosome** is the product of fusing two chromosome fragments, each of which has a centromere. It is unstable and may be broken when the two centromeres are pulled to opposite poles in mitosis.

Direct repeats are identical (or closely related) sequences present in two or more copies in the same orientation in the same molecule of DNA.

Divergence is the percent difference in nucleotide sequence between two related DNA sequences or in amino acid sequences between two proteins.

DNA fingerprinting analyzes the differences between individuals of the fragments generated by using restriction enzymes to cleave regions that contain short repeated sequences or by PCR. The lengths of the repeated regions are unique to every individual, and as a result the presence of a particular subset in any two individuals can be used to define their common inheritance (e.g., a parent–child relationship).

DNA ligase makes a bond between an adjacent 3′-OH and 5′-phosphate end where there is a nick in one strand of duplex DNA.

A **dna mutant** of bacteria is temperature-sensitive; it cannot synthesize DNA at 42°C but can do so at 37°C.

A **DNA polymerase** is an enzyme that synthesizes a daughter strand(s) of DNA (under direction from a DNA template). Any particular enzyme may be involved in repair or replication (or both).

A **DNA replicase** is a DNA-synthesizing enzyme required specifically for replication.

A **domain** of a chromosome may refer *either* to a discrete structural entity defined as a region within which supercoiling is independent of other domains *or* to an extensive region including an expressed gene that has heightened sensitivity to degradation by the enzyme DNAase I.

A **domain** of a protein is a discrete continuous part of the amino acid sequence that can be equated with a particular function.

A **dominant negative** mutation results in a mutant gene product that prevents the function of the wild-type gene product, causing loss or reduction of gene activity in cells containing both the mutant and wild-type alleles. The most common cause is that the gene codes for a homomultimeric protein whose function is lost if only one of the subunits is a mutant.

Dosage compensation describes mechanisms employed to compensate for the discrepancy between the presence of two X chromosomes in one sex but only one X chromosome in the other sex.

A **double-strand break (DSB)** occurs when both strands of a DNA duplex are cleaved at the same site. Genetic recombination is initiated by double-strand breaks. The cell also has repair systems that act on double-strand breaks created at other times.

The **doubling time** is the period (usually measured in minutes) that it takes for a bacterial cell to reproduce.

A **down mutation** in a promoter decreases the rate of transcription.

Downstream identifies sequences proceeding farther in the direction of expression; for example, the coding region is downstream of the initiation codon.

A **DP thymocyte** is a double-positive thymocyte. It is an immature T cell that expresses cell surface CD4 and CD8. Selection of DP thymocytes in the thymus yields mature T cells expressing either CD4 or CD8.

Early genes are transcribed before the replication of phage DNA. They code for regulators and other proteins needed for later stages of infection.

Early infection is the part of the phage lytic cycle between entry and replication of the phage DNA. During this time, the phage synthesizes the enzymes needed to replicate its DNA.

EF-Tu is the elongation factor that binds aminoacyl-tRNA and places it into the A site of a bacterial ribosome.

Elongation is the stage in a macromolecular synthesis reaction (replication, transcription, or translation) when the nucleotide or polypeptide chain is extended by the addition of individual subunits.

Elongation factors (EF in prokaryotes; eEF in eukaryotes) are proteins that associate with ribosomes cyclically during the addition of each amino acid to the polypeptide chain.

Endonucleases cleave bonds within a nucleic acid chain; they may be specific for RNA or for single-stranded or double-stranded DNA.

An **endotoxin** is a toxin that is present on the surface of Gram-negative bacteria (as opposed to exotoxins, which are secreted). LPS is an example of an endotoxin.

An **enhanceosome** is a complex of transcription factors that assembles cooperatively at an enhancer.

An **enhancer** is a *cis*-acting sequence that increases the utilization of (some) eukaryotic promoters, and can function in either orientation and in any location (upstream or downstream) relative to the promoter.

Enzyme turnover is the process through which the enzyme returns to its original shape, enabling the enzyme to catalyze another reaction.

Epigenetic changes influence the phenotype without altering the genotype. They consist of changes in the properties of a cell that are inherited but that do not represent a change in genetic information.

An **episome** is a plasmid able to integrate into bacterial DNA.

An **epitope** is the portion of an antigen that is recognized by the antigen receptor on lymphocytes. It is also called an antigenic determinant.

Error-prone synthesis occurs when DNA incorporates noncomplementary bases into the daughter strand.

Euchromatin comprises most of the genome in the interphase nucleus, is less tightly coiled than heterochromatin, and contains most of the active or potentially active single copy genes.

The **evolutionary clock** is defined by the rate at which mutations accumulate in a given gene.

The **excision** of phage or episome or other sequence describes its release from the host chromosome as an autonomous DNA molecule.

The **excision** step in an excision-repair system consists of removing a single-stranded stretch of DNA by the action of a 5′ to 3′ exonuclease.

Excision repair describes a type of repair system in which one strand of DNA is directly excised and then replaced by resynthesis using the complementary strand as template.

An **exon** is any segment of an interrupted gene that is represented in the mature RNA product.

Exon definition describes the process when a pair of splicing sites are recognized by interactions involving the 5′ site of the intron and also the 5′ site of the next intron downstream.

Exon trapping inserts a genomic fragment into a vector whose function depends on the provision of splicing junctions by the fragment.

Exonucleases cleave nucleotides one at a time from the end of a polynucleotide chain; they may be specific for either the 5′ or 3′ end of DNA or RNA.

Extein sequences remain in the mature protein that is produced by processing a precursor via protein splicing.

The **external domain** is the part of a plasma membrane protein that extends outside of the cell. Upon internalization, the protein's external domain extends into the lumen (the topological equivalent of the outside of the cell) of an organelle.

The **extra arm** of tRNA lies between the TψC and anticodon arms. It is the most variable in length in tRNA, from 3 to 21 bases. tRNAs are called class 1 if they lack it, and class 2 if they have it.

An **extrachromosomal genome** in a bacterium is a self-replicating set of genes that is not part of the bacterial chromosome. In many cases, the genes are necessary for bacterial growth under certain environmental conditions.

Extranuclear genes reside outside the nucleus in organelles such as mitochondria and chloroplasts.

The **F plasmid** is an episome that can be free or integrated in *E. coli*, and which in either form can sponsor conjugation.

Facultative heterochromatin describes the inert state of sequences that also exist in active copies—for example, one mammalian X chromosome in females.

Fixation is the process by which a new allele replaces the allele that was previously predominant in a population.

Footprinting is a technique for identifying the site on DNA bound by some protein by virtue of the protection of bonds in this region against attack by nucleases.

Forward mutations inactivate a wild-type gene.

Frameshifts are mutations caused by deletions or insertions that are not a multiple of three base pairs. They change the frame in which triplets are translated into protein.

A **fully methylated** site is a palindromic sequence that is methylated on both strands of DNA.

A **gain-of-function** mutation usually refers to a mutation that causes an increase in the normal gene activity. It sometimes represents acquisition of certain abnormal properties. It is often, but not always, dominant.

G-bands are generated on eukaryotic chromosomes by staining techniques and appear as a series of lateral striations. They are used for karyotyping (identifying chromosomes and chromosomal regions by the banding pattern).

The **GC box** is a common pol II promoter element consisting of the sequence GGGCGG.

A **gene** is the segment of DNA specifying a polypeptide chain; it includes regions preceding and following the coding region (leader and trailer), as well as intervening sequences (introns) between individual coding segments (exons).

A **gene cluster** is a group of adjacent genes that are identical or related.

Gene conversion is the alteration of one strand of a heteroduplex DNA to make it complementary with the other strand at any position(s) where there were mispaired bases.

A **gene family** consists of a set of genes within a genome that code for related or identical proteins or RNAs. The members were derived by duplication of an ancestral gene followed by accumulation of changes in sequence between the copies. Most often the members of a family are related but not identical.

The **genetic code** is the correspondence between triplets in DNA (or RNA) and amino acids in protein.

The **genome** is the complete set of sequences in the genetic material of an organism. It includes the sequence of each chromosome plus any DNA in organelles.

The **glucocorticoid response element (GRE)** is a sequence in a promoter or enhancer that is recognized by the glucocorticoid receptor, which is activated by glucocorticoid steroids.

GMP-PCP is an analog of GTP that cannot be hydrolyzed. It is used to test which stage in a reaction requires hydrolysis of GTP.

Gratuitous inducers resemble authentic inducers of transcription, but are not substrates for the induced enzymes.

The **GT-AG rule** describes the presence of these constant dinucleotides at the first two and last two positions of introns of nuclear genes.

A **guide RNA** is a small RNA whose sequence is complementary to the sequence of an RNA that has been edited. It is used as a template for changing the sequence of the preedited RNA by inserting or deleting nucleotides.

The **H2 locus** is the mouse major histocompatibility complex, a cluster of genes on chromosome 17. The genes encode proteins for antigen presentation, cytokines, and complement proteins.

The **haplotype** is the particular combination of alleles in a defined region of some chromosome—in effect, the genotype in miniature. Originally used to described combinations of MHC alleles, it now may be used to describe particular combinations of RFLPs, SNPs, or other markers.

A **hapten** is a small molecule that acts as an antigen when conjugated to a protein.

Hb anti-Lepore is a fusion gene produced by unequal crossing-over that has the N-terminal part of β globin and the C-terminal part of δ globin.

Hb Kenya is a fusion gene produced by unequal crossing-over between the between $^{A}\gamma$ and β globin genes.

Hb Lepore is an unusual globin protein that results from unequal crossing-over between the β and δ genes. The genes become fused together to produce a single β-like chain that consists of the N-terminal sequence of δ joined to the C-terminal sequence of β.

HbH disease results from a condition in which there is a disproportionate amount of the abnormal tetramer β_4 relative to the amount of normal hemoglobin ($\alpha_2\beta_2$).

The **headpiece** is the DNA-binding domain of the *lac* repressor.

Heat shock genes are a set of loci activated in response to an increase in temperature (and other abuses to the cell). All organisms have heat shock genes. Their products usually include chaperones that act on denatured proteins.

The **heat shock response element (HSE)** is a sequence in a promoter or enhancer that is used to activate a gene by an activator induced by heat shock.

The immunoglobulin **heavy chain** is one of two types of subunits in an antibody tetramer. Each antibody contains two heavy chains. The N-terminus of the heavy chain forms part of the antigen recognition site, whereas the C-terminus determines the subclass (isotype).

Heavy strands and light strands of a DNA duplex refer to the density differences that result when there is an asymmetry between base representation in the two strands such that one strand is rich in T and G bases and the other is rich in C and A bases. This occurs in some satellite and mitochondrial DNAs.

A **helicase** is an enzyme that uses energy provided by ATP hydrolysis to separate the strands of a nucleic acid duplex.

The **helix-loop-helix (HLH)** motif is responsible for dimerization of a class of transcription factors called HLH proteins. A bHLH protein has a basic DNA-binding sequence close to the dimerization motif.

The **helix-turn-helix** motif describes an arrangement of two α-helices that form a site that binds to DNA, one fitting into the major groove of DNA and other lying across it.

A **helper T cell** is a T lymphocyte that activates macrophages and stimulates B cell proliferation and antibody production. Helper T cells usually express cell surface CD4 but not CD8.

A **helper virus** provides functions absent from a defective virus, enabling the latter to complete the infective cycle during a mixed infection.

Hemimethylated DNA is methylated on one strand of a target sequence that has a cytosine on each strand.

A **hemi-methylated** site is a palindromic sequence that is methylated on only one strand of DNA.

Heterochromatin describes regions of the genome that are highly condensed, are not transcribed, and are late-replicating. Heterochromatin is divided into two types: constitutive and facultative.

Heteroduplex DNA is generated by base pairing between complementary single strands derived from the different parental duplex molecules; it occurs during genetic recombination.

Heterogeneous nuclear RNA (hnRNA) comprises transcripts of nuclear genes made by RNA polymerase II; it has a wide size distribution and low stability.

A **heteromultimer** is a protein that is composed of nonidentical subunits (coded by different genes).

An **Hfr** cell is a bacterium that has an integrated F plasmid within its chromosome. Hfr stands for high frequency recombination, referring to the fact that chromosomal genes are transferred from an Hfr cell to an F^- cell much more frequently than from an F^+ cell.

Histones are conserved DNA-binding proteins that form the basic subunit of chromatin in eukaryotes. Histones H2A, H2B, H3, and H4 form an octameric core around which DNA coils to form a nucleosome. Histone H1 is external to the nucleosome.

Histone acetyltransferase (HAT) enzymes modify histones by addition of acetyl groups; some transcriptional coactivators have HAT activity.

Histone deacetylases (HDAC) remove acetyl groups from histones; they may be associated with repressors of transcription.

The **histone fold** is a motif found in all four core histones in which three α-helices are connected by two loops.

The **HLA** locus is the human major histocompatibility complex, a cluster of genes on chromosome 6. The genes encode proteins for antigen presentation, cytokines, and complement proteins.

An **hnRNP** is the ribonucleoprotein form of hnRNA (heterogeneous nuclear RNA), in which the hnRNA is complexed with proteins. Pre-mRNAs are not exported until processing is complete; thus hnRNPs are found only in the nucleus.

A **Holliday** structure is an intermediate structure in homologous recombination, for which the two duplexes of DNA are connected by the genetic material exchanged between two of the four strands, one from each duplex. A joint molecule is said to be resolved when nicks in the structure restore two separate DNA duplexes.

The RNA polymerase **holoenzyme** is the form that is competent to initiate transcription. It consists of the four subunits of the core enzyme ($\alpha_2\beta\beta'$) and σ factor.

The **homeodomain** is a DNA-binding motif that typifies a class of transcription factors. The DNA sequence that codes for it is called the homeobox.

Homologous recombination involves a reciprocal exchange of sequences of DNA, e.g., between two chromosomes that carry the same genetic loci.

A **homomultimer** is a protein composed of identical subunits.

A **hotspot** is a site in the genome at which the frequency of mutation (or recombination) is very much increased, usually by at least an order of magnitude relative to neighboring sites.

Housekeeping genes are those (theoretically) expressed in all cells because they provide basic functions needed for sustenance of all cell types.

The **humoral response** is an immune response that is mediated primarily by antibodies. It is defined as immunity that can be transferred from one organism to another by serum antibody.

Hybrid dysgenesis describes the inability of certain strains of *D. melanogaster* to interbreed, because the hybrids are sterile (although otherwise they may be phenotypically normal).

Hybridization describes the pairing of complementary RNA and DNA strands to give an RNA–DNA hybrid.

Hybridoma is a cell line produced by fusing a myeloma with a lymphocyte; it continues indefinitely to express the immunoglobulins of both parents.

Hydrops fetalis is a fatal disease resulting from the absence of the hemoglobin α gene.

Hypermutation describes the introduction of somatic mutations in a rearranged immunoglobulin gene. The mutations can change the sequence of the corresponding antibody, especially in its antigen-binding site.

A **hypersensitive site** is a short region of chromatin detected by its extreme sensitivity to cleavage by DNAase I and other nucleases; it comprises an area from which nucleosomes are excluded.

Icosahedral symmetry is typical of viruses that have capsids that are polyhedrons.

IF-1 is a bacterial initiation factor that stabilizes the initiation complex.

IF-2 is a bacterial initiation factor that binds the initiator tRNA to the initiation complex.

IF-3 is a bacterial initiation factor required for 30S subunits to bind to initiation sites in mRNA. It also prevents 30S subunits from binding to 50S subunits.

Immediate early phage genes in phage lambda are equivalent to the early class of other phages. They are transcribed immediately upon infection by the host RNA polymerase.

An **immune response** is an organism's reaction, mediated by components of the immune system, to an antigen.

Immunity in phages refers to the ability of a prophage to prevent another phage of the same type from infecting a cell. It results from the synthesis of phage repressor by the prophage genome.

Immunity in plasmids describes the ability of a plasmid to prevent another of the same type from becoming established in a cell. It results usually from interference with the ability to replicate.

Immunity refers to the ability of certain transposons to prevent others of the same type from transposing to the same DNA molecule.

The **immunity region** is a segment of the phage genome that enables a prophage to inhibit additional phage of the same type from infecting the bacterium. This region has a gene that encodes for the repressor, as well as the sites to which the repressor binds.

An **immunoglobulin** is a protein that is produced by B cells and that binds to a particular antigen.

Imprecise excision occurs when the transposon removes itself from the original insertion site but leaves behind some of its sequence.

Imprinting describes a change in a gene that occurs during passage through the sperm or egg with the result that the paternal and maternal alleles have different properties in the very early embryo. This is caused by methylation of DNA.

***In situ* hybridization** is performed by denaturing the DNA of cells squashed on a microscope slide so that reaction is possible with an added single-stranded RNA or DNA; the added preparation is radioactively labeled and its hybridization is followed by autoradiography.

***In vitro* complementation** is a functional assay used to identify components of a process. The reaction is reconstructed using extracts from a mutant cell. Fractions from wild-type cells are then tested for restoration of activity.

Incision is a step in a mismatch excision repair system. An endonuclease recognizes the damaged area in the DNA and isolates it by cutting the DNA strand on both sides of the damage.

Indirect end labeling is a technique for examining the organization of DNA by making a cut at a specific site and identifying all fragments containing the sequence adjacent to one side of the cut; it reveals the distance from the cut to the next break(s) in DNA.

Induced mutations result from the action of a mutagen. The mutagen may act directly on the bases in DNA or it may act indirectly to trigger a pathway that leads to a change in DNA sequence.

An **inducer** is a small molecule that triggers gene transcription by binding to a regulator protein.

Induction of prophage describes its entry into the lytic (infective) cycle as a result of destruction of the lysogenic repressor, which leads to excision of free phage DNA from the bacterial chromosome.

Induction refers to the ability of bacteria (or yeast) to synthesize certain enzymes only when their substrates are present; applied to gene expression, it refers to switching on transcription as a result of interaction of the inducer with the regulator protein.

Initiation describes the stages of transcription up to synthesis of the first bond in RNA. This includes binding of RNA polymerase to the promoter and melting a short region of DNA into single strands.

The **initiation codon** is a special codon (usually AUG) used to start synthesis of a protein.

An **initiation complex** in bacterial protein synthesis contains a small ribosome subunit, initiation factors, and initiator aminoacyl-tRNA bound to mRNA at an AUG initiation codon.

Initiation factors (IF in prokaryotes; eIF in eukaryotes) are proteins that associate with the small subunit of the ribosome specifically at the stage of initiation of protein synthesis.

Innate immunity is the rapid response mediated by cells with nonvarying (germline-encoded) receptors that recognize pathogen. The cells of the innate immune response act to eliminate the pathogen and initiate the adaptive immune response.

The **Inr** is the sequence of a pol II promoter between –3 and +5 and has the general sequence Py_2CAPy_5. It is the simplest possible pol II promoter.

An **insertion** is the addition of a stretch of base pairs in DNA. Duplications are a special class of insertions.

An **insertion sequence (IS)** is a small bacterial transposon that carries only the genes needed for its own transposition.

An **insulator** is a sequence that prevents an activating or inactivating effect passing from one side to the other.

An **intasome** is a protein-DNA complex between the phage lambda integrase (Int) and the phage lambda attachment site (*attP*).

An **integrase** is an enzyme that is responsible for a site-specific recombination that inserts one molecule of DNA into another.

Integration of viral or another DNA sequence describes its insertion into a host genome as a region covalently linked on either side to the host sequences.

An **intein** is the part that is removed from a protein that is processed by protein splicing.

Interallelic complementation describes the change in the properties of a heteromultimeric protein brought about by the interaction of subunits coded by two different mutant alleles; the mixed protein may be more or less active than the protein consisting of subunits only of one or the other type. For another cause of interallelic complementation see transvection.

Interbands are the relatively dispersed regions of polytene chromosomes that lie between the bands.

The **intercistronic region** is the distance between the termination codon of one gene and the initiation codon of the next gene.

Interspersed repeats were originally defined as short sequences that are common and widely distributed in the genome. They are now known to consist of transposable elements.

Intrinsic terminators are able to terminate transcription by bacterial RNA polymerase in the absence of any additional factors.

An **intron** is a segment of DNA that is transcribed, but later removed from within the transcript by splicing together the sequences (exons) on either side of it.

Intron definition describes the process when a pair of splicing sites are recognized by interactions involving only the 5′ site and the branchpoint/3′ site.

Intron homing describes the ability of certain introns to insert themselves into a target DNA. The reaction is specific for a single target sequence.

Invariant base positions in tRNA have the same nucleotide in virtually all (>95%) tRNAs.

Inverted terminal repeats are the short related or identical sequences present in reverse orientation at the ends of some transposons.

Isoschizomers are restriction enzymes that cleave the same DNA sequence but are affected differently by its state of methylation.

J segments are coding sequences in the immunoglobulin and T cell receptor loci. The J segments are between the variable (V) and constant (C) gene segments.

A **joint molecule** is a pair of DNA duplexes that are connected together through a reciprocal exchange of genetic material.

A **kilobase** is a measure of length and may be used to refer to DNA (1000 base pairs) or to RNA (1000 bases).

Kinetic proofreading describes a proofreading mechanism that depends on incorrect events proceeding more slowly than correct events, so that incorrect events are reversed before a subunit is added to a polymeric chain.

The **kinetochore** is a small organelle associated with the surface of the centromere that attaches a chromosome to the microtubules of the mitotic spindle. Each mitotic chromosome contains two "sister" kinetochores that are positioned on opposite sides of its centromere and face in opposite directions.

Kirromycin is an antibiotic that inhibits protein synthesis by acting on EF-Tu.

A **knot** in the DNA is an entangled region that cannot be resolved without cutting and rearranging the DNA.

Kuru is a human neurological disease caused by prions. It may be caused by eating infected brains.

The **lagging strand** of DNA must grow overall in the 3′ to 5′ direction and is synthesized discontinuously in the form of short fragments (5′–3′) that are later connected covalently.

Lampbrush chromosomes are the extremely extended meiotic bivalents of certain amphibian oocytes.

The **large subunit** of the ribosome (50S in bacteria, 60S in eukaryotes) has the peptidyl transferase active site that synthesizes the peptide bond.

The **lariat** is an intermediate in RNA splicing in which a circular structure with a tail is created by a 5′ to 2′ bond.

Late genes are transcribed when phage DNA is being replicated. They code for components of the phage particle.

Late infection is the part of the phage lytic cycle from DNA replication to lysis of the cell. During this time, the DNA is replicated and structural components of the phage particle are synthesized.

A **lateral element** is a structure in the synaptonemal complex that forms when a pair of sister chromatids condenses on to an axial element.

The **leader** of a protein is a short N-terminal sequence responsible for initiating passage into or through a membrane.

The **leader** of an mRNA is the untranslated sequence at the 5′ end that precedes the initiation codon.

The **leader peptide** is the product that would result from translation of a short coding sequence used to regulate transcription of the tryptophan operon by controlling ribosome movement.

The **leading strand** of DNA is synthesized continuously in the 5′ to 3′ direction.

Leaky mutations leave some residual function—for instance, when the mutant protein is partially active (in the case of a missense mutation), or when read-through produces a small amount of wild-type protein (in the case of a nonsense mutation).

The **leucine zipper** is a dimerization motif that is found in a class of transcription factors.

The **leucine-rich region** is a motif found in the extracellular domains of some surface receptor proteins in animal and plant cells.

A **licensing factor** is located in the nucleus and is necessary for replication; it is inactivated or destroyed after one round of replication. New licensing factors must be provided for further rounds of replication to occur.

The immunoglobulin **light chain (L)** is one of two types of subunits in an antibody tetramer. Each antibody contains two light chains. The N-terminus of the light chain forms part of the antigen recognition site.

Linkage describes the tendency of genes to be inherited together as a result of their location on the same chromosome; measured by percent recombination between loci.

The **linking number** is the number of times the two strands of a closed DNA duplex cross over each other.

A **lipopolysaccharide (LPS)** is a molecule containing both lipid and sugar components. It is present in the outer membrane of Gram-negative bacteria. It is also an endotoxin responsible for inducing septic shock during an infection.

A **locus** is the position on a chromosome at which the gene for a particular trait resides; a locus may be occupied by any one of the alleles for the gene.

The **locus control region (LCR)** that is required for the expression of several genes in a domain.

The **long terminal repeat (LTR)** is the sequence that is repeated at each end of the provirus (integrated retroviral sequence).

A **loop** is a single-stranded region at the end of a hairpin in RNA (or single-stranded DNA); it corresponds to the sequence between inverted repeats in duplex DNA.

A **loose binding site** is any random sequence of DNA that is bound by the core RNA polymerase when it is not engaged in transcription.

A **loss-of-function** mutation eliminates or reduces the activity of a gene. It is often, but not always, recessive.

Luxury genes are those coding for specialized functions synthesized (usually) in large amounts in particular cell types.

Lysis describes the death of bacteria at the end of a phage infective cycle when they burst open to release the progeny of an infecting phage (because phage enzymes disrupt the bacterium's cytoplasmic membrane or cell wall). The same term also applies to eukaryotic cells; for example, when infected cells are attacked by the immune system.

Lysogeny describes the ability of a phage to survive in a bacterium as a stable prophage component of the bacterial genome.

Lytic infection of a bacterium by a phage ends in the destruction of the bacterium with release of progeny phage.

A **maintenance methylase** adds a methyl group to a target site that is already hemimethylated.

The **major groove** of DNA is 22 Å across.

The **major histocompatibility complex (MHC)** is a chromosomal region containing genes that are involved in the immune response. The genes encode proteins for antigen presentation, cytokines, and complement, as well as other functions. The MHC is highly polymorphic. MHC genes and proteins are divided into three classes.

Maternal inheritance describes the preferential survival in the progeny of genetic markers provided by one parent.

A **matrix attachment site (MAR)** is a region of DNA that attaches to the nuclear matrix. It is also known as a scaffold attachment site (SAR).

A **maturase** is a protein coded by a group I or group II intron that is needed to assist the RNA to form the active conformation that is required for self-splicing.

Mediator is a large protein complex associated with yeast bacterial RNA polymerase II. It contains factors that are necessary for transcription from many or most promoters.

A **megabase** is one million base pairs of DNA.

A **memory cell** is a lymphocyte that has been stimulated during the primary immune response to antigen and that is rapidly activated upon subsequent exposure to that antigen. Memory cells respond more rapidly to antigen than naive cells.

Messenger RNA (mRNA) is the intermediate that represents one strand of a gene coding for protein. Its coding region is related to the protein sequence by the triplet genetic code.

A **methyltransferase** is an enzyme that adds a methyl group to a substrate, which can be a small molecule, a protein, or a nucleic acid.

MHC class I proteins are a major type of MHC molecule. In most cases, MHC class I proteins present peptides to $CD8^+$ cytotoxic T lymphocytes. MHC class I-binding peptides are usually produced by proteolytic degradation in the cytosol.

MHC class II proteins are a major type of MHC molecule. In most cases, MHC class II proteins present peptides to $CD4^+$ helper T lymphocytes. MHC class II-binding peptides are usually produced by proteolytic degradation in endosomes and lysosomes.

Micrococcal nuclease is an endonuclease that cleaves DNA; in chromatin, DNA is cleaved preferentially between nucleosomes.

MicroRNAs are very short RNAs that may regulate gene expression.

Microsatellite DNAs consist of repetitions of extremely short (typically <10 bp) units.

A **microtubule organizing center (MTOC)** is a region from which microtubules emanate. In animal cells the centrosome is the major microtubule organizing center.

Middle genes are phage genes that are regulated by the proteins coded by early genes. Some proteins coded by middle genes catalyze replication of the phage DNA; others regulate the expression of a later set of genes.

A **minicell** is an anucleate bacterial (*E. coli*) cell produced by a division that generates a cytoplasm without a nucleus.

Minisatellite DNAs consist of ~10 copies of a short repeating sequence. The length of the repeating unit is measured in 10s of base pairs. The number of repeats varies between individual genomes.

The **minor groove** of DNA is 12 Å across.

Minus strand DNA is the single-stranded DNA sequence that is complementary to the viral RNA genome of a plus strand virus.

A **mismatch** describes a site in DNA where the pair of bases does not conform to the usual G-C or A-T pairs. It may be caused by incorporation of the wrong base during replication or by mutation of a base.

Mismatch repair corrects recently inserted bases that do not pair properly. The process preferentially corrects the sequence of the daughter strand by distinguishing the daughter strand and parental strand, sometimes on the basis of their states of methylation.

Missense mutations change a single codon so as to cause the replacement of one amino acid by another in a protein sequence.

A **missense suppressor** codes for a tRNA that has been mutated to recognize a different codon. By inserting a different amino acid at a mutant codon, the tRNA suppresses the effect of the original mutation.

Modification of DNA or RNA includes all changes made to the nucleotides after their initial incorporation into the polynucleotide chain.

Modified bases are all those except the usual four from which DNA (T, C, A, G) or RNA (U, C, A, G) are synthesized; they result from postsynthetic changes in the nucleic acid.

Monocistronic mRNA codes for one protein.

Monster particles of bacteriophages form as the result of an assembly defect in which the capsid proteins form a head that is much longer than usual.

A plasmid is said to be under **multicopy control** when the control system allows the plasmid to exist in more than one copy per individual bacterial cell.

A **multiforked chromosome** (in a bacterium) has more than one set of replication forks, because a second initiation has occurred before the first cycle of replication has been completed.

A locus is said to have **multiple alleles** when more than two allelic forms have been found. Each allele may cause a different phenotype.

Mutagens increase the rate of mutation by inducing changes in DNA sequence, directly or indirectly.

A **mutator** is a mutation or a mutated gene that increases the basal level of mutation. Such genes often code for proteins that are involved in repairing damaged DNA.

An **N nucleotide** sequence is a short non-templated sequence that is added randomly by the enzyme at coding joints during rearrangement of immunoglobulin and T cell receptor genes. N nucleotides augment the diversity of antigen receptors.

The **n–1 rule** states that only one X chromosome is active in female mammalian cells; any others are inactivated.

A **nascent protein** has not yet completed its synthesis; the polypeptide chain is still attached to the ribosome via a tRNA.

Nascent RNA is a ribonucleotide chain that is still being synthesized, so that its 3′ end is paired with DNA where RNA polymerase is elongating.

Negative complementation occurs when interallelic complementation allows a mutant subunit to suppress the activity of a wild-type subunit in a multimeric protein.

The default state of genes that are controlled by **negative regulation** is to be expressed. A specific intervention is required to turn them off.

A **neutral** mutation has no significant effect on evolutionary fitness and usually has no effect on the phenotype.

Neutral substitutions in a protein cause changes in amino acids that do not affect activity.

N-formyl-methionyl-tRNA ($tRNA_f^{Met}$) is the aminoacyl-tRNA that initiates bacterial protein synthesis. The amino group of the methionine is formylated.

Nick translation describes the ability of *E. coli* DNA polymerase I to use a nick as a starting point from which one strand of a duplex DNA can be degraded and replaced by resynthesis of new material; is used to introduce radioactively labeled nucleotides into DNA *in vitro*.

Nonallelic genes are two (or more) copies of the same gene that are present at *different* locations in the genome (contrasted with alleles, which are copies of the same gene derived from different parents and present at the same location on the homologous chromosomes).

A **nonautonomous controlling element** is a transposon in maize that encodes a non-functional transposase; it can transpose only in the presence of a *trans*-acting autonomous member of the same family.

A **nonhistone** is any structural protein found in a chromosome except one of the histones.

Non-homologous end-joining ligates blunt ends. It is common to many repair pathways and to certain recombination pathways (such as immunoglobulin recombination).

The recombination of V, (D), J gene segments results in a **nonproductive rearrangement** if the rearranged gene segments are not in the correct reading frame. A nonproductive rearrangement occurs when nucleotide addition or subtraction disrupts the reading frame or when a functional protein is not produced.

Nonreciprocal recombination results from an error in pairing and crossing-over in which nonequivalent sites are involved in a recombination event. It produces one recombinant with a deletion of material and one with a duplication.

Nonrepetitive DNA shows reassociation kinetics expected of unique sequences.

Nonreplicative transposition describes the movement of a transposon that leaves a donor site (usually generating a double-strand break) and moves to a new site.

A **nonsense mutation** is any change in DNA that replaces a codon specifying an amino acid with a translation-termination codon (UAG, UGA, or UAA).

A **nonsense suppressor** is a gene coding for a mutant tRNA able to respond to one or more of the termination codons and insert an amino acid at that site.

Nonsense-mediated mRNA decay is a pathway that degrades an mRNA that has a nonsense mutation prior to the last exon.

The **nontranscribed spacer** is the region between transcription units in a tandem gene cluster.

The **nonviral superfamily** of transposons originated independently of retroviruses.

Nopaline plasmids are Ti plasmids of *Agrobacterium tumefaciens* that carry genes for synthesizing the opine, nopaline. They retain the ability to differentiate into early embryonic structures.

The **nucleation center** of TMV (tobacco mosaic virus) is a duplex hairpin where assembly of coat protein with RNA is initiated.

Nucleic acids are molecules that encode genetic information. They consist of a series of nitrogenous bases connected to ribose molecules that are linked by phosphodiester bonds. DNA is deoxyribonucleic acid and RNA is ribonucleic acid.

The **nucleoid** is the structure in a prokaryotic cell that contains the genome. The DNA is bound to proteins and is not enclosed by a membrane.

Nucleolar dominance describes the transcription of rRNA genes inherited from only one parent that occurs in certain crosses.

The **nucleolar organizer** is the region of a chromosome carrying genes coding for rRNA.

The **nucleolus** (*plural:* nucleoli) is a discrete region of the nucleus where ribosomes are produced.

The **nucleosome** is the basic structural subunit of chromatin, consisting of ~200 bp of DNA and an octamer of histone proteins.

Nucleosome positioning describes the placement of nucleosomes at defined sequences of DNA instead of at random locations with regard to sequence.

A **null mutation** completely eliminates the function of a gene.

The **ochre** codon is the triplet UAA, one of the three termination codons that end protein synthesis.

Octopine plasmids of *Agrobacterium tumefaciens* carry genes coding the synthesis of opines of the octopine type. The tumors are undifferentiated.

Okazaki fragments are the short stretches of 1000 to 2000 bases produced during discontinuous replication; they are later joined into a covalently intact strand.

The **opal** codon is the triplet UGA, one of the three termination codons that end protein synthesis. It has evolved to code for an amino acid in a small number of organisms or organelles.

An **open complex** describes the stage of initiation of transcription when RNA polymerase causes the two strands of DNA to separate to form the "transcription bubble."

An **open reading frame** is a sequence of DNA consisting of triplets that can be translated into amino acids starting with an initiation codon and ending with a termination codon.

The **operator** is the site on DNA at which a repressor protein binds to prevent transcription from initiating at the adjacent promoter.

An **operon** is a unit of bacterial gene expression and regulation, including structural genes and control elements in DNA recognized by regulator gene product(s).

An **opine** is a derivative of arginine that is synthesized by plant cells infected with crown gall disease.

The **origin** is a sequence of DNA at which replication is initiated.

Orthologs are corresponding proteins in two species as defined by sequence homology.

A stretch of **overwound** DNA has more base pairs per turn than the usual average (10 bp = 1 turn). This means that the two strands of DNA are more tightly wound around each other, creating tension.

A **P element** is a type of transposon in *D. melanogaster*.

A **P nucleotide** sequence is a short palindromic (inverted repeat) sequence that is generated during rearrangement of immunoglobulin and T cell receptor V, (D), J gene segments. P nucleotides are generated at coding joints when RAG proteins cleave the hairpin ends generated during rearrangement.

The **P site** of the ribosome is the site that is occupied by peptidyl-tRNA, the tRNA carrying the nascent polypeptide chain, still paired with the codon to which it bound in the A site.

The **packing ratio** is the ratio of the length of DNA to the unit length of the fiber containing it.

A **paranemic joint** describes a region in which two complementary sequences of DNA are associated side by side instead of being intertwined in a double-helical structure.

A **parental** strand or duplex of DNA refers to the DNA that will be replicated.

Patch recombinant DNA results from a Holliday junction being resolved by cutting the exchanged strands. The duplex is largely unchanged, except for a DNA sequence on one strand that came from the homologous chromosome.

A pathogen-associated molecular pattern (PAMP) is a molecular structure on the surface of a pathogen. A given PAMP may be conserved across a large number of pathogens. During an immune response, PAMPs may be recognized by receptors on cells that mediate innate immunity.

Peptidyl transferase is the activity of the large ribosomal subunit that synthesizes a peptide bond when an amino acid is added to a growing polypeptide chain. The actual catalytic activity is a property of the rRNA.

Peptidyl-tRNA is the tRNA to which the nascent polypeptide chain has been transferred following peptide bond synthesis during protein synthesis.

The **periplasm** (or periplasmic space) is the region between the inner and outer membranes in the cell envelope of Gram-negative bacteria.

A **periseptal annulus** is a ringlike area where inner and outer membranes appear fused. Formed around the circumference of the bacterium, the periseptal annulus determines the location of the septum.

Peroxins are the protein components of the peroxisome.

The **peroxisome** is an organelle in the cytoplasm enclosed by a single membrane. It contains oxidizing enzymes.

A **phosphorelay** describes a pathway in which a phosphate group is passed along a series of proteins.

Photoreactivation uses a white light-dependent enzyme to split cyclobutane pyrimidine dimers formed by ultraviolet light.

Pilin is the subunit that is polymerized into the pilus in bacteria.

A **pilus** (*plural:* pili) is a surface appendage on a bacterium that allows the bacterium to attach to other bacterial cells. It appears as a short, thin, flexible rod. During conjugation, pili are used to transfer DNA from one bacterium to another.

A **plaque** is an area of clearing in a bacterial lawn. It is created by a single phage particle that has undergone multiple rounds of lytic growth.

A **plasmid** is a circular, extrachromosomal DNA. It is autonomous and can replicate itself.

A **plectonemic** joint is a region that consists of one molecule wound around another molecule, e.g., the DNA strands in a double helix.

Plus strand DNA is the strand of the duplex sequence representing a retrovirus that has the same sequence as that of the RNA.

A **plus strand virus** has a single-stranded nucleic acid genome whose sequence directly codes for the protein products.

A **point mutation** is a change in the sequence of DNA involving a single base pair.

Polarity refers to the effect of a mutation in one gene in influencing the expression (at transcription or translation) of subsequent genes in the same transcription unit.

Poly(A) is a stretch of ~200 bases of adenylic acid that is added to the 3′ end of mRNA following its synthesis.

Poly(A)$^-$ mRNA is mRNA that has does not have a 3′ terminal stretch of poly(A).

Poly(A) polymerase (PAP) is the enzyme that adds the stretch of polyadenylic acid to the 3′ end of eukaryotic mRNA. It does not use a template.

Poly(A)$^+$ mRNA is mRNA that has a 3′ terminal stretch of poly(A).

Poly(A)-binding protein is the protein that binds to the 3′ stretch of poly(A) on a eukaryotic mRNA.

Polycistronic mRNA includes coding regions representing more than one gene.

Polymorphism (more fully genetic polymorphism) refers to the simultaneous occurrence in the population of alleles showing variations at a given position. The original definition applied to alleles producing different phenotypes. Now it is also used to describe changes in DNA affecting the restriction pattern or even the sequence. For practical purposes, to be considered as an example of a polymorphism an allele should be found at a frequency of >1% in the population.

A **polyribosome** is an mRNA that is simultaneously being translated by several ribosomes.

Polytene chromosomes are generated by successive replications of a chromosome set without separation of the replicas.

Position effect variegation (PEV) is silencing of gene expression that occurs as the result of proximity to heterochromatin.

Positional information describes the localization of macromolecules at particular places in an embryo. The localization may itself be a form of information that is inherited.

Postmeiotic segregation describes the segregation of two strands of a duplex DNA that bear different information (created by heteroduplex formation during meiosis) when a subsequent replication allows the strands to separate.

The **postreplication complex** is a protein-DNA–complex in *S. cerevisiae* that consists of the ORC complex bound to the origin.

Posttranslational translocation is the movement of a protein across a membrane after the synthesis of the protein is completed and it has been released from the ribosome.

Precise excision describes the removal of a transposon plus one of the duplicated target sequences from the chromosome. Such an event can restore function at the site where the transposon inserted.

Preinitiation complex in eukaryotic transcription describes the assembly of transcription factors at the promoter before RNA polymerase binds.

Premature termination describes the termination of protein or of RNA synthesis before the chain has been completed. In protein synthesis it can be caused by mutations that create termination codons within the coding region. In RNA synthesis it is caused by various events that act on RNA polymerase.

Pre-mRNA is used to describe the nuclear transcript that is processed by modification and splicing to give an mRNA.

A protein to be imported into an organelle or secreted from bacteria is called a **preprotein** until its signal sequence has been removed.

The **prereplication complex** is a protein–DNA complex at the origin in *S. cerevisiae* that is required for DNA replication. The complex contains the ORC complex, Cdc6, and the MCM proteins.

The **primary immune response** is an organism's immune response upon first exposure to a given antigen. It is characterized by a relatively shorter duration and lower affinity antibodies than in the secondary immune response.

A **primary transcript** is the original unmodified RNA product corresponding to a transcription unit.

The **primase** is a type of RNA polymerase that synthesizes short segments of RNA that will be used as primers for DNA replication.

A **primer** is a short sequence (often of RNA) that is paired with one strand of DNA and provides a free 3′-OH end at which a DNA polymerase starts synthesis of a deoxyribonucleotide chain.

The **primosome** describes the complex of proteins involved in the priming action that initiates replication on φX-type origins. It is also involved in restarting stalled replication forks.

A **prion** is a proteinaceous infectious agent that behaves as an inheritable trait, although it contains no nucleic acid. Examples are PrP^{Sc}, the agent of scrapie in sheep and bovine spongiform encephalopathy, and Psi, which confers an inherited state in yeast.

A **processed pseudogene** is an inactive gene copy that lacks introns, contrasted with the interrupted structure of the active gene. Such genes originate by reverse transcription of mRNA and insertion of a duplex copy into the genome.

Processing of RNA describes changes that occur after its transcription, including modification of the 5′ and 3′ ends, internal methylation, splicing, or cleavage.

Processivity describes the ability of an enzyme to perform multiple catalytic cycles with a single template instead of dissociating after each cycle.

The recombination of V, (D), J gene segments results in a **productive rearrangement** if all the rearranged gene segments are in the correct reading frame.

Programmed frameshifting is required for expression of the protein sequences coded beyond a specific site at which a +1 or –1 frameshift occurs at some typical frequency.

A **promoter** is a region of DNA where RNA polymerase binds to initiate transcription.

Proofreading refers to any mechanism for correcting errors in protein or nucleic acid synthesis that involves scrutiny of individual units *after* they have been added to the chain.

Prophage is a phage genome covalently integrated as a linear part of the bacterial chromosome.

Protein sorting is the direction of different types of proteins for transport into or between specific organelles.

Protein splicing is the autocatalytic process by which an intein is removed from a protein and the exteins on either side become connected by a standard peptide bond.

The **proteome** is the complete set of proteins that is expressed by the entire genome. Some genes code for multiple proteins, and as a result the size of the proteome is greater than the number of genes. Sometimes the term is used to describe the complement of proteins expressed by a cell at any one time.

Provirus is a duplex sequence of DNA integrated into a eukaryotic genome that represents the sequence of the RNA genome of a retrovirus.

PrP is the protein that is the active component of the prion that causes scrapie and related diseases. The form involved in the disease is called PrP^{Sc}.

Pseudogenes are inactive but stable components of the genome derived by mutation of an ancestral active gene. Usually they are inactive because of mutations that block transcription or translation or both.

A **puff** is an expansion of a band of a polytene chromosome associated with the synthesis of RNA at some locus in the band.

Puromycin is an antibiotic that terminates protein synthesis by mimicking a tRNA and becoming linked to the nascent protein chain.

A **pyrimidine dimer** is formed when ultraviolet irradiation generates a covalent link directly between two adjacent pyrimidine bases in DNA. It blocks DNA replication and transcription.

A **quick-stop mutant** is a type of DNA replication temperature-sensitive mutant (*dna*) in *E. coli* that immediately stops DNA replication when the temperature is increased to 42°C.

The **R segments** are the sequences that are repeated at the ends of a retroviral RNA. They are called R-U5 and U3-R.

Random drift describes the chance fluctuation (without selective pressure) of the levels of two alleles in a population.

A **reading frame** is one of three possible ways of reading a nucleotide sequence. Each reading frame divides the sequence into a series of successive triplets. There are three possible reading frames in any sequence, depending on the starting point. If the first frame starts at position 1, the second frame starts at position 2, and the third frame starts at position 3.

Readthrough at transcription or translation occurs when RNA polymerase or the ribosome, respectively, ignores a termination signal because of a mutation of the template or the behavior of an accessory factor.

rec mutations of *E. coli* cannot undertake general recombination.

Recoding events occur when the meaning of a codon or series of codons is changed from that predicted by the genetic code. It may involve altered interactions between aminoacyl-tRNA and mRNA that are influenced by the ribosome.

The **recognition helix** is the one of the two helices of the helix-turn-helix motif that makes contacts with DNA that are specific for particular bases. This determines the specificity of the DNA sequence that is bound.

A **recombinant joint** is the point at which two recombining molecules of duplex DNA are connected (the edge of the heteroduplex region).

Recombination nodules are dense objects present on the synaptonemal complex; they may represent protein complexes involved in crossing-over.

Recombination-repair is a mode of filling a gap in one strand of duplex DNA by retrieving a homologous single strand from another duplex.

Redundancy describes the concept that two or more genes may fulfill the same function, so that no single one of them is essential.

A **regulator gene** codes for a product (typically protein) that controls the expression of other genes (usually at the level of transcription).

A **relaxase** is an enzyme that cuts one strand of DNA and binds to the free 5′ end.

A **release factor (RF)** is required to terminate protein synthesis to cause release of the completed polypeptide chain and the ribosome from mRNA. Individual factors are numbered. Eukaryotic factors are called eRF.

Renaturation describes the reassociation of denatured complementary single strands of a DNA double helix.

Repair of damaged DNA can take place by repair synthesis, when a strand that has been damaged is excised and replaced by the synthesis of a new stretch. It can also take place by recombination reactions, when the duplex region containing the damaged strand is replaced by an undamaged region from another copy of the genome.

Repetitive DNA behaves in a reassociation reaction as though many (related or identical) sequences are present in a component, allowing any pair of complementary sequences to reassociate.

Replacement sites in a gene are those at which mutations alter the amino acid that is coded.

Replication of duplex DNA takes place by synthesis of two new strands that are complementary to the parental strands. The parental duplex is replaced by two identical daughter duplexes, each of which has one parental strand and one newly synthesized strand. Replication is called semiconservative because the conserved units are the single strands of the parental duplex.

A **replication eye** is a region in which DNA has been replicated within a longer, unreplicated region.

A **replication fork** is the point at which strands of parental duplex DNA are separated so that replication can proceed. A complex of proteins including DNA polymerase is found at the fork.

A **replication-defective** virus cannot perpetuate an infective cycle because some of the necessary genes are absent (replaced by host DNA in a transducing virus) or mutated.

Replicative transposition describes the movement of a transposon by a mechanism in which first it is replicated, and then one copy is transferred to a new site.

The **replicon** is a unit of the genome in which DNA is replicated. Each replicon contains an origin for initiation of replication.

The **replisome** is the multiprotein structure that assembles at the bacterial replication fork to undertake synthesis of DNA. It contains DNA polymerase and other enzymes.

Repression describes the ability of bacteria to prevent synthesis of certain enzymes when their products are present; more generally, it refers to inhibition of transcription (or translation) by binding of repressor protein to a specific site on DNA (or mRNA).

A **repressor** is a protein that inhibits expression of a gene. It may act to prevent transcription by binding to an operator site in DNA or to prevent translation by binding to RNA.

Resolution occurs by a homologous recombination reaction between the two copies of the transposon in a cointegrate. The reaction generates the donor and target replicons, each with a copy of the transposon.

Resolvase is the enzyme activity involved in site-specific recombination between two copies of a transposon that has been duplicated.

A **response element** is a sequence in a eukaryotic promoter or enhancer that is recognized by a specific transcription factor.

Restriction endonucleases recognize specific short sequences of DNA and cleave the duplex (sometimes at the target site, sometimes elsewhere, depending on type).

Restriction fragment length polymorphism (RFLP) refers to inherited differences in sites for restriction enzymes (for example, caused by base changes in the target site) that result in differences in the lengths of the fragments produced by cleavage with the relevant restriction enzyme. RFLPs are used for genetic mapping to link the genome directly to a conventional genetic marker.

A **restriction map** is a linear array of sites on DNA cleaved by various restriction enzymes.

A **retroposon** is a transposon that mobilizes via an RNA form; the DNA element is transcribed into RNA, and then reverse-transcribed into DNA, which is inserted at a new site in the genome. The difference from retroviruses is that the retroposon does not have an infective (viral) form.

A **retrovirus** is an RNA virus with the ability to convert its sequence into DNA by reverse transcription.

Reverse transcriptase is an enzyme that uses a template of single-stranded RNA to generate a double-stranded DNA copy.

Reverse transcription is synthesis of DNA on a template of RNA. It is accomplished by the enzyme reverse transcriptase.

Revertants are derived by reversion of a mutant cell or organism to the wild-type phenotype.

RF1 is the bacterial release factor that recognizes UAA and UAG as signals to terminate protein synthesis.

RF2 is the bacterial release factor that recognizes UAA and UGA as signals to terminate protein synthesis.

RF3 is a protein synthesis termination factor related to the elongation factor EF-G. It functions to release the factors RF1 or RF2 from the ribosome when they act to terminate protein synthesis.

Rho factor (ρ) is a protein involved in assisting *E. coli* RNA polymerase to terminate transcription at certain terminators (called rho-dependent terminators).

Rho-dependent terminators are sequences that terminate transcription by bacterial RNA polymerase in the presence of the rho factor.

Ri plasmids are found in *Agrobacterium tumefaciens*. Like Ti plasmids, they carry genes that cause disease in infected plants. The disease may take the form of either hairy root disease or crown gall disease.

Ribonucleases are enzymes that cleave RNA. They may be specific for single-stranded or for double-stranded RNA and may be either endonucleases or exonucleases.

A **ribonucleoprotein** is a complex of RNA with proteins.

Ribosomal RNA (rDNA) is a major component of the ribosome. Each of the two subunits of the ribosome has a major rRNA as well as many proteins.

The **ribosome** is a large assembly of RNA and proteins that synthesizes proteins under direction from an mRNA template. Bacterial ribosomes sediment at 70S, eukaryotic ribosomes at 80S. A ribosome can be dissociated into two subunits.

Ribosome stalling describes the inhibition of movement that occurs when a ribosome reaches a codon for which there is no corresponding charged aminoacyl-tRNA.

A **ribosome-binding site** is a sequence on bacterial mRNA that includes an initiation codon that is bound by a 30S subunit in the initiation phase of protein synthesis.

A **riboswitch** is a catalytic RNA whose activity responds to a small ligand.

A **ribozyme** is an RNA that has catalytic activity.

A helix is said to be **right-handed** if the turns run clockwise along the helical axis.

RNA editing describes a change of sequence at the level of RNA following transcription.

RNA interference describes the technique in which double-strand RNA is introduced into cells to eliminate or reduce the activity of a target gene. It is caused by using sequences complementary to the double-stranded RNA sequences to trigger degradation of the mRNA of the gene.

An **RNA ligase** is an enzyme that functions in tRNA splicing to make a phosphodiester bond between the two exon sequences that are generated by cleavage of the intron.

RNA polymerases are enzymes that synthesize RNA using a DNA template (formally described as DNA-dependent RNA polymerases).

RNA silencing describes the ability of a dsRNA to suppress expression of the corresponding gene systemically in a plant.

RNA splicing is the process of excising introns from RNA and connecting the exons into a continuous mRNA.

The **rolling circle** is a mode of replication in which a replication fork proceeds around a circular template for an indefinite number of revolutions; the DNA strand newly synthesized in each revolution displaces the strand synthesized in the previous revolution, giving a tail containing a linear series of sequences complementary to the circular template strand.

Rotational positioning describes the location of the histone octamer relative to turns of the double helix, which determines which face of DNA is exposed on the nucleosome surface.

An **r-protein** is one of the proteins of the ribosome.

rut is an acronym for rho utilization site, the sequence of RNA that is recognized by the rho termination factor.

The **S domain** is the sequence of 7S RNA of the SRP that is not related to Alu RNA.

S phase is the restricted part of the eukaryotic cell cycle during which synthesis of DNA occurs.

An **S region** is a sequence involved in immunoglobulin class switching. S regions consist of repetitive sequences at the 5′ ends of gene segments encoding the heavy chain constant regions.

Satellite DNA consists of many tandem repeats (identical or related) of a short basic repeating unit.

A chromosome **scaffold** is a proteinaceous structure in the shape of a sister chromatid pair, generated when chromosomes are depleted of histones.

Scarce mRNA consists of a large number of individual mRNA species, each present in very few copies per cell. This accounts for most of the sequence complexity in RNA.

Scrapie is a disease caused by an infective agent made of protein (a prion).

A **secondary attachment site** is a locus on the bacterial chromosome into which phage lambda integrate inefficiently because the site resembles the *att* site.

The **secondary immune response** is an organism's immune response upon a second exposure to a given antigen. This second exposure is also referred to as a "booster." The secondary immune response is characterized by a more rapid induction, greater magnitude, and higher affinity antibodies than the primary immune response.

Second-site reversion occurs when a second mutation suppresses the effect of a first mutation.

A **sector** is a patch of cells made up of a single altered cell and its progeny.

Self-assembly refers to the ability of a protein (or of a complex of proteins) to form its final structure without the intervention of any additional components (such as chaperones). The term can also refer to the spontaneous formation of any biological structure that occurs when molecules collide and bind to each other.

Selfish DNA describes sequences that do not contribute to the genotype of the organism but have self-perpetuation within the genome as their sole function.

Semiconservative replication is accomplished by separation of the strands of a parental duplex, each strand then acting as a template for synthesis of a complementary strand.

A **semiconserved** position is one where comparison of many individual sequences finds the same type of base (pyrimidine or purine) always present.

Semidiscontinuous replication is the mode in which one new strand is synthesized continuously while the other is synthesized discontinuously.

The **septal ring** is a complex of several proteins coded by *fts* genes of *E. coli* that forms at the mid-point of the cell. It gives rise to the septum at cell division. The first of the proteins to be incorporated is FtsZ, which gave rise to the original name of the Z-ring.

A **septum** is the structure that forms in the center of a dividing bacterium, providing the site at which the daughter bacteria will separate. The same term is used to describe the cell wall that forms between plant cells at the end of mitosis.

The **serum response element (SRE)** is a sequence in a promoter or enhancer that is activated by transcription factor(s) induced by treatment with serum. This activates genes that stimulate cell growth.

The **Shine–Dalgarno** sequence is the polypurine sequence AGGAGG centered about 10 bp before the AUG initiation codon on bacterial mRNA. It is complementary to the sequence at the 3′ end of 16S rRNA.

Sigma factor is the subunit of bacterial RNA polymerase needed for initiation; it is the major influence on selection of promoters.

The **sign inversion** model describes the mechanism of DNA gyrase. DNA gyrase binds a positive supercoil (inducing a compensatory negative supercoil elsewhere on the closed circular DNA), breaks both strands in one duplex, passes the other duplex through, and reseals the strands.

A **signal end** is produced during recombination of immunoglobulin and T cell receptor genes. The signal ends are at the termini of the cleaved fragment containing the recombination signal sequences. The subsequent joining of the signal ends yields a signal joint.

Signal peptidase is an enzyme within the membrane of the ER that specifically removes the signal sequences from proteins as they are translocated. Analogous activities are present in bacteria, archaea, and in each organelle in a eukaryotic cell into which proteins are targeted and translocated by means of removable targeting sequences. Signal peptidase is one component of a larger protein complex.

The **signal recognition particle** is a ribonucleoprotein complex that recognizes signal sequences during translation and guides the ribosome to the translocation channel. SRPs from different organisms may have different compositions, but all contain related proteins and RNAs.

A **signal sequence** is a short region of a protein that directs it to the endoplasmic reticulum for cotranslational translocation.

A **silencer** is a short sequence of DNA that can inactivate expression of a gene in its vicinity.

Silencing describes the repression of gene expression in a localized region, usually as the result of a structural change in chromatin.

Silent mutations do not change the sequence of a protein because they produce synonymous codons.

A **silent site** in a coding region is one where mutation does not change the sequence of the protein.

Single copy replication describes a control system in which there is only one copy of a replicon per unit bacterium. The bacterial chromosome and some plasmids have this type of regulation.

Single nucleotide polymorphism describes a polymorphism (variation in sequence between individuals) caused by a change in a single nucleotide. This is responsible for most of the genetic variation between individuals.

The **single X hypothesis** describes the inactivation of one X chromosome in female mammals.

Single-strand assimilation describes the ability of RecA protein to cause a single strand of DNA to displace its homologous strand in a duplex; that is, the single strand is assimilated into the duplex.

The **single-strand binding protein (SSB)** attaches to single-stranded DNA, thereby preventing the DNA from forming a duplex.

Single-strand exchange is a reaction in which one of the strands of a duplex of DNA leaves its former partner and instead pairs with the complementary strand in another molecule, displacing its homologue in the second duplex.

Single-strand passage is a reaction catalyzed by type I topoisomerase in which one section of single-stranded DNA is passed through another strand.

Site-specific recombination occurs between two specific sequences, as in phage integration/excision or resolution of cointegrate structures during transposition.

SL RNA is a small RNA that donates an exon in the *trans*-splicing reaction of trypanosomes and nematodes.

A **slow-stop mutant** is a type of DNA replication temperature-sensitive mutant in *E. coli* that can finish a round of replication at the unpermissive temperature, but cannot start another.

Small cytoplasmic RNAs (scRNA) are present in the cytoplasm (and sometimes are also found in the nucleus).

A **small nuclear RNA (snRNA)** is one of many small RNA species confined to the nucleus; several of the snRNAs are involved in splicing or other RNA processing reactions.

The **small subunit** of the ribosome (30S in bacteria, 40S in eukaryotes) binds the mRNA.

A **somatic mutation** is a mutation occurring in a somatic cell, therefore affecting only its daughter cells; it is not inherited by descendants of the organism.

Somatic recombination describes the process of joining a V gene to a C gene in a lymphocyte to generate an immunoglobulin or T cell receptor.

The **SOS box** is the DNA sequence (operator) of ~20 bp recognized by LexA repressor protein.

An **SOS response** in *E. coli* describes the coordinate induction of many enzymes, including repair activities, in response to irradiation or other damage to DNA; it results from activation of protease activity by RecA to cleave LexA repressor.

A **spacer** is a sequence in a gene cluster that separates the repeated copies of the transcription unit.

Splice recombinant DNA results from a Holliday junction being resolved by cutting the nonexchanged strands. Both strands of DNA before the exchange point come from one chromosome; the DNA after the exchange point come from the homologous chromosome.

Splice sites are the sequences immediately surrounding the exon–intron boundaries.

The **spliceosome** is a complex formed by the snRNPs that are required for splicing together with additional protein factors.

Spontaneous mutations occur in the absence of any added reagent to increase the mutation rate, as the result of errors in replication (or other events involved in the reproduction of DNA) or by environmental damage.

Sporulation is the generation of a spore by a bacterium (by morphological conversion) or by a yeast (as the product of meiosis).

An **SR protein** has a variable length of an Arg-Ser–rich region and is involved in splicing.

An **sRNA** is a small bacterial RNA that functions as a regulator of gene expression.

Startpoint refers to the position on DNA corresponding to the first base incorporated into RNA.

A **stem** is the base-paired segment of a hairpin structure in RNA.

Steroid receptors are transcription factors that are activated by binding of a steroid ligand.

A **stop codon** is one of three triplets (UAG, UAA, or UGA) that causes protein synthesis to terminate. They are also known historically as *nonsense codons*. The UAA codon is called ochre and the UAG codon is called amber, after the names of the nonsense mutations by which they were originally identified.

Strand displacement is a mode of replication of some viruses in which a new DNA strand grows by displacing the previous (homologous) strand of the duplex.

A **structural distortion** is a change in the conformation of DNA caused by bases or base pairs that do not fit into the normal duplex.

A **structural gene** codes for any RNA or protein product other than a regulator.

Structural maintenance of chromosomes (SMC) describes a group of proteins that include the cohesins, which hold sister chromatids together, and the condensins, which are involved in chromosome condensation.

A **subviral pathogen** is an infectious agent that is smaller than a virus, such as a virusoid.

Supercoiling describes the coiling of a closed duplex DNA in space so that it crosses over its own axis.

A **superfamily** is a set of genes all related by presumed descent from a common ancestor, but now showing considerable variation.

Suppression occurs when a second event eliminates the effects of a mutation without reversing the original change in DNA.

A **suppressor** is a second mutation that compensates for or alters the effects of a primary mutation.

A frameshift **suppressor** is an insertion or deletion of a base that restores the original reading frame in a gene that has had a base deletion or insertion.

Surveillance systems check nucleic acids for errors. The term is used in several different contexts. One example is the system that degrades mRNAs that have nonsense mutations. Another is the set of systems that react to damage in the double helix. The common feature is that the system recognizes an invalid sequence or structure and triggers a response.

SWI/SNF is a chromatin remodeling complex; it uses hydrolysis of ATP to change the organization of nucleosomes.

Synapsis describes the association of the two pairs of sister chromatids (representing homologous chromosomes) that occurs at the start of meiosis; the resulting structure is called a bivalent.

The **synaptonemal complex** describes the morphological structure of synapsed chromosomes.

Synonym codons have the same meaning in the genetic code. Synonym tRNAs bear the same amino acid and respond to the same codon.

Synteny describes a relationship between chromosomal regions of different species where homologous genes occur in the same order.

Synthetic genetic array analysis (SGA) is an automated technique in budding yeast whereby a mutant is crossed to an array of approximately 5000 deletion mutants to determine if the mutations interact to cause a synthetic lethal phenotype.

Synthetic lethality occurs when two mutations that by themselves are viable cause lethality when combined.

T cells are lymphocytes of the T (thymic) lineage; they may be subdivided into several functional types. They carry TcR (T cell receptor) and are involved in the cell-mediated immune response.

The **T cell receptor (TCR)** is the antigen receptor on T lymphocytes. It is clonally expressed and binds to a complex of MHC class I or class II protein and antigen-derived peptide.

TAFs are the subunits of $TF_{II}D$ that assist TBP in binding to DNA. They also provide points of contact for other components of the transcription apparatus.

TATA box is a conserved A-T–rich octamer found about 25 bp before the startpoint of each eukaryotic RNA polymerase II transcription unit; it is involved in positioning the enzyme for correct initiation.

The **TATA-binding protein (TBP)** is the subunit of transcription factor $TF_{II}D$ that binds to DNA.

A **TATA-less promoter** does not have a TATA box in the sequence upstream of its startpoint.

T-DNA is the segment of the Ti plasmid of *Agrobacterium tumefaciens* that is transferred to the plant cell nucleus during infection. It carries genes that transform the plant cell.

Telomerase is the ribonucleoprotein enzyme that creates repeating units of one strand at the telomere by adding individual bases to the DNA 3′ end, as directed by an RNA sequence in the RNA component of the enzyme.

A **telomere** is the natural end of a chromosome; the DNA sequence consists of a simple repeating unit with a protruding single-stranded end.

Telomeric silencing describes the repression of gene activity that occurs in the vicinity of a telomere.

A **teratoma** is a growth in which many differentiated cell types—including skin, teeth, bone, and others—grow in a disorganized manner after an early embryo is transplanted into one of the tissues of an adult animal.

A **terminal protein** allows replication of a linear phage genome to start at the very end. The protein attaches to the 5′ end of the genome through a covalent bond, is associated with a DNA polymerase, and contains a cytosine residue that serves as a primer.

A **terminase** enzyme cleaves multimers of a viral genome and then uses hydrolysis of ATP to provide the energy to translocate the DNA into an empty viral capsid starting with the cleaved end.

Termination is a separate reaction that ends a macromolecular synthesis reaction (replication, transcription, or translation), by stopping the addition of subunits, and (typically) causing disassembly of the synthetic apparatus.

A **terminator (t)** is a sequence of DNA that causes RNA polymerase to terminate transcription.

A **terminus** is a segment of DNA at which replication ends.

The **ternary complex** in initiation of transcription consists of RNA polymerase and DNA and a dinucleotide that represents the first two bases in the RNA product.

A **tetrad** describes the four (haploid) spores that result from meiosis in yeast. (The term originally was used to describe the structure found at the beginning of meiosis, now known as a bivalent, that contains all four chromatids, produced by duplication of a homologous chromosome pair.)

$TF_{II}D$ is the transcription factor that binds to the TATA sequence upstream of the startpoint of promoters for RNA polymerase II. It consists of TBP (TATA binding protein) and the TAF subunits that bind to TBP.

Thalassemia is disease of red blood cells resulting from lack of either α or β globin.

Third-base degeneracy describes the lesser effect on codon meaning of the nucleotide present in the third codon position.

The **Ti plasmid** is an episome of the bacterium *Agrobacterium tumefaciens* that carries the genes responsible for the induction of crown gall disease in infected plants.

Tight binding of RNA polymerase to DNA describes the formation of an open complex (when the strands of DNA have separated).

The **TIM complex** resides in the inner membrane of mitochondria and is responsible for transporting proteins from the intermembrane space into the interior of the organelle.

Bacterial transposons carrying markers that are not related to their function, e.g., drug resistance, are named as **Tn** followed by a number.

Tolerance is the lack of an immune response to an antigen (either self antigen or foreign antigen) due to clonal deletion.

A **Toll-like receptor (TLR)** is a plasma membrane receptor that is expressed on phagocytes and other cells and is involved in signaling during the innate immune response. TLRs are related to IL-1 receptors.

The **TOM complex** resides in the outer membrane of the mitochondrion and is responsible for importing proteins from the cytosol into the space between the membranes.

A DNA **topoisomerase** is an enzyme that changes the number of times the two strands in a closed DNA molecule cross each other. It does this by cutting the DNA, passing DNA through the break, and resealing the DNA.

Topological isomers are molecules of DNA that are identical except for a difference in linking number.

A **trailer** is a nontranslated sequence at the 3′ end of an mRNA following the termination codon.

trans configuration of two sites refers to their presence on two different molecules of DNA (chromosomes).

A ***trans*-acting** product can function on any copy of its target DNA. This implies that it is a diffusible protein or RNA.

A **transcript** is the RNA product produced by copying one strand of DNA. It may require processing to generate a mature RNA.

Transcription describes synthesis of RNA on a DNA template.

A **transcription factor** is required for RNA polymerase to initiate transcription at specific promoter(s), but is not itself part of the enzyme.

A **transcription unit** is the sequence between sites of initiation and termination by RNA polymerase; it may include more than one gene.

The **transcriptome** is the complete set of RNAs present in a cell, tissue, or organism. Its complexity is due mostly to mRNAs, but it also includes noncoding RNAs.

A **transducing virus** carries part of the host genome in place of part of its own sequence. The best known examples are retroviruses in eukaryotes and DNA phages in *E. coli*.

A **transesterification** reaction breaks and makes chemical bonds in a coordinated transfer so that no energy is required.

Transfection of eukaryotic cells is the acquisition of new genetic markers by incorporation of added DNA.

The **transfer region** is a segment on the F plasmid that is required for bacterial conjugation.

Transfer RNA (tRNA) is the intermediate in protein synthesis that interprets the genetic code. Each tRNA can be linked to an amino acid. The tRNA has an anticodon sequence that is complementary to a triplet codon representing the amino acid.

Transformation of bacteria is the acquisition of new genetic material by incorporation of added DNA.

Transformation of eukaryotic cells refers to their conversion to a state of unrestrained growth in culture, resembling or identical with the tumorigenic condition.

The **transforming principle** is DNA that is taken up by a bacterium and whose expression then changes the properties of the recipient cell.

A **transgene** is a gene that is introduced into a cell or animal from an external source.

A **transition** is a mutation in which one pyrimidine is replaced by the other, or in which one purine is replaced by the other.

Translation is synthesis of protein on an mRNA template.

Translational positioning describes the location of a histone octamer at successive turns of the double helix, which determines which sequences are located in linker regions.

Translocation describes the stage of nuclear import or export when a protein or RNA substrate moves through the nuclear pore.

Translocation is the movement of the ribosome one codon along mRNA after the addition of each amino acid to the polypeptide chain.

Protein **translocation** describes the movement of a protein across a membrane. This occurs across the membranes of organelles in eukaryotes, or across the plasma membrane in bacteria. Each membrane across which proteins are translocated has a channel specialized for the purpose.

Movement of a protein across a lipid bilayer usually requires a **translocon**, an integral membrane protein that provides a channel for displacement of polypeptide segments across the membrane.

A **transmembrane protein** extends across a lipid bilayer. A hydrophobic region (typically consisting of a stretch of 20 to 25 hydrophobic and/or uncharged amino acids) or regions of the protein resides in the membrane. Hydrophilic regions are exposed on one or both sides of the membrane.

The **transmembrane region** is the part of a protein that spans the membrane bilayer. It is hydrophobic and in many cases contains approximately 20 amino acids that form an α-helix. It is also called the transmembrane domain.

Transplantation antigen is protein coded by a major histocompatibility locus, present on all mammalian cells, involved in interactions between lymphocytes.

A **transposase** is the enzyme activity involved in insertion of transposon at a new site.

Transposition refers to the movement of a transposon to a new site in the genome.

A **transposon** is a DNA sequence able to insert itself (or a copy of itself) at a new location in the genome without having any sequence relationship with the target locus.

A **transversion** is a mutation in which a purine is replaced by a pyrimidine or vice versa.

$tRNA_f^{Met}$ is the special RNA used to initiate protein synthesis in bacteria. It mostly uses AUG, but can also respond to GUG and CUG.

$tRNA_i^{Met}$ is the special tRNA used to respond to initiation codons in eukaryotes.

$tRNA_m^{Met}$ inserts methionine at internal AUG codons.

A **true reversion** is a mutation that restores the original sequence of the DNA.

The **twisting number** of a DNA is the number of base pairs divided by the number of base pairs per turn of the double helix.

A **two hybrid** assay detects interaction between two proteins by means of their ability to bring together a DNA-binding domain and a transcription-activating domain. The assay is performed in yeast using a reporter gene that responds to the interaction.

Ty stands for transposon yeast, the first transposable element to be identified in yeast.

A **type I topoisomerase** is an enzyme that changes the topology of DNA by nicking and resealing one strand of DNA.

A **type II topoisomerase** is an enzyme that changes the topology of DNA by nicking and resealing both strands of DNA.

U3 is the repeated sequence at the 3′ end of a retroviral RNA.

U5 is the repeated sequence at the 5′ end of a retroviral RNA.

A stretch of **underwound** DNA has fewer base pairs per turn than the usual average (10 bp = 1 turn). This means that the two strands of DNA are less tightly wound around each other; ultimately this can lead to strand separation.

Unidirectional replication refers to the movement of a single replication fork from a given origin.

An **uninducible** mutant is one where the affected gene(s) cannot be expressed.

The **unit cell** describes the state of an *E. coli* bacterium generated by a new division. It is 1.7 μm long and has a single replication origin.

An **up mutation** in a promoter increases the rate of transcription.

Upstream identifies sequences in the opposite direction from expression; for example, the bacterial promoter is upstream of the transcription unit, and the initiation codon is upstream of the coding region.

An **upstream activator sequence (UAS)** is the equivalent in yeast of the enhancer in higher eukaryotes and is bound by the GAL4 transcriptional activator proteins.

A **V gene** is sequence coding for the major part of the variable (N-terminal) region of an immunoglobulin chain.

A **variable region (V region)** is an antigen-binding site of an immunoglobulin or T cell receptor molecule. V regions are composed of the variable domains of the component chains. They are coded by V gene segments and vary extensively among antigen receptors as the result of multiple, different genomic copies and of changes introduced during synthesis.

Variegation of phenotype is produced by a change in genotype during somatic development.

The **vegetative phase** describes the period of normal growth and division of a bacterium. For a bacterium that can sporulate, this contrasts with the sporulation phase, when spores are being formed.

The **viral superfamily** comprises transposons that are related to retroviruses. They are defined by sequences that code for reverse transcriptase or integrase.

Virion is the physical virus particle (irrespective of its ability to infect cells and reproduce).

A **viroid** is a small infectious nucleic acid that does not have a protein coat.

Virulent phage mutants are unable to establish lysogeny.

A **virusoid** is a small infectious nucleic acid that is encapsidated by a plant virus together with its own genome.

VNTR (variable number tandem repeat) regions describe very short repeated sequences, including microsatellites and minisatellites.

The **wobble hypothesis** accounts for the ability of a tRNA to recognize more than one codon by unusual (non–G-C, non–A-T) pairing with the third base of a codon.

The **writhing number** is the number of times a duplex axis crosses over itself in space.

The **zinc finger** is a DNA-binding motif that typifies a class of transcription factor.

A **zoo blot** describes the use of Southern blotting to test the ability of a DNA probe from one species to hybridize with the DNA from the genomes of a variety of other species.

Index

Note: *f* = figure; *t* = table

A

B

D

F

G

H

I

J

K

L

M

N

O

P

Q

R

S

T

U

V

W

X

Y

Z